Calculus

INTRODUCTORY EDITION

Volume 1

AL SHENK
University of California- San Diego

with
Lee van de Wetering and Carol Shenk

Addison Wesley

Boston San Francisco New York
London Toronto Sydney Tokyo Singapore Madrid
Mexico City Munich Paris Cape Town Hong Kong Montreal

Senior Acquisitions Editor: Laurie Rosatone
Project Manager: Bess Deck
Assistant Editor: Ellen Keohane
Sr. Production Supervisor: Peggy McMahon
Print Buyer: Evelyn Beaton
Sr. Prepress Supervisor: Caroline Fell
Technical Art Supervisor: Joe Vetere
This book was typeset by the author.

Photo/Illustration Credits
p. 25, From Handbook of Engineering Materials, 1E, Douglas Miner and John Seastone, eds. Copyright © 1955. Reprinted with permission of John Wiley & Sons. p. 40, Library of Congress. p. 220, From Why Buildings Stand Up: The Strength of Architecture, by Mario Salvadori. Copyright © 1980 by Mario Salvadori. Reprinted by permission of W. W. Norton & Company, Inc. p. 390, photo reprinted with permission of Culver Pictures; Graph reprinted with permission of Slim Films.

ISBN 0-201-73037-5

1 2 3 4 5 6 7 8 9 10 — CRS — 03020100

CALCULUS, VOLUME 1

Al Shenk

with Lee Van de Wetering

†These programs and instructions are included in preliminary editions of this text for the convenience of users. In the forthcoming First Edition all programs and implementation instructions for specific calculators or computer software will be in the *Calculator/Computer Supplement* that will accompany the book.

Typeset by $\mathcal{A}\mathcal{M}\mathcal{S}$-TEX

Chapter 3. Derivatives and Applications, II‡

Chapter 4. Integrals and Applications

‡Sections 3.4 through 3.6 can be covered after Sections 4.1 through 4.5 to postpone the study of derivatives of transcendental functions until integrals of other functions have been studied.

Chapter 5. Derivatives and Applications, part III.

Chapter 6. Integrals and applications, part II.

Chapter 7. First-order differential equations.

Chapter 8. Sequences and series.

Chapter 9. Conic sections and polar coordinates.

INTRODUCTION

"Most of calculus is easy, once you figure it out."

The above statement, which is the basic premise of this text, might come as a surprise to you and requires some explanation.

First, the statement does not say that calculus is easy. It is not. Calculus is a complex and abstract subject that has been developed over the past 2500 years and is used to study a wide variety of topics, including the effects of gravity and air resistance on the velocity and acceleration of vehicles and projectiles, the motion of planets and satellites, areas and volumes, probability, chemical reaction rates, air currents and ocean waves, electricity and magnetism, radioactive decay, the effects of supply and demand on market prices, and the growth of organisms and populations.

Second, the statement speaks of "figuring out calculus" and not of "learning calculus" to make an important point. When someone tells you how to get to a restaurant during your first visit to a city, you learn the directions so you can follow them, but you have to figure out how to interpret those directions as you find your way to the restaurant. Moreover, after you are familiar with the city and have been to the restaurant several times, your understanding of how to find it is based more on your experience than on what you were told initially.

The process of mastering mathematics is similar. You come to understand a new mathematical idea by working with it and by figuring out what it means to you based on your experience and on the ways you formulate and remember abstract ideas. You need to do much more than listen to your instructor and study the explanations given in your textbook.

To do your best, you need to approach every new mathematical topic by asking yourself, "What is this really about?" and "How can I explain this so it makes more sense to me?" If you approach the topic as a mystery to be unraveled, rather than just as information and procedures to be learned, you will bring out the best of your intellectual abilities. Then, as you work with the topic and the new concepts and techniques it entails, it will become clearer and more meaningful to you until it is, in fact, easy to understand. At that point your sense of the topic may well differ considerably from the explanations in your textbook or lectures.

This book is designed to help you to figure out calculus—with the assistance of your instructor—so it can be a useful tool in your later work and studies.

You will see that, in addition to solved examples as are found in other mathematics textbooks, most discussions of ideas, procedures, and problems in this book are interspersed with questions for you to answer to supply key steps of reasoning and calculations. The goal is to keep you actively involved in the logical development and to allow you to formulate your own ways of looking at the topics as you study.

To take full advantage of the questions, pause in your reading to spend time on each one. Think carefully about your response. Then, when you are satisfied with your conclusion, compare it with the response given at the end of the chapter. If you have trouble with a question, do not read the response in the book right away. Continue reading the text and return to the question later, when you can approach it with fresh ideas.

The tune-up exercises at the end of each section cover the basic techniques that are needed to apply the ideas and results covered in the section. Be sure you can work each of these types of exercises before you begin the regular problems that follow.

You will also notice that many of the regular problems do not look exactly like the examples in the body of the text, and that some require precalculus procedures, techniques from earlier in the textbook, or other lines of reasoning not covered in the section. This has not been done to make the problems difficult—most of them are not—but to help you develop your overall ability to read, analyze, and solve problems.

Arrange to study with other students in your class. You can solve harder problems and write better solutions by working together, and discussing ideas with others is a natural way to improve your own understanding.

In many cases you will figure out how to solve a problem after you have formulated and tested several strategies for working it. Morover, in addition to thinking about the "big picture," you need to concentrate on the details and pay close attention to the logic of going from each step to the next. As is suggested in the diagram on the right, you need to write down carefully the results of each step and think about what you have achieved and where you hope you are heading before you proceed.†

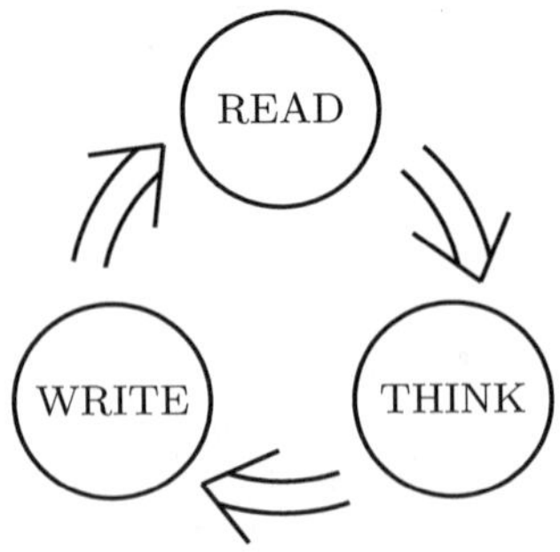

You probably follow this procedure to a certain extent whevever you solve a problem successfully. Suppose, for example, you are asked to find the radius of a circle of area 5π square feet. You might say to yourself, "Because the area of a circle of radius r is πr^2, I need to solve the equation, $5\pi = \pi r^2$." After writing down and reading this equation, you might say, "I need to divide both sides of the equation by π," and write "$5 = r^2$." Reading this might lead you to think, "If I take the square root of both sides to find r, I would have $r = \pm\sqrt{5}$, but a a radius cannot be negative," and you give the answer, "$r = \sqrt{5}$."

It is especially important that you also concentrate on each step when you are working a type of problem that is not familiar to you. Paying close attention to the details and to the logical flow of your reasoning will not only help you avoid mistakes. It can also help you develop a strategy for solving the entire problem as you work on it.

The two basic tools of calculus are the derivative, which is used to determine tangent lines and rates of change, and the integral, which is used to find areas and volumes and to determine changes from rates of change. The derivative is introduced in Chapter 2 of the text and the integral in Chapter 4. Your instructor might choose to cover Sections 4.1 through 4.6 before Section 3.4 through 3.6 to postpone the study of derivatives of exponential, logarithmic, and trigonometric functions until integrals of other functions have been discussed.

†Diagram suggested by Rick LeBorne

Most reflection problems in this book are not technically difficult, but because they ask questions for which there are no matching solved examples in the text, they take students out of a template-application mode and encourage them to think more abstractly and creatively. Many of these problems require additional reasoning or deal with formulas, data, or graphs from other disciplines. Some use precalculus techniques or techniques from previous sections, and others preview concepts to be presented later.

This book has a number of features, in addition to the emphasis on reflection problems, that set it apart from most other calculus texts:

- ***Students need access to graphing calculators or computers.***
  Many of the questions, examples, and problems require a calculator or a computer with calculus software. These remarkable tools make it much easier for students to grasp ideas and make them better problem solvers. Seeing the geometric and numerical aspects of their work helps them avoid errors and leaves them with a much richer understanding of what they are doing.[†]
- ***Questions in the text lead students to be more active learners.***
  Discussions of theory and techniques, sample calculations, and applications are interspersed with questions which students answer to supply details and interpretations. Responses are given at the end of the chapters for them to check the accuracy of their calculations and reasoning. This process focuses their attention on important concepts and techniques and gives them opportunities to see the mathematics from their own points of view, while providing them enough feedback so they can study effectively on their own. Solved examples are used as appropriate to consolidate students' understanding, to introduce new techniques, and to illustrate technical notation and terminology.
- ***Tune-up exercises develop basic skills***
  While the basic goal of the text is to develop students' conceptual understanding and problem-solving abilities, they also need practice with basic operations. Tune-up exercises, which students are expected to master before they attempt the regular problems, cover the easier skills. Other problems provide practice with more difficult procedures, such as differentiation, integration, and the sketching of graphs.
- ***The text gives students extensive resources for studying on their own.***
  By the time students have worked through all of the questions in a section and solved all of that section's tune-up exercises, they have spent a substantial amount of time in structured study. Then they can consolidate and expand their skills and understanding with the problems in the regular problem sets.
- ***Lecture time can be saved by assigning questions as homework.***
  Instructors can have students study each topic before it is covered in class by requiring that they turn in, as part of their homework, detailed responses to some or all of the questions on that topic in the text, and that they begin this work before the class meeting in which the topic is introduced. This leads the students to come to class better prepared to discuss the material with comprehension and allows instructors to spend much less time introducing new material.
- ***Instructors can use problems from the text in their lectures.***
  The preparation of nonroutine problems—especially those that are based on graphs or data from specific applications or that involve calculators or computers—can require a great deal of time and energy. The text contains a wide variety of problems suitable for use in lectures, freeing instructors to spend their class-preparation time on other aspects of teaching, such as planning ways to stimulate student interest and develop their overall understanding. Moreover, if the students can read the problems from their books, instructors do not have to spend class time writing them on the board for the students to copy.

[†]Instructions for using of a number of types graphing calculators and calculus computer software in calculus will be provided in a *Calculator/Computer Supplement* to the forthcoming First Edition of this book.

- ***Instructional notes help instructors plan their classes***
  Footnotes at the beginning of each section indicate, when appropriate, how the section could be divided into smaller teaching units. Lists of the corresponding tune-up exercises and problems are also given with suggestions of questions that might be assigned as homework and of problems that serve well for in-class discussions. In addition, a classification by level of difficulty and topic is given after each tune-up exercise and problem.[†]
- ***The* Instructor's Handbook *is a solution manual and instructor's guide.***
  The *Instructor's Handbook*, which will be supplied to users of the book, will contain answers and outlines of solutions of all tune-up exercises and problems and more detailed, section-by-section suggestions for using the book.
- ***Ample problem sets are included.***
  Substantial numbers of problems are given to meet the requirements of different classes, to give instructors a variety of problems to use in their lectures and in possible group projects, and so instructors can vary their assignments from year to year. The problems are in some cases organized by topic but are generally ordered by level of difficulty, with the most challenging exercises at the end.

This text is the result of eight years of experimenting with a variety of teaching materials and techniques in my large calculus classes at the University of California at San Diego. I found that by using my material, I got much more positive feedback from my students than I had previously. My students said they liked the question-response format and the variety of applications and that they learned more in my course than they had in their previous math classes. I also had much more fun in my lectures, since I spent most of my time discussing problems with the students rather than just presenting material on the board.

In developing this text I have been guided by my students' performance and by their comments and suggestions. In my writing, I have also kept in mind my understanding of other instructors' needs and interests, in the hope that the book will be as flexible and useful as possible.

Al Shenk, Mathematics Department, UCSD
La Jolla, CA 92093 (ashenk@ucsd.edu)

[†]These notes for instructiors are in all copies of this Preliminary Edition because it does not have separate versions for students and instructors. They will only appear in the instructors' desk copies in the forthcoming First Edition.

CHAPTER 1: MATHEMATICAL MODELS, FUNCTIONS, AND GRAPHS

This chapter sets the stage for the study of calculus, which begins in Chapters 2 with the DERIVATIVE and continues in Chapters 4 with the INTEGRAL. Derivatives and integrals are needed for solving problems that cannot be solved using only more elementary mathematics. For instance, if a car moves on a straight road and the forces acting on it are constant, its motion can be studied using only arithmetic and algebra, but if the road is curved and the forces vary, derivatives and integrals are required. Because calculus is needed in almost all studies that involve motion or other types of change or variation, it is the basis of much of higher mathematics and is fundamental in all branches of the physical sciences, including the study of aerodynamics, fluid flow, planetary motion, heat transfer, sound, electricity and magnetism, biological growth, radioactive decay, and chemical reactions. Calculus is also required in the study of probability, in economics, and in certain aspects of the social sciences.

Study this chapter carefully to be sure you are prepared to understand the following chapters. Give full attention to answering the questions in the reading, study the examples, and be sure you can work all of the tune-up exercises and any problems in the regular problem sets that you are assigned. Section 1.1 deals with mathematical models, functions, and graphs. Section 1.2 contains a preliminary analysis of graphs of functions constructed from powers of x and a discussion of infinite limits. Section 1.3 deals with problems involving linear functions and constant rates of change. These serve as a preview of the types of problems you will solve later with other functions by using the tools of calculus.†

Section 1.1

Measuring and modeling: variables and functions

OVERVIEW: *If a problem is to be solved using calculus, the first step is to set up a mathematical model of the situation being studied in which the quantities that vary are represented by variables and functions. The rules and procedures of calculus are then applied to the variables and functions, using whatever arithmetic and algebra are needed to solve the problem. Read this section carefully, so you will know the notation and terminology we will use as we work with variables and functions. This section will also give you practice working narrative problems and solving inequalities.*

Topics:

- ***Variables, functions, and graphs: notation and terminology***
- ***Changes in values of a function***
- ***Percent and relative change***
- ***Finding formulas for mathematical models***
- ***Absolute values and solving inequalities***

Variables and functions

To illustrate how the terms "variable" and "function" are used, take out a quarter or another coin, and use the scale in Figure 1 to measure its diameter in centimeters. Call the diameter d. Calculate the radius of the coin with the formula $r = \frac{1}{2}d$, and use the formula $A = \pi r^2$ for the area of a circle of radius r to find the area of its face (Figure 2).

---

†Study Appendix 1 to prepare yourself for using your calculator or computer with this text.

♦ SUGGESTIONS TO INSTRUCTORS: This section should not be skipped, but can be covered on homework and in a small amount of class time in courses with well-prepared students. Its main topics can be divided into two parts, as follows:
Part 1 (Functions, graphs, mathematical models, change, and relative change; Tune-up Exercises T1–T9, T11–T19, Problems 1-41, 68–72) Have students begin work on some or all of Questions 1 through 23 before class. Problems 1, 7, 12, 17, 21, 33, and 34 are good for in-class discussion.
Part 2 (Absolute values, equations, and inequalities; Tune-up Exercises T10 and T20; Problems 42–67) Have stuents begin work on Question 24 before class. Problems 44, 50, 54, and 55 are good for in-class discussion. (Instructional notes such as this will only be in instructors' desk copies of the forthcoming First Edition.)

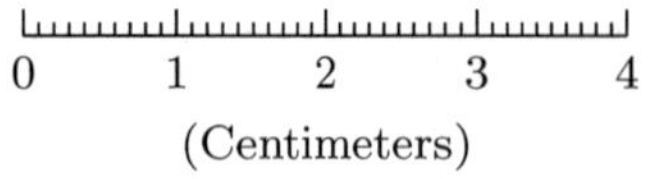

FIGURE 1

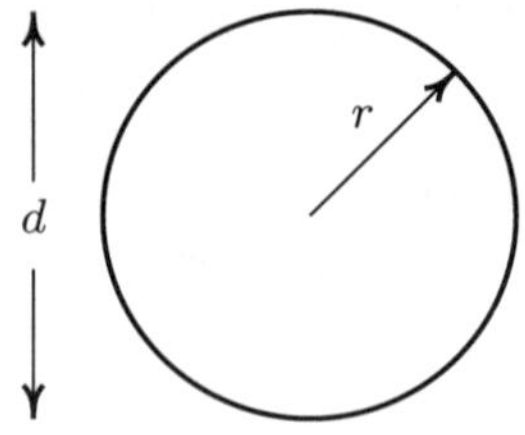

$r = \frac{1}{2}d,\ A = \pi r^2$

FIGURE 2

The letters d, r, and A that represent the diameter, radius, and area are VARIABLES. The rules $r = \frac{1}{2}d$ for finding the radius from the diameter and $A = \pi r^2$ for finding the area from the radius are FUNCTIONS.

In the case of the radius function $r = \frac{1}{2}d$, the variable d is called the INDEPENDENT VARIABLE or the VARIABLE OF THE FUNCTION, the variable r is called the DEPENDENT VARIABLE, and the number $\frac{1}{2}d$ is called the VALUE OF THE FUNCTION at the value d of its variable. The action of the function $r = \frac{1}{2}d$ can be represented by the symbols

$$d \xrightarrow{r=\frac{1}{2}d} r$$

since it converts a diameter d into a radius r.

The area function $A = \pi r^2$ can be represented by

$$r \xrightarrow{A=\pi r^2} A.$$

In this case, r is the independent variable, A is the dependent variable, and the value of the function at r is πr^2.

In the calculation you made with your coin, the diameter that you measured was a value of the independent variable d. You used this to calculate the corresponding value $r = \frac{1}{2}d$, of the radius function, which was a value of the dependent variable r. Then r became the independent variable. You used its value with the function $A = \pi r^2$ to calculate the area, which was a value of the dependent variable A.

The DOMAIN of a function is the set of values of its variable for which it is defined. If we consider all circles of zero or positive diameters, then the domain of the radius function is the interval $d \geq 0$, and the domain of the area function is the interval $r \geq 0$.

The GRAPH of the radius function, $d \to r = \frac{1}{2}d$, consists of the points (d, r) in a dr-plane with d in the domain of the function and $r = \frac{1}{2}d$ the corresponding value of the function. The graph is the line in Figure 3, as you might predict from the following table of values.†

d	0	1	2	3	4
$r = \frac{1}{2}d$	0	$\frac{1}{2}$	1	$\frac{3}{2}$	2

The graph of the area function, $r \to A = \pi r^2$, is the curve in the rA-plane of Figure 4. Its shape is suggested by the values in the next table. We use the symbol $\doteq$ here, and throughout this book, for approximations that come from round-off errors with decimals. We will use $\approx$ for other types of approximations.

r	0	$\frac{1}{2}$	1	$\frac{3}{2}$	2
$A = \pi r^2$	0	$\frac{1}{4}\pi \doteq 0.79$	$\pi \doteq 3.14$	$\frac{9}{4}\pi \doteq 7.07$	$4\pi \doteq 12.6$

†Notice that we use the term "graph" for a curve or other set of points in a coordinate plane, whereas in everyday language, the term "graph" usually refers to an entire drawing.

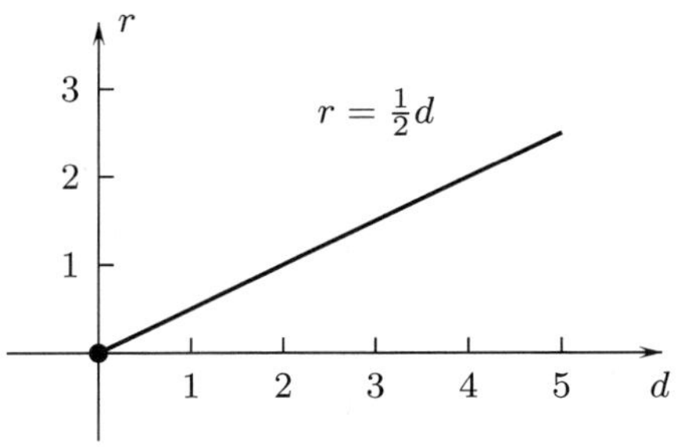

FIGURE 3

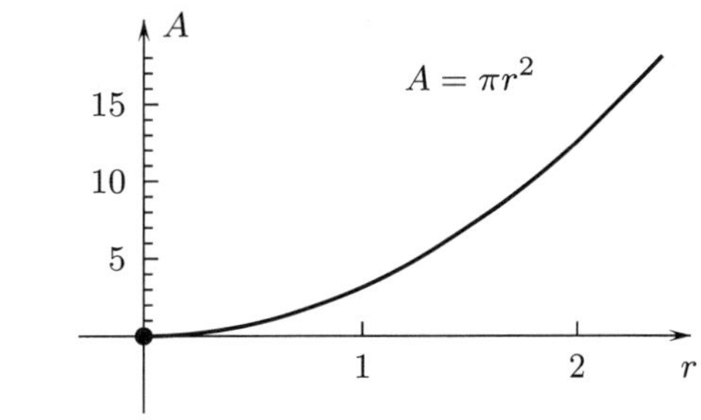

FIGURE 4

The graphs in Figures 3 and 4 do not extend to the left of the vertical axes because the radius and area functions are not defined for negative values of their variables. Dots have been placed at the origins to indicate that we are considering the radius function to be defined at $d = 0$ and the area function to be defined at $r = 0$.

To show how values of functions are found from their graphs, we start with the value $d = 4$ of the diameter. We find this value on the horizontal d-axis in Figure 5, move vertically up to the graph, and then move horizontally to the value $r = \frac{1}{2}(4) = 2$ of the radius on the vertical r-axis. To find the area corresponding to this radius, we find $r = 2$ on the horizontal r-axis in Figure 6, move up to the graph, and then horizontally to the value $A = \pi(2^2) = 4\pi \doteq 12.57$ of the area on the vertical A-axis.

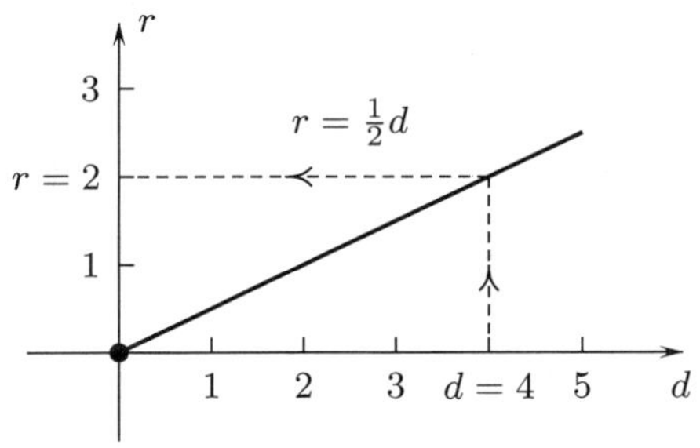

FIGURE 5

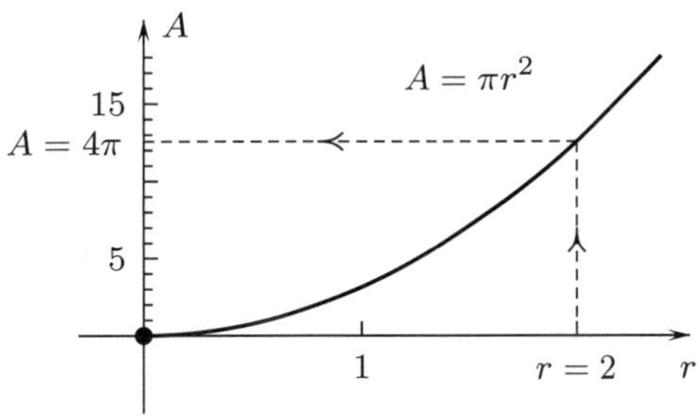

FIGURE 6

The RANGE of a function is the set of its possible values. The range of the radius function of Figure 5 is the interval $r \geq 0$ and the range of the area function of Figure 6 is the interval $A \geq 0$.

In Figures 3 through 6, the formulas $r = \frac{1}{2}d$ and $A = \pi r^2$ are used as names of the functions. Often a single letter is used as the name of a function. Figure 7 shows the graph of a function f with variable x. Its value at x is denoted $f(x)$, as in the representation, $x \xrightarrow{f} f(x)$. The domain of this function is the interval $a < x < b$, and the range is the interval $c < y < d$. The open circles show that the function is not defined at the endpoints $x = a$ and $x = b$ of the interval and, consequently, the endpoints of the curve are not in the graph. Notice that the value $f(x)$ at x can be determined by finding x on the horizontal x-axis, moving up to the graph, and then across to $f(x)$ on the vertical y-axis. The graph is labeled $y = f(x)$ because it consists of all points (x, y) with x in the domain of the function and $y = f(x)$.

Frequently in this book, we will use the symbol $f(x)$ as a name of a function f, to indicate that we are using x as its variable. In such a case the distinction between $f(x)$ as the name of the function and as the value of the function at x will be clear from the context.

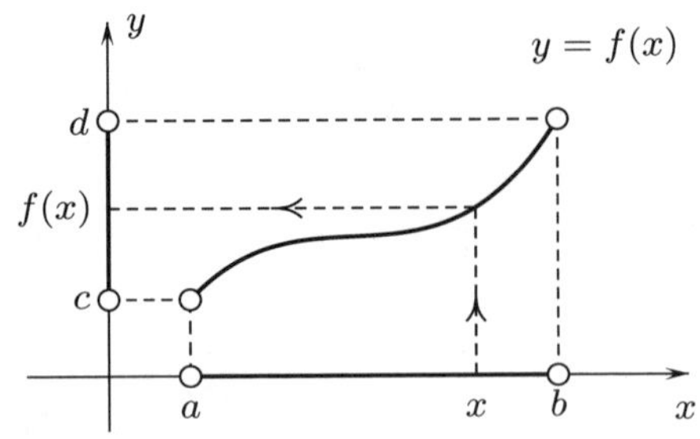

FIGURE 7

Mathematical models

When we use the formula $A = \pi r^2$ to find the area of the face of a coin, we are using a MATHEMATICAL MODEL of the coin's face as a perfect circle. In fact, the face of a coin is not perfectly circular—at least on a microscopic level—but we carry out our calculations as if it were so that we can use the familiar formula for the area of a circle, under the assumption that it yields results that are as accurate as we need.

This mathematical model is used in the following six questions. Take the time to answer each one thoughtfully and then compare your response with that given at the end of the chapter. If you have difficulty with a question, do not look at the response right away. Instead, continue reading and working on other questions and return to those that cause you trouble later, after you have had time to think more about them.

Question 1 What is the area of the face of a coin if its radius is 2 centimeters?

Question 2 The smallest and largest modern Mexican coins are a centavo coined in the 1970's and a ten peso piece from the 1950's (Figure 8).[(1)] The radius of the centavo is 0.65 centimeters and the radius of the ten peso piece is 2 centimeters. Consequently, the radii of all modern Mexican coins lie in the interval $0.65 \le r \le 2$ with r measured in centimeters. What is the smallest interval that contains the areas of the faces of all modern Mexican coins?

FIGURE 8

Question 3 Is the area of a circle increased more by increasing its radius from 1 to 5 feet or by increasing the radius from 12 to 13 feet?

Question 4 How does doubling the radius affect the area of a circle?

[(1)]Data from *A Guide Book of Mexican Coins* by T. Buttrey, Jr. and C. Hubbard, Racine, Wisconsin: Western Publishing Company, Inc., 1969, pp. 173 and 208.

Question 5 How would we change our sketches of the graphs of the radius and area functions in Figures 3 and 4 if in our mathematical model we considered only circles with positive diameters?

Question 6 Find a formula for the area of a circular ring whose inner radius is r and whose outer radius is $r+1$ (Figure 9). This is a LINEAR FUNCTION because its graph is a straight line. Sketch its graph in an rA-plane.

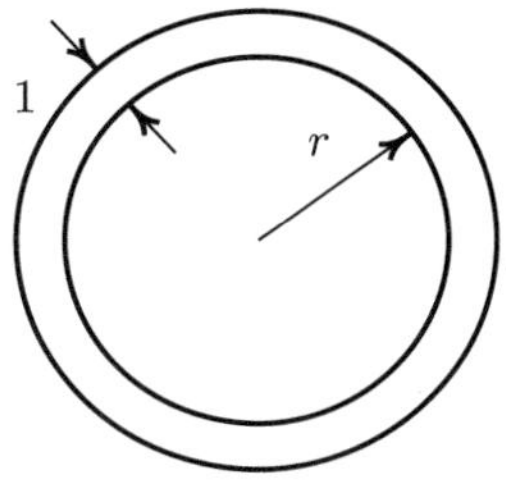

FIGURE 9

Functions given by formulas

If a function, like the function $y = x^2 + 1/x^2$, whose graph is shown in Figure 10, is given by a formula, we will assume that its domain is the set of all values of its variable for which the formula is defined, unless a smaller domain is given explicitly or is determined by the context in which the function is used.

Example 1 **(a)** What is the domain of $y = x^2 + 1/x^2$? **(b)** Use the graph of $y = x^2 + 1/x^2$ in Figure 10 to find its approximate range.

SOLUTION **(a)** Because $x^2 + 1/x^2$ is defined unless $x = 0$, the domain of the function consists of all nonzero x.

(b) The drawing in Figure 10 suggests that the graph extends from its lowest points, at $y = 2$, through all greater values of y. If this is the case, then the range is the interval $y \geq 2$. (With calculus techniques from Chapter 3, we could show this is correct by showing that the lowest points on the graph are, in fact, $(\pm 1, 2)$.) □

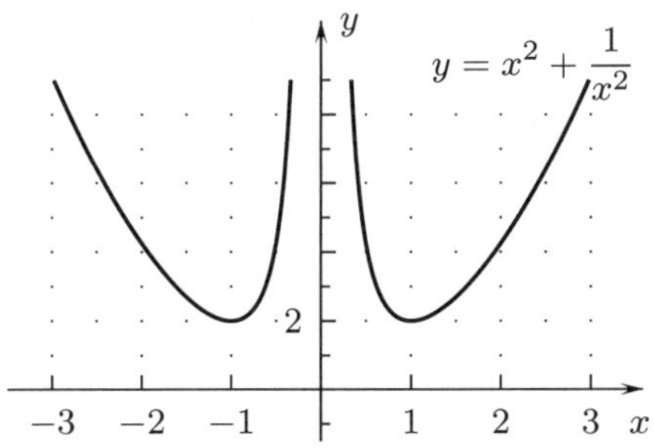

FIGURE 10

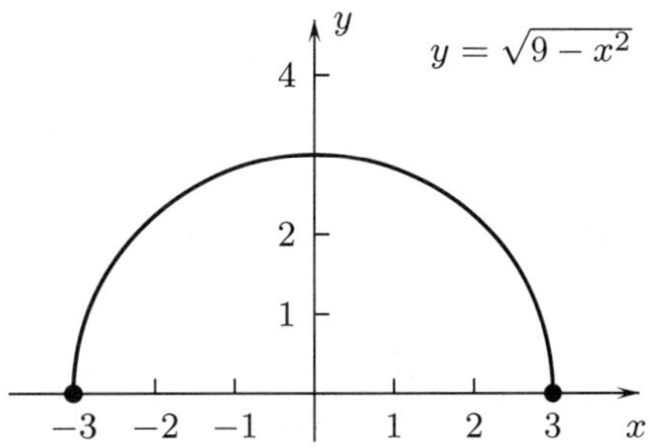

FIGURE 11

Question 7 **(a)** What are the domain and range of $y = \sqrt{9 - x^2}$? **(b)** The graph of $y = \sqrt{9 - x^2}$, drawn in the xy-plane with equal scales on the axes of Figure 11, is the upper half of the circle with the equation $x^2 + y^2 = 9$. Give a formula for the function whose graph is the lower half of the circle.

Functions given by tables of values

Table 1 below gives the amount of corn grown by U. S. farmers each year from 1980 through 1988.[2] We let $B(t)$ denote the number of billions of bushels of corn produced in year t. The table defines the function B for t equal to the nine integers from 1980 to 1988 and with the values $B(1980) = 6.7, B(1981) = 8.1$, etc.

TABLE 1. CORN PRODUCTION BY U.S. FARMERS (BILLIONS OF BUSHELS)

t	1980	1981	1982	1983	1984	1985	1986	1987	1988
$B(t)$	6.7	8.1	8.2	4.2	7.7	8.9	8.2	7.1	4.9

Question 8 What is the value of $B(1986)$ and what does it represent?

Figure 12 is a bar graph of the data in Table 1. To obtain the type of graph that is generally used in calculus, we replace each bar with a dot at its top as in Figure 13. The horizontal axis is broken near the origin in Figures 12 and 13 to indicate that the vertical B-axis is not at $t = 0$.

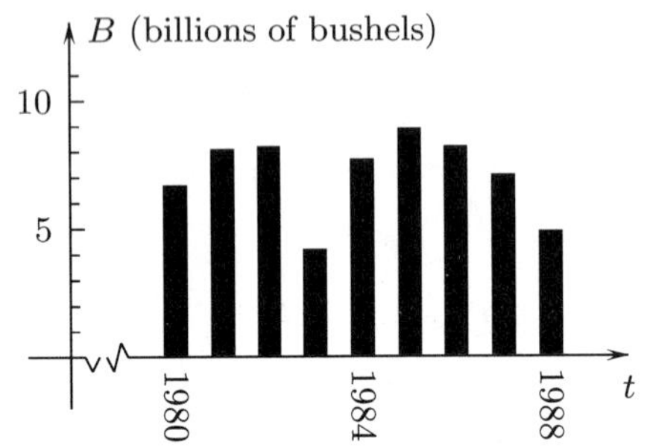

FIGURE 12

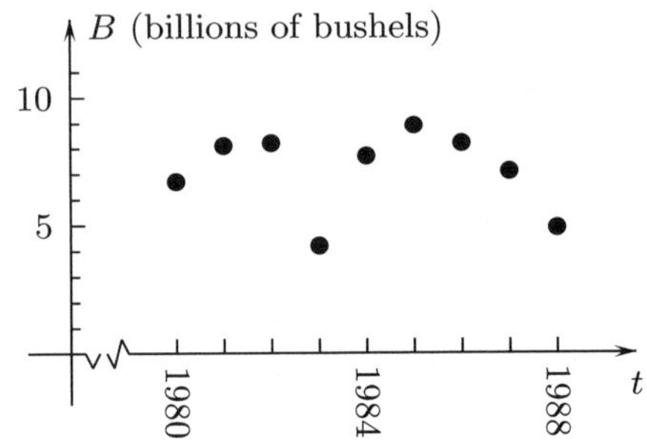

FIGURE 13

Question 9 **(a)** Use Figure 13 to estimate visually the average corn production per year from 1980 through 1988 without using the values in Table 1.
(b) Use Table 1 to calculate the actual average corn production per year from 1980 through 1988.

Question 10 **(a)** In what year from 1980 through 1988 was the most corn produced? How much was produced in that year? **(b)** For what value of t with $1980 \leq t \leq 1988$ does $B(t)$ have its least value? **(c)** What was the total U. S. production of corn in the years 1980 through 1988?

Question 11 Describe one or more ways in which the function $B(t)$ is a mathematical model that does not represent U. S. corn production exactly.

Changes in values of a function

To study the increases and decreases in corn production in Table 1, we consider the change $C(t)$ in production in year t from the production in the previous year. These values for 1981 through 1984 are shown in the Table 2. For instance, from Table 1 we see that the production was 6.7 billion bushels in 1980 and 8.1 billion bushels in 1981, so the change from 1980 to 1981 was $8.1 - 6.7 = 1.4$ billion bushels. The change from 1982 to 1983 equals the 1983 production of 4.2 billion bushels minus the 1982 production of 8.2 billion bushels: $4.2 - 8.2 = -4.0$ billion bushels. This change is negative because production decreased. A graph of the change C in corn production from one year to the next as a function of t is shown in Figure 14.

[2]Data from *Agricultural Statistics, 1991*, Washington, DC: United United States Government Printing Office, 1991, p. 31.

TABLE 2. ANNUAL CHANGES IN CORN PRODUCTION (BILLIONS OF BUSHELS)

t	1980	1981	1982	1983	1984	1985	1986	1987	1988
C(t)		1.4	0.1	−4.0	3.5				

Question 12 **(a)** Complete Table 2 by calculating the change in corn production in the years 1985 through 1988. **(b)** Find the sum of the numbers in the second row of the completed Table 2. **(c)** Show that the sum from part (b) equals $B(1988) - B(1980)$ and explain why this is the case.

Question 13 Why is there no value for 1980 in Table 2 or Figure 14?

Question 14 Give a formula for $C(t)$ in Table 2 in terms of values of the function B defined in Table 1.

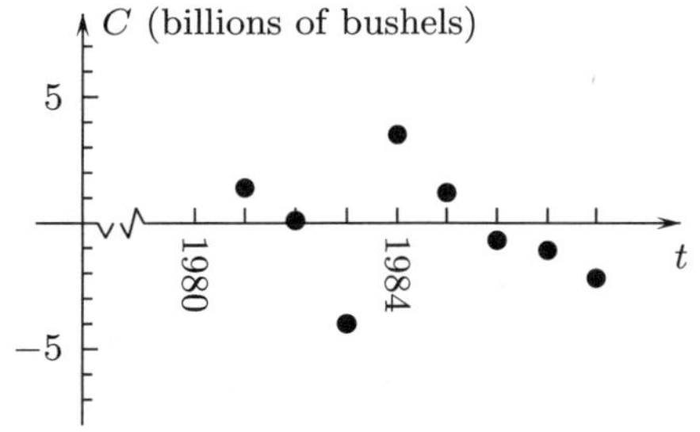

FIGURE 14

Functions given approximately by their graphs

Figure 15 shows the graphs of two functions T and H that were determined from census data collected every ten years by a procedure called LINEAR INTERPOLATION.[3] The functions were defined for all t with $1910 \le t \le 1980$ by plotting the data for 1910, 1920, 1930, ..., 1980 and joining successive points with line segments. In this study, farm workers were divided into two categories: those who were members of the families who owned the farms and those who were hired workers and not family members. In Figure 15, $T(t)$ equals the total number of United States farm workers and $H(t)$ equals the number of those that were hired. Both functions are measured in millions of workers.

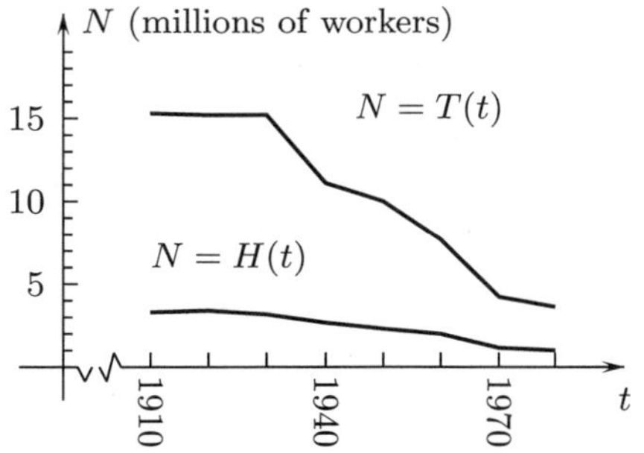

FIGURE 15

Question 15 **(a)** What was the approximate total number of farm workers in the United States in 1940? **(b)** Approximately what percentage of farm workers were hired workers in 1910? **(c)** Approximately how many years did it take after 1910 for the total number of farm workers in the U.S. to be reduced by one-half?

[3] Data adapted from "Labor-intensive Agriculture" by P. Martin, *Scientific American*, New York, NY: Scientific American, Inc., October, 1983, p. 56.

Question 16 Sketch the graph of the number $F(t)$ of millions of family farm workers in the U.S. by plotting approximate values at $t = 1910, 1920, \ldots, 1980$.

Question 17 Describe one or more ways that the functions $T(t)$ and $H(t)$ of Figure 15 are mathematical models that do not represent reality exactly.

Percent and relative changes

Suppose that the price on a TV set you want to buy is raised from \$300 to \$330.

Question 18 What is the percent increase from \$300 to \$330?

You possibly found the answer to Question 18 by subtracting the old price of \$300 from the new price of \$330 to get the change in price, which is \$30. Then you probably divided the change in price by the old price to obtain $\frac{\$30}{\$300} = 0.10$. This is the fraction or proportion by which the price was raised and is known as the RELATIVE CHANGE from \$300 to \$330. Multiplying the relative change of 0.10 by 100 converts it to 10%. These calculations can be made for any change.

Rule 1

$$[\textit{The change from } A \textit{ to } B] = B - A.$$

$$[\textit{The relative change from } A \textit{ to } B] = \frac{B - A}{A}.$$

$$[\textit{The percent change from } A \textit{ to } B] = \frac{100(B - A)}{A}.$$

Example 2 **(a)** What are the relative and percent changes from 10 to 13? **(b)** What are the relative and percent changes from 100 to 103? **(c)** What are the relative and percent changes from 100 to 95?

SOLUTION **(a)** The change from 10 to 13 is 3, so the relative change is $\frac{3}{10} = 0.3$ and the percent change is 30%.
(b) The relative change from 100 to 103 is $\frac{3}{100} = 0.03$; the percent change is 3%.
(c) The relative change from 100 to 95 is -0.05; the percent change is -5%. □

We will often apply Rule 1 to functions, as in the next example.

Example 3 What is the relative change in x^3 when x goes from 2 to 3?

SOLUTION Because x^3 equals 8 at $x = 2$ and equals 27 at $x = 3$, the change is $27 - 8 = 19$, and the relative change is $\frac{19}{8}$. □

Change vs. relative change

The second row of Table 3 lists the amount $U(t)$, measured in billions of dollars, that was invested in utility mutual funds in each of the years from 1986 through 1993.[4] A bar graph of this data is shown in Figure 16. The third row gives the changes in investments from one year to the next, which are plotted in Figure 17.

TABLE 3. UTILITY MUTUAL FUND INVESTMENTS (BILLIONS OF DOLLARS)

Year: t	1986	1987	1988	1989	1990	1991	1992	1993
Investments: $U(t)$	1.6	2.3	2.5	4.9	5.1	7.0	16.3	23.5
Change in $U(t)$		0.7	0.2	2.4	0.2	1.9	9.3	7.2
Relative change in $U(t)$		0.438	0.087	0.960	0.041			
Percent change in $U(t)$		43.8	8.7	96.0	4.1			

[4] *Los Angeles Times*, Los Angeles, CA: The Times Mirror Company, October 4, 1993, source: Lipper Analytical Services.

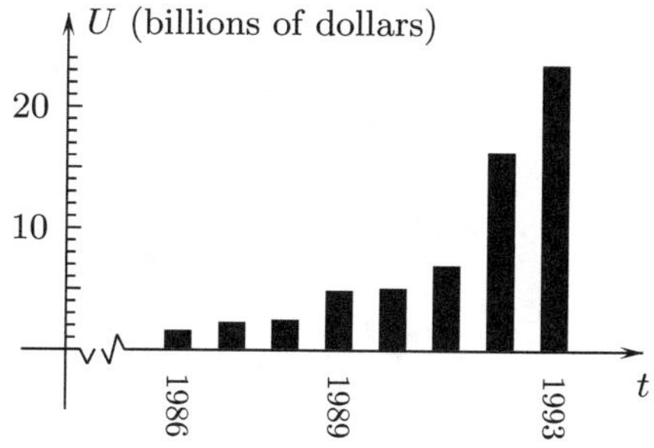

FIGURE 16

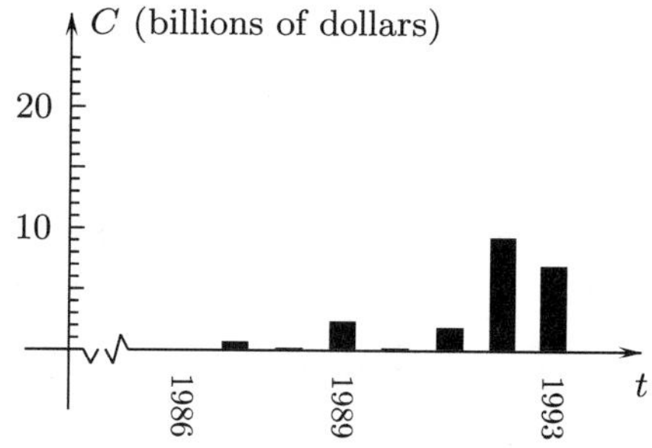

FIGURE 17

Question 19 Complete the last two rows of Table 3 with the relative and percent changes in investments, rounded to three and one decimal places, respectively.

Question 20 Figures 18 and 19 are bar graphs of the relative and percent changes. Why does the overall shape of the bar graph of the change in investments in Figure 17 differ from that of the bar graph of relative change in Figure 18? Why do the bar graphs in Figures 18 and 19 have the same shape?

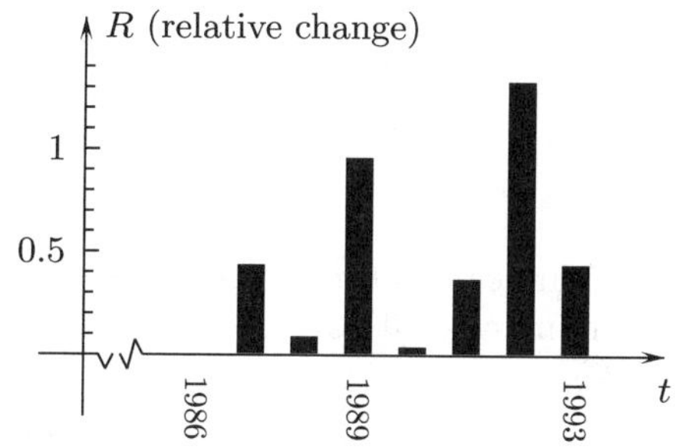

FIGURE 18

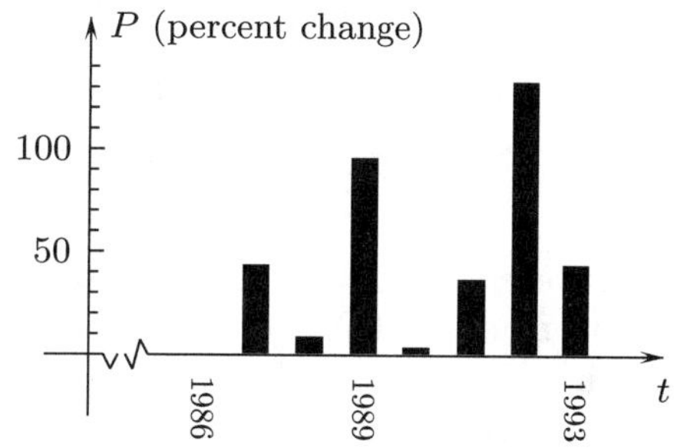

FIGURE 19

Question 21 Give formulas for the relative change $R(t)$ of $U(t)$ and the percentage change $P(t)$ of $U(t)$ as functions of the time t.

Question 22 The investment figures in Table 3 are given in billions of dollars. How would the values of changes and relative changes in investments be affected if the investments were measured in millions of dollars instead?

Finding formulas for mathematical models

One way to obtain a mathematical model is to find a formula that fits or approximately fits empirical data. The next table gives the velocity v (feet per second) of flame through a hydrogen-air mixture as a function of the percent P of hydrogen in the mixture for $P = 5, 15, 25, \ldots, 65$.[(5)] This data is plotted in Figure 20. The velocity is relatively small for $P = 5$ and $P = 15$ because in these cases there is not much hydrogen to burn. The velocity is also relatively small for $P = 65$ because in this case, there is not much oxygen to combine with the hydrogen in the burning process.

TABLE 4. FLAME VELOCITY

P (percent)	5	15	25	35	45	55	65
v (feet per second)	0.8	5.0	11.6	15.7	15.0	8.5	2.9

(5) Data adapted from *Handbook of Engineering Materials*, D. Miner and J. Seastone editors, New York, NY: John Wiley & Sons, 1955, p. 3-322.

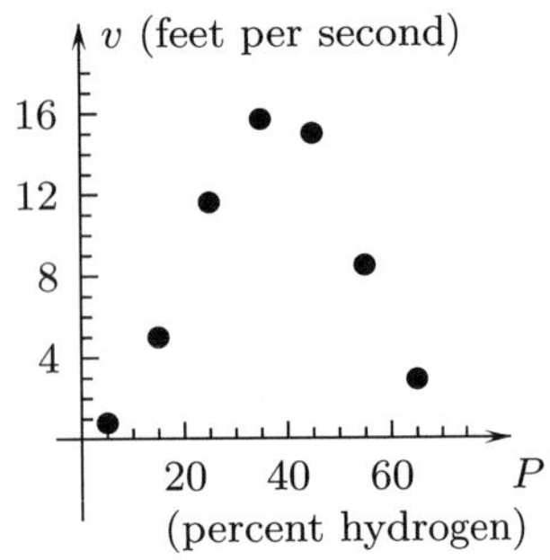

FIGURE 20

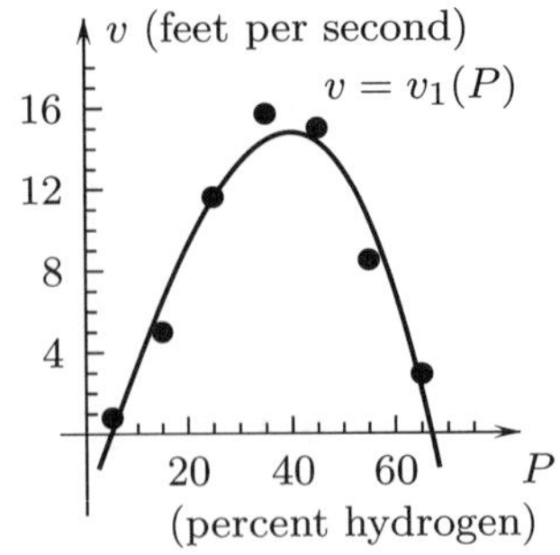

FIGURE 21

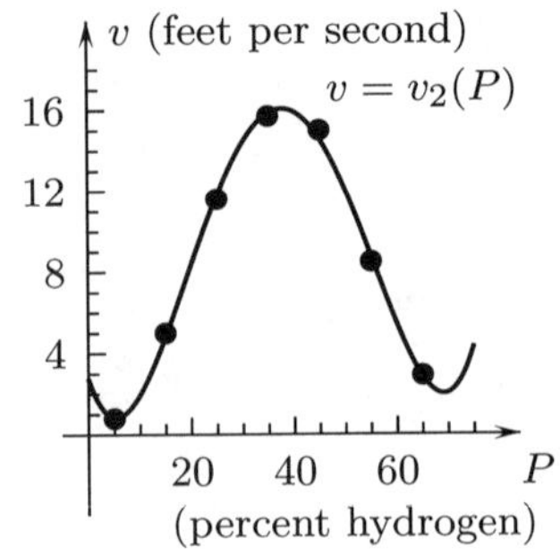

FIGURE 22

If we wanted a formula for calculating approximating flame velocities from percents of hydrogen without referring to the table, we could use a procedure on a graphing calculator or in calculus computer software to find a polynomial that approximates the data in the table. Figures 21 shows, for instance, the graph of a third-degree polynomial,

$$v_1(P) = -(1.3333 \times 10^{-4})P^3 - (7.7381 \times 10^{-4})P^2 + (0.6971)P - 3.3259 \quad \textbf{(1)}$$

and Figure 22 shows the graph of a fourth-degree polynomial

$$\begin{aligned} v_2(P) = \; & (1.4167 \times 10^{-5})P^4 - (2.1167 \times 10^{-3})P^3 + (8.9792 \times 10^{-2})P^2 \\ & - (7.8327 \times 10^{-1})P + 2.7797. \end{aligned} \quad \textbf{(2)}$$

whose coefficients were found with such procedures on a graphing calculator. Notice that the fourth-degree polynomial $v_2(P)$ gives a better approximation of the flame velocity at the values $5, 15, \ldots, 65$ of P in the table than does the third-degree polynomial $v_1(P)$.

We could use either $v_1(P)$ or $v_2(P)$. We might decide to use $v_2(P)$ because it approximates the given data more closely than does $v_1(P)$, or we might use $v_1(P)$ because its formula is simpler and yet it gives results that are as accurate as we need for our particular purposes.

Example 4 **(a)** What is the largest (positive) error that is made if $v_1(P)$ is used to approximate the seven values in Table 4? **(b)** What is the largest error if $v_2(P)$ is used?

SOLUTION **(a)** It appears from Figure 21 that the largest error occurs either at $P = 15, 35$, or 55. To determine which is greatest, we use **(1)** to find the values $v_1(15) \doteq 6.5, v_1(35) \doteq 14.4$, and $v_1(55) \doteq 10.5$ of $v_1(P)$ at the three points and then calculate the absolute values of the differences between these values of $v_1(P)$ and the numbers in Table 4. We obtain

$$\begin{aligned} |v_1(15) - 5.0| &\doteq |6.5 - 5.0| = 1.5 \\ |v_1(35) - 15.7| &\doteq |14.4 - 15.7| = |-1.3| = 1.3 \\ |v_1(55) - 8.5| &\doteq |10.5 - 8.5| = 2.0 \end{aligned} \quad \textbf{(3)}$$

The greatest error using $v_1(P)$ is approximately 2.0 at $P = 55$.

(b) We can see from Figure 22 that the greatest error with $v_2(P)$ is at $P = 45$. Then, since $|v_2(45) - 15.0| \doteq |14.6 - 15.0| = |-0.4| = 0.4$, the greatest error arising from approximating the data in the table by v_2 is about 0.4. □

Question 23 Explain why neither $v_1(P)$ or $v_2(P)$ is a good model of the flame velocity for P close to zero or for $P \geq 70$.

The absolute value function

We used absolute values in each of the calculations in **(3)** because we wanted to determine how far each value of $v_1(P)$ was from the corresponding value in Table 4, without regard to which was greater.

Recall that the ABSOLUTE VALUE FUNCTION can be defined by the formulas,

$$|x| = \begin{cases} x & \text{for} \quad x \geq 0 \\ -x & \text{for} \quad x < 0. \end{cases} \tag{4}$$

The nonnegative number $|x|$ is the distance from x on an x-axis to the origin, and the graph $y = |x|$ has the V-shape in Figure 23. The absolute value can be also be given by the formula,

$$|x| = \sqrt{x^2} \tag{5}$$

since $\sqrt{x^2}$ designates the nonnegative square root of x^2.

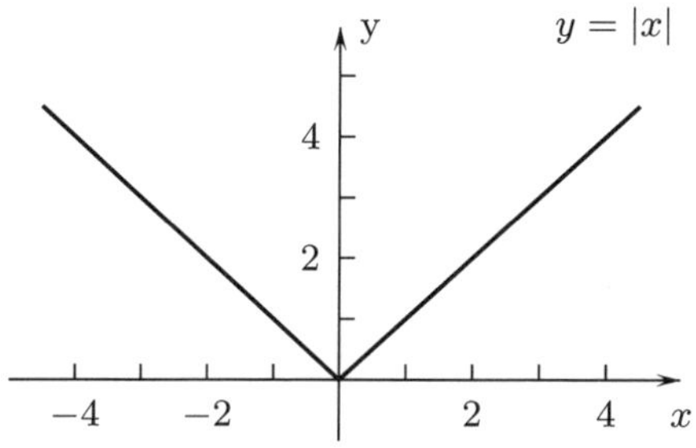

FIGURE 23

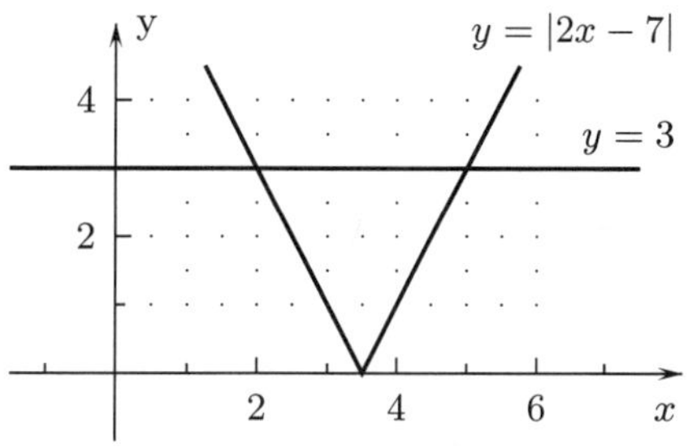

FIGURE 24

Example 5 Solve $|2x - 7| = 3$ for x.

SOLUTION $|2x - 7| = 3$ if $2x - 7 = 3$ or if $2x - 7 = -3$. Adding 7 to both sides of both equations shows that x is a solution if $2x = 10$ or if $2x = 4$, so the solutions are $x = 5$ and $x = 2$. This is illustrated in Figure 24, which shows the curve $y = |2x - 7|$ and the line $y = 3$, which intersect at $x = 2, y = 3$ and at $x = 5, y = 3$. □

Working with inequalities

How does solving inequalities differ from solving equations? We can perform many operations on both sides of an equation without altering its validity. For example, adding a number to both sides of an equation, subtracting a number from both sides, or multiplying or dividing both sides by a nonzero number yields an equivalent equation—one that is valid if and only if the original equation is valid. We have to modify our procedures slightly when we apply these operations on inequalities. Here are some basic rules:

Rule 2 **(a)** *Adding a number to both sides of an inequality, subtracting a number from both sides, or multiplying or dividing both sides by a positive number yields an equivalent inequality.*

(b) *Multiplying or dividing both sides of an inequality by a negative number and reversing the direction of the inequality sign yields an equivalent inequality.*

(c) *Taking an odd power or odd root of both sides of an inequality yields an equivalent inequality.*

(d) *If the numbers on both sides of an inequality are nonnegative, then taking an even power or even root of both sides yields an equivalent inequality.*

Example 6 In answering Question 2 above, you used the fact that circles with larger radii have larger areas to determine the smallest interval that contains the areas of all modern Mexican coins from the smallest interval $0.65 \leq r \leq 2$ containing their radii. Find the same result by using parts (d) and (b) of Rule 2.

SOLUTION Squaring all three positive numbers in the inequalities $0.65 \leq r \leq 2$ gives the equivalent inequalities $(0.65)^2 \leq r^2 \leq 2^2$, which can be rewritten as $0.4225 \leq r^2 \leq 4$. Then multiplying the new numbers by the positive number π yields the equivalent inequalities $0.4225\pi \leq \pi r^2 \leq 4\pi$. We again find that the smallest interval containing the areas A of all modern Mexican coins is $0.4225\pi \leq A \leq 4\pi$. □

Example 7 Solve $6 + 2x \leq 10$ for x (i.e., find an equivalent inequality in which x appears alone and on only one side). Generate $y = 6+2x$ and $y = 10$ together on your calculator or computer in the window $-3 \leq x \leq 3, -3 \leq y \leq 3$ as a partial check of your answer.

SOLUTION Subtracting 6 from both sides of $6 + 2x \leq 10$ gives the equivalent inequality $2x \leq 4$. Then dividing both sides by the positive number 2 yields $x \leq 2$. The solution set is the interval $x \leq 2$. This can be seen in Figure 25, which shows the lines $y = 6 + 2x$ and $y = 10$. □

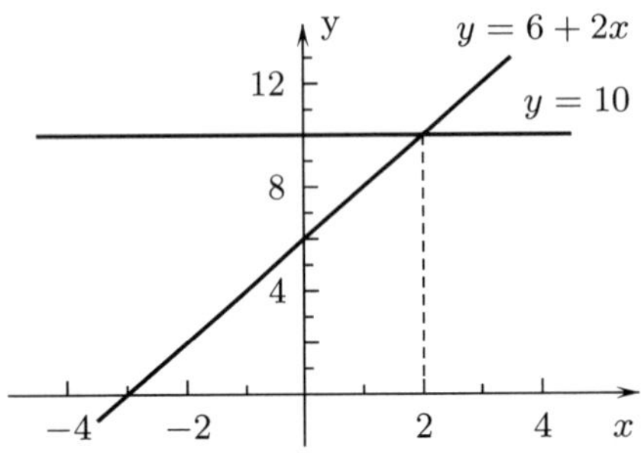

FIGURE 25

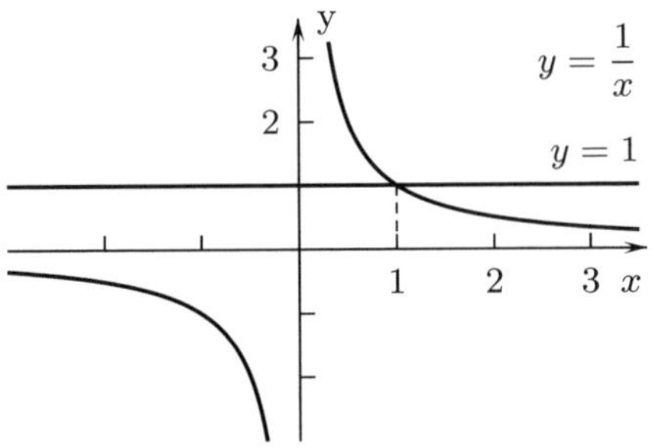

FIGURE 26

Example 8 Solve $\frac{1}{x} \leq 1$ for x. Treat the cases of positive and negative x separately.

SOLUTION First, we observe that $x = 0$ cannot be a solution of the inequality because x cannot be zero in the formula $1/x$. For positive x, multiplying both sides of $\frac{1}{x} \leq 1$ by the positive number x gives the equivalent inequality $1 \leq x$, which we can rewrite as $x \geq 1$. For negative x, we have to reverse the direction of the inequality when we multiply both sides by x, so we obtain $1 \geq x$. This condition automaticallly holds for all $x < 0$. Therefore, the original inequality, $1/x \leq 1$ holds for $x \geq 1$ and for $x < 0$. This can be seen from the graph of $1/x$ and the line $y = 1$ in Figure 26. □

Question 24 Solve the inequalities **(a)** $x^3 > -8$ and **(b)** $x^2 < 16$. (Use the formula $\sqrt{x^2} = |x|$ in part (b).)

The following principles, which can be used to deal with absolute values in many inequalities, are valid because $|f(x)|$ is the distance on a y-axis from $f(x)$ to the origin.

$$\begin{aligned}
&|f(x)| < k \text{ if and only if } -k < f(x) < k \\
&|f(x)| \leq k \text{ if and only if } -k \leq f(x) \leq k \\
&|f(x)| > k \text{ if and only if } f(x) < -k \text{ or } f(x) > k \\
&|f(x)| \geq k \text{ if and only if } f(x) \leq -k \text{ or } f(x) \geq k.
\end{aligned} \qquad \textbf{(6)}$$

Example 9 Solve $|1 - \frac{1}{2}x| > 2$ for x. As a partial check of the answer, generate the curve $y = |1 - \frac{1}{2}x|$ with the line $y = 2$ in the window $-3 \le x \le 9, -0.5 \le y \le 3.5$.

SOLUTION The condition $|1-\frac{1}{2}x| > 2$ holds if and only if $1-\frac{1}{2}x > 2$ or $1-\frac{1}{2}x < -2$. Subtracting 1 from both sides of these inequalities gives the equivalent condition, $-\frac{1}{2}x > 1$ or $-\frac{1}{2}x < -3$. Then multiplying both sides of both inequalities by the negative number -2 and reversing the direction of the inequality shows that x is a solution if $x < -2$ or $x > 6$, as can be seen in Figure 27. We could also work Example 8 by multiplying both sides of $|1 - \frac{1}{2}x| > 2$ by 2 to obtain $|2 - x| > 4$. This is satisfied if and only if the distance between 2 and x is greater than 4. Again we find that x is a solution if $x < -2$ or $x > 6$. □

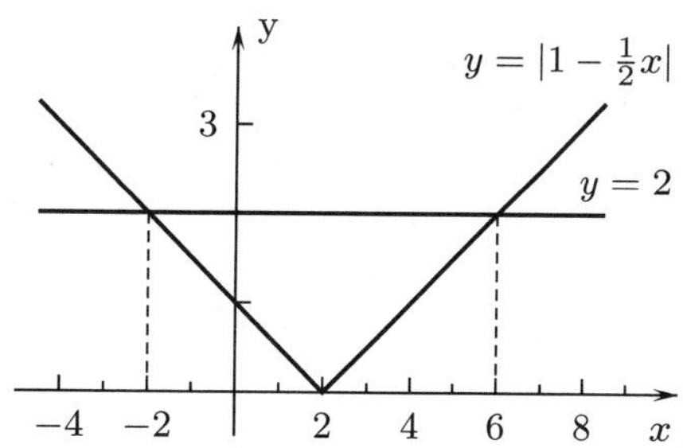

FIGURE 27

Principles and procedures

- The examples in this section illustrate the following formal definition:

Definition 1 *A* FUNCTION *f of x is a rule and a set of numbers. The set of numbers is the* DOMAIN *of the function. The rule assigns a single number, called the* VALUE OF THE FUNCTION *and denoted $f(x)$, to each number x in its domain. The* RANGE *of the function is the set of its values. If we write $y = f(x)$, then x is called the* INDEPENDENT VARIABLE *and y is called the* DEPENDENT VARIABLE. *The* GRAPH *of f in an xy-plane is the curve $y = f(x)$, which is the set of points with coordinates $(x, f(x))$ for all x in the domain.*

- As we have seen, functions can be given by formulas, by lists of values or tables, and approximately by sketches of their graphs.
- We will often use the symbol $f(x)$, which denotes the value of the function f at x, as an alternate name of the function to emphasize that the function's variable is x. In later sections, we will also use the symbol $[f(x)]_{x=a}$ for the value $f(a)$ of f at $x = a$.
- When the formulas in Rule 1 for finding changes, relative changes, and percent changes are applied to values of a function $f(x)$, they read as follows.

$$[\text{The change of } f(x) \text{ from } x_1 \text{ to } x_2] = f(x_2) - f(x_1)$$

$$[\text{The relative change of } f(x) \text{ from } x_1 \text{ to } x_2] = \frac{f(x_2) - f(x_1)}{f(x_1)}$$

$$[\text{The percent change of } f(x) \text{ from } x_1 \text{ to } x_2] = 100\left[\frac{f(x_2) - f(x_1)}{f(x_1)}\right].$$

- Whenever you use functions as models for data and processes, it is important that you recognize how well the models represent the actual data or processes and, in particular, what simplifying assumptions have been made in setting up the mathematical models.

- Equations and inequalities involving an absolute value $|a|$ can usually be handled by considering separately the cases of $a \geq 0$ (where $|a| = a$) and of $a < 0$ (where $|a| = -a$).
- Remember in solving inequalities that you must reverse the direction of the inequality if you multiply or divide both sides by a negative number. If you want to multiply or divide both sides by the variable x or a function $f(x)$, you might have to treat the cases of positive, negative and zero x or $f(x)$ separately.
- Think carefully through the consequences when you consider taking a power or root of an inequality that involves negative numbers. For instance, squaring both sides of the valid inequality $-2 < -1$ would yield the incorrect statement $4 < 0$, and taking square roots of both sides of $-2 < -1$ gives the meaningless statement $\sqrt{-2} < \sqrt{-1}$.
- If, after you have taken square roots or other even roots of both sides of an equation or inequality, your formulas include $\sqrt{x^2}$ or $\sqrt{[f(x)]^2}$ replace $\sqrt{x^2}$ by $|x|$ or $\sqrt{[f(x)]^2}$ by $|f(x)|$.

Tune-Up Exercises 1.1♦

A *Answer provided.* O *Outline of solution provided.*

Be sure you can work the first ten exercises and solve as many of the others as you need for practice. Answers and outlines of solutions are given at the end of the chapter.

T1.A Give the values of $\dfrac{(x-3)(x+4)}{(x+2)(x-1)(x+5)}$ at **(a)** $x = -3$, **(b)** $x = 0$, and **(c)** $x = 3$. **(d)** At what values of x is this function not defined?

♦ Type 1, value of a function

T2.A What are $f(0), f(2), f(-5)$, and $f(4)$ for $f(x) = \dfrac{x^2}{\sqrt{5-2x}}$?

♦ Type 1, value of a function

T3.O What are $K(-2)$, $K(-1)$, $K(0)$, $K(1)$, and $K(100)$ if

$$K(t) = \begin{cases} 2 + 1/t & \text{for } -3 < t < 0 \\ 5 & \text{for } t = 0 \\ -t^2 & \text{for } t > 0? \end{cases}$$

♦ Type 1, value of a function

T4.A **(a)** Give approximate values of $A(1)$ and $A(4)$ for the function $A(x)$ of Figure 28. **(b)** Find the approximate solutions x of $A(x) = 40$ and of $A(x) = 60$ with $0 \leq x \leq 6$. **(c)** Which is greater $[A(2)]^2$ or $A(2^2)$? Explain.

♦ Type 2, values of a function

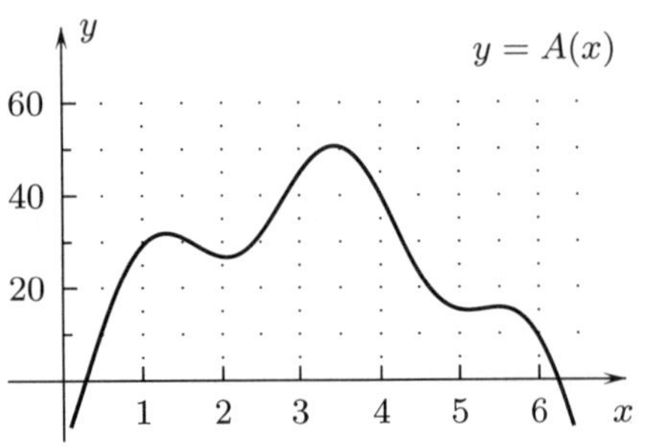

FIGURE 28

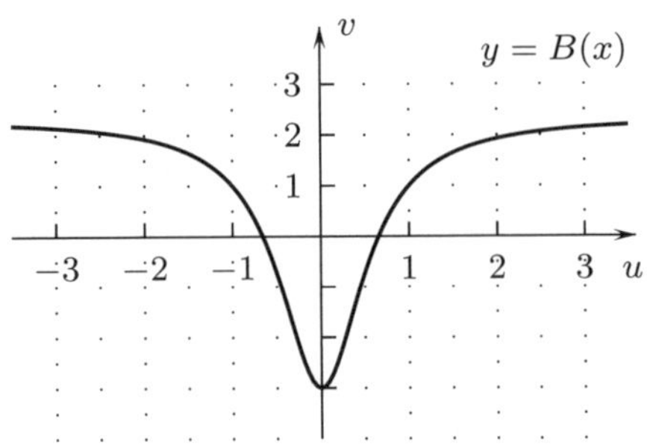

FIGURE 29

♦The Tune-up Exercises and Problems are classified by type and content. The types are (1) basic, reactive; (2) basic reflective; (3) intermediate, reactive; (4) intermediate, reflective; (5) advanced, reactive; (6) advanced, reflective; and (7) advanced, theoretical.

T5.[A] Give a formula without parentheses for $F(-x)$ where $F(t) = \dfrac{1 - t^2 + 3t^3}{10 - t}$.

♦ Type 2, values of a function

T6.[A] Compare $Q(t)$ and $Q(1/t)$ where $Q(x) = x + \dfrac{1}{x} + 2$.

♦ Type 2, values of a function

T7.[A] Give a formula without parentheses for $W(2 + a) - W(2)$ where $W(t) = 5t + 3t^2$.

♦ Type 2, values of a function

T8.[A] What are **(a)** the change, **(b)** the relative change, and **(c)** the percent change from 54 pounds to 60 pounds?

♦ Type 1, change, relative change, percent change

T9.[A] What are **(a)** the change, **(b)** the relative change, and **(c)** the percent change in 2^x from $x = 2$ to $x = 5$?

♦ Type 2, change, relative change, percent change

T10.[O] Solve the inequalities **(a)** $t - 3 \geq 7 - 6t$, **(b)** $|3 - t| \geq 15$, and **(c)** $3t^3 \geq -6$ for t.

♦ Type 3, solving inequalities

More practice

T11.[A] Give **(a)** the approximate values of $B(-2), -2B(-1)$, and $[B(2)]^2$ and **(b)** the approximate solutions u of $B(u) = 1$ for the function B of Figure 29.

♦ Type 1, values of a function

T12.[A] What are $g(-10), g(-3), g(0), g(3)$, and $g(10)$ for $g(x) = -x^2 + 3/x$?

♦ Type 1, values of a function

T13.[A] What are the values of the function h at $x = 1, x = 2$, and $x = 4$, if $h(x) = (4x - x^2)^{-1/2}$?

♦ Type 1, values of a function

T14.[A] What are $f(-33), f(0)$, and $f(78)$ if

$$f(x) = \begin{cases} \sqrt{3 - x} & \text{for} \quad x < 0 \\ \sqrt{3 + x} & \text{for} \quad x \geq 0? \end{cases}$$

♦ Type 1, values of a function

T15.[A] Express $G(t + 1)$ without parentheses, where $G(x) = 8(x - 1)^3 - \dfrac{4}{x - 1} + 10$.

♦ Type 2, value of a function

T16.[A] Express $Z\left(\frac{1}{3}\right)$ as an exact fraction for $Z(x) = x + \dfrac{x}{x + 1}$.

♦ Type 2, value of a function

T17.[A] Give a formula without parentheses for $V(2t^2)$, where $V(x) = x^3 - x^4$.

♦ Type 2, value of a function

T18.[A] What are **(a)** the change, **(b)** the relative change, and **(c)** the percent change from $\frac{3}{10}$ to $\frac{1}{2}$?

♦ Type 1, change, relative change, and percent change

T19.[A] What are **(a)** the change, **(b)** the relative change, and **(c)** the percent change from 1000 to 999?

♦ Type 1, change, relative change, and percent change

T20.[O] Solve the inequalities **(a)** $x + 1 < 5x - 3$, **(b)** $|5x - 4| < 2$, and **(c)** $-x^2 \leq -4$.

♦ Type 3, solving inequalities

Problems 1.1

[A]*Answer provided.* [O]*Outline of solution provided.*

1.[A] The symbols $A(2)$ have one meaning if $A(x)$ is a function of x and another if A represents a constant. Explain.

♦ Type 2, discussion

2.[A] In discussing a function, $x \xrightarrow{f} f(x)$, what symbols are used for **(a)** the variable, **(b)** the values of the function, and **(c)** the coordinates of points on the graph of the function?

♦ Type 2, discussion

3.[A] What are $p(1), p(2)$, and $p(3)$ if the points $(1, 18), (2, -6)$, and $(3, \sqrt{2})$ are on the graph of p?

♦ Type 2, values of a function

4.[A] Which of the following are true for $f(x) = 2x^3$: **(a)** $f(0) = 0$, **(b)** $f(5) = f(-5)$, **(c)** $f(a) + f(-a) = 0$ for all a, and **(d)** $f(a) < 0$ for all $a < 0$?

♦ Type 2, values of a function

5.[A] List in order the steps—adding, multiplying, and finding values of g—that are used to find **(a)** $g(a+b)$, **(b)** $g(a)+g(b)$, **(c)** $g(ab)$, and **(d)** $g(a)g(b)$ from a and b.

♦ Type 2, values of a function

6. How do the steps for calculating $F(a^2)$ for a function F differ from the steps in calculating $[F(a)]^2$?

♦ Type 2, values of a function

7.[A] What is B/A if A and B are positive and the relative change from A to B is 3?

♦ Type 2, relative change

8. Give examples from everyday speech where a change is referred to as "large" or "small" depending on whether a relative change is large or small. (As one example, you might compare inviting ten extra people to a party with increasing the population of a city by ten.)

♦ Type 2, relative change

9.[A] The balance after one year on a deposit of \$1000 that earns $r\%$ annual interest compounded semiannually is $B(r) = 1000 + 10r + \frac{1}{40}r^2$ dollars. **(a)** What is the balance after one year if the interest rate is 8%? **(b)** If the interest rate is 10%? **(c)** Show that $B(r) = 1000(1 + \frac{1}{200}r)^2$.

♦ Type 2, values of a function

10. The density ρ (rho) of dry air at a pressure of one atmosphere and a temperature of T degrees Celsius is $\rho(T) = \dfrac{0.001293}{1 + (0.00367)T}$ grams per milliliter.[6] What is its density **(a)** at 10° Celsius and **(b)** at 50° Celsius? **(c)** Does the density increase or decrease as the temperature increases?

♦ Type 2, values of a function

11.[A] If the radius of a right circular cone is r feet and it is h feet high, then the total surface area of its base and lateral surface is $\pi r^2 + \pi r\sqrt{r^2 + h^2}$ square feet. **(a)** Which part of this formula is the area of the base? **(b)** What is its total surface area if $r = 2$ and $h = 3$? **(c)** Give formulas for the total area $A(h)$ in terms of h for $r = 1, 2$, and 3. Generate the graphs of these three functions on your calculator or computer together in the window $0 \le h \le 10, 0 \le A \le 130$ and copy them on your paper. Describe how increasing r changes the shape of the graph.

♦ Type 2, functions with parameters and graphs

12. As is stated above, the domain of a function given by an expression in x is assumed to consist of all x for which the expression makes sense, unless a different domain is specified. What are the domains of **(a**[A]**)** $y = \sqrt{x+1}$, **(b)** $y = \sqrt[3]{x-1}$, and **(c)** $y = \dfrac{1 - 5x^2}{5x^2}$ and why? Generate the graphs separately in the window $-2 \le x \le 3$, $-2 \le y \le 2$ to guide your reasoning.[†]

♦ Type 2, domains and graphs

Use your calculator or computer as needed in Problems 13 and 14.

13.[A] Find the exact and approximate decimal values of **(a)** $F(10)$, **(b)** $G(3)$, **(c)** $H(2)$, and **(d)** $\sqrt{F(2)}$ where $F(x) = x^4 - 3x^2 + 1$, $G(x) = xF(x)$, and $H(x) = F(\sqrt{x})$.

♦ Type 1, approximate values

14. Find the exact and approximate values of **(a)** $p(3)$, **(b)** $p(3^2)$, and **(c)** $[p(3)]^2$ for $p(x) = \dfrac{\sqrt{x}}{x-2}$. **(d)** What is the domain of p and why? (Generate its graph in the window $-1 \le x \le 5$, $-8 \le y \le 8$.)

♦ Type 2, exact and approximate values and a domain

(6)

---

[6] *C R C Handbook of Chemistry and Physics*, R. Weast and M. Astle, editors, Boca Raton: C R C Press, 1981, p. F-11.

[†] Whenever you are asked in this text to generate a curve, generate it on your calculator or computer.

15.[A] Figure 30 shows the graph of the velocity $v(t)$ of a woman's car t seconds after she leaves a parking space. She stops at a traffic light and then goes on the freeway. Give approximate answers to the following questions: **(a)** How long does she wait at the light? **(b)** The speed limit is 55 miles per hour, and her trip is interrupted when she is stopped and given a speeding ticket. When do you imagine she sees the highway patrolman who stopped her? **(c)** During what time interval does she exceed the speed limit on the freeway? **(d)** During part of the short drive between the parking space and the traffic light, she drives at a constant speed. What is it? How far does she travel during that time? (Remember that an hour consists of $60 \times 60 = 3600$ seconds.)

♦ Type 4, interpreting a graph

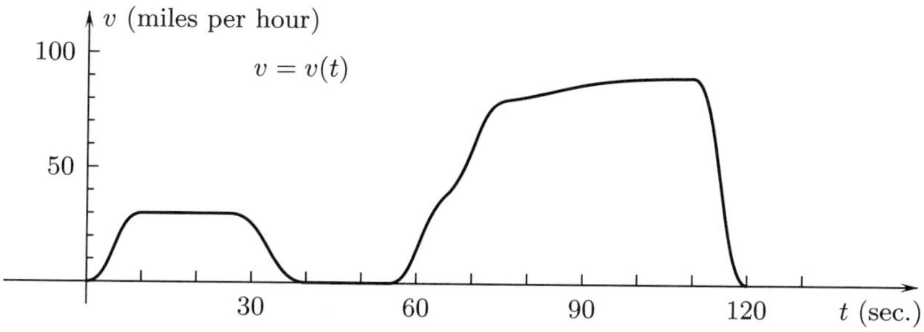

FIGURE 30

16.[A] The table below lists the low and high temperatures in Bismark, North Dakota during five 24-hour periods in May, 1994.[7] Draw a plausible graph of the temperature $T(t)$ in Bismark from 5 PM, May 1, to 5 PM, May 6 as a function of the time t, measured in hours. Suppose that $T(t)$ is defined for all t with $0 \le t \le 120$ (hours) with $t = 0$ at 5 PM, May 1, $t = 24$ at 5 PM, May 2, etc. Your graph should be consistent with the lows and highs in the table and should take into account likely differences in temperature at different times of the day and night.

♦ Type 2, making a mathematical model

TEMPERATURE IN BISMARK, NORTH DAKOTA (°F)

	Low	High
5 PM, May 1 to 5 PM, May 2	28°	56°
5 PM, May 2 to 5 PM, May 3	33°	73°
5 PM, May 3 to 5 PM, May 4	37°	60°
5 PM, May 4 to 5 PM, May 5	39°	58°
5 PM, May 5 to 5 PM, May 6	42°	54°

17.[A] A rubber ball is dropped from a height of 45 inches above a concrete sidewalk. It bounces straight up and falls back down a number of times. Figure 31 shows the graph of the height $h(t)$ of the ball above the sidewalk as a function of the time t (seconds) from $t = 0$, when it is dropped, until it hits the ground for the fourth time. In this mathematical model there is no air resistance and the height that the ball reaches in every bounce is a fixed fraction of the height it reached in the previous bounce. (Energy is lost as heat through the compression and expansion of the ball whenever it hits the sidewalk.) **(a)** By approximately what fraction is the height that the ball reaches decreased with each bounce? **(b)** Approximately when does the ball hit the sidewalk for the fourth time? **(c)** Approximately how far does the ball travel from $t = 0$ to when it hits the sidewalk for the fourth time? **(d)** Describe some likely or possible defects in this mathematical model of a bouncing ball.

♦ Type 4, interpreting a graph

[7] Data from the *Los Angeles Times*, Los Angeles, CA: The Times Mirror Company, May 3, 1994, p. B5.

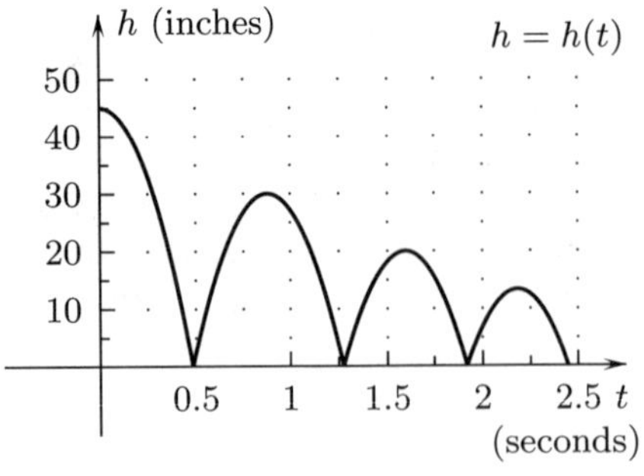

FIGURE 31

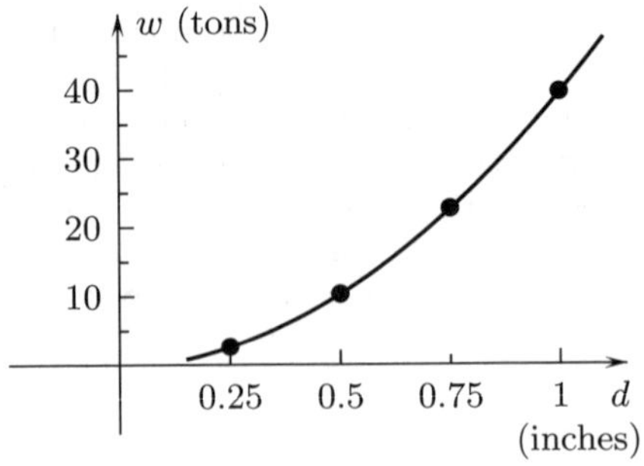

FIGURE 32

18. The table below lists the amount of weight that can be supported by wire ropes of several diameters.[8] This data is plotted as dots in Figure 32, which also shows the graph of a polynomial $w = 37d^2 + 3d - 0.5$ that is to be used as a model for the weight w (tons) the wire can support as a function of its diameter d (inches). **(a)** How large are the errors that would be made if the formula were used instead of the table to determine w for $d = 0.25, 0.5, 0.75$, and 1? **(b)** Does the formula give values for w that are less than or greater than those in the table? **(c)** Why is the polynomial not a good model for the strength of wire ropes of very small diameters?

♦ Type 4, analyzing a mathematical model

Diameter d (inches)	0.25	0.5	0.75	1
Maximum load w (tons)	2.64	10.3	22.7	39.7

19. Tim jumps out of bed at 8:00 AM Saturday morning and rushes into the family room to watch cartoons on television. Every ten minutes or so until 9:00 AM he goes to the kitchen for some food and returns to the TV. At 9:00 he goes back to bed. Draw a possible floor plan for his house and use it to sketch the graph of what might be his distance from bed as a function of time between 8:00 and 9:00.

♦ Type 2, making a mathematical model

20. A small potato at room temperature was put in an oven then taken out and left to cool. The graph of its temperature as a function of time is shown in Figure 33. **(a)** What was the approximate temperature of the room? **(b)** The temperature of the oven was either 120° F, 190° F, or 350° F. Which is possible? Explain. **(c)** When was the potato taken out of the oven? **(d)** Suppose a much larger potato is used. Trace the curve in Figure 21 and sketch in the same drawing a plausible graph of the temperature of the larger potato. Explain how and why the two graphs differ.

♦ Type 4, analyzing a mathmatical model

FIGURE 33

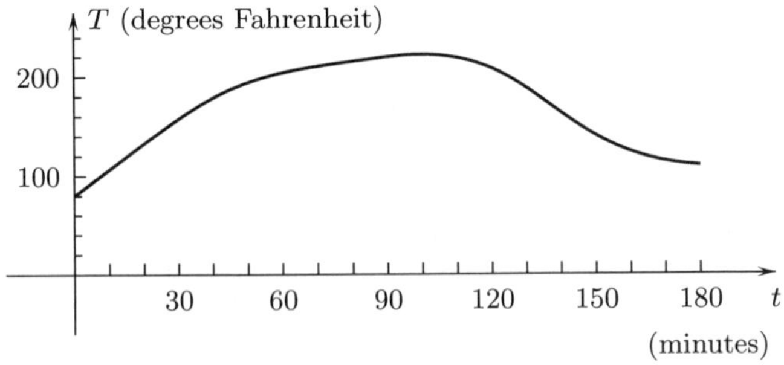

(8) Data from *Handbook of Engineering Materials* by D. Miner and J. Seastone, New York, NY: Wiley, 1955, p. 4-115.

21.[A] The world's human population at the beginning of each century from 100 AD to 1900 AD is given below with the projected population for 2000 AD, all rounded to the nearest 10 million.[9] Figure 34 is a bar graph of the population. **(a)** Calculate the changes and relative changes in population from 200 AD to 1000 AD, from 1000 AD to 1800 AD, and from 1800 AD to 2000 AD. **(b)** Find the change and relative change in population from 200 AD to 2000 AD. **(c)** One of the answers in part (b) is a sum of answers from part (a) and the other is not. Which is which and why? **(d)** In which century did the population decrease and why?

◆ Type 4, sums of change and relative change

WORLD POPULATION (MILLIONS)

Year A.D.	100	200	300	400	500	600	700	800	900	1000
Population	180	190	190	190	190	200	210	220	230	240

Year A.D.	1100	1200	1300	1400	1500	1600	1700	1800	1900	2000
Population	300	360	360	350	415	545	610	900	1630	6150

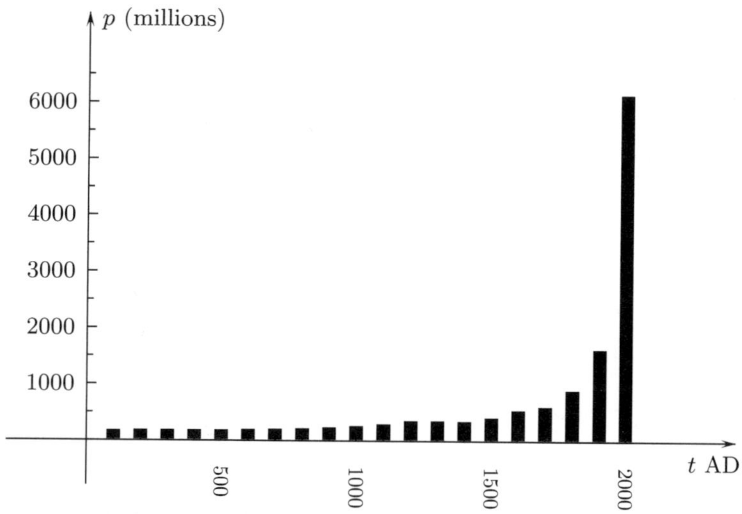

FIGURE 34

22.[A] The next table gives the number of registered political action committees (PAC's) and the amounts they contributed to candidates for the U. S. Congress during the campaigns for the 1974 through 1988 elections.[10] The contributions have been rounded to the nearest thousand dollars. **(a)** What were the least and greatest increases in the number of PAC's from one election to the next during these years? **(b)** What were the least and greatest increases in the amounts contributed by PAC's? **(c)** What was the average amount contributed per PAC in 1974? In 1988?

◆ Type 2, analyzing data in a table

POLITICAL ACTION COMMITTEES, 1974–1988

Year	1974	1976	1978	1980	1982	1984	1986	1988
Number of PAC's	608	1146	1653	2551	3371	4009	4157	4268
Total contributions (million dollars)	12.527	22.572	34.121	55.217	86.620	105.330	132.671	147.898

[9] Data adapted from *World Health and Disease* by A. Gray, Buckingham: Open University Press, 1993, p. 49.
[10] Data adapted from *Vital Statistics on Congress, 1989–1990* by N. Ornstein, T. Mann, and M. Malbin, Washington, DC: Congressional Quarterly Inc., 1990, p. 99.

23. The total value of imports from Mexico to the United States and of exports from the United States to Mexico from 1987 through 1992 are listed below.[11] **(a)** In which of these years did the United States have the greatest trade deficit with Mexico? In which year did Mexico have the greatest trade deficit with the United States? **(b)** Was there a greater increase from 1987 to 1992 in U.S. imports from Mexico or in U.S. exports to Mexico? **(c)** What were the average values of imports per year from Mexico and exports to Mexico in the entire six years?

♦ Type 2, analyzing data in a table

IMPORTS FROM AND EXPORTS TO MEXICO (BILLIONS OF DOLLARS)

Year	1987	1988	1989	1990	1991	1992
U.S. imports from Mexico	20.2	23.2	27.2	30.2	31.2	35.2
U.S. exports to Mexico	14.6	20.6	24.9	28.4	33.3	40.6

24. The table below shows the numbers of games played and won by the major league pitcher Sandy Koufax and his earned-run average (ERA) each year during his major league career with the Brooklyn Dodgers from 1955 through 1957 and with the Los Angeles Dodgers from 1958 through 1966. A pitcher's ERA each year is the average number of earned runs opponents score per nine innings in which he pitched.[12] **(a)** What were Koufax's maximum number of wins and minimum ERA and in what years did they occur? **(b)** What was Koufax's average number of wins per year from 1955 through 1966? **(c)** How many earned runs did opponents score in 1961? In 1962? **(d)** What was Koufax's earned-run average for the two years 1961 and 1962 combined? **(e)** What is the average of Koufax's ERA's for 1961 and his ERA for 1962, and how is it possible that this average is not equal to the result of part (d)?

♦ Type 4, analyzing data and an average of averages

Year	Games	Wins	ERA	Year	Games	Wins	ERA
1955 (NY)	$4\frac{17}{27}$	2	3.02	1961 (LA)	$28\frac{11}{27}$	18	3.52
1956 (NY)	$6\frac{14}{27}$	2	4.91	1962 (LA)	$20\frac{13}{27}$	14	2.54
1957 (NY)	$11\frac{16}{27}$	5	3.88	1963 (LA)	$34\frac{5}{9}$	25	1.88
1958 (LA)	$17\frac{17}{27}$	11	4.48	1964 (LA)	$25\frac{8}{9}$	19	1.74
1959 (LA)	$17\frac{1}{27}$	8	4.05	1965 (LA)	$37\frac{8}{27}$	26	2.04
1960 (LA)	$19\frac{4}{9}$	8	3.91	1966 (LA)	$35\frac{8}{9}$	27	1.73

25.[A] The domain of function $G(x)$ consists of the integers $x = 1, 2, 3, 4$. Its values satisfy the conditions $G(1) = 2, G(2) = 3G(1) + 1, G(1) + G(2) + G(3) = 0$, and $G(4) = 2G(3)$. Find its four values and sketch its graph.

♦ Type 4, values of a function

26.[A] A function Q has the values $Q(5) = 7$ and $Q(10) = 21$. What are **(a)** $10Q(10)$, **(b)** $Q(6 + 3 + 1)$, **(c)** $Q(100/20)$, **(d)** $Q(10)/Q(5)$, and $Q(Q(5) + 3))$?

♦ Type 4, values of a function

[11] Data adapted from "One market for all?" by D. Forde, *Américas*, Vol. 45, Number 5, Washington, DC: Organization of American States,1993, p. 36.

[12] Data adapted from *The Baseball Enclyclopedia*, J. Reichler, editor, New York, NY: Macmillan Publishing Co., Inc., 1979, p. 1829. A starting pitcher records a win if his team goes ahead when he is the active pitcher, if he pitches at least five complete innings, and if his team remains ahead for the rest of the game. A run is earned if it is scored before the defense has the opportunity to retire the batting team. (Major League Baseball Rules 10.18 and 10.19.)

27.[A] Figure 35 is a bar graph of the number of failures $N(t)$ of national savings and loan institutions in year t from 1980 through 1990.[13] **(a)** What was the approximate greatest decrease in $N(t)$ from one year to the next during this time period? **(b)** How many years did it take for $N(t)$ to go from its minimum value to its maximum value?

♦ Type 2, analyzing a bar graph

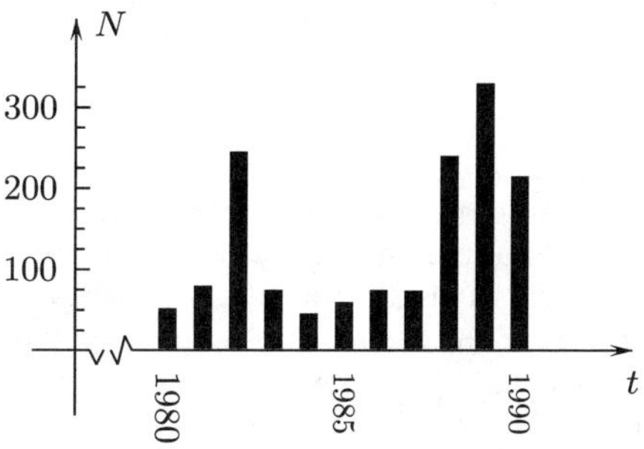

FIGURE 35

28.[A] The graph of a function $W(x)$ is formed by joining the successive points $(1,1)$, $(2,5)$, $(3,2)$, and $(5,2)$ in an xy-plane by straight lines. **(a)** What is the domain of $W(x)$? **(b)** Draw its graph. **(c)** What are $W(1), W(3), W(4)$, and $W(4.5)$?

♦ Type 2, drawing and analyzing a graph

29. The average number of cigarettes $C(t)$ smoked per person in the United States is given below for 1945 and every sixth year until 1987.[14] **(a)** What are the change and relative change of $C(t)$ from 1945 to 1963? **(b)** What are the change and relative change of $C(t)$ from 1963 to 1987? **(c)** $C(t)$ has been defined for all t between 1945 and 1987 in Figure 36 by joining the points given in the table with line segments. Find $C(1953)$. This is the average number of cigarettes smoked per capita in 1953 according to the mathematical model in Figure 36. In fact, the average that year was 3500 cigarettes per person. What error is made by using $C(1953)$? What is the relative error (the error divided by the correct value)? **(d)** In Figure 36 the numbers on the vertical axis run from 3000 to 4500. Figure 37 shows the same graph with all of the vertical axis shown from 0 to 4500. Both graphs give good representations of changes in $C(t)$ but the second graph is better for representing relative changes. Explain.

NUMBER OF CIGARETTES SMOKED PER CAPITA, 1945–1987

Year	1945	1951	1957	1963	1969	1975	1981	1987
Cigarettes smoked	3500	3700	3700	4300	4000	4200	3900	3300

♦ Type 4, analyzing a table and a graph

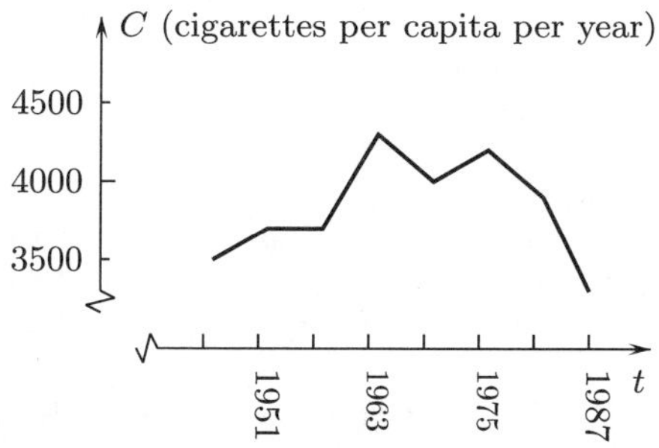

FIGURE 36

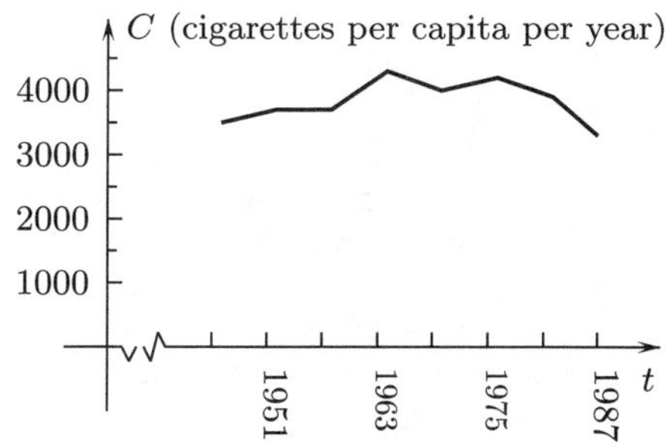

FIGURE 37

[13] Data adapted from *Los Angeles Times*, Los Angeles, CA: The Times Mirror Company, January 2, 1991. Sources: Office of Thrift Supervision and the Resolution Trust Corp.

[14] Data adapted from *Drugs, Society, & Human Behavior* by O. Ray and C. Ksir, St. Louis: Times Mirror/Mosby College Publishing, 1990, p. 200.

30.[A] What is the change from $F(4)$ to $F(10)$ if $F(4) = 6$ and the relative change from $F(4)$ to $F(10)$ is 0.75? What is $F(10)$ under these conditions?

◆ Type 4, change and relative change

31. What is $H(3)$ if $H(0) = 3$, the change in $H(t)$ from $t = 0$ to $t = 1$ is 7, the relative change in $H(t)$ from $t = 1$ to $t = 2$ is -0.5, and the change in $H(t)$ from $t = 2$ to $t = 3$ is -1?

◆ Type 4, change and relative change

32. Suppose that $P(50) = 2P(1)$ and $P(100) = 3P(50)$. How do you obtain $P(100)$ from $P(1)$? What is the relative change from $P(1)$ to $P(100)$?

◆ Type 4, change and relative change

33. The following table gives results from experiments run to determine to what extent expert typists look ahead to characters they have not yet reached as they type.[15] Each typist in the study read a screen which showed a fixed number p of characters beyond the one he or she was typing. The typing speeds were averaged to obtain the values of $t(p)$ in the second row of the table. For instance, $t(2) = 0.450$ because it took the typists an average of 0.450 seconds to type a character when they were shown two characters ahead of the one they were typing. Figure 38 is a graph of t as a function of p. **(a)** Why is it plausible that $p(2) < p(1)$ and $p(3) < p(2)$? **(b)** Why is it plausible that $p(6), p(9), p(12)$, $p(15)$, and $p(18)$ are approximately equal? **(c)** What would be the likely value of $p(25)$? Why? **(d)** Suppose you could use only the table or the graph of typing speeds in a presentation. What would be an advantage of each method of displaying the data?

◆ Type 4, analyzing a table and a graph

TYPING SPEEDS

p (characters seen ahead)	1	2	3	6	9	12	15	18
$t(p)$ (seconds per character)	0.620	0.450	0.280	0.20	0.191	0.189	0.190	0.192

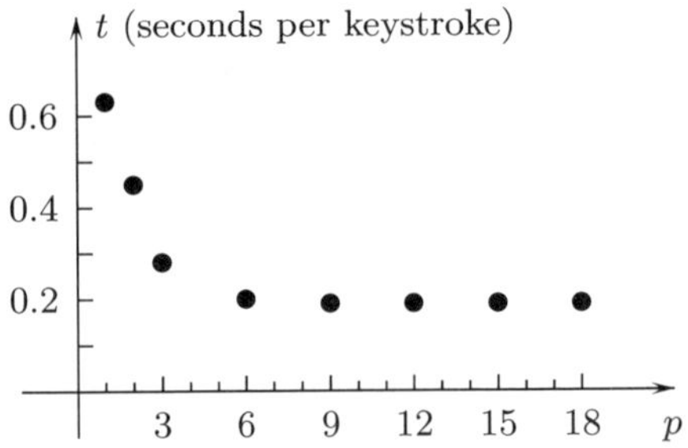

FIGURE 38

34. Your new boss says he will not pay you anything for your first day on the job, but will pay you \$120 for your second day, and in each subsequent day will pay you the average of your previous two days' salaries. Let $S(t)$ be your salary in day t. Find the values of $S(t)$ and sketch its graph for $t = 1, 2, 3, \ldots, 7$.

◆ Type 4, values and graph of a function

35. Light, radio signals, microwaves, X-rays, ultraviolet light, infrared rays, and gamma rays are all electromagnetic waves. An electromagnetic wave with a single frequency ν oscillates ν times each second. In practice, electromagnetic waves with different frequencies occur together with the range of frequencies determined by the source of the electromagnetism. The approximate frequencies of various types of electrogmagnetic waves are listed below.[16] **(a)** What are the change and relative change in frequency from visible red light of frequency 4×10^{14} cycles per second to visible violet light of frequency 7.5×10^{14} cycles per second? **(b)** What are the change and relative change in frequency from radar of frequency 10^{10} cycles per second to X-rays of frequency 1.5×10^{19} cycles per second?

◆ Type 4, analyzing a table

(15) Data from "The Skill of Typing" by T. Salthouse, *Scientific American*, New York, NY: Scientific American, Inc., February, 1984, p. 129.

(16) Data from *Encyclopædia Brittanica*, 15th Edition, Vol. 6, Chicago, IL: Encyclopædia Brittanica, Inc.,1979, p. 645.

FREQUENCIES OF ELECTROMAGNETIC RADIATION

TYPE	FREQUENCY	TYPE	FREQUENCY
Radio waves	10^2–10^8	Visible yellow light	$\approx 5.5 \times 10^{14}$
Radar	10^9–10^{11}	Visible violet light	$\approx 7.5 \times 10^{14}$
Microwaves	10^{11}–10^{13}	Ultraviolet radiation	10^{15}–10^{16}
Infared radiation	10^{13}–10^{14}	X-rays	10^{19}–10^{20}
Visible red light	$\approx 4 \times 10^{14}$	Gamma rays	10^{21}–10^{28}

36. The next table lists the distances ridden by passengers in private automobiles and in trains and busses in Japan in the years 1960, 1965, 1970, 1975, and 1980.[(17)] Let $f(t)$ denote the number of kilometers ridden in private cars and $g(t)$ the number in trains and busses in year t. Calculate $A(t) = \dfrac{100f(t)}{f(t)+g(t)}$ for each of the years in the table. What does $A(t)$ represent?

TRAVEL IN JAPAN, 1960–1980 (HUNDRED MILLION KILOMETERS)

Year	Travel by private cars	Travel by trains and busses
1960	1,610	18,581
1965	4,306	26,355
1970	12,221	28,196
1975	17,681	28,319
1980	23,612	27,900

◆ Type 4, interpreting a table

37.[A] Define $F(0), F(1), F(2), F(3), F(4)$, and $F(5)$ so that the relative change of $F(n)$ to $F(n+1)$ is 3 for $n = 0, 1, 2, 3$, and 4.

◆ Type 4, defining a function from relative changes

38. Define a function $f(x)$ with positive values for $x = 1, 2$, and 3 such that its change from $x = 1$ to $x = 2$ is less than its change from $x = 2$ to $x = 3$, but the relative change in $f(x)$ from $x = 1$ to $x = 2$ is greater than its relative change from $x = 2$ to $x = 3$.

◆ Type 4, defining a function from relative changes

39. Is it possible for a function that has only positive values to have a relative change from one value to another equal to -1.5? Explain.

◆ Type 2, relative change

40.[A] **(a)** You are earning \$1000 per week, when your boss decides to give you a 50% raise. The next day he has second thoughts and gives you a 50% cut in pay, thinking he has restored your earlier salary. What is your actual new salary? **(b)** You complain, and your salary is put back to \$1000 per week. Six months later, business is looking up, so your boss gives you a 20% raise. But then, a week later, he decides that was a mistake and, forgetting his earlier error, thinks he will restore your \$1000 salary by giving you a 20% pay cut. What is your latest salary instead?

◆ Type 2, percent change

41. **(a)** What is your final salary if, starting with \$1000 per week, you get first a $P\%$ raise and then a $P\%$ cut in pay? (Give a formula for your final salary as a function of P.) **(b)** What is your final salary if, starting with \$1000 per week, you get first a $P\%$ pay cut and then a $P\%$ raise? **(c)** Compare the results of parts (a) and (b).

◆ Type 4, percent change

[(17)] Data from "Japan: the late starter who outpaces all her rivals" by K. Shimokawa, *The Economic and Social Effects of the Spread of Motor Vehicles*, London: Macmillan Press, Ltd., 1987, pp. 226–227.

In Problems 42 through 67 find exact solutions algebraically, showing all of your steps. As a partial check of your conclusion generate the graphs of the relevant functions in suitable windows.

42.[A] Find the intersection of the lines $y = x + 1$ and $y = 5 - 2x$.
♦ Type 3, solving an equation

43.[A] Find the solutions x of $|3x + 6| = 12$.
♦ Type 3, solving an equation

44.[A] Solve $x^2 = 4|x|$.
♦ Type 3, solving an equation

45. Find all solutions x of $x^3 = |x|^3$.
♦ Type 3, solving an equation

46. Find all solutions x of $3|x| - 4 = x$.
♦ Type 3, solving an equation

47. Find the intersections of $y = x^3$ and $y = 4x$.
♦ Type 3, solving an equation

48.[O] Solve $-5 < 3x + 4 < 7$ for x.
♦ Type 3, solving an inequality

49.[O] Solve $x + 3 < 6x + 10$ for x.
♦ Type 3, solving an inequality

50. Solve $2 < 5 - 3x \leq 4$.
♦ Type 3, solving an inequality

51.[O] Solve $|6 - 2x| \geq 20$.
♦ Type 3, solving an inequality

52. Solve $|5x - 4| < 2$.
♦ Type 3, solving an inequality

53.[A] Solve $|x + 2| = |3x - 4|$.
♦ Type 3, solving an equation

54. Solve $x^4 - 2x^2 = 2$. (Use the quadratic formula.)
♦ Type 4, solving an equation

55.[A] Solve $\dfrac{3+x}{x} < 2$.
♦ Type 4, solving an inequality

56. Solve $|x^2 + x| = 1$
♦ Type 4, solving an equation

57. For what values of x does $Z(x) = 1 + |x^2 - 1|$ have the value 1?
♦ Type 4, solving an equation

58. For what values of x is $W(x) = x^4 - 4x^2 + 3$ negative?
♦ Type 4, solving an inequality

59.[A] Solve $\dfrac{1}{t^2 - 1} < 1$ for t.
♦ Type 4, solving an inequality

60. Solve $\dfrac{1}{t} \leq t^2$ for t.
♦ Type 4, solving an inequality

61. For what values of x is $|x^2 - 2x - 3|$ nonpositive?
♦ Type 4, solving an inequality

62. Solve $x + 5 < 3x < 4x - 4$.
♦ Type 4, solving an inequality

63. Solve $1 \leq 4 - 2x < 2 + 3x$.
♦ Type 4, solving an inequality

64. Solve **(a)** $16 - 2x^3 > 0$, **(b)** $x^4 < 16$, **(c)** $1 < \sqrt{x} \leq 4$, and **(d)** $\sqrt[3]{x} \geq \frac{1}{2}\sqrt[3]{x} - 5$.
♦ Type 4, solving inequalities

65. Solve $\sqrt{x^2 + 3} \geq 2|x|$.
♦ Type 4, solving an inequality

66. Solve **(a)** $(x+3)^{1/3} > 2$, **(b)** $\sqrt{2x - 1} \geq 2$, and **(c)** $1 < (x + 1)^3 \leq 8$.
♦ Type 4, solving inequalities

67. Solve $x + \dfrac{3}{x} < 4$.
♦ Type 4, solving an inequality

68. Figure 39 shows the graph of a function $y = V(x)$. Draw the graphs of **(a**[A]**)** $y = V(2x)$, **(b)** $y = V(\frac{1}{2}x)$, **(c**[A]**)** $y = V(x + 1)$, **(d)** $y = V(x - 1)$, and **(e)** $y = 2 - V(x + 1) - V(x - 1)$.
♦ Type 4, drawing graphs

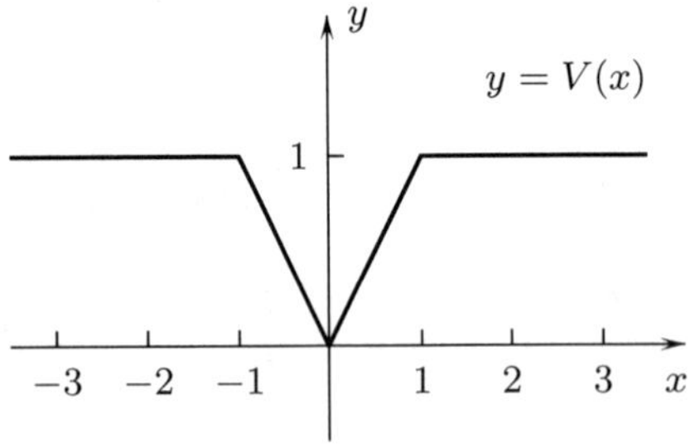

FIGURE 39

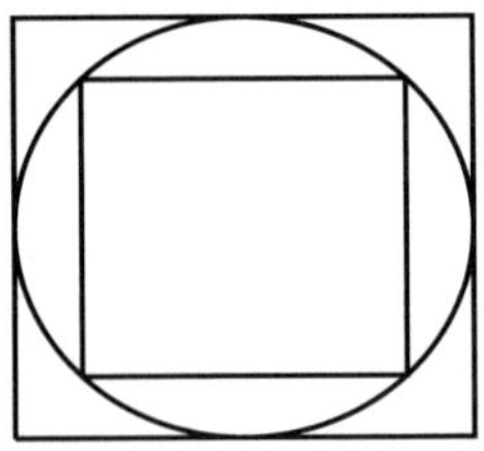

FIGURE 40

69. One square is circumscribed about a circle of radius 1, and another square is inscribed in the circle (Figure 40). **(a)** What percent of the area of the circumscribed square is not in the circle? **(b)** What percent of the area of the circle is not in the inscribed square? **(c)** How are the areas of the two squares related?
♦ Type 4, percent changes

70. If $F(2x) = 10F(x)$ for all $x > 0$ and $F(1) = 5$, what is $F(32)$?

♦ Type 4, value of a function

71. Use your calculator or computer to find the approximate decimal value of $f(f(f(f(f(10)))))$ where $f(x) = \sqrt{x}$.

♦ Type 4, approximate values of a function

72. The upper curve labeled E in Figure 41 was the source of the data on flame velocity in hydrogen/air mixtures that was used in Figure 20 with Example 4 and Question 23. Curve C gives the flame velocities for propane/air mixtures and curve D gives the flame velocities for carbon monoxide/air mixtures. **(a)** How does the maximum flame velocity for hydrogen/air mixtures compare with the maximum flame velocity for carbon monoxide/air and propane/air mixtures? **(b)** Find one quadratic polynomial $v(P) = aP^2 + bP + c$ that approximates the data for propane and another that approximates the data for carbon monoxide. Also give the P-intervals in which each of these functions might be used.

♦ Type 6, making mathematical models

FIGURE 41

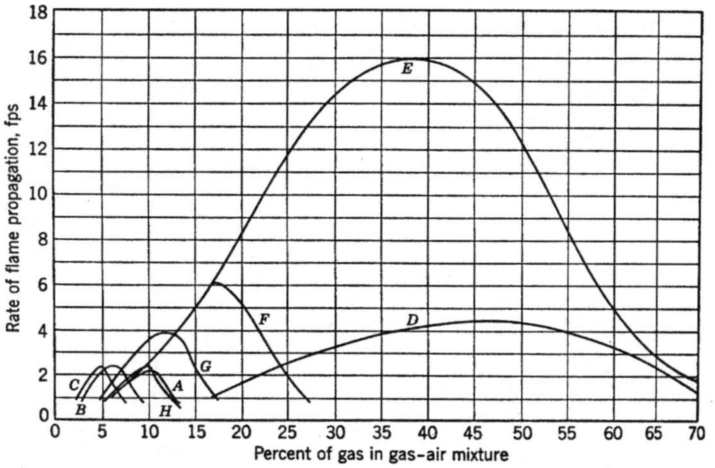

A. Methane.
B. Ethane.
C. Propane.
D. Carbon monoxide.
E. Hydrogen.
F. Coke-oven gas.
G. Refinery gas.
H. Natural gas.

Section 1.2

Functions constructed from powers of x and infinite limits

OVERVIEW: *In order to understand calculus, you need to be able to visualize the graphs of functions given by relatively simple formulas without having to generate them on a calculator or computer. We begin this section with a discussion of lines. Then we consider the curves $y = ax^n + b$ with integers n and define several types of infinite limits. Finally, we present techniques for determining the general shapes, for large and small $|x|$, of graphs of certain other functions constructed from integer powers of x. We will need the formulas for lines later when we study derivatives.*

Topics:

- ***Equations of lines***
- ***The curves $\mathrm{y = ax^n + b}$ with constants a and b and integers n***
- ***Infinite limits***
- ***Predicting shapes of graphs for large and small |x| by recognizing dominant terms***

Slopes and equations of lines

Recall that the SLOPE m of a nonvertical line can be determined from the coordinates of any two points (x_0, y_0) and (x_1, y_1) on it by calculating the RISE $y_1 - y_0$ divided by the RUN $x_1 - x_0$ from one point to the other (Figure 1):

$$\text{Slope} = m = \frac{\text{Rise}}{\text{Run}} = \frac{y_1 - y_0}{x_1 - x_0}. \tag{1}$$

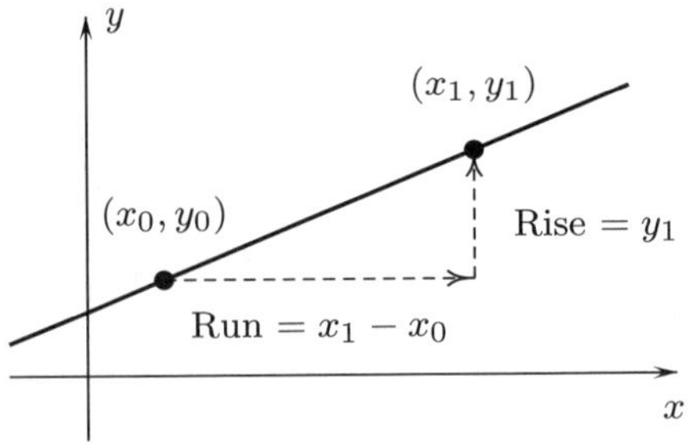

FIGURE 1

Question 1 What is the slope of the line through $(2, 4)$ and $(5, 7)$?

Question 2 What distinguishes the slopes of **(a)** horizontal lines, **(b)** lines that point up to the right, and **(c)** lines that point down to the right?

Question 3 What angles do lines with slopes 1 and -1 make with the x- and y-axes when the scales on the axes are equal?

Question 4 Figure 2 shows the line of slope 2 through the point $(3, 1)$ and a variable point (x, y) on the line. **(a)** Give a formula for the slope of the line in terms of the numbers 3 and 1 and the variables x and y. **(b)** Use the result of part (a) to give an equation of the line.

Question 4 dealt with a specific instance of finding an equation of a line from its slope and the coordinates of a point on the line. Now we will look at the general case.

Question 5 Imagine you want to find the equation of a line when you are given its slope m and the coordinates of a point (x_0, y_0) on it. Let (x, y) represent a variable point on the line, as in Figure 3. **(a)** Express the slope of the line in terms of x_0, y_0, x, and y. **(b)** Use the result of part (a) to give an equation of the line in terms of the constants m, x_0 and y_0 and the variables x and y.

♦ SUGGESTIONS TO INSTRUCTORS: This section should not be skipped, but can be covered on homework and in a small amount of class time in courses with well-prepared students. Have students begin work on some or all of Questions 1 through 18 before class. Tune-up Exercise T12 and Problems 3, 15, 20, 21, 31–33, and 36 are good for in-class discussion.

Typeset by $\mathcal{A}\mathcal{M}\mathcal{S}$-TEX

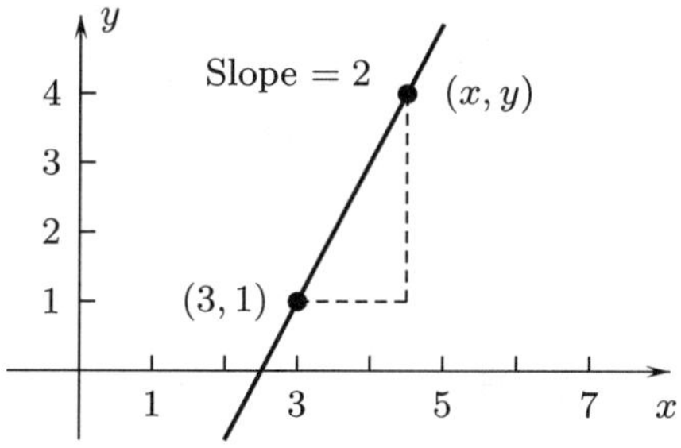

FIGURE 2

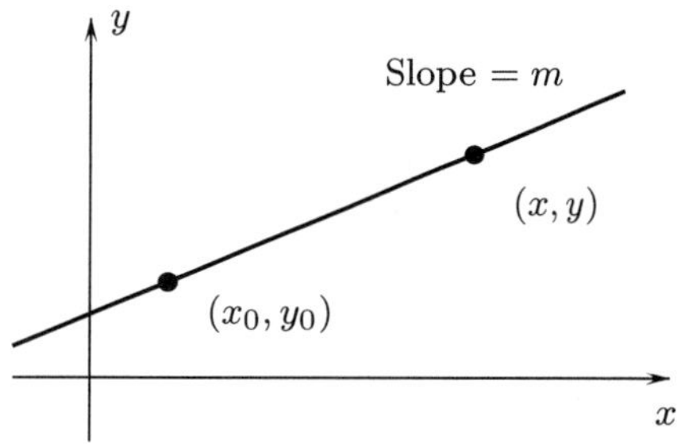

FIGURE 3

This equation you derived in Question 5,

$$y = y_0 + m(x - x_0), \tag{2}$$

is called the POINT-SLOPE EQUATION. Think of it in words: The y-coordinate of the variable point (x, y) on the line equals the y-coordinate of the given point (x_0, y_0), plus the slope m multiplied by the change $x - x_0$ in the x-coordinate from (x_0, y_0) to (x, y). We can use **(2)** to find formulas for linear functions (functions whose graphs are lines).

Example 1 The line in Figure 4 is the graph of a linear function $f(x)$ that has the value 400 at $x = 1$ and the value 200 at $x = 3$. Give a formula for this function. (Notice that there are different scales on the x- and y-axes.)

FIGURE 4

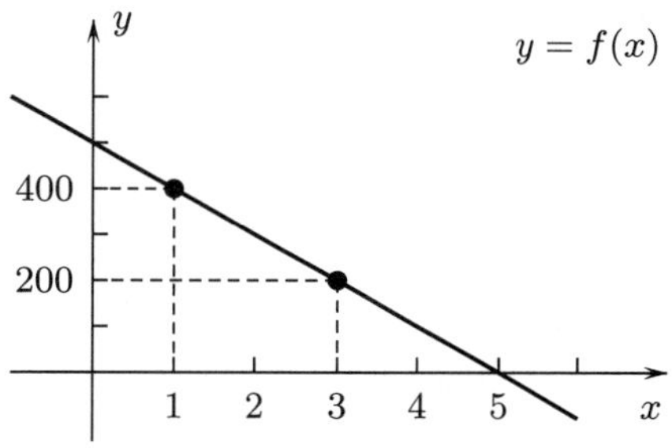

SOLUTION Because the line goes through the points $(1, 400)$ and $(3, 200)$, its slope is

$$m = \frac{[\text{Rise}]}{[\text{Run}]} = \frac{200 - 400}{3 - 1} = -100.$$

We can use **(2)** with $y_0 = 400, m = -100$, and $x_0 = 1$ to obtain the equation $y = 400 - 100(x - 1)$ or $y = 500 - 100x$ for the line. (Using $y_0 = 200$ and $x_0 = 3$ would yield the same result.) The function has the formula, $f(x) = 500 - 100x$. □

The number b in Figure 5 is called the y-INTERCEPT of the line because it is the y-coordinate of the line's intersection with the y-axis. The point-slope formula (2) with $(x_0, y_0) = (0, b)$ gives $y = b + m(x - 0)$ or

$$y = mx + b, \tag{3}$$

as an equation of the line. This is called the SLOPE-INTERCEPT EQUATION. It states that the y-coordinate of the variable point (x, y) on the line equals the y-intercept b, plus the slope multiplied by the x-coordinate of (x, y).

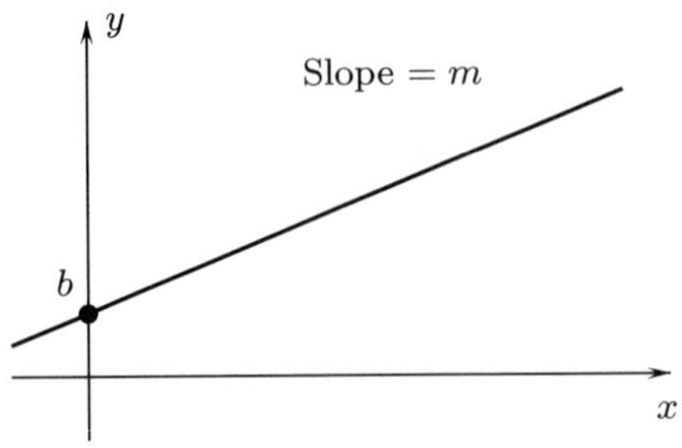

FIGURE 5

Question 6 **(a)** The slope m of the line in Figure 5 is positive. How does the line move if m is increased to a larger value and then decreased to 0, without changing b? **(b)** How does the line in Figure 5 move if m is decreased to a negative value and then increased back to 0, without changing b?

Question 7 **(a)** What would replacing b by $\frac{1}{2}b$ do to the line of Figure 5 if m is not changed? **(b)** What would replacing b by $-2b$ do to the line of Figure 5 if m is not changed?

We cannot define the slope of a vertical line because the run from one point to another on it is zero and calculating the slope would involve dividing by zero. The vertical line through the point (x_0, y_0) has the equation

$$x = x_0 \tag{4}$$

because it consists of all points (x, y) with $x = x_0$. (See Figure 6.)

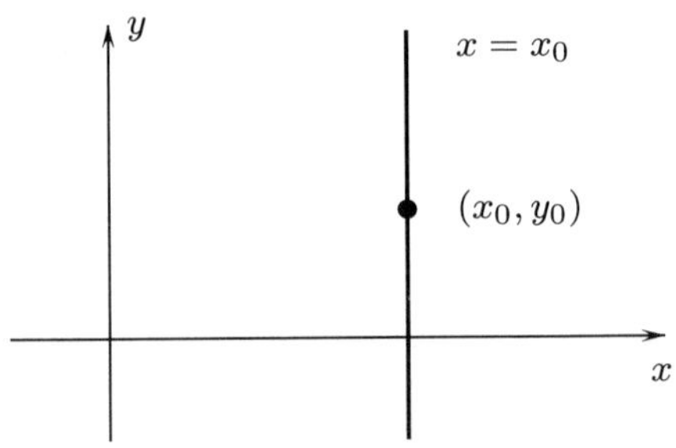

FIGURE 6

Parameters

Constants, such as x_0, y_0, m, and b, in equations for lines, are called PARAMETERS. They have different values for different lines but are fixed for each line.

Question 8 **(a)** Give an equation involving one parameter for all nonvertical lines through the point $(6, -5)$. **(b)** Give an equation for the vertical line through $(6, -5)$.

Slopes of perpendicular lines

Parallel lines have equal slopes. What can we say about slopes of perpendicular lines? Figure 7 shows two perpendicular lines in an xy-plane with equal scales on the axes. One line points up to the right and has positive slope m_1. The other points down to the right and has negative slope m_2. Consider a right triangle with horizontal and vertical legs and hypotenuse along the line of positive slope, such as the triangle in Figure 7. When this triangle is rotated 90° counterclockwise about its left vertex, it becomes the triangle in Figure 8. We can use these two triangles to relate the slopes m_1 and m_2.

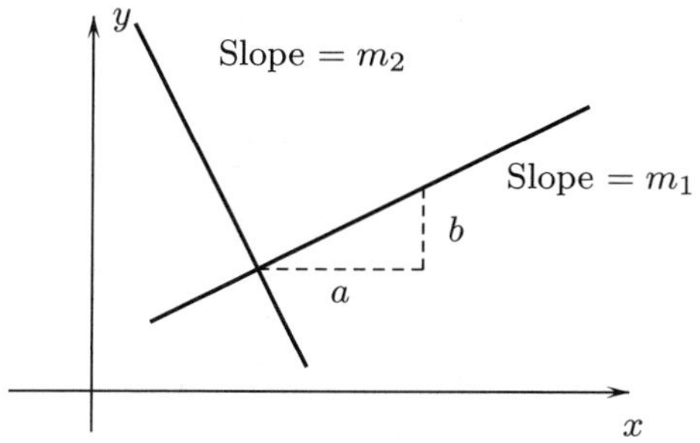

FIGURE 7

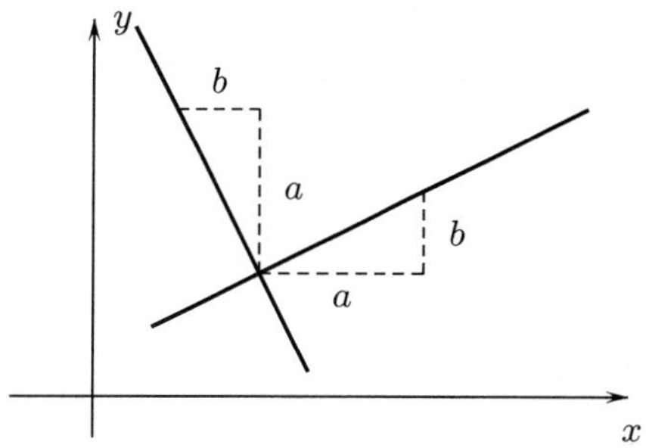

FIGURE 8

Question 9 Express the slopes m_1 and m_2 of the lines in Figures 7 and 8 in terms of a and b.

The formulas, $m_1 = b/a$ and $m_2 = -a/b$, you found in Question 9 show that the slopes of the perpendicular lines are related by the equation,

$$m_1 = -\frac{1}{m_2}. \tag{5}$$

Example 2 Give an equation of the line through $(-5, 2)$ that is perpendicular to the line $y = 3x + 1$.

SOLUTION The line $y = 3x + 1$ has slope 3, so the lines perpendicular to it have slope $-\frac{1}{3}$. The line with this slope that passes through $(-5, 2)$ has the equation $y = 2 - \frac{1}{3}(x - (-5))$, which we rewrite as $y = 2 - \frac{1}{3}(x + 5)$. □

The power functions $y = x^n$ *with integers* $n > 1$

You are probably familiar with the curve $y = x^2$ in Figure 9, which is a type of PARABOLA. The curve $y = x^4$ in Figure 10 is similar but is not a parabola.† How do the equations $y = x^2$ and $y = x^4$ cause the curves to have their particular shapes? One way to approach this question is to examine values of the functions.

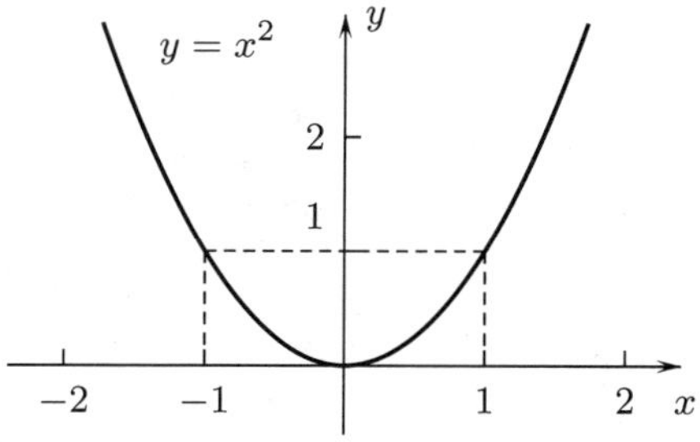

The graph of the function x^2

FIGURE 9

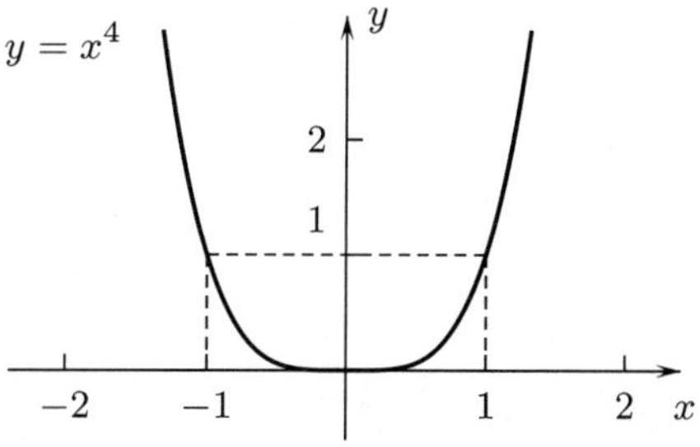

The graph of the function x^4

FIGURE 10

Question 10 **(a)** Calculate the missing values of x^2 and x^4 to complete Table 1. **(b)** Explain why the graph of $y = x^4$ in Figure 10 is closer to the x-axis than the graph of $y = x^2$ in Figure 9 for $-1 < x < 0$ and for $0 < x < 1$. **(c)** Explain why the graph of $y = x^4$ is farther from the x-axis than the graph of $y = x^2$ for $x < -1$ and for $x > 1$.

†The parabolas which are graphs of functions have equations of the form $y = ax^2 + bx + c$ with constants $a \neq 0, b$ and c. The curve $y = x^4$ is called a QUARTIC because x^4 is a fourth-degree polynomial.

TABLE 1

x	-3	-2	-1	$-\frac{1}{2}$	$-\frac{1}{3}$	0	$\frac{1}{3}$	$\frac{1}{2}$	1	2	3
$y=x^2$	9				$\frac{1}{9}$	0	$\frac{1}{9}$				9
$y=x^4$	81				$\frac{1}{81}$	0	$\frac{1}{81}$				81

A function $f(x)$ is called EVEN if $f(-x) = f(x)$ for all x in its domain. This term is used because even powers of x, such as x^2 and x^4, have this property: The graphs of even functions are symmetric about the the y-axis. This can be seen for $y = x^2$ in Figure 11. Because $(-a)^2 = a^2$, the points on the graph with x-coordinates a and $-a$ for any number a are symmetric about the y-axis.

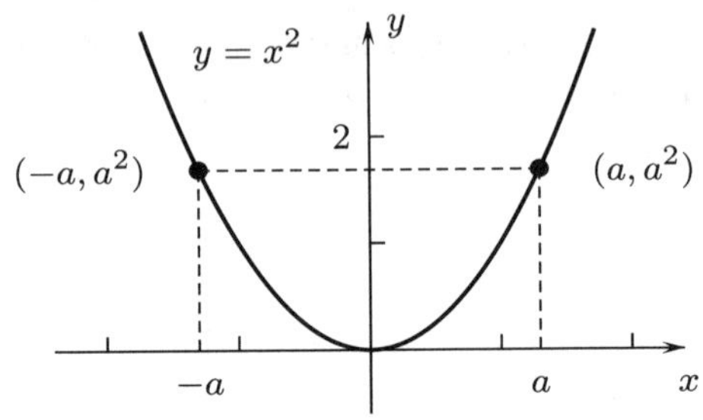

FIGURE 11

As the values in the Table 1 for $x = \pm\frac{1}{2}$ and $x = \pm\frac{1}{3}$ suggest, x^2 and x^4 are extremely small for nonzero x close to 0. This causes their graphs to "hug" the x-axis near $x = 0$. (In Chapter 2, where we define tangent lines to general graphs, we will see that that the curves $y = x^2$ and $y = x^4$ are "tangent" to the x-axis at $x = 0$.) The curves bend away from the x-axis for large positive and negative x because x^2 and x^4 are extremely large for large positive and large negative x.

What about odd powers of x? Table 2 lists values of x^3 and x^5, and Figures 12 and 13 show their graphs.

TABLE 2

x	-2	-1	$-\frac{1}{2}$	0	$\frac{1}{2}$	1	2
$y=x^3$	-8	-1	$-\frac{1}{8}$	0	$\frac{1}{8}$	1	8
$y=x^5$	-32	-1	$-\frac{1}{32}$	0	$\frac{1}{32}$	1	32

Question 11 In what ways are the curves $y = x^3$ and $y = x^5$ similar? How do they differ?

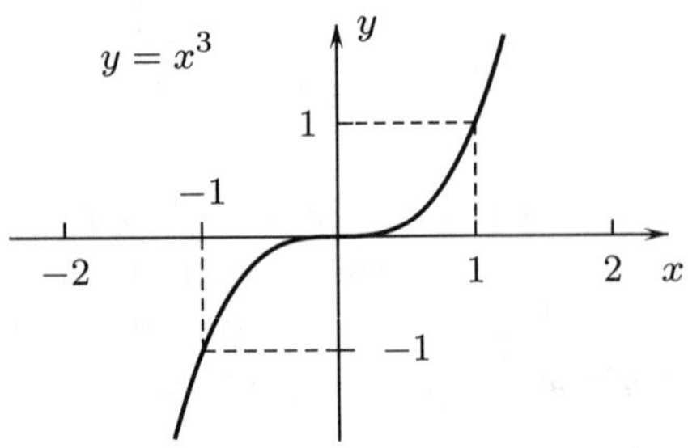

The graph of the function x^3

FIGURE 12

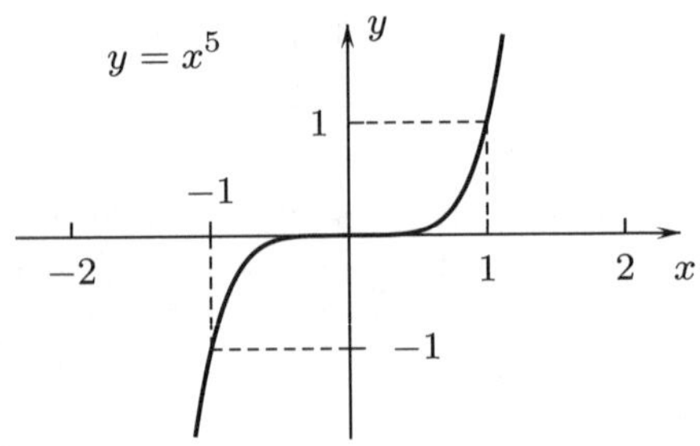

The graph of the function x^5

FIGURE 13

A function $f(x)$ is called ODD if $f(-x) = -f(x)$ for all x in its domain. This term is used because x^3, x^5 and other odd powers of x have this property: $(-x)^3 = -x^3$ and $(-x)^5 = -x^5$ for all x. Graphs of odd functions are symmetric about the origin, as can be seen for $y = x^3$ in Figure 14.

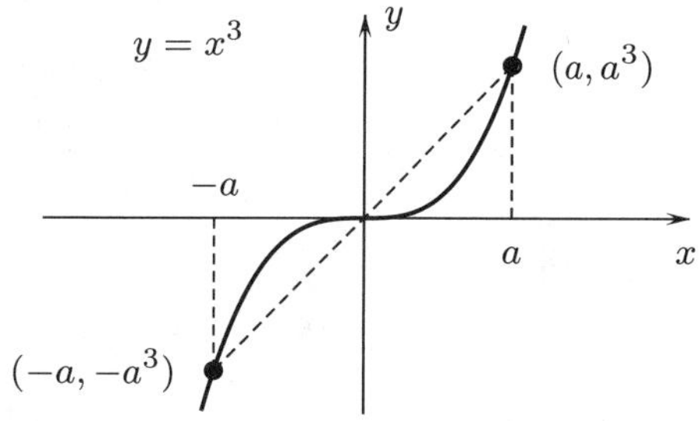

FIGURE 14

Infinite limits as $x \to \pm\infty$

We need terminology and notation to describe the behavior of the graph of $y = x^3$ in Figure 14 as x moves off to the right, increasing through all positive values, and as x moves off to the left, decreasing through all negative values. We start with a calculation.

Example 3 Figure 15 shows the graph of the function $y = x^3$ and horizontal lines at $y = 100$ and $y = -100$. **(a)** Show that the curve is above $y = 100$ for sufficiently large positive x. **(b)** Show that the curve is below $y = -100$ for sufficiently large negative x.

SOLUTION **(a)** Because $x^3 = 100$ at $x = 100^{1/3}$ and x^3 increases as x increases through positive values, x^3 is greater than 100 for $x > 100^{1/3}$. (We could also obtain this result by taking the cube root of both sides of the inequality $x^3 > 100$.)

(b) Because $x^3 = -100$ at $x = (-100)^{1/3} = -100^{1/3}$, and x^3 decreases as x decreases through negative values, x^3 is less than -100 for $x < -100^{1/3}$. □

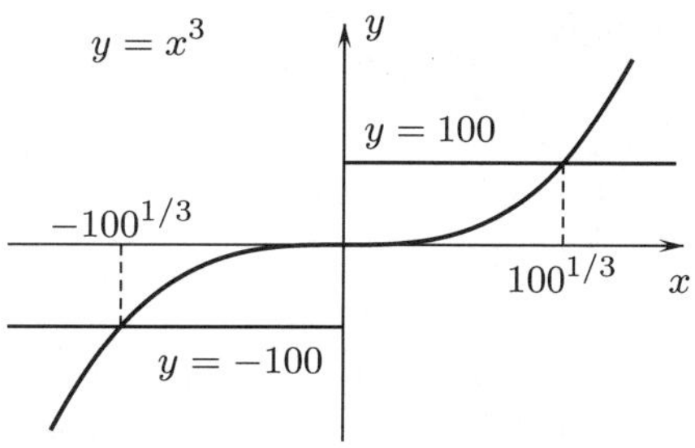

FIGURE 15

We describe the result of Example 3a by saying that x^3 eventually becomes greater than and remains greater than 100 as x increases through all postive numbers, and we describe the result of Example 3b by saying that x^3 eventually becomes less than and remains less than -100 as x decreases through all negative numbers.

We could show with similar calculations that as x increases through all postive numbers, x^3 eventually becomes greater than and remains greater than any positive number Y (such as 1,000 or 1,000,000, etc.) and that as x decreases through all negative numbers, x^3 eventually becomes less than and remains less than any, arbitrarily large, negative number Y. We describe this by saying that x^3 TENDS TO INFINITY as x TENDS TO INFINITY and by saying that x^3 TENDS TO MINUS INFINITY as x TENDS TO MINUS INFINITY. We write $\lim_{x\to\infty} x^3 = \infty$ (or $x^3 \to \infty$ as $x \to \infty$) and $\lim_{x\to-\infty} x^3 = -\infty$ (or $x^3 \to -\infty$ as $x \to -\infty$).

This illustrates two of the following four, almost identical, definitions:

Definition 1 **(a)** $\lim_{x\to\infty} f(x) = \infty$ *if, for every positive number Y, no matter how large, $f(x)$ is eventually greater than and remains greater than Y as x increases through all positive numbers.*

(b) $\lim_{x\to\infty} f(x) = -\infty$ *if, for every negative number Y, $f(x)$ is eventually less than and remains less than Y as x increases through all positive numbers.*

(c) $\lim_{x\to-\infty} f(x) = \infty$ *if, for every positive number Y, $f(x)$ is eventually greater than and remains greater than Y as x decreases through all negative numbers.*

(d) $\lim_{x\to-\infty} f(x) = -\infty$ *if, for every negative number Y, $f(x)$ is eventually less than and remains less than Y as x decreases through all negative numbers.*

In order for parts (a) and (b) of this definition to be applied, $f(x)$ must be defined in (a, ∞) for some number a, and in order for parts (c) and (d) to be applied, $f(x)$ must be defined in $(-\infty, b)$ for some number b.

We present the formal Definition 1 mainly to clarify the meaning of these types of infinite limits. We can generally determine whether such limits occur by visualizing the graphs of the functions being considered.†

Example 4 What are $\lim_{x\to\infty} x^2$ and $\lim_{x\to-\infty} x^2$?

SOLUTION The graph in Figure 9 shows that $\lim_{x\to\infty} x^2 = \infty$ and $\lim_{x\to-\infty} x^2 = \infty$. □

Because part (a) of Definition 1 requires that $f(x)$ remain greater than Y for arbitrarily large Y as x increases through all positive x, it does not apply to functions such as $x \sin x$, whose graph is shown in Figure 16. The number $x \sin x$ is greater than any preassigned positive Y for lots of large positive values of x, but $x \sin x$ does not stay above any positive Y as x increases. Instead, $x \sin x$ oscillates between large positive and negative values. Accordingly, $x \sin x$ does not tend to ∞ as $x \to \infty$. We say that $\lim_{x\to\infty} x \sin x$ does not exist or is not defined.

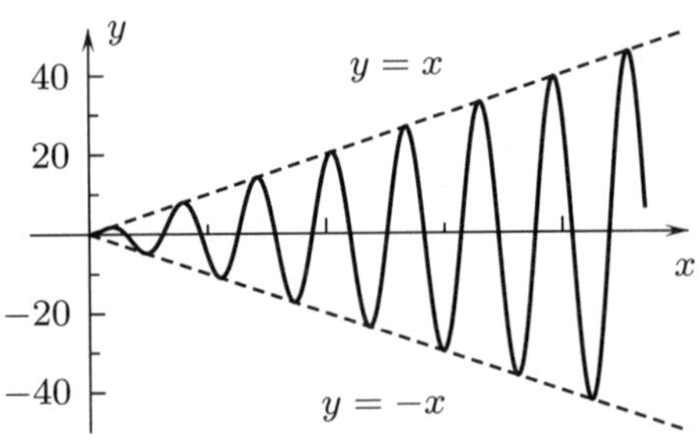

$y = x \sin x$

FIGURE 16

The functions $y = ax^n + b$ *for integers* $n > 1$

You can predict the general shape of the graph of a function of the form $y = ax^n + b$ with constants a and b and positive integers n by imagining how introducing the constants a and b modifies the graph of $y = x^n$. For instance, the graph of $y = 2x^2$ is obtained from the graph of $y = x^2$ by doubling the y-coordinate of every point on the graph (Figure 17). This stretches the curve vertically, making it steeper at each positive y.

---

†Rephrasings of Definition 1 and of the similar Definitions 2 and 3 below in the forms usually used by mathematicians are given with the Principles and Procedures of this section and are illustrated in Problems 51 through 62.

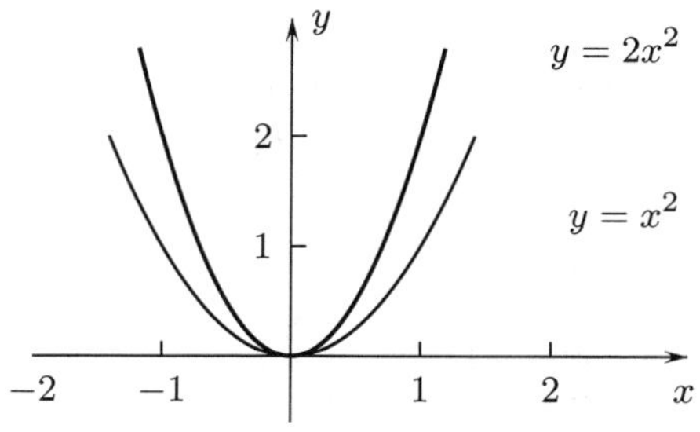

FIGURE 17

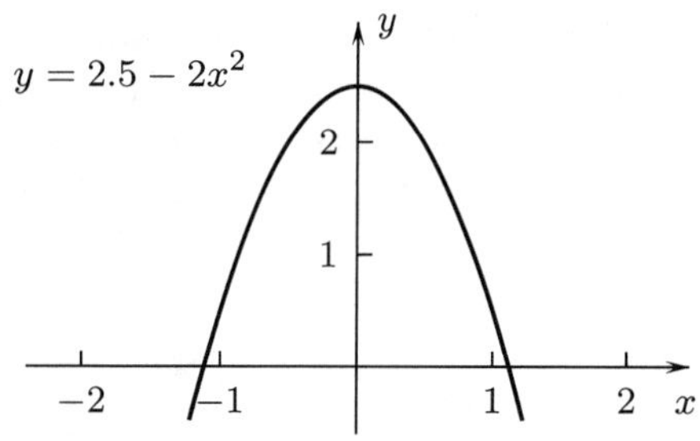

FIGURE 18

Example 5 **(a)** Why does changing $y = 2x^2$ to $y = 2.5 - 2x^2$ transform the upper curve in Figure 17 into the curve in Figure 18? **(b)** What are $\lim_{x\to\infty}(2.5 - 2x^2)$ and $\lim_{x\to-\infty}(2.5 - 2x^2)$?

SOLUTION **(a)** Multiplying the $2x^2$ in $y = 2x^2$ by -1 reflects the curve about the x-axis, and then adding 2.5 raises the curve up 2.5 units.
(b) Figure 18 shows that $\lim_{x\to\infty}(2.5 - 2x^2) = -\infty$ and $\lim_{x\to-\infty}(2.5 - 2x^2) = -\infty$. □

The power functions $y = x^n$ *with negative integers* n

How about negative exponents n? Figures 19 through 22 show the graphs of the functions given by the formulas, $x^{-1} = 1/x$, $x^{-2} = 1/x^2$, $x^{-3} = 1/x^3$, and $x^{-4} = 1/x^4$.†

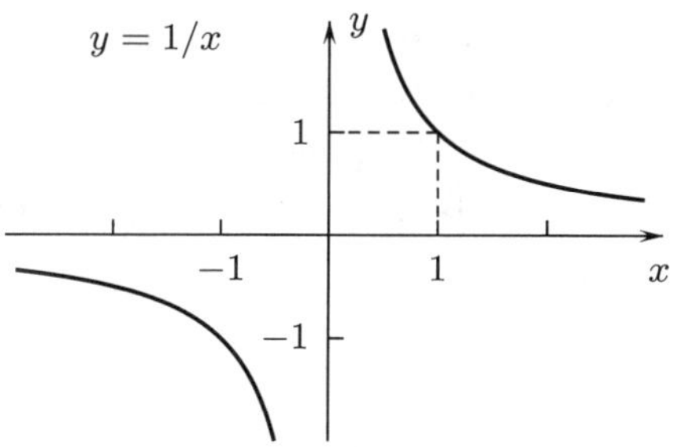

The graph of the function $y = x^{-1}$
FIGURE 19

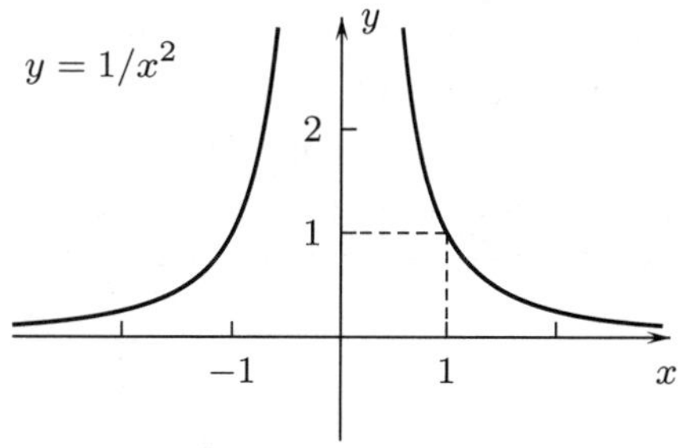

The graph of the function $y = x^{-2}$
FIGURE 20

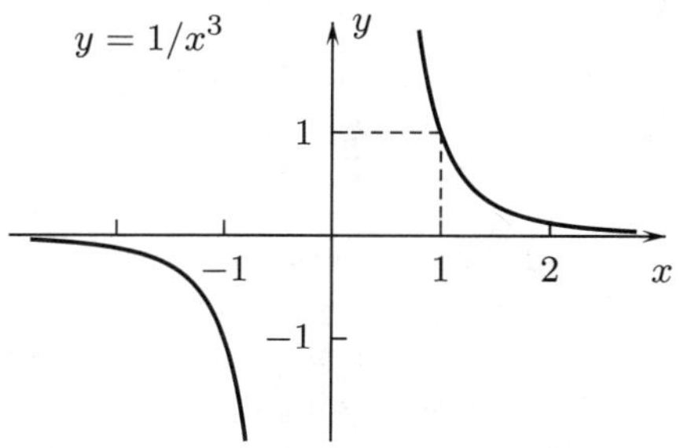

The graph of the function $y = x^{-3}$
FIGURE 21

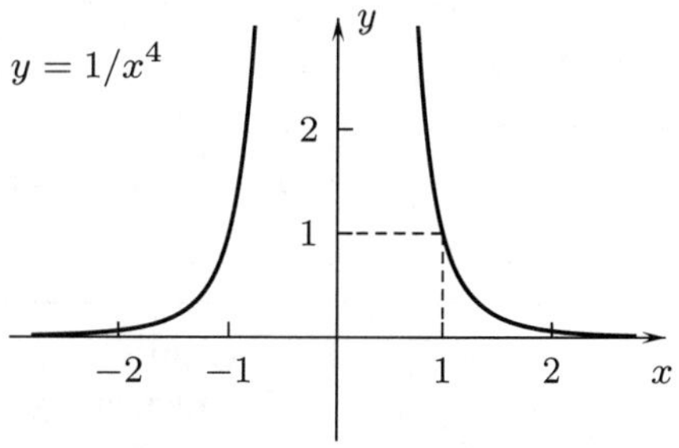

The graph of the function $y = x^{-4}$
FIGURE 22

† $y = x^{-1}$ is a type of HYPERBOLA. The curves $y = x^{-3}$, $y = x^{-2}$, and $y = x^{-4}$ do not have common names.

Question 12 **(a)** Complete Table 3 of values of x^n for $n = -1, -2, -3$, and -4. (The symbol $*$ has been placed under the 0 in the table, because $1/0$ is not defined.) **(b)** Which of the four functions is even? Which is odd? **(c)** Which of the four graphs is closest to the x-axis for $x < -1$ and for $x > 1$? **(d)** Which of the four graphs is closest to the x-axis for $-1 < x < 0$ and for $0 < x < 1$?

TABLE 3

x	-2	-1	$-\frac{1}{2}$	0	$\frac{1}{2}$	1	2
$y = x^{-1}$		-1	-2	$*$		1	$\frac{1}{2}$
$y = x^{-2}$		1	4	$*$		1	$\frac{1}{4}$
$y = x^{-3}$		-1	-8	$*$		1	$\frac{1}{8}$
$y = x^{-4}$		1	16	$*$		1	$\frac{1}{16}$

Finite limits as x $\rightarrow \pm\infty$

Definition 1 dealt with infinite limits of functions as $x \rightarrow \infty$ and as $x \rightarrow -\infty$. Now we will consider finite limits of functions as $x \rightarrow \infty$ and as $x \rightarrow -\infty$. Again we illustrate the ideas with a calculation, this time using the function $1/x$ whose graph is shown in Figure 19.

Question 13 For what values of x does $y = 1/x$ lie between the lines $y = \frac{1}{10}$ and $y = -\frac{1}{10}$? Find the answer algebraically and then check it by generating the curve and lines together in the window $-25 \leq x \leq 25, -0.5 \leq y \leq 0.5$. (Consider the cases of $x > 0$ and $x < 0$ separately.)

According to the result of Question 13, the curve $y = 1/x$ eventually lies between and remains between the lines $y = \frac{1}{10}$ and $y = -\frac{1}{10}$ as x increases through all positive values and as x decreases through all negative values. We could reach the same conclusion with $\frac{1}{10}$ replaced by $\frac{1}{100}$, $\frac{1}{1000}$, or any other small positive number. Because of this, we say that $1/x$ tends to 0 as x tends to infinity and as x tends to minus infinity and we write

$$\lim_{x\rightarrow\infty} \frac{1}{x} = 0 \quad \text{and} \quad \lim_{x\rightarrow-\infty} \frac{1}{x} = 0$$

or "$1/x \rightarrow 0$ as $x \rightarrow \infty$" and "$1/x \rightarrow 0$ as $x \rightarrow -\infty$." This means that the graph of $1/x$ approaches the x-axis when x increases through all positive numbers and when x decreases through all negative numbers, as can be seen in Figure 19.

Here is the general definition of finite limits as $x \rightarrow \pm\infty$:

Definition 2 **(a)** $\lim_{x\rightarrow\infty} f(x) = L$ *with a number* L *if, for every positive number* ϵ*, no matter how small,* $f(x)$ *eventually lies between and remains between* $L - \epsilon$ *and* $L + \epsilon$ *as* x *increases through all positive numbers.*

(b) $\lim_{x\rightarrow-\infty} f(x) = L$ *with a number* L *if, for every positive number* ϵ*,* $f(x)$ *eventually lies between and remains between* $L - \epsilon$ *and* $L + \epsilon$ *as* x *decreases through all negative numbers.*

This definition, like Definition 1, is given primarily to clarify what we mean when we speak of these types of limits. We can usually determine whether the values of a function tend to a finite number or not as x tends to ∞ or to $-\infty$ by just visualizing its graph or examining its formula.

Example 6 What are the limits of $1/x^2$, $1/x^3$, and $1/x^4$ as $x \rightarrow \infty$ and as $x \rightarrow -\infty$?

SOLUTION We can see that $1/x^2$, $1/x^3$, and $1/x^4$ all tend to 0 as $x \rightarrow \pm\infty$ either from the graphs in Figures 20 through 22, or by noting that $|x|, x^2, |x^3|$ and x^4 all tend to ∞ as $x \rightarrow \pm\infty$. □

One-sided and two-sided infinite limits

Now we consider cases where a function has an infinite limit as its variable approaches a finite number.

Figure 23 shows the graph $y = 1/x$ of the function $1/x$ and the horizontal lines $y = Y$ and $y = -Y$, where Y is a positive number. Notice what happens to a point $(x, 1/x)$ on the curve as x decreases toward 0 through positive values. The point on the curve will eventually be above and remain above $y = Y$. Since Y could be arbitrarily large, we say that $1/x$ tends to infinity as x TENDS TO 0 FROM THE RIGHT, and we write

$$\lim_{x\to 0+} \frac{1}{x} = \infty.$$

A plus sign is used in the symbol $x \to 0^+$ for the limit from the right because the positive side of the x-axis is on the right.

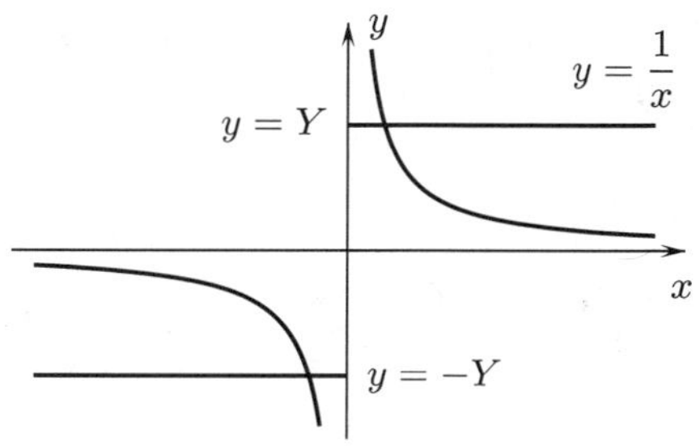

FIGURE 23

Similarly, as x increases toward 0 through negative values, the point on $y = 1/x$ eventually is below and remains below $y = Y$ for any negative number Y. We say that $1/x$ tends to $-\infty$ as x TENDS TO 0 FROM THE LEFT, and we write

$$\lim_{x\to 0-} \frac{1}{x} = -\infty.$$

A minus sign is used in the symbol $x \to 0^-$ for the limit from the left because the negative side of the x-axis is on the left. There are four of this type of infinite limit.

Definition 3 **(a)** $\lim_{x\to x_0^+} f(x) = \infty$ *with a number* x_0 *if, for every positive number* Y, $f(x)$ *eventually is greater than and remains greater than* Y *as* x *decreases toward* x_0 *from the right.*

(b) $\lim_{x\to x_0^-} f(x) = \infty$ *if, for every positive number* Y, $f(x)$ *eventually is greater than and remains greater than* Y *as* x *increases toward* x_0 *from the left.*

(c) $\lim_{x\to x_0^+} f(x) = -\infty$ *if, for every negative number* Y, $f(x)$ *eventually is less than and remains less than* Y *as* x *decreases toward* x_0 *from the right.*

(d) $\lim_{x\to x_0^-} f(x) = -\infty$ *if, for every negative number* Y, $f(x)$ *eventually is less than and remains less than* Y *as* x *increases toward* x_0 *from the left.*

Question 14 What are $\lim_{x\to 0+} \frac{1}{x^2}$ and $\lim_{x\to 0-} \frac{1}{x^2}$?

You saw in Question 14 that $1/x^2$ tends to ∞ as x tends to 0 from the right and as x tends to 0 from the left. Because the limits from the right and from the left are the same, we say that the LIMIT of $1/x^2$ AS x TENDS TO 0 is ∞ and write

$$\lim_{x\to 0} \frac{1}{x^2} = \infty$$

without a superscript + or − on the 0. We sometimes refer to this type of limit, as a TWO-SIDED LIMIT to distinguish it from one-sided limits from the right and from the left. Here are the general definitions:

Definition 4 **(a)** $\lim_{x\to x_0} f(x) = \infty$ *with a number* x_0 *if* $\lim_{x\to x_0^+} f(x) = \infty$ *and* $\lim_{x\to x_0^-} f(x) = \infty$.

(b) $\lim_{x\to x_0} f(x) = -\infty$ *with a number* x_0 *if* $\lim_{x\to x_0^+} f(x) = -\infty$ *and* $\lim_{x\to x_0^-} f(x) = -\infty$.

If $\lim_{x\to x_0^+} f(x)$ and $\lim_{x\to x_0^-} f(x)$ are different, we say that the (two-sided) limit $\lim_{x\to x_0} f(x)$ does not exist or is not defined. For example, $\lim_{x\to 0}(1/x)$ does not exist because $\lim_{x\to 0+}(1/x) = \infty$ and $\lim_{x\to 0^-}(1/x) = -\infty$, as can be seen in Figure 23.

The functions $y = ax^n + b$ *with negative integers* n

We can visualize the graph of $y = ax^n + b$ with negative integers n by imagining how the constants a and b modify the curve $y = x^n$.

Question 15 **(a)** Use properties of $y = \dfrac{1}{x^2}$ to explain the shape of the curve $y = 2 - \dfrac{1}{x^2}$ in Figure 24. **(b)** What are $\lim_{x\to\infty}\left(2 - \dfrac{1}{x^2}\right)$, $\lim_{x\to-\infty}\left(2 - \dfrac{1}{x^2}\right)$, and $\lim_{x\to 0}\left(2 - \dfrac{1}{x^2}\right)$?

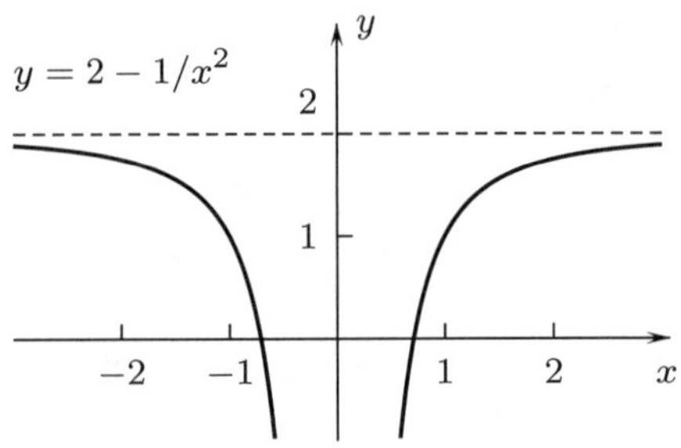

FIGURE 24

Asymptotes

A line is an ASYMPTOTE of a curve if the curve approaches it as the x-coordinate and/or the y-coordinate of a point on the curve tends to ∞ or $-\infty$. Thus, the y-axis is a vertical asymptote and the line $y = 2$ is a horizontal asymptote of $y = 2 - \dfrac{1}{x^2}$ in Figure 24. The x- and y-axes are asymptotes of the curves in Figures 19 through 22.

Comparing $|x^n|$ *and* $|x^m|$

Because many functions that are encountered in calculus are formed by combining powers of x, it is important to be able to recognize the relative sizes of different powers of x for large and small, positive and negative x. Because x^n is negative when x is negative if n is an odd integer, we compare x^n with other functions for odd n and negative x by using absolute values $|x^n|$.

Question 16 How do the values of $x^{-2}, |x^{-1}|, |x|, x^2$, and $|x^3|$ compare **(a)** for large $|x| > 1$ and **(b)** for small positive $|x| < 1$?

You probably found in answering Question 16 that increasing the exponent in $|x^n|$ gives a function that is much larger for large $|x|$ and much smaller for small $|x|$. We state this general principle as a rule.

Rule 1 *For any integers* n *and* m *with* $m > n$, $|x^m|$ *is much larger than* $|x^n|$ *for very large* $|x|$, *and* $|x^m|$ *is much smaller than* $|x^n|$ *for* $|x|$ *very close to zero.*

We describe this rule by saying that for m greater than n, x^m "dominates" x^n when x is a very large positive or negative number and x^n "dominates" x^m when x is very close to 0.

Behavior of more complicated functions for large and small |x|

We will need calculus techniques that will be developed in Chapter 3 to determine the exact shapes of graphs of most functions. However, if the function is constructed from integer powers of x, we can often use Rule 1 to predict the approximate shape of its graph for large positive and negative x and for x near 0 from the shape of $y = x^n$ for the appropriate integers n. The basic principles are illustrated by the graph $y = x - \frac{1}{2}x^3$ in Figure 25.

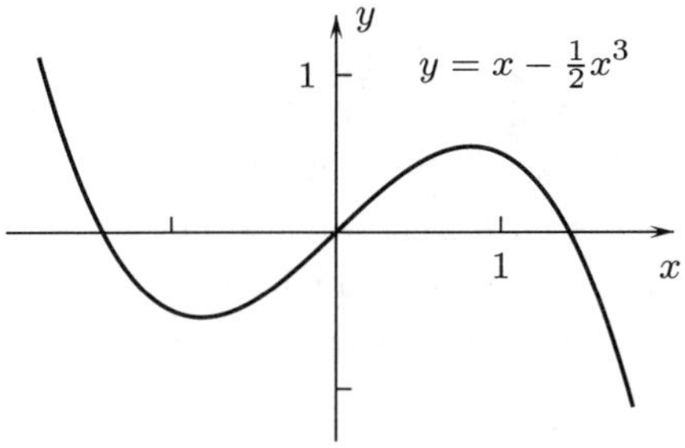

FIGURE 25

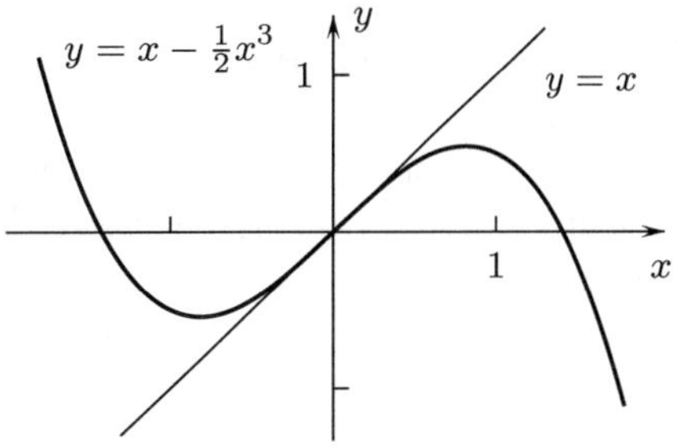

FIGURE 26

To predict the approximate shape of the curve $y = x - \frac{1}{2}x^3$ for x close to 0, we visualize the line $y = x$, whose equation is obtained by taking only the lowest-power term, x, in the formula for the function. Because $\left|\frac{1}{2}x^3\right|$ is much smaller than $|x|$ for x close to zero, the curve $y = x - \frac{1}{2}x^3$ is closely approximated by the line near $x = 0$. This can be seen in Figure 26, which shows the curve $y = x - \frac{1}{2}x^3$ and the line $y = x$.

To predict the approximate shape of $y = x - \frac{1}{2}x^3$ for large positive and large negative x, we visualize the curve $y = -\frac{1}{2}x^3$ whose equation is obtained by taking only the highest-power term $-\frac{1}{2}x^3$ from the original equation. The curve $y = x - \frac{1}{2}x^3$ looks approximately like the curve $y = -\frac{1}{2}x^3$ for large positive and large negative x. You can see this starting to occur in Figure 27. The curve $y = x - \frac{1}{2}x^3$ bends down to the right and up to the left like $y = -\frac{1}{2}x^3$ as soon as x is some distance away from 0.

FIGURE 27

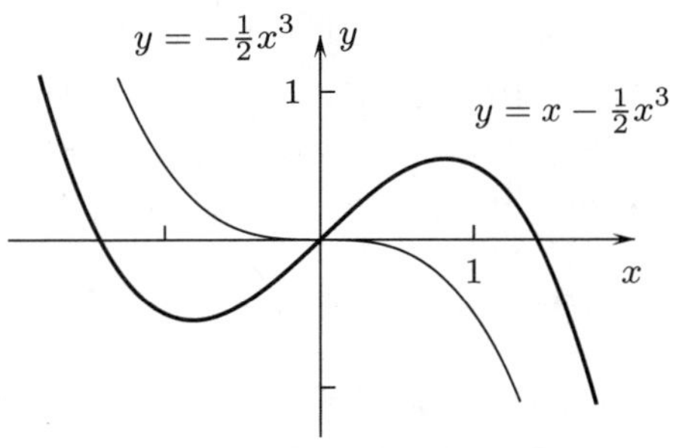

This phenomenon is shown more clearly in the larger windows of Figures 28 and 29, where the curves appear even closer together.

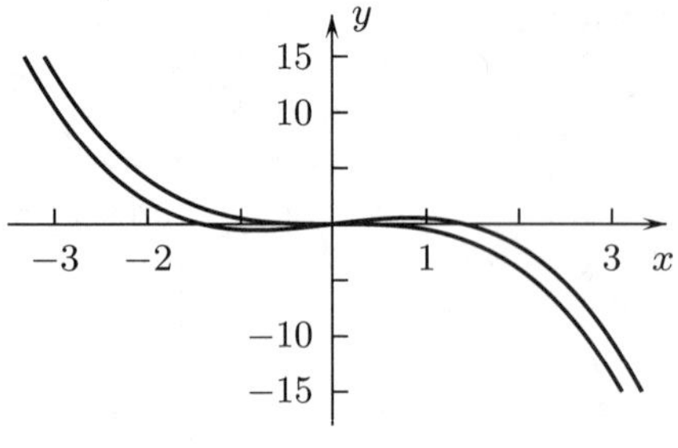

$y = x - \frac{1}{2}x^3$ and $y = -\frac{1}{2}x^3$

FIGURE 28

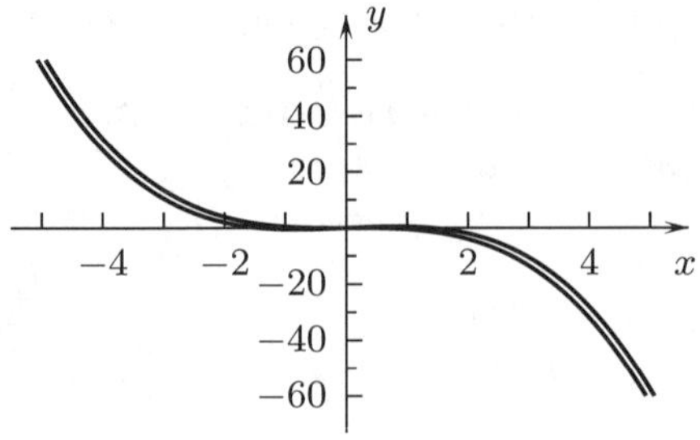

$y = x - \frac{1}{2}x^3$ and $y = -\frac{1}{2}x^3$

FIGURE 29

What is happening here? The curves $y = x - \frac{1}{2}x^3$ and $y = -\frac{1}{2}x^3$ are not close together for large positive or negative x. On the contrary, the vertical distance between the curves at x is $|x|$, which tends to ∞ as $x \to \pm\infty$. However, because x is relatively small compared to $-\frac{1}{2}x^3$ for large $|x|$, the curves appear close together for large $|x|$ in windows that show the curves for a large range of y-values.

Moreover, because $x - \frac{1}{2}x^3$ and $-\frac{1}{2}x^3$ are relatively close for large $|x|$, they have the same limits as $x \to \pm\infty$:

$$\lim_{x\to\infty}(x - \tfrac{1}{2}x^3) = \lim_{x\to\infty}(-\tfrac{1}{2}x^3) = -\infty$$

$$\lim_{x\to-\infty}(x - \tfrac{1}{2}x^3) = \lim_{x\to-\infty}(-\tfrac{1}{2}x^3) = \infty.$$

Thus, we can determine the limits of $x - \frac{1}{2}x^3$ as $x \to \pm\infty$ from the limits of its term $-\frac{1}{2}x^3$ with the highest power of x.

This example illustrates the following general principle.

Rule 2 *A graph, such as $y = x - \frac{1}{2}x^3$, of a sum of constants multiplied by powers of x looks approximately like the graph of the term with the least exponent for x near 0 and looks approximately like the graph of the term with the greatest exponent for large positive and large negative x.*

We can describe this rule by saying that the graph looks approximately like the graph of its dominant term for small and large $|x|$, where the dominant term is the term with the least exponent for small $|x|$ and is the term with the greatest exponent for large $|x|$.

To illustrate Rule 2 with a negative exponent, consider the graph $y = 2x^{-2} + 4x^2 - x^3$ in Figure 30. The exponents in the formula are -2 on $2x^{-2}$, 2 on $4x^2$, and 3 on $-x^3$. The least exponent is -2, so for x near 0, the dominant term is $2x^{-2}$ and the curve looks approximately like $y = 2x^{-2}$ In particular,

$$\lim_{x\to 0}(2x^{-2} + 4x^2 - x^3) = \lim_{x\to 0}(2x^{-2}) = \infty.$$

Because the greatest of the exponents in $2x^{-2} + 4x^2 - x^3$ is 3, its dominant term for large $|x|$ is $-x^3$ and its graph looks approximately like $y = -x^3$ for large positive and large negative x. This can also be seen in Figure 30. It follows that

$$\lim_{x\to\infty}(2x^{-2} + 4x^2 - x^3) = \lim_{x\to\infty}(-x^3) = -\infty$$

$$\lim_{x\to-\infty}(2x^{-2} + 4x^2 - x^3) = \lim_{x\to-\infty}(-x^3) = \infty.$$

The bump in the graph to the right of the y-axis in Figure 30 is caused by the term $4x^2$ in the formula. It can be analyzed by using derivatives, as we will see in Chapter 3.

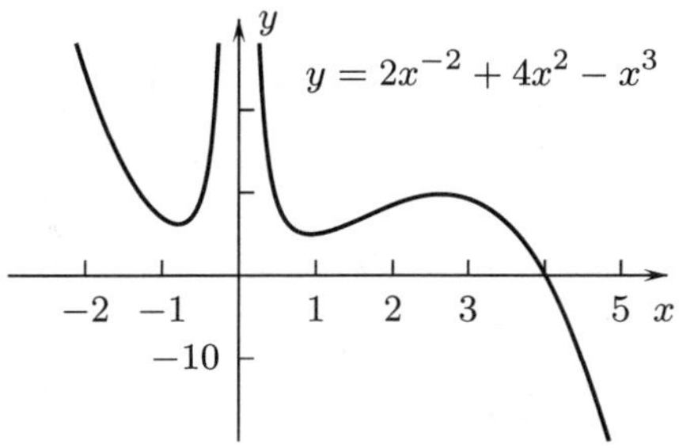

FIGURE 30

Another look at the formulas

We can also give a more algebraic justification of Rule 2 in each case by factoring the dominant term from the original formula for the given function. This shows that the function equals the dominant term multiplied by an expression that is close to 1 for small $|x|$ if we are studying the graphs for x near 0, and close to 1 for large $|x|$ if we are studying the graphs for large $|x|$. Let us apply this technique to the function $y = x - \frac{1}{2}x^3$ of Figure 28.

Because the dominant term in $x - \frac{1}{2}x^3$ for small $|x|$ is x, we write

$$x - \tfrac{1}{2}x^3 = x(1 - \tfrac{1}{2}x^2).$$

The graph $y = x - \frac{1}{2}x^3$ looks approximately like the line $y = x$ for small $|x|$ because the function in parentheses on the right of this equation is close to 1 for small $|x|$.

Similarly, because the dominant term in $x - \frac{1}{2}x^3$ for large $|x|$ is $-\frac{1}{2}x^3$, we factor this expression from the original formula for the function. This requires writing x as $-\frac{1}{2}x^3\left(\frac{-2}{x^2}\right)$. We obtain

$$x - \tfrac{1}{2}x^3 = -\tfrac{1}{2}x^3\left(\frac{-2}{x^2}\right) - \tfrac{1}{2}x^3 = -\tfrac{1}{2}x^3\left(\frac{-2}{x^2} + 1\right).$$

The graph $y = x - \frac{1}{2}x^3$ looks approximately like the graph $y = -\frac{1}{2}x^3$ for large $|x|$ because the function in parentheses on the right of the last equation is close to 1 for large $|x|$.

Products and quotients

Rule 2 can also be applied to certain products and quotients of sums of constants multiplied by powers of x, as in the next example and questions.

Example 7 Use the formula to explain the shape, for large and small $|x|$, of $y = \dfrac{x}{x^2+1}$ in Figure 31.

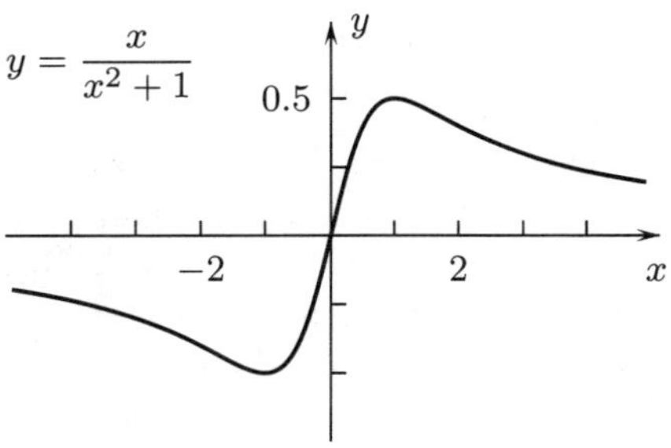

FIGURE 31

SOLUTION For x near 0, the x^2 in the denominator is much smaller than the 1, and $y = \dfrac{x}{x^2+1}$ looks approximately like $y = \dfrac{x}{1}$, which is the line $y = x$. For large positive or negative x, the 1 in the denominator of $\dfrac{x}{x^2+1}$ is much smaller than the x^2 so that $y = \dfrac{x}{x^2+1}$ looks approximately like $y = \dfrac{x}{x^2}$, which is $y = \dfrac{1}{x}$. In particular, $\dfrac{x}{x^2+1} \to 0$ as $x \to \infty$ and as $x \to -\infty$. □

Question 17 **(a)** Generate $y = \dfrac{x^2}{x^2+1}$ in the window $-5 \le x \le 5, -0.25 \le y \le 1.25$ and use the formula to explain the shape of the curve for x near 0 and for large positive and negative x. **(b)** What are $\displaystyle\lim_{x\to\pm\infty} \frac{x^2}{x^2+1}$?

Question 18 **(a)** Generate $y = \left(x^2+\dfrac{1}{x^2}\right)(1-x+x^4)$ in the window $-2 \le x \le 2, -5 \le y \le 25$ with y-scale $= 5$. Use the formula to explain the shape of the curve for small and large $|x|$. **(b)** What are $\displaystyle\lim_{x\to\pm\infty} \left[\left(x^2+\frac{1}{x^2}\right)(1-x+x^4)\right]$ and $\displaystyle\lim_{x\to 0} \left[\left(x^2+\frac{1}{x^2}\right)(1-x+x^4)\right]$?

A bit of history

General algebraic equations were first studied in the sixteenth century, but without the modern convention of using a single letter for the unknown. The Italian physician and algebraist Gerolamo Cardano (1501–1572), for example, used the Latin sentence, *Cubus $\overline{p}$ 6 rebus aequalis 20*, for the equation that we would write $x^3 + 6x = 20$. In Cardano's sentence, the word "cubus" denotes the cube of the unknown, $\overline{p}$ stands for "plus", "rebus" denotes the unknown, and "aequalis" means "equals."

The use of a single letter for the unknown and of exponents for positive integer powers was popularized by a treatise "La Géométrie," written in 1637 by the French philosopher Réné Descartes (1587–1650) as an appendix to a work on the philosophy of science.

Réné Descartes
(1587–1650)

Pierre Fermat
(1601–1665)

Descartes and a French lawyer Pierre Fermat (1601–1665) are considered the inventors of analytic geometry as a tool for giving geometric meaning to aspects of algebra and calculus. Greek mathematicians, including Euclid (ca. 300 BC), Archimedes (287–212 BC), and Apollonius (ca. 225 BC), used the equivalent of coordinate systems with the theory of proportions for studying geometric figures, and algebra was employed in the sixteenth century to solve geometric problems. It was Descartes and Fermat, however, who first studied curves defined by equations as well as by their geometric properties and who made extensive use of the association between the algebra of equations and the geometry of curves. Fermat's work was not published until after his death, more than forty years after the publication of Descartes' "La Géométrie," so Descartes often receives more credit for creating what now is known as analytic or "cartesian" geometry.† Fermat and Decartes generally used only positive coordinates. Negative coordinates were first used systematically by Isaac Newton (1642–1727) in his *Enumeration of Curves of Third Degree* (1676).

Fermat's name has been in the news during the past several years because his "famous last theorem", that $x^n + y^n = z^n$ has no nonzero integer solutions x, y, and z for integers $n > 2$, has finally been proved.

Principles and procedures

- Learn the point-slope equation for a line in the form $y = y_0 + m(x - x_0)$. This is equivalent to the equation $y - y_0 = m(x - x_0)$, which you might have learned before, but is more convenient for calculus. You can also use the slope-intercept equation, $y = mx + b$ instead, but if x_0 is not zero, you have to carry out an extra calculation (and run the risk of introducing an error) to find b.
- You will occasionally need a formula for the distance between two points. To recall it, visualize two points $P(x_0, y_0)$ and $Q(x_1, y_1)$ in an xy-plane with equal scales on the axes, as in Figure 32. The distance $\overline{PQ}$ between the points is the length of the hypotenuse of a right triangle with legs of lengths $|x_1 - x_0|$ and $|y_1 - y_0|$. Consequently,

$$\overline{PQ} = \sqrt{(x_1 - x_0)^2 + (y_1 - y_0)^2}. \qquad \textbf{(6)}$$

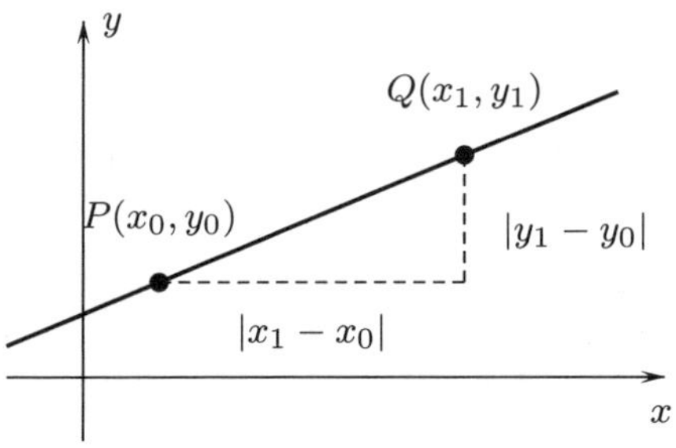

FIGURE 32

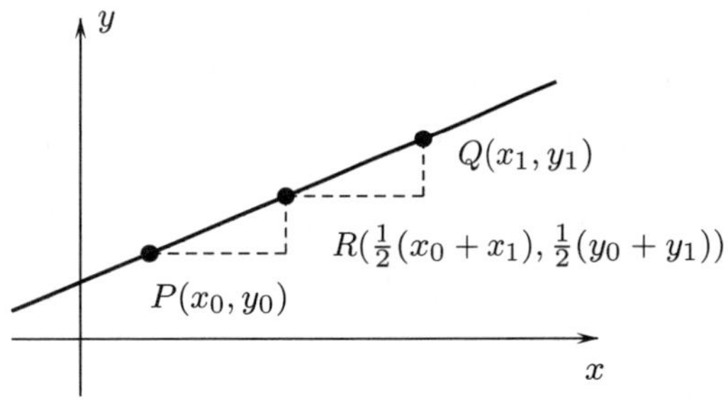

FIGURE 33

- To recall a formula for the coordinates of the midpoint R of a line segment PQ, visualize the triangles in Figure 33. Because the triangles are congruent, the coordinates of R are the averages of the coordinates of P and Q, and

$$R = [\text{Midpoint of } PQ] = \left(\tfrac{1}{2}(x_0 + x_1), \tfrac{1}{2}(y_0 + y_1)\right). \qquad \textbf{(7)}$$

- The general shape of the curve $y = a + bx^n$ with an integer n can be determined by noting whether x^n is positive or negative for positive and negative x and by determining whether $|x^n|$ is large or small for large and small $|x|$.
- To determine the approximate shape, for x near 0 and for large positive or negative x in windows with large ranges of y-values, of the graph of a function constructed from integer powers of x, notice which terms in sums dominate other terms and visualize the graph of the function formed only from the dominant terms.

---

†For an entertaining look at the life and times of the contentious Cardano, read his autobiography, *The Book of My Life* (New York: Dover, 1929) or the biography *Doctor Cardano, Physician Extraordinary* (London: Muller, 1969). See also pp. 74–102 of *History of Analytic Geometry* by C. Boyer (New York: Scripta Mathematica, 1956) and pp. 62–69 on Cardano, pp. 143–150 on Fermat, and pp. 155–157 on Descartes in *A Source Book in Mathematics, 1200–1800* by D. Struik (Cambridge: Harvard University Press, 1969).

- The following rephrasings of Definitions 1, 2 and 3 are generally used by mathematicians when they are studying infinite limits.

Definition 1A (a) $\lim_{x\to\infty} f(x) = \infty$ *if, for every positive number* Y, *there is an* X *such that* $f(x) > Y$ *for* $x > X$.

(b) $\lim_{x\to\infty} f(x) = -\infty$ *if, for every negative number* Y, *there is a number* X *such that* $f(x) < Y$ *for* $x > X$.

(c) $\lim_{x\to-\infty} f(x) = \infty$ *if, for every positive number* Y, *there is a number* X *such that* $f(x) > Y$ *for* $x < X$.

(d) $\lim_{x\to-\infty} f(x) = -\infty$ *if, for every negative number* Y, *there is a number* X *such that* $f(x) < Y$ *for* $x < X$.

Definition 2A (a) $\lim_{x\to\infty} f(x) = L$ *with a number* L *if, for every positive number* ϵ, *there is an* X *such that* $|f(x) - L| < \epsilon$ *for* $x > X$.

(b) $\lim_{x\to-\infty} f(x) = L$ *with a number* L *if, for every positive number* ϵ, *there is an* X *such that* $|f(x) - L| < \epsilon$ *for* $x < X$.

Definition 3A (a) $\lim_{x\to x_0^+} f(x) = \infty$ *with a number* x_0 *if, for every positive number* Y, *there is a positive* δ *such that* $f(x) > Y$ *for* $x_0 < x < x_0 + \delta$.

(b) $\lim_{x\to x_0^-} f(x) = \infty$ *if, for every positive number* Y, *there is a positive* δ *such that* $f(x) > Y$ *for* $x_0 - \delta < x < x_0$.

(c) $\lim_{x\to x_0^+} f(x) = -\infty$ *if, for every negative number* Y, *there is a positive* δ *such that* $f(x) < Y$ *for* $x_0 < x < x_0 + \delta$.

(d) $\lim_{x\to x_0^-} f(x) = -\infty$ *if, for every negative number* Y, *there is a positive* δ *such that* $f(x) < Y$ *for* $x_0 - \delta < x < x_0$.

Tune-Up Exercises 1.2♦

[A]*Answer provided.* [O]*Outline of solution provided.*

Be sure you can work Exercises T1 through T12, and solve as many of Exercises T13 through T19 as you need for practice.

T1.[A] Give equations of **(a)** the line through the point $(3, 5)$ with slope -4 and **(b)** the line through the points $(5, 1)$ and $(-5, 2)$.

♦ Type 1, equations of lines

T2.[A] Give equations of **(a)** the line through $(-4, 7)$ with slope $\frac{1}{2}$ and **(b)** the line through $(-3, -5)$ and $(2, -2)$.

♦ Type 1, equations of lines

T3.[A] Give an equation of **(a)** the line that is parallel to the line $6x - 4y = 7$ and passes through $(1, -1)$ and **(b)** the line that is perpendicular to $3x + y = 0$ and passes through $(4, 7)$.

♦ Type 1, equations of lines

T4.[A] Give an equation of the line with x-intercept 5 and y-intercept -6. What is its slope?

♦ Type 1, equations of lines

♦The Tune-up Exercises and Problems are classified by type and content. The types are (1) basic, reactive; (2) basic reflective; (3) intermediate, reactive; (4) intermediate, reflective; (5) advanced, reactive; (6) advanced, reflective; and (7) advanced, theoretical.

T5.[A] Give equations of the lines through $(-4, 7)$ that are **(a)** parallel to the x-axis and **(b)** parallel to the y-axis.

♦ Type 1, equations of lines

T6.[O] Give an equation of the perpendicular bisector of the line segment between the points $(4, 6)$ and $(6, 0)$. (Use a drawing.)

♦ Type 1, equations of lines

T7.[A] What is the slope of the line $5x + 7y = 35$?

♦ Type 1, slope

T8.[O] Sketch the triangle with vertices $P(-5, 6)$, $Q(4, -3)$ and $R(6, 8)$ and find the lengths of its sides. Is it an isosceles triangle (does it have two equal sides)? Is it an equilateral triangle (are all of its sides equal)?

♦ Type 1, slopes and distances

T9.[O] Find the coordinates of the vertices of the triangle formed by the lines $y = \frac{1}{2}x$, $y = -2x$, and $y = -\frac{1}{3}x + 5$. Then show it is a right triangle and find its area, under the assumption that the scales on the axes are equal.

♦ Type 1, slopes and distances

T10.[A] Without generating the curves on your calculator or computer, describe the shapes of $y = \frac{1}{2}x^2 - 4$ and $y = 10 - 1/x$. Then sketch the two curves.

♦ Type 1, predicting slopes and graphs

T11.[A] What happens to the curve $y = x^{-1}$ if we change it to $y = -3x^{-1}$ and why?

♦ Type 1, predicting slopes and graphs

T12.[A] Each of Figures 34 through 37 shows the graph of $y = ax^2 + b, y = ax^3 + b, y = \dfrac{a}{x} + b$, or $y = \dfrac{a}{x^2} + b$ with integers a and b. Find the equations of the curves.

♦ Type 1, predicting slopes and graphs

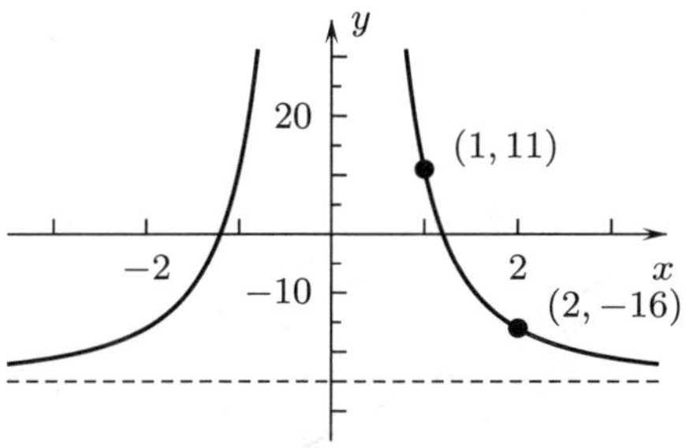

FIGURE 34

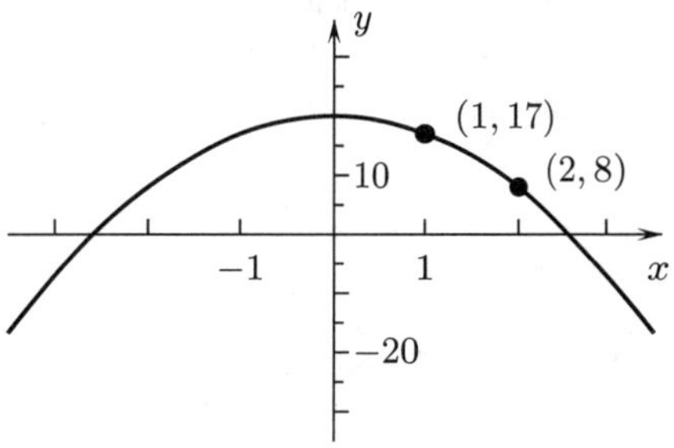

FIGURE 35

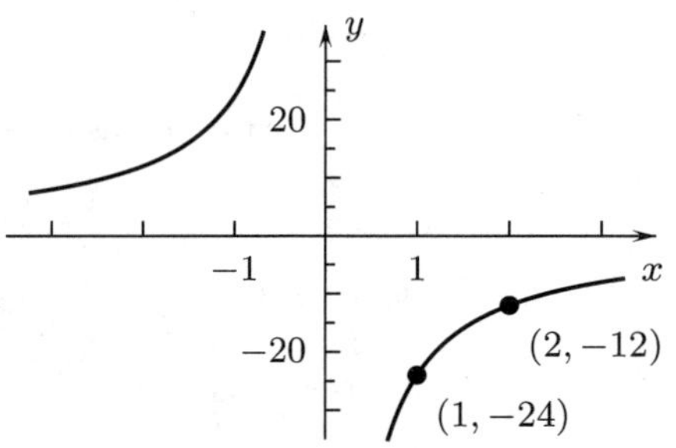

FIGURE 36

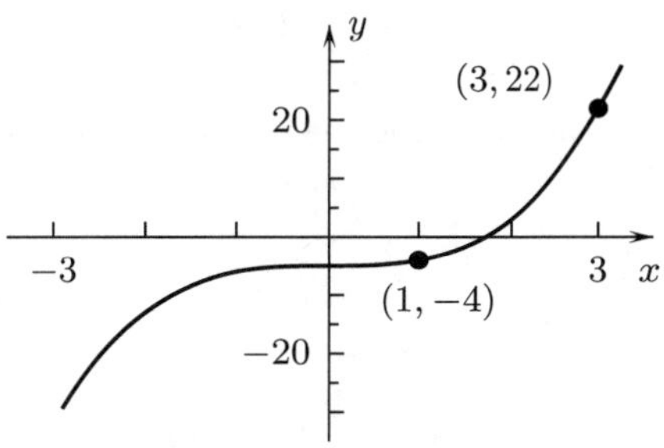

FIGURE 37

More practice

T13.[A] What is the slope of the line $x = 3y + 2$?

♦ Type 1, slopes

T14.[A] What is the slope of the line through $(3, -4)$ and $(-1, 4)$?

♦ Type 1, slopes

T15.[A] Give an equation of the line through the points $(-5, 0)$ and $(6, 10)$.

♦ Type 1, equations of lines

T16.[A] Give an equation of the line through the point $(4, 3)$ and parallel to $5x - 3y = 10$.

♦ Type 1, equations of lines

T17.[A] Give an equation of the perpendicular bisector of the line segment from $(-3, 2)$ to $(7, 6)$.

♦ Type 1, equations of lines

T18.[A] Without generating the curves on your calculator or computer, describe the shapes of $y = 6 - \frac{1}{3}x^3$ and $y = -1/x^2$.

♦ Type 1, predicting shapes and graphs

Problems 1.2

[A]*Answer provided.* [O]*Outline of solution provided.*

1.[A] What do the letters x, y, m, x_0 and y_0 represent in the point-slope equation $y = y_0 + m(x - x_0)$?

♦ Type 2, discussion

2. **(a)** Is the point (x, y) above or below the line $y = x$ if $y > x$? **(b)** How is (x, y) related to the line $y = x$ if $y < x$?

♦ Type 2, discussion

3. The lines $y = x$ and $y = -x$ are perpendicular if the scales on the x- and y-axes are the same. How are they related if the the number of units per inch is greater on the x-axis than on the y-axis?

♦ Type 2, discussion

4.[A] For what values of the parameter a is the curve $y = a/x^2$ entirely above the x-axis? (Answer this without using your calculator or computer.)

♦ Type 2, discussion

5. Give an equation of the line in Figure 38 **(a)** assuming the tick marks are one unit apart on the x- and y-axes and **(b)** assuming the tick marks are one unit apart on the x-axis and ten units apart on the y-axis.

♦ Type 2, equations of lines

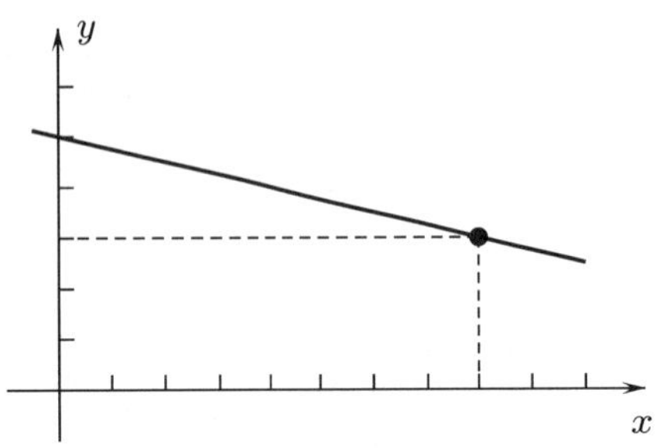

FIGURE 38

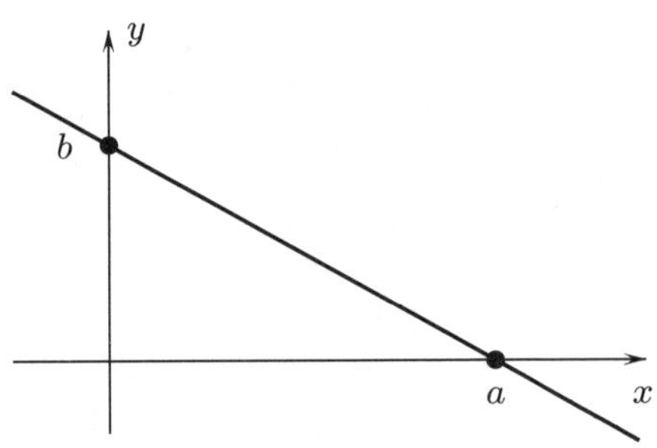

FIGURE 39

6. Show that the line with nonzero x- and y-intercepts a and b, respectively, (Figure 39) has the equation $\frac{x}{a} + \frac{y}{b} = 1$. (This is called the TWO-INTERCEPT EQUATION.)

♦ Type 2, equations of lines

7. For what values of the integer n is the curve $y = 1/x^n$ partially above and partially below the x-axis?

♦ Type 2, predicting the shapes of graphs

8.[A] Which of x^{-2} and x^{-1} dominates the other **(a)** when x is a large positive or large negative number and **(b)** when x is close to 0?

♦ Type 2, recognizing dominating terms

9.[A] Which are the dominant terms in the numerator and in the denominator of $\dfrac{x^3 - 5}{7x^2 + 3x}$ **(a)** when x is a large positive or large negative number and **(b)** when x is close to 0?

♦ Type 2, recognizing dominating terms

10. **(a)** Which of the lines in Figures 40 through 42 has the equation $y = ax + b$ with $a > 0$ and $b > 0$? **(b)** Which has the equation $y = ax + b$ with $a < 0$? Is b positive or negative in this case? **(c)** Which has the equation $y = ax + b$ with $b < 0$? Is a positive or negative in this case?

♦ Type 2, classifying parameters in linear functions

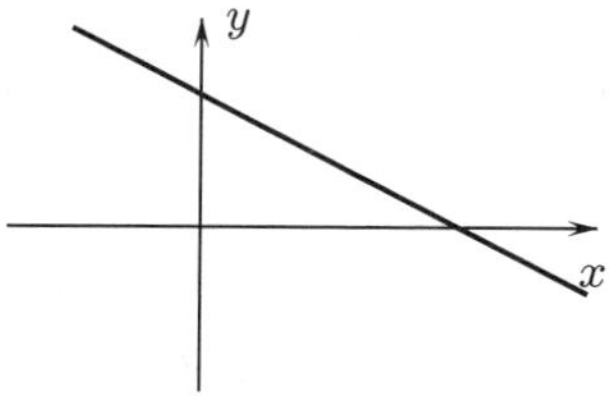

FIGURE 40

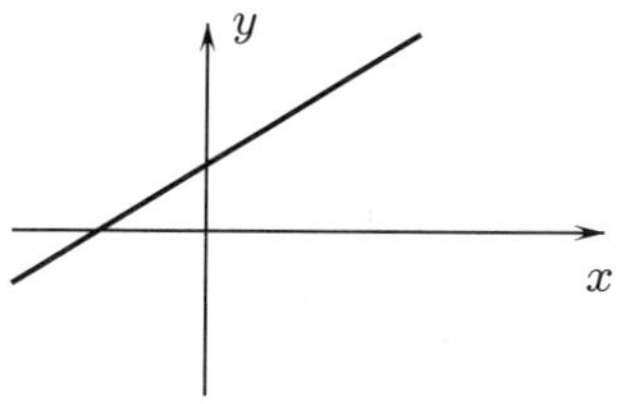

FIGURE 41

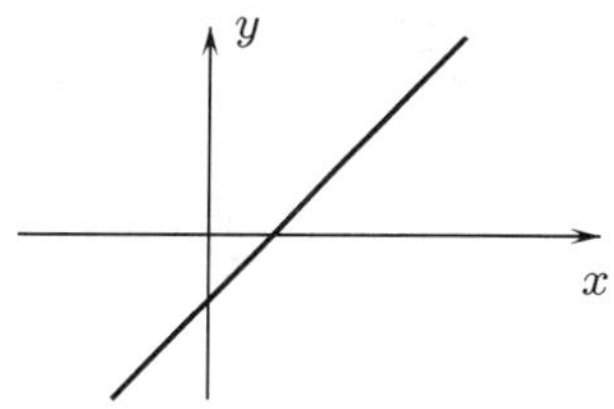

FIGURE 42

11.[A] Give an equation of the line through the origin that has negative slope and bisects the angle between the x- and y-axes in an xy-plane with equal scales on the axes.

◆ Type 2, equation of a line

12. Find the point Q such that the x-axis is the perpendicular bisector of the line segment PQ, where $P = (3, 4)$.

◆ Type 2, a perpendicular bisector

13.[A] Find the points with x-coordinate 4 that are a distance 5 from the origin.

◆ Type 2, using the Pythagorean Theorem

14. Give an equation with one parameter for all lines that do not pass through the origin and are such that their x-intercepts equal their y-intercepts. What are the slopes of these lines?

◆ Type 2, giving a linear function with a parameter

15.[A] For what values of m does the line $y = mx$ pass through the acute angle ABC determined by the points $A(2, 5)$, $B(0, 0)$, and $C(5, 2)$?

◆ Type 4, finding a range of slopes

16. Sketch the quadrilateral $ABCD$ with vertices $A(0, 0)$, $B(4, 1)$, $C(2, 3)$, and $D(-2, 2)$. Is it a parallelogram (are its opposite sides equal)? Is it a rhombus (are all of its sides equal)?

◆ Type 2, parallelograms

17. Sketch the triangle ABC with vertices $A(2, -3)$, $B(-3, -3)$, and $C(3, 2)$. Is it a right triangle? Explain.

◆ Type 2, right triangles

18. Show that the quadrilateral $ABCD$ with vertices $A(0, 0)$, $B(5, 5)$, $C(-2, 6)$, and $D(-7, 1)$ is a rhombus (see Exercise 16). Then show that its diagonals are perpendicular.

◆ Type 2, a rhombus

19.[A] The tangent line at $(-2, -4)$ on the circle $x^2 + y^2 = 20$ is perpendicular to the radius at that point. What is the slope of the tangent line?

◆ Type 2, perpendicular lines

20.[A] **(a)** For what values of x do $y = \frac{1}{4}x^4$ and $y = x^2$ intersect? **(b)** For what values of x is $y = \frac{1}{4}x^4$ above $y = x^2$? **(c)** For what values of x is $y = \frac{1}{4}x^4$ below $y = x^2$? Verify your conclusions algebraically and check by generating $y = \frac{1}{4}x^4$ and $y = x^2$ in the window $-3 \le x \le 3, -1.5 \le y \le 9$.

◆ Type 2, intersections of curves

21. **(a)** Where do $y = x^{-1}$ and $y = 9x^{-3}$ intersect? **(b)** For what values of x is $y = x^{-1}$ above $y = 9x^{-3}$? **(c)** For what values of x is $y = x^{-1}$ below $y = 9x^{-3}$? **(d)** Use the formulas to explain the results of parts (b) and (c). Check your answers by generating $y = x^{-1}$ and $y = 9x^{-3}$ together for $-7 \le x \le 7, -1.5 \le y \le 1.5$.

◆ Type 2, intersections of curves

Generate the curves in Problems 22 through 30 with the given windows and copy them on your paper. Describe in each case the behavior of the graph as $x \to \pm\infty$ and near any vertical asymptotes. Explain how the equation causes the curve to have its particular shape for x near 0 and for large positive and negative x by relating the curve to the appropriate curves $y = ax^n + b$.[†]

[†]If you are using a calculator other than a TI-83 or TI-86, enter $\frac{1}{3}x^3$ in Problems 29 through 30 either as $(1/3)x$^3, x^3/3, or 1/3 * x^3 and not as 1/3x^3 to keep the x^3 from being in the denominator of the fraction.

22.[A] $y = 3x^2 - x^4, -2 \le x \le 2, -3 \le y \le 3$

♦ Type 2, explaining the shapes of graphs

23.[A] $y = 2x + \dfrac{1}{x}, -3 \le x \le 3, -8 \le y \le 8$

♦ Type 2, explaining the shapes of graphs

24. $y = x^2 - \dfrac{1}{x^2}, -3 \le x \le 3, -8 \le y \le 5$

♦ Type 2, explaining the shapes of graphs

25. $y = x^3 - \dfrac{1}{x}, -3 \le x \le 3, -12 \le y \le 12$

♦ Type 2, explaining the shapes of graphs

26.[A] $y = x - \frac{1}{3}x^3, -3 \le x \le 3, -2 \le y \le 2$

♦ Type 2, explaining the shapes of graphs

27. $y = x + \frac{1}{3}x^3, -2 \le x \le 2, -4 \le y \le 4$

♦ Type 2, explaining the shapes of graphs

28. $y = -3x + 2x^2 - \frac{1}{3}x^3, -1 \le x \le 5, -3 \le y \le 3$

♦ Type 2, explaining the shapes of graphs

29. $y = 4 - x^2 + 1/x, -3 \le x \le 3, -4 \le y \le 8$

♦ Type 2, explaining the shapes of graphs

30. $y = (2 + 1/x)(3 - 5/x^2), -4 \le x \le 4, -6 \le y \le 8$

♦ Type 4, explaining the shapes of graphs

31.[A] Which of the curves in Figures 43 through 48 is the graph of $y = a + b/x^2$ with $a > 0$? Is b positive or negative? Give your reasoning.

♦ Type 4, explaining the shapes of graphs

32. Which of the curves in Figures 43 through 48 is the graph of $y = a + bx^3$ for some constants a and b? Is a positive or negative? Is b positive or negative? Give your reasoning.

♦ Type 4, explaining the shapes of graphs

33. Which of the curves in Figures 43 through 48 is the graph of $y = a + b/x$ for some constants a and b? Is a positive or negative. Is b positive or negative? Give your reasoning.

♦ Type 4, explaining the shapes of graphs

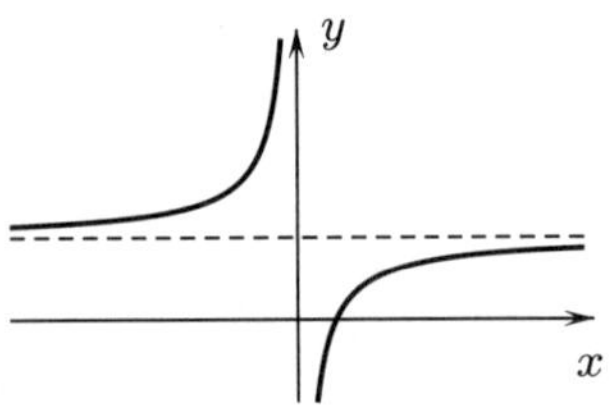

FIGURE 43

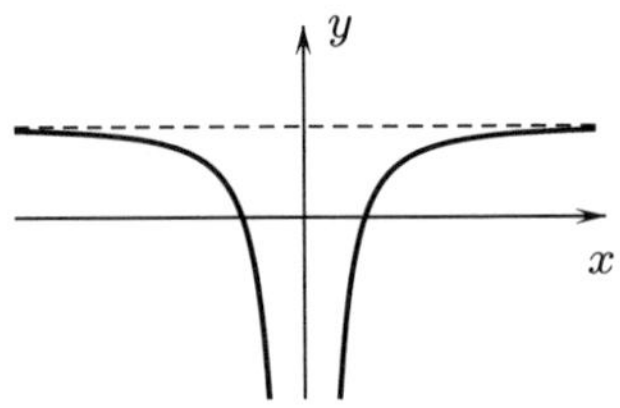

FIGURE 44

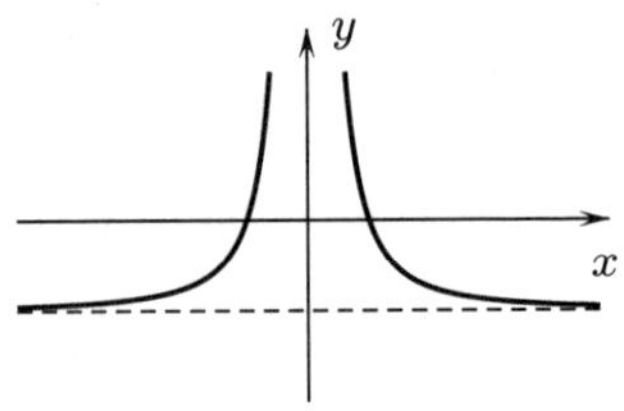

FIGURE 45

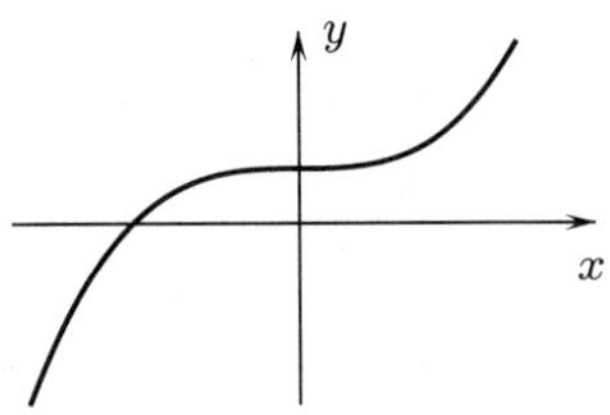

FIGURE 46

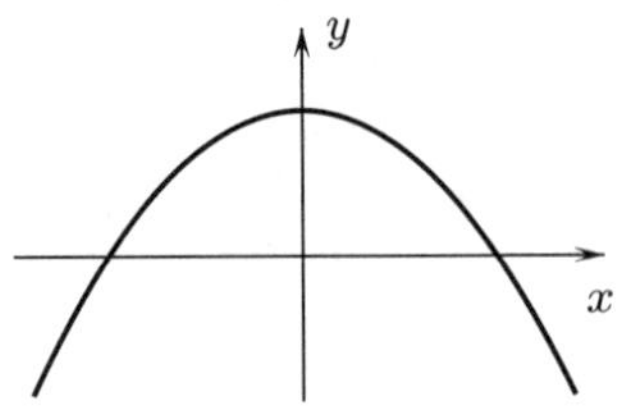

FIGURE 47

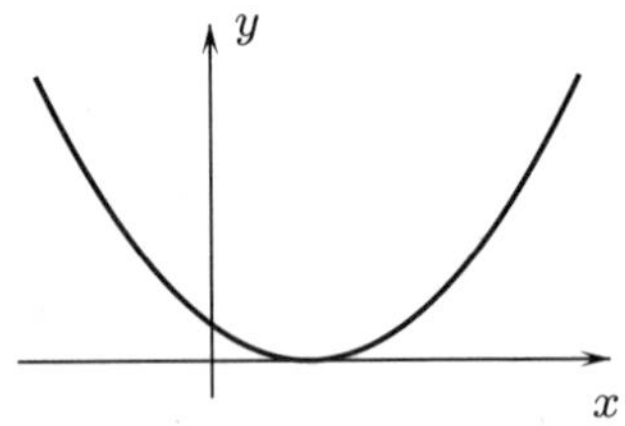

FIGURE 48

34.[A] Are the three points $P = (-3, 6), Q = (3, 2)$, and $R = (6, 0)$ on a line? Explain.

♦ Type 2, slopes

35. **(a)** Show that for any constant k, the points $A(0,0), B(2,1), C(k+2,6)$, and $D(k,5)$ are the vertices of a parallelogram $ABCD$. **(b)** How does the parallelogram move as k is changed? **(c)** For what value of k is it a rectangle?

♦ Type 2, slopes

36. Show that the quadrilateral $ABCD$ with vertices $A(0,0), B(5,5), C(-2,6), D(-7,1)$ is a rhombus (a quadrilateral with sides of equal length). Then show that its diagonals are perpendicular.

♦ Type 2, the Pythagorean Theorem and perpendicular slopes

37. Figure 49 shows a design formed by four lines in an xy-plane with equal scales on the coordinate axes. The line through the origin and the point P has slope less than 1, there is a right angle at P, and the design is symmetric about the y-axis. Find equations of four lines that satisfy these conditions. (There are an infinite number of possible choices.) Check your results by generating the four lines on your calculator or computer.

♦ Type 4, creating a design

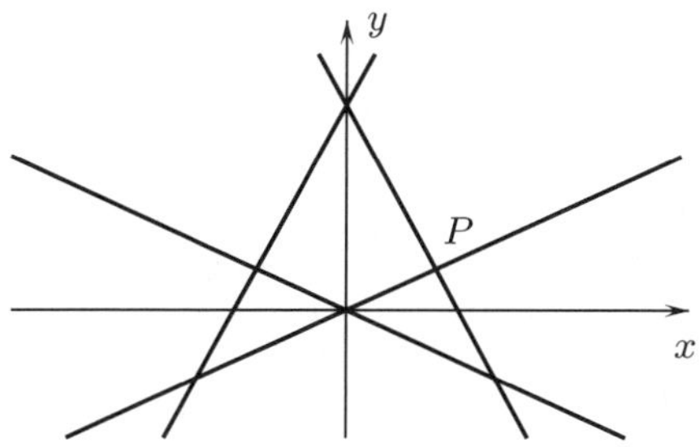

FIGURE 49

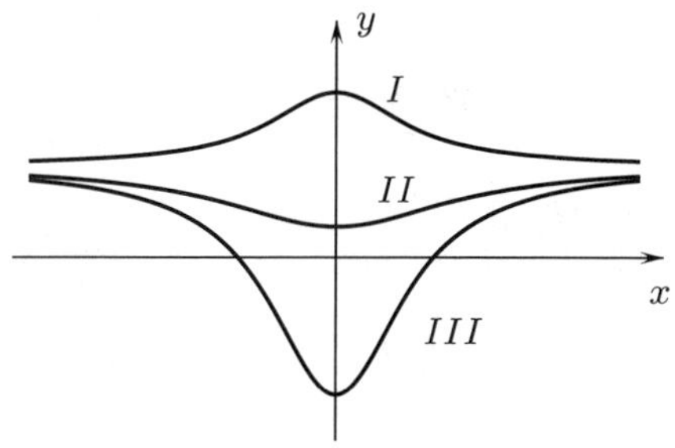

FIGURE 50

38. Which of the curves in Figure 50 has the equation $y = \dfrac{x^2+a}{x^2+b}$ **(a)** with $0 < a < b$, **(b)** with $0 < b < a$, and **(c)** with $a < 0$ and $b > 0$? Give your reasoning.

♦ Type 4, the effects of parameters on the shapes of curves

39. Generate $y = x^2 + ax$ on your calculator or computer with $-4 \le x \le 4, -2 \le y \le 5$, first for $a = 0, 1$ and 2, and then for $a = 0, -1$, and -2. Explain how changing a changes the graph and why.

♦ Type 4, the effects of parameters on the shapes of curves

40. How does changing b change the curve $y = 5x + bx^3$ and why? (Generate the curve for sample values of b in the window $-4 \le x \le 4, -30 \ \le y \le 30$ with y-scale $= 10$.)

♦ Type 4, the effects of parameters on the shapes of curves

41.[A] The tangent line at a point on a circle is perpendicular to the redius at that point. Give the equations of the two tangent lines to the circle $x^2 + y^2 = 25$ that have slope $\frac{3}{4}$.

♦ Type 4, tangent lines of circles

42.[A] The line segment between the points $A(1,3)$ and $B(3,1)$ is a chord of two circles of radius $\sqrt{10}$. Give the coordinates of the centers of the two circles.

♦ Type 4, tangent lines of circles

43.[A] Give an equation for the line that has its x-intercept equal to its y-intercept and passes through $(-4,-2)$.

♦ Type 4, equation of a line

44. Find the two right triangles such that their perpendicular sides are along the x- and y-axes, their longest sides pass through $(2,4)$, and the area of each is 18.

♦ Type 4, right triangles

45. Show that any triangle that is inscribed in the semicircle $y = \sqrt{a^2 - x^2}$ and has one side along the diameter is a right triangle. (Assume that there are equal scales on the coordinate axes and that a is a positive constant.)

♦ Type 4, right triangles

46. Show that for any nonzero constants a, b, and c, the quadrilateral with vertices $A(0,0), B(a,0), C(a+b,c)$, and $D(b,c)$ is a parallelogram. Show it is a rhombus if and only if its diagonals are perpendicular.

♦ Type 4, parallelogram

47.[O] What are $\displaystyle\lim_{x\to 2+} \frac{1-x}{x-2}$ and $\displaystyle\lim_{x\to 2-} \frac{1-x}{x-2}$?

♦ Type 3, limits

48.[A] What are $\displaystyle\lim_{x\to 1} \frac{1}{(x-1)^2}$ and $\displaystyle\lim_{x\to\infty} \frac{1}{(x-1)^2}$?

♦ Type 3, limits

49. What are $\lim_{x\to 5+} \dfrac{1}{x^2-25}$ and $\lim_{x\to 5-} \dfrac{1}{x^2-25}$?

♦ Type 3, limits

50. What are $\lim_{x\to -2} \dfrac{x^5}{(x+2)^4}$ and $\lim_{x\to\infty} \dfrac{x^5}{(x+2)^4}$?

♦ Type 4, limits

Use Definitions 1A, 2A, and 3A (given with the Principles and Procedures) to verify the limits in Problems 51 through 59.

51.[O] $\lim_{x\to\infty} x^3 = \infty$

♦ Type 6, limits

52.[A] $\lim_{x\to -\infty} (\frac{1}{10}x) = -\infty$

♦ Type 6, limits

53. $\lim_{x\to\infty} \sqrt{x} = \infty$

♦ Type 6, limits

54.[O] $\lim_{x\to\infty} \dfrac{3}{x-2} = 0$

♦ Type 6, limits

55.[A] $\lim_{x\to -\infty} \dfrac{8}{x^3} = 0$

♦ Type 6, limits

56. $\lim_{x\to\infty} x^{-1/10} = 0$

♦ Type 6, limits

57. $\lim_{x\to\infty} \sqrt{9+1/x} = 3$

♦ Type 6, limits

58.[A] $\lim_{x\to 0+} \dfrac{1}{x^3} = \infty$

♦ Type 6, limits

59. $\lim_{x\to 5-} \dfrac{1}{\sqrt{5-x}} = \infty$

♦ Type 6, limits

60. $\lim_{x\to 1-} \dfrac{1}{\sqrt[3]{x-1}} = -\infty$

♦ Type 6, limits

61. **(a)** Use Definition 1A to prove that $\lim_{x\to\infty} (1.2x + x\sin x) = \infty$. **(b)** Use Definition 2A to prove that $\lim_{x\to\infty} \left(\dfrac{\sin x}{x}\right) = 0$.

♦ Type 6, limits

62. Use Definitions 1A and 2A to prove that $\lim_{x\to\infty} \cos x$ is not defined.

♦ Type 7, limits

Section 1.3

Linear functions and constant rates of change

OVERVIEW: *In many applications of calculus, the first step is to find formulas for the functions that are to be used in the mathematical models of the situations being studied. This process is illustrated in this section in cases involving linear functions (functions whose graphs are lines) and their constant rates of change. This material also serves as background for the study of the derivative and its applications in Chapters 2 and 3, since the rates of change of nonlinear functions are found by using derivatives.*

Topics:

- ***Constant rates of change of linear functions***
- ***Functions constructed from linear functions***
- ***Fitting data approximately with linear functionns***

Fahrenheit and Celsius

The most commonly used scales for temperature—the Fahrenheit and Celsius scales—are related by linear functions.[†] The graph of degrees Fahrenheit T_F as a function of degrees Celsius T_C is the line in Figure 1, and the graph of degrees Celsius as a function of degrees Fahrenheit is the line in Figure 2. The dots on the graphs correspond to the freezing point of water, which is 0°C and 32°F, and to its boiling point, which is 100°C and 212°F.

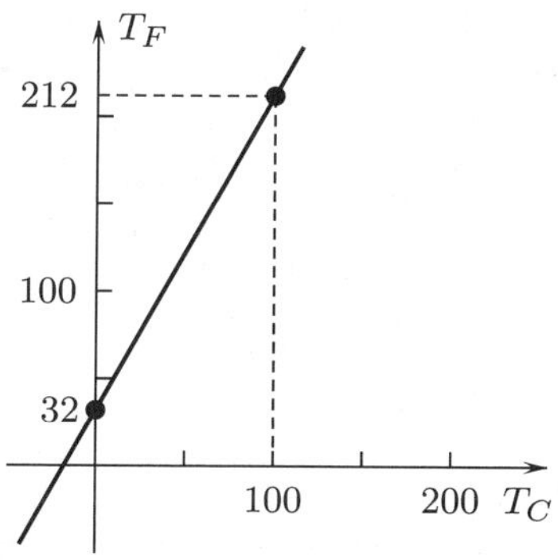

FIGURE 1

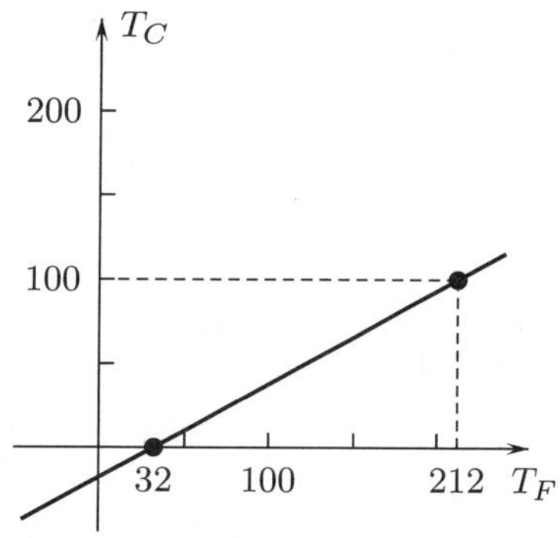

FIGURE 2

Question 1 What are the slopes of the lines in Figures 1 and 2 and how are they related? (Use fractions, not decimals.)

Question 2 Give a formula for T_F (degrees Fahrenheit) as a function of T_C (degrees Celsius). Then give a formula for T_C (degrees Celsius) as a function of T_F (degrees Fahrenheit).

Example 1 What do the expressions **(a)** $T_F(15)$ and **(b)** $T_C(59)$, represent in terms of temperatures and what are their values? What are the values of **(c)** $T_C(T_F(15))$ and **(d)** $T_F(T_C(59))$ and why?

SOLUTION **(a)** $T_F(15) = 32 + \frac{9}{5}(15) = 59$ is the temperature in degrees Fahrenheit that corresponds to 15 degrees Celsius.

(b) $T_C(59) = \frac{5}{9}(59 - 32) = 15$ is the temperature in degrees Celsius corresponding to 59 degrees Fahrenheit.

[†]The Fahrenheit scale was devised by Gabriel Daniel Fahrenheit (1686–1736), a German physicist who was the first to make mercury thermometers. The Celsius or Centigrade scale was introduced by a Swedish astronomer, Anders Celsius (1701–1744).

♦ SUGGESTIONS TO INSTRUCTORS: This section should not be skipped, but can be covered on homework and in a small amount of class time in courses with well-prepared students. Have students begin work on some or all of Questions 1 through 25 before class. Problems 3, 7, 12, 14, 17, 21 are good for in-class discussion.

Typeset by $\mathcal{A}\mathcal{M}\mathcal{S}$-TEX

(c) $T_C(T_F(15)) = T_C(59) = 15°C$ because converting 15°C into degrees Fahrenheit and then back into degrees Celsius gives 15°C.
(d) $T_F(T_C(59)) = T_F(15) = 59°F$ because converting 59°F into degrees Celsius and then back into degrees Fahrenheit gives 59°F. □

Sales, revenue, cost, and profit

Mary pays $250 overhead per day to rent space and hire help for her cotton candy stand at a county fair. She sells the cotton candy for $32 per pound and she has to pay $7 per pound of cotton candy for supplies and the fair management's percent of the revenue.

Question 3 **(a)** How much money would she take in if she made and sold x pounds in a day? This is her revenue $R(x)$, which is a linear function of x. Sketch its graph in an xR-plane.
(b) What would her expenses be if she made and sold x pounds in a day? This is her cost $C(x)$, which is also a linear function of x. Sketch its graph in an xC-plane.
(c) Give a formula for her day's profit $P(x)$ as a function of the number of pounds x of cotton candy she makes and sells in a day, and sketch its graph in an xP-plane.

Question 4 What are the x- and P-intercepts of the graph of $P(x)$ and what do they represent?

Question 5 Describe two or more ways in which this mathematical model of Mary's cotton candy business might not be very realistic.

Constant velocity

If an object moves on a straight path, we can use an s-axis along that path to indicate the object's position. The object's s-coordinate is then a function of the time t.

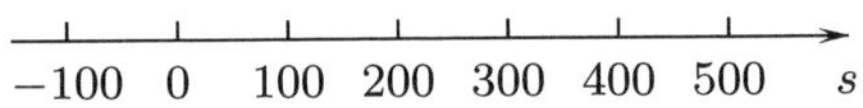

FIGURE 3

Question 6 A moving van is 200 miles east of El Paso, Texas, at noon and is driving east at the constant velocity of 60 miles per hour. How far is the van east of El Paso t hours after noon?

The formula $s = 200 + 60t$ from Question 6 gives the van's distance s east of El Paso as a function of the time t (hours) after noon, and the graph of this function is the line in the ts-plane of Figure 4. Notice that the slope of the graph is the velocity, 60 miles per hour, of the van.

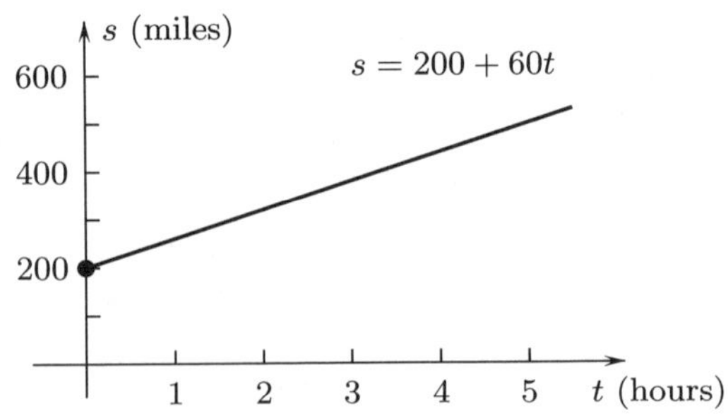

Velocity = Slope
= 60 miles per hour

FIGURE 4

Question 6 illustrates the following basic principle.

Rule 1 *If an object moves at a constant velocity along an s-axis, then its position s is a linear function of the time t. The slope of the graph of this function is the object's velocity in the positive s-direction, with units determined from those used for measuring t and s.*

Example 2 A ball is at $s = 5t + 7$ feet on an s-axis at time t seconds. **(a)** What is its velocity? **(b)** What does the s-intercept 7 of the line $s = 5t + 7$ represent?

SOLUTION **(a)** The velocity of the ball in the positive s-direction is the slope, 5 feet per second, of the line $s = 5t + 7$.
(b) The ball is at $s = 7$ on the s-axis at time $t = 0$. □

Other constant rates of change

If the graph of $f(x)$ is a line of slope m, then the ratio $\dfrac{f(b) - f(a)}{b - a}$ of the rise and the run between any two points on the line equals m (Figure 5). We say that the RATE OF CHANGE of the function with respect to its variable is the constant m:

Rule 2 *If the graph of $f(x)$ is a line, then the slope of that line is the (constant) rate of change of $f(x)$ with respect to x.*

Rate of change

$= \text{Slope}$

$= \dfrac{f(b) - f(a)}{b - a}$

FIGURE 5

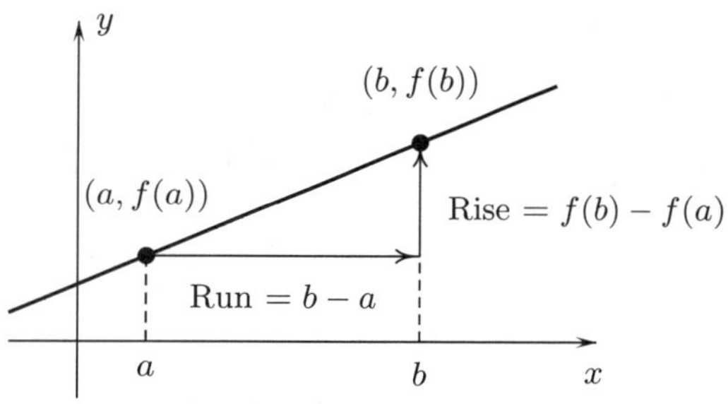

According to Rule 2, the slope $\frac{9}{5}$ of the line $T_F = 32 + \frac{9}{5}T_C$ in Figure 1 is the rate of change of degrees Fahrenheit T_F with respect to degrees Celsius T_C, and the slope $\frac{5}{9}$ of the line $T_C = \frac{5}{9}(T_F - 32)$ in Figure 2 is the rate of change of degrees Celsius with respect to degrees Fahrenheit.

The slopes of the graphs of Mary's revenue $R = 32x$, cost $C = 250 + 7x$, and profit $P = 25x - 250$ from Questions 3 through 5 are 32, 7, and 25, respectively. Consequently, the rate of change of her revenue, cost, and profit with respect to the number of pounds of cotton candy she makes and sells in a day are 32, 7, and 25 dollars per pound.

Question 7 The circumference C of a circle is $C = 2\pi r$ if its radius is r. What is the rate of change of the circumference with respect to the radius if lengths are measured in meters?

Rates of change of functions whose graphs are not lines are found by using derivatives, as we will see in Chapter 2.

Driving on empty

Imagine that you start on a trip with 20 gallons of gas in your tank, that your car consumes gasoline at the constant rate of 0.05 gallons per mile, and that you drive until your tank is empty without buying any more gas.

Question 8 What mileage does your car get on the trip, measured in miles per gallon?

Question 9 Give a formula for the volume V of gasoline remaining in your tank as a function of the distance s that you have traveled on your trip. What is the domain of this function?

Question 10 Suppose you drive at the constant speed of 60 miles per hour. Give a formula for the distance you travel s as a function of the number of hours t that you have traveled. What is its domain?

Question 11 **(a)** Use the formulas from Questions 9 and 10 to give a formula for the volume of gas in your tank as a function of t. What is its domain?
(b) What is the rate of change of this function with respect to t?
(c) How much does the volume of gasoline in the tank change every hour?

The graphs of the linear functions that you found in answering Questions 9 through 11 are shown in Figures 6 through 8. The slope of the line $V = 20 - 0.05s$ in Figure 6 is -0.05 and is the rate of change of the volume of gasoline in your tank with respect to the distance you have traveled. Its units are gallons per mile, and it is negative because the volume decreases as the distance you have traveled increases. The slope of the line $s = 60t$ in Figure 7 is 60 miles per hour and is the rate of change of the distance with respect to time (your velocity). The slope of $V = 20 - 3t$ in Figure 8 is -3 gallons per hour and is the rate of change of the volume of gas with respect to time.

Notice that these rates of change are related by the equation,

$$\left[-0.05\ \frac{\text{gallons}}{\text{mile}}\right]\left[60\ \frac{\text{miles}}{\text{hour}}\right] = -3\ \frac{\text{gallons}}{\text{hour}}.$$

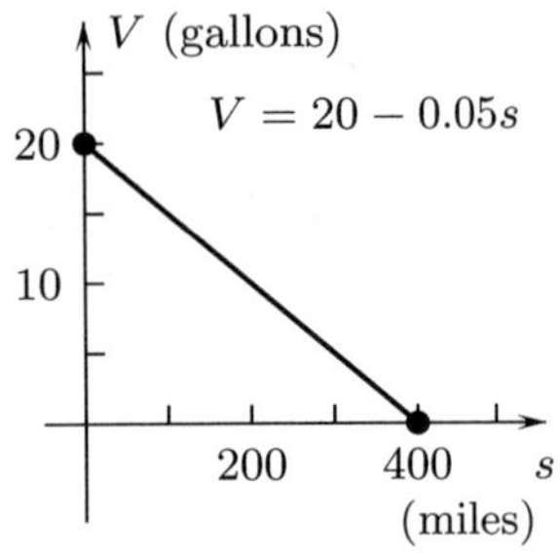

FIGURE 6

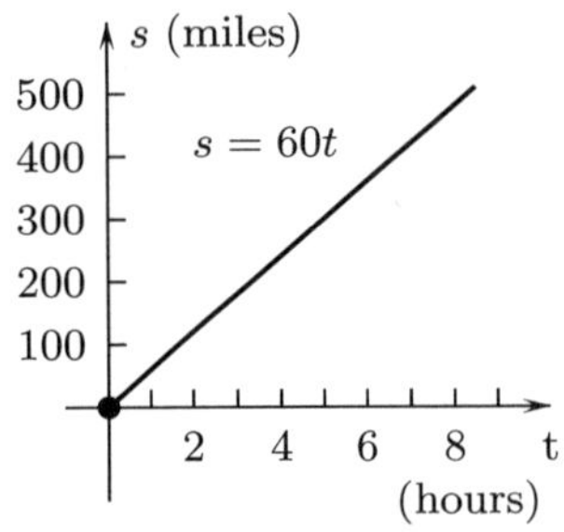

FIGURE 7

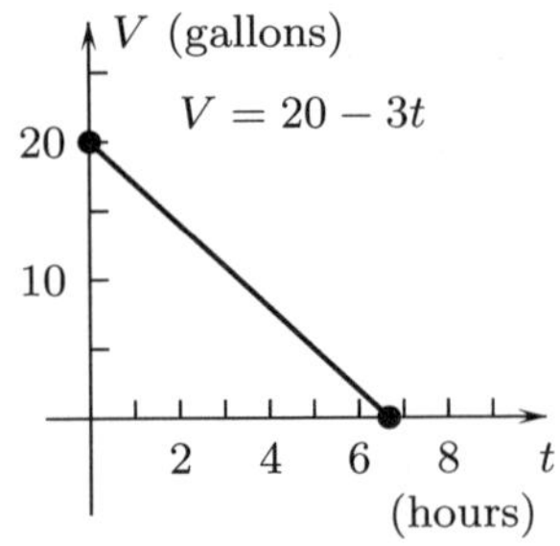

FIGURE 8

In our next application we will find a formula for a linear function by first finding an equation that relates the two variables.

Pot luck

John has \$90 to spend on lima beans and corn to make succotash for a large potluck dinner. The lima beans cost \$3 per pound and the corn costs \$0.75 per pound.

Question 12 **(a)** How much does it cost him to buy L pounds of lima beans?
(b) What does he have to spend to buy C pounds of corn?
(c) What does he spend for L pounds of lima beans and C pounds of corn?

Question 13 Find an equation relating the pounds of lima beans L and pounds of corn C he buys under the assumption that he spends a total of \$90 on both. The graph of this equation in an CL-plane is the line in Figure 9.

Question 14 What is the C-intercept of the graph and what does it represent? What is the L-intercept and what does it represent?

Question 15 Solve the equation in Question 13 to obtain a formula for L as a function of C. The line in Figure 9 is the graph of this function.

Question 16 **(a)** What is the meaning of the function $L(C)$ of Question 15 in terms of pounds of corn and lima beans?
(b) What is the slope of its graph and what does it represent?
(c) Why is the slope negative?

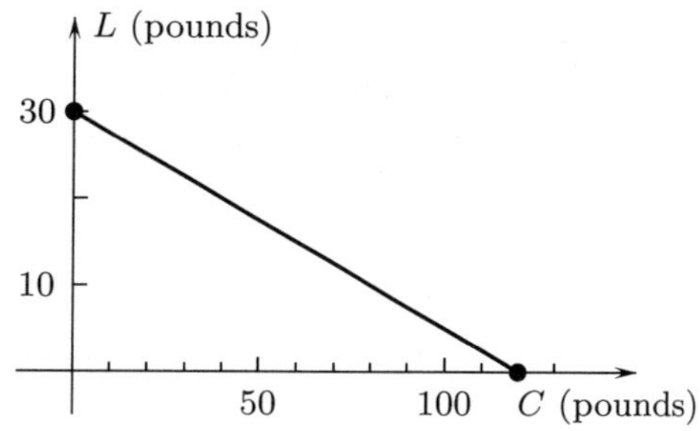

FIGURE 9

Watch your shadow

A six-foot tall man is walking away from a fifteen-foot high lamppost. We want to find the length y of his shadow as a function of his distance x from the lamppost. To do this we use the mathematical model of two right triangles in Figure 10. The lamppost and the man form the vertical sides of the triangles and the ground is their base. We will use the fact that the triangles are similar to find an equation relating x and y.

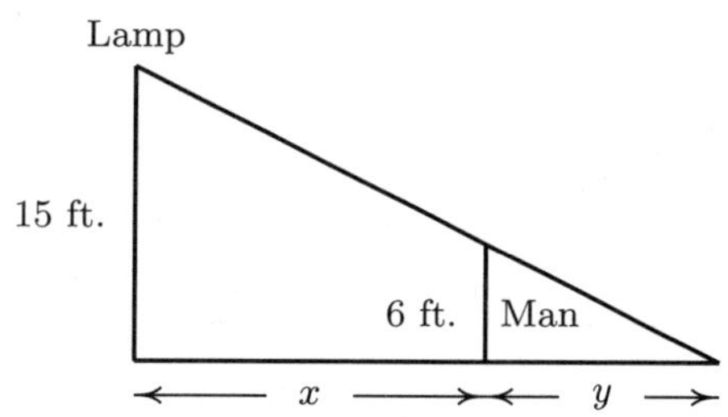

FIGURE 10

Example 3 Use the similar triangles† in Figure 10 to find a formula for the length y (feet) of the man's shadow as a function of his distance x (feet) from the lamppost.

SOLUTION Because the triangles are similar, the length $x + y$ of the base of the larger triangle divided by its height 15 equals the length y of the base of the smaller triangle divided by its height 6: $\dfrac{x+y}{15} = \dfrac{y}{6}$. Multiplying both sides of this equation by 15 and then by 6 gives $6x + 6y = 15y$, from which we obtain $9y = 6x$ and then $y = \frac{2}{3}x$. The graph of this linear function is the line in Figure 11. □

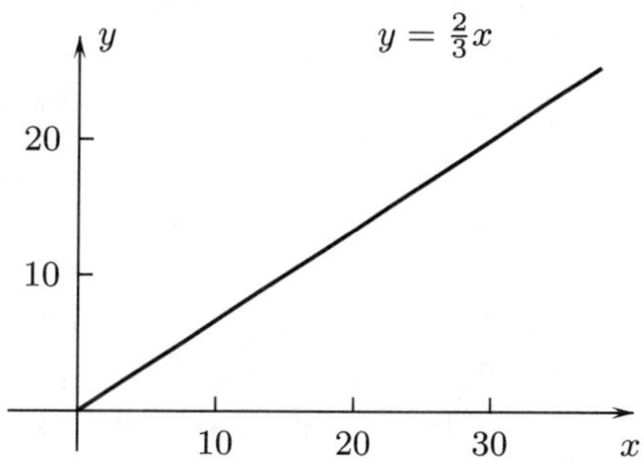

FIGURE 11

†Recall that two triangles are similar if they have the same angles.

Question 17 Imagine that the man in Figure 10 is 10 feet from the lamppost at time $t = 0$ (seconds) and that he walks in a straight path away from it with the constant velocity of 4 feet per second for ten seconds. **(a)** Give a formula for his distance x from the lamppost as a function of t. **(b)** Give a formula for the length of his shadow y as a function of t. **(c)** How fast does the man's shadow grow?

Question 18 Describe two or more ways in which the representation of the lamppost, the man, and his shadow by similar right triangles might not be very accurate.

In the next applications we use linear functions to construct nonlinear functions.

Are they drifting apart?

One Thanksgiving morning, Jack is driving at the constant speed of 70 miles per hour on I-80 west toward Pardoe, Pennsylvania. Jill is driving north on I-79 at the constant speed of 50 miles per hour away from Pardoe (Figure 12). At 10:00 AM Jack is 60 miles east of Pardoe and Jill is 20 miles north of it. We want to know how close they come to each other as they continue to drive.

We imagine there are x- and y-axes along the two freeways, as in Figure 12, and let x be Jack's coordinate on the x-axis and y be Jill's coordinate on the y-axis, both measured in miles. The variable distances x and y are shown in Figure 12, rather than the initial distances of 60 and 20 miles, to help us study the car's motion. We will use formulas for x and y as functions of t to obtain a formula for the distance D between Jack and Jill as a function of t.

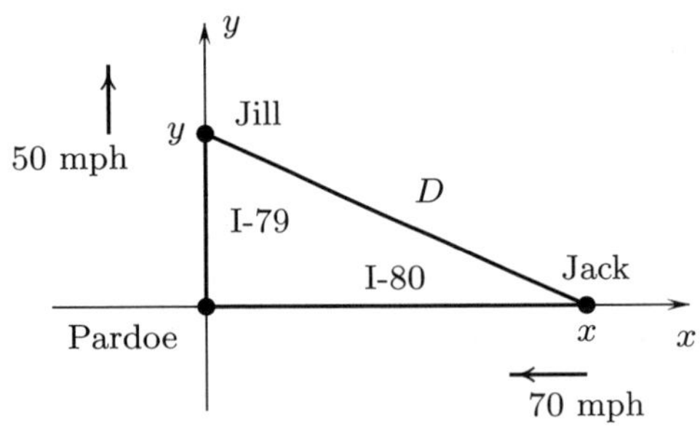

FIGURE 12

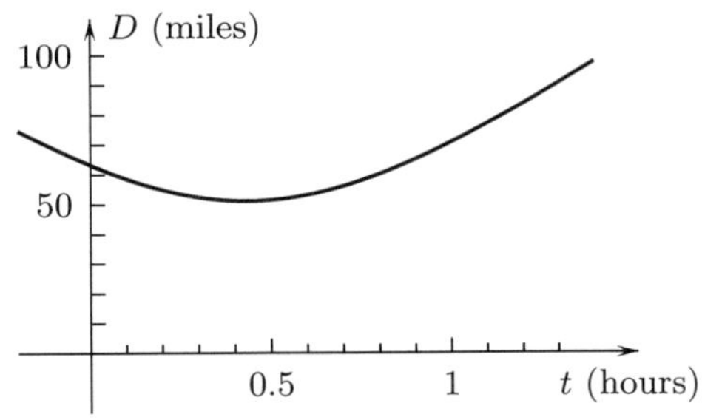

FIGURE 13

Question 19 **(a)** How far is Jack east of Pardoe t hours after 10:00? This formula gives his x-coordinate as a function of t. **(b)** How far is Jill north of Pardoe t hours after 10:00? This gives her y-coordinate as a function of t. **(c)** Use the Pythagorean Theorem to give the distance D between Jack and Jill as a function of t. **(d)** Generate on your calculator or computer the graph of $D(t)$ with $-0.2 \le t \le 1.5$, $-20 \le D \le 110$ that is shown in Figure 13. Use a minimum or trace command or the graph to find approximately how close Jack and Jill come to each other and the time when they are closest.[†]

Average profit

Mary with her cotton candy stand described with Questions 3 through 5 has a profit of $P = 25x - 250$ dollars if she makes and sells x pounds of cotton candy in a day.

Example 4 Give a formula for Mary's average profit per pound A as a function of x.

SOLUTION Her average profit per pound is her profit P divided by the number of pounds x of cotton candy she sells: $A = P/x = 25 - 250/x$ dollars per pound. The graph of this function is the hyperbola in Figure 14. □

Question 20 **(a)** What is the significance of the x-intercept of the curve in Figure 14? **(b)** The formula $A = 25 - 250/x$ shows that A is close to 25 for large x. What is it about the cotton candy business that causes this to occur? **(c)** Why is A negative for x close to 0? **(d)** Why is A a large negative number for x close to zero?

[†]We will be able to find exact answers to questions such as this using the derivative in Chapter 3.

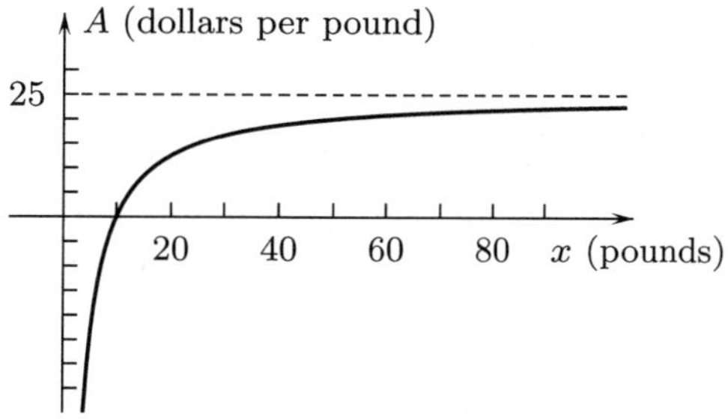

FIGURE 14

In our last examples of this section we use linear functions derived from empirical data.

Energy and aging

Table 1 gives the results of a study of energy expenditure for men of different ages.[(1)] The second row gives the average total daily energy expenditure for each age group; the third row gives the average daily basal energy expenditure (basal metabolism), which is the energy expended by the body apart from that used in activities; and the fourth row gives the average daily energy expenditure due to activity.

TABLE 1. ENERGY EXPENDITURES (CALORIES PER DAY)

Age	30	40	50	60	70	80
Total energy expenditure	2700	2650	2470	2360	2340	2110
Basal energy expenditure	1590	1600	1580	1560	1500	1400
Energy spent in activities	1110	1050	890	800	840	710

Question 21 How are the data in the last three rows of Table 1 related and why?

The data in Table 1 is plotted in Figure 15, where the upper string of dots gives the total energy expenditure, the dots in the middle give the basal energy expenditure, and the dots at the bottom give the energy expended in activities. In Figure 16 lines have been drawn to approximate the dots.

Question 22 Use approximate coordinates of two points on the upper line in Figure 16 to find an approximate equation for it. This gives a linear function $E_T(t)$ that is a model of total daily energy expenditure as a function of age.

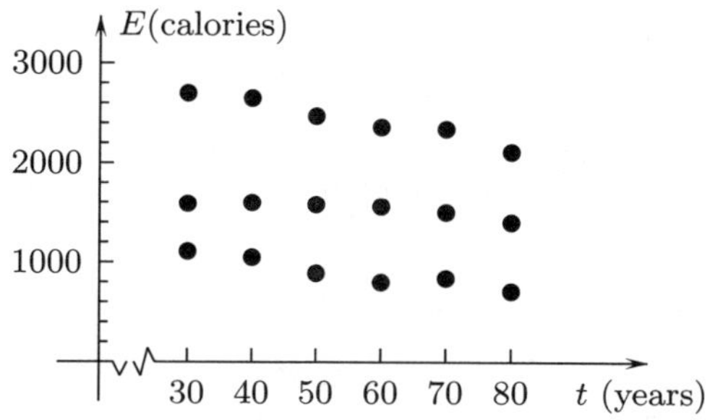

FIGURE 15

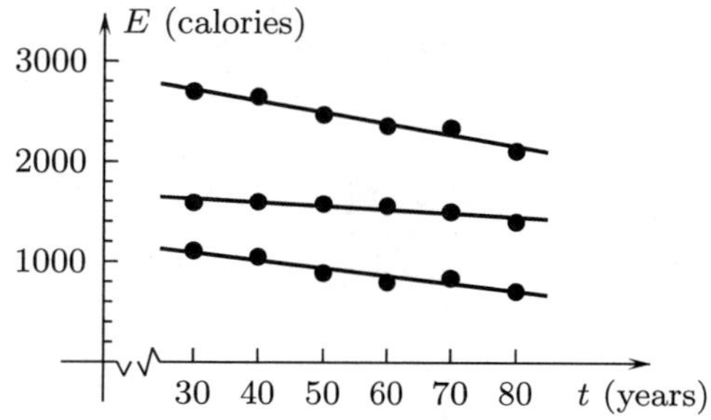

FIGURE 16

Notice that if you extend the top line in Figure 16 to the left, it intersects the E-axis at $E \approx 2950$, but this is not the actual E-intercept of the line because the E-axis is not at $t = 0$.

[(1)]Data adapted from "Nutrition and aging" by C. Finch, *The Handbook of Aging*, New York: Van Nostrand Reinhold Company, Inc., 1985, p. 883.

Question 23 Find an approximate equation of the middle line in Figure 16 to obtain a linear function $E_B(t)$ that models daily basal energy expenditure as a function of t.

Question 24 Use the formulas from Questions 22 and 23 to obtain a model $E_A(t)$ of daily energy expenditure in activity as a function of age.

Question 25 According to these mathematical models, what are the approximate rates of change with respect to age of total energy expenditure, basal energy expenditure, and energy spent on activities?

Feed vs. weight

If you were buying a horse and wanted to have some idea of how much it would cost to feed it, you might consult a table such as Table 2 below, which lists typical daily feed requirements of horses of various weights if they are ridden five hours per day. The graph of this data is shown in Figure 17.[(2)]

TABLE 2. DAILY FEED REQUIREMENTS OF HORSES

Weight w (pounds)	800	900	1000	1100	1200
Feed requirement F (pounds)	18.7	20.5	22.2	23.8	25.4

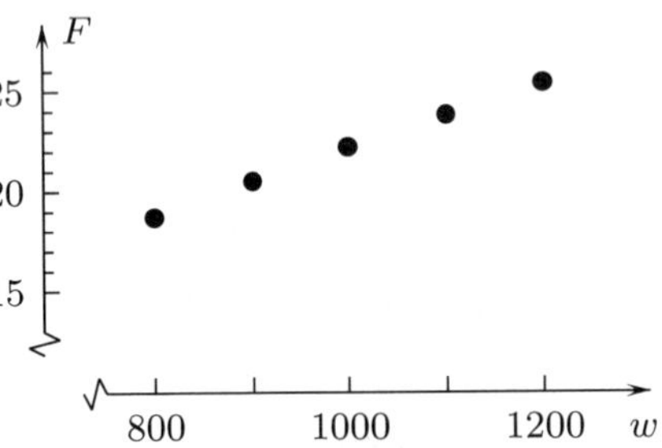

FIGURE 17

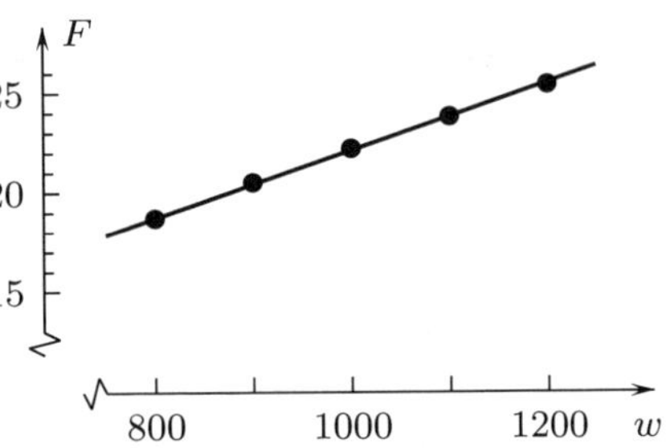

FIGURE 18

Example 5 Use the data in Table 2 to calculate the slopes of the four lines that would go through pairs of adjacent dots in Figure 17.

SOLUTION The 900-pound horse requires 20.5 pounds of food and the 800-pound horse requires 18.7 pounds, so the slope of the line joining the first two dots on the left in Figure 17 is $\dfrac{20.5 - 18.7}{900 - 800} = \dfrac{1.8}{100} = 0.018$. The result of this calculation has been entered in the third column of the next table, and similar calculations would give the other numbers in the table. □

Weight w (pounds)	800	900	1000	1100	1200
Feed requirement F (pounds)	18.7	20.5	22.2	23.8	25.4
Increase		1.8	1.7	1.6	1.6
Slope		0.018	0.017	0.016	0.016

Because the slopes in the last table are approximately equal, we can find a linear function that approximates the data in Table 2 fairly accuratly.

(2) Data adapted from from *The Science of Animal Husbandry*, by J. Blakeky, Reston, VA: Reston Publishing Company, 1974, p. 343.

Question 26 Use a slope that is close to the slopes in Table 2 and a feed requirement in the table to give a linear function $F(w)$ that approximates the data in the table.

Figure 18 shows the graph of the linear function from Question 26 based on a slope of 0.017 and the feed requirement of 18.7 pounds for an 800 pound horse as given in the response at the end of the chapter.

Principles and procedures

- You can find the constant rate of change of a linear function $f(x) = mx + b$ by taking the coefficient m of x in a formula for the function or by using the values of the function at two values of x to calculate the slope of the line that is its graph.
- If you know the constant rate of change m of a linear function $f(x)$ and its value $y_0 = f(x_0)$ at one point, you can use the point-slope equation for a line to give a formula $f(x) = y_0 + m(x - x_0)$ for the function. If x_0 is 0, you use the slope intercept equation of a line to obtain $f(x) = y_0 + mx$.
- If an object's coordinate, as it moves on an s-axis, is given by the linear function $s = mt + b$, then the rate of change m of its s-coordinate with respect to time is its constant velocity in the positive s-direction.
- Linear functions can often be used to construct related nonlinear functions, which then can be analyzed with a calculator or computer—or, as we will see later, with calculus.
- A linear function can be used to approximate data that is given by a finite number of points in a graph by drawing a line that passes through the cluster of points.
- If data given in a table is close to linear, an approximating linear function can be found by using one value in the table and a slope that approximates slopes from consecutive values in the table.[†]

Tune-Up Exercises 1.3[♦]

[A]*Answer provided.* [O]*Outline of solution provided.*

Be sure you can solve Exercises T1 through T7. Work as many of Exercises T8 through T13 as you need for the practice.

T1.[A] What is the constant rate of change of $7 - 6x$ with respect to x?

♦ Type 1, a linear function

T2.[A] Give a formula for the linear function $f(x)$ whose constant rate of change with respect to x is 50 and whose value at $x = 3$ is 10. (Use the point-slope equation for a line.)

♦ Type 1, a linear function

T3.[A] A linear function $g(x)$ has the value 100 at $x = 0$ and its constant rate of change is -40. Give a formula for it. (Use the slope-intercept equation of a line.)

♦ Type 1, a linear function

T4.[A] **(a)** Give a formula for the linear function $h(x)$ whose value at $x = 0.2$ is 6 and whose value at $x = 0.3$ is 2. **(b)** What are the rates of change of $h(x)$ and of $-h(x)$ with respect to x?

♦ Type 2, a linear function

T5.[A] Give a formula for the linear function $k(x)$ whose graph is shown in Figure 19.

♦ Type 2, a linear function

T6.[A] Give a formula for a linear function $y = L(x)$ that approximates the data in Figure 20.

♦ Type 2, a linear function

T7.[A] Give a formula for a linear function $y = M(x)$ that approximates the data in the table below.

x	0	2	4	6	8
y	4.2	8.0	12.0	16.1	19.9

♦ Type 2, a linear function

[†]A technique for finding the line that in a specific sense is the "best" approximation of data in any table will be studied in Chapter 10 and is discussed in statistics classes.

[♦]The Tune-up Exercises and Problems are classified by type and content. The types are (1) basic, reactive; (2) basic reflective; (3) intermediate, reactive; (4) intermediate, reflective; (5) advanced, reactive; (6) advanced, reflective; and (7) advanced, theoretical.

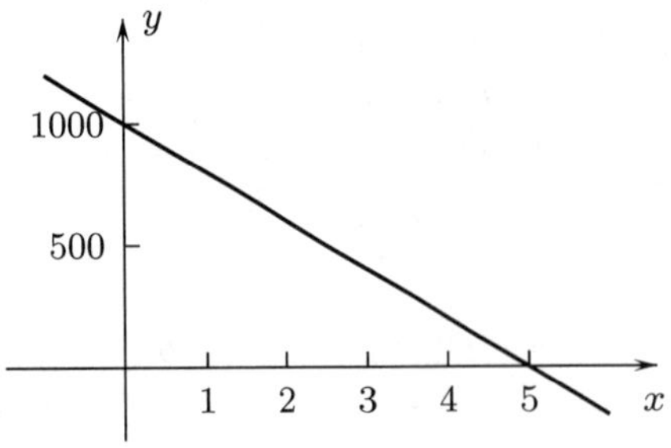

FIGURE 19

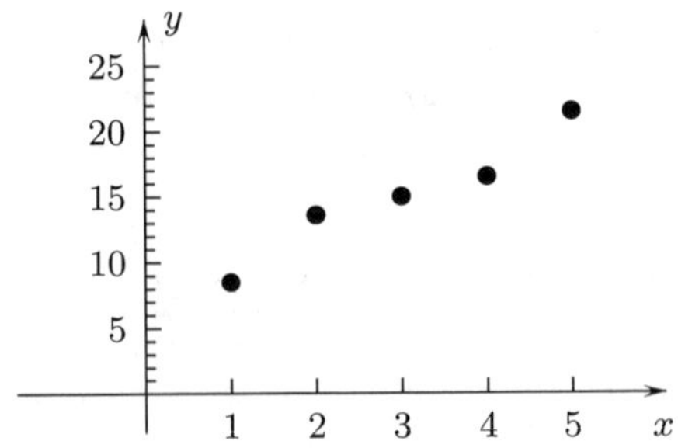

FIGURE 20

More practice

T8.[A] What is the constant rate of change of $470 - \frac{1}{3}z$ with respect to z?

♦ Type 2, a linear function

T9.[A] Give a formula for the linear function $p(w)$ whose constant rate of change with respect to w is -3 and whose value at $w = 10$ is -6.

♦ Type 2, a linear function

T10.[A] Give a formula for the linear function $Z(x)$ such that $Z(0) = 100$ and $Z(5) = 200$.

♦ Type 2, a linear function

T11.[A] Give an approximate formula for the linear function $q(x)$ whose graph is shown in Figure 21.

♦ Type 2, a linear function

T12.[A] Find a formula for the linear function $y = L(x)$ whose graph contains the dots in Figure 22.

♦ Type 2, a linear function

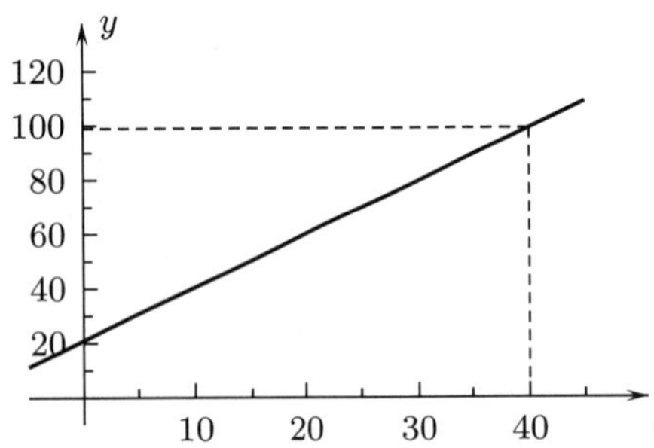

FIGURE 21

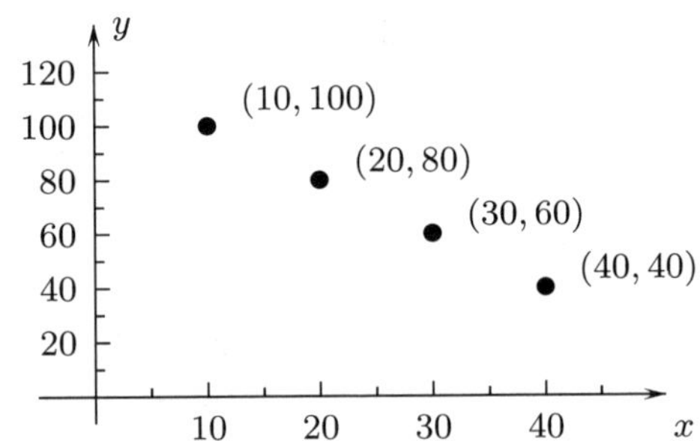

FIGURE 22

T13.[A] Give a formula for the linear function $y = K(x)$ that has the values at $x = 100$, $x = 200$, and $x = 300$ in the next table.

x	100	200	300	400	500
y	50	65	80	96	109

♦ Type 2, a linear function

Problems 1.3

A *Answer provided.* **O** *Outline of solution provided.*

1.[A] The density of sucrose is 1.58 grams per milliliter.[(3)] **(a)** Give a formula for the mass w of a sample of sucrose as a function of its volume V. **(b)** What is the rate of change of the mass of a sample of sucrose with respect to its volume? Give the units in which the rate of change is measured.

♦ Type 2, a linear function

2. The length L of a copper rod is a linear function of the temperature T. The rod is 100 inches long at 50°F and expands 0.093 inches for every degree increase in its temperature.[(4)] **(a)** Give a formula for L as a function of T. **(b)** At what temperature is the rod 100.5 inches long?

♦ Type 2, a linear function

3.[A] At the surface of the ocean, the water pressure equals the air pressure, which is 14.7 pounds per square inch. In one mathematical model, it is assumed that as you descend vertically into the ocean, the pressure increases at the constant rate of 0.44 pounds per square inch per foot.[(5)] **(a)** Give a formula for the water pressure p as a function of the depth h beneath the surface with this mathematical model. **(b)** Based on this model, what is the water pressure at the deepest spot in the oceans, Challenger Deep, which is 36,198 feet beneath the surface?

♦ Type 2, a linear function

4. The Kelvin scale for measuring temperature is calculated by adding a constant to the temperature measured in degrees Celsius, so that absolute zero, which is -273 degrees Celsius, is zero degrees Kelvin. Give formulas for degrees Kelvin T_K, first as a function of degrees Celsius T_C and then as a function of degrees Fahrenheit T_F. (See Question 2.) What is the rate of change of T_K with respect to T_F and how does it relate to the rate of change of T_C with respect to T_F?

♦ Type 2, a linear function

5.[A] A rectangular tank is 2 feet wide, 6 feet long, and 4 feet high. **(a)** What is the area of its base? **(b)** Give a formula for the volume $V(h)$ of water in the tank as a function of the depth of the water h, measured in feet. **(c)** What is the domain of the function $V(h)$ of part (b)? **(d)** What is the rate of change of $V(h)$ with respect to h?

♦ Type 2, a linear function

6. A hiker walks north at the constant speed of 3 miles per hour from 2:00 PM to 5:00 PM. At 2:00 PM he is five miles north of his house. Give his distance from his house as a function of the number of hours t since 2:00 PM for $0 \le t \le 3$.

♦ Type 2, a linear function

7.[A] A car rental agency charges $25 per day plus 15 cents per mile. Let $F_1(x)$ be the charge for renting the car for one day and driving it x miles, let $F_2(x)$ be the charge for renting the car for two days and driving it x miles, and let $F_3(x)$ be the charge for renting the car for three days and driving it x miles. **(a)** Give formulas for $F_1(x)$, $F_2(x)$, and $F_3(x)$ and sketch their graphs in one xF-plane. **(b)** What are the rates of change of $F_1(x), F_2(x)$, and $F_3(x)$ with respect to x?

♦ Type 2, linear functions

8. A parachutist is 300 meters above the ground at time $t = 0$ seconds and descends at the constant rate of 15 meters per second until she hits the ground. **(a)** Give a formula for her height s above the ground as a function of t and sketch its graph. **(b)** When does she reach the ground? **(c)** The slope of the line in part (a) is the parachutist's upward velocity. What is it?

♦ Type 2, a linear function

9.[A] A ball is thrown straight down from a tower with velocity 5 feet per second. We use the mathematical model in which air resistance is disregarded, so that the ball's downward velocity v increases at the constant rate of 32 feet per second per second. **(a)** Give a formula for v as a function of the number of seconds t after the ball was thrown. **(b)** How long does it take for v to increase from 5 feet per second to 37 feet per second? **(c)** When does the ball hit the ground if it hits with velocity 101 feet per second?

♦ Type 2, a linear function

[(3)] *CRC Handbook of Chemistry and Physics*, 62nd Edition, Boca Raton, FL: CRC Press, 1981, p. C-525.

[(4)] *Ibid*, p. D-161.

[(5)] Data adapted from *The World Book Encyclopedia*, Vol. 14, Chicago IL: World Book, Inc., 1978, p.,491.

10. Suppose that the ball in Exercise 9 is thrown upward with an initial upward velocity of 5 feet per second instead of downward. To deal with this case, we let v denote the ball's upward rather than downward velocity. Then (still disregarding air resistance), v decreases at the constant rate of 32 feet per second per second. **(a)** Give a formula for $v(t)$ with t the number of seconds since the ball was thrown. **(b)** When is $v = 0$ and what happens to the ball then?

♦ Type 2, a linear function

11.[A] Suppose you spend exactly \$28 on x pounds of meat that costs \$2 per pound and y pounds of potatoes that cost 70 cents per pound. **(a)** Give a formula for y as a function of x. **(b)** Give a formula for x as a function of y. **(c)** What are the rates of change of y with respect to x and of x with respect to y? **(d)** Why are the rates of change in part (c) both negative? Which is less (more negative) than the other and why?

♦ Type 4, linear functions

12. George and Martha are both driving south from Kansas City on Highway 71. George is $50+60t$ miles from Kansas City and Martha is $25 + 70t$ miles from Kansas City at time t (hours). **(a)** How fast is George driving? **(b)** How fast is Martha driving? **(c)** When and where does Martha catch up to George? **(d)** What is the rate of change of the distance between them before Martha passes George? **(e)** What is the rate of change of the distance between them after Martha passes George?

♦ Type 4, linear functions

13.[A] A furniture manufacturer can build x bookcases in a month for \$3000 overhead and \$25 per bookcase. **(a)** Give a formula for the cost of building x bookcases in a month as a function of x. **(b)** Give a formula for the average cost per bookcase $A(x)$ as a function of x. Generate the graph of $A(x)$ for $-230 \le x \le 300, -8 \le A \le 80$ with x-scale $= 50$ and A-scale $= 10$. Copy it on your paper and explain in reference to the bookcases why the function curves down to the right and has two asymptotes. **(c)** What is the average cost per bookcase if only 30 bookcases are built? **(d)** How many bookcases would have to be built for the average cost to be \$28?

♦ Type 4, a function constructed from a linear function

14.[A] Figure 23 shows the graph of a linear function $O(V)$ that is a fairly good model of a human's oxygen absorption, measured in liters per minute, as a function of the rate of his or her breathing, also measured in liters per minute.[(6)] What is the approximate rate of change of oxygen absorption with respect to the rate of breathing?

♦ Type 2, a linear function

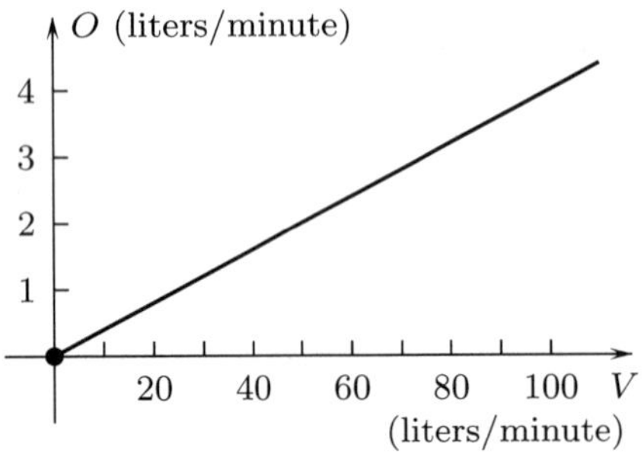

FIGURE 23

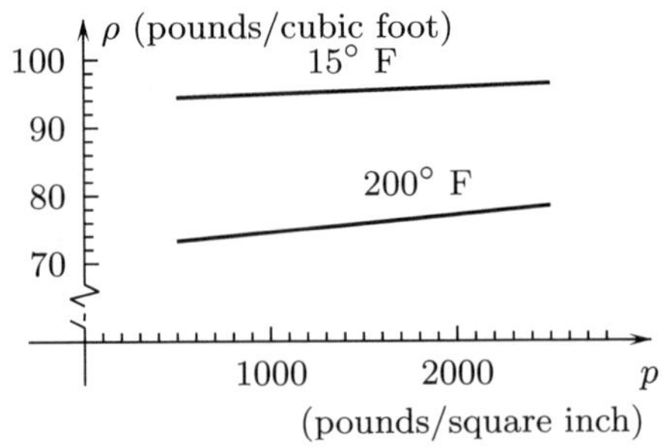

FIGURE 24

15. At each fixed temperature between 15°F and 200°F and for air pressures between 500 and 2500 pounds per square inch, the density of liquid chlorine is approximately a linear function of air pressure. The upper line in Figure 24 gives the density as a function of pressure at 15°F and the lower line gives the density as a function of pressure at 200°F.[(7)] **(a)** Is the rate of change of the density with respect to pressure greater at 15°F or at 200°F? **(b)** By approximately what amount is the density of chlorine decreased if the temperature is raised from 15°F to 200° and the pressure is held at 1000 pounds per square inch? **(c)** By approximately what amount is the density of chlorine increased if the air pressure is raised from 1000 to 2000 pounds per square inch and the temperature is held at 200°F?

♦ Type 4, linear functions

[(6)] Data adapted from *Textbook of Medical Physiology* by A. Guynton, Philadelphia, PA: W. B. Saunders Co., 1976, p. 523.

[(7)] Data adapted from *McGraw Hill Encyclopedia of Science*, Vol. 5, New York, NY: McGraw-Hill, Inc., p. 118.

16. A plastic garden hose has an inner diameter of 1 inch and an outer diameter of 1.2 inches. **(a)** A cross section of the hose is the region between two concentric circles. What is its area? **(b)** Give a formula for the volume of plastic in the hose as a function of its length, measured in inches. **(c)** Suppose the plastic costs 5 cents per cubic inch. What is the rate of change of the cost of the plastic in the hose with respect to its length?

♦ Type 4, a linear function

17. Which is greater, the rate of change of the circumference of a circle with respect to its diameter or the rate of change of the perimeter of a square with respect to its width?

♦ Type 4, a linear function

18. Table 3 gives annual world meat production, measured in millions of metric tons, and average meat production per capita, measured in kilograms, in the years 1950, 1960, 1970, 1980, and 1990.[8] (A metric ton is 1000 kilograms or about 2205 pounds.) **(a)** One of these sets of data can be modeled by a linear function for 1950, 1960, 1970, and 1980. Find its formula. **(b)** What error and relative error are made if this model is used in 1990? (Recall that relative error equals the error divided by the correct value.)

♦ Type 4, a linear function

TABLE 3. WORLD MEAT PRODUCTION

Year	1950	1960	1970	1980	1990
Total (million metric tons per year)	46	68	98	133	171
Per capita (kilograms per year)	18	22	26	30	32

19. **(a)** Plot the total meat production data in Table 3 as five points in a tM-plane. If this data were given by a linear function, the dots would lie on a line, but this is not the case. **(b)** Find slopes of the four lines joining (i) the dot at 1950 to the dot at 1960, (ii) the dot at 1960 to the dot at 1970, (iii) the dot at 1970 to the dot at 1980, and (iv) the dot at 1980 to the dot at 1990. **(c)** What is the average of the slopes from part (b)? **(d)** Show that the answer to part (c) equals the slope of the line joining the dot at 1950 to the dot at 1990.

♦ Type 4, a linear function

20. After a high carbohydrate meal, the amount of glucose in a person's blood rises above its normal level. This causes the pancreas to secrete insulin, which in turn allows the individual's cells to absorb the excess glucose. In a test of the effect of insulin, muscle cells were placed in different concentrations of glucose with and without the presence of insulin, and the glucose concentration in the cells was measured. The results are in Table 4.[9] The numbers in the second row show that in the absence of insulin, the glucose concentration in the cells is barely affected by the increased glucose outside the cells. The glucose concentration in the cells in the presence of insulin is given by the third row of the table. Find a linear function $y(x)$ that approximates the data in the third row as a function of x in the first row. What is the rate of change of $y(x)$ with respect to x, and what does this number tell you about glucose transport into the cells?

♦ Type 4, a linear function

TABLE 4. GLUCOSE TRANSPORT INTO CELLS

Concentration outside the cells $= x$	100	250	500	700
Concentration inside without insulin	0	0	10	15
Concentration inside with insulin	0	100	250	390

21.[A] At time t (hours) for $0 \leq t \leq 2$, a sailor's boat is $3 + 30t$ kilometers directly north of a pier. **(a)** How fast is he sailing? **(b)** The sailor's favorite pub is on the shore 5 kilometers directly west of the pier. Give a formula for the boat's distance from the pub as a function of t. **(c)** When is the boat 13 kilometers from the pub?

♦ Type 4, a function constructed from a linear function

---

[8] Data adapted from *Vital Signs, 1992* by Brown, Flavin, and Kane, New York, NY: W.W. Norton & Company, Inc., 1992, p. 29.

[9] *Textbook of Medical Physiology* by A.C. Guyton, Philadelphia, PA: W.B. Saunders, Co., 1981, p. 961.

22. An uncooked cookie is in the shape of a right circular cylinder 10 centimeters high with horizontal cross section of radius 5 centimeters. For eight minutes after it is put in the oven at $t = 0$ (minutes), the radius expands at the constant rate of one centimeter per minute and the height decreases at the constant rate of one centimeter per minute. **(a)** Give formulas for the radius r, height h, and volume $V = \pi r^2 h$ of the cookie in terms of the time t. **(b)** Generate the graph of $V(t)$ in the window $-1 \le t \le 8, -200 \le V \le 2000$ with V-scale $= 500$ and use a maximum or trace command or the graph to find the approximate maximum volume of the cookie. **(c)** How good is the mathematical model of a cookie in this problem?

♦ Type 4, a function constructed from a linear function

23. In the Shell mileage marathon, specially designed gasoline-powered vehicles compete to see which uses the least amount of gasoline. The winning mileages, rounded to the nearest 100 miles per gallon, for 1974 through 1980 are shown in Table 5.[10] **(a)** Find a linear function that approximates this data. **(b)** What would the winning mileage have been in 1989 if this mathematical model were valid up to then? **(c)** The winning mileage in 1989 was, in fact, 6409 miles per gallon, which is quite close to the result from part (b). Do you think this is due to the accuracy of the mathematical model, or is it just a coincidence? Explain.

TABLE 5. WINNING MILEAGES IN THE SHELL MILEAGE MARATHON

Year	1974	1975	1976	1977	1978	1979	1980
Miles per gallon	400	800	1100	1600	1800	2100	2800

♦ Type 4, a linear function

24.[A] Federal income tax in 1997 on taxable income between \$41,200 and \$99,600 for married couples filing jointly was \$6,180 plus 28% of the amount over \$41,200.[11] **(a)** Give a formula for the tax T on taxable income x for $41{,}200 \le x \le 99{,}600$. **(b)** Imagine that up to August 1, 1997 you and your spouse had earned 1997 taxable income of \$50,000, and that your taxable income increased at the constant rate of \$150 per day through the month. Give a formula for your taxable income x as a function of the time t (days) after the beginning of the month for $0 \le t \le 31$. **(c)** Give a formula for your federal tax as a function of t for $0 \le t \le 31$. **(d)** What are the rates of change of T with respect to x, of x with respect to t, and of T with respect to t, and how are these related?

♦ Type 4, a piecewise linear function

25.[A] A crate is being hauled at the constant rate of 5 feet per minute up a ramp that has slope $\frac{3}{4}$. Let s denote the distance the crate has traveled up the ramp, let x be the corresponding change in the crate's horizontal coordinate, and let y be the change in its vertical coordinate (Figure 25), all measured in feet. **(a)** Find a formula for s as a function of the amount of time t (minutes) the crate has been pulled up the ramp, starting at the bottom of the ramp. **(b)** Find formulas for x and y as functions of s. **(c)** Give formulas for x and y as functions of t. **(d)** What are the rates of change of x and y with respect to t?

♦ Type 4, linear functions

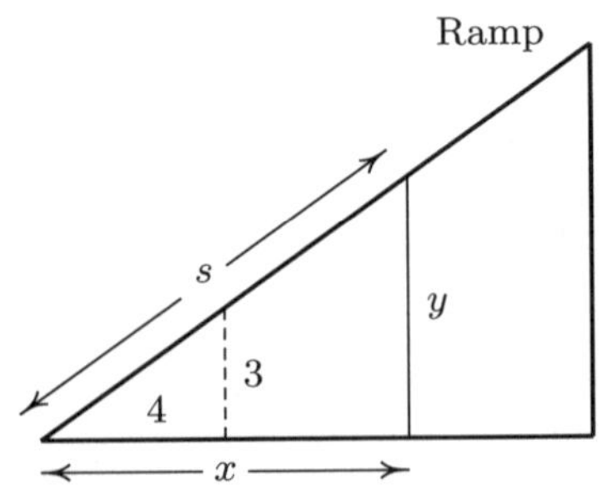

FIGURE 25

[10] Data adapted from "Shell's mileage marathon competition" by D. Blackmore and G. Toft, *Automotive Power Systems—Environment and Conservation*, Beccles, Suffolk: Waveney Print Services Ltd., 1990, p. 257.

[11] Data adapted from *1040 Forms and Instructions*, Bloomington, IL: Department of the Treasury, Internal Revenue Service, 1997, p. 51.

26. A spherical balloon has a three-inch radius at $t = 0$ (minutes) and is inflated in such a way that its radius increases at the constant rate of one-half inch per second, until the radius is six inches. From that time until $t = 9$ the balloon leaks in such a way that its radius decreases at the constant rate of one inch per second. **(a)** What is the greatest volume of the balloon? **(b)** What is the radius at $t = 9$? **(c)** Give formulas for the volume $V(t)$ in terms of t for $0 \leq t \leq 9$ and sketch the graph of V in a tV-plane.

♦ Type 4, linear functions

27. Imagine you have a flock of chickens that on January 1 weighs a total of 1000 pounds and that the flock gains weight at the constant rate of 80 pounds per week. **(a)** Give a formula for the chickens' total weight as a function of the number of weeks t after January 1. **(b)** Suppose that the chickens would have sold for 30 cents per pound on January 1 and that the price drops one cent per pound per week for 15 weeks after that. Give the price of the chickens per pound as a function of t. **(c)** Find a formula for the value $v(t)$ (dollars) of your flock of chickens as a function of t and generate its graph on your calculator or computer with $-1 \leq t \leq 15$, t-scale $= 5$, $-50 \leq v \leq 450$, and v-scale $= 100$. Copy the graph on your paper. **(d)** Use the maximum or trace command or the graph to find the approximate time when the value of your flock of chickens is the greatest.

♦ Type 4, linear functions

RESPONSES TO QUESTIONS IN CHAPTER 1

Responses 1.1

Response 1 $\pi(2)^2 = 4\pi$ square centimeters

Response 2 $0.4225\pi \le A \le 4\pi$ with A measured in square centimeters, since $\pi(0.65)^2 = 0.4225\pi$, $\pi(2)^2 = 4\pi$, and circles with larger radii have larger areas.

Response 3 Increasing the radius from 12 to 13 feet increases the area more since $\pi(5)^2 - \pi(1)^2 = 24\pi$ is less than $\pi(13)^2 - \pi(12)^2 = 25\pi$.

Response 4 The area is quadrupled.

Response 5 The dots at the origin would be replaced by small circles (Figures R5a and R5b).

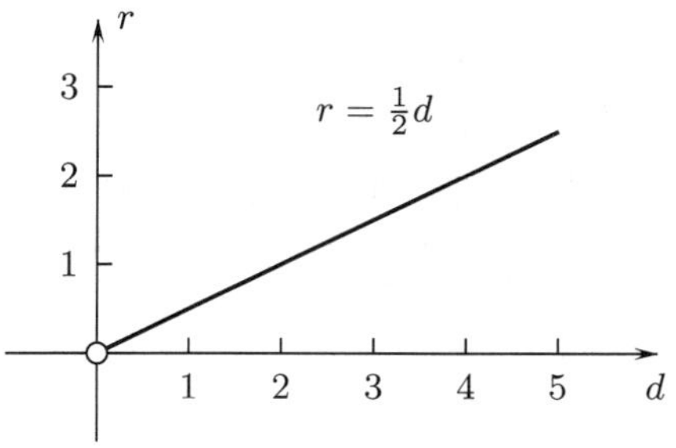

Figure R5a

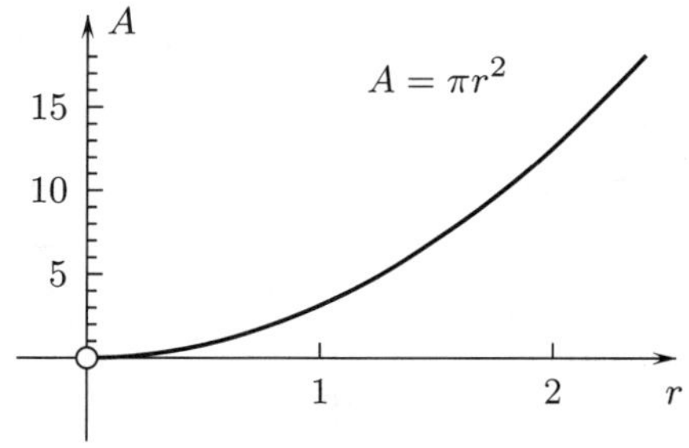

Figure R5b

Response 6 $A = 2\pi r + \pi$, defined for $r \ge 0$ if we consider all circles of zero or positive radii (Use $(r+1)^2 = r^2 + 2r + 1$.) • Figure R6

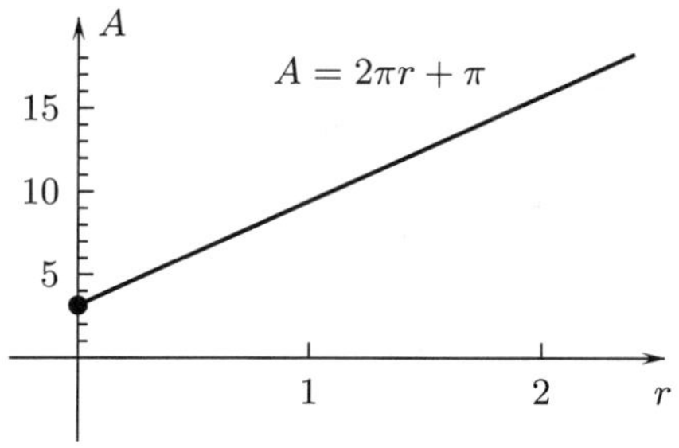

Figure R6

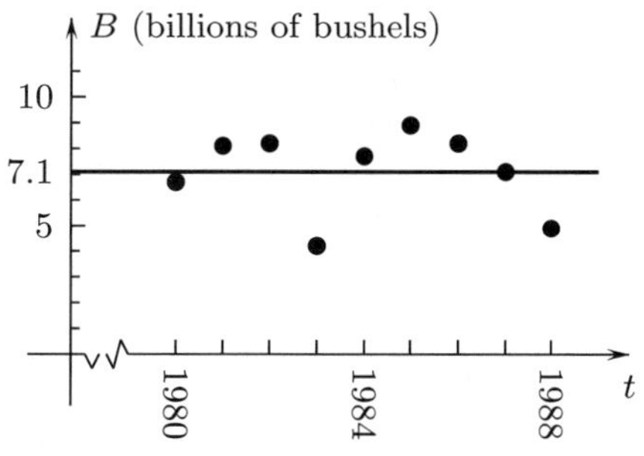

Figure R9

Response 7 **(a)** $\sqrt{9-x^2}$ is defined provided $9-x^2$ is not negative. The domain is the interval $-3 \le x \le 3$. **(b)** The center of the circle is the origin $(0,0)$ and its radius is 3. Since $\sqrt{x^2+y^2}$ is the distance from (x,y) to $(0,0)$, the circle has the equation $x^2+y^2=9$. • Solving for y gives $y = \pm\sqrt{9-x^2}$, so the lower half of the circle is the graph of $y = -\sqrt{9-x^2}$.

Response 8 $B(1986) = 8.2$ billion bushels is the amount of corn produced in the U.S. in 1986.

Response 9 **(a)** One answer: 7 billion bushels, since the horizontal line $B = 7$ passes more or less through the middle of the dots in Figure 13 of Section 1.1 **(b)** The total production in the ten years is 64 billion bushels—the sum of the numbers in the second row of Table 1—so the average for the nine years is $64/9 \doteq 7.1$ billion bushels per year. The line at this value is shown in Figure R9.

Response 10 (a) 1985, when 8.9 billion bushels were produced
(b) $t = 1983$ (the year with the least corn production)
(c) 64 billion bushels (as calculated in answering Question 9)

Response 11 Possible answers: There is no telling how carefully the bushels were measured, and the results have been rounded off to the nearest 100 million bushels.

Response 12a

t	1980	1981	1982	1983	1984	1985	1986	1987	1988
C(t)		1.4	0.1	−4.0	3.5	1.2	−0.7	−1.1	−2.2

Response 12b The sum of the changes is −1.8 billion bushels.

Response 12c $B(1988) - B(1980) = 4.9 - 6.7 = -1.8$ • Explanation: The sum of the changes each year from 1980 to 1988 equals the total change from 1980 to 1988.

Response 13 We do not know $B(1979)$ (the amount of corn produced in 1979).

Response 14 $C(t) = B(t) - B(t-1)$ for $t = 1981, 1982, \ldots, 1988$

Response 15 (a) ≈ 10.5 million (Different readings of the graph yield different values.)
(b) $\approx 3.3/15.5 \doteq 0.21$ or 21% since $T(1910) \approx 15.5$ and $H(1910) \approx 3.3$
(c) ≈ 50 years because $T(1910) \approx 15.3$ and $T(1960) \approx 7.7$

Response 16 $F(1910) = T(1910) - H(1910) \approx 15.3 - 3.3 = 12$ •
$F(1920) = T(1920) - H(1920) \approx 15.2 - 3.4 = 11.8$ •
$F(1930) = T(1930) - H(1930) \approx 15.2 - 3.2 = 12.0$ •
$F(1940) = T(1040) - H(1940) \approx 11.1 - 2.7 = 8.4$ •
$F(1950) = T(1950) - H(1950) \approx 10.0 - 2.3 = 7.7$ •
$F(1960) = T(1960) - H(1960) \approx 7.7 - 2.0 = 5.7$ •
$F(1970) = T(1970) - H(1970) \approx 4.2 - 1.2 = 3$ •
$F(1980) = T(1980) - H(1980) \approx 3.6 - 1.0 = 2.6$ •
Figure R16

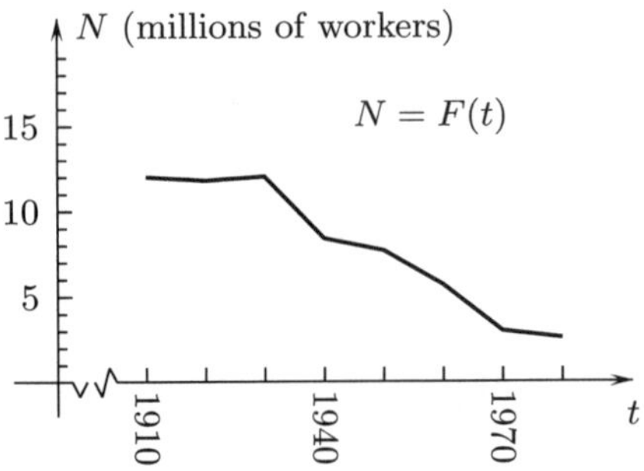

Figure R16

Response 17 Possible answers: The values are interpolated from data collected only every ten years. The graphs do not give exact values.

Response 18 10%

Response 19

Year: t	1986	1987	1988	1989	1990	1991	1992	1993
Investments: $U(t)$	1.6	2.3	2.5	4.9	5.1	7.0	16.3	23.5
Change in $U(t)$		0.7	0.2	2.4	0.2	1.9	9.3	7.2
Relative change in $U(t)$		0.438	0.087	0.960	0.041	0.373	1.329	0.442
Percent change in $U(t)$		43.8	8.7	96.0	4.1	37.3	132.9	44.2

Response 20 Possible explanations: The bar graph of relative change in Figure 18 of Section 1.1 does not look like the bar graph of change in Figure 17 because the relative change equals the change divided by different numbers for different years. • The bar graphs in Figures 18 and 19 of have the same shape because percent change is relative change multiplied by the constant 100, and 1 unit on the vertical axis in Figure 18 corresponds to 100 units on the vertical axis in Figure 19.

Response 21 $R(t) = \dfrac{U(t) - U(t-1)}{U(t-1)}$ and $P(t) = \dfrac{100[U(t) - U(t-1)]}{U(t-1)}$

Response 22 The changes would be multiplied by 1000 and the relative changes would be the same.

Response 23 $v_1(P)$ is negative for small P and for $P > 70$. • $v_2(P)$ is greater for small P than it is at $P = 5$ and greater for $P > 70$ than it is at $P = 65$.

Response 24 **(a)** Take the cube root of both sides: $x > -2$

(b) Take the square root of both sides: $\sqrt{x^2} < 4$ • $|x| < 4$ • $-4 < x < 4$

Responses 1.2

Response 1 [Slope] = [Rise]/[Run] = $(7-4)/(5-2) = 1$

Response 2 **(a)** Horizontal lines have zero slope. **(b)** Lines that point up to the right have positive slopes. **(c)** Lines that point down to the right have negative slopes.

Response 3 If the scales on the axes are equal, then lines with slope 1 or -1 make angles of 45° with the x- and y-axes.

Response 4 **(a)** [Slope] $= \dfrac{y-1}{x-3}$

(b) Set [Slope] = 2 and solve for y: $y = 1 + 2(x-3)$.

Response 5 **(a)** [Slope] $= \dfrac{y - y_0}{x - x_0}$ **(b)** $y = y_0 + m(x - x_0)$

Response 6 **(a)** One description: The line rotates about the point $(0, b)$ on the y-axis, first counterclockwise, without turning so far that it becomes vertical, and then clockwise until it is horizontal. **(b)** One description: The line rotates clockwise until it points downward to the right, without turning so far that it becomes vertical, and then counterclockwise until it is horizontal.

Response 7 **(a)** The line would move down, parallel to its initial position, until its y-intercept is half as far from the origin as it was initially. **(b)** The line would move down, parallel to its initial position, until its y-intercept is below the origin and twice as far from the origin as it was initially.

Response 8 **(a)** $y = -5 + m(x-6)$ (with parameter m) **(b)** $x = 6$

Response 9 When you go from the left vertex to the right vertex of the lower triangle in Figure 8 of Section 1.2, the rise is b and the run is a, so $m_1 = b/a$. Going from the left vertex to the right vertex in the upper triangle, the rise is $-a$ and the run is b, so $m_2 = -a/b$.

Response 10a

x	-3	-2	-1	$-\frac{1}{2}$	$-\frac{1}{3}$	0	$\frac{1}{3}$	$\frac{1}{2}$	1	2	3
$y = x^2$	9	4	1	$\frac{1}{4}$	$\frac{1}{9}$	0	$\frac{1}{9}$	$\frac{1}{4}$	1	4	9
$y = x^4$	81	16	1	$\frac{1}{16}$	$\frac{1}{81}$	0	$\frac{1}{81}$	$\frac{1}{16}$	1	16	81

Response 10b One explanation: For $-1 < x < 0$ and $0 < x < 1$, x^2 is less than 1, so multiplying it by itself to form x^4 gives a smaller number.

Response 10c One explanation: For $x < -1$ and $x > 1$, x^2 is greater than than 1, so multiplying it by itself to form x^4 gives a larger number.

Response 11 One answer: Both curves pass through $(0,0)$, $(1,1)$ and $(-1,-1)$, and both curves hug the x-axis near $x = 0$ and bend away from the x-axis for large positive and large negative x. • $y = x^5$ is closer to the x-axis than $y = x^3$ for $-1 < x < 0$ and $0 < x < 1$ and is farther from the x-axis than $y = x^3$ for $x < -1$ and $x > 1$.

Response 12 **(a)** The completed table is below. **(b)** The functions $y = x^{-2}$ and $y = x^{-4}$ are even. (Their graphs are symmetric about the y-axis.) • The functions $y = x^{-1}$ and $y = x^{-3}$ are odd. (Their graphs are symmetric about the origin.) **(c)** The curve $y = x^{-4}$ is closest to the x-axis for $x < -1$ and for $x > 1$. **(d)** The curve $y = x^{-1}$ is closest to the x-axis for $-1 < x < 0$ and for $0 < x < 1$.

x	-2	-1	$-\frac{1}{2}$	0	$\frac{1}{2}$	1	2
$y = x^{-1}$	$-\frac{1}{2}$	-1	-2	*	2	1	$\frac{1}{2}$
$y = x^{-2}$	$\frac{1}{4}$	1	4	*	4	1	$\frac{1}{4}$
$y = x^{-3}$	$-\frac{1}{8}$	-1	-8	*	8	1	$\frac{1}{8}$
$y = x^{-4}$	$\frac{1}{16}$	1	16	*	16	1	$\frac{1}{16}$

Response 13 The case of $x > 0$: $y = 1/x$ is above the x-axis, and it is between $y = \frac{1}{10}$ and $y = -\frac{1}{10}$ if and only if $\dfrac{1}{x} < \frac{1}{10}$. • Multiply both sides by the positive numbers x and 10 to obtain the equivalent inequality $x > 10$. •
The case of $x < 0$: $y = 1/x$ is below the x-axis, and it is between $y = \frac{1}{10}$ and $y = -\frac{1}{10}$ if and only if $\dfrac{1}{x} > -\frac{1}{10}$. • Multiply both sides by the negative numbers x and -10, reversing the direction of the inequality twice, to obtain the equivalent inequality $x < -10$. • Figure R13

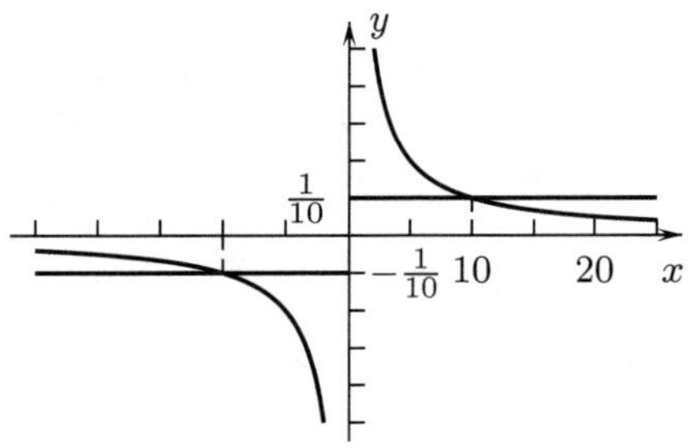

Figure R13

Response 14 $\lim_{x \to 0^+} \frac{1}{x^2} = \infty$ • $\lim_{x \to 0^-} \frac{1}{x^2} = \infty$

Response 15 **(a)** One explanation: $y = -\frac{1}{x^2}$ is the mirror image of $y = \frac{1}{x^2}$ about the x-axis, and $y = 2 - \frac{1}{x^2}$ is obtained by translating (lifting) $y = -\frac{1}{x^2}$ up two units. • Another explanation: $y = 2 - \frac{1}{x^2}$ is symmetric about the y-axis because the function is even. It lies entirely below the line $y = 2$ because $1/x^2$ is always positive. It is close to $y = 2$ for large $|x|$ where $1/x^2$ is small, and is far from it for $|x|$ close to zero, where $1/x^2$ is large.
(b) $\lim_{x \to \infty} \left(2 - \frac{1}{x^2}\right) = 2$ • $\lim_{x \to -\infty} \left(2 - \frac{1}{x^2}\right) = 2$ • $\lim_{x \to 0} \left(2 - \frac{1}{x^2}\right) = -\infty$

Response 16 **(a)** For large $|x| > 1$: x^{-2} and $|x^{-1}|$ are very small, with x^{-2} much smaller than $|x^{-1}|$; while $|x|$, x^2, and $|x^3|$ are large, with $|x^3|$ much larger than x^2 and x^2 much larger than $|x|$.
(b) For small positive $|x| < 1$: x^{-2} and $|x^{-1}|$ are very large, with x^{-2} much larger than $|x^{-1}|$; while $|x|$, x^2, and $|x^3|$ are small, with $|x^3|$ much smaller than x^2 and x^2 much smaller than $|x|$.

Response 17 **(a)** Figure R17 • For x near 0, the x^2 in the denominator is much smaller than 1, and $y = \frac{x^2}{x^2+1}$ looks approximately like the parabola $y = \frac{x^2}{1}$. • For large $|x|$, the 1 in the denominator is much smaller than x^2 and $y = \frac{x^2}{x^2+1}$ looks approximately like $y = \frac{x^2}{x^2}$, which is $y = 1$. **(b)** $\lim_{x \to \pm\infty} \frac{x^2}{x^2+1} = \lim_{x \to \pm\infty} (1) = 1$

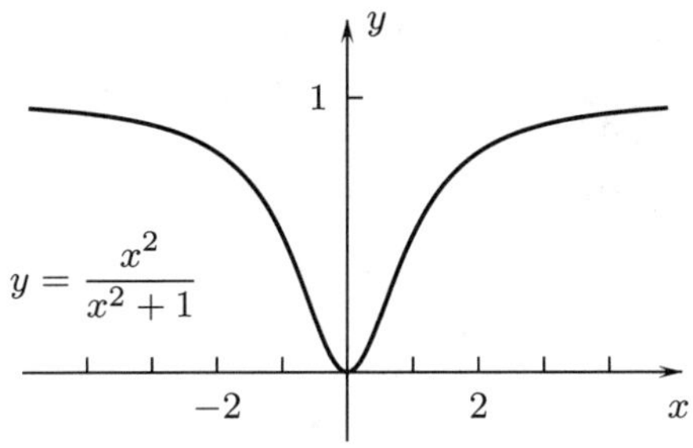

Figure R17

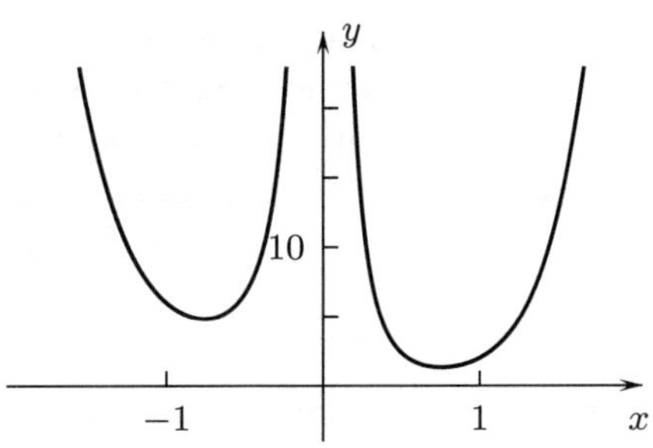

Figure R18

Response 18 **(a)** Figure R18 • For x near 0, the $\frac{1}{x^2}$ dominates the x^2 in the first factor of $(x^2 + \frac{1}{x^2})(1 - x + x^4)$ and 1 dominates the $-x$ and x^4 in the second factor. Therefore, $y = (x^2 + \frac{1}{x^2})(1 - x + x^4)$ looks approximately like $y = (\frac{1}{x^2})(1)$, which is $y = 1/x^2$. • For large $|x|$, the x^2 dominates the $\frac{1}{x^2}$ in the first factor of $(x^2 + \frac{1}{x^2})(1 - x + x^4)$ and x^4 dominates the 1 and $-x$ in the second factor. Consequently, $y = (x^2 + \frac{1}{x^2})(1 - x + x^4)$ looks approximately like $y = (x^2)(x^4)$, which is $y = x^6$.

(b) $\lim_{x\to\pm\infty} (x^2 + \frac{1}{x^2})(1 - x + x^4) = \lim_{x\to\pm\infty} x^6 = \infty$ •

$\lim_{x\to 0} (x^2 + \frac{1}{x^2})(1 - x + x^4) = \lim_{x\to 0} \frac{1}{x^2} = \infty$

Responses 1.3

Response 1 The points $(0, 32)$ and $(100, 212)$ are on the line in Figure 1 of Section 1.3. Its slope is $\frac{9}{5}$. • The points $(32, 0)$ and $(212, 100)$ are on the line in Figure 2. Its slope is $\frac{5}{9}$. • The slopes are reciprocals of each other.

Response 2 $T_F = 32 + \frac{9}{5}T_C$ • $T_C = \frac{5}{9}(T_F - 32)$

Response 3 **(a)** $R(x) = \left[x \text{ pounds}\right]\left[32 \frac{\text{dollars}}{\text{pound}}\right] = 32\,x$ dollars • Figure R3a

(b) $C(x) = [250 \text{ dollars}] + \left[x \text{ pounds}\right]\left[7 \frac{\text{dollars}}{\text{pound}}\right] = 250 + 7x$ dollars • Figure R3b

(c) Profit = Revenue − Cost: $P(x) = R(x) - C(x) = 25x - 250$ dollars • Figure R3c

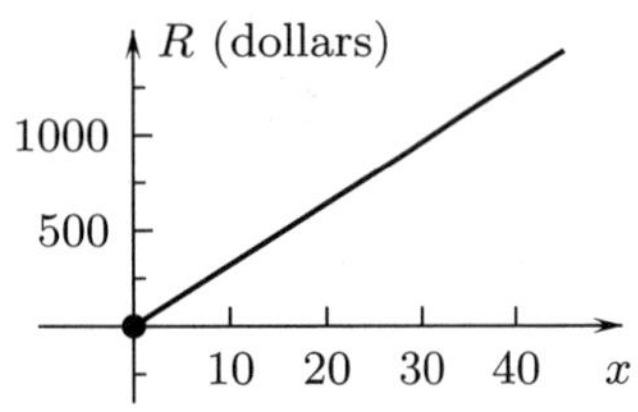

Figure R3a

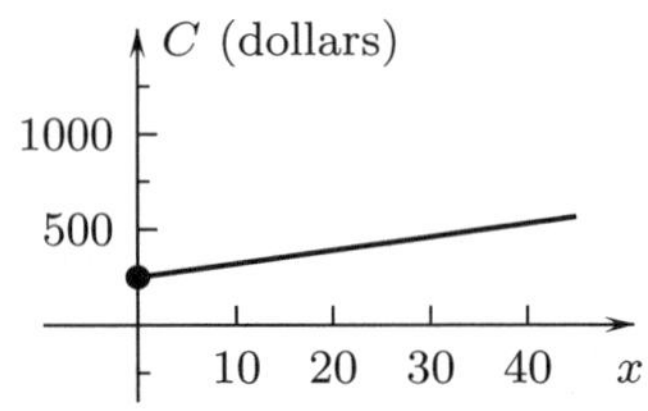

Figure R3b

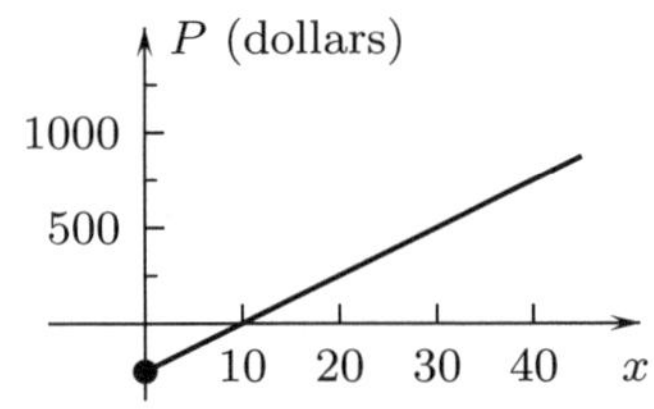

Figure R3c

Response 4 The x-intercept is $250/25 = 10$ pounds and is the amount of cotton candy she would have to sell to break even. • The P-intercept is -250, and is her profit with no sales (when her loss would be \$250).

Response 5 Possible answers: She might have other expenses and probably does not sell all of the cotton candy she makes in a day.

Response 6 $[200 \text{ miles}] + \left[t \text{ hours}\right]\left[60 \frac{\text{miles}}{\text{hour}}\right] = 200 + 60t$ miles

Response 7 2π meters per meter

Response 8 $\frac{1}{0.05 \text{ gallons/mile}} = 20 \frac{\text{miles}}{\text{gallon}}$

Response 9 $V = 20 - \left[s \text{ miles}\right]\left[0.05\, \dfrac{\text{gallons}}{\text{mile}}\right] = 20 - 0.05s$ gallons • The tank is empty when $20 - 0.05s = 0$, which is at $s = 20/0.05 = 400$ miles, so the domain of V as a function of s is the interval $0 \le s \le 400$.

Response 10 $s = [t \text{ hours}][60 \text{ miles per hour}] = 60t$ miles • Since $s = 400$ at $t = 6\frac{2}{3}$, the domain of s as a function of t is $0 \le t \le 6\frac{2}{3}$.

Response 11 **(a)** $V = 20 - 0.05(60t) = 20 - 3t$ (gallons) for $0 \le t \le 6\frac{2}{3}$
(b) -3 gallons per hour **(c)** The volume of gasoline in your tank decreases three gallons every hour.

Response 12 **(a)** $\left[L \text{ pounds}\right]\left[3\, \dfrac{\text{dollars}}{\text{pound}}\right] = 3L$ dollars

(b) $\left[C \text{ pounds}\right]\left[0.75\, \dfrac{\text{dollars}}{\text{pound}}\right] = 0.75C$ dollars

(c) $3L + 0.75C$ dollars

Response 13 $3L + 0.75C = 90$

Response 14 The C-intercept is $C = 90/0.75 = 120$ and is the number of pounds of corn John buys if he spends \$90 and buys no lima beans. • The L-intercept is $L = 90/3 = 30$ and is the number of pounds of lima beans he buys if he spends \$90 and buys no corn.

Response 15 $L = 30 - 0.25C$ for $0 \le C \le 120$

Response 16 **(a)** $L(C)$ is the number of pounds of lima beans that John buys if he spends all \$90 and buys C pounds of corn.
(b) The slope of the line is -0.25 and is the rate of change of pounds of lima beans with respect to pounds of corn.
(c) The slope is negative because John buys fewer pounds of beans if he buys more corn.

Response 17 **(a)** $x = 10 + 4t$ for $0 \le t \le 10$
(b) $y = \frac{2}{3}x = \frac{2}{3}(10 + 4t) = \frac{20}{3} + \frac{8}{3}t$ for $0 \le t \le 10$
(c) $\frac{8}{3}$ feet per second

Response 18 Possible answers: The ground might not be perfectly flat or horizontal. The lamppost might not be perfectly vertical. The top of the man's head is probably not exactly 6 feet above the ground at all times as he walks. The light from the lamp does not all come from one point.

Response 19 **(a)** $x = 60 - 70t$ miles **(b)** $y = 20 + 50t$ miles **(c)** $D = \sqrt{x^2 + y^2}$ $= \sqrt{(60 - 70t)^2 + (20 + 50t)^2}$ **(d)** The minimum is at $t \approx 0.432$ hours when they are $D \approx 51.149$ miles apart.

Response 20 **(a)** As in the response to Question 4, the x-intercept 10 is the number of pounds she must sell in a day to break even.
(b) One explanation: If x is large, then the overhead of \$250 is spread out over a large number of pounds of cotton candy and the average profit is close to the difference, \$25 per pound, between the selling price of \$32 per pound and the \$7 per pound for other expenses.
(c) One explanation: If she sells less than 10 pounds of cotton candy in a day, she has negative profit and consequently negative average profit. **(d)** The average profit tends to $-\infty$ as $x \to 0^+$ because if x is very small, the fixed overhead has to be covered by a very small amount of cotton candy, resulting in a large loss per pound.

Response 21 The second row is the sum of the third and fourth rows because the total energy expenditure is the basal energy expenditure plus the energy expenditure due to activity.

Response 22 The point at $t = 30$ on the top line is at $E \approx 2700$ and the point at $t = 80$ is at $E \approx 2200$. With these coordinates the slope would be $(2200 - 2700)/(80 - 30) = -10$ and the top line would have the equation $E = 2700 - 10(t - 30)$, which can be rewritten $E = 3000 - 10t$. $E_T(t) = 3000 - 10t$. (Different approximations give different answers.)

Response 23 The point at $t = 30$ on the middle line is at $E \approx 1600$ and the point at $t = 80$ is at $E \approx 1400$. With these coordinates, the slope would be $(1400 - 1600)/(80 - 30) = -4$ and the middle line would have the equation $E = 1600 - 4(t - 30)$ or $E = 1720 - 4t$. $E_B(t) = 1720 - 4t$. (Different approximations give different answers.)

Response 24 With the responses above to Questions 22 and 23, $E_A = E_T - E_B$, so $E_A(t) = 1280 - 6t$. (Different approximations give different answers.)

Response 25 With the responses above to Questions 22 through 24, the rate of change of E_T is -10 calories per year, the rate of change of E_B is -4 calories per year, and the rate of change of E_A is -6 calories per year. (Different approximations give different answers.)

Response 26 One answer: We use 0.017 for the slope, since it is close to all of the slopes in the table. With the value of $F = 18.7$ for $w = 800$ in Table 1, we obtain the linear function, $F = 18.7 + 0.017(w - 800)$

CHAPTER 1 ANSWERS

Tune-up exercises 1.1

T1. **(a)** $-\frac{3}{4}$ **(b)** $\frac{6}{5}$ **(c)** 0 **(d)** The function is not defined at $x = -2$, $x = 1$, or $x = -5$, where its denominator is zero.

T2. $f(0) = 0, f(2) = 4$, and $f(-5) = \dfrac{25}{\sqrt{15}}$ • $f(4)$ is not defined because with $x = 4$ there is a negative number inside the square root sign.

T3. $K(-2) = 2 + 1/(-2) = \frac{3}{2}$ since $-3 < -2 < 0$ • $K(-1) = 2 + 1/(-1) = 1$ since $-3 < -1 < 0$ • $K(0) = 5$ • $K(1) = -1^2 = -1$ since $1 > 0$ • $K(100) = -100^2 = -10000$ since $100 > 0$

T4. **(a)** $A(1) \approx 30$ • $A(4) \approx 40$ **(b)** $A(x) = 40$ for $x \approx 2.8$ and $x \approx 4$ • $A(x) = 60$ has no solutions with $1 \leq x \leq 6$. **(c)** $A(2) \approx 26$ • $[A(2)]^2 \approx 26^2 = 676$ • $A(2^2) = A(4) \approx 40$ • $[A(2)]^2$ is greater.

T5. $F(-x) = \dfrac{1 - x^2 - 3x^3}{10 + x}$

T6. $Q(t) = t + \dfrac{1}{t} + 2$ • $Q(1/t) = \dfrac{1}{t} + t + 2$ • $Q(t)$ and $Q(1/t)$ are equal.

T7. $W(2 + a) - W(2) = 17a + 3a^2$

T8. **(a)** 6 **(b)** $\frac{6}{54} = \frac{1}{9}$ **(c)** $\frac{100}{9}\% \doteq 11.11\%$

T9. **(a)** 28 **(b)** 7 **(c)** 700%

T10. **(a)** $t \geq \frac{10}{7}$ **(b)** $t \leq -12$ and $t \geq 18$ **(c)** $t \geq -\sqrt[3]{2}$

T11. **(a)** $B(-2) \approx 1.9$ • $-2B(-1) \approx -2(1) = -2$ • $[B(2)]^2 \approx 2^2 = 4$ **(b)** $u \approx \pm 1$

T12. $g(-10) = -100.3$ • $g(-3) = -10$ • $g(0)$ is not defined. • $g(3) = -8$ • $g(10) = -99.7$

T13. **(a)** $h(1) = 1/\sqrt{3}$ • $h(2) = \frac{1}{2}$ • $h(4)$ is not defined.

T14. $f(-33) = 6$ • $f(0) = \sqrt{3}$ • $f(78) = 9$

T15. $G(t + 1) = 8t^3 - \dfrac{4}{t} + 10$

T16. $Z\left(\frac{1}{3}\right) = \frac{7}{12}$

T17. $V(2t^2) = 8t^6 - 16t^8$

T18. **(a)** $\frac{1}{5}$ **(b)** $\frac{2}{3}$ **(c)** $66\frac{2}{3}\% \doteq 66.67\%$

T19 **(a)** -1 **(b)** -0.001 **(c)** -0.1%

T20. **(a)** $x > 1$ **(b)** $\frac{2}{5} < x < \frac{6}{5}$ **(c)** $x \leq -2$ and $x \geq 2$

Problems 1.1

1. If A is a function, then $A(2)$ is its value at the value 2 of its variable. • If A is a constant, then $A(2)$ is the constant multiplied by 2.

2. **(a)** x is the variable. **(b)** $f(x)$ is the value of the function at x.
(c) $(x, f(x))$ is a point on the graph.

3. $p(1) = 18$ • $p(2) = -6$ • $p(3) = \sqrt{2}$

4. Statements (a), (c), and (d) are true.

5. **(a)** First you add and then you find a value of g.
(b) First you find values of g and then you add.
(c) First you multiply and then you find a value of g.
(d) First you find values of g and then you multiply.

7. $B/A = 4$

9. **(a)** $B(8) = 1000 + 10(8) + \frac{1}{40}(8)^2 = \$1081.60$
(b) $B(10) = 1000 + 10(10) + \frac{1}{40}(10)^2 = \$1102.50$
(c) $1000\left(1 + \frac{1}{200}r\right)^2 = 1000\left[1 + 2\left(\frac{1}{200}r\right) + \left(\frac{1}{200}r\right)^2\right] = 1000 + 10r + \frac{1}{40}r^2 = B(r)$

11. **(a)** [Area of the base] $= \pi r^2$ **(b)** $4\pi + 2\pi\sqrt{13}$ **(c)** $A = \pi + \pi\sqrt{1+h^2}$ for $r = 1$ • $A = 4\pi + 2\pi\sqrt{4+h^2}$ for $r = 2$ • $A = 9\pi + 3\pi\sqrt{9+h^2}$ for $r = 3$ • Figure A11 • One description: Increasing r raises the graph and makes it steeper.

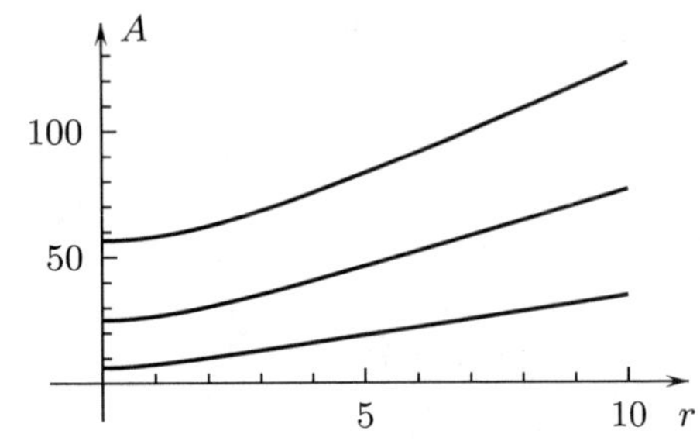

Figure A11

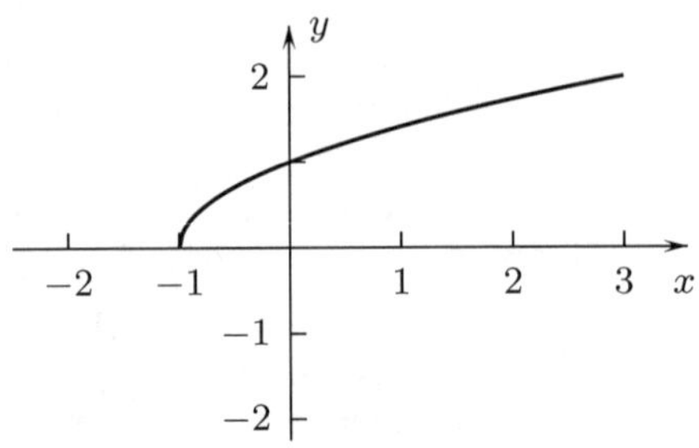

Figure A12a

12. **(a)** The domain is $x \geq -1$ since the $x + 1$ inside the square root sign cannot be negative. • Figure A12a

13. **(a)** $F(10) = 9701$ **(b)** $G(3) = 165$ **(c)** $H(2) = -1$ **(d)** $\sqrt{F(2)} = \sqrt{5} \doteq 2.2361$

15. Possible answers: **(a)** She is stopped from $t \approx 40$ to $t \approx 55$, which is about 15 seconds. **(b)** At $t \approx 110$ **(c)** From $t \approx 70$ to $t \approx 115$ **(d)** From about $t = 7$ to $t = 27$ • She goes about $\left[30\,\frac{\text{miles}}{\text{hour}}\right][20\,\text{seconds}]\left[\frac{1}{3600}\,\frac{\text{hours}}{\text{second}}\right] = \frac{1}{6}$ mile.

16. One possible answer: Figure A16

Figure A16

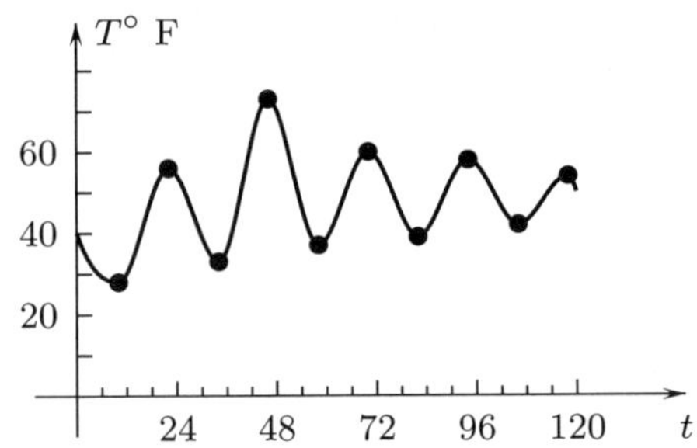

17. **(a)** $\frac{2}{3}$ **(b)** At $t \approx 2.4$ seconds **(c)** ≈ 170 feet **(d)** Possible answers: There is air resistance. The ball maintains contact with the ground for a short time as it contracts and expands before it bounces back up. It is unlikely that the ball would lose exactly the same fraction of its height at each bounce.

21. **(a)** From 200 AD to 1000 AD: [Change] = 50 million, [Relative change] $= \frac{5}{19} \doteq 0.26$ • From 1000 AD to 1800 AD: [Change] = 660 million, [Relative change] $= 2\frac{3}{4} = 2.75$ • From 1800 AD to 2000 AD: [Change] = 5,250 million, [Relative change] $= 5\frac{5}{6} \doteq 5.83$
(b) [Change] = 5960 million, [Relative change] $= 31\frac{7}{19} \doteq 31.37$ **(c)** One answer: The sum of the changes in part (a) equals the change in part (b) because the change from 200 AD to 2000 AD is the sum of the changes from 200 AD to 1000 AD, from 1000 AD to 1800 AD, and from 1800 AD to 2000 AD. The relative changes in part (a) cannot be added to obtain the relative change in part (b) because, while the sum of their numerators is the total change, they have different denominators.
(d) The 14th century • The black (bubonic) plague

22. **(a)** [Least increase in number of PAC's] = 111 (from 1986 to 1988) • [Greatest increase] = 898 (from 1978 to 1980)
(b) [Least increase in contributions] = \$10,045,000 (from 1974 to 1976) • [Greatest increase] = \$31,403,000 (from 1980 to 1982)
(c) \$20,604 per PAC in 1974 and \$34,653 per PAC in 1988 (rounded to the nearest dollar)

25. $G(1) = 2$ • $G(2) = 3(2) + 1 = 7$ • $G(3) = -G(1) - G(2) = -2 - 7 = -9$ • $G(4) = 2(-9) = -18$ • Figure A25

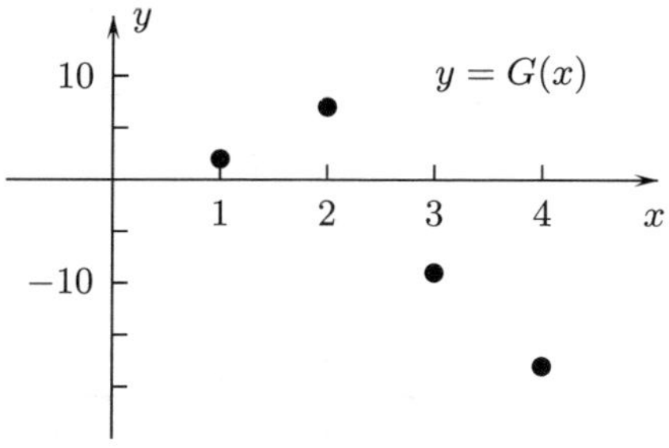

Figure A25

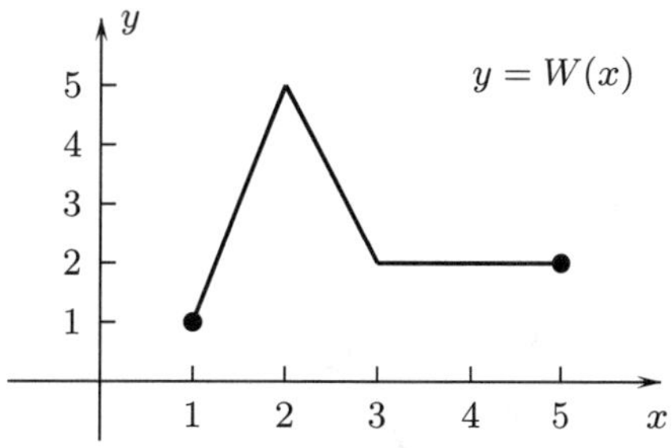

Figure A28

26. **(a)** $10Q(10) = 210$ **(b)** $Q(6+3+1) = 21$ **(c)** $Q(100/20) = 7$ **(d)** $Q(10)/Q(5) = 3$ **(e)** $Q(Q(5)+3) = 21$

27. **(a)** The greatest decrease was about 160 (from 1982 to 1983). **(b)** Five years from 1984 to 1989

28. **(a)** The domain is the interval $1 \leq x \leq 5$ **(b)** Figure A28 **(c)** $W(1) = 1$ • $W(3) = 2$ • $W(4) = 2$ • $W(4.5) = 2$

30. [Change] $=0.75F(4) = 4.5$ • $F(10) = 1.75F(4) = 1.75(6) = 10.5$ or $F(10) = F(4) + 4.5 = 10.5$

37. Possible answer: Set $F(0) = 1$. Then $F(1) = 4F(0) = 4, F(2) = 4F(1) = 16, F(3) = 4F(2) = 64, F(4) = 4F(3) = 256$, and $F(5) = 4F(4) = 1024$.

40. **(a)** \$750 per week **(b)** \$960 per week

42. $\left(\frac{4}{3}, \frac{7}{3}\right)$ • Figure A42

43. $x = 2$ and $x = -6$ • Figure A43

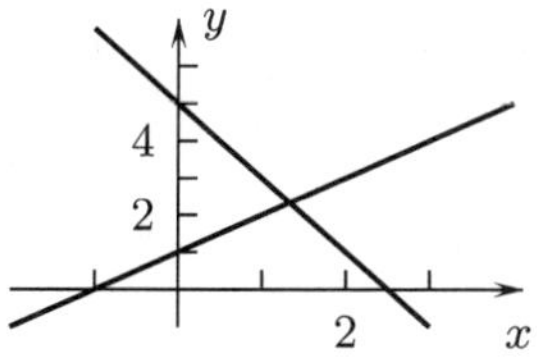

$y = x + 1,\ y = 5 - 2x$

Figure A42

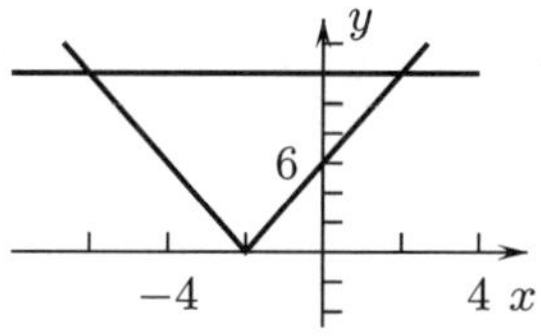

$y = |3x + 6|,\ y = 12$

Figure A43

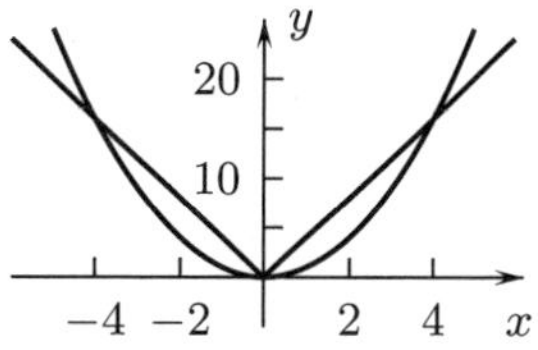

$y = x^2,\ y = 4|x|$

Figure A44

44. $x = -4, 0, 4$

48. x is a solution if $-5 < 3x + 4$ and $3x + 4 < 7$. • $-9 < 3x$ and $3x < 3$ • $-3 < x < 1$ • Figure A48

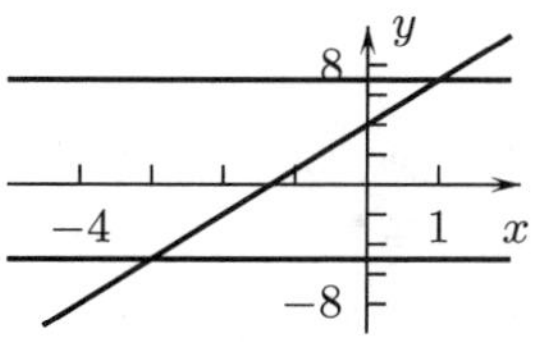

$y = 3x + 4,\ y = -5,\ y = 7$

Figure A48

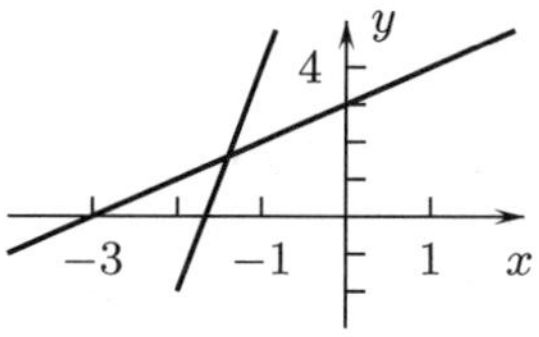

$y = x + 3,\ y = 6x + 10$

Figure A49

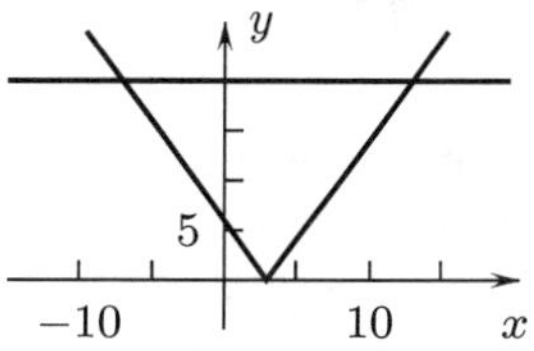

$y = |6 - 2x|,\ y = 20$

Figure A51

49. $-7 < 5x$ • $x > -\frac{7}{5}$ • Figure A49

51. $6 - 2x \geq 20$ or $6 - 2x \leq -20$ • $x \leq -7$ or $x \geq 13$ • Figure A51

53. $x = 3$ and $x = \frac{1}{2}$ • Figure A53

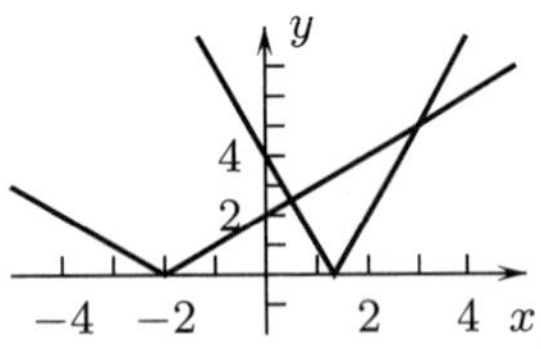

$y = |x + 2|$, $y = |3x - 4|$

Figure A53

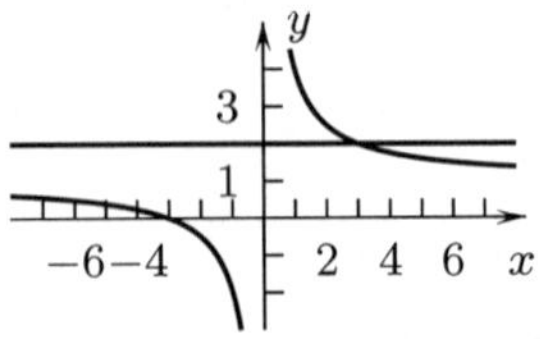

$y = \dfrac{3 + x}{x}$, $y = 2$

Figure A55

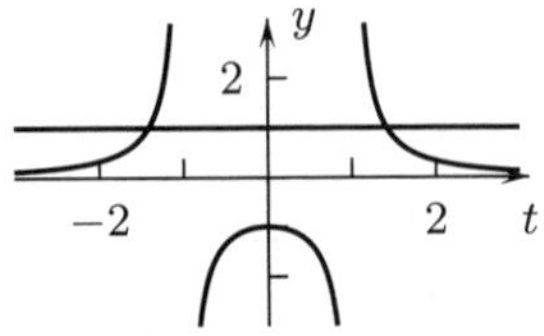

$y = 1/(t^2 - 1)$, $y = 1$

Figure A59

55. $x < 0$ and $x > 3$ • Figure A55

59. $|t| < 1$ and $|t| > \sqrt{2}$ • Figure A59

68. **(a)** Figure A68a **(c)** Figure A68c

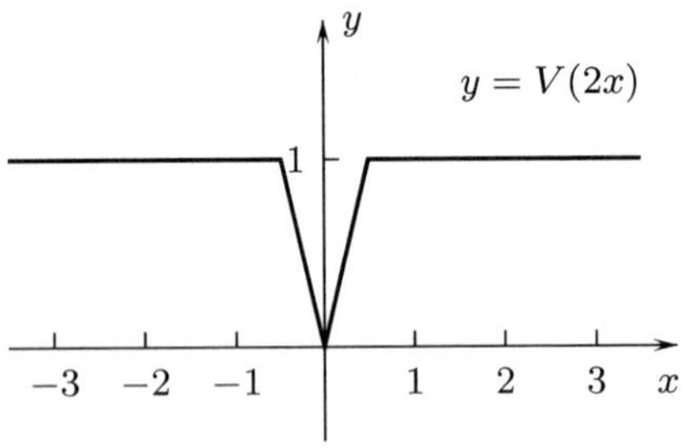

Figure A68a

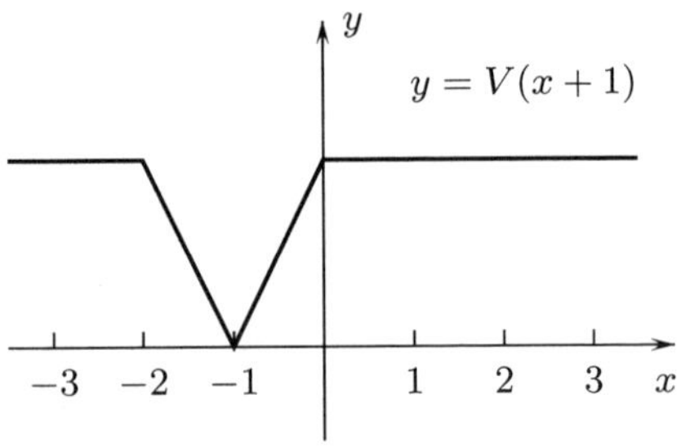

Figure A68b

Tune-up exercises 1.2

T1. **(a)** $y = 5 - 4(x - 3)$ **(b)** $y = 1 - \frac{1}{10}(x - 5)$ or $y = 2 - \frac{1}{10}(x + 5)$

T2. **(a)** $y = 7 + \frac{1}{2}(x + 4)$ **(b)** $y = -5 + \frac{3}{5}(x + 3)$ or $y = -2 + \frac{3}{5}(x - 2)$

T3. **(a)** $y = -1 + \frac{3}{2}(x - 1)$ **(b)** $y = 7 + \frac{1}{3}(x - 4)$

T4. $\frac{1}{5}x - \frac{1}{6}y = 1$ • Slope $= \frac{6}{5}$

T5. **(a)** $y = 7$ **(b)** $x = -4$

T6. The midpoint of the line seqment is $(5, 3)$ and its slope is -3. • The line we want goes through $(5, 3)$ and has slope $\frac{1}{3}$. • $y = 3 + \frac{1}{3}(x - 5)$

T7. $-\frac{5}{7}$

T8. Figure T8 • The triangle has sides of lengths $5\sqrt{5}, 5\sqrt{5}$ and $9\sqrt{2}$ and is isosceles but not equilateral.

T9. The lines $y = \frac{1}{2}x$ and $y = -2x$ intersect at $(0, 0)$. • The lines $y = \frac{1}{2}x$ and $y = -\frac{1}{3}x + 5$ intersect at $(6, 3)$. • The lines $y = -2x$ and $y = -\frac{1}{3}x + 5$ intersect at $(-3, 6)$. • The lines $y = -2x$ and $y = \frac{1}{2}x$ are perpendicular, so the triangle has a right angle. • The lengths of the perpendicular sides are each $\sqrt{45}$, so the area is $\frac{45}{2}$.

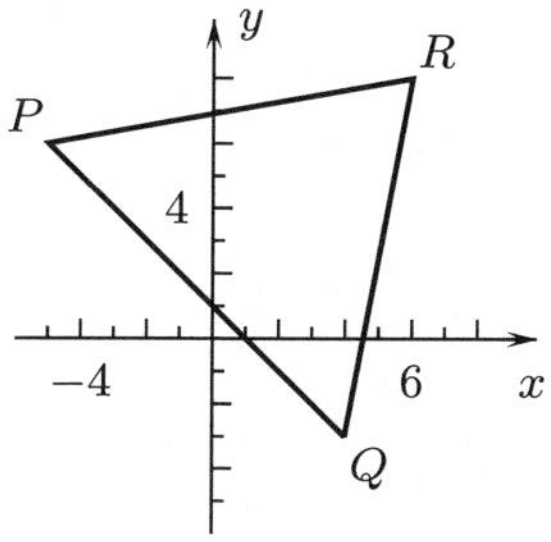

Figure T8

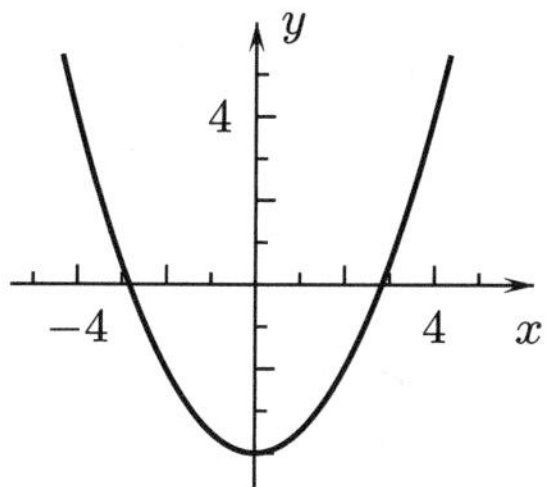

Figure T10a

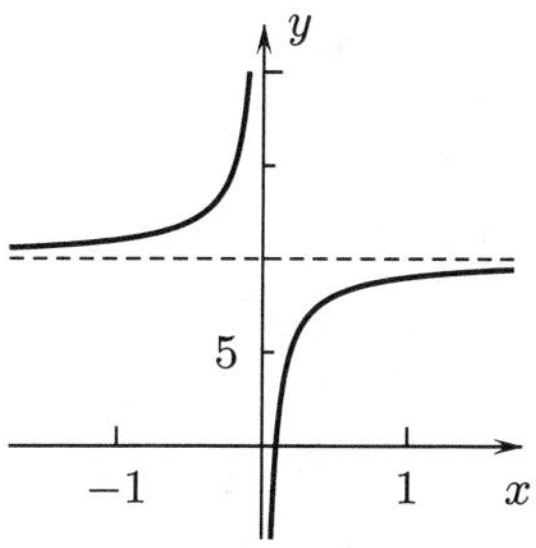

Figure T10b

T10. **(a)** One description: $y = \frac{1}{2}x^2 - 4$ is a parabola with a U-shape that opens upward, is symmetric about the y-axis, and whose lowest point is at $y = -4$ on the y-axis. • Figure 10a **(b)** $y = 10 - 1/x$ is the curve $y = -1/x$ translated 10 units upward. It goes down to the right and up to the left of the y-axis, which is a vertical asymptote. It approaches the horizontal asymptote $y = 10$ as $x \to \pm\infty$ • Figure 10b

T11. One description: Multiplying by -3 reflects the curve about the x-axis and moves each point to three times as far from the x-axis.

T12. Figure 34 of Section 1.2: $y = \dfrac{36}{x^2} - 25$ • Figure 35: $y = 20 - 3x^2$ • Figure 36: $y = -\dfrac{24}{x}$ • Figure 37: $y = x^3 - 5$

T13. $\frac{1}{3}$

T14. -2

T15. $y = \frac{10}{11}(x + 5)$

T16. $y = 3 + \frac{5}{3}(x - 4)$ or $5x - 3y = 11$

T17. $y = 4 - \frac{5}{2}(x - 2)$

T18. One description: $y = 6 - \frac{1}{3}x^3$ crosses the y-axis at $y = 6$ and curves down to the right and curves up to the left. • $y = -1/x^2$ lies entirely below the x-axis and tends to it as $x \to \pm\infty$. It goes down on both sides of the y-axis, which is its vertical asymptote, and it is symmetric about the y-axis.

Problems 1.2

1. x and y are coordinates of the variable point; m is the slope; x_0 and y_0 are coordinates of the given fixed point

4. $a > 0$

8. x^{-1} dominates x^{-2} for large positive and large negative x. x^{-2} dominates x^{-1} for small positive and large negative x.

9. **(a)** x^3 is the dominant term in the numerator and $7x^2$ is the dominant term in the denominator when x is a large positive or negative number. **(b)** -5 is the dominant term in the numerator and $3x$ is the dominant term in the denominator when x is close to zero.

11. $y = -x$

13. $(4, 3)$ and $(4, -3)$

15. $\frac{2}{5} < m < \frac{5}{2}$ or $\frac{2}{5} \le m \le \frac{5}{2}$, depending on how "passing through" is interpreted

18. $-\frac{1}{2}$

20. **(a)** $x = 0, -2, 2$ **(b)** $|x| > 2$ **(c)** $0 < |x| < 2$ • Figure A20

22. Figure A22 • The graph looks like $y = 3x^2$ for small $|x|$ where $3x^2$ is the dominant term and approximately like $y = -x^4$ for large $|x|$ where $-x^4$ is the dominant term. • $y \to -\infty$ as $x \to \pm\infty$

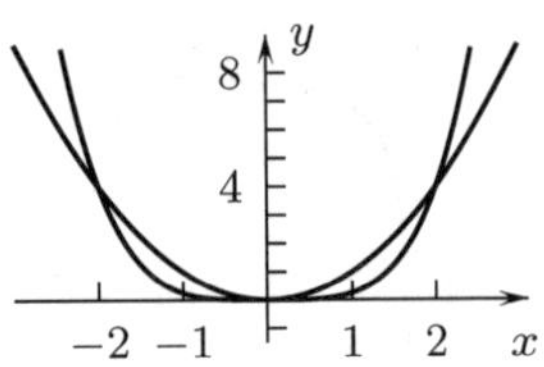

Figure A20

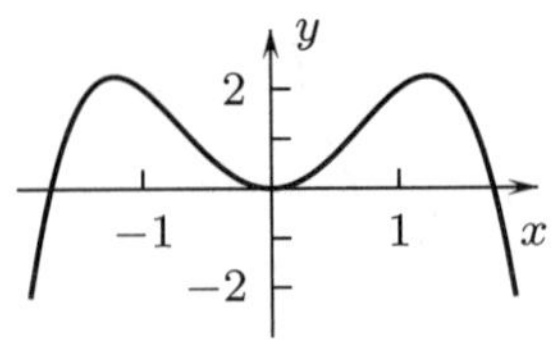

Figure A22

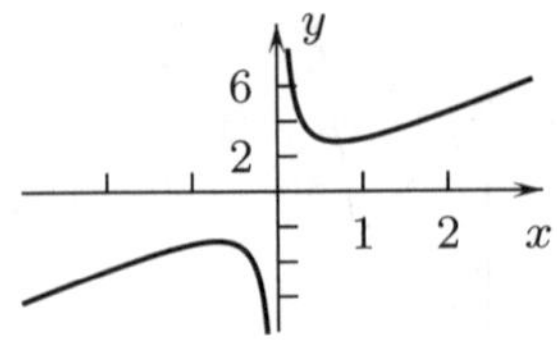

Figure A23

23. Figure A23 • The graph looks like $y = 1/x$ for x near 0, where $1/x$ is the dominant term and like $y = 2x$ for large $|x|$, where $2x$ is the dominant term. • The y-axis is a vertical asymptote. • $y \to \infty$ as $x \to 0^+$ • $y \to -\infty$ as $x \to 0^-$ • $y \to \infty$ as $x \to \infty$ • $y \to -\infty$ as $x \to -\infty$

26. Figure A26 • The graph looks like $y = x$ for small $|x|$, where x is the dominant term and approximately like $y = \frac{1}{3}x^3$ for large $|x|$, where $\frac{1}{3}x^3$ is the dominant term. • $y \to -\infty$ as $x \to \infty$ • $y \to \infty$ as $x \to -\infty$

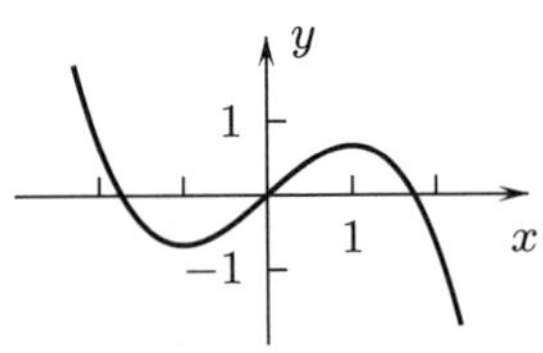

Figure A26

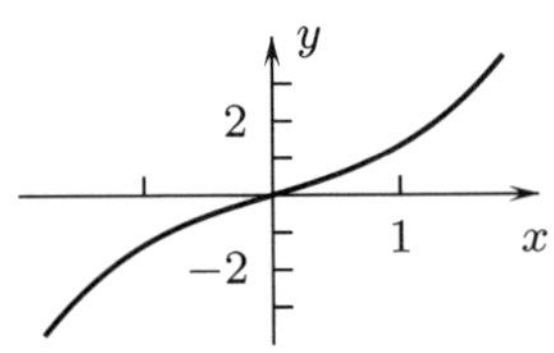

Figure A27

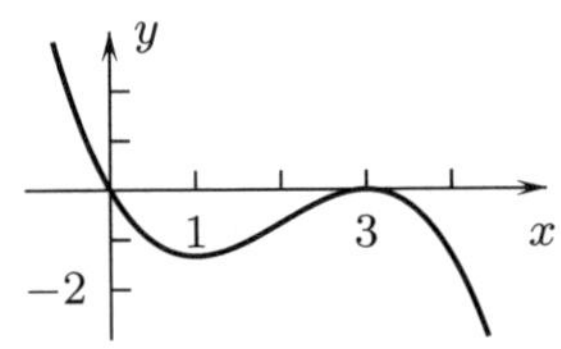

Figure A28

31. Figure 44 in Section 1.2 • One explanation: This is the only curve that has the y-axis as a vertical asymptote and a horizontal asymptote $y = a$ with $a > 0$. • b is negative because the curve goes down on both sides of the y-axis.

34. Yes • One explanation: The slopes of the line segments PQ and QR are both $-\frac{2}{3}$.

41. $y = \frac{3}{4}x + \frac{25}{4}$ and $y = \frac{3}{4}x - \frac{25}{4}$

42. $(0, 0)$ and $(4, 4)$

43. $x + y = -6$

47. $\dfrac{1-x}{x-2} \to -\infty$ as $x \to 2^+$ because $1 - x \to -1$ and $x - 2 \to 0$ as $x \to 2^+$ and $\dfrac{1-x}{x-2} < 0$ for $x > 2$. • $\dfrac{1-x}{x-2} \to \infty$ as $x \to 2^-$ because $1 - x \to -1$ and $x - 2 \to 0$ as $x \to 2^-$ and $\dfrac{1-x}{x-2} > 0$ for $1 < x < 2$.

51. $x^3 > Y$ if and only if $x > Y^{1/3}$. • We can use $X = Y^{1/3}$.

52. $X = 10Y$ will do.

54. For $x < 2$: $\dfrac{3}{x-2} > 0$ and $\dfrac{3}{x-2} < \epsilon$ if and only if $\dfrac{3}{\epsilon} < x - 2$ if and only if $x > \dfrac{3}{\epsilon} + 2$ • We can use $X = \dfrac{3}{\epsilon} + 2$.

55. $X = -\dfrac{2}{\epsilon^{1/3}}$ will do.

58. $\delta = \dfrac{1}{Y^{1/3}}$ will do.

Tune-up exercises 1.3

T1. -6

T2. $f(x) = 10 + 50(x - 3)$

T3. $g(x) = 100 - 40x$

T4. **(a)** The rate of change of $h(x)$ (the slope of its graph) is $(2 - 6)/(0.3 - 0.2) = -40$. • $h(x) = 6 - 40(x - 0.2)$ (Check the values at 0.2 and 0.3 with your calculator.) **(b)** -40 and 40

T5. $k(x) = 1000 - 200x$

T6. One answer: Use the line with y-intercept 7 that goes through the last dot on the right, whose slope is about 3: $k(x) = 7 + 3x$. (Different lines and different approximations of their slopes and y-intercepts yield different answers.)

T7. One answer: Use 2 for the slope and the value 8 at $x = 2$: $L(x) = 8 + 2(x - 2)$.

T8. $-\frac{1}{3}$

T9. $p(w) = -6 - 3(w - 10)$

T10. $Z(x) = 100 + 20x$

T11. One answer: $q(x) = 20 + 2x$

T12. $L(x) = 100 - 2(x - 10)$

T13. $K(x) = 50 + \frac{15}{100}(x - 100)$

Problems 1.3

1. **(a)** $w = 1.58V$ **(b)** 1.58 grams per milliliter

3. **(a)** $p = 0.44h + 14.7$ **(b)** 15941.82 pounds per square inch

5. **(a)** 12 square feet **(b)** $V(h) = 12h$ **(c)** $0 \le h \le 4$ **(d)** 12 cubic feet per foot

7. $F_1(x) = 0.15x + 25$, $F_2(x) = 0.15x + 50$, $F_3(x) = 0.15x + 75$ • Figure A7
(b) 0.15 dollars per mile for all three functions

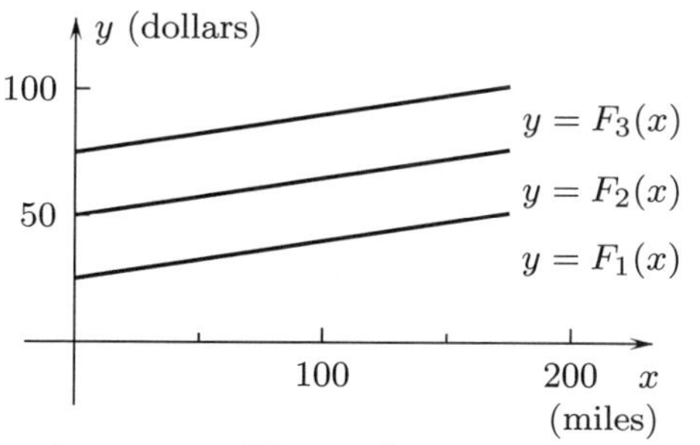

Figure A7

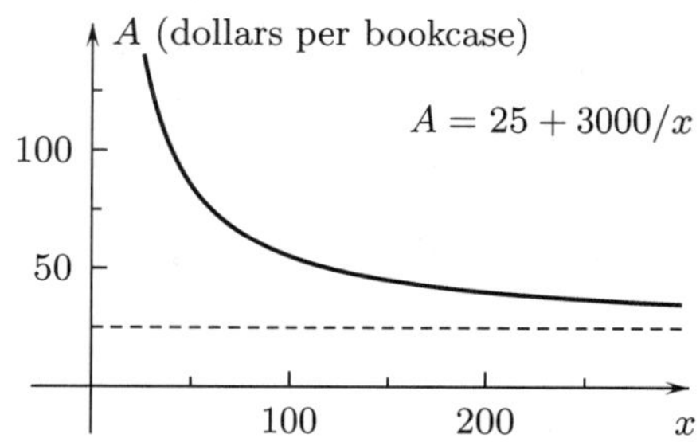

Figure A13

9. **(a)** $v(t) = 32t + 5$ **(b)** One second **(c)** After 3 seconds

11. **(a)** $y = \frac{280}{7} - \frac{20}{7}x$ **(b)** $x = 14 - 0.35y$ **(c)** The rate of change of y with respect to x is $-\frac{20}{7}$ • The rate of change of x with respect to y is $-\frac{7}{20}$. **(d)** One explanation: If you buy more meat you have to buy less potatoes and vice-versa. • The rate of change $-\frac{20}{7}$ of pounds of potatoes with respect to pounds of meat since meat costs more per pound than potatoes.

13. **(a)** $C(x) = 3000 + 25x$ **(b)** $A(x) = \dfrac{3000 + 25x}{x} = \dfrac{3000}{x} + 25$ • Figure A13 • One explanation: $A(x)$ is large if x is small because then the overhead must be paid from sales of only a few bookcases. (The mathematical model is not realistic for $0 < x < 1$.) • $A(x)$ is close to 25 if x is large since then the overhead is spread over many bookcases. **(c)** \$125 **(d)** 1000 bookcases

14. $\frac{1}{25}$ liters per minute per liters per minute

21. **(a)** 30 kilometers per hour **(b)** $D(t) = \sqrt{25 + (3 + 30t)^2} = \sqrt{34 + 180t + 900t^2}$
(c) At $t = 0.3$ hours

24. **(a)** $T(x) = 61800 + 0.28(x - 41200)$ **(b)** $x = 50000 + 150t$ **(c)** $T(t) = 61800 + 0.28(8800 + 150t)$ **(d)** The rate of change of T with respect to x is 0.28 dollars tax per dollar income. • The rate of change of x with respect to t is 150 dollars income per day. • The rate of change of T with respect to t is 42 dollars per day. • [42 dollars tax per day] = [0.28 dollars tax per dollar income] [150 dollars income per day]

25. **(a)** $s = 5t$ **(b)** $x = \frac{4}{5}s$ • $y = \frac{3}{5}s$ **(c)** $x = 4t$ • $y = 3t$ **(d)** The rate of change of x with respect to t is 4. • The rate of change of y with respect to t is 3.

CHAPTER 2
DERIVATIVES AND APPLICATIONS

As you saw in Section 1.3, rates of change of linear functions can be studied using only arithmetic because their graphs are lines and their rates of change are the constant slopes of the lines. Calculus was developed to study rates of change of nonlinear functions that arise in the study air resistance, planetary motion, and a great variety of other applications. We will see in this chapter that the rate of change of a nonlinear function is its DERIVATIVE, which is the slope of a tangent line to its and is found as a finite limit of average rates of change. We study average rates of change in Section 2.1 and finite limits in Section 2.2. These are combined to give the definition of the derivative in Section 2.3, where we derive formulas for exact derivatives of powers of x and of linear combinations of functions and discuss applications that use them. In Section 2.4 we look at more applications, including problems that require finding approximate derivatives from graphs and tables. We derive and use formulas for derivatives of products, quotients, and powers of functions in Section 2.5.

Section 2.1

Average rates of change

OVERVIEW: *This section is background for the introduction of the derivative in Section 2.3. The examples of average velocity and other average rates of change considered here illustrate the types of mathematical models that we will use later with derivatives. Also, the understanding you will gain from this section of how average rates of change relate to secant lines will prepare you to understand the relationship between derivatives and tangent lines in Section 2.3.*

Topics:

- ***Average velocity and other average rates of change***
- ***Estimating velocities with average velocities***

Average velocity

Imagine that a pilot is flying a small plane into a strong headwind toward the west from an airport. As a mathematical model of her flight, we suppose that the plane is $s(t) = t^3 + 30t + 100$ miles from the airport t hours after noon (Figure 1). Note that this function cannot give a perfect description of such a flight, since it would not be possible to know the plane's position with the absolute precision of this formula, and it is unlikely that an actual flight would be even closely described by a formula as simple as this one. We would use this formula —as we use all mathematical models—if it yields results that are as accurate as we need for our particular purposes.

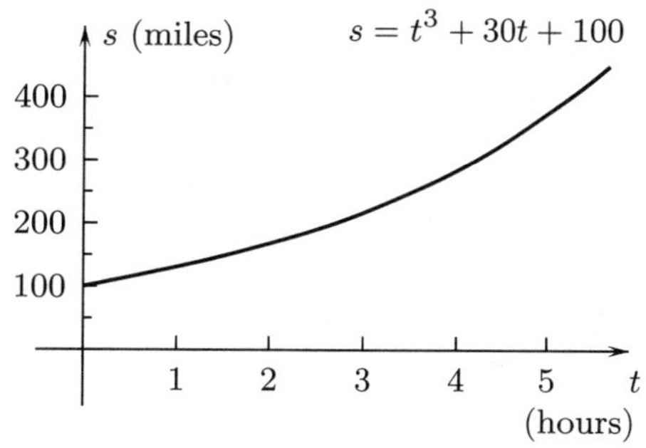

FIGURE 1

Question 1 Complete the following table of values of $s(t) = t^3 + 30t + 100$.

THE PLANE'S DISTANCE FROM THE AIRPORT

t (hours)	0	1	2	3	4	5
$s(t)$ (kilometers)	100	131	168			

♦ SUGGESTIONS TO INSTRUCTORS: Have students begin work on Questions 1 through 15 before class. Problems 2, 3, 9, 14, 17, and 22 are good for in-class discussion.

Question 2 What is the plane's average velocity from $t = 1$ to $t = 5$?

In answering Question 2, you probably first found the distance the plane traveled by subtracting its distance $s(1) = 131$ to the airport at $t = 1$ from its distance $s(5) = 375$ to the airport at $t = 5$, and then divided by the number of hours from $t = 1$ to $t = 5$:

$$\begin{aligned}\text{[The plane's average velocity for } 1 \le t \le 5] &= \frac{[s(5) - s(1)] \text{ miles}}{(5-1) \text{ hours}} \\ &= \frac{(375 - 131) \text{ miles}}{4 \text{ hours}} = \frac{244 \text{ miles}}{4 \text{ hours}} = 61\, \frac{\text{miles}}{\text{hour}}.\end{aligned} \tag{1}$$

Notice that the average velocity **(1)** is the slope of the line in Figure 2 through the points at $t = 1$ and $t = 5$ on the graph of the plane's distance from the airport. The rise is the distance the plane travels, and the run the time it takes to go that distance. Such a line, that passes through two points on a curve, is called a SECANT LINE.†

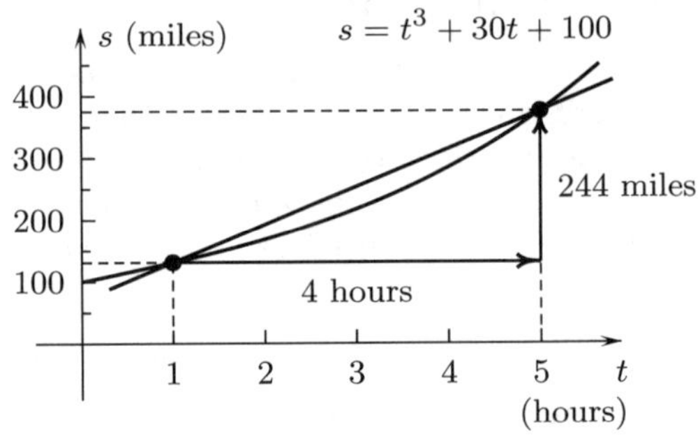

$$\begin{bmatrix}\text{Average} \\ \text{velocity}\end{bmatrix} = \frac{\text{Rise}}{\text{Run}} = \frac{244 \text{ miles}}{4 \text{ hours}}$$

FIGURE 2

Question 3 Calculate the plane's average velocity for $0 \le t \le 4$ and draw the corresponding secant line on the graph of $s(t)$.

Any average velocity can be interpreted as the slope of a secant line:

Definition 1 *If an object is at $s = s(t)$ on an s-axis at time t, then its average velocity in the positive s-direction between times a and b is its change in position, $s(b) - s(a)$, divided by the change in time, $b - a$. This ratio is the slope of the secant line through the points at $t = a$ and $t = b$ on the graph of $s(t)$ (Figure 3):*

$$\begin{aligned}\text{[Average velocity from time } a \text{ to time } b] &= \frac{\text{Change in position}}{\text{Change in time}} \\ &= \frac{s(b) - s(a)}{b - a} \quad (a \ne b).\end{aligned} \tag{2}$$

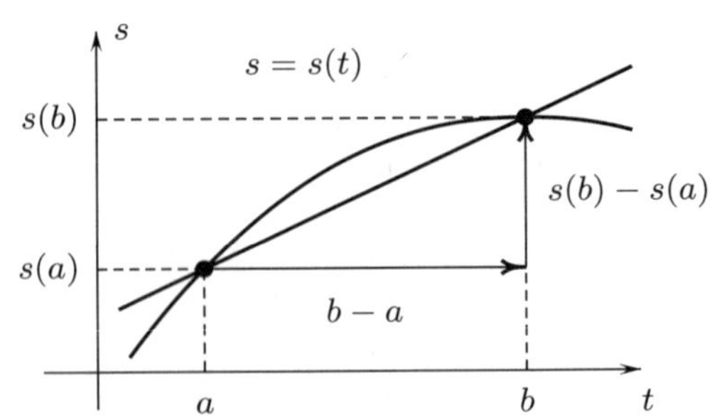

$$\begin{bmatrix}\text{Average} \\ \text{velocity}\end{bmatrix} = \frac{\text{Rise}}{\text{Run}} = \frac{s(b) - s(a)}{b - a}$$

FIGURE 3

†The terminology "secant line" comes from the Latin word *secare*, meaning "to cut." It is used because secant lines generally cut across curves.

Zooming in

The plane's average velocity of 61 miles per hour for $1 \le t \le 5$ that you found in Question 2 differs considerably from its average velocity of 46 miles per hour for $0 \le t \le 4$ that you found in Question 3. The average velocity is different in different time intervals because the function $t^3 + 30t + 100$ that gives the plane's position is not linear.

How fast is the plane flying at $t = 3$? Neither average velocity from Questions 2 and 3 gives a very good estimate because the time intervals $1 \le t \le 5$ and $0 \le t \le 4$ are so long. We can obtain a better estimate from the average velocity in shorter time intervals, as you will see in the next two questions.

Question 4 Generate $s = t^3 + 30t + 100$ in the four windows, **(a)** $2 \le t \le 4, 100 \le s \le 300$, **(b)** $2.5 \le t \le 3.5$, $150 \le s \le 250$, **(c)** $2.99 \le t \le 3.01, 216 \le s \le 218$, and **(d)** $2.999 \le t \le 3.001, 216.9 \le s \le 217.1$. The last curve is shown in Figure 4. Describe the results.

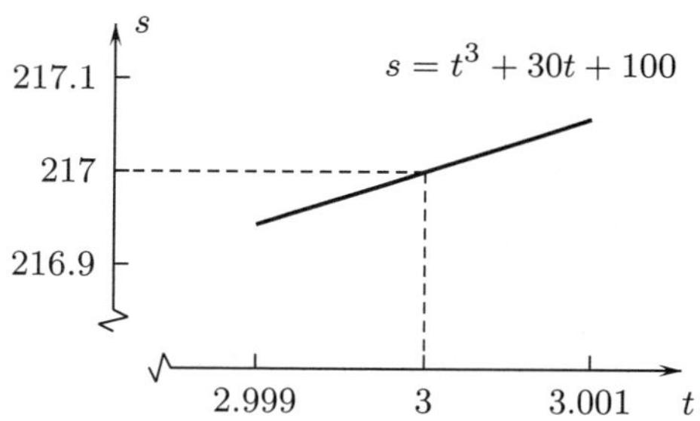

FIGURE 4

Because the portion of the graph in Figure 4 is close to being a straight line, any average velocities over time intervals in that time period will give fairly good estimates of the plane's velocity at $t = 3$.

Question 5 Find the plane's average velocity for $2.999 \le t \le 3$ and for $3 \le t \le 3.001$.

We can conclude from the results of Question 5 that the plane's velocity is approximately 57 miles per hour at $t = 3$. In Section 2.3 we will study differentiation formulas that could be used to show that its velocity is exactly 57 miles per hour at that moment.

What goes up must come down

Imagine that a professional baseball pitcher throws a baseball straight up in the air with a velocity of 80 feet per second ($\doteq$ 54.5 miles per hour), that the ball comes straight down, and that he catches it at the point from which he released it. If we make the simplifying assumption that there is no air resistance and that the only force on the ball is gravity, then, as we will see in Chapter 4, the ball is

$$s(t) = 80t - 16t^2 \text{ feet} \quad \textbf{(3)}$$

above the pitcher's hand t seconds after he throws it, up to the time when he catches it. The graph of $s(t)$ is shown in Figure 5.

Question 6 Use algebra to determine when the pitcher catches the ball.

Question 7 Generate the graph of $s(t)$ in the window $0 \le t \le 5$, $-15 \le s \le 120$ with s-scale $= 20$, as in Figure 5. Use a <maximum> or <trace> command or the graph to determine approximately how high the ball goes and the approximate time when it reaches its highest point.[†] Does the ball spend more time going up or coming down?

[†] See the response to Question 11 in Appendix 1 if you are using a TI-82, TI-83, TI-85, or TI-86 calculator.

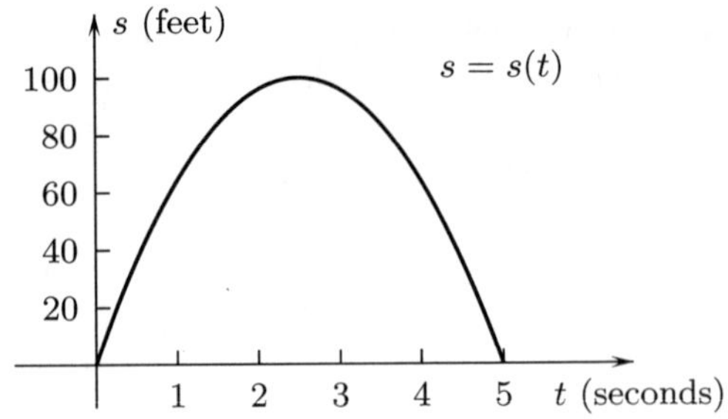

FIGURE 5

Unlike the airplane of Question 1, which went in only one direction (toward the west), the baseball of Figure 5 goes first in one direction (up) and then in the opposite direction (down). In this case, formula **(2)** gives the ball's average velocity in the direction of positive s, which we chose to be the upward direction. The average upward velocity is positive and the secant lines have positive slope for a and b in the time period when the ball is going up. The average upward velocity is negative and the secant lines have negative slope for a and b in the time period when the ball is going down.

Question 8 Find the ball's average upward velocity **(a)** for $0 \le t \le 2$ and **(b)** for $3 \le t \le 4$. Then draw the corresponding secant lines with the graph of $s(t)$.

Question 9 Estimate the ball's upward velocity at $t = 0$, when the pitcher throws the ball, and at $t = 5$, when he catches it, by finding the ball's average upward velocity for $0 \le t \le 0.001$ and for $4.999 \le t \le 5$.

Air resistance

If we include air resistance in the mathematical model of the flight of the baseball, then the ball's height above the pitcher's hand at time t might be given by[†]

$$s_R(t) = 1200 - 1200(0.82)^t - 158.14t \text{ feet.} \tag{4}$$

We use the subscript "R" on the name $s_R(t)$ of this function to suggest "resistance." Its graph is the curve drawn with a thicker line in Figure 6. The curve drawn with a finer line is the graph of the ball's height $s(t)$ with no air resistance given by formula **(3)**.

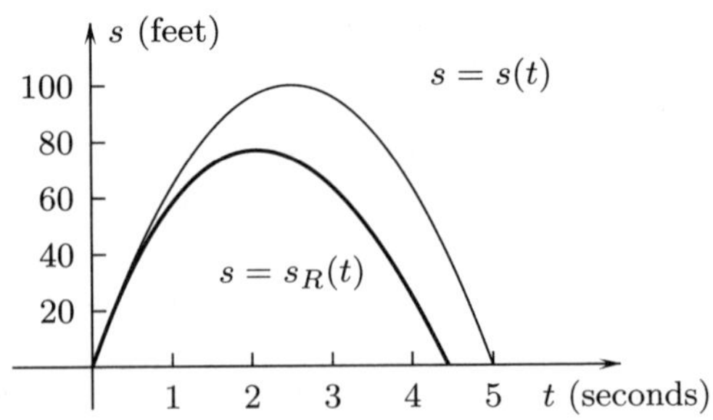

FIGURE 6

Question 10 Generate the graph of $s_R(t)$ in the window $0 \le t \le 6$, $-15 \le s \le 120$, as in Figure 6. Use your calculator or computer to determine, with formula **(4)**, the approximate maximum height of the ball above the pitcher's hand, the approximate time it reaches its highest point, and the approximate time when he catches it.[‡] Does the ball take more time to go up or to come down?

[†]Formula **(4)** comes from a mathematical model in which the force of air resistance is assumed to be proportional to the velocity of the ball. In Chapter 7 you will learn how to obtain such formulas by solving differential equations.

[‡]See the responses to Questions 10 and 11 in Appendix 1 if you are using a TI-82, TI-83, TI-85, or TI-86 calculator.

Question 11 Estimate, with formula **(4)**, the ball's upward velocity when the pitcher throws it and when he catches it, by finding the ball's average upward velocity for $0 \le t \le 0.001$ and for $4.449 \le t \le 4.450$.

Question 12 Describe the effect of air resistance on the flight of the ball.

Other average rates of change

An AVERAGE RATE OF CHANGE of any function is a change in its value, divided by the corresponding change in the value of its variable, and is the slope of a secant line to its graph:

Definition 2 *The* AVERAGE RATE OF CHANGE *of $f(x)$ with respect to x from $x = a$ to $x = b$ is*

$$\frac{\text{Change in } f(x)}{\text{Change in } x} = \frac{f(b) - f(a)}{b - a} \quad (a \ne b) \tag{5}$$

and equals the slope of the secant line through the points at $x = a$ and $x = b$ on the graph of $f(x)$ (Figure 7).

$$\begin{bmatrix}\text{Average rate} \\ \text{of change}\end{bmatrix} = \frac{\text{Rise}}{\text{Run}} = \frac{f(b) - f(a)}{b - a}$$

FIGURE 7

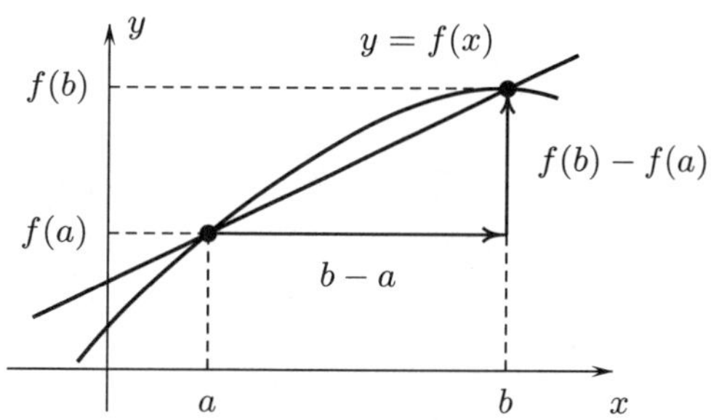

Average velocity is average rate of change of position with respect to time.

Weighing Bossy

If we have only the graph of a function and not an exact formula for its values, we cannot find its exact average rates of change. We can only estimate them by estimating values of the function from the graph.

Question 13 Figure 8 is the graph of a cow's weight as a function of her age t.(1) Find the approximate average rate of change of her weight with respect to time from age two to the time when she weighed the most. Then draw the corresponding secant line on the graph.

FIGURE 8

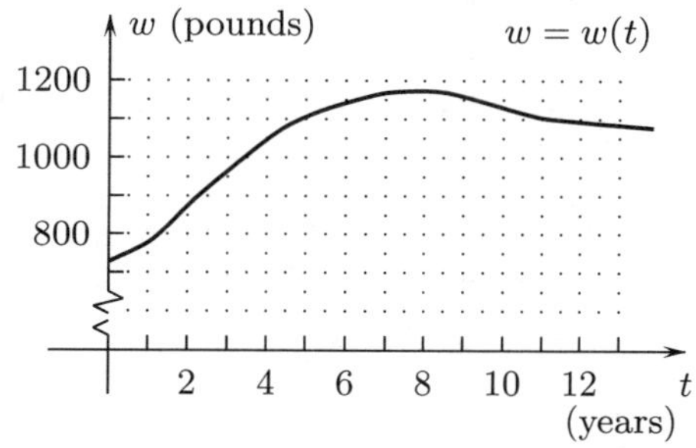

(1) Data adapted from *The Science of Animals that Serve Mankind* by J. Campbell and J. Lasley, New York, NY: McGraw-Hill Book Company, 1975, p. 207. Source: Missouri Agricultural Experimental Station.

The bat is asleep

The next example involves a function given by a table.

Bats are warm-blooded mammals whose body temperatures keep fairly constant when they are awake and active. When a bat is asleep in a cold place, however, it goes into a sort of hybernation and its metabolism (rate of energy expenditure) and body temperature drop. The next table gives the metabolism of a sleeping brown bat as a function of the air temperature around it.[2]

Temperature (°C)	0.5	2.0	10.0	20.0	30.0	37.0	41.5
Metabolic rate (calories/hour)	5.4	1.4	3.4	19.0	96.0	134.0	200.0

Question 14 Describe how the bat's metabolism changes as the temperature drops from 41.5°C to 0.5°C.

Question 15 Find temperatures a and b from the table such that the average rate of change with respect to temperature of the bat's metabolism for $a \le t \le b$ is 7.7 calories per hour per degree.

Principles and procedures

- The average velocity in the positive s-direction for $a \le t \le b$ of an object that is at $s(t)$ on an s-axis at time t is the slope $\dfrac{s(b) - s(a)}{b - a}$ of the secant line through the points $(a, s(a))$ and $(b, s(b))$ at $t = a$ and $t = b$ on the graph of $s(t)$ in a ts-plane.
- An object's velocity at a particular instant can be estimated by using its average velocity in short time intervals including that time.
- The average rate of change of a general function $f(x)$ with respect to x for $a \le x \le b$ is the slope $\dfrac{f(b) - f(a)}{b - a}$ of the secant line through the points at $x = a$ and $x = b$ on the graph of $f(x)$.
- Average velocity is average rate of change of position with respect to time.

Tune-Up Exercises 2.1♦

[A]*Answer provided.* [O]*Outline of solution provided.*

Be sure you can work all of these exercises.

T1.[A] An object is at $s = 50 - 50/t$ (yards) on an s-axis at time t (minutes) for $t \ge 1$. Find its average velocity for $1 \le t \le 5$. Sketch the graph of $s(t)$ without generating it on a calculator or computer and draw the secant line whose slope is the average velocity. (Plot the points on the curve at $t = 1, 2, 4, 5$.)

♦ Type 1, average velocity

T2.[A] Estimate the velocity at $t = 2$ of the object in Exercise T1 by calculating its average velocity for $1.99 \le t \le 2$, for $2 \le t \le 2.01$, and for $1.99 \le t \le 2.01$.

♦ Type 1, average velocity

T3.[A] Values of a function $Q(x)$ are given in the table below. Find the average rate of change of $Q(x)$ with respect to x **(a)** for $0 \le x \le 0.2$, **(b)** for $0.2 \le x \le 0.4$, and **(c)** for $0 \le x \le 0.4$.

♦ Type 1, average velocity

x	0	0.1	0.2	0.3	0.4
$Q(x)$	50.3	52.4	55.0	49.3	50.3

T4.[A] The graph of a function $W(x)$ is shown in Figure 9. Find the average rates of change of $W(x)$ with respect to x **(a)** for $2 \le x \le 5$, **(b)** for $1 \le x \le 3$, and **(c)** for $2.50 \le x \le 2.51$.

♦ Type 1, average velocity

---

[2] Data adapted from *Listening in the Dark, The Acoustic Orientation of Bats and Men* by D. Griffin, Ithaca, NY: Cornell University Press, 1986, p. 40.

♦The Tune-up Exercises and Problems are classified by type and content. The types are (1) basic, reactive; (2) basic reflective; (3) intermediate, reactive; (4) intermediate, reflective; (5) advanced, reactive; (6) advanced, reflective; and (7) advanced, theoretical.

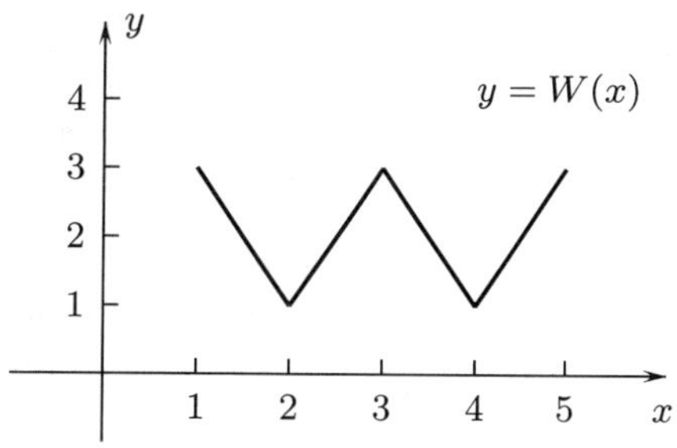

FIGURE 9

Problems 2.1

[A]*Answer provided.* [O]*Outline of solution provided.*

Give units with answers whenever it is appropriate.

1.[A] A motorcyclist, riding away from her home town, is $s(t) = 10t^2$ (miles) from the city limits at time t (hours) for $0 \le t \le 3$. **(a)** What is her average velocity away from the town for $0 \le t \le 3$? **(b)** Draw the graph of $s(t)$ and the secant line whose slope is the average velocity from part (a). (Plot the points at $t = 0, 1, 2, 3$.) **(c)** Does she speed up or slow down during the ride? **(d)** What constant velocity would give her the same average velocity for $0 \le t \le 3$?

♦ Type 2, average velocity

2. A ball rolling toward the edge of a table is $s(t) = 10/t$ centimeters from the edge at time t (seconds) for $t \ge 1$. **(a)** When is it 5 centimeters from the edge? **(b)** When is it 1 centimeter from the edge? **(c)** What is its average velocity away from the edge for $2 \le t \le 10$? **(d)** Draw the graph of $s(t)$ and the secant line whose slope is the average velocity from part (c) (Plot the points at $t = 1, 2, 5, 10$ on the graph of $s(t)$.) **(e)** Does the ball fall off the table?

♦ Type 2, average velocity

3.[A] The following table gives the earth's distance from the sun at three month intervals during one year.[(3)] The distances are not all the same because the earth's orbit is not a circle. The orbit is an ellipse with the sun at one focus. What is the average rate of change with respect to time of the distance from the earth to the sun **(a)** between January 1 and July 1 and **(b)** between July 1 and the following January 1? **(c)** The earth's orbit can be approximated by a circle with the sun at the center. Based on the table, what radius might be used for the circle and what would be the maximum error for the distances in the table with this model?

♦ Type 2, average rates of change

Date	January 1	April 1	July 1	October 1	January 1
Distance ($\times 10^8$ km)	1.4710	1.4949	1.5208	1.4977	1.4710

4. The average number of words in a human's vocabulary at several ages are listed below.[(4)] Which is greater, the average rate of change with respect to age of the size of vocabulary between ages 5 and 6 or between ages 6 and 18?

Age	1	2	3	4	5	6	18
Words in vocabulary	3	300	900	1,500	2,000	2,500	15,000

♦ Type 2, average rates of change

[(3)]Data adapted from *Space Mathematics*, Washington, DC: National Aeronautics and Space Administration, 1985, p. 20.

[(4)]Data adapted from *Great Psychologists* by S. Sargent, New York, NY: Barnes & Noble, Inc., 1958, p. 90.

5.[O] The greatest exchange rate from dollars to Japanese yen from January, 1980 through December, 1998 was 276.74 yen per dollar in November, 1992, and the lowest exchange rate in that period was 83.69 yen per dollar in April, 1995.[(5)] **(a)** What was the percent change in the exchange rate from November, 1992 to April, 1995? **(b)** What was the average rate of change with respect to time of the exchange rate from November, 1992 to April, 1995?

♦ Type 2, percent and average rates of change

6. The deep space probe Pioneer 10 took 21 months to travel the 994 million kilometers from Mars to Jupiter.[(6)] **(a)** What was its average velocity, measured in millions of kilometers per month? **(b)** What was its average velocity, measured in kilometers per hour?

♦ Type 3, average velocity with different units

7. Per capita energy consumption in the United States, Western Europe, Japan, and the formerly communist countries of Eastern Europe are given below for 1968, 1978, and 1988. Energy from all sources is given in terms of the equivalent number of gallons of refined gasoline.[(7)] **(a)** In which of the three years was the ratio of per capita energy consumption in the United States to per capita consumption in Eastern Europe the greatest? **(b)** In which of the four regions was there the greatest and the least change and average rate of change with respect to time of per capita energy consumption from 1968 to 1988?

ANNUAL PER CAPITA ENERGY CONSUMPTION
(GALLONS OF GASOLINE EQUIVALENT)

	United States	Western Europe	Japan	Eastern Europe
1968	1290	430	370	160
1978	1440	550	540	240
1988	1300	540	560	270

♦ Type 2, ratios and rates of change

8.[A] In 1950, a total of 4,718 million pounds of fish were caught by United States fishermen, of which 1,001 million pounds were caught by fishermen sailing from New England and 492 million pounds by fishermen sailing from Alaska. In 1993, 10,436 million pounds were caught, of which 605 million pounds were caught from New England and 5,906 million pounds were caught from Alaska.[(8)] What were **(a)** the percent change from 1950 to 1993 of the total annual catch by U. S. fishermen, **(b)** the average rate of change with respect to time from 1950 to 1993 of the annual catch by fishermen from New England, **(c)** and the fraction of the U. S. catch that was made by Alaska fishermen in 1993?

♦ Type 2, percent change, average rate of change, and a ratio

9.[O] In 1988 there where 26,721 applicants to medical schools in the United States and 15,569 applicants were admitted. There was a 69.77% increase in the number of applicants and a 1.99% increase in the number of admissions from 1988 to 1994.[(9)] What were the average rates of change with respect to time of the numbers of applicants and admissions from 1988 to 1994?

♦ Type 4, average rates of change, etc.

---

(5) Data from the Federal Reserve Bank of St. Louis, http://www.stls.frborg/legal.html.

(6) Data adapted from *Space Mathematics*, Washington, DC: National Aeronautics and Space Administration, 1985, p. 21.

(7) Data adapted from *The Encyclopedia of the Environment*, R. Eblen and W. Eblen, editors, New York, NY: Houghton Mifflin Company, 1994, p. 197.

(8) *Los Angeles Times*, March 29, 1995, Los Angeles, CA: The Times Mirror Company, 1995, p. A5. Source: National Marine Fisheries Service.

(9) *Los Angeles Times*, June 4, 1995 Los Angeles, CA: The Times Mirror Company, 1995, p. D2. Source: Assn. of American Medical Colleges.

10. When a stunt pilot does an inner loop, with his head oriented toward the center of the loop, the centrifugal force causes his blood to concentrate in the vessels of the lower part of his body. This reduces the blood supply to his heart, causing his blood pressure to drop and diminishing the blood supply to his brain. The next table shows the blood pressure of a 150-pound pilot at four values of the centrifugal force on his body.[10] **(a)** What is the pilot's normal blood pressure? **(b)** If his blood pressure drops to 50 millimeters of mercury his vision will black out and then he will quickly fall unconscious. What approximate centrifugal force would cause his blood pressure to drop this low? **(c)** What is the average rate of change of blood pressure with respect to centrifugal force from a centrifugal force of 0 pounds to one of 900 pounds? **(d)** What is the average rate of change of centrifugal force with respect to blood pressure from a blood pressure of 100 millimeters of mercury to one of 30 millimeters of mercury? **(e)** How are the results of parts (b) and (c) related? (Use exact values.)

Centrifugal force (pounds)	0	300	600	900
Blood pressure (mm Hg)	100	80	40	30

♦ Type 4, average rates of change, etc.

11. The average weight of Hereford calves at birth, the average litter size of Poland China hogs, and the average number of eggs laid per year by Leghorn hens are given below as functions of the ages of the cow, sow, and hen, respectively.[11] (There are no entries in the last two columns of the second row because pigs are not generally kept more than seven years.) **(a)** At what age do cows, on average, give birth to the largest calves? **(b)** What is the average rate of change with respect to the sow's age of the average size of a litter from age 1 to the age when the average litter size is greatest? **(c)** During what two-year period does the average egg production decrease the most?

Age of female	1	2	3	4	5	6	7	8	9
Calf weight (lbs.)	370	389	411	427	438	443	442	434	421
Litter size	7.8	9.4	9.8	9.5	9.2	9.2	7.3		
Eggs per year	169	146	124	109	95	86	66	67	51

♦ Type 4, average rates of change, etc.

12.º Figure 10 shows the graphs of annual revenue in the United States from CD-ROM and video games (the black bar graph) and from movie box offices (the gray bar graph) from 1991 through 1994.[12] **(a)** What was the approximate total revenue from CD-ROM and video games in the years 1991 through 1994? What was the approximate total revenue from movie box offices in those four years? **(b)** Give the approximate average rate of change with respect to time of the combined annual revenue from both sources from 1991 to 1994.

♦ Type 2, average rates of change, etc.

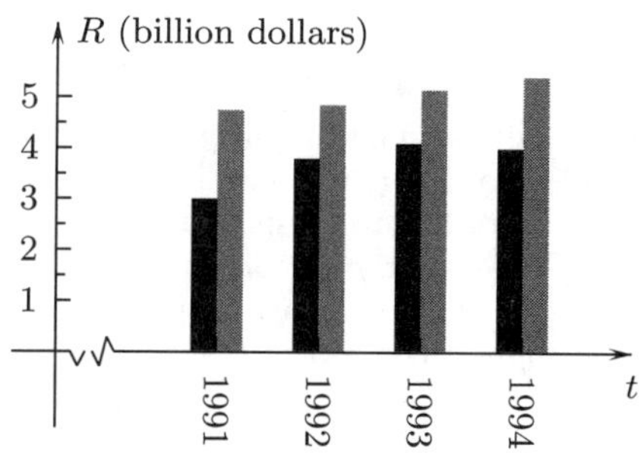

FIGURE 10

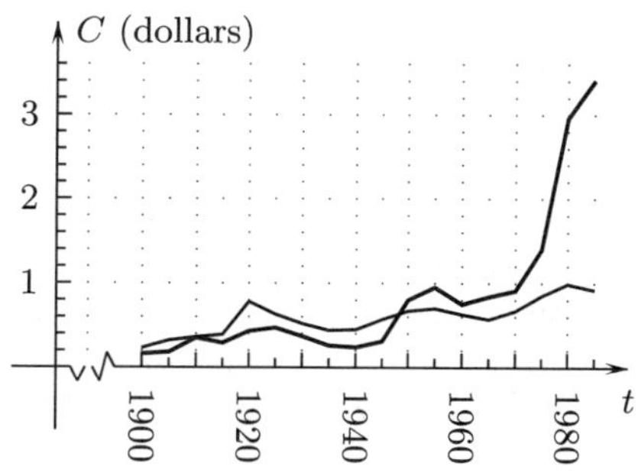

FIGURE 11

[10] Data adapted from *Textbook of Medical Physiology* by A. Guyton, Philadelphia, PA: W. B. Saunders Co., 1981, p. 547.

[11] *The Science of Animals that Serve Mankind* by J. Campbell, New York: McGraw-Hill, Inc., 1975, p. 204.

[12] *Los Angeles Times*, March 23, 1995 Los Angeles, CA: The Times Mirror Company, 1995.

13. The graph drawn with a heavy line in Figure 11 gives the price of a pound of coffee in New York City with the values interpolated from data every five years from 1900 through 1985. The graph drawn with a finer line gives the price of a dozen eggs.[13] **(a)** During which five-year period did the price of coffee increase the most and what was the approximate average rate of change with respect to time of the price of coffee in that period? **(b)** During which fifteen-year period did the price of eggs decrease and what was the approximate average rate of change with respect to time of the price of eggs in that period? **(c)** What was the approximate average rate of change with respect to time from 1900 to 1985 of the total cost of three pounds of coffee and two dozen eggs?

◆ Type 3, average rates of change, etc.

14.[A] A woman's trip is described by the mathematical model in which she is $s(t) = 100 + 50t + 25t^2 - t^4$ miles east of Reno, Nevada on Interstate 80 at time t (hours) for $0 \le t \le 5$. **(a)** Generate the graph of $s(t)$ on your calculator or computer in the window $0 \le t \le 5, -50 \le s \le 500$ with s-scale $= 50$ and use it to describe her trip. **(b)** What is her average velocity toward the east from $t = 0$ to $t = 4$ and from $t = 4$ to $t = 5$? **(c)** Estimate her velocity toward the east at $t = 2$ and $t = 4.5$ by calculating her average velocity toward the east for $2 \le t \le 2.001$, and $4.5 \le t \le 4.501$.

◆ Type 1, average velocity with a graph

15. A tank contains $V(t) = 24t - 3t^2$ gallons of water at time t (hours). **(a)** Generate the graph of $V(t)$ in the window $0 \le t \le 6, -5 \le V \le 55$ with V-scale $= 10$. Use a <maximum> or <trace> command or the graph to find the time T when the volume in the tank is a maximum. (T is an integer.) **(b)** What is the maximum volume? **(c)** Based on the shape of the graph, is water flowing into the tank more rapidly at $t = 1$ or at $t = 3$? Check your answer by estimating the rate of flow at $t = 1$ or at $t = 3$ with average rates of change of the volume with respect to t over short time intervals including $t = 1$ and $t = 3$.

◆ Type 2, average velocity with a graph, etc.

16. The temperature in a room is $T(t) = 85 - 50t^3(3^{-t})$ degrees Fahrenheit at time t (minutes). **(a)** Is the temperature greater at $t = 0$ or at $t = 7$? **(b)** Generate the graph of $T(t)$ for $0 \le t \le 7$, $-10 \le T \le 100$ with T-scale $= 10$. Use a <minimum> or <trace> command or the graph to find the approximate time t when the temperature is a minimum. **(c)** Describe how the temperature varies from $t = 0$ to $t = 7$. **(d)** Based on the graph, at approximately what time is the temperature falling the most rapidly? Use an average rate of change to estimate how fast it is falling at that time.

◆ Type 2, average velocity with a graph, etc.

17. The radius r and area $A = \pi r^2$ of an expanding circle are given as functions of the time t below. What are **(a)** the average rate of change of r with respect to t for $20 \le t \le 60$, **(b)** the average rate of change of A with respect to r for $5 \le r \le 9$, and **(c)** the average rate of change of A with respect to t for $20 \le t \le 60$? **(d)** How are the answers to parts (a), (b), and (c) related? Give units and exact answers, using π as necessary.

t (seconds)	0	20	40	60
r (feet)	2	5	8	9
A (square feet)	4π	25π	64π	81π

◆ Type 4, related average rates of change, etc.

18. An airplane is 3000 feet above the ground three minutes after taking off from sea level. The air pressure is 14.7 pounds per square inch at sea level and 12.9 pounds per square inch at 3000 feet.[14] What are **(a)** the average rate of change of the plane's altitude with respect to time during the first three minutes of the flight, **(b)** the average rate of change of the air pressure with respect to altitude from sea level to 3000 feet, and **(c)** the average rate of change with respect to time of the air pressure on the plane during the first three minutes of flight? **(d)** How are the answers to parts (a), (b), and (c) related? (Give the units.)

◆ Type 4, related average rates of change, etc.

---

(13) Data adapted from *The Value of a Dollar*, S. Derks, editor, Detroit, MI: Gale Research Inc., 1994, pp. 65–546.

(14) Data adapted from *Encyclopædia Britannica*, Vol. I, Chicago, IL: Encyclopædia Britannica, Inc., 1974, p. 627.

19. Mothers of children in kindergarten and the first, fifth, and eleventh grades in Minneapolis, Minnesota and Taipei, Taiwan were asked, (I) "Are you very satisfied with your child's schoolwork?" and, (II) "Do you think your child's school is doing a good job?" The percentages who responded "Yes" to each question are listed below.[15]

Grade of children	K	1st	5th	11th
Mothers in Minneapolis, Question I	57%	49%	36%	33%
Mothers in Taipei, Question I	17%	4%	6%	10%
Mothers in Minneapolis, Question II	93%	91%	87%	78%
Mothers in Taipei, Question II	90%	47%	48%	43%

(a) What were the average rates of change with respect to time from kindergarten to the eleventh grade of the percentages of mothers in the two cities who thought their schools were doing a good job? **(b)** For which age children in each of the two cities was the smallest percentage of mothers "very satisfied" with their children's schoolwork?

◆ Type 4, average rates of change, etc.

20. The next table gives the results of another survey, in which children from Chicago, Illinois and Beijing, China were asked, (I) "Do you consider yourself to be one of the best math students in your class?" and (II) "Do you consider yourself to be one of the students with the most social skills?"[16] **(a)** What were the changes and relative changes from Chicago to Beijing of the percentage of "Yes" responses to the two questions? **(b)** The children were also asked what they would request if a wizard would grant them one wish. The last columns of the table give the percentages who said they would wish for money, objects, education, or other categories of gifts. What are the changes from Chicago to Beijing of the percentage who would wish for money and of the percentage who would wish for education? **(c)** Beijing is 7400 miles from Chicago. What are the average rates of changes with respect to distance from Chicago to Beijing of the the percentage who would wish for money and of the percentage who would wish for education? **(d)** Which are more meaningful, the answer to part (b) or the answer part (c)?

	Question I	Question II	Money	Objects	Education	Other
Chicago	36%	50%	22%	23%	10%	45%
Beijing	15%	35%	1%	10%	67%	22%

◆ Type 4, average rates of change, etc.

21. In 1826, the typical English farmer produced in one day enough food to feed seven people. He obtained 77% of the energy used in the farm from human labor, 21% from animals, and 2% from machines, and his farm produced 40 times the energy that was put into it. The typical English farmer in 1972 produced in one day enough food to feed 200 people with 0.2% of the necessary energy coming from human labor, none from animals, and 99.8% from machines. Because of the energy required for equipment, the farm produced only twice the energy put into it.[17] What were the average rates of change with respect to time from 1826 to 1972 of the number of people one farmer could feed and of the ratio of the energy produced by a farm to the energy put into it?

◆ Type 4, average rates of change, etc.

22. The curve in Figure 12 is the graph of a person's rate of oxygen consumption $R(t)$ (cubic centimeters per minute) t seconds after beginning 40 minutes of meditation.[18] **(a)** What was the approximate average rate of change with respect to time of $R(t)$ from $t = 0$ to $t = 30$? **(b)** Give the approximate value of a time T, such that the average rate of change with respect to time of $R(t)$ from $t = T$ to $t = T + 40$ is zero.

◆ Type 4, average rates of change, etc.

[15] Data adapted from "Learning from Asian Schools" by H. Stevenson, *Scientific American*, Dec., 1992, New York, NY: Scientific American, Inc., 1992, p. 73.

[16] loc. cit., p. 74

[17] Data adapted from *World Health and Disease* by A. Gray and P. Payne, Buckingham: Open University Press, 1993, p. 176.

[18] Data adapted from *Transcendental Meditation* by J. Forem, New York, NY: E.P. Dutton & Co., Inc., 1974, p. 49.

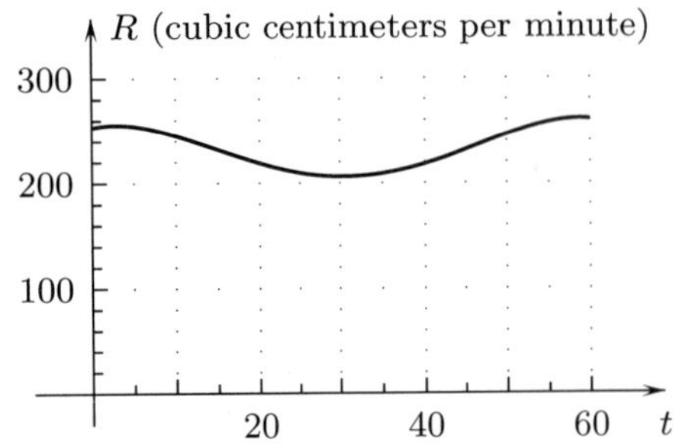

FIGURE 12

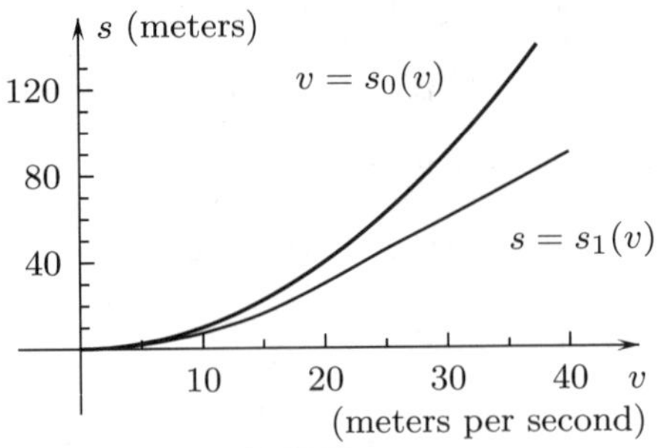

FIGURE 13

23. The upper curve in Figure 13 is the graph of the distance $s_0(v)$ a cricket ball would travel as a function of the velocity v with which it is thrown if there were no air resistance. The lower curve gives the distance $s_1(v)$ the ball would travel with air resistance.[19] In both cases the ball is thrown from the same height at 45° to the horizontal ground and the distance is measured from the point where the ball is thrown to where it hits the ground. **(a)** Why is $s_1(v)$ less than $s_0(v)$ for all $v > 0$? **(b)** What does $D(v) = s_0(v) - s_1(v)$ represent? **(c)** Which is greater, the average rate of change of $D(v)$ with respect to v for $0 \le v \le 20$ or for $0 \le v \le 30$? Justify your conclusion.

♦ Type 4, average rates of change, etc.

24. There were 238 million bushels of wheat grown in Michigan in 1990 and 242 million bushels in 1992. The wheat sold for an average of \$2.21 per bushel in 1990 and \$2.00 per bushel in 1992.[20] **(a)** What was the revenue from wheat in Michigan in 1990? What was it in 1992? **(b)** What was the average rate of change with respect to time from 1990 to 1992 of revenue from wheat in Michigan? **(c)** What would have been the revenue from wheat in Michigan in 1993 if its average rate of change with respect to time was the same from 1992 to 1993 as from 1990 to 1992?

♦ Type 4, average rates of change, etc.

25. The next table gives the number of seconds $t(T)$ of contact with a copper pipe at various temperatures T°C that leads to full-skin-thickness burns.[21] **(a)** For what temperatures a and b in the table is the average rate of change of t with respect to T from $T = a$ to $T = b$ equal to -0.2 seconds per degree? **(b)** What does the inverse function $T(t)$ represent? **(c)** What is the average rate of change of T with respect to t from $t = 2$ to $t = 4$? **(d)** How is the answer to part (c) related to the average rate of change in part (a)? (Give the units.)

Temperature T°C	60	70	80	90	100
Time t (seconds)	100	14	4	2	1

♦ Type 4, average rates of change, etc.

26. A helium balloon is released at time $t = 0$ (seconds). It rises at first but then descends because the helium leaks slowly from it. It is $h = 20 + 10t - t^2$ feet above the ground at time t until it hits the ground. **(a)** Use the Quadratic Formula to find an exact expression for the time when it hits the ground. **(b)** Generate the graph of $h(t)$ in the window $0 \le t \le 12, -5 \le h \le 50$ with h-scale $= 10$, and use the <trace> or <root> command or the graph to check the result of part (a). **(c)** The balloon is highest at an integer value of t. What is it? How high does the balloon go? **(d)** Estimate the balloon's upward velocity at $t = 0$, at $t = 5$, and when it hits the ground by calculating its average velocity in short time intervals.

♦ Type 4, average rates of change, etc., with graphs

27. A fly is $h(t) = 2t - \frac{1}{15}t^2 + 3\sin t$ feet above a picnic table at time t (seconds). **(a)** Generate the graph of $h(t)$ in the window $0 \le t \le 25, -2 \le h \le 50$ with the line $h = 10$ and use an <intersect> or <trace> operation or the graph to find the approximate times a and b when the fly is 10 feet above the table. **(b)** What is the average rate of change with respect to time of the fly's height above the table from time a to time b? **(c)** Estimate the fly's upward velocity at $t = a$ and $t = b$ by calculating average rates of change of $h(t)$ with respect to t in short time intervals.

♦ Type 4, average rates of change, etc., with graphs

(19) Data adapted from *The Human Machine* by R. Alexander, New York, NY: Columbia University Press, 1992, p. 119.

(20) Data adapted from *Statistical Abstract of the United States: 1993*, Washington, DC: U. S. Bureau of the Census, 1993, p. 670.

(21) Data adapted from *Human Thermal Environments* by K. Parsons, London: Taylor & Francis, Ltd., 1993, p. 231.

Section 2.2

Finite limits and continuity

OVERVIEW: *This section contains background material on* FINITE LIMITS *that we will need in Section 2.3, where we will define derivatives as finite limits of average rates of change. We also discuss here the related concept of* CONTINUITY. *This material is fairly technical, and you might not have a very clear understanding of its significance until you have studied derivatives and their applications in this and later chapters. Nevertheless, this section contains notation, terminology, and lines of reasoning that will be important later. Study it carefully.*

Topics:

- ***Closed and open intervals***
- ***One-sided and two-sided finite limits***
- ***Continuity at a point and in an interval***
- ***Limit theorems***
- ***Finding limits by rewriting formulas***
- ***The Intermediate-Value Theorem***

Closed and open intervals

We begin by introducing some terminology and notation concerning intervals that will be used in the rest of the book.

An interval is said to be CLOSED if it includes its endpoints and OPEN if it does not. We write $[a, b]$ for the finite closed interval formed by the numbers x such that $a \leq x \leq b$ with numbers a and b. The symbol (a, b) denotes the finite open interval consisting of all x's with $a < x < b$.† Notice that a square bracket is used when the endpoint is included in the interval and a parenthesis symbol is used when the endpoint is not included. In sketches of intervals, we use dots for endpoints that are included in intervals and small open circles for endpoints that are excluded. This convention is illustrated in Figure 1, which shows the closed interval $[1, 4]$ and the open interval $(1, 4)$.

FIGURE 1

The intervals $a \leq x < b$ and $a < x \leq b$, which are neither open nor closed, are denoted $[a, b)$ and $(a, b]$, respectively. The closed infinite intervals $x \geq a$ and $x \leq b$ are represented by $[a, \infty)$ and $(-\infty, b]$. The open infinite intervals $x > a$ and $x < b$ are denoted (a, ∞) and $(-\infty, b)$. The symbol $(-\infty, \infty)$ represents the infinite interval consisting of all real numbers. It is both open and closed because it has no endpoints to include or exclude.

Question 1 Sketch the intervals $[-5, 2)$ and $(4, \infty)$ on the same x-axis.

---

†Because the symbol (a, b) also denotes the point with horizontal coordinate a and vertical coordinate b, the distinction between open intervals and points has to be determined from the context in which the symbol (a, b) is used.

♦ SUGGESTIONS TO INSTRUCTORS: The main topics in this section can be divided into three parts, as follows:
Part 1 (Finite limits and continuity, apart from the $\epsilon\delta$-definitions; Tune-up Exercises T1–T12; Problems 1–50) Have students begin work on Questions 1 through 3 and 5 through 12 before class. Problems 5, 6, 7, 11, 21, 40 are good for in-class discussion.
Part 2 (The $\epsilon\delta$-definitions of finite limits and continuity; Problems 66-78) Have students prepare a detailed response to Question 4 before class. Problems 66–68, 70–72 are good for in-class discussion.
Part 3 (The Intermediate-Value Theorem; Problems 51–65) Have students begin work on Questios 13 and 14 before class. Problems 51–54, 56, 59, 61, and 65 are good for in-class discussion.

The CLOSURE of a closed interval is the interval itself, and the closure of a non-closed interval is the closed interval obtained by adding its endpoints. The INTERIOR of an open interval is the interval itself, and the interior of a non-open interval is the open interval obtained by removing its endpoints. The interval $[1, 4]$ in Figure 1, for instance, is the closure of the interval $(1, 4)$, and $(1, 4)$ is the interior of $[1, 4]$.

Question 2 What are the closures and interiors of each of the intervals $(-1, 2]$, $(5, 7)$, $[-10, -5]$, and $(-\infty, 3]$?

One-sided finite limits

Figure 2 shows the graph of $G(x)$ defined for all x by

$$G(x) = \begin{cases} 4 - x^2 & \text{for} \quad x < 1 \\ 4 & \text{for} \quad x = 1 \\ 2x & \text{for} \quad x > 1. \end{cases} \qquad \textbf{(1)}$$

This function is defined for all x, but its graph consists of three disconnected pieces—the portion of the parabola $y = 4 - x^2$ for $x < 1$, the point $(1, 4)$, and the portion of the line $y = 2x$ for $x > 1$. A dot has been drawn at $(1, 4)$ to indicate that it is a point on the graph, and open circles have been placed at the right end of the parabola and the left end of the line to show that those points are not part of the graph.

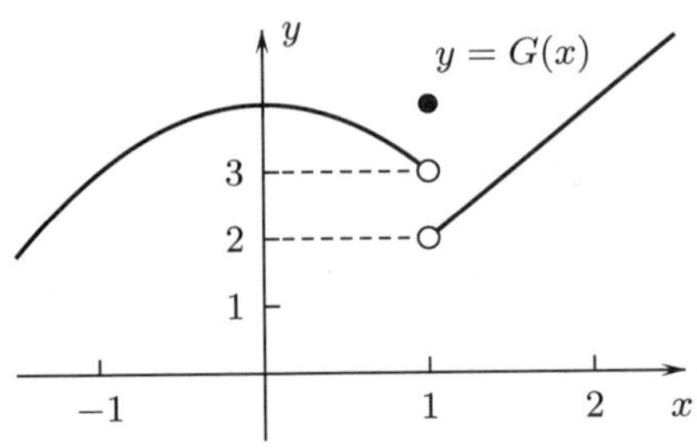

FIGURE 2

We need terminology and notation to describe the behavior of $G(x)$ as x approaches 1 from the right and from the left. We first examine values of $G(x)$ on both sides of $x = 1$.

Question 3 Complete the following tables of values of $G(x)$. The formula $4 - x^2$ is used for $G(x)$ in the table on the left, where $x < 1$, and $2x$ is used in the table on the right, where $x > 1$. Describe the results.

x	$G(x) \doteq$
0.9	3.19
0.99	3.0199
0.999	
0.9999	

x	$G(x) =$
1.1	2.2
1.01	2.02
1.001	
1.0001	

As is suggested by the numbers you found in the table on the right, $G(x)$ approaches 2 as x approaches 1 through values greater than 1 (Figure 3). We say that the LIMIT OF $G(x)$ AS x APPROACHES 1 FROM THE RIGHT is 2 and write[†]

$$\lim_{x \to 1^+} G(x) = 2.$$

[†]The term "limit" in these statements does not mean "barrier" or "boundary" as is generally the case when this term is used in everyday English. The meaning of "limit" here is closer to that of "destination" in everyday speech.

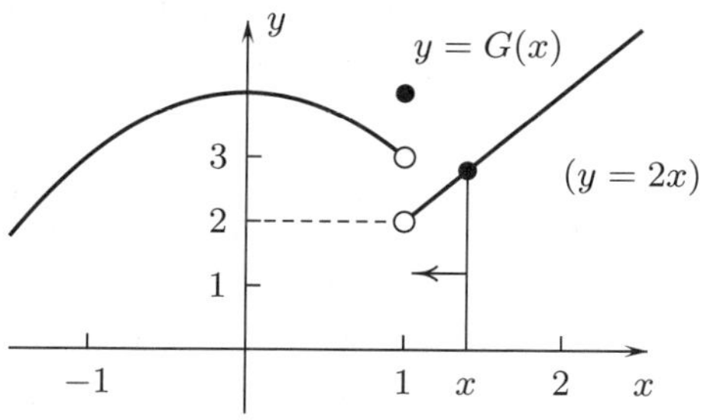

FIGURE 3

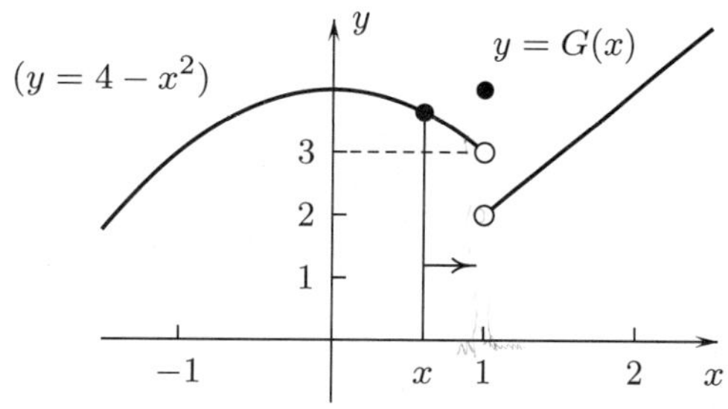

FIGURE 4

The numbers in the completed table on the left suggest that $G(x)$ approaches 3 as x approaches 1 through values less than 1 (Figure 4). We say that the LIMIT OF $G(x)$ AS x APPROACHES 1 FROM THE LEFT is 3 and write

$$\lim_{x \to 1^-} G(x) = 3.$$

As in the notation for infinite limits at vertical asymptotes in Section 1.2, a plus sign is used as a superscript for limits from the right and a minus sign for limits from the left because the positive x-axis is on the right and the negative x-axis is on the left. We can also describe the limits of $G(x)$ as x approaches 1 from the right and left with the notation

$$G(x) \to 2 \text{ as } x \to 1^+ \quad \text{and} \quad G(x) \to 3 \text{ as } x \to 1^-.$$

Notice that the value 4 of $G(x)$ at $x = 1$ is not considered in finding the limits of $G(x)$ as x tends to 1 from the left and from the right. This convention is followed so that limits can be used to distinguish between the behavior of the function near 1 and its value at 1.

A closer look

The intuitive notion of limit that we gave above when we said that $\lim_{x \to 1^+} G(x) = 2$ because "$G(x)$ approaches 2 as x approaches 1 from the right" is adequate for most of our purposes. Sometimes, however, we need a more specific definition—either to deal with cases where the meaning of the term "approaches" might be given more than one interpretation or in proofs. The next example illustrates the type of considerations that lead to the definition we will use.

Example 1 How close does $x > 1$ need to be to 1 in order that $G(x)$ differ from 2 by less than 0.5?

SOLUTION $G(x)$ differs from 2 by less than 0.5 if $1.5 < G(x) < 2.5$, and this is the case if the point on the graph of G at x lies between the horizontal lines $y = 1.5$ and $y = 2.5$. This happens for $1 < x < X$ with X as in Figure 5. Since $G(x) = 2x$ for $x > 1$, this is the X such that $2X = 2.5$, which is $X = 1.25$. Consequently, $G(x)$ differs from 2 by less than 0.5 for $1 < x < 1.25$. □

FIGURE 5

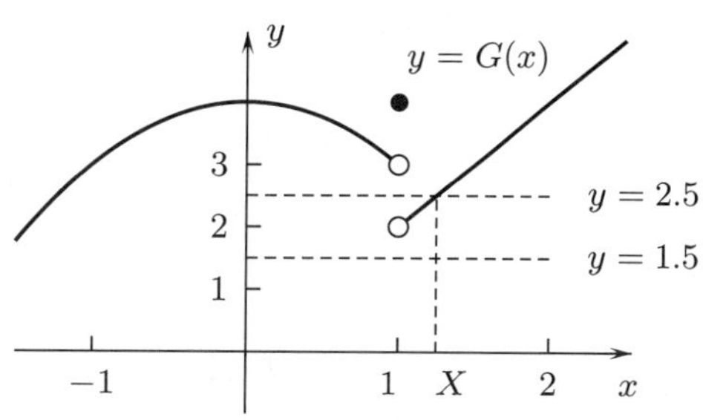

Question 4 **(a)** How close does $x > 1$ need to be to 1 in order that $G(x)$ differ from 2 by less than 0.01? **(b)** How close does $x > 1$ need to be to 1 in order that $G(x)$ differ from 2 by less than ϵ, where ϵ (epsilon) is an arbitrarily small positive number?

In part (b) of Question 4 you showed that for every positive ϵ, no matter how small, $G(x)$ is eventually between and remains between $2-\epsilon$ and $2+\epsilon$ as x decreases toward 1 from the right. This is the phrasing used in part (a) of the following general definition.

Definition 1 *(a)* $\lim_{x \to x_0^+} f(x) = L$ *with numbers* x_0 *and* L *if* $f(x)$ *is defined in an open interval* (x_0, b) *to the right of* x_0 *and if, for every positive* ϵ*, regardless how small,* $f(x)$ *eventually is between and remains between* $L-\epsilon$ *and* $L+\epsilon$ *as* x *decreases toward* x_0 *from the right.*

(b) $\lim_{x \to x_0^-} f(x) = L$ *with numbers* x_0 *and* L *if* $f(x)$ *is defined in an open interval* (a, x_0) *to the left of* x_0 *and if, for every positive* ϵ*,* $f(x)$ *eventually is between and remains between* $L-\epsilon$ *and* $L+\epsilon$ *as* x *increases toward* x_0 *from the left.*

The Greek letter ϵ is used here to suggest "error." The condition $L-\epsilon < f(x) < L+\epsilon$ states that $f(x)$ approximates L with an error less than ϵ. With this point of view, part (a) of Definition 1 states that $f(x) \to L$ as x tends to x_0 from the right if we can cause $f(x)$ to approximate L with arbitrary accuracy by taking $x > x_0$ sufficiently close to x_0. Part (b) of the definition states that $f(x) \to L$ as x tends to x_0 from the left if we can cause $f(x)$ to approximate L with arbitrary accuracy by taking $x < x_0$ sufficiently close to x_0.

Whenever Definition 1 is to be verified algebraically, it is usually rephrased as follows:

Definition 1A *(a)* $\lim_{x \to x_0^+} f(x) = L$ *if* $f(x)$ *is defined in an open interval* (x_0, b) *and if, for every positive* ϵ*, regardless how small, there is a number* $\delta > 0$ *such that* $L-\epsilon < f(x) < L+\epsilon$ *for* $x_0 < x < x_0 + \delta$ *(Figure 6).*

(b) $\lim_{x \to x_0^-} f(x) = L$ *if* $f(x)$ *is defined in an open interval* (a, x_0) *and if, for every positive* ϵ*, there is a number* $\delta > 0$ *such that* $L-\epsilon < f(x) < L+\epsilon$ *for* $x_0 - \delta < x < x_0$ *(Figure 7).*

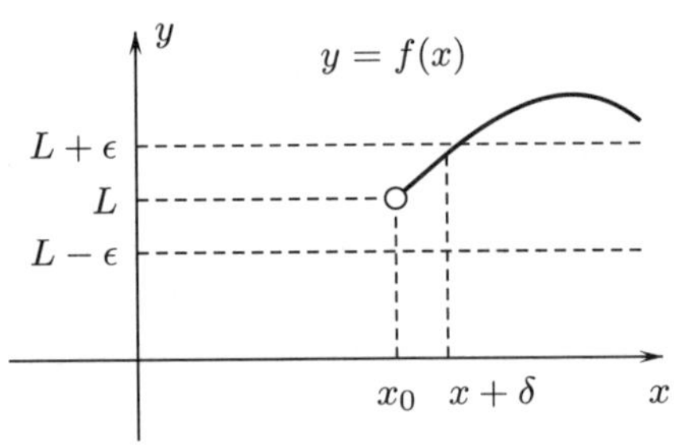

FIGURE 6

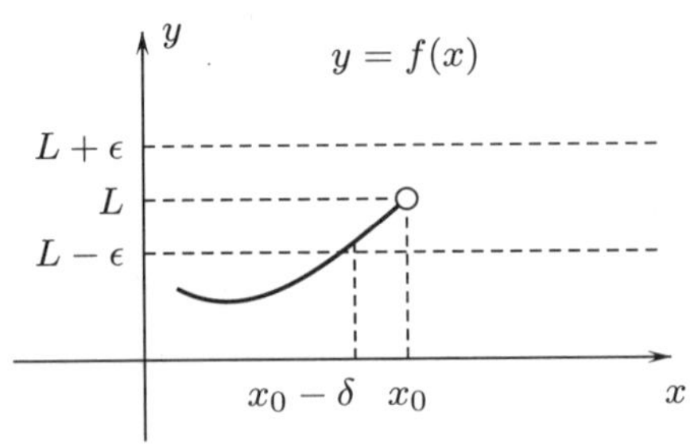

FIGURE 7

To see why part (a) of Definition 1A is stated as it is, imagine a succession of positive ϵ's that are approaching zero. As $\epsilon \to 0$, the horizontal dashed lines at $y = L-\epsilon$ and $y = L+\epsilon$ in Figure 6 squeeze down on the line $y = L$. Definition 1A states that no matter how small ϵ becomes, there is a corresponding positive δ such that the portion of the graph for $x_0 < x < x_0 + \delta$ is between $y = L-\epsilon$ and $y = L+\epsilon$. Because the lines squeeze together as $\epsilon \to 0$, the graph approaches the circle at (x_0, L) and $f(x)$ approaches L as x approaches x_0 from the right. Part (b) of Definition 1A has a similar explanation based on Figure 7.

A function without a limit

The graph $y = \sin(1/x)$ in Figure 8 oscillates infinitely often as x approaches 0 from the right because $1/x$ tends to infinity as x approaches 0 from the right. Accordingly, $\sin(1/x)$ does not have a limit as $x \to 0^+$. The next example shows how Definition 1 can be applied in such a case.

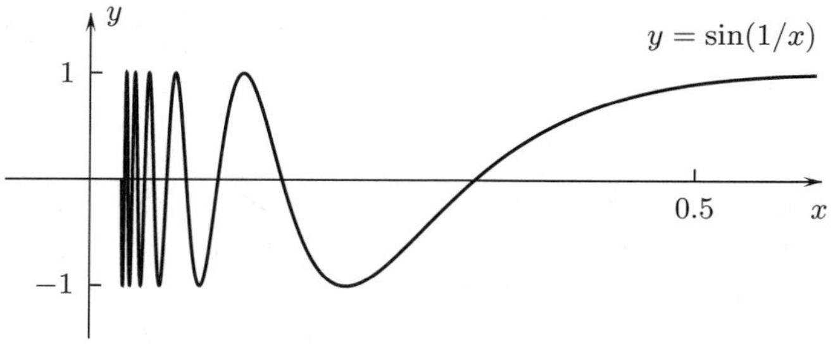

FIGURE 8

Example 2 Use Definition 1 to show that the statement "$\lim_{x\to 0+} \sin(1/x) = 0$" is false.

SOLUTION If the statement "$\lim_{x\to 0+} \sin(1/x) = 0$" were true, then for every $\epsilon > 0$, the number $\sin(1/x)$ would eventually be between and stay between $-\epsilon$ and ϵ as x decreases toward 0 from the right. This is not true for $\epsilon = 0.5$ (or, in fact, for any $\epsilon \le 1$), as can be seen in Figure 8. Therefore, the statement is false. □

We could show that $\lim_{x\to 0+} \sin(1/x)$ does not exist by using Definition 1 to show that the statement "$\lim_{x\to 0+} \sin(1/x) = k$" is false for any number k and by using Definition 3 in Section 1.2 to show that the statements "$\lim_{x\to 0+} \sin(1/x) = \infty$" and "$\lim_{x\to 0+} \sin(1/x) = -\infty$" are also false.[†]

Two-sided finite limits

If we replace the formula $4 - x^2$ by $3 - x^2$ in the first part of definition **(1)** of the function $G(x)$, we obtain the function $H(x)$, defined by

$$H(x) = \begin{cases} 3 - x^2 & \text{for} \quad x < 1 \\ 4 & \text{for} \quad x = 1 \\ 2x & \text{for} \quad x > 1. \end{cases}$$

Its graph is shown in Figure 9.

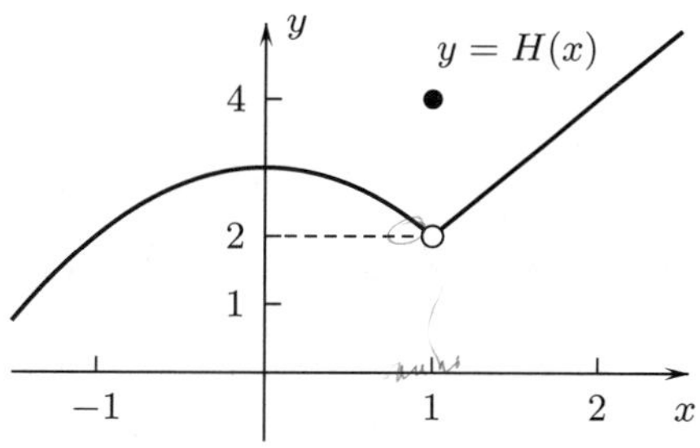

FIGURE 9

As you can see from the graph, $H(x)$ tends to 2 as x tends to 1 from the left and as x approaches 1 from the right. Because the one-sided limits are equal, we say that the LIMIT or the TWO-SIDED LIMIT of $H(x)$ as x approaches 1 is 2. We write

$$\lim_{x\to 1} H(x) = 2 \quad \text{or} \quad H(x) \to 2 \text{ as } x \to 1.$$

[†]Definitions 1 and 1A are further illustrated in Problems 69 through 78.

This is an application of the next definition.

Definition 2 *For a function $f(x)$ defined in open intervals (a, x_0) and (x_0, b) on both sides of x_0, $\lim_{x\to x_0} f(x) = L$ with a number L if and only if $\lim_{x\to x_0^+} f(x) = L$ and $\lim_{x\to x_0^-} f(x) = L$.*

If $\lim_{x\to x_0^+} f(x)$ and $\lim_{x\to x_0^-} f(x)$ are different or one or both do not exist, we say that $\lim_{x\to x_0} f(x) = L$ does not exist or is not defined.

Continuity at a point

In most cases that normally arise in introductory calculus, a function is continuous wherever its graph is not broken. We will use this idea in our formal definitions of continuity, which we first motivate by looking at the function $J(x)$ of Figure 10, which is defined by

$$J(x) = \begin{cases} x + 4 & \text{for} \quad -2 \le x \le 1 \\ \dfrac{3}{x} & \text{for} \quad 1 < x \le 3. \end{cases}$$

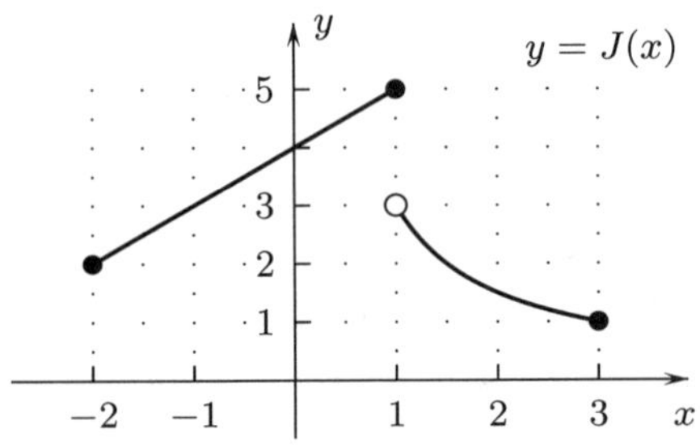

FIGURE 10

Question 5 **(a)** What are $J(-2)$ and $\lim_{x\to -2^+} J(x)$? **(b)** What are $J(3)$ and $\lim_{x\to 3^-} J(x)$? **(c)** What are $J(0)$ and $\lim_{x\to 0} J(x)$?

Notice what the responses to Question 5 say about how the parts of the graph in Figure 10 fit together:

Because $J(-2)$ and $\lim_{x\to -2^+} J(x)$ are equal, the line segment that forms the graph for $-2 < x < 1$ is connected to the dot at $x = -2$. We describe this by saying that $J(x)$ is CONTINUOUS FROM THE RIGHT at $x = -2$.

Because $J(3)$ and $\lim_{x\to 3^-} J(x)$ are equal, the portion of the hyperbola $y = 3/x$ that forms the graph for $1 < x < 3$ is connected to the dot at $x = 3$. We say that $J(x)$ is CONTINUOUS FROM THE LEFT at $x = 3$.

Finally, because $J(0)$ equals the two-sided limit $\lim_{x\to 0} J(x)$, the graph is not broken at $x = 0$. We say that $J(x)$ is CONTINUOUS AT $x = 0$. Here is a general definition of these terms.

Definition 3 *(a) $f(x)$ is continuous from the right at x_0 if it is defined in an interval $[x_0, b)$ containing x_0 and if $\lim_{x\to x_0^+} f(x)$ exists and equals $f(x_0)$.*

(b) $f(x)$ is continuous from the left at x_0 if it is defined in an interval $(a, x_0]$ containing x_0 and if $\lim_{x\to x_0^-} f(x)$ exists and equals $f(x_0)$.

(c) $f(x)$ is continuous at x_0 if it is defined in an open interval containing x_0 and $\lim_{x\to x_0} f(x)$ exists and equals $f(x_0)$.

Notice that a function is continuous at a point if and only if it is continuous from the right and from the left at the point. The function $J(x)$ of Figure 10 is discontinuous at $x = 1$ because it is not continuous from the right at $x = 1$. Its graph is broken there.

Continuity in intervals

Intuitively, a function is continuous in an interval if the portion of its graph for x in the interval is one unbroken piece. We say that $J(x)$ of Figure 10, for example, is continuous in $[-2, 1]$ because the portion of its graph for $-2 \le x \le 1$ is not broken, and that $J(x)$ is continuous in $(1, 3]$ because the part of its graph for $1 < x \le 3$ is not broken. $J(x)$ is also continuous in any smaller intervals contained in $[-2, 1]$ and $(1, 3]$. Here is the general definition.

Definition 4 *$f(x)$ is continuous in an interval if (i) it is continuous at each point interior to the interval, (ii) it is continuous from the right at a if the interval includes a left endpoint a, and (iii) it is continuous from the left at b if the interval includes a right endpoint b.*

Example 3 The function $K(x)$, whose graph is shown in Figure 11, is defined for $-3 \le x \le 3$ by

$$K(x) = \begin{cases} x + 3 & \text{for} \quad -3 \le x \le -1 \\ x^3 + 1 & \text{for} \quad -1 < x < 1 \\ 3 & \text{for} \quad x = 1 \\ 2 & \text{for} \quad 1 < x \le 3. \end{cases}$$

In which of the intervals **(a)** $[-3, -1]$, **(b)** $[-1, 0)$, **(c)** $(-1, 1)$, and **(d)** $(1, 2]$ is $K(x)$ continuous?

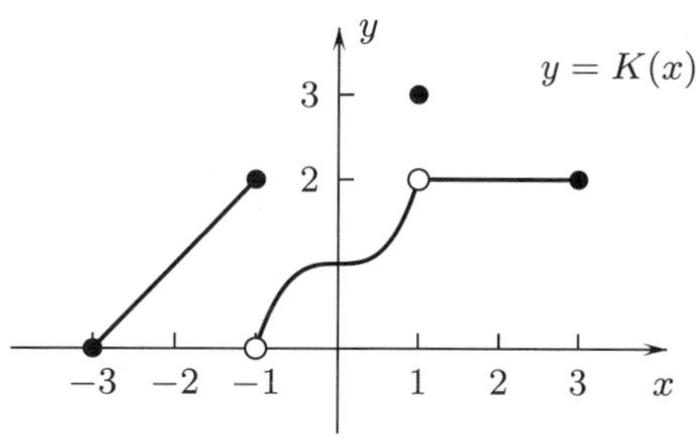

FIGURE 11

SOLUTION **(a)** $K(x)$ is continuous in $[-3, -1]$ because it is continuous at all x with $-3 < x < -1$ and is continuous from the right at -3 and continuous from the left at -1.
(b) $K(x)$ is not continuous in $[-1, 0)$ since it is not continuous from the right at -1.
(c) $K(x)$ is continuous in $(-1, 1)$ because it is continuous at all x with $-1 < x < -1$.
(d) $K(x)$ is continuous in $(1, 2]$ because it is continuous at all x with $1 < x < 2$ and is continuous from the left at 2. □

Continuity of functions given by single formulas

The functions $G(x), H(x), J(x)$, and $K(x)$ above were constructed by using different formulas in different portions of their domains. We will now take a special look at functions given by single formulas, with domains consisting of all values of x for which the formulas are defined.

Example 4 The graphs of $y = \frac{1}{2}x^2$ and $y = 1/x$ are shown in Figures 12 and 13. **(a)** What is the domain of $y = \frac{1}{2}x^2$ and what is the most extensive interval in which it is continuous?[†] **(b)** What is the domain of $y = 1/x$ and what are the most extensive intervals in which it is continuous?

[†] We say that an interval I is the "most extensive" interval or one of the "most extensive" intervals in which a function is continuous if the function is continuous in I and is not continuous in any interval obtained by extending I to the right or to the left.

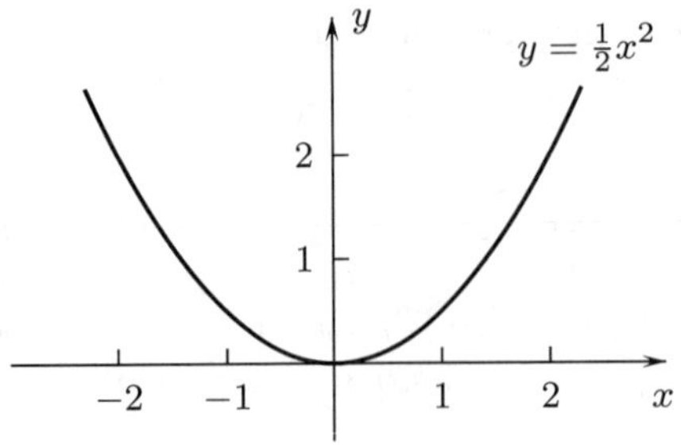

FIGURE 12

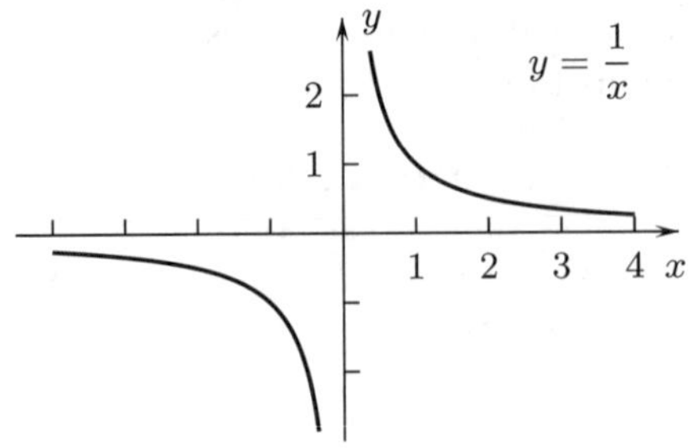

FIGURE 13

SOLUTION (a) $y = \frac{1}{2}x^2$ is defined and continuous in the interval $(-\infty, \infty)$, which cannot be extended.
(b) $y = 1/x$ is defined and continuous in $(-\infty, 0)$ and $(0, \infty)$. These are the most extensive intervals in which $y = 1/x$ is continuous because $1/x$ is not defined at $x = 0$. □

Question 6 (a) The graph of $y = 1 + \sqrt{x-2}$ is shown in Figure 14. What is its domain and why? (b) What is the most extensive interval in which $y = 1 + \sqrt{x-2}$ is continuous?

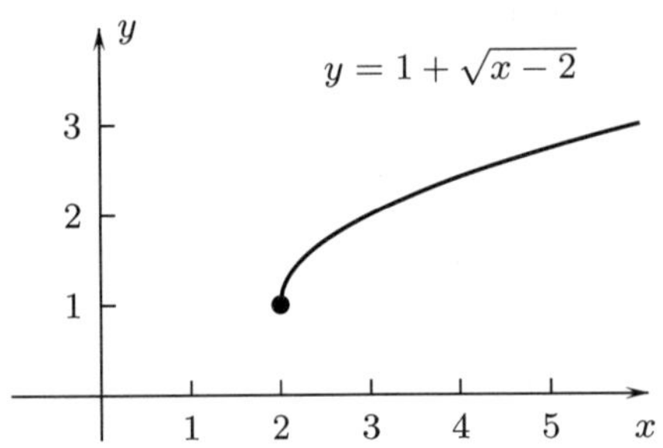

FIGURE 14

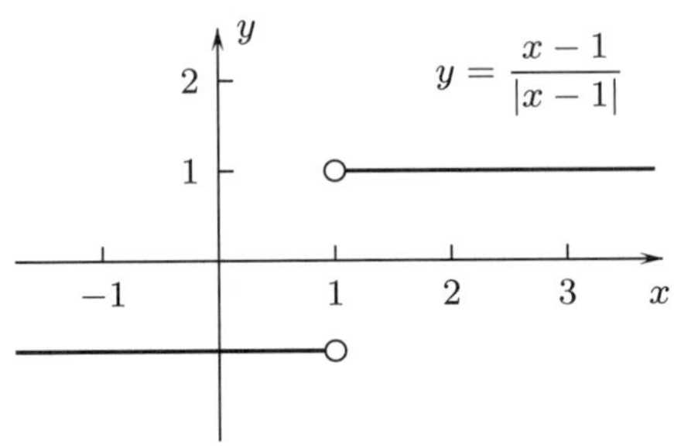

FIGURE 15

Question 7 (a) Explain the shape of the graph $y = \dfrac{x-1}{|x-1|}$ in Figure 15. (Calculate values for x on both sides of 1 to guide your reasoning, if necessary.) (b) What intervals form the domain of $y = \dfrac{x-1}{|x-1|}$? (c) What are the most extensive intervals in which $y = \dfrac{x-1}{|x-1|}$ is continuous?

The functions in Example 4 and Questions 6 and 7 illustrate the following result. It can be proved using Theorem 2 below and similar results that are studied in advanced calculus classes.

Theorem 1 *Any function that is given by one formula formed from the variable and constants by taking sums, differences, products, quotients, powers, or absolute values is continuous in the intervals where it is defined.*†

---

†In Chapter 3 we will use this result with functions given by single formulas that also involve trigonometric functions, exponential functions, or logarithms.

Example 5 Use Theorem 1 to determine the the most extensive intervals in which **(a)** $y = \dfrac{3x^2}{x^2+1}$, **(b)** $y = \sqrt{x^2-1}$, and **(c)** $y = \dfrac{x^2}{x^2-1}$ are continuous. Generate the graphs in the window $-3 \le x \le 3, -3 \le y \le 4$ to guide your reasoning.

SOLUTION **(a)** $y = 3x^2/(x^2+1)$ is defined for all x and, therefore, is continuous in $(-\infty, \infty)$ (Figure 16).

(b) Because of the square root, $y = \sqrt{x^2-1}$ is defined and, consequently, is continuous in $(-\infty, -1]$ and $[1, \infty)$ (Figure 17).

(c) Because $x^2/(x^2-1)$ is defined except at $x = \pm 1$ where its denominator is zero, $y = x^2/(x^2-1)$ is defined and, therefore, continuous in $(-\infty, -1), (-1, 1)$, and $(1, \infty)$ (Figure 18). □

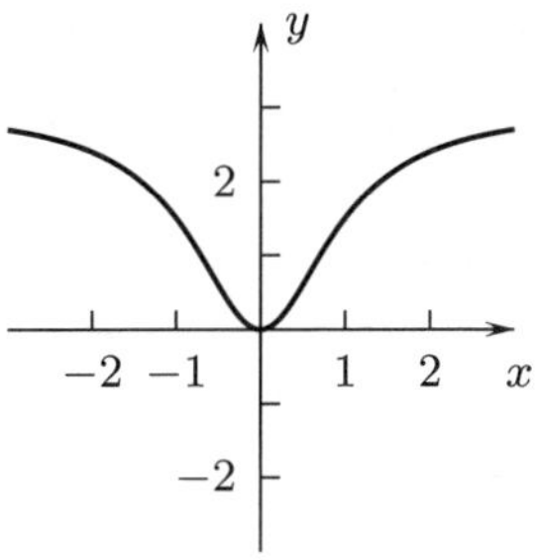

$y = 3x^2/(x^2+1)$

FIGURE 16

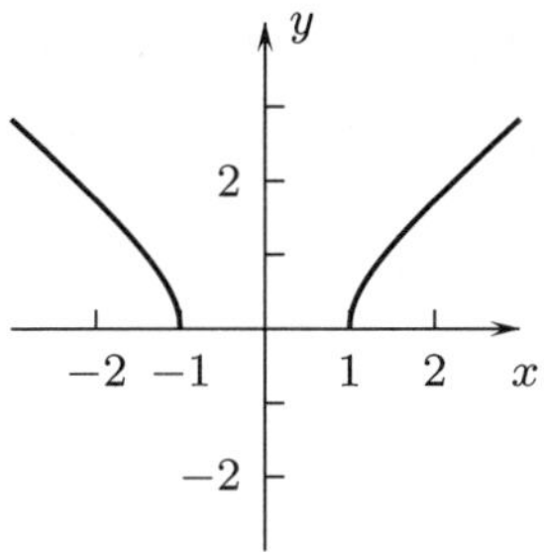

$y = \sqrt{x^2-1}$

FIGURE 17

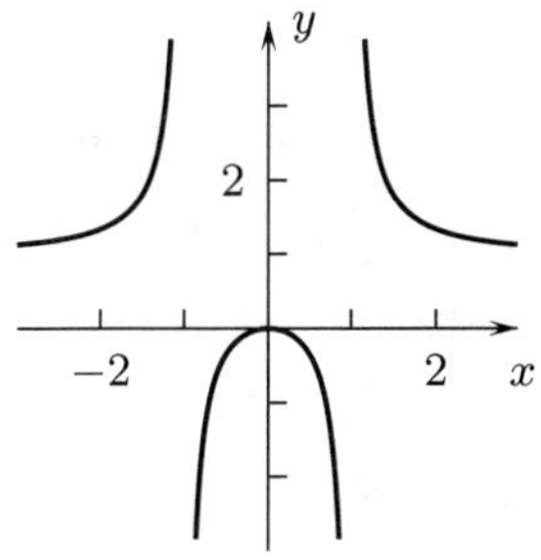

$y = x^2/(x^2-1)$

FIGURE 18

Theorem 1 can often be used to find limits.

Example 6 What is the limit of $\dfrac{5x^2}{x^2+1}$ as x tends to 2?

SOLUTION Because $y = \dfrac{5x^2}{x^2+1}$ is continuous at all x, it is continuous at 2, and its limit as $x \to 2$ equals its value at $x = 2$:

$$\lim_{x \to 2} \frac{5x^2}{x^2+1} = \left[\frac{5x^2}{x^2+1}\right]_{x=2} = \frac{5(2^2)}{2^2+1} = 4. \ \square$$

Example 7 What is the limit of $\sqrt{x^2-1}$ as x tends to 1 from the right?

SOLUTION Because $y = \sqrt{x^2-1}$ is continuous in $[1, \infty)$, it is continuous from the right at 1 and its limit as $x \to 1^+$ equals its value at $x = 1$:

$$\lim_{x \to 1^+} \sqrt{x^2-1} = \left[\sqrt{x^2-1}\right]_{x=1} = \sqrt{1^2-1} = 0. \ \square$$

Question 8 Figure 19 shows the graph of $N(x)$, defined by

$$N(x) = \begin{cases} \frac{1}{4}x^2 & \text{for} \quad x < 3 \\ 7/\sqrt{x} & \text{for} \quad x > 3. \end{cases}$$

What are **(a)** $\lim_{x \to 3^+} N(x)$ and **(b)** $\lim_{x \to 3^-} N(x)$?

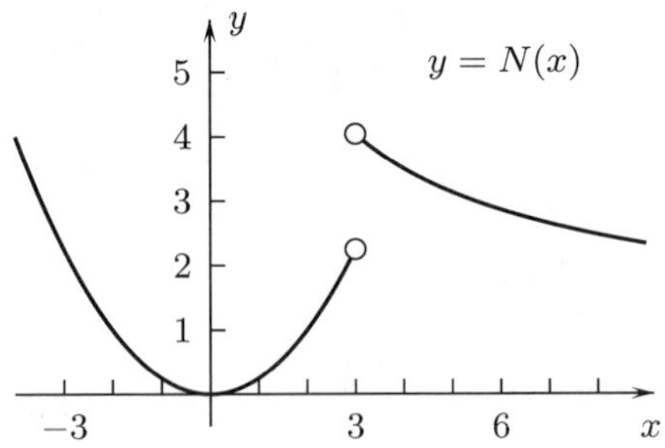

FIGURE 19

Limit theorems

Suppose that $f(x) \to 10$ and $g(x) \to 2$ as $x \to 1$. What happens to $f(x) + g(x)$, $f(x)g(x)$, and $\dfrac{f(x)}{g(x)}$ as $x \to 1$? As you can imagine, $f(x)+g(x)$ tends to $10+2 = 12$, $f(x)g(x)$ tends to $(10)(2) = 20$, and $\dfrac{f(x)}{g(x)}$ tends to $\dfrac{10}{2} = 5$. This illustrates the following rules, which are generally used without referring to them explicitly.

Theorem 2 *If $f(x) \to L$ and $g(x) \to M$ as $x \to x_0$ with finite numbers L and M, then*

$$f(x) + g(x) \to L + M \quad \text{and} \quad f(x)g(x) \to LM \quad \text{as} \quad x \to x_0. \tag{2}$$

If also $M \neq 0$, then

$$\frac{f(x)}{g(x)} \to \frac{L}{M} \quad \text{as} \quad x \to x_0. \tag{3}$$

The two-sided limit in Theorem 2 can be replaced by either type of one-sided limit or by the limit as $x \to \infty$ or as $x \to -\infty$. Remember these results in words: "If two functions have finite limits, then the limit of their sum is the sum of their limits, the limit of their product is the product of their limits, and, if the limit of the denominator is not zero, the limit of their quotient is the quotient of their limits." These results are proved in advanced calculus courses.[†]

Question 9 What is $\lim_{x\to 2^+} \left[A(x)B(x) + \dfrac{C(x)}{D(x)}\right]$ if $\lim_{x\to 2^+} A(x) = 2$, $\lim_{x\to 2^+} B(x) = 5$, $\lim_{x\to 2^+} C(x) = 6$, and $\lim_{x\to 2^+} D(x) = 3$?

Question 10 Use Theorem 2 to prove that if f and g are continuous at x_0, then **(a)** $f + g$ is continuous at x_0, **(b)** fg is continuous at x_0, and **(c)** if $g(x_0) \neq 0$, then f/g is continuous at x_0.

When numerators and denominators both tend to zero

We now look at the type of limit that we will use with the definition of the derivative in Section 2.3.

Does $\dfrac{x^2 - 4}{x - 2}$ have a limit as $x \to 2$, and if so, what is it? We cannot use Theorem 1 because the function is not defined at $x = 2$, and we cannot use Theorem 2 because the denominator $x - 2$ tends to 0 as $x \to 2$. In fact, the numerator $x^2 - 4$ also tends to zero, and if we try to apply rule **(3)** to $\lim_{x\to 2} \left[\dfrac{x^2-4}{x-2}\right]$, we obtain the meaningless expression $\dfrac{0}{0}$.

[†]The proof of the statement concerning sums of functions in Theorem 2 is outlined for limits from the right in Problem 69 at the end of this section.

Before we present an algebraic technique for finding this limit, we will try to predict the result with some calculations.

Question 11 Predict $\lim_{x\to 2} \dfrac{x^2-4}{x-2}$ by completing the following table.

x	$\dfrac{x^2-4}{x-2}$
1.9	3.9
1.99	3.99
1.999	
1.9999	

x	$\dfrac{x^2-4}{x-2}$
2.1	4.1
2.01	4.01
2.001	
2.0001	

We can verify the prediction in Question 11 by rewriting the formula $\dfrac{x^2-4}{x-2}$ for $x \neq 2$ in a form whose limit we can determine.

Example 8 **(a)** Show by factoring the numerator and cancelling one factor with the denominator that $\dfrac{x^2-4}{x-2}$ equals $x+2$ for $x \neq 2$. **(b)** Use this formula to find the limit of $\dfrac{x^2-4}{x-2}$ as $x \to 2$.

SOLUTION **(a)** For $x \neq 2$, $\dfrac{x^2-4}{x-2} = \dfrac{(x+2)(x-2)}{x-2} = x+2$.

(b) $\lim_{x\to 2} \dfrac{x^2-4}{x-2} = \lim_{x\to 2}(x+2) = \Big[x+2\Big]_{x=2} = 2+2 = 4$ because $y = x+2$ is continuous at $x = 2$. □

The graph of $y = \dfrac{x^2-4}{x-2}$ is shown in Figure 20. The calculations in Example 8a show that it consists of the line $y = x+2$ with the point $(2,4)$ removed. This is why its limit as $x \to 2$ is the value 4 of $x+2$ at 2.

FIGURE 20

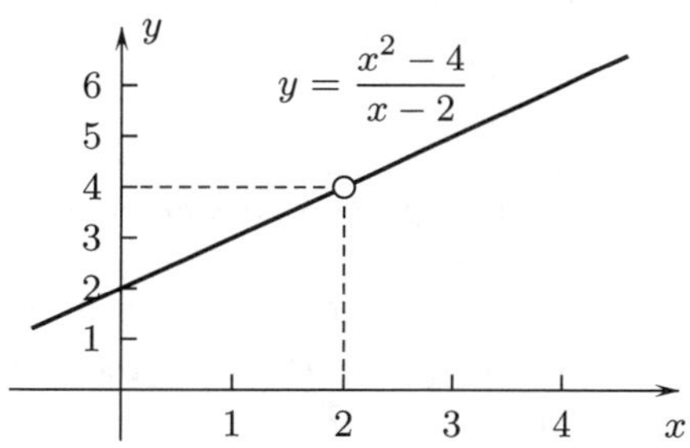

Question 12 Find the limit of $\dfrac{(x+2)^2-4}{(x+1)^2-1}$ as $x \to 0$ by first simplifying the expression for $x \neq 0$: (Expand the squares, factor x from the numerator and denominator, and cancel the x.) Check your answer by generating the graph of $\dfrac{(x+2)^2-4}{(x+1)^2-1}$

The Intermediate-Value Theorem

We close this section with a discussion of the INTERMEDIATE-VALUE THEOREM, which is based on the principle that graphs of continuous functions are "unbroken." It is used to show, without actually solving the equations, that certain equations of the form $f(x) = k$ with constant k have solutions. We will motivate the theorem with the functions $P(x)$, $Q(x)$, and $R(x)$ of Figures 21, 22, and 23.

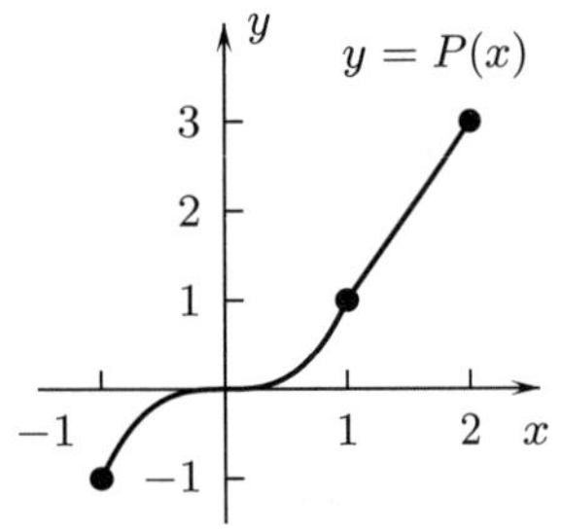

FIGURE 21

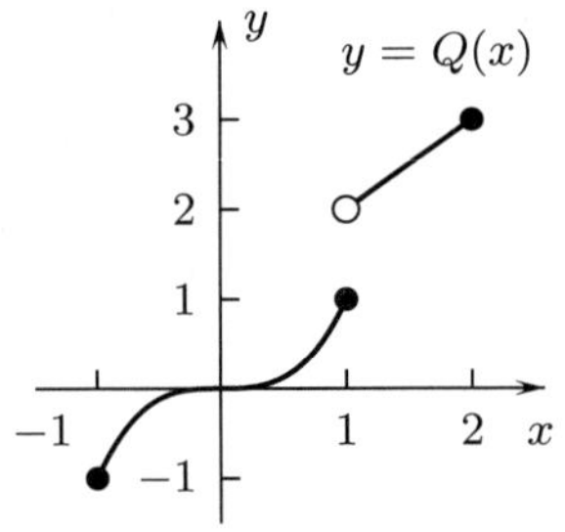

FIGURE 22

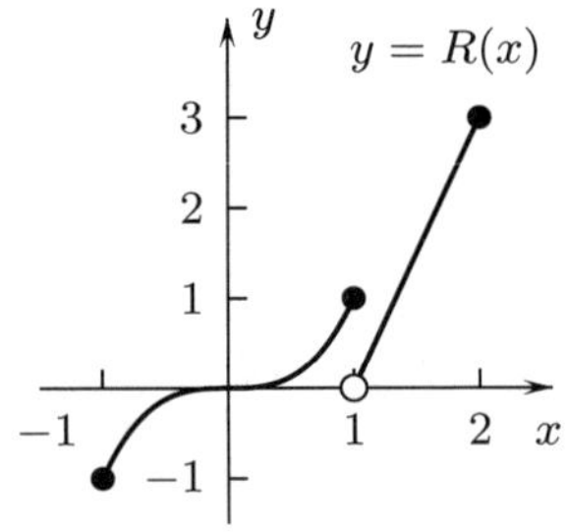

FIGURE 23

These functions are defined by

$$P(x) = \begin{cases} x^3 & \text{for} \quad -1 \le x \le 1 \\ 2x - 1 & \text{for} \quad 1 < x \le 2 \end{cases}$$

$$Q(x) = \begin{cases} x^3 & \text{for} \quad -1 \le x \le 1 \\ x + 1 & \text{for} \quad 1 < x \le 2 \end{cases}.$$

$$R(x) = \begin{cases} x^3 & \text{for} \quad -1 \le x \le 1 \\ 3x - 3 & \text{for} \quad 1 < x \le 2. \end{cases}$$

Notice that each of these functions has the value -1 at $x = -1$ and the value 3 at $x = 2$.

Question 13 **(a)** Which of the three functions, P, Q, and R, is continuous in $[-1, 2]$? **(b)** For what numbers k with $-1 \le k \le 3$ does the equation $P(x) = k$ have a solution x in $[-1, 2]$? **(c)** For what numbers k with $-1 \le k \le 3$ does $Q(x) = k$ have a solution x in $[-1, 2]$? **(d)** For what numbers k with $-1 \le k \le 3$ does $R(x) = k$ have a solution x in $[-1, 2]$?

Part (b) of this question illustrates the next theorem. Because $P(x)$ is continuous in $[-1, 2]$, it takes on all values between $P(-1) = -1$ and $P(2) = 3$ as x increases from -1 to 2, so that the equation $P(x) = k$ has a solution for every k between -1 and 3. Here is the general result, which can be derived from axioms studied in advanced calculus:

Theorem 3 (The Intermediate-Value Theorem) *If $f(x)$ is continuous in a finite closed interval $[a, b]$, then the equation $f(x) = k$ has at least one solution x with $a \le x \le b$ for every number k between $f(a)$ and $f(b)$ (Figure 24).*

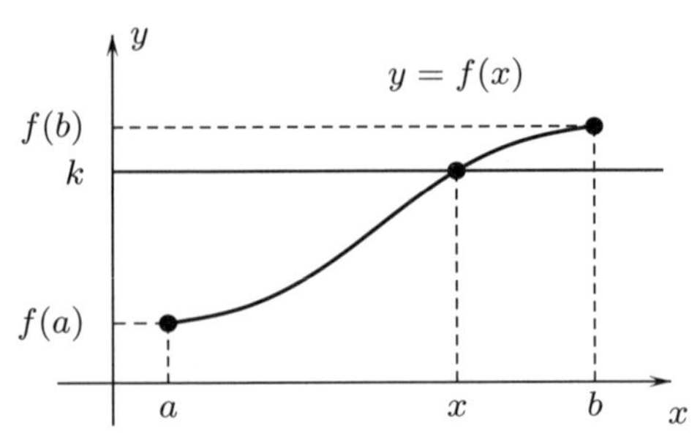

FIGURE 24

Theorem 3, like all theorems, has a HYPOTHESIS ($f(x)$ is continuous in $[a, b]$) and a CONCLUSION ($f(x) = k$ has at least one solution in $[a, b]$ for every k between $f(a)$ and $f(b)$). In each instance where we might use the theorem, there are three possibilities. (i) the hypothesis might be satisfied, in which case the conclusion must hold; (ii) the hypothesis might not be satisfied and the conclusion not hold; or (iii) the hypothesis might not be satisfied but the conclusion hold, nevertheless.

As we said above, part (b) of Question 13 illustrates case (i).

Part (c) of Question 13 illustrates case (ii). The hypothesis of the theorem is not satisfied by $Q(x)$ in $[-1, 2]$ because $Q(x)$ is discontinuous at $x = 1$. The conclusion does not hold because the equation $Q(x) = k$ does not have any solutions for $1 < k \le 2$. The discontinuous $Q(x)$ jumps over values of k as x increases from -1 to 2.

Part (d) of Question 13 illustrates case (iii). The hypothesis of the theorem is not satisfied by $R(x)$ in $[-1, 2]$ because $R(x)$ is discontinuous at $x = 1$. The conclusion, however, does hold because $R(x)$ takes all values between $R(-1) = -1$ and $R(2) = 3$. (In fact, $R(x) = k$ has one solution x for each k with $-1 \le k \le 0$ or $1 < k \le 3$ and two solutions for $0 < k \le 1$, as can be seen from the graph in Figure 23.)

To show how the Intermediate-Value Theorem might be used, let us consider the equation

$$7x - \tfrac{2}{3}x^3 = 5. \tag{4}$$

Example 9 Use the Intermediate-Value Theorem to show that **(4)** has at least one solution.

SOLUTION After some experimentation, we notice that $7x - \frac{2}{3}x^3$ has the value $14\frac{2}{3}$ at $x = -4$ and the value $-14\frac{2}{3}$ at $x = 4$. Since $7x - \frac{2}{3}x^3$ is continuous in the interval $[-4, 4]$ and $-14\frac{2}{3} < 5 < 14\frac{2}{3}$, we conclude from the theorem that **(4)** has at least one solution x in $[-4, 4]$. In fact the equation has three solutions in the interval, as can be seen from the graph of the function in Figure 25. □

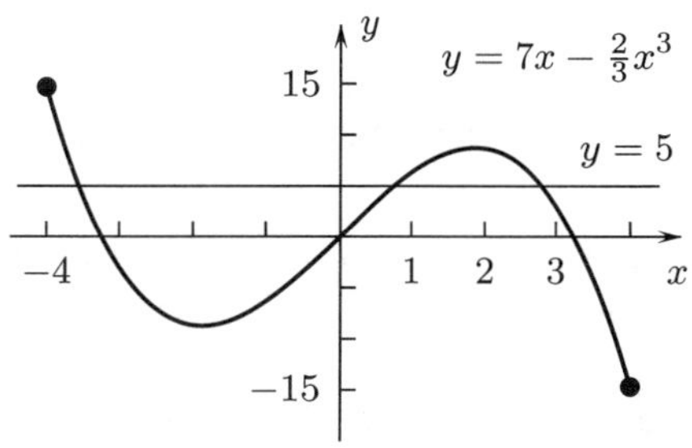

FIGURE 25

Question 14 Use the Intermediate-Value Theorem, without generating any graphs, to show that the equation $x + 2x^2 + 3x^3 + 4x^4 = 10{,}000$ has at least one solution.

Principles and procedures

- If $f(x)$ approaches a number L as x approaches x_0 from the right (Figure 26), we write $f(x) \to L$ as $x \to x_0^+$ or $\lim_{x \to x_0^+} f(x) = L$.
- $f(x)$ is continuous from the right at x_0 if $\lim_{x \to x_0^+} f(x) = f(x_0)$ (Figure 27).

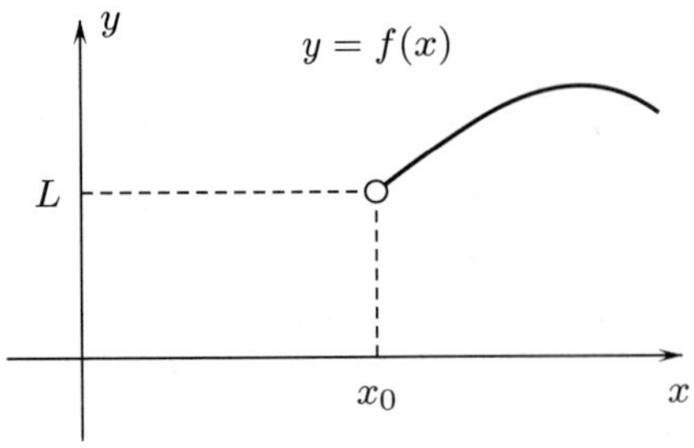

$\lim_{x \to x_0^+} f(x) = L$

FIGURE 26

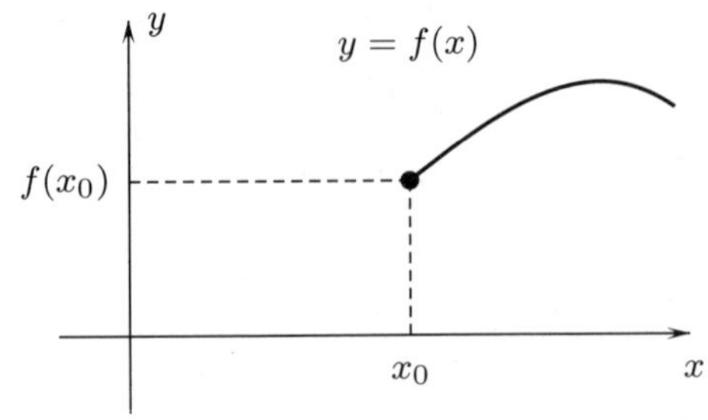

$f(x)$ is continuous from the right at x_0

FIGURE 27

- If $f(x)$ approaches a number L as x approaches x_0 from the left (Figure 28), we write $f(x) \to L$ as $x \to x_0^-$ or $\lim_{x \to x_0^-} f(x) = L$.
- $f(x)$ is continuous from the left at x_0 if $\lim_{x \to x_0^-} f(x) = f(x_0)$ (Figure 29).

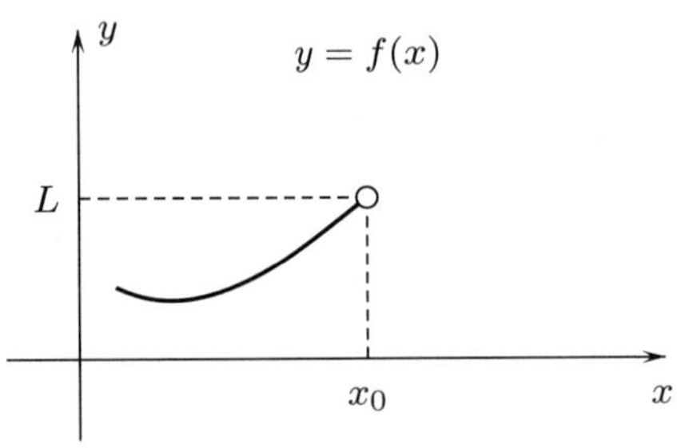

$\lim_{x \to x_0^-} f(x) = L$

FIGURE 28

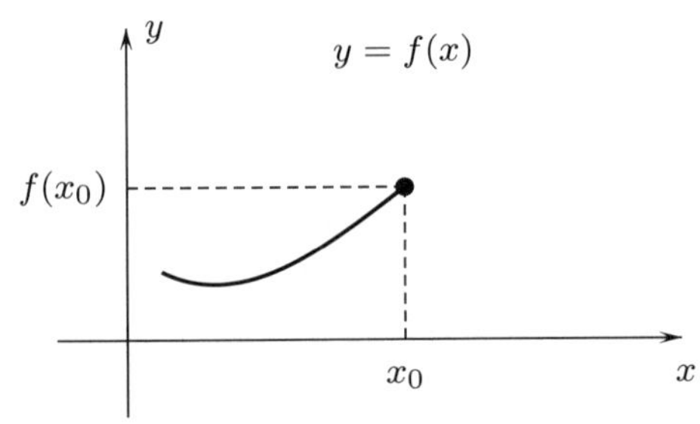

$f(x)$ is continuous from the left at x_0

FIGURE 29

- If $f(x)$ tends to a number L as x approaches x_0 from the left and from the right (Figure 30), we write $f(x) \to L$ as $x \to x_0$ or $\lim_{x \to x_0} f(x) = L$.
- $f(x)$ is continuous at x_0 if $\lim_{x \to x_0} f(x) = f(x_0)$ (Figure 31).

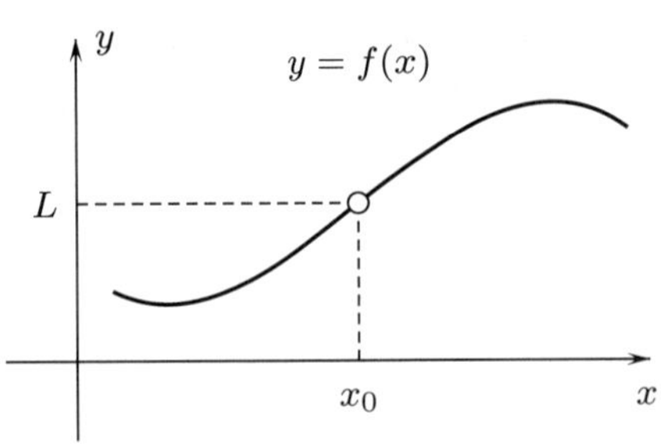

$\lim_{x \to x_0} f(x) = L$

FIGURE 30

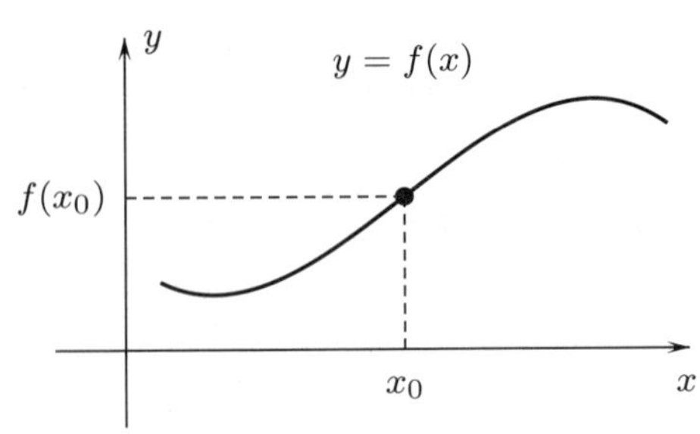

$f(x)$ is continuous at x_0

FIGURE 31

- A function is continuous in an interval where it is defined if (i) it is continuous at each point in the interior of the interval, (ii) it is continuous from the right at a if the interval includes a left endpoint a, and (iii) it is continuous from the left at b if the interval includes a right endpoint b.
- In most cases you can tell whether a function is continuous in an interval by visualizing its graph and noting whether the portion of the graph for x in the interval is one unbroken curve.
- The fact that any function given by a single formula formed from x and constants with sums, differences, products, quotients, powers, and absolute values is continuous in the intervals where it is defined (Theorem 1) can be used, as in Question 8, to study limits of functions that are given by more than one formula, or as in Example 8 and Question 12, to find limits of certain discontinuous functions by rewriting their formulas to show they are equal to continuous functions on one or both sides of the limiting value of the variable.
- The Intermediate-Value Theorem is used to show, without solving the equations, that certain equations involving continuous functions have solutions.

Tune-Up Exercises 2.2[♦]

[A]*Answer provided.* [O]*Outline of solution provided.*

Be sure you can solve Exercises T1 through T6 and work as many of Exercises T7 through T12 as you need for practice.

T1.[A] A function $P(x)$ is defined below. Sketch its graph without using a calculator or computer and find $P(-1)$, $\lim_{x\to-1^-} P(x)$, $\lim_{x\to-1^+} P(x)$, $P(1)$, $\lim_{x\to1^-} P(x)$, $\lim_{x\to1^+} P(x)$, $P(0)$, $\lim_{x\to0} P(x)$, $\lim_{x\to-\infty} P(x)$, and $\lim_{x\to\infty} P(x)$. What are the most extensive intervals in which $P(x)$ is continuous?

$$P(x) = \begin{cases} -3/x & \text{for} \quad x < -1 \\ 1 & \text{for} \quad -1 \le x \le 1 \\ 3/x & \text{for} \quad x > 1. \end{cases}$$

♦ Type 1, limits and continuity

T2.[A] What is the limit of $\sqrt{x} + \dfrac{6}{x+2}$ as $x \to 4$ and why?

♦ Type 1, limits and continuity

T3.[A] What is $\lim_{x\to7} \dfrac{F(x) - x}{F(x) + x}$ if $\lim_{x\to7} F(x) = 20$?

♦ Type 1, limits

T4.[O] Without drawing the graph, find $S(2)$, $\lim_{x\to2^-} S(x)$, $\lim_{x\to2^+} S(x)$, and $\lim_{x\to2} S(x)$ where

$$S(x) = \begin{cases} \sqrt{x+2} & \text{for} \quad x < 2 \\ 14 & \text{for} \quad x = 2 \\ x^8 & \text{for} \quad x > 2. \end{cases}$$

♦ Type 1, limits

T5.[O] Predict $\lim_{x\to1} \dfrac{x^{1.7} - 1}{x - 1}$ by finding at least six values of the function for x near 1.

♦ Type 1, limits

T6.[O] Find the exact value of $\lim_{x\to3} \dfrac{x^2 - 9}{x - 3}$ by rewriting the expression for the function so its denominator does not tend to zero as $x \to 3$.

♦ Type 1, limits

More practice

T7.[A] A function $Q(x)$ is defined below. Sketch its graph without using a calculator or computer. Then find $Q(1)$, $\lim_{x\to1^-} Q(x)$, $\lim_{x\to1^+} Q(x)$, $\lim_{x\to1} Q(x)$, $Q(0)$, $\lim_{x\to0} Q(x)$, $\lim_{x\to-\infty} Q(x)$, and $\lim_{x\to\infty} Q(x)$. What are the most extensive intervals in which $Q(x)$ is continuous?

$$Q(x) = \begin{cases} x^2 - 3 & \text{for} \quad x \le -1 \\ 2x^3 & \text{for} \quad -1 < x < 1 \\ x^2 - 3 & \text{for} \quad x \ge 1. \end{cases}$$

♦ Type 1, limits and continuity

T8.[A] What is the limit of $\sqrt{x} + \dfrac{6}{x+2}$ as $x \to 0^+$ and why?

♦ Type 1, limits and continuity

[♦]The Tune-up Exercises and Problems are classified by type and content. The types are (1) basic, reactive; (2) basic reflective; (3) intermediate, reactive; (4) intermediate, reflective; (5) advanced, reactive; (6) advanced, reflective; and (7) advanced, theoretical.

T9.[A] What is $\displaystyle\lim_{x\to -10+} \frac{xT(x)}{5T(x) - 3x}$ if $\displaystyle\lim_{x\to -10+} T(x) = 2$?

♦ Type 1, limits

T10.[O] Without drawing the graph, find $W(5), W(-5), W(3)$, $\displaystyle\lim_{x\to 3^-} W(x)$, $\displaystyle\lim_{x\to 3^+} W(x)$, and $\displaystyle\lim_{x\to 3} W(x)$ where

$$W(x) = \begin{cases} 9x^2 & \text{for} \quad x < 3 \\ 0 & \text{for} \quad x = 3 \\ x^4 & \text{for} \quad x > 3. \end{cases}$$

♦ Type 1, limits

T11.[O] Use your calculator or computer to predict $\displaystyle\lim_{x\to 2} \frac{x^3 + x^2 - 12}{x^2 - 4}$.

♦ Type 1, limits

T12.[O] Find the exact value of $\displaystyle\lim_{x\to -5} \frac{x^2 - 25}{x + 5}$ by rewriting the expression for the function so its denominator does not tend to zero as $x \to -5$.

♦ Type 1, limits

Problems 2.2

[A]*Answer provided.* [O]*Outline of solution provided.*

1.[A] What is $\displaystyle\lim_{x\to 8} f(x)$ if $f(8) = 5$ and $f(x)$ is continuous at $x = 8$?

♦ Type 2, limits and continuity

2. Sketch the graph of a function $g(x)$ such that $\displaystyle\lim_{x\to 1} g(x) = 2$ and $g(1) = 3$.

♦ Type 2, limits

3. Define a function $h(x)$ such that $\displaystyle\lim_{x\to 0} h(x) = \infty$ and $h(0) = 2$. (Use different formulas for $x = 0$ and for $x > 0$.)

♦ Type 2, limits

4.[A] What is $\displaystyle\lim_{x\to 5} K(x)$ if $\displaystyle\lim_{x\to 5} L(x) = 10$ and $K(x) = L(x)$ for $x \neq 5$?

♦ Type 2, limits

5. Define a function $M(x)$ such that $\displaystyle\lim_{x\to 0+} M(x) = \infty$, $M(0) = 0$, and $\displaystyle\lim_{x\to 0-} M(x) = -2$. (Use different formulas for $x < 0$, for $x = 0$, and for $x > 0$.)

♦ Type 4, limits

6. Sketch the graph of a function $Z(x)$ that is continuous in the interval $[0, 2)$ but is discontinuous in $(-1, 2)$ and in $[0, 2]$.

♦ Type 4, limits

Sketch the required curves in Problems 7 through 11 without generating them on your calculator or computer.

7.[O] **(a)** Sketch the graph of

$$w(x) = \begin{cases} 2/x & \text{for} \quad 0 < x < 2 \\ 2 & \text{for} \quad x > 2. \end{cases}$$

(b) Find the limits of $w(x)$ as $x \to 0^+$, as $x \to 2^-$, as $x \to 2^+$, and as $x \to \infty$. **(c)** Find all solutions x of $w(x) = 3$.

♦ Type 2, limits and solving an equation

8[O] **(a)** Sketch the graph of

$$T(x) = \begin{cases} x & \text{for} \quad x < 1 \\ x - 2 & \text{for} \quad x \geq 1. \end{cases}$$

(b) Find $T(1)$ and the limits of $T(x)$ as $x \to 1^-$ and as $x \to 1^+$.
(c) Where is $T(x)$ discontinuous? **(d)** What are the greatest and least values of $T(x)$ for $0 \leq x \leq 4$ and at what values of x do they occur?

♦ Type 2, limits, continuity, etc.

9.[A] **(a)** Sketch the graph of

$$K(x) = \begin{cases} 2 + x & \text{for} \quad x \leq -1 \\ -x^2 & \text{for} \quad -1 < x < 1 \\ 2 - x & \text{for} \quad x \geq 1. \end{cases}$$

(b) Find the limits of $K(x)$ as $x \to -1^-$, as $x \to -1^+$, and as $x \to -1$. **(c)** Solve $K(x) = 0$. **(d)** For what values of x is $K(x)$ negative? (Give the four open intervals.) **(e)** What are the most extensive intervals in which $K(x)$ is continuous?

♦ Type 2, limits, continuity, etc.

10. **(a)** Sketch the graph of

$$Q(x) = \begin{cases} 2/x & \text{for} \quad x < -1 \\ 2x^3 & \text{for} \quad -1 \le x \le 1 \\ 2/x & \text{for} \quad x > 1. \end{cases}$$

(b) Find $Q(-1)$, $\lim_{x \to -1} Q(x)$, $Q(0)$, $\lim_{x \to 0} Q(x), Q(1)$, and $\lim_{x \to 1} Q(x)$. **(c)** What are the greatest and least values of $Q(x)$ for all x? **(d)** Find the three solutions x of $Q(x) = 2x$. **(e)** At what values of x is $Q(x)$ discontinuous?

♦ Type 2, limits, continuity, etc.

11. **(a)** Sketch the graph of

$$\Lambda(x) = \begin{cases} 2 + 2x & \text{for} \quad -1 \le x < 0 \\ 3 & \text{for} \quad x = 0 \\ 2 - 2x & \text{for} \quad 0 < x \le 1. \end{cases}$$

(b) What are the most extensive intervals in which $\Lambda(x)$ is continuous? **(c)** For what values of x is $\Lambda(x) \ge 2$? **(d)** Does $\Lambda(x)$ increase or decrease as x increases from -1 to 0? **(e)** Does $\Lambda(x)$ increase or decrease as x increases from 0 to 1? **(f)** What one change in the definition of $\Lambda(x)$ would make it continuous for all x?

♦ Type 4, continuity, etc.

12.[A] Without drawing the graph, find $\lim_{x \to 2} g(x)$ and $\lim_{x \to 10} g(x)$ for

$$g(x) = \begin{cases} x^3 + x & \text{for} \quad x < 2 \\ 10 & \text{for} \quad 2 \le x \le 10 \\ 1 + \sqrt{x + 7} & \text{for} \quad x > 10. \end{cases}$$

Justify your answers. What are the most extensive intervals in which $g(x)$ is continuous?

♦ Type 4, limits and continuity

13. Without using a graph, find the limits as $x \to 1^-$ and as $x \to 1^+$ of

$$h(x) = \begin{cases} x^4 - 2x & \text{for} \quad x < 1 \\ 5x^3 - 4x^2 & \text{for} \quad x > 1. \end{cases}$$

Does $h(x)$ have a (two-sided) limit as $x \to 1$? If so, what is it?

♦ Type 4, limits

14. Without using a graph find $\lim_{x \to 10} Z(x)$ and $\lim_{x \to 20} Z(x)$, if they exist, where

$$Z(x) = \begin{cases} x^2 + 900 & \text{for} \quad x < 10 \\ 500 & \text{for} \quad x = 10 \\ x^3 & \text{for} \quad x > 10. \end{cases}$$

15.[A] At what values of x is $f(x)$ discontinuous if

♦ Type 4, limits

$$f(x) = \begin{cases} x^3 - x^4 + 2 & \text{for} \quad x \le -1 \\ 2x^5 + x + 3 & \text{for} \quad -1 < x < 1 \\ 7 & \text{for} \quad x \ge 1? \end{cases}$$

Justify your answer.

♦ Type 4, continuity

16. **(a)** Sketch the graph of

$$M(x) = \begin{cases} 1/x & \text{for} \quad x < 0 \\ x+1 & \text{for} \quad 0 \le x \le 2 \\ 3 & \text{for} \quad x > 2. \end{cases}$$

(b) What are the most extensive intervals in which $M(x)$ is continuous? For what values of k (i.e., in what k-intervals) does the equation $M(x) = k$ have **(c)** one solution x, **(d)** an infinite number of solutions, and **(e)** no solutions? **(f)** Use the graph to find the two solutions of $M(x) = x$.

♦ Type 4, continuity, etc.

17. **(a)** Sketch the graph of

$$A(x) = \begin{cases} -2 - x^2 & \text{for} \quad x < 0 \\ 0 & \text{for} \quad x = 0 \\ 2 + x^2 & \text{for} \quad x > 0. \end{cases}$$

(b) At what values of x is $A(x)$ discontinuous? **(c)** For what values of k does the equation $A(x) = k$ have a solution? **(d)** For what values of x is $A(x) \le A(0)$? **(d)** Does $A(x)$ increase or decrease as x increases?

♦ Type 4, continuity, etc.

18.[A] What are the most extensive intervals in which **(a)** $y = \dfrac{x}{x^2-4}$, **(b)** $y = \dfrac{8}{x^2+4}$, and **(c)** $y = \dfrac{\sqrt{x}}{x-1}$ are continuous and why?

♦ Type 2, continuity

19. What are the most extensive intervals in which **(a)** $y = 1 + \dfrac{1}{x} + \dfrac{1}{x+1} - \dfrac{1}{x-1}$, **(b)** $y = x^2 + x^3 + x^{1/2} + x^{-1/2}$, and **(c)** $y = \dfrac{x^3 - 4}{1 + \sqrt{x}}$ are continuous?

♦ Type 2, continuity

In each of Problems 20 through 23 calculate values of the given function for x near 1 to predict the limit as $x \to 1$.

20.[A] $y = \dfrac{x^3 - 1}{1 - x^4}$

♦ Type 2, predicting limits

21.[A] $y = \dfrac{x^2 + x - 2}{\sqrt{x} - 1}$

♦ Type 2, predicting limits

22.[A] $y = \dfrac{x^2 - 1}{x + 1}$

♦ Type 2, predicting limits

23. $y = \dfrac{x^5 - 1}{\sqrt{2x+7} - 3}$

♦ Type 2, predicting limits

24. Predict the limit of $\dfrac{\sin(kx)}{x}$ as $x \to 0$ for $k = 1, 2, 3$, and 4 by calculating values for x near 0.

♦ Type 2, predicting limits

25.[A] **(a)** What is the limit of $F(x) = \dfrac{(x-1)^2 - 1}{x}$ as $x \to 2$, and why? **(b)** Find the limit of $F(x)$ as $x \to 0$ by rewriting its formula so it does not contain a denominator that tends to 0 as $x \to 0$. Check your answers by generating the graph of $F(x)$ in the window $-2 \le x \le 4, -4 \le y \le 3$.

♦ Type 3, limits

26. **(a)** What is the limit of $G(x) = \dfrac{1 - 2/x}{x - 2}$ as $x \to 1$, and why? **(b)** Find the limit of $G(x)$ as $x \to 2$ by rewriting its formula so it does not contain a denominator that tends to 0 as $x \to 2$. Check your answers by generating the graph of $G(x)$ for $-1 \le x \le 4, -0.5 \le y \le 1.5$.

♦ Type 3, limits

27. Find the limits of $H(x) = \dfrac{x^2 - 1}{x^4 - x^2}$ as $x \to 1$ and as $x \to -1$ by rewriting its formula so its denominator does not tend to 0 as $x \to \pm 1$. Check your answers by generating the graph of $H(x)$ in the window $-2.5 \le x \le 2.5, -0.5 \le y \le 3$. What is $\lim_{x \to 0} H(x)$ and why?

♦ Type 3, limits

28. Find $\lim_{x \to 0^+} \left[\dfrac{\sqrt{x} + x}{\sqrt{x}} \right]$ by rewriting its formula. Check by generating the graph in the window $-0.1 \le x \le 0.5, -0.5 \le y \le 2$.

♦ Type 3, limits

In Problems 29 through 38 find the limits by rewriting the formulas, if necessary, so they do not contain denominators that tend to zero. Check by calculating values of the functions near the limiting values of x.

29.[A] $\lim_{x \to 1} \dfrac{x^2 - 1}{3x - 3}$

♦ Type 1, limits

30. $\lim_{x \to -2} \dfrac{2 + x}{4 - x^2}$

♦ Type 1, limits

31.[A] $\lim_{x \to 2} \dfrac{x^2 - 4}{x^2 + 4}$

♦ Type 1, limits

32. $\lim_{x \to 1} \dfrac{(x - 1)^2}{x^2 - 1}$

♦ Type 1, limits

33.[A] $\lim_{x \to 3} \dfrac{(9/x^2) - 1}{12 - 4x}$

♦ Type 1, limits

34. $\lim_{x \to 2} \dfrac{(1/x) - (1/2)}{x^2 - 4}$

♦ Type 1, limits

35.[A] $\lim_{h \to 0} \dfrac{(3 + h)^2 - 9}{h}$

♦ Type 1, limits

36. $\lim_{x \to 0} \dfrac{x^2 + 4x}{6x - x^3}$

♦ Type 1, limits

37. $\lim_{x \to 2} \dfrac{1 - (2/x)}{x^2 - 4}$

♦ Type 1, limits

38. $\lim_{x \to -2} \dfrac{x^2 + 4}{x^3 - 8}$

♦ Type 1, limits

39.[A] What are $\lim_{x \to 2^-} Z(x)$ and $\lim_{x \to 2^+} Z(x)$ if

$$Z(x) = \begin{cases} kx^2 & \text{for} \quad x < 2 \\ x^3 & \text{for} \quad x > 2? \end{cases}$$

For what values of k does $\lim_{x \to 2} Z(x)$ exist?

♦ Type 2, limits

40. For what two values of a does $\lim_{x \to 5} Y(x)$ exist if

$$Y(x) = \begin{cases} a/x & \text{for} \quad x < 5 \\ -1 & \text{for} \quad x = 5 \\ x/a & \text{for} \quad x > 5? \end{cases}$$

For what one value of a is $Y(x)$ continuous at $x = 5$?

♦ Type 4, limits

41. Find constants a and b such that $g(x)$ is continuous for all x, where

$$g(x) = \begin{cases} x^3 & \text{for} \quad x < -1 \\ ax + b & \text{for} \quad -1 \le x < 1 \\ x^2 + 2 & \text{for} \quad x \ge 1. \end{cases}$$

♦ Type 4, limits

42. Find constants a and b such that $h(x)$ is continuous in $[-3, 0]$, where

$$h(x) = \begin{cases} -x & \text{for} \quad -3 \le x \le -2 \\ ax^2 + b & \text{for} \quad -2 < x < 0 \\ 6 & \text{for} \quad x = 0. \end{cases}$$

♦ Type 4, limits

43. At what values of x is $y = \dfrac{x^2-1}{|x^2-1|}$ discontinuous, and what are its limits from the left and from the right at those points?

♦ Type 4, limits

44. Find **(a)** $\lim\limits_{x\to 2} \dfrac{x^4-16}{x-2}$ and **(b)** $\lim\limits_{x\to 1} \dfrac{x^4+x^2-2}{x-1}$ by factoring the numerators.

♦ Type 4, limits

45. Find $\lim\limits_{x\to 4} \dfrac{\sqrt{x}-2}{x-4}$ by setting $y=\sqrt{x}$ and finding the limit as $y \to 2$.

♦ Type 4, limits

46. Find **(a)** $\lim\limits_{x\to 0} \dfrac{x^2+4}{x^2}$ and **(b)** $\lim\limits_{x\to 0^+} \dfrac{x^3-5}{2x}$.

♦ Type 4, limits

47. The malt for making lager beer is obtained by sprouting and roasting barley seed. The malt is then crushed and mixed with water and heated in what is called "mashing." Enzyme action during this process reduces complex carbohydrates in the "mash" to sugars and converts insoluble proteins to simpler amino acids. The result is rinsed, boiled with hop flowers, and then fermented with yeast to create the beer. The function $T(t)$ below gives the temperature (°F) of the mash during a mashing procedure.[(1)] The time intervals when the mash is allowed to cool slightly are called the "acid rest," the "protein rest," and the "saccharification" period. The mash becomes more acidic during the first of these, the second allows proteins to be broken down, and the third enhances the conversion of starches into sugars. **(a)** Sketch the graph of $T(t)$. **(b)** During which time period does the mash heat the most rapidly? Cool the most rapidly? **(c)** Is $T(t)$ continuous in $[0, 5.25]$?

$$T(t) = \begin{cases} 60+100t & \text{for} \quad 0 \le t \le 0.5 \\ 110-10(t-0.5) & \text{for} \quad 0.5 \le t \le 2 \\ 95+50(t-2) & \text{for} \quad 2 \le t \le 2.5 \\ 120-\frac{40}{3}(t-2.5) & \text{for} \quad 2.5 \le t \le 4 \\ 100+240(t-4) & \text{for} \quad 4 \le t \le 4.25 \\ 160-20(t-4.25) & \text{for} \quad 4.25 \le t \le 5.25. \end{cases}$$

♦ Type 4, continuity, etc.

48. Ten minutes after a hot dog vender opens her stand, she sells one hot dog. Fifteen minutes later she sells four hot dogs. After another half hour she sells three more. During the next thirty five minutes she does not sell any, so she goes home. Let $H(t)$ be the number of hot dogs she sells during the first t minutes that her stand is open. Give formulas for $H(t)$ and sketch its graph. Where is it discontinuous?

♦ Type 4, continuity, etc.

49. A model rocket is fired vertically from the ground at time $t=0$ (seconds). It rises at the constant rate of 50 feet per second until $t=2$, when it runs out of fuel. **(a)** Give its height $h(t)$ above the ground as a function of t for $0 \le t \le 2$. **(b)** Under the assumption that, after the rocket runs out of fuel, there is no air resistance or other force on it apart from gravity, the rocket is $h(t) = -16t^2 + 114t + C$ feet above the ground from $t=2$ until it hits the ground. Explain why it is reasonable to assume that h is continuous at $t=2$ and use this assumption to find the constant C. **(c)** Sketch the graph of h and use the quadratic formula to give an exact expression for the time when the rocket hits the ground.

♦ Type 4, continuity, etc.

50. Figure 32 shows the graph of the density of the earth $\rho(d)$ as a function of the depth d (kilometers) below the surface.[(2)] The dot on the ρ-axis gives the density in the earth's crust, which is about 15 kilometers thick. The line on the left gives the density in the mantle, which extends to 2,900 kilometers. The two lines on the right give the density in the outer liquid core, which extends from 2,900 kilometers to 5,100 kilometers; and in the inner core, which extends to the center of the earth at a depth of 6,370 kilometers. In what ways is it reasonable to represent the density of the earth with the discontinuous function $\rho = \rho(d)$, and in what ways is this model unrealistic?

♦ Type 4, discussing a mathematical model

---

(1) Data adapted from *Brewing Lager Beer* by G. Noonan, Boulder, Colo: Brewers Publications, 1986, p. 104.
(2) Data from *Encyclopædia Brittanica*, Micropædia, Vol, III, p. 749, Chicago: Encyclopædia Brittanica, Inc., 1974.

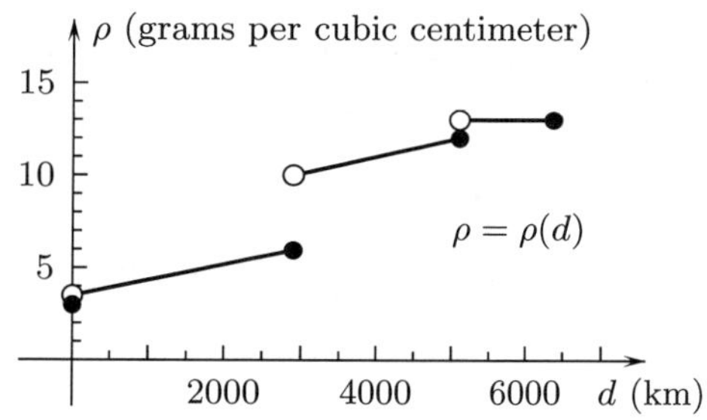

FIGURE 32

51. **(a)** Sketch the graph of

$$W(x) = \begin{cases} -1-x & \text{for} \quad -2 \le x < -1 \\ 1+x & \text{for} \quad -1 \le x < 0 \\ 1-x & \text{for} \quad 0 \le x < 1 \\ -1+x & \text{for} \quad 1 \le x \le 2. \end{cases}$$

(b) Are the hypotheses of the Intermediate-Value Theorem satisfied for $W(x)$ in the interval $[-2, 2]$ with $k = \frac{1}{2}$? Does the conclusion hold in this case?

♦ Type 4, the Intermediate-Value Theorem

52. **(a)** Sketch the graph of

$$C(x) = \begin{cases} -1 & \text{for} \quad x < -1 \\ x^2 & \text{for} \quad -1 \le x \le 1 \\ 1 & \text{for} \quad x > 1. \end{cases}$$

(b) Are the hypotheses of the Intermediate-Value Theorem satisfied for $C(x)$ in the interval $[-2, 2]$ with $k = \frac{1}{2}$? Does the conclusion hold in this case?

♦ Type 4, the Intermediate-Value Theorem

53. **(a)** Sketch the graph of

$$f(x) = \begin{cases} x & \text{for} \quad x < 0 \\ x+1 & \text{for} \quad x \ge 0. \end{cases}$$

(b) Are the hypotheses of the Intermediate-Value Theorem satisfied for $f(x)$ in the interval $[-1, 1]$ with $k = 0.5$? Does the conclusion hold in this case?

♦ Type 4, the Intermediate-Value Theorem

54. **(a)** Sketch the graph of

$$g(x) = \begin{cases} 2/x & \text{for} \quad 0 < x \le 2 \\ -2+x & \text{for} \quad 2 < x \le 4. \end{cases}$$

(b) Are the hypotheses and conclusion of the Intermediate-Value Theorem satisfied for $g(x)$ with $k = 1$ in the interval $[1, 2.5]$? In $(2, 4]$?

♦ Type 4, the Intermediate-Value Theorem

55.[A] **(a)** Use the Intermediate-Value Theorem to show that the equation $12x - 7x^2 + x^3 = 4$ has at least one solution in $[0, 5]$. **(b)** Use your calculator or computer to find approximate values of all solutions of $12x - 7x^2 + x^3 = 4$ in $[0, 5]$.

♦ Type 4, the Intermediate-Value Theorem

56. **(a)** Use the Intermediate-Value Theorem to show that the equation $\dfrac{6 + x^3 - 2x}{x + 3} = 1$ has at least one solution. **(b)** Use your calculator or computer to find approximate values of one solution.

♦ Type 4, the Intermediate-Value Theorem

57. Why does the Intermediate-Value Theorem not apply to $g(x) = \dfrac{x^3 - 3x^2 + 1}{x - 1}$ in $[-2, 3]$? **(b)** Use the graph of $g(x)$, generated on your calculator or computer, to explain why the equation $g(x) = k$ nevertheless has at least one solution in $[-2, 3]$ for each k between $g(-2)$ and $g(3)$.

♦ Type 4, the Intermediate-Value Theorem

58. Show that the equation $F(x) = k$ has at least one solution for every k with $0 \le k \le 200$, where

$$F(x) = \begin{cases} x^3 & \text{for} \quad 0 \le x \le 5 \\ 225 - \dfrac{2500}{x^2} & \text{for} \quad 5 < x \le 10. \end{cases}$$

♦ Type 4, the Intermediate-Value Theorem

59. **(a)** For what values of the constant m does the equation $G(x) = k$ have at least one solution for all numbers x, where

$$G(x) = \begin{cases} 3 - x^2 & \text{for} \quad x < 1 \\ mx & \text{for} \quad x \ge 1? \end{cases}$$

(b) For what values of m is $G(x)$ continuous in $(-\infty, \infty)$?

♦ Type 4, the Intermediate-Value Theorem

Use the Intermediate-Value Theorem in Questions 60 through 64.

60.[A] A sprinter starts from rest and runs around a circular track. He stops when he gets back to the start. Show that at least once during that lap he must have had the same speed at diametrically opposite points.

♦ Type 4, the Intermediate-Value Theorem

61. A rock climber starts to climb a mountain at 8:00 AM on a Saturday and gets to the top at 4:00 PM that afternoon. She camps on top and climbs back down on Sunday, starting at 8:00 AM and getting back to her starting point at 4:00 PM. Show that at some time of day on Sunday she was at the same elevation as she was at the same time on Saturday.

♦ Type 4, the Intermediate-Value Theorem

62. Show that the equation $ax^3 + bx^2 + cx + d = 0$ has at least one solution x for any choice of the constants a, b, c, and d with $a \ne 0$.

♦ Type 4, the Intermediate-Value Theorem

63. Hank weighed one-half stone (seven pounds) at birth and ten stones (140 pounds) at age 15. **(a)** Prove that at some time his weight in stones equalled his age in years. **(b)** Did his weight in pounds have to equal his age in years at some time? **(c)** Did his weight in pounds have to equal his age at some time if his age were measured in months?

♦ Type 4, the Intermediate-Value Theorem

64. Prove the following FIXED-POINT THEOREM: if $f(x)$ is continuous in a finite closed interval and the values $f(x)$ for $a \le x \le b$ all lie in $[a, b]$, then the equation $f(x) = x$ has at least one solution in $[a, b]$.

♦ Type 4, the Intermediate-Value Theorem

65. Give intervals $[a, b]$ and numbers k such that for the function $S(x)$ of Figure 33, **(a)** the hypotheses of the Intermediate-Value are satisfied, **(b)** the hypotheses of the Intermediate-Value are not satisfied and its conclusion does not hold, and **(c)** the hypotheses of the Intermediate-Value Theorem are not satisfied and its conclusion does hold.

♦ Type 4, the Intermediate-Value Theorem

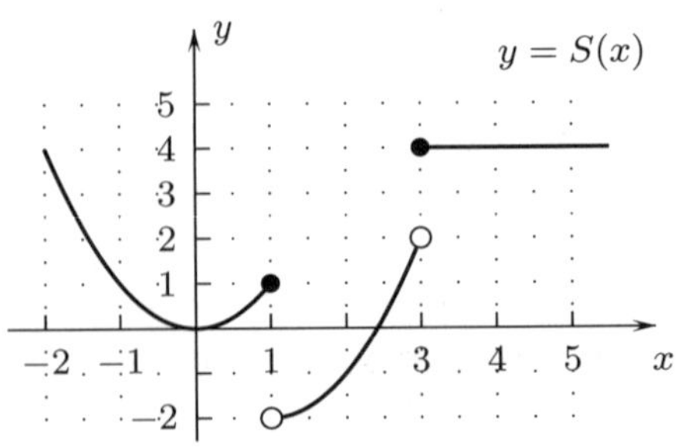

FIGURE 33

66.[O] The radius of a sphere is measured to be one foot and this number is used to calculate the volume. Give a positive number δ such that an error less than δ in the measurement of the radius would lead to an error less than 0.005 cubic feet in the calculated volume of the sphere.

♦ epsilon-delta definition of a limit

67. A sample of clay weighs exactly 100 grams, and its volume is measured to be 50 cubic centimeters. Give a positive number δ such that an error less than δ in the measurement would lead to an error less than 0.003 grams per cubic centimeter in the calculated density $\rho = 100/V$ of the clay.

♦ epsilon-delta definition of a limit

68. A square piece of zinc plate is made from material with a density of four grams per square centimeter. The plate is weighed and found to weigh 36 grams, and this value is used to calculate the width of the plate. Give a positive number δ such that an error less than δ in the measurement of the weight would lead to an error less than 0.01 centimeters in the calculated width of the plate.

♦ epsilon-delta definition of a limit

Use Definitions 1A and 2A of Section 1.2 and Definitions 1A and 2 of this section as needed in Problems 69 through 77.

69.[O] Prove that if $f(x) \to L$ and $g(x) \to M$ as $x \to x_0^+$ with numbers M and N, then $f(x)+g(x) \to L+M$ as $x \to x_0^+$.

♦ epsilon-delta definition of a limit

70.[O] Prove that if $f(x)$ is positive for sufficiently large x and $f(x) \to 0$ as $x \to \infty$, then $1/f(x) \to \infty$ as $x \to \infty$.

♦ epsilon-delta YX-definitions of limits

71. Prove that if $f(x) \to -\infty$ as $x \to x_0^-$, then $1/f(x) \to 0$ as $x \to x_0^-$.

♦ epsilon-delta and Y-delta definitions of limits

72.[A] Prove that the statement $\lim_{x\to 4^+} (5x) = 20.1$ is false. (Show that for some $\epsilon > 0$ there is no suitable δ in Definition 1A.)

♦ epsilon-delta definition of a limit

73. Prove that the statement $\lim_{x\to 3^+} x^2 = 9.2$ is false.

♦ epsilon-delta definition of a limit

74. Prove that $x\sin(1/x)$ has a finite limit as $x \to 0^+$.

♦ epsilon-delta definition of a limit

75. Prove that the statement $\lim_{x\to\infty} \dfrac{2x+10}{x} = \infty$ is false. (Show that for some $Y > 0$ there is no suitable X in Definition 1 of Section 1.2.)

♦ YX-definition of a limit

76. Prove that the statement $\lim_{x\to\infty} x^2 = 5{,}000$ is false.

♦ epsilon-delta definition of a limit

77. Prove that $\lim_{x\to x_0^+} x^n = x_0^n$ for any positive x_0 and any integer n.

♦ epsilon-delta definition of a limit

78. Prove that $\lim_{x\to x_0^+} x^{1/n} = x_0^{1/n}$ for any positive x_0 and any integer n.

♦ epsilon-delta definition of a limit

Section 2.3

Tangent lines, rates of change, and derivatives

OVERVIEW: *In Section 1.3 you saw that the rate of change with respect to x of a linear function $mx + b$ is constant and equals the slope m of the line that is its graph. The rate of change of a general, not necessarily linear, function $f(x)$ is its* DERIVATIVE. *If the function is not linear, then its derivative is a nonconstant function of x. In this section we will describe how derivatives are defined and derive our first formulas for derivatives of functions that are given by exact formulas.*

Topics:

- ***Tangent lines, secant lines, and derivatives***
- ***A secant-line program***
- ***The differentiation operator***
- ***Derivatives of x^n and of linear combinations***
- ***Equations of tangent lines***
- ***Tangent line approximations and differentials***

Finding a rate of change from a tangent line

The function $A(t)$ whose graph is shown in Figure 1 is a mathematical model of the percentage of alcohol in a man's blood t hours after he has drunk eight fluid ounces of whiskey.[(1)] As you can see from the graph, the man's blood-alcohol level rises from 0% to about 0.22% in about two hours, and then drops close to 0% after ten more hours.

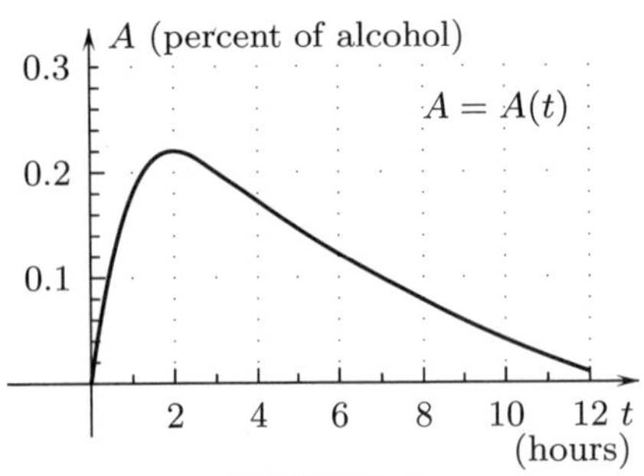

FIGURE 1

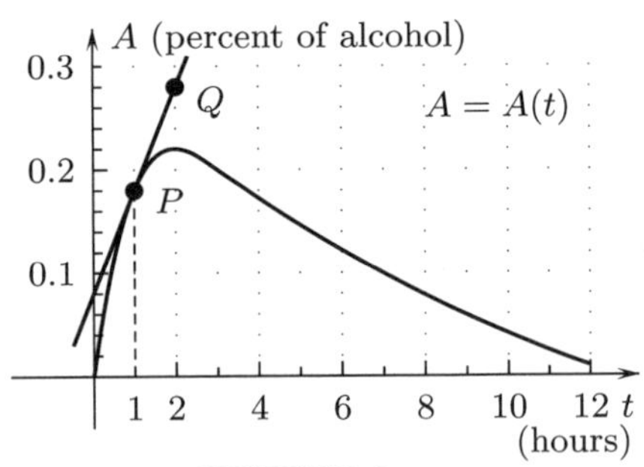

FIGURE 2

How fast is his blood-alcohol level rising after one hour? In other words, what is the rate of change of $A(t)$ with respect to t at $t = 1$? If $A(t)$ were a linear function, its graph would be a line and its rate of change would be the slope of that line. Because $A(t)$ is not linear, we find its rate of change at $t = 1$ from the tangent line to its graph at the point P that is shown in Figure 2.[†]

We will see later in this section how tangent lines are defined. Meanwhile, the geometric idea of a tangent line can be seen in Figure 2. You can think of the tangent line at P as the line that best approximates the graph near P. You can also think of it as the line that has the same direction as the curve at P in the following sense: If the graph were a road and you were driving on it, your car would be pointed in the direction of the tangent line when you reach P. Both interpretations will lead us to define the rate of change of $A(t)$ with respect to t at $t = 1$ to be the slope of the tangent line.

[(1)]Data adapted from *Encyclopædia Britannica*, Vol. 1, Chicago: Encyclopædia Britannica, Inc., 1965, p. 548.

[†]The use of the term "tangent" in "tangent line" comes from the Latin *tangere*, to touch.

◆ SUGGESTIONS TO INSTRUCTORS: The main topics in this section can be divided into two parts, as follows:

Part 1 (Definition of the derivative, the Secant-Line program, and predicting derivatives with values of difference quotients; Tune-up Exercises T1 and T6; Problems 1-6) Have students begin work on Questions 1–3 before class. Problems 1–6 are good for in-class discussion.

Part 2 (Derivatives of x^n and linear combinations, tangent lines, and differentials; Tune-up Exercises T2–T5 and T7–T10;

We can find the slope of the tangent line in Figure 2 from the coordinates of the points P and Q on it in the drawing. In fact, P has coordinates $(1, 0.18)$ and Q has coordinates $(2, 0.28)$, so the tangent line's slope is

$$\frac{\text{Rise}}{\text{Run}} = \frac{[0.28 - 0.18] \text{ percent}}{[2-1] \text{ hour}} = 0.1 \text{ percent per hour.}$$

This is the rate of change of $A(t)$ with respect to t at $t = 1$: One hour after drinking the whiskey, the man's blood-alcohol level is rising 0.1 percent per hour.

Question 1 Find the rate of change of the man's blood-alcohol level eight hours after he drinks the whiskey by calculating the slope of the tangent line at $t = 8$ to the graph of $A(t)$ in Figure 3. Use the coordinates $(4, 0.16)$ and $(8, 0.08)$ of the two points on the tangent line in the drawing. This rate of change is negative because the blood-alcohol level is decreasing at $t = 8$.

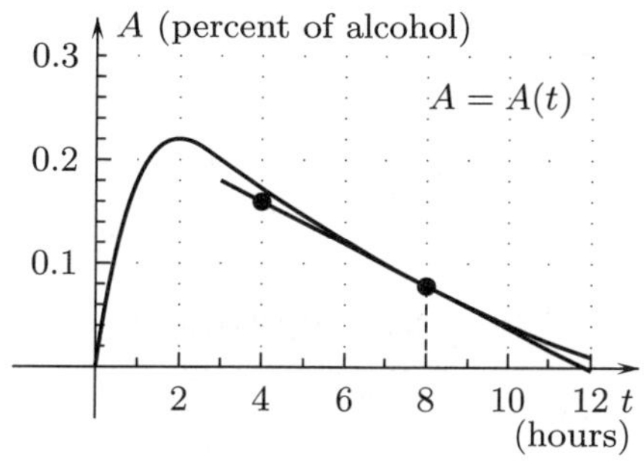

FIGURE 3

Why we cannot use the plane-geometry definition of tangent lines

The tangent line at the point P on the circle in Figure 4 is defined in plane geometry to be the line that intersects the circle only at P.† Tangent lines to many other curves can be characterized as the lines that intersect the curves at only one point, but this is not a suitable general definition. The tangent line to $y = x^3 - 4x + 5$ at $P = (-1, 8)$ in Figure 5, for example, intersects the curve at the second point Q.

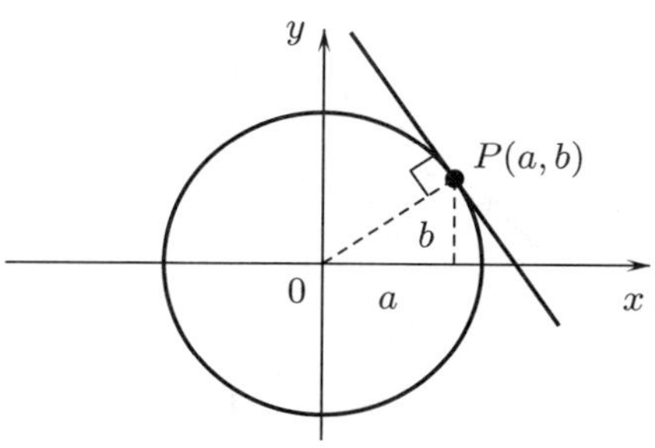

FIGURE 4

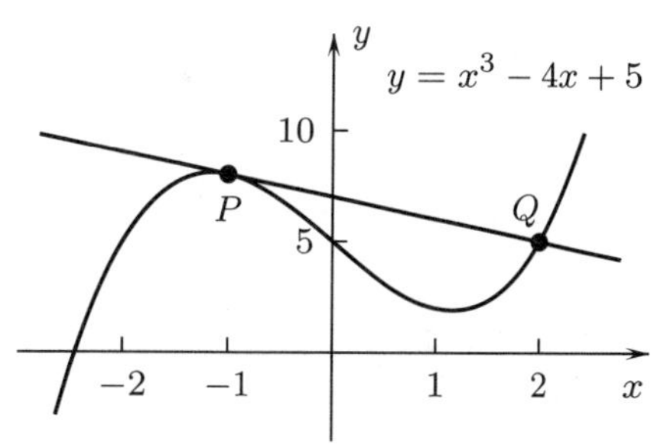

FIGURE 5

Problems 7–75) Have students begin work on Questions 4 through 8 before class; Problems 12, 22, 24, 25, 27, 28, 32–37, 53, 54, and 56 are good for in-class discussion.

†This definition, which was given by Euclid (c. 300 BC), can be used to find the slope of the tangent line to the circle in Figure 4. Because the tangent line intersects the circle only at P, all of the line except the point P is outside the circle, and P is the closest point on the line to the center O of the circle. This implies that the radius OP of the circle is perpendicular to the tangent line, so that, if P has coordinates (a, b) with nonzero constants a and b, then the radius has slope b/a and the tangent line has slope $-a/b$.

This definition would also not work for the tangent line at $x = 0$ to the graph of

$$G(x) = \begin{cases} x^2 + 3x + 1 & \text{for} \quad x \le -1 \\ x & \text{for} \quad -1 < x < 1 \\ -x^2 + 3x - 1 & \text{for} \quad x \ge 1 \end{cases}$$

in Figure 6. The graph consists of the line $y = x$ for $-1 < x < 1$. Therefore, its tangent line at $(0, 0)$ is the line $y = x$ (Figure 7), which intersects the curve at an infinite number of points.

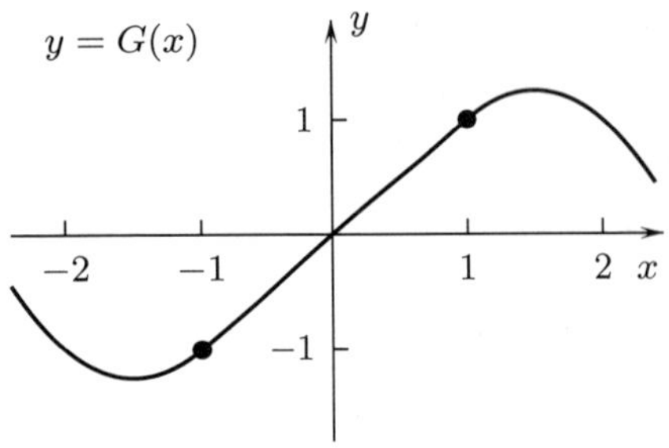

FIGURE 6

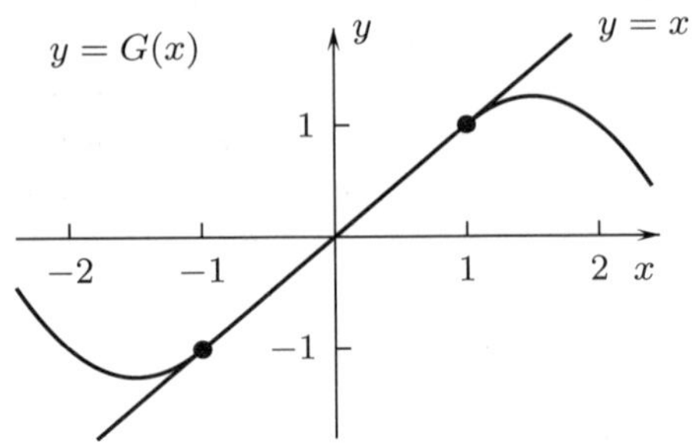

FIGURE 7

Using secant lines and average rates of change: the derivative

To define and find slopes of tangent lines of a wide variety of curves, including those in Figures 5 and 6, we return to the idea of average rate of change from Section 2.1. Recall that the average rate of change with respect to x of $f(x)$ from $x = a$ to $x = b$, with $a \neq b$, equals the change in $f(x)$, divided by the change in x,

$$[\text{Average rate of change of } f(x) \text{ from } a \text{ to } b] = \frac{\text{Rise}}{\text{Run}} = \frac{f(b) - f(a)}{b - a} \tag{1}$$

and that this average rate of change is the slope of the secant line between the points at $x = a$ and $x = b$ on the graph of $f(x)$ (Figure 8).

Secant line of slope

$$\frac{f(b) - f(a)}{b - a}$$

FIGURE 8

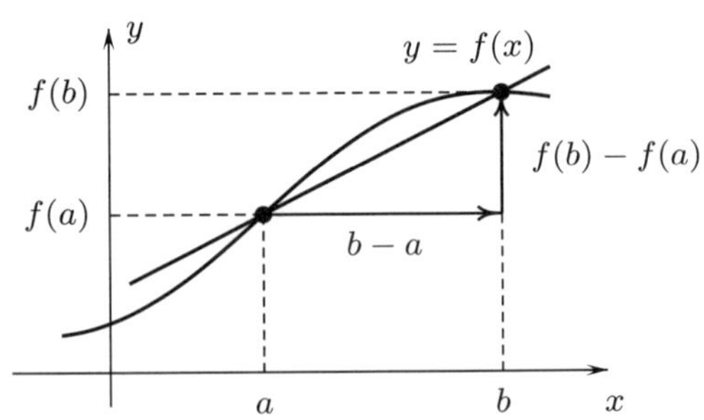

Imagine what happens as b tends to a. If the slope **(1)** of the secant line has a finite limit m, then the secant line turns about the fixed point $(a, f(a))$, as indicated in Figure 9, and approaches the line with slope m—the dashed line in Figure 9—which is the tangent line to the graph of $f(x)$ at $x = a$ that is shown in Figure 10.

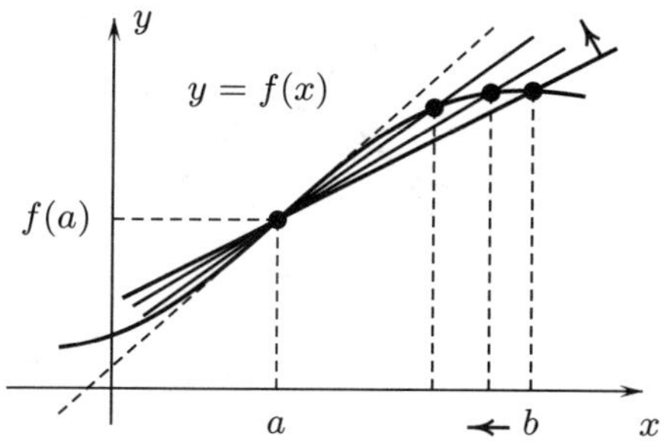

Secant lines

FIGURE 9

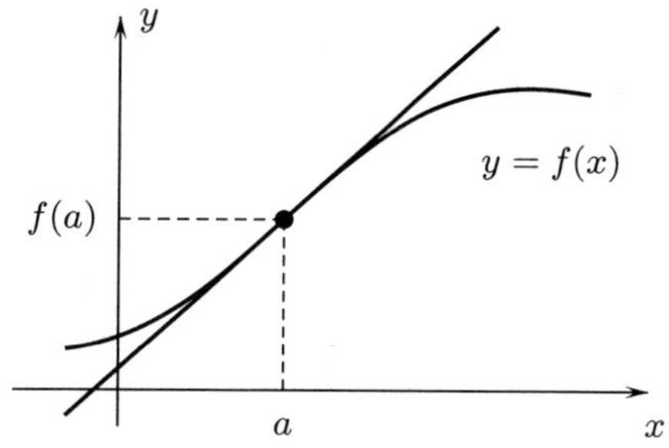

The tangent line

FIGURE 10

The slope of the tangent line is the DERIVATIVE of $f(x)$ at a and is denoted $f'(a)$. Because it is the limit of the average rate of change (1) as $x \to a$, the derivative is also the RATE OF CHANGE of $f(x)$ with respect to x at a:

Definition 1 *The derivative of $f(x)$ at $x = a$ is*

$$f'(a) = \lim_{b \to a} \frac{f(b) - f(a)}{b - a} \tag{2}$$

provided this limit exists and is finite. The derivative $f'(a)$ is the slope of the tangent line to the graph of $f(x)$ at $x = a$ and is the rate of change of $f(x)$ with respect to x at a.

The derivative is occasionally referred to as an "instantaneous rate of change" rather than as a "rate of change" to distinguish it from average rates of change.

The average rate of change **(1)** is also called a DIFFERENCE QUOTIENT because it is a quotient of the differences $f(b) - f(a)$ and $b - a$. We then have three terms for average rate of change and three terms for (instantaneous) rate of change, which are related as shown the following diagram, where the vertical arrows represent the process of having b tend to a:

$\frac{f(b) - f(a)}{b - a}$	=	Average rate of change	=	Slope of the secant line	=	The difference quotient
↓		↓		↓		↓
$f'(a)$	=	Instantaneous rate of change	=	Slope of the tangent line	=	The derivative

Predicting a derivative by calculating difference quotients

Figure 11 shows the graph of $f(x) = 4x - x^2$ and its tangent line at $x = 1$, whose slope equals the derivative $f'(1)$. We can predict the value of this derivative by calculating the slope of the secant line $\frac{f(b) - f(1)}{b - 1}$ for values of b close to 1 (Figure 12).

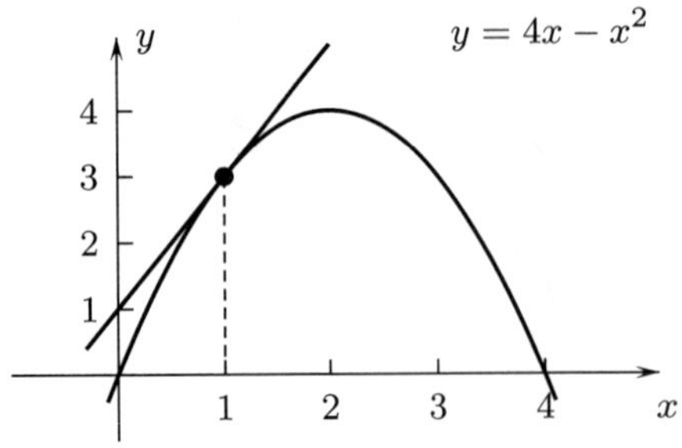

Tangent line

Slope $= f'(1)$

FIGURE 11

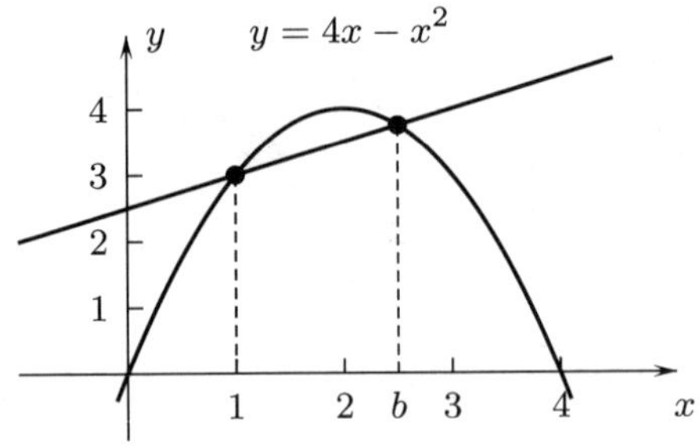

Secant line

$$\text{Slope} = \frac{f(b) - f(1)}{b - 1}$$

FIGURE 12

Question 2 Complete the following table of difference quotients of $f(x) = 4x - x^2$ (slopes of secant lines to $y = 4x - x^2$) and use it to predict the derivative,

$$f'(1) = \lim_{b \to 1} \frac{f(b) - f(1)}{b - 1} = \lim_{b \to 1} \frac{(4b - b^2) - 3}{b - 1}.$$

b	2.5	2	1.5	1.1	1.001	1.00001
$\frac{(4b - b^2) - 3}{b - 1} \doteq$	0.5	1	1.5	1.9		

b	−0.5	0	0.5	0.9	0.999	0.99999
$\frac{(4b - b^2) - 3}{b - 1} \doteq$	3.5	3	2.5	2.1		

A secant-line program

The following program for Texas Instruments TI-82, TI-83, TI-85, and TI-86 graphing calculators draws the secant line through the points at $x = A$ and $x = B$ on the graph of the function Y_1 entered in the $Y =$ menu. It then displays the slope of the secant line. The line numbers are given only for reference and are not used on the calculator. The program for TI-81 calculators and key-by-key instructions for entering the programs in the T1-81, TI-82, TI-83, TI-85, and TI-86 are given at the end of this chapter.†

TI-82, TI-83

```
PROGRAM: SECANT
: ClrDraw
: FnOff
: FnOn 1
: Disp "USE Y1(X)"
: Prompt A, B
```

TI-85, TI-86

```
PROGAM: SECANT
: ClrDrw
: FnOff
: FnOn 1
: Disp "USE y1(x)"
: Prompt A, B
```

†These calculator-specific programs and instructions are included in preliminary editions of this text for the convenience of users. In the first edition only a description of the action of the program on any calculator or computer will be in the text. The programs and implementation instructions for various calculators and computer software will be in the *Calculator/Computer Supplement*.

```
:A → X
:Y1 → W
:B → X
:Y1 → V
:(V − W)/(B − A) → M
:DrawF W + M(X − A)
:Pause
:Disp "SLOPE ="
:Disp M
```

```
:A → x
:y1 → W
:B → x
:y1 → V
:(V − W)/(B − A) → M
:DrawF W + M * (x − A)
:Pause
:Disp "SLOPE ="
:Disp M
```

The function to be studied should be entered as Y_1 in the $Y=$ menu and a suitable window for the graph of the function and the secant line should be chosen. To run the program on a TI-81, TI-82, or TI-83, press [PRGM], the program's number, and [ENTER]. On a TI-85 or TI-86, press [PRGM], find the program under <NAMES>, and press [ENTER]. Give the values of A and B when requested, followed by the [ENTER] key. The curve and the secant line will be drawn. Press [ENTER] to display the slope of the line. You can then press [ENTER] to rerun the program if you do not perform any other operations first.

Run the secant-line program with $Y_1 = 4x - x^2$ and the window $-1 \leq x \leq 5, -1 \leq y \leq 5$ to verify the slopes for $B = 2, 1.5, 1.1, 1.001$, and 1.00001 in the table of Question 2. In the process you should obtain the secant lines in Figures 13 through 18. The secant lines in Figures 17 and 18 are indistinguishable from the tangent line in Figure 11 because B is very close to 1 in these cases.

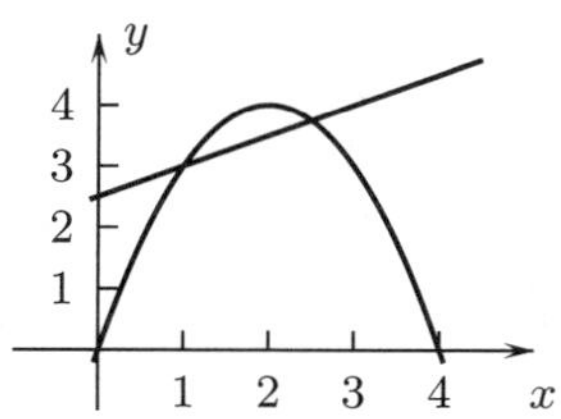

$a = 1, b = 2.5$

FIGURE 13

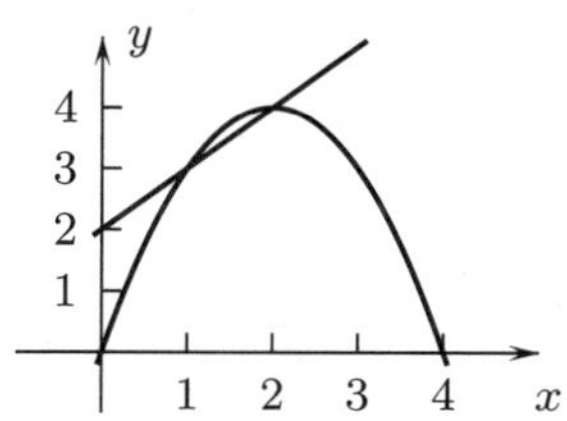

$a = 1, b = 2$

FIGURE 14

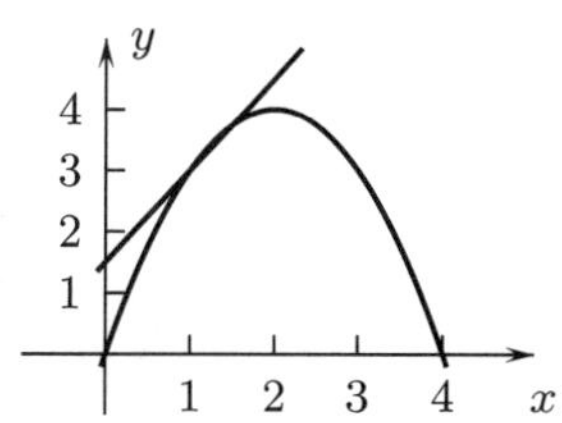

$a = 1, b = 1.5$

FIGURE 15

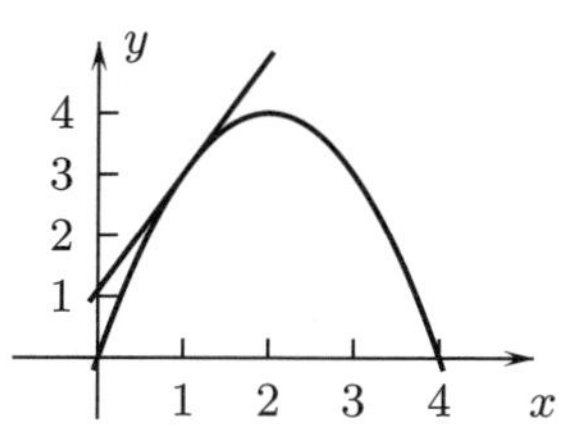

$a = 1, b = 1.1$

FIGURE 16

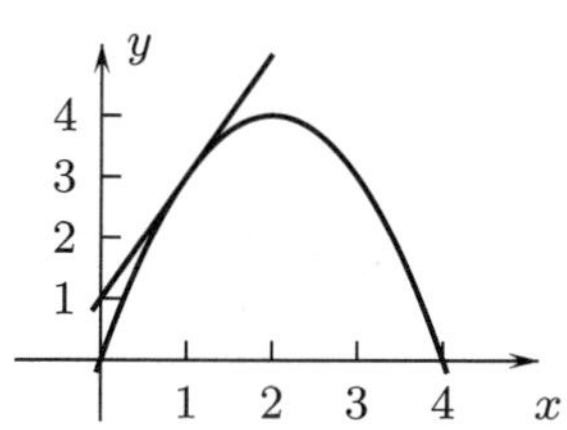

$a = 1, b = 1.001$

FIGURE 17

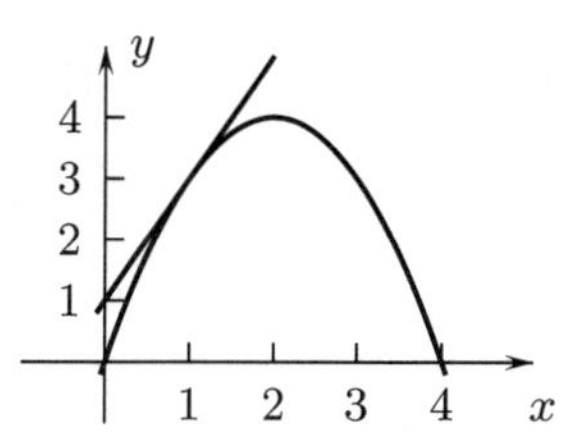

$a = 1, b = 1.00001$

FIGURE 18

Question 3 Use the secant-line program with the window $-1.75 \le x \le 2.75, -0.5 \le y \le 2.5$ or direct calculations to complete the next table with slopes of secant lines to $y = x^2, y = x^3, y = x^{-1}$, and $y = x^{1/2}$ for $a = 1$ and values of b close to 1. Then use these results to predict the derivatives of x^2, x^3, x^{-1}, and $x^{1/2}$ at $x = 1$.

Curve	Slope	$a = 1, b = 1.1$	$a = 1, b = 1.001$	$a = 1, b = 1.00001$
$y = x^2$	$\dfrac{b^2 - 1}{b - 1} \doteq$	2.1		
$y = x^3$	$\dfrac{b^3 - 1}{b - 1} \doteq$	3.31		
$y = x^{-1}$	$\dfrac{b^{-1} - 1}{b - 1} \doteq$	-0.9090909		
$y = x^{1/2}$	$\dfrac{b^{1/2} - 1}{b - 1} \doteq$	0.4880885		

Finding exact derivatives

By Definition 1, the derivative of $f(x) = x^2$ at $x = 1$ is the limit as b tends to 1 of the difference quotient,

$$\frac{f(b) - f(1)}{b - 1} = \frac{b^2 - 1}{b - 1}. \tag{3}$$

We cannot find this limit by taking the limit of the numerator and denominator in **(3)** because they both tend to zero and we would obtain the meaningless expression 0/0. However, as we saw with similar examples in the last section, we can rewrite **(3)** to show that it is equal, for $b \neq 1$, to a continuous function of b whose limit as $b \to 1$ is its value at 1.

Example 1 Rewrite **(3)** by factoring its numerator and cancelling one factor with the denominator. Then use the new expression to find

$$f'(1) = \lim_{b \to 1} \left[\frac{b^2 - 1}{b - 1}\right].$$

SOLUTION For $b \neq 1$, $\dfrac{b^2 - 1}{b - 1} = \dfrac{(b + 1)(b - 1)}{b - 1} = b + 1$. Then because $b + 1$, as a function of b, is continuous at $b = 1$, $b + 1 \to 2$ as $b \to 1$ and $f'(1) = \lim\limits_{b \to 1}(b + 1) = 2$. □

The result of Example 1 probably agrees with your prediction in Question 3. Later in this section we will obtain formulas for exact derivatives of x^3, x^{-1}, and $x^{1/2}$ that can be used to verify your likely other predictions in Question 3.

Leibniz notation

Isaac Newton (1642–1727) and Gottfried Leibniz (1646–1716), who are considered to be the founders of calculus, each introduced notation for the derivative. The PRIME NOTATION $f'(a)$ is similar to that used by Newton, and the symbols used by Leibniz evolved into what is known today as LEIBNIZ NOTATION. In this notation the derivative of $f(x)$ at x is denoted $\dfrac{df}{dx}(x)$, and Definition 1 with x in place of a reads

$$\frac{df}{dx}(x) = \lim_{b \to x} \frac{f(b) - f(x)}{b - x}. \tag{4}$$

The differentiation operator

Mathematicians use the term OPERATOR for a rule that transforms one function into another. The operator that transforms a function into its derivative is called DIFFERENTIATION and is represented by the symbol $\frac{d}{dx}$. If the function is $f(x)$, then

$$\frac{d}{dx}[f(x)] = \frac{df}{dx}(x)$$

or with prime notation for the derivative

$$\frac{d}{dx}[f(x)] = f'(x).$$

Look at these formulas very closely! Especially notice the logical difference between the derivative $\frac{df}{dx}(x) = f'(x)$, which is a function, and the differentiation operator $\frac{d}{dx}$ that transforms the function into its derivative.

The main advantage of Leibniz notation over prime notation is that it includes the name of the variable with respect to which the derivative is being taken. The differentiation operator $\frac{d}{dx}$ also provides a convenient way to express derivatives of functions given by formulas, as in the following discussion of the derivative $\frac{d}{dx}(1)$ of the function $y = 1$, the derivative $\frac{d}{dx}(x)$ of the function $y = x$, and other examples of $\frac{d}{dx}(x^n)$ with constant n.

The derivatives of 1 and x

Because the constant function $y = 1$ and the function $y = x$ are linear, their derivatives are the slopes of the lines that are their graphs (Figures 19 and 20):

$$\frac{d}{dx}(1) = 0 \quad \text{and} \quad \frac{d}{dx}(x) = 1. \tag{5}$$

We can also obtain these formulas from the definition of the derivatives.

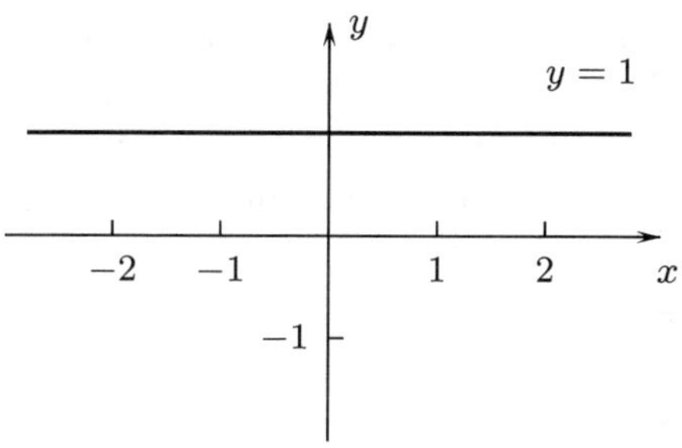

$\frac{d}{dx}(1) = [\text{Slope}] = 0$

FIGURE 19

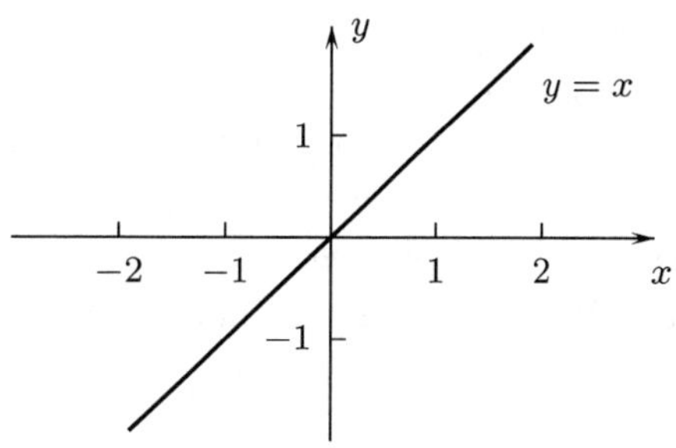

$\frac{d}{dx}(x) = [\text{Slope}] = 1$

FIGURE 20

Example 2 Derive formulas **(5)** from definition **(4)**.

SOLUTION If $f(x) = 1$, then $f(b) = 1$ and

$$\frac{d}{dx}(1) = \lim_{b \to x} \frac{1 - 1}{b - x} = \lim_{b \to x}(0) = 0.$$

For $f(x) = x$, we have $f(b) = b$ and

$$\frac{d}{dx}(x) = \lim_{b \to x} \frac{b - x}{b - x} = \lim_{b \to x}(1) = 1. \square$$

The derivative of the power function x^n *for integers* $n \geq 2$

If we set $f(x) = x^n$ in **(4)** with a constant n, we obtain the definition of its derivative:

$$\frac{d}{dx}(x^n) = \lim_{b \to x} \frac{b^n - x^n}{b - x}. \quad \textbf{(6)}$$

For $n = 2$, we can find this limit as we did in Example 1 with $x = 1$, by factoring $b^2 - x^2$:

$$\frac{d}{dx}(x^2) = \lim_{b \to x} \frac{b^2 - x^2}{b - x} = \lim_{b \to x} \frac{(b + x)(b - x)}{b - x} = \lim_{b \to x} (b + x) = x + x = 2x.$$

We could find the derivative of x^n for any other positive integer n by dividing $b - x$ into $b^n - x^n$ with the notation used for long division of numbers. Here is the calculation for $n = 3$:

$$\begin{array}{r|llll}
 & b^2 & + \; bx & + \; x^2 & \\
\hline
b - x & b^3 & & & - \; x^3 \\
 & b^3 & - \; b^2x & & \\
\cline{2-3}
 & & b^2x & & \\
 & & b^2x & - \; bx^2 & \\
\cline{3-4}
 & & & bx^2 & - \; x^3 \\
 & & & bx^2 & - \; x^3 \\
\cline{4-5}
 & & & & 0
\end{array}$$

In this calculation, we first put b^2 in the quotient (above the horizontal line in the division symbol), so that when the divisor $b - x$ is multiplied by b^2 and subtracted from the dividend $b^3 - x^3$ (under the division sign), the b^3-term is eliminated. Similarly, bx is put in the quotient to eliminate the b^2x below the first horizontal bar, and then x^2 is put in the quotient to eliminate the bx^2 under the second horizontal bar.

Example 3 The calculation above shows that for $b \neq x$,

$$\frac{b^3 - x^3}{b - x} = b^2 + bx + x^2. \quad \textbf{(7)}$$

Use equation **(7)** to find a formula for the derivative of x^3.

SOLUTION As $b \to x$, each of the three terms on the right of **(7)** tends to x^2, so that

$$\frac{d}{dx}(x^3) = \lim_{b \to x} \frac{b^3 - x^3}{b - x} = \lim_{b \to x} (b^2 + bx + x^2) = x^2 + x^2 + x^2 = 3x^2. \square$$

Question 4 Complete the calculation of $\dfrac{b^4 - x^4}{b - x}$ below and use the result to find the derivative of x^4.

$$
\begin{array}{rlllll}
 & b^3 & + \quad & + \quad & + & \\
\cline{2-6}
b - x \,\big) & b^4 & & & & - \; x^4 \\
 & b^4 & - \; b^3x & & & \\
\cline{2-3}
 & & b^3x & & & \\
 & & b^3x & - \; b^2x^2 & & \\
\cline{3-4}
 & & & & & \\
\cline{4-5}
 & & & & & \\
\cline{5-6}
 & & & & & 0
\end{array}
$$

As you can imagine, the long-division calculations above for $n = 3$ and $n = 4$ could be carried out for any integer $n \geq 2$ to yield

$$\frac{b^n - x^n}{b - x} = b^{n-1} + b^{n-2}x + b^{n-3}x^2 + \cdots + x^{n-1}$$

with n terms on the right. Consequently,

$$\frac{d}{dx}(x^n) = \lim_{b \to x} (b^{n-1} + b^{n-2}x + b^{n-3}x^2 + \cdots + x^{n-1}).$$

Since each of the n terms on the right of this equation tends to x^{n-1} as $b \to x$, we obtain the formula

$$\frac{d}{dx}(x^n) = nx^{n-1} \tag{8}$$

for the arbitrary positive integer n.

The case of positive fractions n

Formula **(8)** for positive integers n can now be used to show that it holds for any positive fraction n. To make the ideas in this derivation easier to follow, we will give it only in the special case of $n = \frac{5}{3}$. By definition **(6)** of the derivative with $n = \frac{5}{3}$,

$$\frac{d}{dx}(x^{5/3}) = \lim_{b \to x} \frac{b^{5/3} - x^{5/3}}{b - x}.$$

We set $B = b^{1/3}$ and $X = x^{1/3}$ with $b \neq x$, so that $b = B^3$, $x = X^3$, and

$$\frac{b^{5/3} - x^{5/3}}{b - x} = \frac{B^5 - X^5}{B^3 - X^3} = \frac{\dfrac{B^5 - X^5}{B - X}}{\dfrac{B^3 - X^3}{B - X}}. \tag{9}$$

In the last step we divided the numerator and denominator by the nonzero number $B - X$. Because B tends to X as b tends to x, the numerator and denominator of the last expression in **(9)** tend to the derivatives of X^5 and X^3 with respect to X, which by formula **(8)** with $n = 5$ and $n = 3$ are $5X^4$ and $3X^2$, respectively. Therefore,

$$\frac{d}{dx}(x^{5/3}) = \lim_{B\to X} \frac{\dfrac{B^5 - X^5}{B - X}}{\dfrac{B^3 - X^3}{B - X}} = \frac{\dfrac{d}{dX}(X^5)}{\dfrac{d}{dX}(X^3)} = \frac{5X^4}{3X^2}. \tag{10}$$

Question 5 Rewrite the last expression in **(10)** in terms of x by making the substitution $X = x^{1/3}$. Then use the laws of exponents to show that differentiation formula **(8)** holds for $n = \frac{5}{3}$.

The result of Question 5 shows that formula **(8)** holds for $n = \frac{5}{3}$ and all x. Similar calculations would show that it is valid for any positive fraction n in the open interval(s) where x^{n-1} is defined.

A case of a negative n

How about the derivative of x^n for negative n? Let us consider $n = -4$ and nonzero x. We write

$$\begin{aligned}\frac{d}{dx}(x^{-4}) &= \lim_{b\to x} \frac{b^{-4} - x^{-4}}{b - x} = \lim_{b\to x} \frac{\dfrac{1}{b^4} - \dfrac{1}{x^4}}{b - x}\\ &= \lim_{b\to x} \left\{\frac{1}{b - x}\left[\frac{x^4 - b^4}{b^4x^4}\right]\right\} = \lim_{b\to x} \left\{\frac{-1}{x^4b^4}\left[\frac{b^4 - x^4}{b - x}\right]\right\}.\end{aligned}$$

The goal of all this algebra is to express the difference quotient $\dfrac{b^{-4} - x^{-4}}{b - x}$ for x^{-4} in terms of the difference quotient $\dfrac{b^4 - x^4}{b - x}$ for x^4. When b tends to x, the expression $\dfrac{b^4 - x^4}{b - x}$ tends to the derivative $4x^3$ of x^4 that you found in Question 4. Also, $\dfrac{1}{b^4x^4}$ tends to $\dfrac{1}{x^4x^4} = \dfrac{1}{x^8}$ as $b \to x$. Consequently,

$$\frac{d}{dx}(x^{-4}) = \frac{-1}{x^8}(4x^3). \tag{11}$$

Question 6 Simplify the expression on the right of **(11)** and then express the result in terms of $n = -4$.

The derivative of x^n

Calculations similar to those carried out above would establish the following for all rational (integer or fraction) n. We will derive it for irrational n in Section 3.5 by using logarithms.

Theorem 1 *For any constant n,*

$$\frac{d}{dx}(x^n) = nx^{n-1} \tag{12}$$

in every open interval where x^{n-1} is defined. The formulas

$$\frac{d}{dx}(1) = 0 \tag{13}$$

$$\frac{d}{dx}(x) = 1 \tag{14}$$

which can be considered to be formula **(12)** *for $n = 0$ and $n = 1$, are valid for all x.*

The derivative **(12)** does not exist for $x \le 0$ if n is irrational or if n is a fraction p/q with q an even integer, and does not exist for $x = 0$ if n is < 1.

Example 4 Find formulas for **(a)** $\frac{d}{dx}(x^2)$, **(b)** $\frac{d}{dx}(x^3)$, **(c)** $\frac{d}{dx}(x^{-1})$, and **(d)** $\frac{d}{dx}(x^{1/2})$.

SOLUTION Theorem 1 with $n = 2, 3, -1$, and $\frac{1}{2}$ gives

(a) $\frac{d}{dx}(x^2) = 2x^{2-1} = 2x$,

(b) $\frac{d}{dx}(x^3) = 3x^{3-1} = 3x^2$,

(c) $\frac{d}{dx}(x^{-1}) = (-1)x^{-1-1} = -x^{-2}$, and

(d) $\frac{d}{dx}(x^{1/2}) = \frac{1}{2}x^{(1/2)-1} = \frac{1}{2}x^{-1/2}$. □

Setting $x = 1$ in the formulas from Example 4 gives the values of these derivatives that you probably predicted in Question 3.

Δx and Δf

It is often convenient to give definition **(2)** of the derivative with different notation for the rise and run on the secant line. We denote the change in x from a to b (the run) by

$$\Delta x = b - a$$

where Δ is the Greek capital letter delta. Then $b = a + \Delta x$, and Δx tends to zero as b tends to a. We use Δf for the rise, which is the change in the value of $f(x)$ when x goes from a to $a + \Delta x$:[†]

$$\Delta f = f(a + \Delta x) - f(a) = f(b) - f(a).$$

This notation is illustrated in Figure 21. With it, definition **(2)** becomes

$$\frac{df}{dx}(a) = \lim_{\Delta x \to 0} \frac{\Delta f}{\Delta x} = \lim_{\Delta x \to 0} \frac{f(a + \Delta x) - f(a)}{\Delta x}. \quad \textbf{(15)}$$

Secant line of slope

$$\frac{\Delta f}{\Delta x} = \frac{f(a + \Delta x) - f(a)}{\Delta x}$$

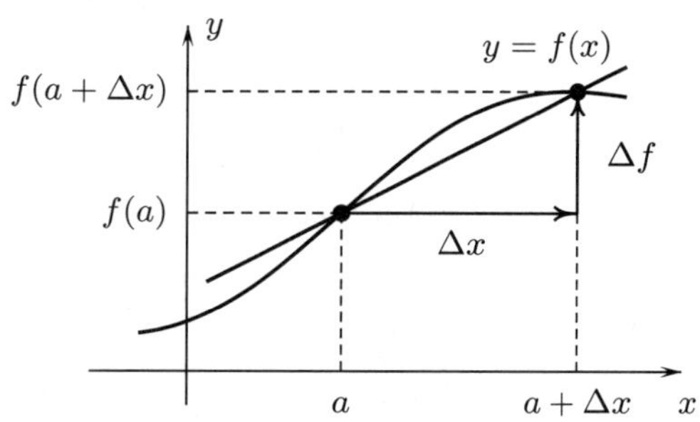

FIGURE 21

We will frequently use definition **(15)** with x in place of a:

$$\frac{df}{dx}(x) = \lim_{\Delta x \to 0} \frac{\Delta f}{\Delta x} = \lim_{\Delta x \to 0} \frac{f(x + \Delta x) - f(x)}{\Delta x}. \quad \textbf{(16)}$$

[†]Notice that each of Δx and Δf denotes a single number and not a product of a Δ and x or f. Think of Δ as "the change in," so that Δx is the change in x and and Δf is the change in f.

Derivatives of linear combinations of functions

A LINEAR COMBINATION of functions $f(x)$ and $g(x)$ is a function of the form $Af(x) + Bg(x)$ with constants A and B. Our next differention rule enables us to find the derivative of any such function if we know the derivatives of $f(x)$ and $g(x)$.

Theorem 2 *If $f(x)$ and $g(x)$ have derivatives at x, then for any constants A and B*

$$\frac{d}{dx}[Af(x) + Bg(x)] = A\frac{df}{dx}(x) + B\frac{dg}{dx}(x) \tag{17a}$$

or, in prime notation,

$$\frac{d}{dx}[Af(x) + Bg(x)] = Af'(x) + Bg'(x). \tag{17b}$$

To establish formula **(17a)**, we use definition **(16)** and write

$$\begin{aligned}\frac{d}{dx}[Af(x) + Bg(x)] &= \lim_{\Delta x \to 0} \frac{[Af(x+\Delta x) + Bg(x+\Delta x)] - [Af(x) + Bg(x)]}{\Delta x} \\ &= \lim_{\Delta x \to 0} \left\{ A\left[\frac{f(x+\Delta x) - f(x)}{\Delta x}\right] + B\left[\frac{g(x+\Delta x) - g(x)}{\Delta x}\right]\right\} \\ &= A\frac{df}{dx}(x) + B\frac{dg}{dx}(x).\end{aligned}$$

Theorems 1 and 2 are often used together, as in the following examples.

Example 5 Find a formula for the derivative of $5x^3 - x^{-2} + 4$.

SOLUTION We write

$$\begin{aligned}\frac{d}{dx}(5x^3 - x^{-2} + 4) &= 5\frac{d}{dx}(x^3) - \frac{d}{dx}(x^{-2}) + 4\frac{d}{dx}(1) \\ &= 5(3x^{3-1}) - (-2x^{-2-1}) + 4(0) = 15x^2 + 2x^{-3}. \square\end{aligned}$$

Derivatives of functions involving roots of x and reciprocals of powers and roots are found by converting to exponential notation:

Example 6 What is $f'(x)$ for $f(x) = 7\sqrt[3]{x} + \dfrac{8}{\sqrt{x}}$?

SOLUTION Since $\sqrt[3]{x} = x^{1/3}$ and $\dfrac{1}{\sqrt{x}} = x^{-1/2}$,

$$\begin{aligned}f'(x) &= \frac{d}{dx}\left(7\sqrt[3]{x} + \frac{8}{\sqrt{x}}\right) = 7\frac{d}{dx}(x^{1/3}) + 8\frac{d}{dx}(x^{-1/2}) \\ &= 7\left(\tfrac{1}{3}x^{(1/3)-1}\right) + 8\left(-\tfrac{1}{2}x^{-(1/2)-1}\right) = \tfrac{7}{3}x^{-2/3} - 4x^{-3/2}. \square\end{aligned}$$

Example 7 A precision heater is controlled by varying the current supplied to it. It produces $Q(I) = 100I^2$ Calories of heat in one second when the current is I amperes. **(a)** What is the average rate of change of the amount of heat Q with respect to the current I for $0 \le I \le 5$? **(b)** Find the instantaneous rate of change of Q with respect to I at $I = 3$. **(c)** Draw the graph of $Q(I)$ for $0 \le I \le 6$ and the secant and tangent lines whose slopes you found in parts (a) and (b).

SOLUTION **(a)** The average rate of change of Q with respect to I for $0 \le I \le 5$ is

$$\frac{[Q(5) - Q(0)] \text{ Calories}}{(5-0) \text{ amperes}} = \frac{2500 - 0}{5 - 0} = \frac{2500}{5} = 500 \text{ Calories per ampere.}$$

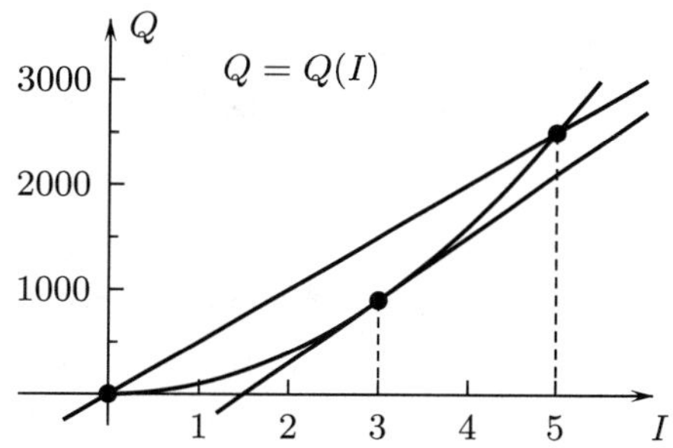

FIGURE 22

(b) The instantaneous rate of change at current I is $\dfrac{dQ}{dI} = \dfrac{d}{dI}(100I^2) = 200I$. Therefore, $\dfrac{dQ}{dI}(3) = 200(3) = 600$ Calories per ampere.

(c) The graph of $Q(I)$ and the secant and tangent lines are shown in Figure 22. □

Velocity

We saw in Section 2.1 that if an object's position on an s-axis is given as a function $s(t)$ of the time, then its average rate of change with respect to t between times t and $t + \Delta t$ is the change in s divided by the change in the time. This equals the difference quotient

$$\frac{\Delta s}{\Delta t} = \frac{s(t+\Delta t) - s(t)}{\Delta t} \tag{18}$$

and is the object's average velocity in the positive direction during the time interval. The limit of expression **(18)** as Δt tends to 0 is the derivative $\dfrac{ds}{dt}(t)$, which is the object's (instantaneous) velocity in the positive s-direction at time t. The velocity is measured in the units used for distance divided by those used for time.

Question 7 A car traveling on a straight road is $5t^2 + 20t + 40$ miles past a gasoline station t hours after noon. What is its velocity at 4:00 PM?

Equations of tangent lines

The tangent line to $y = f(x)$ at $x = x_0$ in Figure 23 passes through the point $(x_0, f(x_0))$ and has slope $f'(x_0)$. By the point-slope formula, it has the equation,

$$y = f(x_0) + f'(x_0)(x - x_0). \tag{19}$$

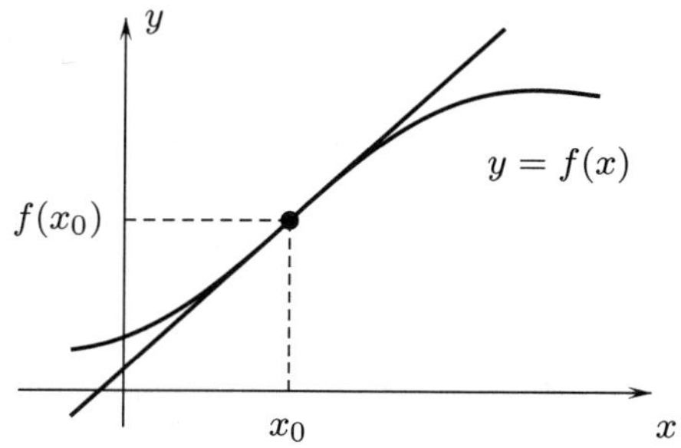

[Slope] $= f'(x_0)$

FIGURE 23

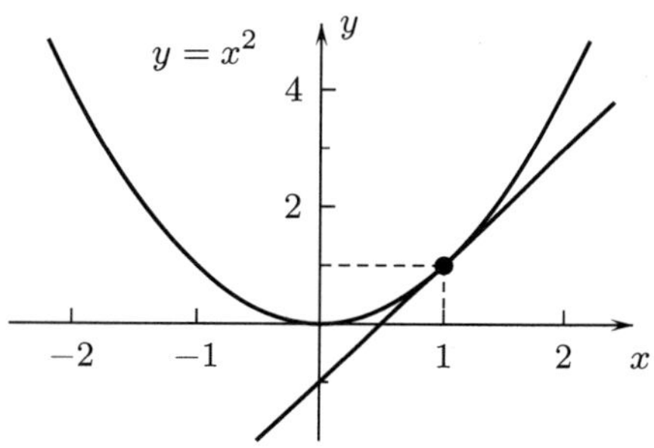

Tangent line at $x = 1$

FIGURE 24

Example 8 Give an equation of the tangent line to $y = x^2$ at $x = 1$.

SOLUTION Set $f(x) = x^2$. Then $f'(x) = 2x$, so that $f(1) = 1$ and $f'(1) = 2$, so that by **(19)** with $x_0 = 1$, the tangent line has the equation

$$y = 1 + 2(x - 1)$$

The curve and tangent line are shown in Figure 24. □

Question 8 Give an equation of the tangent line to $y = x^3 - 4x + 5$ at $x = -1$. Generate the curve and the tangent line in the window $-3 \leq x \leq 3$, $-5 \leq y \leq 15$. The result should look like Figure 5 on the second page of this section.

Using tangent lines to estimate values of functions

The tangent line to $y = f(x)$ at $x = x_0$ approximates the graph near $x = x_0$ and consequently,

$$f(x) \approx f(x_0) + f'(x_0)(x - x_0) \quad \text{for } x \approx x_0. \tag{20}$$

Example 9 One hour after leaving Toulouse, France, on a test flight, the French/British supersonic passenger jet Concord has traveled 975 miles and is flying 1520 miles per hour (twice the speed of sound).[(2)] Approximately how far is it from Toulouse six minutes (0.1 hours) later?

SOLUTION We let $s(t)$ be the Concord's distance from Toulouse t hours after take off. We are given that $s(1) = 975$ miles and $s'(1) = 1520$ miles per hour, and we want to calculate the approximate value of $s(1.1)$. The tangent line to the graph of $s(t)$ at $t = 1$ is

$$s = 975 + 1520(t - 1)$$

so that $s(1.1) \approx 975 + 1520(1.1 - 1) = 1127$ miles. The plane is approximately 1127 miles from Toulouse one hour and six minutes after take off. □

Because the function $f(x_0) + f'(x_0)(x - x_0)$, whose graph is the tangent line to $y = f(x)$ at $x = x_0$, is a linear function that closely approximates $f(x)$ for x near x_0, it is called the LOCAL LINEARIZATION of $f(x)$ at x_0.

Differentials

The symbols dx and df are called DIFFERENTIALS. They were used in the seventeenth and eighteenth centuries to designate "infinitesimal" changes in x and f. Some modern mathematians and many physicists and other scientists adopt this point of view, but in most calculus classes, dx and df are considered to be real, noninfinitesimal numbers corresponding to changes in x and y on a tangent line to the graph of $f(x)$, as shown in Figure 25.

FIGURE 25

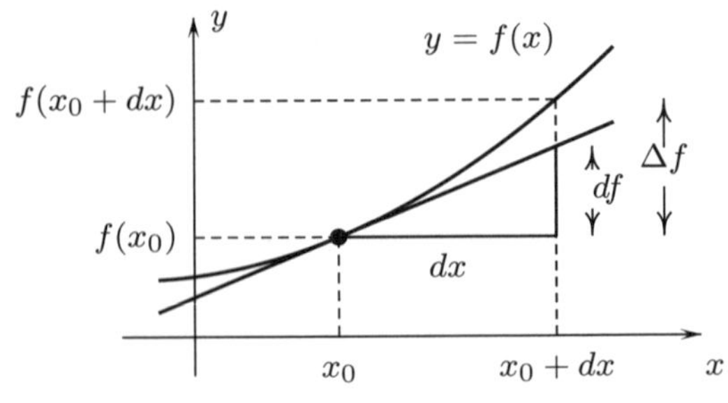

[(2)]Data from "Test flights of supersonic aircraft" by J. F. Renaudie, *Flight Test Instrumentation*, AGARD Conference Proceedings No. 32, London: Technical Editing and Reproduction Ltd., 1967, p. 39.

The differentials dx and df are, for $dx \neq 0$, the run and rise between two points on the tangent line, whose slope is $f'(x_0)$ so that

$$\frac{df}{dx} = f'(x_0). \tag{21}$$

This equation looks like a statement relating Leibniz and prime for the derivative. Here, however, $\frac{df}{dx}$ is an actual ratio of numbers rather than just a symbol for the derivative.

Multiplying both sides of equation **(21)** by dx gives a formula for df in terms of dx:

$$df = f'(x_0)\,dx. \tag{22}$$

Example 10 Give a formula for df in terms of dx for $f(x) = x^3$ at $x = 2$.

SOLUTION $f'(x) = 3x^2$ and $f'(2) = 3(2)^2 = 12$ so at $x = 2, df = 12dx$. □

If we let Δf denote the change $f(x_0 + dx) - f(x_0)$ in the value of $f(x)$ from x_0 to $x_0 + dx$, as in Figure 24, then for small dx,

$$\Delta f = f(x_0 + dx) - f(x_0) \approx df. \tag{23}$$

This statement is convenient for expressing error estimates based on tangent line approximations. Suppose a quantity x is measured to be x_0 and a formula is used to calculate a related quantity $f(x_0)$. If there is an error in the measurement x_0 and the exact value is $x_0 + dx$, then the exact value of $f(x)$ is $f(x_0 + dx)$ and the resulting error in the value of $f(x)$ is $|\Delta f| = |f(x_0 + dx) - f(x_0)|$. Therefore,

$$[\text{The error in } f(x)] = |f(x_0 + dx) - f(x_0)| \approx |df| = |f'(x_0)dx|. \tag{24}$$

The next example illustrates how this notation is used.

Example 11 The width w of a cube is measured to be 10 centimeters with an error ≤ 0.03 centimeters, and the value $w = 10$ is used to find the approximate volume $V = 10^3 = 1000$ cubic centimeters. Use differentials to estimate the maximum error in the calculated volume.

SOLUTION The volume of a cube of width w is $V = w^3$ and $V' = 3w^2$. Therefore the differentials dV and dw are related by the equation, $dV = 3w^2\,dw$. Setting $w = 10$ gives $dV = 300\,dw$. The error from using the measured $w = 10$ instead of the (actual, unknown) width $10 + dw$ is approximately $|dV| = 300|dw|$. Since we are told that $|dw| \leq 0.03$, we conclude that

$$\begin{aligned}[\text{The error in} V] &\approx |dV| = |300\,dx| \\ &\leq 300(0.03) = 9 \text{ cubic centimeters.}\end{aligned}$$

Because the volume of the cube increases when its width increases, the actual maximum error under the conditions of Example 10 would occur if the width of the cube were exactly 0.03 centimeters greater than or less than the measured 10. The error is $|10.03^3 - 10^3| \doteq 9.027$ if the actual width is 10.03 and is $|10^3 - 9.97^3| \doteq 8.973$ if the actual width is 9.97. The actual maximum error is, with three decimal place accuracy, 9.027 cubic centimeters, which is fairly close to the error estimate from Example 11.

Principles and procedures

- You should be completely familiar with prime notation $f'(a)$ and Leibniz notation $\frac{df}{dx}(a)$ for the derivative of $f(x)$ at $x = a$ and with the differentiation operator $\frac{d}{dx}$ so you can understand all discussions where these different styles of notation occur and so you can use them properly in your work.
- Be sure you understand the chart below Definition 1 that shows the relationships among the terms "average rate of change," "slope of a secant line," "difference quotient," "instantaneous rate of change," "slope of a tangent line," and "derivative."
- If a function $f(x)$ is given by an exact formula but you do not know rules for finding a formula for its derivative, you can find the approximate value of the derivative at a specific point a by calculating difference quotients $\dfrac{f(b) - f(a)}{b - a}$ for values of b very close to a, with or without using the secant-line program that is listed above.
- Be sure you know the formula nx^{n-1} for the derivative of x^n and how to apply it to roots and reciprocals of powers and roots of x by first switching to exponential notation. You will often use it in combination with the rule $\frac{d}{dx}[Af(x) + Bg(x)] = A\frac{df}{dx}(x) + B\frac{dg}{dx}(x)$, as in Examples 4 and 5 above.
- Even though the rules $\frac{d}{dx}(1) = 0$ and $\frac{d}{dx}(x) = 1$ can be thought of as $\frac{d}{dx}(x^n) = nx^{n-1}$ in the special cases of $n = 0$ and $n = 1$, it is probably easier to remember these two special cases in terms of slopes of lines: The derivative of 1 is zero because its graph is a horizontal line of zero slope, and the derivative of x is 1 because its graph is a line of slope 1.
- If you know $f(a)$ and $f'(a)$, you can estimate $f(x)$ for x near a with the linear function whose graph is the tangent line to the graph of $f(x)$ at $x = a$:

$$f(x) \approx f(a) + f'(a)(x - a) \quad \text{for } x \text{ close to } a.$$

- The use of differentials dx and df for estimating errors, as in Example 11, may seem confusing at first. Bear in mind that this notation is just a special way of describing approximations by tangent lines.

Tune-Up Exercises 2.3♦

[A] *Answer provided.* [O] *Outline of solution provided.*

Be sure you can work Exercises T1 through T5 and work as many of Exercises T6 through T11 as you need for practice.

T1.[O] Use at least three difference quotients, calculated with or without the secant-line program, to predict the derivative of $\sqrt{x^4 + 3}$ at $x = 1$. For the secant-line program, use the window $-2 \le x \le 2$, $-0.5 \le y \le 5$.

♦ Type 1, predicting a derivative

T2.[O] Find exact formulas for **(a)** $\frac{d}{dx}(5x^3 - 6x^2 + 7x)$, **(b)** $\frac{d}{dx}(6x^{-3} - 4x^{-1})$, and **(c)** $\frac{d}{dt}(t^{3/2} - t^{2/3})$.

♦ Type 1, derivatives

T3.[O] At what values of x is $f'(x) = 2$ if $f(x) = x^3$?

♦ Type 1, a derivative

♦The Tune-up Exercises and Problems are classified by type and content. The types are (1) basic, reactive; (2) basic reflective; (3) intermediate, reactive; (4) intermediate, reflective; (5) advanced, reactive; (6) advanced, reflective; and (7) advanced, theoretical.

T4.[O] Give an exact equation of the tangent line to $y = \sqrt[3]{x}$ at $x = 27$. Generate the curve and tangent line together in the window $-20 \le x \le 70, -3 \le y \le 5$ as a partial check of your answer.

♦ Type 1, an equation of a tangent line

T5.[O] What is the approximate value of $T(9.75)$ if $T(10) = 7$ and $T'(10) = 4$?

♦ Type 1, a tangent-line approximation

More practice

T6.[A] Use at least three difference quotients to predict the derivative of $\dfrac{90x}{x+1}$ at $x = 2$. For the secant-line program use the window $-1 \le x \le 5, -30 \le y \le 100$.

♦ Type 1, predicting a derivative

T7.[A] Find **(a)** $\dfrac{d}{dx}(x^7 + 3x^2 - 15)$, **(b)** $\dfrac{d}{dx}(7x^{-8} - 8x^{-7} + 3)$, and **(c)** $\dfrac{d}{dt}(4t^{-1/3} - 3t^{1/4})$.

♦ Type 1, derivatives

T8.[A] At what values of x is $\dfrac{dw}{dx}(x) \ge 5$ if $w(x) = x^4 + x$?

♦ Type 2, a derivative, etc.

T9.[A] Give an equation of the tangent line to the graph of $y = 10x - x^4$ at $x = 1$. Generate the curve and tangent line together in the window $-1 \le x \le 3, -5 \le y \le 15$ as a partial check of your answer.

♦ Type 1, an equation of a tangent line

T10.[O] What is the approximate value of $Z(u)$ at $u = 1.09$ if $Z(1) = 100$ and $\dfrac{dZ}{du}(1) = -20$?

♦ Type 1, a tangent-line approximation

Problems 2.3

[A]*Answer provided.* [O]*Outline of solution provided.*

Use at least three difference quotients to predict the derivative in each of Problems 1 through 6. The given windows are for the secant-line program.

1.[A] $f'(2)$ for $f(x) = \dfrac{x^2+6}{x^2+1}$
$(-2 \le x \le 4, -1 \le y \le 8)$

♦ Type 1, predicting a derivative

2.[A] $\dfrac{dg}{dx}(2)$ for $g(x) = \dfrac{25}{9-x^2}$
$(-3 \le x \le 3, -1 \le y \le 10)$

♦ Type 1, predicting a derivative

3. $h'(1)$ for $h(x) = (3x+5)^{1/3}$
$(-1.5 \le x \le 6, -0.5 \le y \le 3.5)$

♦ Type 1, predicting a derivative

4.[A] $\dfrac{dL}{dx}(3)$ for $L(x) = \dfrac{4}{(x-1)^2}$
$(-2 \le x \le 6, -0.5 \le y \le 3)$

♦ Type 1, predicting a derivative

5. $p'(2)$ for $p(x) = 10 + (x^2 - 4)^2$
$(-4 \le x \le 5, -5 \le y \le 50)$

♦ Type 1, predicting a derivative

6. $q'(3)$ for $q(x) = \sqrt{25 - x^2}$
$(-6 \le x \le 6, -1 \le y \le 7)$

♦ Type 1, predicting a derivative

Find the derivatives in Problems 7 through 20.

7.[A] $\dfrac{d}{dx}(3x^4 - 6x^2 + 2)$

♦ Type 1, a derivative

8.[A] $\dfrac{d}{dx}(\frac{1}{4}x^4 - 4x^{-1/4})$

♦ Type 1, a derivative

9.[O] $\dfrac{d}{dt}\left[\left(\sqrt[3]{t}\right)^2 + \left(\sqrt{t}\right)^3\right]$

♦ Type 1, a derivative

10.[O] $\dfrac{dy}{dx}(2)$ for $y = \dfrac{3}{x} - \dfrac{4}{x^2}$

♦ Type 1, a derivative

11. $y'(x)$ for $y = 1 - x + (\sqrt{x})^5$

♦ Type 1, a derivative

12. $\dfrac{dy}{dx}$ for $y = x - 3/\sqrt{x}$

♦ Type 1, a derivative

13. $\dfrac{dP}{dt}(t)$ for $P(t) = \dfrac{1}{t} - \dfrac{2}{t^2} + \dfrac{3}{t^3}$

♦ Type 1, a derivative

14.[A] $y'(3)$ for $y = x^{-2} + 3x^2$

♦ Type 1, a derivative

15. $\dfrac{dy}{dx}(1)$ for $y = 1 + x + x^2 + x^3$

♦ Type 1, a derivative

16. $F'(2)$ for $F(x) = 6x^3 - x^{-1}$

♦ Type 1, a derivative

17.[A] $\left[\dfrac{d}{dx}(4x^4 - 4x^{-4})\right]_{x=1}$

♦ Type 1, a derivative

18. $\left[\dfrac{d}{dw}(3w^{4/3} - 6w^{-1/3})\right]_{w=8}$

♦ Type 1, a derivative

19.[O] $f'(10)$ where $f(x) = 3g(x) - 4h(x)$, $g'(10) = 7$, and $h'(10) = -3$

◆ Type 1, a derivative

20. $\left[\dfrac{d}{dt}(5V(t) + 6W(t) - 7Z(t))\right]_{t=0}$ where $V'(0) = -1, W'(0) = 4$, and $Z'(0) = 3$

◆ Type 1, a derivative

In the following problems give the units with which quantities are measured wherever appropriate. In Problems 21 through 26 use differentiation formulas to give exact answers and draw any required curves and lines without generating them on a calculator or computer.

21.[A] A ball rolling straight down a hill is $s = 1 + \frac{1}{2}t^2$ feet from a tree t seconds after it is released. What are **(a)** its average velocity for $0 \le t \le 2$ and **(b)** its instantaneous velocity at $t = 1$? **(c)** Draw the graph of $s(t)$ and the secant and tangent lines whose slopes are the average and instantaneous velocities from parts (a) and (b).

◆ Type 1, average and instantaneous velocity with a graph

22. The temperature in a forest is $T = 10 + \dfrac{16}{t^2}$ degrees Fahrenheit t hours after midnight for $1 \le t \le 5$. Which is greater (less negative), the average rate of change of T with respect to t for $1 \le t \le 4$ or the instantaneous rate of change of T with respect to t at $t = 2$? Draw the graph of $T(t)$ and the secant and tangent lines whose slopes are these average and instantaneous velocities. Plot the points on the curve at $t = 1, 2, 4, 5$.

◆ Type 1, average and instantaneous rates of change with a graph

23.[A] A tank contains $V(t) = 640 - 10t^3$ cubic meters of water at time t (hours) for $0 \le t \le 4$. **(a)** When is water flowing out of the tank at the rate of 270 cubic meters per hour? **(b)** Draw the graph of $V(t)$ with its tangent line at the time from part (a).

◆ Type 2, an instantaneous rate of change with a graph

24. **(a)** What is the rate of change of the volume $V(w) = w^3$ of a cube with respect to the width w of its sides at $w = 3$? Give the units with w measured in yards. **(b)** Are the average rates of change of the volume with respect to the width for $0 \le w \le 3$ and for $1 \le w \le 3$ less than or greater than the instantaneous rate of change from part (a)? **(c)** Illustrate the result of part (b) by drawing the curve $V = w^3$ with the relevant tangent line and two secant lines.

◆ Type 2, average and instantaneous rates of change with a graph

25. A car travels $s = 37t$ miles in t hours $(0 \le t \le 10)$. **(a)** Find its average velocity for $0 \le t \le 10$ and its instantaneous velocity at $t = 5$. **(b)** How are the results of part (a) related and why?

◆ Type 2, average and instantaneous velocity

26. If there is no air resistance, a ball that is dropped from a height of 490 meters is $490 - 4.9t^2$ meters above the ground t seconds later, until it hits the ground. **(a)** When does it hit the ground? **(b)** Draw the graph of the ball's height above the ground as a function of t. **(c)** What is the ball's average upward velocity as it falls? **(d)** What is its average downward velocity as it falls? **(e)** Is it falling faster three seconds or eight seconds after it is dropped? Justify your answer.

◆ Type 2, average and instantaneous velocity

27.[A] A walker has gone $s(t)$ (miles) past Reklaw, Texas, t hours after dawn one summer day. What do each of the following values tell about the hike: **(a)** $s(3) = 7$, **(b)** $\dfrac{ds}{dt}(3) = 3$, **(c)** $\dfrac{ds}{dt}(5) = -0.5$, and **(d)** $\dfrac{s(5) - s(0)}{5} = 2$?

◆ Type 2, average and instantaneous velocity

28.[A] The median price $N(t)$ (thousand dollars) of new houses and $E(t)$ (thousand dollars) of existing houses on the market in the U.S. in year t AD[3] satisfy **(a)** $N(1991) = N(1970) + 97$, **(b)** $N(1991) = 1.25E(1991)$, **(c)** $E(1991) = 5E(1970)$, **(d)** $\dfrac{dE}{dt}(1991) > 0$, and **(e)** $\dfrac{dN}{dt}(1991) = -0.9\dfrac{dE}{dt}(1991)$. Rephrase each of these equations as an English sentence without using the symbols $N(t)$ and $T(t)$ for the functions.

◆ Type 4, change, relative change and instantaneous rates of change

---

[3] Data adapted from *1993 Statistical Abstract of the United States*, Washington, DC: U.S. Department of Commerce, 1993, p. 710.

29. The amount of tar $T(t)$ (milligrams) and the amount of nicotine $N(t)$ (milligrams) in an average cigarette that was manufactured in the United States in year t AD has the properties[(4)] **(a)** $T(1957) = 38$, **(b)** $\frac{dT}{dt}(1957) = -6$, **(c)** $N(1980) - N(1957) = 24$, and **(d)** $\frac{dN}{dt}(1980) = \frac{1}{3}\frac{dN}{dt}(1957)$. Rephrase each of these equations as an English sentence without using the symbols $T(t)$ and $N(t)$ for the functions.

♦ Type 4, change, relative change and instantaneous rates of change

30. A small airplane whose engine is running at 2500 revolutions per minute has air speed $v(h)$ nautical miles per hour (knots) and uses $g(h)$ gallons of gasoline per hour when it is at an altitude of h thousand feet above the ground.[(5)] Express the following statements with equations involving $v(h), g(h)$, and their derivatives, under the assumption that the plane's engine is running at 2500 revolutions per minute. **(a^A)** The Cessna's air speed is 116 nautical miles per hour when it is at an altitude of 4000 feet. **(b^A)** The Cessna's air speed decreases at the rate of 0.5 knots per thousand feet and its rate of gasoline consumption decreases 0.2 gallons per hour per thousand feet when the plane is at an altitude of 6000 feet. **(c)** The rate at which the Cessna uses gasoline at an altitude of 12,000 feet is 81% of the rate at 4000 feet.

♦ Type 4, change, relative change and instantaneous rates of change

31. In a study of vehicle sales in the United States, passenger vehicles were classified into two groups, trucks (including pickups, mini-vans, and sport-utility vehicles) and cars.[(6)] **(a)** Twenty percent of the passenger vehicles that were sold were trucks in 1970. Thirty-eight percent were trucks in 1992. The percent of passenger vehicles sold that were trucks was increasing 0.4 percent per year in 1992. Express these three statements as equations involving values of the percent $T(t)$ of passenger vehicles sold in year t AD that were trucks, the percent $C(t)$ that were cars, and their derivatives. **(b)** What are the values of $C(1970), C(1992)$, and $C'(1992)$?

♦ Type 2, percent and p[ercent change

In Problems 32 through 37 find exact equations of the tangent lines to the curves at the given values of x. Then generate the curves and the tangent lines together in the indicated windows and copy them on your paper.

32.^A The tangent line to $y = 4 - x^2$ at $x = 1$ $(-3 \le x \le 3, -5 \le y \le 8)$

♦ Type 1, tangent lines

33.^A The tangent line to $y = 53 + 7x^{10}$ at $x = 1$ $(-0.5 \le x \le 2, -10 \le y \le 130)$

♦ Type 1, tangent lines

34. The tangent line to $y = x + x^2 + x^3$ at $x = -2$ $(-4 \le x \le 4, -20 \le y \le 20)$

♦ Type 1, tangent lines

35. The tangent line to $y = x^3 - 4x$ at $x = 2$ $(-3 \le x \le 4, -6 \le y \le 6)$

♦ Type 1, tangent lines

36. The tangent line to $y = \frac{2}{\sqrt{x}}$ at $x = 1$ $(-0.5 \le x \le 3, -0.5 \le y \le 5)$

♦ Type 1, tangent lines

37. The tangent line to $y = x^2 + 27/x^2$ at $x = 3$ $(-5 \le x \le 5, -5 \le y \le 30)$

♦ Type 1, tangent lines

Find the derivatives in Problems 38 through 45 by first rewriting the formulas so that Theorems 1 and 2 can be applied.

38.º $\frac{d}{dx}\left(\frac{\sqrt{x} - 3}{x}\right)$

♦ Type 3, a derivative

39.º $\frac{d}{dx}\sqrt{\frac{x^{3/4}}{x^{2/3}}}$

♦ Type 3, a derivative

40.^A $\frac{d}{dx}[(x+3)^2]$

♦ Type 3, a derivative

41. $\frac{d}{dy}[y^2(2y^3 + 3y^2 - 4)]$

♦ Type 3, a derivative

---

(4) Data adapted from *Drugs Society and Human Behavior*, by O. Ray and C. Ksir, Saint Louis, MO: Mosby College Publishing, 1990, p.198.

(5) Data adapted from *Cessna Skyhawk Information Manual*, Wichita, KA: Cessna Aircraft Company, 1978, pp. 5-16.

(6) Data adapted from *The New York Times*, August 4, 1993, New York, NY: The New York Times Company, p. C1. Source: Ford Motor Company.

42.[A] $\dfrac{d}{dx}[(x+2)(3x-4)]$

♦ Type 3, a derivative

43. $\dfrac{d}{dx}\left[\dfrac{2+3x^{30}-5x^{40}}{x^{50}}\right]$

♦ Type 3, a derivative

44. $\dfrac{d}{dx}[x^{1/7}(x^{1/6}-x^{1/5})]$

♦ Type 3, a derivative

45. $\dfrac{d}{dt}\left[\left(\dfrac{t-1}{t}\right)^2\right]$

♦ Type 3, a derivative

46. During a compression stroke in a diesel engine, the pressure p and volume of air V in a cylinder are related by $p = 4000V^{-1.4}$, where the pressure is measured in pounds per square inch and the volume in cubic inches.[(7)] What is the rate of change of the pressure with respect to the volume when the volume is 50 cubic inches? What does the fact that this derivative is negative indicate about the engine?

♦ Type 2, an instantaneous rate of change

47. If a tin rod is 10 meters long at a temperature of 0°C, then its length at a temperature T between 0° and 90° is[(8)]

$$L = 10 + (2.093\times10^{-5})T + (1.75\times10^{-8})T^2 \text{ meters}$$

(a) What is the rate of change of the length with respect to the temperature at a temperature of 20°C? **(b)** Does dL/dT increase or decrease as T increases?

♦ Type 2, an instantaneous rate of change

48. Two positive charges, of one electrostatic unit each, exert a repulsive force of $F = 1/r^2$ dynes on each other when they are r centimeters apart. **(a)** What is the average rate of change of the force with respect to the distance for $5 \le r \le 100$? **(b)** What is the instantaneous rate of change of the force with respect to distance when the charges are 10 centimeters apart? **(c)** How far apart are the charges when dF/dr equals -0.002 dynes per centimeter?

♦ Type 2, an instantaneous rate of change

49. A rock that is propelled straight up into the air at time $t = 0$ (seconds) with no air resistance is $4 + 39.2t - 4.9t^2$ meters above the ground at time t, until it hits the ground. **(a)** From what height is it propelled and with what velocity? **(b)** When is its velocity zero? **(c)** How high does it go?

♦ Type 2, instantaneous velocity

50. An object is at $s = 100 + 24t - 2t^3$ (feet) on an s-axis at time t. **(a)** When is its velocity zero? **(b)** During what time interval is its velocity positive?

♦ Type 2, instantaneous velocity

51. A car is $60 + 120t - 80t^{3/2}$ miles east of a town t hours after noon. **(a)** How fast is it traveling and in what direction is it traveling at 12:15? **(b)** When is its velocity zero and how far is it from the town at that time?

♦ Type 2, instantaneous velocity

52.[A] California's environmental and services industries, which cover work from trash recycling to electric car research, had a total revenue of 17.0 billion dollars at the beginning of 1990 and 18.3 billion dollars at the beginning of 1993.[(9)] Use this data to estimate the revenue and rate of change with respect to time of the revenue of these industries at the beginning of 1992.

♦ Type 2, a tangent-line approximation

[(7)] The mathematical model $p = CV^{-1.4}$ is used with the assumption that no heat enters or leaves the cylinder as the piston is compressed. Such a process with no heat transfer is called ADIABATIC. (*Physical Chemistry* by G. Castillan, Reading, MA: Addison Wesley Publishing Company, 1971, pp. 130–132.)

[(8)] *Handbook of Chemistry and Physics*, Cleveland, OH: The Chemical Rubber Publishing Company, 1960, p. 2246.

[(9)] Data adapted from from *Los Angeles Times*, April 22, 1995, Los Angeles, CA: The Times Mirror Company, 1995 Source: Environmental Business International.

53. At the beginning of 1990 the population of Nevada was 1.28 million and was increasing at the rate of 44 thousand per year.[10] What was the approximate population of Nevada in the middle of 1989?

◆ Type 2, a tangent-line approximation

54. **(a)** Show that the rate of change with respect to the radius of the area of a circle is its circumference, and that the rate of change with respect to the radius of the volume of a sphere is its surface area. Explain why these are plausible results. **(b)** What are the rates of change with respect to the width of the area of a square and of the volume of a cube? Why are they not equal to the perimeter of the square and the surface area of the cube, respectively, as the results of part (a) might suggest?

◆ Type 4, rates of change

55. Find numbers a and b such that the curves $y = \sqrt{x}$ and $y = ax^2 + b$ intersect and are tangent (have the same tangent line) at (1,1).

◆ Type 3, tangent curves

56. Two curves are ORTHOGONAL at a point where they intersect if their tangent lines at that point are perpendicular. Find a constant n such that $y = \frac{1}{2}x^{-2}$ and $y = \frac{1}{2}x^n$ are orthogonal at $(1, \pm\frac{1}{2})$. Then draw the curves in an xy-plane with equal scales on the axes.

◆ Type 3, orthogonal curves

Apply the definition $f'(x) = \lim_{b \to x} \dfrac{f(b) - f(x)}{b - x}$ directly to find the derivatives in Problems 57 through 60.

57.[A] $g'(1)$ for $g(x) = x^2 + 2x$

◆ Type 3, finding a derivative with the definition

58.[A] $G'(4)$ for $G(x) = 1/(5 - 2x)$

◆ Type 3, finding a derivative with the definition

59. $P'(0)$ for $P(x) = (x + 4)/(x + 2)$

◆ Type 3, finding a derivative with the definition

60. $f'(3)$ for $f(x) = 2x + 1/x$

◆ Type 3, finding a derivative with the definition

Use the definition $\dfrac{df}{dx}(x) = \lim_{\Delta x \to 0} \dfrac{f(x + \Delta x) - f(x)}{\Delta x}$ to find the derivatives in Problems 61 through 64.

61.[A] $\dfrac{d}{dx}(2 - 3x^{-2})$

◆ Type 3, finding a derivative with the definition

62.[A] $\dfrac{dy}{dx}(0)$ for $y(x) = x^{4/3}$

◆ Type 3, finding a derivative with the definition

63. $\dfrac{dy}{dx}$ for $y = 8 - 5x$

◆ Type 3, finding a derivative with the definition

64. $\dfrac{dy}{dx}$ for $y = (5 + x)^2$

◆ Type 3, finding a derivative with the definition

65. Find the derivative of $f(x) = \sqrt{x}$ with the definition before Problem 57 by using the substitutions $\sqrt{x} = X$ and $\sqrt{b} = B$ and letting $B \to X$.

◆ Type 3, finding a derivative with the definition

66. Use the quadratic formula to determine the two values of b such that the parabola $y = x^2 + 1$ and the line $y = 2bx$ have exactly one intersection. Then show that $y = 2bx$ is a tangent line to $y = x^2 + 1$ for these two values of b.

◆ Type 4, tangent line

67.[O] How are df and dx related at $x = 10$ if $f(x) = x^3 + 3x^2$?

◆ Type 3, differentials

68.[A] How are dw and dz related at $z = 10$ if $w = 18z^{-1} - 27z^{-2}$?

◆ Type 3, differentials

69.[A] By weighing a square plate of known density, it is determined that its area is 100 square centimeters with an error ≤ 0.05 square centimeters. **(a)** What is the width of the plate if its area is exactly 100 square centimeters? **(b)** Use differentials to estimate the maximum error in the answer to part (a).

◆ Type 3, differentials and an error estimate

70. Exactly ten gallons of soup is measured to weigh 90 pounds with an error ≤ 0.1 pound. **(a)** What is the density ρ of the soup, measured in pounds per gallon, if it weighs exactly 90 pounds? **(b)** Use differentials to estimate the maximum error in the answer to part (a).

◆ Type 3, differentials and an error estimate

---

[10] Data adapted from *1993 Statistical Abstract of the United States*, Washington, DC: U.S. Department of Commerce, 1993, p. 28.

71. A quantity z is measured to be 6 with an error ≤ 0.001, and this measurement is used to calculate another quantity w, where $w = w(z)$ is a function such that $w(6) = 300$ and $\frac{dw}{dz}(6) = 4$. Use differentials to estimate the maximum error in the calculated value of w.

♦ Type 3, differentials and an error estimate

72.[A] Give a formula for dV in terms of r and dr, where V is the volume of a sphere of radius r.

♦ Type 3, differentials

73. **(a[A])** Use differentials to estimate the approximate relative change $\Delta V/V$ in the volume of a sphere that corresponds to the relative change $\Delta r/r$ in its radius. **(b)** Use the formula from part (a) to determine the approximate percent change in the volume that would result in a 6% change in the radius.

♦ Type 3, differentials and error estimates

74. Use differentials to relate the approximate relative change $\Delta A/A$ in the area of an isosceles right triangle to the corresponding relative change $\Delta s/s$ in the lengths of its legs.

♦ Type 3, differentials and error estimates

75. According to Newton's law of gravity, the force of attraction between an object of mass m and an object of mass M is $GmMr^{-2}$, where r is the distance between the objects and G is a constant whose value depends only on the units in which mass, distance, and force are measured. The sun's mass is 2.7×10^7 times that of the moon, and the sun's distance from the earth is 390 times that of the moon. **(a)** How many times greater is the gravitational attraction between an object on the earth and the sun than between the same object and the moon? **(b)** Tides are caused because the gravitational attraction by the sun and the moon on ocean water at the side of the earth closest to the sun or moon is greater than that on water on the far side. Use differentials to explain why the moon has a greater effect on tides than the sun, by comparing the difference in the gravitational attraction by the sun at the near and far sides of the earth to the attraction by the moon on the near and far sides of the earth.

♦ Type 3, differentials and error estimates

Section 2.4

Derivatives as functions and estimating derivatives

OVERVIEW: *In this section we study more examples of derivatives as rates of change and find approximate derivatives from graphs and tables.*

Topics:

- ***The derivative function***
- ***Estimating derivatives from graphs***
- ***Derivatives of piecewise linear functions; step functions***
- ***The term "marginal" in economics***
- ***Estimating derivatives from tables***
- ***The second and higher-order derivatives; acceleration***
- ***Derivatives and continuity***

Filling and draining a tank

Imagine that a tank contains $V(t) = 12t - t^2$ gallons of water at time t (minutes) for $0 \le t \le 12$. As can be seen from the graph of $V(t)$ in the tV-plane of Figure 1, the tank is empty at $t = 0$, water is added for six minutes and then drained out, and the tank is empty again at $t = 12$.

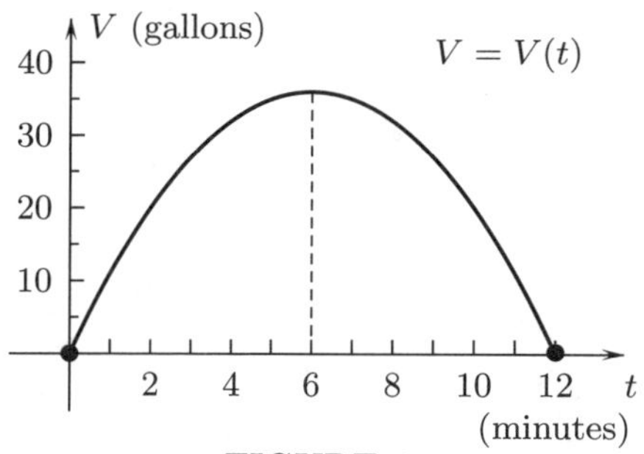

FIGURE 1

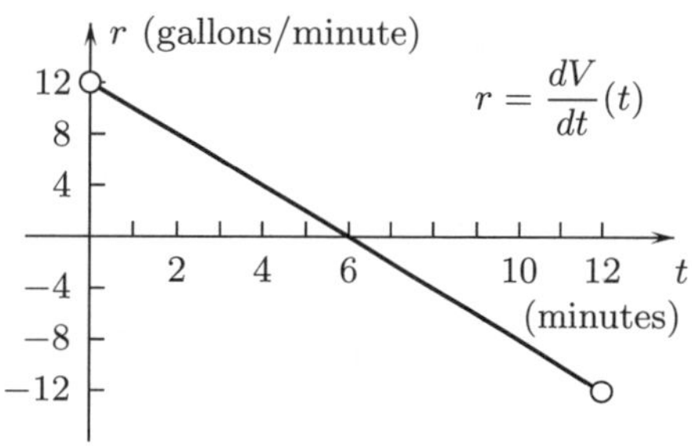

FIGURE 2

Example 1 What is the rate of change with respect to time of the volume of water in the tank at $t = 2$, at $t = 6$, and at $t = 10$?

SOLUTION The rate of change of the volume is the derivative, $\frac{dV}{dt}(t) = \frac{d}{dt}(12t - t^2) = 12 - 2t$. Therefore, $\frac{dV}{dt}(2) = 12 - 2(2) = 8$ gallons per minute, $\frac{dV}{dt}(6) = 12 - 2(6) = 0$ gallons per minute, and $\frac{dV}{dt}(10) = 12 - 2(10) = -8$ gallons per minute. Water is flowing into the tank at the rate of 8 gallons per minute at $t = 2$, water is neither flowing in nor out at $t = 6$, and water is flowing out at the rate of 8 gallons per minute at $t = 10$. □

The derivative $\frac{dV}{dt}(t)$ is defined for $0 < t < 12$. Its graph in Figure 2 is the portion of the line $r = 12 - 2t$ for $0 < t < 12$. The r-coordinate of each point on it is the slope of the tangent line at the corresponding point on the graph of $V(t)$, as is shown in Figures 3 and 4 for $t = 2, 6$, and 10.

◆ SUGGESTIONS TO INSTRUCTORS: Have students begin work on Questions 1 through 6 before class. Problems 4–10, 13, 16, 24, 39, 42, and 43 are good for in-class discussion.

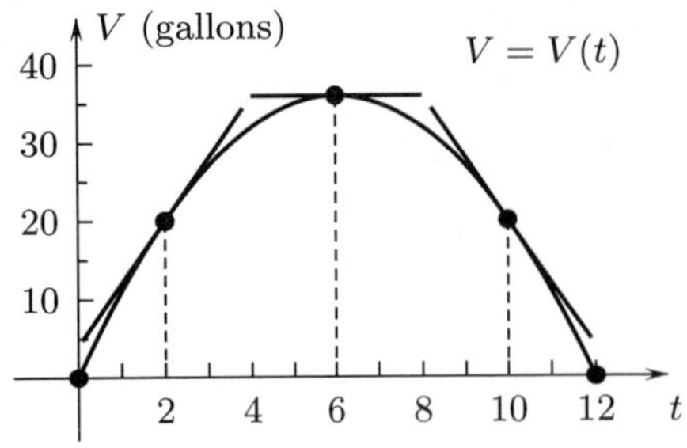

FIGURE 3

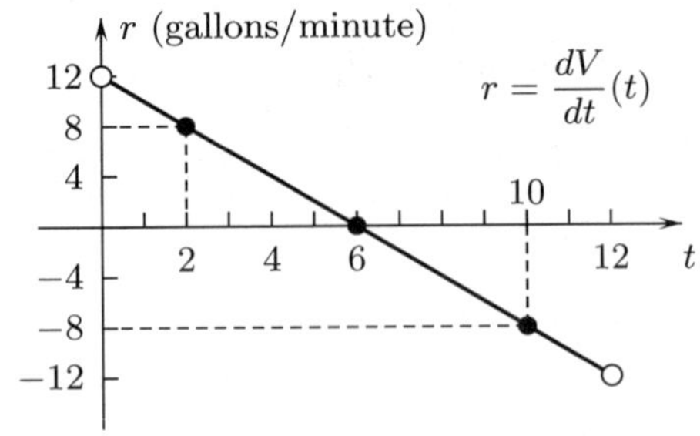

FIGURE 4

Finding approximate derivatives from graphs

If you are given a function only by a drawing of its graph, rather than by an exact formula, you cannot find exact values for its derivative. You can, however, find approximate values of the derivative, as in the next example, by drawing plausible tangent lines to the graph and determining their approximate slopes by estimating the coordinates of pairs of points on them.

Example 2 Figure 5 shows the graph of an object's position $s(t)$ on an s-axis as a function of the time t. What is the object's approximate velocity in the positive s-direction at $t = 2$ and at $t = 6$?

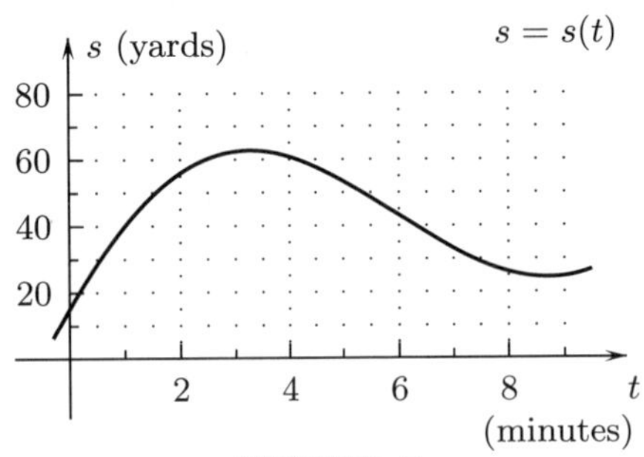

FIGURE 5

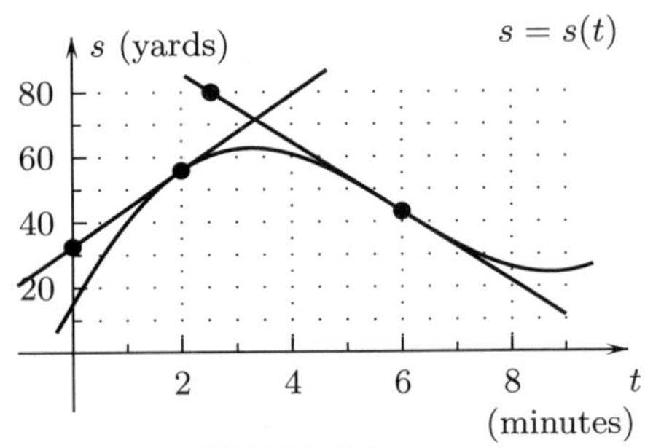

FIGURE 6

SOLUTION We draw plausible tangent lines at $t = 2$ and $t = 6$, as in Figure 6. Their slopes are the approximate velocity at $t = 2$ and $t = 6$. The points shown by dots on the tangent line at $t = 2$ have approximate coordinates $(0, 32)$ and $(2, 55)$, so that

$$[\text{Velocity at } t = 2] = \frac{ds}{dt}(2) \approx \frac{55 - 32}{2 - 0} = 11.5 \text{ yards per minute.}$$

The points on the tangent line at $t = 6$ in Figure 6 have approximate coordinates $(2.5, 80)$ and $(6, 43)$. Therefore,

$$[\text{Velocity at } t = 6] = \frac{ds}{dt}(6) \approx \frac{43 - 80}{6 - 2.5} \doteq -10.6 \text{ yards per minute.} \ \square$$

Stress causes strain

Vertebra in the human spine are separated by fibrous elastic disks that compress when the load on them increases. The load on the disk is called the STRESS, and the percent of compression is called the STRAIN. The function $S(L)$ in Figure 7 gives the strain S of lower back (lumbar) vertebral disks as a function of the stress L.[(1)]

[(1)]Data adapted from *Physics with Illustrative Examples from Medicine and Biology*, Vol. 1, G. Benedeck and F. Villars, Reading, Massachusetts: Addison-Wesley, 1973, p.343.

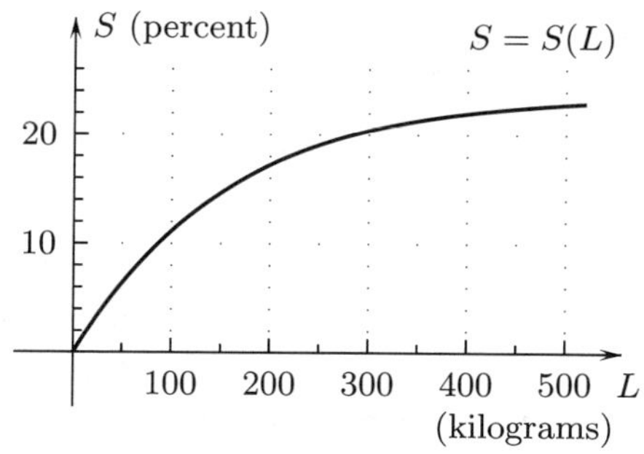

FIGURE 7

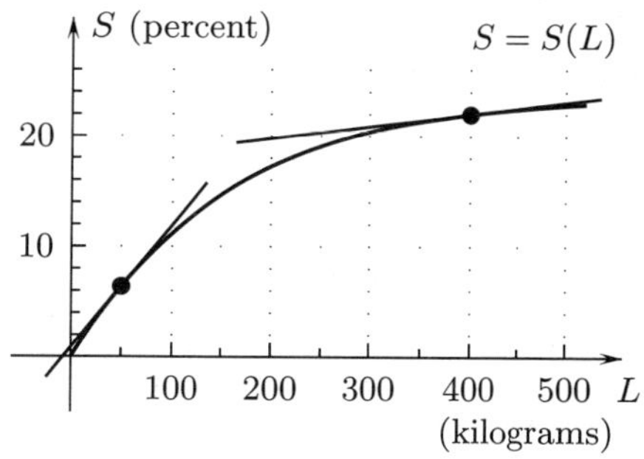

FIGURE 8

Question 1 **(a)** $S(L)$ increases as L increases. What does this mean for the disks in the spine? **(b)** Estimate the rate of change dS/dL for $L = 50$ and $L = 400$ by estimating the slopes of the tangent lines in Figure 8. **(c)** Does dS/dL increase or decrease as L increases? **(d)** Use the information from parts (b) and (c) to sketch the graph of dS/dL.

Income tax: a piecewise linear function

The country of Retro has a regressive tax structure. Each resident pays 20% tax on the first 1000 donas (the local currency) of his or her taxable income, but only 10% on the next 1000 donas, and 5% on any taxable income over 2000 donas. Therefore, the tax $T(x)$ on taxable income x is given by

$$T(x) = \begin{cases} 0.2x & \text{for} \quad 0 \le x \le 1000 \\ 200 + 0.1(x - 1000) & \text{for} \quad 1000 < x \le 2000 \\ 300 + 0.05(x - 2000) & \text{for} \quad x > 2000. \end{cases} \tag{1}$$

$T(x)$ is called PIECEWISE LINEAR because its graph in Figure 11 is formed from segments of lines.

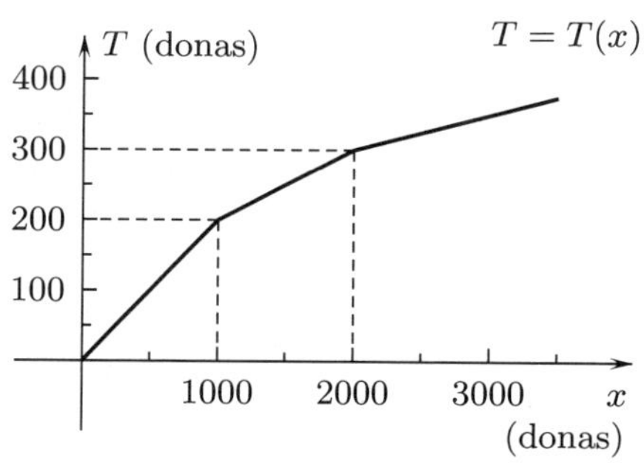

FIGURE 9

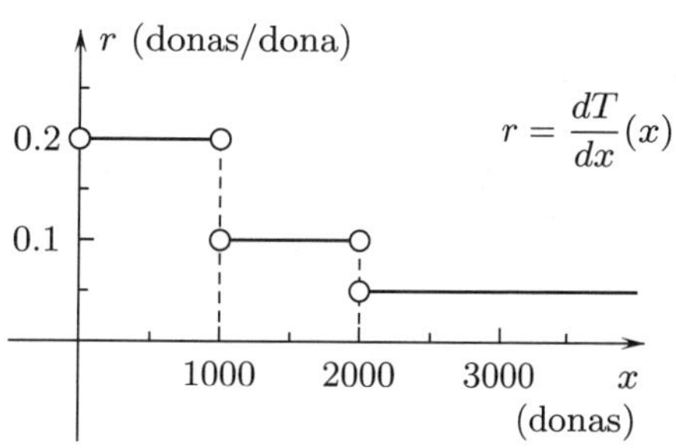

FIGURE 10

Question 2 What is the value of $\dfrac{dT}{dx}(x)$ for $0 < x < 1000$, for $1000 < x < 2000$, and for $x > 2000$?

The derivative dT/dx is called a STEP FUNCTION because its graph in Figure 10 consists of pieces of horizontal lines. Small circles have been drawn at the ends of the lines to emphasize the fact that dT/dx is not defined at $x = 0, x = 1000$, or $x = 2000$. The derivative is not defined at $x = 0$ because $T(x)$ is not defined for $x < 0$. It is not defined at $x = 1000$ because it would be the two-sided limit $\lim_{b \to 1000} \dfrac{T(b) - T(1000)}{b - 1000}$. This limit does not exist because $\dfrac{T(b) - T(1000)}{b - 1000}$ equals 0.2 for $0 < b < 1000$ and equals 0.1 for $1000 < b < 2000$. The derivative is not defined at $t = 2000$ because it would be the two-sided limit of $\dfrac{T(b) - T(2000)}{b - 2000}$ as $b \to 2000$ but $\dfrac{T(b) - T(2000)}{b - 2000}$ equals 0.1 for $1000 < b < 2000$ and equals 0.05 for $b > 2000$.

Marginal quantities in economics

Economists often use the word "marginal" for "derivative of," as in the statement

$$[\text{Marginal } f(x)] = \frac{df}{dx}(x). \quad \textbf{(2)}$$

With this definition applied to the tax function of Figure 9, the marginal tax is given by the derivative from Question 2 and Figure 10. The marginal tax is 0.20 (20%) if the taxable income is less than 1000 donas, is 0.1 (10%) on taxable incomes between 1000 and 2000 donas, and is 0.05 (5%) on taxable incomes greater than 2000 donas. In each case the marginal tax is the tax rate on the highest portion of the taxable income.

Economists also use the term "marginal" for the change in a function when its variable is increased by 1:

$$[\text{Marginal } f(x)] = [f(x+1) - f(x)]. \quad \textbf{(3)}$$

Notice that the quantity in **(3)** is the average rate of change of $f(x)$ from x to $x+1$:

$$f(x+1) - f(x) = \frac{f(x+1) - f(x)}{1}.$$

The definitions in **(2)** and **(3)** give the same number if the function is linear in an open interval containing x and $x+1$, since then the derivative and average rate of change are equal. In particular, they give the same marginal tax on incomes in the country of Retro for integer values of $x \neq 0, 1000$, or 2000.

For a nonlinear $f(x)$, the numbers given by **(2)** and **(3)** are generally different, but are often approximately equal because the average rate of change is close to the derivative.

Question 3 A company makes $R(x) = 2.50x + 0.04x^2$ dollars revenue if they sell x quartz healing crystals in a month. **(a)** Use **(2)** to calculate the marginal revenue when 100 crystals are sold. **(b)** Calculate the marginal revenue on 100 crystals using **(3)**, and compare the result with that from part (a).

Estimating a derivative from a table: Catch a breeze

The next table lists the wind speed at three-hour intervals during a spring day in Dodge City, Kansas.[(2)] The time t is measured in hours with $t = 0$ at midnight. We want to estimate the rate of change $\frac{dW}{dt}$ of the wind's velocity with respect to time at 9:00 AM ($t = 9$).

WIND VELOCITY $W(t)$ (MILES PER HOUR) IN DODGE CITY, KANSAS

t	0	3	6	9	12	15	18	21	24
$W(t)$	12.55	12.40	12.57	15.32	16.00	15.66	14.92	11.30	12.91

Figures 11 through 13, which show the data from the table, illustrate three techniques for estimating the derivative $\frac{dW}{dt}(9)$. We could use the secant line in Figure 11, whose slope is the average rate of change of $W(t)$ with respect to t for $9 \leq t \leq 12$; we could use the secant line in Figure 12, whose slope is the average rate of change for $6 \leq t \leq 9$; or we could use the secant line in Figure 13, whose slope is the average rate of change for $6 \leq t \leq 12$.

[(2)]Data from *Wind Energy Systems*, G. L. Johnson, Englewood Cliffs, N. J: Prentice Hall International, Inc., 1985, p. 47.

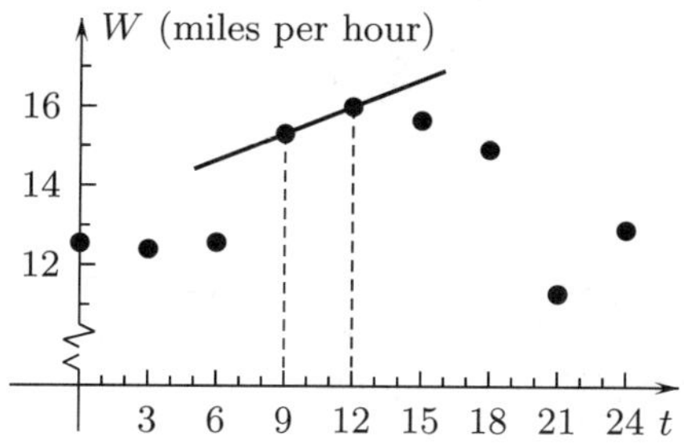

FIGURE 11

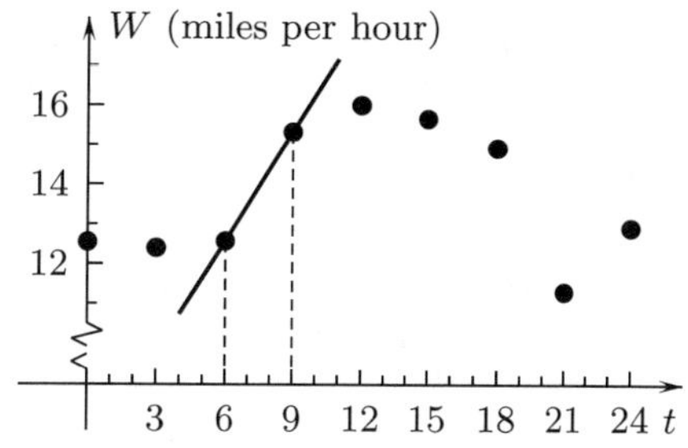

FIGURE 12

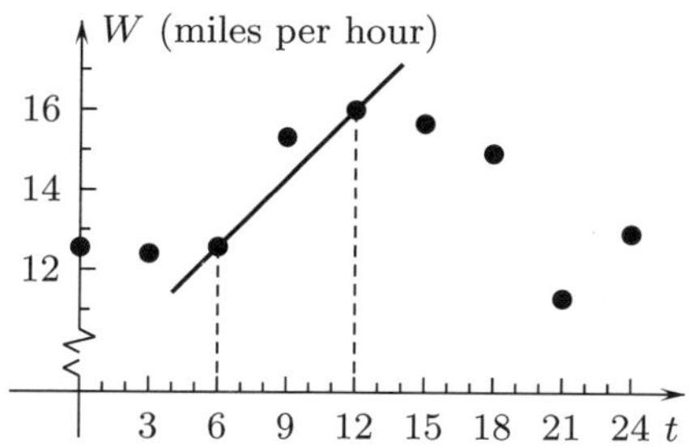

FIGURE 13

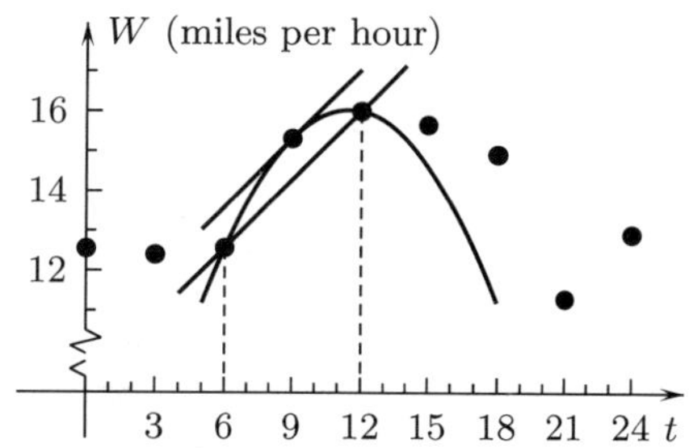

FIGURE 14

Question 4 Give three estimates of $\frac{dW}{dt}(9)$ by finding the slopes of the secant lines **(a)** in Figure 11, **(b)** in Figure 12, and **(c)** in Figure 13. **(d)** Show that the answer to part (c) is the average of the answers to (a) and (b) with the same Δx.

The slopes of the secant lines in Figures 11 through 13 are called a RIGHT-DIFFERENCE QUOTIENT, a LEFT-DIFFERENCE QUOTIENT, and a CENTERED-DIFFERENCE QUOTIENT, respectively. These can be used to approximate a derivative of any function: If $f(x)$ has a derivative at x_0, then for positive Δx,

$$f'(x_0) \approx \frac{f(x_0 + \Delta x) - f(x_0)}{\Delta x} \quad \text{(a right-difference quotient)} \tag{4}$$

$$f'(x_0) \approx \frac{f(x_0) - f(x_0 - \Delta x)}{\Delta x} \quad \text{(a left-difference quotient)} \tag{5}$$

$$f'(x_0) \approx \frac{f(x_0 + \Delta x) - f(x_0 - \Delta x)}{2\Delta x} \quad \text{(a centered-difference quotient).} \tag{6}$$

As was illustrated in Question 4d, the centered difference quotient is the average of the left and right difference quotients with the same Δx. You can see that this is always the case by adding the formulas on the right of **(4)** and **(5)** and dividing the result by 2.

The centered difference quotient is also the slope of the tangent line at $x = x_0$ to the approximation of the graph of $f(x)$ that is obtained by using the parabola that passes through the points at $x = x_0 - \Delta x$, $x = x_0$, and $x = x_0 + \Delta x$ on it. This is the case because the tangent line to the parabola at $x = x_0$ is parallel to the secant line through the points at $x = x_0 - \Delta x$ and $x = x_0 + \Delta x$, as can be seen in Figure 14 for the function $W(t)$ of Question 4.[†]

[†]A short calculation shows that for a function $g(x) = Ax^2 + Bx + C$ whose graph is a parabola, the centered-difference quotient $\frac{g(x_0 + \Delta x) - g(x_0 - \Delta x)}{2\Delta x}$ equals $2Ax_0 + B$, which is $g'(x_0)$.

If we want to estimate the derivative at a value of the variable between the values listed in a table, we can use the slope of the secant line through points on both sides. For example, we can estimate the rate of change of the wind velocity at 10:00 ($t = 10$) from the table above by using the slope 0.23 miles per hour per hour of the secant line in Figure 11 through the points at $t = 9$ and $t = 12$ that you calculated in Question 4a.

Using a derivative to study a function: a sunflower

Now we will use values of a derivative to help us sketch the graph of a function. Imagine that $h(t)$ is the height, measured in centimeters, of a sunflower t days after it begins to grow, that you have the graph of the derivative dh/dt in Figure 15, and that you know how tall the sunflower is after it has grown 10, 35, and 60 days, as given by the three dots in Figure 16.[(3)] Since the values of the derivative are the slopes of the tangent lines to the graph of the function, you can draw tangent lines at the dots and then draw the graph of $h(t)$ to fit them.

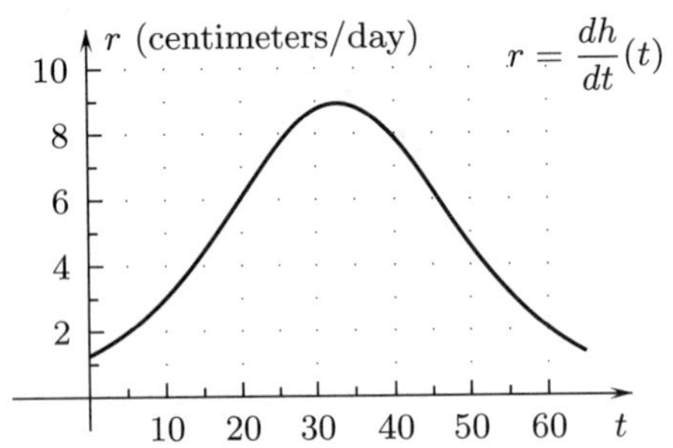

FIGURE 15

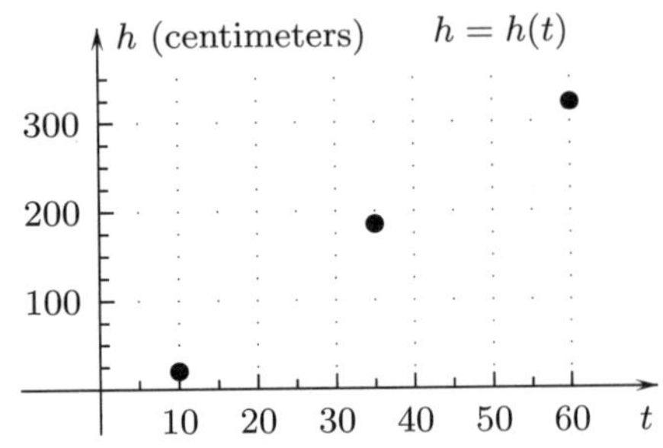

FIGURE 16

Question 5 Find the approximate values of dh/dt at $t = 10, 35$, and 60 from Figure 15 and draw the tangent lines with these slopes at the corresponding points from Figure 16. Then draw a plausible graph of $h(t)$ in Figure 16, with the tangent lines as a guide.

Higher-order derivatives

The second derivative of $f(x)$ is the derivative of its first derivative and is denoted $f''(x)$ or $\frac{d^2 f}{dx^2}(x)$. The third derivative is the derivative of the second derivative and is denoted $f^{(3)}(x)$ or $\frac{d^3 f}{dx^3}(x)$. Higher derivatives are defined similarly. The nth derivative is also called the derivative of ORDER n.

Example 3 What are the first, second, and third derivative of $f(x) = x^5$?

SOLUTION $f'(x) = \frac{d}{dx}(x^5) = 5x^4$, $f''(x) = \frac{d}{dx}(5x^4) = 20x^3$, and $f^{(3)}(x) = \frac{d}{dx}(20x^3) = 60x^2$ □

Acceleration

As we have seen before, if an object's position on an s-axis at time t is given by the function $s(t)$, then the derivative $v(t) = \frac{ds}{dt}(t)$ is the object's velocity in the positive s-direction at time t. The derivative of the velocity, which is the second derivative of the position function, is the object's ACCELERATION in the positive s-direction:

$$a(t) = \frac{dv}{dt}(t) = \frac{d^2 s}{dt^2}(t).$$

Question 6 A car is $s = 10 + 50t - t^3$ miles past a truck stop at the time t (hours) for $0 \le t \le 4$. **(a)** Where is it, **(b)** how fast is it going, and **(c)** how rapidly is it accelerating or decelerating at $t = 2$?

(3) Data adapted from *Technology and Environment*, by J. Ausbel and H. Sladovich, Washington, DC: National Academy of Sciences, 1989, p. 74.

You can see from the graph of $s = 10 + 50t - t^3$ in Figure 17 that the car in Question 6 has positive velocity for $0 < t < 4$ but slows down as t increases from 0 to 4. This can also be seen from the graph of the velocity $v = 50 - 3t^2$ in Figure 18: the velocity is positive for $0 \le t < 4$ and decreases as t increases form 0 to 4. Figure 19 shows the graph of the acceleration $a(t) = -6t$. The car is slowing down for $0 < t < 4$ because the velocity is positive and the acceleration is negative during that time..

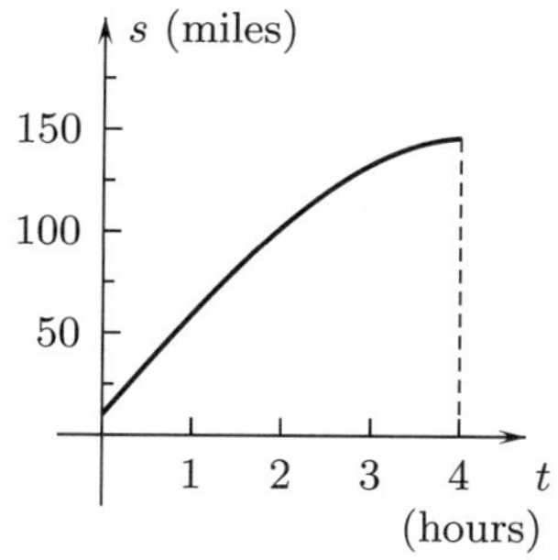

$s = 10 + 50t - t^3$

FIGURE 17

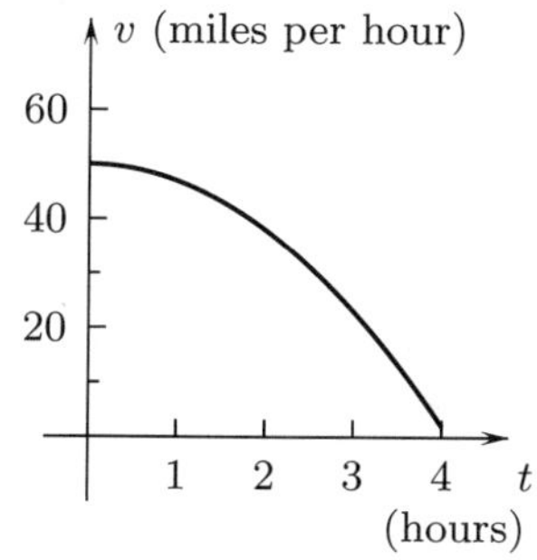

$v = 50 - 3t^2$

FIGURE 18

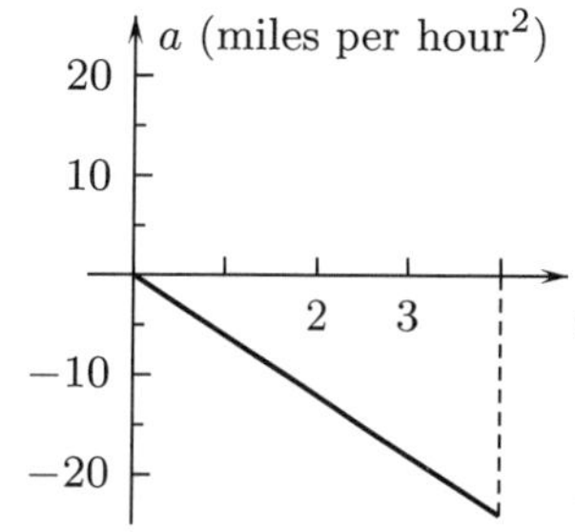

$a = -6t$

FIGURE 19

When was that?

If we are given the graph of a function and we want to determine where its derivative has a specified value, we can look for places on the graph where the tangent line has the given slope.

Example 4 The graph of the upward velocity $v(t)$ of a balloon is shown in Figure 20. Find the approximate values of two times when the balloon's upward acceleration is 20 feet per minute^2.

SOLUTION We draw a line of slope 20, as at the lower right in Figure 21, and use a lined, transparent ruler to determine where the tangent line to the graph of $v(t)$ is parallel to that line. As can be seen in Figure 21, the acceleration dv/dt is 20 feet per minute at $t \approx 2.4$ and $t \approx 7.6$. □

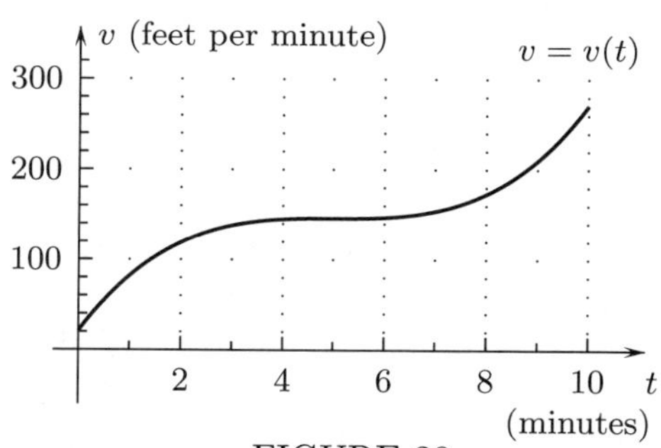

FIGURE 20

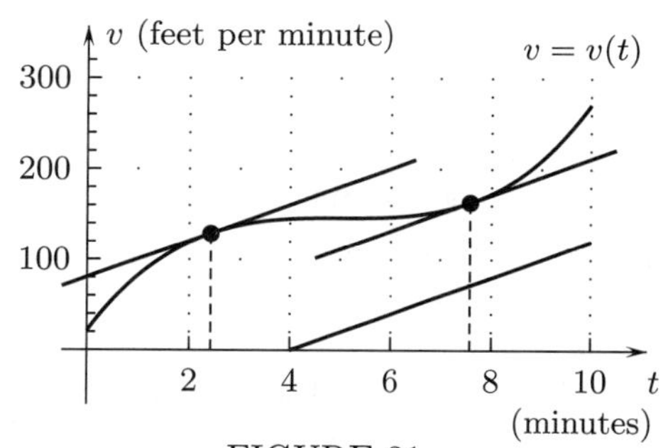

FIGURE 21

Derivatives and continuity

The function

$$F(x) = \begin{cases} x^2 & \text{for} \quad x < 1 \\ 2 & \text{for} \quad x \ge 1 \end{cases}$$

of Figure 22 does not have a derivative at $x = 1$ because it is discontinuous at that point. To see why this is the case, look at the secant line in Figure 23 through the points at $x = 1$ and $x = b$ on the graph with $b < 1$. If $F'(1)$ existed, it would be the limit of the slope of this secant line as $b \to 1^-$. However, as you can see from Figure 23, the slope of the secant line tends to ∞ as $b \to 1^-$.

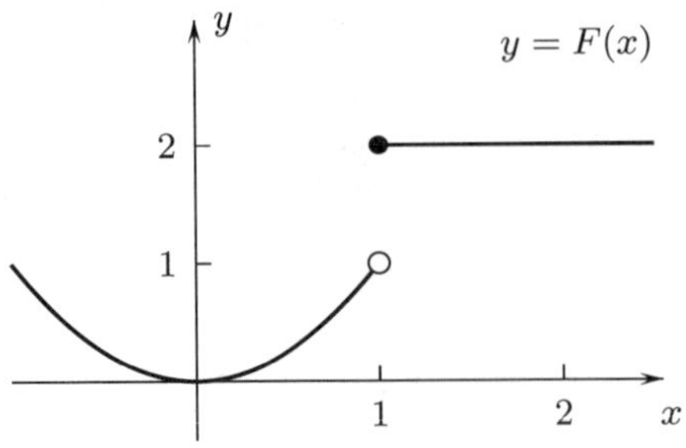

FIGURE 22

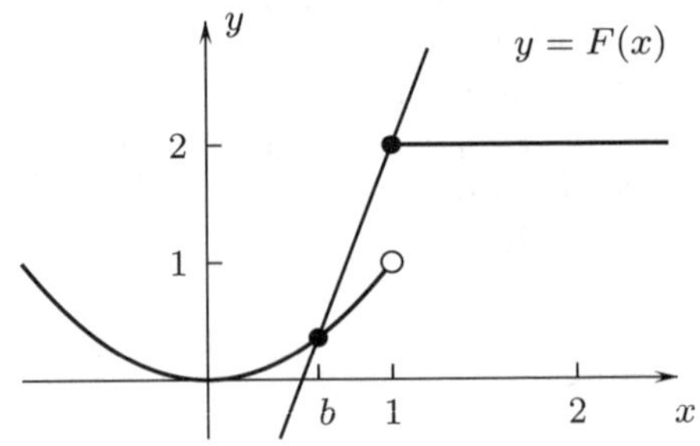

FIGURE 23

Rather than proving that no function can have a derivative at a point where it is not continuous, it is easier to prove the following equivalent result.

Theorem 1 *If $f(x)$ has a derivative at $x = x_0$, then $f(x)$ is continuous at x_0.*

To establish this theorem, we suppose that $f(x)$ has a derivative at x_0. Then

$$f'(x_0) = \lim_{b \to x_0} \frac{f(b) - f(x_0)}{b - x_0}.$$

We write for $b \neq x_0$,

$$f(b) - f(x_0) = (b - x_0)\left[\frac{f(b) - f(x_0)}{b - x_0}\right].$$

As $b \to x_0$, the first factor on the right tends to 0 and the second to the number $f'(x_0)$. Thus $f(b) - f(x_0) \to 0$ and, consequently, $f(b) \to f(x_0)$ as $b \to x_0$. This shows that $f(x)$ is continuous at x_0.

Principles and procedures

- The vertical coordinate at each point on the graph $y = f'(x)$ of the derivative of $f(x)$ is the slope of the tangent line at the corresponding point on the graph of $f(x)$. This principle can be used to gain information about the derivative from a sketch of the graph of the function or about the function from a sketch of the graph of its derivative.
- If $f(x)$ is given by a sketch of its graph rather than by an exact formula, its derivative at x_0 can be estimated by drawing an approximate tangent line to the graph at $x = x_0$ and using approximate coordinates of two points on the tangent line to estimate its slope.
- If $f(x)$ is given only by values in a table that includes the value at x_0, then the derivative $f'(x_0)$ can be estimated by using the right-difference quotient **(4)** with the next value in the table, by using the left-difference quotient **(5)** with the preceding value in the table, or by using the centered-difference quotient **(6)** with the preceding and following values. If x_0 is not in the table, then $f'(x_0)$ can be estimated by using the slope of the secant line through the closest points in the table on both sides of x_0.
- To sketch the graph of a derivative when all you have is a sketch of the graph of the function, first estimate where the tangent lines to its graph are horizontal (where the derivative is zero), where the tangent lines have positive slopes (where the derivative is positive), and where the tangent lines have negative slopes (where the derivative is negative). Then find the approximate value of the derivative near the ends of the given graph and where the graph is steepest by estimating slopes of tangent lines. Plot these values and use the information from the first step to complete the graph.
- If a function is given only by a graph and you want to find where its derivative has a particular value, draw a line of that slope at a convenient place in the drawing and use a lined, transparent ruler to determine where the tangent lines are parallel to that line.

Tune-Up Exercises 2.4[♦]

[A]*Answer provided.* [O]*Outline of solution provided.*

Be sure you can solve Exercises T1 through T8, and work as many of Exercises T9 through T11 as you need for practice.

T1.[O] Use the graph of the function $P(x)$ in Figure 24 **(a)** to estimate $P'(3)$, **(b)** to find the approximate values of x with $0 \le x \le 5$ such that $P'(x) = 0$, and **(c)** to find an approximate value of x such that $P'(x) = 5$.

♦ Type 1, estimating derivatives from graphs

FIGURE 24

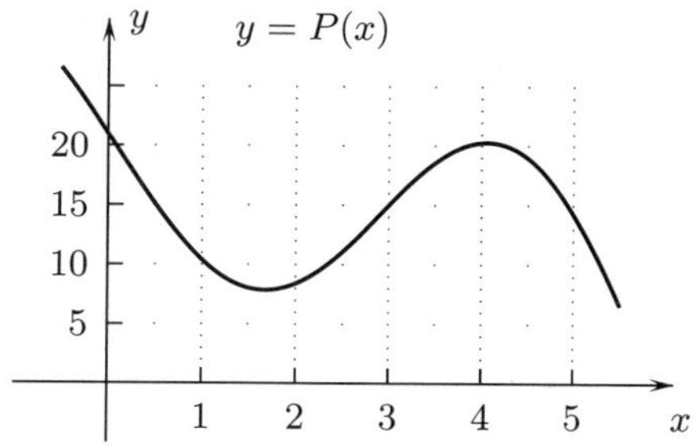

T2.[O] Sketch the graph of the derivative of the piecewise linear function $H(x)$ of Figure 25.

♦ Type 1, drawing the graph of a derivative

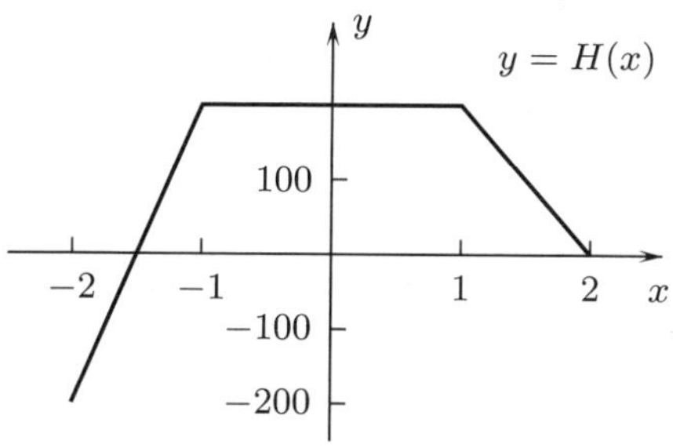

FIGURE 25

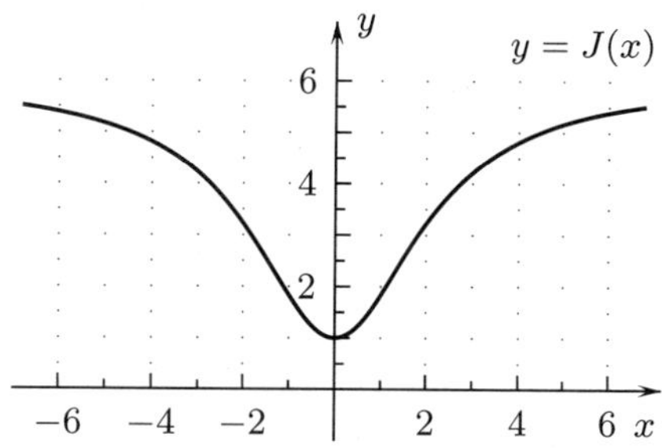

FIGURE 26

T3.[O] Sketch the graph of $J'(x)$ for $J(x)$ of Figure 26.

♦ Type 1, drawing the approximate graph of a derivative

T4.[A] Match **(a)** $P(x)$, **(b)** $Q(x)$, and **(c)** $R(x)$ of Figures 27 through 29 to the graphs of their first derivatives in Figures 30 through 32. Explain your choices.

♦ Type 2, recognizing graphs of derivatives

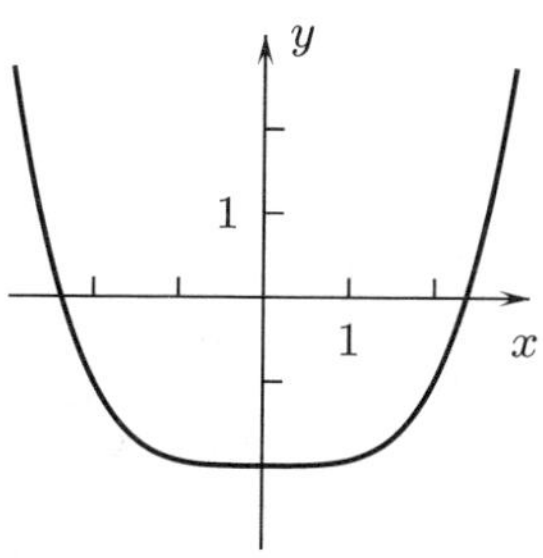

$y = P(x)$

FIGURE 27

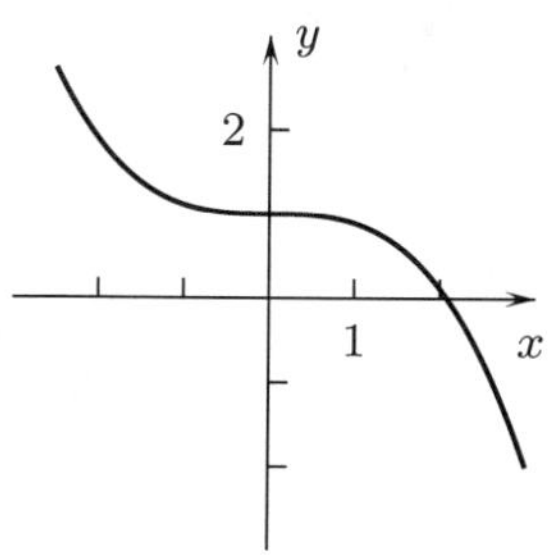

$y = Q(x)$

FIGURE 28

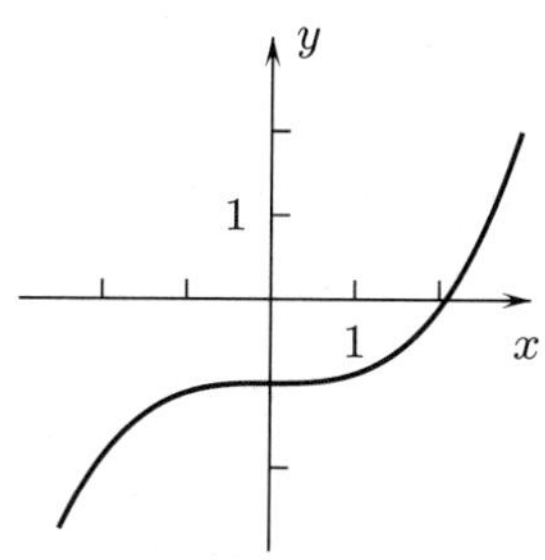

$y = R(x)$

FIGURE 29

[♦]The Tune-up Exercises and Problems are classified by type and content. The types are (1) basic, reactive; (2) basic reflective; (3) intermediate, reactive; (4) intermediate, reflective; (5) advanced, reactive; (6) advanced, reflective; and (7) advanced, theoretical.

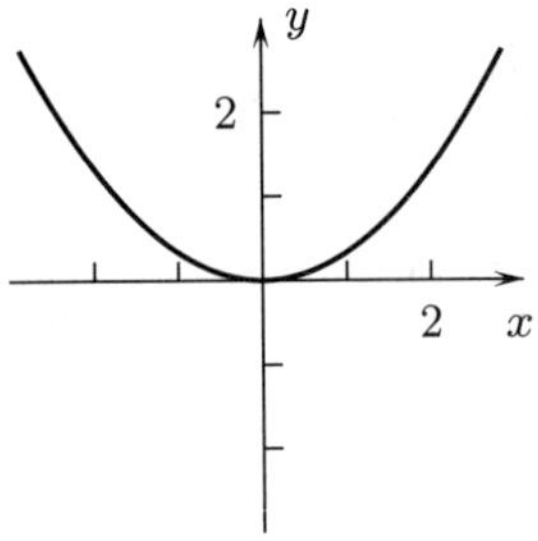

FIGURE 30

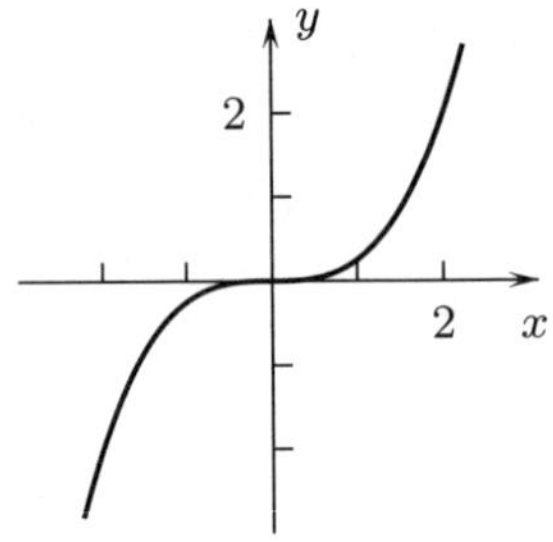

FIGURE 31

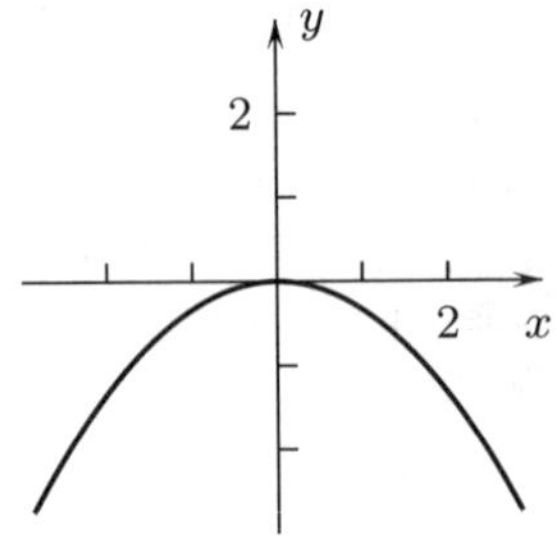

FIGURE 32

T5.[O] Use the values of $Q(x)$ in the next table to estimate $\frac{dQ}{dx}(10)$ and $\frac{dQ}{dx}(10.35)$.

x	9.6	9.8	10.0	10.2	10.4
$Q(x)$	410.9	414.1	417.0	420.2	422.7

◆ Type 1, estimating a derivative from a table

T6.[A] Find the second derivatives of **(a)** $w(x) = 5x^3 - 3x^5$ and **(b)** $y(x) = x^{1/4} + x^{1/3}$.

◆ Type 1, derivatives

T7.[A] What is the fourth derivative of $\frac{1}{90}x^{10} - \frac{1}{56}x^8$?

◆ Type 1, a derivative

T8.[A] Find the first and second derivatives of $y = 6ax^3 - 4bx^5$, where a and b are constant parameters.

◆ Type 1, a derivative

More practice

T9.[A] Use the graph of the function $P(x)$ in Figure 24 **(a)** to estimate $P'(1)$ and **(b)** to find an approximate value of x such that $P'(x) = -5$.

◆ Type 1, estimating derivatives from a graph

T10.[A] Sketch the graph of the function $M(x)$ and of its derivative, where

$$M(x) = \begin{cases} x & \text{for} \quad 0 \le x < 2 \\ 4 - x & \text{for} \quad 2 \le x < 3 \\ x - 2 & \text{for} \quad 3 \le x < 4 \\ 6 - x & \text{for} \quad 4 \le x \le 6. \end{cases}$$

◆ Type 2, drawing the graph of a derivative

T11.[A] What are the second derivatives of **(a)** $t^5 - t^2$ and **(b)** t^{100}?

◆ Type 1, a derivative

Problems 2.4

[A]*Answer provided.* [O]*Outline of solution provided.*

1.[O] Figure 33 shows the graph of the temperature $T(h)$ (degrees Celsius) as a function of the altitude h (feet) above sea level one summer day in Los Angeles.[(4)] This temperature distribution is called an "inversion" because the air is warmer at 3000 feet than at 1000 feet, while normally the air is cooler at higher altitudes. The inversion is caused by warm air from the deserts in the east that blows over the cooler air from the ocean on the west. The temperature inversion traps air pollution near the ground by preventing the upward circulation of warm air. What are the approximate temperature and the approximate rate of change of the temperature with respect to altitude at altitudes of 500, 1500, and 7000 feet?

◆ Type 1, estimating derivatives from graphs

[(4)]Data adapted from "The role of meteorology in the study and control of air pollution" by M. Neiburger, *Bulletin of the American Meteorological Society*, Dec., 1969, Boston, MA: American Meteorological Society, 1969, p. 960.

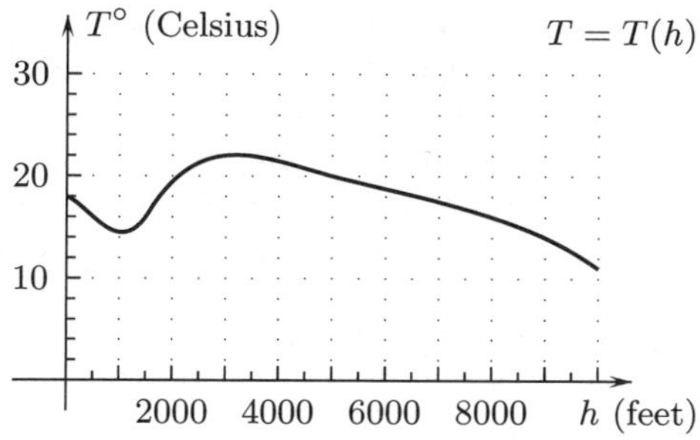

FIGURE 33

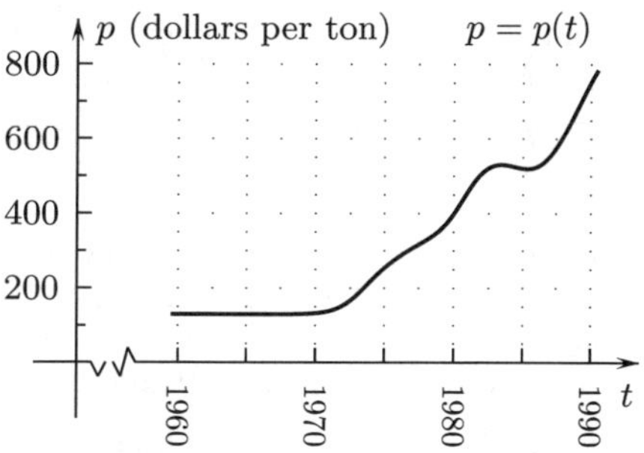

FIGURE 34

2. The graph in Figure 34 gives the price $p(t)$ of newsprint in the U.S. as a function of the year.[(5)] **(a)** During what time period was dp/dt approximately zero? **(b)** When was dp/dt negative? **(c)** What was the approximate percent change in the price of newsprint from 1960 to 1990? **(d)** Approximately how fast was the price of newsprint increasing when it was 400 dollars per ton?

♦ Type 2, estimating derivatives from a graph, etc.

3. The maximum power $p_M(t)$ that the average healthy man can sustain for t minutes is given by the lower curve in Figure 35. The upper curve gives the maximum power $p_A(t)$ for an average first-class athlete.[(6)] **(a)** Why, in terms of the men's activity, do $p_M(t)$ and $p_A(t)$ decrease as t increases? **(b)** Why is $p_M(t)$ less than $p_A(t)$ for $0 < t < 12$? **(c)** The graphs might suggest that $p_M(t) \to \infty$ and $p_A(t) \to \infty$ as $t \to 0^+$. Is this plausible? Explain. **(d)** The graphs might suggest that $p_M(t) \to 275$ and $p_A(t) \to 400$ as $t \to \infty$. Is this plausible? Explain. **(e)** Find the approximate value of t such that $p'_M(t) = p'_A(t)$. Express this equation in terms of rates of change.

♦ Type 2, estimating derivatives from a graph, etc.

FIGURE 35

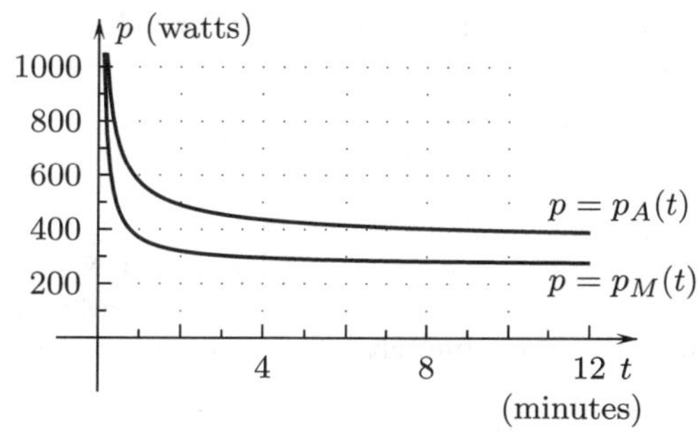

4.[A] Figure 36 shows the graph of a bird's height $h(t)$ above the ground as a function of time. Sketch the graph of its upward velocity dh/dt.

♦ Type 1, drawing the graph of a derivative

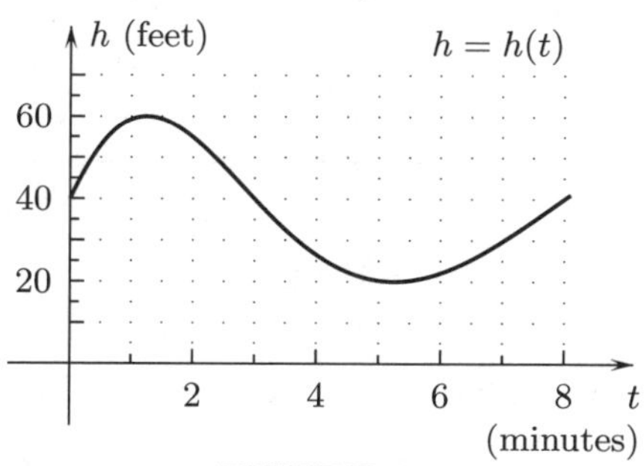

FIGURE 36

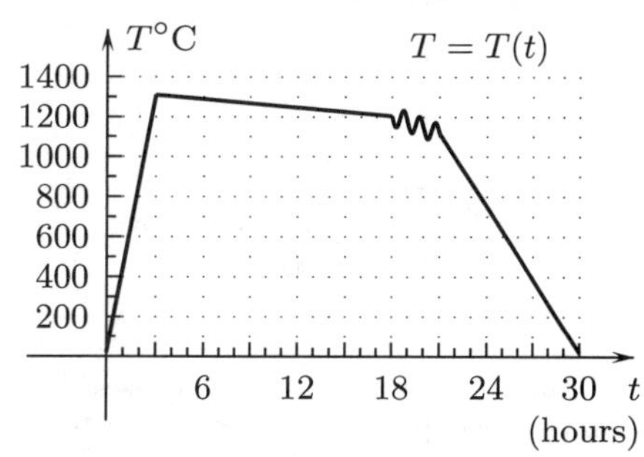

FIGURE 37

[(5)]Data adapted from *Newsprint, Canadian Supply and American Demand* by T. Roach, Durham, NC: Forest History Society, 1994, p. 43.

[(6)]Data adapted from "Human-powered flight" by M. Drela and J. Langford, *Scientific American*, Nov., 1985, New York, NY: Scientific American, Inc., 1985, p. 150.

5.[O] Figure 37 shows the graph of the temperature $T(t)$ as a function of time in the manufacture of glass.[7] Sand and other materials are heated to the melting point and formed into the desired shapes. The glass is cooled slightly over a long period of time, to allow bubbles of air to escape. Then the glass is repeatedly heated and cooled in a process called annealing, which removes internal stress. Finally the glass is cooled. Give approximate formulas for dT/dt, disregarding the annealing period.

◆ Type 2, finding approximate formulas for a derivative

6. Figure 38 shows the graph of the ozone concentration $C(t)$ (parts per million) in Los Angeles t hours after midnight one day, interpolated from measured data given by the dots in the figure.[8] **(a)** $C(t)$ is zero or close to zero at night because ozone (O_3) is an unstable molecule that is created from nitrous oxide with energy from the sun. Could $C(t)$ possibly be negative during another twenty hour period? **(b)** Use the coordinates in the figure to give exact formulas for $\frac{dC}{dt}(t)$.

◆ Type 2, finding formulas for a derivative

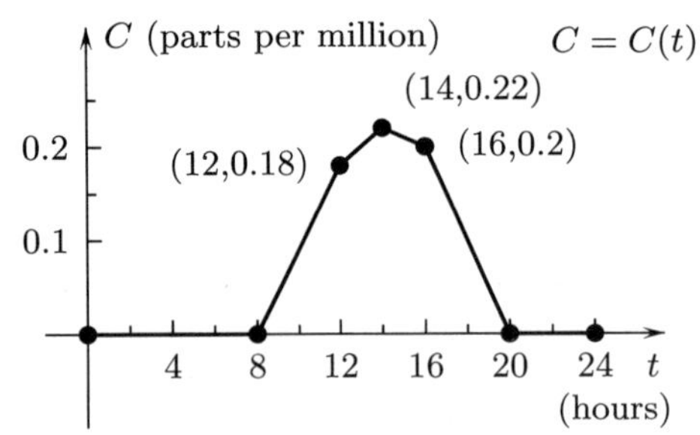

FIGURE 38

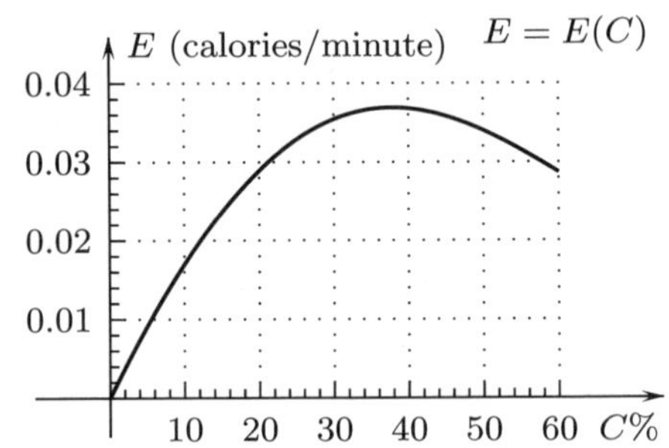

FIGURE 39

7. The rate $E(C)$ at which a gulf fritillary butterfly takes in energy is given as a function of the sugar concentration in the nectar it is sucking by the function in Figure 39.[9] **(a)** Why is $E(C)$ zero at $C = 0$ and small for small C? **(b)** The butterfly has difficulty sucking nectar with high sugar concentration. Explain how this affects the shape of the graph of $E(C)$. **(c)** Find the approximate value of C such that $\frac{dE}{dC}(C) = -\frac{dE}{dC}(50)$.

◆ Type 2, estimating derivatives from a graph, etc.

8.[A] The percent $P(T)$ of water in ice cream that is frozen is given in Figure 40 as a function of the temperature at which the ice cream is stored.[10] **(a)** Why is $P(T)$ a decreasing function for $-20 < T < 32$? Why is $P(T)$ zero for $T \geq 32$? **(b)** Find the approximate value of $P'(10)$ and use it determine approximately how much more water is frozen at 9.99°F than at 10°F.

◆ Type 4, estimating derivatives from a graph, etc.

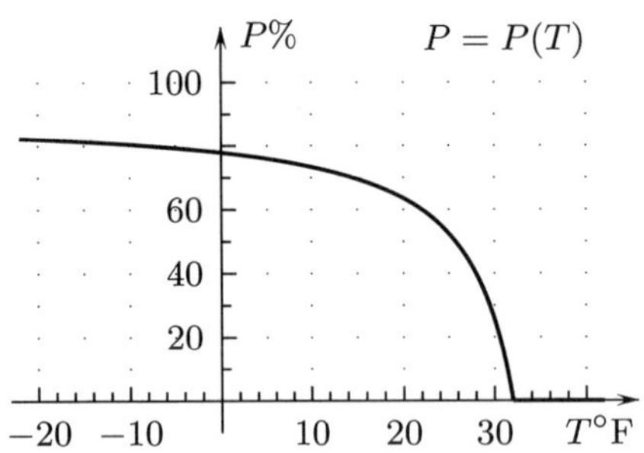

FIGURE 40

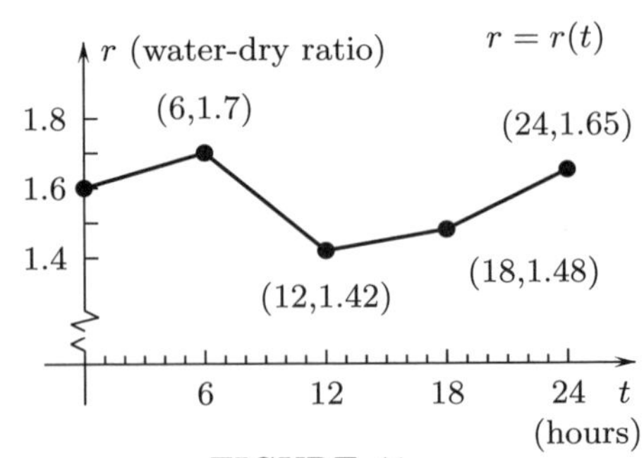

FIGURE 41

---

[7] Data adapted from "Ancient Glazes" by P. Vandiver, *Scientific American*, April, 1990, New York, NY: Scientific American, Inc. 1990, p. 108.

[8] Data adapted from *McGraw-Hill Encyclopedia of Science and Technology*, Vol. vi, New York, NY: McGraw-Hill, Inc., 1992, p. 246.

[9] Data adapted from "Butterfly engineering" by J. Kingsolver, *Scientific American*, August, 1985, New York, NY: Scientific American, Inc., p. 109.

[10] Data adapted from *McGraw-Hill Encyclopedia of Science and Technology*, Vol. vii, New York: NY: McGraw-Hill, Inc., 1992, p. 251.

9. The piecewise linear curve in Figure 41 is the graph of the ratio $r(t)$ of water to dry weight in a holly plant t hours after midnight one day. The function is interpolated from data given by the dots in the figure.[11] **(a)** Notice that $r(t) > 1$ for all t. What does this say about the plant? **(b)** Why is it plausible that $r(6) > r(0)$ and $r(12) < r(6)$? **(c)** Use the coordinates in the figure to give exact formulas for $\frac{dr}{dt}(t)$.

♦ Type 2, exact formulas for a derivative, etc.

10. The function $v(t)$ of Figure 42 is the upward velocity of a World War II V-2 rocket t seconds after it is fired.[12] The rocket's engine exerts a constant force but the rocket becomes lighter as it uses fuel and the air resistance is less at higher altitudes. Explain how this affects the shape of the graph. Then sketch the graph of the rocket's upward acceleration dv/dt.

♦ Type 4, drawing the approximate graph of a derivative, etc.

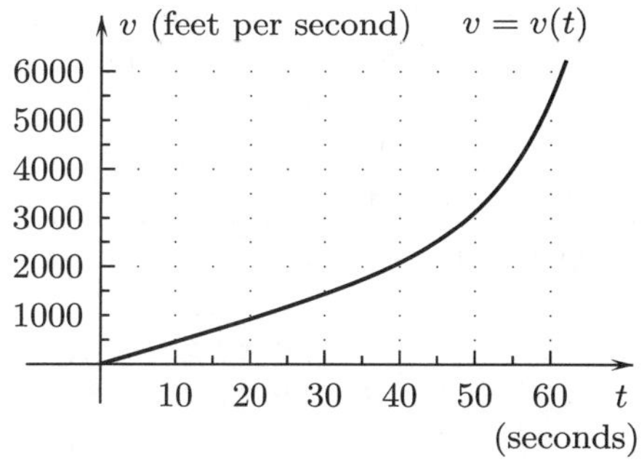

FIGURE 42

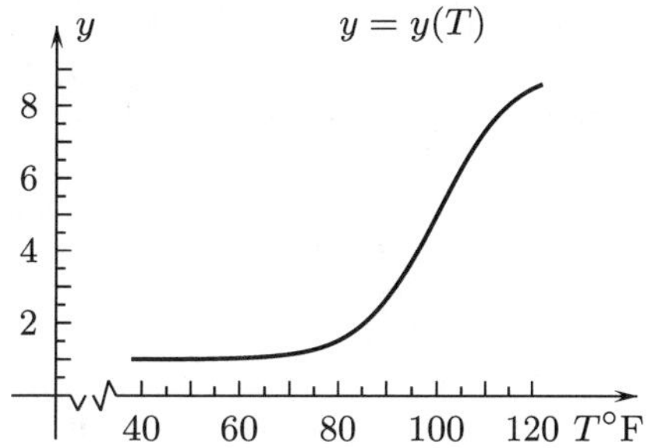

FIGURE 43

11. When the body core is overheated, the blood vessels dilate to allow more blood to reach the skin surface where it can be cooled by perspiration. The function $y(T)$ of Figure 43 gives the rate of flow from the body core to the surface of the skin, as a function of the environmental temperature.[13] The values are given relative to the rate when the temperature is 40°F; i. e., $y(T)$ is the relative change from the rate at 40°F to the rate at temperature T. **(a)** $y'(T)$ is zero or close to zero for $40 \le T \le 65$ and $y'(T)$ is relatively large for $90 \le T \le 110$. What does this indicate about the body's reaction to the environmental temperature? **(b)** Give the approximate value of the temperature T_0 for which $y'(T)$ is a maximum and find the approximate value of $y'(T_0)$. **(c)** Use the result of part (b) to estimate the amount that $y(T)$ increases when T is increased from T_0 to $T_0 + 1$.

♦ Type 4, drawing the approximate graph of a derivative, etc.

12. The table below gives the temperature $T(t)$°F of the radiator in an automobile t minutes after its engine is started.[14] **(a)** Approximately how fast is the temperature of the radiator increasing after 15 minutes? **(b)** After 22 minutes? **(c)** Does the temperature of the radiator increase more rapidly or less rapidly as the car heats up?

t (minutes)	0	5	10	15	20	25
$T(t)$°F	87	110	113	116.5	119	121

♦ Type 2, estimating derivatives from a table, etc.

[11] Data adapted from *Fire in Forestry*, Vol. 1, by I. Chandler and C. Craig, New York, NY: John Wiley & Sons, Inc., 1983, p. 34.

[12] Data adapted from *Introduction to Rocket Technology* by V. Feodosiev and G. Siaiarev, New York, NY: Academic Press, 1959, p. 253.

[13] Data adapted from *Human Adaptation and Accommodation* by A. Frisancho, Ann Arbor, MI.: The University of Michigan Press, 1993, p. 40.

[14] Data adapted from "A Flow Network Approach to Vehicle Underhood Heat Transfer Problem" by P. Li, G. Chui, J. Glidewell, T. Chue, and M. Lai, *1993 Vehicle Thermal Management Systems Proceedings*, Warrendale, PA: Society of Automotive Engineers, 1993, p. 46.

13.º Based on the next table of the number of farms $N(t)$ (millions) and total farm acreage $A(t)$ (million acres) in the United States,[15] what were the approximate rates of change with respect to time of **(a)** the number of farms in 1979 and **(b)** the total farm acreage in 1984? **(c)** What was the average size of an American farm in 1976? In 1991?

t (year)	1976	1979	1982	1985	1988	1991
$N(t)$ (millions)	2.497	2.437	2.407	2.293	2.197	2.105
$A(t)$ (million acres)	1054	1042	1028	1012	995	983

♦ Type 2, estimating derivatives from a table, etc.

14. The following table gives the velocity of sound in water $v(T)$ (feet per second) as a function of the temperature T°F of the water.[16] **(a)** Does the velocity of sound increase or decrease as the temperature increases? **(b)** Based on this data, does the velocity decrease more when the temperature rises from 200°F to 201°F or when it rises from 400°F to 401°F?

T°F	200	240	280	320	360	400	420
$v(T)$ (feet per second)	5079	5000	4879	4724	4537	4331	4081

♦ Type 2, estimating derivatives from a table, etc.

15. The table below lists the percentage of roses sold in the U.S. that were imported in the odd-numbered years from 1981 through 1993.[17] Based on this data, when between 1981 and 1993 was the percentage of imported roses increasing the most rapidly, and how fast was it increasing then?

Year	1981	1983	1985	1987	1989	1991	1993
% of roses imported	14	22	26	31	34	48	58

♦ Type 2, estimating derivatives from a table, etc.

16. The table below gives the average grip strength of men and the rate of change of the grip strength with respect to age at five ages.[18] **(a)** Use the data in the table to estimate the average grip strength of men at ages 12 and 42. **(b)** What are the maximum and minimum of $G(t)$ for $10 \le t \le 50$ if $G'(t)$ is positive for $10 \le t < 30$ and negative for $30 < t \le 50$?

t (years)	10	20	30	40	50
$G(t)$ (kilograms)	15	50	55	51	47
$G'(t)$ (kilograms per year)	1.9	1.5	0	−0.4	−0.3

♦ Type 2, estimating derivatives from a table, etc.

17. After leaving a parking space, a woman drives 400 feet in 20 seconds to a stop sign. She waits 10 seconds at the stop sign and then drives 800 feet more in 20 seconds. Her velocity is 30 feet per second 10 seconds after she drives away from the stop sign. Draw a possible graph of her distance $s(t)$ from the parking space as a function of the length of time since she left it.

♦ Type 4, drawing a graph for a mathematical model

18. A bucket sitting outside contains water one inch deep when it begins to rain. After half an hour of constant rainfall, the water is 1.2 inches deep in the bucket. It then rains at a greater, constant rate for one hour and then stops, adding 0.8 inches of water in the bucket. Draw the graphs of the depth $h(t)$ of the water and of its derivative $h'(t)$ as functions of time in the two hours after the rain starts.

♦ Type 4, drawing a graph for a mathematical model

---

(15) Data from *1993 Statistical Abstract of the United States*, Washington, D.C: U.S. Bureau of the Census, 1993, p. 655.

(16) Data from *CRC Handbook of Chemistry and Physics*, Boca Raton, FL: CRC Press, 1981, p. E-46.

(17) Data adapted from *Los Angeles Times*, January 28, 1995, Los Angeles, CA: The times Mirror Company, 1995, p. D1. Source: Department of Agriculture.

(18) Data adapted from *Physiology of Exercise* by H. deVries, Dubuque, IA: Wm. C. Brown Company, 1980, p. 358.

19. An icicle hanging from a roof is 12 inches long when the sun comes up at 7:00 AM. The temperature is above freezing all morning and the icicle shrinks to a length of 6 inches at noon. Then it turns cold, and the icicle grows to 12 inches at 5:00 PM. During the rest of the afternoon and night it is so cold that no moisture is added to the icicle and it remains 12 inches long. Draw a possible graph of the length $L(t)$ of the icicle t hours after sunrise for $0 \le t \le 15$, so that dL/dt is continuous in $[0, 15]$, is increasing for $0 < t < 7$, and is decreasing for $7 < t < 10$. What is the value of dL/dt for $10 < t < 15$?

♦ Type 4, drawing a graph for a mathematical model

20. A trust fund's investments are worth 10 thousand dollars on January 1 and increase at the constant rate of \$100 per month for three months. The value decreases at the constant rate of \$50 per month for the next six months, and then increases at the constant rate of \$75 per month for three months. Draw the graphs of the value $B(t)$ of the investments t months after January 1 and of its derivative $B'(t)$ for $0 \le t \le 12$.

♦ Type 3, drawing the graph of a piecewise-linear function

21. The percent of water W, measured by weight, in a piece of wood is given as a function of the relative humidity h of the environment by $W(h)$, where $W'(70) = 0.55$.[19] **(a)** Why is it plausible that $W'(70)$ is positive? **(b)** What is the approximate percent of water in the wood at $h = 69$ if $W(70) = 11\%$?

♦ Type 4, a tangent-line approximation, etc.

22. In a study of high school athletes, it is found that the average upper arm circumference of wrestlers at age t (years) is $A_W(t)$ (centimeters), where $A_W(16) = 2.6$ and $A'_W(16) = 0.8$, and that the average upper arm circumference of all boys at age t is $A_B(t)$, where $A_B(t) = 1.16A_W(t)$ for t near 16.[20] What are the values of $A_B - A_W$ and A'_B at $t = 16$ and what do they represent in terms of arm measurements?

♦ Type 2, values and derivatives of related functions

23. The graph in Figure 44 gives the percentage $P(t)$ of the U.S. population whose income was below the government's official poverty level from 1960 to 1990.[21] **(a)** Approximately when was $P(t)$ a minimum for $1960 \le t \le 1990$, and what was dP/dt at that time? **(b)** At what time with $1960 \le t \le 1990$ is dP/dt a minimum (the most negative)? **(c)** What is the approximate minimum value of dP/dt for $1960 \le t \le 1990$?

♦ Type 4, estimating derivatives from graphs, etc.

FIGURE 44

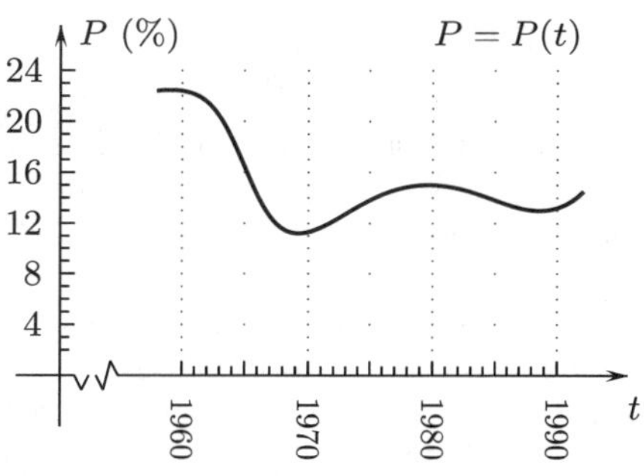

24. Figure 45 shows the velocity $v(t)$ (kilometers per hour) of a car and Figure 46 shows the temperature $T(t)^\circ$C of its brakes t seconds after the car's brakes are applied.[22] **(a)** The values of $v(t)$ decrease and the values of $T(t)$ increase as t increases. Explain why this is the case. **(b)** Approximately how fast is the car decelerating and how rapidly are its brakes heating when it is traveling 30 kilometers per hour?

♦ Type 4, estimating derivatives from graphs, etc.

---

[19] Data adapted from *Handbook of Engineering Materials*, D. Miner and J. Seastone, New York, NY: John Wiley & Sons, 1955, p. 3-22.

[20] Data adapted from "Anthropometric growth patterns of high school wrestlers" by T. Housh, G. Johnson, J. Stout, and D. Housh, *Medicine and Science in Sports and Exercise*, Vol. 25, No.10, American College of Sports Medicine, 1993, p. 1151.

[21] Data adapted from 1993, Statistical Abstract of the United States, Washington,DC: Bureau of the Census, 1993, p. 469.

[22] Data adapted from "Heat flow on disk brakes" by M. Yano and M. Murata, *1993 Vehicle Thermal Management Systems Conference Proceedings*, Worrendale, PA: Society of Automotive Engineers, 1993, pp. 121–122.

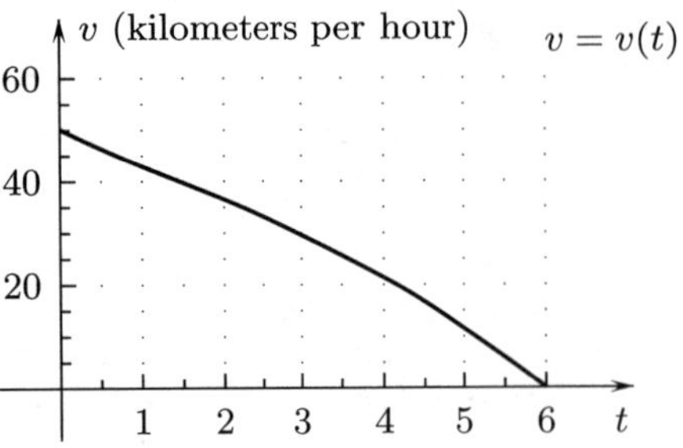

FIGURE 45

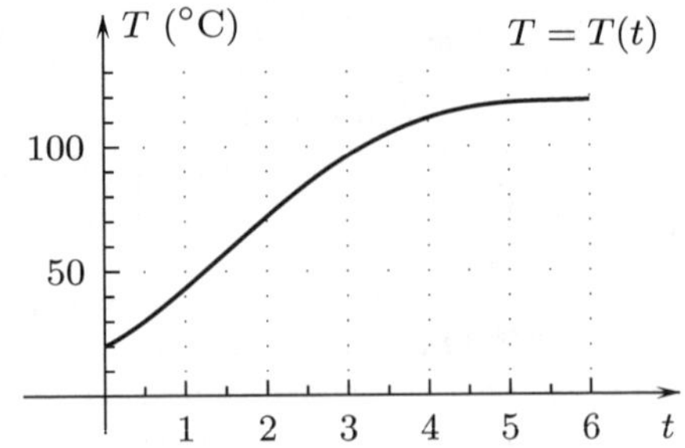

FIGURE 46

25. Figure 47 shows the upward force (lift) $L(x)$ (pounds) of an airplane wing as a function of the angle x (degrees) it makes with the horizontal.[(23)] **(a)** What is the approximate rate of change of the lift with respect to the angle at $x = 0$? At $x = 13$? **(b)** What is the approximate maximum lift of the wing and at approximately what angle does it occur? What is the rate of change of the lift with respect to the angle when the lift is a maximum?

◆ Type 4, estimating derivatives from a graph

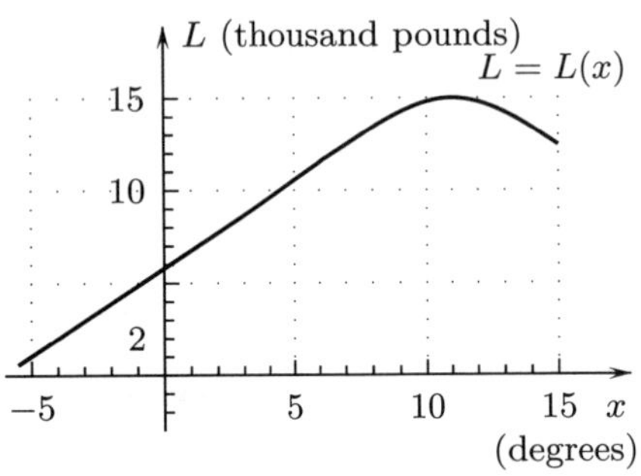

FIGURE 47

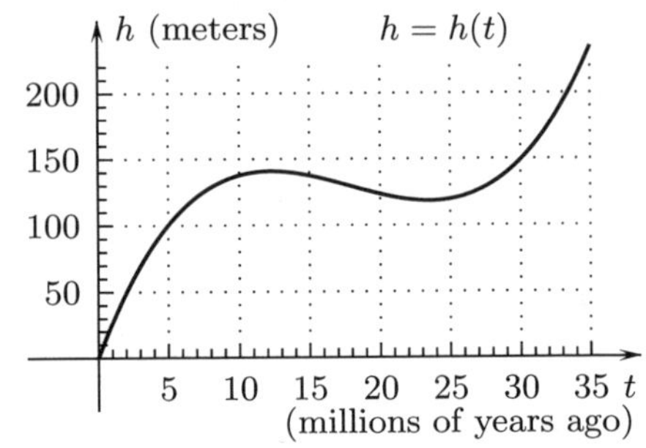

FIGURE 48

26. Figure 48 shows the graph of the average sea level $h(t)$, measured as meters above the present level, t million years ago.[(24)] **(a)** Find the approximate value of $h'(5)$. **(b)** Was the average level of the ocean rising or falling five million years ago? At what approximate rate?

◆ Type 4, estimating derivatives from a graph

27. A gram of ice is initially at $-100°$C and is heated to $200°$C. Figure 49 gives the graph of the temperature $T(Q)$ of the gram of H_2O as a function of the number Q of calories of heat that have been added to it.[†] The ice's temperature increases $2°$C for every calorie of heat until the ice begins to melt at $0°$C. It takes 80 calories to melt the ice into water at $0°$C. The temperature of the gram of water increases 1 degree for every calorie of heat until the water starts to boil at $100°$C. Boiling the water into steam at $100°$C requires 539 calories, and then the steam's temperature increases 2 degrees for every additional calorie of heat. The three values of dT/dQ are the SPECIFIC HEAT of ice, water, and steam, respectively. What are they? Sketch the graph of dT/dQ.

◆ Type 2, finding formulas for a derivative

---

(23) Data adapted from *Boundary Layer Theore* by H. Schlichting, New York, NY: McGraw-Hill, Inc, 1968, p. 21.

(24) Data adapted from "Large Igneous Provinces" by M. Coffin and O. Eldholm, *Scientific American*, October, 1993, New York, NY: Scientific American, Inc., 1993, p. 47.

† A calorie, written with a lower case "c" is the amount of heat required to increase the temperature of a gram of water by one degree Celsius. A Calorie, written with a upper case "C" is the amount of heat required to increase the temperature of a kilogram of water by one degree Celsius.

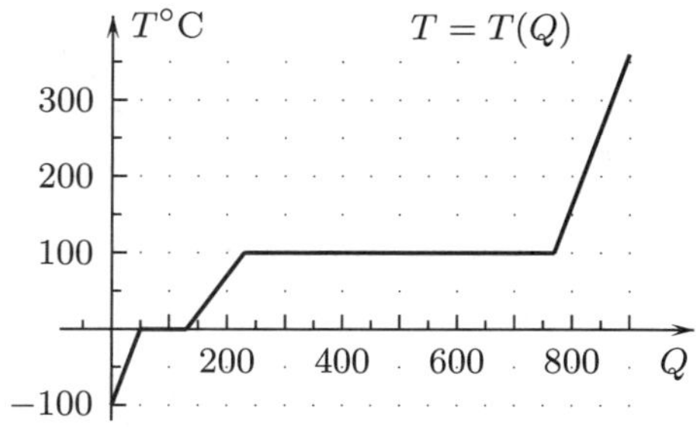

FIGURE 49

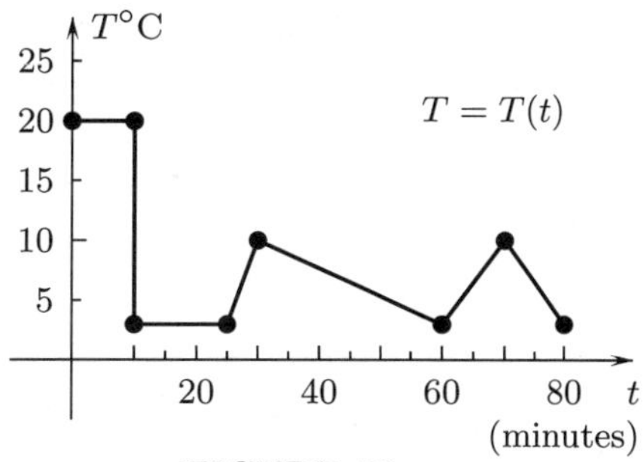

FIGURE 50

28. When a body extremity becomes very cold, the blood vessels either dilate to allow more warming blood to the cold region or the vessels constrict to reduce overall heat loss, depending on the circumstances. Figure 50 shows the graph of the temperature $T(t)$ of a finger which is at 20°C from $t = 0$ to $t = 10$ (minutes), when it is submerged in ice water and held there for 70 minutes. The graph is obtained by interpolating measured data given by the dots.(26) Describe how T and dT/dt vary with time and what this indicates about the finger's reaction to the cold.

♦ Type 4, estimating derivatives from a graph

29. The percent $r(t)$ of the surface of a steel sheet that has rusted t years after it was placed outside is given by the graph in Figure 51.(27) Let $s(t)$ denote the percent that has not rusted. **(a)** How are $s(t)$ and $r(t)$ related? How are $s'(t)$ and $r't)$ related? **(b)** Sketch the graphs of $s(t)$, $r'(t)$ and $s'(t)$.

♦ Type 4, estimating derivatives from a graph

FIGURE 51

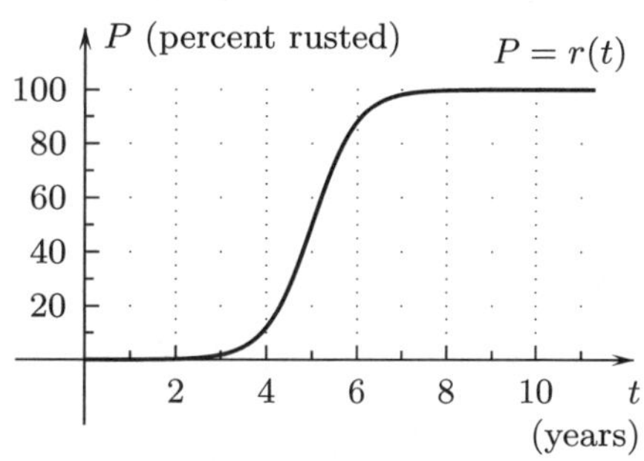

30. Large amounts of organic pollutants in a river can consume so much dissolved oxygen that the water cannot support life. This occurred during the early 1970's in parts of the Rhine River in Germany. Figure 52 is the graph of the average concentration of dissolved oxygen in the Rhine from 1955 to 1990.(28) **(a)** What, according to this mathematical model, were the approximate relative changes in the concentration of dissolved oxygen from 1955 to its minimum value and from its minimum value to its value in 1990? **(b)** Approximately when was the oxygen level in the Rhine decreasing 0.3 milligrams per liter per year? When was it increasing 0.2 milligrams per liter per year?

♦ Type 4, estimating derivatives from a graph

(26) Data adapted from *Handbook of Human Thermal Environments* by K. Parsons, London: Taylor & Francis Ltd., 1993, p. 184.

(27) Data adapted from *Handbook of Engineering Materials* by D. Miner and J. Seastone, New York, NY: John Wiley & Sons, 1955, p. 2-48.

(28) Data adapted from "Cleaning up the river Rhine" by K-G. Malle, *Scientific American*, January, 1996, Washington, DC: Scientific American, Inc., 1996, p. 70.

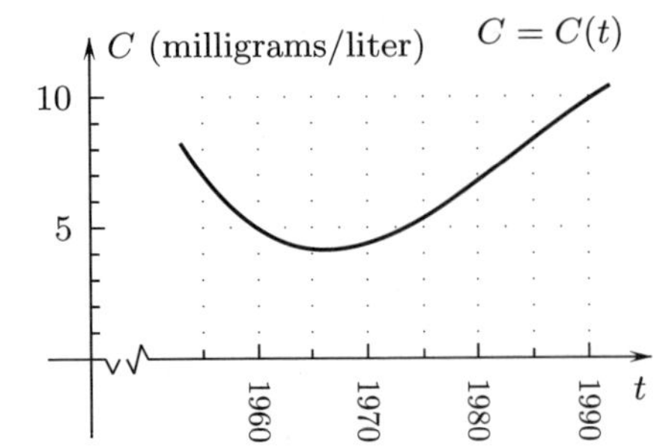

FIGURE 52

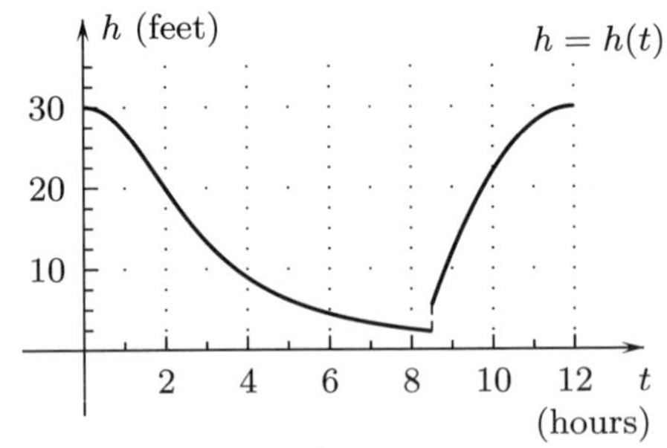

FIGURE 53

31. Tidal bores are waves that move up certain gently sloping rivers when the high tide comes in from the ocean. For this to occur, there must be an exceptionally large difference between the water levels at low and high tide, and the river must have a wide mouth that funnels the water into the narrowing river. Figure 53 gives the depth $h(t)$ of the water at one point on Pettitcodiac River in New Brunswick, Canada as a function of time as a three-foot high tidal bore passes.(29) **(a)** When does the bore arrive? **(b)** Does it take longer for the tide to go from its highest point to its lowest or from its lowest point to its highest? **(c)** According to this mathematical model, what are the approximate rates of change of the depth of the water just before and just after the bore arrives?

♦ Type 4, estimating derivatives from a graph, etc.

32. Figure 54 gives the concentration $S(t)$ of sugar and $A(t)$ of alcohol (ethanol) in a vat during the fermentation of wine.(30) **(a)** Why is $S(t)$ a decreasing function and $A(t)$ an increasing function? **(b)** What are the approximate rates of change of the concentrations at the time when the concentrations are equal? **(c)** Approximately how much sugar does it take to make a gram of alcohol?

♦ Type 4, estimating derivatives from a graph, etc.

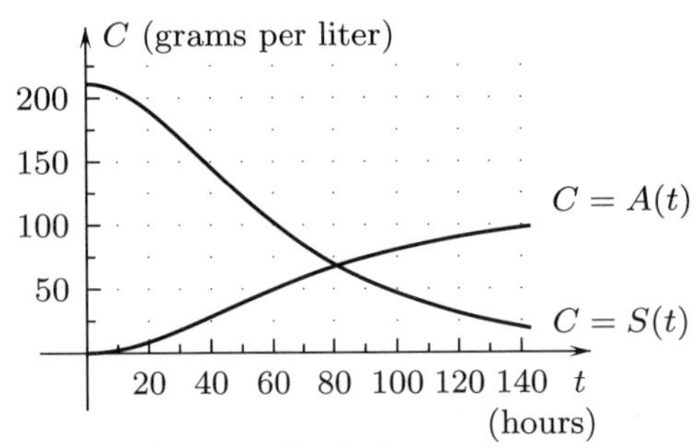

FIGURE 54

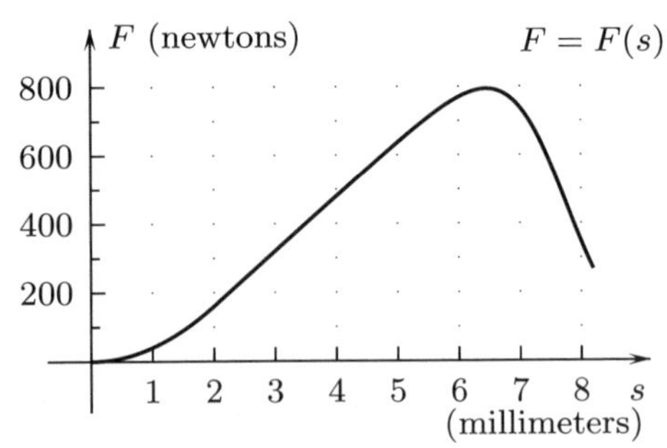

FIGURE 55

33. Figure 55 is the graph of the force $F(s)$ exerted by a monkey ligament as a function of how much it is stretched. The ligament can easily be stretched very small amounts. This gives the portion of the graph for small s the "toe" shape shown in the figure. The portion of the curve that looks like a line is the "linear region," and the rest of the curve on the right is the "fatigue region", where the ligament weakens because increasing numbers of microfibers in it are torn.(31) Describe in a few words how $F(s)$ and $F'(s)$ vary as s increases from 0 to 8.

♦ Type 4, estimating derivatives from a graph, etc.

(29) Data adapted from "Tidal Bores" by D. Lynch, *Scientific American*, October, 1982, Washington, DC: Scientific American, Inc., 1982, p. 150.

(30) Data adapted from *Application à l'Œnologie des Progrès Récent en Microbiologie et en Fermentation* Paris: Office International de la Vigne et du Vin, 1989, p. 236.

(31) Data adapted from *Human Adaptation and Accommodation* by A. Frisancho, Ann Arbor, MI: The University of Michigan Press, 1993, p. 119.

In Problems 34 through 39 use definition **(2)** of "marginal $f(x)$" as the derivative $f'(x)$.

34.[A] It costs a company $C(x) = 2500 + 10x - 30x^{2/3}$ dollars to manufacture x shirts $(100 \le x \le 2000)$. What is the marginal cost at $x = 1000$?

♦ Type 3, marginal cost

35.[A] A company can sell $x(p) = 21000p^{-1/2} - 3000$ jackets if it charges p dollars per jacket $(8 \le p \le 49)$. **(a)** How many will it sell and what revenue will it receive from the jackets if it charges 16 dollars per jacket? **(b)** What revenue will it receive from the jackets if it charges p dollars per jacket? **(c)** Give a formula for the marginal revenue with respect to price as a function of the price.

♦ Type 3, marginal revenue

36. It costs a company $600 + 325x + x^2$ dollars to manufacture x radios in a week and it can sell them for \$500 each. **(a)** What are the revenue and profit of x radios? **(b)** What are the marginal cost, marginal revenue, and marginal profit with respect to x?

♦ Type 3, marginal revenue, cost, and profit

37. It costs \$25 per bookcase plus \$300 overhead to build bookcases. **(a)** What is the average total cost per bookcase if x bookcases are made? **(b)** What is the marginal total cost with respect to x if x bookcases are made?

♦ Type 3, average and marginal cost

38.[A] A company can sell 1000 units of a product if it charges \$30 per unit. For every dollar it raises the price, up to a price of \$50 per unit, it will sell 20 fewer units. **(a)** Give the demand and revenue as functions of price. **(b)** What is the marginal revenue with respect to price at $p = 30, p = 40$, and $p = 48$?

♦ Type 3, demand, revenue, and marginal revenue

39[A] Give a formula in terms of the surface area for the rate of change of the volume of a sphere with respect to its surface area.

♦ Type 4, a rate of change

40. Give a formula in terms of the volume for the rate of change of the surface area of a cube with respect to its volume.

♦ Type 4, a rate of change

41.[A] A car is $60t + \frac{1}{10}t^5$ (kilometers) past a service station at time t (hours). Where is it and how fast is it traveling when it is accelerating 16 kilometers per hour per hour?

♦ Type 2, acceleration

42. A watermelon is dropped from a dormitory window 20 meters above the ground. If there were no air resistance it would fall $4.9t^2$ meters in t seconds. Based on this mathematical model, **(a)** how long does it take the watermelon to hit the ground and **(b)** what are its downward velocity and acceleration when it is ten meters from the ground?

♦ Type 2, velocity and acceleration, etc.

43. A snail is crawling in a straight line toward the north. Figure 56 shows the graph of its velocity $v(t)$ toward the north for $0 \le t \le 3$ **(a)** What were its approximate maximum and minimum velocity for $0 \le t \le 3$? **(b)** Which of the curves in Figures 57 and 58 is the graph of its acceleration $a(t)$ toward the north? **(c)** The number k on the vertical axis with the correct graph of $a(t)$ is either 3, 5, or 10. Which is it? Justify your answers.

♦ Type 4, acceleration

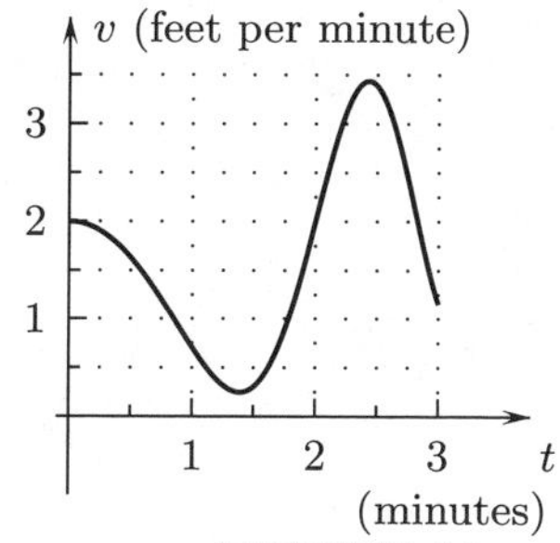

FIGURE 56

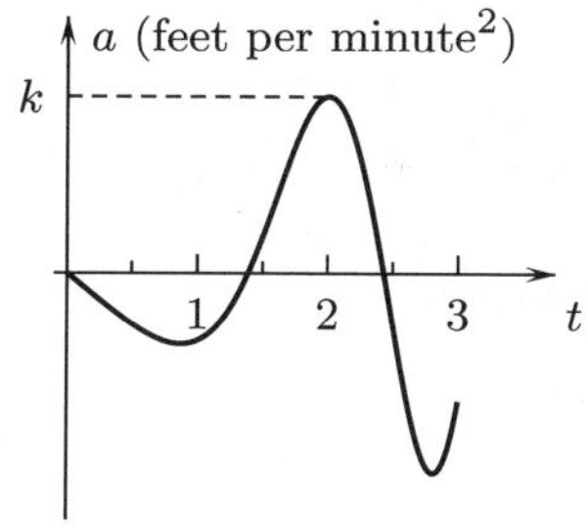

FIGURE 57

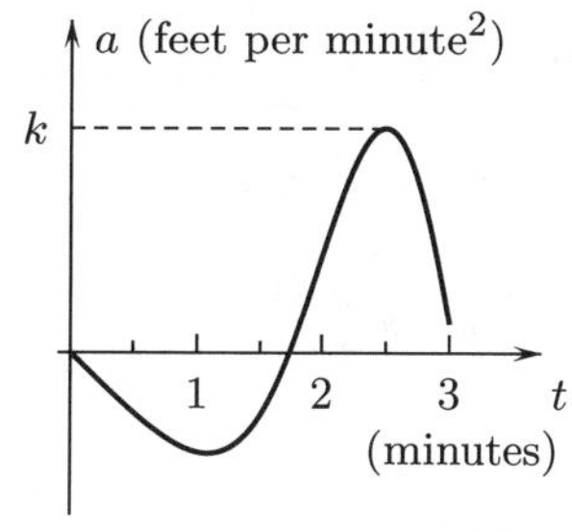

FIGURE 58

44. According to the February 14, 1994 issue of *Newsweek* magazine, "The rate of increase of health costs nudged downward in the last year, a welcome sign. But costs are still increasing." Rephrase these statements in terms of derivatives.

♦ Type 4, translating a quote to mathematics

45. An article in the *Los Angeles Times*, August 26, 1993 had the title "Ozone-depleting emissions drop, study finds." The article, however, stated, "In a paper published in the journal *Nature*, scientists from the Natianal Oceanic and Atmosphere Administration said worldwide emissions of the most widely used chlorofluorocarbons (CFCs) are increasing at substantially reduced rates from previous years." What is wrong with the title of the article?

♦ Type 4, translating a quote to mathematics

46.[A] Draw the graph of a function $F(x)$ such that $F(0) = 1, F'(0) = 0,\ F'(1) = 3,\ F(2) = 5$ and $F'(2) = 0$.

♦ Type 2, drawing graphs with specified derivatives, etc.

47. Draw the graph of a function $H(x)$ such that $H(0) = 0, H'(0) = 0,\ H(2) = 2, H'(2) = 0, H(4) = 0$ and $H'(4) = 0$.

♦ Type 2, drawing graphs with specified derivatives, etc.

48. Draw the graph of $f(x)$ for $0 \le x \le 4$ such that $\dfrac{df}{dx}(1) = \frac{1}{2}$ and $\dfrac{df}{dx}(3) = 1$. Then draw the graph of $g(x) = 2f(x)$. What are $\dfrac{dg}{dx}(1)$ and $\dfrac{dg}{dx}(3)$?

♦ Type 2, drawing graphs with specified derivatives, etc.

49. A mouse running away from its hole is five feet from it at $t = 0$ seconds. Its average velocity away from the hole for $0 \le t \le 3$ is ten feet per second, and its instantaneous velocity away from the hole at $t = 1.5$ is ten feet per second. Draw a possible graph of its distance $s(t)$ from the hole at time t **(a)** with linear $s(t)$ and **(b)** with a nonlinear $s(t)$.

♦ Type 2, drawing graphs with specified derivatives, etc.

50. Sketch the graph of the derivative of $W(x) = ||x| - 1|$. Generate the graph of $W(x)$ in the window $-2.4 \le x \le 2.4, -1 \le y \le 2.2$ to guide your reasoning.

♦ Type 4, drawing the graph of a derivative

51. Sketch the graph of the derivative of $\Lambda(x) = |1 - |x|| - |x|$. Generate the graph of $\Lambda(x)$ in the window $-2.4 \le x \le 2.4, -1.6 \le y \le 1.6$ to guide your reasoning.

♦ Type 4, drawing the graph of a derivative

52.[A] For what values of the parameters A and B is

$$H(x) = \begin{cases} 2 - x + x^2 & \text{for} \quad x \le 0 \\ Ax + B & \text{for} \quad x > 0 \end{cases}$$

(a) continuous for all x? **(b)** differentiable (have a derivative) for all x?

♦ Type 4, continuity and differentiability

53. For what values of the parameter c does $|x^2 - 2x + c|$ have a derivative for all x? Generate sample graphs to guide your reasoning.

♦ Type 4, differentiability

54. The density (grams per cubic centimeter) of pure zinc at T degrees Celsius is[32]

$$\rho(T) = \begin{cases} 7.176 - 0.0008T & \text{for} \quad 0 \le T < 420 \\ 7.04 - 0.001T & \text{for} \quad 420 < T \le 800. \end{cases}$$

Zinc melts at 420°C. **(a)** Is $\rho(T)$ continuous at $T = 420$? **(b)** What are $\lim\limits_{T\to 420-} \rho(T)$ and $\lim\limits_{T\to 420+} \rho(T)$ and what is their significance for zinc? **(c)** What are $\lim\limits_{T\to 420-} \dfrac{d\rho}{dT}(T)$ and $\lim\limits_{T\to 420+} \dfrac{d\rho}{dT}(T)$ and what do they represent?

♦ Type 4, one-sided limits of a derivative, etc.

[32] Data adapted from *Handbook of Engineering Materials* by D. Miner and J. Seastone, New York, NY: John Wiley & Sons, 1955, p. 2-362.

Section 2.5

Derivatives of products, quotients, and powers

OVERVIEW: *In this section, we derive formulas for derivatives of functions that are constructed by taking products, quotients, and powers of other functions, and we use these formulas to study rates of change in a variety of contexts.*

Topics:

- ***The Product Rule***
- ***Related-rate problems***
- ***The Quotient Rule***
- ***Derivatives of powers of functions***

The rate of change of the area of a rectangle

Imagine that the sides of the rectangle in Figure 1 are getting larger. The width w, height h, and area $A = wh$ of the rectangle are functions of the time t. We want a formula for the rate of change of the area in terms of w, h, and their rates of change. We begin by looking at cases where one of the two dimensions is constant.

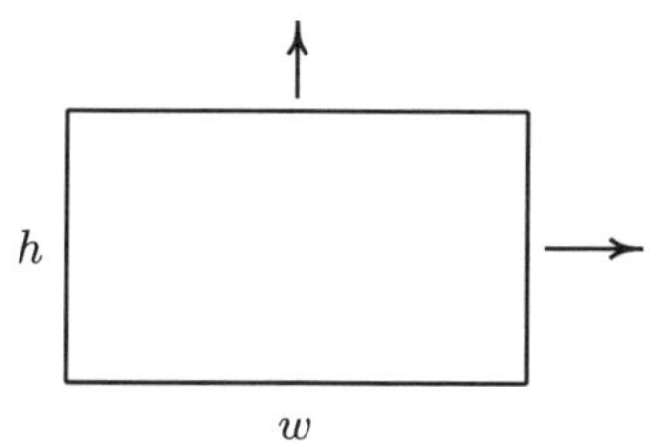

FIGURE 1

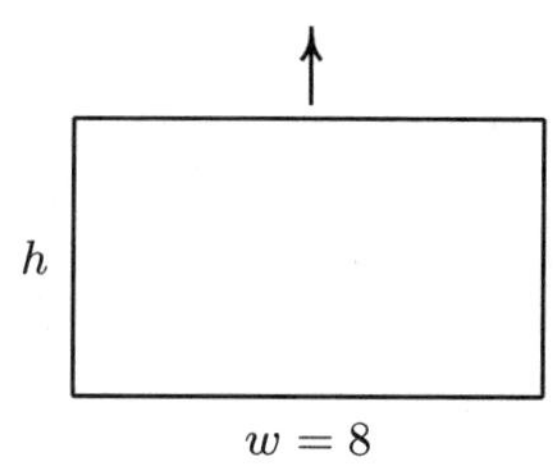

FIGURE 2

Question 1 Suppose that the width of the rectangle is held constant at $w = 8$ inches and the height increases, as in Figure 2. **(a)** Express the area A of the rectangle in terms of h. **(b)** Express the rate of change dA/dt with respect to time of the area in terms of the rate of change dh/dt of the height.

Question 2 Now suppose that the height is held at $h = 6$ inches and the width is increasing, as in Figure 3. **(a)** Express A in terms of w. **(b)** Give dA/dt in terms of dw/dt.

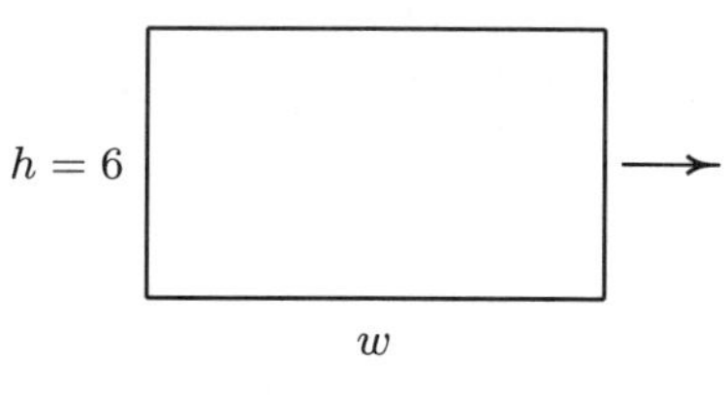

FIGURE 3

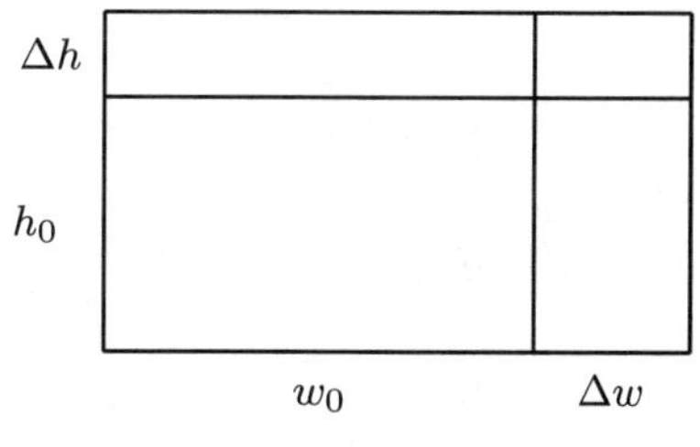

FIGURE 4

♦ SUGGESTIONS TO INSTRUCTORS: The main topics in this section can be divided into two parts, as follows:
Part 1 (The Product and Quotient Rules; Tune-up Exercises T1, T2, T4, T5, T7–T9, T11–T13; Problems 1–35, 58, 59. 67–69. 72–75) Have students begin work on Questions 1 through 6 before class. Problems 17, 20, 22, 23, and 26 are good for in-class discussion.
Part 2 (The Chain Rule for powers; Tune-up Exercises T3, T6, T10, T14–T16; Problems 36–57, 60–66, 70, 71) Have students begin work on Questions 7 through 12 before class. Problems 37, 41, 45, 52, and 57 are good for in-class discussion.

To see what happens if the width and height both change, we suppose at a fixed time t_0, the width is w_0 and the height is h_0. We consider a nonzero change Δt in the time from t_0 to $t_0 + \Delta t$, and let Δw and Δh be the corresponding changes in the width and height, so that at time $t_0 + \Delta t$, the width is $w_0 + \Delta w$ and the height is $h_0 + \Delta h$ (Figure 4). We let ΔA be the change in the area from $t = t_0$ to $t = t_0 + \Delta t$.

Question 3 Suppose that Δw and Δh are both positive, as in Figure 4. Then ΔA is the area of the three small rectangles in that drawing. Express ΔA in terms of $w_0, h_0, \Delta w$ and Δh.

We can also derive the formula from Question 3 algebraically, without the restriction that Δh and Δw be positive. The change in $A = wh$ when w is changed from w_0 to $w_0 + \Delta w$ and h is changed from h_0 to $h_0 + \Delta h$ is

$$\begin{aligned} \Delta A &= (w_0 + \Delta w)(h_0 + \Delta h) - w_0 h_0 \\ &= (w_0 h_0 + h_0 \Delta w + w_0 \Delta h + \Delta w \Delta h) - w_0 h_0 \\ &= w_0 \Delta h + h_0 \Delta w + \Delta w \Delta h. \end{aligned} \tag{1}$$

Now imagine that the graph of $w(t)$ is the curve in Figure 5. The slope of the secant line in that drawing is $\dfrac{\Delta w}{\Delta t}$ and its limit as $\Delta t \to 0$ is the derivative $\dfrac{dw}{dt}(t_0)$, which is the slope of the tangent line in Figure 6:

$$\frac{dw}{dt}(t_0) = \lim_{\Delta t \to 0} \frac{\Delta w}{\Delta t}. \tag{2}$$

Also, because $w(t)$ is continuous at t_0 (by Theorem 1 of Section 2.4),

$$\Delta w \to 0 \text{ as } \Delta t \to 0. \tag{3}$$

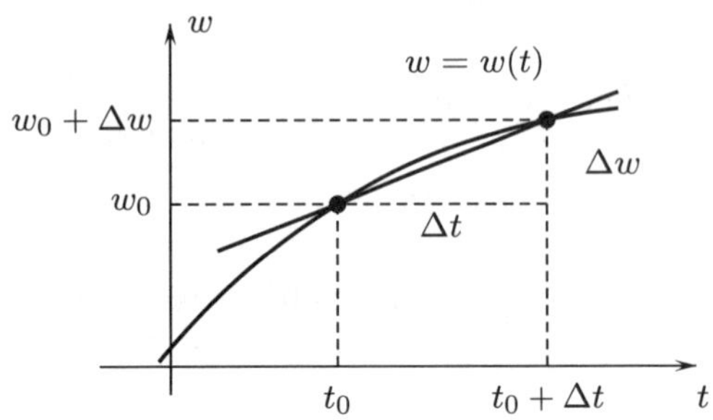

Slope $= \dfrac{\Delta w}{\Delta t}$

FIGURE 5

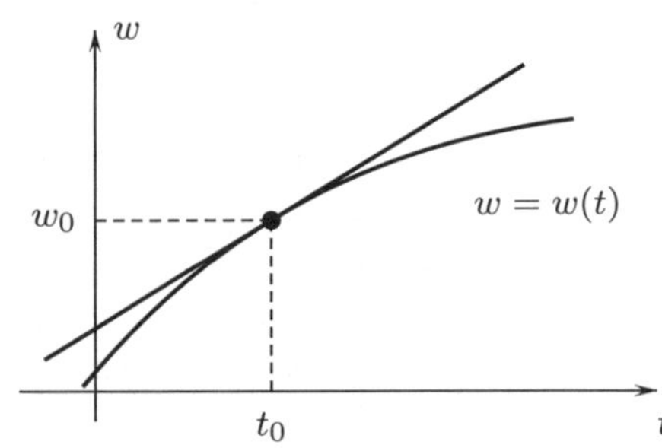

Slope $= \dfrac{dw}{dt}(t_0)$

FIGURE 6

The height $h(t)$ and area $A(t)$ have analogous graphs and

$$\frac{dh}{dt}(t_0) = \lim_{\Delta t \to 0} \frac{\Delta h}{\Delta t} \tag{4}$$

$$\frac{dA}{dt}(t_0) = \lim_{\Delta t \to 0} \frac{\Delta A}{\Delta t}. \tag{5}$$

We will find the rate of change **(5)** of the area by using the last formula in **(1)** for ΔA. We divide both sides by Δt to obtain

$$\frac{\Delta A}{\Delta t} = w_0 \frac{\Delta h}{\Delta t} + h_0 \frac{\Delta w}{\Delta t} + \Delta w \left[\frac{\Delta h}{\Delta t}\right]. \tag{6}$$

Example 1 Use equations **(2)** through **(6)** to find a formula for $\frac{dA}{dt}(t_0)$.

SOLUTION Since w_0 and h_0 are constant, equation **(6)** gives

$$\begin{aligned}\frac{dA}{dt}(t_0) &= \lim_{\Delta t\to 0}\frac{\Delta A}{\Delta t}\\ &= \lim_{\Delta t\to 0}\left(w_0\frac{\Delta h}{\Delta t} + h_0\frac{\Delta w}{\Delta t} + \Delta w\frac{\Delta h}{\Delta t}\right)\\ &= w_0\lim_{\Delta t\to 0}\frac{\Delta h}{\Delta t} + h_0\lim_{\Delta t\to 0}\frac{\Delta w}{\Delta t} + \left(\lim_{\Delta t\to 0}\Delta w\right)\left(\lim_{\Delta t\to 0}\frac{\Delta h}{\Delta t}\right).\end{aligned}$$

Then, because $\frac{\Delta h}{\Delta t}\to\frac{dh}{dt}(t_0)$, $\frac{\Delta w}{\Delta t}\to\frac{dw}{dt}(t_0)$, and $\Delta w\to 0$ as $\Delta t\to 0$,

$$\begin{aligned}\frac{dA}{dt}(t_0) &= w_0\frac{dh}{dt}(t_0) + h_0\frac{dw}{dt}(t_0) + 0\left(\frac{dh}{dt}(t_0)\right)\\ &= w(t_0)\frac{dh}{dt}(t_0) + h(t_0)\frac{dw}{dt}(t_0).\end{aligned} \quad \textbf{(7)}$$

Example 2 Find the rate of change of the area of a rectangle at a moment when the width is 8 inches, the height is 6 inches, the width is increasing 3 inches per minute, and the height is increasing 5 inches per minute.

SOLUTION We use **(7)** with $w = 8, h = 6, \frac{dw}{dt} = 3$, and $\frac{dh}{dt} = 5$. We obtain

$$\begin{aligned}\frac{dA}{dt} &= w\frac{dh}{dt} + h\frac{dw}{dt}\\ &= [8\text{ inches}]\left[5\ \frac{\text{inches}}{\text{minute}}\right] + [6\text{ inches}]\left[3\ \frac{\text{inches}}{\text{minute}}\right]\\ &= 8(5) + 6(3) = 58\ \frac{\text{square inches}}{\text{minute}}.\ \square\end{aligned}$$

The Product Rule

Formula **(7)** is an example of the PRODUCT RULE. For its general statement, we replace t by x, w by f, h by g, and the area A by fg:

Theorem 1 (The Product Rule) *If $f = f(x)$ and $g = g(x)$ have derivatives at x, then so does their product, fg, and*

$$\frac{d}{dx}(fg) = f\frac{dg}{dx} + g\frac{df}{dx} \quad \textbf{(8a)}$$

or in prime notation,

$$\frac{d}{dx}(fg) = fg' + gf'. \quad \textbf{(8b)}$$

Remember the Product Rule in words: The derivative of a product of two functions equals the first function multiplied by the derivative of the second, plus the second function multiplied by the derivative of the first.

Example 3 Find the derivative of $(x^2+3x-1)(4x^{1/2}-6)$.

SOLUTION

$$\frac{d}{dx}[(x^2+3x-1)(4x^{1/2}-6)]$$
$$= (x^2+3x-1)\frac{d}{dx}(4x^{1/2}-6)+(4x^{1/2}-6)\frac{d}{dx}(x^2+3x-1)$$
$$= (x^2+3x-1)(2x^{-1/2})+(4x^{1/2}-6)(2x+3).$$

(We do not simplify the result because we have no further use for it.) □

Related-rate problems

In Example 2 we had an equation, $A = wh$, relating three functions of time, w, h, and A. We differentiated the equation with respect to t to obtain an equation, $\frac{dA}{dt} = w\frac{dh}{dt} + h\frac{dw}{dt}$, which we could use to relate the functions and their derivatives. This sort of problem, in which we study rates of change of related functons, is called a RELATED-RATE PROBLEM. Here is another example.

Question 4 At the beginning of 1990 the total population of the U.S was 248.7 million, of whom 51.3% were women. At that time the total population was increasing at the rate of 3.5 million per year, and the percentage of women was decreasing 0.04% per year.[(1)] We let $P(t)$ be the total population (measured in millions) in year t and let $F(t)$ be the fraction that were women (the percent divided by 100). Then $P(1990) = 248.7, F(1990) = 0.513$, $\frac{dP}{dt}(1990) = 3.5$, and $\frac{dF}{dt}(1990) = -0.0004$.

(a) What was the population of women in the U. S. at the beginning of 1990?

(b) At what rate was the population of women increasing at the beginning of 1990?

(c) At what rate was the population of men increasing at the beginning of 1990?

The Quotient Rule

To derive a formula for differentiating quotients of functions, we suppose that $f(x)$ and $g(x)$ have derivatives at x and that $g(x) \neq 0$. We write f for $f(x)$ and g for $g(x)$; we let Δx denote a small, nonzero change in the variable; and we let $\Delta f = f(x+\Delta x) - f(x)$ and $\Delta g = g(x+\Delta x) - g(x)$ be the resulting changes in f and g. Then at x,

$$\frac{d}{dx}\left(\frac{f}{g}\right) = \lim_{\Delta x\to 0}\frac{\dfrac{f(x+\Delta x)}{g(x+\Delta x)} - \dfrac{f}{g}}{\Delta x} = \lim_{\Delta x\to 0}\frac{\dfrac{f+\Delta f}{g+\Delta g} - \dfrac{f}{g}}{\Delta x}.$$

We replace division by Δx with multiplication by $1/(\Delta x)$ and take a common denominator:

$$\frac{\dfrac{f+\Delta f}{g+\Delta g} - \dfrac{f}{g}}{\Delta x} = \frac{1}{\Delta x}\left(\frac{f+\Delta f}{g+\Delta g} - \frac{f}{g}\right) = \frac{1}{\Delta x}\left(\frac{g\cdot(f+\Delta f) - f\cdot(g+\Delta g)}{(g+\Delta g)g}\right)$$
$$= \frac{1}{(g+\Delta g)g}\left(g\frac{\Delta f}{\Delta x} - f\frac{\Delta g}{\Delta x}\right). \quad \textbf{(9)}$$

Question 5 What are the limits as $\Delta x \to 0$ of $g+\Delta g$, $\Delta f/\Delta x$, and $\Delta g/\Delta x$ and why?

The results of Question 5 and **(9)** give the formula we need.

$$\frac{d}{dx}\left(\frac{f}{g}\right) = \lim_{\Delta x\to 0}\frac{1}{(g+\Delta g)g}\left(g\frac{\Delta f}{\Delta x} - f\frac{\Delta g}{\Delta x}\right) = \frac{g\dfrac{df}{dx} - f\dfrac{dg}{dx}}{g^2}.$$

[(1)]Data adapted from *1993 Statistical Abstract of the United States*, Washington, DC: U. S. Department of Commerce, 1993, p. 15.

We state this formula as a theorem:

Theorem 2 (The Quotient Rule) *At any value of x where $f = f(x)$ and $g = g(x)$ have derivatives and $g(x)$ is not zero, f/g also has a derivative and*

$$\frac{d}{dx}\left(\frac{f}{g}\right) = \frac{g\dfrac{df}{dx} - f\dfrac{dg}{dx}}{g^2} \quad \textbf{(10a)}$$

or with prime notation for the derivatives of f and g,

$$\frac{d}{dx}\left(\frac{f}{g}\right) = \frac{gf' - fg'}{g^2}. \quad \textbf{(10b)}$$

Also remember **(10)** in words: The derivative of a quotient equals the denominator multiplied by the derivative of the numerator, minus the numerator multiplied by the derivative of the denominator, all divided by the square of the denominator.

Example 4 What is $\dfrac{dy}{dx}$ if $y = \dfrac{x^2 + a}{x^4 + b}$ with constants a and b?

SOLUTION For all x such that $x^4 + b \neq 0$,

$$\begin{aligned}\frac{dy}{dx} &= \frac{d}{dx}\left[\frac{x^2 + a}{x^4 + b}\right] \\ &= \frac{(x^4 + b)\dfrac{d}{dx}(x^2 + a) - (x^2 + a)\dfrac{d}{dx}(x^4 + b)}{(x^4 + b)^2} \\ &= \frac{(x^4 + b)(2x) - (x^2 + a)(4x^3)}{(x^4 + b)^2}. \ \square\end{aligned}$$

Question 6 **(a)** Figures 7 and 8 show graphs of the U. S. national debt $D(t)$ and the U. S. population $P(t)$ as functions of time.[2] Find approximate values of $D(1985)$ and $P(1985)$ from the graphs. Then draw approximate tangent lines and estimate their slopes to find approximate values of $\dfrac{dD}{dt}(1985)$ and $\dfrac{dP}{dt}(1985)$. **(b)** Use your estimates from part (a) to give the approximate debt per person and the approximate rate of increase with respect to time of the debt per person at the beginning of 1985.

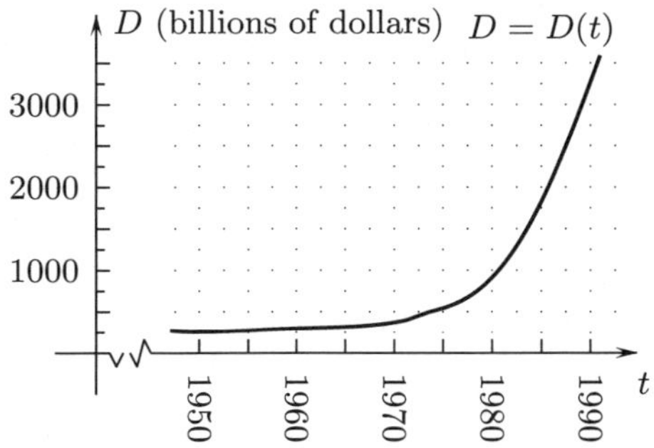

FIGURE 7

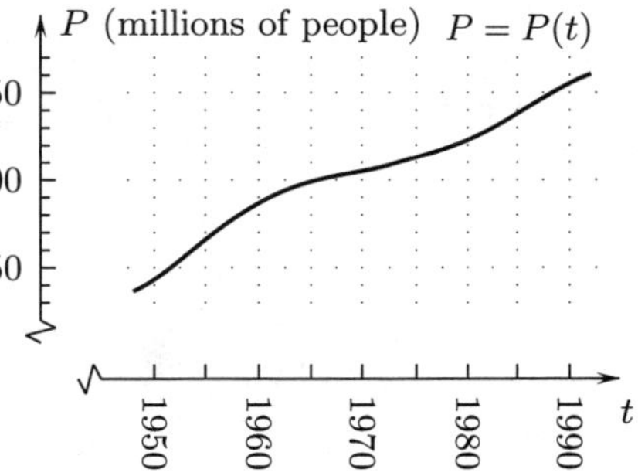

FIGURE 8

---

[2]Data adapted from *1993 Statistical Abstract of the United States*, Washington, DC: U. S. Department of Commerce, 1993, pp. 15 and 328.

Derivatives of powers of functions

The Product Rule can be used to differentiate low integer powers of a differentiable function.

Question 7 **(a)** Find the derivative of $[f(x)]^2$ by writing it as the product of $f(x)$ with itself and using the Product Rule. **(b)** Use the result of part (a) and the Product Rule to find a formula for the derivative of $[f(x)]^3$.

To find a formula for the derivative of $[f(x)]^n$ for other constants n, we start with the rule,

$$\frac{d}{dx}(x^n) = \lim_{\Delta x \to 0} \frac{(x+\Delta x)^n - x^n}{\Delta x} = nx^{n-1} \tag{11}$$

from Section 2.3. We consider a fixed x and a small nonzero Δx and set $f = f(x)$ and $\Delta f = f(x+\Delta x) - f(x)$. Then **(11)** with f in place of x yields

$$\frac{d}{dx}(f^n) = \lim_{\Delta x \to 0} \frac{(f+\Delta f)^n - f^n}{\Delta x}.$$

We assume that Δf is not zero for small nonzero Δx to avoid a technical detail that is needed to deal with the general case. Then we can multiply and divide the difference quotient on the right of the last equation by Δf:

$$\frac{(f+\Delta f)^n - f^n}{\Delta x} = \left[\frac{(f+\Delta f)^n - f^n}{\Delta f}\right]\left[\frac{\Delta f}{\Delta x}\right].$$

Since f is continuous at x (by Theorem 1 of Section 2.4), $\Delta f \to 0$ as $\Delta x \to 0$ and

$$\begin{aligned}\frac{d}{dx}(f^n) &= \left[\lim_{\Delta f \to 0} \frac{(f+\Delta f)^n - f^n}{\Delta f}\right]\left[\lim_{\Delta x \to 0} \frac{\Delta f}{\Delta x}\right] \\ &= \frac{d}{df}(f^n)\,\frac{df}{dx} = nf^{n-1}\frac{df}{dx}.\end{aligned}$$

We have used **(11)** with f in place of x for the limit in the first set of brackets and the definition of the derivative of $f(x)$ in the second set. This formula is called the CHAIN RULE FOR POWERS OF FUNCTIONS. A more general version of the Chain Rule will be presented in Section 3.3.

Theorem 3 (The Chain Rule for powers) *For constant n,*

$$\frac{d}{dx}(f^n) = nf^{n-1}\frac{df}{dx} \tag{12a}$$

provided $\frac{df}{dx} = \frac{df}{dx}(x)$ exists and $f = f(x)$ is in an open interval where $f^{n-1} = [f(x)]^{n-1}$ is defined.

We could include the variable x in **(12a)** by writing

$$\frac{d}{dx}\{[f(x)]^n\} = n[f(x)]^{n-1}\frac{df}{dx}(x) \tag{12b}$$

and use f' in place of $\frac{df}{dx}$ to have

$$\frac{d}{dx}\{[f(x)]^n\} = n[f(x)]^{n-1}f'(x). \tag{12c}$$

Equations **(12a)**, **(12b)**, and **(12c)** are the same formula with different notation. Remember it in words: The derivative of the nth power of a function equals n, multiplied by the $(n-1)$st power of the function, multiplied by the derivative of the function.

Example 5 Find the derivative of $(x^3+1)^5$ with respect to x.

SOLUTION By **(12a)** with $n = 5$ and $f(x) = (x^3+1)^5$,

$$\frac{d}{dx}[(x^3+1)^5] = 5(x^3+1)^4\frac{d}{dx}(x^3+1)$$
$$= 5(x^3+1)^4(3x^2) = 15x^2(x^3+1)^4. \square$$

Question 8 What is the derivative of $[y(x)]^{-2}$ with respect to x at $x = 0$ if $y(0) = 2$ and $\frac{dy}{dx}(0) = -5$?

Question 9 **(a)** Express the rate of change dV/dt of the volume $V = \frac{4}{3}\pi r^3$ of a sphere in terms of the radius r and the rate of change dr/dt of the radius. **(b)** At a particular moment the radius of a sphere is 3 inches and is increasing at rate of $\frac{1}{2}$ inch per second. How fast is the volume of the balloon increasing at that moment?

Notice that in the formula,

$$\frac{dV}{dt} = 4\pi r^2\frac{dr}{dt} \qquad \textbf{(13)}$$

from Question 9 for the t-derivative of the volume $V = \frac{4}{3}\pi r^3$ of a sphere, the quantity $A = 4\pi r^2$ is the surface area. Consequently, the rate of change with respect to time of the volume of the sphere equals its surface area, multiplied by the rate of change of its radius with respect to time.

Example 6 Figure 9 shows the graph of the volume V of air in a spherical balloon as a function of the time t. **(a)** Use the graph to find the approximate volume and rate of change of the volume at $t = 6$. **(b)** Use the formula $V = \frac{4}{3}\pi r^3$ to find the approximate radius of the balloon at $t = 6$. **(c)** Find the approximate rate of change of the radius at $t = 6$.

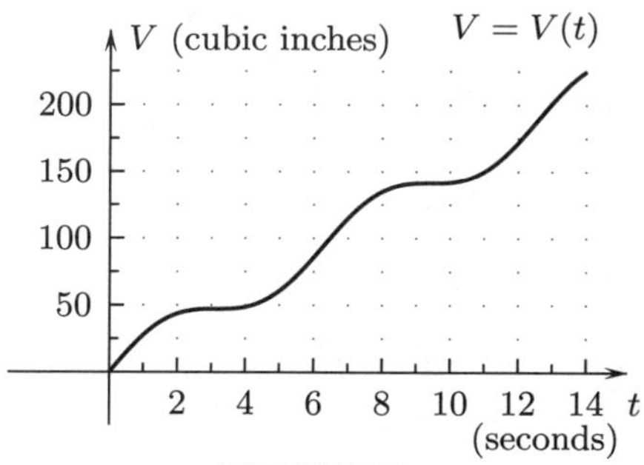

FIGURE 9

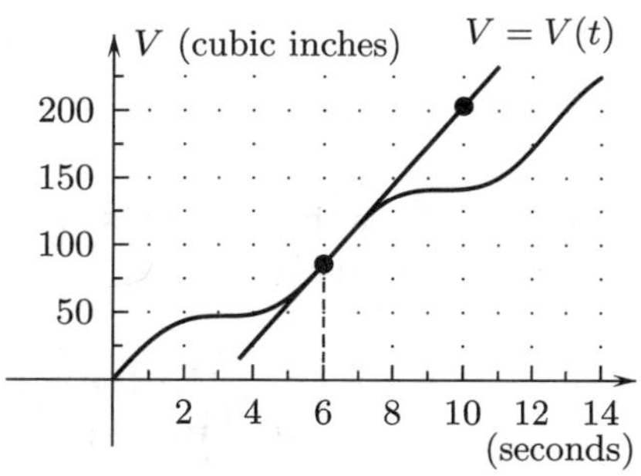

FIGURE 10

SOLUTION **(a)** From the graph we see that $V(6) \approx 85$ cubic inches. To estimate the derivative, we draw an approximate tangent line at $t = 6$ as in Figure 10. The points on the line in the drawing have approximate coordinates $(6, 85)$ and $(10, 205)$. Using these values, we find that the rise ΔV from the first to the second point on the tangent line is approximately $205 - 85 = 120$. The run Δt is 4, so

$$\frac{dV}{dt}(6) \approx \frac{\Delta V}{\Delta t} \approx \frac{120 \text{ cubic inches}}{4 \text{ seconds}} = 30 \text{ cubic inches per second.}$$

(b) We have $\frac{4}{3}\pi r^3 \approx 85$, from which we obtain $r^3 \approx \frac{3(85)}{4\pi}$ and then

$$r \approx \left(\frac{3(85)}{4\pi}\right)^{1/3} \doteq 2.7 \text{ inches.}$$

(c) The estimates from parts (a) and (b), and formula **(13)** give $30 \approx 4\pi(2.7)^2 \frac{dr}{dt}$, and then

$$\frac{dr}{dt} \approx \frac{30}{4\pi(2.7)^2} \doteq 0.33 \text{ inches per second.}$$

(Different estimates of the values of V and of its derivative would yield different answers.) □

On the order of operations

Differentiating a complicated expression may require the Product or Quotient Rule and one or more applications of the Chain Rule. You can determine the order in which to apply these operations by noting the order of the steps used in calculating values of the function. The differentiation proceeds in the reverse order.

Question 10 In what order are the calculations made in finding a value of $f(x) = \left(\frac{x+1}{x-3}\right)^5$?

Because the last step in finding a value of $f(x)$ in Question 10 is the taking of the fifth power, we find its derivative by first using the Chain Rule for differentiating the fifth power of a function:

$$\frac{d}{dx}\left[\left(\frac{x+1}{x-3}\right)^5\right] = 5\left(\frac{x+1}{x-3}\right)^4 \frac{d}{dx}\left(\frac{x+1}{x-3}\right).$$

Since the next-to-last step in calculating $f(x)$ involves division, the Quotient Rule is used next in finding its derivative:

$$\begin{aligned}\frac{d}{dx}\left[\left(\frac{x+1}{x-3}\right)^5\right] &= 5\left(\frac{x+1}{x-3}\right)^4 \frac{d}{dx}\left(\frac{x+1}{x-3}\right)\\ &= 5\left(\frac{x+1}{x-3}\right)^4 \left[\frac{(x-3)\frac{d}{dx}(x+1) - (x+1)\frac{d}{dx}(x-3)}{(x-3)^2}\right].\end{aligned} \quad \textbf{(14)}$$

Finally, because the first steps in evaluating $f(x)$ are to calculate $x+1$ and $x-3$, these functions are differentiated in the last steps:

$$\begin{aligned}\frac{d}{dx}\left[\left(\frac{x+1}{x-3}\right)^5\right] &= 5\left(\frac{x+1}{x-3}\right)^4 \left[\frac{(x-3)-(x+1)}{(x-3)^2}\right]\\ &= 5\left(\frac{x+1}{x-3}\right)^4 \left[\frac{-4}{(x-3)^2}\right].\end{aligned} \quad \textbf{(15)}$$

Be sure to write down all of the steps, as in equations **(14)** and **(15)**,when you perform complex differentiations such as this so you can concentrate on the details and check your work to see that it is correct.

Question 11 In what order would you perform the operations to calculate a value of $g(x) = x^2(x^3+2x)^{10}$? Find $g'(x)$. Do not simplify the answer.

Question 12 Express the derivative of $[x^2 + u(x)]^7$ in terms of x, $u(x)$, and $u'(x)$

A bit of history

Euclid (ca. 300 BC) stated that a line is tangent to a circle if it intersects the curve at one and only one point.[3] This definition was also used for ellipses, but could not be applied without modification to the other conic sections—parabolas and hyperbolas—since there are two such lines at each point on a parabola and three at each point on a hyperbola. For the parabola $y = x^2$ in Figure 11, the two lines are the tangent line and the line parallel to the y-axis. For the hyperbola $y = 1/x$ in Figure 12, the three lines are the tangent line and the lines parallel to the x- and y-axes.

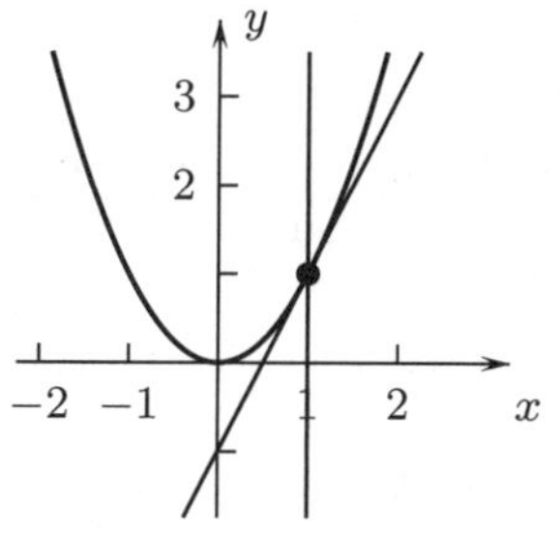

$y = x^2$

FIGURE 11

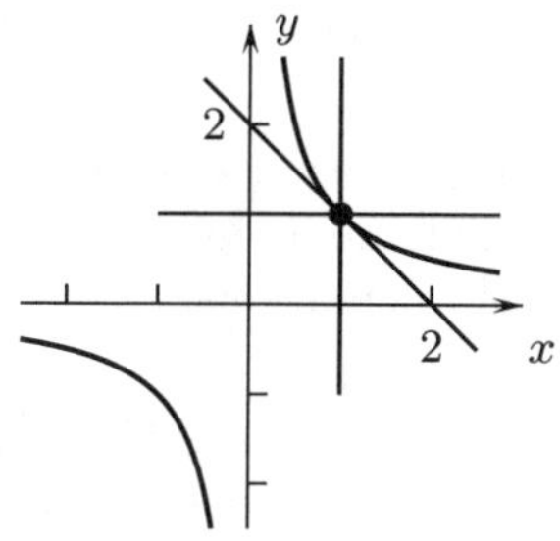

$y = 1/x$

FIGURE 12

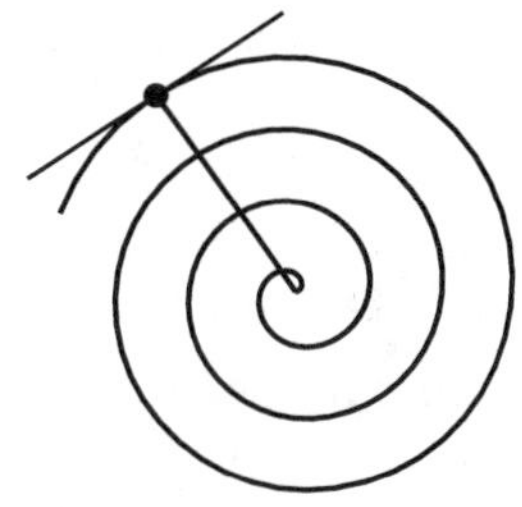

An Archimedian spiral

FIGURE 13

The definition of tangent line that Apollonius (ca. 225 BC) used in his treatise *Conics* could be applied to any conic section. He defined a tangent line to be a line such that no other straight line could fall between it and the curve.[4]

Another early approach to the study of tangent lines was based on the principle that an object's motion at each point as it moves along a curve is in the direction of the tangent line at that point. The Greek scientist and mathematician Archimedes (ca.. 287–222 BC) apparently used this idea to find the lines tangent to a spiral traversed by a point that moves along a ray at a constant speed while the ray rotates about its end at a constant speed (Figure 13). Such a spiral is now known as an Archimedean spiral.

This kinematic approach to tangent lines was also used in the 1640's by a number of scholars, including an Italian physicist and student of Galileo, Evangelista Torricelli (1608–1647). They used this approach to determine tangent lines to a variety of curves, including those given in modern notation by the equations $y = ax^n$ with various rational exponents n.

In the 1630s, Descartes devised the METHOD OF EQUAL ROOTS for finding normal lines to certain curves by studying the solutions of polynomial equations. This approach was later used extensively to find tangent lines (see Problem 75).

The need for mathematical descriptions of velocity contributed to the development of the concept of derivative. The philosopher Aristotle (384–322 BC) believed that an object falling in air would have a constant velocity proportional to its weight. This view was commonly held through the Middle Ages, although it was challenged by various scholars who observed that the velocity is not constant but increases with time. In the fourteenth century, scholastic philosophers at Merton College, Oxford, studied motion with constant acceleration and deduced what is now known as the *Merton rule:* an object with constant acceleration travels the same distance as it would if it had constant velocity equal to the average of its initial and final velocities (see Problem 76). In the seventeenth century, Galileo Galilei (1564–1642) and others discovered that in a void, all falling objects have the same constant acceleration, so that their motion may be determined by using the Merton rule. This result, however did not resolve the question of motion of an object falling in air.

The approaches to the study of tangent lines and velocity that led to the modern concept of the derivative were based on one notion or another of the "infinitely small." Some mathematicians

[3] *The Thirteen Books of Euclid* by T. Heath, Vol. 2, New York, NY: Dover Publications, Inc., 1956, p. 1.
[4] *Apollonius of Perga* by T. Heath, Cambridge: W. Heffer & Sons, 1961, p. 22.

considered a curve to be composed of infinitely short line segments and contended that the tangent line at a point is the extension of the infinitely short line segment at that point. Others stated that tangent lines are the lines through pairs of infinitely close points on the curve. Variable velocities were studied by considering the infinitely small distance that an object travels in an infinitely short period of time.

The algebraic calculations based on these ideas involved INFINITESIMALS, which are quantities that are supposed to be infinitely small and yet not zero. Questions concerning the existence of the infinitely small and the infinitely large had been actively debated by philosophers in previous centuries, and the notion of infinitely small nonzero numbers was questioned to some extent by those who used it. However, the concept of infinitesimals was accepted because of its effectiveness as a mathematical tool.

Fermat used infinitesimals to find tangent lines to curves as early as 1636, and by 1660 a number of others, including the Scottish mathematician James Gregory (1638–75) and the English theologian Isaac Barrow (1630–1677), had done similar calculations.

Techniques using infinitesimals were developed into the general operation now known as differentiation by Barrow's protégé Isaac Newton (1642–1727) in the 1660's and independently by the German philosopher Gottfried Wilhelm Leibniz (1646–1716) in the 1670's. Although Newton did his fundamental work earlier than Liebniz, he made little effort to publicize it and did not publish his results until after 1700. Leibniz published his results promptly and took a keen interest in making his techniques comprehensible and useful to a wide audience. Consequently, he exerted a much greater influence on mathematicians of the next 150 years than did Newton. Leibniz's use of infinitesimals was perhaps related to his philosophy, which contended that the "best of all possible worlds" in which we live consists of infinitesimal, indivisible, and indestructible spiritual atoms called "monads."

Leibniz used what he called "differentials" dx and dy to denote corresponding infinitesimal changes in quantities x and y related by an equation. If the equation were

$$y = x^2 \tag{16}$$

he would write the same equation with x replaced by $x + dx$ and with y replaced by $y + dy$ to obtain $y + dy = (x + dx)^2$. Expanding the square and subtracting the original equation **(16)** would give

$$dy = 2x(dx) + (dx)^2. \tag{17}$$

Leibniz would then drop the square $(dx)^2$ of the differential from equation **(17)**. He would justify this by stating that the square of the differential is negligible in comparison with the differentials themselves (as a point is negligible in comparison with a line, he once said). Without the square, the equation would be

$$dy = 2x(dx) \tag{18}$$

which corresponds to the modern formula $dy/dx = 2x$.

Let us compare this use of infinitesimals with the modern reasoning based on limits. We let Δx denote a nonzero (and noninfinitesimal) change in x and let Δy be the corresponding change in y. By the same algebraic calculations as above, we obtain the equation $\Delta y = 2x(\Delta x) + (\Delta x)^2$. Then, instead of throwing away the $(\Delta x)^2$ term, we divide the equation by Δx and take the limit as $\Delta x \to 0$ to obtain

$$\frac{dy}{dx} = \lim_{\Delta x \to 0} \frac{\Delta y}{\Delta x} = \lim_{\Delta x \to 0} (2x + \Delta x) = 2x.$$

The obvious criticism of the calculations with infinitesimals which led from equation **(17)** to equation **(18)** is that the differential dx is a nonzero quantity, and yet its square, which is dropped from the calculation, is somehow zero. The most famous attack on the use of infinitesimals was published in 1734 by an English bishop, George Berkeley, in a treatise titled *The analyst, or a discourse addressed to an infidel mathematician, wherein it is examined whether the object, principles, and inferences of the modern analyst are more distinctly conceived or more evidently deduced than religious mysteries* *and points of faith.* (The "infidel mathematician" was supposedly the astronomer Edmund Halley.) In this treatise, Berkeley ridiculed infinitesimals as "the ghosts of departed quantities" and discussed the apparent contradictions involved in their use.

At one time or another both Newton and Leibniz gave presentations of their calculus that were fairly close to the modern treatments involving limits. Newton did so, for example, in his theory of prime and ultimate ratios, and Leibniz in a reply to a critic, Nieuwentijt.

Nevertheless, Newton and Leibniz and the other prominent mathematicians of the late seventeenth and the eighteenth centuries generally relied on the use of infinitesimals. Their intuition and insight into the problems they studied enabled them to develop most of what is now known as elementary calculus and a great deal of more advanced mathematics, even though their arguments did not meet modern standards of mathematical analysis.

It was not until the nineteenth century that calculus was widely studied as a systematic application of the concept of limit. This change in point of view was stimulated by the work of the Swiss mathematician L. Euler (1707–1783), the French mathematicians J. Le Rond D'Alembert (1717–1783), S. L'Hullier (1750–1840), A. Cauchy (1789–1857), and the Czechoslovakian priest B. Bolzano (1781–1848).

Principles and procedures

- The Product and Quotient Rules and the Chain Rule for powers (Theorems 1, 2, and 3) are used to express the derivatives of products, quotients and powers of functions in terms of those functions and their derivatives.
- Complex expressions are generally evaluated from the inside out. Differentiation of complex expressions usually goes in the reverse order, from the outside in.
- If two or more functions of one variable are related by an equation, then that equation can be differentiated, using the Product, Quotient, and Chain Rules and other differentiation formulas as necessary, to obtain a new equation relating the functions and their derivatives. Narrative problems that require this technique are called "related rate problems" because they involve rates of change of related functions.
- In your algebra classes, you used letters as symbols to represent unknown numbers. You then performed algebraic operations with the letters to find their values or other properties. In the solutions of most related-rate problems, letters such as y or f are used to represent unknown functions, and you need to use symbols such as y' or df/dt for their derivatives.

Tune-Up Exercises 2.5[♦]

[A]*Answer provided.* [O]*Outline of solution provided.*

Find the derivatives in Exercises T1 through T7. Do not simplify your answers in Exercises T1 or T2.

T1.[A] $\dfrac{d}{dx}[(5x^3 + 2x^2 - 4)(x^7 - 2x^5)]$

♦ Type 1, the Product Rule

T2.[O] $\dfrac{dy}{dx}$ for $y = \dfrac{\sqrt{x}}{x^5 + 1}$

♦ Type 3, the Quotient Rule

T3.[O] $\dfrac{d}{dx}[(a + bx^2 + cx^5)^{10}]$ with constants a, b, c.

♦ Type 1, the Chain Rule for powers

T4.[O] $P'(5)$ where $P(x) = R(x)S(x), R(5) = 3,$ $S(5) = 4, R'(5) = -3,$ and $S'(5) = 10$

♦ Type 1, the Product Rule

T5.[O] $\dfrac{dW}{dx}(4)$ where $W(x) = \dfrac{Y(x)}{Z(x)}, Y(4) = 2,$ $\dfrac{dY}{dx}(4) = 3, Z(4) = 5,$ and $\dfrac{dZ}{dx}(4) = 6$

♦ Type 3, the Quotient Rule

T6.[O] $\dfrac{dG}{dx}(3)$ where $G(x) = [y(x)]^{7/2}, y(3) = 6$ and $\dfrac{dy}{dx}(3) = -5$

♦ Type 1, the Chain Rule for powers

T7.[O] Give an equation of the tangent line to $y = (x^2 + x^3 + x^4)(x^5 + x^6 + x^7)$ at $x = 1$.

♦ Type 1, the Product Rule and a tangent line

[♦]The Tune-up Exercises and Problems are classified by type and content. The types are (1) basic, reactive; (2) basic reflective; (3) intermediate, reactive; (4) intermediate, reflective; (5) advanced, reactive; (6) advanced, reflective; and (7) advanced, theoretical.

T8.^{O} Figures 14 and 15 give the graphs of differentiable functions $A(x)$ and $B(x)$. Give approximate values of AB, A/B, and of their first derivatives at $x = 2$.

♦ Type 1, the Product and Quotient Rules with graphs, etc.

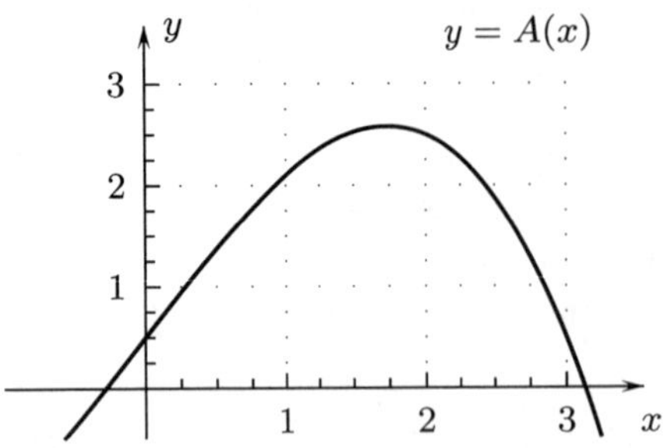

FIGURE 14

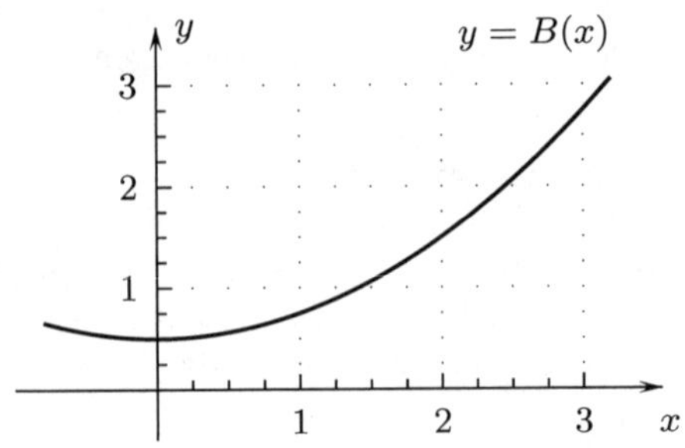

FIGURE 15

More practice

Find the derivatives in Exercises T9 through T14. Do not simplify your answers in Exercises T9 or T12.

T9.^{O} $\dfrac{d}{dx}[(x^2 + 3x - 1)(4x^{1/2} - 6)]$

♦ Type 1, the Product Rule

T10.^{O} $f'(t)$ for $f(t) = \sqrt{t^2 - 4}$

♦ Type 1, the Chain Rule for powers

T11.^{O} $y'(2)$ for $y = x^3 Z(x)$ with $Z(2) = 5$ and $Z'(2) = 10$

♦ Type 2, the Product Rule

T12.^{O} $\dfrac{dy}{dx}$ for $y = \dfrac{ax^2 + 2}{bx^4 + 4}$ with constants a and b

♦ Type 2, the Quotient Rule

T13.^{O} $f'(1)$ for $f(t) = t^2 - \dfrac{t}{t-6}$

♦ Type 3, the Quotient Rule

T14.^{O} $\dfrac{d}{dx}[(x^3 + 1)^5]$

♦ Type 1, the Chain Rule for powers

T15.^{O} Give an equation of the tangent line to $y = (x^4 - 8)^{1/3}$ at $x = 2$.

♦ Type 1, the Chain Rule for powers and a tangent line

T16.^{O} Figure 16 shows the graph of a differentiable function $U(x)$. Give approximate values of **(a)** U, **(b)** $\dfrac{dU}{dx}$, **(c)** $\dfrac{d}{dx}(xU)$, and **(d)** $\dfrac{d}{dx}(U^6)$ at $x = 2$.

♦ Type 2, the Product and Chain Rules with a graph

FIGURE 16

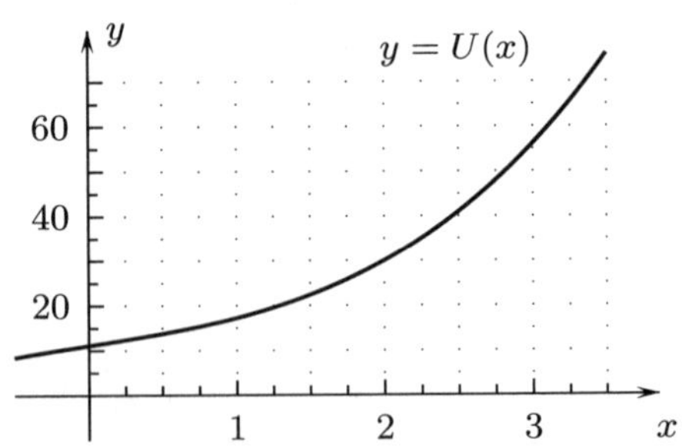

Problems 2.5

[A]*Answer provided.* [O]*Outline of solution provided.*

Find the derivatives in Problems 1 through 18.

1.[O] $\dfrac{d}{dx}[(1+3x-x^2)(x^2-5)]$

♦ Type 1, the Product Rule

2.[A] $\dfrac{d}{dx}[(x-x^2)(1+x^{-1}+x^{-2})]$

♦ Type 1, the Product Rule

3. $\dfrac{d}{dx}[(4x^2+5)(6x^{-2}-3)]$

♦ Type 1, the Product Rule

4.[A] $\dfrac{d}{dx}[(x^2+x^{-2})(2+3x)]$

♦ Type 1, the Product Rule

5.[A] dy/dx for $y = (1+ax^n)(1-bx^n)$ with constants a, b, and n

♦ Type 1, the Product Rule

6. dP/dz for $P = (1-\sqrt{z})(1+\sqrt{z}-3z)$

♦ Type 1, the Product Rule

7.[A] $\dfrac{d}{dx}\left(\dfrac{x^2+x}{x^2-2}\right)$

♦ Type 1, the Quotient Rule

8. $\dfrac{d}{dt}\left(\dfrac{t^{1/3}}{t^3+1}\right)$

♦ Type 1, the Quotient Rule

9.[A] $y'(0)$ for $y = (5+3x+7x^2)(x^3-2x+5)$

♦ Type 1, the Product Rule

10. $v'(1)$ for $v(z) = (2-z^{1/2}+z^{9/4})(z^2-1)$

♦ Type 1, the Product Rule

11.[A] $\dfrac{dy}{dx}(1)$ for $y = \dfrac{1+x+x^2}{3-x^3}$

♦ Type 1, the Quotient Rule

12. $y'(2)$ for $y = \dfrac{3+4/x}{1+8/x}$

♦ Type 1, the Quotient Rule

13. The derivative of $\dfrac{A+Bx^2}{C+Dx^3}$ at $x=1$ with constants a, b, A, B, C and D

♦ Type 1, the Quotient Rule

14. $\dfrac{dy}{dt}(1)$ for $y = \dfrac{1+t^2+t^3}{t^4+10t^3}$

♦ Type 1, the Quotient Rule

15.[A] $f'(10)$ where $f(x) = g(x)h(x)$, $g(10) = -4, h(10) = 560, g'(10) = 0$, and $h'(10) = 35$

♦ Type 1, the Product Rule

16. $y'(5)$ where $y(x) = (x^2+4)z(x), z(5) = -4$, and $z'(5) = 10$

♦ Type 1, the Product Rule

17.[A] $\dfrac{dR}{ds}(1)$ where $R(s) = \dfrac{P(s)}{Q(s)}, P(1) = 13$, $Q(1) = -2, \dfrac{dP}{ds}(1) = 7$, and $\dfrac{dQ}{ds}(1) = -4$

♦ Type 1, the Quotient Rule

18. $y'(-3)$ where $y(x) = \dfrac{z(x)}{1+x^2}, z(-3) = 6$, and $z'(-3) = 15$

♦ Type 3, the Quotient Rule

19.[A] Give an equation of the tangent line to $y = (1+x-x^4)(1-x+x^3)$ at $x=1$. Generate the curve and the line in the window $-1 \le x \le 2, -2 \le y \le 4$ as a partial check of your answer.

♦ Type 1, the Product Rule and a tangent line

20. Give equations of **(a**[A]**)** the tangent line to $y = \dfrac{x^2-1}{x^2+1}$ at $x=3$, **(b)** the tangent line to $y = (1+x+x^2)(1-x^{-1}-x^{-2})$ at $x=1$, **(c)** the tangent line to $y = (x^2-5x+1)(x^3-5x^2+2)$ at $x=5$, and **(d)** the tangent line to $y = \dfrac{x^3}{1-x^2}$ at $x=2$. Generate the curves and the tangent lines in suitable windows as a partial check of your answers.

♦ Type 1, the Product and Quotient Rules with tangent lines

21.[O] At the beginning of 1991 there were 2.1 million farms in the United States with an average size of 467 acres per farm; the number of farms was decreasing 0.035 million farms per year; and the average size was increasing 7 acres per farm per year.[(5)] What was the total acreage of farms and at what rate was it increasing or decreasing at the beginning of 1991?

♦ Type 2, the Product Rule, etc.

22.[O] At what rate is the area of a rectangle increasing or decreasing at a moment when it is 5 meters wide and 7 meters long, its width is increasing 3 meters per second, and its length is decreasing 6 meters per second?

♦ Type 2, the Chain Rule for powers

[(5)]Data adapted from *1993 Statistical Abtstract of the United States*, Washington, DC: U. S. Bureau of the Census, 1993, p. 655.

23. Figure 17 gives the number of inmates $N(t)$ in U. S. federal prisons and Figure 18 gives the the percent $P(t)$ who were incarcerated for drug offences in year t.[(6)] **(a)** Approximately how many more federal prisoners were held for drug offences in 1993 than in 1979? **(b)** What was the approximate rate of change of the number of federal prisoners being held for drug offences at the beginning of 1988?

◆ Type 2, the Product Rule with graphs

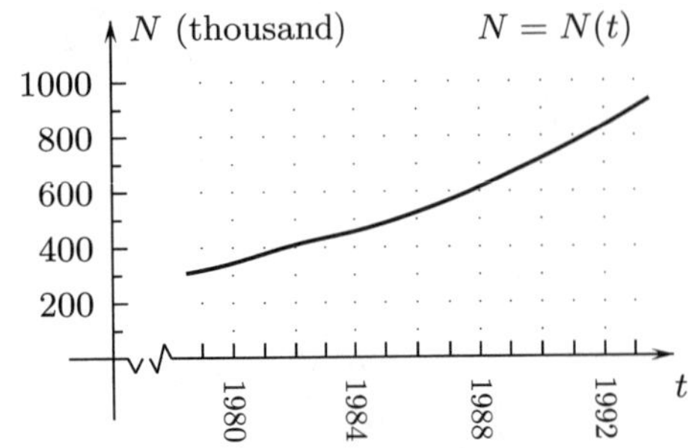

FIGURE 17

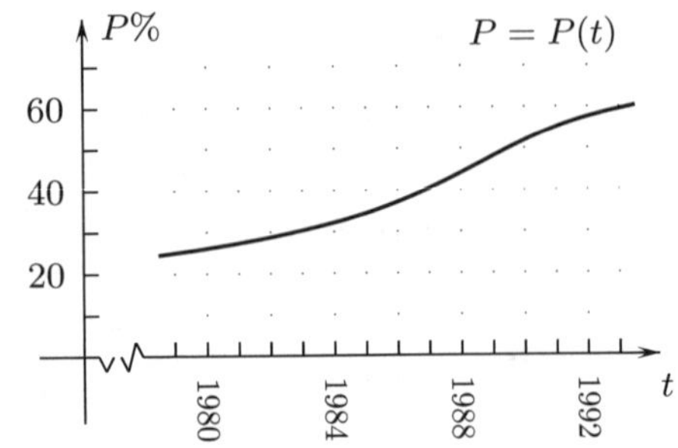

FIGURE 18

24. At the beginning of 1990, annual health care costs in the U. S. were \$2600 per capita and were rising \$260 per capita per year. At that time the population of the U. S. was 250 million and was increasing at the rate of 2.6 million per year.[(7)] At what rate were the annual health costs for the entire country increasing at the beginning of 1990?

◆ Type 2, the Product Rule with graphs

25. In 1950 there were 196 thousand dairy farms and 1,440 thousand dairy cows in the United Kingdom. At the beginning of 1982, there were 52 thousand dairy farms and 3,275 thousand dairy cows; the number of dairy farms was decreasing at the rate of 2 thousand per year; and the number of dairy cows was increasing at the rate of 18 thousand per year.[(8)] Let $A(t)$ denote the average number of dairy cows per farms in t A. D. for $t \geq 1950$. **(a)** Give the values of $A(1950)$ and $A(1986)$. **(b)** What are the values of $\dfrac{A(1986) - A(1950)}{1986 - 1950}$ and $\dfrac{dA}{dt}(1986)$ and what do they represent in terms of rates of change?

◆ Type 2, the Quotient Rule, etc.

26. Figures 19 and 20 give the number $N(t)$ (millions) of MasterCard and Visa accounts and the total outstanding debt $D(t)$ (million dollars) in the U. S. as functions of the year.[(9)] What were the approximate average debt per credit card and the rate of change of the average debt per credit card at the beginning of 1988?

◆ Type 2, the Quotient Rule with graphs

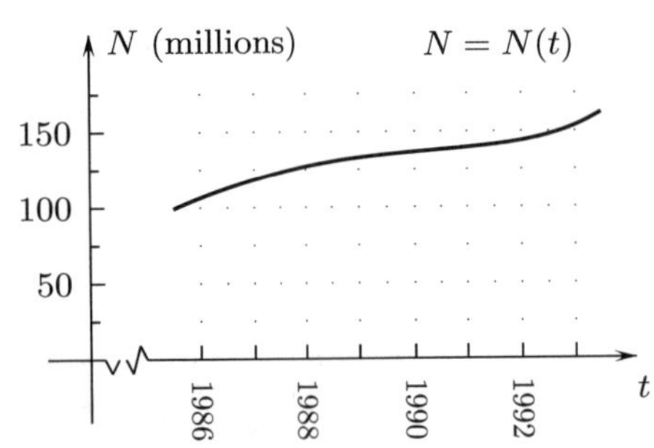

FIGURE 19

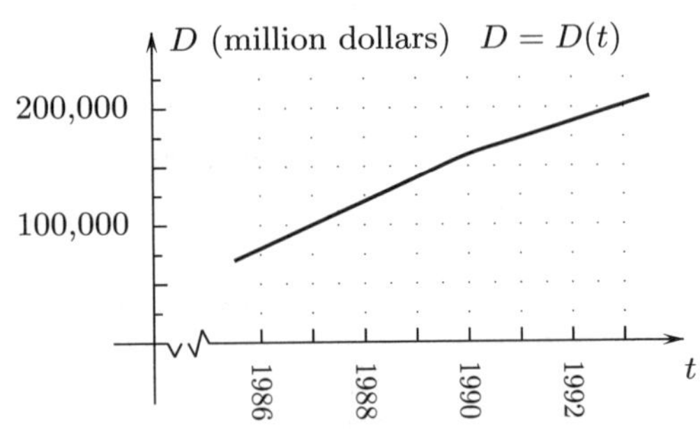

FIGURE 20

---

(6) Data adapted from *Newsweek*, October 17, 1994, New York, NY: Newsweek, Inc., p. 87.

(7) Data adapted from *1993 Statistical Abtstract of the United States*, Washington, DC: U. S. Bureau of the Census, 1993, p. 109.

(8) Data adapted from "Animal behavior and welfare" by P. English and D. MacDonald, *Bioindustrial Ecosystems*, Amsterdam: Elsevier, 1986, p. 90.

(9) Data adapted from *Los Angeles Times*, November 26, 1993, Los Angeles, CA: Times Mirror Company.

27. The upper curve in Figure 21 is the graph of the ratio $r_C(t)$ of median annual earnings of women in the U.S. with 16 or more years of schooling to the median annual earnings of women who had exactly 12 years of schooling. The lower curve is the graph of the ratio $r_P(t)$ of median annual earnings of women in the U.S. with 9–11 years of schooling to the median annual earnings of those who had exactly 12 years of schooling.[10] **(a)** At approximately what value of t is $r_C(t) - r_P(t)$ a maximum? **(b)** What are the approximate values of $\dfrac{r_C(t)}{r_P(t)}$ and $\dfrac{d}{dt}\left[\dfrac{r_C(t)}{r_P(t)}\right]$ at $t = 1985$ and what are the significance of these numbers?

♦ Type 4, the Quotient Rule, etc.

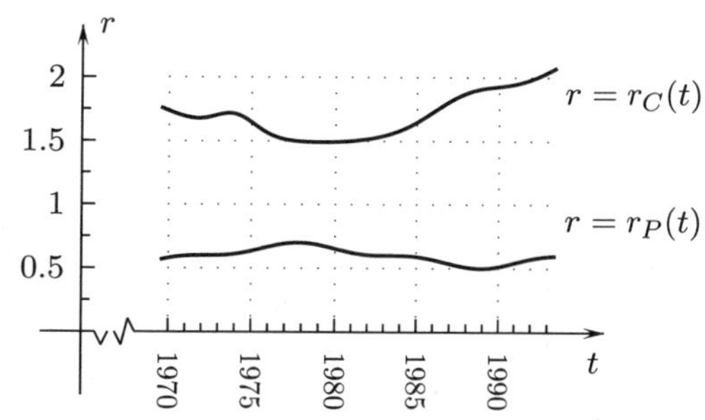

FIGURE 21

28. The table below gives the number $D_1(t)$ of children in the U. S. under the age of 18 for every 100 people between the ages of 18 and 65, and the number $D_2(t)$ of senior citizens over the age of 65 for every 100 people between the ages of 18 and 65 as functions of the year.[11] Figure 22 shows the graphs of $D_1(t)$ and $D_2(t)$, interpolated from the data in the table. For the purpose of this problem, we will consider everyone under 18 and over 65 to be dependents. **(a)** What does $D_1(t) + D_2(t)$ represent and in what years was it a maximum and a minimum? **(b)** What does $F(t) = \dfrac{D_1(t)}{D_1(t) + D_2(t) + 100}$ represent and what is the value of $F(1985)$? **(c)** What does $F'(1985)$ represent and what is its approximate value?

Year $= t$	1900	1910	1920	1930	1940	1950	1960	1970	1980	1990
$D_1(t)$	72.6	65.7	64.0	58.6	48.8	51.1	65.3	61.1	46.2	41.7
$D_2(t)$	7.3	7.5	8.0	9.1	10.9	13.4	16.9	17.6	18.7	20.3

♦ Type 4, the Quotient Rule, etc.

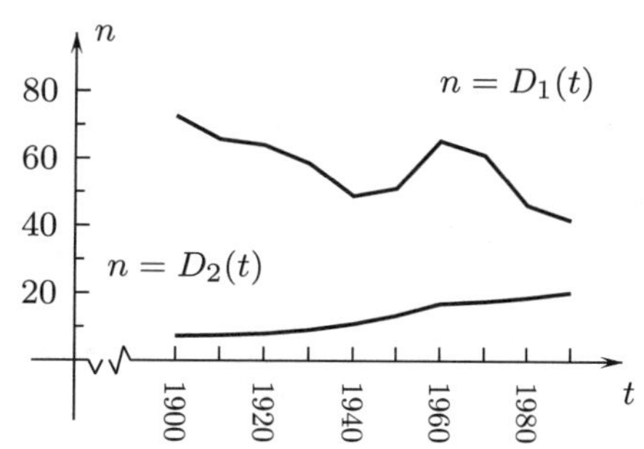

FIGURE 22

29. Figures 23 and 24 give the rate $S(t)$ at which shoes were purchased in the U. S. and the U. S. population $P(t)$ as functions of the year.[12] What do $\dfrac{S(t)}{P(t)}$ and $\dfrac{d}{dt}\left[\dfrac{S(t)}{P(t)}\right]$ represent, and what are their approximate values at $t = 1982$?

♦ Type 2, the Quotient Rule with graphs

[10] Data adapted from *The Cost of Higher Education*, Washington, DC: U. S. Department of Education, 1995, p. 16.

[11] *The Census and You*, Vol. 31, No. 4, Washington, DC: Government Printing Office, 1996, p. 7.

[12] Data adapted from *Technology and Environment* by J. Ausbel and H. Sladovich, Washington, DC: National Academy of Sciences, 1989, p. 53.

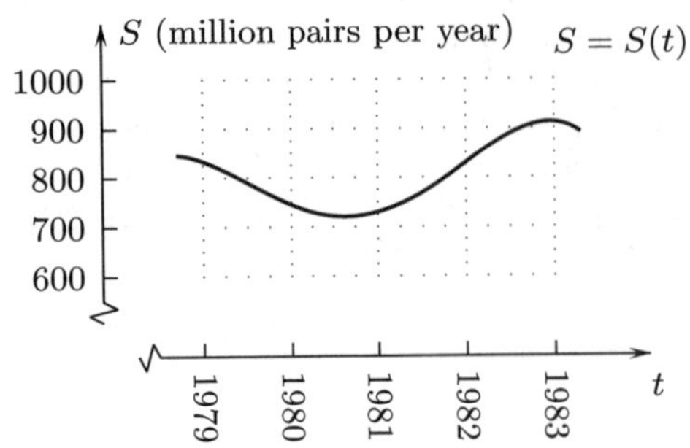

FIGURE 23

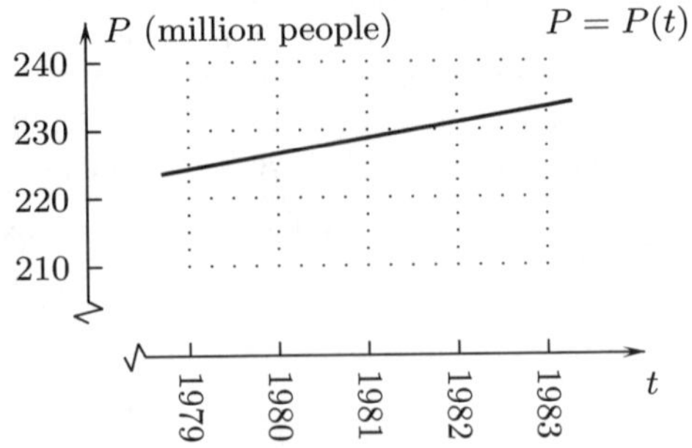

FIGURE 24

30. Imagine that your investments are in the stock market, in real estate, and in livestock. Imagine that on April 15 the value of your stocks is 1.2 million dollars and is rising at the rate of 0.05 million dollars per year; the value of your real estate is 2.1 million dollars and is falling 0.1 million dollars per year; and the value of your livestock is 0.5 million dollars and is rising 0.05 million dollars per year. **(a)** What is the total value of your investments on April 15? **(b)** What percent of your investments is in the stock market on April 15? **(c)** At what rate is the total value of your investments increasing or decreasing on April 15? **(d)** At what rate is the percent of your investments in real estate increasing or decreasing on April 15?

♦ Type 4, the Quotient Rule, etc.

31.[A] What is $P'(3)$ if $P(x) = x^2Q(x), Q(3) = 5$, and $Q'(3) = -6$?

♦ Type 1, the Product Rule

32. What is $\dfrac{dZ}{dy}(9)$ if $Z(y) = \dfrac{R(y)}{S(y)}, R(9) = 2, \dfrac{dR}{dy}(9) = 4, S(9) = 6$, and $\dfrac{dS}{dy}(9) = 8$?

♦ Type 1, the Quotient Rule

33. What is $V'(2)$ where $V(x) = \dfrac{x + U(x)}{x - U(x)}, U(2) = 3$, and $U'(2) = -2$?

♦ Type 2, the Quotient Rule

34. What is the derivative of $\dfrac{G(x)}{x}$ at $x = 2$ if the tangent line to $y = G(x)$ at $x = 2$ has the equation $y = 7 + 2x$?

♦ Type 2, the Quotient Rule

35. What is the derivative of $W(x)$ at $x = 2$ if $W(2) = 6$ and the derivative of $\dfrac{W(x)}{x}$ is 4 at $x = 2$?

♦ Type 2, the Quotient Rule

Find the derivatives in Problems 36 through 46.

36.[A] $\dfrac{d}{dx}[(2 + 3x^2 + 4x^5)^{10}]$

♦ Type 1, the Chain Rule for powers

37. $\dfrac{d}{dt}\sqrt{t^2 + 4}$

♦ Type 1, the Chain Rule for powers

38.[A] dy/dx for $y = (10 - x^{1/2})^{-4}$

♦ Type 1, the Chain Rule for powers

39. $G'(z)$ for $G(z) = \sqrt[3]{100z + 45}$

♦ Type 1, the Chain Rule for powers

40.[A] $y'(3)$ for $y = (x^4 - 3x^3 + 1)^{100}$

♦ Type 1, the Chain Rule for powers

41. $V'(0)$ for $V = (2 - 4t + t^5)^{-3}$

♦ Type 1, the Chain Rule for powers

42.[A] $\dfrac{d}{dx}[(3x - 4)^{-1/3}]$

♦ Type 1, the Chain Rule for powers

43.[A] $\dfrac{d}{dx}[(6\sqrt{x} + 3)^3]$

♦ Type 1, the Chain Rule for powers

44. $\dfrac{d}{dt}[(t^5 - 3t^2 + 1)^{-13}]$

♦ Type 1, the Chain Rule for powers

45.[A] $\dfrac{dW}{du}(0)$ where $W(u) = \sqrt{Z(u)}, Z(0) = 9$, and $\dfrac{dZ}{du}(0) = 10$

♦ Type 1, the Chain Rule for powers

46. $\dfrac{dB}{dv}(9)$ where $B(v) = [A(v)]^7, A(9) = -1$, and $\dfrac{dA}{dv}(9) = 12$

♦ Type 1, the Chain Rule for powers

47. Give equations of **(a[A])** the tangent line to $y = (x^2 + 4x)^2$ at $x = -2$, **(b)** the tangent line to $y = \sqrt{x^4 + 9}$ at $x = 2$, and **(c)** the tangent line to $y = \sqrt{4 - x}$ at $x = 3$. Generate the curves and lines in suitable windows.

♦ Type 1, the Chain Rule for powers and tangent lines

48.[A] The density of dry air at a pressure of one atmosphere and a temperature of $T°$C is $\rho = 1.293(1 + 0.00367T)^{-1}$ grams per liter.[(13)] What is the density and the rate of change of the density with respect to temperature at 20°C?

♦ Type 2, the Chain Rule for powers, etc.

49. A leaky bucket contains $V(t) = (3 - t)^2$ gallons of water from $t = 0$ (hours) until the bucket is empty. **(a)** When does the bucket have 4 gallons of water in it? **(b)** When is the bucket empty? **(c)** What is the rate of flow out of the bucket at $t = 1$?

♦ Type 2, the Chain Rule for powers, etc.

50.[O] What is the rate of change with respect to time of the volume $V = w^3$ of an expanding cubic crystal at a moment when its width is 10 millimeters and its width is increasing 3 millimetes per day?

♦ Type 2, the Chain Rule for powers

51.[A] The one-dimensional density, measured in mass per centimeter, of a rod of mass 160 grams and length L centimeters is $\rho = \dfrac{160}{L}$ grams per centimeter. The rod expands when it is heated. What are **(a)** its density and **(b)** the rate of change with respect to temperature of its density at a moment when it is 40 centimeters long and its length is increasing 0.01 centimeters per degree?

♦ Type 2, the Chain Rule for powers, etc.

52. At what rate is the radius r of a circle decreasing when the area of the circle is 16 square inches if the area is decreasing 3 square inches per minute?

♦ Type 2, the Chain Rule for powers, etc.

53. The weight of an object is the force of gravity on it. If an object weighs 100 pounds on the surface of the earth, then it weighs $w(r) = 100(1 + \frac{1}{4000}r)^{-2}$ pounds at an altitude of r miles above the earth. **(a)** Is $w(r)$ a decreasing or an increasing function for $r > 0$? **(b)** What does the object weigh and how rapidly is its weight decreasing when it is 400 miles above the earth if it is rising 15 miles per second?

♦ Type 2, the Chain Rule for powers, etc.

54. The volume of punch in a hemispherical bowl of radius 10 inches is $10\pi h^2 - \frac{1}{3}\pi h^3$ cubic inches when the punch is h inches deep. **(a)** What are the volume and the rate of change of the volume with respect to h at $h = 5$? **(b)** At what rate is the volume increasing with respect to time at a moment when $h = 5$ inches and h is increasing 6 inches per minute?

♦ Type 2, the Chain Rule for powers, etc.

55. An object moves along the parabola $y = 4x^2$ in such a way that its x-coordinate increases at the constant rate of 5 units per minute. **(a)** What is the object's y-coordinate and how rapidly is it decreasing when its x-coordinate is -2? **(b)** How rapidly is the object's distance to the origin decreasing when its x-coordinate is -2?

♦ Type 2, the Chain Rule for powers, etc.

56. A powerboat with its engine off has velocity $v = \dfrac{200}{t + 20}$ feet per second at time t (seconds) for $t \geq 0$. **(a)** When is the velocity 4 feet per second? **(b)** When is the boat decelerating 0.08 feet per second^2?

♦ Type 2, the Chain Rule for powers, etc.

57.[A] The force of air resistance (drag) on a car is $D = \frac{1}{30}v^2$ pounds when the velocity is v miles per hour.[(14)] The car is accelerating at a constant rate of 500 miles per hour^2. What is the rate of increase with respect to time of the drag when the car is going 50 miles per hour?

♦ Type 2, the Chain Rule for powers, etc.

---

(13) *CRC Handbook of Chemistry and Physics*, R. Weast and ¿ Astle, editors, Boca Raton, FL: CRC Press, Inc., 1981, p. F-11.

(14) Data adopted from *Fluid Dynamic Drag* by S. Hoerner, Published by the author, 1958, p. 12.

58. At the beginning of 1992, the population of Washington, DC, was 589,000 and was dropping 2% per year.[15] **(a)** What was the rate of change with respect to time of the population of Wahington, DC at the beginning of 1990? **(b)** The area of Washington, DC is 68 square miles. At what rate was the population density of Washington, DC, measured in population per square mile, changing at the beginning of 1992?

♦ Type 2, the Product Rule, etc.

59.[A] Figures 25 and 26 show the graphs of the width $w(t)$ (meters) and height $h(t)$ (meters) of a rectangle as functions of the time t. What is the approximate rate of change of the area of the rectangle with respect to t at $t = 30$?

♦ Type 2, the Product Rule, etc.

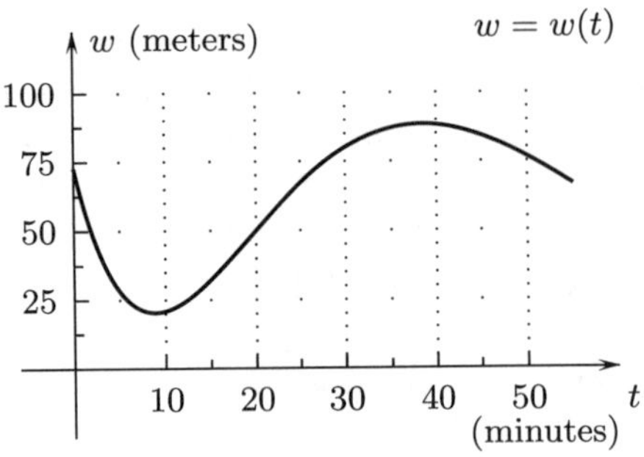

FIGURE 25

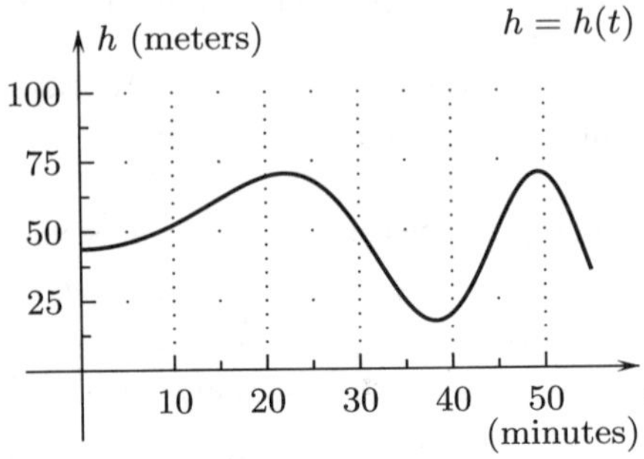

FIGURE 26

60. Imagine a water pipe with a faucet at the end and with the other end connected to a water supply that applies 50 pounds per square inch of water pressure when the faucet is closed. When the valve is opened and water flows through the pipe, the water pressure in the pipe drops. According to an example of BERNOULLI'S LAW, the water pressure in the pipe is $p = 50 - 2v^2$ pounds per square inch when the water is flowing v feet per second. Figure 27 shows the graph of the velocity of water as a function of t (seconds) for $0 \le t \le 200$. **(a)** What is the approximate maximum water pressure in the pipe for $0 \le t \le 200$? **(b)** What is the approximate rate of change of the water pressure with respect to t at $t = 150$?

♦ Type 2, the Chain Rule for powers with a graph

FIGURE 27

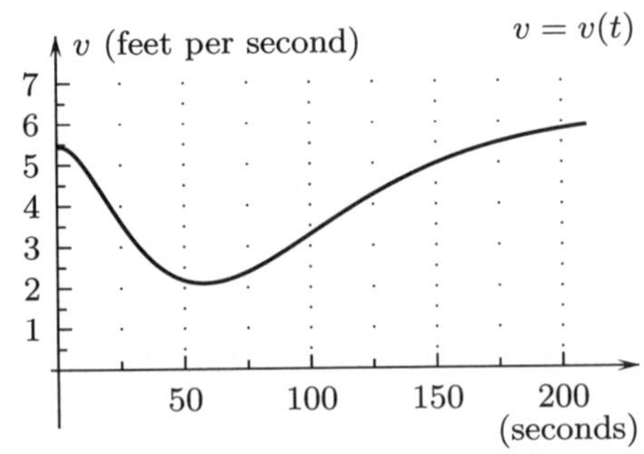

61.[A] Find $\dfrac{dy}{dx}$ for $y = \dfrac{(x+1)^4}{2x+3}$.

♦ Type 3, the Product, Quotient, and Chain Rules

62.[A] Find $y'(x)$ for $y = x(5x+4)^{1/4}$.

♦ Type 3, the Product, Quotient, and Chain Rules

63. Find $\dfrac{dy}{dx}$ for $y = \dfrac{x + \sqrt{5x-2}}{x^2+9}$.

♦ Type 3, the Product, Quotient, and Chain Rules

64. $\dfrac{d}{dx}[(x^2+3)^{1/3}(x^3+2)^{1/2}]$

♦ Type 3, the Product, Quotient, and Chain Rules

65. $\dfrac{d}{dt}(t\sqrt{2t+3})$

♦ Type 3, the Product, Quotient, and Chain Rules

66. Find $y'(t)$ for $y(t) = \dfrac{(t^3+1)^{10}}{t}$.

♦ Type 3, the Product, Quotient, and Chain Rules

---

[15] Data adapted from *1993 Statistical Abtstract of the United States*, Washington, DC: U. S. Bureau of the Census, 1993, p. 28.

67.[A] What is an object's acceleration in the positive s-direction at $t = 1$ (seconds) if it is at $s = \dfrac{t+1}{t+4}$ (centimeters) on an s-axis at time t?

♦ Type 3, the Product, Quotient, and Chain Rules

68.[A] An object is at $s = (3t^2+4)(6t^3-t+1)$ (inches) on an s-axis at time t (hours). Give a formula for its acceleration in the positive s-direction as a function of t.

♦ Type 3, the Product, Quotient, and Chain Rules

69. What is $\dfrac{d^2y}{dx^2}$ for $y = (x^2-1)(x^3+x)$?

♦ Type 3, the Product, Quotient, and Chain Rules

70. The lateral surface area of a right circular cylinder of height h (inches) and base of radius r (inches) is $A = \pi r\sqrt{r^2+h^2}$. Give a formula for $\dfrac{dA}{dt}$ in terms of $r, h, \dfrac{dr}{dt}$, and $\dfrac{dh}{dt}$.

♦ Type 3, the Product, Quotient, and Chain Rules

71. Differentiate $(x^3-5)^2$, first by applying the Chain Rule and then by expanding the square and differentiating the resulting expression.

♦ Type 2, the Chain Rule for powers

72. Derive the Quotient Rule from the Product Rule and the Chain Rule by writing $f(x)/g(x) = f(x)[g(x)]^{-1}$.

♦ Type 4, the Product and Quotient Rules

73. **(a)** Derive the Product Rule $\dfrac{d}{dx}[fgh] = \dfrac{df}{dx}gh + f\dfrac{dg}{dx} + fg\dfrac{dh}{dx}$. Then use this formula to find the derivatives of **(b**[A]**)** $(x^2+1)(x^3+2)(x^4+3)$ and of **(c)** $(x+1)(x^{1/2}+x)(x^{-1}+x-3)$.

♦ Type 4, a generalized Product Rule

74. In this problem we will illustrate Descartes' "method of equal roots" by using it to find the tangent line to $y = x^3$ at $x = 1$. The line with slope m through the point $(1,1)$ on the graph (Figure 28) has the equation $y = 1 + m(x-1)$, and the x-coordinates of the intersections of the curve with the line are the solutions of the third-degree equation,

$$1 + m(x-1) = x^3. \qquad \textbf{(19)}$$

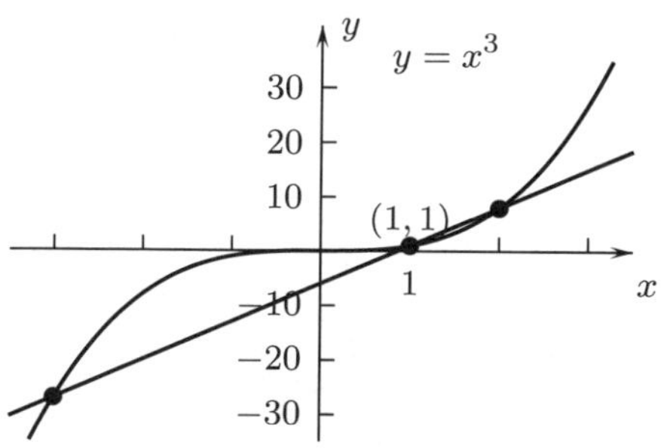

FIGURE 28

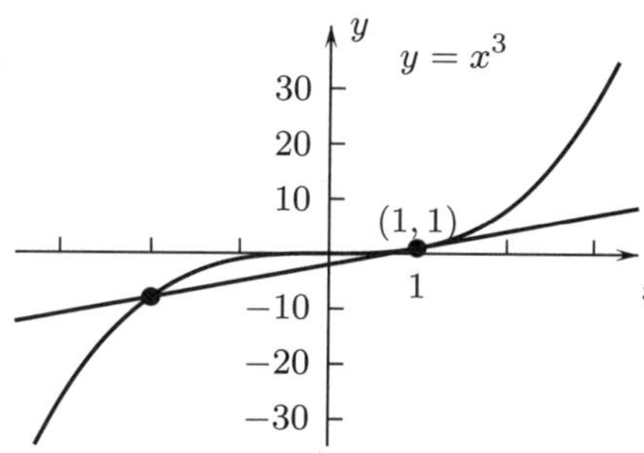

FIGURE 29

For all values of m, $x = 1$ is one solution of **(19)** because the curve and the line intersect at $(1,1)$, and for most values of m there are two more intersections and two more solutions ("roots") of **(19)**. In fact,when we solve **(19)**, we obtain three expressions that give three different values of m. For the value of m corresponding to the tangent line at $x = 1$, however, there are only two intersections and only two solutions of **(19)** (Figure 29). Therefore, we can find the slope of the tangent line by determining the value of m for which two of the three expressions for solutions of **(19)** give $m = 1$, i.e., for which two of the usual three "roots" are equal to 1. **(a)** Carry out this calculation by writing **(19)** first as $(x^3-1) - m(x-1) = 0$ and then as $(x-1)(x^2+x+1-m) = 0$, so you can use the Quadratic Formula. **(b)** Explain why letting $m \to 3$, so that two roots of **(19)** come together, amounts to having a secant line approach the tangent line. **(c)** For what other value of m does **(19)** have two equal roots, and what tangent line does that give?

♦ Type 6, the method of equal roots

75. Derive the Merton rule that is described in the historical notes from the calculus formula $s = \frac{1}{2}at^2 + bt$ for the distance traveled in time t by an object with constant acceleration a.

♦ Type 4, the Merton Rule

RESPONSES TO QUESTIONS IN CHAPTER 2

Responses 2.1

Response 1

t (hours)	0	1	2	3	4	5
$s(t)$ (miles)	100	131	168	217	284	375

Response 2

In the four hours from $t = 1$ to $t = 5$, the plane travels $375 - 131 = 244$ miles. • Its average velocity is $\dfrac{244 \text{ kilometers}}{4 \text{ hours}} = 61$ miles per hour.

Response 3

The plane's average velocity for $0 \le t \le 4$ is $\dfrac{s(4) - s(0)}{4 - 0} = \dfrac{284 - 100}{4} = 46$ miles per hour. • The secant line in Figure R3 passes through the points $(0, 100)$ and $(4, 284)$ on the graph.

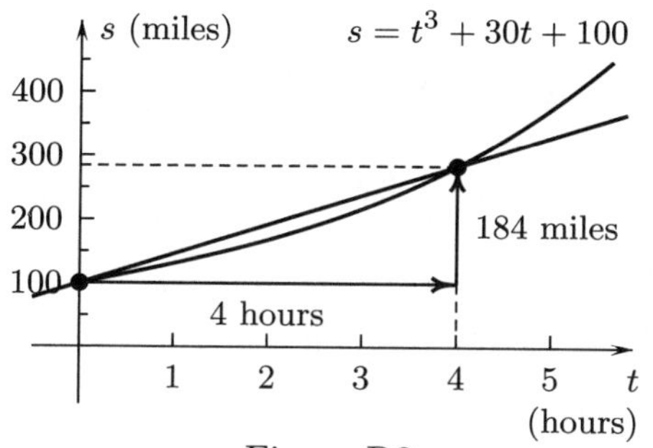

Figure R3

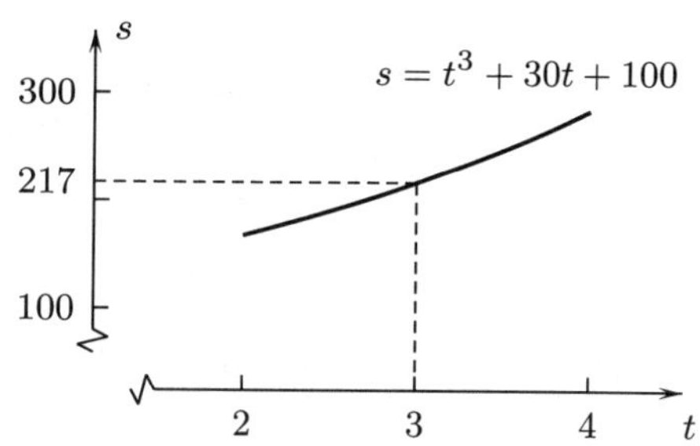

Figure R4a

Response 4

The first three graphs are in Figures R4a,b,c here. The fourth is Figure 4 in Section 2.1. • One description: As you zoom in, the curve looks more and more like a line.

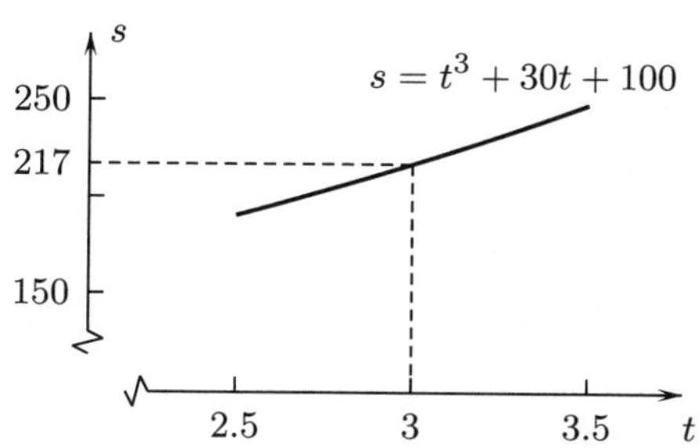

Figure R4b

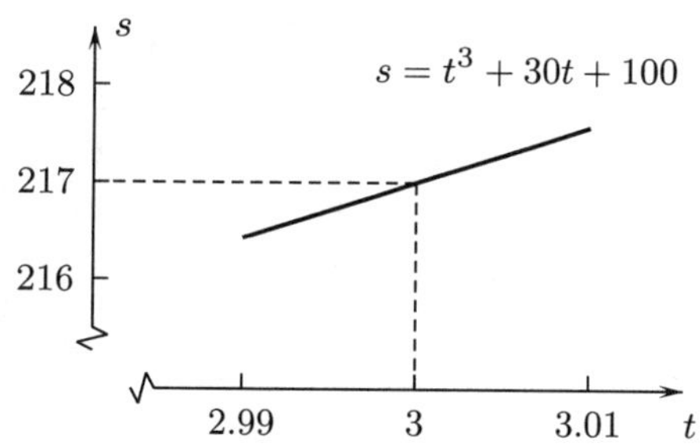

Figure R4c

Response 5

The plane's average velocity for $2.999 \le t \le 3$ is $\dfrac{s(3) - s(2.999)}{3 - 2.999} \doteq \dfrac{217 - 216.943009}{0.001} \doteq 56.991$ miles per hour. •

Its average velocity for $3 \le t \le 3.001$ is $\dfrac{s(3.001) - s(3)}{3.001 - 3} \doteq \dfrac{217.057009 - 217}{0.001} \doteq 57.009$ miles per hour.

Response 6

$80t - 16t^2 = 16t(5 - t)$ is zero at $t = 0$, when the pitcher throws the ball, and at $t = 5$, which is when he catches it.

Response 7 The ball goes approximately 100 feet above the pitcher's hand and reaches that height at approximately $t = 2.5$. (These are, in fact, the exact answers, as you will be able to find using a derivative after you have studied Section 2.4.) • The ball spends the same amount of time going up as coming down.

Response 8 The ball's average upward velocity is **(a)** $\frac{96-0}{2-0} = 48$ feet per second for $0 \le t \le 2$ (Figure R8a) and **(b)** $\frac{64-96}{4-3} = -32$ feet per second for $3 \le t \le 4$ (Figure R8b).

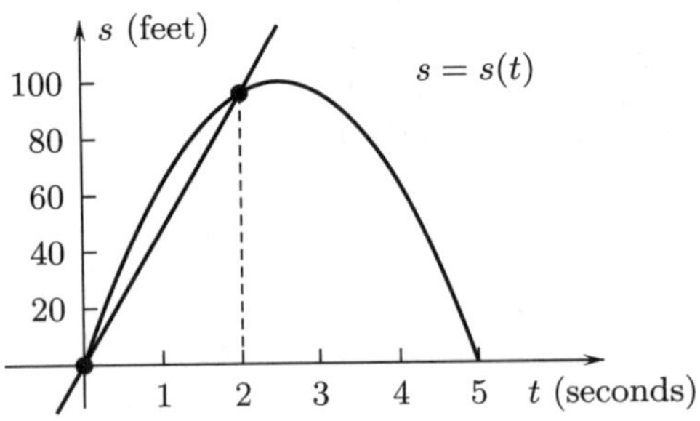

Figure R8a

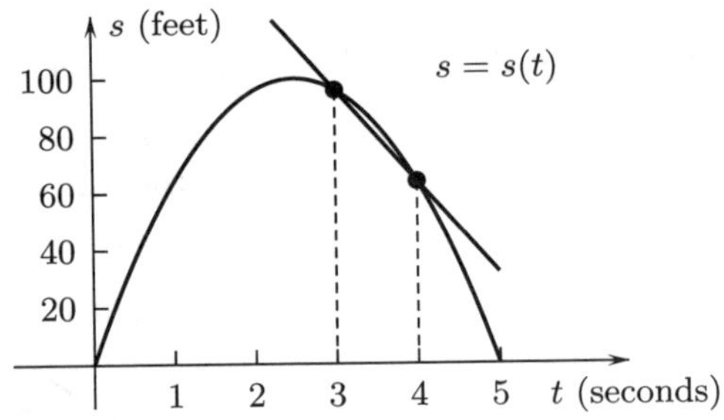

Figure R8b

Response 9 The ball's upward velocity at $t = 0$ is approximately $\frac{s(0.001) - s(0)}{0.001 - 0}$ $= \frac{0.079984}{0.001} = 79.984$ feet per second, and approximately $\frac{s(5) - s(4.999)}{5 - 4.999} = \frac{-0.079984}{0.001} = -79.984$ feet per second at $t = 5$.

Response 10 Possible answers: With air resistance the ball goes about 76.90 feet above the pitcher's hand and takes approximately 2.06 seconds to reach that height, because the maximum value of $s_R(t)$ is approximately 76.90 feet and occurs at $t \approx 2.06$. He catches the ball about 4.45 seconds after he throws it, because $s_R(t) = 0$ at $t \approx 4.45$. • It takes the ball more time to come down than to go up.

Response 11 The ball's upward velocity at $t = 0$ is approximately $\frac{s_R(0.001) - s_R(0)}{0.001 - 0} \doteq \frac{0.07998}{0.001} \doteq 79.98$ feet per second. • The upward velocity is approximately $\frac{s_R(4.450) - s_R(4.449)}{4.450 - 4.449}$ $\doteq \frac{0.08142 - 0.14108}{0.001} \doteq -59.66$ feet per second at $t = 4.45$.

Response 12 Possible answers: With air resistance, the ball does not go as high, it reaches its highest point sooner, and it hits the pitcher's hand sooner than with no air resistance. Also with air resistance, it takes the ball longer to come down than to go up and the ball is going more slowly when it is caught than when it is thrown.

Response 13 The cow weighed the most at about age 7.5, $w(7.5) \approx 1175$, and $w(2) \approx 875$, so her average rate of growth in this time period is approximately $\frac{1175 - 875}{7.5 - 2} \doteq 54.5$ pounds per year. • Figure R13 (Different estimates yield different answers.)

Response 14 One description: The bat's metabolism drops from 200 calories per hour to 1.4 calories per hour as the temperature drops from 41.5°C to 2.0°C and then increases slightly to 5.4 calories per hour as the temperature drops from 2.0°C from 0.5°C (which is close to the freezing temperature of water).

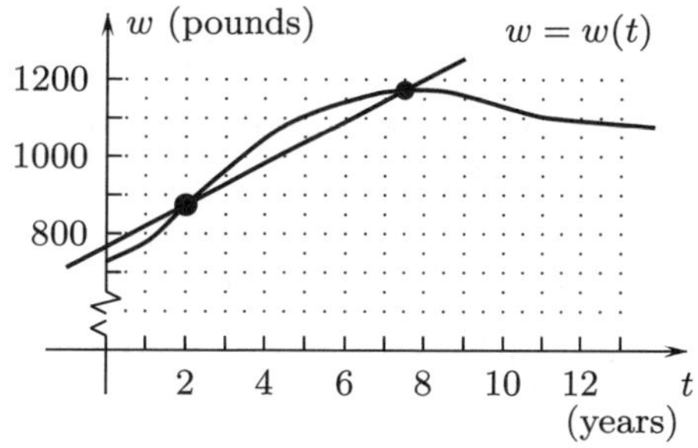

Figure R13

Response 15 Possible answer: $a = 20.0$ and $b = 30.0$, since $\dfrac{96-19}{30-20} = 7.7$

Responses 2.2

Response 1 Figure R1

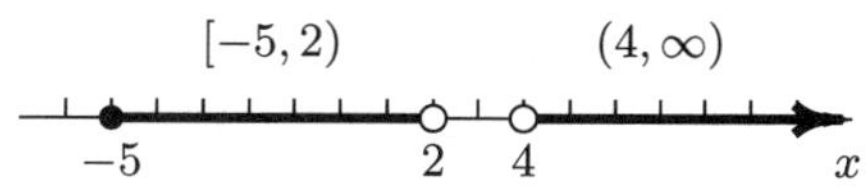

Figure R1

Response 2 The closure of $(-1,2]$ is $[-1,2]$; its interior is $(-1,2)$. • The closure of $(5,7)$ is $[5,7]$; its interior is itself, since it is open. • The closure of $[-10,-5]$ is itself, since it is closed; its interior is $(-10,-5)$. • The closure of $(-\infty,3]$ is itself; its interior is $(-\infty,3)$.

Response 3 The tables are below. • One description: $G(x)$ seems to be approaching 3 in the table on the left and seems to be approaching 2 in the table on the right.

x	$G(x) \doteq$
0.9	3.19
0.99	3.0199
0.999	3.001999
0.9999	3.00019999

x	$G(x) \doteq$
1.1	2.2
1.01	2.02
1.001	2.002
1.0001	2.0002

Response 4 **(a)** For $x > 1$, $G(x) = 2x$ is greater than 2, so it differs from 2 by less than 0.01 if $2x < 2.01$. Dividing this inequality by the positive number 2 yields the equivalent inequality $x < 1.005$. • $G(x)$ differs from 2 by less than 0.01 for $1 < x < 1.005$. **(b)** For $x > 1$, $G(x) = 2x$ differs from 2 by less than ϵ if $2x < 2+\epsilon$. Dividing this inequality by the positive number 2 yields the equivalent inequality $x < 1 + \frac{1}{2}\epsilon$. • $G(x)$ differs from 2 by less than ϵ for $1 < x < 1 + \frac{1}{2}\epsilon$.

Response 5 **(a)** $J(-2) = [x+4]_{x=-2} = 2$ • $\lim_{x\to -2^+} J(x) = 2$

(b) $J(3) = \left[\dfrac{3}{x}\right]_{x=3} = 1$ • $\lim_{x\to 3^-} J(x) = 1$

(c) $J(0) = [x+4]_{x=0} = 4$ • $\lim_{x\to 0} J(x) = 4$

Response 6 **(a)** The domain of $y = 1 + \sqrt{x-2}$ is $[2,\infty)$ where the square root is defined. **(b)** $y = 1 + \sqrt{x-2}$ is continuous in its entire domain, $[2,\infty)$.

Response 7 **(a)** Because $|x-1| = x-1$ for $x > 1$ and $|x-1| = -(x-1)$ for $x < 1$, the graph consists of the portion of the line $y = 1$ for $x > 1$ and the portion of $y = -1$ for $x < 1$.
(b) $(x-1)/|x-1|$ is defined for all $x \neq 1$, so it is defined in the intervals $(-\infty, 1)$ and $(1, \infty)$.
(c) $y = (x-1)/|x-1|$ is continuous in the two intervals $(-\infty, 1)$ and $(1, \infty)$, which cannot be extended because the function is not defined at $x = 1$.

Response 8 $\lim_{x\to 3^+} N(x) = \lim_{x\to 3^+} \frac{7}{\sqrt{x}} = \frac{7}{\sqrt{3}}$ since $N(x) = \frac{7}{\sqrt{x}}$ for $x > 3$ and $\frac{7}{\sqrt{x}}$ is continuous at 3. • $\lim_{x\to 3^-} N(x) = \lim_{x\to 3^-} (\frac{1}{4}x^2) = \frac{1}{4}(3^2) = \frac{9}{4}$ since $N(x) = \frac{1}{4}x^2$ for $x < 3$ and $\frac{1}{4}x^2$ is continuous at 3.

Response 9 $(2)(5) + \frac{6}{3} = 12$

Response 10 If f and g are continuous at x_0, then $\lim_{x\to x_0} f(x) = f(x_0)$ and $\lim_{x\to x_0} g(x) = g(x_0)$, so that by Theorem 2,
(a) $\lim_{x\to x_0} [f(x) + g(x)] = \lim_{x\to x_0} f(x) + \lim_{x\to x_0} g(x) = f(x_0) + g(x_0)$
(b) $\lim_{x\to x_0} [f(x)g(x)] = \left[\lim_{x\to x_0} f(x)\right]\left[\lim_{x\to x_0} g(x)\right] = f(x_0)g(x_0)$ and
(c) if $g(x_0) \neq 0$, $\lim_{x\to x_0} \left[\frac{f(x)}{g(x)}\right] = \frac{\lim_{x\to x_0} f(x)}{\lim_{x\to x_0} g(x)} = \frac{f(x_0)}{g(x_0)}$.
These statements show that $f + g$, fg, and f/g are continuous at x_0.

Response 11 The table is below. • Likely prediction: $\lim_{x\to 2} \frac{x^2-4}{x-2} = 4$

x	$\frac{x^2-4}{x-2}$
1.9	3.9
1.99	3.99
1.999	3.999
1.9999	3.9999

x	$\frac{x^2-4}{x-2}$
2.1	4.1
2.01	4.01
2.001	4.001
2.0001	4.0001

Response 12 For $x \neq 0$, $\frac{(x+2)^2-4}{(x+1)^2-1} = \frac{x^2+4x}{x^2+2x} = \frac{x(x+4)}{x(x+2)} = \frac{x+4}{x+2}$ • $\lim_{x\to 0} \frac{(x+2)^2-4}{(x+1)^2-1} = \lim_{x\to 0} \frac{x+4}{x+2} = \frac{0+4}{0+2} = 2$ since $\frac{x+4}{x+2}$ is continuous at 0. • Figure R12

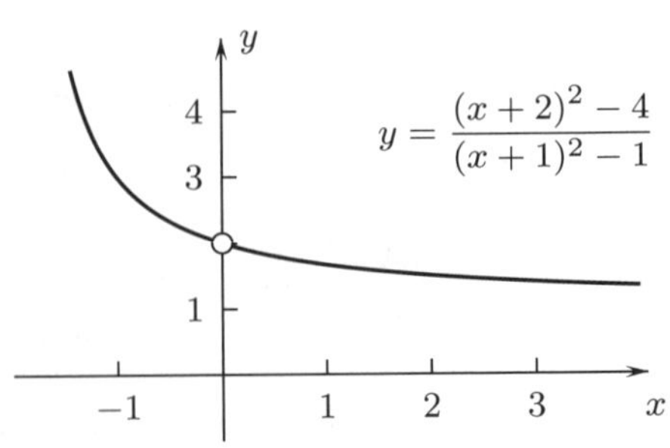

Figure R12

Response 13 (a) Only $P(x)$ is continuous in $[-1, 2]$.
(b) $P(x) = k$ has solutions f or all k with $-1 \le k \le 3$.
(c) $Q(x) = k$ has solutions for all k with $-1 \le k \le 1$ and with $2 < k \le 3$.
(d) $R(x) = k$ has solutions for all k with $-1 \le k \le 3$.

Response 14 One possible solution: After some experimentation we find that $x+2x^2+3x^3+4x^4$ has the value 0 at $x = 0$ and the value 10,738 at $x = 7$. By the Intermediate-Value Theorem, $x + 2x^2 + 3x^3 + 4x^4$ has the value 10,000 at one or more x's in $[0, 7]$ because 10,000 is between 0 and 10,738, and $x + 2x^2 + 3x^3 + 4x^4$ is continuous in $[0, 7]$.

Responses 2.3

Response 1 Because the points $(4, 0.16)$ and $(8, 0.08)$ are on the tangent line, the rate of change is

$$[\text{Slope}] = \frac{\text{Rise}}{\text{Run}} = \frac{[0.08 - 0.16]\ \text{percent}}{[8 - 4]\ \text{hour}} = -0.02 \text{ percent per hour.}$$

Eight hours after drinking the whiskey, the man's blood alcohol level is dropping 0.02 percent per hour.

Response 2 The table is completed below. • Likely prediction: $f'(1) = 2$

b	2.5	2	1.5	1.1	1.001	1.00001
$\frac{(4b - b^2) - 3}{b - 1} \doteq$	0.5	1	1.5	1.9	1.999	1.99999

b	−0.5	0	0.5	0.9	0.999	0.99999
$\frac{(4b - b^2) - 1}{b - 3} \doteq$	3.5	3	2.5	2.1	2.001	2.00001

Response 3 The table is below. • Likely predictions: The derivative of x^2 at 1 is 2, the derivative of x^3 at 1 is 3, the derivative of x^{-1} at 1 is -1, and the derivative of $x^{1/2}$ at $x = 1$ is $\frac{1}{2}$.

Curve	Slope	$a = 1, b = 1.1$	$a = 1, b = 1.001$	$a = 1, b = 1.00001$
$y = x^2$	$\frac{b^2 - 1}{b - 1} \doteq$	2.1	2.001	2.00001
$y = x^3$	$\frac{b^3 - 1}{b - 1} \doteq$	3.31	3.0030010	3.0000300
$y = x^{-1}$	$\frac{b^{-1} - 1}{b - 1} \doteq$	−0.9090909	−0.9990010	−0.9999900
$y = x^{1/2}$	$\frac{b^{1/2} - 1}{b - 1} \doteq$	0.4880885	0.4998751	0.4999988

Response 4 The division is completed below. •
$\frac{d}{dx}(x^4) = \lim_{b\to x} \frac{b^4 - x^4}{b - x} = \lim_{b\to x} (b^3 + b^2x + bx^2 + x^3) = 4x^3$

$$
\begin{array}{rllllllll}
 & b^3 & + & b^2x & + & bx^2 & + & x^3 \\ \hline
b - x \,\big) & b^4 & & & & & - & x^4 \\
 & b^4 & - & b^3x & & & & \\ \cline{2-4}
 & & & b^3x & & & & \\
 & & & b^3x & - & b^2x^2 & & \\ \cline{4-6}
 & & & & & b^2x^2 & & \\
 & & & & & b^2x^2 & - & bx^3 \\ \cline{6-8}
 & & & & & & & bx^3 - x^4 \\
 & & & & & & & bx^3 - x^4 \\ \cline{8-8}
 & & & & & & & 0
\end{array}
$$

Response 5 $\dfrac{5X^4}{3X^2} = \dfrac{5x^{4/3}}{3x^{2/3}} = \frac{5}{3}x^{(4/3)-(2/3)} = \frac{5}{3}x^{2/3}$ and this equals nx^{n-1} with $n = \frac{5}{3}$.

Response 6 $\dfrac{-1}{x^8}(4x^3) = -4x^{-5}$ and this is nx^{n-1} with $n = -4$.

Response 7 The car's velocity t hours after noon is $\dfrac{d}{dt}(5t^2 + 20t + 40) = 10t + 20$ miles per hour. At $t = 4$ it is going 60 miles per hour.

Response 8 $y' = \dfrac{d}{dx}(x^3 - 4x + 5) = 3x^2 - 4$ • $y(-1) = 8$, $y'(-1) = -1$ • Tangent line: $y = 8 - [x - (-1)]$ or $y = 8 - (x + 1)$

Responses 2.4

Response 1 **(a)** The disks compress more when more weight is put on them.
(b) The tangent line at $L = 50$ in Figure 8 of Section 2.4 intersects the y-axis at approximately $(0, 1)$ and intersects the line $L = 100$ at about $(100, 12)$, so its slope is about $(12 - 1)/100 = 0.11$. •

$$\frac{dS}{dL}(50) \approx 0.11 \text{ percent per kilogram.}$$

The tangent line at $L = 400$ intersects $L = 500$ at approximately $(500, 23)$ and intersects $L = 300$ at about $(300, 21)$, so its slope is approximately $(23-21)/200 = 0.01$. •

$$\frac{dS}{dL}(400) \approx 0.01 \text{ percent per kilogram.}$$

(c) $\dfrac{dS}{dL}(L)$ decreases as L increases. **(d)** Figure R1

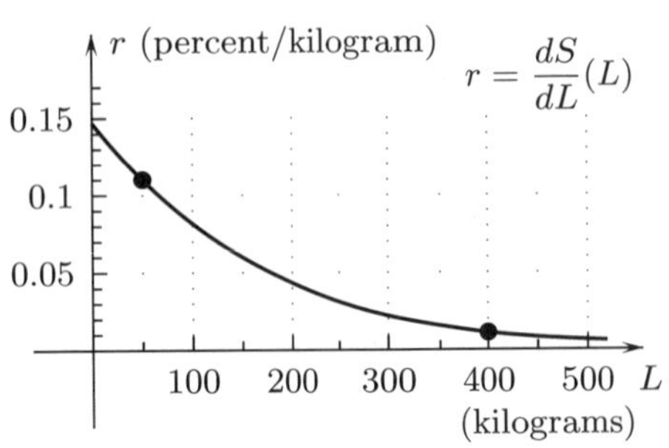

Figure R1

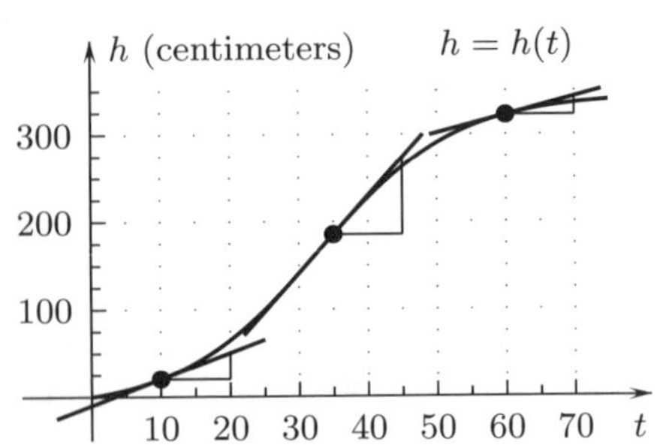

Figure R5

Response 2

$$\frac{dT}{dx}(x) = \begin{cases} 0.2 & \text{for} \quad 0 < x < 1000 \\ 0.1 & \text{for} \quad 1000 < x < 2000 \\ 0.05 & \text{for} \quad x > 2000 \end{cases}$$

Response 3 **(a)** $\frac{dR}{dx} = \frac{d}{dx}(2.50x + 0.04x^2) = 2.50 + 0.08x$ • [Marginal revenue at $x = 100$] $= \frac{dR}{dx}(100) = 2.50 + 0.08(100) = 10.50$ dollars per crystal
(b) [Marginal revenue at 100] $= R(101) - R(100) = 660.54 - 650$ $= 10.54$ dollars per crystal • The two values differ by 4 cents per crystal.

Response 4 **(a)** $W'(9) \approx \dfrac{16.00 - 15.32}{12 - 9} \doteq 0.23$ miles per hour per hour
(b) $W'(9) \approx \dfrac{15.32 - 12.57}{9 - 6} \doteq 0.92$ miles per hour per hour
(c) $W'(9) \approx \dfrac{16.00 - 12.57}{12 - 6} \doteq 0.57$ miles per hour per hour
(d) $\frac{1}{2}\left[\dfrac{16.00 - 15.32}{3} + \dfrac{15.32 - 12.57}{3}\right] = \dfrac{16.00 - 12.57}{6}$

Response 5 By Figure 15 of Section 2.4, $dh/dt \approx 3$ at $t = 10$, $dh/dt \approx 9$ at $t = 35$, and $dh/dt \approx 2$ at $t = 60$. Tangent lines with these slopes have been drawn in Figure R5 (above) and the curve has been drawn tangent to them.

Response 6 **(a)** The car is $s(2) = 10 + 50(2) - 2^3 = 102$ miles past the truck stop at $t = 2$.
(b) $v = \dfrac{ds}{dt} = \dfrac{d}{dt}(10 + 50t - t^3) = 50 - 3t^2$ • $v(2) = 50 - 3(2^2) = 38$ • The car's velocity in the positive s-direction is 38 at $t = 2$: Since the car is at $s = 102$ and the truck stop is at $s = 0$, the car is going 38 miles per hour away from the truck stop at $t = 2$.
(c) $a = \dfrac{d^2s}{dt^2} = \dfrac{dv}{dt} = \dfrac{d}{dt}(50 - 3t^2) = -6t$ • $a(2) = -6(2) = -12$ •
The car's acceleration in the positive s-direction is -12. Since its velocity is positive at $t = 2$, it is decelerating 12 miles per hour $t = 2$.

Responses 2.5

Response 1 **(a)** $A = 8h$ **(b)** $\dfrac{dA}{dt} = 8\dfrac{dh}{dt}$

Response 2 **(a)** $A = 6w$ **(b)** $\dfrac{dA}{dt} = 6\dfrac{dw}{dt}$

Response 3 The change ΔA in the area is the total area of the three small rectangles in Figure 4 of Section 2.5. The area of the rectangle at the upper left is $w_0\Delta h$, the area of the rectangle on the lower right is $h_0\Delta w$, and the area of the rectangle at the upper right is $\Delta w\Delta h$. • $\Delta A = w_0\Delta h + h_0\Delta w + \Delta w\Delta h$

Response 4 **(a)** The population of women at time t is $W(t) = F(t)P(t)$. • $W(1990) = F(1990)P(1990) = (248.7)(0.513) \doteq 127.6$ million
(b) $\dfrac{dW}{dt}(1990) = F(1990)\dfrac{dP}{dt}(1990) + P(1990)\dfrac{dF}{dt}(1990)$ $= (0.513)(3.5) + (248.7)(-0.0004) \doteq 1.7$ million per year
(c) At the beginning of 1990, the rate of change of the population of men was $\doteq 3.5 - 1.7 = 1.8$ million per year.

Response 5 $g + \Delta g \to g(x)$ since by Theorem 1 of Section 2.4, g is continuous at x and $\Delta g \to 0$. • $\dfrac{\Delta f}{\Delta x} \to \dfrac{df}{dx}(x)$ and $\dfrac{\Delta g}{\Delta x} \to \dfrac{dg}{dx}(x)$ by the definition of the derivative

Response 6 (a) $D(1985) \approx 1850$ billion dollars • $P(1985) \approx 240$ million people • The points $(1980, 600)$ and $(1985, 1850)$ are on the approximate tangent line in Figure R6a. • The points $(1970, 180)$ and $(1985, 240)$ are on the approximate tangent line in Figure R6b. • $D'(1985) \approx \dfrac{1850 - 600}{5} \doteq 250$ billion dollars per year • $P'(1985) \approx \dfrac{240 - 180}{15} = 4$ million people per year
(b) [Debt per person] $= D/P \approx$ [1850 billion dollars]/[250 million people] $= 7.4$ thousand dollars per person •
[Rate of change of the debt per person] $= \dfrac{d}{dt}\left[\dfrac{D}{P}\right] = \dfrac{PD' - DP'}{P^2}$
$\approx \dfrac{(240)(250) - (1850)(4)}{240^2} \doteq 0.913$ thousand dollars per person per year. (Different estimates yield different answers.)

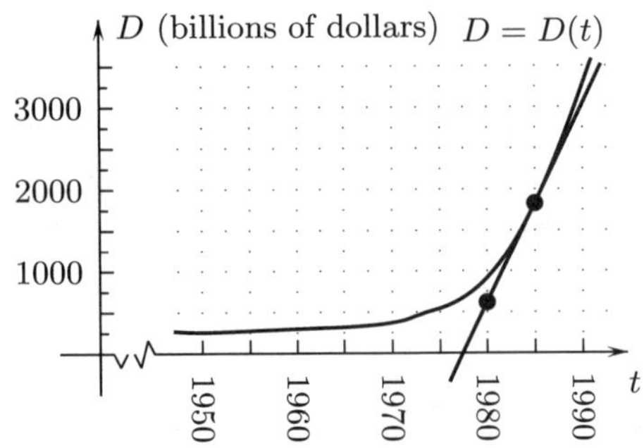

Figure R6a

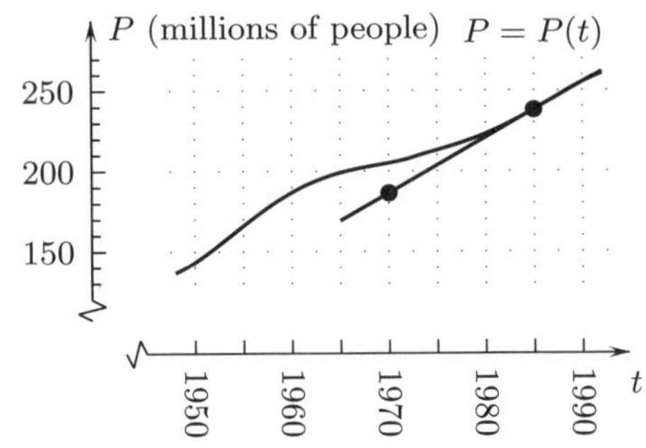

Figure R6b

Response 7 (a) $\dfrac{d}{dx}(f^2) = \dfrac{d}{dx}(f \times f) = f\dfrac{df}{dx} + f\dfrac{df}{dx} = 2f\dfrac{df}{dx}$
(b) $\dfrac{d}{dx}(f^3) = \dfrac{d}{dx}(f^2 \times f) = f^2\dfrac{df}{dx} + f\dfrac{d}{dx}(f^2) = f^2\dfrac{df}{dx} + f(2f)\dfrac{df}{dx} = 3f^2\dfrac{df}{dx}$

Response 8 $\dfrac{d}{dx}(y^{-2}) = -2y^{-3}\dfrac{dy}{dx}$ • Set $y = 2$ and $\dfrac{dy}{dx} = -5$: $\left[\dfrac{d}{dx}(y^{-2})\right]_{x=0}$
$= -2(2^{-3})(-5) = \frac{5}{4}$

Response 9 (a) $\dfrac{dV}{dt} = \frac{4}{3}\pi(3r^2)\dfrac{dr}{dt} = 4\pi r^2\dfrac{dr}{dt}$
(b) $4\pi(3^2)(\frac{1}{2}) = 18\pi$ cubic inches per second

Response 10 You calculate $x + 1$ and $x - 3$, then you divide, and then you take the fifth power.

Response 11 You calculate $x^3 + 2x$ and take its tenth power, you calculate x^2, and then you take the product. •
$g'(x) = \dfrac{d}{dx}[x^2(x^3 + 2x)^{10}] = x^2\dfrac{d}{dx}[(x^3 + 2x)^{10}] + (x^3 + 2x)^{10}\dfrac{d}{dx}(x^2)$
$= x^2[10(x^3 + 2x)^9]\dfrac{d}{dx}(x^3 + 2x) + (x^3 + 2x)^{10}(2x)$
$= 10x^2(x^3 + 2x)^9(3x^2 + 2) + 2x(x^3 + 2x)^{10}$

Response 12 $\dfrac{d}{dx}\{[x^2 + u(x)]^7\} = 7[x^2 + u(x)]^6\dfrac{d}{dx}[x^2 + u(x)] = 7[x^2 + u(x)]^6[2x + u'(x)]$

CHAPTER 2 ANSWERS

Tune-up exercises 2.1

T1. $[\text{Average velocity}] = \dfrac{s(5)-s(1)}{5-1} = \dfrac{40-0}{5-1} = 10$ yards per minute • Figure T1

Figure T1

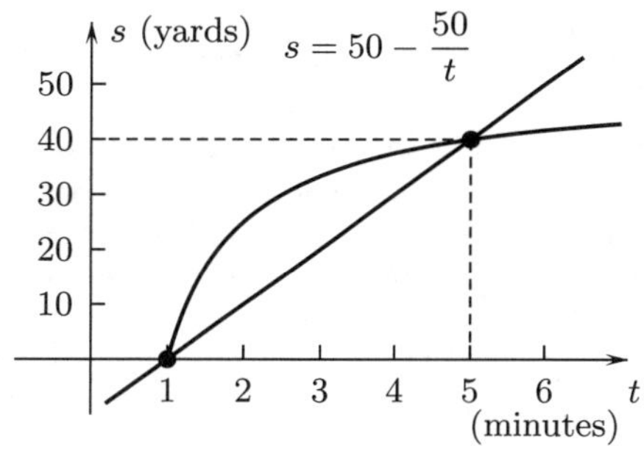

T2. $[\text{Average velocity for } 1.99 \le t \le 2] = \dfrac{s(2)-s(1.99)}{2-1.99} \doteq \dfrac{25-24.8743719}{0.01} \doteq 12.56281$ •
$[\text{Average velocity for } 2 \le t \le 2.01] = \dfrac{s(2.01)-s(2)}{2.01-2} \doteq \dfrac{25.1243781-25}{0.01} \doteq 12.43781$ •
$[\text{Average velocity for } 1.99 \le t \le 2.01] = \dfrac{s(2.01)-s(1.99)}{2.01-1.99} \doteq \dfrac{25.1243781-24.8743719}{0.02} \doteq 12.50031$ •
One answer: The velocity at $t = 2$ is approximately 12.5 yards per minute

T3. **(a)** $[\text{Average rate of change for } 0 \le x \le 0.2] = \dfrac{Q(0.2)-Q(0)}{0.2-0} = \dfrac{55.0-50.3}{0.2} = 23.5$
(b) $[\text{Average rate of change for } 0.2 \le x \le 0.4] = \dfrac{Q(0.4)-Q(0.2)}{0.4-0.2} = \dfrac{50.3-55.0}{0.2} = -23.5$
(c) $[\text{Average rate of change for } 0 \le x \le 0.4] = \dfrac{Q(0.4)-Q(0)}{0.4-0} = \dfrac{50.3-50.3}{0.4} = 0$

T4. **(a)** $[\text{Average rate of change for } 2 \le x \le 5] = \dfrac{W(5)-W(2)}{5-2} = \dfrac{3-1}{3} = \frac{2}{3}$
(b) $[\text{Average rate of change for } 1 \le x \le 3] = \dfrac{W(3)-W(1)}{3-1} = \dfrac{3-3}{2} = 0$
(c) The average rate of change for $2.50 \le x \le 2.51$ is the slope 2 of the line from $(2,1)$ to $(3,3)$.

Problems 2.1

1. **(a)** $\dfrac{s(3)-s(1)}{3-0} = \dfrac{90-0}{3-0} = 30$ miles per hour **(b)** Figure A1 **(c)** She speeds up. **(d)** 30 miles per hour

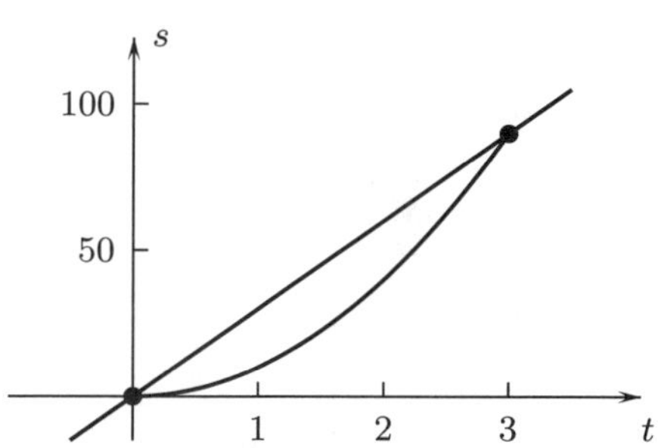

Figure A1

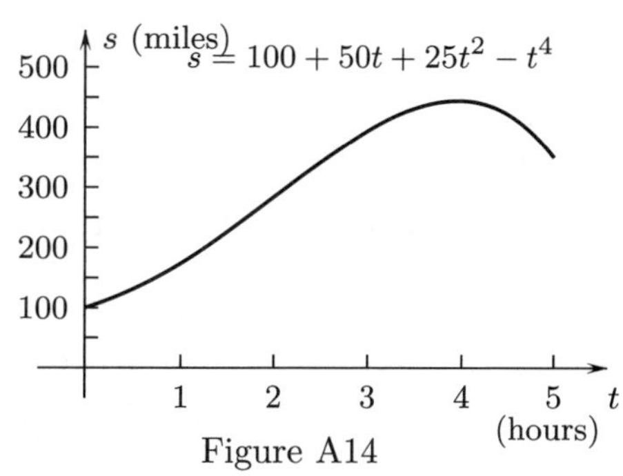

Figure A14

3. **(a)** $0.0083 \times 10^8 = 830{,}000$ kilometers per month **(b)** $-0.0083 \times 10^8 = -830{,}000$ kilometers per month **(c)** One answer: 1.5×10^8 kilometers • $[\text{Maximum error}] = 0.029 \times 10^8 = 2{,}900{,}000$ kilometers

5. **(a)** $100(83.69 - 276.74)/276.74 \doteq -69.8\%$
(b) [-193.05 yen per dollar]/[29 months] $\doteq -6.66$ yen per dollar per month

8. **(a)** $\doteq 121.2\%$ **(b)** $\doteq -9.2$ million pounds per year **(c)** $\doteq 0.566$

9. [Average rate of change of number of applicants] $\doteq (45364 - 26721)/6 \doteq 3107$ applicants per year • [Average rate of change of number of admissions] $\doteq (15878 - 15569)/6 \doteq 52$ admissions per year

12. **(a)** $\approx 3 + 3.8 + 4.1 + 4 = 14.9$ billion dollars from CD-ROM and video games • $\approx 4.7 + 4.9 + 5.2 + 5.4 = 20.2$ billion dollars from movie box offices
(b) $\approx [(4+5.4) - (3+4.7)]/3 \doteq 0.57$ billion dollars per year (Different approximations yield different answers.)

14. **(a)** Figure A14 (above) • One description: She starts 100 miles east of Reno, drives east about four hours, when she is about 440 miles from Reno. Then she drives west for about one hour and ends up about 350 miles east of Reno. **(b)** [Average velocity toward the east for $0 \le t \le 4$] = 86 miles per hour • [Average velocity toward the east for $4 \le t \le 5$] = -94 miles per hour **(c)** [Velocity toward the east at $t = 2$] ≈ 118.0 miles per hour • [Velocity toward the east at $t = 4.5$] ≈ -89.6 miles per hour

Tune-up exercises 2.2

T1. Figure T1 • $P(-1) = 1$ • $\lim_{x\to -1^-} P(x) = 3$ • $\lim_{x\to -1^+} P(x) = 1$ • $P(1) = 1$ • $\lim_{x\to 1^-} P(x) = 1$ • $\lim_{x\to 1^+} P(x) = 3$ • $P(0) = 1$ • $\lim_{x\to 0} P(x) = 1$ • $\lim_{x\to -\infty} P(x) = 0$ • $\lim_{x\to\infty} P(x) = 0$ • $P(x)$ is continuous in $(-\infty, -1)$, $[-1, 1]$, and $(1, \infty)$.

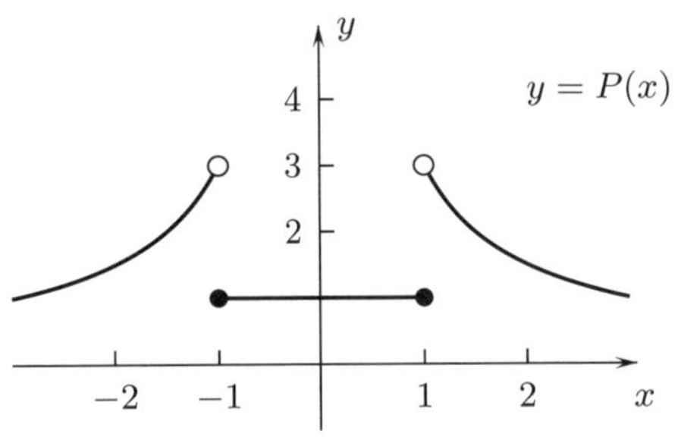

Figure T1

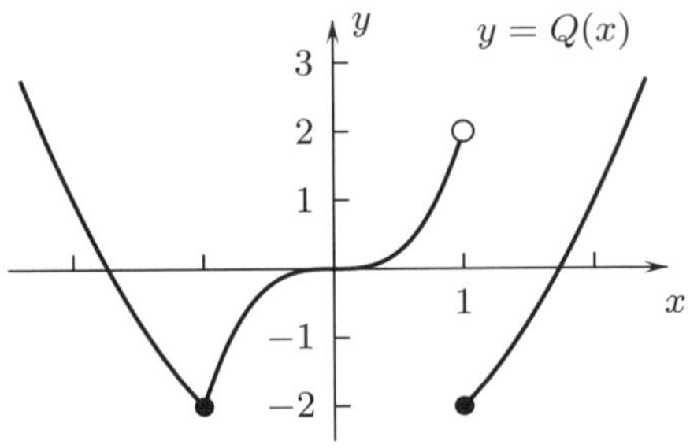

Figure T7

T2. $\sqrt{x} + \dfrac{6}{x+2}$ is defined for $x \ge 0$. • By Theorem 1, it is continuous at $x = 4$, so that $\sqrt{x} + \dfrac{6}{x+2} \to \sqrt{4} + \dfrac{6}{4+2} = 3$ as $x \to 4$.

T3. $\lim_{x\to 7}\left[\dfrac{F(x) - x}{F(x) + x}\right] = \dfrac{20-7}{20+7} = \frac{13}{27}$ since $x \to 7$ and $F(x) \to 20$ as $x \to 7$ (by Theorem 2).

T4. $S(2) = 14$ • $\lim_{x\to 2^-} S(x) = \lim_{x\to 2^-} \sqrt{x+2} = \sqrt{2+2} = 2$ • $\lim_{x\to 2^+} S(x) = \lim_{x\to 2^+} x^8 = 2^8 = 256$ • $\lim_{x\to 2} S(x)$ does not exist because $S(x) \to 2$ as x tends to 2 from the left, $S(x) \to 256$ as x tends to 2 from the right, and these one-sided limits are not equal.

T5. Probable answer: The values in the table below suggest that $\lim_{x\to 1} \dfrac{x^{1.7} - 1}{x - 1} = 1.7$.

x	$\dfrac{x^{1.7}-1}{x-1} \doteq$
0.9	1.639884721
0.99	1.694044031
0.999	1.699404940
0.9999	1.699940499

x	$\dfrac{x^{1.7}-1}{x-1} \doteq$
1.1	1.758923494
1.01	1.705944069
1.001	1.700594941
1.0001	1.700059499

T6. For $x \neq 3$, $\dfrac{x^2-9}{x-3} = \dfrac{(x-3)(x+3)}{(x-3)} = x+3$ • $\lim_{x\to 3}\left[\dfrac{x^2-9}{x-3}\right] = \lim_{x\to 3}(x+3) = 3+3 = 6$

T7. Figure T7 (above) • $Q(1) = -2$ • $\lim_{x\to 1^-} Q(x) = 2$ • $\lim_{x\to 1^+} Q(x) = -2$ • $\lim_{x\to 1} Q(x)$ does not exist. • $Q(0) = 0$ • $\lim_{x\to 0} Q(x) = 0$ • $\lim_{x\to -\infty} Q(x) = \infty$ • $\lim_{x\to\infty} Q(x) = \infty$ • Q is continuous in $(-\infty, 1)$ and $[1,\infty)$.

T8. $\sqrt{x} + \dfrac{6}{x+2}$ tends to 3 as $x \to 0^+$ because it is continuous in $[0,\infty)$ and its value at 0 is 3.

T9. $\dfrac{-10(2)}{5(2)-3(-10)} = -\frac{1}{2}$

T10. $W(5) = 5^4 = 625$ • $W(-5) = 9(-5)^2 = 225$ • $W(3) = 0$ • $\lim_{x\to 3^-} W(x) = \lim_{x\to 3^-}(9x^2) = 81$ • $\lim_{x\to 3^+} W(x) = \lim_{x\to 3^+}(x^4) = 81$ • $\lim_{x\to 3} W(x) = 81$

T11. Likely prediction: 4

T12. For $x \neq -5$, $\dfrac{x^2-25}{x+5} = \dfrac{(x-5)(x+5)}{x+5} = x-5$ • $x-5 \to -10$ as $x \to -5$

Problems 2.2

1. 5

4. 10

7. **(a)** Figure A7 **(b)** $\lim_{x\to 0^+} w(x) = \infty$ • $\lim_{x\to 2^-} w(x) = 1$ • $\lim_{x\to 2^+} w(x) = 2$ • $\lim_{x\to\infty} w(x) = 2$
(c) The line $y = 3$ only intersects the graph of $w(x)$ on the part of $y = 2/x$. • Solve $2/x = 3$: $x = \frac{2}{3}$

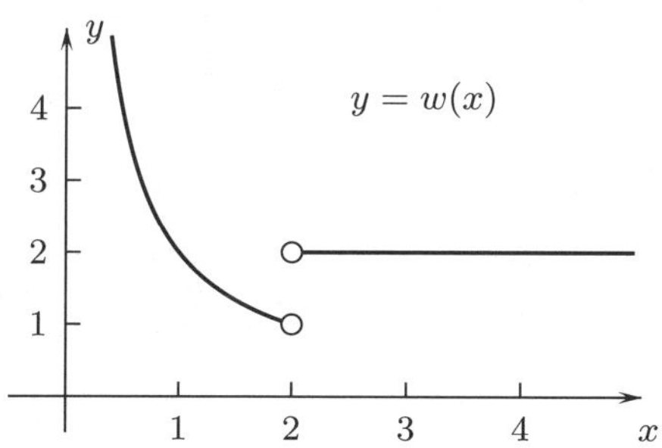

Figure A7

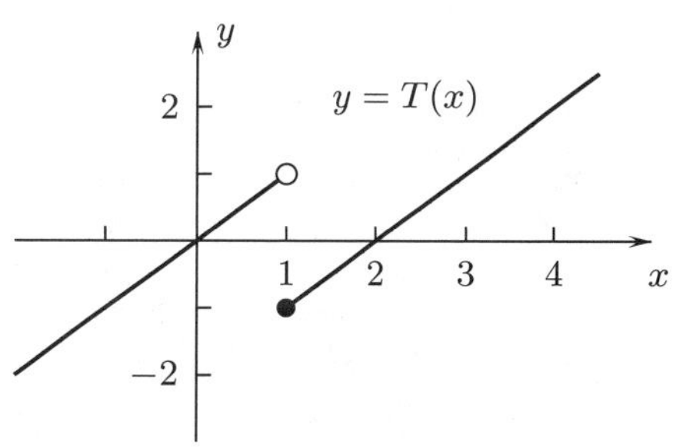

Figure A8

8. **(a)** Figure A8 **(b)** $T(1) = -1$ • $\lim_{x\to 1^-} T(x) = 1$ • $\lim_{x\to 1^+} T(x) = -1$
(c) $T(x)$ is discontinuous at $x = 1$.
(d) The greatest value of $T(x)$ for $0 \le x \le 4$ is its value 2 at $x = 4$. • Its least value for $0 \le x \le 4$ is its value -1 at $x = 1$.

9. **(a)** Figure A9 **(b)** $\lim_{x\to -1^-} K(x) = 1$ • $\lim_{x\to -1^+} K(x) = -1$ • $\lim_{x\to -1} K(x)$ does not exist.
(c) $K(x) = 0$ at $x = 0, -2, 2$ **(d)** $K(x)$ is negative for $x < -2$, for $-1 < x < 0$, for $0 < x < 1$, and for $x > 2$. **(e)** $K(x)$ is continuous in $(-\infty, -1]$, $(-1, 1)$, and $[1, \infty)$.

12. $\lim_{x\to 2^-} g(x) = \lim_{x\to 2^-}(x^3+x) = 10$ • $\lim_{x\to 2^+} g(x) = \lim_{x\to 2^-}(10) = 10$ • $\lim_{x\to 2} g(x) = 10$ • $\lim_{x\to 10^-} g(x) = \lim_{x\to 10^-}(10) = 10$ • $\lim_{x\to 10^+} g(x) = \lim_{x\to 10^-}(1+\sqrt{x+7}) = 1+\sqrt{17}$ • $\lim_{x\to 10} g(x)$ does not exist. • $g(x)$ is continuous in $(-\infty, 10]$ and $(10,\infty)$.

15. $f(x)$ is continuous at all $x \neq \pm 1$ by Theorem 1 of Section 2.2. It is also continuous at $x = -1$ because the three numbers $f(-1) = 0$, $\lim_{x\to -1^-} f(x) = \lim_{x\to 1^-}(x^3 - x^4 + 2) = (-1)^3 - (-1)^4 + 3 = 0$ and $\lim_{x\to -1^+} f(x) = \lim_{x\to -1^+}(2x^5 + x + 3) = 2(-1)^5 + (-1) + 3 = 0$ are equal. • It is discontinuous at $x = 1$ because $\lim_{x\to 1^-} f(x) = \lim_{x\to 1^-}(2x^5 + x + 3) = 2(1)^3 + 1 + 3 = 6$ and $f(1) = 7$ are different.

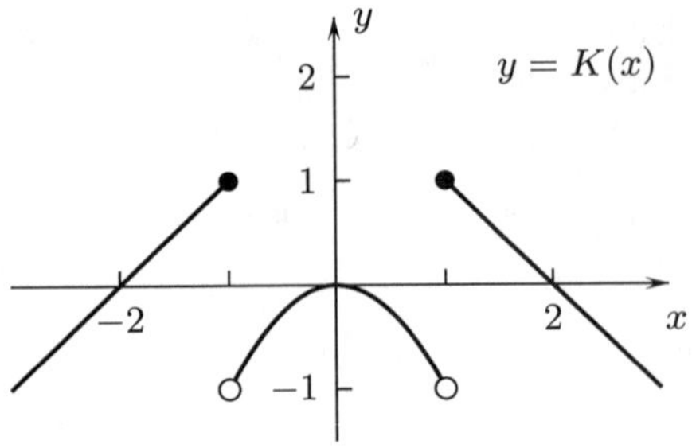

Figure A9

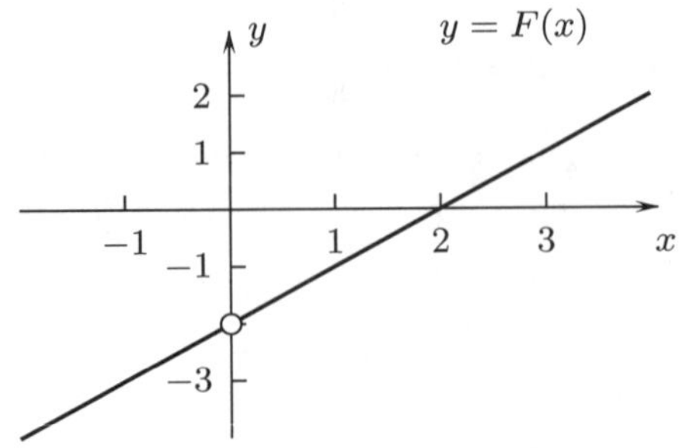

Figure A25

18. **(a)** $\dfrac{x}{x^2-4}$ is continuous in $(-\infty,-2),(-2,2)$, and $(2,\infty)$ because it is given by one formula which is defined for all $x \neq \pm 2$. **(b)** $\dfrac{8}{x^2+4}$ is continuous in $(-\infty,\infty)$ because it is given by one formula which is defined for all x. **(c)** $\dfrac{\sqrt{x}}{x-1}$ is continuous in $[0,1)$ and $(1,\infty)$ because it is given by one formula which is defined for all $x \geq 0$ except $x = 1$.

20. Likely prediction: $-\frac{3}{4}$

21. Likely prediction: 6

22. The limit is $\dfrac{1^2-1}{2} = 0$ because $\dfrac{x^2-1}{x+1}$ is continuous at $x = 1$.

25. **(a)** $\displaystyle\lim_{x\to 2} F(x) = \frac{(2-1)^2-1}{2} = 0$ because $F(x) = \dfrac{(x-1)^2-1}{x}$ is continuous at $x = 2$. **(b)** For $x \neq 0$, $F(x) = \dfrac{(x-1)^2-1}{x} = \dfrac{x^2-2x}{x} = \dfrac{x(x-2)}{x} = x-2$ • $\displaystyle\lim_{x\to 0} F(x) = \lim_{x\to 0}(x-2) = -2$ • Figure A25 (above)

29. $\frac{2}{3}$

31. 0

33. $\frac{1}{6}$

35. 6

39. $\displaystyle\lim_{x\to 2^-} Z(x) = 4k$ • $\displaystyle\lim_{x\to 2^+} Z(x) = 8$ • The two-sided limit exists if $k = 2$.

55. **(a)** One of many possible solutions: Set $f(x) = 12x - 7x^2 + x^3$. • $f(0) = 0$, $f(5) = 10$, $0 < 4 < 10$, and $f(x)$ is continuous in $[0,5]$, so $f(x)$ has the value 4 at one or more points in $[0,5]$. **(b)** $x \approx 0.438447, x = 2$, and $x \approx 4.561553$

60. Let $v(x)$ be his speed after he has run x feet and let L be the circumference of the track, and assume that $v(x)$ is continuous in $[0,L]$. • $v(0) = 0$ and $v(L) = 0$ • $f(x) = v(x+L/2) - v(x)$ gives the difference between his speeds at diametrically opposite points. • $f(0) = v(L/2)$, $f(L/2) = -v(L/2)$, $f(x)$ is continuous in $[0,L/2]$, and 0 is between $v(L/2)$ and $-v(L/2)$, so $f(x) = 0$ for at least one x with $0 \leq x \leq L/2$, for which $v(x+L/2) = v(x)$.

66. The volume $\frac{4}{3}\pi r^3$ is $\frac{4}{3}\pi$ for $r = 1$. • $\frac{4}{3}\pi - 0.005 < \frac{4}{3}\pi r^3 < \frac{4}{3}\pi + 0.005$
$\iff 1 - \dfrac{0.015}{4\pi} < r^3 < 1 + \dfrac{0.015}{4\pi} \iff \left(1-\dfrac{0.015}{4\pi}\right)^{1/3} < r < \left(1+\dfrac{0.015}{4\pi}\right)^{1/3}$ •
$1 - \left(1-\dfrac{0.015}{4\pi}\right)^{1/3} \doteq 3.980 \times 10^{-4}$ and $\left(1+\dfrac{0.015}{4\pi}\right)^{1/3} - 1 \doteq 3.977 \times 10^{-4}$ •
One answer: $\delta = 0.003$ (Any positive $\delta \leq \left(1+\dfrac{0.015}{4\pi}\right)^{1/3} - 1$ would do.)

69. Let ϵ be any positive number. There is a positive δ_1 such that $L - \frac{1}{2}\epsilon < f(x) < L + \frac{1}{2}\epsilon$ for $x_0 < x < x_0 + \delta_1$ and a positive δ_2 such that $M - \frac{1}{2}\epsilon < g(x) < M + \frac{1}{2}\epsilon$ for $x_0 < x < x_0 + \delta_2$. Let δ be the smaller of δ_1 and δ_2. Then (adding the inequalities) $L + M - \epsilon < f(x) + g(x) < L + M + \epsilon$ for $x_0 < x < x_0 + \delta$.

70. For any $Y > 0$, there is an X such that $0 < f(x) < \dfrac{1}{Y}$ for $x > X$. Then $f(x) > Y$ for $x > X$.

72. One answer: $|5x - 20.1| = 5|x - 4.02|$ is greater than $5(0.01) = 0.05$ for $4 < x < 4.01$, so there is no δ for $\epsilon = 0.05$.

Tune-up exercises 2.3

T1. Since $\sqrt{x^4+3}$ equals $\sqrt{1^4+3}=2$ at $x=1$, its derivative at $x=1$ is the limit of $\dfrac{\sqrt{b^4+3}-2}{b-1}$ as $b\to 1$. • Use this formula or the secant-line program with $A=1$ to find values as in the table below. • Likely prediction: The derivative of $\sqrt{x^4+1}$ at $x=1$ is 1. (The graph and its tangent line at $x=1$ are in Figure T1.)

b	1.1	1.001	1.00001	0.9	0.999	0.99999
$\dfrac{\sqrt{b^4+3}-2}{b-1}\doteq$	1.128417	1.001250	1.000013	0.879069	0.998750	0.999988

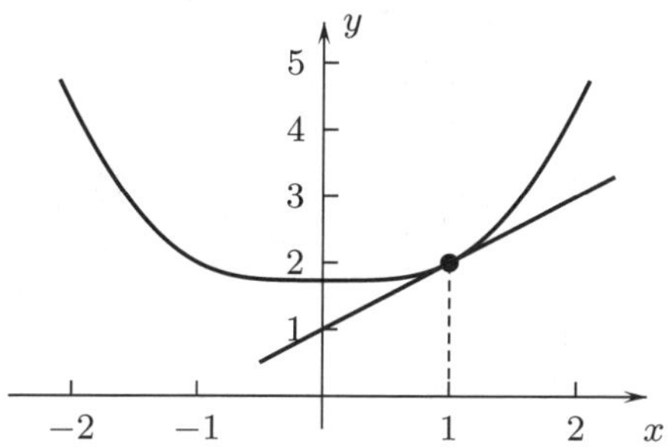

Figure T1

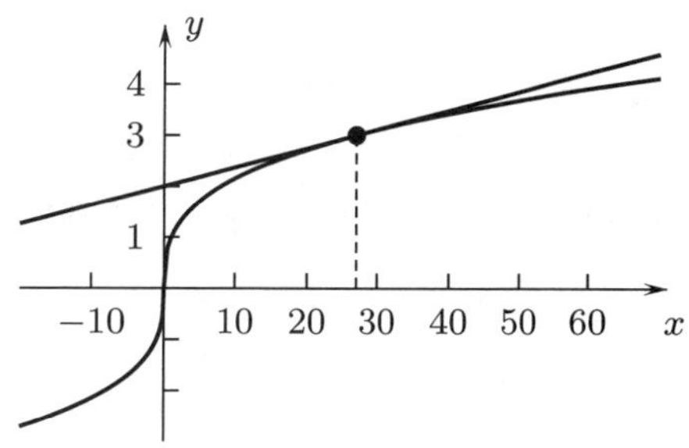

Figure T4

T2. **(a)** $\dfrac{d}{dx}(5x^3-6x^2+7x)=5(3x^{3-1})-6(2x^{2-1})+7(1)=15x^2-12x+7$

(b) $\dfrac{d}{dx}(6x^{-3}-4x^{-1})=6(-3x^{-3-1})-4(-x^{-1-1})=-18x^{-4}+4x^{-2}$

(c) $\dfrac{d}{dt}(t^{3/2}-t^{2/3})=\frac{3}{2}t^{(3/2)-1}-\frac{2}{3}t^{(2/3)-1}=\frac{3}{2}t^{1/2}-\frac{2}{3}t^{-1/3}$

T3. $f'(x)=3x^2$ • Solve $3x^2=2$: $x=\pm\sqrt{2/3}$

T4. The value of $\sqrt[3]{x}$ at $x=27$ is $\sqrt[3]{27}=3$. • $\dfrac{dy}{dx}=\dfrac{d}{dx}(x^{1/3})=\frac{1}{3}x^{-2/3}$, so the slope of the tangent line at $x=27$ is $\frac{1}{3}(27^{-2/3})=\frac{1}{3}\left[\frac{1}{9}\right]=\frac{1}{27}$. • Tangent line: $y=3+\frac{1}{27}(x-27)$ • Figure T4 (above)

T5. Probable answer: The value of $T(x)$ at 9.75 is approximately equal to the value $7+4(-0.25)=6$ given by the linear function $7+4(t-4)$ whose graph is the tangent line.

T6. Likely prediction: 10 • Figure T6

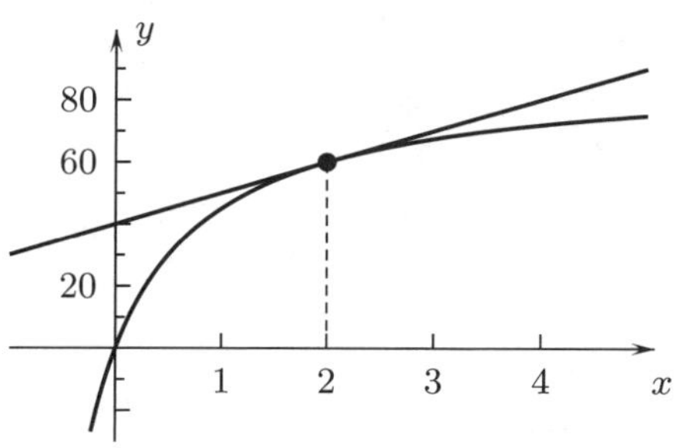

Figure T6

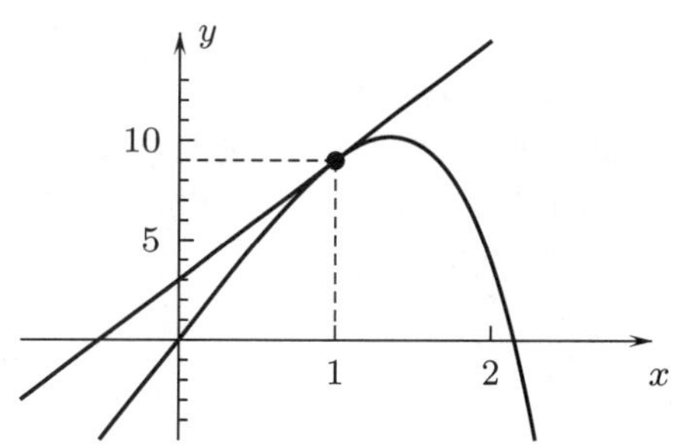

Figure T10

T7. **(a)** $7x^6+6x$ **(b)** $-56x^{-9}+56x^{-8}$ **(c)** $-\frac{4}{3}t^{-4/3}-\frac{3}{4}t^{-3/4}$

T8. $x\ge 1$

T9. $y=9+6(x-1)$ • Figure T10 (above)

T10. $Z(1.09)\approx Z(1)+0.09Z'(1)=100+0.09(-20)=98.2$

Problems 2.3

1. Likely prediction: $f'(2) = -\frac{4}{5}$ • Figure A1

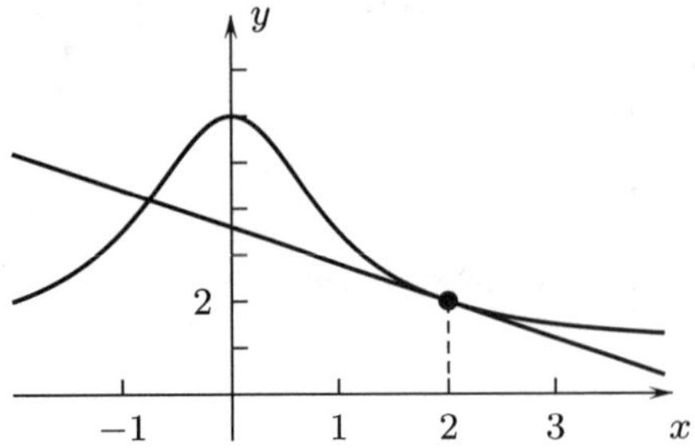

Figure A1

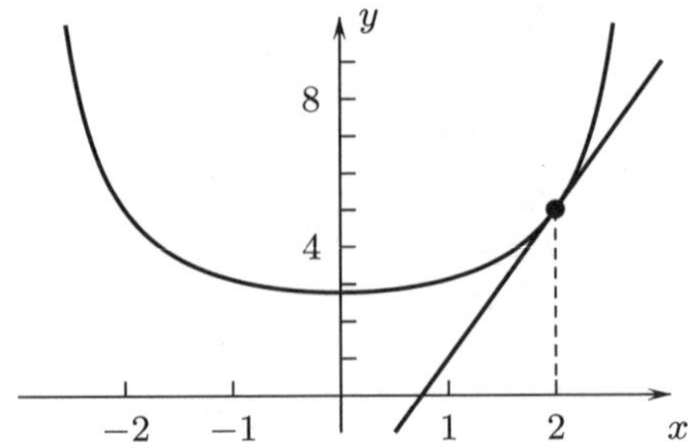

Figure A2

2. Likely prediction: $\frac{dg}{dx}(2) = 4$ • Figure A2

4. Likely prediction: $\frac{dL}{dx}(3) = -1$ • Figure A4

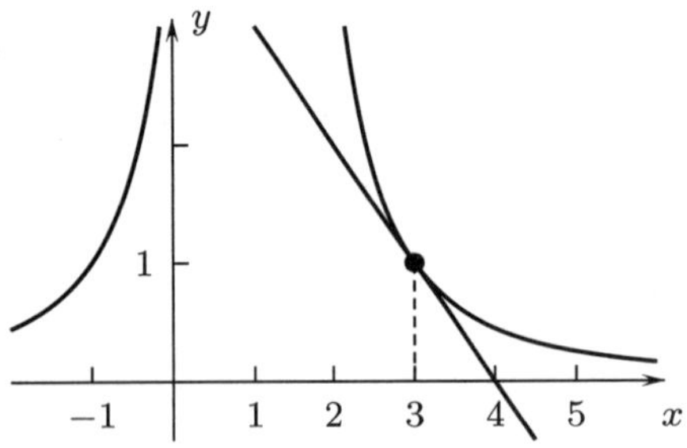

Figure A4

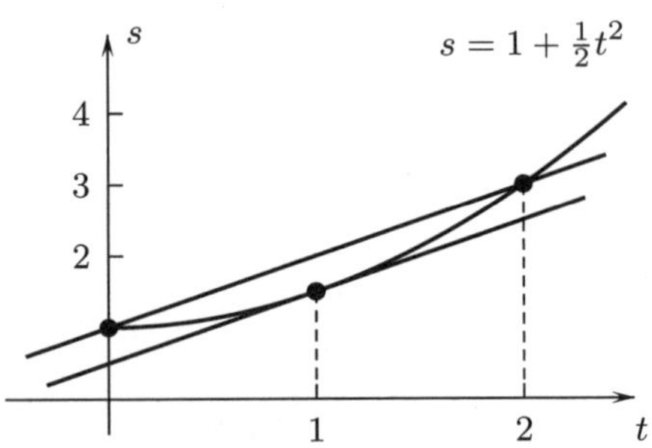

Figure A21

7. $12x^3 - 12x$

8. $x^3 + x^{-5/4}$

9. $\frac{d}{dt}(t^{2/3} + t^{3/2}) = \frac{2}{3}t^{-1/3} + \frac{3}{2}t^{1/2}$

10. $\frac{dy}{dx} = \frac{d}{dx}(3x^{-1} - 4x^{-2}) = -3x^{-2} + 8x^{-3}$ • $\frac{1}{4}$

14. $-\frac{2}{27} + 18 = \frac{484}{27}$

17. 32

19. $f'(10) = 3g'(10) - 4h'(10) = 33$

21. **(a)** 1 foot per second **(b)** 1 foot per second **(c)** Figure A21 (above)

23. **(a)** At $t = 3$ hours **(b)** Figure A23

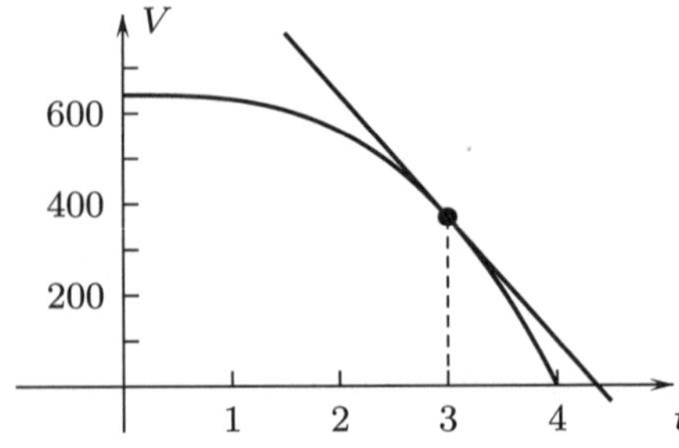

Figure A23

27. **(a)** The walker is 7 miles past Reklaw 3 hours after dawn. **(b)** The walker is walking 3 miles per hour 3 hours after dawn. **(c)** The walker is walking one-half mile per hour toward Reklaw 5 hours after dawn. **(d)** The walker's average velocity for $0 \le t \le 5$ is 2 miles per hour.

28. Possible answers: **(a)** The median price of new houses was \$97,000 more at the beginning of 1991 than at the beginning of 1970. **(b)** The median price of new houses at the beginning of 1991 was 25% more than the median price of existing homes. **(c)** The median price of existing houses at the beginning of 1991 was five times what it was at the beginning of 1970. **(d)** The median price of existing houses was rising at the beginning of 1991. **(e)** The rate of increase of the median price of new houses at the beginning of 1991 was 90% of the rate of decrease of the median price of existing houses.

30. **(a)** $v(4) = 116$ knots **(b)** $\frac{dv}{dh}(6) = -0.5$ knots per thousand feet • $\frac{dg}{dh}(6) = -0.2$ gallons per hour per thousand feet

32. Tangent line: $y = 3 - 2(x-1)$ • Figure A32

33. Tangent line: $y = 60 + 70(x-1)$ • Figure A33

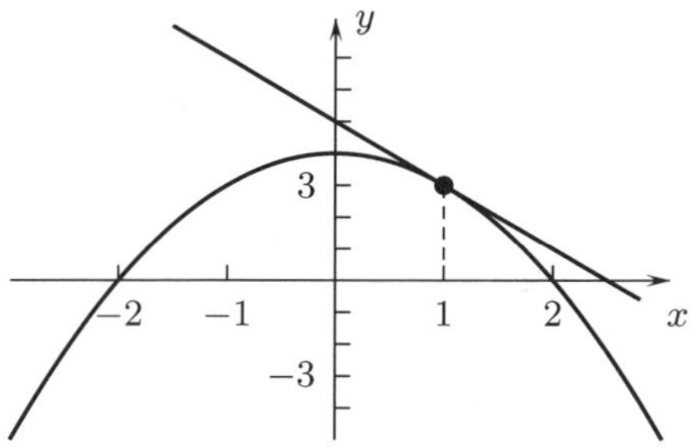

Figure A32

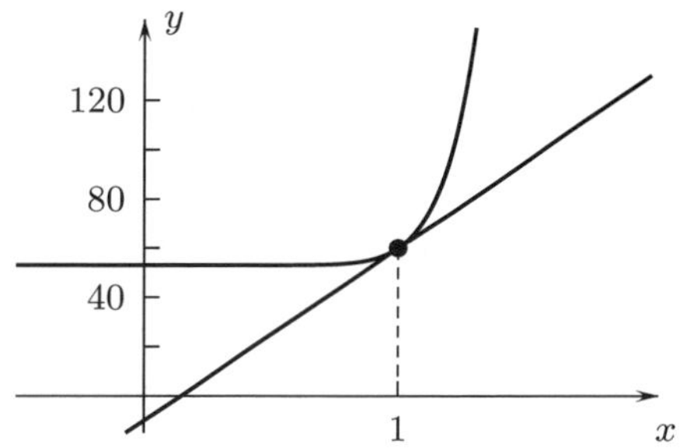

Figure A33

38. $\frac{d}{dx}(x^{-1/2} - 3x^{-1}) = -\frac{1}{2}x^{-3/2} + 3x^{-2}$

39. $\frac{d}{dx}(x^{1/24}) = \frac{1}{24}x^{-23/24}$

40. $2x + 6$

42. $6x + 2$

52. Likely answers: The average rate of change of the revenue from the beginning of 1990 to the beginning of 1993 was $\frac{18.3 - 17.0}{1993 - 1990} \doteq 0.43$ billion dollars per year. The revenue at the beginning of 1992 was approximately $17.0 + 2(0.43) = 17.86$ billion dollars. • The rate of change of the revenue at the beginning of 1992 was approximately 0.43 billion dollars per year.

57. For $b \neq 1$, $\frac{g(b) - g(1)}{b - 1} = \frac{b^2 + 2b - 3}{b - 1} = \frac{(b-1)(b+3)}{b-1} = b + 3$ • $b + 3 \to 4$ as $b \to 1$

58. For $b \neq 4$, $\frac{G(b) - G(4)}{b - 4} = \frac{\frac{1}{5-2b} - \frac{-1}{3}}{b-4} = \frac{-2}{3(5-2b)}$ • $\frac{-2}{3(5-2b)} \to \frac{2}{9}$ as $b \to 4$

61. For $x \neq 0$ and $\Delta x \neq 0$ with $|\Delta x| < |x|$, $\frac{f(x+\Delta x) - f(x)}{\Delta x} = \frac{1}{\Delta x}\left[\frac{-3}{(x+\Delta x)^2} + \frac{3}{x^2}\right]$ $= \frac{3}{x^2(x+\Delta x)^2}(2x + \Delta x) \to \frac{6}{x^3}$ as $\Delta x \to 0$

62. For $\Delta x \neq 0$, $\frac{1}{\Delta x}[(0 + \Delta x)^{4/3} - 0] = (\Delta x)^{1/3}$ • $(\Delta x)^{1/3} \to 0$ as $\Delta x \to 0$

67. $f'(x) = 3x^2 + 6x$ • $f'(10) = 360$ • At $x = 10$, $df = 360\,dx$

68. $dw = -0.126\,dz$

69. **(a)** 10 centimeters **(b)** [Error] $\approx |dw| = 0.05|dA| \le 0.0025$ centimeters

72. $dV = (4\pi r^2)\,dr$

73a. $\frac{dV}{V} = 3\frac{\Delta r}{r}$

Tune-up exercises 2.4

T1. Possible answers: **(a)** The tangent line in Figure T1a passes through the points with approximate coordinates $(3, 15)$ and $(4, 23)$, so $P'(3) \approx \dfrac{23 - 15}{4 - 3} = 8$. **(b)** Figure T1b • $x \approx 1.7, 4.1$
(c) Draw a line of slope 5, as in Figure T1c, and find the points where the tangent line is parallel to it. • $x \approx 2.2$ or $x \approx 3.6$

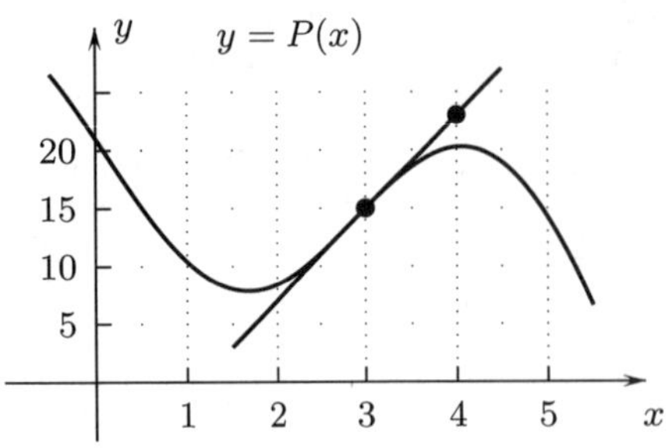

Figure T1a

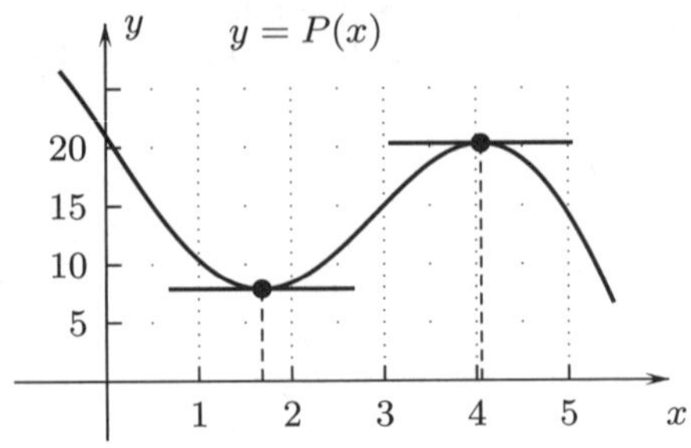

Figure T1b

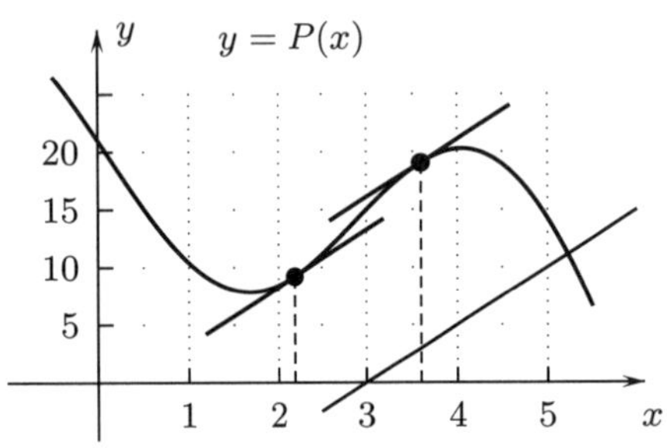

Figure T1c

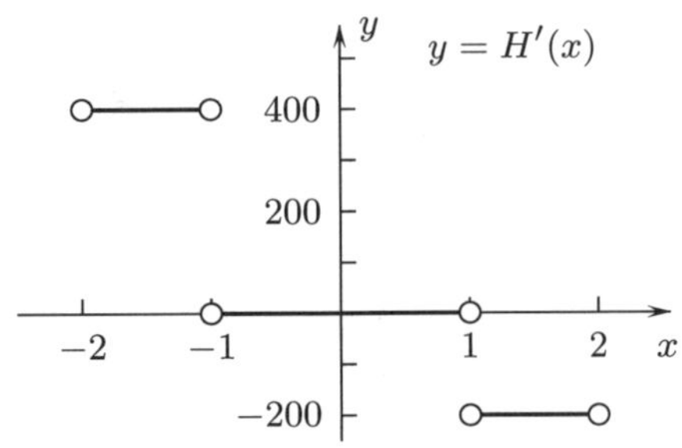

Figure T2

T2. $H'(x) = 400$ for $x < -1$, $H'(x) = 0$ for $-1 < x < 1$, and $H'(x) = -200$ for $1 < x < 2$ • Figure T2

T3. $J'(x)$ is approximately zero at $x = 0$. Its greatest and least values for x in Figure 26 of Section 2.4 are approximately 1.5 at $x = 1.3$ and -1.5 at $x = -1.3$. Its values at $x = \pm 5$ are approximately ± 0.3 (Figure T3a). • The graph of $J'(x)$ is in Figure T3b. (Different approximations yield different answers.)

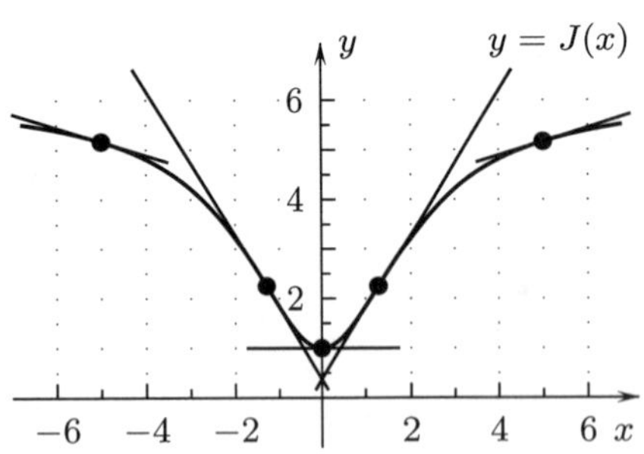

Figure T3a

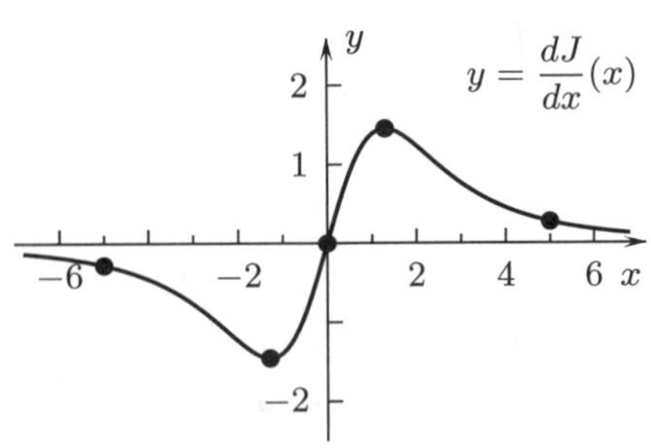

Figure T3b

T4. The graph of $P'(x)$ is in Figure 31 of Section 2.4, the graph of $Q'(x)$ is in Figure 32, and the graph of $R'(x)$ is in Figure 30. • One explanation: $P'(x) < 0$ for $x < 0$ and $P'(x) > 0$ for $x > 0$, $Q'(x) < 0$ for $x \neq 0$, and $R'(x) > 0$ for $x \neq 0$.

T5. $\frac{dQ}{dx}(10) \approx 16$ (right difference quotient), 14.5 (left difference quotient), or 15.25 (centered difference quotient) • $\frac{dQ}{dx}(10.35) \approx 12.5$

T6. **(a)** $30x - 60x^3$ **(b)** $-\frac{3}{16}x^{-7/4} - \frac{2}{9}x^{-5/3}$ **T7.** $56x^6 - 30x^4$

T8. $y' = 18ax^2 - 20bx^4$ • $y'' = 36ax - 80bx^3$

T9. **(a)** $P'(1) \approx -7$ **(b)** $P'(x) = -5$ at $x \approx 1.2$ and $x \approx 4.4$

T10. Figures T10a and T10b **T11.** **(a)** $20t^3 - 2$ **(b)** $9900t^{98}$

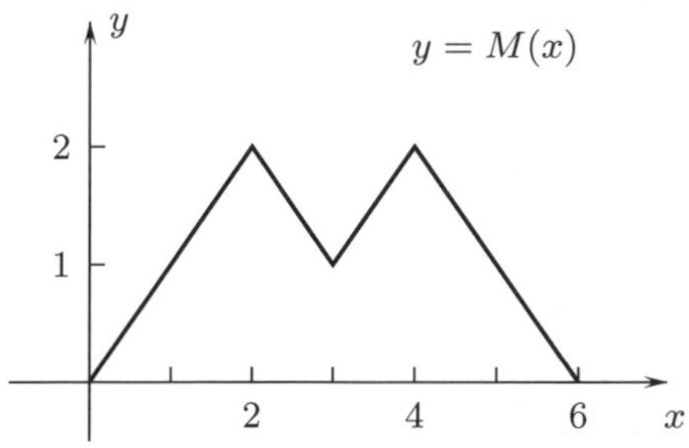

Figure T10a

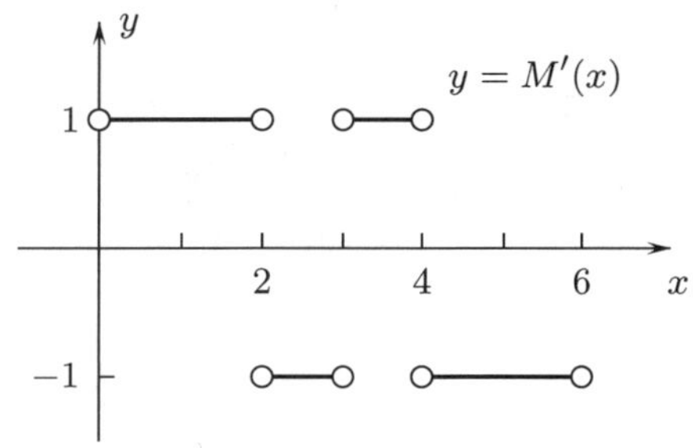

Figure T0b

Problems 2.4

1. Figures A1a,b • Possible answers: $T(500) \approx 16$ degrees Celsius • $\frac{dT}{dh}(500) \approx -0.0045$ degrees Celsius per foot • $T(1500) \approx 16$ degrees Celsius • $\frac{dT}{dh}(1500) \approx 0.01$ degrees Celsius per foot • $T(7000) \approx 17.5$ degrees Celsius • $\frac{dT}{dh}(7000) \approx -0.0015$ degrees Celsius per foot (Different estimates give different answers.)

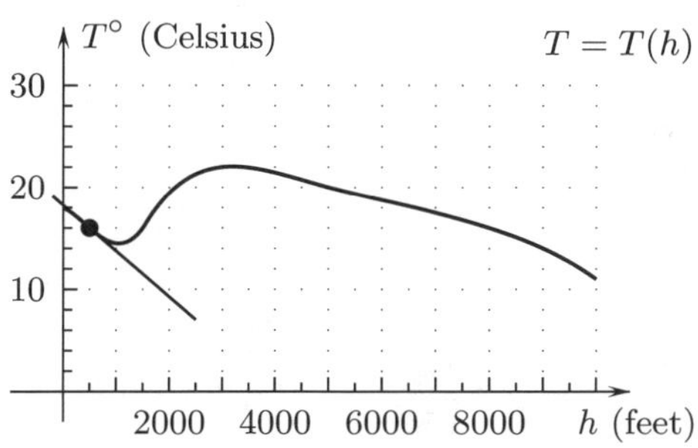

Figure A1a

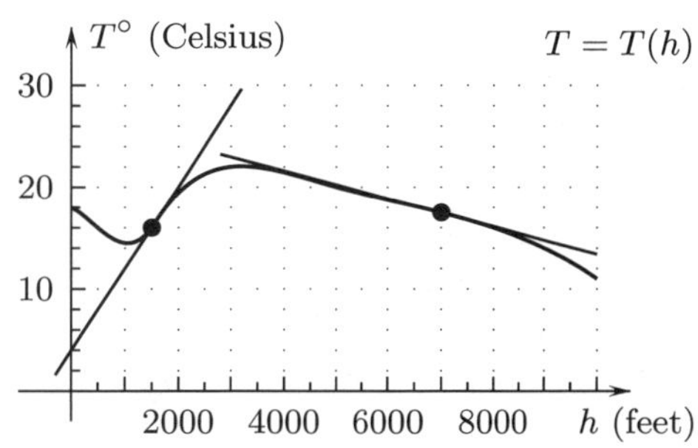

Figure A1b

4. Figure A4

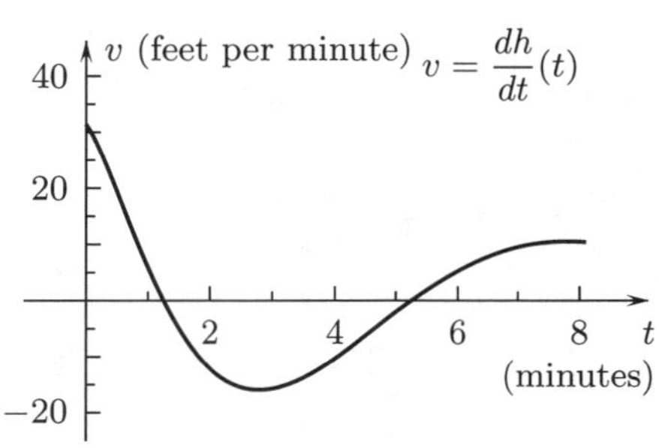

Figure A4

5. One answer using $T(0) \approx 0$, $T(3) \approx 1300$, $T(18) \approx 1200$, $T(21) \approx 1000$, $T(30) \approx 0$: $\frac{dT}{dt} \approx \frac{1300-0}{3-0} \doteq 433$ degrees per hour for $0 < t < 3$ • $\frac{dT}{dt} \approx \frac{1200-1300}{18-3} \doteq -7$ degrees per hour for $3 < t < 18$ • $\frac{dT}{dt} \approx \frac{0-1000}{30-21} \doteq -111$ degrees per hour for $21 < t < 30$

8. **(a)** $P(T)$ is a decreasing function for temperatures above $-20°$F and below freezing because less of the ice cream freezes when the temperature is higher. • $P(T) = 0$ for $T > 32$ because those temperatures are above freezing. **(b)** One answer: $P'(10) \approx -0.6$ percent per degree • $P(T)$ increases about $P'(10)(10 - 9.99) \approx -0.006\%$ (and decreases about 0.006%) when T decreases from 10 to 9.99.

13. **(a)** Likely answers: $\frac{2.407 - 2.437}{1982 - 1979} = -0.01$ or $\frac{2.437 - 2.497}{1979 - 1976} = -0.02$ or $\frac{2.407 - 2.497}{1982 - 1976} = -0.015$ millions per year
(b) Likely answer: $\frac{1012 - 1028}{1985 - 1982} \doteq -5.33$ million acres per year
(c) $A(1976)/N(1976) = 1054/2.497 \doteq 422.1$ acres per farm in 1976 and $A(1991)/N(1991) = 983/2.105 \doteq 467.0$ acres per farm in 1991

34. 8 dollars per shirt

35. **(a)** 2250 jackets • \$36,000 **(b)** $21000p^{1/2} - 3000p$ dollars **(c)** $10500p^{-1/2} - 3000$ dollars per dollar

38. **(a)** [Demand] $= 1600 - 20p$ • [Revenue] $= 1600p - 20p^2$ **(a)** $R'(30) = 400$ dollars per dollar • $R'(40) = 0$ dollars per dollar • $R'(48) = -320$ dollars per dollar

39. $dV/dA = \frac{1}{4}\sqrt{A/\pi}$

41. The car is 123.2 kilometers past the service station and is traveling 68 kilometers per hour.

46. One answer: Figure A46

52. **(a)** $b = 2$, any a **(b)** $a = -1, b = 2$

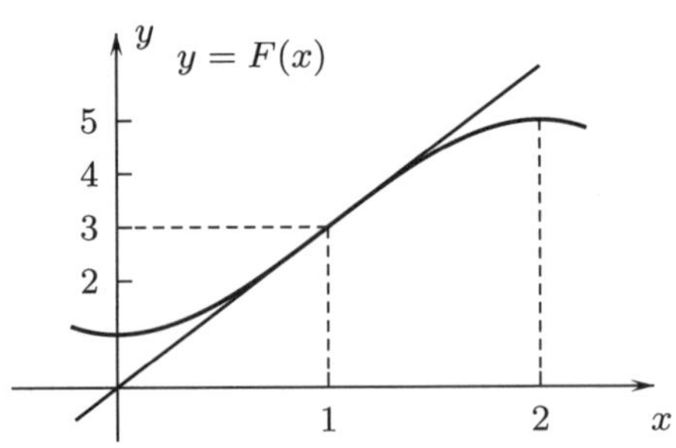

Figure A46

Tune-up exercises 2.5

T1. $(5x^3 + 2x^2 - 4)(7x^6 - 10x^4) + (x^7 - 2x^5)(15x^2 + 4x)$

T2. $\dfrac{(x^5 + 1)(\frac{1}{2}x^{-1/2}) - x^{1/2}(5x^4)}{(x^5 + 1)^2}$

T3. $10(a + bx^2 + cx^5)^9 \frac{d}{dx}(a + bx^2 + cx^5) = 10(a + bx^2 + cx^5)^9(2bx + 5cx^4)$

T4. $P'(5) = R(5)S'(5) + S(5)R'(5) = (3)(10) + (4)(-3) = 18$

T5. $\frac{dW}{dx}(4) = \dfrac{Z(4)\frac{dY}{dx}(4) - Y(4)\frac{dZ}{dx}(4)}{[Z(4)]^2} = \dfrac{(5)(3) - (2)(6)}{5^2} = \frac{3}{25}$

T6. $\frac{dG}{dx} = \frac{7}{2}y^{5/2}\frac{dy}{dx}$ • $\frac{dG}{dx}(3) = \frac{7}{2}[y(3)]^{5/2}\frac{dy}{dx}(3) = \frac{7}{2}(6^{5/2})(-5) = -\frac{35}{2}(6^{5/2})$

T7. $y(1) = 9$ • $y'(x) = (x^2 + x^3 + x^4)(5x^4 + 6x^5 + 7x^6) + (x^5 + x^6 + x^7)(2x + 3x^2 + 4x^3)$ • $y'(1) = 81$ • Tangent line: $y = 9 + 81(x - 1)$

T8. Figures T8a,b • $A(2) \approx 2.5$ • $A'(2) \approx -0.6$ • $B(2) \approx 1.5$ • $B'(2) \approx 1$ • $[AB]_{x=2} \approx (2.5)(1.5) = 3.75$ • $\left[\frac{A}{B}\right]_{x=2} \approx \frac{2.5}{1.5} \doteq 1.7$ •
$\left[\frac{d}{dx}(AB)\right]_{x=2} = A(2)B'(2) + B(2)A'(2) \approx (2.5)(1) + (1.5)(-0.6) = 1.6$ •
$\left[\frac{d}{dx}\left(\frac{A}{B}\right)\right]_{x=2} = \frac{B(2)A'(2) - A(2)B'(2)}{[B(2)]^2} \approx \frac{(1.5)(-0.6) - (2.5)(1)}{1.5^2} \doteq -1.5$
(Different estimates yield different answers.)

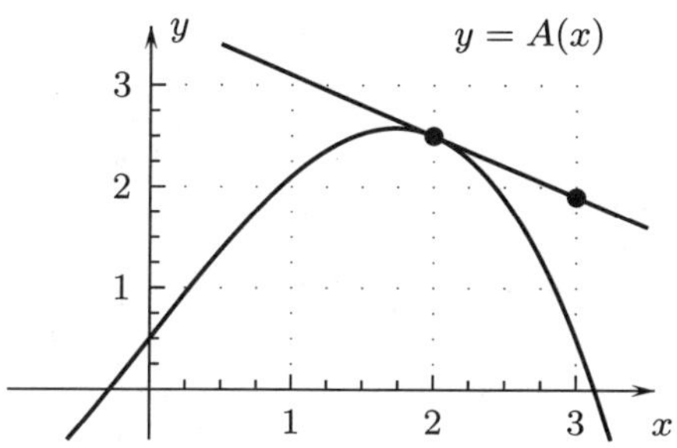

Figure T8a

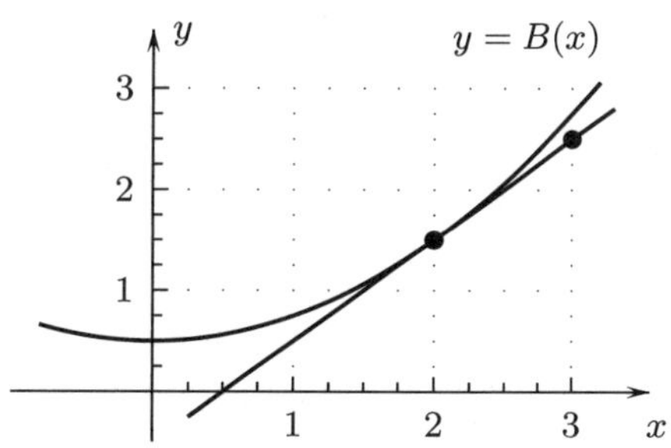

Figure T8b

T9. $(x^2 + 3x - 1)(2x^{-1/2}) + (4x^{1/2} - 6)(2x + 3)$ **T10.** $\frac{1}{2}(t^2 - 4)^{-1/2}\frac{d}{dt}(t^2 - 4) = t(t^2 - 4)^{-1/2}$

T11. $2^3 Z'(2) + 3(2^2)Z(2) = 8(10) + 12(5) = 140$ **T12.** $\frac{(bx^4 + 4)(2ax) - (ax^2 + 2)(4b^3)}{(bx^4 + 4)^2}$

T13. $f'(t) = 2t + \frac{6}{(t - 6)^2}$ • $f'(1) = \frac{56}{25}$

T14. $15x^2(x^3 + 1)^4$ **T15.** $y = 2 + \frac{8}{3}(x - 2)$

T16. Figure T16 **(a)** $U(2) \approx 30$ **(b)** $U'(2) \approx 18$
(c) $\left[\frac{d}{dx}(xU)\right]_{x=2} = [xU'(x) + U(x)]_{x=2} = 2U'(2) + U(2) \approx (2)(18) + 30 = 66$
(d) $\left[\frac{d}{dx}(U^6)\right]_{x=2} = 6[U(2)]^5 U'(2) \approx 6(30^5)(18) \doteq 2.6 \times 10^9$ (Different estimates yield different answers.)

Figure T16

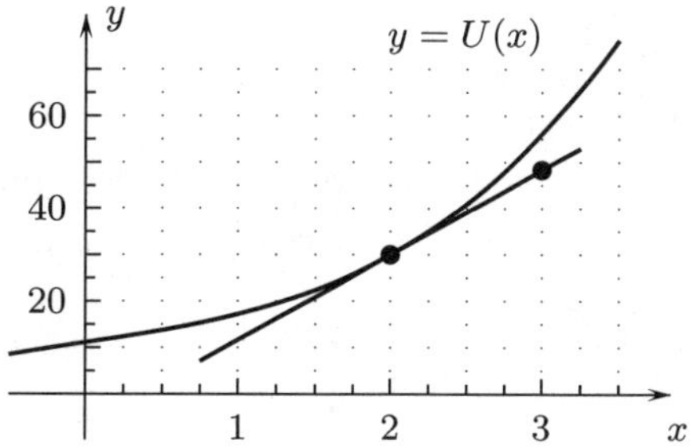

Problems 2.5

1. $(1+3x-x^2)\frac{d}{dx}(x^2-5)+(x^2-5)\frac{d}{dx}(1+3x-x^2)=(1+3x-x^2)(2x)+(x^2-5)(3-2x)$

2. $(x-x^2)(-x^{-2}-2x^{-3})+(1+x^{-1}+x^{-2})(1-2x)$

4. $3(x^2+x^{-2})+(2x-2x^{-3})(2+3x)$

5. $(1+ax^n)(-bnx^{n-1})+(1-bx^n)(nax^{n-1})$

7. $\dfrac{-x^2-4x-2}{(x^2-2)^2}$

9. 5

11. $\frac{15}{4}$

15. $-4(35)+0(560)=-140$

17. $\dfrac{(-2)(7)-(13)(-4)}{(-2)^2}=\frac{19}{2}$

19. $y=1-(x-1)$ • Figure A19

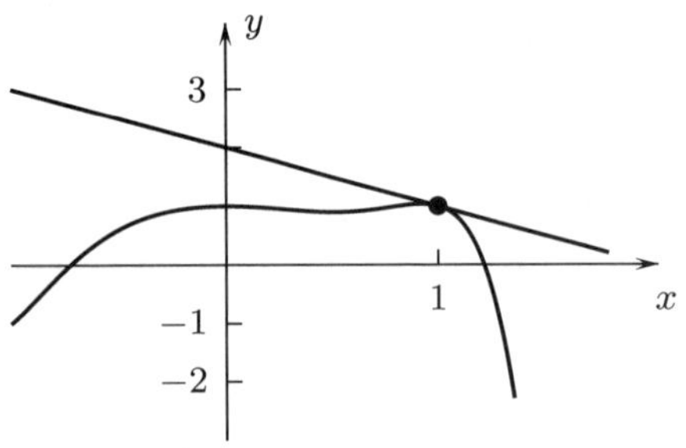

Figure A19

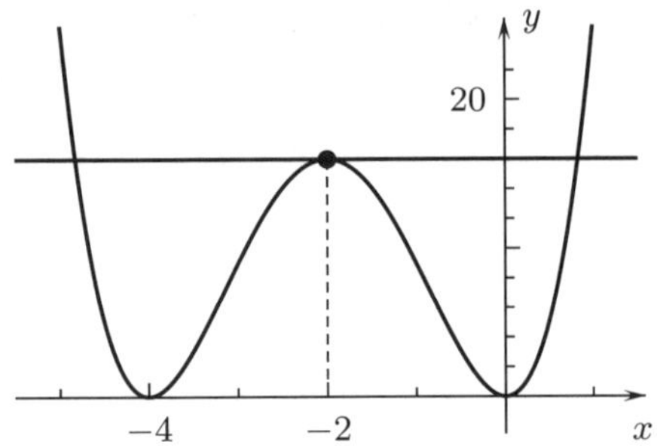

Figure A47a

20. $y=0.8+0.12(x-3)$

21. Let $f(t)$ (acres) be the size of the average farm and $N(t)$ (millions) the number of farms at time t. • [Total acreage] $= A(t) = N(t)f(t)$ million acres • $A(1991)=$[2.105 million farms][467 acres per farm] $\doteq$ 983 million acres • $A'(1991)=N(1991)f'(1991)+N'(1991)f(1991)=$ [2.105 million farms][7 acres per farm per year] + [−0.035 million farms per year][467 acres per farm] = −1.61 million acres per year • The total acreage was decreasing 1.61 million acres per year.

22. $A=wL$ • $\frac{dA}{dt}=w\frac{dL}{dt}+L\frac{dw}{dt}=$ (5 meters)(−6 meters/second) + (7 meters)(3 meters/second) = −9 square meters/second • The area is decreasing 9 square meters/second.

31. -24

36. $10(2+3x^2+4x^5)^9(6x+20x^4)$

38. $2x^{-1/2}(10-x^{1/2})^{-5}$

40. 2700

37. $\frac{5}{3}$

42. $-(3x-4)^{-4/3}$

43. $9x^{-1/2}(6\sqrt{x}+3)^2$

45. $\frac{5}{3}$

47a $y=16$ • Figure A47a (above)

48. $\rho(2)=1.293[1+0.00367(20)]^{-1}\doteq 1.2046$ grams per liter • $\frac{d\rho}{dT}(20)=-1.293[1+0.0036(20)]^{-2}(0.00367)\doteq -0.00412$ grams per liter per degree

50. $V=w^3$ • $\frac{dV}{dt}=3w^2\frac{dw}{dt}$ • $3(10^2)(3)=900$ cubic millimeters per day

51. **(a)** $\rho=4$ grams per centimeter **(b)** -0.01 grams per centimeter per degree

57. $1666\frac{2}{3}$ pounds per hour

59. One answer: -270 meter^2/min.

61. $\dfrac{4(2x+3)(x+1)^3-2(x+1)^4}{(2x+3)^2}$

62. $\frac{5}{4}x(5x+4)^{-3/4}+(5x+4)^{1/4}$

67. $-\frac{6}{125}$ centimeters per second^2

68. $360t^3+126t+6$ inches per hour^2

73b. $(x^2+1)(x^3+2)(4x^2)+(x^2+1)(x^4+3)(3x^2)+(x^2+2)(x^4+3)(2x)$

Secant-line program for TI-81 calculators[†]

The line numbers in the following program are not used on the calculator.

```
1.  Prgm1: SECANT
2.  : ClrDraw
3.  : All Off
4.  : Y1-On
5.  : Disp "USE Y1(X)"
6.  : Disp "B ="
7.  : Input A
8.  : Disp "B ="
9.  : Input B
10. : A → X
11. : Y1 → W
12. : B → X
13. : Y1 → V
14. : (V − W)/(B − A) → M
15. : DrawF  W + M(X − A)
16. : Pause
17. : Disp "SLOPE ="
18. : Disp M
```

To enter the program, press [2nd] [QUIT] to display the home screen and [PRGM] for the menu of programs. Press [▶] so that EDIT is highlighted and press [1] to use program #1 or the number of an empty program if you have already used #1. If you are editing a previously entered program, scroll down to its name and press [ENTER].

Line-by-line instructions for entering the program follow. To make a correction, move the cursor to the error and use [INS] and [DEL] as needed. To return to the program from other screens, press [CLEAR] or follow the steps in the previous paragraph. To interrupt the running of the program, press [ON] and then [2] for QUIT. To delete a program, press [2nd] [MEM] [2] for Delete and [6] for Prgm, put the cursor on the name of the program, and press [ENTER] [2nd] [QUIT].

---

1. Press [S] [E] [C] [A] [N] [T] [ENTER]. The calculator is automatically in alpha mode here. This title identifies the program in the program menu.

---

2. Press [2nd] [DRAW] [1] [ENTER]. ClearDraw clears the graphics screen.

---

3. Press [2nd] [Y-VARS] [◀] [1] [ENTER]. All-Off unselects all the formulas in the Y= menu so their graphs are not generated when the program is run.

---

4. Press [2nd] [Y-VARS] [▶] [2] [ENTER]. This selects Y_1 so its graph will be generated.

---

5. Press [PRGM] to display the CTL (control) commands and then [▶] for I/O (input/output) commands. Press [1] to enter the word Disp and then [2nd] [A-LOCK] to lock the calculator in alpha mode. Press ["] [U] [S] [E] [␣] [2nd] [Y-VARS] [1] [1] [(] [X|T] [)] [ALPHA] ["] [ENTER] with [␣] the space on the zero key.

---

6. Press [PRGM] [▶] [1] to enter the word Disp and then [ALPHA] ["] [ALPHA] [A] [2nd] [TEST] to displaye the TEST menu. Press [1] for the equal sign and [ALPHA] ["] [ENTER]. The prompt "A =?" to appear when the program is run.

---

7. Press [PRGM] [▶] for the I/O menu and [2] [ALPHA] [A] [ENTER]. Here the program pauses for the user to assign a value to A.

[†]These calculator-specific programs are included in preliminary editions of this text for the convenience of users. In the First Edition they will be in the *Calculator/Computer Supplement* that will accompany the final book.

8. Press [PRGM] [▶] [1] [ALPHA] ["] [ALPHA] [B] [2nd] [TEST] [1] [ALPHA] ["] [ENTER]. The prompt "$B =?$" appears when the program is run.

9. Press [PRGM] [▶] [2] [ALPHA] [B] [ENTER]. The user enters a value for B.

10. Press [ALPHA] [A] [STO▶] [$X|T$] [ENTER]. This assigns the value of A to X.

11. Press [2nd] [Y-VARS] [1] [1] [STO▶] [ALPHA] [W] [ENTER]. Here the function Y_1 in the Y= menu is evaluated at $X(= A)$ and the result is stored as W.

12. Press [ALPHA] [B] [STO▶] [$X|T$] [ENTER] to give X the value B.

13. Press [2nd] [Y-VARS] [1] [1] [STO▶] [ALPHA] [V] [ENTER]. $Y_1(B)$ is stored as V.

14. Use [(] [ALPHA] [V] [−] [ALPHA] [W] [)] [÷] [(] [ALPHA] [B] [−] [ALPHA] [A] [)] [STO▶] [ALPHA] [M] [ENTER]. The slope $(V - W)/(B - A)$ of the secant line through $(A, W) = (A, Y_1(A))$ and $(B, V) = (B, Y_1(B))$ on the graph is stored as M.

15. Press [2nd] [DRAW] [6] [ALPHA] [W] [+] [ALPHA] [M] [(] [$X|T$] [−] [ALPHA] [A] [)] [ENTER] This draws the secant line, $y = W + M(x - A)$.

16. Press [PRGM] and then either press [6] [ENTER] or scroll down to <Pause> and press [ENTER] [ENTER] This stops the program until the user presses [ENTER].

17. Press [PRGM] [▶] [3] and then [2nd] [A-LOCK] to lock the calculator in alpha mode. Press ["] [S] [L] [O] [P] [E] [2nd] [TEST] [1] [ALPHA] ["] [ENTER]. SLOPE = appears when the program is run.

18. Press [PRGM] [▶] [3] [ALPHA] [M] [ENTER] to displays the slope of the line on the home screen.

Entering the secant-line program in a TI-82 calculator

Press [2nd] [QUIT] to display the home screen. Then press [PRGM] for the menu of programs. If you are starting a new program, press [▶] twice so that <NEW> is highlighted and press [1] or [ENTER]. If you are editing a previously entered program, press [PRGM] and [▶] to highlight <EDIT>. Then press the number or letter of the program, or scroll down to its name and press [ENTER].

The program is listed in Section 2.3. Line-by-line instructions for entering it follow. To make a correction, move the cursor to the error and use [INS] and [DEL] as needed. To return to the program from other screens, press [CLEAR] or follow the steps in the previous paragraph. To interrupt the running of the program, press [ON] and then [2] for <QUIT>.

To delete a program, press [2nd] [MEM] [2] for <Delete>, [6] for <Prgm>, put the cursor on the name of the program, and press [ENTER] [2nd] [QUIT].

To transfer a program electronically from one TI-82 calculator to another or from a TI-82 to a TI-83, turn both calculators on and connect them with the wire that is provided, pressing the connectors in firmly. On the calculator to receive the program, press [2nd] [LINK], select <Receive>, and press [ENTER]. The screen should display "Waiting. . ." On the transmitting calculator press

[2nd], scroll down to <SelectCurrent>, press [ENTER] and scroll down to the name of the program. Press [ENTER] to select it. The triangle beside the program name should change into a fat arrow. Press [▶] for <Transmit> and press [ENTER] to transmit the program. Select <Overwrite> on the receiving calculator if necessary to replace any old program that has the same name.

---

1. Press [S] [E] [C] [A] [N] [T] [ENTER]. The calculator is automatically in alpha mode here. This title identifies the program in the program menu.

---

2. Press [2nd] [DRAW] [1] [ENTER]. ClearDraw clears the graphics screen.

---

3. Press [2nd] [Y-VARS] [5] [2] [ENTER]. FnOff unselects all the formulas in the Y= menu so their graphs are not generated when the program is run.

---

4. Press [2nd] [Y-VARS] [5] [1] [1] [ENTER]. This command selects Y_1 so its graph will be generated.

---

5. Press [PRGM] to display CTL (control) commands and then [▶] for I/O (input/output) commands. Press [3] to enter the word Disp and then [2nd] [A-LOCK] to lock the calculator in alpha mode. Press ["] [U] [S] [E] [_] [2nd] [Y-VARS] [1] [1] [(] [X, T, θ] [)] [ALPHA] ["] [ENTER] with [_] the space on the zero key.

---

6. Press [PRGM] [▶] [2] for the word Prompt. Press [ALPHA] [A] [,] [ALPHA] [B] [ENTER] to complete the line, which causes the prompts $A =$? and $B =$? to appear when the program is run so the user can supply values of A and B.

---

7. Press [ALPHA] [A] [STO▶] [X, T, θ] [ENTER]. This assigns the value of A to X.

---

8. Press [2nd] [Y-VARS] [1] [1] [STO▶] [ALPHA] [W] [ENTER]. Here the function Y_1 in the Y= menu is evaluated at $X(= A)$ and the result is stored as W.

---

9. Press [ALPHA] [B] [STO▶] [X, T, θ] [ENTER] to give X the value B.

---

10. Press [2nd] [Y-VARS] [1] [1] [STO▶] [ALPHA] [V] [ENTER]. $Y_1(B)$ is stored as V.

---

11. Use [(] [ALPHA] [V] [−] [ALPHA] [W] [)] [÷] [(] [ALPHA] [B] [−] [ALPHA] [A] [)] [STO▶] [ALPHA] [M] [ENTER]. The slope $(V - W)/(B - A)$ of the secant line through $(A, W) = (A, Y_1(A))$ and $(B, V) = (B, Y_1(B))$ on the graph is calculated and stored as M.

---

12. Press [2nd] [DRAW] [6] [ALPHA] [W] [+] [ALPHA] [M] [(] [X, T, θ] [−] [ALPHA] [A] [)] [ENTER] This draws the secant line, $y = W + M(x - A)$.

---

13. Press [PRGM] and then either press [8] [ENTER] or scroll down to <Pause> and press [ENTER] [ENTER] This stops the program until the user presses [ENTER].

---

14. Press [PRGM] [▶] [3] and then [2nd] [A-LOCK] to lock the calculator in alpha mode. Press ["] [S] [L] [O] [P] [E] [2nd] [TEST] [1] [ALPHA] ["] [ENTER]. SLOPE = appears when the program is run.

---

15. Press [PRGM] [▶] [3] [ALPHA] [M] [ENTER] to which displays the slope of the line on the home screen.

Entering the secant-line program in a TI-83 calculator

Press [2nd] [QUIT] to display the home screen. Then press [PRGM] for the menu of programs. If you are starting a new program, press [▶] twice so that NEW is highlighted and press [1] or [ENTER]. If you are editing a previously entered program, press [PRGM] and [▶] to highlight EDIT. Then press the number or letter of the program, or scroll down to its name and press [ENTER].

The program is listed in Section 2.3. Line-by-line instructions for entering it follow. To make a correction, move the cursor to the error and use [INS] and [DEL] as needed. To return to the program from other screens, press [CLEAR] or follow the steps in the previous paragraph. To interrupt the running of the program, press [ON] and then [2] for QUIT.

To delete a program, press [2nd] [MEM] [2] for Delete, [6] for Prgm, put the cursor on the name of the program, and press [ENTER] [2nd] [QUIT].

To transfer a program electronically from one TI-83 calculator to another TI-83 or from a TI-82 to a TI-83, turn both calculators on and connect them with the wire that is provided, pressing the connectors in firmly. On the calculator to receive the program, press [2nd] [LINK], select <RECEIVE>, and press [ENTER]. The screen should display "Waiting. . ." On the transmitting calculator press [2nd] [LINK], scroll down to <Prgm>, press [ENTER], and scroll down to the name of the program. Press [ENTER] to select it. The triangle beside its name should changed to an square. Press [▶] for <Transmit> and [ENTER] to transmit the program. Select <Overwrite> on the receiving calculator if necessary to replace any old program that has the same name.

---

1. Press [S] [E] [C] [A] [N] [T] [ENTER]. The calculator is automatically in alpha mode here. This title identifies the program in the program menu.

---

2. Press [2nd] [DRAW] [1] [ENTER]. ClearDraw clears the graphics screen.

---

3. Press [VARS] [▶] [4] [2] [ENTER]. FnOff unselects all the formulas in the Y= menu so their graphs are not generated when the program is run.

---

4. Press [VARS] [▶] [4] [1] [1] [ENTER]. This command selects Y_1 so its graph will be generated.

---

5. Press [PRGM] to display CTL (control) commands and then [▶] for I/O (input/output) commands. Press [3] to enter the word Disp and then [2nd] [A-LOCK] to lock the calculator in alpha mode. Press ["] [U] [S] [E] [␣] [VARS] [▶] [1] [1] [(] [X, T, θ, n] [)] [ALPHA] ["] [ENTER] with [␣] the space on the zero key.

---

6. Press [PRGM] [▶] [2] for the word Prompt. Press [ALPHA] [A] [,] [ALPHA] [B] [ENTER] to complete the line, which causes the prompts $A = ?$ and $B = ?$ to appear when the program is run so the user can supply values of A and B.

---

7. Press [ALPHA] [A] [STO▶] [X, T, θ, n] [ENTER]. This assigns the value of A to X.

---

8. Press [VARS] [▶] [1] [1] [STO▶] [ALPHA] [W] [ENTER]. Here the function Y_1 in the Y= menu is evaluated at $X(= A)$ and the result is stored as W.

---

9. Press [ALPHA] [B] [STO▶] [X, T, θ, n] [ENTER] to give X the value B.

---

10. Press [VARS] [▶] [1] [1] [STO▶] [ALPHA] [V] [ENTER]. $Y_1(B)$ is stored as V.

---

11. Use [(] [ALPHA] [V] [−] [ALPHA] [W] [)] [÷] [(] [ALPHA] [B] [−] [ALPHA] [A] [)] [STO▶] [ALPHA] [M] [ENTER]. The slope $(V - W)/(B - A)$ of the secant line through $(A, W) = (A, Y_1(A))$ and $(B, V) = (B, Y_1(B))$ on the graph is calculated and stored as M.

---

12. Press [2nd] [DRAW] [6] [ALPHA] [W] [+] [ALPHA] [M] [(] [X, T, θ, n] [−] [ALPHA] [A] [)] [ENTER] This draws the secant line, $y = W + M(x - A)$.

---

13. Press [PRGM] and then either press [8] [ENTER] or scroll down to <Pause> and press [ENTER] [ENTER] This stops the program until the user presses [ENTER].

---

14. Press [PRGM] [▶] [3] and then [2nd] [A-LOCK] to lock the calculator in alpha mode. Press ["] [S] [L] [O] [P] [E] [2nd] [TEST] [1] [ALPHA] ["] [ENTER]. SLOPE = appears when the program is run.

---

15. Press [PRGM] [▶] [3] [ALPHA] [M] [ENTER] to displays the slope of the line on the home screen.

Entering the secant-line program in a TI-85 or TI-86 calculator

Press [2nd] [QUIT] to display the home screen. Then press [PRGM] [F2] for <EDIT> and follow the steps below. If you are editing a previously entered program, select its abbreviated name from the menu and press [ENTER].

The program is listed in Section 2.3. Line-by-line instructions for entering it follow. Commands typed in ***slanted bold type*** can be obtained from the calculator's catalog, which is reached with [2nd] [CATALOG] on a TI-85 and with [2nd] [CATALOG] [F1] on a TI-86. Press the key for the first letter in the command, scroll down to the command if necessary, and press [ENTER] to enter it. For example, to enter ClDrw on line 2 of the program use the keys [2nd] [CATALOG] [C] [▼] [ENTER] [ENTER], and for FnOff on line 3 press [2nd] [CATALOG] [F] followed by [▼] eight times (on a TI-85) or nine times (on a TI-86) and [ENTER] [ENTER]. Most of the commands can also be found in the menus.

To make a correction, move the cursor to the error and use [DEL] and [2nd] [INS] as needed. To return to the program from other screens, press [EXIT] or follow the steps in the first paragraph. To interrupt the running of the program, press [ON] and then [F5] for <QUIT>.

To transfer a program electronically from one TI-85 calculator to another TI-85 or TI-86 or from a TI-86 to another TI-86, turn both calculators on and connect them with the wire that is provided, pressing the connectors in firmly. On the calculator to receive the program, press [2nd] [LINK] ond [F1] for <RECV>. The screen should display "Waiting." On the transmitting calculator press [2nd] [LINK] and [F1] for <Send>, press [F2] for <PRGM>, and scroll down to the name of the program. Press [F2] for <Select>. The triangle beside the program name should be changed to an arrow. Press [F1] for <XMIT> to transmit the program. Select <Overwrite> on the receiving calculator to replace any old program that has the same name.

---

1. Press [S] [E] [C] [A] [N] [T] [ENTER].

---

2. Enter ***ClDrw*** from the catalog and press [ENTER]. This clears the graphics screen.

---

3. Enter ***FnOff*** from the catalog and press [ENTER]. This unselects all the formulas in the $y=$ menu so their graphs are not generated when the program is run.

---

4. Enter ***FnOn*** and press [1] [ENTER]. This selects $y1$ so its graph is generated.

---

5. Enter ***Disp*** and press [2nd] [STRING] [F1] (for "). Press [ALPHA] [ALPHA] to lock the calculator in upper-case alpha mode and [U] [S] [E] [␣] with [␣] the space on the [(-)] key. Press [ALPHA] to leave upper-case alpha mode and [2nd] [alpha] to enter lower-case alpha mode. Press [Y] [1] [(] [x-VAR] [)] [F1] (for ") [ENTER] to finish the line.

---

6. Enter ***Prompt*** and press [ALPHA] [A] [,] [ALPHA] [B] [ENTER]. The prompts $A=?$ and $B=?$ will appear so the user can assign values to A and B.

---

7. Press [ALPHA] [A] [STO▶] [x-VAR] [ENTER] to set $x = A$.

---

8. Press [2nd] [alpha] [Y] [1] [STO▶] [W] [ENTER]. There is no [ALPHA] before [W] because the calculator is locked in upper-case alpha mode after [STO▶]. Here the value at x $(= A)$ of $y1$ in the $y=$ menu is stored as W.

---

9. Press [ALPHA] [B] [STO▶] [x-VAR] [ENTER] to give x the value B.

---

10. Press [2nd] [alpha] [Y] [1] [STO▶] [V] [ENTER]. $y1(B)$ is stored as V.

---

11. Use [(] [ALPHA] [V] [−] [ALPHA] [W] [)] [÷] [(] [ALPHA] [B] [−] [ALPHA] [A] [)] [STO▶] [M] [ENTER]. The slope $(V - W)/(B - A)$ of the secant line through $(A, W) = (A, y1(A))$ and $(B, V) = (B, y1(B))$ on the graph is stored as M.

---

12. Enter ***DrawF*** and press [ALPHA] [W] [+] [ALPHA] [M] [×] [(] [x-VAR] [−] [ALPHA] [A] [)] [ENTER]. The secant line $y = W + M(x - A)$ is drawn.

---

13. Enter ***Pause*** [ENTER] to stop the program until the user presses [ENTER].

---

14. Enter ***Disp*** and press [2nd] [STRING] [F1] (for ") [ALPHA] [ALPHA] [S] [L] [O] [P] [E] [=] [F1] [ENTER]. SLOPE = appears when the program is run.

---

15. Enter ***Disp*** and press [ALPHA] [M] [ENTER] to displays the slope of the line on the home screen.

CHAPTER 3
DERIVATIVES AND APPLICATIONS, II

In Chapter 2 we used formulas for derivatives of powers and approximate derivatives of functions given by graphs and tables to solve problems involving rates of change. We continue our study of derivatives in this chapter by using them to analyze graphs of functions in Section 3.1 and to solve maximum/minimum problems in Section 3.2. We discuss the general Chain Rule in Section 3.3. The TRANSCENDENTAL FUNCTIONS—exponential functions, logarithms, trigonometric functions, and inverse trigonometric functions—and their derivatives are studied and used in applications in Sections 3.4 through 3.6.

Section 3.1

Analyzing graphs of functions

OVERVIEW: *In this section we show how derivatives are used to determine the intervals in which functions are increasing and in which they are decreasing and to determine where their graphs are concave up and concave down. The techniques presented here for analyzing graphs will be used extensively in Section 3.2 and later sections to solve maximum/minumum problems and in other types of applications that require information about graphs.*

Topics:

- ***Local maxima and minima and critical points***
- ***The First-Derivative Test in an interval***
- ***The Mean-Value Theorem***
- ***Global maxima and minima***
- ***The Second-Derivative Test for concavity***

Water flowing in and out of a tank

Because the main concepts of this section—local and global maxima and minima, increasing and decreasing functions in intervals, concavity, and inflection points—are interrelated, we will be able to illustrate them with the same mathematical model. We will use a particular function that describes the changing depth of water in a rectangular tank.

We suppose that the tank is empty at time $t = 0$ (hours) and that water is added until a time t_0, and then drained out, so that the depth is $h(t) = 3t^2 - t^3$ (feet) for $0 \leq t \leq 3$. Figure 1 shows the graph of this function. Since $h(3) = 3(3^2) - 3^3 = 0$, the tank is empty again at $t = 3$. We want to determine the time t_0 when the water is deepest in the tank.

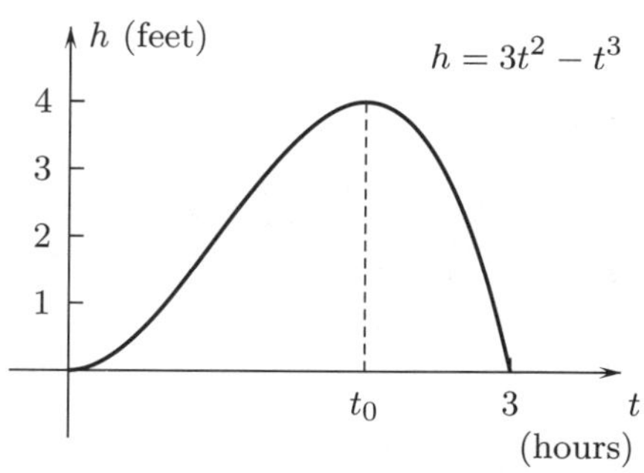

FIGURE 1

Local maxima and minima

The maximum depth at $t = t_0$ is called a LOCAL MAXIMUM of $h(t)$ because it is the greatest value of $h(t)$ in an open interval that includes points on both sides of t_0. Here is a general definition of this term for a function with x as the variable:

♦ SUGGESTIONS TO INSTRUCTORS: Have students begin work on Questions 1 through 9 before class. Problems 1, 9, 12, 19, 20, 22, 27, 29, 45, and 47 are good for in-class discussion.

Definition 1 **(a)** *$f(x)$ has a local maximum at $x = x_0$ if $f(x_0)$ is the greatest value of $f(x)$ in an open interval containing x_0, i.e., if $f(x_0) \geq f(x)$ for all x in the interval.*

(b) *$f(x)$ has a local minimum at $x = x_0$ if $f(x_0)$ is the least value of $f(x)$ in an open interval containing x_0, i.e., if $f(x_0) \leq f(x)$ for all x in the interval.*

Notice that the least value of $h(t)$ for $0 \leq t \leq 3$ in Figure 1 is $h = 0$ at $t = 0$ and $t = 3$, but neither of these values is a local minimum because $h(t)$ is not defined to the left of $t = 0$ or to the right of $t = 3$.

Critical points

How can we find the value of t_0 in Figure 1 where $h(t)$ has the local maximum? We consider the secant line through the points at $t = t_0$ and $t = b$ for values of b close to t_0, as shown in Figure 2 for $b < t_0$ and in Figure 3 for $b > t_0$. Recall that the derivative of $h(t)$ at t_0,

$$h'(t_0) = \left[\frac{d}{dt}(3t^2 - t^3)\right]_{t=t_0} = \left[6t - 3t^2\right]_{t=t_0} = 6t_0 - 3t_0^2 = 3t_0(2 - t_0) \tag{1}$$

is the two-sided limit, as $b \to t_0$, of the slope of the secant line.

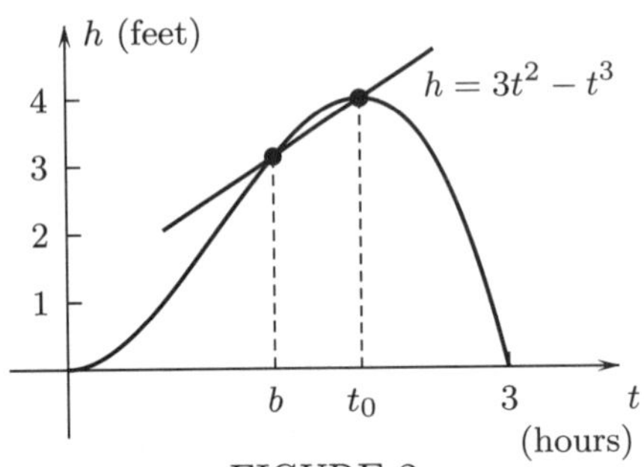

FIGURE 2

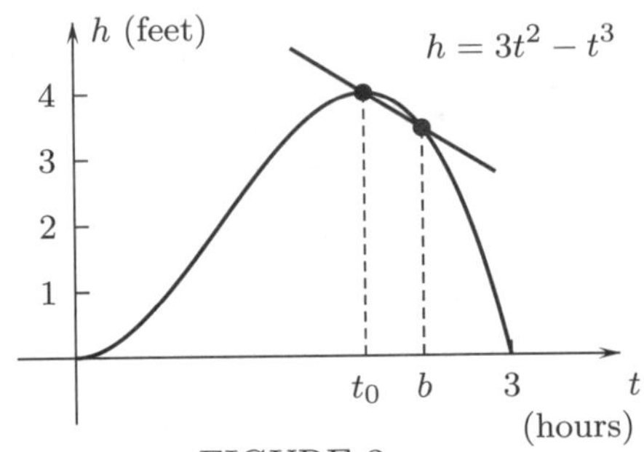

FIGURE 3

Because $h(t_0)$ is greater than all other values of $h(t)$ for $0 \leq t \leq 3$, the slope of the secant line is positive for b just to the left of t_0, as in Figure 2. Consequently, $h'(t_0)$, which is the limit of this slope as $b \to t_0$ from the left, must be positive or zero. On the other hand, the slope of the secant line for b just to the right of t_0, as in Figure 3 is negative, so $h'(t_0)$, which is also the limit of this slope as $b \to t_0$ from the right, has to be negative or zero. Since $h'(t_0)$ is both ≥ 0 and ≤ 0, it is zero, and the tangent line to the graph of $h(t)$ at $t = t_0$ is horizontal, as is shown in Figure 4.

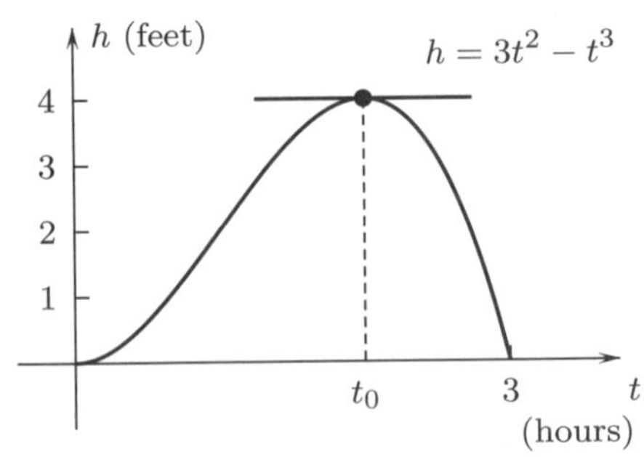

Horizontal tangent line

$h'(t_0) = 0$

FIGURE 4

To find t_0, we set the formula on the right side of **(1)** equal to 0. This gives $3t_0(2 - t_0) = 0$, which shows that $t_0 = 2$. Consequently, the water is deepest at time $t = 2$ hours.

Question 1 What is the maximum depth of the water in the tank?

The line of reasoning we used to show that $h(t)$ has a zero derivative at its local maximum can be applied to any local maximum or minimum and yields the following result.

Theorem 1 *If $f(x)$ has a local maximum or local minimum at $x = x_0$, then either $f'(x_0) = 0$ or $f'(x_0)$ does not exist.*

A value $x = x_0$ that is in the interior of an interval where $f(x)$ is defined is called a CRITICAL POINT of $f(x)$ if $f'(x_0)$ is zero or does not exist. Hence, Theorem 1 can be rephrased by saying that a function must have a critical point wherever it has a local maximum or minimum.

Example 1 Explain how Theorem 1 applies to the function $y = x^2$, which has a local minimum at $x = 0$ (Figure 5).

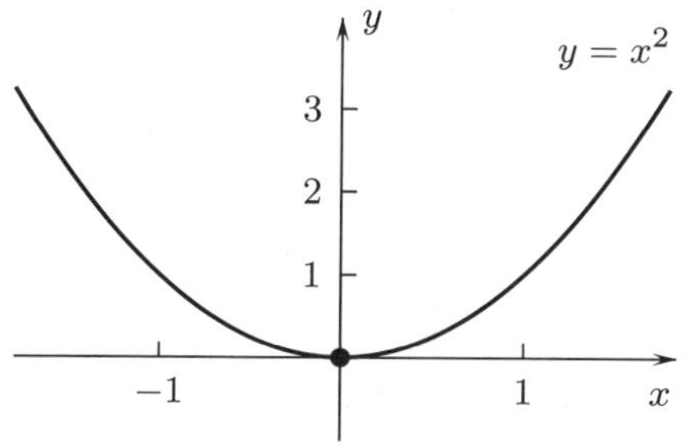

FIGURE 5

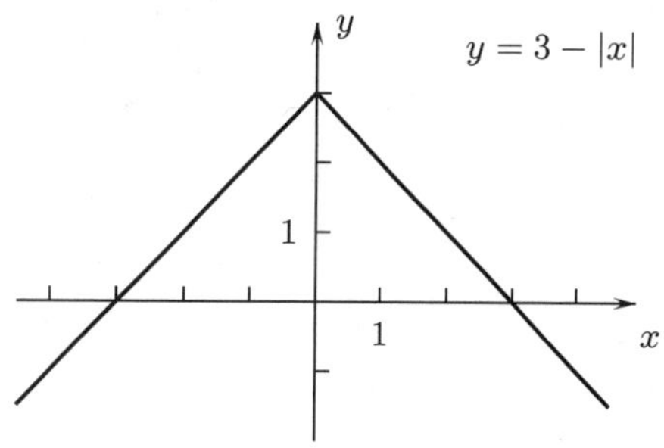

FIGURE 6

SOLUTION According to the theorem, x^2 has a critical point at $x = 0$, where it has a local minimum. This is the case because its derivative $\frac{d}{dx}(x^2) = 2x$ is zero at $x = 0$. □

Question 2 Explain how Theorem 1 applies to $y = 3 - |x|$, which has a local maximum at $x = 0$.

It is important to realize that the converse of Theorem 1 is not valid: A function can have a critical point where it has neither a local maximum nor a local minimum. The function $y = x^3$ of Figure 7, for example, has a critical point at $x = 0$ because its derivative $\frac{d}{dx}(x^3) = 3x^2$ is zero there, but x^3 does not have a local maximum or minimum at $x = 0$ because x^3 is less than its value $0^3 = 0$ at $x = 0$ for $x < 0$ and is greater than its value at $x = 0$ for $x > 0$.

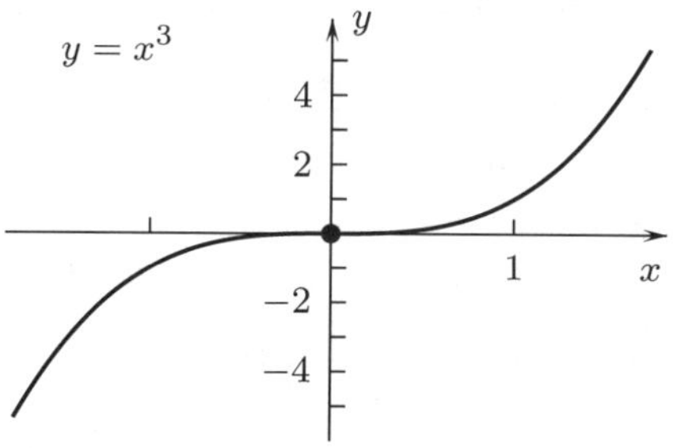

FIGURE 7

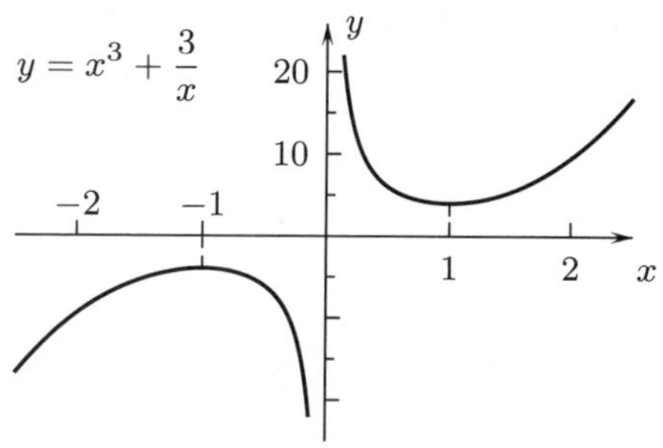

FIGURE 8

Example 2 The function $S(x) = x^3 + 3/x$, whose graph is shown in Figure 8, has a local maximum and a local minimum. Find their values.

SOLUTION The derivative $S'(x) = \frac{d}{dx}(x^3 + 3x^{-1}) = 3x^2 - 3x^{-2} = \frac{3(x^4 - 1)}{x^2}$ exists for $x \neq 0$ and is zero at $x = \pm 1$, where $x^4 = 1$. Consequently $x = \pm 1$ are the two critical points. Since we are told that $S(x)$ has a local maximum and a local minimum, they have to be at $x = 1$ or $x = -1$, by Theorem 1. The graph in Figure 8 shows that the local mimimum is at $x = 1$ and the local maximum is at $x = -1$. Notice that the local minimum $S(1) = 1^3 + 3/1 = 4$ is greater than the local maximum $S(-1) = (-1)^3 + 3/(-1) = -4$. □

Increasing and decreasing functions in intervals

Figure 9 shows the graph of the depth $h(t)$ of the water in the tank that we considered at the beginning of this section. This function is defined to equal $3t^2 - t^3$ for $0 \le t \le 3$, so its derivative $h'(t)$ is $6t - 3t^2$ for $0 < t < 3$. The graph of $h'(t)$ is in Figure 10.

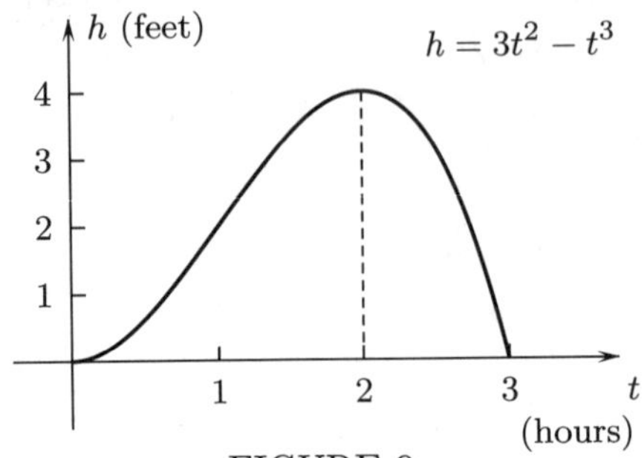

FIGURE 9

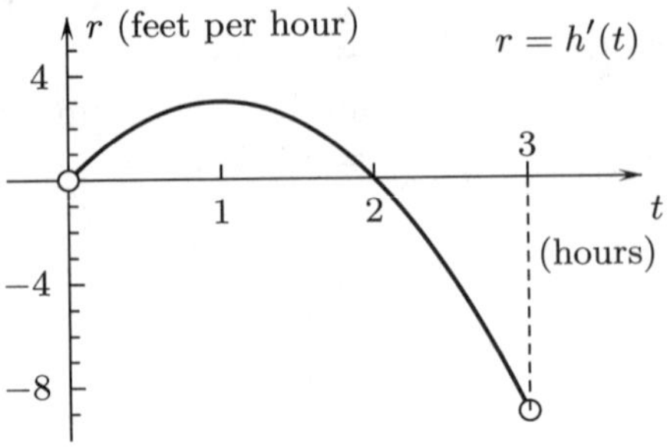

FIGURE 10

We saw above that $h(t)$ has a local maximum and $h'(t)$ is zero at $x = 2$, as is shown in Figures 9 and 10. Figure 10 also shows that $h'(t)$ is positive for $0 < t < 2$ and negative for $2 < t < 3$, so that $h(t)$ is increasing at each t with $0 < t < 2$ and decreasing at each t with $2 < t < 3$.

We can say more about where this function increases and decreases by considering its behavior in intervals. You can see from Figure 9 that its values increase as t increases across the closed interval $[0, 2]$ and its values decrease as t increases across the interval $[2, 3]$ (even though the function is neither increasing nor decreasing at $t = 0, t = 2$, or $t = 3$). We say that $h(t)$ INCREASES IN THE INTERVAL $[0, 2]$ and DECREASES IN THE INTERVAL $[2, 3]$, according to the following definition for a function with variable x.

Definition 2 **(a)** *$f(x)$ is* INCREASING IN AN INTERVAL *if $f(x)$ increases as x increases across the interval. This means that $f(x_1) < f(x_2)$ for all x_1 and x_2 in the interval with $x_1 < x_2$.*

(b) *$f(x)$ is* DECREASING IN AN INTERVAL *if $f(x)$ decreases as x increases across the interval. This means that $f(x_1) > f(x_2)$ for all x_1 and x_2 in the interval with $x_1 < x_2$.*

This definition is further illustrated in the following example and questions.

Example 3 Figure 11 shows the graph of $F(x) = \sqrt{x}$ and Figure 12 shows the graph of its derivative $F'(x) = \dfrac{d}{dx}(x^{1/2}) = \frac{1}{2}x^{-1/2}$. What are the most extensive intervals **(a)** in which $F(x)$ is continuous, **(b)** in which $F'(x)$ is positive, and **(c)** in which $F(x)$ is increasing?

SOLUTION **(a)** $F(x)$ is continuous in $[0, \infty)$, **(b)** $F'(x)$ is positive in $(0, \infty)$, and **(c)** $F(x)$ is increasing in $[0, \infty)$. □

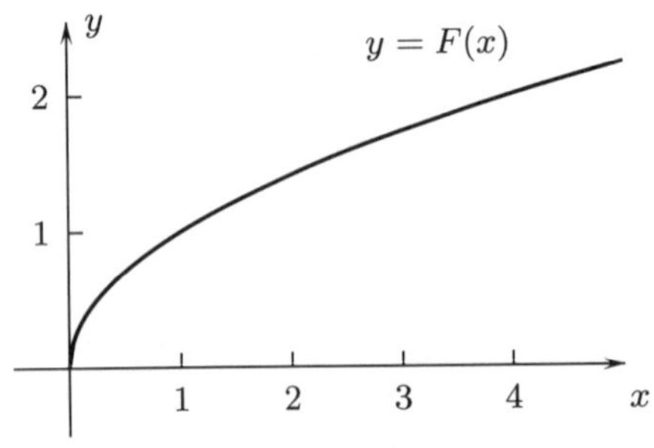

$F(x) = \sqrt{x}$

FIGURE 11

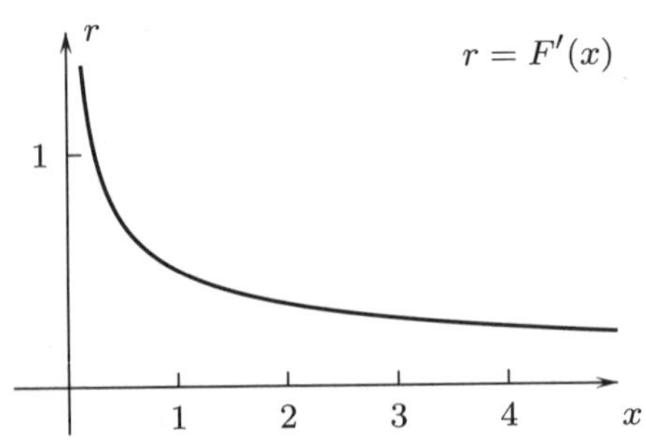

$F'(x) = \frac{1}{2}x^{-1/2}$

FIGURE 12

Question 3 Figures 13 and 14 show the graphs of $G(x) = 1/x$ and of its derivative $G'(x) = \frac{d}{dx}(x^{-1}) = -x^{-2}$. What are the most extensive intervals **(a)** in which $G(x)$ is continuous, **(b)** in which $G'(x)$ is positive and negative, and **(c)** in which $G(x)$ is increasing and decreasing?

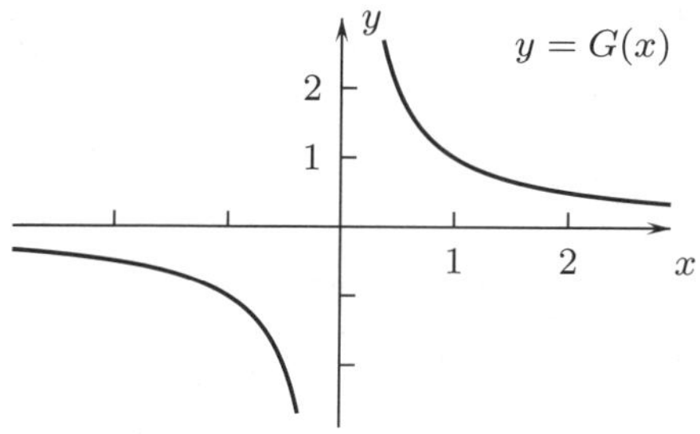

$G(x) = \frac{1}{x}$

FIGURE 13

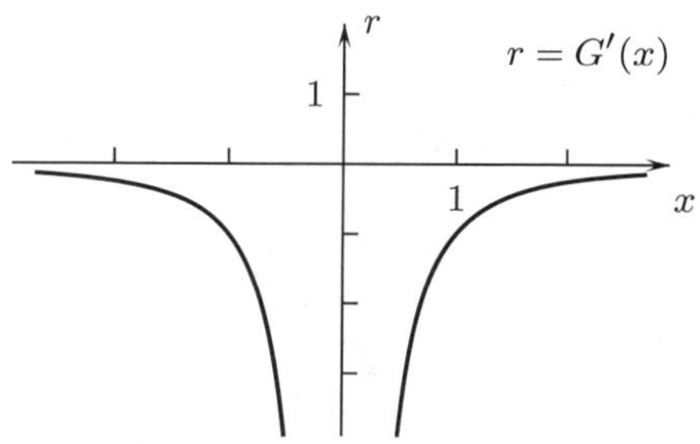

$G'(x) = -\frac{1}{x^2}$

FIGURE 14

Question 4 Figures 15 and 16 show the graphs of

$$H(x) = \begin{cases} 3 - x & \text{for} \quad 0 \le x < 2 \\ x & \text{for} \quad x \ge 2 \end{cases}$$

and of its derivative

$$H'(x) = \begin{cases} -1 & \text{for} \quad 0 < x < 2 \\ 1 & \text{for} \quad x > 2. \end{cases}$$

Find the most extensive intervals **(a)** in which $H(x)$ is continuous, **(b)** in which $H'(x)$ is positive or negative, and **(c)** in which $H(x)$ is increasing or decreasing.

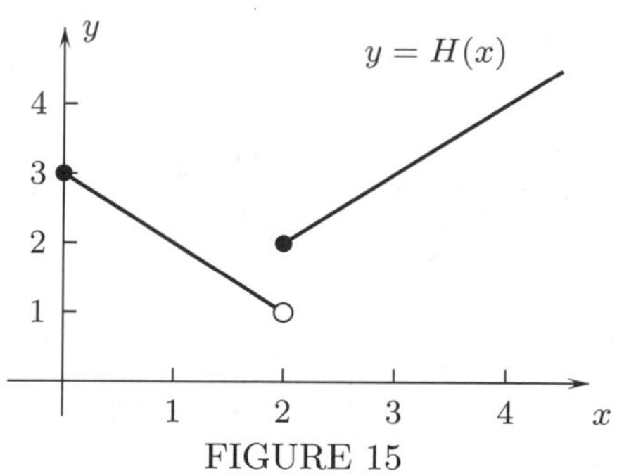

FIGURE 15

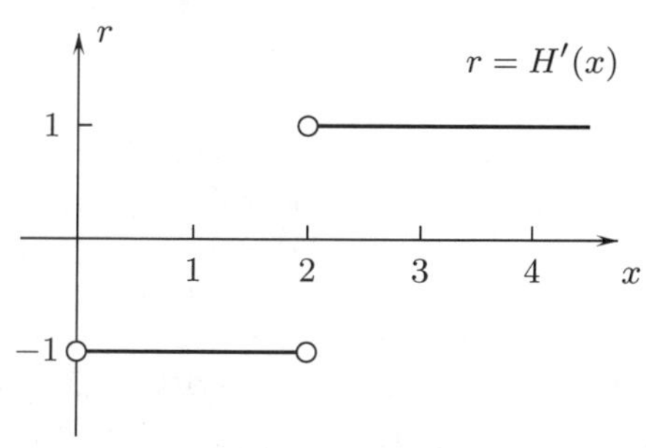

FIGURE 16

The most extensive intervals we found in which the functions $F(x), G(x)$, and $H(x)$ of Example 3 and Questions 3 and 4 are increasing or decreasing have a common property: In each case, the function is continuous in the interval and the derivative is either positive or negative in the interior of the interval. These examples illustrate the following theorem.

Theorem 2 (The First-Derivative Test for Increasing and Decreasing Functions)

(a) *If $f(x)$ is continuous in an interval and $f'(x)$ exists and is positive at all points in the interior of the interval, then $f(x)$ is increasing in the interval.*

(b) *If $f(x)$ is continuous in an interval and $f'(x)$ exists and is negative at all points in the interior of the interval, then $f(x)$ is decreasing in the interval.*

To show how this theorem can be proved, we consider a function $f(x)$ as in Figure 17, which is continuous in the finite closed interval $[a, b]$ and has a positive derivative in the interior (a, b) of that interval.

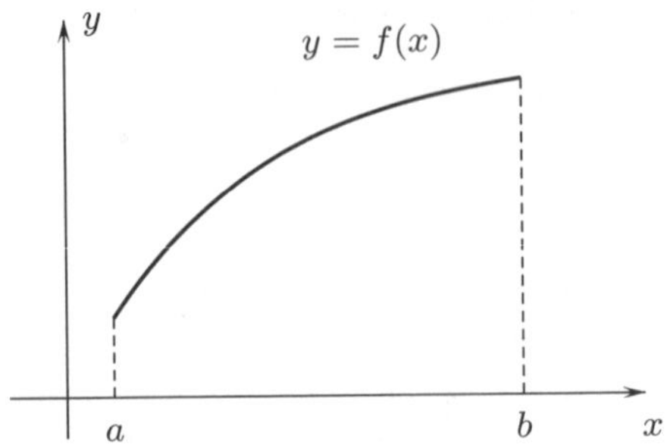

FIGURE 17

We can demonstrate that $f(x)$ is increasing in $[a, b]$ by showing that $f(x_1) < f(x_2)$ for any x_1 and x_2 with $a \le x_1 < x_2 \le b$. To accomplish this, we consider any such points x_1 and x_2, as in Figure 18, and the secant line through the points on the graph at $x = x_1$ and $x = x_2$. We denote the slope of this secant line by m:

$$m = \frac{f(x_2) - f(x_1)}{x_2 - x_1}. \tag{2}$$

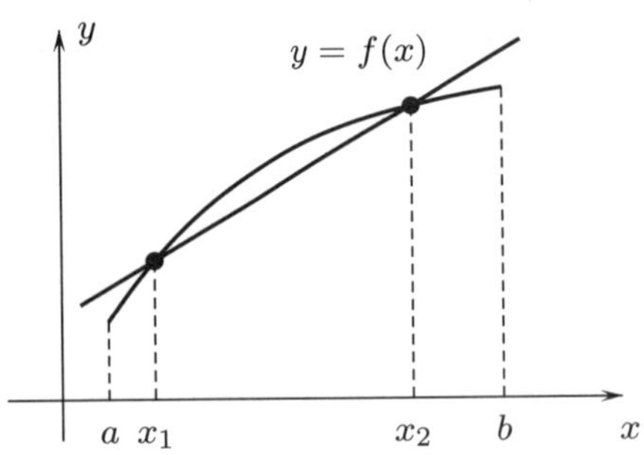

FIGURE 18

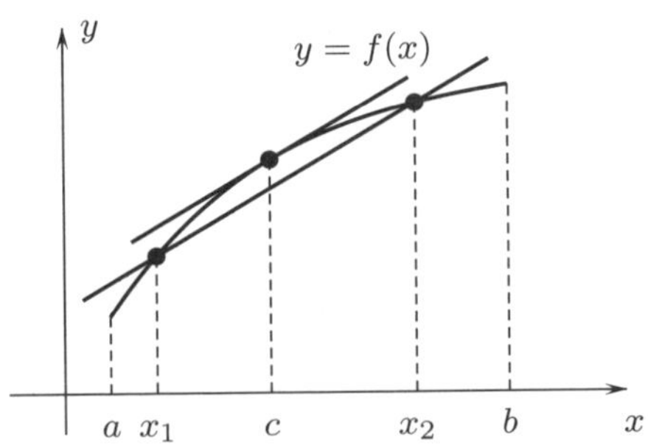

FIGURE 19

As you can see from Figure 18, we can show that $f(x_1)$ is less than $f(x_2)$ by showing that the slope m of the secant line is positive. This requires a new line of reasoning; we used slopes of secant lines to define derivatives, but have not yet used derivatives to study slopes of secant lines. We apply the following result, whose proof is given at the end of this section.

Theorem 3 (The Mean-Value Theorem) *If $f(x)$ is continuous in the finite closed interval $[x_1, x_2]$ and if $f'(x)$ exists for all x with $x_1 < x < x_2$, then there is at least one number c with $x_1 < c < x_2$ such that*

$$f'(c) = \frac{f(x_2) - f(x_1)}{x_2 - x_1}.$$

In the case of the function $f(x)$ of Figure 18, the Mean-Value Theorem implies that there is at least one number c with $x_1 < c < x_2$ such that the tangent line to the graph of $f(x)$ at $x = c$ (with slope $f'(c)$) is parallel to the secant line (with slope m given by **(2)**). Such a tangent line is shown in Figure 19 for the function of Figure 18. Because we are assuming that $f'(c)$ is positive for $a < c < b$, the tangent line has positive slope. Consequently, the slope m of the secant line is also positive, so that $f(x_1)$ is less than $f(x_2)$. Since x_1 and x_2 are arbitrary points in $[a, b]$ with $x_1 < x_2$, this shows that $f(x)$ is increasing in $[a, b]$. This reasoning can be modified to establish Theorem 2 in general.

We now show how Theorem 2 can be applied.

Example 4 Find the most extensive intervals in which $y = x^2$ is increasing and in which it is decreasing.

SOLUTION The function x^2 is continuous for all x, and its derivative, $\frac{d}{dx}(x^2) = 2x$ is negative for $x < 0$ and positive for $x > 0$. By Theorem 2, x^2 is decreasing in the interval $(-\infty, 0]$ and increasing in $[0, \infty)$ (see Figure 5 above).

Question 5 Find the most extensive intervals in which $y = 3 - |x|$ is increasing and in which it is decreasing.

Example 5 Figure 20 shows the graph of the function $R(x) = 30 - 8x + \frac{1}{4}x^4$. Find the most extensive intervals in which it is increasing and in which it is decreasing.

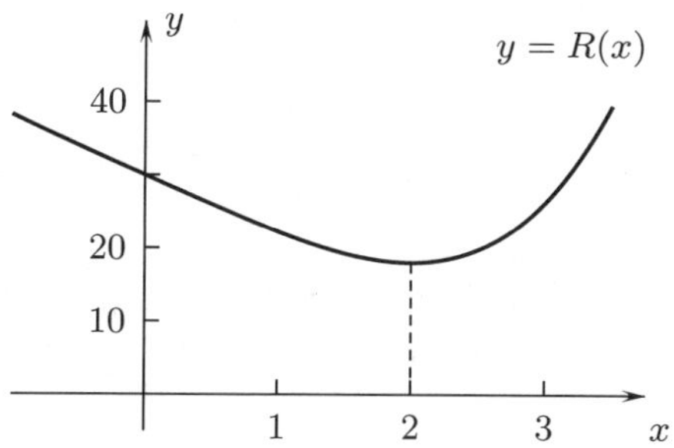

FIGURE 20

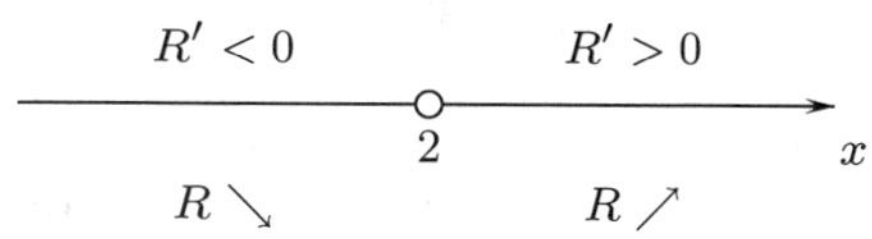

FIGURE 21

SOLUTION Because the derivative $R'(x) = \frac{d}{dx}(30 - 8x + \frac{1}{4}x^4) = -8 + x^3$ exists and is continuous for all x, it must be zero at any point where it changes sign from positive to negative or from negative to positive.[†] Since $R'(x)$ is zero only at $x = 2$, its sign is constant for $x < 2$ and for $x > 2$. We can determine its signs by calculating its value at one point in each of these intervals.

For instance, since $R'(1) = -8 + 1^3 = -7$ is negative, $R'(x)$ is negative for $x < 2$, and because $R'(3) = -8 + 3^3 = 19$ is positive, $R'(x)$ is positive for $x > 2$. We show this information in Figure 21. Then because $R(x)$ is continuous for all x, it is increasing in $[2, \infty)$ and decreasing in $(-\infty, 2]$ and these are the most extensive intervals in which it is increasing and decreasing. □

In the next example we look at a case where applying Theorem 2 to adjacent intervals that include a common endpoint shows that the function is increasing or decreasing in a larger interval.

Example 6 Use Theorem 2 to show that x^3 is increasing in $(-\infty, \infty)$.

SOLUTION The derivative $\frac{d}{dx}(x^3) = 3x^2$ is positive for $x < 0$ and for $x > 0$ and x^3 is continuous for all x. By Theorem 2, x^3 is increasing in $(-\infty, 0]$ and in $[0, \infty)$. Then because these intervals include the common endpoint $x = 0$, the function is actually increasing in $(-\infty, \infty)$ (see Figure 7 above). □

Global maxima and minima

Because the local maximum $h(2) = 4$ (feet) of the depth of water in the tank that you found in Question 1 is the greatest value of $h(t)$ in its domain, we say that 4 is the GLOBAL MAXIMUM or the ABSOLUTE MAXIMUM of $h(t)$.

Similarly, because $R(2) = 18$ for the function of Figure 20 is the least value of $R(x)$ for all x in its domain, 18 is the GLOBAL MINIMUM or the ABSOLUTE MINIMUM of $R(x)$.

[†]This is a consequence of the Intermediate-Value Theorem from Section 2.2 since if $R'(x)$ is positive at one point and negative at another, it must be zero at some point in between.

Example 7 Figure 22 shows the graph of $S(x) = \dfrac{x}{x^2+1}$. **(a)** Use the formula for the function to explain the general shape of the graph for small and large $|x|$. **(b)** Find the global maximum and global minimum values of $S(x)$.

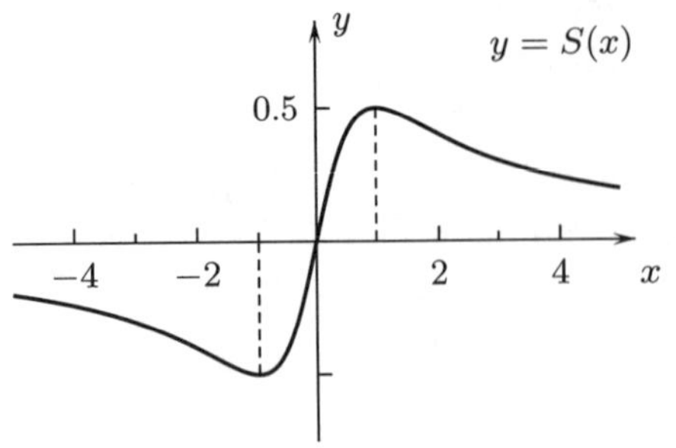

FIGURE 22

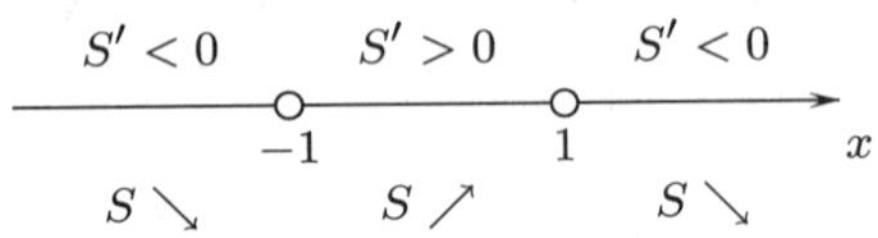

FIGURE 23

SOLUTION **(a)** The graph has the shape in Figure 22 for small and large $|x|$ because $S(x)$ is an odd function that is closely approximated by $x/(0+1) = x$ for x near zero and by $x/x^2 = 1/x$, for large $|x|$.

(b) The derivative

$$S'(x) = \frac{d}{dx}\left[\frac{x}{x^2+1}\right] = \frac{(x^2+1)\frac{d}{dx}(x) - x\frac{d}{dx}(x^2+1)}{(x^2+1)^2}$$
$$= \frac{x^2+1-2x^2}{(x^2+1)^2} = \frac{1-x^2}{(1+x^2)^2}$$

is continuous for all x, so it can change sign only at $x = \pm 1$, where it is zero. Also, $S'(0) = 1$ is positive and $S'(-2) = -\frac{3}{25}$ and $S'(2) = -\frac{3}{25}$ are negative. Therefore, $S'(x)$ is positive for $-1 < x < 1$, and negative for $x < -1$ and for $x > 1$ (Figure 23). Since $S(x)$ is continuous for all x, it is increasing in $[-1, 1]$ and decreasing in $(-\infty, -1]$ and $[1, \infty)$. Then, because $S(x)$ is positive for $x > 0$ and negative for $x < 0$, its global maximum is $S(1) = \frac{1}{2}$ and its global minimum is $S(-1) = -\frac{1}{2}$ as can be seen in Figure 22. □

Question 6 **(a)** Use the formula $f(x) = x^2 + 10/x^2$ for the function of Figure 24 to explain the shape of the graph for small and large $|x|$. **(b)** Find the global minimum of $f(x)$. **(c)** Why does $f(x)$ not have a global maximum?

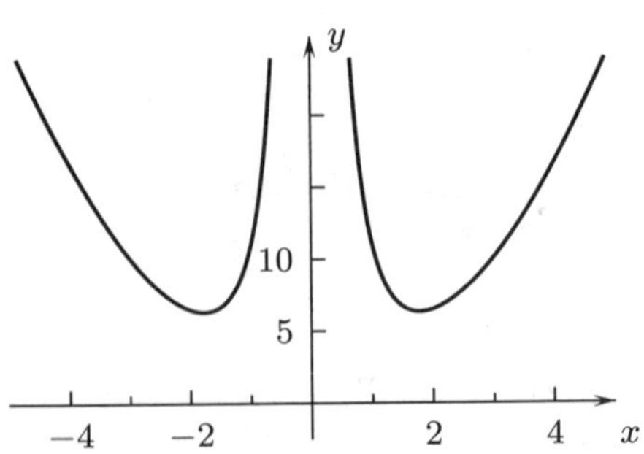

$f(x) = x^2 + \dfrac{10}{x^2}$

FIGURE 24

Concavity and the second derivative

The graph in Figure 25 is said to be CONCAVE UP and the graph in Figure 26 is said to be CONCAVE DOWN, according to the following definition.

Definition 3 **(a)** *If the slope of the tangent line at x to the graph of $f(x)$ increases for x in an open interval, then that portion of the graph of the function is concave up.*

(b) *If the slope of the tangent line at x to the graph of $f(x)$ decreases for x in an open interval, then that portion of the graph of the function is concave down.*

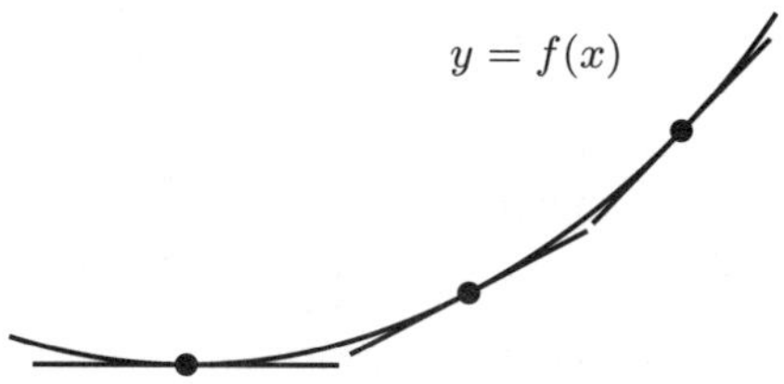

Slope increasing
Graph concave up

FIGURE 25

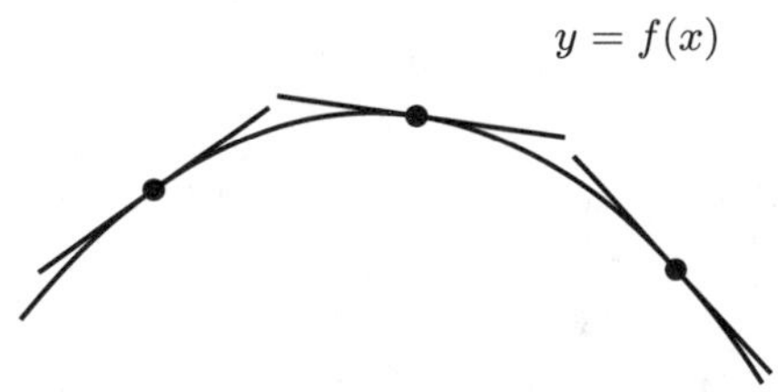

Slope decreasing
Graph concave down

FIGURE 26

We can determine where the first derivative $f'(x)$ of $f(x)$ increases and decreases by studying its derivative, which is the second derivative $f''(x)$ of $f(x)$. We use Theorem 2, applied to $f'(x)$ in place of $f(x)$, and obtain the following:

Theorem 4 (The Second-Derivative Test for concavity) **(a)** *If $f''(x)$ exists and is positive in an open interval, then the graph of $f(x)$ is concave up for x in the interval.*

(b) *If $f''(x)$ exists and is negative in an open interval, then the graph of $f(x)$ is concave down for x in the interval.*

Example 8 Find the most extensive open intervals in which the graph in Figure 1 of the depth $h(t)$ of the water in the tank is concave up and concave down.

SOLUTION We saw above that the first derivative of $h(t) = 3t^2 - t^3$, defined for $0 \le t \le 3$, is $h'(t) = 6t - 3t^2$, defined for $0 < t < 3$. Consequently, the second derivative is $h''(t) = 6 - 6t$ for $0 < t < 3$. It is zero at $t = 1$, positive for $0 < t < 1$, and negative for $1 < t < 3$. By Theorem 4, the graph is concave up in $(0, 1)$ and concave down in $(1, 3)$ (Figure 27). □

Inflection points

The graph of $h(t)$ of Example 8 is shown in Figure 27 with the point plotted at $t = 1$ where the curve switches from concave up to concave down. This point is an INFLECTION POINT of the graph, according to the following definition:

Definition 4 *An* INFLECTION POINT *on the graph of a function is a point where the graph has a tangent line and switches from concave up to concave down or where it switches from concave down to concave up.*

The graph of the derivative $h'(t)$ is shown again in Figure 28. Because $h'(t)$ is increasing for $0 < t < 1$ and decreasing for $1 < t < 3$, it is a maximum at $t = 1$. Consequently, the inflection point on the graph of $h(t)$ in Figure 27 is the point where its tangent line has the greatest slope and is at the time $t = 1$ when water is flowing into the tank the most rapidly.

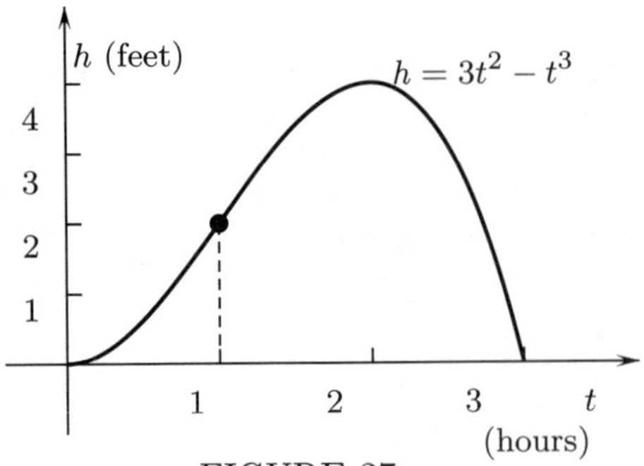

FIGURE 27

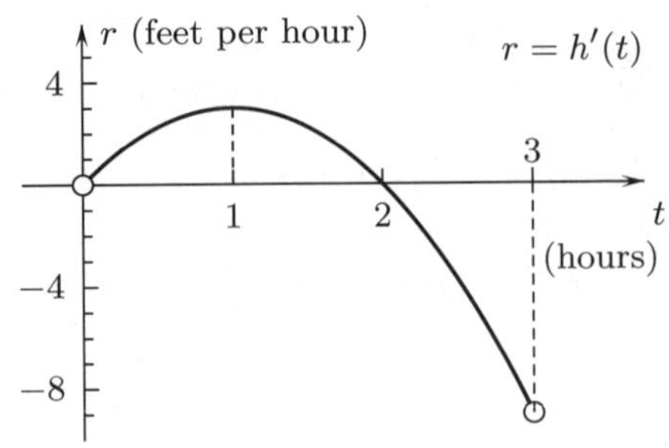

FIGURE 28

Question 7 Find the inflection point of $y = 3 + x^2 - 8/x$. Generate its graph in the window $-5 \leq x \leq x, -15 \leq y \leq 15$ as a partial check of your answer.

The next example gives a summary of the techniques we have developed for analyzing graphs.

Example 9 Use the formula for $f(x) = x^3 - 3x^2 + 4$ to predict the approximate shape of its graph for small and large $|x|$. Then find the most extensive intervals in which $f(x)$ is increasing and decreasing and the most extensive open intervals in which its graph is concave up and concave down. Then draw the graph, showing any local maxima or minima and inflection points.

SOLUTION USING THE FORMULA FOR THE FUNCTION: The lower-power terms, $-3x^2 + 4$, of $f(x)$ dominate for very small $|x|$, so its graph looks approximately like $y = -3x^2 + 4$ near $x = 0$. For large positive and large negative x, x^3 is the dominant term and the graph looks approximately like $y = x^3$ (Figure 29). In particular, $f(x) \to \infty$ as $x \to \infty$ and $f(x) \to -\infty$ as $x \to -\infty$.

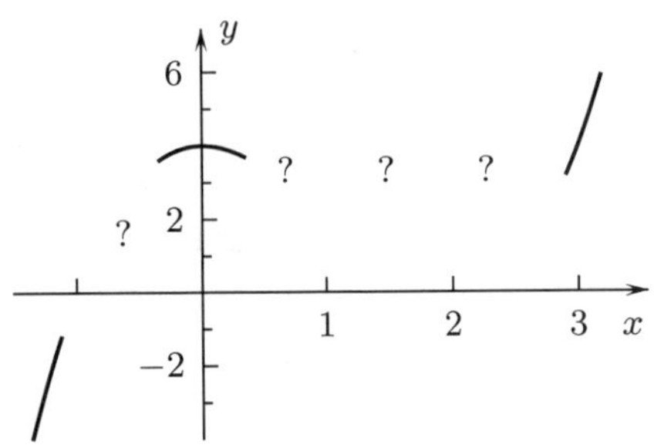

FIGURE 29

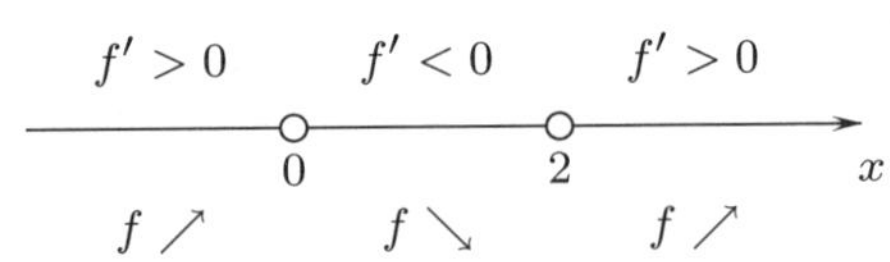

FIGURE 30

USING THE FIRST DERIVATIVE: Because the first derivative,

$$f'(x) = \frac{d}{dx}(x^3 - 3x^2 + 4) = 3x^2 - 6x = 3x(x - 2) \tag{3}$$

is continuous for all x, it can change sign only at $x = 0$ and $x = 2$, where it is zero. These points divide the x-axis into the three open intervals $x < 0, 0 < x < 2$, and $x > 2$ shown in Figure 30.

Examining the formula on the right of equation **(3)** or calculating its values $f'(-1) = 9, f'(1) = -3$, and $f'(3) = 9$ at points in the three intervals shows that the derivative is positive for $x < 0$ and for $x > 2$, and is negative for $0 < x < 2$. Then, because $f(x)$ is continuous for all x, it is increasing in $(-\infty, 0]$ and in $[2, \infty)$ and is decreasing in $[0, 2]$. Hence $f(x)$ has a local maximum of $f(0) = 0^3 - 3(0^2) + 4 = 4$ at $x = 0$ and a local minimum of $f(2) = 2^3 - 3(2^2) + 4 = 0$ at $x = 2$.

USING THE SECOND DERIVATIVE: The second derivative

$$f''(x) = \frac{d}{dx}(3x^2 - 6x) = 6x - 6 = 6(x-1) \qquad \textbf{(4)}$$

can only change sign at $x = 1$, where it is zero. You can see from its formula **(4)** that it is negative for $x < 1$ and positive for $x > 1$. By the Second-Derivative Test (Theorem 4), the graph of $f(x)$ is concave down in $(-\infty, 1)$ and concave up in $(1, \infty)$ and has an inflection point at $\big(1, f(1)\big) = (1, 2)$.

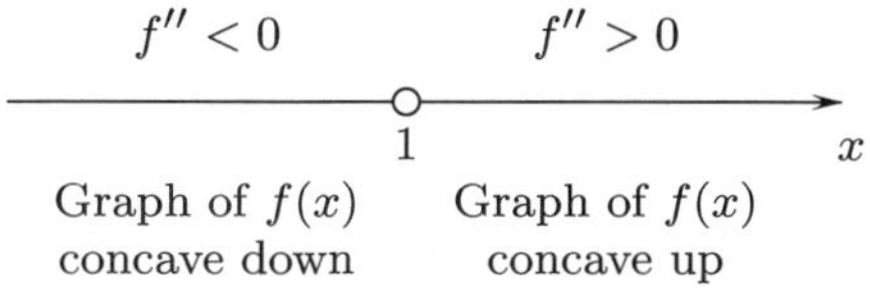

FIGURE 31

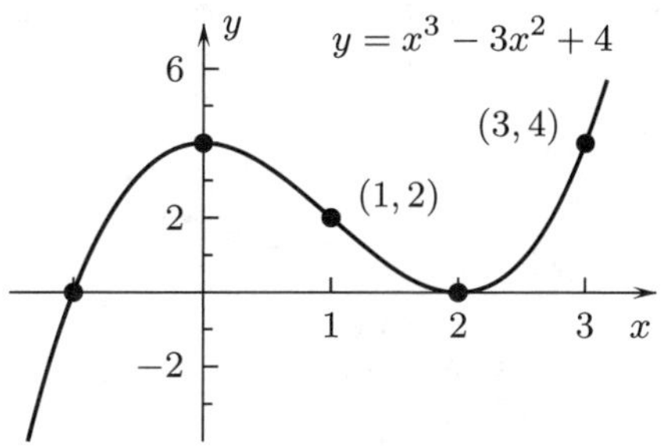

FIGURE 32

COMPLETING THE SKETCH To complete the graph in Figure 32, we use the information in Figures 30 and 31 and plot the points $(0, 4)$ and $(2, 0)$ corresponding to the local maximum and minimum and the inflection point $(1, 2)$. We also plot the points at $x = -2$, where $f(x) = (-2)^3 - 3(-2)^2 + 4 = 0$, and at $x = 3$, where $f(x) = 3^3 - 3(3^2) + 4 = 4$ to position the left and right ends of the curve in our sketch. □

Using the graph of a derivative to study a function

To consolidate your understanding of the ideas we have been discussing, let us now use the graph of the derivative of a function with First- and Second-Derivative Tests to analyze the graph of the function.

Question 8 **(a)** Figure 33 shows the graph of the derivative $g'(x)$ of a function $g(x)$ that is continuous and has continuous first and second derivatives for all x. The first derivative $g'(x)$ is negative for $x < 2$ and for $x > 4$ and is positive for $2 < x < 4$. What does this tell you about $g(x)$? **(b)** The first derivative $g'(x)$ is increasing for $x < 3$ and decreasing for $x > 3$. What does this tell you about the graph of $g(x)$?

FIGURE 33

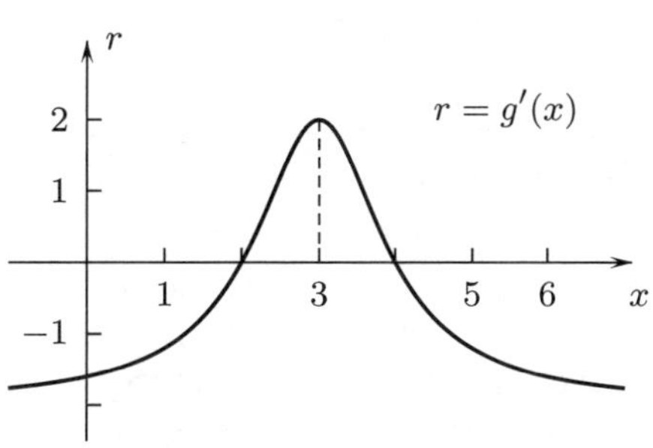

Our last question of this section deals with a function whose formula involves a parameter—a letter representing a constant whose value may change from case to case.

Question 9 **(a)** What is the one critical point of $F(x) = x^4 - 4ax$? **(b)** What is the global minimum of $F(x)$? (Give formulas in terms of the parameter a.)

Proof of the Mean-Value Theorem

To prove Theorem 3 for the function $f(x)$ of Figure 18, which is repeated here as Figure 34, we define a new function $g(x) = f(x) - mx$ with m given by **(2)** (Figure 35). The slope of the secant line through the points at $x = x_1$ and $x = x_2$ on the graph of $g(x)$ is

$$\frac{g(x_2) - g(x_1)}{x_2 - x_1} = \frac{[f(x_2) - mx_2] - [f(x_1) - mx_1]}{x_2 - x_1} = \frac{[f(x_2) - f(x_1)] - [mx_2 - mx_1]}{x_2 - x_1}$$
$$= \frac{f(x_2) - f(x_1)}{x_2 - x_1} - m = 0.$$

This shows that the secant line in Figure 35 is horizontal.

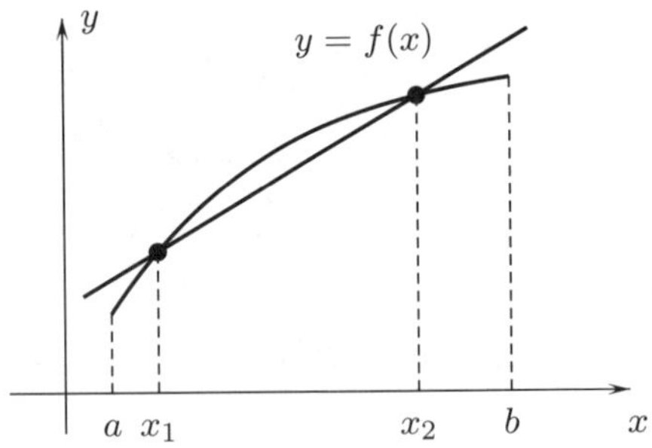

FIGURE 34

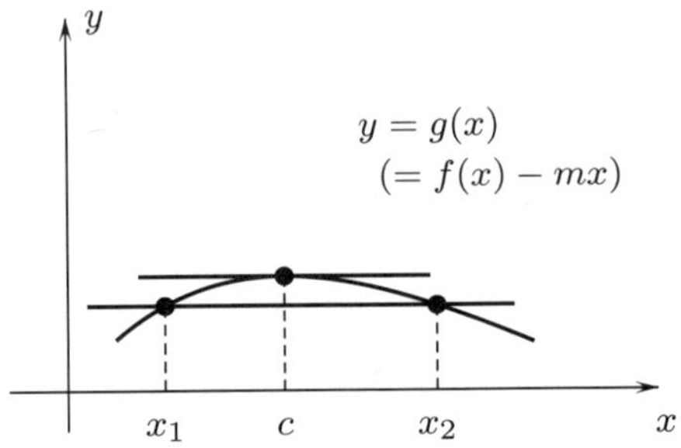

FIGURE 35

Because $g(x)$ is continuous in $[x_1, x_2]$, it has a greatest value in $[x_1, x_2]$. (This is a consequence of the Extreme-Value Theorem, which will be discussed in Section 3.2.) Moreover, because the graph of $g(x)$ is above the secant line for x between x_1 and x_2, the greatest value occurs at a point c with $x_1 < c < x_2$, as shown in Figure 35, so that it is a local maximum of $g(x)$. Then $g'(c) = 0$ by Theorem 1 of this section. Thus, the tangent line in Figure 35 is horizontal and is parallel to the horizontal secant line.

Since $g'(c) = f'(c) - m$ is zero, the slope $f'(c)$ of the tangent line in Figure 19 equals the slope m of the secant line, as is stated in Theorem 3. Similar reasoning would establish Theorem 3 in general.

Principles and procedures

- $f(x)$ has a local maximum at x_0 if $f(x) \leq f(x_0)$ for all x in an open interval containing x_0. $f(x)$ has a local minimum at x_0 if $f(x) \geq f(x_0)$ for all x in an open interval containing x_0.
- x_0 is a critical point of $f(x)$ if $f(x)$ is defined in an open interval containing x_0 and either $f'(x_0)$ is zero or does not exist.
- Any local maxima and minima of a function must occur at critical points, but a function does not have to have a local maximum or minimum at each of its critical point.
- In most cases of a function given by a single formula, you find the most extensive intervals in which it is increasing and decreasing by first finding the points where its derivative is not defined or is zero. These points usually divide the domain of the function into open intervals where the derivative is continuous and nonzero and, therefore, does not change sign. Decide whether the derivative is positive or negative in each of these intervals either by examining its formula or by calculating its value at one point in the interval. Then determine whether the function is continuous in the intervals obtained by adding one or two endpoints to the open intervals where its derivative is positive or negative. The function is increasing in an interval in which it is continuous if its derivative is positive in the interior of the interval, and is decreasing in an interval in which it is continuous if its derivative is negative in the interior of the interval. This information is often used to locate local and global maxima and minima of functions.

- If a function is increasing (or decreasing) in two adjacent intervals, both of which include their common endpoint, then the function is increasing (or decreasing) in the interval obtained by combining the original two intervals. For example, if $f(x)$ is increasing in $(-\infty, 2]$ and in $[2, 10)$, then it is increasing in $(-\infty, 10)$.
- The graph of $f(x)$ is concave up for x in an open interval if $f'(x)$ is increasing in the interval and is concave down if $f'(x)$ is decreasing in the interval. If $f''(x)$ is positive in an open interval, then the graph of $f(x)$ is concave up for x in the interval; if $f''(x)$ is negative in an open interval, then the graph of $f(x)$ is concave down for x in the interval.
- A point on the graph of a function is an inflection point if the graph has a tangent line at that point and switches from concave up to concave down or from concave down to concave up there.

Tune-Up Exercises 3.1♦

[A]*Answer provided.* [O]*Outline of solution provided.*

Be sure you can solve Exercises T1 through T5, and work as many of Exercises T6 through T10 as you need for practice. Justify your answers. Generate relevant graphs as a partial check of your results.

T1.[O] Find the most extensive intervals in which $f(x) = x^4 - 4x^2$ is positive and the most extensive intervals in which it is negative.

♦ Type 3, analyzing a function

T2.[O] What are the most extensive intervals in which $g(x) = 1 + \dfrac{2}{x-2}$ is positive and in which it is negative?

♦ Type 3, analyzing a function

T3.[A] Figure 36 shows the graph of the derivative of a function $G(x)$ such that $G(x)$ and its first and second derivatives are continuous in $[-2, 8]$. **(a)** What are the most extensive subintervals of $[-2, 8]$ in which $G(x)$ is increasing and in which it is decreasing? **(b)** Does $G(x)$ have any local maxima or minima for $-2 < x < 8$? If so, at what values of x? **(c)** What are the most extensive subintervals of $[-2, 8]$ in which the graph of $G(x)$ is concave up and in which it is concave down? **(d)** At what value of x with $-2 < x < 8$ does the graph of $G(x)$ have an inflection point?

♦ Type 3, using the graph of a first derivative

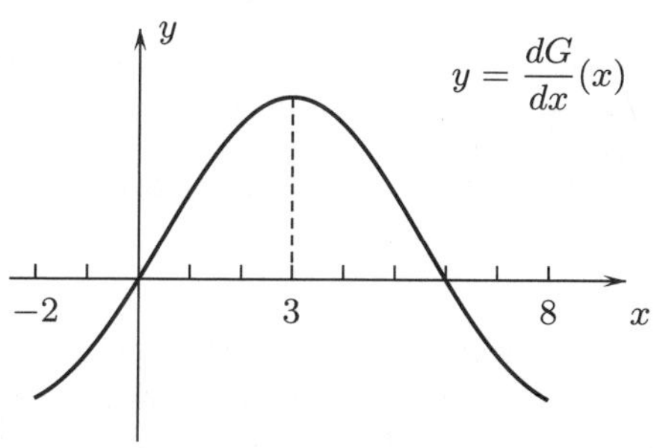

FIGURE 36

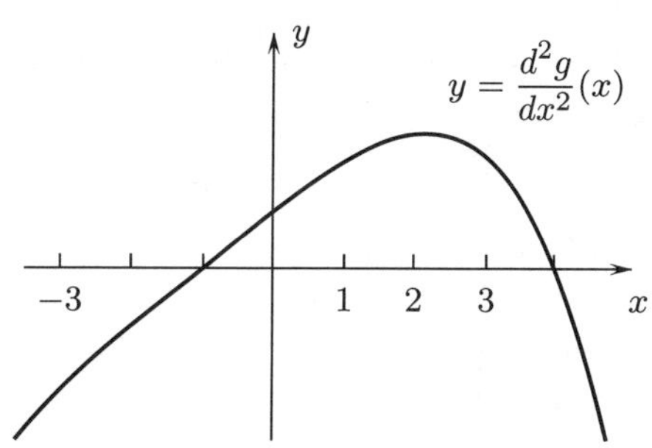

FIGURE 37

T4.[A] Figure 37 shows the graph of the second derivative of a function $g(x)$ such that $g(x)$ and its first and second derivatives are continuous for all x. The second derivative is zero only at $x = -1$ and $x = 4$. **(a)** What are the most extensive open intervals in which the graph of $g(x)$ is concave up and in which it is concave down? **(b)** Does the graph of $g(x)$ have any inflection points? If so, at what values of x?

♦ Type 3, using the graph of a second derivative

T5.[O] Use the First- and Second-Derivative Tests to determine the most extensive intervals in which **(a)** x^{-1} and **(b)** x^{-2} are increasing and decreasing, and the most extensive open intervals where their graphs are concave up and concave down. Check your conclusions by generating the graphs in the window $-2.5 \le x \le 2.5, -4 \le y \le 4$.

♦ Type 1, the First- and Second-Derivative Tests

♦The Tune-up Exercises and Problems are classified by type and content. The types are (1) basic, reactive; (2) basic reflective; (3) intermediate, reactive; (4) intermediate, reflective; (5) advanced, reactive; (6) advanced, reflective; and (7) advanced, theoretical.

More practice

T6.[O] What are the most extensive intervals in which $h(x) = x - 2 + 1/x$ is positive and in which it is negative?

◆ Type 1, the First- and Second-Derivative Tests

T7.[O] Find the most extensive intervals in which $k(x) = x^2 - \dfrac{16}{x^2}$ is positive and in which it is negative.

◆ Type 1, the First- and Second-Derivative Tests

T8.[A] Figure 38 shows the graph of the derivative of a function $H(x)$ that is defined and has continuous first and second derivatives in $[-7, 7]$. **(a)** What are the most extensive open subintervals of $[-7, 7]$ in which $H(x)$ is increasing and in which it is decreasing? **(b)** What are the most extensive subintervals of $[-7, 7]$ in which the graph of $H(x)$ is concave up and in which it is concave down? **(c)** At what values of x with $-7 < x < 7$ does the graph of $H(x)$ have inflection points?

◆ Type 3, using the graph of a first derivative

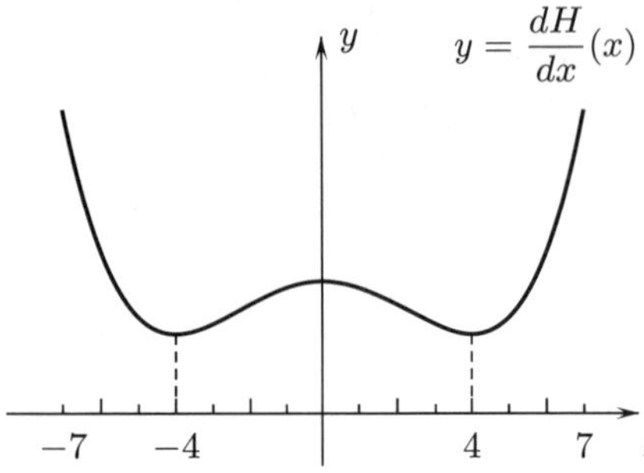

FIGURE 38

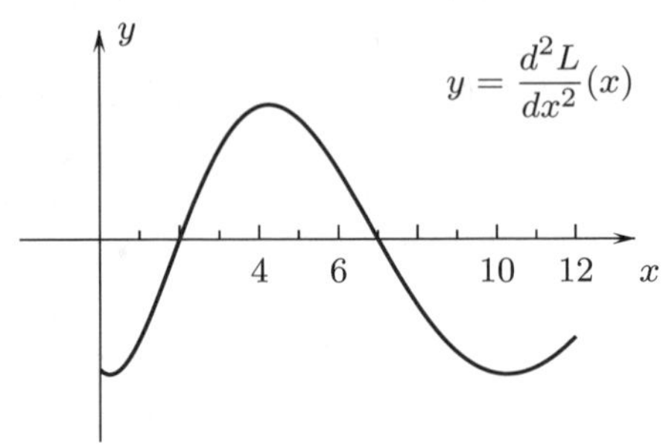

FIGURE 39

T9.[A] Figure 39 shows the graph of the second derivative of a function $L(x)$ that is continuous and has continuous first and second derivatives in $[0, 12]$. **(a)** What are the most extensive open subintervals of $[0, 12]$ in which the graph of $L(x)$ is concave up and in which it is concave down? **(b)** At what values of x with $0 < x < 12$ does the graph of $L(x)$ have inflection points?

◆ Type 3, using the graph of a second derivative

T10.[O] For each of the functions **(a)** $y = x^{1/2}$, **(b)** $y = x^{1/3}$, and **(c)** $y = x^{-1/2}$, determine the most extensive intervals in which the function is defined and continuous, the most extensive intervals in which the function is increasing and decreasing, and the most extensive intervals in which its graph is concave up and concave down. Check your answers by generating the graphs.

◆ Type 1, the First- and Second-Derivative Tests, etc

Problems 3.1

[A]*Answer provided.* [O]*Outline of solution provided.*

In Problems 1 through 18 sketch the graphs of the functions after determining the most extensive intervals in which the functions are increasing and decreasing and the most extensive open intervals in which the graphs are concave up and concave down. Identify any local and global maxima and minima and inflection points. Justify your answers.

1.[A] $y = x^3 - 12x + 7$

◆ Type 1, the First- and Second-Derivative Tests

2.[A] $y = x + x^{-2}$

◆ Type 1, the First- and Second-Derivative Tests

3.[A] $y = 2x^3 - x^4$

◆ Type 1, the First- and Second-Derivative Tests

4.[A] $y = 2x^2 - x^4$

◆ Type 1, the First- and Second-Derivative Tests

5.[A] $y = x^4 - 12x$

◆ Type 1, the First- and Second-Derivative Tests

6. $y = x^3 - 3x^2 + 4$

◆ Type 1, the First- and Second-Derivative Tests

7. $y = x^3 + x - 5$

◆ Type 1, the First- and Second-Derivative Tests

8. $y = x^3 + x^{-3}$

◆ Type 1, the First- and Second-Derivative Tests

9.[A] $y = 4x^{-1} + 4x^{-2}$

◆ Type 1, the First- and Second-Derivative Tests

10. $y = x - 1/x$

◆ Type 1, the First- and Second-Derivative Tests

11. $y = 1 + \dfrac{2}{x} + \dfrac{1}{x^2}$

◆ Type 1, the First- and Second-Derivative Tests

12.[A] $y = 6\sqrt{x} - x$

◆ Type 1, the First- and Second-Derivative Tests

13. $y = x + 16x^{-1/2}$

◆ Type 1, the First- and Second-Derivative Tests

14. $y = x^5 - 80x$

◆ Type 1, the First- and Second-Derivative Tests

15. $y = \frac{1}{4}x^4 + x^3$

◆ Type 1, the First- and Second-Derivative Tests

16 $y = 3x^{-2} - x^2$

◆ Type 1, the First- and Second-Derivative Tests

17. $y = x^3 - 6x^2 + 12x$

◆ Type 1, the First- and Second-Derivative Tests

18. $y = x^2 + 1/x^2$

◆ Type 1, the First- and Second-Derivative Tests

19.[A] The function $f(x)$ is continuous for $x \neq 0$, and $f'(x) = x - 1/x$ for $x \neq 0$. What are the most extensive open intervals in which $f(x)$ is increasing and decreasing and the most extensive open intervals in which its graph is concave up and concave down? Justify your answers.

◆ Type 2, using a formula for the first derivative

20.[A] The function $g(x)$ and its derivatives are continuous for all x, and $g''(x) = x(x-1)^2(x-2)^3$. What are the most extensive open intervals in which the graph of $g(x)$ is concave up and in which it is concave down? At what values of x does it have inflection points? Justify your answers.

◆ Type 2, using a formula for a second derivative

21. $K(x)$ is continuous for all x, and $K'(x) = 4x - x^2$. What are the most extensive intervals in which $K(x)$ is increasing and decreasing? Where is its graph concave up and concave down? Where does $K(x)$ have local maxima or minima and where does its graph have inflection points? Justify your answers.

◆ Type 2, using a formula for a first derivative

22. $L(x)$ and its derivatives are continuous for $x \neq 2$, and $\dfrac{d^2L}{dx^2}(x) = \dfrac{x}{(x-2)^2}$. Does the graph of $L(x)$ have any inflection points? If so, where? Justify your answers.

◆ Type 2, using a formula for a second derivative

23. Figure 40 shows the graph of the temperature in an Algonkian Eskimo home during a 24-hour period with $t = 0$ at midnight.(1) **(a)** What is the approximate minimum temperature and approximately when does it occur? **(b)** At approximately what time with $0 \le t \le 24$ is the rate of change of the temperature equal to zero? **(c)** Approximately when for $0 \le t \le 24$ is the rate of change of the temperature a maximum and when is it a minimum? **(d)** What are the approximate inflection points on the graph of $T(t)$ for $0 \le t \le 24$? **(e)** How are the answers to parts (a) through (d) related?

◆ Type 2, using the graph of a function

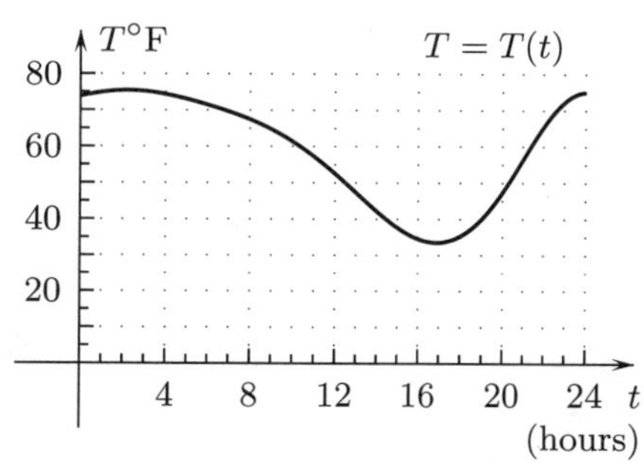

FIGURE 40

24. Draw the graph of a temperature function $T(t)$ for $0 \le t \le 4$ such that $T(t)$ increases for $0 < t < 2$ and decreases for $2 < t < 4$ and the graph of $T(t)$ has inflection points at $t = 1$ and at $t = 3$.

◆ Type 2, drawing the graph of a function with specified properties

25. The formula $M = 70W^{3/4}$ is used to predict the basal metabolic rate (the metabolic rate at rest), measured in Calories per day, of an animal weighing W kilograms.(2) Is the graph of this function concave up or concave down and how does this fact reflect the relationship between basal metabolism and weight?

◆ Type 2, the Second-Derivative Test, etc.

---

(1) *Human Adaptation and Accommodation* by A. Frisancho, Ann Arbor, MI.: The University of Michigan Press, 1993, p.119.

(2) *Animal Agriculture*, H. Cole, San Francisco, CA: W.H.Freeman and Company, 1974, p.534.

26. The braking distance of a particular car (the distance it takes to stop after its brakes have been fully applied) is approximately $s(v) = \frac{1}{48}v^2$ feet if the car is traveling v miles per hour.[(3)] Is the graph of $s(v)$ concave up or concave down and how does this fact relate to how braking distance depends on velocity?

♦ Type 2, the Second-Derivative Test, etc.

27. The following table gives chlorophyll production in a leaf during one year relative to the maximum possible chlorophyll production in the leaf. The numbers in the second row are the the percentages of the maximum production at the beginnings of the months.[(4)] **(a)** The graph of this function of time is either concave up or concave down. Which is it? **(b)** During what month is chlorophyll production a maximum, and why?

Date	June 1	July 1	Aug. 1	Sept. 1	Oct. 1	Nov. 1	Dec.1
Percent	60	77	85	91	95	60	2

♦ Type 2, the Second-Derivative Test, etc.

28. Because concrete is very resistant to compression but not to stretching, reinforcing steel rods are imbedded in any concrete that will be subject to forces that might stretch it. Figure 41 shows the sagging of a horizontal concrete beam that is supported but not fixed at the ends, and Figure 42 shows the sagging of a horizontal concrete beam that is fixed at the ends.[(5)] Extra reinforcing steel is placed near the bottom of beams as in Figure 41. Where is extra reinforcing steel placed in beams as in Figure 42?

♦ Type 2, concavity

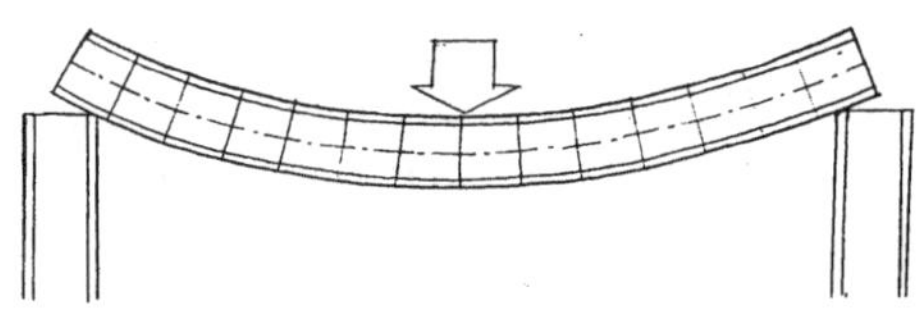

FIGURE 41

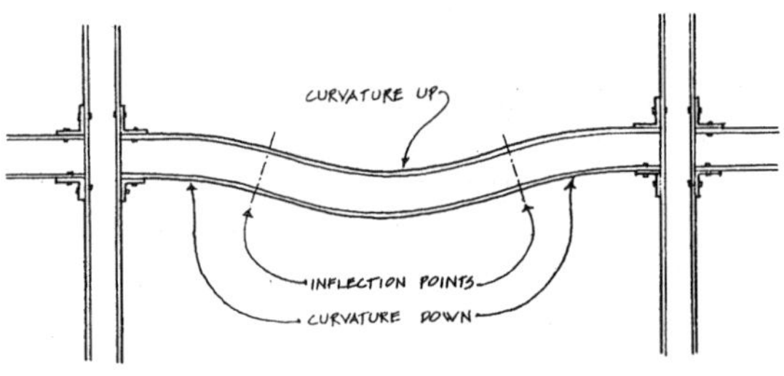

FIGURE 42

29.[A] **(a)** Show that $x = 1$ and $x = 2$ are critical points of $f(x) = \frac{1}{4}x^4 - \frac{5}{3}x^3 + 4x^2 - 4x$. **(b)** The two critical points in part (a) are the only critical points of $f(x)$. What are the most extensive intervals in which $f(x)$ is increasing and in which it is decreasing? Justify your answer.

♦ Type 2, the First-Derivative Test

30. Figure 43 shows the graph of the rate of change dV/dt of the volume of water in a tank for $0 \le t \le 8$. **(a)** What is the maximum rate of flow into the tank? Out of the tank? **(b)** When is the volume of water in the tank increasing? Decreasing? **(c)** When is the rate of flow of water into the tank increasing? Decreasing? **(d)** At what times with $0 < t < 8$ does $V(t)$ have a local maximum? A local minimum?

♦ Type 2, the First-Derivative Test

(3) Data adapted from *A Policy on Geometric Design of Highways and Streets*, Washington, DC: American Association of State Highway and Transportation Officials, 1983, pp. 139–141.

(4) Data adapted from "The Turning of the Leaves", P. Spencer, *Natural History*, 82 (1983), p. 63.

(5) *Why Buildings Stand up*, by M. Salvatori, New York, NY: W.W. Norton & Company, .pp. 79 and 82.

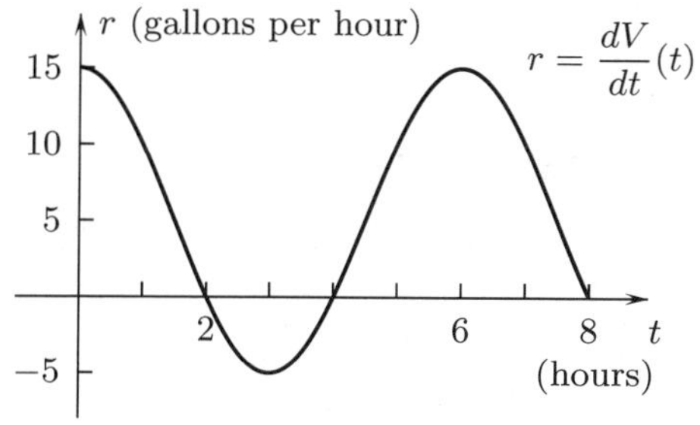

FIGURE 43

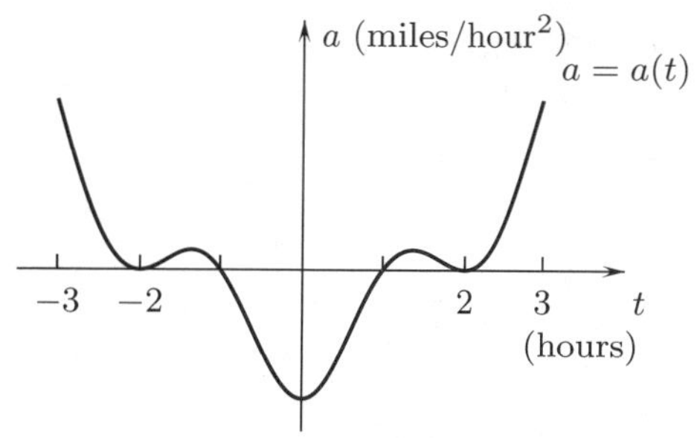

FIGURE 44

31. Figure 44 shows the graph of the acceleration toward the north $a(t)$ of a car on a north-south road for $-3 \le t \le 3$ (hours). **(a)** In which part of the interval $-3 \le t \le 3$ is the car accelerating toward the north? **(b)** When for $-3 \le t \le -1$ is the car's velocity toward the north the greatest?

♦ Type 2, the First-Derivative Test, etc.

32. For what values of the parameter a does $f(x) = ax + 1/x$ have a local maximum and a local minimum? Generate sample graphs in the window $-4 \le x \le 4, -6 \le y \le 6$ to guide your reasoning.

♦ Type 2, the First-Derivative Test, etc.

33. What can you say about $H(x)$ and its graph if $H(x)$ is continuous for all x, $H'(0) = 0$, and $H'(x)$ is increasing for $x < 0$ and decreasing for $x > 0$?

♦ Type 2, the Second-Derivative Test, etc.

34. What can you say about $G(x)$ and its graph if $G(x)$ and its first derivative are negative for all x, $G''(2) = 0$, and $G''(x)$ is increasing for all x?

♦ Type 2, the Second-Derivative Test, etc.

35. For what values of the parameter a does $x^3 - 3ax$ have a local maximum and a local minimum? Give formulas for the local maximum and minimum in terms of a.

♦ Type 2, the First-Derivative Test, etc.

36. For what values of the parameter c does $x^2 + c/x^2$ have an absolute minimum? Give a formula for the absolute minimum as a function of c.

♦ Type 2, the First-Derivative Test, etc.

37. Find a formula in terms of b for the absolute minimum of $x^2 - 2bx$ **(a)** by using calculus and **(b)** by completing the square in the expression $x^2 - 2bx$.

♦ Type 2, the First-Derivative Test, etc.

38.[A] Is it possible that $f'(x)$ is positive for all x and that $f''(x)$ is positive for $x < 0$ and negative for $x > 0$? If so, draw the graph of such a function. If not, explain why not.

♦ Type 4, the Second-Derivative Test, etc.

39. Is it possible that $f''(x)$ is negative for all x and that $f'(x)$ is negative for $x < 0$ and positive for $x > 0$? If so, draw the graph of such a function. If not, explain why not.

♦ Type 4, the Second-Derivative Test, etc.

40. Do the tangent lines at $x = -1$ and $x = 1$ to the graph of $f(x)$ intersect above or below the x-axis if $f(0) = 0$, $f'(0) = 0$, and $f''(x) > 0$ for $-1 \le x \le 1$? Explain.

♦ Type 4, the Second-Derivative Test, etc.

41. Is $P(x)$ less than or greater than $P(6) + P'(6)(x - 6)$ for all $x \ne 6$ if $P''(x) > 0$ for all x? Give a geometric argument.

♦ Type 4, the Second-Derivative Test, etc.

42. Suppose that $g''(x) < 0$ for all x. **(a)** Is it possible that the graph of $g(x)$ be tangent to $y = x$ at $x = -1$ and tangent to $y = -x$ at $x = 1$? **(b)** Is it possible that the graph of $g(x)$ be tangent to $y = -x$ at $x = -1$ and tangent to $y = x$ at $x = 1$? In each case either draw the graph of such a function or explain why there is no such function.

♦ Type 4, the Second-Derivative Test, etc.

In Problems 43 through 50 sketch the graph of the function after determining the most extensive intervals in which it is increasing and decreasing and finding exact values of its local or global maxima or minima. Give your reasoning.

43.[A] $y = \dfrac{8}{x^3 - 3x^2}$

♦ Type 3, the First-Derivative Test

44.[A] $y = \dfrac{x}{x^2 + 1}$

♦ Type 3, the First-Derivative Test

45.[A] $y = \dfrac{x^2}{x - 1}$

♦ Type 3, the First-Derivative Test

46. $y = x\sqrt{4 - x^2}$

47. $y = \sqrt{9 - x^2}$

♦ Type 3, the First-Derivative Test

48. $y = \dfrac{x}{x^2 - 4}$

♦ Type 3, the First-Derivative Test

49.[A] $y = \dfrac{x^2 - 1}{x^2 + 1}$

♦ Type 3, the First-Derivative Test

50. $y = \dfrac{1}{x - x^2}$

♦ Type 3, the First-Derivative Test

In Problems 51 through 58 sketch the graph of the function after determining the most extensive intervals in which it is increasing and decreasing and the most extensive open intervals in which its graph is concave up and concave down. Also find exact values of any local or global maxima or minima of the function, and exact coordinates of any inflection points on its graph. Give your reasoning.

51.[A] $y = \dfrac{1}{(x - 2)^2}$

♦ Type 3, the First- and Second-Derivative Tests

52.[A] $y = \dfrac{5}{x^2 + 1}$

♦ Type 3, the First- and Second-Derivative Tests

53. $y = \dfrac{3x}{x^2 - 4}$

♦ Type 3, the First- and Second-Derivative Tests

54. $y = \dfrac{x^2}{x^2 + 1}$

♦ Type 3, the First- and Second-Derivative Tests

55. $y = \dfrac{x}{x + 1}$

♦ Type 3, the First- and Second-Derivative Tests

56. $y = \dfrac{10}{x^3 + x}$

♦ Type 3, the First- and Second-Derivative Tests

57. $y = \dfrac{1}{\sqrt{x^2 + 8}}$

♦ Type 3, the First- and Second-Derivative Tests

58. $y = \frac{1}{3}x^3 - 10x - 9x^{-1}$

♦ Type 3, the First- and Second-Derivative Tests

59.[A] Use first and second derivatives to analyze $y = \frac{1}{3}x^3 + x^2 + 2x$. Then draw its graph.

♦ Type 3, the First- and Second-Derivative Tests

60.[A] Find the local maximum of $\dfrac{3\sqrt[3]{x}}{x + 2}$. Justify your answer.

♦ Type 3, the First-Derivative Test

61. Find **(a)** the global minimum of $-(2 - |x|)^3$ and **(b)** the inflection points of $y = -(2 - |x|)^3$. (Consider $x \geq 0$ and $x \leq 0$ separately.)

♦ Type 3, the First- and Second-Derivative Tests

62.[A] Find the exact coordinates of the inflection point of $y = \frac{3}{2}x^2 - 2x^{3/2} - 6x$.

♦ Type 4, the Second-Derivative Test

63. **(a)** Show that the one critical point of $\frac{1}{2}x^2 - \frac{2}{3}x^{3/2} - 2x$ is $x = 4$. **(b)** Prove that $\frac{3}{2}x^2 - 2x^{3/2} - 6x$ is ≥ -16 for all $x \geq 0$.

♦ Type 3, the First-Derivative Test

64. Generate $y = \sqrt{1 + 1/x}$ in the window $-3 \leq x \leq 3, -1 \leq y \leq 3$, and use the formulas for the function and its first derivative to explain the shape of the graph.

♦ Type 3, the First-Derivative Test

65. Generate $y = \dfrac{x + 1}{|x|}$ in the window $-3 \leq x \leq 3, -1 \leq y \leq 3$, and use the formulas for the function and its first derivative to explain the shape of the curve.

♦ Type 3, the First-Derivative Test

In Problems 66 through 70 use your calculator or computer to find exact answers or approximate answers with five-decimal-place accuracy.

66.[A] Find the local maximum and minimum of $x^5 + x^3 - 5x^2$ and the coordinates of the inflection point of its graph.

♦ Type 2, the First- and Second-Derivative Tests with a calculator or computer

67. Find the global minimum of $x^4 - 6x^2 - 8x$.

◆ Type 2, the First-Derivative Test with a calculator or computer

68. Find the minimum slope of tangent lines to $y = \frac{1}{4}x^2 - x^{5/4} - \frac{1}{4}x^{3/2}$.

◆ Type 2, the Second-Derivative Test with a calculator or computer

69. What are the coordinates of the inflection point of $y = x^2 + \frac{1}{3}x^4 - \frac{2}{5}x^5$?

◆ Type 2, the Second-Derivative Test with a calculator or computer

70. What is the global maximum of $x - 3x^2 - \frac{1}{4}x^4$?

◆ Type 2, the First-Derivative Test with a calculator or computer

(VERTICAL TANGENT LINES AND CUSPS) The graph of $f(x)$ has a VERTICAL TANGENT LINE at $x = x_0$ if $f(x)$ is continuous at x_0 and its derivative tends to ∞ or to $-\infty$ as $x \to x_0$. The graph has a CUSP at $x = x_0$ if $f(x)$ is continuous at $x = x_0$ and $f'(x)$ tends to ∞ as x tends to x_0 from one side and tends to $-\infty$ as x tends to x_0 from the other side. The curve $y = x^{1/3}$ in Figure 45, for instance, has a vertical tangent line at $x = 0$ because in this case $y' = \frac{1}{3}x^{-2/3} \to \infty$ as $x \to 0$, and $y = x^{2/3}$ in Figure 46 has a cusp at $x = 0$ because in this case $y' = \frac{2}{3}x^{-1/3}$ tends to ∞ as $x \to 0+$ and tends to $-\infty$ as $x \to 0^-$.

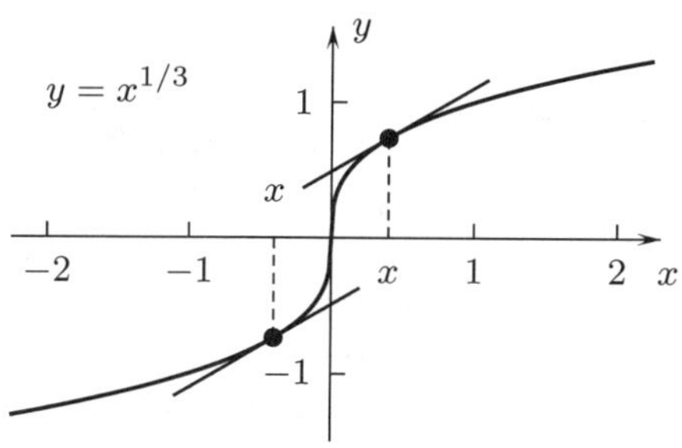

$y' \to \infty$ as $x \to 0$

A vertical tangent line at $x = 0$

FIGURE 45

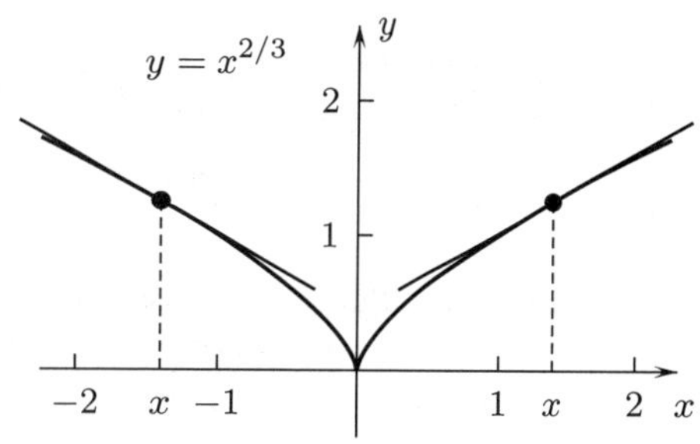

$y' \to \infty$ as $x \to 0^+$

$y' \to -\infty$ as $x \to 0^-$

A cusp at $x = 0$

FIGURE 46

71. Identify any vertical tangent lines and cusps of **(a[A])** $y = 1 - x^{1/3}$, **(b[A])** $y = \frac{2}{3}x - x^{2/3}$, and **(c)** $y = x - 3x^{1/3}$.

◆ Type 4, vertical tangent lines and cusps

72. Where are the vertical tangent lines of $y = \sqrt[3]{x^3 - 4x}$?

◆ Type 4, vertical tangent lines

73. Where are the cusps of $y = (x^2 - 1)^{2/3}$?

◆ Type 4, cusps

74. Show that if $y = f(x)$ has a vertical tangent line at $x = a$ and $f(a) = 0$, then $y = |f(x)|$ has a cusp at $x = a$.

◆ Type 4, vertical tangent lines and cusps

75. Show that if the y-axis is a vertical tangent line for $y = f(x)$, then $y = f(|x|)$ has a cusp at $x = 0$.

◆ Type 4, vertical tangent lines and cusps

76. Use the First- and Second-Derivative Tests to sketch the curve $y = 2x^{1/2} - x$. This graph might be said to have a one-sided vertical tangent line. Explain. (Consider secant lines.)

◆ Type 4, one-sided vertical tangent line

77. The graph of a function $f(x)$ that does not have a derivative is said to be concave up if for all points a, b, and c in the domain of the function with $a < c < b$, the point at $x = c$ on the graph of $f(x)$ is below the secant line through the points on the graph at $x = a$ and $x = b$. The graph is concave down if the point at $x = c$ is above the secant line for all such triples of points. The table below gives the probable average weight $W(n)$ (kilograms) of monsters in Loch Ness as a function of the number n of monsters in the loch.(6) Is the graph of $W(n)$ concave up or concave down? Verify the definition above with two triples of values from the table.

◆ Type 4, concavity

n	5	10	15	20	25	30
$W(n)$	600	350	230	150	100	75

Use the Mean-Value Theorem to justify your answers in Problems 78 through 86.

78.[A] Suppose that $G'(x) > 0$ and $G''(x) < 0$ for $0 \le x \le 10$. Which is the greatest and which is the least of $G'(0)$, $G'(10)$, and $\dfrac{G(10) - G(0)}{10}$?

◆ Type 4, using the Mean-Value Theorem

79. Is it possible that $H''(x) > 0$ for all x and the graph of $H(x)$ is tangent to $y = x$ at $x = 0$ and at $x = 1$? If so, draw the graph of such a function. If not, explain why not.

◆ Type 4, using the Mean-Value Theorem

80. Prove that $(x+1)^{1/4} - x^{1/4}$ tends to zero as $x \to \infty$.

◆ Type 4, using the Mean-Value Theorem

81. **(a)** Suppose that $f(x)$ is continuous in the interval $[x_0, x_0 + 1]$ and that $f'(x)$ has a finite limit m as x tends to x_0 from the right. Show that the ONE-SIDED DERIVATIVE FROM THE RIGHT at x_0, $D^+ f(x_0) = \lim\limits_{b \to x_0^+} \dfrac{f(b) - f(x_0)}{b - x_0}$ exists and equals m. Use the result of part (a) to find the one-sided derivatives from the right of **(b**[A]**)** $|x|$ at $x = 0$ and **(c)** $|x^2 - 9|$ at $x = 3$,.

◆ Type 4, using the Mean-Value Theorem

82. Formulate and prove the result corresponding to Problem 81 for one-sided derivatives from the left.

◆ Type 4, using the Mean-Value Theorem

83. Is it possible that a function $Z(x)$ have the properties that $Z'(x) > 0$ for all x, that $Z(x) \to 10$ as $x \to -\infty$, and that $Z(x) \to 0$ as $x \to \infty$? If so, draw the graph of such a function. If not, explain why not.

◆ Type 4, using the Mean-Value Theorem

84. Consider a function $L(x)$ such that $L'(x) > 0$ for all x. Is it possible that the equation $L(x) = 10$ has **(a)** no solutions x, **(b)** one solution, **(c)** two solutions, **(d)** more than two solutions? In each case draw the graph of such a function if one exists. Otherwise explain why there is no such function.

◆ Type 4, using the Mean-Value Theorem

85. Prove that if $y(x)$ is continuous in $[a, b]$ and has a nonzero derivative for $a < x < b$, then the equation $y(x) = 0$ has at most one solution with $a \le x \le b$.

◆ Type 4, using the Mean-Value Theorem

86. Prove that if $y'(x)$ exists for all x and $y'(x) = 0$ for at most two values of x, then the equation $y(x) = 0$ has at most three solutions.

◆ Type 4, using the Mean-Value Theorem

87. Determine whether the hypotheses of the Mean-Value Theorem are satisfied for each of the following functions in the given intervals. Then determine directly, without using the theorem, whether the conclusion of the theorem holds in each case.
(a[A]**)** $y = \sqrt{x}$ in $[1, 9]$, **(b**[A]**)** $y = \dfrac{x^2}{1 - x^2}$ in $[-2, 2]$, **(c)** $y = x^3 + x + 7$ in $[0, 3]$, **(d)** $y = x^{2/3}$ in $[-1, 1]$, **(e)** $y = \dfrac{1}{x}$ in $[-3, -1]$, **(f)** $y = \sqrt[3]{x}$ in $[-1, 1]$, **(g)** $y = x^{-2}$ in $[-1, 1]$.

◆ Type 4, using the Mean-Value Theorem

(6) Data adapted from "The population density of monsters in Loch Ness" by R. Sheldon and S. Kerr, *Limnology and Oceanography*, Sept., 1972, p. 797.

Section 3.2

Maxima and minima

OVERVIEW: *One of the most important uses of the derivative is in applications that call for finding maximum or minimum values of functions. In this section we use the techniques developed in Section 3.1 for studying increasing and decreasing functions to find maximum and minimum values of areas, volumes, and other quantities. More such* MAXIMUM/MINIMUM PROBLEMS *will be covered in Section 5.1.*

Topics:

- ***Using the derivative in narrative maximum/minimum problems***
- ***Maxima and minima of functions in intervals***
- ***The Extreme-Value Theorem***
- ***The Second-Derivative Test for local maxima and minima***

Grow all you can: finding a maximum area

Imagine you want to enclose a rectangular garden using a dormitory wall as one side and up to 40 feet of fence for the three other sides (Figure 1) and that you want to make the area of the garden as large as possible.

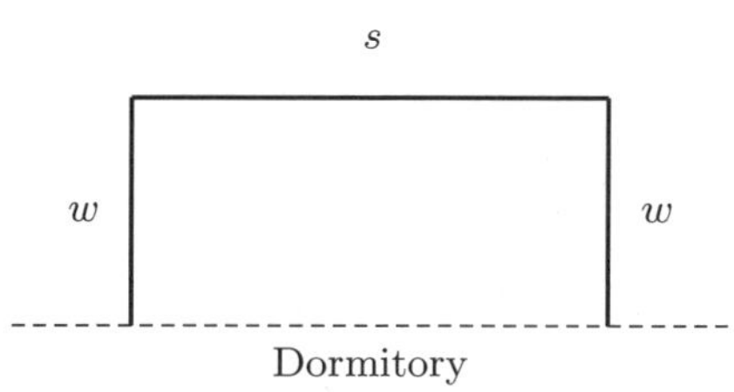

FIGURE 1

Question 1 **(a)** Explain why you would have to use all 40 feet of fence to maximize the area of the garden and why in this case $s = 40 - 2w$. **(b)** Calculate s and the area of the garden for the choices of w in the table below, under the condition that all 40 feet of fence is used. Describe how s and the area change as w runs through the values in the table. **(c)** What is wrong with using $w = -10$ or $w = 30$?

w	0	4	8	12	16	20
$s = 40 - 2w$	40	32				
Area	0	128				

Question 2 **(a)** Give a formula for the area $A(w)$ of the garden as a function of w, under the assumption that all 40 feet of fence is used. What is the domain of this function? (Allow for zero areas.) **(b)** Generate the graph of $A(w)$ in the window $-5 \leq w \leq 25, -50 \leq A \leq 250$ with w-scale = 10 and A-scale = 50, and use a <maximum> or <trace> operation on your calculator or computer to predict the greatest possible area of the garden and the corresponding dimensions.

♦ SUGGESTIONS TO INSTRUCTORS: Have students begin work on Questions 1 through 11 before class. Problems 2, 16, 22, 28, 32, 39, and 53 are good for in-class discussion.

Using the derivative

You probably predicted in Question 2 that the greatest possible area is achieved by using $w = 10$, as is shown in Figure 2. We can use a derivative to verify this result.

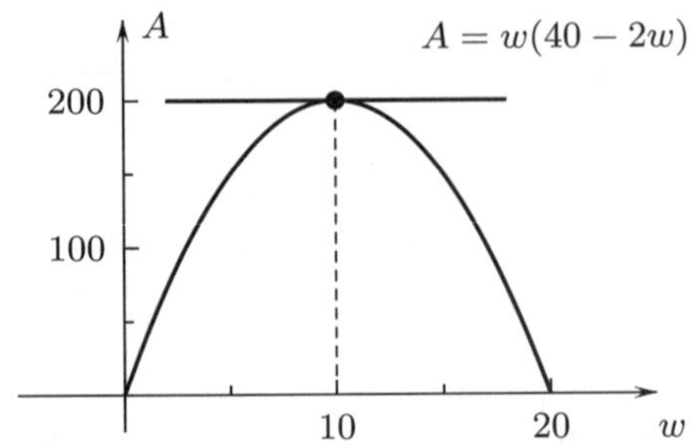

FIGURE 2

Example 1 Use the derivative to find the value of w for which the area of the garden is greatest.

SOLUTION We first eliminate the parentheses to put the formula $A = w(40-2w)$ from Question 2 in the form $A = 40w - 2w^2$. Then we find the derivative:

$$A'(w) = \frac{d}{dw}(40w - 2w^2) = 40 - 4w = 4(10 - w). \quad \textbf{(1)}$$

From formula **(1)** we can see that $A'(w) = 0$ at $w = 10$, that $A'(w) > 0$ for $0 < w < 10$ and $A'(w) < 0$ for $10 < w < 20$. Since $A(w)$ is continuous in $[0, 20]$, it is increasing in $[0, 10]$ and decreasing in $[10, 20]$. As predicted, the area is a maximum for $w = 10$. □

Question 3 What is the length s of the garden in Example 1 that yields the greatest possible area? What is the greatest possible area?

Maxima and minima in intervals

We say that M is the MAXIMUM or MAXIMUM VALUE OF $f(x)$ IN AN INTERVAL if $f(x)$ has the value M at one or more points x in the interval and $f(x) \leq M$ for all other x in the interval. We say that m is the MINIMUM VALUE OF $f(x)$ IN AN INTERVAL if $f(x)$ has the value m at one or more points x in the interval and $f(x) \geq m$ for all other points x in the interval. In Example 1 and Question 3 we found that the maximum of the area function $A(w) = w(40 - 2w)$ for $0 \leq w \leq 20$ is 200 at $w = 10$.

A little theory goes a long way: the Extreme-Value Theorem

In the solution of Example 1 we located the maximum of a function in an interval by finding the points where the derivative was zero and then determining the intervals where the derivative was positive and where it was negative. In many cases, we can use a shorter procedure that only requires finding the critical points (the points where the derivative is zero or does not exist) and comparing values of the function. This shorter procedure uses the following result, which is known as the EXTREME-VALUE THEOREM. It, like the Intermediate-Value Theorem in Section 1.2, is derived in advanced-calculus courses.

Theorem 1 (The Extreme-Value Theorem) *If $f(x)$ is continuous in a finite closed interval $[a, b]$, then $f(x)$ has a maximum and a minimum value in $[a, b]$.*

This result is illustrated in Figure 3 for a particular continuous function $f(x)$ and finite closed interval $[a, b]$. The maximum of $f(x)$ in $[a, b]$ is its value $f(x_1)$ at $x = x_1$, and its minimum in $[a, b]$ is its value $f(x_2)$ at $x = x_2$.

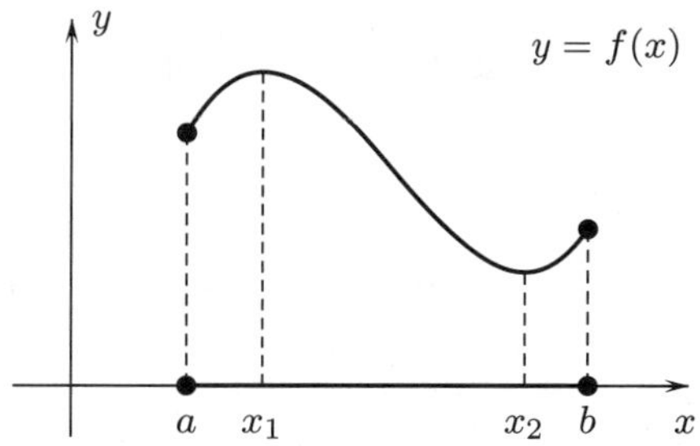

FIGURE 3

Example 2 **(a)** Is the hypothesis of the Extreme-Value Theorem satisfied and does the conclusion hold for the function $y = x^2$ in $[-1, 2]$?
(b) Is the hypothesis of the Extreme-Value Theorem satisfied and does the conclusion hold for $y = 1/x$ in $(0, \infty)$?

SOLUTION **(a)** The function $y = x^2$ is continuous in $[-1, 2]$, and the interval $[-1, 2]$ is closed and finite, so the hypothesis of the theorem is satisfied. As Figure 4 shows, the maximum of x^2 in $[-1, 2]$ is its value $2^2 = 4$ at $x = 2$, and its minimum is its value 0 at $x = 0$, so the conclusion of the theorem holds.
(b) The function $y = 1/x$ is continuous in the interval $(0, \infty)$, but the interval is neither closed nor finite, so the hypothesis of the theorem is not satisfied. The conclusion does not hold because $1/x$ does not have a minimum or minimum in $(0, \infty)$. It does not have a maximum because $1/x \to \infty$ as $x \to 0^+$. It does not have a minimum because $1/x \to 0$ as $x \to \infty$ but $1/x$ never has the value 0 (Figure 5).[†] □

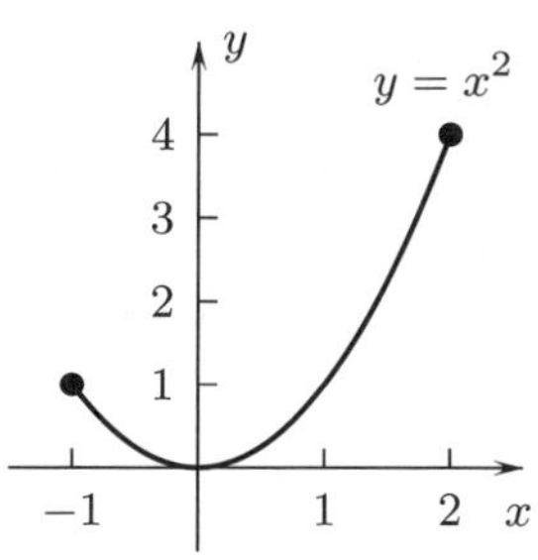

$[-1, 2]$
FIGURE 4

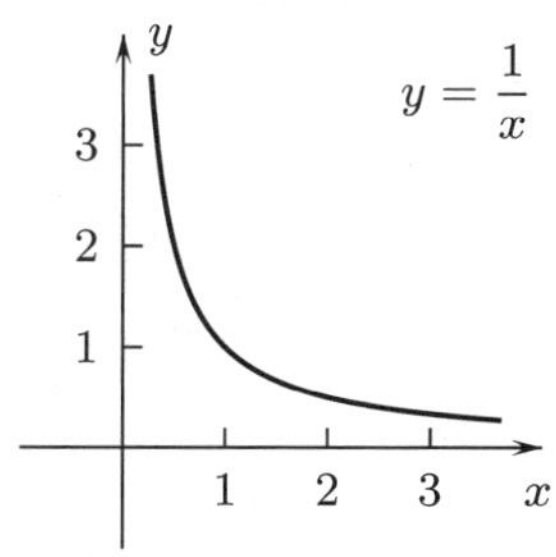

$(0, \infty)$
FIGURE 5

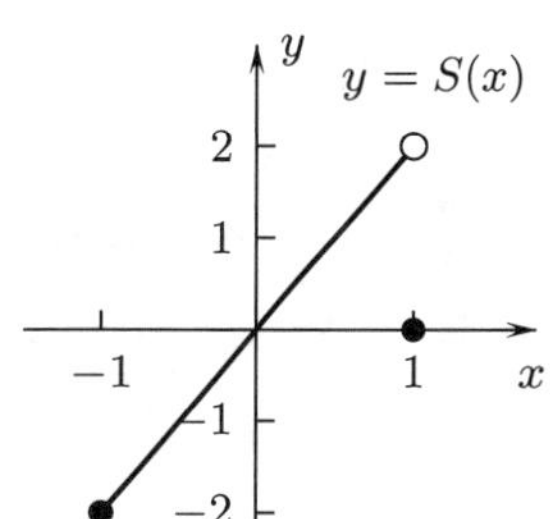

$[-1, 1]$
FIGURE 6

Question 4 The function $S(x)$, whose graph is shown in Figure 6, is defined by

$$S(x) = \begin{cases} 2x & \text{for} \quad -1 \le x < 1 \\ 0 & \text{for} \quad x = 1. \end{cases}$$

(a) Is the hypothesis of the Extreme-Value Theorem satisfied by $S(x)$ in $[-1, 1]$?
(b) Does the conclusion of the Extreme-Value Theorem hold for $S(x)$ in $[-1, 1]$?

[†] The value 0 is called the GREATEST LOWER BOUND of $1/x$ for x in $(0, \infty)$. We will discuss greatest lower bounds in Chapter 8.

Maxima and minima of continuous functions in finite, closed intervals

In problems that call for finding the maximum or minimum value of a continuous function in a finite, closed interval, we can use the following:

Theorem 2 *If $f(x)$ is continuous in a finite closed interval $[a,b]$, then it has a maximum and a minimum value in the interval, and the maximum and minimum occur either at a, at b, or at critical points in the interior of $[a,b]$.*

This theorem is valid because if $f(x)$ is continuous in $[a,b]$, it has a maximum and a minimum value in that interval by the Extreme-Value Theorem. Then the maximum or minimum might occur at an endpoint of the interval. Otherwise, it occurs at a point x_0 in the interior of the interval and is a local maximum or minimum, so that, by Theorem 1 of Section 3.1, x_0 is a critical point of the function.

Example 3 Find the minimum value of $p(x) = 2x^3 - 6x^2 + 12$ for $1 \le x \le 3$.

SOLUTION The polynomial $p(x)$ is continuous in $[1,3]$, and the interval is closed and finite. Therefore, we can apply Theorem 2. The derivative,

$$p'(x) = \frac{d}{dx}(2x^3 - 6x^2 + 12) = 6x^2 - 12x = 6x(x-2)$$

is zero at $x = 0$ and at $x = 2$. By the theorem, the minimum in the interval $[1,3]$ is either at the one critical point $x = 2$ in its interior or at one of the endpoints $x = 1$ or $x = 3$. We calculate

$$\begin{aligned} p(1) &= 2(1^3) - 6(1^2) + 12 = 8 \\ p(2) &= 2(2^3) - 6(2^2) + 12 = 4 \\ p(3) &= 2(3^3) - 6(3^2) + 12 = 12. \end{aligned}$$

Because the least of these is $p(2) = 4$, the minimum of $p(x)$ for $1 \le x \le 3$ is 4 at $x = 2$. This is corroborated by the graph of $p(x)$ in Figure 7. □

Question 5 What is the maximum value of $p(x) = 2x^3 - 6x^2 + 12$ from Example 3 for $1 \le x \le 3$?

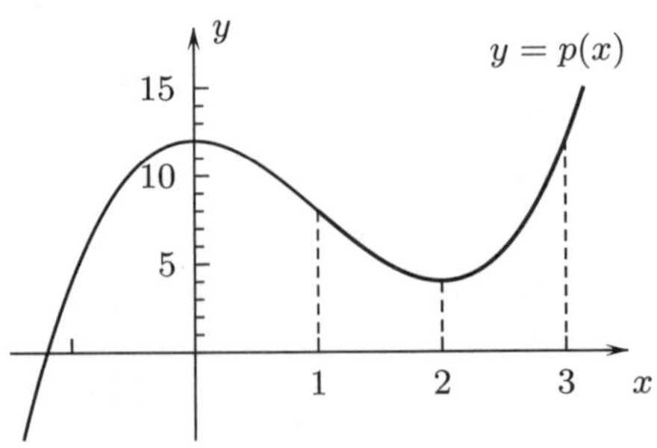

FIGURE 7

Fold a box: maximizing a volume

Example 4 A woman wants to construct a box with no top from a square piece of cardboard 12 inches wide by cutting equal-sized squares from each corner and folding up the sides (Figure 8). The box must be at least 3 inches tall. What size squares should she cut from the cardboard to give the box the greatest possible volume?

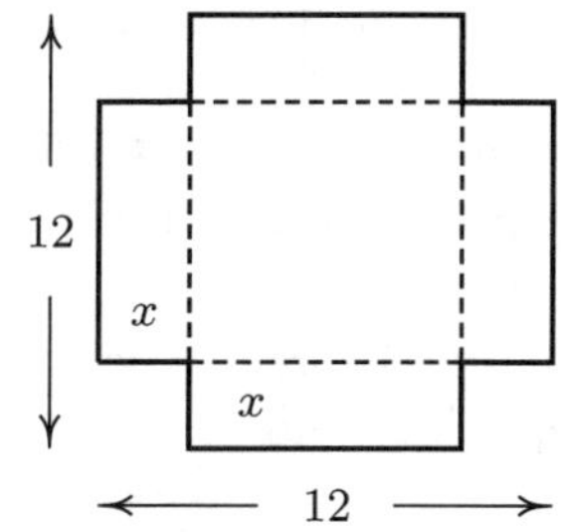

FIGURE 8

SOLUTION We let x denote the width of the four squares to be cut from the corners of the cardboard. The resulting box is x inches high and its base is a square of width $12 - 2x$, so its volume is

$$V(x) = x(12 - 2x)^2 = x(144 - 48x + 4x^2) = 144x - 48x^2 + 4x^3.$$

Because x cannot be greater than 6 inches (half the width of the cardboard) and the box must be at least 3 inches high, we need to find the maximum of $V(x)$ for $3 \le x \le 6$. $V(x)$ has a maximum and a minimum value in this interval because it is continuous for all x and the interval is closed and finite. The derivative is

$$\begin{aligned} \frac{dV}{dx}(x) &= \frac{d}{dx}(144x - 48x^2 + 4x^3) = 144 - 96x + 12x^2 \\ &= 12(x^2 - 8x + 12) \end{aligned}$$

and is zero where $x^2 - 8x + 12 = 0$. We can solve this equation by factoring to obtain $(x-6)(x-2) = 0$ or by using the Quadratic Formula, which gives

$$\begin{aligned} x &= \frac{-(-8) \pm \sqrt{(-8)^2 - 4(1)(12)}}{2(1)} = \frac{8 \pm \sqrt{16}}{2} \\ &= \frac{8 \pm 4}{2} = \begin{cases} \dfrac{8+4}{2} = 6 \\ \dfrac{8-4}{2} = 2. \end{cases} \end{aligned}$$

The critical points are 6 and 2, and neither is in the interior of the interval $[3, 6]$. By Theorem 2, the maximum of $V(x)$ for $3 \le x \le 6$ must occur either at the left endpoint $x = 3$ or at the right endpoint $x = 6$. Since $V(3) = 3\big(12 - 2(3)\big)^2 = 3(6)^2 = 108$ and $V(6) = 0$, the maximum for $3 \le x \le 6$ is at $x = 3$. This is consistent with the graph of $V(t)$ in Figure 9. The box of maximum volume that is at least 3 inches high is obtained by cutting squares 3 inches wide from the corners of the cardboard. □

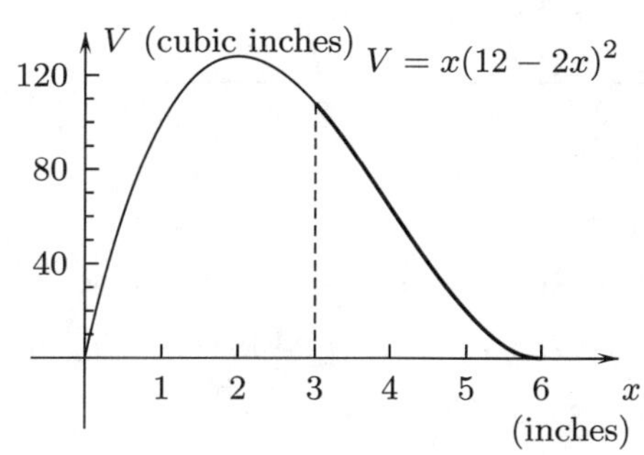

FIGURE 9

Question 6 What dimensions would give the box in Example 4 its maximum volume if the box could have any height between 0 and 6 inches?

An example with an infinite interval: minimizing a distance

Example 5 What point on the curve $y = \sqrt{x^2+1}$ in Figure 10 is closest to the point $(2,0)$ on the x-axis?

SOLUTION By the Pythagorean Theorem, the distance from the point $(x, \sqrt{x^2+1})$ on the curve to the point $(2,0)$ is the length of the line segment in Figure 10, which is

$$\begin{aligned} D(x) &= \sqrt{(x-2)^2 + (\sqrt{x^2+1}-0)^2} = \sqrt{(x^2-4x+4)+(x^2+1)} \\ &= \sqrt{2x^2-4x+5}. \end{aligned} \quad \textbf{(2)}$$

Because x can have any value, we need to find the global minimum of the continuous function $D(x)$ in the infinite interval $(-\infty, \infty)$. We first find its derivative,

$$\begin{aligned} D'(x) &= \frac{d}{dx}[(2x^2-4x+5)^{1/2}] \\ &= \tfrac{1}{2}(2x^2-4x+5)^{-1/2}\frac{d}{dx}(2x^2-4x+5) \\ &= \tfrac{1}{2}(2x^2-4x+5)^{-1/2}(4x-4) = \frac{2(x-1)}{\sqrt{2x^2-4x+5}}. \end{aligned} \quad \textbf{(3)}$$

Notice that the denominator of the last expression in **(3)** is the distance $D(x)$, given by **(2)**. You can see from Figure 10 that this distance is not zero for any x, so the denominator of **(3)** is always positive. Then formula **(3)** shows that $D'(x)$ is negative for $x < 1$ and positive for $x > 1$, so that $D(x)$ is decreasing for $x < 1$ and increasing for $x > 1$. Consequently, $D(x)$ has a global minimum at $x = 1$, and the closest point on the curve is at $x = 1$, where $y = \sqrt{1^2+1} = \sqrt{2}$. The closest point is $(1, \sqrt{2})$.

We could also find this point, with simpler calculations, by minimizing the square $f(x) = [D(x)]^2 = 2x^2 - 4x + 5$ of the distance **(2)**. Because $f'(x) = 4x - 4$ is negative for $x < 1$ and positive for $x > 1$, $f(x)$ has a global minimum at $x = 1$. Then $D(x)$ has a global minimum at the same x because increasing (or decreasing) $f(x)$ increases (or decreases) $D(x)$. □

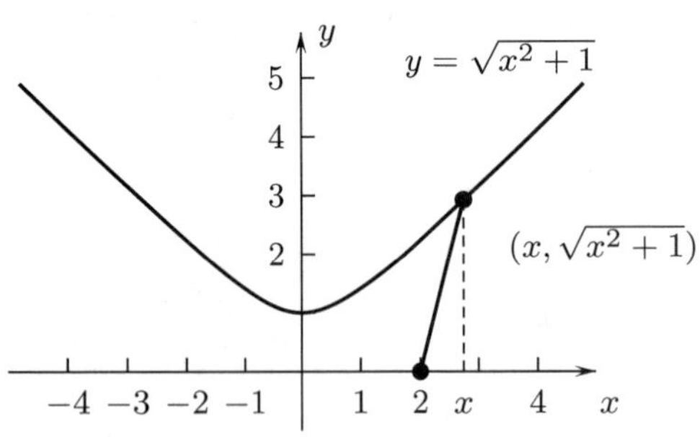

FIGURE 10

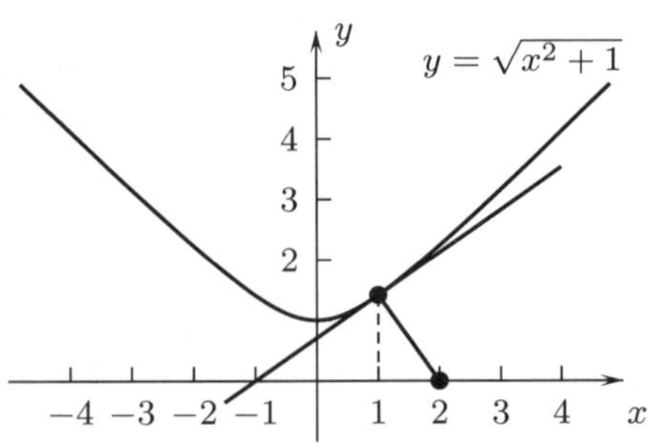

FIGURE 11

Question 7 Show that the line in Figure 11 joining the point $(2,0)$ to the closest point on $y = \sqrt{x^2+1}$ is perpendicular to the tangent line to the curve at that point.

Minimizing a cost

Imagine that we need to make a rectangular box with a square base that has a volume of 20 cubic meters (Figure 12) and that we are to use material that costs 4 dollars per square meter for the base and top and 2 dollars per square meter for the four sides. What dimensions should we give the box to minimize the cost of material? To answer this question, we let w denote the width of the square base and h the height of the box, measured in meters. Then the volume of the box is w^2h, its base and top each has area w^2, and each of the four sides has area wh.

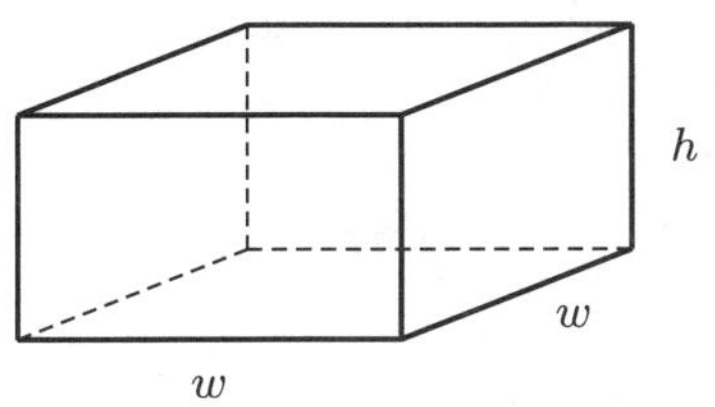

FIGURE 12

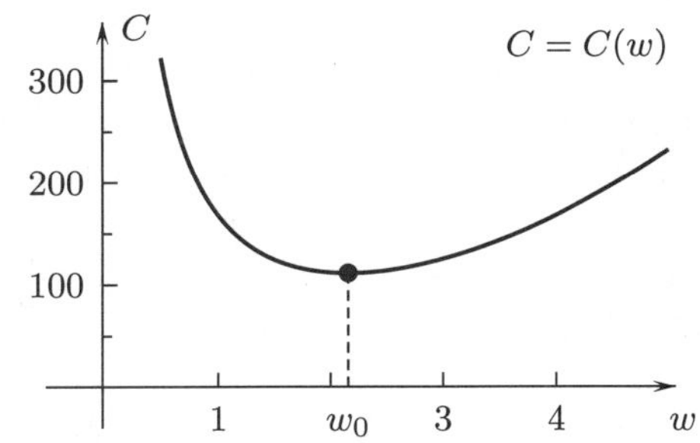

FIGURE 13

Question 8 **(a)** What is the height h of the box if its volume is 20 cubic meters and its base is $\frac{1}{2}$ meter wide? What is the cost C of material for the base, top, and four sides in this case? **(b)** What is the height h and cost C if the volume is 20 cubic meters and the base is 4 meters wide? **(c)** What is the height h and cost C if the volume is 20 cubic meters and the base is w meters wide? This gives formulas for h and C as functions of w. Generate the graph of $C(w)$ for $-0.5 \le w \le 5.5, -35 \le C \le 350$ with A-scale $= 100$. You should obtain the curve in Figure 13.

Question 9 **(a)** What happens to $C(w)$ as $w \to \infty$ and as $w \to 0^+$, and what does this say about the cost of possible boxes? **(b)** Use a minimum or trace operation on your calculator or computer to find the approximate minimum of $C(w)$ and the approximate value of w that gives that area.

Example 6 Find the exact and approximate decimal values of the width of the box that would cost the least.

SOLUTION The derivative of the cost $C(w) = 8w^2 + 160w^{-1}$ from Question 8 is

$$\frac{dC}{dw} = \frac{d}{dw}\left(8w^2 + 160w^{-1}\right) = 16w - 160w^{-2}. \qquad \textbf{(4)}$$

We eliminate the negative exponent in **(4)** by putting w^2 in the denominator and then combine terms by taking the common denominator:

$$\frac{dC}{dw} = 16w - \frac{160}{w^2} = \frac{16w^3}{w^2} - \frac{160}{w^2} = \frac{16(w^3 - 10)}{w^2}.$$

We do not factor the last numerator because we can see from this formula that dC/dw is zero for $w = 10^{1/3}$, is negative for $0 < w < 10^{1/3}$, and is positive for $w > 10^{1/3}$. Consequently, $C(w)$ is decreasing for $0 < w < 10^{1/3}$ and is increasing for $w > 10^{1/3}$. The box that costs the least has width $w_0 = 10^{1/3} \doteq 2.15443$ feet. □

Question 10 What is the height of the box in Example 6 that costs the least and how much does that box cost? Check your answer by comparing it with the results of Question 9.

The Second-Derivative Test at a local maximum or minimum

We usually show that a function $f(x)$ has a local maximum or minimum at a critical point x_0 by determining whether the derivative is positive or negative in open intervals to the left and to the right of x_0. In some cases, however, it is more convenient to calculate the second derivative at x_0. We use the following result.

Theorem 3 (The Second-Derivative Test for a local maximum or minimum)

(a) *If $f'(x_0) = 0$ and $f''(x_0) > 0$, then $f(x)$ has a local minimum at x_0 (Figure 14).*

(b) *If $f'(x_0) = 0$ and $f''(x_0) < 0$, then $f(x)$ has a local maximum at x_0 (Figure 15).*

This theorem is easiest to understand if we make the additional assumption that $f''(x)$ is positive in an open interval containing x_0 for part (a) and that $f''(x)$ is negative in an open interval containing x_0 for part (b). Then in both cases the graph has a horizontal tangent line at x_0 because $f'(x_0) = 0$. In case (a), $f(x)$ has a local minimum at x_0 because its graph is concave up near $x = x_0$ (Figure 14). In case (b), $f(x)$ has a local maximum at x_0 because its graph is concave down near $x = x_0$ (Figure 15).†

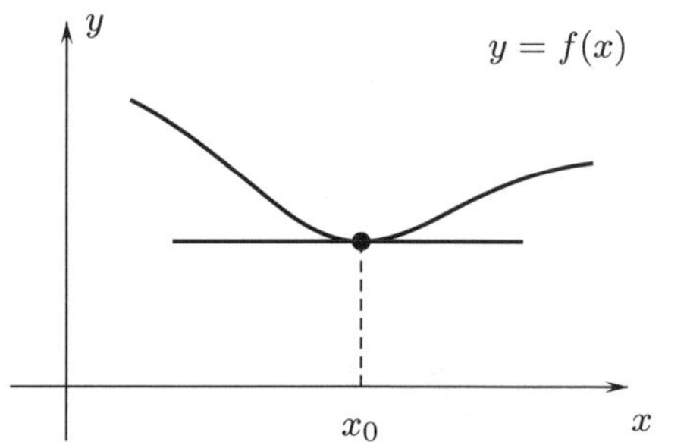

$f'(x_0) = 0$ and $f''(x_0) > 0$

Local minimum

FIGURE 14

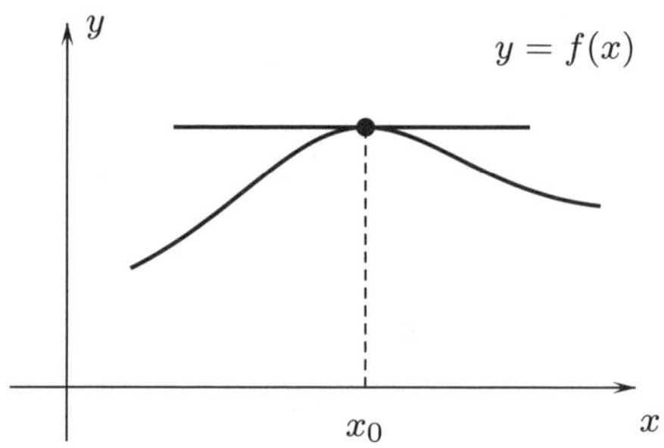

$f'(x_0) = 0$ and $f''(x_0) < 0$

Local maximum

FIGURE 15

Example 7 Use the Second-Derivative Test to show that $V(x) = x(12 - 2x)^2$ from Example 4 has a local maximum at $x = 2$.

SOLUTION In the solution of Example 4 we saw that $V'(x) = 12(x^2 - 8x + 12)$ is zero at $x = 2$. The second derivative $V''(x) = 12(2x-8)$ has the value $V''(2) = 12\big(2(2)-8\big) = -48$, which is negative, so $V(x)$ has a local maximum at $x = 2$. (See Figure 9.) □

Question 11 Use the Second-Derivative Test to show that $C(w) = 8w^2 + 160w^{-1}$ from Example 6 has a local minimum at $w = 10^{1/3}$.

Principles and procedures

- In most maximum/minimum problems, you need to find the maximum or minimum value of a differentiable function in an interval or determine that the function does not have a maximum or minimum in the interval. Usually this can be accomplished by finding where the derivative is zero or does not exist and then finding the open intervals where the derivative is positive and negative—since this shows where the function is increasing and where it is decreasing. In some cases you will have to compare values of the function at critical points and endpoints.

†Part (a) of Theorem 3, as stated, can be established with the following argument: Because $f'(x_0) = 0$, $f''(x_0)$, which is the derivative of $f'(x)$ at x_0, is the limit of $\dfrac{f'(b) - f'(x_0)}{b - x_0} = \dfrac{f'(b)}{b - x_0}$ as $b \to x_0$. Since $f''(x_0)$ is positive, $\dfrac{f'(b)}{b - x_0}$ is positive for b sufficiently close to x_0. Consequently, $f'(b)$ is positive for b sufficiently close to and on the right of x_0, and is negative for b sufficiently close to and on the left of x_0. Hence $f(x)$ is increasing just to the right of x_0 and decreasing just to the left of x_0 and, thus, has a local minimum at x_0. Part (b) can be established by a similar argument.

- If you are looking for the maximum or minimum of a continuous function in a finite closed interval, you can apply Theorem 2 of this section to conclude that the maximum and minimum exist and occur either at endpoints of the interval or at critical points in the interior of the interval. Then you can find the maximum and minimum by comparing the values of the function at the endpoints and critical points.
- You cannot find maxima or minima of a function by just finding its critical points. You either have to determine the intervals where the function is increasing and decreasing or, if appropriate, use Theorem 2 and compare the values of the function at any critical points and endpoints.
- Occasionally it is convenient to use the Second-Derivative Test to see whether the function has a local maximum or minimum at a point where the derivative is zero, but usually you can get all the information you need from the first derivative.
- In some narrative optimization problems, a formula for the function to be maximized or minimized is given in terms of a dimension or other variable. In other cases, the necessary formula has to be found—often from formulas for distances, areas, volumes, and other quantities—by converting a function of two variables with a constraint or a function of three variables with two constraints into a function of one variable. You generally solve the problem by finding the maximum or minimum of the new function in an interval.
- Be sure to check your work by carrying out calculations and generating graphs on your calculator or computer as appropriate.

Tune-Up Exercises 3.2[♦]

[A]*Answer provided.* [O]*Outline of solution provided.*

Be sure you can work all of these exercises.

T1.[A] What are the maximum and minimum values of $f(x)$ for $-5 \leq x \leq 5$ if $f(x)$ is continuous in $[-5, 5]$, if $f'(x)$ is negative for $-5 \leq x < 0$ and positive for $0 < x \leq 5$, and if $f(-5) = 10$, $f(0) = 2$, and $f(5) = 11$?

♦ Type 1, max/min in an interval

T2.[A] What are the maximum and minimum values of $g(x)$ for $0 \leq x \leq 8$ if $g(x)$ is continuous in $[0, 8]$, if $g'(x)$ exists for all x and is zero only at $x = 5$, and if $g(0) = -1$, $g(5) = 3$, and $g(8) = 2$?

♦ Type 1, max/min in an interval

T3.[A] What are the maximum and minimum values of $h(x)$ for $x > 0$ if $h(x)$ is continuous in $(0, \infty)$, if $h(x)$ tends to ∞ as $x \to 0^+$ and as $x \to \infty$, if $h'(x)$ exists for all positive x and is zero only at $x = 2$ and $x = 3$, and if $h(2) = 10$, and $h(3) = 20$?

♦ Type 1, max/min in an interval

T4.[O] **(a)** Give a formula for the perimeter of a rectangle as a function of its width under the condition that its area is 7 square feet. What is the domain of this function? **(b)** Give a formula for the area of a rectangle as a function of its width under the condition that its perimeter is 100 inches. What is the domain of this function?

♦ Type 3, max/min in an interval

T5.[O] A rectangular box has a square base and a square top. **(a)** Give a formula for the volume of the box as a function of the width of its base under the condition that the total area of its six sides is 24 square feet. What is the domain of this function? **(b)** Give a formula for the total area of its six sides as a function of the width of its base under the condition that its volume is 9 cubic meters. What is the domain of this function?

♦ Type 3, max/min in an interval

[♦]The Tune-up Exercises and Problems are classified by type and content. The types are (1) basic, reactive; (2) basic reflective; (3) intermediate, reactive; (4) intermediate, reflective; (5) advanced, reactive; (6) advanced, reflective; and (7) advanced, theoretical.

T6.[A] **(a)** What are the volume V, the area A_B of the base, and the area A_L of the lateral (curved) sides of a right circular cylinder of radius r and height h? (Figure 16) **(b)** What is the total area A of its base, top, and curved sides?

♦ Type 3, max/min in an interval

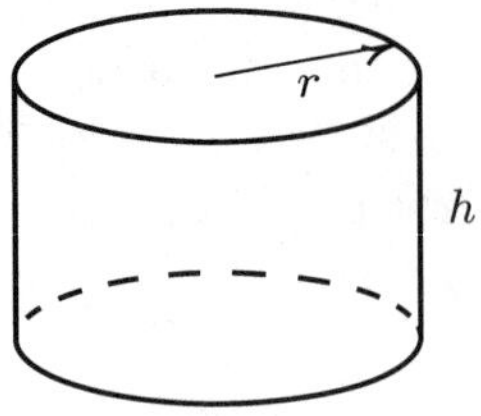

FIGURE 16

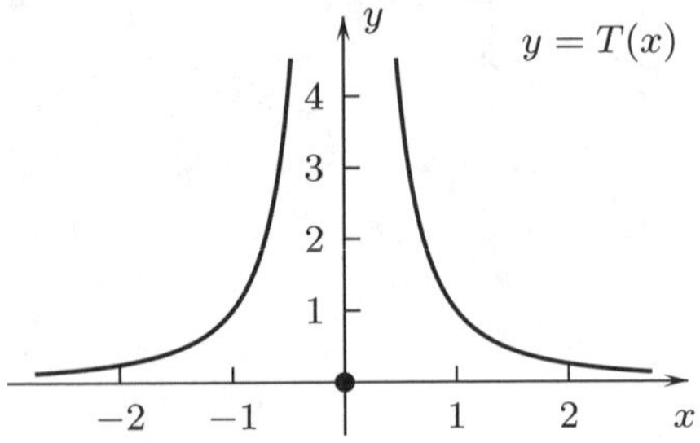

FIGURE 17

T7.[O] A rectangular box with a square bottom and top is to have a volume of 1000 cubic inches. The four sides are to be made from material that costs two cents per square inch and the top and bottom are to be made from material that costs three cents per square inch. Give a formula for the cost of the box as a function of its width.

♦ Type 3, max/min in an interval

T8.[A] The function $y = T(x)$ whose graph is shown in Figure 17 is defined by

$$T(x) = \begin{cases} 1/x^2 & \text{for} \quad x < 0 \\ 0 & \text{for} \quad x = 0 \\ 1/x^2 & \text{for} \quad x > 0. \end{cases}$$

(a) Are the hypotheses of the Extreme-Value Theorem satisfied and does its conclusion hold for $T(x)$ in the interval $[1, 2]$? Explain. **(b)** Are the hypotheses of the Extreme-Value Theorem satisfied and does its conclusion hold for $T(x)$ in the interval $(-2, 2)$? Explain.

♦ Type 3, the Extreme-Value Theorem

Problems 3.2

[A]*Answer provided.* [O]*Outline of solution provided.*

In Problems 1 through 18 find the maximum and minimum values of the functions, if they exist, in the given intervals. Justify your answers. (Generate the graphs as a partial check.)

1.[A] $y = x^3 - 3x + 5$ for $0 \le x \le 2$

♦ Type 1, max/min in an interval

2.[A] $y = x^3 - 6x^2 + 9x$ for $x \ge 0$

♦ Type 1, max/min in an interval

3.[A] $y = x + 4/x^2$ for $x > 0$

♦ Type 1, max/min in an interval

4.[A] $y = x^2 - x^{-2}$ for $1 \le x < 4$

♦ Type 1, max/min in an interval

5.[A] $y = 1/x$ for $x > 0$

♦ Type 1, max/min in an interval

6. $y = x^2 + 1$ for all x

♦ Type 1, max/min in an interval

7. $y = 6x^4 - 4x^6$ for $-2 \le x \le 2$

♦ Type 1, max/min in an interval

8. $y = x^3 - 3x^2 + 3x$ for $0 \le x \le 2$

♦ Type 1, max/min in an interval

9. $y = x + 4/x$ for $1 \le x \le 3$

♦ Type 1, max/min in an interval

10. $y = x^3 + x^5 + x^7$ for $-1 \le x \le 1$

♦ Type 1, max/min in an interval

11. $y = x^{10} - 10x$ for $-1 \le x \le 2$

♦ Type 1, max/min in an interval

12. $y = x^3 + 6x^2 + 1$ for $-1 \le x \le 1$

♦ Type 1, max/min in an interval

13. $y = x^2/(x-1)$ for $-1 \le x \le \frac{1}{2}$

♦ Type 1, max/min in an interval

14. $y = \frac{3}{4}x^4 - 4x^3 + 6x^2 - 5$ for $x \ge -1$

♦ Type 1, max/min in an interval

15.[A] $y = \sqrt{5 - x^2}$ for $-1 \le x \le 2$

♦ Type 1, max/min in an interval

16.[A] $y = \dfrac{x}{x^2 + 9}$ for $x \le 1$

♦ Type 1, max/min in an interval

17. $y = (x^2 + 1)^{-3}$ for $-1 \le x \le 2$

♦ Type 1, max/min in an interval

18. $y = \dfrac{x}{x^2 - 9}$ for $-2 \le x \le 2$

♦ Type 1, max/min in an interval

19.[A] Of all rectangles of perimeter P, which has the maximum area? Justify your answer.

♦ Type 2, a max/min problem

20. Of all rectangles of area A, which has the minimum perimeter? Justify your answer.

♦ Type 2, a max/min problem

21.[A] In order for a package to be mailed by parcel post, its length (the length of its longest side) plus its girth (the perimeter of the cross section perpendicular to its longest side) must be no greater than 108 inches (Figure 18). **(a)** Why does the rectangular box of maximum volume that can be shipped by parcel post have its length plus its girth equal to 108 inches? Why does it have square cross sections? Suppose that a box meets these conditions. **(b)** What is the length of the longest side if the square cross sections are 10 inches wide? What is the volume of the box in this case? **(c)** What is the length of the longest side and the volume V if the square cross sections are w inches wide? **(d)** Find the dimensions of the box of maximum volume. Justify your answer.

♦ Type 2, a max/min problem

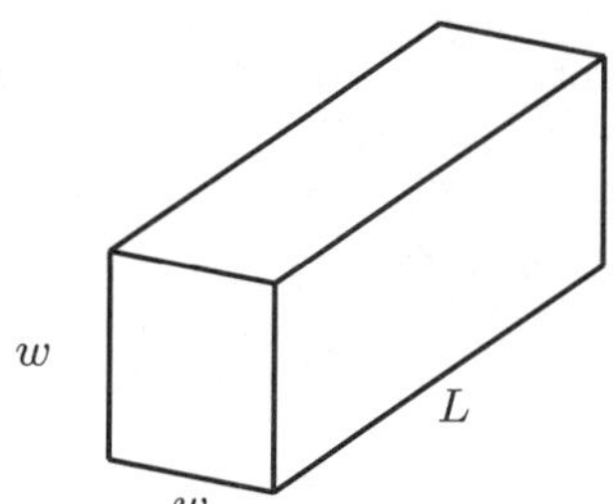

FIGURE 18

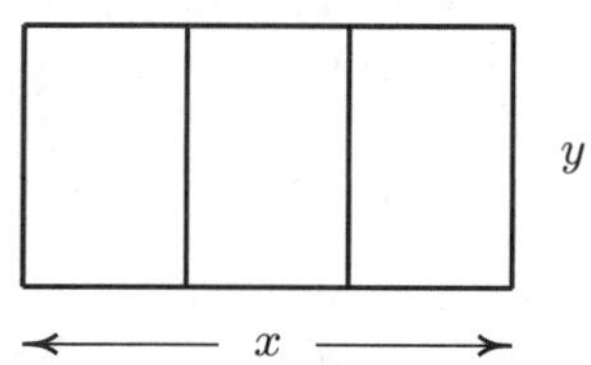

FIGURE 19

22.[A] A farmer uses 80 feet of fence to construct three rectangular pens as in Figure 19. **(a)** What is the depth y of the pens if all of the fence is used and the overall width x across the front of the three pens is 10 feet? What is the total area of the pens in this case? **(b)** What is the depth y of the pens if all the fence is used and $x = 30$ feet? What is the total area of the pens in this case? **(c)** What is the depth y of the pens and their total area $A(x)$ if all of the fence is used and the overall width of the pens is x feet? **(d)** Find the dimensions that give the maximum total area. Justify your answer. As a partial check, generate the graph of $A(x)$ in the window $-5 \le x \le 45, -25 \le A \le 250$ with x-scale $= 10$ and A-scale $= 50$.

♦ Type 2, a max/min problem

23.[A] **(a)** Which positive numbers x are greater than their cubes? **(b)** How much greater is $\frac{1}{2}$ than $\left(\frac{1}{2}\right)^3$? **(c)** How much greater is x than x^3 for $0 < x < 1$? **(d)** What positive number exceeds its cube by the greatest amount? Give the exact answer.

♦ Type 2, a max/min problem

24. What nonnegative number exceeds its fourth power by the greatest amount?

♦ Type 2, a max/min problem

25. What is the width of the cube such that the area of its six sides, measured in square meters, exceeds its volume, measured in cubic meters, by the greatest amount?

♦ Type 2, a max/min problem

26.[A] For what values of x and y with $3x + y = 12$ is $10 + (x - y)^2$ a minimum?

♦ Type 2, a max/min problem

27. For what values of x and y with $y = x - 1$ is $100 - x^2 - y^2$ a maximum?

♦ Type 2, a max/min problem

28. A cylindrical can with no top is to be constructed so its volume is 8π cubic feet. **(a**[A]**)** Give a formula for the total area $A(r)$ of its base and curved sides as a function of the radius of its base. **(b)** Find the dimensions that minimize $A(r)$. Justify your answer. Generate the graph of $A(r)$ in the window $-3 \le r \le 5, -10 \le A \le 100$ with A-scale $= 10$ as a partial check of your answer.

♦ Type 2, a max/min problem

29.[A] A rectangle is to have its base on the x-axis and its upper corners on the parabola $y = 36 - x^2$ as in Figure 20. **(a)** What are the width, height and area of the rectangle if its right side is at $x = 4$? **(b)** For what values of $x \ge 0$ is the area zero? **(c)** What are the width, height and area of the rectangle if its right side is at x with $0 \le x \le 6$? **(d)** Find the dimensions that maximimize the area of the rectangle. **(e)** Of all such rectangles, what are the width and height of those with minimum area? Justify your answer.

♦ Type 2, a max/min problem

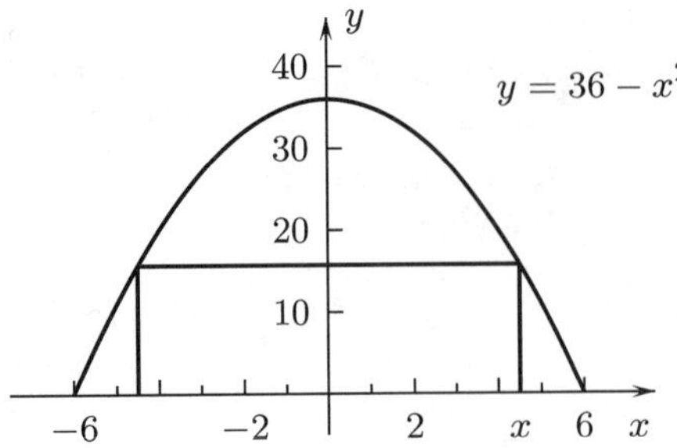

FIGURE 20

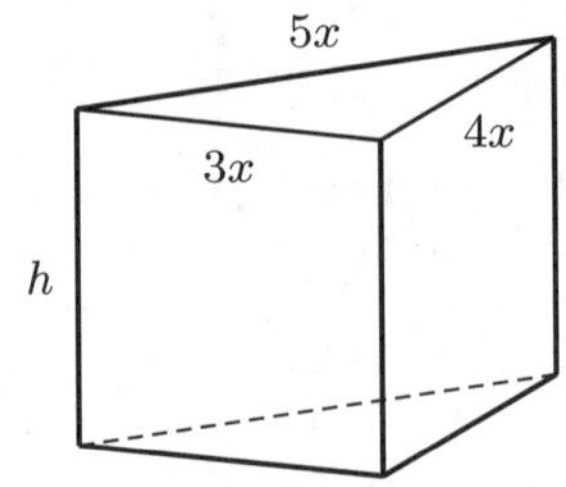

FIGURE 21

30. A triangular prism with vertical sides and with horizontal top and bottom that are right triangles with sides of lengths $3x, 4x$, and $5x$, is to be constructed so its volume is 12 cubic feet (Figure 21). **(a)** Give formulas for its height h and the total area A of its top, bottom, and three sides as functions of x. Justify your answer. **(b)** For what value of x is A a minimum?

♦ Type 2, a max/min problem

31.[A] A rectangular garden is to have an area of 150 square feet. The fencing for three sides costs \$5 per foot and the fencing for the fourth side costs \$10 dollars per foot. **(a)** Give a formula for the cost of the fence as a function of the length of the side that costs \$10 per foot. **(b)** What is the minimum cost of the fence?

♦ Type 2, a max/min problem

32. A rectangular box is to have volume $\frac{8}{3}$ cubic feet and be twice as long as it is wide. What dimensions should it have to minimize the total area of its six sides?

♦ Type 2, a max/min problem

33. A woman wants to build a rectangular garden with an ornamental fence that costs \$2 per foot for three sides and \$3 per foot for the fourth side. She will spend \$120 for the fence. What should the dimensions of the garden be to maximize its area?

♦ Type 2, a max/min problem

34. A rectangular box with no top is to be 5 feet wide and have a volume of 6 cubic feet. **(a)** Give a formula for the total area A of its bottom and four sides as a function of its length. **(b)** How long and how tall should it be to minimize the total area of its four sides and bottom?

♦ Type 2, a max/min problem

35. A rectangular box with a square bottom and no top is to have a volume of six cubic feet. The material for the bottom costs three dollars per square foot and the material for the four sides costs two dollars per square foot. How wide and how tall should it be to minimize its cost?

♦ Type 2, a max/min problem

36. A cylindrical can with no top is to be made from 12π square inches of tin. What should be the height of the can and the radius of its base to maximize its volume?

♦ Type 2, a max/min problem

37. A cylindrical can with a top is to have a volume of 2π cubic meters. What should be the dimensions of the can to minimize the total area of its top, bottom, and curved sides?

♦ Type 2, a max/min problem

38. A printer is to use a page that has an overall area of 80 square inches with margins of one inch at the bottom and sides and $1\frac{1}{2}$ inches at the top. What should be the dimensions of the paper so that the area of the printed portion is a maximum?

♦ Type 2, a max/min problem

39. A boy has enough money to buy 600 square inches of material to make a box with square ends and no top. He wants to paint the ends red, but only has enough paint if they are no larger than 11 inches on each side. What dimensions should he give the box for it to have the maximum volume?

♦ Type 2, a max/min problem

40. A horse breeder wants to use 180 feet of fencing for a rectangular corral that exends at least across the entire front of a 100-foot-long stable, as in Figure 22. What dimensions should he give the corral for it to have the maximum area?

♦ Type 2, a max/minm problem

41. Prove that of all right triangles of area A, the one with the shortest hypotenuse is isosceles.

♦ Type 2, a max/min problem

42. Prove that of all right triangles with hypotenuse of length L, the one of maximum area is isosceles.

♦ Type 2, a max/min problem

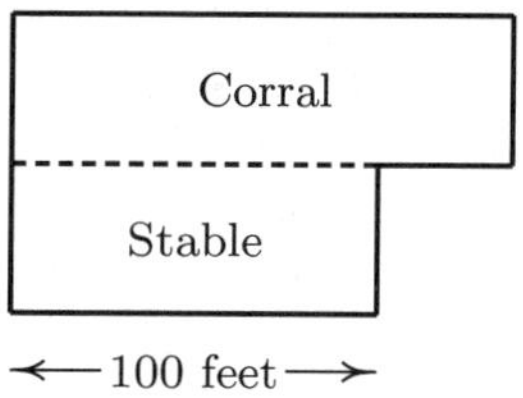

FIGURE 22

In Exercises 43 through 50 use calculus to find the maximum and minimum values of the functions, if they exist, in the given intervals. Use the quadratic formula as necessary and justify your answers. Give exact and approximate decimal answers in cases where the exact answers involve irrational numbers.

43.[O] $y = \frac{1}{2}x^4 + x^3 - x^2$ for $-3 \le x \le 2$

♦ Type 3, max/min in an interval

44.[A] $y = -\frac{1}{3}x^3 + 2x^2 - 2x$ for $0 \le x \le 4$

♦ Type 3, max/min in an interval

45. $y = x^3 - x^2 - x$ for $-1 \le x \le 0$

♦ Type 3, max/min in an interval

46.[A] $y = \frac{1}{4}x^4 - 2x^3 + x^2 + 50$ for all x

♦ Type 3, max/min in an interval

47. $y = x^5 + \frac{7}{3}x^3 + 2x$ for $-3 \le x \le 3$

♦ Type 3, max/min in an interval

48. $y = 2x^3 + x^2 + x$ for all x

♦ Type 3, max/min in an interval

49. $y = \frac{1}{4}x^4 + \frac{1}{3}x^3 + \frac{1}{2}x^2$ for $-1 \le x \le 1$

♦ Type 3, max/min in an interval

50. $y = 2x^{5/2} - 20x^{3/2} + 10x^{1/2}$ for $x \ge 5$

♦ Type 3, max/min in an interval

51. A box with no top is to be made from a rectangular piece of cardboard two feet wide and four feet long by cutting out equal squares from the four corners and folding up the sides (Figure 23). What size squares should be cut out to maximize the volume of the box?

♦ Type 4, a max/min problem

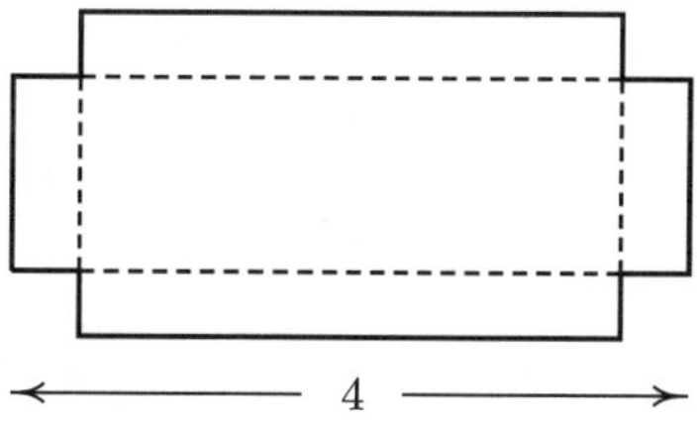

FIGURE 23

52.[A] For what values of x is $3x^5 - 25x^3 + 60x$ negative?

♦ Type 4, a max/min problem

53.[A] For what values of x is $2\sqrt{x} + x^{-1} \ge 3$?

♦ Type 4, a max/min problem

54. Show that $\frac{1}{4}x^4 - 4x^3 + 10x^2 + 501$ is positive for all x.

♦ Type 4, a max/min problem

55. Show without using any calculus that of all trapezoids of given width and area, it is the rectangle that has the minimum perimeter.

♦ Type 4, a max/min problem

56. **(a**[A]**)** Does $x^4 - 4x^3 - 2x^2 + 12x$ have a local maximum or a local minimum at $x = -1$? **(b**[A]**)** Does $x^5 - 4x^4 + 3x^2 - x$ have a local maximum at $x = 3$? Does it have a local minimum at $x = 3$? **(c)** Does $x^5 - 6x^2$ have a local maximum or a local minimum at $x = 0$? **(d)** Does $2x^3 - 25x^2 - 100x$ have a local maximum or a local minimum at $x = 10$? (In each case justify your answer.)

♦ Type 2, the Second-Derivative Test for a local maximum or minimum.

Section 3.3

The general Chain Rule

OVERVIEW: *A function of the form $G(y(x))$, whose value at x is calculated by first finding the value of $y(x)$ at x and then finding the value of $G(y)$ at $y = y(x)$, is called a* COMPOSITE FUNCTION. *Such functions arise in applications where various quantities are considered to be functions of other quantities. In this section we discuss the general* CHAIN RULE *for expressing derivatives of composite functions in terms of the derivatives of the functions that form them. We encountered a special case of this rule in Section 2.5, where we discussed the Chain-Rule formula for the x-derivative of the composite function $[y(x)]^n$ that is obtained by taking the nth power of a function $y = y(x)$.*

Topics:

- ***Rates of change of distance and of gasoline consumption***
- ***The general Chain Rule***

Distance, velocity, and gasoline consumption

Imagine that on a particular trip you travel $s(t)$ miles in t hours and your mobile home uses $G(s)$ gallons of gasoline to travel s miles from your starting point. Then the composite function $G = G(s(t))$ gives the amount of gasoline you use as a function of the time. We want to find a formula for the derivative of $G(s(t))$ in terms of the derivatives of $s(t)$ and $G(s)$. We begin with an example involving linear functions.

Question 1 Suppose that you drive your mobile home at the constant velocity of 50 miles per hour, and that on your drive it consumes gasoline at the constant rate of $\frac{1}{10}$ gallon per mile (10 miles per gallon). **(a)** At what rate does your mobile home consume gasoline, measured in gallons per hour? **(b)** Figure 1 shows the graph of the distance you travel as a function of the amount of time you have driven, Figure 2 gives the graph of the amount of gasoline you use as a function of the distance you have driven, and Figure 3 shows the graph of the amount of gasoline used as a function of the time. Give formulas for these three functions. **(c)** How are the slopes of the lines in Figures 1 through 3 related?

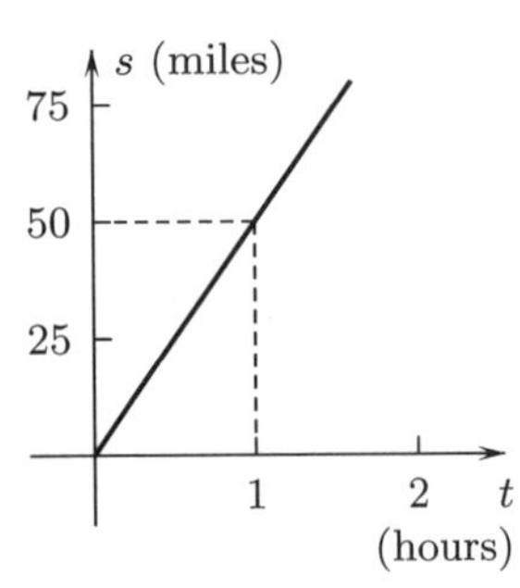

$\text{Slope} = 50 \dfrac{\text{miles}}{\text{hour}}$

FIGURE 1

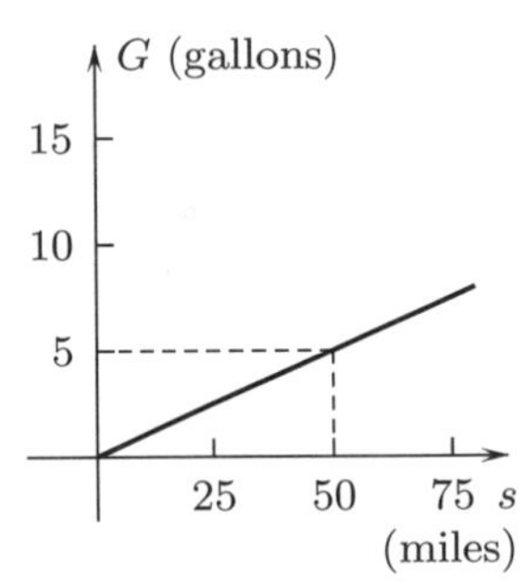

$\text{Slope} = \frac{1}{10} \dfrac{\text{gallon}}{\text{mile}}$

FIGURE 2

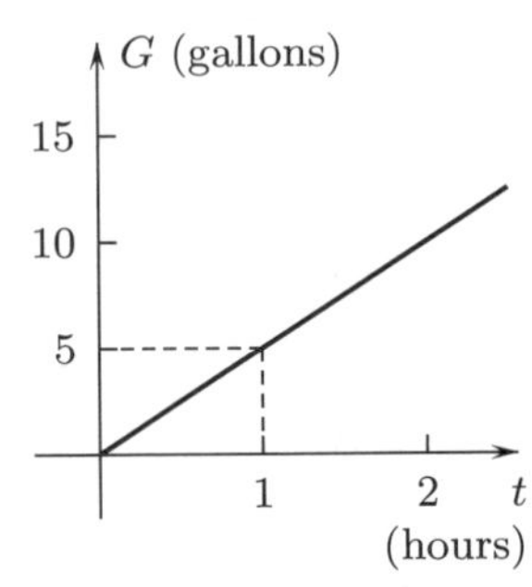

$\text{Slope} = 5 \dfrac{\text{gallons}}{\text{hour}}$

FIGURE 3

Notice that the amount of gasoline $G = 5t$ as a function of time from Question 1 can be obtained from the formulas $s(t) = 50t$ for distance as a function of time and $G(s) = \frac{1}{10}s$ for the amount of gasoline used as a function of distance either with the calculation

♦ SUGGESTIONS TO INSTRUCTORS: Have students begin work on Questions 1 through 5 before class. Problems 2, 5, 7, 8, 10, 21, 22, and 25 are good for in-class discussion.

$$G(s(t)) = \tfrac{1}{10}s(t) = \tfrac{1}{10}(50t) = 5t$$

or with

$$G((s(t)) = G(50t) = \tfrac{1}{10}(50t) = 5t.$$

An example with nonlinear functions

Now suppose that on your trip your distance and gasoline consumption are given instead by the nonlinear functions $s = s(t)$ and $G = G(s)$ of Figures 4 and 5. The resulting composite function $G(s(t))$ of Figure 6 gives the amount of gasoline you use as a function of t. We fix a time t_0 and set $s_0 = s(t_0)$ and $G_0 = G(s_0)$, so that you have driven s_0 miles and used G_0 gallons of gas at time t_0.

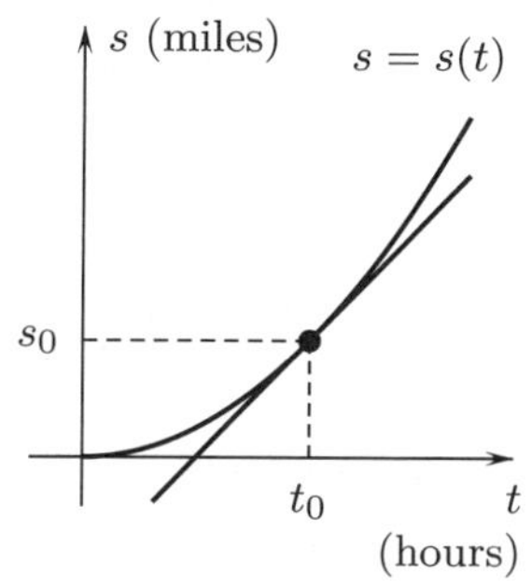

$$\text{Slope} = \left[\frac{ds}{dt}\right]_{t=t_0}$$

FIGURE 4

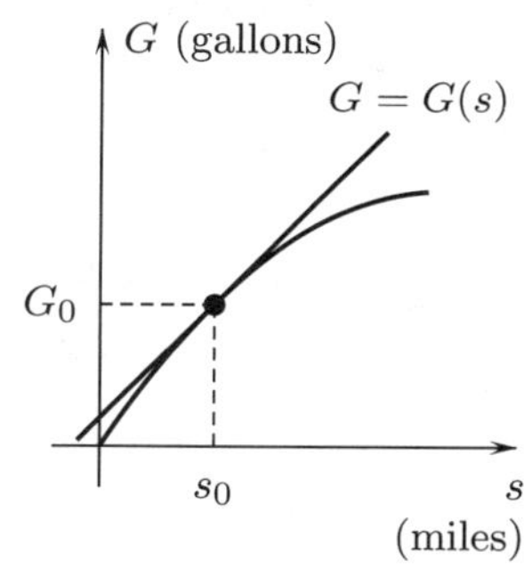

$$\text{Slope} = \left[\frac{dG}{ds}\right]_{s=s_0}$$

FIGURE 5

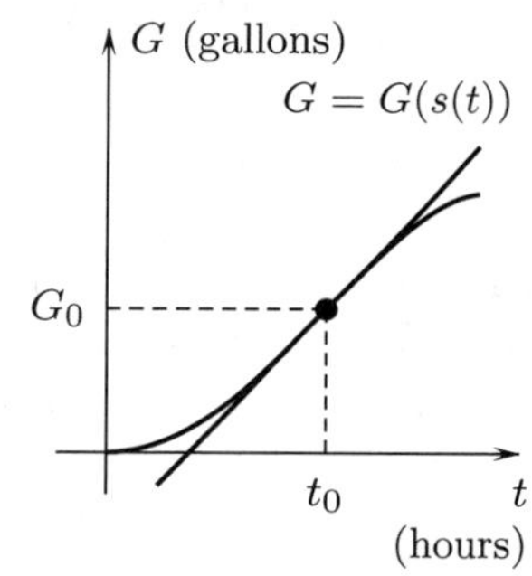

$$\text{Slope} = \left[\frac{dG}{dt}\right]_{t=t_0}$$

FIGURE 6

Figures 4 through 6 also show the tangent lines whose slopes are the velocity ds/dt at time t_0, the rate of gas consumption in gallons per mile dG/ds at distance s_0, and the rate of gas consumption in gallons per hour dG/dt at time t_0. Notice that we use brackets with subscript equations, $t = t_0$ or $s = s_0$, to show where the derivatives are evaluated.

To see how the slopes of these tangent lines are related, we consider a change Δt in the time as in Figure 7. In this time period you would drive Δs miles and use ΔG gallons of gas, with Δs and ΔG as in Figures 7 through 9.

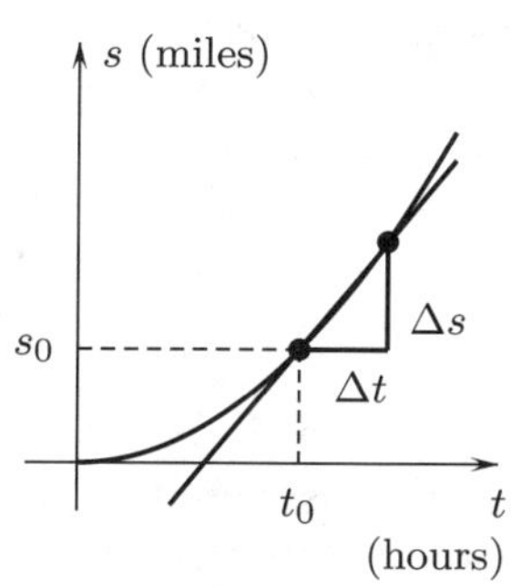

FIGURE 7

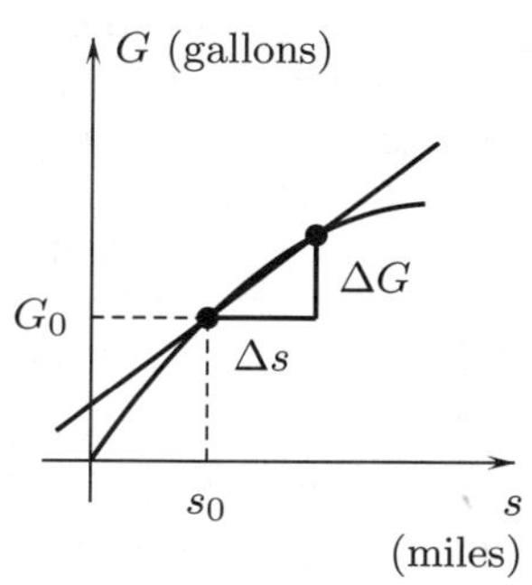

FIGURE 8

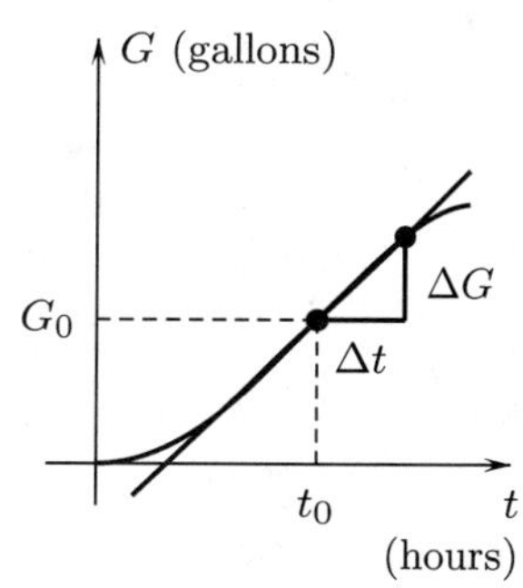

FIGURE 9

Question 2 How are the slopes of the secant lines in Figures 7, 8, and 9 related?

We assume that $s(t)$ has a derivative at t_0 and that $G(s)$ has a derivative at s_0. Then

$$\begin{aligned} \left[\frac{ds}{dt}\right]_{t=t_0} &= \lim_{\Delta t\to 0}\frac{\Delta s}{\Delta t} \\ \left[\frac{dG}{ds}\right]_{s=s_0} &= \lim_{\Delta s\to 0}\frac{\Delta G}{\Delta s} \\ \left[\frac{dG}{dt}\right]_{t=t_0} &= \lim_{\Delta t\to 0}\frac{\Delta G}{\Delta t}. \end{aligned} \tag{1}$$

The change in distance Δs tends to 0 as the change in time Δt tends to 0, since by Theorem 1 of Section 2.4, $s(t)$ is continuous at t_0. To avoid a complication in the general proof, we also assume that Δs is not zero for small nonzero Δt. Then, when we combine the result of Question 2 with the formulas in **(1)**, we obtain

$$\begin{aligned} \left[\frac{dG}{dt}\right]_{t=t_0} &= \lim_{\Delta t\to 0}\frac{\Delta G}{\Delta t} = \lim_{\Delta t\to 0}\left[\frac{\Delta G}{\Delta s}\frac{\Delta s}{\Delta t}\right] \\ &= \left[\lim_{\Delta s\to 0}\frac{\Delta G}{\Delta s}\right]\left[\lim_{\Delta t\to 0}\frac{\Delta s}{\Delta t}\right] = \left[\frac{dG}{ds}\right]_{s=s_0}\left[\frac{ds}{dt}\right]_{t=t_0}. \end{aligned} \tag{2}$$

Thus, the slope of the tangent line to the graph of the composite function in Figure 6 is the product of the slopes of the tangent lines in Figures 4 and 5. To state the result in **(2)** for reference, we replace t by x and, $s(t)$ by $y(x)$:

Theorem 1 (The general Chain Rule) *If $y(x)$ has a derivative with respect to x at x_0 and $G(y)$ has a derivative with respect to y at $y_0 = G(x_0)$, then $G\big(y(x)\big)$ has a derivative with respect to x at x_0, which is given by*

$$\left[\frac{dG}{dx}\right]_{x=x_0} = \left[\frac{dG}{dy}\right]_{y=y_0}\left[\frac{dy}{dx}\right]_{x=x_0}. \tag{3}$$

The form of equation **(3)** is easy to remember if we abbreviate the notation by dropping the references to where the derivatives are calculated. We obtain

$$\frac{dG}{dx} = \frac{dG}{dy}\frac{dy}{dx} \tag{4}$$

where it appears that the symbols dy on the right side would cancel to give the expression on the left. Learn equation **(4)** in words: The derivative of G with respect to x equals the derivative of G with respect to y, multiplied by the derivative of y with respect to x. Or, equivalently, the rate of change of G with respect to x equals the rate of change of G with respect to y, multiplied by the rate of change of y with respect to x.

Example 1 The force F (pounds) exerted by a spring is determined by the distance s (feet) that it is stretched beyond its length when no force is on it. The rate of change of the force with respect to the distance the spring is stretched is 10 pounds per foot when it is stretched 5 feet. What is the rate of change of the force with respect to time t (minutes) at a moment when the spring is stretched 5 feet if it is being lengthened 0.3 feet per minute?

SOLUTION We are given that $\dfrac{dF}{ds} = 10$ pounds per foot and $\dfrac{ds}{dt} = 0.3$ feet per minute at a particular moment. We use equation **(4)** with F instead of G, s instead of y, and t instead of x. At the moment in question, the rate of change of the force with respect to time is

$$\frac{dF}{dt} = \frac{dF}{ds}\frac{ds}{dt} = \left[10\ \frac{\text{pounds}}{\text{foot}}\right]\left[0.3\ \frac{\text{feet}}{\text{minute}}\right] = 3\ \frac{\text{pounds}}{\text{minute}}.\ \square \tag{5}$$

Notice that the units "foot" and "feet" in equation **(5)** appear to cancel to give the units of pounds per minute for the final answer, just the way the symbols "ds" in the numerator and denominator of the second expression in **(5)** appear to cancel.

Version **(4)** of the Chain Rule is easy to remember and is generally fairly easy to apply in narrative problems like Example 1, where the functions represent real quantities, such as time, distance, volume, etc. Sometimes, however, formula **(4)** can be confusing because it does not show the relevant values of the variables. In such cases, it can help to specify the values of the variables completely by writing

$$\frac{d}{dx}\left[G\big(y(x)\big)\right] = \frac{dG}{dy}\big(y(x)\big)\frac{dy}{dx}(x). \tag{6}$$

If we call $G = G(y)$ the outside function and $y = y(x)$ the inside function, we can express equation **(6)** as follows: The x-derivative of the composite function equals the derivative of the outside function with respect to its own variable, evaluated at the inside function, multiplied by the x-derivative of the inside function.

Example 2 What is $\frac{dH}{dx}(5)$ if $H(x) = G\big(y(x)\big)$, $y(5) = 100$, $\frac{dy}{dx}(5) = -20$, and $\frac{dG}{dy}(100) = 7$?

SOLUTION We can use the notation in **(3)** by writing

$$\begin{aligned}\frac{dH}{dx}(5) = \left[\frac{dG}{dx}\right]_{x=5} &= \left[\frac{dG}{dy}\right]_{y=100}\left[\frac{dy}{dx}\right]_{x=5}\\ &= (7)(-20) = -140.\end{aligned}$$

Or, we can use the notation in **(6)** by writing

$$\begin{aligned}\frac{dH}{dx}(5) = \left[\frac{d}{dx}[G\big(y(x)\big)]\right]_{x=5} &= \frac{dG}{dy}\big(y(5)\big)\frac{dy}{dx}(5)\\ &= \frac{dG}{dy}(100)\frac{dy}{dx}(5) = (7)(-20) = -140. \ \square\end{aligned}$$

Question 3 What is the x-derivative of $R(x^3)$ at $x = 10$ if the u-derivative $\frac{dR}{du}(1000)$ of $R(u)$ at $u = 1000$ equals 2?

The Chain Rule in prime notation

The next example uses the Chain Rule with prime notation for the derivatives. Here, as throughout this book, the prime designates the derivative of a function with respect to its own variable.

Example 3 What is $p'(2)$ if $p(x) = q(5x + 3)$ and $q'(13) = -1$?

SOLUTION Because the variable of p is x, $p'(2)$ denotes the value of its x-derivative at 2. In contrast, $q'(13)$ is not an x-derivative. It is the value at 13 of the derivative of q with respect to its own variable, which is not named.

According to the Chain Rule, the x-derivative of the composite function $q(5x + 3)$ equals the derivative of the outside function q with respect to its own variable, evaluated at the value $5x + 3$ of the inside function, multiplied by the x-derivative of the inside function. We obtain

$$\begin{aligned}p'(x) &= \frac{d}{dx}[q(5x+3)] = q'(5x+3)\frac{d}{dx}(5x+3)\\ &= 5q'(5x+3).\end{aligned}$$

Then setting $x = 2$ gives $p'(2) = 5q'\big(5(2)+3\big) = 5q'(13) = 5(-1) = -5.$ $\square$

Question 4 What is $W'(3)$ if $W(x) = [V(x)]^2$, $V(3) = 10$, and $V'(3) = 2$?

Using the Chain Rule with graphs: oxygen deficiency

The crew and passengers in airplanes are supplied pure oxygen to breathe when they fly at high altitudes in unpressurized cabins to maintain enough oxygen in their blood. (A person whose blood-oxygen concentration is 50% of its maximum will lose consciousness in about five minutes and probably go into a coma within half an hour.) The lower graph in Figure 10 gives the percent $P_A(h)$ of a person's maximum oxygen concentration in the passenger's blood when he or she is breathing the ambient air in an unpressurized cabin at an altitude of h thousand feet above sea level. The upper curve gives the percent $P_O(h)$ of the maximum oxygen concentration in the passenger's blood if he or she is breathing pure oxygen at altitude h.(1)

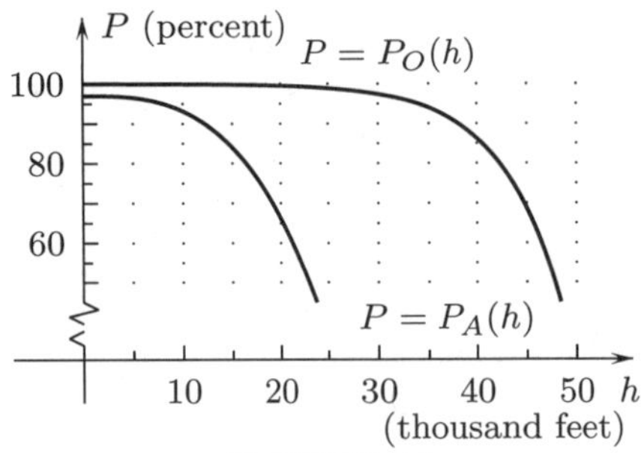

FIGURE 10

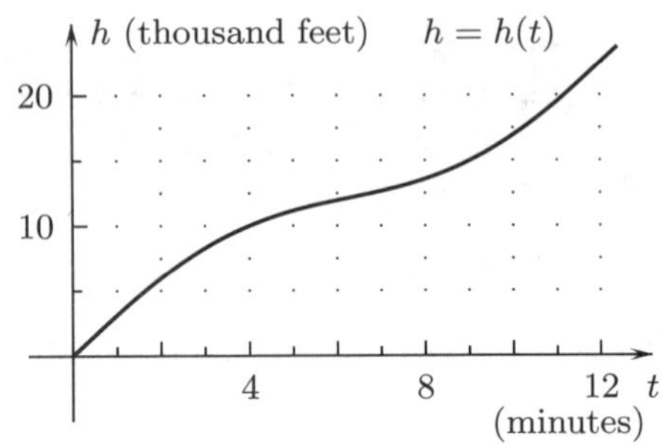

FIGURE 11

Question 5 **(a)** At what altitude is the blood-oxygen concentration of a person breathing pure oxygen in an unpressurized cabin the same as that of a person breathing the ambient air in an unpressurized cabin at 10,000 feet?
(b) Suppose that the graph in Figure 11 gives the height $h(t)$ of an unpressurized plane on a particular trip as a function of time. Approximately how high is the plane at $t = 4$?
(c) What is the percentage P_A at $t = 4$ of the maximum blood-oxygen concentration in the blood of a passenger who is breathing the ambient air?
(d) What is the percentage P_O at $t = 4$ of the maximum blood-oxygen concentration in the blood of a passenger who is breathing pure oxygen?
(e) What are the approximate rates of change of h with respect to t and of P_A with respect to h at $t = 4$? Give the units.

If we make the assumption in our mathematical model of the passenger in Question 5 that the passenger's blood-oxygen concentration changes immediately as the plane rises, then we can use the data in Figures 10 and 11 to estimate how rapidly his or her blood-oxygen concentration falls.

Example 4 What is the approximate rate of change with respect to time of a passenger's blood-oxygen concentration at $t = 4$ in Question 5?

SOLUTION We use the values $h(4) \approx 10$, $\left[\frac{dh}{dt}\right]_{t=4} \approx 1.5 \frac{\text{thousand feet}}{\text{minute}}$, and $\left[\frac{dP_A}{dh}\right]_{h=10} \approx -1.1 \frac{\text{percent}}{\text{thousand feet}}$ from the response to Question 5. The Chain Rule gives

$$\begin{aligned}\left[\frac{dP_A}{dt}\right]_{t=4} &\approx \left[\frac{dP_A}{dh}\right]_{h=10}\left[\frac{dh}{dt}\right]_{t=4} \\ &\approx \left[-1.1\,\frac{\text{percent}}{\text{thousand feet}}\right]\left[1.5\,\frac{\text{thousand feet}}{\text{minute}}\right] \\ &= -1.65\,\frac{\text{percent}}{\text{minute}}.\end{aligned}$$

The blood-oxygen concentration is falling about 1.65 percent per minute. □

(1)Data adapted from *Textbook of Medical Physiology* by G. Guyton, Philadelphia, PA: W. B. Saunders Company, 1981, p. 544.

The Chain Rules for powers revisited

If we use $G(y) = y^n$ with y a function of x and n a constant in the general Chain Rule **(3)**, then $\dfrac{dG}{dy} = ny^{n-1}$, and we obtain

$$\frac{d}{dx}(y^n) = \frac{d}{dy}(y^n)\,\frac{dy}{dx} = ny^{n-1}\,\frac{dy}{dx}$$

which is the Chain Rule for derivatives of powers of functions from Section 2.5.

Principles and procedures

- Be sure you understand the general Chain Rule in the streamlined form $\dfrac{dG}{dx} = \dfrac{dG}{dy}\,\dfrac{dy}{dx}$ and its interpretation in terms of rates of change that is given after **(4)** above. The more complicated looking formulation $\dfrac{d}{dx}\left[G\big(y(x)\big)\right] = \dfrac{dG}{dy}\big(y(x)\big)\,\dfrac{dy}{dx}(x)$ in **(6)** is the same equation with the relevant values of the variables explicitly stated.
- Each of these formulations of the Chain Rule states that the rate of change of $G\big(y(x)\big)$ with respect to x equals the rate of change of $G(y)$ with respect to y at $y(x)$, multiplied by the rate of change of $y(x)$ with respect to x at x. The geometric interpretation is that the slope of the tangent line to the graph of $G\big(y(x)\big)$ at x (Figure 14) is the product of the slope of the tangent line to the graph of $y(x)$ at x (Figure 12) and of the slope of the tangent line to the graph of $G(y)$ at $y = y(x)$ (Figure 13).

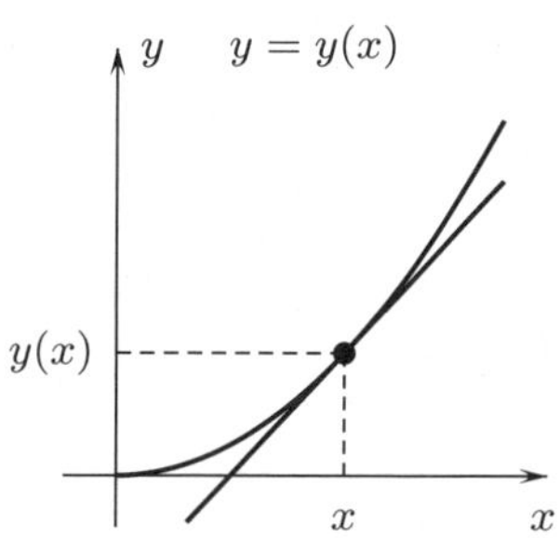

Slope $= \dfrac{dy}{dx}(x)$

FIGURE 12

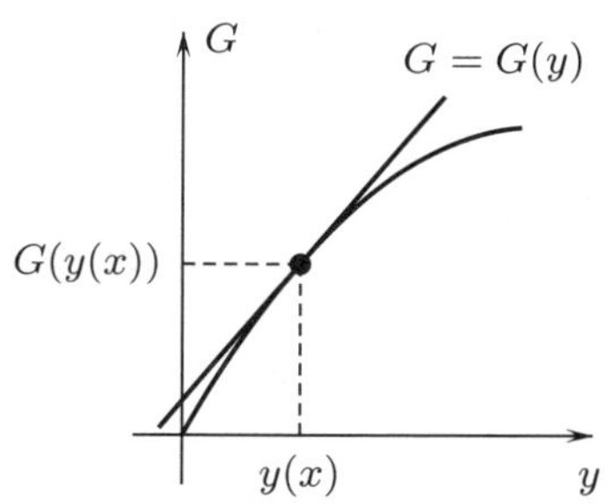

Slope $= \dfrac{dG}{dy}\big(y(x)\big)$

FIGURE 13

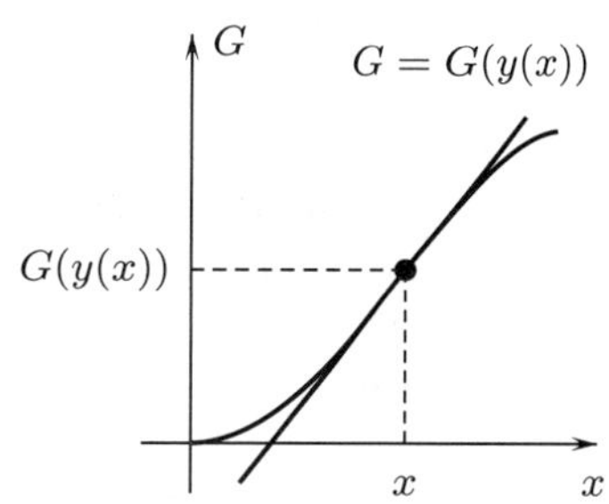

Slope $= \dfrac{d}{dx}[G\big(y(x)\big)]$

FIGURE 14

Tune-Up Exercises 3.3♦

[A]*Answer provided.* [O]*Outline of solution provided.*

T1.[A] What is $A(B(100))$ if $B(100) = 1000$ and $A(1000) = 14$?

♦ Type 1, value of a composite function

T2.[A] Give a formula in terms of x for $U(V(x))$, where $U(V) = V^3/(V - 25)$ and $V(x) = x^2$. What is the domain of $U(V(x))$?

♦ Type 1, value of a composite function

T3.[A] **(a)** Draw the graph of $q(p) = 3 + 2p$ in a pq-plane, the graph of $r(q) = 4 - q$ in a qr-plane, and the graph of $r(q(p))$ in a pr-plane. **(b)** How are the slopes of the three lines related?

♦ Type 2, graphs of functions and their composition

♦The Tune-up Exercises and Problems are classified by type and content. The types are (1) basic, reactive; (2) basic reflective; (3) intermediate, reactive; (4) intermediate, reflective; (5) advanced, reactive; (6) advanced, reflective; and (7) advanced, theoretical.

T4.[A] **(a)** Draw the graph of $y(x) = x^2 + 1$ and its tangent line at $x = 1$ in an xy-plane. **(b)** Draw the graph of $z(y) = 2/y$ and its tangent line at $y = 2$ in a yz-plane. **(c)** Draw the graph of $z(y(x))$ and its tangent line at $x = 1$ in an xz-plane. **(d)** How are the slopes of the tangent lines in parts (a), (b), and (c) related?

♦ Type 2, tangent lines to graphs of functions and their composition

T5.[O] What are the value and the x-derivative of $v(w(x))$ at $x = 0$ if $w(0) = 1$, $\frac{dw}{dx}(0) = 5$, $v(1) = 7$, and $\frac{dv}{dw}(1) = -3$?

♦ Type 3, the general Chain Rule

T6.[O] What are $Q(4)$ and $Q'(4)$ if $Q(x) = R(x^2)$, $R(16) = 10$ and $R'(16) = 6$?

♦ Type 3, the general Chain Rule

T7.[O] Figures 15 and 16 show the graphs of differentiable functions $x(t)$ and $y(x)$. Find the approximate value and the approximate t-derivative of $y(x(t))$ at $t = 3$.

♦ Type 3, the general Chain Rule with graphs

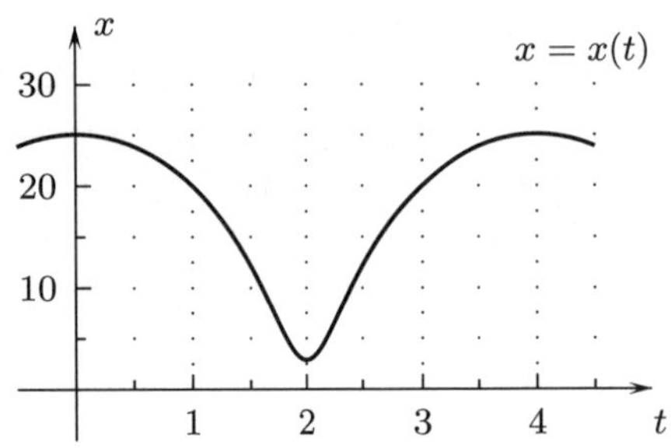

FIGURE 15

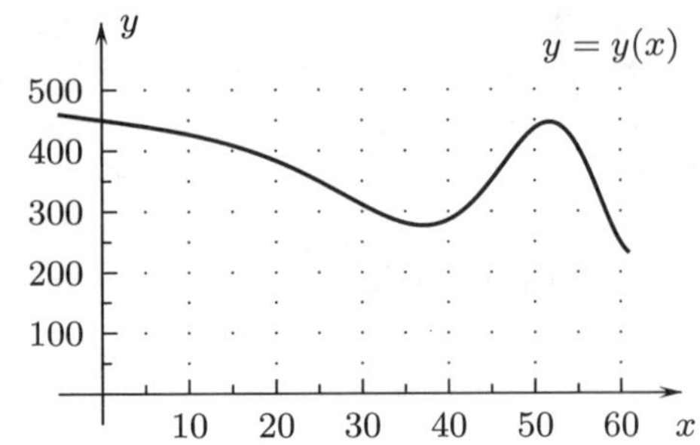

FIGURE 16

Problems 3.3

[A]*Answer provided.* [O]*Outline of solution provided.*

1.[A] After you have driven 100 kilometers, you have 75 liters of gasoline in your car, your car is consuming gasoline at the rate of 0.2 liters per kilometer, and you are traveling 80 kilometers per hour. At what rate, measured in liters per hour, are you using gasoline at that moment?

♦ Type 2, the general Chain Rule, etc.

2.[A] At 12:00 PM a balloon is 200 meters above the ground, it has a volume of 5 liters, and it is rising 3 meters per second. Its volume increases 0.001 liter for every meter it rises. **(a)** At what rate is its volume increasing with respect to time at 12:00 PM? **(b)** Approximately how high is the balloon and what is its approximate volume 15 seconds after 12:00 PM?

♦ Type 2, the general Chain Rule, etc.

3.[A] A store sells $x(t)$ pounds of potatoes in the first t hours of a day ($0 \le t \le 12$), and it makes $P(x)$ dollars profit on x pounds of potatoes. Express the store's profit on potatoes and the rate of change of the profit with respect to time at time t in terms of $x(t)$, $P(x)$, and their derivatives.

♦ Type 2, the general Chain Rule, etc.

4. Water is flowing into a pond at the rate of 300 gallons per minute. What is the rate of change of the depth with respect to time at a time when the rate of change of the depth with respect to the volume is 0.001 feet per gallon?

♦ Type 2, the general Chain Rule, etc.

5. At a temperature of 30°C, the density of mercury is 13.5217 grams per milliliter and the rate of change of the density with respect to temperature is −0.0024 grams per milliliter per degree.[(2)] **(a)** What is the approximate density of mercury at 30.5°C? **(b)** At what rate, with respect to time, does the density of mercury increase or decrease when it is 30°C and the temperature is falling 0.5 degrees per minute?

♦ Type 2, the general Chain Rule, etc.

[(2)]Data adapted from *CRC Handbook of Chemistry and Physics*, Boca Raton, Florida: CRC Press, Inc., 1981, p. F-7.

6. The slope of the graph of the linear function $W = W(x)$ is 5 and the slope of the graph of the linear function $V = V(W)$ is -3. Is the composite function $V = V(W(x))$ a linear function of x? If so, what is the slope of its graph?

♦ Type 4, composing linear functions

7. Figure 17 shows the graph of a mathematical model of the net sales $M(t)$ of Better Mouse Traps in the time period from time 0 up to time t. $M(t)$ is a piecewise linear function that is increasing for $0 \le t \le 2$, when the traps are first sold. It is constant in the time period $2 \le t \le 3$, when one trap is returned for each one that is sold, and it is decreasing for $3 < t < 4$ when returns exceed sales. Figure 18 shows the graph of the mouse-trap designer's income $I(M)$ as a function of the number of traps that are sold. It is piecewise linear because she earns one amount per mouse trap for the first 1000 that are sold and another amount for any more that are sold. **(a)** Draw the graph of $I(M(t))$ in a tI-plane. **(b)** Give formulas for dM/dt and dI/dt as functions of t and for dI/dM as a function of M, and show that the Chain Rule $\dfrac{dI}{dt} = \dfrac{dI}{dM}\dfrac{dM}{dt}$ is satisfied in each of the four open intervals where dI/dt is constant.

♦ Type 4, composing linear functions

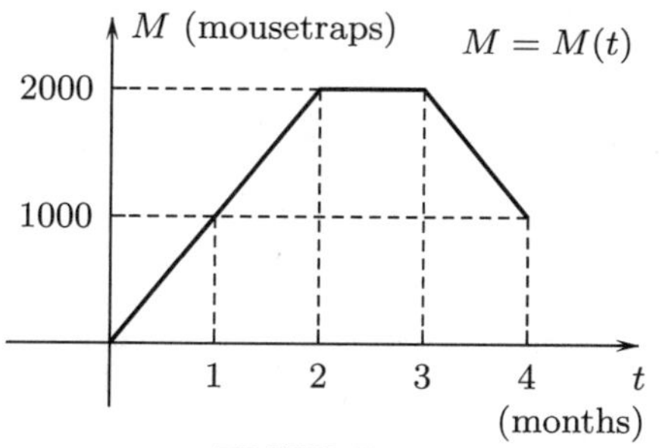

FIGURE 17

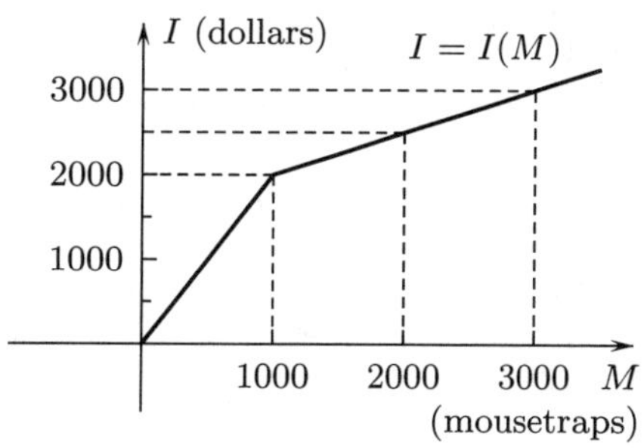

FIGURE 18

8.[A] **(a)** Figures 19 and 20 show the graphs of $y = x^{1/2}$ and $z = \frac{1}{4}y^3$ and their tangent lines at $x = 4$ and $y = y(4) = 2$, respectively. What are the slopes of these tangent lines? **(b)** Figure 21 shows the graph of the composite function $z = z(y(x))$. Give a formula for z in terms of x and use it to find the slope of the tangent line at $x = 4$ in Figure 21. **(c)** Show by direct calculations that the slopes of the three tangent lines in Figures 19, 20, and 21 are related by the Chain Rule, $\dfrac{dz}{dx} = \dfrac{dz}{dy}\dfrac{dy}{dx}$.

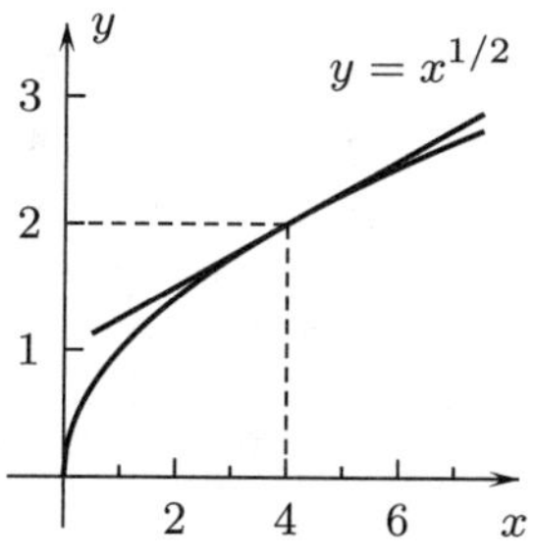

FIGURE 19

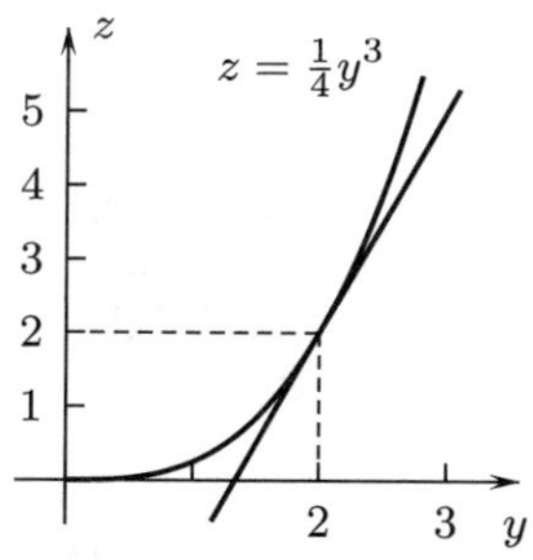

FIGURE 20

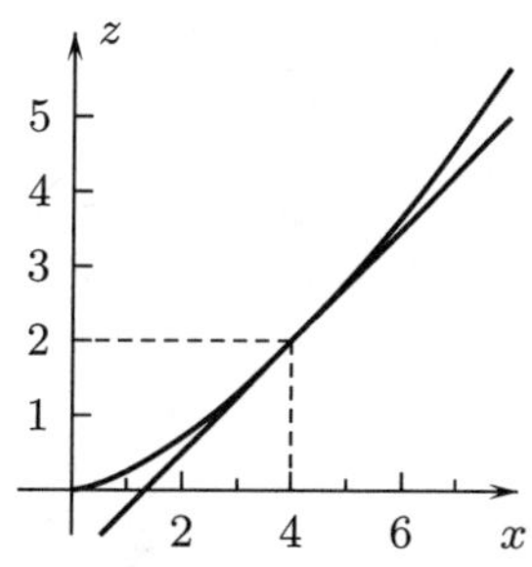

FIGURE 21

(d) Figure 22 shows the graph of $w = 4/z$ and its tangent line at $z = 2$. What is the slope of the tangent line? **(e)** Figure 23 shows the graph of the composite function $w = w(z(y(x)))$, with $z(y)$ and $y(x)$ from Figures 19 and 20. Give a formula for w in terms of x and use it to find the slope of the tangent line in Figure 23. **(f)** Show by direct calculations that the slopes of the tangent lines in Figures 19, 20, 22 and 23 are related by the following equation. (Notice that the right side of this equation ressembles a chain with the two symbols "dz" forming one link and the two symbols "dy" forming a second link. It is this sort of expression that led to the term "Chain Rule.")

$$\frac{dw}{dx} = \frac{dw}{dz}\frac{dz}{dy}\frac{dy}{dx}.$$

♦ Type 4, doubly composed functions

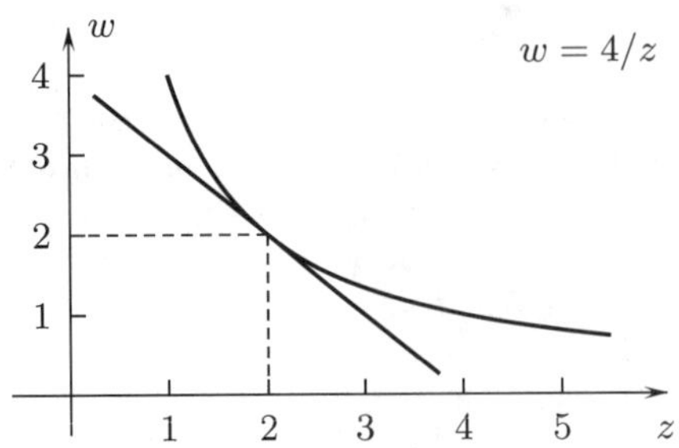

FIGURE 22

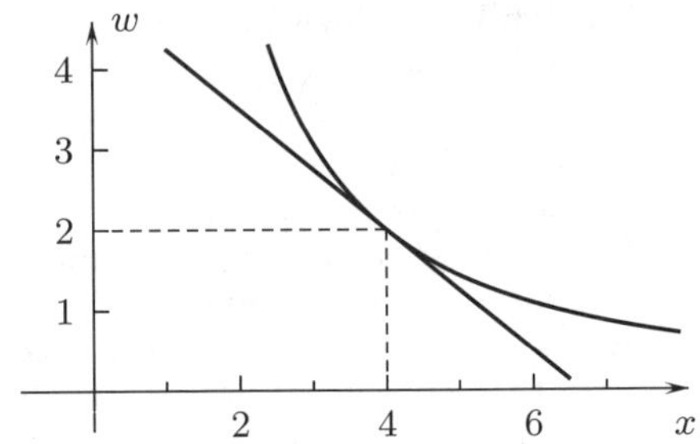

FIGURE 23

9. The pressure P in a balloon is a function of its volume V, the volume is a function of the temperature T, and the temperature is a function of the time. At a particular moment the rate of change of the pressure with respect to the volume is -0.01 atmospheres per liter, the rate of change of the volume with respect to the temperature is 0.2 liters per degree Celsius, and the rate of change of the temperature with respect to the time is 3 degrees Celsius per hour. What is the rate of change of the pressure with respect to time at that moment?

♦ Type 4, doubly composed functions

10.[A] A ball rolls from left to right down the ramp with the shape of the curve $y = y(x)$ in Figure 24. **(a)** When the ball is at $x = 6$ its x-coordinate is increasing 7 feet per second. Approximately how fast is its y-coordinate increasing at that moment? **(b)** When the ball is at $x = 10$ its x-coordinate is increasing 9 feet per second. Approximately how fast is its y-coordinate decreasing at that time?

♦ Type 2, the general Chain Rule with graphs

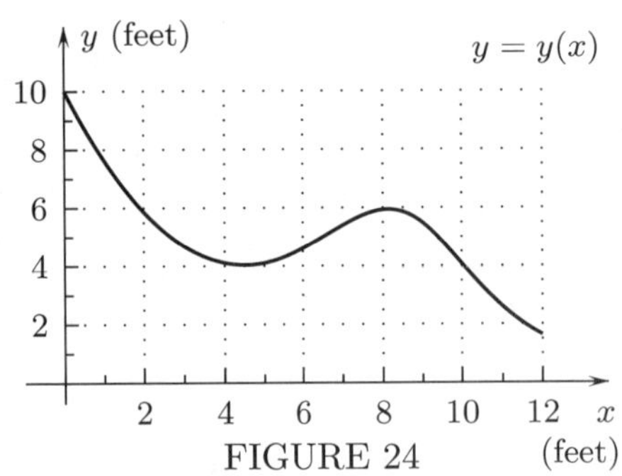

FIGURE 24

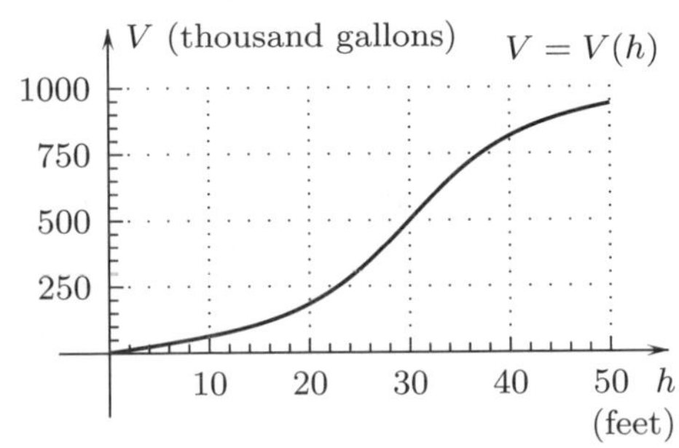

FIGURE 25

11. Figure 25 shows the volume $V(h)$ (thousand gallons) of the water in a lake as a function of the depth h (feet) of the water. Approximately how fast is the volume increasing, measured in gallons per hour, when there are 500,000 gallons of water in the lake if the level of the water is rising at the rate of 3 feet per day?

♦ Type 2, the general Chain Rule with graphs

In Exercises 12 through 19 find the values of the indicated derivatives.

12.[O] $\frac{dA}{dx}(6)$ where $A(x) = B(y(x)), y(6) = 7, \frac{dy}{dx}(6) = 8$, and $\frac{dB}{dy}(7) = 9$

♦ Type 3, the general Chain Rule

13.[A] $W'(100)$ where $W(y) = U(V(y)), V(100) = -3, \frac{dV}{dy}(100) = 7$, and $\frac{dU}{dV}(-3) = 10$

♦ Type 3, the general Chain Rule

14.[A] $T'(10)$ where $T(t) = R((S(t)), S(10) = 4, S'(10) = 5$, and $R'(4) = -6$

♦ Type 3, the general Chain Rule

15.[O] $\frac{dH}{dx}(2)$ where $H(x) = F(x^3)$ and $\frac{dF}{dy}(8) = 10$

♦ Type 3, the general Chain Rule

16.[A] $\dfrac{dG}{dx}(5)$ where $G(x) = R(2x + x^2)$ and $\dfrac{dR}{dy}(35) = 6$

♦ Type 3, the general Chain Rule

17. $Z'(-5)$ where $Z(t) = X(Y(t)), Y(-5) = 6, Y'(-5) = 8$, and $X'(6) = 9$

♦ Type 3, the general Chain Rule

18. $p'(13)$ where $p(t) = q(r(t)), r(13) = 5, r'(13) = -7$, and $q'(5) = 10$

♦ Type 3, the general Chain Rule

19. $\dfrac{dA}{dz}(0)$ where $A(z) = B(C(z)), C(0) = 3, C'(0) = 4$, and $B'(3) = 5$

♦ Type 3, the general Chain Rule

20. It costs a waste disposal plant $C(x)$ dollars to process x tons of garbage in a day, and $x(t)$ tons are processed in the first t hours. **(a)** What does $C(x(t))$ represent? **(b)** What do the derivatives $\dfrac{dx}{dt}(t)$, $\dfrac{dC}{dx}(x(t)g)$, and $\dfrac{d}{dt}[C(x(t))]$ represent, in what units are they measured, and how are they related?

♦ Type 3, the general Chain Rule

21. Increasing the rate of flow of blood from the heart increases the blood pressure in the arteries. Figure 26 shows the graph of arterial blood pressure P, measured in millimeters of mercury, as a function of the rate of blood flow from the heart v, measured in liters per minute.[(3)] Approximately how fast is the blood pressure increasing or decreasing at a moment when the rate of blood flow is 12 liters per minute and is decreasing 0.1 liters per minute per minute?

♦ Type 4, the general Chain Rule with graphs

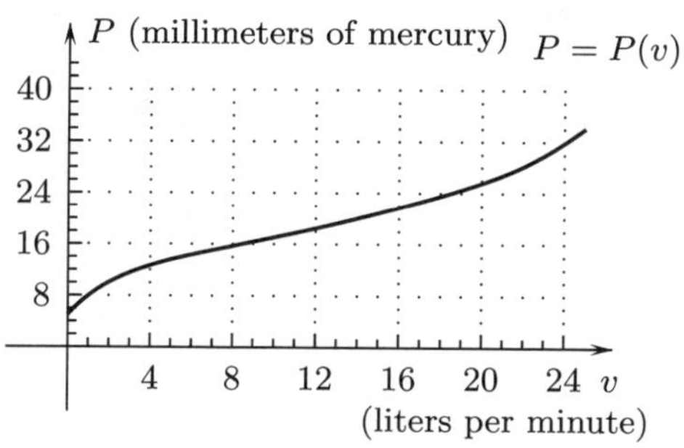

FIGURE 26

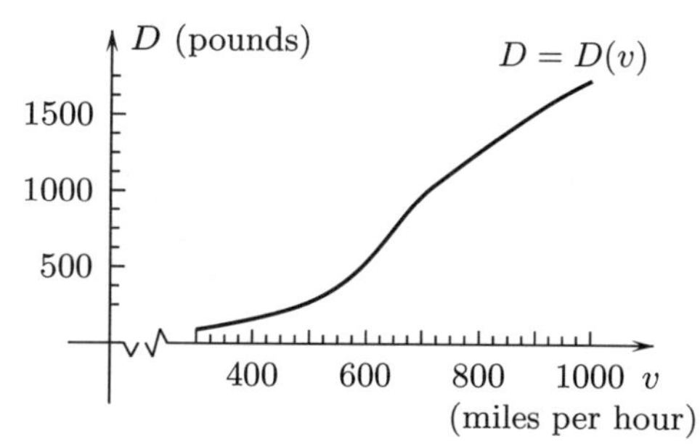

FIGURE 27

22. The force of air resistance on an airplane is called DRAG. Figure 27 gives the drag $D(v)$, measured in pounds, on a particular airplane as a function of its velocity v, measured in miles per hour.[(4)] **(a)** At what approximate velocity is the rate of change of the drag with respect to velocity the greatest and how close is this velocity to the speed of sound, which is 760 miles per hour? **(b)** Suppose the airplane with drag given in Figure 27 is traveling the velocity from part (a) and is accelerating 300 miles per hour^2. What is the approximate rate of increase with respect to time of the drag on the airplane at that moment?

♦ Type 4, the general Chain Rule with graphs

23. Married couples filing joint federal income tax returns for 1996 paid 36% on taxable income between \$21,919 and \$37.667. Suppose that on November 1 a couple's taxable income for 1996 was \$30,000 and was increasing \$500 dollars per day. How fast was their federal tax increasing at that moment?

♦ Type 2, the general Chain Rule

24.[A] What is the approximate value of $\dfrac{dM}{dx}(200)$ if $M(x) = K(y(x))$ and the differentiable functions $y(x)$ and $K(y)$ have the graphs in Figures 28 and 29?

♦ Type 4, the general Chain Rule with graphs

[(3)]Data adapted from *Textbook of Medical Physiology* by A. Guyton, Philadelphia, PA: W. B. Saunders Company, 1981, p. 293.

[(4)]Data adapted from *Fluid-Dynamic Drag*by S. Hoerner, published by the author, 1958, p. 12-7.

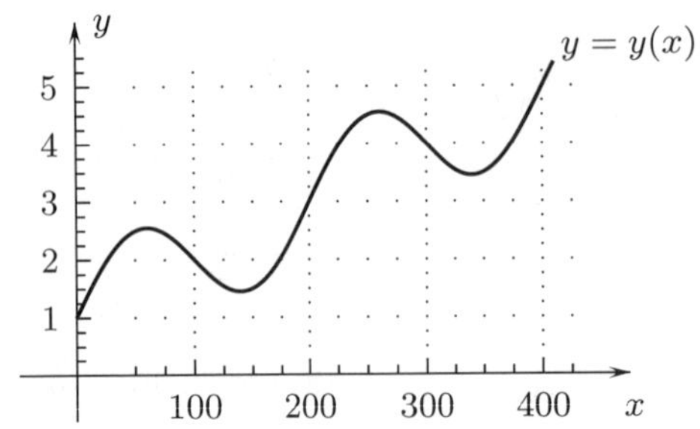

FIGURE 28

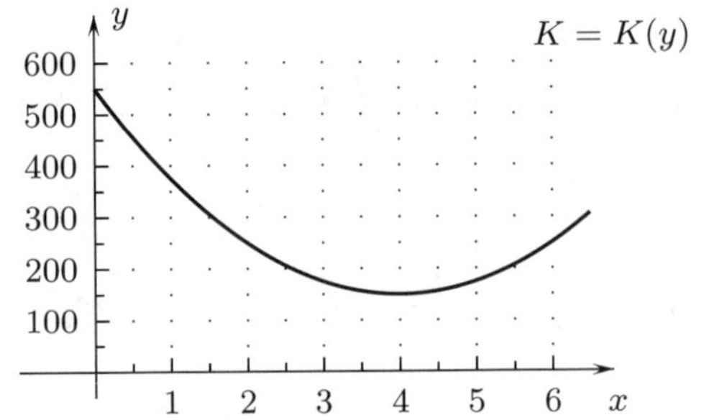

FIGURE 29

25. The curve in Figure 30 gives the power $p(r)$ of the of the engine in an automobile as a function of its speed $r(t)$. The curve in Figure 31 shows the graph of the speed of the engine as a function of the time on one particular trip. **(a)** What are the approximate maximum and minimum power of the engine for $0 \le t \le 50$? **(b)** What is the approximate rate of change of the power with respect to t at $t = 20$? **(c)** What is the approximate rate of change of the power with respect to t at $t = 40$?

◆ Type 4, the general Chain Rule with graphs

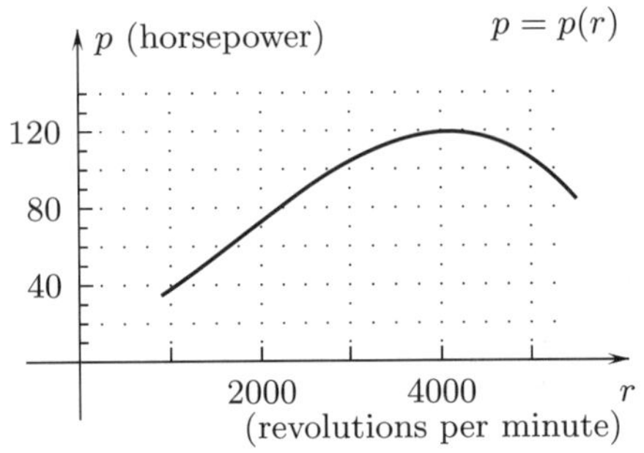

FIGURE 30

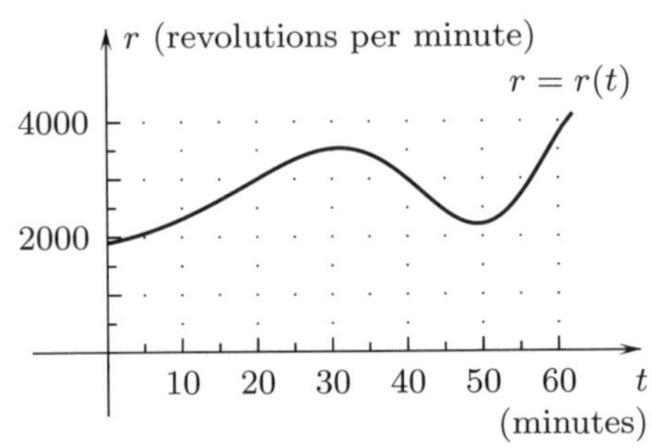

FIGURE 31

Find the derivatives in Problems 26 through 29.

26.[A] $Z'(5)$ where $Z(x) = [F(x^3)]^2$, $F(125) = 10$, $F'(5) = 6$, and $F'(125) = 8$.

◆ Type 4, the general Chain Rule

27.[A] The derivative of $xK(x^2)$ at $x = 6$ where $K(6) = 10$, $K(36) = -5$, $K'(12) = -4$, and $K'(36) = 2$.

◆ Type 4, the general Chain Rule

28. $\frac{dH}{dx}(1)$ where $H(x) = G(x^2)G(x^3)$, $G(1) = 4$, $G(2) = 10$, $G(3) = 1$, $G'(1) = -5$, $G'(2) = 13$, and $G'(3) = 9$.

◆ Type 4, the general Chain Rule

29. $z'(-1)$ where $z(t) = y(t^2 - t^{-2})$ and $y'(0) = 14$.

◆ Type 4, the general Chain Rule

30. Figure 32 shows the graph of a differentiable function $R(x)$. Give approximate values of **(a)** $R(x^2)$, **(b)** $[R(x)]^2$, **(c)** $\frac{d}{dx}[R(x^2)]$, and **(d)** $\frac{d}{dx}\{[R(x)]^2\}$, at $x = 2$

◆ Type 4, the general Chain Rule with graphs

31. Give approximate values of $f(1)$ and of $f'(1)$ where $f(x) = Q(Q(x))$, and the graph of $Q(x)$ is in Figure 33.

◆ Type 4, the general Chain Rule with graphs

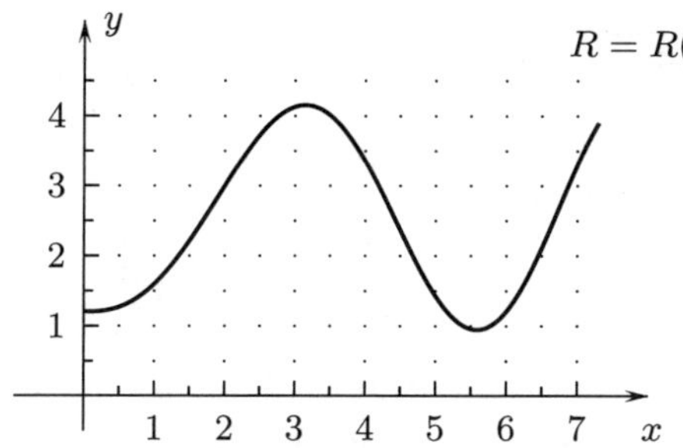

FIGURE 32

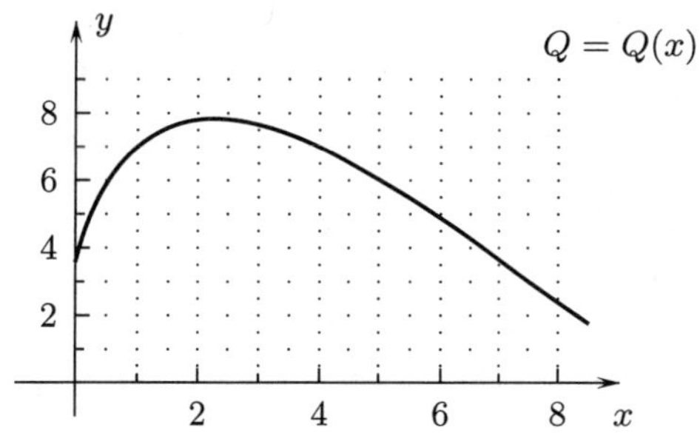

FIGURE 33

32. Give approximate values of $P(2)$ and of $P'(2)$ where $P(x) = G(2x + G(x))$, and the graph of $G(x)$ is in Figure 34.

♦ Type 4, the general Chain Rule with graphs

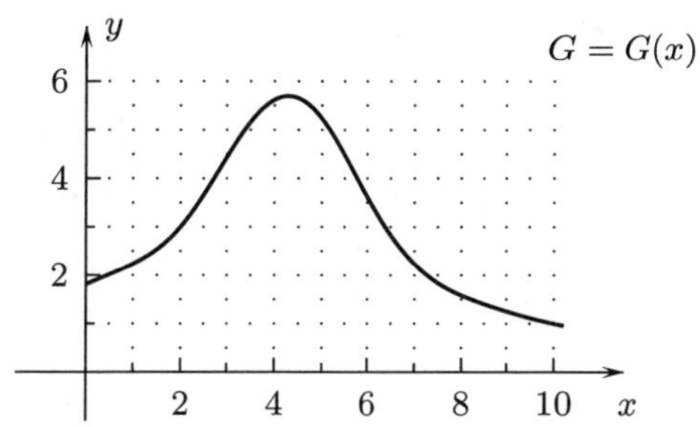

FIGURE 34

33. What are the approximate t-derivatives **(a[A])** of $x(y(t))$ and **(b)** of $y(x(t))$ at $t = 2$ if the differentiable functions $x(t)$ and $y(t)$ have the graphs in Figures 35 and 36?

♦ Type 4, the general Chain Rule with graphs

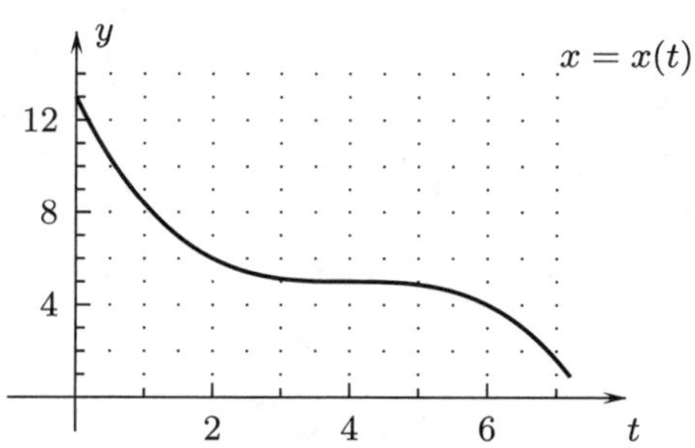

FIGURE 35

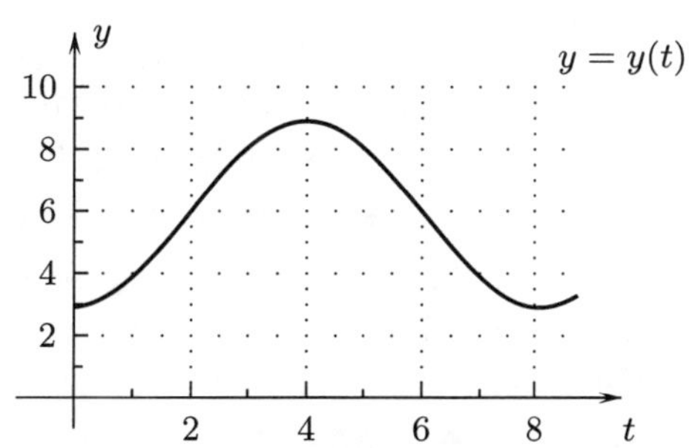

FIGURE 36

34. A function $f(x)$ has a derivative for all x. What special property must $f(x)$ have in order that $f(|x|)$ also have a derivative for all x?

♦ Type 4, the general Chain Rule with the absolute-value function

Section 3.4

Exponential functions and their derivatives

OVERVIEW: *Mathematical models for musical scales, population growth, radioactive decay, compound interest, and a wide variety of other phenomena involve exponential functions of the form Cb^x with positive constants C and b. In this section we first discuss some of these applications and then derive formulas for the derivatives of the exponential functions. Next, we use derivatives to analyze the graphs of formulas constructed from exponential functions. We close the section with a brief discussion of the differential equation $dy/dt = ry$ with constant r, which gives the framework for many of the problems that are modeled with exponential functions. This topic is covered more thoroughly in Section 7.1.*†

Topics:

- ***Growth factors***
- ***Musical pitch and frequency: octaves and the equal-tempered scale***
- ***Exponential growth***
- ***Radioactive decay and half-life***
- ***Interest compounded continuously***
- ***The x-derivatives of*** $\mathbf{e^x, e^y, b^x}$, ***and*** $\mathbf{b^y}$
- ***Graphs of functions constructed from exponential functions***
- ***Relative and percent rates of change***
- ***The differential equation*** $\mathbf{dy/dt = ry}$ ***with constant*** $\mathbf{r}$

Doubling power

Imagine that when you were born, your rich uncle Joe made a big show of setting up a trust fund for your college education. He then infuriated your parents—who never liked him anyway—by putting only twenty-five cents in the trust. What they failed to notice was a provision that he would double the money in the trust every time you had a birthday until you were eighteen, when you would receive the money. Let us see whether your uncle was stingy or generous.

Question 1 Your trust starts with \$0.25, which on your first birthday is doubled to $\$0.25 \times 2 = \$0.50$ and is doubled again to $\$0.25 \times 2^2 = \$1.00$ on your second birthday. **(a)** How much is in the trust on your third and fourth birthdays? **(b)** Express the value of the trust on your tth birthday as a function of t. Use a power of 2.

You would presumably be pleased to receive $\$0.25(2^{18}) = \$65{,}536$ on your eighteenth birthday.

The function $0.25(2^t)$ from Question 1 is a type of EXPONENTIAL FUNCTION, because it is of the form Cb^t with positive constants C and b and the variable t in the exponent. In this section we will study properties and some applications of such functions.

The GROWTH FACTOR from one positive number to another is the new value divided by the old. Because the value $0.25(2^t)$ of your trust fund doubles every year, it has a growth factor of 2 every year. We will now look at another exponential function with a growth factor of 2.

---

†Consult Appendix 2 as necessary to review the definitions and basic properties of exponential functions and logarithms.

♦ SUGGESTIONS TO INSTRUCTORS: The main topics in this section can be divided into three parts, as follows:

Part 1 (Growth factors, exponential growth and decay, and compound interest; Tune-up Exercises T1–T9 and T11–T16; Problems 22, 24–40, 67) Have students begin work on Questions 1 through 9 before class. Problems 27, 28, 30, 31, and 37 are good for in-class discussion.

Part 2 (Derivatives of exponential functions and applications; Tune-up Exercises T10 and T17–T20; Problems 1–19, 21, 23, 41–50, 54–66, 68) Have students begin work on Questions 20 through 25 before class. Problems 17, 23, 41, 43, 44, and 49 are good for in-class discussion.

Part 3 (Relative and percent rates of change and the differential equation $\dfrac{dy}{dt} = ry$ with constant r; Tune-up Exercise T21; Problems 20, 51–53) Have students begin work on Questions 26 through 28 before class.

Musical pitch and frequency: octaves

Sound travels as oscillations in air pressure which cause your eardrum and the bones of your inner ear to vibrate and give you the sensation of the sound. A pure note of frequency ν causes the air pressure to oscillate ν times each second. (ν is the Greek letter nu, which is often used to represent frequencies.) Table 1 lists the frequencies of the note A in different octaves as tuned by most symphony orchestras. The A below middle C is denoted $A^{(0)}$ and is tuned to 220 cycles per second; the A one octave lower is denoted $A^{(-1)}$ and is tuned to 110 cycles per second; the A above middle C is denoted $A^{(1)}$ and is tuned to 440 cycles per second; and so forth. If we represent the octaves by the superscript n in $A^{(n)}$, then the frequency of $A^{(n)}$ is a function $\nu(n)$ of n whose values are in the third row of the table.

TABLE 1. FREQUENCY OF THE NOTE A IN DIFFERENT OCTAVES

Note: $A^{(n)}$	$A^{(-2)}$	$A^{(-1)}$	$A^{(0)}$	$A^{(1)}$	$A^{(2)}$	$A^{(3)}$
Octave: n	-2	-1	0	1	2	3
Frequency: $\nu(n)$	55	110	220	440	880	1760

You can see from the table that raising the pitch by one octave doubles the frequency: the growth factor of the frequency $\nu(n)$ from one n to the next is 2. The growth factor from $\nu(-2) = 55$ to $\nu(-1) = 110$ is $\frac{110}{55} = 2$, the growth factor from $\nu(-1) = 110$ to $\nu(0) = 220$ is $\frac{220}{110} = 2$, and so forth.

Because the function $\nu(n)$, like the value of your Uncle Joe's trust discussed above, grows by the factor of 2 when the variable is increased by 1, it can be represented as the exponential function $\nu(n) = C(2^n)$ for some constant C. We can determine the constant from the value of $\nu(n)$ for any n in Table 1. It is easiest to use the value $\nu(0) = 220$ at $n = 0$ since $2^0 = 1$. We obtain $220 = \nu(0) = C(2^0) = C$, so that for integers n,

$$\nu(n) = 220(2^n). \tag{1}$$

Question 2 $A^{(4)}$ is the highest A and $A^{(-3)}$ is the lowest A on a standard piano. What are the frequencies of these notes?

Question 3 The highest frequency that can be detected by most young people is approximately 24,000 cycles per second. How many octaves above middle C is the highest A that most young people can hear?

The equal-tempered scale

What about the other notes? Counting natural notes (the white keys on a piano) and flats and sharps (the black keys), there are twelve notes to an octave (Figure 1).

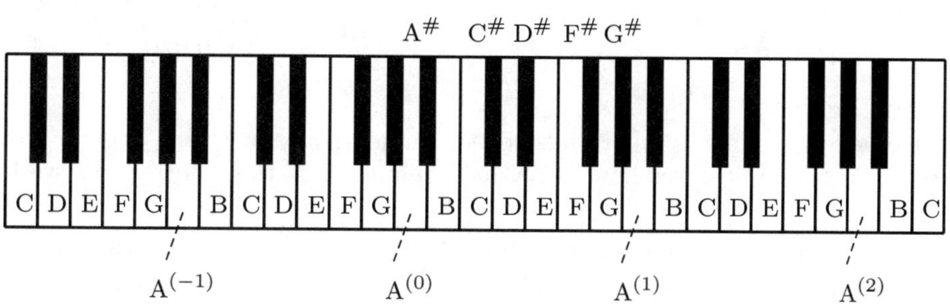

FIGURE 1

The frequencies of the notes between $A^{(0)}$ and $A^{(1)}$ in the EQUAL-TEMPERED SCALE that is generally used on modern instruments are given in Table 2. We label the notes with the variable $m = 0, 1, 2, \ldots, 12$ by associating $A^{(0)}$ with $m = 0$, A# with $m = 1$, B with $m = 2$, C with $m = 3$, and so forth. Then the frequency in the third row is a function $\mu(m)$ of m for $m = 0, 1, 2, \ldots, 12$. (μ is the Greek letter mu.) Notice that, as in Table 1, the frequency of $A^{(1)}$ is twice that of $A^{(0)}$.

TABLE 2. ONE OCTAVE OF THE EQUAL-TEMPERED SCALE

Note	$A^{(0)}$	A#	B	C	C#	D
m	0	1	2	3	4	5
$\mu(m)$	220	$220(2^{1/12})$	$220(2^{2/12})$	$220(2^{3/12})$	$220(2^{4/12})$	$220(2^{5/12})$
	$= 220$	$\doteq 233.1$	$\doteq 247.0$	$\doteq 261.6$	$\doteq 277.2$	$\doteq 293.7$

D#	E	F	F#	G	G#	$A^{(1)}$
6	7	8	9	10	11	12
$220(2^{6/12})$	$220(2^{7/12})$	$220(2^{8/12})$	$220(2^{9/12})$	$220(2^{10/12})$	$220(2^{11/12})$	$220(2)$
$\doteq 311.1$	$\doteq 329.6$	$\doteq 349.2$	$\doteq 370.0$	$\doteq 392.0$	$\doteq 415.3$	$= 440$

Question 4 **(a)** What is the exact growth factor from one note to the next, including sharps and flats, in the equal-tempered scale? **(b)** Give a formula for the frequency $\mu(m)$ in Table 2 as a function of m.

Question 5 Give the exact and approximate decimal values of the frequency of the D above $A^{(1)}$ in the equal-tempered scale.

Figure 2 shows the graph of the frequency $\nu(n)$ of $A^{(n)}$ as a function of n from Table 1. Figure 3 shows the graph of the frequency $\mu(m)$ of the mth note above $A^{(0)}$ in the equal-tempered scale from Table 2. The dots in Figure 3 would fit between the dots at $n = 0$ and $n = 1$ in Figure 2.

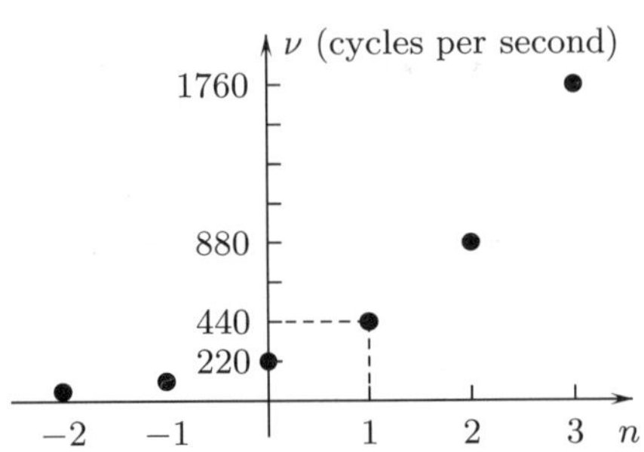

FIGURE 2

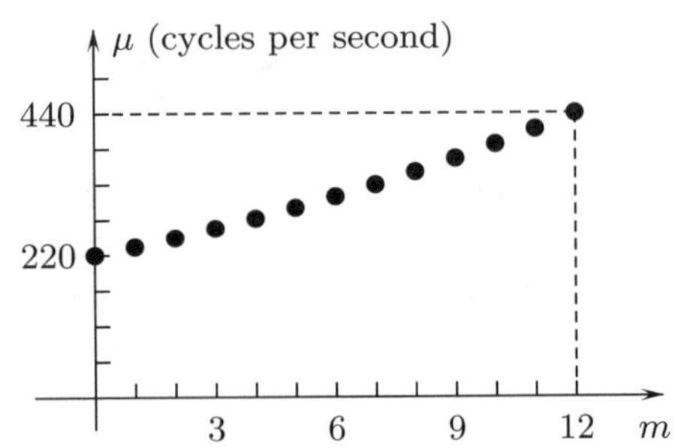

FIGURE 3

Using growth factors

An exponential function of time is determined by its value at any one time and its growth factor over any nonzero time period. To see how to obtain formulas from this information, we look at two examples.

Question 6 The exponential function $g(t)$, with t measured in years, has the value 7 at $t = 0$ and increases by the factor of 10 every 6 years. **(a)** What are $g(6), g(12)$, and $g(18)$? **(b)** Because $g(t)$ is an exponential function, it can be written in the form $g(t) = CB^t$ with positive constants C and B. Use the value of $g(t)$ at 0 to find C. Then use its value at 6 to find B.

Substituting the value of $B = 10^{1/6}$ from Question 6 in the formula $g(t) = 7B^t$ shows that the exponential function that has the value 7 at $t = 0$ and grows by the factor 10 every 6 years is $g(t) = 7(10^{1/6})^t = 7(10^{t/6})$.

Question 7 Suppose that the exponential function $f(t)$ has the value C at $t = 0$ and that it grows by the factor of b whenever t is increased by T. (Here C, B, b, and T are positive constants.) **(a)** What is the value of $f(t)$ at $t = T$? **(b)** Because $f(t)$ is an exponential function and has the value C at $t = 0$, it can be written in the form $f(t) = CB^t$ with a constant B. Use the result of part (a) to give B in terms of b.

The formula $B = b^{1/T}$ from Question 7 gives $f(t) = CB^t = C(b^{1/T})^t = Cb^{t/T}$ and leads us to the following result.

Rule 1 *For any positive constants C, b, and T, the exponential function that has the value C at $t = 0$ and grows by the factor b whenever t is increased by T is*

$$y = Cb^{t/T}. \tag{2}$$

Question 8 An exponential function $h(t)$ has the value 1000 at $t = 0$ (years) and has the growth factor of 1.2 every five years. Give a formula for $h(t)$.

Question 9 Find the exponential function $G(t)$ such that $G(0) = 9$ and $G(5) = 99$. (Use the growth factor from $t = 0$ to $t = 5$.)

Growth factors and relative change

Recall from Section 1.1 that the relative change from one positive number A to another positive number B is the ratio $(B - A)/A$ of the change, $B - A$, from A to B divided by the original value. Since B/A is the growth factor from A to B,

$$\begin{aligned} [\text{The relative change}] &= \frac{B - A}{A} = \frac{B}{A} - 1 \\ &= [\text{The growth factor}] - 1 \end{aligned} \tag{3}$$

and consequently

$$[\text{The growth factor}] = [\text{The relative change}] + 1. \tag{4}$$

Question 10 What is the growth factor from A to B if A and B are positive and B is 37% greater than A?

Question 11 A population that is given by an exponential function is 10,000 at $t = 0$ (years) and increases by 10% every three years. **(a)** What is the growth factor of the population every three years? **(b)** Give the population as a function of $t \geq 0$.

We will now use an exponential function as an approximate mathematical model for empirical data.

Exponential growth of the U.S. population, 1790–1860

The following table gives the population of the United States from the first census in 1790 through the census of 1890.[(1)] To see whether we can model the population accurately with an exponential function, we calculate the growth factor from each census to the next by dividing each population by the population ten years earlier. For instance, the growth factor from 1790 to 1800 was $\frac{5.31}{3.93} \doteq 1.35$. (This means that from 1790 to 1800 the population had a relative change of approximately 0.35 and a percent change of approximately 35%.) The growth factor from 1800 to 1810 was $\frac{7.24}{5.31} \doteq 1.36$, and the growth factor from 1810 to 1820 was $\frac{9.64}{7.24} \doteq 1.33$. These and the growth factors over the other decades, rounded to two decimal places, are listed in the third row of the table.

[(1)] Data adapted from *Historical Statistics of the United States*, Part 1, Washington, DC: U.S. Government Printing Office, 1975, p. 5.

TABLE 3. POPULATION OF THE UNITED STATES (MILLIONS)

Year	1790	1800	1810	1820	1830	1840	1850	1860	1870	1880	1890
Population	3.93	5.31	7.24	9.64	12.87	17.07	23.19	31.44	38.82	50.16	62.95
Growth factor		1.35	1.36	1.33	1.34	1.33	1.36	1.36	1.23	1.29	1.25

If the growth factors were all exactly the same, we could represent the census figures exactly as an exponential function. This is not the case, but since the growth factors every ten years from 1800 through 1860 were close to 1.35, we can approximate that portion of the data fairly well by using an exponential function with 1.35 as base. We let $t = 0$ represent the year 1790. We want the exponential function $P(t)$ that equals 3.93 at $t = 0$ and that grows by the factor 1.35 every ten years. By Rule 1, $P(t) = 3.93(1.35^{t/10})$.

Table 4 lists the census data from Table 3 and values given by this mathematical model. You can see by comparing the last two rows that the approximation is fairly good through 1860 but not very accurate in 1870, 1880, or 1890. This can also be seen in Figure 4, which shows the graph of the exponential model $P = 3.93(1.35^{t/10})$ and the eleven dots with the actual census data. The exponential model based on data from the early censuses is poor for the later decades because the Civil War and other events caused the growth factors to be smaller then.

TABLE 4. POPULATION OF THE UNITED STATES (MILLIONS)

Year	1790	1800	1810	1820	1830	1840	1850	1860	1870	1880	1890
t (years)	0	10	20	30	40	50	60	70	80	90	100
Population	3.93	5.31	7.24	9.64	12.87	17.07	23.19	31.44	38.82	50.16	62.95
$3.93(1.35^{t/10})$	3.93	5.31	7.16	9.67	13.05	17.62	23.79	32.12	43.36	58.53	79.02

Question 12 The U.S. population in the 1990 census was 248.7 million. What would it have been in 1990 if the population was 3.93 million in 1790 and its growth factor was 1.35 every decade until now?

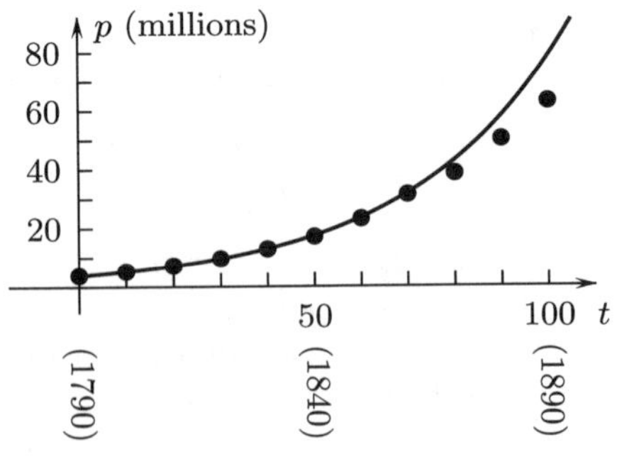

FIGURE 4

Radioactive decay and half life

If the positive number B is less than the number A, then the change from A to B involves shrinking or decay rather than growth. Nevertheless, we call the ratio B/A the "growth factor" from A to B. If there is growth and B is greater than A, then the growth factor from A to B is greater than 1. If there is decay and B is less than A, then the growth factor from A to B is a positive number less than 1.

Radioactive decay is modeled by exponential functions with growth factors that are less than 1. Radium, for example, has a HALF LIFE of 1620 years. This means that a sample of radium decomposes to half its weight or mass every 1620 years. In other words, the weight or mass of a sample of radium has the growth factor of $\frac{1}{2}$ every 1620 years. By Rule 1, the mass or weight is given as a function of t by

$$C\left(\tfrac{1}{2}\right)^{t/1620} \tag{5}$$

where C is the weight or mass at $t = 0$. Figure 5 shows the graph with $C = 4$.

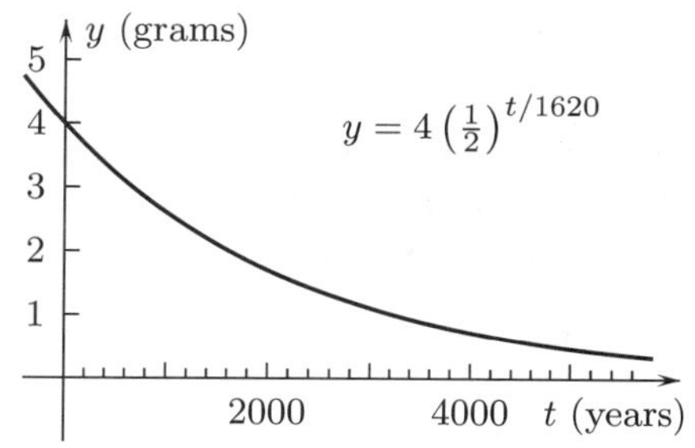

FIGURE 5

Question 13 If a sample of radium has a mass of 4 grams now, what will its mass be in 1620 years? In 3240 years? In 5000 years?

Question 14 A sample of radioactive Carbon-14 has a mass of $0.05\left(\frac{1}{2}\right)^{t/5700}$ kilograms at time t (years). What is its half life?

Compound interest

Formulas for compound interest can also be found from Rule 1. Suppose that on your eighteenth birthday you put \$50,000 from your Uncle Joe's trust in a savings account that paid 12% annual interest compounded quarterly, and that you left it there without making any withdrawals or further deposits. Every quarter $\frac{1}{4}(12\%) = 3\%$ interest is paid on the current balance.

Question 15 **(a)** What is the growth factor of the balance in your savings account every quarter?
(b) Use Rule 1 to give a formula for the balance after t years.

Now suppose the annual interest rate is $P\%$. We let $r = P/100$ denote the fraction or decimal corresponding to $P\%$. If the interest is compounded n times a year, then every pay period the relative change in the balance is r/n and the growth factor is $1 + r/n$. Rule 1 yields the following:

Rule 2 *C dollars deposited at time $t = 0$ in a savings account that pays $P\%$ annual interest compounded n times a year grows to*

$$B = C\left(1 + \frac{r}{n}\right)^{nt} \tag{6}$$

at $t = \frac{1}{n}, \frac{2}{n}, \frac{3}{n}, \ldots,$ *where* $r = \frac{P}{100}$ *is the decimal or fraction corresponding to* $P\%$.

Question 16 According to legend, the Dutch colonist Peter Minuit paid \$24 to buy Manhattan from the indians in 1626. How much would this investment have become as of 1996 if it had been kept in a savings account paying 6% annual interest compounded semi-annually?

The restriction of t to integer multiples of $1/n$ is necessary in **(6)** because the formula is only valid immediately after interest payments have been made. The balance is a step function, which is constant between payments. This is illustrated in Figure 6, which shows the graph of the balance in an account with an initial balance of \$100 and an annual interest rate of 50%, compounded annually.

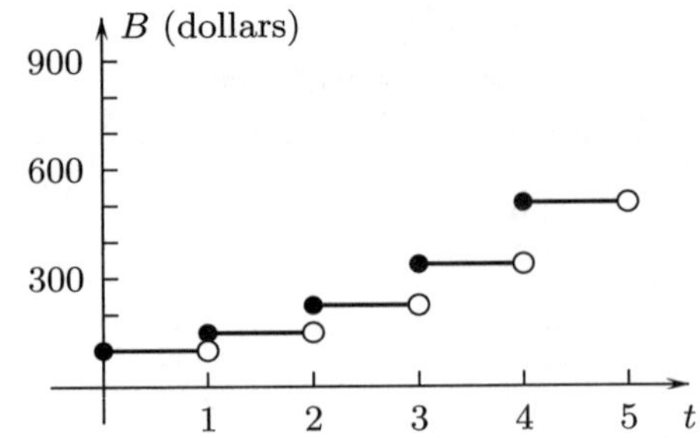

FIGURE 6

Interest compounded continuously and the number e

The limit of the balance **(6)** as $n \to \infty$ is the balance with interest COMPOUNDED CONTINUOUSLY. The formula for the limit is based on the definition

$$e = \lim_{n\to\infty}\left(1+\frac{1}{n}\right)^n = 2.718281828\ldots \tag{7}$$

of the number e and on the related formula,

$$\lim_{n\to\infty}\left(1+\frac{x}{n}\right)^n = e^x \tag{8}$$

for e^x with any number x.

Combining **(8)** with formula **(6)** gives

$$\lim_{n\to\infty}\left[C\left(1+\frac{r}{n}\right)^{nt}\right] = C\left\{\lim_{n\to\infty}\left[\left(1+\frac{r}{n}\right)^n\right]\right\}^t = C(e^r)^t = Ce^{rt}$$

and leads to the following:

Rule 3 *A deposit of C dollars that earns $P\%$ annual interest compounded continuously grows to*

$$B = Ce^{rt} \tag{9}$$

after t years, where $r = P/100$ is the decimal or fraction that corresponds to $P\%$.

Figure 7 shows the graph of the step function from Figure 6 of the balance in an account containing \$100 initially that earns 50% annual interest compounded annually ($n = 1$) and the graph of the balance $100e^{0.5t}$ with interest compounded continuously. Figure 8 shows the graphs of the balances with interest compounded semi-annually ($n = 2$) and continuously. Notice that the graph of the step function in Figure 8 is closer to the curve for interest compounded continuously than is the step function in Figure 7. The graph for interest compounded more than two times a year would consist of more, shorter line segments and be even closer to the exponential curve.

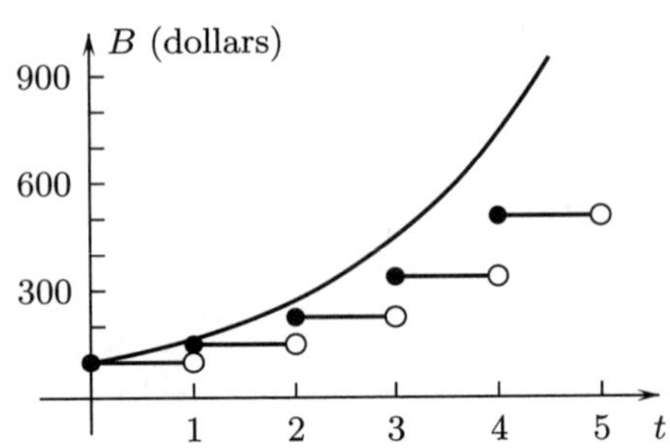

FIGURE 7

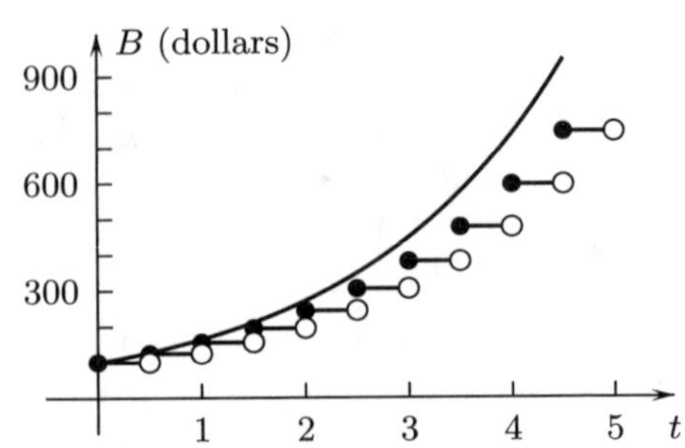

FIGURE 8

Question 17 How much would the \$24 paid in 1626 for Manhattan have yielded in 1996 if it earned 6% annual interest compounded continuously?

Question 18 What is the annual interest rate if the balance in an account with interest compounded continuously is $1000e^{2t}$ dollars after t years?

Simple interest

Simple (noncompounded) interest is calculated with linear functions.

Question 19 You loan your cousin \$1000 at 3% annual interest, not compounded. **(a)** How much does she owe you after one year? **(b)** After t years?

What if an initial balance of C dollars earns $P\%$ simple interest each year? As above, we let r denote the fraction $r = P/100$ corresponding to $P\%$. Then in one year the money earns rC dollars interest and the balance is increased to $C + rC$ dollars. In t years it earns rCt dollars interest to become

$$B = C + rCt \quad \text{dollars.} \tag{10}$$

The graph of the balance is the line with B-intercept C and slope rC (Figure 9).

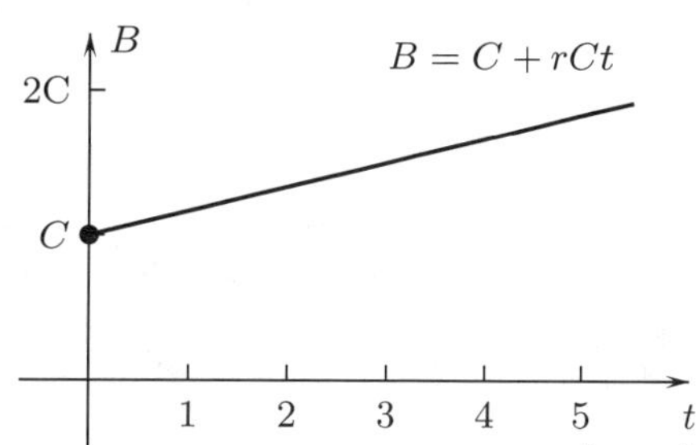

FIGURE 9

The derivative of e^x

To study rates of change of functions given by exponential functions, we need formulas for their derivatives. We begin by finding the derivative of the natural exponential function $y = e^x$. As you could predict from the graph in Figure 10, the derivative of e^x is positive for all x and tends to 0 as $x \to -\infty$ and to ∞ as $x \to \infty$. What is its formula?

$$[\text{Slope}] = \frac{d}{dx}(e^x)$$

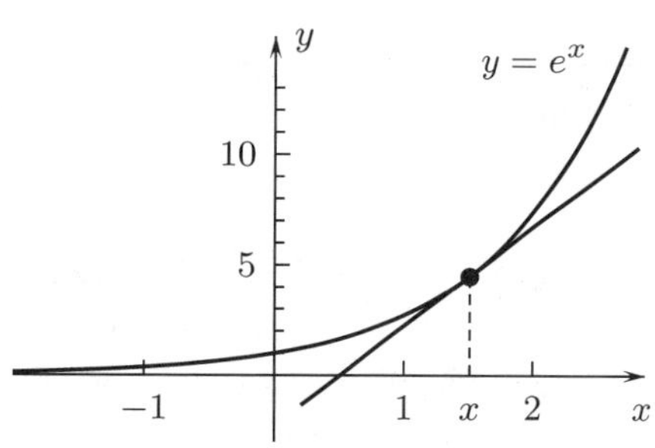

FIGURE 10

Question 20 Complete the following table of values of e^x and of its approximate derivative, and use the results to predict a formula for the derivative as a function of x. Use the secant-line program of Section 2.3 with the window $-1 \le x \le 12$, $-3 \le y \le 3$. (The secant line is not on the screen for $x = 10$.)

TABLE 5

x	$e^x \doteq$	$\dfrac{e^{(x+10^{-6})} - e^x}{10^{-6}} \doteq$
-1	0.367879441	0.367879630
0	1	1.000000500
1		
2		
3		
10		

You probably predicted in Question 20 that e^x is its own derivative. Why is this true?

Question 21 Use the property $e^{x_1+x_2} = e^{x_1}e^{x_2}$ of the exponential function to show that for $\Delta x \neq 0$,

$$\frac{e^{(x+\Delta x)} - e^x}{\Delta x} = e^x \left[\frac{e^{\Delta x} - 1}{\Delta x}\right]. \tag{11}$$

The limit of the difference quotient on the left of **(11)** is the derivative of e^x at x. If we set $x = 0$ in this difference quotient, we obtain the expression in square brackets on the right, whose limit is the derivative at $x = 0$. Then, assuming that the limits exist and are finite, equation **(11)** yields

$$\frac{d}{dx}(e^x) = \lim_{\Delta x \to 0} \frac{e^{x+\Delta x} - e^x}{\Delta x} = e^x \left[\lim_{\Delta x \to 0} \frac{e^{\Delta x} - 1}{\Delta x}\right] = e^x \left[\frac{d}{dx}(e^x)\right]_{x=0}. \tag{12}$$

This shows that the derivative of e^x at any x equals e^x multiplied by its derivative at 0. It can be shown—as you might predict from the entry for $x = 0$ in Table 5 above—that the dervative of e^x at $x = 0$ is 1. This gives the next result.†

Theorem 1 *e^x is its own derivative:*

$$\frac{d}{dx}(e^x) = e^x \quad \text{for all } x. \tag{13}$$

Example 1 Show that $y(x) = 25x - e^x$ has a global maximum and find its exact value.

SOLUTION The derivative,

$$y'(x) = \frac{d}{dx}(25x - e^x) = 25 - e^x$$

exists for all x and is zero where $e^x = 25$, which is at $x = \ln(25) \doteq 3.22$.

Since the derivative $y' = 25 - e^x$ is continuous for all x and is zero only at $x = \ln(25)$, it is of constant sign for $x < \ln(25)$ and for $x > \ln(25)$. The derivative equals $25 - e^0 = 24$ at $x = 0$ and equals $25 - e^5 \doteq -123$ at $x = 5$. Therefore, it is positive for $x < \ln(25)$ and negative for $x > \ln(25)$. Because $y(x)$ is continuous for all x, it is increasing in $(-\infty, \ln(25)]$ and decreasing in $[\ln(25), \infty)$ (Figure 11). Its global maximum is at $x = \ln(25)$ and is $y(\ln(25)) = 25\ln(25) - e^{\ln(25)} = 25\ln(25) - 25$. □

---

†In Section 3.5, where we study the derivatives of logarithms, we will see how the formula for the derivative of e^x is related to the definition **(7)** of the number e.

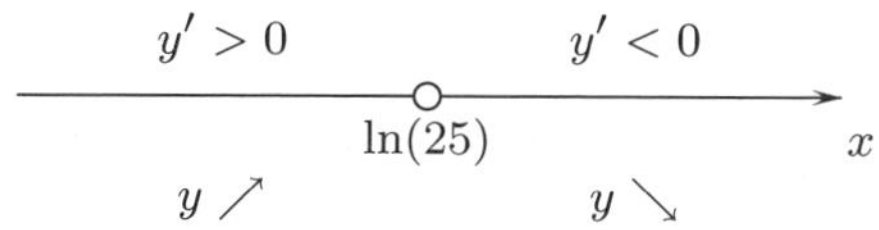

FIGURE 11

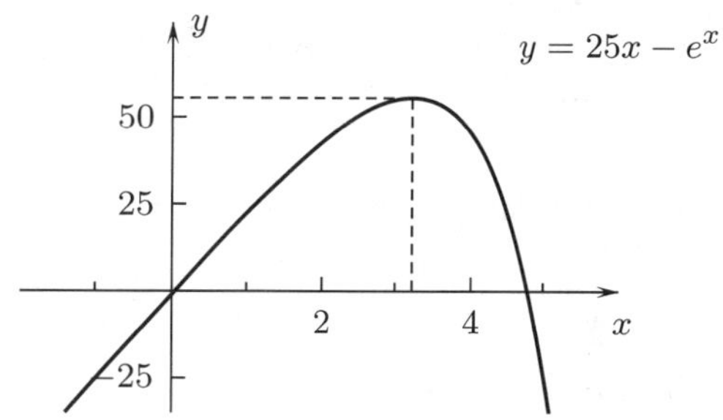

FIGURE 12

The conclusion of Example 1 is corroborated by the graph $y = 25x - e^x$ in Figure 12, since the global maximum, $25\ln(25) - 25$, of $y = 25x - e^x$ has the approximate decimal value 55.47.

The derivative of $e^{y(x)}$

To derive a formula for the x-derivative of e^y with a differentiable function $y = y(x)$ we use the Chain-Rule from Section 3.3:

Theorem 2 *If $y(x)$ has a derivative at x, then $e^{y(x)}$ has a derivative at x, given by*

$$\frac{d}{dx}(e^y) = \frac{d}{dy}(e^y)\,\frac{dy}{dx} = e^y\,\frac{dy}{dx} \tag{14a}$$

or with prime notation,

$$\frac{d}{dx}[e^{y(x)}] = e^{y(x)}\,y'(x). \tag{14b}$$

Remember **(14)** in words: The x-derivative of e raised to the power $y(x)$ for a function $y(x)$ equals e raised to the function, multiplied by the derivative of the function.

Example 2 What are the x-derivatives of **(a)** e^{x^3} and **(b)** $2x + 3e^{4x}$?

SOLUTION **(a)** $\frac{d}{dx}(e^{x^3}) = e^{x^3}\frac{d}{dx}(x^3) = 3x^2e^{x^3}$

(b) $\frac{d}{dx}(2x + 3e^{4x}) = 2 + 3e^{4x}\frac{d}{dx}(4x) = 2 + 12e^{4x}$ □

Question 22 Figure 13 shows the graph of $g(x) = e^{-x^2/2}$. **(a)** Why is the graph symmetric about the y-axis? **(b)** What is its y-intercept? **(c)** Why does the graph approach the x-axis as $x \to \pm\infty$? **(d)** Find the exact coordinates of the two inflection points of the graph that are shown in Figure 13.†

FIGURE 13

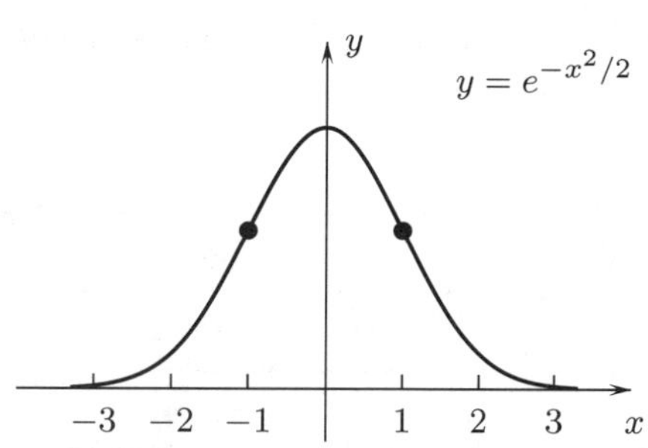

†The graph of $\frac{1}{\sqrt{2\pi}}g(x) = \frac{1}{\sqrt{2\pi}}e^{-x^2/2}$ is the "bell curve" from probability.

The derivative of b^x

The exponential function $y = b^x$ for any positive constant b can be expressed with base e by using the equation $b = e^{\ln b}$ to write

$$b^x = (e^{\ln b})^x = e^{x \ln b}.$$

This formula then can be used to find the derivative of b^x:

$$\frac{d}{dx}(b^x) = \frac{d}{dx}(e^{x \ln b}) = e^{x \ln b}\frac{d}{dx}(x \ln b) = (\ln b)b^x.$$

This gives part (a) of the following theorem, and part (b) follows from the Chain Rule.

Theorem 3 **(a)** *For any positive constant b,*

$$\frac{d}{dx}(b^x) = (\ln b)b^x \quad \text{for all } x. \tag{15}$$

(b) *At any value of x where $y = y(x)$ has a derivative,*

$$\frac{d}{dx}(b^y) = \frac{d}{dy}(b^y)\,\frac{dy}{dx} = (\ln b)b^y\,\frac{dy}{dx}. \tag{16}$$

Example 3 Based on the mathematical model $P = 3.93(1.35^{t/10})$ from Table 4 for the population of the United States from 1790 ($t = 0$) to 1860 ($t = 70$), what was the rate of change of the population with respect to time **(a)** at the beginning of 1800 ($t = 10$) and **(b)** at the beginning of 1860 ($t = 70$)? Give the exact and approximate decimal values.

SOLUTION By formula **(16)** with $b = 1.35$, $y = t/10$, and t in place of x,

$$\begin{aligned}\frac{dP}{dt} &= \frac{d}{dt}[3.93(1.35^{t/10})] = 3.93\ln(1.35)(1.35^{t/10})\frac{d}{dt}\left(\tfrac{1}{10}t\right)\\ &= \tfrac{1}{10}\ln(1.35)[3.93(1.35^{t/10})]\end{aligned}$$

so that **(a)** $\frac{dP}{dt}(10) = \frac{1}{10}\ln(1.35)[3.93(1.35^1) \doteq 0.15922$ million per year, and **(b)** $\frac{dP}{dt}(70) = \frac{1}{10}\ln(1.35)[3.93(1.35^7) \doteq 0.96383$ million per year. □

Question 23 **(a)** What was the exact growth factor of the population in Example 3 from the beginning of 1800 to the beginning of 1860? **(b)** What was the exact growth factor of the rate of change of the population from the beginning of 1800 to the beginning of 1860? **(c)** How are the answers to parts (a) and (b) related?

Comparing b^x *and* x^n *for large positive and large negative* x

If b is a constant > 1, then $b^x \to \infty$ as $x \to \infty$ (Figure 14). Moreover, as x increases through all positive values, b^x is eventually much larger than x^n for any fixed number n.[†] This is illustrated for 3^x and x^3 in the following table.

x	x^3	3^x
4	$4 \cdot 4 \cdot 4 = 64$	$3 \cdot 3 \cdot 3 \cdot 3 = 81$
6	$6 \cdot 6 \cdot 6 = 216$	$3 \cdot 3 \cdot 3 \cdot 3 \cdot 3 \cdot 3 = 729$
8	$8 \cdot 8 \cdot 8 = 512$	$3 \cdot 3 \cdot 3 \cdot 3 \cdot 3 \cdot 3 \cdot 3 \cdot 3 = 6{,}561$
10	$10 \cdot 10 \cdot 10 = 1000$	$3 \cdot 3 \cdot 3 \cdot 3 \cdot 3 \cdot 3 \cdot 3 \cdot 3 \cdot 3 \cdot 3 = 59{,}049$

[†] We will establish this fact in Section 5.5 by using L'HÔPITAL'S RULE

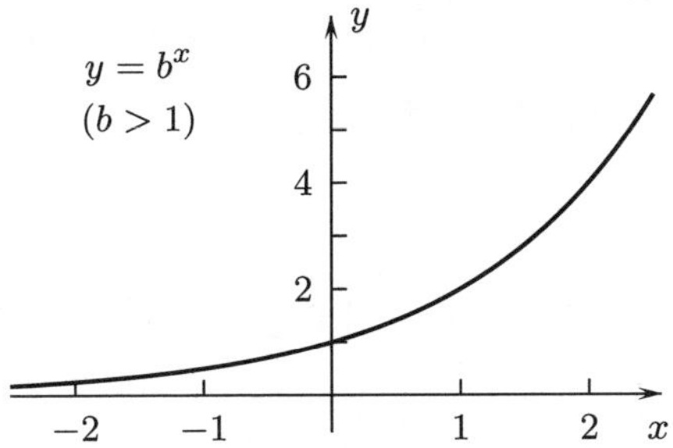

FIGURE 14

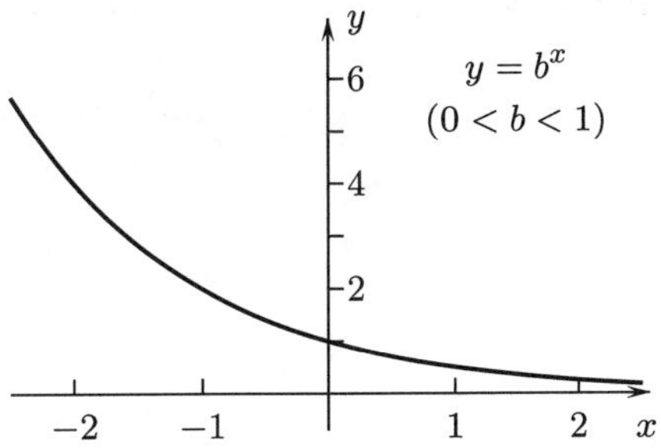

FIGURE 15

For $0 < b < 1$, the expression b^x can be written $(1/b)^{-x}$ with $1/b > 1$. Consequently, b^x tends to ∞ as $x \to -\infty$ (Figure 15), and b^x is much larger than any power of $|x|$ for sufficiently large negative x. We state these facts as a rule:

Rule 4 *(i) For any constant $b > 1$ and any constant n, b^x is much larger than x^n for sufficiently large positive x.*

(ii) For any constant b with $0 < b < 1$ and any constant n, b^x is much larger than $|x^n|$ for sufficiently large negative x.

We describe this rule by saying that for $b > 1$, b^x dominates x^n for sufficiently large positive x. and that for $0 < b < 1$, b^x dominates x^n for sufficiently large negative x. These principles enable us to predict the shapes of graphs of certain functions constructed from b^x and powers of x for large positive and large negative x.

Consider, for example the graph of $e^x - 4x$ in Figure 16. It is closely approximated by the line $y = -4x$ for large negative x, where e^x is very small (Figure 17). For large positive x, on the other hand, e^x dominates $4x$. Consequently, $e^x - 4x \to \infty$ as $x \to \infty$, and $4x$ is small relative to e^x for large positive x, so that $y = e^x - 4x$ appears close to $y = e^x$ in any window with a large y-range, as in Figure 18.

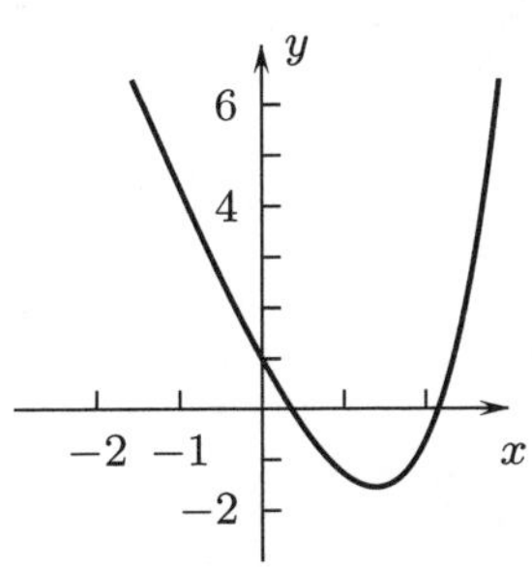

$y = e^x - 4x$

FIGURE 16

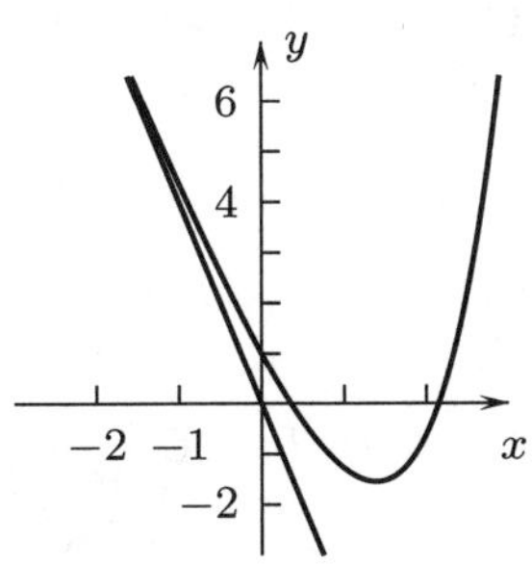

$y = e^x - 4x$
and $y = -4x$

FIGURE 17

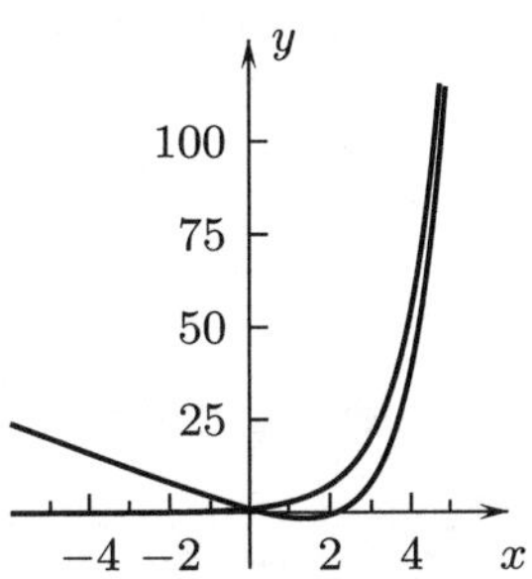

$y = e^x - 4x$
and $y = e^x$

FIGURE 18

Example 4 Find the global minimum of $y = e^x - 4x$.

SOLUTION The derivative $y' = e^x - 4$ is zero at $x = \ln(4)$, is negative for $x < \ln(4)$, and is positive for $x > \ln(4)$. Because $y = e^x - 4x$ is continuous for all x, it is decreasing in $(-\infty, \ln(4)]$ and increasing in $[\ln(4), \infty)$ and has a global minimum of $e^{\ln(4)} - 4\ln(4) = 4 - 4\ln(4)$ at $x = \ln(4)$. □

Question 24 Figure 19 shows the graph of $y = \dfrac{2^{-x}}{x^2}$. **(a)** What are $\lim_{x\to-\infty}\left[\dfrac{2^{-x}}{x^2}\right]$, $\lim_{x\to 0}\left[\dfrac{2^{-x}}{x^2}\right]$ and $\lim_{x\to\infty}\left[\dfrac{2^{-x}}{x^2}\right]$ and why? **(b)** What is the one critical point of $\dfrac{2^{-x}}{x^2}$?

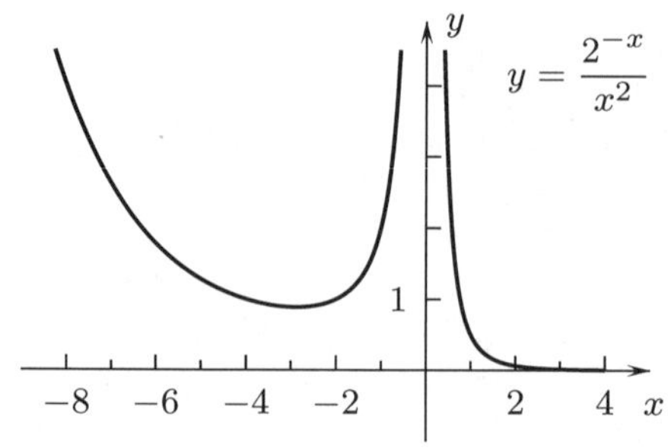

FIGURE 19

Question 25 Find $\lim_{x\to\infty}(x^4 - 2x - 3^x)$ and $\lim_{x\to-\infty}(x^4 - 2x - 3^x)$.

Relative and percent rates of change

The RELATIVE RATE OF CHANGE of a nonzero, differentiable function $f(t)$ is defined to be the ratio of the value of the derivative and the value of the function:

$$\begin{bmatrix}\text{Relative rate of change of}\\ f(t)\text{ with respect to }t\end{bmatrix} = \frac{f'(t)}{f(t)}. \tag{17}$$

The PERCENT RATE OF CHANGE is the relative rate of change multiplied by 100.

Example 5 What is the relative rate of change of t^4 at $t = 10$?

SOLUTION The derivative of $f(t) = t^4$ is $f'(t) = 4t^3$. Therefore, $f(10) = 10^4 = 10{,}000$, $f'(10) = 4(10^3) = 4000$, and the relative rate of change is $\dfrac{f'(10)}{f(10)} = \dfrac{4000}{10{,}000} = 0.4$. □

Example 6 Show that for constants $C \neq 0$ and $b > 0$, the relative rate of change of $y = Cb^t$ is constant, and find its value.

SOLUTION The derivative of Cb^t is

$$\begin{aligned}\frac{d}{dt}(Cb^t) &= \frac{d}{dt}(Ce^{\ln(b^t)}) = \frac{d}{dt}(Ce^{t\ln b}) = Ce^{t\ln b}\frac{d}{dt}(t\ln b)\\ &= Ce^{t\ln b}(\ln b) = Cb^t(\ln b).\end{aligned}$$

Consequently, the relative rate of change of Cb^t is

$$\frac{\dfrac{d}{dt}(Cb^t)}{Cb^t} = \frac{Cb^t(\ln b)}{Cb^t} = \ln b. \ \square$$

The result of Example 6 gives another way of contrasting linear functions $mt+b$ and exponential functions Cb^t: linear functions have constant rates of change; exponential functions have constant relative rates of change. This anology between linear and exponential functions is explained by the results of the next question.

Question 26 **(a)** Show that $\ln[E(t)]$ is a linear function of t if $E(t) = Cb^t$ with positive constants b and C. **(b)** What are the slope and the y-intercept of the graph of $y = \ln[E(t)]$ in a ty-plane?

The differential equation dy/dt = ry

A variety of important applications use the mathematical model that the relative rate of change of an unknown function $y = y(t)$ is equal to a constant r. This gives the equation $\dfrac{dy/dt}{y} = r$, which can be written as $\dfrac{dy}{dt} = ry$. The latter is called a DIFFERENTIAL EQUATION because it involves the derivative dy/dt of the unknown function y. The equation $\dfrac{dy}{dt} = ry$ can also be described as the requirement that dy/dt be proportional to y with constant of proportionality r. As you might expect from Example 6, the solutions of this differential equation are exponential functions. To find them we use the base e and apply the next theorem.

Theorem 4 *The solutions of the differential equation*

$$\frac{dy}{dx} = ry \tag{18}$$

with a constant r are

$$y = Ce^{rt} \tag{19}$$

with constants C.

Any function of the form $y = Ce^{rt}$ is a solution of **(18)** because with $y = Ce^{rt}$,

$$\frac{dy}{dt} = \frac{d}{dt}(Ce^{rt}) = Ce^{rt}\frac{d}{dt}(rt) = Cre^{rt} = ry.$$

To show that all solutions of **(18)** are given by **(19)**, we suppose that y is any solution. By the Product Rule, $\dfrac{d}{dt}(ye^{-rt}) = \dfrac{dy}{dt}e^{-rt} - rye^{-rt} = \left(\dfrac{dy}{dt} - ry\right)e^{-rt} = 0$ for all t. This shows that ye^{-rt} equals a constant C, so that $y = Ce^{rt}$.

The value of the constant C in **(19)** can be determined from a value of y at one value of t. This is called an INITIAL CONDITION. This is especially easy to carry out if the value at $t = 0$ is given since $e^0 = 1$. The differential equation with an initial condition is called an INITIAL-VALUE PROBLEM.

Example 7 Find the solution of the initial-value problem $\dfrac{dy}{dt} = 2y,\ y(0) = 4$.

SOLUTION The solutions of the differential equation are $y = Ce^{2t}$, by Theorem 4 with $r = 2$. To find the constant C, we set $t = 0$ and $y = 4$ to obtain $4 = Ce^0 = C$. The solution is $y = 4e^{2t}$.

Question 27 Show that formula $B = Ce^{rt}$ in Rule 3 for the balance in an account that pays interest compounded continuously can be obtained from the assumption that the rate of growth dB/dt of the balance is at all times proportional to the balance B.

Question 28 Because Carbon-14 has a half-life of 5700 years, a sample with mass C at $t = 0$ (years) has mass $y = C\left(\frac{1}{2}\right)^{t/5700}$ in year t. The rate of decay dy/dt of the mass is always proportional to the mass y at that moment. What is the constant of proportionality?

Principles and procedures

- The growth factor from a positive number A to a positive number B is B/A. The growth factor is greater than 1 if B is greater than A and is between 0 and 1 if B is less than A. The growth factor equals the relative change plus 1.
- The exponential function that grows by the factor b when t is increased by T and that has the value C at $t = 0$ is $Cb^{t/T}$.

- Formulas for exponential functions in terms of their half lives, doubling times, tripling times, etc., are given by $Cb^{t/T}$ with $b = \frac{1}{2}$ for half lives, $b = 2$ for doubling times, $b = 3$ for tripling times, etc. For example, the exponential function that triples every T years and has the value C at $t = 0$ is $C(3^{t/T})$.
- The balance in a savings account that pays $P\%$ annual interest compounded n times a year has the growth factor $(1+r/n)$ every $1/n$ years, where $r = P/100$ is the fraction corresponding to $P\%$. Consequently, if the balance is C at $t = 0$, it is $C(1 + r/n)^{nt}$ at time t for $t = 1/n, 2/n, 3/n, \ldots$.
- The number e is defined as the limit $e = \lim_{n\to\infty}\left(1 + \frac{1}{n}\right)^n$, and e^x for any number x can be expressed as the similar limit $e^x = \lim_{n\to\infty}\left(1 + \frac{x}{n}\right)^n$.
- The balance in a savings account with interest compounded continuously is the limit as $n \to \infty$ of the balance compounded n times a year. If the annual interest rate is $P\%$ and the balance is C dollars at $t = 0$ (years), then the balance at time t is Ce^{rt} with $r = P/100$ the fraction corresponding to $P\%$.
- The balance in a savings account that pays $P\%$ annual interest without compounding (i.e., with simple interest) is given by the linear function $B = C + rCt = C(1 + rt)$ with $r = P/100$.
- The behavior for large positive and large negative x of functions constructed from powers and exponential functions can often be determined by using the facts that for $b > 1$, b^x tends to ∞ and dominates all powers of x as $x \to \infty$ and b^x tends to 0 as $x \to -\infty$, while for $0 < b < 1$, b^x tends to ∞ and dominates all powers of x as $x \to -\infty$ and b^x tends to 0 as $x \to \infty$.
- The fact that $b^0 = 1$ can often be used to find the y-intercept of the graph of a function constructed from b^x or to determine how the graph relates to the y-axis if the y-axis is a vertical asymptote.
- To remember that e^{rx} is an increasing function of x if $r > 0$ and a decreasing function if $r < 0$, note that $e^{rx} = (e^r)^x$ and that $e^r > 1$ for $r > 0$ and $0 < e^r < 1$ for $r < 0$.
- The linear function $L(t) = mt + b$ has the property that increasing the value of the variable by a fixed amount Δt from t_0 to $t_0 + \Delta t$ increases the value of the function by the same amount $m\Delta t$, regardless of the value of t_0. This is because $L(t_0 + \Delta t) = m(t_0 + \Delta t) + b = mt_0 + b + m\Delta t = L(t_0) + m\Delta t$ (Figure 20).

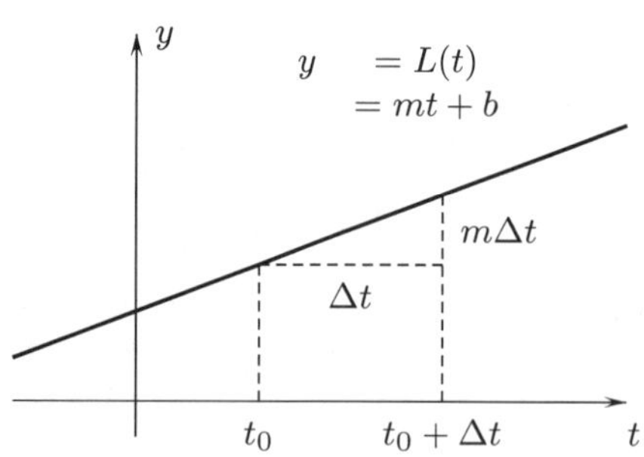

FIGURE 20

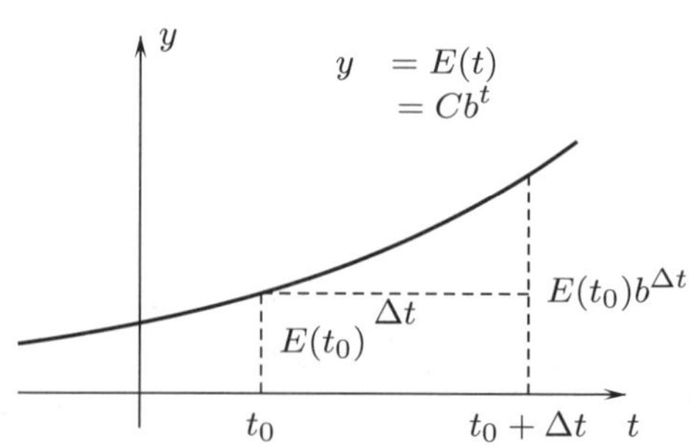

FIGURE 21

- For an exponential function $E(t) = Cb^t$ with positive constants b and C, increasing the value of the variable by a fixed amount Δt from t_0 to $t_0 + \Delta t$ multiplies the value of the function by the same factor $b^{\Delta t}$, regardless of the value of t_0. This is because $E(t_0 + \Delta t) = Cb^{t_0+\Delta t} = Cb^{t_0}b^{\Delta t} = E(t_0)b^{\Delta t}$ (Figure 21).
- Any exponential function b^x with constant $b > 0$ can be expressed in terms of the natural exponential function by writing $b^x = (e^{\ln b})^x = e^{x \ln b}$. The base can be changed to any positive $c \neq 1$ by a similar calculation.

- Here, as in earlier sections, differentiation rules are applied in the reverse order of the corresponding steps used to evaluate a function.
- Think of the differential equation $\frac{dy}{dt} = ry$ either as the requirement that the relative rate of change of $y = y(t)$ be equal to the constant r or that the rate of change of y is proportional to y with constant of proportionality r.

Tune-Up Exercises 3.4♦

[A]*Answer provided.* [O]*Outline of solution provided.*

T1.[A] **(a)** What is the growth factor from 10 to 15? **(b)** What is the relative change from 10 to 15? **(c)** What is the growth factor from 10 to 5? **(d)** What is the relative change from 10 to 5?

♦ Type 1, growth factor and relative change

T2.[A] What are the growth factors of $4(5^t)$ when t **(a)** increases by 1 and **(b)** increases by 4?

♦ Type 1, growth factors

T3.[A] What are the growth factors of $7\left(\frac{1}{10}\right)^{t/4}$ when t increases **(a)** by 1 and **(b)** by 4?

♦ Type 1, growth factors

T4.[A] **(a)** By what amount does the linear function $2x+7$ increase when x is increased by 3? **(b)** By what factor does the exponential function $7(2^x)$ increase when x is increased by 3?

♦ Type 2, growth factors, etc.

T5.[A] An exponential function $f(x)$ has the values in the next table. Give formulas for the function and its x-derivative.

x	0	1	2	3	4
$f(x)$	5	15	45	135	405

♦ Type 2, fitting an exponential function to data, etc.

T6.[A] **(a)** Give a formula for the exponential function $g(t)$ with the values in the next table. **(b)** Is the ratio $g'(t)/g(t)$ positive or negative? What is its value?

t	0	1	2	3	4
$g(t)$	64	16	4	1	0.25

♦ Type 2, fitting an exponential function to data, etc.

T7.[A] **(a)** Give a formula for the exponential function $h(t)$ such that $h(0) = 1000, h(5) = 800, h(10) = 640,$ and $h(15) = 512$. **(b)** Give exact and approximate decimal values of $h(-3)$ and $h'(-3)$.

♦ Type 2, fitting an exponential function to data, etc.

T8.[A] **(a)** Find a formula for the exponential function $K(y)$ such that $K(-200) = 400$, $K(-100) = 600, K(0) = 900$, and $K(100) = 1350$. **(b)** What are the values of $\frac{K(y+100)}{K(y)}$ and $\frac{K'(y+100)}{K'(y)}$?

♦ Type 2, fitting an exponential function to data, etc.

T9[A] Give a formula for the exponential function whose graph is shown in Figure 22. Then generate the graph of the function with the graph of its derivative.

♦ Type 2, fitting an exponential function to data, etc.

♦The Tune-up Exercises and Problems are classified by type and content. The types are (1) basic, reactive; (2) basic reflective; (3) intermediate, reactive; (4) intermediate, reflective; (5) advanced, reactive; (6) advanced, reflective; and (7) advanced, theoretical.

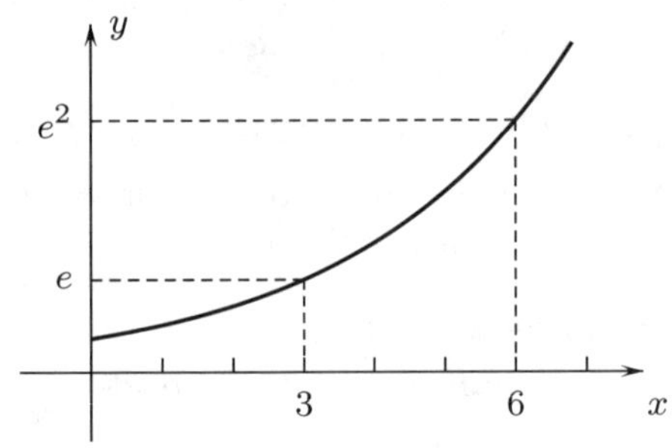

FIGURE 22

T10.[A] Generate the graphs of $10 + e^{-x}$ and of its x-derivative.

♦ Type 1, a derivative

T11.[A] **(a)** What is the half-life of $1500\left(\frac{1}{2}\right)^{t/8}$? **(b)** What is the doubling time of $1500(2)^{t/8}$?

♦ Type 2, half-life and doubling time

T12.[A] An initial deposit of \$3000 is made in a savings account that pays 9.6% annual interest compounded monthly. No other deposits or withdrawals are made. How much is in the account after 4 years?

♦ Type 1, compound interest

T13.[A] An initial deposit of \$100 is made in a savings account and subsequently no other deposits or withdrawals are made. What would the account balance be after 9.5 years at 6% annual interest if **(a)** the interest is compounded annually, **(b)** if the interest is compounded continuously, and **(c)** if the interest is not compounded?

♦ Type 1, compound interest

T14.[A] What are $\lim_{x\to\infty}\left[\left(\frac{1}{6}\right)^x - x^2\right]$ and $\lim_{x\to-\infty}\left[\left(\frac{1}{6}\right)^x - x^2\right]$ and why?

♦ Type 2, limits involving exponential functions

T15.[A] Find either a linear function or an exponential function $R(t)$ with the values in the next table.

t	1	2	3	4	5
$R(t)$	95	90	85	80	75

♦ Type 2, linear and exponential functions

T16.[A] Find the linear function $A(t)$ and the exponential function $B(t)$ such that $A(0) = 20$, $A(1) = 10, B(0) = 20$, and $B(1) = 10$.

♦ Type 2, linear and exponential functions

Find the derivatives in Exercises T17 through T20.

T17.[O] $\dfrac{d}{dx}(e^{2x})$

♦ Type 1, derivatives involving exponential functions

T18.[O] $\dfrac{d}{dx}(e^{x^2+3x})$

♦ Type 1, derivatives involving exponential functions

T19.[A] y' for $y = 2^x + 3^x$

♦ Type 1, derivatives involving exponential functions

T20.[O] $f'(x)$ for $f(x) = (1 + e^x)^4$

♦ Type 1, derivatives involving exponential functions

T21.[A] Solve the initial-value problems, **(a)** $\dfrac{dy}{dt} = -4y,\ y(0) = 6$ and **(b)** $\dfrac{dy}{dt} = \ln(3)y,\ y(2) = 9.$

♦ Type 1, an initial value problem with an exponential function

Problems 3.4

[A]*Answer provided.* [O]*Outline of solution provided.*

Find the (exact) derivatives in Problems 1 through 16.

1.[O] $\dfrac{d}{dx}(e^{6x})$

♦ Type 1, a derivative

2.[A] $\dfrac{d}{dx}(e^x + x^e)$

♦ Type 1, a derivative

3.[O] $\dfrac{d}{dx}\left[\dfrac{x}{1+e^x}\right]$

♦ Type 1, a derivative

4.[A] $\dfrac{d}{dx}(\sqrt[3]{1+e^x})$

♦ Type 1, a derivative

5.[O] $\dfrac{dp}{dt}$ for $p(t) = \dfrac{t^2+1}{e^t}$

♦ Type 1, a derivative

6.[A] $\dfrac{d}{dx}(1 - e^4)$

♦ Type 1, a derivative

7.[A] $\left[\dfrac{d}{dx}(e^{100x-x^2})\right]_{x=0}$

♦ Type 1, a derivative

8.[A] $\dfrac{d}{dx}(xe^{-x})$

♦ Type 1, a derivative

9.[A] $\dfrac{d}{dx}(\sqrt{1+e^x})$

♦ Type 1, a derivative

10. $\dfrac{d}{dx}[(1-e^x)^{10}]$

♦ Type 1, a derivative

11.[A] $q'(2)$ for $q(x) = 3^x + x^3$

♦ Type 1, a derivative

12.[A] $\dfrac{d}{dx}(5^{x/3})$

♦ Type 1, a derivative

13. $\dfrac{d}{dx}(e^{-1/x})$

♦ Type 1, a derivative

14. $\dfrac{d}{dx}(\dfrac{e^x}{x+3})$

♦ Type 1, a derivative

15. $y'(x)$ for $y = x3^x$

♦ Type 1, a derivative

16. $\dfrac{d}{dx}[(1+x^2)^\pi]$

♦ Type 1, a derivative

Give exact answers and their approximate decimal values, with units, wherever appropriate in the following problems.

17.[A] In an exponential model of atmospheric pressure, it is assumed that the air pressure is 1035 grams per square centimeter on the surface of the earth and is halved for every 5.2 kilometers of vertical ascent. **(a)** Give a formula for air pressure $p(h)$ (grams per square centimeter) with this model as a function of the height h (kilometers) above the earth. **(b)** What is $\dfrac{dp}{dh}(10.4)$?

♦ Type 2, exponential decay and a derivative

18.[A] If a population that grows exponentially is 500 initially and doubles every three years, at what rate is it growing after t years?

♦ Type 2, exponential growth

19. A power pack on a satellite produces $P = 50(\frac{1}{2})^{t/173}$ watts t years after the satellite is put in orbit. What is the half life of dP/dt?

♦ Type 2, exponential decay

20.[A] A population was 2000 at the beginning of 1980 and increased 1.5% every month. What was its relative rate of change with respect to time, measured in years?

♦ Type 2, exponential growth and a relative rate of change

21. Uranium-238 has a half life of 4.5×10^9 years. **(a)** How long does it take for a five-gram sample of Uranium-238 to decay to one-fourth its original weight? **(b)** At what rate is it decaying at that time?

♦ Type 2, exponential decay and a derivative

22. The number of bacteria in a culture grows exponentially. At 12 PM there are 1000 bacteria in the culture and at 5 PM there are 1500. When are there 2250 bacteria in the culture?

♦ Type 2, exponential growth y

23. A man has 10 milligrams of lead per liter in his blood from breathing polluted air. His body eliminates the lead with a half-life of approximately 16 days. If the half-life is exactly 16 days and the man is not exposed to more lead pollution, what is the lead concentration in his blood and how rapidly is it decaying 48 days later?

♦ Type 2, exponential decay and a derivative

24. What percentage of a five-gram sample of radium decomposes in 15 years? (The half life of radium is 1620 years.)

♦ Type 2, half-life

25. The number of bacteria in a test tube triples every 10 hours. How many were there and at what rate were they increasing initially if 20 hours later there were 9000 bacteria in the test tube?

♦ Type 2, exponential growth and a derivative

26. Do you earn more each year with a five-year savings certificate that pays 12% annual interest compounded quarterly or with one that pays 12.5% annual interest compounded annually?

♦ Type 2, compound interest

27. **(a^{O})** What rate of simple interest has the same yield after one year as 5% compounded continuously? **(b)** What rate of simple interest has the same yield after ten years as 5% compounded continuously?

♦ Type 2, compound interest

28. You, your sister, and your brother each put \$5,000 for three years in savings accounts that pay 8% annual interest. Your sister's bank compounds interest quarterly, your bank compounds interest daily (365 times each year), and your brother's bank pays simple interest without compounding. How much more money is in your account than in your sister's and brother's after the three years?

♦ Type 2, compound interest

29. A retiree puts \$2000 each year for three years in a savings account that pays 6% annual interest compounded annually. Then he closes that account, spends \$1000 of the proceeds, and puts the rest for three more years in an account that pays 8% annual interest compounded daily. What is the balance at the end of that time period?

♦ Type 2, compound interest

30. The half-life of plutonium-242 is 3.79×10^5 years. What percentage of a sample of plutonium decomposes in ten thousand years?

♦ Type 2, exponential decay

31. What are the growth factors every 30 years of the balance in a savings account that pays 5% annual interest **(a)** if interest is compounded annually, **(b)** if interest is compounded continuously, and **(c)** if simple interest is paid?

♦ Type 2, exponential growth and growth factors

32.º Carbon-14 (^{14}C) is formed in the upper atmosphere from nitrogen-14 (^{14}N) by the bombardment of neutrons in cosmic rays. The ^{14}C then disintegrates back into ^{14}N with the emission of beta particles and with a half-life of 5700 years. The result is an equilibrium of one ^{14}C atom per 10^{12} atoms of carbon in the atmosphere and in living things. When a plant or animal dies, it is no longer supplied with ^{14}C at the equilibrium ratio, so the amount of ^{14}C in a sample diminishes with time. (The amount of ^{14}C in a sample is determined by counting the beta particles with a Geiger counter.) Tests show that 20% of the ^{14}C in a piece of charcoal from an archaeological site has decomposed since the charcoal was a living tree. How old is the charcoal?

♦ Type 2, radioactive dating

33. Lava contains Uranium-238 (^{238}U), which is constantly decomposing into lead-206 (^{206}Pb) with a half-life of 4.5 billion years. In a volcanic eruption, the lead is removed from the lava, leaving pure uranium. If a sample of lava has one molecule of ^{206}Pb for each 99 molecules of ^{238}U, when did the volcanic eruption that formed it occur?

♦ Type 2, radioactive dating

34. Find the linear function $L(x)$ and the exponential function $E(x)$ such that $L(0) = 4$, $L(2) = 16$, $E(0) = 4$, and $E(2) = 16$. Draw their graphs together for $0 \le x \le 3.5$. In which of the three intervals, $x < 0$, $0 < x < 2$, and $x > 2$, is $E(x)$ less than $L(x)$, and in which is $E(x)$ greater than $L(x)$?

♦ Type 2, linear and exponential functions

35. Some prefixes for powers of 10 are listed below. **(a)** How many nanometers are there in a gigameter? **(b)** How many centimeters are there in an attometer?

Power of 10	10^{12}	10^9	10^6	10^3	10^2	10	10^{-1}
Prefix	tera	giga	mega	kilo	hecto	deka	deci

Power of 10	10^{-2}	10^{-3}	10^{-6}	10^{-9}	10^{-12}	10^{-15}	10^{-18}
Prefix	centi	milli	micro	nano	pico	femto	atto

♦ Type 1, powers of 10

36.[A] The values of the function $H(t)$ in the next table can be represented by an exponential function with errors that are less than 0.5. Find such a function and determine the largest error in its approximation of the values in the table.

t	0	1	2	3	4
$H(t)$	5.00	6.25	7.79	9.34	12.20

♦ Type 2, modeling with an exponential function, etc.

37. The table below lists the percentages of U. S. residents 5 to 18 years of age that were enrolled in school as determined in the censuses of 1870 through 1910. **(a)** What was the growth factor of the percentage from 1870 to 1880, from 1880 to 1890, from 1890 to 1900, and from 1900 to 1910? **(b)** What was the growth factor from 1870 to 1910. **(c)** How can the answer to part (b) be calculated from the answer to part (a)?

Year	1870	1880	1890	1900	1910
Percent enrolled	57.0	65.5	68.6	72.4	72.2

♦ Type 2, modeling with an exponential function, etc.

38. The next table lists the number of bank employees, the number of checks written, and the number of electronic transfers made in banking in the years 1950, 1960, 1970, and 1980,[(2)] **(a)** One of these sets of data can be modelled closely by an exponential function. Find an exponential function that fits this data with less than 2% error for each value. **(b)** The growth factors from one decade to the next increase significantly with time for another set of the data in the table and decrease for the third. Which is which?

Year	1950	1960	1970	1980
Bank employees (thousands)	600	820	1100	1510
Checks written (billions)	6.0	11.4	21.1	31.6
Electronic transfers (millions)	1.3	2.8	6.8	40.8

♦ Type 2, modeling with an exponential function, etc.

39. **(a)** How close is $(1+1/n)^n$ to $e = e^1$ for $n = 1$, 10, 100, and 1000? **(b)** Find by trial and error an integer n such that $(1+5/n)^n$ differs from e^5 by less than 3 but more than 2.

♦ Type 2, the definition of e

40. An initial balance of k dollars is made to a savings account that pays 5% annual interest. No further deposits or withdrawals are made, and the balance after t years is $B(t)$ dollars. What is $B'(9.5)$ **(a)** if the interest is compounded continuously, **(b)** if the interest is compounded annually, and **(c)** if simple interest is paid?

♦ Type 2, compound and simple interest

41.[A] **(a)** Give an equation of the tangent line to $y = 1 + e^{x^3 - x}$ at $x = 1$. Generate the curve and the tangent line in the window $-3 \le x \le 2, -1 \le y \le 5$ and copy them on your paper. **(b)** What is the vertical distance between the curve and the tangent line at $x = 0$ and at $x = 2$? (Give approximate decimal values.)

♦ Type 2, tangent lines, etc.

42. Give an equation of the tangent line to $y = 2 + e^{-(x-3)^2}$ at $x = 3$. Use the formulas to explain why the curve is below the tangent line for $x \neq 3$.

♦ Type 2, tangent lines, etc.

---

(2) "The Mechanization of Commerce," M. L. Ernst, *Scientific American*, Sept., 1982, New York, NY: Scientific American, Inc, 1982, p. 143.

43. **(a)** Use properties of x^3 and of e^x to find $\lim_{x\to\infty} e^{x^3}$ and $\lim_{x\to-\infty} e^{x^3}$. **(b)** Give an equation of the tangent line to $y = e^{x^3}$ at $x = 0$. Generate the curve and the tangent line as a partial check of your answers.

◆ Type 2, tangent lines, etc.

44.[A] Give an equation of the tangent line to $y = 3 + 2x - e^x$ at its highest point.

◆ Type 2, tangent lines, etc.

45. **(a)** What is the global minimum of $e^x - 5x$? **(b)** Give the equation of the tangent line to $y = e^x - 5x$ that has slope 1.

◆ Type 2, tangent lines, etc.

46.[A] What are the exact maximum and minimum values of x^2e^x for $-5 \le x \le 1$?

◆ Type 2, max/min in an interval

47. Use the formulas for the function and its first and second derivatives to analyze the graph of $y = 3 - e^{-x^2/2}$.

◆ Type 2, the First- and Second-Derivation Tests

48. Give an equation of the tangent line to $y = xe^{-x}$ at its inflection point.

◆ Type 2, an inflection point, etc.

49. What are the exact maximum and minimum values of xe^{2x} for $0 \le x \le 1$?

◆ Type 2, the First- and Second-Derivation Tests

50. What are the exact maximum and minimum values of $x - e^x$ for $-2 \le x \le 2$?

◆ Type 2, the First- and Second-Derivation Tests

51.[A] Solve **(a)** $\frac{dy}{dt} = \frac{1}{2}y,\ y(0) = 4$ and **(b)** $\frac{dy}{dt} = -y,\ y(5) = 1$.

◆ Type 2, initial-value problem with an exponential function

52. Solve **(a)** $\frac{dy}{dt} = 3y,\ y(0) = 7$ and **(b)** $\frac{dy}{dt} = 0.01y,\ y(100) = 20$.

◆ Type 2, initial-value problem with an exponential function

53. Solve **(a**[A]**)** $\frac{dy}{dt} = \ln(2)y,\ y(0) = 300$ and **(b)** $\frac{dy}{dt} = -\ln(2)y,\ y(0) = 7$.

◆ Type 2, initial-value problem with an exponential function

54. A vertical beam of light which has intensity I_0 when it hits the ocean surface has intensity $I = I_0e^{-1.4y}$ at a depth of y meters. **(a)** At what depth is the light intensity equal to 1% of its value at the surface? **(b)** What is the rate of change of the intensity with respect to the depth at the depth from part (a)?

◆ Type 2, exponential decay

55.[O] What is the x-derivative of e^{-2Z} at $x = 0$ if $Z = Z(x)$ such that $Z(0) = 3$ and $\frac{dZ}{dx}(0) = -1$?

◆ Type 3, the Chain Rule with an exponential function

56. What is $\frac{dR}{dx}(5)$ if $R(x) = xe^{s(x)}$, $s(5) = -4$, and $\frac{ds}{dx}(5) = 3$?

◆ Type 3, the Chain Rule with an exponential function

57.[A] Which rectangle with its base on the x-axis and upper corners on $y = e^{-x^2}$ has maximum area? Give its exact area.

◆ Type 2, a max/min problem

58. A store will sell $5000e^{-x/20}$ shirts if they are sold for x dollars each ($5 \le x \le 50$). What price would maximize the total revenue from the shirts?

◆ Type 2, a max/min problem

59. Find the exact x-coordinates of the three inflection points of $y = x^3e^x$.

◆ Type 3, an inflection point, etc.

60. Prove that the tangent line to $y = 1 + xe^{-x^2}$ at $x = 0$ intersects the curve only at the point of tangency.

◆ Type 4, a tangent line, etc.

61. What is the range of $(1 - e^x)^{-1}$? Justify your answer.

◆ Type 4, the range of a function

62. Prove that the equation $xe^{-x^2} = 10^{-10}$ has one and only one solution with $x > 1$. Use your calculator or computer to find the solution with two-decimal-place accuracy.

♦ Type 4, analyzing the solutions of an equation, etc.

63. Prove that **(a)** $2^x + 3^{-x} < e^x + e^{-x}$ for $x > 0$ and **(b)** $2^x + 3^{-x} > e^x + e^{-x}$ for $x < 0$.

♦ Type 4, deriving inequalities

64. What is the maximum of $x^{20}e^{-x}$ for $x \geq 0$?

♦ Type 4, a max/min problem

65. Find the most extensive open intervals in which $\frac{1}{2}e^{2x} - 6e^x + 5x$ is increasing and decreasing and in which its graph is concave up and concave down.

♦ Type 4, the First- and Second-Derivation Tests

66. What is the minimum value of $2^x + 4^{-x}$?

♦ Type 4, a max/min problem

67. Find the three solutions of $x^2 = 1.5^x$.

♦ Type 4, solving an equation

68. Find the x-coordinate of the inflection point of $y = x^2 - 2^x$.

♦ Type 4, an inflection point

Section 3.5

Logarithms and their derivatives

OVERVIEW: *Logarithms are used most frequently as tools in calculations with exponential functions, but they are also used directly in mathematical models, including the* DECIBEL LEVELS *of sounds and radio and TV signals and the* PH *levels of ionized solutions. In this short section we first look at two of these applications and illustrate how* LOGARITHMIC SCALES *are used in graphs. Next, we show how the number e, defined to be the limit of* $(1 + 1/n)^n$ *as* $n \to \infty$*, arises in the formulas for derivatives of logarithms and why the natural logarithm with base e is the most convenient logarithm for calculus. We then use these differentiation formulas in applications and to analyze the graphs of functions constructed from logarithms.*†

Topics:

- ***Decibels and pH***
- ***Logarithmic scales***
- ***The derivative of*** $\log_b x$ ***and the number*** e
- ***The derivatives of*** $\ln x$ ***and*** $\ln(y(x))$
- ***Graphs of functions constructed from logarithms***

Decibels

Sounds come to you as oscillations in air pressure which cause your eardrum and the bones of the ear to vibrate, giving you the sensation of the sound. The loudness of the sound is often measured in DECIBELS.‡ Table 1 lists the variation P in air pressure for sounds at various decibel levels D.[(1)] The decibel scale is designed so that a change of one decibel is the approximate smallest change in loudness that most humans can detect. A whisper might be at 10 decibels, and the music from a rock band might be as loud as 120 decibels.

TABLE 1

D (decibels)	0	20	40	60	80	100	120
P (grams per square meter)	0.002	0.02	0.2	2	20	200	2000

Question 1 The pressure P in Table 1 is an exponential function of the decibel level D. Give a formula for this function.

Because the pressure is an exponential function of the decibel level, the decibel level is expressed in terms of the pressure by using a logarithm.

Example 1 Give a formula for D in terms of P.

SOLUTION We begin with the formula $P = 0.002(10^{D/20})$ from Question 1. Since $1/0.002 = 500$, we first multiply both sides by 500 to obtain $500P = 10^{D/20}$. Then applying the common logarithm to both sides gives $\log_{10}(500P) = D/20$, and we finally obtain $D = 20\log_{10}(500P)$. □

The pressure in Table 1 is a measure of the physical STIMULUS of the sound and the decibel level is a measure of the SENSATION that we feel. In Section 7.1 we will see, by solving the corresponding differential equation, why sensations are often measured by calculating logarithms of the variables that measure stimuli. We will also see in Section 7.1 that logarithms are needed to solve other types of differential equations.

---

†Consult Appendix 2, as necessary, as you study this Section.

‡One BELL, named in honor of Alexander Graham Bell (1847–1922), equals 10 decibels.

(1)Data adapted from *McGraw-Hill Encyclopedia of Science & Technology*, Vol. 5, New York: McGraw-Hill, Inc. 1992, pp. 65–66.

♦ SUGGESTIONS TO INSTRUCTORS: Have students begin work on Questions 1 through 12 before class. Problems 11, 12, 18, 19, 26, 31, 35, and 36 are good for in-class discussion.

pH of ionized solutions

In chemistry, the pH of a solution is defined by $\text{pH} = -\log_{10}([\text{H}^+])$, where $[\text{H}^+]$ is the concentration in the solution of hydrogen ions, measured in moles per liter. One mole is 6×10^{23} molecules.[(2)]

Example 2 Pure water has a pH of 7. What is the hydrogen ion concentation of pure water?

SOLUTION The hydrogen ion concentration in pure water satisfies the equation $-\log_{10}([\text{H}^+]) = 7$ or $\log_{10}([\text{H}^+]) = -7$. Applying the function 10^x to both sides gives $[\text{H}^+] = 10^{-7}$, so the hydrogen ion concentration of pure water is 10^{-7} moles per liter. □

Question 2 Solutions with pH less than 7 are acidic, and those with pH greater than 7 are alkaline or basic. How can you distinguish acidic from alkaline solutions if you know their hydrogen ion concentrations?

Question 3 What is the growth factor of $[\text{H}^+]$ from a solution with $\text{pH} = 8$ to a solution with $\text{pH} = 7$?

Logarithmic scales

If the values of a variable or function to be plotted are too spread out to make a clear sketch, the common logarithms of the numbers are often plotted instead by using a LOGARITHMIC SCALE. Table 2 gives for each of the nine planets, the average r of of its closest and farthest distances from the sun, and its period (the length of its year) T measured in earth years.[(3)]

TABLE 2

Planet	Mercury	Venus	Earth	Mars	Jupiter
r (million kilometers)	58	108	147	227	778
T (earth years)	0.241	0.616	1	1.88	11.9

Planet	Saturn	Uranus	Neptune	Pluto
r (million kilometers)	1428	2875	4498	5914
T (earth years)	29.5	84.1	165	248

When this data is plotted in an rT-plane, as in Figure 1, the values for the planets nearest to the sun cannot be distinguished. To show these values more clearly, we can use logarithmic scales on the r- and T-axes in Figure 2, where the values of $\log_{10} r$ in Table 3 are plotted on the horizontal axis and values of $\log_{10} T$ on the vertical axis.

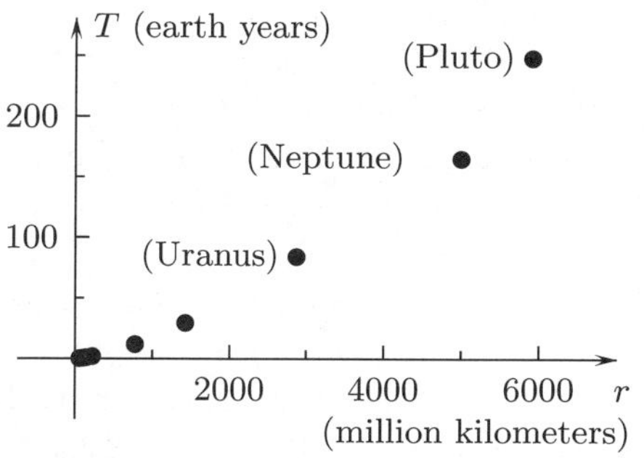

FIGURE 1

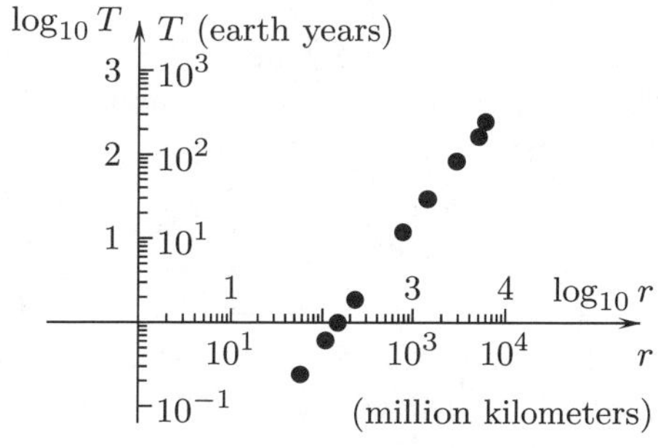

FIGURE 2

(2) Data from *McGraw-Hill Encyclopedia of Science & Technology*, Vol. 1, New York, NY: McGraw-Hill, Inc. 1992, pp. 52-53.

(3) Data adapted from *The World Book Encyclopedia*, Vo. 15, Chicago, IL: Childcraft International, Inc., 1977, pp. 466–467.

TABLE 3

Planet	Mercury	Venus	Earth	Mars	Jupiter
$\log_{10} r$	1.76	2.03	2.17	2.36	2.89
$\log_{10} T$	−0.618	−0.210	0	0.274	1.08

Planet	Saturn	Uranus	Neptune	Pluto
$\log_{10} r$	3.16	3.46	3.65	3.77
$\log_{10} T$	1.47	1.92	2.22	2.39

In Figure 2, the values of r are given by the numbers below the horizontal axis and the values of $\log_{10} r$ by the numbers above the horizontal axis. Similarly, the values of T are given on the right of the vertical axis and the values of $\log_{10} T$ on the left.

Question 4 **(a)** At what values of r and T do the axes in Figure 2 intersect ? **(b)** The short tick marks between the r-axis and $T = 10$ on the vertical axis in Figure 2 are at $T = 2, 3, 4, 5, 6, 7, 8$, and 9. What are the values of T at the other short tick marks on the T-axis?

The points in Figure 2 appear very close to the line shown in Figure 3. What does this tell us about the relationship between the average r of the planets' least and greatest distance from the sun and their periods T?

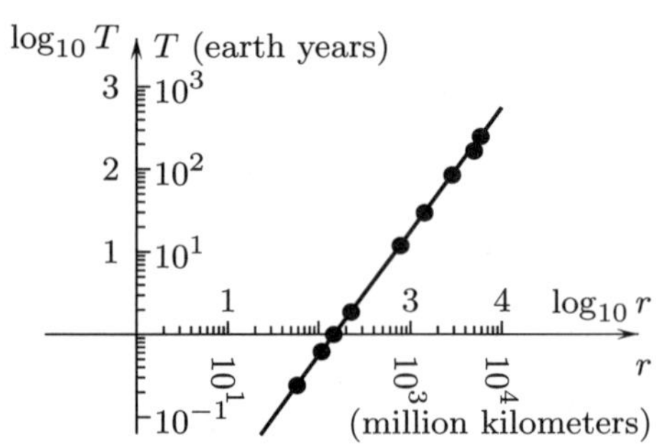

FIGURE 3

Question 5 Use the values of $\log_{10} r$ and $\log_{10} T$ at the dots corresponding to Mercury and Pluto in Figure 3 to estimate the slope of the line relative to $\log_{10} r$ and $\log_{10} T$ coordinates.

Question 6 The line in Figure 3 actually has slope 1.5. Consequently it has the equation $\log_{10} T = 1.5 \log_{10} r + b$ with a constant b. Solve this equation to obtain a formula relating T and r.

The mathematical model $T = Cr^{1.5}$ from Question 6 that relates T and r for the nine planets is the third of Kepler's laws of planetary motion. The German astronomer Johann Kepler (1571–1630) discovered his three laws between 1605 and 1618 after years of studying astronomical data. His first law states that each planet moves in an elliptical path with the sun at one focus. The second states that each planet moves so that the line segment between it and the sun sweeps out equal areas in equal times.†

†In Chapter 9, we will show how Kepler's laws can be derived from Newton's law of gravity, as was demonstrated by Isaac Newton in 1687.

The derivative of $\log_b x$ *and the number* e

As you can predict from the graph in Figure 4, the derivative of $\log_b x$ for $b > 1$ is positive for all $x > 0$ and tends to ∞ as $x \to 0^+$ and to 0 as $x \to \infty$. We want a formula for that derivative. We begin by experimenting with the common logarithm $\log_{10} x$ and the natural logarithm $\ln x = \ln_e x$.

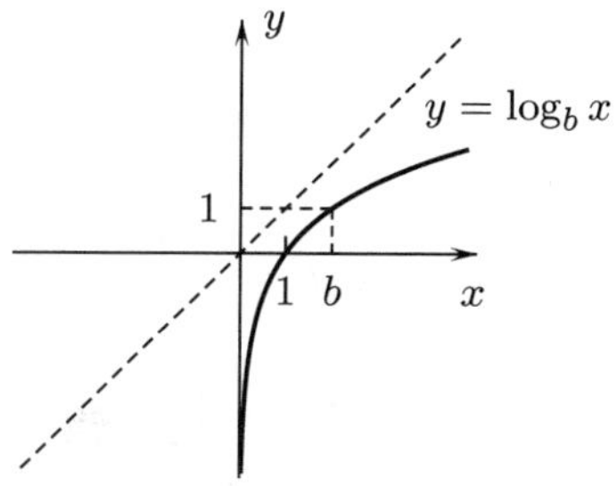

FIGURE 4

Question 7 Complete the following tables of approximate derivatives of $\log_{10} x$ and $\ln x$ and use it to predict formulas for the derivatives as functions of x. Use the secant-line program of Section 2.3 with $A = x, B = x + 0.00001$ if it is available to you in the window $-1 \leq x \leq 6$, $-1.5 \leq y \leq 2$ for $x = 1, 2$ in the window $-3 \leq x \leq 30$, $-1.5 \leq y \leq 2$ for $x = 10, 20$, and in the window $-15 \leq x \leq 150$, $-1 \leq y \leq 3$ for $x = 100$.

x	$\dfrac{\log_{10}(x + 10^{-5}) - \log_{10} x}{10^{-6}} \doteq$
1	0.434294
2	0.217147
10	
20	
100	

x	$\dfrac{\ln(x + 10^{-6}) - \ln x}{10^{-6}} \doteq$
1	0.9999995
2	0.49999989
10	
20	
100	

We now come to one of the most remarkable aspects of calculus: the fact that the exact formula for the derivative of $\log_b x$ for any $b > 1$ requires the number e, defined by

$$e = \lim_{n\to\infty} \left(1 + \frac{1}{n}\right)^n. \tag{1}$$

In this derivation we will use the following properties of the logarithm:

$$\log_b(y/z) = \log_b y - \log_b z \quad \text{for positive } y \text{ and } z \tag{2}$$

$$\log_b(y^n) = n \log_b y \quad \text{for positive } y \text{ and any } n \tag{3}$$

We start with the definition of the derivative. For fixed $x > 0$,

$$\frac{d}{dx}(\log_b x) = \lim_{\Delta x\to 0} \frac{\log_b(x + \Delta x) - \log_b x}{\Delta x}. \tag{4}$$

Question 8 Use identity **(2)** to show that, for nonzero Δx,

$$\frac{\log_b(x+\Delta x) - \log_b x}{\Delta x} = \frac{1}{\Delta x}\log_b\left(1 + \frac{\Delta x}{x}\right). \tag{5}$$

Next we set $\Delta x = \dfrac{x}{n}$ with a positive integer n, so that

$$\frac{1}{\Delta x} = \frac{n}{x} \quad \text{and} \quad \frac{\Delta x}{x} = \frac{1}{n}. \tag{6}$$

Question 9 Use formulas **(4)**, **(5)**, and **(6)** to show that

$$\frac{\log_b(x+\Delta x) - \log_b x}{\Delta x} = \frac{1}{x}\log_b\left[\left(1 + \frac{1}{n}\right)^n\right]. \tag{7}$$

Since x is fixed and positive, we can cause Δx to tend to 0 by having n tend to ∞. We assume that the derivative exists, so that $\log_b x$ is continuous for $x > 0$. Formula **(7)** then yields

$$\begin{aligned}\frac{d}{dx}(\log_b x) &= \lim_{\Delta x \to 0} \frac{\log_b(x+\Delta x) - \log_b x}{\Delta x} \\ &= \lim_{n\to\infty} \frac{1}{x}\log_b\left[\left(1+\frac{1}{n}\right)^n\right] \\ &= \frac{1}{x}\log_b\left[\lim_{n\to\infty}\left(1+\frac{1}{n}\right)^n\right].\end{aligned}$$

The last equation with **(1)** gives

$$\frac{d}{dx}(\log_b x) = \frac{\log_b e}{x}. \tag{8}$$

Formula **(8)** shows that, no matter what base b we use for the logarithm, the number e is involved in the formula for its derivative. Moreover, because $\log_e e = 1$, we have the following result for the natural logarithm, $\ln x = \log_e x$, which you probably predicted in Question 7.

Theorem 1 For $x > 0$,

$$\frac{d}{dx}(\ln x) = \frac{1}{x}. \tag{9}$$

The logarithm with base e is called the "natural" logarithm because its derivative **(9)** has the simplest formula.

To put formula **(8)** for the derivative of $\log_b x$ in a more useful form, we can use the rule $\log_b e = 1/(\log_e b) = 1/(\ln b)$ to obtain

$$\frac{d}{dx}(\log_b x) = \frac{1}{x(\ln b)}. \tag{10}$$

We could also derive **(10)** from **(9)** by employing the identity $\log_b x = (\ln x)/(\ln b)$. We will use **(10)** only with the common logarithm, for which $b = 10$.

Question 10 Find an exact equation of the tangent line to $y = \ln x$ at $x = 2$. Generate the graph and the tangent line together as a partial check of your answer.

Question 11 **(a)** What are $\lim_{x\to 0+} [(\ln x)^2]$ and $\lim_{x\to\infty} [(\ln x)^2]$ and why? (Use properties of $\ln x$.) **(b)** Find, without using any derivatives, the minimum value of $(\ln x)^2$ and the value of x where it occurs. **(c)** Use the first derivative to determine the most extensive open intervals in which $(\ln x)^2$ is increasing and in which it is decreasing. **(d)** Use the second derivative to find the exact coordinates of the inflection point of $y = (\ln x)^2$. Generate the graph as a partial check of your answers.

Derivatives of natural logarithms of functions

The Chain Rule from Section 3.3 and formula **(9)** for the x-derivative of $\ln x$ yield the folowing formula for the x-derivative of $\ln y$ for positive differentiable functions $y = y(x)$.

Theorem 2 *If $y = y(x)$ is positive and has a derivative at x, then $\ln(y(x))$ has a derivative at x, which is given by*

$$\frac{d}{dx}(\ln y) = \frac{d}{dy}(\ln y)\,\frac{dy}{dx} = \frac{1}{y}\frac{dy}{dx} \tag{11a}$$

or, with prime notation,

$$\frac{d}{dx}[\ln y(x)] = \frac{1}{y(x)}\,y'(x). \tag{11b}$$

Learn **(11)** in words: The derivative of the natural logarithm of a function equals 1 divided by the function, multiplied by the derivative of the function.

Example 3 **(a)** What is the domain of $\ln(x^2+1)$? **(b)** What is its x-derivative?

SOLUTION **(a)** $\ln(x^2+1)$ is defined for all x because x^2+1 is positive for all x and $\ln y$ is defined for $y > 0$. **(b)** We use differentiation formula **(11a)** with $y = x^2+1$:

$$\frac{d}{dx}[\ln(x^2+1)] = \frac{1}{x^2+1}\frac{d}{dx}(x^2+1) = \frac{2x}{x^2+1}. \ \square$$

Question 12 What are the maximum and minimum values of $\ln(x^2+1)$ for $-3 \le x \le 3$? Generate the graph for $-3 \le x \le 3, -0.5 \le y \le 3$ as a partial check of your answers.

The next example requires **(11)** and the Product Rule.

Example 4 What is the derivative of $x^2\ln(3x+1)$?

SOLUTION As we saw in earlier sections, if values of a function are calculated by a sequence of operations, then in most cases its derivative is found by applying the corresponding differentiation formulas in the reverse order. In this case we would calculate $x^2\ln(3x+1)$ from x by first calculating $3x+1$, then $\ln(3x+1)$ and x^2, and finally the product $x^2\ln(3x+1)$. Consequently, to find the derivative, we first use the Product Rule, then differentiate x^2 and the logarithm, and finally differentiate $3x+1$:

$$\begin{aligned}
\frac{d}{dx}[x^2\ln(3x+1)] &= \ln(3x+1)\frac{d}{dx}(x^2) + x^2\frac{d}{dx}[\ln(3x+1)] \\
&= 2x\ln(3x+1) + x^2\frac{1}{3x+1}\frac{d}{dx}(3x+1) \\
&= 2x\ln(3x+1) + \frac{3x^2}{3x+1}. \ \square
\end{aligned}$$

The Chain Rule combined with formula **(10)** for the x-derivative of $\log_b x$ for positive $b \ne 1$, gives the following differentiation formula for $\log_b y$ with y a differentiable function of x:

$$\frac{d}{dx}(\log_b y) = \frac{1}{y(\ln b)}\frac{dy}{dx}. \tag{12}$$

Example 5 An acidic solution contains 5×10^{-9} moles of hydrogen ions per liter at a time when its hydrogen-ion concentration is decreasing 0.4×10^{-9} moles per liter per hour. At what rate is the pH of the solutions increasing or decreasing at that moment?

SOLUTION We let h be the hydrogen-ion concentration and let $p = -\log_{10} h$ be the pH. Then by **(12)** with t as the independent variable,

$$\frac{dp}{dt} = \frac{d}{dt}(-\log_{10} h) = -\frac{1}{h\ln(10)}\frac{dh}{dt}.$$

Setting $h = 5 \times 10^{-9}$ and $dh/dt = -0.4 \times 10^{-9}$ in the last formula yields

$$\frac{dp}{dt} = -\frac{1}{(5 \times 10^{-9})\ln(10)}(-0.4 \times 10^{-9}) = \frac{0.08}{\ln(10)}.$$

The pH of the solution is increasing $\dfrac{0.08}{\ln(10)} \doteq 0.03474$ units per hour. □

Comparing ln x *and* x^n *for large positive* x

The values of $\ln x$ tend to ∞ as $x \to \infty$, but very slowly compared to x^n with any positive n, as is illustrated in the next table, which compares values of $\ln x$ with those of $x^{1/3}$.

x	10	100	10,000	1,000,000
$x^{1/3} \doteq$	2.15	4.64	21.54	100
$\ln x \doteq$	2.30	4.60	9.21	13.82

We state this principle as a rule:†

Rule 1 *For any positive constant n, x^n is much larger than $\ln x$ for sufficiently large positive x.*

Example 6 What are **(a)** $\lim_{x\to\infty}(3+2\ln x - x)$ and **(b)** $\lim_{x\to 0^+}(3+2\ln x - x)$ and why? **(c)** What are the inflection points of $y = 3 + 2\ln x - x$?

SOLUTION **(a)** $\lim_{x\to\infty}(3+2\ln x - x) = -\infty$ because x dominates $\ln x$ as $x \to \infty$ and $-x \to -\infty$.

(b) $\lim_{x\to 0^+}(3+2\ln x - x) = -\infty$ because $3 - x \to 0$ and $\ln x \to -\infty$ as $x \to 0^+$.

(c) $\dfrac{dy}{dx} = \dfrac{d}{dx}(3 + 2\ln x - x) = 2x^{-1} - 1$ and $\dfrac{d^2y}{dx^2} = \dfrac{d}{dx}(2x^{-1} - 1) = -2x^{-2}$. The second derivative is negative in the entire domain $(0, \infty)$ of the function, so the entire graph is concave down and has no inflection points. (These results are corroborated by the graph in Figure 5.) □

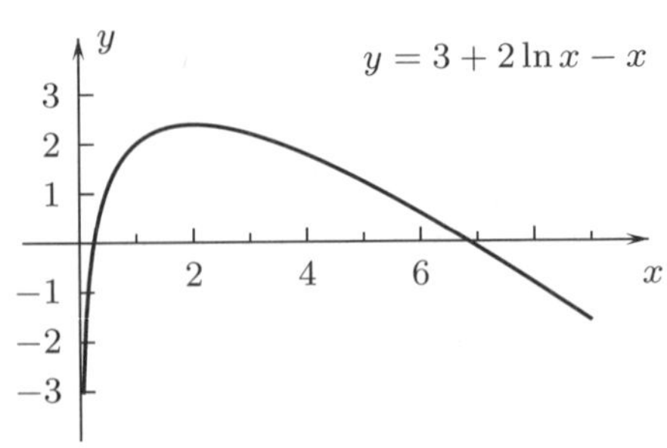

FIGURE 5

†We will establish this result in Section 5.7 by using L'HÔPITAL'S RULE.

Principles and procedures

- Logarithms and their derivatives are used directly in applications but are more frequently needed as tools for dealing with exponential functions and, as we will see in Chapters 6 and 7, for evaluating integrals and solving differential equations.
- Use the formula $\log_b x = (\ln x)/(\ln b)$ to calculate values and derivatives of $\log_b x$ from values and derivatives of $\ln x$. This formula also shows that the graphs of $\log_b x$ for various values of b have the same general shape but are closer to the x-axis for larger values of b.
- The logarithm to the base 10 is called the "common" logarithm because it is the most commonly used outside of calculus and physics. The logarithm to the base e is called the "natural" logarithm because it is has the simplest formula for its derivative.
- Learn formulas **(9)** and **(11)** for the derivatives of the natural logarithm and of the natural logarithm of functions.
- For any positive constant n, x^n dominates $\ln x$ and $\log_{10} x$ as $x \to \infty$ in the sense that x^n is much larger than $\ln x$ and $\log_{10} x$ for sufficiently large positive x.
- The common logarithm $\log_{10} x$ is used to measure decibels and pH scales and with logarithmic scales on coordinate axes, in part, because large ranges of x result in small ranges of $\log_{10} x$.

Tune-Up Exercises 3.5♦

[A]*Answer provided.* [O]*Outline of solution provided.*

T1.[A] Solve **(a)** $\log_{10}(x^2) = -3.5$, **(b)** $10^{2x} = 50$, **(c)** $\sqrt{\ln x} = 2$, and **(d)** $e^{\sqrt{x}} = 0.001$. Give exact answers and approximate decimal values, and check your answers with your calculator or computer.

♦ Type 1, solving equations

T2.[A] Find the number (I)—(VIII) in the second column below that equals each of the numbers (a)–(h) in the first column. Use properties of the logarithm to show that the different expressions represent the same numbers. Check by evaluating the expressions with your calculator or computer.

(a) $\log_{10}(300)$	(I) $2\log_{10}(3) + 2\log_{10}(4)$
(b) $\log_{10}(10)$	(II) $5\log_{10}(2)$
(c) $\log_{10}(25)$	(III) $2\log_{10}(3)$
(d) $\log_{10}(30) - \log_{10}(3)$	(IV) 1
(e) $\log_{10}(1)$	(V) $2\log_{10}(5)$
(f) $\sqrt{\log_{10}(81)}$	(VI) $2 + \log_{10}(3)$
(g) $\log_{10}(3^2 + 4^2)$	(VII) 0
(h) $\log_{10}[9(16)]$	(VIII) $2\sqrt{\log_{10}(3)}$

♦ Type 1, properties of logarithms

T3.[A] **(a)** Give an equation of the tangent line to $y = \ln x$ at $x = 1$. **(b)** Give an equation of the tangent line to $y = e^x$ at $x = 0$. Generate the graphs of the two functions and the two tangent lines together in a window with equal scales on the axes. Explain the symmetry of the result.

♦ Type 1, a tangent line

Find the derivatives in Exercises T4 through T7.

T4.[A] $\dfrac{d}{dx}(3\ln x - 4e^x)$

♦ Type 1, a derivative

T5.[O] $\dfrac{d}{dx}[\ln(x^2 + 3x)]$

♦ Type 1, a derivative

T6.[O] $f'(5)$ for $f(x) = \sqrt{\ln x}$

♦ Type 1, a derivative

T7.[A] $\dfrac{d}{dx}\left(\dfrac{\ln x}{\log_{10} x}\right)$

♦ Type 1, a derivative

♦The Tune-up Exercises and Problems are classified by type and content. The types are (1) basic, reactive; (2) basic reflective; (3) intermediate, reactive; (4) intermediate, reflective; (5) advanced, reactive; (6) advanced, reflective; and (7) advanced, theoretical.

T8.[O] What are the derivatives of $\ln y$ and e^y with respect to x at $x = 7$ if y is a function of x such that $y(7) = 10$ and $y'(7) = 5$?

◆ Type 1, derivatives

Problems 3.5

[A]*Answer provided.* [O]*Outline of solution provided.*

Find the derivatives in Problems 1 through 10.

1.[A] $\frac{d}{dx}(3x - 4\ln x)$

◆ Type 1, a derivative

2.[A] $L'(300)$ for $L(x) = \ln x$

◆ Type 1, a derivative

3.[O] $\frac{d}{dx}[(\ln x)^7]$

◆ Type 1, a derivative

4.[O] $y'(x)$ for $y(x) = x^2 \ln x$

◆ Type 1, a derivative

5.[A] $\frac{d}{dx}[\ln(1 - e^x)]$

◆ Type 1, a derivative

6. $F'(3)$ for $F(x) = x \ln x$

◆ Type 1, a derivative

7. $\frac{d}{du}\left(\frac{\ln u}{u}\right)$

◆ Type 1, a derivative

8. $\frac{d}{dt}[\ln(6t^5 - 5t^6)]$

◆ Type 1, a derivative

9.[A] $\frac{d}{dx}[\ln(\ln x)]$

◆ Type 1, a derivative

10. $\frac{d}{dx}(e^{1+\ln x})$

◆ Type 1, a derivative

11.[A] Give an equation of the tangent line to $y = x - 5\ln x$ at $x = 5$. Generate the curve and the tangent line in the window $-2 \le x \le 16, -5 \le y \le 5$ and copy them on your paper.

◆ Type 1, a tangent line

12.[A] **(a)** What is the domain of $y = \sqrt{\ln x}$ and why? **(b)** Give an equation of the tangent line to the curve $y = \sqrt{\ln x}$ at $x = e$. **(c)** Does the tangent line of part (b) approximate the curve better at $x = 2.5$ or at $x = 2.8$? Give your calculations.

◆ Type 1, a tangent line , etc.

13. **(a)** What is the domain of $x \ln x$? **(b)** What is $\lim_{x\to\infty}(x \ln x)$ and why? **(c)** Use your calculator or computer to predict $\lim_{x\to 0^+}(x \ln x)$. **(d)** Analyze the graph of $x \ln x$ by studying its first and second derivatives. Then sketch the graph.

◆ Type 1, limits and the First- and Second-Derivative Tests

14. **(a)** How are the curves $y = \frac{1}{2}\ln(x^2)$ and $y = \ln x$ related? **(b)** Give equations of the tangent lines to $y = \frac{1}{2}\ln(x^2)$ at the points where it intersects the x-axis.

◆ Type 2, tangent lines, etc.

15. How do the graphs of $e^{\ln x}$ and $\ln(e^x)$ differ and why?

◆ Type 2, domains

16.[A] Find an equation of the tangent line to $y = \ln x + 1/x$ at its inflection point.

◆ Type 2, an inflection point and a tangent line

17. Where is $y = e^x - \ln x$ concave up and where is it concave down? Justify your answer.

◆ Type 2, concavity

18. What are the exact maximum and minimum values of $(\ln x)^2$ for $0.1 \le x \le 10$?

◆ Type 2, maximum and minimum in an interval

19.[A] What are the maximum and minimum values of $\frac{\ln x}{\sqrt{x}}$ for $1 \le x \le 20$?

◆ Type 2, maximum and minimum in an interval

20. The following table lists the number n of television stations and the number m of television sets in mainland China in five years.(4) **(a)** Plot this data in the tn- and tm-planes with logarithmic scales on the vertical axes in Figures 6 and 7. **(b)** What does the quantity $10^{\log_{10} m - \log_{10} n}$ represent?

NUMBER OF TELEVISION STATIONS AND SETS IN MAINLAND CHINA

Year	1958	1978	1983	1988	1992
n = Number of TV stations	3	32	52	422	591
m = Number of TV sets	200	3×10^6	36×10^6	125×10^5	200×10^6

♦ Type 3, a logarithmic scale

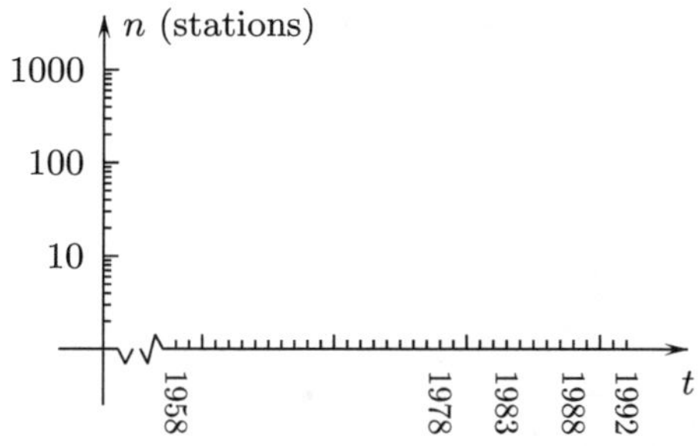

FIGURE 6

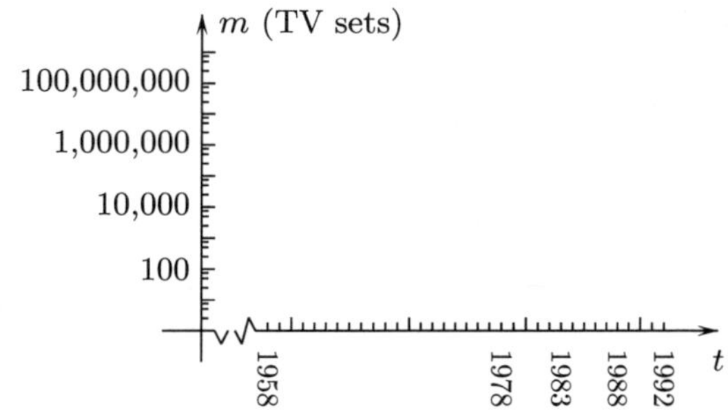

FIGURE 7

21. Draw the graph of pH $= -\log_{10}[\mathrm{H}^+]$ in a $[\mathrm{H}^+]$pH-plane with a logarithmic scale on the horizontal $[\mathrm{H}^+]$-axis as in Figure 8.

♦ Type 3, a logarithmic scale

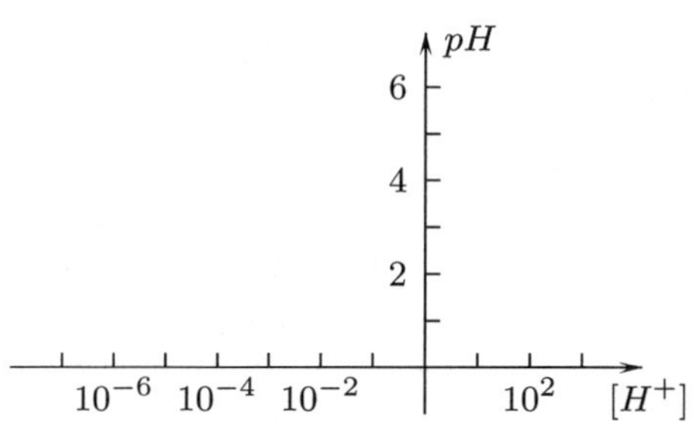

FIGURE 8

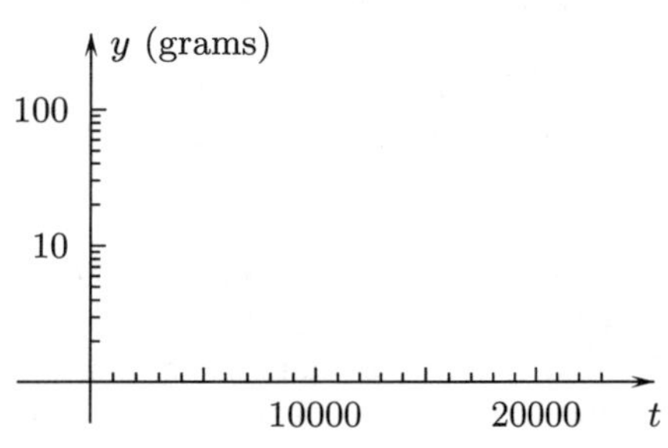

FIGURE 9

22. The next table gives the mass of a sample of radioactive carbon-14 at 5,000 year intervals. **(a)** Plot these values with a logarithmic scale on the vertical axis as in Figure 9. **(b)** Explain the shape of the graph.

t (years)	0	5,000	10,000	15,000	20,000	25,000
y(grams)	100	54.4	29.6	16.1	8.8	4.8

♦ Type 3, a logarithmic scale

(4) Data from "Peaceful evolution: the case of television reform in post-Mao China", *Media, Culture, and Society*, Vol. 16, April, 1994, Thousand Oaks, CA: SAGE Publications, p. 222.

23. Electromagnetic radiation with frequency ν cycles per second has WAVELENGTH λ (lambda) meters, where $\lambda = [3 \times 10^8]/\nu$. Here 3×10^8 is the velocity of light, measured in meters per second. **(a)** The frequencies of five types of electromagnetic radiation are given below. Calculate the corresponding wavelengths and plot this data using logarithmic scales on the ν- and λ-axes as in Figure 10. **(b)** The points you plot appear to be on a line. Are they? Explain.

	Frequency ν	Wavelength λ
An AM radio station (600 KHz)	6×10^5	
An FM radio station (100 MHz)	10^8	
Microwaves	10^4	
Thermal (infrared) radiation	2×10^{14}	
Orange visible light	6×10^{16}	
X-rays	10^{20}	

♦ Type 3, logarithmic scales

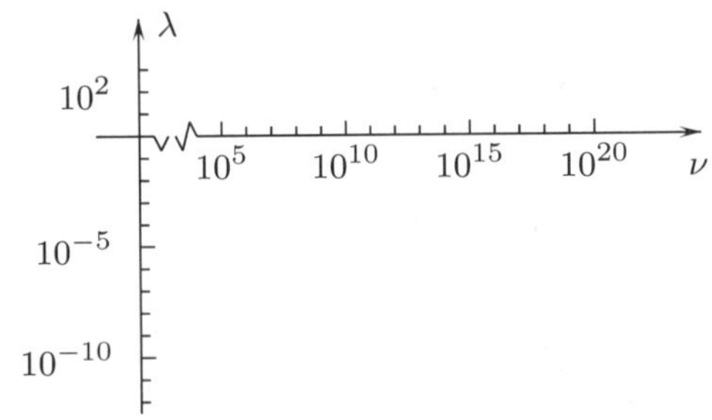

FIGURE 10

24. The magnitude of a star can be defined by the formula $M = -2.5 \log_{10}(kI)$, where k is a constant and I is the intensity of the light from the star. **(a)** What is the ratio of the intensities of light from the brightest star Sirius, which has a magnitude of -1.6 and from the star Betelgeuse, which has a magnitude of 0.9? **(b)** Show that dM/dI is independent of the constant k.

♦ Type 2, a ratio and a derivative

25. The magnitude M of an earthquake in the Richter scale can be defined by the formula $M = \frac{2}{3} \log_{10} E + c$, where E is the total energy of the earthquake, measured in hundreds of thousands of kilowatt-hours, and c is a constant. **(a)** What are the growth factors in the total energy from an earthquake of magnitude 5 to an earthquake of magnitude 6, from an earthquake of magnitude 6 to an earthquake of magnitude 7, and from an earthquake of magnitude 7 to an earthquake of magnitude 8? **(b)** Give a formula for dM/dE in terms of E.

♦ Type 2, a growth factor and a derivative

26. Based on the fomula $D = 20 \log_{10}(500P)$ from Example 1 for the decibel level of a sound in terms of the variation P in air pressure caused by the sound, how rapidly is the decibel level of a scream increasing when the variation in air pressure from the scream is 10^{-4} pounds per square inch and is increasing 10^{-3} pounds per square inch per second?

♦ Type 2, decibels

27. Decibels are used in describing the effect of antennas, amplifiers, and transmission lines on the power of radio signals. The decibel gain from each element of a circuit is defined by the formula $dB = \log_{10}\left(\frac{P_{\text{out}}}{P_{\text{in}}}\right)$, where P_{in} is the power of the radio signal entering the element and P_{out} is the power of the signal leaving it. A transmission line contains two amplifiers that cause 6 and 10 decibel gains and an antenna that causes a 3 decibel gain. If a signal with 2 watts of power is fed into the initial amplifier, what is the power of the resulting signal leaving the antenna?

♦ Type 4, decibels

28. Based on the definition $-\log_{10}[\mathrm{H}^+]$ of pH as a function of the concentration $[\mathrm{H}^+]$ of hydrogen ions in a solution, how rapidly is the hydrogen-ion concentration rising or falling in a solution in which the pH is 8 units and is rising 0.1 units per hour?

♦ Type 2, the Chain Rule with a logarithm

29. A PRIME NUMBER is an integer greater than 1 that cannot be written as a product of smaller integers greater than 1. The first six prime numbers are 2, 3, 5, 7, 11, and 13. According to the Prime-Number Theorem, there are approximately $n/\ln n$ prime numbers less than n for large n. Based on this result, approximately how many prime numbers should there be that are **(a)** less than 1000, **(b)** less than 10,000, and **(c)** less than 100,000? **(d)** There are, in fact, 168 prime numbers less than 1000, 1229 prime numbers less than 10,000, and 9592 prime numbers less than 100,000.[6] Recall that the relative error is the error divided by the actual value. Calculate the relative errors in the values from parts (a) through (c). (As these numbers suggest, the relative error tends to 0 as $n \to \infty$.)

♦ Type 4, counting prime numbers

30. What is the t-derivative of $\ln(1+w)$ at $t = 1$ if $w = w(t)$ and its derivative have the values $w(1) = 4$ and $w'(1) = 5$?

♦ Type 2, the Chain Rule with a logarithm

31. The radius of a spherical balloon is $\ln(5 + 6t - t^2)$ meters at time t (hours) for $0 \le t \le 6$. **(a)** What are the maximum and minimum volume of the balloon for $0 \le t \le 6$? **(b)** How fast is the volume of the balloon increasing at $t = 1$?

♦ Type 2, the Chain Rule with a logarithm

32. Use the formula for the function and its first derivative to analyze the graph of $y = \dfrac{x^2}{\ln x}$.

♦ Type 3, analyzing a graph

33. Find the exact global minimum of $\ln x + (\ln x)^2$.

♦ Type 3, a global minimum

34. What is the x-coordinate of the inflection point of $y = (\ln x)^{10}$? Give the exact and approximate decimal values.

♦ Type 3, an inflection point

35. **(a)** Find the constant n such that $y = (\ln x)^n$ has an inflection point at $x = e^2$. **(b)** Where is the other inflection point of the curve in part (a)?

♦ Type 4, an inflection point

36. Find the exact maximum and minimum of $\ln x + (\ln x)^2$ for $1 \le x \le 3.2166$.

♦ Type 2, max/min in an interval

37. What are the exact maximum and minimum of $3 + x^2 - 8\ln x$ for $1 \le x \le 3.2166$?

♦ Type 2, max/min in an interval

38. Show that the global maximum M of $\dfrac{\ln x}{x}$ occurs at $x = \dfrac{1}{M}$.

♦ Type 2, a global maximum

Use the formula $f(x)^{g(x)} = e^{g(x)\ln[f(x)]}$ to find the derivatives in Problems 39 through 43.

39.[O] $\dfrac{d}{dx}(x^x)$

♦ Type 3, a derivative

40.[A] $\dfrac{d}{dx}(x^{\ln x})$

♦ Type 3, a derivative

41. $\dfrac{d}{dx}[(3+x)^{3-x}]$

♦ Type 3, a derivative

42. $\dfrac{d}{dx}(x^{x^2})$

♦ Type 3, a derivative

43. Give an equation of the tangent line to $y = x^x$ at $x = e$.

♦ Type 3, a derivative

44. Use the YX-definition of $\lim_{x\to\infty} f(x) = \infty$ to show that $\ln x \to \infty$ as $x \to \infty$.

♦ Type 7, the YX-definition

---

[6] *Introduction to Number Theory*, T. Nagell, Chelsea Publishing Co., N.Y., N.Y,, 1981, p. 54.

Section 3.6

Trigonometric functions and their derivatives

OVERVIEW: *The subject that we now call trigonometry began long ago. Its earliest traces are found in Egyptian tables from the 13th Century BC which relate the lengths of shadows to the time of day. The concepts and techniques of trigonometry were developed over the centuries, initially for astronomy and geography and later for the study of mechanics, circular motion, light, sound, electric circuits, and many other phenomena. In this section we first consider some problems that are modeled with trigonometric functions. Then we derive and use formulas for the derivatives of the trigonometric and inverse trigonometric functions.*†

Topics:

- ***Circular motion: A Ferris wheel***
- ***A shadow function***
- ***Derivatives of the sine and cosine functions***
- ***Sinusoidal functions***
- ***Derivatives of the tangent, cotangent, secant, and cosecant functions***
- ***Derivatives of the inverse trigonometric functions***

Circular motion: George's biggest wheel

The largest Ferris wheel of all time was designed by George Ferris and built for the 1893 World's Columbian Exposition in Chicago. It was 250 feet in diameter and could carry 2160 passengers.(1) Assume that the center of the wheel was 150 feet above the ground and that the wheel turned counterclockwise as viewed in the schematic sketch in Figure 1.

Question 1 The dot in Figure 1 represents one of the cars on Ferris' wheel. Let $h(\theta)$ be its height (feet) above the ground when it is at the angle θ (radians) in Figure 1. **(a)** Why are $h(\theta + 2\pi)$ and $h(\theta)$ equal for all θ? **(b)** Which of the curves in Figures 2 through 4 is the graph of $h(\theta)$ and why?

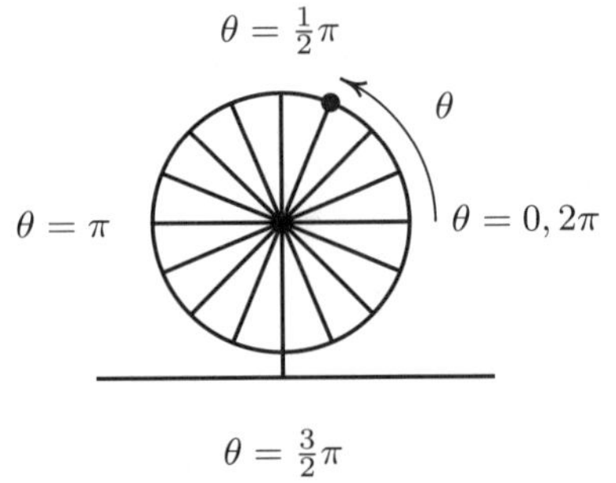

FIGURE 1

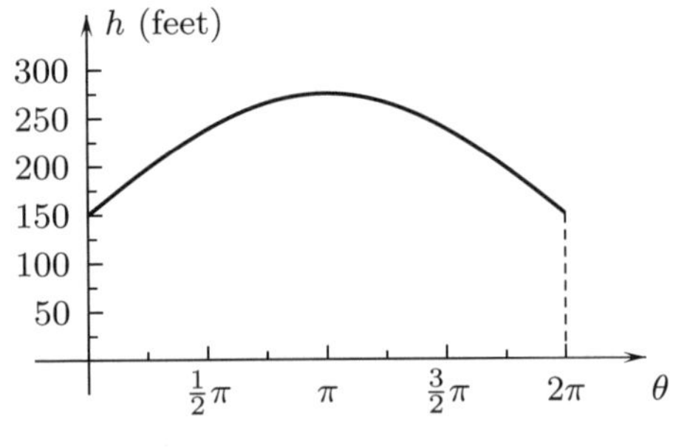

FIGURE 2

---

†Read Appendix 3 on the definitions and basic properities of trigonometric and inverse trigonometric functions before you study this section and consult it as necessary as you study this section.

(1) "The Ferris wheel" by H. Petrovski, *American Scientist*, May-June, 1993, Phyladelphia, PA: Sigma Xi, 1993 pp. 216–222.

♦ SUGGESTIONS TO INSTRUCTORS: The main topics on this section can be divided into two parts, as follows: **Part 1** (Angles, triangles, the sine and cosine function and their derivatives; Tune-up Exercises T1-T4, T6, T9, T11, T13, T14, T16, T19, T20; Problems 1–2, 5, 8, 12, 14–17, 19–21, 29–333, 38–41, 43, 44, 48–50, 52–54, 57–61, 65, 66, 68, 69, 71, 72, 75–77, 80–85, 88, 89, 91, 92, 94, 95, 100, 101) Have students begin work on Questions 1 through 22 before class. Problems 1, 3, 43, 44, 48, 52, 59, 60, 68, 83, and 88 are good for in-class discussion.
Part 2 (The tangent, cotangent, secant, cosecant and inverse trigonometric functions and their derivatives; Tune-up Exercises T5, T7, T8, T10, T15, T17, T18, T21–T23; Problems 4, 6, 7, 9–11, 13, 18, 22–28, 34–37, 42, 45–47, 55, 56, 62–64, 67, 70, 73, 74, 78, 79, 86, 87, 90, 93, 95–99) Have students begin work on Questions 17 through 22 before class. Problems 7, 13, 42, 46, 51, 70, 87, 90, 95, and 96 are good for in-ckass discussion.

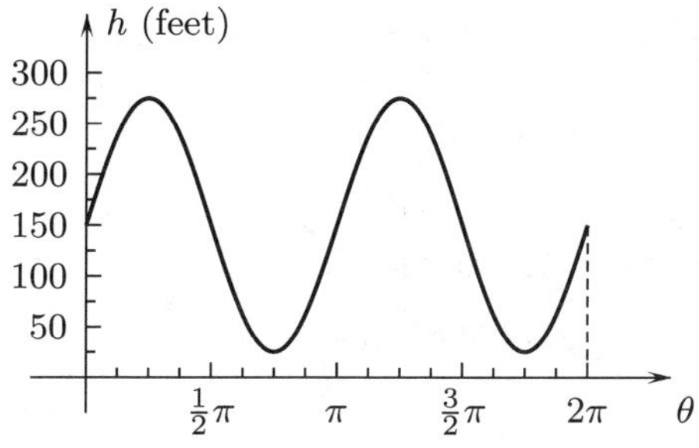

FIGURE 3

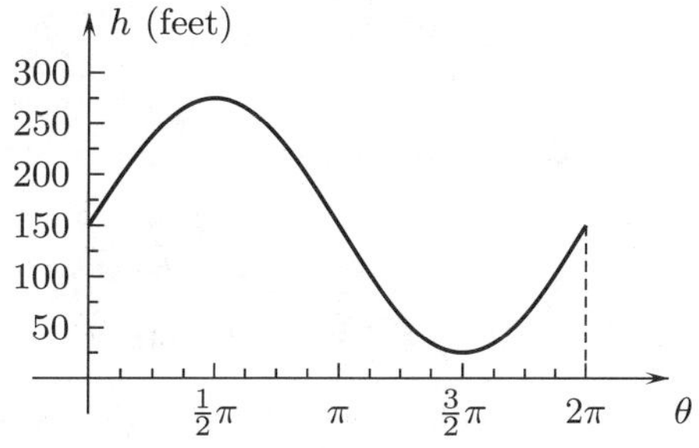

FIGURE 4

Example 1 Give a formula for the height of the Ferris wheel in Figure 1 as a function of θ.

SOLUTION We introduce auxiliary uv-coordinates with origin at the center of the wheel as in Figure 5. Then the Ferris-wheel car, represented by the dot with coordinates (u, v) in Figure 5, moves on the circle of radius $r = 125$ with its center at the origin and by the definition of the sine function (formula **(7)** in Appendix 3),

$$\sin\theta = \frac{v}{125}.$$

Then $v = 125\sin\theta$, and, since the center of the Ferris wheel is 150 feet above the ground, the car's height at angle θ radians is $h = 150 + v = 150 + 125\sin\theta$ feet. As you probably predicted in Question 1, the graph of this function is in Figure 4. □

FIGURE 5

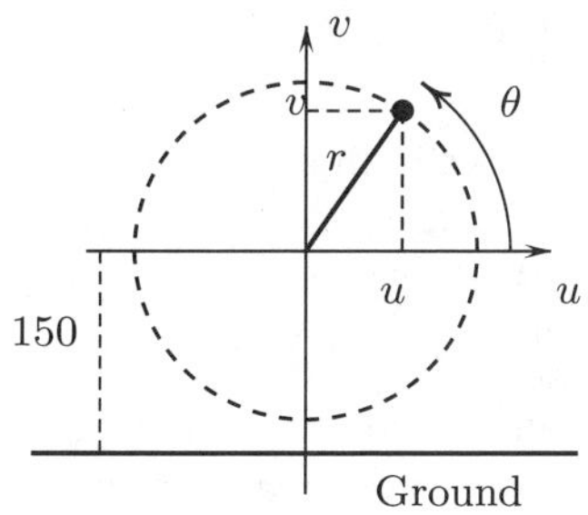

Question 2 Find a formula for the horizontal position u of the Ferris wheel as a function of the angle θ and generate its graph.

A shadow function

Figure 6 shows a vertical pole 10 feet high and its shadow cast on the horizontal ground by the sun. Figure 7 shows the graph of the length $s(\theta)$ of the shadow as a function of the angle θ between the rays of the sun and the pole, measured in radians.

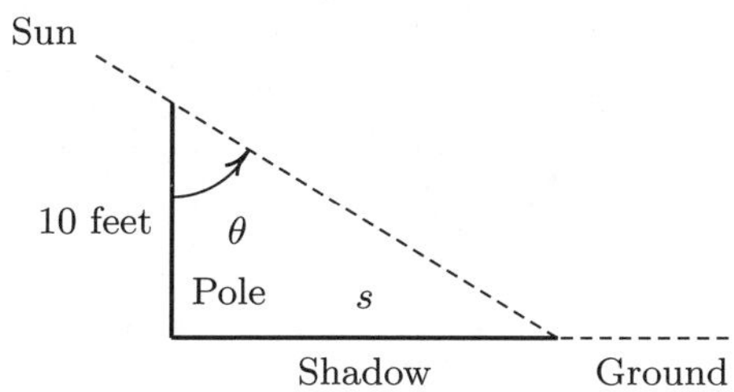

FIGURE 6

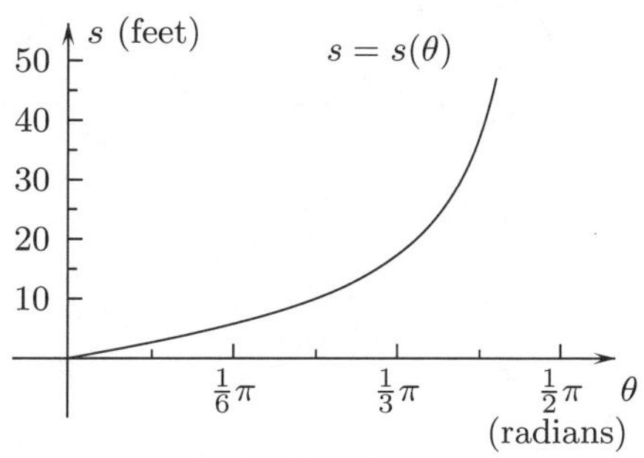

FIGURE 7

Question 3 **(a)** What is $s(0)$ and why? **(b)** Why is $s(\theta)$ an increasing function of θ for $0 \le \theta < \frac{1}{2}\pi$? **(c)** What is $s(\frac{1}{4}\pi)$ and why? **(d)** What is the limit of $s(\theta)$ as $\theta \to \frac{1}{2}\pi^-$?

Example 2 Give a formula for the length s of the shadow in Figures 6 and 7.

SOLUTION Because the vertical leg of the right triangle in Figure 6 is formed by the pole and does not change as the sun moves, we use the tangent function to express the length s of the side oposite the angle θ as a function of θ. We have $\tan\theta = \dfrac{\text{Opposite}}{\text{Adjacent}} = \dfrac{s}{10}$ and therefore $s = 10\tan\theta$. □

Predicting the derivatives of sin x *and* cos x

The graphs of $\sin x$ and $\cos x$ are shown in Figures 8 and 9. We can use these graphs to predict the general shapes of the graphs of their derivatives.

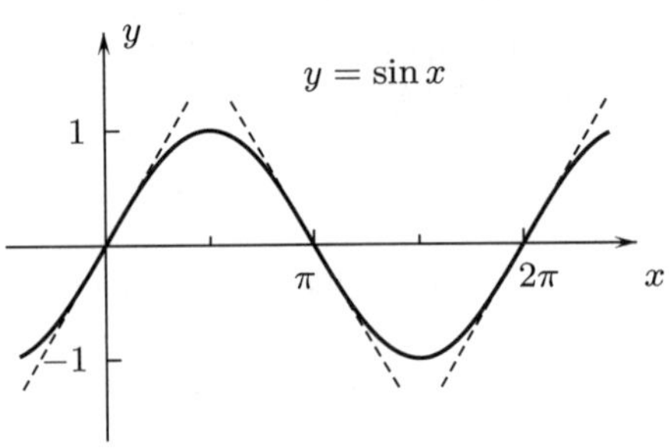

FIGURE 8

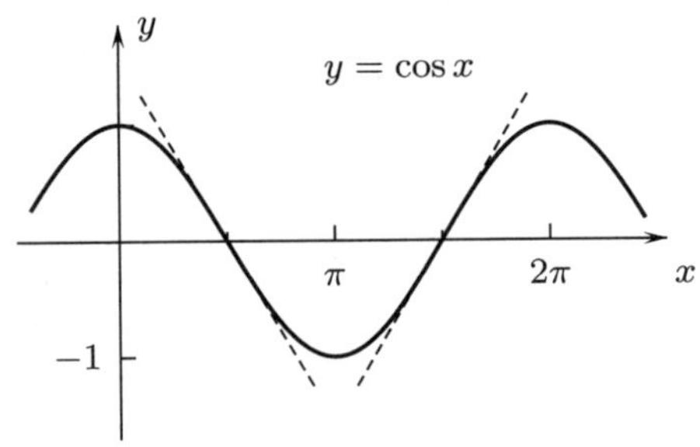

FIGURE 9

Question 4 The three tangent lines to the graph of $\sin x$, drawn with dashed lines in Figure 8, have slopes 1 and -1. Use this information to describe how the derivative of $\sin x$ changes **(a)** as x goes from 0 to $\frac{1}{2}\pi$, **(b)** as x goes from $\frac{1}{2}\pi$ to π, **(c)** as x goes from π to $\frac{3}{2}\pi$, and **(d)** as x goes from $\frac{3}{2}\pi$ to 2π.

Question 5 The two tangent lines to the graph of $\cos x$ in Figure 9 also have slopes 1 and -1. Describe how the derivative of $\cos x$ changes **(a)** as x goes from 0 to $\frac{1}{2}\pi$, **(b)** as x goes from $\frac{1}{2}\pi$ to π, **(c)** as x goes from π to $\frac{3}{2}\pi$, and **(d)** as x goes from $\frac{3}{2}\pi$ to 2π.

We will show below that the derivative of $\sin x$ is $\cos x$ and the derivative of $\cos x$ is $-\sin x$, so that the derivatives have the graphs in Figures 10 and 11. These drawings should corroborate your responses to Questions 4 and 5.

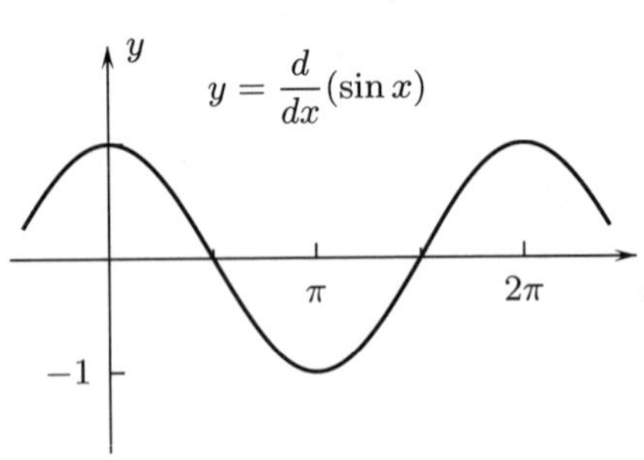

FIGURE 10

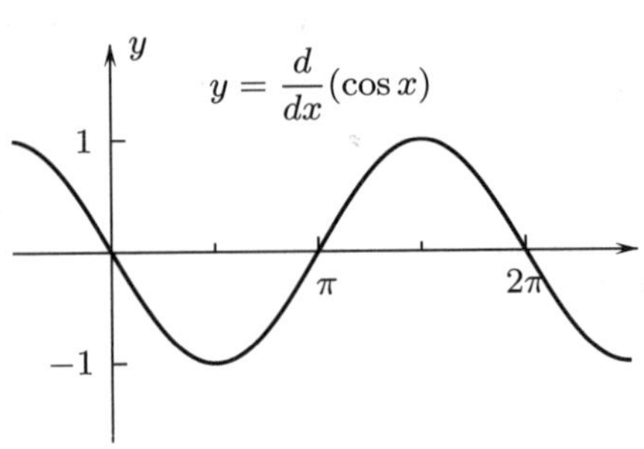

FIGURE 11

Finding the derivatives of sin x *and* cos x

Because $\sin x$ and $\cos x$ are given by distances, while the angle x is given as the length of an arc on a circle, we will need to relate the chord length $\overline{PQ}$ to the arclength $\overset{\frown}{PQ}$ for nearby points P and Q on a circle as in Figure 12 in order to find formulas for the derivatives of $\sin x$ and $\cos x$.

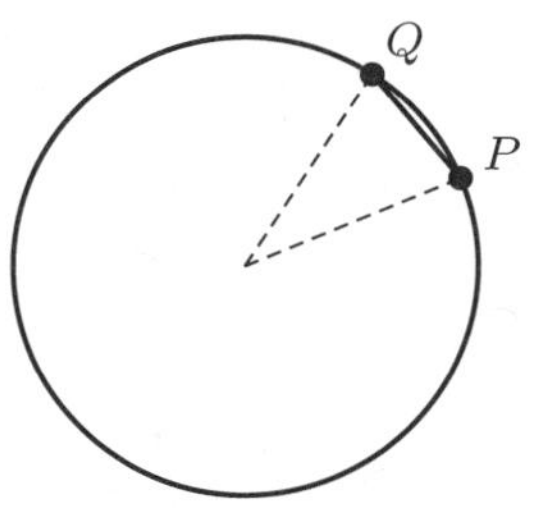

FIGURE 12

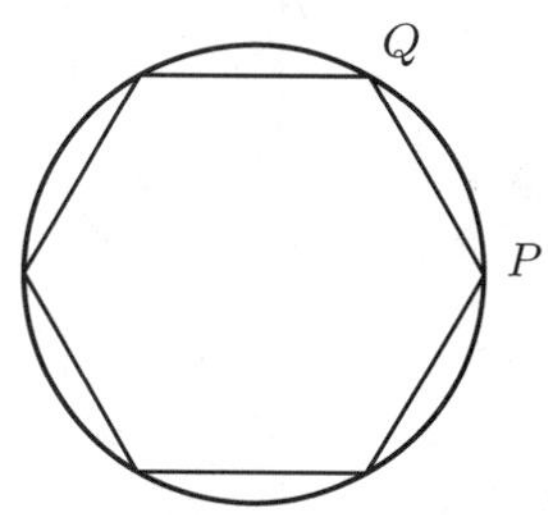

FIGURE 13

We consider the special case where P and Q are adjacent vertices in an equilateral polygon of n sides that is inscribed in the circle, as is illustrated in Figure 13 for $n = 6$ (a regular hexagon). The length $\overline{PQ}$ of the chord in the polygon with n equal sides is the perimeter of the entire polygon divided by n, and the length $\overset{\frown}{PQ}$ of the arc is the circumference of the circle divided by n. Consequently,

$$\begin{aligned}\frac{\overline{PQ}}{\overset{\frown}{PQ}} &= \frac{[\text{Perimeter of the polygon}]/n}{[\text{Circumference of the circle}]/n} \\ &= \frac{\text{Perimeter of the polygon}}{\text{Circumference of the circle}}.\end{aligned}$$

In the last step we cancelled the n's. When $n \to \infty$, Q approaches P, and the perimeter of the n-sided polygon tends to the circumference of the circle, so that $\overline{PQ}/\overset{\frown}{PQ}$ tends to 1. This leads to the following, which can be established for arbitrary points P and Q on the circle.

Proposition 1 *The ratio of the chord length $\overline{PQ}$ between points P and Q on a circle to the arclength $\overset{\frown}{PQ}$ between the points tends to 1 as the points move together:*

$$\lim_{Q \to P} \frac{\overline{PQ}}{\overset{\frown}{PQ}} = 1.$$

Now we can find formulas for the derivatives of $\sin x$ and $\cos x$ from Proposition 1 and the definitions of the sine and cosine. According to the definition of the derivative from Section 2.3,

$$\begin{aligned}\frac{d}{dx}(\sin x) &= \lim_{b \to x} \frac{\sin b - \sin x}{b - x} \\ \frac{d}{dx}(\cos x) &= \lim_{b \to x} \frac{\cos b - \cos x}{b - x}.\end{aligned} \qquad \mathbf{(1)}$$

In order to visualize the geometric meanings of the numerators and denominators of the difference quotients in **(1)**, we consider a fixed angle x and a variable angle b with $0 < x < b < \frac{1}{2}\pi$ and the points $P(\cos x, \sin x)$ and $Q(\cos b, \sin b)$ at those angles on a unit circle, as in Figure 14.

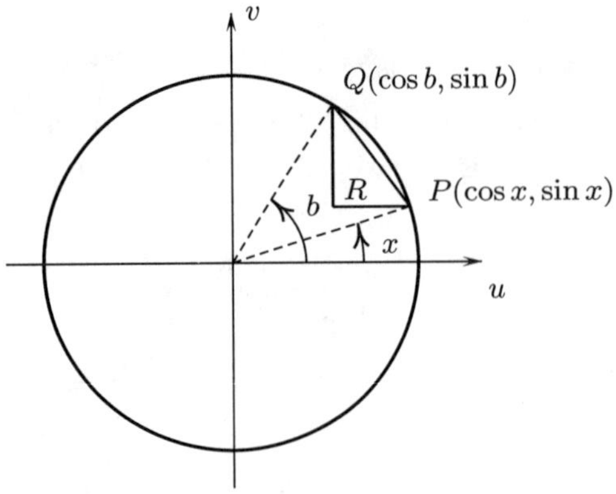

FIGURE 14

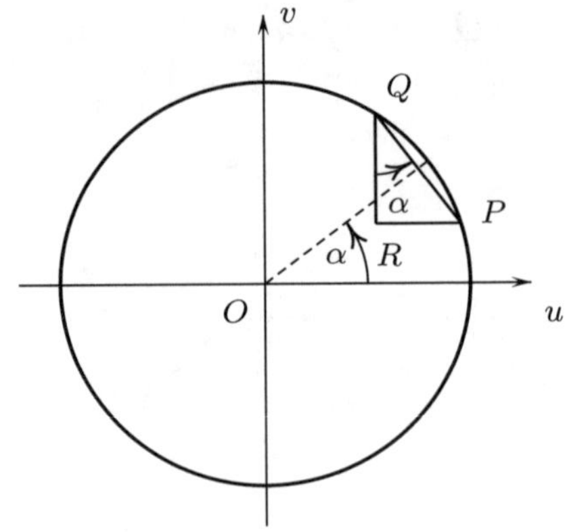

FIGURE 15

Because the circle has radius 1, the difference $b - x$ between the angles equals the arclength from P to Q, and the denominator in **(1)** is

$$b - x = \overset{\frown}{PQ}. \tag{2}$$

Also, by the unit-circle definitions of the sine and cosine, the lengths $\overline{RQ}$ and $\overline{RP}$ of the legs of the right triangle PRQ are $\overline{RQ} = \sin b - \sin x$ and $\overline{RP} = \cos x - \cos b$, so that

$$\begin{aligned} \frac{\sin b - \sin x}{b - x} &= \frac{\overline{RQ}}{\overset{\frown}{PQ}} \\ \frac{\cos b - \cos x}{b - x} &= -\frac{\overline{RP}}{\overset{\frown}{PQ}}. \end{aligned} \tag{3}$$

Next, we consider Figure 15, where the dashed line is the perpendicular bisector of the chord PQ. The two angles labeled α in this drawing are equal because their sides are perpendicular. Therefore, the ratios of the lengths of the legs and hypotenuse of this triangle are given by the sine and cosine of α:

$$\begin{aligned} \frac{\overline{RQ}}{\overline{PQ}} &= \cos\alpha \\ \frac{\overline{RP}}{\overline{PQ}} &= \sin\alpha. \end{aligned} \tag{4}$$

If b is very close to x, then the points P and Q in Figures 14 and 15 are very close together, so that $\alpha \approx x$ and

$$\cos\alpha \approx \cos x, \ \sin\alpha \approx \sin x, \ \overset{\frown}{PQ} \approx \overline{PQ}.$$

The last statement is a consequence of Proposition 1. Combining **(3)** and **(4)** then gives, for b very close to x,

$$\begin{aligned} \frac{\sin b - \sin x}{b - x} &= \frac{\overline{RQ}}{\overset{\frown}{PQ}} \approx \frac{\overline{RQ}}{\overline{PQ}} = \cos\alpha \approx \cos x \\ \frac{\cos b - \cos x}{b - x} &= -\frac{\overline{RP}}{\overset{\frown}{PQ}} \approx -\frac{\overline{RP}}{\overline{PQ}} = -\sin\alpha \approx -\sin x. \end{aligned} \tag{5}$$

The approximations in **(5)** become increasingly accurate as $b \to x^+$, and with similar considerations for $b \to x^-$ we would obtain

$$\begin{aligned} \frac{d}{dx}(\sin x) &= \lim_{b \to x} \frac{\sin b - \sin x}{b - x} = \cos x \\ \frac{d}{dx}(\cos x) &= \lim_{b \to x} \frac{\cos b - \cos x}{b - x} = -\sin x. \end{aligned}$$

Slight modifications of this reasoning would yield this result for arbitrary x to establish the following:

Theorem 1 *Sin x and $\cos x$ have derivatives for all x, which are given by*

$$\frac{d}{dx}(\sin x) = \cos x \quad \textbf{(6)}$$

$$\frac{d}{dx}(\cos x) = -\sin x. \quad \textbf{(7)}$$

Combining these formulas with Theorem 2 of Section 2.3,

$$\frac{d}{dx}[Af(x) + Bg(x)] = A\frac{df}{dx}(x) + B\frac{dg}{dx}(x) \quad \textbf{(8)}$$

enables us to differentiate linear combinations of the sine and cosine.

Example 3 What is the derivative of $\frac{d}{dx}(5\sin x - 6\cos x)$?

SOLUTION We use **(6)**, **(7)**, and **(8)** with $A = 5$ and $B = -6$ to obtain

$$\frac{d}{dx}(5\sin x - 6\cos x) = 5\frac{d}{dx}(\sin x) - 6\frac{d}{dx}(\cos x) = 5\cos x + 6\sin x. \ \square$$

Question 6 Give an equation of the tangent line to the graph of $f(x) = 5\sin x - 6\cos x$ at $x = 2$. Generate the graph and the tangent line in the window $-3 \leq x \leq 8$, $-12 \leq y \leq 12$ as a partial check of your answer.

Pick your angle: a maximization problem

Of all the right triangles with hypotenuse of length 10 feet (see Figure 16), which has the largest area? To answer this question, we first need to express the width w and height h of the triangle in terms of the angle x.

Question 7 Use the sine and cosine to express h, w, and the area A of the triangle in Figure 16 in terms of x.

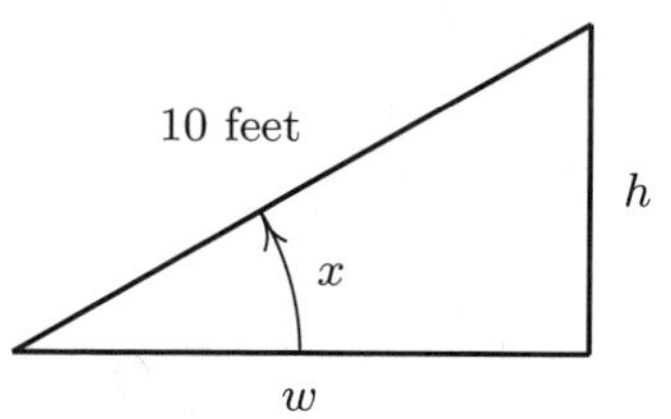

FIGURE 16

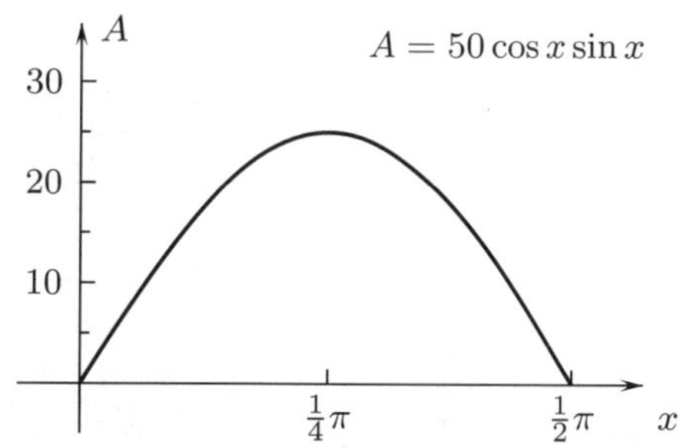

FIGURE 17

Question 8 The graph of the area $A(x) = 50\cos x \sin x$ of the triangle from Question 9 is shown in Figure 17. **(a)** Use Figure 16 to explain why $A(0) = 0$ and $A\left(\frac{1}{2}\pi\right) = 0$. **(b)** Use a derivative to find the exact value of x for which the area is a maximum.

The Chain Rule for sines and cosines of functions

We can now give formulas for the x-derivatives of $\sin y$ and $\cos y$, where $y(x)$ is a differentiable function of x. The general Chain rule in Section 3.3 combined with Theorem 1 yields the folowing:

Theorem 2 *If $y = y(x)$ is a differentiable function of x, then*

$$\frac{d}{dx}(\sin y) = \frac{d}{dy}(\sin y)\,\frac{dy}{dx} = \cos y\,\frac{dy}{dx} \tag{9a}$$

$$\frac{d}{dx}(\cos y) = \frac{d}{dy}(\cos y)\,\frac{dy}{dx} = -\sin y\,\frac{dy}{dx} \tag{10a}$$

or with prime notation,

$$\frac{d}{dx}\{\sin[y(x)]\} = \cos[y(x)]\,y'(x) \tag{9b}$$

$$\frac{d}{dx}\{\cos[y(x)]\} = -\sin[y(x)]\,y'(x). \tag{10b}$$

Learn formulas **(9)** and **(10)** in words: The derivative of the sine of a function equals the cosine of the function, multiplied by the derivative of the function, and the derivative of the cosine of a function equals the negative of the sine of the function, multiplied by the derivative of the function.

Example 4 Give formulas for the derivatives of **(a)** $\sin(x^3 - 4)$ and **(b)** $\cos(x^{-1})$.

SOLUTION **(a)** $\frac{d}{dx}[\sin(x^3 - 4)] = \cos(x^3 - 4)\frac{d}{dx}(x^3 - 4) = 3x^2\cos(x^3 - 4)$

(b) $\frac{d}{dx}[\cos(x^{-1})] = -\sin(x^{-1})\frac{d}{dx}(x^{-1}) = -\sin(x^{-1})(-x^{-2})$

$= x^{-2}\sin(x^{-1})$. □

We can also use the formulas in Theorems 1 and 2 with the Chain Rule for powers of functions (Theorem 3 of Section 2.5),

$$\frac{d}{dx}(y^n) = \frac{d}{dy}(y^n)\,\frac{dy}{dx} = ny^{n-1}\,\frac{dy}{dx}. \tag{11}$$

Example 5 What is $f'(x)$ for $f(x) = \sqrt{\cos x}$?

SOLUTION We use **(11)** with $y = \cos x$ and $n = \frac{1}{2}$ and then **(7)**:

$$f'(x) = \frac{d}{dx}(\sqrt{\cos x}) = \frac{d}{dx}[(\cos x)^{1/2}] = \tfrac{1}{2}(\cos x)^{-1/2}\frac{d}{dx}(\cos x)$$
$$= -\tfrac{1}{2}\sin x\,(\cos x)^{-1/2} = \frac{-\sin x}{2\sqrt{\cos x}}. \ \square$$

Question 9 What is the rate of change of $\sin^5(3x)$ with respect to x at $x = 10$?

The results of the next question show why we measure angles in radians instead of degrees in calculus.

Question 10 The sine and cosine of x degrees can be defined by $S(x) = \sin\left(\frac{1}{180}\pi x\right)$ and $C(x) = \cos\left(\frac{1}{180}\pi x\right)$. Use these formulas to express $S'(x)$ in terms of $C(x)$ and $C'(x)$ in terms of $S(x)$.

A sliding ladder

The following related-rate problem, like the optimization problem in Question 8, uses the sine and cosine functions because it involves a right triangle whose angles change but whose hypotenuse is fixed.

Suppose that the right triangle in Figure 16 represents a ten-foot-long ladder whose top is sliding down a vertical wall while its base is sliding away from the wall along the horizontal ground. Recall from Question 7 that $h = 10\sin x$ and $w = 10\cos x$.

Question 11 At a particular time the angle x in Figure 16 is 1.1 radians and is decreasing 0.3 radians per minute. How fast is the top of the ladder falling and how rapidly is its base moving at that moment?

Sinusoidal functions

The study of vibrating springs, light, sound, electric circuits, and many other phenomena involving periodic oscillations employ SINUSOIDAL FUNCTIONS, which are functions of the form

$$f(x) = A\sin(\omega x + \psi) + C \qquad \textbf{(12)}$$

with arbitrary constants ψ (psi) and C and nonzero constants A and ω (omega). Figure 18 shows the graph of such a function with positive ω, ψ, A, and C.

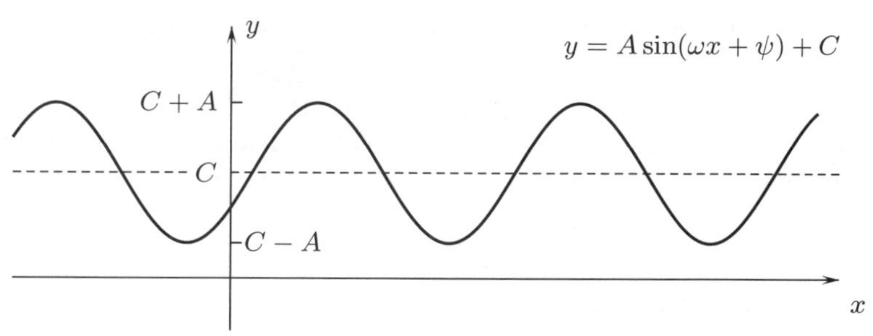

FIGURE 18

Because the values of the sine function oscillate between 1 and -1, the values of $A\sin(\omega x+\psi)+C$ oscillate between $C+A$ and $C-A$. The positive number $|A|$ is called the AMPLITUDE of the function defined by **(12)**.

A vibrating spring

Visualize a weight that moves up and down, subject to gravity and the force of a spring from which it is suspended (Figure 19). We measure the location of the weight on a vertical s-axis, as in Figure 19. We put $s = 0$ at the weight's REST POSITION, which is the position where the upward force of the spring balances the downward force of gravity and the weight can remain stationary.

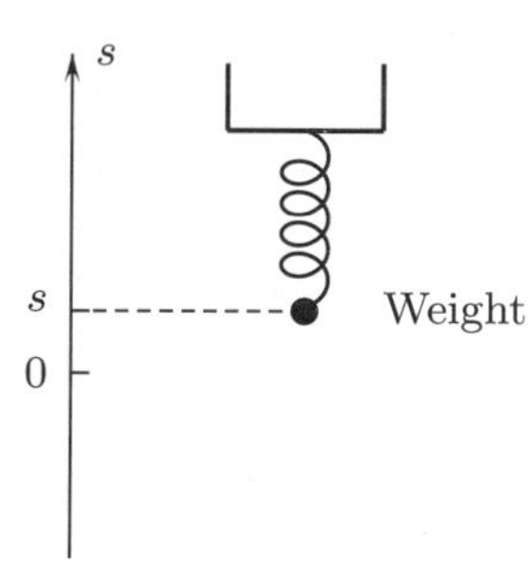

FIGURE 19

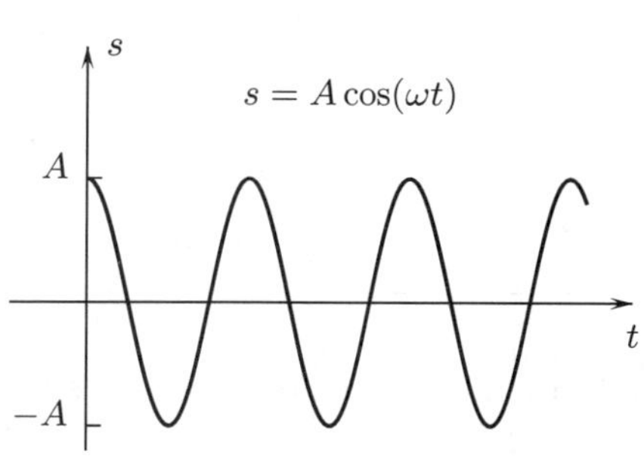

FIGURE 20

Imagine that the weight has been set in motion by moving it a short distance above its rest position and letting it go at time $t = 0$ (seconds). If there is no air resistance so that the only forces on the weight are gravity and the spring, and if the force of the spring is proportional to the displacement of the weight from its rest position, then the position of the weight is given as a function of t by the sinusoidal function

$$s(t) = A\cos(\omega t) \qquad \textbf{(13)}$$

(Figure 20). The positive constant ω is determined by the size of the weight and the strength of the spring. The amplitude of the oscillation A is determined by how far the weight is raised initally.

The function **(13)** is periodic of period $2\pi/\omega$ since increasing t by $2\pi/\omega$ increases ωt by 2π.

Question 12 How many times per second does the weight with position given by **(13)** move from its lowest point to its highest point and back to its lowest point again? This number is called the FREQUENCY of the motion.

Question 13 Making the weight heavier without changing the spring makes the constant ω smaller. How does this affect the motion of the weight?

Question 14 Making the spring stronger without changing the weight makes ω larger. How does this affect the motion of the weight?

Question 15 The derivative $\frac{ds}{dt}(t)$ of **(13)** is the weight's upward velocity. It is also a sinusoidal function. How do its amplitude and period relate to those of $s(t)$?

Question 16 How is the upward acceleration of the weight $\frac{d^2s}{dt^2}(t)$ related to $s(t)$?

Derivatives of the other four trigonometric functions

Now we can use the formulas for the derivatives of the sine and cosine to derive formulas for the derivatives of the tangent, cotangent, secant, and cosecant functions.[†]

Example 6 Find a formula for the derivative of $\tan x = \dfrac{\sin x}{\cos x}$.

SOLUTION By the Quotient Rule,

$$\frac{d}{dx}(\tan x) = \frac{d}{dx}\left[\frac{\sin x}{\cos x}\right] = \frac{\cos x \frac{d}{dx}(\sin x) - \sin x \frac{d}{dx}(\cos x)}{\cos^2 x}$$
$$= \frac{\cos^2 x + \sin^2 x}{\cos^2 x} = \frac{1}{\cos^2 x} = \sec^2 x. \ \square$$

This gives the first statement in the following:

Theorem 3 *At all values of x where the denominators are not zero,*

$$\frac{d}{dx}(\tan x) = \sec^2 x \qquad \textbf{(14)}$$
$$\frac{d}{dx}(\cot x) = -\csc^2 x \qquad \textbf{(15)}$$
$$\frac{d}{dx}(\sec x) = \sec x \, \tan x \qquad \textbf{(16)}$$
$$\frac{d}{dx}(\csc x) = -\csc x \, \cot x. \qquad \textbf{(17)}$$

[†]This section and Appendix 3 with the definitions and basic properties of the trigonometric functions contain a lot of facts and information because they deal with all six trigonometric functions. We would have a lot less to learn and still be able to handle all applications with only the sine. We could use $\pm\sqrt{1-\sin^2 x}$ in place of $\cos x$, $(\sin x)/(\cos x)$ in place of $\tan x$, $(\cos x)/(\sin x)$ instead of $\cot x$, $1/(\cos x)$ instead of $\sec x$, and $1/(\sin x)$ instead of $\csc x$. But then we would often end up with very complicated formulas, and it is easier in the long run to use all of the trigonometric functions.

Question 17 Use Theorem 1 and the Quotient and Chain Rules to derive formulas **(15)**, **(16)**, and **(17)** for the derivatives of **(a)** $\cot x$, **(b)** $\sec x$ and **(c)** $\csc x$.

Example 7 **(a)** What is the minimum slope of the tangent lines to the graph $y = \tan x$ in Figure 21? **(b)** What happens to the slope of the tangent line to $y = \tan x$ as $x \to \frac{1}{2}\pi$?

SOLUTION **(a)** The minimum slope of the tangent lines to $y = \tan x$ in Figure 21 is 1 because, according to formula **(14)**, the slope at x is $\sec^2 x = 1/\cos^2 x$ and $\cos^2 x$ oscillates between 0 and 1. **(b)** As $x \to \frac{1}{2}\pi$, the slope $\sec^2 x = 1/\cos^2 x$ of the tangent line at x tends to ∞ because $\cos^2 x$ is positive and tends to zero. □

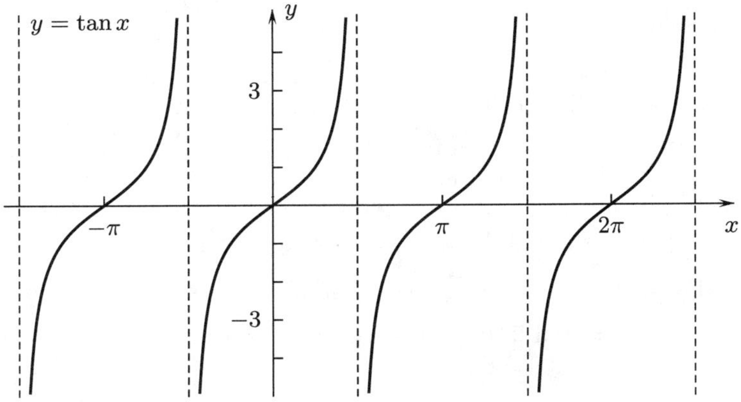

FIGURE 21

Question 18 Give an equation of the tangent line to $y = \tan x$ at $x = 0$. Then generate the curve and the tangent line together in the window $-6 \le x \le 6, -4 \le y \le 4$ with x-scale$= \frac{1}{2}\pi$ as a partial check of your answer.

Notice how formulas **(15)**, **(16)**, and **(17)** relate to the graphs of $\cot x$, $\sec x$, and $\csc x$ in Figures 22, 23, and 24: $\cot x$ is decreasing in the open intervals where it is defined because its derivative $-\csc^2 x$ is negative wherever it is defined. In contrast, $\sec x$ and $\csc x$ increase and decrease in alternating intervals because their derivatives, $\sec x \tan x = \dfrac{\sin x}{\cos^2 x}$ and $-\csc x \cot x = -\dfrac{\cos x}{\sin^2 x}$, are positive and negative in alternating intervals.

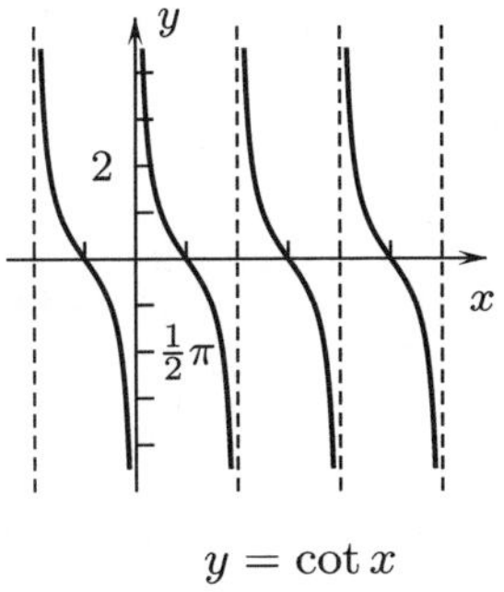

FIGURE 22

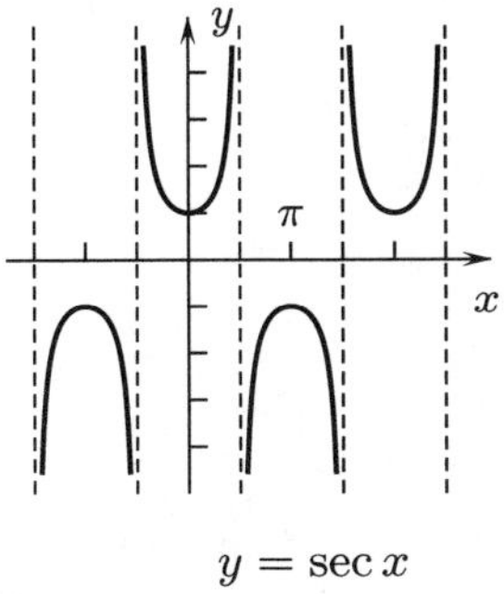

FIGURE 23

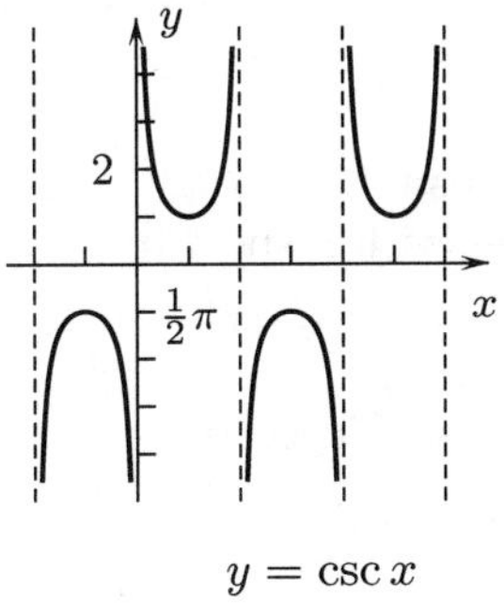

FIGURE 24

Chain Rule formulas for tangents, cotangents, secants, and cosecants

Applying the Chain Rule to the formulas in Theorem 2 yields the following formulas for derivatives of tangents, cotangents, secants, and cosecants of differentiable functions $y = y(x)$.

$$\frac{d}{dx}(\tan y) = \frac{d}{dy}(\tan y)\,\frac{dy}{dx} = \sec^2 y\,\frac{dy}{dx} \tag{18}$$

$$\frac{d}{dx}(\cot y) = \frac{d}{dy}(\cot y)\,\frac{dy}{dx} = -\csc^2 y\,\frac{dy}{dx} \tag{19}$$

$$\frac{d}{dx}(\sec y) = \frac{d}{dy}(\sec y)\,\frac{dy}{dx} = \sec y\,\tan y\,\frac{dy}{dx} \tag{20}$$

$$\frac{d}{dx}(\csc y) = \frac{d}{dy}(\csc y)\,\frac{dy}{dx} = -\csc y\,\cot y\,\frac{dy}{dx}. \tag{21}$$

The tangent or cotangent is often used in related-rate problems involving right triangles with varying angles and one leg of fixed length, as in the next question.

Question 19 A lighthouse is two miles from a pier on a line perpendicular to the straight shore (Figure 25). **(a)** Give a formula for the distance s from the pier to the place where the beam of light hits the shore in terms of the angle θ in Figure 25. **(b)** The beam of light from the lighthouse rotates at a constant rate and makes one full counterclockwise revolution, as viewed from above, every three minutes. What is $\frac{d\theta}{dt}$ if the time t is measured in minutes? **(c)** How fast is the beam of light moving along the shore when $\theta = 0.9$ radians?

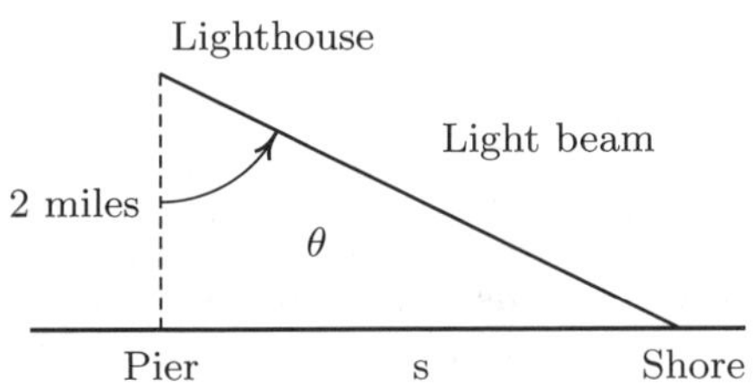

FIGURE 25

Derivatives of the inverse sine, cosine, tangent, and cotangent[†]

In many applications, the angles that occur in the mathematical models are expressed as inverse trigonometric functions, and we need formulas for their derivatives to solve the problems. To find a formula for the derivative of $\sin^{-1} x$, we start with the identity,

$$\sin(\sin^{-1} x) = x$$

which holds for $-1 \le x \le 1$. Assuming that $\sin^{-1} x$ has a derivative for $-1 < x < 1$, we differentiate both sides with respect to x:

$$\frac{d}{dx}[\sin(\sin^{-1} x)] = \frac{d}{dx}(x).$$

Then we use Chain-Rule formula **(9b)** to obtain

$$\cos(\sin^{-1} x)\frac{d}{dx}(\sin^{-1} x) = 1. \tag{22}$$

For $0 < x < 1$, $\sin^{-1} x$ is the angle opposite the leg of length x in a right triangle with hypotenuse of length 1 (Figure 26), and $\cos(\sin^{-1} x)$ is the length of the other leg. Therefore,

$$\cos(\sin^{-1} x) = \sqrt{1 - x^2}.$$

[†]We do not study the inverse secant and cosecant here because they are seldom used.

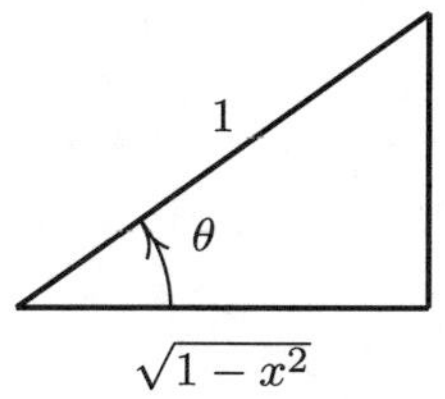

$\theta = \sin^{-1} x$

$\cos\theta = \sqrt{1 - x^2}$

FIGURE 26

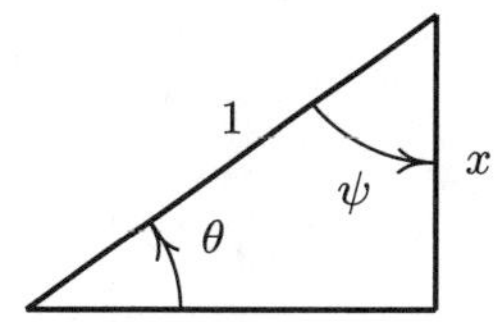

$\theta = \sin^{-1} x$

$\psi = \cos^{-1} x$

FIGURE 27

The last formula also holds for $-1 < x \leq 0$, and with **(22)** gives the first formula in the next theorem.

Theorem 4 *For $-1 < x < 1$,*

$$\frac{d}{dx}(\sin^{-1} x) = \frac{1}{\sqrt{1 - x^2}} \tag{23}$$

$$\frac{d}{dx}(\cos^{-1} x) = \frac{-1}{\sqrt{1 - x^2}} \tag{24}$$

and for all x,

$$\frac{d}{dx}(\tan^{-1} x) = \frac{1}{1 + x^2} \tag{25}$$

$$\frac{d}{dx}(\cot^{-1} x) = \frac{-1}{1 + x^2}. \tag{26}$$

To derive **(24)**, we notice that for $0 < x < 1$, $\sin^{-1} x$ and $\cos^{-1} x$ are the acute angles in a right triangle (Figure 27). Since the sum of the acute angles is $\frac{1}{2}\pi$,

$$\cos^{-1} x = \tfrac{1}{2}\pi - \sin^{-1} x. \tag{27}$$

This identity also holds for $x = 1$ and $-1 \leq x < 0$. We obtain **(24)** by differentiating both sides of **(27)** and using **(23)**.

To derive **(25)**, we differentiate both sides of the identity,

$$\tan(\tan^{-1} x) = x$$

which is valid for all x. We obtain

$$\frac{d}{dx}[\tan(\tan^{-1} x)] = \frac{d}{dx}(x)$$

and then by the Chain-Rule formula **(18)**

$$\sec^2(\tan^{-1} x)\frac{d}{dx}(\tan^{-1} x) = 1. \tag{28}$$

For $x > 0$, $\tan^{-1} x$ is the angle opposite the side of length x in a right triangle whose other leg has length 1 (Figure 28). Consequently, $\sec(\tan^{-1} x)$ is the length of the hypotenuse and $\sec(\tan^{-1} x) = \sqrt{1 + x^2}$. This formula also holds for $x \leq 0$, and with **(28)**, it gives **(25)**.

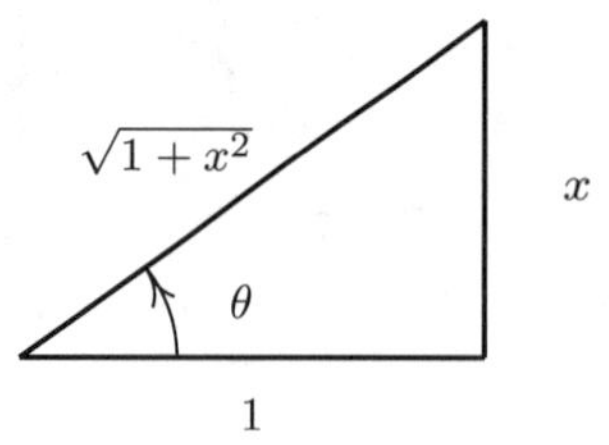

$\theta = \tan^{-1} x$

$\sec\theta = \sqrt{1+x^2}$

FIGURE 28

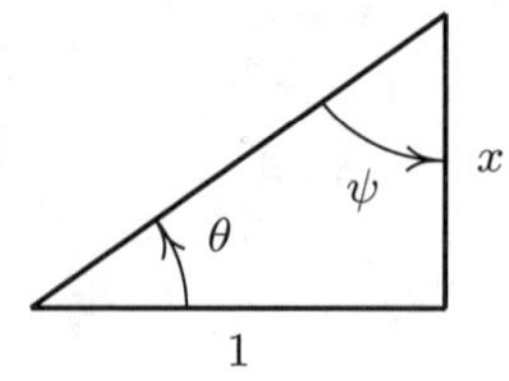

$\theta = \tan^{-1} x$

$\psi = \cot^{-1} x$

FIGURE 29

Formula **(26)** follows from **(25)** and the identity (see Figure 29),

$$\cot^{-1} x = \tfrac{1}{2}\pi - \tan^{-1} x. \qquad \textbf{(29)}$$

Question 20 How do the formulas for the derivatives in Theorem 4 relate to the graphs of **(a)** $\sin^{-1} x$, **(b)** $\cos^{-1} x$, **(c)** $\tan^{-1} x$, and **(d)** $\cot^{-1} x$ in Figures 30 through 33?

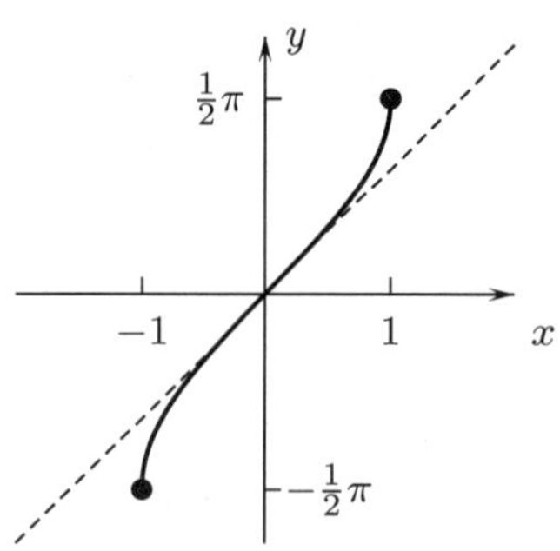

$y = \sin^{-1} x$

FIGURE 30

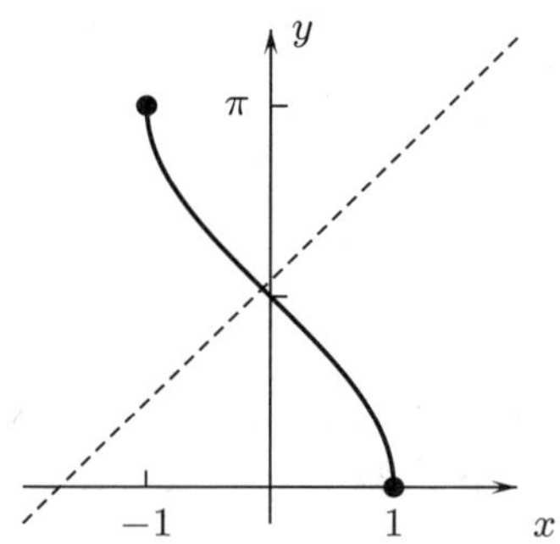

$y = \cos^{-1} x$

FIGURE 31

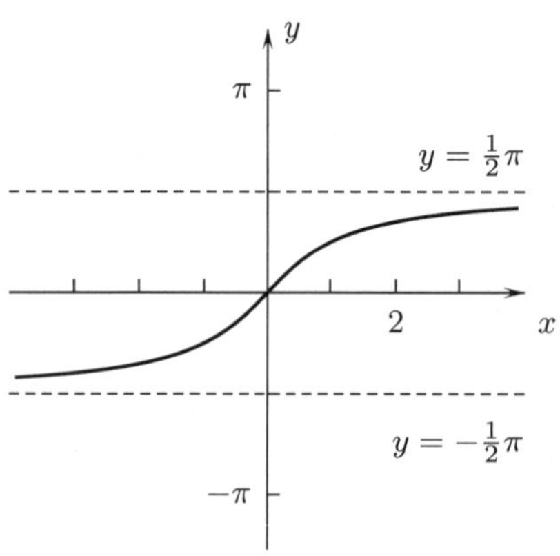

$y = \tan^{-1} x$

FIGURE 32

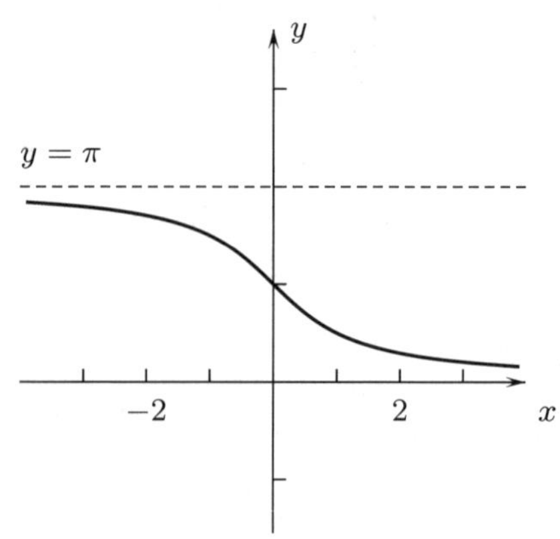

$y = \cot^{-1} x$

FIGURE 33

Chain-Rule formulas for the inverse trigonometric functions

With the Chain-Rule, Theorem 4 gives for differentiable functions $y = y(x)$,

$$\frac{d}{dx}(\sin^{-1} y) = \frac{d}{dy}(\sin^{-1} y)\,\frac{dy}{dx} = \frac{1}{\sqrt{1-y^2}}\,\frac{dy}{dx} \tag{30}$$

$$\frac{d}{dx}(\cos^{-1} y) = \frac{d}{dy}(\cos^{-1} y)\,\frac{dy}{dx} = \frac{-1}{\sqrt{1-y^2}}\,\frac{dy}{dx} \tag{31}$$

$$\frac{d}{dx}(\tan^{-1} y) = \frac{d}{dy}(\tan^{-1} y)\,\frac{dy}{dx} = \frac{1}{1+y^2}\,\frac{dy}{dx}. \tag{32}$$

$$\frac{d}{dx}(\cot^{-1} y) = \frac{d}{dy}(\cot^{-1} y)\,\frac{dy}{dx} = \frac{-1}{1+y^2}\,\frac{dy}{dx}. \tag{33}$$

Formulas **(30)** and **(31)** hold provided that $-1 < y(x) < 1$, and $y(x)$ can have any values in **(32)** and **(33)**.

Question 21 Generate $y = \tan^{-1}(x^2 - 4)$ in the window $-8 \le x \le 8, -2 \le y \le 2$. Then use the formulas for the function and its first derivative to explain the shape of the graph.

Question 22 A woman is walking at the constant speed of three feet per second toward one post of a four-foot-wide gate in a tall fence. Her path is perpendicular to the fence (Figure 34). At what rate is her horizontal angle of vision θ through the gate increasing when she is eight feet from the fence?

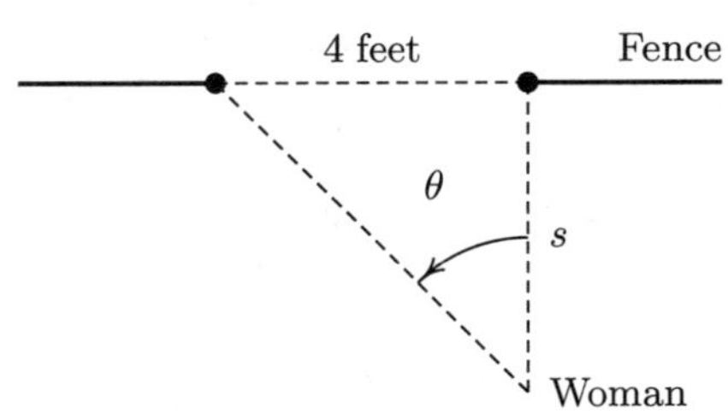

FIGURE 34

Formulas for the derivatives of the inverse secant and inverse cosecant functions can be derived from the definitions $\sec^{-1} x = \cos^{-1}(1/x)$ and $\csc^{-1} x = \sin^{-1}(1/x)$, but are seldom used.

Principles and procedures

- You can recall the shapes of the graphs $y = \sin x$ and $y = \cos x$ in Figures 8 and 9 by visualizing the unit circle in Figure 1 and noticing that as x increases from 0 to $\frac{1}{2}\pi$, $\sin x$ increases from 0 to 1 and $\cos x$ decreases from 1 to 0.
- Remembering that $\sin x$ increases and $\cos x$ decreases as x increases from 0 to $\frac{1}{2}\pi$ will also remind you where the minus sign goes in the differentiation formulas $\frac{d}{dx}(\sin x) = \cos x$ and $\frac{d}{dx}(\cos x) = -\sin x$, since $\sin x$ and $\cos x$ are both positive for $0 < x < \frac{1}{2}\pi$.
- Listing the formulas for the derivatives of the six trigonometric functions in two columns as below, with the "functions" $\sin x, \tan x$, and $\sec x$ on the left and the "cofunctions" $\cos x, \cot x$, and $\csc x$ on the right shows a pattern that can help you remember the formulas. Given the formulas on the left, the formulas on the right can be obtained by replacing each "function" by the corresponding "cofunction" and vice versa, and by inserting minus signs.

$$\frac{d}{dx}(\sin x) = \cos x \qquad \frac{d}{dx}(\cos x) = -\sin x$$
$$\frac{d}{dx}(\tan x) = \sec^2 x \qquad \frac{d}{dx}(\cot x) = -\csc^2 x$$
$$\frac{d}{dx}(\sec x) = \sec x\,\tan x \qquad \frac{d}{dx}(\csc x) = -\csc x\,\cot x$$

- To help remember the formulas for the derivatives $\sin^{-1} x, \tan^{-1} x, \cos^{-1} x$, and $\cot^{-1} x$ below, recall that $\sin^{-1} x$ is not defined if $|x| > 1$, while $\tan^{-1} x$ is defined for all x, that $\sin^{-1} x$ and $\tan^{-1} x$ are increasing functions, and that $\cos^{-1} x = \frac{1}{2}\pi - \sin^{-1} x$ and $\cot^{-1} x = \frac{1}{2}\pi - \tan^{-1} x$, so their derivatives are the negatives of the derivatives of $\sin^{-1} x$ and $\tan^{-1} x$, respectively.

$$\frac{d}{dx}(\sin^{-1} x) = \frac{1}{\sqrt{1-x^2}} \qquad \frac{d}{dx}(\tan^{-1} x) = \frac{1}{1+x^2}$$
$$\frac{d}{dx}(\cos^{-1} x) = \frac{-1}{\sqrt{1-x^2}} \qquad \frac{d}{dx}(\cot^{-1} x) = \frac{-1}{1+x^2}$$

- Some problems in this section may seem more difficult than those you encountered in earlier sections, until you have had practice working with trigonometric and inverse trigonometric functions and their derivatives. Then you will see that the problems here use the same ideas and procedures as similar problems that you have worked before, and are not, in principle, much more difficult.

Tune-Up Exercises 3.6♦

[A] *Answer provided.* [O] *Outline of solution provided.*

T1.[A] Give the radian measures of **(a)** $-60°$, **(b)** $135°$, and **(c)** $450°$. Then sketch each angle in a unit circle and find its sine, cosine, and tangent.

♦ Type 1, radians and degrees

T2.[A] What is the length of the arc subtended on a circle of radius 14 meters by an angle of $\frac{7}{5}\pi$ radians with its vertex at the center of the circle? What is the area of the sector of the circle inside the angle?

♦ Type 1, arc length on a circle

T3.[A] Find $\sin\theta$, $\cos\theta$, and $\tan\theta$ for the angle θ in Figure 35.

♦ Type 1, values of trigonometric functions

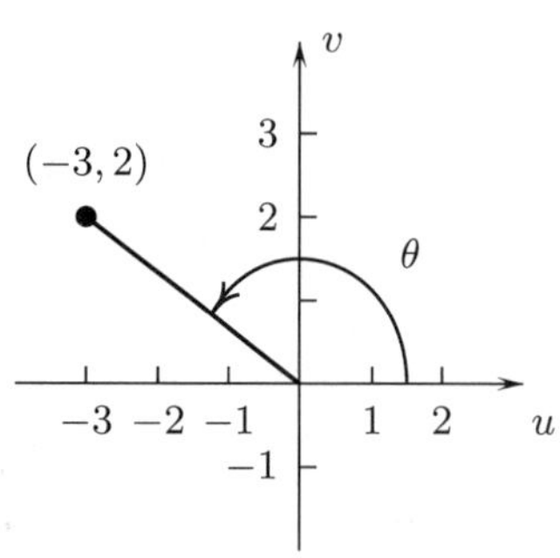

FIGURE 35

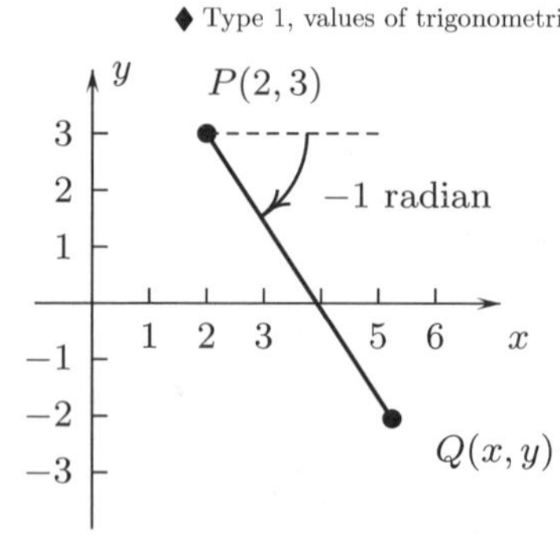

FIGURE 36

T4.[O] What are the coordinates (x, y) of the point Q in Figure 36 if the points P and Q are 6 units apart?

♦ Type 1, using a sine and a cosine

T5.[O] What is the angle γ in the right triangle of Figure 37?

♦ Type 1, an inverse trigonometric function

♦The Tune-up Exercises and Problems are classified by type and content. The types are (1) basic, reactive; (2) basic reflective; (3) intermediate, reactive; (4) intermediate, reflective; (5) advanced, reactive; (6) advanced, reflective; and (7) advanced, theoretical.

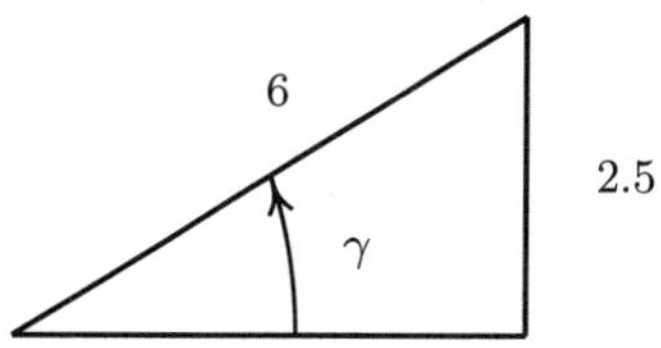

FIGURE 37

T6.[A] Use $\sqrt{2}$ and $\sqrt{3}$ as needed to give the exact values of **(a)** $\sin(\frac{7}{4}\pi)$, **(b)** $\cos(\frac{2}{3}\pi)$, **(c)** $\tan(-3\pi)$, **(d)** $\sec(\frac{4}{3}\pi)$, **(e)** $\cot(\frac{1}{6}\pi)$, **(f)** $\csc(-\frac{3}{4}\pi)$, **(g)** $\tan(-\frac{5}{6}\pi)$, and **(h)** $\sec(-\frac{3}{2}\pi)$.

♦ Type 1, values of trigonometric functions

T7.[O] Use π as necessary to give exact values of **(a)** $\sin^{-1}(1/\sqrt{2})$, **(b)** $\tan^{-1}(-1)$, **(c)** $\sin^{-1}(-1)$, **(d)** $\tan^{-1}(-1/\sqrt{3})$, **(e)** $\sin^{-1}(\frac{1}{2}\sqrt{3})$, and **(f)** $\tan^{-1}(0)$.

♦ Type 1, values of inverse trigonometric functions

T8.[O] Give the exact and approximate decimal values of the angle α in Figure 38.

♦ Type 1, approximate values of an inverse trigonometric function

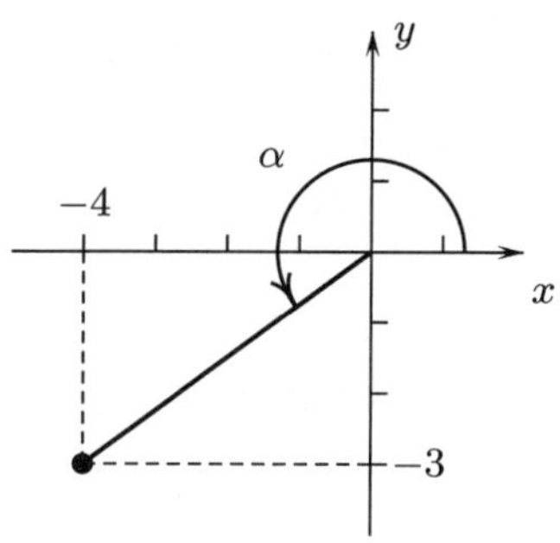

FIGURE 38

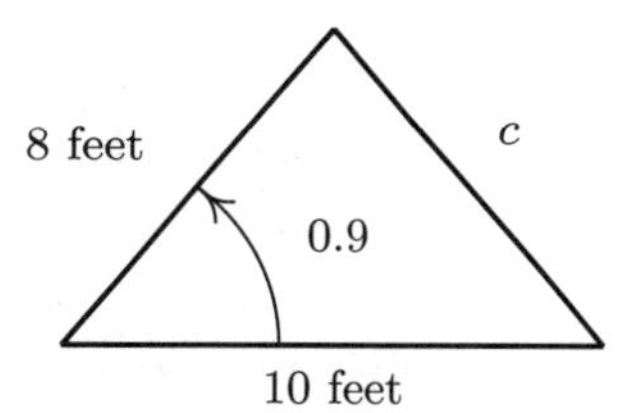

FIGURE 39

T9.[O] The angle between the sides of lengths 8 and 10 feet in the triangle of Figure 39 is 0.9 radians. Find the length c of the other side.

♦ Type 3, the Law of Cosines

T10.[O] What is the smallest angle in the right triangle with sides of lengths 3, 4, and 5 meters?

♦ Type 3, the Law of Cosines and an inverse trigonometric function

Find the derivatives in Exercises T11 through T18.

T11.[A] $\dfrac{d}{dx}(5\sin x - 6\cos x)$

♦ Type 1, a derivative

T12.[A] $y'(\pi)$ for $y = 3\tan x$

♦ Type 1, a derivative

T13.[A] $\dfrac{d}{dx}(\sqrt{\cos x})$

♦ Type 1, a derivative

T14.[A] The derivative of $\sin(x^3 - 4)$

♦ Type 1, a derivative

T15.[O] $\left[\dfrac{d}{dx}[\tan(2x)]\right]_{x=0.6}$

♦ Type 1, a derivative

T16.[A] dy/dx for $y = x^2 \sin x$

♦ Type 1, a derivative

T17.[O] $\dfrac{d}{dx}[\tan^{-1}(2x)]$

♦ Type 1, a derivative

T18.[A] The x-derivative of $\sin^{-1}(x^2)$

♦ Type 1, a derivative

T19.[O] **(a)** Express h in the right triangle of Figure 40 in terms of θ. **(b)** What is $\dfrac{dh}{dt}$ at a moment when $\theta = \frac{1}{3}\pi$ radians and $\dfrac{d\theta}{dt} = 0.2$ radians per minute?

♦ Type 3, a related-rate problem with a trigonometric function

T20.[A] What is $\dfrac{d\psi}{dt}$ in Figure 40 at a moment when $\dfrac{d\theta}{dt} = 0.2$ radians per minute?

◆ Type 3, a related-rate problem with an inverse trigonometric function

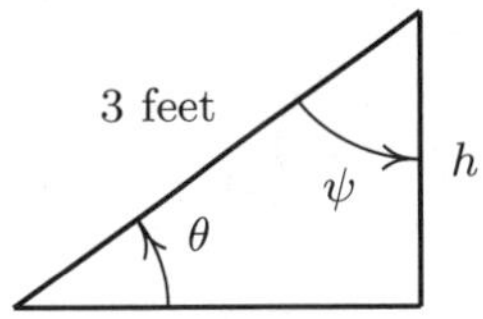

FIGURE 40

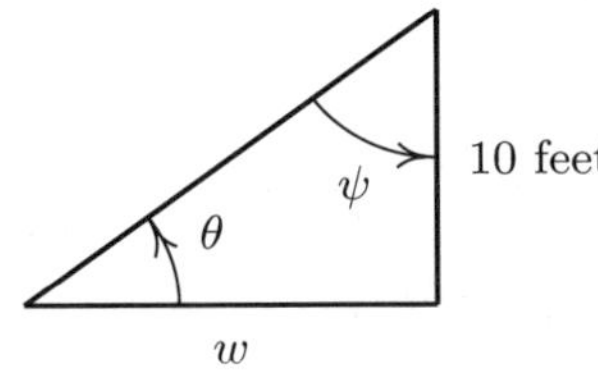

FIGURE 41

T21.[O] **(a)** Express ψ in Figure 40 in terms of h. **(b)** What is $\dfrac{d\psi}{dt}$ at a moment when $h = 2$ feet and $\dfrac{dh}{dt} = -1$ foot per minute?

◆ Type 3, a related-rate problem with an inverse trigonometric function

T22.[O] **(a)** Express w in the right triangle of Figure 41 in terms of θ. **(b)** What is $\dfrac{dw}{dt}$ at a moment when $\theta = \frac{1}{4}\pi$ and $\dfrac{d\theta}{dt} = 0.2$?

◆ Type 3, a related-rate problem with trigonometric functions

T23.[O] **(a)** Express ψ in Figure 41 in terms of w. **(b)** What is $\dfrac{d\psi}{dt}$ at a moment when $w = 8$ and $\dfrac{dw}{dt} = 4$?

◆ Type 3, a related-rate problem with an inverse trigonometric function

Problems 3.6

[A]*Answer provided.* [O]*Outline of solution provided.*

Give exact and approximate decimal answers, as appropriate, to the following problems.

1.[A] The earth's orbit about the sun is approximately a circle of radius 93 million miles. Approximately how far does the earth move in its orbit in a day?

◆ Type 2, arclength on a circle

2.[A] The angle α in Figure 42 is $\frac{1}{5}\pi$ and the angle β is $\frac{3}{8}\pi$. What is the angle γ?

◆ Type 1, angles in a triangle

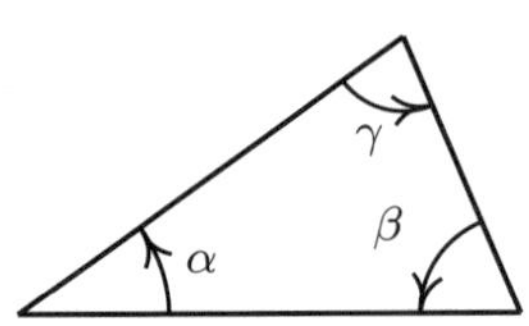

FIGURE 42

3.[A] The direction from Helsinki, Finland toward St. Petersburg, Russia is 4.8° south of east, and Helsinki and St. Petersburg are 274 kilometers apart (Figure 43). How far is St. Petersburg south and how far is it east of Helsinki?

◆ Type 2, using trigonometric functions

4. Ravenna, Italy is 70 kilometers north and 70 kilometers east of Florence (Figure 44). How far is Ravenna from Florence? What is the direction from Florence toward Ravenna?

◆ Type 2, using an inverse trigonometric function, etc.

5. The Eiffel tower was designed by Alexandre Gustave Eiffel for the Paris exposition of 1889. Its base is a horizontal square 100 meters wide and a line from a corner of its base to its pointed top makes an angle of 1.34 radians. **(a)** How long are the diagonals of the base? **(b)** How high is the Eiffel tower?

◆ Type 2, using trigonometric functions

FIGURE 43

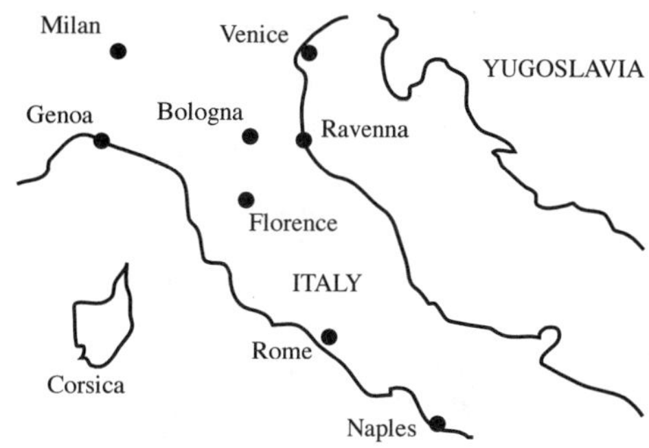

FIGURE 44

6.[A] Panama City is 480 kilometers east and 90 kilometers south of San Jose, Costa Rica (Figure 45). **(a)** How far is San Jose from Panama City? **(b)** What is the direction from Panama City toward San Jose?

◆ Type 2, using an inverse trigonometric function, etc.

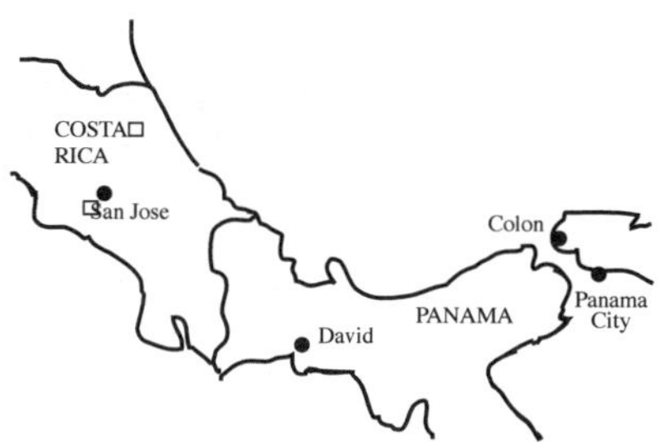

FIGURE 45

FIGURE 46

7. Beirut, Lebanon and Haifa, Israel are 56 miles apart, and Haifa is 21 miles west of Beirut (Figure 46). **(a)** How far is Haifa south of Beirut? **(b)** What is the direction from Beirut toward Haifa?

◆ Type 2, using an inverse trigonometric function, etc.

8. Draw an angle of 1 radian with its vertex at the center of a circular arc of radius 1 and compare the angular sector with an equilateral triangle to explain why 1 radian is slightly less than 60°.

◆ Type 2, an angle of one radian

9.[A] After an argument with Martha, George stomps out of the house and walks one-half mile east on one road and then one-quarter mile to the north on another road to a coffee shop. What are the angles in the triangle with vertices at his house, the coffee shop, and the intersection of the two roads?

◆ Type 2, using an inverse trigonometric function

10.[A] One leg of a right triangle is 8 feet long and the hypotenuse is 10 feet long. What is the smallest angle in the triangle?

◆ Type 2, using an inverse trigonometric function

11. What is the obtuse angle in the triangle with vertices $(1,0)$, $(0,0)$, and $(-1,1)$ in an xy-plane with equal scales on the axes?

◆ Type 2, using an inverse trigonometric function

12.[A] An angle of 4 radians with its vertex at the center of a circle subtends an arc 10 feet long on the circle. What is the radius of the circle?

◆ Type 2, arclength on a circle

13.[A] When the Great Pyramid of Egypt was built in the fourth dynasty (c. 2600–2500 BC), it was a right pyramid 481 feet high with a square base 776 wide. What angle did its sides make with the ground? (Calculate the angle at the middle of one side of the base.)

◆ Type 2, using an inverse trigonometric function

Find the derivatives in Problems 14 through 37.

14.[A] $y'(x)$ for $y = 2\tan x - 3\sin x$

♦ Type 1, a derivative

15.[A] dy/dx for $y = 3\sin x + 5\cos x$

♦ Type 1, a derivative

16.[A] dy/dx for $y = (\sin x)/x$

♦ Type 1, a derivative

17.[O] $\dfrac{d}{dt}(\sin(4t) - 4\sin t)$

♦ Type 1, a derivative

18.[A] $\dfrac{d}{dx}(x\tan x)$

♦ Type 1, a derivative

19.[A] $y'(1)$ for $y = \cos^4 x$

♦ Type 1, a derivative

20.[A] $y'(\frac{1}{4}\pi)$ for $y(x) = x\sin x$

♦ Type 1, a derivative

21.[A] $\left[\dfrac{d}{dx}[\sin(\pi x)]\right]_{x=1}$

♦ Type 1, a derivative

22.[O] $\dfrac{d}{dt}[\tan(4t)]$

♦ Type 1, a derivative

23.[A] $\left[\dfrac{d}{dx}(\sin^{-1} x)\right]_{x=0.5}$

♦ Type 1, a derivative

24.[A] $\left[\dfrac{d}{dx}(\tan^{-1} x)\right]_{x=3}$

♦ Type 1, a derivative

25.[A] $\dfrac{d}{dx}(x\tan^{-1} x)$

♦ Type 1, a derivative

26.[O] $\dfrac{d}{dx}[(\sin^{-1} x)^2]$

♦ Type 1, a derivative

27. $\dfrac{d}{dx}(x\sin^{-1} x)$

♦ Type 1, a derivative

28.[A] y' for $y = \sin^{-1}\left(\dfrac{x}{3}\right)$

♦ Type 1, a derivative

29.[A] $\dfrac{d}{dx}[\sin(\ln x)]$

♦ Type 1, a derivative

30.[A] $\dfrac{d}{dx}[\ln(\sin x)]$

♦ Type 1, a derivative

31. $\dfrac{d}{dx}(e^{\cos x})$

♦ Type 1, a derivative

32.[A] $\dfrac{d}{dx}[(\sin x)^\pi]$

♦ Type 1, a derivative

33. $\dfrac{d}{dx}[\sin(\cos x)]$

♦ Type 1, a derivative

34.[A] $\left[\dfrac{d}{dx}\csc x\right]_{x=\pi/3}$

♦ Type 1, a derivative

35.[A] $\dfrac{d}{dx}\left(\dfrac{\cot x}{x}\right)$

♦ Type 1, a derivative

36 $\dfrac{d}{dx}(\cot^2 x)$

♦ Type 1, a derivative

37. $\dfrac{d}{dx}[\sec(3x+1)]$

♦ Type 1, a derivative

In Problems 38 through 47, give equations of the tangent lines at the indicated points on the graphs. Check by generating the curves and the tangent lines together in suitable windows.

38.[A] $y = \sin x$ at $x = 3$

♦ Type 1, equation of a tangent line

39.[A] $y = x\cos x$ at $x = \pi$

♦ Type 1, equation of a tangent line

40. $y = \sin(x^2)$ at $x = 3$

♦ Type 1, equation of a tangent line

41 $y = \sin x + \cos x$ at $x = \frac{1}{2}\pi$

♦ Type 1, equation of a tangent line

42. $y = \tan^{-1} x$ at $x = 1$

♦ Type 1, equation of a tangent line

43.[A] $y = e^{\sin x}$ at $x = 4$.

♦ Type 1, equation of a tangent line

44. $y = \sin(\ln x)$ at $x = e$.

♦ Type 1, equation of a tangent line

45. $y = \sin^{-1} x$ at $x = 0.9$

♦ Type 1, equation of a tangent line

46.[A] $y = \sec x$ at $x = \pi/4$

♦ Type 1, equation of a tangent line

47. $y = \tan(\pi x)$ at $x = 1$

♦ Type 1, equation of a tangent line

48.[A] The voltage in an electrical circuit is $220\sin(120\pi t)$ volts at time t (seconds). **(a)** What is the frequency of the current? **(b)** What is the rate of change of the voltage with respect to time at $t = 45$?

♦ Type 2, a related-rate problem

49.[A] A right triangle has a hypotenuse of fixed length 10 feet and one of its acute angles is θ. Give the rates of change with respect to θ of its area and perimeter at $\theta = \pi/8$.

◆ Type 2, a related-rate problem

50. An acute angle x in a right triangle is changing while the leg adjacent to x has the constant length of 10 feet. Give a formula for the rate of change of the area of the triangle with respect to x.

◆ Type 2, a related-rate problem

51. The top of a 13-foot-long ladder is sliding down a tall vertical wall while its base is sliding away from the wall along the horizontal ground. When the base of the ladder is five feet from the wall, it is moving away from the wall at the rate of two feet per second. **(a)** What is the angle between the ground and the ladder at that time? **(b)** At what rate is the angle decreasing at that moment?

◆ Type 2, a related-rate problem

52. An object moving in an xy-plane has coordinates $x = 2\cos t, y = 3\sin t$ at time t. **(a)** Show that its path is on the ellipse $9x^2 + 4y^2 = 36$. **(b)** At what rate is its distance to the origin increasing at $t = \frac{1}{4}\pi$?

◆ Type 2, a related-rate problem

53. The top of a twelve-foot-long ladder is sliding down a vertical wall as its bottom is sliding on the horizontal ground. How fast are **(a)** its top and **(b)** its bottom moving when the angle between the ladder and the ground is 0.4 radians if the angle is decreasing 0.05 radians per second?

◆ Type 2, a related-rate problem

54.[A] A five-meter-tall vertical post casts a shadow on the horizontal ground. What is the rate of change of the length of the shadow with respect to the angle x between the sun's rays and the ground when the angle is 0.8 radians?

◆ Type 2, a related-rate problem

55.[A] An airplane is flying at an altitude of 400 feet at the speed of 200 feet per second directly away from a searchlight on the ground. At what rate is the angle between the ray of light and ground changing when the angle is $30°$?

◆ Type 2, a related-rate problem

56.[A] A man is watching a helium balloon rise vertically over his daughter's head. She released the balloon from a point 5 feet above the ground and it rises at the constant rate of 3 feet per second. Her father is 30 feet from her and his eyes are 5 feet above the ground. Give a formula for the rate of change with respect to time of the angle between his line of sight and the horizontal as a function of the time (seconds) since the release of the balloon.

◆ Type 2, a related-rate problem

57. An aviarist is watching a bird that is flying directly south from the aviary at a constant height of 100 feet above the aviarist's eye level. When the aviarist's line of vision makes an angle of $\frac{1}{3}\pi$ radians with the horizontal, the angle is decreasing at the rate of $\frac{1}{4}$ radians per second. How fast is the bird flying at that moment?

◆ Type 2, a related-rate problem

58. A lighthouse is 2 miles offshore and 3 miles down the straight shoreline from a farmhouse. The light from the lighthouse shining on the farmhouse passes it at the rate of 1 mile per second. How many seconds does it take for the light to make a complete revolution if it turns at a constant rate?

◆ Type 2, a related-rate problem

59. A searchlight in a lighthouse 15 miles off a straight shore is turning at the rate of 2 revolutions per minute. At what speed does the beam of light pass a point 20 miles down the shore from the lighthouse?

◆ Type 2, a related-rate problem

60. Find the maximum and minimum values of $3\sin x - 4\cos x$ for $0 \le x \le 1$.

◆ Type 2, a max/min problem

61. Of all the right triangles with hypotenuse of length 4, which has the minimum perimeter and why?

◆ Type 2, a maximum/minimum problem

62. What is $\dfrac{d^2U}{dw^2}$ for $U = w^2 \cot w$?

◆ Type 1, a second derivative

63.[A] What is $\dfrac{dZ}{dx}(1)$ where $Z(x) = \tan[Y(x)], Y(1) = \frac{1}{3}\pi$, and $\dfrac{dY}{dx}(1) = -5$?

◆ Type 2, the Chain Rule with trigonometric functions

64. Find $L'(\frac{1}{4}\pi)$ where $L(x) = M(\tan x), M(1) = 5$ and $M'(1) = 8$.

♦ Type 2, the Chain Rule with trigonometric functions

65. Generate $y = \frac{1}{6}x^2 \sin(2\pi x)$ and $y = \dfrac{\sin(2\pi x)}{x^2+1}$ separately for $-3 \le x \le 3$, $-1.3 \le y \le 1.3$. **(a)** Why do the two curves have the same x-intercepts? **(b)** Explain the overall shapes of the two curves without using any derivatives.

♦ Type 2, explaining the shape of a graph

66. Sound reaches your ears as oscillations of air pressure. An electronic synthesizer could produce a pure tone at a frequency of 220 cycles per second (the C below middle C on a piano) by causing the oscillations of air pressure to be $p = k\sin(110\pi t)$ at time t (seconds) where k is a positive constant. When this note is played on a piano, the HARMONIC NOTES with frequencies $440, 660, 880, \ldots$ (the integer multiples of 220) are also created. If the variation in the air pressure were $p = k\sin(110\pi t) + \frac{1}{4}k\sin(330\pi t)$ the sound would be a combination of C and its second harmonic, which is the E above middle C. Generate the graph of $p = k\sin(110\pi t) + \frac{1}{4}k\sin(330\pi t)$ with $k = 1$ in the window $0 \le t \le 0.04, -1.5 \le p \le 1.5$ and explain its shape.

♦ Type 4, explaining the shape of a graph

67. Generate $y = \cos x$ and $y = \sqrt{\sec x}$ together in the window $-8\pi \le x \le 8$, $-1.5 \le y \le 3$. Use the shape of $y = \cos x$ to explain the shape of $y = \sqrt{\sec x}$.

♦ Type 4, explaining the shape of a graph

68. Figures 47 through 49 show $y = \dfrac{A}{2+\cos x}$, $y = \dfrac{B}{1-\cos x}$ and $y = \dfrac{C}{1-2\cos x}$ with positive constants A, B, and C. **(a)** Which curve is which? **(b)** What are A, B and C?

♦ Type 2, analyzing shapes of graphs

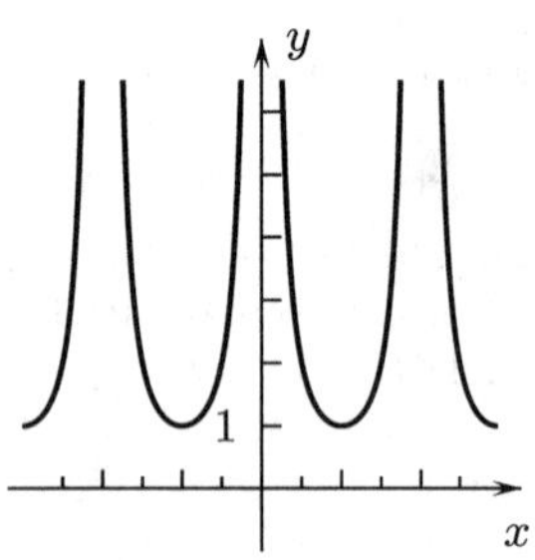

FIGURE 47

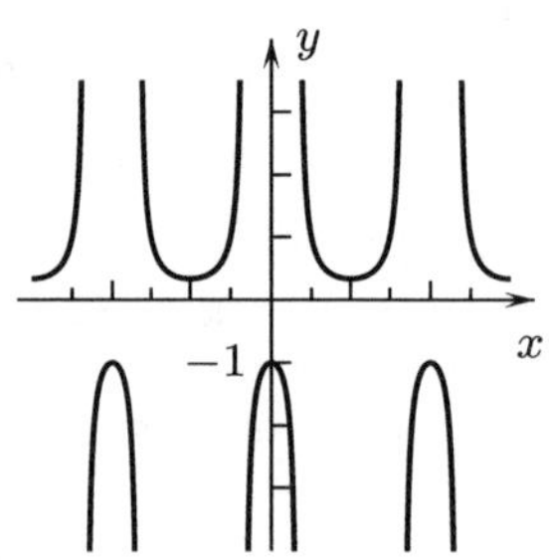

FIGURE 48

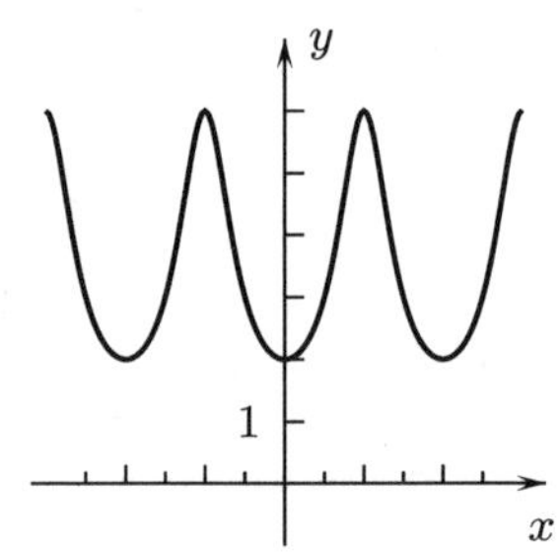

FIGURE 49

69. Find, without using derivatives, the maximum and minimum values of $\sin(x^2)$ for $-1 \le x \le 1$.

♦ Type 3, max/min in an interval

70.[A] Find the maximum and minimum values of $x + \tan x$ for $-\frac{1}{4}\pi \le x \le \frac{1}{4}\pi$.

♦ Type 3, max/min in an interval

71.[A] What are the maximum and minimum values of $\dfrac{\cos x}{x}$ for $0.1 \le x \le 1.2$?

♦ Type 3, max/min in an interval

72. **(a)** Find the most extensive open intervals in which $\cos\left(\dfrac{\pi}{x^2+1}\right)$ is increasing and decreasing. **(b)** What are $\lim_{x\to\infty} \cos\left(\dfrac{\pi}{x^2+1}\right)$ and $\lim_{x\to-\infty} \cos\left(\dfrac{\pi}{x^2+1}\right)$ and why?

♦ Type 2, increasing and decreasing functions and limits

73. Give approximate values of **(a)** $f'(2)$ and **(b)** $g'(2)$ where $f(x) = \sin(h(x))$, $g(x) = \tan(h(x))$, and $h(x)$ is the differentiable function whose graph is shown in Figure 50.

♦ Type 3, the Chain Rule with trigonometric functions and a graph

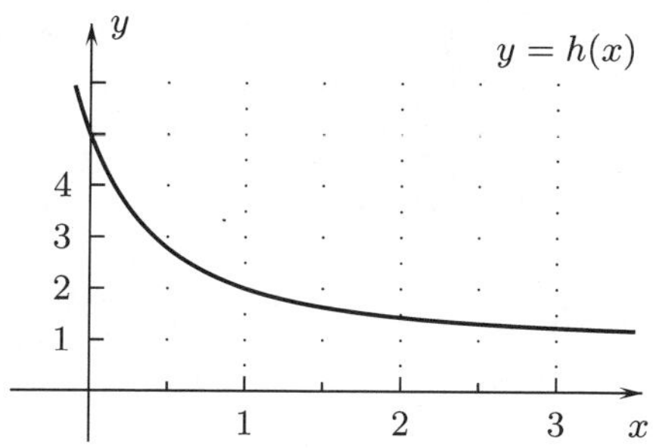

FIGURE 50

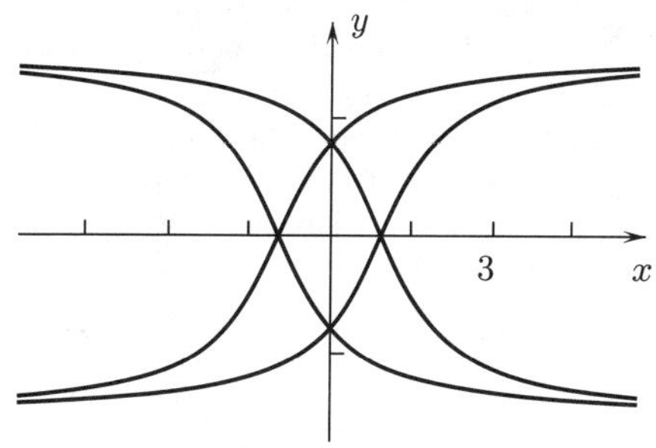

FIGURE 51

74. Figure 51 shows the graphs of four functions given by $\tan^{-1}(Ax+B)$ with integer values of A and B. What are the four functions? Check your answer by generating the four graphs on your calculator or computer for $-6 \le x \le 6, -1.75 \le y \le 1.75$.

♦ Type 4, the inverse tangent function

75.[O] Two sides of a triangle are 5 and 7 feet long and the angle between them is $\frac{1}{3}\pi$. How long is the other side?

♦ Type 4, the Law of Cosines, etc.

76.[A] The town of Jalapeño is 20 kilometers north-northeast of the town of Ancho, and the town of Habanero is 30 kilometers southwest of Ancho. How far is it from Habanero to Jalapeño?

♦ Type 4, the Law of Cosines, etc.

77. The angle γ in the isosceles triangle of Figure 52 is $\frac{3}{4}\pi$. **(a)** What are the other angles? **(b)** How long is the unequal side?

♦ Type 4, the Law of Cosines, etc.

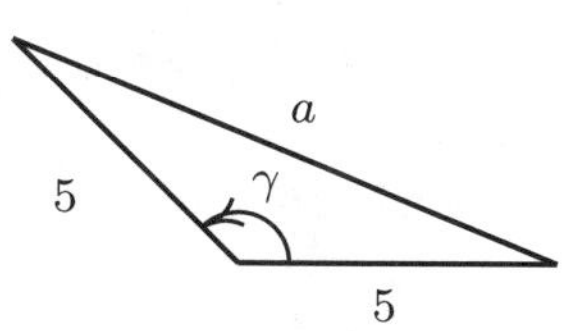

FIGURE 52

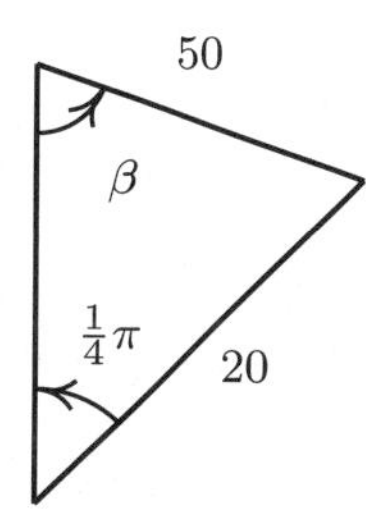

FIGURE 53

78.[A] What is the angle β in Figure 53?

♦ Type 4, the Law of Cosines, etc.

79. The Bermuda Triangle is an area of the Atlantic Ocean where many ships and airplanes have disappeared. The vertices of the triangle are in Bermuda, in Puerto Rico, and at Melbourne, Florida.[2] Melbourne is approximately 1175 miles from Bermuda and approximately 1000 miles from Puerto Rico, and it is about 975 miles from Bermuda to Puerto Rico. What is the smallest angle in the Bermuda Triangle?

♦ Type 4, the Law of Cosines, etc.

80. The moon's diameter is approximately 2,000 miles and its distance from the earth varies between 226,000 and 252,000 miles. **(a)** What are the approximate maximum and minimum angles of vision subtended by the moon when viewed from the earth? **(b)** As viewed from the earth the moon moves relative to the stars because its phases have a period of 29.53 days. By approximately what multiple of its diameter does the moon move every hour relative to the stars?

♦ Type 2, arclength on a circle

---

[2] *The World Book Encyclopedia*, Volume 2, Chicago, Il: : World Book-Childcraft International, Inc., 1978, p. 208.

81. The phases of the moon have a period of 29.53 days; there is a full moon every 29.53 days. **(a)** Is there at least one full moon every month? Explain. **(b)** If there is a second full moon in any month, it is called a "blue moon." The table below lists the dates of full moons in the years 2015 through 2020.[(3)] Which are the blue moons? What are the greatest and the least time spans between these blue moons? **(c)** It is rare for there to be two blue moons in the same year, and most years with blue moons are separated by one or two years. Explain why blue moons are never as much as three years apart.

◆ Type 2, blue moons

FULL MOONS, 2015–2020

2015	2016	2017	2018	2019	2020
Jan. 5	Jan. 24	Jan. 12	Jan. 2	Jan. 21	Jan. 10
Feb. 3	Feb. 22	Feb. 11	Jan. 31	Feb. 19	Feb. 9
Mar. 5	Mar. 23	Mar. 12	Mar. 2	Mar. 21	Mar. 9
Apr. 4	Apr. 22	Apr. 11	Mar. 31	Apr. 19	Apr. 8
May 4	May 21	May 10	Apr. 30	May 18	May 7
Jun. 2	Jun. 20	Jun. 9	May 29	Jun. 17	Jun. 5
Jul. 2	Jul. 19	Jul. 9	Jun. 28	Jul. 16	Jul. 5
Jul. 31	Aug. 18	Aug. 7	Jul. 27	Aug. 15	Aug. 3
Aug. 29	Sept. 16	Sept. 6	Aug. 26	Sept. 14	Sept. 2
Sept. 28	Oct. 16	Oct. 5	Sept. 25	Oct. 13	Oct. 1
Oct. 27	Nov. 14	Nov. 4	Oct. 24	Nov. 12	Oct. 31
Nov. 25	Dec. 14	Dec. 3	Nov. 23	Dec. 12	Nov. 30
Dec. 25			Dec. 22		Dec. 30

82. A gear of radius 20 inches is engaged with a gear of radius 30 inches. The larger gear is turning at the constant rate of 3 revolutions per minute. How fast is the smaller gear turning?

◆ Type 4, arclengths on circles

83. **(a)** Generate $y = x + \frac{1}{100}\cos x$ in the window $-30 \le x \le 30, -30 \le y \le 30$ and explain its shape. **(b)** The curve $y = x + \frac{1}{100}\cos x$ has an infinite number of inflection points. Find the two that are closest to the origin.

◆ Type 4, inflection points, etc.

84. Generate $y = \sin(\cos x)$ and $y = \cos(\sin x)$ together in the window $-2\pi \le x \le 2\pi$, $-1.5 \le y \le 1.5$ **(a)** What are the (smallest) periods of each of these two functions? **(b)** Find the exact global maximum and global minimum of both functions. (Look at $0 \le x \le 2\pi$.)

◆ Type 4, composite functions, etc.

85.[A] Find the maximum and minimum of $\sqrt{3}\cos(2x) - \sin(2x)$ for $0 \le x \le 4\pi$.

◆ Type 3, max/min in an interval

86. What are the maximum and minimum of $\tan x - \sqrt{2}\sec x$ for $0 \le x \le \frac{1}{4}\pi$?

◆ Type 3, max/min in an interval

87. Find the minimum value of $y = \tan x + \dfrac{1}{\tan x}$ for $0 < x < \frac{1}{2}\pi$.

◆ Type 3, max/min in an interval

88. **(a)** Match $y = \sin^2 x$, $y = |\sin x|$, $y = \sin|x|$, and $y = \sqrt{\sin x}$ to their graphs in Figures 54 through 57. **(b)** What happens to the first derivatives of each of these functions as $x \to 0^+$?

◆ Type 2, recognizing graphs

(3) *Moon Tables for Times Past, Present, and Future*, by R. Brahde, Bergen, Norway: Norlanger Forlag, 1977, p. 9.

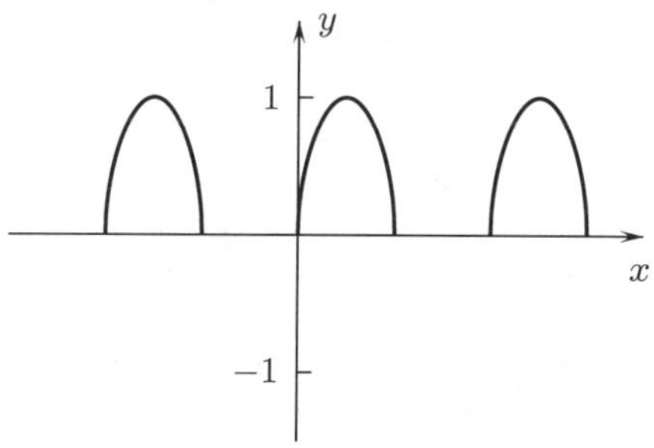

FIGURE 54

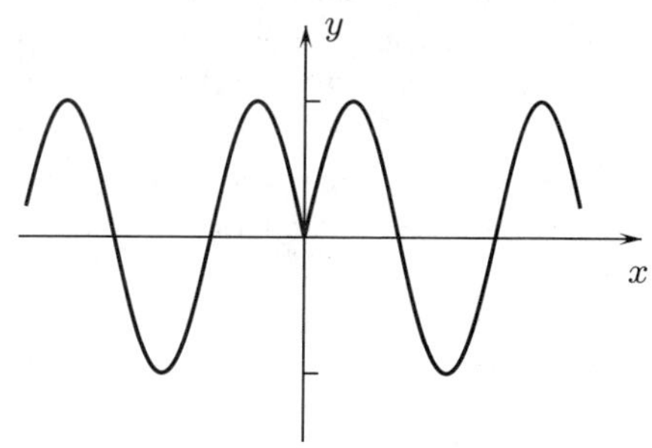

FIGURE 55

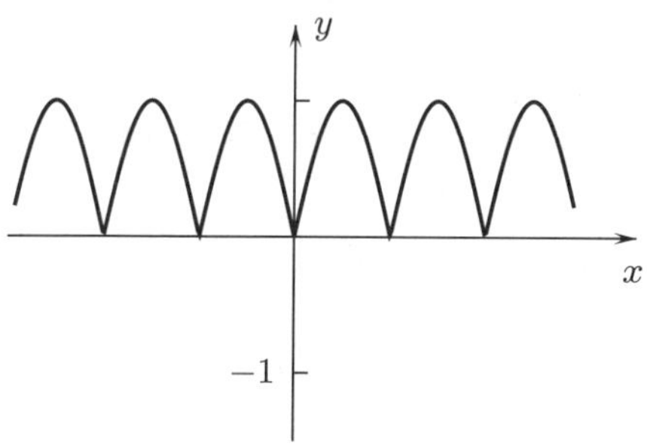

FIGURE 56

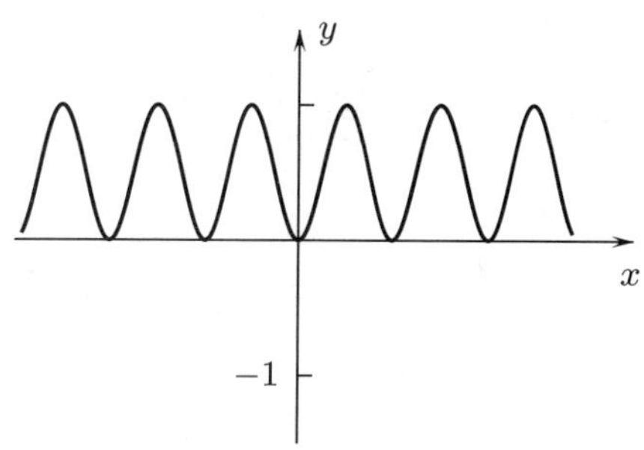

FIGURE 57

89. **(a)** Match $y = \cos^3 x$, $y = \cos^{1/3} x$, and $y = \cos(x^3/\pi^3)$ to their graphs in Figures 58 through 60. **(b)** What happens to the first derivatives of each of these functions as $x \to \frac{1}{2}\pi$?

♦ Type 4, recognizing graphs and limits of derivatives

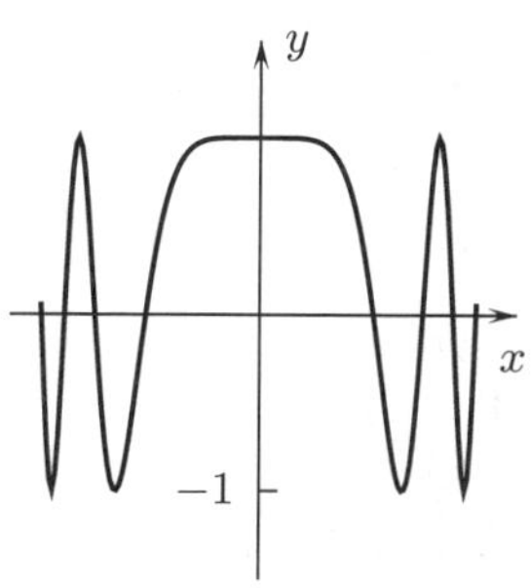

FIGURE 58

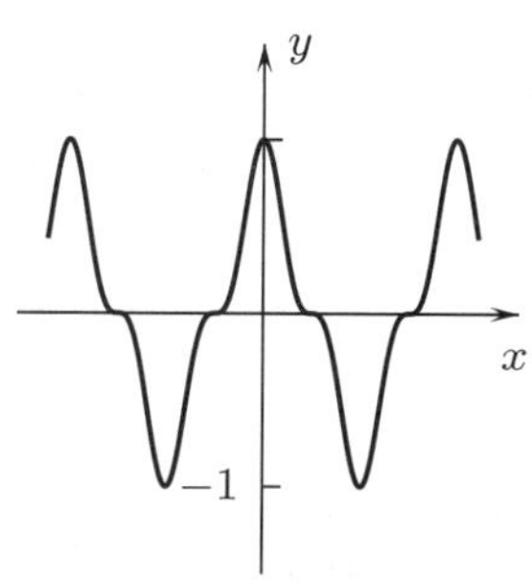

FIGURE 59

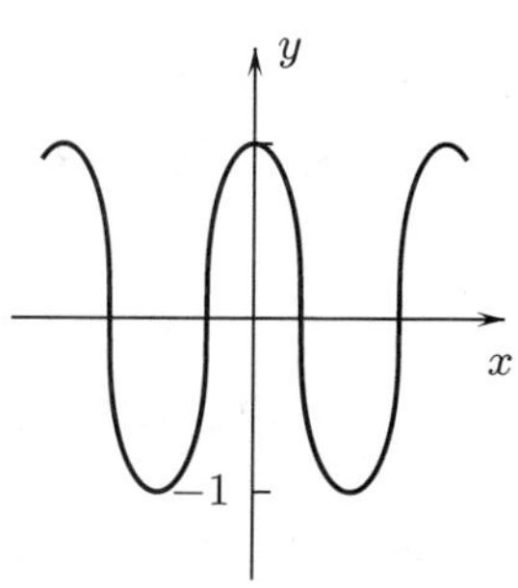

FIGURE 60

90. Generate the graph of $f(x) = \dfrac{4}{\tan^2 x + 1}$ in the window $-2\pi \le x \le 2\pi, -1 \le y \le 5$. **(a)** At what values of x with $-2\pi \le x \le 2\pi$ is $f(x)$ not defined? **(b)** Use the first derivative to determine the most extensive subintervals of $[2\pi, 2\pi]$ in which $f(x)$ is increasing and in which it is decreasing.

♦ Type 4, the First-Derivative Test, etc.

91. Find a constant k such that $y = k\sin x$ and $y = k\cos x$ intersect at right angles if they are drawn in an xy-plane with equal scales on the coordinate axes.

♦ Type 4, perpendicular curves

92. A 20-foot-high vertical pole is 20 feet east of a vertical wall. The sun is setting in the west and its rays make an angle α with the pole $(0 < \alpha < \frac{1}{2}\pi)$. **(a)** For what values of α does the shadow reach the wall? **(b)** How fast is the shadow moving up the wall when $\alpha = 1$ radian if α is increasing at the rate of $\frac{1}{12}\pi$ radians per hour?

♦ Type 4, a related-rate problem, etc.

93. What are $\lim_{x\to\infty} \tan^{-1}(\ln x)$ and $\lim_{x\to 0^+} \tan^{-1}(\ln x)$?

♦ Type 3, limits

94. The least force acting at an angle θ with the floor which will pull an object of weight W along a floor is given by the formula $F = \mu W/(\cos\theta + \mu\sin\theta)$. Here μ is a positive constant called the COEFFICIENT OF STATIC FRICTION of the object. What angle requires the least force?

♦ Type 4, a max/min problem

95. Show that the curves $y = \sec^2 x$ and $y = 2\tan x$ intersect at only one point for $-\pi/2 < x < \pi/2$ and that they are tangent there.

♦ Type 4, tangent curves

96. **(a)** Match $y = \sin(\sin^{-1} x)$, $y = \sin^{-1}(\sin x)$, $\cos(\sin^{-1} x)$, $y = \tan(\tan^{-1} x)$, and $y = \tan^{-1}(\tan x)$ to their graphs in Figures 61 through 65. **(b)** What are the derivatives of these functions at $x = 0$?

♦ Type 4, composite functions and their derivatives

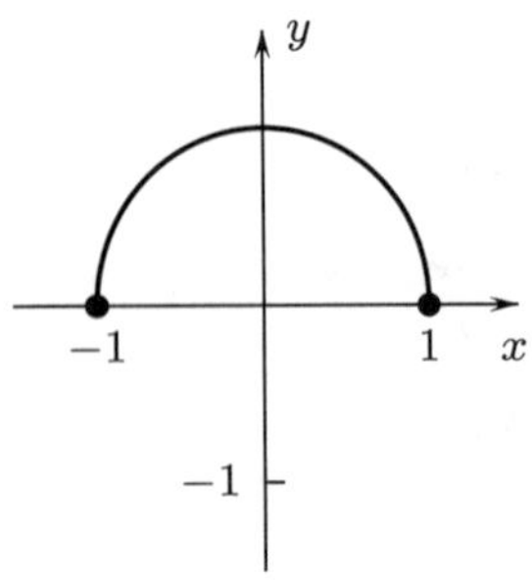

FIGURE 61

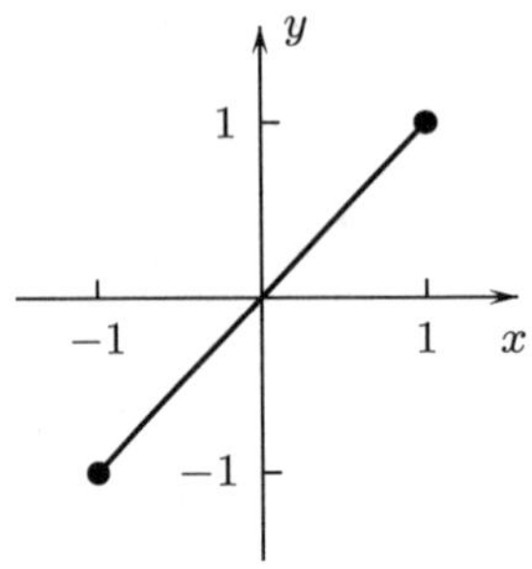

FIGURE 62

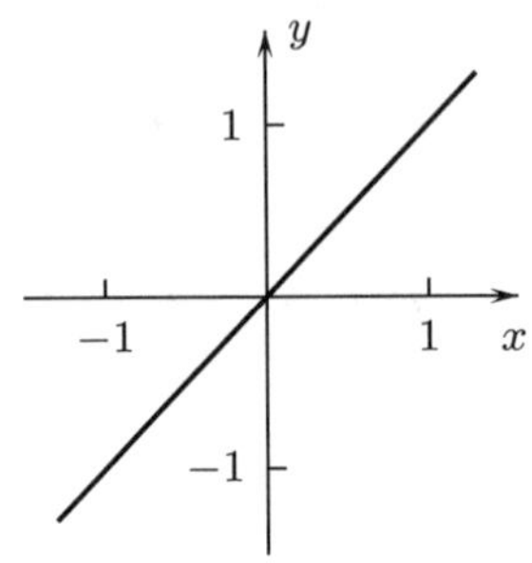

FIGURE 63

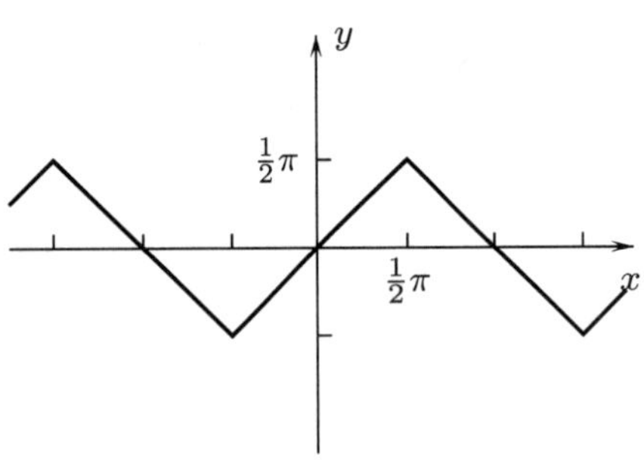

FIGURE 64

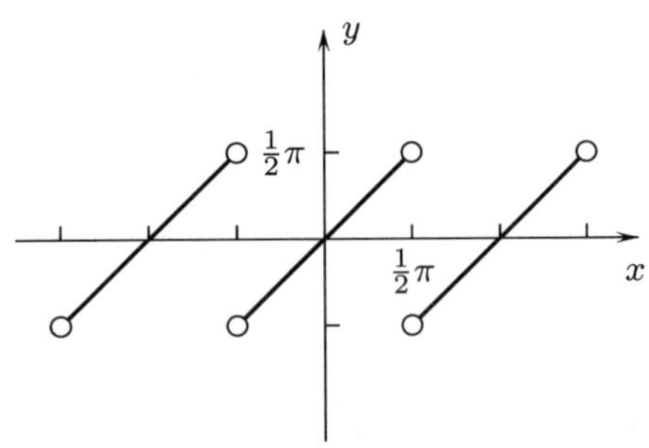

FIGURE 65

97. **(a)** Generate the curve $y = \sin^{-1}(\cos x)$ in the window $-6 \le x \le 6, -4 \le y \le 4$ and explain its shape. **(b)** Give a formula for the derivative of $\sin^{-1}(\cos x)$.

♦ Type 4, a composite function and its derivative

98. **(a)** Generate the curve $y = \tan(\sin^{-1} x)$ in the window $-2 \le x \le 2, -2 \le y \le 2$ and explain its shape. **(b)** What is $\lim_{x\to 1^-} \tan(\sin^{-1} x)$? Give a geometric explanation using a right triangle. **(c)** Give a formula for $\tan(\sin^{-1} x)$ without any trigonometric or inverse trigonometric functions.

♦ Type 4, a composite function

99. What are the exact ranges of **(a)** $\sin^{-1}(x^2)$ and **(b)** $(\sin^{-1} x)^2$?

♦ Type 4, ranges of composite functions

100. Use your calculator or computer to find the approximate coordinates of the first inflection point of $y = \sin(\sqrt{x})$ to the right of the y-axis.

♦ Type 4, an inflection point

101. Find the limits as $x \to 0$ of $x^2 \sin(1/x)$ and $x^3 \sin(1/x)$ and of their first derivatives or explain why the limits do not exist.

♦ Type 4, limits of derivatives

RESPONSES TO QUESTIONS IN CHAPTER 3

Responses 3.1

Response 1 The maximum depth is $h(2) = 3(2^2) - 2^3 = 12 - 8 = 4$ feet.

Response 2 According to Theorem 1 of Section 3.1, $3 - |x|$ should have a critical point at $x = 0$ since is a local maximum there. This is the case because $3 - |x|$ does not have a derivative at $x = 0$. (Its graph has a corner at $x = 0$.)

Response 3 **(a)** $G(x)$ is continuous in $(-\infty, 0)$ and $(0, \infty)$.
(b) $G'(x)$ is negative in $(-\infty, 0)$ and $(0, \infty)$.
(c) $G(x)$ is decreasing in $(-\infty, 0)$ and $(0, \infty)$.

Response 4 **(a)** $H(x)$ is continuous in $[0, 2)$ and $[2, \infty)$.
(b) $H'(x)$ is negative in $(0, 2)$ and positive in $(2, \infty)$.
(c) $H(x)$ is decreasing in $[0, 2)$ and increasing in $[2, \infty)$.

Response 5 $3 - |x|$ is continuous for all x, and its derivative is $\frac{d}{dx}(3 - x) = -1$ for $x > 0$ and $\frac{d}{dx}(3 + x) = 1$ for $x < 0$. • $3 - |x|$ is decreasing in $[0, \infty)$ and increasing in $(-\infty, 0]$ (see Figure 6 in Section 3.1)

Response 6 **(a)** The graph of $f(x) = x^2 + 10/x^2$ looks approximately like $y = 10/x^2$ for x near zero, where the term $10/x^2$ dominates, and looks approximately like $y = x^2$ for large positive and negative x, where x^2 dominates.
(b) Find the derivative:

$$\begin{aligned} f(x) &= \frac{d}{dx}\left(x^2 + \frac{10}{x^2}\right) = \frac{d}{dx}(x^2 + 10x^{-2}) \\ &= 2x - 20x^{-3} = 2x - \frac{20}{x^3} = \frac{2x^4 - 20}{x^3} = \frac{2(x^4 - 10)}{x^3}. \end{aligned} \tag{1}$$

$f'(x)$ is continuous for $x \neq 0$ and is zero only at $x = \pm 10^{1/4}$ and, therefore, can change sign only at $x = 0, 10^{1/4}$ or $-10^{1/4}$. • For large negative x, the denominator x^3 in **(1)** is negative and the numerator $2(x^4 - 10)$ is positive, so $f'(x)$ is negative. For small negative x the denominator and numerator are negative, so $f'(x)$ is positive. For small positive x the denominator is positive and the numerator is negative, so $f'(x)$ is negative. For large positive x the denominator and numerator are positive, so $f'(x)$ is positive. • $f(x)$ is continuous in $(-\infty, 0)$ and $(0, \infty)$. • $f(x)$ is decreasing in $(-\infty, -10^{1/4}]$ and $(0, 10^{1/4}]$, increasing in $[-10^{1/4}, 0)$ and $[10^{1/4}, \infty)$, and has local minima at $x = \pm 10^{1/4}$. • Because $f(\pm 10^{1/4}) = (\pm 10^{1/4})^2 + \frac{10}{(\pm 10^{1/4})^2} = \sqrt{10} + 10/\sqrt{10} = 2\sqrt{10}$, the global minimum is $2\sqrt{10}$ at $x = \pm 10^{1/4}$.
(c) $f(x)$ has no global maximum because it tends to ∞ as $x \to 0$ and as $x \to \pm\infty$.

Response 7 $y = 3 + x^2 - 8x^{-1}$, $y' = 2x + 8x^{-2}$, and $y'' = 2 - 16x^{-3} = \frac{2(x^3 - 8)}{x^3}$ •
y'' is continuous for $x \neq 0$, is zero at $x = 2$, where $x^3 = 8$, y'' is positive for $x < 0$ and for $x > 2$, and is negative for $0 < x < 2$. • The graph is concave up in $(-\infty, 0)$ and $(2, \infty)$ and is concave down in $(0, 2)$. • The inflection point is $(2, 3)$ at $x = 2$. (There is no inflection point at $x = 0$ because the function is not defined there.) • Figure R7

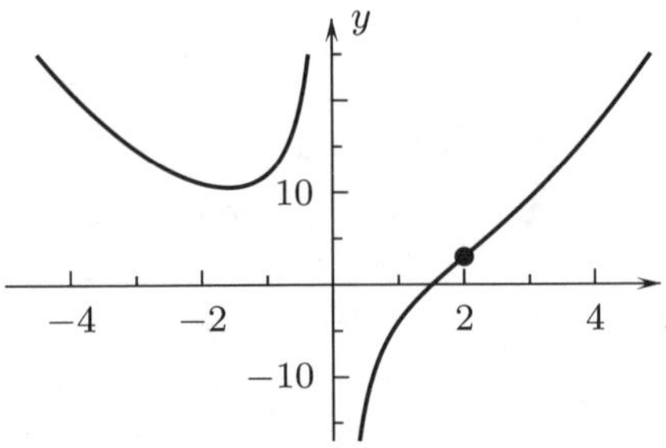

Figure R7

Response 8 **(a)** $g(x)$ is decreasing in $(-\infty, 2]$ and in $[4, \infty)$ and is increasing in $[2, 4]$. **(b)** The graph of $g(x)$ is concave up for $x < 3$ and concave down for $x > 3$ and has an inflection point at $x = 3$.

Response 9 **(a)** $F'(x) = 4x^3 - 4a$ is defined and continuous for all x and is zero where $x^3 = a$. • The one critical point is $x = \sqrt[3]{a}$. **(b)** $F'(x) < 0$ for $x < \sqrt[3]{a}$ and $F'(x) > 0$ for $x > \sqrt[3]{a}$ • $F(x)$ is decreasing in $(-\infty, \sqrt[3]{a}]$ and increasing in $[\sqrt[3]{a}, \infty)$, so its global minimum is at $x = \sqrt[3]{a}$ and is $(\sqrt[3]{a})^4 - 4a\sqrt[3]{a} = -3a^{4/3}$.

Responses 3.2

Response 1 **(a)** If you did not use all the fence, you could give the garden a larger area by using the left-over fence to lengthen the sides. • The perimeter is $P = s + 2w$ feet. • if $P = 40$, then $s + 2w = 40$ and $s = 40 - 2w$. **(b)** The table is below. • One description: As w increases, s decreases and the area first increases and then decreases. **(c)** w cannot be negative or be greater than half the length of available fence.

w	0	4	8	12	16	20
$s = 40 - 2w$	40	32	24	16	8	0
Area	0	128	192	192	128	0

Response 2 **(a)** $A = w(40 - 2w)$ square feet for $0 \le w \le 20$ **(b)** Figure R2 • Likely prediction: The maximum area is obtained with $w = 10$ feet, for which $s = 40 - 2w = 20$ feet and $A = 200$ square feet.

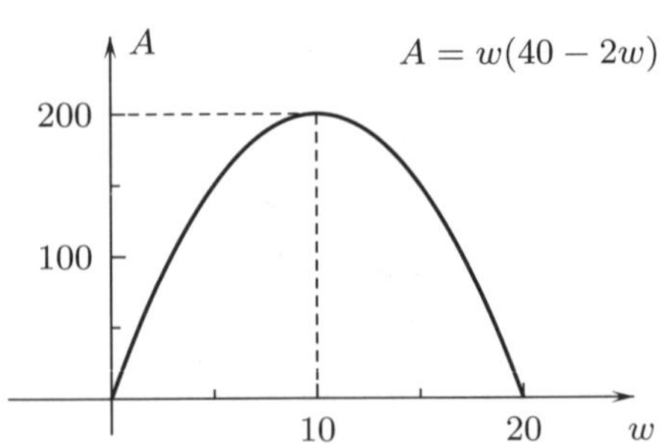

Figure R2

Response 3 For $w = 10$: $s = 40 - 2w = 20$ feet and $A = ws = 10(20) = 200$ square feet

Response 4 Part of the hypothesis of the Extreme-Value Theorem is satisfied in this case since the interval $[-1, 1]$ is closed and finite. The entire hypothesis is, however, not satisfied because $S(x)$ is not continuous in the interval. • Part of the conclusion holds because $S(x)$ has a minimum $S(-1) = -2$ in the interval, but the entire conclusion does not hold because $S(x)$ does not have a maximum in the interval: The maximum is not 2 even though $S(x) < 2$ for all x in the interval and $S(x) \to 2$ as $x \to 1^-$, because $S(x)$ is never equal to 2.†

Response 5 Because $p(x)$ is continuous in $[1, 3]$ and its one critical point in this interval is $x = 2$, its maximum in the interval is the greatest of $p(1) = 8, p(2) = 4$, and $p(3) = 12$. The maximum is 12 at the right endpoint, $x = 3$.

Response 6 With no restrictions on the width x of the box, x can take any value in the interval $[0, 6]$. • $V(x)$ is continuous in the finite closed interval $[0, 6]$, with the critical point $x = 2$ in its interior. • $V(0) = 0$, $V(2) = 128$, and $V(6) = 0$ • The maximum volume is $V(2) = 128$ cubic inches with $x = 2$.

Response 7 The slope $\dfrac{dy}{dx} = \dfrac{x}{\sqrt{x^2+1}}$ of the tangent line at $x = 1$ is $\dfrac{1}{\sqrt{2}}$. • The line segment from $(1, y(1)) = (1, \sqrt{2})$ to $(2, 0)$ has slope $\dfrac{0-\sqrt{2}}{2-1} = -\sqrt{2}$. • Since $\dfrac{1}{\sqrt{2}}$ is the negative reciprocal of $-\sqrt{2}$, the line segment is perpendicular to the tangent line.

Response 8 **(a)** If the volume is 20 cubic meters and the width is $\frac{1}{2}$ meter, then $\left(\frac{1}{2}\right)^2 h = 20$, so that $h = 80$. • Each of the top and bottom has area $\frac{1}{4}$ square meters and costs

$$[\tfrac{1}{4}\text{ square meter}]\left[\frac{4\text{ dollars}}{\text{square meter}}\right] = 1\text{ dollar.}\ \bullet$$

Each of the four sides has area $\frac{1}{2}(80) = 40$ square meters and costs

$$[40\text{ square meters}]\left[\frac{2\text{ dollars}}{\text{square meter}}\right] = 80\text{ dollars.}\ \bullet$$

[Total cost] $= C = 2(1) + 4(80) = 322$ dollars

(b) If the volume is 20 cubic feet and $w = 4$, then $h = \frac{5}{4}$, each of the top and bottom has area 16 square meters and costs 64 dollars, each side has area 5 square meters and costs 10 dollars. • [Total cost] $C = 2(64) + 4(10) = 168$ dollars

(c) $h = 20/w^2$ • The top and bottom each has area w^2 square meters and costs $4w^2$ dollars. • Each of the four sides has area $wh = 20/w$ and costs $40/w$ dollars. • [Total cost] $= C(w) = 2(4w^2) + 4\left(\dfrac{40}{w}\right) = 8w^2 + 160w^{-1}$ dollars

Response 9 **(a)** $C(w) \to \infty$ as $w \to \infty$ and as $w \to 0^+$ • The box would cost a great deal if you made it extremely wide or extremely narrow. **(b)** The minimum cost is approximately \$111.40 for a box with $w \approx 2.15$ feet.

Response 10 Since the box that costs the least has width $w = 10^{1/3}$, its height is $20/w^2 = 2(10^{1/3}) \doteq 4.30887$ meters and it costs $C(w) = 8w^2 + 160/w = (10^{1/3})^2 + 160/(10^{1/3}) = 24(10^{2/3}) \doteq 111.40$ dollars (in agreement with the result in Question 9b above).

Response 11 As we saw in Example 6, $C'(w) = 16w - 160w^{-2}$ is zero at $w = 10^{1/3}$. • $C''(w) = 16 + 320w^{-3}$ is positive for all $w > 0$, so $A(w)$ has a local minimum at $w = 10^{1/3}$.

---

†The number 2 is the LEAST UPPER BOUND of the values of $S(x)$ in the interval $[-1, 1]$.

Responses 3.3

Response 1 (a) $\left[\frac{1}{10}\frac{\text{gallons}}{\text{mile}}\right]\left[50\frac{\text{miles}}{\text{hour}}\right] = 5\,\frac{\text{gallons}}{\text{hour}}$ (b) $s = 50t$ • $G = \frac{1}{10}s$ • $G = 5s$
(c) The slope 5 of the line in Figure 3 of Section 3.3 is the product of the slope 50 of the line in Figure 1 and the slope $\frac{1}{10}$ of the line in Figure 2, as in the calculation from part (a).

Response 2 $[\text{Slope in Figure 9}] = \frac{\Delta G}{\Delta t} = \left[\frac{\Delta G}{\Delta s}\right]\left[\frac{\Delta s}{\Delta t}\right]$
$= [\text{Slope in Figure 8}]\times[\text{Slope in Figure 7}]$

Response 3 $\frac{d}{dx}[R(x^3)] = \frac{dR}{du}(x^3)\frac{d}{dx}(x^3) = 3x^2\frac{dR}{du}(x^3)$ •
$\left[\frac{d}{dx}[R(x^3)]\right]_{x=10} = 3(10^2)\frac{dR}{du}(10^3) = 300\frac{dR}{du}(1000) = 300(2) = 600$

Response 4 $W'(x) = \frac{d}{dx}\{[V(x)]^2\} = 2V(x)\frac{d}{dx}[V(x)] = 2V(x)V'(x)$ •
$W'(3) = 2V(3)V'(3) = 2(10)(2) = 40$

Response 5 Possible answers: **(a)** About 37 thousand feet **(b)** $h \approx 10$ thousand feet **(c)** $P_A \approx 93\%$ **(d)** $P_O \approx 100\%$ **(e)** Using the points on the approximate tangent lines in Figures R5a and R5b, we obtain

$$\left[\frac{dh}{dt}\right]_{t=4} \approx \frac{16-10}{8-4} = 1.5\ \frac{\text{thousand feet}}{\text{minute}}$$
$$\left[\frac{dP_A}{dh}\right]_{h=10} \approx \frac{82-93}{20-10} = -1.1\ \frac{\text{percent}}{\text{thousand feet}}$$

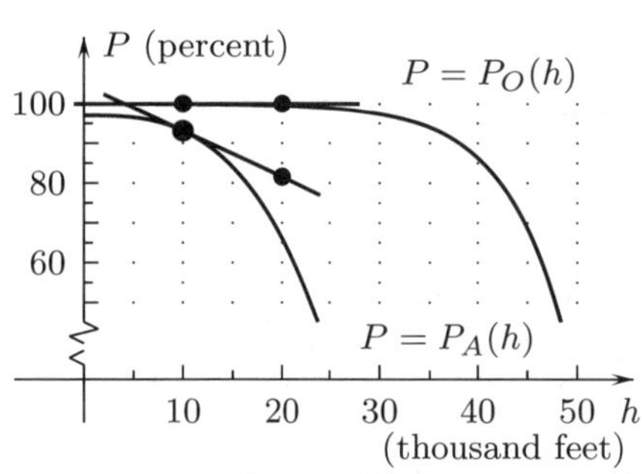

Figure R5a

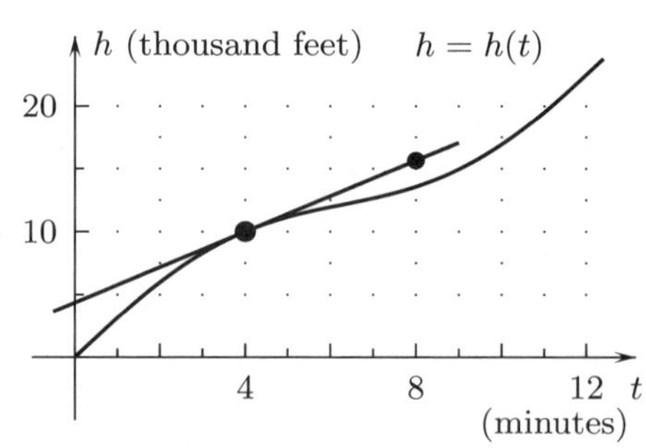

Figure R5b

Responses 3.4

Response 1 **(a)** Your trust fund is increased to $0.25 \times 2^3 = 2$ dollars on your third birthday and to $0.25 \times 2^4 = 4$ dollars on your fourth birthday.
(b) The trust increases $0.25(2^t)$ dollars on your tth birthday for $t = 1, 2, 3, \ldots$.

Response 2 The frequency of $A^{(4)}$ is $\nu(4) = 220(2^4) = 3520$ cycles per second. • The frequency of $A^{(-3)}$ is $\nu(-3) = 220(2^{-3}) = 27.5$ cycles per second.

Response 3 $\nu(6) = 220(2^6) = 14{,}080$ • $\nu(7) = 220(2^7) = 28{,}160$ • The highest A most young people can hear is $A^{(6)}$.

Response 4 **(a)** $2^{1/12}$ **(b)** $\mu(m) = 220(2^{m/12})$

Response 5 Since the D is five notes above $A^{(1)}$, counting sharps and flats, its frequency in the equal-tempered scale is $440(2^{5/12}) \doteq 587.33$ cycles per second.

Response 6 **(a)** $g(6) = 10g(0) = 10(7) = 70$ • $g(12) = 10g(6) = 10(70) = 700$ • $g(18) = 10g(12) = 10(700) = 7000$ **(b)** $7 = g(0) = C(B^0) = C$ and $g(t) = 7B^t$ • $70 = g(6) = 7B^6 \Longrightarrow B^6 = 10$ and $B = 10^{1/6}$

Response 7 **(a)** $f(T) = bf(0) = bC$ **(b)** $f(T) = CB^T$ because $f(t) = CB^t$ and $f(T) = bC$ by part (a) • $CB^T = bC$, $B^T = b$, and $B = b^{1/T}$

Response 8 Use Rule 1 with $C = 1000$, $b = 1.2$, and $T = 5$: $h(t) = 1000(1.2^{t/5})$

Response 9 $G(t)$ grows from 9 to 99, which is by the factor of 11, when t is increased by 5 from $t = 0$ to $t = 5$, and $G(0) = 9$. • $G(t) = 9(11^{t/5})$

Response 10 The relative change from A to B is 0.37. • The growth factor is 1.37.

Response 11 The relative change is 0.1 every three years, so the growth factor is 1.1 every three years. • Use Rule 1 with $C = 10,000$, $b = 1.1$, and $T = 3$: $P = 10{,}000(1.1)^{t/3}$

Response 12 $\left[3.93(1.35^{t/10})\right]_{t=200} = 3.9(1.35^{20}) \doteq 1588.80$ million, or approximately 1.5888 billion

Response 13 $4\left(\frac{1}{2}\right)^{1620/1620} = 2$ grams in 1620 years • $4\left(\frac{1}{2}\right)^{3240/1620} = 1$ gram in 3240 years • $4\left(\frac{1}{2}\right)^{5000/1620} \doteq 0.47$ grams in 5000 years

Response 14 5700 years

Response 15 **(a)** 1.03 **(b)** Use $C = 50,000, b = 1.03$, and $T = \frac{1}{4}$: [Balance after t years] $= 50,000(1.03^{t/(1/4)}) = 50,000(1.03)^{4t}$ dollars

Response 16 Use Rule 2 with $C = 24, r = 0.06, n = 2$, and $t = 370$:
$24(1 + 0.06/2)^{2(370)} = 24(1.03^{740}) \doteq 7.58 \times 10^{10}$ or 75.8 billion dollars

Response 17 $24e^{0.06(370)} \doteq 1.05 \times 10^{11}$ or approximately 105 billion dollars

Response 18 $2(100\%) = 200\%$

Response 19 **(a)** 1000 dollars principal plus $0.03(1000) = 30$ dollars interest • \$1030
(b) $1000 + 1000(0.03t) = 1000 + 30t$ dollars

Response 20 The table is below. • Likely prediction: $\frac{d}{dx}(e^x) = e^x$

x	$e^x \doteq$	$\frac{e^{(x+10^{-6})} - e^x}{10^{-6}} \doteq$
-1	0.367879441	0.367879630
0	1	1.000000500
1	2.718281828	2.718283200
2	7.389056099	7.389059700
3	20.08553692	20.08554700
10	22026.46158	22026.47600

Response 21 $$\frac{e^{x+\Delta x} - e^x}{\Delta x} = \frac{e^x e^{\Delta x} - e^x}{\Delta x} = \frac{e^x(e^{\Delta x} - 1)}{\Delta x} = e^x\left[\frac{e^{\Delta x} - 1}{\Delta x}\right]$$

Response 22 **(a)** The graph is symmetric about the y-axis since $g(-x) = e^{-(-x)^2/2}$ $= e^{-x^2/2} = g(x)$ for all x.
(b) The y-intercept is $g(0) = e^0 = 1$.
(c) $g(x) \to 0$ as $x \to \pm\infty$ because $-x^2/2 \to -\infty$ as $x \to \pm\infty$ and $e^t \to 0$ as

$t \to -\infty$.

(d) Use the Chain Rule: $g'(x) = \dfrac{d}{dx}(e^{-x^2/2}) = e^{-x^2/2}\dfrac{d}{dx}(-\tfrac{1}{2}x^2) = -xe^{-x^2/2}$ •

Use the Product and Chain Rules: $g''(x) = \dfrac{d}{dx}(-xe^{-x^2/2})$

$= -e^{-x^2/2}\dfrac{d}{dx}(x) - x\dfrac{d}{dx}(e^{-x^2/2}) = (x^2-1)e^{-x^2/2}$ • Since $e^{-x^2/2}$ is positive for all x, $g''(x)$ is positive for $|x| > 1$ and negative for $|x| < 1$, and the inflection points are $(1, e^{-1/2})$ and $(-1, e^{-1/2})$.

Response 23 1.35^{60} **(b)** 1.35^{60} **(c)** They are the same.

Response 24 **(a)** $\dfrac{2^{-x}}{x^2} \to \infty$ as $x \to -\infty$ because 2^{-x} goes to ∞ much faster than x^2. •

$\dfrac{2^{-x}}{x^2} \to \infty$ as $x \to 0$ because $2^{-x} \approx 1$ for $x \approx 0$ and $1/x^2 \to \infty$ as $x \to 0$. •

$\dfrac{2^{-x}}{x^2} \to 0$ as $x \to \infty$ because both 2^{-x} and $1/x^2$ tend to 0.

(b) $\dfrac{d}{dx}\left(\dfrac{2^{-x}}{x^2}\right) = \dfrac{x^2\dfrac{d}{dx}(2^{-x}) - 2^{-x}\dfrac{d}{dx}(x^2)}{x^4} = \dfrac{x^2(2^{-x})(-\ln(2)) - 2x(2^{-x})}{x^4}$

$= \dfrac{x2^{-x}(-x\ln(2) - 2)}{x^4}$ • The critical point is $x = -2/\ln(2) \doteq -2.88539$.

Response 25 $\lim\limits_{x\to\infty}(x^3 - 2x - 3^x) = -\infty$ because 3^x is much larger than both $2x$ and x^3 for large positive x and 3^x tends to ∞ as $x \to \infty$. •

$\lim\limits_{x\to-\infty}(x^3 - 2x - 3^x) = -\infty$ because x^3 is much larger than $2x$ for large negative x, $x^3 \to -\infty$ as $x \to \infty$, and $3^x \to 0$ as $x \to -\infty$.

Response 26. **(a)** $\ln[E(t)] = \ln(Cb^t) = \ln(C) + \ln(b^t) = \ln(C) = t\ln(b)$
(b) [Slope] $= \ln(b)$ • [y-intercept] $= \ln(C)$

Responses 3.5

Response 1 $P(D)$ grows by the factor 10 when D is increased by 20, so by Rule 1 of Section 3.4, $P = C(10^{D/20})$. • $P(0) = 0.002 \implies C = 0.002$ • $P = 0.002(10^{D/20})$

Response 2 For acidic solutions, $-\log_{10}([\mathrm{H}^+]) < 7$, so $\log_{10}([\mathrm{H}^+]) > -7$ and $[\mathrm{H}]^+ > 10^{-7}$. • For alkaline solutions, $[\mathrm{H}]^+ < 10^{-7}$

Response 3 The growth factor from $[\mathrm{H}^+] = 10^{-8}$ for a solution with pH $= 8$ to $[\mathrm{H}^+] = 10^{-7}$ for a solution with pH $= 7$ is 10.

Response 4 **(a)** The axes intersect at $r = 1$ and $T = 1$ because $\log_{10} r = 0$ and $\log_{10} T = 0$.
(b) The other short tick marks on the vertical axis are at $0.2, 0.3, 0.4, \ldots 0.9$, at $20, 30, 40, \ldots, 90$, and at $200, 300, 400, \ldots, 900$.

Response 5 [Slope] $\approx \dfrac{2.39 - (-0.618)}{3.77 - 1.76} \doteq 1.4965$

Response 6 $\log_{10} T = 1.5\log_{10} r + b \iff 10^{\log_{10} T} = 10^{1.5\log_{10} r + b} \iff T = 10^b(r^{1.5})$ •
$T = Cr^{1.5}$ with $C = 10^b$

Response 7 The table is completed below. • Likely predictions: $\dfrac{d}{dx}(\log_{10} x) \approx \dfrac{0.43429}{x}$ and : $\dfrac{d}{dx}(\ln x) \approx \dfrac{1}{x}$

x	$\dfrac{\log_{10}(x+10^{-6})-\log_{10}x}{10^{-6}} \doteq$
1	0.434294
2	0.217147
10	0.043429
20	0.0217147
100	0.0043429

x	$\dfrac{\ln(x+10^{-6})-\ln x}{10^{-6}} \doteq$
1	0.9999995
2	0.49999989
10	0.0999999
20	0.0500000
100	0.0099999

Response 8

$$\begin{aligned}\frac{\log_b(x+\Delta x)-\log_b x}{\Delta x} &= \frac{1}{\Delta x}[\log_b(x+\Delta x)-\log_b x]\\ &= \frac{1}{\Delta x}\log_b\left(\frac{x+\Delta x}{x}\right) = \frac{1}{\Delta x}\log_b\left(1+\frac{\Delta x}{x}\right)\end{aligned}$$

Response 9

$$\begin{aligned}\frac{\log_b(x+\Delta x)-\log_b x}{\Delta x} &= \frac{1}{\Delta x}\log_b\left(1+\frac{\Delta x}{x}\right)\\ &= \frac{n}{x}\log_b\left(1+\frac{1}{n}\right) = \frac{1}{x}\log_b\left[\left(1+\frac{1}{n}\right)^n\right]\end{aligned}$$

Response 10 For $y(x)=\ln x$: $y'(x)=1/x$, $y(2)=\ln(2)$, and $y'(2)=\frac{1}{2}$ • Tangent line: $y=\ln(2)+\frac{1}{2}(x-2)$ • Figure R10

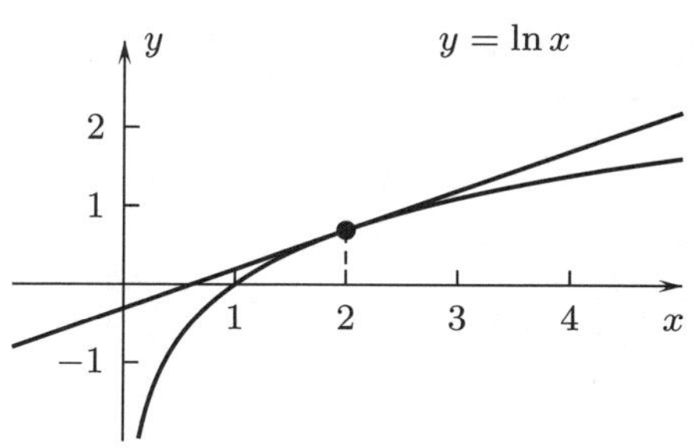

Figure R10

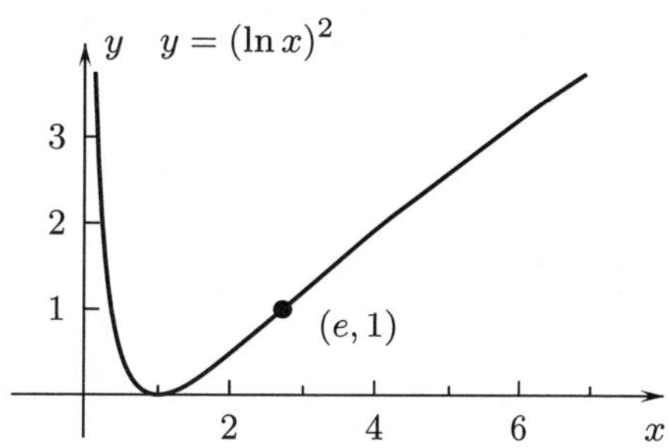

Figure R11

Response 11 **(a)** $\lim_{x\to 0^+}(\ln x)^2 = \infty$ and $\lim_{x\to\infty}(\ln x)^2 = \infty$ since $\ln x \to -\infty$ as $x \to 0^+$ and $\ln x \to \infty$ as $x \to \infty$. **(b)** Because $(\ln x)^2$ is 0 at $x = 1$ and positive for all positive $x \neq 1$, its minimum value is 0 at $x = 1$.
(c) Set $f(x) = (\ln x)^2$ • Use the Chain Rule for differentiating powers of functions from Section 2.5: $f'(x) = \frac{d}{dx}[(\ln x)^2] = 2\ln x \frac{d}{dx}(\ln x) = \frac{2\ln x}{x}$ • $f'(x) < 0$ for $0 < x < 1$ and $f'(x) > 0$ for $x > 1$. • $f(x)$ is continuous in $(0,\infty)$, decreasing in $(0,1]$, and increasing in $[1,\infty)$.
(d) Use the Quotient Rule:

$$f''(x) = \frac{d}{dx}\left(\frac{2\ln x}{x}\right) = \frac{x\frac{d}{dx}(2\ln x) - 2\ln x\frac{d}{dx}(x)}{x^2}$$
$$= \frac{2 - 2\ln x}{x^2} = \frac{2(1-\ln x)}{x^2}$$

$f''(x) = 0 \iff \ln x = 1 \iff x = e$ • $f''(x) > 0 \iff \ln x < 1$, which is for $0 < x < e$ • $f''(x) < 0 \iff \ln x > 1$, which is for $x > e$ • The graph is concave up for $0 < x < e$ and concave down for $x > e$. • Inflection point: $(e, (\ln e)^2) = (e, 1)$ • Figure R11

Response 12 $\frac{d}{dx}[\ln(x^2+1)] = \frac{2x}{x^2+1}$ is positive for $x > 0$ and negative for $x < 0$. • $\ln(x^2+1)$ is continuous in $(-\infty,\infty)$, decreasing in $(-\infty, 0]$, and increasing in $[0,\infty)$. • The minimum of $\ln(x^2+1)$ for $-3 \le x \le 3$ is its value $\ln(0^2+1) = \ln(1) = 0$ at $x = 0$ and the maximum is the value $\ln(3^2+1) = \ln(10)$ at $x = \pm 3$. Figure R12

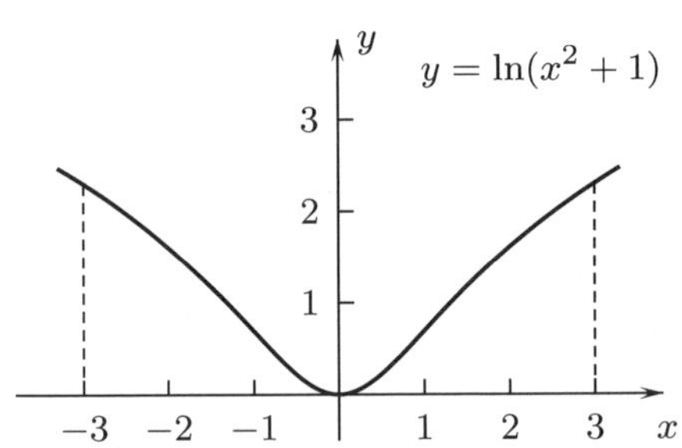

Figure R12

Responses 3.6

Response 1 **(a)** $h(\theta + 2\pi) = h(\theta)$ because the Ferris wheel is at the same location when the angle is θ as it is $\theta + 2\pi$.
(b) Figure 4 in Section 3.6 because the car goes up to a height of 275 feet, down to 25 feet, and back up to 150 feet in one revolution.

Response 2 By the definition of the cosine in Appendix 3, $\cos\theta = u/125$, so $u = 125\cos\theta$.

Response 3 **(a)** $s(0) = 0$ because the sun is directly overhead when $\theta = 0$.
(b) The length $s(\theta)$ of the shadow increases as θ increases from 0 toward $\frac{1}{2}\pi$ because then the sun is descending.
(c) $s(\frac{1}{4}\pi) = 10$ because in this case the triangle in Figure 6 is isosceles.
(d) $s(\theta) \to \infty$ as $\theta \to \frac{1}{2}\pi^-$

Response 4 $\frac{d}{dx}(\sin x)$ **(a)** decreases from 1 to 0 as x goes from 0 to $\frac{1}{2}\pi$,
(b) decreases from 0 to -1 as x goes from $\frac{1}{2}\pi$ to π,
(c) increases from -1 to 0 as x goes from π to $\frac{3}{2}\pi$, and
(d) increases from 0 to 1 as x goes from $\frac{3}{2}\pi$ to 2π.

Response 5 $\frac{d}{dx}(\cos x)$ **(a)** decreases from 0 to -1 as x goes from 0 to $\frac{1}{2}\pi$, **(b)** increases from -1 to 0 as x goes from $\frac{1}{2}\pi$ to π, **(c)** increases from 0 to 1 as x goes from π to $\frac{3}{2}\pi$, and **(d)** decreases from 1 to 0 as x goes from $\frac{3}{2}\pi$ to 2π.

Response 6 $f(2) = 5\sin(2) - 6\cos(2)$ • $f'(2) = 5\cos(2) + 6\sin(2)$ •
Tangent line: $y = [5\sin(2) - 6\cos(2)] + [5\cos(2) + 6\sin(2)](x-2)$ • Figure R5

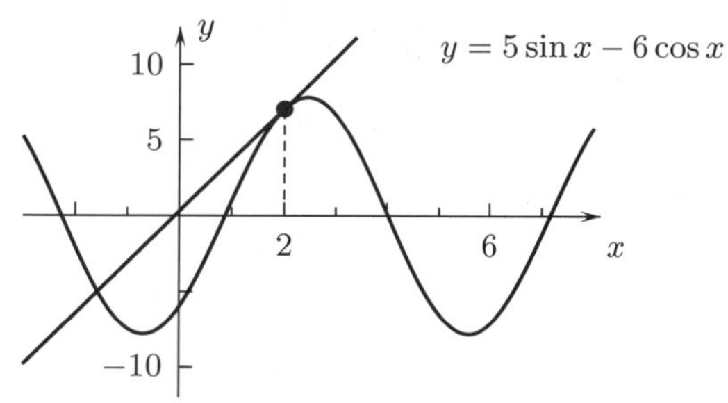

Figure R6

Response 7 $h = 10\sin x$ • $w = 10\cos x$ • $A = \frac{1}{2}wh = 50\cos x\sin x$

Response 8 **(a)** The height of the triangle is 0 if $x = 0$, and the width is zero if $x = \frac{1}{2}\pi$.
(b) Product rule: $\frac{dA}{dx} = \frac{d}{dx}(50\cos x\sin x) = 50[\cos x\frac{d}{dx}(\sin x) + \sin x\frac{d}{dx}(\cos x)]$ $= 50(\cos^2 x - \sin^2 x)$ • Solve $dA/dx = 0$ with $0 \le x \le \frac{1}{2}\pi$: $\cos^2 x = \sin^2 x$ • $\cos x = \sin x$ • $x = \frac{1}{4}\pi$. • Since $A(x)$ is continuous in $[0, \frac{1}{2}\pi]$, $A(0) = 0$, and $A\left(\frac{1}{2}\pi\right) = 0$, the maximum is at $x = \frac{1}{4}\pi$ (which is for the isosceles triangle, as you might have predicted from Figure 17 of Section 3.6).

Response 9 $\frac{d}{dx}[\sin^5(3x)] = 5\sin^4(3x)\frac{d}{dx}[\sin(3x)] = 5\sin^4(3x)\cos(3x)\frac{d}{dx}(3x)$
$= 15\sin^4(3x)\cos(3x)$ • At $x = 10$, this equals $15\sin^4(30)\cos(30)$.

Response 10 Chain-Rule formulas **(9)** and **(10)** in Section 3.6 give

$$\frac{dS}{dx}(x) = \frac{d}{dx}\left[\sin\left(\tfrac{1}{180}\pi x\right)\right] = \cos\left(\tfrac{1}{180}\pi x\right)\frac{d}{dx}\left(\tfrac{1}{180}\pi x\right) = \tfrac{1}{180}\pi C(x)$$
$$\frac{dC}{dx}(x) = \frac{d}{dx}\left[\cos\left(\tfrac{1}{180}\pi x\right)\right] = -\sin\left(\tfrac{1}{180}\pi x\right)\frac{d}{dx}\left(\tfrac{1}{180}\pi x\right) = -\tfrac{1}{180}\pi S(x)$$

Response 11 $h = 10\sin x$ • $\frac{dh}{dt} = 10\cos x\frac{dx}{dt} = 10\cos(1.1)(-0.3) = -3\cos(1.1)$ • The top of the ladder is falling $3\cos(1.1) \doteq 1.36$ feet per minute. • $w = 10\cos x$ • $\frac{dw}{dt} = -10\sin x\frac{dx}{dt} = -10\sin(1.1)(-0.3) = 3\sin(1.1)$ • The base of the ladder is moving $3\sin(1.1) \doteq 2.67$ feet per minute away from the wall.

Response 12 It takes $2\pi/\omega$ seconds for the weight to move from its rest position at $s = 0$ down to $s = -A$, up through its rest position to $s = A$ and back down to its rest position. Therefore, it oscillates $\frac{1}{2\pi/\omega} = \frac{\omega}{2\pi}$ times per second.

Response 13 The weight oscillates more slowly.

Response 14 The weight oscillates more rapidly.

Response 15 $\dfrac{ds}{dt} = \dfrac{d}{dt}[A\cos(\omega t)] = -A\sin(\omega t)\dfrac{d}{dt}(\omega t) = -A\omega\sin(\omega t)$ • The amplitude ωA of the upward velocity equals ω times the amplitude A of $s(t)$. • The velocity and position functions have the same period since $\sin(\omega t)$ and $\cos(\omega t)$ have the same period.

Response 16 $\dfrac{d^2s}{dt^2} = \dfrac{d}{dt}[-\omega A\sin(\omega t)] = -\omega A\cos(\omega t)\dfrac{d}{dt}(\omega t) = -\omega^2 A\cos(\omega t)$ • The upward acceleration equals $s(t)$ multiplied by $-\omega^2$.

Response 17 **(a)** $\dfrac{d}{dx}(\cot x) = \dfrac{d}{dx}\left[\dfrac{\cos x}{\sin x}\right] = \dfrac{-\sin^2 x - \cos^2 x}{\sin^2 x} = \dfrac{-1}{\sin^2 x} = -\csc^2 x$

(b) $\dfrac{d}{dx}(\sec x) = \dfrac{d}{dx}[(\cos x)^{-1}] = -(\cos x)^{-2}(-\sin x) = \dfrac{\sin x}{\cos^2 x} = \sec x\,\tan x$

(c) $\dfrac{d}{dx}(\csc x) = \dfrac{d}{dx}[(\sin x)^{-1}] = -(\sin x)^{-2}(\cos x) = -\dfrac{\cos x}{\sin^2 x} = -\csc x\,\cot x$

Response 18 **(a)** $y(0) = \tan(0) = 0$ • $y' = \sec^2 x$ • $y'(0) = \sec^2(0) = 1/\cos^2(0) = 1$ • Tangent line: $y = x$ • Figure R18

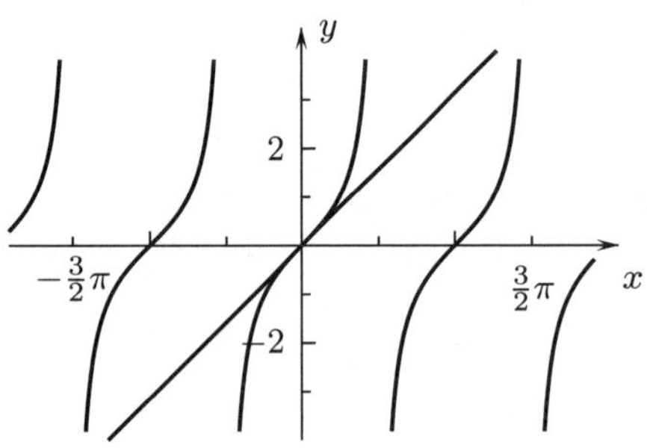

$y = \tan x$

Figure R18

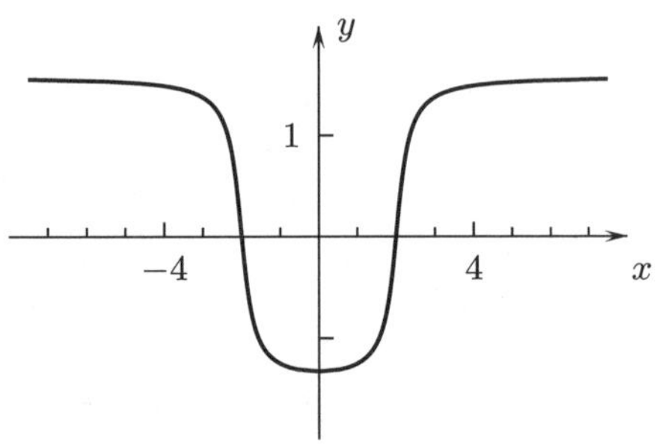

$y = \tan^{-1}(x^2 - 4)$

Figure R21

Response 19 **(a)** $\tan\theta = \dfrac{s}{2}$ • $s = 2\tan\theta$ miles

(b) $\dfrac{d\theta}{dt} = \dfrac{2\pi \text{ radians}}{3 \text{ minutes}} = \frac{2}{3}\pi$ radians per minute

(c) $\dfrac{ds}{dt} = 2\sec^2\theta\,\dfrac{d\theta}{dt} = \frac{4}{3}\pi\sec^2(0.9) \doteq 10.84$ miles per minute

Response 20 Possible answers: **(a)** The derivative of $\sin^{-1} x$ is $\dfrac{1}{\sqrt{1-x^2}}$, which equals 1 at $x = 0$, is > 1 for $-1 < x < 0$ and $0 < x < 1$, and tends to ∞ as $x \to -1^+$ and as $x \to 1^-$. • The tangent line to the graph of $\sin^{-1} x$ has slope 1 at $x = 0$, has slope > 1 for $-1 < x < 0$ and $0 < x < 1$, and its slope tends to ∞ as $x \to -1^+$ and as $x \to 1^-$.

(b) The derivative of $\cos^{-1} x$ is $\dfrac{-1}{\sqrt{1-x^2}}$, which equals -1 at $x = 0$, is < -1 for $-1 < x < 0$ and $0 < x < 1$, and tends to $-\infty$ as $x \to -1^+$ and as $x \to 1^-$. • The tangent line to the graph of $\cos^{-1} x$ has slope -1 at $x = 0$, has slope < -1 for $-1 < x < 0$ and $0 < x < 1$, and its slope tends to $-\infty$ as $x \to -1^+$ and as $x \to 1^-$.

(c) The derivative of $\tan^{-1} x$ is $\dfrac{1}{x^2+1}$, which equals 1 at $x = 0$, is positive and

< 1 for $x \neq 0$, and tends to 0 as $x \to -\infty$ and as $x \to \infty$. • The tangent line to the graph of $\tan^{-1} x$ has slope 1 at $x = 0$, has positive slope < 1 for $x \neq 0$, and its slope tends to 0 as $x \to \infty$ and as $x \to -\infty$.

(d) The derivative of $\cot^{-1} x$ is $\dfrac{-1}{x^2+1}$, which equals -1 at $x = 0$, is negative and > -1 for $x \neq 0$, and tends to 0 as $x \to -\infty$ and as $x \to \infty$. • The tangent line to the graph of $\cot^{-1} x$ has slope -1 at $x = 0$, has negative slope > -1 for $x \neq 0$, and its slope tends to 0 as $x \to \infty$ and as $x \to -\infty$.

Response 21 Figure R21 (above) • One explanation: $\tan^{-1}(x^2 - 4)$ is even, so its graph is symmetric about the y-axis. • The y-intercept of the graph is $\tan^{-1}(-4) \doteq -1.33$. • The x-intercepts of the graph are ± 2. • $\tan^{-1}(x^2-4) \to \frac{1}{2}\pi \doteq 1.57$ as $x \to \pm\infty$. • $\dfrac{d}{dx}[\tan^{-1}(x^2-4)] = \dfrac{1}{1+(x^2-4)^2}\dfrac{d}{dx}(x^2-4) = \dfrac{2x}{1+(x^2-4)^2}$ is zero at $x = 0$, is negative for $x < 0$, is positive for $x > 0$, and tends to 0 as $x \to \pm\infty$. • $\tan^{-1}(x^2-4)$ is continuous in $(-\infty, \infty)$, decreasing in $(-\infty, 0)$ and increasing in $(0, \infty)$. • The slope of the tangent line to its graph is 0 at $x = 0$ and tends to 0 as $x \to \pm\infty$.

Response 22 $\cot\theta = s/4$ and $\theta = \cot^{-1}\left(\frac{1}{4}s\right)$ (The inverse tangent can also be used but leads to more complicated formulas.) •

$$\frac{d\theta}{dt} = \frac{-1}{(\frac{1}{4}s)^2+1}\frac{d}{dt}\left(\tfrac{1}{4}s\right) = \frac{-1}{(\frac{1}{4}s)^2+1}\left(\tfrac{1}{4}\right)\frac{ds}{dt}$$ • Set $s = 4$ and $ds/dt = -3$:

$$\frac{d\theta}{dt} = -\tfrac{1}{4}\frac{1}{\left(\frac{1}{4}(4)\right)^2+1}(-3) = \tfrac{3}{8} \text{ radians per second}$$

CHAPTER 3 ANSWERS

Tune-Up Exercises 3.1

T1. [O] $f(x) = x^4 - 4x^2 = x^2(x^2 - 4)$ • $f(x) > 0 \iff x^2 > 4$ and $x \neq 0$ • $f(x) < 0 \iff x^2 < 4$ and $x \neq 0$ • $f(x) > 0$ in $(-\infty, -2)$ and $(2, \infty)$ • $f(x) < 0$ in $(-2, 0)$ and $(0, 2)$ 4.11.98A

T2. [O] $g(x) = 1 + \dfrac{2}{x-2} = \dfrac{x}{x-2}$ is positive when x and $x - 2$ have the same sign and negative when they have opposite signs. • $g(x)$ is positive in $(-\infty, 0)$ and $(2, \infty)$ and negative in $(0, 2)$. 4.11.98B

T3. [A] **(a)** Increasing in $[0, 6]$ and decreasing in $[-2, 0]$ and $[6, 8]$ **(b)** Yes, a local minimum at $x = 0$ and a local maximum at $x = 6$. **(c)** Concave up in $(-2, 3)$ • Concave down in $(3, 8)$ **(d)** At $x = 3$ 4.11.98C

T4. [A] **(a)** Concave down in $(-\infty, -1)$ and $(4, \infty)$ • Concave up in $(-1, 4)$ **(b)** Inflection points at $x = -1, 4$ 4.11.98D

T5. [O] **(a)** $y = x^{-1}$ • $y' = -x^{-2}$ • $y'' = 2x^{-3}$ • y is continuous and decreasing in $(-\infty, 0)$ and $(0, \infty)$. • The graph is concave down in $(-\infty, 0)$ and concave up in $(0, \infty)$. • Figure T5a
(b) $y = x^{-2}$ • $y' = -2x^{-3}$ • $y'' = 6x^{-4}$ • y is continuous in $(-\infty, 0)$ and $(0, \infty)$, increasing in $(-\infty, 0)$, and decreasing in $(0, \infty)$. • The graph is concave up in $(-\infty, 0)$ and $(0, \infty)$. • Figure T5b 8.29.96G

Jun6.96G May9.95B

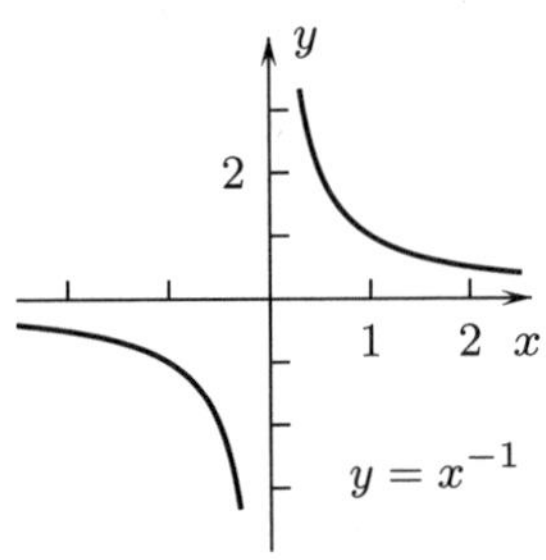

Figure T5a

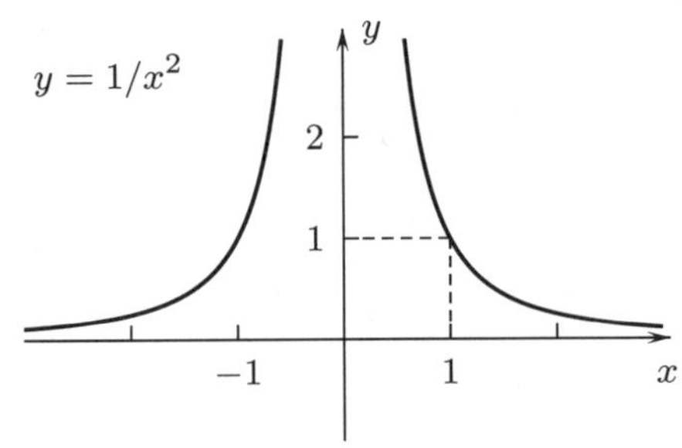

Figure T5b

T6. [O] $y = x - 2 + 1/x = (x^2 - 2x + 1)/x = (x-1)^2/x$ is positive in $(0, 1)$ and $(1, \infty)$ and negative 1n $(-\infty, 0)$. 4.11.98E

T7. [O] $k(x) = x^2 - 16/x^2 = (x^4 - 16)/x^2$ is positive in $(-\infty, -2)$ and $(2, \infty)$ and negative in $(-2, 0)$ and $(0, 2)$. 4.11.98F

T8. [A] **(a)** $H(x)$ is Increasing in $[-7, 7]$.
(b) The graph is concave up in $(-4, 0)$ and $(4, 7)$ and concave down in $(-7, -4)$ and $(0, 4)$
(c) $x = -4, 0, 4$ 4.11.98G

T9. [A] **(a)** Concave up in $(2, 7)$ and concave down in $(0, 2)$ and $(7, 12)$ **(b)** The inflection points are at $x = 2, 7$. 4.11.98H

T10. [O] **(a)** $y = x^{1/2}$ is defined and continuous in $[0, \infty)$ • $y' = \frac{1}{2}x^{-1/2}$ • y is increasing in $[0, \infty)$ • $y'' = -\frac{1}{4}x^{-3/2}$ • The graph is concave down in $(0, \infty)$. • Figure T10a
(b) $y = x^{1/3}$ is defined and continuous in $(-\infty, \infty)$ • $y' = \frac{1}{3}x^{-2/3}$ is positive for $x \neq 0$. • y is increasing in $(-\infty, 0]$ and $[0, \infty)$ and therefore in $(-\infty, \infty)$. • $y'' = -\frac{2}{9}x^{-5/3}$ • The graph is concave up in $(-\infty, 0)$ and concave down in $(0, \infty)$. • Figure T10b
(c) $y = x^{-1/2}$ is defined and continuous in $(0, \infty)$ • $y' = -\frac{1}{2}x^{-3/2}$ • y is decreasing in $(0, \infty)$ • $y'' = \frac{3}{4}x^{-5/2}$ • The graph is concave up in $(0, \infty)$ • Figure T10c

4.11.98I

Apr11.98B

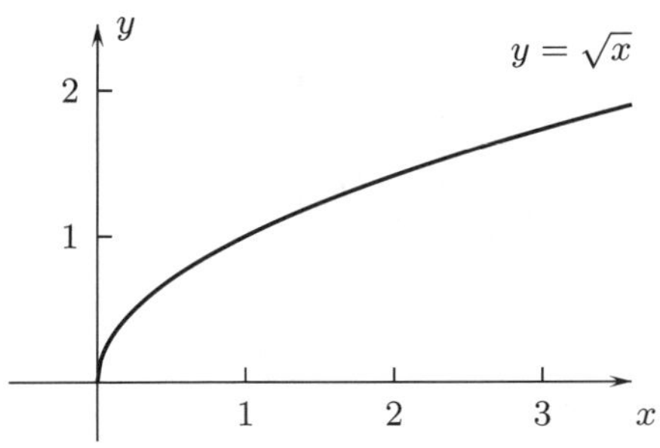

Figure T10a

Apr11.98D

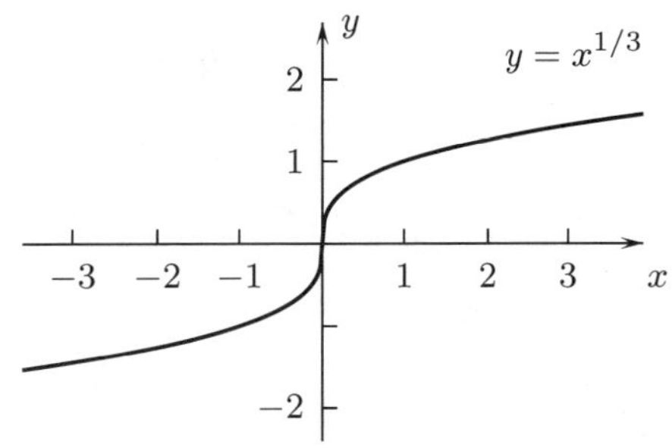

Figure T10b

Apr11.98C

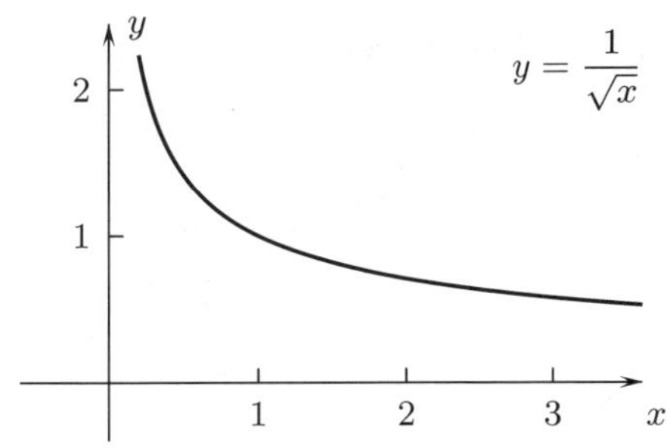

Figure T10c

Problems 3.1

1. $y = x^3 - 12x + 7$ • $y' = 3x^2 - 12 = 3(x^2 - 4)$ • $y'' = 6x$ • y is increasing in $(-\infty, -2]$ and $[2, \infty)$ and decreasing in $[-2, 2]$. • The graph is concave up for $x > 0$ and concave down for $x < 0$ • y has a local maximum of 23 at at $x = -2$ and a local minimum of -9 at $x = 2$. • The graph has an inflection point at $x = 0,\ y = 7$. • Figure A1

4.12.98A

Feb8.99A

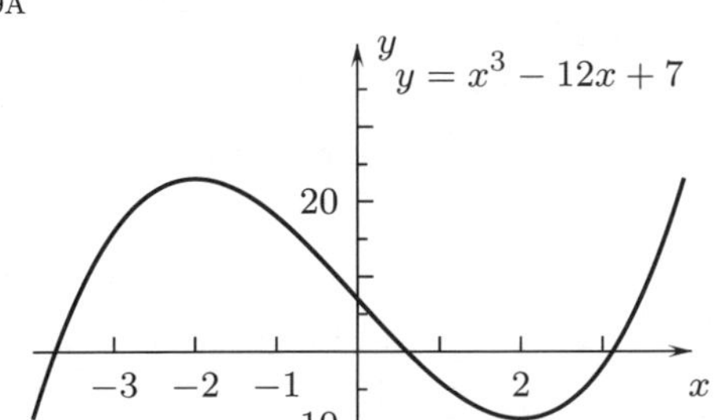

Figure A1

iFeb8.99B

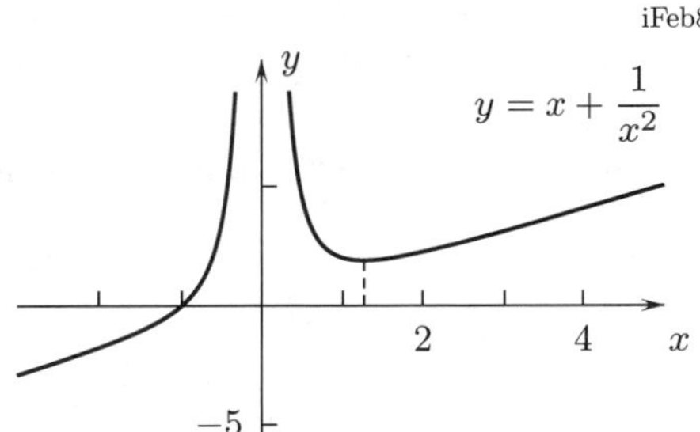

Figure A2

2. $y = x + x^{-2}$ • $y' = 1 - 2x^{-3} = (x^3 - 2)/x^3$ • $y' = 0 \iff x = \sqrt[3]{2}$ • $y'' = 6x^{-4}$ • y is increasing in $(-\infty, 0)$ and $[\sqrt[3]{2}, \infty)$ and decreasing in $(0, \sqrt[3]{2}]$. • The graph is concave up for $x < 0$ and for $x > 0$ • y has a local minimum of $2^{1/3} + 2^{-2/3} \doteq 1.8899$ at $x = \sqrt[3]{2}$. • Figure A2

4.12.98B

3. $y = 2x^3 - x^4$ • $y' = 6x^2 - 4x^3 = 2x^2(3 - 2x)$ • $y'' = 12x - 12x^2 = 12x(1 - x)$ • y is increasing in $(-\infty, \frac{3}{2}]$, decreasing in $[\frac{3}{2}, \infty)$, and has a global maximum of $1\frac{11}{16}$ at $x = \frac{3}{2}$. • The graph is concave for $0 < x < 1$ and concave down for $x < 0$ and for $x > 1$. • The graph has inflection points at $x = 0,\ y = 0$ and $x = 1,\ y = 1$. • Figure A3

4.12.98C

4. $y = 2x^2 - x^4$ • $y' = 4x - 4x^3 = 4x(1 - x^2)$ • $y'' = 4 - 12x^2 = 4(1 - 3x^2)$ • y is increasing in $(-\infty, -1]$ and $[0, 1]$, decreasing in $[-1, 0]$ and $[1, \infty)$, has a global maximum of 1 at $x = \pm 1$, and has a local minimum of 0 at $x = 0$ • The graph is concave up in $(-\sqrt{3}/3, \sqrt{3}/3)$ and concave down in $(-\infty, -\sqrt{3}/3)$ and $(\sqrt{3}/3, \infty)$ • The graph has inflection points at $x = \pm\sqrt{3}/3,\ y = \frac{5}{9}$. • Figure A4

4.12.98D

Jun8.99A iFeb8.99D

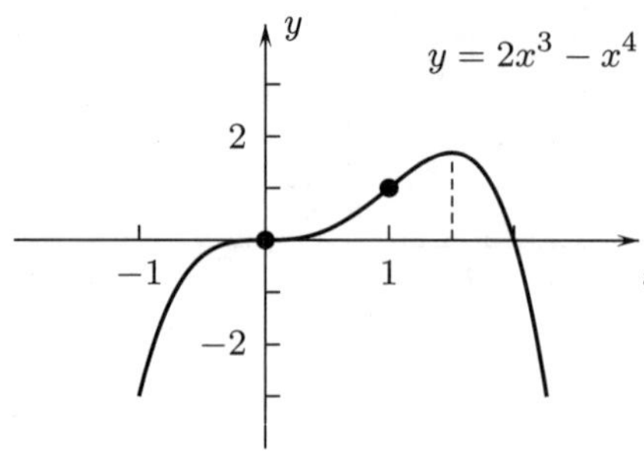

Figure A3

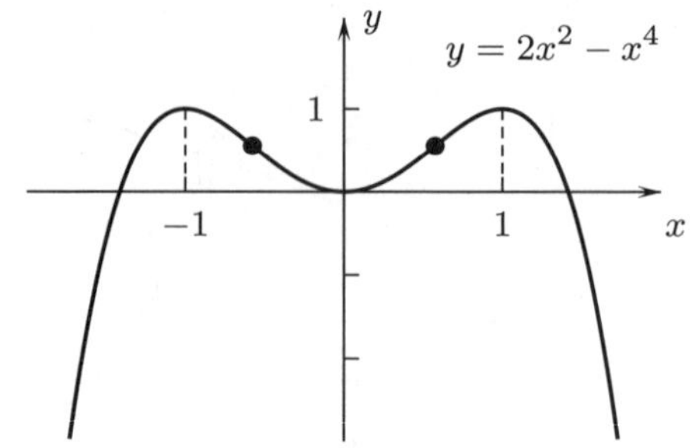

Figure A4

5. $y = x^4 - 12x$ • $y' = 4x^3 - 12 = 4(x^3 - 3)$ • $y'' = 12x^2$ • y is increasing in $[\sqrt[3]{3}, \infty)$, decreasing in $(-\infty, \sqrt[3]{3}]$, and has a global minimum of $-9(\sqrt[3]{3})$ at $x = \sqrt[3]{3}$. • The graph is concave up for all x. • Figure A5

4.12.98E

iFeb8.99E Jun8.99C

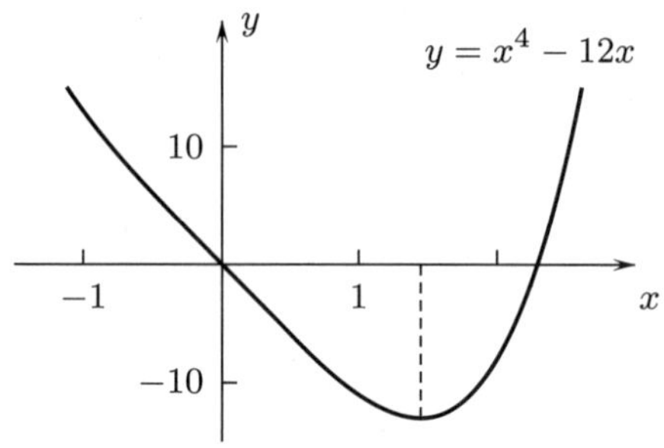

Figure A5

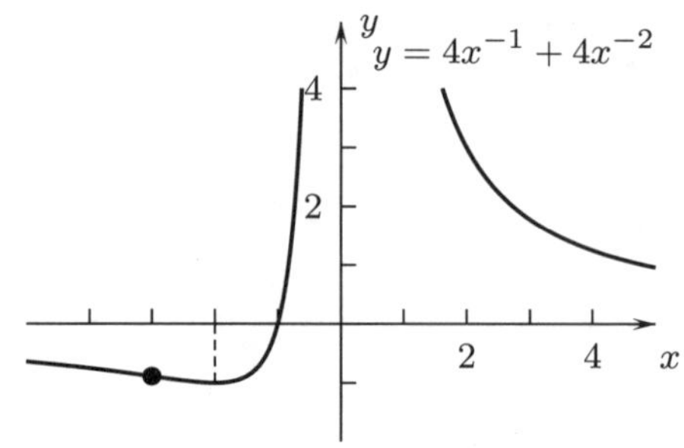

Figure A9

9. $y = 4x^{-1} + 4x^{-2}$ • $y' = -4x^{-2} - 8x^{-3} = -4x^{-3}(x+2)$ • $y'' = 8x^{-3} + 24x^{-4} = 8x^{-4}(x+3)$ • y is increasing in $[-2, 0)$, decreasing in $(-\infty, -2]$ and $(0, \infty)$, and has a global minimum of -1 at at $x = -2$. • The graph is concave up in $(-3, 0)$ and $(0, \infty)$ and concave down in $(-\infty, -3)$ and has an inflection point at $x = -3$, $y = -8/9$.

12. $y = 6x^{1/2} - x$ • $y' = 3x^{-1/2} - 1 = x^{-1/2}(3 - x^{1/2})$ • $y'' = -(1/2)x^{-3/2}$ • y is increasing in $[0, 9]$ and decreasing in $[9, \infty)$ and has a global maximum of 9 at $x = 9$. • The graph is concave down in $(0, \infty)$ • Figure A12

iFeb8.99G Feb25.99C

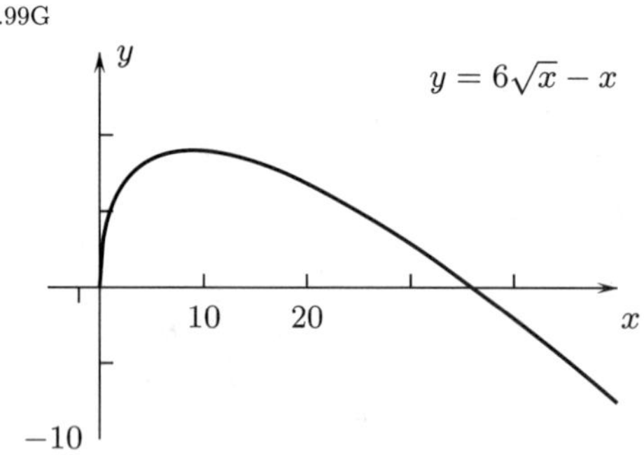

Figure A12

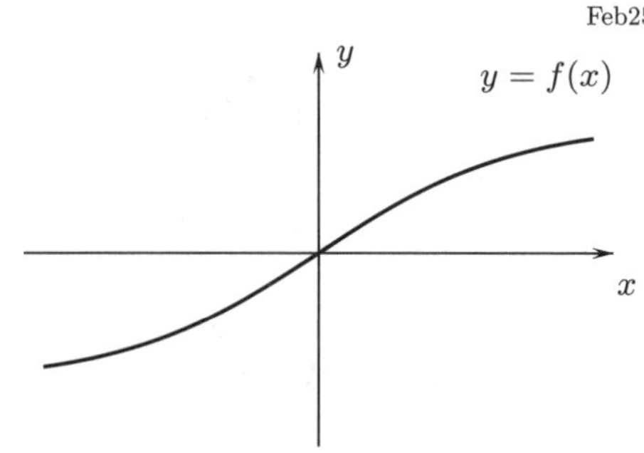

Figure A38

19. $f(x)$ is increasing in $[-1, 0)$ and $[1, \infty)$ and decreasing in $(-\infty, -1]$ and $(0, 1]$. • The graph of $f(x)$ is concave up for $x < 0$ and for $x > 0$.

4.13.98A

20. The graph of $g(x)$ is concave up in $(-\infty, 0)$ and $(2, \infty)$, is concave down in $(0, 1)$ and $(1, 2)$, and has inflection points at $x = 0,\ 2$.

10.3.98A

29. **(a)** $f(x) = x^3 - 5x^2 + 8x - 4$ • $f'(1) = 0$ and $f'(2) = 0$
(b) $f'(x)$ is continuous for all x and can change sign only at $x = 1$ or $x = 2$. • $f'(0) = -4$ is negative, so $f'(x) < 0$ for $x < 1$ and $f(x)$ is decreasing in $(-\infty, 1]$. • $f'(1.5) = -0.125$ is positive, so $f'(x) > 0$ for $1 < x < 2$ and $f(x)$ is increasing in $[1, 2]$. • $f'(3) = 2$ is positive, so $f'(x) > 0$ for $x > 2$ and $f(x)$ is increasing in $[2, \infty)$ • $f(x)$ is decreasing in $(-\infty, 1]$ and increasing in $[1, \infty)$.

4.12.98W, 3.3.15, 16, 17, 19

38.[A] Yes • Figure A38

43. $y = \dfrac{8}{x^3 - 3x^2}$ • $y' = -\dfrac{8(3x^2 - 6x)}{(x^3 - 3x^2)^2} = -\dfrac{24(x-2)}{x^3(x-3)^2}$ • y is increasing in $(0, 2]$ and decreasing in $(-\infty, 0)$, $[2, 3)$ and $(3, \infty)$ • y has a local maximum of -2 at $x = 2$. • Figure A43

Jun8.99D 4.13.98A, 3.4.13 Jun29.99M

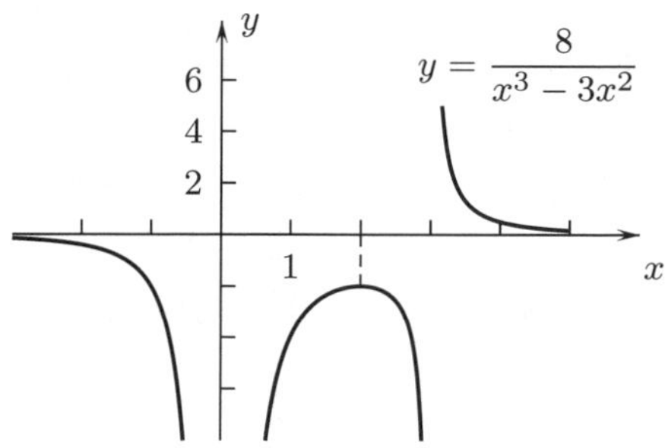

Figure A43

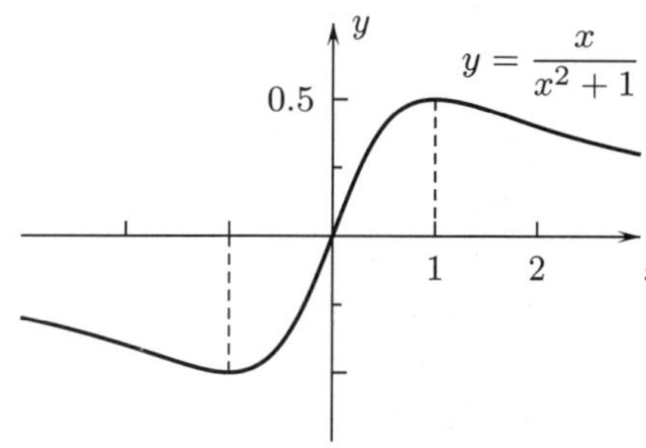

Figure A44

44. $y = \dfrac{x}{x^2 + 1}$ • $y' = \dfrac{1 - x^2}{(x^2 + 1)^2}$ • y is increasing in $[-1, 1]$, decreasing in $(-\infty, -1]$ and $[1, \infty)$, and has a global maximum of $\frac{1}{2}$ at $x = 1$ and a global minimum of $-\frac{1}{2}$ at $x = -1$. • Figure A44

4.13.98B

45. $y = \dfrac{x^2}{x - 1}$ • $y' = \dfrac{x(x-2)}{(x-1)^2}$ • y is increasing in $(-\infty, 0]$ and $[2, \infty)$ and decreasing in $[0, 1)$ and $(1, 2]$ and has a local maximum of 0 at $x = 0$ and a local minimum of 4 at $x = 2$. • Figure A45

Jun8.99F 4.13.98M Jun8.99G

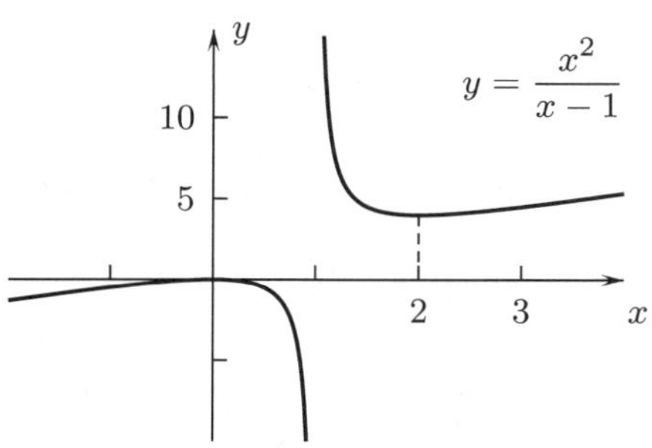

Figure A45

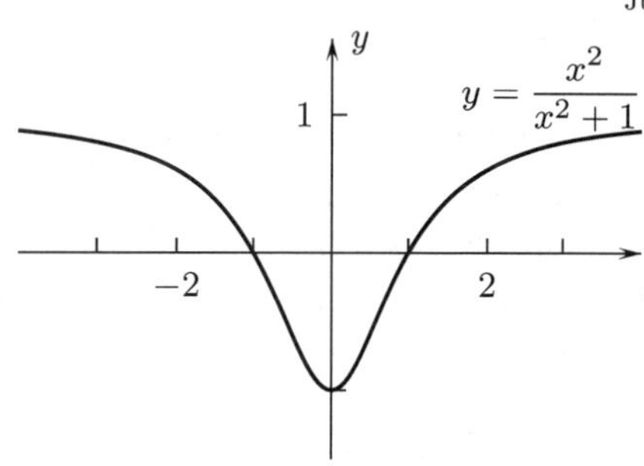

Figure A49

49. $y = \dfrac{x^2 - 1}{x^2 + 1}$ • $y' = \dfrac{4x}{(x^2 + 1)^2}$ • y is increasing in $[0, \infty)$ and decreasing in $(-\infty, 0]$ and has a global minimum of -1 at $x = 0$. • Figure A49

4.13.98Q

51. $y = \dfrac{1}{(x-2)^2}$ • $y' = \dfrac{-2}{(x-2)^3}$ • $y'' = 6(x-2)^{-4}$ • y is increasing in $(-\infty, 2)$ and decreasing in $(2, \infty)$ • The graph is concave up for $x < 2$ and for $x > 2$. • y has no local maxima or minima and its graph has no inflection points.

4.13.98S,3.4.3

52. $y = \dfrac{5}{x^2 + 1}$ • $y' = \dfrac{-10x}{(x^2 + 1)^2}$ • $y'' = \dfrac{10(3x^2 - 1)}{(x^2 + 1)^3}$ • y is increasing in $(-\infty, 0]$ and decreasing in $[0, \infty)$ and has a global maximum of 5 at $x = 0$. • The graph is concave up for $x < -1/\sqrt{3}$ and for $x > 1/\sqrt{3}$ and concave down for $-1/\sqrt{3} < x < 1/\sqrt{3}$ and has inflection points at $x = -1/\sqrt{3}$, $y = 15/4$ and $x = 1/\sqrt{3}$, $y = 15/4$.

4.13.98T,3.4.2

59. $y = \frac{1}{3}x^3 + x^2 + 2x$ • $y' = x^2 + 2x + 2$ • $y'' = 2x + 2$ • Since $(2)^2 - 4(1)(2) = -4 < 0$, $y' > 0$ and y is increasing for all x. • The graph is concave down in $(-\infty, -1)$ and concave up in $(-1, \infty)$. • Figure A59

4.14.98A

iFeb8.99M

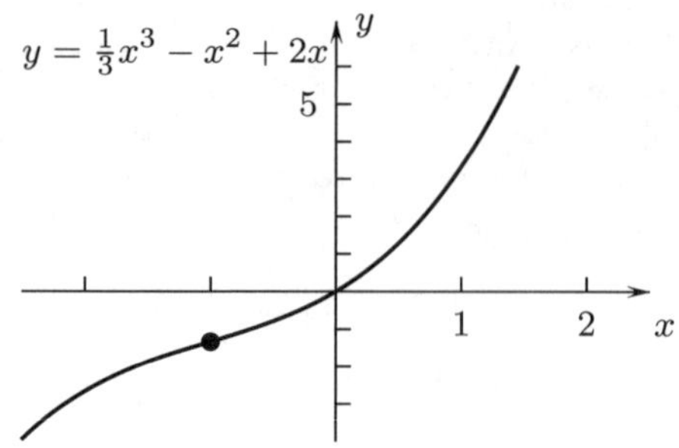

Figure A59

60. $y' = \dfrac{2(x^{-2/3} - x^{1/3})}{(x+2)^2}$ is positive for $x < -2, -2 < x < 0$ and $0 < x < 1$ and negative for $x > 1$. • y and has a local maximum of 1 at $x = 1$.

4.14.98B

62. Inflection point: $(\frac{1}{4}, -\frac{53}{32})$

4.14.98D

66. Local maximum of 0 at $x = 0$ • Local minimum ≈ -3.10854 at $x \approx 1.10212$. • Inflection point at $x \approx 0.66889,\ y \approx -1.80407$

4.14.98H

71. **(a)** Vertical tangent line: $x = 0$ **(b)** Cusp: $(0, 0)$

78. $G'(10) < \dfrac{G(10) - G(0)}{10} < G'(0)$

9.1.96N

81b. $D^+(|x|) = 1$ at $x = 0$

87. **(a)** $f(x) = \sqrt{x}$ is continuous in $[1, 9]$ and the interval is closed and finite, so the hypotheses are satisfied. • $\dfrac{f(9) - f(1)}{9 - 1} = f'(c)$ at $c = 4$, so the conclusion holds. **(b)** $f(x) = \dfrac{x^2}{1 - x^2}$ is not defined at $x = \pm 1$, so the hypotheses are not satisfied. • $\dfrac{f(2) - f(-2)}{2 - (-2)} = f'(c)$ at $c = 0$, so the conclusion holds.

Tune-Up Exercises 3.2

T1. $f(x)$ is decreasing in $[-5, 0]$ and increasing in $[0, 5]$. • $g(-5) = 10$ is less than $g(5) = 11$. • [Minimum] $= f(0) = 2$ • [Maximum] $= f(5) = 11$

6.7.97A

T2. The maximum and minimum are the greatest and least of $g(0) = -1, g(5) = 3$ and $g(8) = 2$ • [Minimum] $= g(0) = -1$ • [Maximum] $= g(5) = 3$

6.7.97B

T3. The minimum is the least of $h(2) = 10$ and $h(3) = 20$. • [Minimum] $= h(2) = 10$ • There is no maximum since $h(x) \to \infty$ as $x \to 0^+$ and as $x \to \infty$.

8.13.98A

T4. **(a)** [Area] $= A = wh$ • $A = 7 \Rightarrow h = 7/w$ • [Perimeter] $= P = 2w + 2h = 2w + 14/w$, defined for $w > 0$
(b) $P = 100 \Rightarrow h = 50 - w$ and $A = w(50 - w)$, defined for $0 \le w \le 50$

4.23.98A

T5. **(a)** $V = w^2h$ and $A = 4wh + 2w^2$ • $A = 24 \Rightarrow h = \frac{1}{2}(12 - w^2)/w$ and $V = \frac{1}{2}w(12 - w^2)$, defined for $0 < w \le \sqrt{12}$
(b) $V = 9 \Rightarrow h = 9/w^2$ and $A = 36/w + 2w^2$, defined for $w > 0$

4.23.98B

T6. **(a)** $V = \pi r^2 h$ • $A_B = \pi r^2$ • $A_L = 2\pi rh$
(b) $A = 2\pi r^2 + 2\pi rh$

4.23.98C

T7. [Cost] $= C = (0.02)(4wh) + 0.03(2w^2) = 0.08wh + 0.06w^2$ • $V = w^2h$ • $V = 1000 \Rightarrow h = 1000/w^2$ and $C = 80/w + 0.06w^2$, defined for $w > 0$

10.3.98C

T8. **(a)** The hypotheses are satisfied in $[1, 2]$ because $T(x)$ is continuous in the interval, which is closed and finite. The conclusion holds because the hypotheses are satisfied or because the maximum of $T(x)$ in $[1, 2]$ is $T(1) = 1$ and the minimum is $T(2) = \frac{1}{4}$. **(b)** The hypotheses are not satisfied in $[-2, 2]$ because $T(x)$ is discontinuous at $x = 0$. The conclusion does not hold because $T(x)$ has no maximum in $[-2, 2]$. (Its minimum is $T(0) = 0$.)

6.9.99B

Problems 3.2

1. $y = x^3 - 3x + 5$ is continuous in $[0, 2]$. • $y' = 3x^2 - 3$ is zero at $x = \pm 1$. • The maximum and minimum in $[0, 2]$ are the greatest and least of $y(0) = 5, y(1) = 3$, and $y(2) = 7$. • [Maximum] $= y(2) = 7$ • [Minimum] $= y(1) = 3$

9.3.96A,3.7.1

2. $y = x^3 - 6x^2 + 9x$ is continuous in $[0, \infty)$ and tends to ∞ as $x \to \infty$. • $y' = 3x^2 - 12x + 9 = 3(x-1)(x-3)$ is positive for $0 \le x < 1$ and for $x > 3$ and is negative for $1 < x < 3$. • y is increasing in $[0, 1]$ and $[3, \infty)$ and decreasing in $[1, 3]$. • $y(0) = 0, y(3) = 0$ • [Minimum] $= y(3) = 0$. • There is no maximum since $y \to \infty$ as $x \to \infty$.

9.3.96B,3.8 Example 1

3. $y = x + 4x^{-2}$ is continuous in $(0, \infty)$ and tends to ∞ as $x \to 0^+$ and as $x \to \infty$. • $y' = \dfrac{x^3 - 8}{x^3}$ is negative for $0 < x < 2$ and positive for $x > 2$. • y is decreasing in $(0, 2]$ and increasing in $[2, \infty)$. • [Minimum] $= y(2) = 3$. • There is no maximum.

9.3.96C,3.8.10

4. $y = x^2 + x^{-2}$ is continuous in $[1, 4)$ • $y' = 2x + 2x^{-3}$ is positive for $1 \le x < 4$. • y is increasing in $[1, 4)$. • [Minimum] $= y(1) = 0$ • There is no maximum (because 4 is not in the interval).

9.3.96D,3.8.11

5. $y = x^{-1}$ is continuous in $(0, \infty)$, $y \to \infty$ as $x \to 0^+$, and $y \to 0$ as $x \to \infty$. • $y' = -x^{-2}$ is negative for $x > 0$. • There is no maximum or minimum.

9.3.96E

15. $y = \sqrt{5 - x^2}$ is continuous in $[-1, 2]$. • $y' = -\dfrac{x}{\sqrt{5 - x^2}}$ is 0 at $x = 0$. • The maximum and minimum in $[-1, 2]$ are the greatest and least of $y(-1) = 2, y(0) = \sqrt{5}$, and $y(2) = 1$. • [Maximum] $= y(0) = \sqrt{5}$ • [Minimum] $= y(2) = 1$

6.7.97F, 3.7.35

16. $y = \dfrac{x}{x^2 + 9}$ is continuous in $(-\infty, 1]$. • $y' = \dfrac{9 - x^2}{(x^2+9)^2}$ is negative for $x < -3$ and positive for $-3 < x \le 1$. • y is decreasing in $(-\infty, -3]$ and increasing in $[-3, 1]$. [Minimum] $= y(-3) = -\frac{1}{6}$ • [Maximum] $= y(1) = \frac{1}{10}$ since $y < 0$ for $x < 0$.

6.7.97G, 3.8.4

19. $P = 2w + 2h \implies h = \frac{1}{2}P - w$ and $A = wh = w(\frac{1}{2}P - w) = \frac{1}{2}wP - w^2$ for $0 \le w \le \frac{1}{2}P$ • $\dfrac{dA}{dw} = \frac{1}{2}P - 2w$ is positive for $0 < w < \frac{1}{4}P$ and negative for $w > \frac{1}{4}P$. • A is increasing in $[0, \frac{1}{4}P]$ and decreasing in $[\frac{1}{4}P, \frac{1}{2}P]$. • The maximum area is at $w = \frac{1}{4}P$, which is for a square.

9.3.96Q

21. **(a)** If the length L plus the girth G is less than 108, the volume can be increased by increasing L or G until $L + G = 108$. • The volume is LA where A is the cross sectional area, and for a fixed perimeter of the rectangular cross section, A is a maximum if the cross section is a square by Problem 19.
(b) $L = 68$ and $V = 6800$ cubic inches
(c) $L = 108 - 4w$ and $V = 4w^2(27 - w)$ for $0 \le w \le 27$ **(d)** $V'(w) = 12w(18 - w)$ is zero at $w = 0, 18$. • $V(0) = 0, V(27) = 0$ • A is a maximum for $w = 18, L = 36$ inches (when $L = 2w$).

8.13.98B

22. **(a)** $y = 15$ and $A = 150$ **(b)** $y = 5$ and $A = 150$ **(c)** $y = \frac{1}{2}(40 - x)$ and $A(x) = \frac{1}{2}x(40 - x)$ **(d)** The total area is a maximum for $x = 20$, $y = 10$ feet.

6.9.99A, 3.7.17 modified

23. **(a)** Numbers x with $0 < x < 1$ **(b)** $\frac{3}{8}$ **(c)** $x - x^3$ **(d)** $x = 1/\sqrt{3}$

3.8.97A

26. $x = 3, y = 3$

4.23.98D

28a $A = \pi r^2 + 16\pi r^{-1}$

4.24.98P

29. **(a)** $w = 8$, $h = 20$, $A = 160$ **(b)** $x = 0$ and $x = 6$
(c) $w = 2x$, $h = 36 - x^2$, $A = 2x(36 - x^2)$ for $0 \le x \le 6$
(d) A is continuous in $[0, 6]$. • $A' = 6(12 - x^2)$ is zero at $x = \sqrt{12} = 2\sqrt{3}$ • A is a maximum at $x = 2\sqrt{3}$, for which $w = 4\sqrt{3}$, $h = 24$
(e) $w = 0$ and $h = 36$ or $w = 6$ and $h = 0$

6.7.97B

31. **(a)** $C = 15x + 1500/x$ **(b)** \$300

6.7.97C

43. $y = \frac{1}{2}x^4 + x^3 - x^2$ is continuous in $[-3, 2]$. • $y' = x(12x^2 + 3x - 2)$ is zero at $x = 0, \frac{1}{2}, -2$. • $y(-3) = 4.5, y(-2) = -4, y(0) = 0, y(\frac{1}{2}) = -\frac{3}{32}, y(2) = 12$. • The minimum is $y(-2) = -4$ and the maximum is $y(2) = 12$.

3.4.97D

44. $y = -\frac{1}{3}x^3 + 2x^2 - 2x$ is continuous in $[0, 4]$. • $y' = -x^2 + 4x - 2$ is zero at $x = 2 \pm \sqrt{2}$. • $y(0) = 0, y(2 - \sqrt{2}) \doteq -0.5523, y(2 + \sqrt{2}) \doteq 3.219, y(4) = \frac{8}{3} \doteq 2.667$. • The minimum is $y(2 - \sqrt{2}) = -\frac{1}{3}(2 - \sqrt{2})^3 + 2(2 - \sqrt{2})^2 - 2(2 - \sqrt{2}) \doteq -0.5523$ and the maximum is $y(2 + \sqrt{2}) = -\frac{1}{3}(1 + \sqrt{2})^3 + 2(2 + \sqrt{2})^2 - 2(2 + \sqrt{2})) \doteq 3.219$.

3.4.97E

46. $y = \frac{1}{4}x^4 - 2x^3 + x^2 + 50$ is continuous for all x and tends to ∞ as $x \to \pm\infty$. • $y' = x(x^2 - 6x + 2)$

is zero at $x = 0, 3 \pm \sqrt{7}$. • $y(0) = 50, y(3-\sqrt{7}) \doteq 50.041, y(3+\sqrt{7}) \doteq -24.041$. • The minimum is $y(3+\sqrt{7}) = \frac{1}{4}(3+\sqrt{7})^4 - 2(3+\sqrt{7})^3 + (3+\sqrt{7})^2 + 50 \doteq -24.041$. • There is no maximum.

9.3.96K

52. For $x < 0$ 6.9.99C

53. For $x > 0$ 6.9.99D

65. **(a)** Local minimum because $y'(-1) = 0$ and $y''(-1) = 21$ is positive
(b) Neither a local maximum nor a local minimum because $y'(3) = 5$ is not zero

4.12.98W, 3.3.15, 16, 17, 19

Tune-Up Exercises 3.3

T1.[A] $A(B(100)) = 14$

T2.[A] $U(V(x)) = x^6/(x^2-25)$ • Domain: $x \neq \pm 5$ 5.5.98A,2.6.4

T3.[A] **(a)** Figures T3a,b,c **(b)** The slope -2 of $r = 1-2p$ is the product of the slope -1 of $r = 4-q$ and the slope 2 of $q = 3+2p$. 8.30.98A

5.6.98A

Jun9.99A Jun9.99B Jun9.99C

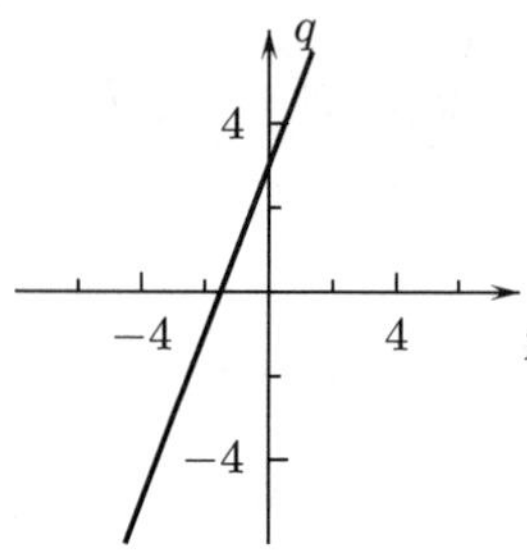

$q = 3 + 2p$

Figure T3a

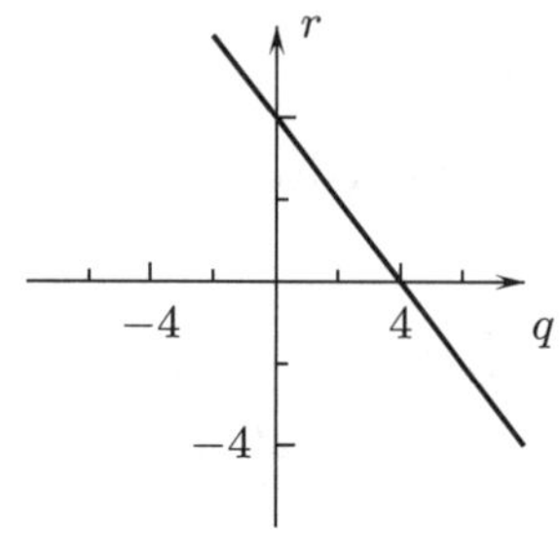

$r = 4 - q$

Figure T3b

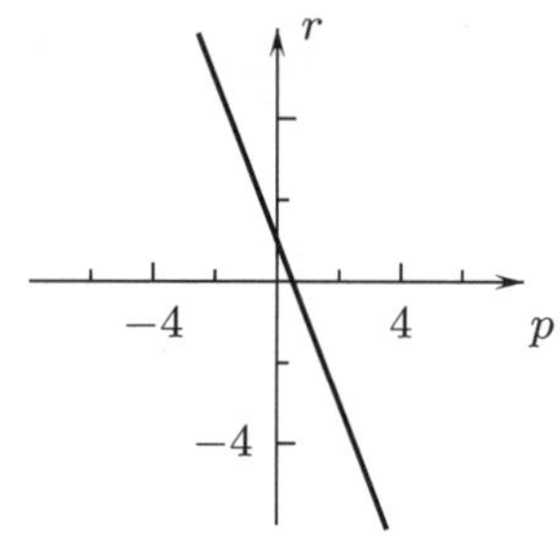

$r = 1 - 2p$

Figure T3c

T4.[A] **(a)** Tangent line: $y = 2 + 2(x-1)$ • Figure T4a
(b) Tangent line: $z = 1 - \frac{1}{2}(y-2)$ • Figure T4b
(c) Tangent line: $z = 1 - (x-1)$ • Figure T4c
(d) The slope -1 of the tangent line in part (c) is the product of the slopes 2 and $-\frac{1}{2}$ of the tangent lines in parts (a) and (b).

5.6.98D

Jun9.99D Jun9.99E Jun9.99F

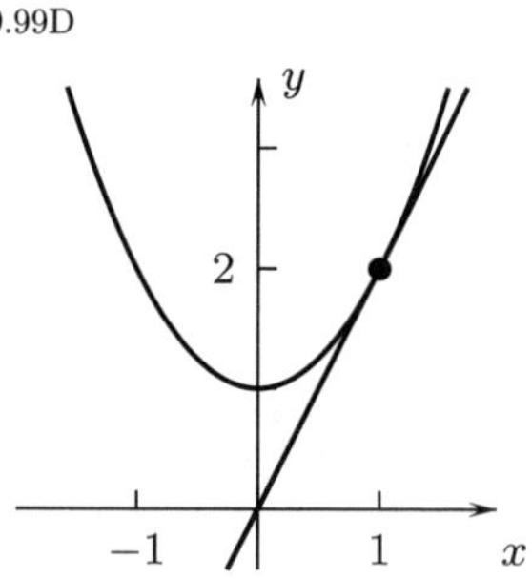

$y = x^2 + 1$

Figure T4a

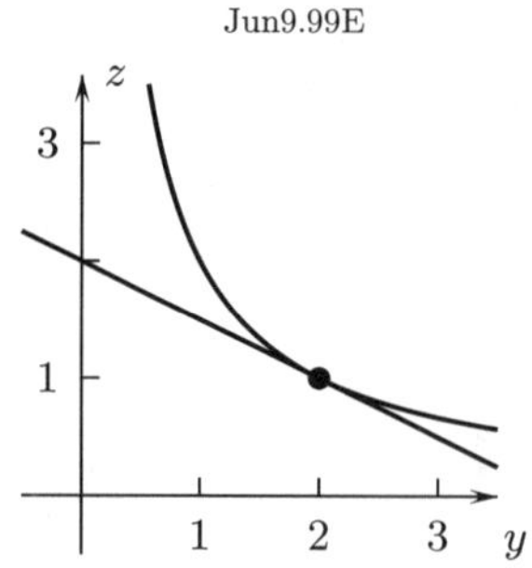

$z = 2/y$

Figure T4b

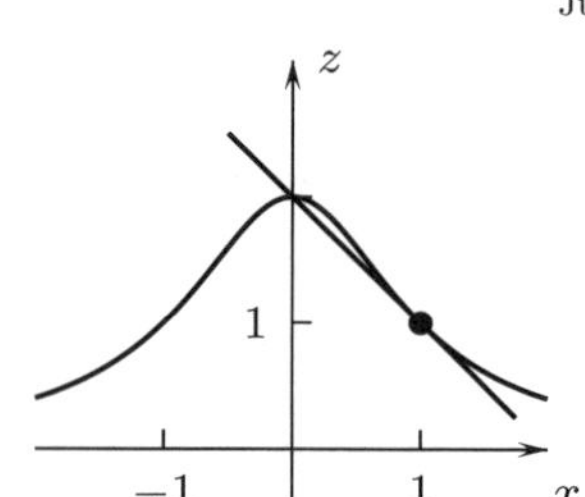

$z = 2/(x^2+1)$

Figure T4c

T5.[O] $v(w(0)) = v(1) = 7$ • $\frac{dv}{dx}(0) = \frac{dv}{dw}(w(0))\frac{dw}{dx}(0) = \frac{dv}{dw}(1)\frac{dw}{dx}(0) = (-3)(5) = -15$

T6.[O] $Q(4) = R(16) = 10$ • $Q'(x) = R'(x^2)\frac{d}{dx}(x^2) = 2xR'(x^2)$ • $Q'(4) = 2(4)R'(16) = 48$ 8.30.98C

T7.[O] $x(3) \approx 20$ • $y(x(3)) \approx y(20) \approx 380$ • $\frac{dx}{dt}(3) \approx 6, \frac{dy}{dx}(20) \approx -10$ • 8.30.98B

$$\left[\frac{dy}{dt}\right]_{t=3} = \frac{dy}{dx}(x(3))\,\frac{dx}{dt}(3) \approx \frac{dy}{dx}(20)\,\frac{dx}{dt}(3) \approx 6(-10) = -60$$

5.9.98A

Problems 3.3

1. $\frac{dG}{dt} = \frac{dG}{ds}\frac{ds}{dt} = \left[0.2\,\frac{\text{liters}}{\text{kilometer}}\right]\left[80\,\frac{\text{kilometers}}{\text{hour}}\right] = 16\,\frac{\text{liters}}{\text{hour}}$

2. **(a)** Set $t = 0$ at 12:00 PM. • $\frac{dV}{dt}(0) = \frac{dV}{dh}(h(0))\frac{dh}{dt}(0) = \left[0.001\,\frac{\text{liters}}{\text{meter}}\right]\left[3\,\frac{\text{meters}}{\text{second}}\right] = 0.003\,\frac{\text{liters}}{\text{second}}$ 5.6.98L
(b) $h \approx 200 + 15(3) = 245$ meters • $V \approx 5 + 15(0.003) = 5.045$ liters

5.7.98M,2.6,Part (a) is Example 9 modified

3. $P(x(t))$ dollars • $\frac{d}{dt}[P(x(t))] = \left[\frac{dP}{dx}(x(t))\,\frac{\text{dollars}}{\text{pound}}\right]\left[\frac{dx}{dt}(t)\,\frac{\text{pounds}}{\text{hour}}\right] = \frac{dP}{dx}(x(t))\,\frac{dx}{dt}(t)\,\frac{\text{dollars}}{\text{hour}}$

5.6.98L,2.6 SP 8

8. **(a)** The tangent line to $y = x^{1/2}$ at $x = 4$ has slope $\frac{1}{4}$. • The tangent line to $z = \frac{1}{4}y^3$ at $y = 2$ has slope 3.
(b) $z = \frac{1}{4}x^{3/2}$ • The tangent line to $z = \frac{1}{4}x^{3/2}$ at $x = 4$ has slope $\frac{3}{4}$
(c) $\frac{dz}{dy}\frac{dy}{dx} = 3(\frac{1}{4}) = \frac{3}{4} = \frac{dz}{dx}$
(d) The tangent line to $w = 4/z$ at $z = 2$ has slope -1.
(e) $w = 16x^{-3/2}$ • The tangent line to $w = 16x^{-3/2}$ at $x = 4$ has slope $-\frac{3}{4}$.
(f) $\frac{dw}{dz}\frac{dz}{dy}\frac{dy}{dx} = (-1)(3)(\frac{1}{4}) = -\frac{3}{4} = \frac{dw}{dx}$

10. **(a)** $\frac{dy}{dx}(6) \approx 0.875$ • At this value of t, $\frac{dy}{dt} = \frac{dy}{dx}\frac{dx}{dt} \approx (0.875)(7) = 6.125$ feet per second 10.26.97F
(b) $\frac{dy}{dx}(10) \approx -1$ • At this value of t, $\frac{dy}{dt} = \frac{dy}{dx}\frac{dx}{dt} \approx (-1)(9) = -9$ feet per second

8.31.98D

12. $\frac{dA}{dx}(6) = \frac{dB}{dy}(y(6))\frac{dy}{dx}(6) = \frac{dB}{dy}(7)\frac{dy}{dx}(6) = (9)(8) = 72$

5.7.98A

13. $W'(100) = 70$

5.7.98B

14. $T'(10) = -30$

5.7.98C

15. $\frac{dH}{dx} = F'(x^3)\frac{d}{dx}(x^3) = 3x^2F'(x^3)$ • $\frac{dH}{dx}(2) = 3(2^2)F'(2^3) = 12F'(8) = (12)(10) = 120$

5.7.98D

16. $\frac{dG}{dx}(5) = (12)(6) = 72$

5.7.98E

24 $\frac{dM}{dx}(200) \approx (-50)(0.04) = -2$

6.9.99d

26. 12,000

5.10.98L, 2.6.26

27. 139

5.10.98M, 2.6.30 modified

33a. $\frac{d}{dt}[x(y(t))]\Big|_{t=2} \approx (2.3)(-1.5) = -3.5$

5.10.98H

Tune-up Exercises 3.4

T1. **(a)** 1.5 **(b)** 0.5 **(c)** $\frac{1}{2}$ **(d)** $-\frac{1}{2}$

T2. **(a)** 5 **(b)** $5^4 = 625$

T3. **(a)** $(\frac{1}{10})^{1/4}$ **(b)** $\frac{1}{10}$

T4. **(a)** 6 **(b)** 8

T5. $f(x) = 5(3^x)$ • $f'(x) = 5\ln(3)(3^x)$

T6. **(a)** $g(t) = 64(\frac{1}{4})^t$ **(b)** $g'(t)/g(t)$ is negative and equals $\ln\left(\frac{1}{4}\right)$.

T7. **(a)** $h(t) = 1000(0.8)^{t/5}$
(b) $h(-3) = 1000(0.8)^{-3/5} \doteq 1143.26$ • $h'(-3) = 200\ln(0.8)(0.8)^{-3/5} \doteq -51.02$

T8. **(a)** $K(y) = 900(1.5)^{y/100}$ **(b)** $K(y+100)/K(y) = 1.5$ and $K'(y+100)/K'(y) = 1.5$

T9. $y = e^{x/3}$ • $y' = \frac{1}{3}e^{x/3}$ • Figure T9

T10. $y' = -e^{-x}$ • Figure T10

T11. **(a)** 8 **(b)** 8

T12. $\$3000\left(1 + \frac{0.096}{12}\right)^{4(12)} = \$3000(1.008)^{48} \doteq \$4397.71$

Jun9.99H

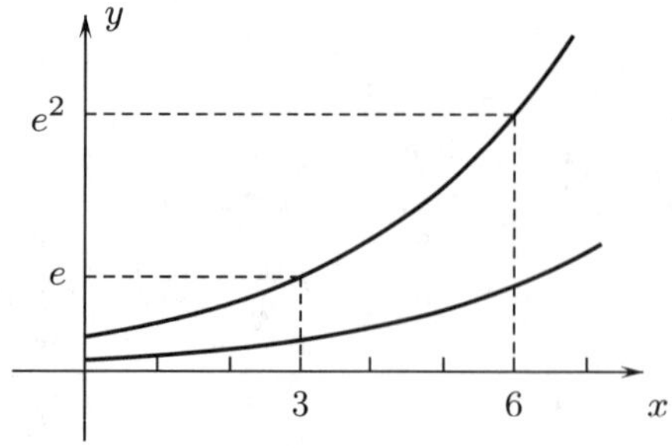

Figure T9

Jun9.99I

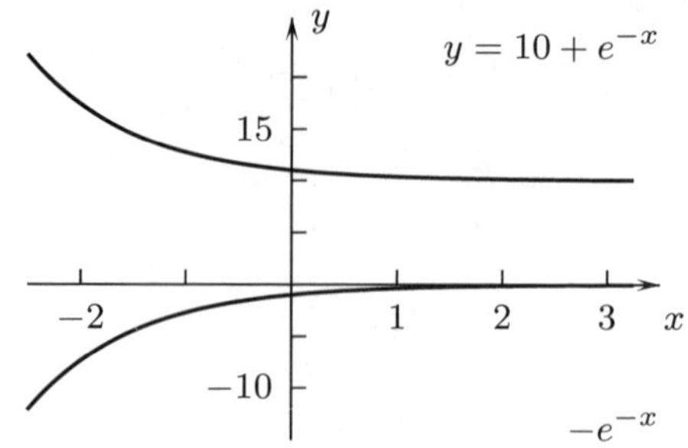

Figure T10

T13. **(a)** $\$100(1.06)^{10} \doteq \$179.08$ **(b)** $\$100e^{0.06(10)} = \$100e^{0.6} \doteq \$182.21$ **(c)** $160

T14. $\lim_{x\to\infty} \left[\left(\frac{1}{6}\right)^x - x^2\right] = -\infty$ because $\left(\frac{1}{6}\right)^x \to 0$ and $x^2 \to \infty$ as $x \to \infty$ • $\lim_{x\to-\infty} \left[\left(\frac{1}{6}\right)^x - x^2\right] = \infty$ because $\left(\frac{1}{6}\right)^x \to \infty$ as $x \to -\infty$ and $\left(\frac{1}{6}\right)^x$ is much larger than x^2 for large negative x.

T15. $R(t) = 95 - 5(t-1)$

T16. $A(t) = 20 - 10t$ • $B(t) = 20\left(\frac{1}{2}\right)^t$

T17. $\frac{d}{dx}(e^{2x}) = e^{2x}\frac{d}{dx}(2x) = 2e^{2x}$

10.23.97C

T18. $\frac{d}{dx}[e^{x^2+3x}] = e^{x^2+3x}\frac{d}{dx}(x^2+3x) = (2x+3)e^{x^2+3x}$

10.23.97E

T19. $y' = \ln(2)(2^x) + \ln(3)(3^x)$

5.10.99G

T20 $f'(x) = \frac{d}{dx}[(1+e^x)^4] = 4(1+e^x)^3\frac{d}{dx}(1+e^x) = 4e^x(1+e^x)^3$

5.10.99H

T21. **(a)** $y = 6e^{-4t}$ **(b)** $y(t) = 3^t$

8.23.98F

Problems 3.4

1. $\frac{d}{dx}(e^{6x}) = e^{6x}\frac{d}{dx}(6x) = 6e^{6x}$

2.28.98A, 6.3.21

2. $\frac{d}{dx}(e^x + x^e) = e^x + ex^{e-1}$

2.28.98B, 6.4.6

3. $\frac{d}{dx}\left[\frac{x}{1+e^x}\right] = \frac{(1+e^x) - xe^x}{(1+e^x)^2}$

10.23.97G

4. $\frac{d}{dx}(\sqrt[3]{1+e^x}) = \frac{1}{3}e^x(1+e^x)^{-2/3}$

10.24.97G

5. $\frac{d}{dt}[e^{-t}(t^2+1)] = e^{-t}\frac{d}{dt}(t^2+1) + (t^2+1)\frac{d}{dt}(e^{-t}) = -e^{-t}(t-1)^2$

10.24.97I

6. $\frac{d}{dx}(1-e^4) = 0$

10.24.97J

7. $\left[\frac{d}{dx}(e^{100x-x^2})\right]_{x=0} = 100$

10.24.97O

8. $\frac{d}{dx}(xe^{-x}) = e^{-x}(1-x)$

10.24.97P,6.3.23

9. $\frac{d}{dx}(\sqrt{1+e^x}) = \frac{1}{2}e^x(1+e^x)^{-1/2}$

10.24.97Q,6.3.24

11. $q'(2) = 9\ln(3) + 12$

5.10.99I

12. $\frac{d}{dx}(5^{x/3}) = \frac{1}{3}\ln(5)(5^{x/3})$

5.10.99J

17. **(a)** $p(h) = 1035(\frac{1}{2})^{h/5.2}$ **(b)** $\frac{dp}{dh}(10.4) = \frac{1035}{20.8}\ln(0.5) \doteq -34.49$ grams per square centimeter per kilometer

5.1099L

18. $\frac{500}{3}\ln(2)(2^{t/3})$

20. $12\ln(1.015)$

27a. Let C be the initial balance and r the fraction corresponding to the simple interest rate. • Solve $C(1+r) = Ce^{0.05}$ for r: $r = e^{0.05} - 1 \doteq 0.05127$ or $\doteq 5.127\%$

32. $5700\left(\dfrac{\ln(4/5)}{\ln(1/2)}\right) \doteq 1835$ years

36. One answer: Use the growth factor of 1.25 from $t=0$ to $t=1$ and the value 5 at $t=0$: $H(t) \approx 5(1.25)^t$ • Maximum error $= |5(1.25)^3 - H(3)| = |5(1.25)^3 - 9.34| \doteq 0.426$ 6.6.98P

41. **(a)** $y = 2x$ • Figure A41 **(b)** At $x=0$, the distance is 2. • At $x=2$, the distance is $1 + e^6 - 4 \doteq 400.43$

6.4.98D

Jun29.99Q

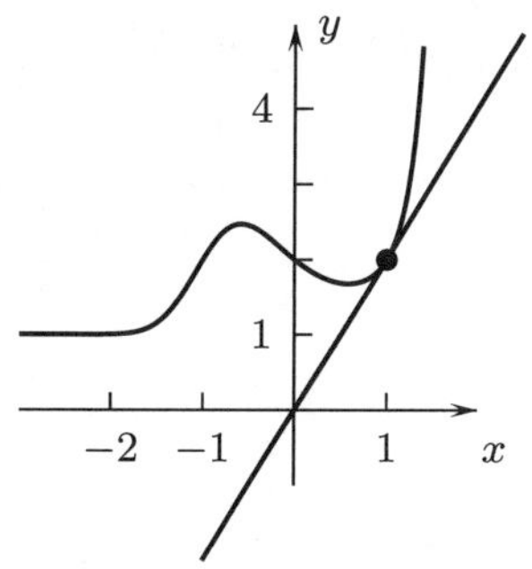

FIGURE A41

44. Tangent line $y = 1 + 2\ln(2)$

46. [Maximum] $= y(1) = e$ • [Minimum] $= y(0) = 0$ 6.4.98H

51. **(a)** $y = 4e^{t/2}$ **(b)** $y = e^{-t+5}$ 6.4.98P

53a $y = 300(2^t)$ 8.26.98A

55. $2e^{-6}$ 5.19.99A

57. The rectangle with corners at $x = \pm\frac{1}{2}\sqrt{2}$ on the x-axis • [Maximum area] $= \sqrt{2}e^{-1/2}$ 6.6.98W

6.7.98D, 6.3.67

Tune-up Exercises 3.5

T1. **(a)** $x = \pm 10^{-1.75} \doteq \pm 0.017783$ **(b)** $x = \frac{1}{2}\log_{10}(50) \doteq 0.849485$
(c) $x = e^4 \doteq 54.59815$ **(d)** There are no solutions because $\sqrt{x}$ is ≥ 0 and so $e^{\sqrt{x}} \geq 1$. (To check an answer, store its exact expression as x in your calculator or computer and evaluate the given expression.)

T2. Write log for $\log_{10}$.
(a) (VI) since $\log(300) = 2 + \log(3)$
(b) (IV) since $\log(10) = 1$
(c) (V) since $\log(25) = 2\log(5)$
(d) (IV) since $\log(30) - \log(3) = \log(10)$
(e) (VII) since $\log(1) = 0$
(f) (VIII) since $\log(81) = \log(3^4) = 4\log(3)$
(g) (V) since $\log(3^2 + 4^2) = \log(5^2) = 2\log(5)$
(h) (I) since $\log[9(16)] = 2\log(3) + 2\log(4)$

T3. **(a)** $y(1) = 0$ • $y'(1) = 1$ • Tangent line: $y = x - 1$
(b) $y(0) = 1$ • $y'(0) = 1$ • Tangent line: $y = x + 1$ • Figure T3

T4. $\dfrac{d}{dx}(3\ln x - 4e^x) = \dfrac{3}{x} - 4e^x$ 10.23.97A

T5. $\dfrac{d}{dx}[\ln(x^2+3x)] = \left(\dfrac{1}{x^2+3x}\right)\dfrac{d}{dx}(x^2+3x) = \dfrac{2x+3}{x^2+3x}$ 10.23.97B

T6. $f'(x) = \frac{1}{2}(\ln x)^{-1/2}\dfrac{d}{dx}(\ln x) = \dfrac{1}{2x\sqrt{\ln x}}$ • $f'(5) = \dfrac{1}{10\sqrt{\ln(5)}}$ 10.23.97D

T7. 0 10.23.97F

Jun9.99K

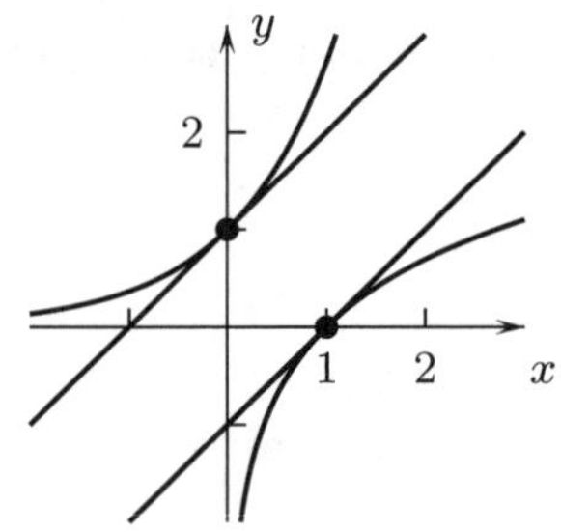

Figure T3

T8. $\left[\frac{d}{dx}(\ln y)\right]_{x=7} = \frac{1}{y(7)}\, y'(7) = \frac{1}{2}$ • $\left[\frac{d}{dx}(e^y)\right|_{x=7} = e^{y(7)}\, y'(7) = 5e^{10}$

10.23.97H

Problems 3.5

1. $\frac{d}{dx}(3x - 4\ln x) = 3 - \frac{4}{x}$

10.24.97A

2. $L'(300) = \frac{1}{300}$

10.24.97E, 6.1.1

3. $\frac{d}{dx}[(\ln x)^7] = 7(\ln x)^6 \frac{d}{dx}(\ln x) = \frac{7(\ln x)^6}{x}$

10.24.97C

4. $y'(x) = \frac{d}{dx}(x^2 \ln x) = x^2 \frac{d}{dx}(\ln x) + \ln x \frac{d}{dx}(x^2) = x + 2x\ln x$

10.24.97D

5. $\frac{d}{dx}[\ln(1 - e^x)] = \frac{-e^x}{1 - e^x}$

10.24.97H,6.1.2

9. $\frac{d}{dx}[\ln(\ln x)] = \frac{1}{x \ln x}$

10.24.97N,6.1.10

11. Tangent line: $y = 5 - 5\ln(5)$ • Figure A11

10.25.97A

Jun9.99L

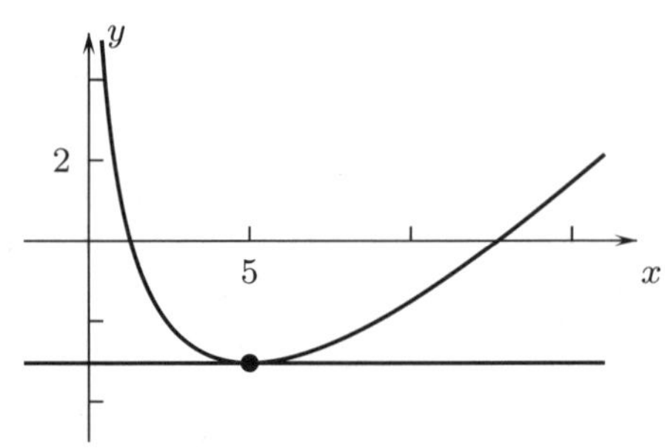

Figure A11

12. **(a)** Domain: $[1, \infty)$ since $\ln x \geq 0$ for $x \geq 1$ **(b)** Tangent line: $y = 1 + \frac{1}{2e}(x - e)$
(c) $\sqrt{\ln 2.5} - [1 + \frac{1}{2e}(2.5 - e)] \doteq -0.0026$ • $\sqrt{\ln 2.8} - [1 + \frac{1}{2e}(2.8 - e)] \doteq -0.00033$ •
The approximation is better at $x = 2.8$.

6.4.98G

16. Tangent line: $y = \ln(2) + \frac{1}{4}x$

6.4.98O

19. [Minimum] $= 0$ at $x = 1$ • [Maximum] $= 2/e$ at $x = e^2$

6.5.98E

39. $\frac{d}{dx}(x^x) = \frac{d}{dx}(e^{x\ln x}) = x^x(\ln x + 1)$

6.10.98R

40. $\frac{d}{dx}(x^{\ln x}) = 2(\ln x)(x^{\ln x - 1})$

6.10.98S,6.M.7

Tune-up Exercises 3.6

T1. **(a)** -60 degrees is $-\frac{1}{3}\pi$ radians • Figure T1a • $\sin(-\frac{1}{3}\pi) = -\frac{1}{2}\sqrt{3}$ • $\cos(-\frac{1}{3}\pi) = \frac{1}{2}$ • $\tan(-\frac{1}{3}\pi) = -\sqrt{3}$
(b) 135 degrees is $\frac{3}{4}\pi$ radians • Figure T1b • $\sin(\frac{3}{4}\pi) = \frac{1}{2}\sqrt{2}$ • $\cos(\frac{3}{4}\pi) = -\frac{1}{2}\sqrt{2}$ • $\tan(\frac{3}{4}\pi) = -1$
(c) 450 degrees is $\frac{5}{2}\pi$ radians • Figure T1c • $\sin(\frac{5}{2}\pi) = 1$ • $\cos(\frac{5}{2}\pi) = 0$ • $\tan(\frac{5}{2}\pi)$ is not defined.

Dec3.97B

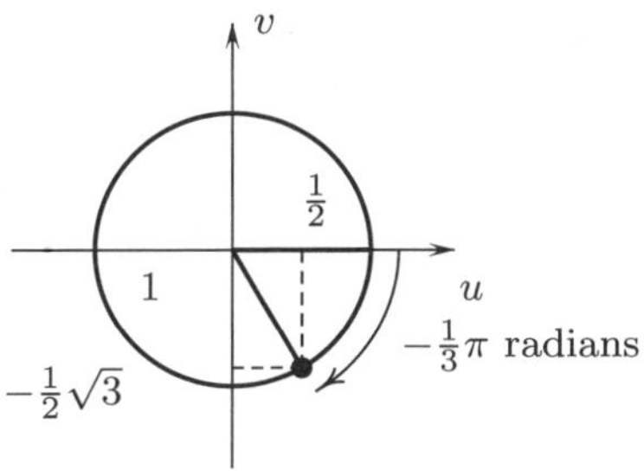

FIGURE T1a

Dec3.97C

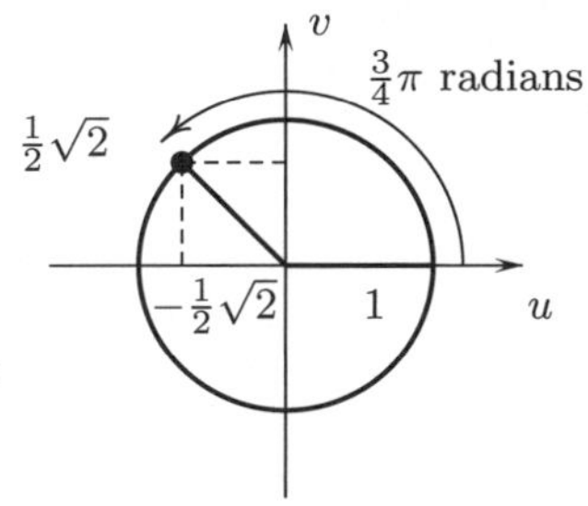

FIGURE T1b

Dec3.97D

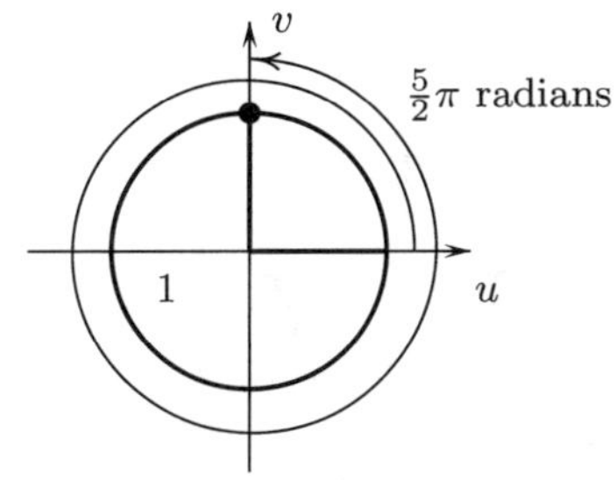

FIGURE T1c

T2. [Arclength] $= \frac{7}{5}\pi(14) = 19.6\pi$ meters • [Area] $= (\frac{1}{2})(\frac{7}{5}\pi)(14)^2 = 137.2\pi$ square meters

T3. $r = \sqrt{13}$ • $\sin\theta = \dfrac{2}{\sqrt{13}}$ • $\cos\theta = -\dfrac{3}{\sqrt{13}}$ • $\tan\theta = -\frac{2}{3}$

T4. Use uv-coordinates with origin at P. • The u-coordinate of Q is $6\cos(-1)$ and the v-coordinate of Q is $6\sin(-1)$ • $x = 2 + 6\cos(-1) \doteq 5.242$ • $y = 3 + 6\sin(-1) \doteq -2.049$

T5. $\sin\gamma = 2.5/6$ and $0 < \gamma < \frac{1}{2}\pi$ • $\gamma = \sin^{-1}(2.5/6) \doteq 0.4298$ radians

T6. **(a)** $-1/\sqrt{2}$ **(b)** $-\frac{1}{2}$ **(c)** 0 **(d)** -2 **(e)** $\sqrt{3}$ **(f)** $-\sqrt{2}$ **(g)** $1/\sqrt{3}$ **(h)** Undefined

T7. **(a)** $\frac{1}{4}\pi$ radians **(b)** $-\frac{1}{4}\pi$ radians **(c)** $-\frac{1}{2}\pi$ radians **(d)** $-\frac{1}{6}\pi$ radians
(e) $\frac{1}{3}\pi$ radians **(f)** 0

T8. $\alpha = \tan^{-1}(\frac{3}{4}) + \pi \doteq 3.7851$

T9. $c = \sqrt{8^2 + 10^2 - 2(8)(10)\cos(0.9)} = \sqrt{164 - 160\cos(0.9)} \doteq 8.0338$ feet

T10. The smallest angle θ is opposite the side of length 3. • The hypotenuse has length $\sqrt{3^2 + 4^2} = 5$ • $\sin\theta = \frac{3}{5} = 0.6$ • $\theta = \sin^{-1}(0.6) \doteq 0.6435$ radians

T11. $\dfrac{d}{dx}(5\sin x - 6\cos x) = 5\cos x + 6\sin x$

T12. $y'(x) = 3\sec^2 x$ • $y'(\pi) = 3(-1)^2 = 3$ 12.31.97A

T13. $\dfrac{d}{dx}\sqrt{\cos x} = \frac{1}{2}(\cos x)^{-1/2}\dfrac{d}{dx}\cos x = -\frac{1}{2}(\cos x)^{-1/2}\sin x$ 12.31.97B

T14. $\dfrac{d}{dx}[\sin(x^3 - 4)] = \cos(x^3 - 4)\dfrac{d}{dx}(x^3 - 4) = 3x^2\cos(x^3 - 4)$ 12.31.97C

T15. $\left[\dfrac{d}{dx}[\tan(2x)]\right]_{x=0.6} = \left[2\sec^2(2x)\right]_{x=0.6} = \dfrac{2}{\cos^2(1.2)} \doteq 15.23$ 12.31.97D

T16. $\dfrac{d}{dx}(x^2\sin x) = x^2\cos x + 2x\sin x$ 12.31.97E

T17. $\dfrac{d}{dx}[\tan^{-1}(2x)] = \dfrac{1}{1 + (2x)^2}\dfrac{d}{dx}(2x) = \dfrac{2}{1 + 4x^2}$ 12.31.97F

T18. $\dfrac{d}{dx}[\sin^{-1}(x^2)] = \dfrac{1}{\sqrt{1 - (x^2)^2}}\dfrac{d}{dx}(x^2) = \dfrac{2x}{\sqrt{1 - x^4}}$ 12.31.97G

T19. **(a)** $h = 3\sin\theta$ **(b)** $\dfrac{dh}{dt} = \dfrac{d}{dt}(3\sin\theta) = 3\cos\theta\dfrac{d\theta}{dt}$ • 12.31.97H

At the time in question, $\dfrac{dh}{dt} = 3(\tfrac{1}{2})(0.2) = 0.3$ feet per minute

T20. $\psi + \theta = \frac{1}{2}\pi$ • $\dfrac{d\psi}{dt} + \dfrac{d\theta}{dt} = 0$ • $\dfrac{d\psi}{dt} = -0.2$ radians per minute 8.18.98A

T21. **(a)** $\psi = \cos^{-1}(h/3)$ **(b)** $\dfrac{d\psi}{dt} = \dfrac{-1/3}{\sqrt{1-(h/3)^2}}\dfrac{dh}{dt} = \dfrac{-1}{\sqrt{9-h^2}}\dfrac{dh}{dt}$ • 8.18.98B

At the time in question, $\dfrac{d\psi}{dt} = \dfrac{-1}{\sqrt{9-2^2}}(-1) = \frac{1}{\sqrt{5}}$ radians per minute

T22. **(a)** $w = 10\cot\theta$ **(b)** $\dfrac{dw}{dt} = -10\csc^2\theta\dfrac{d\theta}{dt}$ • At the time in question, $\dfrac{dw}{dt} = -(10)(2)(0.2) = -4$ 8.18.98C

T23. **(a)** $\tan\psi = w/10 \Rightarrow \psi = \tan^{-1}(w/10)$ 8.18.98D

(b) $\dfrac{d\psi}{dt} = \dfrac{1/10}{1+(w/10)^2}\dfrac{dw}{dt} = \dfrac{10}{100+w^2}\dfrac{dw}{dt}$ • At the time in question,

$\dfrac{d\psi}{dt} = \dfrac{10}{100+64}(4) = \frac{10}{41}$ radians per minute

8.18.98E

Problems 3.6

1. $\frac{1}{365}(2\pi)(93) \doteq 1.6009$ million miles

2. $\gamma = \frac{17}{40}\pi$

3. $274\sin(4.8\pi/180) \doteq 22.93$ kilometers south • $274\cos(4.8\pi/180) \doteq 273.04$ kilometers east

6. **(a)** $\sqrt{480^2+90^2} \doteq 488.36$ kilometers **(b)** $\tan^{-1}(90/480) \doteq 0.1853$ radians south of east

9. $\tan^{-1}(\frac{1}{2}) \doteq 0.464$ radians and $\tan^{-1}(2) \doteq 1.107$ radians

10. $\cos^{-1}(\frac{4}{5}) \doteq 0.6435$ radians

12. 2.5 feet

13. $\tan^{-1}(\frac{481}{388}) \doteq 0.8920$ radians

14. $y'(x) = 2\sec^2 x - 3\cos x$ 12.31.97I

15. $\dfrac{dy}{dx} = 3\cos x - 5\sin x$ 8.18.97A

16. $\dfrac{dy}{dx} = \dfrac{x\cos x - \sin x}{x^2}$ 12.31.97J

17. $\dfrac{d}{dt}[\sin(4t) - 4\sin t] = \cos(4t)\dfrac{d}{dt}(4t) - 4\cos t = 4\cos(4t) - 4\cos t$ 12.31.97K

18. $\dfrac{d}{dx}(x\tan x) = x\sec^2 x + \tan x$ 1.2.97B,2.R.20

19. $y'(1) = -4\sin(1)\cos^3(1)$ 12.31.97O,2.8.6

20. $y'(\frac{1}{4}\pi) = \frac{\pi}{8}\sqrt{2} + \frac{1}{2}\sqrt{2}$ 12.31.97P,2.8.7

21. $\left[\dfrac{d}{dx}[\sin(\pi x)]\right]_{x=1} = -\pi$ 12.31.97Q,2.8.8

22. $\dfrac{d}{dt}[\tan(4t)] = \sec^2(4t)\dfrac{d}{dt}(4t) = 4\sec^2(4t)$ 12.31.97S,2.8.12

23. $\left[\dfrac{d}{dx}(\sin^{-1} x)\right]_{x=0.5} = \dfrac{1}{\sqrt{0.75}}$ 8.19.98B

24. $\left[\dfrac{d}{dx}(\tan^{-1} x)\right]_{x=3} = \frac{1}{10}$ 3.1.98E

25. $\dfrac{d}{dx}(x\tan^{-1} x) = \dfrac{x}{1+x^2} + \tan^{-1} x$ 3.1.98G

26. $\dfrac{d}{dx}(\sin^{-1} x)^2 = 2(\sin^{-1} x)\dfrac{d}{dx}(\sin^{-1} x) = \dfrac{2\sin^{-1} x}{\sqrt{1-x^2}}$ 3.1.98F

28. $y'(x) = \dfrac{1}{\sqrt{9-x^2}}$ 8.19.98D

29. $\dfrac{d}{dx}[\sin(\ln x)] = \dfrac{\cos(\ln x)}{x}$ 8.23.98A

30. $\dfrac{d}{dx}[\ln(\sin x)] = \dfrac{\cos x}{\sin x}$

32. $\dfrac{d}{dx}[(\sin x)^{\pi}] = \pi(\sin x)^{\pi-1}\cos x$ 8.23.98B

34. $\left[\dfrac{d}{dx}\csc x\right]_{x=\pi/3} = -\frac{2}{3}$ 8.23.98A

35. $\dfrac{d}{dx}\left(\dfrac{\cot x}{x}\right) = \dfrac{-x\csc^2 x - \cot x}{x^2}$ 8.19.98F,2.8.28

38. $y = \sin(3) + \cos(3)(x-3)$ 8.19.98G,2.8.11

39. $y'(x) = -x\sin x + \cos x$ • $y'(\pi) = -1,\ y(\pi) = -\pi$ • $y = -\pi - (x-\pi)$ (or $y = -x$) 2.7.98R, 2.8.41

43. $y = e^{\sin(4)} + \cos(4)e^{\sin(4)}(x-4)$ 2.2.98S,2.8.42

46. $y = \sqrt{2} + \sqrt{2}(x - \frac{1}{4}\pi)$ 8.24.98B

48. **(a)** [Frequency]$= \frac{1}{30}\pi$ cycles per second **(b)** $v'(t) = \frac{22}{3}\pi\cos(\frac{1}{30}\pi t)$ • $v'(45) = 0$ 8.19.98K, 2.R.54

49. $\dfrac{dA}{d\theta}(\frac{1}{8}\pi) = -50\sin^2(\frac{1}{8}\pi) + 50\cos^2(\frac{1}{8}\pi) \doteq 35.36$ square feet per radian • 8.19.98M,2.8.50

$\dfrac{dP}{d\theta}(\frac{1}{8}\pi) = 10\cos(\frac{1}{8}\pi) - 10\sin(\frac{1}{8}\pi) \doteq 5.41$ feet per radian

54. $-5\csc^2(0.8) \doteq -9.7163$ meters per radian 3.1.98O,3.R.44

55. $-\frac{1}{8}$ radians per second 1.1.8D,2.8.ExA8

56. $\dfrac{d\theta}{dt} = \dfrac{10}{100+t^2}$ 8.19.98Q, 3.6.26

63. -20 9.27.98A

70. [Maximum] $= \frac{1}{4}\pi + 1$ • [Minimum] $= -\frac{1}{4}\pi - 1$ 8.20.98M

71. [Maximum] $= \dfrac{\cos(0.1)}{0.1} \doteq 9.95$ • [Minimum] $= \dfrac{\cos(1.2)}{1.2} \doteq 0.302$ 8.20.98B

75. $\sqrt{39}$ 8.20.98C

76. $\sqrt{20^2 + 30^2 - 2(20)(30)\cos(\frac{9}{8}\pi)} \doteq 49.078$ kilometers

87. $\sin^{-1}(\frac{1}{5}\sqrt{2}) \doteq 0.2868$ radians

CHAPTER 4
INTEGRALS AND APPLICATIONS, PART I

The derivative, studied in Chapters 2 and 3, and the INTEGRAL, which is introduced in this chapter, are the two main tools of calculus. The integral is used to determine values of quantities from their rates of change, to find areas, volumes, weights, average values, lengths of curves, and in many other contexts. In Section 4.1 we illustrate the types of applications that generally call for integrals by looking at special cases requiring only arithmetic. The DEFINITE INTEGRAL is defined in Section 4.2. In Section 4.3 we derive the FUNDAMENTAL THEOREM OF CALCULUS, which relates definite integrals and derivatives, and use it to obtain a formula for definite integrals of power functions x^n with constant $n \neq -1$. We also introduce in Section 4.3 the term INDEFINITE INTEGRALS for antiderivatives and use integrals to study areas and rates of change. Section 4.4 deals with estimating definite integrals and Section 4.5 with INTEGRATION BY SUBSTITUTION. In Section 4.6 we use integrals to study weight and density, arclength, average value, and NEWTON'S LAW OF MOTION. We use the Fundamental Theorem to derive basic integration formulas involving exponential functions, logarithms, and trigonometric functions in Section 4.7.

Section 4.1

Problems whose solutions are given by areas of rectangles

OVERVIEW: *In this section we illustrate the types of problems that usually require integrals with special examples that can be solved using only products and sums. These problems provide background for the definition of the integral in Section 4.2 and will help you understand how integrals are used in applications.*

Topics:

- ***Finding position from velocity***
- ***Finding volumes from rates of flow***
- ***Finding weight from one-dimensional density***
- ***Finding volumes from cross-sectional areas***
- ***Average values of functions***
- ***Length of a graph***

How far?

We start with two problems that use the formula for distance traveled at constant velocity,

$$[\text{Distance traveled}] = [\text{Velocity}] \times [\text{Time}].$$

Example 1 Imagine that one day you drive west on Interstate 94 from Fargo, North Dakota and that your velocity toward the west as a function of time is given by the function $v(t)$ of Figure 1. As can be seen from the graph, you travel 50 miles per hour for two hours, 25 miles per hour for one hour, and 75 miles per hour for three more hours. **(a)** How far do you travel in the entire six hours? **(b)** How is the answer to part (a) related to the area of the rectangles in Figure 2?[†]

SOLUTION **(a)** In the first two hours you travel [50 miles per hour]×[2 hours] = 100 miles; in the next hour you travel [25 miles per hour]×[1 hour] = 25 miles; and in the last three hours you go [75 miles per hour]×[3 hours] = 225 miles, for a total distance of $100 + 25 + 225 = 350$ miles.

(b) The distance traveled equals the area of the three rectangles in Figure 2. □

[†]We consider the area of a rectangle with sides parallel to the coordinate axes to be the product of the lengths of its sides even if different scales are used on the two axes.

♦ SUGGESTIONS TO INSTRUCTORS: Have students begin work on Questions 1 through 7 and possibly some of the Tune-up Exercises in this section before the class in which integrals are introduced. Problems 2, 4, 5, 6, 8, and 9 are good for in-class discussion.

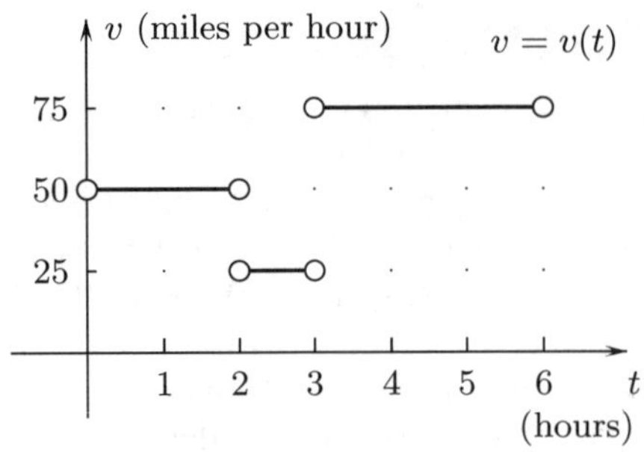

FIGURE 1

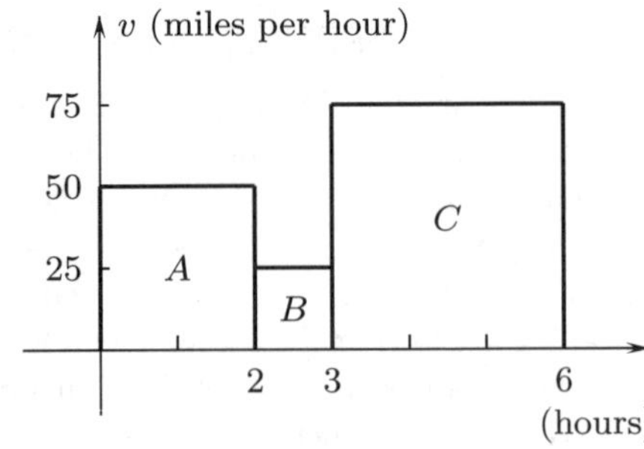

FIGURE 2

Question 1 At 10:00 AM one morning a truck driver is 50 miles east of Ozona, Texas. He drives 75 miles per hour toward the east for two hours to make a delivery outside San Antonio. Next, he drives west at 50 miles per hour for two hours to make another delivery and then drives east at 50 miles per hour for two more hours. Figure 3 is the graph of his velocity toward the east with $t = 0$ at 10 AM. **(a)** How far is he from Ozona at $t = 6$? **(b)** How is the answer to part (a) related to the areas of the rectangles in Figure 4?

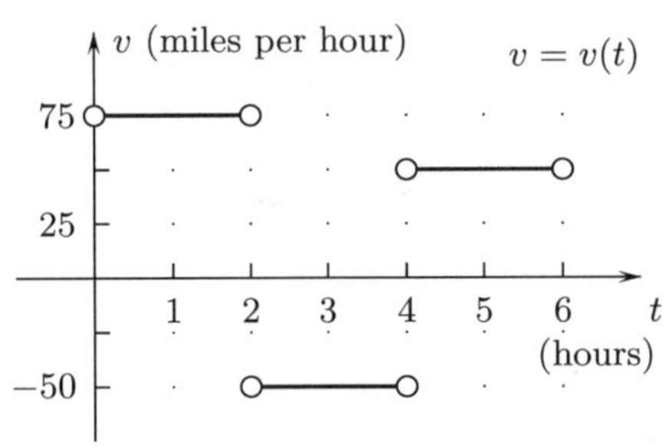

FIGURE 3

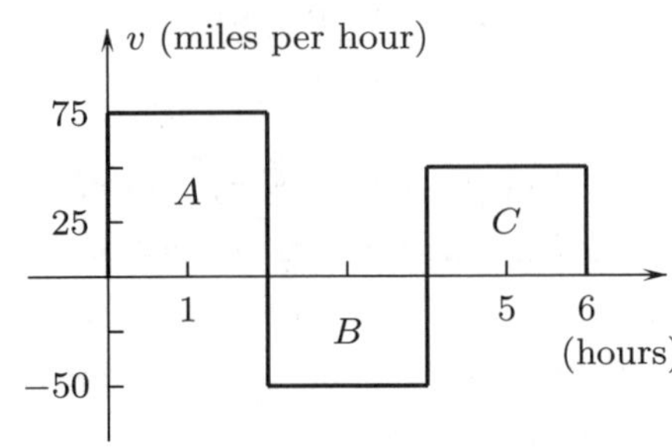

FIGURE 4

The position functions

The function $v(t)$ of Figure 1 that gives your velocity during your drive from Fargo is called a STEP FUNCTION because its graph consists of horizontal line segments and looks like steps. Because your velocity is constant in each of the three intervals where it is defined, the graph of your distance from Fargo consists of three line segments, whose slopes are the three velocities. We can use formulas for equations of lines to find formulas for your distance as a function of time.

Since you drive 50 miles per hour for the first two hours, you are $s(t) = 50t$ miles from Fargo for $0 \le t \le 2$ and are 100 miles from Fargo after two hours. During the next hour you drive 25 miles per hour, so you are $s(t) = 100 + 25(t - 2)$ miles from Fargo for $2 \le t \le 3$. After three hours you are 125 miles from Fargo, and you drive 75 miles per hour for the next three hours. Consequently, you are $s(t) = 125 + 75(t - 3)$ miles from Fargo for $3 \le t \le 6$.

The graph of your distance $s(t)$ from Fargo is given in Figure 5. Notice that the portion of the graph for $0 < t < 2$ has slope 50, the portion for $2 < t < 3$ has slope 25, and the portion for $3 < t < 6$ has slope 75, so that $\frac{ds}{dt}(t)$ is, as you would expect, the velocity function of Figure 1.

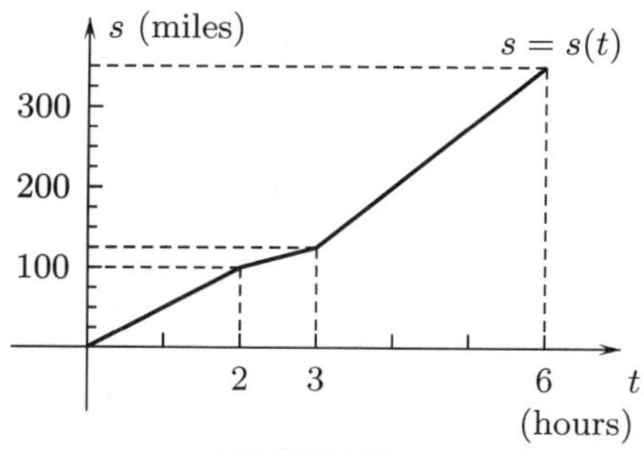

FIGURE 5

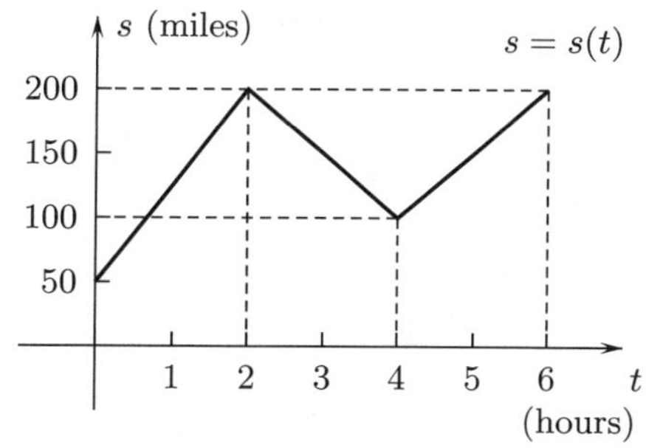

FIGURE 6

Question 2 Figure 6 shows the graph of the truck driver's distance from Ozona $s(t)$, as determined from the information in Question 1. **(a)** What are the slopes of the three line segments that form the graph in Figure 6 and how do they relate to Figure 3? **(b)** Give formulas for $s(t)$ in Figure 6.

Fill and drain

Suppose that a water tank contains 300 gallons of water at time $t = 0$ (minutes), and the rate of flow $r(t)$ (gallons per minute) into the tank for $0 \leq t \leq 70$ is the step function in Figure 7.

Question 3 **(a)** How much water flows into the tank in the time period $0 \leq t \leq 10$? **(b)** How much water flows out of the tank from $t = 10$ to $t = 60$? **(c)** How much water flows into the tank from $t = 60$ to $t = 70$? **(d)** How much water is in the tank at $t = 70$, and how is this number related to the areas of the four rectangles in Figure 8?

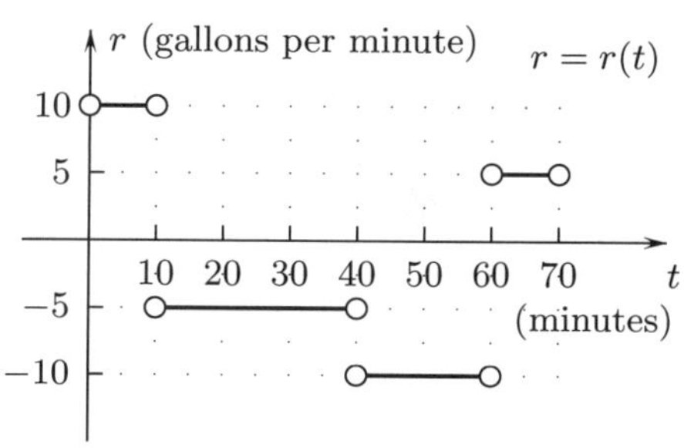

FIGURE 7

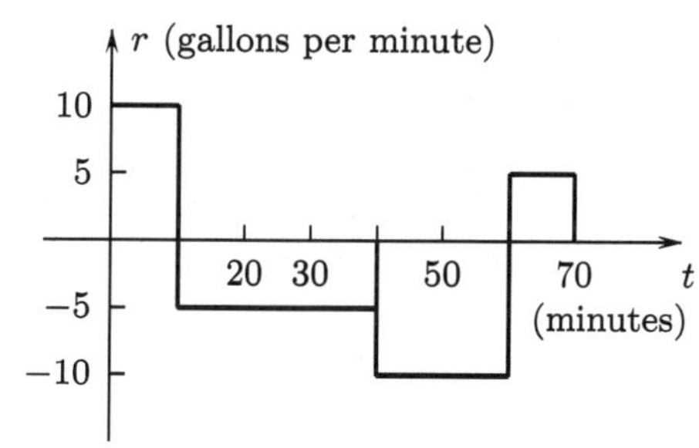

FIGURE 8

Question 4 Give formulas for the volume $V(t)$ of water in the tank for $0 \leq t \leq 70$ and draw its graph in a tV-plane.

The weight of a core sample

A 100-centimeter-long cylindrical core sample, drilled from the earth, consists of two different types of granite at the ends and sandstone in the middle. The density $\rho(x)$ (grams per centimeter) as a function of the distance x (centimeters) from one end (Figure 9) is[(1)]

$$\rho(x) = \begin{cases} 200 & \text{for} \quad 0 < x < 30 \\ 160 & \text{for} \quad 30 < x < 80 \\ 220 & \text{for} \quad 80 < x < 100. \end{cases}$$

Question 5 **(a)** What is the weight of the entire core sample? **(b)** How is the weight of the core sample related to the areas of the rectangles in Figure 10?

[(1)]Data adapted from *Encyclopædia Britannica*, Vol. 15, Chicago: Encyclopædia Britannica, Inc., 1979, p. 953.

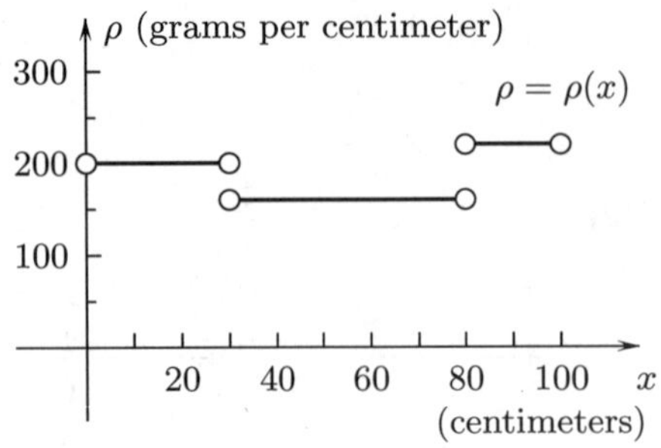

FIGURE 9

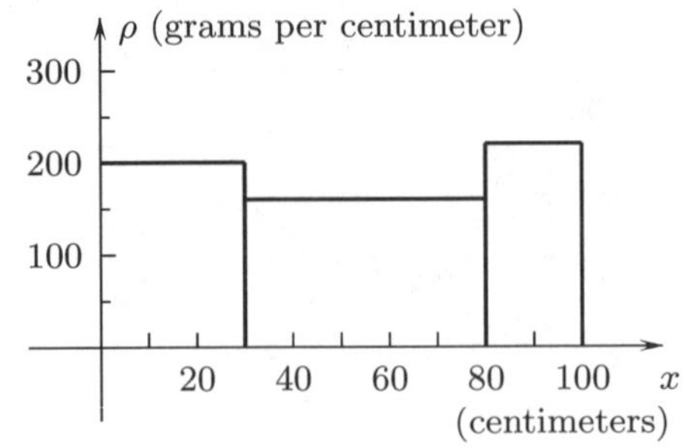

FIGURE 10

The volume of a spa

Figure 11 shows a home spa in the shape of a right circular cylinder of radius 3 feet and height 4 feet beneath a right circular cylinder of radius 5 feet and height 2 feet.

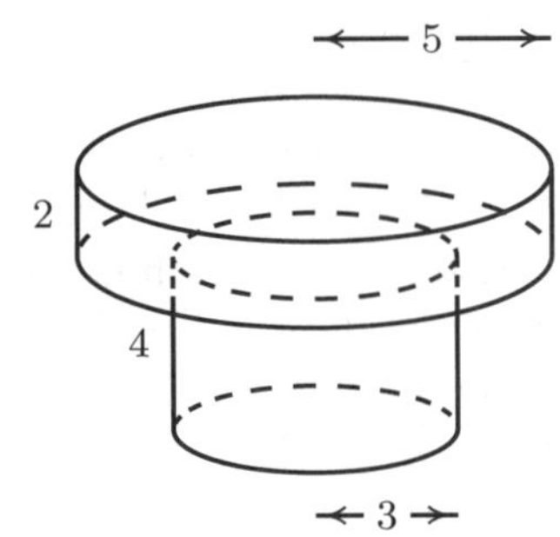

FIGURE 11

Question 6 **(a)** What is the total volume of the spa? (The volume of a cylinder with base a circle of radius r and height h is $\pi r^2 h$.) **(b)** Figure 12 shows the graph of the area $A(x)$ of the horizontal cross section of the jacuzzi as a function of the distance x of the cross section above the bottom. Because the cross section is a circle of radius 3 for $0 < x < 4$ and a circle of radius 5 for $4 < x < 6$, $A(x) = \pi(3)^2 = 9\pi$ for $0 < x < 4$ and $A(x) = \pi(5)^2 = 25\pi$ for $4 < x < 6$. How is the volume of the spa from part (a) related to the areas of the two rectangles in Figure 13?

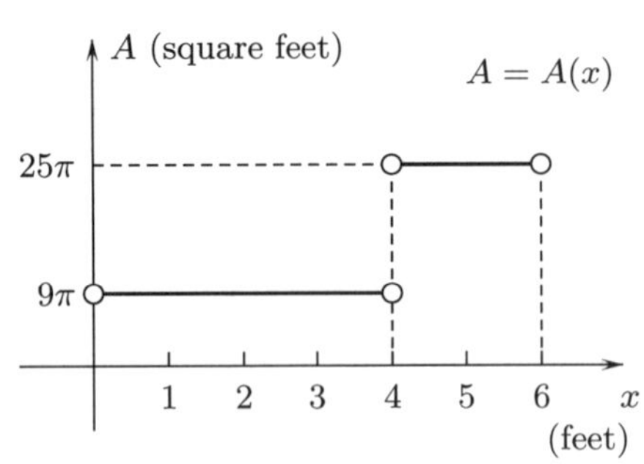

FIGURE 12

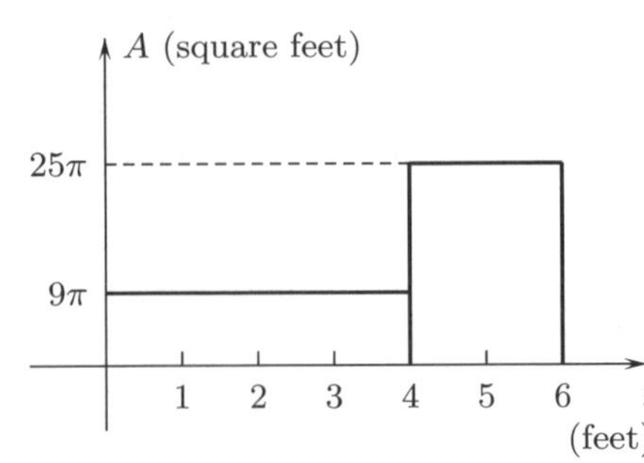

FIGURE 13

An average maximum daily temperature

The graph of the maximum daily temperature $T(t)$ in a northern U. S. city during a week in the winter is shown in Figure 14. The time t is measured in days with $t = 0$ at 12:00 midnight Saturday night. As you can see from the graph, the maximum daily temperature was 20°F on Sunday and Monday, 10°F on Tuesday, −20°F on Wednesday and Thursday, and 40°F on Friday and Saturday.

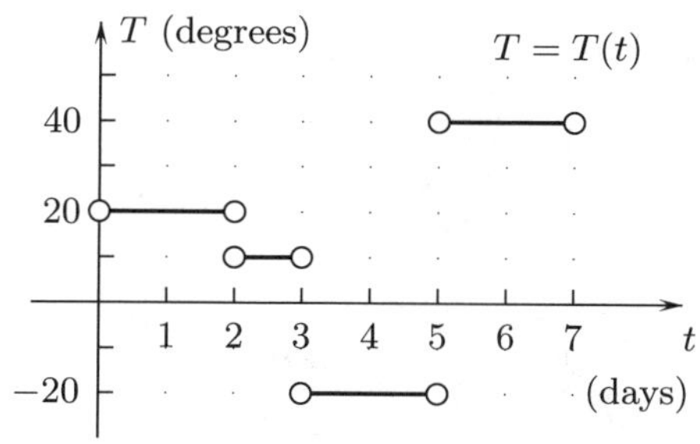

FIGURE 14

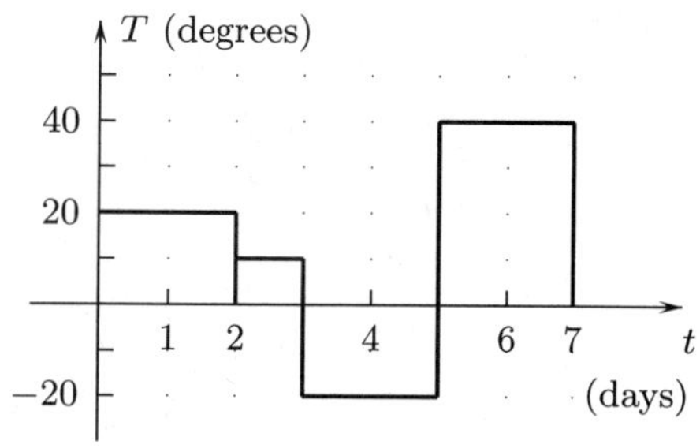

FIGURE 15

The maximum daily temperature was constant during each day. It was 20°F for two days, 10°F for one day, −20°F for two days, and 40°F for two days, so the average maximum temperature over the seven days was

$$\tfrac{1}{7}[20(2) + 10(1) + (-20)(2) + 40(2)] = 12\tfrac{6}{7} \text{ degrees Fahrenheit.}$$

Question 7 How is the average maximum daily temperature during the week related to the areas of the four rectangles in Figure 15?

The length of a curve

Because the graph of the piecewise linear function $f(x)$ in Figure 16 consists of three line segments, we can determine its total length by using the Pythagorean Theorem to find the lengths of each of the pieces.

FIGURE 16

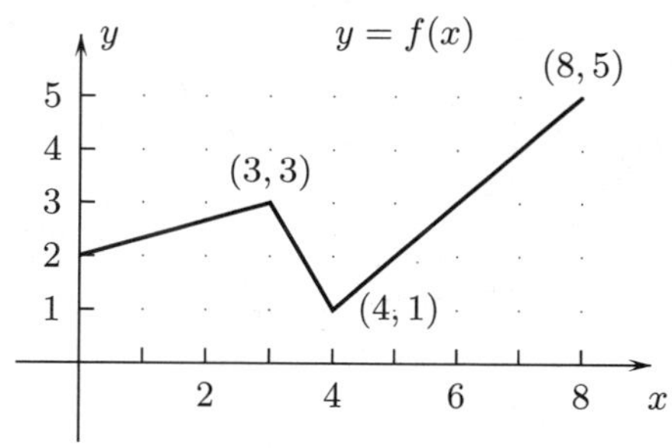

Example 2 What is the length of the graph of $f(x)$ in Figure 16?

SOLUTION Because the first line segment runs from $(0, 2)$ to $(3, 3)$, its length is $\sqrt{(3-0)^2 + (3-2)^2} = \sqrt{10}$. The second line segment runs from $(3, 3)$ to $(4, 1)$, and its length is $\sqrt{(4-3)^2 + (1-3)^2} = \sqrt{5}$. The third line segment goes from $(4, 1)$ to $(8, 5)$, and its length is $\sqrt{(8-4)^2 + (5-1)^2} = \sqrt{32} = 4\sqrt{2}$. Therefore, the total length of the graph of $f(x)$ in Figure 16 is $\sqrt{10} + \sqrt{5} + 4\sqrt{2}$. □

To have the solution of Example 2 fit the pattern of the other applications in this section, we need to express the answer in terms of areas of rectangles. We can accomplish this by expressing the lengths of the line segments in terms of their slopes.

Consider a line segment of slope m that extends from $x = a$ to $x = b$, as in Figure 17. The run from its left end to its right end is $\Delta x = b - a$ and the rise is $\Delta y = m\Delta x$, so its length is

$$\begin{aligned}\sqrt{(\Delta x)^2 + (\Delta y)^2} &= \sqrt{(\Delta x)^2 + (m\Delta x)^2} = \sqrt{(\Delta x)^2(1 + m^2)} \\ &= \sqrt{1 + m^2}\,\Delta x.\end{aligned} \tag{1}$$

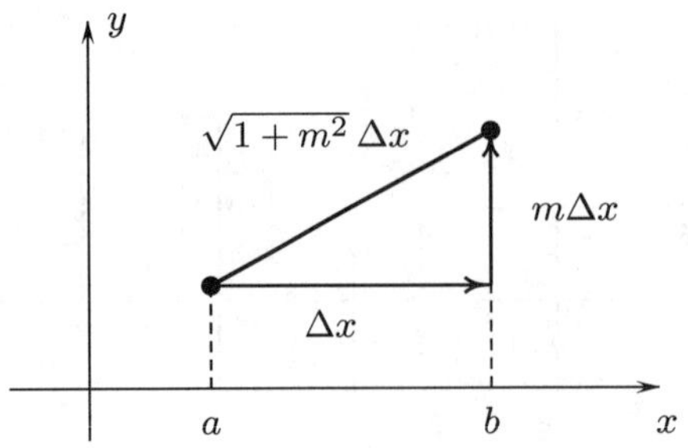

[Length] $= \sqrt{1+m^2}\,\Delta x$

FIGURE 17

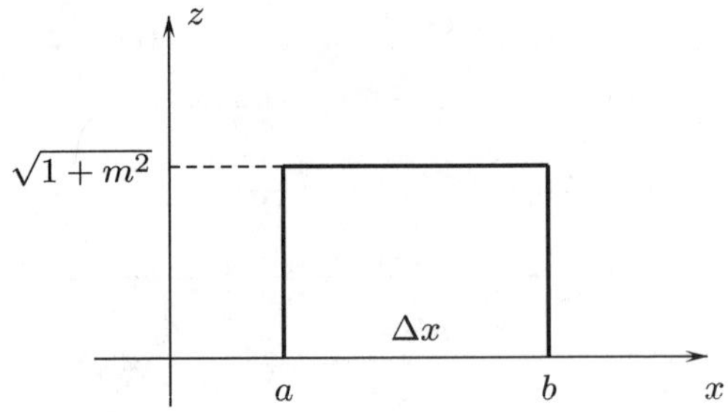

[Area] $= \sqrt{1+m^2}\,\Delta x$

FIGURE 18

Then, because the length **(1)** is the product of $\sqrt{1+m^2}$ and Δx, it is the area of the rectangle in Figure 18 whose height is $\sqrt{1+m^2}$ and whose width is Δx. This principle is applied in the next example to the graph of $f(x)$ in Figure 16, where the slope of each piece of the graph is the derivative $f'(x)$ for the corresponding values of x.

Example 3 **(a)** Find the values of $f'(x)$ for $0 < x < 3$, for $3 < x < 4$, and for $4 < x < 8$ for $f(x)$ of Figure 16. Then show that the values of $\sqrt{1+[f'(x)]^2}$ are the numbers on the z-axis with its graph in Figure 19. **(b)** How is the length $\sqrt{10}+\sqrt{5}+4\sqrt{2}$ of the graph of $f(x)$ from Example 2 related to the three rectangles in Figure 20?

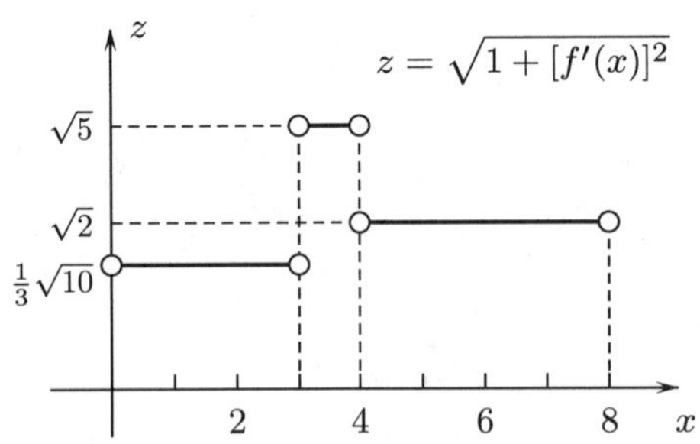

FIGURE 19

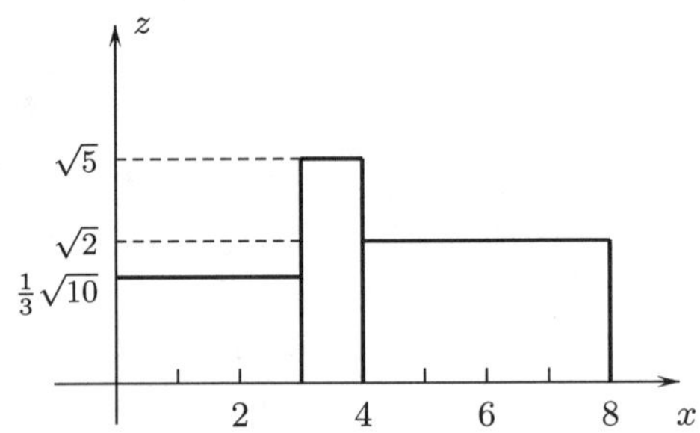

FIGURE 20

SOLUTION **(a)** For the function $f(x)$ of Figure 16,

$$f'(x) = \begin{cases} \frac{1}{3} & \text{for} \quad 0 < x < 3 \\ -2 & \text{for} \quad 3 < x < 4 \\ 1 & \text{for} \quad 4 < x < 8 \end{cases}$$

and

$$\sqrt{1+[f'(x)]^2} = \begin{cases} \sqrt{1+\left(\frac{1}{3}\right)^2} = \frac{1}{3}\sqrt{10} & \text{for} \quad 0 < x < 3 \\ \sqrt{1+(-2)^2} = \sqrt{5} & \text{for} \quad 3 < x < 4 \\ \sqrt{1+1^2} = \sqrt{2} & \text{for} \quad 4 < x < 8. \end{cases}$$

(b) The total area of the three rectangles is

$$\left(\tfrac{1}{3}\sqrt{10}\right)(3) + (\sqrt{5})(1) + (\sqrt{2})(4) = \sqrt{10}+\sqrt{5}+4\sqrt{2}$$

and equals the length of the graph of $f(x)$. □

Principles and procedures

- This section is a preview of the applications to be discussed in later sections, where we will see that the types of problems that are solved here by finding areas of rectangles are usually solved with integrals. This section also provides background for the definition in Section 4.2 of the DEFINITE INTEGRAL, which is based on approximations of curved regions by collections of rectangles. Keep this section in mind as you study the rest of this chapter, and review this material later to clarify your understanding of the underlying ideas.
- If $f(t)$ is continuous and piecewise linear for $a \le t \le b$, then $f'(t)$ is a step function, and the change in $f(t)$ from a to b equals the total area of the rectangles below the graph of $f'(t)$ and above the t-axis in the portion of $[a, b]$ where $f'(t)$ is positive, minus the total area of the rectangles above the graph of $f'(t)$ and below the t-axis where $f'(t)$ is negative.
- If an object, such as a stick or a rod, extends from a to b on a x-axis and its density per unit length $\rho(x)$ is a step function for $a \le x \le b$, then the object's weight equals the total area of the rectangles between the x-axis and the graph of the density for $a \le x \le b$.
- If a solid extends from $x = a$ to $x = b$ on an x-axis with $a < b$ and if the area $A(x)$ of its cross section perpendicular to the x-axis at x is a step function, then the volume of the solid equals the total area of the rectangles between the graph of $A(x)$ and the x-axis for $a \le x \le b$.
- To find the average value of a step function in an interval, multiply each of its values by the length of the subinterval where it has that value, add the results, and divide by the length of the entire interval being considered.
- If $f(x)$ is piecewise linear, then $\sqrt{1 + [f'(x)]^2}$ is a step function, and the length of the graph of $f(x)$ for $a \le x \le b$ equals the area of the rectangles between the graph of $\sqrt{1 + [f'(x)]^2}$ and the x-axis for $a \le x \le b$.

Tune-Up Exercises 4.1♦

[A]*Answer provided.* [O]*Outline of solution provided.*

T1.[O] The step function $v(t)$ of Figure 21 gives velocity in the positive direction at time t of an object as it moves on an s-axis. The scale on the s-axis is given in feet, t is measured in minutes, and v is measured in feet per minute. The object is at $s = 300$ at $t = 10$. Where is it at $t = 60$?

♦ Type 1, distance from a velocity step function

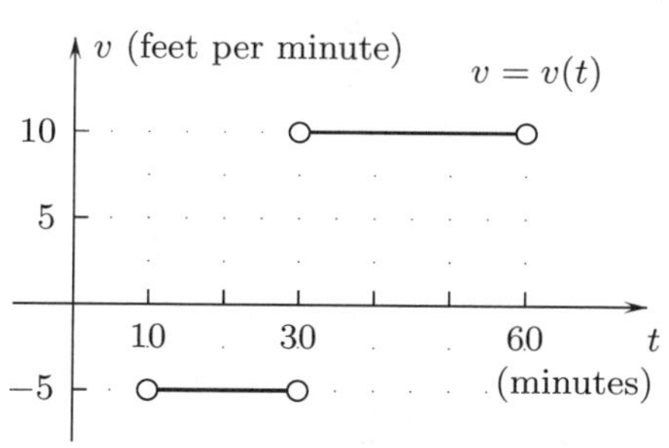

FIGURE 21

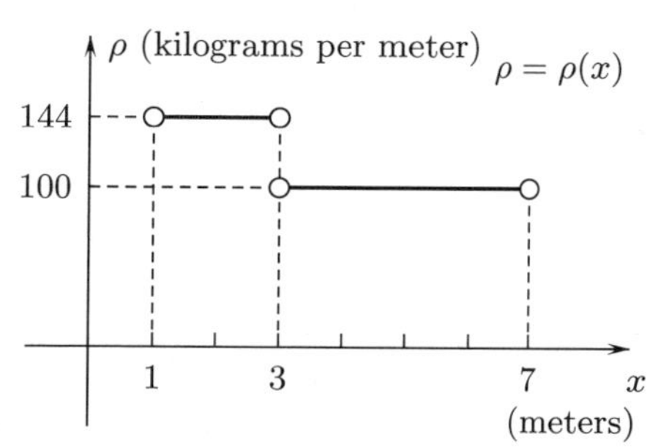

FIGURE 22

T2.[O] A rod extends from $x = 1$ to $x = 7$ (meters) on an x-axis, and its density $\rho(x)$ (kilograms per meter) at x is given by the step function of Figure 22. **(a)** How much does the rod weigh, measured in kilograms? **(b)** What is its average density for $1 \le x \le 7$?

♦ Type 1, weight from a density step function

♦The Tune-up Exercises and Problems are classified by type and content. The types are (1) basic, reactive; (2) basic reflective; (3) intermediate, reactive; (4) intermediate, reflective; (5) advanced, reactive; (6) advanced, reflective; and (7) advanced, theoretical.

T3.[A] A sculpture consists of two rectangular blocks of granite, one on top of the other. The lower block is five meters high and has a square base three meters wide. The upper block is six meters high and has a square base two meters wide. The graph of the area $A(x)$ (square meters) of the horizontal cross section of the sculpture as a function of its distance x above the ground is the step function of Figure 23. **(a)** What are the values of the numbers a and b on the A-axis in Figure 23? **(b)** What is the total volume of the sculpture?

♦ Type 1, volume from a step function of cross-sectional area

T4.[O] Figure 24 shows the graph of a step function $f(x)$. What is its average value for $100 \le x \le 800$?

♦ Type 1, average value of a step function

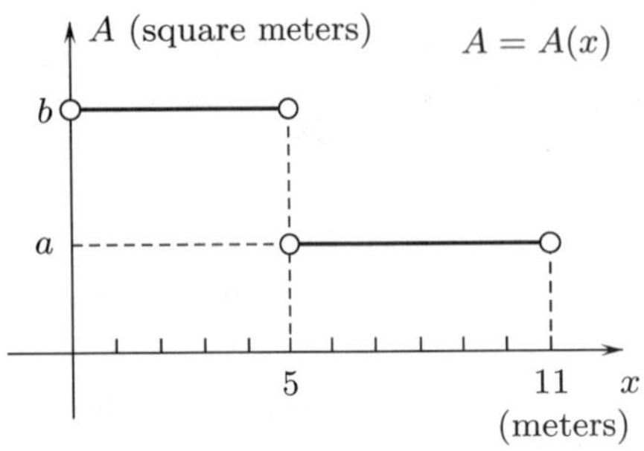

FIGURE 23

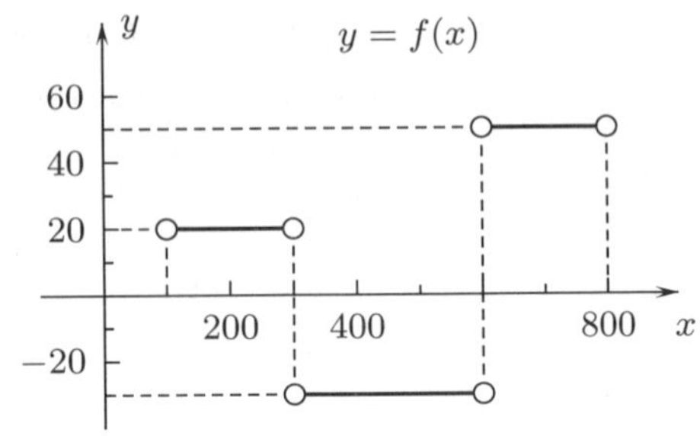

FIGURE 24

T5.[O] The graph of a function $f(x)$ consists of a line of slope 5 for $1 \le x \le 6$ and a line of slope -3 for $6 \le x \le 9$. The graph of $\sqrt{1 + [f'(x)]^2}$ is the step function of Figure 25. What are the numbers a and b on the z-axis? What is the exact length of the graph of $f(x)$ for $1 \le x \le 9$?

♦ Type 1, length of a piecewise-linear graph

FIGURE 25

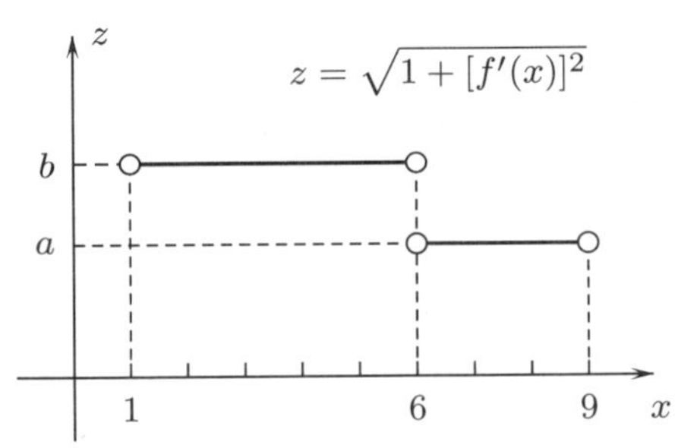

Problems 4.1

[A] *Answer provided.* [O] *Outline of solution provided.*

1.[O] World-class jogger Ranier Schein runs eight miles between 8:30 AM and 10:00 AM. He then runs 6 miles per hour between 10:00 AM and 12:00 PM and 4 miles per hour between 12:00 PM and 1:30 PM. **(a)** How far does he run between 8:30 AM and 1:30 PM? **(b)** What is his average velocity from 8:30 AM to 1:30 PM?

♦ Type 1, distance from a velocity step function

2.[A] The step function of Figure 26 is a mathematical model of the rate of rainfall $r(t)$ (inches per year) in Los Angeles from the beginning of 1881 to the beginning of 1886.[(2)] **(a)** Based on this data, what was the relative change in annual rainfall in Los Angeles from the year 1883 to year 1884 and from the year 1884 to the year 1885? **(b)** What was the total rainfall in Los Angeles in the years 1881 through 1885? **(c)** What was the average annual rainfall in Los Angeles in the years 1881 through 1885 and how did it compare with the average annual rainfall of 14.9 inches per year from 1878 through 1992?

♦ Type 1, total change from a rate-of-change step function

[(2)] Data adapted from *Los Angeles Times*, January 7, 1993, Source: National Weather Service.

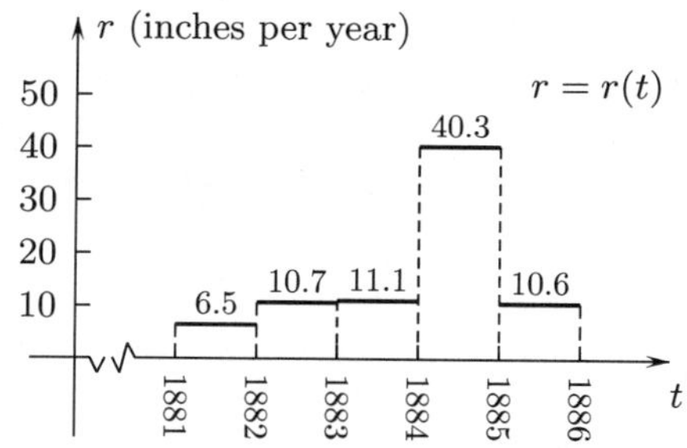

FIGURE 26

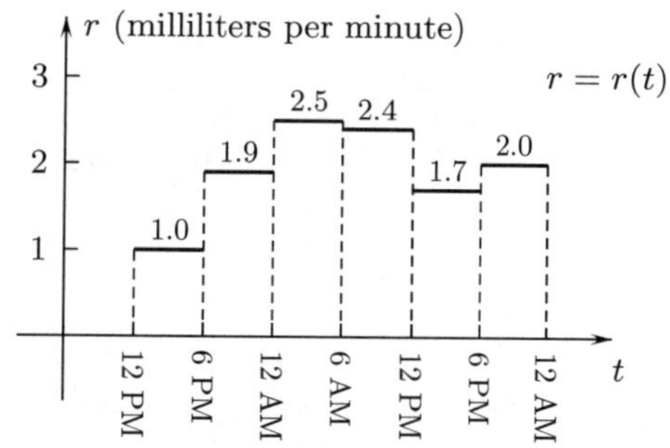

FIGURE 27

3. A few plants, including jack in the pulpit, skunk cabbage, and sacred lotus, can create heat from oxygen and other nutrients. The step function in Figure 27 is a model of the rate of consumption of oxygen by a sacred lotus plant from noon one day to midnight 36 hours later. The flower kept its temperature between 30°C and 37°C as the air temperature varied between 10°C and 35°C by consuming more oxygen in the colder periods.[3] How much oxygen did the flower consume and what was its average rate of consumption during the 36 hours? (Be careful with the units.)

♦ Type 1, total change from a-rate-of-change step function

4. A kite that is initially 110 feet above the ground is allowed to rise at the constant rate of 60 feet per minute for ninety seconds and then is pulled down so its height decreases at the constant rate of 100 feet per minute until it hits the ground. **(a)** Sketch the graph of the kite's upward velocity, measured in feet per minute, as a function of time, measured in minutes. **(b)** What is its maximum height above the ground? **(c)** When does it hit the ground?

♦ Type 1, distance from a velocity step function

5.[A] Figure 28 is a sketch of an office building, and Figure 29 shows the graph of the area $A(x)$ of the east-west, vertical cross section of the building x feet from the south end. Notice that the vertical cross section is a 30-foot-wide square for $0 < x < 60$, a 60-foot-wide square for $60 < x < 100$, and an 80-foot-wide square for $100 < x < 160$. **(a)** What is the total volume of the building. **(b)** Give the values a, b, and c of $A(x)$ in Figure 29. **(c)** The region between the graph of $A(x)$ and the x-axis for $0 \le x \le 160$ consists of three rectangles. How are their areas related to the volume of the building?

♦ Type 1, volume from a step function of cross-sectional area

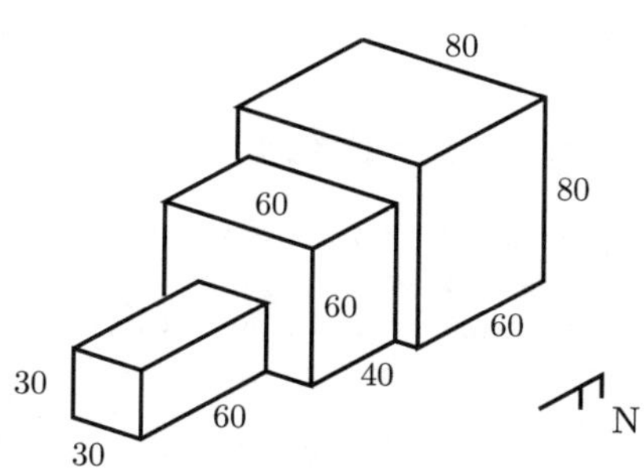

FIGURE 28

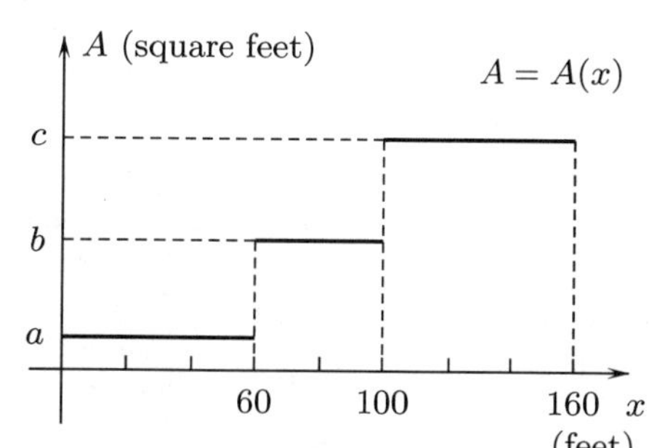

FIGURE 29

6.[A] Madeline earned \$40,000 per year for six months, \$45,000 per year for eighteen months, and then \$50,000 per year for two years. What was her average salary over the entire four year period?

♦ Type 1, average value of a step function

[3] Data adapted from "Plants that warm themselves" by R. Seymour, *Scientific American*, March, 1997.

7. The weights at the ends of a bar bell are 10-inch long, circular cylinders of four-inch radius. The bar between them is 30 inches long and has a one-inch radius. Consequently, the cross section of the bar bell at a distance x inches from one end is a circle with a four-inch radius for $0 < x < 10$, a circle with a one-inch radius for $10 < x < 50$, and a circle with a four-inch radius for $50 < x < 60$ **(a)** Draw the graph of the area $A(x)$ of the cross section of the bar bell as a function of the distance x from one end. **(b)** What is the total volume of the bar bell and how is it related to the graph from part (a)? **(c)** How much does the bar bell weigh if it is made of steel with density 0.273 pounds per cubic inch?[4]

♦ Type 1, weight from a density step function

8. The GRADE of a highway is its slope given as a percentage. A highway has a constant grade of 5% and extends for 10 miles, measured horizontally. How long is it, measured along its surface?

♦ Type 1, length of a road from a grade step function

9.[A] The graph in Figure 30 consists of a line of slope m_1 for $0 \le x \le 2$ and a line of slope m_2 for $2 \le x \le 5$. Express the length of the graph in terms of m_1 and m_2.

♦ Type 1, length of a piecewise-linear graph

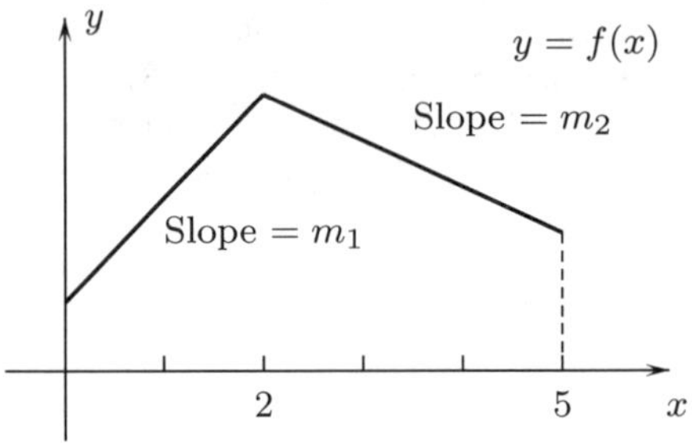

FIGURE 30

10. The step function $r(t)$ defined below gives the approximate rate of U.S. gasoline consumption, measured in millions of barrels per day, from the beginning of 1975 to the beginning of 1995.[5] Based on this data, how much more gasoline was consumed in the U.S in the time period $1985 \le t \le 1995$ than in the time period $1975 \le t \le 1985$? (Disregard leap years and be careful with the units.)

$$\mathrm{r(t)} = \begin{cases} 6.2 & \text{for} \quad 1975 \le t < 1980 \\ 5.8 & \text{for} \quad 1980 \le t < 1985 \\ 6.3 & \text{for} \quad 1985 \le t < 1990 \\ 6.5 & \text{for} \quad 1990 \le t < 1995 \end{cases}$$

♦ Type 1, total change from a rate-of-change step function

11. The volcano Mount Tambara in the Dutch East Indies (now Indonesia) erupted in 1815. The resulting cloud of volcanic dust gradually shadowed the higher latitudes and caused the summer of 1816 to be exceptionally cold in New England. It snowed in June and frosts continued through August. Figure 31 is a mathematical model of the average daily temperature in New Haven, Connecticut from the beginning of June 1 through the end of June 22, 1816.[6] What was the average temperature in New Haven during those 22 days?

♦ Type 1, average value of a step function

12. The step function $p(t)$ of Figure 32 is a mathematical model of the power produced in the entire world by wind-driven generators as a function of the year t.[7] A kilowatt of electrical power yields one kilowatt-hour of energy in one hour. Based on the graph, how much energy (kilowatt-hours) was generated by wind-driven generators from the beginning of 1981 to the beginning of 1992, and what was the average power they generated over that time period? Be careful with the units and disregard leap years in your calculation.

♦ Type 1, average value of a step function

---

(4) Data adapted from *Handbook of Engineeering Materials* by D. Miner and J. Seastone, New York, N. Y.: John Wiley & Sons, 1955, p. 2-144.

(5) Data adapted from "Improving automotive efficiency" by J. DeCicco and M. Ross, *Scientific American*, December, 1994, p. 56.

(6) Data adapted from "The year without a summer" by H. Stommel and E. Stommel, *Scientific American*, June, 1979, p. 177.

(7) Data adapted from *Vital Signs*, p. 51.

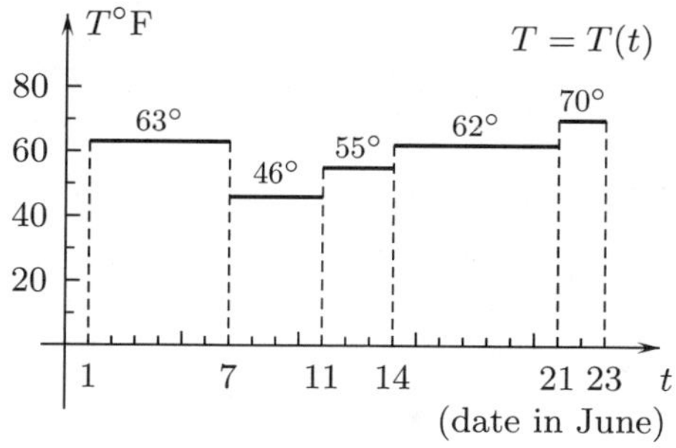

FIGURE 31

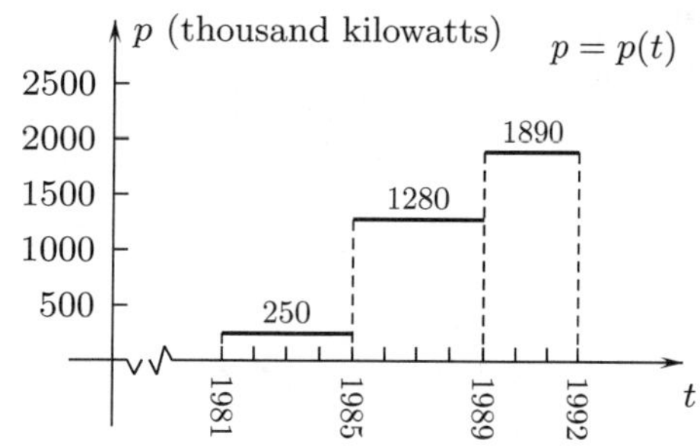

FIGURE 32

13. The functions $r_I(t)$ and $r_E(t)$ of Figures 33 and 34 give the approximate rates of merchandise imports to and exports from the United States as functions of the year t.[8] **(a)** Based on this data, what was the merchandise trade deficit for the time period from 1950 through 1954? (The merchandise trade deficit in a time period is the total value of imports minus the total value of exports during that time.) **(b)** What was the merchandise trade deficit from 1990 through 1994? **(c)** What was the ratio of the merchandise trade deficit to the total value of merchandise imports in the period from 1950 through 1954? **(d)** What was the ratio of the merchandise trade deficit to the total value of merchandise imports in the period from 1990 through 1994?

◆ Type 2, changes from rate-of-change step functions, etc.

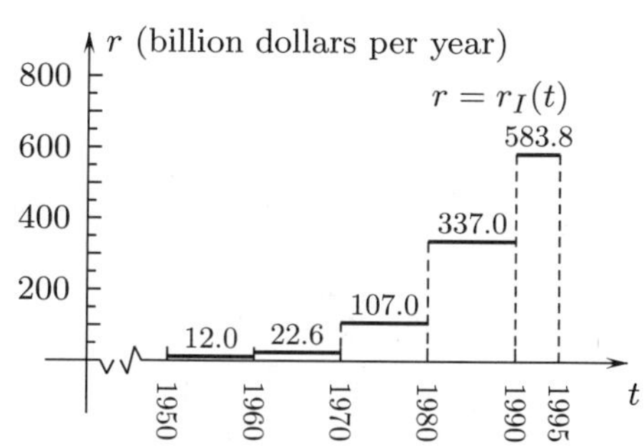

Merchandise imports

FIGURE 33

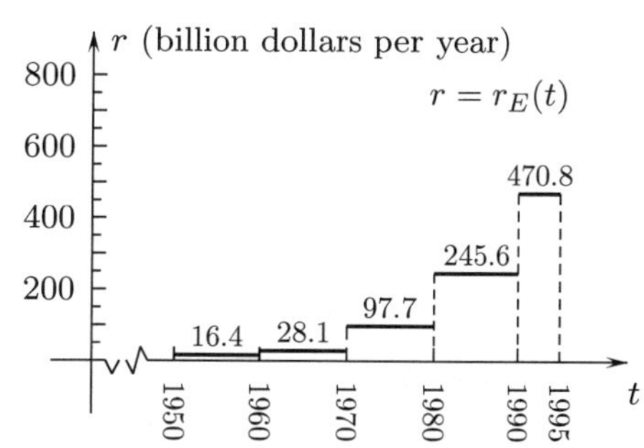

Merchandise exports

FIGURE 34

14. Timber is treated with creosote or other preservatives by first putting the wood in a vacuum cylinder to draw out moisture. Then the cylinder is filled with preservative and put under pressure to force the chemicals into the wood. Finally the preservative is drained from the cylinder and the vacuum reestablised to extract excess preservative from the wood. Figure 35 is the graph of the pressure during one such procedure, with the surrounding atmospheric pressure taken to be zero.[9] What is the average pressure in the cylinder for $0 \le t \le 8$?

◆ Type 1, average value of a step function

[8] Data adapted from *Historical Statistics of the United States*, Part 2, Washington, D. C.: U. S. Department of Commerce, p. 886 and *Statistical Abstracts of the United States, 1996*, Washington, D. C.: U. S. Department of Commerce, p. 796.

[9] Data adapted from *Preservative Treatment of Hardwoods* by W. Thompson and P. Koch, Washington, D. C.: United States Department of Agriculture, 1981, p. 15.

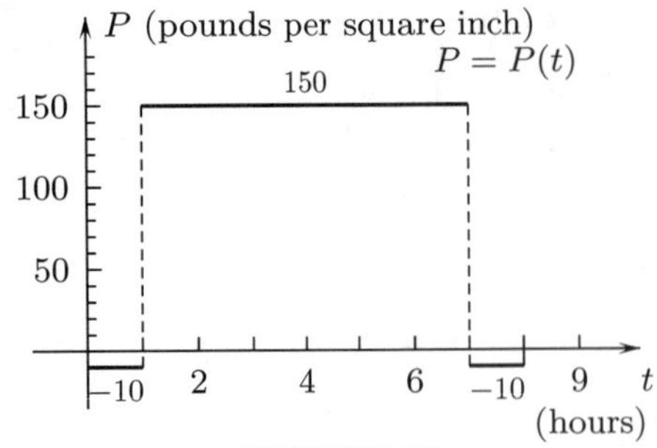

FIGURE 35

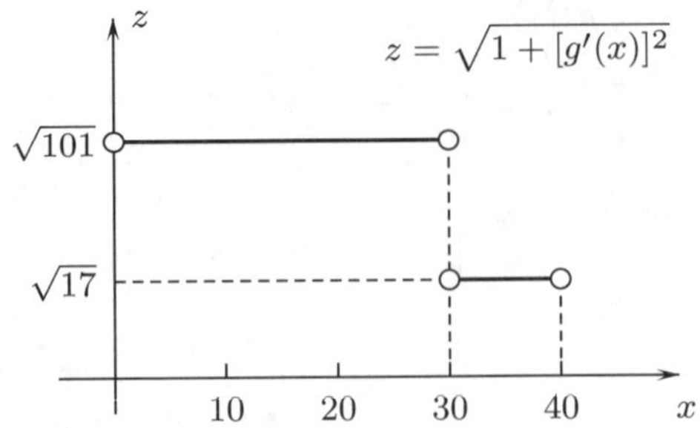

FIGURE 36

15. Figure 36 shows the graph of $\sqrt{1+[g'(x)]^2}$ for a function $g(x)$ whose graph consists of one line segment from $x = 0$ to $x = 30$ and a second line segment from $x = 30$ to $x = 40$. What are the possible slopes of the line segments and how long is the graph of $g(x)$?

◆ Type 1, length of a piecewise-linear graph

16. Figure 37 shows the graph of the rate of flow of water into a tank. The tank contains 100 gallons of water at $t = 0$. When for $0 \le t \le 8$ does it contain 75 gallons?

◆ Type 2, total change from a rate-of-change step function

FIGURE 37

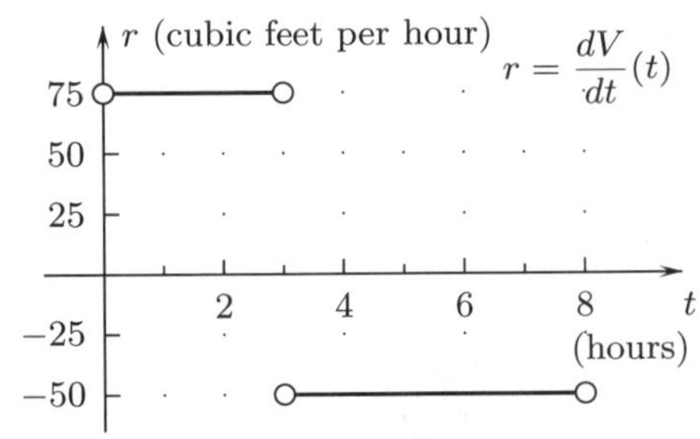

17. The four tiers of a wedding cake are five-inch-high circular cylinders with 3.5 inch, 5 inch, 7 inch, and 9 inch radii. **(a)** What is the (exact) total area of the four horizontal surfaces of the cake that need to be frosted? **(b)** What is the total area of the four curved vertical surfaces of the cake? **(c)** The circumference $C(h)$ (inches) of the circular horizontal cross section as a function of the distance h (inches) from the bottom of the cake is a step function. Sketch its graph. **(d)** The region between the graph of $C(h)$ and the h-axis consists of four rectangles. How is their total area related to the answers to part (b)?

◆ Type 1, volume from a step function of cross-sectional area

Section 4.2

The definite integral

OVERVIEW: *We saw in Section 4.1 how the change in an object's position during a time interval can be calculated from its velocity in cases where the velocity is a step function. In this section we first show how these calculations can be carried out for other types of functions by using step-function approximations. The resulting formula gives a preview of the* FUNDAMENTAL THEOREM OF CALCULUS *and leads to the definition of the* DEFINITE INTEGRAL. *The definition and basic properties of the definite integral are discussed in the remainder of this section. The Fundamental Theorem will be established in Section 4.3.*

Topics:

- ***Finding distance from velocity: a preview of the Fundamental Theorem***
- ***Riemann sums and the definite (Riemann) integral***
- ***The definite integral and areas***
- ***Definite integrals of linear combinations and over adjacent intervals***

Finding distance from velocity: a preview

Because a complete discussion of the definitions and results of this section requires considerable special notation and terminology, we begin with a preview of the main ideas.

Question 1 of Section 4.1 dealt with a truck moving on an east-west road whose velocity toward the east was given by the step function $v(t)$ of Figure 1. The truck travels seventy-five miles per hour to the east from $t = 0$ to $t = 2$, fifty miles per hour to the west from $t = 2$ to $t = 4$, and fifty miles per hour to the east from $t = 4$ to $t = 6$. As we saw in Section 4.1, the change in the truck's position from $t = 0$ to $t = 6$ is

$$\left[75\,\frac{\text{miles}}{\text{hour}}\right][2\text{ hours}] - \left[50\,\frac{\text{miles}}{\text{hour}}\right][2\text{ hours}] + \left[50\,\frac{\text{miles}}{\text{hour}}\right][2\text{ hours}]$$
$$= 75(2) - 50(2) + 50(2) = 150 \text{ miles.} \tag{1}$$

At $t = 6$, the truck is 150 miles to the east of its position at $t = 0$.

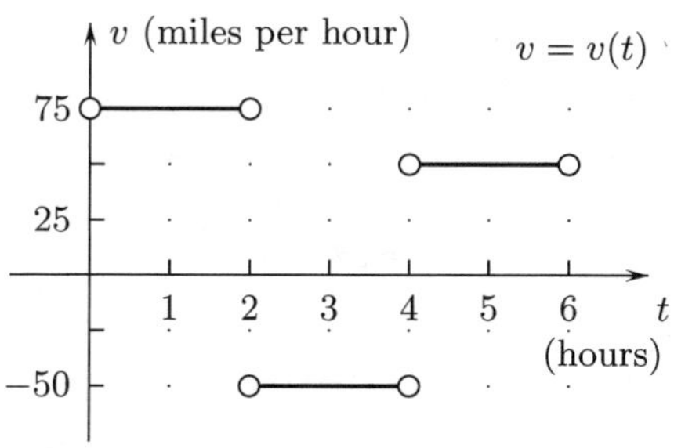

FIGURE 1

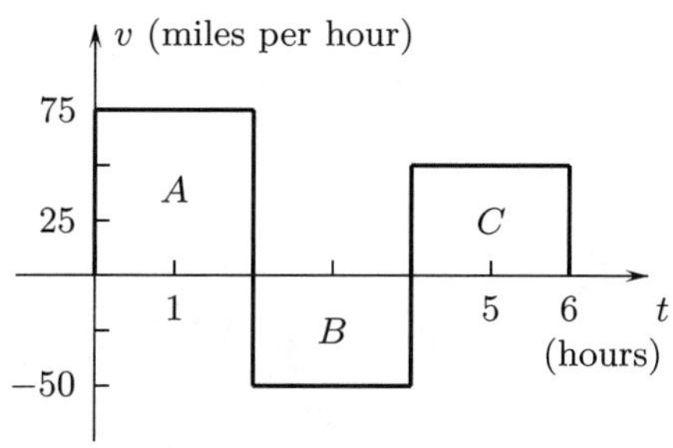

FIGURE 2

♦ SUGGESTIONS TO INSTRUCTORS: The main topics in this section can be divided into two parts, as follows:
Part 1 (Summation notation, Riemann sums, the definition of the definite integral, definite integrals in terms of areas; Tune-up Exercises T1 through T7, Problems 1–15) Have students begin work on Questions 1 through 4 before class. Problems 3, 4, 8, 9, 12, and 14 are good for in-class discussion.
Part 2 (The Riemann-sum program, integrals over adjacent intervals and of linear combinations, evaluating integrals with summation formulas; Tune-up Exercises T8–T10, Problems 16–31) Have students begin work on Questions 5 through 8 before class. Problems 16, 19, 22, 24, and 27 are good for in-class discussion.

The calculations in **(1)** show that the change in the truck's position equals the area of rectangles A and C above the t-axis in Figure 2, minus the area of rectangle B below the t-axis. We express this result as follows:

$$\begin{aligned}&[\text{The change in the truck's position}]\\ &\qquad = \begin{bmatrix}\text{The area between the}\\ \text{graph of } v(t) \text{ and the}\\ t\text{-axis where } v(t) > 0\end{bmatrix} - \begin{bmatrix}\text{The area between the}\\ \text{graph of } v(t) \text{ and the}\\ t\text{-axis where } v(t) < 0\end{bmatrix}.\end{aligned} \qquad \textbf{(2)}$$

As you saw with other examples in Section 4.1, similar calculations would lead to equation **(2)** whenever the velocity is a step function. What happens if the truck's velocity, like the continuous function $v(t)$ of Figure 3, is not a step function? In this case, we can determine the change in the truck's position by a process of approximating $v(t)$ by step functions.

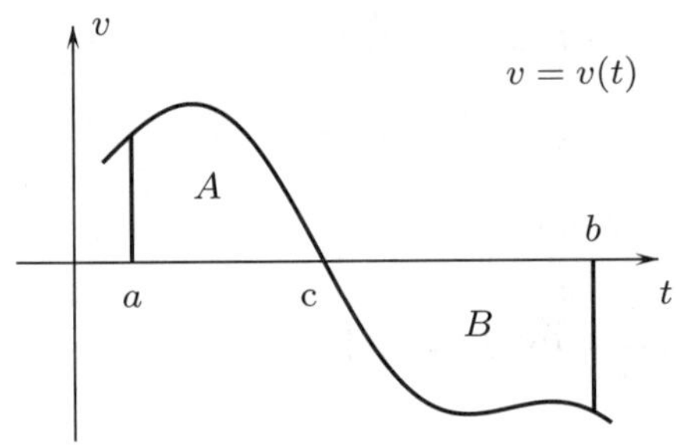

FIGURE 3

Figure 4 shows rectangles that are obtained when we approximate $v(t)$ by a step function whose graph consists of five horizontal lines of equal length that intersect the graph of $v(t)$ at their midpoints. The two rectangles above the t-axis approximate the region A in Figure 3 corresponding to the time period $a \le t \le c$ when the truck's velocity is positive and the three rectangles below the t-axis approximate the region B corresponding to the time period $c \le t \le b$ when the velocity is negative. Because the step function approximates $v(t)$, the change in the truck's position with velocity $v(t)$ is approximately equal to the change with velocity given by the step function. The latter equals the area of the rectangles above the t-axis minus the area of the rectangles below the t-axis, so that

$$\begin{aligned}&[\text{The change in the truck's position from time } a \text{ to time } b]\\ &\qquad \approx \begin{bmatrix}\text{The area of}\\ \text{the rectangles}\\ \text{above the } t\text{-axis}\end{bmatrix} - \begin{bmatrix}\text{The area of}\\ \text{the rectangles}\\ \text{below the } t\text{-axis}\end{bmatrix}.\end{aligned} \qquad \textbf{(3)}$$

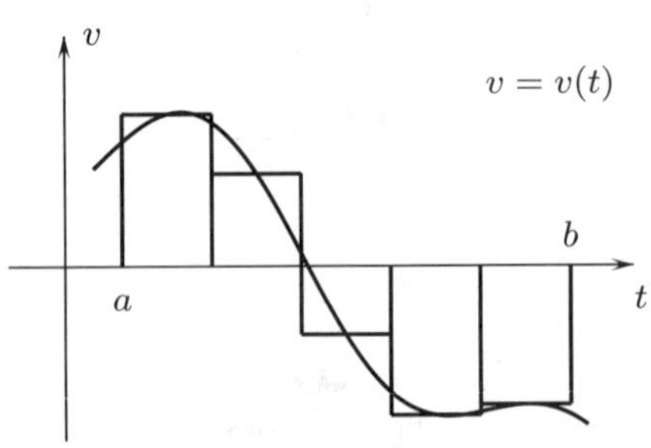

FIGURE 4

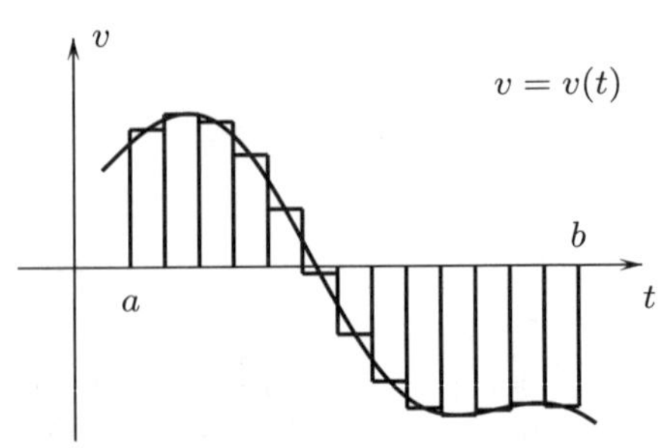

FIGURE 5

Figure 5 shows an approximation by a step function that gives rise to thirteen rectangles, where the five rectangles above the t-axis approximate region A and the eight rectangles below the t-axis approximate region B. This gives a closer approximation of $v(t)$ than that in Figure 4, so estimate **(3)** is more accurate with this step function than with the step function of Figure 4.

To obtain exact results, we let the number of rectangles tend to infinity and their widths tend to zero. As you can imagine, the combined areas of the rectangles determined by the graph of the continuous function $f(x)$ approach the areas of regions A and B in Figure 3, and the change in position with velocity given by the step function approaches the change in position of the truck with velocity given by $v(t)$. Consequently, we can expect that

$$\begin{aligned}&[\text{The change in the truck's position from time } a \text{ to time } b]\\ &= \lim\left\{\begin{bmatrix}\text{The area of}\\ \text{the rectangles}\\ \text{above the } t\text{-axis}\end{bmatrix} - \begin{bmatrix}\text{The area of}\\ \text{the rectangles}\\ \text{below the } t\text{-axis}\end{bmatrix}\right\}\\ &= \begin{bmatrix}\text{The area of region}\\ A \text{ in Figure 3}\end{bmatrix} - \begin{bmatrix}\text{The area of region}\\ B \text{ in Figure 3}\end{bmatrix}\end{aligned} \quad \textbf{(4)}$$

where the limit is taken as the number of rectangles in the approximation tends to ∞ and their widths tend to 0. We will give a more detailed proof of this result in Section 4.3

Later in this section we will define the DEFINITE INTEGRAL of $v(t)$ from $t = a$ to $t = b$, which is denoted by the symbol $\int_a^b v(t)\,dt$, to be equal to the difference of areas on the last line of **(4)**. With this notation, **(4)** gives

$$[\text{The change in the truck's position from time } a \text{ to time } b] = \int_a^b v(t)\,dt. \quad \textbf{(5)}$$

Equation **(5)** is an example of the FUNDAMENTAL THEOREM OF CALCULUS, which we will return to in Section 4.3. Here we now introduce terminology and notation to describe the approximations by step functions that will be used in the general definition of the definite integral.

Partitions, Riemann sums, and the definite integral

We will use $f(x)$ in place of $v(t)$ in defining the definite integral because we will use integrals for many purposes in addition to studying velocity. We want to define the integral from a to b of the function $f(x)$ of Figure 6 so that it is equal to the area of region A minus the area of region B in the drawing. We will accomplish this by defining the integral to be the limit of the area of rectangles above the x-axis that approximate A, minus the area of rectangles below the x-axis that approximate B.

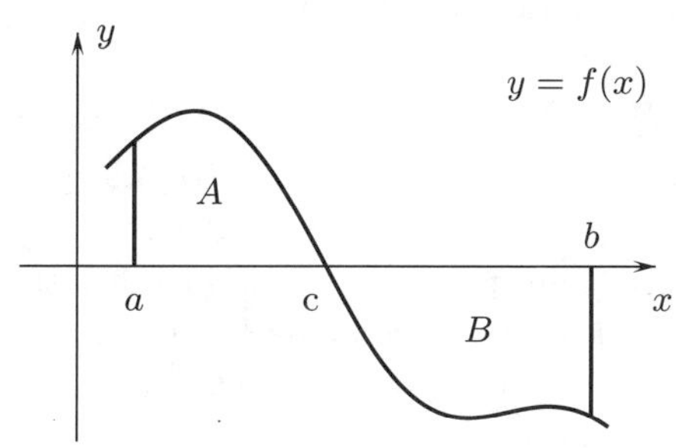

FIGURE 6

The notation we will use in the definition is illustrated in Figure 7, which shows five rectangles determined by a step function that approximates $f(x)$ for $a \le x \le b$. We let $x_0, x_1, x_2, x_3, x_4, x_5$ denote the x-coordinates of the sides of the rectangles, labeled from left to right, as in Figure 7. Then $x_0 = a$, $x_5 = b$, and the other numbers are such that

$$a = x_0 < x_1 < x_2 < x_3 < x_4 < x_5 = b. \quad \textbf{(6)}$$

The collection of points **(6)** is called a PARTITION of the interval $[a, b]$ into five subintervals.

Question 1 What are the five subintervals of the partition **(6)**, and how wide is each of them?

We can refer to the numbers $x_0, x_1, x_2, x_3, x_4, x_5$ in the partition **(6)** as the numbers x_j, where the subscript j is a variable that takes the values $0, 1, 2, 3, 4, 5$. Then the results of Question 1 show that for $j = 1, 2, 3, 4, 5$, the jth subinterval is $[x_{j-1}, x_j]$ and its width is $x_j - x_{j-1}$. We denote this width by the symbol Δx_j:

$$[\text{The width of the } j\text{th subinterval}] = \Delta x_j = x_j - x_{j-1}. \tag{7}$$

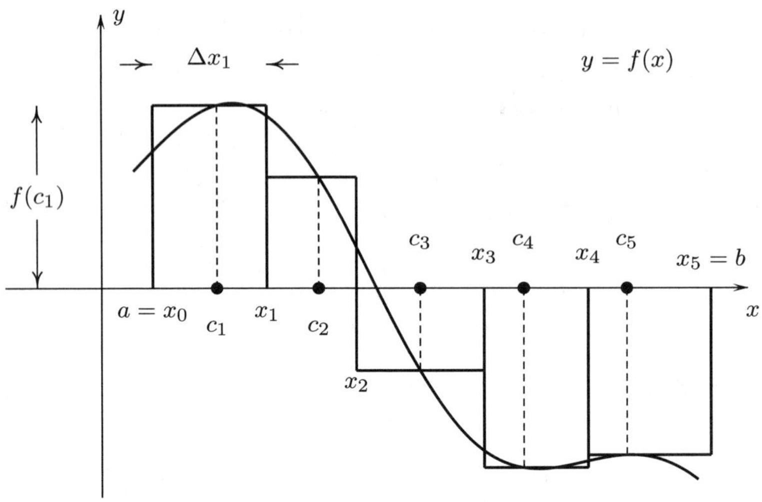

FIGURE 7

As is shown in Figure 7, we let c_1, c_2, c_3, c_4, and c_5 denote values of x where the graph of the step function intersects the graph of $f(x)$. Then the height of the first rectangle in Figure 7 is $f(c_1)$ and its width is Δx_1, so its area is $f(c_1)\Delta x_1$.

Question 2 Give formulas for the areas of the other four rectangles in Figure 7. (The answers should all be positive.)†

The approximation of the integral $\int_a^b f(x)\, dx$ that is given by the step function in Figure 7 equals the area of the two rectangles above the x-axis, minus the areas of the three rectangles below the x-axis. With the formulas for these areas from Question 2 and the statement before it, we find that the approximation of the integral is

$$\begin{aligned}&[f(c_1)\Delta x_1 + f(c_2)\Delta x_2] - [(-f(c_3)\Delta x_3) + (-f(c_4)\Delta x_4) + (-f(c_5)\Delta x_5)]\\ &\qquad = f(c_1)\Delta x_1 + f(c_2)\Delta x_2 + f(c_3)\Delta x_3 + f(c_4)\Delta x_4 + f(c_5)\Delta x_5.\end{aligned} \tag{8}$$

Notice that there are no minus signs in the last formula. The quantity **(8)** equals the areas of the first two rectangles minus the areas of the last three rectangles because $f(c_1)$ and $f(c_2)$ are positive whereas $f(c_3), f(c_4)$, and $f(c_5)$ are negative.

The sum **(8)** is called the RIEMANN SUM‡ for the integral $\int_a^b f(x)\, dx$ corresponding to the partition **(6)** and the choice of the numbers c_j in Figure 7.

†Some calculus textbooks speak of "signed areas," which can be positive, negative, or zero. We prefer to consider all areas to be ≥ 0.

‡Named after the German mathematician G. F. B. Riemann (1826–1866), who formulated the modern definition of the integral.

To use N rectangles, where N is an arbitrary positive integer, we let

$$a = x_0 < x_1 < x_2 < \cdots < x_{N-1} < x_N = b \tag{9}$$

be the partition that determines the sides of the rectangles. Then the Riemann sum is

$$f(c_1)\Delta x_1 + f(c_2)\Delta x_2 + \cdots + f(c_{N-1})\Delta x_{N-1} + f(c_N)\Delta x_N \tag{10}$$

where $x_{j-1} \le c_j \le x_j$ for $j = 1, 2, 3, \ldots, N$.

Look at the the pattern in **(10)**. All the terms are the same, except for the subscripts, which run from 1 to N. Consequently, we can express the sum more concisely by using SUMMATION NOTATION. The following example and question show how this notation works.

Example 1 **(a)** Write out the sum $\sum_{j=1}^{6} j^2$. Do not perform the calulations.

(b) Express $\frac{1}{3}+\frac{1}{4}+\frac{1}{5}+\cdots+\frac{1}{N}$ with summation notation, where N is an unspecified integer ≥ 3.

SOLUTION **(a)** $\sum_{j=1}^{6} j^2$ denotes the sum of the numbers j^2 with $j = 1, 2, 3, 4, 5,$ and 6. Consequently, it equals $1^2 + 2^2 + 3^2 + 4^2 + 5^2 + 6^2$.

(b) $\frac{1}{3}+\frac{1}{4}+\frac{1}{5}+\cdots+\frac{1}{N} = \sum_{j=3}^{N} \frac{1}{j}$. □

Question 3 **(a)** Write out the sum $\sum_{j=1}^{4} \sqrt{x_j}\,\Delta x_j$. **(b)** Express $6(x_1)^3\Delta x_1 + 6(x_2)^3\Delta x_2 + 6(x_3)^3\Delta x_3 + \cdots + 6(x_N)^3\Delta x_N$ with summation notation.

Since the Riemann sum **(10)** is the sum of the quantities $f(c_j)\Delta x$ with j running from 1 to N, we are led to the following definition.

Definition 1 *For a function $f(x)$ defined at all or at all but a finite number of points in the interval $[a, b]$, a Riemann sum for $\int_a^b f(x)\,dx$ corresponding to the partition $a = x_0 < x_1 < x_2 < \cdots < x_{N-1} < x_N = b$ is a sum*

$$\sum_{j=1}^{N} f(c_j)\Delta x_j = f(c_1)\Delta x_1 + f(c_2)\Delta x_2 + \cdots + f(c_{N-1})\Delta x_{N-1} + f(c_N)\Delta x_N \tag{11}$$

where for $j = 1, 2, 3, \ldots, N$, $\Delta x_j = x_j - x_{j-1}$ is the width of the jth subinterval, and c_j is a point in the jth subinterval such that $f(c_j)$ is defined.

If the numbers c_j are the left endpoints x_{j-1} of the subintervals, then **(11)** is the LEFT RIEMANN SUM for the integral corresponding to the partition (Figure 8), if c_j are the right endpoints x_j of the subintervals, then **(11)** is the RIGHT RIEMANN SUM (Figure 9), and if c_j are the midpoints $\frac{1}{2}(x_{j-1}+x_j)$, then **(11)** is the MIDPOINT RIEMANN SUM (Figure 10). Midpoint Riemann sums with equal subintervals are called MIDPOINT-RULE approximations of the integrals.

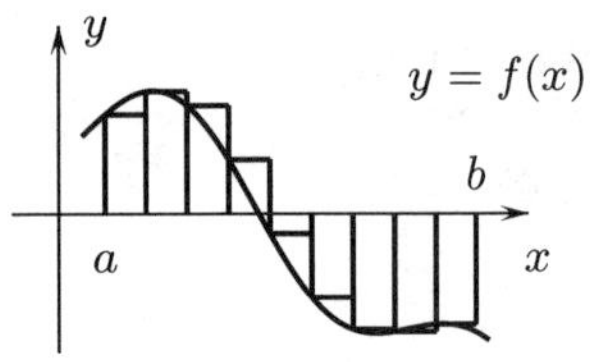

FIGURE 8

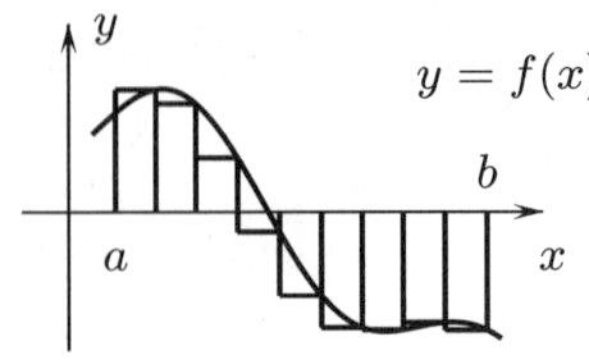

FIGURE 9

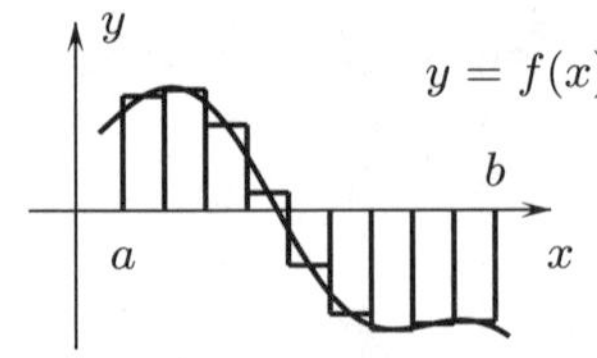

FIGURE 10

When we use a partition into subintervals of equal widths we will write Δx intead of Δx_j for their width.

The integral $\int_a^b f(x)\,dx$ is defined to be the limit of the Riemann sums **(11)**:

Definition 2 *The definite (Riemann) integral* $\int_a^b f(x)\,dx$ *is the limit of the Riemann sums* **(11)** *as the number N of subintervals in the partition tends to infinity and their widths tend to zero,*

$$\int_a^b f(x)\,dx = \lim_{N\to\infty} \sum_{j=1}^{N} f(c_j)\Delta x_j \tag{12}$$

provided the limit exists and is finite.[†]

The numbers a and b in **(12)** are called the LIMITS OF INTEGRATION,[‡] $[a, b]$ is the INTERVAL OF INTEGRATION, and $f(x)$ is the INTEGRAND. The letter "x" is called the DUMMY VARIABLE in the integral because the integral is a single number and not a variable function of x and the symbol "dx" is only part of the notation. (In early versions of calculus, which were based on the use of "infinitesimals" instead of limits as in modern calculus, "dx" represented the "infinitely small width" of an "infinitely narrow rectangle." See the historical notes at the end of Section 4.7.)

What functions have Riemann integrals?

As you have seen, many functions that arise in calculus do not have derivatives at all values of their variables. In particular, a function does not have a derivative at any point where it is discontinuous or where its graph has a corner.

In contrast, the Riemann integral **(12)** exists for all functions $f(x)$ that are normally encountered in introductory calculus provided the function is BOUNDED in the interval of integration, i.e., provided there is a constant M such that $|f(x)| \le M$ for all x in the interval where it is defined. This is because the bounded functions that are normally encountered in calculus are PIECEWISE CONTINUOUS, as defined below, and—as is established in advanced calculus classes—the integral **(12)** is defined if $f(x)$ piecewise continuous in $[a, b]$.

Definition 3 *The function $f(x)$ is* PIECEWISE CONTINUOUS *in $[a, b]$ if there is a fixed partition of the interval such that $f(x)$ is defined and continuous in the interior of each subinterval and has a finite limit from the right at the left endpoint of each subinterval and from the left at the right endpoint of each subinterval.*

[†]This type of limit has an $\epsilon\delta$-formulation: The integral is the limit of the Riemann sums if for every positive ϵ, no matter how small, there is a positive δ such that the Riemann sum differs from the integral by less than ϵ for all partitions into subintervals of widths less than δ and all choices of the numbers c_j in the subintervals.

[‡]The term "limit" in the phrase "limits of integration" just refers to the endpoints of the interval of integration and has nothing to do with the limit in Definition 2 or other types of limits elsewhere in calculus.

According to this definition, a piecewise continuous function may not be defined and may be discontinuous at a finite number of points in the interval, but it must be bounded in the interval because it has finite one-sided limits at any points where it is not defined or is discontinuous.

Example 2 **(a)** Where is the function $g(x)$ of Figure 11 defined? **(b)** Explain why $g(x)$ is bounded in $[0,7]$. **(c)** Why is $g(x)$ piecewise continuous in $[0,7]$?

SOLUTION **(a)** As can be seen in Figure 11, $g(x)$ is defined for all x in $[0,7]$ except at $x = 3$.
(b) $g(x)$ is bounded in $[0,7]$ because $|g(x)| \le 8$ for $0 \le x < 3$ and $3 < x \le 7$.
(c) $g(x)$ is piecewise continuous in $[0,7]$ because it is continuous in the open intervals $(0,3)$, $(3,5)$, and $(5,7)$ and it has finite limits as $x \to 0^+$, as $x \to 3^-$, as $x \to 3^+$, as $x \to 5^-$, as $x \to 5^+$, and as $x \to 7^-$. □

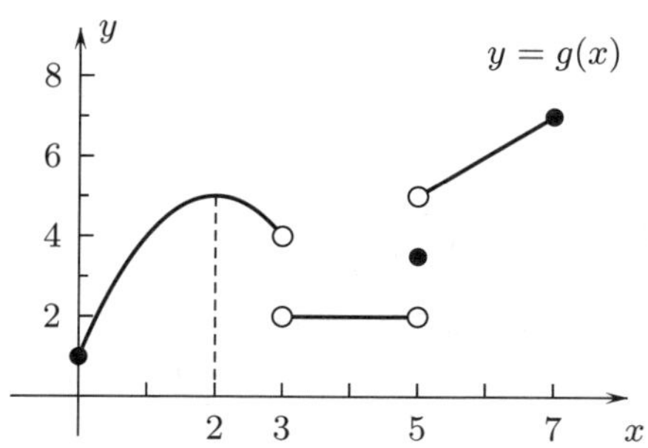

FIGURE 11

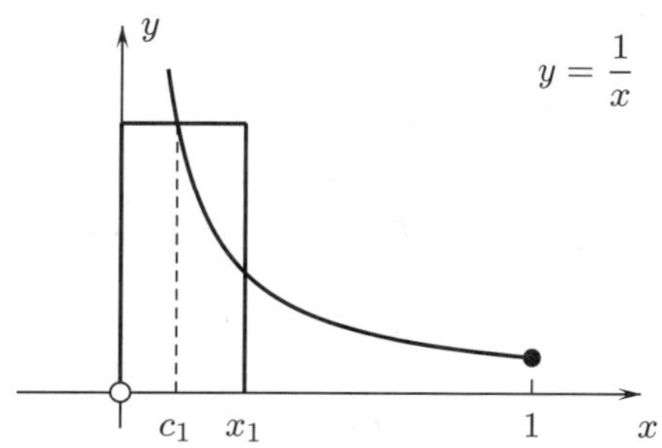

FIGURE 12

The Riemann integral is not defined if the function is not bounded. For example, $\int_0^1 \frac{1}{x}\,dx$ is not defined as a Riemann integral because $1/x \to \infty$ as $x \to 0^+$ and consequently $1/x$ is not bounded in the interval of integration $[0,1]$ (Figure 12). The Riemann sums for this integral cannot have a finite limit because we could construct an arbitrarily large Riemann sum for any partition by choosing c_1 sufficiently close to 0.†

Question 4 Suppose that $x_1 = \frac{1}{3}$ in Figure 12. Find values of c_1 such that the rectangle in Figure 12 has area 10, has area 100, and has area 1000.

Integrals and areas

We defined the integral so that, for the function $f(x)$ of Figure 6, $\int_a^b f(x)\,dx$ is equal to the area of region A minus the area of region B in that drawing. This illustrates the following rule, which can be used to define the areas of regions concerned.

Rule 1 *Suppose that the integral $\int_a^b f(x)\,dx$ exists with $a < b$.*

(a) *If $f(x) \ge 0$ for $a \le x \le b$, then the integral equals the area of the region between the graph of $f(x)$ and the x-axis for $a \le x \le b$ (Figure 13).*

(b) *If $f(x) \le 0$ for $a \le x \le b$, then the integral equals the negative of the area of the region between the graph of $f(x)$ and the x-axis for $a \le x \le b$ (Figure 14).*

(c) *If $f(x)$ has positive and negative values for $a \le x \le b$, then the integral equals the area of the region above the x-axis and below the graph where $f(x)$ is positive, minus the area of the region below the x-axis and above the graph where $f(x)$ is negative. (Figure 15).*

---

†We will see in Section 6.3 that integrals of certain unbounded functions can be defined as IMPROPER INTEGRALS.

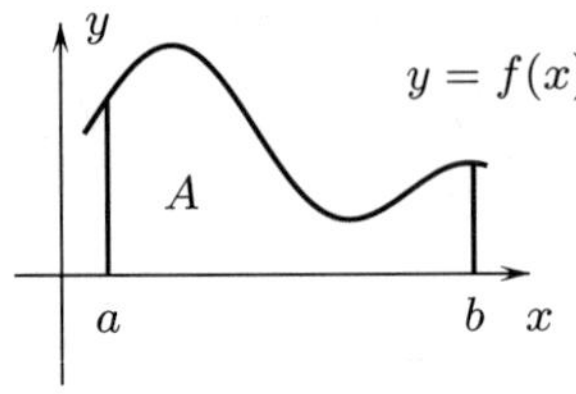

$$\int_a^b f(x)\,dx = [\text{Area } A]$$

FIGURE 13

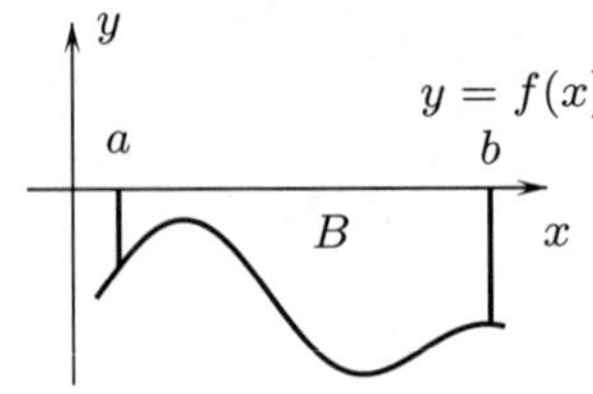

$$\int_a^b f(x)\,dx = -\ [\text{Area } B]$$

FIGURE 14

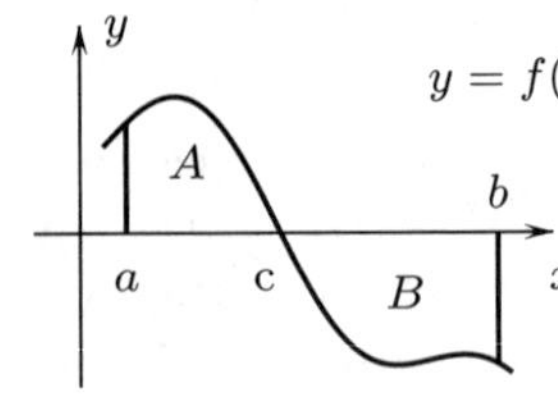

$$\int_a^b f(x)\,dx$$
$$= [\text{Area } A] - [\text{Area } B]$$

FIGURE 15

Example 3 Use the formula for the area of a triangle to evaluate $\int_{-3}^{3} (x+1)\,dx$.

SOLUTION The graph of $x+1$ is the line $y = x+1$ in Figure 16, which intersects the x-axis at $x = -1$. By Rule 1, the integral equals the area of triangle B above the x-axis in Figure 13, minus the area of triangle A below the x-axis. Since triangle B is 4 units wide and 4 units high, its area is $\frac{1}{2}(4)(4) = 8$, and since triangle A is 2 units wide and 2 units high, its area is $\frac{1}{2}(2)(2) = 2$. Therefore,

$$\int_{-3}^{3} (x+1)\,dx = [\text{Area } B] - [\text{Area } A] = 8 - 2 = 6. \ \square$$

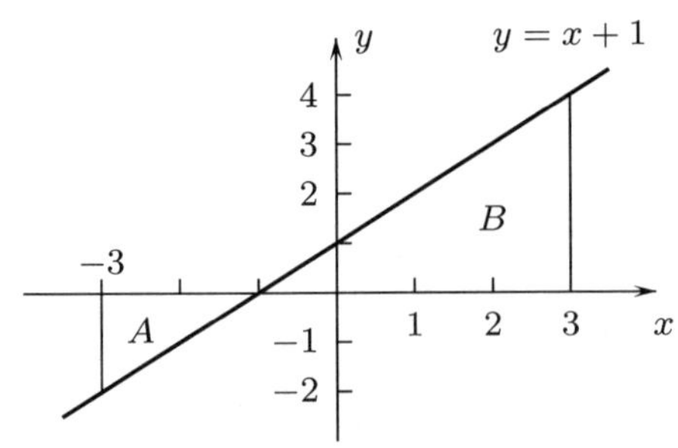

FIGURE 16

Calculating Riemann sums

To find values of integrals that we cannot evaluate with formulas from elementary geometry, we need to study their Riemann sums.

Example 4 Calculate the right Riemann sum for $\int_0^1 x^2\,dx$ corresponding to the partition of $[0,1]$ into five equal subintervals.

SOLUTION Since $[0,1]$ has width 1, each of the five subintervals has width $1/5 = 0.2$, and the partition is $0 < 0.2 < 0.4 < 0.6 < 0.8 < 1$. The rectangles are shown in Figure 17. Their heights are the values of x^2 at the right endpoints, which are $0.2, 0.4, 0.6, 0.8$, and 1. Because the rectangles are above the x-axis, the Riemann sum equals their total area, which is

$$(0.2)^2(0.2) + (0.4)^2(0.2) + (0.6)^2(0.2) + (0.8)^2(0.2) + 1^2(0.2)$$
$$= [(0.2)^2 + (0.4)^2 + (0.6)^2 + (0.8)^2 + 1^2](0.2) = 0.44. \ \square$$

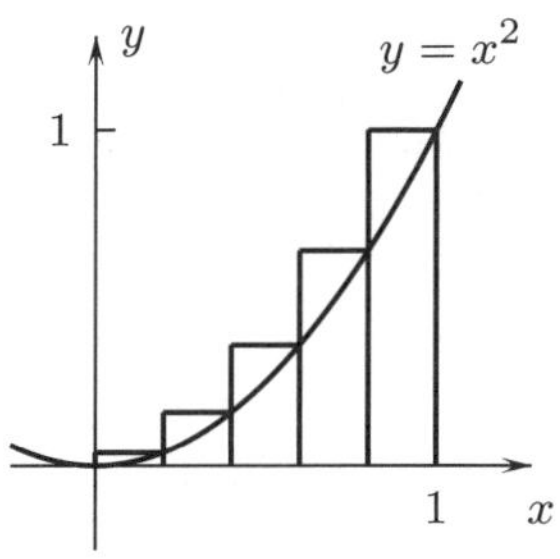

FIGURE 17

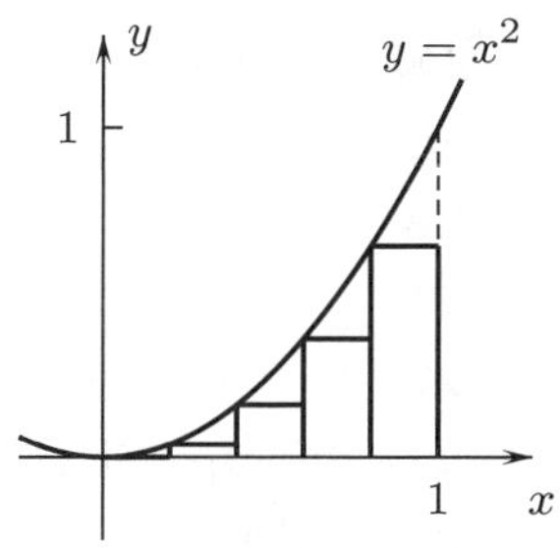

FIGURE 18

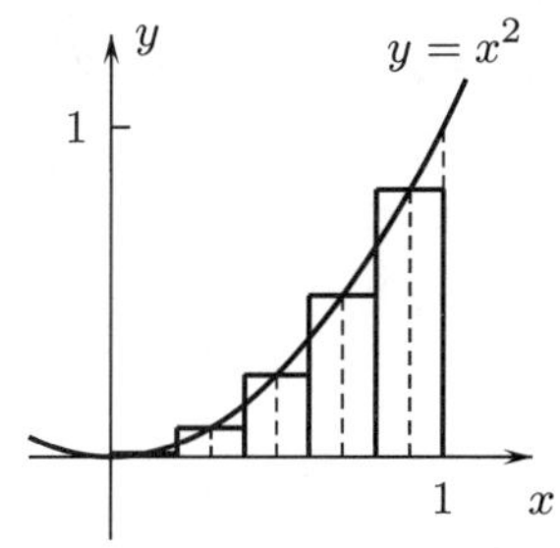

FIGURE 19

Question 5 Calculate **(a)** the left Riemann sum and **(b)** the midpoint Riemann sum for $\int_0^1 x^2\,dx$ with the partition of $[0,1]$ into five equal subintervals. The corresponding rectangles are shown in Figures 18 and 19.

Question 6 Use Figures 17 through 19 to explain why the Riemann-sum approximation from Example 4 is greater than the integral, why the Riemann sum from Question 5a is less than the integral, and why the Riemann sum from Question 5b is the most accurate of the three approximations of the integral.

A Riemann-sum program

The following program for Texas Instruments TI-82, TI-83, TI-85, and TI-86 calculators draws the graph and rectangles and calculates Riemann sums for $\int_a^b Y_1(x)\,dx$ with $Y_1(x)$ entered in the Y= menu. The user gives $a = A$, $b = B$, the number N of subintervals, and a number T between 0 and 1. The rectangles are of equal widths, and the program gives the left Riemann sum if $T = 0$, the right Riemann sum if $T = 1$, and the midpoint Riemann sum if $T = 0.5$.†

TI-82, TI-83

```
PROGRAM: RIEMANN
: ClrDraw
: FnOff
: FnOn 1
: Disp "USE Y1(X)"
: Prompt A, B, N
: Disp "T = ∅ FOR LEFT"
: Disp "T = 1 FOR RIGHT"
: Disp "T = ∅.5 FOR MID"
: Input T
: (B − A)/N → D
: ∅ → S
: For(J, 1, N, 1)
: A + (J − 1)D → U
: U + TD → X
: Y1 → W
```

TI-85, TI-86

```
PROGRAM: RIEMANN
: ClrDrw
: FnOff
: FnOn 1
: Disp "USE y1(x)"
: Prompt A, B, N
: Disp "T = ∅ FOR LEFT"
: Disp "T = 1 FOR RIGHT"
:Disp "T = ∅.5 FOR MID"
: Input T
: (B − A)/N → D.
: ∅ → S
: For(J, 1, N, 1)
: A + (J − 1) * D → U
: U + T * D → x
: y1 → W
```

†In this preliminary edition of this book key-by-key instructions for entering these programs and the program for TI-81 calculators are given at the end of this chapter. In the Preliminary and First Editions all such calculator- or computer-specific instructions will be in the *Calculator/Computer Supplement*.

```
17. : Line(U, ∅, U, W)
18. : Line(U, W, U + D, W)
19. : Line(U + D, W, U + D, ∅)
20. : S + DW → S
21. : End
22. : Pause
23. : Disp "SUM ="
24. : Disp S
```

```
17. : Line(U, ∅, U, W)
18. : Line(U, W, U + D, W)
19. : Line(U + D, W, U + D, ∅)
20. : S + D * W → S
21. : End
22. : Pause
23. : Disp "SUM ="
24. : Disp S
```

The symbol ∅ in lines 7, 9, 12, 17, and 19 of the program denotes zero. Be sure to use – and not (-) for the subtraction in lines 11 and 14.

Before this program is run, the function to be studied should be entered as Y_1 in the $Y=$ menu and a suitable window for the graph of the function should be chosen. To run the program on a TI-82 or TI-83 calculator, press PRGM, the program's number, and ENTER. On a TI-85, press PRGM, find the program under <NAMES>, and press ENTER. Give the limits of integration, A and B, the number of subintervals, N, and T $(= 0, 1$ or $0.5)$ when requested, followed by the ENTER key. The curve and the rectangles will be drawn. Press ENTER to display the value of the Riemann sum. You can then press ENTER to rerun the program if you do not perform any other operations first.

Example 5 **(a)** Use the Riemann-sum program to find the midpoint Riemann sum for $\int_0^1 x^2\,dx$ with partitions of $[0, 1]$ into 10, 20, 50, and 100 equal subintervals. **(b)** Use the results of part (a) to predict the exact value of $\int_0^1 x^2\,dx$.

SOLUTION **(a)** On a Texas Instruments calculator, we set $Y_1 = x^2$ in the $Y=$ menu and use the window $-0.25 \le x \le 1.25, -0.25 \le y \le 1.25$. In each case, we run the program and give $A = 0$, $B = 1$, the number of subintervals for N, and $T = 0.5$. The rectangles for $N = 10$ and 20 are shown in Figures 20 and 21. The Riemann sum equals 0.3325 for $N = 10$, 0.333125 for $N = 20$, 0.3333 for $N = 50$, and 0.333325 for $N = 100$. **(b)** It appears that the Riemann sums are approaching $\frac{1}{3}$, so we predict that $\int_0^1 x^2\,dx = \frac{1}{3}$. (We will see that this prediction is correct in Section 4.3.) □

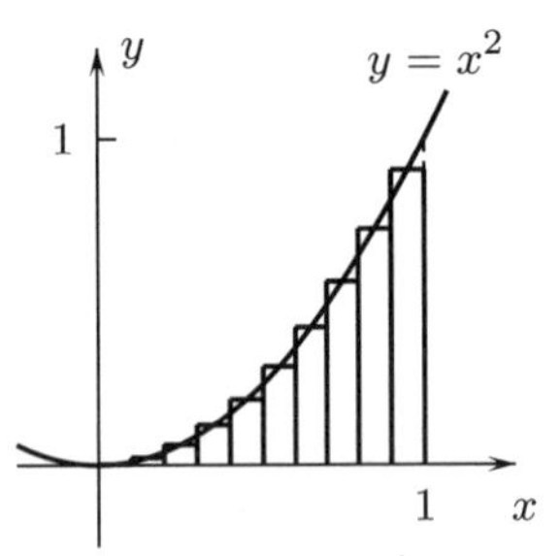

FIGURE 20

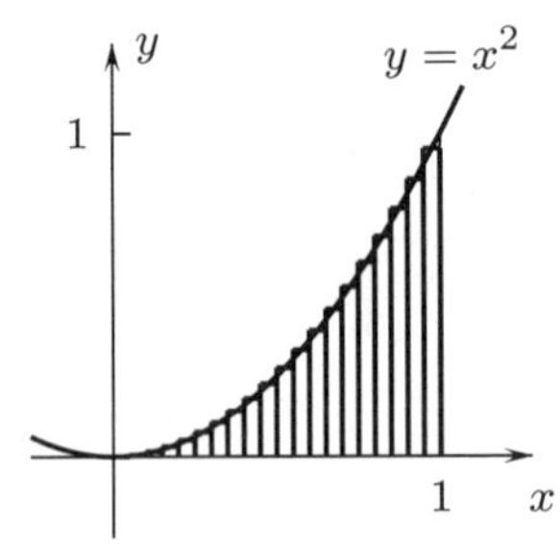

FIGURE 21

Question 7 Use a Riemann-sum program and the procedure of Example 5 to predict the values of $\int_0^1 x^3\,dx$, $\int_0^1 x^4\,dx$, and $\int_0^1 x^{1/4}\,dx$.

Integrals from a *to* b *with* a ≥ b

Because regions of zero width have zero area, integrals with equal limits of integration are defined to be zero. Also, for convenience in many applications, an integral from b to a with $a < b$ is defined to be the negative of the integral from a to b:

Definition 4 **(a)** *For any function $f(x)$,*

$$\int_a^a f(x)\,dx = 0. \tag{13}$$

(b) *If the integral of $f(x)$ from a to b is defined with $a < b$, then*

$$\int_b^a f(x)\,dx = -\int_a^b f(x)\,dx. \tag{14}$$

Rules **(13)** and **(14)** are illustrated by the equations

$$\int_8^8 f(x)\,dx = 0 \quad \text{and} \quad \int_{10}^0 f(x)\,dx = -\int_0^{10} f(x)\,dx.$$

Integrals over adjacent intervals

Figure 22 shows the graph of a positive function $f(x)$ for $a \le x \le c$ and a number b between a and c. Since the area between the graph and the x-axis for $a \le x \le c$ is the sum of the areas between the graph and the x-axis for $a \le x \le b$ and for $b \le x \le c$, the integral of $f(x)$ from a to c is the sum of its integrals from a to b and from b to c:†

$$\int_a^b f(x)\,dx + \int_b^c f(x)\,dx = \int_a^c f(x)\,dx. \tag{15}$$

This formula also holds if $f(x)$ has negative values or positive and negative values and for any choices of a, b, and c such that the integrals are defined.

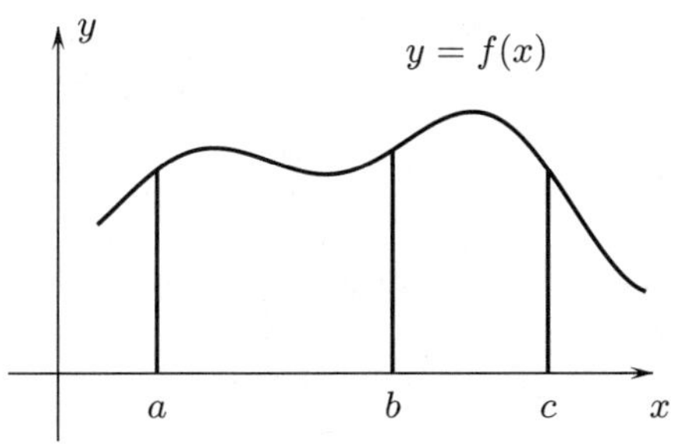

FIGURE 22

Question 8 What is the integral of $g(x)$ from 1 to 8 if its integral from 1 to 5 is -7 and its integral from 5 to 8 is 9?

---

†We could prove this result with Definition 2 by using the fact that a Riemann sum for the integral from a to b plus a Riemann sum for the integral from b to c is a Riemann sum for the integral from a to c.

Integrals of linear combinations of functions

If the integrals of $f(x)$ and $g(x)$ from a to b exist, then for any constants A and B, any Riemann sum for $Af(x) + Bg(x)$ equals A multiplied by a Riemann sum for the integral of $f(x)$, plus B multiplied by a Riemann sum for the integral of $g(x)$:

$$\sum_{j=1}^{N}[Af(c_j) + Bg(c_j)]\Delta x = A\sum_{j=1}^{N} f(c_j)\Delta x + B\sum_{j=1}^{N} g(c_j)\Delta x.$$

Since the integrals are the limits of their Riemann sums when we let $N \to \infty$, we obtain the formula,

$$\int_a^b [Af(x) + Bg(x)]\,dx = A\int_a^b f(x)\,dx + B\int_a^b g(x)\,dx. \qquad \textbf{(16)}$$

Question 9 What is $\displaystyle\int_{-35}^{35} [2p(x) - 3q(x)]\,dx$ if $\displaystyle\int_{-35}^{35} p(x)\,dx = 10$ and $\displaystyle\int_{-35}^{35} q(x)\,dx = 20$?

Principles and procedures

- A Riemann sum for the definite (Riemann) integral $\displaystyle\int_a^b f(x)\,dx$ with $a < b$ is a sum $\displaystyle\sum_{j=1}^{N} f(c_j)\Delta x_j$, where $a = x_0 < x_1 < x_2 < \cdots < x_{N-1} < x_N = b$ is a partition of $[a, b]$ and for each $j = 1, 2, 3, \ldots, N$, $\Delta x_j = x_j - x_{j-1}$ is the width of the jth subinterval in the partition and c_j is a point in the jth subinterval where $f(x)$ is defined. If the subintervals are of equal width, we write Δx instead of Δx_j for their widths, which are then $(b - a)/N$.
- The definite (Riemann) integral $\displaystyle\int_a^b f(x)\,dx$ is the finite limit of the Riemann sums as the number of subintervals tends to ∞ and their widths tend to zero.
- The integral is defined for all functions that are normally encountered in calculus, unless the function is not bounded, i.e., unless there are values of the function for x in the interval of integration that tend to ∞ or $-\infty$. If the function is not bounded, then the Riemann integral is not defined.
- If $\displaystyle\int_a^b f(x)\,dx$ with $a < b$ is defined, then it equals the area of the region or regions between the graph of $f(x)$ and the x-axis for $a \leq x \leq b$ where $f(x) \geq 0$, minus the area of the region or regions between the graph of $f(x)$ and the x-axis for $a \leq x \leq b$ where $f(x) < 0$.
- Any definite integral $\displaystyle\int_a^a f(x)\,dx$ with equal limits of integration is zero, and for any a and b, $\displaystyle\int_b^a f(x)\,dx$ is the negative of $\displaystyle\int_a^b f(x)\,dx$.
- Definite integrals over adjacent intervals can be combined by using the rule,

$$\int_a^b f(x)\,dx + \int_b^c f(x)\,dx = \int_a^c f(x)\,dx.$$

- Definite integrals of linear combinations of functions can be written as linear combinations of the integrals: For constants A and B,

$$\int_a^b [Af(x) + Bg(x)]\,dx = A\int_a^b f(x)\,dx + B\int_a^b g(x)\,dx.$$

Tune-Up Exercises 4.2♦

[A]*Answer provided.* [O]*Outline of solution provided.*

T1.[O] Use the formula for the area of a triangle to find the value of $\int_0^3 (x-1)\,dx$.

♦ Type 1, using an area formula to evaluate an integral

T2.[A] Use the fact that the curve $y = \sqrt{16 - x^2}$ is the upper half of the circle $x^2 + y^2 = 16$ of radius 4 with its center at the origin to find the exact value of $\int_{-4}^{0} \sqrt{16 - x^2}\,dx$.

♦ Type 2, using an area formula to evaluate an integral

T3.[O] Use areas to evaluate $\int_1^6 P(x)\,dx$ where$P(x) = \begin{cases} -1 & \text{for} \quad 1 < x < 3 \\ 3 & \text{for} \quad 3 < x < 4 \\ 2 & \text{for} \quad 4 < x < 6. \end{cases}$

♦ Type 2, using an area formula to evaluate an integral

T4.[A] Calculate $\sum_{j=1}^{6} j(j-1)(j-2)$.

♦ Type 1, summation notation

T5.[A] Express $1^2 + 2^2 + 3^2 + \cdots + 99^2$ with summation notation.

♦ Type 1, summation notation

T6.[O] Calculate directly, without using the Riemann-sum program, the right Riemann sum for $\int_0^2 (4x - x^2)\,dx$ corresponding to the partition $0 < 0.5 < 1 < 1.5 < 2$. Then draw the curve $y = 4x - x^2$ and the rectangles whose area equals the Riemann sum.

♦ Type 2, a Riemann sum

T7.[A] Follow the instructions in Exercise T6, but with the left Riemann sum.

♦ Type 2, a Riemann sum

T8.[A] Predict the value of $\int_0^2 (4x - x^2)\,dx$ by using the Riemann-sum program to calculate midpoint Riemann sums with 20, 50, and 200 equal subintervals. Use the window $0 \le x \le 2, -1 \le y \le 5$.

♦ Type 1, using the Riemann-sum program

T9.[A] What is $\int_{-7}^{7} [6f(x) + 3g(x)]\,dx$ if $\int_{-7}^{7} f(x)\,dx = 4$ and $\int_{-7}^{7} g(x)\,dx = -5$?

♦ Type 2, an integral of a linear combination

T10.[A] What is $\int_1^6 h(x)\,dx$ if $\int_1^2 h(x)\,dx = 10$ and $\int_2^6 h(x)\,dx = 7$?

♦ Type 2, integrals over adjacent intervals

Problems 4.2

[A]*Answer provided.* [O]*Outline of solution provided.*

In Problems 1 through 9, sketch the graphs of the integrands. Then use formulas for areas of rectangles, triangles, and circles to find the values of the integrals.

1.[O] $\int_{-2}^{1} (-2x)\,dx$

♦ Type 1, using an area formula to evaluate an integral

2.[A] $\int_{-2}^{4} (\frac{1}{2}x - 2)\,dx$

♦ Type 1, using an area formula to evaluate an integral

3. $\int_{-10}^{5} 6\,dx$

♦ Type 1, using an area formula to evaluate an integral

4. $\int_{-3.96}^{5.72} 0\,dx$

♦ Type 1, using an area formula to evaluate an integral

♦The Tune-up Exercises and Problems are classified by type and content. The types are (1) basic, reactive; (2) basic reflective; (3) intermediate, reactive; (4) intermediate, reflective; (5) advanced, reactive; (6) advanced, reflective; and (7) advanced, theoretical.

5.[A] $\displaystyle\int_{-1}^{0} \sqrt{1-x^2}\,dx$

♦ Type 1, using an area formula to evaluate an integral

6. $\displaystyle\int_{-5}^{5} \sqrt{25-x^2}\,dx$

♦ Type 1, using an area formula to evaluate an integral

7.[A] $\displaystyle\int_{3}^{1} 7\,dx$

♦ Type 1, using an area formula to evaluate an integral

8. $\displaystyle\int_{5}^{0} x\,dx$

♦ Type 1, using an area formula to evaluate an integral

9.[A] $\displaystyle\int_{-2}^{3} H(x)\,dx$ where $H(x) = \begin{cases} \sqrt{4-x^2} & \text{for } -2 \le x \le 0 \\ -\sqrt{9-x^2} & \text{for } 0 < x \le 3 \end{cases}$

♦ Type 2, using an area formula to evaluate an integral

10. Calculate **(a**[A]**)** $\displaystyle\sum_{j=1}^{5}\left(\frac{j}{12}\right)$, **(b**[A]**)** $\displaystyle\sum_{j=1}^{5}\left(\frac{12}{j}\right)$, **(c)** $\displaystyle\sum_{j=1}^{5}\left(1-\frac{j}{5}\right)$, and **(d)** $\displaystyle\sum_{j=1}^{6}\left[\frac{j}{6}-\left(\frac{j}{6}\right)^2\right]\left(\frac{1}{6}\right)$.

♦ Type 1, summation notation

11. Express **(a**[A]**)** $\frac{1}{3}+\frac{2}{4}+\frac{3}{5}+\frac{4}{6}+\cdots+\frac{175}{177}$, **(b)** $5+\frac{5}{4}+\frac{5}{9}+\frac{5}{16}+\cdots+\frac{5}{121}$, and **(c)** $2^3+\left(\frac{2}{2}\right)^3+\left(\frac{2}{3}\right)^3+\left(\frac{2}{4}\right)^3+\cdots+\left(\frac{2}{N}\right)^3$ with summation notation.

♦ Type 1, summation notation

12.[A] Calculate the left Riemann sum for $\displaystyle\int_{1}^{8}\frac{1}{x}\,dx$ for the partition $1 < 2 < 4 < 8$ of $[1, 8]$. Then draw the curve $y = 1/x$ with the rectangles whose areas equal the Riemann sum. (Notice the unequal subintervals.)

♦ Type 1, a Riemann sum

13. What is the right Riemann sum for $\displaystyle\int_{0}^{10} x^2\,dx$ for the partition $0 < 5 < 9 < 10$ of $[0, 10]$. Draw the curve $y = x^2$ with the rectangles whose areas equal the Riemann sum. (Notice the unequal subintervals.)

♦ Type 1, a Riemann sum

14.[A] Calculate the left and the right Riemann sums for $\displaystyle\int_{0}^{2} Q(x)\,dx$ for the partition of $[0, 2]$ into four equal subintervals, where $Q(x)$ is the function whose graph is shown in Figure 23.

♦ Type 1, a Riemann sum

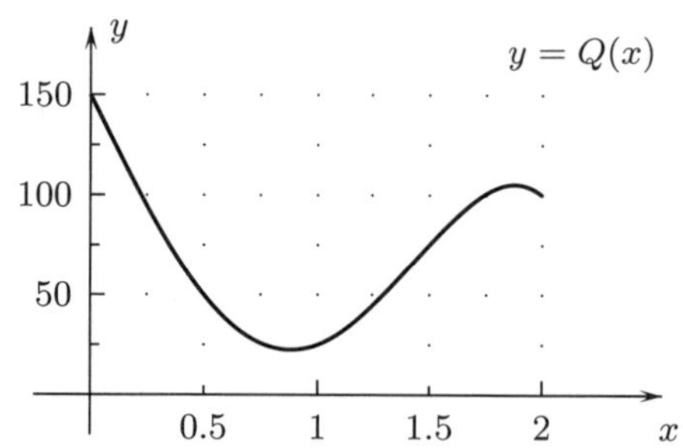

FIGURE 23

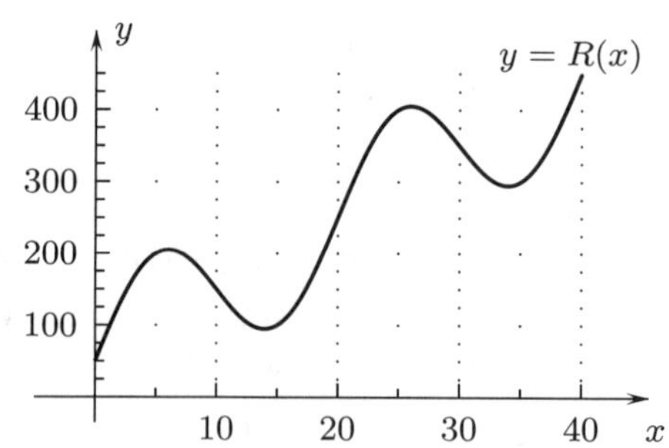

FIGURE 24

15. The graph of $R(x)$ is shown in Figure 24. What is the midpoint Riemann sum for $\displaystyle\int_{0}^{40} R(x)\,dx$ corresponding to the partition of $[0, 40]$ into four equal subintervals?

♦ Type 1, a Riemann sum

16. How much larger is the right Riemann sum than the left Riemann sum for $\displaystyle\int_{0}^{2}(3x^2-x^3)\,dx$ **(a**[A]**)** with the partition into four equal subintervals, **(b)** with the partition of $[0, 2]$ into 20 equal subintervals, and **(c)** with the partition of $[0, 2]$ into 80 equal subintervals? Use the Riemann-sum program with the window $0 \le x \le 2, 0 \le y \le 5$ in your calculations.

♦ Type 2, Riemann sums, etc.

Use at least three Riemann sums, calculated with the Riemann-sum program, to predict the values of the integrals in Exercises 17 through 20. Be sure your calculator is in radian mode for Problem 18.

17.[A] $\int_1^4 (3x - x^2)\, dx$

♦ Type 1, using the Riemann-sum program

19 $\int_1^4 \frac{1}{x^2}\, dx$

♦ Type 1, using the Riemann-sum program

18.[A] $\int_0^1 \pi \sin(\pi x)\, dx$

♦ Type 1, using the Riemann-sum program

20. $\int_0^5 (100x - 4x^3)\, dx$

♦ Type 1, using the Riemann-sum program

21.[A] What is $\int_3^1 F(x)\, dx$ if $\int_1^4 F(x)\, dx = -4$ and $\int_3^4 F(x)\, dx = 6$?

♦ Type 1, integrals over adjacent intervals

22. What is $\int_3^0 Y(x)\, dx$ if $\int_0^4 Y(x)\, dx = 100$ and $\int_4^3 Y(x)\, dx = -25$?

♦ Type 1, integrals over adjacent intervals

23. What is $\int_1^{10} P(x)\, dx$ if $\int_{10}^5 P(x)\, dx = 100$ and $\int_1^5 P(x)\, dx = -25$?

♦ Type 1, integrals over adjacent intervals

24. Use the formulas for areas of rectangles and circles to evaluate $\int_0^2 (5 - 3\sqrt{4 - x^2})\, dx$.

♦ Type 2, using an area formula to evaluate an integral

25.[A] The integral $\int_{0.5}^4 \left(8x + \frac{4}{x^2}\right) dx$ has the value 70. Use the Riemann-sum program and trial and error to find a positive integer N such that the midpoint Riemann sum with N equal subintervals for the integral differs from the integral by less than 0.1 and more than 0.09.

♦ Type 2, using the Riemann-sum program

26. **(a)** Find the value of $\int_0^2 (1+2x)\, dx$ from the formula for the area of a trapezoid or from the formulas for areas of rectangles and triangles. **(b)** Use the Riemann-sum program to calculate midpoint Riemann sums for $\int_0^2 (1 + 2x)\, dx$ with N equal subintervals for various positive integers N. All such Riemann sums give the exact value of the integral. Explain..

♦ Type 2, using the Riemann-sum program

27. **(a)** Explain why $\int_{-2}^2 x^3\, dx$ equals zero. **(b)** Use the Riemann-sum program to calculate midpoint Riemann sums for $\int_{-2}^2 x^3\, dx$ with N equal subintervals for various positive integers N. Why do all such Riemann sums give the exact value of the integral?

♦ Type 2, using the Riemann-sum program

28. Use the Riemann-sum program and trial and error to find a positive integer N such that the left and right Riemann sums with N equal subintervals for $\int_{0.3}^4 \sqrt{1 + x^3}\, dx$ differ by more than 1 and less than 1.2.

♦ Type 2, using the Riemann-sum program

29. **(a)** Show, by writing the sum twice, once forward and once backwards, and adding corresponding terms, that for any positive integer N,

$$1 + 2 + 3 + \cdots + N = \tfrac{1}{2}N(N + 1). \qquad \textbf{(17)}$$

(b) Explain why $\sum_{j=1}^{N} \frac{j}{N}\left(\frac{1}{N}\right)$ is the right Riemann sum for $\int_0^1 x\, dx$ with N equal subintervals.

(c) Use the results of parts (a) and (b) and Definition 1 of the integral to find the value of $\int_0^1 x\, dx$.

♦ Type 6, deriving an integration formula

30. **(a)** Derive the summation formula,

$$1^3 + 2^3 + 3^3 + \cdots + N^3 = \left[\tfrac{1}{2}N(N+1)\right]^2 \qquad \textbf{(18)}$$

from **(17)** by the following geometric argument:† The square of width $1^3 + 2^3 + 3^3 + \cdots + N^3$ can be divided into a square of width 1 and L-shaped GNOMONS of widths $2, 3, \ldots, N$ as in Figure 25. Show that the jth gnomen has area j^3 (Figure 26), so that the area of the large square is the sum in **(18)**.

(b) Explain why $\displaystyle\sum_{j=1}^{N} \frac{j^3}{N^3}\left(\frac{1}{N}\right)$ is the right Riemann sum for $\displaystyle\int_0^1 x^3\,dx$ with N equal subintervals.

(c) Use the results of parts (a) and (b) and Definition 1 of the integral to find the value of $\displaystyle\int_0^1 x^3\,dx$.

◆ Type 6, exact integration with a summation formula

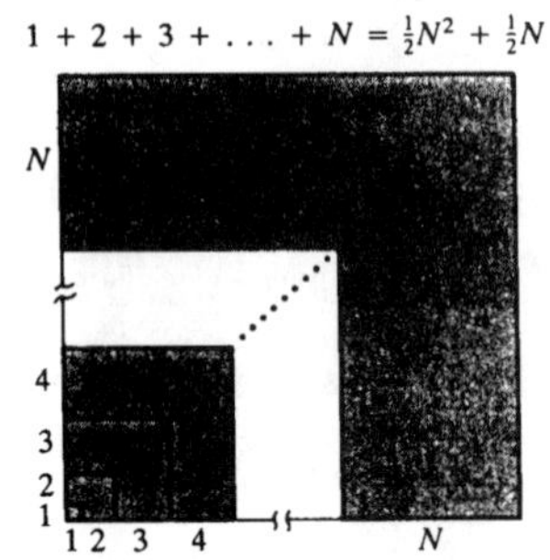

FIGURE 25

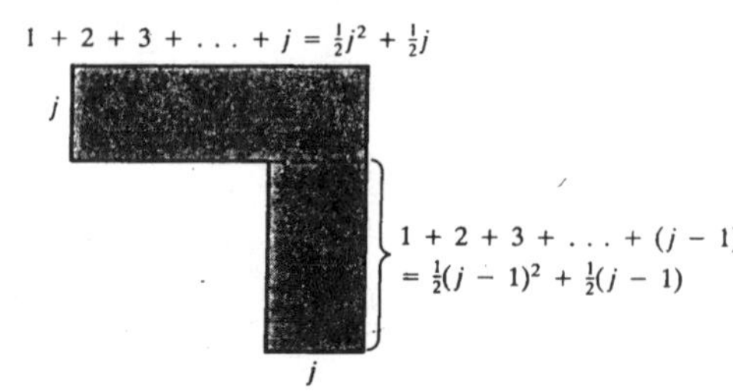

FIGURE 26

31. **(a)** Derive the formula,

$$1^2 + 2^2 + 3^2 + \cdots + N^2 = \tfrac{1}{3}N^3 + \tfrac{1}{2}N^2 + \tfrac{1}{6}N \qquad \textbf{(19)}$$

from **(17)** and **(18)** by writing the sum $\displaystyle\sum_{j=1}^{N}[(j+1)^3 - j^3]$ in two ways and equating the results. First, show that $\displaystyle\sum_{j=1}^{N}[(j+1)^3 - j^3] = (N+1)^3 - 1$ because the negative part of each term except the last cancels with the positive part of the following term. Then use the identity $(j+1)^3 - j^3 = 3j^2 + 3j + 1$ and put the expression on the right in the sum. **(b)** Explain why $\displaystyle\sum_{j=1}^{N}\frac{j^2}{N^2}\left(\frac{1}{N}\right)$ is the right Riemann sum for $\displaystyle\int_0^1 x^2\,dx$ with N equal subintervals. **(c)** Use the results of parts (a) and (b) and Definition 1 of the integral to find the value of $\displaystyle\int_0^1 x^2\,dx$.

◆ Type 6, exact integration with a summation formula

†This derivation was known to Arab mathematicians in the eleventh century (see *Extrait du Fakhrî par Alkarkhî* by F. Woepcke, Paris: l'Imprimerie Imperériale, 1853, p. 61).

Section 4.3

The Fundamental Theorem and integrals of x^n with $n \neq -1$

OVERVIEW: *In this section we study the Fundamental Theorem of Calculus, which shows how the derivative, defined as the slope of a tangent line, is related to the definite integral, defined in terms of areas, so that integration and differentiation are, in a sense, inverse operations. Next, we use this result to derive a formula for definite integrals of the power functions x^n with constants $n \neq -1$. Then we introduce the term* INDEFINITE INTEGRALS *for antiderivatives and look at applications of definite and indefinite integrals involving areas and rates of change.*

Topics:

- ***The Fundamental Theorem of Calculus***
- ***Integrals of x^n with $n \neq -1$***
- ***The indefinite integral***
- ***Areas of regions between the graphs of functions and the x-axis***
- ***Integrating rates of change***

The Fundamental Theorem of Calculus

At the beginning of Section 4.2, we considered an example where a truck's velocity in the positive direction on an s-axis was given by the function $\frac{ds}{dt}(t)$ of Figure 1. We saw that we could estimate the change $s(b) - s(a)$ in the truck's position from time a to time b by approximating $\frac{ds}{dt}(t)$ by step functions and approximating the regions between its graph and the t-axis by rectangles, as in Figure 2. If the truck's velocity were given by this step function, then the area of each rectangle above the t-axis in Figure 2 would be a distance the truck would travel in the positive direction and the area of each rectangle below the t-axis would be a distance the truck would travel in the negative direction. This yields the following estimate of the truck's actual change in position:

$$s(b) - s(a) \approx \begin{bmatrix} \text{The area of} \\ \text{the rectangles} \\ \text{above the } t\text{-axis} \end{bmatrix} - \begin{bmatrix} \text{The area of} \\ \text{the rectangles} \\ \text{below the } t\text{-axis} \end{bmatrix}. \qquad (1)$$

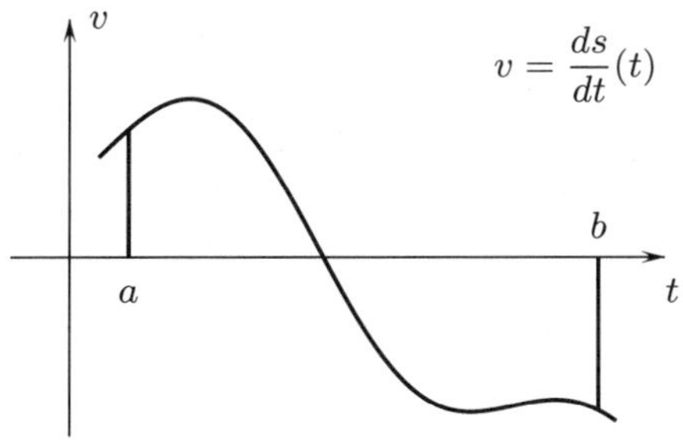

FIGURE 1

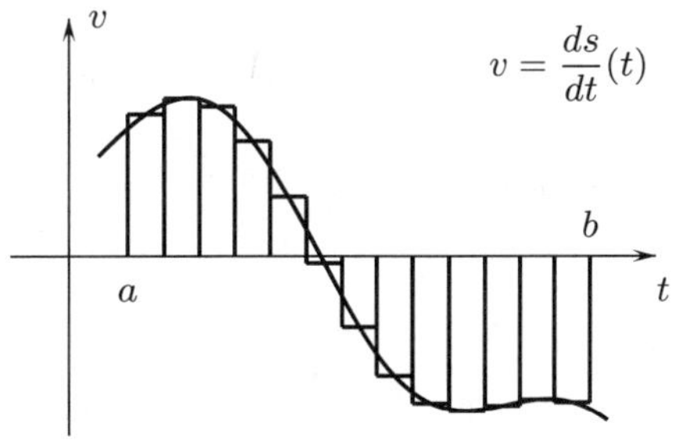

FIGURE 2

♦ SUGGESTIONS TO INSTRUCTORS: The main topics in this section can be divided into two parts, as follows: **Part 1** (The Fundamental Theorem for integrals of derivatives; Tune-up Exercise T1; Problems 1–7) Have students begin work on Questions 1 and 2 before class. Problems 2, 5, and 7 are good for in-class discussion.
Part 2 (Indefinite integrals and integrals of x^n for $n \neq -1$; Tune-up Exercises T2–T6; Problems 8–57) Have students begin work on Questions 3 through 5 before class. Problems 11, 16, 19, 26, 30, 32, 37, 38, 39, and 52 are good for in-class discussion.

If we let the number of rectangles tend to infinity and their widths tend to zero, we can expect that the change in position determined by the areas of the rectangles would approach the change in position of the truck with velocity given by $v(t)$. On the other hand, the limit of the difference of areas on the right of **(1)** is the definite integral $\int_a^b \frac{ds}{dt}(t)\,dt$, as defined in Section 4.2. Therefore, we can expect that

$$s(b) - s(a) = \int_a^b \frac{ds}{dt}(t)\,dt. \tag{2}$$

To prove this result, we study the distance function $s(t)$ of Figure 3, whose derivative is the velocity function of Figures 1 and 2. We consider the partition,

$$a = t_0 < t_1 < t_2 < t_3 < t_4 < t_5 = b \tag{3}$$

of the interval $[a, b]$ into five subintervals, and the secant lines through the points on the graph at $t = t_0, t = t_1, \cdots, t = t_5$ in Figure 3.

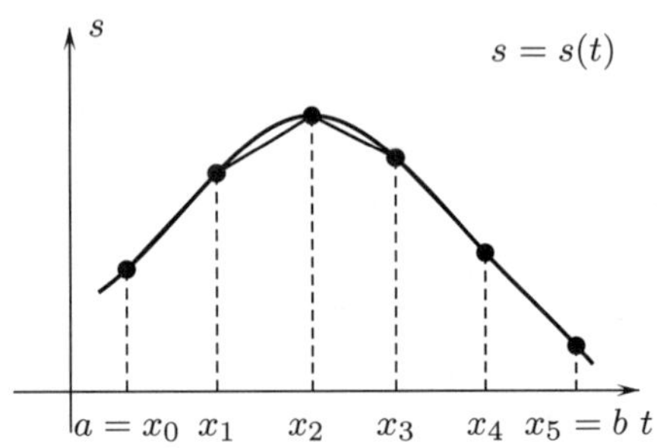

FIGURE 3

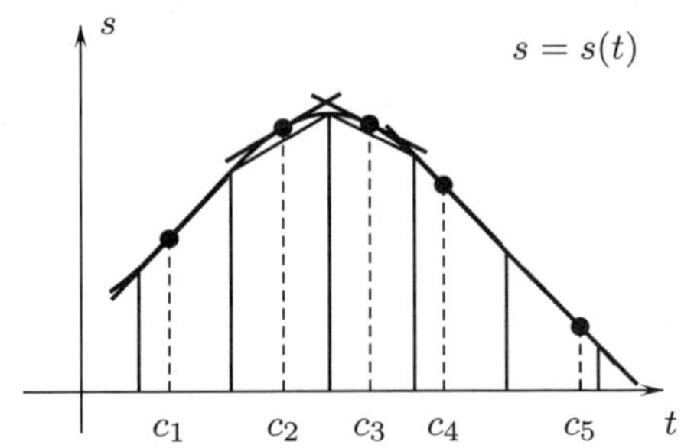

FIGURE 4

We assume that $s(t)$ is continuous in $[a, b]$ and that $\frac{ds}{dt}(t)$ exists for $a < t < b$. Then by the Mean-Value Theorem of Section 3.2, there is, for each $j = 1, 2, 3, 4, 5$, a number c_j in the jth subinterval such that the tangent line to the graph at $t = c_j$ is parallel to the secant line over that subinterval, as shown in Figure 4.

Because each tangent line is parallel to the corresponding secant line, the slope $\frac{s(t_j) - s(t_{j-1})}{\Delta t_j}$ of the secant line is equal to the slope $\frac{ds}{dt}(c_j)$ of the tangent line:

$$\frac{s(t_j) - s(t_{j-1})}{\Delta t_j} = \frac{ds}{dt}(c_j) \quad \text{for } j = 1, 2, 3, 4, 5.$$

Here $\Delta t_j = t_j - t_{j-1}$.

We multiply both sides of the last equation by Δt_j to obtain

$$s(t_j) - s(t_{j-1}) = \frac{ds}{dt}(c_j)\Delta t_j \quad \text{for } j = 1, 2, 3, 4, 5. \tag{4}$$

Thus, the change $s(t_j) - s(t_{j-1})$ in the object's position across the jth subinterval is equal to its velocity at $t = c_j$, multiplied by the amount of time Δt_j corresponding to the jth subinterval.

Equations **(4)** for $j = 1, 2, 3, 4, 5$ imply that the sum

$$[s(t_1) - s(t_0)] + [s(t_2) - s(t_1)] + [s(t_3) - s(t_2)] + [s(t_4) - s(t_3)] + [s(t_5) - s(t_4)] \quad \textbf{(5)}$$

of the changes in the object's position across the five subintervals is equal to the sum

$$\frac{ds}{dt}(c_1)\Delta t_1 + \frac{ds}{dt}(c_2)\Delta t_2 + \frac{ds}{dt}(c_3)\Delta t_3 + \frac{ds}{dt}(c_4)\Delta t_4 + \frac{ds}{dt}(c_5)\Delta t_5. \quad \textbf{(6)}$$

Sum **(6)** is the Riemann sum for the integral $\int_a^b \frac{ds}{dt}(t)\, dt$ corresponding to the rectangles in Figure 5.

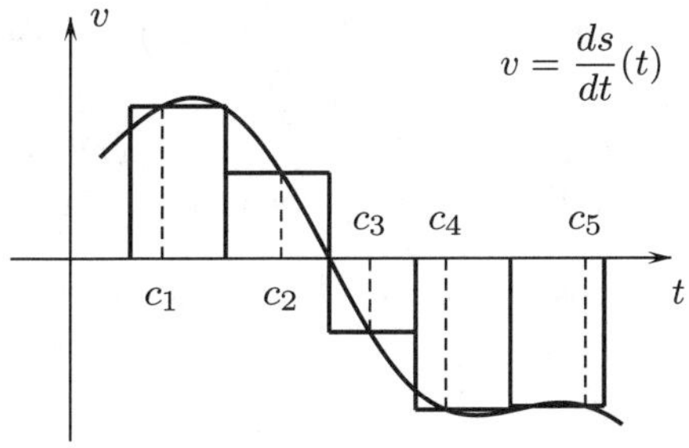

FIGURE 5

As you might expect, the sum **(5)** of the changes in $s(t)$ across the five subintervals is equal to its change from $t = a$ to $t = b$. We can also reach this conclusion from the fomulas.

Question 1 Show that the sum **(5)** is equal to the change $s(t_5) - s(t_0) = s(b) - s(a)$ in $s(t)$ from $t = a$ to $t = b$.

Because the sum **(5)** is equal to the Riemann sum **(6)** and is equal to $s(b) - s(a)$, the Riemann sum equals $s(b) - s(a)$.

Similar reasoning would show that for each partition of $[a, b]$ there is at least one Riemann sum that equals $s(b) - s(a)$. Then, since the integral $\int_a^b \frac{ds}{dt}(t)\, dt$ is the limit of all of the Riemann sums as the number of subintervals in the partitions tend to ∞ and their widths tend to zero, the integral equals $s(b) - s(a)$ and we have established formula **(2)**.

Equation **(2)** is an example of the FUNDAMENTAL THEOREM OF CALCULUS. Here is its statement for a function $F(x)$:†

Theorem 1 (The Fundamental Theorem of Calculus) *Suppose that $F(x)$ is continuous in an interval containing a and b, that $\frac{dF}{dx}(x)$ exists for all or all but a finite number of points in the interval, and that the Riemann integral of $\frac{dF}{dx}(x)$ from a to b is defined. Then*

$$\int_a^b \frac{dF}{dx}(x)\, dx = F(b) - F(a). \quad \textbf{(7a)}$$

or in prime notation

$$\int_a^b F'(x)\, dx = F(b) - F(a) \quad \textbf{(7b)}$$

---

†This version of the Fundamental Theorem deals with definite integrals of derivatives. A second version that involves derivatives of integrals will be discussed in Section 6.3.

We include in Theorem 1 the case where $\dfrac{dF}{dx}(x)$ does not exist at a finite number of points in the interval to allow for functions $F(x)$, as in the next example, whose graphs have corners.[‡]

Example 1 **(a)** Show that the hypotheses of Theorem 1 are satisfied for

$$F(x) = \begin{cases} 3x & \text{for} \quad 0 \le x \le 1 \\ 4 - x & \text{for} \quad 1 < x \le 3 \end{cases}$$

in the interval $[0, 3]$. **(b)** Show, without using the theorem, that the conclusion holds in this case by using areas to evaluate $\displaystyle\int_0^3 F'(x)\,dx$.

SOLUTION **(a)** The graph of $F(x)$ is shown in Figure 6. Its derivative $F'(x)$, whose graph is shown in Figure 7, equals $\dfrac{d}{dx}(3x) = 3$ for $0 < x < 1$ and equals $\dfrac{d}{dx}(4 - x) = -1$ for $1 < x < 3$. The hypotheses of the theorem are satisfied (i) because $F(x)$ is continuous in $[0, 3]$, (ii) because $F'(x)$ is defined at all x in the interval except 0, 1, and 3, and (iii) because $F'(x)$ is piecewise continuous in $[0, 3]$ and consequently the integral $\displaystyle\int_0^3 F'(x)\,dx$ is defined.

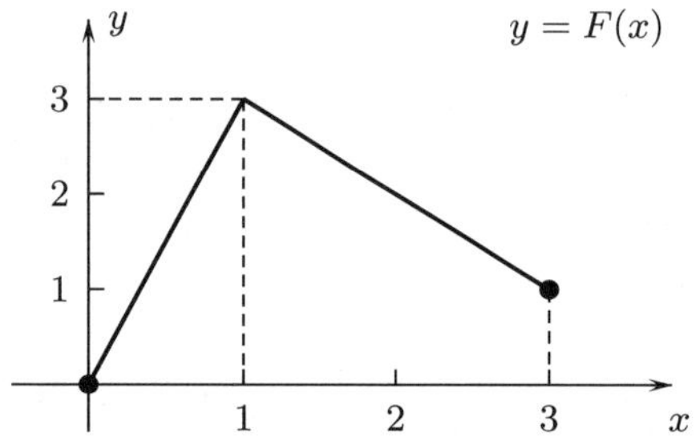

FIGURE 6

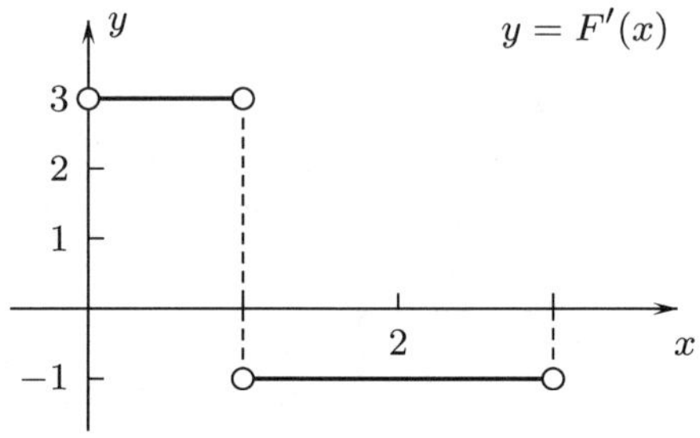

FIGURE 7

(b) We will show that the conclusion of the theorem holds by calculating both sides of the equation $\displaystyle\int_0^3 F'(x)\,dx = F(3) - F(0)$. The integral of $F'(x)$ equals the area of the rectangle above the x-axis in Figure 7, minus the area of the rectangle below the x-axis:

$$\int_0^3 F'(x)\,dx = (1)(3) - (2)(1) = 1.$$

$F(3) - F(0)$ also equals 1 because $F(3) = 1$ and $F(0) = 0$. □

[‡]A proof in this case can be obtained by applying the reasoning above to partitions that include all points where the derivative is not defined. This does not affect the existence of the limit defining the integral because adding partition points where the derivative is not defined changes the Riemann sums by amounts that tend to zero as the widths of the subintervals tend to zero.

Example 2 What is the value of $\displaystyle\int_0^2 \frac{d}{dx}(x^4+5)\,dx$?

SOLUTION By formula **(7b)** with $F(x) = x^4 + 5$,

$$\int_0^2 \frac{d}{dx}(x^4+5)\,dx = \Big[x^4+5\Big]_{x=2} - \Big[x^4+5\Big]_{x=0}$$
$$= [2^4+5] - [0^4+5] = 2^4 = 16. \square$$

Question 2 What is the value of $\displaystyle\int_4^9 F'(x)\,dx$ if $F(x) = \sqrt{x}$?

Example 3 Figure 8 shows the graph of the derivative $\dfrac{dG}{dx}(x)$ of a continuous function $G(x)$. Region A in the drawing has area 89 and region B has area 62. What is $G(6) - G(1)$?

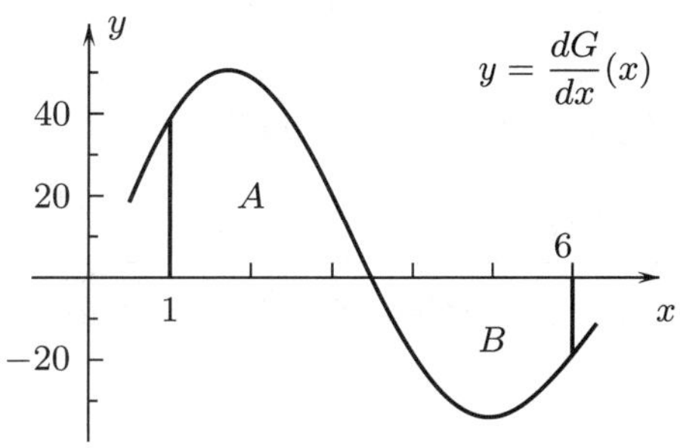

FIGURE 8

SOLUTION The Fundamental Theorem **(7a)** with $G(x)$ in place of $F(x)$ gives

$$G(6) - G(1) = \int_1^6 \frac{dG}{dx}(x)\,dx.$$

By Rule 1 of Section 4.2, the integral $\displaystyle\int_1^6 \frac{dG}{dx}(x)\,dx$ is equal to the area of region A in Figure 8 between the graph and the x-axis for $1 \le x \le 6$ where $\dfrac{dG}{dx}(x)$ is positive, minus the area of region B between the graph and the x-axis for $1 \le x \le 6$, where $\dfrac{dG}{dx}(x)$ is negative. Therefore,

$$G(6) - G(1) = \text{[Area of } A] - \text{[Area of } B] = 89 - 62 = 27. \square$$

Evaluating definite integrals with antiderivatives

How can we evaluate definite integrals if we do not know the areas of the regions set off by the graph of the function? The most important technique for making such calculations is based on the Fundamental Thereom, which enables us to derive integration formulas from differentiation formulas.

A function $F(x)$ is called an ANTIDERIVATIVE of a function $f(x)$ in an open interval if $f(x)$ is the derivative of $F(x)$ there; i.e., if

$$F'(x) = f(x).$$

If we know the antiderivative $F(x)$, we can evaluate definite integrals of $f(x)$ because, by the Fundamental Theorem 1,

$$\int_a^b f(x)\,dx = \int_a^b F'(x)\,dx = F(b) - F(a).$$

In applying this equation, we will use the symbol $\left[F(x)\right]_a^b$ for $F(b)-F(a)$ and write

$$\int_a^b f(x)\,dx = \int_a^b F'(x)\,dx = \Big[F(x)\Big]_a^b = F(b) - F(a). \tag{8}$$

Thus, to evaluate a definite integral of a power function x^n with constant n, we need to find a function whose derivative is x^n.

We notice that in finding the derivative $\frac{d}{dx}(x^n) = nx^{n-1}$, we first multiply x^n by the exponent n and then decrease the exponent by 1. Accordingly, we should be able to find an antiderivative of x^n by performing the opposite steps in the opposite order. We raise the exponent in x^n to $n+1$ and then divide by it to obtain $\frac{1}{n+1}x^{n+1}$. This is, in fact, an antiderivative of x^n: for $n \neq -1$ and for x in any open interval where x^n is defined,

$$\frac{d}{dx}\left(\frac{1}{n+1}x^{n+1}\right) = \frac{1}{n+1}[(n+1)x^n] = x^n.$$

We cannot carry out this calculation if $n = -1$ because then $n+1 = 0$ and we cannot divide by zero.

Setting $f(x) = x^n$ and $F(x) = \frac{1}{n+1}x^{n+1}$ in **(8)** gives the following integration formula:

Theorem 2 *For any constant $n \neq -1$ and for a and b in any interval where x^n is defined,*

$$\int_a^b x^n\,dx = \left[\frac{1}{n+1}x^n\right]_a^b = \frac{1}{n+1}b^{n+1} - \frac{1}{n+1}a^{n+1}. \tag{9}$$

Notice that equation **(9)** gives $\int_a^b 1\,dx = b - a$ for $n = 0$ and $\int_a^b x\,dx = \frac{1}{2}b^2 - \frac{1}{2}a^2$ for $n = 1$. In Section 4.7 we will see that integrals of x^{-1} are given by logarithms.

Example 4 What is the exact value of the integral $\int_0^1 x^2\,dx$?

SOLUTION By **(9)** with $n = 2$, for which $n+1 = 3$,

$$\int_0^1 x^2\,dx = \left[\tfrac{1}{3}x^3\right]_0^1 = \tfrac{1}{3}(1^3) - \tfrac{1}{3}(0^3) = \tfrac{1}{3}. \ \square$$

Question 3 Find exact values of **(a)** $\int_0^1 x^3\,dx$, **(b)** $\int_0^1 x^4\,dx$, and **(c)** $\int_0^1 x^{1/4}\,dx$.

Using integrals to find areas

Before the development of calculus exact areas could be determined only for regions bounded by line segments and for a few special types of curves with curved boundaries such as circles and ellipses, whose areas are given by formulas from geometry. With the integral we can find exact areas of regions bounded by the graphs of any functions whose definite integrals we can evaluate.

In particular, if $f(x)$ is nonnegative for $a \le x \le b$, then, by Rule 1 of Section 4.2, the region between its graph and the x-axis for $a \le x \le b$ is $\int_a^b f(x)\,dx$.

Example 5 Find the area of the region bounded by the graph of $3x^2 - x^3$ and the x-axis in Figure 9.

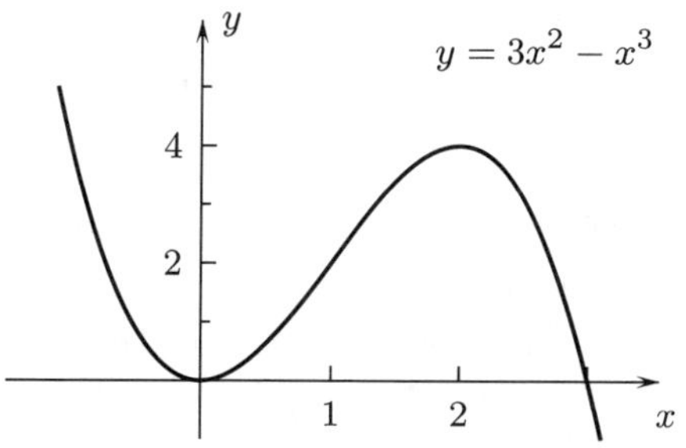

FIGURE 9

SOLUTION We first need to find the x-intercepts of the graph to determine the values of x at the left and right sides of the region. To solve the equation $3x^2 - x^3 = 0$, we first factor the polynomial to have $x^2(3-x) = 0$. This has the solutions $x = 0$ and $x = 3$, so the region extends for $0 \le x \le 3$, and its area is

$$\int_0^3 (3x^2 - x^3)\,dx = 3\int_0^3 x^2\,dx - \int_0^3 x^3\,dx.$$

Applying **(9)** with $n = 2$ and $n + 1 = 3$ to the first integral and with $n = 3$ and $n + 1 = 4$ to the second shows that the area equals

$$\begin{aligned}\int_0^3 (3x^2 - x^3)\,dx &= 3\left[\tfrac{1}{3}x^3\right]_0^3 - \left[\tfrac{1}{4}x^4\right]_0^3 \\ &= 3[\tfrac{1}{3}3^3 - \tfrac{1}{3}0^3] - [\tfrac{1}{4}3^4 - \tfrac{1}{4}0^4] \\ &= 3(9) - \tfrac{81}{4} = 6\tfrac{3}{4}. \ \square\end{aligned}$$

If $f(x)$ is ≤ 0 for $a \le x \le b$, then $\int_a^b f(x)\,dx$ is equal to the negative of the area region between the graph of $f(x)$ and the x-axis for $a \le x \le b$. Consequently, the area of the region is the negative of the integral.

Question 4 What is the area of the region between $y = -6x^{-2}$ and the x-axis for $1 \le x \le 3$ in Figure 10?

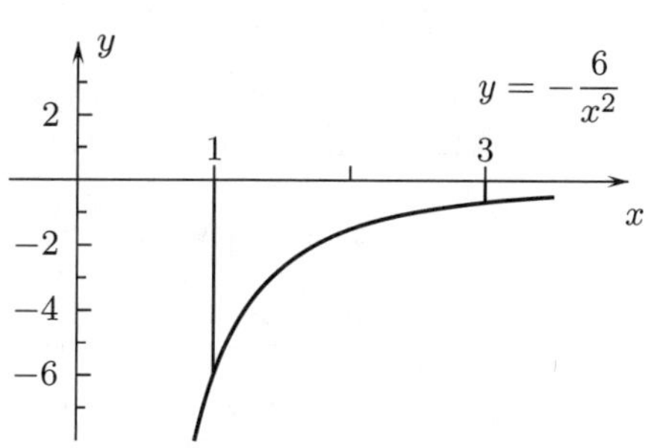

FIGURE 10

Indefinite integrals

We can use the Fundamental Theorem to find values $F(b)$ of a function at points b from its derivative and its value $F(a)$ at one particular point by using the formula,

$$F(b) = F(a) + \int_a^b \frac{dF}{dx}(x)\,dx \tag{10}$$

which is obtained by adding $F(a)$ to both sides of equation **(7a)**.

Equation **(10)** shows that the process of finding definite integrals is, in a sense, the inverse of the process of differentiation and leads us to think of integration as antidifferentiation. In particular, we use the term INDEFINITE INTEGRALS for antiderivatives and use the integration symbol to represent them, as in the following definition

---

Definition 1 *The indefinite integral* $\int f(x)\,dx$ *is an antiderivative of* $f(x)$.

We saw in equation **(9)** that $\frac{1}{1+n}x^{n+1}$ is an antiderivative of x^n for $n \neq -1$. The function $\frac{1}{1+n}x^{n+1} + C$ is also an antiderivative of x^n for any constant C because the derivative of the constant is zero. We express this result as follows.

Theorem 3 *If* $n \neq -1$*, then in any open interval where* x^n *is defined,*

$$\int x^n\,dx = \frac{1}{n+1}x^{n+1} + C \tag{11}$$

where C *is an unspecified constant.*

Notice that equation **(11)** gives $\int 1\,dx = x + C$ for $n = 0$ and $\int x\,dx = \frac{1}{2}x^2 + C$ for $n = 1$. The integral in **(10)** is called "indefinite" because the constant C, which is called the CONSTANT OF INTEGRATION, can be any number.

The differentiation formula $\frac{d}{dx}[AF(x) + BG(x)] = A\frac{dF}{dx}(x) + B\frac{dG}{dx}(x)$ translates into a similar formula for indefinite integrals:

$$\int [Af(x) + Bg(x)]\,dx = A\int f(x)\,dx + B\int g(x)\,dx. \tag{12}$$

We usually use **(12)** without referring to it explicitly, as in the next example.

Example 6 Find the antiderivative $\int \left(3\sqrt{x} + \frac{4}{x^2} - 3\right)\,dx$.

SOLUTION We switch the first two terms to exponential notation by writing

$$\int \left(3\sqrt{x} + \frac{4}{x^2} - 3\right)\,dx = \int (3x^{1/2} + 4x^{-2} - 3)\,dx.$$

Then we use Theorem 3 with $n = \frac{1}{2}$ and $n + 1 = \frac{1}{2} + 1 = \frac{3}{2}$ to find an antiderivative of $x^{1/2}$ and with $n = -2$ and $n + 1 = -1$ to find an antiderivative of x^{-2}, and use the antiderivative $3x$ of 3. We obtain

$$\int \left(3\sqrt{x} + \frac{4}{x^2} - 3\right) dx = \int (3x^{1/2} + 4x^{-2} - 3)\, dx$$
$$= 3\left[\frac{1}{3/2}x^{3/2}\right] + 4\left[\frac{1}{-1}x^{-1}\right] - 3x + C$$
$$= 2x^{3/2} - 4x^{-1} - 3x + C. \ \square$$

We can verify a formula for an indefinite integral by differentiating the result.

Question 5 Check the result of Example 5 by differentiation.

Integrating a rate of change

If a rate of change in a problem is given by an exact formula and we can find a formula for its antiderivative, then we can usually study the antiderivative by using either a definite or an indefinite integral, as we will illustrate by giving two solutions of the next example.

Example 7 Suppose that the temperature in a room is 50°F at time $t = 0$ (hours) and that the rate of change of the temperature is $12t^2 - 4t^3$ degrees per hour for $0 \leq t \leq 2$. What is the temperature at $t = 2$?

SOLUTION (USING A DEFINITE INTEGRAL) We let $T(t)$ denote the temperature at time t. Then $T(0) = 50$ and $T'(t) = 12t^2 - 4t^3$. We want to find $T(2)$.

We use the Fundamental Theorem to write $T(2) - T(0) = \int_0^2 T'(t)\, dt$ and then

$$T(2) = T(0) + \int_0^2 T'(t)\, dt = 50 + \int_0^2 (12t^2 - 4t^3)\, dt$$
$$= 50 + \left[12(\tfrac{1}{3}t^3 - 4(\tfrac{1}{4}t^4)\right]_0^2 = 50 + \left[4t^3 - t^4\right]_0^2 \qquad \textbf{(13)}$$
$$= 50 + [4(2^3) - 2^4] - [4(0^3) - 0^4]$$
$$= 50 + 32 - 16 = 66°\text{F}.$$

(USING AN INDEFINITE INTEGRAL) Since $T(t)$ is an antiderivative of its derivative,

$$T(t) = \int T'(t)\, dt = \int (12t^2 - 4t^3)\, dt$$
$$= 12(\tfrac{1}{3}t^3) - 4(\tfrac{1}{4}(t^4) + C = 4t^3 - t^4 + C. \qquad \textbf{(14)}$$

The constant of integration C in **(14)** can be determined from the given temperature of 50° at $t = 0$. Setting $t = 0$ and $T(0) = 50$ in **(8)** gives $50 = 4(0^4) - 0^3 + C$, so that $C = 50$ and by **(14)**, $T(t) = 4t^3 - t^4 + 50$. Finally, we set $t = 2$ to obtain $T(2) = 4(2^3) - 2^4 + 50 = 32 - 16 + 50 = 66°\text{F}$. This is the result we obtained in **(13)** with a definite integral. □

When is one approach better than the other?

It is probably a matter of taste whether the first or the second solution of Example 7 is preferable. There are, however, some cases where it is definitely more convenient to use either an indefinite or a definite integral. If, for instance, we want to obtain a formula for a function from a formula for its derivative and its value at one point, it is generally easiest to use an indefinite integral, as in the next question.

Question 6 Use an indefinite integral to find a formula for the x-antiderivative $F(x)$ of $10+3\sqrt{x}$ that has the value 25 at $x = 1$.

On the other hand, if we want to use a function's derivative to find how much the function changes from one particular point to another, without knowing its value at any point, it is best to use a definite integral.

Question 7 The derivative of $G(t)$ is $4t^{-3}$ for $t > 0$. Use a definite integral to determine how much $G(t)$ increases or decreases when t increases from 1 to 10.

Principles and procedures

- If $F(x)$ is continuous in an interval containing a and b, if $F'(x)$ exists at all but possibly a finite number of points between a and b, and if the Riemann integral of $F'(x)$ from a to b is defined, then

$$\int_a^b F'(x)\,dx = \Big[F(x)\Big]_a^b = F(b) - F(a).$$

- For $n \neq -1$ and a and b in any interval where x^n is defined,

$$\int_a^b x^n\,dx = \left[\frac{1}{n+1}x^n\right]_a^b = \frac{1}{n+1}b^{n+1} - \frac{1}{n+1}a^{n+1}.$$

- The indefinite integral $\int f(x)\,dx$ represents all of the antiderivatives of $f(x)$. If $F(x)$ is one antiderivative of $f(x)$ in any open interval, then $\int f(x)\,dx = F(x) + C$ in the interval.
- For any constant $n \neq -1$ and in any open interval where x^n is defined,

$$\int x^n\,dx = \frac{1}{n+1}x^{n+1} + C.$$

- Indefinite integrals of linear combinations of functions can be written as linear combinations of the integrals: For constants A and B,

$$\int [Af(x) + Bg(x)]\,dx = A\int f(x)\,dx + B\int g(x)\,dx.$$

- If you can integrate the formula for the derivative of a function, then you can obtain a formula for the function by finding the indefinite integral and determining the constant of integration from a value of the function at one point
- You can usually use either an indefinite or a definite integral to find a particular value of a function from its derivative and its value at another point.
- Check all of your formulas for integrals with your calculator or computer. To check the value of $\int_a^b f(x)\,dx$ on a Texas Instruments TI-82, TI-83, TI-85, or TI-86 calculator enter fnInt$(f(x), x, a, b)$ for the approximate value of the integral of $f(x)$ with respect to x from a to b. On a TI-81, use the Riemann sum program with $N = 50$ or $N = 100$ and $T = 0.5$. To check a formula for an indefinite integral $\int f(x)\,dx = F(x) + C$, check that $F'(x) = f(x)$ by setting $Y_1 = \text{nDeriv}(F(x), x, x)$, and $Y_2 = f(x)$. Then confirm that Y_1 and Y_2 are approximately equal by comparing their values in a table (TI-82 or TI-83) or by generating their graphs together and using the trace if the graphs are not on the screen (TI-81, TI-85, or TI-86).

Tune-Up Exercises 4.3♦

[A]*Answer provided.* [O]*Outline of solution provided.*

T1.[O] What is $f(2)$ if $f(x)$ is continuous in $[2,6]$, $\displaystyle\int_2^6 \frac{df}{dx}(x)\,dx = 5$, and $f(6) = 8$?

♦ Type 1, the Fundamental Theorem

T2.[O] Perform the integrations **(a)** $\displaystyle\int_1^9 \frac{1}{\sqrt{x}}\,dx$ and **(b)** $\displaystyle\int (3 + 4x - 5x^2)\,dx$.

♦ Type 1, evaluating an integral

T3.[O] Evaluate $\displaystyle\int_0^4 K(x)\,dx$ where $K(x) = \begin{cases} x^3 & \text{for } x < 2 \\ x^{-2} & \text{for } x > 2. \end{cases}$

♦ Type 3, evaluating an integral

T4.[O] Find the exact area of the region between $y = 1 + x^2 + x^4$ and the x-axis for $0 \le x \le 1$.

♦ Type 1, area

T5.[O] Give a formula for $W(x)$ where $W'(x) = 3x^5$ and $W(1) = 2$.

♦ Type 2, the Fundamental Theorem

T6.[O] What is $Z(4) - Z(3)$ if $Z'(t) = 4t^3$?

♦ Type 2, the Fundamental Theorem

Problems 4.3

[A]*Answer provided.* [O]*Outline of solution provided.*

Evaluate the integrals in Problems 1 through 4.

1.[O] $\displaystyle\int_{-5}^5 \frac{d}{dx}\left(\frac{x}{x^2+1}\right)dx$

♦ Type 1, the Fundamental Theorem

2.[A] $\displaystyle\int_0^5 \frac{d}{dx}\left(\sqrt{x^2+144}\right)dx$

♦ Type 1, the Fundamental Theorem

3. $\displaystyle\int_1^4 \frac{d}{dx}(2^x)\,dx$

♦ Type 1, the Fundamental Theorem

4. $\displaystyle\int_0^1 \frac{d}{dx}[(x^3+1)^4]\,dx.$

♦ Type 1, the Fundamental Theorem

5. What is $W(0)$ if $W(x)$ is continuous in $[0,300]$, $W(300) = 7$, and $\displaystyle\int_0^{300} W'(x)\,dx = 10$?

♦ Type 2, the Fundamental Theorem

6. What is $G(4)$ if $G(x)$ is continuous in $[-4,4]$, $G(-4) = 3$, and $\displaystyle\int_{-4}^4 \frac{dG}{dx}(x)\,dx = 0$?

♦ Type 2, the Fundamental Theorem

7. What is $Q(10) - Q(-10)$ if $Q(x)$ is continuous for all x and $\displaystyle\int_{-10}^{10} \frac{dQ}{dx}(x)\,dx = 50$?

♦ Type 2, the Fundamental Theorem

Find exact formulas for the indefinite integrals and exact values of the definite integrals in Problems 8 through 29.

8.[A] $\displaystyle\int (4 + 7x^3 - x^5)\,dx$

♦ Type 1, evaluating an integral

9.[A] $\displaystyle\int (9x^8 - 8x^7)\,dx$

♦ Type 1, evaluating an integral

10.[A] $\displaystyle\int (x^{1/3} + x^{1/4} + x^{1/5})\,dx$

♦ Type 1, evaluating an integral

11. $\displaystyle\int (x - \sqrt{x} + \sqrt[3]{x})\,dx$

♦ Type 1, evaluating an integral

12.[A] $\displaystyle\int_1^3 (x^2 + 10x)\,dx$

♦ Type 1, evaluating an integral

13. $\displaystyle\int (7x^{5/7} + 5x^{7/5})\,dx$

♦ Type 1, evaluating an integral

♦The Tune-up Exercises and Problems are classified by type and content. The types are (1) basic, reactive; (2) basic reflective; (3) intermediate, reactive; (4) intermediate, reflective; (5) advanced, reactive; (6) advanced, reflective; and (7) advanced, theoretical.

14.[A] $\int_8^0 (4x^{2/3} - 3)\, dx$

♦ Type 1, evaluating an integral

15. $\int_0^1 (1 - 6x^7)\, dx$

♦ Type 1, evaluating an integral

16. $\int_1^3 (x^{-2} - 2x)\, dx$

♦ Type 1, evaluating an integral

17. $\int_0^1 x^{99}\, dx$

♦ Type 1, evaluating an integral

18.[A] $\int_1^2 \left(\frac{1}{t^2} - \frac{t^2}{4}\right) dt$

♦ Type 1, evaluating an integral

19.[A] $\int_{-4}^{-1} \left(\frac{2}{y^3} - \frac{3}{y^4}\right) dt$

♦ Type 1, evaluating an integral

20. $\int_{96}^{97} 1\, dt$

♦ Type 1, evaluating an integral

21.[O] $\int (x+3)^2\, dx$

♦ Type 2, evaluating an integral

22.[A] $\int_2^1 \frac{1-w}{w^3}\, dw$

♦ Type 2, evaluating an integral

23. $\int_1^3 \left(1 + \frac{3}{x^2}\right)^2 dx$

♦ Type 2, evaluating an integral

24. $\int_0^1 (1 + 3t^2)^2\, dt$

♦ Type 2, evaluating an integral

25. $\int \frac{x - x^3}{\sqrt{x}}\, dx$

♦ Type 2, evaluating an integral

26.[O] $\int_{-8}^{9} P(t)\, dt$ where $P(t)$ equals $t^{1/3}$ for $t < 0$ and equals $t^{1/2}$ for $t > 0$.

♦ Type 2, evaluating an integral

27. $\int_0^3 G(x)\, dx$ where $G(x)$ equals x for $x < 2$ and equals $3x^2$ for $x \geq 2$.

♦ Type 2, evaluating an integral

28. $\int_{-30}^{30} H(x)\, dx$ where $H(x)$ equals 100 for $x < 0$ and equals -100 for $x > 0$.

♦ Type 2, evaluating an integral

29. $\int_0^2 S(t)\, dt$ where $S(t)$ equals $5t^4$ for $t < 1$ and equals $3t^{-4}$ for $t > 1$

♦ Type 2, evaluating an integral

30.[O] Find the exact area of the region between $y = 3x^2 + 90x^{-2}$ and the x-axis for $1 \leq x \leq 6$.

♦ Type 1, area

31.[A] What is the exact area of the region between $y = 1 - x^4$ and the x-axis for $0 \leq x \leq 1$?

♦ Type 1, area

32. What fraction of the area of the rectangle $0 \leq x \leq 9, 0 \leq y \leq 3$ is occupied by the region between $y = \sqrt{x}$ and the x-axis for $0 \leq x \leq 9$?

♦ Type 2, area

33.[A] What is the exact area of the region between $y = \sqrt{x} - x^2$ and the x-axis?

♦ Type 2, area

34. What is the total area of the two regions bounded by $y = x - 8x^{1/3}$ and $y = -x$?

♦ Type 2, area

35.[O] What is the area of the region bounded by the curve $y = x^3 - 1$, the x-axis and the line $x = -2$?

♦ Type 2, area

36. Find the area of the region bounded by the curve $y = x^3$, by the x-axis, and by the lines $x = -2$ and $x = -1$.

♦ Type 2, area

37.[O] A store takes in money at the rate of $10t - t^2$ dollars per hour t hours after it has opened. **(a)** How much money does it take in during its 10 hour day? **(b)** When is it taking in money at the greatest rate?

♦ Type 2, integral of a rate of change, etc.

38.[A] A ball that is rolling down a grass-covered hill has velocity $5 + \frac{3}{2}\sqrt{t}$ feet per second t seconds after it was thrown. **(a)** What is its initial velocity (its velocity at $t = 0$)? **(b)** When is its velocity 11 feet per second? **(c)** How far does it go in 16 seconds?

◆ Type 2, integral of a rate of change, etc.

39.[A] A ship is 5 nautical miles north of a lighthouse at midnight one morning and is traveling north with a velocity of $20 + \frac{1}{5}t$ knots (nautical miles per hour) at t AM. **(a)** When is its velocity 21 knots? **(b)** How far is it from the lighthouse at 10 AM?

◆ Type 2, integral of a rate of change, etc.

40. Air is pumped into a balloon at the rate of $3\sqrt{t}$ cubic feet per minute from time $t = 0$ (minutes) until it breaks. Its volume is one cubic foot at $t = 0$ and it breaks when its volume is 17 cubic feet. When does it break?

◆ Type 2, integral of a rate of change, etc.

41. Water flows into a tank at the rate of $10t^4 + 100t + 10$ cubic feet per day at time t (days) and leaks out at the constant rate of 6 cubic feet per day. The tank contains 25 cubic feet of water at $t = 0$. How much does it contain a day later?

◆ Type 2, integral of a rate of change, etc.

42.[A] A balloon's upward velocity is $9t - t^3$ feet per minute t minutes after it is released from six feet above the ground. **(a)** What is its upward acceleration after two minutes? **(b)** How high is it above the ground after three minutes?

◆ Type 2, integral of a rate of change, etc.

43. A small airplane is flying north with the constant air speed of 125 miles per hour. The wind is blowing south at the speed of t^3 miles per hour at time t (hours) for $0 < t \leq 6$. Where is the plane at $t = 6$ relative to its position at $t = 0$?

◆ Type 2, integral of a rate of change, etc.

44.[A] An airplane wing is to be tested in a wind tunnel. Its vertical cross section at its widest point has the shape of the region bounded by the curves $y = \pm(0.3\sqrt{x} - 0.13x - 0.35x^2 + 0.28x^3 - 0.1x^4)$ for $0 \leq x \leq 1$ in an xy-plane with distance measured in meters.[(1)] **(a)** Draw the cross section. **(b)** What is its exact area?

◆ Type 2, area

45.[O] Find the total area of the two regions bounded by the curve $y = x^{1/3}$, by the x-axis, and by the lines $x = -8$ and $x = 1$.

◆ Type 2, area

46. Compute the total area of the three regions bounded the parabola $y = x^2 - 1$, the x-axis, and the lines $x = -2$ and $x = 2$.

◆ Type 2, area

47. The curve $y = (x^2 - 1)(x - 2)$ and the x-axis form the boundaries of two regions. Find their exact total area.

◆ Type 2, area

48.[A] A ball is rolling in a straight line toward the west on a horizontal sidewalk. Its velocity toward the west is $t^3 - 4t$ feet per second at time t (seconds) for $-3 \leq t \leq 1$. **(a)** How far is it at $t = 1$ from its position at $t = -3$? **(b)** What is the total distance it travels from $t = -3$ to $t = 1$?

◆ Type 4, total and net distance from velocity

49.[A] What is the total distance an object travels between times $t = 0$ and $t = 5$ (minutes) if its velocity is $3t - t^2$ meters per minute at time t?

◆ Type 4, total and net distance from velocity

50. Find the constant k such that the region between $y = \sqrt[3]{x}$ and the x-axis for $0 \leq x \leq k$ has area 60.75.

◆ Type 4, area

---

(1) Data adapted from *McGraw Hill Encyclopedia of Science*, Vol. 1, p. 283.

51. What is the value of $\int_0^2 (ax^4 - 20x)\,dx$ if the minimum of $ax^4 - 20x$ occurs at $x = 1$?

♦ Type 4, area

52. Evaluate $\int_{-5}^{5} |x^2 - 9|\,dx$.

♦ Type 4, area

53. A car's velocity is $a + bt^2$ kilometers per hour at time t (hours). If it travels 57 kilometers from $t = 0$ to $t = 2$ and 69 kilometers from $t = 2$ to $t = 3$, what are the positive constants a and b?

♦ Type 4, distance and velocity

54. Give formulas for the antiderivative $V(x)$ of x^{-2} such that $V(1) = 2$ and $V(-1) = 3$.

♦ Type 4, an antiderivative

55. Give formulas for the function $W(x)$, continuous for all x, such that $W(0) = 3, W'(x) = 2x$ for $x < 0$, and $W'(x) = -1$ for $x > 0$.

♦ Type 4, an antiderivative

56. What is $Z(15) - Z(-15)$ if $\frac{dZ}{dx}(x)$ is an odd function that is continuous for all x?

♦ Type 4, an antiderivative

57. Which is greater, $P(x)$ or $-P(-x)$, if $P(0) = 0$ and $P'(x)$ is an even function that is continuous for all x?

♦ Type 4, an antiderivative

Section 4.4

Estimating integrals

In Section 4.3 we used integration formulas in problems where we had exact formulas for the functions involved. In this section we will deal with similar applications in cases where the data, given by graphs or tables, is not exact or where we do not have techniques for finding exact integrals of the relevant functions.

Topics:

- ***Finding approximate integrals from graphs***
- ***Finding approximate integrals from tables***
- ***The Trapezoid rule***
- ***Upper and lower sums***
- ***Simpson's rule***

A man, a plan, a dam, . . . a disaster

The St. Francis dam, constructed in 1928 northeast of the present Magic Mountain near Los Angeles, was designed by William Mulholland with the plans he had used for the Mulholland dam that still supports the Hollywood Reservoir. The sides of the canyon where the St. Francis dam was built had geological flaws that were not recognized at the time. When the reservoir was filled for the first time, the dam broke, flooding the San Francisquito and Santa Clara River valleys and drowning 450 people. This tragedy ended the previously glamorous career of the self-educated engineer Mulholland, who had been the chief architect of the Los Angeles–Owens River Aquaduct that supplies much of the water to Los Angeles.

Example 1 Figure 1 shows the graph of the rate of flow of water from the St. Francis dam in a 90-minute period starting 15 minutes before it broke.[(1)] Estimate the total volume of water that flowed from the dam for $0 \le t \le 90$ by approximating the area between the graph and the t-axis by six rectangles of equal width.

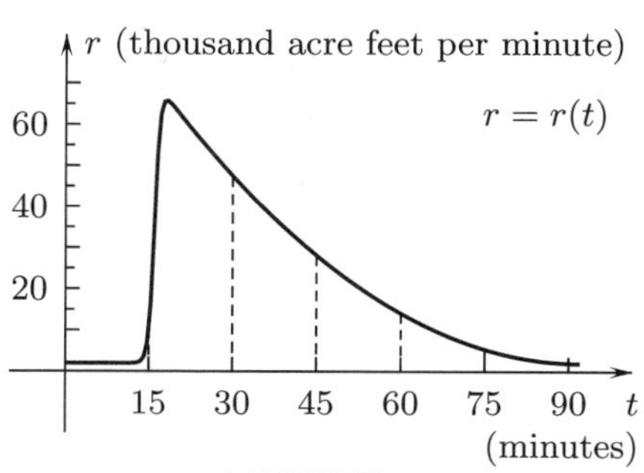

FIGURE 1

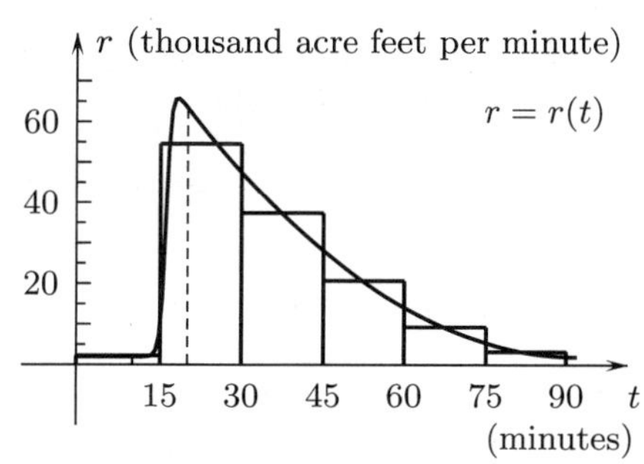

FIGURE 2

SOLUTION By the Fundamental Theorem of Calculus (Theorem 1 of Section 4.3), the total volume of water that flowed from the dam from $t = 0$ to $t = 90$ is the integral $\int_0^{90} r(t)\, dt$ of the rate of flow and equals the area of the region between its graph and the t-axis for $0 \le t \le 90$ in Figure 1.

[(1)]Data adapted from "A man, a dam and a disaster: Mulholland and the St. Francis Dam" by J. Rogers, *Ventura County Historical Society Quarterly*, Vol. 77, Ventura California: Ventura County Historical Society, 1995, p. 76.

♦ SUGGESTIONS TO INSTRUCTORS: The main topics in this section can be divided into two parts, as follows:
Part 1 (Estimating definite integrals from graphs and tables, the Trapezoid Rule, upper and lower Riemann sums; Tune-up Exercises T1–T4b; Problems 1–24, 27, 35, 36) Have students begin work on Questions 1 through 4 before class. Problems 2, 4, 5, 7, 8, 12, 15, 21, 35, 36 are good for in-class discussion.
Part 2 (Simpson's Rule and error estimates; Tune-up Exercise T4c; Problems 25, 26, 28–34) Have students begin work on Problems 25 and 31–33 before class. Problems 25 and 31–33 are good for in-class discussion

We will estimate this area visually. The vertical lines at $t = 0, 15, 30, 45, 60, 75$, and 90 in Figure 1 slice this region into six sections of equal width. We approximate each section by a rectangle, as in Figure 2, with the tops chosen to have the area of each rectangle appear approximately equal to the area of the corresponding portion of the region under the curve. The width of each rectangle is 15. We use the sketch to estimate the heights of the rectangles to be 2, 55, 40, 20, 10, and 3, so that

$$\begin{aligned}[\text{Total volume}] &\approx (2)(15) + (55)(15) + (40)(15) + (20)(15) \\ &\quad + (10)(15) + (3)(15) = 1950 \text{ thousand acre-feet.}\end{aligned}$$

The areas of the rectangles are in units of [thousand acre-feet per minute]×[minutes] = thousand-acre feet, because the heights are measured in thousand acre-feet per minute and the widths in minutes. (An acre-foot is the volume of an acre of water one foot deep.) □

Question 1 What was the approximate percent increase in the amount of water that flowed from the dam for $0 \le t \le 90$ over what would have flowed from it if it had not broken?

Using a table: Spending during the Depression

The table below lists the rate $r(t)$ at which residents of the U.S. spent money on commodities and services, as measured on January 1 every other year just before and during the Great Depression.[(2)] This data is plotted in Figures 3 and 4. The total spent from the beginning of 1929 to the beginning of 1939 is given by the integral of $r(t)$ from 1929 to 1939.

TABLE 1. RATE OF SPENDING (BILLION DOLLARS PER YEAR)

t	1929	1931	1933	1935	1937	1939
$r(t)$	77.2	60.5	45.8	55.7	66.5	66.8

Question 2 Find the approximate total amount that was spent in the U. S. on commodities and services from the beginning of 1929 to the beginning of 1939 **(a)** by calculating the left Riemann sum for $\int_{1929}^{1939} r(t)\,dt$ that is equal to the area of the rectangles in Figure 3, and **(b)** by calculating the right Riemann sum that is equal to the area of the rectangles in Figure 4.

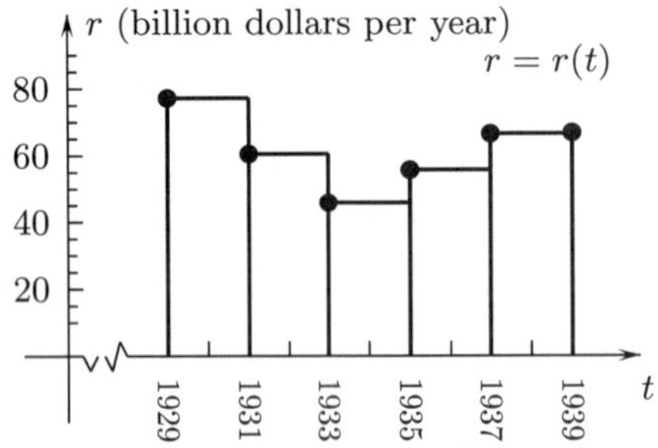

FIGURE 3

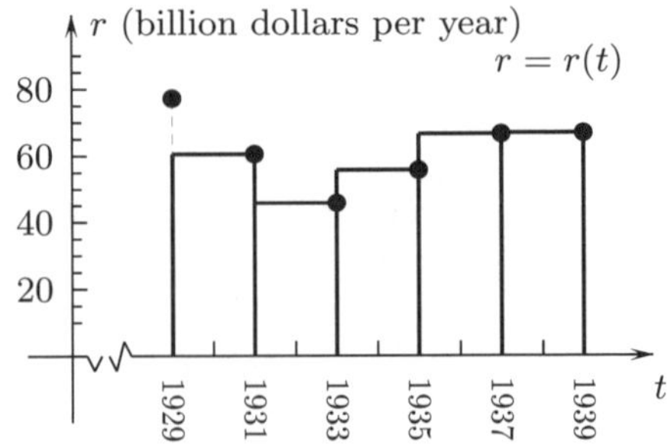

FIGURE 4

[(2)]Data adapted from *Historical Statistics of the United States*, Washington, DC: U.S. Department of Commerce, p. 319.

The Trapezoid rule

The rectangles in Figures 3 and 4 probably do not give very accurate estimates of the actual amount that was spent in the United States during the ten years from 1929 through 1938 because the rate of spending $r(t)$ was probably not constant during any of the two-year periods in the decade. The rate of spending $r(t)$ probably changed gradually from the beginning of one year to the beginning of the next. We would expect to obtain a better estimate by using the region in Figure 5 that is obtained by joining the given points on the graph with line segments. Because this region consists of five trapezoids, we can find its area with the formula for the area of a trapezoid.

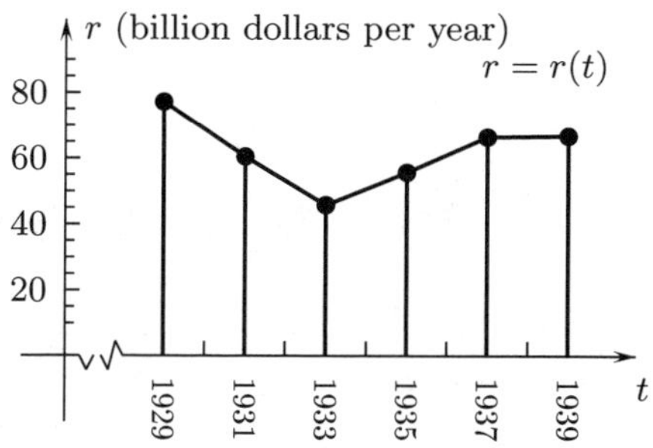

FIGURE 5

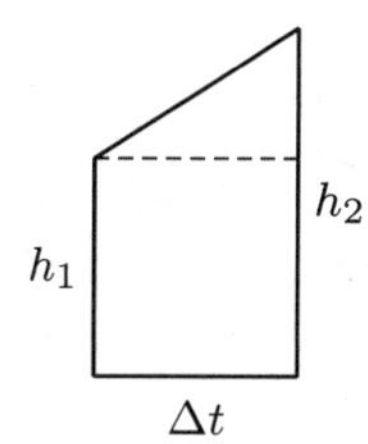

FIGURE 6

Question 3 Let h_1 and h_2 denote the height of the vertical sides of a trapezoid with $h_2 \geq h_1$ as in Figure 6, and let Δt be its width. Use the formulas for the areas of rectangles and triangles to give a formula for the area of the trapezoid.

To remember the formula from the response to Question 3, notice that it expresses the area as the average $\frac{1}{2}(h_1 + h_2)$ of the heights of the sides of the trapezoid, multiplied by its width:

$$[\text{Area of the trapezoid in Figure 6}] = \tfrac{1}{2}(h_1 + h_2)\Delta t.$$

Example 2 Find the approximate total amount that was spent in the U. S. on commodities and services from the beginning of 1929 to the beginning of 1939 by finding the areas of the six trapezoids in Figure 5.

SOLUTION With the values Table 1 and the formula for the area of a trapezoid from Question 1, we obtain

$$\begin{aligned}[\text{Total spent for } 1929 \leq t \leq 1939] &= \int_{1929}^{1939} r(t)\,dt \\ &\approx \tfrac{1}{2}(77.2 + 60.5)(2) + \tfrac{1}{2}(60.5 + 45.8)(2) + \tfrac{1}{2}(45.8 + 55.7)(2) \\ &\quad + \tfrac{1}{2}(55.7 + 66.5)(2) + \tfrac{1}{2}(66.5 + 66.8)(2) \\ &= 601 \text{ billion dollars } \square.\end{aligned}$$

Notice that the width 2 of the trapezoids is a factor in every term of the last sum and that all of the values of the function from the table, except the first and last, appear in it twice. Consequently, we can rewrite the combined area of the five trapezoids in the form,

$$\begin{aligned}&[\tfrac{1}{2}(77.2 + 60.5) + \tfrac{1}{2}(60.5 + 45.8) + \tfrac{1}{2}(45.8 + 55.7) \\ &\quad + \tfrac{1}{2}(55.7 + 66.5) + \tfrac{1}{2}(66.5 + 66.8)](2) \\ &= [\tfrac{1}{2}(77.2) + 60.5 + 45.8 + 55.7 + 66.5 + \tfrac{1}{2}(66.8)](2).\end{aligned} \quad \textbf{(1)}$$

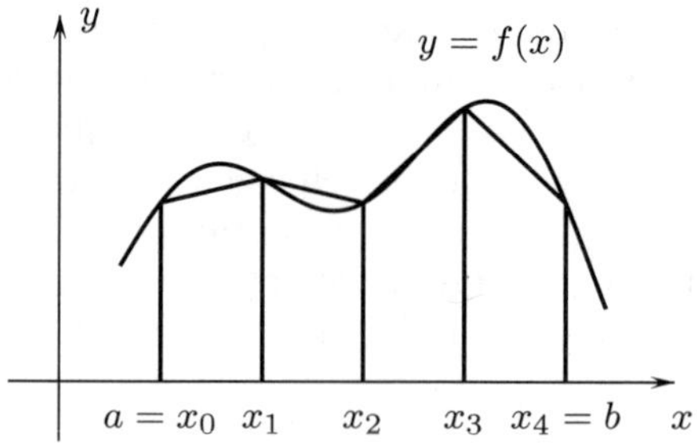

FIGURE 7

To use areas of trapezoids to approximate a general integral $\int_a^b f(x)\,dx$ with $a < b$, we take a partition $a = x_0 < x_1 < x_2 < \cdots < x_{N-1} < x_N = b$ of the interval of integration into N subintervals of equal length Δx, as indicated for the positive function in Figure 7 with $N = 4$. As shown in **(1)** for the case of Example 2, the combined areas of the trapezoids equals $\frac{1}{2}f(x_0)$, plus the sum of $f(x_j)$ for $j = 1, 2, \ldots, N-1$, plus $\frac{1}{2}f(x_N)$, all multiplied by the width Δx of the trapezoids:

Rule 1 (The Trapezoid Rule) *The Trapezoid-Rule approximation of* $\int_a^b f(x)\,dx$ *corresponding to the partition of* $[a, b]$ *into* N *subintervals of equal width* Δx *is*

$$\int_a^b f(x)\,dx \approx \left[\tfrac{1}{2}f(x_0) + f(x_1) + f(x_2) + \cdots + f(x_{N-1}) + \tfrac{1}{2}f(x_N)\right]\Delta x. \tag{2}$$

Formula **(2)** can also be used if $f(x)$ has negative values at some or all points in the partition.

Question 4 Calculate the Trapezoid-Rule approximation of $\int_0^3 (-8x^3 + 24x^2 - 16)\,dx$ with six subintervals. Draw the graph of the integrand for $0 \le x \le 3$ and the six trapezoids that form the approximation and compare the Trapezoid-Rule approximation with the exact value of the integral.

Upper and lower Riemann sums: Russia's oil production

If $f(x)$ has a maximum value in each subinterval of the partition $a = x_0 < x_1 < \cdots < x_N = b$, then the UPPER RIEMANN SUM for the partition is

$$\int_a^b f(x)\,dx \approx \sum_{j=1}^{N} f(c_j)\Delta x_j = f(c_1)\Delta x_1 + f(c_2)\Delta x_2 + \cdots + f(c_{N-1})\Delta x_{N-1} + f(c_N)\Delta x_N$$

where for each j, $f(c_j)$ is the greatest value of $f(x)$ in the jth subinterval. The upper Riemmnn sum is the greatest of all Riemann sums for $\int_a^b f(x)\,dx$ coresponding to that partition. We call it an UPPER APPROXIMATION or UPPER ESTIMATE of the integral because it is either greater than or equal to it. Figure 8 shows the rectangles corresponding to the upper Riemann sum for the function and partition of Figure 7.

We obtain the LOWER RIEMANN SUM from the same formula by choosing $f(c_j)$ to be the least value of $f(x)$ in the jth subinterval for each j (Figure 8). The lower Riemann sum is the least of all Riemann sums for the partition. It is a LOWER APPROXIMATION or LOWER ESTIMATE of the integral.

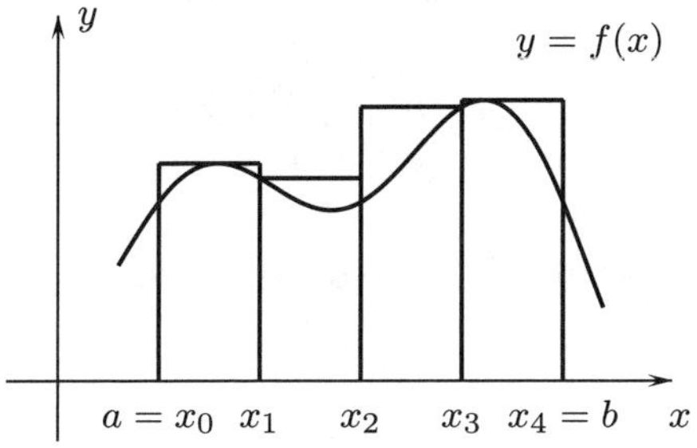

FIGURE 8

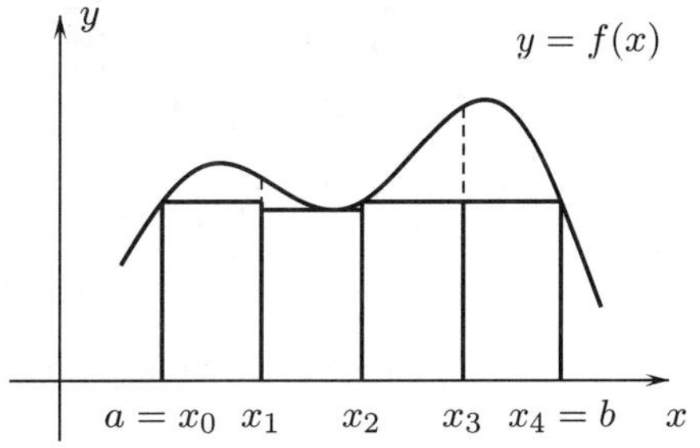

FIGURE 9

Upper and lower Riemann sums are easiest to find if, as in the next example, the function is increasing or decreasing, since then the maxima and minima occur at endpoints of the subintervals.

Example 3 The following table gives the rate of oil production, measured in million barrels per day, in Russia at the beginning of each year from 1990 through 1995.[3] The rate of production decreased during this time period. Find upper and lower estimates of the total production for $1990 \le t \le 1996$, by calculating the upper and lower Riemann sums for the integral corresponding to the partition into five equal subintervals. (Use 365 as the number of days in a year.)

TABLE 2. RATE OF OIL PRODUCTION IN RUSSIA (MILLION BARRELS PER DAY)

t	1990	1991	1992	1993	1994	1995
$r(t)$	10.2	9.0	7.8	6.8	6.0	5.7

SOLUTION We multiply $r(t)$ by the number of days in a year to obtain the rate of production measured in million barrels per year. The total production from the beginning of 1990 to the beginning of 1996 is

$$\int_{1990}^{1996} 365\, r(t)\, dt = 365 \int_{1990}^{1996} r(t)\, dt \quad \text{million barrels.}$$

The data from Table 2 is plotted in Figures 10 and 11. Because $r(t)$ is decreasing, the complete graph of $r(t)$ could look like the dashed curve in the drawings. The upper Riemann sum for $\int_{1990}^{1996} r(t)\, dt$ is the left Riemann sum and equals the area of the five rectangles in Figure 10.

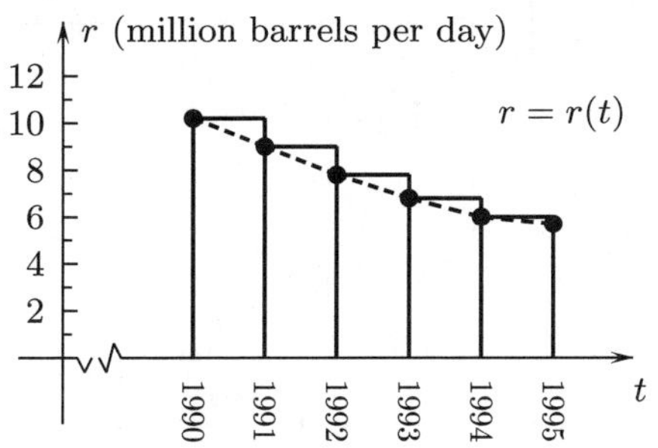

FIGURE 10

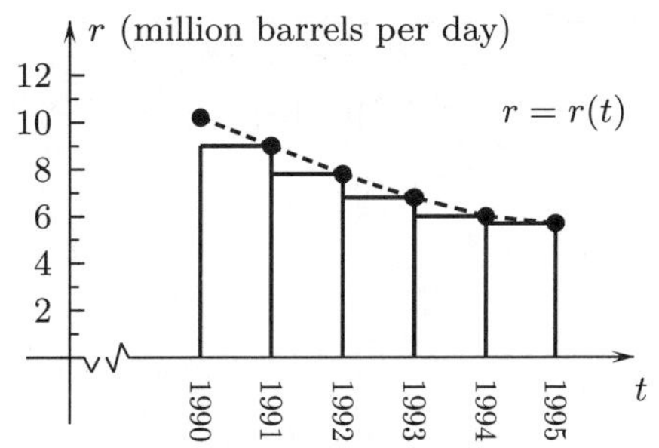

FIGURE 11

(3) Data adapted from *Time*, May 27, 1996, p. 52, Source: PlanEcon.

The lower Riemann sum is the right Riemann sum and equals the area of the five rectangles in Figure 11.

Using the values in Table 2, we see that the upper Riemann sum for $\int_{1990}^{1996} r(t)\,dt$ is

$$[(10.2)(1) + (9.0)(1) + (7.8)(1) + (6.8)(1) + (6.0)(1)] = 39.8.$$

The lower Riemann sum is

$$(9.0)(1) + (7.8)(1) + (6.8)(1) + (6.0)(1) = 35.3.$$

We conclude that, based on the data in Table 2, the total production of oil in Russia for $1990 \le t \le 1996$ was less than $365(39.8) = 14,527$ million barrels and greater than $365(35.3) = 12,884.5$ million barrels. □

Simpson's Rule

If you are using a calculator or computer program to estimate a definite integral of a function given by an exact formula, you can usually achieve all the accuracy you require with Riemann sums or the Trapezoid Rule by using a sufficiently large number of subintervals. If, however, you have to use a fixed number of subintervals because the values of the function are given only in a table or for some other reason, you can generally obtain greater accuracy by using SIMPSON'S RULE,† with which portions of the graph are approximated by parabolas.

To find a Simpson-Rule approximation of $\int_a^b f(x)\,dx$ for $a < b$, we use a partition $a = x_0 < x_1 < x_2 < \cdots < x_{2M} = b$ into an even number of subintervals of equal width, where M is a positive integer. We approximate the graph of $f(x)$ over the first two subintervals by the parabola through the points on the graph at $x = x_0, x = x_1$, and $x = x_2$; we approximate the graph over the next two subintervals by the parabola through the points at $x = x_2, x = x_3$, and $x = x_4$; and we continue in this manner until the entire graph has been approximated.

Figures 12 and 13 illustrate the case of $M = 3$, for which there are 6 subintervals. The values of a function $f(x)$ at the seven points $a = x_0, x_1, x_2, \ldots, x_6 = b$ are plotted in Figure 12. Figure 13 shows the three portions of parabolas that are used to approximate the entire graph of $f(x)$.

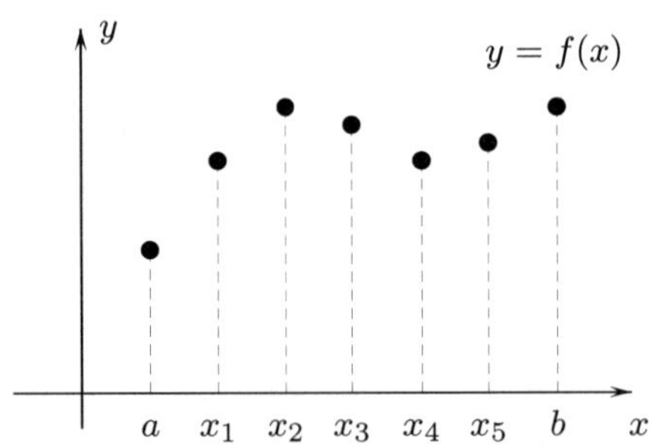

FIGURE 12

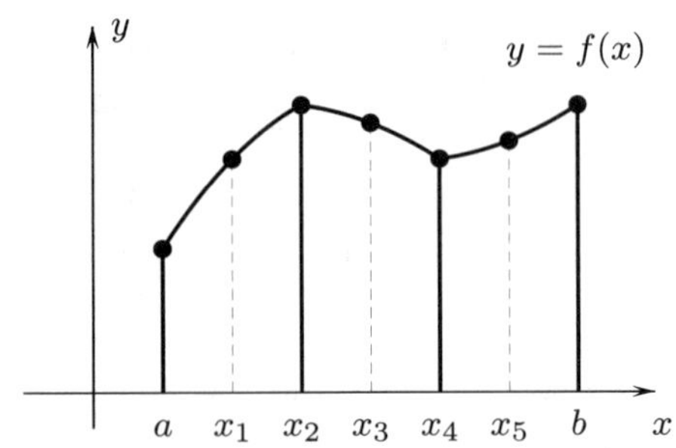

FIGURE 13

†Named after the English mathematician, Thomas Simpson (1710–1761).

To find a formula for Simpson's Rule in terms of values of the function $f(x)$, we need to find formulas for the quadratic polynomials whose graphs are used to approximate the graph of $f(x)$ and then integrate those formulas. The somewhat lengthy calculations that this requires are outlined in Problem 34 at the end of this section. The result is that the integral of the kth approximating quadratic polnomial from $x = x_{2k-2}$ to $x = x_{2k}$ is

$$\tfrac{1}{3}[f(x_{2k-2}) + 4f(x_{2k-})) + f(x_{2k})]\Delta x \quad (k = 1, 2, \ldots, M)$$

where $\Delta x = (b-a)/(2M)$ is the width of the subintervals. Adding these expressions gives the following rule:

Rule 2 (Simpson's Rule) *The Simpson-Rule approximation of* $\int_a^b f(x)\,dx$ *with* $2M$ *subintervals is*

$$\begin{aligned}\int_a^b f(x)\,dx &\approx \sum_{k=1}^{M} \tfrac{1}{3}[f(x_{2k-2})) + 4f(x_{2k-1}) + f(x_{2k})]\Delta x \\ &= \tfrac{1}{3}[f(x_0) + 4f(x_1) + 2f(x_2) + 4f(x_3) + 2f(x_4) + 4f(x_5) + 2f(x_6) + 4f(x_7) \\ &\qquad + \cdots + 2f(x_{2M-2}) + 4f(x_{2M-1}) + f(x_{2M})]\Delta x\end{aligned} \quad \textbf{(3)}$$

where $a = x_0 < x_1 < x_2 < \cdots < x_{2M-1} < x_{2M} = b$ *is the partition of* $[a, b]$ *into* $2M$ *subintervals of width* $\Delta x = (b-a)/(2M)$.

Notice that inside the square brackets on the right of **(3)**, all of the numbers $f(x_j)$ with an odd subscript j are multiplied by 4 and that, for even j, the numbers $f(x_j)$ except the first and the last ($f(x_0)$ and $f(x_{2M})$) are multiplied by 2.

Example 4 Calculate the Simpson-Rule approximation of $\int_0^1 \frac{1}{x^2+1}\,dx$ with four subintervals.

SOLUTION We first calculate the values of $f(x) = 1/(x^2+1)$ at the points in the partition $0 < 0.25 < 0.5 < 0.75 < 1$ of $[0, 1]$:

x	0	0.25	0.5	0.75	1
$\frac{1}{x^2+1} \doteq$	1	0.94117647	0.8	0.64	0.5

Since the width Δx of the subintervals is 0.25, the Simpson-Rule approximation is

$$\begin{aligned}\int_0^1 \frac{1}{x^2+1}\,dx &\approx \tfrac{1}{3}[f(0) + 4f(0.25) + 2f(0.5) + 4f(0.75) + f(1)](0.25) \\ &\doteq \tfrac{1}{3}[1 + 4(0.94117468) + 2(0.8) + 4(0.64) + 0.5](0.25) \\ &\doteq 0.78539216. \ \square\end{aligned}$$

Principles and procedures

- If $f(x)$ is given only by its graph, a fairly good approximation of $\int_a^b f(x)\,dx$ with $a < b$ can often be obtained with a few rectangles by choosing each rectangle so it appears to have approximately the same area as the portion of the region between the graph of $f(x)$ and the x-axis that it is approximating. The Trapezoid Rule can also be used to approximate definite integrals of functions given by graphs.

- Approximate values of definite integrals of functions given by tables can be found with Riemann sums or by using the Trapezoid Rule.
- If you have a formula for $f(x)$, you can use the Riemann-sum program of Section 4.2 to calculate Trapezoid-rule approximations of $\int_a^b f(x)\,dx$ as the average of left and right Riemann sums (see Problem 27 below).
- If $f(x)$ has a maximum and a minimum value in each subinterval of a partition of $[a, b]$, then the upper Riemann sum for $\int_a^b f(x)\,dx$ with that partition is the Riemann sum using the maximum values, and the lower Riemann sum is the Riemann sum using the minimum values. The upper Riemann sum is the greatest and the lower Riemann sum is the least of all possible Riemann sums for the particular partition.
- If $f(x)$ is either nonincreasing or nondecreasing in the entire interval of integration, then upper and lower Riemann sums are obtained by evaluating $f(x)$ at either the left or right endpoints of all the subintervals.
- Simpson's Rule often gives considerably more accurate estimates of integrals than Riemann sums or the Trapezoid rule with the same number of subintervals. For Simpson's Rule, the interval of integration is partitioned into an even number of subintervals and the graph of the integrand in pairs of subintervals is approximated by parabolas.
- Most graphing calculators and computer calculus software have procedures for finding approximate values of definite integrals of functions given by formulas that are much more accurate than those obtained from Riemann sums, the Trapezoid Rule, or Simpson's Rule with small numbers of subintervals. The calculator and computer procedures are generally fairly complicated but, like the Trapezoid Rule and Simpson's Rule, use approximate functions whose integrals can be calculated exactly.

Tune-Up Exercises 4.4♦

[A]*Answer provided.* [O]*Outline of solution provided.*

T1.[O] **(a)** Use six rectangles of equal width with the graph of $g(x)$ in Figure 14 to find the approximate value of $\int_0^3 g(x)\,dx$. **(b)** Use the Trapezoid Rule to estimate the integral of part (a) .

♦ Type 1, a Riemann sum and the Trapezoid Rule

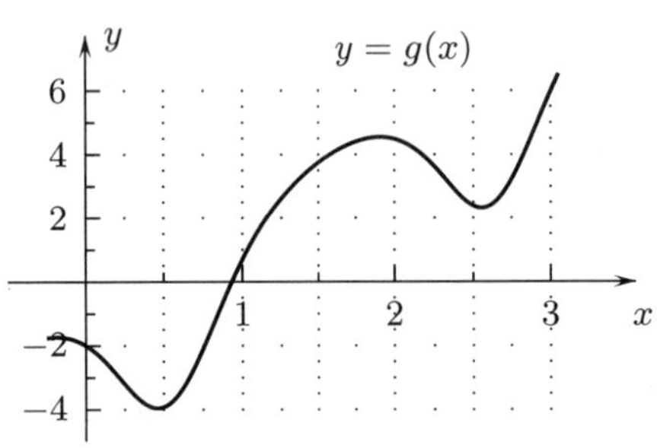

FIGURE 14

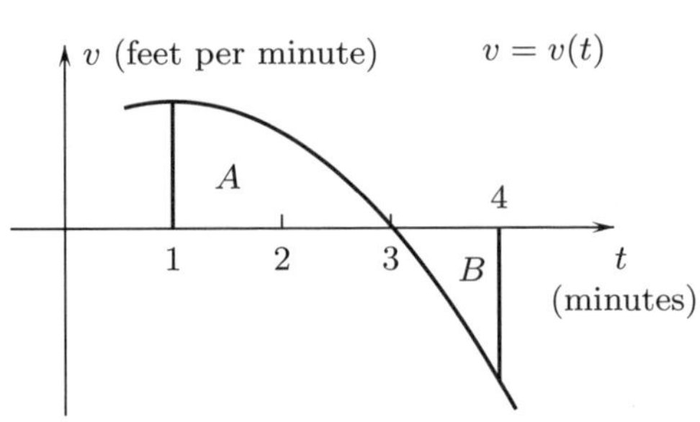

FIGURE 15

♦The Tune-up Exercises and Problems are classified by type and content. The types are (1) basic, reactive; (2) basic reflective; (3) intermediate, reactive; (4) intermediate, reflective; (5) advanced, reactive; (6) advanced, reflective; and (7) advanced, theoretical.

T2.^{O} An object moving on an s-axis with coordinates given in feet is at $s = 10$ at time $t = 1$. The graph of its velocity in the positive s-direction as a function of t is shown in Figure 15. The area of region A in the drawing is 16.5 and the area of region B is 6.75. Where is the object at $t = 4$?

◆ Type 1, an integral in terms of areas

T3.^{O} Values of an increasing function $J(x)$ are given below. Calculate **(a)** the Trapezoid Rule approximation, **(b)** the lower Riemann sum approximation, and **(c)** the upper Riemann sum approximation of $\int_1^2 J(x)\,dx$ with five equal subintervals.

◆ Type 1, the Trapezoid Rule and upper and lower sums

x	1.0	1.2	1.4	1.6	1.8	2.0
$J(x)$	30	32	33	38	40	45

T4.^{O} Calculate **(a)** the Midpoint-Rule, **(b)** the Trapezoid-Rule, and **(c)** the Simpson-Rule approximations of $\int_1^9 x^{1/3}\,dx$ with four subintervals.

◆ Type 1, Midpoint, Trapezoid, and Simpson's Rules

Problems 4.4

^{A}*Answer provided.* ^{O}*Outline of solution provided.*

1.^{O} Chlorofluorocarbons (CFC's) that are used as coolants in refrigerators and air conditioners and for other purposes are catalysts in chemical processes that destroy ozone in the upper atmosphere. Figure 16 shows the graph of the rate of world production of CFC's from 1950 to 1990.[(4)] **(a)** What was the approximate percent increase in the rate of CFC production from 1950 to its highest value? **(b)** Estimate the total production of CFC's from the beginning of 1950 to the beginning of 1990.

◆ Type 1, an integral of a rate of change from a graph

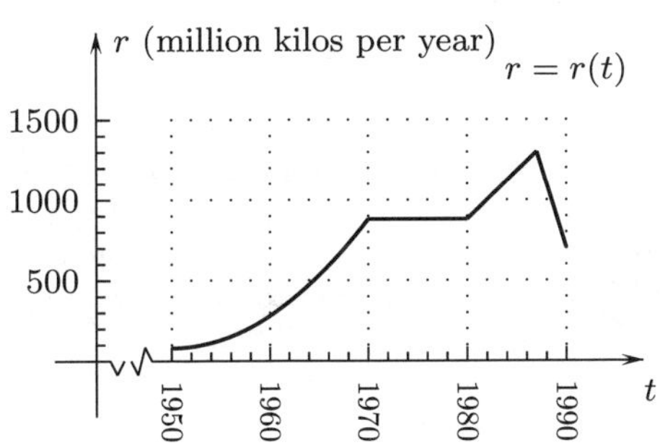

FIGURE 16

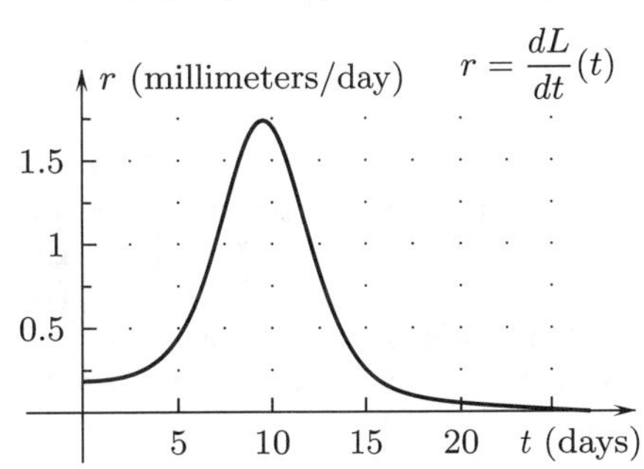

FIGURE 17

2. The graph of the rate of regeneration of a tadpole's lost tail is given in Figure 17. What is the approximate final length of the tail?

◆ Type 1, an integral of a rate of change from a graph

3.^{A} Figure 18 shows the graph of the rate $r(t)$ (milliliters per hour) at which oxygen is taken into a kiwi egg during the 40 days between when the egg is laid and when it is hatched.[(5)] **(a)** Approximately how much oxygen is taken during the entire time? (Be careful with the units.) **(b)** Approximately when is the rate of intake of oxygen increasing the most rapidly?

◆ Type 1, an integral of a rate of change from a graph

[(4)]Data adapted from "Making environmental treaties work" by H. French, *Scientific American, December, 1994, San Francisco: W. H. Freeman and Company, p. 97.*

[(5)]Data adapted from "The Kiwi" by W. Calder III, *Scientific American*, July, 1978, New York: NY: Scientific American, Inc., p.140.

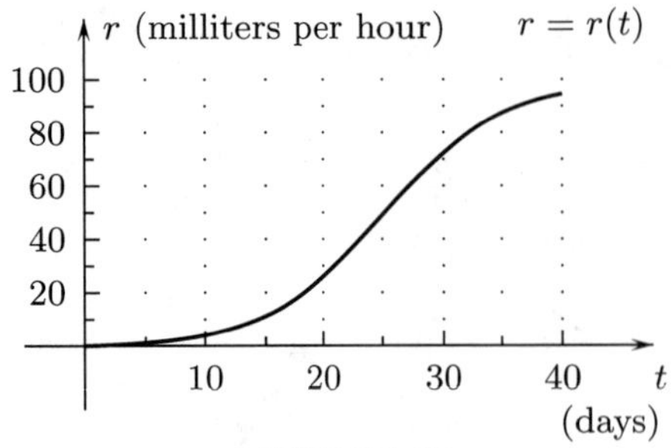

FIGURE 18

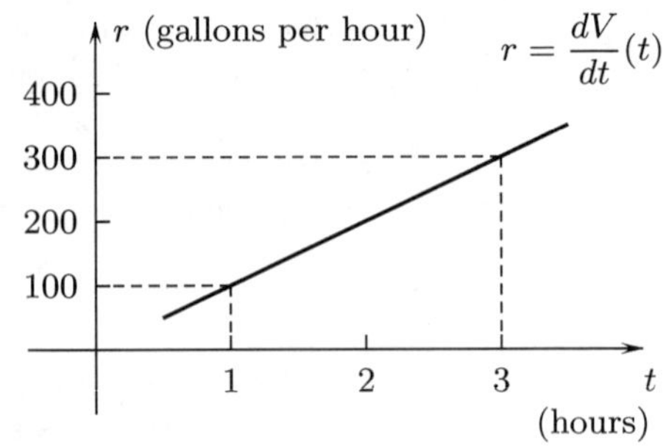

FIGURE 19

4.[A] A pool contains 100 gallons of water at $t = 1$ (hours), and the rate of flow of water into it is the linear function whose graph is the line in Figure 19. **(a)** Find an equation for the line. **(b)** Use the result of part (a) to find a formula for the volume of water in the pool as a function of t.

♦ Type 2, a linear rate of change

5.[A] The rate at which previously owned houses were being sold in the U. S. at the beginning of 1972, 1977, 1982, 1987, and 1992 is given in the table below.[(6)] Use the Trapezoid Rule to estimate the total number of sales for $1972 \le t \le 1992$. Then draw the four trapezoids whose areas give the approximation.

♦ Type 2, the Trapezoid Rule

RATE OF SALE
(MILLION HOMES PER YEAR)

t	1972	1977	1982	1987	1992
$r(t)$	2.2	4.0	2.0	3.5	3.3

6. The dots in Figure 20 give the rate at which world airlines were carrying passengers at the beginning of 1975, 1980, 1985, 1990, and 1995.[(7)] **(a)** Draw a line that approximates this data, and find an approximate formula for it. **(b)** The formula from part (a) gives a linear function $L(t)$ that approximates the rate of passenger use. Use it to give an estimate of the total number of airline passengers that were carried from the beginning of 1975 to the beginning of 1995.

♦ Type 2, a linear model of a rate of change

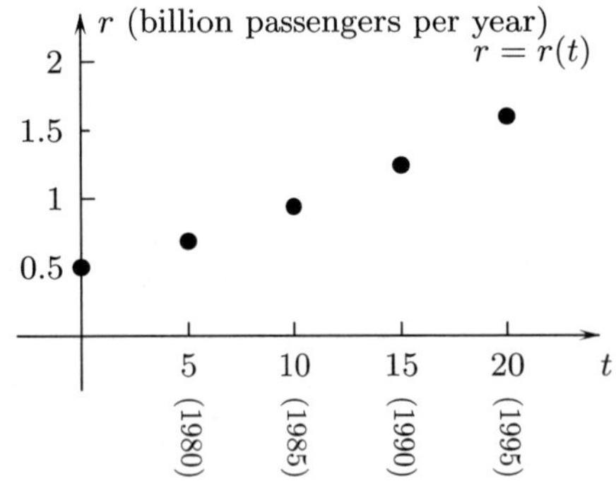

FIGURE 20

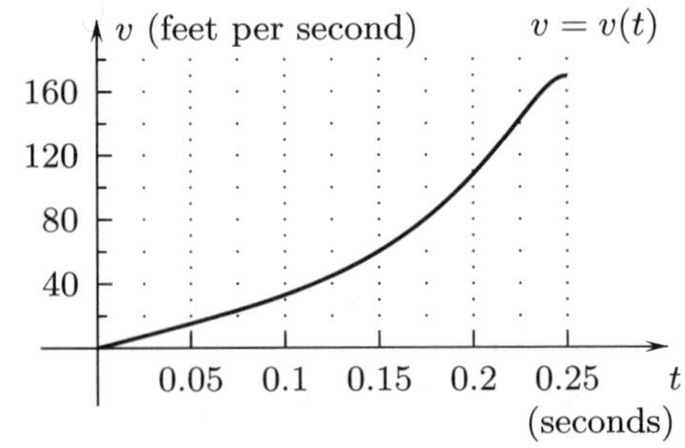

FIGURE 21

[(6)]Data adapted from *1993 Statistical Abstract of the United States*, Washington, DC: U. S. Department of Commerce, 1993, p. 710.

[(7)]Data adapted from *New York Times*, December 9, 1996. New York, NY: The New York Times Company, Source: Airclaims Limited

7. A golf club is swung at a golf ball at time $t = 0$ (seconds) and hits the ball at $t = 0.25$. The curve in Figure 21 is the graph of the derivative $v(t)$ of the distance that the head of the club travels up to time t.[8] Approximately how long does it take for the club head to travel half the distance that it goes from $t = 0$ to $t = 0.25$?

◆ Type 2, an integral from a graph

8. The next table gives the rates at which three-piece steel and two-piece aluminum beverage cans were being produced in the U. S. at the beginnings of six years.[9] **(a)** What were the average rates of change with respect to time of the rates of production of both types of cans from the beginning of 1970 to the beginning of 1980? **(b)** Use Trapeizoid-Rule approximations to estimate how many more aluminum cans were produced than steel cans during the 35 years from the beginning of 1965 to the beginning of 1990.

RATE OF PRODUCTION (BILLION CANS PER YEAR)

t	1965	1970	1975	1980	1985	1990
Steel	18	28	22	4	0	0
Aluminum	1	4	17	42	63	92

◆ Type 2, the Trapezoid Rule, etc.

9. A tank contains 1000 gallons of oil at $t = 0$ (hours). Figure 22 shows the graph of the rate of change of the volume for $0 \leq t \leq 50$. **(a)** When is oil flowing in and when is it flowing out of the tank? **(b)** Approximately how much oil is in the tank at $t = 50$? **(c)** What is the approximate maximum volume of oil in the tank?

◆ Type 2, an integral in terms of area, etc.

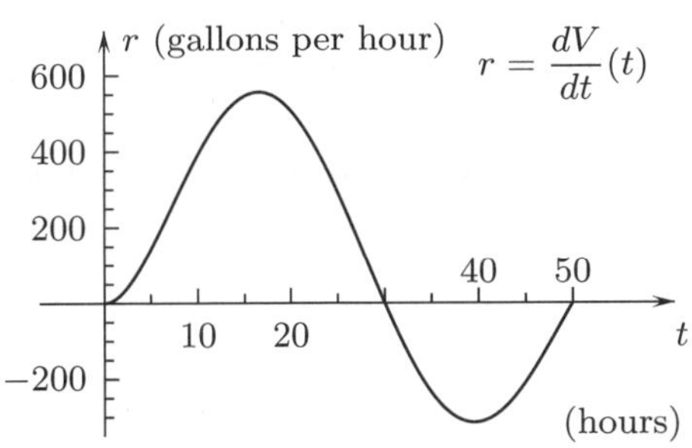

FIGURE 22

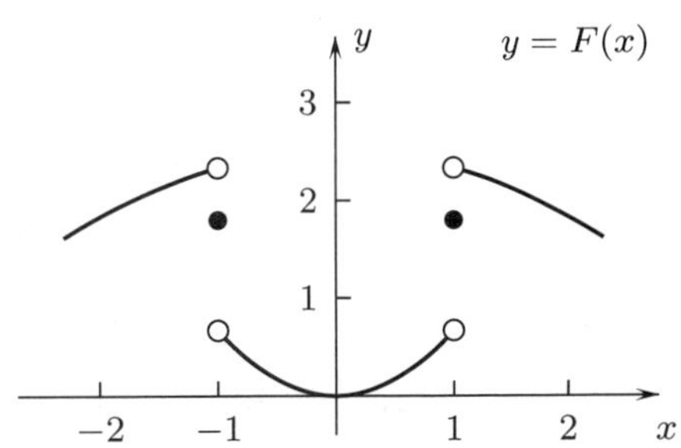

FIGURE 23

10. What is the approximate value of $\int_{-2}^{2} F(x)\,dx$ for the function $F(x)$ of Figure 23?

◆ Type 2, an integral in terms of area

11. The graph of $G(x)$ in Figure 24 consists of three semicircles with the same radius. What is their radius if $\int_{0}^{b} G(x)\,dx = 8 + 2\pi$? (The scales are equal on the axes.)

◆ Type 2, an integral in terms of area

12. The function of Figure 25 gives the rate of oxygen intake by a man who bicycles strenuously for 20 minutes and then rests.[10] Approximately what percent of the oxygen that he breathes in for $0 \leq t \leq 40$ does he consume while bicycling?

◆ Type 2, an integral in terms of area

[8] Data adapted from *The Physics of Golf* by T. Jorgensen, Springer-Verlag, 1994, p. 15.

[9] Data adapted from "The aluminum beverage can" by W. Hosford and J. Duncan, *Scientific American*, Sept.,1994, New ew York, NY: Scientific American, Inc., p. 52.

[10] Data adapted from *Handbook of Physiology*, W. Fenn and H. Rahn editors, Washington, DC: American Physiological Society, 1965, p. 945.

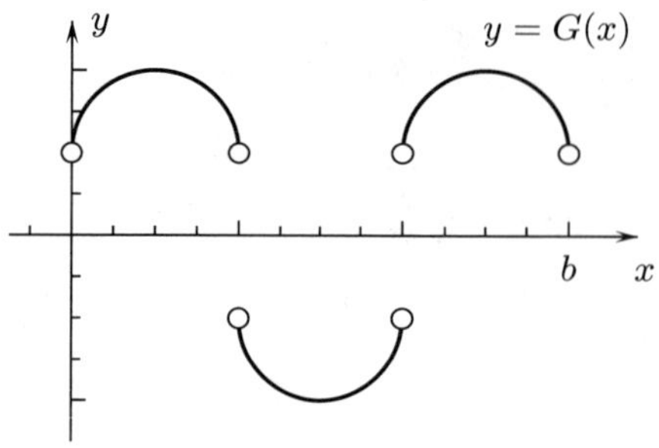

FIGURE 24

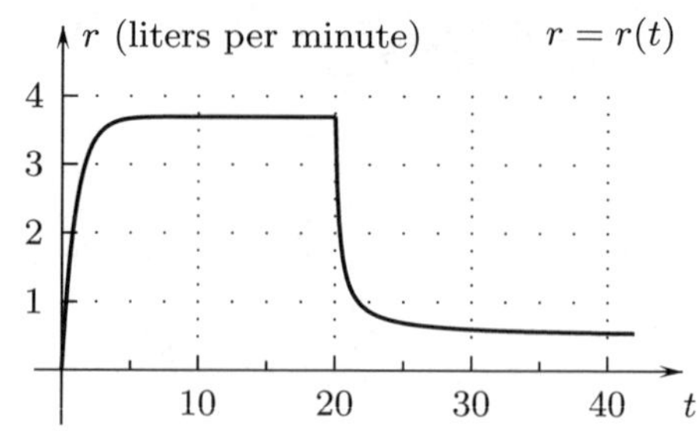

FIGURE 25

13.º Find the exact upper and lower Riemann sums for $\int_0^2 (x^3 - 3x + 3)\, dx$ corresponding to the partition of $[0, 2]$ into six equal subintervals.

♦ Type 2, upper and lower Riemann sums

14. Calculate with four-decimal-place accuracy, the upper and lower Riemann sums for $\int_{-1}^{1} \frac{1}{(x^2+1)}\, dx$ corresponding to the partition of $[-1, 1]$ into six equal subintervals.

♦ Type 2, upper and lower Riemann sums

15. Calculate with four-decimal-place accuracy, the upper and lower Riemann sums for $\int_1^3 (3 - \frac{2}{x^3})\, dx$ with four equal subintervals. Which is closer to the exact value of the integral?

♦ Type 2, upper and lower Riemann sums

16. As we will see in Section 4.7, the integral $\int_1^5 \frac{1}{x}\, dx$ equals $\ln(5)$, where $\ln(x)$ is the natural logarithm of x. Calculate the upper and lower Riemann sums for $\int_1^5 \frac{1}{x}\, dx$ corresponding to the partition into four equal subintervals. Which is closer to the exact value of the integral?

♦ Type 2, upper and lower Riemann sums

17. Use the Riemann-sum program of Section 4.2 to find a positive integer N such that the integral $\int_0^3 (5x^4 - 12x^2)\, dx$ differs form its Midpoint-Rule approximation with N equal subintervals by less than 0.5 and more than 0.1.

♦ Type 2, using the Riemann-sum program

18. The table below lists the per-capita rate of consumption of wood $r(t)$ (cubic feet per person per year) and the population (millions) in the United States at the beginnings of 1900, 1920, 1940, and 1960.[11] Based on this data, **(a)** what was the percent increase or decrease in the rate of consumption for the entire country from the beginning of 1900 to the beginning of 1960, and **(b)** approximately how much wood was consumed by the entire country from the beginning of 1900 to the beginning of 1960? (Use the Trapezoid Rule in part (b).)

♦ Type 2, approximating an integral from a table, etc.

Year	1900	1920	1940	1960
$r(t)$	157	113	85	63
Population	76	106	132	178

[11] Data adapted from *Historical Statistics of the United States, Bicentennial Edition, Part 2*, Washington, DC: U. S. Bureau of the Census, 1975, p. 540.

19. The next table gives the rate $r(t)$ (million metric tons per year) at which grain was produced in the world at ten-year intervals from 1950 to 1990.[12] The linear function $G(t) = 581 + 29(t - 1950)$ approximates this data. **(a)** For what year does the value of $G(t)$ differ the most from values in the table? **(b)** Use $G(t)$ to estimate the total world grain production from the beginning of 1950 to the beginning of 1990. **(c)** Use the values in the table and the Trapezoid Rule to estimate the total world grain production from the beginning of 1950 to the beginning of 1990.

♦ Type 2, approximating an integral from a table, etc.

Year	1950	1960	1970	1980	1990
$r(t)$	631	847	1096	1447	1780

20. Figure 26 shows the upward acceleration of a space ship during the three stages of its launch.[13] The first and second parts of the curve correspond to the time periods when the space ship is being propelled by its first and second booster rockets, which are jettisoned. After that, the space ship is propelled by its own rocket. The acceleration is measured in g's (multiples of the acceleration due to gravity). **(a)** What is the approximate maximum acceleration of the space ship? **(b)** What is space ship's approximate upward velocity five minutes after take-off?

♦ Type 2, an integral in terms of area, etc.

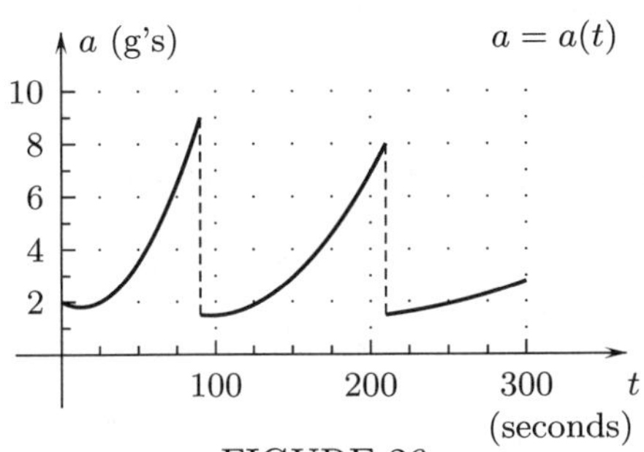

FIGURE 26

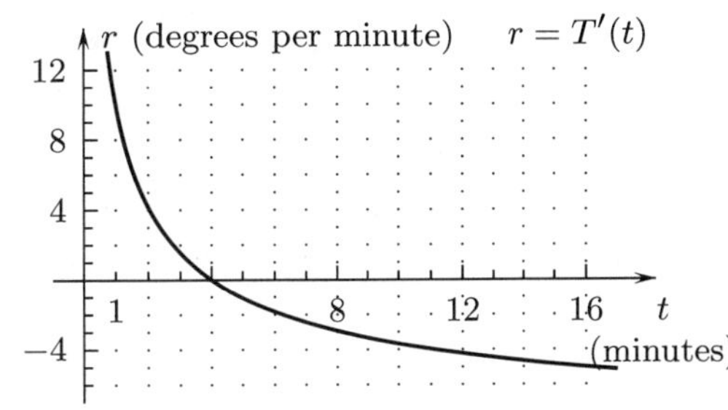

FIGURE 27

21. The graph of the rate of change of the temperature in a freezer as a function of time is given in Figure 27. The temperature is 25°F at $t = 1$ (minutes). **(a)** When for $1 \le t \le 16$ is the temperature increasing and when is it decreasing? **(b)** Is the graph of the temperature concave up or concave down for $1 \le t \le 16$ and why? **(c)** Sketch the graph of the temperature for $1 \le t \le 16$ by plotting the point at $t = 1$ and the approximate points at $t = 16$ and at the maximum temperature and by using the results of parts (a) and (b).

♦ Type 2, an integral in terms of area, etc.

22. A 1793 epidemic of yellow fever killed approximately 10% of the population of Philadelphia, Pennsylvania. No accurate health records were kept, but a historian, Mattew Carey, estimated the death rate by counting new graves. His data is shown in Figure 28 which also shows Carey's picture.[14] From this information he determined that approximately 4,000 died in the time period shown in the graph. What units are used on the vertical axis in the drawing? (The numbers on the vertical scale run from 0 to 120.)

♦ Type 4, an integral in terms of area, etc.

---

(12) Data adapted from *Vital Signs*, 1992, p. 25. Source: USDA, *World Grain Database*.

(13) Data adapted from *Textbook of Medical Physiology* by A. Guyton, Philadelphia, PA: W. B..Saunders Co., 1981, p. 548.

(14) "The Philadelphia yellow fever epidemic of 1793" by K. Foster, M. Jenkins, and A. Toogood, *Scientific American*, August, 1988, New York, NY: Scientific American, Inc., p 93.

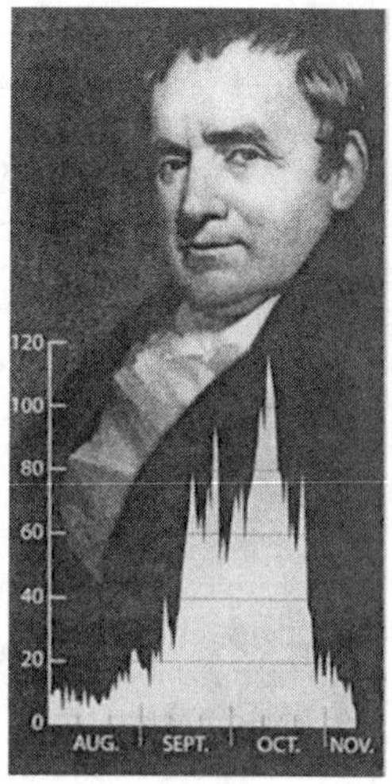

FIGURE 28

FIGURE 29

23. Figure 29 shows a frog who is about to try to catch a cockroach. The function $s(t)$ drawn with a heavy line in Figure 30 gives the frog's distance, as a function of time, from the initial location of the cockroach—which turns and escapes.[15] The dip in the graph between $t = 0.15$ and $t = 0.25$ occurs when the frog is sticking out its tongue to catch the cockroach, and the distance from the frog's mouth to the initial location of the cockroach during that period is given by the curve that is drawn with a finer line. Figure 31 shows the graphs of the t-derivatives of the functions in Figure 31. **(a)** Approximately how far does the frog stick out its tongue? **(b)** Why do the two regions between the two curves in Figure 31 have the same area?

♦ Type 4, an integral in terms of area, etc.

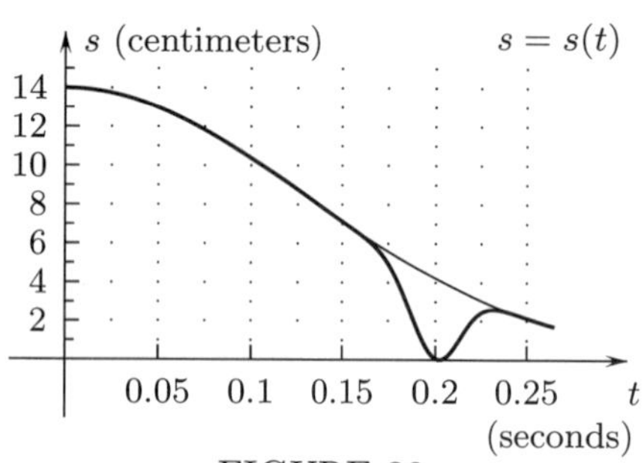

FIGURE 30

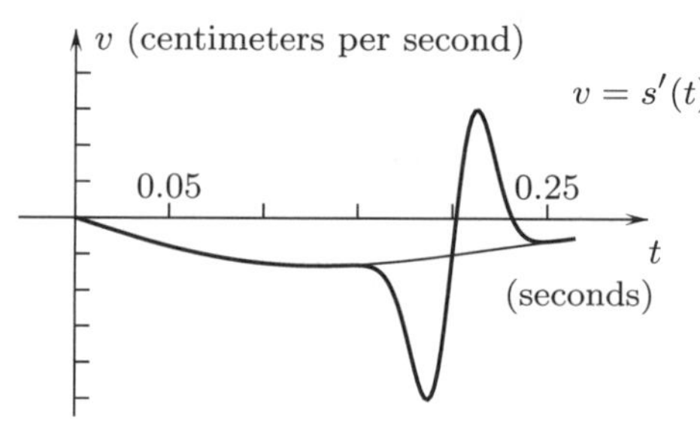

FIGURE 31

24. Determine which is the greatest and which is the least of the three integrals $\int_2^5 (x^2+1)\,dx$, $\int_2^5 (x^3+x)\,dx$, and $\int_2^5 \frac{9}{x}\,dx$ without evaluating them. (Compare the integrands.)

♦ Type 4, comparing integrals

25.[A] Use **(a)** the Midpoint Rule, **(b)** the Trapezoid Rule, and **(c)** Simpson's Rule with eight subintervals to estimate $\int_0^4 \sqrt{x}\,dx$. Do your calculations with four-decimal place accuracy. Then find the exact value of the integral and calculate the errors in the three approximations.

♦ Type 3, the Midpoint-, Trapezoid- and Simpson's Rules

26. Carry out the instructions of Problem 25 for $\int_0^1 x^3\,dx$ with four subintervals.

♦ Type 3, the Midpoint-, Trapezoid- and Simpson's Rules

[15] Drawing and data adapted from "The escape system of the cockroach" by J. Cambi, *Scientific American*, December, 1980, New York: NY: Scientific Americam, Inc., p. 166.

27. Show that any Trapezoid-Rule approximation of an integral is the average of the left and right Riemann sums with the same partition.

♦ Type 3, the Trapezoid Rule and Riemann sums

28. (ERROR ESTIMATES WITH THE MIDPOINT AND TRAPEZOID RULES) It is shown in advanced courses that if $f(x)$ has a continuous second derivative in $[a, b]$ and M_2 is the maximum of $|f''(x)|$ for $a \le x \le b$, then the Midpoint- and Trapezoid-rule approximations with N equal subintervals of length $\Delta x = (b - a)/N$ satisfy

$$\left| \int_a^b f(x)\,dx - \begin{bmatrix} \text{The Midpoint-rule} \\ \text{approximation} \end{bmatrix} \right| \le \tfrac{1}{24}(b-a)M_2(\Delta x)^2 \tag{4}$$

$$\left| \int_a^b f(x)\,dx - \begin{bmatrix} \text{The Trapezoid-rule} \\ \text{approximation} \end{bmatrix} \right| \le \tfrac{1}{12}(b-a)M_2(\Delta x)^2. \tag{5}$$

(a[A]) Calculate $\int_0^1 (x^4 + 2x)\,dx$ and its Midpoint-Rule approximation with four equal subintervals. Then compare the error in the approximation with the estimate given by formula **(4)**.

(b[A]) Calculate $\int_1^4 x^{3/2}\,dx$ and its Trapezoid-Rule approximation with six equal subintervals. Then compare the error in the approximation with the estimate given by formula **(5)**.

(c) Calculate $\int_1^{0.8} x^{3/4}\,dx$ and its Midpoint-Rule approximation with eight equal subintervals. Then compare the error in the approximation with the estimate given by formula **(4)**.

♦ Type 3, error estimates

29. (ERROR ESTIMATES WITH SIMPSON'S RULE) If $f(x)$ has a continuous fourth derivative in $[a, b]$ and M_4 is the maximum of $|f^{(iv)}(x)|$ for $a \le x \le b$, then the Simpson-Rule approximation with N equal subintervals of length $\Delta x = (b - a)/N$ satisfies

$$\left| \int_a^b f(x)\,dx - \begin{bmatrix} \text{The Simpson-rule} \\ \text{approximation} \end{bmatrix} \right| \le \tfrac{1}{180}(b-a)M_4(\Delta x)^4. \tag{6}$$

(a[A]) Calculate $\int_1^5 \sqrt{x}\,dx$ and its Simpson-Rule approximation with eight equal subintervals. Then compare the error in the approximation with the estimate obtained from **(6)**.

(b) As we will be able to determine from an integration formula in Section 4.7, the integral $\int_0^1 \frac{1}{x+1}\,dx$ has the value $\ln(2)$, where $\ln x$ is the natural logarithm. Calculate the Simpson-Rule approximation of the integral with four equal subintervals. Then compare the error in the approximation with the estimate given by **(10)**.

♦ Type 3, error estimates

30. Use estimate **(6)** to show that all Simpson-Rule approximations of polynomials of degree ≤ 3 are exact.

♦ Type 3, Simpson's Rule

31.[A] Based on estimate **(4)**, how many subintervals would be required with the Midpoint Rule to find the value of $\int_0^2 \frac{1}{3-x}\,dx$ with an error less than 10^{-5}?

♦ Type 3, error estimates

32. Based on estimate **(5)**, how many subintervals would be required with the Trapezoid Rule to find the value of $\int_0^3 (1+x^2)^{2.5}\,dx$ with an error less than 10^{-8}?

♦ Type 3, error estimates

33. Based on estimate **(6)**, how many subintervals would be required with Simpson's Rule to find the value of $\int_0^{10} x^6\,dx$ with an error less than 10^{-6}?

♦ Type 3, error estimates

34. (SIMPSON'S RULE AND LAGRANGE INTERPOLATING POLYNOMIALS)

To study Simpson's Rule applied to $\int_a^b f(x)\,dx$, we only need to look at the first two subintervals $x_0 < x_1 < x_2$ of any partition, because we could then modify the notation and add the results for the other pairs of subintervals. The Simpson-Rule approximation for the first two subintervals is $\int_{x_0}^{x_2} S(x)\,dx$, where $S(x)$ is the second-degree polynomial that has the same values as $f(x)$ at x_0, x_1, and x_2.

(a) Explain why

$$S(x) = f(x_0)P_0(x) + f(x_1)P_1(x) + f(x_2)P_2(x) \tag{7}$$

where P_0, P_1, and P_2 are the second-degree polynomials such that

$$\begin{array}{lll} P_0(x_0) = 1, & P_0(x_1) = 0, & P_0(x_2) = 0 \\ P_1(x_0) = 0, & P_1(x_1) = 1, & P_1(x_2) = 0 \\ P_2(x_0) = 0, & P_2(x_1) = 0, & P_2(x_2) = 1. \end{array}$$

(b) Show that $P_0(x) = \dfrac{(x-x_1)(x-x_2)}{(x_0-x_1)(x_0-x_2)}$, $P_1(x) = \dfrac{(x-x_0)(x-x_2)}{(x_1-x_0)(x_1-x_2)}$, and $P_2(x) = \dfrac{(x-x_0)(x-x_1)}{(x_2-x_0)(x_2-x_1)}$.

(c) Use the substitution $u = x - x_1$ to derive formula **(3)** for $M = 1$ by showing that $\int_{x_0}^{x_2} P_j(x)\,dx$ equals $\frac{1}{3}\Delta x$ for $j = 0$ and $j = 2$ and equals $\frac{4}{3}\Delta x$ for $j = 1$.

◆ Type 7, a proof

35. **(a)** Suppose that $f(x)$ has a maximum and a minimum value in every subinterval of $[a, b]$. Explain why adding a new point to a partition for $\int_a^b f(x)\,dx$ either leaves the lower sum unchanged or raises it, and either leaves the upper sum unchanged or lowers it. **(b)** Use the result of part (a) to show that all lower sums for $\int_a^b f(x)\,dx$ are $\le$ all of its upper sums (Figure 32). (Compare the upper and lower sums for any two partitions with the upper and lower sums for the third partition obtained by combining the points in the original two.)

◆ Type 7, a proof

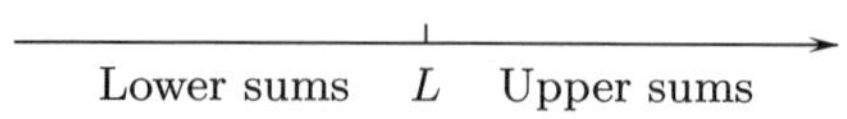

FIGURE 32

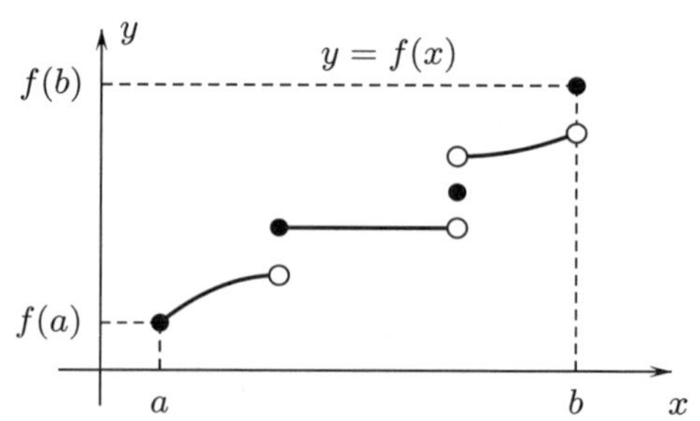

FIGURE 33

36. (BOUNDED NONDECREASING FUNCTIONS HAVE RIEMANN INTEGRALS) Suppose that $f(x)$ is defined at all but possibly a finite number of points in $[a,b]$ and is bounded and nondecreasing there. Define $f(x)$ at the missing points, if necessary, so it is nondecreasing in the entire interval. (Extending the definition of $f(x)$ allows for more Riemann sums, but the proof below shows that all such Riemann sums have the same finite limit.) **(a)** Explain why for any partition of $[a,b]$,

$$[\text{The upper sum}] - [\text{The lower sum}] \le \delta[f(b) - f(a)] \tag{8}$$

where δ is the width of the widest subinterval in the partition. (The ideas are illustrated in Figures 34 through 37 for the function of Figure 33 and a partition into four subintervals. The upper sum is equal to the area of the four rectangles in Figure 34 and the lower sum is equal to the area of the four rectangles in Figure 35. The difference between the upper and lower sums is equal to the area of the four rectangles in Figure 36, and these could be stacked as in Figure 37 to show that inequality **(8)** is valid.)

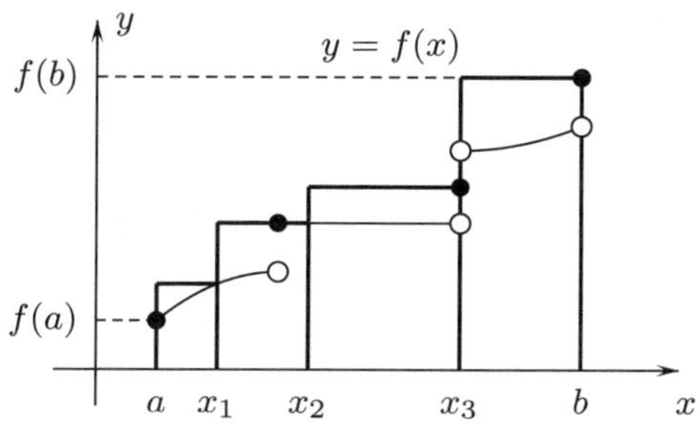

[Area] = [Upper sum]

FIGURE 34

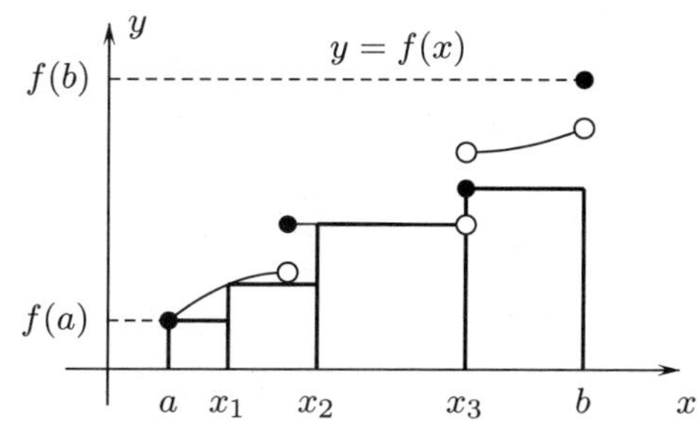

[Area] = [Lower sum]

FIGURE 35

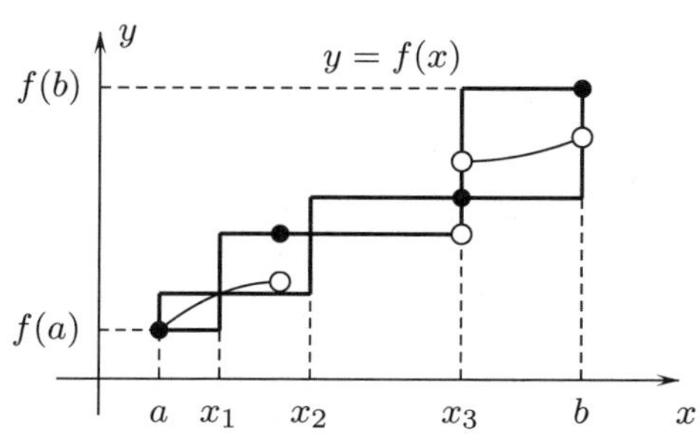

[Area] = [Upper sum] − [Lower sum]

FIGURE 36

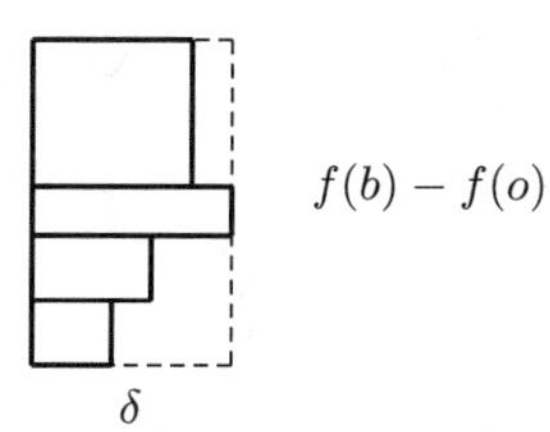

[Area] = [Upper sum] − [Lower sum]

FIGURE 37

(b) Use the results of part (a) and Problem 35 to show that the Riemann integral $\int_a^b f(x)\,dx$ is defined. (Hint: All of the lower sums are less than all of the upper sums by Problem 35. Use **(8)** to show that there is exactly one number L that is $\le$ all upper sums and $\ge$ all lower sums and that all of the upper and lower Riemann sums tend to L as the number of subintervals in the partition tend to ∞ and their widths tend to zero. Finally, explain why all Riemann sums tend to L, so that L is the value of the integral.)

♦ Type 7, a proof

Section 4.5

Integration by substitution with powers

Recall from Chapter 2 that we first derived a formula for derivatives of powers x^n of the variable x, and then extended this with the Chain Rule to enable us to find derivatives of powers $[y(x)]^n$ of functions. In this section we discuss the technique of INTEGRATION BY SUBSTITUTION, *which is obtained from this Chain Rule, and use it to find exact integrals of certain functions involving powers of functions.*

Topics:

- ***Integration by substitution***
- ***Changing limits of integration in definite integrals***
- ***Integrals of $\mathrm{x^n(ax+b)^r}$ with small positive integers n***

Integration by substitution

The Chain-Rule formula from Section 2.5 for the derivative of powers of functions gives

$$\frac{d}{dx}\left(\frac{1}{n+1}u^{n+1}\right)dx = \frac{1}{n+1}\left[(n+1)u^n\frac{du}{dx}\right] = u^n\,\frac{du}{dx}$$

for any constant $n \neq -1$ and any differentiable function $u = u(x)$. This equation translates into the integration (antidifferentiation) formula,

$$\int u^n\,\frac{du}{dx}\,dx = \frac{1}{n+1}\,u^{n+1} + C \quad (n \neq -1) \tag{1}$$

which we can use, along with the corresponding formula for definite integrals, to integrate powers of fuctions multiplied by their derivatives. This technique for changing variables in integrals is called INTEGRATION BY SUBSTITUTION. In carrying out such calculations, we take advantage of Leibniz notation by writing

$$du = \frac{du}{dx}\,dx. \tag{2}$$

In this symbolic equation, $\dfrac{du}{dx}$ denotes the derivative of the function u, dx is the symbol that is used in indefinite integrals with x as variable (which are x-antiderivatives), and du is the symbol that is used in indefinite integrals with u as variable (which are u-antiderivatives). Then instead of referring to the Chain Rule to obtain formula **(1)**, we use the symbolic equation **(2)** and write

$$\int u^n\,\frac{du}{dx}\,dx = \int u^n\,du = \frac{1}{n+1}u^n + C \quad (n \neq -1). \tag{3}$$

Here the first equation relates the antiderivative $\displaystyle\int u^n\,\frac{du}{dx}\,dx$ with respect to x to the antiderivative $\displaystyle\int u^n\,du$ with respect to u, and the second equation is the integration formula from Section 4.2 with u as variable.

Example 1 Find a formula for the antiderivative $\displaystyle\int (x^2+1)^5(2x)\,dx$.

SOLUTION We set $u = x^2 + 1$. Then

$$du = \frac{du}{dx}\,dx = \frac{d}{dx}(x^2+1)\,dx = 2x\,dx$$

♦ SUGGESTIONS TO INSTRUCTORS: Have students begin Questions 1 through 5 before class. Problems 1, 6, 9, 10, 11, 16, and 17 are good for in-class discussion

so that

$$\int (x^2+1)^5(2x)\,dx = \int u^5 \frac{du}{dx}\,dx = \int u^5\,du$$
$$= \tfrac{1}{6}u^6 + C = \tfrac{1}{6}(x^2+1)^6 + C. \;\Box$$

We can check the result of Example 1 by differentiating the result:

$$\frac{d}{dx}[\tfrac{1}{6}(x^2+1)^6] = \tfrac{1}{6}[6(x^2+1)^5\frac{d}{dx}(x^2+1) = (x^2+1)^5(2x).$$

The result is the original integrand, so the integration is correct.

Adjusting a constant in the integrand

Usually when we carry out an integration by substitution with formula **(3)**, we have to adjust a constant in the integrand to construct du.

Example 2 Perform the integration $\int x^3\sqrt{x^4+16}\,dx$.

SOLUTION Because of the square root of x^4+16 in the integral, we use the substitution $u = x^4+16$, for which $\dfrac{du}{dx} = \dfrac{d}{dx}(x^4+16) = 4x^3$ and

$$du = \frac{du}{dx}dx = 4x^3\,dx.$$

We use this substitution by constructing du from the dx and other elements of the integral. First, we move the x^3 next to the dx to have

$$\int x^3\sqrt{x^4+16}\,dx = \int \sqrt{x^4+16}\,(x^3\,dx).$$

The integral on the right is similar to $\int \sqrt{u}\,du$, but du equals $4x^3\,dx$ and not $x^3\,dx$. However, since the missing factor 4 is a constant, we can insert it to construct the correct du if we compensate by dividing the entire integral by 4, as in the following calculation:

$$\begin{aligned}\int x^3\sqrt{x^4+16}\,dx &= \int \sqrt{x^4+16}\,(x^3\,dx)\\ &= \tfrac{1}{4}\int \sqrt{x^4+16}\,(4x^3\,dx)\\ &= \tfrac{1}{4}\int \sqrt{u}\,du = \tfrac{1}{4}\int u^{1/2}\,du \qquad \textbf{(4)}\\ &= \tfrac{1}{4}\left[\frac{1}{3/2}u^{3/2}\right] + C = \tfrac{1}{4}(\tfrac{2}{3})u^{3/2} + C\\ &= \tfrac{1}{6}u^{3/2} + C.\end{aligned}$$

We put the result in terms of the original variable x by setting $u = x^4+16$ and obtain our final result:

$$\int x^3\sqrt{x^4+16}\,dx = \tfrac{1}{6}(x^4+16)^{3/2} + C. \;\Box$$

Look closely at the transition from the first line to the second line in equations **(4)**. This step involves a line of reasoning that you may not have encountered very often. Most of your work with mathematics has involved simplifying expressions. Here we put the integral $\int \sqrt{x^4+16}\,(x^3\,dx)$ in the more complicated form $\frac{1}{4}\int \sqrt{x^4+16}\,(4x^3\,dx)$ in order to construct $du = 4x^3\,dx$ from the terms $x^3\,dx$ in the original integral.

Question 1 Check Example 2 by differentiating the answer.

Question 2 The substitution $u = 1+5x$ can be used with $\int \frac{1}{(1+5x)^4}\,dx$. **(a)** What is du in this case? **(b)** Carry out the integration.

We we were able to insert the factor 4 in the integrand of Example 2 and the factor 5 in the integrand of Question 2 and compensate by dividing the entire integral by 4 and 5, respectively, because 4 and 5 are constants. We cannot, however, perform this sort of manipulations with x or any other nonconstant expressions, as is illustrated in the next question.

Question 3 Show that $\int x^4\,dx = \int x \cdot x^3\,dx$ is not equal to $x\int x^3\,dx$.

Substitution in definite integrals

In the next example we will illustrate two procedures for making substitutions in definite integrals.

Example 3 Figure 1 shows the region between $y = \frac{10x}{(x^2+1)^2}$ and the x-axis for $0 \le x \le 3$. Find its exact area.

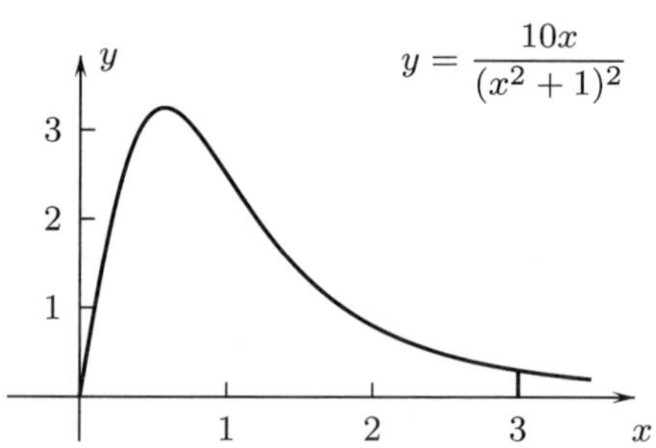

FIGURE 1

SOLUTION Because $\frac{10x}{(x^2+1)^2}$ is positive for $0 \le x \le 3$, the area of the region is $\int_0^3 \frac{10x}{(x^2+1)^2}\,dx$. To find its value, we make the substitution $u = x^2+1$, for which $du = \frac{d}{dx}(x^2+1)\,dx = 2x\,dx$. To construct du from $x\,dx$, we write $10 = 5(2)$ and put the 2 with the x and dx to obtain

$$\begin{aligned}\int_0^3 \frac{10x}{(x^2+1)^2}\,dx &= \int_{x=0}^{x=3} \frac{5}{(x^2+1)^2}\,(2x\,dx)\\ &= \int_{x=0}^{x=3} \frac{5}{u^2}\,du = \int_{x=0}^{x=3} 5u^{-2}\,du.\end{aligned} \tag{5}$$

We have written the equations $x = 0$ and $x = 3$ as the limits of integration in the last three integrals because these are values of x and not of the new variable u.

There are two ways to deal with the last integral in **(5)**. In the first procedure, we carry out the u-integration (find the u-antiderivative) and then switch the variable back to x:

$$\int_0^3 \frac{10x}{(x^2+1)^2}\,dx = \int_{x=0}^{x=3} 5u^{-2}\,du = \Big[-5u^{-1}\Big]_{x=0}^{x=3}$$
$$= \Big[-5(x^2+1)^{-1}\Big]_{x=0}^{x=3} = \left[\frac{-5}{x^2+1}\right]_{x=0}^{x=3}$$
$$= \left[\frac{-5}{3^2+1}\right] - \left[\frac{-5}{0^2+1}\right] = -\tfrac{1}{2} + 5 = 4.5.$$

In the alternate procedure, we express the limits of integration on the right of **(5)** in terms of u instead of switching back to x. Because x runs from 0 to 3 in the original integral, we write

$$0 \underset{x}{\longrightarrow} 3.$$

The formula $u = x^2 + 1$ shows that as x runs from 0 to 3, u runs from $0^2 + 1 = 1$ to $3^2 + 1 = 10$, so we write

$$1 \underset{u}{\longrightarrow} 10.$$

This shows that, in terms of u, the lower limit of integration in the last integral of **(5)** should be $u = 1$ and the upper limit should be $u = 10$, so that

$$\int_0^3 \frac{10x}{(x^2+1)^2}\,dx = \int_{x=0}^{x=3} 5u^{-2}\,du = \int_{u=1}^{u=10} 5u^{-2}\,du$$
$$= \Big[-5u^{-1}\Big]_{u=1}^{u=10} = \left[\frac{-5}{u}\right]_{u=1}^{u=10}$$
$$= -\tfrac{5}{10} - (-\tfrac{5}{1}) = -\tfrac{1}{2} + 5 = 4.5. \ \square$$

Question 4 Find the exact value of $\displaystyle\int_1^2 (1-2x)^3\,dx$.

A special technique

An integral of $x^n(ax+b)^r$ with a small, positive integer n and a larger, negative, fractional, or irrational r with constants a and b, can be found by using the substitution $u = ax + b$.

Example 4 Perform the integration $\displaystyle\int x(2x+1)^6\,dx$.

SOLUTION We set $u = 2x + 1$, so that $(2x+1)^6$ becomes u^6. For this type of integral, we first solve for x in terms of u and then express dx in terms of du. We obtain $2x = u - 1$ and $x = \frac{1}{2}(u-1)$. This gives $dx = \frac{1}{2}\,du$, and making these substitutions in the integral yields

$$\int x(2x+1)^6\,dx = \int \tfrac{1}{2}(u-1)(u^6)(\tfrac{1}{2}\,du)$$
$$= \tfrac{1}{4}\int (u^7 - u^8)\,du$$
$$= \tfrac{1}{4}(\tfrac{1}{8}u^6) - \tfrac{1}{4}(\tfrac{1}{7}u^7) + C$$
$$= \tfrac{1}{32}(2x+1)^8 - \tfrac{1}{28}(2x+1)^7 + C. \ \square$$

Question 5 **(a)** Express $\displaystyle\int_0^1 x\sqrt{1-x}\,dx$ as a definite integral with variable $u = 1 - x$.

(b) Evaluate the integral in part (a).

Principles and procedures

- Notice that in the equation $du = \frac{du}{dx}\,dx$, which enables us to carrry out integrations by substitution without referring explicitly to the Chain Rule, du is the symbol in integrals with u as variable, dx is the symbol in integrals with x as variable, and du/dx is the x-derivative of the function u.
- In applying the formula $\int u^n \frac{du}{dx}\,dx = \int u^n\,du = \frac{1}{n+1}u^{n+1} + C$ for integration by substitution, the given integrand can be multiplied by a constant factor to construct du if the entire integral is divided by that constant to preserve its value.
- Substitution in definite integrals can be carried out, after the u-antiderivative has been determined, either by switching back to x as variable for the final evaluation or by converting the limits of integration to the appropriate values of u.
- An integral of a small, positive, integer power of x, multiplied by any power of a linear function $ax + b$ can be found by using the substitution $u = ax + b$.
- Remember to check the values of all definite integrals with a procedure of approximate integration on your calculator or computer, and check formulas for indefinite integrals by differentiating the results.

Tune-Up Exercises 4.5[♦]

[A]*Answer provided.* [O]*Outline of solution provided.*

T1.[O] Find the antiderivatives $\int x^2\sqrt{x^3 - 1}\,dx$ and check the result by differentiation.

♦ Type 1, an integration

T2.[A] Perform the integration $\int \frac{1}{(3x+4)^2}\,dx$ and check the result by differentiating it.

♦ Type 1, an integration

T3.[O] Find the exact value of $\int_0^1 \frac{1}{(3-x)^3}\,dx$.

♦ Type 1, an integration

T4.[A] Perform the integration $\int \frac{x}{\sqrt{x+3}}\,dx$.

♦ Type 1, an integration

T5.[O] Evaluate $\int_0^2 [(1+x)^3 - (1-x)^4]\,dx$ by writing it as the difference of two integrals and using one substitution in one and another substitution in the other.

♦ Type 1, an integration

Problems 4.5

[A]*Answer provided.* [O]*Outline of solution provided.*

Perform the integrations in Problems 1 through 8.

1.[A] $\int \frac{x}{(x^2+4)^3}\,dx$

♦ Type 1, an integration

2.[A] $\int \frac{x^8}{(2+x^9)^3}\,dx$

♦ Type 1, an integration

3.[O] $\int_0^1 x\sqrt{x^2+9}\,dx$

♦ Type 1, an integration

4. $\int \frac{x^3}{(5-x^4)^2}\,dx$

♦ Type 1, an integration

♦The Tune-up Exercises and Problems are classified by type and content. The types are (1) basic, reactive; (2) basic reflective; (3) intermediate, reactive; (4) intermediate, reflective; (5) advanced, reactive; (6) advanced, reflective; and (7) advanced, theoretical.

5. $\int_0^4 (2x+1)^{-1/2}\,dx$

♦ Type 1, an integration

6. $\int_1^{10} \sqrt{2x+1}\,dx$

♦ Type 1, an integration

7. $\int (4x+1)^{1/3}\,dx$

♦ Type 1, an integration

8. $\int \sqrt{5x+1}\,dx$

♦ Type 1, an integration

9.[O] Find the exact area of the region bounded by the x-axis, the curve $y = \sqrt{1+5x}$, and the lines $x = 1$ and $x = 6$.

♦ Type 1, area

10.[A] An object's velocity in the positive s-direction on an s-axis is $v = t^2(2+t^3)^3$ feet per minute for $0 \le t \le 1$. It is at $s = 3$ (feet) at $t = 0$ (minutes). Where is it at $t = 1$?

♦ Type 1, an integral of velocity

11.[A] What is the total area of the regions bounded by $y = x\sqrt{9-x^2}$ and the x-axis?

♦ Type 1, area

12. What is the exact area of the region bounded by the curve $y = \sqrt{6-x}$ and by the x- and y-axes?

♦ Type 1, area

13. Find the exact area of the region bounded by $y = x^2\sqrt{8-x^3}$ and the x-axis.

♦ Type 1, area

Perform the integrations in Problems 14 through 21.

14.[O] $\int \frac{x}{\sqrt{1-x}}\,dx$

♦ Type 3, integration by substitution

15.[A] $\int \frac{y}{(3y-4)^3}\,dy$

♦ Type 3, integration by substitution

16. $\int \frac{t}{\sqrt{t+10}}\,dt$

♦ Type 3, integration by substitution

17.[A] $\int \frac{(\sqrt{x}+9)^9}{\sqrt{x}}\,dx$

♦ Type 3, integration by substitution

18. $\int \frac{x+1}{\sqrt{x^2+2x+5}}\,dx$

♦ Type 3, integration by substitution

19. $\int_0^1 x^{1/2}(1-x^{3/2})\,dx$

♦ Type 3, integration by substitution

20. $\int_0^1 x(x-1)^{3/2}\,dx$

♦ Type 3, integration by substitution

21. $\int_0^8 x^{2/3}(10+x^{5/3})^{10}\,dx$

♦ Type 3, integration by substitution

22. What is the area of the region between $y = 1 + x(x-1)^4$ and the x-axis for $0 \le x \le 1$?

♦ Type 3, area

23. Give a formula for $\int \frac{x}{\sqrt{x+a}}\,dx$ in terms of the parameter a.

♦ Type 3, an integral with a parameter

24. Evaluate $\int_0^1 (1+\sqrt{x})^3\,dx$ by using the change of variables $u = 1+\sqrt{x}$.

♦ Type 4, a special substitution

25.[A] Find the positive constant b such that the region between $y = \frac{x}{(1+x^2)^2}$ and the x-axis for $0 \le x \le b$ has area $\frac{1}{4}$.

♦ Type 2, area

26. **(a)** Find the positive number k such that $\int_3^k x\sqrt{x^2-9}\,dx = \frac{64}{3}$. **(b)** Why is there no negative k that satisfies the equation in part (a)?

♦ Type 4, determining a parameter

Section 4.6

Area, density, arclength, average value, and acceleration

In this section we use ideas and techniques from Sections 4.1 through 4.4 to study areas, density and centers of gravity of rods, lengths of curves, and average values of functions. We then discuss mass, acceleration and Newton's law of motion.

Topics:

- ***Area between two curves***
- ***One-dimensional density***
- ***Centers of gravity of rods***
- ***Lengths of curves***
- ***Average value of a function***
- ***Force, mass, acceleration, and Newton's law***

The area between two curves

We saw in Section 4.2 that the area of the region between the graph of a positive function and the x-axis for $a \leq x \leq b$ equals the integral of the function from a to b. How can we find the area of a region between the graphs of two positive functions $f(x)$ and $g(x)$, as in Figure 1? We assume that the graph of $g(x)$ is below the graph of $f(x)$, and that the Riemann integrals of both functions from a to b are defined. The existence of the integrals implies that the the functions are bounded for $a \leq x \leq b$.

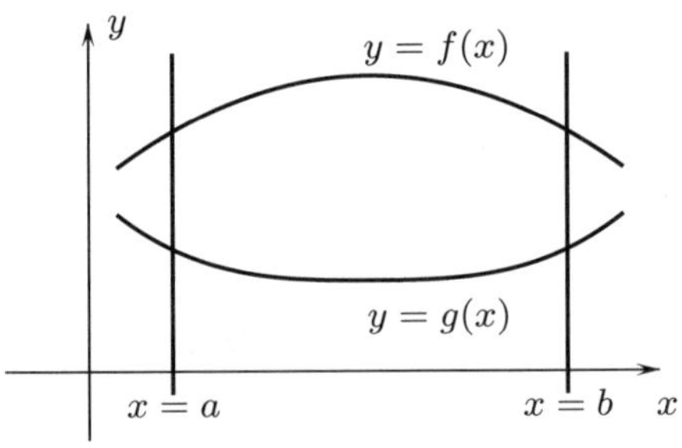

FIGURE 1

Because the region between the upper curve $y = f(x)$ and the lower curve ok $y = g(x)$ for $a \leq x \leq b$ in Figure 1 is obtained by removing the region between the lower curve and the x-axis from the region between the upper curve and the x-axis, the area of the region between $y = g(x)$ and $y = f(x)$ equals

$$\int_a^b f(x)\,dx - \int_a^b g(x)\,dx = \int_a^b [f(x) - g(x)]\,dx.$$

This gives the following rule for the case of positive functions:

♦ SUGGESTIONS TO INSTRUCTORS: The main topics in this section can be divided into five parts, as follows:
Part 1 (Area; Tune-up Exercises T1, T2; Problems 1–8, 22, 23, 32–46, 50–53, 64, 66, 67, 70, 71) Have students begin work on Question 1 before class. Problems 22, 33, 37, 40 and 43 are good for in-class discussion.
Part 2 (Density, weight, and centers of gravity of rods; Tune-up Exercise T4, Problems 17–19, 68) Have students begin work on Questions 2 through 4 before class. Problems 17–19 are good for in-class discussion.
Part 3 (Lengths of curves; Tune-up Exercise T5; Problems 20, 21, 47, 65, 69, 72–74) Have students begin work on Question 5 before class. Problems 47 and 65 are good for in-class discussion.
Part 4 (Average value; Tune-up Exercise T3; Problems 9–16, 48, 54) Have students begin work on Question 6 before class. Problems 12-16 and 48 are good for in-class discussion.
Part 5 (Force, mass, velocity, acceleration, and Newton's Law; Tune-up Exercises T6, T7; Problems 24–31, 49, 55–63, 75) Have students begin work on Questions 7 through 10 before class. Problems 27–31, 56, 58, and 61 are good for in-class discussion.

Rule 1 *If $g(x) \le f(x)$ for $a \le x \le b$, then the area of the region between their graphs for $a \le x \le b$ is given by*

$$[\text{Area}] = \int_a^b [f(x) - g(x)]\, dx. \qquad \textbf{(1)}$$

If $g(x)$ has any negative values, as in Figure 2, we pick a positive constant K so that the graph of $g(x)+K$ is above the x-axis. Then the region between $y = g(x)+K$ and $y = f(x)+K$ for $a \le x \le b$ in Figure 3 has the same shape and the same area as the region between $y = g(x)$ and $y = f(x)$ for $a \le x \le b$ in Figure 2. Consequently, the area of the region in Figure 2 equals

$$\int_a^b [f(x) + K]\, dx - \int_a^b [g(x) + K]\, dx$$
$$= \int_a^b \{[f(x) + K] - [g(x) + K]\}\, dx = \int_a^b [f(x) - g(x)]\, dx$$

and we again obtain equation **(1)**.

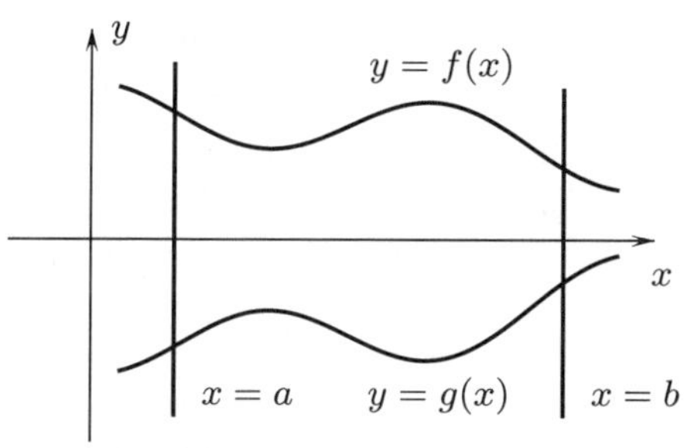

FIGURE 2

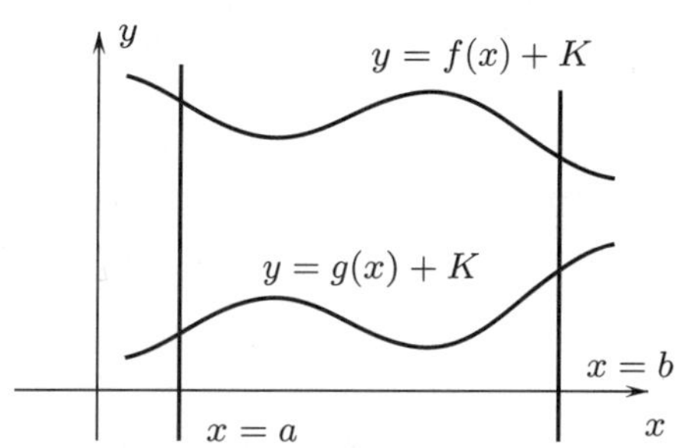

FIGURE 3

Example 1 Figure 4 shows the region bounded by the curves $y = x^2$ and $y = -x^2$ and by the vertical line $x = 2$. Find its area.

SOLUTION Because the curves intersect at the origin, the region extends from $x = 0$ to $x = 2$. Its top is formed by $y = x^2$ and its bottom by $y = -x^2$. By Rule 1, its area is

$$\int_0^2 [x^2 - (-x^2)]\, dx = \int_0^2 2x^2\, dx = \left[\tfrac{2}{3}x^3\right]_0^2 = \left[\tfrac{2}{3}(2^3)\right] - \left[\tfrac{2}{3}(0^3)\right] = \tfrac{16}{3}. \ \square$$

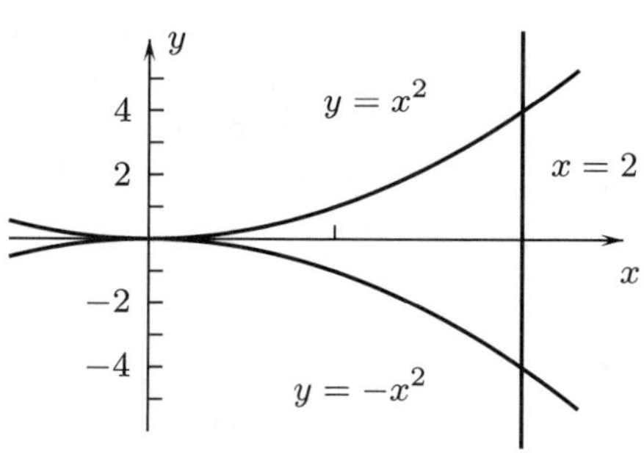

FIGURE 4

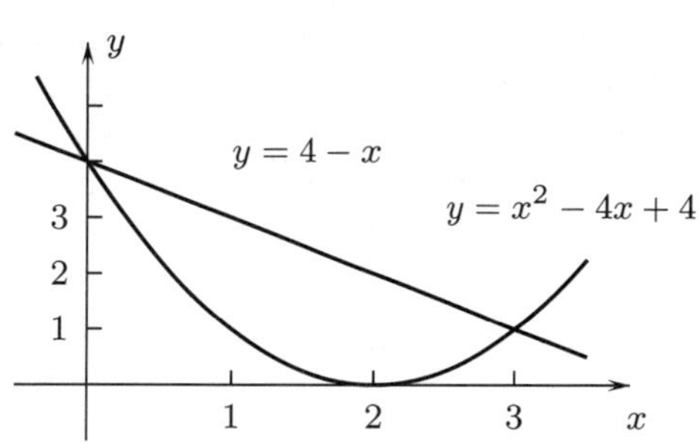

FIGURE 5

Notice that the x-and y-axes, except for the point at the origin, are not part of the boundary of the region in Figure 4, whose area is calculated in Example 1. Also, the regions between the curves for $x < 0$ and for $x > 3$ are not included because those regions extend infinitely far to the left and right and are not bounded by the two curves.

Question 1 **(a)** Find the x-coordinates of the two points of intersection of $y = 4 - x$ and $y = x^2 - 4x + 4$ in Figure 5. **(b)** What is the area of the region bounded by the two curves in Figure 5?

One-dimensional density

The density of a three-dimensional object is usually given in units of weight or mass per volume, and we will study this topic by using integrals of functions of more than one variable in Chapter 11. Here we consider one-dimensional densities that are given as weight or mass per unit length.

In Question 4 of Section 4.1 we calculated the mass of a core sample drilled from the earth whose density $\rho(x)$, measured in grams per centimeter, is the step function in Figure 6. The core sample extends from $x = 0$ to $x = 100$. Its density is 200 grams per centimeter for $0 < x < 30$, 160 grams per centimeter for $30 < x < 80$, and is 220 grams per centimeter for $80 < x < 100$. Therefore, its total mass is

$$\left[200\,\frac{\text{grams}}{\text{centimeter}}\right][30\text{ centimeters}] + \left[160\,\frac{\text{grams}}{\text{centimeter}}\right][50\text{ centimeters}] + \left[220\,\frac{\text{grams}}{\text{centimeter}}\right][20\text{ centimeters}] = 18{,}400\text{ grams.}$$

This sum equals the area of the three rectangles between the graph of $\rho(x)$ and the x-axis for $0 \le x \le 100$ in Figure 7, and therefore equals the integral of $\rho(x)$ from 0 to 100. Similar calculations would show that the mass or weight of any object with a step-function density is given by the integral of the density.

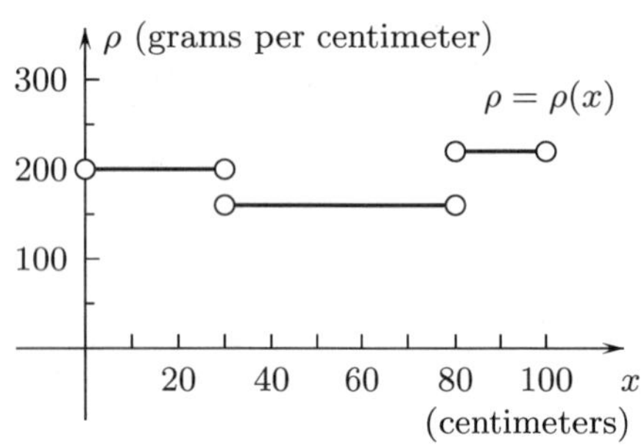

FIGURE 6

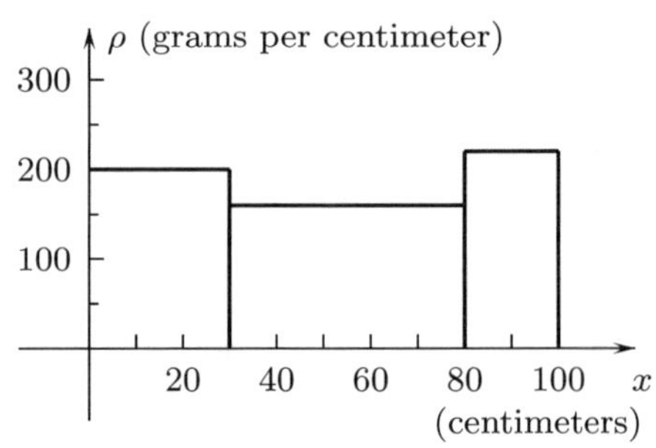

FIGURE 7

What if the density, as in Figure 8, is not a step function? In this case, the object extends from $x = a$ to $x = b$ (feet), and its density at x is $\rho(x)$ pounds per foot. For each partition $a = x_0 < x_1 < \cdots < x_N = b$ of $[a, b]$, we approximate $\rho(x)$ by the step function whose value in the interior of the jth subinterval for each j is $\rho(c_j)$ with c_j a point in the jth subinterval. Figure 9 shows the graph of such a step function for a partition into six subintervals.

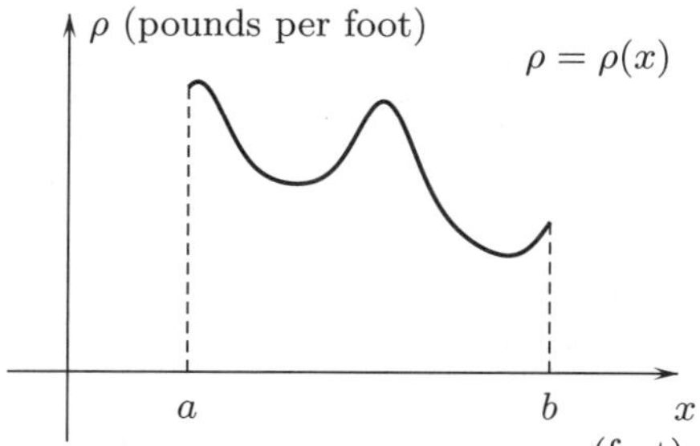

FIGURE 8

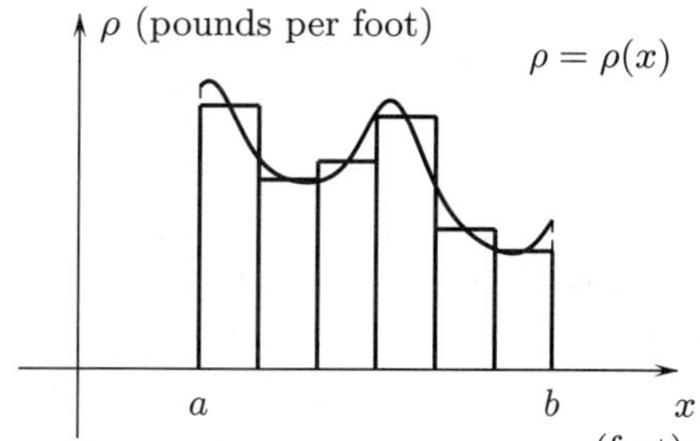

FIGURE 9

If the object had density given by the step function, its weight would equal the area of the N rectangles,

$$\sum_{j=1}^{N}\left[\rho(c_j)\ \frac{\text{pounds}}{\text{foot}}\right]\left[\Delta x_j \text{ feet}\right] = \sum_{j=1}^{N}\rho(c_j)\Delta x_j \text{ pounds} \tag{2}$$

where, as usual, Δx_j denotes the width of the jth subinterval. The sum **(2)** is approximately equal to the actual weight of the rod and is a Riemann sum for the integral of $\rho(x)$ from a to b. We assume that the integral is defined. Then the Riemann sum tends to the integral as the number of subintervals tend to ∞ and their widths tend to zero, we are led to the following rule.

Rule 2 *If an object extends from a to b for $a < b$ and its density at x is $\rho(x)$, measured in weight (or mass) per unit length, then its weight (or mass) is $\int_a^b \rho(x)\,dx$.*

Question 2 The density of a two-foot-long rod is $2+6x$ pounds per foot at a distance x feet from one end. How much does it weigh?

Centers of gravity of rods

The CENTER OF GRAVITY (or CENTER OF MASS) of a rod is the point where it would balance on your finger (Figure 10). If the rod has constant density, its center of gravity is at its midpoint. To see how to find centers of gravity in other cases, we first consider the case of a large man and a small boy who are balanced on a see-saw.

We represent the see-saw as a weightless x-axis with its fulcrum (supporting point) at $x = L$ (feet) and represent the man and the boy by POINT MASSES, that is, by weights concentrated at points. We suppose that the man is at $x = c_1$ and weighs w_1 pounds and that the boy is at $x = c_2$ and weighs w_2 pounds, as in Figure 11.

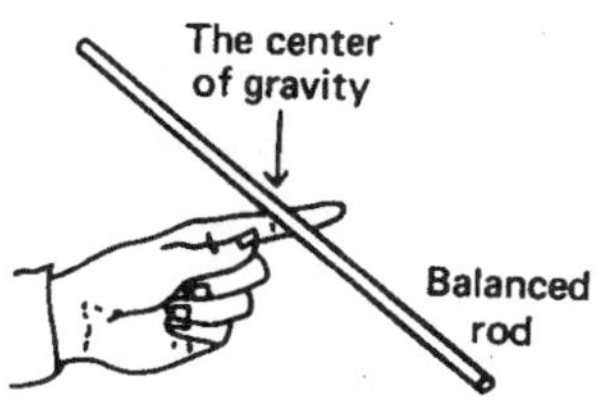

FIGURE 10

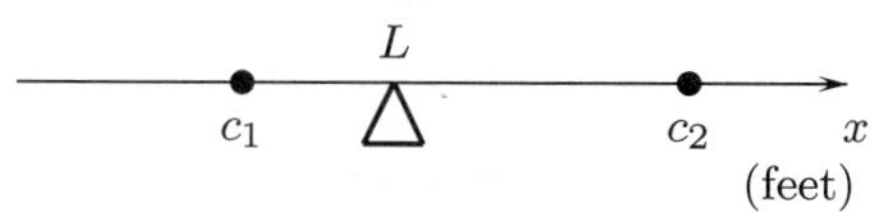

FIGURE 11

Example 2 The see-saw in Figure 11 balances if the boy's weight multiplied by his distance to the fulcrum equals the man's weight multiplied by his distance from the fulcrum. Express this condition as an equation involving the symbols w_1, w_2, c_1, c_2, and L.

SOLUTION Since the distance from c_1 to L is $L - c_1$ and the distance from c_2 to L is $c_2 - L$, the man and boy are balanced if $w_1(L - c_1) = w_2(c_2 - L)$. □

The condition from Example 2 can be written as $-w_1(c_1 - L) = w_2(c_2 - L)$ and then as

$$w_1(c_1 - L) + w_2(c_2 - L) = 0.$$

The expression on the left of this equation is called the MOMENT OF INERTIA of the man and boy about the point L. The center of gravity of the two people is at L, and the see-saw balances with its fulcrum at L if this moment of inertia is zero.

If there were N point masses on the see-saw with weight w_j at $x = c_j$ for $j = 1, 2, \ldots, N$, then their total moment of inertia about L would be $w_1(c_1 - L) + w_2(c_2 - L) + w_3(c_3 - L) + \cdots + w_N(c_N - L)$. We express this with summation notation by writing

$$[\text{Moment of inertia of the } N \text{ weights about } L] = \sum_{j=1}^{N} w_j(c_j - L). \tag{3}$$

The center of gravity of the weights is at L and the weights balance with the fulcrum at L if this moment of inertia is zero.

To find the moments of inertia and center of gravity of a rod as in Figure 12, we approximate it by point masses. As in the discussion of the rod's weight leading to Rule 2, we suppose that the density at x, measured in weight per unit length, is given by the function $\rho(x)$ of Figure 8. In finding the weight of the rod, we approximated $\rho(x)$ by step function, as in Figure 9, with the value $\rho(c_j)$ in the interior of the jth subinterval. Since this subinterval has length Δx_j, the portion of the rod in the jth subinterval weighs approximately $\rho(c_j)\Delta x_j$. In this case, we approximate the rod by N point masses with the point at c_j having weight $\rho(c_j)\Delta x_j$ for $j = 1, 2, \ldots, N$. (See Figure 13 for $N = 6$.).

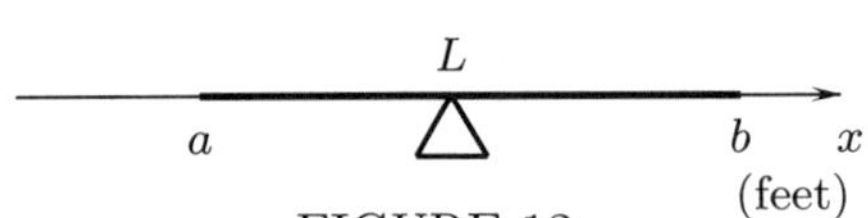

FIGURE 12

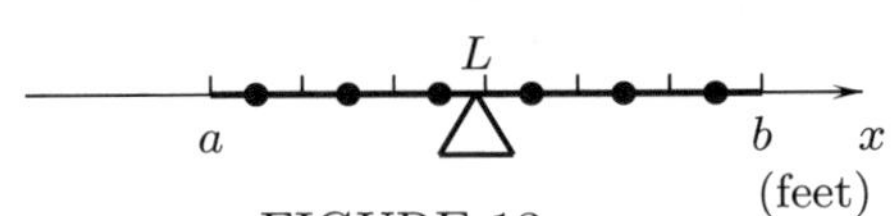

FIGURE 13

According to formula **(2)**, the moment of the N point masses about $x = L$ is

$$\sum_{j=1}^{N} \rho(c_j)(c_j - L)\Delta x_j. \tag{4}$$

We assume that the integral $\int_a^b \rho(x)(x - L)\,dx$ is defined. Then the Riemann sum **(3)** tends to this integral as the number of subintervals tends to ∞ and their widths tend to zero, and we are led to the following:

Rule 3 *If an object extends from a to b for $a < b$ and its density at x is $\rho(x)$, measured in weight or mass per length, then its moment about $x = L$ is $\int_a^b \rho(x)(x - L)\,dx$.*

The number L is the center of gravity of the rod of Rule 3 if the moment

$$\int_a^b \rho(x)(x - L)\,dx = \int_a^b x\rho(x)\,dx - L\int_a^b \rho(x)\,dx \tag{5}$$

is zero.

Question 3 Find a formula for the center of gravity by setting the expression on the right of **(5)** equal to zero and then solving for L.

The formula from Question 3 is stated in the next rule, in which we use the symbol $\overline{x}$ for the center of gravity.

Rule 4 *If an object extends from a to b for $a < b$ and its density at x is $\rho(x)$, measured in weight or mass per unit length, then its center of gravity is*

$$\overline{x} = \frac{\displaystyle\int_a^b x\rho(x)\,dx}{\displaystyle\int_a^b \rho(x)\,dx}.$$

Notice that the center of gravity of the rod is its moment about $x = 0$ (the origin), divided by its weight. To remember this formula, notice that there is an extra x in the numerator on the right.

Question 4 Find the center of gravity of the rod from Question 3 with density $2 + 6x$ pounds per foot for $0 \le x \le 2$.

The length of a curve

In Section 4.1 we used the Pythagorean Theorem to find the length of the graph of the piecewise linear function $f(x)$ in Figure 14. We found that the length is equal to the area between the graph of $\sqrt{1 + [f'(x)]^2}$ and the x-axis in Figure 15. Consequently, the length equals the integral of $\sqrt{1 + [f'(x)]^2}$ from 1 to 8.

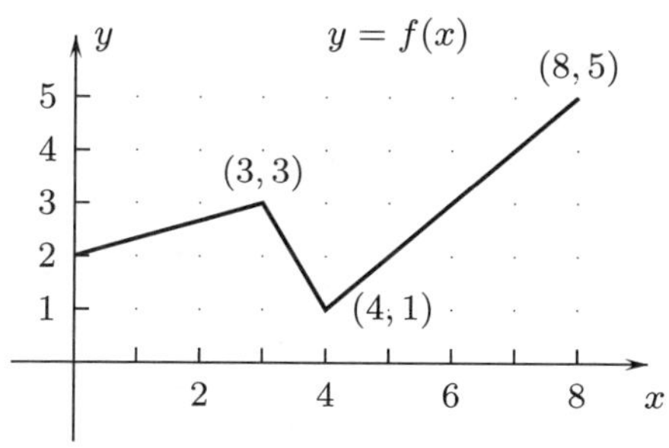

FIGURE 14

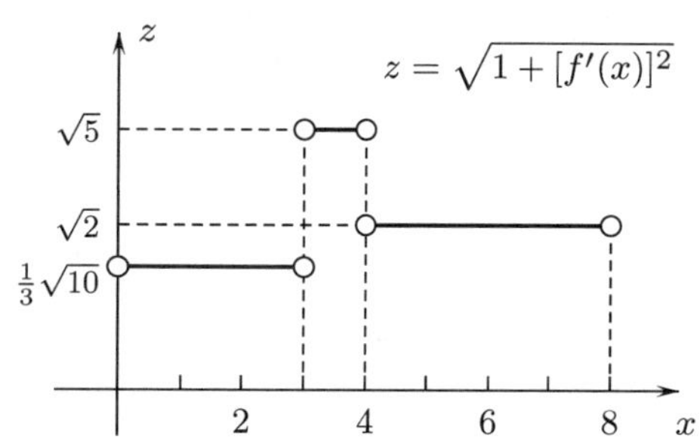

FIGURE 15

Suppose we want to find the length of the graph of a function $f(x)$, as in Figure 16, that is not piecewise linear. We assume that $f(x)$ has a continuous derivative for $a \le x \le b$. For the partition $a = x_0 < x_1 \cdots < x_{N-1} < x_N = b$ of $[a, b]$ into N equal subintervals of width Δx, we approximate the graph by the N line segments that join the points $(x_j, f(x_j))$ for $j = 0, 1, 2, \ldots, N$, as shown in Figure 17 for $N = 6$.

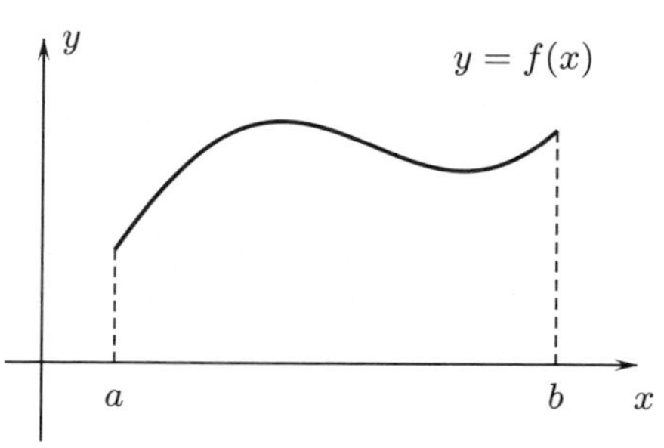

FIGURE 16

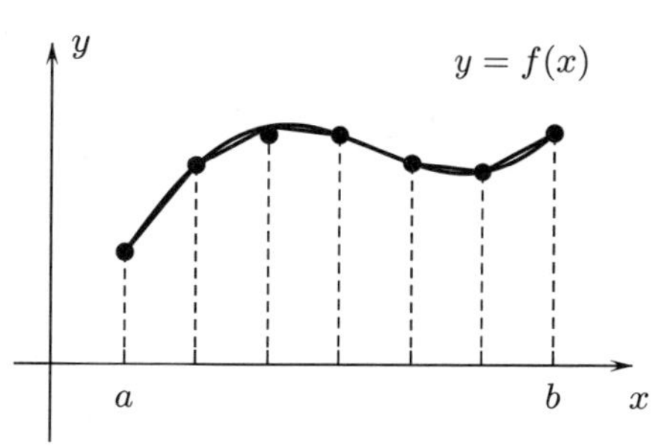

FIGURE 17

Because the run in the x-direction from the left to the right end of each line segment is Δx, the rise along the jth line segment is $m_j \Delta x$, with m_j its slope, and by the Pythagorean Theorem the length of the jth line segment is $\sqrt{[\Delta x]^2 + [m_j \Delta x]^2} = \sqrt{1 + (m_j)^2}\,\Delta x$. Since the jth line segment is on a secant line to the graph of $f(x)$, there is, by the Mean-Value Theorem (Theorem 2 of Section 3.1), a point c_j in the interior of the jth subinterval such that $m_j = f'(c_j)$. Consequently the length of the curve is approximately

$$\sum_{j=1}^{N} \sqrt{1 + [f'(c_j)]^2}\,\Delta x.$$

Because this sum is equal to the area of N rectangles of heights $\sqrt{1 + [f'(c_j)]^2}$ for $j = 1, 2, \ldots, N$, and width Δx, it is a Riemann sum for the integral $\int_a^b \sqrt{1 + [f'(x)]^2}\,dx$ and tends to this integral as $N \to \infty$. Since the collection of secant lines approaches the graph of $f(x)$ as $N \to \infty$,we are led to the following:

Rule 5 *The length of the graph of $f(x)$ for $a \le x \le b$ is* $\int_a^b \sqrt{1 + [f'(x)]^2}\,dx.$

We can often use this rule to find lengths of graphs with discontinuities or corners by applying it to the pieces that are graphs of functions with continuous derivatives and adding the results.

Because of the square root in the formula of Rule 5, we cannot find exact lengths of most graphs that are not piecewise linear, but we can find their approximate lengths with procedures for estimating integrals on calculator and computers.

Example 3 Express the length of $y = x^2$ for $-1 \le x \le 1$ as a definite integral and use your calculator or computer to find its approximate value.

SOLUTION For $f(x) = x^2$, $f'(x) = 2x$, so the length of the curve is

$$\int_{-1}^{1} \sqrt{1 + [f'(x)]^2}\,dx = \int_{-1}^{1} \sqrt{1 + (2x)^2}\,dx.$$

A procedure for finding approximate values of integrals on a calculator or computer shows that the length is approximately 2.958. □

The next question deals with one of the few types of curves whose lengths can be found by exact integration.

Question 5 How long is $y = 2x^{3/2}$ for $0 \le x \le 1$?

Average values

In Question 6 of Section 4.1 we considered the step function $T(t)$ of Figure 18, which gives the maximum daily temperature in a city as a function of the time during one week. Because this function has the value 20°F for two days, the value 10°F for one day, the value −20°F for two days, and the value 40°F for two days, we concluded that its average value during the seven-day period is $\frac{2}{7}(20) + \frac{1}{7}(10) + \frac{2}{7}(-20) + \frac{2}{7}(40) = 12\frac{6}{7}$°F, which is equal to the area of the three rectangles above the t-axis in Figure 19, minus the area of the rectangle below the t-axis, all divided by 7. Since the difference of the areas of the rectangles is the integral of $T(t)$ from 0 to 7,

$$[\text{The average value of } T(t) \text{ for } 0 \le t \le 7] = \frac{1}{7}\int_0^7 T(t)\,dt.$$

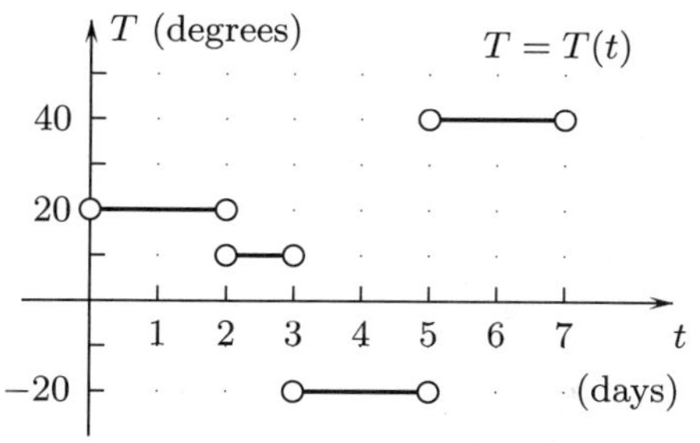

FIGURE 18

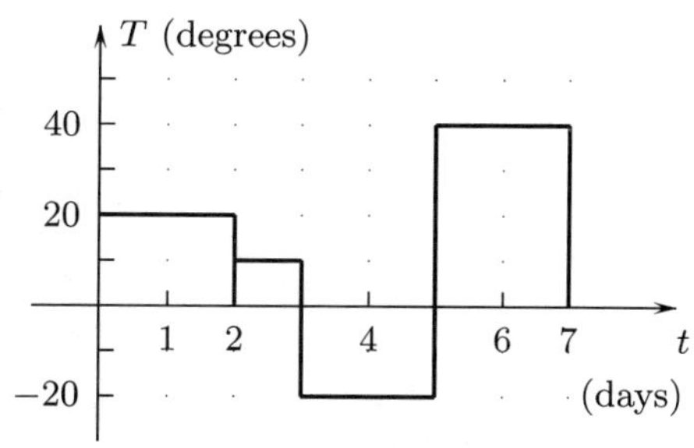

FIGURE 19

How can we find the average values of $f(x)$ in an interval $a \le x \le b$ if $f(x)$ is not a step function? We consider a partition $a = x_0 < x_1 < x_2 < \cdots < x_{N-1} < x_N = b$ of $[a, b]$ into N subintervals, and pick a point c_j in the jth subinterval for each j. We approximate $f(x)$ by the step function that equals $f(c_j)$ in the interior of the jth subinterval for each j. If, as usual, we denote the width of the jth subinterval by Δx_j, then the jth subinterval is the fraction $(\Delta x_j)/(b-a)$ of the entire interval $[a, b]$, and the average value of the step function for $a \le x \le b$ is

$$\sum_{j=1}^{N} \frac{\Delta x_j}{b-a} f(c_j) = \frac{1}{b-a} \sum_{j=1}^{N} f(c_j)\Delta x_j.$$

Since the Riemann sum on the right of this equation tends to the integral $\int_a^b f(x)\, dx$ as the number of subintervals tends to infinity and their widths tend to zero, we are led to the next rule:

Rule 6 *If the integral of $f(x)$ from a to b is defined, then*

$$[\text{The average value of } f(x) \text{ for } a \le x \le b] = \frac{1}{b-a} \int_a^b f(x)\, dx.$$

Example 4 What is the average value of $3x^2 + 3$ for $0 \le x \le 2$?

SOLUTION The integral of $3x^2 + 3$ from 0 to 2 is

$$\int_0^2 (3x^2 + 3)\, dx = \Big[x^3 + 3x\Big]_0^2 = [2^3 + 3(2)] - [0] = 14.$$

By Rule 2, the average value is $\frac{1}{2} \int_0^2 (3x^2 + 3)\, dx = \frac{1}{2}(14) = 7.$ □

Geometric interpretations of average value

Figure 20 shows the graph of $3x^2 + 3$ from Example 4 and the horizontal line $y = 7$ at its average value. We will use this drawing to illustate two geometric interpretations of average value.

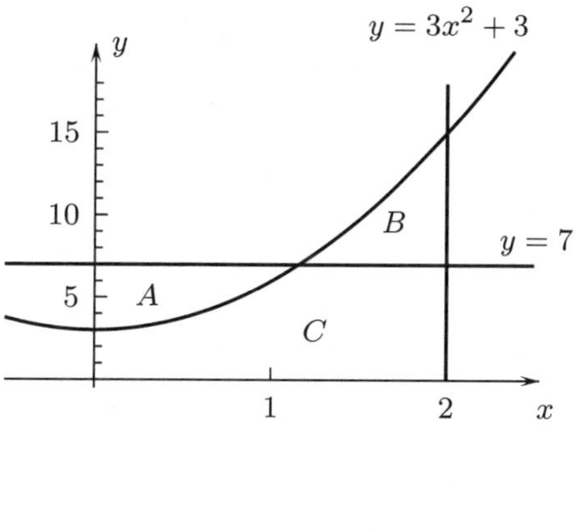

FIGURE 20

Example 5 **(a)** Show that the integral from 0 to 2 of the average value of $3x^2+3$ for $0 \le x \le 2$ equals the integral of $3x^2+3$ from 0 to 2. **(b)** Use part (a) to show that the regions A and B in Figure 20 have the same area.

SOLUTION **(a)** The integral of the averge value from 0 to 2 is $\int_0^2 7\,dx = 7(2) = 14$, which equals the integral of $3x^2+3$ that we calculated in Example 4.

(b) The integral of the average value is equal to the area of the rectangle in Figure 20, and the integral of $3x^2+3$ is equal to the area of the region between the curve $y = 3x^2+3$ and the x-axis for $0 \le x \le 2$. Since these are equal, the combined area of regions A and C in Figure 20 equals the combined area of regions B and C, and regions A and B have the same area. □

The reasoning in Example 5 can be applied to any average value to give the following:

Rule 7 **(a)** *The average value of $f(x)$ for $a \le x \le b$ is the constant L whose integral from a to b equals the integral of $f(x)$ from a to b.*

(b) *If the average value of $f(x)$ for $a \le x \le b$ is L, then the line $y = L$ splits the graph of $f(x)$ equally in the sense that the area of the region between $y = f(x)$ and $y = L$ for $a \le x \le b$ where $f(x) > L$ equals the area of the region between $y = f(x)$ and $y = L$ for $a \le x \le b$ where $f(x) < L$ (Figure 21).*

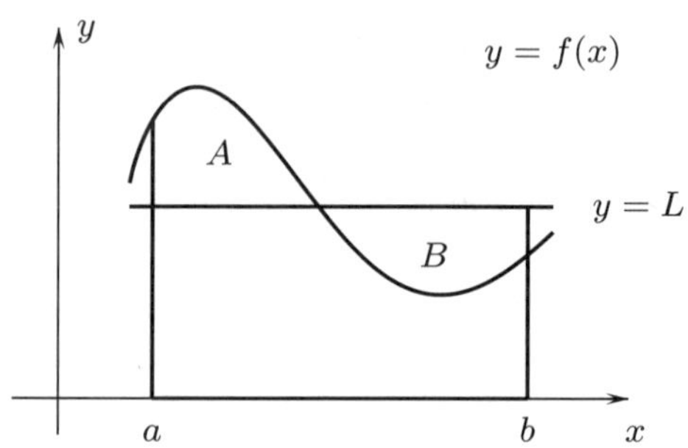

FIGURE 21

Question 6 Find the average value L of $4x^{-2}$ for $1 \le x \le 4$ and show that it satisfies Rule 7b by calculating the area of the region between $y = 4x^{-2}$ and $y = L$ for $1 \le x \le 4$ where $4x^{-2} \le L$ and the area of the region between $y = 4x^{-2}$ and $y = L$ for $1 \le x \le 4$ where $4x^{-2} \ge L$.

The beat goes on

Figure 22 shows the graph of blood pressure $p(t)$ at the opening of the aorta leading from a person's heart during about two and a half heart beats. The aorta carries blood from the left ventricle of the heart to the arteries that lead to all of the body except the lungs and the heart itself (Figure 23). The rises in pressure in Figure 22 occur when the ventricle contracts to force blood into the aorta. The notches in the graph after the points of highest pressure occur when the ventricle relaxes and a small backflow of blood occurs. This backflow causes the valve between the ventricle and the aorta to close and prevent more blood from flowing back into the heart. The function $p(t)$ of Figure 22 is periodic of period 0.8 seconds because the pressure pattern from its minimum at $t = 0.2$ to its next minimum at $t = 1.0$ repeats every 0.8 seconds.

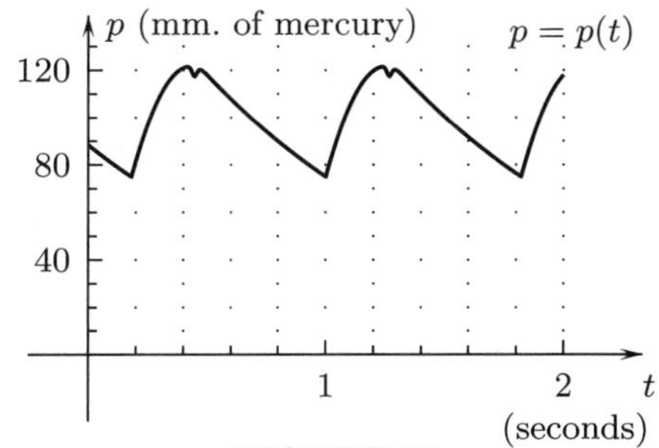

FIGURE 22

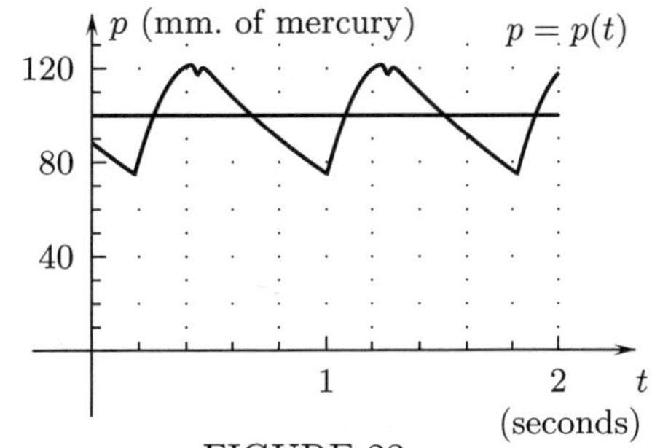

FIGURE 23

Example 6 **(a)** What are the approximate systolic (maximum) and diastolic (minimum) blood pressures in Figure 22?† **(a)** Use Rule 7 to estimate the average blood pressure in Figure 22.

SOLUTION **(a)** The systolic pressure in Figure 22 is approximately 121 millimeters of mercury, and the diastolic pressure is approximately 75 millimeters of mercury. (A pressure of p millimeters of mercury will support a column of mercury p millimeters high.)
(b) We consider the blood pressure during the period $0.2 \le t \le 1.0$, and draw a horizontal line, as in Figure 23, such that the area of the region for $0.2 \le t \le 1.0$ between the line and the graph where the graph is below the line appears to have about the same area as the region for $0.2 \le t \le 1.0$ between the line and the graph where the graph is above the line. The line is at $p \approx 100$, so we conclude that the average pressure is about 100 millimeters of mercury. □

Integrating acceleration

If an object's position on an s-axis is given as a function of the time t by $s(t)$, then its velocity in the positive direction on the s-axis is $v(t) = \frac{ds}{dt}(t)$ and its acceleration in the positive direction is $a(t) = \frac{dv}{dt}(t) = \frac{d^2s}{dt^2}(t)$. Consequently, velocity is an antiderivative of acceleration,

$$v(t) = \int a(t)\,dt \tag{6}$$

and position is an antiderivative of velocity,

$$s(t) = \int v(t)\,dt. \tag{7}$$

We can express the relationship between position and velocity and between velocity and acceleration by writing

$$s(t) \xrightarrow{\frac{d}{dt}} v(t) \xrightarrow{\frac{d}{dt}} a(t)$$

and

$$a(t) \xrightarrow{\int dt} v(t) \xrightarrow{\int dt} s(t).$$

The two constants of integration which come from the indefinite integrals **(6)** and **(7)** can be determined from values of the velocity and position at a particular time.

†The terms "systolic" and "diastolic" come from the Greek words for "contraction" and "dilation." They refer to the contraction and dilation of the left ventricle that cause the blood pressure to rise and fall.

Example 7 A car's position as it travels on a road is given by the variable s, measured in miles. At time $t = 0$ the car is at $s = 100$ and has velocity 40 miles per hour in the positive s-direction. It is accelerating $6t$ miles per hour per hour in the positive s-direction at time t for $0 \le t \le 3$. Give formulas for the car's velocity and position as functions of t for $0 \le t \le 3$.

SOLUTION Since the car's acceleration a is $6t$, its velocity is

$$v = \int a\,dt = \int (6t)\,dt = 6(\tfrac{1}{2}t^2) + C = 3t^2 + C.$$

To find the constant C of integration, we set $t = 0$ and $v = 40$: $6(0) + C = 40$. We conclude that $C = 40$ and $v = 3t + 40$ miles per hour.

We integrate again to find the car's position.

$$s = \int v\,dt = \int (3t^2 + 40)\,dt = 3(\tfrac{1}{3}t^3) + 40t + C = t^3 + 40t + C.$$

To determine the new constant of integration, we set $t = 0$ and $s = 100$ in the last equation. We obtain $0^3 + 40(0) + C = 100$, so that $C = 100$ and $s = t^3 + 40t + 100$ miles. □

Question 7 A car that is stopped at $t = 0$ (minutes) accelerates at the constant rate of 1000 feet per minute^2 down a straight road for $0 \le t \le 5$. **(a)** Give a formula for its velocity down the road as a function of t for $0 \le t \le 5$. **(b)** How far does it travel from $t = 0$ to $t = 5$?

Weight and mass

The WEIGHT of an object, which is the force of gravity on it, is not an intrinsic characteristic because it depends on the object's location. On the moon, for example, an object weighs approximately one-sixth of what it weighs on the surface of the earth. To obtain an intrinsic property, we divide the object's weight w by the acceleration g due to gravity at its location. This ratio $m = w/g$ is called the object's MASS. It is constant because if the weight changes by a certain factor when the object is moved from one place to another, then the acceleration due to gravity changes by the same factor.

In English units, time is measured in seconds, distance in feet, force in pounds, and mass is in SLUGS (a term with the same etymology as the word "sluggish"). On the surface of the earth, the acceleration due to gravity is 32 feet per second^2, and we calculate mass in slugs by dividing weight in pounds by 32. In metric units, the acceleration due to gravity on the surface of the earth is 9.8 meters per second^2 or 980 centimeters per second^2. Therefore, we find mass in kilograms by dividing weight in NEWTONS by 9.8 and find mass in grams by dividing weight in DYNES by 980. This terminology is listed in the next table for reference.†

Time	Distance	g	Weight (force) w	Mass $m = w/g$
seconds	feet	$32\,\dfrac{\text{feet}}{\text{second}^2}$	pounds	slugs
seconds	meters	$9.8\,\dfrac{\text{meters}}{\text{second}^2}$	newtons	kilograms
seconds	centimeters	$980\,\dfrac{\text{centimeters}}{\text{second}^2}$	dynes	grams

†In countries which employ the metric system, grams and kilograms are also used as units of weight. Then one gram weight is 980 dynes, or the weight of an object of one gram mass, and one kilogram weight is 9.8 newtons, or the weight of an object of one kilogram mass.

Question 8 The heaviest primate ever kept in captivity was a gorilla named N'gagi in the San Diego Zoo in the early 1940's. His greatest weight was 683 pounds.[1] What was his greatest mass in slugs?

Newton's law of motion

NEWTON'S LAW OF MOTION,

$$F = ma \tag{8}$$

states that for an object moving on an s-axis, the total force F applied to it in the positive s-direction at any particular time is equal to the object's mass m, multiplied by the object's acceleration in the positive s-direction at that time. Here time, distance, force, and mass should be measured with the units in the same row of the table above. The formula $m = w/g$ relating weight and mass, that was discussed above, can be written $w = mg$. Then it is a special case of **(8)** because the weight w of an object of mass m is the downward force on it caused by the downward acceleration g of gravity.

Question 9 An object weighing 19.6 newtons is moving on an s-axis subject to a force of $20 - 12t^{-1/2}$ newtons in the positive s-direction at time t (seconds) for $1 \leq t \leq 10$. At $t = 1$, The onject is at $s = 50$ (meters) and its velocity in the positive direction is 12 meters per second. Give formulas for its acceleration, velocity, and position as functions of t for $1 \leq t \leq 10$.

Constant acceleration

In the important special case of constant acceleration, the formulas for velocity and position are especially simple.

Example 8 A pickup truck weighing 4000 pounds is 500 feet past an on-ramp and is traveling 20 feet per second ($\doteq$ 13.6 miles per hour) down a straight freeway at time $t = 0$ (seconds). The total force on the truck from its engine and air and rolling resistance is the constant 625 pounds, directed down the freeway, for $0 \leq t \leq 20$. Give formulas for the truck's acceleration, velocity, and distance from the on-ramp for $0 \leq t \leq 20$.

SOLUTION With time measured in seconds and distance in feet, the acceleration due to gravity at the surface of the earth is $g = 32$ feet per second^2. The truck's weight is $w = 4000$ pounds, so its mass is $m/g = 4000/32 = 125$ slugs. By Newton's law, the truck's acceleration is $a = F/m = 625/125 = 5$ feet per second^2. Its velocity is

$$v = \int a\,dt = \int 5\,dt = 5t + v_0.$$

We use the symbol v_0 here for the constant of integration because it is the velocity at $t = 0$. Therefore, $v_0 = 20$ and $v = 5t + 20$ feet per second.

The truck's distance from the on-ramp is

$$s = \int v\,dt = \int (5t + 20)\,dt = 2.5t^2 + 20t + s_0.$$

Here s_0 is the distance at $t = 0$, so $s_0 = 500$, and we obtain $s = 2.5t^2 + 20t + 500$. □

Question 10 Figures 24 and 25 show the graphs of the truck's velocity and position from Example 8. **(a)** What property of the acceleration causes the graph of the velocity to be a line that slopes up to the right? **(b)** What property of the truck's velocity causes the graph of its position to be concave up?

---

[1] *The Guinness Book of World Records, 1998*, Stamford, CT: Guinness Media, Inc., 1997, p. 159.

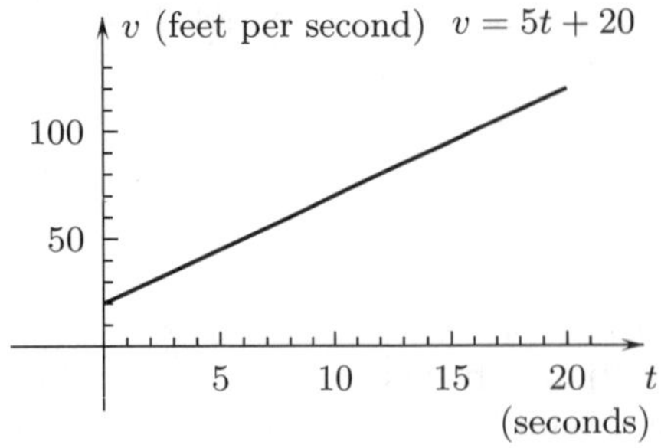

FIGURE 24

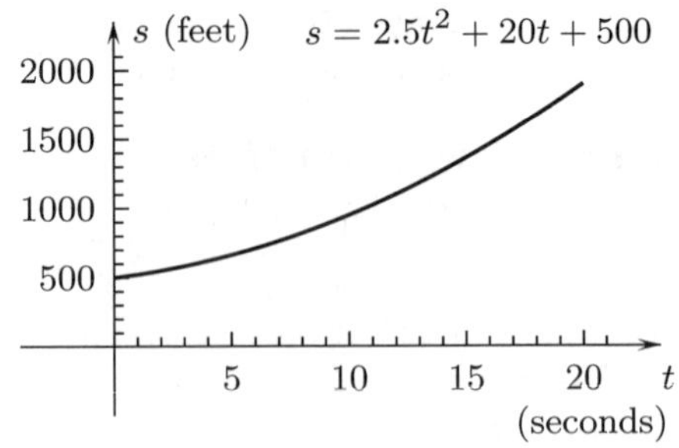

FIGURE 25

Principles and procedures

- To find the area of a region between the graphs of functions $f(x)$ and $g(x)$, determine which curve forms the top and which forms the bottom, or, if the graphs cross, determine the intervals in which one graph is above the other. If the left or right side is not given but is an intersection of the graphs, find it by solving the equation $f(x) = g(x)$. Then find the area of the region, or of each piece of the region, by integrating the top function minus the bottom function from the left side to the right side.
- The weight (or mass) of an object lying along the interval $[a, b]$ on an x-axis is the integral from a to b of its one-dimensional density $\rho(x)$, measured in units of weight (or mass) per length. The moment of the rod about the origin is the integral from a to b of $x\rho(x)$. Its center of gravity $\overline{x}$ is its moment about the origin divided by its weight (or mass).
- The length of the graph of $f(x)$ for $a \leq x \leq b$ is the integral from a to b of $\sqrt{1 + [f'(x)]^2}$. Because of the square root, most integrals giving lengths of curves cannot be found exactly with techniques of introductory calculus.
- The average value of a function in an interval $[a, b]$ is the integral of the function over the interval, divided by the width of the interval. Average values have the two geometric interpretations described in Rule 7.
- Because velocity is the derivative of position and acceleration is the derivative of velocity, velocity can be found by integrating acceleration and position can be found by integrating velocity.
- The total acceleration a of an object is calculated from the total force F on it by using Newton's law $F = ma$ with m the object's mass in the appropriate units.
- If an object's acceleration is a constant a, then its velocity $v = \int a\,dt = at + v_0$ is linear, and its position $s = \int v\,dt = \frac{1}{2}at^2 + v_0 t + s_0$ is a quadratic function of t, where v_0 and s_0 are the object's velocity and position at $t = 0$ or are constants that can be calculated from the values of v and s at some other time.

Tune-Up Exercises 4.6♦

[A]*Answer provided.* [O]*Outline of solution provided.*

Find exact answers in Exercises T1 through T4. Check the values of integrals with your calculator or computer.

T1.[O] Find the area of the region between $y = 2 + x^2$ and $y = 1$ for $0 \le x \le 2$.

♦ Type 1, area

T2.[O] Find the area of the region bounded by $y = 3 - x^2$ and $y = -2x$.

♦ Type 1, area

T3.[O] What is the average value of $6x^{-2}$ for $1 \le x \le 3$?

♦ Type 1, average value

T4.[O] Find the weight, moment of inertia about the origin, and center of gravity of a rod that extends from $x = 0$ to $x = 1$ (feet) on an x-axis and whose density at x is $\rho = 12x^2$ pounds per foot.

♦ Type 1, weight and center of gravity

T5.[A] Give an integral that equals the length of $y = x^4 + 1$ for $0 \le x \le 2$. Then use your calculator or computer to find its approximate value.

♦ Type 1, length of a graph

T6.[O] What is $g(0)$ if $\dfrac{d^2g}{dx^2}(x) = 6x$ for all x, $\dfrac{dg}{dx}(1) = 2$, and $g(1) = -1$?

♦ Type 1, using a second derivative

T7.[O] An object with constant acceleration of 2 yards per minute^2 in the positive direction on an s-axis. is at $s = 16$ (yards) and is traveling 4 yards per minute in the negative s-direction at $t = 0$ (minutes). When is it at $s = 12$?

♦ Type 1, acceleration, velocity, and position

Problems 4.6

[A]*Answer provided.* [O]*Outline of solution provided.*

Fxact answers in the following problems whenever possible. In Problems 1 through 8 sketch the regions and find their areas.

1.[O] The region bounded by $y = x^2$ and $y = x^{1/2}$.

♦ Type 1, area

2.[O] The region bounded by $y = 5 - x^2$ and $y = x^2 - 3$.

♦ Type 1, area

3. The region bounded by $y = x^2$ and $y = 4$.

♦ Type 1, area

4.[A] The region bounded by $y = 3 - x^2$ and $y = 3 - 2x$.

♦ Type 1, area

5. The region bounded by $y = 4x^{-2}, x = 1$ and $y = 1$.

♦ Type 1, area

6.[A] The region bounded by $y = x^2$ and $y = x^3$.

♦ Type 1, area

7. The region bounded by $y = 1 - x^2$ and $y = -3$.

♦ Type 1, area

8. The region bounded by $y = x^2$ and $y = 2x^2 - 9$.

♦ Type 1, area

9.[O] What is the average value of $12x^{-2}$ for $1 \le x \le 4$?

♦ Type 1, average value

10.[A] What is the average value of $2x - x^2$ for $0 \le x \le 2$?

♦ Type 1, average value

♦The Tune-up Exercises and Problems are classified by type and content. The types are (1) basic, reactive; (2) basic reflective; (3) intermediate, reactive; (4) intermediate, reflective; (5) advanced, reactive; (6) advanced, reflective; and (7) advanced, theoretical.

11. Find the average value of x^3 for $-2 \le x \le 2$.

♦ Type 1, average value

12. **(a)** If a falling object has velocity $32t + 10$ feet per second at time t (seconds), what is its average velocity for $0 \le x \le 3$? **(b)** How far does it travel for $0 \le x \le 3$?

♦ Type 2, average velocity, etc.

13.[A] The table below gives the temperature $T(t)$°F in a room at 10 minute intervals in a forty-minute time period. **(a)** Express the average temperature for $0 \le t \le 40$ as an integral. **(b)** Use the Trapezoid Rule to estimate the average temperature.

t	0	10	20	30	40
$T(t)$	50°	53°	57°	55°	51°

♦ Type 2, average value and the Trapezoid Rule

14. Figure 26 shows the graphs of the volume V (liters) of oxygen in a diver's lungs during a breath-hold dive as a function of her depth h (feet). Approximately how much oxygen is in her lungs **(a)** when she starts her dive, **(b)** when she is at the deepest point of her dive, and **(c)** at the end of her dive? **(d)** Use the graphs to estimate the average volume in her lungs for $0 \le h \le 90$ on her descent.

♦ Type 2, average value, etc.

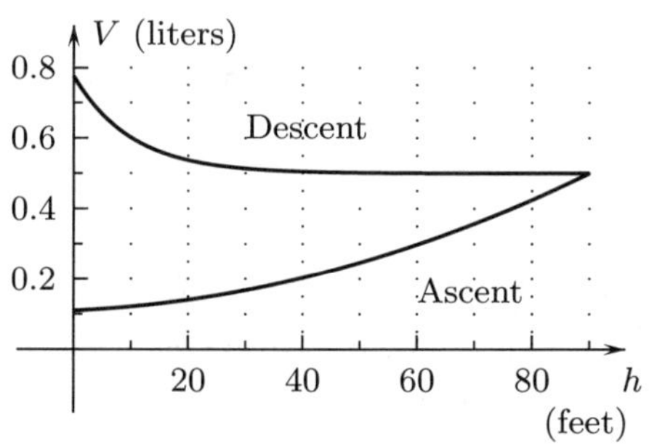

FIGURE 26

15. The table below gives the rate $r(t)$ (eggs per month) of egg production by a British chicken at the beginning of each month in 1931, with $t = 0$ at the beginning of January 1.[2] Express her average egg production during the year as an integral. Then use the Trapezoid Rule to estimate its value.

t	0	2	4	6	8	10	12
$r(t)$	14	21	20	16	10	9	13

♦ Type 2, average value and the Trapezoid Rule

16. The next table gives the price of gasoline in Italy at the beginnings of 1982, 1984, 1986, 1988, and 1990.[3] Use the Trapezoid Rule to estimate the average price of gasoline in Italy for $1982 \le t \le 1990$.

t (year)	1982	1984	1986	1988	1990
Price (dollars per gallon)	3.75	3.40	3.75	4.30	4.35

♦ Type 2, average value and the Trapezoid Rule

17.[A] Find the weight and center of gravity of a rod that extends from 0 to 1 on an x-axis with the scale given in inches and whose density at x is $4 + 4x^{1/3}$ pounds per inch.

♦ Type 1, weight and center of gravity

[2] Data adapted from *Hammond Farm Animals* by J. Hammond, Jr., J. Bowman, and T. Robinson, London: Edward Arnold Publishers, 1983, p. 174.

[3] Data adapted from "Energy for Motor Vehicles" by D. Bleviss and P. Walzer, *Scientific American* September, 1990, New York, NY: Scientific American, Inc., p. 107.

18. Find the weight and center of gravity of a rod that extends from 1 to 2 on an x-axis with the scale given in meters and whose density at x is $2 + 8x^{-3}$ kilograms per meter.

♦ Type 1, weight and center of gravity

19. Where is the center of gravity of a rod that extends from 0 to 1 on an x-axis with the scale given in feet and whose density at x is $x^{1/3}$ pounds per foot.

♦ Type 1, weight and center of gravity

20.[O] Give an integral that equals the length of $y = x^3 + 3x^2 - 4$ for $-10 \le x \le 10$, and use your calculator or computer to find the approximate value of the integral.

♦ Type 1, length of a graph

21. Use your calculator or computer to determine whether the portion of $y = x^{-2}$ for $0.1 \le x \le 1$ is shorter or longer than the portion for $1 \le x \le 100$. Give the results of your calculations.

♦ Type 2, length of a graph, etc.

22.[A] In the creation of a company's logo, the two regions bounded by $y = x$ and $y = x^3$ in an xy-plane with distances measured in yards are to be painted with paint that costs five dollars per square yard. How much does the paint cost?

♦ Type 2, area

23. A metal plate weighing 15 kilograms per square meter has the shape of the region bounded by $y = x$ and $y = \frac{1}{8}x^4$ in an xy-plane with distances measured in meters. How much does the plate weigh?

♦ Type 2, weight

24.[A] An object is moving on an s-axis. Its acceleration in the positive s-direction is t feet per minute^2 at time t minutes for $0 \le t \le 10$. At $t = 0$ it is at $s = -2$ and its velocity in the positive direction is 3 feet per minute. Give a formula for its s-coordinate as function of t.

♦ Type 1, acceleration and position

25.[A] An object's acceleration in the positive direction on an s-axis is $3t^2$ inches per hour^2 at time t (hour) for $t \ge 0$. It is at $s = 5$ and has velocity 4 inches per hour in the positive direction at $t = 0$. Give a formula for its position as a function of t for $t \ge 0$.

♦ Type 1, acceleration, velocity, and position

26. An object's velocity in the positive direction on an s-axis is $5 - t^{-2}$ meters per minute at time t (minutes) for $t \ge 1$. It is at $s = 0$ at $t = 1$. Give formulas for its position and acceleration as functions of t for $t > 0$.

♦ Type 1, acceleration, velocity, and position

27.[A] A ball is thrown down from a height of 32 feet above the ground with a velocity of 16 feet per second. There is no air resistance. **(a)** How long does it take to hit the ground? **(b)** What is its velocity when it hits the ground?

♦ Type 2, acceleration, velocity, and position

28. Figures 27 through 29 show the graphs of a car's acceleration, velocity, and position as functions of time. Find formulas for the three functions.

♦ Type 4, acceleration, velocity, and position

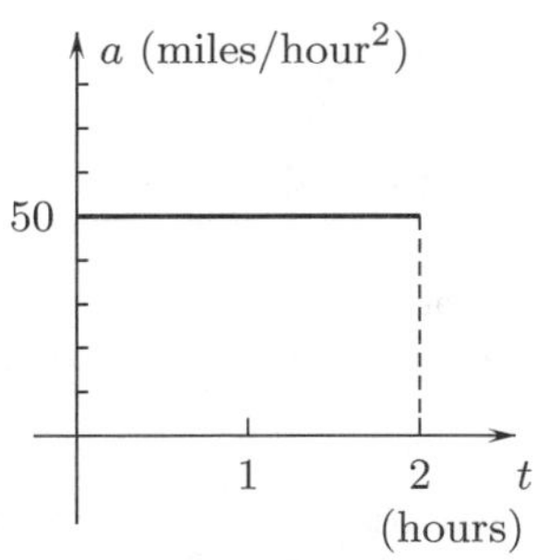

FIGURE 27

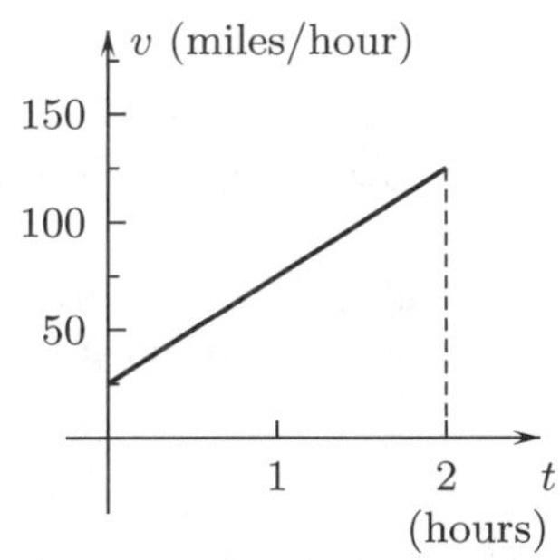

FIGURE 28

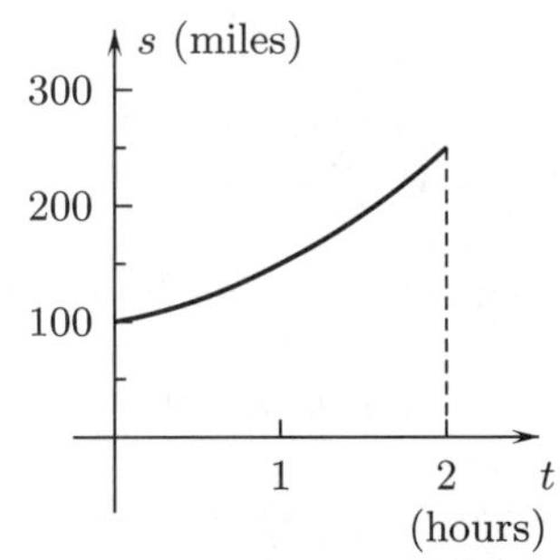

FIGURE 29

29.[O] A golf ball that is dropped from a window hits the ground in three seconds. There is no air resistance. How high is the window?

♦ Type 2, acceleration, velocity, and position

30. A suitcase is dropped from a height of 100 meters. How long does it take to hit the ground if there is no air resistance?

♦ Type 2, acceleration, velocity, and position

31. A ball that is thrown into the air from four feet above the ground is 20 feet above the ground one-half second later. What was the ball's initial velocity if there was no air resistance?

♦ Type 2, acceleration, velocity, and position

32 Find the total area of the two regions bounded by $y = x^3 + 1$ and $y = 4x + 1$.

♦ Type 4, area

33. **(a)** Show that the lines $y = 2x$ and $y = -2x$ are tangent to the parabola $y = x^2 + 1$. **(b)** Find the area of the region bounded by the curve and tangent lines from part (a).

♦ Type 4, area

34.[A] Give a formula in terms of the parameter a for the area of the region between $y = x - 1$ and $y = x^2$ for $0 \le x \le a$ with $a > 0$.

♦ Type 4, area

35. Find the total area of the two regions bounded by $y = x^{2/3}$ and $y = |x|$.

♦ Type 4, area

36. What is the exact area of the regions bounded by $y = x^2$ and $y = \dfrac{2x}{\sqrt{x^2+3}}$?

♦ Type 4, area

37. What is the exact area of the region bounded by $\sqrt{x} + \sqrt{y} = 1$ and the x- and y-axes?

♦ Type 4, area

38. The curves $y = x^3$ and $y = x^{1/3}$ bound two regions. Use integrals to show that they have the same area.

♦ Type 4, area

39.[A] Give a formula in terms of the positive parameter b for the area of the region bounded by $y = x^2$ and $y = b$.

♦ Type 4, area

40.[O] Use integrals to find the area of the triangle with vertices $(0, 0), (2, 4)$, and $(3, 3)$.

♦ Type 4, area

41. Use integrals to find the area of the triangle formed by the lines $y = 4 - x, y = -5x$, and $y = x$.

♦ Type 4, area

42. Use an integral to find the area of the triangle with vertices $(0, 0), (b, 0)$, and (b, h) for positive b and h.

♦ Type 4, area

43. Find the value of the parameter $b > 0$ such that the region between $y = x^2$ and $y = -x^2$ for $0 \le x \le b$ has area 18.

♦ Type 4, area

44. A region is bounded by $y = 1 - x + x^2$, by its tangent line at $x = 1$, and by the line $x = 3$, What is its area?

♦ Type 4, area, etc.

45. **(a)** Show that the tangent line to $y = 10x^4 - 20x^3 + 10x^2 + x$ at $x = 0$ is also its tangent line at $x = 1$. **(b)** Find the area of the region bounded by the curve and line from part (a).

♦ Type 4, area, etc.

46. **(a)** Find the intersections of $y = 3x^{2/3}$ and $y = 2 + x^2$ by generating their graphs. Check your answers. **(b)** Show that the curves from part (a) are tangent where they intersect. **(c)** Find the area of the region bounded by the curves from part (a).

♦ Type 4, area, etc.

47. **(a)** Generate the curves $y = x^n$ for $0 \le x \le 1$ with $n = 1, 2, 5, 10, 20$, and 100 and copy them on your paper. **(b)** Use your calculator or computer to find the approximate lengths of the curves from part (a). **(c)** Predict the limit as $n \to \infty$ of the length of $y = x^n$ for $0 \le x \le 1$ and give a geometric explanation of your prediction.

♦ Type 4, area, etc.

48.[A] **(a)** Find the average value L of $\sqrt{x}$ for $0 \le x \le 9$. **(b)** The curve $y = \sqrt{x}$ and the line $y = L$ for $0 \le x \le 9$ with L from part (a) bound two regions. According to Rule 3, the two regions should have the same area. Show that this is the case by finding the two areas.

♦ Type 3, area

49. The California Overland Express took 25 days to travel from San Francisco to St. Louis in November of 1858. It averaged 5.8 miles per hour in the 80 hours from San Francisco to Los Angeles, 3.5 miles per hour in the 144 hours from Los Angeles to Tucson, 3.9 miles per hour in the 274 hours from Tucson to Red River, and 6.7 miles per hour in the 102 hours from Red River to St. Louis.[(4)] **(a)** Express this data in terms of the express's velocity $v(t)$ (miles per hour) after t hours on the trip $(0 \le t \le 600)$. **(b)** What was the total distance of the trip from San Francisco to St. Louis?

♦ Type 4, velocity and position

50.[O] **(a)** Use your calculator or computer to find, with three-decimal-place accuracy, the x-coordinates of the two intersections of $y = 2x + 1/x^2$ and $y = 8$ to the right of the y-axis. **(b)** Use an integral and the values from part (a) to find the approximate area of the region bounded by $y = 2x + 1/x^2$ and $y = 8$.

♦ Type 3, area with approximate limits of integration

51. Use graphs generated on your calculator or computer to find the exact x-coordinates of the points of intersection of $y = x^2 + x^4 + x^6$ and $y = x^2 + 2$. Check your answers. Then use these numbers to find the exact area of the region bounded by the two curves.

♦ Type 2, area

52. Use your calculator or computer to find the approximate area of the region bounded by $y = \sqrt{25 - x^2}$ and $y = x^4$.

♦ Type 3, area with approximate limits of integration

53.[A] Use your calculator or computer to find the approximate area of the region bounded by $y = x^6$ and $y = \dfrac{6}{x^4 - 2x + 2}$.

♦ Type 3, area with approximate limits of integration

54. Find the number $b > 1$ such that the average value of x^3 for $0 \le x \le b$ is 250.

♦ Type 3, average value

55.[A] An object weighing 98 dynes is at $s = -10$ centimeters on an s-axis and has a velocity of 20 centimeters per second in the positive direction at $t = 0$. It is subject to a force of $0.3\sqrt{t}$ dynes in the positive direction on an s-axis at time t. Where is it at $t = 4$?

♦ Type 3, force and position

56. A 128-pound object is subject to a force of $24t + 96$ pounds in the positive direction on an s-axis at time t (seconds). At $t = 0$ it is at rest. How far does it travel for $0 \le t \le 10$?

♦ Type 3, force and position

57. An arrow that is shot into the air is 45 meters above the ground after four seconds and is 7 feet above the ground after six seconds. There is no air resistance. What was the arrow's initial velocity and from what height was it shot?

♦ Type 3, acceleration and velocity

58. If a car goes from zero to eighty miles per hour in six seconds with constant acceleration, what is its acceleration? (Be careful with the units.)

♦ Type 3, acceleration and velocity

59. A ball that is rolling up a hill has constant deceleration of eight yards per second^2 until it stops and starts rolling back down. How far does it go up the hill if its initial velocity is five yards per second?

♦ Type 4, acceleration and position

60. An object is at the origin on an s-axis at the time $t = 0$ and has acceleration $a(t) = -3t^2$ in the positive direction for $t > 0$. Its greatest s-coordinate is $s = 3/4$. What is its velocity at $t = 0$?

♦ Type 4, acceleration and position

61. An object weighing 49 newtons is at rest at $s = 0$ on an s-axis at time $t = 0$ (seconds). It is subject to a force of $10 - 5t^{1/2}$ newtons in the positive direction at time t for $t \ge 0$. **(a)** What is its greatest s-coordinate? **(b)** At what time does it pass again through the origin?

♦ Type 4, force and position

[(4)]Data adapted from "The California Overland Express, *The Quarterly Publication of the Historical Society of Southern California*, September, 1935, p. 107.

62. A rocket weighing 16 pounds is shot up vertically from the ground at $t = 0$ (seconds). The upward force of its engine is $47 - 21t^2$ pounds until $t = 1$, when it shuts off. There is no air resistance. **(a)** How high is the rocket and what is its velocity at $t = 1$? **(b)** What is its maximum upward velocity? **(c)** How high does it go? **(d)** When does it hit the ground? (Use the quadratic formula.)

◆ Type 4, force and position

63. A rod that extends from $x = 0$ to $x = 1$ (meters) on an x-axis has density x^n newtons per meter at x, and its center of gravity is at $x = 0.6$. What is the constant n?

◆ Type 4, center of gravity

64. A valentine box is two inches deep and has vertical sides. Its horizontal top and bottom have the shape of the region between $y = |2x| - 10$ and $y = |x| - \frac{1}{5}x^2$ in an xy-plane with distances measured in inches. Generate the curves in the window $-12 \le x \le 12, -12 \le y \le 4$. Then draw the box and find its exact volume.

◆ Type 4, volume

65. What is the exact length of the portion of the curve $y = 2000 - 2x^{1.5}$ that is above the x-axis?

◆ Type 3, length of a graph

66. **(a)** Give a formula in terms of the positive parameter m for the area of the region bounded by $y = \sqrt{x}$ and $y = mx$. **(b)** What is the limit of this area as $m \to 0^+$?

◆ Type 4, area, etc.

67. **(a)** Give a formula in terms of the positive parameter k for the area $A(k)$ of the region bounded by $y = x^2$ and $y = k$. **(b)** Show that $\frac{dA}{dk}(k)$ is the width of the top of the region for any choice of $k > 0$.

◆ Type 4, area, etc.

68. A rod that extends from $x = 0$ to $x = k$ (inches) with a positive constant k has density $x^{3/2}$ pounds per inch at x and its center of gravity is at $x = 5$. How long is the rod?

◆ Type 4, center of gravity

69. Find the exact length of $y = \frac{1}{3}(x^2 + 2)^{3/2}$ for $-2 \le x \le 3$.

◆ Type 3, length of a graph

70. Find the values of the positive constant k such that $y = x^4 + 1$ and $y = kx^2$ **(a)** do not intersect, **(b)** intersect at four different points, and **(c)** intersect at two points. **(d)** Show that with the choice of k from part (c), the curves are tangent where they intersect. **(e)** Find the area of the region bounded by the curves with k the value from part (c).

◆ Type 4, area, etc.

71. Use your calculator or computer to find the approximate value of the constant k such that portion of the curve $y = x^2$ for $0 \le x \le k$ has length 25.

◆ Type 4, area, etc.

72. Suppose you attempt to find the length of $y = f(x)$ for $a \le x \le b$ by approximating $f(x)$ by step functions as in the study of areas, density, and average values. What (incorrect) formula would you obtain for the length?

◆ Type 4, definition of arclength

73. **(a)** Give a formula for the function $f(x)$ whose graph is the upper half of the ASTROID $x^{2/3}+y^{2/3} = 1$. What is its domain? **(b)** The integral in Rule 5 of this section cannot be used to find the length of the graph of $f(x)$ for $0 \le x \le 1$ because $f'(x) \to -\infty$ as $x \to 0^+$. Find the length by determining the limit as $a \to 0^+$ of the length for $a \le x \le 1$.

◆ Type 4, length of a graph

74. Find the exact length of $y = \frac{4}{5}x^{5/4}$ for $0 \le x \le 9$ by using the substitution $u = 1+\sqrt{x}$ in the integral.

◆ Type 3, length of a graph

75. A boy drops a rock in a river from a bridge. He hears the rock hit the water $4\frac{3}{8}$ seconds later. There is no air resistance. How high is the bridge above the water? (The speed of sound is 1089 feet per second.)

◆ Type 5, acceleration, velocity, and position

Section 4.7

Integrals involving transcendental functions

In this section we will use the Fundamental Theorem of Calculus and differentiation formulas from Sections 3.3 and 3.4 to derive integration formulas that involve logarithms, exponential functions, and trigonometric functions. You will probably find these formulas fairly easy to learn because most of them are closely related to differentiation formulas you have already learned.

Topics:

- ***Integrals of*** $\mathbf{x^{-1}}$
- ***Integrals of*** $\mathbf{e^x}$ ***and*** $\mathbf{b^x}$
- ***Present value and income streams***
- ***Integrals of*** $\cos \mathbf{x}$ ***and*** $\sin \mathbf{x}$
- ***Integrals of*** $\sec^2 \mathbf{x}$, $\csc^2 \mathbf{x}$, $\sec \mathbf{x} \tan \mathbf{x}$, ***and*** $\csc \mathbf{x} \cot \mathbf{x}$
- ***Integrals of*** $\tan \mathbf{x}$, $\cot \mathbf{x}$, $\sec \mathbf{x}$, ***and*** $\csc \mathbf{x}$
- ***Integrals of*** $\dfrac{1}{\sqrt{\mathbf{a^2 - x^2}}}$ ***and*** $\dfrac{1}{\mathbf{a^2 + x^2}}$

The curious integrals of $\mathbf{x^{-1}}$

The integration formula

$$\int_a^b x^n \, dx = \frac{1}{n+1} b^{n+1} - \frac{1}{n+1} a^{n+1}$$

that we studied in Section 4.3 does not give integrals of $x^{-1} = 1/x$ because we cannot set $n = -1$ in the fraction $1/(n+1)$. We begin out study of this special case by examining Riemann sums for integrals of $1/x$.

Question 1 Use the Riemann-sum program from Section 4.2 to find the left Riemann sums for **(a)** $\displaystyle\int_1^2 \frac{1}{x}\,dx$, **(b)** $\displaystyle\int_2^4 \frac{1}{x}\,dx$ and **(c)** $\displaystyle\int_3^6 \frac{1}{x}\,dx$ partitions of the intervals of integration into three equal subintervals. Use the window $-0.5 \le x \le 4.5$, $-0.67 \le y \le 3$ in parts (a) and (b) and the window $2.5 \le x \le 7.5, -0.67 \le y \le 3$ in part (c) The program generates the rectangles in Figures 1 through 3.

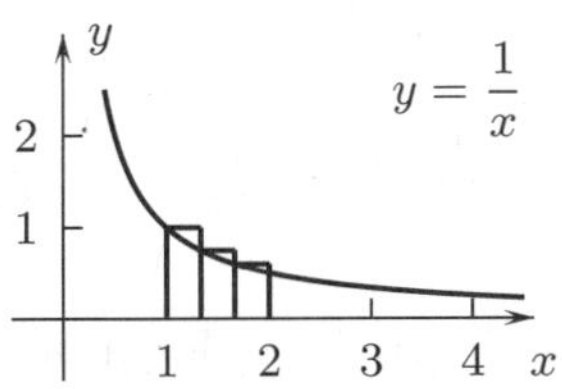

FIGURE 1

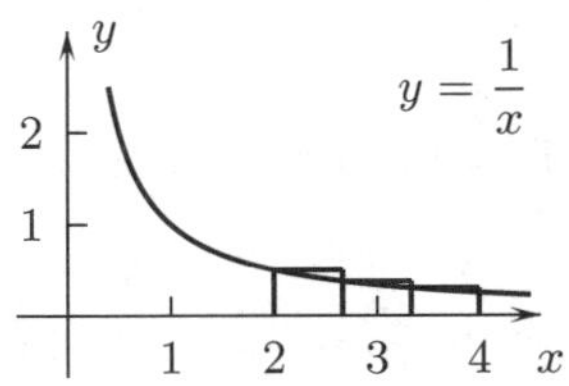

FIGURE 2

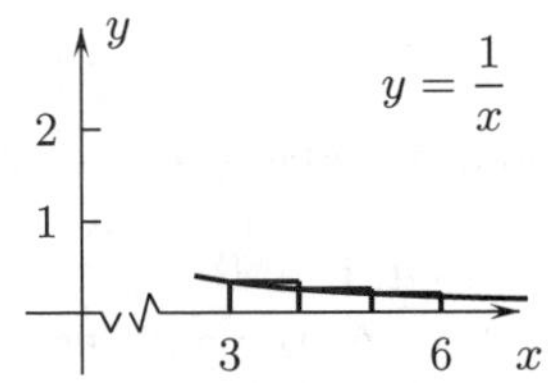

FIGURE 3

♦ SUGGESTIONS TO INSTRUCTORS: The main topics in this section can be divided into two parts, as follows:
Part 1 (Integrals involving powers, logarithms and exponential functions, present value, and income streams; Tune-up Exercises T1–T6, T8; Problems 3, 4, 7, 10, 12–14, 16–19, 21, 27–30, 32, 33, 36–39, 42, 44, 48, 51–57, 61–67, 70–83, 87, 90–93, 96, 99) Have students begin work on Questions 1 through 5 before class. Problems 37, 44, 52, 56, 67, 72, 75, 80, and 82 are good for in-class discussion.
Part 2 (Integrals involving trigonometric and inverse-trigonometric functions; Tune-up Exercises T7, T9, T11–T16; Problems 1, 2, 5, 6, 8, 9, 11, 15, 20, 22–26, 31, 34, 35, 40, 41, 43, 45–47, 49, 50, 58–60, 62–65, 68, 69, 84–86, 88, 89, 94, 95, 97, 98) Have students begin work on Questions 6 through 10 before class. Problems 43, 45, 46, 60, 62 and 85 are good for in-class discussion.

Question 2 Explain, by comparing the widths and heights of respective rectangles, why the Riemann sums corresponding to the rectangles in Figures 1 through 3 are equal.

The reasoning in the response to Question 2 could be used to show that the left Riemann sums for $\int_1^2 \frac{1}{x}\,dx$, $\int_2^4 \frac{1}{x}\,dx$, and $\int_3^6 \frac{1}{x}\,dx$ are equal for any partitions into equal subintervals. Since the integrals are the limits of these Riemann sums as the number of subintervals in the partitions tend to ∞, the three integrals are equal:

$$\int_1^2 \frac{1}{x}\,dx = \int_2^4 \frac{1}{x}\,dx = \int_3^6 \frac{1}{x}\,dx.$$

Similar reasoning would show that for any constants $b > 1$ and $c > 1$,

$$\int_1^b \frac{1}{x}\,dx = \int_c^{bc} \frac{1}{x}\,dx. \qquad \textbf{(1)}$$

We can explain this result by finding a formula for definite integrals of $1/x$.

The differentiation formula from Section 2.3 for the natural logarithm

$$\frac{d}{dx}(\ln x) = \frac{1}{x} \quad \text{for } x > 0$$

translates into the antidifferentiation formula

$$\int \frac{1}{x}\,dx = \ln x + C \quad \text{for } x > 0. \qquad \textbf{(2)}$$

This, with the Fundamental Theorem, gives

$$\int_a^b \frac{1}{x}\,dx = \ln(b) - \ln(a) \quad \text{for positive } a \text{ and } b \qquad \textbf{(3)}$$

When we apply this formula to the integrals in **(1)**, we obtain

$$\int_1^b \frac{1}{x}\,dx = \ln(b) - \ln(1)$$
$$\int_c^{bc} \frac{1}{x}\,dx = \ln(bc) - \ln(c).$$

These equations give **(1)** since $\ln(b) - \ln(1) = \ln(b)$ and $\ln(bc) - \ln(c) = \ln(bc/c) = \ln(b)$.

Integrals of $1/x$ *for negative* x

Formula **(2)** gives the antiderivatives of $1/x$ for positive x. What are its antiderivatives for negative x, where $\ln x$ is not defined? To answer this, we look at the graph of $\ln x$ and its tangent line at $x = a$ for a positive constant a, as in Figure 4. The slope of the tangent line to $y = \ln x$ at $x = a$ is $1/a$. The graph of an antiderivative of $1/x$ for negative x has to be such that its tangent line at $x = -a$ has slope $1/(-a) = -1/a$. We can achieve this by having the graph of the antiderivative for $x < 0$ be the mirror image in the y-axis of the graph of $\ln x$, as in Figure 5, because taking the mirror image of the tangent line multiplies the run between two points on the line by -1, without changing the rise, so the slope is multiplied by -1.

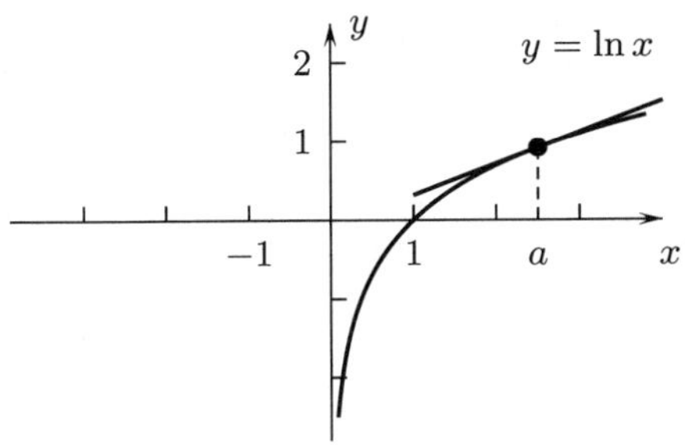

Slope $\frac{1}{a}$

FIGURE 4

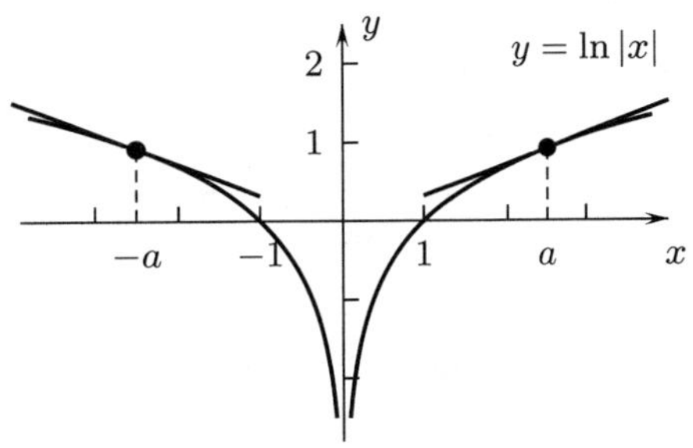

Slopes $\frac{1}{-a}$ and $\frac{1}{a}$

FIGURE 5

The function that equals $\ln x$ for $x > 0$ and whose graph for $x < 0$ is the mirror image of the graph of $\ln x$ is $\ln|x|$. Consequently, $\frac{d}{dx}(\ln|x|) = \frac{1}{x}$ for all $x \neq 0$, and we have the following:

Theorem 1 **(a)** *In any interval not containing zero,*

$$\int \frac{1}{x}\,dx = \ln|x| + C. \tag{4}$$

(b) *If a and b are either both positive or negative, then*

$$\int_a^b \frac{1}{x}\,dx = \Big[\ln|x|\Big]_a^b = \ln|b| - \ln|a|. \tag{5}$$

Example 1 Imagine you have \$100,000 in investments at the beginning of the year 2000, that your investments do not change until the beginning of 2001, and that thereafter your investments are increasing at the rate of $500/t$ dollars per year t years after the beginning of 2000. What would be the value of your investments at the beginning of 2005, 2025, 2050, and 2075?

SOLUTION Let $I(t)$ be the value of your investments t years after the beginning of 2000. Then $dI/dt = 500/t$ for $t \geq 1$, so that for $t \geq 1$,

$$I(t) = \int \frac{500}{t}\,dt = 500\ln|t| + C.$$

SOLUTION Because $I(1) = 100{,}000$, the constant C of integration here is 100,000 and $I(t) = 100{,}000 + 500\ln|t|$. Your investments would be worth $I(25) \doteq 101{,}609.44$ dollars at the beginning of 2025, $I(50) \doteq 101{,}956.01$ dollars at the beginning of 2050, and $I(75) \doteq 102{,}258.74$ dollars at the beginning of 2075. (The value of your investments would tend to ∞ as $t \to \infty$, since $\ln x \to \infty$, but would grow very slowly.) □

In the next question we examine an integral of $1/x$ with negative values of x.

Question 3 Give the exact value of $\displaystyle\int_{-7}^{-2} \frac{1}{x}\,dx$ Then give a geometric explanation of why the integral is negative.

Integrals of 1/u *with substitutions*

Any function that can be put in the form $\frac{1}{u}\frac{du}{dx}$ with a function $u(x)$ by adjusting a constant factor can be integrated by making this substitution and applying **(4)** or **(5)**.

Example 2 A rod that extends from $x = 0$ to $x = 4$ (feet) on an x-axis has one-dimensional density $1/(3x+8)$ pounds per foot at x. How much does it weigh?

SOLUTION By Rule 4 of Section 4.5, the rod's weight is the integral of $1/(3x+8)$ from 0 to 4. We make the substitutions

$$u = 3x + 8, \quad du = \frac{d}{dx}(3x+8)\,dx = 3\,dx$$

in the integral and obtain

$$\begin{aligned}
[\text{Weight}] &= \int_{x=0}^{x=4} \frac{1}{3x+8}\,dx = \tfrac{1}{3}\int_{x=0}^{x=4}\frac{1}{3x+8}(3\,dx) \\
&= \tfrac{1}{3}\int_{x=0}^{x=4}\frac{1}{u}\,du = \tfrac{1}{3}\Big[\ln|u|\Big]_{x=0}^{x=4} \\
&= \tfrac{1}{3}\Big[\ln|3x+8|\Big]_{x=0}^{x=4} \\
&= \tfrac{1}{3}[\ln|3(4)+8|] - \tfrac{1}{3}[\ln|3(0)+8|] \\
&= \tfrac{1}{3}[\ln(20) - \ln(8)].
\end{aligned}$$

We could also have changed the limits of integration to values of u instead of switching back to x for the evaluations: since $u = 3x + 8$ runs from 8 to 20 when x goes from 0 to 4, we would have

$$\begin{aligned}
[\text{Weight}] &= \tfrac{1}{3}\int_{x=0}^{x=4}\frac{1}{u}\,du = \tfrac{1}{3}\int_{u=8}^{u=20}\frac{1}{u}\,du \\
&= \tfrac{1}{3}\Big[\ln|u|\Big]_{u=8}^{u=20} = \tfrac{1}{3}[\ln(20) - \ln(8)].\ \Box
\end{aligned}$$

Integrals of e^x

The differentiation formula from Section 3.3

$$\frac{d}{dx}(e^x) = e^x$$

gives our next integration formulas.

Theorem 2 **(a)** *For all* x,

$$\int e^x\,dx = e^x + C. \qquad \textbf{(6)}$$

(b) *For any* a *and* b,

$$\int_a^b e^x\,dx = \Big[e^x\Big]_a^b = e^b - e^a. \qquad \textbf{(7)}$$

Question 4 Find a formula for the function $g(x)$ such that $g'(x) = e^x$ for all x and $g(2) = 10$.

We can also use **(6)** and **(7)** with integration by substitution.

Example 3 Imagine your investments at the beginning of 2000 are worth \$100,000 and that that are increasing at the rate of $500e^{t/5}$ dollars per year t years after the beginning of 2000. What would be the value of your investments at the beginning of 2025 and 2050?

SOLUTION Again let $I(t)$ be the value of your investments t years after the beginning of 2000. Then, since $\frac{dI}{dt} = 500e^{t/5}$ for $t \geq 0$,

$$\begin{aligned} I(t) &= \int 500e^{t/5}\,dt = 5\int 500e^{t/5}\,\tfrac{1}{5}\,dt \\ &= 2500\int e^u\,du = 2500e^u + C = 2500e^{t/5} + C. \end{aligned}$$

We used here the substitution $u = \frac{1}{5}t, du = \frac{1}{5}\,dt$. Since $I(0) = 2500 + C$ is equal to 100,000 the constant of integration C is 97,500 and $I(t) = 2500e^{t/5} +$ 97,500. Your investments would be worth $I(25) \doteq$ 468,532.90 dollars at the beginning of 2025 and $I(50) \doteq$ 55,163,664.49 dollars at the beginning of 2050. (The value of your investments would grow more and more rapidly as time passes and tend to ∞ as $t \to \infty$.) □

Example 4 Figure 6 shows the region between $y = e^x$ and the x-axis for $-1 \leq x \leq 1$. **(a)** What is its exact area? **(b)** What is the exact average value of $f(x) = 2xe^{-x^2} + 1$ for $0 \leq x \leq 3$? (See Figure 7.)

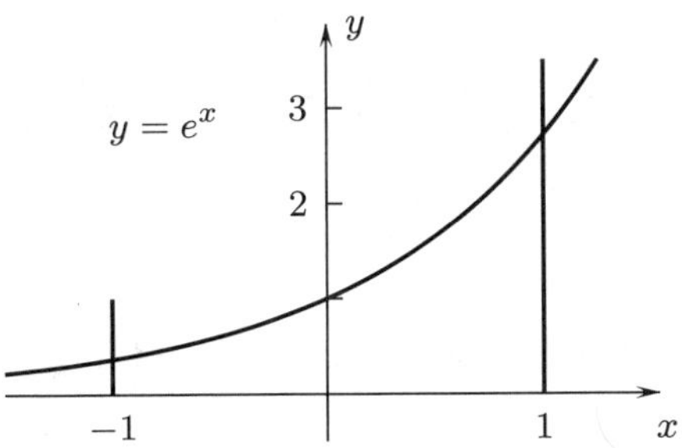

FIGURE 6

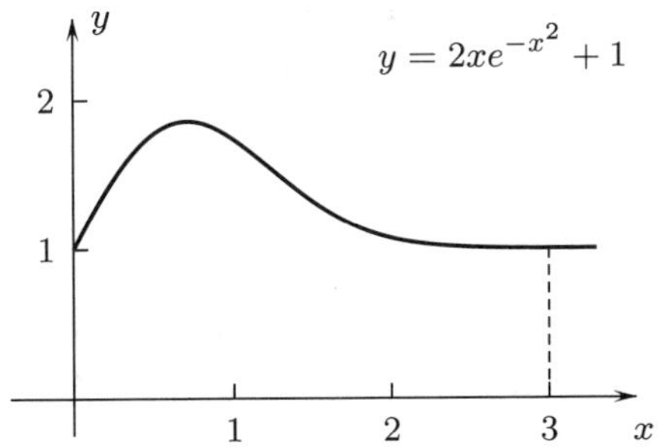

FIGURE 7

SOLUTION **(a)** Because e^x is positive for $-1 \leq x \leq 1$, the area of the region is $\int_{-1}^{1} e^x\,dx$ $= \Big[e^x\Big]_{-1}^{1} = e^1 - e^{-1}.$

(b) The interval $0 \leq x \leq 3$ is three units long. Therefore, the average value is

$$\tfrac{1}{3}\int_{x=0}^{x=3} f(x)\,dx = \tfrac{1}{3}\int_{x=0}^{x=3} (2xe^{-x^2} + 1)\,dx = \tfrac{1}{3}\int_{x=0}^{x=3} 2xe^{-x^2}\,dx + \tfrac{1}{3}\int_0^3 1\,dx$$

We evaluate the second integral and make the substitution $u = -x^2$ in the first. Then $du = -2x\,dx$, and the average value is

$$\begin{aligned} -\tfrac{1}{3}\int_{x=0}^{x=3} e^{-x^2}\,(-2x\,dx) + \tfrac{1}{3}\Big[x\Big]_{x=0}^{x=3} &= -\tfrac{1}{3}\int_{x=0}^{x=3} e^u\,du + 1 \\ &= -\tfrac{1}{3}\Big[e^u\Big]_{x=0}^{x=3} + 1 = -\tfrac{1}{3}\Big[e^{-x^2}\Big]_{x=0}^{x=3} + 1 \\ &= -\tfrac{1}{3}(e^{-9} - 1) + 1 = \tfrac{4}{3} - \tfrac{1}{3}e^{-9}. \ \square \end{aligned}$$

Integrals of b^x

As we have seen before, b^x with a positive $b \neq 1$ can be expressed in terms of e^x by writing $b = e^{\ln b}$ to have

$$b^x = (e^{\ln b})^x = e^{x \ln b}. \tag{8}$$

We used this formula in Section 3.3 to derive the differentiation formula $\frac{d}{dx}(b^x) = (\ln b)b^x$, which yields the integration formula

$$\int b^x\, dx = \frac{1}{\ln b} b^x + C. \tag{9}$$

Example 5 Evaluate $\int_0^2 3^x\, dx$.

SOLUTION Formula **(9)** and the Fundamental Theorem give

$$\int_0^2 3^x\, dx = \frac{1}{\ln(3)}\left[3^x\right]_0^2 = \frac{1}{\ln(3)}(3^2 - 3^0) = \frac{8}{\ln(3)} \square$$

Present value

The PRESENT VALUE of future income is the amount that would have to be invested today to yield the income at a specified future date. Present value depends on the interest rate and the method of compounding used.

Recall that C dollars invested at $100r\%$ annual interest grows to $C(I + r/N)^{Nt}$ dollars after t years if the interest is compounded N times a year, and grows to Ce^{rt} dollars after t years if the interest is compounded continuously.

Consequently to find the present value at $100r\%$ annual interest compounded n times a year of an income of I dollars t years from now, we solve the equation $I = C(1 + r/n)^{nt}$ for the present value C. We obtain

$$[\text{Present value}] = I\left(1 + \frac{r}{n}\right)^{-nt}. \tag{10}$$

To find the present value at $100r\%$ annual interest compounded continuously of an income of I dollars t years from now, we solve $I = Ce^{rt}$ for C:

$$[\text{Present value}] = Ie^{-rt}. \tag{11}$$

Example 6 Imagine that you win a six million dollar lottery and that you will receive two million dollars right away, two million in five years, and two million in ten years. What is the present value of the prize at 6% annual interest compounded annually?

SOLUTION The present value of the first payment is 2 million dollars, the present value of the payment to be made in five years is $2(1 + 0.06)^{-5}$, and the present value of the last payment is $2(1 + 0.06)^{-10}$. The present value of the entire six million is

$$2 + 2(1 + 0.06)^{-5} + 2(1 + 0.06)^{-10} \doteq 4.6113 \text{ million dollars.} \square$$

Question 5 What is the present value of the lottery prize at 6% annual interest compounded continuously?

Income streams

In Examples 1 and 3 above we studied incomes that were received continuously, rather than in a succession of individual payments. Such a mathematical model is called an INCOME STREAM and is described by giving the rate $R(t)$ at which the income is received as a continuous or piecewise continuous function of the time t.

If an income stream at the rate of $R(t)$ dollars per year at t years is received during the time interval $T_0 \le t \le T_1$, then the total amount received in that period is

$$[\text{Total income}] = \int_{T_0}^{T_1} R(t)\,dt \text{ dollars.} \tag{12}$$

To find the formula for the present value of this income stream at $100r\%$ annual interest, compounded continuously we first review a derivation of formula **(12)**. We use partitions $T_0 = t_0 < t_1 < t_2 < \cdots < t_N = T_1$ to approximate the income stream with lump sum payments. The rate of flow of income in the jth subinternal, $t_{j-1} \le t \le t_j$, is approximately $R(t_j)$ dollars per year, so the amount received in that time period is approximately $R(t_j)\Delta t_j$ dollars, where Δt_j is the length of the jth subinterval. The total income is approximately

$$\sum_{j=1}^{N} R(t_j)\Delta t_j.$$

This Riemann sum tends to the integral as the number of subintervals in the partition tends to ∞ and their widths tend to 0.

Similarly, a lump payment of $R(t_j)\Delta t_j$ dollars made at time t_j has present value $e^{-rt_j}R(t_j)\Delta t_j$ dollars at time t_j, so the present value at time T_1 of all N payments is

$$\sum_{j=1}^{N} e^{-rt_j}R(t_j)\Delta t_j.$$

This is a Riemann sum for the integral that gives the present value:

$$[\text{Present value}] = \int_{T_0}^{T_1} e^{-rt}R(t)\,dt. \tag{13}$$

Example 7 Your investments in Example 3 had an income stream of $500e^{t/5}$ dollars per year t years after the beginning of 2000 and, as we saw in solving Example 3, would earn \$368,532.90 by the beginning of 2025 (raising the value of your investments from \$100,000 at the beginning of 2000 to \$468,532.90 at the beginning of 2025.) What is the present value of this income at the beginning of 2000 at 10% annual interest compounded continuously?

SOLUTION We again set $t = 0$ (years) at the beginning of 2000 and write $500e^{t/5}$ as $500e^{0.2t}$. Then by **(13)**, the present value is

$$\begin{aligned}
\int_0^{25} 500e^{0.2t}e^{-0.1t}\,dt &= \int_0^{25} 500e^{0.1t}\,dt = \frac{500}{0.1}\int_{t=0}^{t=25} e^{0.1t}\,(0.1\,dt)\\
&= 5000\int_{t=0}^{t=25} e^{u}\,du = 5000\Big[e^{u}\Big]_{t=0}^{t=25}\\
&= 5000\Big[e^{0.1t}\Big]_0^{25} = 5000(e^{2.5}-1) \doteq 55{,}912.47 \text{ dollars.} \ \square
\end{aligned}$$

Integrals of cos x and sin x

The differentiation formula

$$\frac{d}{dx}(\sin x) = \cos x$$

translates into the first integration formula in the next theorem.

Theorem 3 **(a)** *For any x,*

$$\int \cos x\, dx = \sin x + C \tag{14}$$

$$\int \sin x\, dx = -\cos x + C. \tag{15}$$

(b) *For any constants a and b,*

$$\int_a^b \cos x\, dx = \Big[\sin x\Big]_a^b = \sin b - \sin a \tag{16}$$

$$\int_a^b \sin x\, dx = \Big[-\cos x\Big]_a^b = -\cos b + \cos a. \tag{17}$$

Formula **(15)** comes from

$$\frac{d}{dx}(\cos x) = -\sin x$$

when we multiply both sides by -1. Formulas **(16)** and **(17)** are consequences of **(14)**,**(15)**, and the Fundamental Theorem.

Example 8 A car is 30 miles north of a town at time $t = 0$ (hours) and its velocity toward the north is $60 + 5\cos t + 8\sin t$ miles per hour for $0 \le t \le 6$. Give a formula in terms of t for its distance from the town for $0 \le t \le 6$.

SOLUTION The car's distance north of the town is

$$s = \int v\, dt = \int (60 + 5\cos t + 8\sin t)\, dt = 60t + 5\sin t - 8\cos t + C.$$

With this formula, we obtain $s(0) = 60(0)+5\sin(0)-8\cos(0)+C = -8+C$. To satisfy the condition $s(0) = 30$, we take $C = 38$, which gives $s = 60t+5\sin t-8\cos t+38$. □

Question 6 Now imagine your investments at the beginning of 2000 are worth \$100,000 and that that are increasing at the rate of $500\pi\cos\left(\frac{1}{50}\pi t\right)$ dollars per year t years after the beginning of 2000. What would be the value of your investments at the beginnings of 2025, 2050 and 2075 in this case?

Example 9 What is the average value of $x\sin(x^2)$ for $-3 \le x \le 3$? Generate the graph of the function for $-3 \le x \le 3$ to check your answer.

SOLUTION The interval $[-3, 3]$ has width 6, so by Rule 2 of Section 4.3, the average value is $\frac{1}{6}\int_{-3}^{3} x\sin(x^2)\, dx$. We use $u = x^2$, for which $du = 2x\, dx$.

$$\begin{aligned}[\text{Average value}] &= \left(\tfrac{1}{6}\right)\left(\tfrac{1}{2}\right)\int_{x=-3}^{x=3} \sin(x^2)\,(2x\, dx) = \tfrac{1}{12}\int_{x=-3}^{x=3} \sin u\, du \\ &= \tfrac{1}{12}\Big[-\cos u\Big]_{x=-3}^{x=3} = \tfrac{1}{12}\Big[-\cos(x^2)\Big]_{x=-3}^{x=3} \\ &= \tfrac{1}{12}[-\cos(9) + \cos(-9)] = 0.\end{aligned}$$

The graph is in Figure 8. The average value of $x\sin(x^2)$ for $-3 \le x \le 3$ is zero because the function is odd ($-x\sin\left((-x)^2\right) = -x\sin(x^2)$ for all x.) □

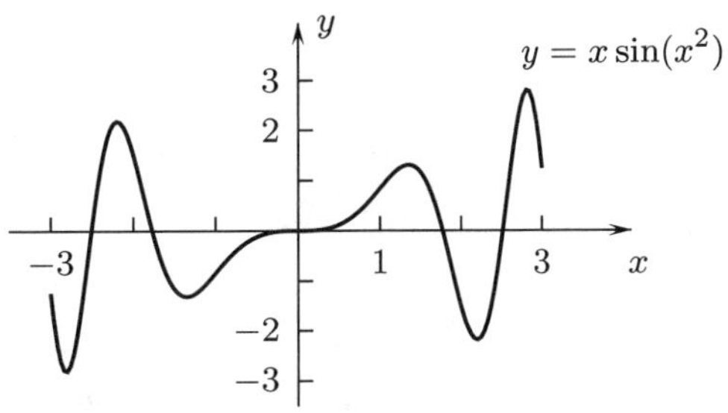

FIGURE 8

Integrals of sec²x, csc²x, sec x tan x and csc x cot x

The differentiation formulas from Section 3.3,

$$\frac{d}{dx}(\tan x) = \sec^2 x$$
$$\frac{d}{dx}(\cot x) = -\csc^2 x$$
$$\frac{d}{dx}(\sec x) = \sec x \tan x$$
$$\frac{d}{dx}(\csc x) = -\csc x \cot x$$

translate into the integration formulas in the next theorem.

Theorem 4 *For all x where the functions are defined,*

$$\int \sec^2 x\, dx = \tan x + C \tag{18}$$

$$\int \csc^2 x\, dx = -\cot x + C \tag{19}$$

$$\int \sec x \tan x\, dx = \sec x + C \tag{20}$$

$$\int \csc x \cot x\, dx = -\csc x + C \tag{21}$$

Example 10 What is the area of the region between $y = \sec^2 x$ and the x-axis for $-1 \le x \le 1$?

SOLUTION The region is shown in Figure 9. With **(18)**, we see that its area is

$$\int_{-1}^{1} \sec^2 x\, dx = \Big[\tan x\Big]_{-1}^{1} = \tan(1) - [-\tan(1)] = 2\tan(1). \ \square$$

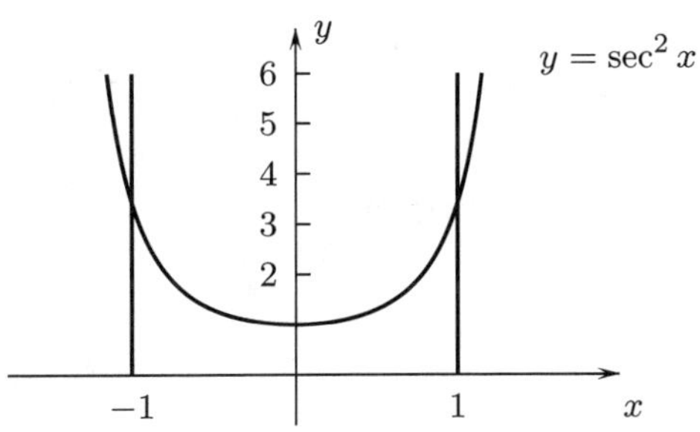

FIGURE 9

Integrals of $\tan$x*,* $\cot$x*,* $\sec$x*, and* $\csc$x

Integral of $\tan x, \cot x, \sec x$ and $\csc x$ are obtained from

$$\int \frac{1}{u}\, du = \ln|u| + C$$

with the appropriate substitutions.

To integrate $\tan x$, we write $\tan x = (\sin x)/(\cos x)$ and use the substitution $u = \cos x, du = -\sin x\, dx$:

$$\begin{aligned}\int \tan x\, dx &= \int \frac{\sin x}{\cos x}\, dx = -\int \frac{1}{u}\, du = -\ln|u| + C \\ &= -\ln|\cos x| + C.\end{aligned}$$

This is the first formula in our next theorem.

Theorem 5 *For all x where the functions are defined,*

$$\int \tan x\, dx = -\ln|\cos x| + C \tag{22}$$

$$\int \cot x\, dx = \ln|\sin x| + C \tag{23}$$

$$\int \sec x\, dx = \ln|\sec x + \tan x| + C \tag{24}$$

$$\int \csc x\, dx = -\ln|\csc x + \cot x| + C \tag{25}$$

Question 7 Derive formula **(23)** by using the substitution $u = \sin x$.

To derive **(24)**, we use a special technique. We multiply and divide $\sec x$ by $\sec x + \tan x$ and use the substitution $u = \sec x + \tan x, du = (\sec x \tan x + \sec^2 x)\, dx$:

$$\begin{aligned}\int \sec x\, dx &= \int \frac{\sec x(\sec x + \tan x)}{\sec x + \tan x}\, dx = \int \frac{\sec^2 x + \sec x \tan x}{\sec x + \tan x}\, dx \\ &= \int \frac{1}{u}\, du = \ln|u| + C = \ln|\sec x + \tan x| + C.\end{aligned}$$

Question 8 Derive formula **(25)** with the substitution $u = \csc x + \cot x$.

Example 11 Find the antiderivatives $\int x\cot(x^2)\, dx$.

SOLUTION We can use $u = x^2, du = 2x\, dx$ to obtain with **(22)**

$$\begin{aligned}\int x\cot(x^2)\, dx &= \tfrac{1}{2}\int \cot(x^2)(2x\, dx) = \int \cot u\, du \\ &= \tfrac{1}{2}\ln|\sin u| + C = \tfrac{1}{2}\ln|\sin(x^2)| + C. \square\end{aligned}$$

Example 12 Evaluate $\displaystyle\int_0^{\pi/4} \sec x\, dx$.

SOLUTION With **(24)** we obtain

$$\begin{aligned}\int_0^{\pi/4} \sec x\, dx &= \Big[\ln|\sec x + \tan x|\Big]_0^{\pi/4} \\ &= \ln|\sec(\tfrac{1}{4}\pi) + \tan(\tfrac{1}{4}\pi)| - \ln|\sec(0) + \tan(0)| \\ &= \ln(\sqrt{2} + 1) - \ln(1) = \ln(\sqrt{2} + 1). \square\end{aligned}$$

Integrals of $1/\sqrt{a^2 - x^2}$ *and* $1/\sqrt{a^2 + x^2}$

The differentiation formulas from Section 3.3

$$\frac{d}{dx}(\sin^{-1} x) = \frac{1}{\sqrt{1 - x^2}}$$

$$\frac{d}{dx}(\tan^{-1} x) = \frac{1}{1 + x^2}$$

translate into the integration formulas

$$\int \frac{1}{\sqrt{1 - x^2}}\, dx = \sin^{-1} x + C \qquad \textbf{(26)}$$

$$\int \frac{1}{1 + x^2}\, dx = \tan^{-1} x + C. \qquad \textbf{(27)}$$

Learn the generalization of **(26)** and **(27)** that are stated in the next theorem.

Theorem 6 *For any positive constant a*

$$\int \frac{1}{\sqrt{a^2 - x^2}}\, dx = \sin^{-1}\left(\frac{x}{a}\right) + C \quad \text{for } -a < x < a \qquad \textbf{(28)}$$

$$\int \frac{1}{a^2 + x^2}\, dx = \frac{1}{a}\tan^{-1}\left(\frac{x}{a}\right) + C \quad \text{for all } x. \qquad \textbf{(29)}$$

To derive **(28)** from **(26)** we use the substitution $u = x/a$, for which $x = au$ and $dx = a\, du$:

$$\int \frac{1}{\sqrt{a^2 - x^2}}\, dx = \int \frac{1}{\sqrt{a^2 - (au)^2}}\,(a\, du) = \int \frac{1}{a\sqrt{1 - u^2}}\,(a\, du)$$

$$= \int \frac{1}{\sqrt{1 - u^2}}\, du = \sin^{-1} u + C = \sin^{-1}\left(\frac{x}{a}\right) + C$$

Question 9 Derive **(29)** from **(27)**.

Example 13 Give a formula for $\displaystyle\int \frac{x}{9 + x^4}\, dx$.

SOLUTION We use **(29)** with $a = 3$ and the substitution $u = x^2, du = 2x\, dx$:

$$\int \frac{x}{9 + x^4}\, dx = \tfrac{1}{2}\int \frac{1}{9 + (x^2)^2}\, 2x\, dx = \tfrac{1}{2}\int \frac{1}{9 + u^2}\, du$$

$$= \tfrac{1}{2}\left[\tfrac{1}{3}\tan^{-1}\left(\tfrac{1}{3}u\right)\right] + C = \tfrac{1}{6}\tan^{-1}\left(\tfrac{1}{3}x^2\right) + C. \square$$

Question 10 Figure 10 shows the graph of $y = \dfrac{25}{\sqrt{100 - x^2}}$ for $-9 \le x \le 9$. **(a)** Based on the drawing, which positive integer is closest to the average value L of $\dfrac{5}{\sqrt{100 - x^2}}$ for $-9 \le x \le 9$? **(b)** Find the exact average value and its approximate decimal value.

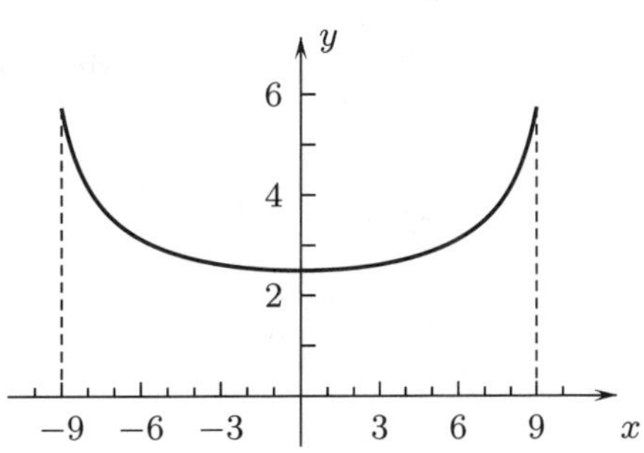

$y = \dfrac{25}{\sqrt{100 - x^2}}$

FIGURE 10

A bit of history

Archimedes (ca. 287–212 B.C.) was the first known mathematician to determine the exact area of a region with a curved boundary by an approximation procedure. He demonstrated, in particular, that the area of a region bounded by a parabola and by a straight line as in Figure 11 has area equal to four thirds of the area of the inscribed triangle whose vertex is at the point where the tangent line is parallel to the top of the region. (The Greeks did not think of areas as numbers; they only compared areas of related regions.) We describe two ways Archimedes obtained this result in the problems sets of Sections 7.3 and 11.1.

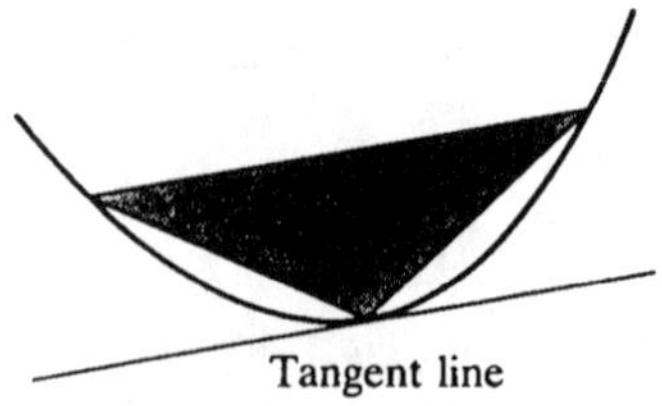

FIGURE 11

The problem of determining the areas of regions bounded by the curves $y = x^n$ for general rational constants n did not arise until the study of analytic geometry in the seventeenth century led to the consideration of these curves. Results amounting to the integration formula

$$\int_a^b x^n \, dx = \frac{1}{n+1} b^{n+1} - \frac{1}{n+1} a^{n+1} \quad (0 < a < b, n \neq -1) \tag{30}$$

for particular positve integers n appeared in various forms during the years 1635-1660 in the work of the two Italian students of Galileo, Bonaventura Cavalieri (1598-1647) and Evangelista Torricelli (1608-1647A), a Frenchman, Gilles Personne de Roberval (1602-1675), and the philosopher Blaise Pascal (1623-1662). The French lawyer Pierre Fermat (1601-1665) is usually given credit for first establishing integration formula **(30)** for arbitrary rational numbers $n \neq -1$. One of his procedures for doing this is described in the problem set for Section 7.3. Pascal and others also derived results which, with modern definitions and notation, would read

$$\int_a^b \cos x \, dx = \sin(b) - \sin(a)$$

$$\int_a^b \sin x \, dx = -\cos(b) + \cos(a).$$

Most early derivations of integration formulas employed the concept of INDIVISIBLES. In the context of area, indivisibles are infintely narrow rectangles of infinitesimal area. They are called "indivisibles" because they are assumed to be so small that they cannot be divided into narrower rectangles. A region as in Figure 12 was considered to consist of an infinite number of parallel indivisibles such as is represented by the line in the sketch. The area of each indivisible was considered to be its length multiplied by its infinitesimal width, and the area of the region was considered to be the sum of the areas of the infinite number of indivisibles in it. Leibniz used the symbol dx to denote the width of an indivisible, so that the area of an indivisible of length y was given by the product $y\,dx$. He then intrduced the symbol $\int y \, dx$ for the sum or "integral" of the areas of the indivisibles that gives the area of the region.

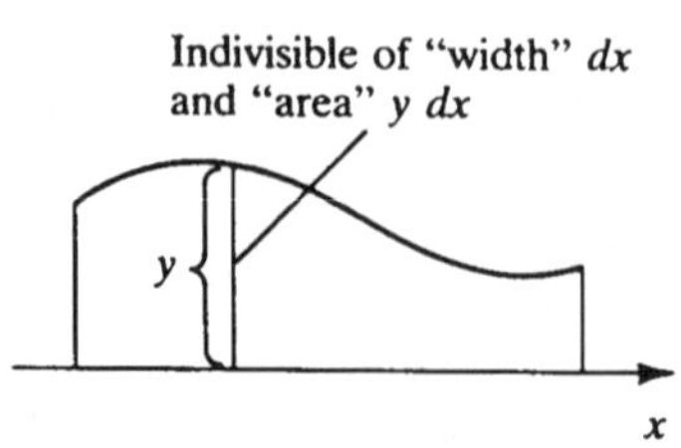

FIGURE 12

The use of indivisibles in integration, like the use of infinitesimals in early treatments of the derivative, has always been subject to criticism. The French writer Voltaire, for example, once complained that calculus was "the Art of numbering and measuring exactly a Thing whose Existence cannot be conceived."[(1)] Yet even those who argued that there were logical difficulties with the use of infinitesimals and indivisibles could not deny the importance of the mathematical tools to which these concepts led. By the second half of the seventeenth century, indivisibles had been used extensively in determining the areas, volumes, and centers of gravity of a large number of geometric figures.

The mathematicians who used indivisibles to make these calculations were schooled in the classical Greek mathematics of Euclid and Archimedes. At first, they supplied rigorous proofs, based on approximation procedures, of any formulas they derived using indivisibles. But, eventually, as more results were obtained, many felt that supplying the rigorous proof to meet the standards of classical mathematics was a waste of time and omitted this final step. The German astronomer Johann Kepler (1571-1630) took this point of view in his popular book *Nova Stereometria Doliorum Binariorum (New Solid Geometry of Wine Barrels)*, in which he computed volumes of a large number of solids of revolution. Referring to the absense of classical proofs in the book, he wrote in its preface, "We could obtain absolute and in all respects perfect demonstrations from the books of Archimedes themselves, were we not repelled by the thorny reading thereof."[(2)]

Special cases of the fundamental theorem of calculus were discovered long before its full significance and generality were recognized. The fourteenth-century French bishop Nicole Oresme (ca. 1323–1382) showed in special cases that if a time interval were represented on a horizontal line and if the corresponding variable velocity were represented by vertical line segments along it, then the distance traveled would be equal to the area on the resulting region. Galileo used the same idea in his study of falling bodies. Fermat used a primitive version of the derivative to solve problems that we solve today with integrals, and he solved certain tangent line problems by using area.

It was the English theologian Isaac Barrow (1630-1677), however, who was the first to clearly recognize that the finding of tangent lines and the computing of areas are inverse operations. This insight contributed to the work of his protégé Newton, and perhaps also that of Leibniz, who purchased a copy of Barrow's *Lectures* on a trip to London in 1673. The Fundamental Theorem played a central role in the rules and procedures of calculus as developed, independently, by Newton and Leibniz.

The point of view that integrals are given geometrically through the use of indivisibles prevailed until the early nineteenth century. The definition of the integral of a continuous function as a limit of finite sums was given by Augustin Cauchy (1789-1857) in the 1820's. Thirty years later, G.F.B. Riemann (1826-1866) realized that Cauchy's definition could be extended to apply to certain discontinuous functions. He gave the definition of the definite integral that we use today.

Principles and procedures

- Most of the formulas for indefinite integrals in this section do not have to be memorized directly since they all come from differentiation formulas from Sections 3.5 and 3.6. The exceptions are formulas **(22)** through **(25)** for integrals of $\tan x, \cos x, \sec x$, and $\csc x$, which are obtained from the integral of $1/u$ with special substitutions, and formulas **(27)** and **(28)** for integrals of $\dfrac{1}{\sqrt{a^2 - x^2}}$ and $\dfrac{1}{a^2 + x^2}$, which are generalizations of the integration formulas that come from the formulas for the derivatives of $\sin^{-1} x$ and $\tan^{-1} x$.
- As always, check your formulas for indefinite integrals by differentiating the results. Check your values of definite integrals with a procedure of approximate integration on your calculator or computer.

---

(1) *History of the Calculus* by C. Boyer, New York, NY: Dover Publications, Inc., 1959, p.238

(2) *A source Book of Mathematics* by D. Struik, Cambridge, MA: Harvard University Press, 1969, p. 188.

Tune-Up Exercises 4.7♦

^A *Answer provided.* ^O *Outline of solution provided.*

T1.^A Find the exact value of $\int_1^6 \frac{1}{x}\, dx$.

♦ Type 1, an integral

T2.^A What is the exact value of $\int_{-6}^{-1} \frac{1}{x}\, dx$? Give a geometric explanation of why this integral is negative.

♦ Type 1, an integral

T3.^O Show that $\int_0^5 \frac{x}{x^2+1}\, dx = \frac{1}{2}\ln(26)$.

♦ Type 1, an integral

Use integration formulas to carry out exact integrations in Exercises T4 through T16.

T4.^O $\int \frac{\sqrt{\ln x}}{x}\, dx$

♦ Type 1, an integral

T5.^A $\int_{-4}^{4} e^x\, dx$

♦ Type 1, an integral

T6.^A $\int (x + e^x)\, dx$

♦ Type 1, an integral

T7.^A $\int (\sin x - 5\cos x)\, dx$

♦ Type 1, an integral

T8.^O $\int x^2 e^{x^3}\, dx$

♦ Type 1, an integral

T9.^O $\int_5^0 \cos x\, dx$

♦ Type 1, an integral

T10.^O $\int_{3\pi/4}^{\pi/4} \sin x\, dx$

♦ Type 1, an integral

T11.^O $\int e^t \cos(e^t)\, dt$

♦ Type 1, an integral

T12.^A $\int \frac{1}{16+x^2}\, dx$

♦ Type 1, an integral

T13.^O $\int_0^4 \frac{1}{\sqrt{25-x^2}}\, dx$

♦ Type 1, an integral

T14.^O $\int_0^1 \tan x\, dx$

♦ Type 1, an integral

T15.^A $\int \sec^2(3x)\, dx$

♦ Type 1, an integral

T16.^A $\int_0^{0.5} \sec x \tan x\, dx$

♦ Type 1, an integral

Problems 4.7

^A *Answer provided.* ^O *Outline of solution provided.*

1.^A Find a formula for $\int (\frac{1}{x} + 2e^x + 3\sin x + 4\cos x)\, dx$.

♦ Type 1, an integral

Use integration formulas to carry out exact integrations in Problems 2 through 35.

2.^A $\int_1^{10} (\frac{2}{x} - 5\sin x)\, dx$

♦ Type 1, an integral

3.^A $\int \frac{x}{(x^2+3)^2}\, dx$

♦ Type 1, an integral

4.^A $\int_0^2 e^{-3t}\, dt$

♦ Type 1, an integral

5.^A $\int \cos x\sqrt{\sin x}\, dx$

♦ Type 1, an integral

♦The Tune-up Exercises and Problems are classified by type and content. The types are (1) basic, reactive; (2) basic reflective; (3) intermediate, reactive; (4) intermediate, reflective; (5) advanced, reactive; (6) advanced, reflective; and (7) advanced, theoretical.

6. $\int_{-\pi}^{\pi} x\sin(x^2)\,dx$

◆ Type 1, an integral

7.[A] $\int_{-1}^{0} \frac{2}{6x-1}\,dx$

◆ Type 1, an integral

8.[A] $\int e^{-x}\sin(e^{-x})\,dx$

◆ Type 1, an integral

9. $\int \frac{1+\cos x}{x+\sin x}\,dx$

◆ Type 1, an integral

10.[A] $\int_{1}^{2} \frac{4x^3+1}{1-x-x^4}\,dx$

◆ Type 1, an integral

11.[A] $\int_{0}^{2} \sin x\cos x\,dx$

◆ Type 1, an integral

12.[A] $\int \frac{\ln x}{x}\,dx$

◆ Type 1, an integral

13. $\int \frac{1}{x\sqrt{\ln x}}\,dx$

◆ Type 1, an integral

14. $\int_{0}^{1} \frac{x^2}{4-x^3}\,dx$

◆ Type 1, an integral

15.[A] $\int \frac{\sin(\ln x)}{x}\,dx$

◆ Type 1, an integral

16.[A] $\int_{1}^{9} \frac{1}{\sqrt{x}}e^{\sqrt{x}}\,dx$

◆ Type 1, an integral

17. $\int x\sqrt{x^2+6}\,dx$

◆ Type 1, an integral

18.[A] $\int \frac{e^y}{1+e^y}\,dy$

◆ Type 1, an integral

19. $\int_{0}^{1} \frac{x}{x^2-4}\,dx$

◆ Type 1, an integral

20. $\int_{0}^{1} \frac{\sin x}{\sqrt{\cos x}}\,dx$

◆ Type 1, an integral

21. $\int_{0}^{1} \frac{x}{\sqrt{4-x^2}}\,dx$

◆ Type 1, an integral

22.[A] $\int_{0}^{\pi/2} \frac{\sin x}{(1+\cos x)^4}\,dx$

◆ Type 1, an integral

23.[A] $\int \cot(3x)\,dx$

◆ Type 1, an integral

24.[A] $\int_{0}^{7} \frac{3}{4+x^2}\,dx$

◆ Type 1, an integral

25.[O] $\int_{-1}^{1} \frac{1}{\sqrt{9-4x^2}}\,dx$

◆ Type 1, an integral

26.[A] $\int \frac{\sin^{-1} x}{\sqrt{1-x^2}}\,dx$

◆ Type 1, an integral

27. $\int_{0}^{1} \frac{x^2}{6+x^3}\,dx$

◆ Type 1, an integral

28. $\int \frac{\tan^{-1} x}{1+x^2}\,dx$

◆ Type 1, an integral

29.[A] $\int \frac{x^2}{9+x^6}\,dx$

◆ Type 1, an integral

30. $\int \frac{1}{\sqrt{1-9x^2}}\,dx$

◆ Type 1, an integral

31. $\int e^x\sqrt{2+e^2}\,dx$

◆ Type 1, an integral

32. $\int_{0}^{1} \frac{x^3}{\sqrt{4-x^8}}\,dx$

◆ Type 1, an integral

33. $\int \frac{(\tan^{-1} x)^3}{1+x^2}\,dx$

◆ Type 1, an integral

34. $\int \tan(6x)\,dx$

◆ Type 1, an integral

35. $\int \csc(6x)\,dx$

◆ Type 1, an integral

36.[A] Find the area of the region bounded by $y=\sin x+3$, and the lines $y=0$, $x=0$, and $x=5$.

◆ Type 1, area

37.[A] What is the exact average value of $2x-3+4x^{-1}$ for $-3\le x\le -1$?

◆ Type 1, average value

38.[A] What is the exact area of the region bounded by the curve $y = \dfrac{x^2}{x^3+2}$, the x-axis, and the line $x = 2$?

◆ Type 1, area

39.[A] Find the exact area of the region bounded by the curve $y = e^{2x}$, the x-axis, and the lines $x = -1$ and $x = 1$.

◆ Type 1, area

40.[A] Sketch the region bounded by the x-axis, by the lines $x = -3$ and $x = 3$, and by the graph $y = (x^2+1)^{-1}$. Then find the exact area of the region.

◆ Type 1, area

41. Draw the region between $y = \dfrac{1}{\sqrt{4-x^2}}$ and the x-axis for $0 \le x \le 1.98$ and find its exact area.

◆ Type 1, area

42. What is the exact area of the region between the x-axis and the curve $y = xe^{-x^2}$ for $0 \le x \le 4$?

◆ Type 1, area

43. What is the exact area of the region bounded by the x-axis, the lines $x = -1$ and $x = 1$, and the curve $y = \sec^2 x$?

◆ Type 1, area

44. A rod that extends form $x = 0$ to $x = 10$ (inches) on the x-axis has density $1/(30-2x)$ ounces per inches at x. How much does it weigh?

◆ Type 1, density and weight

45. Find the exact area of the region between $y = \dfrac{1}{1+x^2}$ and $y = \dfrac{1}{\sqrt{1-x^2}}$ for $0 \le x \le 0.9$

◆ Type 1, area

46.[A] Find the value of the parameter $a > 0$ such that the area between $y = \dfrac{a}{a^2+x^2}$ and the x-axis for $0 \le x \le 1$ is $\frac{1}{3}\pi$.

◆ Type 2, area

47. An object is at $s = -1$ (feet) on an s-axis at $t = 0$ (minutes), and its velocity in the positive s-direction is $\sin t$ feet per minute for $t \ge 0$. Give a formula for its s-coordinate as a function of t.

◆ Type 1, velocity and position

48.[A] What is the exact area of the region between $y = (e^x+1)^2$ and the x-axis for $0 \le x \le 5$?

◆ Type 2, area

49.[A] Give the exact value of $\displaystyle\int_0^{\pi/4} (\cos x - 3)\, dx$ in terms of π and $\sqrt{2}$.

◆ Type 2, an integral

50. Express $\displaystyle\int_0^{\pi/2} \cos(3x)\, dx$ as a fraction.

◆ Type 2, an integral

51. Find exact formulas for **(a)** $\displaystyle\int \frac{x^2+1}{x}\, dx$, **(b)** $\displaystyle\int (e^{3x}+1)^2\, dx$, and **(c)** $\displaystyle\int \frac{e^x+1}{e^x}\, dx$, by putting the integrands in forms to which the techniques of this section can be applied.

◆ Type 2, an integral

52. Sketch the region bounded by $y = e^x, y = e^{-x}$, and $y = 4$ and find its exact area. Use two integrals or the symmetry of the region. (You need to use $\ln x$ to express the limits of integration.)

◆ Type 3, area

53. Find the exact value of one positive number and one negative number a such that the region bounded by the x- and y-axes, by the curve $y = e^{-x}$, and by the line $x = a$ has area 0.99.

◆ Type 3, area

54.[A] A rod extends from $x = 1$ to $x = 4$ on an x-axis with its scale measured in feet. Its density at x is $2/x$ pounds per foot.Where is its center of gravity?

◆ Type 1, center of gravity

55. **(a)** Sketch the region bounded by $y = \dfrac{6}{6-x}$, by the line $y = 3$, and by the y-axis. **(b)** Use the formulas for the area of a square and the area of a triangle to show that the area of the region of part (a) is greater than 4 and less than 8. **(c)** Show that the conclusion of part (b) is correct by using calculus to find the exact area of the region.

♦ Type 2, area, etc.

56. Find the exact area of the region bounded by the line $x + y = 3$ and the hyperbola $y = 2/x$.

♦ Type 2, area

57. Give formulas for the antiderivatives of **(a)** $\dfrac{\sqrt{\ln x}}{x}$, **(b)** $\dfrac{1}{x \ln x}$, and **(c)** $\dfrac{e^{\ln x}}{x}$ that have the value 2 at $x = e$.

♦ Type 2, antiderivatives

58. Give a formula for $\displaystyle\int_0^{\pi/k} \sin(kx)\,dx$ in terms of the positive constant k.

♦ Type 3, an integral

59. Determine by comparing the integrands which is the greatest and which is the least of $\displaystyle\int_0^1 \sin x\,dx$, $\displaystyle\int_0^1 \sin^2 x\,dx$ and $\displaystyle\int_0^1 \sqrt{\sec x}\,dx$.

♦ Type 3, comparing integrals

60. **(a)** Generate the graph of $3 + (\sin x)e^{\cos x}$ and the line $y = 3$ for $0 \le x \le 6\pi$ and copy them on your paper. **(b)** Use the drawing from part (a) to predict the average value of $3 + (\sin x)e^{\cos x}$ for $0 \le x \le 6\pi$. **(c)** Find the average value of $3 + (\sin x)e^{\cos x}$ for $0 \le x \le 6\pi$ by exact integration.

♦ Type 2, average value, etc.

61. Find the antiderivative $f(x)$ of xe^{-x^2} such that $\displaystyle\lim_{x\to\pm\infty} f(x) = 10$.

♦ Type 4, antiderivative, etc.

62. Find the exact area of the region bounded by $y = \sin x$ and $y = x^2 - \pi x$.

♦ Type 3, area

63. What is the area of the region between $y = \cos x$ and the x-axis for x between successive x-intercepts of the curve?

♦ Type 2, area

64. Find the exact area of the region between $y = x$ and $y = x\cos(x^2)$ for $0 \le x \le \sqrt{4\pi}$.

♦ Type 2, area

65. **(a^A)** What are the x-intercepts of $y = \dfrac{\sin(\ln x)}{x}$? **(b)** Find the area of the region between $y = \dfrac{\sin(\ln x)}{x}$ and the x-axis for x between 1 and the first x-intercept that is greater than 1.

♦ Type 3, area

66. **(a)** Generate $y = \dfrac{\ln x}{x}$ in the window $0 \le x \le e^3, -0.2 \le y \le 0.5$ and copy it on your paper. Use the formulas for the function and its derivative to explain the shape of the curve. **(b)** What is the exact global maximum of $\dfrac{\ln x}{x}$? **(c)** Find the exact area between $y = \dfrac{\ln x}{x}$ and the x-axis for $1 \le x \le e^3$.

♦ Type 3, area

67. Find the exact value of $k > 0$ such that $\displaystyle\int_0^1 \frac{1}{x+k}\,dx = 0.9$.

♦ Type 3, an integral

68.^A Choose k so that $\displaystyle\int_0^k \frac{1}{\sqrt{4-x^2}}\,dx = \tfrac{1}{6}\pi$.

♦ Type 3, an integral

69.^A A rod that extends from $x = 0$ to $x = 4$ (inches) on an x-axis has density $(9 + x^2)^{-1}$ ounces per inch at x. What does it weigh and where is its center of gravity?

♦ Type 3, center of gravity

70.[A] What is the present value at 8% interest compounded annually of a payment of \$75,000 to be made in 12 years?

♦ Type 1, present value

71.[A] What is the present value at 9% annual interest compounded quarterly of a gift of \$5000 to be made in three years?

♦ Type 1, present value

72.[A] What is the present value at 7% annual interest compounded continuously of an income of \$35,000 to be received in six years?

♦ Type 1, present value

73. What is the total present value at 8% interest compounded quarterly of gifts of \$1000 each to be made one, two, and three years from now?

♦ Type 1, present value

74. What is the total present value January 1, 2007 at 6% interest compounded continuously of \$3,000 to be earned January 1, 2010, \$5,000 to be earned July 1, 2020, and \$10,000 to be earned January 1, 2021?

♦ Type 1, present value

75.[A] **(a)** What is the total income from an income stream of \$5,000 per year from the beginning of 2005 to the beginning of 2035? **(b)** What is the present value of the income at the beginning of 2005 at 8% annual interest compounded continuously?

♦ Type 1, income stream

76. **(a)** What is the total income from a constant income stream of $\$7{,}500e^{-t/10}$ per year for $5 \le t \le 20$ (years)? **(b)** What is the present value of the income at $t = 0$ with 10% annual interest compounded continuously?

♦ Type 1, income stream

77.[A] **(a)** What is the total income from an income stream of $10{,}000+500t$ dollars per year for $0 \le t \le 10$ (years)? **(b)** What is the present value at $t = 0$ of the income at 7% annual interest compounded continuously? Use the integration formula $\int te^{-rt}\,dt = -(\frac{t}{r} + \frac{1}{r^2})e^{-rt} + C$.

♦ Type 3, income stream

78. **(a)** What is the total income from an income stream of $3{,}000+1000\sin(\pi t)$ dollars per year for $10 \le t \le 20$ (years)? **(b)** What is the present value at $t = 0$ of the income if the interest rate is 10% per year compounded continuously? Use the integration formula $\int e^{-rt}\sin(\pi t)\,dt = \frac{-e^{-rt}}{r^2+\pi^2}[r\sin(\pi t) + \pi\cos(\pi t)] + C$.

♦ Type 3, income stream

79. If a gift includes future income, gift tax is computed on the present value when the gift is made. Suppose a gift consists of an income stream at the contant rate of \$10,000 per year for ten years after the gift is made plus \$25,000 to be paid at the end of 10 years. What is the present value of the gift at the time it is given if the interest rate is 5% per year compounded continuously?

♦ Type 3, income stream

80. What is $\lim_{x\to -3} F(x)$ if $F(x)$ is an antiderivative of $f(x) = \dfrac{1}{2x+6}$?

♦ Type 4, limit of an antiderivative

81. An object is at the origin on an s-axis and its velocity $v(t)$ is zero at time $t = 1$ (seconds). Its acceleration $a(t)$ in the positive direction is t^{-2} feet per second per second at time t for $t \ge 1$. **(a)** Give a formula for its position $s(t)$ (feet) for $t \ge 1$. Then generate the graph of $s(t)$ graph for $1 \le t \le 10$ and copy it on your paper. **(b)** What are $\lim_{t\to\infty} a(t)$, $\lim_{t\to\infty} v(t)$, and $\lim_{t\to\infty} s(t)$, and how are these facts shown on the graph of $s(t)$?

♦ Type 2, acceleration, velocity, and position, etc.

82.[A] Find the exact value of the constant a such that the region lying between $y = e^x$ and the x-axis for $0 \le x \le a$ has the same area as the region between $y = 3^x$ and the x-axis for $0 \le x \le 3$.

♦ Type 3, area

83. An object that is at $s = 10$ (feet) on an s-axis and has zero velocity at $t = 0$ (seconds) has acceleration $t/(1+t^2)^2$ feet per second^2 in the positive direction at time t. Give its position as a function of t.

♦ Type 3, acceleration and position

84. Sketch the curve $y = 1/\sqrt{1-x^2}$ and find the exact area of the region bounded by this curve and the lines $y = 0, x = \frac{1}{2}$, and $x = -\frac{1}{2}$.

♦ Type 3, area

85.[A] Find the exact area of one of the regions bounded by $y = (\sin x)e^{\cos x}$ and the x-axis.

♦ Type 3, area

86. What is the exact area of the region bounded by $y = 0$, $x = \frac{1}{2}$, and $y = x\sec^2(x^2)$?

♦ Type 3, area

87. Find the number k such that $\int_0^k \frac{1}{1+x}\,dx = 20$. Give the exact and decimal values.

♦ Type 3, an integral

88. Choose k so that $\int_0^k \frac{1}{1+x^2}\,dx = \frac{1}{4}\pi$.

♦ Type 3, an integral

89. Find b such that $\int_0^b \frac{1}{1+x^2}\,dx$ is greater than 1.57.

♦ Type 3, an integral

Perform exact integrations in Problems 90 through 93.

90.[A] $\int_0^2 (2^x + x^2)\,dx$

♦ Type 3, an integral

91. $\int_0^e (x^e + e^x)\,dx$

♦ Type 3, an integral

92. $\int_0^{1/2} \frac{4^x}{4^x - 4}\,dx$

♦ Type 3, an integral

93. $\int_{-1}^1 |e^x - 1|\,dx$

♦ Type 3, an integral

94. **(a)** The voltage in a circuit with alternating current is $V(t) = V_{max}\cos(\omega t)$ volts and the current is $i(t) = i_{max}\cos(\omega t)$ amperes at time t (seconds). The PERIOD of the current is $T = (2\pi/\omega)$ seconds. Give a formula for the AVERAGE POWER of the circuit, which is the average value of $V(t)i(t)$ (watts) in an interval of lenth T. **(b)** The voltage and current in part (a) are IN PHASE because they peak at the same times. Give a formula for the average power for $V(t) = V_{max}\cos(\omega t + \theta)$ volts and $i(t) = i_{max}\cos(\omega t + \psi)$ with $\theta \neq \psi$, so the current and voltage are out of phase. (Use the trigonometric identity $\cos\alpha\cos\beta = \frac{1}{2}[\cos(\alpha-\beta) + \cos(\alpha+\beta)]$.)

♦ Type 3, average power

95. A man is trying to drive his car at a constant velocity, but because of a badly tuned engine his acceleration is $100\cos(100t)$ miles per hour^2 at time t (hours). At $t = 0$ his velocity is 55 miles per hour. **(a)** What are the maximum and minimum velocities? **(b)** How far does he travel in the time period $0 \le t \le 2$?

♦ Type 2, velocity and position, etc.

96. A stand of trees will yield $\$5000e^{\sqrt{t}}$ profit if it is cut in t years. When should the trees be cut to maximize the present value of the profit at $t = 0$ if the interest is 10% per year compounded continuously?

♦ Type 4, present value, etc.

97. If an object moving on an s-axis has acceleration $a = \cos t$ meters per minute^2 at time t (minutes) and has velocity 0.5 meters per minute at time $t = 0$, what is the total distance it travels in the time interval $0 \le t \le 2\pi$?

♦ Type 4, acceleration and position

98. A country has enough resources to last 700 years at the present rate of consumption. If the population increases at the continuous rate of 5% per year and if the rate of consumption per capita does not change, how long will the resources last?

♦ Type 4, exponential growth, etc.

99. Derive the formula $\int_{-1}^1 x^2\,dx = \frac{2}{3}$ from the formulas for areas of triangles and rectangles and Archimedes' result concerning areas of regions bounded by parabolas that is stated with Figure 11.

♦ Type 4, area

RESPONSES TO QUESTIONS IN CHAPTER 4

Responses 4.1

Response 1 **(a)** The truck driver is 50 miles east of Ozona at 10:00 AM. He drives [75 miles per hour]×[2 hours] = 150 miles farther to the east, then [50 miles per hour]×[2 hours] = 100 miles to the west, and then [50 miles per hour]×[2 hours] = 100 miles to the east. At $t = 6$ he is $50 + 150 - 100 + 100 = 200$ miles east of Ozona.
(b) His distance east of Ozona at $t = 6$ equals his distance at $t = 0$, plus the sum of the areas of rectangles A and C above the t-axis in Figure 4 of Section 4.1, minus the area of rectangle B below the t-axis.

Response 2 **(a)** The slope of the graph in Figure 6 of Section 4.1 is $(200 - 50)/2 = 75$ miles per hour for $0 < t < 2$, $(100 - 200)/2 = -50$ miles per hour for $2 < t < 4$, and $(200 - 100)/2 = 50$ miles per hour for $4 < t < 6$, and these are the values of the step function in Figure 3.
(b)

$$s(t) = \begin{cases} 50 + 75t & \text{for} \quad 0 \le t \le 2 \\ 200 - 50(t-2) & \text{for} \quad 2 \le t \le 4 \\ 100 + 50(t-4) & \text{for} \quad 4 \le t \le 6. \end{cases}$$

Response 3 **(a)** [10 gallons/minute][10 minutes] = 100 gallons flow in from $t = 0$ to $t = 10$.
(b) [5 gallons/minute][30 minutes] + [10 gallons/minute][20 minutes] = 350 gallons flow out from $t = 10$ to $t = 60$.
(c) [5 gallons/minute][10 minutes] = 50 gallons flow in from $t = 60$ to $t = 70$.
(d) There are $300 + 100 - 350 + 50 = 100$ gallons in the tank at $t = 70$. • The final volume is the initial volume plus the areas of the two rectangles above the t-axis in Figure 8 of Section 4.1, minus the areas of the two rectangles below the t-axis.

Response 4 $V(t) = 300 + 10t$ for $0 \le t \le 10$ • $V(10) = 400$ • $V(t) = 400 - 5(t - 10)$ for $10 \le t \le 40$ • $V(40) = 250$ • $V(t) = 250 - 10(t - 40)$ for $40 \le t \le 60$ • $V(60) = 50$ • $V(t) = 50 + 5(t - 60)$ for $60 \le t \le 70$ • Figure R4

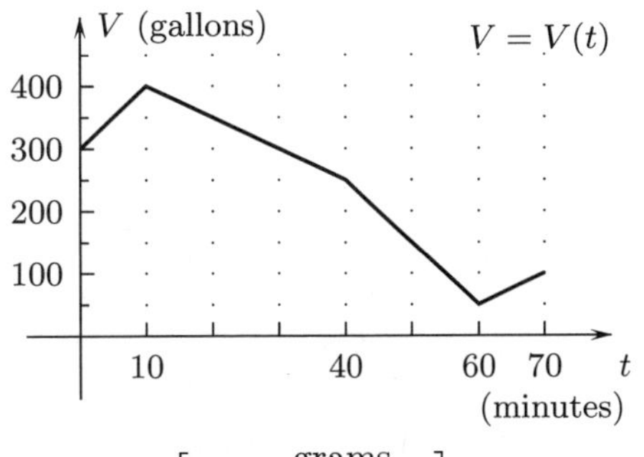

Figure R4

Response 5 **(a)** $\left[200\,\dfrac{\text{grams}}{\text{centimeter}}\right]$[30 centimeters] + $\left[160\,\dfrac{\text{grams}}{\text{centimeter}}\right]$[50 centimeters] + $\left[220\,\dfrac{\text{grams}}{\text{centimeter}}\right]$[20 centimeters] = 18,400 grams
(b) The total weight equals the area of the three rectangles below the graph of the density and above the x-axis in Figure 10 of Section 4.1.

Response 6 **(a)** [Volume] $= \pi(3^2)(4) + \pi(5^2)(2) = 86\pi$ cubic feet
(b) The volume equals the area of the two rectangles in Figure 13 of Section 4.1.

Response 7 The average maximum temperature equals the area of the three rectangles above the t-axis in Figure 15 of Section 4.1, minus the area of the rectangle below the t-axis, all divided by 7.

Responses 4.2

Response 1 The first subinterval is $[x_0, x_1]$ and is $x_1 - x_0$ units wide. • The second subinterval is $[x_1, x_2]$ and is $x_2 - x_1$ units wide. • The third subinterval is $[x_2, x_3]$ and is $x_3 - x_2$ units wide. • The fourth subinterval is $[x_3, x_4]$ and is $x_4 - x_3$ units wide. • The fifth subinterval is $[x_4, x_5]$ and is $x_5 - x_4$ units wide.

Response 2 The second rectangle has height $f(c_2)$, width Δx_2, and area $f(c_2)\Delta x_2$. • Because $f(c_3)$ is negative, the height of the the third rectangle is $-f(x_3)$ and its area is $-f(c_3)\Delta x_3$. • Because $f(c_4)$ is negative, the height of the the fourth rectangle is $-f(x_4)$ and its area is $-f(c_4)\Delta x_4$. • Because $f(c_5)$ is negative, the height of the the fifth rectangle is $-f(x_5)$ and its area is $-f(c_5)\Delta x_5$.

Response 3 **(a)** $\sum_{j=1}^{4} \sqrt{x_j}\,\Delta x_j = \sqrt{x_1}\,\Delta x_1 + \sqrt{x_2}\,\Delta x_2 + \sqrt{x_3}\,\Delta x_3 + \sqrt{x_4}\,\Delta x_4$

(b) $6(x_1)^3\Delta x_1 + 6(x_2)^3\Delta x_2 + 6(x_3)^3\Delta x_3 + \cdots + 6(x_N)^3\Delta x_N = \sum_{j=1}^{N} 6(x_j)^3\Delta x_j$

Response 4 The height of the rectangle in Figure 12 of Section 4.2 is $\dfrac{1}{c_1}$, so its area is $\dfrac{1}{c_1}x_1 = \dfrac{1}{3c_1}$. • The area of the rectangle is 10 if $c_1 = \frac{1}{30}$, is 100 if $c_1 = \frac{1}{300}$, and is 1000 if $c_1 = \frac{1}{3000}$.

Response 5 **(a)** As in Example 4 of Section 4.2, the partition is $0 < 0.2 < 0.4 < 0.6 < 0.8 < 1$ and the width of the rectangles is 0.2. The left endpoints are $0, 0.2, 0.4, 0.6$, and 0.8, so in this case the Riemann sum is

$$\begin{aligned}(0^2)(0.2) + (0.2^2)(0.2) + (0.4^2)(0.2) + (0.6^2)(0.2) + (0.8^2)(0.2))\\ = (0^2 + 0.2^2 + 0.4^2 + 0.6^2 + 0.8^2)(0.2) = 0.24.\end{aligned}$$

(b) The midpoints of the subintervals are $0.1, 0.3, 0.5, 0.7$, and 0.9. The Riemann sum is

$$\begin{aligned}(0.1^2)(0.2) + (0.3^2)(0.2) + (0.5^2)(0.2) + (0.7^2)(0.2) + (0.9^2)(0.2))\\ = (0.1^2 + 0.3^2 + 0.5^2 + 0.7^2 + 0.9^2)(0.2) = 0.33.\end{aligned}$$

Response 6 One answer: The Riemann sum in Example 4 of Section 4.2 is greater than the integral because the region formed by the rectangles in Figure 17 contains the region between the curve $y = x^2$ and the x-axis, whose area equals the integral.

The Riemann sum in Question 5a is less than the integral because the rectangles in Figure 18 are contained in the region between the curve and the x-axis.

The Riemann sum in Question 5b is the closest of the three to the integral because each rectangle in Figure 19 is partially outside and partially inside the region between the curve and the x-axis.

Response 7 The table below gives the values, rounded to six digits, of the Riemann sums for the three integrals with $N = 10, 20, 50$, and 100 equal subintervals and the function evaluated at the midpoints. The rectangles for $N = 10$ are shown in Figures R7a, R7b, and R7c. • Likely predictions: $\int_0^1 x^3\,dx = \frac{1}{4}$, $\int_0^1 x^4\,dx = \frac{1}{5}$, and $\int_0^1 x^{1/4}\,dx = \frac{4}{5}$

N	10	20	50	100
$\int_0^1 x^3\,dx$	0.248750	0.249688	0.249950	0.249988
$\int_0^1 x^4\,dx$	0.198336	0.199584	0.199933	0.199983
$\int_0^1 x^{1/4}\,dx$	0.802763	0.801179	0.800379	0.800160

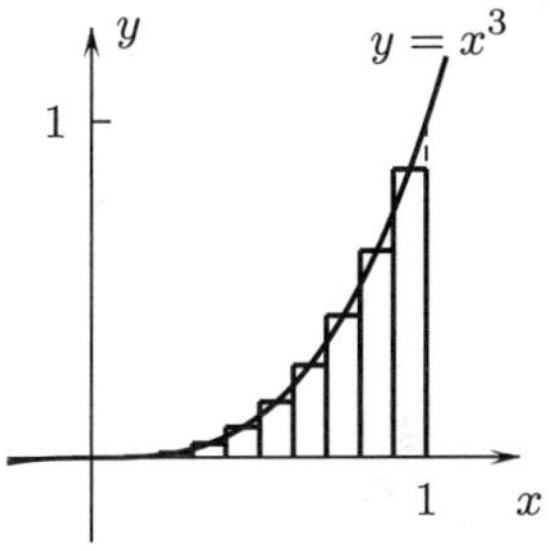

Figure R7a

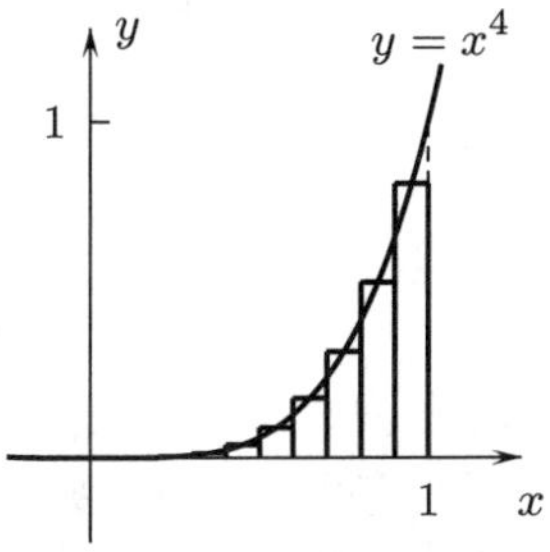

Figure R7b

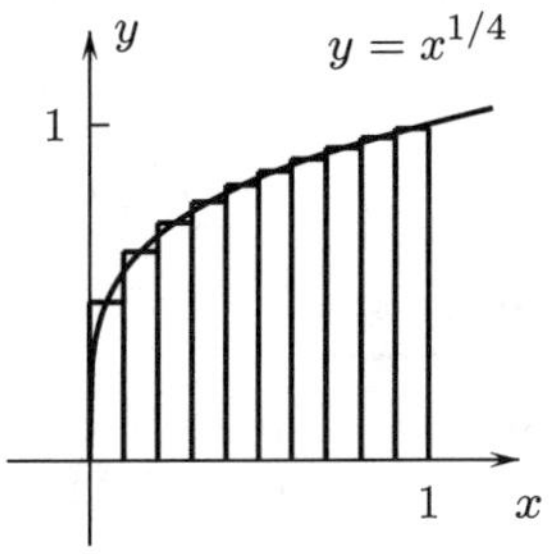

Figure R7c

Response 8 $\int_1^8 g(x)\,dx = \int_1^5 g(x)\,dx + \int_5^8 g(x)\,dx = -7 + 9 = 2$

Response 9 $\int_{-35}^{35} [2p(x) - 3q(x)]\,dx = 2\int_{-35}^{35} p(x)\,dx - 3\int_{-35}^{35} q(x)\,dx = 2(10) - 3(20) = -40$

Responses 4.3

Response 1 $[s(t_1) - s(t_0)] + [s(t_2) - s(t_1)] + [s(t_3) - s(t_2)] + [s(t_4) - s(t_3)] + [s(t_5) - s(t_4)]$ $= -s(t_0) + s(t_5) = s(b) - s(a)$ because all of the other terms cancel when the square brackets are removed.

Response 2 $\int_4^9 F'(x)\,dx = F(9) - F(4) = \sqrt{9} - \sqrt{4} = 1$

Response 3 **(a)** Here $n = 3$ and $n + 1 = 4$:

$$\int_0^1 x^3\,dx = \left[\tfrac{1}{4}x^4\right]_0^1 = \tfrac{1}{4}(1^4) - \tfrac{1}{4}(0^4) = \tfrac{1}{4}$$

(b) Here $n = 4$ and $n + 1 = 5$:

$$\int_0^1 x^4\,dx = \left[\tfrac{1}{5}x^5\right]_0^1 = \tfrac{1}{5}(1^5) - \tfrac{1}{5}(0^5) = \tfrac{1}{5}$$

(c) In this case, $n = \frac{1}{4}$ and $n + 1 = \frac{1}{4} + 1 = \frac{5}{4}$:

$$\int_0^1 x^{1/4}\,dx = \left[\frac{1}{5/4}x^{5/4}\right]_0^1 = \left[\tfrac{4}{5}x^{5/4}\right]_0^1 = \tfrac{4}{5}(1^{5/4}) - \tfrac{4}{5}(0^{5/4}) = \tfrac{4}{5}$$

Response 4 $[\text{Area}] = -\int_1^3 (-6x^{-2})\,dx = \int_1^3 6x^{-2}\,dx$ • $n = -2$ and $n + 1 = -1$ •

$[\text{Area}] = \left[\frac{1}{-1}6x^{-1}\right]_1^3 = \left[-\frac{6}{x}\right]_1^3 = -\frac{6}{3} + \frac{6}{1} = 4$

Response 5 $\frac{d}{dx}(2x^{3/2} - 4x^{-1} - 3x) = 2\left(\frac{3}{2}x^{1/2}\right) - 4(-x^{-2}) - 3 = 3x^{1/2} + 4x^{-2} - 3$ (We drop the $+C$ because the derivative of the constant is zero.) • The formula for the indefinite integral is correct because its derivative is the original integrand.

Response 6 $F(x) = \int (10 + 3\sqrt{x})\,dx = \int (10 + 3x^{1/2})\,dx = 10x + 3\left(\frac{2}{3}x^{3/2}\right) + C$ $= 10x + 2x^{3/2} + C$ • Set $F(1) = 25$: $25 = 10(1) + 2(1)^{3/2} + C$ • $C = 13$ and $F(x) = 10x + 2x^{3/2} + 13$

Response 7 $G(10) - G(1) = \int_1^{10} G'(t)\,dt = \int_1^{10} 4t^{-3}\,dt = \left[4\left(-\frac{1}{2}t^{-2}\right)\right]_1^{10}$ $= \left[\frac{-2}{t^2}\right]_1^{10} = \left[\frac{-2}{10^2}\right] + \left[\frac{-2}{1^2}\right] = 1.98$

Responses 4.4

Response 1 If, as it appears in Figure 1 of Section 4.4, water was flowing at the approximate rate of 2 thousand cubic feet per minute before the dam broke, then $2 \times 90 = 180$ thousand cubic feet would have flowed from the dam if it had not broken. According to the solution of Example 1, about 1950 thousand feet of water crossed it. • The increase was about 983%.

Response 2 **(a)** From the table we see that the heights of the rectangles in Figure 3 of Section 4.4 are 77.2, 60.5, 45.8, 55.7, and 66.5. They are 2 units wide, so their total area is $(77.2)(2) + (60.5)(2) + (45.8)(2) + (55.7)(2) + (66.5)(2) = 611.4$ billion dollars.

(b) The heights of the rectangles in Figure 4 are 60.5, 45.8, 55.7, 66.5, and 66.8. Their total area is $(60.5)(2) + (45.8)(2) + (55.7)(2) + (66.5)(2) + (66.8)(2) = 590.6$ billion dollars.

Response 3 $[\text{Area of the rectangle}] = h_1(\Delta t)$ • $[\text{Area of the triangle}] = \frac{1}{2}(h_2 - h_1)\Delta t$ • $[\text{Area of the trapezoid}] = h_1(\Delta t) + \frac{1}{2}(h_2 - h_1)\Delta t = \frac{1}{2}(h_1 + h_2)\Delta t$

Response 4 $\Delta x = 0.5$ • Partition: $0 < 0.5 < 1 < 1.5 < 2 < 2.5 < 3$ •
Set $f(x) = -8x^3 + 24x^2 - 16$. • [Trapezoid-Rule approximation]
$= [\frac{1}{2}f(0) + f(0.5) + f(1) + f(1.5) + f(2) + f(2.5) + \frac{1}{2}f(3)](0.5)$
$= [\frac{1}{2}(-16) - 11 + 0 + 11 + 16 + 9 + \frac{1}{2}(-16)](0.5) = 4.5$ • $\int_0^3 (-8x^3 + 24x^2 - 16)\,dx$
$= 6$ • The error with the Trapezoid-Rule approximation is 1.5. • Figure R4

Figure R4

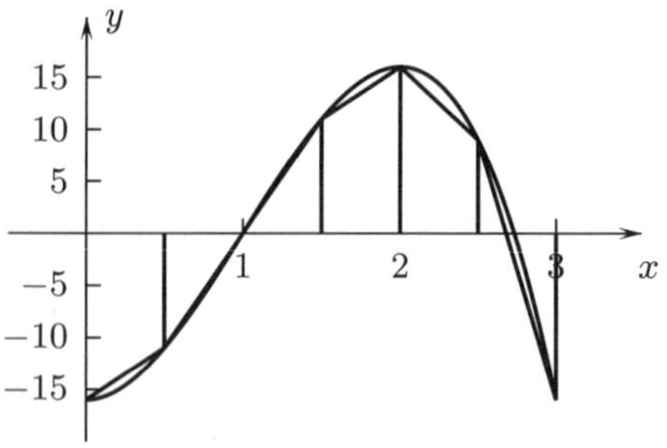

Responses 4.5

Response 1 $\frac{d}{dx}[\frac{1}{6}(x^4+16)^{3/2}] = \frac{1}{6}[\frac{3}{2}(x^4+16)^{1/2}]\frac{d}{dx}(x^4+16) = \frac{1}{4}\sqrt{x^4+16}\,(4x^3)$
$= x^3\sqrt{x^4+16}$ (The Chain Rule is needed to check the answer because the integral was found by substution.)

Response 2 **(a)** $du = \frac{du}{dx}\,dx = [\frac{d}{dx}(1+5x)]\,dx = 5\,dx$

(b) $\int \frac{1}{(1+5x)^4}\,dx = \frac{1}{5}\int (1+5x)^{-4}(5\,dx) = \frac{1}{5}\int u^5\,du = \frac{1}{5}(\frac{1}{6}u^6) + C$
$= \frac{1}{30}(1+5x)^6 + C$

Response 3 $\int x^4\,dx = \frac{1}{5}x^5 + C$ but $x\int x^3\,dx = x(\frac{1}{4}x^4 + C) = \frac{1}{4}x^5 + Cx$

Response 4 Set $u = 1-2x$ • $du = [\frac{d}{dx}(1-2x)]\,dx = -2\,dx$ •

$$\int_1^2 (1-2x)^3\,dx = -\frac{1}{2}\int_{x=1}^{x=2} (1-2x)^3(-2\,dx) = -\frac{1}{2}\int_{x=1}^{x=2} u^3\,du$$

One approach. Perform the integration with respect to u and put the result in terms of x for the final evaluations:

$$\begin{aligned}\int_1^2 (1-2x)^3\,dx &= -\frac{1}{2}\int_{x=1}^{x=2} u^3\,du = -\frac{1}{2}\left[\frac{1}{4}u^4\right]_{x=1}^{x=2} \\ &= -\frac{1}{2}\left[\frac{1}{4}(1-2x)^4\right]_{x=1}^{x=2} = -\frac{1}{8}\left[(1-2x)^4\right]_{x=1}^{x=2} \\ &= -\frac{1}{8}\{[(-3)^4] - [(-1)^4]\} = -10\end{aligned}$$

Another approach. Switch the limits of integration to values of $u = 1-2x$, which equals -1 when $x = 1$ and equals -3 when $x = 2$, so that $-1 \underset{u}{\longrightarrow} -3$ when $1 \underset{x}{\longrightarrow} 2$:

$$\begin{aligned}\int_1^2 (1-2x)^3\,dx &= -\frac{1}{2}\int_{u=-1}^{u=-3} u^3\,du = -\frac{1}{2}\left[\frac{1}{4}u^4\right]_{u=-1}^{u=-3} \\ &= -\frac{1}{8}\left[u^4\right]_{u=-1}^{u=-3} = -\frac{1}{8}\{[(-3)^4] - [(-1)^4]\} = -10\end{aligned}$$

Response 5 **(a)** For $u = 1-x$: $x = 1-u$, $dx = -du$, and $0 \underset{x}{\longrightarrow} 1$, so that $1 \underset{u}{\longrightarrow} 0$ •

$\int_0^1 x\sqrt{1-x}\,dx = \int_1^0 (1-u)\sqrt{u}\,(-du) = -\int_1^0 (u^{1/2} - u^{3/2})\,du$

(b) $\int_0^1 x\sqrt{1-x}\,dx = -\left[\frac{2}{3}u^{3/2} - \frac{2}{5}u^{5/2}\right]_1^0 = \frac{4}{15}$

Responses 4.6

Response 1 **(a)** Solve the equations $y = 4-x, y = x^2 - 4x + 4$ for x:
$4 - x = x^2 - 4x + 4$ • $x^2 - 3x = 0$ • $x(x-3) = 0$ • $x = 0, 3$

(b) $[\text{Area}] = \int_0^3 [(4-x) - (x^2 - 4x + 4)]\,dx = \int_0^3 (3x - x^2)\,dx$
$= \left[\frac{3}{2}x^2 - \frac{1}{3}x^3\right]_0^3 = 4\frac{1}{2}$

Response 2 $[\text{Weight}] = \int_0^2 (2+6x)\left(\frac{\text{pounds}}{\text{foot}}\right) dx\ (\text{feet}) = \int_0^2 (2+6x)\, dx = \left[2x+3x^2\right]_0^2$ $= 16$ pounds

Response 3 $\int_a^b x\rho(x)\, dx - L\int_a^b \rho(x)\, dx = 0$ • $\int_a^b x\rho(x) dx = L\int_a^b \rho(x)\, dx$ •

$$L = \frac{\int_a^b x\rho(x)\, dx}{\int_a^b \rho(x)\, dx}.$$

Response 4 The weight of the rod is 16 pounds, from Question 2. •

$[\text{Moment about } x=0] = \int_0^2 x\rho(x)\, dx = \int_0^2 x(2+6x)\, dx$

$= \int_0^2 (2x+6x^2)\, dx = \left[x^2+2x^3\right]_0^2 = [2^2+2(2^3)] - [0] = 20$ foot-pounds •

$\overline{x} = [\text{Moment about } x=0]/[\text{Weight}] = 20/16 = 1.25$ feet

Response 5 For $y = 2x^{3/2}$, $y' = 3x^{1/2}$ • $[\text{Length}] = \int_0^1 \sqrt{1+[f'(x)]^2}\, dx = \int_0^1 \sqrt{1+9x}\, dx$ • Set $u = 1+9x, du = 9\, dx$ • $[\text{Length}] = \frac{1}{9}\int_0^1 \sqrt{1+9x}(9\, dx) = \frac{1}{9}\int_{x=0}^{x=1} \sqrt{u}\, du$ $= \frac{2}{27}\left[(1+9x)^{3/2}\right]_{x=0}^{x=1} = \frac{2}{27}(10^{3/2}-1)$

Response 6 $[\text{Average value}] = \frac{1}{3}\int_1^4 4x^{-2}\ dx = 1$ • $4x^{-2} = 1$ at $x = 2$ • Figure R6 • $\int_1^2 (4x^{-2}-1)\, dx = 1$ • $\int_2^4 (1-4x^{-2})\, dx = 1$

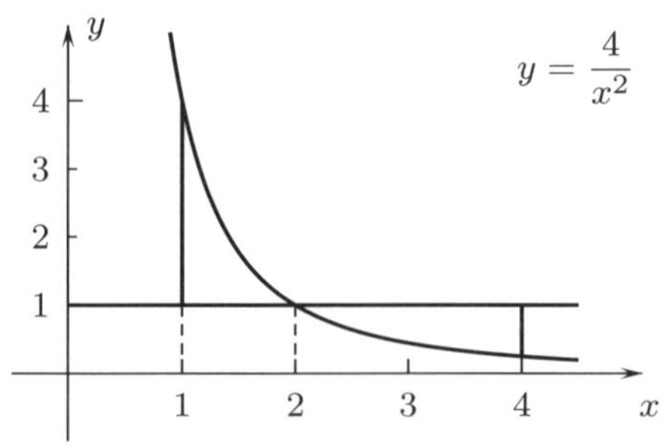

Figure R6

Response 7 $v = \int a\ dt = \int 1000\ dt = 1000t + C$ • $v(0) = 0$ • $v = 1000t$ • Because $v(t)$ is ≥ 0 for $0 \leq t \leq 5$, the distance traveled by the car for $0 \leq t \leq 5$ is $\int_0^5 1000t\ dt = \left[500t^2\right]_0^5 = 12{,}500$ feet.

Response 8 N'gagi's greatest mass was $683/32 \doteq 21.34$ slugs.

Response 9 Mass $= m = w/g = 19.6/9.8 = 2$ slugs • $F = ma$ •

$a = F/m = (20-12t^{-1/2})/2 = 10 - 6t^{-1/2}$ feet per second^2 •

$v = \int (10-6t^{-1/2})\, dt = 10t - 12t^{1/2} + v_0$ •

Set $t = 1, v = 8$: $v_0 = 10$ • $v = 10t - 12t^{1/2} + 10$ •

$$s = \int (10t - 12t^{1/2} + 10)\, dt = 5t^2 - 8t^{3/2} + 10t + s_0 \quad \bullet$$

Set $t = 1, s = 50$: $s_0 = 43$ • $s = 5t^2 - 8t^{3/2} + 43$

Response 10 **(a)** The graph of the velocity is a line of positive slope because its derivative, the acceleration, is a positive constant.
(b) The graph of the position is concave up because its derivative, the velocity, is an increasing function.

Responses 4.7

Response 1 All three Riemann sums equal $\frac{47}{60} \doteq 0.78333333$.

Response 2 One explanation: The rectangles in Figure 2 of Section 4.7 are twice as wide as those in Figure 1; the rectangles in Figure 3 are three times as wide as those in Figure 1; the rectangles in Figure 2 are half as high as those in Figure 1; and the rectangles in Figure 3 are one-third the height of those in Figure 1. Therefore, each rectangle in Figures 2 and 3 has the same area as one in Figure 1.

Response 3 $\int_{-7}^{-2} \frac{1}{x}\, dx = \Big[\ln|x|\Big]_{-7}^{-2} = \ln|-2| - \ln|-7| = \ln(2) - \ln(7) = \ln\left(\frac{2}{7}\right) \doteq -1.2527630$
• The integral is negative because $-7 < -2$ and $1/x < 0$ for $-7 \le x \le -2$.

Response 4 $g(x) = \int e^x\, dx = e^x + C$ • $g(2) = e^2 + C$ • Set $g(2) = 10$: $e^2 + C = 10$ •
$C = 10 - e^2$ • $g(x) = e^x + 10 - e^2$

Response 5 [Present value at 6% annual interest compounded continuously]
$= 2 + 2e^{-5(0.06)} + 2e^{-10(0.06)} = 2 + e^{-0.3} + 2e^{-0.6} \doteq 4.5792$ million dollars

Response 6 With $I(t)$ the value of your investments t years after the beginning of 2000,
$I(t) = 500\pi \int \cos(\frac{1}{50}\pi t)\, dt = 25{,}000 \sin(\frac{1}{50}\pi t) + C$ • $I(0) = 100{,}000 \implies$
$C = 100{,}000$ • $I(t) = 25{,}000 \sin(\frac{1}{50}\pi t) + 100{,}000$ • At the beginnings of 2025, 2050, and 2075, your investments are worth $I(25) = 125{,}000$, $I(50) = 100{,}000$, and $I(75) = 75{,}000$, respectively. (Because of the sine function, your investments would oscillate for ever between \$125,000 and \$75,000.)

Response 7 $u = \sin x, du = \cos x\, dx$ • $\int \cot x\, dx = \int \frac{\cos x}{\sin x}\, dx = \int \frac{1}{u}\, du = \ln|u| + C$
$= \ln|\sin x| + C$

Response 8 $u = \csc x + \cot x,\ du = (-\csc x \cot x - csc^2 x)\, dx$ •
$\int \csc x\, dx = \int \frac{\csc x(\csc x + \cot x)}{\csc x + \cot x}\, dx = -\int \frac{1}{u}\, du = -\ln|u| + C$
$= -\ln|\csc x + \cot x| + C.$

Response 9 Use $u = x/a, x = au$, and $dx = a\, du$ • $\int \frac{1}{a^2 + x^2}\, dx = \int \frac{1}{a^2 + (au)^2}(a\, du)$
$= \frac{1}{a}\int \frac{1}{1 + u^2}\, du = \frac{1}{a}\tan^{-1} u + C = \frac{1}{a}\tan^{-1}\left(\frac{x}{a}\right) + C$

Response 10 **(a)** [Likely prediction]: $L = 3$
(b) Use **(28)** from Section 4.8 with $a = 10$:
[Average value] $= \frac{1}{18}\int_{-9}^{9} \frac{25}{\sqrt{100 - x^2}}\, dx = \frac{25}{18}\Big[\sin^{-1}(\frac{1}{10}x)\Big]_{-9}^{9}$
$= \frac{25}{18}[\sin^{-1}\left(\frac{9}{10}\right) - \sin^{-1}\left(-\frac{9}{10}\right)] = \frac{25}{9}\sin^{-1}(0.9) \doteq 3.110$

CHAPTER 4 ANSWERS

Tune-up exercises 4.1

T1. At $t = 60$, $s = 300$ feet $- \left[5 \dfrac{\text{feet}}{\text{minute}}\right]$ [20 minutes] $+ \left[10 \dfrac{\text{feet}}{\text{minute}}\right]$ [30 minutes] $= 500$ feet

T2. **(a)** [Weight] $= \left[144 \dfrac{\text{kilograms}}{\text{meter}}\right]$ [2 meters] $+ \left[100 \dfrac{\text{kilograms}}{\text{meter}}\right]$ [4 meters] $= 688$ kilograms
(b) [Average density] $=$ [688 kilograms]/[6 meters] $= 114\frac{2}{3}$ kilograms per meter

T3. **(a)** $a = 4, b = 9$ **(b)** [Volume] $=$ [9 square meters][5 meters] $+$ [4 square meters][6 meters] $= 69$ cubic meters

T4. [Average value] $= \frac{1}{700}[(20)(200) + (-30)(300) + (50)(200)] = 7\frac{1}{7}$

T5. $f'(x) = 5$ and $\sqrt{1 + [f'(x)]^2} = \sqrt{26}$ for $1 < x < 6$ • $a = \sqrt{26}$
$f'(x) = -3$ and $\sqrt{1 + [f'(x)]^2} = \sqrt{10}$ for $6 < x < 9$ • $b = \sqrt{10}$
[Length] $= 5\sqrt{26} + 3\sqrt{10}$

Problems 4.1

1. **(a)** $8 + \left[6 \dfrac{\text{miles}}{\text{hour}}\right]$ [2 hours] $+ \left[4 \dfrac{\text{miles}}{\text{hour}}\right]$ [1.5 hours] $= 26$ miles
(b) 5.2 miles per hour

2. **(a)** $\doteq 2.63$ from 1883 to 1884 and $\doteq -0.74$ from 1984 to 1885 **(b)** 79.2 inches **(c)** [Average] $= 15.84$ inches per year • One answer: 0.94 inches • Another answer: 6.3% more

5. **(a)** 582,000 cubic feet **(b)** $a = 900, b = 3600, c = 6400$ **(c)** The volume equals the area of the three rectangles.

6. $46,875

9. $2\sqrt{1 + m_1^2} + 3\sqrt{1 + m_2^2}$.

Tune-up exercises 4.2

T1. Figure T1 • The area of the triangle above the x-axis is $\frac{1}{2}(2)(2) = 2$, and the area of the triangle below the x-axis is $\frac{1}{2}(1)(1) = \frac{1}{2}$. • $\displaystyle\int_0^3 (x-1)\,dx = 2 - \frac{1}{2} = \frac{3}{2}$

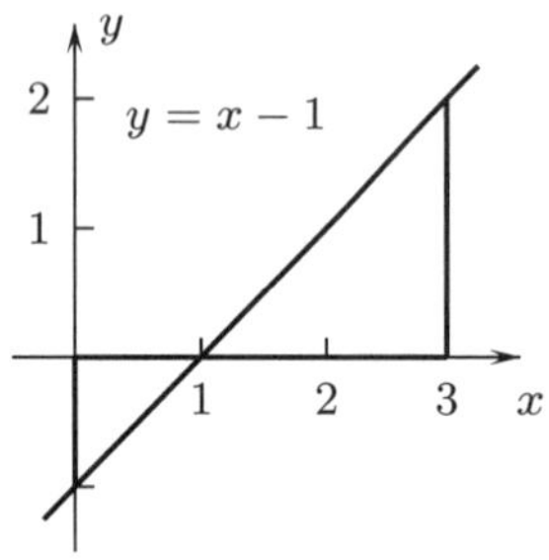

Figure T1

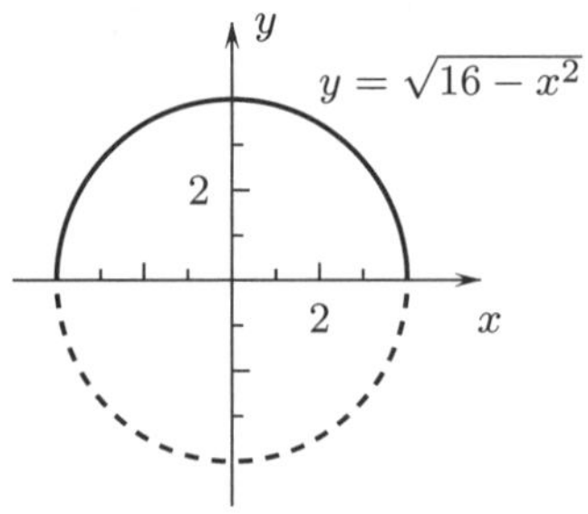

Figure T2

T2. $\displaystyle\int_{-4}^0 \sqrt{16 - x^2}\,dx$ is the area of the region between $y = \sqrt{16 - x^2}$ and the x-axis for $-4 \le x \le 0$, which is one fourth of the circle. • $\displaystyle\int_{-4}^0 \sqrt{16 - x^2}\,dx = \frac{1}{4}\pi(4^2) = 4\pi$

T3. Figure T3 • Use the areas of the three rectangles: $\displaystyle\int_1^6 P(x)\,dx = -(1)(2) + (3)(1) + (2)(2) = 5.$

T4. 210

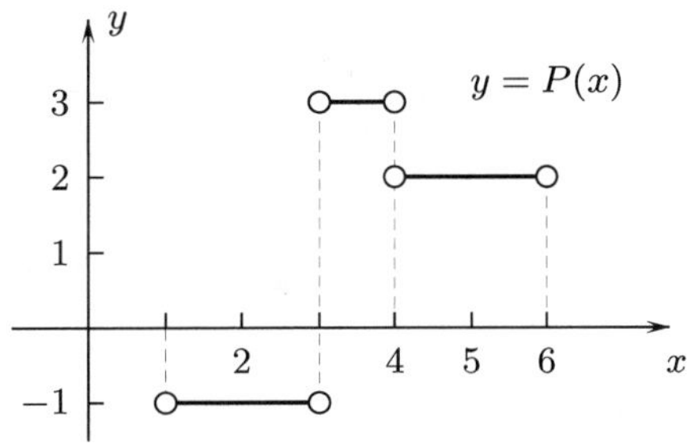

Figure T3

T5. $\sum_{j=1}^{99} j^2$

T6. Figure T6 • $[f(0.5) + f(1) + f(1.5) + f(2)](0.5) = (1.75 + 3 + 3.75 + 4)(0.5) = 6.25$

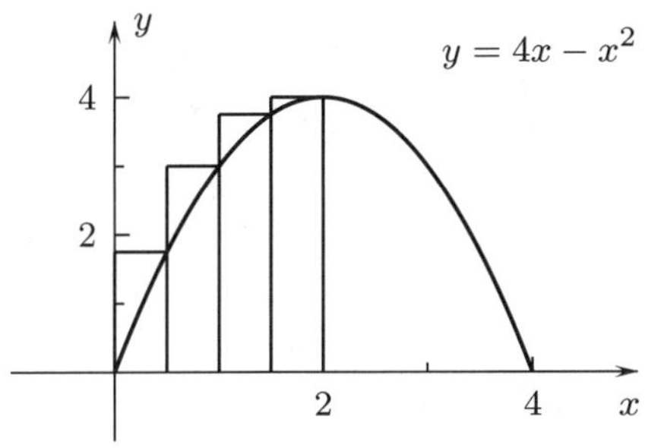

Figure T6

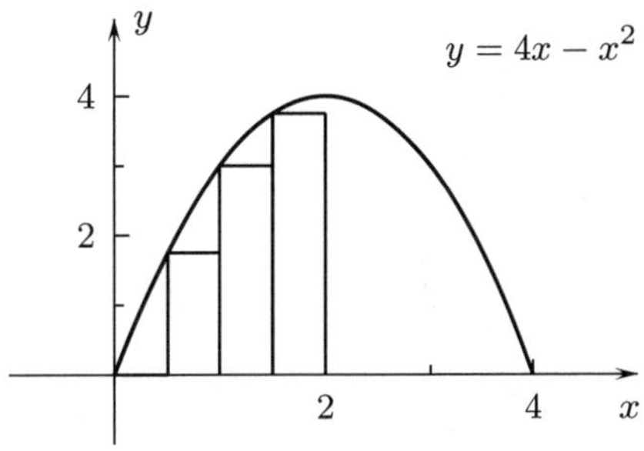

Figure T7

T7. Figure T7 • $[f(0) + f(0.5) + f(1) + f(1.5)](0.5) = (0 + 1.75 + 3 + 3.75)(0.5) = 4.25$

T8. $N = 20$: 5.335 • $N = 50$: 5.3336 • $N = 200$: 5.3335 • One prediction: The integral equals $5\frac{1}{3}$.

T9. 9

T10. 17

Problems 4.2

1. $\frac{1}{2}(2)(4) - \frac{1}{2}(1)(2) = 3$

2. -9

5. $\frac{1}{4}\pi$

7. -14

9. $-\frac{5}{4}\pi$

10 **(a)** $\frac{5}{4}$ **(b)** $27\frac{2}{5}$

11a $\sum_{j=1}^{175} \frac{j}{j+2}$

12. Figure A12 • $1(1) + \frac{1}{2}(2) + \frac{1}{4}(4) = 3$

14. [Left Riemann sum] = 150 • [Right Riemann sum] = 125

16a 2

17. Likely prediction: 1.5

18. Likely prediction: 2

21. 10

25. $N = 18$

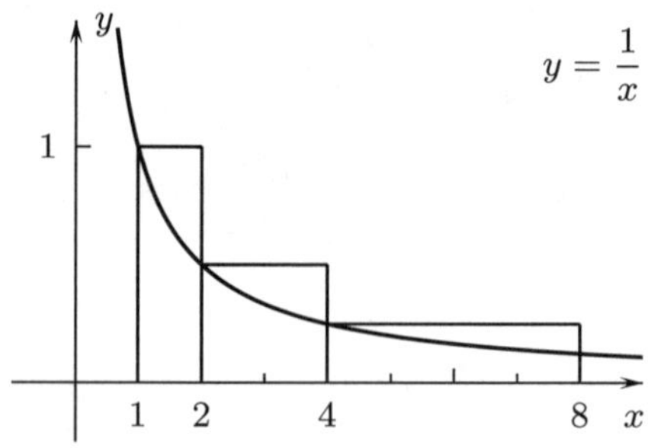

Figure A12

Tune-up exercises 4.3

T1 $f(6) - f(2) = \int_2^6 \frac{df}{dx}(x)\,dx$ • $8 - f(2) = 5$ • $f(2) = 3$

T2. **(a)** $\int_1^9 x^{-1/2}\,dx = \left[2x^{1/2}\right]_1^9 = 2(9)^{1/2} - 2(1)^{1/2} = 4$

(b) $\int (3 + 4x + 5x^2)\,dx = 3x + 4(\frac{1}{2}x^2) - 5(\frac{1}{3}x^3) + C = 3x + 2x^2 - \frac{5}{3}x^3 + C$

T3. $\int_0^4 K(x)\,dx = \int_0^2 x^3\,dx + \int_2^4 x^{-2}\,dx = 4 + \frac{1}{4} = 4.25$

T4. $[\text{Area}] = \int_0^1 (1 + x^2 + x^4)\,dx = \left[x + \frac{1}{3}x^3 + \frac{1}{5}x^5\right]_0^1 = \frac{23}{15}$

T5. $W(x) = \int 3x^5\,dx = \frac{1}{2}x^6 + C$ • $W(1) = 2 \Longrightarrow C = \frac{3}{2}$ • $W(x) = \frac{1}{2}x^6 + \frac{3}{2}$.

T6 $Z(4) - Z(3) = \int_3^4 4t^3\,dt = \left[t^4\right]_3^4 = 175$

Problems 4.3

1. $\int_{-5}^5 \frac{d}{dx}\left(\frac{x}{x^2+1}\right)dx = \left[\frac{x}{x^2+1}\right]_{-5}^5 = \frac{5}{26} - (-\frac{5}{26}) = \frac{5}{13}$

2 1

8. $4x + \frac{7}{4}x^4 - \frac{1}{6}x^6 + C$

9. $x^9 - x^8 + C$

10. $\frac{3}{4}x^{4/3} + \frac{4}{5}x^{5/4} + \frac{5}{6}x^{6/5} + C$

12. $48\frac{2}{3}$

14. $-52\frac{4}{5}$

18. $-\frac{1}{12}$

19. $-1\frac{59}{64}$

21. $\int (x^2 + 6x + 9)\,dx = \frac{1}{3}x^3 + 3x^2 + 9x + C$

22. $\frac{1}{8}$

26. $\int_{-8}^0 t^{1/3}\,dt + \int_0^9 t^{1/2}\,dt = 6$

30. $[\text{Area}] = \int_1^6 (3x^2 + 90x^{-2})\,dx = 290$

31. $\frac{4}{5}$

33. $\frac{1}{3}$

35. $-\int_{-2}^{1}(x^3-1)\,dx = \frac{27}{4}$

37. **(a)** $\int_0^{10}(10t-t^2)\,dt = 166\frac{2}{3} \doteq 166.67$ dollars **(b)** $\frac{d}{dt}(10t-t^2) = 10-2t$ is >0 for $0<t<5$ and is <0 for $5<t<10$. • After 5 hours

38. **(a)** 5 feet per second **(b)** At $t=16$ **(c)** 144 feet

39. **(a)** At $t=5$ **(b)** 215 nautical miles

42. **(a)** -3 feet per second^2 **(b)** 26.25 feet

44. **(a)** Figure A44 **(b)** $\frac{41}{300} \doteq 0.1367$ square meters

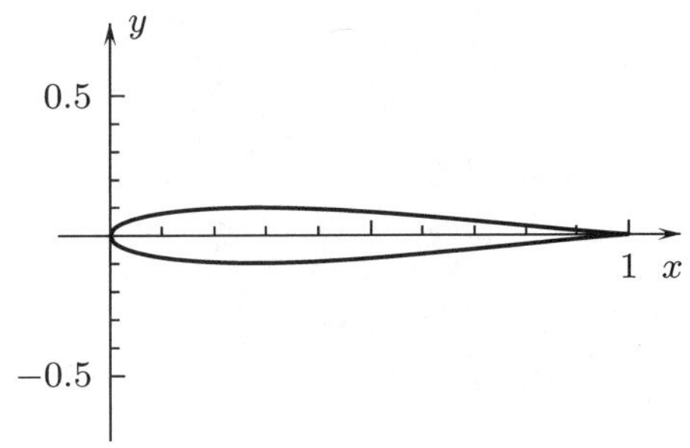

Figure A44

45. $-\int_{-8}^{0} x^{1/3}\,dx + \int_0^1 x^{1/3}\,dt = 12.75$

48. **(a)** 4 feet **(b)** 12 feet

49. $13\frac{1}{6}$ meters

Tune-up exercises 4.4

T1. **(a)** One answer: $(-3)(0.5)+(-2)(0.5)+2.5(0.5)+4.3(0.5)+3.5(0.5)+3.5(0.5) = 4.4$
(b) One answer: $[\frac{1}{2}(-2)-4+0.7+3.7+4.5+2.4+\frac{1}{2}(6)](0.5) = 4.65$

T2. $s(14) = 10 + \int_1^4 v(t)\,dt = 10+16.5-6.75 = 19.75$

T3. **(a)** Trapezoid Rule: $[-\frac{1}{2}(30)+32+33+38+40+\frac{1}{2}(45)](0.2) = 36.1$
(b) Lower sum: $(30+32+33+38+40)(0.2) = 34.6$
(c) Upper sum: $(32+33+38+40+45)(0.2) = 37.6$

T4. **(a)** Midpoint Rule: $[2^{1/3}+4^{1/3}+6^{1/3}+8^{1/3}](2) \doteq 13.3289$
(b) Trapezoid Rule: $[\frac{1}{2}(1^{1/3})+3^{1/3}+5^{1/3}+7^{1/3}+\frac{1}{2}(9^{1/3}](2) \doteq 13.2104$
(c) Simpson's Rule: $\frac{1}{3}[1^{1/3}+4(3^{1/3})+2(5^{1/3})+4(7^{1/3})+(9^{1/3}](2) \doteq 13.2805$

Problems 4.4

1. **(a)** One answer: $[(1300-80)/80](100) = 1525\%$
(b) One answer: $\int_{1950}^{199} r(t)\,dt \approx (145+550+875+1065)(10) = 26{,}350$ million kilograms

3. **(a)** One answer: 35,700 milliliters **(b)** One answer: At $t=25$ days

4. **(a)** $r=100t$ **(b)** $V=50t^2+50$ gallons

5. 61.25 million

13. [Upper sum:] $= (3+\frac{55}{27}+\frac{35}{27}+\frac{37}{27}+\frac{71}{27}+5)(\frac{1}{3}) = 5\frac{1}{9}$ •
[Lower sum $= (\frac{55}{27}+\frac{35}{27}+1+1+\frac{37}{27}+\frac{71}{27})(\frac{1}{3}) = 3\frac{1}{9}$

25. **(a)** Midpoint Rule: 5.3523 **(b)** Trapezoid Rule: 5.2650 **(c)** Simpson's Rule: 5.3046 •

$\int_0^4 \sqrt{x}\,dx = \frac{16}{3}$ • [Error in (a)] $\doteq 1.9 \times 10^{-2}$ • [Error in (b)] $\doteq 6.8 \times 10^{-2}$ • [Error in (c)] $\doteq 2.9 \times 10^{-2}$

28. **(a)** Midpoint Rule: 1.896973 • $\int_0^1 (x^4 + 2x)\,dx = \frac{6}{5}$ • [Error] $\doteq 1.0 \times 10^{-2}$ • [Error estimate] $= \frac{1}{24}(1)(12)(\frac{1}{4})^2 \doteq 3.1 \times 10^{-2}$

(b) Trapezoid Rule: 12.431222 • $\int_1^4 x^{3/2}\,dx = \frac{62}{5}$ • [Error] $\doteq 3.1 \times 10^{-2}$ • [Error estimate] $= \frac{1}{12}(3)(\frac{3}{4})(\frac{1}{2})^2 \doteq 4.7 \times 10^{-2}$

29a. Simpson's rule: 6.78678795 • $\int_1^5 \sqrt{x}\,dx = \frac{2}{5}(5^{3/5} - 1) \doteq 6.78689326$ • [Error] $\doteq 1.1 \times 10^{-4}$ • [Error estimate] $= \frac{1}{180}(4)(\frac{15}{16})(\frac{1}{2})^4 \doteq 1.3 \times 10^{-3}$

31. More than 258 subintervals

Tune-up exercises 4.5

T1. $u = x^3 - 1, du = 3x^2\,dx$ • $\int x^2\sqrt{x^3-1}\,dx = \frac{1}{3}\int \sqrt{x^3-1}(3x^2\,dx) = \frac{1}{3}\int u^{1/2}\,du$
$= \frac{1}{3}(\frac{2}{3})u^{3/2} + C = \frac{2}{9}(x^3-1)^{3/2} + C$ •
Check: $\frac{d}{dx}[\frac{2}{9}(x^3-1)^{3/2}] = \frac{2}{9}(\frac{3}{2})(x^3-1)^{1/2}\frac{d}{dx}(x^3-1) = x^2(x^3-1)^{1/2}$

T2. $-\frac{1}{3}(3x+4)^{-1} + C$ • Check: $\frac{d}{dx}[-\frac{1}{3}(3x+4)^{-1}] = \frac{1}{3}(3x+4)^{-2}\frac{d}{dx}(3x+4) = (3x+4)^{-2}$

T3. $u = 3 - x, du = -dx$ • $\int_0^1 (3-x)^{-3}\,dx = -\int_{x=0}^{x=1} (3-x)^{-3}(-dx) = -\int_{x=0}^{x=1} u^{-3}\,du$
$= \left[\frac{1}{2}u^{-2}\right]_{x=0}^{x=1} = \left[\frac{1}{2}(3-x)^{-2}\right]_{x=0}^{x=1} = \frac{5}{72}$

T4. $u = x + 3, x = u - 3, du = dx$ • $\int \frac{x}{\sqrt{x+3}}\,dx = \int \frac{u-3}{\sqrt{u}}\,du$
$= \int (u^{1/2} - 3u^{-1/2})\,du = \frac{2}{3}u^{3/2} - 6u^{1/2} + C = \frac{2}{3}(x+3)^{3/2} - 6(x+3)^{1/2} + C$

T5. Use $u = 1 + x, du = dx$ in the first integral and $v = 1 - x, du = -dx$ in the second. •
$\int_{x=0}^{x=2} u^3\,du + \int_{x=0}^{x=2} v^4\,dv = \left[\frac{1}{4}(1+x)^4\right]_0^2 + \left[\frac{1}{5}(1-x)^5\right]_0^2 = 19.6$

Problems 4.5

1. $-\frac{1}{4}(x^2+4)^{-2} + C$

2. $-\frac{1}{18}(2+x^9)^{-2} + C$

3. $\frac{1}{2}\int_{x=0}^{x=1} u^{1/2}\,du = \left[\frac{1}{3}(x^2+9)^{3/2}\right]_0^1 = \frac{1}{3}(10)^{3/2} - 9$

9. $\int_1^6 \sqrt{1+5x}\,dx = \frac{2}{15}[(31)^{3/2} - (6)^{3/2}]$

10. $s = 8\frac{5}{12}$ feet

11. 18

14. $u = 1 - x, x = 1 - u, dx = -du$ • $-\int (u^{-1/2} - u^{1/2})\,du = -2(1-x)^{1/2} + \frac{2}{3}(1-x)^{3/2} + C$

15. $-\frac{1}{9}(3y-4)^{-1} - \frac{2}{9}(3y-4)^{-2} + C$

17. $\frac{1}{5}(\sqrt{x}+9)^{10} + C$

25. $b = 1$

Tune-up exercises 4.6

T1. Figure T1 • [Area] $= \int_0^2 [(2+x^2)-1]\,dx = \frac{14}{3}$

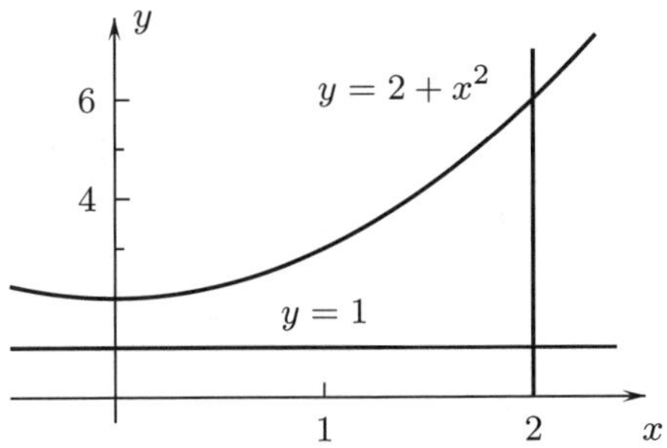

Figure T1

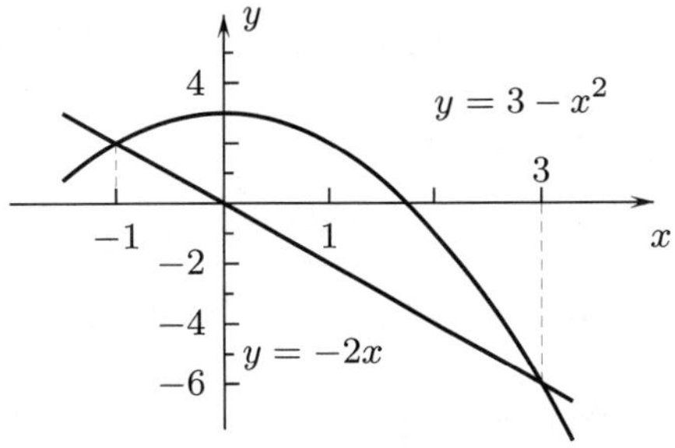

Figure T2

T2. Figure T2 • [Area] $= \int_{-1}^3 [(3-x^2)-(-2x)]\,dx = \frac{32}{3}$

T3. [Average value] $= \frac{1}{2}\int_1^3 6x^{-2}\,dx = 2$

T4. [Weight] $= \int_0^1 12x^2\,dx = 4$ • [Moment of inertia about 0] $= \int_0^1 12x^3\,dx = 3$ •
[Center of gravity] $= \overline{x} = \frac{3}{4}$

T5. [Length] $= \int_0^2 \sqrt{1+(4x^3)^2}\,dx \approx 16.65$

T6. $g'(x) = \int 6x\,dx = 3x^2 + C$ • $g'(1) = 2 \implies C = -1$ • $g'(x) = 3x^2 - 1$ •
$g(x) = \int (3x^2-1)\,dx = x^3 - x + C$ • $g(1) = -1 \implies C = -1$ • $g(x) = x^3 - x - 1$ • $g(0) = -1$

T7. $a = 2$ • $v = \int 2\,dt = 2t + v_0$ • $v(0) = -4 \implies v_0 = -4$ • $v = 2t - 4$ •
$s = \int (2t-4)\,dt = t^2 - 4t + s_0$ • $s(0) = 16 \implies s_0 = 16$ • $s = t^2 - 4t + 16$ •
$s = 12 \iff t^2 - 4t + 4 = 0 \iff (t-2)^2 = 0 \iff t = 2$

Problems 4.6

1. Figure A1 • $x^{1/2} = x^2 \iff x = 0$ or (cancelling $x^{1/2}$) $1 = x^{3/2} \iff x = 0, 1$ •
[Area] $= \int_0^1 (x^{1/2} - x^2)\,dx = \frac{1}{3}$

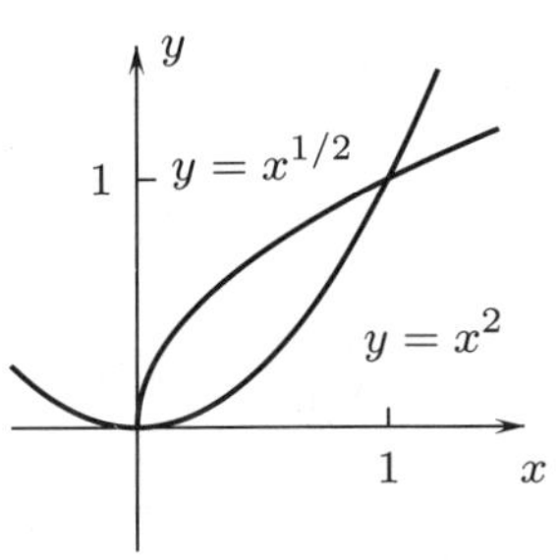

Figure A1

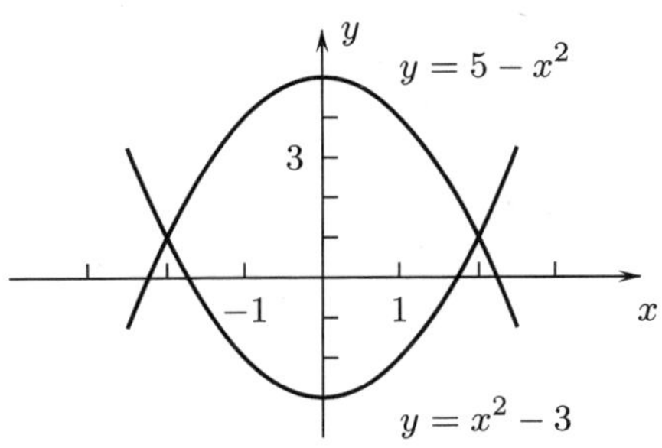

Figure A2

2. Figure A2 • $5-x^2 = x^2-3 \iff x^2 = 4 \iff x = \pm 2$ • $[\text{Area}] = \int_{-2}^{2} [(5-x^2)-(x^2-3)\,dx = \frac{64}{3}$

4. $[\text{Area}] = \frac{4}{3}$

6. $[\text{Area}] = \frac{1}{12}$

9. $[\text{Average value}] = \frac{1}{3}\int_1^4 12x^{-2}\,dx = 3$

10. $[\text{Average value}] = \frac{2}{3}$

13. **(a)** $[\text{Average value}] = \frac{1}{40}\int_0^{40} T(t)\,dt$ **(b)** $[\text{Average value}] \approx 2155/40 = 53.875°\text{F}$

17. $[\text{Weight}] = 7$ pounds • $[\text{Center of gravity}] = \frac{26}{49}$ inches

20. $[\text{Length}] = \int_{-10}^{10} \sqrt{1+(3x^2+6x)^2}\,dx \approx 2009.2$

22. $[\text{Cost}] = \$2.50$

24. $s = \frac{1}{6}t^3 + 3t - 2$

25. $s = \frac{1}{4}t^4 + 4t + 5$

27. **(a)** One second **(b)** Its downward velocity is 48 feet/second.

29. $a = -32$ feet/second^2 • $v = -32t + v_0$ • Set $t = 0$ when the ball is dropped. Then $v(0) = 0 \implies v_0 = 0$ • $v = -32t$ feet per second • $s = -16t^2 + s_0$ • $s(3) = 0 \implies s_0 = s(0) = 144$ feet

34. $[\text{Area}] = \frac{1}{3}a^3 - \frac{1}{2}a^2 + a$

39. $[\text{Area}] = \frac{4}{3}b^{3/2}$

40. $[\text{Area}] = \int_0^2 (2x-x)\,dx + \int_2^3 (6-x-x)\,dx = 3$

48. **(a)** $L = 2$ **(b)** Both areas are $\frac{8}{3}$.

50. **(a)** $y = 2x + 1/x^2$ and $y = 8$ intersect at $x \approx 0.371$ and 3.968. • $[\text{Area}] \approx \int_{0.37}^{3.97} [8-(2x+x^{-2})]\,dx$
$= \Big[8x - x^2 + 1/x\Big]_{0.37}^{3.97} \doteq 10.73$

53. $[\text{Area}] \approx 7.759$

55. $s = 95.6$ centimeters

Tune-up Exercises 4.7

T1. $\ln(6) - \ln(1) = \ln(6)$

T2. $-\ln(6)$ • The integral is negative because $1/x < 0$ for $-6 \le x \le -1$.

T3. Use $u = x^2+1, du = 2x\,dx$. • $\int_0^5 \frac{x}{x^2+1}\,dx = \frac{1}{2}\int_{x=0}^{x=5} \frac{1}{u}\,du = \frac{1}{2}\Big[\ln|u|\Big]_{x=0}^{x=5}$
$= \frac{1}{2}\Big[\ln|x^2+1|\Big]_{x=0}^{x=5} = \frac{1}{2}\ln(26)$

T4. $u = \ln x, du = \frac{1}{x}\,dx$ • $\int \frac{\sqrt{\ln x}}{x}\,dx = \int u^{1/2}\,du = \frac{2}{3}u^{3/2} + C = \frac{2}{3}(\ln x)^{3/2} + C$

T5. $e^4 - e^{-4}$

T6. $\frac{1}{2}x^2 + e^x + C$

T7. $-\cos x - 5\sin x + C$

T8. $u = x^3, du = 3x^2\,dx$ • $\int x^2 e^{x^3}\,dx = \frac{1}{3}\int e^u\,du = \frac{1}{3}e^u + C$
$= \frac{1}{3}e^{x^3} + C$

T9. $\displaystyle\int_5^0 \cos x\, dx = \Big[\sin x\Big]_5^0 = -\sin(5)$

T10. $\displaystyle\int_{3\pi/4}^{\pi/4} \sin x\, dx = \Big[-\cos x\Big]_{3\pi/4}^{\pi/4} = -\cos(\tfrac{1}{4}\pi) + \cos(\tfrac{3}{4}\pi) = -\sqrt{2}$

T11. $u = e^t, du = e^t\, dt$ • $\displaystyle\int e^t \cos(e^t)\, dt = \int \cos u\, du = \sin u + C = \sin(e^t) + C$

T12. $\frac{1}{4}\tan^{-1}(\frac{1}{4}x) + C$

T13. $\Big[\sin^{-1}(\tfrac{1}{5}x)\Big]_0^4 = \sin^{-1}(0.8)$

T14. Use $u = \cos x, du = -\sin x\, dx.$ • $\Big[-\ln|\cos x|\Big]_0^1 = -\ln(\cos(1))$

T15. $\frac{1}{3}\tan(3x) + C$

T16. $\sec(0.5) - 1$

Problems 4.7

1. $\ln|x| + 2e^x - 3\cos x + 4\sin x + C$

2. $2\ln(10) + 5\cos(10) - 5\cos(1)$

3. $-\frac{1}{2}(x^2+3)^{-1} + C$

4. $-\frac{1}{3}(e^{-6} - 1)$

5. $\frac{2}{3}(\sin x)^{3/2} + C$

7. $-\frac{1}{3}\ln(7)$

8. $\cos(e^{-x}) + C$

10. $-\ln(17)$

11. $\frac{1}{2}\sin^2(2)$

12. $\frac{1}{2}(\ln x)^2 + C$

15. $-\cos(\ln x) + C$

16. $2(e^3 - e)$

18. $\ln(1 + e^y) + C$

22. $\frac{7}{24}$

23. $\frac{1}{3}\ln|\sin(3x)| + C$

24. $\frac{3}{2}\tan^{-1}(3.5)$

25. Use $u = 2x, du = 2\,dx.$ • $\displaystyle\frac{1}{2}\int_{x=-1}^{x=1} \frac{1}{\sqrt{9-u^2}}\, du = \sin^{-1}(\tfrac{2}{3})$

26. $\frac{1}{2}(\sin^{-1} x)^2 + C$

29. $\frac{1}{9}\tan^{-1}(\frac{1}{3}x^3) + C$

36. Figure A36 • $16 - \cos(5)$

37. [Average value] $= -7 - 2\ln(3)$

38. $\frac{1}{3}\ln(5)$

39. $\frac{1}{2}(e^2 - e^{-2})$

40. Figure A40 • $2\tan^{-1}(3)$

46. $1/\sqrt{3}$

48. $\frac{1}{2}e^{10} + 2e^5 + \frac{5}{2}$

49. $\frac{1}{2}\sqrt{2} - \frac{3}{4}\pi$

54. $\overline{x} = 3/(\ln(4))$

65a. $x = e^{n\pi}$ for integers $n = 0, \pm 1, \pm 2, \ldots$

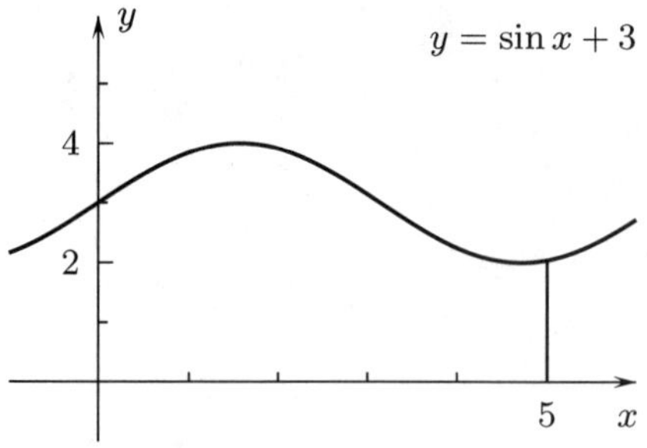

Figure A36

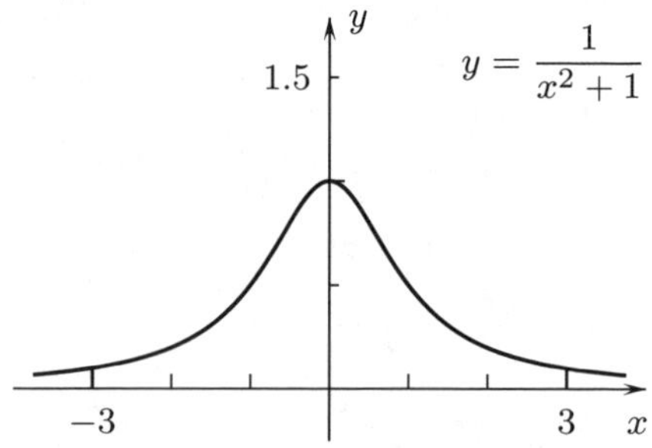

Figure A40

68. $k = 1$

69. $[\text{Weight}] = \frac{1}{3}\tan^{-1}(\frac{4}{3})$ • $\overline{x} = \dfrac{3\ln(25/9)}{2\tan^{-1}(4/3)} \doteq 1.6256$

70. \$29,783.53

71. \$3,828.34

72. \$22,996.64

75. **(a)** \$150,000 **(b)** \$56,830.13

77. **(a)** \$125,000 **(b)** \$87,814.85

82. $b = \ln[1 + 26/\ln(3)]$

85. $e - 1/e$

90. $3/[\ln(2)] + \frac{8}{3}$

Riemann sum program for TI-82 calculators

Line-by-line instructions are given here for entering the program in Section 4.2. Press [2nd] [QUIT] [PRGM] [▶] [▶] [1] to start a new program. See the discussion of the Secant line program for instructions on editing and deleting programs.

---

1. Press [R] [I] [E] [M] [A] [N] [N] [ENTER]. The calculator is automatically in alpha mode here. This title identifies the program.

---

2. Press [2nd] [DRAW] [1] [ENTER]. ClrDraw clears the graphics screen.

---

3. Press [2nd] [Y-Vars] [5] [2] [ENTER] to unselect all the functions in the $Y=$ menu.

---

4. Press [2nd] [Y-VARS] [5] [1] [1] [ENTER] to select Y_1.

---

5. Press [PRGM] [▶] [3] for the word Disp and then [2nd] [A-LOCK] [“] [U] [S] [E] [␣] [2nd] [Y-VARS] [1] [1] [(] [X,T,θ] [)] [ALPHA] [“] [ENTER]. The space [␣] is on the key with 0.

---

6. Press [PRGM] [▶] [2] [ALPHA] [A] [,] [ALPHA] [B] [,] [ALPHA] [N] [ENTER] to cause the prompts $A=$?, $B=$?, and $N=$? to appear when the program is run so the user can supply values of A, B, and N.

---

7. Press [PRGM] [▶] [3] [ALPHA] [“] [ALPHA] [T] [2nd] [TEST] [1] [Ø] [2nd] [A-LOCK] [␣] [F] [O] [R] [␣] [L] [E] [F] [T] [“] [ENTER]. This and the next two lines explain the role of T.

---

8. Press [PRGM] [▶] [3] [ALPHA] [“] [ALPHA] [T] [2nd] [TEST] [1] [1] [2nd] [A-LOCK] [␣] [F] [O] [R] [␣] [R] [I] [G] [H] [T] [“] [ENTER].

---

9. Press [PRGM] [▶] [3] [ALPHA] [“] [ALPHA] [T] [2nd] [TEST] [1] [Ø] [.] [5] [2nd] [A-LOCK] [␣] [F] [O] [R] [␣] [M] [I] [D] [“] [ENTER].

---

10. Press [PRGM] [▶] [1] [ALPHA] [T] [ENTER]. The user picks T.

---

11. Press [(] [ALPHA] [B] [−] [ALPHA] [A] [)] [÷] [ALPHA] [N] [STO▶] [ALPHA] [D] [ENTER]. The width D of the rectangles is calculated and stored.

---

12. Press [Ø] [STO▶] [ALPHA] [S] [ENTER]. The Riemann sum is stored as S during the calculations. Here S is set equal to 0 to erase its previous value.

---

13. Press [PRGM] [4] [ALPHA] [J] [,] [1] [,] [ALPHA] [N] [,] [1] [)] [ENTER]. The commands between here and the End on line 21 will be executed for each integer J from 1 to N.

---

14. Press [ALPHA] [A] [+] [(] [ALPHA] [J] [−] [1] [)] [ALPHA] [D] [STO▶] [ALPHA] [U] [ENTER]. U is the x-coordinate of the left side of the rectangle.

---

15. Press [ALPHA] [U] [+] [ALPHA] [T] [ALPHA] [D] [STO▶] [X,T,θ] [ENTER]. Here X is the x-coordinate of the left edge of the rectangle if $T=0$, the right edge if $T=1$, and the midpoint if $T=0.5$.

---

16. Press 2nd Y-vars 1 1 STO▶ ALPHA W ENTER. The value at X of the function Y_1 in the Y= menu is calculated and stored as W.

17. Press 2nd DRAW 2 ALPHA U , Ø , ALPHA U , ALPHA W) ENTER. The left side of the rectange is drawn.

18. Press 2nd DRAW 2 ALPHA U , ALPHA W , ALPHA U + ALPHA D , ALPHA W) ENTER. The top of the rectange is drawn.

19. Press 2nd DRAW 2 ALPHA U + ALPHA D , ALPHA W , ALPHA U + ALPHA D , Ø) ENTER. The right side of the rectangle is drawn.

20. Press ALPHA S + ALPHA D ALPHA W STO▶ ALPHA S ENTER. The area of the rectangle is added to S if W is positive and is subtracted from S if W is negative.

21. Press PRGM 7 ENTER.

22. Press PRGM 8 ENTER The program pauses to show the graph and rectangles.

23. Press PRGM ▶ 3 2nd A-LOCK " S U M 2nd TEST 1 ALPHA " ENTER . The symbols SUM = are displayed.

24. Press PRGM ▶ 3 ALPHA S ENTER . The value of the Riemann sum is displayed.

Riemann sum program for TI-83 calculators

Line-by-line instructions are given here for entering the program in Section 4.2. Press 2nd QUIT PRGM ▶ ▶ 1 to start a new program. See the discussion of the Secant line program for instructions on editing and deleting programs.

1. Press R I E M A N N ENTER. The calculator is automatically in alpha mode here. This title identifies the program.

2. Press VARS ▶ 4 2 ENTER. ClrDraw clears the graphics screen.

3. Press VARS ▶ 4 2 ENTER to unselect all the functions in the Y= menu.

4. Press VARS ▶ 4 1 1 ENTER to select Y_1.

5. Press PRGM ▶ 3 for the word Disp and then 2nd A-LOCK " U S E ␣ VARS ▶ 1 1 (X,T,θ,n) ALPHA " ENTER. The space ␣ is on the key with 0.

6. Press PRGM ▶ 2 ALPHA A , ALPHA B , ALPHA N ENTER to cause the prompts $A=$?, $B=$?, and $N=$? to appear when the program is run so the user can supply values of A, B, and N.

7. Press PRGM ▶ 3 ALPHA " ALPHA T 2nd TEST 1 Ø 2nd A-LOCK _ F O R _ L E F T " ENTER. This and the next two lines explain the role of T.

8. Press PRGM ▶ 3 ALPHA " ALPHA T 2nd TEST 1 1 2nd A-LOCK _ F O R _ R I G H T " ENTER.

9. Press PRGM ▶ 3 ALPHA " ALPHA T 2nd TEST 1 Ø . 5 2nd A-LOCK _ F O R _ M I D " ENTER.

10. Press PRGM ▶ 1 ALPHA T ENTER. The user picks T.

11. Press (ALPHA B − ALPHA A) ÷ ALPHA N STO▶ ALPHA D ENTER. The width D of the rectangles is calculated and stored.

12. Press Ø STO▶ ALPHA S ENTER. The Riemann sum is stored as S during the calculations. Here S is set equal to 0 to erase its previous value.

13. Press PRGM 4 ALPHA J , 1 , ALPHA N , 1) ENTER. The commands between here and the End on line 21 will be executed for each integer J from 1 to N.

14. Press ALPHA A + (ALPHA J − 1) ALPHA D STO▶ ALPHA U ENTER. U is the x-coordinate of the left side of the rectangle.

15. Press ALPHA U + ALPHA T ALPHA D STO▶ X, T, θ, n ENTER. Here X is the x-coordinate of the left edge of the rectangle if $T = 0$, the right edge if $T = 1$, and the midpoint if $T = 0.5$.

16. Press VARS ▶ 1 1 STO▶ ALPHA W ENTER. The value at X of the function Y_1 in the Y= menu is calculated and stored as W.

17. Press 2nd DRAW 2 ALPHA U , Ø , ALPHA U , ALPHA W) ENTER. The left side of the rectange is drawn.

18. Press 2nd DRAW 2 ALPHA U , ALPHA W , ALPHA U + ALPHA D , ALPHA W) ENTER. The top of the rectange is drawn.

19. Press 2nd DRAW 2 ALPHA U + ALPHA D , ALPHA W , ALPHA U + ALPHA D , Ø) ENTER. The right side of the rectangle is drawn.

20. Press ALPHA S + ALPHA D ALPHA W STO▶ ALPHA S ENTER. The area of the rectangle is added to S if W is positive and is subtracted from S if W is negative.

21. Press PRGM 7 ENTER.

22. Press PRGM 8 ENTER The program pauses to show the graph and rectangles.

23. Press [PRGM] [▶] [3] [2nd] [A-LOCK] [“] [S] [U] [M] [2nd] [TEST] [1] [ALPHA] [“] [ENTER] . The symbols SUM = are displayed.

24. Press [PRGM] [▶] [3] [ALPHA] [*S*] [ENTER]. The value of the Riemann sum is displayed.

Riemann sum program for TI-85 calculators

Line-by-line instructions are given here for entering the program in Section 4.2. Press [2nd] [QUIT] [PRGM] [F2] to start a new program. See the discussion of the Secant line program for instructions on editing, and deleting programs.

1. Press [R] [I] [E] [M] [A] [N] [N] [ENTER].

2. Enter ***ClDrw*** and press [ENTER] to clear the graphics screen.

3. Enter ***FnOff*** and press [ENTER]. This command unselects the functions in the $y=$ menu so their graphs will not be drawn when the program is run.

4. Enter ***FnOn*** and press [1] [ENTER]. This command selects $y1$ so its graph is generated as soon as a drawing command is encountered in the program.

5. Enter ***Disp*** and press [2nd] [STRING] [F1] for the quotation marks. Press [ALPHA] [ALPHA] to lock the calculator in upper-case alpha mode and [*U*] [*S*] [*E*] [␣] using the space key with [(–)] . Press [ALPHA] [2nd] [alpha] to switch to lower-case alpha mode and [*Y*] [1] [(] [*x*-VAR] [)] [F1] [ENTER].

6. Enter ***Prompt*** and press [ALPHA] [*A*] [,] [ALPHA] [*B*] [,] [ALPHA] [*N*] [ENTER]. This line causes prompts “$A =$”, “$B =$”, and “$N =$” to appear so the user can enter values for these parameters.

7. Enter ***Disp*** and press [2nd] [STRING] [F1] (for “) [ALPHA] [*T*] [ALPHA] [=] [∅] [ALPHA] [ALPHA] [␣] [F] [O] [R] [␣] [L] [E] [F] [T] [F1] [ENTER]. This and the next two lines explain the role of T.

8. Enter ***Disp*** and press [2nd] [STRING] [F1] (for “) [ALPHA] [*T*] [ALPHA] [=] [1] [ALPHA] [ALPHA] [␣] [F] [O] [R] [␣] [R] [I] [G] [H] [T] [F1] [ENTER].

9. Enter ***Disp*** and press [2nd] [STRING] [F1] (for “) [ALPHA] [*T*] [ALPHA] [=] [0] [.] [5] [ALPHA] [ALPHA] [␣] [F] [O] [R] [␣] [M] [I] [D] [F1] [ENTER].

10. Enter ***Input*** and press [ALPHA] [T] [ENTER]. The user picks T.

11. Press [(] [ALPHA] [*B*] [–] [ALPHA] [*A*] [)] [÷] [ALPHA] [*N*] [STO▶] [*D*] [ENTER]. The width D of the rectangles is calculated and stored.

12. Press [∅] [STO▶] [S] [ENTER]. The Riemann sum is stored as S during the calculations. Here S is set equal to 0 to erase its previous value.

13. Enter ***For(*** and press [ALPHA] [J] [,] [1] [,] [ALPHA] [N] [,] [1] [)] [ENTER]. This is the beginning of a loop where the commands up to line 21 are carried out succesively as J runs from 1 to N in steps of 1, i.e., for $J = 1, 2, \ldots, N$.

14. Press [ALPHA] [A] [+] [(] [ALPHA] [J] [−] [1] [)] [×] [ALPHA] [D] [STO▶] [U] [ENTER]. U is the x-coordinate of the left side of the rectangle.

15. Press [ALPHA] [U] [+] [ALPHA] [T] [×] [ALPHA] [D] [STO▶] [x-VAR] [ENTER]. Here x is the x-coordinate of the left edge of the rectangle if $T = 0$, the right edge if $T = 1$, and the midpoint if $T = 0.5$.

16. Press [2nd] [alpha] [Y] [1] [STO▶] [W] [ENTER]. The value at x of the function $y1$ in the y= menu is calculated and stored as W.

17. Enter ***Line(*** and press [ALPHA] [U] [,] [∅] [,] [ALPHA] [U] [,] [ALPHA] [W] [)] [ENTER]. The left side of the rectange is drawn.

18. Enter ***Line(*** and press [ALPHA] [U] [,] [ALPHA] [W] [,] [ALPHA] [U] [+] [ALPHA] [D] [,] [ALPHA] [W] [)] [ENTER]. The top of the rectange is drawn.

19. Enter ***Line(*** and press [ALPHA] [U] [+] [ALPHA] [D] [,] [ALPHA] [W] [,] [ALPHA] [U] [+] [ALPHA] [D] [,] [∅] [)] [ENTER]. The right side of the rectangle is drawn.

20. Press [ALPHA] [S] [+] [ALPHA] [D] [×] [ALPHA] [W] [STO▶] [S] [ENTER]. The area of the rectangle is added to S if W is positive and is subtracted from S if W is negative.

21. Enter ***End*** and press [ENTER]. This is the end of the loop that begins on line 13.

22. Enter ***Pause*** and press [ENTER].

23. Enter ***Disp*** and press [2nd] [STRING] [F1] (for ") [ALPHA] [ALPHA] [S] [U] [M] [=] [F1] [ENTER]. The symbols SUM = are displayed.

24. Enter ***Disp*** and press [ALPHA] [S] [ENTER]. The value of the Riemann sum is displayed.

CHAPTER 5
DERIVATIVES AND APPLICATIONS, PART III

In this chapter we continue our study of the derivative. Section 5.1 deals with related-rate and maximum/minimum problems that use techniques from Sections 2.5 and 3.2 in other contexts. In Section 5.2, we discuss a technique called IMPLICIT DIFFERENTIATION for finding derivatives of functions $y(x)$ that are defined by equations involving x and y, rather than by formulas involving only x, as has been the case for all the functions we have studied up to now. NEWTON'S METHOD, which uses tangent line approximations to produce accurate approximate solutions of equations of the form $F(x) = 0$, is presented in Section 5.3. In Section 5.4 we discuss L'HOPITAL'S RULE, which uses derivatives to determine limits of certain quotients, products, and sums that cannot be found directly.

Section 5.1

More related-rate and maximum/minimum problems

OVERVIEW: *In Sections 2.5 and 3.2 we studied related-rate and maximum/minimum problems in cases where the necessary functions were given or could be easily obtained from familar formulas for areas and volumes. Most of the Problems in this section involve functions whose formulas must be derived by more extensive analysis of the information that is provided. Once the necessary formulas have been found, the solutions are obtained with the techniques used in Sections 2.5 and 3.2.*

Topics:

- ***Using similar triangles***
- ***Using the Pythagorean Theorem***
- ***Dimension analysis***
- ***Finding a maximum by finding a minimum***

Using similar triangles: the height of a shadow

Imagine that a light on the horizontal ground is casting the shadow of a 10-foot-high fence on the side of a building. Suppose that the fence is 8 feet from the building as in Figure 1, and that the light is being moved along the ground away from the wall.

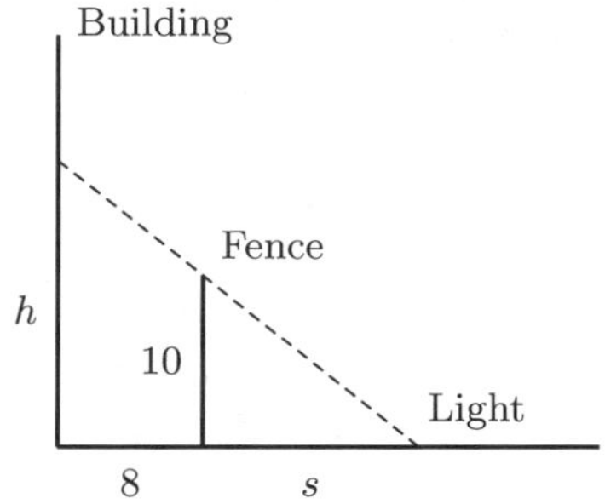

FIGURE 1

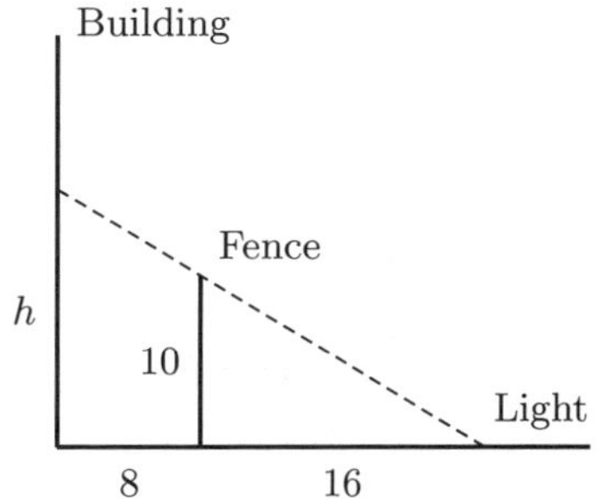

FIGURE 2

Question 1 **(a)** Does the height h of the shadow in Figure 1 increase or decrease as the distance s from the lamp to the fence increases? **(b)** What are the limits of h as $s \to \infty$ and as $s \to 0^+$ (assuming that the building is infinitely tall)?

Notice that the two right triangles in Figure 1 are similar for any positive value of s. We can use this fact to find an equation relating s to the height h of the shadow.

Question 2 **(a)** How high is the shadow if the lamp is 16 feet from the fence as in Figure 2? **(b)** Give a formula for the height h of the shadow as a function of the distance s, as in Figure 1, of the lamp from the fence. Generate its graph with the line $h = 10$. The graph should confirm the responses to Question 1.

♦ SUGGESTIONS TO INSTRUCTORS: Have students begin work on Questions 1 through 10 before class. Problems 7, 13, 25, 42, 48, 58, 60, 64, and 67 are good for in-class discussion.

Example 1 Suppose that when the light in Figure 1 is 16 feet from the fence it is moving away from the fence at the rate of 12 feet per minute. At what rate is the height of the shadow increasing or decreasing at that moment?

SOLUTION We differentiate the equation $h = 80/s + 10$ from Question 2 with respect to t, using the Chain Rule for derivatives of powers of functions. We obtain

$$\frac{dh}{dt} = \frac{d}{dt}(80s^{-1} + 10) = -80s^{-2}\frac{ds}{dt} = \frac{-80}{s^2}\frac{ds}{dt}.$$

At the moment in question, $s = 16$ and $\dfrac{ds}{dt} = 12$, so that

$$\frac{dh}{dt} = \frac{-80}{16^2}(12) = -3.75.$$

The shadow is shrinking 3.75 feet per minute at the moment under consideration. □

Using the Pythagorean Theorem: minimizing a distance

Imagine that a car is being driving east, away from an intersection of two highways, while a truck is being driving south toward the intersection on the other road (Figure 3). Suppose we know how far they are from the intersection and how fast they are driving at one particular time. We want to determine the rate of change with respect to time of the distance between them at that moment. Notice that for any position of the car and truck in Figure 3, the distances x, y, and D are the lengths of the sides of a right triangle.

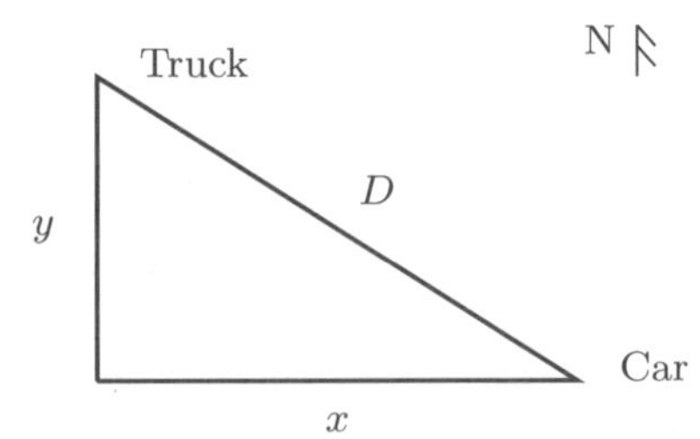

FIGURE 3

Question 3 Find an equation that relates the car's distance x from the intersection, the truck's distance y from the intersection, and the distance D between the car and the truck.

Example 2 Suppose that when the car is 4 miles east of the intersection, it is going 50 miles per hour away from it, the truck is 3 miles north of the intersection and is traveling 70 miles per hour toward it. At what rate is the distance between the car and the truck increasing or decreasing at that moment?

SOLUTION We need more than the equation,

$$D^2 = x^2 + y^2 \tag{1}$$

that you found in Question 3 because it does not involve the rates of change of x, y, and D. To obtain an equation that involves the rates of change, we differentiate both sides of **(1)** with respect to t using the Chain Rule, $\dfrac{d}{dt}(f^2) = 2f\,\dfrac{df}{dt}$. We obtain

$$2D\frac{dD}{dt} = 2x\,\frac{dx}{dt} + 2y\,\frac{dy}{dt}$$

and, after cancelling the 2's,

$$D\,\frac{dD}{dt} = x\,\frac{dx}{dt} + y\,\frac{dy}{dt}. \tag{2}$$

Notice that **(1)** and **(2)** are two equations with the six unknowns, x, y, D, $dx/dt, dy/dt$ and dD/dt. We can use these equations to find any two of the unknowns from values of the other four. In this case, we use $x, y, \frac{dx}{dt}$ and $\frac{dy}{dt}$ to find D and $\frac{dD}{dt}$. We set $x = 4$, $y = 3$ in **(1)** to see that $D = 5$. Next, we set $x = 4, y = 3$, $D = 5, \frac{dx}{dt} = 50$ and $\frac{dy}{dt} = -70$ in **(2)**. This gives $5\frac{dD}{dt} = (4)(50) + (3)(-70) = -10$, so that $\frac{dD}{dt} = -2$. The distance between the car and the truck is decreasing 2 miles per hour at the moment in question. □

Using the Pythagorean Theorem: a cone in a sphere

A right circular cone is INSCRIBED in a sphere if it is contained in the sphere and touches the sphere at its vertex and along the circumference of its base. We position such a cone so its base is horizontal and its vertex is at the highest point on the sphere, as in Figure 4. Let r be the radius of the cone's base and h its height. As you can visualize, the volume of the cone would be very small if the base of the cone were very close to the bottom or to the top of the sphere. Therefore, there must be an inscribed cone of maximum volume whose base is near the center of the sphere. To find it, we first derive a formula for the volume of the cone as a function of the position of its base. We suppose that the radius of the sphere is three meters.

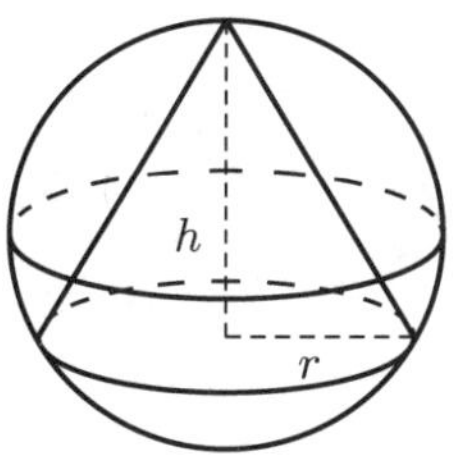

FIGURE 4

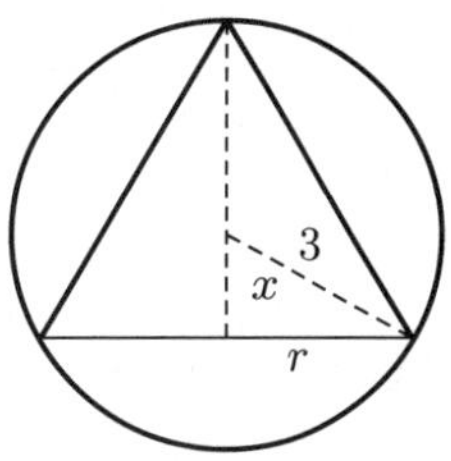

FIGURE 5

Question 4 Consider the vertical cross section through the centers of the sphere and of the cone in Figure 5. Let x be the distance of the center of the sphere above the base of the cone, with x positive if the base of the cone is below the center, as in Figure 5, and negative if the base of the cone is above the center of the sphere. **(a)** Use the Pythagorean Theorem to give a formula, in terms of x, for the radius r of the base of the cone. **(b)** Use the formula $V = \frac{1}{3}\pi r^2 h$ for the volume of the cone to give a formula for the volume as a function of x.

Example 3 How far below or above the center of a sphere of radius 3 meters is the base of the inscribed right circular cone of maximum volume?

SOLUTION The volume function $V(x) = \frac{1}{3}\pi(9 - x^2)(3 + x)$, whose formula you found in Question 4, is defined, continuous, and differentiable in the interval $[-3, 3]$, and it is zero at the endpoints $x = -3$ and $x = 3$. It follows (by Theorem 3 of Section 3.2) that the maximum of $V(x)$ must occur at a point in the interior of the interval where its derivative is zero.

To avoid having to use the Product Rule in finding the derivative, we first expand the product in the formula for $V(x)$:

$$V(x) = \tfrac{1}{3}\pi(9 - x^2)(3 + x) = \tfrac{1}{3}\pi(27 + 9x - 3x^2 - x^3).$$

Then we obtain

$$\begin{aligned}\frac{dV}{dx} &= \frac{d}{dx}\left[\tfrac{1}{3}\pi(27+9x-3x^2-x^3)\right]\\ &= \tfrac{1}{3}\pi(9-6x-3x^2) = \pi(3-2x-x^2)\\ &= -\pi(x^2+2x-3).\end{aligned} \tag{3}$$

The critical points of $V(x)$ are the zeros of x^2+2x-3, which we can find with the Quadratic Formula. We obtain

$$\begin{aligned}x = \frac{-2\pm\sqrt{2^2-4(1)(-3)}}{2(1)} &= \frac{-2\pm\sqrt{4+12}}{2}\\ &= \frac{-2\pm 4}{2} = \begin{cases}\dfrac{-2+4}{2} = 1\\ \dfrac{-2-4}{6} = -3.\end{cases}\end{aligned}$$

(We could also find the zeros of x^2+2x-3 by factoring it in the form $(x-1)(x+3)$.) Since $x=1$ is the only critical point in $[-3,3]$, the maximum of $V(x)$ is at $x=1$. The cone of maximum volume has its base 1 meter below the center of the circle. □

Question 5 Generate the graph of $V(t)$ from Example 3. How does it corroborate the solution?

Dimension analysis: allocating resources

A grower who is planning an apple orchard in a one-acre field expects that each tree would yield 7 boxes of apples if she plants 30 trees, and that for every tree over 30 that she plants, the yield from the entire orchard would decrease 0.15 box. For example, if she planted 32 trees, each tree would produce 0.3 box fewer apples and the yield would be $7-0.3=6.7$ boxes per tree. In this case, the yield from the entire orchard would be

$$\left[6.7\ \frac{\text{boxes}}{\text{tree}}\right]\left[32\ \text{trees}\right] = 214.4\ \text{boxes}. \tag{4}$$

As a partial check of formula **(4)**, we note that the units "tree" in the denominator on the left of the equation cancel with the units "trees" in the numerator, leaving the units "boxes," which match the units of "boxes" on the right. This sort of reasoning, which you have seen before in dealing with rates of change, is known as DIMENSION ANALYSIS.

The grower wants to decide how many trees to plant to maximize the yield from the orchard.

Question 6 **(a)** What would be the yield per tree if 40 trees were planted? What would be the yield from the entire orchard with 40 trees? **(b)** Give formulas in terms of x for the yield per tree and the total yield $Y(x)$ from the orchard under the condition that $30+x$ trees are planted.

Example 4 How many trees should the grower plant to maximize the total yield from the orchard? Give an integer answer.

SOLUTION We expand the product in the formula for the yield from Question 6:

$$\begin{aligned}Y(x) &= (7-0.15x)(30+x)\\ &= 210+7x-4.5x-0.15x^2 = 210+2.5x-0.15x^2.\end{aligned}$$

We want to find the maximum value of $Y(x)$ for integer values of x. We start by finding its derivative:

$$Y'(x) = \frac{d}{dx}(210+2.5x-0.15x^2) = 2.5-0.3x.$$

$Y(x)$ is continuous for all x, and its derivative is zero if $0.3x = 2.5$, which is at $x = 8\frac{1}{3}$. The derivative is positive for $0 < x < 8\frac{1}{3}$ and negative for $x > 8\frac{1}{3}$. Consequently, $Y(x)$ is increasing in $[0, 8\frac{1}{3}]$ and decreasing in $[8\frac{1}{3}, \infty)$. Its maximum for $x \geq 0$ is at $x = 8\frac{1}{3}$, as is illustrated in Figure 6. However, because $8\frac{1}{3}$ is not an integer, we need to study the function further to find the best integer answer.

The maximum of $Y(x)$ for integer values of x is either at $x = 8$ or at $x = 9$ (Figure 7). Since $Y(8) = 210 + 2.5(8) - 0.15(8^2) = 220.4$ is greater than $Y(9) = 210 + 2.5(9) - 0.15(9^2) = 220.35$, the maximum is at $x = 8$. The grower should plant $30 + 8 = 38$ trees. □

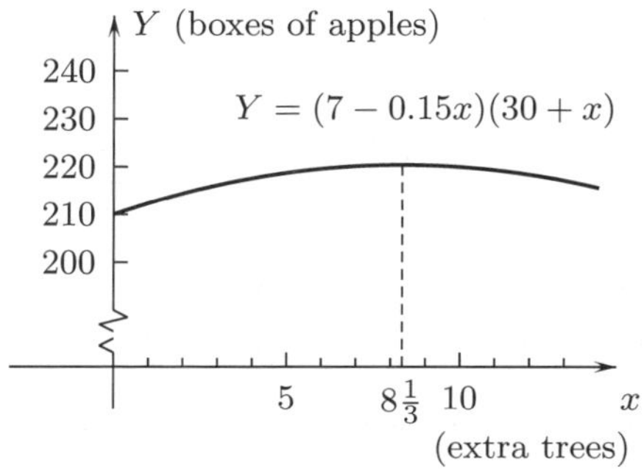

FIGURE 6

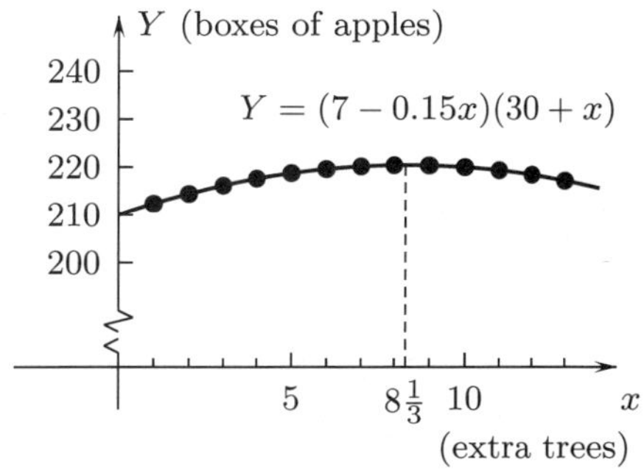

FIGURE 7

Dimension analysis: demand and revenue

Imagine that you have two thousand bushels of soybeans, and that if you charged p dollars per bushel for them you would sell

$$x(p) = \frac{200}{p^3 + 100} \text{ thousand bushels.} \tag{5}$$

A function such as $x(p)$ that gives the amount of sales as a function of the price is called a DEMAND FUNCTION. The amount of money you would receive from sales is the REVENUE.

Question 7 **(a)** How many bushels of soybeans would you sell and what would be your revenue if you charged 2 dollars per bushel? **(b)** How many bushels of soybeans would you sell and what would be your revenue receive if you charged 10 dollars per bushel? **(c)** Give a formula for your revenue $R(p)$, measured in thousands of dollars, in terms of the price p (dollars per bushel) that you charge. Check the formula by analyzing the dimensions.

The graphs of $x(p)$ and $R(p)$ are shown in Figures 8 and 9.

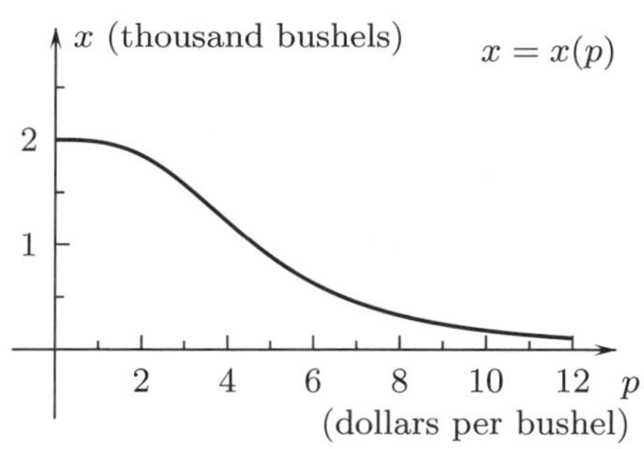

The demand function
FIGURE 8

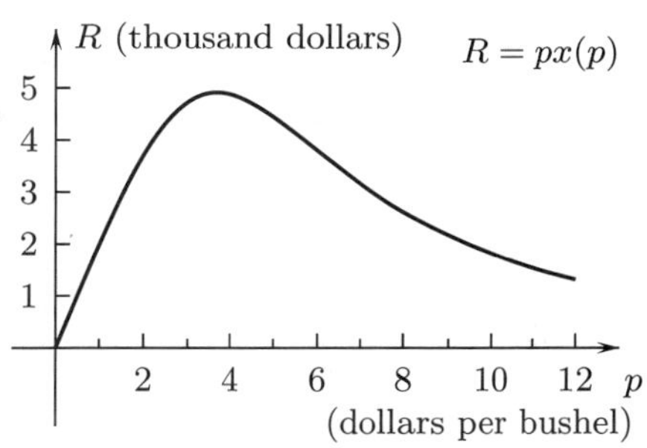

The revenue function
FIGURE 9

The result of Question 7 illustrates the general principle that revenue as a function of price p is the product $R(p) = px(p)$ of price p and demand $x(p)$.

Question 8 Why is it plausible, in the context of your soybean holdings, that the revenue is small for small p and for large p?

Example 5 Find the price (to the nearest cent) that would maximize the revenue from your soybeans.

SOLUTION We use the Quotient Rule to find the derivative of the revenue $R(p) = \dfrac{200p}{p^3 + 100}$ from Question 7:

$$\begin{aligned}\frac{dR}{dp}(p) &= \frac{d}{dp}\left[\frac{200p}{p^3+100}\right] \\ &= \frac{(p^3+100)\dfrac{d}{dp}(200p) - 200p\dfrac{d}{dp}(p^3+100)}{(p^3+100)^2} \qquad \textbf{(6)} \\ &= \frac{200(p^3+100)) - 200p(3p^2)}{(p^3+100)^2} = \frac{200(100-2p^3)}{(p^3+100)^2}.\end{aligned}$$

The derivative is zero at $p = \pm 50^{1/3}$. The negative value of p does not apply, and from formula **(6)**, you can see that dR/dp is positive for $0 < p < 50^{1/3}$ and negative for $p > 50^{1/3}$. Therefore, $R(p)$ is is increasing for $0 < p < 50^{1/3}$ and decreasing for $p > 50^{1/3}$, and its maximum is at $p = 50^{1/3}$.

Because $50^{1/3} \doteq 3.68403$, the optimum price must be either \$3.68 or \$3.69 per bushel. A calculator or computer gives $R(3.68) \doteq 4.91204$ thousand dollars (\$4,912.04) and $R(3.69) \doteq 4.91203$ thousand dollars (\$4,912.03), so the optimum price is \$3.68 per bushel—although charging \$3.69 per bushel would give only one cent less revenue. □

Question 9 Suppose that when the price of soybeans from Example 5 is five dollars per bushel, it is rising 27 cents per bushel per day. At what rate is the potential revenue from your soybeans increasing or decreasing at that moment?

Finding a maximum by finding a minimum

Imagine that you want to roll a large pane of glass on dollies around a corner from one hallway that is 10 feet wide to another hallway that is 15 feet wide, as in Figure 10. What is the widest pane of glass that would fit? We answer this question by finding the minimum of the distance between the points P and Q in Figure 10 as a function of the angle θ in the drawing. We assume in our mathematical model that the glass has zero thickness to simplify the calculations.

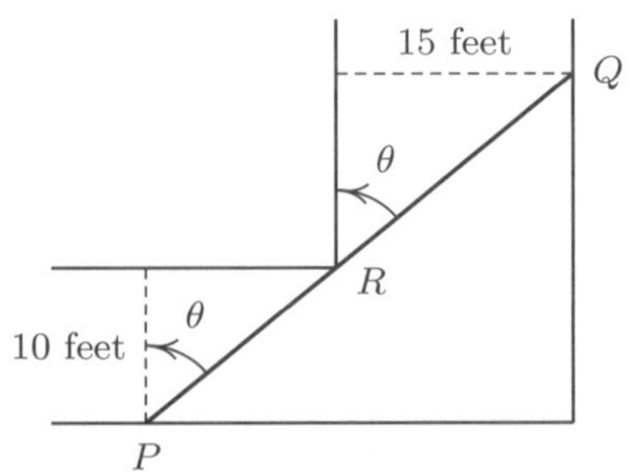

FIGURE 10

Question 10 **(a)** Give formulas for the distances $\overline{PR}$ and $\overline{RQ}$ in Figure 10 in terms of θ. **(b)** Give a formula for $L(\theta) = \overline{PQ}$ and find its one critical point θ with $0 < \theta < \frac{1}{2}\pi$.

Example 6 How wide is the widest pane of glass that could be moved around the corner in Figure 10?

SOLUTION Because L is a continuous function of θ for $0 < \theta < \frac{1}{2}\pi$ which tends to ∞ as θ tends to 0 from the right and as θ tends to $\frac{1}{2}\pi$ from the left, the minimum of L is at the critical point $\theta = \tan^{-1}(1.5) \doteq 0.852770877$ that you found in Question 10. The value of L at this point is $10\sec\theta + 15\csc\theta = \dfrac{10}{\cos\theta} + \dfrac{15}{\sin\theta} \doteq 35.12$. The widest pane of class that would fit around the corner would be approximately 35.12 feet wide. □

Principles and practice

- Most of the related-rate and maximum/minimum problems in Sections 2.5 and 3.2 dealt with functions given by graphs or tables or whose formulas were given in the statements of the problems or were fairly easy to derive from formulas for areas, volumes, and other familiar quantities. Many of the problems in this section require the use of similar triangles, the Pythagorean Theorem, or formulas relating other quantities such as velocity, demand, price, revenue, cost, and profit.
- As in Section 2.5, most of the related-rate problems in this section can be solved by considering all of the varying quantities to be functions of time t and differentiating equations relating those quantities with respect to t to obtain equations relating the quantities and their rates of change. The Chain Rule $\dfrac{dF}{dt} = \dfrac{dF}{dx}\dfrac{dx}{dt}$ can be used instead if the quantities involved are $x = x(t)$ and $F = F(x)$.
- In most of the maximum/minimum problems here, as with the problems in Section 3.2, a formula is found for the quantity to be maximized or minimized as a function of a dimension or other parameter, and the maximum or minimum value of this function in the relevant interval is determined by using its derivative.
- Many of the problems here ask for the rate of change or maximum or minimum value of a quantity that is given initially by a formula with two variables. This formula is converted to the formula for a function of one variable by the use of a constraint equation that relates those variables.
- The Tune-up Exercises in this section are not themselves related-rate or maximum/minimum problems. Instead they cover some of the techniques that are used to set up the mathematical models of the problems that follow. The emphasis is placed on setting up the models because once the necessary formulas have been obtained, the procedures required to finish the Problems generally do not differ very much from those you used in Sections 2.5 and 3.2.

Tune-Up Exercises 5.1[♦]

[A]*Answer provided.* [O]*Outline of solution provided.*

T1.[A] Figure 11 shows a right triangle with legs of lengths 6 and 4 and a smaller similar triangle with legs of lengths x and y. Give a formula for y in terms of x.

♦ Type 1, similar triangles

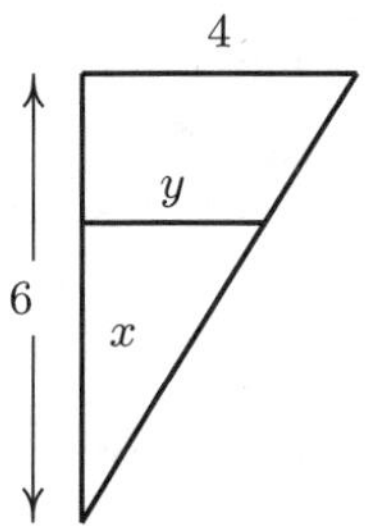

FIGURE 11

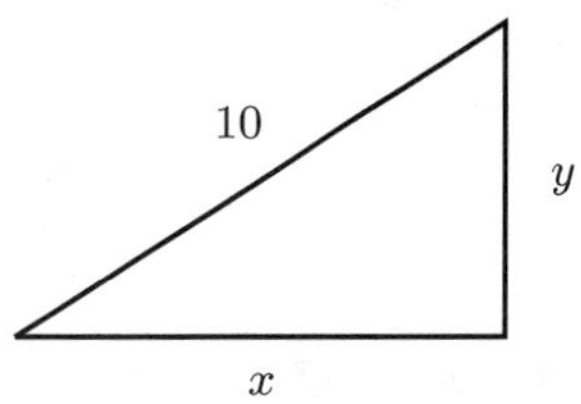

FIGURE 12

[♦]The Tune-up Exercises and Problems are classified by type and content. The types are (1) basic, reactive; (2) basic reflective; (3) intermediate, reactive; (4) intermediate, reflective; (5) advanced, reactive; (6) advanced, reflective; and (7) advanced, theoretical.

T2.[O] Express the length of one leg of the right triangle of Figure 12 in terms of the length of the other leg.

♦ Type 1, the Pythagorean Theorem

T3.[O] You can sell 50 shirts if you sell them for \$2.50 each, and for every five cents you lower the price, down to \$2.00, you can sell four more. Give formulas for the number of sales and total revenue as functions of the amount x (dollars) that you lower the price.

♦ Type 1, price, sales, and revenue

T4.[O] Figure 13 shows an isosceles right triangle with legs of length w. Give formulas for the length s of the hypotenuse and for the area A of the triangle in terms of w.

♦ Type 1, an isosceles right triangle

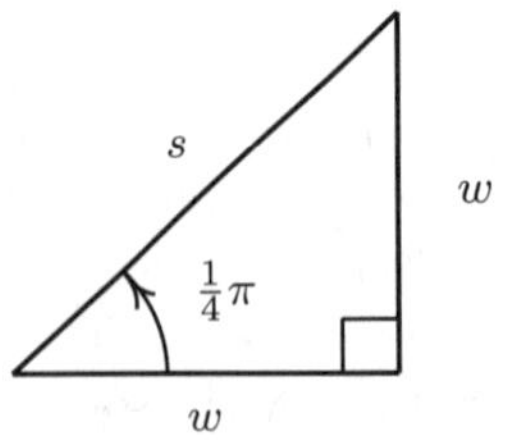

FIGURE 13

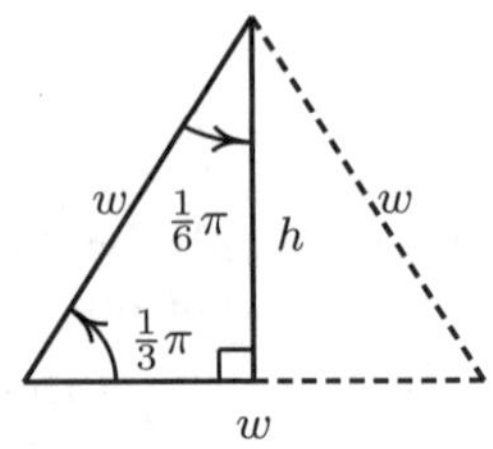

FIGURE 14

T5.[O] Figure 14 shows an equilateral triangle and the 30°-60°-90° triangle that is its left half. Give formulas for the height and areas of both triangles in terms of w.

♦ Type 1, a 30 degree-60 degree-right triangle

T6.[O] Imagine that you run at the constant speed of 8 miles per hour from point A to point B in Figure 15 and then you walk at the constant rate of 3 miles per hour from point B to point C. Express the amount of time it takes you to go from A to B in terms of the distance x in the drawing, where distances are given in miles.

♦ Type 1, the Pythagorean Theorem, distance, time, and velocity

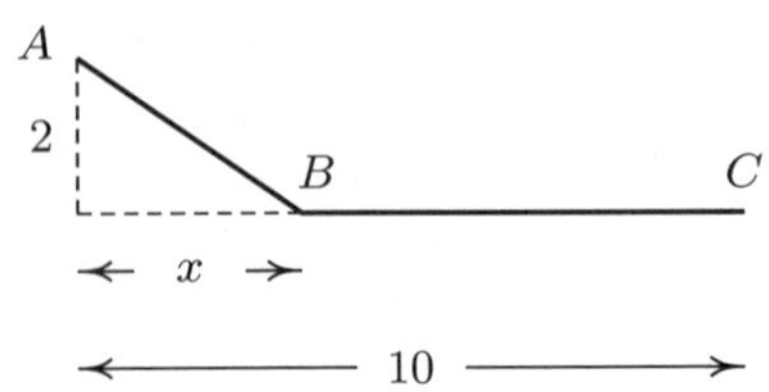

FIGURE 15

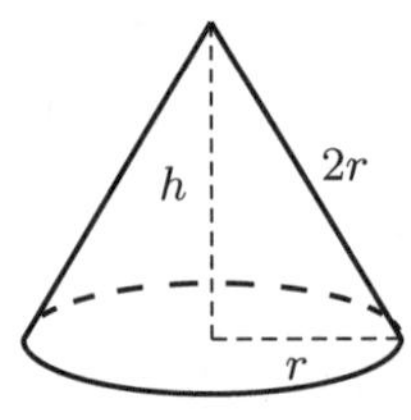

FIGURE 16

T7.[A] The right circular cone in Figure 16 has a 60° angle at its vertex. Give a formula for its volume as a function of its height.

♦ Type 1, volume of a cone

T8.[O] The large triangle in Figure 17 is a right triangle with base of length x and height 8. Use the inverse cotangent to give formulas for the angles θ and ψ in terms of x.

♦ Type 3, angles in triangles

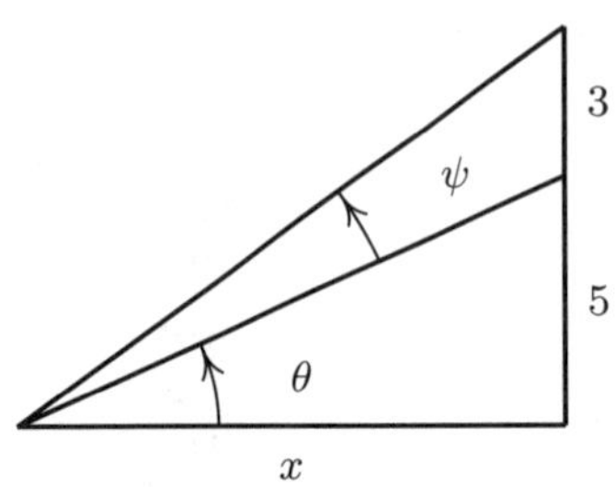

FIGURE 17

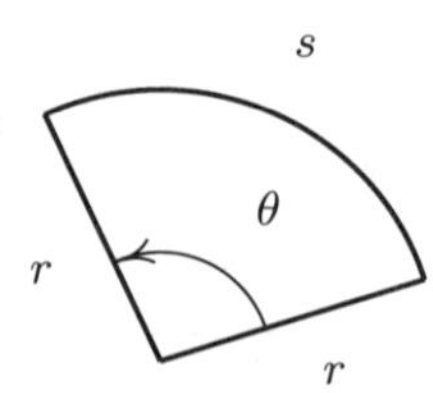

FIGURE 18

T9.[O] A SEGMENT OF A CIRCLE is the portion between two radii. **(a)** Express the length s of the curved boundary of the segment shown Figure 18 in terms of the angle θ (in radians) and the radius r of the circle. **(b)** Give a formula for the area of the segment of Figure 18 in terms of θ and r.

◆ Type 1, area and arclength

T10.[A] Express the distance from the point $(1,3)$ to a point on the parabola $y = x^2$ in terms of the x-coordinate of the point on the curve.

◆ Type 1, distance

Problems 5.1

[A]*Answer provided.* [O]*Outline of solution provided.*

1.[A] A man seven feet tall is walking toward a vertical 28-foot-high lamppost. **(a)** Give a formula for the length y of his shadow on the horizontal ground as a function of his distance x from the lamppost. **(b)** How fast is his shadow shrinking when he is 20 feet from the lamppost if he is walking toward it at the rate of 4 feet per second?

◆ Type 2, a related-rate problem

2.[O] An airplane, flying at the constant velocity of 600 feet per second in a straight path at an altitude of 3000 feet, passes directly over the pilot's house. How rapidly is the distance from the plane to the house increasing 12 seconds later?

◆ Type 2, a related-rate problem

3. An airplane flying in a straight path with constant velocity at a constant altitude of 3000 feet passes directly over a radar station. A few seconds later, the station determines that the plane is 5000 feet away and that its distance from the station is increasing at the rate of 700 feet per second. What is the plane's velocity?

◆ Type 2, a related-rate problem

4.[A] A car that is 25 miles north of an intersection at $t = 0$ hours is traveling south with the constant speed of 40 miles per hour. Another car, traveling east at the constant speed of 30 miles per hour on the cross road is at the intersection at $t = 0$. **(a)** Give a formula for the distance between the cars as a function of t. **(b)** When is the distance between the cars a minimum? What is the minimum distance?

◆ Type 2, a max/min problem

5. One bicyclist is 4 miles east of an intersection and is traveling toward the intersection at the rate of 9 miles per hour. A the same time a second bicyclist is 3 miles south of the intersection and is traveling away from the intersection at the rate of 10 miles per hour. Is the distance between the bicycles increasing or decreasing at that moment? At what rate?

◆ Type 2, a related-rate problem

6. At the end of their conversation two people walk off in different directions. One walks due north at the constant speed of 3 miles per hour and the other walks due east at the constant speed of 4 miles per hour. How fast does the distance between them increase?

◆ Type 2, a related-rate problem

7. The point P is moving on the x-axis and the point Q is moving on the y-axis in an xy-plane with distances measured in meters. When P is at $(5,0)$, it is moving two meters per second in the positive x-direction, Q is at $(0,12)$, and the distance between them is increasing ten meters per second. How fast is Q moving at that moment?

◆ Type 2, a related-rate problem

8.[A] Water boils at 100°C if the barometric pressure (air pressure) is one atmosphere—the normal air pressure on the surface of the earth. For barometric pressures p between 0.92 and 1.05 atmospheres, the boiling point of water is given with an accuracy of one hundredth of a degree by $T = 100 + 28.012(p-1) - 11.64(p-1)^2 + 7.1(p-1)^3$ degrees Celsius.[(1)] **(a)** Based on this formula, what is the average rate of change of the boiling point of water with respect to the barometric pressure from $p = 0.9$ to $p = 1.0$ atmospheres? **(b)** How fast is the boiling point of water dropping if the barometric pressure is 0.9 atmospheres and is dropping at the rate of 0.01 atmospheres per minute?

◆ Type 2, a related-rate problem

[(1)]Data adapted from *CRC Handbook of Chemistry and Physics*, Boca Raton, FL: CRC Press, Inc., 1981, p. D-160.

9.[A] The THIN LENS EQUATION $\frac{1}{x}+\frac{1}{y}=\frac{1}{f}$ relates the distances x and y from an object and its image to the lens. The constant f is called the FOCAL LENGTH of the lens. An object is moving away from a lens with a focal length of 20 centimeters at the rate of 10 centimeters per minute. How fast is its image moving when the object is 60 centimeters from the lens?

♦ Type 2, a related-rate problem

10. An object with mass m grams moving with velocity v centimeters per second has KINETIC ENERGY $E=\frac{1}{2}mv^2$ dyne-centimeters. The kinetic energy of an object of mass 100 grams is increasing at the constant rate of 750 dyne-centimeters per second. What is the object's acceleration when it is traveling 10 centimeters per second?

♦ Type 2, a related-rate problem

11 Two light sources are located 12 feet apart. One of them is eight times as bright as the other. The intensity of the light from both sources at a point on the line between them and x feet from the brightest one is given by the formula $I=8x^{-2}+(12-x)^{-2}$. At what point between the light sources is the light intensity the least?

♦ Type 4, a max/min problem

12. The STRENGTH OF THE REACTION to a particular medical drug is given by $R=kD^2-\frac{1}{3}D^3$, where D is the amount of the drug that is administered, k is a positive constant, and $0\le D\le 2k$. The derivative dR/dD is the SENSITIVITY of the drug as a function of the dosage. **(a)** For what value of D with $0\le D\le 2k$ is the strength of the reaction a maximum? **(b)** For what value of D is the sensitivity the greatest? **(c)** Draw the graph of $R(D)$ for $0\le D\le 2k$ and explain the relevance of the answer from part (b) to the graph. **(d)** Why is it appropriate to call dR/dD the "sensitivity" of the drug?

♦ Type 4, a max/min problem

13. Water is the only common liquid whose greatest density occurs at a temperature above its freezing point. This favors the survival of aquatic life by preventing ice from forming in the bottom of lakes. One kilogram of water occupies a volume of $V=1-AT+BT^2-CT^3$ liters at $T°$C for $0\le T\le 30$ where $A=6.42\times10^{-5}$, $B=8.51\times10^{-6}$, and $C=6.79\times10^{-8}$.(2) Use the Quadratic Formula to find, with two-decimal-place accuracy, the temperature at which the density of water is greatest. (Minimize the volume of one kilogram of water.)

♦ Type 4, a max/min problem

14.[A] A woman is walking away from a building toward a light that is shining up from the ground. The light is 20 feet from the building and she is six feet tall. At what rate is the height of her shadow on the building increasing when she is 8 feet from the light if she walks 5 feet per second?

♦ Type 2, a related-rate problem

15.[A] A swimming pool with triangular vertical cross sections, as shown in Figure 19, is 45 feet long, 10 feet wide, and 15 feet deep. **(a)** Give a formula for the length s of the rectangular surface of the water as a function of the depth h (feet) of the water. **(b)** Give a formula for the volume of the water in the pool as a function of its depth. **(c)** When the water is six feet deep, water is being added at the rate of 180 cubic feet per hour. How fast is the depth of the water increasing at that moment?

♦ Type 4, a related-rate problem

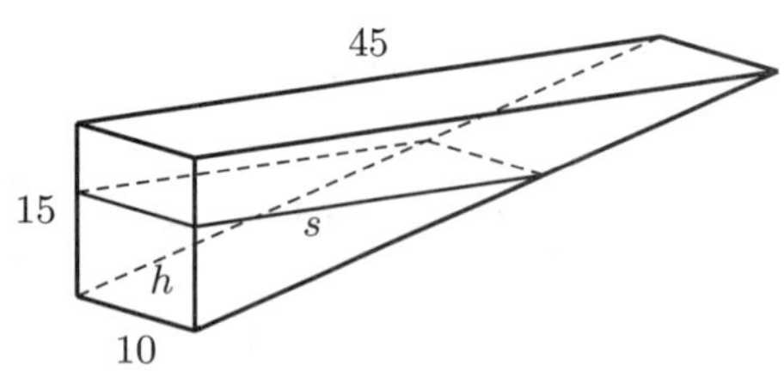

FIGURE 19

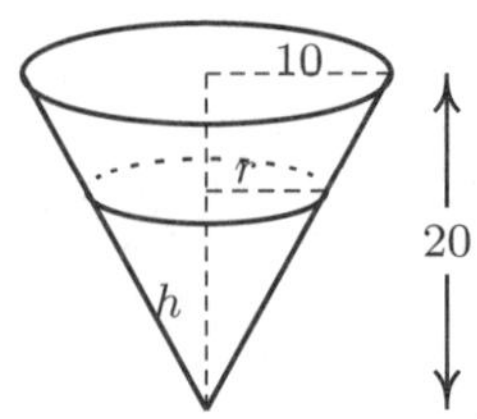

FIGURE 20

(2)Data adapted from *Handbook of Chemistry and Physics*, Cleveland OH: Chemical Rubber Publishing Company, 1960, p. 2248.

16. Water is flowing into a tank that has the shape of a right circular cone with its vertex pointed down (Figure 20). The radius of its top is 10 feet and it is 20 feet high. At what rate is water being added when it is 5 feet deep if, at that moment, the water level is rising at the rate of 4 feet per hour?

♦ Type 4, a related-rate problem

17.[A] A conical tank with its vertex at the bottom has vertical cross sections that are equilateral triangles. At what rate is the volume of water in the tank increasing when the water is 6 feet deep if the depth of the water is increasing 2 feet per hour?

♦ Type 4, a related-rate problem

18.[A] Coffee mugs can be manufactured for 60 cents each. At a price of one dollar each, 1000 can be sold, and for each penny the price is lowered, 50 more mugs will be sold. **(a)** What would be the revenue and profit from the mugs if they were sold at one dollar each? **(b)** What would be the revenue and profit from the mugs if the price were lowered two cents? **(c)** How much should the price be lowered to maximize the profit?

♦ Type 4, a max/min problem

19. A store can sell 275 cameras if it sells them for 75 dollars each. For every dollar the price is raised, three sales are lost. What price (whole dollars) should be charged to maximize revenue?

♦ Type 4, a max/min problem

20. A train has an average of 1200 passengers per day when its fare is \$2.00 per person. Suppose that for every ten cents that the fare is decreased, down to \$1.50 per person, the train will average 100 more passengers each day. What fare should be charged to maximize the average daily revenue?

♦ Type 4, a max/min problem

21. A store can sell 300 radios if it sells them for \$120 each. For every dollar the price is raised, two sales are lost. What is the marginal revenue with respect to price (the derivative of the revenue with respect to price) at a price of \$130?

♦ Type 4, marginal revenue

22.[A] It costs a company $x^3 + 100x + 1500$ dollars to produce a batch of x gallons of a chemical that sells for 400 dollars per gallon. **(a)** What size batch should it produce to maximize the profit on each batch? **(b)** What is the maximum profit?

♦ Type 2, a max/min problem

23. A company uses 50 units of an item each day. It costs 20 dollars each time an order is placed for the item and costs 40 cents to store a unit of the item for any part of one day. **(a)** Show that if they order every x days and have no remaining stock when each order arrives, then they have, on average, $25(x+1)$ units to store each day. **(b)** How frequently should they order to minimize the total annual costs of ordering and storage? (Give an integer answer.)

♦ Type 4, a max/min problem

24.[O] Find the point on the parabola $y = x^2$ that is closest to the point $(4, \frac{1}{2})$.

♦ Type 4, a max/min problem

25. Find the point on the curve $y = \sqrt{x}$ that is closest to the point $(1, 0)$.

♦ Type 4, a max/min problem

26. **(a)** Use calculus to find the point Q on the line $y = 4x + 3$ that is closest to the point $P(2, -6)$. **(b)** Show that Q is the intersection of the line $y = 4x + 3$ and the perpendicular line through P.

♦ Type 4, a max/min problem

27. At a certain moment the radius of the base of a right circular cone is 4 inches and is increasing at the rate of 5 inches per second, while the height of the cone is 3 inches and is decreasing at the rate of 6 inches per second. Is the volume of the cone increasing or decreasing at that moment? At what rate?

♦ Type 2, a related-rate problem

28. A farmer has 40 tons of cattle that he could sell for a profit of 200 dollars per ton. For every week he waits to sell the cattle, their total weight will increase by one ton, but his profit will fall by four dollars per ton. **(a**[A]**)** What will be his profit if he sells in two weeks? **(b)** When should he sell to maximize his profit?

♦ Type 4, a max/min problem, etc.

29.[A] A store will sell $\dfrac{7500}{50+p^2}$ T-shirts if they charge p dollars per shirt $(5 \le p \le 15)$. What is the marginal revenue with respect to price from the T-shirts **(a)** at $p = 5$ and **(b)** at $p = 15$? **(c)** What price would maximize the revenue?

♦ Type 4, a max/min problem, etc.

30. If $1000e^{-p/10}$ units of a product will sell at p dollars per unit, what price would maximize the revenue from the product?

♦ Type 4, a max/min problem, etc.

31. The demand for an item is $\dfrac{a}{b+cx^3}$ units if it is sold for x dollars per unit with positive constants a, b, and c. What price would maximize the revenue from the sales of the item?

♦ Type 4, a max/min problem, etc.

32.[A] If the marginal revenue in producing x items is $50 - \frac{1}{80}x^2$ dollars per unit and the marginal cost is $10 + \frac{1}{80}x^2$ dollars per unit, how many units should be produced to maximize the profit?

♦ Type 4, a max/min problem, etc.

33. Show that if profit is calculated as revenue minus cost and all three are differentiable functions of a variable x, then the marginal revenue with respect to x equals the marginal cost with respect to x at any x where the profit has a local maximum or minimum.

♦ Type 4, a max/min problem, etc.

34. Suppose that the profit on x units of an item is $P(x)$ dollars. Show that if the average profit per unit $\dfrac{P(x)}{x}$ is a local maximum or a local minimum at x_0, then it equals the marginal profit per unit $\dfrac{dP}{dx}$ at x_0 and the tangent line to the graph of $P(x)$ at x_0 passes through the origin.

♦ Type 4, a max/min problem, etc.

35.[O] Karen has landed her glider in a field one mile from a straight road and walked to a general store three miles down the road from the point on the road that is closest to the plane. Now, she wants to get back to the glider as soon as possible by walking back down the road and cutting across the field. She can walk three miles per hour on the road and two miles per hour in the field. How far should she walk on the road before cutting across the field?

♦ Type 4, a max/min problem, etc.

36. A powerhouse is on one side of a 200-foot-wide straight river, and a factory is across the river and 300 feet downstream. An electric cable is to be strung from the powerhouse to the factory. It costs 5 dollars for each foot of cable in the water and 3 dollars for each foot of cable on land. No more than 100 feet of cable can be run on land by the powerhouse and none can be run on land by the factory. What path would minimize the cost of the cable?

♦ Type 4, a max/min problem, etc.

37. A fisherman in a rowboat on a lake is 2 miles from the shore when a storm hits. His cabin is 3 miles down the (straight) shore. He can row 5 miles per hour and run 13 miles per hour. To what point on the shore should he row to get to his cabin as quickly as possible?

♦ Type 4, a max/min problem, etc.

38 A rope is to run from a point 20 feet from a fence, to the fence, and then to a point 30 feet from the fence as in Figure 21. To what point on the fence should the rope be run to minimize the amount of rope that is used?

♦ Type 4, a max/min problem, etc.

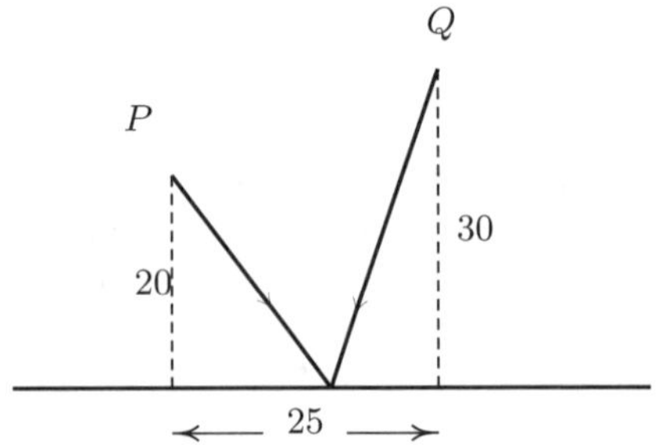

FIGURE 21

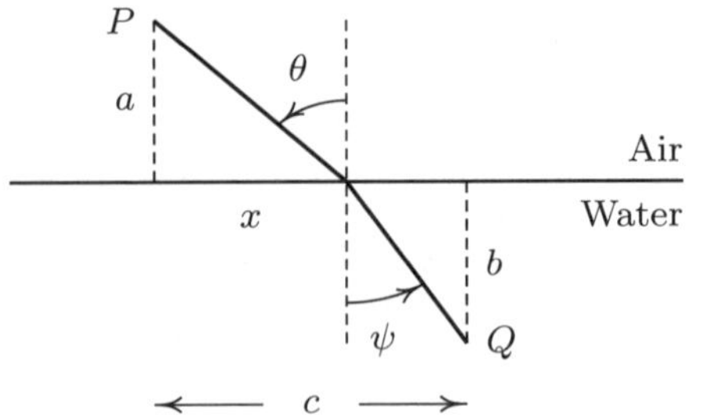

FIGURE 22

39. According to FERMAT'S PRINCIPLE in optics, a ray of light that goes from a point P above the water to a point Q below it, as in Figure 22, will take a path that requires the least time. In air and in water, where the speed of light is constant, the light goes in straight lines, but the path bends at the surface of the water, as shown in the drawing, because light travels more slowly in water. Suppose that the speed of light in air is v_A and its speed in water is v_W, with both measured in meters per second. **(a)** Express the amount of time T it would take light to go from point P to point Q along the path in Figure 22 in terms of the distance x (meters) and the distances a, b, and c. **(b)** Show that the time T is minimized if the angles θ and ψ in Figure 22 satisfy SNELL'S LAW, which states that $\dfrac{\sin\theta}{\sin\psi} = \dfrac{v_A}{v_W}$. **(c)** The ratio $\dfrac{v_A}{v_W}$ is approximately 1.33. Suppose that light takes the path in Figure 22 and that $\theta = \frac{1}{4}\pi$. Give the value of ψ with two-decimal-place accuracy.

♦ Type 4, a max/min problem, etc.

40.[A] It costs 96 dollars per hour plus $\frac{1}{2} + \frac{1}{60}v^{3/2}$ dollars per mile to run a boat v miles per hour. **(a)** What is the cost per mile at v miles per hour? **(b)** At what speed is the cost per mile a minimum?

♦ Type 4, a max/min problem, etc.

41.[A] Two identical lamps are at the tops of 3-foot and 6-foot vertical posts that are 12 feet apart. The intensity of the light from each lamp is Kr^{-2} foot-candles at a distance of r feet. A cat runs on the ground from the 3-foot post to the 6-foot post with constant velocity of 5 feet per second. What is the rate of change with respect to time of the total intensity of light from both lamps on the cat when it has gone 4 feet?

♦ Type 4, a related-rate problem

42. Rectangular beams are to be cut from logs with circular cross sections with 12 inch diameters. The STRENGTH of the beam is equal to a constant times h^2w, where h and w are the height and width of its cross section. What are the height and width of the beams with maximum strength?

♦ Type 4, a max/min problem

43.[A] Sand is falling in a conical pile at the rate of 12 cubic feet per minute. The vertical cross section of the pile is always an isoceles right triangle. **(a)** How fast is the height of the pile increasing when the height is 4 feet? **(b)** In what ways is this mathematical model of a pile of sand not very accurate?

♦ Type 4, a related-rate problem

44. The ends of a trough are equilateral triangles with one vertex pointing down. The trough is 9 feet long. When the water in the trough is 2 feet deep, its depth is increasing at the rate of $\frac{1}{2}$ foot per minute. At what rate is water flowing into the trough at that moment?

♦ Type 4, a related-rate problem

45.[A] A light is at the top of a 48-foot pole. A ball is dropped from the same height from a point 10 feet away. After one second, the ball has fallen 16 feet and is falling at the rate of 32 feet per second. How fast is the shadow of the ball moving on the ground at that moment?

♦ Type 4, a related-rate problem

46.[A] A window is to be made in the shape of a rectangle with a semicircle over it. What dimensions should it have to maximize its area if its perimeter is to be 7 feet?

♦ Type 4, a max/min problem

47. As a 20-foot-long ladder is resting on the top of a 2-foot-high fence, its bottom is being pulled along the ground away from the fence at the rate of 0.5 feet per second. At what rate is the top of the ladder falling when the bottom is 10 feet from the fence?

♦ Type 4, a related-rate problem

48. Show that the least possible volume of a right circular cone circumscribed about a sphere is twice that of the sphere.

♦ Type 4, a max/min problem

49. A fish uses kv^3 ergs per second of energy to swim at a speed of v centimeters per second relative to the speed of the stream. Show that in order to minimize his total energy used in swimming upstream a given distance, the fish should swim at a speed of $v = \frac{3}{2}c$, relative to the water, where c is the speed of the stream.

♦ Type 4, a max/min problem

50. A right circular cone with height h and base of radius r has volume $V = \frac{1}{3}\pi r^2 h$ and lateral surface area $\pi r\sqrt{r^2 + h^2}$. Of all such cones with volume $\frac{2}{3}\pi$, what are the dimensions of the one with minimum lateral surface area?

♦ Type 4, a max/min problem

51. Find the dimensions of the right circular cylinder of greatest volume that can be inscribed in a hemisphere with a three meter radius.

◆ Type 4, a max/min problem

52.[A] Find the dimensions of the right circular cylinder of greatest volume that can be inscribed in a right circular cone of height 24 inches and with a base of radius 10 inches.

◆ Type 4, a max/min problem

53. Show that the distance from the point (x_0, y_0) to the line $Ax + By + C = 0$ is $\dfrac{|Ax_0 + By_0 + C|}{\sqrt{A^2 + B^2}}$. (Write the equation of the line in the form $v = Cu + E$ with $u = x - x_0$ and $v = y - y_0$.)

◆ Type 4, a max/min problem

54. Prove that if a function f has a derivative for all x, then the shortest distance to the graph of f from a point (x_0, y_0) not on the graph is along a line perpendicular to the graph.

◆ Type 4, a max/min problem

55. Use the result of Problem 54 to find the point on the line $y = 2x - 11$ that is closest to the parabola $y = x^2$.

◆ Type 4, a max/min problem

56. **(a)** Find the points P and Q on the line $y = x + \dfrac{\sqrt{2}}{x}$ that is closest to the origin. **(b)** Show that the tangent lines to $y = x + \dfrac{\sqrt{2}}{x}$ at the points P and Q from part (a) are perpendicular to the line through the origin and the corresponding point P or Q.

◆ Type 4, a max/min problem

57. In sixteenth and seventeenth century Austria, wine barrels were measured by how far a bung stick could be inserted through the bung hole in the middle of the side of the barrel (Figure 23). What shaped cylindrical wine barrel with a volume of 8π has the smallest bung stick measurement? (This is the approximate shape of Austrian wine barrels from that time.)

◆ Type 4, a max/min problem

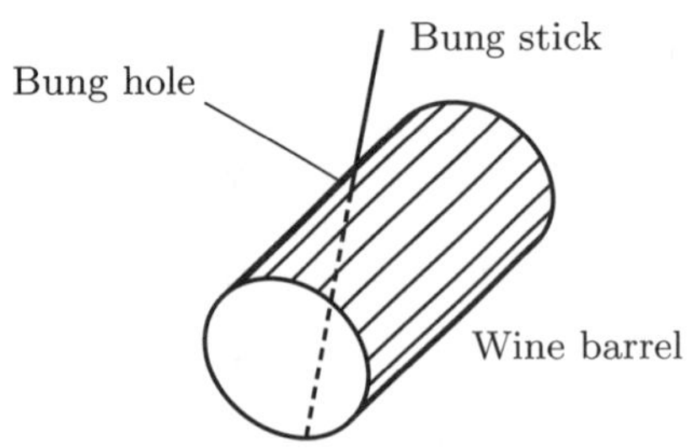

FIGURE 23

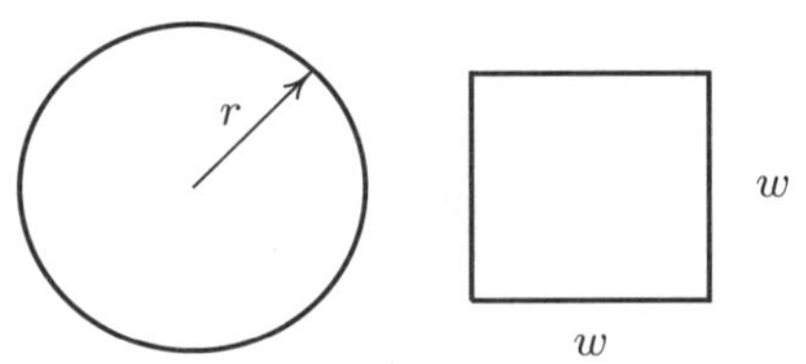

FIGURE 24

58. Your boss at Circle in the Square Pizza has decided to celebrate his sixty-fourth birthday with a promotion, during which he will sell one square and one round pizza for \$9.98, with the total perimeter of the two pizzas equal to 64 inches (Figure 24). **(a**[A]**)** Give a formula for the total area of the two pizzas as a function of the width of the square one. What is the domain of this function? **(b)** Your boss wants to maximize the profit on each order by minimizing the total area of the two pizzas. Show that the minimum area is achieved by having the diameter of the circle equal the width of the square. (This involves some relatively difficult algebra.) **(c)** What sized pizzas would maximize their total area?

◆ Type 4, a max/min problem

59. From theoretical considerations a researcher expects pairs (x, y) of data from her experiment to be on a line $y = mx$. She wants to determine the constant m. In four runs of the experiment she obtains the data $x_1 = 1, y_1 = 1.1; x_2 = 2, y_2 = 2.3; x_3 = 3, y_3 = 3.8$; and $x_4 = 4, y_4 = 5.0$. Compute the value of m that minimizes the quantity $E(m) = (y_1 - mx_1)^2 + (y_2 - mx_2)^2 + (y_3 - mx_3)^2 + (y_4 - mx_4)^2$. ($E(m)$ is the sum of the squares of the vertical distances from the points to the line. Consequently, this technique of choosing the constant m is called the METHOD OF LEAST SQUARES.)

◆ Type 4, a max/min problem

60. (**a**[A]) Of all pie slices of perimeter 10 inches (other than a whole pie), which has the maximum area? (**b**) How much less is the area of the pie slice from part (a) than the area of a whole pie of circumference 10 inches?

♦ Type 4, a max/min problem

61.[A] A cone is to be formed from a circular piece of paper with a five-inch radius by cutting it along two radii, as in Figure 25 and joining its edges. What is the maximum volume of a cone that can be formed in this way?

♦ Type 4, a max/min problem

62. A billboard on a building is 10 feet high and its bottom is 23 feet above the ground. A man whose eyes are five feet above the ground is walking toward the building (Figure 26). How far is he from the building when the billboard subtends the greatest vertical angle at his eyes?

♦ Type 4, a max/min problem

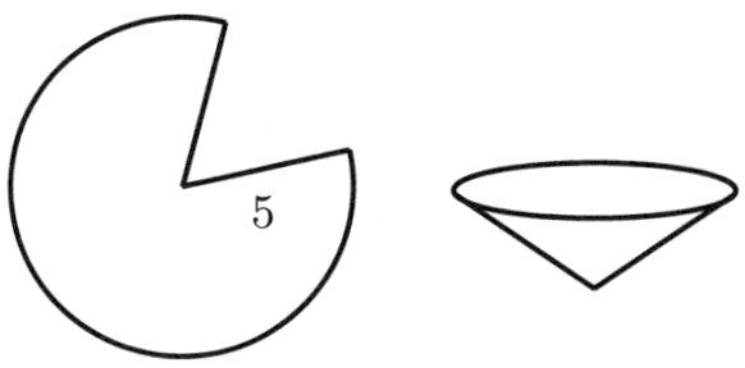

FIGURE 25

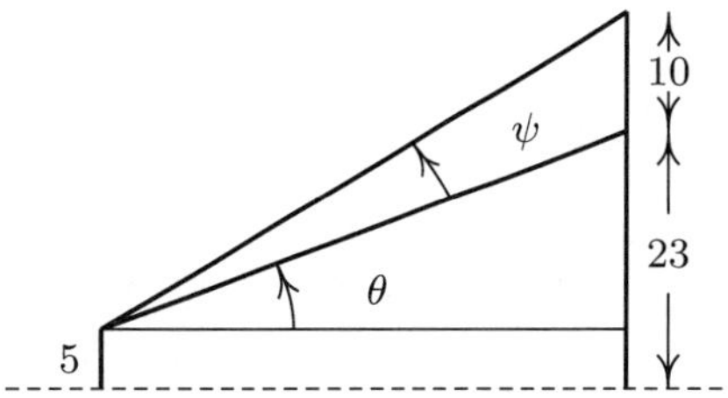

FIGURE 26

63. What is the length of the shortest ladder that will reach from the ground, over a six-foot-high fence, and touch a building that is five feet behind the fence? Give (**a**) an approximate decimal answer and (**b**) an exact expression for the length.

♦ Type 4, a max/min problem

64. The minute hand on a clock is 11 inches long, and the hour hand is 10 inches long. How rapidly is the distance between the tips of the hands increasing at 8:00 AM?

♦ Type 4, a related-rate problem

65. A gutter is to be made from a long 18-inch-wide strip of galvanized steel by bending it in thirds and folding up the sides. (Figure 27 shows a cross section.) At what angle should the metal be bent to maximize the volume of the gutter?

♦ Type 4, a related-rate problem

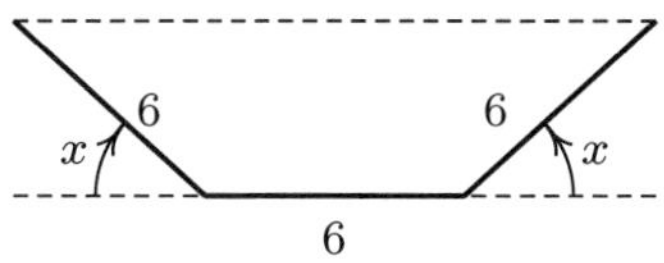

FIGURE 27

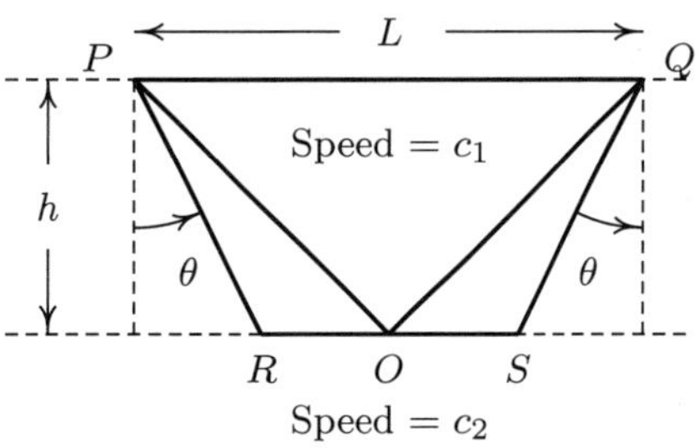

FIGURE 28

66. The speeds of sound c_1 and c_2 (kilometers per second) in an upper and a lower horizontal layer of rock and the thickness h (kilometers) of the upper layer can often be determined by seismic exploration techniques if the speed of sound is significantly greater in the lower layer (Figure 28). A dynamite charge is set off at point P and the transmitted signals are recorded at a second point Q on the surface of the earth. Typically, the first signal to arrive travels in a straight line along the surface from P to Q. Suppose it takes T_1 seconds. The second signal to arrive goes on straight paths from P to R at the top of the lower layer, along the top of the lower layer to S and then to Q. Suppose it takes T_2 seconds. The third signal goes from P to O, where it reflects off the lower layer, and then to Q. Suppose it takes T_3 seconds. **(a**[A]**)** Give formulas for T_1, T_2, and T_3 in terms of L, h, c_1, c_2 and the angle θ in Figure 28. **(b)** Show by minimizing T_2 that $\sin\theta = c_1/c_2$. **(c)** Suppose that $L = \frac{1}{4}$ kilometers, $T_1 = \frac{1}{8}$ second, $T_2 = \frac{1}{6}$ second, and $T_3 = \frac{3}{40}\sqrt{5}$ second. Find c_1 and then h. If you have access to a calculator or computer algebra system, use it to find two possible values of c_2.

♦ Type 4, a related-rate problem

67. The top of a 10-foot-long ladder is leaning against a vertical wall and its base is being pulled away from the wall along the ground at the constant rate of 0.2 feet per second. **(a)** Suppose we assume that the ladder, wall, and ground form a right triangle. Show that the rate of change of the height of the top of the ladder is then $\dfrac{dh}{dt} = \dfrac{-0.2x}{\sqrt{100 - x^2}}$. **(b)** According to the formula in part (a), $\dfrac{dh}{dt} \to \infty$ as $x \to 10^-$, so the top of the ladder hits the ground with infinite velocity. What is wrong with the mathematical model in this case? What actually happens?

♦ Type 4, a related-rate problem

Section 5.2

Implicit differentiation and derivatives of inverse functions

OVERVIEW: *Up to now, we have only considered functions $y(x)$ whose exact values are given by formulas in terms of the variable x or whose approximate value at each x can be determined directly from graphs or tables. Such functions are said to be* DEFINED EXPLICITLY. *A function $y(x)$ is said to be* DEFINED IMPLICITLY *if it is defined by an equation involving x and y that has to be solved to obtain values of the function. In this section we discuss a procedure called* IMPLICIT DIFFERENTIATION *for finding derivatives of implicitly defined functions. The calculations required to carry out this procedure are similar to those used to solve related-rate problems in Sections 2.5 and 5.1.*

Topics:

- ***Derivatives of implicitly defined functions***
- ***Tangent lines to curves that are not graphs of functions***
- ***Derivatives of inverse functions***

Derivatives of implicitly defined functions

The equation $y^3 - x = 0$ defines $y(x)$ IMPLICITLY as a function of x because we have to solve the equation to determine values of the function. To solve the equation, we write first $y^3 = x$ and then $y = x^{1/3}$. This is an EXPLICIT formula for the function.

In many cases, one equation in x and y defines more than one function $y(x)$. For example, if we solve $x^2 + y^2 = 1$ for y, we write first $y^2 = 1 - x^2$ and then $y = \pm\sqrt{1-x^2}$. Thus, the equation $x^2 + y^2 = 1$ defines implicitly two functions with the explicit formulas $y_1(x) = \sqrt{1-x^2}$ and $y_2(x) = -\sqrt{1-x^2}$. The graph of $x^2 + y^2 = 1$ is the circle in Figure 1, and the graphs of $y_1(x)$ and $y_2(x)$ are the upper and lower semicircles in Figures 2 and 3.

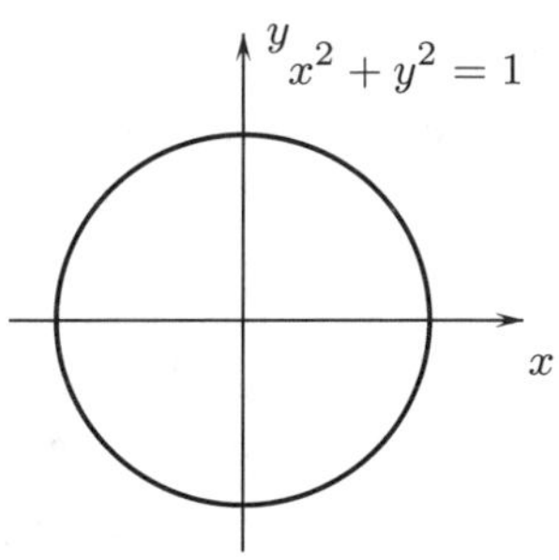

FIGURE 1

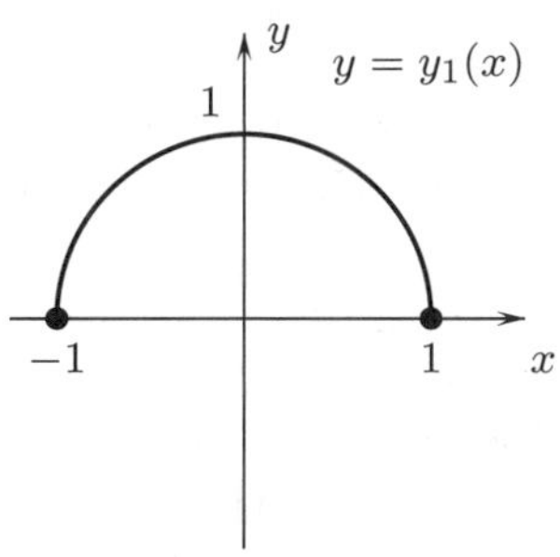

FIGURE 2

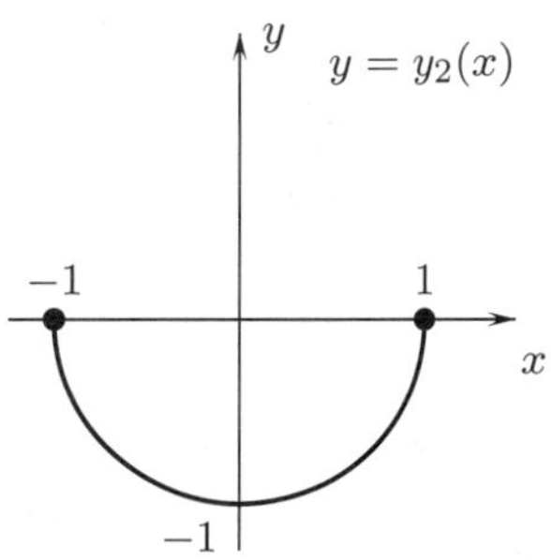

FIGURE 3

If an equation defines more than one function, as in the case of $x^2 + y^2 = 1$, we can often specify one of the functions by giving its value at one value of x. For example, the function $y_1(x) = \sqrt{1-x^2}$ of Figure 2 can be distinguished from the function $y_2(x) = -\sqrt{1-x^2}$ of Figure 3 by requiring that $y_1(0) = 1$.

Differentiation of implicitly defined functions

If we cannot solve the equation that defines a function $y(x)$ implicitly to obtain an explicit formula for the function, we can generally find its derivative at any point x_0 where we know its value $y(x_0)$. We take the x-derivative of both sides of the equation that the function satisfies, evaluate the result at x_0 and solve for $y'(x_0)$. This process is called IMPLICIT DIFFERENTIATION.

♦ SUGGESTIONS TO INSTRUCTORS: Have students begin work on Questions 1 through 4 before class. Problems 19, 21, 22, 25, 30, 34, and 38 are good for in-class discussion.

Example 1 The function $y(x)$, defined implicitly by

$$x^3 + y^3 = 9xy \tag{1}$$

and the condition

$$y(4) = 2 \tag{2}$$

has a derivative at $x = 4$. What is its value?

SOLUTION We differentiate both sides of equation **(1)** with respect to x, treating y as a function of x and using, in this case, the Product and Chain Rules. This yields

$$\frac{d}{dx}(x^3) + \frac{d}{dx}(y^3) = 9y\frac{dx}{dx} + 9x\frac{dy}{dx}$$

and then

$$3x^2 + 3y^2\frac{dy}{dx} = 9y + 9x\frac{dy}{dx}.$$

We factor 3 from both sides and divide both sides by 3 to obtain

$$x^2 + y^2\frac{dy}{dx} = 3y + 3x\frac{dy}{dx}. \tag{3}$$

To satisfy **(2)**, we set $x = 4$ and $y = 2$ to have

$$4^2 + 2^2\frac{dy}{dx} = 3(2) + 3(4)\frac{dy}{dx}. \tag{4}$$

Here, as with any implicit differentiation, it is easy to solve for dy/dx because it appears as a single factor in the terms containing it. Equation **(4)** simplifies to $8\frac{dy}{dx} = 10$, so that $\frac{dy}{dx}(4) = \frac{10}{8} = \frac{5}{4}$. □

What is the geometric meaning of Example 4? The graph of $x^3 + y^3 = 9xy$ is the FOLIUM OF DESCARTES† shown in Figure 4. In order for the function $y(x)$ to be defined implicitly by this equation, its graph must be part of the curve, and the requirement that $y(4) = 2$ means that the point $(4, 2)$ must be on the graph of $y(x)$.

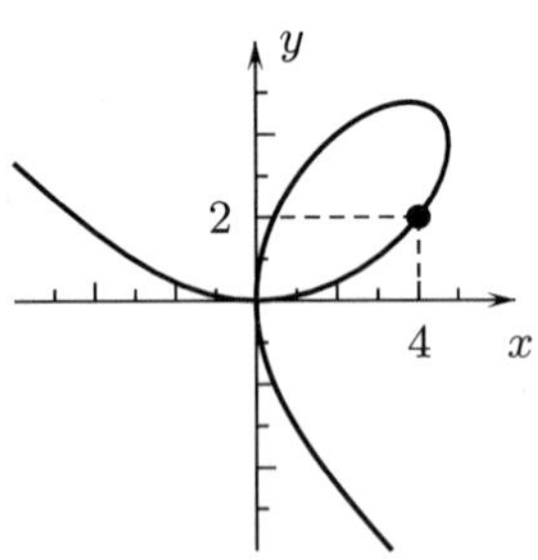

$x^3 + y^3 = 9xy$

FIGURE 4

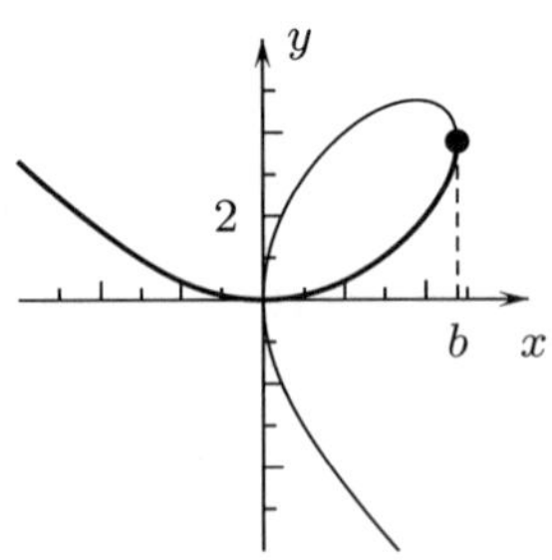

$y = y(x)$

FIGURE 5

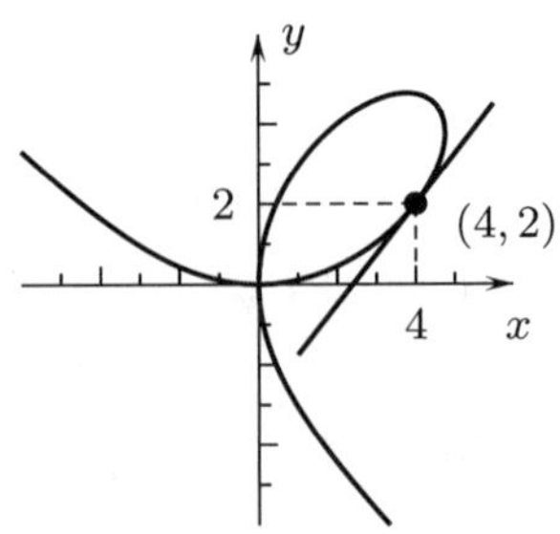

$x^3 + y^3 = 9xy$

FIGURE 6

†Discussed by René Descartes in 1638. *A Catalog of Special Plane Curves* by J. Lawrence, New York: Dover Publications, 1972, p. 106.

The curve drawn with a heavier line in Figure 5 is the graph of the continuous function $y(x)$, defined implicitly by $x^3 + y^3 = 9xy$ with the condition $y(4) = 2$ that has the most extensive possible interval domain. Its domain is a closed infinite interval $(-\infty, b]$ for a positive constant b between 4 and 5, as indicated in Figure 5. (In fact, $b = 3(2^{2/3})$, but we do not need this fact.)

Tangent lines to curves

Figure 6 shows the tangent line to the curve $x^3 + y^3 = 9xy$ at $(4, 2)$. This line is also the tangent line at $(4, 2)$ to the graph of the implicitly defined function $y(x)$ that we studied in Example 1, so its slope is equal to the derivative $\frac{dy}{dx}(4) = \frac{5}{4}$ that we calculated in that example.

Example 2 Give an equation for the tangent line to the curve $x^3 + y^3 = 9xy$ in Figure 6.

SOLUTION The tangent line has slope $\frac{5}{4}$ and passes through $(4, 2)$, so it has the equation $y = 2 + \frac{5}{4}(x - 4)$. □

Question 1 Find an equation of the tangent line to $x^3 + y^3 = 9xy$ at $(2, 4)$. Then draw the curve and the tangent line.

Giving derivatives in terms of x *and* y

If we do not know the value of the implicitly defined function $y(x)$ we can express its derivative in terms of x and the value of y of the function.

Example 3 A function $y(x)$ is defined implicitly by the equation $x^4 - e^y = 15 + y^2$. Give a formula for $\frac{dy}{dx}$ in terms of x and $y = y(x)$.

SOLUTION We differentiate both sides of $x^4 - e^y = 15 + y^2$ with respect to x, viewing y as a function of x and using the Chain Rule: We write

$$\frac{d}{dx}(x^4 - e^y) = \frac{d}{dx}(15 + y^2)$$

and obtain

$$4x^3 - e^y \frac{dy}{dx} = 2y \frac{dy}{dx}.$$

Collecting the terms with the derivative gives $(e^y + 2y) \frac{dy}{dx} = 4x^3$, so that $\frac{dy}{dx} = \frac{4x^3}{2y + e^y}$. □

Question 2 A function $y(x)$ is defined implicitly by the equation $x^4 - e^y = 15 + y^2$ from Example 3 and the condition $y(2) = 0$. What is $y'(2)$?

Inverse functions

We can relate the derivatives of functions and their inverses by implicit differentiation. We assume that the function $y = Y(x)$ has a positive or negative derivative in an open interval. Then, as is proved in courses on advanced calculus, $y = Y(x)$ has a differentiable inverse $x = X(y)$, defined in an open y-interval, which we can consider to be defined implicitly by the equation

$$Y\big(X(y)\big) = y.$$

We differentiate both sides of this equation with respect to y, using the Chain Rule, to obtain $\frac{dY}{dx}\big(X(y)\big)\frac{dX}{dy}(y) = 1$, so that

$$\frac{dX}{dy}(y) = \frac{1}{\frac{dY}{dx}\big(X(y)\big)}. \tag{5}$$

Formula in **(5)** is easiest to understand with a geometric explanation. Suppose that the graph of $y = Y(x)$ is the curve in the xy-plane with equal scales on the coordinate axes of Figure 7. Then the graph of $x = X(y)$ in the yx-plane of Figure 8 is obtained by flipping Figure 7 about the line $y = x$ to interchange the x- and y-axes. We suppose that $Y(x_0) = y_0$, as in Figure 7, so that $X(y_0) = x_0$.

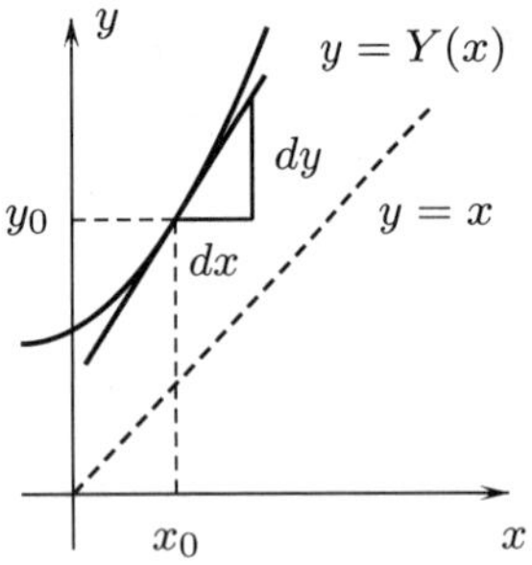

FIGURE 7

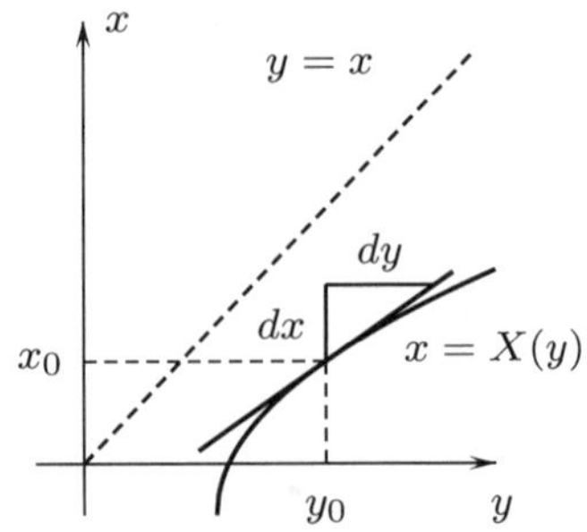

FIGURE 8

The nonzero run dx and the corresponding rise dy along the tangent line at $x = x_0$ to the graph of $y = Y(x)$ in Figure 7 transform into a rise dx and run dy along the tangent line at $y = y_0$ to the graph of $x = X(y)$ in Figure 8. Since $\dfrac{dy}{dx} = \dfrac{dY}{dx}(x_0)$ and $\dfrac{dx}{dy} = \dfrac{dX}{dy}(y_0)$, we have

$$\frac{dX}{dy}(y_0) = \frac{dx}{dy} = \frac{1}{\dfrac{dy}{dx}} = \frac{1}{\dfrac{dY}{dx}(x_0)} \tag{6}$$

which is **(5)** for $x = x_0, y = y_0$.

We have used capital letters for the functions in these calculations to help distinguish them from the variables. If we simplify the notation by using "x" and "y" as the names of the functions and deleting the references to where the functions are evaluated, then **(6)** reads

$$\frac{dx}{dy} = \frac{1}{\dfrac{dy}{dx}}. \tag{7}$$

This form is easy to remember and shows another advantage of Leibniz notation.

Equation **(5)** can also be written with prime notation as

$$X'(y) = \frac{1}{Y'(X(y))}$$

where the prime on X denotes a y-derivative and the prime on Y denotes an x-derivative.

Question 3 Equation **(7)** looks just like the innermost equation in **(6)**, but they are not the same because the symbols have different meanings in the two equations. Explain.

Example 4 What is $\dfrac{dX}{dy}(10)$ if $X(y)$ is the inverse of $Y(x)$, $Y(2) = 10$, and $\dfrac{dY}{dx}(2) = \frac{1}{3}$?

SOLUTION Since $X(10) = 2$, **(6)** with $x_0 = 2$ and $y_0 = 10$ gives

$$\frac{dX}{dy}(10) = \frac{1}{\dfrac{dY}{dx}(2)} = \frac{1}{1/3} = 3. \ \square$$

Question 4 Suppose that your truck uses $G(s)$ gallons of gasoline to go s miles on a trip and that when you have gone 50 miles it has used 3 gallons of gasoline and is consuming gasoline at the rate of $\frac{1}{15}$ gallons per mile. Let $s(G)$ denote the inverse of $G(s)$. Give the values of $G(50), s(3), \frac{dG}{ds}(50)$, and $\frac{ds}{dG}(3)$ with units.

Principles and procedures

- Implicit differentiation is not a new kind of differentiation. It is just a procedure for finding derivatives of functions that are defined implicitly. The techniques it requires are similar to those used in solving related-rate problems. You are given an equation that is satisfied by an unknown function $y(x)$, and you differentiate both sides of the equation with respect to x, viewing y as a function of x, to obtain an equation relating the values of the function and its derivative.
- If you know the value of the implicitly defined function at a particular value of x, then implicit differentiation gives the value of the derivative there. Otherwise, implicit differentiation gives a formula for the derivative in terms of x and the value y of the function.
- Implicit differentiation can often be used to find the slope of the tangent line at a point on a curve that is given by an equation in x and y if both coordinates (x_0, y_0) of the point are known. We consider a portion of the curve including (x_0, y_0) to be the graph of an implicitly defined function and find its derivative at x_0, which is the slope of the tangent line.
- The relationship between derivatives of a function $y = Y(x)$ and its inverse $x = X(y)$ is easiest to understand as $\frac{dx}{dy} = \frac{1}{dy/dx}$. To show where the derivatives are evaluated, you can use the more complete formulation $\frac{dX}{dy}(y) = \frac{1}{\frac{dY}{dx}(X(y))}$ of the same result.

Tune-Up Exercises 5.2[♦]

[A]*Answer provided.* [O]*Outline of solution provided.*

T1.[A] A function $y(x)$ is defined implicitly by the curve in Figure 9 and the condition $y(4) = 5$. The domain of $y(x)$ is the most extensive possible interval in which it is continuous. **(a)** What is its domain? **(b)** Trace its graph from Figure 9. **(c)** What is its approximate derivative at $x = 4$?

♦ Type 1, implicit derivative from a graph

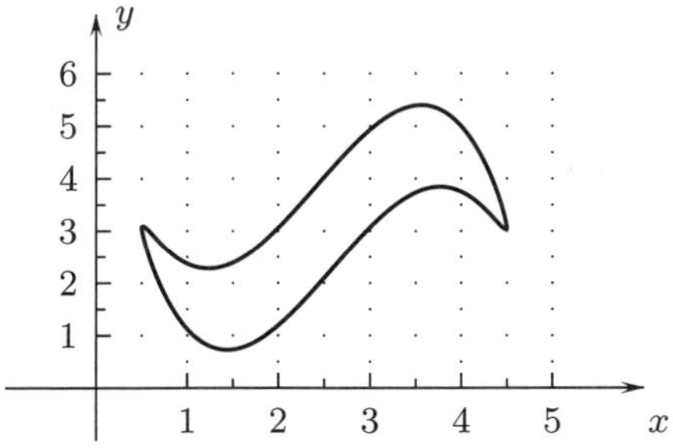

FIGURE 9

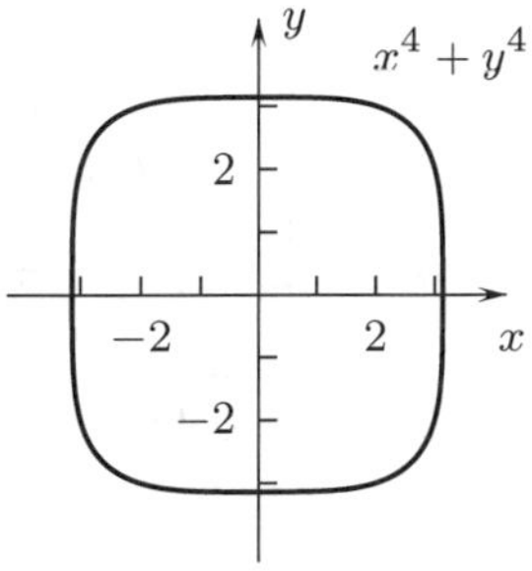

FIGURE 10

T2.[O] What is $y'(3)$ if $y(x)$ is defined implicitly by $xy^3 + xy = 6$ and $y(3) = 1$?

♦ Type 1, implicit differentiation

T3.[O] The curve in Figure 10 is the SUPERCIRCLE $x^4 + y^4 = 97$. Use implicit differentiation to find an equation of its tangent line at $(2, 3)$. Then draw the tangent line with the curve.

♦ Type 1, implicit differentiation and a tangent line

[♦]The Tune-up Exercises and Problems are classified by type and content. The types are (1) basic, reactive; (2) basic reflective; (3) intermediate, reactive; (4) intermediate, reflective; (5) advanced, reactive; (6) advanced, reflective; and (7) advanced, theoretical.

T4.[A] The function $z(x)$ is defined implicitly by the equation $z^5 + z = x^2$. Give a formula for $\frac{dz}{dx}$ in terms of x and z.

♦ Type 1, implicit differentiation

T5.[A] The function $Y(x)$ has a positive derivative for all x. Also, $Y(3) = 100$ and $\frac{dY}{dx}(3) = 8$. What are $X(100)$ and $\frac{dX}{dy}(100)$ if $X(y)$ is the inverse of $Y(x)$?

♦ Type 1, derivative of an inverse function

T6.[A] Figure 11 shows the graph of $y = f(x)$ and its tangent line at $x = 2$. What are $g(3)$ and $\frac{dg}{dy}(3)$ if $x = g(y)$ is the inverse of of $y = f(x)$?

♦ Type 1, derivative of an inverse function from a graph

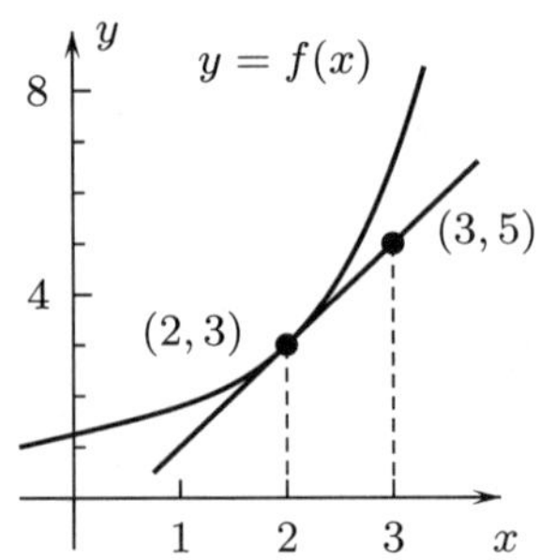

FIGURE 11

Problems 5.2

[A]*Answer provided.* [O]*Outline of solution provided.*

Assume that all implicitly defined functions and inverse functions in the following problems have derivatives wherever they are defined.

1.[O] The function $y(x)$ is defined implicitly by the equation $xy + y - 6x = 0$ and the condition $y(2) = 4$. Find $\frac{dy}{dx}(2)$ **(a)** by implicit differentiation and **(b)** by first solving for y as a function of x and then using this formula to find the derivative.

♦ Type 1, implicit differentiation

2.[A] What is $\frac{dy}{dx}(3)$ if $y(3) = 2$ and $y(x)$ is such that $x^2 - 2xy(x) + [y(x)]^3 = 5$ near $x = 3$?

♦ Type 1, implicit differentiation

3. Find $h'(2)$ where $h(2) = 1$ and $h(t)$ is such that $t[h(t)]^2 = 1 + \sqrt{h(t)}$ near $t = 2$.

♦ Type 1, implicit differentiation

4.[A] What is the slope of the tangent line to the graph of $y(x)$ at $x = 0$ if $y(x)$ is defined implicitly by the equation $xy^{3/2} + 4 = 2x + y$ and the condition $y(0) = 4$?

♦ Type 1, implicit differentiation

5.[O] What is $\frac{dy}{dx}(0)$ if $y(x)$ is defined implicitly by the equation $\sin x + \sin y = 5x - y$ and the condition $y(0) = 0$?

♦ Type 1, implicit differentiation

6. Find $y'(1)$ where $y(x)$ is defined implicitly by $\ln x + \ln y + x^2 - y^2 = 0$ and the condition $y(1) = 1$.

♦ Type 1, implicit differentiation

In Problems 7 through 11 functions $y(x)$ are defined implicitly by the given equations. Give formulas for dy/dx in terms of x and y.

7.[O] $y(x)$ defined by $x^2y^3 - x^2 + y^2 = 1$

♦ Type 1, implicit differentiation

8.[A] $y(x)$ defined by $2y + \cos y = x$

♦ Type 1, implicit differentiation

9.[A] $y(x)$ defined by $x^2 - 5xy + y^2 = 4$

♦ Type 1, implicit differentiation

10. $y(x)$ defined by $y^5 + x^2y^2 = 10x$

♦ Type 1, implicit differentiation

11. $y(x)$ defined by $e^x + e^y = x - y$

♦ Type 1, implicit differentiation

12. $y(x)$ defined by $\frac{y}{y^4 + 1} = x$

♦ Type 1, implicit differentiation

13. **(a)** Explain why the curve in Figure 12 is the graph of the equation

$$(x^2+y^2-1)(x^2+y^2-49)[(x-3)^2+(y-3)^2-1] \times [(x+3)^2+(y-3)^2-1][y+\sqrt{16-x^2}]=0.$$

(b) Give formulas for the nine continuous functions $y(x)$, with the most extensive possible interval domains, that are defined implicitly by this equation.

♦ Type 2, implicitly defined functions

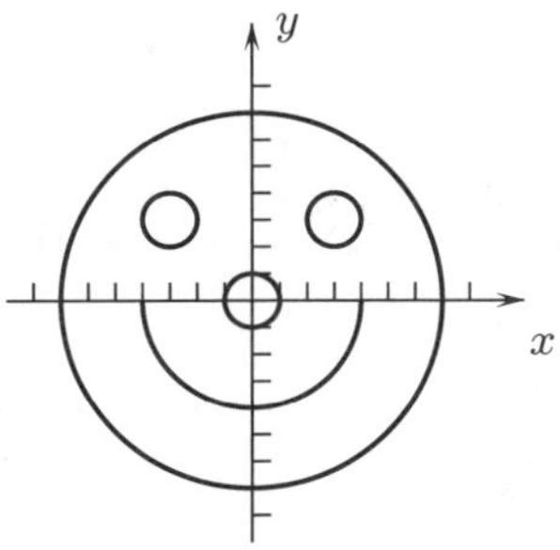

FIGURE 12

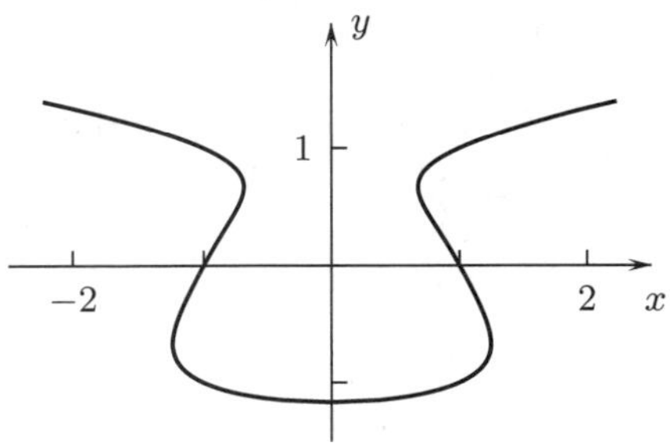

FIGURE 13

14. **(a^A)** Figure 13 shows the curve $y^5-y-x^2+1=0$. A continuous function $y(x)$ with the most extensive possible interval domain is defined implicitly by this equation and the condition $y(1)=1$. Draw the graph of $y(x)$. Then find an equation for the tangent line to the graph at $x=1$ and add the tangent line to your drawing. **(b^A)** Follow the instructions in part (a) with the condition $y(1)=0$. **(c)** Follow the instructions in part (a) with the condition $y(1)=-1$.

♦ Type 2, implicit differentiation and tangent lines

In Problems 15 through 18 find equations for the tangent lines to the curves at the given points.

15.º The tangent line to $xy+x^2y^2=8$ at $(1,2)$

♦ Type 2, implicit differentiation and tangent lines

16.^A The tangent line to $e^y-e^x=6x$ at $(0,0)$

♦ Type 2, implicit differentiation and tangent lines

17. The tangent line to $y^3+xy^2-x=11$ at $(1,2)$

♦ Type 2, implicit differentiation and tangent lines

18. The tangent line to $y^3+y^5+x^4+x^6=0$ at $(1,-1)$

♦ Type 2, implicit differentiation and tangent lines

19.º The curve in Figure 14 is known as TSCHIRNHAUSEN'S CUBIC. It has the equation $9y^2=\frac{1}{3}x^3+3x^2$. **(a)** Give an equation for its tangent line at $(-3,\sqrt{2})$. **(b)** At what two points on the curve is its tangent line horizontal?

♦ Type 2, implicit differentiation and tangent lines

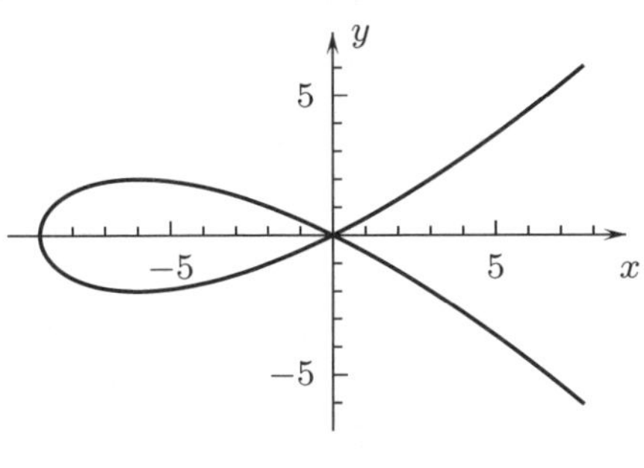

FIGURE 14

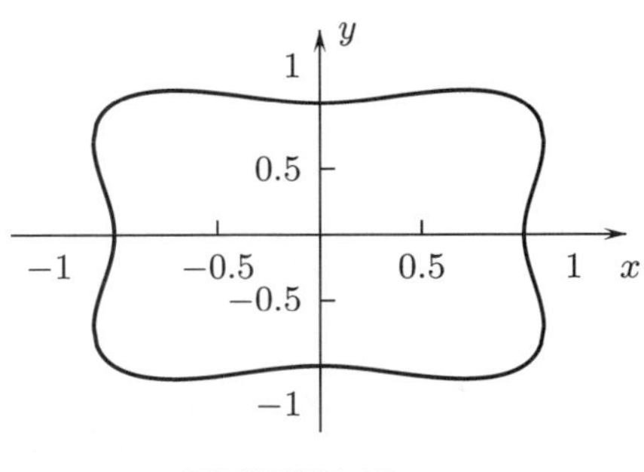

FIGURE 15

20.[A] Figure 15 shows the curve $y^4 - y^2 + x^4 - x^2 = 0$. Find the slopes of its tangent lines at $(1,1), (1,-1), (-1,1)$ and $(-1,-1)$. Then draw the tangent lines with the curve.

♦ Type 2, implicit differentiation and tangent lines

21. **(a)** Draw the graph of $y(x) = x^3$ and its tangent line at $x = 1$ in an xy-plane. **(b)** Draw the graph of $z(y) = y^{1/3}$ and its tangent line at $y = 1$ in a yz-plane **(c)** Draw the graph of $z(y(x))$ and its tangent line at $x = 1$ in an xz-plane **(d)** How are the tangent lines in parts (a), (b), and (c) related? What result in this section does this illustrate?

♦ Type 2, implicit differentiation and tangent lines

22.[A] What is the rate of change of the depth of water in a lake with respect to the volume of water in the lake at a moment when the rate of change of the volume with respect to depth is 200,000 gallons per foot?

♦ Type 2, derivative of an inverse function

23. What is the rate of change of the voltage (volts) in a light bulb with respect to the current (amperes) when the rate of change of the current with respect to voltage is 0.0005 amperes per volt?

♦ Type 2, derivative of an inverse function

24.[A] What is $\frac{dg}{dy}$ if $x = g(y)$ is the inverse of $y = f(x)$ with $f(x) = 10x - 6$?

♦ Type 2, derivative of an inverse function

25.[O] What is $\frac{dr}{dy}(3)$ if $x = r(y)$ is the inverse of $y = q(x)$ with $q(x) = x^2 - 2/x$ for $x > 0$? Notice that $q(2) = 3$.

♦ Type 2, derivative of an inverse function

26. What is $\frac{dV}{dy}(16)$ if $x = V(y)$ is the inverse of $y = U(x)$ and $U(x) = 20 - 3x^2 - x^4$? Notice that $U(1) = 16$.

♦ Type 2, derivative of an inverse function

27.[A] What is $\frac{dK}{dy}(0)$ if $x = K(y)$ is the inverse of $y = J(x)$, where $\frac{dJ}{dx}(x)$ exists and is negative for all x, $J(0) = 16, J(4) = 0, \frac{dJ}{dx}(0) = -3, \frac{dJ}{dx}(4) = -5$ and $\frac{dJ}{dx}(16) = -30$?

♦ Type 2, derivative of an inverse function

28. The function $x = G(y)$ is the inverse of $y = F(x)$, $\frac{dF}{dx}(x)$ exists and is positive for all x, $F(12) = 3, F(0) = 12, \frac{dF}{dx}(12) = 6, \frac{dF}{dx}(3) = 4$ and $\frac{dF}{dx}(0) = 2$. What is $\frac{dG}{dy}(12)$?

♦ Type 2, derivative of an inverse function

29. What is the y-derivative of $y^2 + Q(y)$ at $y = 10$ if $x = Q(y)$ is the inverse of $y = P(x)$, $\frac{dp}{dx}(x)$ exists and is positive for all x, $P(3) = 101$ and $\frac{dP}{dx}(3) = \frac{1}{7}$?

♦ Type 2, derivative of an inverse function

30. The function $y = F(x)$, whose graph is in Figure 16, is defined and has a positive derivative for $x > 0$; $x = G(y)$ is its inverse. Which is greater, $\frac{dG}{dy}(4)$ or $\frac{dG}{dy}(7)$? Explain.

♦ Type 2, derivative of an inverse function

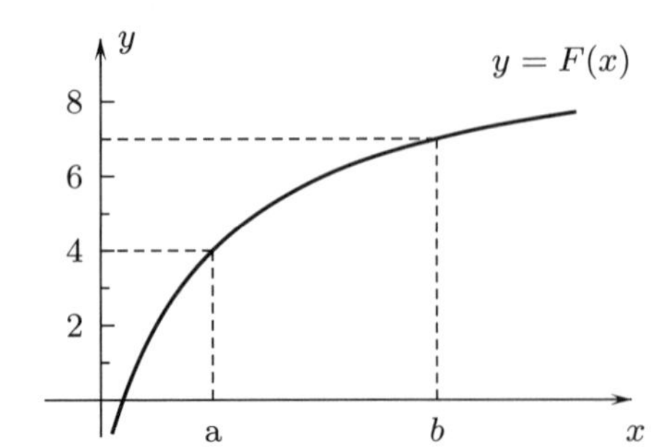

FIGURE 16

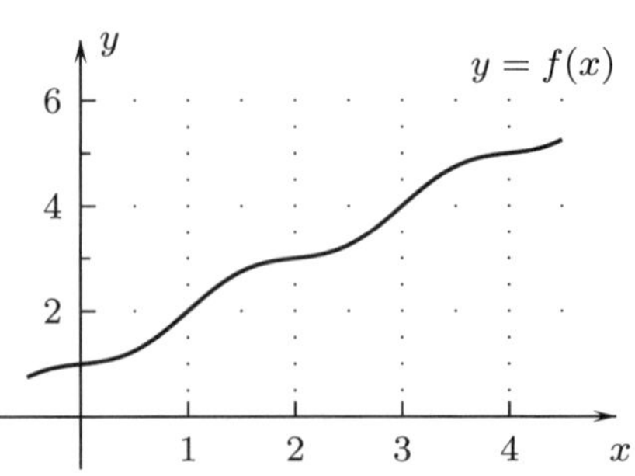

FIGURE 17

31.[A] Figure 17 shows the graph of $y = f(x)$ with $f(x) = x + \frac{1}{4}\sin(\pi x) + 1$. Give the exact value of $\frac{dg}{dy}(4)$ where $x = g(y)$ is the inverse of $y = f(x)$.

♦ Type 2, derivative of an inverse function

32. Figure 18 shows the LEMNISCATE OF BERNOULLI, $3(x^2+y^2)^2 = 25(x^2-y^2)$.[(1)] Show that if the point $(a,b) \neq (0,0)$ is on the curve, then $(-a,-b)$ is also on the curve and the tangent lines at the two points are parallel.

♦ Type 4, implicit differentiation and tangent lines

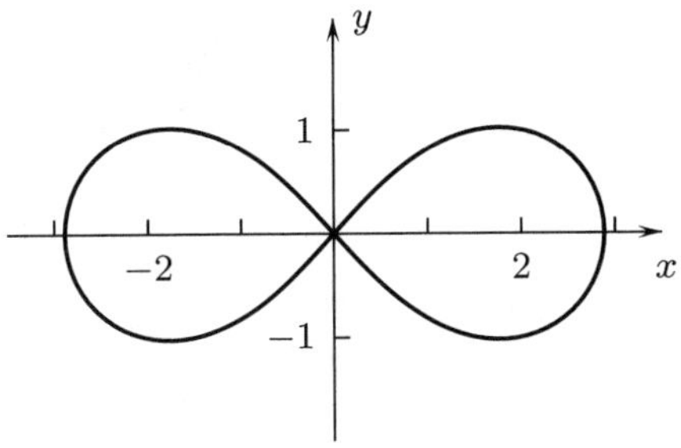

FIGURE 18

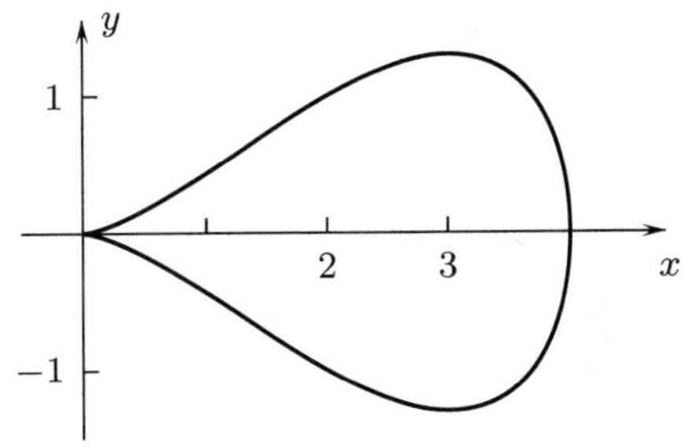

FIGURE 19

33.[A] The PIRIFORM in Figure 19 has the equation $16y^2 - 4x^3 + x^4 = 0$. Find equations of the tangent lines at the points on the curve where it intersects the line $x = 2$.

♦ Type 4, implicit differentiation and tangent lines

34. The ASTROID $x^{2/3} + y^{2/3} = 2$ in Figure 20 has two tangent lines of slope 1 and two tangent lines of slope -1. Use the symmetry of the curve to predict their locations. Then use derivatives to verify your predictions.

♦ Type 4, implicit differentiation and tangent lines

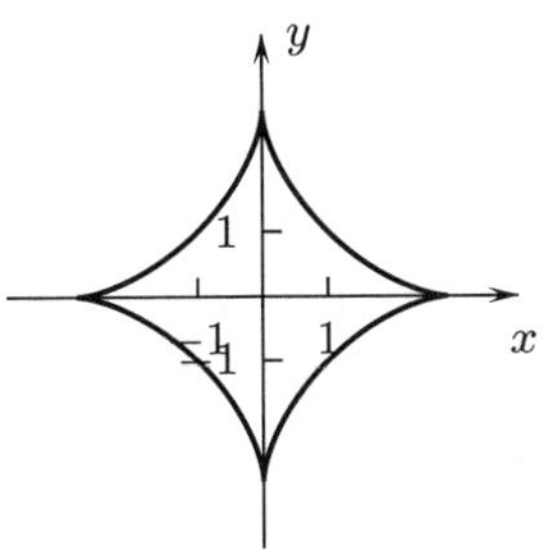

FIGURE 20

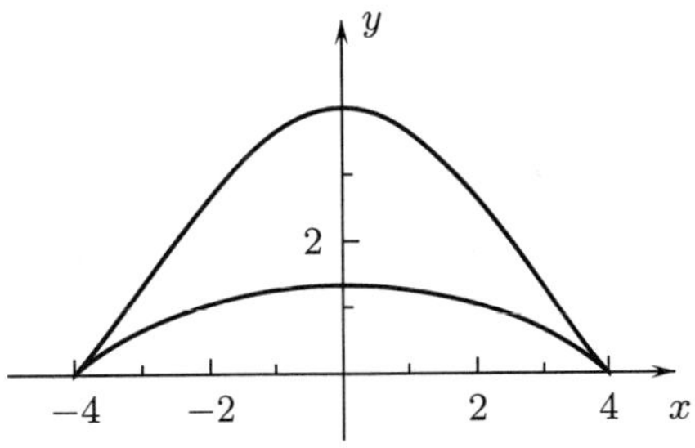

FIGURE 21

35. The curve in Figure 21 is called a BICORN. It has the equation $(x^2 + 8y - 16)^2 = y^2(16 - x^2)$. Show that its tangent lines are horizontal where it crosses the y-axis.

♦ Type 4, implicit differentiation and tangent lines

36.[O] The function $x = N(y)$ is the inverse of $y = M(x)$ with $M(x) = x^2 + 1$. **(a)** Give a formula for $\frac{dM}{dx}$ and use it to find $\frac{dN}{dy}(5)$. **(b)** Find a formula for $N(y)$ and use it to find $\frac{dN}{dy}(5)$.

♦ Type 4, derivative of an inverse function

37. The function $y = Y(x)$ is defined by $Y(x) = 5 - x^{-2}$ for $x > 0$. Its inverse is $x = X(y)$. Determine the value of $\frac{dX}{dy}(4)$, first by using $\frac{dY}{dx}$ and then by finding a formula for $X(y)$.

♦ Type 4, derivative of an inverse function

38. The tangent line to $y = f(x)$ at $x = 1$ is $y = 1 + 2(x-1)$. Give equations of the tangent lines to the graphs of $f^{-1}(x)$, $f(x^3)$ and $[f(x)]^3$ at $x = 1$.

♦ Type 4, tangent lines

[(1)]Studied by Bernoulli in 1694, Ibid, p. 121.

SECOND DERIVATIVES OF IMPLICITLY DEFINED FUNCTIONS In Problems 39 through 42 find the second derivatives of the implicitly defined functions by differentiating the given equations twice.

39.[O] Find $y'(2)$ and $y''(2)$ with $y(x)$ defined by $x^3 - y^3 + x^2 + y^2 = 12$ and $y(2) = 1$.

♦ Type 3, second derivatives of implicitly defined functions

40.[A] Find $y'(1)$ and $y''(1)$ where $y(x)$ is defined by $y^3 + 4y^2 = 11x^2$ and $y(1) = 1$.

♦ Type 3, second derivatives of implicitly defined functions

41. What are $y'(1)$ and $y''(1)$ if $y(x)$ is defined by $2\sin^2 x = x^3$ and $y(1) = \frac{1}{4}\pi$.

♦ Type 3, second derivatives of implicitly defined functions

42. Find $y'(1)$ and $y''(1)$ where $x^3 + xy + y^4 = 1$ and $y(1) = -1$.

♦ Type 3, second derivatives of implicitly defined functions

43. The function $G(t)$ is defined by $G(t) = F(x(t))$, where $x(0) = 3, F(3) = 9, \frac{dx}{dt}(0) = 5$, and $\frac{dF}{dx}(3) = 10$. The inverse of $y = G(t)$ is $t = H(y)$. What is the value of $\frac{dH}{dy}(9)$?

♦ Type 3, second derivatives of implicitly defined functions

Section 5.3

Newton's method

OVERVIEW: *Many applications of calculus and more advanced mathematics involve solving equations that are in the form $F(x) = 0$ or that can be put in this form. In sketching the graph $y = f(x)$ of a function, for example, we solve the equation $f(x) = 0$ to find the x-intercepts, we solve $f'(x) = 0$ to find critical points, and we solve $f''(x) = 0$ to look for possible inflection points. We also solve equations of the form $f'(x) = 0$ in working maximum/minimum problems. Algebraic techniques such as factoring can often be used to solve the necessary equations. In other cases, exact solutions cannot be determined but approximate solutions can be found by a number of methods with a calculator or computer. In this section we discuss* NEWTON'S METHOD, *in which approximate solutions of $F(x) = 0$ are found by using tangent lines to the graph of $F(x)$. With it, highly accurate solutions can usually be found quite quickly with a calculator or computer. Newton's method is also the basis of many procedures for finding accurate approximate solutions of equations that are built into graphing calculators and calculus computer software.*

Topics:

- ***Tangent-line approximations and Newton's method***
- ***How Newton's method can fail***

Tangent-line approximations and Newton's method

To find an approximate solution of $F(x) = 0$, we first pick a value x_0 that we expect to be fairly close to a solution and we calculate $F(x_0)$. If $F(x_0) = 0$, we have our desired solution. Otherwise we consider the tangent line to the graph $y = F(x)$ at $x = x_0$. If $F'(x_0) \neq 0$, the tangent line is not horizontal and we let x_1 denote the value of x where it intersects the x-axis (Figure 1).

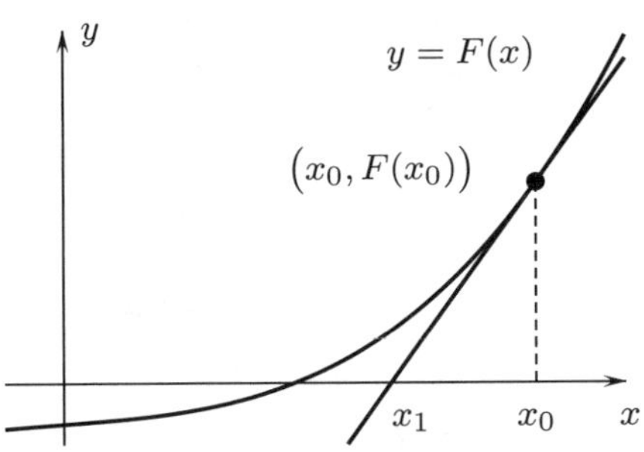

FIGURE 1

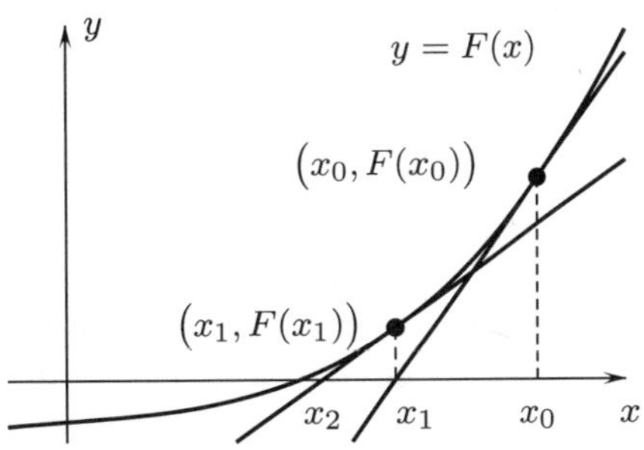

FIGURE 2

Question 1 Find a formula for x_1 in Figure 1 in terms of $x_0, F(x_0)$, and $F'(x_0)$.

If the number x_1 from Question 1 is such that $F(x_1) = 0$, then we have a solution of the equation. Otherwise, if $F'(x_1) \neq 0$, we let x_2 be the x-intercept of the tangent line to the graph of $F(x)$ at x_1 (Figure 2). Continuing in this fashion, we obtain a succession of numbers $x_0, x_1, x_2, \ldots$, and in many cases, the number x_n for a small integer n gives an highly accurate approximate solution of $F(x) = 0$.

A formula for each $x_{n+1} (n = 1, 2, \ldots)$ in terms of x_n can be found by repeating the calculation from Question 1 with x_n in place of x_0: the tangent line to $y = F(x)$ at $x = x_n$ has the equation $y = F(x_n) + F'(x_n)(x - x_n)$ (Figure 3). Setting $y = 0$ and $x = x_{n+1}$ gives $0 = F(x_n) + F'(x_n)(x_{n+1} - x_n)$ and then, if $F'(x_n) \neq 0, x_{n+1} = x_n - \dfrac{F(x_n)}{F'(x_n)}$, and we have the next rule.

♦ SUGGESTIONS TO INSTRUCTORS: Have students begin work on Questions 1 through 3 before class. Problems 5, 8, 11, 12, 13, 16, 22, and 27 are good for in-class discussion.

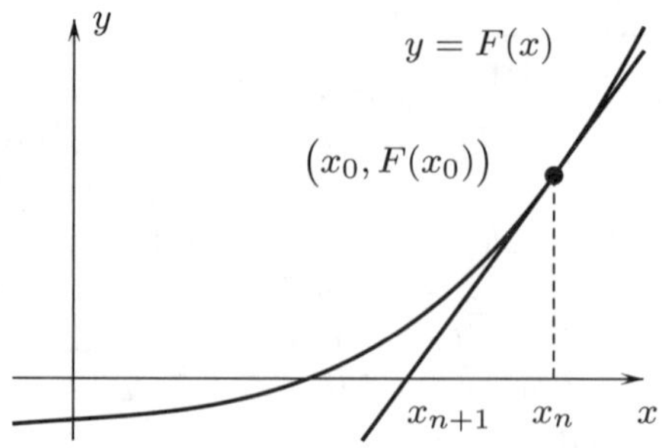

FIGURE 3

Rule 1 (Newton's method) *In Newton's method for finding an approximate solution of* $F(x) = 0$, *an initial approximation* x_0 *is chosen. Then the approximation* x_{n+1} *for* $n = 0, 1, 2 \ldots$ *is calculated from the previous approximation* x_n *by*

$$x_{n+1} = x_n - \frac{F(x_n)}{F'(x_n)}. \tag{1}$$

In successful applications of this rule, the number $F(x_n)$ is so close to zero for a sufficiently large n that x_{n+1}, as displayed on the calculator or computer being used, equals x_n. We define $x^* \doteq x_n$ and use this approximate solution of $F(x) = 0$.

A Newton's-method program

Newton's method can be applied with the following program for Texas Instruments TI-82, TI-83, TI-85, TI-86, TI-89, and TI-92 graphing calculators. The use it, $F(x)$ is first entered as Y_1, its derivative $F'(x)$ is entered as Y_2, and a suitable window is chosen. The initial approximate solution x_0 is entered when the program is run. Then the program displays the curve $y = Y_1(x)$ and the successive tangent lines as it calculates and displays the values of $x_1, x_2, \ldots$ and of $Y_1(x_0), Y_1(x_1), \ldots$. The program is terminated, when appropriate, by pressing the <ON> key.[†]

TI-82, TI-83	**TI-85, TI-86**	**TI-89, TI-92**
PROGRAM: NEWTON	PROGRAM: NEWTON	: newton()
: ClrDraw	: ClrDrw	: Prgm
: FnOff	: FnOff	: ClrDrw
: FnOn 1	: FnOn 1	: FnOff
: Disp "SOLVE $Y_1(X) = 0$"	: Disp "SOLVE $y1(x) = 0$"	: FnOn 1
: Disp "SET $Y_2(X) = DY_1(x)$"	: Disp "SET $y2(x) = Dy1(x)$"	: Disp "solve $y1(x) = 0$"
: Disp "PRESS ON 1"	: Disp "PRESS ON F5"	: Disp "s $y2(x) = dy1(x)$"
: Disp "TO QUIT"	: Disp "TO QUIT "	: Disp "press on esc to quit"
: Disp "$X0 =$?"	: Disp "$x0 =$ "	: Disp "$x0 =$ "
: Input Z	: Input Z	: Input z
: $\emptyset \to J$	: $\emptyset \to J$	: $\emptyset \to j$
: Lbl 1	: Lbl A	: Lbl a
: $Z \to X$	: $Z \to x$	: $z * 0.1 \to x$

[†]These calculator-specific programs are included in preliminary editions of this text for the convenience of users. In the first edition only a description of the action of the program on any calculator or computer will be in the text. The programs, key-by-key instructions for entering them, and implementation instructions for various calculators and computer software will be in the *Calculator/Computer Supplement* of later editions.

```
:Y1 → W
:Y2 → M
:Line(Z,∅,Z,W)
:Pause
:Disp "J,XJ,Y1(XJ)="
:Disp J
:Disp Z
:Disp W
:Pause
:DrawF W+M(X−Z)
:If M=∅
:Goto 2
:Z−W/M → Z
:J+1 → J
:Goto 1
:Lbl 2
:Pause
:Disp "SLOPE = ∅"
:Disp "THE METHOD FAILS"
```

```
:y1 → W
:y2 → M
:Line(Z,∅,Z,W)
:Pause
:Disp "J,xJ,y1(xJ)="
:Disp J
:Disp Z
:Disp W
:Pause
:DrawF W+M(x−Z)
:If M==∅
:Goto B
:Z−W/M → Z
:J+1 → J
:Goto A
:Lbl B
:Pause
:Disp "SLOPE = ∅"
:Disp "THE METHOD FAILS"
```

```
:y1(x) → w
:y2(x) → m
:Line z,∅,z,w)
:Pause
:Disp "j,xj,y1(xj)="
:Disp j
:Disp z
:Disp w
:Pause
:DrawFunc w+m(x−z)
:If m=∅
:Goto b
:z−w/m → z
:j+1 → j
:Goto a
:Lbl b
:Pause
:Disp "slope = ∅,
      the method fails"
:EndPrgm
```

Example 1 Use Newton's method to find the approximate x-intercept of the curve $y = \frac{1}{5}x^5 + \frac{1}{3}x^3 + x - 4$ that is shown in Figure 4.

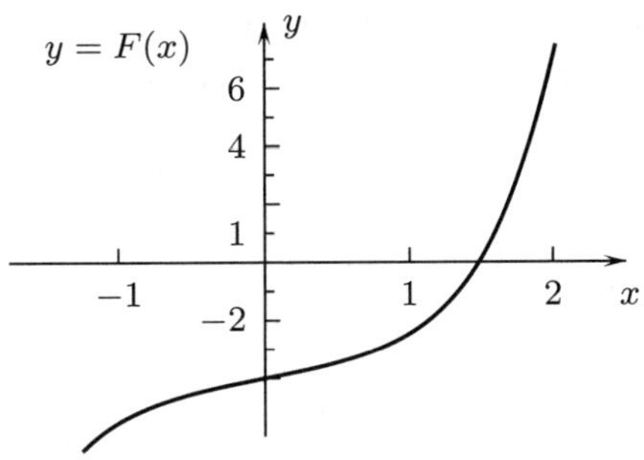

FIGURE 4

SOLUTION Because we want to determine the approximate solution of $F(x) = 0$ with $F(x) = \frac{1}{5}x^5 + \frac{1}{3}x^3 + x - 4$, we find the derivative $F'(x) = \frac{d}{dx}(\frac{1}{5}x^5 + \frac{1}{3}x^3 + x - 4) = x^4 + x^2 + 1$ and define

$$N(x) = x - \frac{F(x)}{F'(x)} = x - \frac{\frac{1}{5}x^5 + \frac{1}{3}x^3 + x - 4}{x^4 + x^2 + 1}. \tag{2}$$

This function is named $N(x)$ for "Newton." As you can see from **(1)**, this is the function that yields x_1 from x_0, x_2 from x_1, etc. Because the x-intercept in Figure 4 is close to 1.5, we could use $x_0 = 1.5$ as our initial approximation. We will use $x_0 = 1$ instead, so that the resulting pictures show more clearly how Newton's method works.

(USING THE PROGRAM) To run the Newton's-method program on a TI-82, TI-83, TI-85, TI-86, TI-89, or TI-92 calculator, we use the window $-1.75 \le x \le 2.5$, $-6 \le y \le 8$ and set

$$Y_1 = x \wedge 5/5 + x \wedge 3/3 + x - 4$$
$$Y_2 = x \wedge 4 + x^2 + 1.$$

We enter 1 for x_0 at the prompt. We press [ENTER] to see the curve and [ENTER] again to see the values $J = 0$, $xJ = x_0 = 1$, and $F(XJ) = F(x_0) = -2.4666666667$. We press [ENTER] again to yield Figure 5, which shows the tangent line at $x = x_0$. The program uses the function **(2)** to calculate $x_1 = N(x_0) = N(1) \doteq 1.8222215836$, which is the x-intercept of the tangent line. The program next displays x_1 and $F(x_1) \doteq 3.857352506$ and then draws the tangent line at $x = x_1$ that is shown in Figure 6.

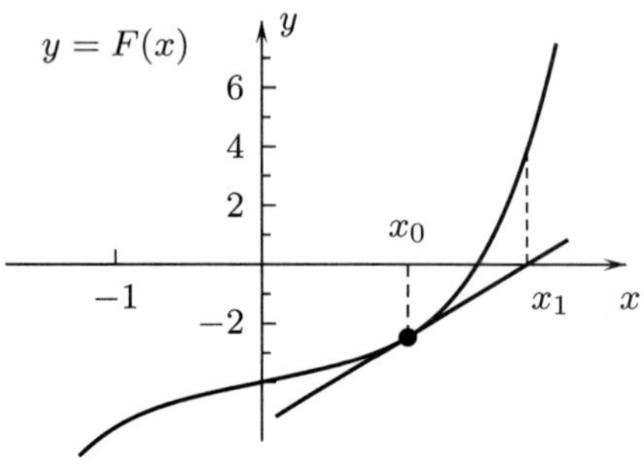

FIGURE 5

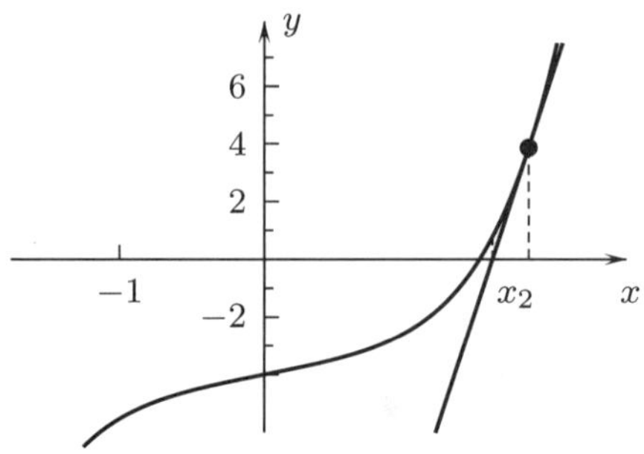

FIGURE 6

We continue to obtain the remaining numbers in the following table, and then press [ON] [1] to terminate the program. We stop with x_7 because its displayed value is the same as that of x_6.

$$\begin{aligned}
x_0 &= 1\\
x_1 &= N(x_0) \doteq 1.822222222\\
x_2 &= N(x_1) \doteq 1.570865594\\
x_3 &= N(x_2) \doteq 1.489665542\\
x_4 &= N(x_3) \doteq 1.482454829\\
x_5 &= N(x_3) \doteq 1.482402580\\
x_6 &= N(x_3) \doteq 1.482402577\\
x_7 &= N(x_3) \doteq 1.482402577.
\end{aligned}$$

(WITHOUT USING THE PROGRAM) If you do not run the program, use formula **(2)** to calculate $x_1 = N(x_0) = N(1), x_2 = N(x_1)$, and so forth, and complete the table above.[†]

[†]If you have a TI-82, TI-83, TI-85, TI-86, TI-89, or TI-92 calculator and do not want to use the program to display the tangent lines, store $N(x)$, given by **(2)**, as Y_3, use $1 \to X$ to store x_0 as X, and type $Y_3 \to X$ in the home screen. Press [ENTER] for x_1, and press [2nd] [ENTRY] [ENTER] repeatedly for the values of $x_2, x_3, \ldots$.

GIVING THE SOLUTION: We denote by x^* the last two numbers in the table, rounded to nine decimal places, and conclude that $x^* = 1.482402577$ is the approximate solution of $\frac{1}{5}x^5 + \frac{1}{3}x^3 + x - 4 = 0$. □

Question 2 Figure 7 shows the curves $y = x^2$ and $y = \cos x$. Use Newton's method to find the approximate x-coordinates of their intersections. (Solve $x^2 - \cos x = 0$.)

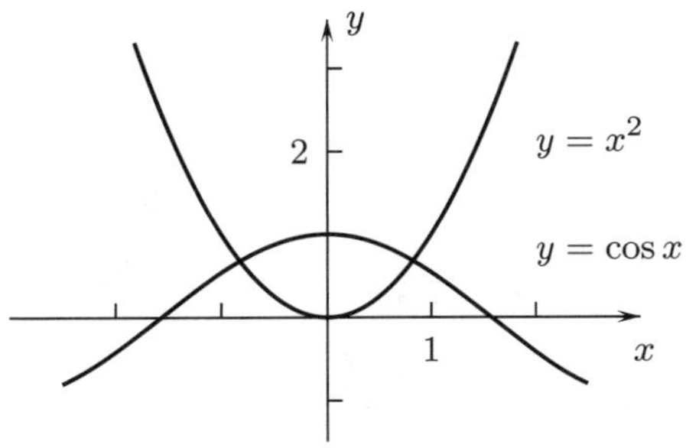

FIGURE 7

Example 2 Use Newton's method to find the approximate coordinates of the point on the curve $y = x^3$ that is closest to the point $(3, 2)$ (see Figure 8).

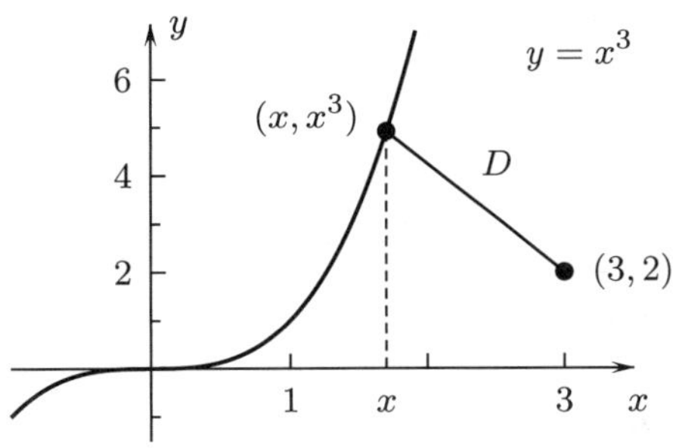

FIGURE 8

SOLUTION The distance from the point $(3, 2)$ to the point (x, x^3) on the curve $y = x^3$ is $D(x) = \sqrt{(x-3)^2 + (x^3-2)^2}$. We minimize $D(x)$ by minimizing its square, which is

$$
\begin{aligned}
A(x) &= [D(x)]^2 = (x-3)^2 + (x^3-2)^2 \\
&= x^2 - 6x + 9 + x^6 - 3x^3 + 4 \\
&= x^6 - 4x^3 + x^2 - 6x + 13.
\end{aligned}
$$

To find the minimum of $A(x)$, we need to find where its derivative $A'(x)$ is zero, so we set $F(x) = A'(x) = 6x^5 - 12x^2 + 2x - 6$ and solve $F(x) = 0$ for x. We either use the Newton's-method program or the function,

$$
\begin{aligned}
N(x) &= x - \frac{F(x)}{F'(x)} = x - \frac{A'(x)}{A''(x)} \\
&= x - \frac{6x^5 - 12x^2 + 2x - 6}{30x^4 - 24x + 2}.
\end{aligned}
$$

Figure 8 shows that the closest point on the curve to the point $(3, 2)$ is fairly close to the point at $x = 2$, so we take $x_0 = 2$ as our initial approximation. This yields $x_1 \doteq 1.672811060, x_2 \doteq 1.457533587, x_3 \doteq 1.351197153, x_4 \doteq 1325075515$, $x_5 \doteq 1.323625823$, and $x_6 \doteq 1.323621525$. The next approximation x_7 has the same approximate decimal value as x_6, so we set $x^* = 1.3236215225$, for which $y = (x^*)^3 \doteq 2.31895042$. Rounding the decimals to three places, we conclude that the closest point has the approximate coordinates $(1.324, 2.319)$. □

How Newton's method can fail

In applying Newton's method it is important to choose the initial approximations x_0 that is fairly close to the anticipated exact solution. Good initial approximations can often be found from graphs or from data in the problem that is being solved.

If the initial approximation is not accurate enough, Newton's method might yield an approximation of a different solution than the one that is sought, or might not work at all. The latter case is illustrated by the function $F(x) = \dfrac{5x}{2 + x^3} + 1$ of Figure 9. The equation,

$$\frac{5x}{2 + x^3} + 1 = 0 \tag{3}$$

has a solution between -1 and 0, as can be seen in Figure 9. As you will see in the next Question, Newton's method yields this solution if we take $x_0 = 0$ but not with $x_0 = 1$.

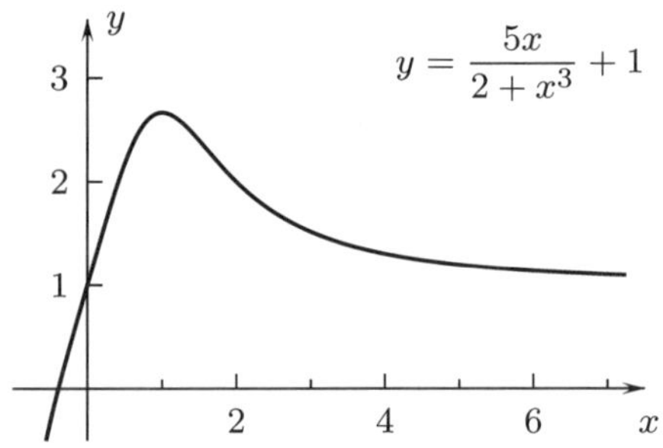

FIGURE 9

Question 3 Apply Newton's method to equation **(3)** **(a)** with $x_0 = 0$ and **(b)** with $x_0 = 1$.

Other types of situations where Newton's method can fail are described in Problems 11 through 14.

Principles and procedures

- You can use Newton's method, as described in this section, to find accurate approximate solutions of equations that you cannot solve exactly if your calculator or computer software does not have a built-in procedure for this purpose.
- Because many of the built-in procedures on calculators and computers for finding accurate approximate solutions of equations with one or more variables are based on or employ Newton's method, understanding how Newton's method works—and how it might fail—can help you use the other calculator/computer procedures more effectively.

Tune-Up Exercises 5.3♦

[A]*Answer provided.* [O]*Outline of solution provided.*

T1[O] Sketch the graph of $F(x) = x^2 + 4x - 2$ and use Newton's method to find approximate solutions of $F(x) = 0$ with whatever accuracy your calculator or computer provides. Check your results by using the quadratic formula to find the exact solutions.

♦ Type 1, Newton's method

T2.[O] Sketch $y = x^3 - 3x^2 + 5$ and use Newton's method to find its x-intercepts.

♦ Type 1, Newton's method

T3.[A] **(a)** What happens if you take $x_0 = 2$ as the initial approximation in Tune-up Exercise T2? **(b)** What happens if you use $x_0 = 1.9$? **(c)** What happens if you use $x_0 = -10$?

♦ Type 2, Newton's method

Problems 5.3

[A]*Answer provided.* [O]*Outline of solution provided.*

In Problems 1 through 6 and 11 through 26, use Newton's method to find approximate solutions of the given equations with whatever accuracy your calculator or computer provides. Generate together the graphs of the functions on both sides of each equation to pick initial approximate solutions x_0. Give your choices of x_0 and list the values of $x_1, x_2, \ldots$ that you use.

1.[O] Find the x-coordinate of the intersection of $y = x^3$ and $y = \cos x$.

♦ Type 1, Newton's method

2.[A] Find the solutions x of $\sin x = 2\cos x$ with $-\pi \leq x \leq \pi$.

♦ Type 1, Newton's method

3. Find the solutions x of $\sin x = 1/x$ with $|x| < \pi$. (Use $F(x) = x \sin x - 1$.)

♦ Type 1, Newton's method

4.[A] Solve $\frac{1}{4}x^4 + \frac{1}{3}x^3 + 3x - 1 = 0$ for x.

♦ Type 1, Newton's method

5. Solve $e^x = 10/x$.

Type 1, Newton's method

6.[A] Solve $3 + x - \frac{1}{60}x^4 = 0$.

♦ Type 2, Newton's method

In Problems 7 through 10, use Newton's method to find approximate solutions of the given equations. Then use the Quadratic Formula to find exact solutions and compare the results.

7.[A] $x^2 + 6x + 4 = 0$

♦ Type 1, Newton's method

8.[O] $x = 4 + x^{1/2}$

♦ Type 2, Newton's method

9. $x^4 + x^2 - 6 = 0$

♦ Type 1, Newton's method

10. $\dfrac{3}{x} = x - 1$ (Use $F(x) = x^2 - x - 3$).

♦ Type 1, Newton's method

11.[A] Newton's method applied to the function $F(x) = x^{1/3}$ and with any $x_0 \neq 0$ fails to give numbers that approach the solution $x = 0$. Explain.

♦ Type 4, a failure of Newton's method

12. Explain why Newton's method with any $x_0 \neq 0$ fails to give numbers that approach the solution of $F(x) = 0$ with $F(x) = \begin{cases} \sqrt{x} & \text{for} \quad x \geq 0 \\ -\sqrt{-x} & \text{for} \quad x < 0. \end{cases}$

♦ Type 4, a failure of Newton's method

13. Newton's method with $F(x) = 2x^3 - 3x^2 + 2$ and $x_0 = 1$ fails to give numbers that approach a solution of the equation $2x^3 - 3x^2 + 2 = 0$. Explain.

♦ Type 4, a failure of Newton's method

♦The Tune-up Exercises and Problems are classified by type and content. The types are (1) basic, reactive; (2) basic reflective; (3) intermediate, reactive; (4) intermediate, reflective; (5) advanced, reactive; (6) advanced, reflective; and (7) advanced, theoretical.

14. **(a)** Sketch the curve $y = \frac{17}{16} + 2x^2 - x^4$ to show that the equation $\frac{17}{16} + 2x^2 - x^4 = 0$ has two solutions. Show algebraically that Newton's method with $x_0 = \frac{1}{2}$ fails to lead to either solution. **(b)** Show by carrying out the calculations that Newton's method for the equation of part (a) can also fail for x_0 close to $\frac{1}{2}$.

♦ Type 4, a failure of Newton's method

Use graphs and Newton's method to find approximate solutions of Problems 15 through 26.

15.[O] If B dollars are borrowed at 100r% interest per month, the loan will be paid off in k months with monthly payments of $P = \dfrac{rB}{1-(1+r)^{-k}}$ dollars. At what annual interest rate would a loan of \$10,000 be paid off in 60 months with monthly payments of \$200 per month?

♦ Type 4, Newton's method, etc.

16.[A] The total area of the base and lateral surface of a right circular cone one meter high is $\pi r^2 + \pi r(1+r^2)^{1/2}$ square meters, where r is the radius of the base. What choice of r would give a total area of 6π square meters?

♦ Type 4, Newton's method, etc.

17.[A] Find the global maximum of $\sin(\pi x) - \frac{1}{4}x^2$.

♦ Type 4, Newton's method, etc.

18. Find the global minimum of $e^x - x^3$.

♦ Type 4, Newton's method, etc.

19. Find the critical points of $\frac{1}{4}x^4 + x^3 - x$.

♦ Type 4, Newton's method, etc.

20. Find the maximum and minimum of $5x - \frac{5}{4}x^4 - x^5$ for $-0.3 \le x \le 1.3$.

♦ Type 4, Newton's method, etc.

21. Find the maximum and minimum of $\sin x - \frac{1}{10}x^2$ for $0 \le x \le \pi$.

♦ Type 4, Newton's method, etc.

22. Find the global maximum and minimum of $\dfrac{\sin x}{1+x^2}$.

♦ Type 4, Newton's method, etc.

23. Find the maximum and minimum of $x^{-1} + 2x^{-2} - 3x^{-4}$ for $x \ge 1$.

♦ Type 4, Newton's method, etc.

24.[A] Find the global minimum of $x^2 + x \ln x$.

♦ Type 4, Newton's method, etc.

25. Find the inflection point of $y = 8 + \ln x - 6x^2 + x^4$.

♦ Type 4, Newton's method, etc.

26.[A] One object is on the x-axis at $x = 2 - \sin t$ at time t, when another object is at $y = t^2$ on the y-axis. What is the minimum distance between the two objects and at what times are they that close?

♦ Type 4, Newton's method, etc.

27. Find the point on the curve $y = x + x^3$ that is closest to the point $(5, 0)$.

♦ Type 4, Newton's method, etc.

28. Use the Intermediate-Value Theorem and the first derivative to prove that the equation $2^x + 3^x = 4$ has exactly one solution. Then find its approximate value.

♦ Type 4, Newton's method, etc.

Section 5.4

L'Hopital's Rule

OVERVIEW: *We saw in Section 3.2 that if $f(x) \to L$ and $g(x) \to M$ as $x \to x_0$, where L and M are finite numbers, and if M is not zero, then the limit, as $x \to x_0$, of the quotient $f(x)/g(x)$ equals the quotient L/M of the limits. If L and M are both zero, it is more difficult to determine whether $f(x)/g(x)$ has a limit. In this case, the quotient $f(x)/g(x)$ is called* INDETERMINATE. *Its limit as $x \to x_0$ might be a finite number, ∞, or $-\infty$ or it might not be defined, depending on the signs and relative sizes of $f(x)$ and $g(x)$ near x_0. In this section we discuss* L'HOPITAL'S RULE, *which uses derivatives to find limits of this and other types of indeterminate forms.*

Topics:

- ***Indeterminate forms of type 0/0***
- ***L'Hopital's Rule***
- ***Indeterminate forms of type ∞/∞***
- ***Indeterminate forms of type $0\infty, 0^0, 1^\infty$, and ∞^0***
- ***The Generalized Mean-Value Theorem***
- ***A partial proof of l'Hopital's Rule***

Indeterminate forms of type 0/0

The rule,

$$\lim_{x\to x_0}\frac{f(x)}{g(x)} = \frac{\lim\limits_{x\to x_0} f(x)}{\lim\limits_{x\to x_0} g(x)} \tag{1}$$

from Section 2.2, cannot be used to find the limit

$$\lim_{x\to 0}\frac{x^2-2x}{x^3+x} \tag{2}$$

because $x^2 - 2x$ and $x^3 + x$ both tend to zero as $x \to 0$. This can be seen from their formulas or from their graphs in Figures 1 and 2. Because its numerator and denominator both tend to 0 as $x \to 0$, the quotient $(x^2 - 2x)/(x^3 + x)$ is called INDETERMINATE OF TYPE $0/0$ as $x \to 0$. Its graph in Figure 3 suggests that it has a finite limit as $x \to 0$. In the following example, we find the limit by factoring the numerator and denominator.

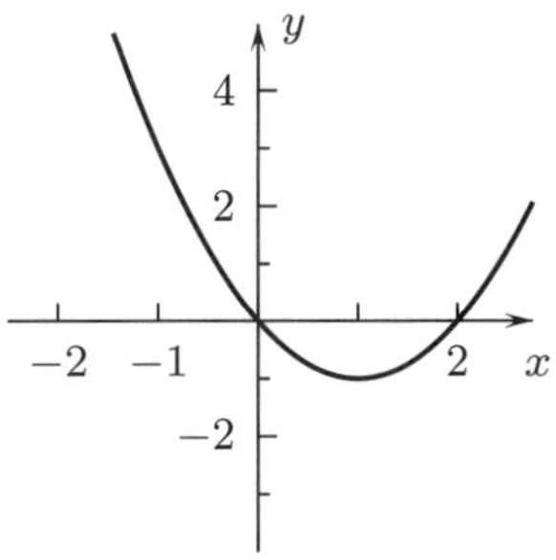

$y = x^2 - 2x$

FIGURE 1

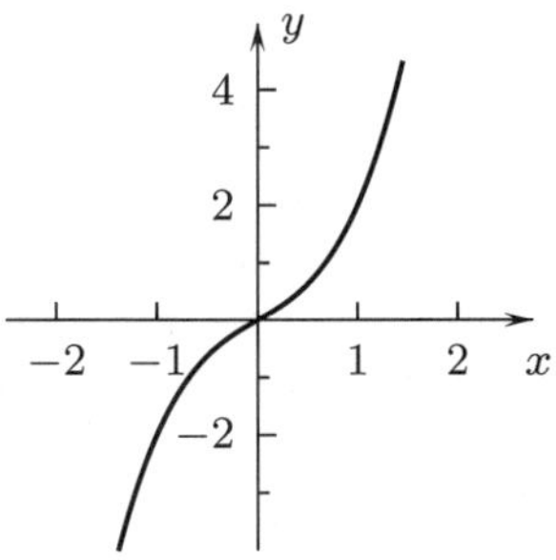

$y = x^3 + x$

FIGURE 2

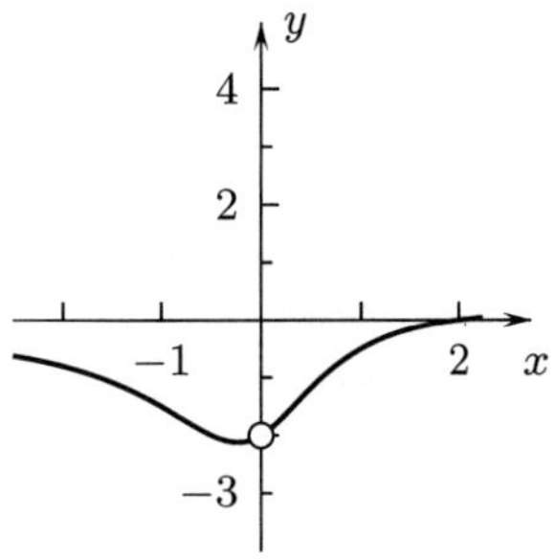

$y = \dfrac{x^2 - 2x}{x^3 + x}$

FIGURE 3

Example 1 Find the limit $\lim_{x\to 0} \dfrac{x^2 - 2x}{x^3 + x}$ in **(2)**.

SOLUTION Factoring x from the numerator and denominator of $\dfrac{x^2-2x}{x^3+x}$ and then cancelling it gives, for $x \neq 0$,

$$\frac{x^2-2x}{x^3+x} = \frac{x(x-2)}{x(x^2+1)} = \frac{x-2}{x^2+1}.$$

We can apply rule **(1)** to the last quotient because its denominator does not tend to zero as $x \to 0$. Its denominator tends to 1 and its numerator tends to -2, so that

$$\lim_{x\to 0} \frac{x^2-2x}{x^3+x} = \lim_{x\to 0} \frac{x-2}{x^2+1} = \frac{-2}{1} = -2.$$

This result is corroborated by the graph of $\dfrac{x^2-2x}{x^3+x}$ in Figure 3. □

Using tangent lines

We solved Example 1 by factoring the polynomials. What can we do when the functions involved are not polynomials or we cannot factor the polynomials that are used? Consider the functions $y = \sin(2\pi x)$ and $y = x^2 - 1$, whose graphs are given in Figures 4 and 5. Notice that both functions are zero at $x = 1$ and tend to zero as $x \to 1$, so that their quotient $\dfrac{\sin(2\pi x)}{x^2-1}$ is an indeterminate form of type 0/0 at 1. The graph of the quotient is shown in Figure 6. We want to find its limit as $x \to 1$. We begin with some numerical experiments.

Question 1 Use your calculator or computer to predict the value of $\lim_{x\to 1} \dfrac{\sin(2\pi x)}{x^2-1}$.

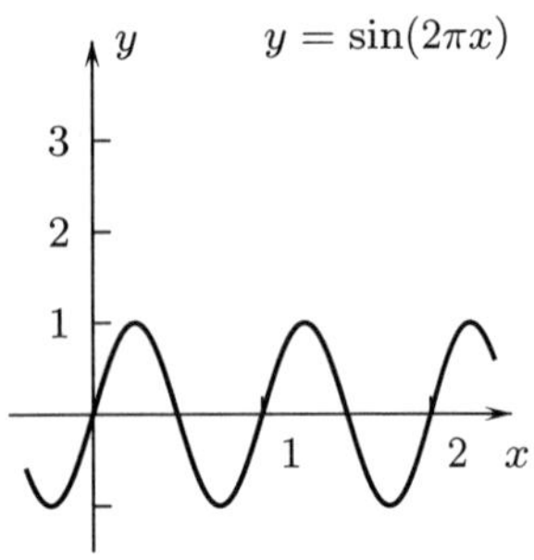

FIGURE 4

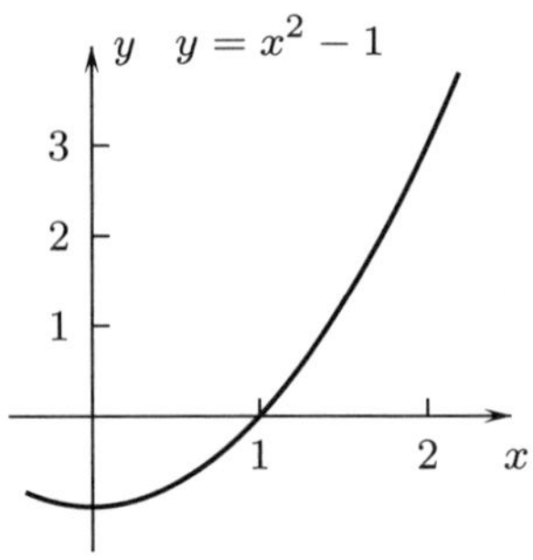

FIGURE 5

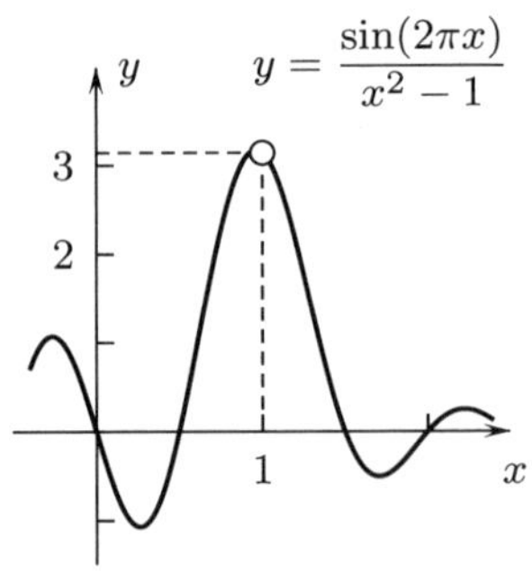

FIGURE 6

We can also predict the limit in Question 1 with greater accuracy by using tangent-line approximations. The linear functions whose graphs are the tangent lines to $y = \sin(2\pi x)$ and $y = x^2-1$ at $x = 1$ in Figures 7 and 8 approximate $\sin(2\pi x)$ and $x^2 - 1$ near $x = 1$, and we can study their quotient to predict the limit of $\dfrac{\sin(2\pi x)}{x^2-1}$ as $x \to 1$.

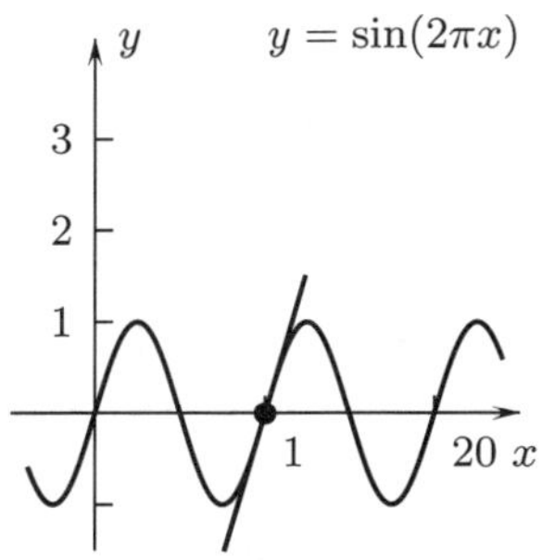

FIGURE 7

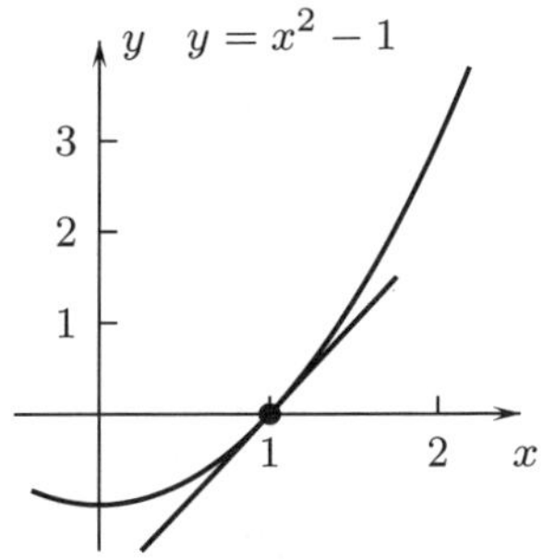

FIGURE 8

Question 2 Give equations of the tangent lines in Figures 7 and 8.

Example 2 Use the formulas for the tangent lines from Question 2 to predict the value of $\lim_{x \to 1} \dfrac{\sin(2\pi x)}{x^2 - 1}$.

SOLUTION Because the tangent lines from Question 2 closely approximate the graphs near $x = 1$, we conclude that, for x near 1,

$$\begin{cases} \sin(2\pi x) \approx 2\pi(x - 1) \\ x^2 - 1 \approx 2(x - 1). \end{cases}$$

This suggests that

$$\frac{\sin(2\pi x)}{x^2 - 1} \approx \frac{2\pi(x-1)}{2(x-1)} = \pi \quad \text{for } x \text{ near } 1$$

so that $\lim_{x \to 1} \dfrac{\sin(2\pi x)}{x^2 - 1} = \pi$. This is probably consistent with your prediction in Question 1 since $\pi \doteq 3.14159$. □

We use tangent lines to predict a limit of another indeterminate form in the following Question.

Question 3 Figures 9 through 11 show the graphs of $y = x^2 - 2x$, $y = e^x - 1$, and their quotient $y = \dfrac{x^2 - 2x}{e^x - 1}$. **(a)** Find equations of the tangent lines at $x = 0$ in Figures 9 and 10. **(b)** Use the formulas for the tangent lines from part (a) to predict the limit of $\dfrac{x^2 - 2x}{e^x - 1}$ as $x \to 0$.

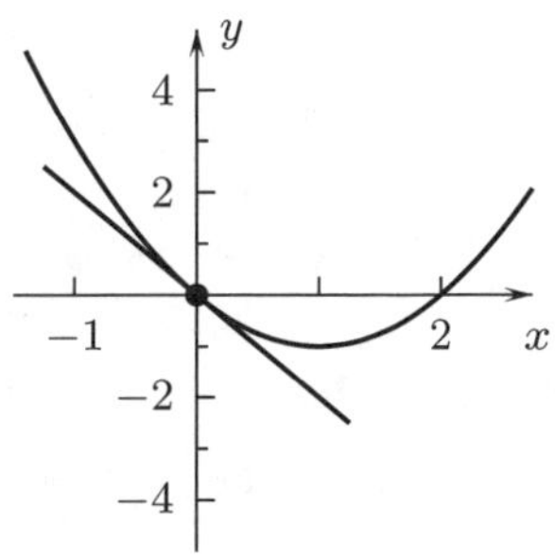

$y = x^2 - 2x$

FIGURE 9

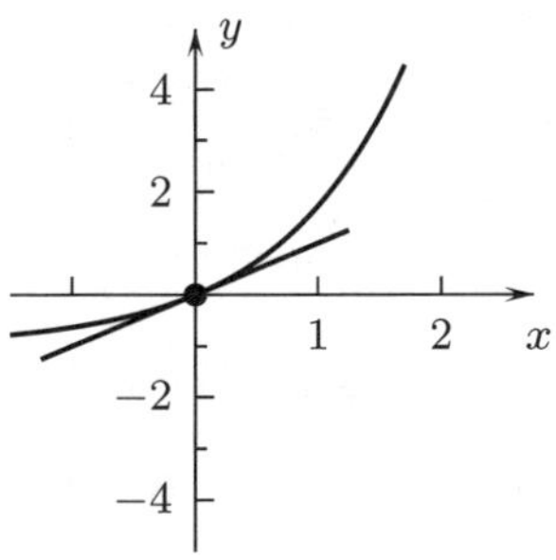

$y = e^x - 1$

FIGURE 10

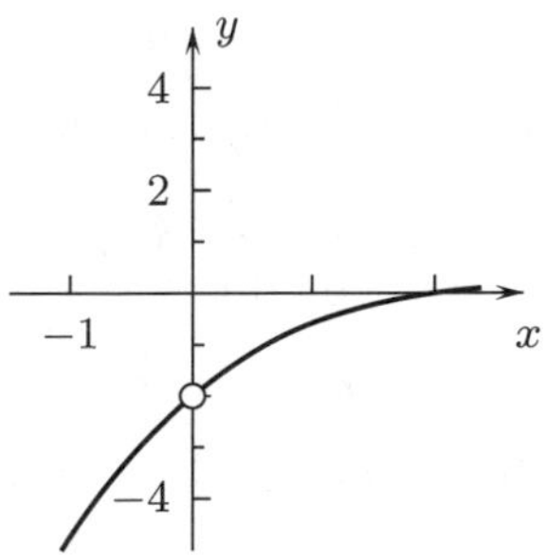

$y = \dfrac{x^2 - 2x}{e^x - 1}$

FIGURE 11

We will see below that the predictions in Example 2 and Question 3 are correct. Notice that in both cases the limit of the indeterminate form equals the quotient of the slopes of the tangent lines, which equals the quotient of the derivatives of the numerator and denominator at the limiting value of x.

l'Hopital's Rule

The technique used in Example 2 and Question 3 to predict limits does not work if the derivative of the denominator is zero at the point where the quotient is indeterminate. Consider, for example, the functions $y = x^2 - 2x + 1$ and $y = \ln x - x + 1$, whose graphs are in Figures 12 and 13, and their quotient $\dfrac{x^2 - 2x + 1}{\ln x - x + 1}$, whose graph is in Figure 14. The quotient is indeterminate of type $0/0$ at $x = 1$ because the numerator and denominator tend to zero as $x \to 1$. Moreover, their derivatives, $\dfrac{d}{dx}(x^2 - 2x + 1) = 2x - 2$ and $\dfrac{d}{dx}(\ln x - x + 1) = x^{-1} - 1$, are also zero at $x = 1$, so the tangent lines are the x-axis, which has the equation $y = 0$. The resulting tangent-line approximations

$$\begin{cases} x^2 - 2x + 1 \approx 0 \\ \ln x - x + 1 \approx 0 \end{cases}$$

for x near 1, do not help us predict the limit of the quotient as x tends to 1 because they only lead to the meaningless statement,

$$\frac{x^2 - 2x + 1}{\ln x - x + 1} \approx \frac{0}{0} \quad \text{for } x \text{ near } 1.$$

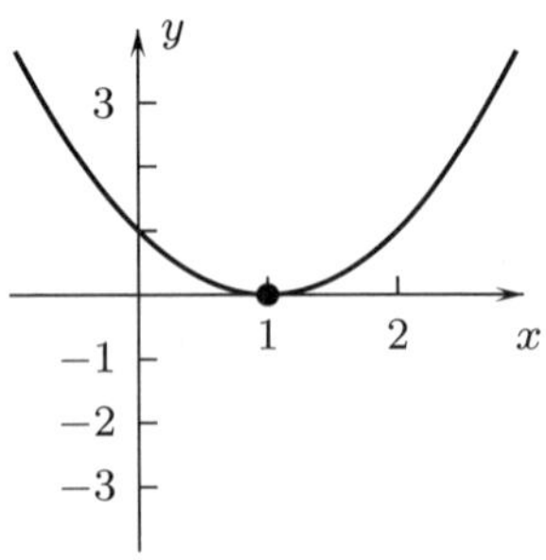

$y = x^2 - 2x + 1$

FIGURE 12

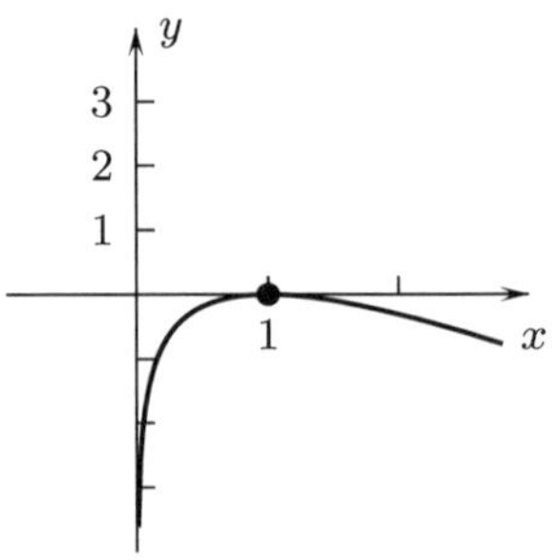

$y = \ln x - x + 1$

FIGURE 13

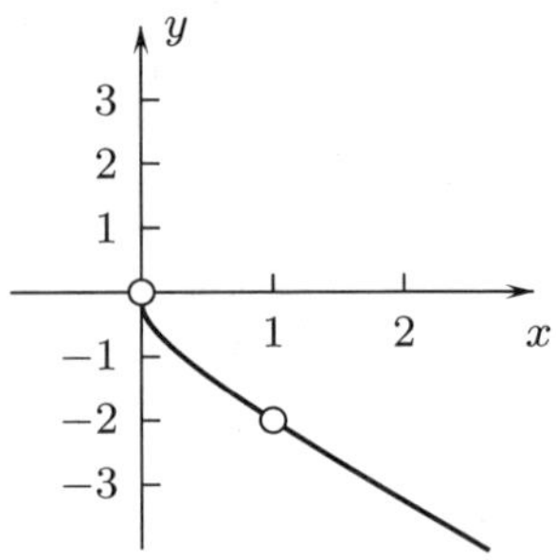

$y = \dfrac{x^2 - 2x + 1}{\ln x - x + 1}$

FIGURE 14

To deal with limits of indeterminate forms, $\lim\limits_{x \to x_0} \dfrac{f(x)}{g(x)}$, such as this, where the limit is not given by the quotient $\dfrac{f'(x_0)}{g'(x_0)}$ of the derivatives at $x = x_0$, we study instead the limit, as $x \to x_0$, of the quotient $\dfrac{f'(x)}{g'(x)}$ of the derivatives at x. This gives the following result, which also applies to other types of indeterminate forms and which, rather than just giving predictions of limits, can be used to prove that the limits are defined.

Theorem 1 (L'Hopital's Rule[†]) *Suppose that $f'(x)$ and $g'(x)$ exist with $g'(x) \neq 0$ for $x \neq x_0$ in an open interval I containing x_0, that either $f(x)$ and $g(x)$ both tend to 0 or $f(x) \to \pm\infty$ and $g(x) \to \pm\infty$ as $x \to x_0$, and that $f'(x)/g'(x) \to L$ as $x \to x_0$. Then*

$$\lim_{x \to x_0} \frac{f(x)}{g(x)} = \lim_{x \to x_0} \frac{f'(x)}{g'(x)} = L. \qquad \textbf{(3)}$$

Here, L may be a number or $\pm\infty$. Furthermore, the condition $x \to x_0$ may be replaced by $x \to x_0^+$ with $I = (x_0, b)$ for some $b > x_0$; by $x \to x_0^-$ with $I = (a, x_0)$ for some $a < x_0$: by $x \to \infty$ with $I = (a, \infty)$ for some a; or by $x \to -\infty$ with $I = (-\infty, b)$ for some b.

We give a partial proof of this theorem at the end of the section.

Example 3 Find the limit $\displaystyle\lim_{x \to 0} \frac{\sin(2\pi x)}{x^2 - 1}$ from Example 2.

SOLUTION Because $\sin(2\pi x)$ and $x^2 - 1$ both tend to 0 as $x \to 1$, the quotient $\dfrac{\sin(2\pi x)}{x^2 - 1}$ is an indeterminate form of type $0/0$ as $x \to 1$, so we look for the limit of the quotient of the derivatives. Because

$$\lim_{x \to 1} \frac{\dfrac{d}{dx}[\sin(2\pi x)]}{\dfrac{d}{dx}[x^2 - 1]} = \lim_{x \to 1} \frac{2\pi \cos(2\pi x)}{2x} = \frac{2\pi}{2} = \pi$$

we can apply l'Hopital's Rule and obtain

$$\lim_{x \to 1} \frac{\sin(2\pi x)}{x^2 - 1} = \pi.$$

This confirms the prediction from Example 2. □

To deal with the limit from Figure 14, we need to apply L'Hopital's Rule twice.

Example 4 Find $\displaystyle\lim_{x \to 1} \frac{x^2 - 2x + 1}{\ln x - x + 1}$.

SOLUTION The quotient $(x^2 - 2x + 1)/(\ln x - x + 1)$ is an indeterminate form of type $0/0$ as $x \to 1$ since $x^2 - 2x + 1$ and $\ln x - x + 1$ tend to zero as $x \to 1$. By **(3)**,

$$\lim_{x \to 1} \frac{x^2 - 2x + 1}{\ln x - x + 1} = \lim_{x \to 1} \frac{\dfrac{d}{dx}(x^2 - 2x + 1)}{\dfrac{d}{dx}(\ln x - x + 1)} = \lim_{x \to 1} \frac{2x - 2}{x^{-1} - 1}. \qquad \textbf{(4)}$$

provided the limit on the right of **(4)** is defined.

The expression on the right of **(4)** is also indeterminant of type $0/0$ as $x \to 1$ because $2x - 2 \to 0$ and $x^{-1} - 1 \to 0$ as $x \to 1$. We could find its limit by simplfying and factorization, but it is easier in this case to apply l'Hopital's Rule again. We obtain

[†]This result is called L'Hopital's rule because it was first published in *Analyse des infiniments petits* by the Marquis G. F. A. de L'Hopital (1661-1704). This book was one of the first textbooks on differential calculus. The result, however, was not proved by de l'Hopital but by another mathematician, Jean Bernoulli, who received a regular salary from the marquis in exchange for the rights to Bernoulli's mathematical discoveries.

$$\lim_{x\to 1} \frac{2x-2}{x^{-1}-1} = \lim_{x\to 1} \frac{\dfrac{d}{dx}(2x-2)}{\dfrac{d}{dx}(x^{-1}-1)} = \lim_{x\to 1} \frac{2}{-x^{-2}} = -2$$

This limit is corroborated by the graph in Figure 14. □

In the next example, L'Hopital's Rule does not apply, and using it inappropriately would lead to an incorrect answer.

Example 5 What is $\lim\limits_{x\to 0} \dfrac{\sin(3x)}{1+\sin(4x)}$?

SOLUTION The numerator $\sin(3x)$ tends to 0, but the denominator $1+\sin(4x)$ tends to the nonzero number 1, so the quotient is not indeterminant. Its limit equals the quotient $0/1 = 0$ of the limits. □

If we made the mistake of applying l'Hopital's Rule to the quotient in Example 5, we would obtain the incorrect result,

$$\lim_{x\to 0} \frac{\dfrac{d}{dx}[\sin(3x)]}{\dfrac{d}{dx}[1+\sin(4x)]} = \lim_{x\to 0} \frac{3\cos(3x)}{4\cos(4x)} = \tfrac{3}{4}.$$

.

Indeterminate forms of type ∞/∞

If the numerator and denominator of $f(x)/g(x)$ tend to ∞ or to $-\infty$, then the quotient is called an INDETERMINATE FORM OF TYPE ∞/∞. L'Hopital's Rule can also be used in this case.

Example 6 Find $\lim\limits_{x\to\infty} \dfrac{e^x}{x}$.

SOLUTION The expression e^x/x is of type ∞/∞ as $x \to \infty$ because $e^x \to \infty$ and $x \to \infty$ as $x \to \infty$. By L'Hopital's Rule,

$$\lim_{x\to\infty} \frac{e^x}{x} = \lim_{x\to\infty} \frac{\dfrac{d}{dx}(e^x)}{\dfrac{d}{dx}(x)} = \lim_{x\to\infty} \frac{e^x}{1} = \infty. \ \square$$

Example 6 establishes a fact that we used occasionally in earlier chapters, namely, that e^x grows more rapidly than x as $x \to \infty$.

Question 4 Show that $\dfrac{\ln x}{\sqrt[n]{x}} \to 0$ as $x \to \infty$ for every positive integer n.

The conclusion of Question 4 shows that for any positive integer n, $\sqrt[n]{x}$ grows more rapidly than $\ln x$ as $x \to \infty$.

Indeterminate forms of type $0 \cdot \infty$

The product $f(x)g(x)$ of a function $f(x)$ that tends to 0 and a function $g(x)$ that tends to ∞ is called an INDETERMINATE FORM OF TYPE $0 \cdot \infty$. Such expressions can be rewritten as forms of type $0/0$ or ∞/∞ to which L'Hopital's Rule can often be applied.

Question 5 Find $\lim\limits_{x\to 0^+} [\sqrt{x}\,(\ln x)]$ by writing $\sqrt{x}\,(\ln x)$ as $\dfrac{\ln x}{x^{-1/2}}$, which is an indeterminate form of type ∞/∞ as $x \to 0^+$.

We could attempt to find the limit in Question 5 by writing the given expression in the form $\dfrac{\sqrt{x}}{(\ln x)^{-1}}$, which is of type $0/0$ as $x \to 0^+$. Then, however, we cannot use l'Hopital's Rule because the quotient of the derivatives is

$$\frac{\dfrac{d}{dx}(x^{1/2})}{\dfrac{d}{dx}[(\ln x)^{-1})]} = \frac{\frac{1}{2}x^{-1/2}}{-(\ln x)^{-2}x^{-1}} = \frac{-\sqrt{x}}{2(\ln x)^{-2}}$$

which is of type $0/0$, and we cannot tell what its limit is as $x \to 0^+$.

Indeterminate forms of types $0^0, 1^\infty$ and ∞^0

The expression $f(x)^{g(x)}$ is INDETERMINATE OF TYPE 0^0 if $f(x) \to 0$ and $g(x) \to 0$; it is OF TYPE 1^∞ if $f(x) \to 1$ and $g(x) \to \pm\infty$; and it is OF TYPE ∞^0 if $f(x) \to \infty$ and $g(x) \to 0$. The limits of such expressions can often by found by applying l'Hopital's Rule to their logarithms.

Example 7 Find the limit of x^{-x} as $x \to 0^+$.

SOLUTION The expression x^{-x} is of type 0^0 as $x \to 0^+$. We first find the limit of its logarithm, which is indeterminate of type 0∞:

$$\begin{aligned}\lim_{x\to 0^+} \ln(x^{-x}) &= \lim_{x\to 0^+} (-x\ln x) = \lim_{x\to 0^+} \frac{\ln x}{x^{-1}} \\ &= \lim_{x\to 0^+} \frac{\dfrac{d}{dx}(-\ln x)}{\dfrac{d}{dx}(x^{-1})} = \lim_{x\to 0^+} \frac{-x^{-1}}{-x^{-2}} = \lim_{x\to 0^+} x = 0.\end{aligned}$$

Because its logarithm tends to 0, the function $x^{-x} = e^{\ln(x^{-x})}$ tends to 1 as $x \to 0^+$. □

Question 6 What are the limits of $(1+x)^{1/x^2}$ **(a)** as $x \to 0^+$ and **(b)** as $x \to 0^-$? Use a calculator or computer to corroborate your answers.

Indeterminate forms of type $\infty - \infty$

If $f(x)$ and $g(x)$ both tend to ∞ as $x \to a$ then so does their sum $f(x) + g(x)$. Their difference, $f(x) - g(x)$, however, might tend to ∞, to $-\infty$, or to a finite number, or it might not have a limit as $x \to a$. We say that it is INDETERMINATE OF TYPE $\infty - \infty$. The limits of such expressions can often be found by combining $f(x)$ and $g(x)$ into another expression, as in Example 8 below, or by first factoring the more rapidly growing function from the difference, as in Question 7.

Example 8 Find the limit of $\dfrac{x^2+2}{1+x} - x$ as $x \to \infty$.

SOLUTION The given expression is of type $\infty - \infty$ since both $\dfrac{x^2+2}{1+x}$ and x tend to ∞ as $x \to \infty$. We rewrite it by taking the common denominator:

$$\frac{x^2+2}{1+x} - x = \frac{x^2+2-x(1+x)}{1+x} = \frac{x^2+2-x-x^2}{1+x} = \frac{2-x}{1+x}.$$

The new expression is of type ∞/∞ as $x \to \infty$. We could find its limit by l'Hopital's Rule, but it is easier to write, for $x \neq 0$,

$$\frac{x^2+2}{1+x} - x = \frac{2-x}{1+x} = \frac{2x^{-1}-1}{x^{-1}+1}.$$

The last expression is not indeterminate as $x \to \infty$. Its numerator tends to -1 and its denominator tends to to 1. Therefore,

$$\lim_{x\to\infty}\left(\frac{x^2+2}{1+x} - x\right) = \lim_{x\to\infty} \frac{2x^{-1}-1}{x^{-1}+1} = -1. \ \square$$

Question 7 Find the limit of $5\ln x - 7x$ as $x \to \infty$ by writing it in the form $x\left(\frac{5\ln x}{x} - 7\right)$ and using the result of Question 5.

A partial proof of l'Hopital's Rule

l'Hopital's Rule can be proved with a generalization of the Mean-Value Theorem from Section 3.2:

Theorem 2 (The Generalized Mean-Value Theorem) *Suppose that $g(x)$ and $h(x)$ are continuous in $[a,b]$ and that $g'(x)$ and $h'(x)$ exist with $h'(x) \neq 0$ for all x with $a < x < b$. Then there is at least one number c with $a < c < b$ such that*

$$\frac{g(b)-g(a)}{h(b)-h(a)} = \frac{g'(c)}{h'(c)}. \tag{5}$$

Recall that the Mean-Value Theorem states that if $f(x)$ is continuous in a finite closed interval $[a,b]$ and if $f'(x)$ exists for all x with $a < x < b$, then there is at least one number c with $a < c < b$ such that

$$\frac{f(b)-f(a)}{b-a} = f'(c). \tag{6}$$

This is the special case of **(5)** for $g(x) = f(x)$ and $h(x) = x$.

We can establish Theorem 2 by applying the Mean-Value Theorem to the function

$$F(x) = [g(b)-g(a)][h(x)-h(a)] - [h(b)-h(a)][g(x)-g(a)]$$

which is continuous in $[a,b]$ and has the derivative

$$F'(x) = [g(b)-g(a)]h'(x) - [h(b)-h(a)]g'(x). \tag{7}$$

Since $F(a)$ and $F(b)$ are both zero, there is, by the Mean-Value Theorem, at least one number c with $a < c < b$ such that $F'(c) = 0$. Then **(7)** with $x = c$ gives

$$[g(b)-g(a)]h'(c) - [h(b)-h(a)]g'(c) = 0. \tag{8}$$

Applying the Mean-Value Theorem now to $h(x)$ shows that there is a number c_1 with $a < c_1 < b$ such that $h(b) - h(a) = h'(c_1)(b-a)$. Because $h'(c_1)$ is not zero, $h(b) - h(a)$ is also not zero, and we can rewrite **(8)** as statement **(6)** in the Generalized Mean-Value Theorem.

To establish l'Hopital's Rule **(3)** in the case where $f(x) \to 0$ and $g(x) \to 0$ as $x \to x_0^+$, we set $a = x_0$ in **(5)** and define $f(x_0)$ and $g(x_0)$ to be zero so that $f(x)$ and $g(x)$ are continuous in $[x_0, b]$. Then by the Generalized Mean-Value Theorem in the interval $[x_0, x]$, there is for each x with $x_0 < x < b$ a number c with $x_0 < c < x$ such that

$$\frac{f(x)}{g(x)} = \frac{f(x)-f(x_0)}{g(x)-g(x_0)} = \frac{f'(c)}{g'(c)}.$$

As x tends to x_0 from the right, the number c also tends to x_0 from the right, so $f'(c)/g'(c) \to L$ and consequently $f(x)/g(x) \to L$, as stated in Theorem 1. A similar argument yields the same result in the case of $x \to x_0^-$, and the result for $x \to x_0$ then follows.

In the case where $f(x) \to 0$ and $g(x) \to 0$ as $x \to \infty$, we set $x = 1/t$, so that $x \to \infty$ as $t \to 0^+$, and we define $F(t) = f(1/t), G(t) = g(1/t)$. Then $F(t)$ and $G(t)$ tend to 0 as $t \to 0^+$, we obtain

$$\lim_{x\to\infty}\frac{f(x)}{g(x)} = \lim_{t\to 0+}\frac{F(t)}{G(t)} = \lim_{t\to 0+}\frac{F'(t)}{G'(t)}$$

$$= \lim_{t\to 0+}\frac{f'\left(\frac{1}{t}\right)(-t^{-2})}{g'\left(\frac{1}{t}\right)(-t^{-2})} = \lim_{t\to 0+}\frac{f'\left(\frac{1}{t}\right)}{g'\left(\frac{1}{t}\right)} = \lim_{x\to\infty}\frac{f'(x)}{g'(x)}.$$

We omit the proofs in the cases when $f(x) \to \pm\infty$ and $g(x) \to \pm\infty$.

Principles and Procedures

- To apply l'Hopital's Rule correctly, you need to show all of your calculations in a way that displays how and why the rule applies.
- Before you apply l'Hopital's Rule to a quotient $f(x)/g(x)$, be sure that it is indeterminate of type $0/0$ or ∞/∞. If not, L'Hopital's Rule does not apply but you can probably find its limit by studying directly the behavior of $f(x)$ and $g(x)$. For example, if the numerator $f(x)$ tends to 0 and the denominator $g(x)$ tends to a nonzero number or $\pm\infty$ as $x \to x_0$, then the quotient tends to 0; and if $f(x)$ has a nonzero, finite limit while $g(x)$ tends to 0, then $|f(x)/g(x)| \to \infty$ as $x \to x_0$ and $f(x)/g(x)$ may tend to ∞ or to $-\infty$ or not be defined, depending on whether $f(x)$ and $g(x)$ are positive or negative near x_0.
- If a quotient $f(x)/g(x)$ is indeterminate of type $0/0$ or ∞/∞, try first to rewrite it in a new form that is not indeterminate, perhaps by factoring the numerator and denominator or by dividing the numerator and denominator by the appropriate function. The quotient $\dfrac{4e^{2x}+e^x}{1+3e^{2x}}$, for example is indeterminate of type ∞/∞ as $x \to \infty$, but dividing the numerator and denominator by the fastest growing term e^{2x} in each of them puts it in the form $\dfrac{4+e^{-x}}{e^{-2x}+3}$, which is not indeterminate. Its limit is $\frac{4}{3}$ because e^{-x} and e^{-2x} tend to 0 as $x \to \infty$.
- If $f(x)/g(x)$ is indeterminate of type $0/0$ or ∞/∞ and you cannot write it in a form that is not indeterminate, see if $f'(x)/g'(x)$ has a limit so you can apply l'Hopital's Rule.
- An indeterminate form $f(x)g(x)$ of type 0∞, where $f(x) \to 0$ and $g(x) \to \pm\infty$, can be written as $\dfrac{f(x)}{1/g(x)}$, which is indeterminate of type $0/0$, or as $\dfrac{g(x)}{1/f(x)}$, which is indeterminate of type ∞/∞. Usually only one of these two forms can be successfully studied with l'Hopital's Rule.
- Expressions of the form $f(x)^{g(x)}$ can be indeterminate of type $0^0, 1^\infty$, or ∞^0 depending on the behavior of $f(x)$ and $g(x)$. Most of these types of indeterminate forms are analyzed by studying their logarithms, $\ln[f(x)^{g(x)}] = g(x)\ln[f(x)]$.
- Check limits as $x \to x_0$, as $x \to x_0^+$, and as $x \to x_0^-$ by calculating values of the functions for x near x_0, and check limits as $x \to \infty$ or as $x \to -\infty$ by calculating values of the functions for large positive or negative x. Generating graphs can also help you verify results and lines of reasoning.

Tune-Up Exercises 5.4♦

[A]*Answer provided.* [O]*Outline of solution provided.*

Be sure you can work all of the exercises. Show your calculations and give your reasoning. If any graphs are required, include them with your solutions.

T1.[O] Find $\lim_{x\to 0} \dfrac{3x^4+6x}{x^2+5x}$ **(a)** by factoring the numerator and denominator and **(b)** by using l'Hopital's Rule.

♦ Type 2, l'Hopital's Rule, etc.

T2.[O] **(a)** Give equations of the tangent lines to $y = \sin(3x)$ and to $y = \sin(2x)$ at $x = 0$. Generate the graphs with their tangent lines in the window $-3 \le x \le 3, -3 \le y \le 3$ and copy them on your paper. **(b)** Use approximations by the tangent lines in part (a) to predict $\lim_{x\to 0} \dfrac{\sin(3x)}{\sin(2x)}$.

(c) Use l'Hopital's Rule to find the limit in part (b).

♦ Type 2, l'Hopital's Rule, etc.

♦The Tune-up Exercises and Problems are classified by type and content. The types are (1) basic, reactive; (2) basic reflective; (3) intermediate, reactive; (4) intermediate, reflective; (5) advanced, reactive; (6) advanced, reflective; and (7) advanced, theoretical.

T3.[O] Predict $\lim_{x\to 1} \dfrac{x+x^2+x^3-3}{x^2+x-2}$ **(a)** by calculating values of the function for x near 1 and **(b)** by generating the graph in the window $-1 \le x \le 3, -1 \le y \le 4.5$. **(c)** Use l'Hopital's Rule to find the limit in part (a).

♦ Type 2, l'Hopital's Rule, etc.

T4.[O] Predict $\lim_{x\to 0} \dfrac{x}{\cos x}$ **(a)** by calculating values of the function for x near 0 and **(b)** by generating the graph in the window $-2 \le x \le 2, -3 \le y \le 3$. **(c)** Find the limit in part (a).

♦ Type 2, l'Hopital's Rule, etc.

T5.[O] Find $\lim_{x\to 0} \dfrac{1-\cos x}{x^2}$. Use the graph $y = \dfrac{1-\cos x}{x^2}$ in the window $-5 \le x \le 5, -0.25 \le y \le 1$ to guide your reasoning.

♦ Type 2, l'Hopital's Rule, etc.

T6.[O] **(a)** Predict $\lim_{x\to\infty} \dfrac{4x^2+2x}{5x^2+3}$ by calculating the values of the function at $x = 10, 100, 1,000$, and 10,000. **(b)** Find the limit in part (a).

♦ Type 2, l'Hopital's Rule, etc.

T7.[O] Find $\lim_{x\to\infty} \dfrac{e^x+x}{\ln x + x}$.

♦ Type 1, l'Hopital's Rule

T8.[O] Find **(a)** $\lim_{x\to\infty} (xe^{-x})$, **(b)** $\lim_{x\to 0} (xe^{-x})$, and **(c)** $\lim_{x\to -\infty} (xe^{-x})$. Generate the graph in the window $-1 \le x \le 4, -2 \le y \le 1$ to guide your reasoning.

♦ Type 1, l'Hopital's Rule

T9.[O] What is the limit of $G(x) = \dfrac{5x^2+1}{x+4} - 5x$ as $x \to \infty$?

♦ Type 1, l'Hopital's Rule

T10.[O] What is $\lim_{x\to 0^+} (\sqrt{x}\ln x)$?

♦ Type 1, l'Hopital's Rule

Problems 5.4

[A]*Answer provided.* [O]*Outline of solution provided.*

Show your calculations and give your reasoning in the following problems. If any graphs are required, include them with your solutions.

1.[O] Find $\lim_{x\to 3} \dfrac{x-3}{x^2-9}$ **(a)** by factoring the denominator and **(b)** by using l'Hopital's Rule.

♦ Type 2, l'Hopital's Rule, etc.

2. Find $\lim_{x\to 0} \dfrac{x^3+6x^2-x}{x^4+3x}$ **(a)** by factoring the numerator and denominator and **(b)** by using l'Hopital's Rule.

♦ Type 2, l'Hopital's Rule, etc.

3. **(a)** Find $\lim_{x\to 3} \dfrac{x^2-6x+9}{x-3}$. **(b)** Generate the graph of $\dfrac{x^2-6x+9}{x-3}$, copy it on your paper, and explain its shape.

♦ Type 2, l'Hopital's Rule, etc.

4. **(a)** Give equations of the tangent lines to $y = \ln x$ and to $y = 1 - x^2$ at $x = 1$. Generate the graphs with their tangent lines in the window $0 \le x \le 3, -2 \le y \le 2$ and copy them on your paper. **(b)** Use approximations by the tangent lines in part (a) to predict $\lim_{x\to 1} \dfrac{\ln x}{1-x^2}$. **(c)** Use l'Hopital's Rule to find the limit in part (b).

♦ Type 2, l'Hopital's Rule, etc.

Find the limits in Problems 5 through 10. Calculate values of the functions for x near its limiting values and generate the graphs in the given windows to guide your reasoning and corroborate your answers.

5.[O] $\lim_{x\to 0} \dfrac{1-e^{2x}}{\sin(4x)}$ (Window: $-\frac{1}{4}\pi \le x \le \frac{1}{4}\pi, -4 \le y \le 2$)

♦ Type 2, l'Hopital's Rule, etc.

6. $\lim_{x\to 0} \frac{\sin(3x)}{6-x^2}$ (Window: $-2 \le x \le 2, -0.5 \le y \le 0.5$)

♦ Type 2, l'Hopital's Rule, etc.

7.º $\lim_{x\to 1} \frac{x^2-2x+1}{\ln x}$ and $\lim_{x\to 0+} \frac{x^2-2x+1}{\ln x}$ (Window: $0 \le x \le 3, -0.5 \le y \le 1.5$)

♦ Type 2, l'Hopital's Rule, etc.

8.º $\lim_{x\to 0} \frac{e^{6x}-1}{\tan(3x)}$ (Window: $-1 \le x \le 1, -4 \le y \le 6$)

♦ Type 2, l'Hopital's Rule, etc.

9. $\lim_{x\to 10} \frac{x+x^2+x^3-1110}{\ln x - \ln(10)}$ (Window: $0 \le x \le 14, 0 \le y \le 5500$ with y-scale $= 1000$)

♦ Type 2, l'Hopital's Rule, etc.

10. $\lim_{x\to 1} \frac{\ln(x^{10})}{x-1}$ (Window: $0 \le x \le 2, -5 \le y \le 25$ with y-scale $= 5$)

♦ Type 2, l'Hopital's Rule, etc.

11. **(a)** Find $\lim_{x\to 0} \frac{\sin x}{\tan x}$. **(b)** Generate the graph $y = \frac{\sin x}{\tan x}$ in the window $-1 \le x \le 12, -1.5 \le y \le 1.5$ and explain its shape.

♦ Type 2, l'Hopital's Rule, etc.

12.º Find **(a)** $\lim_{x\to 0} \frac{\sin(x^2)}{x^2}$ and **(b)** $\lim_{x\to\infty} \frac{\sin(x^2)}{x^2}$. Generate the graph $y = \frac{\sin(x^2)}{x^2}$ in the window $-4 \le x \le 4, -0.5 \le y \le 1.25$ to corroborate your answers.

♦ Type 2, l'Hopital's Rule, etc.

Find the limits in Problems 13 through 18.

13.º $\lim_{x\to 0} \frac{x^3}{\sin x - x}$

♦ Type 1, l'Hopital's Rule

14.º $\lim_{x\to 1} \frac{(\ln x)^2}{x^3-2x^2+x}$

♦ Type 1, l'Hopital's Rule

15. $\lim_{x\to 0} \frac{x^2}{1-e^x}$

♦ Type 1, l'Hopital's Rule

16.^{A} $\lim_{y\to 0} \frac{\cos(6y)}{\cos(5y)}$

♦ Type 1, l'Hopital's Rule

17. $\lim_{x\to 0+} \frac{\ln x}{x^2-1}$

♦ Type 1, l'Hopital's Rule

18. $\lim_{x\to 0} \frac{\sin^{-1} x}{\tan^{-1} x}$

♦ Type 1, l'Hopital's Rule

19.º Find the limits of $\frac{e^x+x}{1-e^x}$ **(a)** as $x \to 0^-$, **(b)** as $x \to 0^+$, and **(c)** as $x \to \infty$. Generate the graph in the window $-3 \le x \le 8, -3 \le x \le 3$ to guide your reasoning.

♦ Type 2, l'Hopital's Rule, etc.

20. Find the limits of $\frac{10+e^{x+1}}{e^x+1}$ **(a)** as $x \to -\infty$, **(b)** as $x \to 0$, and **(c)** as $x \to \infty$. Generate the graph in the window $-15 \le x \le 15, -2 \le x \le 13$ to guide your reasoning.

♦ Type 2, l'Hopital's Rule, etc.

21. Predict the limits $\lim_{x\to\infty} \frac{3\ln x+5}{\ln x+1}$ and $\lim_{x\to 0+} \frac{3\ln x+5}{\ln x+1}$ by calculating the values of the function at $x = 10^2, 10^4, 10^8, 10^{16}, 10^{32}$ and at $x = 10^{-2}, 10^{-4}, 10^{-8}, 10^{-16}, 10^{-32}$. **(b)** Find the limits in part (a).

♦ Type 2, l'Hopital's Rule, etc.

Find the limits in Problems 22 through 29. Show your calculations and give your reasoning.

22.º $\lim_{x\to\infty} \frac{x+\ln x}{x}$

♦ Type 1, l'Hopital's Rule

23. $\lim_{x\to\infty} \frac{x}{e^{2x}}$

♦ Type 1, l'Hopital's Rule

24.º $\lim_{x\to\infty} \frac{\tan^{-1}(3x)}{\tan^{-1}(-x)}$

♦ Type 1, l'Hopital's Rule

25.º $\lim_{x\to\infty} \frac{\sin x}{\ln x}$

♦ Type 1, l'Hopital's Rule

26. $\lim_{x\to\infty} \frac{e^x+5}{e^{-x}+10}$

♦ Type 1, l'Hopital's Rule

27. $\lim_{x\to\infty} \frac{\sin(1/x)}{\cos(2/x)}$

♦ Type 1, l'Hopital's Rule

28. $\lim\limits_{x\to-\infty} \dfrac{e^x}{x^2+10}$

♦ Type 1, l'Hopital's Rule

29. $\lim\limits_{x\to\infty} \dfrac{2e^{-x}+e^{-2x}}{3e^{-x}+e^{-3x}}$

♦ Type 1, l'Hopital's Rule

30.º What are **(a)** $\lim\limits_{x\to0+} \dfrac{\ln x}{x}$ and **(b)** $\lim\limits_{x\to\infty} \dfrac{\ln x}{x}$? **(c)** What is the maximum value of $\dfrac{\ln x}{x}$?

♦ Type 2, l'Hopital's Rule, etc.

31. **(a)** What are $\lim\limits_{x\to0} \dfrac{\cos(\pi x)-1}{x}$ and $\lim\limits_{x\to\pm\infty} \dfrac{\cos(\pi x)-1}{x}$? **(b)** In what intervals is $\dfrac{\cos(\pi x)-1}{x} \geq 0$? **(c)** Use your calculator or computer to find the approximate maximum and minimum values of $\dfrac{\cos(\pi x)-1}{x}$ and the approximate values of x where they occur. Generate the graph in the window $-7 \leq x \leq 7, -3 \leq y \leq 3$ to guide your reasoning.

♦ Type 2, l'Hopital's Rule, etc.

32. Find **(a)** $\lim\limits_{x\to0} \dfrac{\ln x}{\ln x + x^2 - x}$ and **(b)** $\lim\limits_{x\to1} \dfrac{\ln x}{\ln x + x^2 - x}$. Generate the graph in the window $0 \leq x \leq 4, -0.25 \leq y \leq 1.5$ to guide your reasoning.

♦ Type 2, l'Hopital's Rule, etc.

33. What are **(a)** $\lim\limits_{x\to0+} \dfrac{x}{\ln x}$, **(b)** $\lim\limits_{x\to1-} \dfrac{x}{\ln x}$, **(c)** $\lim\limits_{x\to1+} \dfrac{x}{\ln x}$, and **(d)** $\lim\limits_{x\to\infty} \dfrac{x}{\ln x}$? **(e)** What are the most extensive intervals in which $\dfrac{x}{\ln x}$ is increasing and decreasing? Generate the graph in the window $0 \leq x \leq 6, -8 \leq y \leq 8$ to guide your reasoning.

♦ Type 2, l'Hopital's Rule, etc.

Find the limits in Problems 34 through 51.

34.º $\lim\limits_{x\to0} \dfrac{\sin(ax)}{\sin(bx)}$ with nonzero constants a and b

♦ Type 2, l'Hopital's Rule, etc.

35. $\lim\limits_{x\to1} \dfrac{\ln(x^n)}{\ln(x^m)}$ with nonzero constants n and m

♦ Type 2, l'Hopital's Rule, etc.

36. $\lim\limits_{x\to k} \dfrac{x^3 - kx^2}{e^x - e^k}$ with constant k

♦ Type 2, l'Hopital's Rule, etc.

37. $\lim\limits_{x\to0} \dfrac{\sin(bx)}{x^2}$ with positive constant b

♦ Type 2, l'Hopital's Rule, etc.

38. $\lim\limits_{x\to0} \dfrac{\sqrt{x^2+kx+1}-1}{x}$ with positive constant k

♦ Type 2, l'Hopital's Rule, etc.

39.º $\lim\limits_{x\to0+} x(\ln x)^{1/3}$

♦ Type 3, l'Hopital's Rule

40. $\lim\limits_{x\to-\infty} xe^x$

♦ Type 3, l'Hopital's Rule

41.º $\lim\limits_{x\to0} (1+3x+x^2)^{1/x}$

♦ Type 3, l'Hopital's Rule

42.º $\lim\limits_{x\to\infty} (x\ln x - 20x^2)$

♦ Type 3, l'Hopital's Rule

43. $\lim\limits_{x\to0+} [(\ln x)^2 + 100\ln x]$

♦ Type 3, l'Hopital's Rule

44. $\lim\limits_{x\to\infty} \left(\dfrac{x^2}{x-1} - \dfrac{x^2}{x+1}\right)$

♦ Type 3, l'Hopital's Rule

45.º $\lim\limits_{x\to0-} (1-4x+5x^2)^{1/x^2}$

♦ Type 3, l'Hopital's Rule

46. $\lim\limits_{x\to0+} (1-4x+5x^2)^{1/x^2}$

♦ Type 3, l'Hopital's Rule

47. $\lim\limits_{t\to\infty} (1+t^{-2})^t$

♦ Type 3, l'Hopital's Rule

48. $\lim\limits_{x\to0} (1-3x)^{1/x}$

♦ Type 3, l'Hopital's Rule

49. $\lim\limits_{x\to0+} (1-\sqrt{x})^{1/x}$

♦ Type 3, l'Hopital's Rule

50. $\lim\limits_{x\to\infty} (2+1/x)^x$

♦ Type 3, l'Hopital's Rule

51. $\lim_{x\to\infty} \left(1 - \frac{a}{x}\right)^{-x}$ with constant a

♦ Type 4, l'Hopital's Rule

52. Show that for every positive constant n, **(a)** e^x tends to ∞ faster than x^n by showing that $e^x/x^n \to \infty$ as $x \to \infty$. **(b)** Show that e^{-x} tends to zero faster than x^{-n} as $x \to \infty$. (Use logarithms.)

♦ Type 4, l'Hopital's Rule, etc.

53. Show that for every positive integer n, **(a)** $|\ln x|$ tends to ∞ more slowly than $x^{-1/n}$ as $x \to 0^+$.

♦ Type 4, l'Hopital's Rule, etc.

54. What is $\lim_{x\to\pi/4} \dfrac{\ln(\tan x)}{\sin x - \cos x}$? Generate the graph to guide your reasoning.

♦ Type 3, l'Hopital's Rule, etc.

55. **(a)** Find $\lim_{x\to 0} \dfrac{x \sin x}{\sin(x^2)}$. **(b)** Use your calculator or computer to find the approximate minimum value of $\dfrac{x\sin x}{\sin(x^2)}$ for $-1.75 \le x \le 1.75$ and the approximate values of x where it occurs.

♦ Type 4, l'Hopital's Rule, etc.

56.[A] Use l'Hopital's Rule and the first derivative to analyze and sketch the graph of $y = x\ln x$.

♦ Type 3, l'Hopital's Rule, etc.

57. Use l'Hopital's Rule and the first derivative to analyze and sketch the graph of $y = x^x$.

♦ Type 3, l'Hopital's Rule, etc.

58. Use properties of $y = 2 - x^2$ and $y = x\ln x$ to explain the shape of the curve $y = (2 - x^2)\ln(2 - x^2)$ in Figure 15 and to show that the minimum value of $(2 - x^2 x)\ln(2 - x^2)$ is $-e^{-1}$.

♦ Type 4, l'Hopital's Rule, etc.

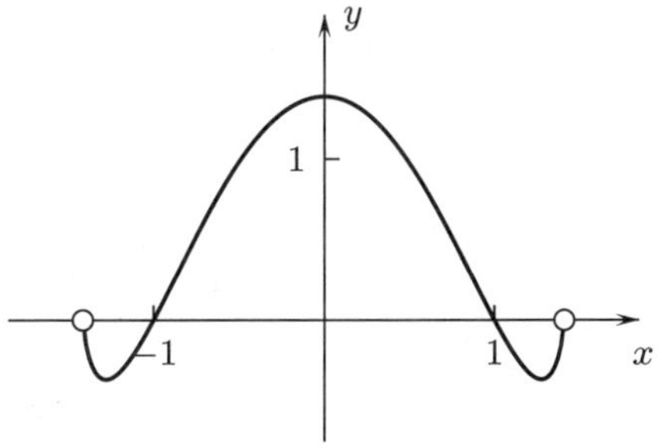

FIGURE 15

59. **(a)** Find $\lim_{x\to 1} \dfrac{\ln x}{x - 1}$. **(b)** Show that $1 - \dfrac{1}{x} - \ln x < 0$ for $0 < x < 1$ and for $x > 1$. **(c)** Give the most extensive intervals in which $\dfrac{\ln x}{x - 1}$ is increasing and decreasing.

♦ Type 4, l'Hopital's Rule, etc.

60. Which of Figures 16 and 17 shows the curve $y = f(g(x))$ and which shows the curve $y = g((f(x))$ for $f(x) = x^x$ and $g(x) = \sin x$? Use properties of $f(x)$ and $g(x)$ to explain the shapes of the curves.

♦ Type 4, l'Hopital's Rule, etc.

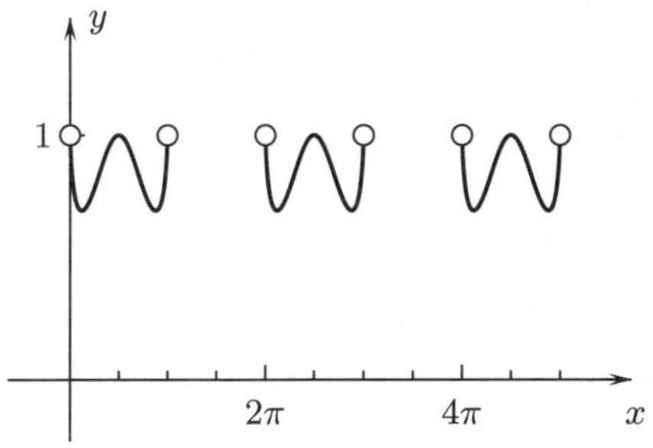

FIGURE 16

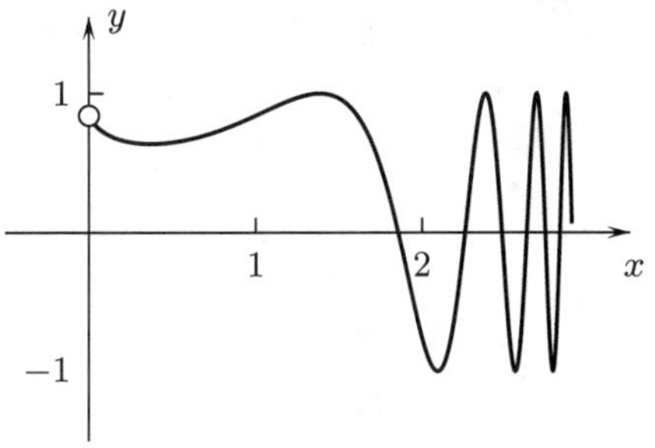

FIGURE 17

61. Generate $y = \dfrac{x}{2x+\sin(\pi x)}$ in the window $-6 \le x \le 6, -0.25 \le y \le 1.25$. **(a)** Why is the graph symmetric about the y-axis? **(b)** What is $\lim_{x\to 0} \dfrac{x}{2x+\sin(\pi x)}$? **(c)** Show that $\lim_{x\to\infty} \dfrac{x}{2x+\sin(\pi x)}$ exists even though $\dfrac{x}{2x+\sin(\pi x)}$ is of type $0/0$ as $x \to \infty$ and $\lim_{x\to 0} \dfrac{\frac{d}{dx}(x)}{\frac{d}{dx}(2x+\sin(\pi x))}$ is not defined.

♦ Type 6, l'Hopital's Rule, etc.

62. A function $y = y(x)$ is ALGEBRAIC if it satisfies an equation of the form $a_n(x)y^n + a_{n-1}(x)y^{n-1} + \cdots + a1(x)y + a_0(x) = 0$, where n is a positive integer and $a_0(x), a_1(x), \ldots a_n(x)$ are polynomials with $a_n(x)$ not the zero function. Use the result of Problem 52 to prove that $y = e^x$ is not an algebraic function.

♦ Type 7, l'Hopital's Rule, etc.

RESPONSES TO QUESTIONS IN CHAPTER 5

Responses 5.1

Response 1 **(a)** The height h of the shadow decreases as s increases.
(b) $h \to 10$ as $s \to \infty$ • $h \to \infty$ as $s \to 0^+$

Response 2 **(a)** Use the similar triangles in Figure 2: $\dfrac{h}{8+16} = \dfrac{10}{16}$ • $h = 15$ feet
(b) Use the similar triangles in Figure 1: $\dfrac{h}{8+s} = \dfrac{10}{s}$ • $sh = 10(8+s)$ • $h = \dfrac{80}{s} + 10$ feet • Figure R2

Jul30.96B

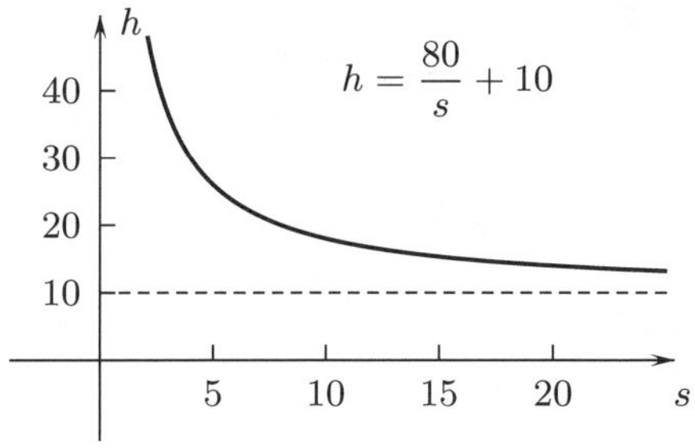

Figure R2

Aug4.99A

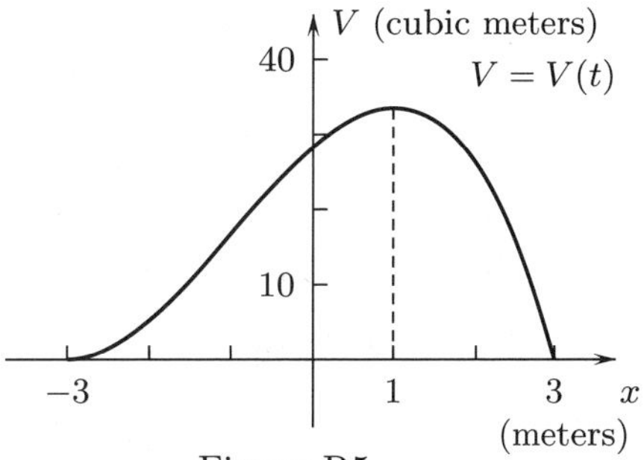

Figure R5

Response 3 $D^2 = x^2 + y^2$

Response 4 **(a)** The right triangle on the lower right of the cross section of the sphere in Figure 5 has hypotenuse equal to the radius 3 of the sphere, one leg of length equal to the radius r of the base of the cone, and the other leg equal to the distance $|x|$ from the center of the sphere to the base of the cone. • The height of the cone is $h = 3+x$, and by the Pythagorean Theorem, its radius is $r = \sqrt{3^2 - x^2} = \sqrt{9 - x^2}$.
(b) The volume is

$$V = \tfrac{1}{3}\pi r^2 h = \tfrac{1}{3}\pi\left[\sqrt{9-x^2}\,\right]^2(3+x) = \tfrac{1}{3}\pi(9-x^2)(3+x).$$

Response 5 The graph of $V(x) = \frac{1}{3}\pi(9-x^2)(3+x)$ is in Figure R5. • The graph shows that the maximum is at $x \approx 1$, as determined in Example 3.

Response 6 **(a)** To have 40 trees in each acre, 10 more would be planted per acre, reducing the yield to $7 - 1.5 = 5.5$ boxes of apples per tree. • The total yield would be [5.5 boxes per tree][40 trees] = 220 boxes.
(b) With $30+x$ trees, the yield would be $7 - 0.15x$ boxes per tree and the overall yield from the orchard would be
$Y(x) =$ [$7 - 0.15x$ boxes per tree][$30 + x$ trees]$= (7 - 0.15x)(30 + x)$ boxes.

Response 7 **(a)** At 2 dollars per bushel, you would sell $x(2) = \dfrac{200}{2^3 + 100} = \frac{200}{108}$ $\doteq 1.85$ thousand bushels and receive$\left[\frac{200}{108}\text{ thousand bushels}\right]\left[2\ \dfrac{\text{dollars}}{\text{bushel}}\right] = \frac{400}{108}$ $\doteq 3.70$ thousand dollars.
(b) At 10 dollars per bushel, you would sell $x(10) = \dfrac{200}{10^3 + 100} = \frac{200}{1100} \doteq 0.18$ thousand bushels and receive$\left[\frac{200}{1100}\text{ thousand bushels}\right]\left[10\ \dfrac{\text{dollars}}{\text{bushel}}\right] = \frac{2000}{1100}$ $\doteq 1.82$ thousand dollars.

(c) At p dollars per bushel, your revenue would be

$$R(p) = \left[\frac{200}{p^3+100} \text{ thousand bushels}\right]\left[p\,\frac{\text{dollars}}{\text{bushel}}\right] = \frac{200p}{p^3+100} \text{ thousand dollars.}$$

Response 8 If you charge very little, you sell a lot but do not receive much money because the price is so low. • If you charge a lot, you do not receive much money because you sell so little.

Response 9 This is a related-rate problem. • Use the Chain Rule:

$$\frac{dR}{dt}\,\frac{\text{thousand dollars}}{\text{day}} = \left[\frac{dR}{dp}\,\frac{\text{thousand dollars}}{\text{dollars per bushel}}\right]\left[\frac{dp}{dt}\,\frac{\text{dollars per bushels}}{\text{day}}\right]$$

At the moment in question,

$$\frac{dR}{dp} = \frac{200(100-2(5^3))}{(5^3+100)^2} = -\tfrac{16}{27}\,\frac{\text{thousand dollars}}{\text{dollar per bushel}}$$

and $\dfrac{dp}{dt} = 0.27$ dollars per day, so that $\dfrac{dR}{dt} = -\frac{16}{27}(0.27) = -0.16$ thousand dollars per day (or -160 dollars per day). • The revenue is decreasing 160 dollars per day.

Response 10 **(a)** $\cos\theta = \dfrac{10}{\overline{PR}}$ and $\sin\theta = \dfrac{15}{\overline{RQ}}$ • $\overline{PR} = 10\sec\theta$ and $\overline{RQ} = 15\csc\theta$

(b) $L = \overline{PQ} = 10\sec\theta + 15\csc\theta$ • $\dfrac{dL}{d\theta} = 10\sec\theta\tan\theta - 15\csc\theta\cot\theta$ $= \dfrac{10\sin\theta}{\cos^2\theta} - \dfrac{15\cos\theta}{\sin^2\theta}$ • $\dfrac{dL}{d\theta} = 0 \iff \dfrac{10\sin\theta}{\cos^2\theta} = \dfrac{15\cos\theta}{\sin^2\theta} \iff \tan^3\theta = 1.5 \iff$ $\tan\theta = 1.5^{1/3}$ • The critical point with $0 < \theta < \frac{1}{2}\pi$ is $\theta = \tan^{-1}(1.5^{1/3})$.

Responses 5.2

Response 1 As in the derivation of **(3)** in Section 5.2, differentiate both sides of $x^3 + y^3 = 9xy$ with respect to x, considering y to be a function of x, and divide by 3 to obtain $x^2 + y^2\dfrac{dy}{dx} = 3y + 3x\dfrac{dy}{dx}$. • Set $x = 2, y = 2$: $4 + 16\dfrac{dy}{dx} = 12 + 6\dfrac{dy}{dx}$ • $\dfrac{dy}{dx} = \frac{4}{5}$ • Tangent line: $y = 4 + \frac{4}{5}(x-2)$ • Figure R1

Oct6.98C

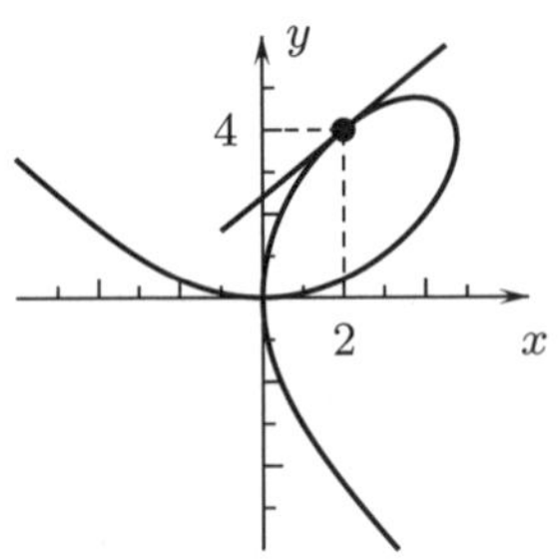

Figure R1

Response 2 Set $x = 2$ and $y = 0$ in the formula $\dfrac{dy}{dx} = \dfrac{4x^3}{2y + e^y}$ from Example 3 of Section 5.2:

$$\frac{dy}{dx}(2) = \frac{4(2^3)}{2(0)+e^0} = -32$$

Response 3 dy and dx in the two interior terms in **(6)** of Section 5.2 are numbers. They are rise and run along the tangent line in Figures 7 and 8 of Section 5.2. • dy/dx and dx/dy in equation **(7)** denote the derivatives $y'(x)$ and $x'(y)$.

Response 4 $G(50) = 3$ gallons • $s(3) = 50$ miles • $\dfrac{dG}{ds}(50) = \frac{1}{15}$ gallons per mile • $\dfrac{ds}{dG}(3) = 15$ miles per gallon

Responses 5.3

Response 1 Tangent line: $y = F(x_0) + F'(x_0)(x - x_0)$ • Solve $F(x_0) + F'(x_0)(x_1 - x_0) = 0$ to find the x-intercept x_1 of the tangent line: $x_1 - x_0 = F(x_0)/F'(x_0)$ • $x_1 = x_0 + F(x_0)/F'(x_0)$

Response 2 Set $F(x) = x^2 - \cos x$ and $F'(x) = 2x + \sin x$. To use the Newton's-method program, set $Y_1 = x^2 - \cos x$ and $Y_2 = x^2$ and use the window $-2 \le x \le 2, -1.5 \le y \le 3.5$. If you do not use the program, set

$$N(x) = x - \frac{F(x)}{F'(x)} = x - \frac{x^2 - \cos x}{2x + \sin x}$$

and calculate $x_1 = N(x_0) = N(1)$, $x_2 = N(x_1), \ldots$ • With $x_0 = 0.5$ we obtain

$$\begin{aligned}
x_1 &= N(x_0) \doteq 0.924296927\\
x_2 &= N(x_1) \doteq 0.829105756\\
x_3 &= N(x_2) \doteq 0.824126132\\
x_4 &= N(x_3) \doteq 0.824132312\\
x_5 &= N(x_4) \doteq 0.824132312
\end{aligned}$$

$x^* = 0.824132312$ is the approximate positive solution of $x^2 - \cos x = 0$ and is the x-coordinate of the intersection of $y = x^2$ and $y = \cos x$ that is to the right of the y-axis. • Because the curves are symmetric about the y-axis, the negative solution is approximately $-x^* = -0.824132312$.

Response 3

$$F'(x) = \frac{d}{dx}\left[\frac{5x}{2 + x^3}\right] = \frac{(2 + x^3)\dfrac{d}{dx}(5x) - 5x\dfrac{d}{dx}(2 + x^3)}{(2 + x^3)^2} = \frac{10 - 10x^3}{(2 + x^3)^2} \quad \bullet$$

To use the Newton's-method program, set $Y_1 = F(x) = 5x/(2 + x^3) + 1$ and $Y_2 = F'(x) = (10 - 10x^3)/(2 + x^3)^2$ and use the window $-1 \le x \le 7, -1 \le y \le 3.5$. If you do not use the program use $N(x) = x - F(x)/F'(x)$.

$$\begin{aligned}
x_0 &= 0\\
x_1 &= N(x_0) \doteq -0.033057851\\
x_2 &= N(x_1) \doteq -0.388354887\\
x_3 &= N(x_2) \doteq -0.388291443\\
x_4 &= N(x_3) \doteq -0.388291441\\
x_5 &= N(x_4) \doteq -0.388291441
\end{aligned}$$

The approximate solution is $x^* = -0.388291441$.

(b) With $x_0 = 1$, $F'(x_0) = 0$, the tangent line is horizontal and does not intersect the x-axis. • Newton's method fails to give a solution.

Responses 5.4

Response 1 Likely prediction: The values in the table below suggest that $\lim_{x\to 1}\dfrac{\sin(2\pi x)}{x^2-1}\approx 3.14$.

x	$\dfrac{\sin(2\pi x)}{x^2-1} \doteq$	x	$\dfrac{\sin(2\pi x)}{x^2-1} \doteq$
0.9	3.093606591	1.1	2.798977392
0.999	3.143143555	1.001	3.140001982
0.99999	3.141608361	1.00001	3.141576945

Response 2 Figure 7: $y=\sin(2\pi x), y'=\dfrac{d}{dx}[\sin(2\pi x)]]=\cos(2\pi x)\dfrac{d}{dx}(2\pi x)=2\pi\cos(2\pi x)$ • $y(1)=0, y'(1)=2\pi$ • Tangent line: $y=2\pi(x-1)$ •
Figure 8: $y=x^2-1, y'=2x$ • $y(1)=0, y'(1)=2$ • Tangent line: $y=2\pi(x-1)$

Response 3 **(a)** Figure 9: $y=x^2-2x, y'=2x-2$ • $y(0)=0, y'(0)=-2$ • Tangent line: $y=-2x$ •
Figure 10: $y=e^x-1, y'=e^x$ • $y(0)=0, y'(0)=1$ • Tangent line: $y=x$ •
(b) For x near 0, $x^2-2x\approx -2x$ and $e^x-1\approx x$ • Likely prediction:
For x near 0, $\dfrac{x^2-2x}{e^x-1}\approx\dfrac{-2x}{x}=-2$ and therefore $\lim_{x\to 0}\dfrac{x^2-2x}{e^x-1}=-2$.

Response 4 $\dfrac{\ln x}{\sqrt[n]{x}}=\dfrac{\ln x}{x^{1/n}}$ is indeterminate of type ∞/∞ as $x\to\infty$. •

$$\frac{\frac{d}{dx}(\ln x)}{\frac{d}{dx}(x^{1/n})}=\frac{x^{-1}}{\frac{1}{n}x^{1/n-1}}=\frac{1}{nx^{1/n}}\to 0 \text{ as } x\to\infty$$

• Therefore, $\lim_{x\to\infty}\dfrac{\ln x}{x^{1/n}}=0$.

Response 5

$$\frac{\frac{d}{dx}(\ln x)}{\frac{d}{dx}(x^{-1/2})}=\frac{x^{-1}}{-\frac{1}{2}x^{-3/2}}=-2x^{1/2}\to 0 \text{ as } x\to 0^+$$

• $\lim_{x\to 0^+}[\sqrt{x}(\ln x)]=0$

Response 6 **(a)** $\ln[(1+x)^{1/x^2}]=\dfrac{\ln(1+x)}{x^2}$ is of type $0/0$ as $x\to 0$. •

$$\frac{\frac{d}{dx}[\ln(x+1)]}{\frac{d}{dx}(x^2)}=\frac{1/(x+1)}{2x}=\frac{1}{2x(x+1)}$$

tends to ∞ as $x\to 0^+$ and tends to $-\infty$ as $x\to 0^-$. • $(1+x)^{1/x^2}=e^{\ln[(1+x)^{1/x^2}]}$ tends to ∞ as $x\to 0^+$ and tends to 0 as $x\to 0^-$.
(b) A calculator gives the table below. (The value at $x=0.001$ is not given because it is too large for the calculator.)

x	$(1+x)^{1/x^2}\doteq$	x	$(1+x)^{1/x^2}\doteq$
0.1	13780.612	−0.1	2.65×10^{-5}
0.01	1.635×10^{44}	−0.01	2.25×10^{-44}
0.001	* * *	−0.001	0

Question 7 $5\ln x-7x=x\left(\dfrac{5\ln x}{x}-7\right)$ tends to $-\infty$ as $x\to\infty$ because $x\to\infty$ and $\dfrac{5\ln x}{x}-7\to -7$ as $x\to\infty$ by Question 4.

CHAPTER 5 ANSWERS

Tune-Up Exercises 5.1

T1. $\frac{y}{4} = \frac{x}{6}$ • $y = \frac{2}{3}x$

T2. $x^2 + y^2 = 10^2$ • $y = \sqrt{100 - x^2}$ or $x = \sqrt{100 - y^2}$

T3. [Number of shirts] $= 50 + \dfrac{4x}{0.05} = 50 + 80x$ • [Price] $= 2.50 - x$ • [Revenue] $= (2.50 - x)(50 + 80x)$ for $0 \le x \le 2.5$

T4. $w^2 + w^2 = s^2$ • $s = \sqrt{2}w$ • $A = \frac{1}{2}w^2$

T5. $(\frac{1}{2}w)^2 + h^2 = w^2$ • $h^2 = \frac{3}{4}w^2$ • $h = \frac{1}{2}\sqrt{3}w$ •
Equilateral triangle: [Height] $= \frac{1}{2}\sqrt{3}w$, [Width] $= w$, [Area] $= \frac{1}{2}(w)(\frac{1}{2}\sqrt{3}w) = \frac{1}{4}\sqrt{3}w^2$
$30°$-$60°$-$90°$ triangle: [Height] $= \frac{1}{2}\sqrt{3}w$, [Width] $= \frac{1}{2}w$, [Area] $= \frac{1}{2}(\frac{1}{2}w)(\frac{1}{2}\sqrt{3}w) = \frac{1}{8}\sqrt{3}w^2$

T6. $\overline{AB} = \sqrt{2^2 + x^2}$ miles • $\overline{BC} = 10 - x$ miles •
[Time] $= \dfrac{\sqrt{4 + x^2}\text{ miles}}{8\text{ miles/hour}} + \dfrac{10 - x\text{ miles}}{3\text{ miles/hour}} = \frac{1}{8}\sqrt{4 + x^2} + \frac{1}{3}(10 - x)$ hours

T7. $(2r)^2 = h^2 + r^2$ • $r = \frac{1}{\sqrt{3}}h$ • $V = \frac{1}{3}\pi r^2 h = \frac{1}{9}\pi h^3$

T8. $\cot\theta = \dfrac{x}{5}$ • $\cot(\theta+\psi) = \dfrac{x}{8}$ • Since θ and $\theta+\psi$ are acute angles, $\theta = \cot^{-1}(\frac{1}{5}x)$ and $\theta+\psi = \cot^{-1}(\frac{1}{8}x)$ • $\psi = \cot^{-1}(\frac{1}{8}x) - \cot^{-1}(\frac{1}{5}x)$

T9. **(a)** $\theta = \dfrac{s}{r}$ • $s = r\theta$ **(b)** $A = \dfrac{\theta}{2\pi}(\pi r^2) = \frac{1}{2}\theta r^2$

T10. [Distance] $= \sqrt{(x - 1)^2 + (x^2 - 3)^2}$

Problems 5.1

1. **(a)** $y = \frac{1}{3}x$ **(b)** The shadow is shrinking $\frac{4}{3}$ feet per second.

2. After t seconds, the plane has traveled $600t$ feet and its distance from the house is $D = \sqrt{(600t)^2 + (3000)^2} = 600\sqrt{t^2 + 25}$ • $\dfrac{dD}{dt} = \dfrac{600t}{\sqrt{t^2 + 25}}$ • After 12 seconds, $\dfrac{dD}{dt} = \dfrac{600(12)}{\sqrt{12^2 + 25}} = 553\frac{11}{13}$ feet per second

4. **(a)** $D =$ [Distance] $= \sqrt{(30t)^2 + (25 - 40t)^2} = \sqrt{2500t^2 - 2000t + 625}$ miles
(b) D is a minimum at $t = \frac{2}{5}$. • [Minimum distance] $= D(\frac{2}{5}) = \sqrt{225} = 15$ miles

8. **(a)** $\dfrac{T(1) - T(0.9)}{1 - 0.9} = 29.247$ degrees per atmosphere
(b) The boiling point of water is dropping 0.30553 degrees per minute.

9. The image is moving toward the lens $\frac{5}{2}$ centimeters per minute.

14. $\frac{75}{8}$ feet per second

15. **(a)** $s = 3h$ **(b)** $V = 15h^2$ **(c)** 1 foot per second

17. 24π cubic feet per hour

18. **(a)** [Revenue] = 1000 dollars • [Cost] = 600 dollars • [Profit] = 400 dollars **(b)** [Revenue] = 1078 dollars • [Cost] = 660 dollars • [Profit] = 478 dollars
(c) Lower the price to \$0.90.

22. **(a)** P is a maximum for $x = 10$ dollars. **(b)** [Maximum profit] = 500 dollars.

24. Minimize the square $f(x) = (x - 4)^2 + (x^2 - \frac{1}{4})^2$ of the distance from $(4, \frac{1}{2})$ to (x, x^2). • $f(x) = 4x^4 - 8x + \frac{65}{4}$ • $f'(x) = 4(x^3 - 2)$ is negative for $x < \sqrt[3]{2}$ and positive for $x > \sqrt[3]{2}$. • The minimum of $f(x)$ is at $x = \sqrt[3]{2}$, and the closest point is $(2^{1/3}, 2^{2/3})$.

28a. [Profit] = 8064 dollars

29. **(a)** $\dfrac{dR}{dp}(5) = 33\frac{1}{3}$ dollars **(b)** $\dfrac{dR}{dp}(15) = -17\frac{43}{121}$ dollars per dollar **(c)** \$7.07 or \$7.08 per shirt

32. 40 units

35. Figure A35 • $T=$ [Time] $= \dfrac{(3-x)\text{miles}}{3 \text{ miles per hour}} + \dfrac{\sqrt{1+x^2}\text{ miles}}{2 \text{ miles per hour}} = \frac{1}{3}(3-x) + \frac{1}{2}\sqrt{1+x^2}$ hours • $T(x)$ is continuous in [0,3]. • $\dfrac{dT}{dx} = -\frac{1}{3} + \frac{1}{2}\dfrac{x}{\sqrt{1+x^2}}$ • $\dfrac{dT}{dx} = 0 \iff$ $\frac{1}{3} = \frac{1}{2}\dfrac{x}{\sqrt{1+x^2}} \iff 3x = 2\sqrt{1+x^2} \iff 9x^2 = 4 + 4x^2 \iff 5x^2 = 4 \iff x = \sqrt{0.8}$ • $T(0) = 1.5$, $T(\sqrt{0.8}) \doteq 1.378$, $T(3) \doteq 1.58$. • The minimum of $T(x)$ for $0 \le x \le 3$ is at $x = \sqrt{0.8} \doteq 0.89$ miles.

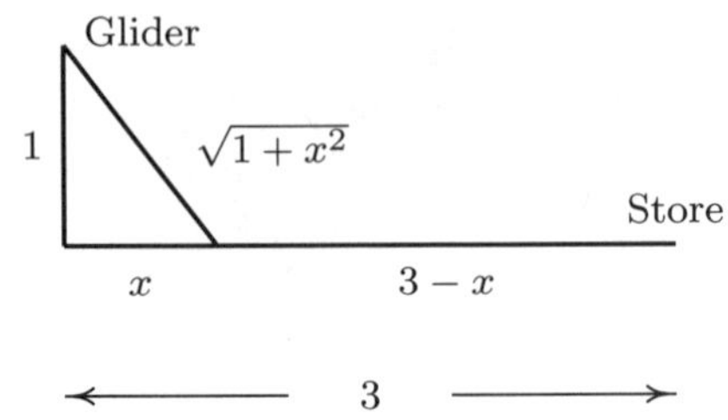

Figure A35

40. **(a)** $96v^{-1} + \frac{1}{2} + \frac{1}{60}v^{3/2}$ dollars per mile **(b)** $v = (3840)^{2/5} \doteq 27.15$ miles per hour

41. $-\frac{7}{125}K$ foot-candles per second

43. **(a)** $\dfrac{3}{4\pi}$ feet per minute **(b)** One answer: Sand does not form a perfect cone.

45. The shadow is moving 60 feet per second

46. $r = \dfrac{7}{4+\pi}$ feet • $h = \dfrac{7}{4+\pi}$ feet

52. [Radius of the base] $= \frac{20}{3}$ inches • [Height] = 8 inches

60a. [Radius of the pie] $= \frac{5}{2}$ inches • [Angle of the slice] = 2 radians

61. [Maximum volume] $= \frac{250}{27}\sqrt{3}\pi$ cubic inches

Tune-Up Exercises 5.2

T1. **(a)** Domain: $[a,b]$ with $a \approx 0.5$ and $b \approx 4.5$ **(b)** Figure T1 **(c)** $\dfrac{dy}{dx} \approx \dfrac{3-5}{5-4} = -2$

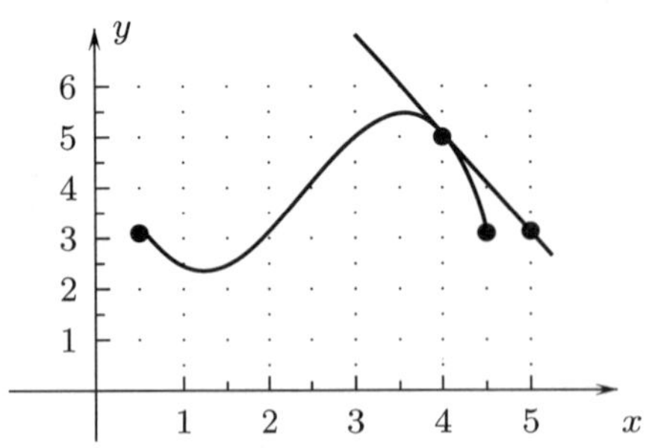

Figure T1

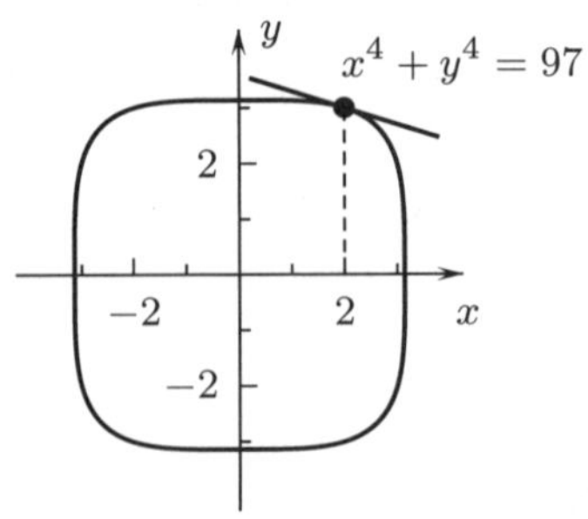

Figure T3

T2. $\dfrac{d}{dx}(xy^3 + xy) = \dfrac{d}{dx}(6)$ • $x\dfrac{d}{dx}(y^3) + y^3\dfrac{d}{dx}(x) + x\dfrac{dy}{dx} + y\dfrac{dx}{dx} = 0$ • $x(3y^2\dfrac{dy}{dx}) + y^3 + x\dfrac{dy}{dx} + y = 0$ • $(3xy^2 + x)\dfrac{dy}{dx} + y^3 + y = 0$ • Set $x = 3$, $y = 1$: $12\dfrac{dy}{dx} + 2 = 0$ • $\dfrac{dy}{dx}(3) = -\frac{1}{6}$

T3. $\frac{d}{dx}(x^4 + y^4) = \frac{d}{dx}(97)$ • $4x^3 + 4y^3\frac{dy}{dx} = 0$ • $x^3 + y^3\frac{dy}{dx} = 0$ • Set $x = 2,\ y = 3 : 27\frac{dy}{dx} + 8 = 0, \frac{dy}{dx}(2) = -\frac{8}{27}$ • $y(2) = 3$ • Tangent line: $y = 3 - \frac{8}{27}(x - 2)$ • Figure T3

T4. $\frac{d}{dx}(z^5 + z) = \frac{d}{dx}(x^2)$ • $(5z^4 + 1)\frac{dz}{dx} = 2x$ • $\frac{dz}{dx} = \frac{2x}{5z^4 + 1}$

T5. $X(100) = 3$ • $\frac{dX}{dy}(100) = \frac{1}{\frac{dY}{dx}(3)} = \frac{1}{8}$

T6. $\frac{df}{dx}(2) = \frac{5-3}{3-2} = 2$ • $f(2) = 3$ • $\frac{dg}{dy}(3) = \frac{1}{\frac{df}{dx}(2)} = \frac{1}{2}$

Problems 5.2

1. **(a)** $\frac{d}{dx}(xy + y - 6x) = \frac{d}{dx}(0)$ • $x\frac{dy}{dx} + y\frac{dx}{dx} + \frac{dy}{dx} - 6\frac{dx}{dx} = 0$ • $(x - 1)\frac{dy}{dx} + y - 6 = 0$ • Set $x = 2,\ y = 4 : 3\frac{dy}{dx} - 2 = 0$ • $\frac{dy}{dx} = \frac{2}{3}$

(b) $y = \frac{6x}{x+1}$ • $\frac{dy}{dx} = \frac{(x+1)\frac{d}{dx}(6x) - 6x\frac{d}{dx}(x+1)}{(x+1)^2} = \frac{6}{(x+1)^2}$ • $\frac{dy}{dx}(2) = \frac{6}{(2+1)^2} = \frac{2}{3}$

2. $\frac{dy}{dx}(3) = -\frac{1}{3}$

4. [Slope] $= \frac{dy}{dx}(0) = 6$

5. $\frac{d}{dx}(\sin x + \sin y) = \frac{d}{dx}(5x - y)$ • $\cos x + \cos y\frac{dy}{dx} = 5 - \frac{dy}{dx}$ • Set $x = 0,\ y = 0 : 1 + \frac{dy}{dx} = 5 - \frac{dy}{dx}$ • $\frac{dy}{dx}(0) = 2$

7. $\frac{d}{dx}(x^2y^3 - x^2 + y^2) = \frac{d}{dx}(1)$ • $2xy^3 + 3x^2y^2\frac{dy}{dx} - 2x + 2y\frac{dy}{dx} = 0$ • $(3x^2y^2 + 2y)\frac{dy}{dx} = 2x - 2xy^3$ • $\frac{dy}{dx} = \frac{2x - 2xy^3}{3x^2y^2 + 2y}$

8. $\frac{dy}{dx} = \frac{1}{2 - \sin y}$

9. $\frac{dy}{dx} = \frac{5y - 2x}{2y - 5x}$

14. **(a)** The heavy curve in Figure A14a • Tangent line: $y = 1 + \frac{1}{2}(x - 1)$ **(b)** The heavy curve in Figure A14b • Tangent line: $y = -2(x - 1)$

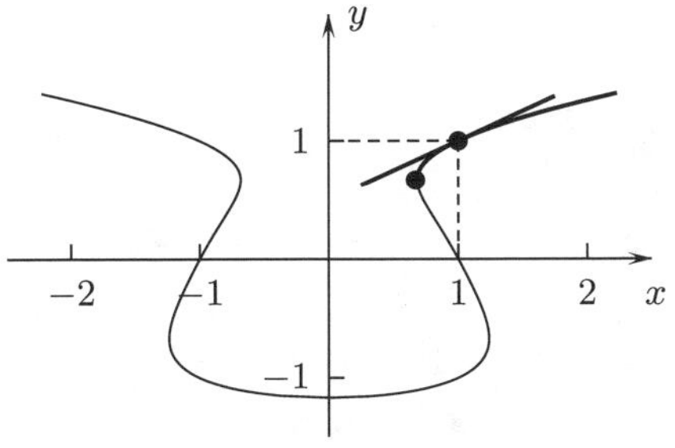

Figure A14a

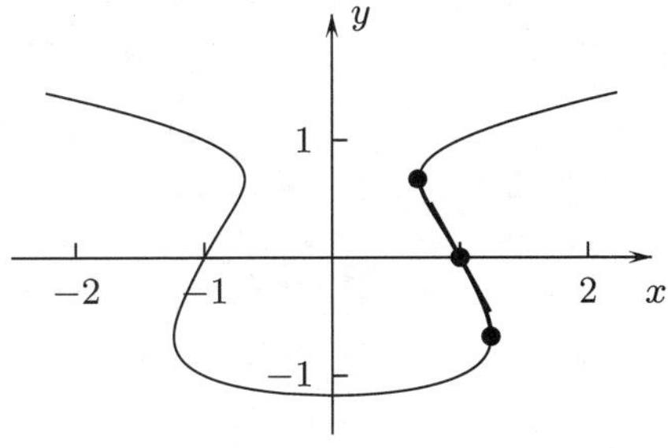

Figure A14b

15. $\frac{d}{dx}(xy+x^2y^2) = \frac{d}{dx}(6)$ • $x\frac{dy}{dx}+y+2xy^2+2x^2y\frac{dy}{dx} = 0$ • $\frac{dy}{dx} = -\frac{2xy^2+y}{2x^2y+x}$ • $\frac{dy}{dx}(1) = -\frac{10}{5} = -2$ • $y(1) = 2$ • Tangent line: $y = 2 - 2(x-1)$

16. Tangent line: $y = 7x$

19. **(a)** Tangent line: $y = \sqrt{2} - \frac{1}{2\sqrt{2}}(x+3)$ **(b)** Horizontal tangent lines at $(-6, 2)$ and $(-6, -2)$

20. $\frac{dy}{dx} = 1$ at $(-1, 1)$, $(1, -1)$ • $\frac{dy}{dx} = -1$ at $(1, 1)$, $(-1, -1)$ • Figure A20

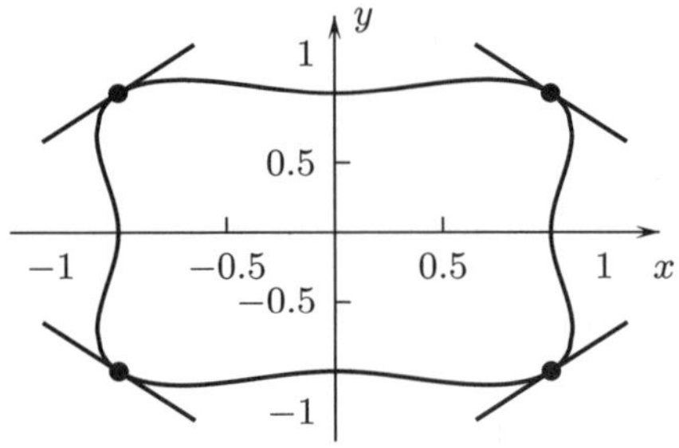

Figure A20

22. $\frac{dh}{dV} = 5 \times 10^{-6}$ feet per gallon

24. $\frac{dg}{dy} = \frac{1}{10}$

25. $\frac{dr}{dy}(3) = \frac{2}{9}$

27. $\frac{dK}{dy}(0) = -\frac{1}{5}$

31. $f'(x) = 1 + \frac{\pi}{4}\cos(\pi x)$ • $f(3) = 4$ • $\frac{dg}{dy}(4) = 1/\frac{df}{dx}(3) = \frac{1}{1-\pi/4}$

33. Tangent lines: $y = 1 + \frac{1}{2}(x-2)$ and $y = -1 - \frac{1}{2}(x-2)$

36. **(a)** $M(2) = 5$ • $M'(x) = 2x$ • $\frac{dN}{dy}(5) = 1/\frac{dM}{dx}(2) = \frac{1}{4}$

(b) $N(y) = \sqrt{y-1}$ • $\frac{dN}{dy} = \frac{1}{2}(y-1)^{-1/2}$ • $\frac{dN}{dy}(5) = \frac{1}{4}$

39. $\frac{d}{dx}(x^3 - y^3 + x^2 + y^2) = \frac{d}{dx}(12)$ • $3x^2 - 3y^2y' + 2x + 2yy' = 0$ • $y'(2) = 16$ • $6x - 6y(y')^2 - 3y^2y'' + 2 + 2(y')^2 + 2yy'' = 0$ • Set $x = 2, y = 1$: $6(2) - 6(1)(16)^2 - 3(1)^2y'' + 2 + 2(16)^2 + 2(1)y'' = 0$ • $y''(2) = -1010$

40. $y''(1) = -\frac{34}{11}$

Tune-Up Exercises 5.3

T1. $F = x^2 + 4x - 2$ • $F' = 2x + 4$ is negative for $x < -2$ and positive for $x > -2$. • F has a global minimum of -6 at $x = -2$ • Figure T1 • To run the program set $Y_1 = x^2 + 4x - 2$, $Y_2 = 2x + 4$, and $-6 \le x \le 3$, $-8 \le y \le 8$. • To make direct calculations use $N(x) = x - \frac{x^2+4x-2}{2x+4}$. • For the poitive solution set $x_0 = 1$. • $x_1 = 0.5$, $x_2 = 0.45$, $x_3 \doteq 0.449489796$, $x_4 \doteq x_5 \doteq x^* = 0.449489743$ • For the negative solution set $x_0 = -5$ • $x_1 = -4.5$, $x_2 = -4.45$, $x_3 \doteq -4.4498489796$, $x_4 \doteq x_5 \doteq x^* = -4.449489743$ • Quadratic formula: $x = \frac{-4 \pm \sqrt{4^2 - 4(-2)(1)}}{2(1)} = -2 \pm \sqrt{6} \doteq 0.449489743$ and -4.449489743

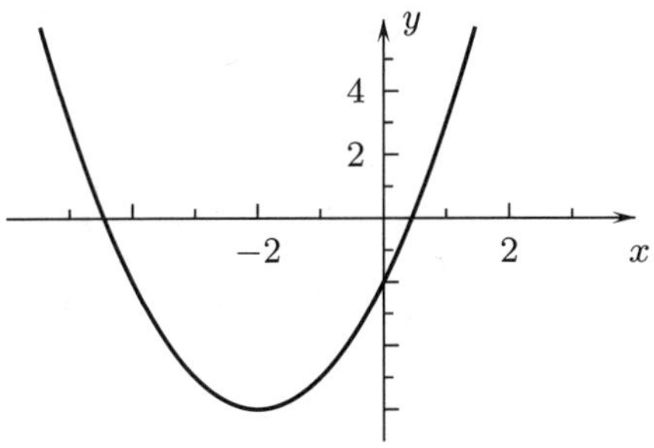

Figure T1

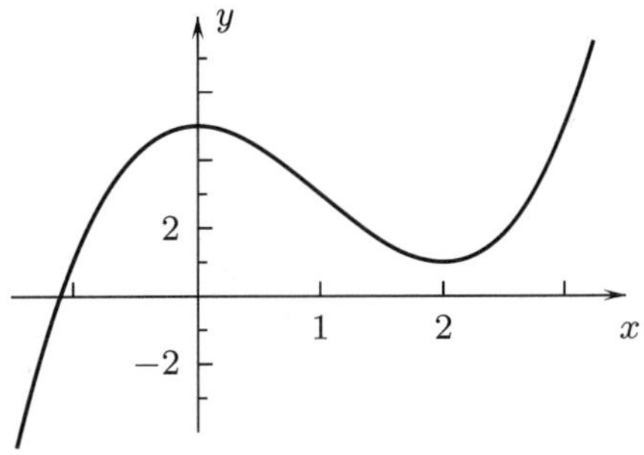

Figure T2

T2. $F = x^3 - 3x^2 + 5$ • $F' = 3x^2 - 6x = 3x(x-2)$ is positive for $x < 0$ and $x > 2$ and negative for $0 < x < 2$. • Local maximum of $F(0) = 5$ at $x = 0$ and local minimum of $F(2) = 1$ at $x = 2$ • Figure T2 • With $x_0 = -1$: $x_1 \doteq -1.111111111$ $x_2 \doteq -1.103835979$, $x_3 \doteq x_4 \doteq x^* = -1.103803403$

T3. **(a)** With $x_0 = 2$, x_1 is not defined because $F'(2) = 0$ and the tangent line at $x = x_1 = 2$ does not intersect the x-axis. **(b)** With $x_0 = 1.9$, Newton's method works but takes 13 steps. • $x_{12} \doteq x_{13} \doteq x^* = -1.103803403$ **(c)** With $x_0 = -10$, Newton's method works but takes 11 steps. • $x_{10} \doteq x_{11} = x^* = -1.103803403$

Problems 5.3

1. Figure A1 • $F = x^3 - \cos x$ • $F' = 3x^2 + \sin x$ • $N(x) = x - \dfrac{x^3 - \cos x}{3x^2 + \sin x}$ • With $x_0 = 1 : x_1 \doteq 0.880332900$, $x_2 \doteq 0.865684163$, $x_3 \doteq 0.865479076$, $x_4 \doteq x_5 \doteq x^* = 0.8654740033$

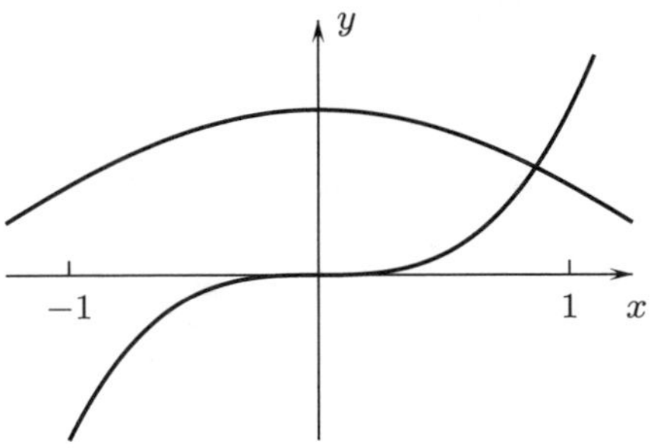

Figure A1

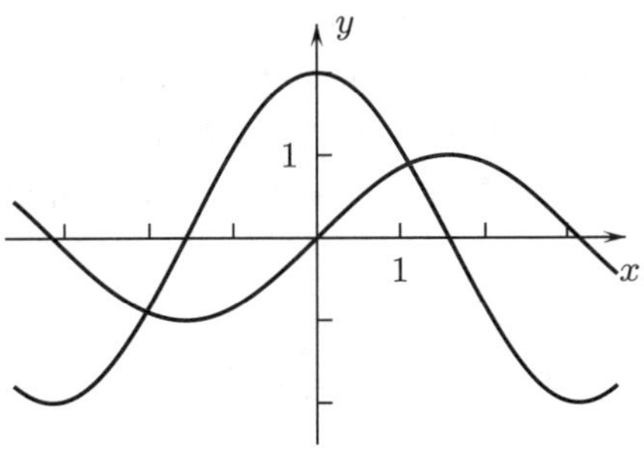

Figure A2

2. Figure A2 • With $x_0 = 1$: $x_1 \doteq 1.107560662$, $x_2 \doteq 1.1077148718$, $x_3 \doteq x_4 = x^* \doteq 1.107148718$ • With $x_0 = -1$: $x_1 \doteq -2.682136248$, $x_2 \doteq -1.925566804$, $x_3 \doteq -2.034876204$, $x_4 \doteq x_5 \doteq x^* = -2.3443936$ • $x \approx 1.107148718, -2.03443936$

4. Figure A4 • With $x_0 = -3$: $x_1 \doteq -2.916666667$, $x_2 \doteq -2.911301343$, $x_3 \doteq -2.911279905$, $x_4 \doteq x_5 \doteq x^* = -2.911279904$ • With $x_0 = 0$: $x_1 \doteq 0.333333333$, $x_2 \doteq 0.328427575$, $x_3 \doteq x_4 \doteq x^* = 0.328427575$ • $x \approx -2.911279904, 0.328427575$

6. Figure A6 • With $x_0 = -1$: $x_1 \doteq -2.859375000$, $x_2 \doteq -2.478887480$, $x_3 \doteq -2.425196558$, $x_4 \doteq -2.424301624$, $x_5 \doteq x_6 \doteq x^* = -2.424301383$ • With $x_0 = 3$: $x_1 \doteq 8.8125$, $x_2 \doteq 6.824709485$, $x_3 \doteq 5.520609273$ $x_4 \doteq 4.839349228$ $x_5 \doteq 4.640784735$ $x_6 \doteq 4.624921380$ $x_7 \doteq 4.624824736$ $x_8 \doteq x_9 \doteq x^* = 4.624824733$ • $x \approx -2.424301383, 4.624824733$

7. With $x_0 = 0$: $x_1 \doteq -0.666666667$, $x_2 \doteq -0.761904762$, $x_3 \doteq -0.763931104$, $x_4 \doteq x_5 \doteq x^* = -0.763932023$ • With $x_0 = -6$: $x_1 \doteq -5.333333333$, $x_2 \doteq -5.238095238$, $x_3 \doteq -5.236068896$, $x_4 \doteq x_5 \doteq x^* = -5.236067978$ •

Quadratic fomula: $x = \dfrac{-6 \pm \sqrt{20}}{2} \doteq -0.763932023, -5.236087978$

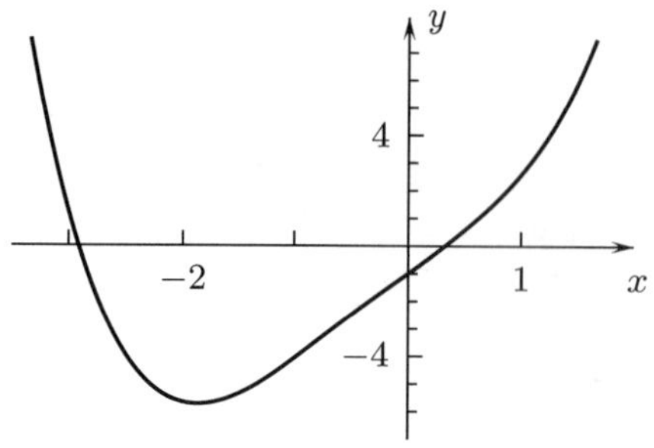

Figure A4

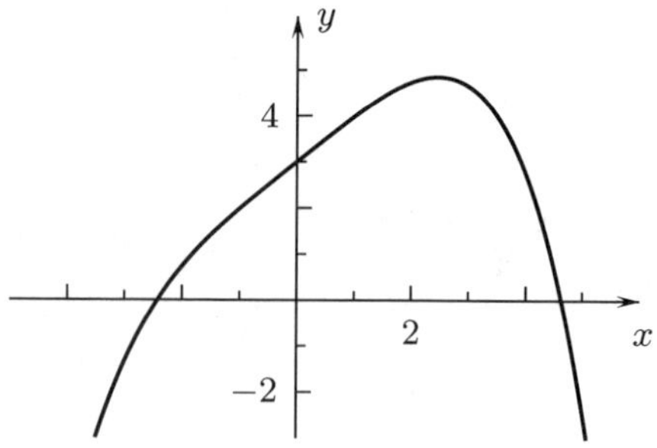

Figure A6

8. $F = x - x^{1/2} - 4$ • $F' = 1 - \frac{1}{2}x^{-1/2}$ • With $x_0 = 3$: $x_1 \doteq 6.840791937$, $x_2 \doteq 6.562240432$, $x_3 \doteq 6.561552817$, $x_4 \doteq 6.561552813$, $x_5 \doteq x_6 \doteq x^* = 6.561552813$ • Quadratic formula; $\sqrt{x} = \dfrac{1 \pm \sqrt{1+16}}{2}$ • The minus sign does not apply because $\sqrt{x} \geq 0$. • $x = (\dfrac{1+\sqrt{1+16}}{2})^2 \doteq 6.561552813$

11. For $F = x^{1/3}, F' = \frac{1}{3}x^{-2/3}$ and $N(x) = x - \dfrac{x^{1/3}}{\frac{1}{3}x^{-2/3}} = -2x$ • $x_1 = -2x_0$, $x_2 = -2x_1$, etc. • $x_n = (-2)^n x_0$ • For $x_0 \neq 0, |x_n| \to \infty$ as $n \to \infty$ and the x_n are alternately positive and negative.

15. $200 = \dfrac{10000r}{1-(1+r)^{-60}} \iff 1 - (1+r)^{-60} = 50r$ • $F = 50r + (1+r)^{-60} - 1$ • $N(r) = r - \dfrac{50r + (1+r)^{-60} - 1}{50 - 60(1+r)^{-61}}$ • (Use $0 \leq r \leq 0.01$, $-0.2 \leq y \leq 0.01$ with the program) • With $r_0 = 0.005$: $r_1 \doteq 0.0065033706$, $r_2 \doteq 0.0061969608$, $r_3 \doteq 0.0061834376$, $r_4 \doteq r_5 \doteq r^* \doteq 0.0061834132$ • $r \doteq 0.618\%$ per month or 7.416% per year

16. With $r_0 = 2$: $r_1 \doteq 1.870057685$, $r_2 \doteq 1.75268570$, $r_3 \doteq 1.664137066$, $r_4 \doteq r_5 = r^* \doteq 1.664100589$

17. $x^* = 0.4758690835$ • [Maximum] $\approx y(x^*) \doteq 0.9405149889$

24. $x^* = 0.231527567$ • [Minimum] $\approx y(x^*) \doteq -0.285132859$

26. $f = [\text{Distance}]^2$ • $t^* = 0.774980736$ • [Minimum] $\approx \sqrt{f(t^*)} \doteq 1.432361853$

Tune-Up Exercises 5.4

T1. **(a)** For $x \neq 0$, $\dfrac{3x^4+6x}{x^2+5x} = \dfrac{x(3x^3+6)}{x(x+5)} = \dfrac{3x^3+6}{x+5}$ • $\dfrac{3x^3+6}{x+5} \to \frac{6}{5}$ as $x \to 0$

(b) $\lim\limits_{x\to 0} \dfrac{3x^4+6x}{x^2+5x}$ (Type $\dfrac{0}{0}$) $= \lim\limits_{x\to 0} \dfrac{\frac{d}{dx}(3x^4+6x)}{\frac{d}{dx}(x^2+5x)} = \lim\limits_{x\to 0} \dfrac{12x^3+6}{2x+5} = \frac{6}{5}$

T2. **(a)** $\sin(3x) = 0$ at $x = 0$ • $\dfrac{d}{dx}(\sin(3x)) = \cos(3x)\dfrac{d}{dx}(3x) = 3\cos(3x) = 3$ at $x = 0$ • Tangent line: $y = 3x$ • Figure T2a • $\sin(2x) = 0$ at $x = 0$ • $\dfrac{d}{dx}(\sin(2x)) = \cos(2x)\dfrac{d}{dx}(2x) = 2\cos(2x) = 2$ at $x = 0$ • Tangent line: $y = 2x$ • Figure T2b

(b) Prediction: $\dfrac{\sin(3x)}{\sin(2x)} \approx \dfrac{3x}{2x} = \frac{3}{2}$ for x near 0 so $\dfrac{\sin(3x)}{\sin(2x)} \to \frac{3}{2}$ as $x \to 0$

(c) $\lim\limits_{x\to 0} \dfrac{\sin(3x)}{\sin(2x)}$ (Type $\dfrac{0}{0}$) $= \lim\limits_{x\to 0} \dfrac{\frac{d}{dx}\sin(3x)}{\frac{d}{dx}\sin(2x)} = \lim\limits_{x\to 0} \dfrac{3\cos(3x)}{2\cos(2x)} = \frac{3}{2}$

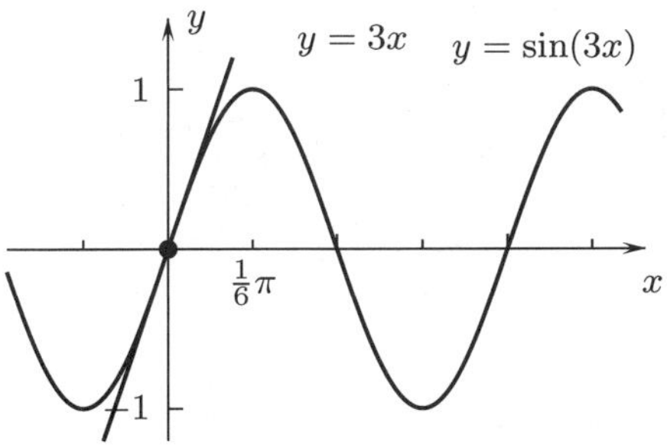

Figure T2a

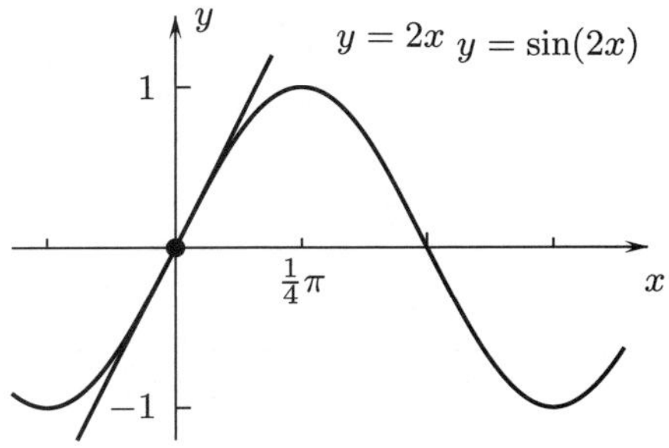

Figure T2b

T3. **(a)** A table of values is below. • Prediction: $\lim\limits_{x\to 1} \dfrac{x+x^2+x^3-3}{x^2+x-2} = 2$

x	1.1	1.001	1.00001	0.9	0.999	0.99999
$\dfrac{x+x^2+x^3-3}{x^2+x-2} \doteq$	2.0667741	2.000667	2.000007	1.934483	1.993333	1.999993

(b) Figure T3 • Prediction: $\lim\limits_{x\to 1} \dfrac{x+x^2+x^3-3}{x^2+x-2} = 2$

(c) $\lim\limits_{x\to 1} \dfrac{x+x^2+x^3-3}{x^2+x-2}$ (Type $\dfrac{0}{0}$) $= \lim\limits_{x\to 1} \dfrac{\dfrac{d}{dx}(x+x^2+x^3-3)}{\dfrac{d}{dx}(x^2+x-2)} = \lim\limits_{x\to 1} \dfrac{1+2x+3x^2}{2x+1} = \frac{6}{3} = 2$

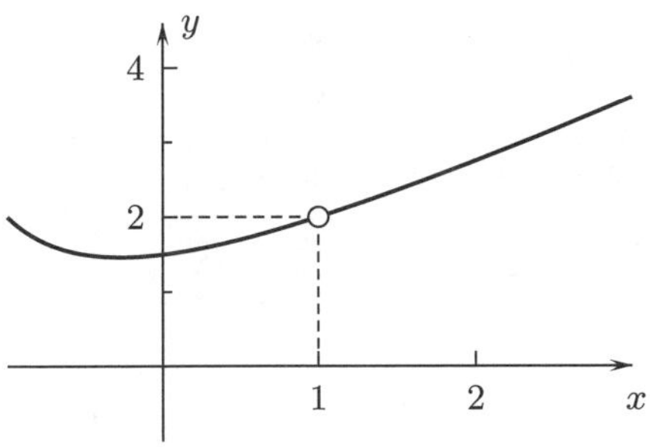

Figure T3

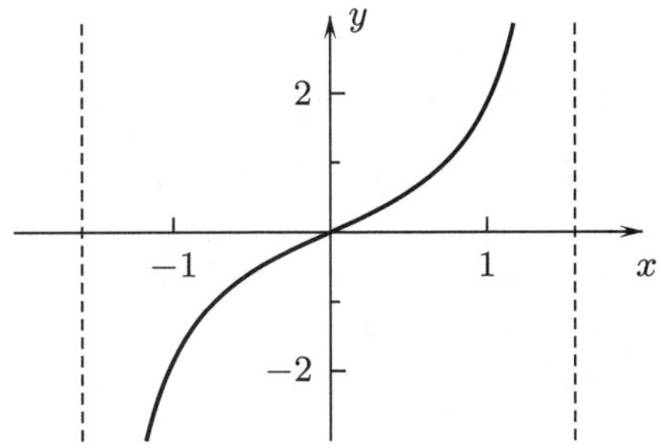

Figure T4

T4. **(a)** A table of values is below. • Prediction: $\lim\limits_{x\to 0} \dfrac{x}{\cos x} = 0$

x	0.1	0.001	0.00001	−0.1	−0.001	−0.00001
$\dfrac{x}{\cos x} \doteq$	0.100503	0.001000	0.000010	−0.100503	−0.001000	−0.000010

(b) Figure T4 • Prediction: $\lim\limits_{x\to 0} \dfrac{x}{\cos x} = 0$

(c) $\lim\limits_{x\to 0} \dfrac{x}{\cos x} = \frac{0}{1} = 0$ since $x \to 0$ and $\cos x \to 1$ as $x \to 0$

T5. Figure T5 • $\lim\limits_{x\to 0} \dfrac{1-\cos x}{x^2}$ (Type $\dfrac{0}{0}$) $= \lim\limits_{x\to 0} \dfrac{\dfrac{d}{dx}(1-\cos x)}{\dfrac{d}{dx}(x^2)} = \lim\limits_{x\to 0} \dfrac{\dfrac{d}{dx}(1-\cos x)}{\dfrac{d}{dx}(x^2)}$

$= \lim\limits_{x\to 0} \dfrac{\sin x}{2x}$ (Type $\dfrac{0}{0}$) $= \lim\limits_{x\to 0} \dfrac{\dfrac{d}{dx}(\sin x)}{\dfrac{d}{dx}(2x)} = \lim\limits_{x\to 0} \dfrac{\cos x}{2} = \frac{1}{2}$

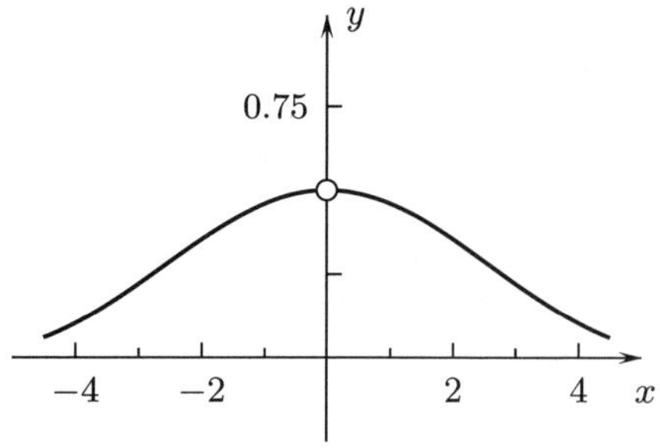

Figure T5

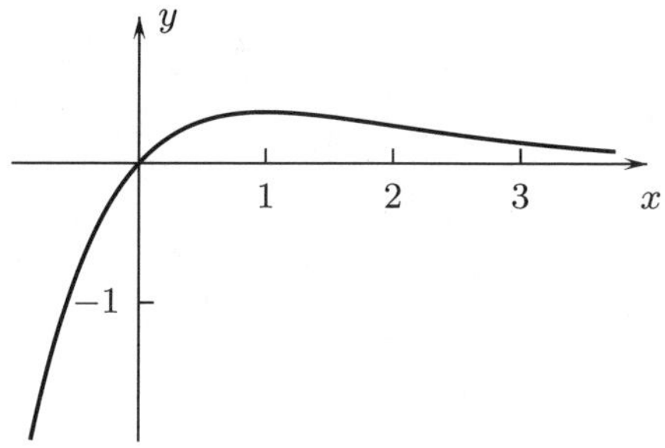

Figure T8

T6 **(a)** The table is below. • Prediction: $\lim\limits_{x\to\infty}\dfrac{4x^2+2x}{5x^2+3}=\frac{4}{5}$

x	10	1000	1,000	10,000	100,000
$\dfrac{4x^2+2x}{5x^2+3}\doteq$	0.834990	0.803952	0.800400	0.800040	0.800004

(b) $\lim\limits_{x\to\infty}\dfrac{4x^2+2x}{5x^2+3}=\lim\limits_{x\to\infty}\dfrac{4+2/x}{5+3/x}=\frac{4}{5}$

T7. $\lim\limits_{x\to\infty}\dfrac{e^x+x}{\ln x+x}$ (Type $\dfrac{\infty}{\infty}$) $=\lim\limits_{x\to\infty}\dfrac{\frac{d}{dx}(e^x+x)}{\frac{d}{dx}(\ln x+x)}=\lim\limits_{x\to\infty}\dfrac{e^x+1}{\frac{1}{x}+1}=\infty$

T8. Figure T8 **(a)** $\lim\limits_{x\to\infty}(xe^{-x})$ (Type $\infty\cdot 0$) $=\lim\limits_{x\to\infty}\dfrac{x}{e^x}$ (Type $\dfrac{\infty}{\infty}$) $=\lim\limits_{x\to\infty}\dfrac{\frac{d}{dx}(x)}{\frac{d}{dx}(e^x)}=\lim\limits_{x\to\infty}\dfrac{1}{e^x}=0$

(b) $\lim\limits_{x\to 0}(xe^{-x})=\Big[xe^{-x}\Big]_{x=0}=0$

(c) $\lim\limits_{x\to-\infty}(xe^{-x})=-\infty$ since $x\to-\infty$ and $e^{-x}\to\infty$ as $x\to-\infty$

T9. For $x\neq -4: G(x)=\dfrac{5x^2+1}{x+4}-5x=\dfrac{5x^2+1-5x(x+4)}{x+4}=\dfrac{1-20x}{x+4}=\dfrac{x^{-1}-20}{1+4x^{-1}}$ • $\dfrac{x^{-1}-20}{1+4x^{-1}}\to -20$ as $x\to\infty$

T10. $\lim\limits_{x\to 0+}(\sqrt{x}\ln x)$ (Type $0\cdot\infty$) $=\lim\limits_{x\to 0+}\dfrac{\ln x}{x^{-1/2}}$ (Type $\dfrac{\infty}{\infty}$) $=\lim\limits_{x\to 0+}\dfrac{x^{-1}}{-\frac{1}{2}x^{-3/2}}=\lim\limits_{x\to 0+}(-2x^{1/2})=0$

Problems 5.4

1. **(a)** $\lim\limits_{x\to 3}\dfrac{x-3}{x^2-9}=\lim\limits_{x\to 3}\dfrac{x-3}{(x-3)(x+3)}=\lim\limits_{x\to 3}\dfrac{1}{x+3}=\frac{1}{6}$

(b) $\lim\limits_{x\to 3}\dfrac{x-3}{x^2-9}=\lim\limits_{x\to 3}\dfrac{\frac{d}{dx}(x-3)}{\frac{d}{dx}(x^2-9)}=\lim\limits_{x\to 3}\dfrac{1}{2x}=\frac{1}{6}$

5. Figure A5 • $\lim\limits_{x\to 0}\dfrac{1-e^{2x}}{\sin(4x)}$ (Type $\dfrac{0}{0}$) $=\lim\limits_{x\to 0}\dfrac{\frac{d}{dx}(1-e^{2x})}{\frac{d}{dx}(\sin(4x))}=\lim\limits_{x\to 0}\dfrac{-2e^{2x}}{4\cos(4x)}=-\frac{2}{4}=-\frac{1}{2}$

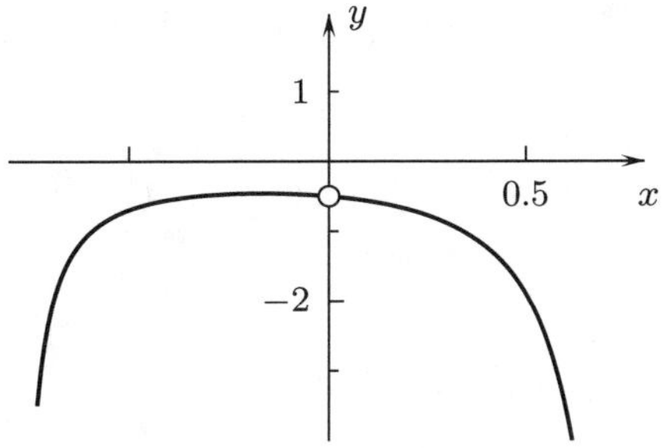

Figure A5

7. Figure A7 • $\lim_{x\to 1} \frac{x^2-2x+1}{\ln x}$ (Type $\frac{0}{0}$) $= \lim_{x\to 1} \frac{\frac{d}{dx}(x^2-2x+1)}{\frac{d}{dx}(\ln x)} = \lim_{x\to 1} \frac{2x-2}{1/x} = 0$ •

$\lim_{x\to 0^+} \frac{x^2-2x+1}{\ln x} = 0$ since $x^2-2x+1 \to 1$ and $\ln x \to -\infty$ as $x \to 0^+$.

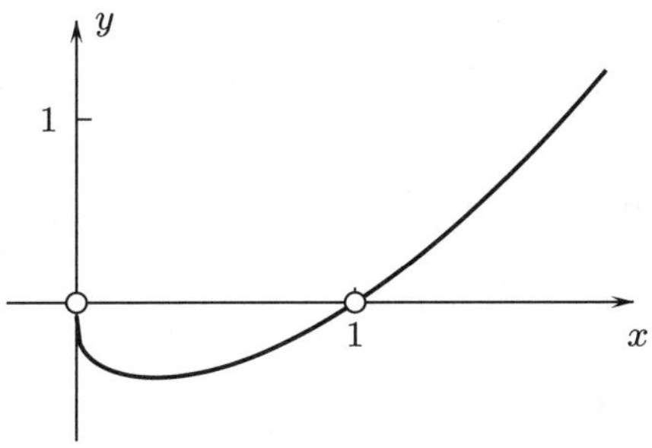

Figure A7

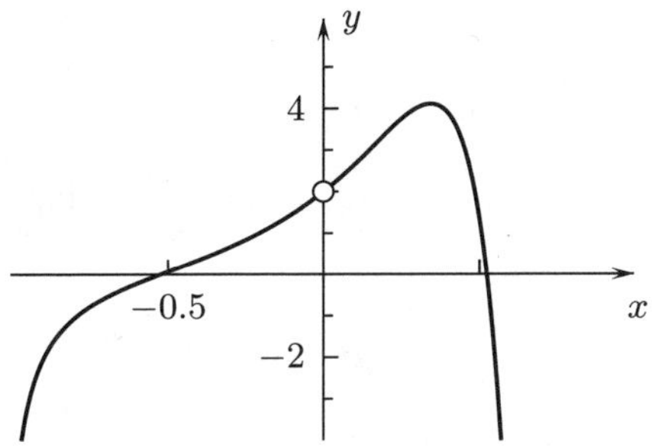

Figure A8

8. Figure A8 • $\lim_{x\to 0} \frac{e^{6x}-1}{\tan(3x)}$ (Type $\frac{0}{0}$) $= \lim_{x\to 0} \frac{\frac{d}{dx}(e^{6x}-1)}{\frac{d}{dx}(\tan(3x))} = \lim_{x\to 0} \frac{6e^{6x}}{3\sec^2(3x)} = \frac{6}{3} = 2$

12. **(a)** $\lim_{x\to 0} \frac{\sin(x^2)}{x^2}$ (Type $\frac{0}{0}$) $= \lim_{x\to 0} \frac{\frac{d}{dx}(\sin(x^2))}{\frac{d}{dx}(x^2)} = \lim_{x\to 0} \frac{2x\cos(x^2)}{2x} = \lim_{x\to 0} \cos(x^2) = 1$

(b) $\lim_{x\to\infty} \frac{\sin(x^2)}{x^2} = 0$ since $x^2 \to \infty$ as $x \to \infty$ and $|\sin(x^2)| \le 1$ for all x. • Figure A12

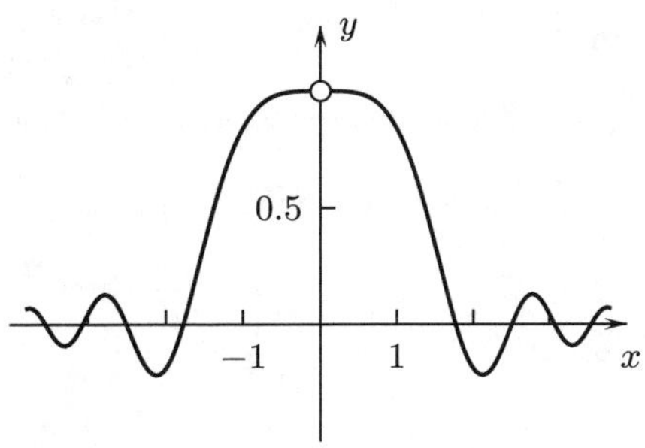

Figure A12

13. $\lim\limits_{x\to 0}\dfrac{x^3}{\sin x - x}$ (Type $\frac{0}{0}$) $= \lim\limits_{x\to 0}\dfrac{\frac{d}{dx}(x^3)}{\frac{d}{dx}(\sin x - x)} = \lim\limits_{x\to 0}\dfrac{3x^2}{\cos x - 1}$ (Type $\frac{0}{0}$)

$= \lim\limits_{x\to 0}\dfrac{\frac{d}{dx}(3x^2)}{\frac{d}{dx}(\cos x - 1)} = \lim\limits_{x\to 0}\dfrac{6x}{-\sin x}$ (Type $\frac{0}{0}$) $= \lim\limits_{x\to 0}\dfrac{\frac{d}{dx}(6x)}{\frac{d}{dx}(-\sin x)} = \lim\limits_{x\to 0}\dfrac{6}{-\cos x} = -6$

14. $\lim\limits_{x\to 1}\dfrac{(\ln x)^2}{x^3 - 2x^2 + x}$ (Type $\frac{0}{0}$) $= \lim\limits_{x\to 1}\dfrac{\frac{d}{dx}[(\ln x)^2]}{\frac{d}{dx}(x^3 - 2x^2 + x)} = \lim\limits_{x\to 1}\dfrac{2\ln x\frac{d}{dx}(\ln x)}{3x^2 - 4x + 1} =$

$\lim\limits_{x\to 1}\dfrac{2(\ln x)/x}{3x^2 - 4x + 1} = \lim\limits_{x\to 1}\dfrac{2\ln x}{3x^3 - 4x^2 + x}$ (Type $\frac{0}{0}$) $=$

$\lim\limits_{x\to 1}\dfrac{\frac{d}{dx}(2\ln x)}{\frac{d}{dx}(3x^3 - 4x^2 + x)} = \lim\limits_{x\to 1}\dfrac{2/x}{9x^2 - 8x + 1} = \dfrac{2}{9 - 8 + 1} = 1$

16. $\lim\limits_{y\to 0}\dfrac{\cos(6y)}{\cos(5y)} = 1$

19. Figure A19 • **(a)** $\lim\limits_{x\to 0^-}\dfrac{e^x + x}{1 - e^x} = \infty$ since $e^x + x \to 1$ and $1 - e^x \to 0$ as $x \to 0^-$ and $1 - e^x > 0$ for $x < 0$

(b) $\lim\limits_{x\to 0^+}\dfrac{e^x + x}{1 - e^x} = -\infty$ since $e^x + x \to 1$ and $1 - e^x \to 0$ as $x \to 0^+$ and $1 - e^x < 0$ for $x > 0$

(c) $\lim\limits_{x\to\infty}\dfrac{e^x + x}{1 - e^x}$ (Type $\frac{\infty}{\infty}$) $= \lim\limits_{x\to\infty}\dfrac{\frac{d}{dx}(e^x + x)}{\frac{d}{dx}(1 - e^x)} = \lim\limits_{x\to\infty}\dfrac{e^x + 1}{-e^x} = \lim\limits_{x\to\infty}(-1 - e^{-x}) = -1$

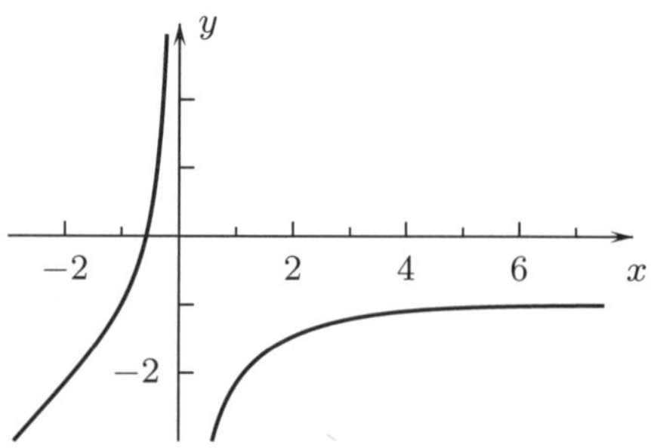

Figure A19

22. $\lim\limits_{x\to\infty}\dfrac{x + \ln x}{x}$ (Type $\frac{\infty}{\infty}$) $= \lim\limits_{x\to\infty}\dfrac{\frac{d}{dx}(x + \ln x)}{\frac{d}{dx}(x)} = \lim\limits_{x\to\infty}\dfrac{1 + 1/x}{1} = 1$

24. $\lim\limits_{x\to\infty}\dfrac{\tan^{-1}(3x)}{\tan^{-1}(-x)} = \dfrac{\frac{1}{2}\pi}{-\frac{1}{2}\pi} = -1$

25. $\lim\limits_{x\to\infty}\dfrac{\sin x}{\ln x} = 0$ since $\ln x \to \infty$ as $x \to \infty$ and $|\sin x| \le 1$ for all x

30. **(a)** $\lim\limits_{x\to 0^+}\dfrac{\ln x}{x} = -\infty$ since $\ln x \to -\infty$ as $x \to 0^+$

(b) $\lim\limits_{x\to\infty}\dfrac{\ln x}{x}$ (Type $\frac{\infty}{\infty}$) $= \lim\limits_{x\to\infty}\dfrac{\frac{d}{dx}(\ln x)}{\frac{d}{dx}(x)} = \lim\limits_{x\to\infty}\dfrac{1/x}{1} = 0$

(c) For $y = \dfrac{\ln x}{x}$: $\dfrac{dy}{dx} = \dfrac{x\frac{d}{dx}(\ln x) - \ln x\frac{d}{dx}(x)}{x^2} = \dfrac{1 - \ln x}{x^2}$ is positive for $0 < x < e$ and negative for $x > e$ • $y(x)$ is increasing in $(0, e]$ and decreasing in $[e, \infty)$ • [Maximum]$= \dfrac{\ln e}{e} = \dfrac{1}{e}$ at $x = e$

34. $\lim_{x\to 0}\dfrac{\sin(ax)}{\sin(bx)}$ (Type $\dfrac{0}{0}$) $= \lim_{x\to 0}\dfrac{\dfrac{d}{dx}\sin(ax)}{\dfrac{d}{dx}\sin(bx)} = \lim_{x\to 0}\dfrac{\cos(ax)\dfrac{d}{dx}(ax)}{\cos(bx)\dfrac{d}{dx}(bx)} = \lim_{x\to 0}\dfrac{a\cos(ax)}{b\cos(bx)} = \dfrac{a}{b}$

39. $\lim_{x\to 0^+} x(\ln x)^{1/3}$ (Type $0\cdot\infty$) $= \lim_{x\to 0^+}\dfrac{\ln x}{x^{-1}}$ (Type $\dfrac{\infty}{\infty}$) $= \lim_{x\to 0^+}\dfrac{\dfrac{d}{dx}(\ln x)^{1/3}}{\dfrac{d}{dx}(x^{-1})} =$

$\lim_{x\to 0^+}\dfrac{\frac{1}{3}(\ln x)^{-2/3}\dfrac{d}{dx}(\ln x)}{-x^{-2}} = \lim_{x\to 0^+}\dfrac{-x}{3(\ln x)^{2/3}} = 0$ since $-x\to 0$ and $(\ln x)^{2/3}\to\infty$ as $x\to 0^+$

41. Type 1^∞ • $\lim_{x\to 0}[\ln(1+3x+x^2)^{1/x}] = \lim_{x\to 0}\dfrac{\ln(1+3x+x^2)}{x}$ (Type $\dfrac{0}{0}$) $=$

$\lim_{x\to 0}\dfrac{\dfrac{d}{dx}(\ln(1+3x+x^2))}{\dfrac{d}{dx}(x)} = \lim_{x\to 0}\dfrac{\dfrac{1}{1+3x+x^2}\dfrac{d}{dx}(1+3x+x^2)}{1} = \lim_{x\to 0}\dfrac{3+2x}{1+3x+x^2} = 3$ •

$\lim_{x\to 0}(1+3x+x^2)^{1/x} = e^3$

42. $\lim_{x\to\infty}(x\ln x - 20x^2)$ (Type $\infty-\infty$) $= \lim_{x\to\infty} x^2(\dfrac{\ln x}{x} - 20) = -\infty$ since

$\lim_{x\to\infty}\dfrac{\ln x}{x}$ (Type $\dfrac{\infty}{\infty}$) $= \lim_{x\to\infty}\dfrac{\dfrac{d}{dx}(\ln x)}{\dfrac{d}{dx}(x)} = \lim_{x\to\infty}\dfrac{1}{x} = 0$

45. Type 1^∞ • $\lim_{x\to 0^-}[\ln(1-4x+5x^2)^{1/x^2}] = \lim_{x\to 0^-}\dfrac{\ln(1-4x+5x^2)}{x^2}$ (Type $\dfrac{0}{0}$) $=$

$\lim_{x\to 0^-}\dfrac{\dfrac{d}{dx}(\ln(1-4x+5x^2))}{\dfrac{d}{dx}(x^2)} = \lim_{x\to 0^-}\dfrac{\dfrac{1}{1-4x+5x^2}\dfrac{d}{dx}(1-4x+5x^2)}{2x} =$

$\lim_{x\to 0^-}\dfrac{10x-4}{2x(1-4x+5x^2)} = \infty$ • $\lim_{x\to 0^-}(1-4x+5x^2)^{1/x^2} = \infty$

56. $y = x\ln x$ is defined for $x>0$ • $\lim_{x\to 0^+} x\ln x$ (Type $0\cdot\infty$)

$= \lim_{x\to 0^+}\dfrac{\ln x}{x^{-1}}$ (Type $\dfrac{\infty}{\infty}$) $= \lim_{x\to 0^+}\dfrac{\dfrac{d}{dx}(\ln x)}{\dfrac{d}{dx}(x^{-1})} = \lim_{x\to 0^+}\dfrac{x^{-1}}{-x^{-2}} = \lim_{x\to 0^+}(-x) = 0$ •

$\dfrac{d}{dx}(x\ln x) = x\dfrac{d}{dx}(\ln x) + \ln x\dfrac{d}{dx}(x) = 1+\ln x$ is negative for $0<x<e^{-1}$ and positive for $x>e^{-1}$ • $x\ln x$ is decreasing in $(0,e^{-1}]$ and increasing in $[e^{-1},\infty)$ • [Minimum of $x\ln x$] $= e^{-1}\ln(e^{-1}) = -e^{-1}$ at $x=e^{-1}$ • Figure A56

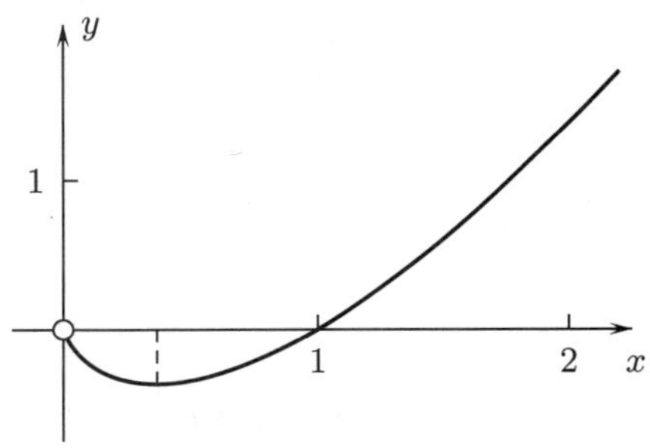

Figure A56

CHAPTER 6
INTEGRALS AND APPLICATIONS, PART II

This chapter is a continuation of Chapter 4 on the integral and its applications. In Section 6.1 we use integrals to find volumes of solids and areas of curved surfaces. Several special integration techniques are presented in Sections 6.2 and 6.3, and Section 6.4 provides practice in the use of integral tables and calculator or computer algebra systems to obtain exact values of definite integrals and exact formulas for indefinite integrals. Distribution densities and the Fundamental Theorem of Calculus for derivatives of integrals are discussed in Section 6.5. Improper integrals of unbounded functions and of integrals with infinite intervals of integration are studied in Section 6.6. Integrals are used to study work, energy, and forces of fluids in Section 6.7, and the hyperbolic functions are introduced in Section 6.8.

Section 6.1
Volume and surface area

OVERVIEW: *In this section we first see how volumes of solids can be calculated by integrating the areas of their parallel planar cross sections. We apply this technique, in particular, to solids that are obtained by rotating regions between graphs of functions about lines parallel to the x-axis. Next, we show how volumes of solids that are generated by rotating such regions about lines parallel to the y-axis can be calculated by integrating areas of their concentrlc cylindrical sections. Finally we show how integrals can be used to find the areas of surfaces that are obtained by rotating graphs of functions about lines parallel to the x-axis.*

Topics:

- ***Volumes by slicing***
- ***Volumes by cylindical sections***
- ***Areas of surfaces of revolution***

Volumes by slicing

We begin by reviewing the discussion in Question 6 of Section 4.1 concerning the volume of the spa in Figure 1 and how the volume is related to the areas of horizontal cross sections.

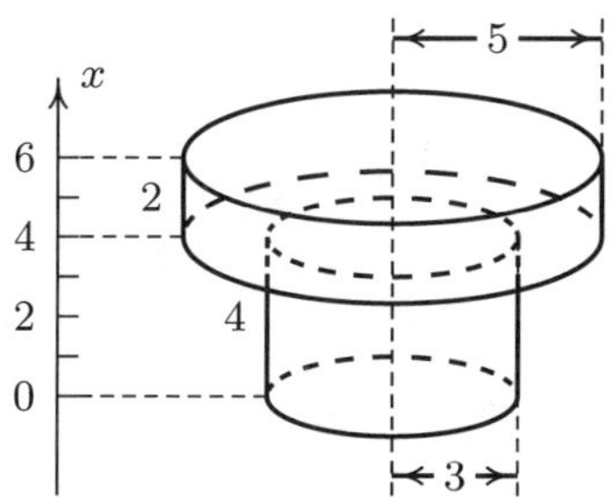

FIGURE 1

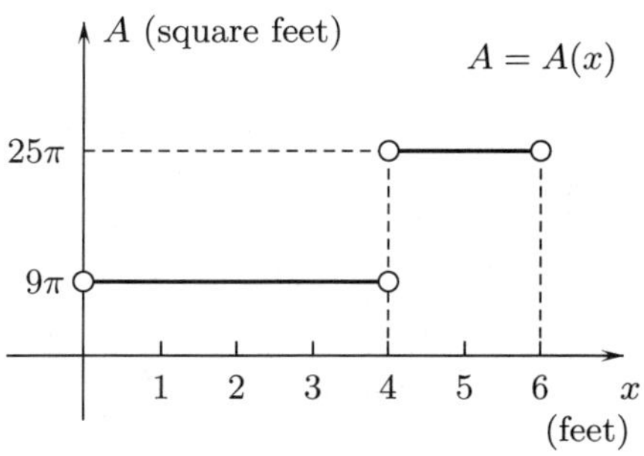

FIGURE 2

♦ SUGGESTIONS TO INSTRUCTORS: The main topics in this section can be divided into three parts, as follows:
Part 1 (Volumes by slicing; Tune-up Exercises T1, T2, T4; Problems 1–14, 16, 17, 19, 22, 26, 27, 30, 31, 32, 34, 35, 39, 42, 43, 47, 48, 50–58, 60) Have students begin work on Questions 1 through 6 before class. Problems 1, 2, 3, 4, 14, 27, 32, 43, and 50 are good for in-class discussion.
Part 2 (Volumes by cylindrical sections; Tune-up Exercise T3; Problems 15, 18, 20, 21, 28, 29, 33, 34, 37, 40, 41, 44, 45, 49) Have students begin work on Question 7 before class. Problems 15, 18, 19, 29, 33, 37, and 41 are good for in-class discussion.
Part 3 (Areas of surfaces of revolution; Tune-up Exercise T5; Problems 23–25, 46, 59) Have students begin work on Questions 8 and 9 before class. Problem 46 is good for in-class discussion.

The spa is in the shape of a cylinder of radius 3 feet and height 4 feet under a cylinder of radius 5 feet and height 2 feet. Because the volume of a cylinder of radius r and height h is equal to the area πr^2 of its base multiplied by its height h, the volume of the spa is

$$(\pi 3^2)(4) + (\pi 5^2)(2) = (9\pi)(4) + (25\pi)(2) = 86\pi \text{ square feet.} \quad \textbf{(1)}$$

The horizontal cross sections of the spa are circles. To identify them, we introduce a vertical x-axis with $x = 0$ at the bottom of the spa, as in Figure 1, and we refer to the horizontal cross section at height x as the CROSS SECTION AT x. As you can see from Figure 1, the cross section at x for $0 < x < 4$ is a circle of radius 3 feet and area $\pi 3^2 = 9\pi$, and the cross section at x for $4 < x < 6$ is a circle of radius 5 feet and area $\pi 5^2 = 25\pi$. Consequently, the area $A(x)$ of the cross section is the step function of Figure 2. Its integral from 0 to 6 is equal to the sum of the areas of the two rectangles between the graph and the x-axis:

$$\int_0^6 A(x)\,dx = 9\pi(4) + 25\pi(2) = 86\pi.$$

This equals the volume **(1)** of the cylinder.

We will see that volumes of solids whose cross-sectional areas are not step functions can also be found by integrating the areas. We begin by considering the solid in Figure 3, whose base is the region between the curves $y = \frac{1}{2}x^2$ and $y = 1 - \frac{1}{2}x^2$ for $0 \le x \le 1$ in Figure 4, and whose cross sections perpendicular to the x-axis are squares.

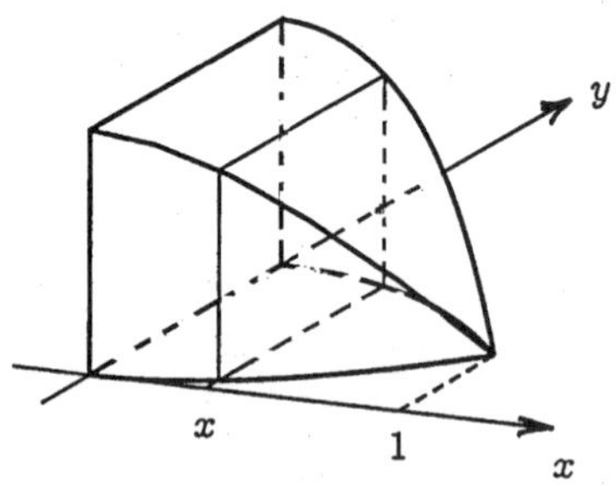

FIGURE 3

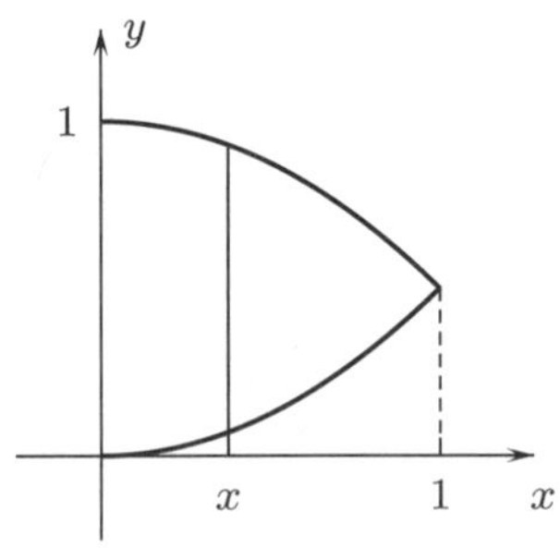

FIGURE 4

The vertical line across the region at x in Figure 4 is the base of the square cross section at x in Figure 3. The cross section at $x = 0$ is a square of side 1 and the cross section at $x = 1$ is a square of side 0 (a point). You can visualize that as x increases from 0 to 1, the cross sections generate (sweep out) the solid and that the cross-sectional areas $A(x)$ decrease continuously from 1 to 0.

Question 1 Find a formula for the area $A(x)$ of the cross section at x and generate its graph. You should obtain the curve in Figure 5.

FIGURE 5

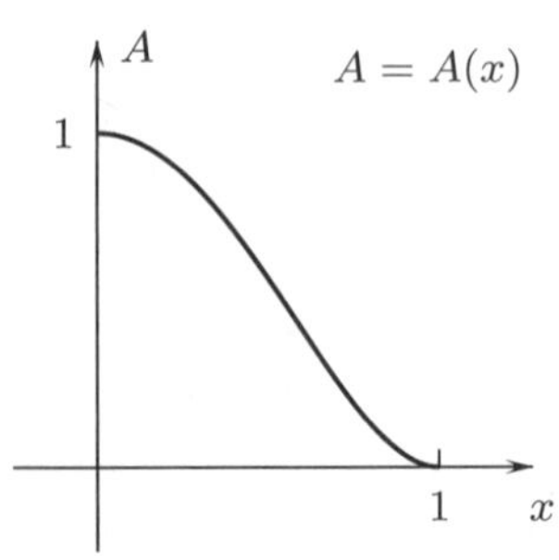

We will find the volume of the solid in Figure 3 by slicing it into pieces and approximating each piece by a box and find an expression for the total area of the boxes. To show the ideas, we will first make the calculations in the case where the solid is sliced into four regions as in Figure 6.

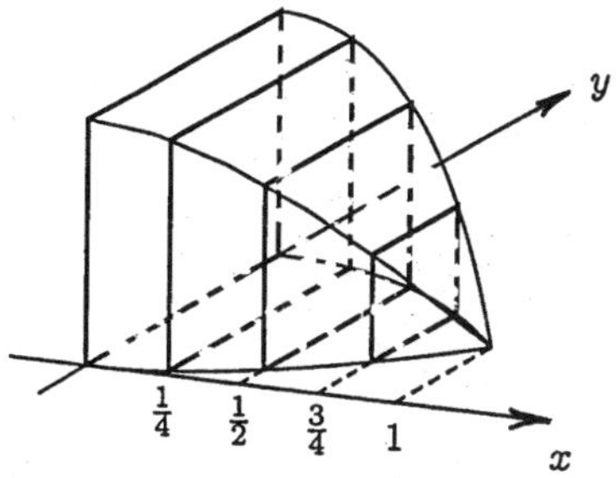

FIGURE 6

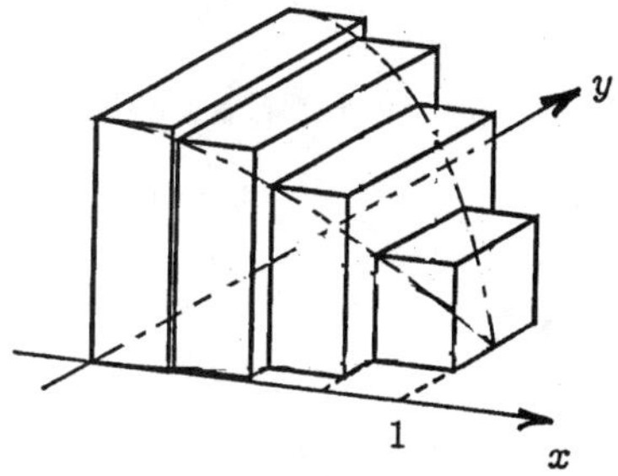

FIGURE 7

In Figure 6 the solid has been sliced into four pieces of width $\frac{1}{4}$ by the planes perpendicular to the x-axis at $x = \frac{1}{4}, \frac{1}{2}$, and $\frac{3}{4}$. We approximate each piece by a box of width $\frac{1}{4}$ whose vertical sides have the shape of the cross section of the solid at their left sides, as shown in Figure 7.

Example 1 Calculate the total volume of the four boxes in Figure 8.

SOLUTION The volume of each box equals its width $\frac{1}{4}$, multiplied by the area of its square sides, which we can calculate with the formula $A(x) = (1 - 2x^2)^2$ that you derived in Question 1.

The sides of the first box, on the left, have area $A(0) = (1 - 0^2)^2$, the sides of the second box have area $A(\frac{1}{4}) = \left(1 - (\frac{1}{4})^2\right)^2$, the sides of the third box have area $A(\frac{1}{2}) = \left(1 - (\frac{1}{2})^2\right)^2$, and the sides of the fourth box, on the right, have area $A(\frac{3}{4}) = \left(1 - (\frac{3}{4})^2\right)^2$. Therefore, the total volume of the boxes is

$$[(1-0^2)^2 + \left(1-(\tfrac{1}{4})^2\right)^2 + \left(1-(\tfrac{1}{2})^2\right)^2 + \left(1-(\tfrac{3}{4})^2\right)^2](\tfrac{1}{4}) \qquad \textbf{(2)}$$

A calculator gives the value $\frac{337}{512} = 0.6582\ldots$ for this area □

You can imagine that we would obtain better approximations of the original solid if we used more, thinner boxes, and that the volume of the approximations by boxes approaches the exact volume of the solid as the number of boxes tends to ∞.

To give a formula for the approximation by N boxes, where N is any positive integer, we use the partition

$$0 = x_0 < x_1 < x_2 < \cdots < x_N = 1$$

of the interval $[0, 1]$ into N equal subintervals of width $\Delta x = 1/N$ and slice the solid into N pieces with the planes perpendicular to the x-axis at $x = x_1, x_2, \ldots, x_{N-1}$. As in Figure 7 for the case of $N = 4$, we approximate each piece by a box of width Δx, whose left side is the cross section of the solid at its left side. Then the area of the side of the jth box is $A(x_{j-1}$, its volume is $A(x_{j-1}\Delta x$ and

$$[\text{The total area of the } N \text{ boxes}] = \sum_{j=1}^{N} A(x_{j-1})\,\Delta x = \sum_{j=1}^{N} \left(1-(x_{j-1})^2\right)^2 \Delta x. \qquad \textbf{(3)}$$

Notice that the sum on the right of **(3)** gives **(2)** for $N = 4$.

Question 2 What integral has the Riemann sums on the right of **(3)**?

To find the exact volume of the original solid, we let the number N of subintervals and boxes tend to ∞. The corresponding collections of boxes approach the original solid in Figure 4 and their total volume tends to its volume. On the other hand, the Riemann sum **(4)** tends to the integral $\int_0^1 (1-x^2)^2\,dx$ that you found in Question 3. Therefore,

$$\begin{aligned}[\text{Volume of the solid in Figure 4}] &= \int_0^1 (1-x^2)^2\,dx \\ &= \int_0^1 (1-2x^2+x^4)\,dx = \left[x - \tfrac{2}{3}x^3 + \tfrac{1}{5}x^5\right]_0^1 \\ &= [1-\tfrac{2}{3}+\tfrac{1}{5}] - [0] = \tfrac{8}{15}\end{aligned} \tag{4}$$

Question 3 Why is the approximate volume $\frac{337}{512} = 0.6582\ldots$ in **(2)** greater than the exact volume $\frac{8}{15} = 0.5333\ldots$ in **(4)**?

To explain how the procedure leading to integral **(4)** can be applied to other solids, we consider the solid of Figure 8. Its cross section at x is its intersection with the plane perpendicular to the x-axis that intersects the x-axis at x. We suppose that the region extends from its cross section at $x = a$ to its cross section at $x = b$ with $a < b$ and that we have a formula for the areas $A(x)$ of the cross sections

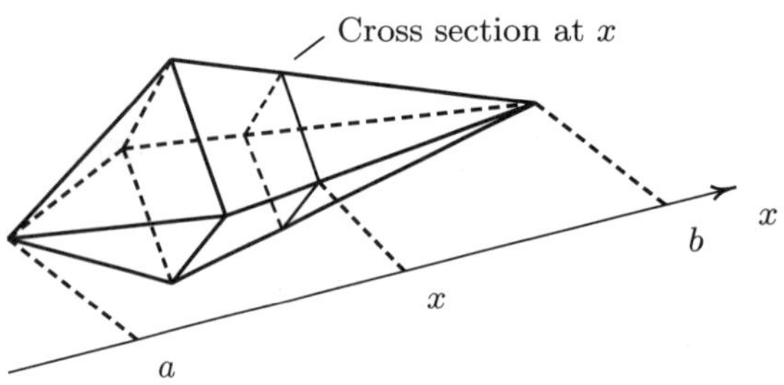

Cross section of area $A(x)$

FIGURE 8

The cross sections at x_j for $j = 0, 1, \ldots, N$ in a partition $a = x_0 < x_1 < x_2 < \cdots < x_N = b$ of $[a, b]$ into N equal subintervals of width Δx divide the solid into N slices of thickness Δx. We approximate the jth slice for each j by a wafer-shaped region whose face is the cross section of the solid at x_{j-1}, whose sides are perpendicular to its ends, and whose thickness is $\Delta x = x_j - x_{j-1}$ (Figure 9).

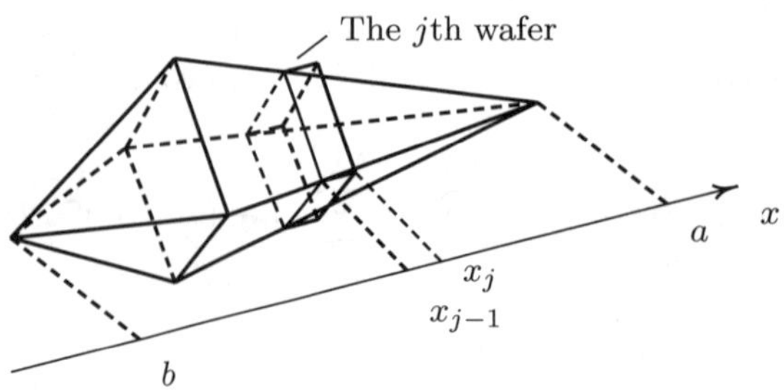

FIGURE 9

Question 4 What is the volume of the jth wafer in Figure 9?

Because the original solid is approximated by the N wafers, its volume is approximately equal to their total volume

$$\sum_{j=1}^{N} A(x_j)\,\Delta x. \tag{5}$$

We assume that the volume is defined and that the integral of $A(x)$ from a to b exists. Then as $N \to \infty$, the collections of wafers approximate the original solid with increasing accuracy and the Riemann sum in **(5)** tends to the integral of $A(x)$ from a to b. This leads us to the following rule:

Rule 1 *If a solid extends from its cross section perpendicular to an x-axis at a to its cross section at b with $a < b$ and if the area of its cross section at x is $A(x)$, then the volume of the solid is* $\int_a^b A(x)\,dx$.

Question 5 The intersection of a solid with an xy-plane is the region between the curves $y = x^{1/3}$ and $y = \frac{4}{3}x^{1/3}$ for $0 \le x \le 1$ in Figure 10. The cross sections of the solid perpendicular to the x-axis are circles with diameters in the xy-plane. **(a)** Describe the solid. **(b)** Find its volume. Distances are measured in feet.

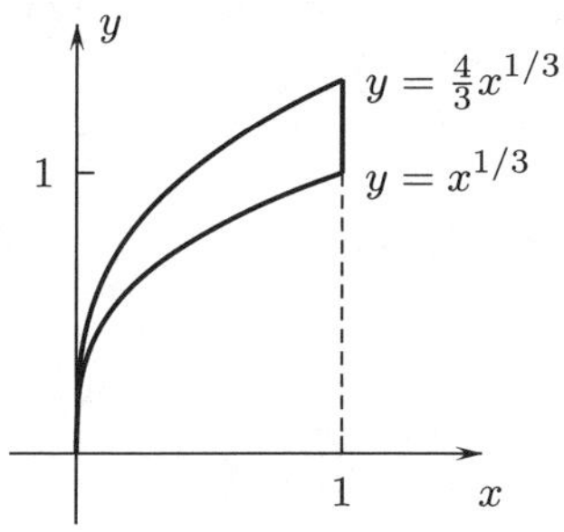

FIGURE 10

Volumes by slicing: solids of revolution

Rule 1 can often be used to find volumes of solids that are generated by regions between the graphs of functions and the x-axis or, as in the next Example, between the graphs of two functions when they are rotated about the x-axis or other lines parallel to the x-axis.

Example 2 Find the volume of the solid that is generated when the triangle bounded by $y = \frac{1}{2}x$ and $y = 2x$ for $0 \le x \le 2$ in Figure 11 is rotated about the x-axis.

SOLUTION We consider the vertical line across the triangle at x for $0 \le x \le 2$ (Figure 11). When the triangle is rotated about the x axis, this line generates the washer-shaped region in Figure 12, which is the cross section of the solid at x.

The outer radius of the cross section in Figure 12 is equal to the distance $2x$ from the upper line to the x-axis in Figure 11, and its inner radius is equal to the distance $\frac{1}{2}x$ from the lower line to the x-axis. The area of the cross section at x is equal to the area of the larger circle, minus the area of the smaller circle:

$$A(x) = \pi(2x)^2 - \pi\left(\tfrac{1}{2}x\right)^2 = 4\pi x^2 - \tfrac{1}{4}\pi x^2 = \tfrac{15}{4}\pi x^2.$$

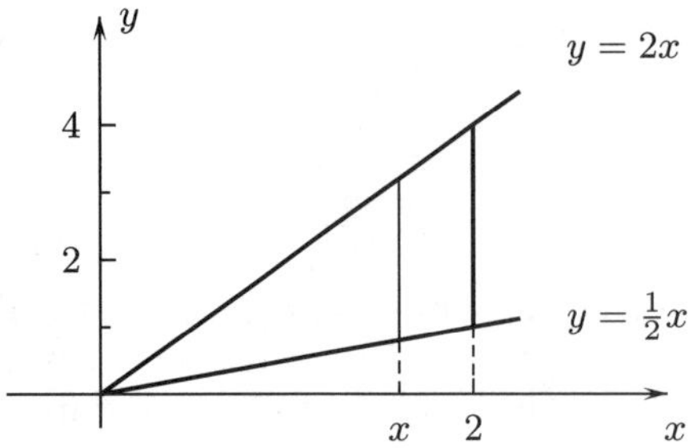

The triangle that is rotated

FIGURE 11

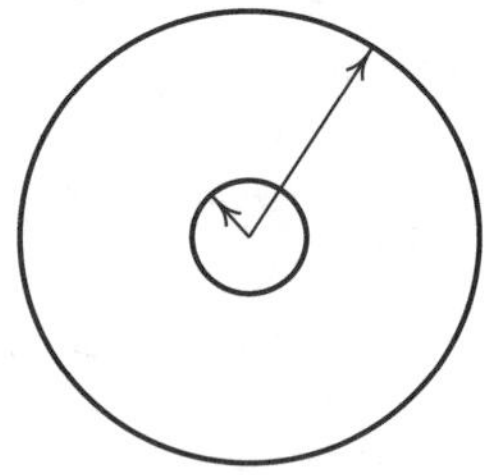

The cross section at x

FIGURE 12

Because the solid extends from its cross section at $x = 0$ on the left to its cross section at $x = 2$ on the right, its volume is

$$\int_0^2 A(x)\,dx = \int_0^2 \tfrac{15}{4}\pi x^2\,dx = \tfrac{15}{4}\pi\int_0^2 x^2\,dx = \tfrac{15}{4}\pi\left[\tfrac{1}{3}x^3\right]_0^2$$
$$= \tfrac{15}{4}\pi[\tfrac{1}{3}(2^3) - \tfrac{1}{3}(0^3)] = 10\pi.\ \square$$

Question 6 The region bounded by the curve $y = x - x^4$ and the x-axis in Figure 13 is rotated about the x-axis. The cross section of the solid at x is a circle. **(a)** Give a formula for the area of the cross section in terms of x. **(b)** Find the volume of the solid.

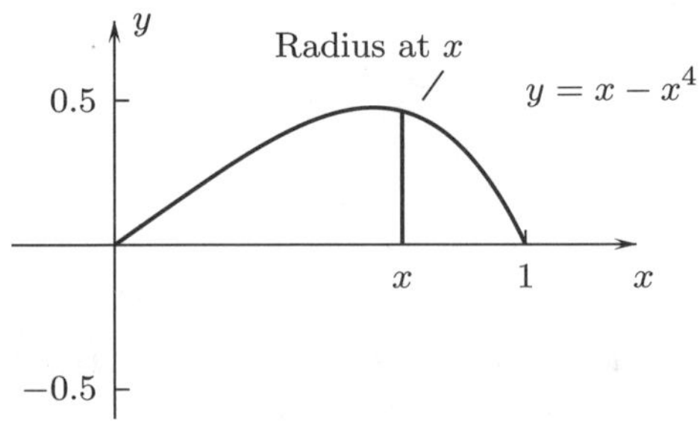

The region that is rotated

FIGURE 13

Finding volumes of solids of revolution from cylindrical sections

Figure 14 shows a region between the graphs of two positive functions $f(x)$ and $g(x)$ and and Figure 15 shows the solid that is generated when the region is rotated about the y-axis. How can we use formulas for the functions to calculate the volume of the solid?

When the vertical line segment extending across the region at the value x in Figure 14 is rotated about the y-axis, it generates the cylinder on the left of Figure 16, which we call the CYLINDRICAL SECTION of the solid at x or the CYLINDER AT x. You can imagine that as x increases from a at the left side of the region to b at the right side in Figure 14, the cylindrical section expands from the inside to the outside of the solid in Figure 15.

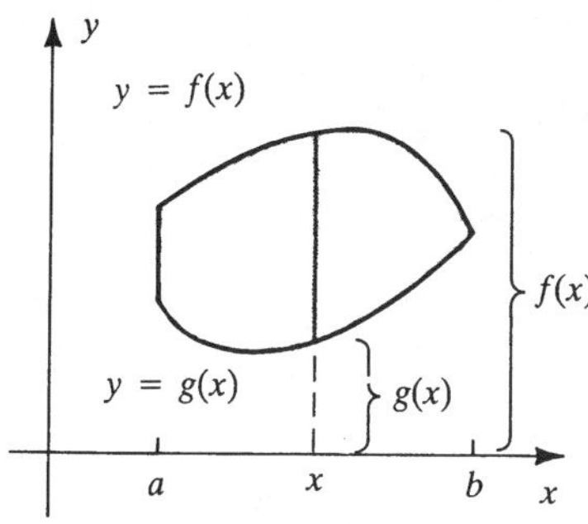

FIGURE 14

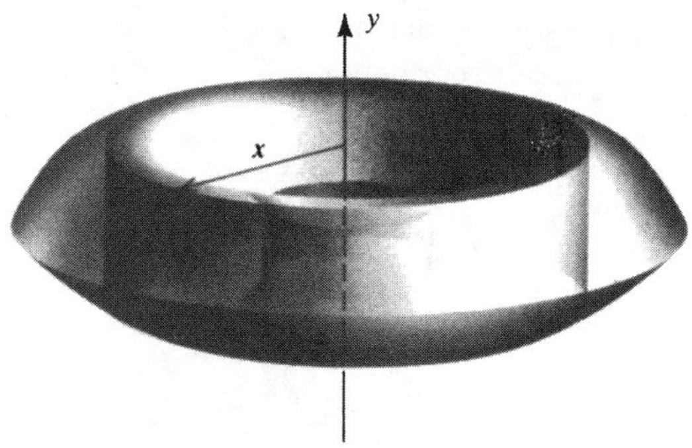

FIGURE 15

The radius of the cylinder in Figure 15 is the distance x from the line in Figure 14 to the y-axis, and its height is the length $f(x) - g(x)$ of that line. If we slice the cylinder vertically and flatten it out, it forms the rectangle on the right of Figure 16, whose height is equal to the height $f(x) - g(x)$ of the cylinder and whose width is equal to its perimeter $2\pi x$. Consequently, the area of the cylindrical section at x is

$$A(x) = [\text{Circumference}][\text{Height}] = 2\pi x[f(x) - g(x)]. \quad \textbf{(6)}$$

We will now show that the volume of the solid is equal to the integral of $A(x)$ from a to b.

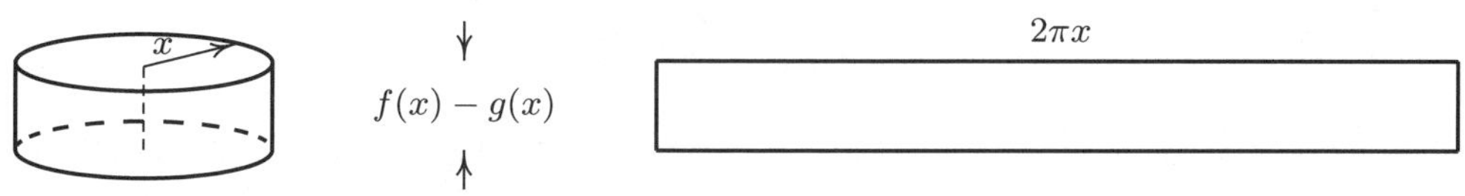

FIGURE 16

We consider a partition $a = x_0 < x_1 < \cdots < x_N = b$ of the interval $[a, b]$ into N equal subintervals of width Δx and use it to approximate the region in Figure 14 by rectangles with their sides at the values of x in the partition and such that their tops and bottoms intersect $y = f(x)$ and $y = g(x)$ at the midpoints c_j of the subintervals, as is shown in Figure 17 for $N = 4$.

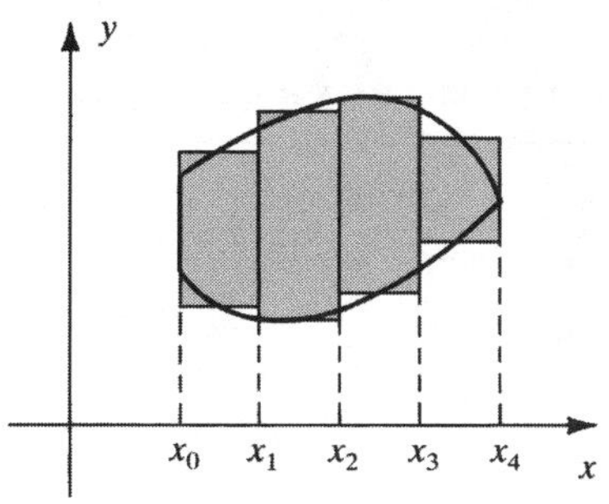

FIGURE 17

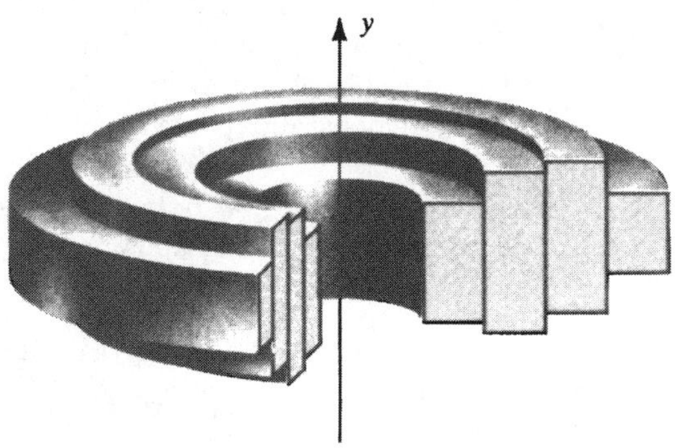

FIGURE 18

When the rectangles are rotated about the y-axis, they generate a solid consisting of N concentric cylindrical shells that is an approximation of the solid whose volume we want to find. The collection of cylindrical shells for the rectangles of Figure 17 is shown in Figure 18.

Figure 19 is a schematic sketch of the jth shell. Its height is the distance $f(c_j) - g(c_j)$ between the two graphs at $x = c_j$, and its base is the washer-shaped region in Figure 20. Because its outer radius is x_j and its inner radius is x_{j-1}, the area of the base is

$$\begin{aligned}\pi(x_j)^2 - \pi(x_{j-1})^2 &= \pi(x_j + x_{j-1})(x_j - x_{j-1}) \\ &= 2\pi[\tfrac{1}{2}(x_j + x_{j-1})](x_j - x_{j-1}) \\ &= 2\pi c_j \Delta x.\end{aligned} \tag{7}$$

The last equation in **(7)** holds because $\Delta x = x_j - x_{j-1}$ is the width of the subinterals and the midpoint of the jth subinterval is $c_j = \frac{1}{2}(x_j + x_{j-1})$.

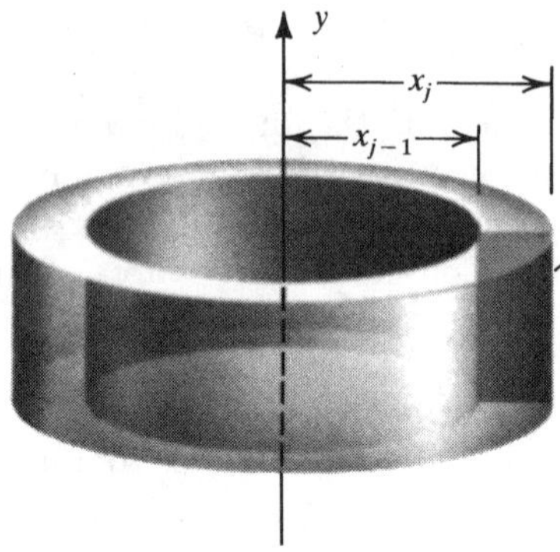

FIGURE 19

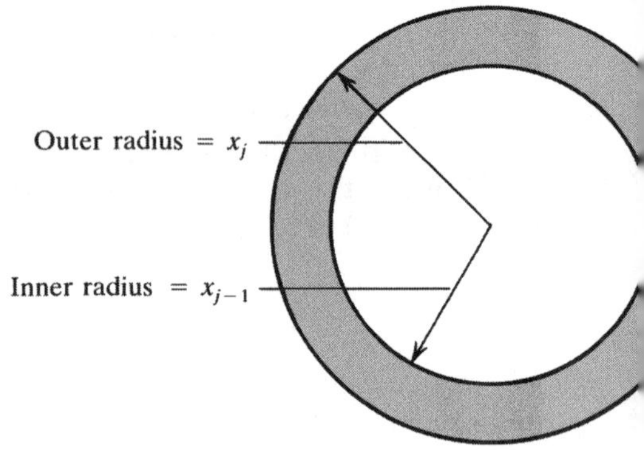

FIGURE 20

The volume of the jth shell is its height $f(c_j) - g(c_j)$ multiplied by the area **(7)** of its base, so its volume is $2\pi c_j[f(c_j) - g(c_j)]\Delta x$, and the total volume of the N shells is $\sum_{j=1}^{N} 2\pi c_j[f(c_j) - g(c_j)]\Delta x$. We assume that the volume of the solid generated by rotating the original region and the integral

$$\int_a^b 2\pi x[f(x) - g(x)]\, dx \tag{8}$$

with these Riemann sums are defined. Then the integral is the limit of the Riemann sums as $N \to \infty$, and we conclude that the volume of the solid is equal to the integral. Notice that **(8)** is the integral of the area **(6)** of the cylindrical section of the solid at x. Similar reasoning for solids obtained by rotating regions about a vertical lines that are not necessarily the y-axis leads to the following rule.

Rule 2 *If the region between the graphs of two functions for $a \le x \le b$ does not lie on both sides of a vertical line, then when the region is rotated about the line, the volume of the solid it generates is*

$$\begin{aligned}[\text{Volume}] &= \int_a^b A(x)\, dx \\ &= \int_a^b [\text{Area of the cylinder at } x\,]\, dx \\ &= \int_a^b 2\pi \begin{bmatrix}\text{Radius of the} \\ \text{cylinder at } x\end{bmatrix} \begin{bmatrix}\text{Height of the} \\ \text{cylinder at } x\end{bmatrix} dx.\end{aligned} \tag{9}$$

Example 3 The region between $y = (x^2+1)^{1/2}$ and the x-axis for $1 \le x \le 3$ in Figure 21 is rotated about the y-axis. What is the volume of the solid it generates?

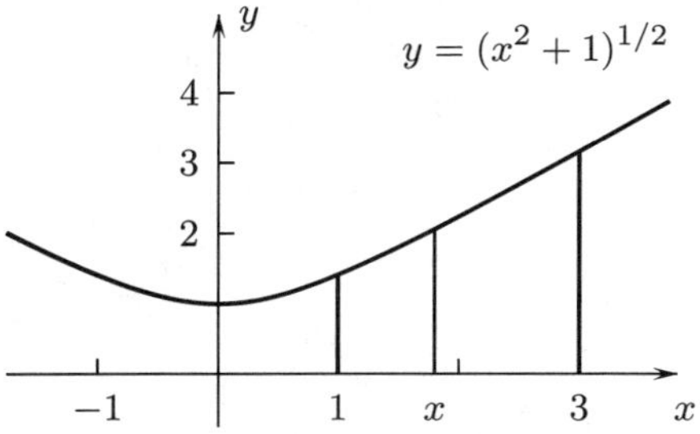

FIGURE 21

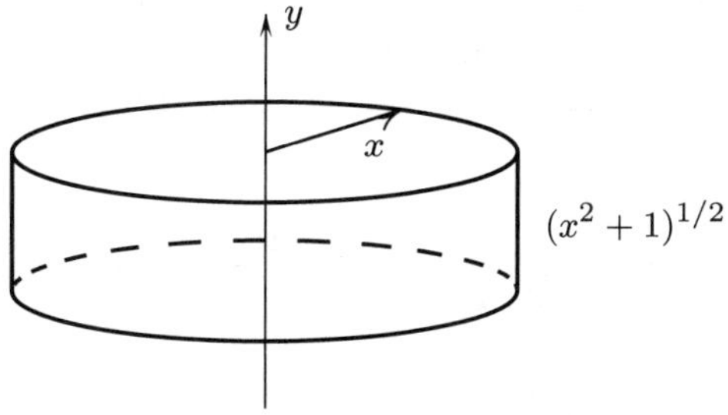

FIGURE 22

SOLUTION When the vertical line segment across the region at x in Figure 21 is rotated about the y-axis, it generates the cylindrical section of the solid in Figure 22. The radius of the cylindrical section is x, its circumference is $2\pi x$, and its height is $(x^2+1)^{1/2}$, so its area is $A(x) = 2\pi x(x^2+1)^{1/2}$. Because the solid extends from its innermost cylindrical section at $x = 1$ to its outermost cylindrical section at $x = 3$, its volume is

$$\int_1^3 A(x)\,dx = 2\pi\int_1^3 x(x^2+1)^{1/2}\,dx.$$

We use the substitution $u = x^2+1, du = 2x\,dx$ in the integral and obtain

$$\begin{aligned}[\text{Volume}] &= 2\pi\int_1^3 x(x^2+1)^{1/2}\,dx = \pi\int_{x=1}^{x=3}(x^2+1)^{1/2}(2x\,dx)\\ &= \pi\int_{x=1}^{x=3} u^{1/2}\,du = \left[\tfrac{2}{3}\pi u^{3/2}\right]_{x=1}^{x=3}\\ &= \left[\tfrac{2}{3}\pi(x^2+1)^{3/2}\right]_{x=1}^{x=3} = \tfrac{2}{3}\pi(10^{3/2} - 2^{3/2}).\ \square\end{aligned}$$

Question 7 What is the volume of the solid that is generated when the region bounded by the curve $y = 4/x$ and the lines $y = x$ and $x = 1$ in Figure 23 is rotated about the y-axis?

FIGURE 23

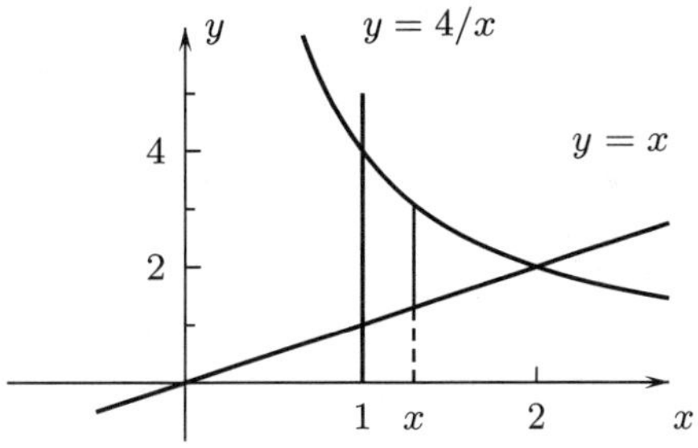

The area of a lampshade

We now want to see how to find the area of a surface that is generated by rotating the graph of s posiitve function about the x-axis. To accomplish this, we need to derive a formula for the area of a surface like the lampshade in Figure 24. The top of the lampshade is a circle of radius 3 inches and the bottom is a circle of radius 6 inches. The distance along the side of the shade from the bottom to the top is called its SLANT HEIGHT, which in this case is 5 inches. The shade could be formed by cutting off the top of the right circular cone in Figure 25. Such a surface is called a FRUSTUM of a cone.

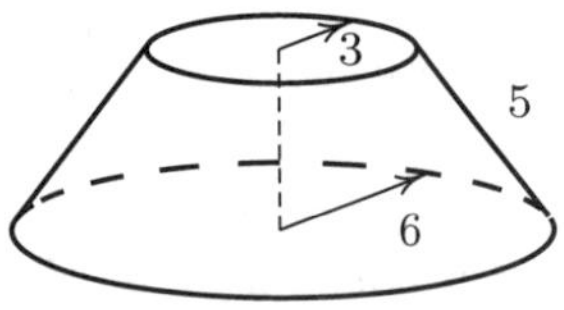

FIGURE 24

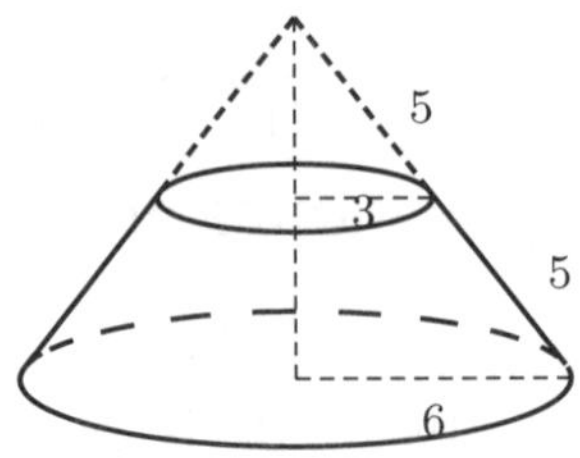

FIGURE 25

The two right triangles in Figure 25 with bases of lengths 3 inches and 6 inches are similar. Therefore, because the base of the larger triangle is twice the base of the smaller triangle, the hypotenuse of the larger triangle is twice that of the smaller, so that the hypotenuse of the smaller triangle is 5 inches and the hypotenuse of the larger is 10 inches, as is shown in Figure 25.

To find the area of the lamp shade, we imagine that the cone in Figure 25 is cut along a straight line through its vertex and it is spread open to form part of a circle, as shown (reduced) in Figure 26. As you can see by comparing Figures 25 and 26, the radius of the circle is 10 inches, and the outer circumference of the flattened lamp shade is the circumference 12π of the base of the shade.

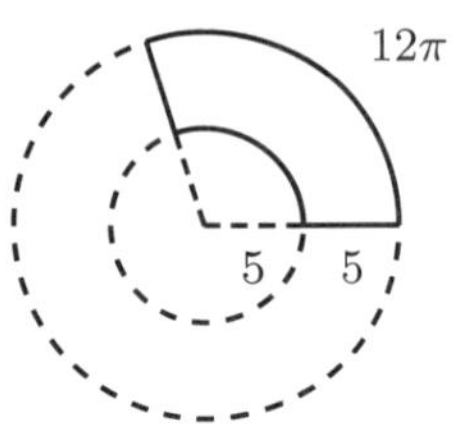

FIGURE 26

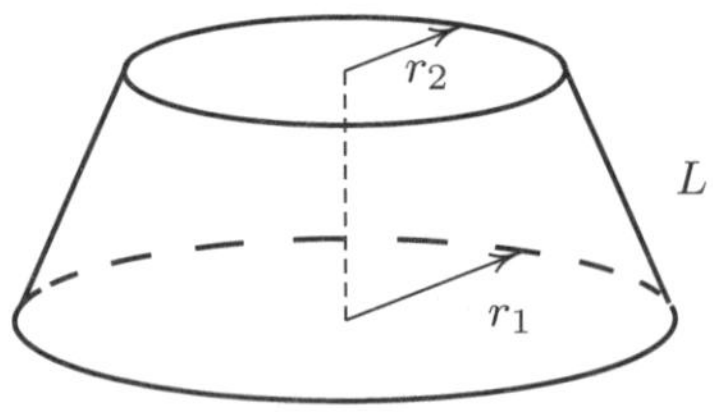

FIGURE 27

Question 8 **(a)** What is the area of the entire washer in Figure 26 between the circle of radius 10 inches and the circle of radius 5 inches? **(b)** Why is the area of the lampshade the fraction $\dfrac{12\pi}{20\pi} = \frac{3}{5}$ of the entire washer?

Combining the results of parts (a) and (b) of Question 6 shows that

$$[\text{The area of the lampshade}] = \tfrac{3}{5}(75\pi) = 45\pi. \qquad \textbf{(10)}$$

Similar calculations would show that for the frustum of the right circular cone in Figure 27, whose base and top have radii r_1 and r_2 and whose slant height is L,

$$[\text{Area of the frustum}] = \pi(r_1 + r_2)L \qquad \textbf{(11)}$$

(see Problem 59). Notice that area in **(9)** is equal to the average $\frac{1}{2}(2\pi r_1 + 2\pi r_2)$ of the circumferences of the top and bottom of the frustum, multiplied by its slant height L.

Question 9 Show that formula **(11)** yields the value **(10)** for the area of the lampshade in Figure 24.

Areas of surfaces of revolution

Now we can apply **(11)** to find the area of the surface in Figure 29 that is generated when the portion for $a \le x \le b$ of the graph of the nonnegative function $f(x)$ in Figure 28 is rotated about the x-axis. We assume that the derivative $f'(x)$ exists and is continuous in $[a, b]$.

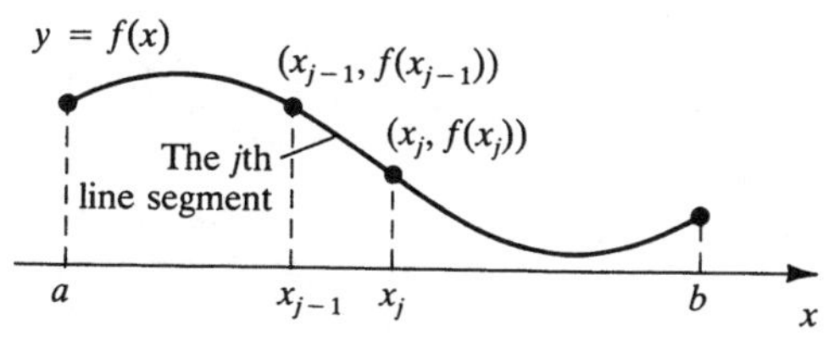

FIGURE 28

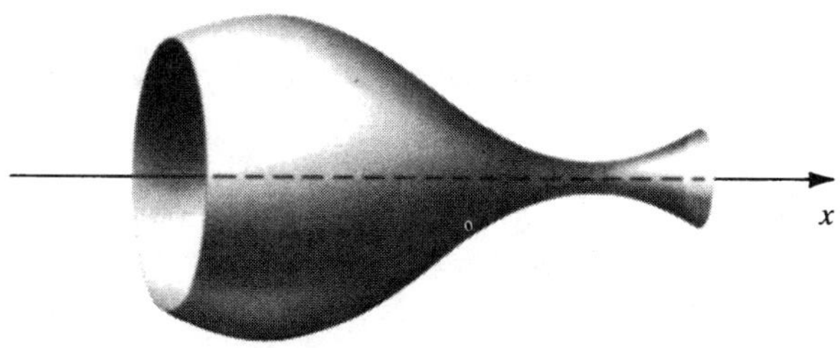

FIGURE 29

We use a partition $a = x_0 < x_1 < \cdots < x_N = b$ of $[a, b]$ into N equal subintervals of width Δx to approximate the curve by N line segments as we did in Section 4.5 to find the formula

$$\int_a^b \sqrt{1 + [f'(x)]^2}\, dx$$

for the length of the curve. When the jth line segment is rotated about the x-axis, it generates the frustum of a cone shown in Figure 30.

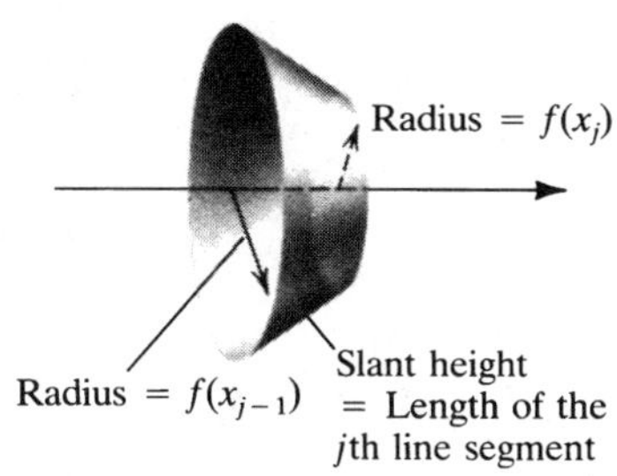

FIGURE 30

The radii of the ends of the frustum in Figure 30 are $r_1 = f(x_{j-1})$ and $r_2 = f(x_j)$ and its slant height is the length $L = \sqrt{(x_j - x_{j-1})^2 + [f(x_j) - f(x_{j-1})]^2}$ of the jth line segment. We write $x_j - x_{j-1} = \Delta x$ and use the Mean-Value Theorem, as we did in Section 4.6 in studying lengths of graphs, to write $f(x_j) - f(x_{j-1}) = f'(c_j)\Delta x$ and then

$$L = \sqrt{(x_j - x_{j-1})^2 + [f(x_j) - f(x_{j-1})]^2} = \sqrt{(\Delta x)^2 + [f'(c_j)]^2(\Delta x)^2} = \sqrt{1 + [f'(c_j)]^2}\,\Delta x$$

with c_j a point between x_{j-1} and x_j. Then, by formula **(11)**, the area of the jth frustum is

$$\pi\big[f(x_{j-1}) + f(x_j)\big]\sqrt{1 + [f'(c_j)]^2}\,\Delta x.$$

The frustums generated by all N line segments combine to form an approximation of the surface in Figure 30, and their total surface area

$$\sum_{j=1}^{N} \pi[f(x_{j-1}) + f(x_j)]\sqrt{1 + [f'(c_j)]^2}\Delta x \tag{12}$$

approximates its surface area.

The sum **(12)** is not a Riemann sum because x_{j-1}, x_j, and c_j are different points in the jth subinterval. Nevertheless, it can be shown that **(12)** tends to the integral

$$\int_a^b \pi[f(x) + f(x)]\sqrt{1 + [f'(x)]^2}\, dx = \int_a^b 2\pi f(x)\sqrt{1 + [f'(x)]^2}\, dx$$

as the number of subintervals in the partition tends to ∞. This leads us to the following rule.

Rule 3 **(a)** *If $f(x)$ is nonnegative and has a continuous derivative in $[a, b]$, then the surface generated by rotating the portion of its graph for $a \le x \le b$ about the x-axis has area*

$$\int_a^b 2\pi f(x)\sqrt{1 + [f'(x)]^2}\, dx. \tag{13}$$

Example 4 When the portion of the curve $y = 2.4 + 2\cos x$ for $0 \le x \le 2\pi$ in Figure 31 is rotated about the x-axis, it generates the hour-glass-shaped surface in Figure 32. Give an integral that equals its surface area. Then use a calcuator or computer to find the approximate decimal value of the integral.

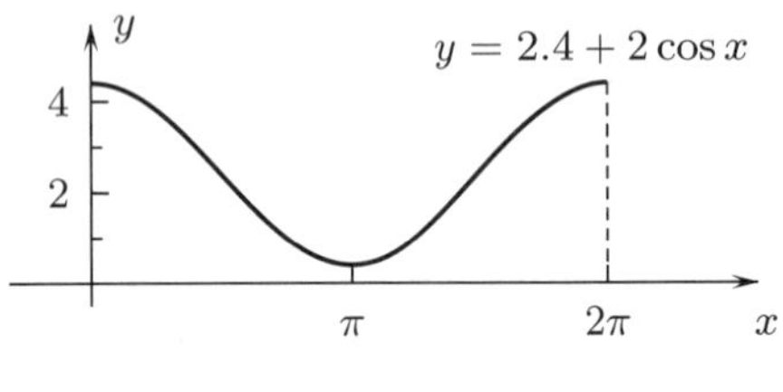

FIGURE 31

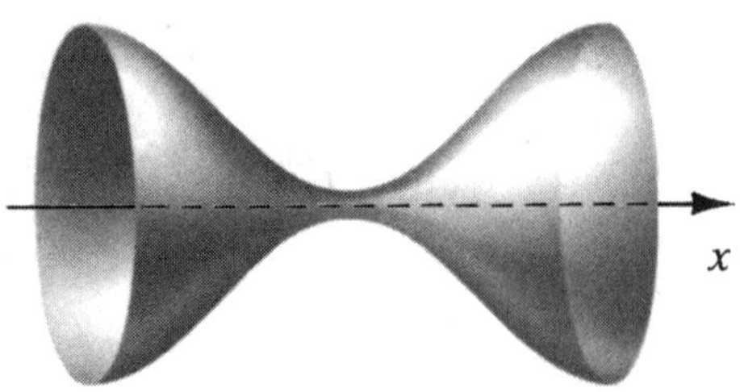

FIGURE 32

SOLUTION Set $f(x) = 2.4 + 2\cos x$. Then $f'(x) = -2\sin x$ and with **(13)** we find that the surface area equals

$$\int_0^{\pi} 2\pi(2.4 + 2\cos x)\sqrt{1 + 4\sin^2 x}\; dx.$$

A calculator gives the approximate value 192.024 for this integral. □

Question 9 Give an integral that equals area of the surface that is generated when the portion of $y = 2x^{3/2}$ for $0 \le x \le 5$ is rotated about the x-axis, and use a calculator or computer to find its approximate value.

Using differentials

Frequently mathematicians and other scientists solve problems that require integrals without going through the process, used in this text, of finding approximate solutions in the form of Riemann sums and then taking limits to obtain the required integrals. Instead they carry out analogous reasoning and calculations with differentials dx, thought of as "infinitesimal" widths and thicknesses. Let us see how this works for areas, volumes of solids of revolution, lengths of curves, and areas of surfaces of revolution.

To find the area of the region between the graphs of the two functions in Figure 33, for example, the vertical line segment across the region is considered to have the infinitesimal "width" dx, and its infinitesimal "area" is considered to be the product $[f(x) - g(x)]\,dx$ of its length and width. Then the area of the region is the "sum" or "integral" of the "areas" of all the vertical line segments in the region:

$$[\text{Area}] = \int_a^b [f(x) - g(x)]\,dx$$

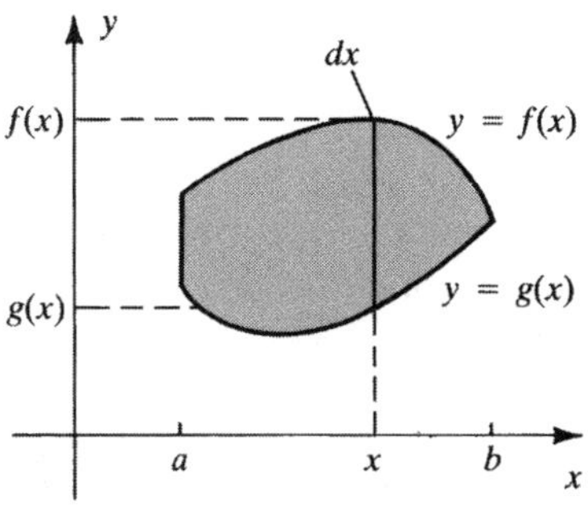

Line of "width" dx

FIGURE 33

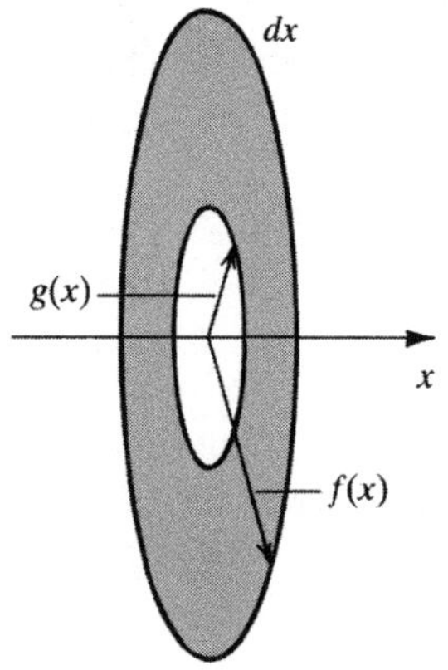

Washer of "thickness" dx

FIGURE 34

To determine the volume generated by rotating the region of Figure 33 about the x-axis, the washer-shaped cross section in Figure 34 generated by rotating the line segment at x is considered to have "volume" equal to the area $\pi[f(x)]^2 - \pi[g(x)]^2$ of its face, multiplied by its "thickness" dx. The volume of the solid is then the "sum" of the infinitesimal "volumes" of all the cross sections:

$$[\text{Volume}] = \int_a^b \{\pi[f(x)]^2 - \pi[g(x)]^2\}\,dx.$$

If the region in Figure 33 is rotated about the y-axis, then the vertical line segment across the region generates a cylindical section of the solid whose radius is x, whose height is $f(x) - g(x)$, and whose "thickness" is dx (Figure 35). The "volume" of the cylindrical section is its circumference multiplied by its height and "thickness," and the volume of the solid is the "sum" of the "volumes" of the cylindrical sections:

$$[\text{Volume}] = \int_a^b 2\pi x[f(x) - g(x)]\,dx.$$

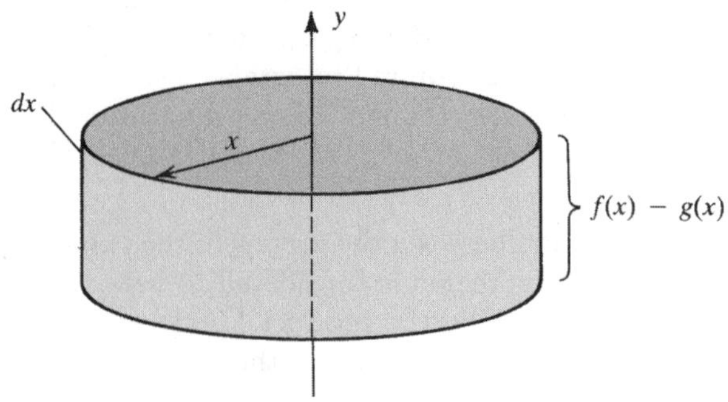

Cylinder of
"thickness" dx
FIGURE 35

Because the tangent line at x to the graph of $f(x)$ has slope $f'(x)$, the infinitesimal piece of the graph at x of "width" dx is considered to have "length" $\sqrt{1+[f'(x)]^2}\,dx$, and the length of the graph of $f(x)$ for $a \le x \le b$ is the "sum" of the "lengths" of the infinite number of its infinitesimal pieces:

$$[\text{Length}] = \int_a^b \sqrt{1+[f'(x)]^2}\,dx.$$

Finally, when the graph of $f(x)$ is rotated about the x-axis, the infinitesimal piece of the graph at x generates an infinitesimal frustum of a cone of circumference $2\pi f(x)$ and "slant height" $\sqrt{1+[f'(x)]^2}\,dx$. The "area" of the frustum at x is $2\pi f(x)\sqrt{1+[f'(x)]^2}\,dx$, and the area of the entire surface is

$$[\text{Area}] = \int_a^b 2\pi f(x)\sqrt{1+[f'(x)]^2}\,dx.$$

These lines of reasoning are similar to those used by mathematicians in the sixteenth, seventeenth, and early eighteenth centuries, before the definition of integrals as limits of Riemann sums was formulated. We use approximate solutions and Riemann sums in this text, rather than infinitesimals, because we prefer to describe explicitly how basic arithmetic formulas, such as those for areas of rectangles and volumes of cylinders, are combined with the taking of limits to find the relevant integrals.

Principles and procedures

- If $A(x)$ is the area of the cross section of a solid perpendicular to an x-axis at x and if the solid extends from its cross section at $x = a$ to its cross section at $x = b$ with $a < b$, then the volume of the solid is equal to the integral of $A(x)$ from a to b (Rule 1).
- If the region between the graphs of $f(x)$ and $g(x)$ for $a \le x \le b$ is rotated about a horizontal line $x = L$ and the region does cross the line, then the cross sections are disks or washers,whose areas $A(x)$ can be expressed in terms of x by using $f(x)$ and $g(x)$ with the formula πr^2 for the area of a circle of radius r.
- If a region between the graphs $y = g(x)$ and $y = f(x)$ for $a \le x \le b$ is rotated about the y-axis and the region does not extend to both sides of the y-axis, then the volume of the solid generated can be calculated from the areas $A(x)$ of cylindrical sections (Rule 2). The area $A(x)$ of the cylindrical section can be expressed in terms of x by using $f(x)$ and $g(x)$ and the formula $2\pi rh$ for the area of a right circular cylinder of radius r and height h.
- The surface area of a surface that is generated by rotating a bounded portion of the graph of a function with a continuous derivative about the x-axis can be expressed as an integral involving the function and its derivative (Rule 3). The integrals can be evaluated by exact integration only in special cases, but their approximate values can always be found with integration algorithms on calculators or computers.

Tune-Up Exercises 6.1[♦]

[A]*Answer provided.* [O]*Outline of solution provided.*

T1.[O] The base of a solid is the region between $y = x^3$ and the x-axis for $0 \le x \le 1$ and its cross sections perpendicular to the x-axis are squares with one side on the xy-plane. Find its volume.

♦ Type 1, volume by slicing

T2.[O] The region between $y = x^3$ and the x-axis for $0 \le x \le 1$ is rotated about the x-axis. What is the volume of the solid that is generated?

♦ Type 1, volume by slicing

T3.[O] What is the volume of the solid that is generated when the region between $y = x^3$ and the x-axis for $0 \le x \le 1$ is rotated about the y-axis?

♦ Type 1, volume by cylindrical section

T4.[O] The region between $y = x^3$ and the x-axis for $0 \le x \le 1$ is rotated about the line $y = -1$. **(a)** The cross section at x is a washer-shaped region. What are its inner and outer radii? **(b)** What is the volume of the solid that is generated?

♦ Type 1, volume by slicing,etc

T5.[O] The portion of the curve $y = x^3$ for $0 \le x \le 1$ is rotated about the x-axis. Give an integral that equals the surface area of the surface that is generated. Then use a procedure of approximate integration on your calculator or computer to find the approximate value of the integral.

♦ Type 1, surface area

Problems 6.1

[A]*Answer provided.* [O]*Outline of solution provided.*

1.[O] Find the volume of a solid whose base is bounded by $y = x^2$ and $y = \sqrt{x}$ in an xy-plane and whose cross sections perpendicular to the x-axis are rectangles twice as high as they are wide.

♦ Type 1, volume by slicing

2. Find the volume of a solid whose base is bounded by $y = x^2$ and $y = 2 - x^2$ in an xy-plane and whose cross sections perpendicular to the x-axis are semicircles with diameters in the xy-plane.

♦ Type 1, volume by slicing

3.[A] The base of a solid is the region between $y = 5x - 4x^2 + x^3$ and $y = 3 + 5x - 4x^2 + x^3$ in an xy-plane for $0 \le x \le 3$. Its cross sections perpendicular to the x-axis are squares with one edge on the xy-plane. What is the volume of the solid?

♦ Type 1, volume by slicing

4. The Greek bowl from the 5th century B. C. in Figure 36 shows Heracles and Apollo fighting. If we disregard the handles and base, the bowl is nine inches high and its horizontal cross sections are circles. The table below gives the radii $r(h)$ (inches) of the cross sections at seven distances h (inches) from the base.[(1)] **(a)** Express the volume of the bowl, not including the handles and base, as an integral involving $r(h)$. **(b)** Use the Trapezoid Rule and the values in the table to estimate the value of the integral.

♦ Type 2, volume by slicing and the Trapezoid Rule

h	$r(h)$
0	1.3
1.5	2.4
3	3.0
4.5	3.6
6	4.1
7.5	4.6
9	5.3

FIGURE 36

[♦]The Tune-up Exercises and Problems are classified by type and content. The types are (1) basic, reactive; (2) basic reflective; (3) intermediate, reactive; (4) intermediate, reflective; (5) advanced, reactive; (6) advanced, reflective; and (7) advanced, theoretical.

[(1)]Photo copied from *Encyclopædia Brittanica*, Vol. 18, 1949, p. 342.

5. The great pyramid of Cheops at Giza in Egypt is 150 meters high and has square horizontal cross sections. The width of the cross section h meters from the top is $1.5h$ meters. What is the volume of the pyramid?

♦ Type 2, volume by slicing

6.[A] Figure 37 shows the graph of the area $A(h)$ of the surface of a lake as a function of the depth of the water in it. What is the approximate volume of the water in the lake when the water in it is 40 feet deep?

♦ Type 2, volume by slicing with a graph

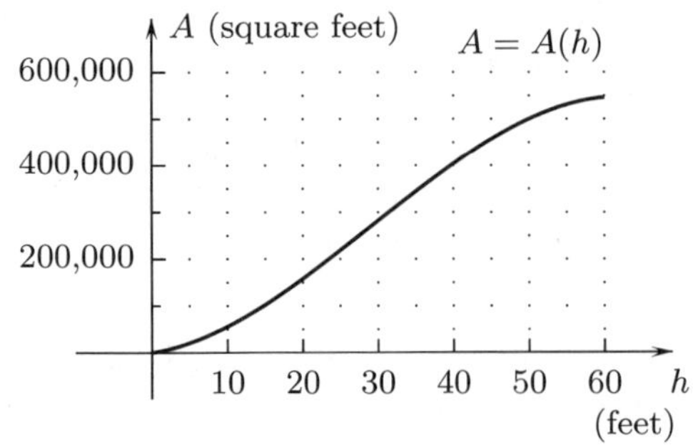

FIGURE 37

7.[O] The region between $y = x^{1/3}$ and the x-axis for $0 \le x \le 8$ is rotated about the x-axis. What is the volume of the solid it generates?

♦ Type 1, volume by slicing

8.[O] Find the volume of the solid that is generated when the region between $y = 1 - \frac{1}{2}x^2$ and the x-axis is rotated about the x-axis. (Use the symmetry to simplify the calculation.)

♦ Type 1, volume by slicing

9. Find the volume of the solid that is generated when the region between $y = 2/x$ and the x-axis for $1 \le x \le 4$ is rotated about the x-axis.

♦ Type 1, volume by slicing

10.[A] The base of a solid is the region between $y = x^3$ and $y = 2 - x^3$ for $x \ge 0$ and its cross sections perpendicular to the x-axis are squares with one edge on the xy-plane. What is the volume of the solid?

♦ Type 1, volume by slicing

11. When the region bounded by $y = x^{5/2}$ and $y = k^2\sqrt{x}$ with a positive constant k is rotated about the x-axis, it generates a solid of volume 243π. What is the value of k?

♦ Type 2, volume by slicing

12.[A] The region between $y = x$ and $y = \sqrt{4 + x^2}$ for $1 \le x \le 2$ is rotated about the x-axis. What is the volume of the solid that is generated?

♦ Type 1, volume by slicing

13. The cross sections perpendicular to the x-axis of a solid are circles whose diameters extend from $y = x^3$ to $y = x^2$ for $0 \le x \le 1$ in an xy-plane. Find the volume of the solid.

♦ Type 1, volume by slicing

14. What is the volume of the solid that is generated when the region between $y = 2/x, x = 2$, and $y = 2$ is rotated about the x-axis?

♦ Type 1, volume by slicing

15.[O] The region between $y = x$ and $y = 3x - x^2$ is rotated about the y-axis. What is the volume of the solid that is generated?

♦ Type 1, volume by slicing

16.[A] The intersection of a solid with an xy-plane is the region between the x-axis and the parabola $y = x^2$ for $0 \le x \le 2$. The cross sections of the solid perpendicular to the x-axis are squares with one diagonal in the xy-plane. What is the volume of the solid?

♦ Type 1, volume by slicing

17. Show that when the region bounded by $y = 8/x^2, x = 5$, and $y = 2$ is rotated about the x-axis it generates a solid of volume $\frac{3564}{375}\pi$.

♦ Type 1, volume by slicing

18.[A] The region between $y = x^{-1/2}$ and $y = -x^{-1/2}$ for $1 \le x \le k$ with $k > 1$ is rotated about the y-axis. Find the volume of the solid that is generated.

◆ Type 1, volume by cylindrical sections

19. What is the volume of the solid that is generated when the region between the x-axis and $y = ax - x^6$ with $a > 0$ is rotated about the y-axis?

◆ Type 1, volume by cylindrical sections

20. Does the region between $y = x$ and $y = \frac{1}{3}x^2$ generate a solid of greater volume when it is rotated about the x-axis or when it is rotated about the y-axis?

◆ Type 1, volume by cylindrical sections

21. Show that the region bounded by $y = \frac{1}{4}x^2$ and $y = 2\sqrt{x}$ generates a solid of the same volume whether it is rotated about the x-axis or about the y-axis.

◆ Type 1, volume by cylindrical sections

22.[A] The base of a solid is the triangle with vertices $(0,0), (1,0)$, and $(0,1)$ in an xy-plane. Its vertical cross sections perpendicular to the x-axis are semicircles with diameters on the xy-plane. Find its volume.

◆ Type 3, volume by slicing

23.[A] Give an integral that equals the area of the surface that is generated when the portion of the curve $y = 1/x$ for $1 \le x \le 4$ is rotated about the x-axis. Then use your calculator or computer to find the approximate value of the integral.

◆ Type 2, surface area

24. Give an integral that equals the area of the surface that is generated when the portion of the curve $y = x^{1/3}$ for $1 \le x \le 10$ is rotated about the x-axis. Then use your calculator or computer to find the approximate value of the integral.

◆ Type 2, surface area

25. The surface that is generated when the portion of $y = x^2 + 1$ for $-1 \le x \le 1$ is rotated about the x-axis is contained in the cylindrical surface that is generated when the line segment from $(-1,2)$ to $(1,2)$ is rotated about the x-axis. Which of these two surfaces has the greater area?

◆ Type 2, surface area

26.[O] Find the volume of the solid whose base is the semicircle between $y = \sqrt{4 - x^2}$ and the x-axis in an xy-plane and whose cross sections perpendicular to the x-axis are equilateral triangles with one side on the xy-plane.

◆ Type 3, volume by slicing

27. Show that when the region bounded by $y = 1 + 3x - x^2$, and $y = 1$ is rotated about the x-axis it generates a solid of volume 17.1π.

◆ Type 3, volume by slicing

28.[A] The region between $y = x^2$ and $y = mx$ with $m > 0$ is rotated about the y-axis. What is the volume of the solid it generates?

◆ Type 3, volume by slicing

29. What the volume of the solid generated by rotating the region between $y = \dfrac{1}{x^2 + 4}$ and $y = \frac{1}{8}$ is rotated half way about the y-axis?

◆ Type 3, volume by slicing

30. Find the volume of the solid whose base is the region bounded by $y = 1 - x$ and $y = 1 - x^4$ in an xy-plane and whose cross sections perpendicular to the x-axis are equilateral triangles with one side on the xy-plane.

◆ Type 3, volume by slicing

31.[A] A church steeple is five feet high and its horizontal cross section x feet from the top is a square with diagonal of length $\frac{2}{5}x^2$. What is its volume?

◆ Type 3, volume by slicing

32. The region bounded by $y = x^{1/4}$, the x-axis, and $x = k$ is rotated about the x-axis. The resulting solid has volume 12π. What is the constant k?

◆ Type 3, volume by slicing

33. Derive the formula $\frac{4}{3}\pi R^3$ for the volume of a sphere of radius R from the formula $x^2 + y^2 = R^2$ for a circle **(a)** by slicing and **(b)** by cylindrical sections.

◆ Type 1, volume by slicing and by cylindrical sections

34. The intersection of a solid with the xy-plane is the region between $y = x^2$ and $y = a^2$ with a positive constant a. The cross section perpendicular to the x-axis at x is a circle with diameter in the xy-plane and the volume of the solid is $\frac{16}{15}\pi\sqrt{2}$. What is the value of a?

♦ Type 2, volume by slicing

35.[A] Find the volume of the solid that is generated when the region between $y = e^{x^2}$ and the x-axis for $0 \le x \le 2$ is rotated about the y-axis.

♦ Type 2, volume by cylindrical sections

36. What is the volume of the solid that is generated when the region between $y = 1/\sqrt{x}$ and the x-axis for $a \le x \le b$ with $0 < a < b$ is rotated about the x-axis?

♦ Type 2, volume by slicing

37. Find the volume of the solids that are generated when the region bounded by $y = 1/\sqrt{x^2+1}$ and $y = 2$ for $0 \le x \le 2$ is rotated **(a)** about the y-axis and **(b)** about the x-axis.

♦ Type 2, volume by slicing and by cylindrical sections

38. Show that when the region between $y = 1 + \cos(x^2)$ and the x-axis for $0 \le x \le \sqrt{\pi}$ is rotated about the y-axis it generates a solid of volume π^2.

♦ Type 2, volume by slicing

39. The region between $y = \sqrt{\sin x}$ and the x-axis consists of an infinite number of congruent, disjoint pieces. Find the volume of the solid that is generated when one of the pieces is rotated about the x-axis.

♦ Type 2, volume by slicing

40.[A] Find a formula in terms of the positive constant k for the volume of the solid that is generated when the triangle with vertices $(0,0)$, (k,k), and $(k, \frac{1}{2}k)$ is rotated about the y-axis.

♦ Type 3, volume by cylindrical sections

41. Give definite integrals that equal **(a[A])** the area of the region between $y = e^{3-x}$ and $y = e^{2/x}$, **(b)** the volume of the solid generated when the region of part (a) is rotated about the x-axis, and **(c)** the volume of the solid generated when the region of part (a) is rotated about the y-axis. Then use an algorithm on your calculator or computer to find the approximate values of the three integrals.

♦ Type 3, volume by slicing and by cylindrical sections

42.[A] The base of a solid is the region between $y = x^2$ and $y = \cos x$ and its cross sections perpendicular to the x-axis are squares with one side on the xy-plane. Use your calculator or computer to find the approximate volume of the solid.

♦ Type 3, volume by slicing

43. Show that the region bounded by $y = x^2$ and $y = x$ generates solids with the same volume when it is rotated about the line $x = -1$ and when it is rotated about the line $x = 3$.

♦ Type 3, volume by cylindrical sections

44. Use your calculator or computer to find the approximate volumes of the solids that are generated when the region between $y = 3\cos x$ and $y = e^{-x^2} + 1$ are rotated **(a)** about the x-axis and **(b)** about the y-axis.

♦ Type 2, volume by slicing and by cylindrical sections

45. Use your calculator or computer to find the approximate volumes of the solids that are generated when the region between $y = \frac{1}{4}x^2$ and $y = \tan^{-1} x$ are rotated **(a)** about the line $y = -1$ and **(b)** about the $y = -2$.

♦ Type 2, volumes by cylindrical sections

46. When the portion of $y = x^3$ for $0 \le x \le 2$ is rotated about the x-axis it generates a surface. Show by evaluating the integral that the surface has area $\frac{1}{27}\pi(145^{3/2} - 1)$.

♦ Type 3, surface area

In Problems 47 through 49 use ARCHIMEDES' PRINCIPLE that a floating object displaces its weight in water, and use 62.4 pounds per cubic foot for the density of water.

47.[A] The vertical cross sections of a six-foot-long boat have the shape of the region bounded by $y = x^2$ and $y = 1$ in an xy-plane with coordinates measured in feet. How deep does the boat settle in the water if it weighs 200 pounds?

♦ Type 4, Archimedes' Principle

48.[A] The bottom of a sphere with a one-foot radius sinks four inches below the surface when the sphere is placed in water. How much does the sphere weigh?

♦ Type 4, Archimedes' Principle

49. When the region bounded by $y = x^{1/3}$, $y = 1$, and the y-axis in an xy-plane with distances measured in feet is rotated about the y-axis, it generates a top-shaped solid. A top with this shape and a weighted tip to keep it upright is placed in a swimming pool. The top weighs one pound. How far does is sink below the surface?

♦ Type 4, Archimedes' Principle

50.[A] Find the volume common to two cylinders with a four-foot radius whose axes intersect at right angles (Figure 38). (The cross sections parallel to the axes are squares.)

♦ Type 4, volume by slicing

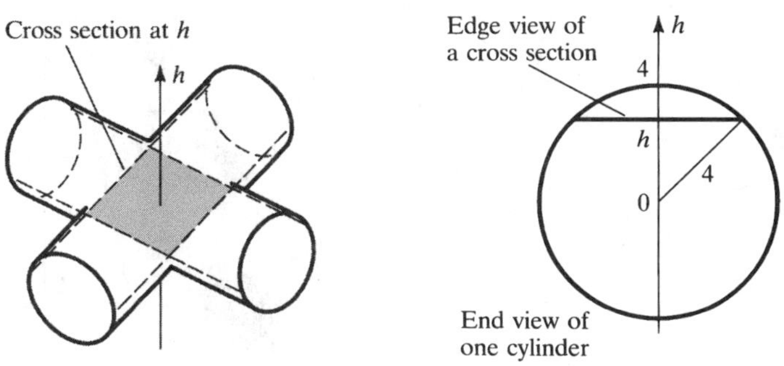

FIGURE 38

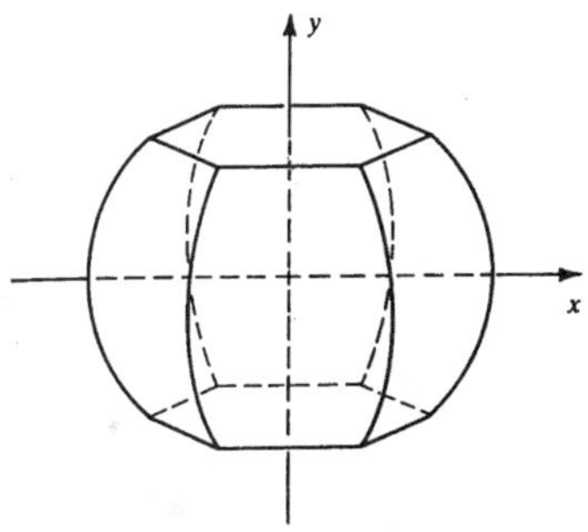

FIGURE 39

51. The intersection of a stool with a vertical xy-plane is the portion of the circle $x^2 + y^2 \leq 1$ with $-1 \leq y \leq 1$ and its cross sections perpendicular to the y-axis are regular hexagons with diagonals in the xy-plane (Figure 39). What is its volume?

♦ Type 4, volume by slicing

52. A right circular cylindrical hole is drilled through the center of a sphere of radius $r > 1$ The radius of the hole is chosen so that the hole remaining in the sphere is two feet long. Show that the volume of resulting solid is the same for all $r > 1$ and equals the volume of a sphere of radius 1.

♦ Type 4, volume by slicing

53. Use an integral to derive the formula $\frac{1}{3}\pi r^2 h$ for the volume of a right circular cylinder of height h and with base of radius r.

♦ Type 4, volume by slicing

54.[A] Find the volume of a slice cut from a cylinder of cheese ten inches in diameter and ten inches high by a plane that passes through a diameter of the circular base of the cylinder and that makes an angle of 45° with the base.

♦ Type 4, volume by slicing

55. A vertical plane a distance a to the right of the center of a sphere of radius r with $0 \leq a \leq r$ cuts the sphere in two pieces. Show that the volume of the portion on the right is $\pi(\frac{2}{3}r^3 - r^2 a + \frac{1}{3}a^3)$.

♦ Type 4, volume by slicing

56.[A] **(a)** Use the formula for the area of a circle to find the exact value of $\int_0^a \sqrt{a^2 - x^2}\, dx$ in terms of the positive constant a. Then use the result of part (a) **(b)** to find the area inside the ellipse $\frac{x^2}{a^2} + \frac{y^2}{b^2} = 1$ with positive constants a and b and **(c)** to find the volume of the torus that is generated when the ellipse in part (b) is rotated about the line $y = c$ with a constant $c > b$.

♦ Type 4, volume by slicing, etc.

57. A general cone with a bounded region in a horizontal plane as base and vertex P above the plane, consists of all line segments between P and points in the base. Imagine a vertical x-axis with its origin at P and its positive side pointing down. Suppose that P is h units above the plane (Figure 40). **(a)** Show that if S and S_1 are points in the base and T that T_1 are the points on the line segments PS and PS_1, respectively, that are vertical distance x below P, then $\overline{TT_1} = (x/h)\overline{PP_1}$ (Figure 41).

(b) Imagine that R is approximated by rectangles. Then the lines through P and the corners of the rectangles intersect the cross section of the cone at x in the corners of rectangles in an approximation of that cross section. The result of part (b) shows that the length of each side of each rectangle in the cross section equals x equals x/h multiplied by the length of the corresponding side of a rectangle in the base, so the area of the rectangle in the cross section is $(x/h)^2$ multiplied by the area of the rectangle in the base. Since the two regions can be approximated by collections of rectangles with whatever accuracy we want, the area of the cross section at x equals $(x/h)^2 A$ with A the area of the base. Use this formula to show that the volume of the cone is $\frac{1}{3}Ah$.

♦ Type 4, volume by slicing, etc.

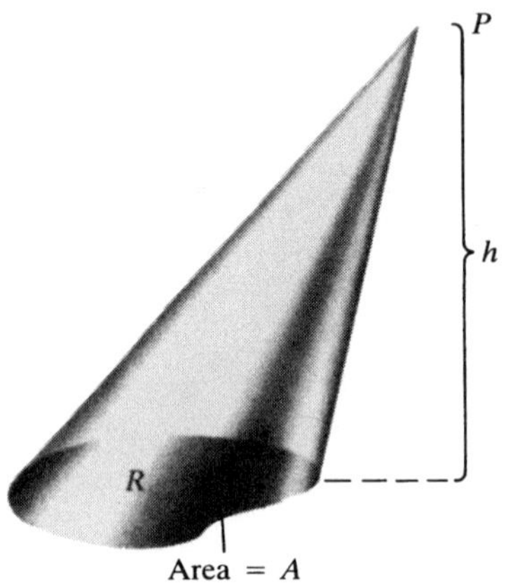

FIGURE 40

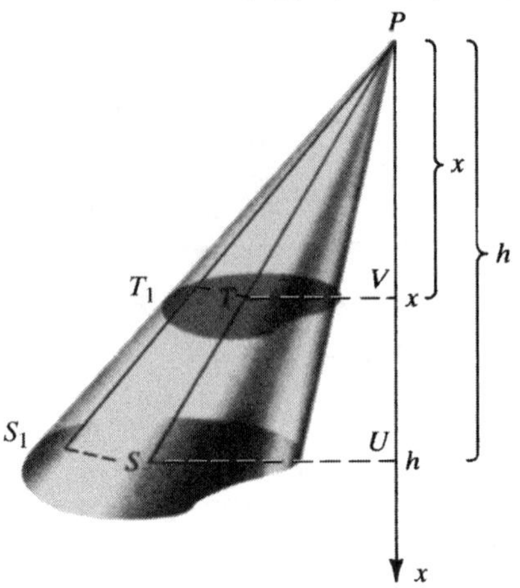

FIGURE 41

58. When a right circular cone is cut on a straight line from a point on the base to the vertex, it can be flattened into segment of a circle. Use this fact to derive the formula $A = \pi rs$ for the surface area of a right circular cone whose base has radius r and whose slant height is s.

♦ Type 4, area of a cone

59. Use the result of Problem 58 to derive the formula

$$\pi(s_1 - s_2)(r_1 + r_2) = 2\pi(s_1 - s_2)\left[\tfrac{1}{2}(r_1 + r_2)\right]$$
$$= 2\pi[\text{Slant height}][\text{Average of the two radii}]$$

for the surface area of the frustum in Figure 42 with outer radius r_1, inner radius r_2, and slant height $s_1 - s_2$.

♦ Type 4, area of a frustum

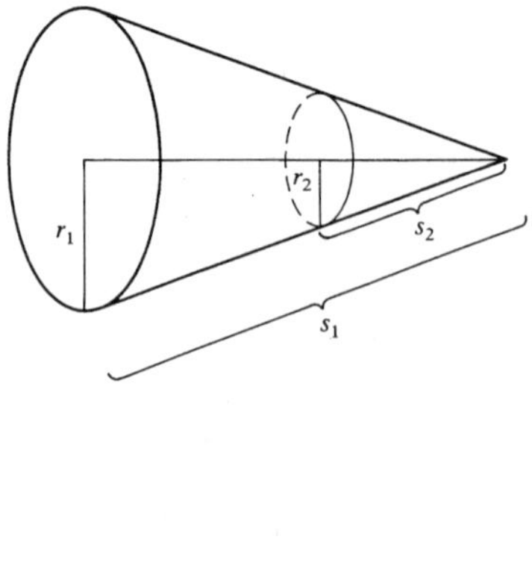

FIGURE 42

60. A glass has the shape of a right circular cone $\sqrt{3}$ inches high whose top is a circle with a one-inch radius. The glass is full of cola, and a spherical ball is to be placed in the glass, if it is small, and resting on the rim if it is larger. What sized ball would make the biggest mess?

♦ Type 4, a maximum/minimum problem

Section 6.2

Integration by parts and by partial fractions

OVERVIEW: *In Chapter 4 we derived basic integration formulas from formulas for the derivatives of x^n, exponential functions, logarithms, trigonometric functions, and inverse trigonometric functions. Then we saw that exact formulas and values of other integrals could be found in cases where the integrals could be put in one of the basic forms by a substitution. In this section we study two other integration techniques:* INTEGRATION BY PARTS, *which is a consequence of the Product Rule for derivatives, and* INTEGRATION BY PARTIAL FRACTIONS, *in which quotients of polynomials are expressed as sums of simpler terms which can be integrated with basic formulas.*

Topics:

- ***Integration by parts***
- ***Decomposing rational functions***
- ***Factoring polynomials***
- ***Partial-fraction decompositions***
- ***Integration by partial fractions***

Integration by parts

The area of the region in Figure 1 between the curve $y = x\sin x$ and the x-axis for $0 \le x \le b$, for a constant b between 0 and π, is given by the integral

$$[\text{Area for } 0 \le x \le b] = \int_0^b x\sin x\,dx \tag{1}$$

which we cannot integrate by substitution with the basic integration formulas from Chapter 4.

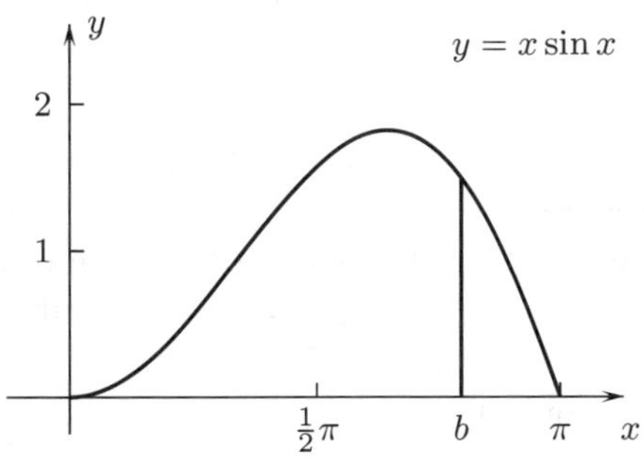

FIGURE 1

Suppose that we need values of this integral for different values of b. We could, as with any Riemann integral of an elmentary function, use a calculator or computer to find its approximate values as in Table 1 for as many particular values of b as we need.

TABLE 1

b	0.5	1	1.5	2	2.5	3
$\int_0^b x\sin x\,dx \approx$	0.040634	0.301169	0.891389	1.741591	2.601331	3.111097

♦ SUGGESTIONS TO INSTRUCTORS: The main topics in this section can be divided into two parts, as follows:
Part 1 (Integration by parts and more elementary integration techniques; Tune-up Exercises T1–T4; Problems 1–26, 59, 61–63) Have students begin work on Questions 1 and 2 before class. Problems 61–63 are good for in-class discussion.
Part 2 (Integration by partial fractions and more elementary integration techniques) Tune-up Exercises T5–T12; Problems 27–58, 60, 64–67) Have students begin work on Questions 3 through 12 before class. Problems 64–67 are good for in-class discussion.

It would be preferable, however, to have an exact formula for the area **(1)** in terms of the parameter b. We can find such a formula by INTEGRATION BY PARTS.

To derive the necessary procedure, we write the Product Rule,

$$\frac{d}{dx}(uv) = u\frac{dv}{dx} + v\frac{du}{dx}$$

in the form

$$u\frac{dv}{dx} = \frac{d}{dx}(uv) - v\frac{du}{dx}.$$

We take antiderivatives of both sides of this equation to obtain

$$\int u\frac{dv}{dx}\,dx = \int \frac{d}{dx}(uv)\,dx - \int u\frac{dv}{dx}\,dx.$$

Then, since uv is an antiderivative of its derivative, we have $\int \frac{d}{dx}(uv)\,dx = uv + C$, and we are led to the first formula in the following theorem.

Theorem 1 (Integration by parts) **(a)** *At any value of x where $u = u(x)$ and $v = v(x)$ have derivatives,*

$$\int u\frac{dv}{dx}\,dx = uv - \int v\frac{du}{dx}\,dx. \qquad \textbf{(2)}$$

(b) *If a and b are in an interval where $u(x)$ and $v(x)$ are continuous and are such that the definite integrals of $u\frac{dv}{dx}$ and $v\frac{du}{dx}$ from a to b are defined, then*

$$\int_a^b u\frac{dv}{dx}\,dx = \Big[uv\Big]_a^b - \int_a^b v\frac{du}{dx}\,dx. \qquad \textbf{(3)}$$

We do not need to include a $+C$ on the right of **(2)** because the indefinite integral on the right already involves an arbitrary constant. Formula **(3)** is a consequence of **(2)** and the Fundamental Theorem of Calculus from Section 4.2.

To apply formula **(2)** or **(3)** to an integral, we view the integrand as the product of two functions, u and dv/dx. Formula **(2)** or **(3)** then expresses the given integral, on the left, in terms of a new integral on the right. In the new integral, u has been differentiated and dv/dx has been integrated. This is useful if the new integral is easier to integrate than the one we started with. Let us use this technique with integral **(1)** for the area described above.

Example 1 Evaluate the integral $\int_0^b x \sin x\,dx$.

SOLUTION To integrate by parts, we could either integrate x and differentiate $\sin x$ or differentiate x and integrate $\sin x$. We need to visualize the results of both procedures to choose which to use. If we integrate x (yielding $\frac{1}{2}x^2$) and differentiate $\sin x$ (yielding $\cos x$), we end up an integral of the product $\frac{1}{2}x^2 \cos x$. On the other hand, if we differentiate x (yielding 1) and integrate $\sin x$ (yielding $-\cos x$), we end up an integral of $(1)(-\cos x)$. We use the latter procedure because it leaves us with an integral of $-\cos x$, which we can evaluate directly.

Because we want to differentiate x and integrate $\sin x$, we set

$$u = x \quad\text{and}\quad \frac{dv}{dx} = \sin x.$$

Then $du/dx = 1$, and to have $dv/dx = \sin x$, we use $v = -\cos x$[†], so that **(3)** yields

$$\begin{aligned}\int_0^b x\sin x\,dx &= \int_0^b u\frac{dv}{dx}\,dx = \Big[uv\Big]_0^b - \int_0^b v\frac{du}{dx}\,dx \\ &= \Big[x(-\cos x)\Big]_0^b - \int_0^b -(\cos x)(1)\,dx \\ &= \Big[-x\cos x\Big]_0^b + \int_0^b \cos x\,dx \\ &= [-\cos b - 0] + \Big[\sin x\Big]_0^b = -b\cos b + \sin b \ \square\end{aligned}$$

We will use integration by parts primarily with two types of integrals in this section. The first type, which is illustrated in Example 1 above and Question 1 below, involves a small integer power of x multiplied by $\sin(ax), \cos(ax)$, or e^{ax} with a constant a. In this case, we integrate by parts one or more times, differentiating the power of x each time until it is eliminated.

Question 1 Perform the integration $\int x^2 e^x\,dx$ by integrating by parts twice.

The second main type of integral to which we will apply integration by parts is illustrated in Example 2 and Question 2 below. These integrals involve logarithms, inverse sine functions, or inverse tangent functions in forms that cannot be integrated by substitution. In this case, we eliminate the logarithm, inverse sine, or inverse tangent by differentiating it, yielding a new integral which in many cases can be integrated by other means.

Example 2 Evaluate $\int_1^4 x^2 \ln x\,dx$.

SOLUTION We differentiate the logarithm and integrate x^2 by setting

$$u = \ln x, \quad \frac{dv}{dx} = x^2$$

for which

$$\frac{du}{dx} = \frac{1}{x}, \quad v = \tfrac{1}{3}x^3.$$

Then, with the integration-by-parts formula **(3)**, we obtain

$$\begin{aligned}\int_1^4 x^2\ln x\,dx &= \int_1^4 u\frac{dv}{dx}\,dx = \Big[uv\Big]_1^4 - \int_1^4 v\frac{du}{dx}\,dx \\ &= \Big[\tfrac{1}{3}x^3\ln x\Big]_1^4 - \int_1^4 (\tfrac{1}{3}x^3)\Big(\frac{1}{x}\Big)\,dx \\ &= [\tfrac{1}{3}(4)^3\ln(4)] - [\tfrac{1}{3}(1)^3\ln(1)] - \tfrac{1}{3}\int_1^4 x^2\,dx \\ &= \tfrac{64}{3}\ln(4) - \tfrac{1}{3}\Big[\tfrac{1}{3}x^3\Big]_1^4 = \tfrac{64}{3}\ln(4) - \tfrac{1}{9}(4^3 - 1^3) \\ &= \tfrac{64}{3}\ln(4) - 7. \ \square\end{aligned}$$

Question 2 Find the antiderivative $\int \tan^{-1} x\,dx$ by viewing the integrand as the product of 1 and $\tan^{-1} x$ and then integrating 1 and differentiating $\tan^{-1} x$.

[†]We use $v = -\cos x$ instead of $-\cos x + C$ term because we need only one antiderivative of $\sin x$.

Decomposing rational functions

Polynomials, quotients of polynomials, and sums of polynomials and quotients of polynomials are called RATIONAL FUNCTIONS.[†] We saw in Chapter 4 how to find formulas for indefinite integrals of polynomials and of certain quotients of polynomials given by simple expressions, such as

$$\int \frac{1}{x}\,dx = \ln|x| + C \quad \text{and} \quad \int \frac{1}{x^2+1} = \tan^{-1} x + C.$$

Other quotients of polynomials can be integrated either by substitution or, as we will see below, by the technique of PARTIAL FRACTIONS, in which a given quotient of polynomials is written as a sum of functions that can be integrated with our basic formulas. Before we discuss this new procedure, we will illustrate in the next three Questions how quotients of polynomials given by relatively complicated expressions can be written as sums of functions given by simpler expressions.

Question 3 The function $\dfrac{2}{x^2-1}$, whose graph is shown in Figure 2, is the sum of two of the functions $\dfrac{1}{x-1}, \dfrac{3}{x-2}, \dfrac{1}{2x}$, and $\dfrac{-1}{x+1}$. Which two are they? Use properties of the graph to justify your choices. Then check by combining the appropriate sum into a single quotient by rewriting the terms so they have the same denominator.

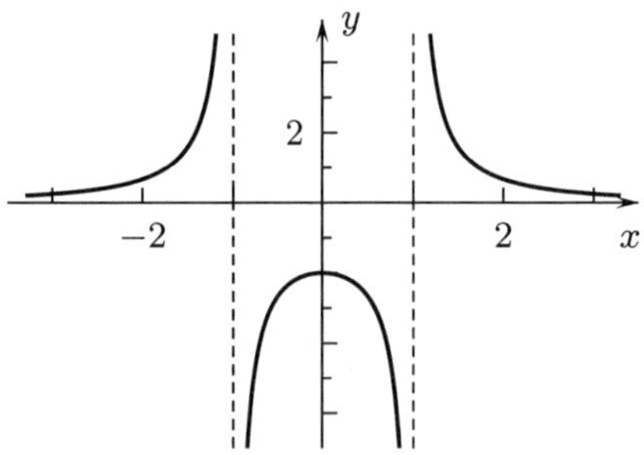

$y = \dfrac{2}{x^2-1}$

FIGURE 2

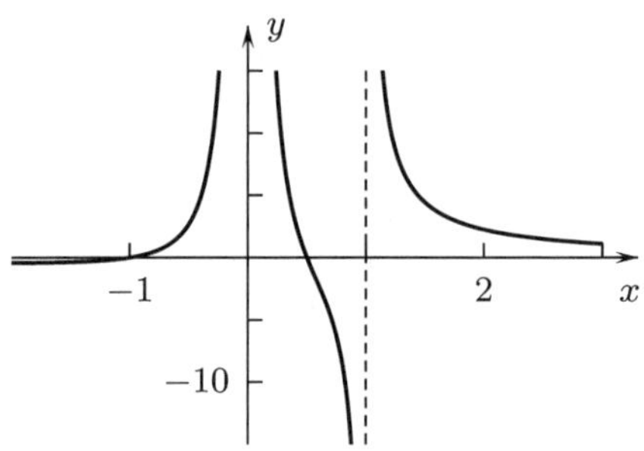

$y = \dfrac{2x^2+x^2-1}{x^3-x^2}$

FIGURE 3

Question 4 Use properties of the graph of $\dfrac{2x^2+x-1}{x^3-x^2}$ in Figure 3 to determine whether $\dfrac{2x^2+x-1}{x^3-x^2}$ is equal to $\dfrac{1}{x} + \dfrac{2}{x-1}$ or to $\dfrac{1}{x^2} + \dfrac{2}{x-1}$. To check your conclusion, combine the appropriate sum into a single quotient.

Question 5 The function $\dfrac{x^3+x-2}{x^2+1}$ of Figure 4 can be written in the form $Ax + \dfrac{B}{x^2+1}$ for certain constants A and B. The graph has y-intercept -2 and has the line $y = x$ as an asymptote. Find A and B and check your results.

[†]These are called rational functions because they are formed from polynomials by taking quotients and sums, just as the rational numbers (fractions) are formed from the integers by taking quotients and sums.

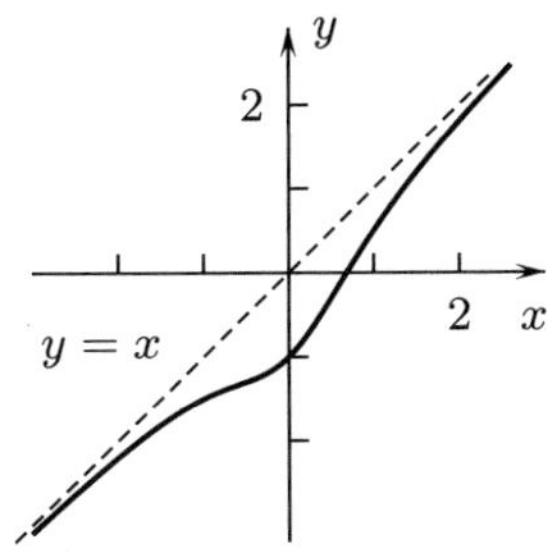

$y = \dfrac{x^3 + x - 2}{x^2 + 1}$

FIGURE 4

The results of Questions 3 through5 can be written

$$\frac{2}{x^2 - 1} = \frac{1}{x - 1} - \frac{1}{x + 1} \tag{4}$$

$$\frac{2x^2 + x - 1}{x^3 - x^2} = \frac{1}{x^2} + \frac{2}{x - 1} \tag{5}$$

$$\frac{x^3 + x - 2}{x^2 + 1} = x - \frac{2}{x^2 + 1}. \tag{6}$$

The expressions on the right of these equations are called the PARTIAL FRACTION DECOMPOSITIONS of the expressions on the left and can be used to find formulas for their integrals.

Example 3 Use **(4)** to perform the integration $\displaystyle\int \frac{2}{x^2 - 1}\, dx$.

SOLUTION We write

$$\begin{aligned}
\int \frac{2}{x^2 - 1}\, dx &= \int \frac{1}{x - 1}\, dx - \int \frac{1}{x + 1}\, dx \\
&= \int \frac{1}{u}\, du - \int \frac{1}{v}\, dv \\
&= \ln|u| - \ln|v| + C \\
&= \ln|x - 1| - \ln|x + 1| + C
\end{aligned}$$

where we have used the substitution $u = x - 1$, $du = dx$ in the first integral and $v = x + 1$, $dv = dx$ in the second. □

Question 6 Use **(5)** to find a formula for the antiderivatives of $\dfrac{2x^2 + x - 1}{x^3 - x^2}$.

Question 7 Use **(6)** to evaluate $\displaystyle\int_0^3 \frac{x^3 + x - 2}{x^2 + 1}\, dx$.

Factoring polynomials

In each of Questions 4 through 6 you used properties of the graphs of rational functions to select their partial-fraction decompositions from several given choices. How can we find the partial-fraction decomposition of a quotient $p(x)/q(x)$ of polynomials if all we have is formulas for $p(x)$ and $q(x)$? Because the first step is to factor the denominator if it is not already given in factored form, we need to review the basic principles of factoring.

The solutions (real or complex) of the equation

$$ax^2 + bx + c = 0 \tag{7}$$

with real constants $a \neq 0, b$, and c are given by the Quadratic Formula

$$x = \frac{-b \pm \sqrt{b^2 - 4ac}}{2a}. \quad \mathbf{(8)}$$

The quantity $b^2 - 4ac$ in **(8)** is known as the DISCRIMINANT of $ax^2 + bx + c$. As formula **(8)** shows, equation **(7)** has two, different real solutions if the discriminant is positive; **(7)** has one real solution if the discriminant is zero; and **(7)** has no real solutions if the discriminant is negative. Accordingly, the form of the factorization of $ax^2 + bx + c$ also depends on whether $b^2 - 4ac$ is positive, zero, or negative.

If $b^2 - 4ac$ is positive, the factorization is

$$ax^2 + bx + c = a(x - r_1)(x - r_2) \quad \mathbf{(9)}$$

where r_1 and r_2 are the numbers given by **(8)**. If $b^2 - 4ac$ is zero, then **(8)** yields only one number r and[†]

$$ax^2 + bx + c = a(x - r)^2. \quad \mathbf{(10)}$$

Finally, if $b^2 - 4ac$ is negative, then the polynomial cannot be factored using only real numbers and is said to be an IRREDUCIBLE QUADRATIC POLYNOMIAL.

The FUNDAMENTAL THEOREM OF ALGEBRA states that every polynomial with real coefficients can, in principle, be factored into real linear (first degree) and irreducible quadratic factors. Finding the factors can be difficult, but in many cases they can be determined by inspection or by using the Quadratic Formula.

Example 4 Factor $x^3 - 2x^2 - 3x$ into linear and irreducible quadratic factors.

SOLUTION **(a)** We first factor an x from $x^3 - 2x^2 - 3x$ to obtain

$$x^3 - 2x^2 - 3x = x(x^2 - 2x - 3).$$

We might be able to factor the polynomial by inspection to have $x^2 - 2x - 3 = (x-3)(x+1)$ since its zeros are small positive integers. Or, we could use the Quadratic Formula, which shows that the solutions of $x^2 - 2x - 3$ are

$$\begin{aligned} x &= \frac{-(-2) \pm \sqrt{(-2)^2 - 4(1)(-3)}}{2(1)} \\ &= \frac{2 \pm \sqrt{16}}{2} = 3, -1. \end{aligned}$$

Then applying **(9)** with $a = 1, r_1 = 3$, and $r_2 = -1$ gives $x^2 - 6x - 3 = (x-3)(x+1)$.

With either approach to factoring the quadratic polynomial, we obtain

$$x^3 - 2x^2 - 3x = x(x^2 - 2x - 3) = x(x - 3)(x + 1). \ \square$$

Example 5 Factor $x^4 + 2x^2 + 1$ into linear and irreducible quadratic factors.

SOLUTION Since $a^2 + 2ab + b^2 = (a + b)^2$ for any numbers a and b, we have

$$x^4 + 2x^2 + 1 = (x^2 + 1)^2. \quad \mathbf{(11)}$$

The quadratic polynomial $x^2 + 1$ in **(11)** is irreducible because it is not zero for any real x (and because its discriminant $0^2 - 2(1)(1) = -2$ is negative). Therefore, **(11)** is the factorization we are seeking. $\square$

[†]To remember **(9)**, notice that the right side is zero at $x = r_1$ and $x = r_2$ and that if it were expanded, the coefficient of x^2 would be a. To remember **(10)**, notice that the right side has a double zero at $x = r$ and that if it were expanded, the coefficient of x^2 would be a.

Proper rational functions

An expression $n + p/q$ for a positive rational number (fraction), with nonnegative integers n and p and a positive integer q, is called PROPER if $p < q$.

By analogy, a sum of a polynomial $n(x)$ plus a quotient of polynomials $p(x)/q(x)$ is called a PROPER RATIONAL FUNCTION and $p(x)/q(x)$ is called a PROPER QUOTIENT if the degree of the numerator $p(x)$ is less than the degree of the denominator $q(x)$. The formula, $6x + \dfrac{3x-1}{x^2-1}$, for instance, is a proper expression of a rational function because the numerator in the ratio has degree 1 and this is less than the degree of the denominator, which is 2.

On the other hand, $\dfrac{x^2}{x^2+1}$ is not proper because its numerator and denominator have the same degree, and $\dfrac{x^5-x}{x^4-10}$ is not proper because the degree of its numerator is greater than the degree of its denominator.

Partial-fraction decompositions with nonrepeated linear factors

If the denominator of a proper quotient of polynomials is a product,

$$a(x-r_1)(x-r_2)\cdots(x-r_N) \tag{12}$$

of N distinct, linear factors, then its partial-fraction decomposition can be found with the next rule.

Rule 1 *If $r_1, r_2, \cdots, r_N$, are N different numbers and the degree of the polynomial $p(x)$ is less than N, then there are numbers $A_1, A_2, \cdots, A_N$ such that*

$$\frac{p(x)}{(x-r_1)(x-r_2)\cdots(x-r_N)} = \frac{A_1}{x-r_1} + \frac{A_2}{x-r_2} + \cdots + \frac{A_N}{x-r_N}. \tag{13}$$

To remember this rule, notice that the graphs of both sides of **(13)** have vertical asymptotes at $x = r_1, r_2, \cdots, r_N$ and that the denominator on the left is the least common denominator of the terms on the right.

Two techniques for finding the constants $A_1, A_2, \cdots, A_N$ in Rule 1 are illustrated in the next example.

Example 6 Use Rule 1 to find the partial-fraction decomposition of $\dfrac{2}{x^2-1} = \dfrac{2}{(x+1)(x-1)}$ from Question 3.

SOLUTION Because $(x+1)(x-1)$ consists of distinct linear factors, Rule 1 can be applied and the partial-fraction decomposition is of the form,

$$\frac{2}{(x+1)(x-1)} = \frac{A}{x+1} + \frac{B}{x-1} \tag{14}$$

with constants A and B that we need to determine. To find them, we first give the terms on the right the same denominator by multiplying and dividing the first term by $(x-1)$ and multiplying and dividing the second term by $(x+1)$. This gives

$$\begin{aligned}\frac{2}{(x+1)(x-1)} &= \frac{A(x-1)}{(x+1)(x-1)} + \frac{B(x+1)}{(x+1)(x-1)}\\ &= \frac{A(x-1)+B(x+1)}{(x+1)(x-1)}.\end{aligned}$$

Because the first and last functions in these equations have the same denominators, their numerators must be equal wherever the functions are defined. Consequently,

$$2 = A(x-1) + B(x+1) \tag{15}$$

for $x \neq \pm 1$. This equation must also hold at $x = \pm 1$ since both sides of it are continuous for all x.

One way to solve **(15)** for A and B is to multiply out the terms on the right to have $2 = Ax - A + Bx + B$ and then write

$$2 = (A+B)x + (B-A).$$

In order for this equation to hold for all x, the coefficient of x and the constant term must be the same on both sides, so that $A + B = 0$ and $B - A = 2$. Solving these equations would give $A = -1$ and $B = 1$.

There is, however, an easier approach that uses **(15)** as it is. We first set $x = 1$, for which the first term on the right is zero. This gives $2 = A(0) + B(1+1)$, showing that $2 = 2B$ and $B = 1$. Next we set $x = -1$ to make the second term zero. This gives $2 = A(-1-1)$ or $2 = -2A$ so that $A = -1$.

Having found A and B by either approach, we substitute their values in **(14)** to obtain the partial-fraction decomposition that we examined in Question 3:

$$\frac{2}{x^2-1} = \frac{-1}{x+1} + \frac{1}{x-1}. \ \square$$

Question 8 **(a)** Find the partial-fraction decomposition of the proper quotient $\dfrac{x+2}{x^3-x}$.

(b) Use the result of part (a) to perform the integration $\displaystyle\int \frac{x+2}{x^3-x}\,dx$.

Partial-fraction decompositions: the general case

The factorization of the denominator of a proper quotient of polynomials might include repeated linear or irreducible quadratic factors rather than only nonrepeated linear factors, as in Rule 1. Then the form of the partial-fraction decomposition is determined by the following rule, which includes the case covered by Rule 1.

Rule 2 *To form the partial-fraction decomposition of a proper quotient $\dfrac{p(x)}{q(x)}$ of polynomials, include one or more terms for each factor of the denominator, according to the following prescriptions.*

(a) *For every nonrepeating linear factor $x - r$, add a term of the form $\dfrac{A}{x-r}$.*

(b) *If the linear factor $x - r$ occurs to the kth power with $k \geq 2$, then add instead k terms of the form*

$$\frac{A_1}{x-r} + \frac{A_2}{(x-r)^2} + \cdots + \frac{A_k}{(x-r)^k}.$$

(c) *For each nonrepeating irreducible quadratic factor $ax^2 + bx + c$, add a term $\dfrac{Ax+B}{ax^2+bx+c}$*

(d) *If the irreducible quadratic factor $ax^2 + bx + c$ occurs to the kth power with $k \geq 2$, add instead k terms of the form*

$$\frac{A_1x+B_1}{ax^2+bx+c} + \frac{A_2x+B_2}{(ax^2+bx+c)^2} + \cdots + \frac{A_kx+B_k}{(ax^2+bx+c)^k}.$$

Terms of the types in (a) and (b) in Rule 2 can be integrated with the substitution $u = x - r$, and terms of the type in (c) can be integrated by using an inverse tangent if $A = 0$ and by using the substitution $u = ax^2 + bx + c$ and possibly an inverse tangent if $A \neq 0$.

We will not consider in this section functions that require terms of the type in (d) because their integrals often require inverse tangent substitutions, which we will not discuss until Section 6.3.

Example 7 Give the forms of the partial-fraction decompositions of **(a)** $\dfrac{2x-1}{(x-1)^2}$ and **(b)** $\dfrac{3x^2-4x+1}{(x-2)(x^2+1)}$.

SOLUTION **(a)** This rational function is proper and its denominators is $x - 2$ is raised to the second power so its decomposition has the form

$$\frac{2x-1}{(x-1)^2} = \frac{A}{x-1} + \frac{B}{(x-1)^2}. \tag{16}$$

(b) Here the factors in the denominator are the linear function $x - 2$ and the irreducible quadratic $x^2 + 1$, and the decomposition has the form

$$\frac{3x^2-4x+1}{(x-2)(x^2+1)} = \frac{A}{x-2} + \frac{Bx+C}{x^2+1}. \ \square \tag{17}$$

One constant in the partial fraction decomposition can be found for each linear factor by the second technique described in the solution of Example 8. Any other constants are then determined by comparing the coefficients of powers of x and solving the resulting equations.

Example 8 Find the partial-fraction decomposition of $\dfrac{2x-1}{(x-1)^2}$.

SOLUTION Multiplying the numerator and denominator of the first term on the right of **(16)** by $x - 1$ to give both terms the same denominator yields

$$\frac{2x-1}{(x-1)^2} = \frac{A(x-1)+B}{(x-1)^2}.$$

In order for the last equation to hold for $x \neq 1$, we must have

$$2x - 1 = A(x-1) + B. \tag{18}$$

This equation then holds for all x because both sides are continuous for all x. We set $x = 1$ to make the first term on the right of **(18)** be zero. This gives $2(1) - 1 = A(0) + B$, showing that $B = 1$.

With $B = 1$, equation **(18)** becomes $2x - 1 = Ax + A - 1$ which simplifies to $2x - 1 = Ax - A + 1$. Comparing coefficients gives the equations $A = 2$ and $-1 = -A + 1$, both of which show that $A = 2$. (We have here two equations with one unknown because we have already found B.) With $B = 1$ and $A = 2$, **(18)** yields

$$\frac{2x-1}{(x-1)^2} = \frac{2}{x-1} + \frac{1}{(x-1)^2}. \ \square \tag{19}$$

Question 9 Use the result of Example 8 to evaluate $\displaystyle\int_2^4 \frac{2x-1}{(x-1)^2}\,dx$. Check the result with your calculator or computer.

Question 10 Find the constants A, B and C in the partial-fraction decomposition **(17)**.

Example 9 Evaluate $\displaystyle\int_0^1 \frac{3x^2 - 4x + 1}{(x-2)(x^2+1)}\,dx.$

SOLUTION With the result of Question 10, we obtain

$$\int_0^1 \frac{3x^2 - 4x + 1}{(x-2)(x^2+1)}\,dx = \int_0^1 \left(\frac{1}{x-2} + \frac{2x}{x^2+1}\right)dx.$$

We use the substitution $u = x - 2, du = 2\,dx$ in the first integral and $v = x^2 + 1$, $dv = 2x\,dx$ in the second to have

$$\begin{aligned}\int_0^1 \frac{3x^2 - 4x + 1}{(x-2)(x^2+1)}\,dx &= \int_{x=0}^{x=1} \frac{1}{u}\,du + \int_{x=0}^{x=1} \frac{1}{v}\,dv \\ &= \Big[\ln|u|\Big]_{x=0}^{x=1} + \Big[\ln|v|\Big]_{x=0}^{x=1} \\ &= \Big[\ln|x-2|\Big]_{x=0}^{x=1} + \Big[\ln(x^2+1)\Big]_{x=0}^{x=1} \\ &= [\ln(1) - \ln(2)] + [\ln(2) - \ln(1)] = 0. \ \square\end{aligned}$$

Improper quotients of polynomials

Converting an improper quotient of polynomials into the sum of a polynomial and a proper quotient is analogous to converting an improper fraction into the sum of an integer and a proper fraction.

To express the improper fraction $\frac{23}{5}$, for example, as an integer plus a proper fraction, we divide 5 into 23, as shown in Figure 5. This calculation yields 4 with a remainder of 3, so that $\frac{23}{5} = 4 + \frac{3}{5}$.

$$\begin{array}{r|l} & \ \ 4 \\ \hline 5 & 23 \\ & 20 \\ \hline & \ \ 3 \end{array}$$

FIGURE 5

$$\begin{array}{r|rrrrr} & 3x & & & & \\ \hline x^2+1 & 3x^3 & + & 4x & + & 2 \\ & 3x^3 & + & 3x & & \\ \hline & & & x & + & 2 \end{array}$$

FIGURE 6

Improper quotients of polynomials can be expressed as polynomials plus proper quotients of polynomals by similar calculations.

Example 10 Express $\dfrac{3x^3 + 4x + 2}{x^2 + 1}$ as a polynomial plus a proper quotient of polynomials.

SOLUTION In the calculation of Figure 5, we put the number 4 above the horizontal line in the division symbol because $4 \times 5 = 20$ is the largest multiple of 5 that is ≤ 23. Then we multiplied the divisor 5 by 4 and subtracted the result from 23 to obtain the remainder of 3.

The calculation in Figure 6 that yields a proper expression for $\dfrac{3x^3 + 4x + 2}{x^2 + 1}$ is similar. We put $3x$ above the division symbol because multiplying the divisor $x^2 + 1$ by $3x$ yields the first term $(3x^3)$ in the dividend $3x^3 + 4x + 1$. We mulitply the divisor by $3x$ and subtract the result $(3x^3 + 3x)$ from the dividend. The remainder is $x + 2$. Therefore, dividing $3x^3 + 4x + 2$ by $x^2 + 1$ yields $3x$ with a remainder of $x + 2$, and

$$\frac{3x^3+4x+2}{x^2+1} = 3x + \frac{x+2}{x^2+1} \tag{20}$$

This is a proper expression of the rational function because the degree of the numerator in the quotient is less than the degree of the denominator. □

Question 11 Check **(20)** by combining the terms on the right into a single quotient.

Question 12 Use the result of Example 10 to perform the integration $\int \frac{3x^3+4x+2}{x^2+1}\,dx$.

Calculations of the type in Figure 5 can be used to express any improper quotient of polynomials as a polynomial plus a proper quotient of polynomals. The partial-fraction decomposition of the proper quotient can then be found if necessary by using Rule 2. The sum of the polynomial and the partial-fraction decomposition is then an expression for the original function that can be integrated.

Principles and procedures

- To integrate $x^n\sin(ax), x^n\cos(ax)$, or x^ne^{ax} with a small positive integer n and constant a, integrate by parts n times, differentiating the power of x each time.
- To integrate $\ln x, \sin^{-1} x$, or $\tan^{-1} x$ or a product of one of these functions and another function, first try the substitution $u = \ln x, u = \sin^{-1} x$, or $u = \tan^{-1} x$, respectively. If this does not work, try integration by parts where you differentiate $\ln x, \sin^{-1} x$, or $\tan^{-1} x$. This may (or may not) give you a new integral which you can integrate.
- To integrate a rational function, first use division and other algebra, if necessary, to express it as a polynomial plus a proper quotient of polynomials. Then factor the denominator. If it consists only of distinct linear factors, use Rule 1 to determine the form of the partial-fraction decomposition of the quotient. If the denominnator has repeated linear or irreducible quadratic factors, use Rule 2. Next, use the algebraic procedures described in the examples above to find the values of the constants involved and integrate the result.

Tune-Up Exercises 6.2[♦]

[A]*Answer provided.* [O]*Outline of solution provided.*

Carry out the integration in Exercises T1 through T4.

T1.[O] $\int_1^5 \ln(2x)\,dx$

♦ Type 1, integration by parts

T2.[O] $\int x^2e^{3x}\,dx$

♦ Type 1, integration by parts

T3.[O] $\int \frac{\tan^{-1} x}{1+x^2}\,dx$

♦ Type 1, integration

T4.[O] $\int \frac{1+2x}{x+x^2}\,dx$

♦ Type 1, integration

Give the partial fraction decomposition of the rational functions in Exercises T5 through T8.

T5.[O] $\frac{5x-7}{x^2-3x+2}$

♦ Type 1, partial-fraction decomposition

T6.[O] $\frac{x^3}{x^2-4}$

♦ Type 1, partial-fraction decomposition

T7.[O] $\frac{9}{x^3-6x^2+9x}$

♦ Type 1, partial-fraction decomposition

T8.[O] $\frac{x^4}{x^2-4x+5}$

♦ Type 1, partial-fraction decomposition

[♦]The Tune-up Exercises and Problems are classified by type and content. The types are (1) basic, reactive; (2) basic reflective; (3) intermediate, reactive; (4) intermediate, reflective; (5) advanced, reactive; (6) advanced, reflective; and (7) advanced, theoretical.

Perform the integration in Exercises T9 through T12. (Use the results of Exercises T5–T8.)

T9.[O] $\displaystyle\int \frac{5x-7}{x^2-3x+2}$

♦ Type 1, integration by partial-fractions

T10.[O] $\displaystyle\int_6^{10} \frac{x^3}{x^2-4}\,dx$

♦ Type 1, integration by partial-fractions

T11.[O] $\displaystyle\int \frac{9}{x^3-6x^2+9x}\,dx$

♦ Type 1, integration by partial-fractions

T12.[O] $\displaystyle\int \frac{x^4}{x^2-4x+5}\,dx$

♦ Type 1, integration by partial-fractions

Problems 6.2

[A]*Answer provided.* [O]*Outline of solution provided.*

Perform the integration in Exercises 1 through 20.

1.[O] $\displaystyle\int x\cos(5x)\,dx$

♦ Type 1, integration

2.[A] $\displaystyle\int xe^{4x}\,dx$

♦ Type 1, integration

3.[A] $\displaystyle\int x\sin(x^2)\,dx$

♦ Type 1, integration

4. $\displaystyle\int x\sin(2x)\,dx$

♦ Type 1, integration

5.[A] $\displaystyle\int x\ln x\,dx$

♦ Type 1, integration

6. $\displaystyle\int_1^2 x^3\ln x\,dx$

♦ Type 1, integration

7.[A] $\displaystyle\int \frac{1}{x\ln x}\,dx$

♦ Type 1, integration

8. $\displaystyle\int \sin^{-1}(2x)\,dx$

♦ Type 1, integration

9.[A] $\displaystyle\int \frac{\sin^{-1}(2x)}{\sqrt{1-4x^2}}\,dx$

♦ Type 1, integration

10. $\displaystyle\int \tan^{-1}x\,dx$

♦ Type 1, integration

11.[A] $\displaystyle\int e^x\sin x\,dx$

♦ Type 1, integration

12. $\displaystyle\int x\sin(3x)\,dx$

♦ Type 1, integration

13.[A] $\displaystyle\int x^4\ln x\,dx$

♦ Type 1, integration

14. $\displaystyle\int_0^{10} xe^{5x}\,dx$

♦ Type 1, integration

15.[A] $\displaystyle\int xe^{x^2}\,dx$

♦ Type 1, integration

16. $\displaystyle\int \frac{(\ln x)^4}{x}\,dx$

♦ Type 1, integration

17. $\displaystyle\int \ln(5x)\,dx$

♦ Type 1, integration

18.[A] $\displaystyle\int_0^{\pi/2} x^2\sin x\,dx$

♦ Type 1, integration

19. $\displaystyle\int e^x\cos x\,dx$

♦ Type 1, integration

20. $\displaystyle\int \frac{\ln x}{x}\,dx$

♦ Type 1, integration

21.[A] Find the area of the region bounded by the x-axis, the line $x = 11$, and the curve $y = \ln x$.

♦ Type 1, area

22. Sketch the region bounded by the curves $y = xe^x, y = e$, and $x = -2$ and find its area.

♦ Type 1, area

23.[A] A rod extends from $x = 0$ to $x = 3$ (centimeters) and its density at x is e^{-x} grams per centimeter. **(a)** What is its mass? **(b)** Where is its center of gravity?

♦ Type 1, weight and center of gravity

24. A rod extends from $x = -\pi$ to $x = \pi$ (inches) and its density at x is $5 + \sin x$ ounces per inch. **(a)** How much does it weigh? **(b)** Where is its center of gravity?

♦ Type 1, weight and center of gravity

25.[A] The region R is bounded by the curves $x = 1, y = e^3$ and $y = e^x$. Compute the volume of the solid generated by rotating R about the y-axis.

♦ Type 1, volume

26. Compute the volume of the solid generated by rotating the region between $y = \ln x$ and the x-axis for $1 \le x \le 11$ about the y-axis.

♦ Type 1, volume

Give the partial fraction decompositions of the rational functions in Exercises 27 through 38.

27.[A] $\dfrac{13x+6}{x^3+5x^2+6x}$

♦ Type 1, partial-fraction decomposition

28.[A] $\dfrac{2x^2+x-22}{x^2-4}$

♦ Type 1, partial-fraction decomposition

29.[O] $\dfrac{x-2}{x^2+2x+1}$

♦ Type 1, partial-fraction decomposition

30. $\dfrac{x^2+3x+1}{x^3+2x^2+x}$

♦ Type 1, partial-fraction decomposition

31.[A] $\dfrac{x^5}{x^2+16}$

♦ Type 1, partial-fraction decomposition

32. $\dfrac{3x^3+5x^2-x-1}{x^3+x^2}$

♦ Type 1, partial-fraction decomposition

33.[A] $\dfrac{x}{(x+1)^3}$

♦ Type 1, partial-fraction decomposition

34. $\dfrac{2x^2+1}{x^3+x}$

♦ Type 1, partial-fraction decomposition

35. $\dfrac{x^2+x+1}{2x^3-x^2-x}$

♦ Type 1, partial-fraction decomposition

36. $\dfrac{4x}{x^2-4x-12}$

♦ Type 1, partial-fraction decomposition

37. $\dfrac{x^2-2x}{(x-1)(x^2+2)}$

♦ Type 1, partial-fraction decomposition

38. $\dfrac{x^3}{x^2+2x-3}$

♦ Type 1, partial-fraction decomposition

Perform the integration in Exercises 39 through 59.

39.[A] $\displaystyle\int \frac{3x}{x^2+4x+4}\,dx$

♦ Type 1, integration by partial-fractions

40.[A] $\displaystyle\int \frac{5x-1}{x^2+x}\,dx$

♦ Type 1, integration by partial-fractions

41. $\displaystyle\int \frac{2x+3}{x-1}\,dx$

♦ Type 1, integration by partial-fractions

42.[A] $\displaystyle\int \frac{2x+3}{x^2+3x+2}\,dx$

♦ Type 1, integration by partial-fractions

43. $\displaystyle\int \frac{x^3}{x^2-9}\,dx$

♦ Type 1, integration by partial-fractions

44. $\displaystyle\int \frac{x-14}{x^2-4}\,dx$

♦ Type 1, integration by partial-fractions

45.[O] $\displaystyle\int \frac{x+3}{x^2+4}\,dx$

♦ Type 1, integration by partial-fractions

46.[A] $\displaystyle\int \frac{x^2+13x}{x^2+7}\,dx$

♦ Type 1, integration by partial-fractions

47.[A] $\displaystyle\int \frac{x^2+3x+1}{x^3+2x^2+x}\,dx$

♦ Type 1, integration by partial-fractions

48. $\displaystyle\int \frac{2x^2-x-1}{x^3+x^2}\,dx$

♦ Type 1, integration by partial-fractions

49. $\displaystyle\int_1^5 \frac{x^2-3}{x^3+x}\,dx$

♦ Type 1, integration by partial-fractions

50. $\displaystyle\int \frac{x+1}{x^2+2x+10}\,dx$

♦ Type 1, integration by partial-fractions

51. $\displaystyle\int \frac{x^4+3x^2+3}{x^2+1}\,dx$

♦ Type 1, integration by partial-fractions

52. $\displaystyle\int \frac{x}{x^2-2x+10}\,dx$

♦ Type 1, integration by partial-fractions

53.[A] $\displaystyle\int \frac{\cos x}{\sin^2 x-1}\,dx$

♦ Type 1, integration by partial-fractions

54. $\displaystyle\int \frac{x}{x^2+2x+5}\,dx$

♦ Type 1, integration by partial-fractions

55. $\displaystyle\int \frac{x^3+9x+4}{x^4+2x^3+x^2}\,dx$

♦ Type 1, integration by partial-fractions

56.[A] $\displaystyle\int_{-10}^{-1} \frac{2x^2-9x+4}{x^3-4x^2+4x}\,dx$

♦ Type 1, integration by partial-fractions

57. $\displaystyle\int \frac{x^5-35x}{x^2+6}\,dx$

♦ Type 1, integration by partial-fractions

58. $\displaystyle\int_0^{10} \frac{x^2-1}{x^2+1}\,dx$

♦ Type 1, integration by partial-fractions

59. Find a formula for $\displaystyle\int (\ln x)^2\,dx$ by integrating by parts twice.

♦ Type 3, integration by parts

60.[A] Give the exact value of the number b with $0 < b < 1$ such that $\displaystyle\int_0^b \frac{1}{1-x^2}\,dx = 2.$

♦ Type 3, integration by partial-fractions

61. **(a)** Use an interpretation of definite integrals in terms of areas to explain why $\displaystyle\int_0^1 xe^x\,dx$ is less than $\displaystyle\int_0^1 e^x\,dx$. **(b)** Evaluate both integrals in part (a).

♦ Type 2, integration by parts, etc.

62. Use an interpretation of definite integrals in terms of areas to explain why $\displaystyle\int_\pi^{2\pi} x\sin x\,dx$ is negative. **(b)** Evaluate the integral in part (a).

♦ Type 2, integration by parts, etc.

63. Does the region bounded by $y = e - e^x$ and the x-axis generate a greater volume when it is rotated about the x-axis or when it is rotated about the y-axis?

♦ Type 2, volume

64.[A] Find the area of the region bounded by $y = \dfrac{24}{12+4x-x^2}$ and the line $y = 2$.

♦ Type 4, area

65. What is the area of the region bounded by the x-axis, the curve $y = \dfrac{x^2}{x^2+1}$, and the line $x = 3$?

♦ Type 4, area

66. When the region between $y = \dfrac{5}{x+1}$ and the x-axis for $0 \le x \le 4$ is rotated about the y-axis it generates a solid in the shape of a curved cone. What is its volume?

♦ Type 2, volume

67.[A] What is the average value of $\dfrac{3x^2-1}{x^3-x}$ for $2 \le x \le 12$?

♦ Type 2, average value

Section 6.3

More integrals involving trigonometric functions

OVERVIEW: *In this section we discuss two integration techniques that involve trigonometric and inverse trigonometric functions. We first use integration by parts and trigonometric identities to integrate products and powers of* $\sin x$ *and* $\cos x$*. Then we show how substitutions with inverse trigonometric functions can be applied to integrals involving odd powers of* $\sqrt{a^2+x^2}$, $\sqrt{a^2-x^2}$, *and* $\sqrt{x^2-a^2}$ *with positive constants* a.

Topics:

- ***Integrals of powers and products of sin*** **x** ***and cos*** **x**
- ***Integrals of products of sin*** **(ax)** ***and cos*** **(bx)**
- ***Inverse trigonometric substitutions***
- ***Completing squares***

Integrals of powers and products of sin x *and cos* x

Many applications involve products or positive integer powers of functions. We saw, for example, when we studied the method of slicing in Section 6.1 that if a region between the graph of a positive function $f(x)$ and the x-axis, as in Figures 1 and 2 is rotated about the x-axis, it generates a solid whose volume is given by an integral of $\pi[f(x)]^2$.

In the case of Figure 1, the solid that is obtained by rotating the region about the x-axis has the shape of a truncated cone with curved sides. The vertical line at x from the x-axis to the curve in Figure 1 generates the cross section perpendicular to the x-axis at x, which is a circle with radius $r = e^x$ feet and area $A(x) = \pi r^2 = \pi(e^x)^2$ square meters. Therefore, the volume is

$$[\text{Volume}] = \int_0^1 A(x)\,dx = \pi\int_0^1 (e^x)^2\,dx.\text{ cubic meters.} \qquad \textbf{(1)}$$

We can obtain an exact formula for this volume by substitution because of a special property of the exponential function. We can write $(e^x)^2 = e^{2x}$, so that with the subtitution $u = 2x, du = 2\,dx$ we obtain

$$\begin{aligned}[\text{Volume}] &= \pi\tfrac{1}{2}\int_0^1 e^{2x}\,(2\,dx) = \tfrac{1}{2}\pi\int_{x=0}^{x=1} e^u\,du \\ &= \tfrac{1}{2}\pi\Big[e^u\Big]_{x=0}^{x=1} = \tfrac{1}{2}\pi\Big[e^{2x}\Big]_{x=0}^{x=1} = \tfrac{1}{2}(e^2-1)\text{ cubic meters.}\end{aligned}$$

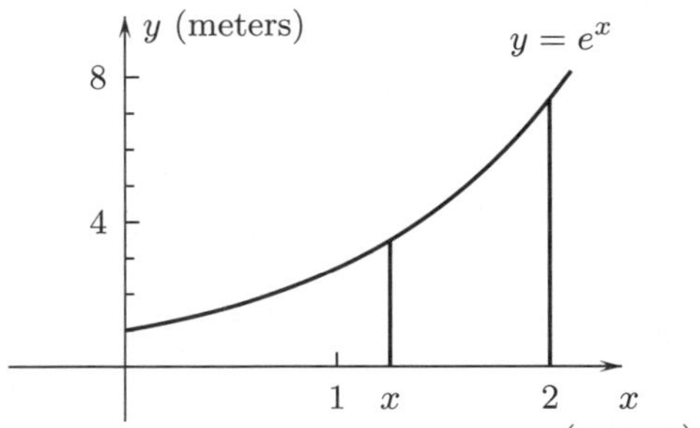

FIGURE 1

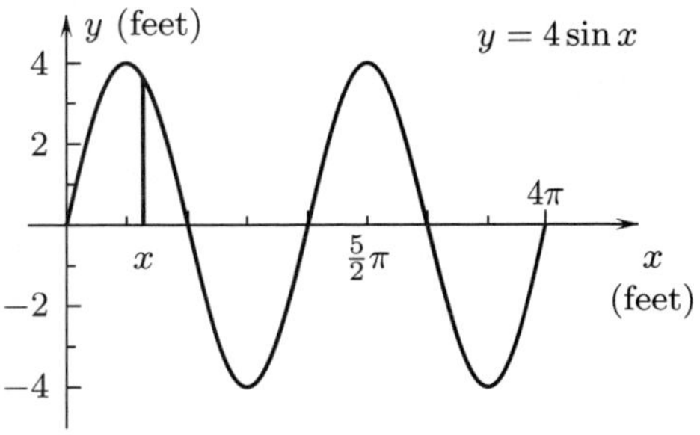

FIGURE 2

♦ SUGGESTIONS TO INSTRUCTORS: Have students begin work on Questions 1 through 6 before class. Problems 21–26 are good for in-class discussion.

Question 1 **(a)** Describe the solid that is generated when the four regions between $y = 4\sin x$ and the x-axis in Figure 2 for $0 \le x \le 4\pi$ are rotated about the x-axis. **(b)** Show that the volume of the solid is

$$16\pi \int_0^{4\pi} \sin^2 x \, dx \text{ cubic feet.} \qquad \textbf{(2)}$$

We could perform the integration in **(1)** by using the special property of the exponential function that $(e^x)^2 = e^{2x}$. Similarly, we can evaluate **(2)** by using the first of the DOUBLE-ANGLE FORMULAS:[†]

$$\sin^2\theta = \tfrac{1}{2}[1 - \cos(2\theta)] \qquad \textbf{(3)}$$

$$\cos^2\theta = \tfrac{1}{2}[1 + \cos(2\theta)]. \qquad \textbf{(4)}$$

Example 1 Use a double angle formula to calculate the volume **(2)**.

SOLUTION We use **(3)** with x in place of θ to have

$$16\pi \int_0^{4\pi} \sin^2 x \, dx = 16\pi \int_0^{4\pi} \tfrac{1}{2}[1 - \cos(2x)] \, dx = 8\pi \int_0^{4\pi} 1 \, dx - 8\pi \int_0^{4\pi} \cos(2x) \, dx.$$

Then, with the substitution $u = 2x, du = 2\,dx$ in the second integral, we obtain

$$\begin{aligned} 16\pi \int_0^{4\pi} \sin^2 x \, dx &= (8\pi)(4\pi) - 4\pi \int_{x=0}^{x=4\pi} \cos(2x)(2\,dx) = 32\pi^2 - 8\pi \int_{x=0}^{x=4\pi} \cos u \, du \\ &= 32\pi^2 - 8\pi \Big[\sin u\Big]_{x=0}^{x=4\pi} = 32\pi^2 - 8\pi \Big[\sin(2x)\Big]_{x=0}^{x=4\pi} \\ &= 32\pi^2 - 8\pi[\sin(8\pi) + \sin(0)] = 32\pi^2. \ \square \end{aligned}$$

We can also evaluate **(2)** by integration by parts with the Pythagorean Identity

$$\sin^2\theta + \cos^2\theta = 1. \qquad \textbf{(5)}$$

Example 2 Use integration by parts to calculate the volume **(2)**.

SOLUTION We set $\left\{\begin{matrix} u = \sin x \\ dv/dx = \sin x \end{matrix}\right\}$ for which $\left\{\begin{matrix} du/dx = \cos x \\ v = -\cos x \end{matrix}\right\}$ to obtain

$$\begin{aligned} 16\pi \int_0^{4\pi} \sin^2 x \, dx &= 16\pi \int_0^{4\pi} u \frac{dv}{dx} \, dx \\ &= 16\pi \Big[uv\Big]_0^{4\pi} - 16\pi \int_0^{4\pi} v \frac{du}{dx} \, dx \\ &= 16\pi \Big[-\sin x \cos x\Big]_0^{4\pi} - 16\pi \int_0^{4\pi} \cos x(-\cos x) \, dx \\ &= 16\pi \int_0^{4\pi} \cos^2 x \, dx. \end{aligned}$$

Substituting $\cos^2 x = 1 - \sin^2 x$ from the Pythagorean Theorem **(5)** into the last integral shows that

[†]To remember which of identities **(3)** and **(4)** has the plus sign and which has the minus sign, notice that there is a minus sign in **(3)** because $\sin^2(0) = 0$ and $\frac{1}{2}[1 - \cos(0)] = 0$, and there is a plus sign in **(4)** because $\cos^2(0) = 1$ and $\frac{1}{2}[1 + \cos(0)] = 1$.

$$16\pi \int_0^{4\pi} \sin^2 x\, dx = 16\pi \int_0^{4\pi} (1 - \sin^2 x)\, dx$$
$$= 64\pi^2 - 16\pi \int_0^{4\pi} \sin^2 x\, dx.$$

The last equation gives first $32\pi \int_0^{4\pi} \sin^2\, dx = 64\pi^2$, which then shows the volume $16\pi \int_0^{4\pi} \sin^2\, dx$ equals $32\pi^2$ cubic feet, as we saw in Example 1. □

Integrals $\int \sin^n x \cos^m x\, dx$, where either n or m is an odd positive integer and the other is any constant, can be found by using the Pythagopream identity and a substitution, as is illustrated in the next example.

Example 3 Perform the integration $\int \sin^2 x \cos^3 x\, dx$.

SOLUTION Because the integrand includes the cube (an odd power) of $\cos x$, we use the Pythagorean Identity **(1)** with $\theta = x$ in the form $\cos^2 x = 1 - \sin^2 x$ to express all but one factor $\cos x$ in terms of $\sin x$. We obtain

$$\int \sin^2 x \cos^3 x\, dx = \int \sin^2 x (1 - \sin^2 x) \cos x\, dx$$
$$= \int (\sin^2 x - \sin^4 x) \cos x\, dx.$$

The substitution $u = \sin x, du = \cos x\, dx$ then gives the result

$$\int \sin^2 x \cos^3 x\, dx = \int \sin^2 x (1 - \sin^2 x) \cos x\, dx$$
$$= \int (u^2 - u^4)\, du = \tfrac{1}{3}u^3 - \tfrac{1}{5}u^5 + C$$
$$= \tfrac{1}{3}\sin^3 x - \tfrac{1}{5}\sin^5 x + C. \ \square$$

Question 2 Evaluate $\int_0^{\pi/2} \sin^3 x\, dx$.

Integrals of $\sin^n x$ and $\cos^n x$ with positive even integers n and of $\sin^n x \cos^m x$ with positive, even integers n and m by be found by the techniques of Examples 1 and 2 above. We can either integrate by parts and use the Pythagorean identity **(1)**, as in Example 1, or apply the double-angle formulas **(3)** and **(4)**, as in Example 2. In the next Example we use double-angle formulas.

Example 4 Perform the integration $\int \sin^2 x \cos^2 x\, dx$.

SOLUTION Because the integrand involves even powers of $\sin x$ and $\cos x$, we first use the double-angle formulas **(3)** and **(4)**, with $\theta = x$ to obtain

$$\int \sin^2 x \cos^2 x\, dx = \int \left[\tfrac{1}{2}(1 - \cos(2x))\right]\left[\tfrac{1}{2}(1 + \cos(2x))\right] dx$$
$$= \tfrac{1}{4}\int \left[1 - \cos^2(2x)\right] dx.$$

Then the double-angle formula **(4)** with $\theta = 2x$ gives

$$\int \sin^2 x \cos^2 x\, dx = \tfrac{1}{4}\int [1 - \tfrac{1}{2} - \tfrac{1}{2}\cos(4x)]\, dx$$
$$= \tfrac{1}{8}\int dx - \tfrac{1}{8}\int \cos(4x)\, dx.$$

The last integral can be found with the substitution $u = 4x, du = 4\,dx$:

$$\int \sin^2 x \cos^2 x\, dx = \tfrac{1}{8}x - \tfrac{1}{32}\int \cos(4x)(4\,dx) = \tfrac{1}{8}x - \tfrac{1}{32}\int \cos u\, du$$
$$= \tfrac{1}{8}x - \tfrac{1}{32}\sin u + C = \tfrac{1}{8}x - \tfrac{1}{32}\sin(4x) + C. \square$$

The identities

$$\sin(2\theta) = 2\sin\theta\cos\theta \qquad \textbf{(6)}$$
$$\cos(2\theta) = 1 - 2\sin^2\theta \qquad \textbf{(7)}$$

could be used, if necessary, to express the result of the last Example in terms of $\sin x$ and $\cos x$ instead of $\sin(4x)$.

Integrals of products of sin (ax) *and cos* (bx)

Integrals of $\cos(ax)\cos(bx)$, $\cos(ax)\sin(bx)$, and $\sin(ax)\sin(bx)$, with constants a and b are most easily evaluated by using the PRODUCT IDENTITIES

$$\sin\theta\sin\psi = \tfrac{1}{2}[\cos(\theta - \psi) - \cos(\theta + \psi)] \qquad \textbf{(8)}$$
$$\cos\theta\cos\psi = \tfrac{1}{2}[\cos(\theta - \psi) + \cos(\theta + \psi)] \qquad \textbf{(9)}$$
$$\sin\theta\cos\psi = \tfrac{1}{2}[\sin(\theta - \psi) + \sin(\theta + \psi)]. \qquad \textbf{(10)}$$

If necessary, the results can then be expressed in terms of $\sin(ax)$ and $\cos(bx)$ by using the SUM AND DIFFERENCE IDENTITIES

$$\cos(\theta \pm \psi) = \cos\theta\cos\psi \mp \sin\theta\sin\psi \qquad \textbf{(11)}$$
$$\sin(\theta \pm \psi) = \sin\theta\cos\psi \pm \cos\theta\sin\psi. \qquad \textbf{(12)}$$

Example 5 Perform the integration $\int \sin(5x)\cos(2x)\, dx$.

SOLUTION We use **(10)** with $\theta = 5x$ and $\psi = 2x$ to write

$$\int \sin(5x)\cos(2x)\, dx = \int \tfrac{1}{2}[\sin\big((5-2)x\big) + \sin\big((5+2)x\big)\, dx$$
$$= \tfrac{1}{2}\int \sin(3x)\, dx + \tfrac{1}{2}\int \sin(7x)\, dx.$$

Then we use the substitutions $u = 2x, du = 2\,dx$ in the first integral and $v = 8x$, $dv = 8\,dx$ in the second to obtain

$$\int \sin(5x)\cos(2x)\, dx = \tfrac{1}{6}\int \sin(3x)(3\,dx) + \tfrac{1}{14}\int \sin(7x)(7\,dx)$$
$$= \tfrac{1}{6}\int \sin u\, du + \tfrac{1}{14}\int \sin v\, dv$$
$$= -\tfrac{1}{6}\cos u - \tfrac{1}{14}\cos v + C$$
$$= -\tfrac{1}{6}\cos(3x) - \tfrac{1}{14}\cos(7x) + C. \square$$

Question 3 Find the antiderivative $F(x)$ of $\sin(5x)\cos(2x)$ that has the value 1 at $x = 0$..

Integrals involving $\sqrt{a^2 - x^2}$

Integrals of integer powers of x multiplied by odd powers of either $\sqrt{a^2 - x^2}, \sqrt{a^2 + x^2}$ or $\sqrt{x^2 - a^2}$ arise in many contexts because of the Pythagorean Theorem and other formulas. We will now see that such integrals can be transformed into integrals involving integer powers of trigonometric functions with no square roots by using inverse sine, inverse tangent, or inverse secant substitutions. The resulting integrals can then be found by direct substitution or other techniques.

An integral of an integer power of x multiplied by an odd power of $\sqrt{a^2 - x^2}$ with constant $a > 0$ can be transformed into one involving integer powers of trigonometric functions with the substitution $x = a\sin\theta$, which is achieved by introducing the new variable

$$\theta = \sin^{-1}\left(\frac{x}{a}\right). \tag{13}$$

Figure 3 shows how x and θ are related for $0 \le x \le a$. Since $\sin\theta = x/a$, we have

$$x = a\sin\theta, \quad dx = a\cos\theta\, d\theta \tag{14}$$

and these substitutions in the given integral yield an integral involving powers of $\sin x$ and $\cos x$. Once the θ-integration has been carried out to give a formula involving $\theta, \sin\theta$ and $\cos\theta$, the result can be expressed in terms of x by using formulas **(13)** and **(14)** with the identity

$$\cos\theta = \frac{\sqrt{a^2 - x^2}}{a} \tag{15}$$

as required. Formula **(14)** follows from Figure 2 for $0 \le x \le a$ and also holds for $-a \le x < 0$.

$x = a\sin\theta$

$\sin\theta = \dfrac{x}{a}$

$\theta = \sin^{-1}\left(\dfrac{x}{a}\right)$

$\cos\theta = \dfrac{\sqrt{a^2 - x^2}}{a}$

FIGURE 3

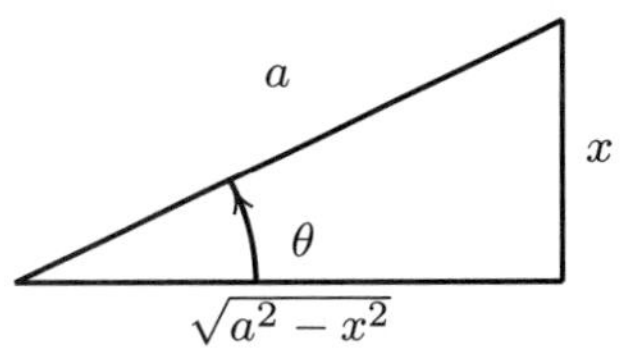

To keep the solutions of examples and problems in this section from being excessively long, we will use the table of integrals formulas for the integrals of trigonometric functions given with the list of principles and procedures at the end of the section, except in cases where this integration can be carried out by direct substitution.

Question 4 What are $x, dx, \sqrt{1 - x^2}, \sin\theta$, and $\cos\theta$ if $\theta = \sin^{-1} x$?

Example 6 Perform the integration

$$\int \frac{\sqrt{1 - x^2}}{x^2}\, dx. \tag{16}$$

SOLUTION We use the substitution

$$\theta = \sin^{-1} x \tag{17}$$

for which, as you saw in Question 4,

$$x = \sin\theta,\ dx = \cos\theta\, d\theta,\ \sqrt{1 - x^2} = \cos\theta \tag{18}$$

With these substitutions **(16)** becomes

$$\int \frac{\sqrt{1-x^2}}{x^2}\,dx = \int \frac{\cos\theta}{\sin^2\theta}\cos\theta\,d\theta = \int \cot^2\theta\,d\theta.$$

Formula **(32)** of the table of integrals given with the Principles and Procedures of this section reads

$$\int \cot^2\theta\,d\theta = -\cot\theta - \theta + C.$$

Consequently,

$$\int \frac{\sqrt{1-x^2}}{x^2}\,dx = -\cot\theta - \theta + C.$$

Finally, to express this result in terms of x, we write $\cot\theta = \cos\theta/\sin\theta$ and use **(17)** and **(18)** to obtain

$$\begin{aligned}\int \frac{\sqrt{1-x^2}}{x^2}\,dx &= -\frac{\cos\theta}{\sin\theta} - \theta + C\\ &= -\frac{\sqrt{1-x^2}}{x} - \sin^{-1}x + C. \ \square\end{aligned}$$

Question 5 Perform the integration $\displaystyle\int \frac{1}{x\sqrt{1-x^2}}\,dx$.

Integrals involving $\sqrt{a^2+x^2}$

If the integrand is the product of an integer power of x and an odd power of $\sqrt{a^2+x^2}$, we use $x = a\tan\theta$ by defining

$$\theta = \tan^{-1}\left(\frac{x}{a}\right) \quad \textbf{(19)}$$

(Figure 4), for which

$$x = a\tan\theta, \quad dx = \frac{d}{d\theta}[a\tan\theta]\,d\theta = a\sec^2\theta\,d\theta \quad \textbf{(20)}$$

and

$$\sqrt{a^2+x^2} = \sqrt{a^2(1+\tan^2\theta)} = a\sec\theta. \quad \textbf{(21)}$$

$x = a\tan\theta$

$\tan\theta = \dfrac{x}{a}$

$\theta = \tan^{-1}\left(\dfrac{x}{a}\right)$

$\sin\theta = \dfrac{x}{\sqrt{a^2+x^2}}$

$\cos\theta = \dfrac{a}{\sqrt{a^2+x^2}}$

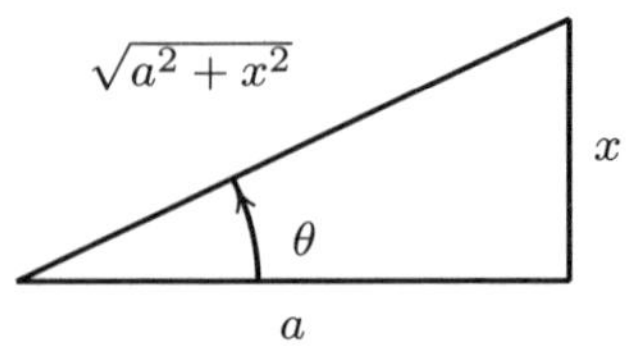

FIGURE 4

Example 7 Find the exact length of the parabola $y = \frac{1}{2}x^2$ for $0 \le x \le 3$.

SOLUTION By Rule 5 in Section 4.6 with $f(x) = \frac{1}{2}x^2$ and $f'(x) = x$,the length is

$$\int_0^3 \sqrt{1+[f'(x)]^2}\,dx = \int_0^3 \sqrt{1+x^2}\,dx. \tag{22}$$

We make the substitutions $\theta = \tan^{-1} x, x = \tan\theta, dx = \sec^2 d\theta$, and $\sqrt{1+x^2} = \sec\theta$ in the indefinite integral to obtain

$$\int \sqrt{1+x^2}\,dx = \int \sqrt{1+\tan^2\theta}\,(\sec^2\theta\,d\theta) = \int \sec^3\theta\,d\theta$$

Formula **(33)** in the table is

$$\int \sec^3\theta\,d\theta = \tfrac{1}{2}\tan\theta\sec\theta + \tfrac{1}{2}\ln|\sec\theta + \tan\theta| + C$$

so that

$$\begin{aligned}\int \sqrt{1+x^2}\,dx &= \tfrac{1}{2}\tan\theta\sec\theta + \tfrac{1}{2}\ln|\sec\theta+\tan\theta| + C\\ &= \tfrac{1}{2}x\sqrt{1+x^2} + \tfrac{1}{2}\ln|\sqrt{1+x^2}+x| + C\end{aligned}$$

We used here formulas **(19)**, **(20)**, and **(21)** with $a = 1$.The length **(22)** is

$$\begin{aligned}\int_0^3 \sqrt{1+x^2}\,dx &= \left[\tfrac{1}{2}x\sqrt{1+x^2} + \tfrac{1}{2}\ln|\sqrt{1+x^2}+x|\right]_0^3\\ &= \tfrac{3}{2}\sqrt{10} + \tfrac{1}{2}\ln(\sqrt{10}+3).\ \square\end{aligned}$$

Integrals involving $\sqrt{\mathbf{x^2 - a^2}}$

If the integrand is a product of an integer power of x and an odd power of $\sqrt{x^2-a^2}$, we use $x = a\sec\theta$. This requires the inverse secant function, which is a function we have not needed before. It can be defined by

$$\sec^{-1} x = \cos^{-1}\left(\frac{1}{x}\right) \quad \text{for } |x| \ge 1 \tag{23}$$

and then is an angle θ with $0 \le \theta < \frac{1}{2}\pi$ if $x \ge 1$ and with $\frac{1}{2}\pi < \theta \le \pi$ if $x \le -1$. We set

$$\theta = \sec^{-1}\left(\frac{x}{a}\right). \tag{24}$$

Then $x = a\sec\theta$ and

$$dx = \frac{d}{d\theta}[a\sec\theta]d\theta = a\sec\theta\tan\theta\,d\theta \tag{25}$$

$$\sqrt{x^2-a^2} = \sqrt{a^2(\sec^2\theta - 1)} = a|\tan\theta| \tag{26}$$

Absolute values are required on $|\tan\theta|$ in **(26)** because, for $x \le -a$, $-\frac{1}{2}\pi \le \theta < 0$, $\tan\theta \le 0$, and $\sqrt{x^2-a^2} \ge 0$. Figure 5 shows the geometric interpretation of **(24)** and **(26)** for $x > a$.

$$x = a\sec\theta$$
$$\sec\theta = \frac{x}{a}$$
$$\theta = \sec^{-1}\left(\frac{x}{a}\right)$$
$$|\tan\theta| = \frac{\sqrt{x^2 - a^2}}{a}$$
$$\sin\theta = \frac{\sqrt{x^2 - a^2}}{|x|}$$
$$\cos\theta = \frac{a}{x}$$

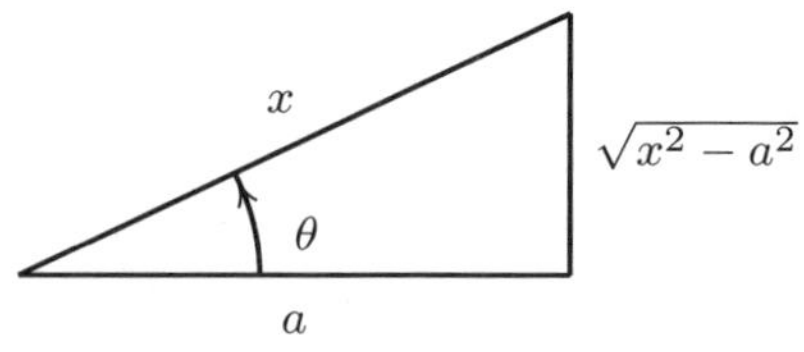

FIGURE 5

Question 6 What is the average value of $\dfrac{1}{x^2\sqrt{x^2-1}}$ for $2 \le x \le 4$?

Completing squares

Many integrals involving integer powers of x and odd-integer powers of $\sqrt{ax^2 + bx + c}$ with nonzero constants a, b and c can be evaluated by completing the square to rewrite $ax^2 + bx + c$ in the form $a(x^2 + 2rx + r^2) + s = a(x+r)^2 + s$. This is accomplished by first factoring out the coefficient a of x^2 from the first two terms and then adding and subtracting the square of half the resulting coefficient of x.

Example 8 Complete the square in the expression $4x^2 + 8x + 5$.

SOLUTION Factoring the coefficient 4 of x^2 from the first two terms gives

$$4x^2 + 8x + 5 = 4(x^2 + 2x) + 5.$$

Because the coefficient of x inside the parentheses is 2 and half of that coefficient is 1, we add and subtract $1^2 = 1$ inside the parentheses to obtain

$$\begin{aligned} 4x^2 + 8x + 5 &= 4(x^2 + 2x + 1 - 1) + 5 \\ &= 4(x^2 + 2x + 1) - 4 + 5 \\ &= 4(x+1)^2 + 1. \;\square \end{aligned}$$

Once the square has been completed in $ax^2 + bx + c$, the integral involving integer powers of x and $\sqrt{ax^2 + bx + c}$ can be found directly or by substitution, using inverse trigonometric functions if necessary.

Example 9 Perform the integration $\displaystyle\int \frac{1}{4x^2 + 8x + 5}\,dx$.

SOLUTION We write the result of Example 8 in the form $4(x+1)^2 + 1 = [2(x+1)]^2 + 1$ and use the substitution $u = 2(x+1), du = 2\,dx$:

$$\begin{aligned} \int \frac{1}{4x^2 + 8x + 5}\,dx &= \tfrac{1}{2}\int \frac{1}{[2(x+1)]^2 + 1}(2\,dx) = \tfrac{1}{2}\int \frac{1}{u^2+1}\,du \\ &= \tfrac{1}{2}\tan^{-1}(u) + C \\ &= \tfrac{1}{2}\tan^{-1}[2(x+1)] + C. \;\square \end{aligned}$$

Principles and procedures

- To integrate $\int \sin^n x \cos^m x \, dx$ with n an odd positive integer and m any constant, use the Pythagorean Identity **(2)** to express all but one $\sin x$ in terms of $\cos x$ and use the substitution $u = \cos x$.
- To integrate $\int \sin^n x \cos^m x \, dx$ with m an odd positive integer and any constant n, use the Pythagorean Identity **(2)** to express all but one $\cos x$ in terms of $\sin x$ and use the substitution $u = \sin x$.
- To find $\int \sin^n x \cos^m x \, dx$ with n and m both nonnegative, even integers, use the double-angle formulas **(3)** and **(4)**—as many times as necessary—to express the integrand as a linear combination of $\cos(kx)$ with one or more positive, even integers k.
- An integral of an integer power of x multiplied by a odd power of $\sqrt{a^2 - x^2}$ with constant $a > 0$ can be converted to an integral of products of powers of $\cos\theta$ and $\sin\theta$ with the substitution $\theta = \sin^{-1}(x/a)$ and the other formulas listed with Figure 2.
- For an integer power of x multiplied by an odd power of $\sqrt{a^2 + x^2}$, use $\theta = \tan^{-1}(x/a)$ and the other formulas listed with Figure 3.
- For an integer power of x multiplied by an odd power of $\sqrt{x^2 - a^2}$, use $\theta = \sec^{-1}(x/a)$ and the other formulas listed with Figure 4.
- Many integrals involving odd powers of $\sqrt{ax^2 + bx + c}$ can be integrated by completing the square to put $\sqrt{ax^2 + bx + c}$ in the form $\sqrt{a(x + r)^2 + s}$.
- The trigonometric integrals that arise in the Tune-up Exercises and Problems of this section from inverse trigonometic substitutions can be found either by direct substitution or by using the formulas in the following table.

$$\int \sec\theta \, d\theta = \ln|\sec\theta + \tan\theta| + C \qquad \textbf{(27)}$$

$$\int \csc\theta \, d\theta = -\ln|\csc\theta + \cot\theta| + C \qquad \textbf{(28)}$$

$$\int \sin^2\theta \, d\theta = \tfrac{1}{2}(\theta - \sin\theta\cos\theta) + C \qquad \textbf{(29)}$$

$$\int \cos^2\theta \, d\theta = \tfrac{1}{2}(\theta + \sin\theta\cos\theta) + C \qquad \textbf{(30)}$$

$$\int \tan^2\theta \, d\theta = \tan\theta - \theta + C \qquad \textbf{(31)}$$

$$\int \cot^2\theta \, d\theta = -\cot\theta - \theta + C \qquad \textbf{(32)}$$

$$\int \sec^3\theta \, d\theta = \tfrac{1}{2}\tan\theta\sec\theta + \tfrac{1}{2}\ln|\sec\theta + \tan\theta| + C \qquad \textbf{(33)}$$

$$\int \sec^4\theta \, d\theta = \tfrac{1}{3}\sec^2\theta\tan\theta + \tfrac{2}{3}\tan\theta + C \qquad \textbf{(34)}$$

- A special substitution u for converting any rational combination of $\sin x$ and $\cos x$ into a rational function of u is described before Problem 33 below. The rational function of u can then be integrated directly or by using partial fractions. This technique is illustrated in Problems 34 through 43.

Tune-Up Exercises 6.3♦

[A]*Answer provided.* [O]*Outline of solution provided.*

Perform the integration in Exercises T1 through T6. Use formulas **(27)** through **(33)** as needed.

T1.[O] $\displaystyle\int \sin^3(2x)\cos^5(2x)\,dx$

♦ Type 1, using trigonometric identities

T2.[O] $\displaystyle\int \sin^2 x\cos^2 x\,dx$

♦ Type 1, using trigonometric identities

T3.[O] $\displaystyle\int \cos(3x)\cos x\,dx$

♦ Type 1, using trigonometric identities

T4.[O] $\displaystyle\int x^3\sqrt{4-x^2}\,dx$

♦ Type 3, using inverse trigonometric substitutions

T5.[O] $\displaystyle\int_0^1 \sqrt{x^2+1}\,dx.$

♦ Type 3, using inverse trigonometric substitutions

T6.[O] $\displaystyle\int \frac{1}{\sqrt{12-4x-x^2}}\,dx.$

♦ Type 3, using inverse trigonometric substitutions

Problems 6.3

[A]*Answer provided.* [O]*Outline of solution provided.*

Use formulas **(27)** through **(33)** as needed to perform the integrations in Exercises 1 through 22.

1.[O] $\displaystyle\int \sin^3(4x)\,dx$

♦ Type 1, using trigonometric identities

2. $\displaystyle\int \cos^3(5x)\,dx$

♦ Type 1, using trigonometric identities

3.[O] $\displaystyle\int (\sin x)^{\sqrt{2}}\cos^3 x\,dx$

♦ Type 1, using trigonometric identities

4. $\displaystyle\int \sin^{5/2} x\cos^3 x\,dx$

♦ Type 1, using trigonometric identities

5.[A] $\displaystyle\int \sin^4 x\,dx$

♦ Type 1, using trigonometric identities

6. $\displaystyle\int \cos^4(3x)\,dx$

♦ Type 1, using trigonometric identities

7.[A] $\displaystyle\int \sin(3x)\cos x\,dx$

♦ Type 1, using trigonometric identities

8. $\displaystyle\int \sin(3x)\sin(5x)\,dx$

♦ Type 1, using trigonometric identities

9.[A] $\displaystyle\int_0^{\pi} \sin^2 x\,dx$

♦ Type 1, using trigonometric identities

10. $\displaystyle\int \cos^2(3x)\,dx$

♦ Type 1, using trigonometric identities

11.[O] $\displaystyle\int \frac{x^3}{\sqrt{x^2-1}}\,dx$

♦ Type 3, using inverse trigonometric substitutions

12.[A] $\displaystyle\int \frac{1}{x^2\sqrt{1-x^2}}\,dx$

♦ Type 3, using inverse trigonometric substitutions

13. $\displaystyle\int x^2\sqrt{25-x^2}\,dx$

♦ Type 3, using inverse trigonometric substitutions

14.[O] $\displaystyle\int \frac{x^2}{\sqrt{x^2-4}}\,dx \quad (x>2)$

♦ Type 3, using inverse trigonometric substitutions

15.[A] $\displaystyle\int \frac{\sqrt{16-x^2}}{x^2}\,dx$

♦ Type 3, using inverse trigonometric substitutions

16. $\displaystyle\int \frac{x^2}{(1-x^2)^{3/2}}\,dx$

♦ Type 3, using inverse trigonometric substitutions

17. $\displaystyle\int \frac{x^2}{\sqrt{4-x^2}}\,dx$

♦ Type 3, using inverse trigonometric substitutions

18. $\displaystyle\int \frac{1}{(x^2+1)^2}\,dx$

♦ Type 3, using inverse trigonometric substitutions

19.[A] $\displaystyle\int \frac{1}{x\sqrt{9-x^2}}\,dx$

♦ Type 3, using inverse trigonometric substitutions

20. $\displaystyle\int \frac{1}{(4-x^2)^{3/2}}\,dx$

♦ Type 3, using inverse trigonometric substitutions

♦The Tune-up Exercises and Problems are classified by type and content. The types are (1) basic, reactive; (2) basic reflective; (3) intermediate, reactive; (4) intermediate, reflective; (5) advanced, reactive; (6) advanced, reflective; and (7) advanced, theoretical.

21.[A] What is the volume of the solid generated when the region bounded by $y = (1-x^2)^{1/4}$ and the x-axis is rotated about the x-axis?

♦ Type 3, volume

22. Evaluate $\int_0^2 \sqrt{4-x^2}\,dx$ **(a)** with an inverse trigonometric substitution and **(b)** by using the formula for the area of a circle.

♦ Type 3, using inverse trigonometric substitutions

23.[A] What is the area of the region between $y = \dfrac{1}{\sqrt{x^2+9}}$ and the x-axis for $0 \le x \le 4$?

♦ Type 3,area

24. What is the average value of $x^3(1-x^2)^{3/2}$ for $0 \le x \le 1$?

♦ Type 3, average value

25.[A] What is the exact area of the region bounded by the hyperbola $y^2 = x^2 + 1$ and the line $y = 2$?

♦ Type 3, area

26. Give a formula in terms of the parameter a for the area of the portion of the circle $x^2 + y^2 \le 1$ for $-a \le x \le a$.

♦ Type 3, area

Perform the integration in Problems 29 through 34.

27.[A] $\displaystyle\int \frac{1}{\sqrt{x^2-8x+25}}\,dx$

♦ Type 4, using inverse trigonometric substitutions

28.[A] $\displaystyle\int \frac{1}{x^2-6x+25}\,dx$

♦ Type 4, using inverse trigonometric substitutions

29. $\displaystyle\int \frac{1}{9x^2+6x+2}\,dx$

♦ Type 4, using inverse trigonometric substitutions

30.[A] $\displaystyle\int \frac{1}{\sqrt{84-8x-x^2}}\,dx$

♦ Type 4, using inverse trigonometric substitutions

31. $\displaystyle\int \frac{1}{(2x-x^2)^{3/2}}\,dx$

♦ Type 3, using inverse trigonometric substitutions

32. $\displaystyle\int \frac{x-1}{x^2-2x}\,dx$

♦ Type 4, using inverse trigonometric substitutions

INTEGRALS OF RATIONAL FUNCTIONS OF SINES AND COSINES A function that is formed from $\sin x$ and $\cos x$ by taking linear combinations, products, and quotients is called a rational function of $\sin x$ and $\cos x$. Any integral of such a function can be converted into a rational function of u with the substitution

$$u = \tan\left(\tfrac{1}{2}x\right) \qquad \textbf{(35)}$$

for which

$$\sin x = \frac{2u}{1+u^2} \qquad \textbf{(36)}$$

$$\cos x = \frac{1-u^2}{1+u^2} \qquad \textbf{(37)}$$

$$dx = \frac{2}{1+u^2}\,du. \qquad \textbf{(38)}$$

33. Derive formulas **(36)**. **(37)**, and **(38)** from formula **(35)**.

♦ Type 7, a derivation

Carry out the integrations in Problems 34 through 43 by using substitutions **(33)** through **(36)**.

34.[O] $\displaystyle\int \frac{1}{\tan x - 1}\,dx$

♦ Type 4, rational functions of sines and cosines

35.[O] $\displaystyle\int \frac{1}{1+\cos x}\,dx$

♦ Type 4, rational functions of sines and cosines

36.[A] $\displaystyle\int \frac{1}{5+3\cos x}\,dx$

♦ Type 4, rational functions of sines and cosines

37. $\displaystyle\int \frac{1}{\tan x + \sin x}\,dx$

♦ Type 4, rational functions of sines and cosines

38.[A] $\displaystyle\int \frac{1}{\tan x - \sin x}\,dx$

♦ Type 4, rational functions of sines and cosines

39. $\displaystyle\int \frac{1-\cos x}{1+\cos x}\,dx$

♦ Type 4, rational functions of sines and cosines

40.[A] $\displaystyle\int_0^{\pi/3} \frac{1}{5\sec x - 3}\,dx$

♦ Type 4, rational functions of sines and cosines

41. $\displaystyle\int_0^{\pi/4} \frac{1}{1-\sin x}\,dx$

♦ Type 4, rational functions of sines and cosines

42. $\displaystyle\int_0^{\pi/2} \frac{\sin x + 1}{\cos x + 1}\,dx.$

♦ Type 4, rational functions of sines and cosines

43. $\displaystyle\int \frac{1}{\sin x + \cos x + 1}\,dx.$

♦ Type 4, rational functions of sines and cosines

Section 6.4

Using integral tables and calculator/computer-algebra systems

OVERVIEW: *Many integrals that would require lengthy calculations can be found more easily and with greater accuracy by using integral tables with other integration techniques as needed, or by using calculator or computer algebra systems which produce exact formulas for antiderivatives and exact values of definite integrals. The Examples, Questions, Tune-up Exercises, and Problems in this section provide practice with a table of integrals and with such calculator or computer software.*

Topics:

- ***The direct use of formulas in a table of integrals***
- ***Using substitutions with a table of integrals***
- ***Using a calculator or computer algebra system***

The direct use of formulas in a table of integrals

In the next two Examples we perform integrations by using the table of integrals at the end of this section.[†] The table gives formulas for a large number of indefinite integrals that arise in applications.

Example 1 What is the area of the region between $y = \dfrac{1}{x(3x+6)}$ and the x-axis for $1 \le x \le 3$?

SOLUTION Because $\dfrac{1}{x(3x+6)}$ is positive for $1 \le x \le 3$, the area is given by the integral $\displaystyle\int_1^3 \frac{1}{x(3x+6)}\,dx$, which could be evaluated by partial fractions. Instead we use formula **(79)** in the table at the end of the section, which reads

$$\int \frac{1}{(ax+b)(cx+d)}\,dx = \frac{1}{ad-bc}\ln\left|\frac{ax+b}{cx+d}\right| + C \quad \text{for } a \neq 0, c \neq 0, ad-bc \neq 0.$$

With the values $a = 1, b = 0, c = 3$, and $d = 6$, of the parameters, this formula becomes

$$\int \frac{1}{x(3x+6)}\,dx = \tfrac{1}{6}\ln\left|\frac{x}{3x+6}\right| + C$$

so that

$$\begin{aligned}[\text{Area}] &= \int_1^3 \frac{1}{x(3x+6)}\,dx = \left[\tfrac{1}{6}\ln\left|\frac{x}{3x+6}\right|\right]_1^3 \\ &= \tfrac{1}{6}\ln\left|\frac{3}{3(3)+6}\right| - \tfrac{1}{6}\ln\left|\frac{1}{3(1)+6}\right| \\ &= \tfrac{1}{6}[\ln(\tfrac{1}{5}) - \ln(\tfrac{1}{9})] = \tfrac{1}{6}\ln(\tfrac{9}{5}). \ \square\end{aligned}$$

Example 2 Use the table of integrals to find a formula for the function $F(x)$ such that $F'(x) = x^2e^x$ for all x and $F(0) = 3$.

SOLUTION Formula **(72)** in the table with $a = 1$ gives

$$F(x) = \int x^2e^x\,dx = (x^2 - 2x + 2)e^x + C.$$

Then $F(0) = 2 + C$, so to have $F(0) = 3$ we must take $C = 1$. The function is $F(x) = (x^2 - 2x + 2)e^x + 1$. □

[†]The table will be placed at the end of the text in the First Edition.

◆ SUGGESTIONS TO INSTRUCTORS: Have students begin work on Questions 1 and 2 before class. Problems 16, 20, 22, 27, 28, 32, and 36 are good for in-class discussion.

Using substitutions with an integral table

The table of integrals can be used to find a formula for any indefinite integral that can be put in the form of one in the table with an appropriate substitution.

Example 3 Use the table of integrals to perform the integration $\int x^5 \sin(x^2)\, dx$.

SOLUTION There are several integrals in the table involving $\sin x$ but none with $\sin(x^2)$, so we try the substitution $u = x^2, du = 2x\, dx$. We obtain

$$\int x^5 \sin(x^2)\, dx = \tfrac{1}{2}\int (x^2)^2 \sin(x^2)(2x\, dx) = \tfrac{1}{2}\int u^2 \sin(u)\, du. \qquad \textbf{(1)}$$

This is a good approach because formula **(66)** in the table reads

$$\int x^2 \sin(ax)\, dx = \frac{2}{a^2}\, x\sin(ax) - \left(\frac{1}{a}x^2 - \frac{2}{a^3}\right)\cos(ax) + C.$$

Using **(1)** and the last formula with $a = 1$ and u in place of x gives

$$\begin{aligned}\int x^5 \sin(x^2)\, dx &= \tfrac{1}{2}\int u^2 \sin u\, du \\ &= \tfrac{1}{2}[2u\sin u - (u^2 - 2)\cos u] + C \\ &= x^2\sin(x^2) - \tfrac{1}{2}(x^4 - 2)\cos(x^2) + C. \ \square\end{aligned}$$

Question 1 Use the table of integrals to evaluate $\displaystyle\int_0^{\pi/2} \frac{\sin x}{\sqrt{1+\cos^2 x}}\, dx$.

Using a calculator or computer algebra system

The above Examples and Question can be worked quickly using a calculator or computer algebra system, with appropriate interpretations of the results given by the machines.

Example 4 Work **(a)** Example 1, **(b)** Example 2, **(c)** Example 3, and **(d)** Question 1 with a calculator or computer algebra system.

SOLUTION **(a)** A calculator gives $\displaystyle\int_1^3 \frac{x}{3x+6}\, dx = \frac{-\ln(5/9)}{6}$, which is the same as the answer $\frac{1}{6}\ln(\frac{9}{5})$ in the solution to Example 1 because $\ln(\frac{5}{9}) = -\ln(\frac{9}{5})$.

(b) A calculator gives $\int F(x)\, dx = \int x^2 e^x\, dx = (x^2 - 2x + 2)e^x$. When we include the constant of integration, we obtain

$$F(x) = (x^2 - 2x + 2)e^x + C.$$

As above, setting $x = 0$ and $F(0) = 3$ gives $3 = 2 + C$, so that $C = 1$ and $F(x) = (x^2 - 2x + 2)e^x + 1$.

(c) A calculator gives

$$\int x^5 \sin(x^2)\, dx = \frac{-((x^4 - 2)\cos(x^2) - 2x^2\sin(x^2))}{2}.$$

We simplify and add the constant of integration to obtain the result $\int x^5 \sin(x^2)\, dx = x^2 \sin(x^2) - \frac{1}{2}(x^4 - 2)\cos(x^2) + C$ from Example 3.

(d) The answer $\displaystyle\int_0^{\pi/2} \frac{\sin x}{\sqrt{1+\cos^2 x}}\, dx = \ln(\sqrt{2}+1)$ obtained with a calculator is the same as is given in the Response to Question 1. $\square$

The next example requires calculations that would be difficult to do by hand, but are suitable for a calculator or computer algebra system.

Example 5 Figure 1 shows the region between the curve $y = x^4 - 6x^3 + 11x^2 - 6x + 1$ and the x-axis. Find its exact area.

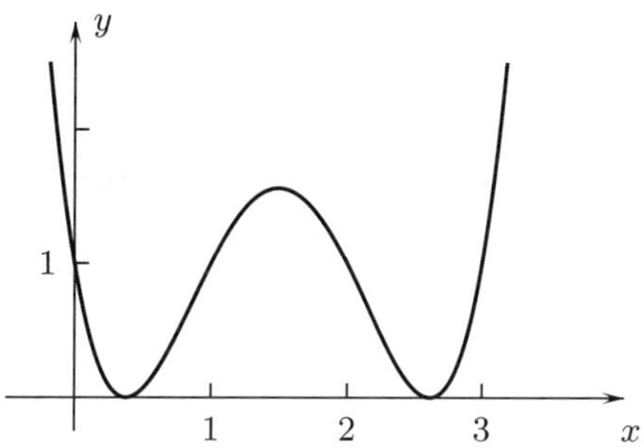

$y = x^4 - 6x^3 + 11x^2 - 6x + 1$
FIGURE 1

SOLUTION The curve intersects the x-axis at the solutions of $x^4 - 6x^3 + 11x^2 - 6x + 1 = 0$, which a calculator shows are $x = \frac{1}{2}(3 \pm \sqrt{5})$. Therefore, the area is

$$\int_{(3-\sqrt{5})/2}^{(3+\sqrt{5})/2} (x^4 - 6x^3 + 11x^2 - 6x + 1)\,dx.$$

A calculator gives the value $\frac{5}{6}\sqrt{5}$ for this integral. □

Example 6 **(a)** Rain is falling at the rate of $1/(t^2+3t+2)$ inches per hour t hours after midnight for $0 \le t \le 10$. Give a formula for the amount of rain that falls from midnight to time t for $0 \le t \le 10$. **(b)** At what time has $\frac{1}{2}$ inches of rain fallen? Give an exact answer.

SOLUTION **(a)** The number of inches $I(t)$ of rain that falls from midnight to time t is an antiderivative of the rate of rainfall:

$$I(t) = \int \left(\frac{1}{t^2+3t+2}\,\frac{\text{inches}}{\text{hour}}\right)(dt\text{ hours}) = \int \frac{1}{t^2+3t+2}\,dt\ \text{inches}.$$

A calculator algebra system gives the formula, $\ln\left(\dfrac{|t+1|}{|t+2|}\right)$, for this integral, to which we add a constant of integration to obtain

$$I(t) = \int \frac{1}{t^2+3t+2}\,dt = \ln\left(\frac{|t+1|}{|t+2|}\right) + C.$$

$I(0) = 0$ because no rain falls for $0 \le t \le 0$. Since $I(0) = \ln(\frac{1}{2}) + C = -\ln(2) + C$, we will have $I(0) = 0$ if $-\ln(2) + C = 0$. Solving this equation gives $C = \ln(2)$ and then $I(t) = \ln\left(\dfrac{|t+1|}{|t+2|}\right) + \ln(2)$ inches.

(b) The solutions of $\ln\left(\dfrac{|t+1|}{|t+2|}\right) + \ln(2) = \frac{1}{2}$ given by a calculator are $t_1 = \dfrac{2\sqrt{e}-2}{2-\sqrt{e}} \doteq 3.6935$ and $t_2 = \dfrac{-2\sqrt{e}-2}{2+\sqrt{e}} \doteq -1.4519$. Only the positive value applies in this case. One-half inch of rain has fallen after $\dfrac{2\sqrt{e}-2}{2-\sqrt{e}}$ hours. □

Question 2 Find the exact value of the positive constant k such that the average value of $f(x) = \dfrac{1}{(x^2+k^2)^2}$ for $-k \le x \le k$ is 10.

Question 3 An object's acceleration in the positive direction as it moves on a s-axis with distances measured in meters is t^2e^{-t} meters per minute at time $t \geq 0$. Give a formula with two parameters for its position in terms of t,

Principles and procedures

- To perform an integration using an integral table, look first for a formula that gives the integral with an appropriate choice of parameter or parameters. If there is no such formula, try substitutions that would put in one of the forms in the table. In many cases the appropriate substitution simplifies the most complicated part of the integrand you are considering.
- Most calculator and computer algebra systems do not include constants of integration in formulas for indefinite integrals. Be sure to add them as needed. Also, many systems do not include absolute values in formulas involving logarithms. They might give, for example, the formulas

$$\int \frac{1}{x}\,dx = \ln(x) \quad \text{and} \quad \int \frac{1}{x(x+1)(x+2)}\,dx = \tfrac{1}{2}\ln\left(\frac{x(x+2)}{(x+1)^2}\right).$$

These should be rewritten as

$$\int \frac{1}{x}\,dx = \ln|x| + C \quad \text{and} \quad \int \frac{1}{x(x+1)(x+2)}\,dx = \tfrac{1}{2}\ln\left(\frac{|x(x+2)|}{(x+1)^2}\right) + C$$

before they are used.

Tune-Up Exercises 6.4[♦]

[A]*Answer provided.* [O]*Outline of solution provided.*

Use the table of integrals at the end of this section or a calculator- or computer-algebra system to perform the integrations in Exercises T1 through T5. Give exact answers.

T1.[O] $\displaystyle\int_0^1 \frac{1}{\sqrt{x^2+5}}\,dx$

♦ Type 1, using an integral table or a CAS

T2.[O] $\displaystyle\int_0^{\pi/6} \sin^2(3x)\cos^2(3x)\,dx$

♦ Type 1, using an integral table or a CAS

T3.[O] $\displaystyle\int \cos x\,\sin(3x)\,dx$

♦ Type 1, using an integral table or a CAS

T4.[O] $\displaystyle\int_0^2 \frac{1}{x^2+x+1}\,dx$

♦ Type 1, using an integral table or a CAS

T5.[O] $\displaystyle\int \frac{e^x}{\sqrt{e^{2x}-1}}\,dx$

♦ Type 1, using an integral table or a CAS

T6.[O] $\displaystyle\int_0^{\pi/3} \sin x\,\ln(\cos x))\,dx$

♦ Type 1, using an integral table or a CAS

[♦]The Tune-up Exercises and Problems are classified by type and content. The types are (1) basic, reactive; (2) basic reflective; (3) intermediate, reactive; (4) intermediate, reflective; (5) advanced, reactive; (6) advanced, reflective; and (7) advanced, theoretical.

Problems 6.4

[A] *Answer provided.* [O] *Outline of solution provided.*

Use the table of integrals at the end of this section or a calculator- or computer-algebra system as required to perform the integrations in Problems 1 through 14 and to solve Problems 15 through 40. Give exact answers.

1.[O] $\displaystyle\int_0^{12} \sqrt{169 - x^2}\, dx$

♦ Type 1, using an integral table or a CAS

2.[A] $\displaystyle\int_{-1}^{1} \frac{1}{9 - x^2}\, dx$

♦ Type 1, using an integral table or a CAS

3.[O] $\displaystyle\int x\sqrt{1 - x^4}\, dx$

♦ Type 1, using an integral table or a CAS

4.[A] $\displaystyle\int \frac{x}{(x+1)(x+2)}\, dx$

♦ Type 1, using an integral table or a CAS

5.[A] $\displaystyle\int x \sin^3(x^2)\, dx$

♦ Type 1, using an integral table or a CAS

6. $\displaystyle\int_{-1}^{1} \tan^2 x\, dx$

♦ Type 1, using an integral table or a CAS

7.[O] $\displaystyle\int_{-1/4}^{1/4} \sec(\pi x)\, dx$

♦ Type 1, using an integral table or a CAS

8.[A] $\displaystyle\int \frac{\sin x}{\cos^2 x + \cos x}\, dx$

♦ Type 1, using an integral table or a CAS

9. $\displaystyle\int \arctan(3x)\, dx$

♦ Type 1, using an integral table or a CAS

10. $\displaystyle\int_0^1 \frac{1 + e^x}{x + e^x}\, dx$

♦ Type 1, using an integral table or a CAS

11.[A] $\displaystyle\int x 3^{x^2}\, dx$

♦ Type 1, using an integral table or a CAS

12.[A] $\displaystyle\int \frac{1}{x^2 - 4x + 13}\, dx$

♦ Type 1, using an integral table or a CAS

13. $\displaystyle\int \frac{1}{x^2 - 4x + 4}\, dx$

♦ Type 1, using an integral table or a CAS

14. $\displaystyle\int x \sin(2x)\, dx$

♦ Type 1, using an integral table or a CAS

15.[A] Sketch the region bounded by $y = xe^x$, the x-axis, and the line $x = 1$ and find its area.

♦ Type 1, area

16.[A] What is the average value of $\dfrac{1}{x^2 - 4x + 13}$ for $0 \le x \le 2$?

♦ Type 1, average value

17.[A] Find the length of the curve $y = \frac{1}{2}x^2$ for $0 \le x \le 3$.

♦ Type 1, length of a curve

18.[A] Find the area of the region between $y = x^2 \sin x$ and the x-axis for x between 0 and the smallest positive x-intercept of the curve.

♦ Type 2, area

19.[A] An object moving on an s-axis has velocity $\sqrt{1 + t^2}$ yards per hour in the positive s-direction for all time t (hours). At $t = 0$ it is at $s = 5$ (yards). Give a formula for its s-coordinate as a function of t.

♦ Type 2, velocity and position

20. The acceleration in the positive direction of an object moving on an s-axis is $t^2 e^{-t}$ meters per second^2 for all time t (seconds). At $t = 0$ it is at $s = 10$ (meters) and its velocity in the positive direction is one meter per second. Give a formula for its s-coordinate as a function of t.

♦ Type 2, acceleration, velocity and position

21. Sketch the region bounded by $y = 1/(x^2 - 4)$, the x-axis, and the lines $x = \pm 1$. Find its area.

♦ Type 1, area

22.[A] Sketch the region bounded by $y = 2 - \ln x$ and $y = 2(\ln x)^2$ and find its area. (To find the intersections, solve $2t^2 + t - 2 = 0$ first.)

♦ Type 4, area

23. What is the area of the region bounded by $y = x\cos^{-1} x$ and $y = \frac{1}{4}\pi x$?

♦ Type 4, area

24.[A] Find the volumes of the solids that are generated when the region bounded by $y = \ln x$, $y = 2$, and the x-axis is rotated **(a)** about the x-axis and **(b)** about the y-axis.

◆ Type 3, volume

25. Derive the formula for the area of the circle $x^2 + y^2 \leq r^2$ by evaluating an integral.

◆ Type 2, area

26.[A] Water is flowing out of a pipe at the rate of $\tan^{-1} t$ gallons per minute at time t for $0 \leq t \leq 60$ (minutes). How much water flows out of the pipe during the hour?

◆ Type 2, an integral of a rate of change

27. What is the volume of the solid that is generated when the region between $y = \dfrac{1}{x^2+1}$ and the x-axis for $-1 \leq x \leq 1$ is rotated about the x-axis?

◆ Type 1, volume

28. A rod extends from $x = 1$ to $x = 10$ (inches) on an x-axis, and its density at x is $\ln x$ pounds per inch. How much does it weigh and where is its center of gravity?

◆ Type 2, weight and center of gravity

29. Find a formula for the function $F(x)$ such that $F'(x) = \sin x \cos(5x)$ and $F(0) = 1$.

◆ Type 2, finding an antiderivative

30. What is the area between $y = \dfrac{1}{(2x+3)(4x+5)}$ and the x-axis for $-1 \leq x \leq 1$?.

◆ Type 1, area

31.[A] **(a)** Find a formula for the function $G(x)$ such that $\dfrac{dG}{dx}(x) = \dfrac{1}{x^2(x+1)}$ for $x > 0$ and $G(1) = 1$. **(b)** Is $G(x)$ increasing or decreasing for $x > 0$?

◆ Type 2, finding an antiderivative

32. Give an equation of the tangent line at $x = \frac{1}{4}\pi$ to the graph of the function $f(x)$ such that $f'(x) = \cos x \sin(3x)$ for all x and $f(\frac{1}{4}\pi) = \frac{1}{4}$. Generate the curve and the tangent line together as a partial check of your answer.

◆ Type 2, a tangent line

33. Find the constant k such that the region between $y = \dfrac{kx}{1+x^2}$ and the x-axis for $0 \leq x \leq 3$ has area 10.

◆ Type 4, area

34. Find the area of the region bounded by $y = \tan^{-1} x$, $y = \frac{1}{4}\pi$ and the y-axis.

◆ Type 4, area

35. Find the area of the region bounded by the graph of $f(x) = -x^4 + 8x^3 - 23x^2 + 28x - 10$ and the horizontal line that touches the graph at its highest points.

◆ Type 4, area

36. What is the total area of the two regions bounded by the graph of $g(x) = -36x - 28x^2 - 9x^3 - x^4$ and the line $y = 16$?

◆ Type 4, area

37.[A] The curves $y = 2\sin x$ and $y = \cos x$ bound an infinite number of regions with the same area. What is that area?

◆ Type 4, area

38. Find the value of the parameter k such that the area bounded by $y = x^2$ and $y = (x+k) - (x+k)^2$ has the greatest possible area. Explain the result with a drawing.

◆ Type 4, area etc.

39. For what value of the parameter a with $0 \leq a \leq 1$ is the area bounded by $y = x^2$ and $y = a^2 - ax + x^2$ a maximum?

◆ Type 4, area etc.

40. Prove that $\displaystyle\int_0^a \frac{x^2}{(x+a)(x+2a)}\,dx$ is a linear function of a.

◆ Type 4, an integral

A Table of Integrals[†]

Basic Formulas

(1) $\int x^n\,dx = \frac{1}{n+1}x^{n+1} + C \quad \text{for } n \neq -1$

(2) $\int \frac{1}{x}\,dx = \ln|x| + C$

(3) $\int \cos x\,dx = \sin x + C$

(4) $\int \sin x\,dx = -\cos x + C$

(5) $\int \sec^2 x\,dx = \tan x + C$

(6) $\int \csc^2 x\,dx = -\cot x + C$

(7) $\int \sec x \tan x\,dx = \sec x + C$

(8) $\int \csc x \cot x\,dx = -\csc x + C$

(9) $\int \frac{1}{\sqrt{a^2 - x^2}}\,dx = \arcsin\left(\frac{x}{a}\right) + C$

(10) $\int \frac{1}{x^2 + a^2}\,dx = \frac{1}{a}\arctan\left(\frac{x}{a}\right) + C$

(11) $\int \frac{1}{|x|\sqrt{x^2 - a^2}}\,dx = \frac{1}{a}\operatorname{arcsec}\left(\frac{x}{a}\right) + C$

(12) $\int e^x\,dx = e^x + C$

(13) $\int b^x\,dx = \frac{1}{\ln b}b^x + C$

(14) $\int \sinh x\,dx = \cosh x + C$

(15) $\int \cosh x\,dx = \sinh x + C$

(16) $\int \tan x\,dx = -\ln|\cos x| + C$

(17) $\int \cot x\,dx = \ln|\sin x| + C$

(18) $\int \sec x\,dx = \ln|\sec x + \tan x| + C = \ln\left|\tan\left(\frac{x}{2} + \frac{\pi}{4}\right)\right| + C$

(19) $\int \csc x\,dx = -\ln|\csc x + \cot x| + C = \ln\left|\tan\left(\frac{x}{2}\right)\right| + C$

[†]This table will be at the back of the text in the First Edition. $\arcsin x$ is used here for $\sin^{-1} x$ and $\arctan x$ is used for $\tan^{-1} x$.

Other Integrals of Basic Functions

(20) $\int \ln x\,dx = x\ln x - x + C$

(21) $\int \arcsin(ax)\,dx = x\arcsin(ax) + \frac{1}{a}\sqrt{1-(ax)^2} + C$

(22) $\int \arctan(ax)\,dx = x\arctan(ax) - \frac{1}{2a}\ln(1+(ax)^2) + C$

Powers of Trigonometric Functions

(23) $\int \sin^2 x\,dx = -\frac{1}{2}\cos x\sin x + \frac{1}{2}x + C = -\frac{1}{4}\sin(2x) + \frac{1}{2}x + C$

(24) $\int \sin^3 x\,dx = -\frac{1}{3}\cos x\sin^2 x - \frac{2}{3}\cos x + C$

(25) $\int \sin^4 x\,dx = -\frac{1}{4}\cos x\sin^3 x - \frac{3}{8}\cos x\sin x + \frac{3}{8}x + C$

(26) $= -\frac{1}{4}\sin(2x) + \frac{1}{32}\sin(4x) + \frac{3}{8}x + C$

(27) $\int \sin^n x\,dx = -\frac{1}{n}\cos x\sin^{n-1}x + \frac{n-1}{n}\int \sin^{n-2}x\,dx$ for $n = 2, 3, 4, \ldots$

(28) $\int \cos^2 x\,dx = \frac{1}{2}\sin x\cos x + \frac{1}{2}x + C = \frac{1}{4}\sin(2x) + \frac{1}{2}x + C$

(29) $\int \cos^3 x\,dx = \frac{1}{3}\sin x\cos^2 x + \frac{2}{3}\sin x + C$

(30) $\int \cos^4 x\,dx = \frac{1}{4}\sin x\cos^3 x + \frac{3}{8}\sin x\cos x + \frac{3}{8}x + C$

(31) $= \frac{1}{4}\sin(2x) + \frac{1}{32}\sin(4x) + \frac{3}{8}x + C$

(32) $\int \cos^n x\,dx = \frac{1}{n}\sin x\cos^{n-1}x + \frac{n-1}{n}\int \cos^{n-2}x\,dx$ for $n = 2, 3, 4, \ldots$

(33) $\int \tan^2 x\,dx = \tan x - x + C$

(34) $\int \tan^3 x\,dx = \frac{1}{2}\tan^2 x + \ln|\cos x| + C$

(35) $\int \tan^n x\,dx = \frac{1}{n-1}\tan^{n-1}x - \int \tan^{n-2}x\,dx$ for $n = 2, 3, 4, \ldots$

(36) $\int \cot^2 x\,dx = -\cot x - x + C$

(37) $\int \cot^3 x\,dx = -\frac{1}{2}\cot^2 x - \ln|\sin x| + C$

(38) $\int \cot^n x\,dx = -\frac{1}{n-1}\cot^{n-1}x - \int \cot^{n-2}x\,dx$ for $n = 2, 3, 4, \ldots$

(39) $\int \sec^2 x\,dx = \tan x + C$

(40) $\int \sec^n x\,dx = \frac{1}{n-1}\sec^{n-2}x\tan x + \frac{n-2}{n-1}\int \sec^{n-2}x\,dx$ for $n = 2, 3, 4, \ldots$

(41) $\int \csc^2 x\,dx = -\cot x + C$

(42) $\int \csc^n x\,dx = -\frac{1}{n-1}\csc^{n-2}x\cot x + \frac{n-2}{n-1}\int \csc^{n-2}x\,dx$ for $n = 2, 3, 4, \ldots$

Products of Trigonometric Functions

(43) $\int \sin(ax)\sin(bx)\,dx = \frac{\sin[(a-b)x]}{2(a-b)} - \frac{\sin[(a+b)x]}{2(a+b)} + C$

(44) $= \frac{1}{a^2-b^2}[b\sin(ax)\cos(bx) - a\cos(ax)\sin(bx)] + C$ for $a^2 \neq b^2$

(45) $\int \cos(ax)\cos(bx)\,dx = \frac{\sin[(a-b)x]}{2(a-b)} + \frac{\sin[(a+b)x]}{2(a+b)} + C$

(46) $= \frac{1}{a^2-b^2}[a\sin(ax)\cos(bx) - b\cos(ax)\sin(bx)] + C$ for $a^2 \neq b^2$

(47) $\int \sin(ax)\cos(ax)\,dx = \frac{1}{2a}\sin^2(ax) + C = -\frac{1}{2a}\cos^2(ax) + C$

(48) $\int \sin(ax)\cos(bx)\,dx = -\frac{\cos[(a-b)x]}{2(a-b)} - \frac{\cos[(a+b)x]}{2(a+b)} + C$

(49) $= \frac{1}{a^2-b^2}[-a\cos(ax)\cos(bx) - b\sin(ax)\sin(bx)] + C$ for $a^2 \neq b^2$

(50) $\int \sin^2 x\cos^2 x\,dx = -\frac{1}{8}\sin x\cos^3 x + \frac{1}{8}\sin^3 x\cos x + \frac{1}{8}x + C$

(51) $= -\frac{1}{32}\sin(4x) + \frac{1}{8}x + C$

(52) $\int \sin^3 x\cos^2 x\,dx = -\frac{1}{3}\cos^3 x + \frac{1}{5}\cos^5 x + C$

(53) $\int \sin^2 x\cos^3 x\,dx = \frac{1}{3}\sin^3 x - \frac{1}{5}\sin^5 x + C$

(54) $\int \sin^3 x\cos^3 x\,dx = \frac{1}{4}\sin^4 x - \frac{1}{6}\sin^6 x + C$

(55) $\int \sin^4 x\cos^3 x\,dx = \frac{1}{5}\sin^5 x - \frac{1}{7}\sin^7 x + C$

(56) $\int \sin^3 x\cos^4 x\,dx = \frac{1}{7}\cos^7 x - \frac{1}{5}\cos^5 x + C$

(57) $\int \sin^n x \cos^m x \, dx = \frac{1}{n+m} \sin^{n+1}x \cos^{m-1}x + \frac{m-1}{n+m} \int \sin^n x \cos^{m-2}x \, dx$ for $n = 1, 2, 3, \ldots; m = 2, 3, 4, \ldots$

(58) $= -\frac{1}{n+m} \sin^{n-1}x \cos^{m+1}x + \frac{n-1}{n+m} \int \sin^{n-2}x \cos^m x \, dx$ for n 2, 3, 4 . . . ; $m = 1, 2, 3, \ldots$

(59) $\int \frac{\cos^m x}{\sin^n x} dx = -\frac{\cos^{m+1}x}{(n-1)\sin^{n-1}x} - \frac{m-n+2}{n-1} \int \frac{\cos^m x}{\sin^{n-2}x}$ for $n = 2, 3, 4, \ldots; m = 1, 2, 3, \ldots$

(60) $= \frac{\cos^{m-1}x}{(m-n)\sin^{n-1}x} + \frac{m-1}{m-n} \int \frac{\cos^{m-2}x}{\sin^n x} dx$ for $n = 1, 2, 3, \ldots; m = 2, 3, 4, \ldots$

(61) $\int \frac{\sin^m x}{\cos^n x} dx = \frac{\sin^{m+1}x}{(n-1)\cos^{n-1}x} - \frac{m-n+2}{n-1} \int \frac{\sin^m x}{\cos^{n-2}x} dx$ for $n = 2, 3, 4, \ldots; m = 0, 1, 2, \ldots$

(62) $= -\frac{\sin^{m-1}x}{(m-n)\cos^{n-1}x} + \frac{m-1}{m-n} \int \frac{\sin^{m-2}x}{\cos^n x} dx$ for $n = 0, 1, 2, \ldots; m = 2, 3, 4, \ldots$

(63) $\int \frac{1}{\sin^m x \cos^n x} dx = -\frac{1}{(m-1)\sin^{m-1}x \cos^{n-1}x} + \frac{m+n-2}{m-1} \int \frac{1}{\sin^{m-2}x \cos^n x} dx$
for $n = 1, 2, 3, \ldots; m = 2, 3, 4, \ldots$

(64) $= \frac{1}{(n-1)\sin^{m-1}x \cos^{n-1}x} - \frac{m+n-2}{n-1} \int \frac{1}{\sin^m x \cos^{n-2}x} dx$
for $n = 2, 3, 4, \ldots; m = 1, 2, 3, \ldots$

Integrals of $x^n f(x)$

(65) $\int x \sin(ax) \, dx = \frac{1}{a^2} \sin(ax) - \frac{1}{a} x \cos(ax) + C$

(66) $\int x^2 \sin(ax) \, dx = \frac{2}{a^2} x \sin(ax) - \left(\frac{1}{a}x^2 - \frac{2}{a^3}\right) \cos(ax) + C$

(67) $\int x^n \sin(ax) \, dx = -\frac{1}{a}x^n \cos(ax) + \frac{n}{a} \int x^{n-1} \cos(ax) \, dx$ for $n = 1, 2, 3, \ldots$

(68) $\int x \cos(ax) \, dx = \frac{1}{a^2} \cos(ax) + \frac{1}{a} x \sin(ax) + C$

(69) $\int x^2 \cos(ax) \, dx = \frac{2}{a^2} x \cos(ax) + \left(\frac{1}{a}x^2 - \frac{2}{a^3}\right) \sin(ax) + C$

(70) $\int x^n \cos(ax) \, dx = \frac{1}{a}x^n \sin(ax) - \frac{n}{a} \int x^{n-1} \sin(ax) \, dx$ for $n = 1, 2, 3, \ldots$

(71) $\int x e^{ax} \, dx = \left(\frac{1}{a}x - \frac{1}{a^2}\right)e^{ax} + C$

(72) $\int x^2 e^{ax} \, dx = \left(\frac{1}{a}x^2 - \frac{2}{a^2}x + \frac{2}{a^3}\right)e^{ax} + C$

(73) $\int x^n e^{ax} \, dx = \frac{1}{a}x^n e^{ax} - \frac{n}{a} \int x^{n-1} e^{ax} \, dx$ for $n = 1, 2, 3, \ldots$

(74) $\int x^n \ln x \, dx = \frac{x^{n+1}}{n+1} \ln x - \frac{x^{n+1}}{(n+1)^2} + C$ for $n \neq -1$

Other Products and Powers

(75) $\displaystyle\int e^{ax}\sin(bx)\,dx = \frac{e^{ax}}{a^2+b^2}[a\sin(bx) - b\cos(bx)] + C$

(76) $\displaystyle\int e^{ax}\cos(bx)\,dx = \frac{e^{ax}}{a^2+b^2}[a\cos(bx) + b\sin(bx)] + C$

(77) $\displaystyle\int (\ln x)^2\,dx = x(\ln x)^2 - 2x\ln x + 2x + C$

(78) $\displaystyle\int (\ln x)^n\,dx = x(\ln x)^n - n\int(\ln x)^{n-1}\,dx \quad \text{for} \quad n = 1, 2, 3, \ldots$

Integrals Involving $ax + b$ and $cx + d$

(79) $\displaystyle\int \frac{1}{(ax+b)(cx+d)}\,dx = \frac{1}{ad-bc}\ln\left|\frac{ax+b}{cx+d}\right| + C \quad \text{for} \quad ad - bc \neq 0$

(80) $\displaystyle\int \frac{x}{(ax+b)(cx+d)}\,dx = \frac{1}{ad-bc}\left[\frac{d}{c}\ln|cx+d| - \frac{b}{a}\ln|ax+b|\right] + C \quad \text{for} \quad a \neq 0, c \neq 0, ad - bc \neq 0$

(81) $\displaystyle\int \frac{1}{(ax+b)(cx+d)^2}\,dx = \frac{1}{(ad-bc)(cx+d)} + \frac{a}{(ad-bc)^2}\ln\left|\frac{ax+b}{cx+d}\right| + C \quad \text{for} \quad ad - bc \neq 0$

Integrals Involving $a^2 \pm x^2$, $x^2 - a^2$, or $ax^2 + bx + c$

(82) $\displaystyle\int \frac{1}{a^2+x^2}\,dx = \frac{1}{a}\arctan\left(\frac{x}{a}\right) + C$

(83) $\displaystyle\int \frac{1}{a^2-x^2}\,dx = \frac{1}{2a}\ln\left|\frac{x+a}{x-a}\right| + C$

(84) $\displaystyle\int \frac{x}{a^2+x^2}\,dx = \frac{1}{2}\ln(a^2+x^2) + C$

(85) $\displaystyle\int \frac{x}{a^2-x^2}\,dx = -\frac{1}{2}\ln|a^2-x^2| + C$

(86) $\displaystyle\int \sqrt{x^2+a^2}\,dx = \frac{1}{2}x\sqrt{x^2+a^2} + \frac{1}{2}a^2\ln(x+\sqrt{x^2+a^2}) + C$

(87) $\displaystyle\int \sqrt{x^2-a^2}\,dx = \frac{1}{2}x\sqrt{x^2-a^2} - \frac{1}{2}a^2\ln|x+\sqrt{x^2-a^2}| + C$

(88) $\displaystyle\int \sqrt{a^2-x^2}\,dx = \frac{1}{2}x\sqrt{a^2-x^2} + \frac{1}{2}a^2\arcsin\left(\frac{x}{|a|}\right) + C$

(89) $$\int \frac{1}{\sqrt{x^2 + a^2}}\,dx = \ln(x + \sqrt{x^2 + a^2}) + C$$

(90) $$\int \frac{1}{\sqrt{x^2 - a^2}}\,dx = \ln|x + \sqrt{x^2 - a^2}| + C$$

(91) $$\int \frac{1}{\sqrt{a^2 - x^2}}\,dx = \arcsin\left(\frac{x}{a}\right) + C$$

(92) $$\int \frac{1}{x\sqrt{a^2 + x^2}}\,dx = -\frac{1}{a}\ln\left|\frac{a + \sqrt{a^2 + x^2}}{x}\right| + C$$

(93) $$\int \frac{1}{|x|\sqrt{x^2 - a^2}}\,dx = \frac{1}{a}\,\text{arcsec}\left(\frac{x}{a}\right) + C$$

(94) $$\int \frac{1}{x\sqrt{a^2 - x^2}}\,dx = -\frac{1}{a}\ln\left|\frac{a + \sqrt{a^2 - x^2}}{x}\right| + C$$

(95) $$\int \frac{1}{ax^2 + bx + c}\,dx = \frac{1}{\sqrt{b^2 - 4ac}}\ln\left|\frac{2ax + b - \sqrt{b^2 - 4ac}}{2ax + b + \sqrt{b^2 - 4ac}}\right| + C \quad \text{for} \quad a \neq 0,\ b^2 > 4ac$$

(96) $$= \frac{2}{\sqrt{4ac - b^2}}\arctan\left(\frac{2ax + b}{\sqrt{4ac - b^2}}\right) + C \quad \text{for} \quad a \neq 0,\ b^2 < 4\,ac$$

(97) $$= \frac{-2}{2ax + b} + C \quad \text{for} \quad a \neq 0,\ b^2 = 4ac$$

(98) $$\int \frac{1}{(x^2 + a^2)^n}\,dx = \frac{1}{2a^2(n - 1)}\frac{x}{(x^2 + a^2)^{n-1}} + \frac{2n - 3}{2a^2(n - 1)}\int \frac{1}{(x^2 + a^2)^{n-1}}\,dx \quad \text{for} \quad n = 2, 3, 4, \ldots$$

Section 6.5

Distribution densities and the Fundamental Theorem, II

OVERVIEW: *In Section 4.6 we discussed one-dimensional densities of rods and other objects. We saw that if an object's density, measured in weight or mass per length, is $\rho(x)$ at point x, then the weight or mass of the portion of the object for $a \le x \le b$ is given by the integral $\int_a^b \rho(x)\,dx$. We open this section with a discussion of another type of density, called a* DISTRIBUTION DENSITY, *and we show how such densities are used to model statistical data, such as the distribution of population in a country by age or income. We then introduce the* CUMULATIVE DISTRIBUTION *of a density $f(x)$, which is defined to be a definite integral $F(X) = \int_{x_0}^{X} f(x)\,dx$ with a variable upper limit of integration. Finally, we explain why the X-derivative of the cumulative distribution $F(X)$ is equal to the density $f(X)$ for X at any point where $f(x)$ is continuous. This result, like the Fundamental Theorem of Calculus that we discussed in Section 4.3, relates integrals and derivatives and is considered to be another part of the Fundamental Theorem.*

Topics:

- ***Distribution densities***
- ***Cumulative distributions***
- ***The Fundamental Theorem for derivatives of integrals***
- ***Integrals with functions as limits of integration***

Density of an age distribution

The second column of Table 1 lists the 1990 population of United States residents, divided in 15-year age brackets.[(1)] We use this data to form a mathematical model of the type we will discuss in this section by making the simplifying assumption that the age distribution was uniform in each of the age brackets. Then the number of residents $p(t)$ per year of age is obtained by dividing the numbers in the second column of the table by 15. These numbers are given in the third column, rounded to two decimal places to simplify later calculations. The graph of the step function $p(t)$ is shown in Figure 1.

TABLE 1. U.S. POPULATION BY AGE IN 1990

Age	Population (millions)	$p(t)$ (millions per year of age)
$0 < t < 15$	52.8	3.52
$15 \le t < 30$	58.8	3.92
$30 \le t < 45$	59.2	3.95
$45 \le t < 60$	35.5	2.37
$60 \le t < 75$	28.6	1.91
$75 \le t \le 90$	13.3	0.89

[(1)]Data adapted from *Statistical Abstract of the United States, 1993*, Washington, D.C.: U.S. Department of Commerce, 1994, p.15.

♦ SUGGESTIONS TO INSTRUCTORS: Have students begin work on Questions 1 through 6 before class. Problems 1, 2, 4, 7, 8, 27, 29, and 33 are good for in-class discussion.

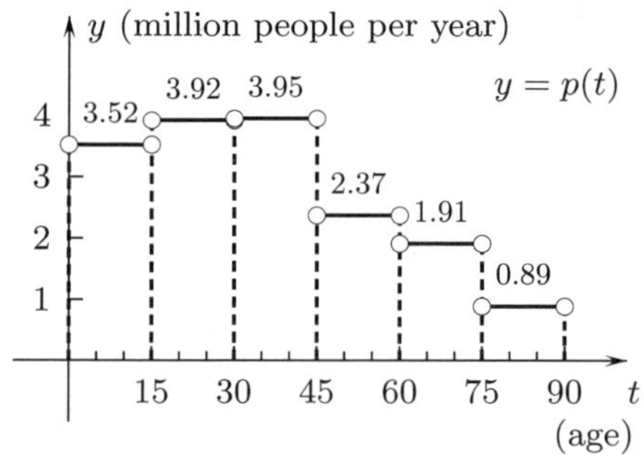

FIGURE 1

Example 1 **(a)** Based on the data in Figure 1, how many U.S. residents in 1990 were less than 20 years old? **(b)** How many were at least 20 but less than 90 years old? **(c)** What percent of the population younger than 90 was younger than 20?

SOLUTION **(a)** Because $p(t)$ equals 3.52 million per year for $0 < t < 15$ and $p(t) = 3.92$ million per year for $15 < t < 20$, the population of age t with $0 \le t < 20$ was

$$\left[3.52\ \frac{\text{million}}{\text{year}}\right][15 \text{ years}] + \left[3.92\ \frac{\text{million}}{\text{year}}\right][5 \text{ years}] = 72.4 \text{ million}. \quad \textbf{(1)}$$

(b) Similarly, because $p(t)$ equals 3.92 for $20 \le t < 30$, equals 3.95 for $30 < t < 45$, equals 2.37 for $45 < t < 60$, equals 1.91 for $60 < t < 75$, and equals 0.89 for $75 < t < 90$, the population of age t with $20 \le t < 90$ was

$$3.92(10) + +3.95(15) + 2.37(15) + 1.91(15) + 0.89(15) = 176 \text{ million} \quad \textbf{(2)}$$

(c) Since the total population younger than 90 was $72.4 + 176 = 248.4$ million, the fraction less than 20 was $72.4/248.4 \doteq 0.291$ and the percent less than 20 was $\doteq 29.1\%$. □

Notice that the number calculated in **(1)** is the area of the region between the graph of $p(t)$ and the t-axis for $0 \le t \le 20$ in Figure 1, so that this area equals the integral $\int_0^{20} p(t)\,dt$. Similarly, the number calculated in **(2)** equals $\int_{20}^{90} p(t)\,dt$. In general, the number of U.S. residents of age t with $a \le t \le b$ for any a and b with $0 \le a < b \le 90$ was $\int_a^b p(t)\,dt$. The function $p(t)$ is called the DENSITY of the age distribution by analogy with the case of weight, where the integral $\int_a^b \rho(x)\,dx$ of the one-dimensional density of an object equals the weight of the portion of the object for $a \le x \le b$.

Question 1 Figure 2 shows the graph of the density $f(x)$ of the age distribution of the 1988 population of the Marshall Islands in the South Pacific.[(2)] **(a)** What are the values of $\int_0^{20} f(t)\,dt$ and $\int_{20}^{90} f(t)\,dt$ and what do they represent? **(b)** What percentage of the population less than 90 years old in the Marshall Islands in 1990 was less than 20?

[(2)] Data adapted from "The evolution of regional demography in the Marshall Islands" by L. Gorenflo and M. Levin, *Pacific Studies*, Honolulu: Brigham Young University Press (verify the publisher), March, 1994, pp. 94, 132.

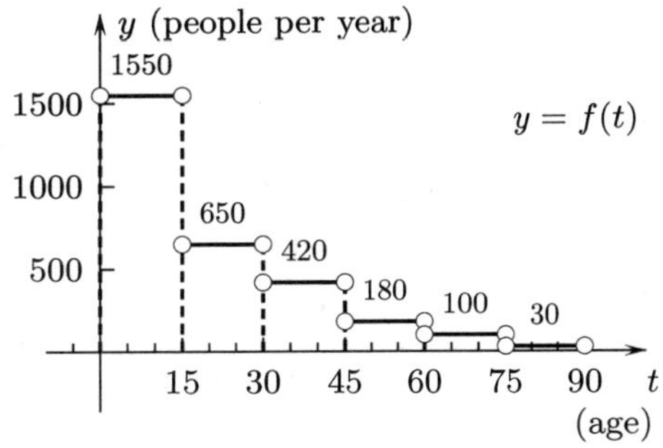

FIGURE 2

Continuous distribution densities

Table 1 above gives the 1990 U.S. population divided into 15-year brackets. More detail is shown in Table 2, which gives the same population divided into five-year brackets.[(3)]. The values of the corresponding density function $p_1(t)$ in the third columns of Table 2 are obtained by dividing the numbers in the second columns by 5. The function $p_1(t)$, like the function $p(t)$ of Table 1, is measured in millions of people per year of age.

TABLE 2. U.S. POPULATION BY AGE IN 1990

Age	Population (millions)	$p_1(t)$ (millions/year)	Age	Population (millions)	$p_1(t)$ (millions/year)
$0 < t < 5$	18.8	3.76	$45 \le t < 50$	13.7	2.74
$5 \le t < 10$	18.0	3.60	$50 \le t < 55$	11.3	2.26
$10 \le t < 15$	17.0	3.40	$55 \le t < 60$	10.5	2.10
$15 \le t < 20$	17.9	3.58	$60 \le t < 65$	10.6	2.12
$20 \le t < 25$	19.1	3.82	$65 \le t < 70$	9.0	1.80
$25 \le t < 30$	21.3	4.26	$70 \le t < 75$	9.0	1.80
$30 \le t < 35$	21.8	4.36	$75 \le t < 80$	5.3	1.06
$35 \le t < 40$	19.8	3.96	$80 \le t < 85$	5.0	1.06
$40 \le t < 45$	17.6	3.52	$85 \le t < 90$	3.0	0.60

The graphs of the density $p_1(t)$ from Table 2 is given in Figure 3, and Figure 4 shows the graph of the density function $p_2(t)$ that might be obtained if had more information and could divide the population into one-year age brackets. We have omitted the open circles at the ends of the horizontal lines in these last drawings because they would obscure the graphs.

[(3)]See footnote (1).

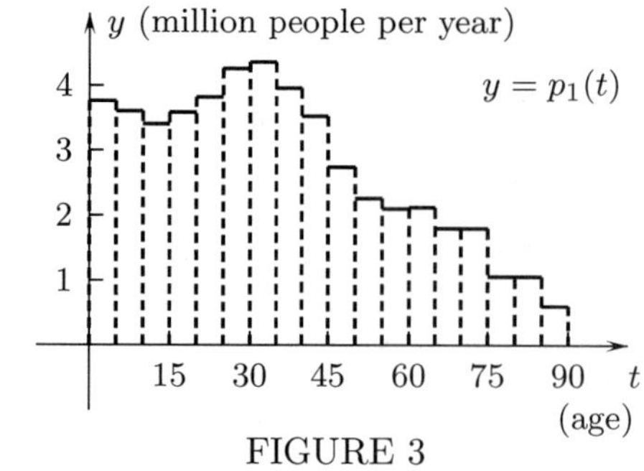

FIGURE 3

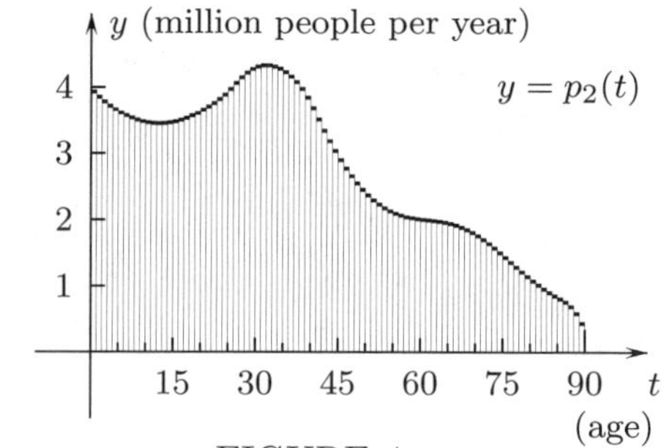

FIGURE 4

The density functions of Figures 1 through 4 are all discontinuous step functions. We could also model the age distribution in the United States in 1990 with a continuous density, such as the function $p_3(t)$ of Figure 5. In each model, the population between ages a and b with $a < b$ is given by the integral of the density from a to b.

FIGURE 5

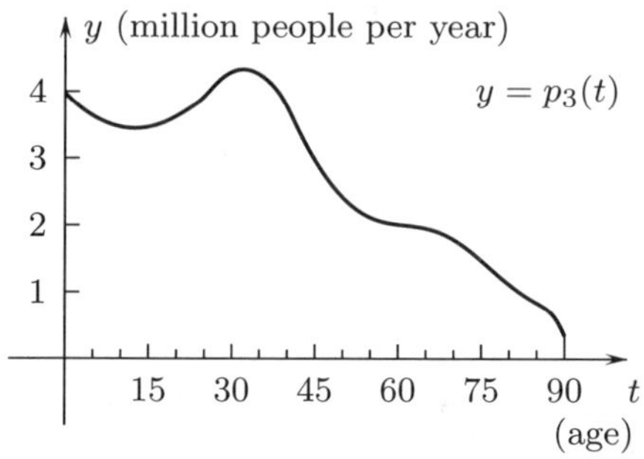

The next example deals with another continuous density function.

Example 2 Figure 6 shows the graph of the density $p(r)$ of the distribution of the radii r (millimeters) of grains of typical beach sand.[(4)] The density is measured in percent per millimeter. Based on the graph, what approximate percent of grains of beach sand have radii between 0.2 and 0.25 millimeter?

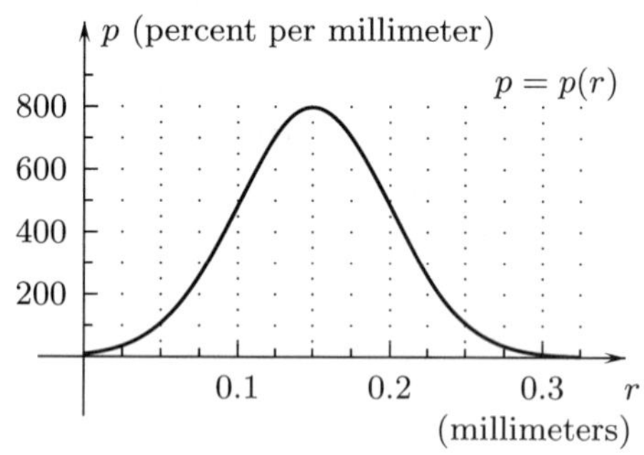

FIGURE 6

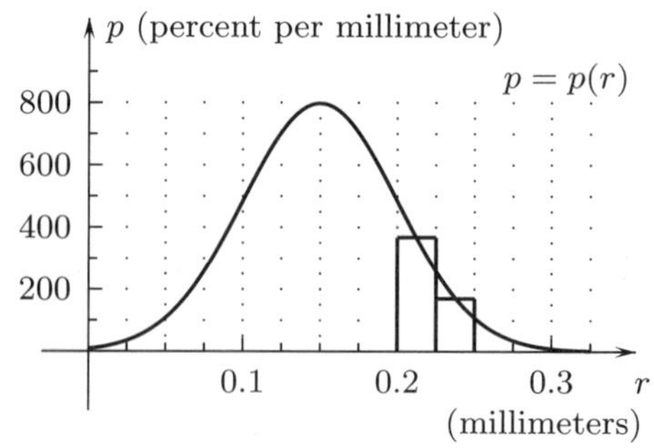

FIGURE 7

[(4)]Data adopted from *Encyclopaedia Britannica, Macropaedia*, Vol. 16, Chicago: Encyclopaedia Britannica, Inc., 1979, p. 468.

SOLUTION The percent of grains of sand with radii r between 0.2 and 0.25 millimeters is given by the integral of $p(r)$ from 0.2 to 0.25 and equals the area of the region between the graph and the r-axis for $0.2 \le r \le 0.25$. Because the horizontal distance between the grid points in Figure 6 is 0.025 millimeters, we approximate this region by two rectangles as in Figure 7, with their sides on the grid and their tops chosen so that each has approximately the same area as the corresponding portion of the region. The height of the first rectangle is approximately 370 percent per millimeter, and the height of the second is about 175 percent per milllimeter. Since the width of the rectangles is 0.025 millimeters, their combined area is approximately $370(0.025) + 175(0.025) = 13.625$ percent, and we conclude that approximately 14 percent of grains of typical beach sand have radii between 0.2 and 0.25 millimeters. □

Question 2 What is the approximate value of $\int_0^{0.3} p(r)\,dr$ for the density function $p(r)$ of Figure 6 and why is this plausible?

Cumulative distributions

Figure 8 is a duplicate of Figure 1, showing the graph of the density $p(t)$ of the age distribution of the United States population in 1990 grouped in fifteen-year intervals. Based on this data, the number of U.S. residents in 1990 who were less than T years old for $0 \le t \le 90$ was

$$P(T) = \int_0^T p(t)\,dt. \tag{3}$$

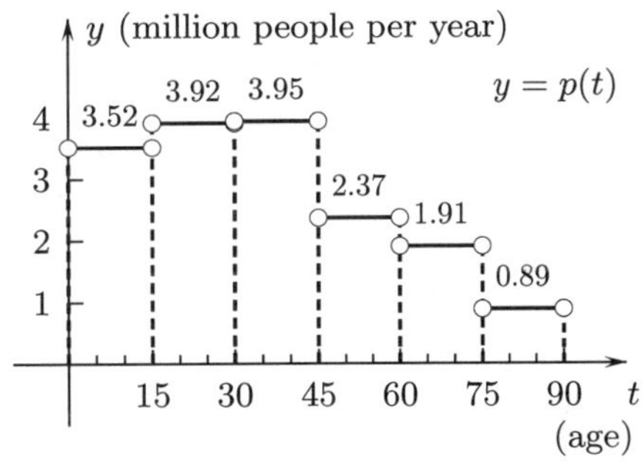

FIGURE 8

Example 3 Give formulas for the values of the function $P(T)$ of **(3)** for $0 \le T \le 15$ and $15 \le T \le 30$.

SOLUTION As is shown in Figure 8, $p(t) = 3.52$ for $0 \le t < 15$, so that

$$P(T) = \int_0^T 3.52\,dt = 3.52T \quad \text{for} \quad 0 \le T \le 15.$$

This gives $P(15) = (3.52)(15) = 52.8$, and since $p(t) = 3.92$ for $15 \le t < 30$, we obtain

$$P(T) = 52.8 + \int_{15}^T (3.92)\,dt = 52.8 + 3.92(T - 15) \quad \text{for } 15 \le T \le 30. \;\square$$

Question 3 Give a formula for $P(T)$ in the interval $30 \le T \le 45$.

With calculations similar to those in Example 3 and Question 3 we could complete the following collection of formulas for $P(T)$.

$$P(T) = \begin{cases} 3.52T & \text{for} \quad 0 \le T \le 15 \\ 52.8 + 3.92(T - 15) & \text{for} \quad 15 \le T \le 30 \\ 111.6 + 3.95(T - 30) & \text{for} \quad 30 \le T \le 45 \\ 170.85 + 2.37(T - 45) & \text{for} \quad 45 \le T \le 60 \\ 206.4 + 1.91(T - 60) & \text{for} \quad 60 \le T \le 75 \\ 235.05 + 0.89(T - 75) & \text{for} \quad 75 \le T \le 90. \end{cases} \tag{4}$$

The graph of $P(T)$ is shown in Figure 9. Notice that this function is piecewise linear and continuous. It is called the CUMULATIVE DISTRIBUTION corresponding to the density function $p(t)$ with base point $t = 0$. A general definition of such functions follows.

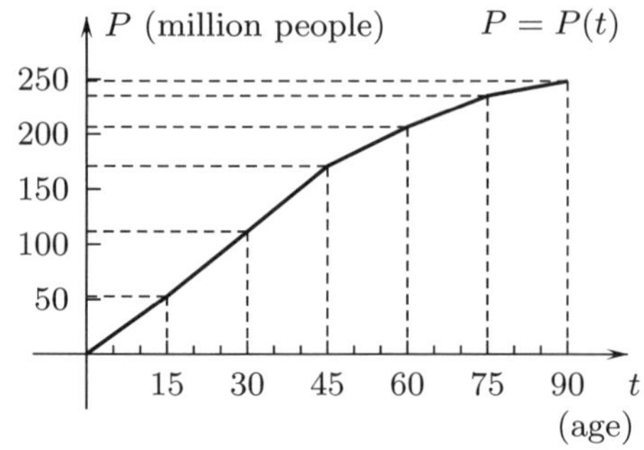

FIGURE 9

Definition 1 *The* CUMULATIVE DISTRIBUTION *of a density function $f(x)$ with basepoint x_0 is the function with variable X defined by*

$$F(X) = \int_{x_0}^{X} f(x)\, dx. \tag{5}$$

Example 4 What is the cumulative distribution $F(X)$ with basepoint 0 of the distribution $f(x) = x^2$?

SOLUTION By **(4)** with $x_0 = 0$, the cumulative distribution is

$$F(X) = \int_0^X x^2\, dx = \left[\tfrac{1}{3}x^3\right]_0^X = \tfrac{1}{3}X^3. \square$$

Finding a distribution from a cumulative distribution

Often data is given in terms of a cumulative distribution, rather than a distribution density. Then the corresponding population density can be found as the the derivative of the cumulative distribution at all points where the distribution density is continuous. Before we derive this general result we will look at three cases where we can verify it directly.

First, we consider the cumulative distribution **(4)**, whose value $P(T)$ is the number of people in the United States in 1990 who were of age $\le T$ years. Its derivative can be found from **(4)**. We obtain

$$\frac{dP}{dT}(T) = \begin{cases} 3.52 & \text{for} \quad 0 < T < 15 \\ 3.92 & \text{for} \quad 15 < T < 30 \\ 2.95 & \text{for} \quad 30 < T < 45 \\ 2.37 & \text{for} \quad 45 < T < 60 \\ 1.91 & \text{for} \quad 60 < T < 75 \\ 0.89 & \text{for} \quad 75 < T < 90. \end{cases} \tag{6}$$

Comparing the values in **(6)** with the values of $p(t)$ in Figure 8 shows that $\frac{dP}{dT}(T) = p(T)$ at the points T where $p(t)$ is continuous.

Example 5 Use the formula from Example 4 for the cumulative distribution $F(X) = \int_0^X x^2\,dx$ with density x^2 to show that $F'(X) = X^2$ for all X.

SOLUTION We saw in Example 4 that $F(X) = \frac{1}{3}X^3$ for all X. Therefore,

$$\frac{dF}{dX}(X) = \frac{d}{dX}(\tfrac{1}{3}X^3) = X^2 \quad \text{for all } X. \square$$

Question 4 Show that the X-derivative of the cumulative distribution $S(X) = \int_0^X \sin x\,dx$ with the density function $\sin x$ equals $\sin(X)$ for all X .

In Example 5 and Question 4 the derivative of the cumulative distribution is the distribution density (with x replaced by X) because the density of continuous in $(-\infty, \infty)$. These and calculations **(6)** illustrate the folowing result.

Theorem 1 (The Fundamental Theorem for derivatives of integrals) *Suppose that the definite integral*

$$F(X) = \int_{x_0}^X f(x)\,dx \tag{7}$$

is defined for all X in an interval I containing the point x_0. Then

$$\frac{dF}{dX}(X) = \frac{d}{dX}\int_{x_0}^X f(x)\,dx = f(X) \tag{8}$$

for all X in the interior of I where $f(X)$ is continuous.

To see why this theorem holds, we consider the case of the positive continuous function $f(x)$ in Figure 10. For $X > x_0$, $F(X) = \int_{x_0}^X f(x)\,dx$ is the area between the graph of $f(x)$ and the x-axis for $x_0 \le x \le X$ that is shown in Figure 10. For small, positive ΔX, $F(X+\Delta X) = \int_{x_0}^{X+\Delta X} f(x)\,dx$ is the area between the graph and the x-axis for $x_0 \le x \le X + \Delta X$, so that the difference

$$F(X+\Delta X) - F(X) = \int_X^{X+\Delta X} f(x)\,dx$$

is the area between the graph and the x-axis for $X \le x \le X + \Delta X$ that is shown in Figure 11.

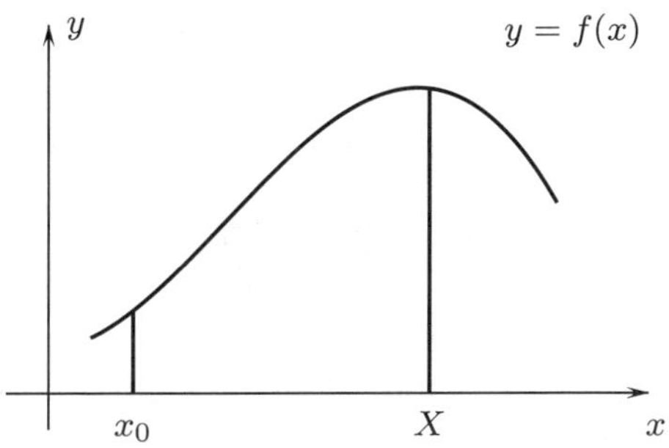

Region of area $F(X)$

FIGURE 10

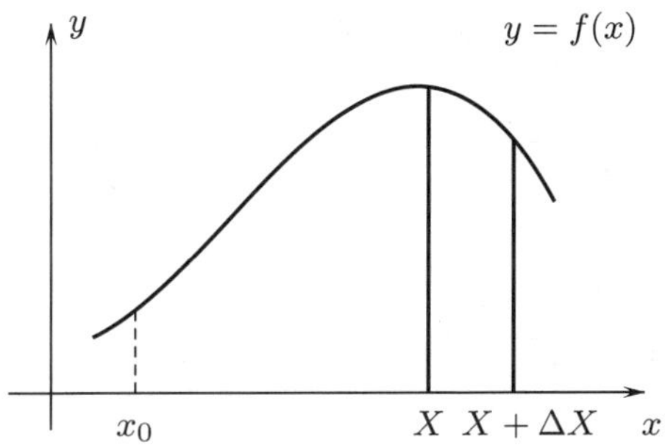

Region of area $F(X+\Delta X) - F(X)$

FIGURE 11

Because $f(x)$ is continuous in the finite closed interval $X \le x \le X + \Delta X$, it has—by the Extreme-Value Theorem of Section 3.2—a minimum value m (Figure 12) and a maximum value M (Figure 13) in the interval.

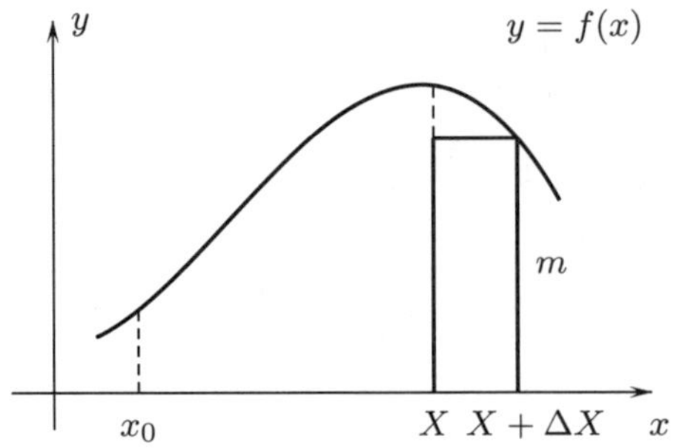

Region of area $m\Delta X$

FIGURE 12

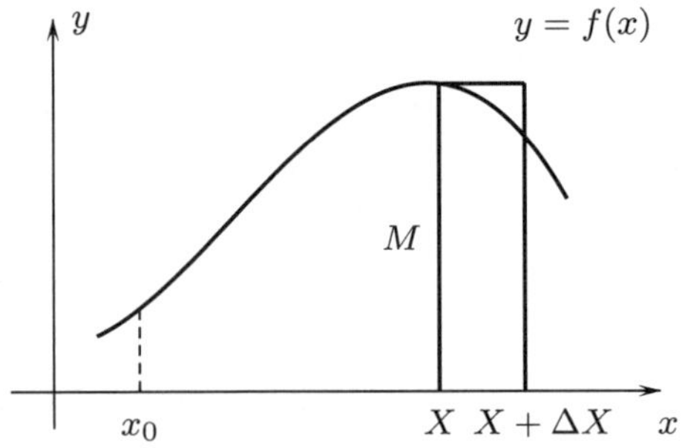

Region of area $M\Delta X$

FIGURE 13

The rectangle in Figure 12 has area $m\Delta X$ and is contained in the region of Figure 11, while the rectangle in Figure 13 has area $M\Delta X$ and contains the region of Figure 11. Therefore,

$$m\Delta X \le F(X + \Delta X) - F(X) \le M\Delta X.$$

Dividing all the expressions in these inequalities by the positive number ΔX yields

$$m \le \frac{F(X + \Delta X) - F(X)}{\Delta X} \le M. \qquad \textbf{(9)}$$

Similar reasoning would give **(9)** for small, negative ΔX. The numbers M and m are values of $f(x)$ in the interval between X and $X + \Delta X$. Because we are assuming that $f(x)$ is continuous at X, m and M both tend to $f(X)$ as ΔX tends to 0, and **(9)** shows that

$$\frac{dF}{dX}(X) = \lim_{\Delta X \to 0} \frac{F(X + \Delta X) - F(X)}{\Delta X} = f(X).$$

This gives formula **(8)**.

The Fundamental Theorem **(8)** has a geometric interpretation for a positive continuous function $f(x)$, which is illustrated in Figure 10. The rate of change with respect to X of the area $F(X)$ between the graph of $f(x)$ and the x-axis for $x_0 \le x \le X$ equals the height $f(X)$ of the region at its right side.

Example 6 Find a formula for $\dfrac{d}{dX}\displaystyle\int_1^X \sqrt{x^4 + 7}\, dx$.

SOLUTION Because $\sqrt{x^4 + 7}$ is continuous for all x, formula **(8)** shows that

$$\frac{d}{dX}\int_1^X \sqrt{x^4 + 7}\, dx = \sqrt{X^4 + 7} \quad \text{for all } X. \square$$

Example 7 Imagine you are interested in an experiment that has been run hundreds of times and whose outcome can be any number x between 0 and 10. Suppose that the number of experiments $N(X)$ for which the outcome x is in the interval $0 \le x \le X$ is given by the cumulative distribution

$$N(X) = X^5 e^{-X/2} \quad \text{for } 0 \le X \le 10 \qquad \textbf{(10)}$$

whose graph is in Figure 14. The corresponding density is the function $n(x)$ such that

$$N(X) = \int_0^X n(x)\,dx \quad \text{for } 0 \le X \le 10 \tag{11}$$

Find a formula for $n(x)$ in **(11)**, under the assumption that it is continuous in $[0, 10]$, and generate its graph.

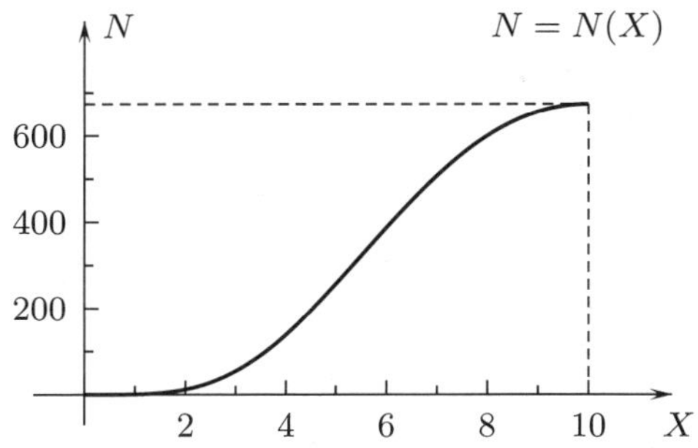

FIGURE 14

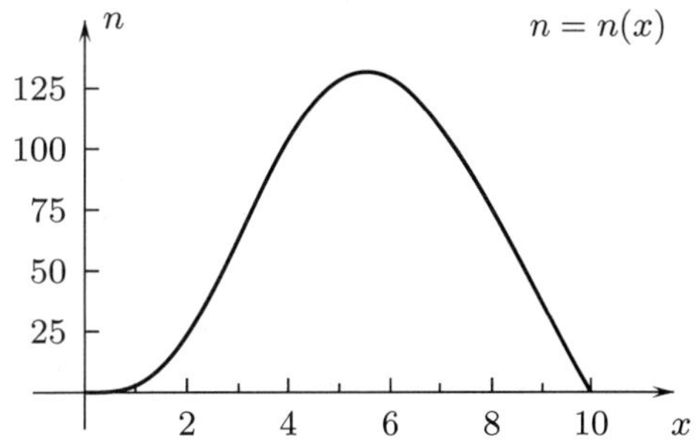

FIGURE 15

SOLUTION By the Fundamental Theorem 1, $n(X)$ is the X-derivative of $N(X)$, which we can find with the Product and Chain Rules:

$$\begin{aligned} n(X) &= \frac{d}{dX}N(X) = \frac{d}{dX}(X^5e^{-X/2}) \\ &= X^5\frac{d}{dX}(e^{-X/2}) + e^{-X/2}\frac{d}{dX}(X^5) \\ &= X^5e^{-X/2}\frac{d}{dX}(-\tfrac{1}{2}X) + 5X^4e^{-X/2} \\ &= (-\tfrac{1}{2}X^5 + 5X^4)e^{-X/2}. \end{aligned}$$

Therefore, $n(x) = (-\frac{1}{2}x^5 + 5x^4)e^{-x/2}$. Its graph is in Figure 15. □

Question 5 **(a)** How many times was the experiment run that led to the cumulative distribution **(10)**? **(b)** How many of the experiments had results between 5 and 8? **(c)** How are the responses to parts (a) and (b) related to the graphs of the functions $N(X)$ and $n(x)$ in Figures 14 and 15?

Other derivatives with variable limits of integration

We can use Theorem 1 to find the X-derivatives of integrals whose upper limit of integration are constant and whose lower limit of integration is X and of integrals where either or both limits of integration are functions of X.

To apply Theorem 1 when the upper limit of integration is constant and the lower limit is X, we first interchange the limits of integration and, to avoid changing its value, multiply the integral by -1.

Example 8 What is the X-derivative of $\int_X^5 e^{\sqrt{x}}\,dx$ at $X = 4$?

SOLUTION We write

$$\frac{d}{dX}\int_X^5 e^{\sqrt{x}}\,dx = \frac{d}{dX}\left[-\int_5^X e^{\sqrt{x}}\,dx\right] = -e^{\sqrt{X}}.$$

The derivative at $X = 4$ is $-e^{\sqrt{4}} = -e^2$. □

If the upper limit of integration is a function $b(X)$, we need the Chain Rule to find the derivative.

Example 9 What is the X-derivative of $\int_0^{X^3} \cos x \, dx$?

SOLUTION We set $F(Y) = \int_0^Y \cos x \, dx$, so that

$$\int_0^{X^3} \cos x \, dx = F(X^3).$$

Since $\cos x$ is continuous for all x, then by **(8)**

$$\frac{dF}{dY}(Y) = \frac{d}{dY}\int_0^Y \cos x \, dx = \cos Y$$

and by the Chain Rule,

$$\begin{aligned}\frac{d}{dX}\int_0^{X^3} \cos x \, dx &= \frac{d}{dX}[F(X^3)] \\ &= \frac{dF}{dY}(X^3)\frac{d}{dX}(X^3) = 3X^2\cos(X^3). \ \square\end{aligned}$$

Question 6 Find the result of Example 9 by first evaluating the integral—using the Fundamental Theorem of Section 4.3—and then differentiating the result, instead of using the Fundamental Theorem of this section.

If the lower limit of integration is one function $a(x)$ and the upper limit of integration is another function $b(x)$, we write the integral as the integral from $a(x)$ to c plus the integral from c to $b(x)$ with c chosen so that both integrals are defined.

Example 10 What is the X-derivative of $\int_{X^4}^{\sin X} e^{x^2} \, dx$?

SOLUTION Since e^{x^2} is defined and continuous for all x, we can use any constant c. We choose $c = 0$ and write

$$\begin{aligned}\frac{d}{dX}\int_{X^4}^{\sin X} e^{x^2} \, dx &= \frac{d}{dX}\left[\int_{X^4}^0 e^{x^2} \, dx + \int_0^{\sin X} e^{x^2} \, dx\right] \\ &= \frac{d}{dX}\left[-\int_0^{X^4} e^{x^2} \, dx + \int_0^{\sin X} e^{x^2} \, dx\right] \\ &= -e^{(X^4)^2}\frac{d}{dX}(X^4) + e^{\sin^2 X}\frac{d}{dX}(\sin X) \\ &= -4X^3e^{X^8} + \cos x \, e^{\sin^2 X}. \ \square\end{aligned}$$

Principles and procedures

- A density function $f(x)$ and the corresponding cumulative distribution $F(X)$ are used to describe data associated with a parameter x.
- In most cases, the amount or number of data for which $a \le x \le b$ is given by the integral $\int_a^b f(x) \, dx$ of the density.
- The cumulative distribution $F(X)$ for a density $f(x)$ with base x_0 is the amount of data for which $x_0 \le x \le X$ and equals $\int_{x_0}^X f(x) \, dx$.

- Densities can be calculated from the corresponding cumulative distributions by using Theorem 1 of this section, which states that at the points where $f(x)$ is continuous,

$$\frac{d}{dX}F(X) = \frac{d}{dX}\int_{x_0}^{X} f(x)\,dx = f(X).$$

- To find a derivative $\frac{d}{dX}\int_a^{b(X)} f(x)\,dx$ with a differentiable function $b(X)$, we use Theorem 1 and the Chain Rule. If $f(x)$ is continuous at $b(X)$, then

$$\frac{d}{dX}\int_a^{b(X)} f(x)\,dx = f\big(b(X)\big)\frac{db}{dX}(X).$$

- To find $\frac{d}{dX}\int_X^{b} f(x)\,dx$ with a constant b, we interchange the limits of integration and multiply the integral by -1 to preserve its value. At the points X where $f(x)$ is continuous,

$$\frac{d}{dX}\int_X^{b} f(x)\,dx = -\frac{d}{dX}\int_b^{X} f(x)\,dx = -f(X).$$

- To find $\frac{d}{dX}\int_{a(X)}^{b(X)} f(x)\,dx$ with differentiable functions $a(x)$ and $b(x)$, we write, with a suitable constant c

$$\begin{aligned}\frac{d}{dX}\int_{a(X)}^{b(X)} f(x)\,dx &= \frac{d}{dX}\left[\int_{a(x)}^{c} f(x)\,dx + \int_c^{b(X)} f(x)\,dx\right]\\ &= -\frac{d}{dX}\int_c^{a}(X) f(x)\,dx + \frac{d}{dX}\int_c^{b(X)} f(x)\,dx\\ &= -f\big(a(X)\big)\frac{da}{dX}(X) + f\big(b(X)\big)\frac{db}{dX}(X).\end{aligned}$$

Tune-Up Exercises 6.5♦

A *Answer provided.* O *Outline of solution provided.*

T1.O Figure 16 shows the approximate distribution $y(x)$ (students per score) of the scores on an exam taken by 30,000 students. Approximately how many students received scores x with $40 \le x \le 80$?

♦ Type 1, integral of a distribution from areas

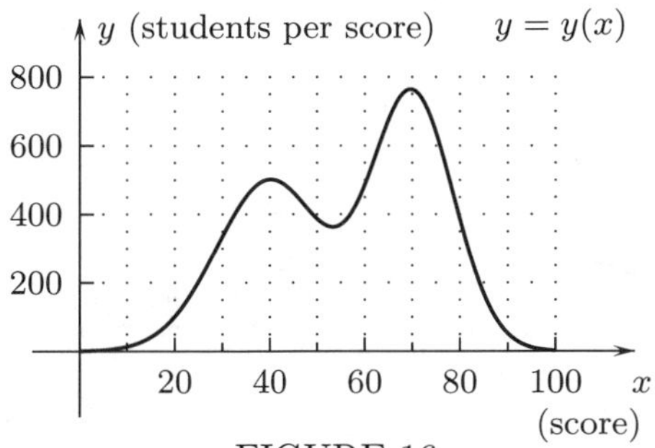

FIGURE 16

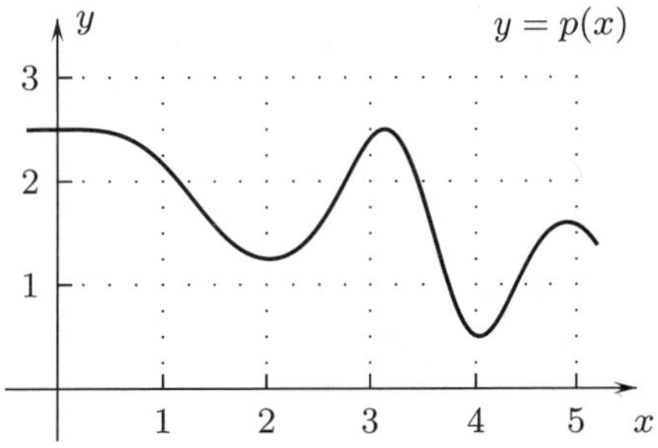

FIGURE 17

♦The Tune-up Exercises and Problems are classified by type and content. The types are (1) basic, reactive; (2) basic reflective; (3) intermediate, reactive; (4) intermediate, reflective; (5) advanced, reactive; (6) advanced, reflective; and (7) advanced, theoretical.

T2.[A] **(a)** Give formulas for $G(X) = \int_0^X g(x)\,dx$ for $0 \le X \le 5$, where

$$g(x) = \begin{cases} 1 & \text{for} \quad 0 \le x \le 2 \\ -\frac{1}{2} & \text{for} \quad 2 < x \le 5. \end{cases}$$

(b) At what values of X is $G'(X)$ defined? **(c)** Give formulas for $G'(X)$.

◆ Type 3, derivative of an integral

T3.[A] Draw the graph of $\frac{d}{dX}\int_0^X p(x)\,dx$, where $p(x)$ is the continuous function of Figure 17.

◆ Type 3, derivative of an integral, etc.

T4.[A] What is the X-derivative of $\int_1^X \sqrt{x^2+10}\,dx$ at $X = 3$?

◆ Type 3, derivative of an integral

T5.[A] Find the X-derivative of $\int_X^{10} (2x+\sin x)\,dx$ **(a)** by performing the integration (using the Fundamental Theorem of Section 4.2) and then differentiating the result, and **(b)** by using the Fundamental Theorem of this section.

◆ Type 3, derivative of an integral, etc.

T6.[O] Find the X-derivative of $\int_1^{\ln X} \frac{1}{1+x^4}\,dx$.

◆ Type 3, derivative of an integral,etc.

Problems 6.5

[A] *Answer provided.* [O] *Outline of solution provided.*

1.[A] The table below gives the distribution by age of deaths of U.S. residents in 1990, divided into four age groups.[(5)] Based on this data, **(a)** how many U.S. residents under 100 died in 1990 and **(b)** what percent of those under 100 who died were less than 25 years old?

◆ Type 2, integral of a distribution step function, etc.

Age	$0 < t < 25$	$25 \le t < 50$	$50 \le t < 75$	$75 \le t \le 100$
Deaths per year	3,600	8,200	31,600	42,600

2.[A] The function in Figure 18 gives the distribution by date of the average number of tornados recorded in the U.S. from 1950 through 1982, with $t = 0$ on January 1 and t measured in months.[(6)] **(a)** Is $n(t)$ a density function or a cumulative distribution for the average number of tornados? **(b)** What does the integral $\int_0^6 n(t)\,dt$ represent? **(c)** What was the approximate total number of tornados during the month of June in all 33 years from 1950 though 1982?

◆ Type 2, integral of a distribution from areas, etc.

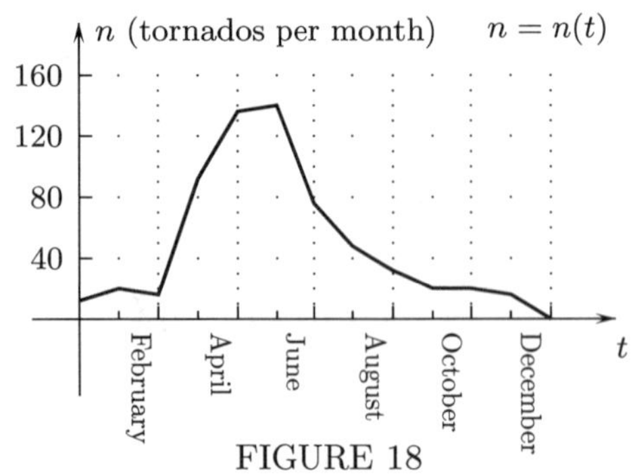

FIGURE 18

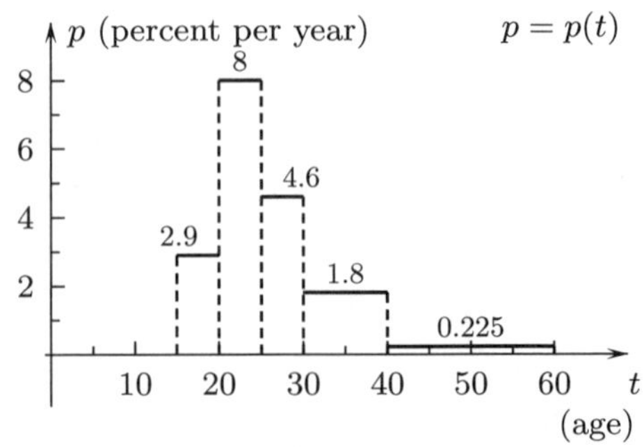

FIGURE 19

[(5)] Data adapted from *Statistical Abstract of the United States, 1993*, Washington, D.C.: U.S. Department of Commerce, 1994, p. 92.

[(6)] Data adapted from "The Tornado" by J. Snow, *Scientific American*, April, 1984, p. 95.

3. Figure 19 gives the density $p(t)$ (percent per year) of the distribution by age of the Pennsylvania Regiment in 1775.[7] **(a)** What percent of the regiment was less than 25 years old? **(b)** What does the cumulative distribution $P(T) = \int_0^T p(t)\, dt$ represent? **(c)** What is the slope of the graph of $P(T)$ for $25 \le T < 30$?

◆ Type 2, integral of a distribution step function, etc.

4. Figure 20 shows the graph of the percent $P(H)$ of the earth's surface area that is at an elevation $\le H$ kilometers above sea level.[8] **(a)** What percent of the surface area of the earth is covered by oceans more than two kilometers deep? **(b)** In this matematical model, the function $P(H)$ is piecewise linear, and, consequently, the corresponding density function $p(h)$ is a step function. Give its approximate values. **(c)** In what units is $p(h)$ measured?

◆ Type 2, a cumulative distribution , etc.

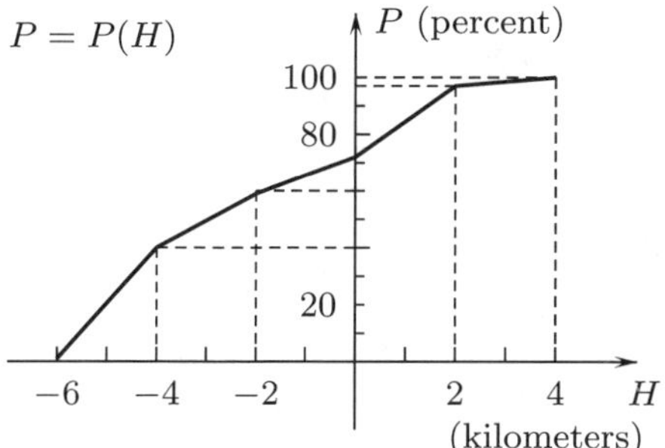

FIGURE 20

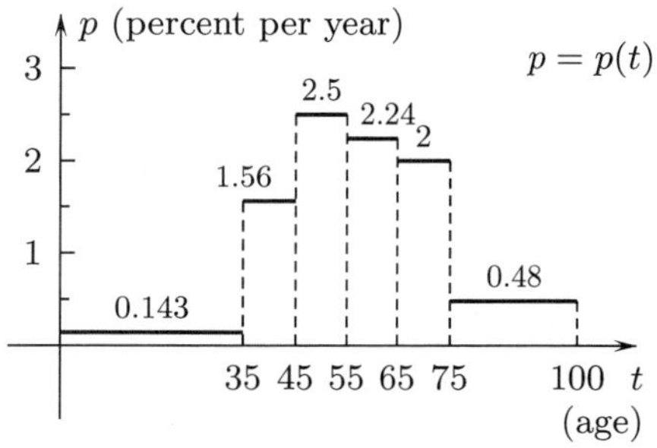

FIGURE 21

5. The graph of the density $p(t)$ of the distribution of wealth in the U.S. in 1995, measured according to the age t of the family head, is shown in Figure 21.[9] **(a)** Draw the graph of the cumulative distribution $P(T) = \int_0^T p(t)\, dt$. **(b)** What does the number $P(T)$ represent?

◆ Type 2, a cumulative distribution , etc.

6. Figure 22 shows results from a large diet study in Framingham, Massachusetts. The functions $p_W(C)$ and $p_M(C)$ give the distribution of women and men in a women with respect to their daily Calorie intake.[10] **(a)** What does $\int_{2000}^{2001} p_M(C)\, dC$ represent and what is its approximate value? **(b)** Approximately what percent of the women had Calorie intake between 2000 and 2001 Calories?

◆ Type 2, integral of a distribution from areas, etc.

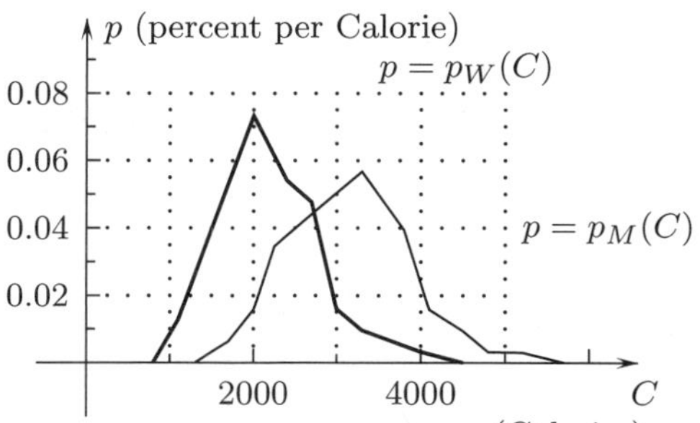

FIGURE 22

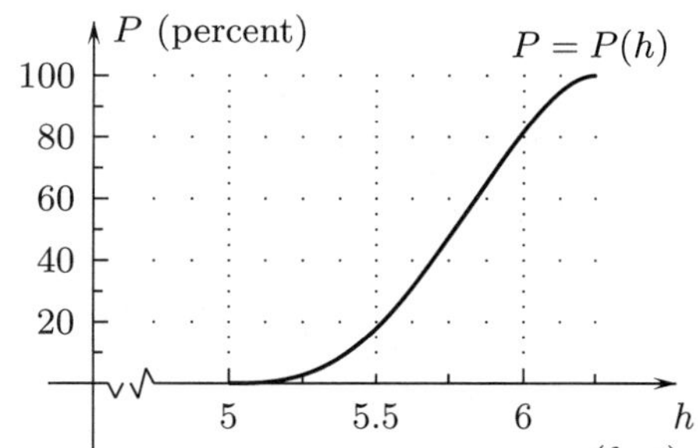

FIGURE 23

---

[7] Data adapted from "An army of servants: The Pennsylvania Regiment during the Seven Years' War" by M. Ward, *The Pennsylvania Magazine of History and Biography*, January/April, 1995, p. 87.

[8] Data adapted from *McGraw-Hill Encyclopedia of Science and Technology*, Vol. 5,1982, p. 484.

[9] Data adapted from *Statistical Abstract of the United States, 1998*, Washington, D.C.: U.S. Department of Commerce, 1998, p. 482.

[10] Data adapted from *The Framingham Study*, W. Kannel and T. Gordon editors, Section 31, Washington, DC: Department of Health, Education, and Welfare, 1976, p.???

7. The function $P(h)$ in Figure 23 is the cumulative distribution $P(h)$ of the heights of men 20–29 years old in 1991.[11] **(a)** What percent of the men were ≤ 6 feet tall? **(b)** What is the approximate value of $\frac{dP}{dh}(5.75)$ and what does it represent?

♦ Type 2, a cumulative distribution , etc.

8.[A] Figure 24 shows the graph of the density function $p(h)$ of the distribution of larch trees in a 200-square-mile region in Siberia, classified by their height.[12] **(a)** Approximately what percent of the larch trees in this study were less than six meters tall? **(b)** What is the maximum value of the cumulative distribution $P(H) = \int_0^H p(h)\,dh$? **(c)** What are the approximate coordinates of the inflection point of the graph of $P(H)$?

♦ Type 2, integral of a distribution from areas, etc.

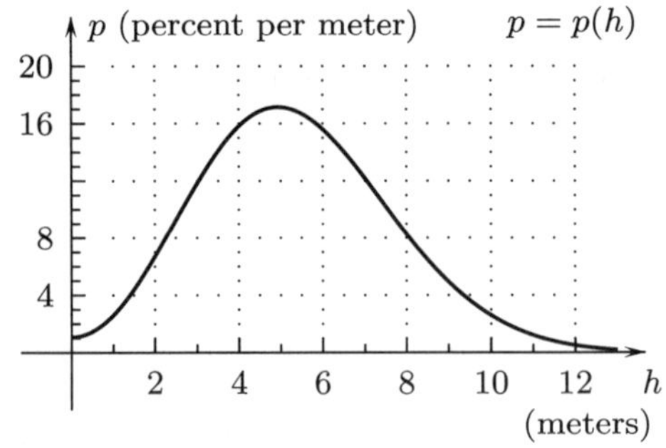

FIGURE 24

9.[A] The cumulative world industrial-energy use since 1850 is given approximately as a function of the year t by $E(t) = 5 + \frac{1}{2}(t - 1900) + \frac{1}{60}(t - 1900)^2 + \frac{1}{3600}(t - 1900)^3$ terrawatt-years.[13] Based on this mathematical model, what was the percent increase in the rate of world industrial-energy use from 1900 to 1990?

♦ Type 4, a cumulative distribution , etc.

10. The table below gives in seven age groups the numbers of Ph.D. recipients in mathematics from U.S. universities from July, 1997, through June, 1998.[14] **(a)** Sketch the graph of the number $N(T)$ of Ph.D. recipients of age $\le T$ with the mathematical model that the distribution $n(t)$ of Ph.D. recipients per year is constant in the interiors of the age intervals in the table. **(b)** What is the value of $\frac{dN}{dt}(33)$?

♦ Type 2, integral of a distribution step function, etc.

Age	20–24	25–29	30–34	35–39	40-44	45–49	50–54
Number	6	257	233	77	39	7	9

Find the derivatives in Problems 11 and 12 **(a)** by using the Fundamental Theorem of this section and **(b)** by first using techniques of Chapter 4 to perform the integration and then finding the derivative.

11.[O] $\frac{d}{dX}\int_2^X \sqrt{x+3}\,dx$

♦ Type 3, derivative of an integral

12. $\frac{d}{dY}\int_3^Y \frac{y}{y^2+4}\,dy$

♦ Type 3, derivative of an integral

Find the derivatives in Problems 13 through 24.

13.[A] $\frac{d}{dX}\int_{-1}^X x\sin x\,dx$

♦ Type 3, derivative of an integral

14. $\frac{d}{dT}\int_0^T (e^{t^2}+1)\,dt$

♦ Type 3, derivative of an integral

(11) Data adapted from *Statistical Abstract of the United States, 1998*, Washington, DC: Bureau of the Census, 1998, p. 153.

(12) Data adapted from "Growth dynamics and biodiversity of larch forest after wildfire at the north of central Siberia", *World Resource Review*, Napervill IL: World Resource Review, Vol. 8, No. 2, 1996, p. 221.

(13) Adapted from "Energy in transition" by J. Holdren, *Scientific American*, September, 1990, p.160.

(14) Data adapted from "1998 Annual survey of the mathematical sciences", *Notices of the American Mathematical Society*, Vol. 46, Number 8, Providence, RI: American Mathematical Society, September, 1999, p. 899.

15.[A] $\dfrac{d}{dX}\displaystyle\int_X^2 t\sin(3t)\,dt$

♦ Type 3, derivative of an integral

16. $\dfrac{d}{dY}\displaystyle\int_Y^0 y^{8/3}\,dy$

♦ Type 3, derivative of an integral

17.[O] $\dfrac{d}{dX}\displaystyle\int_1^{X^2} \sqrt{x+1}\,dx$

♦ Type 3, derivative of an integral

18.[A] $\dfrac{d}{dX}\displaystyle\int_1^{3X} \sin^5 x\,dx$

♦ Type 3, derivative of an integral

19. $\dfrac{d}{dZ}\displaystyle\int_1^{\sin Z} \frac{z}{z^4+1}\,dz$

♦ Type 3, derivative of an integral

20. $\dfrac{d}{dX}\displaystyle\int_{X^3}^{10} \sqrt[3]{\cos z}\,dz$

♦ Type 3, derivative of an integral

21. $\dfrac{d}{dX}\displaystyle\int_{\cos X}^{100} \frac{1}{x^2+1}\,dx$

♦ Type 3, derivative of an integral

22.[O] $\dfrac{d}{dX}\displaystyle\int_{X^2}^{X^3} (1-x^2)^{1/3}\,dx$

♦ Type 3, derivative of an integral

23.[A] $\dfrac{d}{dX}\displaystyle\int_{4X}^{5X} x\sin x\,dx$

♦ Type 3, derivative of an integral

24. $\dfrac{d}{dX}\displaystyle\int_{\sin X}^{\cos X} \sqrt{x^2+1}\,dx$

♦ Type 3, derivative of an integral

25.[A] Evaluate $\displaystyle\int_0^1 \frac{d}{dx}(x+\cos(x^2))\,dx$

♦ Type 3, integral of a derivative

26. Evaluate $\displaystyle\int_{-X}^{X} \frac{d}{dx}[(x^2+1)^6]\,dx.$

♦ Type 3, integral of a derivative

27.[A] For $X > 0$, the area of the region between the graph of the positive continuous function $f(x)$ and the x-axis for $0 \le x \le X$ is $5X^5 + X$. What is the function $f(x)$?

♦ Type 3, derivative of an integral, etc.

28. A rod extends from $x = 0$ to $x = 10$ (feet) on an x-axis. The portion of the rod for $0 \le X \le 10$ for $0 \le X \le 10$ weighs $6X^2 + 3X$ pounds. What is the one-dimensional density of the rod?

♦ Type 3, derivative of an integral, etc.

29. A function $f(x)$ has a continuous derivative for $x \ge 0$ and the length of its graph for $0 \le x \le X$ is $\frac{2}{3}(1+X)^{3/2} - \frac{2}{3}$. What is the function $f(x)$?

♦ Type 3, derivative of an integral, etc.

30. A region is bounded by the graph of a positive continuous function $f(x)$, by the x- and y-axes, and by the line $x = a$ with a positive constant a. When the region is rotated about the x-axis it generates a solid of volume $\pi(\frac{1}{5}a^5 + \frac{2}{3}a^3 + a)$. What is the function $f(x)$?

♦ Type 4, derivative of an integral, etc.

31.[A] Find a function $y(x)$ such that its graph contains the point $(0, 1)$ and such that the length of its graph from $x = 0$ to $x = X$ is $e^X - y(X)$ for $X > 0$.

♦ Type 4, derivative of an integral, etc.

32. Suppose that $H(X) = \displaystyle\int_0^X G(x)\,dx$, where $G(x)$ is positive and continuous for all x and $G(0) = 10$. What is the Y-derivative of the inverse function $H^{-1}(Y)$ at $Y = 0$?

♦ Type 4, derivative of an integral, etc.

33. Komodo dragons are large monitor lizards that live on Komodo Island and a few other islands in Indonesia, where they survive by eating deer, wild boars, as well as smaller animals and carrion. A study of the ages of deer eaten by the dragons resulted in the table below.[(15)] Let $P(T)$ denote the percent of the deer that were of age $t \le T$ years when they were eaten and let $p(t)$ be the corresponding density. **(a)** What is $P(2)$ and what does it represent? **(b)** What is $P^{-1}(59)$ and what does it represent? **(c)** Under the assumption that $p(t)$ is constant in the intervals of the table, in which interval is $p(t)$ equal to 46 percent per year?

Age t of deer	$0 < t < 0.5$	$0.5 \le t < 1$	$1 \le t < 1.5$	$1.5 \le t < 2$	$2 \le t < 6$	$6 \le t \le 8$
Percent	27	32	23	4	3	11

♦ Type 4, integral of a distribution step function, etc.

[(15)] Data adapted from *The Behaviorial Ecology of the Komodo Monitor* by W. Auffenberg, Gainsville FL: University Presses of Florida, 1981, p. 282. (See also, "The Komodo dragon" by C. Coifi, *Scientific American*, March 1999, pp. 84–91.)

Section 6.6

Improper integrals

OVERVIEW: *The definition of the Riemann integral* $\int_a^b f(x)\,dx$ *as a finite limit of Riemann sums that we gave in Section 4.2 can only be used if the interval of integration is finite and* $f(x)$ *is bounded there. In this section we show how certain integrals over infinite intervals and integrals of unbounded functions can be defined as* IMPROPER INTEGRALS, *which are limits of Riemann integrals as defined in Section 4.3.*

Topics:

- ***The area of an infinite region***
- ***Improper integrals with infinite limits of integration***
- ***An improper integral of an unbounded function***
- ***Improper integrals involving more than one limit***
- ***Normal probability distributions***

The area of an infinite region

What is the area of the region between the curve $y = x^{-2}$ and the x -axis for $x \geq 1$ that is shown in Figure 1? To answer this question, we have to extend our concept of "area" since the region in Figure 1 extends infinitely far to the right.

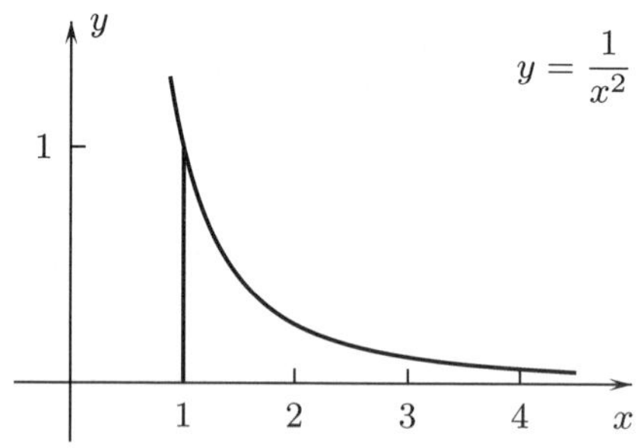

FIGURE 1

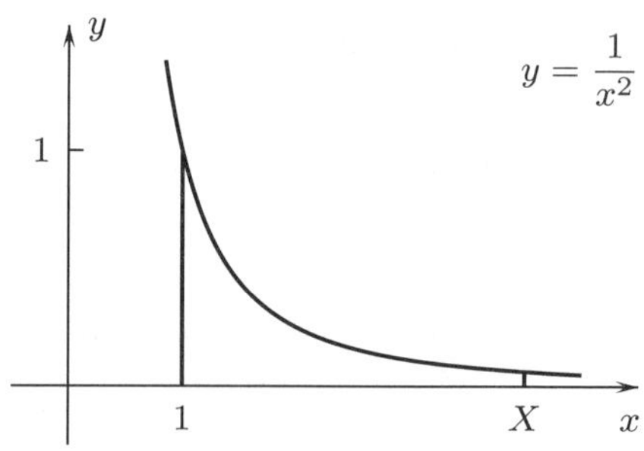

FIGURE 2

We calculate first the area of the region between $y = x^{-2}$ and the x -axis for $1 \leq x \leq X$ for fixed $X > 1$ (Figure 2). This area is given by the Riemann integral $\int_1^X \frac{1}{x^2}\,dx$, which we can evaluate with techniques of Chapter 4:

$$\begin{aligned}[\text{Area for } 1 \leq x \leq X] &= \int_1^X \frac{1}{x^2}\,dx = \int_1^X x^{-2}\,dx \\ &= \Big[-x^{-1}\Big]_1^X = \left[\frac{-1}{x}\right]_1^X = \frac{-1}{X} + 1.\end{aligned} \tag{1}$$

Question 1 What is the limit of area **(1)** as $X \to \infty$?

We define the area of the infinite region in Figure 1 for $x \geq 1$ to be the limit of the area for $1 \leq x \leq X$ that you found in Question 1, and we denote this limit by the symbol $\int_1^\infty \frac{1}{x^2}\,dx$. Then

---

♦ SUGGESTIONS TO INSTRUCTORS: The main topics in this section can be divided into two parts, as follows:
Part 1 (Improper integrals; Tune-up Exercises T1–T7; Problems 1–35, 40–43, 45–52) have students begin work on Questions 1–4 before class. Problems 26, 27, 35, 39 are good for in-class discussion.
Part 2 (Probability and normal distributions; Problems 36–39, 44) Have students begin work on Questions 5 through 7 before class. Examples 6 through 8 and Problems 36, 37, and 44 are good for in-class discussion .

we have

$$[\text{Area for } x \geq 1] = \int_1^\infty \frac{1}{x^2}\, dx = \lim_{X\to\infty} \int_1^X \frac{1}{x^2}\, dx$$
$$= \lim_{X\to\infty} [\frac{-1}{X} + 1] = 1. \qquad \textbf{(2)}$$

The integral $\int_1^\infty \frac{1}{x^2}\, dx$ in **(2)** is called an IMPROPER INTEGRAL because it is not a Riemann integral defined directly as a limit of Riemann sums.

We obtain a different sort of result in the following example.

Example 1 What is the area of the infinite region between $y = 1/x$ and the x-axis for $x \geq 1$, which is shown in Figure 3?

SOLUTION The area of the portion of the region for $1 \leq x \leq X$ that is shown in Figure 4 is given by the integral

$$\int_1^X \frac{1}{x}\, dx = \Big[\ln|x|\Big]_1^X = \ln(X) - \ln(1) = \ln(X).$$

Because $\ln(X) \to \infty$ as $X \to \infty$, this integral tends to ∞ as $X \to \infty$. We say that the area of the infinite region and the improper integral $\int_1^\infty \frac{1}{x}\, dx$ are both ∞ and write

$$[\text{Area for } x \geq 1] = \int_1^\infty \frac{1}{x}\, dx = \lim_{X\to\infty} \int_1^X \frac{1}{x}\, dx = \lim_{X\to\infty} \ln(X) = \infty. \square \qquad \textbf{(3)}$$

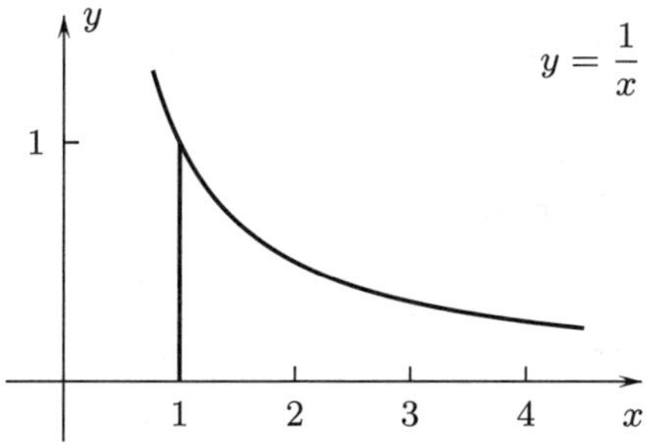

FIGURE 3

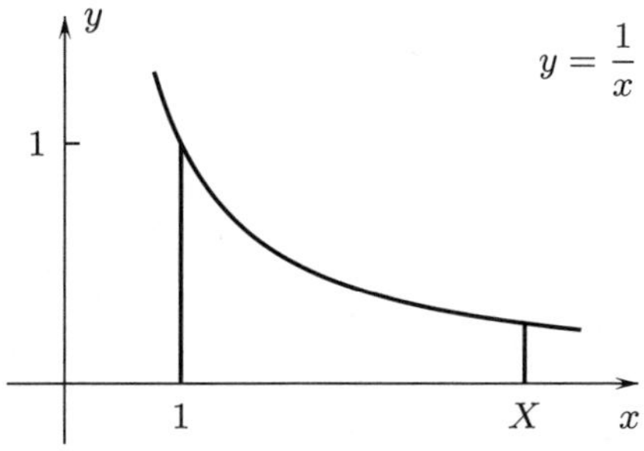

FIGURE 4

An improper integral of an unbounded function

The region between $y = 1/x^2$ and the x-axis for $0 < x \leq 1$ in Figure 5 extends infinitely high to the right of the y-axis. Its area is given by the integral $\int_0^1 \frac{1}{x^2}\, dx$, which is improper because $\frac{1}{x^2} \to \infty$ as $x \to 0^+$. To evaluate this integral, we first find the area of the bounded portion of the region with $X \leq x \leq 1$ for X between 0 and 1 that is shown in Figure 6. The area of this region is

$$\int_X^1 \frac{1}{x^2}\, dx = \int_X^1 x^{-2}\, dx = \Big[-x^{-1}\Big]_X^1 = \Big[\frac{-1}{x}\Big]_X^1 = -1 + \frac{1}{X}.$$

The area of the entire region is the limit of this area as $X \to 0^+$:

$$\begin{aligned}[\text{Area for } 0 < x \le 1] &= \int_0^1 \frac{1}{x^2}\, dx \\ &= \lim_{X\to 0+} \int_X^1 \frac{1}{x^2}\, dx = \lim_{X\to 0+} \left[\frac{-1}{x}\right]_X^1 \\ &= \lim_{X\to 0+} \left[-1 + \frac{1}{X}\right] = \infty.\end{aligned} \tag{4}$$

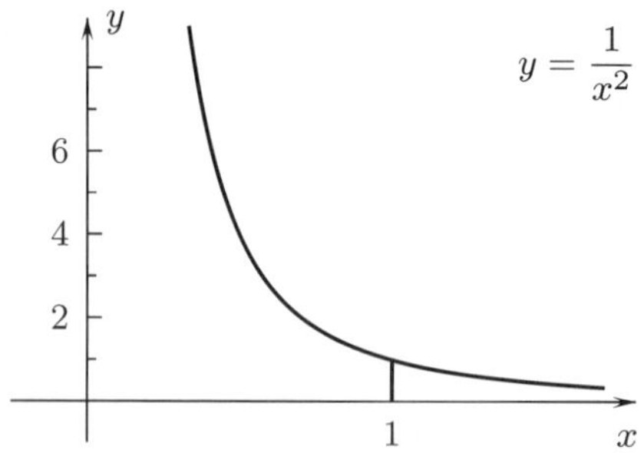

FIGURE 5

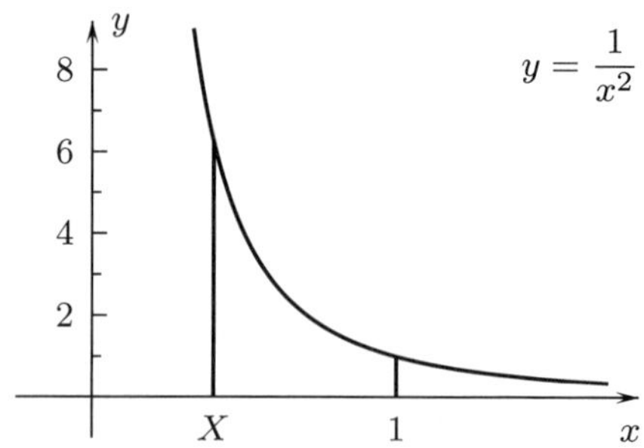

FIGURE 6

We say that the improper integral $\int_0^1 \frac{1}{x^2}\, dx$ equals ∞ and that the region in Figure 5 has infinite area.

Question 2 What is the area of the region between $y = 1/x$ and the x-axis for $0 < x \le 1$?

The formulas in **(2)**, **(3)**, and **(4)** and in the response to Question 2 illustrate the following general definition, where a and b are any finite numbers.

Definition 1 **(a)** *If the Riemann integral* $\int_a^X f(x)\, dx$ *is defined for all* $X > a$ *then*

$$\int_a^\infty f(x)\, dx = \lim_{X\to\infty} \int_a^X f(x)\, dx \tag{5}$$

provided the limit is defined, either as a finite number or as $\pm\infty$.

(b) *If the Riemann integral* $\int_X^b f(x)\, dx$ *is defined for all* $X < b$, *then*

$$\int_{-\infty}^b f(x)\, dx = \lim_{X\to-\infty} \int_X^b f(x)\, dx \tag{6}$$

provided the limit is defined, either as a finite number or as $\pm\infty$.

(c) *If* $\int_X^b f(x)\, dx$ *is defined as a Riemann integral for all* X *with* $a < X < b$ *but* $\int_a^b f(x)\, dx$ *is not, then*

$$\int_a^b f(x)\, dx = \lim_{X\to a+} \int_X^b f(x)\, dx \tag{7}$$

provided the limit is defined, either as a finite number or as $\pm\infty$.

(d) *If* $\int_a^X f(x)\,dx$ *is defined as a Riemann integral for all* X *with* $a < X < b$ *but* $\int_a^b f(x)\,dx$ *is not, then*

$$\int_a^b f(x)\,dx = \lim_{X\to b^-} \int_a^X f(x)\,dx \qquad \textbf{(8)}$$

provided the limit is defined, either as a finite number or as $\pm\infty$.

If the limit in **(5)** through **(8)** is a finite number, we say that the improper integral CONVERGES. The improper integral is said to DIVERGE TO ∞ if the limit is ∞ and to DIVERGE TO $-\infty$ if the limit is $-\infty$. If the limit is not defined, we say that the improper integral DIVERGES WITHOUT A LIMIT.

Example 2 Does $\int_0^\infty \cos x\,dx$ converge or diverge?

SOLUTION The improper integral $\int_0^\infty \cos x\ dx$ equals $\lim_{X\to\infty}\int_0^X \cos x\ dx$ if it is defined. However, because

$$\int_0^X \cos x\,dx = \Big[\sin x\Big]_0^X = \sin(X) - \sin(0) = \sin(X)$$

oscillates between -1 and 1 and does not have a limit as $X\to\infty$, $\int_0^\infty \cos x\ dx$ is not defined. It diverges without a limit. □

Improper integrals involving more than one limit

If the definition of an improper integral would require finding more than one limit, we write the integral as the sum of integrals, each of which involves only one limit and add the results, using the following rules if any of the integrals are ∞ or $-\infty$:

$$\begin{aligned} \infty + \infty &= \infty \\ -\infty - \infty &= -\infty \\ \infty + [\text{A finite number}] &= \infty \\ -\infty + [\text{A finite number}] &= -\infty. \end{aligned} \qquad \textbf{(9)}$$

The expression $\infty - \infty$ is not defined. Consequently, the sum of one improper integral that equals ∞ and another that equals $-\infty$ is not defined.

Example 3 What is the area of the infinite region between $y = \dfrac{1}{x^2+1}$ and the x-axis that is shown in Figure 7?

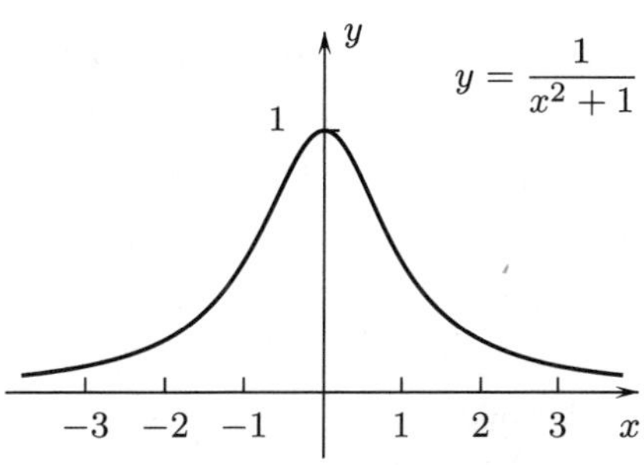

FIGURE 7

SOLUTION The area is given by the improper integral $\int_{-\infty}^{\infty} \frac{1}{x^2+1}\, dx$, which involves two limits because of the two infinite limits of integration. We first study separately the improper integrals from $-\infty$ to 0 and from 0 to ∞. (We could use any finite number instead of 0 here.) Because the antiderivative of $\frac{1}{x^2+1}$ is $\tan^{-1} x + C$ and $\tan^{-1} x$ tends to $\frac{1}{2}\pi$ as $x \to \infty$ and tends to $-\frac{1}{2}\pi$ as $x \to -\infty$, we obtain

$$\int_0^{\infty} \frac{1}{x^2+1}\, dx = \lim_{X\to\infty} \int_0^X \frac{1}{x^2+1}\, dx = \lim_{X\to\infty} \Big[\tan^{-1} x\Big]_0^X$$
$$= \lim_{X\to\infty} \tan^{-1}(X) = \tfrac{1}{2}\pi$$

and

$$\int_{-\infty}^{0} \frac{1}{x^2+1}\, dx = \lim_{X\to-\infty} \int_X^0 \frac{1}{x^2+1}\, dx = \lim_{X\to-\infty} \Big[\tan^{-1} x\Big]_X^0$$
$$= \lim_{X\to-\infty} [-\tan^{-1}(X)] = -(-\tfrac{1}{2}\pi) = \tfrac{1}{2}\pi.$$

Since both of these improper integrals converge, so does their sum and

$$\int_{-\infty}^{\infty} \frac{1}{x^2+1}\, dx = \int_{-\infty}^{0} \frac{1}{x^2+1}\, dx + \int_0^{\infty} \frac{1}{x^2+1}\, dx$$
$$= \tfrac{1}{2}\pi + \tfrac{1}{2}\pi = \pi. \ \square$$

Example 4 Does $\int_0^{\infty} \frac{1}{x}\, dx$ converge or diverge?

SOLUTION Evaluating this improper integral requires two limits because of the infinite limit of integration and because $1/x \to \infty$ as $x \to 0^+$. We consider it to be the sum of the integrals from 0 to 1 and 1 to ∞. We saw in Example 1 that $\int_1^{\infty} \frac{1}{x}\, dx = \infty$ and in Question 2 that $\int_0^1 \frac{1}{x}\, dx = \infty$. Consequently,

$$\int_0^{\infty} \frac{1}{x}\, dx = \int_0^1 \frac{1}{x}\, dx + \int_1^{\infty} \frac{1}{x}\, dx = \infty + \infty = \infty.$$

The integral from 0 to ∞ diverges to ∞. $\square$

Question 3 Is $\int_{-1}^{1} \frac{1}{x}\, dx$ defined? If so, what is its value?

The average value of an unbounded function

We saw above how improper integrals can be used to find areas. They are also used in other applications where we have previously used Riemann integrals.

Example 5 What is the average value of $x^{-1/3}$ for $0 \le x \le 8$?

SOLUTION By Rule 6 of Section 4.6, the average value is

$$\tfrac{1}{8}\int_0^8 x^{-1/3}\, dx. \tag{10}$$

Because $x^{-1/3} \to \infty$ as $x \to 0^+$, the integral is improper. It is equal to the limit of the integral from X to 8 as $X \to 0^+$:

$$\begin{aligned}[\text{Average value}] &= \lim_{X\to 0+} \tfrac{1}{8}\int_X^8 x^{-1/3}\,dx = \tfrac{1}{8}\lim_{X\to 0+}\left[\tfrac{3}{2}x^{2/3}\right]_X^8 \\ &= \tfrac{1}{8}\lim_{X\to 0+}\left(\tfrac{3}{2}8^{2/3} - \tfrac{3}{2}X^{2/3}\right) \\ &= \tfrac{1}{8}\left(\tfrac{3}{2}\right)\left(8^{2/3}\right) = \tfrac{3}{4}. \square\end{aligned}$$

Flipping coins and the normal probability distribution

One important application of improper integrals is in the analysis of continuous PROBABILITY DISTRIBUTIONS.To explain why improper integrals are needed in this context, we will consider first a few examples of probabilities of possible outcomes from the repeated flipping of a coin.

Example 6 Suppose a coin with a head side and a tail side is flipped two times. What are the possible outcomes? Assuming that the coin is fair, so that a head or a tail occur with equal probability, what is the probability that no heads are thrown, that one head and one tail is thrown, or that two heads are thrown?

SOLUTION The possible outcomes of flipping the coin twice are

{T, T} {T, H} {H, T} {H, H}

where **T** represents a tail, **H** represents a head, the first letter gives the result of the first throw, and the second letter gives the result of the second throw.

Because each of the four possible outcomes has equal probability there is one chance in four, or a probability of $\frac{1}{4}$ (25%), that no heads (two tails) occur, two chances in four, or a probability of $\frac{2}{4} = \frac{1}{2}$ (50%) , that one head and one tail occur, and one chance in four, or a probability of $\frac{1}{4}$ (25%), that two heads occur. □

Notice that the sum of the probabilities from Example 6 is 1: $\frac{1}{4} + \frac{1}{2} + \frac{1}{4} = 1$. The percentages are plotted in Figure 8. Their sum is 100.

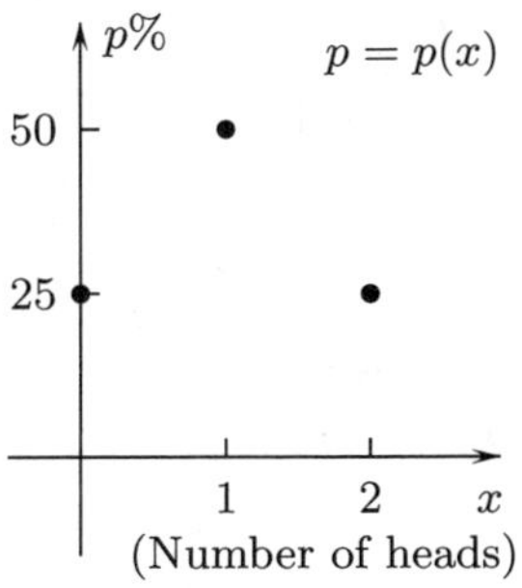

Two flips

FIGURE 8

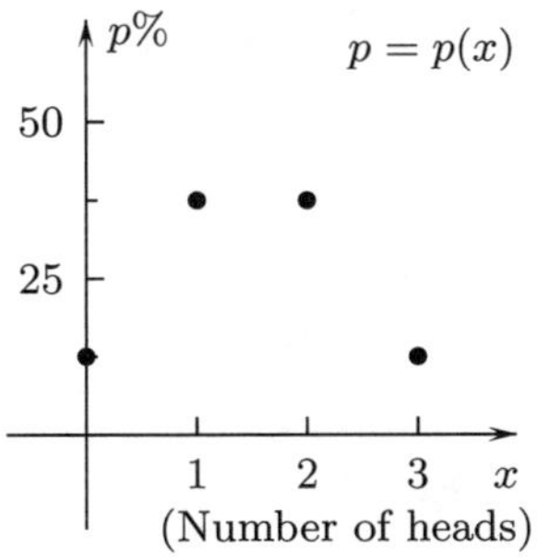

Three flips

FIGURE 9

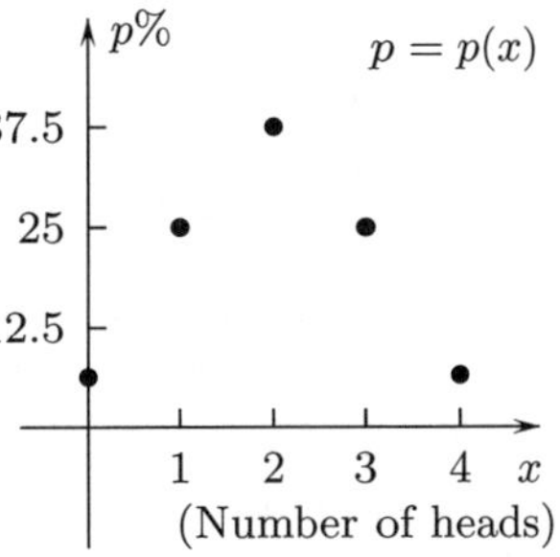

Four flips

FIGURE 10

Question 4 Figure 9 shows the graph of the probability that x heads occur when a fair coin is flipped three times. Find the four values of this function. What is their sum?

With four flips of the coin there would be sixteen possible outcomes:

{T, T, T, T}

{T, T, T, H} {T, T, H, T} {T, H, T, T} {H, T, T, T}

{T, T, H, H} {T, H, T, H} {T, H, H, T} {H, T, T, H} {H, T, H, T} {H, H, T, T}

{H, H, H, T} {H, H, T, H} {H, T, H, H} {T, H, H, H}

{H, H, H, H}

In this case, there is one chance in sixteen (6.25%) that no heads occur, four chances in sixteen (25%) that exactly one head occurs, six chances in sixteen (37.5%) that exactly two heads tail occur, four chances in sixteen (25%) that exactly three heads occur, and one chance in sixteen (6.25%) that four heads occur. Consequently the function of Figure 10 has the values $p(0) = 6.25, p(1) = 25, p(2) = 37.5, p(3) = 25$, and $p(4) = 6.25$. Notice that the sum of these percentages is also 100.

Figures 11 and 12 show the graphs of the probability of getting x heads with sixteen flips and twenty-five flips of the coin, respectively.

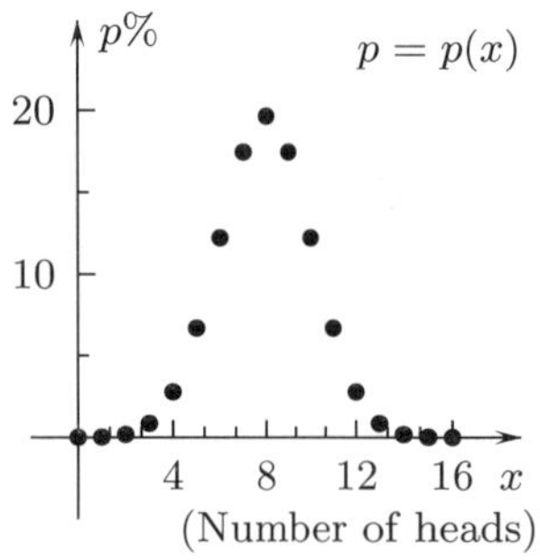

Sixteen flips

FIGURE 11

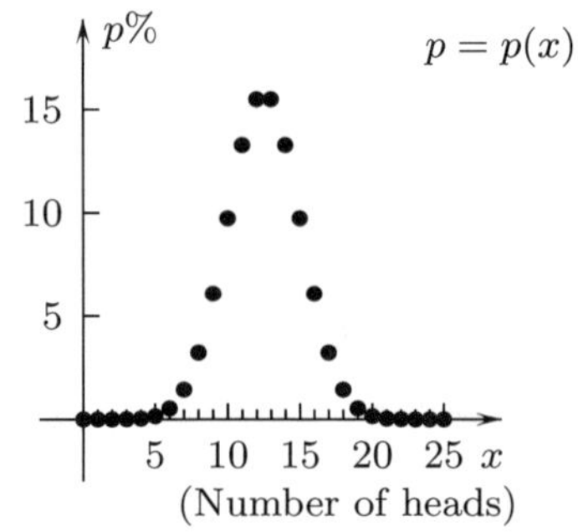

Twenty-five flips

FIGURE 12

Formulas for probabilities such as in Figures 11 and 12 are discussed in courses in probability and statistics. Here we are mainly interested in the approximations given by continuous curves as in Figures 13 through 15 for four, sixteen, and twenty-five flips of the coin.

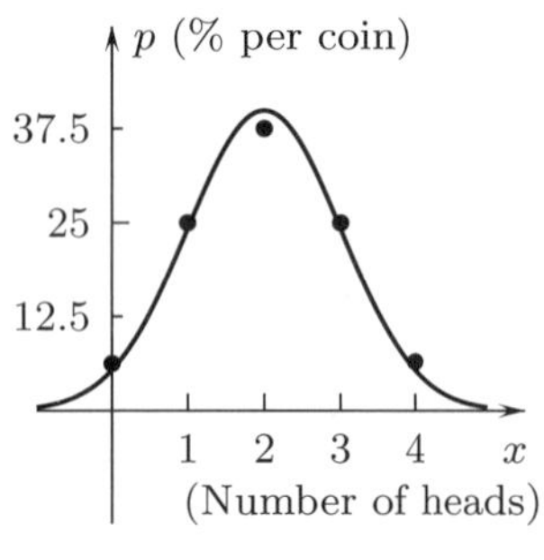

Four flips

FIGURE 13

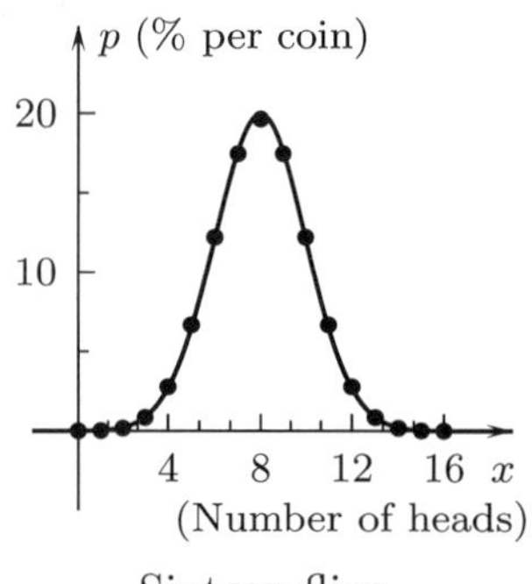

Sixteen flips

FIGURE 14

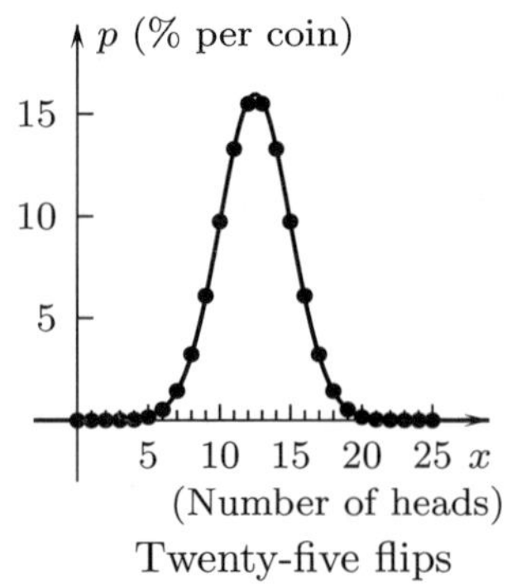

Twenty-five flips

FIGURE 15

The units of p on the vertical axes in Figures 13 through 15 are percent per coin, rather than percent, as in Figures 10 through 12, to allow for the continuous distributions. This does not change the meanings of the dots because, in each case, there is one dot for each integer value of x.

The curves in Figures 13 through 15 are graphs of the function

$$n(x) = \frac{1}{\sigma\sqrt{2\pi}} e^{-(x-\mu)^2/(2\sigma^2)} \tag{11}$$

for different choices of the constants μ (mu) and σ (sigma). This function is the NORMAL PROBABILITY DISTRIBUTION and its graph is the "bell curve" from probability.

Notice that the normal distribution in each case is defined for all x and that its approximation of the probabilities from coin tosses is more accurate for sixteen and twenty-five throws than for four throws. In fact, the accuracy of such approximations increases as the number of throws increases.

The value μ in formula **(11)** is the value of x at the highest point on the graph because $e^{-(x-\mu)^2/2}$ has the value 1 at $x = \mu$ and is less than 1 for $x \neq \mu$. The number μ is the AVERAGE or MEAN of the distribution. Thus, $\mu = 2$ for the distribution in Figure 13, $\mu = 8$ in Figure 14, and $\mu = 12.5$ in Figure 15. As is illustrated in Figures 13 through 15, the graph is symmetric about the vertical line $x = \mu$. The function $n(x)$ is positive for all x, but tends to zero very fast as $x \to \pm\infty$.

The number σ is called the STANDARD DEVIATION of the distribution. It determines how spread out its graph is. The graph is steeper and narrower for small σ and is flatter and more spread out for large σ. In every case, about 68% of the area under the graph is for x within a standard deviation of the mean, i. e., for $\mu - \sigma < x < \mu + \sigma$. The standard deviation also has a special geometric property, which is examined in the next question.

Question 5 Figure 16 shows the graph of the normal distribution

$$n(x) = \frac{1}{\sigma\sqrt{2\pi}} e^{-x^2/(2\sigma^2)} \tag{12}$$

with mean $\mu = 0$ and standard deviation σ. Find its points of inflection.

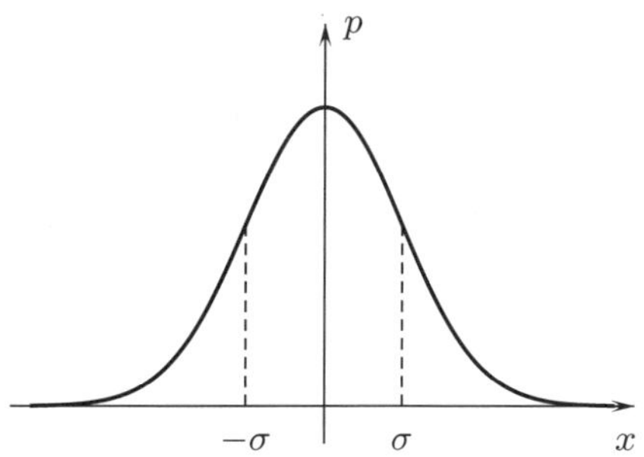

$p = n(x)$

$= \frac{1}{\sigma\sqrt{2\pi}} e^{-x^2/(2\sigma^2)}$

FIGURE 16

Probabilities are determined from probability distributions by using the following rule:

Rule 1 *If the outcome x of an experiment has the probability distribution $p(x)$, then the probablity that the outcome x is in the interval $[a, b]$ is*

$$[\text{Probability that } a \leq x \leq b] = \int_a^b p(x)\, dx.$$

Example 7 The normal distribution in Figure 14 is given by **(11)** with $\mu = 8$ and $\sigma = 2$. Based on this approximation, what is the probablility that in sixteen throws of the dice four or fewer heads occur?

SOLUTION Since $\mu = 8, \sigma = 2$, and $2\sigma^2 = 8$, the approximate probability is

$$\int_0^4 n(x)\, dx = \frac{1}{2\sqrt{2\pi}} \int_0^4 e^{-(x-8)^2/8}\, dx.$$

An integration algorithm on a calculator gives the approximate value 0.023 for this integral. There is about a 2.3% probability that four or fewer heads occur in 16 throws of the coin. □

Now we can explain why improper integrals are needed in studying the normal distribution and other probablity distributions. The probability that one of all possible results of an experiment occurs is 1. Therefore, because the normal distribution $n(x)$ is defined for all x, we should have

$$\int_{-\infty}^{\infty} n(x)\, dx = 1. \tag{13}$$

We cannot verify this result with the tools at our disposal, so will just verify it approximately for $\mu = 0$ and two values of σ.

Question 6 Use an integration algorithm on your calculator to find the approximate values of

(a) $\int_{-3}^{3} n(x)\,dx$ and $\int_{-5}^{5} n(x)\,dx$ for the normal distribution **(12)** with $\sigma = 1$ and

(b) $\int_{-15}^{15} n(x)\,dx$ and $\int_{-25}^{25} n(x)\,dx$ for the normal distribution **(12)** with $\sigma = 5$.

The calculations in the Response to Question 6 corroborate that in both cases the improper integral from $-\infty$ to ∞ is 1.

Modeling with a normal distribution

Figure 17 shows the distribution according to blood-cholesterol level of the participants in study of diet and cardiovascular disease in the United States. The graph is obtained from data at $x = 140, 180, 200, \ldots, 240$ by linear interpolation: points on the graph at these values of x are joind by line segments.[(1)] Figure 18 shows an approximation of the data in Figure 17 by a normal distribution, which can be used as a model of the original data.

Example 8 The inflection points of the normal curve in Figure 18 are shown by two dots. **(a)** Use the graph to determine the approximate mean and standard deviation of this distribution. **(b)** Many U.S. doctors recommend that their patients keep the blood-cholersterol levels below 200 milligrams per deciliter. Based on the normal distribution, what approximate percentage of the participants in the Framingham study had blood-cholersterol levels below 200?

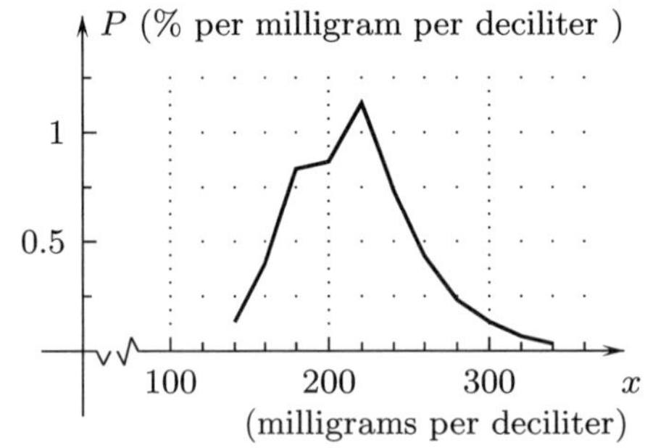

FIGURE 17

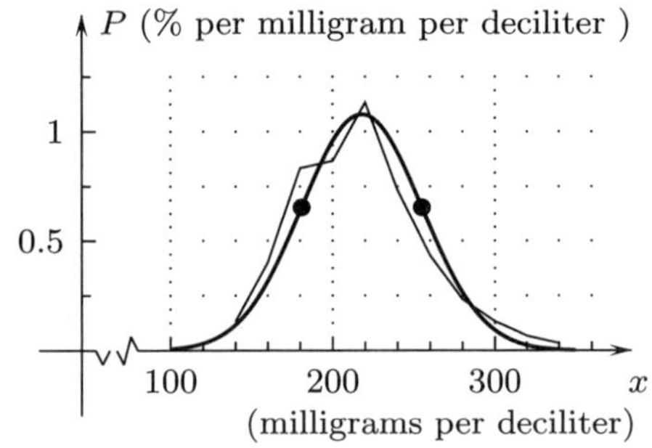

FIGURE 18

SOLUTION **(a)** The highest point on the normal curve is Figure 18 is at $x = \mu \approx 218$ and the horizontal distance from $x = \mu$ to the inflection points is $\sigma \approx 37$.
(b) Based on the normal distribution with mean 218 and standard deviation 37, the fraction with blood-cholesterol less than 200 milligrams per deciliter is

$$\frac{1}{37\sqrt{2\pi}} \int_{-\infty}^{200} e^{-(x-218)^2/(2(37)^2)}\,dx \approx \frac{1}{37\sqrt{2\pi}} \int_{-100}^{200} e^{-(x-218)^2/(2(37)^2)}\,dx \doteq 0.313.$$

Approximately 31.3% of the participants in the study had blood-cholesterol levels less than 200 milligrams per deciliter. □

[(1)]Data adapted from *The Framingham Study: An Epidemiological Investigation of Cardiovascular Disease*, Section 31W. Kannel and T. Gordon, editors, Washington, DC: U.S. Department of Health, Education, and Welfare, 1976, p.???.

Question 7 The blood-cholesterol level of men in a rural area of northwestern China was measured to have a mean of 148 milligrams per deciliter with a standard deviation of 33 milligrams per deciliter.[(2)] **(a)** Give a formua for this distribution under the assumption that it was normal and generate its graph. **(b)** Based on this mathematical model, what percentage of the men in the in the study had blood-cholesterol levels below 200?

Principles and procedures

- If an integral of $f(x)$ is not defined directly as a limit of Riemann sums, either because the interval of integration is infinite or because $f(x)$ is not bounded as x approaches a finite endpoint of the interval of integration, the integral might be defined as an improper integral involving one limit of a Riemann integral by applying Definition 1. In other cases, the integral can be defined by writing it as the sum of two or more such improper integrals and applying rules **(9)**.
- An improper integral is said to converge if it defined and finite, to diverge to ∞ is it is defined to be ∞, to diverge to $-\infty$ is it is defined to be $-\infty$, and to diverge without a limit is it is not defined.
- Improper integrals can be used to find areas, volumes, and average values and in other applications where we have used Riemann integrals previously.
- The "bell curves" of probability are the graphs of the normal probability distributions

$$n(x) = \frac{1}{\sigma\sqrt{2\pi}}e^{-(x-\mu)^2/(2\sigma^2)}.$$

The parameter μ here is the average or mean of the distribution and is the value of x at the center and highest point of the graph. The distribution is concentrated near the mean if the standard deviation σ is small and is spread out if σ is large. The constant $\dfrac{1}{\sigma\sqrt{2\pi}}$ is included in the definition of $n(x)$ so that $\int_{-\infty}^{\infty} n(x)\,dx = 1$.

- Normal distrbutions can be shown to provide very accurate approximations for distribution of special types of probabilities, such as arise from flipping coins or rolling dice, and are used as models for many other types of phenenoma.

Tune-Up Exercises 6.6[♦]

[A]*Answer provided.* [O]*Outline of solution provided.*

T1.[O] Does $\displaystyle\int_3^{\infty} \frac{1}{x^3}\,dx$ converge, diverge to ∞, diverge to $-\infty$ or diverge without a limit? If it converges, give its value.

♦ Type 1, an improper integral

T2.[O] Does $\displaystyle\int_0^{3} \frac{1}{x^3}\,dx$ converge, diverge to ∞, diverge to $-\infty$ or diverge without a limit? If it converges, give its value.

♦ Type 1, an improper integral

T3.[O] Evaluate $\displaystyle\int_{-\infty}^{\infty} xe^{-x^2}\,dx$.

♦ Type 1, an improper integral involving two limits

[(2)]Data adaped from "Nutritional habits and serum lipid levels in a low-fat intake Chinese population sample", J. Xie, L, Liu, J Huang, H. Hu, and H. Kesteloot, sl Acta Cardiologica, No. 6, Brussels: Belgian Society of Cardiology, December, 1998, p. 361.

[♦]The Tune-up Exercises and Problems are classified by type and content. The types are (1) basic, reactive; (2) basic reflective; (3) intermediate, reactive; (4) intermediate, reflective; (5) advanced, reactive; (6) advanced, reflective; and (7) advanced, theoretical.

T4.[O] Evaluate $\int_0^9 \frac{1}{\sqrt{x}}\, dx$.

♦ Type 1, an improper integral

T5.[O] What is the area of the region between $y = 3x^{-2/3}$ and the x-axis for $-8 \le x \le 27$?

♦ Type 1, area

T6.[O] Does $\int_{-\infty}^{\infty} e^x\, dx$ converge, diverge to ∞, diverge to $-\infty$ or diverge without a limit? If it converges, give its value.

♦ Type 1, an improper integral involving two limits

T7.[A] Suppose that $\int_{-\infty}^{-5} g(x)\, dx = 7$, $\int_{-5}^{5} g(x)\, dx = -\infty$, $\int_{5}^{10} g(x)\, dx = 21$, and $\int_{10}^{\infty} g(x)\, dx = \infty$. Which of $\int_{-\infty}^{10} g(x)\, dx$ and $\int_{-5}^{\infty} g(x)\, dx$ is defined?

♦ Type 1, an improper integral, etc.

Problems 6.6

[A]Answer provided. [O]Outline of solution provided.

In Problems 1 through 24 determine whether the given improper integrals converge, diverge to ∞, diverge to $-\infty$, or diverge without a limit. Find the values of those that converge.

1.[A] $\int_1^{\infty} e^{-10x}\, dx$

♦ Type 1, an improper integral

2. $\int_{-\infty}^{0} xe^{-x^2}\, dx$

♦ Type 1, an improper integral

3.[A] $\int_0^{\infty} \frac{x}{1+x^2}\, dx$

♦ Type 1, an improper integral

4.[A] $\int_5^{\infty} \sin x\, dx$

♦ Type 1, an improper integral

5. $\int_{-\infty}^{1} \frac{1}{x^2+1}\, dx$

♦ Type 1, an improper integral

6.[A] $\int_0^{27} x^{-1/3}\, dx$

♦ Type 1, an improper integral

7. $\int_0^{\pi/4} \tan x\, dx$

♦ Type 1, an improper integral

8. $\int_{-\infty}^{\infty} x^3\, dx$

♦ Type 3, an improper integral involving two limits

9.[A] $\int_1^{2} \frac{x}{\sqrt{x^2-1}}\, dx$

♦ Type 1, an improper integral

10.[A] $\int_{-32}^{0} x^{-4/5}\, dx$

♦ Type 1, an improper integral

11. $\int_{-8}^{27} x^{-1/3}\, dx$

♦ Type 3, an improper integral involving two limits

12.[A] $\int_{-5}^{5} x^{-2}\, dx$

♦ Type 3, an improper integral involving two limits

13. $\int_{-2}^{2} x^{-5}\, dx$

♦ Type 3, an improper integral involving two limits

14.[A] $\int_{-\infty}^{\infty} \frac{e^x}{(1+e^x)^2}\, dx$

♦ Type 3, an improper integral involving two limits

15. $\int_{-\infty}^{\infty} \frac{e^{-x}}{1+e^{-x}}\, dx$

♦ Type 3, an improper integral involving two limits

16. $\int_0^{1} \frac{1}{(1-x)^{1/3}}\, dx$

♦ Type 1, an improper integral

17.[A] $\int_0^{4} (4-x)^{-1/2}\, dx$

♦ Type 1, an improper integral

18. $\int_0^{1} (1-x^2)^{-1/2}\, dx$

♦ Type 1, an improper integral

19.[A] $\int_0^{4} \frac{1}{\sqrt{16-x^2}}\, dx$

♦ Type 1, an improper integral

20. $\int_{-\infty}^{\infty} \frac{1}{x^2+25}\, dx$

♦ Type 3, an improper integral involving two limits

21. $\displaystyle\int_{10}^{\infty} (x-2)^{-4/3}\,dx$

♦ Type 1, an improper integral

22. $\displaystyle\int_{0}^{2} (x-2)^{-4/3}\,dx$

♦ Type 1, an improper integral

23. $\displaystyle\int_{e}^{\infty} \frac{1}{x(\ln x)^3}\,dx$

♦ Type 1, an improper integral

24. $\displaystyle\int_{0}^{1/e} \frac{\ln x}{x}\,dx$

♦ Type 1, an improper integral

25.[O] Sketch the region between $y = \frac{1}{3}x$ and $y = \frac{1}{3}x + 5/x^2$ for $x \geq 1$ and find its area.

♦ Type 1, area

26.[A] Give a formula for the area between $y = x^{-3}$ and the x-axis for $x \geq a$ with $a > 0$.

♦ Type 1, area

27. Sketch the region above $y = x$ and below $y = 8/\sqrt{x}$ for $x > 0$ and find its area.

♦ Type 1, area

28. Find the area of the region between the curves $y = x^{-1/2}$ and $y = -x^{-1/2}$ for $x > 9$.

♦ Type 1, area

29. What is the area of the region between $y = \dfrac{\cos x}{\sqrt{\sin x}}$ and the x-axis for $0 < x \leq \frac{1}{2}\pi$?

♦ Type 1, area

30.[A] What is the area of the region between $y = \dfrac{1}{1+x^2}$ and $y = \dfrac{1}{4+x^2}$?

♦ Type 1, area

31. Find the area of the region above $y = x^2$ and below $y = x^{-2/3}$.

♦ Type 1, area

32. What is the area of the region between the graph of $\tan x$ and the x-axis for $0 \leq x < \frac{1}{2}\pi$?

♦ Type 3, area

33.[A] If we take the radius of the earth to be 4000 miles, then the force of the earth's gravity on a rocket ship that weighs 1000 pounds on the surface of the earth is $16r^{-2}$ thousand pounds when the ship is r thousand miles above the center of the earth ($r \geq 4$). In order for the ship to escape the earth's gravity (i. e., for it to be lifted an infinite distance above the earth), its engines must to $\displaystyle\int_4^{\infty} \frac{16}{r^2}\,dr$ million mile-pounds of work against the earth's gravity. What is the value of this integral?

♦ Type 2, improper integral, etc.

34. Find $k > 0$ such that the area of the region between $y = \dfrac{1}{x-1}$ and $y = \dfrac{1}{x}$ for $x \leq k$ is $\ln(1.25)$.

♦ Type 3, area

35.[A] Find two values of the number k such that $\displaystyle\int_k^{\infty} \frac{x}{x^4+1}\,dx = \frac{1}{8}\pi$. Give a geometric explanation of why there are two answers.

♦ Type 4, improper integral, etc.

36.[A] In a mathematical model of the failure of a computer component, it is assumed that the probability that the component will fail after a years and before b years ($0 \leq a < b$) is $\displaystyle\int_a^b ke^{-kx}\,dx$ with a positive constant k. **(a)** With this model, what is the probability that the component will eventually fail? **(b)** What is the constant k if there is a one-fifth chance that the component will fail during the first year? **(c)** What is the probability, under the conditions of part (b), that the component will fail during the second year?

♦ Type 4, improper integral, etc.

37. A die is a cube with the numbers 1 through 6 on its six sides. It is fair if each side is equally likely to be rolled (be on top) when the die is thrown. In this case, the probability of rolling a 6 is $\frac{1}{6}$ and the probability not rolling a 6 is $\frac{5}{6}$. **(a**[A]**)** What is the distribution of probabilities that zero, one, or two 6's will be rolled when a fair die is rolled two times? (The possible outcomes are {**not 6, not 6**} {**6, not 6**} {**not 6, 6**} and {**6, 6**}.) **(b)** The dots in Figure 19 show the probability distribution that x 6's are rolled when a fair die is rolled 24 times. The curve is an approximating normal curve with mean 4 and standard deviation $\sqrt{5}$. Based on this approximation, what is the probability of rolling eight or more 6's in twenty-four rolls of a fair die?

♦ Type 4, a normal distribution, etc.

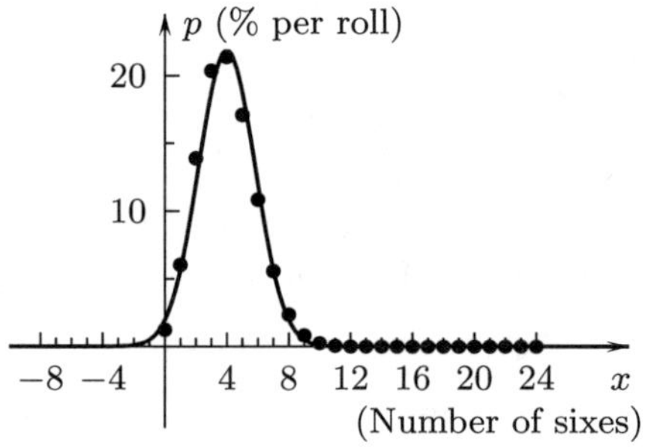

FIGURE 19

38. Figure 20 shows the graph of a CAUCHY DISTRIBUTION $n(x) = \frac{a}{\pi}\frac{1}{x^2 + a^2}$ with mean 0. The positive parameter a determines how spread out the curve is. **(a)** What is $\int_{-\infty}^{\infty} n(x)\,dx$? **(b)** Figure 21 shows the graph of the Cauchy distribution with the normal curve with mean 0 and standard deviation 1. The area between the normal curve and the x-axis for $-1 \le x \le 1$ is approximatelly 68.3% of the area under the entire curve. As is suggested in Figure 21, the constant a in the Cauchy distribution has been chosen so that the portion of the area between it and the x-axis for $-1 \le x \le 1$ is also 68.3%. What is the approximate value of the constant a in this case. **(c)** Describe how the Cauchy distribution differs from the normal distribution.

♦ Type 4, a Cauchy distribution.

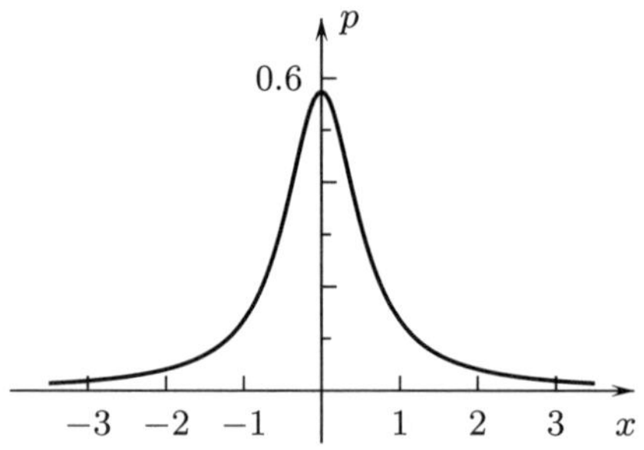

FIGURE 20

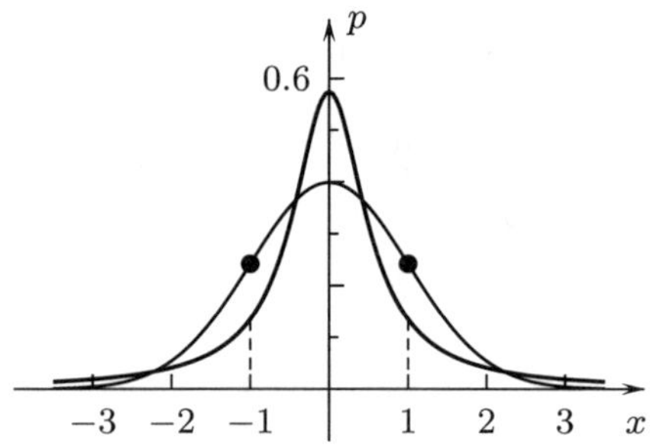

FIGURE 21

39. A solid is obtained by rotating about the x-axis the region between $y = 1/x$ and the x-axis for $x \ge 1$. **(a)** Show that the solid has finite volume. **(b)** Use the inequality $1 + x^{-2} \ge 1$ for $x \ge 1$ to show that the solid has infinite surface area. **(c)** The results of parts (a) and (b) seem contradictory because they suggest you could fill the solid but not paint its surface with a finite amount of paint. Explain.

♦ Type 4, improper integral, etc.

40.[A] For what values of k does $\int_1^{\infty} \frac{\ln x}{x^k}\,dx$ converge and for what values does it diverge to ∞?

♦ Type 3, improper integral, etc.

41.[A] For what values of k does $\int_0^1 x^k \ln x\,dx$ converge and for what values does it diverge to ∞?

♦ Type 3, improper integral, etc.

42. Find the smallest positive integer X such that $\int_0^X \frac{1}{x^2 + 16}\,dx > 0.3925$.

♦ Type 3, improper integral, etc.

43. Find $k > e$ such that $\int_e^k \frac{1}{x\sqrt{\ln x - 1}}\,dx = \frac{2}{3}$.

♦ Type 3, improper integral, etc.

44. In a study of diets of U.S. residents it is determined that the distribution of fat intake by the men was 136 grams per day with a standard deviation of 37 grams per day and the distribution of fat intake by the women was 96 grams per day with a standard deviation of 31 grams per day. Under the assumption that the distributions were normal, what percentage of the women ate more fat per day than the average daily fat intake for men?[(3)]

♦ Type 4, a normal distribution, etc.

Work Problems 45 through 49 by using integration techniques from Sections 6.2 and 6.3, a table of integrals, or a calculator- or computer-algebra system.

45.[O] Does $\int_1^2 \frac{1}{\sqrt{x^2-1}}\, dx$ converge? If so, give its value.

♦ Type 1 with a CAS or integral table, type 3 otherwise, improper integral

46.[A] Find two values of k such that $\int_0^\infty (4x - x^2)e^{-kx}\, dx = 2$.

♦ Type 2 with a CAS or integral table, type 4 otherwise, improper integral

47. Give a formula for the value of $\int_0^\infty \sin x\, e^{-kx}\, dx$ for the values of k such that the integral converges, and show that it diverges for the other values of k.

♦ Type 1 with a CAS or integral table, type 3 otherwise, improper integral

48. Show that $\int_0^\infty \frac{1}{(x^2+1)^{3/2}}\, dx = 1$ by finding an antiderivative of $\frac{1}{(x^2+1)^{3/2}}$.

♦ Type 1 with a CAS or integral table, type 3 otherwise, improper integral

49. Evaluate $\int_{13}^\infty \frac{1}{x^3 - 9x^2 + 27x - 27}\, dx$.

♦ Type 2 with a CAS or integral table, type 4 otherwise, improper integral

50. Prove that $\lim_{X\to\infty} \int_{-X}^{X} \sin x\, dx = 0$ although $\int_{-\infty}^{\infty} \sin x\, dx$ is not defined.

♦ Type 3, improper integral, etc.

51. Show that the change of variables $u = 1/x$ transforms the improper integral $\int_2^\infty x^{-3}\, dx$ into a Riemann integral and evaluate both integrals.

♦ Type 3, improper integral, etc.

52. Show that the change of variables $u = x^{1/3}$ transforms the improper integral $\int_0^8 x^{-1/3}\, dx$ into a Riemann integral. Evaluate both integrals.

♦ Type 3, improper integral, etc.

---

(3) Data adapted from *The Framingham Study: An Epidemiological Investigation of Cardiovascular Disease*, Section 24, W. Kannel and T. Gordon, editors, Washington, DC: U.S. Department of Health, Education, and Welfare, 1976, p.???.

Section 6.7

Work, energy, and forces of fluids

OVERVIEW: *In this section, we first discuss the concept of* WORK, *as used by physicists to describe transfers of energy. Then we study how work is used to describe changes in* KINETIC *and* POTENTIAL ENERGY *of moving objects. At the end of the section, we use integrals to calculate forces caused by the pressure of water or other fluids against flat surfaces such as dams, sides of containers, and sides of submerged objects.*

Topics:

- ***Work done by a constant force***
- ***Work done by a variable force***
- ***Work lifting water***
- ***Kinetic energy of point masses***
- ***Potential energy***
- ***Forces on dams***

Work done by a constant force

For a rocket to reach a point in its upward flight, energy is required (i) to lift it above the earth, (ii) to give it upward velocity, and (iii) to overcome the forces of air resistance. As we will see, lifting the rocket increases its POTENTIAL ENERGY and giving it an upward velocity increases its KINETIC ENERGY. The energy used to overcome air resistance is dissipated as heat. These transfers of energy are described by saying that the rocket's engine does WORK on the rocket and that the rocket does work against the forces of gravity and air resistance.

We begin by defining the work done an object that moves on an s-axis subject to a constant force exerted parallel to the axis. Then we will use this definition and the definition of the Riemann integral to determine the integrals that give work under nonconstant forces.

Definition 1 *The* WORK *done by a constant force F in the positive s-direction on an object as it moves from $s = a$ to $s = b$ on an s-axis is*

$$[\text{Work}] = [\text{Force}][\text{Net distance moved}] = F(b - a). \qquad \textbf{(1)}$$

In applying Definition 1, we represent forces in the negative s-direction by negative numbers. For example, a force $F = -10$ pounds in the positive direction represents a force of 10 pounds in the negative direction. The net distance moved, $b - a$, is positive if $b > a$ and is negative if $b < a$. Consequently, if the force and motion are in the same direction, then F and $b - a$ are both positive or both negative and the work is positive. If the motion and force are in opposite directions, then F and $b - a$ have opposite signs and the work is negative.

Notice that in Definition 1, there is no reference to the nature of the object, to the nature of the force, or to the amount of time it takes for the motion to occur. Also, the force being considered might be just one of a number of forces applied to the object during the motion.

The units for measuring work are those used for force multiplied by those used for distance. If the force is given in pounds and the distance in feet, work is given in foot-pounds. If the force is given in newtons and the distance in meters, work is given in newton-meters or JOULES. For force given in dynes and distance in centimeters, work is given in dyne-centimeters or ERGS.

♦ SUGGESTIONS TO INSTRUCTORS: The main topics in this section can be divided into three parts, as follows:
Part 1 (Work; Tune-up Exercises T1–T4; Problems 1–13, 15, 16, 19–27) Have students begin work on Questions 1 through 3 before class. Problems 3, 7, 15, and 24 are good for in-class discussion.
Part 2 (Kinetic and potential energy; Tune-up Exercises T5 and T6; Problems 14, 17, 18, 28–34, 53–55) Have students begin work on Question 4 before class. Problem 17 and 29 are good for in-class discussion.
Part 3 (Forces of fluids; Tune-up Exercise T7; Problems 35–52) Have students study Examples 6 and 7 before class. Problems 35 and 40 are good for in-class discussion.

In studying motion near the surface of the earth, as in the next example, we consider the acceleration due to gravity to be the constant g. Then the force of gravity on an object of mass m is its constant weight $w = mg$.†

Example 1 A ball weighing one-half pound is thrown 20 feet straight up in the air and then is caught at the point from which it was thrown. What is the work done by gravity on the ball as it rises?

SOLUTION We use a vertical s-axis with distances measured in feet, with the positive side pointing up, and with the origin where the ball is thrown and caught. The upward force of gravity F is is the negative of the ball's weight. Since the ball weighs one-half pound, $F = -\frac{1}{2}$ pounds. As the ball rises from $s = 0$ to $s = 20$, the work done by gravity on it is

$$[\text{Force}][\text{Net distance}] = [-\tfrac{1}{2} \text{ pounds}][(20-0) \text{ feet}] = -10 \text{ foot-pounds.} \ \square$$

Question 1 **(a)** How much work does the force of gravity do on the ball in Example 1 as it falls back to the point from which it was thrown? **(b)** What is the total work done by gravity on the ball as it rises and falls?

We define the work AGAINST a force to be the negative of the work done by a force. Thus, the ball in Example 1 does 10 feet-pounds of work against gravity as it rises and does -10 foot-pounds of work against gravity as it falls.

Work done by a variable force

Now suppose that an object is moving on an s-axis and that it is subject to a force $F(s)$ in the positive direction when it is at s, where $F(s)$ is the function of Figure 1.

To determine the work done by the force as the object moves from $s = a$ to $s = b$, with $a < b$, we approximate $F(s)$ by step functions.

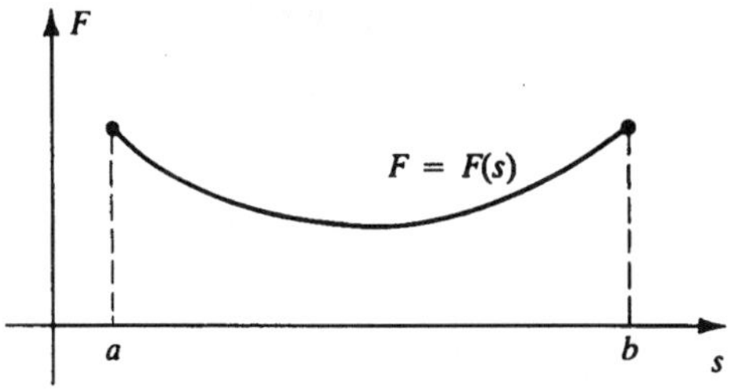

FIGURE 1

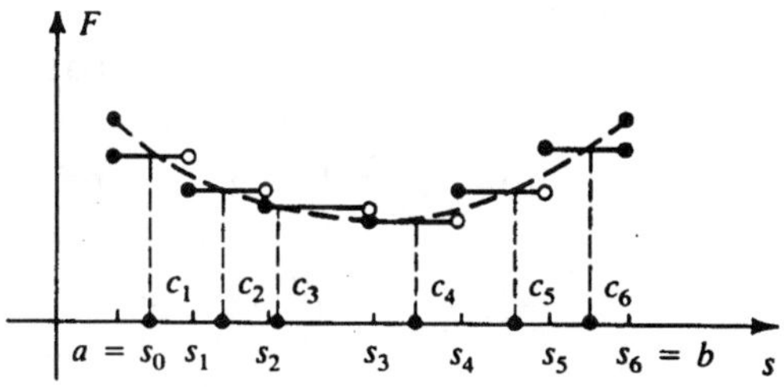

FIGURE 2

For each partition $a = s_0 < s_1 < s_2 < \cdots < s_N = b$ of $[a, b]$, we choose a number c_j in the jth subinterval for each j and approximate $F(s)$ in the jth subinterval by the constant $F(c_j)$. Figure 2 shows the graph of such an approximation of the function $F(s)$ of Figure 1.

According to Definition 1, the work done by the approximating force as the object goes from s_{j-1} to s_j across the jth subinterval is [Force][Net distance] $= F(c_j)\Delta s_j$ where Δs_j is the width of the subinterval. The total work done by the approximating force as the object goes from a to b is

$$\sum_{j=1}^{N} F(c_j)\Delta s_j. \tag{2}$$

The limit of this Riemann sum as the number of subintervals in the partition tends to ∞ and their widths tend to zero is the integral $\int_a^b F(s)\,ds$. We obtain the following rule in the case of $a < b$.

---

†Recall that $g \doteq 32$ feet/second^2 if mass is measured in slugs and weight in pounds, $g \doteq 9.8$ meters/second^2 if mass is measured in kilograms and weight in newtons; and $g \doteq 980$ centimeters/second^2 if mass is measured in grams and force in dynes.

Rule 1 *If an object is subject to a force $F(s)$ in the positive s-direction when it is at s on an s-axis, then the* WORK *done by the force on the object as it moves from $s = a$ to $s = b$ is*

$$[\text{Work}] = \int_a^b F(s)\,ds. \tag{3}$$

In applying this rule, we assume that the integral is defined. Notice that the work **(3)** is zero if $a = b$. Also, interchanging a and b multiplies the integral in **(3)** by -1. Accordingly, the work done as the object goes from $s = b$ to $s = a$ is the negative of the work done as the object goes from $s = a$ to $s = b$.

Example 2 The engine of a bus exerts a force of $3000 - 2s$ pounds on the bus when it has gone s feet from its starting point, for $0 \le s \le 1000$. How much work does the engine do on the bus as it travels the 1000 feet?

SOLUTION By **(3)** with $F(s) = 3000 - 2s$, the work is

$$\begin{aligned}\int_0^{1000} (3000 - 2s)\ (\text{pounds})\ ds\ (\text{feet}) &= \Big[3000s - s^2\Big]_0^{1000}\\ &= [3000(1000) - (1000)^2] - [0] = 2{,}000{,}000 \text{ foot-pounds.} \ \square\end{aligned}$$

To study the force of gravity at large distances from the surface of the earth, we have to use Newton's law, which states that the force of the earth's gravity on an object above or on the surface of the earth is kr^{-2}, where r is the distance from the object to the center of the earth, and the constant k is chosen so that on the surface of the earth the force of gravity is the object's weight. In the next Question, we use 6500 kilometers as the radius of the earth.

Question 2 **(a)** What is the weight on the surface of the earth of a ball with a mass of 10 kilograms? **(b)** What is the downward force of the earth's gravity on such a ball when it is $r \ge 6500$ kilometers from the center of the earth?

Example 3 How much work against gravity is required to lift the ten-kilogram ball of Question 2 from the surface of the earth to 1000 kilometers above it?

SOLUTION We assume that the ball moves on an r-axis with distances measured in kilometers and with its origin at the center of the earth. Gravity pulls the ball down, so that by the result of Question 2, the force of gravity on the ball in the positive r-direction is $F = -98(6500)^2r^{-2}$ newtons. The work done by gravity when the ball moves from $r = 6500$ on the surface of the earth to $r = 7500$ a thousand kilometers above it is

$$\begin{aligned}\int_{6500}^{7500} F\,dr &= \int_{6500}^{7500} (-98(6500)^2r^{-2})\,dr\\ &= 98(6500)^2\Big[r^{-1}\Big]_{6500}^{7500} = 98(6500)^2\left[\tfrac{1}{7500} - \tfrac{1}{6500}\right]\\ &= -84{,}933\tfrac{1}{3} \text{ newton-kilometers.}\end{aligned}$$

The work done against gravity is the negative of this amount or $84{,}933\frac{1}{3}$ newton-kilometers. $\square$

Springs

Imagine a spring that is fastened to a rigid object at its left end and whose right end is free to move in a straight line (Figure 3). We say that the spring is at its REST POSITION and has its NATURAL LENGTH when it is neither stretched nor compressed and exerts no force in either direction. We indicate the location of the free end of the spring on an s-axis with the positive side pointing to the right and the origin at the rest position of the end of the spring.

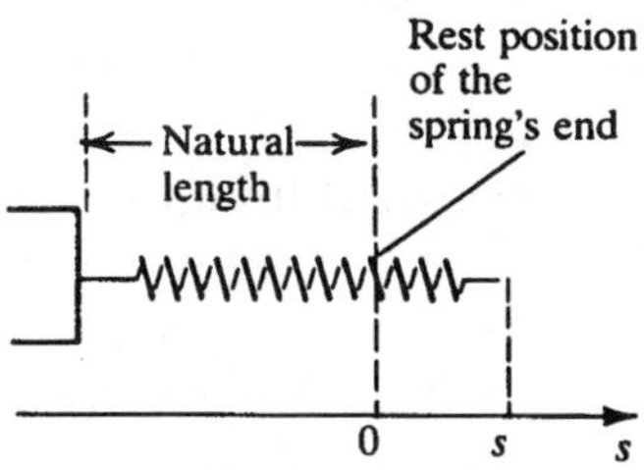

FIGURE 3

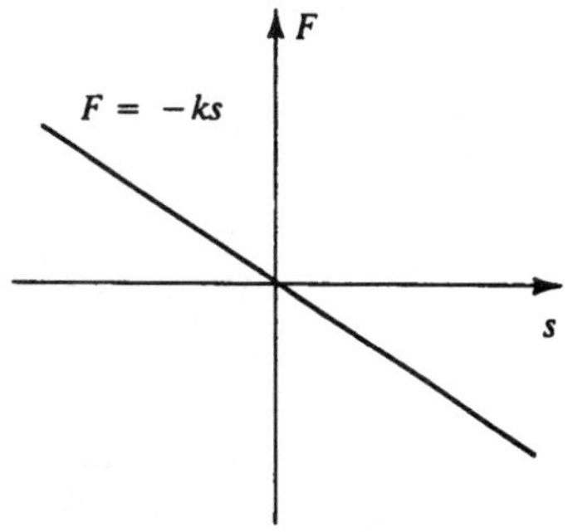

FIGURE 4

We let $F(s)$ denote the force that the spring exerts toward the right when its free end is at s. $F(s)$ is 0 when s is 0 and the spring is in its rest position. $F(s)$ is positive when s is negative because then the spring is compressed and exerts a force toward the right. $F(s)$ is negative when s is positive because then the spring is stretched and exerts a force to the left, which is in the negative direction on the s-axis.

If $F(s)$ has the form

$$F(s) = -ks \tag{4}$$

with a positive constant k, the spring is said to satisfy HOOKE'S LAW. With this model, the graph of $F(s)$ is a line through the origin as in Figure 4. Hooke's law is not valid for any spring if s is very large, but is reasonably accurate under the normal operation of most springs.

Example 4 A spring that satisfies Hooke's law has a natural length of 10 feet and exerts a force of 3 pounds when it is compressed to a length of 8 feet. Give a formula for the force exerted by the spring when it is stretched s feet beyond its natural length.

SOLUTION By Hooke's law **(4)**, the force exerted by the spring is $F(s) = -ks$ with some constant k. Because the natural length of the spring is 10 feet, $s = -2$ when it is compressed to a length of 8 feet and $F(-2) = 2k$ equals 3 pounds. Therefore, $k = \frac{3}{2}$ and $F(s) = -\frac{3}{2}s$. □

Question 3 How much work must be done against the spring of Example 4 to stretch it from its natural length to a length of 15 feet?

Work lifting water

We will solve the next example by using an approximation procedure to find the necessary integral rather than by applying Rule 1.

Example 5 A full water tank in the shape of a right circular cone with its vertex pointing downward is 3 feet high and its top has a 1-foot radius (Figure 5). How much work against gravity is needed to pump the water to the top of the tank? (Use 62.4 pounds/foot^3 for the density of water.)

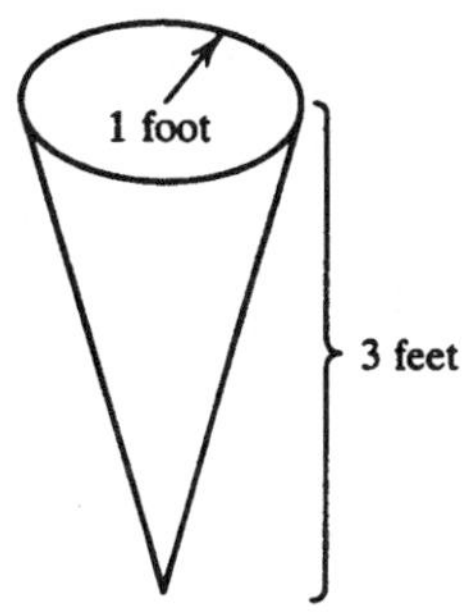

FIGURE 5

SOLUTION We cannot apply Definition 1 or Rule 1 directly because we are not dealing with a single object or with a finite number of objects that are moved a fixed distance. The tank of water is a continuum, and the water at different depths is lifted different amounts. We find the integral that gives the work by approximating the water by collections of point masses.

We let s feet be the vertical distance measured from the bottom of the tank and use similar triangles to see that the circular cross section of the tank at height s has radius $\frac{1}{3}s$ (Figure 6). For a partition $0 = s_0 < s_1 < s_2 < \cdots < s_N = 3$, the horizontal planes at $s = s_j$ slice the tank into N pieces. We first approximate the jth piece by a disk of radius $\frac{1}{3}s_j$ equal to the radius of the cross section at s_j and of thickness equal to the width Δs_j of the jth subinterval (Figure 7). The volume of the jth disk is $\pi(\frac{1}{3}s_j)^2\Delta s_j$ feet^3, and the water in it would weigh $\left[\pi(\frac{1}{3}s_j)^2\Delta s_j\right.$ $\left.\text{feet}^3\right]\left[62.4 \text{ pounds/foot}^3\right] = 62.4\pi(\frac{1}{3}s_j)^2\Delta s_j$ pounds. We approximate this volume of water by a point mass of weight $62.4\pi(\frac{1}{3}s_j)^2\Delta s_j$ pounds positioned at $s = s_j$ at the top of the disk.

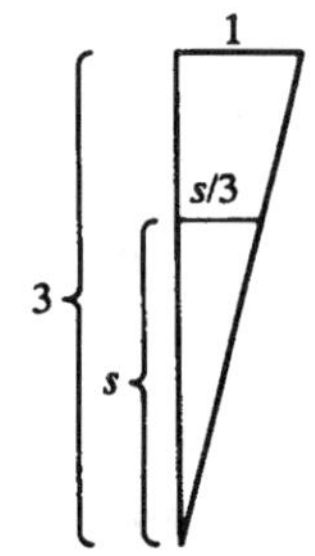

FIGURE 6

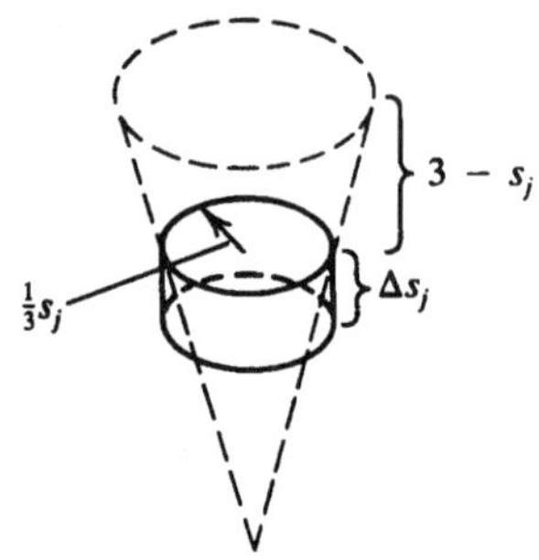

FIGURE 7

Because the jth point mass is $(3 - s_j)$ feet from the top of the tank, the work to lift the N point masses to the top is

$$\sum_{j=1}^{N} 62.4\left[\frac{\text{pounds}}{\text{cubic foot}}\right]\pi(\tfrac{1}{3}s_j \text{ (feet)})^2(3 - s_j) \text{ (feet)}\Delta s_j \text{ (feet)}$$

$$= \sum_{j=1}^{N} 62.4\pi(\tfrac{1}{3}s_j)^2(3 - s_j)\Delta s_j \text{ foot-pounds}$$

and the work to pump the water to the top of the tank is the corresponding integral

$$\begin{aligned}\int_0^3 (62.4)\pi(\tfrac{1}{3}s)^2(3 - s)\,ds &= (62.4)\pi\int_0^3 (\tfrac{1}{3}s^2 - \tfrac{1}{9}s^3)\,ds \\ &= (62.4)\pi\left[\tfrac{1}{9}s^3 - \tfrac{1}{36}s^4\right]_0^3 \\ &= (62.4)\pi\left\{\left[\tfrac{1}{9}(3)^2 - \tfrac{1}{36}(3)^4\right] - \left[\tfrac{1}{9}(0)^2 - \tfrac{1}{36}(0)^4\right]\right. \\ &= (62.4)\pi(\tfrac{3}{4}) = 46.8\pi \text{ foot-pounds. } \square\end{aligned}$$

Kinetic energy of point masses

The KINETIC ENERGY of a point mass with mass m which is moving on an s-axis with velocity v is defined by

$$[\text{Kinetic energy}] = \tfrac{1}{2}mv^2. \tag{5}$$

This quantity, like all types of energy, has the same units as work. For instance, if mass is given in slugs, distance in feet, time in seconds, and force in pounds, the units in **(5)** are [slugs-feet^2/second^2], which are the same as foot-pounds, since by Newton's Law:

$$F \text{ pounds} = [m \text{ slugs}][a \text{ feet/second}^2]$$

and [pounds] = [slug-feet/second^2].

Let us see how kinetic energy is related to work.

Suppose that a car of mass m moving on an s-axis, is at $s(t)$ for $\alpha \le t \le \beta$ and that, on this particular trip, the total of all forces on the car when it is at s is $F(s)$. The car goes from $s = s(\alpha)$ to $s(\beta)$ as t increases from α to β, so the total work done by all forces on the car during the trip is

$$[\text{Work by all forces}] = \int_{s(\alpha)}^{s(\beta)} F(s)\, ds. \tag{6}$$

The car is at $s = s(t)$ and its acceleration is $\frac{d^2s}{dt^2}(t)$ at time t, so that, by Newton's Law, $F(s) = m\frac{d^2s}{dt^2}(t)$. We make this substitution in integral **(6)** and use the change of variables $s = s(t), ds = \frac{ds}{dt}(t)\, dt$ to obtain

$$\begin{aligned}[\text{Work by all forces}] &= \int_{s(\alpha)}^{s(\beta)} m\frac{d^2s}{dt^2}\, ds \\ &= \int_{\alpha}^{\beta} m\frac{d^2s}{dt^2}\frac{ds}{dt}\, dt.\end{aligned} \tag{7}$$

The Chain Rule gives

$$\frac{d}{dt}\left[\tfrac{1}{2}m\left(\frac{ds}{dt}\right)^2\right] = m\frac{d^2s}{dt^2}\frac{ds}{dt}$$

so that, by **(6)** and the Fundamental Theorem of Calculus,

$$\begin{aligned}[\text{Work by all forces}] &= \int_{\alpha}^{\beta} \frac{d}{dt}\left[\tfrac{1}{2}m\left(\frac{ds}{dt}\right)^2\right] dt \\ &= \left[\tfrac{1}{2}m\left(\frac{ds}{dt}\right)^2\right]_{t=\alpha}^{t=\beta} \\ &= \left[\tfrac{1}{2}m\left(\frac{ds}{dt}\right)^2\right]_{t=\beta} - \left[\tfrac{1}{2}m\left(\frac{ds}{dt}\right)^2\right]_{t=\alpha}.\end{aligned} \tag{8}$$

By definition **(5)**, the quantity on the right of **(8)** is the change in the car's kinetic energy from time α to time β. Equations **(8)** illustrate the next result.

Rule 2 *If, as an object moves on a straight line, all forces on it are parallel to the line, then the work done by all forces on the object is equal to the change in the object's kinetic energy.*

Question 4 How much work is done by all forces on a 4000-pound truck as it speeds up from a velocity of 20 feet per second to 80 feet per second?

Potential energy

The forces acting on an object might depend on the object's position, the object's velocity, the time, or other factors. Some forces, however, such as the force of gravity and the force of a spring, depend only on the object's position. In these cases, the object's POTENTIAL ENERGY $U(s)$ relative to that force is defined by the formula,

$$[\text{Potential energy}] = U(s) = U_0 - \int_{s_0}^{s} F(s)\,ds. \tag{9}$$

where U_0 and s_0 are an arbitrary constant. Because $-F(s)$ is the force by the object against the force, the potential energy **(9)** equals U_0 plus the work done against the force as the object goes from s_0 to s. Consequently, the increase in the potential energy from s_0 to s equals the work done against the force.

In the examples and exercises dealing with kinetic and potential energy in this section, we will assume that the only forces acting on an object are (i) an external force, (ii) the force of gravity or a spring, and (iii) the force of air resistance. Then the total force on the object is the sum of three forces, and as the object moves from $s = a$ to $s = b$,

$$\begin{bmatrix}\text{The increase in}\\ \text{the kinetic energy}\end{bmatrix} = \begin{bmatrix}\text{The work done by}\\ \text{all forces}\end{bmatrix}$$
$$= \begin{bmatrix}\text{The work done by}\\ \text{the external force}\end{bmatrix} + \begin{bmatrix}\text{The work done by}\\ \text{the spring or gravity}\end{bmatrix} + \begin{bmatrix}\text{The work done by}\\ \text{air resistance}\end{bmatrix}.$$

To examine the work done by the external force, we rewrite the last equation in the form

$$\begin{bmatrix}\text{The work done by}\\ \text{the external force}\end{bmatrix}$$
$$= \begin{bmatrix}\text{The increase in}\\ \text{kinetic energy}\end{bmatrix} - \begin{bmatrix}\text{The work done by}\\ \text{the spring or gravity}\end{bmatrix} - \begin{bmatrix}\text{The work done by}\\ \text{air resistance}\end{bmatrix}.$$

Then since the work done by the spring or gravity is the negative of the increase in the object's potential energy; the work done by air resistance is the negative of the work done against air resistance; and the work done by air resistance equals the loss of energy due to air resistance we find that

$$\begin{bmatrix}\text{The work done by}\\ \text{the external force}\end{bmatrix}$$
$$= \begin{bmatrix}\text{The increase in}\\ \text{kinetic energy}\end{bmatrix} + \begin{bmatrix}\text{The increase in}\\ \text{potential energy}\end{bmatrix} - \begin{bmatrix}\text{The energy lost to}\\ \text{air resistance}\end{bmatrix}. \tag{10}$$

Example 5 A ten-gram ball is at the end of a spring that exerts a force of $-2s$ dynes when the object is at s centimeters on an s-axis as in Figure 3. An external force does 110 ergs of work on the ball to move it from $s = 0$, where its velocity is 1 centimeter/second to $s = 5$, where its velocity is 4 centimeters/second. How much energy is lost due to air resistance?

SOLUTION The ball's mass is ten grams and its velocity changes from 1 centimeter per second to 4 centimeters per second, so the change in its kinetic energy is $\frac{1}{2}(10)(4^2) - \frac{1}{2}(10)(1^2) = 75$ ergs. The increase in potential energy is the work against the spring, which is

$$-\int_0^5 (-2s)\,ds = \left[s^2\right]_0^5 = 25 \text{ ergs}.$$

By **(10)**, the energy lost due to air resistance equals the work by the external force minus the increases in kinetic and potential energy, or $110 - 75 - 25 = 10$ ergs. □

Forces of fluids

If a surface is horizontal, the force of a fluid is equal to the weight of the fluid above it. Thus, if a surface has area A square feet and is h feet beneath the surface of the fluid, and if the fluid has constant density ρ pounds per cubic foot, then the force on the surface is

$$[\text{Force}] = \left[\rho \, \frac{\text{pounds}}{\text{cubic foot}}\right]\left[hA \text{ cubic feet}\right] = \rho h A \,\text{pounds}. \tag{11}$$

This formula shows that the force at a fixed depth is proportional to the area of the surface. The constant of proportionality ρh is the PRESSURE at that depth:

$$P = [\text{Pressure at depth } h] = \rho h. \tag{12}$$

Pressure is measured in units of force per area. For instance, if force is measured in pounds and distance in feet, pressure is given in pounds per square foot. Combining **(11)** and **(12)** shows that the force on a horizontal surface is equal to the product

$$[\text{Force}] = [\text{Pressure}][\text{Area}] = PA \tag{13}$$

of its area and the pressure at that depth.

If a flat surface submerged in a fluid is not horizontal, then the pressure on it is not constant, and we have to use an integral to find the force of the fluid. We put xy-coordinates on the plane containing the surface, with the x-axis horizontal at constant depth. If the surface is vertical, as in Figure 9, then the y-axis is vertical. If the surface is at an angle, then the y-axis is not vertical. The positive y-direction can be up or down, whichever is more convenient.

We suppose that the surface occupies the region in the xy-plane between two curves $x = g(y)$ and $x = f(y)$ for $a \le y \le b$. We let $W(y)$ denote the width of the surface measured on the horizontal line through y on the y-axis (Figure 8), and let $D(y)$ denote the depth of that line beneath the surface of the fluid. If the surface is vertical, $D(y)$ is measured along the y-axis. Otherwise it must be measured along a vertical line.

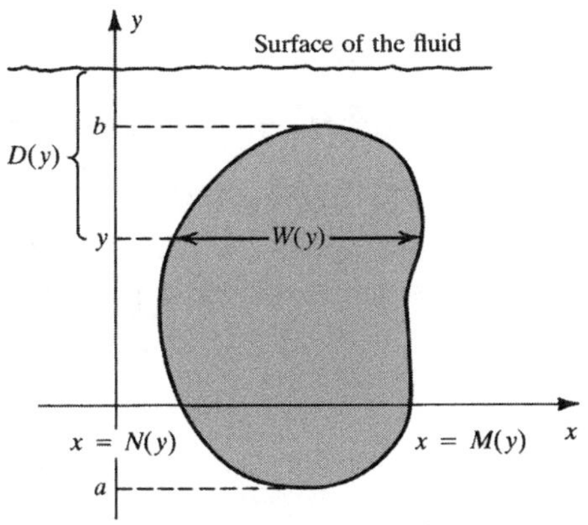

FIGURE 8

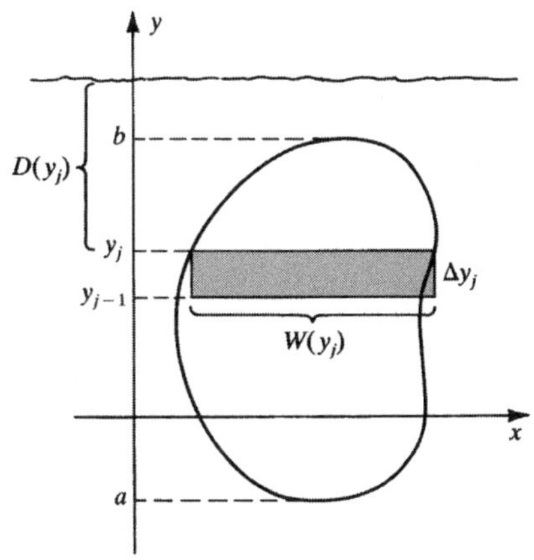

FIGURE 9

We consider a partition $a = y_0 < y_1 < y_2 < \cdots < y_N = b$ of the interval $[a, b]$ and approximate the portion of the surface between the horizontal lines $y = y_{j-1}$ and $y = y_j$ by a rectangle of width $W(y_j)$ and height $\Delta y_j = y_j - y_{j-1}$ (Figure 9). If the pressure on the rectangle were constant, the force on it would be the product of its area $W(y_j)\Delta y_j$ and the pressure. The pressure is not constant, but is approximately equal to the pressure $\rho D(y_j)$ at the top of the rectangle. Therefore, the force on the jth rectangle is approximately $\rho D(y_j)W(y_j)\Delta y_j$ with ρ the density of the fluid. The sum

$$\sum_{j=1}^{N} \rho D(y_j)W(y_j)\Delta y_j$$

is the approximate force on the entire surface and is a Riemann sum for the integral that gives the exact force.

Rule 3 *If a flat plate, submerged in a fluid of constant density ρ, measured in weight per volume, occupies the region between curves $x = g(y)$ and $x = f(y)$ for $a \le y \le b$ in an xy-plane with a horizontal x-axis, then the total force on the surface is*

$$\int_a^b \rho D(y)W(y)\,dy \tag{14}$$

where $W(y)$ and $D(y)$ are the width of the plate at y and $D(y)$ is the depth of the fluid at y.

The force **(14)** is directed perpendicular to the surface. If the plate is submerged with fluid on both sides, then the forces on the two sides are equal and in opposite directions and balance each other.

Example 6 A dam has the shape of the region $\frac{1}{16}x^2 \le y \le 16$ with distances on the coordinate axes measured in feet and with the y-axis pointing upward. What is the total force of water on the dam when the water is nine feet deep? (Use 62.4 pounds per cubic foot for the density of water.)

SOLUTION The dam is sketched in Figure 10. Solving the equation $y = \frac{1}{16}x^2$ for x gives $x = \pm\sqrt{y}$. This shows that the dam is $8\sqrt{y}$ feet wide at height y. The points at height y are $9 - y$ feet below the surface of the water. By formula **(13)** with $\rho = 62.4, D(y) = 9 - y, W(y) = 8\sqrt{y}, a = 0$, and $b = 9$, the total force of the water on the dam is

$$\int_0^9 \left[62.4\,\frac{\text{pounds}}{\text{cubic feet}}\right][(9-y)\text{ (feet)}]\,[8\sqrt{y}\text{ (feet)}]\,[dy\text{ (feet)}]$$

$$= \int_0^9 (62.4)(9-y)(8\sqrt{y})\,dy = (62.4)(8)\int_0^9 (9y^{1/2} - y^{3/2})\,dy$$

$$= (62.4)(8)\left[6y^{3/2} - \tfrac{2}{5}y^{5/2}\right]_0^9 = (62.4)(8)[6(9^{3/2}) - \tfrac{2}{5}(9^{5/2})]$$

$$= (62.4)(518.4) = 32{,}348.16 \text{ pounds.}$$

The units are included in the first of these integrals as a partial check that the formula is correct. □

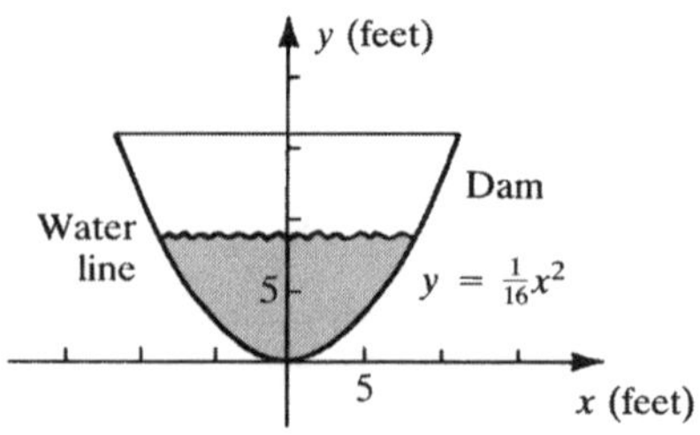

FIGURE 10

In the next example, the surface is in a nonvertical plane. We use a horizontal x-axis, but the y-axis is not vertical.

Example 7 A plate has the shape of the region in Figure 11 bounded by the parabola $x = y^2$, the line $y = 1$, and the y-axis in an xy-plane with distances measured in meters. The plate is submerged in water with the line $y = 1$ along the surface of the water and with the y-axis making an angle of 45° with the vertical. What is the force of water pressure on each side of the plate? (Use 9800 newtons per cubic meter as the density of water.)

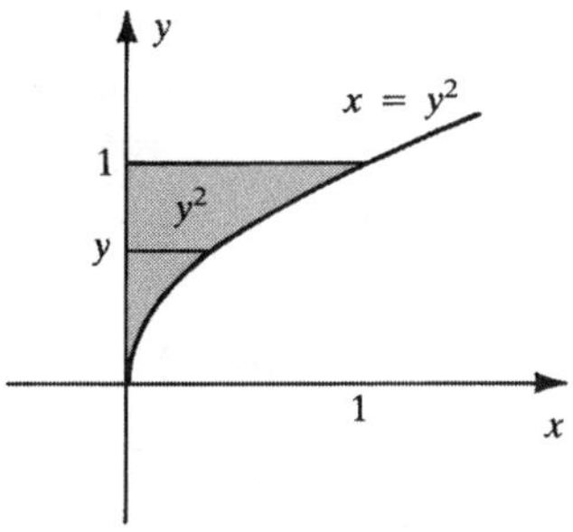

FIGURE 11

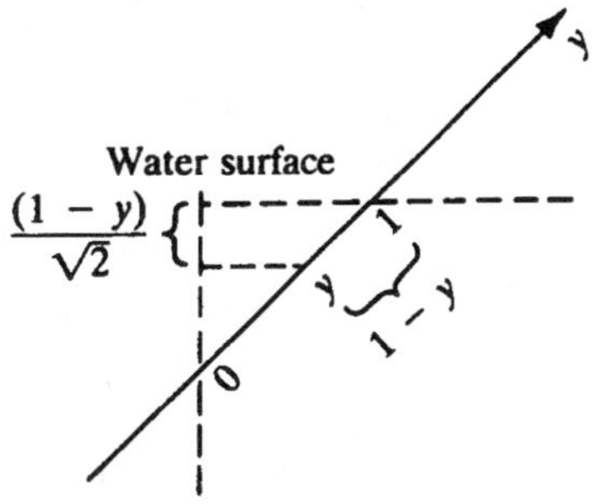

FIGURE 12

SOLUTION From the drawing of the plate in Figure 11, we see that its width at y is $W(y) = y^2$ meters. Because the point at y on the y-axis is $1 - y$ meters, measured along that axis, from the surface of the water, and because the axis makes an angle of 45° with the vertical, the depth at y is $D(y) = (1 - y)/\sqrt{2}$ (Figure 12). By Rule 3 the force on one side of the plate is

$$\int_0^1 9800[(1-y)/\sqrt{2}](y^2)\,dy = \frac{9800}{\sqrt{2}}\int_0^1 (y^2 - y^3)\,dy$$

$$= \frac{9800}{\sqrt{2}}\left[\tfrac{1}{3}y^3 - \tfrac{1}{4}y^4\right]_0^1 = \frac{9800}{\sqrt{2}}\left[\tfrac{1}{3} - \tfrac{1}{4}\right] = \frac{2450}{3\sqrt{2}} \text{ newtons. } \square$$

Principles and procedures

- The use of term "work" in physics and mathematics has little similarity to how it is used in everyday English. In physics and mathematics, work is a quantity used to describe transfers of energy.
- The work done by a constant force in the direction of an object's motion is the the force multiplied by the net distance traveled by the object (Definition 1). To allow for forces and motion in opposite directions, the force, the next distance, and the work can be positive, negative or zero.
- We consider the force applied to an object to be the negative of the force applied by the object, and we say that the work done by a force on an object is the negative of the work done by the object against the force.
- If the force is not constant but is given as a function of the object's position, then the work is calculated as an integral using Rule 1.
- If a force depends only on the object's position, then a potential energy associated with the force can be defined, and the work done by or against the force results in decreases or increases in the potential energy.
- The work done by all forces on an object equals its change in kinetic energy.
- Equation **(10)** for the case of motion under the effect of gravity or a spring should be easy to remember. It states that the work done by an external force on an object equals the increase in the kinetic energy plus the increase in potential energy minus the loss in energy die to air resistance.

Tune-Up Exercises 6.7♦

[A]*Answer provided.* [O]*Outline of solution provided.*

T1. [O] If a rubber band exerts a force of $50s$ dynes when it is stretched s centimeters, how much work must be done against it to stretch it 10 centimeters?

♦ Type 1, work

T2. [O] A spring as in Figure 3 exerts a force of $-ks$ newtons toward the right when it is stretched s meters. Its natural length is one meter, and it requires 3 joules of work against the spring to stretch it from a length of 1.2 meters to a length of 1.4 meters. What is the constant k?

♦ Type 2, work, etc.

T3. [O] When a spring that satisfies Hooke's law is stretched 10 centimeters beyond its natural length, it exerts a force of 500 dynes. What work against the spring is required to compress it 8 centimeters from its natural length?

♦ Type 2, work, etc.

T4. [O] A 40-foot-long chain weighing 5 pounds per foot is held by one end at the edge of a 60-foot-high roof. How much work must be done against gravity to pull the chain onto the roof?

♦ Type 3, work

T5. [A] What are the increases in the potential energy due to gravity of **(a)** a ten-pound box that is lifted 500 feet, **(b)** a cat with a mass of 5 kilograms that climbs 3 meters down a telephone pole, and **(c)** a mouse with a mass of 25 grams that climbs up a 210-centimeter-high clock and then back down again?

♦ Type 1, potential energy

T6. [O] An eight-pound ball that is initially at rest is lifted straight up in the air by an upward force equal to $25 - 3\sqrt{s}$ pounds when the ball has been lifted s feet. What is the ball's velocity after it has been lifted 25 feet if 200 foot-pounds of work is done against air resistance in the motion?

♦ Type 3, kenetic and potential energy

T7. [O] A vertical plate with its top 3 feet beneath the surface of a lake has the shape of an isosceles trapezoid 10 feet wide at the top, 6 feet wide at the bottom, and 3 feet high. What is the force of water pressure on each side of the plate? (Use 62.4 pounds per cubic foot as the density of water.)

♦ Type 4, force of fluid

Problems 6.7

[A]*Answer provided.* [O]*Outline of solution provided.*

1. [A] A locomotive diesel engine supplies a force of $400s^{1/2}$ pounds when the train has traveled s feet. How much work does the engine do on the train to move it 3600 feet?

♦ Type 1, work

2. [A] A woman pushing her stalled car exerts a force of $200(1+s)^{-1/2}$ newtons on it when she has pushed it s meters. The engine starts and she stops pushing it when she has pushed 24 meters. How much work does she do pushing the car?

♦ Type 1, work

3. A boy rolling a large boulder exerts $10 + 5\sin s$ pounds of force on it when he has rolled it s feet. How much work does he do in rolling it 30 feet?

♦ Type 1, work

4. An accelerating car's engine exerts a force of $\frac{1}{6}s^{1/9}$ newtons on it after the car has traveled s meters. How much work does the engine do on the car in the first 800 meters?

♦ Type 1, work

5. [A] The repulsive force between two negatively-charged particles is equal to a constant divided by the square of the distance between them (Coulomb's law). Two negative charges exert a repulsive force of 2 dynes on each other when they are 10 centimeters apart. How much work against the force is required to move one of them from a distance of 10 centimeters to a distance of 5 centimeters from the other?

♦ Type 2, work

♦The Tune-up Exercises and Problems are classified by type and content. The types are (1) basic, reactive; (2) basic reflective; (3) intermediate, reactive; (4) intermediate, reflective; (5) advanced, reactive; (6) advanced, reflective; and (7) advanced, theoretical.

6. The force of gravity on a meteorite that would weigh 1 pound on the surface of the earth is $(16 \times 10^6)s^{-2}$ pounds when it is s miles from the center of the earth. How much work does gravity do on such a meteorite as it falls from 96,000 miles above the earth to 1000 miles above the earth? (Take 4000 miles as the radius of the earth.)

♦ Type 2, work

7. O A spring exerts a force of 6 pounds when it is stretched from its natural length of 15 inches to a length of 16 inches. How much work against the spring is required to compress it from a length of 14 inches to a length of 12 inches?

♦ Type 2, work

8. A spring with a natural length of 10 feet exerts a force of 12 pounds when it is stretched to 12 feet. How much work against the spring does it take to stretch it from its natural length to a length of 14 feet?

♦ Type 2, work

9. If a spring exerts two pounds of force when it is stretched half a foot from its rest position, how much work must be done against it to compress it a foot and a half from its rest position?

♦ Type 2, work

10. It takes 6 ergs of work against a spring to compress it 2 centimeters from its natural length. What force does the spring exert when it is compressed 2 centimeters?

♦ Type 2, work

11. A spring with a natural length of 4 meters exerts a force of 0.4 newtons when it is compressed to a length of 3.8 meters. What is the increase in the potential energy due to the spring of an object at the end of it when the spring is stretched from a length of 3.8 meters to a length of 4.3 meters?

♦ Type 2, work

12. How much work against a spring is required to stretch it 5 centimeters from its rest position and then return it to its rest position?

♦ Type 2, work

13. A five-pound monkey climbs a 50-foot-long dangling rope that weighs a half pound per foot. How much work does the monkey do against gravity?

♦ Type 2, work

14. The potential energy due to a spring of an object at the end of the spring increases 20 foot-pounds when the spring is stretched from its natural length of 6 feet to a length of 6.5 feet. What force does the spring exert when its length is 6.5 feet?

♦ Type 1, potential energy

15. The force of air resistance on a small car is $0.63v^2$ newtons when the car is traveling v meters per second. How much work must be done by the car against air resistance in traveling 1000 meters if its velocity is $\frac{1}{3}s^{2/3}$ meters per second when it has traveled s meters?

♦ Type 4, work

16. A The weight of an object on the surface of the earth is the force of gravity exerted on it. As the object is lifted from the surface of the earth, the force is equal to a constant times r^{-2} where r is the distance to the center of the earth (Newton's law). How much work against gravity does it take to lift a 100-pound satellite 300 miles above the surface of the earth? (Use 4000 miles as the radius of the earth.)

♦ Type 4, work

17. O A 10-kilogram ball, moving vertically, is at $s = \frac{1}{4}t^3$ (meters) above the ground at time t (seconds) for $0 \le t \le 4$. During the motion 1000 joules of work are done against air resistance. (a) What is the increase in the ball's potential energy? (b) What is its final kinetic energy? (c) How much work is done by the force lifting the ball?

♦ Type 4, kenetic and potential energy

18. A A 64-pound weight at the end of a spring is moved from $s = 0$ to $s = 4$ (feet) by the force $2 + 3\sqrt{s}$ pounds toward the right at s. There is no air resistance and the weight's final velocity is 5 feet per second. What is its initial velocity?

♦ Type 4, work and energy

19. A A bag of sand is lifted at the constant rate of 2 feet per second for 10 seconds. Initially the bag contains 100 pounds of sand, but the sand leaks out at the constant rate of 3 pounds per second. How much work is done against gravity in lifting the bag?

♦ Type 4, work

20. (a) How much work against gravity is required to pump the water out of a full hemispherical fish bowl of radius 1 foot to a point 3 feet above the top of the bowl? (b) A one-tenth horsepower pump does 55 foot-pounds of work per second. How long would it take to empty the fishbowl? (The density of water is 62.4 pounds per cubic foot.)

♦ Type 4, work

21. The vertical cross sections of a ten-meter-long trough are semicircles of radius two meters, and the trough is full of water weighing 9800 newtons per cubic meter. (a) How much work against gravity does it take to pump the water to the top of the trough? (b) How much work does it take to pump the water to a point fifteen meters above the top of the trough? (Use the result from part (a).)

♦ Type 4, work

22. [A] A rectangular tank 3 feet wide, 6 feet long, and 4 feet deep contains 2 feet of linseed oil weighing 58.5 pounds per cubic foot on top of 2 feet of water weighing 62.4 pounds per cubic foot. How much work must be done against gravity to pump both liquids to a pipe 10 feet above the top of the tank?

♦ Type 4, work

23. [A] A 30-foot-long rope weighing 0.2 pounds per foot is hanging from a roof, and a 20-foot-long chain weighing 3 pounds per foot is hanging from the end of the rope. How much work must be done against gravity to pull the rope and chain up to the roof?

♦ Type 4, work

24.[A] A cement block is at the end of a 20-foot-long chain hanging from the edge of a roof. How much does the block weigh if the chain weighs 5 pounds per foot and it takes 5,000 foot-pounds of work against gravity to pull the chain and block to the roof?

♦ Type 4, work

25. Sand leaks out of a bag at a constant rate. As the bag is lifted at the constant rate of 3 feet per minute for 6 minutes, the amount of sand in the bag decreases from 500 pounds to 250 pounds. How much work is done against gravity lifting the bag?

♦ Type 4, work

26. An underground chamber has the shape obtained when the region bounded by $y = x^2/500$ and $y = 500$ (meters) in an xy-plane with the y-axis pointing downward is rotated about the y-axis. The chamber is full of oil weighing 8350 newtons per cubic meter. The top of the chamber is 200 meters below the surface of the ground. How much work must be done against gravity to pump all the oil to the surface?

♦ Type 4, work

27. A tank in the shape of an inverted cone is 85 centimeters high. The shape of the base of the cone is not known, but the area of the cross section s centimeters above the bottom is proportional to s^2. The area of the base is 200 square centimeters. The tank is full of glycerine weighing 1235 dynes per cubic centimeter. How much work must be done against gravity to pump the glycerine to the top of the tank?

♦ Type 4, work

28.[A] It requires 180 foot-pounds of work to lift an eight-pound ball ten feet and increase its upward velocity from 8 feet per second to 24 feet per second. What are the increases in the ball's potential and kinetic energy and how much energy is lost due to air resistance?

♦ Type 4, work and energy

29. How much work is required to lift a 10-kilogram rocket 1000 meters if it is initially at rest, if its velocity at 1000 meters is 40 meters per second, and if 5,000 joules of energy are dissipated due to air resistance? Assume g is constant.

♦ Type 4, work and energy

30. A spring exerts a force of $-2s$ pounds when its end is at s (feet) on an s-axis. How much work is required to move a 16-pound object at its end from $s = 1$ to $s = 3$ if 2 foot-pounds of energy are dissipated due to air resistance, and if the object's speed is 3 feet per second at $s = 1$ and 2 feet per second at $s = 3$?

♦ Type 4, work and energy

31. [A] A one-kilogram object is at $s = t^2$ (meters) on a vertical s-axis at time t (seconds) for $0 \le t \le 10$. During the motion, the upward force that is lifting it does 3000 joules of work on the object. How much work is done against air resistance?

♦ Type 4, work and energy

32. A 4-pound weight that has an initial upward velocity of 5 feet per second is lifted 12 feet. The total work done on it by the upward force is 200 foot-pounds, of which 50 foot-pounds of work is done against air resistance. What is the weight's final velocity?

◆ Type 4, work and energy

33. A ten-gram weight at the end of a spring is subject to a constant force of 500 dynes toward the right as it moves from $s = 0$ to $s = 200$ (centimeters). The weight has zero velocity at $s = 0$ and velocity 20 centimeters per second at $s = 200$; there is no air resistance or other external force; and the force of the spring is given by Hooke's law $F = ks$. What is the constant k?

◆ Type 4, work and energy

34. A 64-pound box is lifted from $s = 0$ to $s = 10$ (feet) on a vertical s-axis by an upward force of $200 - 13.6s$ pounds at s. The box's initial and final velocities are equal. How much work is done against air resistance during the motion?

◆ Type 4, work and energy

35. The top of a trough is 10 feet long and 4 feet wide. The ends are vertical equilateral triangles with a vertex pointing downward. What is the total force on one end of the trough when it was full of water?

◆ Type 4, force of a fluid

36. A pool is 20 yards long, 8 yards wide, and 4 yards deep. What is the total force on the four (rectangular) sides and the bottom when it is full of water?

◆ Type 4, force of a fluid

37.[A] The back of a rowboat is a vertical trapezoid which is 2 feet high, 4 feet wide at the bottom, and 6 feet wide at the top. What is the total force of the water on the back of the boat when the bottom of the boat is 1 foot beneath the surface of the water?

◆ Type 4, force of a fluid

38.[O] An open cylindrical can with radius 4 inches and height 10 inches is full of paint weighing 110 pounds per cubic foot. (a) What is the force of the paint on the bottom of the can? (b) What is the force on the side?

◆ Type 4, force of a fluid

39.[A] A rectangular can with a 2-foot square base contains water 1 foot deep with six inches of oil on top of the water. What is the total force on the sides of the can if the oil has density of 59.3 pounds per cubic foot?

◆ Type 4, force of a fluid

40. A plate in the shape of a right triangle with legs of lengths 5 feet and 8 feet is submerged under water so that it is vertical and so that its 8-foot-long side is horizontal and is 16 feet below the surface of the water. What is the force of water on one side of the plate?

◆ Type 4, force of the fluid

41. One side of a plate in the shape of an equilateral triangle is resting on the bottom of a pool 20 feet deep. The tip is resting on the side of the pool so that the plate makes an angle of 45° with the vertical. Each side of the plate is 1 foot long. What is the force of the water on one side of the plate?

◆ Type 4, force of a fluid

42. The ends of a box are equilateral triangles and the three sides are squares. Each edge of the box is 4 feet long. What is the force on each of the ends when one side of the box lies on the horizontal bottom of a 10-foot-deep pool? What is the force on each of the three sides?

◆ Type 4, force of a fluid

43. What is the force of the water on one side of a vertical semicircular disk of radius 2 feet whose flat edge lies along the surface of the water?

◆ Type 4, force of a fluid

44. A vertical plate has the shape of an isosceles trapezoid with vertices $(-3,0), (5,0), (2,3)$, and $(0,3)$ in an xy-plane with its scales given in centimeters and its y-axis pointing upward. The plate is submerged in water at 20° Celsius and of density 0.988 grams per cubic centimeter. The surface of the water is at $y = 4$. What is the force of the water on each side of the plate?

◆ Type 4, force of a fluid

45. A plate bounded by the curve $y = x^{2/3}$ and the line $y = 1$ is submerged in benzene of density 0.879 grams per cubic centimeter. The plate forms an angle of 30° with the surface of the liquid and its straight edge lies along the surface. What is the force on each side of the plate?

◆ Type 4, force of a fluid

46.[A] The horizontal top of a trough is a rectangle, and the ends are semicircles with one-meter radius which make an angle of 60° with the top of the trough. What is the force on each end when the trough is full of milk weighing 10,100 newtons per cubic meter?

♦ Type 4, force of a fluid

47.[A] A plate in the shape of an isosceles right triangle with a 10-foot-long hypotenuse is suspended in water at its right angle with its hypotenuse 15 feet below the surface. What is the force of water pressure on each side of it?

♦ Type 4, force of a fluid

48.[A] A square swimming pool is 20 meters wide and its bottom is a slanted rectangle. The pool is one meter deep at one end and three meters deep at the other end. What is the force on the bottom of the pool when it is full of water weighing 9800 newtons per cubic meter?

♦ Type 4, force of a fluid

49. A boat 16 feet from a pier is moved to a distance of 9 feet by a sailor pulling on a rope with the constant force of 20 pounds. The rope passes through a pulley at the end of the pier and is fastened to the bow of the boat. The pulley is 12 feet above the level of the boat. How much work does the sailor do on the boat?

♦ Type 4,work

50. A right circular cone of radius 6 centimeters and height 10 centimeters is submerged in water with its rim at the surface. What is the force on the sides of the cone due to water pressure?

♦ Type 6, force of a fluid

51. A right triangle with sides of lengths 5 feet, 4 feet, and 3 feet is placed in water with its hypotenuse on the surface. What is the force of water pressure on one side of the triangle if it makes an angle of 30° with the surface of the water?

♦ Type 6, force of a fluid

52. A cylindrical can of radius 2 feet and of length 5 feet is full of carbon tetrachloride that weighs 100 pounds per cubic foot. What is the force of the oil on one end of the can when it is lying on its side?

♦ Type 6, force of a fluid

53. A child drinks a half pound of milk through a 10-inch straw that she keeps vertical and with its end on the bottom of her cylindrical glass. The depth of the milk in the glass changes from 6 inches to 2 inches as she drinks. How much work does she do against gravity?

♦ Type 4,work

54. A cylinder with a piston in one end contains air at a pressure of 100 pounds per square foot when the piston is 1 foot from the closed end of the cylinder. The area of the end of the cylinder is one square foot. The pressure p and the volume V satisfy an equation of the form $pV^{1.4} =$ constant. How much work is done by the air on the piston as the piston moves to a distance of two feet from the end of the cylinder?

♦ Type 4,work

55. It takes twice as much work against a spring to compress it from a length of 5 inches to a length of 4 inches as it takes to compress it from a length of 6 inches to a length of 5 inches. The spring satisfies Hooke's law. What is its natural length?

♦ Type 4,work

Section 6.8

The hyperbolic functions

OVERVIEW: *The* HYPERBOLIC FUNCTIONS *are combinations of exponential functions that are used to simplify formulas and calculations in the study of air resistance , heat, electrical circuits, ocean waves, and other topics in science and engineering. The hyperbolic functions are given names similar to those of the trigonometric functions because the two collections of functions have some similar properties.*

Topics:

- ***The hyperbolic sine and cosine***
- ***Hyperbolic functions and hyperbolas***
- ***The hyperbolic tangent***
- ***The other hyperbolic functions***
- ***Inverse hyperbolic functions***

The hyperbolic sine and cosine

The HYPERBOLIC SINE is denoted $\sinh x$, and the HYPERBOLIC COSINE is denoted $\cosh x$.[†] These functions are defined as follows.

Definition 1 *For any number x,*

$$\sinh x = \tfrac{1}{2}(e^x - e^{-x}) \quad \text{and} \quad \cosh x = \tfrac{1}{2}(e^x + e^{-x}). \tag{1}$$

The letter "h" in the symbols for these functions stands for "hyperbola."

Most properties of the hyperbolic functions can be determined easily from properties of exponential functions. For instance, $\sinh(0) = \frac{1}{2}(e^0 - e^{-0}) = 0$ and $\cosh(0) = \frac{1}{2}(e^0 + e^{-0}) = 1$ so that $\sinh x$, like $\sin x$, equals 0 at $x = 0$ and $\cosh x$, like $\cos x$, equals 1 at $x = 0$.

Definitions **(1)** also show that $\sinh x$ and $\cosh x$ have the derivatives

$$\begin{aligned}\frac{d}{dx}(\sinh x) &= \frac{d}{dx}\left[\tfrac{1}{2}(e^x - e^{-x})\right] = \tfrac{1}{2}\left[e^x - e^{-x}\frac{d}{dx}(-x)\right] \\ &= \tfrac{1}{2}(e^x + e^{-x}) = \cosh x\end{aligned}$$

and

$$\begin{aligned}\frac{d}{dx}(\cosh x) &= \frac{d}{dx}\left[\tfrac{1}{2}(e^x + e^{-x}\right] = \tfrac{1}{2}\left[e^x + e^{-x}\frac{d}{dx}(-x)\right] \\ &= \tfrac{1}{2}(e^x - e^{-x}) = \sinh x.\end{aligned}$$

These are the first pair of formulas in the next theorem. The second pair follows with the Chain Rule.

Theorem 1 **(a)** *For any x,*

$$\frac{d}{dx}(\sinh x) = \cosh x \quad \text{and} \quad \frac{d}{dx}(\cosh x) = \sinh x. \tag{2}$$

(b) *For any differentiable function $y = y(x)$,*

$$\frac{d}{dx}(\sinh y) = \cosh y\,\frac{dy}{dx} \quad \text{and} \quad \frac{d}{dx}(\cosh y) = \sinh y\,\frac{dy}{dx}. \tag{3}$$

[†]Pronounced "sinch" and "cosh.".

♦ SUGGESTIONS TO INSTRUCTORS: Have students begin work on Questions 1 through 7 before class.Problems 18, 21, 22, 23, 24, and 29 are good for in-class discussion.

Theorem 1 shows another similarity between hyperbolic functions and trigonometric functions. Formulas **(2)** can be obtained from the differentiation formulas, $\frac{d}{dx}(\sin x) = \cos x$ and $\frac{d}{dx}(\cos x) = -\sin x$, for the sine and cosine, by replacing "sin" by "sinh," "cos" by "cosh," and deleting the minus sign.

The graphs $y = \sinh x$ and $y = \cosh x$ are shown in Figures 1 and 2. Their particular shapes can be determined from definitions **(1)** and differentiation formulas **(2)**.

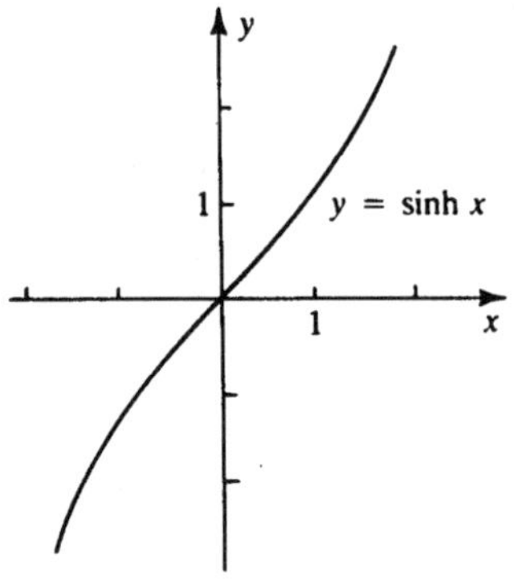

FIGURE 1

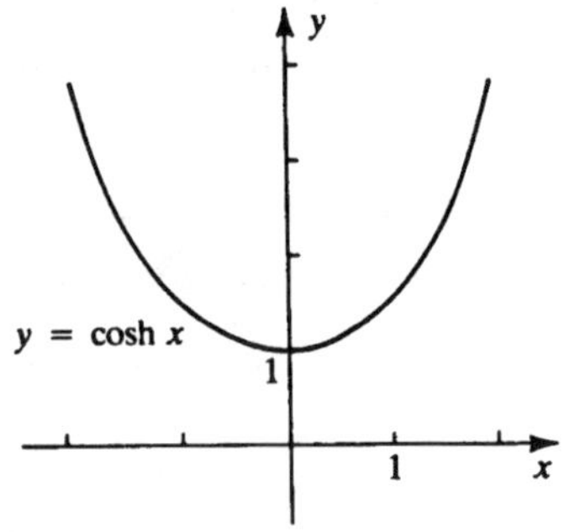

FIGURE 2

Question 1 **(a)** Why does the graph of $\sinh x$ in Figure 1 pass through the origin, and why does the graph of $\cosh x$ in Figure 2 have y-intercept equal to 1? **(b)** Why is the graph of $\sinh x$ symmetric about the origin and the graph of $\cosh x$ symmetric about the y-axis? **(c)** Why does $\sinh x$ tend to $-\infty$ as $x \to -\infty$ and tend to ∞ as $x \to \infty$? Why does $\cosh x$ tend to ∞ as $x \to \pm\infty$?

Question 2 Why is $\sinh x$ increasing for all x, and why is $\cosh x$ decreasing in $(-\infty, 0]$ and increasing in $[0, \infty)$?

Question 3 Why is the graph of $\sinh x$ concave down for $x < 0$ and concave up for $x > 0$? Why is the graph of $\cosh x$ concave up for all x?

We use differentiation formulas **(3)** in the next example and question.

Example 1 What is the x-derivative of $\sinh(6x^7 - 5x)$?

SOLUTION We use the first formula in **(3)** with $y = 6x^7 - 5x$:

$$\begin{aligned}\frac{d}{dx}[\sinh(6x^7 - 5x)] &= \cosh(6x^7 - 5x)\frac{d}{dx}(6x^7 - 5x)\\ &= \cosh(6x^7 - 5x)(42x^6 - 5). \ \square\end{aligned}$$

Question 4 Give an equation of the tangent line to $y = 3x^2 - \cosh x$ at $x = 4$. Generate the graph and the tangent line together as a partial check of your answer.

The differentiation formulas in Theorem 1 translate into the following integration rules.

Theorem 2 *For any x,*

$$\int \sinh x\, dx = \cosh x + C \tag{4}$$

$$\int \cosh x\, dx = \sinh x + C \tag{5}$$

and for any differentiable function $u = u(x)$,

$$\int \sinh u \frac{du}{dx}\, dx = \int \sinh u\, du = \cosh u + C \tag{6}$$

$$\int \cosh u \frac{du}{dx}\, dx = \int \cosh u\, du = \sinh u + C. \tag{7}$$

Example 2 What is the area of the region bounded by the curve $y = x^2\cosh(x^3)$ and the line $y = \cosh(1)$?

SOLUTION The region is shown in Figure 3. Because $x^2\cosh(x^3)$ equals $\cosh(1)$ at $x = \pm 1$, the curve and line intersect at $x = \pm 1$. Then, since the region is bounded on the bottom by $y = x^2\cosh(x^3)$ and on the top by $y = \cosh(1)$, its area is

$$[\text{Area}] = \int_{-1}^{1} [\cosh(1) - x^2\cosh(x^3)]\, dx. \tag{8}$$

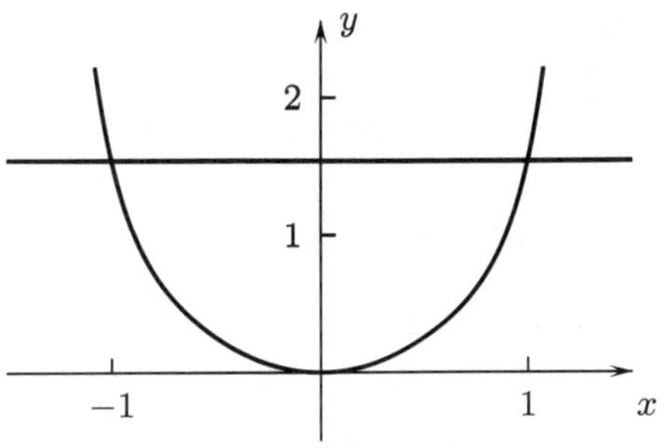

FIGURE 3

We separate (8) into two integrals and use the substitutions $u = x^3, du = 3x^2\, dx$ in the second :

$$\begin{aligned}
[\text{Area}] &= \int_{-1}^{1} \cosh(1)\, dx - \tfrac{1}{3}\int_{-1}^{1} \cosh(x^3)(3x^2\, dx) \\
&= \Big[x\cosh(1)\Big]_{-1}^{1} - \tfrac{1}{3}\int_{x=-1}^{x=1} \cosh u\, du \\
&= [\cosh(1) - (-1)\cosh(1)] - \tfrac{1}{3}\Big[\sinh u\Big]_{x=-1}^{x=1} \\
&= 2\cosh(1) - \tfrac{1}{3}\Big[\sinh(x^3)\Big]_{x=-1}^{x=1} \\
&= 2\cosh(1) - \tfrac{1}{3}[\sinh(1) - \sinh(-1)] \\
&= 2\cosh(1) - \tfrac{2}{3}\sinh(1). \ \square
\end{aligned}$$

The hyperbolic functions and hyperbolas

What do $\sinh x$ and $\cosh x$ have to do with hyperbolas and how are their definitions related to the definitions of $\sin x$ and $\cos x$?

Recall that $\sin x$ and $\cos x$ can be defined as the vertical, and horizontal coordinates of the intersection of the terminal side of the angle x with the unit circle $u^2+v^2=1$ in a uv-plane, as shown for a positive acute angle x in Figure 4. The point $(\cos x, \sin x)$ is on the circle because the sine and cosine satisfy the Pythagorean Identity

$$\cos^2 x + \sin^2 x = 1$$

for all values of x. As you will see in answering the next Question, the hyperbolic sine and cosine satisfy the Hyperbolic Identity,†

$$\cosh^2 x - \sinh^2 x = 1. \qquad \textbf{(9)}$$

Consequently, for each x the point $(\cosh x, \sinh x)$ is on the hyperbola $u^2 - v^2 = 1$ in Figure 5.

Question 5 Use definitions **(1)** to derive identity **(9)**.

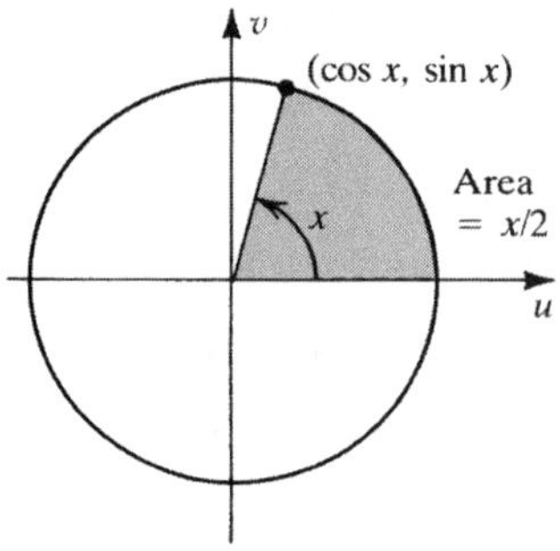

FIGURE 4

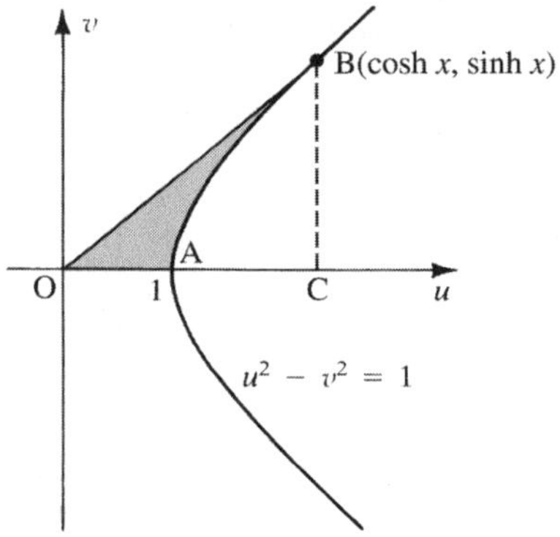

FIGURE 5

In the definition of $\sin x$ and $\cos x$ for a positive x as in Figure 4, x is the angle from the positive u-axis to the line through the point $(\cos x, \sin x)$ on the unit circle. To relate this to the definition of the hyperbolic sine and cosine in terms of the hyperbola in Figure 5, we need to express x in terms of an area.

Because x is measured in radians, the shaded area in Figure 4 is the fraction $\dfrac{x}{2\pi}$ of the area π of the entire circle and equals $\pi(\dfrac{x}{2\pi}) = \frac{1}{2}x$. Consequently, $(\cos x, \sin x)$ is the point on the circle such that the shaded area in Figure 4 has area $\frac{1}{2}x$.

The analogous definition of $\sinh x$ and $\cosh x$ is given by the following rule

---

Rule 1 *For positive x, $\cosh x$ and $\sinh x$ are the u- and v-coordinates of the point on the hyperbola $u^2 - v^2 = 1$ such that the shaded region in Figure 5 has area $\frac{1}{2}x$.*

To verify this rule, we will find the area of the shaded region in Figure 5 by finding first the areas of the triangle OBC and of the region ABC between the curve and the u-axis.

Question 6 What is the area of the triangle OBC? Express the answer in terms of e^{2x} and e^{-2x}.

We find the area of the region ABC by integration. We solve $u^2 - v^2 = 1$ for v by writing $v^2 = u^2 - 1$ and then $v = \pm\sqrt{u^2-1}$. This shows that the upper half of the curve in Figure 5 is the graph $v = \sqrt{u^2-1}$. Then because the u-coordinate of the point A is 1 and the u-coordinate of C is $\cosh x$,

$$[\text{Area of region } ABC] = \int_1^{\cosh x} \sqrt{u^2-1}\, du.$$

†To recall where the minus sign occurs in **(9)**, use the facts that $\cosh(0) = 1$ and $\sinh(0) = 0$.

We use the substitution $u = \cosh t$ for which $du = \sinh t\,dt$. Then, because $\sinh x$ is positive, identity **(9)** gives $\sqrt{u^2 - 1} = \sqrt{\cosh^2 t - 1} = \sinh t$. Also $1 \underset{u}{\longrightarrow} \cosh x$ as $0 \underset{t}{\longrightarrow} x$, so that

$$\begin{aligned}[\text{Area of region } ABC] &= \int_0^x \sinh^2 t\,dt \\ &= \int_0^x \left[\tfrac{1}{2}(e^t - e^{-t})\right]^2 dt = \tfrac{1}{4}\int_0^x (e^{2t} - 2 + e^{-2t})\,dt \\ &= \tfrac{1}{4}\left[\tfrac{1}{2}e^{2t} - 2t - \tfrac{1}{2}e^{-2t}\right]_0^x = \tfrac{1}{8}(e^{2x} - e^{-2x}) - \tfrac{1}{2}x.\end{aligned} \tag{10}$$

Combining **(10)** with the result of Question 5 shows that

$$[\text{Area of region } OAB] = [\text{Area of triangle } OCB] - [\text{Area of region } ABC] = \tfrac{1}{2}x.$$

This establishes Rule 1.

The hyperbolic tangent

The definition of the hyperbolic tangent function $\tanh x$ in terms of $\sinh x$ and $\cosh x$ is similar to the definition of $\tan x$ in terms of $\sin x$ and $\cos x$:

Definition 2 *For any x,*

$$\tanh x = \frac{\sinh x}{\cosh x} = \frac{e^x - e^{-x}}{e^x + e^{-x}}. \tag{11}$$

We can find the derivative of $\tanh x$ from the derivative of $\sinh x$ and $\cosh x$ by using Theorem 1 with the Quotient Rule. We obtain, with identity **(9)**

$$\begin{aligned}\frac{d}{dx}(\tanh x) &= \frac{d}{dx}\left(\frac{\sinh x}{\cosh x}\right) \\ &= \frac{\cosh x \dfrac{d}{dx}(\sinh x) - \sinh x \dfrac{d}{dx}(\cosh x)}{\cosh^2 x} \\ &= \frac{(\cosh x)(\cosh x) - (\sinh x)(\sinh x)}{\cosh^2 x} \\ &= \frac{\cosh^2 x - \sinh^2 x}{\cosh^2 x} = \frac{1}{\cosh^2 x}.\end{aligned}$$

With the Chain Rule, we obtain the following.

Theorem 3 *For any x,*

$$\frac{d}{dx}(\tanh x) = \frac{1}{\cosh^2 x} \tag{12}$$

and for any differentiable function $y = y(x)$,

$$\frac{d}{dx}(\tanh y) = \frac{1}{\cosh^2 y}\frac{dy}{dx}. \tag{13}$$

The graph of $\tanh x$ is shown in Figure 6. This function is zero at $x = 0$ because $\sinh(0) = 0$ and is increasing for all x because its derivative **(12)** , is positive for all x.

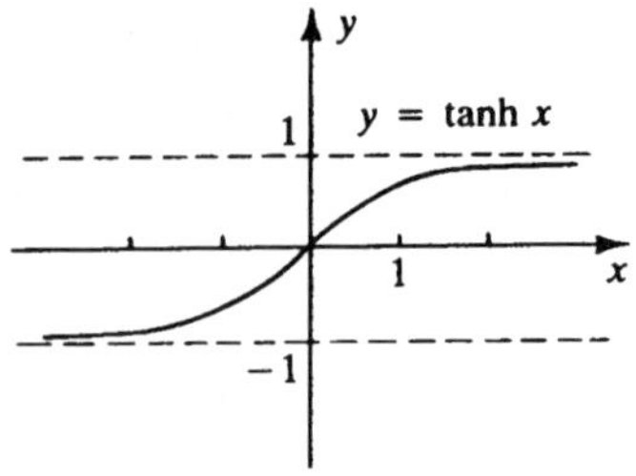

FIGURE 6

Question 7 Use definition **(11)** to show that **(a)** $\lim_{x\to\infty} \tanh x = 1$ and **(b)** $\lim_{x\to-\infty} \tanh x = -1$, as is suggested by Figure 6.

The other hyperbolic functions

The hyperbolic cotangent, secant, and cosecant can be given in terms of $\sinh x$ and $\cosh x$ by formulas similar to the definitions of the trigonometric functions $\cot x, \sec x$, and $\csc x$ in terms of $\sin x$ and $\cos x$.

$$\coth x = \frac{\cosh x}{\sinh x}, \quad \operatorname{sech} x = \frac{1}{\cosh x}, \quad \operatorname{csch} x = \frac{1}{\sinh x}.$$

With these definitions, the derivative of $\tanh x$ in **(12)** can be expressed in terms of $\operatorname{sech} x$, and the derivatives of the other functions can be calculated by the Quotient and Chain Rules (see Problem 51):

Theorem 4 *For all x except where denominators are zero,*

$$\frac{d}{dx}(\tanh x) = (\operatorname{sech} x)^2 \tag{14}$$

$$\frac{d}{dx}(\coth x) = -(\operatorname{csch} x)^2 \tag{15}$$

$$\frac{d}{dx}(\operatorname{sech} x) = -\operatorname{sech} x \tanh x \tag{16}$$

$$\frac{d}{dx}(\operatorname{csch} x) = -\operatorname{csch} x \coth x. \tag{17}$$

We list formulas **(14)** through **(17)**. primarily to show their similarities with corresesponding formulas for the trigonometric functions.

Inverse hyperbolic functions

Because the hyperbolic functions are defined in terms of e^x, the inverse hyperbolic sine $\sinh^{-1} x$ and other inverse hyperbolic functions can be expressed in terms of the natural logarithm.

Example 3 Give a formula for $\sinh^{-1} x$ in terms of $\ln x$.

SOLUTION $\sinh^{-1} x$ is defined for all x because $\sinh x$ is increasing for all x and because $\sinh x \to \infty$ as $x \to \infty$ and $\sinh x \to -\infty$ as $x \to -\infty$.

We set $x = \sinh y = \frac{1}{2}(e^y - e^{-y})$ and solve for y. Multiplying both sides of the equation by $2e^y$ gives $2xe^y = e^{2y} - 1$, which we rewrite as $(e^y)^2 - 2x(e^y) - 1 = 0$. Then, by the Quadratic Formula.

$$e^y = \frac{2x \pm \sqrt{4x^2 + 4}}{2} = x \pm \sqrt{x^2 + 1}.$$

We take the plus sign here because e^y is positive. Applying the natural logarithm to both sides of this equation gives

$$\sinh^{-1} x = y = \ln(x + \sqrt{x^2 + 1}). \square$$

Principles and procedures

- The hyperbolic functions are special combinations of e^x and e^{-x} which are given their own names because they arise in many applications. Their names are chosen to be similar to those of the trigonometric functions because they have many analogous properties. They are called "hyperbolic" functions because $\cosh x$ and $\sinh x$ can be defined in terms of the hyperbola $u^2 - v^2 = 1$, as described in Rule 1.
- You should know the definitions of $\cosh x$ and $\sinh x$ in terms of exponential functions, the shapes of their graphs, and the formulas for their derivatives. You should also know the defintion of $\tanh x$ and the shape of its graph.
- If at any time you cannot proceed with calculations using hyperbolic functions, use the definitions to express the hyperbolic functions in terms of exponential functions and then continue.
- Because the hyperbolic functions are constructed from e^x and e^{-x}, the inverse hyperbolic functions can all be expressed in terms of the natural logarithm.
- We will see in Section 13.2 that the sine and cosine functions can be written

$$\sin x = \frac{e^{ix} - e^{-ix}}{2i} \quad \text{and} \quad \cos x = \frac{e^{ix} + e^{-ix}}{2}$$

  where i is the symbol that represents the "imaginary" square root of -1. Most of the similarities between trigonometric and hyperbolic functions can be attributed to the similarity of these formulas with definitions **(1)** of $\sinh x$ and $\cosh x$.

Tune-Up Exercises 6.8♦

[A]*Answer provided.* [O]*Outline of solution provided.*

T1.[A] Use definition **(1)** to explain why $4\sinh x + 6\cosh x$ is positive for all x.

♦ Type 2,hyperbolic function

T2.[A] What is the derivative of $4\sinh x + 6\cosh x$?

♦ Type 1, a derivative

T3.[O] What is the derivative of $\sinh(6x)$?

♦ Type 1, a derivative

T4.[O] Evaluate $\displaystyle\int_0^3 \sinh(2x)\,dx$.

♦ Type 1, an integral

T5.[O] **(a)** For what values of x is $\dfrac{\cosh x}{\sqrt{\sinh x}}$ defined? **(b)** Find its antiderivatives.

♦ Type 2, a derivative, etc.

Problems 6.8

[A]*Answer provided.* [O]*Outline of solution provided.*

1.[A] Find all solutions x of **(a)** $\sinh(\ln x) = 0$ and **(b)** $\cosh(x^2 + 1) = 1$.

♦ Type 2, solving equations

Find formulas for the derivatives in Problems 2 through 9.

♦The Tune-up Exercises and Problems are classified by type and content. The types are (1) basic, reactive; (2) basic reflective; (3) intermediate, reactive; (4) intermediate, reflective; (5) advanced, reactive; (6) advanced, reflective; and (7) advanced, theoretical.

2.[O] $\frac{d}{dx}[3\sinh(4x) - 4\cosh(3x)]$

♦ Type 1, a derivative

3.[A] $\frac{d}{dx}(x^2 \sinh x)$

♦ Type 1, a derivative

4.[A] dy/dx for $y = \ln(\cosh x)$

♦ Type 1, a derivative

5. y' for $y = \sinh(e^{-x})$

♦ Type 1, a derivative

6. $\frac{d}{dx}[\cos(\cosh x)]]$

♦ Type 1, a derivative

7.[A] $\frac{d}{dt}\left[\frac{\cosh t}{t}\right]$

♦ Type 1, a derivative

8. $\frac{d}{dx}(\cosh^6 x)$

♦ Type 1, a derivative

9 $\frac{d}{dx}[\sinh(\ln x)]$

♦ Type 1, a derivative

Perform the integration in Problems 10 through 17.

10.[O] $\int_0^{10} \cosh(3x)\,dx$

♦ Type 1, an integral

11. $\int \sinh x \cosh x\,dx$

♦ Type 1, an integral

12.[O] $\int_1^3 \frac{\sinh x}{1+\cosh x}\,dx$

♦ Type 1, an integral

13[A] $\int \frac{\sinh\sqrt{x}}{\sqrt{x}}\,dx$

♦ Type 1, an integral

14.[A] $\int \tanh x\,dx$

♦ Type 1, an integral

15. $\int \sinh(\sin x)\cos x\,dx$

♦ Type 1, an integral

16. $\int_{-1}^{1} e^{3x}\cosh(e^{3x})\,dx$

♦ Type 1, an integral

17. $\int_1^{10} \frac{1}{x}\cosh(\ln x)\,dx$

♦ Type 1, an integral

18 **(a)** Use l'Hopital's Rule to find $\lim_{x\to 0}\left(\frac{\sinh x}{x}\right)$. **(b)** Give an equation of the tangent line to $y = \frac{\sinh x}{x}$ at $x = 1$. Generate the curve and the tangent line and copy them on your paper.

♦ Type 2, a derivative

19. Show that the derivative of $\sqrt{\sinh^2 x + 1}$ is $\sinh x$.

♦ Type 2, l'Hopital's Rule and a tangent line

20. Derive formulas **(15)** through **(17)** for the derivatives of $\coth x$, $\operatorname{sech} x$, and $\operatorname{csch} x$.

♦ Type 3, deriving differentiation formulas

21. A rectangular metal plate occupies the region $0 \le x \le \pi, 0 \le y \le 3$ in an xy-plane with distances measured in feet (Figure 7). The temperature at each point on the plate is kept constant by keeping the temperature constant at each point along its four sides. The temperature at (x, y) is $T(x,y) = \frac{20\sin x \sinh y}{\sinh(3)}$ degrees Celsius. **(a)** What is the temperature of the plate along its four edges? **(b)** What is the maximum temperature of the plate and at what point does it have that temperature? **(c)** How much cooler is the center of the plate than the maximum temperature? **(d)** Show that $T(x, y)$ satisfies LAPLACE'S EQUATION, $T_{xx}(x,y) + T_{yy}(x,y) = 0$.

♦ Type 3, hyperbolic functions with maximum

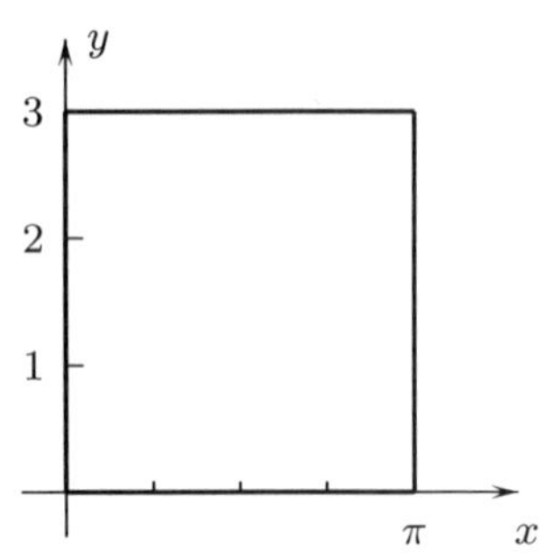

FIGURE 7

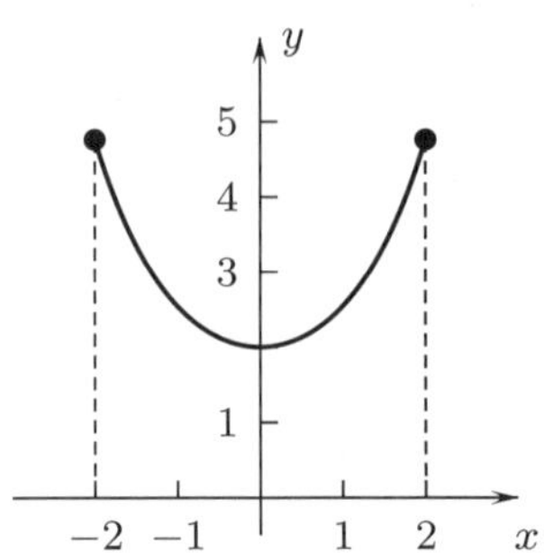

FIGURE 8

22. When a cable of uniform density is hung from the two points in Figure 8, it takes the shape of the CATENARY, $y = \cosh x + 1$. How long is the cable? Give the exact and approximate decimal values.

♦ Type 3, length of a graph

23. A ball is thrown straight up in the air from a window of a low-flying helicopter. Because of the combined forces of gravity and air resistance, the ball's upward velocity $v(t)$ at time t is $16\tan(1-2t)$ feet per second for $0 \le t \le 0.5$ and $16\tanh(1-2t)$ for $t \ge 0.5$ until the ball hits the ground. The graph of this function is shown in Figure 9.† **(a)** What is the ball's initial upward velocity? **(b)** At what time does the ball reach its highest point? **(c)** What is the ball's upward acceleration at its highest point? **(d)** What would be the limit as $t \to \infty$ of the ball's downward velocity if it could fall forever?

♦ Type 4, a problem requiring the tangent and the hyperbolic tangent

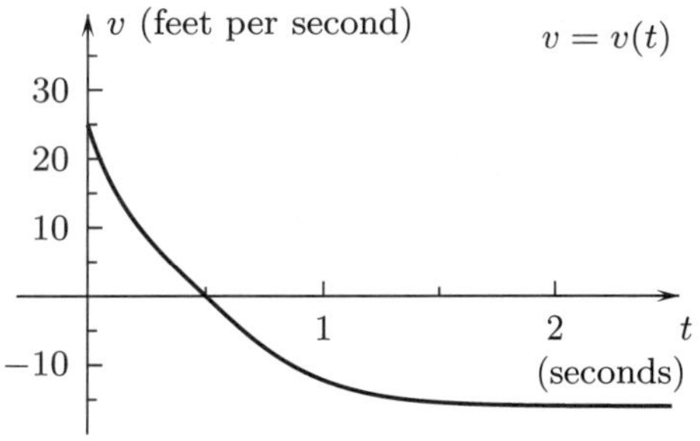

FIGURE 9

24.[A] When the catenary $y = \cosh x$ for $-1 \le x \le 1$ is rotated about the x-axis, it generates a surface called a CATENOID, which has the least area of all surfaces generated by curves having the same endoints. What is its area?

♦ Type 3, surface area

25. The WAVELENGTH L(meters) of a sequence of ocean waves is the distance between crests and the FREQUENCY ν (seconds^{-1}) is the number of crests that pass a fixed point each second. In shallow water h meters deep, the frequency and wavelength are related by the equation $\nu^2 = \dfrac{g}{2\pi L}\tanh\left(\dfrac{2\pi h}{L}\right)$, where $g \doteq 9.8$ meters per second^2 is the acceleration due to gravity.(1) **(a)** Show, without using any derivatives that for fixed h the frequency decreases when the wavelength increases. **(b)** Suppose your boat is anchored where the ocean is ten meters deep and a wave crest passes every 12 seconds. Use an equation-solving algorithm on your calculator or computer to find the approximate wavelength and speed of the waves at your boat. **(c)** Use the tangent line to $y = \tanh x$ at $x = 0$ to give an approximate formula for ν^2 in terms of g, h, and L for L much larger than h. **(d)** Give an approximate formula for ν^2 for L much smaller than h.

♦ Type 4, tangent lines, etc.

26. Show that the global minimum of $\sinh x + 2\cosh x$ is $\sqrt{3}$. Generate the graph to guide your reasoning.

♦ Type 3, a global maximum

27. Find **(a)** the exact area of the region between $y = \sinh x\, dx$ and the x-axis for $0 \le x \le \ln(10)$ and **(b)** the exact volume of the solid that is generated when the region of part (a) is rotated about the x-axis.

♦ Type 3, an area and volume

28. What is the global minimum of $(1 + \cosh x)^8$?

♦ Type 3, a global minimum

29. The inverse hyperbolic tangent function is the solution $y = y(x)$ of the equation $\tanh y = x$ for $-1 < x < 1$. **(a)** Solve this equation to obtain a formula for $\tanh^{-1} x$ in terms of the natural logarithm. **(b)** Use the result of part (a) to give a formula for the derivative of the inverse hyperbolic tangent, $\tanh^{-1} x$. How does this formula differ from that for the derivative of the inverse tangent function, $\tan^{-1} x$?

♦ Type 3, the inverse hyperbolic tangent

30. Find the antiderivative of $x\sinh x$ that has the value 10 at $x = 0$.

♦ Type 3, an antiderivative

---

†In this mathematical model the air resistance is assumed to be proprotional to the square of the ball's velocity.

(1) *Wind Waves. Their Generation and Propagation on the Ocean Surface* by B. Kinsman, New York, NY: Dover, 1984, p. 125.

RESPONSES TO QUESTIONS IN CHAPTER 6

Responses 6.1

Response 1 The length of the base of the square cross section at x in Figure 4 is the distance $L = (1 - \frac{1}{2}x^2) - \frac{1}{2}x^2 = 1 - x^2$ between the curves $y = 1 - \frac{1}{2}x^2)$ and $y = \frac{1}{2}x^2$. • The area of the cross section is $A(x) = L^2 = (1 - x^2)^2$.

Response 2 The integrand is the function $(1 - x^2)^2$ obtained by replacing x_{j-1} with x in formula **(3)**. • The limits of integration are 0 and 1 because the partition is of the interval $[0, 1]$. • The integral is $\int_0^1 (1 - x^2)^2\,dx$.

Response 3 The volume of the four boxes in Figure 7 is greater than the volume of the solid in Figure 6 because the solid is contained in the boxes.

Response 4 The volume of the jth wafer equals the area $A(x_{j-1})$ of its faces multiplied by its thickness Δx. • $[\text{Volume}] = A(x_{j-1})\,\Delta x$

Response 5 **(a)** The solid has the shape of a curved horn with its tip at the origin in the xy-plane and with a circular end. (Figure R5)
(b) $[\text{Diameter of the cross section at } x] = \frac{4}{3}x^{1/3} - x^{1/3} = \frac{1}{3}x^{1/3}$ •
$[\text{Radius}] = \frac{1}{6}x^{1/3}$ • $A(x) = \pi(\frac{1}{6}x^{1/3})^2 = \frac{1}{36}\pi x^{2/3}$ •
$[\text{Volume}] = \int_0^1 A(x)\,dx = \int_0^1 \frac{1}{36}\pi x^{2/3}\,dx = \left[\frac{1}{60}\pi x^{5/3}\right]_0^1 = \frac{1}{60}\pi$ cubic feet

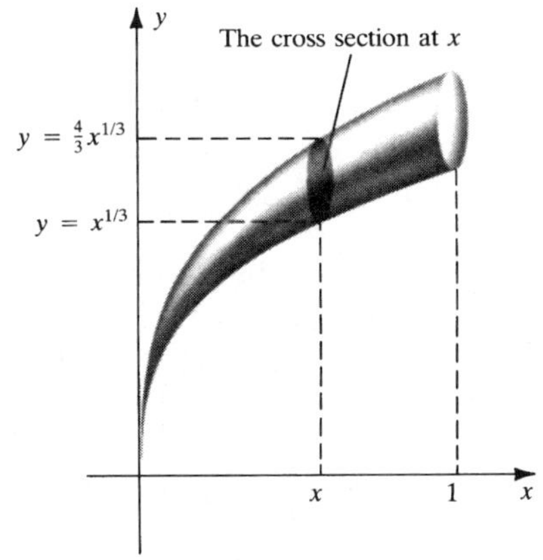

Figure R5

Response 6 **(a)** The cross section at x is a circle of radius $x - x^4$ and area $A(x) = \pi(x - x^4)^2 = \pi(x^2 - 2x^5 + x^8)$ for $0 \le x \le 1$. •
(b) $[\text{Volume}] = \int_0^1 A(x)\,dx = \int_0^1 \pi(x^2 - 2x^5 + x^8)\,dx$
$= \pi\left[\frac{1}{3}x^3 - \frac{2}{6}x^6 + \frac{1}{9}x^9\right]_0^1 = \pi[\frac{1}{3}(1^3) - \frac{2}{6}(1^6) + \frac{1}{9}(1^9)] - \pi[0] = \frac{1}{9}\pi$

Response 7 Find the intersection of the curves: $4/x = x \iff x^2 = 4 \iff x = \pm 2$ • The region extends from $x = 1$ to $x = 2$. • When the vertical line across the region at x is rotated about the y-axis, it generates a cylinder of radius x, height $\frac{4}{x} - x$, and area $A(x) = 2\pi x(\frac{4}{x} - x) = 2\pi(4 - x^2)$. •
$[\text{Volume}] = \int_1^2 A(x)\,dx = 2\pi\int_1^2 (4 - x^2)\,dx = 2\pi\left[4x - \frac{1}{3}x^3\right]_1^2 = \frac{10}{3}\pi$

Response 8 **(a)** [Area of the entire washer] $= \pi(10^2) - \pi(5^2) = 75\pi$

(b) The cut and flattened lampshade in Figure 26 of Section 6.1 has outer circumference 12π and the outer circumference of the entire washer is 20π, so the lampshade is the fraction $\dfrac{12\pi}{20\pi} = \frac{3}{5}$ of the entire washer.

Response 9 For the lampshade in Figure 24, $r_1 = 6, r_2 = 3$, and $L = 5$, so that [Area] $= \pi(r_1 + r_2)L = \pi(6+3)(5) = 45\pi$

Response 10 $f(x) = 2x^{3/2}$ • $f'(x) = 3x^{1/2}$ •

$$[\text{Surface area}] = 2\pi\int_0^5 2x^{3/2}\sqrt{1+9x}\ dx \approx 1596.7$$

Responses 6.2

Response 1 First integrate by parts, differentiating x^2 and integrating e^x:

Set $\left\{\begin{matrix} u = x^2 \\ dv/dx = e^x \end{matrix}\right\}$ for which $\left\{\begin{matrix} du/dx = 2x \\ v = e^x \end{matrix}\right\}$:

$\displaystyle\int x^2e^x\ dx = x^2e^x - \int (2x)e^x\ dx$ •

Then differentiate $2x$ and integrate e^x:

Set $\left\{\begin{matrix} u = 2x \\ dv/dx = e^x \end{matrix}\right\}$ for which $\left\{\begin{matrix} du/dx = 2 \\ v = e^x \end{matrix}\right\}$:

$\displaystyle\int x^2e^x\ dx = x^2e^x - [2xe^x - \int 2e^x\ dx] = x^2e^x - [2xe^x - 2e^x] + C$

$= x^2e^x - 2xe^x + 2e^x + C$

Response 2 Set $\left\{\begin{matrix} u = \tan^{-1} x \\ dv/dx = 1 \end{matrix}\right\}$ for which $\left\{\begin{matrix} du/dx = 1/(x^2+1) \\ v = x \end{matrix}\right\}$:

$\displaystyle\int \tan^{-1} x\ dx = \int (\tan^{-1} x)(1)\ dx = x\tan^{-1} x - \int \frac{x}{x^2+1}\ dx$ •

Use the substitution $u = x^2 + 1, du = 2x\ dx$:

$\displaystyle\int \tan^{-1} x\ dx = x\tan^{-1} x - \tfrac{1}{2}\int \frac{1}{u}\ dx = x\tan^{-1} x - \tfrac{1}{2}\ln|u| + C$

$= x\tan^{-1} x - \frac{1}{2}\ln(x^2+1) + C$

Response 3 In order for the graph of $\dfrac{2}{x^2-1}$ to have asymptotes at $x = 1$ and $x = -1$ as in Figure 2 of Section 6.2, it must be the sum of $\dfrac{1}{x-1}$ and $\dfrac{-1}{x+1}$. •

Check: $\dfrac{1}{x-1} + \dfrac{-1}{x+1} = \dfrac{1}{x-1} - \dfrac{1}{x+1}$

$= \dfrac{x+1}{(x-1)(x+1)} - \dfrac{x-1}{(x-1)(x+1)} = \dfrac{(x+1)-(x-1)}{x^2-1} = \dfrac{2}{x^2-1}$

Response 4 According to Figure 3 of Section 6.2, $\dfrac{2x^2+x-1}{x^3-x^2} \to \infty$ as $x \to 0$, so it must be equal to $\dfrac{1}{x^2} + \dfrac{2}{x-1}$ rather than $\dfrac{1}{x} + \dfrac{2}{x-1}$. •

Check: $\dfrac{1}{x^2} + \dfrac{2}{x-1} = \dfrac{x-1}{x^2(x-1)} + \dfrac{2x^2}{x^2(x-1)} = \dfrac{2x^2+x-1}{x^3-x^2}$

Response 5 Since $Ax + \dfrac{B}{x^2+1} = B$ at $x = 0$, $B = -2$. •

Because $Ax + \dfrac{B}{x^2+1} \approx Ax$ for large $|x|$, $A = 1$.

Check: $x + \dfrac{-2}{x^2+1} = \dfrac{x(x^2+1)}{x^2+1} - \dfrac{2}{x^2+1} = \dfrac{x^3+x-2}{x^2+1}$

Response 6 $\displaystyle\int \frac{2x^2+x-1}{x^3-x^2}\,dx = \int [\frac{1}{x^2} + \frac{2}{x-1}]\,dx = -x^{-1} + 2\ln|x-1| + C$

Response 7 $\displaystyle\int_0^3 \frac{x^3+x-2}{x^2+1}\,dx = \int_0^3 \left[x - \frac{2}{x^2+1}\right]\,dx = \left[\tfrac{1}{2}x^2 - 2\tan^{-1}x\right]_0^3$
$= [\frac{9}{2} - 2\tan^{-1}(3)] - [0] = \frac{9}{2} - 2\tan^{-1}(3)$

Response 8 (a) $\dfrac{x+2}{x^3-x} = \dfrac{x+2}{x(x-1)(x+1)} = \dfrac{A}{x} + \dfrac{B}{x-1} + \dfrac{C}{x+1}$ •

$\dfrac{x+2}{x^3-x} = \dfrac{A(x-1)(x+1) + Bx(x+1) + Cx(x-1)}{x^3-x}$ •

$x+2 = A(x-1)(x+1) + Bx(x+1) + Cx(x-1)$ •

Set $x = 0$: $-A = 2 \implies A = -2$

Set $x = 1$: $3 = 2B \implies B = \frac{3}{2}$

Set $x = -1$: $1 = 2C \implies C = \frac{1}{2}$

$\dfrac{x+2}{x^3-x} = -\dfrac{2}{x} + \frac{3}{2}(\dfrac{1}{x-1}) + \frac{1}{2}(\dfrac{1}{x+1})$

(b) $\displaystyle\int \frac{x+2}{x^3-x}\,dx = \int \left[-\frac{2}{x} + \tfrac{3}{2}(\frac{1}{x-1}) + \tfrac{1}{2}(\frac{1}{x+1})\right]\,dx$

$\displaystyle = -\int \frac{1}{x}\,dx + \tfrac{3}{2}\int \frac{1}{u}\,du + \tfrac{1}{2}\int \frac{1}{v}\,dv$

(with $u = x-1, du = dx$ and $v = x+1, dv = dx$)

$= -2\ln|x| + \frac{3}{2}\ln|x-1| + \frac{1}{2}\ln|x+1| + C$

Response 9 $\displaystyle\int_2^4 \frac{2x-1}{(x-1)^2}\,dx = \int_2^4 \left[\frac{2}{x-1} + (x-1)^{-2}\right]\,dx = \left[2\ln|x-1| - (x-1)^{-1}\right]_2^4$
$= [2\ln(3) - 2\ln(1)] - [\frac{1}{3} - 1] = 2\ln(3) + \frac{2}{3}$ •

Check: With a procedure of approximate integration, a calculator gives the value 2.86389124 for the integral and for $2\ln(3) + \frac{2}{3}$.

Response 10 $\dfrac{3x^2-4x+1}{(x-2)(x^2+1)} = \dfrac{A(x^2+1) + (x-2)(Bx+C)}{(x-2)(x^2+1)}$ •

$3x^2 - 4x + 1 = A(x^2+1) + (Bx+C)(x-2)$ •

Set $x = 2$: $3(2^2) - 4(2) + 1 = (2^2+1)A \implies A = 1$ •

$$\begin{aligned} 3x^2 - 4x + 1 &= x^2 + 1 + Bx^2 - 2Bx + Cx - 2C \\ &= (1+B)x^2 + (-2B+C)x + (1-2)C \end{aligned}$$

$$\left\{\begin{array}{c} 1 + B = 3 \\ -2B + C = -4 \\ 1 - 2C = 1 \end{array}\right\} \implies B = 2,\ C = 0 \bullet$$

$\dfrac{3x^2-4x+1}{(x-2)(x^2+1)} = \dfrac{1}{x-2} + \dfrac{2x}{x^2+1}$

Response 11 $3x + \dfrac{x+2}{x^2+1} = \dfrac{3x(x^2+1)}{x^2+1} + \dfrac{x+2}{x^2+1} = \dfrac{3x(x^2+1) + (x+2)}{x^2+1} = \dfrac{3x^3+4x+2}{x^2+1}$

Response 12 $\displaystyle\int \frac{3x^3+4x+2}{x^2+1}\,dx = \int [3x + \frac{x}{x^2+1} + \frac{2}{x^2+1}]\,dx$
$= \frac{3}{2}x^2 + \frac{1}{2}\ln(x^2+1) + 2\tan^{-1}x + C$

Responses 6.3

Response 1 **(a)** One decription: The solid looks like four lens-shaped beads on a wire.
(b) The vertical line from the x-axis to the curve at x in Figure 2 is of length $4|\sin x|$ feet and when it is rotated about the x-axis, it generates a circle of area $A(x) = \pi(4\sin^x)^2 = 16\pi\sin^2 x$ square feet. • [Volume] $= \int_0^{4\pi} A(x)\,dx = 16\pi\int_0^{4\pi} \sin^2 x\,dx$ cubic meters.

Response 2 Use the Pythgorean Identity $\sin^2 x = 1 - \cos^2 x$ to replace two factors of $\sin x$ by $\cos x$:
$\int_0^{\pi/2} \sin^3 x\,dx = \int_0^{\pi/2} (1-\cos^2 x)\sin x\,dx$ •
Use $u = \cos x, du = -\sin x\,dx$:
$\int_0^{\pi/2} \sin^3 x\,dx = -\int_{x=0}^{x=\pi/2} (1-u^2)\,du = \left[\frac{1}{3}u^3 - u\right]_{x=0}^{x=\pi/2}$
$= \left[\frac{1}{3}\cos^3 x - \cos x\right]_0^{\pi/2} = 0 - \left[\frac{1}{3} - 1\right] = \frac{2}{3}$

Response 3 By Example 5, $F(x) = \int \sin(5x)\cos(2x)\,dx = -\frac{1}{6}\cos(3x) - \frac{1}{14}\cos(7x) + C$ •
$F(0) = -\frac{1}{6} - \frac{1}{14} = -\frac{5}{21}$ • Use $F(0) = 1 \iff C = \frac{26}{21}$ •
$F(x) - \frac{1}{6}\cos(3x) - \frac{1}{14}\cos(7x) + \frac{26}{21}$

Response 4 $x = \sin\theta, dx = \cos\theta\,d\theta$, and $\cos\theta = \sqrt{1-x^2}$

Response 5 Use $x = \sin^{-1}\theta, \sqrt{1-x^2} = \cos\theta, x = \sin\theta$, and $dx = \cos\theta\,d\theta$:
$\int \frac{1}{x\sqrt{1-x^2}}\,dx = \int \frac{1}{\sin\theta\cos\theta}\cos\theta\,d\theta = \int \csc\theta\,d\theta$ • Use formula **(35)**:
$-\ln|\csc\theta + \cot\theta| + C = -\ln\left|\frac{1}{\sin\theta} + \frac{\cos\theta}{\sin\theta}\right| + C = -\ln\left|\frac{1+\sqrt{1-x^2}}{x}\right| + C$

Response 6 [Average value] $= \frac{1}{4-2}\int_2^4 \frac{1}{x^2\sqrt{x^2-1}}\,dx = \frac{1}{2}\int_2^4 \frac{1}{x^2\sqrt{x^2-1}}\,dx.$ •
Use $\theta = \sec^{-1} x$, for which $x = \sec\theta, dx = \sec\theta\,\tan\theta\,dx$, and (since $x > 0$ in the integral) $\sqrt{x^2-1} = \sqrt{\sec^2\theta - 1} = \tan\theta$:
[Average value] $= \frac{1}{2}\int_{x=2}^{x=4} \frac{1}{(\sec\theta)^2\tan\theta}(\sec\theta\,\tan\theta\,d\theta)$
$= \frac{1}{2}\int_{x=2}^{x=4} \cos\theta\,d\theta = \frac{1}{2}\left[\sin\theta\right]_{x=2}^{x=4} = \frac{1}{2}\left[\frac{\sqrt{x^2-1}}{x}\right]_{x=2}^{x=4} = \frac{1}{8}\sqrt{15} - \frac{1}{4}\sqrt{3}$

Responses 6.4

Response 1 Use $u = \cos x, du = -\sin x\,dx$:
$\int_0^{\pi/2} \frac{\sin x}{\sqrt{1+\cos^2 x}}\,dx = -\int_{x=0}^{x=\pi/2} \frac{1}{\sqrt{1-\cos^2 x}}(-\sin x\,dx)$
$= -\int_{x=0}^{x=\pi/2} \frac{1}{\sqrt{1-u^2}}\,dx$ • Use formula **(89)** in the table with u in place of x:
$\int_0^{\pi/2} \frac{\sin x}{\sqrt{1+\cos^2 x}}\,dx = -\left[\ln(u + \sqrt{u^2+1})\right]_{x=0}^{x=\pi/2}$
$= -\left[\ln(\cos x + \sqrt{\cos^2 x + 1})\right]_{x=0}^{x=\pi/2}$

$= -[\ln(\cos(\frac{1}{2}\pi) + \sqrt{\cos^2(\frac{1}{2}\pi) + 1})] + [\ln(\cos(0) + \sqrt{\cos^2(0) + 1}) = \ln(1 + \sqrt{2})$

Response 2 Because the interval $[-k, k]$ has length $2k$, the average value is $\frac{1}{2k}\int_{-k}^{k} \frac{1}{(x^2 + k^2)^2}\,dx$. • A calculator gives the value $\frac{\pi + 2}{8k^4}$ for this integral and then the solutions $k = \pm\frac{1}{10}5^{3/4}(\pi + 2)^{1/4}$ of the equation $\frac{\pi + 2}{8k^4} = 10$. • Only the positive value $k = \frac{1}{10}5^{3/4}(\pi + 2)^{1/4}$ applies.

Response 3 The object's velocity is an antiderivative $v = \int t^2 e^{-t}\,dt$ of its acceleration. • A calculator gives $\int t^2 e^{-t}\,dt = (-t^2 - 2t - 2)e^{-t}$. • Add a constant of integration: $v = (-t^2 - 2t - 2)e^{-t} + C$ • The object's s-coordinate is an antiderivative of its velocity. • A calculator gives $\int [(-t^2 - 2t - 2)e^{-t} + C]\,dt = (t^2 + 4t + 6)e^{-t} + Ct$. • Add another constant of integration: $s = (t^2 + 4t + 6)e^{-t} + Ct + C_1$

Responses 6.5

Response 1 **(a)** $\int_0^{20} f(t)\,dt = 1550(15) + 650(5) = 26{,}500$ (people) •

$\int_{20}^{90} f(t)\,dt = 650(10) + 420(15) + 180(15) + 100(15) + 30(15) = 17{,}450$

(b) The fraction of the population less than 20 was $\frac{26{,}500}{26{,}500 + 17{,}450} \doteq 0.603$. • About 60.3%

Response 2 We approximate the area between the graph of $p(r)$ and the r-axis by six rectangles, as in Figure R2. • $\int_0^{0.3} p(r)\,dr \approx 40(0.05) + 270(0.05) + 680(0.05) + 680(0.05) + 270(0.05) + 40(0.05) = 99\%$ • The value 99% for the integral is plausible because, according to Figure 6 in Section 6.5, a high percentage of the grains of sand have radii between 0 and 3 millimeters.

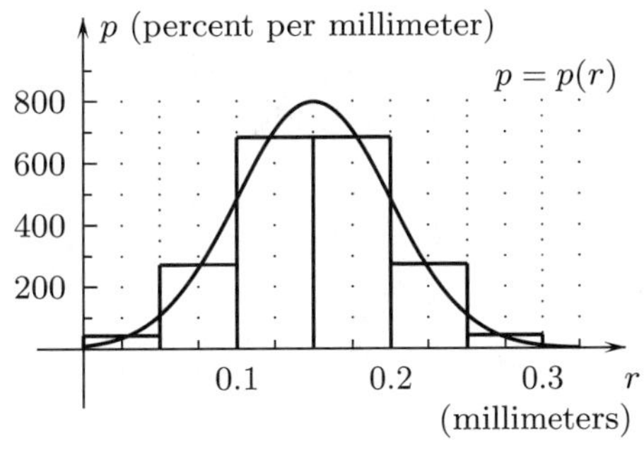

Figure R2

Response 3 The formula $P(T) = 52.8 + 3.92(T - 15)$ for $15 \le T \le 30$ from Example 3 in Section 6.5 gives $P(30) = 111.6$. • $P(T) = 111.6 + 2.95(T - 30)$ for $30 \le T \le 45$

Response 4 $\int_0^X \sin x\,dx = \Big[-\cos x\Big]_0^X = -\cos X + 1$ •

$\frac{d}{dX}\int_0^X \sin x\,dx = \frac{d}{dX}(-\cos X + 1) = \sin(X)$

Response 5 **(a)** Since the outcomes of all the experiments were ≤ 10, the total number of experiments was $N(10) = 10^5 e^{-10/2} \doteq 674$. **(b)** The number of experiments for which the outlcome was between 5 and 8 was
$N(8) - N(5) = 8^5 e^{8/2} - 5^5 e^{5/2} \doteq 600 - 257 = 343$.
(c) $N(10)$ is the vertical coordinate of the point on the graph of $N(X)$ at $X = 10$ and is the area between the graph of $n(x)$ and the x-axis. • $N(8) - N(5)$ is the difference between the vertical coordinates of the point on the graph of $N(X)$ at $X = 8$ and $X = 5$ (Figure R5a) and is the area between the graph of $n(x)$ and the x-axis for $5 \leq x \leq 8$ (Figure R5b).

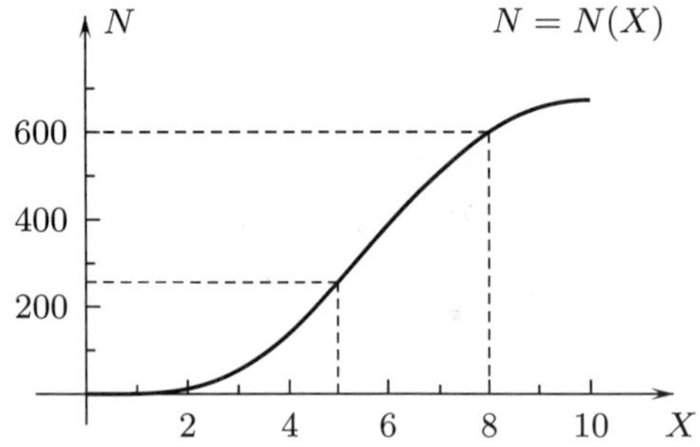

Figure R5a

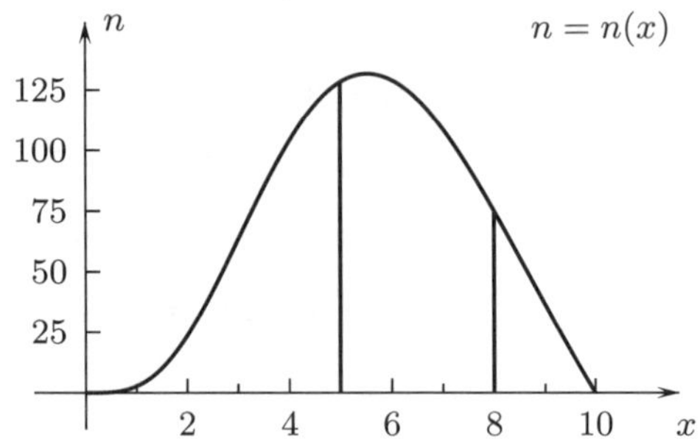

Figure R5b

Response 6 $\displaystyle\int_0^{X^3} \cos y \, dy = \sin(X^3)$ •

$$\frac{d}{dX}\int_0^{X^3} \cos y \, dy = \frac{d}{dX}[\sin(X^3)] = \cos(X^3)\frac{d}{dX}(X^3) = 3X^2\cos(X^3)$$

Responses 6.6

Response 1 $\displaystyle\lim_{X\to\infty} [\text{Area for } 1 \leq x \leq X] = \lim_{X\to\infty}\left[\frac{-1}{X} + 1\right] = 1$

Response 2 The area of the region (Figure R2a) is given by the integral $\displaystyle\int_0^1 \frac{1}{x}\,dx$, which is improper because $\dfrac{1}{x} \to \infty$ as $x \to 0^+$. •

For $0 < X < 1$, $\displaystyle\int_X^1 \frac{1}{x}\,dx = \Big[\ln|x|\Big]_X^1 = \ln(1) - \ln(X) = -\ln(X)$ is the area of the region in Figure R2b. • Because $\ln(X) \to -\infty$ as $X \to 0^+$,

$$[\text{Area for } 0 < x \leq 1] = \int_0^1 \frac{1}{x}\,dx = \lim_{X\to 0+}\int_X^1 \frac{1}{x}\,dx = \lim_{X\to 0+}[-\ln(X)] = \infty$$

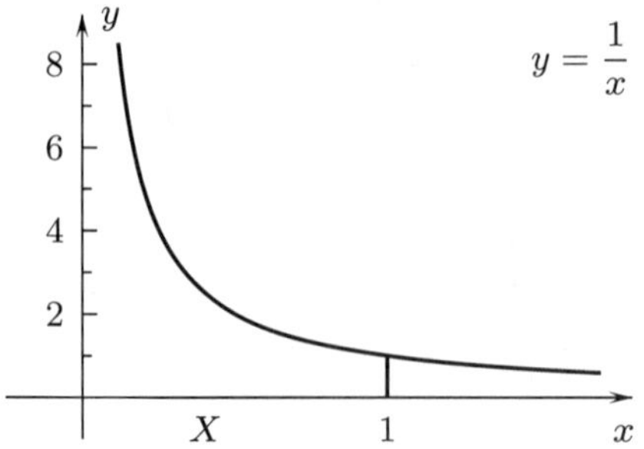

Figure R2a

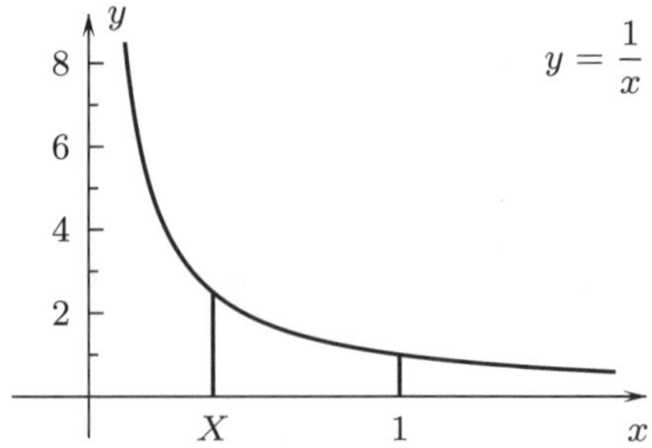

Figure R2b

Response 3 $\int_{-1}^{1} \frac{1}{x}\,dx = \int_{-1}^{0} \frac{1}{x}\,dx + \int_{0}^{1} \frac{1}{x}\,dx$ if it is defined (Figure R3). •

$\int_{0}^{1} \frac{1}{x}\,dx = \infty$ from Question 2 •

$$\int_{-1}^{0} \frac{1}{x}\,dx = \lim_{X\to 0^-} \int_{-1}^{X} \frac{1}{x}\,dx = \lim_{X\to 0^-} \Big[\ln|x|\Big]_{-1}^{X} = \lim_{X\to 0^-} [\ln|X| - \ln|-1|]$$

$= \lim_{X\to 0^-} \ln|X| = -\infty$ • $\int_{-1}^{1} \frac{1}{x}\,dx$ is not defined because it would equal $-\infty+\infty$. It diverges without a limit.

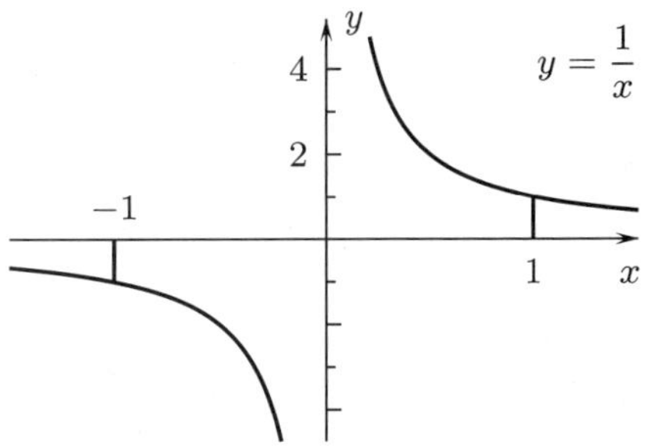

Figure R3

Response 4 With three flips there are eight possible outcomes:

{T, T, T}
{T, T, H} {T, H, T} {H,T, T}
{T, H, H} {H, ,T, H} {H, H, T}
{H,H, H}

There is one chance in eight (12.5%) that no heads (three tails) occur, three chances in eight (37.5%) that one head and two tails occur, three chances in eight (37.5%) that two heads and one tail occur, and one chance in eight (12.5%) that two heads occur. • The function in Figure 8 has the values $p(0) = 12.5, p(1) = 37.5, p(2) = 37.5$, and $p(3) = 12.5$. •
The sum of these values is $12.5 + 37.5 + 37.5 + 12.5 = 100$.

Response 5 To simplify the calculations, we delete the constant $1/(\sigma\sqrt{2\pi})$ from formula **(12)** and find the inflection points of $y = f(x)$ with $f(x) = e^{-x^2/(2\sigma^2)}$. •

$$f'(x) = \frac{d}{dx}[e^{-x^2/(2\sigma^2)}] = e^{-x^2/(2\sigma^2)}\frac{d}{dx}\left[\frac{-x^2}{2\sigma^2}\right] = -\frac{1}{\sigma^2}\,x\,e^{-x^2/(2\sigma^2)} \quad \bullet$$

By the Product Rule,

$$f''(x) = -\frac{1}{\sigma^2}\left[x\frac{d}{dx}[e^{-x^2/(2\sigma^2)}] + e^{-x^2/(2\sigma^2)}\frac{d}{dx}(x)\right]$$

$$= -\frac{1}{\sigma^2}\left[\frac{-x^2}{\sigma^2}e^{-x^2/(2\sigma^2)} + e^{-x^2/(2\sigma^2)}\right] = \frac{1}{\sigma^2}e^{-x^2/(2\sigma^2)}\left(\frac{x^2}{\sigma^2} - 1\right) \bullet$$

Because $\frac{1}{\sigma^2}e^{-x^2/(2\sigma^2)}$ is positive for all x, $f''(x)$ is positive for $x < -\sigma$, negative for $-\sigma < x < \sigma$, and positive for $x > \sigma$. • The inflection points of the graph are at $x = \pm\sigma$, as is suggested by Figure 15 in Section 6.6.

Response 6 A calculator gives **(a)** $\int_{-3}^{3} n(x)\,dx \approx 0.9973$ and $\int_{-6}^{6} n(x)\,dx \approx 0.99999994$ for $\sigma = 1$ and **(b)** $\int_{-15}^{15} n(x)\,dx \approx 0.9973$ and $\int_{-25}^{25} n(x)\,dx \approx 0.99999994$ for $\sigma = 5$.

Response 7 **(a)** $n(x) = \dfrac{1}{33\sqrt{2\pi}} e^{-(x-148)^2/(2(33)^2)}$ • Figure R7 •

(b) [Fraction with blood-cholesterol levels < 200] $= \displaystyle\int_{-\infty}^{200} n(x)\,dx$
$\approx \displaystyle\int_{-100}^{200} p(x)\,dx \approx 0.94$ • [Percent with blood-cholesterol levels < 200] $\approx 94\%$

Figure R7

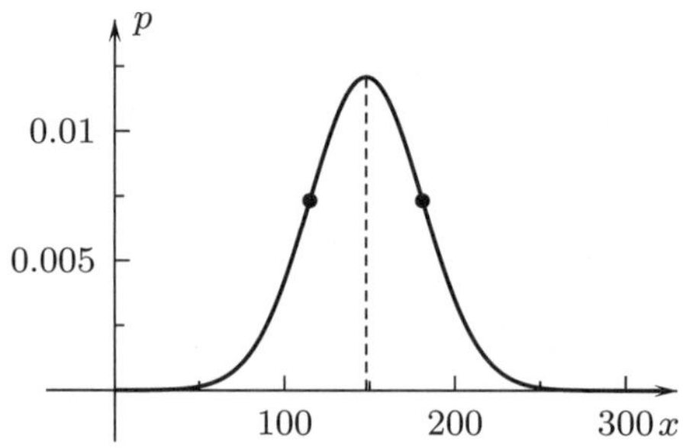

Responses 6.7

Response 1 **(a)** On the way down the ball goes from $s = 20$ to $s = 0$, so the work done by the constant force of gravity on the ball is

$$[-0.5 \text{ pounds}][(0-20) \text{ feet}] = 10 \text{ foot-pounds.}$$

(b) The total work done by gravity is the the work done as the ball rises plus the work done as the ball falls. The results of part (a) and Example 1 give

$$[\text{Total work}] = -10 + 10 = 0 \text{ foot-pounds.}$$

Response 2 **(a)** The force of gravity on the ball, measured in newtons, is its weight, which is equal to its its mass multiplied by the acceleration due to gravity. • On the surface of the earth, the acceleration due to gravity is $g = 9.8$ meters per second^2 and the ball weighs

$$[10 \text{ kilograms}][9.8 \text{ meters per second}^2] = 98 \text{ newtons.}$$

(b) The force of gravity r meters from the center of the earth is $F = \dfrac{k}{r^2}$ with some constant k. • On the surface of the earth, $r = 6500$ and $F = 98$. •
$98 = \dfrac{k}{6500^2} \iff k = 98(6500)^2$ • $F = \dfrac{98(6500)^2}{r^2}$ newtons

Response 3 When the spring is stretched s feet, it exerts a force of $-\frac{3}{2}s$ and the force on it is $\frac{3}{2}s$. • s goes from 0 to 5 as the spring is stretched from 10 feet to 15 feet. •
[Work done against the spring] $= \displaystyle\int_0^5 \tfrac{3}{2}s\,ds = \left[\tfrac{3}{4}s^2\right]_0^5 = \tfrac{3}{4}(5^2) = 18\tfrac{3}{4}$ foot-pounds

Response 4 [Acceleration due to gravity] $= 32$ feet per second^2 •
[The truck's mass] $= m = 4000/32 = 125$ slugs •
[Initial kinetic energy] $= \frac{1}{2}(125)(20^2) = 25{,}000$ foot-pounds •
[Final kinetic energy] $= \frac{1}{2}(125)(80^2) = 400{,}000$ foot-pounds •
[Work by all forces] $= 400{,}00 - 25{,}000 = 375{,}000$ foot-pounds

Responses 6.8

Response 1 **(a)** $y = \sinh x$ passes through the origin because $\sinh(0) = 0$. • $y = \cosh x$ has y-intercept 1 because $\cosh(0) = 1$.
(b) $y = \sinh x$ is symmetric about the origin because $\sinh(-x) = \frac{1}{2}(e^{-x} - e^{x}) = -\sinh x$. • $y = \cosh x$ is symmetric about the y-axis because $\cosh(-x) = \frac{1}{2}(e^{-x} + e^{x}) = \cosh x$.
(c) $\sinh x \to \infty$ and $\cosh x \to \infty$ as $x \to \infty$ because $e^{x} \to \infty$ and $e^{-x} \to 0$ as $x \to \infty$ • $\sinh x \to -\infty$ and $\cosh x \to \infty$ as $x \to -\infty$ because $e^{x} \to 0$ and $e^{-x} \to \infty$ as $x \to -\infty$

Response 2 $\sinh x$ is increasing for all x because $\dfrac{d}{dx}(\sinh x) = \cosh x$ and $\cosh x$ is positive for all x. • $\cosh x$ is decreasing in $(-\infty, 0]$ and increasing in $[0, \infty)$ because it is continuous for all x and its derivative $\sinh x$ is negative for $x < 0$ and positive for $x > 0$.

Response 3 The graph of $\sinh x$ is concave down for $x < 0$ and concave up for $x > 0$ because its second derivative is $\sinh x$ which is negative for $x < 0$ and positive for $x > 0$. • The graph of $\cosh x$ is concave up for all x because its second derivative $\cosh x$ is positive for all x.

Response 4 $y = 3x^2 - \cosh x$ • $y' = 6x - \sinh x$ • $y(4) = 48 - \cosh(4)$ • $y'(4) = 24 - \sinh(4)$ • Tangent line: $y = 48 - \cosh(4) + (24 - \sinh(4))(x - 4)$ • Figure R4

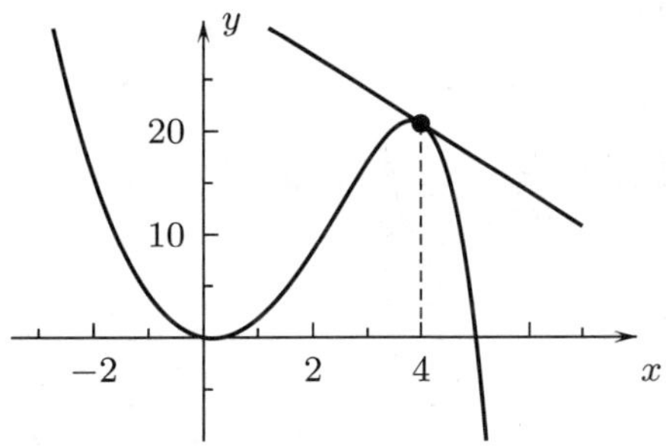

Figure R4

Response 5 $\cosh^2 x - \sinh^2 x = \left[\frac{1}{2}(e^{x} + e^{-x})\right]^2 - \left[\frac{1}{2}(e^{x} - e^{-x})\right]^2$
$= \frac{1}{4}\left[(e^{x})^2 + 2e^{x}e^{-x} + (e^{-x})^2\right] - \left[(e^{x})^2 - 2e^{x}e^{-x} + (e^{-x})^2\right] = \frac{1}{4}(4e^{x}e^{-x}) = 1$

Response 6 The base of the triangle is $b = \cosh x$ and its height is $h = \sinh x$, so its area is

$$\begin{aligned}[\text{Area}] &= \tfrac{1}{2}bh = \tfrac{1}{2}\cosh x \sinh x = \tfrac{1}{2}(\tfrac{1}{2})(e^{x} - e^{-x})(\tfrac{1}{2})(e^{x} + e^{-x}] \\ &= \tfrac{1}{8}[(e^{x})^2 - (e^{-x})^2) = \tfrac{1}{8}(e^{2x} - e^{-2x}).\end{aligned}$$

Response 7 **(a)** Divide the numerator and denominator in the definition of $\tanh x$ by e^{x}:

$$\tanh x = \frac{e^{x} - e^{-x}}{e^{x} + e^{-x}} = \frac{1 - e^{-2x}}{1 + e^{-2x}} \to 1 \text{ as } x \to \infty$$

(b) Divide the numerator and denominator in the definition of $\tanh x$ by e^{-x}:

$$\tanh x = \frac{e^{x} - e^{-x}}{e^{x} + e^{-x}} = \frac{e^{2x} - 1}{e^{2x} + 1} \to -1 \text{ as } x \to -\infty$$

CHAPTER 6 ANSWERS

Tune-Up Exercises 6.1

T1. Base: Figure T1a • The cross section at x is a square of width $w = x^3$. • Figure T1b • $A(x) = w^2 = x^6$ for $0 \le x \le 1$ • [Volume]= $\int_0^1 A(x)\,dx = \int_0^1 (x^3)^2\,dx = \int_0^1 x^6\,dx = \left[\frac{1}{7}x^7\right]_0^1 = \frac{1}{7}$

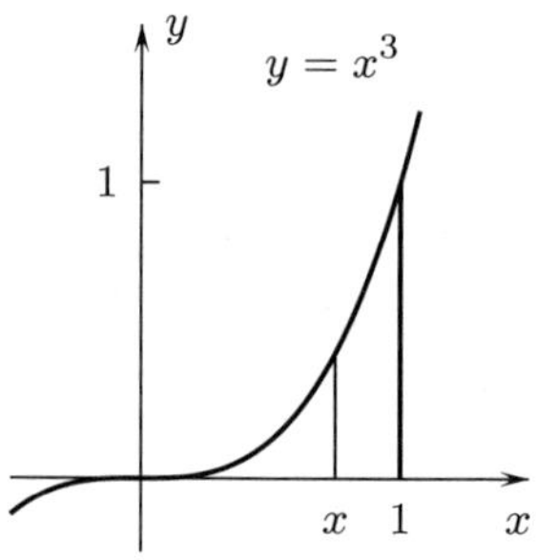

Figure T1a

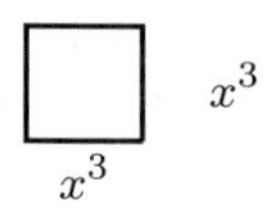

Figure T1b

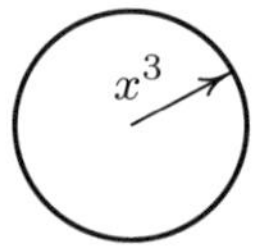

Figure T2

T2. The vertical line across the region at x in Figure T1a generates the cross section at x in Figure T2, which is a circle of radius $r = x^3$. • $A(x) = \pi r^2 = \pi x^6$ •
[Volume]= $\int_0^1 A(x)\,dx = \int_0^1 \pi(x^3)^2\,dx = \pi\int_0^1 x^6\,dx = \left[\frac{1}{7}\pi x^7\right]_0^1 = \frac{1}{7}\pi$

T3. When the vertical line at x in Figure T1 is rotated about the y-axis, it generates the cylinder at x in Figure T3 with radius $r = x$ and height $h = x^3$ • $A(x) = 2\pi rh = 2\pi x^4$ • [Volume]= $\int_0^1 A(x)\,dx = \int_0^1 2\pi x(x^3)\,dx = 2\pi\int_0^1 x^4\,dx = \left[\frac{2}{5}\pi x^5\right]_0^1 = \frac{2\pi}{5}$

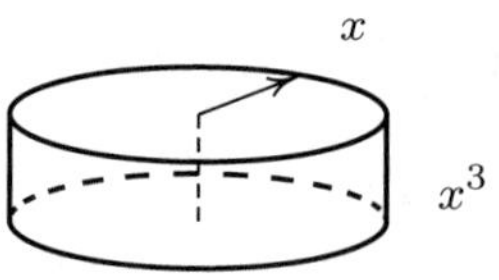

Figure T3

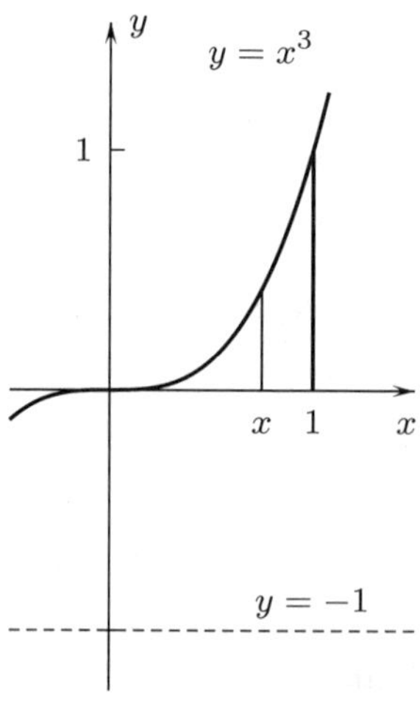

Figure T4a

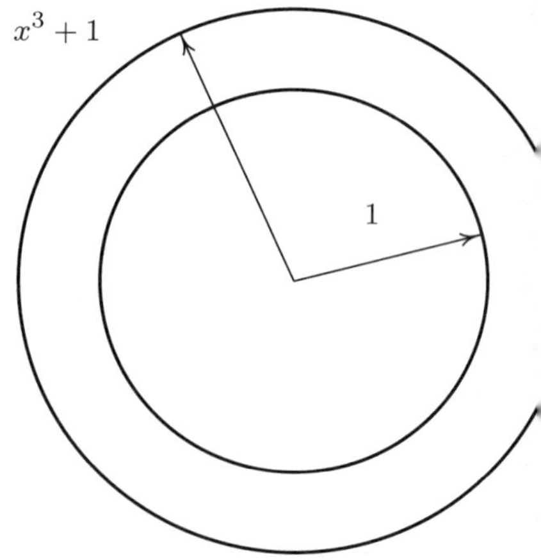

Figure T4b

T4. **(a)** When the cross section at x in Figure T4a is rotated about the line $y = -1$, it generates the cross section at x in Figure T4b, which is a ring with outer radius $r_1 = x^3 + 1$ and inner radius $r_0 = 1$
(b) $A(x) = \pi(r_1)^2 - \pi(r_0)^2 = \pi(x^3+1)^2 - \pi(1)^2 = \pi(x^6 + 2x^3)$ • [Volume]= $\int_0^1 A(x)\,dx$
$= \int_0^1 \pi[(x^3+1)^2 - 1]\,dx = \pi\int_0^1 (x^6 + 2x^3)\,dx = \pi\left[\frac{1}{7}x^7 + \frac{1}{2}x^4\right]_0^1 = \pi(\frac{1}{7} + \frac{1}{2}) = \frac{9}{14}\pi$

T5. $y = x^3$ • $y' = 3x^2$ • [Surface area]$= \int_0^1 2\pi y\sqrt{1+(y')^2}\,dx = \int_0^1 2\pi x^3\sqrt{1+(3x^2)^2}\,dx$
$= 2\pi\int_0^1 x^3\sqrt{1+9x^4}\,dx \approx 3.563121052$

Problems 6.1

1. $w = \sqrt{x} - x^2$ and $h = 2w$ • $A(x) = 2(\sqrt{x}-x^2)^2$ • [Volume] $= \int_0^1 A(x)\,dx$
$= \int_0^1 2(\sqrt{x}-x^2)^2\,dx = \int_0^1 (2x - 4x^{5/2} + 2x^4)\,dx = \frac{9}{35}$

3. 27

6. One answer: $\approx$ 7,000,000 cubic feet

7. $\pi\int_0^8 (x^{1/3})^2\,dx = \frac{96}{5}\pi$

8. $\pi\int_{-\sqrt{2}}^{\sqrt{2}} (1-\frac{1}{2}x^2)^2\,dx = 2\pi\int_0^{\sqrt{2}} (1-\frac{1}{2}x^2)^2\,dx = 2\pi\int_0^{\sqrt{2}} (\frac{1}{4}x^4 - x^2 + 1)\,dx = \frac{16}{15}\sqrt{2}\,\pi$

10. $4\left[x - \frac{1}{2}x^4 + \frac{1}{7}x^7\right]_0^1 = 4[1 - \frac{1}{2} + \frac{1}{7}] - 4[0] = \frac{18}{7}$

12. 4π

15. $\int_0^2 2\pi x(3x - x^2 - x)\,dx = 2\pi\int_0^2 (2x^2 - x^3)\,dx = \frac{8}{3}\pi$

16. $\frac{16}{5}$

18. $\frac{8}{3}\pi(k^{3/2} - 1)$

22. $\frac{1}{24}\pi$

23. $2\pi\int_1^4 x^{-1}\sqrt{1+x^{-4}}\,dx \approx 9.4172$

26. $\frac{1}{4}\sqrt{3}\int_{-2}^2 (4-x^2)\,dx = \frac{8}{3}\sqrt{3}$

28. $2\pi\left[\frac{1}{3}mx^3 - \frac{1}{4}x^4\right]_0^m = \frac{1}{6}\pi m^4$

31. 50 cubic feet

35. $\pi\left[e^{x^2}\right]_0^2 = \pi(e^4 - 1)$

40. The cylindrical section at x has radius x and height $x - \frac{1}{2}x = \frac{1}{2}x$ • $A(x) = 2\pi(\frac{1}{2}x) = \pi x^2$ •
[Volume]$= \int_0^k \pi x^2\,dx = \left[\frac{1}{3}\pi x^3\right]_0^k = \frac{1}{3}\pi k^3$

41a. [Area]$= \int_1^2 (e^{3-x} - e^{2/x})\,dx \approx 0.505033633$

42. $x^2 = \cos x$ at $x = k \approx \pm 0.824132312$ • [Volume]$= \int_{-k}^k (\cos x - x^2)^2\,dx \approx 0.874405088$

47. $h = \left(\frac{25}{62.4}\right)^{2/3} \doteq 0.5435$ feet

48. [Weight]$= (62.4)(\frac{8}{81}\pi) \doteq 19.36$ pounds

50. [Volume]$= 8\int_0^4 (16 - x^2)\,dx = 8\left[16x - \frac{1}{3}x^3\right]_0^4 = 8\left[16(4) - \frac{1}{3}(4)^3\right] = 341\frac{1}{3}$

54. $[\text{Volume}] = -\left[\frac{2}{3}(25-x^2)^{3/2}\right]_0^5 = \frac{2}{3}(25)^{3/2} = \frac{250}{3}$

56. **(a)** $\frac{1}{4}\pi a^2$ **(b)** πab **(c)** $2\pi^2 abc$

Tune-Up Exercises 6.2

T1. $u = \ln(2x),\ \dfrac{dv}{dx} = 1$ • $\dfrac{du}{dx} = \dfrac{1}{x},\ v = x$ •

$$\int_1^5 \ln(2x)\,dx = \Big[x\ln(2x)\Big]_1^5 - \int_1^5 x\left(\frac{1}{x}\right)dx$$

$$= \Big[x\ln(2x) - x\Big]_1^5 = 5\ln(10) - \ln(2) - 4$$

T2. $u = x^2,\ \dfrac{dv}{dx} = e^{3x}$ • $\dfrac{du}{dx} = 2x,\ v = \frac{1}{3}e^{3x}$ •

$$\int x^2e^{3x}\,dx = \tfrac{1}{3}x^2e^{3x} - \int 2x(\tfrac{1}{3}e^{3x})\,dx = \tfrac{1}{3}x^2e^{3x} - \tfrac{2}{3}\int xe^{3x}\,dx \ \bullet$$

$u = x,\ \dfrac{dv}{dx} = e^{3x}$ • $\dfrac{du}{dx} = 1, v = \frac{1}{3}e^{3x}$ •

$$\int x^2e^{3x}\,dx = \tfrac{1}{3}x^2e^{3x} - \tfrac{2}{3}\int xe^{3x}\,dx = \tfrac{1}{3}x^2e^{3x} - \tfrac{2}{3}\left[\tfrac{1}{3}xe^{3x} - \int \tfrac{1}{3}e^{3x}\,dx\right]$$

$$= \tfrac{1}{3}x^2e^{3x} - \tfrac{2}{9}xe^{3x} + \tfrac{2}{27}e^{3x} + C$$

T3. Use substitution, not integration by parts: $u = \tan^{-1}x, du = (1+x^2)^{-1}dx$ •

$$\int \frac{\tan^{-1}x}{1+x^2}\,dx = \int u\,du = \tfrac{1}{2}u^2 + C = \tfrac{1}{2}(\tan^{-1}x)^2 + C$$

T4. Use substitution, not partial fractions: $u = x + x^2,\ du = (1+2x)\,dx$. •

$$\int \frac{1+2x}{x+x^2}\,dx = \int \frac{1}{u}\,du = \ln|u| + C = \ln|x+x^2| + C$$

T5. $x^2 - 3x + 2 = (x-1)(x-2)$ by inspection or the Quadratic Formula. •

$\dfrac{5x-7}{x^2-3x+2} = \dfrac{5x-7}{(x-1)(x-2)} = \dfrac{A}{x-1} + \dfrac{B}{x-2}$ • $\dfrac{5x-7}{(x-1)(x-2)} = \dfrac{A(x-2)+B(x-1)}{(x-1)(x-2)}$ •

$5x - 7 = A(x-2) + B(x-1)$ • Set $x = 2$ • $5(-2) - 7 = B(2-1)$ • $3 = B$

Set $x = 1$ • $5(1) - 7 = A(1-2)$ • $A = 2$ • $\dfrac{5x-7}{x^2-3x+2} = \dfrac{2}{x-1} + \dfrac{3}{x-2}$

T6. Figure T6 • $\dfrac{x^3}{x^2-4} = x + \dfrac{4x}{x^2-4}$ • $\dfrac{4x}{x^2-4} = \dfrac{4x}{(x+2)(x-2)} = \dfrac{A}{x+2} + \dfrac{B}{x-2}$ •

Multiply by $(x+2)(x-2)$: $4x = A(x-2) + B(x+2)$ • Set $x = 2$: $4(2) = B(4)$ • $B = 2$ •

Set $x = -2$: $4(-2) = A(-4)$ • $A = 2$ • $\dfrac{4x}{x^2-4} = \dfrac{2}{x+2} + \dfrac{2}{x-2}$ • $\dfrac{x^3}{x^2-4} = x + \dfrac{2}{x+2} + \dfrac{2}{x-2}$

$$\begin{array}{r|l} & x \\ \hline x^2-4 & x^3 \\ & \underline{x^3 - 4x} \\ & 4x \end{array}$$

Figure T6

$$\begin{array}{r|l} & x^2 + 4x + 11 \\ \hline x^2 - 4x + 5 & x^4 \\ & \underline{x^4 - 4x^3 + 5x^2} \\ & 4x^3 - 5x^2 \\ & \underline{4x^3 - 16x^2 + 20x} \\ & 11x^2 - 20x \\ & \underline{11x^2 - 44x + 55} \\ & 24x - 55 \end{array}$$

Figure T8

T7. $\dfrac{9}{x^3-6x^2+9x} = \dfrac{9}{x(x-3)^2} = \dfrac{A}{x} + \dfrac{B}{x-3} + \dfrac{C}{(x-3)^2}$ • Multiply by $x(x-3)^2$: $9 = A(x-3)^2 + Bx(x-3) + Cx$ • Set $x=0$: $9 = 9A$ • $A = 1$ • Set $x=3$: $9 = 3C$ • $C = 3$ • $9 = (x-3)^2 + Bx(x-3) + 3x = (1+B)x^2 + (-3-3B)x + 9$ • $1+B = 0, -3-3B = 0$ • $B = -1$ • $\dfrac{9}{x^3-6x^2+9x} = \dfrac{1}{x} - \dfrac{1}{x-3} + \dfrac{3}{(x-3)^2}$

T8. Figure T8 • $\dfrac{x^4}{x^2-4x+5} = x^2 + 4x + 11 + \dfrac{24x-55}{x^2-4x+5}$ • x^2-4x+5 is irreducible because here $b^2-4ac = (-4)^2 - 4(5) = -4$ is negative.

T9. By Exercise T5: $\displaystyle\int \frac{5x-7}{x^2-3x+2}\,dx = \int \frac{2}{x-1}\,dx + \int \frac{3}{x-2}\,dx = 2\int \frac{1}{v}\,dv + 3\int \frac{1}{u}\,du$ with $u = x-2, du = dx$ and $v = 1-x, dv = dx$ • $\displaystyle\int \frac{5x-7}{x^2-3x+2}\,dx = 3\ln|u| + 2\ln|v| + C = 3\ln|x-2| + 2\ln|x-1| + C$

T10. By Exercise T6: $\displaystyle\int \frac{x^3}{x^2-4}\,dx = \int \left(x + \frac{2}{x+2} + \frac{2}{x-2}\right)dx = \tfrac{1}{2}x^2 + 2\ln|x+2| + 2\ln|x-2| + C$ • $\displaystyle\int_6^{10} \frac{x^3}{x^2-4}\,dx = \left[\tfrac{1}{2}x^2 + 2\ln|x+2| + 2\ln|x-2|\right]_6^{10}$ $= [\tfrac{1}{2}(10)^2 + 2\ln(12) + 2\ln(8)] \quad -[\tfrac{1}{2}(6)^2 + 2\ln(8) + 2\ln(4)] = 32 + \ln(9)$

T11. By Exercise T7: $\displaystyle\int \frac{9}{x^3-6x^2+9x}\,dx = \int \left[\frac{1}{x} - \frac{1}{x-3} + \frac{3}{(x-3)^2}\right]dx$ $= \ln|x| - \ln|x-3| - 3(x-3)^{-1} + C$

T12. Use the result of Exercise T8 and complete the square in the irreducible quadratic: $x^2-4x+5 = (x-2)^2 + 1$. • $u = x-2, x = u+2, du = dx$:
$\displaystyle\int \frac{x^4}{x^2-4x+5}\,dx = \int (x^2+4x+11)\,dx + \int \frac{24x-55}{(x-2)^2+1}\,dx = \tfrac{1}{3}x^3 + 2x^2 + 11x + \int \frac{24u-7}{u^2+1}\,du$
$= \tfrac{1}{3}x^3 + 2x^2 + 11x + 12\ln(u^2+1) - 7\tan^{-1}u + C$
$= \tfrac{1}{3}x^3 + 2x^2 + 11x + 12\ln(x^2-4x+5) \quad - 7\tan^{-1}(x-2) + C$

Answers to Exercises 6.2

1. $u = x$, $\dfrac{dv}{dx} = \cos(5x)$ • $du/dx = 1$ $v = \tfrac{1}{5}\sin(5x)$ • $\displaystyle\int x\cos(5x)\,dx$
$\displaystyle = \tfrac{1}{5}x\sin(5x) - \tfrac{1}{5}\int \sin(5x)\,dx = \tfrac{1}{5}x\sin(5x) + \tfrac{1}{5}(\tfrac{1}{5})\cos(5x) + C = \tfrac{1}{5}x\sin(5x) + \tfrac{1}{25}\cos(5x) + C$

2. $\tfrac{1}{4}xe^{4x} - \tfrac{1}{16}e^{4x} + C$

3. $-\tfrac{1}{2}\cos(x^2) + C$

5. $\tfrac{1}{2}x^2\ln x - \tfrac{1}{4}x^2 + C$

7. $\ln|\ln x| + C$

9. $\tfrac{1}{4}[\sin^{-1}(2x)]^2 + C$

11. $\tfrac{1}{2}(e^x\sin x - e^x\cos x) + C$

13. $\tfrac{1}{5}x^5\ln x - \tfrac{1}{25}x^5 + C$

15. $\tfrac{1}{2}e^{x^2} + C$

18. $\pi - 2$

21. $11\ln(11) - 10$

23. **(a)** [Mass] $= 1 - e^{-3}$ grams **(b)** [Moment about $x = 0$] $= 1 - 4e^{-3}$ gram-centimeters • [Center of gravity] $= \bar{x} = \dfrac{1-e^{-3}}{1-4e^{-3}}$ centimeters

25. $4\pi e^3$

27. $\dfrac{13x+6}{x^3+5x^2+6x} = \dfrac{1}{x} + \dfrac{10}{x+2} - \dfrac{11}{x+3}$

28. $\dfrac{2x^2+x-22}{x^2-4} = 2 + \dfrac{4}{x+2} - \dfrac{3}{x-2}$

29. $\dfrac{x-2}{x^2+2x+1} = \dfrac{x-2}{(x+1)^2} = \dfrac{A}{x+1} + \dfrac{B}{(x+1)^2}$ • $x-2 = A(x+1)+B$ • $x-2 = Ax+(1+B)$ • $A=1,\ 1+B=-2$ • $B=-3$ • $\dfrac{x-2}{x^2+2x+1} = \dfrac{1}{x+1} - \dfrac{3}{(x+1)^2}$

31. $\dfrac{x^5}{x^2+16} = x^3 - 16x + \dfrac{256x}{x^2+16}$

33. $\dfrac{x}{(x+1)^3} = \dfrac{1}{(x+1)^2} - \dfrac{1}{(x+1)^3}$

39. $3\ln|x+2| + \dfrac{6}{x+2} + C$

40. $-\ln|x| + 6\ln|x+1| + C$

42. $\ln|x+2| + \ln|x+1| + C = \ln|x^2+3x+2| + C$

45. Use $u = x^2+4$ • $\displaystyle\int \frac{x+3}{x^2+4}\,dx = \tfrac{1}{2}\int \frac{2x}{x^2+4}\,dx + 3\int \frac{1}{x^2+4}\,dx = \tfrac{1}{2}\int \frac{1}{u}\,du + \tfrac{3}{2}\tan^{-1}(\tfrac{1}{2}x)$
$= \tfrac{1}{2}\ln|u| + \tfrac{3}{2}\tan^{-1}(\tfrac{1}{2}x) + C = \tfrac{1}{2}\ln(x^2+4) + \tfrac{3}{2}\tan^{-1}(\tfrac{1}{2}x) + C$

46. $x + \tfrac{13}{2}\ln(x^2+7) - \sqrt{7}\tan^{-1}(\dfrac{x}{\sqrt{7}}) + C$

47. $\ln|x| - \dfrac{1}{x+1} + C$

53. $\ln|\sin x - 1| - \ln|\sin x + 1| + C$

56. $-\tfrac{3}{4} - \ln 40$

60. $\displaystyle\int_0^b \frac{1}{1-x^2}\,dx = \tfrac{1}{2}\ln\left|\frac{1+b}{1-b}\right|$ • $\tfrac{1}{2}\ln\left|\dfrac{1+b}{1-b}\right| = 2 \iff \dfrac{1+b}{1-b} = e^4$ (since $0<b<1$) $\iff b = \dfrac{e^4-1}{e^4+1}$

64. $8 - 6\ln(3)$

67. $\tfrac{1}{10}\ln(286)$

Tune-up Exercises 6.3

T1. $\displaystyle\int \sin^3(2x)\cos^5(2x)\,dx = \int \sin(2x)[1-\cos^2(2x)]\cos^5(2x)\,dx = \int \sin(2x)[\cos^5(2x) - \cos^7(2x)]\,dx$
• Let $u = \cos(2x), du = -2\sin(2x)\,dx$ •
$\displaystyle\int \sin^3(2x)\cos^5(2x)\,dx = -\tfrac{1}{2}\int (u^5-u^7)\,du = -\tfrac{1}{2}(\tfrac{1}{6}u^6 - \tfrac{1}{8}u^8)+C = -\tfrac{1}{12}\cos^6(2x)+\tfrac{1}{16}\cos^8(2x)+C$

T2. $\displaystyle\int \sin^2 x\cos^2 x\,dx = \int [\tfrac{1}{2} - \tfrac{1}{2}\cos(2x)][\tfrac{1}{2} + \tfrac{1}{2}\cos(2x)]\,dx = \int [\tfrac{1}{4} - \tfrac{1}{4}\cos^2(2x)]\,dx$
$\displaystyle= \int (\tfrac{1}{4} - \tfrac{1}{4}[\tfrac{1}{2} - \tfrac{1}{2}\cos(4x)]\,dx = \int [\tfrac{1}{8} - \tfrac{1}{8}\cos(4x)]\,dx = \tfrac{1}{8}x - \tfrac{1}{32}\sin(4x) + C$

T3. $\displaystyle\int \cos(3x)\cos x\,dx = \tfrac{1}{2}\int [\cos(2x) + \cos(4x)]\,dx = \tfrac{1}{4}\sin(2x) + \tfrac{1}{8}\sin(4x) + C$

T4. Let $x = 2\sin\theta,\ dx = 2\cos\theta\ d\theta,\ \sqrt{4-x^2} = 2\cos\theta$ • $\int x^3\sqrt{4-x^2}\ dx$

$= \int (2\sin\theta)^3(2\cos\theta)(2\cos\theta)\ d\theta = 32\int \sin^3\theta\cos^2\theta\ d\theta$ • Use $\sin^2\theta = 1-\cos^2\theta$ •

$\int x^3\sqrt{4-x^2}\ dx = 32\int \sin\theta\cos^2\theta\ d\theta - 32\int \sin\theta\cos^4\theta\ d\theta = 32(-\frac{1}{3}\cos^3\theta + \frac{1}{5}\cos^5\theta + C_1)$

$= -\frac{32}{3}\left[\frac{1}{2}\sqrt{4-x^2}\right]^3 + \frac{32}{5}\left[\frac{1}{2}\sqrt{4-x^2}\right]^5 + C = -\frac{4}{3}(4-x^2)^{3/2} + \frac{1}{5}(4-x^2)^{5/2} + C$

T5. Let $x = \tan\theta,\ dx = \sec^2\theta\ d\theta,\ \sqrt{x^2+1} = \sec\theta$ • $\int_0^1 \sqrt{x^2+1}\ dx = \int_{x=0}^{x=1} \sec\theta\sec^2\theta\ d\theta$

$= \int_{x=0}^{x=1} \sec^3\theta\ d\theta$ • Use (33). • $\int_0^1 \sqrt{x^2+1}\ dx = \left[\frac{1}{2}\tan\theta\sec\theta + \frac{1}{2}\ln|\sec\theta + \tan\theta|\right]_{x=0}^{x=1}$

$= \left[\frac{1}{2}x\sqrt{x^2+1} + \frac{1}{2}\ln|\sqrt{x^2+1}+x|\right]_0^1 = \frac{1}{2}\sqrt{2} + \frac{1}{2}\ln(\sqrt{2}+1)$

T6. $12 - 4x - x^2 = 16 - (x+2)^2$ • $\int \frac{1}{\sqrt{12-4x-x^2}}\ dx = \int \frac{1}{\sqrt{16-(x+2)^2}}\ dx$ •

Use $u = x+2,\ du = dx$ • $\int \frac{1}{\sqrt{12-4x-x^2}}\ dx = \int \frac{1}{\sqrt{16-u^2}}\ du$

$= \sin^{-1}(\frac{u}{4}) + C = \sin^{-1}(\frac{x+2}{4}) + C$

Problems 6.3

1. $\int \sin^3(4x)\ dx = \int \sin(4x)[1-\cos^2(4x)]\ dx =$ • Let $u = \cos(4x), du = -4\sin(4x)\ dx$ •

$\int \sin^3(4x)\ dx = -\frac{1}{4}\int (1-u^2)\ du = -\frac{1}{4}(u - \frac{1}{3}u^3) + C = -\frac{1}{4}\cos(4x) + \frac{1}{12}\cos^3(4x) + C$

3. $\int (\sin x)^{\sqrt{2}}(1-\sin^2 x)\cos\ dx$ • Set $u = \sin x,\ du = \cos x dx$ • $\int u^{\sqrt{2}}(1-u^2)\ du$

$= \int (u^{\sqrt{2}} - u^{2+\sqrt{2}})\ du = \frac{u^{1+\sqrt{2}}}{1+\sqrt{2}} - \frac{u^{3+\sqrt{2}}}{3+\sqrt{2}} + C = \frac{(\sin x)^{1+\sqrt{2}}}{1+\sqrt{2}} - \frac{(\sin x)^{3+\sqrt{2}}}{3+\sqrt{2}} + C$

5. $\frac{3}{8}x - \frac{1}{4}\sin(2x) + \frac{1}{32}\sin(4x) + C$

7. $-\frac{1}{4}\cos(2x) - \frac{1}{8}\cos(4x) + C$

9. $\frac{1}{2}\pi$

11. $\int \frac{x^3}{\sqrt{x^2-1}}\ dx$ • Set $\theta = \sec^{-1}x$. • $x = sec\theta$ • $dx = \sec\theta\tan\theta\ d\theta$ •

$\sqrt{x^2-1} = \tan\theta.$ • $\int \frac{x^3}{\sqrt{x^2-1}}\ dx = \int \sec^4\theta\ d\theta$

$= \frac{1}{3}\sec^2\theta\tan\theta + \frac{2}{3}\tan\theta + C = \frac{1}{3}x^2\sqrt{x^2-1} + \frac{2}{3}\sqrt{x^2-1} + C$

12. $-\frac{\sqrt{1-x^2}}{x} + C$

14. Set $x = 2\sec\theta,\ dx = 2\sec\theta\tan\theta\ d\theta,\ \sqrt{x^2-4} = 2\tan\theta$ •

$\int \frac{x^2}{\sqrt{x^2-4}}\ dx = \int \frac{4\sec^2\theta}{2\tan\theta} 2\sec\theta\tan\theta\ d\theta = 4\int \sec^3\theta\ d\theta = 4[\frac{1}{2}\sec\theta\tan\theta + \frac{1}{2}\ln|\sec\theta + \tan\theta] + C$

$= \frac{1}{2}x\sqrt{(x^2-4)} + 2\ln|\frac{1}{2}x + \frac{1}{2}\sqrt{x^2-4}| + C = \frac{1}{2}x\sqrt{(x^2-4)} + 2\ln|x + \sqrt{x^2-4}| + C_1$

15. $-\dfrac{\sqrt{16-x^2}}{x} - \sin^{-1}(\dfrac{x}{4}) + C$

19. $-\frac{1}{3}\ln\left|\dfrac{3}{x} + \dfrac{\sqrt{9-x^2}}{x}\right| + C$

21. $\frac{1}{4}\pi^2$

23. $\ln(3)$

25. $2\sqrt{3} - \ln(\sqrt{3}+2)$

27. $\ln\left|\frac{1}{3}\sqrt{x^2-8x+25} + \frac{1}{3}(x-4)\right| + C$

28. $\frac{1}{4}\tan^{-1}[\frac{1}{4}(x-3)] + C$

30. $\sin^{-1}[\frac{1}{10}(x+4)] + C$

34. $\displaystyle\int \frac{1}{\tan x - 1}\,dx = -2\int \frac{(u^2-1)}{(1+u^2)(u^2+2u-1)}\,du$ • $\dfrac{(u^2-1)}{(1+u^2)(u^2+2u-1)}$
$= \dfrac{(Au+B)}{1+u^2} + \dfrac{(Cu+D)}{u^2+2u-1}$ • $A = \frac{1}{2},\ B = \frac{1}{2},\ C = -\frac{1}{2},\ D = -\frac{1}{2}$ •
$\displaystyle -2\int \frac{(u^2-1)}{(1+u^2)(u^2+2u-1)}\,du = -2\int\left(\frac{(\frac{1}{2}u+\frac{1}{2})}{1+u^2} + \frac{(-\frac{1}{2}u-\frac{1}{2})}{u^2+2u-1}\right)du$
$= -\frac{1}{2}\ln(1+u^2) - \tan^{-1}u + \frac{1}{2}\ln|u^2+2u-1| + C = \frac{1}{2}\ln|\sin x - \cos x| - \frac{1}{2}x + C$

35. $\displaystyle\int \frac{1}{1+\cos x}\,dx = \int \frac{1+u^2}{1+u^2+1-u^2}(\frac{2}{1+u^2})\,du = u + C = \tan(\tfrac{1}{2}x) + C$

36. $\frac{1}{2}\tan^{-1}(\frac{1}{2}\tan(\frac{1}{2}x)) + C$

38. $-\frac{1}{4}[\tan(\frac{x}{2})]^{-2} - \frac{1}{2}\ln|\tan(\frac{x}{2})| + C = -\frac{1}{2}\ln\left|\dfrac{\sin x}{1+\cos x}\right| - \frac{1}{4}\left(\dfrac{1+\cos x}{1-\cos x}\right) + C$

40. $\frac{5}{6}\tan^{-1}\left(\dfrac{2}{\sqrt{3}}\right) - \frac{1}{9}\pi$

Tune-Up Exercises 6.4

T1. (WITH A CALCULATOR) $\frac{1}{2}\ln(\frac{2}{5}\sqrt{6}+\frac{7}{5})$
(WITH THE TABLE) Formula (89) is $\displaystyle\int \frac{1}{\sqrt{x^2+a^2}}\,dx = \ln(x+\sqrt{x^2+a^2}) + C$ •
Set $a^2 = 5$: $\displaystyle\int_0^1 \frac{1}{\sqrt{x^2+5}}\,dx = \left[\ln(x+\sqrt{x^2+5})\right]_0^1 = \ln(1+\sqrt{6}) - \ln(\sqrt{5})$

T2. (WITH A CALCULATOR) $\frac{1}{48}\pi$
(WITH THE TABLE) Use $u = 3x$, $du = 3dx$ and formula (51) with x replaced by u:
$\displaystyle\int_0^{\pi/6} \sin^2(3x)\cos^2(3x)\,dx = \frac{1}{3}\int_{x=0}^{x=\pi/6} \sin^2(u)\cos^2(u)\,du = \frac{1}{3}\left[-\frac{1}{32}\sin(4u) + \frac{1}{8}u\right]_{x=0}^{x=\pi/6}$
$\displaystyle = \frac{1}{3}\left[-\frac{1}{32}\sin(12x) + \frac{3}{8}x\right]_0^{\pi/6} = \frac{1}{48}\pi$

T3. (WITH A CALCULATOR) $-\frac{1}{4}\cos(2x) - \frac{1}{8}\cos(4x) + C$
(WITH THE TABLE) Formula (48) with $a = 3, b = 1$:
$\displaystyle\int \cos x\sin(3x)\,dx = -\frac{\cos[(3-1)x]}{2(3-1)} - \frac{\cos[(3+1)x]}{2(3+1)} + C = -\frac{1}{4}\cos(2x) - \frac{1}{8}\cos(4x) + C$

T4. (WITH A CALCULATOR) $\frac{2}{3}\sqrt{3}\tan^{-1}(\frac{5}{3}\sqrt{3}) - \frac{1}{9}\pi\sqrt{3}$
(WITH THE TABLE) Formula (96) with $a = 1, b = 1, c = 1$ gives
$\displaystyle\int_0^2 \frac{1}{x^2+x+1}\,dx = \left[\frac{2}{3}\sqrt{3}\tan^{-1}(\frac{2x+1}{\sqrt{3}})\right]_0^2 = \frac{2}{3}\sqrt{3}\left[\tan^{-1}(\tfrac{5}{3}\sqrt{3}) - \tan^{-1}(\tfrac{1}{\sqrt{3}})\right]$

T5. (WITH A CALCULATOR) $\displaystyle\int \frac{e^x}{\sqrt{x^{2x}-1}}\,dx = \ln(e^x + \sqrt{x^{2x}-1}) + C$
(WITH THE TABLE) Use $u = e^x$, $du = e^x dx$ and (90) with $a = 1$:

$$\int \frac{e^x}{\sqrt{x^{2x}-1}}\,dx = \int \frac{1}{\sqrt{u^2-1}}\,du = \ln(u+\sqrt{u^2-1}) + C = \ln(e^x+\sqrt{x^{2x}-1}) + C$$

T6. (WITH A CALCULATOR) $\int_0^{\pi/3} \sin x \ln(\cos x)\,dx = -\frac{1}{2}\ln(2) - \frac{1}{2}$

(WITH THE TABLE) Use $u = \cos x, du = -\sin x dx = \int_0^{\pi/3} \sin x \ln(\cos x)\,dx$

$$= -\int_{x=0}^{x=\pi/3} \ln(u)\,du = -\left[u\ln u - u\right]_{x=0}^{x=\pi/3} = -\left[\cos x\ln(\cos x) - \cos x\right]_0^{\pi/3}$$

$$= [\tfrac{1}{2}(1-\ln(\tfrac{1}{2}))] - (1) = -\tfrac{1}{2}\ln(\tfrac{1}{2}) - \tfrac{1}{2}$$

Problems 6.4

1. (WITH A CALCULATOR) $30 + \frac{169}{2}\sin^{-1}(\frac{12}{13})$

(WITH THE TABLE) Formula (88) with $a = 13$ gives

$$\int_0^{12} \sqrt{169-x^2}\,dx = \left[\tfrac{1}{2}x\sqrt{169-x^2} + \tfrac{1}{2}(169)\sin^{-1}(\tfrac{1}{13}x)\right]_0^{12} = 30 + \tfrac{169}{2}\sin^{-1}(\tfrac{12}{13})$$

2. (WITH A CALCULATOR) $\frac{1}{3}\ln(2)$

(WITH THE TABLE) $\frac{1}{3}\ln(2)$

3. (WITH A CALCULATOR) $\frac{1}{4}x^2\sqrt{1-x^4} + \frac{1}{4}\sin^{-1}(x^2) + C$

(WITH THE TABLE) Use $u = x^2$, $du = 2x dx$, and formula (88) with $a = 1$:

$$\int x\sqrt{1-x^4}\,dx = \tfrac{1}{2}\int \sqrt{1-u^2}\,du = \tfrac{1}{4}u\sqrt{1-u^2} + \tfrac{1}{4}\sin^{-1}u + C = \tfrac{1}{4}x^2\sqrt{1-x^4} + \tfrac{1}{4}\sin^{-1}(x^2) + C$$

4. (WITH A CALCULATOR) $\ln\left(\frac{(x+2)^2}{|x+1|}\right) + C$

(WITH THE TABLE) $2\ln|x+2| - \ln|x+1| + C$

5. (WITH A CALCULATOR) $-\frac{1}{6}\cos(x^2)\sin^2(x^2) - \frac{1}{3}\cos(x^2) + C$

(WITH THE TABLE) Use $u = x^2$, $du = 2x\,dx$, and (24): $\int x\sin^3(x^2)\,dx = \frac{1}{2}\int \sin^3 u\,du$

$$= \tfrac{1}{2}(-\tfrac{1}{3}\cos u\sin^2 u - \tfrac{2}{3}\cos u) + C = -\tfrac{1}{6}\cos(x^2)\sin^2(x^2) - \tfrac{1}{3}\cos(x^2) + C$$

7. (WITH A CALCULATOR) $\int_{-1/4}^{1/4} \sec(\pi x)\,dx = \int_{-1/4}^{1/4} \frac{1}{\cos(\pi x)}\,dx = -\frac{1}{\pi}\ln(\frac{1}{2}(\sqrt{2}-2)^2)$

(WITH THE TABLE) Use $u = \pi x$, $du = \pi\,dx$, and formula (18):

$$\int_{-1/4}^{1/4} \sec(\pi x)\,dx = \frac{1}{\pi}\int_{x=-1/4}^{x=1/4} \sec(u)\,du = \frac{1}{\pi}\left[\ln|secu + \tan u|\right]_{x=-1/4}^{x=1/4}$$

$$= \frac{1}{\pi}\left[\ln|\sec(\pi x) + \tan(\pi x)|\right]_{-1/4}^{1/4} = \frac{1}{\pi}[\ln(\sqrt{2}+1) - \ln(\sqrt{2}-1)]$$

8. (WITH A CALCULATOR) $\ln\left|\frac{\cos x + 1}{\cos x}\right| + C$

(WITH THE TABLE) Use $u = \cos x, du = -\sin x dx$, and (95) with $a = 1, b = 1, c = 0$:

$$\int \frac{\sin x}{\cos^2 x + \cos x}\,dx = \int \frac{1}{u^2+u}\,du = -\ln\left|\frac{2u}{2u+2}\right| + C = -\ln\left|\frac{\cos x}{\cos x + 1}\right| + C$$

11. (WITH A CALCULATOR) $\frac{1}{2\ln(3)}3^{x^2} + C$

(WITH THE TABLE) $\int x3^{x^2}\,dx = \frac{1}{2}\int 3^u\,du = \frac{1}{2\ln(3)}3^{x^2} + C$

12. (WITH A CALCULATOR) $\int \frac{1}{x^2-4x+13}\,dx = \frac{1}{3}\tan^{-1}(\frac{1}{3}(x-2)) + C$

(WITH THE TABLE) $\int \frac{1}{x^2-4x+13}\,dx = \frac{2}{\sqrt{36}}\tan^{-1}(\frac{2x-4}{\sqrt{36}}) + C = \frac{1}{3}\tan^{-1}(\frac{1}{3}(x-2)) + C$

15. $[\text{Area}] = \int_0^1 xe^x\,dx$

(WITH A CALCULATOR) $\int_0^1 xe^x\,dx = 1$

(WITH THE TABLE) $= \Big[(x-1)e^x\Big]_0^1 = 1$

16. $[\text{Average value}] = \frac{1}{2}\int_0^2 \frac{1}{x^2-4x+13}\,dx$

(WITH A CALCULATOR) $\frac{1}{2}\int_0^2 \frac{1}{x^2-4x+13}\,dx = \frac{1}{6}\tan^{-1}(\frac{2}{3})$

(WITH THE TABLE) $\frac{1}{2}\int_0^2 \frac{1}{x^2-4x+13}\,dx = \frac{1}{2}\left[\frac{1}{3}\tan^{-1}(\frac{1}{3}(x-2))\right]_0^2 = \frac{1}{6}\tan^{-1}(\frac{2}{3})$

17. $[\text{Length}]= \int_0^3 \sqrt{1+x^2}\,dx$

(WITH A CALCULATOR) $[\text{Length}] = \frac{1}{2}\ln(\sqrt{10}+3) + \frac{3}{2}\sqrt{10}$

(WITH THE TABLE) Formula (86) with $a = 1$: $[\text{Length}]= \left[\frac{1}{2}x\sqrt{1+x^2} + \frac{1}{2}\ln(x+\sqrt{1+x^2})\right]_0^3$
$= \frac{1}{2}\ln(\sqrt{10}+3) + \frac{3}{2}\sqrt{10}$

18. $[\text{Volume}]= \int_0^\pi x^2\sin x\,dx$

(WITH A CALCULATOR) $[\text{Volume}]= \pi^2 - 4$

(WITH THE TABLE) $[\text{Volume}]= \Big[2x\sin x - (x^2-2)\cos x\Big]_0^\pi = \pi^2 - 4$

19. $s = \frac{1}{2}\ln(\sqrt{1+t^2}+t) + \frac{1}{2}t\sqrt{1+t^2} + 5$ yards

22. Limits of integration: $a = e^{\frac{-1-\sqrt{17}}{4}}, b = e^{\frac{-1+\sqrt{17}}{4}}$ • $[\text{Area}]= \int_a^b (2 - \ln x - 2(\ln x)^2)\,dx =$
$(\sqrt{17}-4)e^{\frac{-1+\sqrt{17}}{4}} + (\sqrt{17}+4)e^{\frac{-1-\sqrt{17}}{4}} \doteq 2.5255336$

24. **(a)** Rotated about the x-axis: $[\text{Volume}]= \pi\int_1^{e^2} (2^2 - (\ln x)^2)\,dx = 2\pi(e^2-1)$

(b) Rotated about the y-axis: $[\text{Volume}]= 2\pi\int_1^{e^2} x(2-\ln x)\,dx = \frac{1}{2}\pi(e^4-5)$

26. $[\text{Total flow}] = \int_0^{60} \tan^{-1} t\,dt$

(WITH A CALCULATOR) $[\text{Total flow}]= -\frac{1}{2}\ln\left(\frac{1}{3601}\right) - 60\tan^{-1}\left(\frac{1}{60}\right) + 30\pi$ gallons

(WITH THE TABLE) Formula (22): $[\text{Total flow}]= \left[t\tan^{-1}t - \frac{1}{2}\ln(1+t^2)\right]_0^{60}$
$= 60\tan^{-1}(60) - \frac{1}{2}\ln(3601)$ gallons (The answers are the same because $\tan^{-1}(1/k) = \frac{1}{2}\pi - \tan^{-1}(k)$.)

31. **(a)** $G(x) = \ln\left(\frac{x+1}{x}\right) - \frac{1}{x} + 2 - \ln(2)$ **(b)** Increasing since $\frac{dG}{dx} = \frac{1}{x^2(x+1)} > 0$ for $x > 0$

37. $[\text{Area}]= \int_{\tan^{-1}(\frac{1}{2})}^{\tan^{-1}(\frac{1}{2})+\pi} (2\sin x - \cos x)\,dx = 2\sqrt{5}$

Tune-up Exercises 6.5

T1. One answer with the Riemann sum corresponding to the rectangles in Figure T1:

$$\int_{40}^{80} y(x)\,dx \approx 425(10) + 400(10) + 630(10) + 610(10) = 20{,}650$$

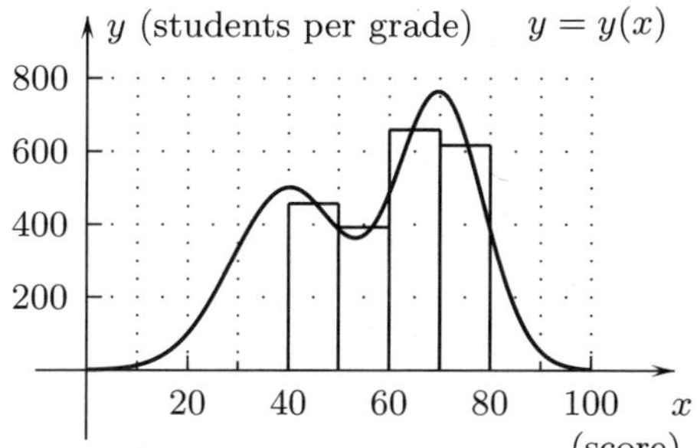

Figure T1

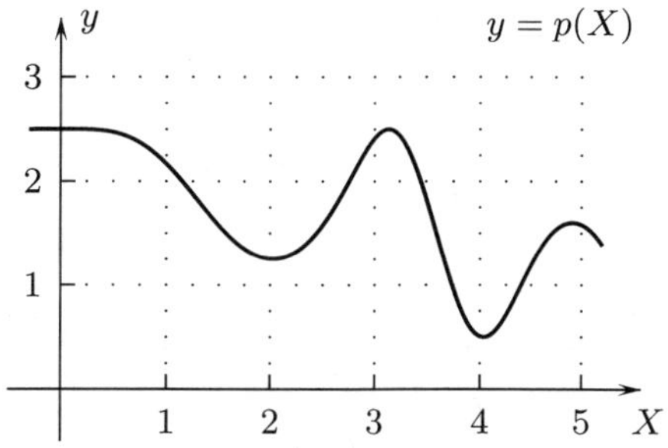

Figure T3

T2. **(a)** $G(X) = \begin{cases} X & \text{for } 0 \le X \le 2 \\ 2 - \frac{1}{2}(X-2) & \text{for } 2 \le X \le 5 \end{cases}$

(b) $G'(X)$ is defined for $0 < X < 2$ and for $2 < X < 5$.

(c) $G'(X) = \begin{cases} 1 & \text{for } 0 < X < 2 \\ -\frac{1}{2} & \text{for } 2 < X < 5 \end{cases}$

T3. $\dfrac{d}{dX}\displaystyle\int_0^X p(x)\,dx = p(X)$, so the graph of $p(X)$ in the Xy-plane of Figure T3 is the same as the graph of $p(x)$ in the xy-plane of Figure 17.

T4. $\left[\dfrac{d}{dX}\displaystyle\int_0^X \sqrt{x^2+10}\,dx\right]_{X=3} = \left[\sqrt{X^2+10}\right]_{X=3} = \sqrt{19}$

T5. **(a)** $\dfrac{d}{dX}\displaystyle\int_X^{10} (2x + \sin x)\,dx = \dfrac{d}{dX}\left[x^2 - \cos x\right]_{x=X}^{x=10} = \dfrac{d}{dX}[\{10^2 - \cos(10)\} - \{X^2 - \cos X\}]$
$= -2X - \sin X$

(b) $\dfrac{d}{dX}\displaystyle\int_X^{10} (2x + \sin x)\,dx = \dfrac{d}{dX}\left[-\displaystyle\int_{10}^{X} (2x + \sin x)\,dx\right] = -2X - \sin X$

T6. $\dfrac{d}{dX}\displaystyle\int_1^{\ln X} \dfrac{1}{1+x^4}\,dx = \dfrac{1}{1+(\ln X)^4}\,\dfrac{d}{dX}(\ln X) = \dfrac{1}{X[1+(\ln X)^4]}$

Problems 6.5

1. **(a)** 2,150,000 **(b)** $\doteq 4.18\%$

2. **(a)** $n(t)$ is a density function. **(b)** The average number of tornados occuring in the first half of the year from 1950 to 1982. **(c)** $\approx 108(33) = 3564$ tornados

8. **(a)** One estimate: 62% **(b)** 100% **(c)** Approximately (5,17)

9. 79123%

11. (a) $\sqrt{X+3}$ (b) $\dfrac{d}{dX}\displaystyle\int_2^X \sqrt{x+3}\,dx = \dfrac{d}{dX}\int_2^X (x+3)^{1/2}\,dx = \dfrac{d}{dX}\left[\tfrac{2}{3}(x+3)^{3/2}\right]_2^X =$
$\dfrac{d}{dX}[\tfrac{2}{3}(X+3)^{3/2} - \tfrac{2}{3}(5)^{3/2}] = \sqrt{X+3}$

13. $X \sin X$

15. $-X\sin(3X)$

17. Set $G(u) = \displaystyle\int_1^u \sqrt{x+1}\,dx$ • $G'(u) = \sqrt{u+1}$ • $\dfrac{d}{dX}G(X^2) = G'(X^2)\dfrac{d}{dX}(X^2) = 2X\sqrt{X^2+1}$

18. $3\sin^5(3X)$

22. Set $G(u) = \displaystyle\int_0^u (1-x^2)^{1/3}\,dx$ • $\displaystyle\int_{X^2}^{X^3} (1-x^2)^{1/3}\,dx = G(X^3) - G(X^2)$ •
$\dfrac{d}{dX}[G(X^3) - G(X^2)] = 3X^2G'(X^3) - 2xG'(X^2) = 3X^2(1-X^6)^{1/3} - 2X(1-X^4)^{1/3}$

23. $25X\sin(5X) - 16X\sin(4X)$

25. $\cos(1)$

27. $25X^4 + 1$

31. $y(x) = \frac{1}{2}(e^x + e^{-x}) = \cosh x$

Tune-up Exercises 6.6

T1. $\displaystyle\int_3^\infty \frac{1}{x^3}\,dx = \lim_{X\to\infty}\int_3^X x^{-3}\,dx = \lim_{X\to\infty}\left[-\tfrac{1}{2}x^{-2}\right]_0^X = \lim_{X\to\infty}\left[-\frac{1}{2x^2}\right]_0^X$
$\displaystyle = \lim_{X\to\infty}\left[-\frac{1}{2X^2} + \frac{1}{18}\right] = \tfrac{1}{18}$

T2. $\displaystyle\int_0^3 \frac{1}{x^3}\,dx = \lim_{X\to0+}\int_X^3 x^{-3}\,dx = \lim_{X\to0+}\left[-\tfrac{1}{2}x^{-2}\right]_X^3 = \lim_{X\to0+}\left[-\frac{1}{2x^2}\right]_X^3$
$\displaystyle = \lim_{X\to0+}\left[-\frac{1}{2X^2} + \frac{1}{18}\right] = \infty$

T3. $\displaystyle\int_{-\infty}^\infty xe^{-x^2}\,dx = \int_{-\infty}^0 xe^{-x^2}\,dx + \int_0^\infty xe^{-x^2}\,dx = \lim_{X\to-\infty}\int_X^0 xe^{-x^2}\,dx + \lim_{X\to\infty}\int_0^X xe^{-x^2}\,dx$ •
Set $u = -x^2$, $du = -2xdx$ • $\displaystyle\lim_{X\to-\infty}\left[-\tfrac{1}{2}\int_X^0 e^u\,du\right] + \lim_{X\to\infty}\left[-\tfrac{1}{2}\int_0^X e^u\,du\right]$
$\displaystyle = \lim_{X\to-\infty}\left[-\tfrac{1}{2}e^u\right]_{x=X}^{x=0} + \lim_{X\to\infty}\left[-\tfrac{1}{2}e^u\right]_{x=0}^{x=X} = \lim_{X\to-\infty}\left[-\tfrac{1}{2}e^{-x^2}\right]_X^0 + \lim_{X\to\infty}\left[-\tfrac{1}{2}e^{-x^2}\right]_0^X$
$\displaystyle = \lim_{X\to-\infty}\left[-\tfrac{1}{2} + \tfrac{1}{2}e^{-X^2}\right] + \lim_{X\to\infty}\left[\tfrac{1}{2} - \tfrac{1}{2}e^{-X^2}\right] = -\tfrac{1}{2} + \tfrac{1}{2} = 0$

T4. $\displaystyle\int_0^9 \frac{1}{\sqrt{x}}\,dx = \lim_{X\to0+}\int_X^9 x^{-1/2}\,dx = \lim_{X\to0+}\left[2x^{1/2}\right]_X^9 = \lim_{X\to0+}\left[2\sqrt{9} - 2\sqrt{X}\right] = 6$

T5. $\displaystyle\int_{-8}^{27} 3x^{-2/3}\,dx = \lim_{X\to0-}\int_{-8}^0 3x^{-2/3}\,dx + \lim_{X\to0+}\int_0^{27} 3x^{-2/3}\,dx$
$\displaystyle = \lim_{X\to0-}\int_{-8}^X 3x^{-2/3}\,dx + \lim_{X\to0+}\int_X^{27} 3x^{-2/3}\,dx = \lim_{X\to0-}\left[9x^{1/3}\right]_{-8}^X + \lim_{X\to0+}\left[9x^{1/3}\right]_X^{27}$
$\displaystyle = \lim_{X\to0-}\left[9(X)^{1/3} - 9(-8)^{1/3}\right]_X^3 + \lim_{X\to0+}\left[9(27)^{1/3} - 9(X)^{1/3}\right] = -9(-2) + 9(3) = 45$

T6. $\int_{-\infty}^{\infty} e^x\,dx = \int_{-\infty}^{0} e^x\,dx + \int_{0}^{\infty} e^x\,dx = \lim_{X\to-\infty}\int_X^0 e^x\,dx + \lim_{X\to\infty}\int_0^X e^x\,dx$
$= \lim_{X\to-\infty}(e^0 - e^X) + \lim_{X\to\infty}(e^X - e^0) = 1 + \infty = \infty$

T7. $\int_{-\infty}^{10} g(x)\,dx = \int_{-\infty}^{-5} g(x)\,dx + \int_{-5}^{5} g(x)\,dx + \int_{5}^{10} g(x)\,dx = 7 - \infty + 21 = -\infty$

Problems 6.6

1. Converges to $\frac{1}{10}e^{-10}$

3. Diverges to ∞

4. Diverges without a limit

6. Converges to $\frac{27}{2}$

9. Converges to $\sqrt{3}$

10. Converges to 10

12. Diverges to ∞

14. Substitution: $u = 1 + e^x, du = e^x\,dx$ • $\int_0^{\infty} \frac{e^x}{(1+e^x)^2}\,dx = \lim_{X\to\infty}\int_0^X \frac{e^x}{(1+e^x)^2}\,dx$
$= \lim_{X\to\infty}\int_{x=0}^{x=X} u^{-2}\,dx = \lim_{X\to\infty}\left[-u^{-1}\right]_{x=0}^{x=X} = \lim_{X\to\infty}\left[\frac{-1}{1+e^x}\right]_0^X = \lim_{X\to\infty}\left[\frac{-1}{1+e^X} + \frac{1}{2}\right] = \frac{1}{2}$ •
$\int_{-\infty}^{0} \frac{e^x}{(1+e^x)^2}\,dx = \lim_{X\to-\infty}\int_X^0 \frac{e^x}{(1+e^x)^2}\,dx = \lim_{X\to-\infty}\int_{x=X}^{x=0} u^{-2}\,dx = \lim_{X\to\infty}\left[-u^{-1}\right]_{x=X}^{x=0}$
$= \lim_{X\to-\infty}\left[\frac{-1}{1+e^x}\right]_X^0 = \lim_{X\to-\infty}\left[-\frac{1}{2} + \frac{1}{1+e^X}\right] = \frac{1}{2}$ • $\int_{-\infty}^{\infty} \frac{e^x}{(1+e^x)^2}\,dx = \frac{1}{2} + \frac{1}{2} = 1$ •
Converges to 1

17. Diverges to ∞

19. Converges to $\frac{1}{2}\pi$

25. Figure A25 • [Area]$= \int_1^{\infty} [(\frac{1}{3}x + \frac{5}{x^2}) - \frac{1}{3}x]\,dx = \int_1^{\infty} 5x^{-2}\,dx = \lim_{X\to\infty}\int_5^X x^{-2}\,dx =$
$\lim_{X\to\infty}\left[-\frac{5}{x}\right]_1^X = \lim_{X\to\infty}\left[-\frac{5}{X} + 5\right] = 5$

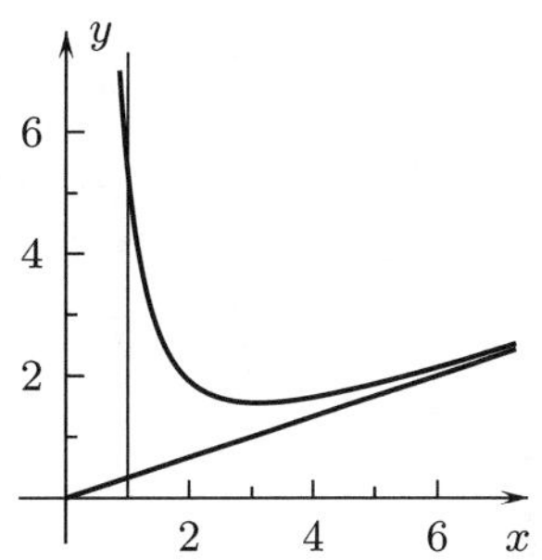

Figure A25

30. [Area] $= \frac{1}{2}\pi$

33. 4000 mile-pounds $\doteq 2.112 \times 10^7$ foot-pounds

35. $k = \pm 1$ • one explanation: The integral from -1 to 1 is zero, because the graph is symmetric about the origin.

36. **(a)** [Probability] = 1 **(b)** $k = -\ln(0.8)$ **(c)** 0.16

37a. $P(0) = (\frac{5}{6})^2 = \frac{25}{36}$ • $P(1) = (\frac{1}{6})(\frac{5}{6}) + (\frac{5}{6})(\frac{1}{6}) = \frac{10}{36}$ • $P(2) = (\frac{1}{6})^2 = \frac{1}{36}$

40. Converges for $k > 1$ • Diverges for $k \le 1$

41. Converges for $k > -1$ • Diverges for $k \le -1$

45. $\displaystyle\int_1^2 \frac{1}{\sqrt{x^2-1}}\,dx = \lim_{X\to 1+}\int_X^2 \frac{1}{\sqrt{x^2-1}}\,dx = \lim_{X\to 1+}\Big[\ln(x+\sqrt{x^2-1}\,\Big]_X^2$
$\displaystyle = \lim_{X\to 1+}\Big[\ln(2+\sqrt{3}) - \ln(X+\sqrt{X^2-1})\Big] = \ln(2+\sqrt{3})$ • The integral converges.

46. $k = 1,\ k = \frac{1}{2}(\sqrt{5}-1)$

Tune-up Exercises 6.7

T1. $\displaystyle[\text{Work}] = \int_0^{10} 50s\,ds = \Big[25s^2\Big]_0^{10} = 25(10)^2 - 25(0)^2 = 2500$ ergs

T2. $\displaystyle[\text{Work}] = \int_{0.2}^{0.4} ks\,ds = \Big[\tfrac{1}{2}ks^2\Big]_{0.2}^{0.4} = \tfrac{1}{2}k[(0.4)^2 - (0.2)^2] = 0.06k$ • $0.06k = 3 \Longrightarrow k = 50$

T3. [Force by the spring toward its free end] $= F = -ks$ dynes • Set $s = 10, F = -500 : -500 = -k(10)$ • $k = 50$ • [Force against the spring toward its free end] $= 50s$ dynes • [Work against the spring to compress its end from $s = 0$ to $s = -8$]
$\displaystyle = \int_0^{-8} 50s\,ds = \Big[25s^2\Big]_0^{-8} = 25(-8)^2 - 25(0)^2 = 1600$ ergs

T4. Put a vertical s-axis along the chain with the bottom at $s = 0$ and the top at $s = 40$ (feet). • For a partition $0 = s_0 < s_1 < s_2 \cdots < s_N = 40$, the part of the chain in the jth subinterval weighs $5\Delta s_j$ pounds and is lifted approximately $40 - s_j$ feet. • [Approximate work] $\displaystyle = \sum_{j=1}^{N} 5(40-s_j)\,\Delta s_j$ •

$\displaystyle[\text{Work}] = \int_0^{40} 5(40-s)\,ds = \Big[200s - \tfrac{5}{2}s^2\Big]_0^{40} = 200(40) - \tfrac{5}{2}(40)^2 = 4000$ foot-pounds

T5. **(a)** (10 pounds)(500 feet) = 5000 foot-pounds
(b) [Cat's weight] $= mg =$ [5 kilograms][9.8 meters/second^2] = 49 newtons •
[Work] = [49 newtons][−3 meters] = −147 joules
(c) [25 grams][210 centimeters] + [25 grams][−210 centimeters] = 0 dynes

T6. Let v = [Final upward velocity] (feet/second) • [Mass] $= \frac{1}{4}$ slug • $\frac{1}{2}(\frac{1}{4})v^2 =$
[Increase in kinetic energy] = [Work done by the upward force] − [Work done against air resistance]
$\displaystyle = \int_0^{25} (25 - 3s^{1/2})\,ds - 200 = \Big[25s - 2s^{3/2}\Big]_0^{25} - 200 = [625 - 250] - 200 = 175$ foot-pounds •
$v^2 = 8(175) = 1400$ • $v = 10\sqrt{14}$ feet per second

T7. Use an xy-plane with the x-axis along the bottom of the plate and the positive y-axis pointing up, as in Figure T7. • Similar triangles: [Width at y] $= W(y) = (6 + \frac{4}{3}y)$ feet • [Depth at y] $= (6 - y)$ feet • $\displaystyle[\text{Force}] = \int_0^3 62.4(6-y)(6+\tfrac{4}{3}y)\,dy = 62.4\int_0^3 (36 + 2y - \tfrac{4}{3}y^2)\,dy = 62.4\Big[36y + y^2 - \tfrac{4}{9}y^3\Big]_0^3$
$= 6552$ pounds

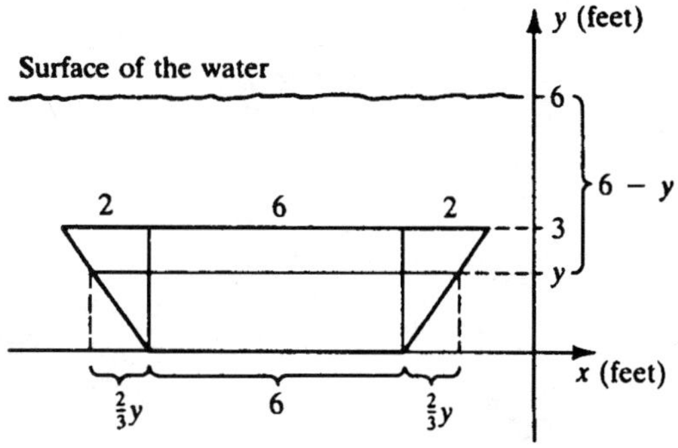

Figure T7

Problems 6.7

1. [Work] = 57,600,000 foot-pounds

2. [Work] = 1600 joules

5. [Work] = 20 ergs

7. $F(s) = -ks$ • $-6 = -ks \implies k = 6$ • $[\text{Work}] = \int_{-3}^{-1} (-6s)\, ds = \left[-3s^2\right]_{-3}^{-1} = 24$ inch-pounds

16. $[\text{Work}] = \frac{12}{43} \times 10^5 \doteq 2.7907 \times 10^4$ mile-pounds

17. **(a)** [Weight of the ball] = [10 kilograms][9.8 meters per second^2] = 98 newtons •
[Change in height] = $s(4) - s(0) = 16$ meters •
[Change in potential energy] = [98 newtons][16 meters] = 1568 joules
(b) $v = \frac{ds}{dt} = \frac{3}{4}t^2$ • $v(4) = 12$ • [Final kinetic energy] $= \frac{1}{2}mv^2 = \frac{1}{2}(10)(12)^2 = 720$ joules
(c) $v(0) = 0$ • [Initial kinetic energy] = 0 joules •
[Total work] = [Increase in potential energy] + [Increase in potential energy]
+ [Work against air resistance] = 1568 + (720 − 0) + 1000 = 3288 joules

18. 9 feet per second

19. 1700 foot-pounds

22. 52,3669.2 foot-pounds

23. 2490 foot-pounds

24. [Weight] = 200 pounds

28. 80 foot-pounds • 64 foot-pounds • 36 foot-pounds

31. 1820 joules

37. 135.2 pounds

38. **(a)** [Radius of the base] = $\frac{1}{3}$ foot • [Height] = $\frac{5}{6}$ foot •
[Volume of the can] $= \pi r^2 h = \pi(\frac{1}{3})^2(\frac{5}{6}) = \frac{5}{54}\pi$ cubic feet •
[Force on the bottom of the can] = [110 pounds per cubic foot][$\frac{5}{54}\pi$ cubic feet] = $\frac{275}{27}\pi$ pounds
(b) Let the bottom of the can be at $y = \frac{5}{6}$ and the top at $y = 0$ on a vertical y-axis. •
For a partition $0 = y_0 < y_1 < \cdots < y_N = \frac{5}{6}$, the side of the can is formed by rings of area $2\pi(\frac{1}{3})\Delta y_j$ square feet, where the pressure is approximately $110y_j$ pounds per square foot. •
[Approximate force on the side of the can]
$= \sum_{j=1}^{N}$ [$110y_j$ pounds per square foot][$(\frac{2}{3})\pi\Delta y_j$ square feet] $= \sum_{j=1}^{N} \frac{200}{3}\pi y_j \Delta y_j$ pounds •
[Force on the side of the can] $= \int_0^{5/6} \frac{220}{3}\pi y\, dy = \left[\frac{110}{3}\pi y^2\right]_0^{5/6} = \frac{1375}{54}\pi$ pounds

39. 273.05 pounds

46. $\frac{1}{3}(10{,}100\sqrt{3})$ newtons

47. 20,800 pounds

48. $784{,}000\sqrt{101}$ newtons

Tune-Up Exercises 6.8

T1. $4\sinh x + 6\cosh x = 4(\frac{1}{2})(e^x - e^{-x}) + 6(\frac{1}{2})(e^x + e^{-x}) = 5e^x + e^x$ is positive for all x because e^x and e^{-x} are positive for all x.

T2. $\frac{d}{dx}[4\sinh x + 6\cosh x] = 4\frac{d}{dx}(\sinh x) + 4\frac{d}{dx}(\cosh x) = 4\cosh x + 6\sinh x$

T3. $\frac{d}{dx}[\sinh(6x)] = \cosh(6x)\frac{d}{dx}(6x) = 6\cosh(6x)$

T4. Substitution: $u = 2x,\ du = 2dx$: $\int_0^3 \sinh(2x)\,dx = \frac{1}{2}\int_{x=0}^{x=3} \sinh u\,du$

$= \frac{1}{2}\left[\cosh u\right]_{x=0}^{x=3} = \frac{1}{2}\left[\cosh(2x)\right]_0^3 = \frac{1}{2}[\cosh(6) - 1]$

T5. **(a)** $\dfrac{\cosh x}{\sqrt{\sinh x}}$ is defined where $\sinh x > 0$, which is for $x > 0$.

(b) Substitution: $u = \sinh x,\ du = \cosh x$ •

$\int \frac{\cosh x}{\sqrt{\sinh x}}\,dx = \int u^{-1/2}\,du = 2u^{1/2} + C = 2\sqrt{\sinh x} + C$

Problems 6.8

1. **(a)** $x = 1$ **(b)** There are no solutions.

2. $\frac{d}{dx}[3\sinh(4x) - 4\cosh(3x)] = 3\cosh(4x)\frac{d}{dx}(4x) - 4\sinh(3x)\frac{d}{dx}(3x) = 12\cosh(4x) - 12\sinh(3x)$

3. $x^2\cosh x + 2x\sinh x$

4. $\tanh x$

7. $\dfrac{t\sinh t - \cosh t}{t^2}$

10. Substitution: $u = 3x,\ du = 3dx$: $\int_0^{10} \cosh(3x)\,dx = \frac{1}{3}\int_{x=0}^{x=10} \sinh u\,du$

$= \frac{1}{3}\left[\sinh u\right]_{x=0}^{x=10} = \frac{1}{3}\left[\sinh(3x)\right]_0^{10} = \frac{1}{3}\sinh(30)$

12. Substitution: $u = \sinh x + 1,\ du = \cosh x\,dx$ • $\int \frac{\cosh x}{\sinh x + 1}\,dx = \int u^{-1}\,du$

$= \ln|u| + C = \ln|\sinh x + 1| + C$

13. $\cosh(x^{1/2}) + C$

14. $\ln(\cosh x) + C$

24. $\pi\sinh(2) + 2\pi$

CHAPTER 7
FIRST-ORDER DIFFERENTIAL EQUATIONS

A DIFFERENTIAL EQUATION is an equation that involves an unknown function and one or more of its derivatives. The highest order of the derivatives is called the ORDER of the differential equation. In this chapter we consider first-order equations in the form $\frac{dy}{dx} = f(x,y)$, where the derivative $\frac{dy}{dx}$ of the unknown function is required to equal a given function $f(x,y)$ of the variable x and values y of the unknown function. The differential equation is SEPARABLE if $f(x,y)$ is the product of a function of x and a function of y. We study first-order separable equations and their their applications in Sections 7.1 and 7.2. Section 7.3 deals with methods for finding approximate solutions of more general first-order equations with a calculator or computer and their applications.[†]

Section 7.1

First-order, separable differential equations

OVERVIEW: *We begin this section with a discussion of* SLOPE FIELDS, *which give geometric representations of first-order differential equations. Next, we present techniques for solving* SEPARABLE *differential equations and then use these techniques in applications.*

Topics:

- ***Slope (direction) fields***
- ***The differential equation* dy/dx = ry *with constant* r**
- ***Linear separable differential equations***
- ***General separable differential equations***
- ***Applications to growth and decay***
- ***The Logistic Equation of population growth***

Slope fields

The differential equation $\frac{dy}{dx} = f(x,y)$ requires that at each point (x,y) on the graph of a solution the tangent line has slope $f(x,y)$. To visualize this condition, imagine a short line segment of slope $f(x,y)$ centered a point (x,y) (Figure 1). The differential equation is satisfied by the function $y(x)$ at x if its graph is tangent to the slope line at $\big(x, y(x)\big)$ (Figure 2).

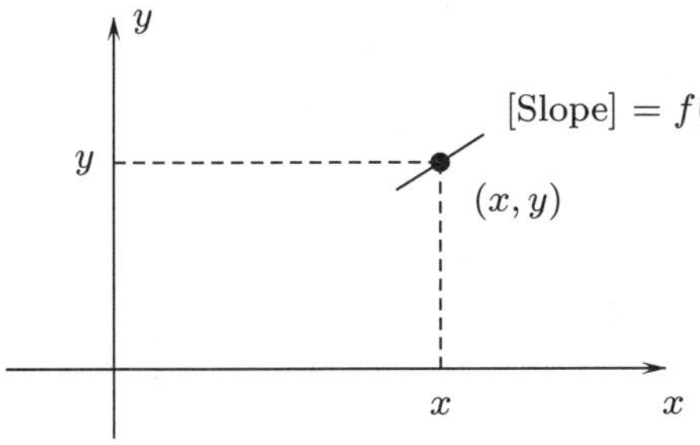

The slope line at (x,y)

FIGURE 1

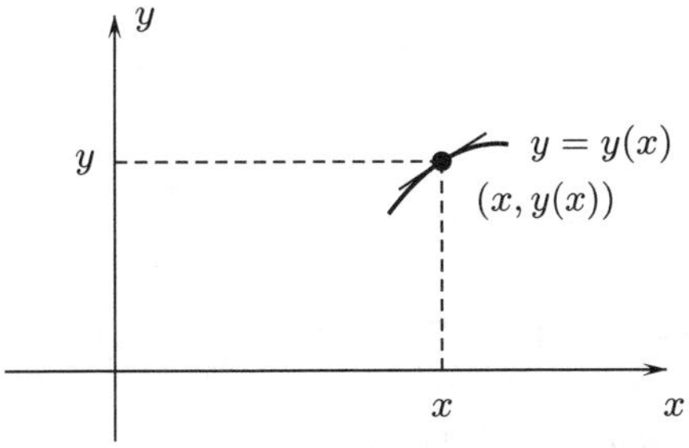

Graph of a solution tangent to the slope line

FIGURE 2

[†]Procedures for finding exact solutions of other types of first-order equations will be covered in Section 13.1.

♦ SUGGESTIONS TO INSTRUCTORS: Have students begin work on Questions 1 through 12 before class. Problems 2, 21, 25, 30, 37, 39,45, and 57 are good for in-class discussion.

A collection of slope lines in the portion of the xy-plane where $f(x, y)$ is defined is called a SLOPE FIELD or DIRECTION FIELD for the differential equation $\frac{dy}{dx} = f(x, y)$. A slope field is displayed by drawing slope lines at an array of points, as in the next Example.

Example 1 **(a)** Draw slope lines for the differential equation $\frac{dy}{dx} = \frac{1}{2}(x-y)$ at the twenty points with coordinates $x = 0, 1, 2, 3, 4$ and $y = 0, 1, 2, 3$ in Figure 3. **(b)** Describe how changing x and y changes the orientation of the slope line at (x, y).

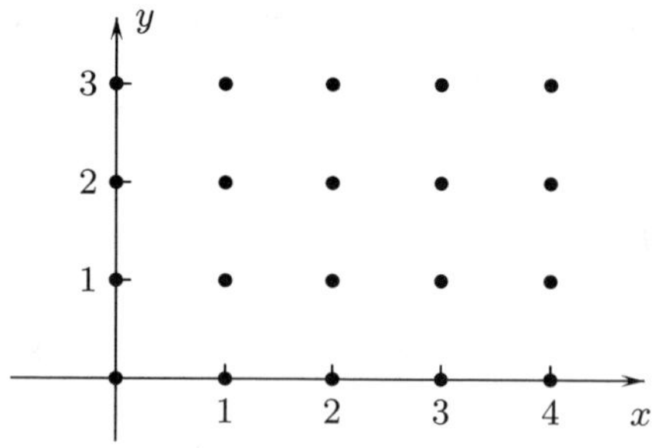

FIGURE 3

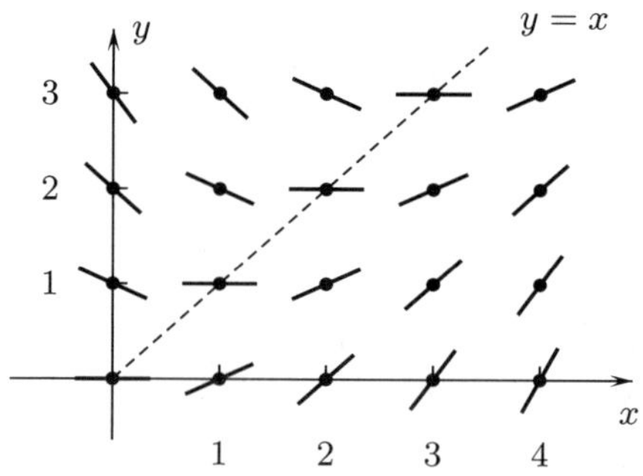

FIGURE 4

SOLUTION **(a)** We calculate the value $\frac{1}{2}(x - y)$ of the slope at each point (x, y) in the table below, and then draw a short line segment with its center at each point and slope given in the table, as shown in Figure 4.

(b) One description: the slope lines are horizontal on the line $y = x$ where $\frac{1}{2}(x - y)$ is zero, point up to the right under the line $y = x$ where $y < x$ and $\frac{1}{2}(x - y)$ is positive, point down to the right above the line $y = x$ where $y > x$ and $\frac{1}{2}(x - y)$ is negative, and become steeper as they move away from the line $y = x$. □

[SLOPE] $= \frac{1}{2}(x - y)$

	$x = 0$	$x = 1$	$x = 2$	$x = 3$	$x = 4$
$y = 3$	-1.5	-1	-0.5	0	0.5
$y = 2$	-1	-0.5	0	0.5	1
$y = 1$	-0.5	0	0.5	1	1.5
$y = 0$	0	0.5	1	1.5	2

Question 1 Draw the lines in the slope field for $\frac{dy}{dx} = \frac{1}{10}x^2$ at the points in Figure 3.

Question 2 Draw the lines in the slope field for $\frac{dy}{dx} = 2 - y$ at the points in Figure 3.

The slope field can give us a general idea of the shapes of graphs of solutions of a differential equation because the graphs are tangent to the lines in the slope field. Figure 5, for example, shows the slope field for

$$\frac{dy}{dx} = 1 \tag{1}$$

which consists of line segments of slope 1, and Figure 6 shows the graphs of seven solutions of the differential equation.

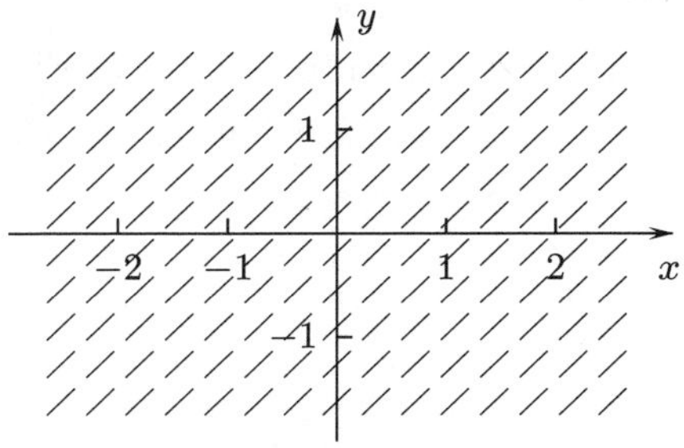

FIGURE 5

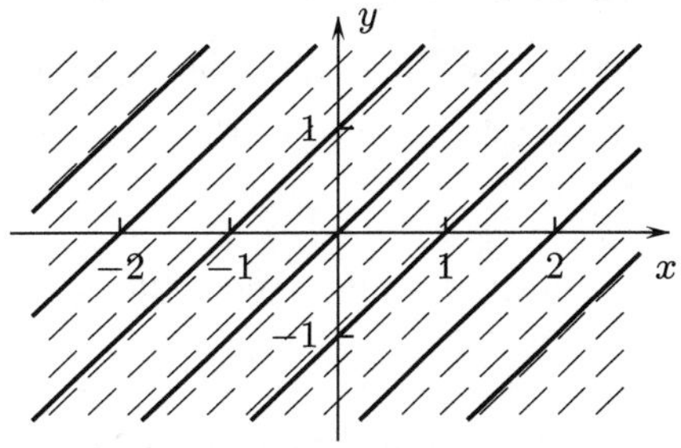

FIGURE 6

Example 2 Find a formula for all solutions of **(1)**.

SOLUTION Because there is no y on the right side of the differential equation, we can solve it by direct x-integration. The solutions are given by

$$y = \int \frac{dy}{dx}\, dx = \int 1\, dx = x + C$$

with an arbitrary constant C. The graphs of the solutions of **(1)** in Figure 6 are parallel lines $y = x + C$ of slope 1. □

Figure 7 shows the slope field for

$$\frac{dy}{dx} = x. \qquad \textbf{(2)}$$

The lines in each vertical column, where x is constant, are parallel because there is no y on the right side of the differential equation. The lines are horizontal along the y-axis, where $x = 0$. They have positive slope to the right of the y-axis, where x is positive, and have negative slope to the left of the y-axis, where x is negative. Also, the lines get steeper as x moves away from 0 in either direction.

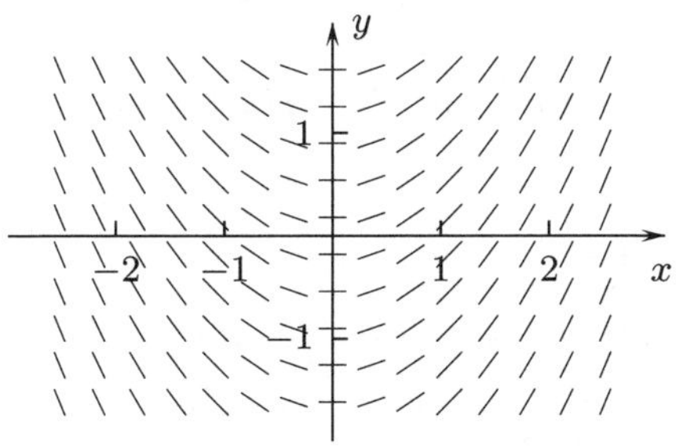

FIGURE 7

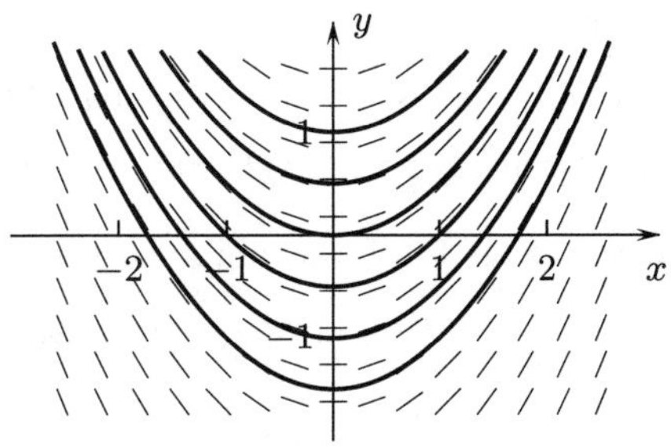

FIGURE 8

The graphs of six solutions of the differential equation **(2)** are shown in Figure 8. Notice that they follow the directions of the lines in the slope field.

Question 3 Find a formula for the solutions of **(2)**.

The differential equation dy/dx = ry *with constant* r

We cannot solve the differential equation $\frac{dy}{dx} = ry$ by direct x-integration, as we did in Example 1 and Question 1, because of the y on the right side. Instead we use the following result, which was discussed in Section 3.6.

Rule 1 *For constant r, the solutions of*

$$\frac{dy}{dx} = ry \qquad \textbf{(3)}$$

are $y = Ce^{rx}$ with arbitrary constants C.

Figure 8 shows the slope field of

$$\frac{dy}{dx} = y \qquad \textbf{(4)}$$

which is of type **(3)** with $r = 1$.

Question 4 **(a)** Use the differential equation **(4)** to explain the pattern of its slope lines in Figure 9. **(b)** Give formulas for the six solutions of **(4)** whose graphs are in Figure 10.

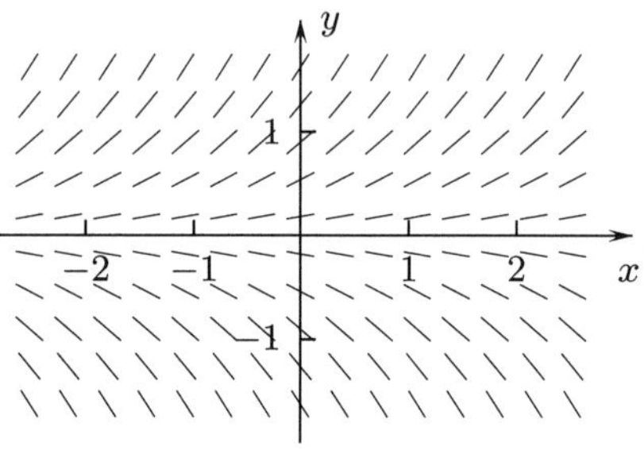

FIGURE 9

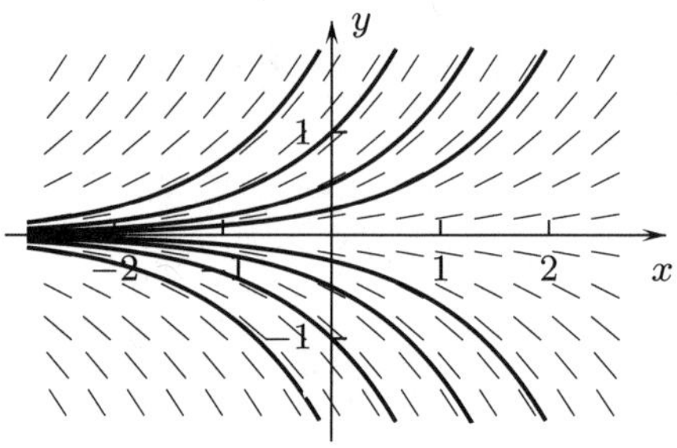

FIGURE 10

Rule 1 can also be used to solve differential equations of the form $\frac{dy}{dx} = r(y+k)$ with constants r, and k by applying it to the function $y + k$.

Example 3 Find a formula for the solutions of $\frac{dy}{dx} = 40 - 4y$.

SOLUTION We first factor the coefficient -4 of y from the right side of the differential equation to have $\frac{dy}{dx} = -4(y - 10)$. Then we rewrite this equation in the form

$$\frac{d}{dx}(y - 10) = -4(y - 10).$$

The last change is permissible because the constant -10 has zero derivative. By Rule 1 with $r = -4$ and $y - 10$ in place of y,

$$y - 10 = Ce^{-4x}$$

with an arbitrary constant C. Consequently, $y = 10 + Ce^{-4x}$. □

Linear separable differential equations

A first-order differential equation is called SEPARABLE if it can be written in the form

$$\frac{dy}{dx} = r(x)s(y) \tag{5}$$

where the right side is the product of a function of x and a function of y. The separable differential equation is also LINEAR if $s(y) = y$ and the equation is

$$\frac{dy}{dx} = r(x)y. \tag{6}$$

In the case of a linear separable equation **(6)**, we look for solutions of the form $y = Ce^{R(x)}$ with a constant C and a function $R(x)$ of x. By the Chain Rule,

$$\frac{dy}{dx} = \frac{d}{dx}(Ce^{R(x)}) = Ce^{R(x)}R'(x) = y\,R'(x).$$

The last equation shows that $y = Ce^{R(x)}$ is a solution of **(6)** if $R'(x) = r(x)$, which means that $R(x)$ is an antiderivative of $r(x)$. It can also be shown that all solutions have this form,† so we have the following generalization of Rule 1.

Rule 2 *The solutions of* $\frac{dy}{dx} = r(x)y$ *are* $r = Ce^{R(x)}$ *where* $R(x) = \int r(x)\,dx$ *is one antiderivative of* $r(x)$ *and* C *is an arbitrary constant.*

This rule is a generalization of Rule 1 because if r is a constant, we can use $R(x) = rx$ as its antiderivative.

In each case, Rules 1 and 2 give an infinite number of different solutions for different values of the constant C. This constant is usually determined from the value of the solution at one value of x as an INITIAL CONDITION. The differential equation with an initial condition is called an INITIAL-VALUE PROBLEM.

Example 4 Find the solution of the initial-value problem $y' = 2y\cos x$, $y(0) = 4$.

SOLUTION Here $r(x) = 2\cos x$, so $R(x) = \int 2\cos x\,dx = 2\sin x + c$. (We use a lower-case "$c$" here to reserve an upper-case "C" for the application of Rule 2.) Because we need only one function $R(x)$, we take $c = 0$ and $R(x) = 2\sin x$. By Rule 2, the solutions of the differential equation are $y = Ce^{2\sin x}$.

The initial condition $y(0) = 4$ determines the value of C. We set $x = 0$ and $y = 4$ in the last formula to obtain

$$4 = y(0) = Ce^{2\sin(0)} = Ce^0 = C.$$

This shows that $C = 4$, so that the solution is $y = 4e^{2\sin x}$. □

Question 5 Check the result of Example 4 by verifying that the solution satisfies the initial condition and the differential equation.

---

†If y is a solution in an open interval, then $\frac{dy}{dx} - r(x)y = 0$ there. Multiplying both sides of this equation by $e^{-R(x)}$ shows, with the Product and Chain Rules, that $\frac{d}{dx}[ye^{-R(x)}] = \frac{dy}{dx}e^{-R(x)} + y\frac{d}{dx}(e^{-R(x)}) = \left[\frac{dy}{dx} - r(x)y\right]e^{-R(x)}$ is zero. Therefore, $ye^{-R(x)}$ is equal to a constant C and $y = Ce^{R(x)}$ in the interval.

General separable differential equations

To see how a general separable differential equation $\frac{dy}{dx} = r(x)s(y)$ can be solved, we consider the example,

$$\frac{dy}{dx} = -2xy^2. \tag{7}$$

First, we find the values of y for which the function on the right is zero. In this case there is one such value: $y = 0$. This implies that the constant function $y(x) = 0$ is a solution of the differential equation, since with this function, $\frac{dy}{dx}$ and $-2xy^2$ are both zero for all x and equation **(7)** is satisfied.

We will describe two procedures for finding the other solutions. The first is a longer, logically correct approach. The second is a short-cut procedure that is generally used.

To carry out the longer procedure, we divide both sides of **(7)** by y^2 to put the unknown function y and its derivative on the left side of the equation, leaving only a function of x on the right:

$$\frac{1}{y^2}\frac{dy}{dx} = -2x.$$

We write $1/y^2$ as y^{-2} and take the x-antiderivative of both sides:

$$\int y^{-2}\frac{dy}{dx}\,dx = \int -2x\,dx$$

Next, we make the substitution $y = y(x), dy = \frac{dy}{dx}\,dx$ in the integral on the left to obtain

$$\int y^{-2}\,dy = \int -2x\,dx. \tag{8}$$

Then we integrate with respect to y on the left and with respect to x on the right, yielding

$$-y^{-1} = -x^2 + C. \tag{9}$$

We might have put one constant of integration C_1 on the left and another C_2 on the right, but then we would have obtained $-y^{-1} = -x^2 + (C_2 - C_1)$, which is equivalent to **(9)**, since the difference of two arbitrary constants is just another arbitrary constant. We multiply by -1 and take reciprocals to obtain

$$y = \frac{1}{x^2 - C} \tag{10}$$

where C is an arbitrary constant.

In the short-cut procedure for solving such differential equations, we view the symbol dy/dx as a fraction and put all the y's and dy on the left and all the x's and dx on the right. For the differential equation **(7)**, we divide both sides by y^2 and multiply both sides by the symbol dx to have

$$y^{-2}\,dy = -2x\,dx.$$

Then we add integral signs to obtain **(8)**, which leads to the solutions **(10)**, as above. This technique is called SEPARATION OF VARIABLES.[†]

Remembering that we found the constant solution $y = 0$ before separating variables, we see that the solutions of the differential equation **(7)** are

$$y = 0 \quad \text{and} \quad y = \frac{1}{x^2 - C} \quad \text{with any constant } C. \tag{11}$$

[†]This technique can also be used with linear separable equations, but it is easier to apply Rules 1 and 2 in these cases.

We can check the second formula in **(11)** by differentiation. For $y = (x^2 - C)^{-1}$, the Chain-Rule formula for derivatives of powers gives

$$\frac{dy}{dx} = \frac{d}{dx}[(x^2 - C)^{-1}] = -(x^2 - C)^{-2}\frac{d}{dx}(x^2 - C) = -2x(x^2 - C)^{-2} \qquad \textbf{(12)}$$

while

$$-2xy^2 = -2x[(x^2 - C)^{-1}]^2 = -2x(x^2 - C)^{-2}. \qquad \textbf{(13)}$$

Since the expressions on the right of **(12)** and **(13)** are the same, $\frac{dy}{dx}$ equals $-2xy^2$ for all x where $y(x)$ is defined, and the differential equation is satisfied.

We examine the slope field of this differential equation in the next Question.

Question 6 **(a)** Figure 11 shows the slope field of the differential equation **(7)**. Use the differential equation to explain the pattern of slope lines. **(b)** The graphs of the solutions with the initial conditions $y(0) = 1$ and $y(0) = -\frac{1}{4}$ are given in Figure 12. Find formulas for them.

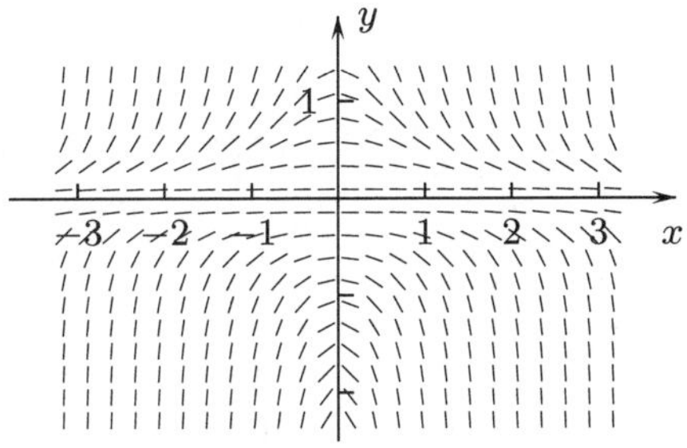

Slope field of $\frac{dy}{dx} = -2xy^2$

FIGURE 11

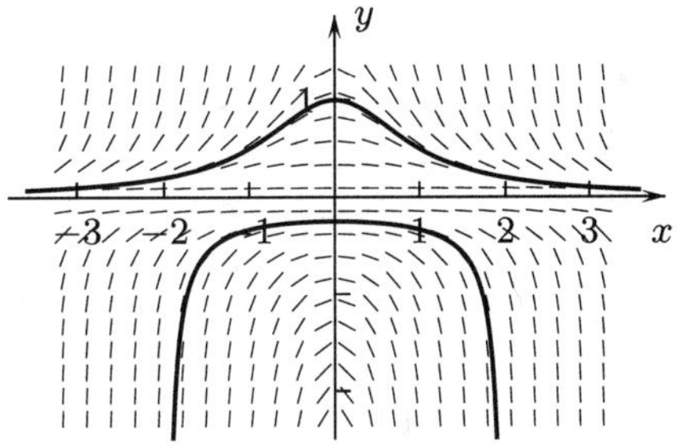

Solutions with $y(0) = 1$ and $y(0) = -\frac{1}{4}$

FIGURE 12

Four differential equations of growth and decay

We now look at four problems of growth and decay that are modeled with first-order differential equations.

Question 7 Each of problems **(A)** through **(D)** below leads to a differential equation for the function $y(t)$. Match the four problems to differential equations **(I)** through ((**IV**) that follow.

(A) The thickness of the ice on a lake grows at a rate that is proportional to the reciprocal of its thickness. Find the thickness $y(t)$ as a function of the time t.

(B) A population grows at a rate proportional to its size. Find the population $y(t)$ as a function of the time t.

(C) A hot potato is taken out of the oven at time $t = 0$ into a kitchen that is at 20° Celsius. The rate of change of the potato's temperature is proportional to the difference between its temperature and that of the kitchen. Find the temperature $y(t)$ of the potato as a function of t.

(D) Find a function $y(t)$ whose rate of change with respect to t is $-2t$.

(I) $\frac{dy}{dt} = 0.2y$ **(II)** $\frac{dy}{dt} = \frac{20}{y}$

(III) $\frac{dy}{dt} = -2t$ **(IV)** $\frac{dy}{dt} = 2(20 - y)$

Question 8 Match the differential equations in **(I)** through **(IV)** with the slope fields in Figures 13 through 16. Explain how you make your choices.

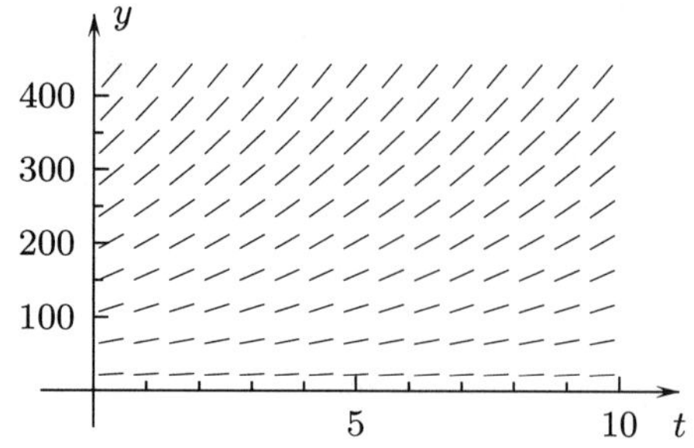

FIGURE 13

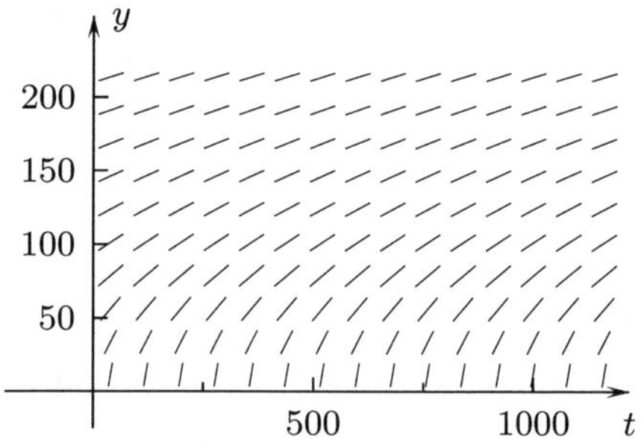

FIGURE 14

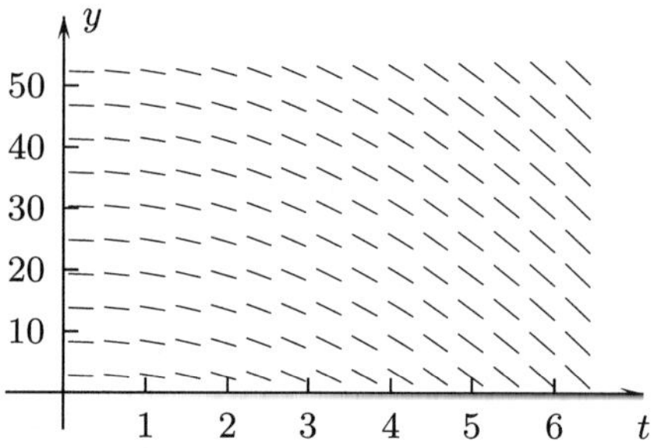

FIGURE 15

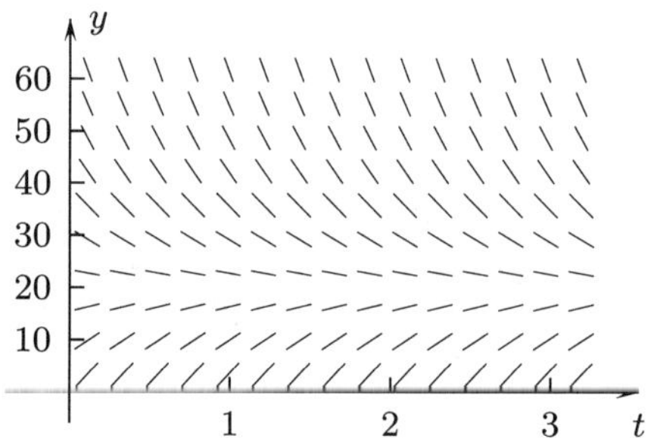

FIGURE 16

Question 9 Solve each of the differential equations in Question 7 with the initial condition $y(0) = 50$. Generate the graph of each solution and draw it with the corresponding slope field. Then determine what happens to each solution as $t \to \infty$.

Population growth in a limited environment: the Logistic Equation

Imagine that rabbits are introduced on an island where they have no natural enemies, and that the island can support a stable population of two thousand rabbits. A mathematical model of the growth of the population of rabbits might given by the LOGISTIC DIFFERENTIAL EQUATION

$$\frac{dP}{dt} = \tfrac{1}{2}P(2 - P) \qquad \textbf{(14)}$$

where $P = P(t)$ is population, measured in thouands, at time t measure in years. Figure 17 shows the slope field of this differential equation.

FIGURE 17

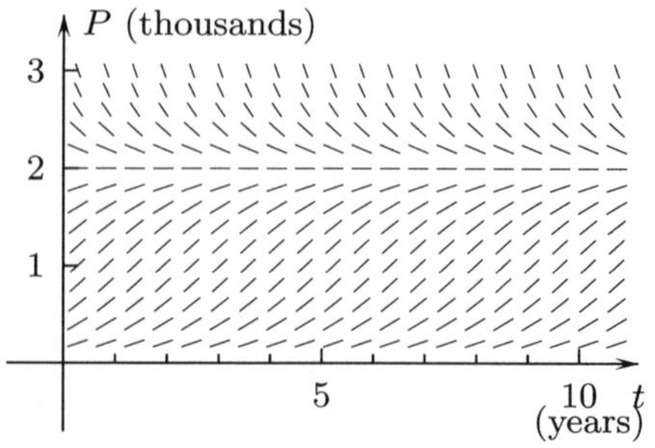

Question 10 Describe the patterns of the slope lines in Figure 16 and explain how those patterns are caused by the differential equation **(14)**.

Example 5 Find the solutions of **(14)**.

SOLUTION Because $\frac{1}{2}P(2-P)$ is zero for $P=0$ and $P=2$, the constant solutions are $P=0$ and $P=2$. To find the other solutions, we separate the variables to obtain

$$\frac{dP}{P(2-P)} = \tfrac{1}{2}dt$$

and then

$$\int \frac{dP}{P(2-P)} = \tfrac{1}{2}\int dt. \tag{15}$$

Next we use the identity

$$\frac{1}{P(L-P)} = \frac{1}{L}\left(\frac{1}{P} + \frac{1}{L-P}\right) \tag{16}$$

which can be derived by the Method of Partial Fractions of Section 6.2 and can be verified by combining the two terms on the right into a single quotient. Equation **(14)** and **(16)** with $L=2$ give

$$\tfrac{1}{2}\left(\int \frac{1}{P}\,dP + \int \frac{1}{2-P}\,dP\right) = \tfrac{1}{2}\int dt.$$

We multiply both sides of this equation by 2 and integrate, with the substitution $u = 2-P, du = -dP$ in the second of integrals to obtain

$$\ln|P| - \ln|2-P| = t + C_1. \tag{17}$$

Question 11 Solve **(17)** for the population P of rabbits.

To study the behavior as $t \to \infty$ of the function $P = \dfrac{2Ce^t}{Ce^t+1}$ from Question 11, we first divide the numerator and denominator by Ce^t to obtain $P = \dfrac{2}{1+1/(Ce^t)}$. Then, we set $c = 1/C$ to have

$$P = \frac{2}{1+ce^{-t}} \text{ thousands.} \tag{18}$$

This formula for the population shows that it tends to 2 thousand as $t \to \infty$ for any choice of c.

All solutions of **(14)**, except the constant solution $P=0$, are of the form **(18)** for different values of the constant c. The constant can be determined by an initial condition, as in the next example.

Example 6 Suppose that 100 rabbits are introduced initially. Based on the logistic differential equation **(14)**, what is the population t years later?

SOLUTION Since P is measured in thousands of rabbits, $P = 0.1$ when the population is 100. We set $t=0$ and $P=0.1$ in **(18)** to have $0.1 = 2/(1+c)$. This gives $1+c=20$ and then $c=19$. In this case, there are $\dfrac{2}{1+19e^{-t}}$ thousand rabbits on the island after t years. □.

Figure 18 shows the graph of the solution from Example 6, corresponding to an initial population of 100 rabbits. Figure 19 shows what happens if initially there are 3000 rabbits on the island—which is more than the island can support.

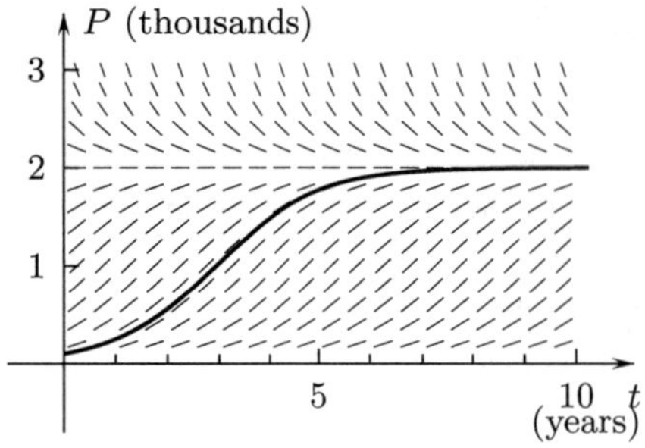

FIGURE 18

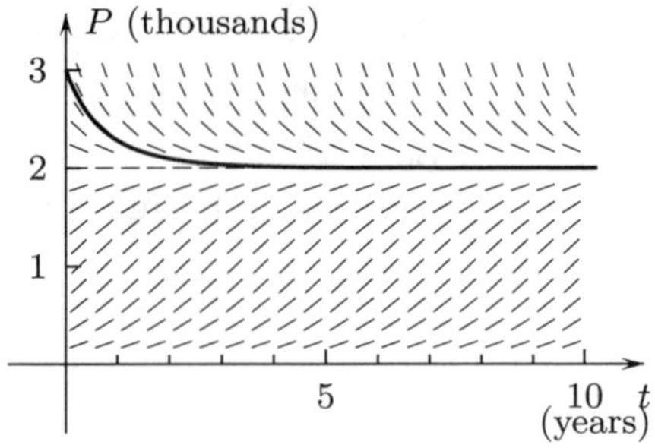

FIGURE 19

Question 12 What is the population of rabbits after t years if the initial population is 3000?

Principles and procedures

- When you are working with applications involving first-order differential equations, visualize their slope fields and notice how they relate to the problems, and when you have found solutions of the differential equations be sure their graphs follow the directions of the lines in the slope fields.†
- When you have a separable differential equation to solve, check first to see if it is linear, because in this case it is easier to apply Rule 1 or 2 above than it is to separate variables.
- You can check a formula for a solution $y(x)$ of the differential equation $\frac{dy}{dx} = f(x, y)$ by finding the derivative of the function, substituting the formula for the function in the expression for $f(x, y)$, and comparing the results. This process often, however, requires more difficult algebra than solving the differential equation.
- When you have solved an initial-value problem, check that your solution satisfies the initial condition. This generally takes only a few seconds and it is easy to make minor mistakes in matching initial conditions.

Tune-Up Exercises 7.1♦

[A] *Answer provided.* [O] *Outline of solution provided.*

T1.[O] Draw the slope lines for the differential equation $\frac{dy}{dx} = \frac{y}{x}$ at the sixteen points in Figure 20. Then explain how the orientation of the lines varies as x and y are changed.

♦ Type 1, sketching a slope field

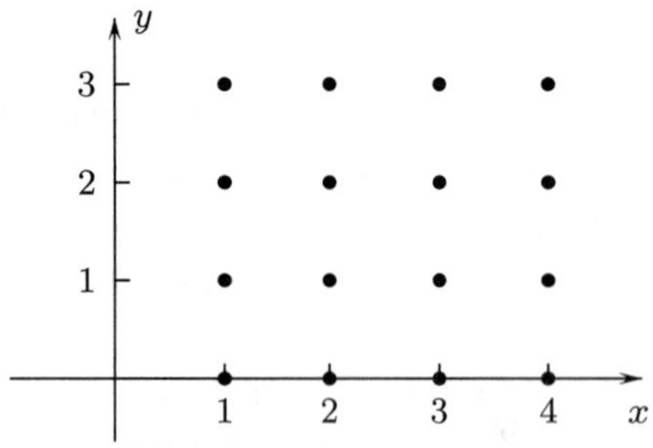

FIGURE 20

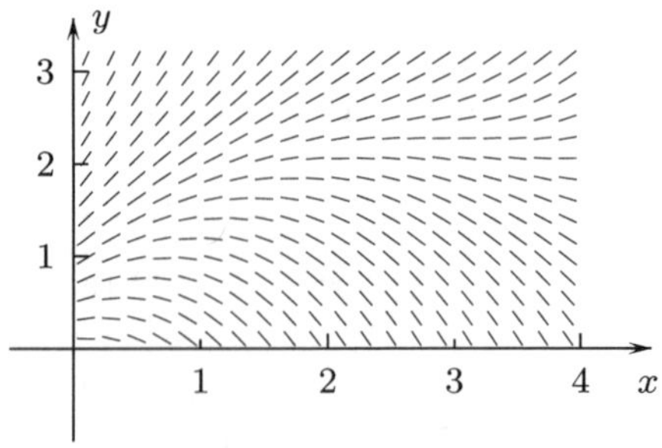

FIGURE 21

T2.[O] Figure 21 shows the slope field for the differential equation $\frac{dy}{dx} = f(x, y)$. Draw the approximate graph of the solution which satisfies the initial condition $y(0) = 1$.

♦ Type 1, using a slope field

---

†Programs for generating slope fields on calculators will be given in the *Calculator/Computer Supplement.*

♦The Tune-up Exercises and Problems are classified by type and content. The types are (1) basic, reactive; (2) basic reflective; (3) intermediate, reactive; (4) intermediate, reflective; (5) advanced, reactive; (6) advanced, reflective; and (7) advanced, theoretical.

Solve differential equations and initial-value problems T3 through T6.

T3.[O] $\frac{dy}{dx} = -4y,\ y(0) = -3$

♦ Type 1, a linear, separable differential equation

T4.[O] $\frac{dy}{dt} = 4t^3y,\ y(0) = 10$

♦ Type 1, a linear, separable differential equation

T5.[O] $f'(x) = -3[f(x)]^4$

♦ Type 1, a separable differential equation

T6.[O] $K'(x) = \sqrt{xK(x)},\ K(1) = 1$

♦ Type 1, a separable differential equation

Problems 7.1

[A]*Answer provided.* [O]*Outline of solution provided.*

1.[A] Figure 22 shows the slope field for $\frac{dy}{dx} = g(x, y)$. Draw the approximate graph of the solution which satisfies the initial condition $y(0) = 2$.

♦ Type 1, using a slope field

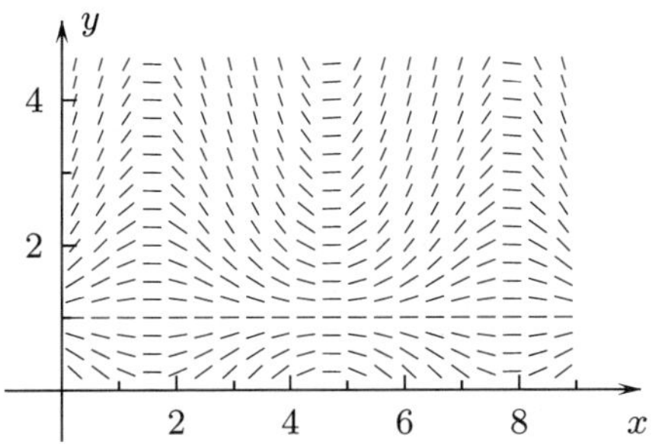

FIGURE 22

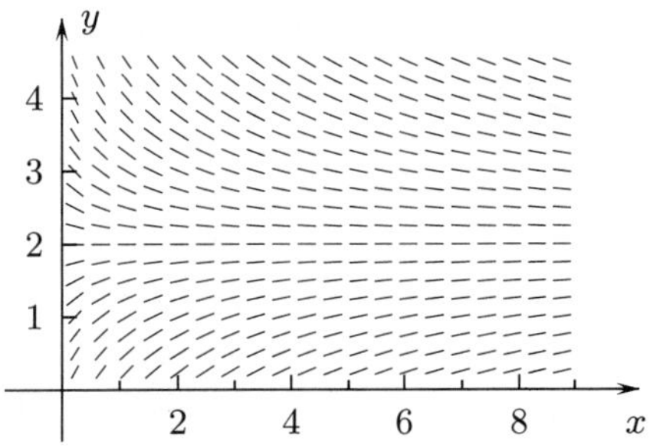

FIGURE 23

2.[A] The slope field for $\frac{dy}{dx} = h(x, y)$ is in Figure 23. Draw the approximate graphs of the solutions with the initial values **(a)** $y(0) = 1$ and **(b)** $y(0) = 4$.

♦ Type 1, using a slope field

3. Draw the approximate graph of the solution of the initial-value problem $\frac{dy}{dx} = K(x, y), y(0) = 2$, where the slope field for the differential equation is Figure 24.

♦ Type 1, using a slope field

FIGURE 24

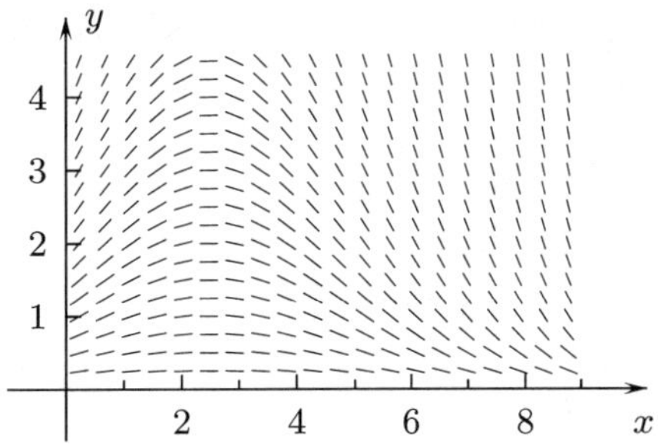

Solve differential equations and initial-value problems in Problems 4 through 17.

4.[A] $\frac{dy}{dt} = 8y,\ y(0) = 4$

♦ Type 1, an initial-value problem

5.[A] $\frac{dy}{dt} = -y,\ y(4) = 10$

♦ Type 1, an initial-value problem

6.[O] $\frac{dz}{dx} = (\ln 4)z,\ z(2) = 15$

♦ Type 1, an initial-value problem

7. $\frac{dy}{dt} = (0.006)y,\ y(50) = 3$

♦ Type 1, an initial-value problem

8.[O] $\frac{dv}{dt} = 5 - v,\ v(0) = 10$

♦ Type 1, an initial-value problem

9.[O] $\frac{dQ}{dx} = -3Q^{1/4}$

♦ Type 1, a differential equation

10.[O] $\frac{dy}{dx} = -y\cos x$

♦ Type 1, a differential equation

11.[A] $Q'(t) = t[Q(t)]^{1/3},\ Q(0) = 1$

♦ Type 1, an initial-value problem

12. $\dfrac{dy}{dx} = -2xy$

♦ Type 1, a differential equation

13.[O] $\dfrac{dy}{dx} = -3y\cos(3x), y(\pi) = 6$

♦ Type 1, an initial-value problem

14. $\dfrac{dy}{dx} = x\sin(x^2)y$

♦ Type 1, a differential equation

15.[A] $\dfrac{dy}{dx} = 3x^2y^2$

♦ Type 1, a differential equation

16. $\dfrac{dy}{dx} = 3\sqrt{xy}, y(1) = 1$

♦ Type 1, an initial-value problem

17. $\dfrac{dy}{dt} = 4yt^3, y(0) = -10$

♦ Type 1, an initial-value problem

18.[A] Figure 25 shows the slope field of $\dfrac{dy}{dx} = \frac{1}{4}y^2\sin x$. Find the solutions that satisfy **(a)** $y(0) = 2$ and **(b)** $y(0) = -2$. Then generate the graphs of the solutions and copy them with the slope field.

♦ Type 1, initial-value problems, etc.

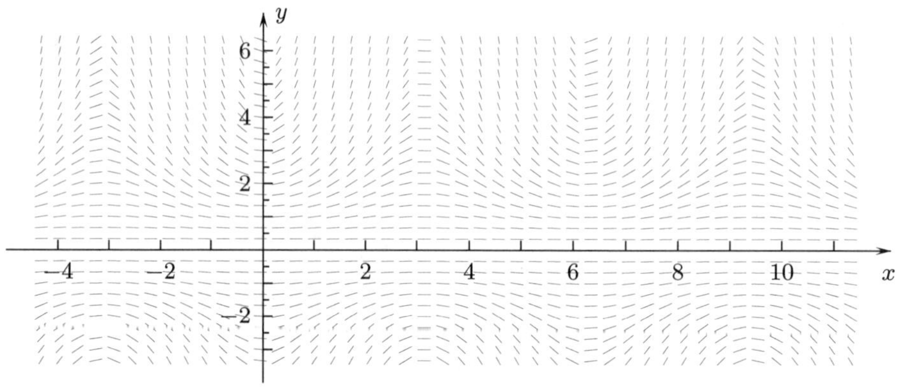

FIGURE 25

19. The slope field of $\dfrac{dy}{dx} = \frac{1}{9}x^2y^2$ is in Figure 26. Find the solution with $y(-2) = 1$. Then generate its graph and copy it with the slope field.

♦ Type 1, an initial-value problem, etc.

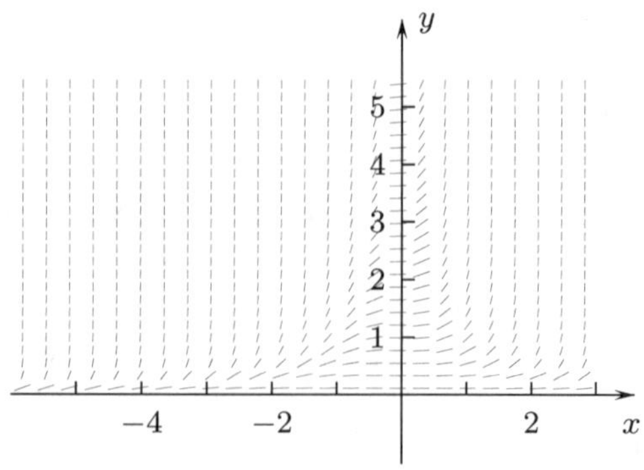

FIGURE 26

In the following Problems give exact answers and, where appropriate, their approximate decimal values.

20.[O] The relative rate of growth of a culture of bacteria is the constant 0.03 day^{-1}.[†] The mass of the culture is two grams initially. What is its mass one week later?

♦ Type 2, an initial-value problem

21.[A] The rate of increase of the number of flies in a colony is proportional to the number in the colony and the number increases by 50% every two weeks. What is the constant of proportionality?

♦ Type 2, a differential equation

22. What is $y(40)$ if $y(0) = 200$ and the relative rate of decrease of $y(t)$ is 0.25 per year, with t measured in years?

♦ Type 2, an initial-value problem

23. If the percentage rate of increase of the price of potatoes is the constant 5% per year and potatoes cost \$0.20 per pound at the beginning of 1998, what would they cost at the beginning of the year 2021?

♦ Type 2, an initial-value problem

24. What is $P(t)$ for $t \geq 100$, with t measured in months, if $P(100) = 1000$ and the percentage rate of increase of $P(t)$ is 5% per month?

♦ Type 2, an initial-value problem

25. What is the ratio $Z(50)/Z(10)$ if the relative rate of increase of $Z(t)$ is 0.1% per year and t is measured in years?

♦ Type 2, a differential equation

26. If the buying power $B(t)$ of a currency decreases at the rate $0.05B(t)$ at time t (years), how long does it take for $B(t)$ to be cut in half?

♦ Type 2, a differential equation

27. The air pressure $P(y)$ (kilograms per square meter) at an altitude of y kilometers above the surface of the earth satisfies $\frac{dP}{dy} = -0.12P$. The air pressure at the surface of the earth is 10^4 kilograms per square meter. What is it at an altitude of 10 kilometers?

♦ Type 2, an initial-value problem

28. If the source of direct current is removed at time $t = 0$ (seconds) from a electrical circuit consisting of one loop with (constant) resistance R ohms and (constant) inductance L henries, then for $t > 0$, the current J in the circuit satisfies the differential equation $L\frac{dJ}{dt} + RJ = 0$. **(a**[A]**)** Give a formula for the current $J(t)$ in terms of R, L, and the size J_0 of the current at $t = 0$. **(b)** How would doubling J_0 affect $J(t)$? **(c)** How would increasing the resistance affect $J(t)$ for the same nonzero values of L and J_0? **(d)** How would increasing the inductance affect $J(t)$ for the same nonzero values of R and J_0?

♦ Type 4, a differential equation, etc.

29. A leaky bucket contains water to a depth of 10 inches at time $t = 0$ (second). The depth decreases at the rate $\frac{1}{3}h$ when it is h inches deep. How long does it take for the depth to reduce to 1 inch?

♦ Type 2, a differential equation

30.[A] The population of the United States was 4 million in 1790 and 249 million in 1990. If the rate of growth of the population had been always proportional to the population during the two-hundred years from 1790 to 1990, what would have been the constant of proportionality?

♦ Type 2, a differential equation

31. Due to a curse imposed by a neighboring tribe, the members of a village are driven to murder each other. The rate of change of the population is $-2\sqrt{P}$ people per month when the population is P. When the curse is made, the population is 1600. When will there be only one person left in the village?

♦ Type 2, an initial-value problem

32. Twenty-thousand dollars is invested at time $t = 0$ (years) and, because of steadily declining interest rates, the relative rate of change of the investment is $(0.12 - 0.01t)$ year^{-1} at time t for $0 \leq t \leq 3$. What is the value of the investment at $t = 3$?

♦ Type 2, an initial-value problem

[†]Recall from Section 3.4 that the relative rate of change of $f(x)$ (with respect to x) is $f'(x)/f(x)$ and its percent rate of change is $100f'(x)/f(x)$.

33. Due to periodic changes in the environment, the rate of change of the population P of a bacteria at time t (years) is $(2 + \sin t)P$ bacteria per year. The population is 100,000 at $t = 0$. Give a formula for $P(t)$ for $t \geq 0$.

◆ Type 2, an initial-value problem

34. If at $t = 0$ (seconds), the source of current in removed from a circuit consisting only os resistance R ohms and inductance L henries, then for $t > 0$ the current I (amps) satisfies the diffrential equation $L\dfrac{dI}{dt} + RI = 0$. What is the current at $t = 6$ if $R = 2, L = 3$, and the current is 5 amps at $t = 0$?

◆ Type 2, an initial-value problem

35. Nitrogen pentoxide (N_2O_5) is a solid that decomposes into the gases nitrogen dioxide (NO_2) and oxygen (O_2). If $N(t)$ denotes the amount of N_2O_5 present in a sample of time t seconds, then[(1)] $\dfrac{dN}{dt} = -0.0005N$. What fraction of the N_2O_5 in a sample decomposes in 1000 seconds?

◆ Type 2, an initial-value problem

36. Because of natural growth and immigration, the rate of increase of a country's population is $0.05P + 3000$ people per year when the population is P. If the country has 124,000 inhabitants now, how many will it have in five years?

◆ Type 2, an initial-value problem

37. **(a)** In a mathematical model of a melting spherical snowball, it is assumed that the rate of change of its volume is proportional to its surface area. Use the formulas $V = \frac{4}{3}\pi r^3$ and $A = 4\pi r^2$ for the volume and surface area of a sphere of radius r to show that this leads to the differential equation $\dfrac{dV}{dt} = -kV^{2/3}$ for the volume with a positive constant k. **(b)** Suppose that the volume V (cubic feet) of a snowball satisfies the differential equation $\dfrac{dV}{dt} = -3V^{2/3}$ with t measured in hours, and that $V(0) = 8$ cubic feet. Give formulas for the volume of the snowball for $t \geq 0$. When does it disappear?

◆ Type 2, an initial-value problem, etc.

38. The moss on a hillside increases at the rate of $\frac{1}{12}M$ pounds per month if there are M pounds of it present on April 1 and decreases at the same rate if there are M pounds present on October 1. As a mathematical model of the growth and decline of the moss, it is assumed that the mass at time t satisfies $\dfrac{dM}{dt} = \frac{1}{12}\sin(\frac{1}{6}\pi t)M$, where t measured in months and $t = 0$ is at the beginning of the year. Based on this model, how much moss will be on the hill at the beginning of October if there are 600 pounds on January 1?

◆ Type 2, an initial-value problem

39.[O] A box at a temperature of 10°C is placed in water at the constant temperature 0°C. The rate of change of the temperature of the box is proportional to the difference between its temperature and the temperature of the water. The temperature of the box is falling at the rate of 7 degrees per minute when it is 4°C. At what rate is the temperature falling when it is 1°C?

◆ Type 2, Newton's law of cooling

40. An orange at a temperature of 50°F is placed in a room at a constant temperature of 70°F, and two minutes later its temperature is 60°F. The rate of change of the temperature of the orange is proportional to the difference between its temperature and the temperature of the room. What is its temperature five minutes after it is placed in the room?

◆ Type 2, Newton's law of cooling

41. After a cake is removed from an oven at 300°F, the rate of change of its temperature is proportional to the difference between its temperature and the constant temperature 65°F of the room. One minute later the temperature of the cake is 200°F. How rapidly is it cooling off at that moment?

◆ Type 2, Newton's law of cooling

42.[A] **(a)** A population P that satisfies a Logistic Equation $\dfrac{dP}{dt} = kP(L-P)$ has an equilibrium population of 100 and is such that $P(0) = 10$ and $\dfrac{dP}{dt}(0) = 9$. What are the constants L and k? **(b)** Find a formula for the population from part (a) as a function of $t \geq 0$.

◆ Type 2, an initial-value problem, etc.

[(1)]Data adapted from *Physical Chemistry* by W. Moore, 3rd ed., Englewood Cliffs, NJ: Prentice Hall, Inc., 1962, p. 261.

43. The mass M of bacteria in a culture satisfies a Logistic Equation $\frac{dM}{dt} = kM(L - M)$. The rate of change of M is 180 grams per day when M is 10 grams and is 320 grams per day when M is 20. **(a)** What is the equilibrium mass of the bacteria in the culture? **(b)** What is the mass t days after it is 10 grams?

♦ Type 2, an initial-value problem, etc.

44. Let $P(t)$ be the number of people in a population of P_1 who know a rumor at t days. Suppose that the rate at which P increases is jointly proportional to the product of the number of people who know the rumor and of the number who do not. Give a differential equation for $P(t)$ and describe the nature of its solutions

♦ Type 4, a differential equation

45. If a nonconstant population P satisfies the Logistic Equation $\frac{dP}{dt} = kP(L-P)$, what is the population when its rate of growth is a maximum?

♦ Type 4, a differential equation

46. **(a)** When a powder is dissolved in a liquid, the rate at which the weight $x(t)$ of dissolved substance increases satisfies a Logistic Equation $\frac{dx}{dt} = kx(L - x)$ with positive constants k and L. **(a)** Sketch the slope field for this differential equation and discuss the nature of its solutions. **(b)** What interpretation does the constant L have? **(c)** For what values of x is the solution "supersaturated"?

♦ Type 4, a differential equation

Solve the differential equations and initial-value problems in Problems 47 through 56.

47.[O] $\frac{dF}{dx} = 4 + F^2, F(0) = 2$

♦ Type 3, an initial-value problem,

48.[A] $\frac{dx}{dt} = t^2\sqrt{1 - x^2}, x(0) = 0$

♦ Type 3, an initial-value problem

49.[A] $\frac{dQ}{dt} = 6t(Q^2 + 4Q + 4)$

♦ Type 3, a differential equation

50.[A] $\frac{dy}{dx} = \frac{\sin x}{\cos y}, y(0) = 0$

♦ Type 3, an initial-value problem

51. $\frac{dy}{dx} = \frac{\sec y}{x^3}, y(1) = \frac{1}{6}\pi$

♦ Type 3, an initial-value problem

52. $\frac{dy}{dx} = \cos^2 y$

♦ Type 3, a differential equation

53. $\frac{dy}{dx} = \frac{4x^3 + 1}{y^2 + 1}, y(0) = -1$

♦ Type 3, an initial-value problem

54.[A] $\frac{dy}{dx} = (1 + \ln x)y, y(1) = 1$

♦ Type 3, an initial-value problem

55. $\frac{dy}{dx} = y^2 - 2y - 3, y(0) = 4$

♦ Type 3, an initial-value problem

56. $\frac{dy}{dx} = -\sin x \cos^2 y, y(0) = \frac{1}{4}\pi$

♦ Type 3, an initial-value problem

57. A 120-gallon tank is full of water in which twenty pounds of salt is dissolved. Starting at $t = 0$ (minutes), a salt solution containing one-half pound of salt per gallon is added to the tank at the constant rate of six gallons per minute. Assume that the two solutions mix instantly and the excess drains off. **(a)** How much salt is in the tank at time $t > 0$? **(b)** What is the limit as $t \to \infty$ of the concentration of salt in the tank?

♦ Type 4, an initial-value problem, etc.

58. Someone puts five pounds of soap in a water fountain contaning 300 gallons of water. When the soap is discovered, clean water is run into the fountain at the rate of ten gallons per minute and the excess drains off. Under the assumption that the solutions mix instantly, when is there only one pound of soap left in the water?

♦ Type 4, an initial-value problem, etc.

Section 7.2

More applications of separable differential equations

OVERVIEW: *In this section we discuss other applications of separable, first-order differential equations, beginning with problems involving motion on a straight path subject to air resistance. Next, we describe how curves can be considered as solutions of first-order differential equations and look at the* LOTKA-VOLTERRA *differential equation that arises in a mathematical model of the interaction between predators and their prey. Then we discuss the differential equation associated with* WEBER'S LAW *that relates scales for measuring stimulus and sensation, and show how differential equations are used in studying chemical reaction rates and orthogonal trajectories.*

Topics:

- ***Velocity with air resistance***
- ***Curves as solutions of differential equations***
- ***Predator-prey differential equations***
- ***Stimulus and sensation***
- ***Orthogonal trajectories***
- ***Chemical reaction rates***

Velocity with air resistance

As we have discussed before, if an object moving on an s-axis has s-coordinate $s(t)$ at time t, then its velocity $v(t)$ in the positive s-direction is the first derivative of $s(t)$ and its acceleration $a(t)$ in the positive s-direction is the derivative of the velocity, which is the second derivative of $s(t)$:

$$v(t) = \frac{ds}{dt}(t) \quad \text{and} \quad a(t) = \frac{dv}{dt}(t) = \frac{d^2s}{dt^2}(t).$$

According to Newton's law, the total F of all forces on the object in the positive s-direction equals the object's MASS multiplied by its acceleration in the positive s-direction:

$$F = ma. \tag{1}$$

We will generally measure the time in seconds when we apply this formula. Then if distance is measured in centimeters and mass in grams, the force is measured in DYNES. This is called the c-g-s SYSTEM of units—for centimeter-gram-second. If distance is measured in meters and mass in kilograms, the force is measured in NEWTONS. This is the m-k-s SYSTEM—for meter-kilogram-second. In the BRITISH ENGINEERING SYSTEM, distance is measured in feet, mass in slugs, and force in pounds.

Recall also, that if an object is moving vertically on an s-axis with its positive side pointing down, and if the only force on it is gravity, then its downward acceleration is the constant g, which on or near the surface of the earth has the approximate values 9.8 meters per second^2, 980 centimeters per second^2, and 32 feet per second^2. If the s-axis points up, then the upward acceleration of gravity is $-g$.

♦ SUGGESTIONS TO INSTRUCTORS: The main material in this section can be divided into five parts, as follows:
Part 1 (Velocity and air resistance; Tune-up Exercises T1–T3; Problems 1–7, 20, 37, 38) Have students begin work on Questions 1 through 6 before class; Questions 1–6 and Problems 1–3 are good for in-class discussion.
Part 2 (Curves as solutions of differential equations; Tune-up Exercise T4; Problems 8–19) Have students begin work on Question 7 before class. Problems 8–10, 18, and 19 are good for in-class discussion.
Part 3 (Predator-prey differential equations; Problems 21 and 22) Have students begin work on Question 8 before class. Example 2 and Problems 21 and 22 are good for in-class discussion.
Part 4 (Orthogonal trajectories; Tune-up Exercise T5; Problems 39–42) Have students begin work on Question 10 before class. Example 4 and Problems 39–42 are good for in-class discussion.
Part 5 (Chemical reaction rates; Problems 34–36) Example 5 and Problem 35 are good for in-class discussion.

Because an object's weight is the force of gravity on it, Newton's law **(14)** with $a = g$ is used to convert between weight w and mass m:

$$w = mg. \tag{2}$$

In the special cases where the force is a function of only velocity and time (and, in particular, is independent of the object's position), the acceleration dv/dt depends only on the velocity and the time, and we obtain a first-order differential equation for the velocity:

$$\frac{dv}{dt} = f(v, t).$$

The next Questions deal with six such differential equations.

Question 1 Match problems **(A)**, **(B)**, and **(C)** below with differential equations **(I)**, **(II)**, and **(III)** given after the problems.

(A) v is the downward velocity, measured in meters per second, of a ball that is falling under the force of gravity but with no air resistance or other forces on it. The time t is measured in seconds and the acceleration due to gravity is 9.8 meters per second^2.

(B) v is the downward velocity, measured in meters per second, of a suitcase that is falling under the force of gravity with air resistance that is proportional to its velocity. The suitcase's downward acceleration is zero when its downward velocity is 49 meters per second. (This is called the EQUILIBRIUM VELOCITY because it is the velocity at which the upward force of air resistance equals the downward force of gravity. It is also referred to as the TERMINAL VELOCITY because it is the limit of the suitcase's velocity as $t \to \infty$.)

(C) v is the downward velocity, measured in meters per second, of a rock that is falling under the force of gravity with air resistance that is proportional to its velocity. Its equilibrium velocity is 98 meters per second.

$$\textbf{(I)}\ \frac{dv}{dt} = 9.8 \quad \textbf{(II)}\ \frac{dv}{dt} = \tfrac{1}{10}(98 - v) \quad \textbf{(III)}\ \frac{dv}{dt} = \tfrac{1}{5}(49 - v).$$

Question 2 Match the differential equations in Question 1 to the slope fields in Figures 1 through 3.

Question 3 Find the solutions of the differential equations in Question 2 with the initial condition $v(0) = 0$, and draw their graphs with the corresponding slope fields.

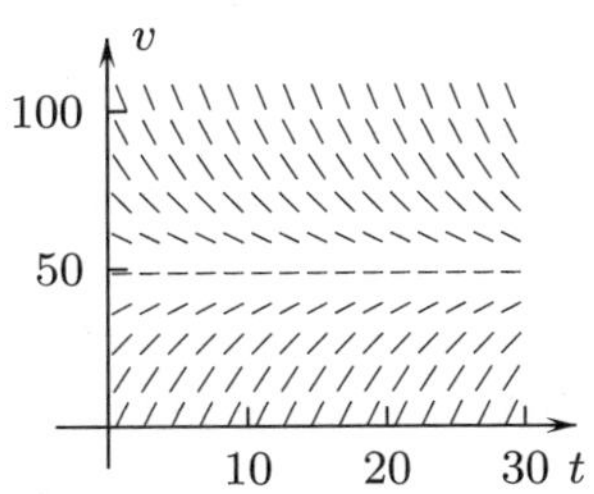

FIGURE 1

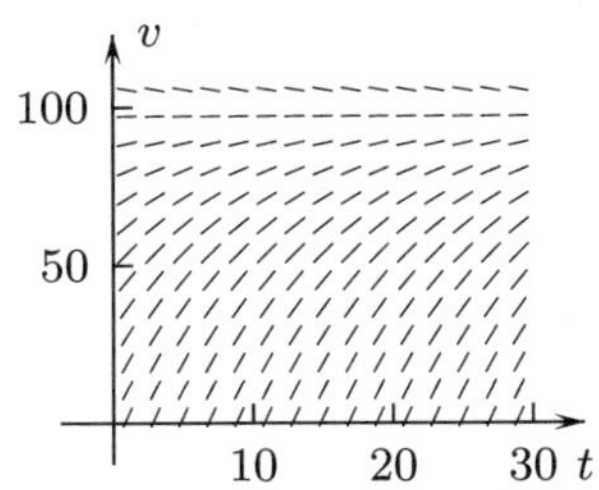

FIGURE 2

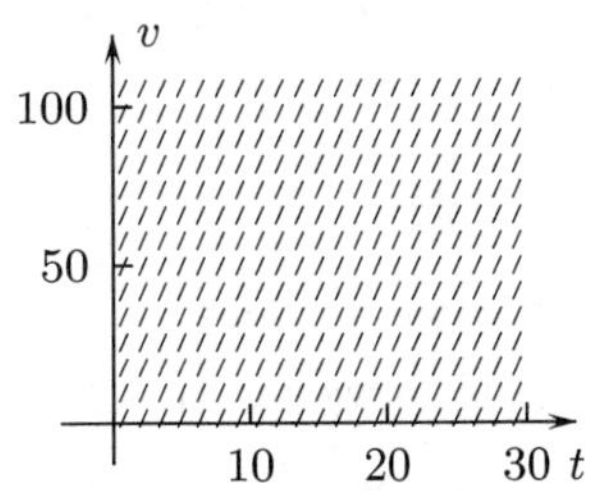

FIGURE 3

Question 4 Match problems **(D)**, **(E)**, and **(F)** below with differential equations **(IV)**, **(V)**, and **(VI)** that follow.

(D) v is the horizontal velocity, measured in miles per hour, of a model car whose acceleration, due to its faltering engine, oscillates between 0 and 90 miles per hour^2. There are no other forces on the car and time is measured in hours.

(E) v is the horizontal velocity, measured in feet per minute, of a motor boat that has its engine turned off and is slowing down because of water and air resistance. The resistance is proportional to the square of the boat's velocity, and there are no other forces on it. Time is measured in minutes.

(F) v is the horizontal velocity, measured in miles per hour, of a truck whose engine exerts a constant force. The truck is subject to air and rolling resistance that is proportional to its velocity. Time is measured in hours.

$$\textbf{(IV)}\ \frac{dv}{dt} = 200 - 5v, \quad \textbf{(V)}\ \frac{dv}{dt} = 45 + 45\cos(15t), \quad \textbf{(VI)}\ \frac{dv}{dt} = -\tfrac{1}{8}v^2$$

Question 5 Match the differential equations in Question 4 to the direction fields in Figures 4 through 6.

Question 6 Find the solutions of differential equations **(IV)** and **(V)** with $v(0) = 0$ and of **(VI)** with $v(0) = 40$. Draw the graphs of the solutions with the corresponding slope fields.

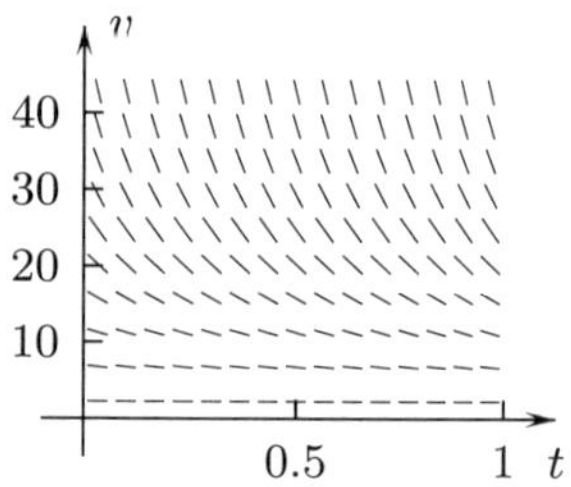

FIGURE 4

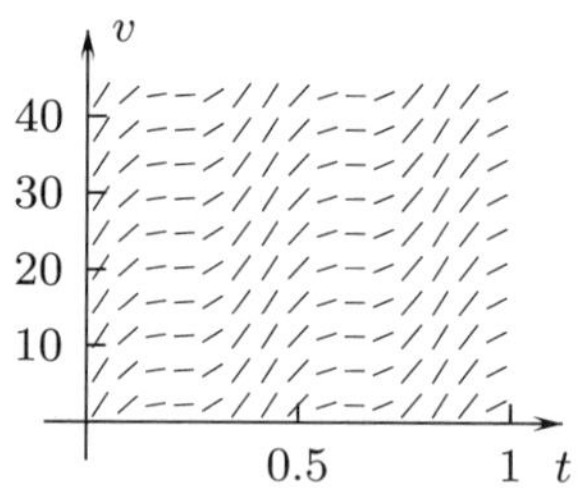

FIGURE 5

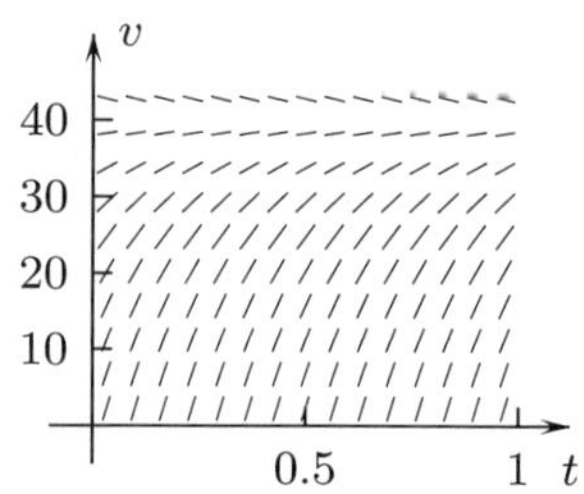

FIGURE 6

Curves as solutions of differential equations

First-order differential equations are often given in the form

$$P(x, y)\, dx + Q(x, y)\, dy = 0. \tag{3}$$

We can think of **(3)** as an alternate way of writing

$$P(x, y) + Q(x, y)\frac{dy}{dx} = 0$$

which is what we obtain if we divide both sides of **(3)** by the symbol dx.

There is, however, another way to interpret **(3)** that is convenient in many applications. We think of dx and dy not just an symbols that arise from the formula $\dfrac{dy}{dx}$ for a derivative, but as actual numbers, which we call DIFFERENTIALS.

To see how this is done, we first look at the graph of a function $y(x)$, as in Figure 7, which shows the tangent line at a point x_0. We let dx denote a nonzero change in x and let dy be the corresponding change in y along the tangent line, as shown in the drawing. Then the slope of the tangent line is $\dfrac{dy}{dx}$ and this expression has the form of the familiar notation for the derivative.

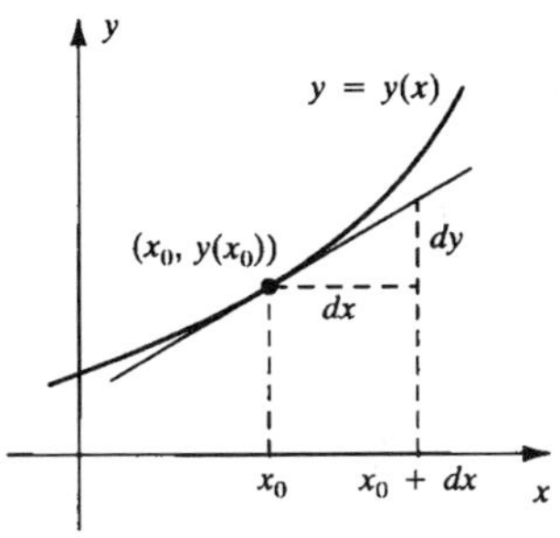

FIGURE 7

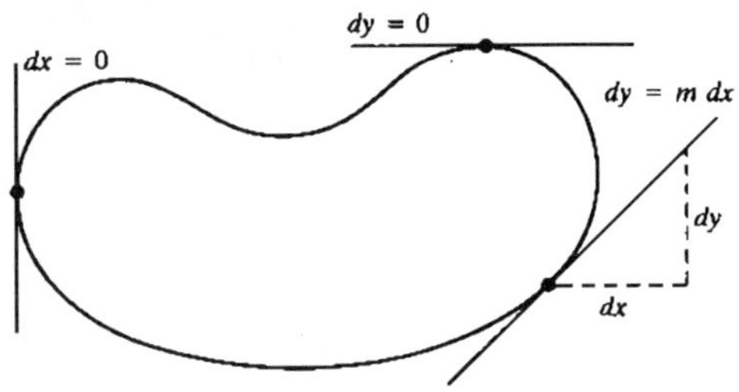

FIGURE 8

We can define differentials to any curve with tangent lines, whether the curve is the graph of a function or not, by using the following definition:

Definition 1 *A pair of differentials dx and dy to a curve at a point where it has a tangent line are the changes in x and y between two points on the tangent line at that point.*

Figure 8 shows how this definition works for curves that are not graphs of functions. The tangent line on the lower right has positive slope m and the differentials at that point satisfy $dy = m\,dx$. The tangent line on the top is horizontal, so at this point $dy = 0$ for all choices of dx. Finally, the tangent line on the left is vertical, and at this point $dx = 0$ for all choices of dy.

With this definition of dx and dy, we can interpret **(3)** as a differential equation whose solutions are curves instead of functions. Moreover, if $P(x, y)$ and $Q(x, y)$ are each products of functions of x and y, then **(3)** is separable and we can solve it with the techniques described above. Notice that the horizontal line $y = k$ is a solution of **(3)** if $P(x, k) = 0$ for all x and the vertical line $x = k$ is a solution if $Q(k, y) = 0$ for all y.

Example 1 Figure 9 shows the slope field of the differential equation

$$4\cos x\; dx - y\; dy = 0. \tag{4}$$

Find an equation for the curve that satisfy this differential equation and passes through the point $(0, 3)$.

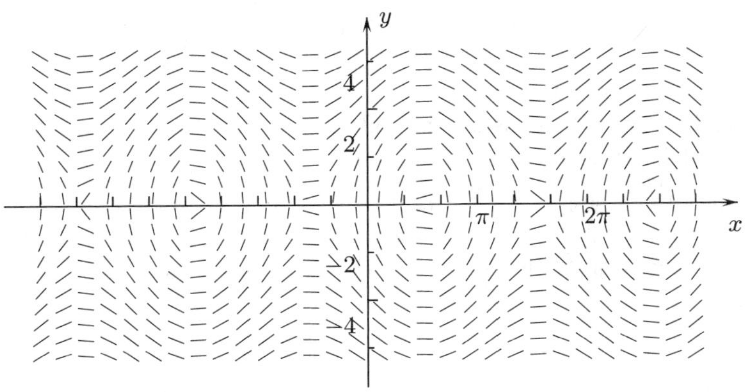

FIGURE 9

SOLUTION We first look for solutions that are horizontal or vertical lines, since such solutions might not be found when we separate variables. A horizontal-line solution $y = k$ with a constant k would have $dy = 0$ for all x and dx. There are no such solutions of **(4)** because there are no constant y's such that $4\cos x$ is zero for all x.

Vertical-line solutions are given by $x = k$ and their differentials satisfy $dx = 0$ for all y and dy. There are also no such solutions of **(4)**, since there is no value of x for which $-y$ is zero for all y.

The variables are separated in **(4)**. Adding integral signs gives

$$\int 4\cos x\, dx = \int y\, dy$$

and when we carry out the two integrations we obtain

$$4\sin x = \tfrac{1}{2}y^2 + C. \tag{5}$$

To find the solution curve through $(0, 3)$, we set $x = 0$ and $y = 3$ in **(5)**. This gives $0 = \frac{1}{2}(3)^2 + C$ so that $C = -\frac{9}{2}$ and the curve is $4\sin x = \frac{1}{2}y^2 - \frac{9}{2}$ or

$$y^2 = 8\sin x + 9.$$

Solving for y gives $y = \pm\sqrt{8\sin x + 9}$. The solution is $y = \sqrt{8\sin x + 9}$, which is the upper curve in Figure 10, because $(0, 3)$ is on the solution curve. The curve $y = -\sqrt{8\sin x + 9}$ below the x-axis is not part of the solution because it is not connected to the point $(0, 3)$. □

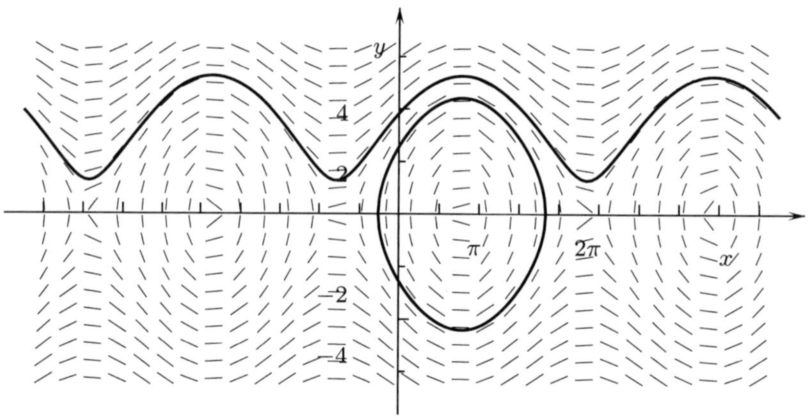

FIGURE 10

Question 7 The lower curve in Figure 10 satisfies the differential equation and passes through $(0, 2)$. Show that it has the equation $y^2 = 8\sin x + 4$.

The graphs of the two functions, $y = \sqrt{8\sin x + 4}$ and $y = -\sqrt{8\sin x + 4}$ combine to form the egg-shaped solution curve $y^2 = 8\sin x + 4$ in Figure 10 because they meet and have vertical tangent lines at the x-axis. The differential equation **(4)** is satisfied at those points because $y = 0$ and $dx = 0$ there.

Predator-prey differential equations

The Austrian mathematician A. J. Lotka (1880–1949) and the Italian mathematician Vito Volterra (1860–1940) proposed the following model for the fluxuations of populations of a predator and its prey. We let $x(t)$ denote the population of predators and $y(t)$ the population of prey at time t. The predator might be a type of insect and the prey a type of scale that is its food, or the predator might be a type of bird and the prey a type of insect that is its diet.

The population $y(t)$ of prey increases by reproduction and is decreased by the action of the prey. In the Lotka-Volterra model it is assumed that $y(t)$ increases at a rate αy proportional to their number and decreases at a rate βxy proportional to the number of predators multiplied by the number of prey. This gives the differential equation $\dfrac{dy}{dt} = \alpha y - \beta xy$ or

$$\frac{dy}{dt} = (\alpha - \beta x)y. \tag{6}$$

It is also assumed that the population $x(t)$ of predators decreases by natural attrition and is increased by the presence of the prey as food in such a way that $x(t)$ decreases at a rate γx proportional to its number and increases at a rate δxy proportional to the product of the number of predators and the number of prey, so that $\dfrac{dx}{dt} = -\gamma x + \delta xy$ or

$$\frac{dx}{dt} = (-\gamma + \delta y)x. \tag{7}$$

The parameters α, β, γ, and δ in **(6)** and **(7)** are positive constants.

Equations **(6)** and **(7)** are the LOTKA-VOLTERRA system of differential equations. We cannot solve them for $x(t)$ and $y(t)$ with the techniques covered in this textbook, but dividing **(6)** by **(7)** yields a separable differential equation involving only x and y:

$$\frac{dy}{dx} = \frac{\dfrac{dy}{dt}}{\dfrac{dx}{dt}} = \frac{(\alpha - \beta x)y}{(-\gamma + \delta y)x}.$$

We rewrite the last equation in the form

$$(\alpha - \beta x)y\ dx + (-\gamma + \delta y)x\ dy = 0. \tag{8}$$

Example 2 Suppose that the number $x(t)$ of predators and the number $y(t)$ of prey satisfy the Lotka-Volterra differential equations

$$\begin{aligned} \frac{dy}{dt} &= (3 - x)y \\ \frac{dx}{dt} &= x(-4 + 2y) \end{aligned} \tag{9}$$

where x is measured in hundreds and y in thousands. Find an equation that relates x and y under the requirement that $x = 1$ when $y = 1$.

SOLUTION Dividing the first of equations **(9)** by the second gives

$$\frac{dy}{dx} = \frac{\dfrac{dy}{dt}}{\dfrac{dx}{dt}} = \frac{(3 - x)y}{x(-4 + 2y)}.$$

Then separating variables yields $\left(\frac{2y-4}{y}\right) dy = \left(\frac{3-x}{x}\right) dx$ or

$$\left(2 - \frac{4}{y}\right) dy = \left(\frac{3}{x} - 1\right) dx \tag{10}$$

or, with integral signs

$$\int \left(2 - \frac{4}{y}\right) dy = \int \left(\frac{3}{x} - 1\right) dx.$$

Carrying out the integrations gives $2y - 4\ln|y| = 3\ln|x| - x + C$. We drop the absolute-value signs because x and y are positive. Then setting $x = 1$ and $y = 1$ gives $C = 3$, so the equation relating x and y is

$$2y - 4\ln y = 3\ln x - x + 3. \ \square \tag{11}$$

Figure 11 shows the slope field of the differential equation **(10)** and the curve **(11)**.

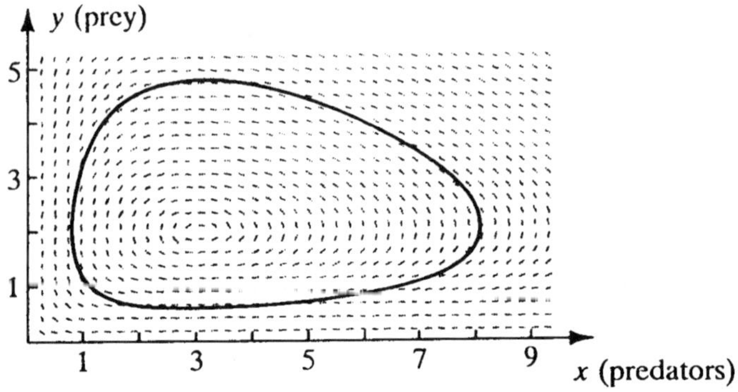

FIGURE 11

Question 8 Use differential equations **(9)** to determine whether the point $(x(t), y(t))$ traverses the curve in Figure 11 clockwise or counterclockwise as t increases.

Figure 11 shows that the variations in the populations are cyclic. The population of prey rises when $(x(t), y(t))$ is on the left side of the curve and the population of predators in relatively low. Then along the top of the curve, the number of predator rises because their is an abundance of prey. Next, along the right side of the curve the number of prey decreases because of the large number of predators.

Stimulus and sensation

Sound travels as oscillating air pressure, which causes the drum and bones of the listener's ear to vibrate. The vibration sends signals to the brain, which the listener perceives as sound. In studying sound and hearing, physiologists refer to the oscillating air pressure as the STIMULUS and the listener's resulting mental activity as the SENSATION of the sound, and they use different scales to measure the two phenomena.

The stimulus is described by measuring the variation in air pressure P in convenient units, such as pounds per square inch. The sensation D is measured in DECIBELS.† The decibel scale was developed from experiments in which participants were asked whether they could distinguish between two sounds of almost the same "loudness." The result was that for sounds in the audible range, most individuals can distinguish between two sounds only if the air-pressure variation P in the oscillations causing the louder sound is more than 12% greater than P for the quieter sound. The decibel scale is defined so that sounds differ by one decibel if there is a 12% difference in the air-pressure variation for the two sounds. With this definition, most people can distinguish between two sounds only if one

†One BEL, named after Alexander Graham Bell, is ten decibels.

is one decibel "louder" than the other. This relationship between the stimulus P and the sensation D is achieved by using the differential equation

$$dD = 8.69\,\frac{dP}{P}. \tag{12}$$

Example 3 **(a)** Solve **(12)** to give a formula for D as a function of P with the initial condition that a sound is at zero decibels if $P = 2.9 \times 10^{-9}$ pounds per square inch. **(b)** How are the values of P related for sounds that are one decibel apart?

SOLUTION **(a)** Differential equation **(12)** is separated. Adding integral signs gives

$$\int dD = 8.69 \int \frac{dP}{P}$$

and then, since P is positive,

$$D = 8.69 \ln P + C.$$

Setting $P = 2.9 \times 10^{-9}$ and $D = 0$ gives $C = -8.69 \ln(2.9 \times 10^{-9})$ so that

$$D = 8.68 \ln P - 8.69 \ln(2.9 \times 10^{-9}) = 8.69 \ln(P/(2.9 \times 10^{-9})).$$

(b) If $D = D_1$ for one sound and $D = D_1 + 1$ for another and P_1 and P_2 are the corresponding values of P, then

$$1 = 8.69 \ln(P_2/(2.9 \times 10^{-9})) - 8.69 \ln(P/(2.9 \times 10^{-9})) = 8.69 \ln(P_2/P_1)$$

so that $\ln(P2/P1) = 1/8.69$ and $P_2/P_1 = e^{1/8.69} \doteq 1.12$ and P_2 is approximately 12% greater than P_1. □

Equation **(12)** is an example of WEBER'S LAW,[†]

$$dS = k\,\frac{ds}{s} \tag{13}$$

in which a small change ds in the stimulus from s to $s + ds$ causes a change dS in the sensation that is proportional to the relative change ds/s of the stimulua. This law which is used to define sensation scales for the lifting of weights and other phenomena as well as in studying sound.

In describing Weber's law, we have followed a common practice of referring to the differentials ds and dS as small changes in the related quantities s and S. We can interpret this as meaning that ds and dS are corresponding changes along a tangent line to a graph relating s and S, as in Definition 1 above. Then, when dS and dS are small, they will be approximately equal to actual changes in s and S. We can also interpret **(12)** as an alternate way of writing the differential equation

$$\frac{dS}{ds} = \frac{k}{s}.$$

[†]Formulated in 1846 by the German psychologist E. H. Weber.

Orthogonal trajectories

An equation involving the variables x and y and a parameter c gives a FAMILY OF CURVES in an xy-plane. Their ORTHOGONAL TRAJECTORIES are a second family of curves such that at each point where two curves from both families meet, they have orthogonal (perpendicular) tangent lines. (For the families to appear orthogonal in drawings, equal scales must be used on the x- and y-axes.)

Figure 12 shows four curves in the family

$$y = ce^{-x/4}. \tag{14}$$

Question 9 What are the values of c for the curves in Figure 12?

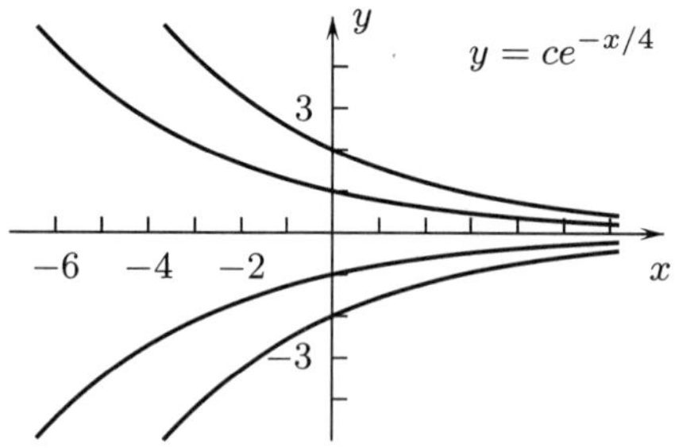

FIGURE 12

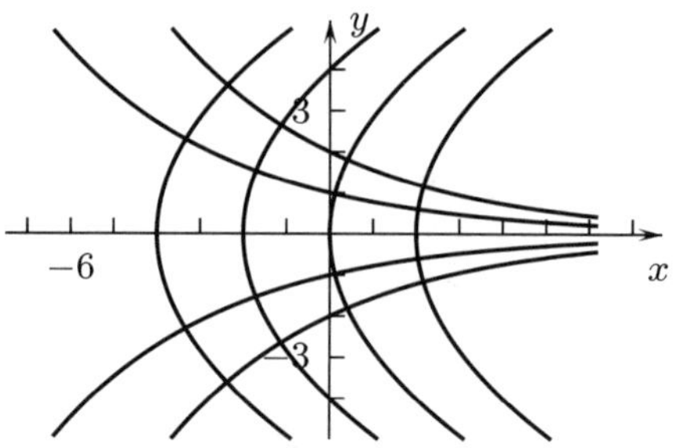

FIGURE 13

Example 4 Find a differential equation for the family of orthogonal trajectories of family **(14)**. (Four of the orthogonal trajectories are shown in Figure 13.)

SOLUTION We begin by finding a differential equation for the curves in Figure 12. Differentiating both sides of **(14)** with respect to x gives $\dfrac{dy}{dx} = ce^{-x/4}\dfrac{d}{dx}(-\tfrac{1}{4}x)$ and then

$$\frac{dy}{dx} = -\tfrac{1}{4}ce^{-x/4}. \tag{15}$$

This equation gives the slopes of tangent lines to the given family of curves in terms of x and the parameter c that identifies the curves. We need to have the slope in terms of x and y. We achieve this by solving **(14)** for c in terms of x and y and substituting the result in **(15)**.

Formula **(14)** gives $c = ye^{x/4}$, which with **(15)** yields $\dfrac{dy}{dx} = -\tfrac{1}{4}ye^{x/4}e^{-x/4}$ or

$$\frac{dy}{dx} = -\tfrac{1}{4}y. \tag{16}$$

This shows that the slope of the tangent line at (x, y) to the curve in the original family **(14)** that passes through that point has slope $-\tfrac{1}{4}y$. The slope of the tangent line to the orthogonal trajectory through (x, y) is the negative recprical $4/y$, and the orthogonal trajectories satisfy the differential equation,

$$\frac{dy}{dx} = \frac{4}{y}. \;\square \tag{17}$$

Question 10 **(a)** Solve **(17)** to find an equation for the family of orthogonal trajectories in terms of a parameter C and describe these curves. **(b)** Find the values of C for the four orthogonal trajectories in Figure 13.

Chemical reaction rates

The numbers of molecules involved in chemical reactions are measured in MOLES, where one mole is 6×10^{23} molecules. The concentration of molecules of type A in a solution is measured in moles per liter and is denoted by the symbol $[\,A\,]$.

The rates of chemical reactions are often studied through differential equations in which the rate of change $\frac{d}{dt}[\,A\,]$ of the concentration of one type of molecule in a reaction is expressed in terms of the concentrations of one or more molecules.

In most cases, the rate at which a chemical reaction proceeds depends at least in part on how frequently the component molecules collide. In the next Example it is assumed that the rate at which the concentration of one component decreases is proportional to the product of the concentrations of the two components.

Example 5 The chemical reaction $CH_3COCH_3 + I_2 \rightarrow CH_3COCOCH_2I + HI$, in which acetone ($CH_3COCH_3$) and iodine ($I_2$) combine, is determined by the differential equation,[(1)]

$$\frac{d}{dt}[\,I_2\,] = -0.0004[\,CH_3COCH_3\,][\,I_2\,]$$

where t is the time, measured in seconds. At $t = 0$ the concentration of acetone and iodine are both a moles per liter with $a \neq 0$. **(a)** Give a formula for the concentration of iodine as a function of t. **(b)** What is the limit of the iodine concentration as $t \to \infty$?

SOLUTION **(a)** We let $y(t)$ be the concentration $[\,I_2\,]$ of iodine at time t. Because there are equal numbers of acetone and iodine molecules initially and equal numbers are consumed in the reaction, the concentration $[CH_3COCH_3]$ of acetone is also $y(t)$, and the differential equation is

$$\frac{dy}{dt} = -0.0004y^2. \tag{18}$$

The constant solution $y = 0$ of **(18)** does not satisfy the initial condition $y(0) = a$, so we separate variables to obtain

$$\int y^{-2}\,dy = -0.0004\int dt.$$

Performing the integrations gives $-y^{-1} = -0.004t + C$. Then setting $t = 0$ and $y = a$ shows that $C = -a^{-1}$, so that $y^{-1} = 0.004t + a^{-1}$ and $y = \dfrac{1}{0.004t + 1/a}$.

(b) The concentration of iodine tends to zero as $t \to \infty$. □

Principles and procedures

- The first step in studying motion on a line subject to air or rolling resistance is to pick the positive direction of the s-axis on which the object under consideration moves.. Then velocity, force, and acceleration are positive in the positive s-direction and negative in the negative s-direction. For an object moving vertically, for instance, you could have the positive s-axis point up or down. Either choice will work, but you have to stick to one throughout your calculations.
- Two models of air and rolling resistance are commonly used. In one it is assumed that the force of resistance F_R on an object is proportional to its velocity v, so $F_R = -kv$ with a positive constant k. There is a minus sign in this formula because the resistance is in the negative direction when v is positive and the motion is in the positive direction and the resistance is in the positive direction when v is negative and the motion is in the negative direction.

[(1)]Data adapted from *Physical Chemistry* by G. Castellan, Reading, MA: Addison-Wesley Publishing Company, Inc., 1971, p. 769.

- In the second model it is assumed that the force of resistance is proportional to the square of the velocity. Because v^2 is always ≥ 0, this requires one formula when v is positive and the motion is in the positive direction and another formula when v is negative and the motion is in the negative direction. We use $F_R = -kv^2$ when v is positive and $F_R = kv^2$ when v is negative with positive constants k.
- If the only force on the object is resistance, then Newton's law, $F = m\dfrac{dv}{dt}$ with m the object's mass in appropriate units, gives $m\dfrac{dv}{dt} = -kv$ in the first type of model for resistance and $m\dfrac{dv}{dt} = \pm kv^2$ in the second type. Both of these differential equations are separable. The first is linear and has solutions $v = Ce^{-kt}$. The second is also relatively easy to solve.
- In many cases, there is a constant force F_0 , such as the force of gravity or an engine, in addition to resistance. Then we obtain $m\dfrac{dv}{dt} = F_0 - kv$ in the first type of model for resistance and $m\dfrac{dv}{dt} = F_0 \pm kv^2$ in the second. Again both differential equations are separable. The first can be rewritten as a linear equation with a new unknown function $v + v_0$. Techniques for solving the second are described in the outlines of solutions of Problem 37.
- Formulas for an object's velocities in terms of t can be integrated to obtain formulas for their positions in terms of the time.
- The term "differential equation" is used for equations involving an unknown function and its derivative because Leibniz and his successors who developed the basic theory of these equations wrote them in terms of differentials as in the equation, $P(x,y)\,dx + Q(x,y)\,dy = 0$.

Tune-Up Exercises 7.2[♦]

[A]*Answer provided.* [O]*Outline of solution provided.*

T1.[O] (Air resistance) **(a)** Solve the initial-value problem $\dfrac{dv}{dt} = -6v, v(0) = 50$ for an object of mass 1 that is subject only to a force of air resistance $-6v$, proportional to its velocity. **(b)** What is the object's terminal velocity (the limit of its velocity as $t \to \infty$)?

♦ Type 2, an initial-value problem, etc.

T2.[O] (Air resistance) **(a)** Solve the initial-value problem $\dfrac{dv}{dt} = -6v^2, v(0) = 50$ for an object of mass 1 that is moving in the positive direction, subject only to a force of air resistance $-6v^2$, proportional to the square of its velocity. **(b)** What is the object's terminal velocity?

♦ Type 2, an initial-value problem, etc.

T3.[O] (Air resistance) Solve the initial-value problem $\dfrac{dv}{dt} = -32 - 32v, v(0) = 0$ for an object of mass 1 that is subject to a constant force -32 (a force of 32 in the negative direction) and a force of air resistance $-32v$, proportional to its velocity. **(b)** What is the object's terminal velocity?

♦ Type 2, an initial-value problem, etc.

T4.[O] Find equations for and describe the curves that satisfy the differential equations **(a)** $y\,dx + x\,dy = 0$ and **(b)** $x\,dx + y\,dy = 0$.

♦ Type 1, curves as solutions of differential equations

T5.[O] (Orthogonal trajectories) Figure 14 shows four curves in the family $y = cx^4$ and three of their orthogonal trajectories. Find an equation with parameter C for all of the curves in the family of orthogonal trajectories.

♦ Type 1, orthogonal trajectories

[♦]The Tune-up Exercises and Problems are classified by type and content. The types are (1) basic, reactive; (2) basic reflective; (3) intermediate, reactive; (4) intermediate, reflective; (5) advanced, reactive; (6) advanced, reflective; and (7) advanced, theoretical.

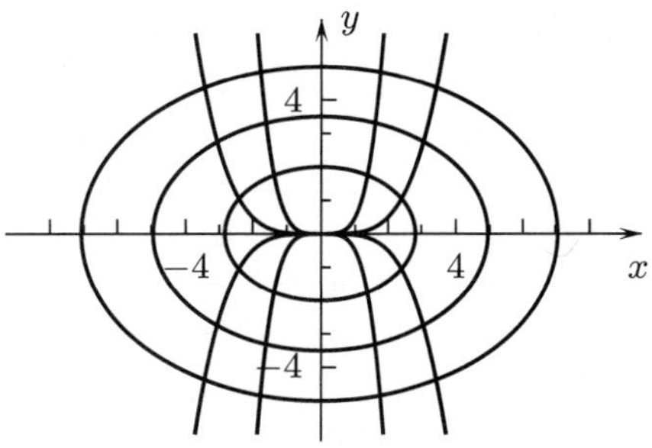

FIGURE 14

Problems 7.2

[A]*Answer provided.* [O]*Outline of solution provided.*

Give exact answers and their approximate decimal values, as appropriate, in the following Problems.

1.[O] (Air resistance) The force of air resistance on a two-kilogram rock is $-0.98v$ newtons when its downward velocity is v meters per second. The rock is falling at the rate of 15 meters per second at $t = 0$ (seconds). **(a)** What is its downward velocity ten seconds later? **(b)** What is its terminal velocity?

◆ Type 2, an initial-value problem, etc.

2.[O] (Air and rolling resistance) A truck with a mass of eight thousand kilograms is coasting on a level road subject to a force of $8v^2$ newtons of air and rolling resistance when its velocity is v meters per second. If the truck's velocity is twenty-five meters per second initially, what is its velocity fifty seconds later?

◆ Type 2, an initial-value problem

3.[A] (Air resistance) A baseball weighing 5.12 ounces is thrown straight up in the air and then caught at the position from which it is thrown. It is thrown with a velocity of 80 feet per second, and is subject to the force of gravity and to the upward force $-\frac{1}{500}v$ pounds of air resistance when its upward velocity is v feet per second. **(a)** What is the ball's mass? (There are sixteen ounces in a pound.) **(b)** Give a formula for the ball's upward velocity up as a function of the time t (seconds) it is in the air. **(c)** Give a formula for the ball's height above the point from which it is thrown, up to when it is caught and generate its graph in a suitable window. **(d)** Use your calculator or computer to find approximately how long the ball is in the air.[†]

◆ Type 4, an initial-value problem, etc.

4. (Air resistance) An object with a mass of two slugs is dropped from a height of 500 feet above the ground. Air resistance exerts a force of $-\frac{1}{2}v$ pounds on the object when its velocity is v feet per second. **(a)** Give a formula for the object's upward velocity as a function of the time (seconds) until it hits the ground. **(b)** Give a formula for the ball's height above the ground, up to when it hits the ground and generate the graph of this function in a suitable window. **(c)** Use your calculator or computer to find the object's approximate velocity when it hits the ground.

◆ Type 4, an initial-value problem, etc.

5. (Air resistance) An eight-pound shot is thrown straight down from a stationary helicopter with a downward velocity of 15 feet per second at time $t = 0$ (seconds). The upward force of air resistance is $\frac{1}{4}v$ pounds when the shot's downward velocity is v feet per second. What is its downward velocity at $t = 1$? (The mass of the shot is $8/32 = \frac{1}{4}$ slug.)

◆ Type 4, an initial-value problem, etc.

6. (Air resistance) A ten-kilogram ball is thrown straight up into the air from the ground with upward velocity 10 meters per second. When does it reach its highest point and how high does it go if the downward force of air resistance on it is $2v$ newtons when its upward velocity is v meters per second?

◆ Type 4, an initial-value problem, etc.

[†]This Problem is fairly close to the mathematical model of a baseball's flight in Question 10 of Section 2.1.

7.[A] (Air resistance) What is the equilibrium (terminal) velocity of a 400-gram rock if the upward force of air resistance is $80v$ dynes when its downward velocity is v centimeters per second?

◆ Type 2, a differential equation

Find equations of the curves that satisfy the differential equations in Problems 8 through 17. If a point is given find an equation of the particular solution through that point.

8.[O] $(x+1)y\,dx + x(y-1)\,dy = 0$

◆ Type 1, curves as solutions of differential equations

9.[A] $y^2\,dx + x^3\,dy = 0$; the point $(2,-2)$

◆ Type 1, curves as solutions of differential equations

10. $x(4-y^2)\,dx + y\,dy = 0.$

◆ Type 1, curves as solutions of differential equations

11. $(x-1)\,dx - 3xy^2\,dy = 0$; the point $(1,1)$

◆ Type 1, curves as solutions of differential equations

12. $2x^3(y^4+1)\,dx + y\,dy = 0$; the point $(1,0)$

◆ Type 1, curves as solutions of differential equations

13.[A] $(y^2+1)\,dx + (2x^2y + 2y)\,dy = 0$

◆ Type 1, curves as solutions of differential equations

14. $y^2\sin x\,dx + 2\,dy = 0$

◆ Type 1, curves as solutions of differential equations

15. $y^2\,dx + x\,dy = 0$; the point $(1,2)$

◆ Type 1, curves as solutions of differential equations

16. $\sqrt{y}\,dx + \,dy = 0$; the point $(0,4)$

◆ Type 1, curves as solutions of differential equations

17. $e^x(y^2+1)\,dx + 2y\,dy = 0$; the point $(0,0)$

◆ Type 1, curves as solutions of differential equations

18.[A] Figure 15 shows the slope field of $7\cos x\,dx - y^2\,dy = 0$. Find the solutions through the points **(a)** $(0,3)$ and **(b)** $(0,-1)$. Then generate the graphs of the solutions and copy them with the slope field.

◆ Type 2, curves as solutions of differential equations, etc.

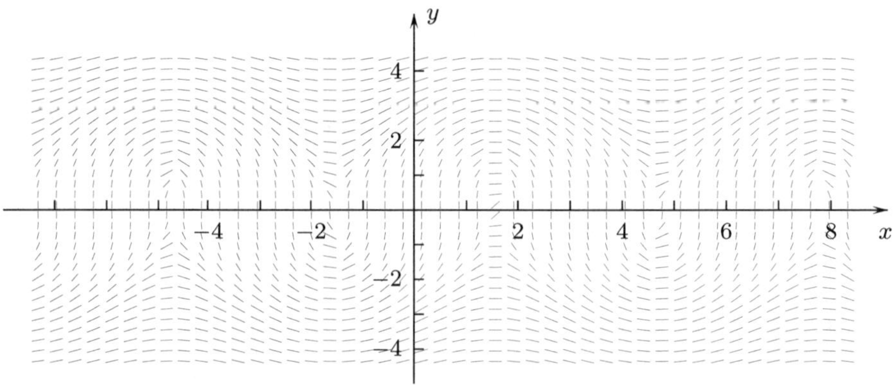

FIGURE 15

19. Figure 16 shows the slope field of $(x^3 - 6x)\,dx + 2y\,dy = 0$. **(a)** Find the solution that intersects the y-axis at $y = 3$. Then generate its graph and copy it with the slope field. (The curve is the graphs of two functions.) **(b)** What is its other y-intercept?

◆ Type 2, curves as solutions of differential equations, etc.

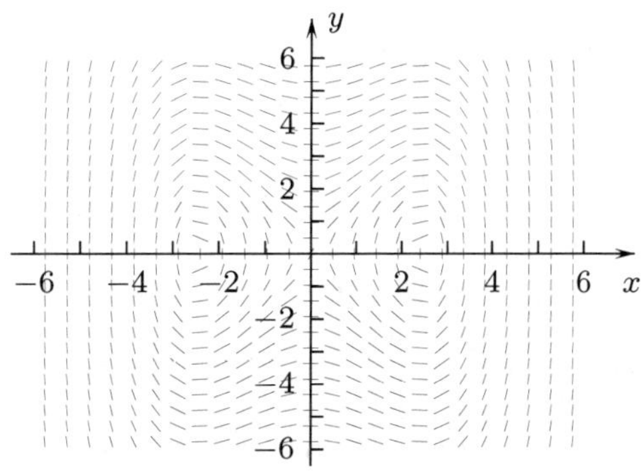

FIGURE 16

20. (Air and water resistance) As the yacht *Gimcrack* is coasting with its sails down and with no forces on it other than air and water resistance, it slows from a speed of 10 feet per second to 5 feet per second in 20 seconds. The *Gimcrack* weighs 6,560 pounds and the force of air and water resistance on it is kv^2 pounds when it is moving v feet per second.[(2)] What is the constant k?

♦ Type 2, a differential equation, etc.

21.[A] (Predator-prey) Find an equation relating the solutions x and y of the Lotka-Volterra differential equations, $\dfrac{dy}{dt} = 100y - 5xy$, $\dfrac{dx}{dt} = -60x + 30xy$, under the assumption that $x = 5$ and $y = 1$ at one time.

♦ Type 3, an initial-value problem

22. (Predator-prey) Give an equation relating x and y, where $x(t)$ and $y(t)$ are the solutions of the Lotka-Volterra differential equations, $\dfrac{dy}{dt} = 1200y - 5xy$, $\dfrac{dx}{dt} = -60x + 30xy$ with the initial conditions $x(0) = 2, y(0) = 2$.

♦ Type 3, an initial-value problem

23.[A] (Stimulus and sensation) Most people can distinguish between two weights only if one weighs at least 5% more than the other. Use Weber's law to define a measure S of apparent weight such that a 5% increase in the actual weight s corresponds to one unit of change in S and so that $S = 1$ for a weight of one pound.

♦ Type 3, an initial-value problem

24. (Stimulus and sensation) In some case the scales for measuring stimuli s are related to the scales for measuring the corresponding sensations S through differential equations of the form $\dfrac{dS}{S} = \dfrac{k\,ds}{s}$. What is the form of the equations relating S and s?

♦ Type 3, a differential equation

25. (Ice growth) The thickness $x(t)$ (inches) of ice forming on a lake satisfies the differential equation $x'(t) = 3/x(t)$. At $t = 0$ (days) the ice is one inch thick. When is it two inches thick?

♦ Type 2, an initial-value problem

26.[A] (Population growth) A lake contains 400 fish at time $t = 0$ (months). Thereafter, the population increases at the rate of $\sqrt{P}$ fish per month when there are P fish in the lake. Give a formula for the number of fish in the lake as a function of t.

♦ Type 2, an initial-value problem

27. (Orthogonal trajectories) Find the orthogonal trajectories of the family $x^2 - y^2 = c$ of hyperbolas. Then draw three curves from each family in an xy-plane with equal scales on the axes. (Each hyperbola is the graph of two functions.)

♦ Type 1, orthogonal trajectories

28.[A] (Orthogonal trajectories) Find the orthogonal trajectories of the family $x^2 + 2y^2 = c$ of ellipses. Then draw three curves from each family in an xy-plane with equal scales on the axes. (Each ellipse is the graph of two functions.)

♦ Type 1, orthogonal trajectories

[(2)] Data adapted from C.A. Marchaj, *Sailing Theory and Practice*, New York: Dodd, Mead, 1964, p 244.

29. (Orthogonal trajectories) Find the orthogonal trajectories of the family $y = cx^3$ of curves. Then draw three curves from each family in an xy-plane with equal scales on the axes.

♦ Type 1, orthogonal trajectories

30. (Orthogonal trajectories) Use the techniques of this section to show that the orthogonal trajectories of the lines through the origin are the circles centered at the origin.

♦ Type 1, orthogonal trajectories

31. (Orthogonal trajectories) Draw in an xy-plane with equal scales on the axes four curves of the family $y = \sqrt[3]{x+c}$ and four curves in the family of orthogonal trajectories.

♦ Type 1, orthogonal trajectories

32. (Diffusion) Suppose that a cell in an organism has fixed volume V (cubic centimeters) and fixed surface area A (square centimeters) and that c_1 is the constant concentration of a substance (grams per cubic centimeter) outside the cell. Let $C(t)$ denote the concentration of the same substance inside the cell at time t. FICK'S LAW states that the rate at which the substance enters the cell is proportional to A and to the difference in concentrations inside and outside. The constant k of proportionality depends on the nature of the substance and the cell. **(a)** Express the mass m of the substance inside the cell in terms of V and $c(t)$. **(b)** Give a differential equation for $c(t)$. **(c)** Solve the differential equation to express $c(t)$ in terms of V, A, k, c_1, t, and $c_0 = c(0)$.

♦ Type 3, an initial-value problem

33. (Bernoulli's equation of physical and moral wealth) In the 1738 *Commentari* of the St. Petersberg Academy of Sciences, Daniel Bernoulli proposed a mathematical model to relate moral fortune, which we might think of as happiness, to physical fortune, or wealth. He suggested that an increase in a person's wealth from M to $M + dM$ causes an increase dH in his happiness which is proportional, not to the amount dM by which his or her wealth is increased but by the fraction dM/M by which it is increased. Use this principle, with dM and dH viewed as differentials, to express a person's happiness H as a function of her or her wealth M, measured in dollars.

♦ Type 4, a differential equation

34. (Chemical reaction rates) The inversion of sucrose into glucose and fructose $C_{12}H_{22}O_{11} + H_2O \to HOCH_2(CHOH)_4CHO + HOCH_2(CHOH)_3COCH_2OH$ is governed by the differential equation,[(3)] $\frac{d}{dt}\,[C_{12}H_{22}O_{11}] = -5.7 \times 10^{-5}\,[C_{12}H_{22}O_{11}]$. How long does it take for 75% of the sucrose in a solution to be inverted?

♦ Type 4, an initial-value problem

35. (Chemical reaction rates) The decomposition of acetaldehyde $CH_3CHO \to CH_4 + CO$ is governed by the differential equation, $\frac{d}{dt}\,[CH_4] = k\,[CH_3CHO]^{3/2}$. If $[CH_3CHO]$ is 0.01 mole per liter at $t = 0$, what is it for $t > 0$?

♦ Type 4, an initial-value problem

36. (Chemical reaction rates) In an AUTOCATALYIC REACTION, $A + \cdots \to B + \cdots$, the product [B] hastens the reaction. In many such cases, the reaction is described by a differential equation of the form $\frac{d}{dt}[A] = -k[A][B]$. Show that this leads to a Logistic Differential Equation.

♦ Type 4, an initial-value problem

37. (Air resistance) A 320 pound object is propelled straight up in the air and reaches his highest point 3 seconds later. The downward force of air resistance on the object is $0.1v$ pounds when its upward velocity is v feet per second. What was its initial upward velocity?

♦ Type 4, an initial-value problem

38. A 128-pound shell from a six-inch World War II naval gun from is fired horizontally with a muzzle velocity of 2500 feet per second. We disregard gravity, so that the path of the shell is a horizontal straight line. Then the force of air resistance on the shell is $(7 \times 10^{-5})\,v^2$ pounds when its velocity is v feet per second. What is its velocity four seconds after it is fired?

♦ Type 4, an initial-value problem

Find equations for the orthogonal trajectories of the families of curves in Problems 39 through 42.

39. $x^2 + 8x + y^2 - 2y = c$

♦ Type 3, orthogonal trajectories

40. $x^4 + y^4 = c$

♦ Type 1, orthogonal trajectories

41. $y = cx + 3$

♦ Type 1, orthogonal trajectories

42. $y = e^{cx}$

♦ Type 3, orthogonal trajectories

---

(3) Adapted from *Physical Chemistry*, Englewood Cliffs, NJ: Pretice-Hall, Inc., 1962, p. 316.

Section 7.3

Approximate solutions of first-order differential equations

OVERVIEW: *If an exact formula for the solution of an initial-value problem for a first-order differential equation cannot be found, a calculator or computer can be used to construct approximate solutions. In this section we discuss* EULER'S METHOD, *a* TRAPEZOID METHOD, *and a* RUNGE-KUTTA METHOD *for finding such approximate solutions. In the Problems we use the Runge-Kutta method in applications involving first-order differential equations that we cannot solve exactly.*

Topics:

- ***Euler's Method***
- ***A Trapezoid Method with Euler's Method as predictor***
- ***A Runge-Kutta Method***
- ***Euler/Runge-Kutta calculator programs***
- ***Slope-field calculator programs***

Euler's Method

We begin our discussion of techniques for finding approximate solutions of differential equations by considering the initial-value problem

$$\frac{dy}{dx} = (1-x)y, \ y(0) = 1. \tag{1}$$

The slope field of the differential equation is shown in Figure 1. The graph of the solution $y = y(x)$ of **(1)** has to be tangent to slope lines to satisfy the differential equation and has to start, on the left, at the point $(0, 1)$ to satisfy the initial condition.

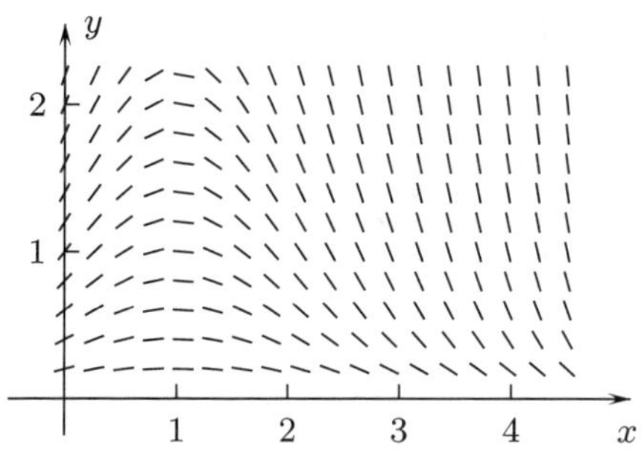

FIGURE 1

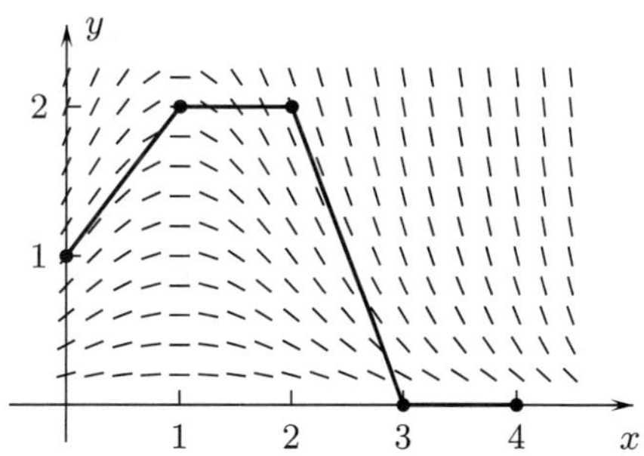

FIGURE 2

Figure 2 shows the graph of the approximate solution of **(1)** in the interval $0 \le x \le 4$ that is given by EULER'S METHOD with four subintervals. The graph consists of four connected line segments. The first line segment, on the left, starts at the point $(0, 1)$ so that the approximate solution satisfies the initial condition, and the slope of each line segment is the slope $(1-x)y$ given by the differential equation at its left end.

Example 1 Calculate the values at $x = 1, 2, 3, 4$ of the Euler's-method approximate solution in Figure 2.

SOLUTION The left end of the first line segment in the graph is $(0, 1)$ so its slope is

$$\Big[(1-x)y\Big]_{x=0,y=1} = (1-0)(1) = 1.$$

The change in x (the run) from the left to the right end of this line sequent is 1, so its right end is one unit higher than the left end and is $(1, 2)$.

♦ SUGGESTIONS TO INSTRUCTORS: Have students work Question 1 before class. Problems 1 and 16–19 are good for in-class discussion.

The slope of the second line segment in Figure 2 is the value

$$\Big[(1-x)y\Big]_{x=1,y=2} = (1-1)(2) = 0$$

given by the differential equation at its left end. Consequently, this line segment is horizontal and its right end is $(2,2)$. Similarly, the slope of the third line segment is

$$\Big[(1-x)y\Big]_{x=2,y=2} = (1-2)(2) = -2$$

so its right end is $(3,0)$, and the slope of the last line segment is

$$\Big[(1-x)y\Big]_{x=3,y=0} = (1-3)(0) = 0.$$

so the last line segment is horizontal and its right end is $(4,0)$. □

Question 1 Figure 3 shows the slope field of the differential equation $\frac{dy}{dx} = x - \frac{1}{2}y$, and Figure 4 shows the graph of the approximate solution that is obtained by Euler's method with four subintervals of $[0,4]$ and the initial condition $y(0) = 2$. The left end of the first line segment is at $(0,2)$ to satisfy the initial condition, and the slope of each line segment is the value $x - \frac{1}{2}y$ given by the differential equation at its left end. Find the values of the approximation at $x = 1, 2, 3, 4$.

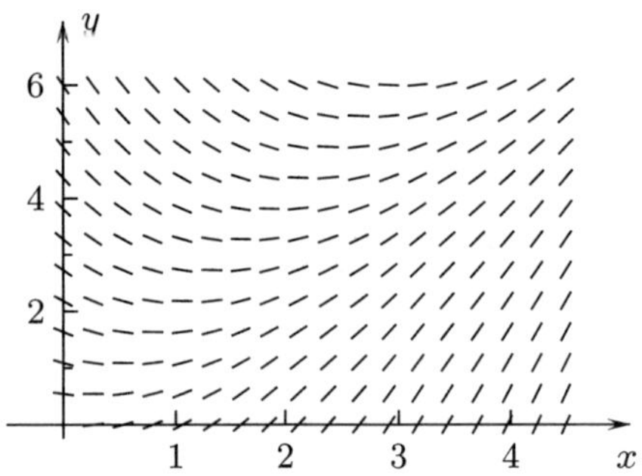

FIGURE 3

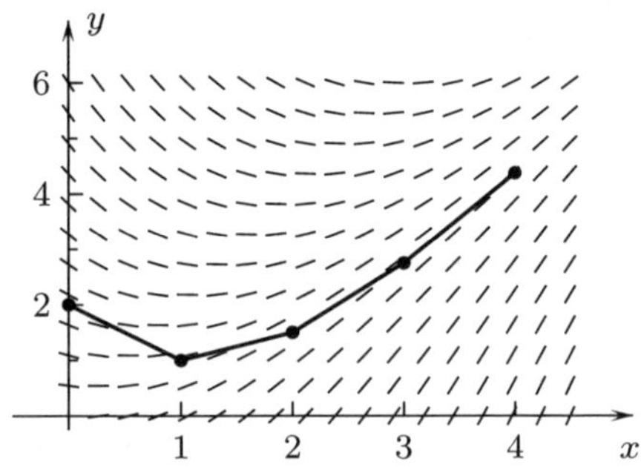

FIGURE 4

Euler's method for constructing the approximate solution in Example 1 and Question 1 can be applied to the general initial-value problem

$$\frac{dy}{dx} = f(x,y), \quad y(a) = y_0 \tag{2}$$

in an interval $[a,b]$. If we want the graph of the approximation to consist of N line segments, we use the partition

$$a = x_0 < x_1 < x_2 < \cdots < x_{N-1} < x_N = b \tag{3}$$

of $[a,b]$ into N equal subintervals of width $\Delta x = (b-a)/N$. Values $Y_0, Y_1, Y_2, \ldots, Y_N$ of the approximate solution at the points in the partition are determined and the points $(x_0, Y_0), (x_1, Y_1), \ldots,$ (x_N, Y_N) are connected by line seqments to form the graph of the appoximation.

To satisfy the initial condition, Y_0 is set equal to y_0, and the other numbers $Y_1, Y_2, \ldots, Y_N$ are chosen so that eqch line segment has the slope $f(x, y)$ given by the differential equation at its left end. Figure 5 is a schematic drawing of the first two line segments. The line segment on the left extends from (x_0, Y_0) to (x_1, Y_1), with Y_0 set equal to y_0 to satisfy the initial condition in **(2)**, and with $Y_1 = Y_0 + f(x_0, y_0)\Delta x$, so that the line segment has slope $f(x_0, Y_0)$.

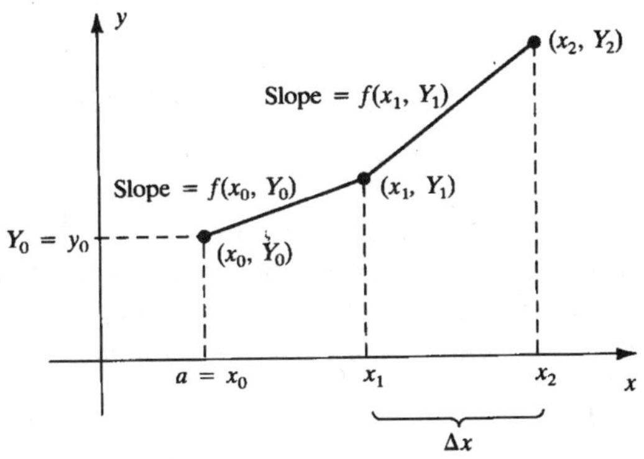

FIGURE 5

The second line segment, on the right in Figure 3, has slope $f(x_1, Y_1)$. This is achieved by setting $Y_2 = Y_1 + f(x_1, Y_1)\Delta x$. Continuing in this fashion, we construct the graph of the approximate solution for $a \le x \le b$ from N connected line segments, each with slope given by the value of $f(x, y)$ at its left endpoint. This leads to the following rule.[†]

Rule 1 (Euler's Method) *The Euler's-Method approximate solution of* **(2)** *corresponding to the partition* **(3)** *of* $[a, b]$ *into* N *subintervals of width* $\Delta x = (b - a)/N$ *is obtained by defining*

$$Y_0 = y_0 \tag{4}$$

and

$$Y_{j+1} = Y_j + f(x_j, Y_j)\Delta x \tag{5}$$

for $j = 0, 1, 2, \ldots, N - 1$, *and then joining successive points* (x_j, Y_j) *by line segments.*

We chose a linear separable differential equation in Example 1 so we could find the exact solution and check the accuracy of the approximation. By Rule 2 of Section 7.1, the solutions of the differential equation $\dfrac{dy}{dx} = (1 - x)y$ in **(1)** are

$$y = Ce^{R(x)}$$

where $R(x)$ is an antiderivative of $r(x) = 1 - x$. Since $\displaystyle\int (1 - x)\, dx = x - \tfrac{1}{2}x^2 + c$, we can take $c = 0$ and $R(x) = x - \frac{1}{2}x^2$, to put the solutions in the form $y = Ce^{1-x^2/2}$. Because $y(0) = Ce^0 = C$, the initial condition, $y(0) = 1$, is satisfied with $C = 1$, and the solution of the initial-value problem is $y = e^{x-x^2/2}$. Figure 6 shows the graph of this exact solution with the graph of the Euler's-method approximation from Example 1.

[†]Instructions and programs for generating approximate solutions with Euler's Method and the Runge-Kutta Method described below are given for Texas Instruments calculators at the end of this section.

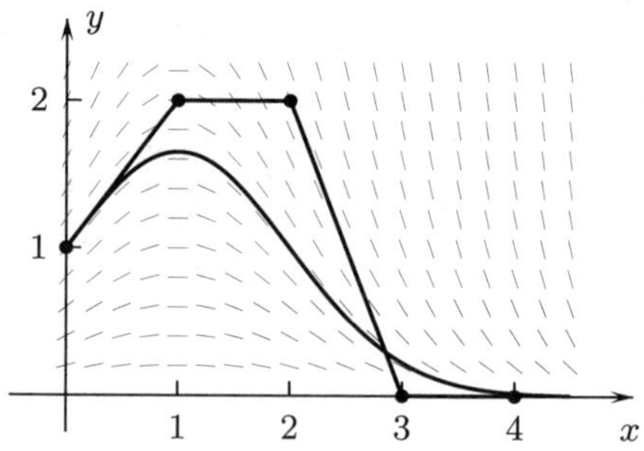

FIGURE 6

The approximate solution is not very accurate, in part, because only four line segments are used in constructing its graph, but also because the slope of each of the line segments is equal to the slope $(1-x)y$ given by the differential equation at its left end and $(1-x)y$ changes considerably as (x,y) moves from the left to the right end of each line segment.

A Trapezoid Method

More accurate methods of constructing approximate solutions are obtained by taking into account the values of $f(x,y)$ at more than one point in calculating the slope of each line segment. TRAPEZOID METHODS use a line segment over the subinterval $x_j \le x \le x_{j+1}$ with slope equal to the average of the values of $f(x,y)$ at one point with $x = x_j$ and a second point with $x = x_{j+1}$. We will use a Trapezoid Method that uses Euler's method to pick the point at x_{j+1}. This technique is called the TRAPEZOID RULE WITH EULER'S METHOD AS PREDICTOR.

As with Euler's Method, we start with a partition **(3)** and set $Y_0 = y_0$ to have the approximation satisfy the initial condition . Then we calculate the slope $m^{(1)} = f(x_0, Y_0)$ given by the differential equation at the initial point. If we were using Euler's method, the next point would be (x_1, Y_1) with $Y_1 = Y_0 + m^{(1)}\Delta x$. Instead, we calculate the slope $m^{(2)} = f(x_1 + m^{(1)}\Delta x)$ at the point given by Euler's method and use the average $m = \frac{1}{2}(m^{(1)} + m^{(2)})$ of the slopes $m^{(1)}$ and $m^{(2)}$ as the slope of the line segment for $x_0 \le x \le x_1$. Thus, we set $Y_1 = y_0 + \frac{1}{2}(m^{(1)} + m^{(2)})\Delta x$ (Figure 7).

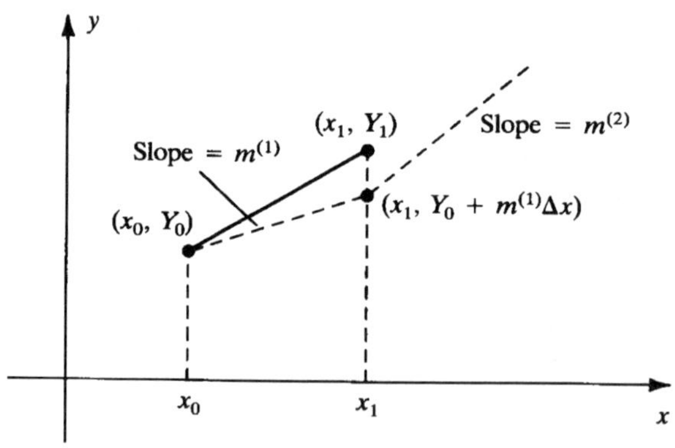

Line with slope $\frac{1}{2}(m_1 + m_2)$

FIGURE 7

We follow the same procedure for each subinterval and obtain the following rule.

Rule 2 (A Trapezoid Method) *The Trapezoid-Method approximate solution of* **(1)** *corresponding to the partition* **(2)** *of* $[a, b]$ *into* N *subintervals of width* $\Delta x = (b - a)/N$ *is obtained by defining*

$$Y_0 = y_0$$

and, for $j = 0, 1, 2, \ldots, N - 1$,

$$\begin{aligned} m^{(1)} &= f(x_j, Y_j) \\ m^{(2)} &= f(x_{j+1}, y_j + m^{(1)}\Delta x) \\ Y_{j+1} &= Y_j + \tfrac{1}{2}(x^{(1)} + m^{(2)})\Delta x \end{aligned}$$

and then joining successive points (x_j, Y_j) *by line segments.*

The graph of the Trapezoid-Method approximate solution of **(1)** is shown in Figure 8 with the graph of the exact solution. You can see by comparing Figures 6 and 8 that the Trapezoid Method is considerably more acccurate than Euler's method in this case.

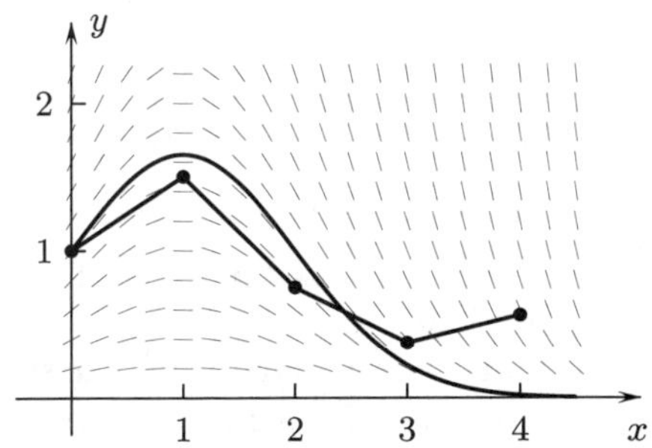

A Trapezoid-Method approximation

FIGURE 8

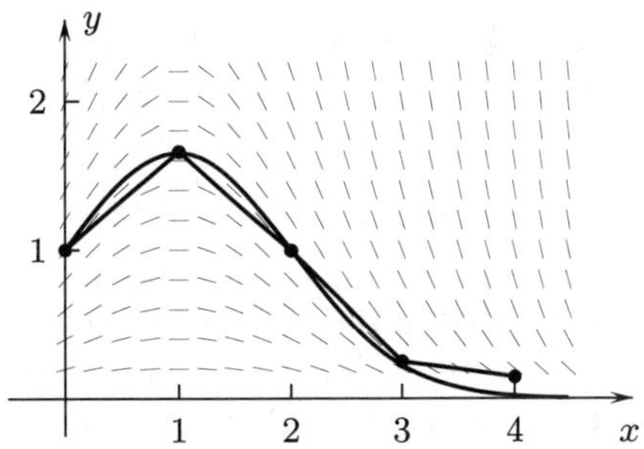

A Runge-Kutta-Method approximation

FIGURE 9

A Runge-Kutta Method

We obtain approximate solutions that are generally much more accurate than either Euler's Method or the Trapezoid Method with a RUNGE-KUTTA METHOD in which the slope of each line segment is a weighted average of $f(x, y)$ at four points. The graph of the Runge-Kutta approximate solution of **(1)** with four subintervals is in Figure 9.

As with the other methods, a partition of $[a, b]$ into N subintervals of width Δx is used and Y_0 is set equal to y_0. We then calculate

$$\begin{aligned} m^{(1)} &= f(x_0, Y_0) \\ m^{(2)} &= f(x_0 + \tfrac{1}{2}\Delta x, Y_0 + \tfrac{1}{2}m^{(1)}\Delta x) \\ m^{(3)} &= f(x_0 + \tfrac{1}{2}\Delta x, Y_0 + \tfrac{1}{2}m^{(2)}\Delta x) \\ m^{(4)} &= f(x_1, Y_0 + m^{(3)}\Delta x). \end{aligned}$$

Here $m^{(1)}$ is the value of $f(x, y)$ at the initial point; $m^{(2)}$ is the value at a point found using the slope $m^{(1)}$ as predictor; $m^{(3)}$ is a value found by using $m^{(2)}$ as predictor, and $m^{(4)}$ is a value found by using $m^{(3)}$ as predictor.

The graph of the approximate solution for $x_0 \le x \le x_1$ is a line of slope equal to the weighted average

$$m = \tfrac{1}{6}(m^{(1)} + 2m^{(2)} + 2m^{(3)} + m^{(4)}).$$

Here is the general rule:

Rule 3 (A Runge-Kutta Method) *The Runge-Kutta-Method approximate solution of* **(1)** *corresponding to the partition* **(2)** *of $[a, b]$ into N subintervals of width $\Delta x = (b - a)/N$ is obtained by defining $Y_0 = y_0$ and, for $j = 0, 1, 2, \ldots, N - 1$,*

$$\begin{aligned}
m^{(1)} &= f(x_j, Y_j)\\
m^{(2)} &= f(x_j + \tfrac{1}{2}\Delta x, Y_j + \tfrac{1}{2}m^{(1)}\Delta x)\\
m^{(3)} &= f(x_j + \tfrac{1}{2}\Delta x, Y_j + \tfrac{1}{2}m^{(2)}\Delta x)\\
m^{(4)} &= f(x_1, Y_j + m^{(3)}\Delta x).\\
Y_{j+1} &= Y_j + \tfrac{1}{6}(m^{(1)} + 2m^{(2)} + 2m^{(3)} + m^{(4)})\Delta x.
\end{aligned}$$

The graph is formed by joining successive points (x_j, Y_j) with line segments.

An Euler/Runge-Kutta program†

The following program for Texas Instruments TI-82, TI-83, TI-85, and TI-86 calculators uses either Euler's method or a Runge-Kutta method with N subintervals to generate the graph of an approximate solution of $\dfrac{dv}{dx} = Y_1(x, v)$ in the interval $A \le x \le B$ with initial condition $v(A) = v_0$. It then displays the value of the approximation at $x = B$. The variable v is used in place of y because y cannot be used in the Y= menu. The interval $A \le x \le B$ is partitioned into N equal subintervals of length $D = (B - A)/N$, and the graph of the approximate solution is constructed from line segments over the N subintervals. The left end of the first line segment (on the left) is put at (A, v_0) so the approximation satisfies the initial condition $v(A) = v_0$.

If the Runge-Kutta method is used, (C, Z) denotes the coordinates of the left end of the line segment that is being constructed. Initially C is set equal to A and Z is set equal to v_0. The value of $Y_1(x, v)$ at (C, Z) is calculated and stored as P. A line of this slope determines the point $(C + D/2, Z + DP/2)$, and the value of $Y_1(x, v)$ at this point is stored as Q. This slope and the point (C, Z) are used to determine the point $(C + D/2, Z + DQ/2)$. The value of $Y_1(x, v)$ at this point is stored as R, and this slope and the point (C, Z) are used to determine the point $(C + D, Z + DR)$. Finally, the value of $Y_1(x, v)$ at this point is stored as S.

The weighted average of the values P, Q, R, and S of $Y_1(x, v)$ is calculated and stored as M. This slope is used with the point (C, Z) to determine the point $(T, V) = (C + D, Z + DM)$, which is taken as the right endpoint of the line segment in the graph of the approximate solution. For Euler's method, M is redefined to equal P. If this is not the last segment of the graph, then C is set equal to T, Z is set equal to V, and the next segment is drawn.

The screen should be cleared before running the program unless you want the graph of the approximate solution to be shown with a previously generated picture. Other graphs can be added with the DrawF command after the program has run.

---

†These calculator-specific programs and instructions are included in preliminary editions of this text for the convenience of users. In the first edition only a description of the action of the program on any calculator or computer will be in the text. The programs and implementation instructions for various calculators and computer software will be in the *Calculator/Computer Supplement*.

TI-82, TI-83

```
PROGRAM: RUNGE
: FnOff
: Disp "USE "
: Disp "DV/DX = Y1(X,V)"
: Disp "GIVE 1 FOR EULER"
: Disp "2 FOR RUNGEKUTTA"
: Input L
: Prompt A,B,N
: Disp "VA = "
: Input V
: (B − A)/N → D
: For(J,1,N,1)
: A + (J − 1)D → C
: V → Z
: C → X
: Y1 → P
: C + D/2 → X
: Z + DP/2 → V
: Y1 → Q
: Z + DQ/2 → V
: Y1 → R
: C + D → X
: Z + DR → V
: Y1 → S
: (P + 2Q + 2R + S)/6 → M
: If L = 1
: P → M
: X → T
: Z + DM → V
: Line(C,Z,T,V)
: End
: Pause
: Disp "THE METHOD GIVES "
: Disp "V = "
: Disp V
: Disp "AT X = "
: Disp B
```

TI-85, TI-86

```
PROGRAM: RUNGE
: FnOff
: Disp "USE"
: Disp "DV/DX = y1(x,V) "
: Disp "GIVE 1 FOR EULER"
: Disp "2 FOR RUNGEKUTTA "
: Input L
: Prompt A,B,N
: Disp "VA = "
: Input V
: (B − A)/N → D
: For(J,1,N,1)
: A + (J − 1) * D → C
: V → Z
: C → x (Use x not X.)
: y1 → P (Use y not Y.)
: C + D/2 → x
: Z + D * P/2 → V
: y1 → Q
: Z + D * Q/2 → V
: y1 → R
: C + D → x
: Z + D * R → V
: y1 → S
: (P + 2 * Q + 2 * R + S)/6 → M
: If L == 1
: P → M
: x → T
: Z + D * M → V
: Line(C,Z,T,V)
: End
: Pause
: Disp "THE METHOD GIVES"
: Disp "V = "
: Disp V
: Disp "AT x = "
: Disp B
```

Slope-field calculator programs

The following program draws a slope field of the differential equation $\frac{dv}{dx} = Y_1(x, v)$ with the function $Y_1(x, v)$ in the Y =menu. The variable v is used instead of y because y cannot be used in the Y =menu. The user chooses the number of columns of lines to be drawn. The command DrawF in the Draw menu can be used to add graphs without erasing the slope field.

TI-82, TI-83

```
PROGRAM: SLOPEFLD
: ClrDraw
: FnOff
: Disp "USE "
: Disp "DV/DX = Y1(X,V)"
: Disp "HOW MANY COLUMNS"
: Input N
: (Xmax − Xmin)/N → D
: 3(Ymax − Ymin)/(2N) → H
: Ymin + H/2 → V
: Lbl 1
: Xmin + D/2 → U
: Lbl 2
: U → X
: Y1 → M
: Ø.35D → W
: If abs M > (H/D)
: Ø.35H/abs M → W
: Line(U−W, V−MW, U+W, V+MW)
: U + D → U
: If U < Xmax
: Goto 2
: V + H → V
: If V < Ymax
: Goto 1
```

TI-85, TI-86

```
PROGRAM: SLOPEFLD
: ClrDraw
: FnOff
: Disp "USE ‘"
: Disp "DV/DX = y1(x,V)"
: Disp "HOW MANY COLUMNS"
: Input N
: (Xmax − Xmin)/N → D
: 3(Ymax − Ymin)/(2N) → H
: Ymin + H/2 → V
: Lbl L
: Xmin + D/2 → U
: Lbl LL
: U → X
: Y1 → M
: Ø.35 * D → W
: If abs M > (H/D)
: Ø.35 * H/abs M → W
: Line(U−W, V−MW, U+W, V+MW)
: U + D → U
: If U < Xmax
: Goto LL
: V + H → V
: If V < Ymax
: Goto L
```

Principles and procedures

- Euler's Method, the Trapezoid Method, and the Runge-Kutta Method discussed in this section are all ways of finding approximate solutions of initial-value problems for first-order differential equations for which exact solutions cannot be found. Euler's Method and the Trapezoid Method are discussed to show more clearly the ideas behind the Runge-Kutta Method, which we use primarily in applications because it is more accurate.

Tune-Up Exercises 7.3[♦]

[A]*Answer provided.* [O]*Outline of solution provided.*

T1.[A] Carry out the calculations required in Euler's method with four subintervals to find an approximate solution of the initial value problem $\frac{dy}{dt} = \frac{1}{9}x^2y^2$, $y(-2) = 1$ for $-2 \le x \le 2$. Draw the graph of the approximation with the graph of the exact solution $y = 27/(19 - x^3)$.

♦ Type 2, Euler's method

T2.[A] Use a Runge-Kutta program on your calculator or computer with four subintervals to find an approximate solution of the initial-value problem in Exercise T1. Compare the accuracy of the two approximations at $x = 2$.

♦ Type 1, using the Runge-Kutta program

Problems 7.3

[A]*Answer provided.* [O]*Outline of solution provided.*

In Problems 1 through 5, find **(a)** the exact solutions of the initial-value problems in the given intervals and **(b)** the approximate values Y_j obtained by Euler's method with four subintervals. **(c)** Compare the accuracy of the approximations at the right endpoints of the subintervals. **(d)** Draw the grphs of the exact and approximate solutions.

1.[O] $\frac{dy}{dx} = \frac{1}{4}y^2 \sin x$, $y(0) = 2$, $0 \le x \le 2$

♦ Type 1, Euler's method, etc.

2. $\frac{dy}{dx} = -y^{3/2}$, $y(1) = 4$, $1 \le x \le 3$

♦ Type 1, Euler's method, etc.

3.[A] $\frac{dy}{dx} = 1+y^2$, $y(-\frac{1}{4}\pi) = -1$, $-\frac{1}{4}\pi \le x \le \frac{1}{4}\pi$

♦ Type 1, Euler's method, etc.

4.[A] $\frac{dy}{dx} = \frac{(1-x)y}{x}$, $y(1) = 3$, $1 \le x \le 5$

♦ Type 1, Euler's method, etc.

5. $\frac{dy}{dx} = \frac{2y}{x}$, $y(1) = 5$, $1 \le x \le 5$

♦ Type 1, Euler's method, etc.

6. $\frac{dy}{dx} = y \cos x$, $y(0) = 1$, $0 \le x \le 4$

♦ Type 1, Euler's method, etc.

In each of Problems 7 through 12, **(a)** find the exact solution of the initial-value problem. **(b)** Generate the graph of the Runge-Kutta approximate solution with four subintervals of the given subinterval and determine how much it differs from the exact solution at the right endpoint. **(c)** Add the graph of the exact solution and copy the two curves on your paper.

7.[A] $\frac{dy}{dx} = y^2$, $y(1) = -1$, $1 \le x \le 2$

♦ Type 1, using the Runge-Kutta program, etc.

8.[A] $\frac{dy}{dx} = \frac{3y}{2x}$, $y(1) = 0$, $1 \le x \le 6$

♦ Type 1, using the Runge-Kutta program, etc.

9. $\frac{dy}{dx} = \frac{2y+6}{x}$, $y(1) = -2$, $1 \le x \le 3$

♦ Type 1, using the Runge-Kutta program, etc.

10. $\frac{dy}{dx} = \frac{x}{y}$, $y(0) = 4$, $0 \le x \le 4$

♦ Type 1, using the Runge-Kutta program, etc.

11. $\frac{dy}{dx} = xy^2$, $y(1) = -1$, $1 \le x \le 13$

♦ Type 1, using the Runge-Kutta program, etc.

12. $\frac{dy}{dx} = x \cos^2 y$, $y(0) = 0$, $0 \le x \le 4$

♦ Type 1, using the Runge-Kutta program, etc.

In Problems 13 and 14 generate graphs of Runge-Kutta approximate solutions of the initial-value problems with twelve subintervals of the given intervals and copy them on your paper.

13.[A] $\frac{dy}{dx} = x^2 + y^2$, $y(0) = -3$, $0 \le x \le 2$

♦ Type 1, using the Runge-Kutta program

14. $\frac{dy}{dx} = \frac{1}{2}(x^2 - y^2)$, $y(0) = \frac{1}{2}$, $0 \le x \le 4$

♦ Type 1, using the Runge-Kutta program

[♦]The Tune-up Exercises and Problems are classified by type and content. The types are (1) basic, reactive; (2) basic reflective; (3) intermediate, reactive; (4) intermediate, reflective; (5) advanced, reactive; (6) advanced, reflective; and (7) advanced, theoretical.

15. **(a)** Generate together in the window $0 \le x \le 4, 0 \le y \le 4$ the Runge-Kutta approximate solutions with twelve subintervals of

$$\frac{dy}{dx} = \frac{x^2 - y}{\sqrt{x^2 + y^2 + 1}}, \quad y(0) = k \tag{6}$$

for $k = 1, 2, 3$, and 4. Then add the curve $y = x^2$ to the drawing, and copy the four curves on your paper. **(b)** Use a calculus principle to explain why the lowest point on the graphs of solutions of **(6)** are on the curve $y = x^2$.

♦ Type 1, using the Runge-Kutta program, etc.

16. The initial value problem,

$$\frac{dT}{dt} = -k(T - 100), \quad T(0) = 0 \tag{7}$$

with a positive constant k, is a mathematical model of the temperature $T(t)°$F as a function of the time t (days) in a building that is partially insulated from the outside. The temperature outside is constant. **(a)** What is the temperature inside at $t = 0$, and what is the temperature outside? **(b)** Find formulas for the exact solutions of **(7)** with $k = 1, 2$, and 5 and generate them together with the line $T = 100$ in the window $0 \le t \le 3, 0 \le T \le 150$ with T-scale $= 10$ Copy the three curves and the line on your paper. **(c)** What happens to the temperature in the room as $t \to \infty$? **(d)** Do large and small values of k correspond to a well or poorly insulated building? **(e)** What value of k would give a model for a perfectly insulated building, and what is the solution of **(7)** in this case? Explain this result.

♦ Type 2, an initial-value problem, etc.

17. The initial value problem,

$$\frac{dT}{dt} = -k(T - 100 + 25\cos(2\pi t)), \quad T(0) = 0 \tag{8}$$

with a positive constant k, is a modification of **(7)** that deals with the case where the temperature outside is $100 - 25\cos(2\pi t)$ at time t with $t = 0$ at midnight one day. **(a)** What are the maximum and minimum temperatures outside and when do they occur? **(b)** Generate, in the window from Problem 16, the Runge-Kutta approximate solution of **(8)** with $k = 10$. Add the graph of the outside temperature and copy the two curves together on your paper. Describe how the temperature inside is affected by the temperature outside. **(c)** Follow the instructions in part (b) with $k = 2$. **(d)** How does increasing or decreasing the amount of insulation on the building affect how the eventual oscillation of the temperature in the building relates to the oscillation of the temperature outside? **(e)** How is the solution of **(8)** changed for $k = 2$ and $k = 10$ if the initial condition is changed to $T(0) = 150$? (Compare the graphs.)

♦ Type 4, an initial-value problem, etc.

18. As was explained in Section 7.1, the Logistic Differential Equation

$$\frac{dP}{dt} = \tfrac{1}{2}P(4 - P) \tag{9}$$

is a mathematical model for a population $P(t)$ in a limited environment with equilibrium population $P = 4$. **(a)** Generate the slope field for **(9)** in the window $0 \le t \le 5, 0 \le P \le 6.5$ and use the differential equation to explain the pattern of the lines. **(b)** Generate in the same window the following solutions of **(9)** and copy them on your paper:

$$P_1(t) = 4,\ P_2(t) = \frac{4}{1 + 3e^{-2t}},\ P_3(t) = \frac{4}{1 + e^{-2t}},\ P_4(t) = \frac{4}{1 - \frac{1}{3}e^{-2t}}.$$

(c) What is the value of each solution at $t = 0$ and what happens to each solution as $t \to \infty$?

♦ Type 4, an initial-value problem, etc.

19. Carry out the following steps for each modification **(a)** through **(g)** below of the Logistic Differential Equation **(9)**.
(i) Match the problem to one of differential equations **(I)** through **(VII)** listed below the problems.
(ii) Generate the slope field of the differential equation in the window $0 \le t \le 5, 0 \le P \le 6.5$ and copy it on your paper. Explain how its pattern is determined by the differential equation.
(iii) Generate the graph of a Runge-Kutta approximate solution of the differential equation for the differential equation with fifty subintervals and the initial condition,

$$P(0) = 1.$$

Add the graph of the solution $P = \dfrac{4}{1 + 3e^{-2t}}$ of **(9)** with the same initial condition and copy the two curves together on your paper.†
(iv) Describe how the solution of the modified problem differs from the solution of the Logistic Equation and explain how the changes in the solution relate to the change in the mathematical model.

(a) The Logistic Equation **(9)** modified to include a constant rate of immigration.

(b) The Logistic Equation **(9)** modified to include a constant rate of emmigration.

(c) The Logistic Equation **(9)** modified to include an increasing rate of immigration.

(d) The Logistic Equation **(9)** modified to include a sinusoidal variation between immigration and emmigration with the greatest rate of emmigration at the beginning of each year and the greatest rate of immigration at the middle of each year.

(e) The Logistic Equation **(9)** modified to have a sinusoidal variation in the equilibrium population with the least equilibrium population at the beginning of each year and the greatest equilibrium population at the middle of each year.

(f) The Logistic Equation **(9)** with the constant $\frac{1}{4}$ replaced by a sinusoidal function of time to incorporate the fact that changes in population affect the rate of change of the population more during part of the year than at other times.

(g) The Logistic Equation **(9)** modified to include the effect of a short epidemic that kills some of the population.

$$\frac{dP}{dt} = \tfrac{1}{2}P(4 - P) - 2\cos(2\pi t) \qquad \textbf{(I)}$$

$$\frac{dP}{dt} = \tfrac{1}{2}P(4 - P) + t \qquad \textbf{(II)}$$

$$\frac{dP}{dt} = \tfrac{1}{2}P(4 - P) - 1 \qquad \textbf{(III)}$$

$$\frac{dP}{dt} = \tfrac{1}{2}P(4 - P) - f(t) \quad \text{with} \quad f(t) = \begin{cases} 0 & \text{for} \quad 0 \le t < 2.9 \\ 4 & \text{for} \quad 2.9 \le t \le 3.1 \\ 0 & \text{for} \quad t > 3.1. \end{cases} \qquad \textbf{(IV)}$$

$$\frac{dP}{dt} = \left(\tfrac{1}{2} - 0.4\cos(2\pi t)\right) P(4 - P) \qquad \textbf{(V)}$$

$$\frac{dP}{dt} = \tfrac{1}{2}P(4 - P) + 1 \qquad \textbf{(VI)}$$

$$\frac{dP}{dt} = \tfrac{1}{2}P\left(4 - \cos(2\pi t) - P\right) \qquad \textbf{(VII)}$$

♦ Type 4, modifications of a Logistic Differential Equation

†On a Texas Instruments calculator, enter $f(t)$ in **(IV)** as $4(\text{abs}(X - 3) \le 0.1)$. (The function $(\text{abs}(X - 3) \le 0.1)$ equals 1 if the inequality holds and 0 otherwise.)

RESPONSES TO QUESTIONS IN CHAPTER 7

Responses 7.1

Response 1 The values of the slope $\frac{dy}{dx} = \frac{1}{10}x^2$ are shown in the table below and the slope lines are in Figure R1. • One description: The lines are parallel in each vertical column because the slope $\frac{1}{10}x^2$ is independent of the value of y. The slopes are 0 and the lines are horizontal along the y-axis. The lines point up to the right for positive x and become steeper as x moves away from 0.

$[\text{SLOPE}] = \frac{1}{10}x^2$

	$x = 0$	$x = 1$	$x = 2$	$x = 3$	$x = 4$
$y = 3$	0	0.1	0.4	0.9	1.6
$y = 2$	0	0.1	0.4	0.9	1.6
$y = 1$	0	0.1	0.4	0.9	1.6
$y = 0$	0	0.1	0.4	0.9	1.6

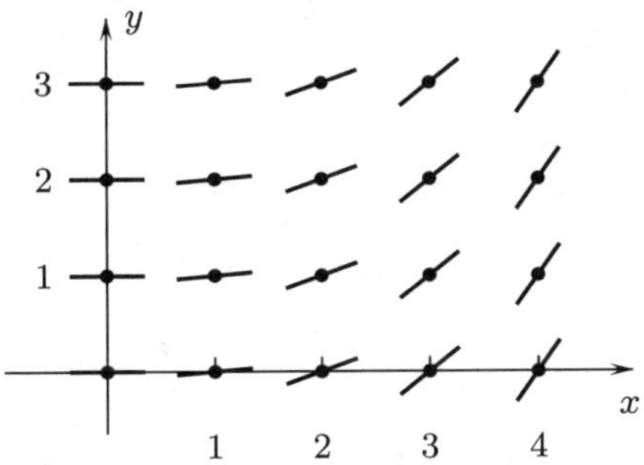

Figure R1

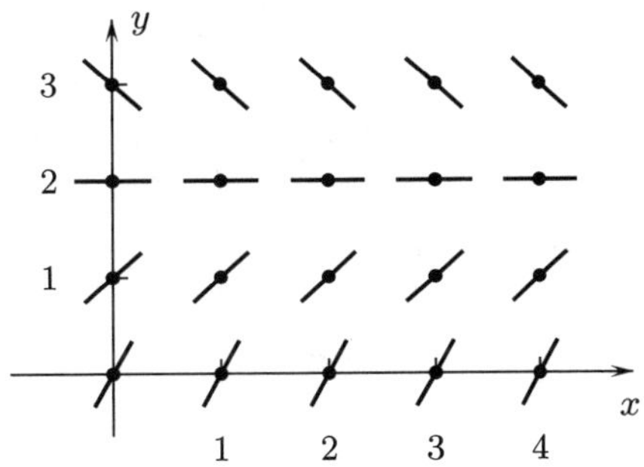

Figure R2

Response 2 The values $\frac{dy}{dx} = 2 - y$ are in the table below and the slope lines are in Figure R2. • One description: The lines are parallel in each horizontal column because the slope $y - 2$ is independent of the value of x. The slopes are 0 and the lines are horizontal along the line $y = 2$. The lines point up to the right for $y < 2$ and down to the right for $y > 2$ and become steeper as y moves away from 2.

$[\text{SLOPE}] = 2 - y$

	$x = 0$	$x = 1$	$x = 2$	$x = 3$	$x = 4$
$y = 3$	-1	-1	-1	-1	-1
$y = 2$	0	0	0	0	0
$y = 1$	1	1	1	1	1
$y = 0$	2	2	2	2	2

Response 3 Because there is no y on the right side of the differential equation, we can solve it by direct x-integration: $y = \int \frac{dy}{dx}\,dx = \int x\,dx = \frac{1}{2}x^2 + C$ • The graphs of solutions are the parabolas $y = \frac{1}{2}x^2 + C$ with arbitrary constants C.

Response 4 **(a)** One explanation: The lines in the slope field of $\frac{dy}{dx} = y$ in Figure 8 have the same slope along each horizontal line because the formula on the right does not involve x. The lines are horizontal along the x-axis where $y = 0$, have positive slopes above the x-axis where $y > 0$, and have negative slopes below the x-axis where $y < 0$. They become steeper as y increases through positive values or decreases through negative values.
(b) The solutions, by Rule 1 with $r = 1$, are $y = Ce^x$ with a parameter C.
• Because the y-intercept of $y = Ce^x$ is C, the curves in Figure 9 are $y = Ce^x$ with $C = \pm\frac{1}{4}, \pm\frac{1}{2}, \pm 1, \pm 2$.

Response 5 With $y = 4e^{2\sin x}$, $y(0) = 4e^{2\sin(0)} = 4$ • The initial condition is satisfied. •
$\frac{dy}{dx} = \frac{d}{dx}(4e^{2\sin x}) = 4e^{2\sin x}\frac{d}{dx}(2\sin x) = 4e^{2\sin x}(2\cos x) = 8(\cos x)e^{2\sin x}$ •
$2y\cos x = 2(4e^{2\sin x})\cos x = 8(\cos x)e^{2\sin x}$ •
$\frac{dy}{dx}$ equals $2y\cos x$ and the differential equation is also satisfied.

Response 6 **(a)** One explanation: The slope lines in Figure 10 are horizontal on the x- and y-axes, where $-2xy^2$ is zero. The lines have negative slopes on the right where $x > 0$, $y \neq 0$, and $-2xy^2$ is negative; they have positive slopes on the left where $x < 0$, $y \neq 0$, and $-2xy^2$ is positive. They become steeper as x and y move away from 0 and $|-2xy^2|$ increases.
(b) The constant solution $y = 0$ does not satisfy either initial condition. •
$y = \frac{1}{x^2 - C}$ • $y(0) = \frac{1}{0^2 - C} = \frac{1}{-C}$. •
$y(0) = 1 \implies C = -1$ and $y = \frac{1}{x^2+1}$, which is the upper curve in Figure 11. •
$y(0) = -\frac{1}{4} \implies C = 4$ and $y = \frac{1}{x^2-4}$, which is the lower curve in Figure 11.

Response 7 Problem **(A)** goes with equation **(II)**, Problem **(B)** goes with equation **(I)**, Problem **(C)** goes with equation **(IV)**, and Problem **(D)** goes with equation **(III)**.

Response 8 **(I)** Equation **(I)** goes with the slope field in Figure 13.
One explanation: There is no t on the right side of the equation and the lines have the same slope on each horizontal line. Also the slopes are small and positive for small positive y and large and positive for large positive y.

(II) Equation **(II)** goes with the slope field in Figure 14.
One explanation: There is no t on the right side of the equation and the lines have the same slope on each horizontal line. Also the slopes are large and positive for small positive y and small and positive for large, positive y.

(III) Equation **(III)** goes with the slope field in Figure 15.
One explanation: There is no y on the right side of the equation and the lines have the same slope on each vertical line. Also the slopes are small and negative for small, positive t and large and negative for large, positive t. (The lines do not appear very steep in Figure 14 because of the different scales on the axes.)

(IV) Equation **(IV)** goes with the slope field in Figure 16.
One explanation: There is no t on the right side of the equation and the lines have the same slope on each horizontal line. Also the slopes are small for y close to 20, positive for $0 < y < 20$ and negative for $y > 20$.

Response 9 **(I)** $\frac{dy}{dt} = 0.2y$ • $y = Ce^{0.2t}$ by Rule 1 with $r = 0.2$ and t as variable • $y(0) = 50 \implies C = 50$ • $y = 50e^{0.2t}$ • Figure R9a • The population y tends to ∞ as $t \to \infty$.

(II) $\frac{dy}{dt} = \frac{20}{y}$ • Separate variables: $y\,dy = 20\,dt$ • $\int y\,dy = \int 20\,dt$ • $\frac{1}{2}y^2 = 20t + C$ • $y(0) = 50 \implies C = 1250$ • $y^2 = 40t + 2500$ • $y = \sqrt{40t + 2500}$ • Figure R9b • The thickness y of the ice tends to ∞ as $t \to \infty$.

(III) $\frac{dy}{dt} = -2t$ • $y = C - t^2$ • $y(0) = 50 \implies C = 50$ • $y = 50 - t^2$ • Figure R9c • $y \to -\infty$ as $t \to \infty$

(IV) $\frac{dy}{dt} = 2(20 - y)$ • $\frac{dy}{dt} = -2(y - 20)$ • $\frac{d}{dt}(y - 20) = -2(y - 20)$ • $y - 20 = Ce^{-2t}$ by Rule 1 • $y(0) = 50 \implies C = 30$ • $y = 20 + 30e^{-2t}$ • Figure R9d • The potato's temperature tends to 20°C as $t \to \infty$.

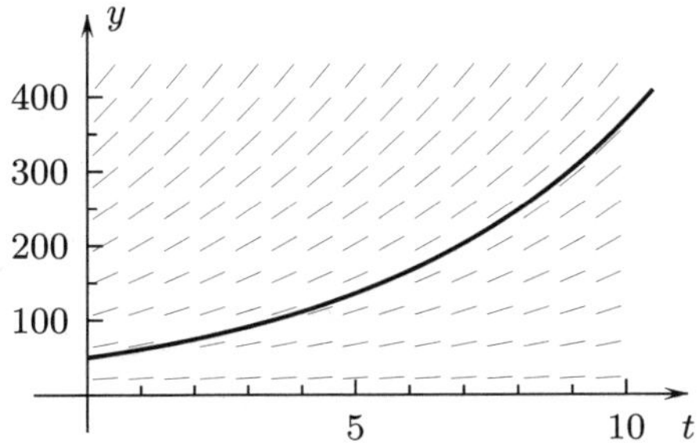

Figure R9a

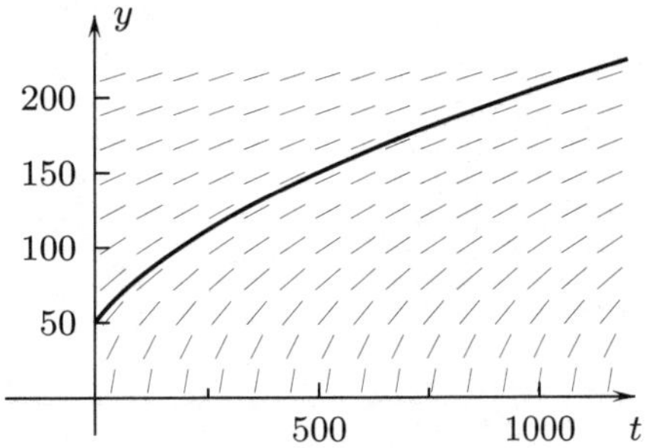

Figure R9b

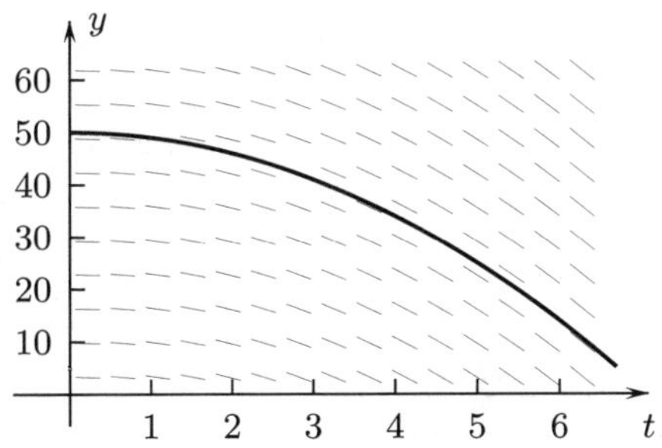

Figure R9c

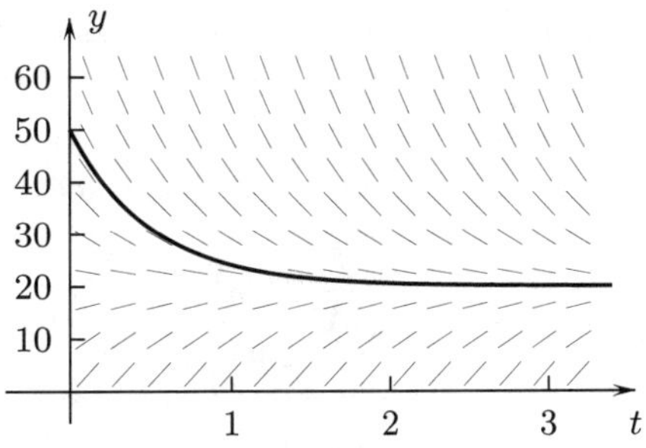

Figure R9d

Response 10 One description: The lines are parallel in each vertical column because there is no t on the right side of the equation. • The lines are horizontal at $P = 2$ where $\frac{1}{2}P(2 - P)$ is 0, point up to the right for $0 < P < 2$ where $\frac{1}{2}P(2 - P)$ is positive, and point down to the right for $P > 2$ where $\frac{1}{2}P(2 - P)$ is negative.

Response 11 $\ln|P| - \ln|2 - P| = t + C_1$ • $\ln\left|\frac{P}{2 - P}\right| = t + C_1$ • $\frac{P}{2 - P} = \pm e^{t + C_1}$ • $\frac{P}{2 - P} = \pm e^{C_1}e^t$ • $\frac{P}{2 - P} = Ce^t$ with $C = \pm e^{C_1}$ • $P = 2Ce^t - PCe^t$ • $(Ce^t + 1)P = 2Ce^t$ • $P = \frac{2Ce^t}{Ce^t + 1}$

Response 12 $P = \frac{2}{1 + ce^{-t}}$ • $P(0) = 3$ thousand $\implies 3 = \frac{2}{1 + c}$ • $3 + 3c = 2$ • $c = -\frac{1}{3}$ • $P = \frac{2}{1 - \frac{1}{3}e^{-t}}$

Responses 7.2

Response 1 Problem **(A)** goes with differential equation **(I)**, problem **(B)** goes with equation **(III)**, and problem **(C)** goes with equation **(II)**.

Response 2 Differential equation **(I)** goes with the slope field in Figure 3, equation **(II)** goes with the slope field in Figure 2, and equation **(III)** goes with the slope field in Figure 1.

Response 3 **(I)** $\dfrac{dv}{dt} = 9.8$ • $v = \displaystyle\int 9.8\,dt = 9.8t + C$ • $v(0) = 0 \implies C = 0$ • $v = 9.8t$ • Figure R3c

(II) $\dfrac{dv}{dt} = \frac{1}{10}(98 - v)$ • $\dfrac{dv}{dt} = -\frac{1}{10}(v - 98)$ $\dfrac{d}{dt}(v - 98) = -\frac{1}{10}(v - 98)$ • $v - 98 = Ce^{-t/10}$ • $v(0) = 0 \implies C = -98$ • $v = 98 - 98e^{-t/10}$ • Figure R3b

(III) $\dfrac{dv}{dt} = \frac{1}{5}(49 - v)$ • $\dfrac{dv}{dt} = -\frac{1}{5}(v - 49)$ $\dfrac{d}{dt}(v - 49) = -\frac{1}{5}(v - 49)$ • $v - 49 = Ce^{-t/5}$ • $v(0) = 0 \implies C = -49$ • $v = 49 - 49e^{-t/5}$ • Figure R3a

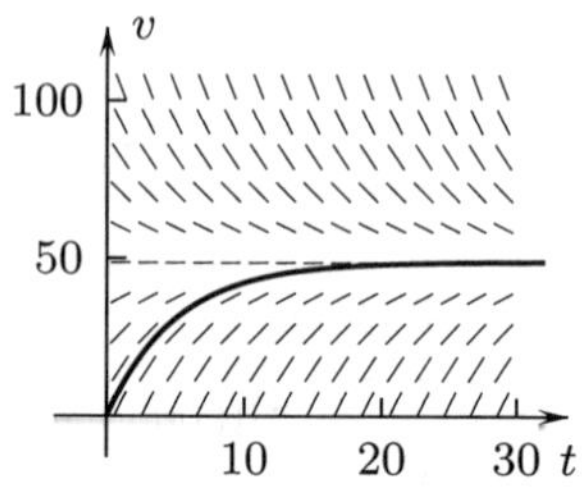

Figure R3a

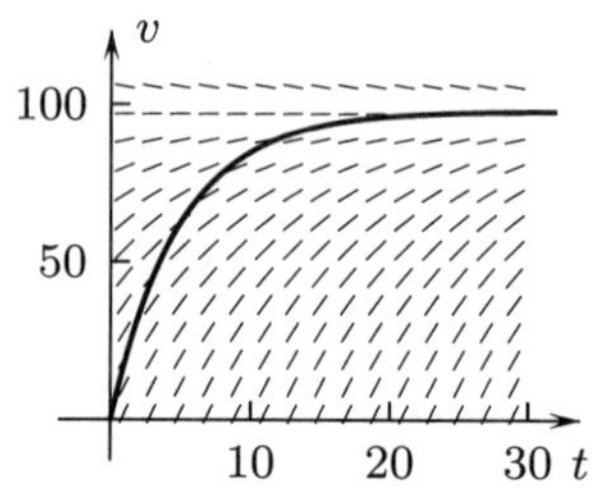

Figure R3b

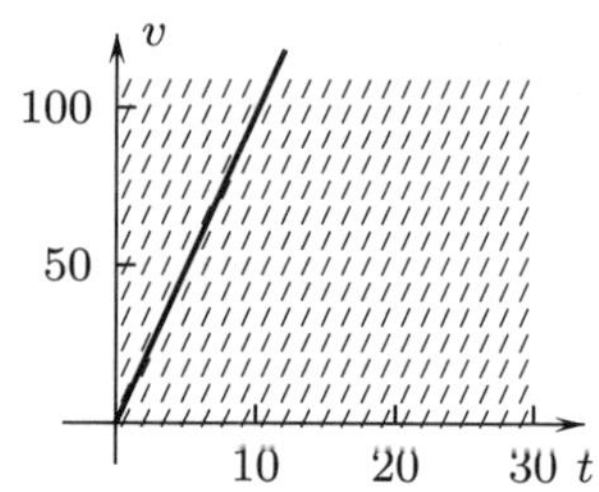

Figure R3c

Response 4 Problem **(D)** goes with differential equation **(V)** , Problem **(E)** goes with equation **(VI)** , and Problem **(F)** goes with equation **(IV)**.

Response 5 Differential equation **(IV)** goes with the slope field in Figure 6, equation **(V)** goes with the slope field in Figure 5, and equation **(VI)** goes with the slope field in Figure 4.

Response 6 **(IV)** $\dfrac{dv}{dt} = 200 - 5v$ • $\dfrac{dv}{dt} = -5(v - 40)$ • $\dfrac{d}{dt}(v - 40) = -5(v - 40)$ • $v - 40 = Ce^{-5t}$ • $v(0) = 0 \implies C = -40$ • $v = 40 - 40e^{-5t}$ • Figure R6a

(V) $\dfrac{dv}{dt} = 45 + 45\cos(15t)$ • $v = \displaystyle\int [45 + 45\cos(15t)]\,dt = 45t + 3\sin(15t) + C$ (using $u = 15t, du = 15\,dt$) • $v(0) = 0 \implies C = 0$ • $v = 45t + 3\sin(15t)$ • Figure R6b •

(VI) $\dfrac{dv}{dt} = -\frac{1}{8}v^2$ • Separate variables: $\displaystyle\int v^{-2}\,dv = -\frac{1}{8}\int dt$ • $-v^{-1} = -\frac{1}{8}t + C_1$ • $v^{-1} = \frac{1}{8}t + C$ • $v(0) = 40 \implies C = \frac{1}{40}$ • $v = \dfrac{1}{\frac{1}{8}t + \frac{1}{40}} = \dfrac{40}{1 + 5t}$ • Figure R6c

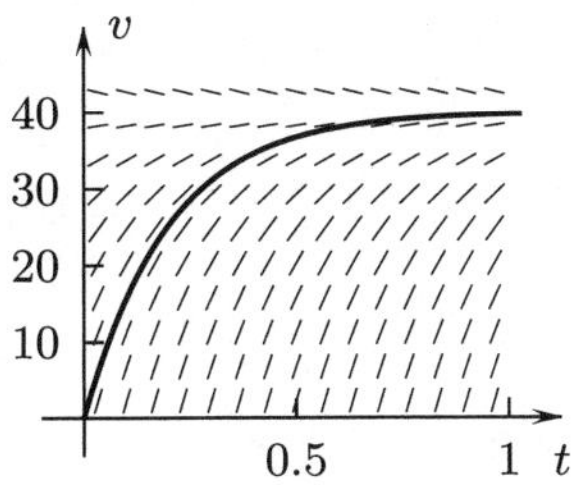

Figure R6a

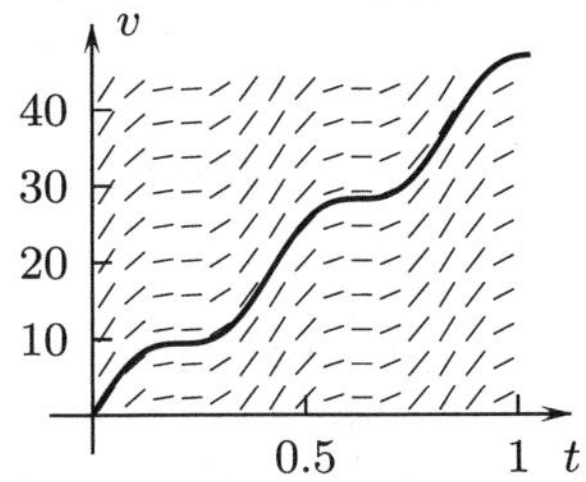

Figure R6b

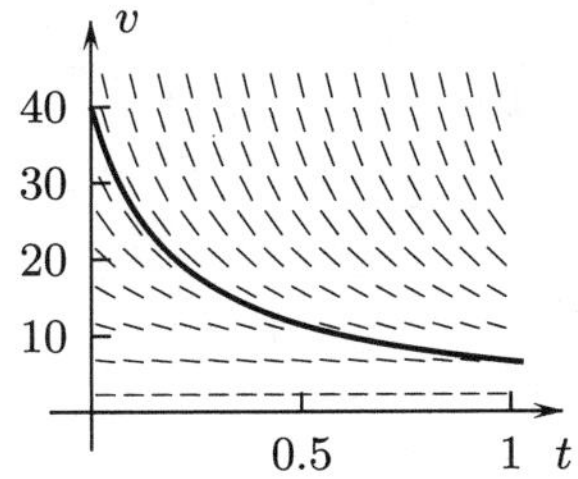

Figure R6c

Response 7 $\int 4\cos x\, dx = \int y\, dy$ • $4\sin x = \frac{1}{2}y^2 + C$ • Set $x = 0, y = 2$: $\frac{1}{2}4^2 + C = 0$ • $C = -2$ • $4\sin x = \frac{1}{2}y^2 - 2$ • $y^2 = 8\sin x + 4$

Response 8 $(x(t), y(t))$ moves clockwise around the curve as t increases. • One explanation: On the left of the curve, $x < 3$, $y > 0$, and, by the first equation in **(9)**, $dy/dt > 0$ so $y(t)$ is increasing and $(x(t), y(t))$ moves up the curve. •
On the right, $x > 3$, $y > 0$, and $dy/dt < 0$ so $y(t)$ is decreasing and $(x(t), y(t))$ moves down the curve. •
On the top, $y > 2$, $x > 0$, and, by the second equation in **(9)**, $dx/dt > 0$ so that $x(t)$ is increasing and $(x(t), y(t))$ moves to the right •
On the bottom, $y < 2$, $x > 0$, and $dx/dt < 0$ so $x(t)$ is decreasing and $(x(t), y(t))$ moves to the left.

Response 9 $c = 1, 2, -1, -2$

Response 10 **(a)** $\int y\, dy = 4\int dx$ • $\frac{1}{2}y^2 = 4x + C_1$ or $x = \frac{1}{8}y^2 + C$ • For each C, the curve $x = \frac{1}{8}y^2 + C$ is a parabola that opens to the right and whose vertex is at $x = C$ on the x-axis.
(b) $C = -4, -2, 0, 2$

Responses 7.3

Response 1 The slope of the first line segment is $\left[x - \frac{1}{2}y\right]_{x=0,y=2} = 0 - \frac{1}{2}(2) = -1$ and the y-coordinate of its right end is $2 + (-1)(1) = 1$. •
The slope of the second line segment is $\left[x - \frac{1}{2}y\right]_{x=1,y=1} = 1 - \frac{1}{2}(1) = 0.5$ and the y-coordinate of its right end is $1 + \frac{1}{2}(1) = 1.5$. •
The slope of the third line segment is $\left[x - \frac{1}{2}y\right]_{x=2,y=1.5} = 2 - \frac{1}{2}(1.5) = 1.25$ and the y-coordinate of its right end is $1.5 + (1.25)(1) = 2.75$. •
The slope of the fourth line segment is $\left[x - \frac{1}{2}y\right]_{x=3,y=2.75} = 3 - \frac{1}{2}(2.75) = 1.625$ and the y-coordinate of its right end is $2.75 + (1.625)(1) = 4.375$. •
Values:

x	0	1	2	3	4
Euler's-method approximation	2	1	1.5	2.75	4.375

CHAPTER 7 ANSWERS

Tune-up Exercises 7.1

T1. The slopes are given in the table below and the slope lines in Figure T1.

[SLOPE] $= y/x$

	$x=1$	$x=2$	$x=3$	$x=4$
$y=3$	3	$\frac{3}{2}$	1	$\frac{3}{4}$
$y=2$	2	1	$\frac{2}{3}$	$\frac{1}{2}$
$y=1$	1	$\frac{1}{2}$	$\frac{1}{3}$	$\frac{1}{4}$
$y=0$	0	0	0	0

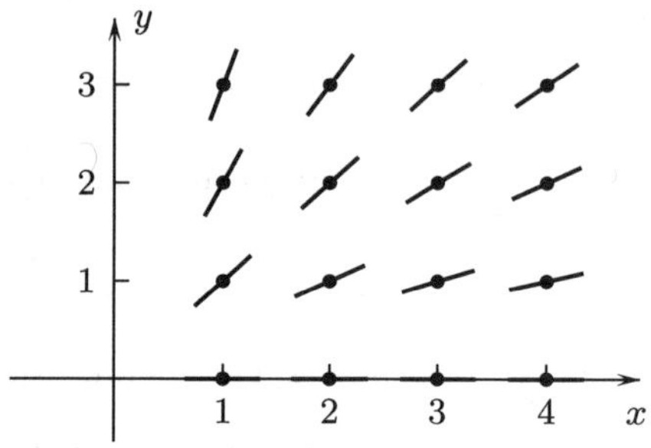

Figure T1

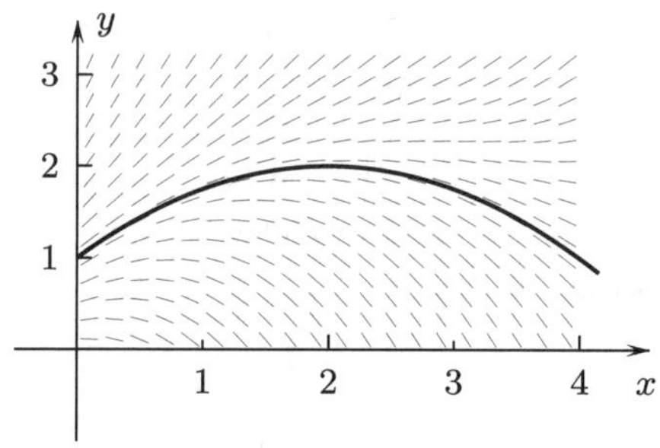

Figure T2

T2. Figure T2

T3. $y = Ce^{-4x}$ • Set $x = 0, y = -3$ • $C = -3$ • $y = -3e^{-4x}$

T4. $r(t) = 4t^3$ • $R(t) = \int 4t^3\,dt = t^4 + c$ • Use $c = 0$ • $y = Ce^{t^4}$ • Set $t = 0, y = 10$: $C = 10$ • $y = 10e^{t^4}$

T5. Write $f'(x)$ as $\dfrac{df}{dx}$ • $\int f^{-4}\,df = \int -3\,dx$ • $-\frac{1}{3}f^{-3} = -3x + C_1$ • $f^{-3} = 9x + C$ • $f = (9x + C)^{-1/3}$

T6. $\int K^{-1/2}\,dK = \int x^{1/2}\,dx$ • $K^{1/2} = \frac{1}{3}x^{3/2} + C$ • Set $x = 1, K = 1$: $C = \frac{2}{3}$ • $K = (\frac{1}{3}x^{2/3} + \frac{2}{3})^2$

Problems 7.1

1. Figure A1

2. Figure A2

4. $y = 4e^{8t}$

5. $y = 10e^4e^{-t} = 10e^{4-t}$

6. $z = Ce^{\ln(4)x} = C(e^{\ln(4)})^x = C4^x$ • Set $x = 2, z = 15 : 15 = 16C$ • $C = \frac{15}{16}$ • $z = \frac{15}{16}4^x$

8. $\dfrac{dv}{dt} = -(v-5)$ • $\dfrac{d}{dt}(v-5) = -(v-5)$ • $v - 5 = Ce^{-t}$ • $v = 5 + Ce^{-t}$ •
Set $t = 0, v = 10 : C = 5$ • $v = 5 + 5e^{-t}$

9. $\int Q^{-1/4}\,dQ = -3\int dx$ • $\frac{4}{3}Q^{3/4} = -3x + C_1$ • $Q^{3/4} = -\frac{9}{4}x + C$ • $Q = (C - \frac{9}{4}x)^{4/3}$

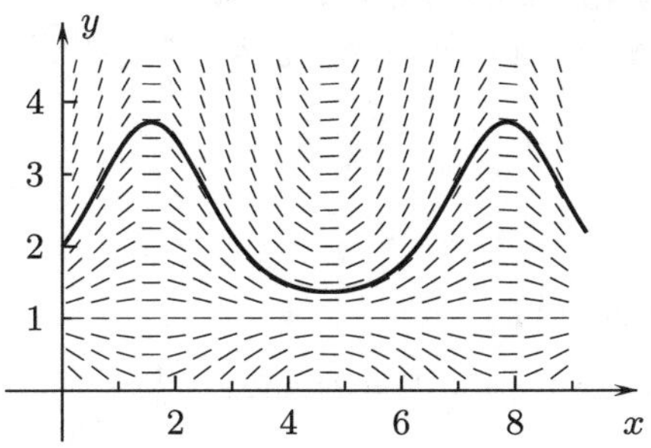

Figure A1

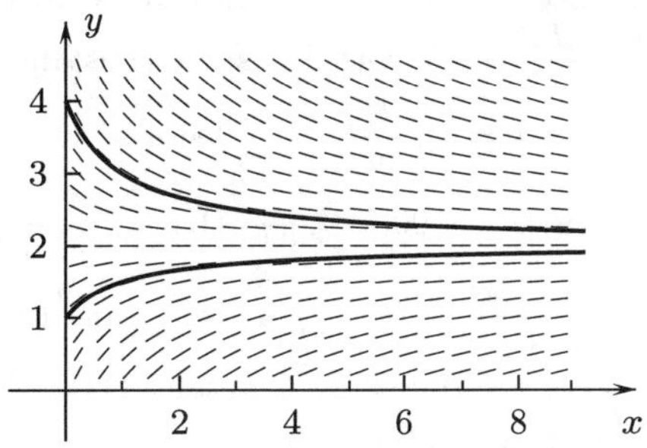

Figure A2

10. $r(x) = -\cos x$ • $R(x) = \int (-\cos x)\,dx = -\sin x + c$ • Take $c = 0$ • $y = Ce^{R(x)} = Ce^{-\sin x}$

11. $Q = (\frac{1}{3}t^2 + 1)^{3/2}$

13. $r(x) = -3\cos(3x)$ • $R(x) = \int (-3\cos(3x)\,dx = -\sin(3x) + c$ • Take $c = 0$ •
$y = Ce^{R(x)} = Ce^{-sin(3x)}$ • Set $x = \pi, y = 6 : 6 = Ce^{-\sin(3\pi)} = C$ • $y = 6e^{-\sin(3x)}$

15. $y = \dfrac{1}{C - x^3}$

18. **(a)** $y = \dfrac{4}{\cos x + 1}$ • The upper curve in Figure A18

(b) $y = \dfrac{4}{\cos x - 3}$ • The lower curve in Figure A18

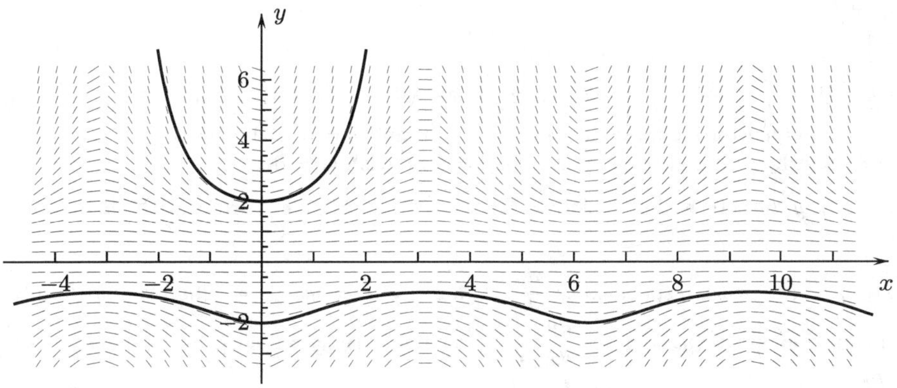

Figure A18

20 Let $y(t)$ be the mass of the culture (grams) at time t (days) with $t = 0$ initially. • $\dfrac{dy/dt}{y} = 0.03$ •
$\dfrac{dy}{dt} = 0.03y$ • $y = Ce^{0.03t}$ • Set $t = 0, y = 2 : C = 2$ • $y = 2e^{0.03t}$ • $y(7) = 2e^{0.21} \doteq 2.467$ grams

21. $\frac{1}{2}\ln(1.5) \doteq 0.2027$

30. $\frac{1}{200}\ln(60.25)$

39. Let $T(t)$ be the temperature °C at time t (minutes) with $t = 0$ when $T = 10$. • $\dfrac{dT}{dt} = -kT$ with a positive constant k • Set $\dfrac{dT}{dt} = -7, T = 4 : -7 = k(4)$ • $k = \frac{7}{4}$ • $\dfrac{dT}{dt} = -\frac{7}{4}T$ • Set $T = 1 : \dfrac{dT}{dt} = -\frac{7}{4}$ degrees per minute

42. **(a)** $L = 100, k = \frac{1}{100}$ **(b)** $P = \dfrac{100}{1 + 9e^{-t}}$

47. $\displaystyle\int \frac{1}{4 + F^2}\, dF = \int dx$ • $\frac{1}{2}\tan^{-1}(\frac{1}{2}F) = x + C_1$ • $\tan^{-1}(\frac{1}{2}F) = 2x + C$ • Set $x = 0, F = 2 : C = \tan^{-1}(1) = \frac{1}{4}\pi$ • $\tan^{-1}(\frac{1}{2}F) = 2x + \frac{1}{4}\pi$ • $F = 2\tan(2x + \frac{1}{4}\pi)$

48. $x = \sin(\frac{1}{3}t^3)$

49. $Q = \dfrac{-1}{3t^2 + C} - 2$

50. $y = \sin^{-1}(1 - \cos x)$

54. $y = x^x$

Tune-up Exercises 7.2

T1. **(a)** $v = Ce^{-6t}$ • Set $t = 0, v = 50 : C = 50$ • $v = 50e^{-6t}$
(b) $v \to 0$ as $t \to \infty$

10.27.99A

T2. **(a)** $v = 0$ or $\displaystyle\int v^{-2}\, dv = -6\int dt$ • $v = 0$ or $-v^{-1} = -6t + C_1$ • $v = 0$ or $\dfrac{1}{v} = 6t + C$ • Set $t = 0, v = 50 : C = \frac{1}{50}$ • $\dfrac{1}{v} = 6t + \frac{1}{50}$ • $v = \dfrac{1}{6t + \frac{1}{50}}$
(b) $v \to 0$ as $t \to \infty$

10.27.99B

T3. **(a)** $\dfrac{dv}{dt} = -32(v + 1)$ • $\dfrac{d}{dt}(v + 1) = -32(v + 1)$ • $v + 1 = Ce^{-32t}$ • Set $t = 0, v = 0 : C = 1$ • $v = -1 + e^{-32t}$
(b) $v \to -1$ as $t \to \infty$

10.27.99C

T4. **(a)** $x = 0, y = 0$ or $\displaystyle\int \frac{1}{x}\, dx + \int \frac{1}{y}\, dy = 0$ • In the last case: $\ln|x| + \ln|y| = C_1$ • $\ln|xy| = C_1$ • $xy = C$ (The last equation includes the cases of $x = 0$ and $y = 0$ with $C = 0$.) • The lines $x = 0$ and $y = 0$ and the hyperbolas $y = C/x$ with $C \neq 0$
(b) $\displaystyle\int x\, dx + \int y\, dy = 0$ • $\frac{1}{2}x^2 + \frac{1}{2}y^2 = C_1$ • $x^2 + y^2 = C$ • Circles with their centers at the origin

10.27.99D

T5. Differentiate $y = cx^4$ with respect to x: $\dfrac{dy}{dx} = 4cx^3$ • $c = yx^{-4}$ • The original family has the differential equation $\dfrac{dy}{dx} = 4yx^{-4}x^3$ or $\dfrac{dy}{dx} = \dfrac{4y}{x}$ • The orthogonal trajectories have the differential equation $\dfrac{dy}{dx} = \dfrac{-x}{4y}$ • $x\, dx + 4y\, dy = 0$ • $\displaystyle\int x\, dx + \int 4y\, dy = 0$ • $\frac{1}{2}x^2 + 2y^2 = C$ (ellipses)

10.27.99E

Problems 7.2

1. **(a)** With a downward pointing s-axis with distances measured in meters: [Weight] $= mg = 2(9.8) = 19.6$ newtons • $2\dfrac{dv}{dt} = 19.6 - 0.98v$ • $\dfrac{dv}{dt} = -0.49(v - 20)$ • $\dfrac{d}{dt}(v - 20) = -0.49(v - 20)$ • $v - 20 = Ce^{-0.49t}$ • Set $t = 0, v = 15 : C = -5$ • $v = 20 - 5e^{-0.49t}$ • $v(10) = 20 - 5e^{-4.9} \doteq 19.96$ meters per second
(b) $v \to 20$ as $t \to \infty$

2. With an s-axis in the direction of the motion and distances measured in meters: $F = ma$ • $-8v^2 = 8000\dfrac{dv}{dt}$ • $-\displaystyle\int v^{-2}\, dv = \frac{1}{1000}\int dt$ • $v^{-1} = (0.001)t + C$ Set $t = 0, v = 25 : C = 0.04$ • $v = \dfrac{1}{(0.001)t + 0.04}$ •

$$v(50) = \frac{1}{0.09} = 11\tfrac{1}{9} \text{ meters per second}$$

3. **(a)** [Mass] $= \frac{1}{100}$ slugs **(b)** $v = 240e^{-t/5} - 160$ feet per second

7. 4900 centimeters per second

8. $x = 0, y = 0$, or $\displaystyle\int \frac{x+1}{x}\,dx + \int \frac{y-1}{y}\,dy = C$ • In the last case:
$\displaystyle\int \left(1+\frac{1}{x}\right)dx + \int \left(1-\frac{1}{y}\right)dy = C$ • $x = 0, y = 0$ or $x - \ln|x| + y - \ln|y| = C$

9. $y = \dfrac{-8x^2}{4+3x^2}$

13. $\tan^{-1} x + \ln(y^2+1) = C$

18. **(a)** $y = \sqrt[3]{21\sin x + 27}$ **(b)** $y = \sqrt[3]{21\sin x - 1}$ • Figure A18

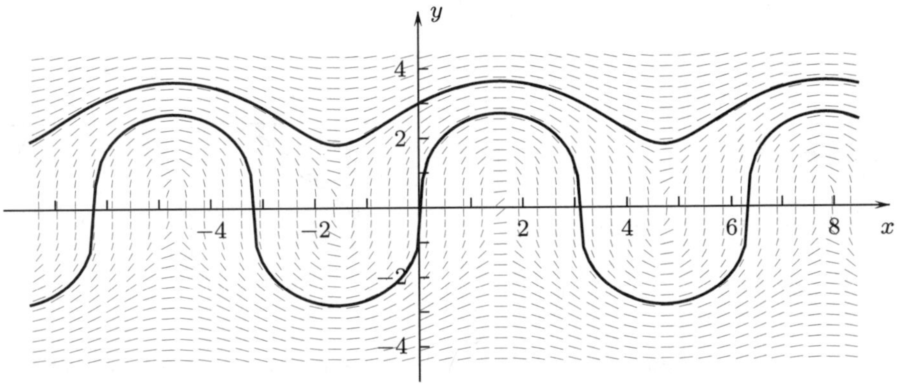

Figure 18

21. $y - 2\ln y - \frac{10}{3}\ln x + \frac{1}{6}x = \frac{11}{6} - \frac{10}{3}\ln(5)$

23. $S = \dfrac{\ln s}{\ln(1.05)} + 1$

26. $P = (\frac{1}{2}t + 20)^2$

28. $y = Cx^2$

Tune-up Exercises 7.3

T1. Partition: $-2 < -1 < 0 < 1 < 2$ • Euler's-method approximation: $Y(-2) = 1$, $Y(-1) \doteq 1.44444, Y(0) \doteq 1.676269, Y(1) \doteq 1.676269, Y(2) \doteq 1.988478$ • Figure T1

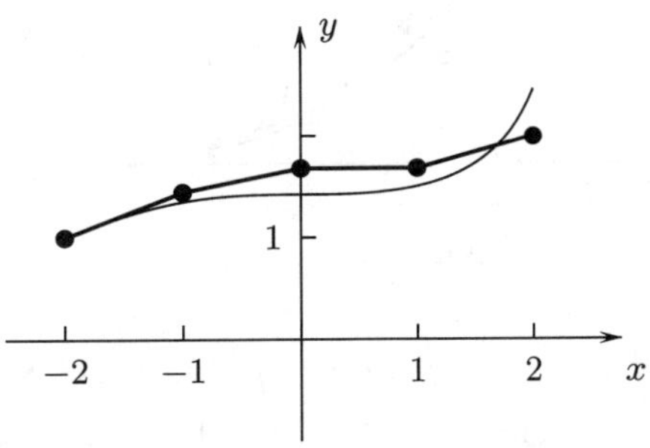

Figure T1

T2. Partition: $-2 < -1 < 0 < 1 < 2$ • Run the Runge-Kutta program in $[-2, -1]$ with one subinterval to calculate $Y(-1)$, in $[-2, 0]$ with two subintervals to calculate $Y(0)$, etc. • Runge-Kutta-method approximation: $Y(-2) = 1, Y(-1) \doteq 1.349774, Y(0) \doteq 1420615, Y(1) \doteq 1.499236, Y(2) \doteq 2.443469$ • $[\text{Error at } x = 2] = |Y(2) - 27/(19 - 2^3)| \doteq 0.011$

Problems 7.3

1. **(a)** $\int y^{-2}\,dy = \frac{1}{4}\int \sin x\,dx$ • $-y^{-1} = -\frac{1}{4}\cos x + C$ • $\frac{1}{y} = \frac{1}{4}\cos x + C_1$ • $y(0) = 2 \implies C_1 = \frac{1}{4}$ • $y = \dfrac{4}{1 + \cos x}$
(b) Partition: $0 < 0.5 < 1 < 0.75 < 1$ • Euler's-method approximation:
$Y(0) = 2,\ Y(0.5) = 2, Y(1) \doteq 2.239713, Y(0.75) \doteq 2.767348, Y(2) \doteq 3.722227$ •
(c) $[\text{Error}] = |y(2) - Y(2)| \doteq 3.13$ **(d)** Figure A1

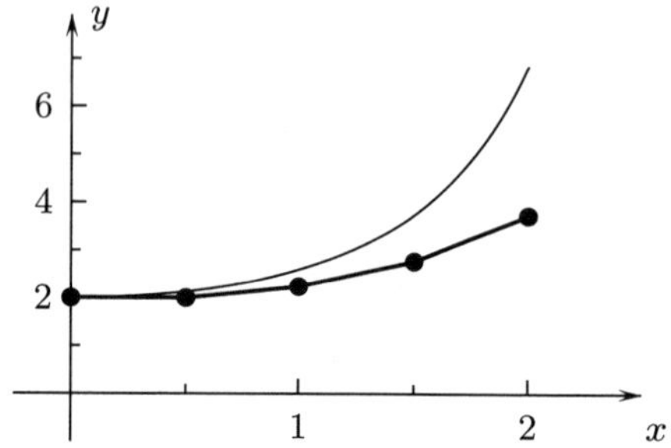

Figure A1

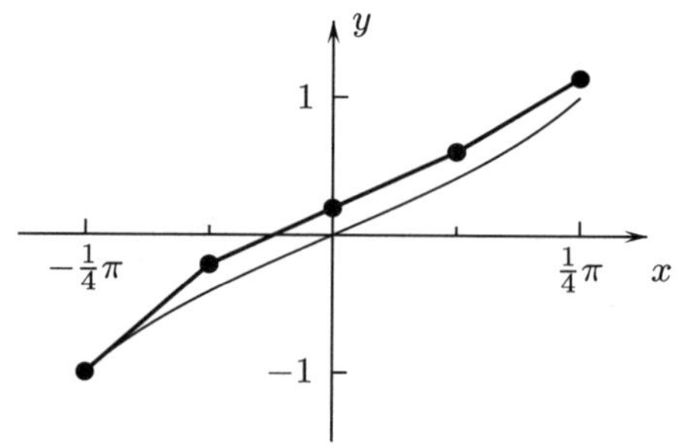

Figure A3

3. **(a)** $\int \frac{1}{y^2 + 1}\,dy = \int dx$ • $\tan^{-1} y = x + C$ • $y(-\frac{1}{4}\pi) = -1 \implies C = 0$ • $y = \tan x$
(b) Partition: $-\frac{1}{4}\pi < -\frac{1}{8}\pi < 0 < \frac{1}{8}\pi < \frac{1}{4}\pi$ • Euler's-method approximation: $Y(-\frac{1}{4}\pi) = -1$, $Y(-\frac{1}{8}\pi) \doteq -0.214602, Y(0) \doteq 0.196183, Y(\frac{1}{8}\pi) \doteq 0.603996, Y(\frac{1}{4}\pi) \doteq 1.139956$ •
(c) $[\text{Error}] = |y(\frac{1}{4}\pi) - Y(\frac{1}{4}\pi)| \doteq 0.14$ **(d)** Figure A2

4. **(a)** $\int \frac{1}{y}\,dy = \int \left(\frac{1}{x} - 1\right)\,dx$ • $\ln|y| = \ln|x| - x + C_1$ • $y = Cxe^{-x}$ • $y(1) = 3 \implies C = 3e$ • $y = 3xe^{1-x}$
(b) Partition: $1 < 2 < 3 < 4 < 5$ • Euler's-method approximation:
$Y(1) = 3, Y(2) = 3, Y(3) = 1.5, Y(4) = 0.5, Y(5) = 0.125$ •
(c) $[\text{Error}] = |y(5) - Y(5)| \doteq 0.15$ **(d)** Figure A4

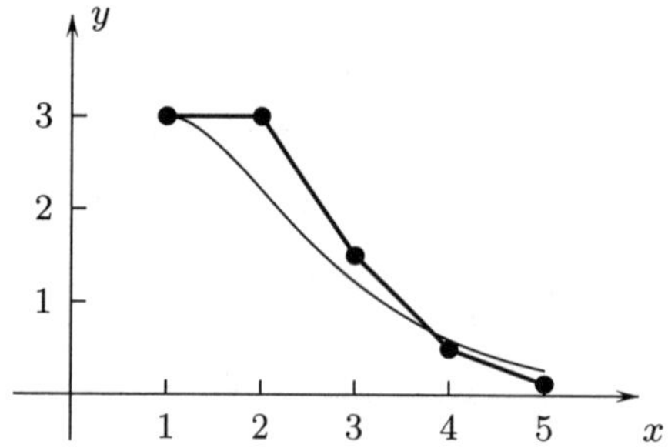

Figure A4

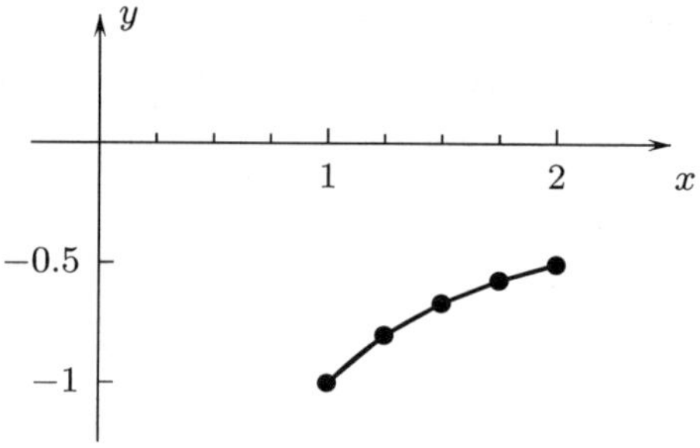

Figure A7

7. **(a)** $\int y^{-2}\,dy = \int dx$ • $-y^{-1} = x + C$ • $y(1) = -1 \implies C = 0$ • $y = -1/x$
(b) Partition: $1 < 1.25 < 1.5 < 1.75 < 2$ • Runge-Kutta-method approximation:
$Y(1) = -1, Y(1.25) \doteq -0.800014, Y(1.5) \doteq -0.666681, Y(1.75) = -571441, Y(2) \doteq -0.5000010$ •
(c) $[\text{Error}] = |y(2) - Y(2)| \doteq 1.00 \times 10^{-5}$ **(c)** Figure A7

CHAPTER 8: SEQUENCES AND SERIES

Most of the applications we have considered up to now have involved continuous mathematical models using continuous functions or functions that are continuous in subintervals of their domains. In Chapter 7, for example, we used differential equations to study population growth by viewing the populations as continuous functions $P(t)$ of the time t. Another approach is to use DISCRETE mathematical models in which phenomena are considered to change by stages rather than continuously. In the case of population growth, we might study only the size of a population on the same date each year and denote the population in year n for integers $n \geq 0$ by P_n. The resulting string of numbers $P_0, P_1, P_2, P_3, \ldots$ is called a SEQUENCE. In Section 8.1 we discuss basic properties of sequences and illustrate their use in studying population growth. Section 8.2 deals with special sums called GEOMETRIC SERIES and their use in other types of discrete mathematical models The main objectives of the remaining four sections of the chapter are to study approximation of continuous functions by polynomials and representations of functions by infinite sums constructed from powers of x. Criteria for determining whether infinite sums are defined are discussed in Sections 8.3 and 8.4; approximation of functions by TAYLOR POLYNOMIALS are discussed in Section 8.5; and Section 8.6 deals with representing functions by TAYLOR SERIES.

Section 8.1

Infinite sequences

OVERVIEW: *In this section we discuss basic properties of sequences, as illustrated by sequences that are defined through explicit formulas. Then we look at sequences that are constructed by* ITERATION *and look at how these are used in in the study of population growth through* DISCRETE LOGISTIC EQUATIONS.

Topics:

- ***Convergence and divergence of sequences***
- ***Bounded and monotone sequences***
- ***Least upper bounds and greatest lower bounds of sequences***
- ***Iterations***
- ***Discrete logistic equations***

Convergence and divergence of infinite sequences

An INFINITE SEQUENCE is an infinite string of numbers

$$a_{n_0},\ a_{n_0+1},\ a_{n_0+2},\ a_{n_0+3}, \cdots \qquad (1)$$

with one number a_n for each integer $n \geq n_0$ with some integer n_0. The variable subscript n is called the INDEX of the sequence and the numbers a_n are its VALUES or TERMS. We also denote the sequence (1) with the symbols $\{a_n\}_{n=n_0}^{\infty}$. We can consider a sequence $\{a_n\}_{n=n_0}^{\infty}$ to be a function with variable n defined on the set of integers $\geq n_0$ with the value a_n at n. Then its graph for finite collections of n's in an ny-plane consists of a finite number of points.

Imagine that a nonfunctioning freezer is at 30°C at time $n = 0$, when it is turned on and that its temperature n hours later is $\dfrac{30}{n+1}$. Figure 1 shows the graph of the sequence $\left\{\dfrac{30}{n+1}\right\}_{n=0}^{\infty}$. Because the temperature $\dfrac{30}{n+1}$ tends to 0 as $n \to \infty$, we say that the infinite sequence CONVERGES to 0.

♦ SUGGESTIONS TO INSTRUCTORS: The main topics in this section can be divided in two parts, as follows:
Part 1 (Convergence and divergence of infinite sequences, the ϵN- and YN-definitions, bounded and monotone sequences, and least upper and greatest lower bounds; Tune-up Exercises T1 through T5; Problems 1–21, 27–54) Have students begin work on Questions 1 through 8 before class. Problems 1–3, 8, 11, 40, 41, and 55 are good for in-class discussion.
Part2 (Iteration, web plots, and the discrete logistic equation; Tune-up Exercises T6 and T7; Problems 22–26) Have students begin work on Questions 9 through 14 before class. Problems 24–26 are good for in-class discussion.

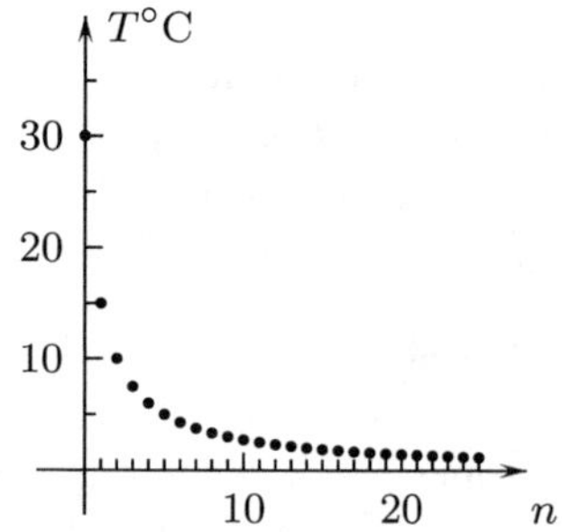

$T = \dfrac{30}{n+1}$

FIGURE 1

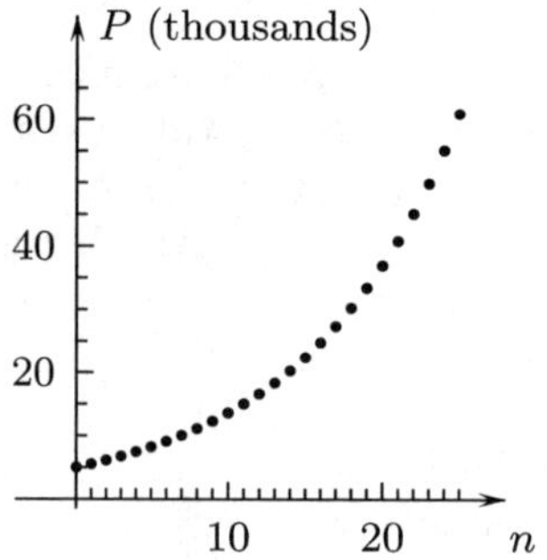

$P = 5e^{n/10}$

FIGURE 2

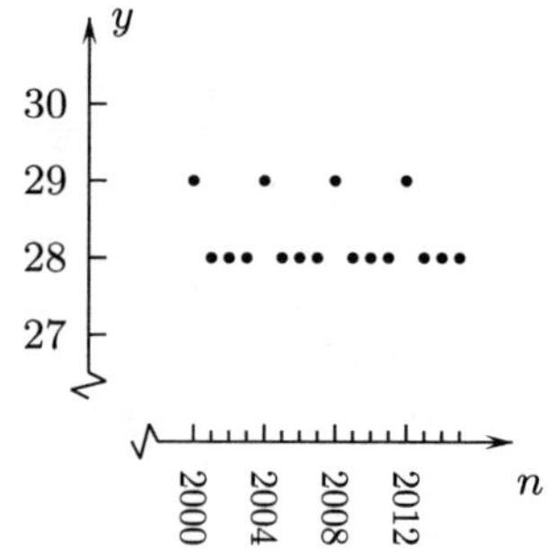

$y = d_n$

FIGURE 3

Figure 2 shows the graph of the sequence $\{5e^{n/10}\}_{n=0}^{\infty}$ that gives the population on day n of a colony of bacteria that consists of 5000 bacteria at $n = 0$ and grows exponentially. Because $5e^{n/10} \to \infty$ as $n \to \infty$, we say that this sequence DIVERGES TO ∞.

If the terms in a sequence tend to $-\infty$ as $n \to \infty$, we say that the sequence DIVERGES TO $-\infty$.

Question 1 **(a)** Figure 3 shows the graph of the sequence $\{d_n\}_{n=2000}^{\infty}$, where d_n is defined to be 29 if $n/4$ is an integer and to be 28 otherwise. Complete Table 1 below of values of the sequence. **(b)** What might this sequence represent?

TABLE 1

n	d_n	n	d_n
2000	29	2004	
2001	28	2005	
2002		2006	
2003		2007	

Because d_n equals 28 for every three out of four integers n and and 29 for the fourth, it does not have a limit as $n \to \infty$. We say that $\{d_n\}_{n=2000}^{\infty}$ DIVERGES WITHOUT A LIMIT.

Calculators and computers with calculus software can be used to plot finite numbers of terms of infinite sequences. We will explain next how this can be done on Texas Instruments calculators in cases where, as in Figures 1 through 3, the terms of the sequences are given by explicit formulas.†

Plotting and calculating explicit sequences on a TI-82 calculator

To generate the graph of $\{a_n\}$ for $n_1 \le n \le n_2$ on a TI-82 calculator with $n_1 \ge 0$‡, where the numbers a_n are given by an explicit formula in terms of n, first press [MODE], highlight <Seq> on the fourth row, and press [ENTER] to put the calculator in sequence mode. Then highlight <Dot> on the fifth row and press [ENTER] so the points on the graph will not be connected by line segments. Press [Y =] and enter the formula for a_n as Un. The variable n is obtained with [2nd] [n] (above [9]). Press [WINDOW]. Set nStart = 0, nMin = n_1, and nMax = n_2. If $n_1 = 0$, also set UnStart = a_{n_1}. Pick Xmin, Xmax, Ymin, and Ymax for a suitable viewing window. Press [GRAPH] to generate the

†These calculator-specific instructions are included in preliminary editions of this text for the convenience of uses. They will be in the *Calculator/Computer Supplement* of the First Edition.

‡This calculator does not allow negative values of n in sequence mode. Any terms a_n with $n < 0$ must be calculated and plotted by hand.

graph. To see the values of the sequence, press TRACE and use ► and ◄ to show different values of n.

To show the values in a table, press 2nd TblSet, set TblMin $= n_1$ and ΔTbl = 1, and select <Auto> on the lines with <Indpnt> and <Depend>. Then press 2nd TABLE and scroll down, as necessary, to see the values.

Plotting and calculating explicit sequences on a TI-83 calculator

To generate the graph of $\{a_n\}$ for $n_1 \le n \le n_2$ on a TI-83 calculator with $n_1 \ge 0$[†] where the numbers a_n are given by an explicit formula in terms of n, press MODE, highlight <Seq> on the fourth row, and press ENTER to put the calculator in sequence mode. Then highlight <Dot> on the fifth row and press ENTER so the points on the graph will not be connected by line segments. Press $Y=$, set nMin equal to n_1, enter the formula for a_n as $u(n)$. The variable n is obtained with X, T, θ, n. If $n_1 = 0$, set $u(n\text{Min})$ equal to a_{n_0}. Press WINDOW, set nMin = 0, nMax $= n_2$, PlotStart = 1, and PlotStep = 1. Pick Xmin, Xmax, Ymin, and Ymax for a suitable viewing window, and press GRAPH to generate the graph. To see the values of the sequence, press TRACE and use ► and ◄ to show different values of n.

To show the values in a table, press 2nd TblSet, set TblMin $= n_1$and ΔTbl = 1, and select <Auto> on the rows with <Indpnt> and <Depend>. Then press 2nd TABLE and scroll down, as necessary, to see the values.

Plotting and calculating explicit sequences on a TI-89 or TI-92 calculator

To generate the graph of $\{a_n\}$ for $n_1 \le n \le n_2$, where the numbers a_n are given by an explicit formula in terms of n and with $n_1 \ge 1$ on a TI-89 or TI-92 calculator,[‡] press MODE and ► for the menu of graphing options, highlight <SEQUENCE> on the fourth row, and press ENTER twice to put the calculator in sequence mode. Press ♦ $Y=$ and enter the formula for a_n as $u1$. The variable n is obtained with alpha N on a TI-89 and N on a TI-92. Set $ui1 = a_{n_1}$. Press ♦ WINDOW, set $n$min = 0, and PlotStrt = 1, $n$max $= n_2$, and PlotStep = 1. Pick $x$min, $x$max, $y$min, and $y$max for a suitable viewing window, and press ♦ GRAPH to generate the graph. To see the values of the sequence, press F3 for <Trace> and use ► and ◄ to show different values of n.

To show the values in a table, press ♦ TblSet, set tblStart $= n_1$and Δtbl = 1, select <OFF> on the line with <Graph <-> Table>, and select <Auto> on the line with <Independent>. Then press ENTER twice and ♦ TABLE and scroll down, as necessary, to see the values.

Plotting and calculating explicit sequences on a TI-85 or TI-86 calculator

The TI-85 and TI-86 calculators do not have a sequence mode, but graphs of sequences given by explicit formulas in terms of n can be plotted on these calculators with the following program. Key-by-key instructions for entering it are given at the end of the chapter.

1. PROGRAM: SEQUENCE
2. : ClrDraw
3. : FnOff
4. : Disp "USE $AN = y1(N)$"
5. : Disp "$N\emptyset =$"
6. : Input A
7. : Disp "MaxN ="
8. : Input B
9. : For$(N, A, B, 1)$
10. : $y1 \to Z$
11. : PtOn(N, Z)
12. : End

[†]This calculator does not allow negative values of n in sequence mode. Any terms a_n with $n < 0$ must be calculated and plotted by hand.

[‡]These calculators allow only positive values of n in sequence mode. Any values of a_n with $n \le 0$ must be calculated and plotted by hand.

To generate the graph of the portion of $\{a_n\}$ for $n_1 \le n \le n_2$, press 2nd QUIT GRAPH F1 and enter the formula for a_n as $y1(N)$ in the $y(x)=$ menu. The N is found with ALPHA N. Press F2 or M2 for <Window> and choose values of $x$min, $x$max, $y$min, and $y$max for a suitable viewing window. Then run the program, entering n_1 for $N\emptyset$ and n_2 for MaxN when requested.

To calculate the approximate decimal value of a_n for n equal to a particular integer k, press 2nd QUIT to go to the home screen. Press k STO▶ N 2nd : 2nd alpha Y ALPHA ALPHA 1 to have the screen read $k \to N : y1$. Press ENTER. To calculate another term, press ENTRY to recall the last string of commands, change k to the new value, and press ENTER again.

Question 2 Generate in suitable windows plots of the first twenty-five terms of the sequences **(a)** $\{n + 2(-1)^n\}_{n=0}^{\infty}$ and **(b)** $\{21n - n^2\}_{n=1}^{\infty}$. **(c)** Which of these sequences diverges to ∞ and which diverges to $-\infty$?

Question 3 **(a)** Explain the pattern of the points in the graph of $\left\{\cos\left(\frac{1}{3}n\pi\right)\right\}_{n=1}^{\infty}$ in Figure 4. **(b)** Does this sequence converge, diverge to ∞, diverge to $-\infty$, or diverge without a limit?

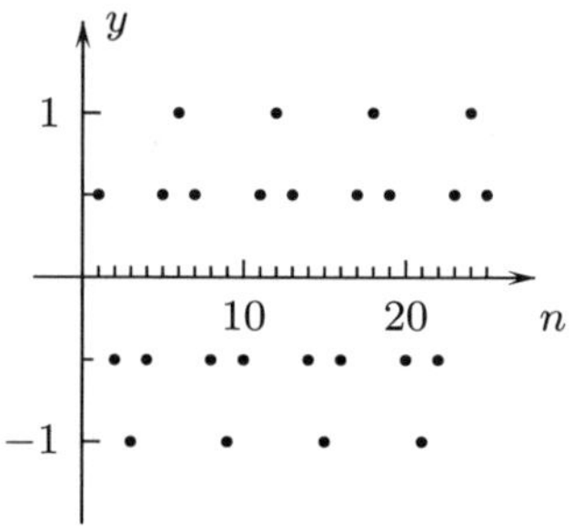

$y = \cos\left(\frac{1}{3}n\pi\right)$

FIGURE 4

ϵN- and YN-definitions

The following formal definitions of convergence and divergence to $\pm\infty$ of infinite sequences are modifications of the definitions of limits of functions $f(x)$ as $x \to \infty$ from Sections 1.2 and 3.2. These definitions play a much more important role here than they did in earlier chapters, where their main purpose was to clarify the meanings of the various types of limits which we found by examining formulas for the functions involved. In this chapter the definitions are needed to derive results that we will use to prove that limits exist in cases where we cannot determine directly from the formulas whether the limits are defined.

Definition 1 **(a)** *The sequence* $\{a_n\}_{n=n_0}^{\infty}$ *converges to the number* L *if for every positive* ϵ, *no matter how small, there is a number* N *such that*

$$|a_n - L| < \epsilon \quad \text{for } n > N.$$

(b) $\{a_n\}_{n=n_0}^{\infty}$ *diverges to* ∞ *if for every positive* Y, *no matter how large, there is a number* N *such that*

$$a_n > Y \quad \text{for } n > N.$$

(c) $\{a_n\}_{n=n_0}^{\infty}$ *diverges to* $-\infty$ *if for every negative* Y, *there is an integer* N *such that*

$$a_n < Y \quad \text{for } n > N.$$

An infinite sequence that does not converge, diverge to ∞, or diverge to $-\infty$ is said to DIVERGE WITHOUT A LIMIT or DIVERGE BY OSCILLATION.

Example 1 **(a)** Use Definition 1a to show that $\{(0.5)^n\}_{n=1}^{\infty}$ converges to 0. **(b)** Give the approximate decimal value of a suitable N in the definition for $\epsilon = 10^{-10}$. **(c)** Calculate the approximate decimal value of $(0.5)^n$ for the first integer n greater than the number N from part (b).

SOLUTION **(a)** Let ϵ be any positive number. We need to find a number N such that

$$|(0.5)^n| < \epsilon \quad \text{for } n > N. \tag{2}$$

We can solve this inequality using logarithms. Because $\ln x$ is an increasing function for all $x > 0$, inequality **(2)** is equivalent to the condition $\ln[(0.5)^n] < \ln(\epsilon)$, which is the same as $n \ln(0.5) < \ln(\epsilon)$. Dividing both sides of this inequality by the negative number $\ln(0.5)$ and reversing the direction of the inequality sign gives the equivalent inequality $n > \ln(\epsilon)/\ln(0.5)$. Therefore, **(2)** holds for $n > \ln(\epsilon)/\ln(0.5)$ and Definition 1a is satisfied with $N = \ln(\epsilon)/\ln(0.5)$.

(b) For $\epsilon = 10^{-10}$, we can use $N = \ln(10^{-10})/\ln(0.5) \doteq 33.22$.

(c) The first integer greater than the N from part (a) is $n = 34$, for which $(0.5)^{34} \doteq 5.8 \times 10^{-11}$. This is less than $\epsilon = 10^{-10}$, as required in **(2)**. □.

Question 4 **(a)** Use Definition 1b to show that $\{3^n\}_{n=1}^{\infty}$ diverges to ∞. **(b)** Give the approximate decimal value of a suitable N in the definition for $Y = 10^{10}$.
(c) Calculate the approximate value of 3^n for an integer n greater than the number N from part (b).

Bounded and monotone sequences

A sequence $\{a_n\}_{n=n_0}^{\infty}$ is said to be BOUNDED if there is a constant M such that $|a_n| \leq M$ for $n \geq n_0$. The sequences of Figures 1, 3, and 4 are bounded. The sequence of Figure 2 is not.

When we speak of an INTERVAL OF INTEGERS, we will mean an interval that contains at least two integers.

A sequence $\{a_n\}_{n=n_0}^{\infty}$ is said to be INCREASING for n in an interval of integers if $a_{n+1} > a_n$ for n in the interval; it is DECREASING for n in an interval of integers if $a_{n+1} < a_n$ for n in the interval; it is NONDECREASING for n in an interval of integers if $a_{n+1} \geq a_n$ for n in the interval; and it is NONINCREASING for n in an interval of integers if $a_{n+1} \leq a_n$ for n in the interval. The sequence of Figure 1 is decreasing for $n \geq 0$, and the sequence of Figure 2 is increasing for $n \geq 0$.

We say that $\{a_n\}_{n=n_0}^{\infty}$ is increasing, decreasing, nondecreasing, or nonincreasing if it has that property for all integers in its domain. A sequence that is either nonincreasing or nondecreasing is said to be MONOTONE.

Example 2 **(a)** What is the most extensive interval of integers n in which the sequence $\{21n - n^2\}_{n=1}^{\infty}$ of Figure 5 is decreasing? **(b)** What is the most extensive interval of integers in which is it nonincreasing?

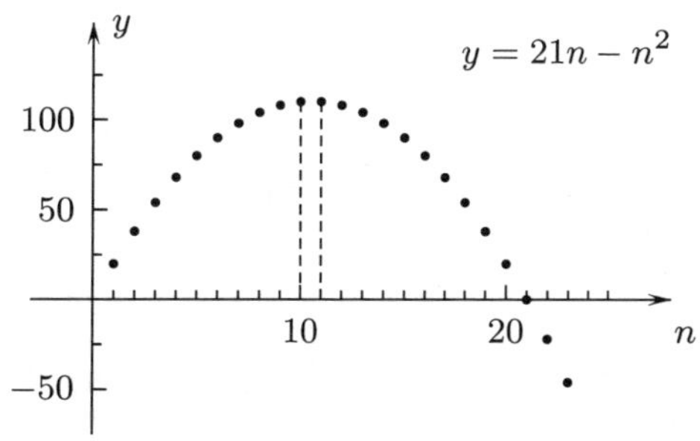

FIGURE 5

SOLUTION **(a)** To find where $21n - n^2$ is increasing and decreasing, we study the derivative of the function $f(x) = 21x - x^2$, defined for all x, whose value at $x = n$ is the value of the given sequence. The derivative

$$f'(x) = \frac{d}{dx}(21x - x^2) = 21 - 2x = 2(10.5 - x)$$

is zero at $x = 10.5$, is positive for $x < 10.5$, and is negative for $x > 10.5$. By the First-Derivative Test, $f(x)$ is increasing in $(-\infty, 10.5]$ and decreasing in $[10.5, \infty)$, so that the sequence $\{21n - n^2\}_{n=1}^{\infty}$ is decreasing for $n \geq 11$.

(b) Because $f(10) = 21(10) - 10^2 = 110$ and $f(11) = 21(11) - 11^2 = 110$ are equal, the sequence is nonincreasing for $n \geq 10$. □

Question 5 What are the most extensive integer intervals in which the sequence $\{ne^{-n^2/50}\}_{n=0}^{\infty}$ of Figure 6 is nonincreasing and in which it is nondecreasing?

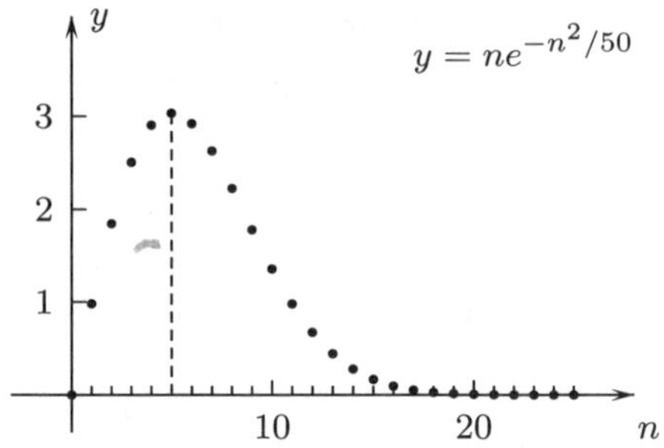

FIGURE 6

Least-upper and greatest-lower bounds

An UPPER BOUND of a sequence is a number M that is $\geq$ all of the terms of the sequence. A LOWER BOUND is a number m that is $\leq$ all of the terms of the sequence. M is the LEAST UPPER BOUND if it is the least of all upper bounds, and m is the GREATEST LOWER BOUND if it is the greatest of all lower bounds.

A basic axiom of mathematics states that if a set of real numbers, containing at least one number, has an upper bound, then it has a least upper bound and if it has a lower bound, then it has a greatest lower bound.[†]

Example 3 What are the least upper bound and the greatest lower bound of the sequence $\{30/(n+1)\}_{n=0}^{\infty}$ whose graph is in Figure 7? (Figure 7 is a duplicate of Figure 1.)

SOLUTION Because the sequence is decreasing for $n \geq 0$ and converges to 0, its least upper bound is its first term $30/(0+1) = 30$, and its greatest lower bound is its limit 0. □

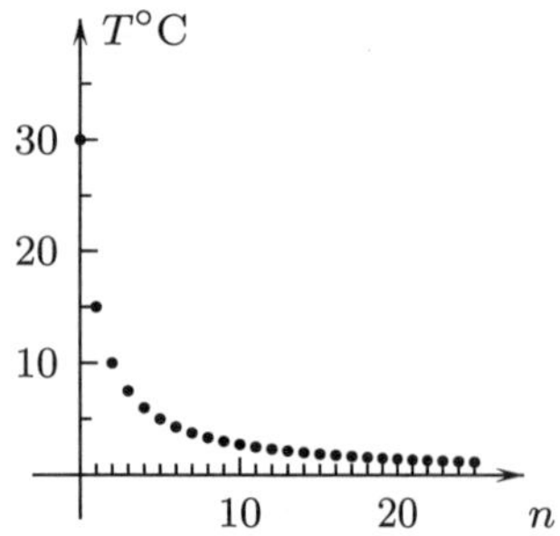

FIGURE 7

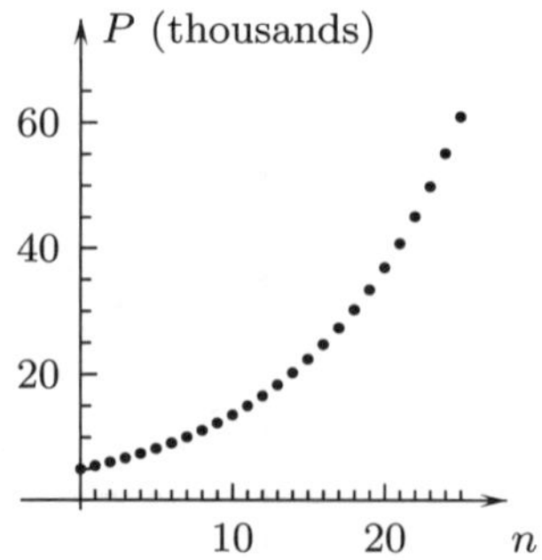

FIGURE 8

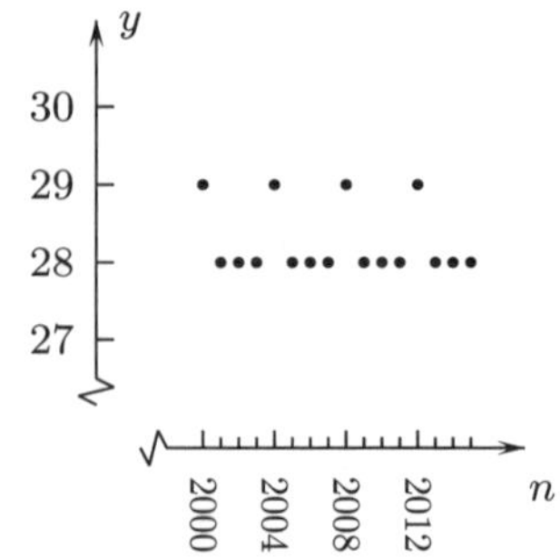

FIGURE 9

[†]This axiom is called the COMPLETENESS AXIOM for the real numbers.

Question 6 What are the least upper and greatest lower bounds of **(a)** $\{5e^{n/10}\}_{n=0}^{\infty}$, whose graph is in Figure 8 and of **(b)** $\{d_n\}_{n=1}^{\infty}$, defined in Question 1, whose graph is in Figure 9? (Figure 8 is a duplicate of Figure 2, and Figure 9 is a duplicate of Figure 3.)

The next Question combines the definitions of nonincreasing and of least upper bounds.

Question 7 Suppose that the sequence $\{a_n\}_{n=n_0}^{\infty}$ is bounded and nondecreasing and its least upper bound is M. Let ϵ denote an arbitrary positive number. **(a)** Why is $a_n \leq M$ for all $n \geq n_0$? **(b)** Why is there an integer N such that $a_N > M - \epsilon$? **(c)** Why, with the number N from part (b), do the inequalities $M - \epsilon < a_n \leq M$ hold for all $n > N$?

The response to Question 7 proves that if $\{a_n\}_{n=n_0}^{\infty}$ is bounded and nondecreasing, then it converges to its least upper bound. In the next Question we consider unbounded, nondecreasing sequences.

Question 8 Suppose that the sequence $\{a_n\}_{n=n_0}^{\infty}$ is nondecreasing but not bounded. Let Y denote an arbitrary positive number. **(a)** Why is there an integer N such that $a_N > Y$? **(b)** Why does the inequality $a_n > Y$ hold for all $n > N$?

The response to Question 8 proves that if $\{a_n\}_{n=n_0}^{\infty}$ is unbounded and nondecreasing, then it diverges to ∞, so with the response to Question 7, we have established part (a) of the following Theorem. Part (b) can be proved similarly. This Theorem will be the basis of many other results in later sections of this chapter.

Theorem 1 **(a)** *A nondecreasing sequence is either bounded and converges to its least upper bound or is unbounded and diverges to* ∞.
(b) *A nonincreasing sequence is either bounded and converges to its greatest lower bound or is unbounded and diverges to* $-\infty$.

Two cases of this theorem are illustrated by the sequence $\{6/(n+1)\}_{n=1}^{\infty}$ of Figure 7, which is decreasing and bounded and converges to its greastest lower bound 0, and by the sequence $\{n^{1/3}\}_{n=1}^{\infty}$ of Figure 8, which is increasing and unbounded and diverges to ∞. The theorem can also be applied in cases where a sequence $\{a_n\}_{n=0}^{\infty}$ is nondecreasing or nonincreasing for $n > n_1$ with some integer $n_1 > n_0$ since the values of a finite number of the terms does not affect the convergence or divergence of the sequence.

Iteration

Up to now we have considered only sequences defined by explicit formulas in terms of the index n. The sequences that will arise in the discrete models of population growth that we will discuss below will be defined instead by ITERATION. The first term of the sequence is given along with a procedure for calculating each subsequent term from the previous one. Before we discuss these applications, however, we will look at some examples of iteration. We start with a definition.

Definition 2 *The iterates of the number* x_0 *under the action of the function* $g(x)$ *are the numbers* $x_1, x_2, x_3, \ldots$ *defined by*

$$x_{n+1} = g(x_n) \quad \text{for } n = 0, 1, 2, \cdots. \tag{3}$$

The set of iterates, $x_0, x_1, x_2, \ldots$ *is called the* ORBIT *of* x_0 *under the action of* $g(x)$.

The procedures described above for calculating and plotting explicit sequences on Texas Instruments calculators can be modified as follows to calculate and plot sequences defined by iteration.

Plotting and calculating iterative sequences on a TI-82 calculator

To calculate and plot the iterates $x_0, x_1, \ldots, x_M$ of x_0 under the action of the function $g(x)$, follow the instructions given above to put the calculator in sequence and dot modes. Press Y = and enter the formula for $x_n = g(x_{n-1})$ as Un = $g(\mathrm{U}_{n-1})$. The symbol U_{n-1} is obtained with 2nd U_{n-1} (above 7). Press WINDOW. Set UnStart = x_0, nStart = 0, nMin = 0, and nMax = M. Pick Xmin, Xmax, Ymin, and Ymax for a suitable viewing window. Press GRAPH to generate the graph. The values can be seen with the <TRACE> command or in a table, as described above.

Plotting and calculating iterative sequences on a TI-83 calculator

To calculate and plot the iterates $x_0, x_1, \ldots, x_M$ of x_0 under the action of the function $g(x)$, follow the instructions given above to put the calculator in sequence and dot modes. Press Y = and set nMin = 0. Enter the formula for $x_n = g(x_{n-1})$ as $u(n) = g(u(n-1))$. The symbol u is obtained with 2nd u (above 7) and n is on the main variable ket X, T, θ, n. Set $u(n\text{Min}) = x_0$. Press WINDOW. Set nMin = 0, nMax = M. PlotStart = 1, and PlotStep = 1. Pick Xmin, Xmax, Ymin, and Ymax for a suitable viewing window. Press GRAPH to generate the graph. The values can be seen with the <TRACE> command or in a table, as described above.

Plotting and calculating iterative sequences on a TI-89 or TI-92 calculator

To calculate and plot the iterates $x_0, x_1, \ldots, x_M$ of x_0 under the action of the function $g(x)$, follow the instructions given above to put the calculator in sequence mode. Press ♦ Y = and enter the formula for $x_n = g(x_{n-1})$ as $u1 = g(u1(n-1))$. The symbol $u1$ is obtained with alpha u 1 and n with with alpha n . Set $ui1 = x_0$. Press ♦ WINDOW. Set nMin = 0, nMax = M. PlotStart = 1, and PlotStep = 1. Pick Xmin, Xmax, Ymin, and Ymax for a suitable viewing window. Press ♦ GRAPH to generate the graph. The values can be seen with the <TRACE> command or in a table, as described above.

Plotting and calculating iterative sequences on a TI-85 or TI-86 calculator

Sequences defined by iteration can be plotted and calculated on these calculators with the following program. Key-by-key instructions for entering it are given at the end of the chapter.

```
PROGRAM: ITERATION
:ClrDraw
:FnOff
:Disp "USE y1 = g(x)"
:Disp "x(Ø) ="
:Input Z
:Disp "MAXN ="
:Input B
:Disp "SHOW AFTER"
:Input M
:For(N,0,B,1)
:If N ≤ M
:Goto L
:Disp "N ="
:Disp N
:Disp "xN ="
:Disp Z
:Pause
:PtOn(N,Z)
:Pause
:Lbl L
:Z → x
:y1 → Z
:End
```

To make the calculations set $y1(x) = g(x)$ in the $y =$ menu and pick a suitable window. Run the program, giving the value of x_0, and a value of MAXN for the calculations to stop. If you want to plot and see all of the values, enter 0 after the prompt "SHOW AFTER." Give a larger integer to show only x_n for higher values of n. Press ENTER to see the successive values of n and x_n.

We consider first iterates under the square-root function $g(x) = \sqrt{x}$ with an initial value a_0 that is greater than 1.

Question 9 A sequence $\{a_n\}_{n=0}^{\infty}$ is defined by setting $a_0 = 25, a_1 = \sqrt{a_0} = \sqrt{25} = 5$ and, in general, for $n = 0, 1, 2, 3, \cdots$,

$$a_{n+1} = \sqrt{a_n}\,. \tag{4}$$

(a) Complete Table 2 below of values of a_n, defined by **(4)**. **(b)** What appears to happen to a_n as n increases?

TABLE 2

n	$a_n \doteq$	n	$a_n \doteq$	n	$a_n \doteq$
0	25	4	1.22284	8	
1	5	5		9	
2	2.23607	6		10	
3	1.49535	7		11	

The next Question deals with iteration by the square root function with an initial value between 0 and 1.

Question 10 Define the sequence $\{b_n\}_{n=1}^{\infty}$ by setting $b_0 = 0.16, b_1 = \sqrt{b_0} = 0.4$, and in general,

$$b_{n+1} = \sqrt{b_n} \quad \text{for } n = 0, 1, 2, \cdots. \tag{5}$$

(a) Complete Table 3 of values of b_n. **(b)** What appears to happen to b_n as n increases?

TABLE 3

n	$b_n \doteq$	n	$b_n \doteq$	n	$b_n \doteq$
0	0.16	4	0.89178	8	
1	0.4	5		9	
2	0.63246	6		10	
3	0.79527	7		11	

We can verify the predictions in Question 9 and 10 by finding explcit formulas for a_n and b_n in terms of n.

Example 4 Give a formula for a_n from Question 9 in terms of n and use it to show that $\{a_n\}_{n=0}^{\infty}$ converges to 1.

SOLUTION We use exponential notation and write

$$\begin{aligned}
a_0 &= 25 = 25^1 \\
a_1 &= \sqrt{a_0} = \sqrt{25} = 25^{1/2} \\
a_2 &= \sqrt{a_1} = \sqrt{25^{1/2}} = (25^{1/2})^{1/2} = 25^{1/4} \\
a_3 &= \sqrt{a_2} = \sqrt{25^{1/4}} = (25^{1/4})^{1/2} = 25^{1/8} \\
a_4 &= \sqrt{a_3} = \sqrt{a_3} = \sqrt{25^{1/8}} = (25^{1/8})^{1/2} = 25^{1/16}.
\end{aligned} \tag{6}$$

To find a general formula for a_n, we notice that the exponents on the right of the equations in **(5)** are $1 = 1/2^0$, $\frac{1}{8} = 1/2^3$, and $\frac{1}{16} = 1/2^4$. Consequently, we might expect that

$$a_n = 25^{1/2^n} \quad \text{for } n = 0, 1, 2, \cdots. \tag{7}$$

This is correct because, to calculate a_n, we start with 25 and take n square roots.

Finally, because $1/2^n \to 0$ as $n \to \infty$ and 25^x is a continuous function of x, it follows from formula **(7)** that

$$\lim_{n\to\infty} a_n = \lim_{n\to\infty} (25^{1/2^n}) = 25^0 = 1. \square$$

Question 11 Give a formula for the numbers b_n from Question 10 and use it to show that $\{b_n\}_{n=0}^{\infty}$ also converges to 1.

Web plots

The iteration process has in many cases a geometric interpretation which can help explain how the iterates are related and whether they have a limit as $n \to \infty$. We consider first the iterative sequence $\{b_n\}_{n=0}^{\infty}$ of Question 10, for which $g(x) = \sqrt{x}$, $a_0 = 0.16$, and $a_{n+1} = \sqrt{a_n}$ for $n \geq 0$.

To apply the function $y = \sqrt{x}$ to $b_0 = 0.16$, we plot b_0 on the x-axis. Then $b_1 = \sqrt{b_0} = 0.4$ is the y-coordinate of the corresponding point on the curve $y = \sqrt{x}$ (Figure 10).

Next, to apply the square root function to b_1 we need to plot it on the x-axis. We do this by moving horizontally from the point on the graph in Figure 10 to the line $y = x$ and then down to the x-axis as in Figure 11. Then $b_2 = \sqrt{b_1}$ is found by moving up to the curve $y = \sqrt{x}$, as in Figure 12.

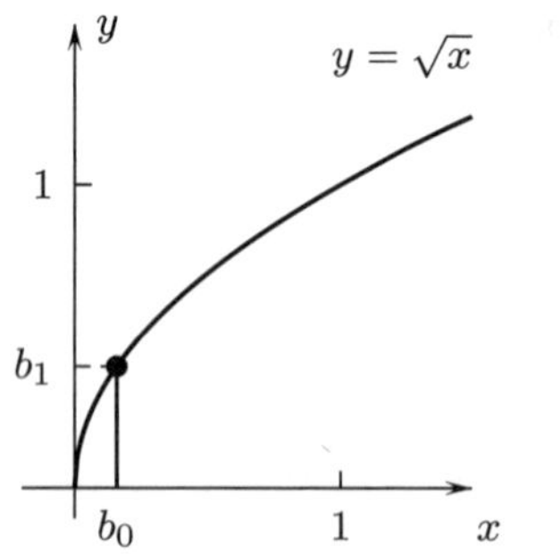

$b_0 = 0.16, b_1 = \sqrt{b_0}$

FIGURE 10

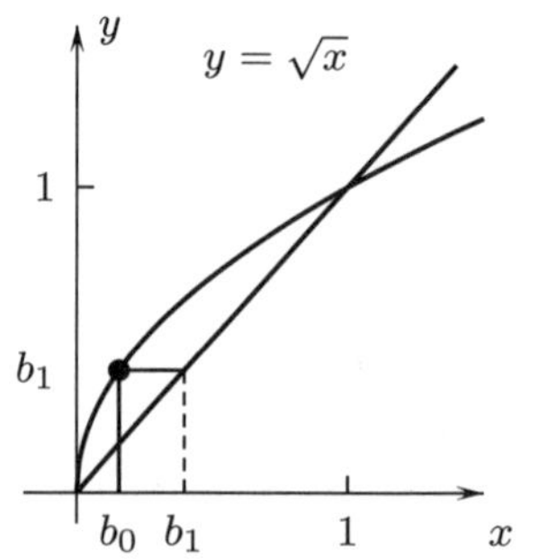

FIGURE 11

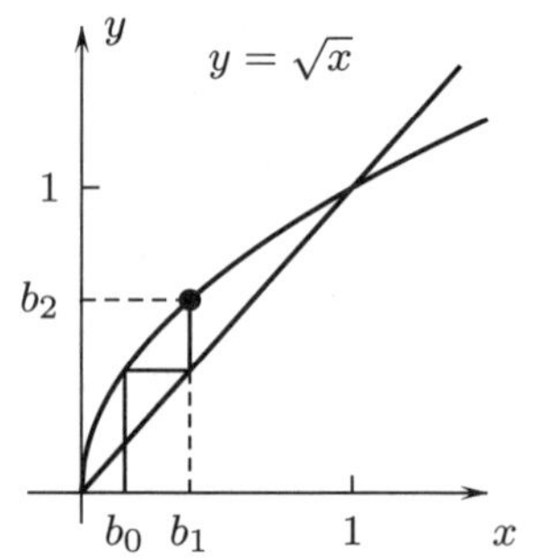

$b_2 = \sqrt{b_1}$

FIGURE 12

Figure 13 shows how b_3 is found from b_2. As the process continues, the zig-zag curve between $y = \sqrt{x}$ and $y = x$ approaches the intersection $(1, 1)$ of the two curves (Figure 14). This is why b_n tends to 1 as $n \to \infty$. The drawings in Figures 10 through 14 are often called WEB PLOTS or WEB DIAGRAMS.[†]

[†]Calculator instructions and programs for generating web plots willl be given in the *Calculator/Computer Supplement* of the First Edition.

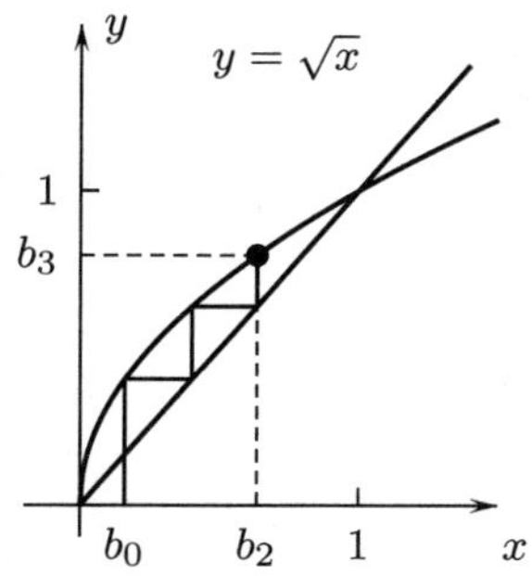

$b_3 = \sqrt{b_2}$

FIGURE 13

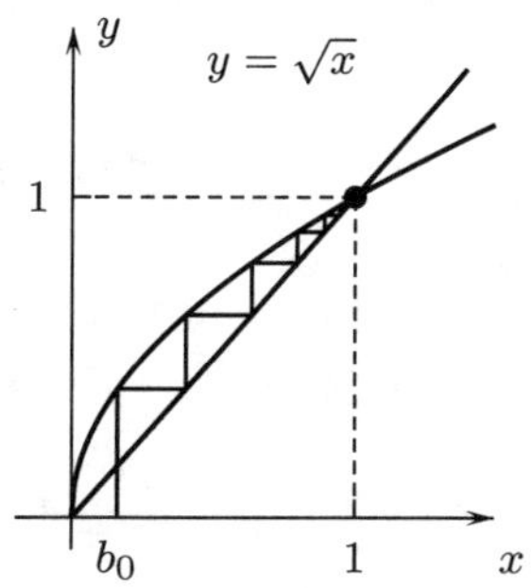

$b_n \to 1$ as $n \to \infty$

FIGURE 14

Question 9 dealt with the sequence $\{a_n\}_{n=0}^{\infty}$ obtained by taking repeated square roots starting with $a_0 = 25$. The web plot in Figure 15 shows how $a_2 = \sqrt{5}$ is obtained from $a_1 = \sqrt{25} = 5$, and Figure 16 shows how a_3 is found. This zig-zag curve would also approach $(1,1)$, as in Figure 17 as a_n tends to 1.

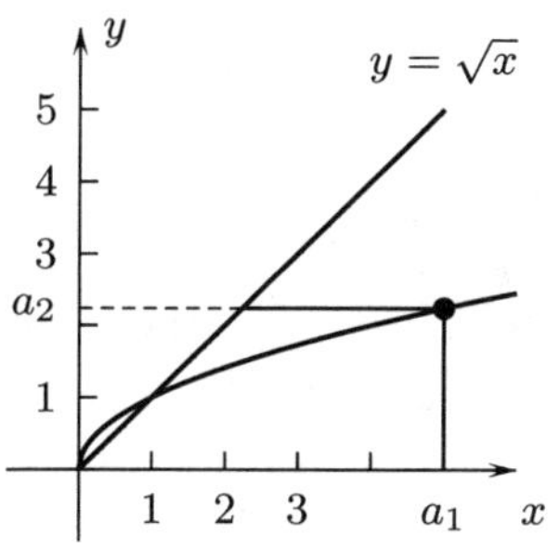

$a_1 = 5, a_2 = \sqrt{a_1}$

FIGURE 15

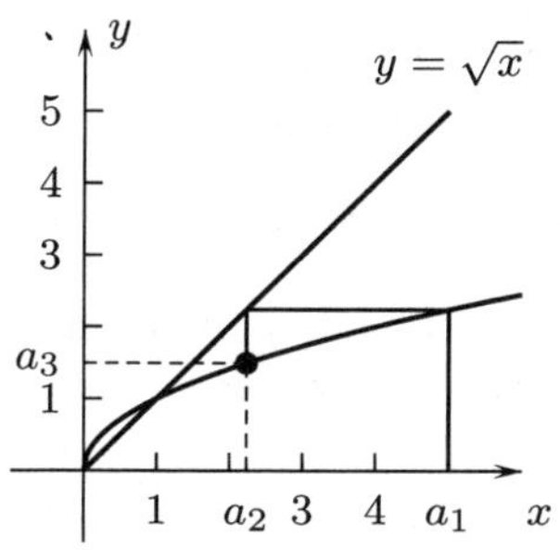

$a_3 = \sqrt{a_2}$

FIGURE 16

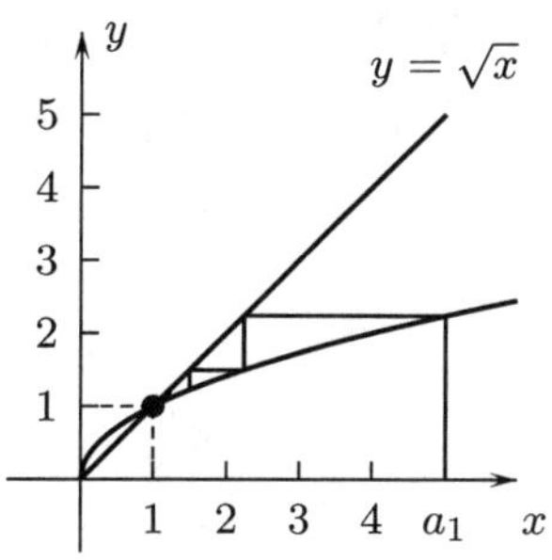

$a_n \to 1$ as $n \to \infty$

FIGURE 17

As you can see in Figures 10 through 17, the equation $\sqrt{x} = x$ has two solutions $x = 1$ and $x = 0$. These are called FIXED POINTS under iteration of the function $\sqrt{x}$ because all the iterates of $x_0 = 1$ are 1 and all iterates of $x_0 = 0$ are 0. We have just seen that the sequences $\{a_n\}_{n=0}^{\infty}$ and $\{b_n\}_{n=0}^{\infty}$, which are defined by iterating $a_0 = 25$ and $b_0 = 0.16$ approach the fixed point $x = 1$. Why do they not approach the other fixed point $x = 0$? This question is partially answered by the following theorem, whose proof is described at the end of the section

---

Theorem 2 *Suppose that $g(x)$ is continuous in the finite closed interval $[\alpha, \beta]$, such that*

$$\alpha \le g(x) \le \beta \quad \text{for } \alpha \le x \le \beta \tag{8}$$

and that there is a constant r with $0 < r < 1$ such that

$$|g'(x)| \le r \quad \text{for } \alpha \le x \le \beta. \tag{9}$$

Then for any choice of x_0 with $\alpha \le x \le \beta$ the sequence $\{x_n\}_{n=0}^{\infty}$ of iterates of x_0 defined by $x_{n+1} = g(x_n)$ for $n \ge 0$, converges to a solution x^* of $g(x) = x$, and this is the only solution with $\alpha \le x^* \le \beta$.

For the function $g(x) = \sqrt{x}$ that was used to construct the sequences $\{a_n\}$ and $\{b_n\}$ above,

$$g'(x) = \frac{d}{dx}(x^{1/2}) = -\tfrac{1}{2}x^{-1/2} = \frac{-1}{2\sqrt{x}} \quad \textbf{(10)}$$

and consequently, $|g'(x)| \approx \frac{1}{2}$ near $x = 1$ and we can apply the theorem there. In fact, $|g'(0.3)| = 1/(2\sqrt{0.3}) \doteq 0.9129$ and $|g'(x)|$ is a decreasing function of x, so that

$$|g'(x)| \leq 0.92 \quad \text{for } x \geq 0.35.$$

We can apply Theorem 2 with $g(x) = \sqrt{x}$ and $r = 0.92$ in the interval $[0.35, 25]$ to show that the sequences $\{a_n\}_{n=0}^{\infty}$ and $\{b_n\}_{n=1}^{\infty}$ both converge to the equilibrium point $x = 1$. It is called an ASYMPTOTICALLY STABLE fixed point because all iterates of numbers near it approach it.[†]

We cannot apply Theorem 2 with $g(x) = \sqrt{x}$ near the second equilibrium point $x = 0$, however, because, as **(10)** shows, $|g'(x)|$ is very large near $x = 0$. In fact, the only number whose iterates tend to 0 under the iteration is $x = 0$ itself. Consequently, the number 0 is called an ASYMPTOTICALLY UNSTABLE fixed point of the iteration.

Before we look at applications of these ideas to population growth, we will consider another example of an asymptotically stable fixed point.

Example 5 **(a)** Use graphs to show that the function

$$g(x) = \cos x \quad \textbf{(11)}$$

has exactly one fixed point. **(b)** Show that Theorem 2 can be applied in the interval $[0, 1]$ to prove that iterates $x_0, x_1, x_2, x_3, \cdots$ of $x_0 = 0.2$ under this function converge to the equibrium point.

SOLUTION **(a)** Figure 18 shows the curves $y = \cos x$ and $y = x$. They intersect only at one point because $\cos x$ decreases from 1 to -1 as x increases from 0 to π.

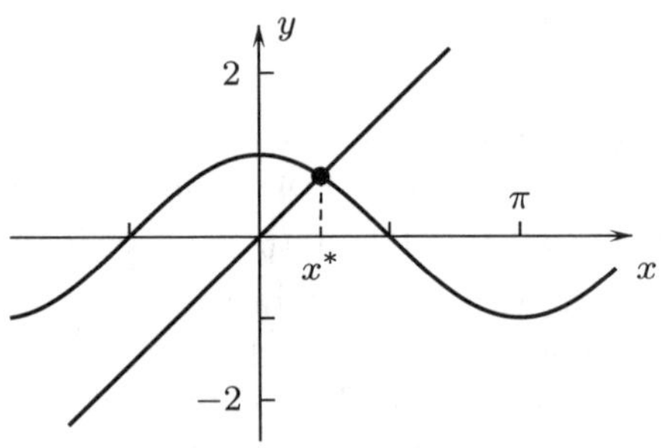

$y = \cos x$ and $y = x$
FIGURE 18

(b) We can use $\alpha = 0$, and $\beta = 1$ with $g(x) = \cos x$ in the theorem because $0 \leq \cos x \leq 1$ for $0 \leq x \leq 1$ (Figure 19). Also, for $0 \leq x \leq 1$, $|g'(x)| = |-\sin x| = \sin(x)$ is less than $\sin(1) \doteq 0.84$, so we can use $r = \sin(1)$. According to Theorem 2, if we set $x_0 = 0.2$ and $x_{n+1} = \cos(x_n)$ for $n \geq 0$, then the sequence $\left\{x_n\right\}_{n=0}^{\infty}$ converges to the fixed point x^*. □

[†]The term "asymptotically" is used here because we are studying the behavior of the iterations as $n \to \infty$.

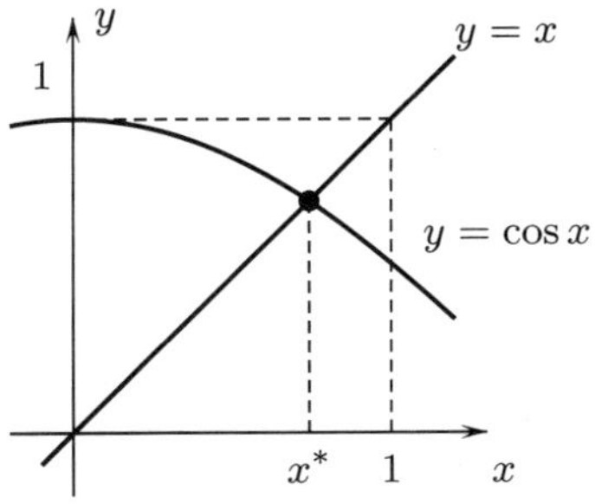

FIGURE 19

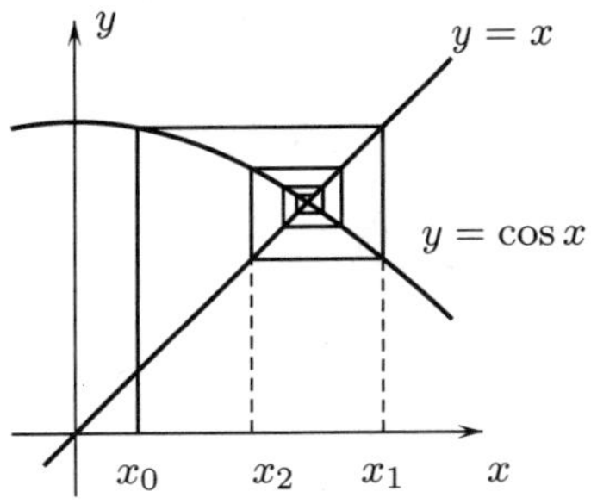

FIGURE 20

Figure 20 shows a web diagram for the iterates, and the table below give some values of their values. A calculator or computer can be used to show that the fixed point x^* has the approximate decimal value 0.73908513.

n	$x_n \doteq$
0	0.2
1	0.98007
2	0.55697
3	0.84886

n	$x_n \doteq$
10	0.73198
11	0.74385
12	0.73586
13	0.74125

n	$x_n \doteq$
20	0.73895
30	0.73908
40	0.73909
50	0.73909

Question 12 **(a)** What are the fixed points under iteration by the function $g(x) = x^2$? **(b)** Complete the tables below of values of the iterates with $x_0 = 0.9$ and $x_0 = 1.2$, respectively and use the results to predict which of the fixed points is asymptotically stable and which is asymptotically unstable. **(c)** How do the results of part (a) relate to Theorem 2?

n	$x_n \doteq$
0	0.9
1	0.81
2	0.6561

n	$x_n \doteq$
3	0.43047
4	
5	

n	$x_n \doteq$
6	
7	
8	

n	$x_n \doteq$
0	1.2
1	1.44
2	2.0736

n	$x_n \doteq$
3	
4	
5	

n	$x_n \doteq$
6	
7	
8	

A web diagram for the iteration in Question 12 with $x_0 = 0.9$ is shown in Figure 21, and a web diagram for $x_0 = 1.2$ is shown in Figure 22.

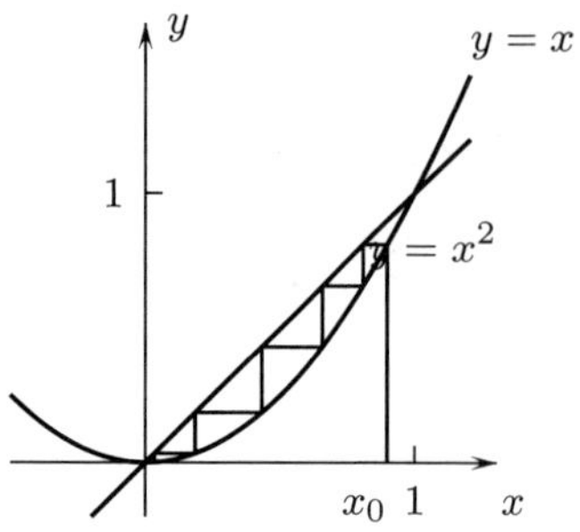

FIGURE 21

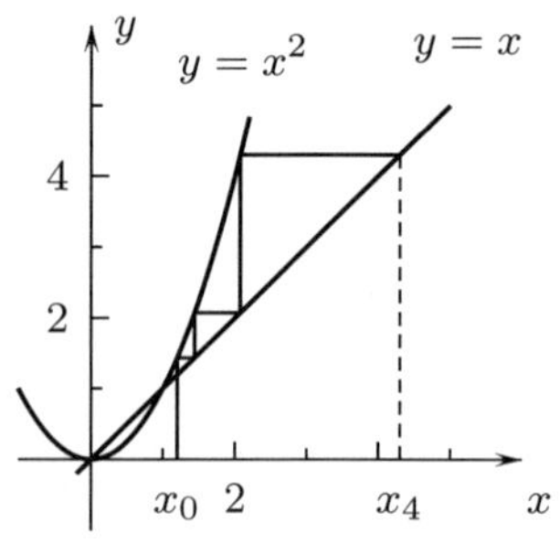

FIGURE 22

The discrete logistic equation

As we saw in Section 7.1, the LOGISTIC DIFFERENTIAL EQUATION,

$$\frac{dP}{dt} = kP(L - P) \tag{12}$$

with positive constants k and L is a mathematical model for a population $P = P(t)$ in a limited environment. Since the right side of **(12)** is zero for $P = L$, the constant population L is a solution of the differential equation and L is called a FIXED population. $P = 0$ is the other fixed population.

The logistic differential equation **(12)** can be solved by separating variables with integration by partial fractions to show that its nonzero solutions are

$$P = \frac{L}{1 + ce^{-kLt}} \tag{13}$$

with a parameter c determined by the population at one particular time. Equation **(13)** shows that every nonzero solution tends to L as $t \to \infty$, and that for fixed L and initial condition, increasing the constant k causes the solution to approach L more rapidly. This is illustrated in Figure 23 which shows the graphs of the solutions with $L = 2, P(0) = 0.1$ for $k = 0.25$ and $k = 1$.

FIGURE 23

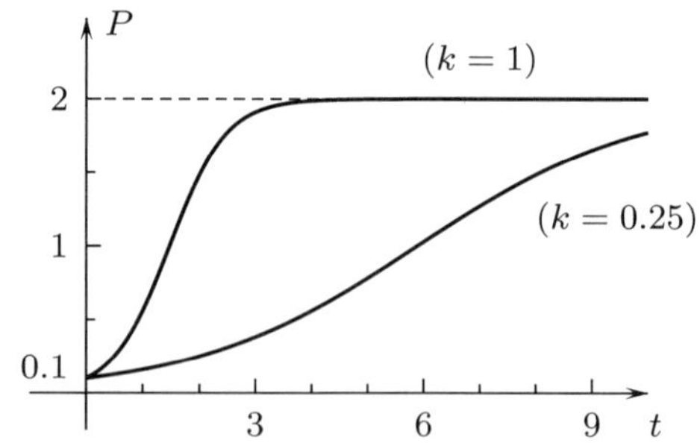

In **(12)**, the rate of change $\frac{dP}{dt}$ of the population is assumed to be proportional to the product $P(L - P)$ of the population P and the difference $L - P$ between the population and the equilibrium value L. We obtain a different sort of mathematical model if we look at how the population changes at fixed time increments rather than studying the population as a continuous function defined for all t in a time interval.

We might, for example, consider only how the population changes from one date in one year to the same date in the next year. To do this, we have P_0 denote the initial population, let P_1 be the population after one year, and, in general let P_n denote the population after n years. This yields a sequence $\{P_n\}_{n=0}^{\infty}$. Then, instead of assuming that the rate of change of the population P at each moment is proportional to $P(L-P)$, we assume that the change $P_{n+1}-P_n$ in the population from the nth to the $n+1$st year is proportional to the product $P_n(L-P_n)$ of the population P_n after n years and the difference $L-P_n$ between P_n and the fixed value. This gives the equation

$$P_{n+1}-P_n = kP_n(L-P_n)$$

which we write in the form

$$P_{n+1} = P_n + kP_n(L-P_n). \tag{14}$$

Here, as in **(13)**, k is a positive constant. Notice that in **(14)**, $P_{n+1}=g(P_n)$ with

$$g(x) = x + kx(L-x) \tag{15}$$

so the numbers $P_1, P_2, P_3, \cdots$ are obtained from P_0 by iterating this function $g(x)$.

Equation **(15)** is called the DISCRETE LOGISTIC EQUATION because it involves sequences rather than differentiable functions as in the Logistic Differential Equation. In our first example of the discrete equation, the population tends to a fixed population.

Question 13 A population that is P_0 (millions) initially and $P_1, P_2, P_3, \cdots$ in subsequent years, satisfies the discrete logistic equation

$$P_{n+1} = P_n + 0.5P_n(2-P_n). \tag{16}$$

(a) What is the sequence $P_0, P_1, P_2, \cdots$ if $P_0 = 2$? **(b)** Complete the following table with values of P_n with the initial value $P_0 = 0.45$.

TABLE 4

n	$P_n \doteq$	n	$P_n \doteq$
0	0.45	5	
1	0.79875	6	
2	1.278499	7	
3		8	
4		9	

Figure 24 shows how the values of P_n in Table 4 are calculated in a web plot.

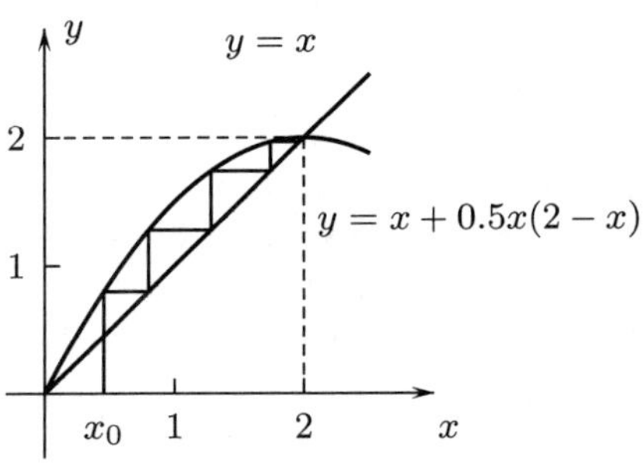

FIGURE 24

Now we come to something of a surprise. Increasing the constant k in the logistic differential equation **(12)** only causes the solutions $P(t)$ to approach the equilibrium population L more rapidly. Increasing the value of k in the discrete logistic equation **(14)**, however, leads to solutions with very different behavior. We first change the value of k in **(14)** from 0.5 to 1.2 to have

$$P_{n+1} = P_n + 1.2P_n(2 - P_n) \tag{17}$$

and we use the same intial condition as in Question 13.

Question 14 Complete Table 5 of values of P_n, defined by **(17)** with the intial value $P_0 = 0.45$.

TABLE 5

n	$P_n \doteq$	n	$P_n \doteq$
0	0.45	5	
1	1.287	6	
2	2.388157	7	
3		8	
4		9	

Figure 25 shows a web plot of the sequence in Table 5 by iteration. As Table 5 suggests, the numbers P_n for odd n approach one number P^* (whose value is $\doteq 1.280563$) while the numbers P_n for even n approach another number ($P^{**} \doteq 2.386104$). This means that as time goes on, the population determined by the mathematical model **(17)** jumps back and forth between about 1.28 million in odd years and about 2.39 million in even years, instead of approaching, the fixed population of 2 million, as would be the case with the logistic differential equation. We say that the orbit of $P_0 = 0.45$ approaches the two values P^* and P^{**}.

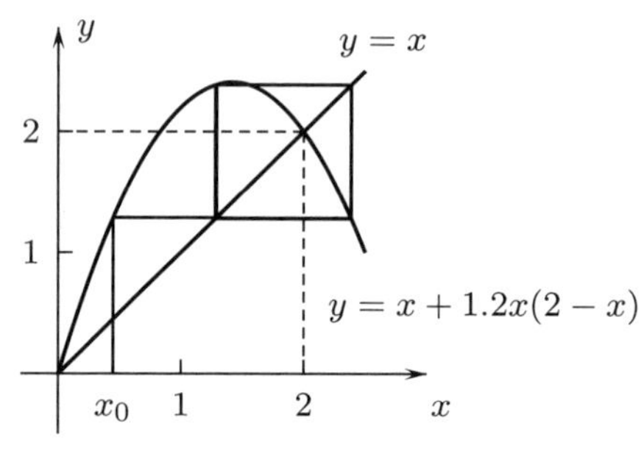

FIGURE 25

If k is increased somewhat more, the orbits approach 4 points, then 8 points, then 16 points, etc., until eventually, when k is large enough, the orbits are CHAOTIC and have very complex and unpredictable behavior.

Proof of Theorem 2

Theorem 1 on the convergence of certain sequences constructed by iteration can be established by using the Intermediate-Value Theorem of Section 2.2 and the Mean-Value Theorem of Section 3.1. Under the conditions of the theorem, $g(x)$ and therefore, $h(x) = g(x) - x$ are continuous in $[\alpha, \beta]$. Also, since $\alpha \le g(x) \le \beta$ in this interval, $h(\alpha) = g(\alpha) - \alpha$ is ≥ 0 and $h(\beta) = g(\beta) - \beta$ is ≤ 0, so that, by the Intermediate-Value Theorem, $h(x^*) = 0$ for some x^* with $\alpha \le x^* \le \beta$. Then $g(x^*) = x^*$.

To show that $x_n \to x^*$, we use the equation $x^* = g(x^*)$ and the definition $x_{n+1} = g(x_n)$ of x_{n+1}. We obtain

$$x_{n+1} - x^* = g(x_n) - g(x^*). \tag{18}$$

By the Mean-Value Theorem, there is a number c between x_n and x^* such that $g(x_n) - g(x^*) = g'(c)(x_n - x^*)$. With **(18)**, this gives

$$|x_{n+1} - x^*| = |g'(c)||x_n - x^*|$$

and since by the hypothesis of the theorem $g'(c) \le r$ with a positive constant $r < 1$, we obtain

$$|x_{n+1} - x^*| \le r|x_n - x^*|. \tag{19}$$

Applying **(19)** repeatedly shows that $|x_1 - x^*| \le r|x_0 - x^*|$, $|x_2 - x^*| \le r|x_1 - x^*| \le r^2|x_0 - x^*|, |x_3 - x^*| \le r|x_2 - x^*| \le r^3|x_0 - x^*|$, and in general,

$$|x_n - x^*| \le r^n|x_0 - x^*| \quad \text{for } n \ge 1. \tag{20}$$

Since $r^n \to 0$ as $n \to \infty$ and $|x_0 - x^*|$ is a fixed number, inequality **(20)** shows that $x_n \to x^*$ as $n \to \infty$. The number x^* is the only solution of $g(x) = x$ in $[a, b]$ because if x^{**} is another solution in the interval, we would have

$$\begin{aligned} |x^{**} - x^*| &= |g(x^{**}) - g(x^*)| = |g'(c)||x^{**} - x^*| \\ &\le r|x^{**} - x^*|. \end{aligned} \tag{21}$$

Here we have used the Mean-Value Theorem again. Inequalities **(21)** imply that $x^{**} - x^* = 0$ and $x^{**} = x^*$ since r is less than 1. This completes the proof.

Principles and procedures

- The most important result of this section is Theorem 1, which states that nondecreasing sequences either converge to diverge to ∞ and that nonincreasing sequences either converge or diverge to $-\infty$. This principle will be the basis of many of the convergence tests for infinite sums in Sections 8.3 and 8.4.
- The terms "nondecreasing," "nonincreasing," "least upper bound," and "greatest lower bound" can be confusing because you have to read them carefully to see their actual meanings. For instance, increasing sequences are nondecreasing, and the "least upper bound" of a set of numbers might sound like a number that is less than the numbers in the set, while in fact it is an upper bound and is $\ge$ all numbers in the set.
- The Discrete Logistic Equation and other mathematical models involving iteration and the asssociated problems involving "chaos" consititute a large branch of mathematics that we can only illustrate here with a few examples.

Tune-Up Exercises 8.1[♦]

[A]*Answer provided.* [O]*Outline of solution provided.*

In Exercises T1 through T4, **(a)** calculate approximate decimal values of the first six terms of the sequence, **(b)** generate the graph of the first twenty-five terms in a suitable window and copy it on your paper and **(c)** determine whether the sequence converges, diverges to ∞, diverges to $-\infty$ or diverges without a limit. If it converges, give its limit.

T1.[A] $\left\{2 + \dfrac{(-1)^n}{n}\right\}_{n=1}^{\infty}$.

♦ Type 1, convergence, etc.

T2.[A] $\left\{\dfrac{n^2 - 5n}{n^2 + 1}\right\}_{n=1}^{\infty}$.

♦ Type 1, convergence, etc.

T3.[A] $\left\{n + (-1)^n\sqrt{n}\right\}_{n=1}^{\infty}$.

♦ Type 1, convergence, etc.

T4.[A] $\left\{n(-1)^n\right\}_{n=1}^{\infty}$.

♦ Type 1, convergence, etc.

[♦]The Tune-up Exercises and Problems are classified by type and content. The types are (1) basic, reactive; (2) basic reflective; (3) intermediate, reactive; (4) intermediate, reflective; (5) advanced, reactive; (6) advanced, reflective; and (7) advanced, theoretical.

T5.[O] **(a)** Generate the graph of the first thirteen terms of $\{9n-n^2\}_{n=1}^{\infty}$ in a suitable window and copy it on your paper. **(b)** Determine the most extensive intervals of integers where the sequence is increasing, nondecreasing, decreasing, and nonincreasing. **(c)** What are the least upper bound and greatest lower bound of the sequence?

◆ Type 2, monotonicity, least upper bound, and greatest lower bound

T6.[O] **(a)** Calculate approximate decimal values of the first seven iterates of $a_0 = 1$ under the function $g(x) = 1 + 1/x$. **(b)** Does the sequence of iterates seem to converge?

◆ Type 1, iteration

T7.[O] **(a)** Calculate approximate decimal values of the first five iterates of $a_0 = 1$ under the function $g(x) = 4x^2$. **(b)** Does the sequence of iterates seem to converge?

◆ Type 1, iteration

Problems 8.1

[A]*Answer provided.* [O]*Outline of solution provided.*

In Problems 1 through 6, **(a)** calculate approximate decimal values of the first six terms of the sequence, **(b)** generate the graph of the first twenty-five terms in a suitable window and copy it on your paper and **(c)** determine whether the sequence converges, diverges to ∞, diverges to $-\infty$ or diverges without a limit. If it converges, give its limit.

1.[O] $\left\{\dfrac{n^2+1}{5n}\right\}_{n=1}^{\infty}$

◆ Type 1, convergence, etc.

2.[O] $\left\{\dfrac{(-1)^n}{n}\right\}_{n=1}^{\infty}$

◆ Type 1, convergence, etc.

3. $\left\{\dfrac{n-n^3}{2n^3+1}\right\}_{n=1}^{\infty}$

◆ Type 1, convergence, etc.

4. $\left\{\dfrac{1.5^n+1}{1.5^n}\right\}_{n=1}^{\infty}$

◆ Type 1, convergence, etc.

5. $\left\{\dfrac{100n}{50+n^3}\right\}_{n=1}^{\infty}$

◆ Type 1, convergence, etc.

6. $\left\{2+\sin(\tfrac{1}{2}n\pi)\right\}_{n=1}^{\infty}$

◆ Type 1, convergence, etc.

7. Are the sequences **(a**[A]**)** $\left\{\tan^{-1} n\right\}_{n=1}^{\infty}$ and **(b)** $\left\{\dfrac{10-n}{1+n}\right\}_{n=1}^{\infty}$ increasing or decreasing?

◆ Type 1, monotonicity

8.[A] What are the most extensive intervals of integers in which $\left\{n+\dfrac{36}{n}\right\}_{n=1}^{\infty}$ is increasing and in which it is decreasing?

◆ Type 1, monotonicity

9. What are the most extensive intervals of integers in which $\left\{n^5e^{-n}\right\}_{n=1}^{\infty}$ is increasing and in which it is decreasing?

◆ Type 1, monotonicity

Find the the least upper bounds and greatest lower bounds of the sequences in Problems 10 through 17.

10.[O] $\left\{(-1)^n(1-e^{-n/10})\right\}_{n=1}^{\infty}$

◆ Type 1, least upper and greatest lower bounds

11.[A] $\left\{(-1)^n+\dfrac{1}{n}\right\}_{n=1}^{\infty}$

◆ Type 1, least upper and greatest lower bounds

12. $\left\{1+\dfrac{1}{n}\right\}_{n=1}^{\infty}$

◆ Type 1, least upper and greatest lower bounds

13. $\left\{-2-\dfrac{3}{n}\right\}_{n=1}^{\infty}$

◆ Type 1, least upper and greatest lower bounds

14. $\left\{n+(-1)^n\right\}_{n=1}^{\infty}$

◆ Type 1, least upper and greatest lower bounds

15. $\left\{\dfrac{1+(-1)^n}{n}\right\}_{n=1}^{\infty}$

◆ Type 1, least upper and greatest lower bounds

16.[A] $\left\{\tan^{-1} n\right\}_{n=1}^{\infty}$

◆ Type 1, least upper and greatest lower bounds

17. $\left\{(-n)^n\right\}_{n=1}^{\infty}$

◆ Type 1, least upper and greatest lower bounds

18. Predict the limit of $\left\{\dfrac{(n!)^2}{(n^2)!}\right\}_{n=1}^{\infty}$ by calculating the approximate decimal values of the first five terms.

◆ Type 1, predicting a limit

19. The sequence $\left\{\dfrac{(2n)!}{(n!)^2}\right\}_{n=1}^{\infty}$ is increasing and diverges to ∞. How large must n be for the *nth* term to be greater than 10^{10}?

♦ Type 1, the YN-definition of divergence to infinity

20.[O] Find the limits of **(a)** $\{a_n\}_{n=3}^{\infty}$ **(b)** $\{b_n\}_{n=3}^{\infty}$, and **(c)** $\{a_n + b_n\}_{n=3}^{\infty}$ where $a_n = \dfrac{6n}{2n+1}$ and $b_n = \dfrac{3n}{2-n}$.

♦ Type 1, linear combinations of sequences

21. Find the limits of **(a)** $\{a_n\}_{n=3}^{\infty}$ **(b)** $\{b_n\}_{n=3}^{\infty}$, and **(c)** $\{a_n - b_n\}_{n=3}^{\infty}$ where $a_n = \dfrac{2n^2}{n+3}$ and $b_n = \dfrac{4n^2}{2n+1}$.

♦ Type 1, linear combinations of sequences

22.[A] **(a)** What are the equilibrium points under iteration by the function $g(x) = 4x^{1/3}$? **(b)** Calculate the first ten iterates of $x_0 = 0.01$ under the function of part (a). **(c)** Which equilibrium points of $g(x)$ seem to be asymptotically stable?

♦ Type 2, stable and unstable equilibrium points

23.[A] **(a)** What is the positive equilibrium point under iteration by the function $g(x) = 1 + 2/x$? **(b)** Calculate the first nine iterates of $x_0 = 1$ under the function of part (a). **(c)** Does the equilibrium point of $g(x)$ from part (a) seem to be asymptotically stable?

♦ Type 2, stable and unstable equilibrium points

24. **(a)** What are the two equilibrium points under iteration by the function $g(x) = \frac{1}{2}xe^x$? **(b)** Calculate the first few iterates of $x_0 = 1$ under the function of part (a). **(c)** Does the positive equilibrium point of $g(x)$ from part (a) seem to be asymptotically stable? **(d)** Calculate the first nine iterates of $x_0 = 0.5$ under the function of part (a). **(c)** Does 0 seem to be an asymptotically stable equilibrium point of $g(x)$?

♦ Type 2, stable and unstable equilibrium points

25. Show with numerical calculations that the orbit of $x_0 = 0.5$ under $g(x) = kx(1-x)$ seems to approach **(a)** one point if $k = 2.8$, **(b)** two points if $k = 3.2$, and **(c)** four points if $k = 3.5$.

♦ Type 2, orbits, etc.

26. What are the orbits of **(a)** $x_0 = 2$, **(b)** $x_0 = 4$, and **(c)** $x_0 = 1$ under $g(x) = 5 - x$. **(d)** Explain the results of parts (a), (b) and (c) by analyzing the composite function $g(g(x))$.

♦ Type 2, orbits, etc.

Use logarithms and l'Hopital's Rule as required in Problems 27 through 36 to determine whether the sequences converge and to find their limits if they do converge.

27.[O] $\left\{\left(\dfrac{1}{n}\right)^n\right\}_{n=1}^{\infty}$

♦ Type 3, convergence

28.[O] $\left\{\left(3 - \dfrac{1}{n}\right)^n\right\}_{n=1}^{\infty}$

♦ Type 3, convergence

29.[O] $\left\{(1+n)^{1/n}\right\}_{n=1}^{\infty}$

♦ Type 3, convergence

30. $\left\{\left(\dfrac{1}{n}\right)^{1/n}\right\}_{n=1}^{\infty}$

♦ Type 3, convergence

31.[O] $\left\{n \sin\left(\dfrac{1}{n}\right)\right\}_{n=1}^{\infty}$

♦ Type 3, convergence

32. $\left\{n^2 \sin\left(\dfrac{1}{n}\right)\right\}_{n=1}^{\infty}$

♦ Type 3, convergence

33. $\left\{ne^{-n}\right\}_{n=1}^{\infty}$

♦ Type 3, convergence

34. $\left\{\dfrac{\ln n}{n}\right\}_{n=1}^{\infty}$

♦ Type 3, convergence

35. $\left\{n \sin\left(\dfrac{5}{n}\right)\right\}_{n=1}^{\infty}$

♦ Type 3, convergence

36. $\left\{\dfrac{e^n}{n^2}\right\}_{n=1}^{\infty}$

♦ Type 3, convergence

37. What are the most extensive intervals of integers in which **(a)** $\{n\ln(1/n)\}_{n=1}^{\infty}$ and **(b)** $\left\{\dfrac{\ln n}{n}\right\}_{n=1}^{\infty}$ are decreasing?

♦ Type 3, intervals of increase and decrease

38.[O] What is the limit of $\left\{\left(1-\frac{2}{n}\right)^n\right\}_{n=1}^{\infty}$?

♦ Type 3, a limit

39 What is the limit of $\left\{\left(\frac{1+5/n}{1-6/n}\right)^n\right\}_{n=1}^{\infty}$?

♦ Type 3, a limit

40. **(a**[A]**)** What is the perimeter P_n of a polygon with n equal sides that is inscribed in a circle of radius R? (Divide the polygon into $2n$ congruent right triangles with vertices at the center of the circle.) **(b)** Show that the limit of $\{P_n\}$ is the circumference of the circle.

♦ Type 3, a limit, etc.

41. **(a)** What is the area A_n of the polygon in Problem 40? **(b)** Show that the limit of $\{A_n\}$ is the area of the circle.

♦ Type 3, a limit, etc.

42. Prove that $\left\{n \sin\left(\frac{1}{n}\right)\right\}_{n=1}^{\infty}$ is an increasing sequence. (First show that $x\cos x - \sin x < 0$ for $0 < x \le \frac{1}{2}\pi$ and then that $(\sin x)/x$ is decreasing for $0 < x \le \frac{1}{2}\pi$.)

♦ Type 4, an increasing sequence

43.[O] The limit of $n^{1/n}$ as $n \to \infty$ is 1. Find a positive integer n such that $n^{1/n}$ differs from its limit as $n \to \infty$ by less than 0.01 and by more than 0.001.

♦ Type 3, the epsilon-N-definition of convergence

44. The limit of e^{n^2} as $n \to \infty$ is ∞. Find the smallest positive integer n such that $e^{n^2} > 10^{10}$.

♦ Type 3, the YN-definition of divergence to infinity

Use Definition 1 to verify the limits in Problems 45 through 48.

45.[A] $\lim_{n\to\infty} \frac{1}{n^2} = 0$

♦ Type 5, the epsilon-N-definition of convergence

46.[A] $\lim_{n\to\infty} e^{1/n} = 1$

♦ Type 5, the epsilon-N-definition of convergence

47. $\lim_{n\to\infty} \left(1-\frac{1}{n}\right)^{100} = 1$

♦ Type 5, the epsilon-N-definition of convergence

48. $\lim_{n\to\infty} (-\sqrt{n}) = -\infty$

♦ Type 5, a YN-definition of divergence

49. $\lim_{n\to\infty} \frac{2n^2}{n^2+1} = 2$

♦ Type 5, the epsilon-N-definition of convergence

50. $\lim_{n\to\infty} \frac{\ln n}{\ln n - 1} = 1$

♦ Type 5, the epsilon-N-definition of convergence

51. Use Definition 1a with $\epsilon = 1$ to prove that if an infinite sequence converges, then it is bounded.

♦ Type 5, the epsilon-N-definition of convergence

52.[A] Complete the following sentence: "The statement $\lim_{n\to\infty} a_n = L$ with a number L is false if and only if there is a positive number ϵ such that for every positive number N . . . ."

♦ Type 5, the epsilon-N-definition of convergence

53.[A] Use the criterion in Problem 52 to prove that the statement $\lim_{n\to\infty} \left(1+\frac{1}{n}\right) = 1.0001$ is false.

♦ Type 5, the epsilon-N-definition of convergence

54.[A] Use the criterion in Problem 52 to prove that the statement $\lim_{n\to\infty} (-1)^n = 1$ is false.

♦ Type 5, the epsilon-N-definition of convergence

55. The FIBONACCI SEQUENCE $\{F_n\}_{n=1}^{\infty}$ is defined by setting $F_1 = 1, F_2 = 1$, and defining each F_n for $n \ge 3$ to be the sum of the previous two terms in the sequence: $F_n = F_{n-2} + F_{n-1}$. **(a)** Calculate the first ten Fibonacci numbers. **(b)** Show that the ratio $r_n = \frac{F_n}{F_{n-1}}$ of successive Fibonacci numbers $(n = 2, 3, 4, \ldots)$ can be obtained from $r_2 = 1$ by iteration under the function $g(x) = 1 + 1/x$. **(c)** Express $r_{n+2} - r_n$ in terms of $r_{n+1} - r_{n-1}$ and then in terms of $r_n - r_{n-2}$. **(d)** Use the results of part (c) to show that the numbers r_n with even n form an increasing bounded sequence and those with odd n form a decreasing bounded sequence so both of these subsequences converge. **(e)** Show that r_n with even and odd n have the same limit. **(f)** A GOLDEN RECTANGLE is one of sides w and h with $w > h$ such that it is congruent to the rectangle with sides h and $w - h$ obtained by removing a square. The ratio w/h of the sides of a golden rectangle is the GOLDEN MEAN. Show that the golden mean is $\frac{1}{2}(1+\sqrt{5})$ and that it is the limit of the ratio r_n of the Fibonacci numbers.

♦ Type 4, Fibonacci sequence

Section 8.2

Geometric series

OVERVIEW: *In many contexts the terms "sequence" and "series" are interchangeable in everyday English. They both denote a succession of things or events, such as a sequence or series of calamities, payments, or publications. These terms are, however, given very different meanings in calculus, where "sequence" denotes a finite or infinite string of numbers and where a "series" is a finite or infinite sum. In this section we look at a particular type of series, called* GEOMETRIC SERIES, *which arises frequently in applications.*

Topics:

- ***Finite geometric series***
- ***Infinite geometric series***
- ***Infinite repeating decimals***

Finite geometric series

We start with three Questions that show the patterns in finite geometric series.

Question 1 Suppose you want to go from a point A toward a second point B two miles away. First you go one mile. Then you go a half mile further for a total of $1 + \frac{1}{2}$ miles. Next you go half that distance, and so forth, so that at each stage you go half as far as you did in the previous stage. **(a)** Complete Table 1 below by calculating how far you go in three, five, and ten stages. **(b)** How far do you think you would go in an infinite number of stages?

TABLE 1

Stages	Total distance	Decimal approximation
1	1	1.0
1–2	$1 + \frac{1}{2}$	1.5
1–3	$1 + \frac{1}{2} + (\frac{1}{2})^2$	1.75
1–5	$1 + \frac{1}{2} + (\frac{1}{2})^2 + (\frac{1}{2})^3 + (\frac{1}{2})^4$	
1–10		

You could anticipate the results of Question 1 by visualizing the distances on an s-axis with point A at the origin and point B at $s = 2$, as in Figure 1. In the first step you go one mile from A to the dot at $s = 1$ in Figure 1, then you go one-half mile farther to the second dot at $s = 1.5 = 1\frac{1}{2}$ in Figure 2, and so forth. Figure 3 shows dots corresponding to the next three stages, when you are at $s = 1.75 = 1\frac{3}{4}, 1.875 = 1\frac{7}{8}$, and $1.9375 = 1\frac{15}{16}$. At each stage, you go half the remaining distance to B, so that you approach B as the number of stages tends to ∞.

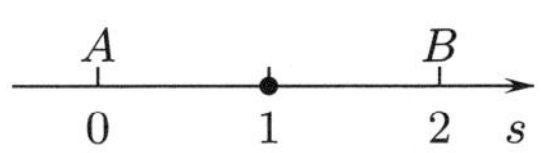

FIGURE 1

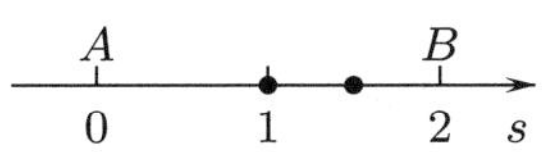

FIGURE 2

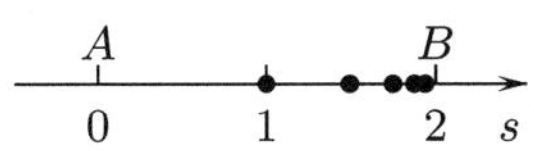

FIGURE 3

♦ SUGGESTIONS TO INSTRUCTORS: Have students begin work on Questions 1 through 13 before class. Problems 23, 28, 30, and 33–37 are good for in-class discussion.

Question 2 Now suppose you go one mile in the first stage, one-fourth mile in the second stage, and in each subsequent stage one-fourth as far as the previous stage. **(a)** Complete Table 2 to show how far you go in three, five, and eight stages. Use powers of $\frac{1}{4}$ in the sums. **(b)** How far do you think you would go in an infinite number of stages?

TABLE 2

Stages	Distance	Decimal approximation
1	1	1.0
1–2	$1 + \frac{1}{4}$	1.25
1–3	$1 + \frac{1}{4} + (\frac{1}{4})^2$	1.3125
1–5	$1 + \frac{1}{4} + (\frac{1}{4})^2 + (\frac{1}{4})^3 + (\frac{1}{4})^4$	
1–8		

Question 3 This time suppose you go one mile from A toward B in the first stage, but then $\frac{1}{2}$ mile backwards toward A in the second stage, $(\frac{1}{2})^2 = \frac{1}{4}$ mile forward in the third stage, $(\frac{1}{2})^3 = \frac{1}{8}$ mile backward in the fourth stage, and so on. **(a)** Complete the following table to show how far you go in three, five and ten stages. **(b)** How far do you think you would go in an infinite number of stages?

TABLE 3

Stages	Total distance	Decimal approximation
1	1	1.0
1-2	$1 - \frac{1}{2}$	0.5
1-3	$1 - \frac{1}{2} + (\frac{1}{2})^2$	0.75
1–5	$1 - \frac{1}{2} + (\frac{1}{2})^2 - (\frac{1}{2})^3 + (\frac{1}{2})^4$	
1-10		

Notice that the numbers in the completed Table 1 are of the form

$$s_n = 1 + r + r^2 + r^3 + \cdots + r^n \tag{1}$$

with $r = \frac{1}{2}$ and $n = 0, 1, 2, 4$ and 9. Similarly, the numbers in Table 2 are of the form **(1)** with $r = \frac{1}{4}$, and the numbers in Table 3 are of the form **(1)** with $r = -\frac{1}{2}$. The sum **(1)** is a FINITE GEOMETRIC SERIES. This series is very easy to study because there are concise fomulas for its values in terms of the constant r and the nonnegative integer n.

Question 4 What is the value of s_n in **(1)** if $r = 1$?

To find a formula for the finite geometric series **(1)** when $r \neq 1$, we use the notation of long division to write $\dfrac{1}{1-r}$ as the sum of powers of r plus a remainder. We first put 1 in the quotient above the horizontal line in the division sign, as in Figure 4, multiply the divisor $1 - r$ by 1, and subtract the result from the dividend 1 to see that $\dfrac{1}{1-r}$ equals 1 with a remainder r, so that

$$\frac{1}{1-r} = 1 + \frac{r}{1-r}. \tag{2}$$

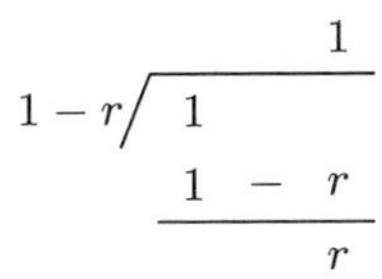

FIGURE 4

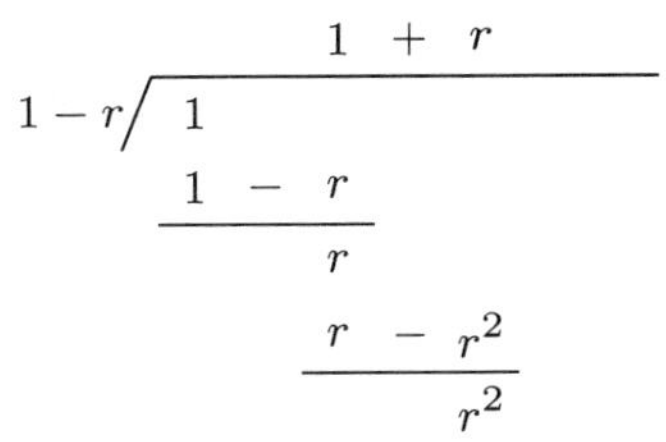

FIGURE 5

Figure 5 shows the result of carrying out a second step of division. We put $+r$ in the dividend, multiply the divisor by r and subtract the result $r - r^2$, leaving r^2. This shows that $\dfrac{1}{1-r}$ equals $1 + r$ with a remainder r^2 and, consequently,

$$\frac{1}{1-r} = 1 + r + \frac{r^2}{1-r}. \tag{3}$$

Carrying out this procedure a third time gives Figure 6, which shows that $\dfrac{1}{1-r}$ equals $1 + r + r^2$ with a remainder r^3, so that

$$\frac{1}{1-r} = 1 + r + r^2 + \frac{r^3}{1-r}. \tag{4}$$

FIGURE 6

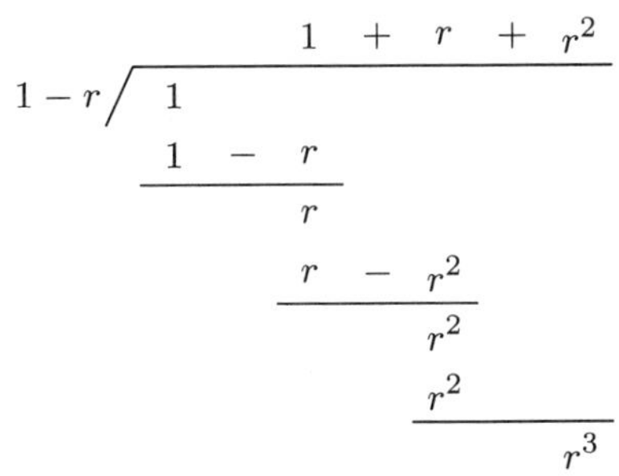

Question 5 Carry out one more step of the long division starting with Figure 6.

The division process leading to **(2)**, **(3)**, **(4)**, and the result of Question 5 can be carried out any number of times to show that for any positive integer n,

$$\frac{1}{1-r} = 1 + r + r^2 + +r^3 + \cdots + r^n + \frac{r^{n+1}}{1-r}.$$

Subtracting $\dfrac{r^{n+1}}{1-r}$ from both sides of this equation then shows that

$$1 + r + r^2 + +r^3 + \cdots + r^n = \frac{1 - r^{n+1}}{1-r}.$$

With the result of Question 4 we obtain the following formulas for the value of the Geometric Series **(1)**:

$$1 + r + r^2 + r^3 + \cdots + r^n = \begin{cases} n+1 & \text{for } r = 1 \\ \dfrac{1 - r^{n+1}}{1-r} & \text{for } r \neq 1. \end{cases} \tag{5}$$

Example 1 Use **(5)** to check the value from Question 1 for the sum in the last row of Table 1.

SOLUTION The sum is

$$1 + \tfrac{1}{2} + (\tfrac{1}{2})^2 + (\tfrac{1}{2})^3 + (\tfrac{1}{2})^4 + (\tfrac{1}{2})^5 + (\tfrac{1}{2})^6 + (\tfrac{1}{2})^7 + (\tfrac{1}{2})^8 + (\tfrac{1}{2})^9$$

which is of the form in **(5)** with $r = \frac{1}{2}$ and $n = 9$. Therefore, it equals

$$\frac{1 - (\frac{1}{2})^{10}}{1 - \frac{1}{2}} = 2\left(1 - (\tfrac{1}{2})^{10}\right) \doteq 1.998047. \square$$

Question 6 Use **(5)** to check the values from Questions 2 and 3 of the sums in the last rows of **(a)** Table 2 and **(b)** Table 3.

A loan balance

One of the most important practical applications of finite geometric series is in the study of payments and loan balances in installment loans and mortgages. Suppose, for example, you take out a student loan of \$10,000 at an interest rate of 12% per year (1% per month) and you make payments of \$200 per month starting one month after you take out the loan. The \$10,000 in the fourth column for month 0 in Table 4 is the amount of the loan. The interest after one month is 1% of \$10,000 or \$100, so the balance just before the first payment is the \$10,000 + \$100 = \$10,100 in the second column of the row for month 1. The third column of the row for month 2 shows the balance \$10,100 − \$200 = \$9,900 after the first payment. The last number in the fourth column is the loan balance, rounded to the nearest dollar, after the fifth payment

TABLE 4

Month	Balance with interest before the payment	Payment	Balance after the payment
0			\$10,000
1	\$10,100	\$200	\$9,900
2	\$9,999	\$200	
3		\$200	
4		\$200	
5		\$200	\$9,490

Question 7 Calculate the missing numbers in Table 4. Round the numbers to the nearest dollar after each calculation.

The calculation of the loan balance after any payment in Table 4 can be found directly, without making the calculations for the earlier months, by using the finite-geometric-series formula **(5)**.

To show the pattern in the calculations more clearly, we write P for the monthly payment and B_n for the loan balance just after the nth monthly payment has been made. Then $P = 200$ and $B_0 = 10{,}000$.

Just before the first monthly payment is made, 1% interest has accrued, raising the balance from B_0 to $B_0 + 0.01B_0 = (1.01)B_0$. Then a payment of P dollars is made, so that

$$B_1 = 1.01B_0 - P. \tag{6}$$

Then, just before the second payment, another 1% interest has accrued, raising the balance from B_1 to $B_1 + 0.01B_1 = (1.01)B_1$. Then a payment of P dollars is made, so that $B_2 = (1.01)B_1 - P$. With formula **(6)** for B_1, we find that

$$B_2 = (1.01)(1.01B_0 - P) - P = (1.01)^2B_0 - 1.01P - P \tag{7}$$

Question 8 **(a)** Give formulas for B_3, first in terms of B_2 and then in terms of B_0 and P. **(b)** Give formulas for B_4, first in terms of B_3 and then in terms of B_0 and P.

The calculations in Question 8 could be extended to show that the loan balance after n months for any positive integer n is

$$B_n = (1.01)^n B_0 - P - 1.01P - (1.01)^2 P - (1.01)^3 P - \cdots - (1.01)^{n-1} P. \tag{8}$$

Example 2 **(a)** Use formula **(5)** for the finite geometric series to simplify expression **(8)** for the loan balance after n months. **(b)** Use the resulting formula to check the value from Question 7 for the loan balance after the fifth payment.

SOLUTION **(a)** We first factor P from all but the first term in **(8)** to have

$$B_n = (1.01)^n B_0 - P\big(1 + 1.01 + (1.01)^2 + (1.01)^3 + \cdots + (1.01)^{n-1}\big).$$

Then by **(5)** with $r = 1.01$ and with n replaced by $n - 1$

$$1 + 1.01 + (1.01)^2 + (1.01)^3 + \cdots + (1.01)^{n-1} = \frac{1 - (1.01)^n}{1 - (1.01)} = \frac{(1.01)^n - 1}{0.01}$$

so that

$$B_n = (1.01)^n B_0 - P\left[\frac{(1.01)^n - 1}{0.01}\right]. \tag{9}$$

(b) To find the balance after the fifth payment, we set $B_0 = 10{,}000$, $P = 200$, and $n = 5$ in **(9)**. We obtain

$$B_5 = (1.01)^5(10{,}000) - 200\left[\frac{(1.01)^5 - 1}{0.01}\right] = 9489.89$$

which, rounded to the nearest dollar, is the loan balance \$9,490 in Table 4. □

Question 9 What would be the balance due on the loan after 60 payments (five years)?

Example 3 How long does it take for the loan to be paid off?

SOLUTION We substitute $B_0 = 10{,}000$, and $P = 200$ in **(9)** to obtain

$$\begin{aligned} B_n &= 10{,}000(1.01)^n - 200\left[\frac{(1.01)^n - 1}{0.01}\right] \\ &= 10{,}000(1.01)^n - 20{,}000[(1.01)^n - 1] \\ &= -10{,}000(1.01)^n + 20{,}000. \end{aligned}$$

This shows that $B_n = 0$ if $(1.01)^n = 2$. Using logarithms to solve the last equation gives $n \ln(1.01) = \ln(2)$ and then $n = (\ln(2))/(\ln(1.01)) \doteq 69.66$. Since loan payments are made only every month, the loan will be paid off with the 70th payment. □

Question 10 What would be the 70th payment in Example 9?

St. Ives

Expanding an eighteenth-century nursery rhyme leads to the following puzzle: "As I was going to St. Ives, I met a man with seven wives. Every wife had seven sacks, every sack had seven cats, every cat had seven kitts, every kitt had seven mice, and every mouse had seven lice. Lice, mice, kitts, cats, sacks, wives—how many were going to St. Ives?"†

Question 11 **(a)** Answer the question. **(b)** How many—lice, mice, kitts, cats, sacks, wives, and the man—did the narrator meet?

Infinite geometric series

An INFINITE SERIES is an expression of the form $\sum_{j=j_0}^{\infty} a_j$, where a_j is a real number for each integer $j \geq j_0$. We give meaning to this symbol by considering finite PARTIAL SUMS.

$$s_n = \sum_{j=j_0}^{n} a_j = a_{j_0} + a_{j_0+1} + a_{j_0+2} + \cdots + a_n. \quad \textbf{(10)}$$

The partial sums form a sequence $\{s_n\}_{n=j_0}^{\infty}$ which we use in the following definition.

Definition 1 **(a)** *The infinite series* $\sum_{j=j_0}^{\infty} a_j$ CONVERGES *if the partial sums* $\{s_n\}_{n=j_0}^{\infty}$ *have a finite limit L as* $n \to \infty$. *In this case, we say that L is the* SUM *or* VALUE *of the infinite series and we write* $\sum_{j=j_0}^{\infty} a_j = L$.

(b) *If* $\lim_{n\to\infty} s_n = \infty$, *we say that the the series* DIVERGES TO ∞ *and write* $\sum_{j=j_0}^{\infty} a_j = \infty$.

(c) *If* $\lim_{n\to\infty} s_n = -\infty$, *we say that the the series* DIVERGES TO $-\infty$ *and write* $\sum_{j=j_0}^{\infty} a_j = -\infty$.

(d) *If the partial sums* $\{s_n\}_{n=j_0}^{\infty}$ *do not have a finite limit and do not tend to* ∞ *or to* $-\infty$, *we say that the the series* $\sum_{j=j_0}^{\infty} a_j$ DIVERGES WITHOUT A LIMIT.

Notice that deleting, adding, or changing a finite number of terms in an infinite series does not affect its convergence. Also, you can see from the definition that if the series $\sum_{j=j_0}^{\infty} a_j$ and $\sum_{j=j_0}^{\infty} b_j$ converge, then for any constants A and B, the series $\sum_{j=j_0}^{\infty} (Aa_j + Bb_j)$ converges and

$$\sum_{j=j_0}^{\infty} (Aa_j + Bb_j) = A\sum_{j=j_0}^{\infty} a_j + B\sum_{j=j_0}^{\infty} b_j.$$

The series $\sum_{j=0}^{\infty} r^j$ with a constant r is an INFINITE GEOMETRIC SERIES. To study its convergence, we write formula **(5)** for a finite geometric series with summation notation as follows:

$$s_n = 1 + r + r^2 + r^3 + \cdots + r^n = \sum_{j=0}^{n} r^j = \begin{cases} n+1 & \text{for } r = 1 \\ \dfrac{1-r^{n+1}}{1-r} & \text{for } r \neq 1. \end{cases} \quad \textbf{(11)}$$

---

†The traditional nursery rhyme does not include mice or lice.

If $r = 1$, then upper formula on the right of **(11)** shows that $s_n = n + 1$, so that $s_n \to \infty$ as $n \to \infty$ and $\sum_{j=0}^{\infty} r^j = \infty$. We will use the results of the next Question to study this series for other values of r.

Question 12 Use logarithms to show that $|r|^n$ **(a)** tends to 0 as $n \to \infty$ if $-1 < r < 1$ and **(b)** tends to ∞ if $|r| > 1$.

For $|r| < 1$, the lower formula on the right of **(11)** applies, and the result of part (a) of Question 12 shows that r^{n+1} tends to 0 as $n \to \infty$. Consequently, the geometric series converges and has the value $\sum_{j=0}^{\infty} r^j = \dfrac{1}{1-r}$.

If $r > 1$, then by part (b) of Question 11 and **(10)**, $s_n = \dfrac{r^{n+1} - 1}{r - 1} \to \infty$ as $n \to \infty$ and the geometric series series diverges to ∞. If $r \le -1$, then $\lim_{n\to\infty} s_n$ is not defined and the geometric series diverges without a limit.

We summarize these results in the next theorem:

Theorem 1 *The geometric series* **(10)** *converges and has the value*

$$\sum_{j=0}^{\infty} r^j = \frac{1}{1-r} \quad \text{for } |r| < 1. \qquad \textbf{(12)}$$

It diverges to ∞ *for* $r \ge 1$ *and diverges without a limit for* $r \le -1$.

Example 4 Use Theorem 1 to verify the prediction in Question 1.

SOLUTION If you go one mile, then half a mile, then half that distance, etc. then the total distance you travel approaches

$$1 + \tfrac{1}{2} + \left(\tfrac{1}{2}\right)^2 + \left(\tfrac{1}{2}\right)^3 + \cdots = \sum_{j=0}^{\infty} \left(\tfrac{1}{2}\right)^j = \frac{1}{1 - \frac{1}{2}} = 2 \text{ miles. } \Box$$

Question 13 Use Theorem 1 to check the predictions **(a)** in Question 2 and **(b)** in Question 3.

Example 5 Does $\sum_{j=0}^{\infty} (-\tfrac{3}{4})^j = 1 - \tfrac{3}{4} + (\tfrac{3}{4})^2 - (\tfrac{3}{4})^3 + \cdots$ converge? If so, what is its value?

SOLUTION By Theorem 1 with $r = -\tfrac{3}{4}$, for which $|r| = \tfrac{3}{4} < 1$, this Geometric Series converges and has the value

$$\sum_{j=0}^{\infty} (-\tfrac{3}{4})^j = \frac{1}{1 - (-\frac{3}{4})} = \frac{1}{1 + \frac{3}{4}} = \tfrac{4}{7}. \ \Box$$

Example 6 Does $\sum_{j=0}^{\infty} (-5)^j = 1 - 5 + 5^2 - 5^3 + 5^4 - 5^5 + \cdots$ converge?

SOLUTION This series diverges without a limit because it is the Geometric Series with $r = -5$, which is ≤ -1. $\Box$

In the next example, we use **(12)** to find the sum of a Geometric Series that starts at $j = 2$ rather than at $j = 0$.

Example 7 What is the exact value of $\sum_{j=2}^{\infty}(0.7)^j = (0.7)^2 + (0.7)^3 + (0.7)^4 + (0.7)^5 + \cdots$?

SOLUTION We factor the first term $(0.7)^2$ from each term of the series and obtain

$$\begin{aligned}\sum_{j=2}^{\infty}(0.7)^j &= (0.7)^2 + (0.7)^3 + (0.7)^4 + (0.7)^5 + \cdots \\ &= (0.7)^2[1 + 0.7 + (0.7)^2 + (0.7)^3 + \cdots] \\ &= (0.7)^2\sum_{j=0}^{\infty}(0.7)^j \\ &= (0.7)^2\frac{1}{1-(0.7)} = \frac{0.49}{0.3} = 1\tfrac{19}{30}. \ \square\end{aligned}$$

Infinite repeating decimals

Every rational number (integer or fraction) can be expressed as an infinite repeating decimal and every infinite repeating decimal represents a rational number. The decimal can be obtained from a fraction by long division. The fraction can be obtained from the decimal by using a Geometric Series.

Example 8 Express the infinite repeating decimal $0.002222\cdots$ as a fraction.

SOLUTION We write

$$\begin{aligned}0.002222\cdots &= \frac{2}{10^3} + \frac{2}{10^4} + \frac{2}{10^5} + \frac{2}{10^6} + \cdots \\ &= \frac{2}{10^3}\left(1 + \frac{1}{10} + \frac{1}{10^2} + \frac{1}{10^3} + \cdots\right) \\ &= \tfrac{1}{500}\left[1 + \tfrac{1}{10} + \left(\tfrac{1}{10}\right)^2 + \left(\tfrac{1}{10}\right)^3 + \cdots\right] \\ &= \tfrac{1}{500}\sum_{j=0}^{\infty}\left(\tfrac{1}{10}\right)^j = \tfrac{1}{500}\left(\frac{1}{1-\frac{1}{10}}\right) = \left(\tfrac{1}{500}\right)\left(\tfrac{10}{9}\right) = \tfrac{1}{450}. \ \square\end{aligned}$$

Principles and Procedures

- To recall the formula in **(5)** for the value of a Finite Geometric Series with $r \neq 1$, notice that, for $n = 1$, it is $1 + r = \dfrac{1-r^2}{1-r}$, a formula which comes from the factorization $1 - r^2 = (1+r)(1-r)$.
- The Infinite Geometric Series $\sum_{j=0}^{\infty} r^j$ arises in many applications and will serve as a point of reference as you study properties of other types of infinite series in the following sections of this chapter. Its convergence properties, as stated in Theorem 1, are easy to remember, especially if you know formula **(5)** for a Finite Geometric Series.
- If you are ever confused by summation notation, write out some of the terms in the sums, as in the solution of Example 7 above, to see the patterns in the formulas and calculations.

Tune-Up Exercises 8.2◆

^A *Answer provided.* ^O *Outline of solution provided.*

T1.^O **(a)** Give the approximate decimal value of the Finite Geometic Series $\sum_{j=0}^{100}(0.95)^j$. **(b)** Does the Infinite Geometric Series $\sum_{j=0}^{\infty}(0.95)^j$ converge? If so, give its value.

◆ Type 1, finite and infinite geometric series

T2.^O Give the exact value of the Finite Geometic Series $3^3 + 3^4 + 3^5 + 3^6 + 3^7 + 3^8 + 3^9$.

◆ Type 1, a finite geometric series

T3.^O Give the exact value of the Infinite Geometic Series $\sum_{j=3}^{\infty}(-\frac{3}{4})^j$.

◆ Type 1, an infinite geometric series

T4.^O **(a)** Does $\sum_{j=0}^{\infty}(-4)^j$ converge, diverge to ∞, diverge to $-\infty$, or diverge without a limit?

(b) Give the exact values of $\sum_{j=0}^{n}(-4)^j$ for $n = 10$ and $n = 11$.

◆ Type 1, finite and infinite geometric series

Problems 8.2

^A *Answer provided.* ^O *Outline of solution provided.*

Give exact or approximate decimal values of the Finite Geometric Series in Problems 1 through 12.

1.^O $\sum_{j=0}^{12}(\frac{1}{2})^j$

◆ Type 1, a finite geometric series

2.^A $\sum_{j=0}^{8}(-\frac{1}{4})^j$

◆ Type 1, a finite geometric series

3.º $\sum_{j=1}^{31}(\frac{3}{4})^j$

◆ Type 1, a finite geometric series

4. $\sum_{j=2}^{10}5^j$

◆ Type 1, a finite geometric series

5.^A $6 + 0.6 + 0.06 + 0.006 + 0.0006 + 0.00006$

◆ Type 1, a finite geometric series

6. $\sum_{j=1}^{511}(0.99)^j$

◆ Type 1, a finite geometric series

7. $\sum_{j=1}^{100}1^j$

◆ Type 1, a finite geometric series

8. $(1.1)^3 + (1.1)^4 + (1.1)^5 + \cdots + (1.1)^{34}$

◆ Type 1, a finite geometric series

9. $\sum_{j=10}^{25}2^j$

◆ Type 1, a finite geometric series

10. $\sum_{j=0}^{20}5(0.8)^j$

◆ Type 1, a finite geometric series

11.^A $\sum_{j=0}^{400}(-1)^j$

◆ Type 1, a finite geometric series

12. $\sum_{j=2}^{7}11^j$

◆ Type 1, a finite geometric series

◆The Tune-up Exercises and Problems are classified by type and content. The types are (1) basic, reactive; (2) basic reflective; (3) intermediate, reactive; (4) intermediate, reflective; (5) advanced, reactive; (6) advanced, reflective; and (7) advanced, theoretical.

Determine whether the Infinite Geometric Series in Problems 13 through 20 converge, diverge to ∞, diverge to $-\infty$ or diverge without a limit. Give the values of those that converge.

13.[A] $\sum_{j=0}^{\infty} (\frac{1}{7})^j$

♦ Type 1, an infinite geometric series

14.[A] $\sum_{j=0}^{\infty} (1.001)^j$

♦ Type 1, an infinite geometric series

15.[A] $\sum_{j=0}^{\infty} (-2)^j$

♦ Type 1, an infinite geometric series

16.[A] $\sum_{j=2}^{\infty} (0.99)^j$

♦ Type 1, an infinite geometric series

17. $\sum_{j=4}^{\infty} (-\frac{3}{4})^j$

♦ Type 1, an infinite geometric series

18. $\sum_{j=4}^{\infty} 10(\frac{4}{5})^j$

♦ Type 1, an infinite geometric series

19. $\sum_{j=10}^{\infty} \frac{1}{100}(\frac{9}{8})^j$

♦ Type 1, an infinite geometric series

20. $\sum_{j=5}^{\infty} 9(\frac{1}{10})^j$

♦ Type 1, an infinite geometric series

21.[A] What is the exact value of $\sum_{j=0}^{\infty} [4(\frac{1}{3})^j - 2(\frac{1}{9})^j]$?

♦ Type 2, an infinite geometric series

22. What is the exact value of $\sum_{j=0}^{\infty} [(0.9)^j + (0.8)^j + (0.7)^j]$?

♦ Type 2, an infinite geometric series

Express the repeating decimals in Problems 23 through 26 as fractions.

23.[O] 3.42121212121...

♦ Type 2, repeating decimal

24.[A] 4.123123123123...

♦ Type 2, repeating decimal

25. 0.1534343434...

♦ Type 2, repeating decimal

26. 5.99999999...

♦ Type 2, repeating decimal

27. Determine the values of x such that $\sum_{j=0}^{\infty} 10(\frac{1}{5}x)^j$ converges and give a concise formula for the value of the sum for such x.

♦ Type 2, an infinite geometric series, etc.

28.[O] Find the smallest positive integer n such that $\sum_{j=0}^{n} (\frac{3}{4})^j$ differs from $\sum_{j=0}^{\infty} (\frac{3}{4})^j$ by less than 0.004.

♦ Type 4, the epsilon-N-definition of convergence

29. How close is $\sum_{j=0}^{40} (0.9)^j$ to $\sum_{j=0}^{\infty} (0.9)^j$?

♦ Type 2, finite and infinite geometric series

30.[A] A woman opens a new savings account and then deposits \$5000 in it on January 1 each year for twenty years. The account pays 8% annual interest compounded annually and she does not make any additional withdrawals or deposits. On January 1 of the twentieth year she closes the account. How much is in it at that time?

♦ Type 4, a finite geometric series, etc.

31. If the U.S. government sets the RESERVE REQUIREMENT for banks at 20%, then the banks have to leave 20% of every deposit in reserve and can loan only 80% of it to other customers.Suppose one customer deposits \$1000 in a bank. How much would the bank loan others, based on this deposit, under the following circumstances? The bank loans the maximum \$800 of the initial deposit allowable to a second customer, who deposits itall back in the bank. The bank then loans the maximum \$640 of that amount to a third customer, who deposits it in the bank. The bank loans the maximum of that to a fourth customer, who deposits it in the bank, and so forth, until the thirtieth customer (whose loan would be less than a dollar).

◆ Type 4, a finite geometric series, etc.

32. Suppose that B_0 dollars are borrowed on January 1 from a bank at $100r\%$ interest per month, compounded monthly, and that loan payments of P are made at the first of each month, starting the following February 1. **(a)** Show that the balance after n months is $B_n = \frac{P}{r} - (1+r)^n\left(\frac{P}{r} - B_0\right)$. **(b)** What happens to the loan if $P = rB_0$? **(c)** What happens to the loan if $P < rB_0$?

◆ Type 4, a finite geometric series, etc.

33. Zeno's paradox "The Achilles"† asserts that if Achilles were racing a tortoise that is given a headstart, he could never catch up to it because when he reaches the tortoise's initial position, it will have advanced to another position, and when Achilles reaches that position, the tortoise will have moved on, and so forth. The flaw in this reasoning can be explained with geometric series. Suppose that Achilles runs at the constant speed of v_A meters per second and that the tortoise crawls at the constant speed of v_T meters per second. Suppose the race is on an x-axis, with Achilles starting at $x = x_0$ (meters) and the tortoise at $x_1 > x_0$ at time t_0 (seconds). Let t_1 be the time when Achilles reaches x_1 and let x_2 be the tortoise's position at time t_1; let t_2 be the time when Achilles reaches x_2 and x_3 the tortoise's position at time t_2; and so forth. **(a)** Show that $t_{j+1} - t_j = \frac{x_{j+1} - x_j}{v_A}$ and $x_{j+1} - x_j = v_T(t_j - t_{j-1})$ for $j \geq 1$. **(b)** Show that $t_{j+1} - t_j = \left(\frac{v_T}{v_A}\right)^j (t_1 - t_0)$ and $x_{j+1} - x_j = \left(\frac{v_T}{v_A}\right)^j (x_1 - x_0)$ for $j \geq 1$. **(c)** Write $t_n = t_0 + \sum_{j=0}^{n-1}(t_{j+1} - t_j)$ and $x_n = x_0 + \sum_{j=0}^{n-1}(x_{j+1} - x_j)$ to prove that the sequences $\{t_n\}$ and $\{x_n\}$ converge and Achilles does catch up to the tortoise if he runs facter than it does.

◆ Type 4, a finite geometric series, etc.

34. Consider a triangle inscribed in a parabola as in Figure 7 with vertex at the point C such that the tangent line at C is parallel to the line segment AB. Suppose that the parabola is $y = ax^2$ in an xy-plane and that the x-coordinate of A is x_0 and the x-coordinate of B is x_1. **(a)** Show that the x-coordinate of C is the average $\frac{1}{2}(x_0 + x_1)$ of the x-coordinates of the other points. **(b)** Use the formula for the area of a trapezoid to show that triangle ABC has area $\frac{1}{8}a(x_1 - x_0)^3$. **(c)** Let D be the point on the parabola where the tangent line is parallel to AC and let E be the point where the tangent line is parallel to BC as in Figure 8. Use the formula from part (b) to prove that the combined area of triangles ADC and CEB equals one-fourth the area of triangle ABC.

◆ Type 4, a finite geometric series, etc.

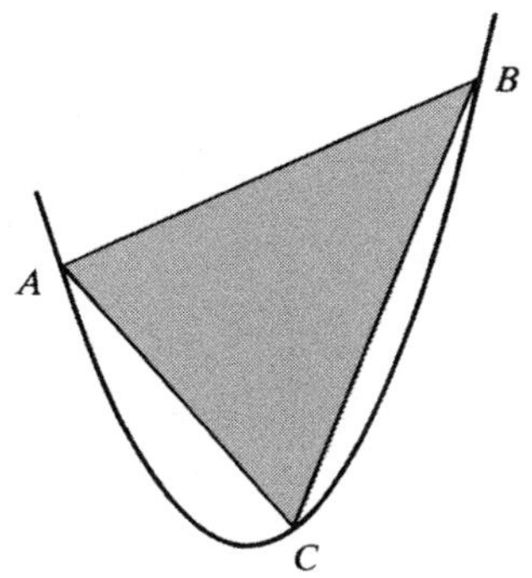

FIGURE 7

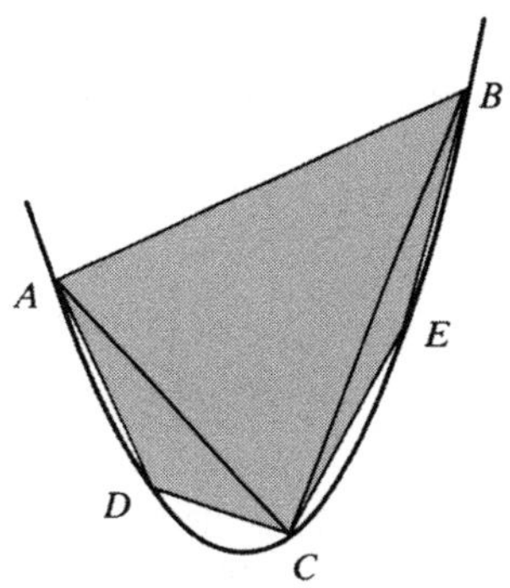

FIGURE 8

†Proposed by Zeno of Elia, ca. 450 BC. See *A History of Mathematics* by C. Boyer, New York, NY: John Wiley & Sons, 1968, p. 82.

35. Archimedes used the Finite-Geometric-Series formula and the result of Problem 34 (proved by studying the conic section directly, rather than by analytic geometry) to show that the region R of Figure 9 bounded by the parabola and the line segment AB has area equal to four-thirds that of the triangle ABC of Figure 7. He used an approximation procedure, starting with the triangle ABC. The second approximation is obtained by adding the triangles ADC and CEB of Figure 4. For the third approximation four triangles are constructed on the sides AD, DC, CE, and EB just as ADC and CEB were constructed from ABC, so the tangent lines to the parabola at the third vertices are parallel to those sides. In the fourth approximation eight similarly constructed triangles are added, and so forth. Derive Archimedes' result by finding the limit of the areas of the approximations.‡

◆ Type 4, a finite geometric series, etc.

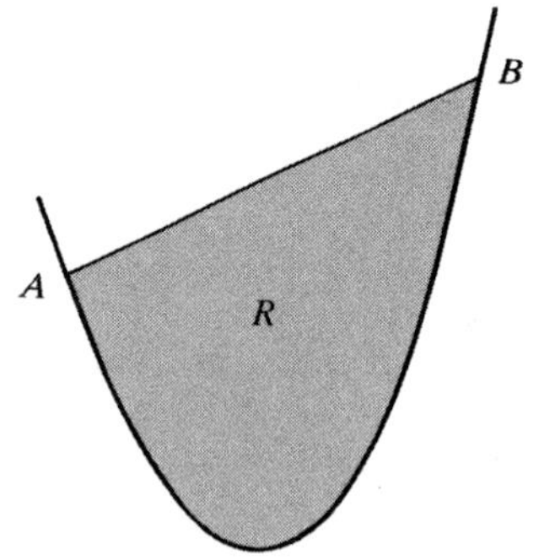

FIGURE 9

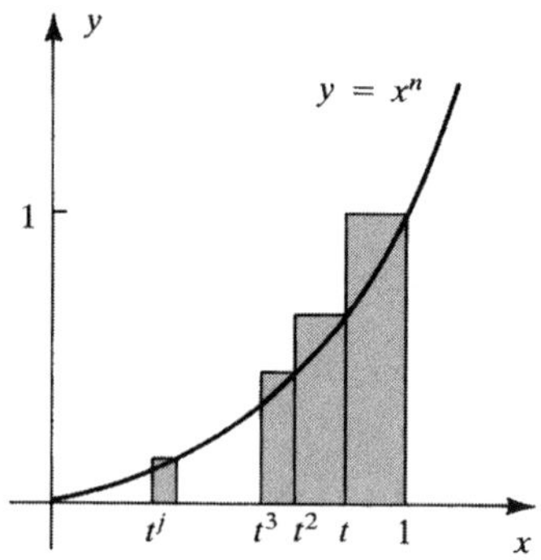

FIGURE 10

36. One technique Fermat (1601–1665) used to find exact areas of regions bounded by the curves $y = x^n$ for rational constants $n \neq -1$ was to approximate the region by infinite collections of rectangles. To determine the area of the region bounded by $y = x^n$ with $n > 0$, by the x-axis, and by the line $x = 1$, as in Figure 9, he would consider, for each t with $0 < t < 1$, the approximation by rectangles with lower right corners at $x = 1, x = t, x = t^2, x = t^3, \ldots$ and upper-right corners on the curve, as in Figure 10. **(a)** Prove that the total area of the infinite number of rectangles in Figure 10 is $\dfrac{1-t}{1-t^{n+1}}$. **(b)** Show that for positive integers n the area of the triangles is $\dfrac{1}{1+t^2+t^3+\cdots+t^n}$ by dividing $1-t$ into $1-t^{n+1}$. **(c)** What happens to the rectangles as $t \to 0^+$? **(d)** Find the area of the curved region as the limit of formula in part(b) as $t \to 1^-$.

◆ Type 4, a finite geometric series, etc.

37. Because a cube has six sides, when a fair die is rolled, there is one chance in six that the $\boxed{6}$ comes up (is on the top) and five chances in six that another number comes up. Imagine that a fair die is rolled an infinite number of times. The probability that the first $\boxed{6}$ comes up on the second roll is $\frac{1}{6}(\frac{5}{6})$ because another number must come up in the first roll and a $\boxed{6}$ on the second roll. **(a)** What is the probability that the first $\boxed{6}$ comes up on the third roll? On the fourth roll? On the jth roll with $j \geq 2$? **(b)** What probability is represented by the infinite series $\sum_{j=1}^{\infty} \frac{1}{6}(\frac{5}{6})^{j-1}$ and what is its value?

◆ Type 4, a finite geometric series, etc.

‡Archimedes did not refer spefically to limits. He completed his demonstrations with arguments by contradiction, a line of reasoning known as THE METHOD OF EXHAUSTION and attributed to Eudoxus (ca. 408–355 B.C.)

Section 8.3

The Integral and Comparison Tests

OVERVIEW: *In Section 8.2 we saw that the Geometric Series $\sum_{j=0}^{\infty} r^j$ converges if $|r| < 1$ and diverges if $|r| \geq 1$. In this section we look at several tests for the convergence and divergence of other infinite series. We begin with the* INTEGRAL TEST, *in which partial sums of infinite series are compared with improper integrals. Next, we present a basic divergence test. Then we discuss the* COMPARISON *and* LIMIT-COMPARISON TESTS, *which are used to determine the convergence of series by comparing them with other series that are known to converge or diverge.*

Topics:

- ***Convergence of series with nonnegative terms***
- ***The Integral Test***
- ***The p-series***
- ***A divergence test***
- ***Absolute convergence***
- ***The Comparison Test***
- ***The Limit Comparison Test***

Convergence and divergence of series with nonnegative terms

An infinite series $\sum_{j=j_0}^{\infty} a_j$ with positive terms a_j has increasing partial sums because each partial sum $s_n = \sum_{j=j_0}^{n} a_j$ is obtained from the previous one $s_{n-1} = \sum_{j=j_0}^{n-1} a_j$ by adding the positive number a_n. Similarly, if all of the terms of a series are nonnegative (≥ 0), then the sequence of partial sums $\{s_n\}_{n=j_0}^{\infty}$ is nondecreasing.

We saw in Theorem 1 of Section 8.1 that a nondecreasing sequence either is bounded and converges are is unbounded an diverges to ∞. Applying this result to the partial sums of a series with nonegative terms establishes the following theorem, which will be the basis of other convergence tests that we will study here and in Section 8.4.

Theorem 1 *An infinite series with nonnegative terms either converges or diverges to ∞. The series converges if its partial sums are bounded and diverges to ∞ if its partial sums are not bounded.*

The Integral Test

As our first application of Theorem 1, we will show how the convergence or divergence of certain series can be determined by studying related improper integrals. We suppose that the terms in the series are given by

$$|a_j| = f(j) \quad \text{for } j \geq J \tag{1}$$

where J is a positive integer and $f(x)$ is a nonincreasing positive function for $x \geq J$. To illustrate the ideas, we suppose that $f(x)$ is the function of Figure 1.

♦ SUGGESTIONS TO INSTRUCTORS: Have students begin work on Questions 1 through 5 before class. Problems 17 and 18–24 are good for in-class discussion.

$$\text{Area} = \int_J^n f(x)\, dx$$

FIGURE 1

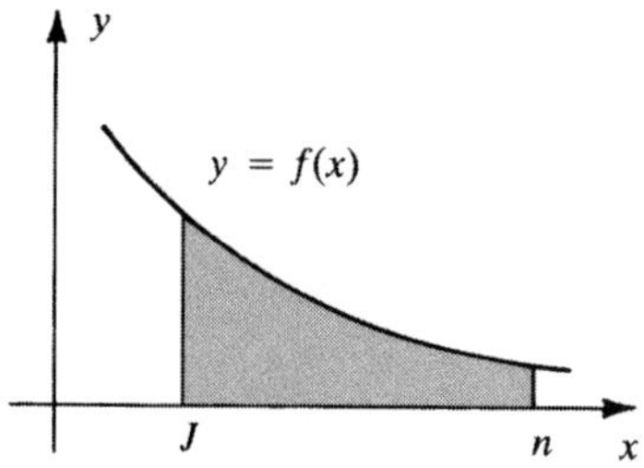

Consider an integer $n > J$. Because $f(x)$ is positive, the integral

$$\int_J^n f(x)\, dx$$

is equal to the area of the region under the graph of $f(x)$ for $J \le x \le n$. We compare this area to the area of the collections of rectangles in Figure 2 and 3.

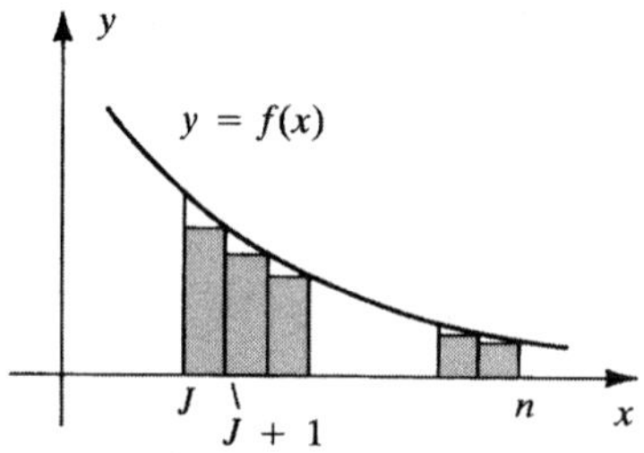

$$\text{Area} = \sum_{j=J+1}^{n} f(j) = \sum_{j=J+1}^{n} |a_j|$$

FIGURE 2

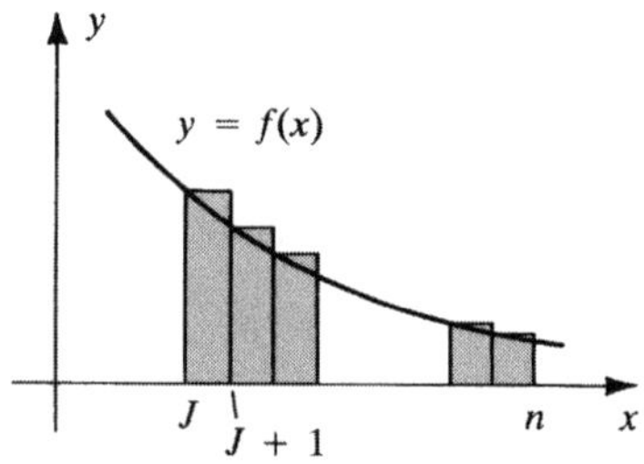

$$\text{Area} = \sum_{j=J}^{n-1} f(j) = \sum_{j=J}^{n-1} |a_j|$$

FIGURE 3

Each of the rectangles in Figures 2 and 3 has its base on the x-axis, has width 1, and has height and area equal to $f(j)$ for an integer j. Therefore, the rectangles in Figure 2 have area

$$\sum_{j=J+1}^{n} f(j) = \sum_{j=J+1}^{n} |a_j| \tag{2}$$

and the rectangles in Figure 3 have area

$$\sum_{j=J}^{n-1} f(j) = \sum_{j=J}^{n-1} |a_j|. \tag{3}$$

Because $f(x)$ is nonincreasing, the region of Figure 1 contains the rectangles of Figure 2 and is contained in the rectangles of Figure 3. Accordingly, the areas **(1)**, **(2)**, **(3)** are related by the inequalities

$$\sum_{j=J+1}^{n} |a_j| \le \int_J^n f(x)\, dx \le \sum_{j=J}^{n-1} |a_j|. \tag{4}$$

Now, we consider the improper integral $\int_J^\infty f(x)\, dx$. If it converges, then the integrals $\int_J^n f(x)\, dx$ are bounded and the inequality on the left of **(4)** shows that the partial sums of $\sum_{j=j_0}^{\infty} |a_j|$ are also bounded and, consequently, the series $\sum_{j=j_0}^{\infty} |a_j|$ converges.

On the other hand, if $\int_J^\infty f(x)\,dx$ diverges to ∞, then the numbers $\int_J^n f(x)\,dx$ tend to ∞ as $n \to \infty$. Then, because of inequality on the right of **(4)**, the partial sums $\sum_{j=j_0}^{n} |a_j|$ tend to ∞ as $n \to \infty$ and $\sum_{j=j_0}^{\infty} |a_j|$ does not converge. This establishes the next theorem.

Theorem 2 (The Integral Test) *Suppose that the series*

$$\sum_{j=j_0}^{\infty} a_j \tag{5}$$

is such that $|a_j| = f(j)$ for integers $j \geq J$, where $f(x)$ is a positive, nonincreasing function for $j \geq J$ with some J. Suppose also that the Rieman integral $\int_J^X f(x)\,dx$ is defined for all $X > J$. Then **(5)** *converges absolutely if and only if the improper integral*

$$\int_J^\infty f(x)\,dx \tag{6}$$

converges.

Example 1 Does $\sum_{j=2}^{\infty} \frac{1}{j(\ln j)^2}$ converge or does it diverge to ∞?

SOLUTION The series either converges or diverges to ∞ because its terms are positive. We can apply the Integral Test to it because $f(x) = \frac{1}{x(\ln x)^2}$ is positive and decreasing for $x \geq 2$.

To determine whether the improper integral of $f(x)$ converges, we use the substitution $u = \ln x$, $du = (1/x)\,dx$ to obtain for $X > 2$,

$$\begin{aligned} \int_2^X \frac{1}{x(\ln x)^2}\,dx &= \int_{x=2}^{x=X} \frac{1}{u^2}\,du = \Big[-u^{-1}\Big]_{x=2}^{x=X} \\ &= \left[-\frac{1}{\ln x}\right]_{x=2}^{x=X} = -\frac{1}{\ln X} + \frac{1}{\ln(2)}. \end{aligned}$$

Therefore,

$$\begin{aligned} \int_2^\infty \frac{1}{x(\ln x)^2} &= \lim_{X\to\infty} \int_2^X \frac{1}{x(\ln x)^2} \\ &= \lim_{X\to\infty} \left[-\frac{1}{\ln X} + \frac{1}{\ln(2)}\right] = \frac{1}{\ln(2)}. \end{aligned}$$

Since the improper integral converges, the infinite series also converges. □

The procedures described in Section 8.1 for generating graphs of finite numbers of terms in infinite sequences can be modified to plot partial sums of infinite series.

Plotting and calculating partial sums on Texas Instruments calculators

Partial sums of an infinite series $\sum_{j=j_0}^{\infty} a_j$ with $j_0 \geq 0$ can be plotted and calculated on most Texas Instruments calculators by making two modifications in the procedures given in Section 7.1. for iterative sequences.

On a TI-82, set $U_n = U_{n-1} + a_n$ instead of $U_n = g(U_{n-1})$ and $U n\text{Start} = a_{j_0}$ instead of $U n\text{Start} = x_0$.

On a TI-83, set $u(n) = u(n-1) + a_n$ instead of $u(n) = g(u(n-1))$ and $u(n\text{Min}) = a_{j_0}$ instead of $u(n\text{Min}) = x_0$.

On a TI-89 or TI-92, set $u1 = u1(n-1) + a_n$ instead of $u1 = g(u1(n-1))$ and $ui1 = a_{j_0}$ instead of $ui1 = x_0$

On a TI-85 or TI-86, use the following program with $y1 = a_N$ in the $y =$ menu and a suitable window.

```
PROGRAM: SERIES
: ClrDraw
: FnOff
: Disp "USE y1 = AN"
: Disp "JØ ="
: Input A
: Disp "MAXN ="
: Input B
: Disp "SHOW AFTER"
: Input M
: 0 → S
: For(N, A, B, 1)
: y1 + S → S
: PtOn(N, S)
: If N ≤ M
: Goto L
: Pause
: Disp "N ="
: Disp N
: Disp "SUM ="
: Disp S
: Pause
: Lbl L
: End
```

Question 1 **(a)** Plot the partial sums $\sum_{j=2}^{n} \frac{1}{j(\ln j)^2}$ for $n = 2, 3, \ldots, 25$ of the convergent infinite series of Example 1. Use the window $-1 \leq n \leq 26, -0.25 \leq y \leq 2.25$. **(b)** Give the approximate decimal values of the partial sums for $n = 23, 24$, and 25.

Question 2 Use Theorem 1 to determine whether $\sum_{j=1}^{\infty} \frac{1}{\sqrt{j}}$ converges or diverges to ∞. Then generate a plot of first 25 partial sums in the window $-1 \leq n \leq 26, -1 \leq y \leq 10$.

The p-series

The infinite series $\sum_{j=1}^{\infty} \frac{1}{\sqrt{j}}$ of Question 2 is the special case with $p = \frac{1}{2}$ of the p-SERIES

$$\sum_{j=1}^{\infty} \frac{1}{j^p} = 1 + \frac{1}{2^p} + \frac{1}{3^p} + \frac{1}{4^p} + \cdots \tag{7}$$

This series diverges if $p \leq 0$ since then all of its terms are ≥ 1, and you saw in Question 2 that it diverges for $p = \frac{1}{2}$. We can determine whether it converges for other positive values of p with the Integral Test because, for $p > 0$, the function $f(x) = 1/x^p$ is positive and decreasing for $x \geq 1$.

For $p = 1$ we have

$$\int_1^\infty \frac{1}{x^p}\,dx = \lim_{X\to\infty}\int_1^X \frac{1}{x}\,dx = \lim_{X\to\infty}\Big[\ln|x|\Big]_1^X$$
$$= \lim_{X\to\infty} \ln X = \infty.$$

In this case, the improper integral and the p-series both diverge to ∞.

For $p \neq 1$

$$\int_1^\infty \frac{1}{x^p}\,dx = \lim_{X\to\infty}\int_1^X x^{-p}\,dx = \lim_{X\to\infty}\left[\frac{1}{1-p}x^{1-p}\right]_1^X$$
$$= \lim_{X\to\infty}\left[\frac{1}{1-p}X^{1-p} - \frac{1}{1-p}\right].$$

If $p > 1$, then this limit is $-\dfrac{1}{1-p} = \dfrac{1}{p-1}$ and the improper integral and series converge. If $0 < p < 1$, then the limit is ∞, so that the improper integral and series diverge. We summarize these results in the next theorem.

Theorem 3 *The p-series*

$$\sum_{j=1}^\infty \frac{1}{j^p} = 1 + \frac{1}{2^p} + \frac{1}{3^p} + \frac{1}{4^p} + \cdots \tag{8}$$

converges if $p > 1$ and diverges to ∞ if $p \leq 1$.

The divergent p-series

$$\sum_{j=1}^\infty \frac{1}{j} \tag{9}$$

with $p = 1$ is known as the HARMONIC SERIES.

A divergence test

If you examine the convergent series $\sum_{j=j_0}^\infty a_j$ that we have considered up to now, you would see that the terms a_j tend to 0 as $j \to \infty$. This is true in general, as we will show by proving the following theorem.

Theorem 4 (A divergence test) *If the infinite series $\sum_{j=j_0}^\infty a_j$ converges, then $a_j \to 0$ as $j \to \infty$. Accordingly, if a_j does not tend to 0 as $j \to \infty$, then the infinite series diverges.*

To prove this theorem, we let s_n denote the partial sum

$$s_n = \sum_{j=j_0}^n a_j \quad \text{for } n \geq j_0.$$

Then for $n > j_0$,

$$a_n = s_n - s_{n-1}. \tag{10}$$

If the series converges, then the partial sums have a finite limit L and by **(10)**,

$$\lim_{n\to\infty} a_n = \lim_{n\to\infty}(s_n - s_{n-1}) = \lim_{n\to\infty} s_n - \lim_{n\to\infty} s_{n-1} = L - L = 0$$

as is stated in the theorem.

Example 2 Show that $\sum_{j=1}^{\infty} \frac{j}{j+1}$ diverges.

SOLUTION The series diverges because its terms $a_j = \frac{j}{j+1} = \frac{1}{1+1/j}$ tend to 1 as $j \to \infty$, rather than to 0 as is required for the series to converge □

Question 3 Apply Theorem 3 to the series $\sum_{j=1}^{\infty} j(-1)^{j+1}$. Generate a plot of the first 25 partial sums in the window $-1 \le n \le 26, -15 \le y \le 18$ to corroborate your answer.

Absolute convergence

If some of the terms a_j in an infinite series $\sum_{j=j_0}^{\infty} a_j$ are positive and some are negative, and if $a_j \to 0$ as $j \to \infty$, then usually the next step in studying its convergence is to study the series $\sum_{j=j_0}^{\infty} |a_j|$ obtained by taking the absolute values of all of the terms.

Example 3 **(a)** Generate in the window $-1 \le n \le 26, -0.25 \le y \le 3.5$ plots of the first 25 partial sums of $\sum_{j=1}^{\infty} \frac{(-1)^{j+1}}{j^{1.25}}$ and of $\sum_{j=1}^{\infty} \left|\frac{(-1)^{j+1}}{j^{1.25}}\right| = \sum_{j=1}^{\infty} \frac{1}{j^{1.25}}$. **(b)** Why do the two graphs contain the same point for $n = 1$? **(c)** Why are the points giving the other partial sums for $\sum_{j=1}^{\infty} \frac{(-1)^{j+1}}{j^{1.25}}$ closer to the n-axis than those for $\sum_{j=1}^{\infty} \frac{1}{j^{1.25}}$?

SOLUTION **(a)** The partial sums are plotted in Figures 4 and 5.
(b) The partial sums for $n = 1$ are equal to their first terms, $(-1)^{1+1}/1^{1.25}$ and $1/1^{1.25}$, which are both equal to 1.
(c) The other partial sums for $\sum_{j=1}^{\infty} \frac{(-1)^{j+1}}{j^{1.25}}$ are closer to the n-axis than those for $\sum_{j=1}^{\infty} \frac{1}{j^{1.25}}$ because there is cancellation of positive and negative terms in the first series. □

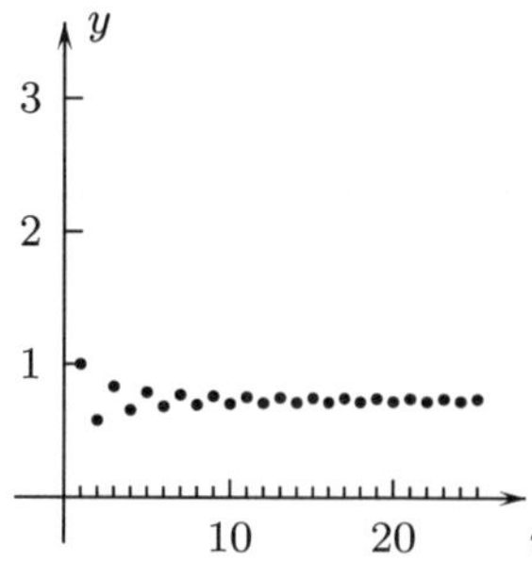

Partial sums of $\sum_{j=1}^{\infty} \frac{(-1)^{j+1}}{j^{1.25}}$

FIGURE 4

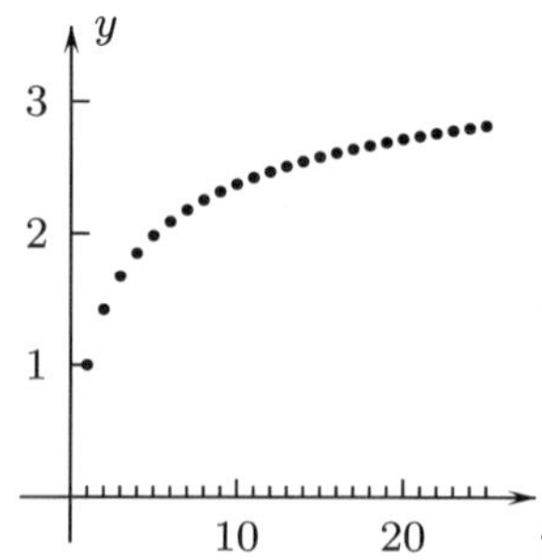

Partial sums of $\sum_{j=1}^{\infty} \frac{1}{j^{1.25}}$

FIGURE 5

The series $\sum_{j=1}^{\infty} \frac{1}{j^{1.25}}$ converges because it is a p-series with $p = 1.25 > 1$. This means that its terms tend to zero rapidly enough that their sum is a finite number. It should not be surprising that the series $\sum_{j=1}^{\infty} \frac{(-1)^{j+1}}{j^{1.25}}$, which involves the same numbers but with some added and some subtracted, also converges. This is established in the next theorem.

Theorem 5 *If* $\sum_{j=j_0}^{\infty} |a_j|$ *converges, then so does the series* $\sum_{j=j_0}^{\infty} a_j$.

If $\sum_{j=j_0}^{\infty} |a_j|$ converges, then the original series $\sum_{j=j_0}^{\infty} a_j$ is said to CONVERGE ABSOLUTELY. With this terminology, Theorem 5 can be rephrased as the statement that any series that converges absolutely, in fact, converges.

To establish Theorem 5, we suppose that $\sum_{j=j_0}^{\infty} |a_j|$ converges. Then its partial sums are bounded. We define numbers b_j and c_j by

$$b_j = \begin{cases} a_j & \text{if } a_j \geq 0 \\ 0 & \text{if } a_j < 0 \end{cases} \quad \text{and} \quad c_j = \begin{cases} 0 & \text{if } a_j \geq 0 \\ -a_j & \text{if } a_j < 0. \end{cases}$$

The numbers b_j and c_j are nonnegative and satisfy $b_j \leq |a_j|$ and $c_j \leq |a_j|$ for $j \geq j_0$. Consequently, the partial sums for $\sum_{j=j_0}^{\infty} b_j$ and $\sum_{j=j_0}^{\infty} c_j$ are also bounded and these series converge. Furthermore, $a_j = b_j - c_j$ for each j so that for each integer $n \geq j_0$

$$\sum_{j=j_0}^{n} a_j = \sum_{j=j_0}^{n} b_j - \sum_{j=j_0}^{n} c_j.$$

Since the partial sums on the right of this equation have finite limits as $n \to \infty$ so does the partial sum on the left and $\sum_{j=j_0}^{\infty} a_j$ converges, as stated in the theorem.

The Comparison Test

In many cases we can determine whether an infinite series converges absolutely or not by comparing it with a simpler series that we know converges or diverges.

Example 4 Figures 6 and 7 show the graphs of the first 26 partial sums of $\sum_{j=0}^{\infty} \frac{(0.6)^j}{j+1}$ and $\sum_{j=0}^{\infty} (0.6)^j$, respectively. **(a)** Why are all the partial sums of Figure 6 less than or equal to the corresponding partial sums of Figure 7? **(b)** Why does the series of Figure 7 converge? **(c)** Why does $\sum_{j=0}^{\infty} \frac{(0.6)^j}{j+1}$ converge?

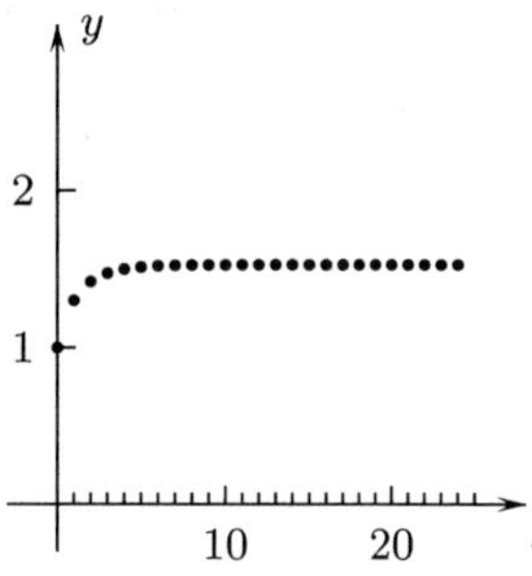

Partial sums of $\sum_{j=0}^{\infty}\frac{(0.6)^j}{j+1}$

FIGURE 6

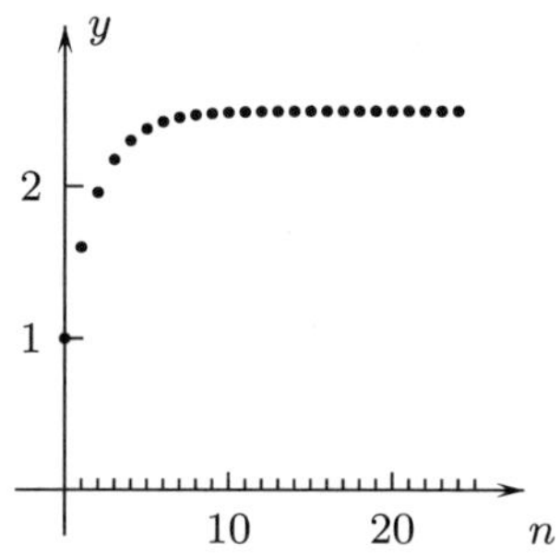

Partial sums of $\sum_{j=0}^{\infty}(0.6)^j$

FIGURE 7

SOLUTION (a) Each partial sum for $\sum_{j=0}^{\infty}\frac{(0.6)^j}{j+1}$ are less than or equal to the corresponding partial sum for $\sum_{j=0}^{\infty}(0.6)^j$ because $\frac{(0.6)^j}{j+1} \le (0.6)^j$ for $j \ge 0$.

(b) $\sum_{j=0}^{\infty}(0.6)^j$ converges because it is the Geometric Series $\sum_{j=0}^{\infty} r^j$ with $r = 0.6$.

(c) $\sum_{j=0}^{\infty}\frac{(0.6)^j}{j+1}$ converges because it has positive terms and its partial sums are bounded. This is the case because its partial sums are no greater than those for the convergent $\sum_{j=0}^{\infty}(0.6)^j$, whose partial sums are bounded. □

The reasoning in the solution of Example 4 with Theorem 5 can be generalized to establish part (a) of the following theorem.

Theorem 6 (The Comparison Test) **(a)** *If $\sum_{j=j_0}^{\infty} b_j$ is a convergent series with nonnegative terms and if there are constants M and J such that $|\, a_j| \le M\, b_j$ for $j \ge J$, then $\sum_{j=j_0}^{\infty} a_j$ converges absolutely.*

(b) *If $\sum_{j=j_0}^{\infty} b_j$ has nonnegative terms and diverges to ∞ and if there are positive constants m and J such that $|\, a_j| \ge m\, b_j$ for $j \ge J$, then $\sum_{j=j_0}^{\infty} a_j$ does not converge absolutely.*

To establish part **(b)** of the theorem, we suppose its hypotheses are satisfied. Then

$$\sum_{j=J}^{n} |\, a_j| \ge m \sum_{j=J}^{n} b_j.$$

The partial sums on the left of this inequality tend to ∞ as $n \to \infty$ because $\sum_{j=j_0}^{\infty} b_j$ has positive terms and diverges to ∞ Consequently, the partial sums of $\sum_{j=j_0}^{\infty} |\, a_j|$ also tend to ∞ and it diverges.

In the case of part (b) of Theorem 7, we cannot conclude that $\sum_{j=j_0}^{\infty} a_j$ diverges because, as we will see in Section 8.4, there are infinite series that converge even though they do not converge absolutely.

Example 5 Show that $\sum_{j=0}^{\infty} \frac{4+\sin j}{2^j}$ converges.

SOLUTION Because $|\sin j| \leq 1$ for all j, the terms in the series are positive and we have

$$\frac{4+\sin j}{2^j} \leq \frac{5}{2^j} \leq 5(\tfrac{1}{2})^j \quad \text{for all } j.$$

Therefore, the given series converges by the Comparison Test with the convergent geometric series $\sum_{j=0}^{\infty} (\frac{1}{2})^j$. □

Example 6 Does

$$\sum_{j=1}^{\infty} \frac{j+9}{3j^2} \tag{11}$$

converge or does it diverge to ∞?

SOLUTION For large j, we have

$$\frac{j+9}{3j^2} = \frac{1+9/j}{3j} \approx \frac{1}{3j}$$

so the terms in **(11)** are approximately equal to $\frac{1}{3}$ multiplied by the terms in the divergent Harmonic Series, and we can expect **(1)** to diverge. To show that if it does, we note that $j+9 > j$, and hence

$$\frac{j+9}{3j^2} \geq \frac{j}{3j^2} = \tfrac{1}{3}(\frac{1}{j}) \quad \text{for } j \geq 1.$$

The series diverges by the Comparison Test with the divergent harmonic series $\sum_{j=1}^{\infty} \frac{1}{j}$.

□

Question 4 Show that $\sum_{j=1}^{\infty} (-1)^j \frac{10\sqrt{j}}{j^2+1}$ converges absolutely.

The Limit Comparison Test

For large j, the terms in the series

$$\sum_{j=1}^{\infty} \frac{j-9}{3j^2} \tag{12}$$

are very close to those in series **(11)** of Example 6. In fact, $\frac{j-9}{3j^2} = \frac{1-9/j}{3j} \approx \frac{1}{3j^2}$ for large j, so we can anticipate that **(12)**, like **(11)**, diverges to ∞. However, because of the minus sign in **(12)**, it is somewhat difficult to apply Theorem 6. Instead we use the following result.

Theorem 7 (The Limit Comparison Test) *Suppose that*

$$\lim_{j\to\infty} \frac{|a_j|}{b_j} = L \tag{13}$$

with the b_j's positive and with L either a nonnegative number or ∞.

(a) *If L is finite and positive, then $\sum_{j=j_0}^{\infty} a_j$ converges absolutely, if and only if $\sum_{j=j_0}^{\infty} b_j$ converges.*

(b) *If $L = 0$ and $\sum_{j=j_0}^{\infty} b_j$ converges, then $\sum_{j=j_0}^{\infty} a_j$ converges absolutely.*

(c) *If $L = \infty$ and $\sum_{j=j_0}^{\infty} b_j$ diverges to ∞, then $\sum_{j=j_0}^{\infty} a_j$ does not converge absolutely.*

To prove part (a) of the theorem, we suppose that **(13)** holds with L a positive finite number. Then by Definition 1 of Section 7.2 with $\epsilon = \frac{1}{2}L$,

$$\left|\frac{|a_j|}{b_j} - L\right| < \tfrac{1}{2}L$$

for sufficiently large j. Solving this inequality for $|a_j|$ gives first $\frac{1}{2}L < |a_j|/b_j < \frac{3}{2}L$ and then

$$\tfrac{1}{2}Lb_j < |a_j| < \tfrac{3}{2}Lb_j$$

for sufficiently large j, so that, by the Comparison Test (Theorem 5), $\sum_{j=j_0}^{\infty} a_j$ converges absolutely if and only if $\sum_{j=j_0}^{\infty} b_j$ converges. The proofs of parts (b) and (c) are similar.

Example 7 Show that the series **(12)** diverges to ∞.

SOLUTION We compare $a_j = \dfrac{j-9}{3j^2}$ with the harmonic series $b_j = \dfrac{1}{j}$ by finding the limit

$$\begin{aligned}\lim_{j\to\infty} \frac{|a_j|}{b_j} &= \lim_{j\to\infty} \frac{(j-9)/(3j^2)}{1/j} = \lim_{j\to\infty} \frac{j(j-9)}{3j^2} \\ &= \lim_{j\to\infty} \frac{1-9/j}{3} = \tfrac{1}{3}.\end{aligned} \tag{14}$$

Because the limit **(14)** is nonzero and finite and the harmonic series diverges, the given series does not converge absolutely. Because its terms are positive, it diverges to ∞. □

Example 8 Does $\displaystyle\sum_{j=1}^{\infty} \frac{(-2)^j + 10}{5^j}$ converge absolutely?

SOLUTION For large j, we have

$$\left|\frac{(-2)^j + 10}{5^j}\right| = \frac{2^j|(-1)^j + 10/2^j|}{5^j} \approx (\tfrac{2}{5})^j$$

so we use the Limit Comparison Test with the convergent geometric series $\sum_{j=0}^{\infty} (\frac{2}{5})^j$.

We find that as $j \to \infty$,

$$\frac{\left|\dfrac{(-2)^j+10}{5^j}\right|}{\dfrac{2^j}{5^j}} = \frac{5^j|(-2)^j+10|}{2^j(5^j)} = |(-1)^j + 10/2^j| \to 1.$$

The given series converges absolutely because the geometric series converges and this limit is finite. □

Question 5 Does $\displaystyle\sum_{j=1}^{\infty} \frac{3j^2+7j}{j^5-j^2+5}$ converge or diverge? Justify your answer.

Principles and Procedures

- The integral Test and Comparison Test reflect the general principle that a series with nonnegative terms converges if and only if its terms tend to zero rapidly enough so they add up to a finite number.
- Theorem 3 with its criteria for the convergence of the p-series is the most important application of the Integral Test.
- The Limit-Comparison Test is generally easier to use than the Comparison Test, unless there is an obvious way to compare the terms of the given series with those of a series that you know converges or diverges.

Tune-Up Exercises 8.3♦

[A]*Answer provided.* [O]*Outline of solution provided.*

Determine whether the infinite series in Exercises T1 through T6 converge or diverge. For those convergent series that have positive and negative terms, determine whether they converge absolutely or not. Justify your answers.

T1.[O] $\displaystyle\sum_{j=0}^{\infty} \frac{9^j}{5+10^j}$

♦ Type 3, convergence

T2.[O] $\displaystyle\sum_{j=0}^{\infty} \frac{25}{\sqrt{j^2+9}}$

♦ Type 3, convergence

T3.[O] $\displaystyle\sum_{j=2}^{\infty} \frac{1}{j\ln j}$

♦ Type 3, convergence

T4.[O] $\displaystyle\sum_{j=1}^{\infty} (-1)^j \tan^{-1} j$

♦ Type 3, convergence

T5.[O] $\displaystyle\sum_{j=1}^{\infty} \frac{j^2}{\sqrt{j^5+7}}$

♦ Type 3, convergence

T6.[O] $\displaystyle\sum_{j=1}^{\infty} \frac{(-1)^j}{2j^2-15}$

♦ Type 3, convergence

♦The Tune-up Exercises and Problems are classified by type and content. The types are (1) basic, reactive; (2) basic reflective; (3) intermediate, reactive; (4) intermediate, reflective; (5) advanced, reactive; (6) advanced, reflective; and (7) advanced, theoretical.

Problems 8.3

[A]*Answer provided.* [O]*Outline of solution provided.*

Determine whether the infinite series in Problems 1 through 16 converge or diverge. For those convergent series that have positive and negative terms, determine whether they converge absolutely or not. Justify your answers.

1.[O] $\displaystyle\sum_{j=1}^{\infty} \frac{\cos j}{j^{3/2}}$

♦ Type 3, convergence

2.[O] $\displaystyle\sum_{j=1}^{\infty} \frac{4^j}{3^j - 10}$

♦ Type 3, convergence

3.[O] $\displaystyle\sum_{j=1}^{\infty} \frac{1}{\ln(\ln j)}$

♦ Type 3, convergence

4.[A] $\displaystyle\sum_{j=1}^{\infty} \frac{100}{3^j - 4j}$

♦ Type 3, convergence

5. $\displaystyle\sum_{j=0}^{\infty} \frac{2^j + 3}{4^j + 7}$

♦ Type 3, convergence

6. $\displaystyle\sum_{j=5}^{\infty} \left(\frac{20 + j}{3j + 10}\right)^j$

♦ Type 3, convergence

7. $\displaystyle\sum_{j=10}^{\infty} \frac{2j - 5}{10 + j^2}$

♦ Type 3, convergence

8. $\displaystyle\sum_{j=10}^{\infty} \frac{2 + \cos j}{3^j}$

♦ Type 3, convergence

9. $\displaystyle\sum_{j=2}^{\infty} \frac{1}{2^j}\left(50 + \frac{2}{j}\right)$

♦ Type 3, convergence

10. $\displaystyle\sum_{j=2}^{\infty} \left(\frac{8}{2^j - 7}\right)^2$

♦ Type 3, convergence

11. $\displaystyle\sum_{j=2}^{\infty} \frac{\ln j}{\ln j + 5}$

♦ Type 3, convergence

12. $\displaystyle\sum_{j=9}^{\infty} \sqrt{j}$

♦ Type 3, convergence

13. $\displaystyle\sum_{j=9}^{\infty} \frac{j^2 + 1}{j^3 - 12j}$

♦ Type 3, convergence

14. $\displaystyle\sum_{j=0}^{\infty} e^{j-10}$

♦ Type 3, convergence

15. $\displaystyle\sum_{j=2}^{\infty} \frac{1}{j\sqrt{\ln j}}$

♦ Type 3, convergence

16. $\displaystyle\sum_{j=0}^{\infty} \frac{(-1)^j}{1 + e^j}$

♦ Type 3, convergence

17. For what values of the constant k does the infinite series $\displaystyle\sum_{j=1}^{\infty} \frac{50j^k}{100 + j^9}$ converge?

♦ Type 3, convergence

Use a calculator or computer algorithm for finding approximate values of partial sums as required in the following Problems.

18.[A] Approximately how close is $\displaystyle\sum_{j=10}^{\infty} \frac{1}{j^2}$ to $\displaystyle\sum_{j=1}^{\infty} \frac{1}{j^2}$?

♦ Type 3, values of partial sums

19. Compare the values of $\displaystyle\sum_{j=1}^{10} \frac{1}{2^j + 4^j}$ and $\displaystyle\sum_{j=1}^{50} \frac{1}{2^j + 4^j}$ and explain the result.

♦ Type 3, values of partial sums

20. Compare the values of $\displaystyle\sum_{j=2}^{100} \frac{1}{\ln j}$ and $\displaystyle\sum_{j=2}^{200} \frac{1}{\ln j}$ and explain the result.

♦ Type 3, values of partial sums

21. Compare the values $\sum_{j=1}^{100} \frac{10}{j^3}$ and $\sum_{j=1}^{100} \frac{1}{j^3}$ and explain the result.

♦ Type 3, values of partial sums

22. Compare the values $\sum_{j=101}^{110} \tan^{-1} j$ and $\sum_{j=111}^{120} \tan^{-1} j$ and explain the result.

♦ Type 3, values of partial sums

23. The infinite series $\sum_{j=1}^{\infty} \frac{1}{j^2}$ has the value $\frac{1}{6}\pi^2$. Find a positive integer n such that $\sum_{j=1}^{n} \frac{1}{j^2}$ differs from $\frac{1}{6}\pi^2$ by less than 0.001 and more than 0.0005.

♦ Type 3, the epsilon-N-definition of convergence

24. The infinite series $\sum_{j=1}^{\infty} \frac{(-1)^{j+1}}{j^4}$ has the value $\frac{7}{720}\pi^4$. Find a positive integer n such that $\sum_{j=1}^{\infty} \frac{(-1)^{j+1}}{j^4}$ differs from $\frac{7}{720}\pi^4$ by less than 10^{-7} and more than 10^{-8}.

♦ Type 3, the epsilon-N-definition of convergence

Section 8.4

The Ratio and Alternating-Series Tests

OVERVIEW: *In this section we discuss the* RATIO *and* ALTERNATING-SERIES TESTS. *With the Ratio Test an infinite series is shown to converge or diverge by studying the ratios of its successive terms. The Alternating-Series Test deals with certain series whose terms are alternately positive and negative and which in many cases converge, even though they because there is cancellation between their positive and negative terms. We also show how the Alternating-Series Test can be used to estimate the errors that are made when partial sums are used to approximate convergent infinite series.*

Topics:

- ***The Ratio Test***
- ***The Alternating-Series Test***
- ***Conditional convergence***
- ***Approximating alternating series with partial sums***

The Ratio Test

When we use the Comparison or Limit-Comparison Test to study the convergence of an infinite series, we compare its terms with the terms in a series that is known to converge or diverge. In the Ratio Test, we study instead the ratios of succesive terms in the series under consideration. We begin with an example.

Question 1 Complete Table 1 below with partial sums of the infinite series $\sum_{j=1}^{\infty} j\left(\frac{2}{3}\right)^j$ with positive terms. Does the completed table suggest that the infinite series converges or diverges?

TABLE 1

n	$\sum_{j=1}^{n} j\left(\frac{2}{3}\right)^j \doteq$	n	$\sum_{j=1}^{n} j\left(\frac{2}{3}\right)^j \doteq$	n	$\sum_{j=1}^{n} j\left(\frac{2}{3}\right)^j \doteq$
1	0.6667	4	3.2346	15	
2	1.5556	5	3.8930	20	
3	2.4444	10		25	

The plot in Figure 1 of partial sums of $\sum_{j=1}^{\infty} j\left(\frac{2}{3}\right)^j$, like their values in the completed Table 1, suggest that the infinite series converges. How can we be sure? We might try to use the Comparison or Limit-Comparison Test to compare it with the convergent geometric series $\sum_{j=1}^{\infty}\left(\frac{2}{3}\right)^j$, but this will not work because the factor j in the jth term of $\sum_{j=1}^{\infty} j\left(\frac{2}{3}\right)^j$ tends to infinity.

♦ SUGGESTIONS TO INSTRUCTORS: Have students begin work on Questions 1 through 8 before class. Problems 1–4 and 21–22 are good for in-class discussion.

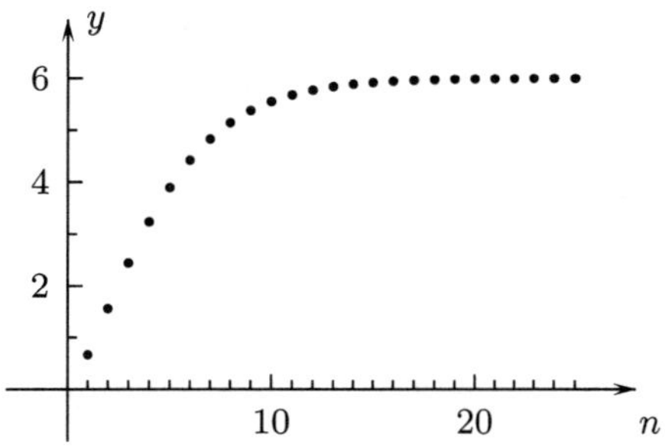

FIGURE 1

Instead we look at the limit of the ratio of successive terms in the series. We set $a_j = j\left(\frac{2}{3}\right)^j$ so that the series is $\sum_{j=1}^{\infty} a_j$ and the ratio of the $(j+1)$st and jth terms is

$$\frac{a_{j+1}}{a_j} = \frac{(j+1)\left(\frac{2}{3}\right)^{j+1}}{j\left(\frac{2}{3}\right)^j} = \frac{j+1}{j}\left(\frac{2}{3}\right)$$
$$= \frac{2}{3}\left(1+\frac{1}{j}\right). \tag{1}$$

The ratio **(1)** is greater than $\frac{2}{3}$ for all $j \geq 1$ but tends to $\frac{2}{3}$ as $j \to \infty$. Consequently, for any number r with $\frac{2}{3} < r < 1$, the ratio is less than r for sufficiently large j. We will use $r = \frac{3}{4}$.

Question 2 For what integer values of j is the ratio **(1)** $\leq \frac{3}{4}$?

Formula **(1)** with the result of Question 1 shows that $\dfrac{a_{j+1}}{a_j} \leq \frac{3}{4}$ for $j \geq 8$, so that

$$a_{j+1} \leq \tfrac{3}{4}\, a_j \quad \text{for } j \geq 8. \tag{2}$$

Question 3 Use inequality **(2)** **(a)** to compare a_9 with a_8, **(b)** to compare a_{10} first with a_9 and then with a_8, and **(c)** to compare a_{11} first with to a_{10} and then with a_8.

Continuing the calculations in Question 3 would show that $a_j \leq \left(\frac{3}{4}\right)^{j-8} a_8$ for all $j \geq 9$, so that

$$a_j \leq \left[\left(\tfrac{3}{4}\right)^{-8} a_8\right]\left(\tfrac{3}{4}\right)^j \quad \text{for } j \geq 9. \tag{3}$$

Because $\left[\left(\frac{3}{4}\right)^{-8} a_8\right]$ is a constant and the numbers a_j are positive, inequalities **(3)** show that the series $\sum_{j=1}^{\infty} a_j = \sum_{j=1}^{\infty} j\left(\frac{2}{3}\right)^j$ converges by the Comparison Test with the convergent Geometric Series $\sum_{j=0}^{\infty} \left(\frac{3}{4}\right)^j$.

This reasoning can be generalized to establish part (a) of the following theorem.

Theorem 1 (The Ratio Test) *Suppose that a_j are nonzero numbers for $j \geq j_0$ and that*

$$\lim_{j\to\infty} \frac{|a_{j+1}|}{|a_j|} = \rho \tag{4}$$

where ρ is a nonnegative number or ∞.

(a) *If $\rho < 1$, then $\sum_{j=j_0}^{\infty} a_j$ converges absolutely.*

(b) *If ρ is a number greater than 1 or is ∞, then $|a_j| \to \infty$ as $j \to \infty$ and the series diverges.*

(c) *If $\rho = 1$, the test gives no information: the series might converge absolutely or not.*

If $\rho = 1$, as in part (c) of this theorem, we say that the Ratio Test does not apply.

The general proof of part (b) of Theorem 2 is very short. If ρ defined by **(4)** is ∞ or is a number greater than 1, then by Definition 1 of Section 8.1, there is a number J such that $\dfrac{|a_{j+1}|}{|a_j|} > 1$ for $j > J$. Then $|a_{j+1}| > |a_j|$ for $j > J$, so the terms of the infinite series do not tend to zero as $j \to \infty$ and the series diverges.

Part (c) of the theorem is illustrated by the p-series $\sum_{j=1}^{\infty} \dfrac{1}{j^p}$. For any constant p,

$$\rho = \lim_{j\to\infty} \frac{1/(j+1)^p}{1/j^p} = \lim_{j\to\infty} \frac{j^p}{(j+1)^p} = \lim_{j\to\infty} \frac{1}{(1+1/j)^p} = 1$$

and the Ratio Test does not apply, but the series converges if $p > 1$ and diverges if $p \leq 1$.

The Ratio Test is often used with series whose terms involve factorials, such as $j!$, or powers, such as a^j or j^j.[†]

Example 1 Does $\sum_{j=0}^{\infty} \dfrac{5^j}{j!}$ converge or diverge?

SOLUTION Since the terms in the series are positive, it either converges absolutely or diverges to ∞. Because of the factorial $j!$ and the power 5^j in the series, we try the Ratio Test. We set $a_j = 5^j/j!$, so that $a_{j+1} = 5^{j+1}/(j+1)!$ and

$$\frac{|a_{j+1}|}{|a_j|} = \frac{\dfrac{5^{j+1}}{(j+1)!}}{\dfrac{5^j}{j!}} = \frac{j!}{5^j}\frac{5^{j+1}}{(j+1)!} = \frac{5}{j+1}.$$

The limit ρ of this ratio as $j \to \infty$ is 0, which is less than 1. By part (a) of Theorem 1, the series converges absolutely. □

[†]The FACTORIAL FUNCTION $j!$ equals 1 for $j = 0$ and $j = 1$ and, for integers $j > 1$, equals the product $1 \cdot 2 \cdot 3 \cdots j$ of the positive integers $\leq j$. Thus, $2! = 2 \cdot 1 = 2$; $3! = 3 \cdot 2 \cdot 1 = 6$; $4! = 4 \cdot 3 \cdot 2 \cdot 1 = 24$; and so forth.

Example 2 Does $\sum_{j=0}^{\infty} \frac{(-2)^j}{j^3}$ converge absolutely or diverge?

SOLUTION We set $a_j = (-2)^j/j^3$ and calculate the ratio

$$\frac{|a_{j+1}|}{|a_j|} = \frac{\left|\frac{(-2)^{j+1}}{(j+1)^3}\right|}{\left|\frac{(-2)^j}{j^3}\right|} = \frac{j^3}{(j+1)^3}\frac{2^{j+1}}{2^j} = \frac{2}{(1+1/j)^3}.$$

The limit ρ of this ratio as $j \to \infty$ is 2, which is greater than 1. The series diverges by part (b) of Theorem 2. □

Example 3 **(a)** Apply the Ratio Test to $\sum_{j=1}^{\infty} \frac{1}{j^3+1}$. **(b)** Use the Comparison Test to determine whether the series of part (a) converges or diverges.

SOLUTION **(a)** With $a_j = 1/(j^3+1)$, we obtain

$$\frac{|a_{j+1}|}{|a_j|} = \frac{1/[(j+1)^3+1]}{1/(j^3+1)} = \frac{j^3+1}{(j+1)^3+1} = \frac{1+1/j^3}{(1+1/j)^3+1/j^3}.$$

This ratio tends to $\rho = 1$ as $j \to \infty$, so the Ratio Test does not apply.

(b) Since $\frac{1}{j^3+1} < \frac{1}{j^3}$ for $j \geq 1$, the given series converges by the Comparison Test with the convergent p-series $\sum_{j=1}^{\infty} \frac{1}{j^3}$. □

Question 4 Does $\sum_{j=1}^{\infty} \frac{j!}{j^3}$ converge or diverge?

Question 5 Does $\sum_{j=1}^{\infty} \frac{j^3}{j!}$ converge or diverge?

Alternating Series

When the Comparison, Limit-Comparison or Integral Test of the last section is used to show that an infinite series $\sum_{j=j_0}^{\infty} a_j$ converges, the terms a_j tend to zero fast enough so that the sum $\sum_{j=j_0}^{\infty} |a_j|$ of their absolute values is finite and the series converges absolutely. The ALTERNATING-SERIES TEST deals with series of the form

$$\sum_{j=0}^{\infty} (-1)^j c_j \tag{5}$$

with positive numbers c_j. We call such a series an ALTERNATING SERIES. We begin again with an example.

Question 6 **(a)** Complete Table 2 below of partial sums of the divergent Harmonic Series $\sum_{j=1}^{\infty} \frac{1}{j}$ and of the alternating series $\sum_{j=1}^{\infty} \frac{(-1)^{j+1}}{j}$. **(b)** Describe how the partial sums of the alternating series change as n increases.

TABLE 2

n	$\sum_{j=1}^{n}\frac{1}{j} \doteq$	$\sum_{j=1}^{n}\frac{(-1)^{j+1}}{j} \doteq$	n	$\sum_{j=1}^{n}\frac{1}{j} \doteq$	$\sum_{j=1}^{n}\frac{(-1)^{j+1}}{j} \doteq$
1	1	1	5		
2	1.5	0.5	6		
3	1.8333	0.8333	7		
4	2.0833	0.5833	8		

Partial sums of the two series in Question 6 are plotted in Figures 2 and 3. The Harmonic Series of Figure 2 diverges to ∞, but series **(5)** of Figure 3 converges, as is shown by the next theorem.

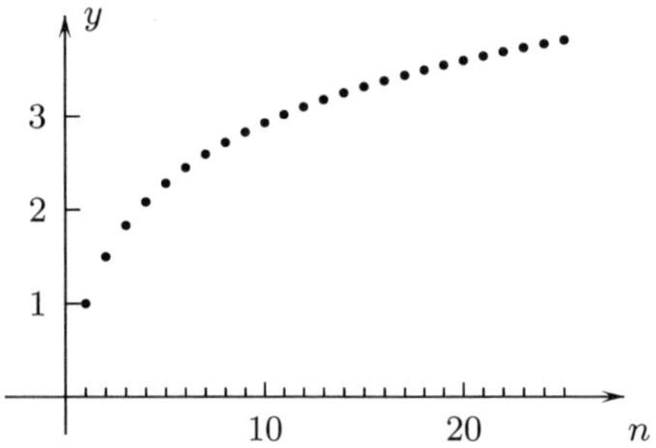

Partial sums of $\sum_{j=1}^{\infty}\frac{1}{j}$

FIGURE 2

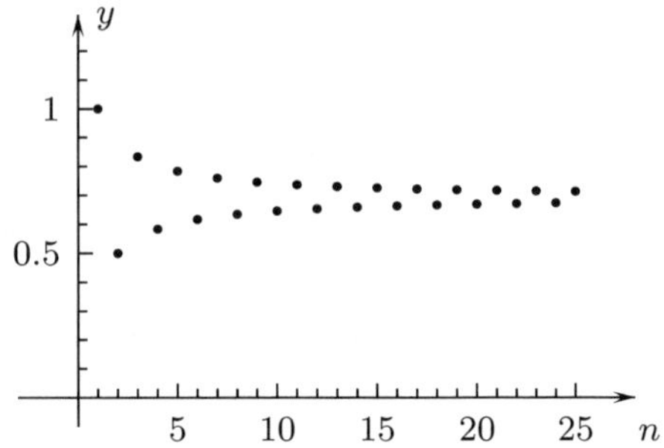

Partial sums of $\sum_{j=1}^{\infty}\frac{(-1)^{j+1}}{j}$

FIGURE 3

Theorem 2 (The Alternating-Series Test) *If the sequence $\{c_j\}_{j=j_0}^{\infty}$ of positive numbers tends to 0 as $j \to \infty$ and is nonincreasing for $j \geq J$ with some J, then the alternating series*

$$\sum_{j=j_0}^{\infty}(-1)^j c_j \tag{6}$$

converges.

To see why this theorem holds, imagine that the partial sums

$$s_n = \sum_{j=0}^{n}(-1)^j c_j = c_0 - c_1 + c_2 - \cdots + (-1)^n c_n \tag{7}$$

are plotted on an s-axis. The partial sum $s_0 = c_0$ is positive (Figure 4) and the inequalities $0 < c_1 \leq c_0$ imply that $s_1 = c_0 - c_1$ lies between 0 and s_0 (Figure 5).

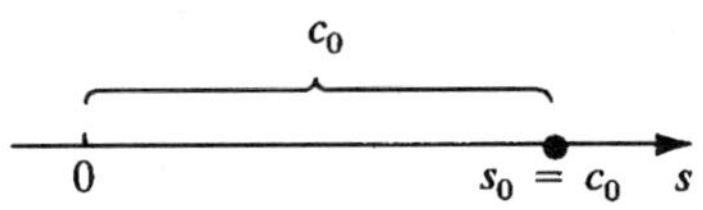

FIGURE 4

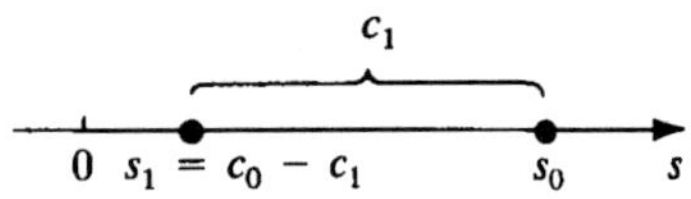

FIGURE 5

The partial sum $s_2 = c_o - c_1 + c_2$ lies between s_0 and s_1 because $0 < c_2 \leq c_1$ (Figure 6), and the next partial sum $s_3 = c_0 - c_1 + c_2 - c_3$ is between s_1 and s_2 because $0 < c_3 \leq c_2$ (Figure 7). Similarly, each subsequent partial sum lies between the previous two. Moreover, as can be seen in Figure 7 for the first few partial sums, the sequence $s_0, s_2, s_4, \cdots$ of the partial sums s_n with even integers n is nonincreasing, whereas the sequence $s_1, s_3, s_5, \cdots$ with odd n is nondecreasing.

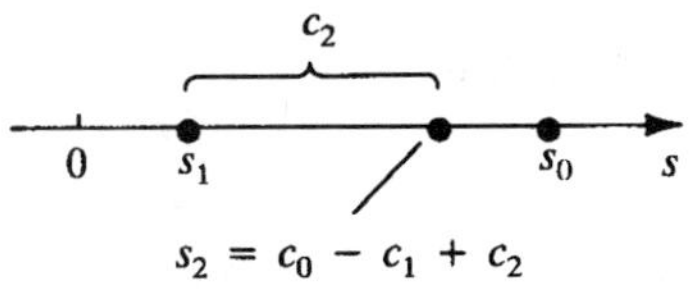

FIGURE 6

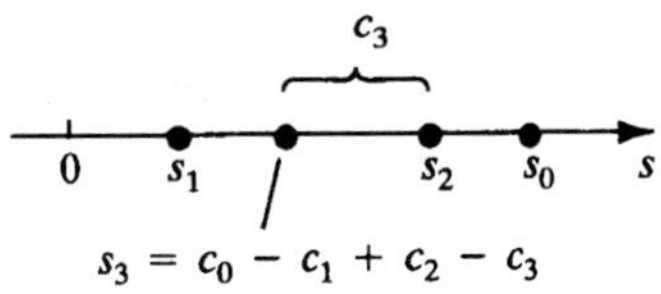

FIGURE 7

The monotone sequences $\{s_{2k}\}_{k=0}^{\infty}$ and $\{s_{2k+1}\}_{k=0}^{\infty}$ both have finite limits because they are between 0 and c_0 and consequently are bounded. The limits must be the same because $|s_{2k+1} - s_{2k}| = c_{2k+1}$ tends to 0 as $k \to \infty$. Thus, all the partial sums have the same finite limit and the series **(1)** converges. This establishes Theorem 2 for the case of $j_0 = 0$

Notice that under the conditions of Theorem 2, the series $\sum\limits_{j=j_0}^{\infty} (-1)^{j+1} c_j$ also converges since it equals series **(2)** multiplied by -1.

Example 4 Show that the series $\sum\limits_{j=1}^{\infty} \frac{(-1)^{j+1}}{\sqrt{j}}$ converges. Does it converge absolutely?

SOLUTION The series converges by the Alternating-Series test because the sequence $c_j = 1/\sqrt{j}$ is positive and decreasing and tends to 0 as $j \to \infty$. The series does not converge absolutely because replacing the terms $(-1)^{j+1}/\sqrt{j}$ by their absolute values $1/\sqrt{j}$ gives a divergent p-series with $p = \frac{1}{2}$. □

A series such as in Example 4, which converges but does not converge absolutely, is said to CONVERGE CONDITIONALLY.

Question 7 Does the alternating series $\sum\limits_{j=1}^{\infty} (-1)^{j+1}(1/j^2)$ converge conditionally, converge absolutely, or diverge? Plot its first 10 partial sums.

Question 8 Does the alternating series $\sum\limits_{j=1}^{\infty} (-1)^{j+1}\sqrt{j}$ converge condtionally, converge absolutely, or diverge? Plot its first 10 partial sums.

Example 5 Does

$$\sum_{j=0}^{\infty} (-1)^j \frac{j}{j^2 + 20} \tag{8}$$

converge absolutely, converge conditionally, or diverge?

SOLUTION We first test for absolute convergence. Because

$$\frac{j}{j^2+20} = \frac{j}{j^2(1+20/j^2)} = \frac{1}{j(1+20/j^2)} \approx \frac{1}{j}$$

for large j, we use the limit comparison test with the divergent harmonic series $\sum 1/j$. We find that

$$\frac{j/(j^2+20)}{1/j} = \frac{j^2}{j^2+20} = \frac{1}{1+20/j^2} \to 1 \quad \text{as} \quad j \to \infty.$$

Since this limit is positive, **(8)** does not converge absolutely.

Because of the factor $(-1)^j$ in the series, we try the Alternating Series test. The numbers $c_j = \dfrac{j}{j^2+20}$ are positive and tend to 0. We need to see if they are nonincreasing for large j. We consider the function $f(x) = x/(x^2+20)$ whose value at $x = j$ is the number c_j. The derivative

$$\begin{aligned} f'(x) = \frac{d}{dx}\left(\frac{x}{x^2+20}\right) &= \frac{(x^2+20)\dfrac{d}{dx}x - x\dfrac{d}{dx}(x^2+20)}{(x^2+20)^2} \\ &= \frac{(x^2+20) - x(2x)}{(x^2+20)^2} = \frac{20-x^2}{(x^2+20)^2} \end{aligned}$$

is negative for $x > \sqrt{20}$. Therefore, the sequence c_j is decreasing for $j \geq 5$, and **(8)** converges by the alternating series test. Because it converges but does not converge absolutely, it converges conditionally. □

Approximating alternating series with partial sums

Suppose L is the sum $\sum\limits_{j=j_0}^{\infty} (-1)^j c_j$ of an alternating series that satisfies the hypotheses of Theorem 2. Then the reasoning we used to establish the theorem shows that L lies between the partial sums $s_{N-1} = \sum\limits_{j=j_0}^{N-1} (-1)^j c_j$ and $s_N = \sum\limits_{j=j_0}^{N} (-1)^j c_j$ for each N. Because $|s_N - s_{N-1}| = c_N$, we conclude that the error made in approximating the sum of the infinite series by the partial sums satifies

$$[\text{Error}] = |L - s_N| \leq c_N. \qquad \textbf{(9)}$$

Example 6 **(a)** Use **(9)** to find a number N such that the partial sum $s_N = \sum\limits_{j=1}^{N} (-1)^{j+1}(1/j)$ differs from the infinite series $\sum\limits_{j=1}^{\infty} (-1)^{j+1}(1/j)$ by less than 0.01. **(b)** Use a calculator or computer to calculate the partial sum from part (a). **(c)** As we will see in Section 8.6, the alternating series from part (a) equals ln(2). How closely does the partial sum from part (b) approximate this value?

SOLUTION **(a)** In this case, $c_n = 1/n$ so we pick n so that $\dfrac{1}{n} \leq 0.01$. We can use $n = 1/0.01 = 100$. **(a)** A procedure for calculating partial sums on a calculator or computer gives $s_{100} = \sum\limits_{j=1}^{100} (-1)^{j+1}(1/j) \doteq 0.688172$, so the error made in using this to approximate L is $|L - s_{100}|$, which has the approximate decimal value $|0.693147 - 0.688172| \doteq 0.00497$. □

Principles and procedures

- The integral Test and Comparison Test of the last section are based on the general principle that a series with nonnegative terms converges if its terms tend to zero rapidly enough. The Alternating-Series Test is based on a different idea. Under its conditions, the partial sums of the series have a finite limit because each partial sum lies between the previous two and the difference between successive partial sums tends to zero as $n \to \infty$.
- The techniques developed in the last three sections for determining whether an infinite series $\sum_{j=j_0}^{\infty} a_j$ converges or diverges can be summarized as follows.

  STEP 1. Try to find the limit of a_j as $j \to \infty$. If the limit does not exist or exists and is not 0, then the series diverges. If the limit is 0 or cannot be determined, try other convergence or divergence test.

  STEP 2. If $|a_j|$ resembles Cr^j or C/j^p for large j with constants C and r or p, try the Comparison Test or Limit-Comparison Test with the geometric series $\sum r^j$ or the p-series $\sum 1/j^p$, respectively.

  STEP 3. If $|a_j| = f(j)$ for $j \geq J$, where $f(x)$ is a positive nonincreasing function that can be readily integrated, try the Integral Test.

  STEP 4. If other tests show that the series does not converge absolutely and if $a_j = (-1)^j c_j$ or $a_j = (-1)^{j+1} c_j$ with positive numbers c_j, try the Alternating-Series Test.

  STEP 5. If a_j involves powers such as r^j or j^j or factorials such as $j!$, try the Ratio Test.

Tune-Up Exercises 8.4[♦]

[A]*Answer provided.* [O]*Outline of solution provided.*

T1.[O] **(a)** Calculate approximate decimal values of the first eight partial sums of $\sum_{j=0}^{\infty} \frac{3^j}{j!}$.

(b) Generate a plot of the first twenty-five partial sums in the window $-1 \leq x \leq 26, -2.5 \leq y \leq 25$.

(c) Determine whether the infinite series converges or diverges.

(d) Compare approximate decimal values of $\sum_{j=0}^{16} \frac{3^j}{j!}$ and $\sum_{j=0}^{17} \frac{3^j}{j!}$ and explain the result.

♦ Type 3, convergence, etc.

T2.[O] **(a)** Calculate approximate decimal values of the first eight partial sums of $\sum_{j=0}^{\infty} \frac{j!}{7^j}$.

(b) Generate a plot of the first twenty-five partial sums in the window $-1 \leq x \leq 26, -1 \leq y \leq 10$.

(c) Determine whether the infinite series converges or diverges.

(d) Compare approximate decimal values of $\sum_{j=0}^{24} \frac{j!}{7^j}$ and $\sum_{j=0}^{25} \frac{j!}{7^j}$ and explain the result.

♦ Type 3, convergence, etc.

[♦]The Tune-up Exercises and Problems are classified by type and content. The types are (1) basic, reactive; (2) basic reflective; (3) intermediate, reactive; (4) intermediate, reflective; (5) advanced, reactive; (6) advanced, reflective; and (7) advanced, theoretical.

T3.[O] **(a)** Calculate approximate decimal values of the first eight partial sums of $\sum_{j=2}^{\infty} \frac{(-1)^j}{\ln j}$.
(b) Generate a plot of the first twenty-five partial sums in the window $-1 \le x \le 26, -0.2 \le y \le 2$.
(c) Determine whether the infinite series converges absolutely, converges conditionally, or diverges.
(d) Compare approximate decimal values of $\sum_{j=2}^{20} \frac{(-1)^j}{\ln j}$, $\sum_{j=2}^{21} \frac{(-1)^j}{\ln j}$, and $\sum_{j=2}^{22} \frac{(-1)^j}{\ln j}$ and explain the result.

♦ Type 3, convergence, etc.

T4.[O] **(a)** Calculate approximate decimal values of the first eight partial sums of $\sum_{j=1}^{\infty} \frac{10j(-1)^j}{j+5}$.
(b) Generate a plot of the first twenty-five partial sums in the window $-1 \le x \le 26, -6 \le y \le 6$.
(c) Determine whether the infinite series converges absolutely, converges conditionally, or diverges.
(d) Compare approximate decimal values of $\sum_{j=1}^{19} \frac{10j(-1)^j}{j+5}$ and $\sum_{j=1}^{20} \frac{10j(-1)^j}{j+5}$ and explain the result.

♦ Type 3, convergence, etc.

Problems 8.4

[A] *Answer provided.* [O] *Outline of solution provided.*

In Problems 1 through 20 determine whether the series converges or diverges. If it converges and has some negative terms, determine whether it converges absolutely or conditionally. Justify your answer. In Problems 1 through 6 generate the graph of the first 25 partial sums in the indicated window and copy it on your paper.

1.[A] $\sum_{j=1}^{\infty} \frac{j^2(-1)^{j+1}}{2^j}$ $(-1 \le x \le 26, -0.75 \le y \le 0.75)$

♦ Type 3, convergence, etc.

2.[A] $\sum_{j=1}^{\infty} \frac{(-1.5)^j}{j^2}$ $(-1 \le x \le 26, -10 \le y \le 10)$

♦ Type 3, convergence, etc.

3. $\sum_{j=0}^{\infty} \frac{j!}{(2j)!}$ $(-1 \le x \le 26, -0.25 \le y \le 2.5)$

♦ Type 3, convergence, etc.

4. $\sum_{j=0}^{\infty} \frac{10(-1)^j}{\sqrt{j^2+1}}$ $(-1 \le x \le 26, -1.5 \le y \le 12)$

♦ Type 3, convergence, etc.

5. $\sum_{j=2}^{\infty} (-1)^j \sqrt{\ln j}$ $(-1 \le x \le 26, -1.5 \le y \le 2)$

♦ Type 3, convergence, etc.

6.[A] $\sum_{j=0}^{\infty} \frac{10}{8-1.1^j}$ $(-1 \le x \le 26, -30 \le y \le 75)$

♦ Type 3, convergence, etc.

7.[A] $\displaystyle\sum_{j=0}^{\infty} \frac{(-1)^j}{(2j)!}$

♦ Type 3, convergence, etc.

8. $\displaystyle\sum_{j=1}^{\infty} j^{10}(0.9)^j$

♦ Type 3, convergence, etc.

9. $\displaystyle\sum_{j=1}^{\infty} \left(\frac{-10}{j}\right)^j$

♦ Type 3, convergence, etc.

10.[A] $\displaystyle\sum_{j=1}^{\infty} \left(\frac{-j}{10}\right)^j$

♦ Type 3, convergence, etc.

11. $\displaystyle\sum_{j=1}^{\infty} (-1)^j \frac{j!}{j^j}$

♦ Type 3, convergence, etc.

12. $\displaystyle\sum_{j=0}^{\infty} \frac{j^4}{4^j}$

♦ Type 3, convergence, etc.

13. $\displaystyle\sum_{j=1}^{\infty} (-1)^j e^{1/j}$

♦ Type 3, convergence, etc.

14.[A] $\displaystyle\sum_{j=0}^{\infty} \left(\frac{-e}{\pi}\right)^j$

♦ Type 3, convergence, etc.

15. $\displaystyle\sum_{j=0}^{\infty} \frac{10+j^4}{j!}$

♦ Type 3, convergence, etc.

16. $\displaystyle\sum_{j=0}^{\infty} \frac{3+5^j}{2-6^j}$

♦ Type 3, convergence, etc.

17. $\displaystyle\sum_{j=1}^{\infty} \left(\frac{10-4j}{7j}\right)^j$

♦ Type 3, convergence, etc.

18. $\displaystyle\sum_{j=1}^{\infty} \left(\frac{10-7j}{4j}\right)^j$

♦ Type 3, convergence, etc.

19. $\displaystyle\sum_{j=0}^{\infty} \frac{j^2}{j!-20}$

♦ Type 3, convergence, etc.

20. $\displaystyle\sum_{j=0}^{\infty} \frac{j!}{(j+2)!}$

♦ Type 3, convergence, etc.

In Problems 21 through 23 find the approximate values of the infinite series with the given degree of accuracy by using a calculator or computer to find the approximate value of the partial sum with N chosen by estimate **(9)**.

21. $\displaystyle\sum_{j=0}^{\infty} \frac{(-1)^j}{\sqrt{1+j^3}}$; [Error] $\leq 10^{-3}$

♦ Type 4, applying error estimates

22. $\displaystyle\sum_{j=0}^{\infty} (-1)^j \frac{100}{\sqrt{j+3^j}}$; [Error] $\leq 10^{-7}$

♦ Type 4, applying error estimates

23. $\displaystyle\sum_{j=2}^{\infty} (\frac{(-1)^j}{(\ln j)^8)}$; [Error] $\leq 10^{-6}$

♦ Type 4, applying error estimates

24. $\displaystyle\sum_{j=1}^{\infty} \frac{(-1)^{j+1}}{j^{3/4}}$; [Error] ≤ 0.02

♦ Type 4, applying error estimates

Section 8.5

Taylor Polynomials and Taylor's Theorem

OVERVIEW: *In this section we discuss approximations of functions by* TAYLOR POLYNOMIALS *and* TAYLOR'S THEOREM, *which is used to estimate errors that are made with such approximations.*

Topics:

- ***Taylor polynomials***
- ***Taylor's Theorem***

Taylor polynomials

As we saw in Chapter 2, the tangent line $y = f(x_0) + f'(x_0)(x - x_0)$ to the graph of a differentiable function $y = f(x)$ at $x = x_0$ approximates the graph near $x = 0$. Consequently, the first-degree polynomial $P_1(x) = f(a) + f'(a)(x - a)$, whose graph is the tangent line, approximates $f(x)$ near $x = x_0$.

For example, $f(x) = \sin x$ has the value 0 at $x = 0$ and its derivative $f'(x) = \cos x$ equals 1 at $x = 0$, so the tangent line to $y = \sin x$ at $x = 0$ is the line $y = x$ shown in Figure 1, and the first-degree polynomial x

$$P_1(x) = x. \tag{1}$$

approximates $\sin x$ closely near $x = 0$ is

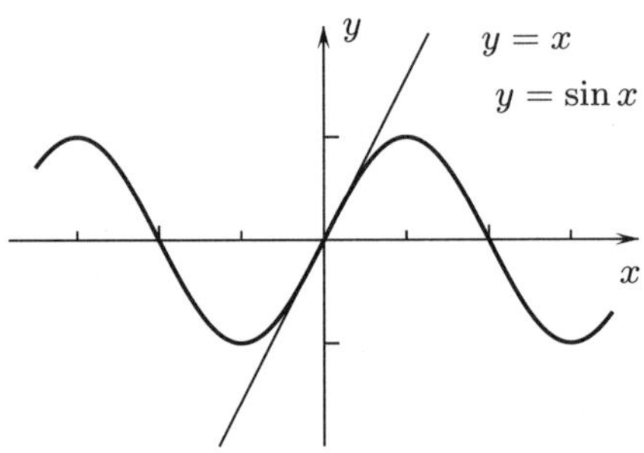

$P_1(x) = x$
FIGURE 1

The polynomial **(1)** can be characterized as the first-degree polynomial that has the same value and the same first derivative as $\sin x$ at $x = 0$. We can obtain better approximations of $\sin x$ near $x = 0$ by using polynomials of higher degree whose higher derivatives have the same values at $x = 0$ as the corresponding derivatives of $\sin x$.

Example 1 Find the constants a_2 and a_3 such that $P_3(x) = x + a_2x^2 + a_3x^3$ has the same second and third derivatives as $\sin x$ at $x = 0$.

SOLUTION With $P_3(x) = x + a_2x^2 + a_3x^3$,‡

$$P_3'(x) = 1 + 2a_2x + 3a_3x^2$$
$$P_3''(x) = 2a_2 + 6a_3x$$
$$P_3^{(3)}(x) = 6a_3.$$

Therefore $P_3''(0) = 2a_2$ and $P_3^{(3}(0) = 6a_3$. The second derivative of $\sin x$ is $\frac{d}{dx}(\cos x) = -\sin x$, which equals 0 at $x = 0$. Its third derivative is $\frac{d}{dx}(-\sin x) = -\cos x$, which equals -1 at $x = 0$. For $P_3(x)$ and $\sin x$ to have the same second and third derivatives at $x = 0$ we must have $2a_2 = 0$ and $6a_3 = -1$, which implies that $a_2 = 0$ and $a_3 = -\frac{1}{6}$. □.

‡Recall that $f^{(j)}(x)$ denotes the jth derivative of $f(x)$.

◆ SUGGESTIONS TO INSTRUCTORS: Have students begin work on Questions 1 through 4 before class.Problems 15, 16, 19, 24, 25, 29, and 35 are good for in-class discussion.

Figure 2 shows the graphs of $\sin x$ and of the third-degree polynomial

$$P_3(x) = x - \tfrac{1}{6}x^3 \tag{2}$$

whose coefficients were found in Example 1. Notice that $P_3(x)$ approximates $\sin x$ better near $x = 0$ than does $P_1(x) = x$ of Figure 2.

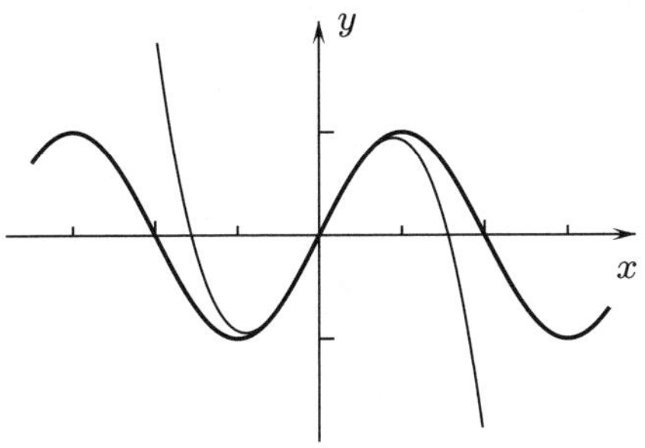

$P_3(x) = x - \frac{1}{6}x^3$

FIGURE 2

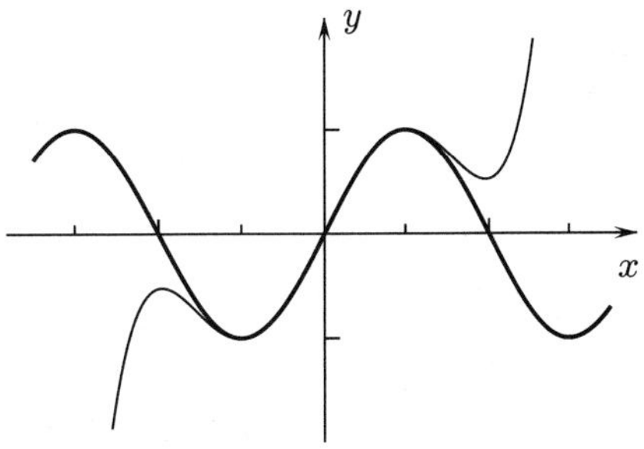

$P_5(x) = x - \frac{1}{6}x^3 + \frac{1}{120}x^5$

FIGURE 3

Question 1 Find constants a_4 and a_5 such that $P_5(x) = x - \frac{1}{6}x^3 + a_3x^3 + a_4x^4 + a_5x^5$ has the same fourth and fifth derivatives as $\sin x$ at $x = 0$.

The fifth-degree polynomial

$$P_5(x) = x - \tfrac{1}{6}x^3 + \tfrac{1}{120}x^5 \tag{3}$$

from Question 1 has the same value and the same first five derivatives as $\sin x$ at $x = 0$. Its graph is is Figure 3. Notice that it approximates $\sin x$ near $x = 0$ even better than does $P_3(x)$ of Figure 2.

We can also compare the approximations of $\sin x$ by $P_1(x)$, $P_3(x)$, and $P_5(x)$ by calculating how close each of the polynomials is to $\sin x$ for specific values of x.

Question 2 Complete Table 1 below with values of $|\sin x - P_1(x)|$, $|\sin x - P_3(x)|$, and $|\sin x - P_5(x)|$ and describe the results.

TABLE 1

x	$\lvert\sin x - P_1(x)\rvert \doteq$	$\lvert\sin x - P_3(x)\rvert \doteq$	$\lvert\sin x - P_5(x)\rvert \doteq$
2	1.09070	0.24263	0.02404
1	0.15853	0.00814	1.96×10^{-4}
0.1			
0.01			

The polynomial $P_n(x)$ of degree n that has the same value and the same first n derivatives as a function $f(x)$ at $x = x_0$ is called the nth degree TAYLOR POLYNOMIAL of $f(x)$ CENTERED at $x = x_0$.[†] The polynomials $P_1(x), P_3(x)$, and $P_5(x)$, defined in **(1)**, **(2)**, and **(3)**, are the first-, third-, and fifth-degree Taylor polynomials of $\sin x$ centered at $x = 0$.

We need to find concise formulas for the Taylor polynomials of a general function $f(x)$ in terms of values of the function and its derivatives. We start with the case of fourth-degree polynomials.

[†]Taylor polynomials are named after the British mathematician Brook Taylor (1685–1731).

If $f(x)$ has derivatives up to the fourth order at x_0, then its fourth-degree Taylor polynomial centered at x_0 can be written in the form

$$P_4(x) = c_0 + c_1(x - x_0) + c_2(x - x_0)^2 + c_3(x - x_0)^3 + c_4(x - x_0)^4 \tag{4}$$

with constants c_0, c_1, c_2, c_3, and c_4.

Example 2 Give formulas in terms of $f(x)$ and its derivatives for c_0, c_1, c_2, c_3, and c_4 such that **(4)** and $f(x)$ have the same values and the same first four derivatives at $x = x_0$.

SOLUTION We calculate

$$\begin{aligned}
P_4'(x) &= c_1 + 2c_2(x - x_0) + 3c_3(x - x_0)^2 + 4c_4(x - x_0)^3 \\
P_4''(x) &= 2c_2 + 3(2)c_3(x - x_0) + 4(3)(x - x_0)^2 \\
P_4^{(3)}(x) &= 3(2)c_3 + 4(3)(2)c_4(x - x_0) \\
P_4^{(4)}(x) &= 4(3)(2)c_4.
\end{aligned}$$

We have not carried out the multiplications in these formulas because we want to show the patterns of the numbers. These formulas with **(4)** show that

$$P_4(x_0) = c_0, P_4'(x_0) = c_1, P_4''(x_0) = 2c_2, P_4^{(3)}(x_0) = 3(2)c_3, P_4^{(4)}(x_0) = 4(3)(2)c_4.$$

In order for these to equal the values of $f(x)$ and its first four derivatives at x_0, we must have $c_0 = f(x_0), c_1 = f'(x_0), 2c_2 = f''(x_0), 3(2)c_3 = f^3(x_0)$, and $4(3)2f^{(4)}(x_0)$. We write $2 = 2!, 3(2) = 3!$ and $4(3)(2) = 3!$ and solve for the numbers c_j to obtain

$$c_0 = f(x_0), c_1 = f'(x_0), c_2 = \frac{1}{2!}f''(x_0), c_3 = \frac{1}{3!}f^{(3)}(x_0), c_4 = \frac{1}{4!}f^{(4)}(x_0). \ \square$$

Substituting the expressions from Example 2 in **(4)** gives a formula for the fourth-degree Taylor polynomial of $f(x)$ centered at x_0:

$$\begin{aligned}
P_4(x) = f(x_0) &+ f'(x_0)(x - x_0) + \frac{1}{2!}f^{(2)}(x_0)(x - x_0)^2 \\
&+ \frac{1}{3!}f^{(3)}(x_0)(x - x_0)^3 + \frac{1}{4!}f^{(4)}(x_0)(x - x_0)^4.
\end{aligned}$$

Similar calculations for Taylor polynomials of other degrees lead to the following rule.

Rule 1 *Suppose that $f(x)$ has n derivatives at $x = x_0$ with n a nonnegative integer. The nth-degree Taylor approximation of $f(x)$ centered at x_0 is*

$$\begin{aligned}
P_n(x) =& f(x_0) + f'(x_0)(x - x_0) + \frac{1}{2!}f''(x_0)(x - x_0)^2 \\
&+ \frac{1}{3!}f^{(3)}(x_0)(x - x_0)^3 + \frac{1}{4!}f^{(4)}(4)(x_0)(x - x_0)^4 \\
&+ \cdots + \frac{1}{(n-1)!}f^{(n-1)}(x_0)(x - x_0)^{n-1} + \frac{1}{n!}f^{(n)}(x_0)(x - x_0)^n.
\end{aligned} \tag{5}$$

We say that the Taylor polynomial **(5)** is CENTERED at x_0 because it is a linear combination of powers of $x - x_0$. Notice that the Taylor polynomial $P_n(x)$ of degree n is obtained from the polynomial $P_{n-1}(x)$ of degree $n - 1$ by adding the term $\frac{1}{n!}f^{(n)}(a)(x - x_0)^n$.

Example 3 Give the first-, second, and third-degree Taylor Polynomials of $\ln x$ centered at $x = 1$. Generate the graph of each with the graph of $\ln x$.

SOLUTION We set $f(x) = \ln x$ and find that $f'(x) = x^{-1}, f''(x) = -x^{-2}$ and $f^{(3)}(x) = 2x^{-3}$. The values at $x = 1$ are $f(1) = \ln(1) = 0, f'(1) = 1, f''(1) = -1$, and $f^{(3)}(1) = 2$. By **(5)** with $x_0 = 1$ and $n = 1, 2$, and 3,

$$\begin{aligned} P_1(x) &= f(1) + f'(1)(x-1) = x - 1 \\ P_2(x) &= f(1) + f'(1)(x-1) + \frac{1}{2!}f''(1)(x-1)^2 = (x-1) - \tfrac{1}{2}(x-1)^2 \\ P_3(x) &= f(1) + f'(1)(x-1) + \frac{1}{2!}f''(1)(x-1)^2 + \tfrac{1}{3!}f^{(3)}(1)(x-1)^3 \\ &= (x-1) - \tfrac{1}{2}(x-1)^2 + \tfrac{1}{3}(x-1)^3. \end{aligned}$$

The graphs of three polynomials are shown in Figures 4 through 6. □

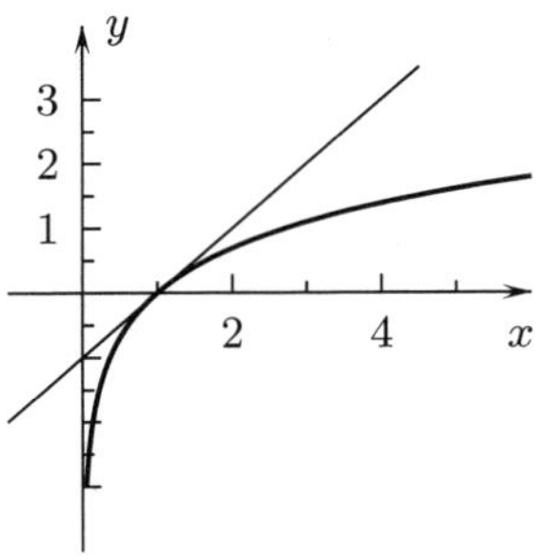

$P_1(x) = x - 1$

FIGURE 4

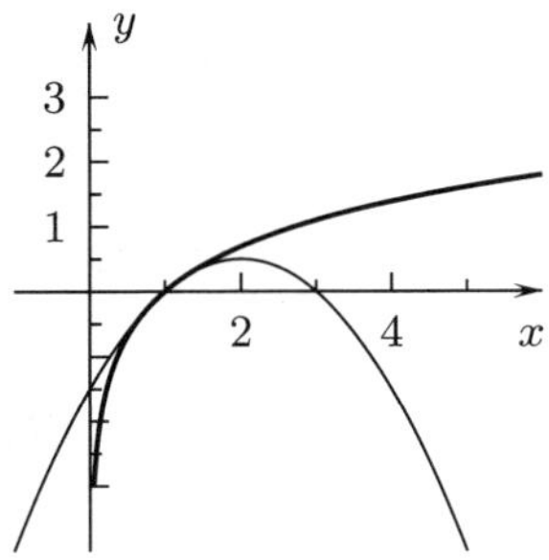

$P_2(x) = (x-1) - \frac{1}{2}(x-1)^2$

FIGURE 5

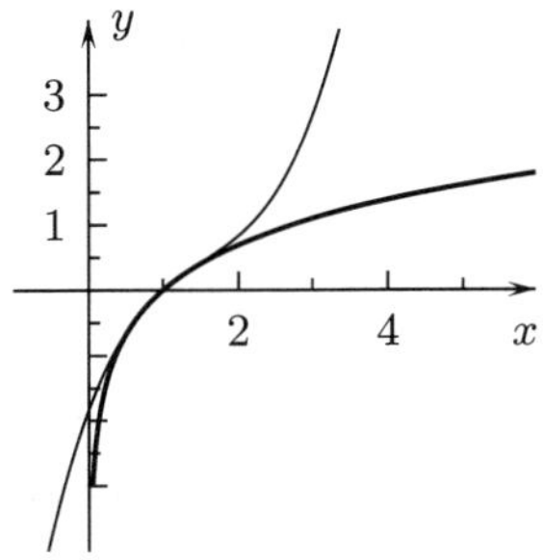

$P_3 = (x-1) - \frac{1}{2}(x-1)^2 + \frac{1}{3}(x-1)^3$

FIGURE 6

Question 3 **(a)** Find a formula for the third-degree Taylor Polynomial of e^x centered at $x = 0$ and generate its graph with the graph of e^x. **(a)** How accurately does the polynomial from part (a) approximate e^x at $x = 1$?.

Taylor's Theorem

To determine the accuracy of the approximation of $f(x)$ by a Taylor polynomial $P_n(x)$, we study the difference

$$R_n(x) = f(x) - P_n(x) \tag{6}$$

between the function and the approximation. Equation **(6)** can be written in the form

$$f(x) = P_n(x) + R_n(x) \tag{7}$$

and for this reason, $R_n(x)$ is called the REMAINDER in the approximation. In many cases, we can estimate the size and determine the sign of the remainder by using the next result, which is known as TAYLOR'S THEOREM.

Theorem 1 (Taylor's Theorem) *Suppose that $f(x)$ has $n+1$ derivatives in an open interval I and that x_0 is a point in I. For each x in I, there is at least one number c between x and x_0 such that the remainder in the approximations of $f(x)$ by $P_n(x)$ has the representation,*

$$R_n(x) = \frac{1}{(n+1)!}f^{(n+1)}(c)(x - x_0)^{n+1}. \tag{8}$$

When we say that c lies between x and x_0, we mean that $x_0 < c < x$ if $x > x_0$ and $x < c < x_0$ if $x < x_0$. We describe the proof of the theorem later in the section.

We can combine definition **(5)** of $P_n(x)$ with formula **(8)** for $R_n(x)$ and write

$$\begin{aligned} f(x) &= P_n(x) + R_n(x) \\ &= f(x_0) + f'(x_0)(x - x_0) + \frac{1}{2!}f''(x_0)(x - x_0)^2 + \frac{1}{3!}f^{(3)}(x_0)(x - x_0)^3 \\ &+ \cdots + \frac{1}{n!}f^{(n)}(x_0)(x - x_0)^n + \frac{1}{(n+1)!}f^{(n+1)}(c)(x - x_0)^{n+1}. \end{aligned} \tag{9}$$

Notice that the remainder, which is the last term in **(9)**, has the same form as the term $\frac{1}{(n+1)!}f^{(n+1)}(a)(x - a)^{n+1}$ that would be added to $P_n(x)$ to form $P_{n+1}(x)$, except that in the remainder, the derivative $f^{(n+1)}$ is evaluated at the unknown point c between x and a rather than at x_0.

Because the remainder $R_n(x)$ is the difference between $f(x)$ and $P_n(x)$, the absolute value of the remainder is the error in approximating $f(x)$ by $P_n(x)$. With expression **(8)** for the remainder we find that

$$\begin{bmatrix} \text{The error in} \\ \text{approximating} \\ f(x) \text{ by } P_n(x) \end{bmatrix} = |R_n(x)| = \frac{1}{(n+1)!}|f^{(n+1)}(c)||x - x_0|^{n+1} \tag{10}$$

for some c between x_0 and x. As is illustrated in the next example, we can often estimate the error by replacing $|f^{n+1}(c)|$ in **(10)** with its maximum for $x_0 \le c \le x$ if $x > x_0$ and with its maximum for $x \le c \le x_0$ if $x < x_0$.

Example 4 **(a)** Use the third-degree Taylor polynomial for $\ln x$ centered at $x = 1$ from Example 3 to give the approximate value of $\ln(1.1)$. **(b)** Use Taylor's theorem to determine whether the approximation is greater than or less than the exact value and to estimate the error.

SOLUTION **(a)** The Taylor polynomial is

$$P_3(x) = (x - 1) - \tfrac{1}{2}(x - 1)^2 + \tfrac{1}{3}(x - 1)^3$$

and its value at $x = 1.1$ is

$$\begin{aligned} P_3(1.1) &= (1.1 - 1) - \tfrac{1}{2}(1.1 - 1)^2 + \tfrac{1}{3}(1.1 - 1)^3 \\ &= 0.1 - \tfrac{1}{2}(0.01) + \tfrac{1}{3}(0.001) = 0.095333333. \end{aligned}$$

(b) Because the fourth derivative of $\ln x$ is $-6x^{-4}$, formula **(8)** with $n = 3$ and $x_0 = 1$ gives

$$R_3(x) = \frac{1}{4!}f^{(4)}(c)(x - 1)^4 = \frac{1}{4!}(-6c^{-4})(x - 1)^4 = -\frac{1}{4c^4}(x - 1)^4$$

with c an unknown number between 1 and x. Setting $x = 1.1$ shows that

$$R_3(1.1) = -\frac{1}{4c^4}(1.1 - 1)^4 = -\frac{1}{4c^4}(10^{-4}) \tag{11}$$

where c is an unknown number between 1 and 1.1. Because the remainder **(11)** is negative, the equation $\ln(1.1) = P_3(1.1) + R_3(1.1)$ shows that the approximation $P_3(1.1)$ is greater than the exact value of $\ln(1.1)$.

Equation **(11)** also shows that $|R_3(1.1)| = 10^{-4}/(4c^4)$. Therefore, since the maximum of $1/(4c^4)$ for $1 \le c \le 1.1$ is its value $\frac{1}{4} = 0.25$ at $c = 1$, the error $|R_3(1.1)|$ is less than $0.25(10^{-4}) = 2.5 \times 10^{-5}$. □

We can verify the conclusion of Example 4 by comparing the approximation $P_3(1.1) = 0.095333333$ with the approximate decimal value of $\ln(1.1) = 0.095310179$ that can be obtained with a calculator. We find that the approximation is greater than the actual value and that the error is $|\ln(1.1) - P_3(1.1)| = |0.095310179 - 0.095333333| = 2.3 \times 10^{-5}$ and is, in fact, less than the estimate 2.5×10^{-5} given by Taylor's theorem.

Question 4 **(a)** Use the third degree Taylor polynomial for $x^{1/3}$ centered at $x = 1$ to give the approximate value of $1.4^{1/3}$. **(b)** Estimate the error using Taylor's Theorem and then check the result by using the value $1.4^{1/3}$ given by a calculator.

A proof of Taylor's Theorem

We suppose that $f(x)$ satisfies the hypothesis of Theorem 1 above for a nonnegative n. We let $P_n(x)$ be its nth-degree Taylor polynomial centered at x_0 and $R_n(x) = f(x) - P_n(x)$ the corresponding remainder. We set $g(x) = R_n(x)$ and $h(x) = (x - x_0)^{n+1}$. Because $f(x)$ and $P_n(x)$ have the same value and the same first n derivatives at $x = x_0$, the value and first n derivatives of $g(x)$ at $x = x_0$ are 0. A short calclulation shows that $h(x)$ and its first n derivatives are 0 at $x = x_0$ and at no other points.

Consider x_1 in the interval I with $x_1 > x_0$. (A similar argument would handle the case of $x_1 < x_0$.) By the Generalized Mean-Value Theorem of Section 5.3, there is a number c_1 with $x_0 < c_1 < x_1$ such that

$$\frac{g(x_1)}{h(x_1)} = \frac{g(x_1) - g(x_0)}{h(x_1) - h(x_0)} = \frac{g'(c_1)}{h'(c_1)}.$$

Then, by the same theorem applied to the functions $g'(x)$ and $h'(x)$, there is a number c_2 with $x_0 < c_2 < c_1$ such that

$$\frac{g(x_1)}{h(x_1)} = \frac{g'(c_1)}{h'(c_1)} = \frac{g'(c_1) - g'(x_0)}{h'(c_1) - h'(x_0)} = \frac{g''(c_2)}{h''(c_2)}.$$

We continue in this fashion, applying the Generalized Mean-Value Theorem to successive derivatives of $g(x)$ and $h(x)$ until we obtain the equation

$$\frac{g(x_1)}{h(x_1)} = \frac{g^{(n+1)}(c_{n+1})}{h^{(n+1)}(c_{n+1})} \qquad \textbf{(12)}$$

where $x_0 < c_{n+1} < c_n < \cdots < c_2 < c_1 < x_1$. Because $P_n(x)$ is a polynomial of degree n, its $(n+1)$th derivative is 0 and hence $g^{(n+1)}(c_{n+1}) = f^{(n+1)}(c_{n+1})$. The $(n+1)$th derivative of $h(x)$ is the constant $(n+1)!$. Therefore, when we make the substitutions $g(x_1) = R_n(x_1)$ and $h(x_1) = (x_1 - x_0)^{n+1}$ in **(12)**, we obtain

$$\frac{R_n(b)}{(x_1 - x_0)^{n+1}} = \frac{1}{(n+1)!} f^{(n+1)}(c_{n+1}),$$

which, if we write c for c_{n+1} and replace x_1 with x, yields **(8)** to prove the theorem.

Principles and Procedures

- The nth-degree Taylor polynomial approximation $P_n(x)$ of $f(x)$, centered at x_0 is the nth-degree polynomial that has the same value and first n derivatives as $f(x)$ at x_0.
- Taylor's Theorem is used to estimate the error that is made when $P_n(x)$ is used to approxiamte $f(x)$ at any particular x.
- $P_n(x)$ is the nth-degree polynomial that best approximates $f(x)$ near $x = x_0$ in the sense that it is the only n-degree polynomial that satisfies the error estimates in Taylor's Theorem.

Tune-Up Exercises 8.5♦

[A]*Answer provided.* [O]*Outline of solution provided.*

T1.[O] (a) Give the fourth-degree Taylor Polynomial $P_4(x)$ of $2 + xe^x$ centered at $x = 0$. **(b)** Generate the graphs of xe^x and $P_4(x)$ together in the window $-4 \leq x \leq 3, -3 \leq y \leq 15$. **(c)** Calculate the values of $xe^x - P_4(x)$ at $x = -2, -1, -0.1, 0.1, 1$ and 2 and describe the results.

♦ Type 2, a Taylor polynomial, etc.

T2.[O] (a) Give the third-degree Taylor Polynomial $P_3(x)$ of $\cos x$ centered at $x = 3$. **(b)** Generate the graphs of $\cos x$ and $P_3(x)$ together in the window $-2 \leq x \leq 7, -1.5 \leq y \leq 15$. **(c)** Calculate the values of $\cos x - P_3(x)$ at $x = 1, 2, 2.9, 3.1, 4$ and 5 and describe the results.

♦ Type 2, a Taylor polynomial, etc.

T3.[O] What is the third-degree Taylor Polynomial $P_3(x)$ of $f(x) = 1 + 2x + 3x^2 + 4x^3 + 5x^4 + 6x^5$ centered at $x = 0$ and how is it related to $f(x)$?

♦ Type 2, a Taylor polynomial, etc.

T4.[O] (a) Use the estimate in Theorem 1 to estimate the error that is made when the fourth-degree Taylor Polynomial $P_4(x)$ of e^x centered at $x = 0$ is used to approximate $e^{0.5}$ **(b)** What is the actual error?

♦ Type 2, an error estimate, etc.

Problems 8.5

[A]*Answer provided.* [O]*Outline of solution provided.*

In Problems 1 through 14 find the indicated Taylor Polynomial of the given function. Then generate the graph of the function and the Taylor Polynomial together in a suitable window and copy them on your paper.

1.[O] The sixth-degree Taylor polynomial of $\cos x$ centered at $x = 0$

♦ Type 2, a Taylor polynomial, etc.

2.[A] The fourth-degree Taylor polynomial of $\tan^{-1} x$ centered at $x = 0$

♦ Type 2, a Taylor polynomial, etc.

3.[O] The second-degree Taylor polynomial of $\sqrt{x}$ centered at $x = 9$

♦ Type 2, a Taylor polynomial, etc.

4.[A] The third-degree Taylor polynomial of $\ln x$ centered at $x = 10$

♦ Type 2, a Taylor polynomial, etc.

5. The fourth-degree Taylor polynomial of $1 + e^{3x}$ centered at $x = 0$

♦ Type 2, a Taylor polynomial, etc.

6.[A] The second-degree Taylor polynomial of $\sin^{-1} x$ centered at $x = 0$

♦ Type 2, a Taylor polynomial, etc.

7.[A] The fourth-degree Taylor polynomial of $\ln(\cos x)$ centered at $x = 0$

♦ Type 2, a Taylor polynomial, etc.

8. The ninth-degree Taylor polynomial of $\sin x$ centered at $x = 0$

♦ Type 2, a Taylor polynomial, etc.

9. The third-degree Taylor polynomial of $(1 + x)^{1/3}$ centered at $x = 0$

♦ Type 2, a Taylor polynomial, etc.

10. The third-degree Taylor polynomial of $e^{\sin x}$ centered at $x = 0$

♦ Type 2, a Taylor polynomial, etc.

11. The third-degree Taylor polynomial of $\tan x$ centered at $x = \frac{1}{4}\pi$

♦ Type 2, a Taylor polynomial, etc.

12. The third-degree Taylor polynomial of $(1 + e^x)^{-1}$ centered at $x = 0$

♦ Type 2, a Taylor polynomial, etc.

13.[A] The second-degree Taylor polynomial of e^{x^2} centered at $x = 0$

♦ Type 2, a Taylor polynomial, etc.

14. The second-degree Taylor polynomial of x^5 centered at $x = 0$

♦ Type 2, a Taylor polynomial, etc.

♦The Tune-up Exercises and Problems are classified by type and content. The types are (1) basic, reactive; (2) basic reflective; (3) intermediate, reactive; (4) intermediate, reflective; (5) advanced, reactive; (6) advanced, reflective; and (7) advanced, theoretical.

15. How do the second-degree Taylor polynomials centered at $x = 0$ of $\sin x$ and $\sinh x$ differ?

♦ Type 3, sine and hyperbolic sine

16. **(a)** Find the fourth-degree Taylor Polynomial $P_4(x)$ of x^4 centered at $x = 10$. **(b)** How well do $P_4(11)$ and $P_4(20)$ approximate 11^4 and 20^4, respectively? Explain the results.

♦ Type 2, a Taylor polynomial, etc.

17.[A] Which approximates $e^{1/2}$ better, the value of the third-degree Taylor Polynomial of e^x centered at $x = 0$ or the third-degree Taylor Polynomial of e^x at $x = 1$?

♦ Type 2, a Taylor polynomial, etc.

18. Which approximates $\sin(\pi)$ better, the value of the third-degree Taylor Polynomial of $\sin x$ centered at $x = 0$ or the third-degree Taylor Polynomial of $\sin x$ centered at $x = 2\pi$?

♦ Type 2, a Taylor polynomial, etc.

19. Generate in one window the graphs of $\ln x$ and of its third-degree Taylor Polynomials centered at $x = 1, 2$, and 3. Copy the drawing on your paper.

♦ Type 2, a Taylor polynomial, etc.

20. Generate in one window the graphs of $\cos x$ and of its third-degree Taylor Polynomials centered at $x = 0, \pi$, and 2π. Copy the drawing on your paper.

♦ Type 2, a Taylor polynomial, etc.

21. Give a formula in terms of the parameter k for the fourth-degree Taylor Polynomial of e^{kx} centered at $x = 0$.

♦ Type 3, Taylor polynomial with parameters

22. Give a formula in terms of the positive parameter k for the fourth-degree Taylor Polynomial of $\ln(kx)$ centered at $x = 1$.

♦ Type 3, Taylor polynomial with parameters

23. Give a formula in terms of the nonzero parameter k for the third-degree Taylor Polynomial of $\dfrac{1}{k - x}$ centered at $x = 0$.

♦ Type 3, Taylor polynomial with parameters

24. Match the functions $\sin x, \cos x, \sinh x, \cosh x, e^x, 2^x \tan^{-1} x$, and $\ln(1 + x)$ to the Taylor Polynomials below by comparing values and derivatives at $x = 0$: **(a**[A]**)** $1 + x + \frac{1}{2}x^2 + \frac{1}{6}x^3 + \frac{1}{24}x^4$, **(b**[A]**)** $1 - \frac{1}{2}x^2 + \frac{1}{24}x^4 - \frac{1}{720}x^6$, **(c)** $x - \frac{1}{6}x^3 + \frac{1}{120}x^5 - \frac{1}{5040}x^7$, **(d**[A]**)** $x + \frac{1}{6}x^3 + \frac{1}{120}x^5 + \frac{1}{5040}x^7$, **(e)** $1 + \frac{1}{2}x^2 + \frac{1}{24}x^4 + \frac{1}{720}x^6$, **(f)** $1 + \ln(2)x - \frac{1}{2}[\ln(2)]^2x^2 + \frac{1}{6}[\ln(2)]^3x^3$, **(g)** $x - \frac{1}{2}x^2 + \frac{1}{3}x^3 - \frac{1}{4}x^4$, **(h)** $x - \frac{1}{3}x^3 + \frac{1}{5}x^5 - \frac{1}{7}x^7$

♦ Type 2, recognizing Taylor polynomials

25.[A] The fifth-degree Taylor Polynomial of $f(x)$ centered at $x = 0$ is $P_5(x) = x - 4x^4 + 3x^5$. What are **(a)** $f'(0)$, **(b)** $f''(0)$, **(c)** $f^{(4)}(0)$, and **(d)** $f^{(5)}(0)$?

♦ Type 2, derivatives of Taylor polynomials

26. The sixth-degree Taylor Polynomial of $g(x)$ centered at $x = 5$ is $P_6(x) = 10 - (x - 5)^3 + 7(x - 5)^4 - (x - 5)^6$. What are **(a)** $g(5)$, **(b)** $g^{(3)}(5)$, **(c)** $g^{(5)}(5)$, and **(d)** $g^{(6)}(5)$?

♦ Type 2, derivatives of Taylor polynomials

27.[A] Use the error representation in Theorem 1 to determine whether the second-degree Taylor Polynomial $P_2(x)$ of $f(x) = x^{2/3}$ centered at $x = 1$ is less than or greater than $f(x)$ at $x = 1.03$. Then verify your answer by calculating approximate decimal values of $f(1.03)$ and $P_2(1.03)$.

♦ Type 2, an error estimate, etc.

28. Use the error representation in Theorem 1 to determine whether the first-degree Taylor Polynomial $P_1(x)$ of $f(x) = \sin^{-1} x$ centered at $x = 0$ is less than or greater than $f(x)$ at $x = 0.5$. Then verify your answer by calculating approximate decimal values of $f(0.5)$ and $P_1(0.5)$.

♦ Type 2, an error estimate, etc.

29.[A] Use the error representation in Theorem 1 to estimate the error when the fifth-degree Taylor Polynomial $P_5(x)$ of $f(x) = \sin x$ centered at $x = 0$ is used to approximate $\sin(\frac{1}{10}\pi)$. Then verify your answer by calculating approximate decimal values of $\sin(\frac{1}{10}\pi)$ and $P_5(\frac{1}{10}\pi)$.

♦ Type 2, an error estimate, etc.

30. How much closer at $x = 0.5$ is the third-degree Taylor Polynomial of $\sqrt{1 + x}$ centered at 0 than the second -degree Taylor Polynomial of $\sqrt{1 + x}$ centered at 0?

♦ Type 2, a Taylor polynomial, etc.

31. Use the error representation in Theorem 1 for the first-degree Taylor Polynomial of $\tan^{-1} x$ centered at $x = 0$ to solve the inequality $x < \tan^{-1} x$. Check your answer by generating the graphs of $\tan^{-1} x$ and the Taylor polynomial in the same window.

♦ Type 4, an error estimate, etc.

32. **(a)** Use the error representation in Theorem 1 for the fifth-degree Taylor Polynomial $P_5(x)$ of e^x centered at $x = 0$ to find numbers a and b such that $|P_5(x) - e^x| < 1$ for $a < x < b$. **(b)** Use a formula for $R_5(x)$ and your calculator or computer to find an interval (a, b) such that $|P_5(x) - e^x| < 1$ for $a < x < b$.

♦ Type 4, an error estimate, etc.

33. How accurate is the estimate in Theorem 1 for predicting the error that is made when the fourth-degree Taylor Polynomial of $\ln x$ centered at $x = 1$ is used to approximate $\ln(2)$?

♦ Type 4, an error estimate, etc.

34. Why can the remainder $R_{2k}(x)$ be used instead of $R_{2k-1}(x)$ for positive integers k to estimate errors made when $\sin x$ is approximated by its $(2k - 1)$st Taylor Polynomial centered at $x = 0$?

♦ Type 4, an error estimate, etc.

35. Based on Theorem 1 and the estimate $0 < e < 3$, what degree Taylor Polynomial of e^x centered at $x = 0$ would be required to estimate **(a)** e^{10} and **(b)** e^{-10} within 0.1?

♦ Type 4, an error estimate, etc.

36. Based on Theorem 1, what degree Taylor Polynomial of $\sin x$ centered at $x = 0$ would be required to estimate $\sin x$ within 0.01 for $|x| \leq 2\pi$?

♦ Type 4, an error estimate, etc.

37. Based on Theorem 1, what is the largest value of k such that the 20th-degree Taylor Polynomial of $\cos x$ centered at $x = 0$ approximates $\cos x$ with an error < 0.1 for $|x| \leq k$?

♦ Type 4, an error estimate, etc.

38. Based on Theorem 1, what is the largest value of k such that the 50th-degree Taylor Polynomial of e^x centered at $x = 0$ approximates e^x with an error < 0.01 for $-k < x \leq 0$?

♦ Type 4, an error estimate, etc.

Section 8.6

Power Series

OVERVIEW: *The Taylor Polynomials we used in Section 8.5 to approximate functions are finite linear combinations of powers $(x - x_0)^n$ of $x - x_0$ with constants x_0 and nonnegative integers n. Infinite linear combinations of such powers are called* POWER SERIES *or* TAYLOR SERIES. *In this section we show how power series can be used to represent functions. We then explain how properties of the power functions $(x - x_0)^n$ can then often be used to determine properties of the functions represented by the power series.*

Topics:

- ***Taylor and MacLaurin Series***
- ***Intervals of convergence***
- ***Proving that functions equal their Taylor Series***
- ***Operations with power series***

Taylor and MacLaurin series

Recall that the nth-degree Taylor Polynomial of $f(x)$ centered at x_0 is the polynomial

$$P_n(x) = \sum_{j=0}^{n} \frac{1}{j!} f^{(j)}(x_0)(x - x_0)^j = f(x_0) + f'(x_0)(x - x_0) + \frac{1}{2!} f''(x_0)(x - x_0)^2$$
$$+ \frac{1}{3!} f^{(3)}(x_0)(x - x_0)^3 + \cdots + \frac{1}{n!} f^{(n)}(x_0)(x - x_0)^n.$$

The corresponding infinite series

$$\sum_{j=0}^{\infty} \frac{1}{j!} f^{(j)}(x_0)(x - x_0)^j = f(x_0) + f'(x_0)(x - x_0) + \frac{1}{2!} f''(x_0)(x - x_0)^2$$
$$+ \frac{1}{3!} f^{(3)}(x_0)(x - x_0)^3 + \frac{1}{4!} f^{(4)}(x_0)(x - x_0)^4 + \cdots \quad \textbf{(1)}$$

is called the TAYLOR SERIES of $f(x)$ centered at x_0, or the MACLAURIN Series of the function if $x_0 = 0$. An infinite series of the form **(1)** is also called a POWER SERIES. We begin our discussion of Taylor Series by seeing how they can be used to construct a solution of a differential equation.

We consider the differential equation with initial condition

$$f'(x) = f(x), \; f(0) = 1. \quad \textbf{(2)}$$

We know from Sections 3.4 and 7.1 that the solution of **(2)** is $f(x) = e^x$. How could we find the solution if we had not yet studied the exponential function? One approach is to use conditions **(2)** to construct a MacLaurin Series that gives the solution.

Example 1 Find the MacLaurin Series of the solution $f(x)$ of the initial-value problem **(2)**.

SOLUTION By **(1)** with $x_0 = 0$, the MacLaurin series of $f(x)$ centered at 0 is

$$\sum_{j=0}^{\infty} \frac{1}{j!} f^{(j)}(0)x^j = f(0) + f'(0)x + \frac{1}{2!} f''(0)x^2 + \frac{1}{3!} f^{(3)}(0)x^3 + \cdots. \quad \textbf{(3)}$$

The initial condition in **(2)** gives the value $f(0) = 1$ of the solution at $x = 0$. We need to find the values of all of its derivatives at $x = 0$ to use in **(3)**.

♦ SUGGESTIONS TO INSTRUCTORS: Have students begin work on Questions 1 through 4 before class. Problems 1–4, 9, 12–14, 20, and 22 are good for in-class discussion.

Equation **(2)** implies that $f'(0) = f(0) = 1$, and differentiating the differential equation in **(2)** gives

$$f''(x) = f'(x)$$

so that $f''(0) = f'(0) = 1$. In fact, differentiating n times with any positive integer n yields

$$f^{(n+1)}(x) = f^{(n)}(x).$$

Consequently, $f^{(n+1)}(0) = f^{(n)}(0)$ for all integers $n \geq 0$, so that $f^{(n)}(0) = 1$ for $n \geq 0$, and the MacLaurin Series **(3)** is

$$\sum_{j=0}^{\infty} \frac{1}{j!} x^j = 1 + x + \frac{1}{2!} x^2 + \frac{1}{3!} x^3 + \cdots . \square \tag{4}$$

Before we can use the MacLaurin Series **(4)** to define the solution of **(2)**, we have to determine whether it converges.

Example 2 For what values of x does the MacLaurin series **(4)** converge?

SOLUTION Because the terms of **(4)** involve factorials and powers, we use the Ratio Test of Section 8.4. We used it there only for infinite series with constant terms, whereas the terms $\frac{1}{j!} x^j$ in **(4)** include the variable x. Nevertheless, we can use the Ratio Test for each fixed x. We set $a_j = \frac{1}{j!} x^j$ and study the ratio of the absolute values of successive terms,

$$\frac{|a_{j+1}|}{|a_j|} = \frac{\frac{1}{(j+1)!}|x|^{j+1}}{\frac{1}{j!}|x|^j} = \frac{j!}{(j+1)!}|x| = \frac{|x|}{j+1}.$$

The limit ρ of this ratio as $j \to \infty$ is 0 for any fixed x. By the Ratio Test, the MacLaurin series **(4)** converges absolutely for all x. $\square$

Because the MacLaurin Series **(4)** converges for all x, we could use it to define the function e^x by writing

$$e^x = \sum_{j=0}^{\infty} \frac{1}{j!} x^j = 1 + x + \frac{1}{2!} x^2 + \frac{1}{3!} x^3 + \cdots . \tag{5}$$

Later in the section we will show that the function e^x defined by **(5)** is the natural exponential function that we have been using since Chapter 3.

In Chapter 13 we will apply the principles used in the solution of Example 1 to construct Taylor-Series solutions of other differential equations. Our main goal in this section is to study Taylor Series of familar functions. We begin with the function $\ln(x + 1)$.

Question 1 **(a)** What are the first five derivatives of $f(x) = \ln(1+x)$? **(b)** What are the values of $f(x)$ and of its first five derivatives at $x = 0$?

Example 3 Give the MacLaurin Series of $\ln(1+x)$.

SOLUTION The calculations in Question 1 show that for $f(x) = \ln(1+x)$,

$$\begin{aligned}
f'(x) &= (1+x)^{-1} \\
f''(x) &= -(1+x)^{-2} \\
f^{(3)}(x) &= 2(1+x)^{-3} \\
f^{(4)}(x) &= -(3)(2)(1+x)^{-4} \\
f^{(5)}(x) &= (4)(3)(2)(1+x)^{-5}.
\end{aligned}$$

As is illustrated by these calculations, taking the nth derivative of $\ln(1+x)$ for $n \geq 2$ multiplies the formula for the $(n-1)$st derivative by $-(n-1)$ and lowers the exponent on $(1+x)$ from $-(n-1)$ to $-n$. Consequently, for $j \geq 1$,

$$f^{(j)}(x) = (-1)^{j-1}(j-1)!(1+x)^{-j} \tag{6}$$

so that $f^{(j)}(0) = (-1)^{j-1}(j-1)!$ for $j \geq 1$. Since $\ln(1+x)$ equals 0 at $x = 0$, the MacLaurin Series **(3)** of $f(x) = \ln(1+x)$ is

$$\begin{aligned}
\sum_{j=0}^{\infty} \frac{1}{j!} f^{(j)}(0)\, x^j &= \sum_{j=1}^{\infty} \frac{1}{j!}(-1)^{j-1}(j-1)!\, x^j = \sum_{j=1}^{\infty} \frac{(-1)^{j-1}}{j} x^j \\
&= x - \tfrac{1}{2}x^2 + \tfrac{1}{3}x^3 - \tfrac{1}{4}x^4 + \cdots . \ \square
\end{aligned} \tag{7}$$

Example 4 For what values of x does the MacLaurin Series **(7)** of $\ln(1+x)$ converge?

SOLUTION We begin with the Ratio Test. Set $a_j = \dfrac{(-1)^{j-1}}{j} x^j$. Then for $j \geq 1$,

$$\frac{|\,a_{j+1}|}{|\,a_j|} = \frac{\dfrac{1}{j+1}|x|^{j+1}}{\dfrac{1}{j}|x|^j} = \frac{j}{(j+1)}\,|x| = \frac{|x|}{1+1/j}.$$

The limit ρ of this ratio as $j \to \infty$ is $|x|$. By the Ratio Test, the series converges absolutely if $|x| < 1$, for which $\rho < 1$, and diverges if $|x| > 1$, for which $\rho > 1$.

This leaves the case of $|x| = 1$, for which $x = -1$ or $x = 1$. For $x = -1$, the series is $\displaystyle\sum_{j=1}^{\infty} \frac{(-1)^j}{j}(-1)^j = \sum_{j=1}^{\infty} \frac{1}{j}$, which is the divergent Harmonic Series. For $x = 1$, the series is $\displaystyle\sum_{j=1}^{\infty} \frac{(-1)^j}{j}(1)^j = \sum_{j=1}^{\infty} \frac{(-1)^{j-1}}{j}$, which converges by the Alternating-Series Test since $1/(j+1) < 1/j$ and $1/j \to 0$ as $j \to \infty$. The Taylor series converges for $-1 < x \leq 1$. $\square$

Intervals of convergence

The power series (Taylor series)

$$\sum_{j=0}^{\infty} c_j(x-x_0)^j = c_0 + c_1(x-x_0) + c_2(x-x_0)^2 + \cdots \tag{8}$$

converges if $x = x_0$ because then all of its terms except possibly c_0 are zero. We will now show that if it converges for any $x \neq x_0$, then it converges for all x in an interval containing x_0.

Suppose that **(8)** converges for some $x_1 \neq x_0$. Then by Theorem 4 of Section 8.3, the terms of the series $c_j(x_1 - x_0)^j$ tend to 0 and there is a constant M such that $|c_j(x_1 - x_0)^j| \le M$ for $j \ge 0$. Consequently, for any x

$$|c_j(x - x_0)^j| \le |c_j(x_1 - x_0)^j| \left|\frac{x - x_0}{x_1 - x_0}\right|^j \le M \left|\frac{x - x_0}{x_1 - x_0}\right|^j. \tag{9}$$

For any number $x \neq x_0$ that is closer to x_0 than x_1, we have $|x - x_0| < |x_1 - x_0|$ so that the positive number,

$$r = \left|\frac{x - x_0}{x_1 - x_0}\right| \tag{10}$$

is less than 1. Also **(9)** gives

$$|c_j(x - x_0)^j| \le Mr^j$$

so that the power series **(8)** converges absolutely by the Comparison Test with the convergent geometric series $\sum_{j=0}^{\infty} r^j$. This gives the next result.

Lemma 1 *If the power series* **(8)** *converges for some* $x_1 \neq x_0$*, then the series converges absolutely for all* x *with* $|x - x_0| < |x_1 - x_0|$.

Lemma 1 is the basis of the next theorem.

Theorem 1 *For each power series* **(8)** *there are three possibilities. Either* **(a)** *the series converges only for* $x = x_0$*,* **(b)** *it converges absolutely for all* x*, or* **(c)** *there is a positive number* R *such that it converges absolutely for* $|x - x_0| < R$ *and diverges for* $|x - x_0| > R$.

If R is positive, then the series converges for x in an interval, which is called its INTERVAL OF CONVERGENCE. In case **(c)**, R is called the RADIUS OF CONVERGENCE, and the series may converge absolutely, converge conditionally, or diverge at the endpoints $x_0 \pm R$ of the interval of convergence. In case **(a)**, we say that the radius of convergence is 0, and in case **(b)** we say it is ∞.

To prove Theorem 1 we only need to deal with the case where the series converges at one value $x_1 \neq x_0$ and diverges at another value x_2. We let I be the set of numbers x for which it converges. Lemma 1 implies that $|x - x_0| \le |x_2 - x_0|$ for all x in I, so I is bounded. We define R to be the least upper bound of the numbers $|x - x_0|$ for x in I. R is positive because the series converges at x_1. The series diverges for $|x - x_0| > R$ by the definitions of I and R. To show that it converges absolutely for $|x - x_0| < R$, we consider any such x. Because R is the least upper bound of $|x - x_0|$ for x in I, there is an x_2 in I with $|x - x_0| < |x_2 - x_0| \le R$, and by Lemma 1, the series converges absolutely at x.

Example 5 Determine the radius of convergence and interval of convergence of the power series

$$\sum_{j=1}^{\infty} \frac{1}{j}\left(\frac{x}{3}\right)^j. \tag{11}$$

SOLUTION In this case, as in studying the convergence of most power series, we start with the Ratio Test. We set $a_j = \frac{1}{j}\left(\frac{x}{3}\right)^j$ with x a fixed but arbitrary nonzero number. The ratio

$$\frac{|a_{j+1}|}{|a_j|} = \frac{\left|\frac{1}{j+1}\left(\frac{x}{3}\right)^{j+1}\right|}{\left|\frac{1}{j}\left(\frac{x}{3}\right)^j\right|} = \frac{1}{1 + \frac{1}{j}}\left|\frac{x}{3}\right|$$

tends to $\rho = \frac{1}{3}|x|$ as $j \to \infty$. For $|x| < 3$, ρ is < 1 and the power series **(11)** converges absolutely. For $|x| > 3$, ρ is > 1 and the power series diverges. The radius of convergence R is 3.

We still have to study the convergence of the power series at the endpoints ± 3 of the interval of convergence. For $x = 3$, the power series **(11)** is the divergent Harmonic Series

$$\sum_{j=1}^{\infty} \frac{1}{j}\left(\frac{3}{3}\right)^j = \sum_{j=1}^{\infty} \frac{1}{j}.$$

For $x = -3$, it is the series

$$\sum_{j=1}^{\infty} \frac{1}{j}\left(\frac{-3}{3}\right)^j = \sum_{j=1}^{\infty} \frac{1}{j}(-1)^j$$

which converges conditionally by the Alternating-Series Test. Thus, the interval of convergence is $[-3, 3)$. □

Question 2 What is the radius of convergence of $\displaystyle\sum_{j=0}^{\infty} \frac{1}{(2j)!} x^j$?

Question 3 What is the radius of convergence of $\displaystyle\sum_{j=0}^{\infty} j^j x^j$?

Proving that functions equal their Taylor Series

In Section 8.5 on Taylor-polynomial approximations, we defined the remainder $R_n(x)$ to be the error in approximating the function $f(x)$ by its Taylor polynomial $P_n(x)$ centered at a:

$$f(x) = P_n(x) + R_n(x).$$

Taylor's Theorem then stated that the remainder can be written in the form

$$R_n(x) = \frac{1}{(n+1)!} f^{(n+1)}(c)(x-a)^{n+1} \tag{12}$$

where c is an unknown number between x and a. In many cases, we can use this formula to show that for x in the interval of convergence of the Taylor series, the remainder tends to 0 as $n \to \infty$ and, hence, the function is equal to its Taylor series. Completing this proof for $e^x, \sin x, \cos x$, and related functions requires the following lemma.

Lemma 2 *For any fixed x the quantity $\dfrac{x^n}{n!}$ tends to 0 as $n \to \infty$.*

This lemma follows from the response to Question 2. The terms $x^j/j!$ in the power series $\displaystyle\sum_{j=0}^{\infty} \frac{x^j}{j!}$ tend to zero as $j \to \infty$ for any x because the series converges for all x.

Example 6 Prove that the Maclaurin series **(5)** has the value e^x for all x.

SOLUTION In the case of $f(x) = e^x$, we have $f^{(j)}(x) = e^x$ for all x and all $j \geq 0$. Therefore, the Taylor polynomial for e^x centered at $x = 0$ is $P_n(x) = \displaystyle\sum_{j=0}^{n} \frac{1}{j!} x^j$ and

$$e^x = P_n(x) + R_n(x) \tag{13}$$

where $R_n(x) = \sum_{j=0}^{n} \frac{1}{(n+1)!} e^c x^{n+1}$ with c is an unknown number between 0 and x.

Because $e^c \le e^{|x|}$, we have

$$|R_n(x)| = |\frac{1}{(n+1)!} e^c x^{n+1}| \le e^{|x|} \frac{|x|^{n+1}}{(n+1)!}.$$

For any x, this expression tends to 0 as $n \to \infty$ by Lemma 2, so that by **(13)**, $e^x = \lim_{n\to\infty} \sum_{j=0}^{n} \frac{1}{j!} x^j = \sum_{j=0}^{\infty} \frac{1}{j!} x^j$, as stated in **(5)** □

Operations with power series

Power series are useful because in the interiors of their intervals of convergence they can be added, multiplied, differentiated and integrated term by term as if they were polynomials. With some restrictions, they can also be divided and composed. We will describe these results in three rules, which are established in courses on advanced calculus. To simplify the notation, we consider only Maclaurin Series.

Rule 1 *Suppose that $f(x)$ and $g(x)$ are equal to the power series*

$$f(x) = \sum_{j=0}^{\infty} a_j x^j \quad \text{and} \quad g(x) = \sum_{j=0}^{\infty} b_j x^j \tag{14}$$

in the open interval $(-R, R)$, where R is a positive number or ∞. Then adding and multiplying the two power series in **(14)** *term by term yields power series for $f(x) + g(x)$ and $f(x)g(x)$ in $(-R, R)$. If $g(x)$ is not 0 for $|x| < R$, then $f(x)/g(x)$ has a power series representation in $(-\rho, \rho)$ for some ρ with $0 < \rho \le \infty$, any number of whose terms can be found from* **(14)** *by division.*

The Examples, Tune-up Exercises, and Problems that use Rule 1 and the two rules that follow generally involve the following MacLaurin series.

$$e^x = \sum_{j=0}^{\infty} \frac{1}{j!} x^j = 1 + x + \frac{1}{2} x^2 + \frac{1}{3!} x^3 + \frac{1}{4!} x^4 + \cdots \tag{15}$$

$$\sin x = \sum_{j=0}^{\infty} \frac{(-1)^j}{(2j+1)!} x^{2j+1} = x - \frac{1}{3!} x^3 + \frac{1}{5!} x^5 - \cdots \tag{16}$$

$$\cos x = \sum_{j=0}^{\infty} \frac{(-1)^j}{(2j)!} x^{2j} = 1 - \frac{1}{2} x^2 + \frac{1}{4!} x^4 + \cdots \tag{17}$$

$$\sinh x = \sum_{j=0}^{\infty} \frac{1}{(2j+1)!} x^{2j+1} = x + \frac{1}{3!} x^3 + \frac{1}{5!} x^5 + \cdots \tag{18}$$

$$\cosh x = \sum_{j=0}^{\infty} \frac{1}{(2j)!} x^{2j} = 1 + \frac{1}{2} x^2 + \frac{1}{4!} x^4 + \cdots \tag{19}$$

$$\frac{1}{1-x} = \sum_{j=0}^{\infty} x^j = 1 + x + x^2 + x^3 + \cdots \tag{20}$$

$$\ln(1+x) = \sum_{j=1}^{\infty} \frac{(-1)^{j+1}}{j} x^j = x - \tfrac{1}{2} x^2 + \tfrac{1}{3} x^3 - \tfrac{1}{4} x^4 + \cdots \tag{21}$$

Example 7 What is the MacLaurin series of $\sin x + \sin(3x)$?

SOLUTION We have by **(16)**, for all x,

$$\sin x = \sum_{j=0}^{\infty} (-1)^j \frac{1}{(2j+1)!} x^{2j+1} = x - \frac{1}{3!}x^3 + \frac{1}{5!}x^5 - \cdots$$

$$\begin{aligned}\sin(3x) &= \sum_{j=0}^{\infty} (-1)^j \frac{1}{(2j+1)!} (3x)^{2j+1} \\ &= \sum_{j=0}^{\infty} (-1)^j \frac{3^{2j+1}}{(2j+1)!} (x)^{2j+1} \\ &= 3x - \frac{3^3}{3!}x^3 + \frac{3^5}{5!}x^5 - \cdots\end{aligned}$$

and consequently

$$\begin{aligned}\sin x + \sin(3x) &= \sum_{j=0}^{\infty} (-1)^j \left\{\frac{1+3^{2j+1}}{(2j+1)!}\right\} x^{2j+1} \\ &= 4x - \left[\frac{1+3^3}{3!}\right]x^3 + \left[\frac{1+3^5}{5!}\right]x^5 - \cdots . \square\end{aligned}$$

Example 8 Give the MacLaurin series of $x^2 \ln(1+x)$.

SOLUTION Multipling **(21)** by x^2 gives

$$x^2 \ln(1+x) = \sum_{j=1}^{\infty} \frac{(-1)^{j+1}}{j} x^{j+2} = x^3 - \tfrac{1}{2}x^4 + \tfrac{1}{3}x^5 - \tfrac{1}{4}x^6 + \cdots$$

for $-1 < x \le 1$. $\square$

Question 4 Give the terms up to degree 5 in the MacLaurin series for $\sin x \cos x$.

Example 9 Give the terms up to degree 6 in the MacLaurin series for $\dfrac{x^2}{\cos x}$.

SOLUTION We divide x^2 by the power series

$$\cos x = 1 - \frac{1}{2!}x^2 + \frac{1}{4!}x^4 - \cdots = 1 - \tfrac{1}{2}x^2 + \tfrac{1}{24}x^4 - \cdots$$

as in Figure 1, keeping track of only those terms that contribute terms of degree ≤ 6 in the dividend. This calculation shows that $x^2/(\cos x) = x^2 + \frac{1}{2}x^4 + \frac{5}{24}x^6 + \cdots$. $\square$

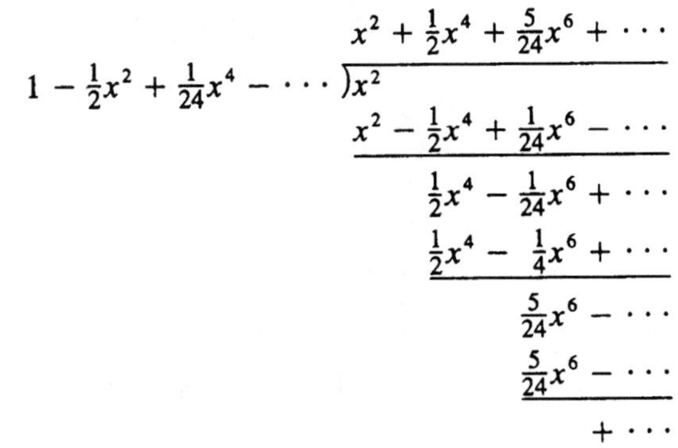

FIGURE 1

We now consider the compositions of power series.

Rule 2 *If $y(x)$ has a convergent MacLaurin series in $(-R, R)$ and $F(y)$ has a convergent power series in an open interval containing the values of $y(x)$ for x in $(-R, R)$, then $F\big(y(x)\big)$ has a convergent MacLaurin series for x in $(-R, R)$.*

We will apply Rule 2 only if $y(0) = 0$ because in other cases, each coefficient in the MacLaurin series for $F\big(y(x)\big)$ is itself an infinite series of constants.

Example 10 Give the terms of degree ≤ 4 in the MacLaurin series for $\sinh[\ln(1-x)]$.

SOLUTION We use the two Maclaurin series

$$\begin{aligned}\sinh y &= y + \frac{1}{3!}y^3 + \frac{1}{5!}y^5 + \cdots = y + \tfrac{1}{6}y^3 + \tfrac{1}{120}y^5 + \cdots \\ \ln(1-x) &= -x - \tfrac{1}{2}x^2 - \tfrac{1}{3}x^3 - \tfrac{1}{4}x^4 - \cdots\end{aligned}$$

the first of which converges for all y and the second for $-1 \le x < 1$. We obtain for $-1 < x < 1$

$$\begin{aligned}\sinh[\ln(1-x)] &= \ln(1-x) + \tfrac{1}{6}[\ln(1-x)]^3 + \tfrac{1}{120}[\ln(1-x)]^5 + \cdots \\ &= (-x - \tfrac{1}{2}x^2 - \tfrac{1}{3}x^3 - \tfrac{1}{4}x^4 - \cdots) \\ &\quad + \tfrac{1}{6}(-x - \tfrac{1}{2}x^2 - \tfrac{1}{3}x^3 - \cdots)^3 \\ &\quad + \tfrac{1}{120}(-x - \tfrac{1}{2}x^2 - \cdots)^5 + \cdots \\ &= -x - \tfrac{1}{2}x^2 - \tfrac{1}{3}x^3 - \tfrac{1}{4}x^4 - \cdots \\ &\quad - \tfrac{1}{6}(x^3 + \tfrac{3}{2}x^4 + \cdots) + \cdots \\ &= -x - \tfrac{1}{2}x^2 - \tfrac{1}{2}x^3 - \tfrac{1}{2}x^4 + \{\text{Terms of degree} \ge 5\}.\end{aligned}$$

In the next to last step of this calculation, we used the formula $(a+b)^3 = a^3 + 3a^2b + 3ab^2 + b^3$. □

The last rule of this section deals with differentiating and integrating Maclaurin series.

Rule 3 *If a MacLaurin series$f(x) = \sum_{j=0}^{\infty} a_j x^j$ has a positive or infinite radius of convergence R, then the derivatives and antiderivatives of $f(x)$ for x in $(-R, R)$ and definite integrals of $f(x)$ over closed finite intervals $[a, b]$ contained in $(-R, R)$ may be determined by differentiating and integrating the power series term by term:*

$$\frac{df}{dx}(x) = \sum_{j=0}^{\infty} \frac{d}{dx}(a_j x^j) = \sum_{j=1}^{\infty} j a_j x^{j-1} \tag{22}$$

$$\int f(x)\,dx = \sum_{j=0}^{\infty} \int a_j x^j\,dx = \left(\sum_{j=0}^{\infty} \frac{1}{j+1} a_j x^{j+1}\right) + C. \tag{23}$$

$$\int_a^b f(x)\,dx = \sum_{j=0}^{\infty} \int_a^b a_j x^j\,dx = \sum_{j=0}^{\infty} \frac{1}{j+1} a_j (b^{j+1} - a^{j+1}). \tag{24}$$

The series **(24)** *of numbers converge and the power series* **(22)** *and* **(23)** *have the same radius of convergence as the series for $f(x)$.*

Example 11 Use the geometric series **(20)** to find the MacLaurin series for $(1-x)^{-2}$. What is its radius of convergence?

SOLUTION We write

$$(1-x)^{-2} = \frac{d}{dx}[(1-x)^{-1}] = \frac{d}{dx}\sum_{j=0}^{\infty} x^j = \sum_{j=0}^{\infty} \frac{d}{dx}(x^j) = \sum_{j=1}^{\infty} jx^{j-1}$$
$$= 1 + 2x + 3x^2 + 4x^3 + \cdots . \qquad \textbf{(25)}$$

We start the summation at $j = 1$ in the last sum because the term for $j = 0$ is 0. The series **(25)**, like the geometric series, has a radius of convergence 1. Both series converge for $|x| < 1$. □

Example 12 Use the geometric series to find the MacLaurin series **(21)** for $\ln(1+x)$.

SOLUTION Because the derivative of $\ln(1+x)$ is $1/(1+x)$, we have for $x > -1$

$$\int_0^x \frac{1}{1+t}\,dt = \Big[\ln(1+t)\Big]_{t=0}^{t=x} = \ln(1+x) - \ln(1) = \ln(1+x).$$

We obtain the Maclaurin series for $1/(1+t)$ by replacing x by $-t$ in the geometric series **(20)**. Then we integrate that series term by term to have

$$\ln(1+x) = \int_0^x \frac{1}{1+t}\,dt = \int_0^x \left(\sum_{j=0}^{\infty} (-1)^j t^j\right) dt = \sum_{j=0}^{\infty} \int_0^x (-1)^j t^j\,dt$$
$$= \sum_{j=0}^{\infty} \frac{(-1)^j}{j+1} x^{j+1}. \;\square$$

Principles and Procedures

- The most important convergence test for power (Taylor or MacLaurin) series is the Ratio Test, because it determines the radius of convergence. Other convergence tests are only needed in cases where the radius of convergence is nonzero and finite and the Ratio Test cannot be used to test the convergence at one or both endpoints of the interval of convergence.
- Most of the Tune-up Exercises and Problems in this section deal with MacLaurin series—Taylor series centered at the origin—to simplify the calculations.
- Rules 1 through 3 are illustrations of the basic principle that many operations of arithmetic and calculus can be performed on Taylor series in the interiors of their intervals of convergence as if the infinite series were polynomials.

Tune-Up Exercises 8.6[♦]

[A]*Answer provided.* [O]*Outline of solution provided.*

T1.[O] A function $f(x)$ that is equal to its MacLaurin series centered at $x = 0$ is such that $f(0) = 1$, $f'(0) = 5, f''(0) = 5^2$, and, in general, $f^{(j)}(0) = 5^j$ for $j = 0, 1, 2, 3, \ldots$. Find the Taylor series for $f(x)$ and determine its radius of convergence.

♦ Type 2, recognizng a MacLaurin series

T2.[O] Use the MacLaurin series for e^x to give the MacLaurin series for $e^x + e^{2x}$.

♦ Type 2, a MacLaurin series

T3.[O] Use the MacLaurin series for e^x and $\cos x$ to give the terms up to degree 4 in the MacLaurin series for $e^x \cos x$.

♦ Type 2, terms in a MacLaurin series

[♦]The Tune-up Exercises and Problems are classified by type and content. The types are (1) basic, reactive; (2) basic reflective; (3) intermediate, reactive; (4) intermediate, reflective; (5) advanced, reactive; (6) advanced, reflective; and (7) advanced, theoretical.

T4.[O] Use the MacLaurin series for $\sin x$ to give the MacLaurin series for the function that equals $\dfrac{\sin x}{x}$ for $x \neq 0$ and has the value 1 at $x = 0$

♦ Type 3, a MacLaurin series

T5.[O] Use the Geometric Series and the MacLaurin series for $\sin x$ to give the terms up to degree 4 in the MacLaurin series $\dfrac{1}{1 - \sin x}$.

♦ Type 3, a MacLaurin series

T6.[O] Use the Geometric Series to give the MacLaurin series for $\dfrac{1}{1+t^2}$ and integrate the result from $t = 0$ to $t = x$ to give the MacLaurin series for $\tan^{-1} x$.

♦ Type 3, a MacLaurin series

T7.[O] Use the MacLaurin series for e^x and $\sin x$ to give the terms up to degree 3 in the MacLaurin series for $e^{\sin x}$.

♦ Type 3, a MacLaurin series

Problems 8.6

[A]*Answer provided.* [O]*Outline of solution provided.*

Carry out the following steps in Problems 1 through 8: **(a)** Find the radius of convergence of the power series. **(b)** Determine whether the power series converges or diverges at the endpoints of its interval of convergence. **(c)** if the radius of convergence is nonzero and finite, calculate partial sums on your calculator or computer to predict its value at a value of x where it converges and to illustrate its divergence at a value of x where it diverges. Justify your application of all convergence and divergence tests.

1.[O] $\displaystyle\sum_{j=1}^{\infty} \frac{x^j}{j}$

♦ Type 3, convergence of a power series

2.[A] $\displaystyle\sum_{j=1}^{\infty} \sqrt{j}\, x^j$

♦ Type 3, convergence of a power series

3. $\displaystyle\sum_{j=1}^{\infty} \frac{x^j}{j^3 3^j}$

♦ Type 3, convergence of a power series

4. $\displaystyle\sum_{j=0}^{\infty} (j+1)!\, x^{j+1}$

♦ Type 3, convergence of a power series

5. $\displaystyle\sum_{j=1}^{\infty} \frac{5^j x^j}{\sqrt{j}}$

♦ Type 3, convergence of a power series

6. $\displaystyle\sum_{j=1}^{\infty} \frac{x^{2j}}{2j}$

♦ Type 3, convergence of a power series

7. $\displaystyle\sum_{j=1}^{\infty} \frac{(-1)^j x^{2j}}{j 4^j}$

♦ Type 3, convergence of a power series

8. $\displaystyle\sum_{j=1}^{\infty} \frac{j!\, x^{2j+1}}{j^{2j}}$

♦ Type 3, convergence of a power series

Justify the application of all convergence and divergence tests in the following Problems.

9.[O] A function $f(x)$ that is equal to its MacLaurin series for all x is such that the value $f^{(j)}(0)$ at $x = 0$ of its jth derivative is j^3 for $j \geq 0$. Give its MacLaurin series and use your calculator or computer to predict its value at $x = 1$.

♦ Type 4, a MacLaurin series, etc.

10. A function $g(x)$ that is equal to its MacLaurin series for all x is such that $g^{(j)}(0) = \sin j$ for $j \geq 0$. Give its MacLaurin series and use your calculator or computer to predict its value at $x = 5$.

♦ Type 4, a MacLaurin series, etc.

11. Use your calculator or computer to predict the value at $x = 1$ of the function $h(x)$ that is equal to its MacLaurin series for all x, that has the value 0 at $x = 0$, and that satisfies $h^{(j)}(0) = \ln j$ for $j \geq 1$.

♦ Type 4, a MacLaurin series, etc.

In Problems 12 through 17 determine the values of x where the series converges and where it diverges.

12.[A] $\sum_{j=0}^{\infty} \frac{j+3}{j^2+1}\left(\frac{x-3}{2}\right)^j$

♦ Type 3, convergence of a power series

13. $\sum_{j=0}^{\infty} \frac{x^{2j+1}}{(j!)^2}$

♦ Type 3, convergence of a power series

14. $\sum_{j=2}^{\infty} \frac{x^j}{\ln j}$

♦ Type 3, convergence of a power series

15. $\sum_{j=1}^{\infty} \sqrt{\frac{j}{j^4+1}}\, x^{2j}$

♦ Type 3, convergence of a power series

16. $\sum_{j=1}^{\infty} \frac{x^j}{(j+1)^j}$

♦ Type 3, convergence of a power series

17. $\sum_{j=1}^{\infty} j^{2j} x^j$

♦ Type 3, convergence of a power series

18.[O] Give all of the terms in the MacLaurin series for $(1+x)^5$. What is its radius of convergence?

♦ Type 4, a MacLaurin series, etc.

19.[A] A function $f(x)$ that equals its MacLaurin series near $x = 0$ is such that $f^{(j)}(0) = j!$ for $j \geq 0$. What is the function?

♦ Type 4, a MacLaurin series, etc.

20. A function $g(x)$ that equals its MacLaurin series for all x is such that $g^{(j)}(0) = j(-1)^{j+1}$ for $j \geq 0$. What is the function?

♦ Type 4, a MacLaurin series, etc.

In Problems 21 through 36 use the MacLaurin Series for $e^x, \sin x, \cos x, (1-x)^{-1}$, and $\ln(1-x)$ as appropriate to give the MacLaurin series of the given functions.

21.[O] $\ln\left(\frac{1+x}{1-x}\right)$.

♦ Type 3, a MacLaurin series

22.[A] $3e^{2x} + 4e^{5x}$.

♦ Type 3, a MacLaurin series

23. $x^2 e^{-x}$.

♦ Type 3, a MacLaurin series

24. $\frac{1+x}{1-x}$.

♦ Type 3, a MacLaurin series

25. $\sin(x^3)$.

♦ Type 3, a MacLaurin series

26.[O] $(1-x)^{-3}$.

♦ Type 3, a MacLaurin series

27.[A] $\frac{d^3}{dx^3}(e^{x^2})$.

♦ Type 3, a MacLaurin series

28. $\frac{1}{1+10x}$

♦ Type 3, a MacLaurin series

29. $\ln(1-x^3)$.

♦ Type 3, a MacLaurin series

30. $(1+x^5)^{-1}$.

♦ Type 3, a MacLaurin series

31.[O] $\frac{d^2}{dx^2}(x^{30}\sin x)$.

♦ Type 3, a MacLaurin series

32.[A] $\frac{d^3}{dx^3}\left(\frac{x^{10}}{1-x}\right)$.

♦ Type 3, a MacLaurin series

33. $\frac{d}{dx}[(1+x^3)^{-1}])$.

♦ Type 3, a MacLaurin series

34.[O] $\int_0^x e^{t^2}\, dt$.

♦ Type 3, a MacLaurin series

35.[A] $\int_0^x \sin(t^3)\, dt$.

♦ Type 3, a MacLaurin series

36. $\int_0^x t^2\cos(t^2)\, dt$.

♦ Type 3, a MacLaurin series

In Problems 37 through 44 find the first four nonzero terms in the MacLaurin Series for the given functions.

37.[O] $e^x \sin x$.

◆ Type 4, terms in a MacLaurin series

38.[A] $[\ln(1-x)]^2$.

◆ Type 4, terms in a MacLaurin series

39. $\sin x \cos(3x)$.

◆ Type 4, terms in a MacLaurin series

40. $\sin x \cos(3x)$.

◆ Type 4, terms in a MacLaurin series

41.[O] $\sec x$.

◆ Type 4, terms in a MacLaurin series

42.[A] $\dfrac{\ln(1+x)}{\cos x}$.

◆ Type 4, terms in a MacLaurin series

43. $\tan(x^2)$.

◆ Type 4, terms in a MacLaurin series

44.[O] $\cos(\sin x)$.

◆ Type 4, terms in a MacLaurin series

45.[A] Use a MacLaurin series to find $f^{(10)}(0)$ for $f(x) = x^7 e^{3x}$.

◆ Type 4, derivatives for MacLaurin series

46. Use a MacLaurin series to find $g^{(15)}(0)$ for $g(x) = \dfrac{x^5}{1-x}$.

◆ Type 4, derivatives from MacLaurin series

47. What is the value of the twentieth derivative of $\sin(x^3)$ at $x = 0$?

◆ Type 4, derivatives from MacLaurin series

48. Which derivative of $\ln(1+x^2)$ has the value $-\frac{1}{6}(12!)$ at $x = 0$?

◆ Type 4, derivatives from MacLaurin series

RESPONSES TO QUESTIONS IN CHAPTER 8

Responses 8.1

Response 1 **(a)** The table is completed below. **(b)** One answer: d_n is the number of days in February of year n AD.

n	d_n	n	d_n
2000	29	2004	29
2001	28	2005	28
2002	28	2006	28
2003	28	2007	28

Response 2 **(a)** TI-82 CALCULATORS: Follow the instructions in the section to put the calculator in sequence and dot modes. • Press [Y =] and set $U_n = n+2(-1)\wedge n$. • Press [WINDOW] and set U_nStart equal to 2 (the value of $a_0 = 0 + 2(-1)^0$), set nStart and nMin equal to 0, and set nMax = 25, Xmin = 0, Xmax = 26, Xscale = 1, Ymin = −5 , Ymax = 30, and Yscale = 5. • Then press [GRAPH]. • Figure R2a

TI-83 CALCULATORS: Follow the instructions in the section to put the calculator in sequence and dot modes. • Press [Y =] and set nMin equal to 1, $u(n) = n + 2(-1) \wedge n$, and $u(n\text{Min})$ equal to 2 (the value of $a_0 = 0 + 2(-1)^0$). • Press [WINDOW] and set nMin = 0, nMax = 25, PlotStart = 1, PlotStep = 1, Xmin = 0, Xmax = 26, Xscale = 1, Ymin = −5 , Ymax = 30, and Yscale = 5. • Then press [GRAPH]. • Figure R2a

TI-85 AND TI-86 CALCULATORS: Enter the Sequence Program. • Press [GRAPH] and [F1] and set $y1 = N + 2(-1) \wedge N$, • Select <WINDOW> and set Xmin = 0, Xmax = 26, Xscale = 1, Ymin = −5, Ymax = 30, and Yscale = 5. • Run the program with $N0 = 0$ and MAXN = 25. • Figure R2a

TI-89 AND TI-92 CALCULATORS: Follow the instructions in the section to put the calculator in sequence mode. • Press [♦] [Y =] and set $u1 = n+2*(-1)\wedge n$. (No value is needed for $ui1$.) • Press [♦] [WINDOW] and set nmin = 1, nmax = 25, PlotStrt = 1, PlotStep = 1, Xmin = 0, Xmax = 26, Xscale = 1, Ymin = −5 , Ymax = 30, and Yscale = 5. • Then press [GRAPH]. • Figure R2a

(b) Use the window $0 \le x \le 26, -110 \le y \le 100$ • Figure R2b

(c) $\{n + 2(-1)^n\}_{n=0}^{\infty}$ diverges to ∞ and $\{21n - n^2\}_{n=1}^{\infty}$ diverges to $-\infty$.

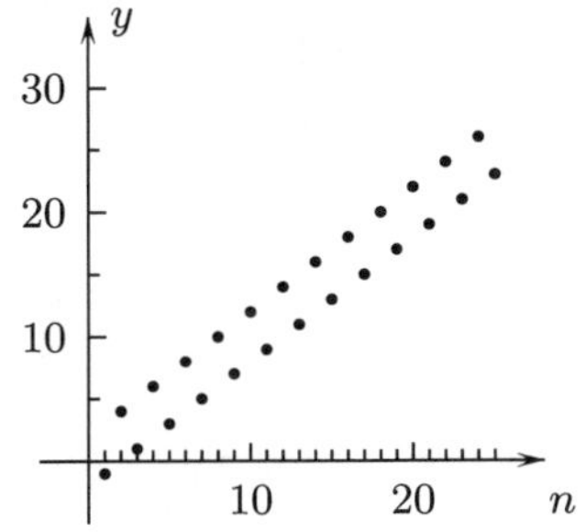

Figure R2a

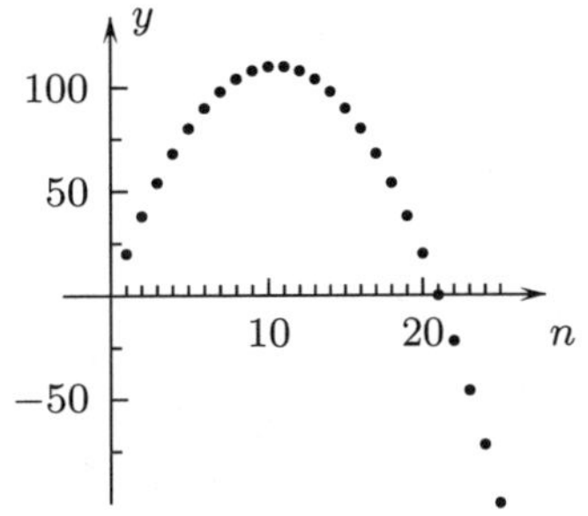

Figure R2b

Response 3 **(a)** Set $a_n = \cos\left(\frac{1}{3}n\pi\right)$. • $a_1 = \cos\left(\frac{1}{3}\pi\right) = \frac{1}{2}$, $a_2 = \cos\left(\frac{2}{3}\pi\right) = -\frac{1}{2}$, $a_3 = \cos(\pi) = -1$, $a_4 = \cos\left(\frac{4}{3}\pi\right) = -\frac{1}{2}$, $a_5 = \cos\left(\frac{5}{3}\pi\right) = \frac{1}{2}$, $a_6 = \cos(2\pi) = 1$, and this pattern repeats every time n increases by 6. •
(b) The sequence diverges without a limit.

Response 4 Let Y be an arbitrary positive number. • $3^n > Y \iff \ln(3^n) > \ln(Y) \iff n\ln(3) > \ln(Y) \iff n > \ln(Y)/\ln(3)$ • We can use $N = \ln(Y)/\ln(3)$.
(b) For $Y = 10^{10}$, $N = \ln(10^{10})/\ln(3) \doteq 20.96$
(c) With $N = 21$, we have $3^{21} \doteq 1.05 \times 10^{10}$, which, as required, is greater than 10^{10}).

Response 5 Set $f(x) = xe^{-x^2/50}$. • Product Rule: $f'(x) = x\frac{d}{dx}(e^{-x^2/50}) + e^{-x^2/50}\frac{d}{dx}(x)$
$= xe^{-x^2/50}\frac{d}{dx}(-\frac{1}{50}x^2) + e^{-x^2/50} = e^{-x^2/50}(-\frac{1}{25}x^2 + 1) = \frac{1}{25}e^{-x^2/50}(25 - x^2)$ •
$f'(x)$ is positive for $0 \le x < 5$ and negative for $x > 5$. • $f(x)$ is increasing in $[0, 5]$ and decreasing in $[5, \infty)$. •
$\left\{ne^{-n^2/50}\right\}_{n=0}^{\infty}$ is increasing and, therefore, nondecreasing for $0 \le n \le 5$. It is decreasing and, therefore, nonincreasing for $n \ge 5$.

Response 6 **(a)** Because $\left\{5e^{n/10}\right\}_{n=0}^{\infty}$ is increasing for $n \ge 0$, its greatest lower bound is its first term $5e^0 = 5$. • Because $5e^{n/10} \to \infty$ as $n \to \infty$, the sequence has no upper bounds.
(b) Since $\left\{d_n\right\}_{n=2000}^{\infty}$ has only the values 28 and 29, its least upper bound is 29 and its greatest lower bound is 28.

Response 7 **(a)** $a_n \le M$ for $n \ge n_0$ because M is an upper bound of the sequence.
(b) $a_N > M - \epsilon$ for some N because M is the least upper bound of the sequence and, consequently, the smaller number $M - \epsilon$ is not an upper bound.
(c) $a_n \ge a_N > M - \epsilon$ for $n > N$ because the sequence is nondecreasing.

Response 8 **(a)** $a_N > Y$ for some N because the sequence is unbounded and, consequently, Y cannot be an upper bound.
(b) $a_n \ge a_N > Y$ for $n > N$ because the sequence is nondecreasing.

Response 9 **(a)** The table is completed below. **(b)** Likely response: The sequence seems to be approaching 1.

n	$a_n \doteq$
0	25
1	5
2	2.23607
3	1.49535

n	$a_n \doteq$
4	1.22284
5	1.10582
6	1.05158
7	1.02547

n	$a_n \doteq$
8	1.01265
9	1.00631
10	1.00315
11	1.00157

Response 10 **(a)** The completed table is below. **(b)** Likely response: b_n seems to tend to 1.

n	$a_n \doteq$
0	0.16
1	0.4
2	0.63246
3	0.79527

n	$a_n \doteq$
4	0.89178
5	0.94434
6	0.97177
7	0.98578

n	$a_n \doteq$
8	0.99287
9	0.99643
10	0.99821
11	0.99911

Response 11 $b_n = (0.16)^{1/2^n}$ • $\lim_{n\to\infty} b_n = \lim_{n\to\infty} (0.16)^{1/2^n} = (0.16)^0 = 1$

Response 12 **(a)** $x^2 = x \iff x = 0$ or $x = 1$
(b) The table is completed below. It appears that 0 is asymptotically stable and 1 is not.
(c) The derivative $g'(x) = 2x$ is close to 0 for x near 0 and close to 2 for x near 1.

n	$x_n \doteq$
0	0.9
1	0.81
2	0.6561

n	$x_n \doteq$
3	0.43047
4	0.18530
5	0.03434

n	$x_n \doteq$
6	0.00118
7	1.4×10^{-6}
8	1.9×10^{-12}

n	$x_n \doteq$
0	1.2
1	1.44
2	2.0736

n	$x_n \doteq$
3	4.2998
4	18.4884
5	341.821

n	$x_n \doteq$
6	116842
7	1.36×10^{10}
8	1.86×10^{20}

Response 13 **(a)** If $P_0 = 2$, then $P_n = 2$ for $n \geq 1$ since 2 is a fixed point.
(b) The table is completed below.

n	$P_n \doteq$
0	0.45
1	0.79875
2	1.27850
3	1.73972
4	1.96613

n	$P_n \doteq$
5	1.99943
6	2.00000
7	2.00000
8	2.00000
9	2.00000

Response 14 The table is completed below.

n	$P_n \doteq$
0	0.45
1	1.287
2	2.38816
3	1.27578
4	2.38451

n	$P_n \doteq$
5	1.28426
6	2.38730
7	1.27779
8	2.38519
9	1.28269

Responses 8.2

Response 1 **(a)** The table is completed below.
(b) Likely answer: The total distance seems to be approaching 2 miles.

Stages	Total distance	Decimal approximation
1	1	1.0
1–2	$1+\frac{1}{2}$	1.5
1–3	$1+\frac{1}{2}+(\frac{1}{2})^2$	1.75
1–5	$1+\frac{1}{2}+(\frac{1}{2})^2+(\frac{1}{2})^3+(\frac{1}{2})^4$	1.9375
1–10	$1+\frac{1}{2}+(\frac{1}{2})^2+(\frac{1}{2})^3+(\frac{1}{2})^4+(\frac{1}{2})^5$ $+(\frac{1}{2})^6+(\frac{1}{2})^7+(\frac{1}{2})^8+(\frac{1}{2})^9$	1.998047

Response 2 **(a)** The table is completed below.
(b) Likely answer: The total distance seems to be approaching $1\frac{1}{3}$ miles.

Stages	Total distance	Decimal approximation
1	1	1.0
1–2	$1+\frac{1}{4}$	1.25
1–3	$1+\frac{1}{4}+(\frac{1}{4})^2$	1.3125
1–5	$1+\frac{1}{4}+(\frac{1}{4})^2+(\frac{1}{4})^3+(\frac{1}{4})^4$	1.332031
1–8	$1+\frac{1}{4}+(\frac{1}{4})^2+(\frac{1}{4})^3+(\frac{1}{4})^4$ $+(\frac{1}{4})^5+(\frac{1}{4})^6+(\frac{1}{4})^7$	1.333313

Response 3 **(a)** The table is completed below.
(b) Likely answer: The total distance seems to be approaching $\frac{2}{3}$ mile.

Stages	Total distance	Decimal approximation
1	1	1.0
1–2	$1 - \frac{1}{2}$	0.5
1–3	$1 - \frac{1}{2} + (\frac{1}{2})^2$	0.75
1–5	$1 - \frac{1}{2} + (\frac{1}{2})^2 - (\frac{1}{2})^3 + (\frac{1}{2})^4$	0.6875
1–10	$1 - \frac{1}{2} + (\frac{1}{2})^2 - (\frac{1}{2})^3 + (\frac{1}{2})^4 - (\frac{1}{2})^5$ $+(\frac{1}{2})^6 - (\frac{1}{2})^7 + (\frac{1}{2})^8 - (\frac{1}{2})^9$	0.666016

Response 4 If $r = 1$, then $s_n = 1 + 1^2 + 1^3 + \cdots + 1^n = 1 + 1 + 1 + \cdots + 1$ with $n+1$ terms and equals $n+1$.

Response 5 Put $+r^3$ in the dividend and subtract r^3 times the divisor, as in Figure R5 to see that $\dfrac{1}{1-r} = 1 + r + r^2 + r^3 + \dfrac{r^4}{1-r}$.

$$\begin{array}{r|l}
 & 1 + r + r^2 + r^3 \\
1-r & 1 \\
 & \underline{1 - r} \\
 & \quad r \\
 & \quad \underline{r - r^2} \\
 & \qquad r^2 \\
 & \qquad \underline{r^2 - r^3} \\
 & \qquad\quad r^3 \\
 & \qquad\quad \underline{r^3 - r^4} \\
 & \qquad\qquad r^4
\end{array}$$

FIGURE R5

Response 6 **(a)** The sum in the last row of completed Table 2 is

$$1 + \tfrac{1}{4} + (\tfrac{1}{4})^2 + (\tfrac{1}{4})^3 + \cdots (\tfrac{1}{4})^7 = \frac{1 - (\frac{1}{4})^8}{1 - \frac{1}{4}} = \tfrac{4}{3}[1 - (\tfrac{1}{4})^8] \doteq 1.333313.$$

(b) The sum in the last row of completed Table 3 is

$$1 + (-\tfrac{1}{2}) + (-\tfrac{1}{2})^2 + (-\tfrac{1}{2})^3 + \cdots (-\tfrac{1}{2})^9 = \frac{1 - (-\frac{1}{2})^{10}}{1 - (-\frac{1}{2})} = \tfrac{2}{3}[1 - (\tfrac{1}{2})^{10}] \doteq 0.666016.$$

Response 7 The completed table is below.

TABLE 4

Month	Balance with interest before the payment	Payment	Balance after the payment
0			\$10,000
1	\$10,100	\$200	\$9,900
2	\$9,999	\$200	\$9,799
3	\$9897	\$200	\$9,697
4	\$9794	\$200	\$9,594
5	\$9690	\$200	\$9,490

Response 8 **(a)** $B_3 = 1.01B_2 - P$ • Use **(7)**:
$B_3 = 1.01[(1.01)^2 B_0 - 1.01P - P] - P = (1.01)^3 B_0 - (1.01)^2 P - 1.01P - P$
(b) $B_4 = 1.01B_3 - P$ • Use the result of part (a):
$B_4 = 1.01[(1.01)^3 B_0 - (1.01)^2 P - 1.01P - P] - P$
$= (1.01)^4 B_0 - (1.01)^3 P - (1.01)^2 P - 1.01P - P$

Response 9 $B_{60} = (1.01)^{60} B_0 - P\left[\dfrac{(1.01)^{60} - 1}{0.01}\right] = (1.01)^{60} 10{,}000 - 200\left[\dfrac{(1.01)^{60} - 1}{0.01}\right]$
$\doteq \$1833.03$

Response 10 The balance due at 70 months is $B_{69} = (1.01)^{69} 10{,}000 - 200\left[\dfrac{(1.01)^{69} - 1}{0.01}\right] \doteq$ \$131.06 plus one month's interest. • The balance due is (1.01)\$131.06 $\doteq$ \$132.37

Response 11 **(a)** Presumably only the narrator was going to St. Ives.
(b) The narrator met the man, his 7 wives, their 7^2 sacks, the 7^3 cats, the 7^4 kitts, the 7^5 mice, and the 7^6 lice for a total of

$$1 + 7 + 7^2 + 7^3 + 7^4 + 7^5 + 7^6 = \frac{1 - 7^7}{1 - 7} = 137{,}257.$$

Response 12 **(a)** if $r = 0$, then $|r|^n = 0$ for all n. • If $0 < |r| < 1$, then $\ln |r| < 0$, so that $\ln |r^n| = n \ln |r|$ tends to $-\infty$ as $n \to \infty$ and $|r|^n = e^{\ln(|r|^n)} \to 0$.
(b) If $|r| > 1$, then $\ln |r| > 0$, so that $\ln |r^n| = n \ln |r|$ tends to ∞ as $n \to \infty$ and $|r|^n = e^{\ln(|r|^n)} \to \infty$

Response 13 **(a)** If you go one mile, then one-fourth mile, then one-fourth that distance, etc. forever, you go the total distance

$$1 + \tfrac{1}{4} + (\tfrac{1}{4})^2 + (\tfrac{1}{4})^3 + \cdots = \sum_{j=0}^{\infty} (\tfrac{1}{4})^j = \frac{1}{1 - \frac{1}{4}} = \frac{1}{\frac{3}{4}} = \tfrac{4}{3} \text{ miles.}$$

(b) If you go one mile forward, then one-half mile backwards, then one-half that distance forward, then one-half that distance backward, etc. forever, you go the total distance

$$1 - \tfrac{1}{2} + (\tfrac{1}{2})^2 - (\tfrac{1}{2})^3 + \cdots = \sum_{j=0}^{\infty} (-\tfrac{1}{2})^j = \frac{1}{1 + \frac{1}{2}} = \frac{1}{\frac{3}{2}} = \tfrac{2}{3} \text{ miles.}$$

Responses 8.3

Response 1 **(a)** Figure R1

(b) With $s_n = \sum_{j=2}^{n} \frac{1}{j(\ln j)^2}$: $s_{23} \doteq 1.792998788$, $s_{24} \doteq 1.797124190$, $s_{25} \doteq 1.800984761$

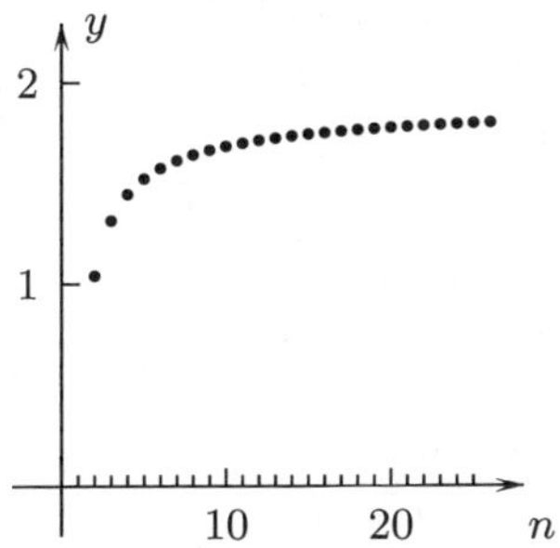

Figure R1

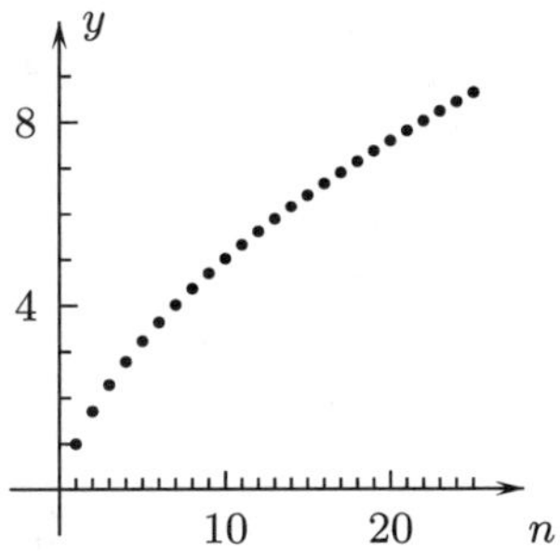

Figure R2

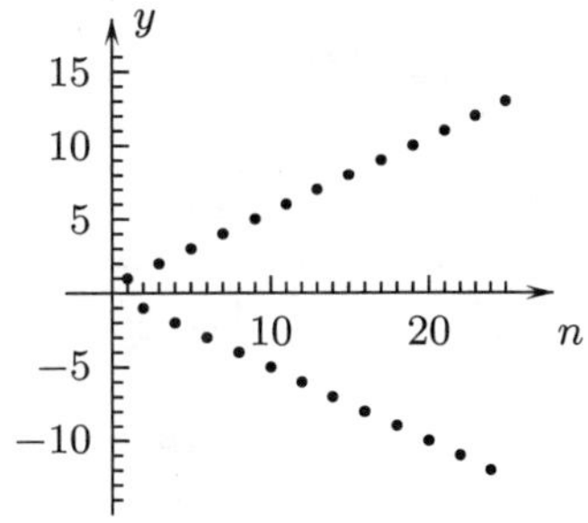

Figure R3

Response 2 $f(x) = \frac{1}{\sqrt{x}}$ is positive and decreasing for $x \geq 1$ and $f(j) = \frac{1}{\sqrt{j}}$. • For $X > 1$,
$\int_1^X \frac{1}{\sqrt{x}}\, dx = \int_1^X x^{-1/2}\, dx = \left[2x^{1/2}\right]_1^X = 2\sqrt{X} - 2$ •
$2\sqrt{X} - 2 \to \infty$ as $X \to \infty$ •
$\int_1^\infty \frac{1}{\sqrt{x}}\, dx = \infty$, so $\sum_{j=1}^{\infty} \frac{1}{\sqrt{j}}$ diverges by the Integral Test. • Figure R2

Response 3 The series diverges because $j(-1)^{j+1}$ does not tend to 0 as $j \to \infty$. (In fact $j(-1)^{j+1}$ tends to ∞ for odd j and tends to $-\infty$ for even j.) • Figure R3

Response 4 Because $\left|(-1)^j \frac{10\sqrt{j}}{j^2+1}\right| = \frac{10\sqrt{j}}{j^2+1} < \frac{10\sqrt{j}}{j^2} = 10\left(\frac{1}{j^{3/2}}\right)$ for $j \geq 1$, the series converges absolutely by the Comparison Test with the convergent p-series $\sum_{j=1}^{\infty} \frac{1}{j^{3/2}}$.

Response 5 Because $\frac{3j^2+7j}{j^5-j^2+5}$ behaves like $\frac{3j^2}{j^5} = \frac{3}{j^3}$ for large j, we compare with the convergent p-series $\sum_{j=1}^{\infty} \frac{1}{j^3}$. • For $j \geq 1$, $\dfrac{\frac{3j^2+7j}{j^5-j^2+5}}{\frac{1}{j^3}} = \frac{3j^5+7j^4}{j^5-j^2+5}$
$= \frac{3+7/j}{1-1/j^3+5/j^5}$ and $\frac{3+7/j}{1-1/j^3+5/j^5} \to 3$ as $j \to \infty$ •
The series converges by the Limit Comparison Test with $\sum_{j=1}^{\infty} \frac{1}{j^3}$.

Responses 8.4

Response 1 The completed table below suggests that the series $\sum_{j=1}^{n} j\left(\frac{2}{3}\right)^j$ converges to a number close to 6.

n	$\sum_{j=1}^{n} j\left(\frac{2}{3}\right)^j \doteq$
1	0.6667
2	1.5556
3	2.4444

n	$\sum_{j=1}^{n} j\left(\frac{2}{3}\right)^j \doteq$
4	3.2346
5	3.8930
10	5.5491

n	$\sum_{j=1}^{n} j\left(\frac{2}{3}\right)^j \doteq$
15	5.9178
20	5.9862
25	5.9978

Response 2 $\frac{2}{3}\left(1+\frac{1}{j}\right) \le \frac{3}{4} \iff 1+\frac{1}{j} \le \frac{3}{2}\left(\frac{3}{4}\right) \iff 1+\frac{1}{j} \le \frac{9}{8} \iff \frac{1}{j} \le \frac{1}{8} \iff j \ge 8$

Response 3 **(a)** $a_9 \le \frac{3}{4} a_8$

(b) Use part (a): $a_{10} \le \frac{3}{4} a_9 \le \frac{3}{4}\left(\frac{3}{4} a_8\right) = \left(\frac{3}{4}\right)^2 a_8$

(c) Use part (b): $a_{11} \le \frac{3}{4} a_{10} \le \frac{3}{4}\left[\left(\frac{3}{4}\right)^2 a_8\right] = \left(\frac{3}{4}\right)^3 a_8$

Response 4 Set $a_j = \dfrac{j!}{j^3}$ • $\dfrac{|a_{j+1}|}{|a_j|} = \dfrac{\dfrac{(j+1)!}{(j+1)^3}}{\dfrac{j!}{j^3}} = \dfrac{j^3}{(j+1)^3}\,\dfrac{(j+1)!}{j!} = \dfrac{j+1}{(1+1/j)^3}$ tends to $\rho = \infty$ as $j \to \infty$ • The series diverges by the Ratio Test.

Response 5 Set $a_j = \dfrac{j^3}{j!}$ • $\dfrac{|a_{j+1}|}{|a_j|} = \dfrac{\dfrac{(j+1)^3}{(j+1)!}}{\dfrac{j^3}{j!}} = \dfrac{(j+1)^3}{j^3}\,\dfrac{j!}{(j+1)!} = \dfrac{(1+1/j)^3}{j+1}$ tends to $\rho = 0$ as $j \to \infty$ • The series converges absolutely by the Ratio Test.

Response 6 **(a)** The completed table is below.

(b) One description: The partial sums go up and down with each one lying between the previous two.

n	$\sum_{j=1}^{n} \frac{1}{j} \doteq$	$\sum_{j=1}^{n} \frac{(-1)^{j+1}}{j} \doteq$
1	1	1
2	1.5	0.5
3	1.8333	0.8333
4	2.0833	0.5833

n	$\sum_{j=1}^{n} \frac{1}{j} \doteq$	$\sum_{j=1}^{n} \frac{(-1)^{j+1}}{j} \doteq$
5	2.2833	0.7833
6	2.45	0.6167
7	2.5929	0.7595
8	2.7119	0.6345

Response 7 Because $\left|\dfrac{(-1)^{j+1}}{j^2}\right| = \dfrac{1}{j^2}$, which is a term in a convergent p-series, the given series converges absolutely. • Figure R7

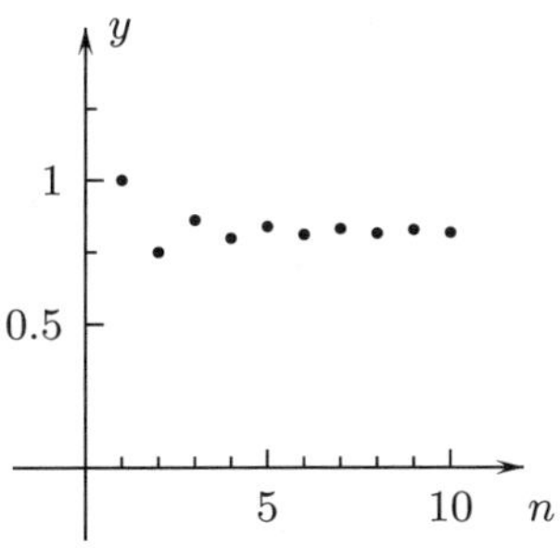

Figure R7

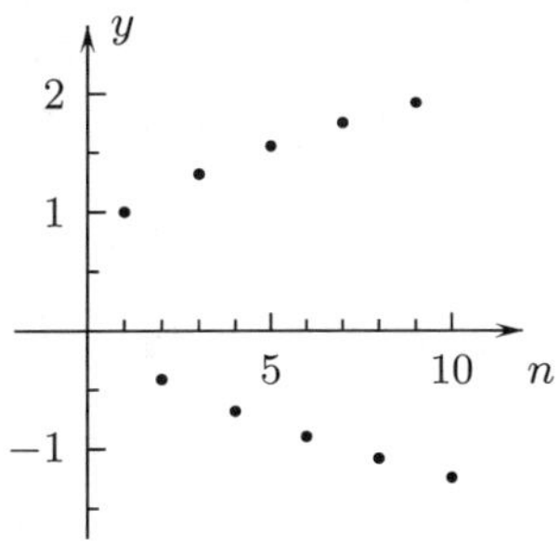

Figure R8

Response 8 The series diverges because $\sqrt{j} \to \infty$ as $j \to \infty$ and the terms do not tend to 0. • Figure R8

Responses 8.5

Response 1 With $P_5(x) = x - \frac{1}{6}x^3 + a_3x^3 + a_4x^4 + a_5x^5$,

$$\begin{aligned}
P_5'(x) &= 1 - \tfrac{1}{2}x^2 + 4a_4x^3 + 5a_5x^4 \\
P_5''(x) &= -x + 12a_3x^2 + 20a_5x^3 \\
P_5^{(3)}(x) &= -1 + 24a_4x + 60a_5x^2 \\
P_5^{(4)}(x) &= 24a_4 + 120a_5x \\
P_5^{(5)}(x) &= 120a_5
\end{aligned}$$

$P_5^{(4)}(0) = 24a_4$ and $P_5^{(5)}(0) = 120a_5$ • The fourth derivative of $\sin x$ is $\dfrac{d}{dx}(-\cos x) = \sin x$, which equals 0 at $x = 0$, and its fifth derivative is $\dfrac{d}{dx}(\sin x) = \cos x$, which equals 1 at $x = 0$. • $24a_4 = 0$ and $120a_5 = 1$ • $a_4 = 0$ and $a_5 = \frac{1}{120}$

Response 2 The table is completed below. • One description: For values of x near 0, $P_3(x)$ is a better approximation of $\sin x$ than $P_1(x)$ and $P_5(x)$ is a better approximation of $\sin x$ than $P_3(x)$, and each approximation improves as x gets closer to 0.

TABLE 1

x	$\lvert \sin x - P_1(x)\rvert \doteq$	$\lvert \sin x - P_3(x)\rvert \doteq$	$\lvert \sin x - P_5(x)\rvert \doteq$
2	1.09070	0.24263	0.02404
1	0.15853	0.00814	1.96×10^{-4}
0.1	1.67×10^{-4}	8.33×10^{-8}	1.98×10^{-11}
0.01	1.67×10^{-7}	8.33×10^{-13}	2×10^{-18}

Response 3

(a) All the derivatives of $f(x) = e^x$ are e^x, so $f^{(j)}(x) = e^x$ and $f^{(j)}(0) = 1$ for all nonnegative integers j. Therefore, the third-degree Taylor polynomial is $P_3(x) = 1 + x + \dfrac{1}{2!}x^2 + \dfrac{1}{3!}x^3$. • Figure R3

(b) $P_3(1) = 1 + 1 + \dfrac{1}{2!}1^2 + \dfrac{1}{3!}1^3 = 2 + \frac{1}{2} + .\frac{1}{6} = \frac{8}{3}$ •

$[\text{Error}] = |e^1 - \frac{8}{3}| \doteq 0.05156$

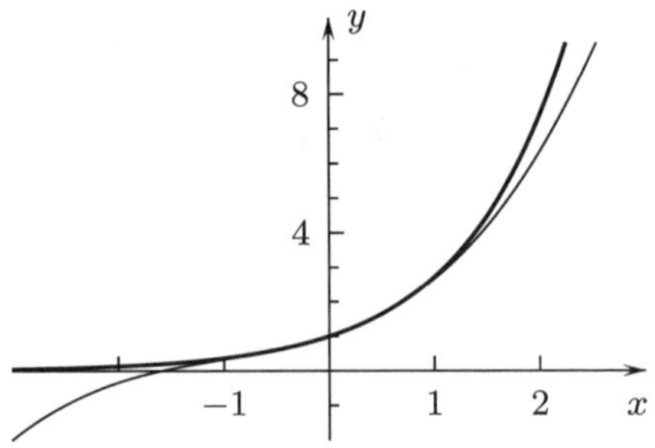

e^x and
$P_3(x) = 1 + x + \dfrac{1}{2!}x^2 + \dfrac{1}{3!}x^3$
Figure R3

Response 4

(a) For $f(x) = x^{1/3} : f'(x) = \frac{1}{3}x^{-2/3}, f''(x) = -\frac{2}{9}x^{-5/3}, f^{(3)}(x) = \frac{10}{27}x^{-8/3}$ • $f(1) = 1, f'(1) = \frac{1}{3}, f''(1) = -\frac{2}{9}, f^{(3)}(1) = \frac{10}{27}$ • $P_3(x) = 1 + \frac{1}{3}(x-1) + \frac{1}{2}\left(-\frac{2}{9}(x-1)^2\right) + \frac{1}{6}\left(\frac{10}{27}(x-1)^3\right) = 1 + \frac{1}{3}(x-1) - \frac{1}{9}(x-1)^2 + \frac{5}{81}(x-1)^3$

(b) $1.4^{1/3} \approx P_3(1.4) = 1 + \frac{1}{3}(0.4) - \frac{1}{9}(0.4)^2 + \frac{5}{81}(0.4)^3 = \frac{2267}{2025}$ •

$f^{(4)}(x) = -\frac{80}{81}x^{-11/3}$ •

$[\text{Error estimate}] = \left|\dfrac{1}{4!}\left(-\frac{80}{81}c^{-11/3}(0.4)^4\right)\right|$ with $1 < c < 1.4$ •

$[\text{Error estimate}] \leq= \frac{10}{243}(0.4)^4 \doteq 1.053 \times 10^{-3}$ •

$[\text{Error}] = |(1.4)^{1/3} - P_3(1.4)| \doteq 8.172 \times 10^{-4}$

Responses 8.6

Response 1

(a) $f(x) = \ln(1+x)$ •

$f'(x) = \dfrac{d}{dx}[\ln(1+x)] = \dfrac{1}{1+x}\dfrac{d}{dx}(1+x) = (1+x)^{-1}$ •

$f''(x) = \dfrac{d}{dx}[g'(x)] = \dfrac{d}{dx}[(1+x)^{-1}] = -(1+x)^{-2}\dfrac{d}{dx}(1+x) = -(1+x)^{-2}$ •

$f^{(3)}(x) = \dfrac{d}{dx}[g''(x)] = \dfrac{d}{dx}[-(1+x)^{-2}] = (-2)[-(1+x)^{-3}] = 2(1+x)^{-3}$ •

$f^{(4)}(x) = \dfrac{d}{dx}[g^{(3)}(x)] = \dfrac{d}{dx}[2(1+x)^{-3}] = -(3)(2)(1+x)^{-4}$ •

$f^{(5)}(x) = \dfrac{d}{dx}[g^{(4)}(x)] = \dfrac{d}{dx}[-(3)(2)(1+x)^{-4}] = (4)(3)(2)(1+x)^{-5}$ •

(b) $f(0) = \ln(1+0) = 0$ •

$f'(0) = (1+0)^{-1} = 1$ •

$f''(0) = -(1+0)^{-2} = -1$ •

$f^{(3)}(0) = 2(1+0)^{-3} = 2$ •

$f^{(4)}(0) = -(3)(2)(1+0)^{-4} = -(3)(2) = -3!$ •

$f^{(5)}(0) = (4)(3)(2)(1+0)^{-5} = (4)(3)(2) = 4!$

Response 2 $$\frac{|a_{j+1}|}{|a_j|} = \frac{\frac{1}{(2j+2)!}|x|^{2j+2}}{\frac{1}{(2j)!}|x|^{2j}} = \frac{(2j)!}{(2j+2)!}|x|^2 = \frac{|x|^2}{(2j+1)(2j+2)}$$

• For any x $\dfrac{|x|^2}{(2j+1)(2j+2)} \to \rho = 0$ as $j \to \infty$ • The series converges absolutely for all x. • The radius of convergence is ∞.

Response 3 $$\frac{|a_{j+1}|}{|a_j|} = \frac{(j+1)^{j+1}|x|^{j+1}}{j^j|x|^j} = (j+1)\left(\frac{j+1}{j}\right)^j |x| = (j+1)\left(1+\frac{1}{j}\right)^j |x|$$

• For any $x \neq 0$, $(j+1)\left(1+\dfrac{1}{j}\right)^j |x| \to \rho = \infty$ as $j \to \infty$ • The series diverges for all $x \neq 0$. • The radius of convergence is 0.

Response 4 $\sin x \cos x = (x - \frac{1}{3!}x^3 + \frac{1}{5!}x^5 - \cdots)(1 - \frac{1}{2}x^2 + \frac{1}{4!}x^4 - \frac{1}{6!}x^6 + \cdots)$
$= x - \frac{2}{3}x^3 + \frac{2}{15}x^5 + \cdots$

CHAPTER 8 ANSWERS

Tune-Up Exercises 8.1

T1. **(a)** $a_n = 2 + \frac{(-1)^n}{n}$ • The table is below.

n	1	2	3	4	5	6
$a_n = 2 + \frac{(-1)^n}{n} \doteq$	1	2.5	1.667	2.25	1.8	2.167

(b) Figure T1

(c) $\lim_{n\to\infty} (2 + \frac{(-1)^n}{n}) = 2$ • The sequence converges to 2.

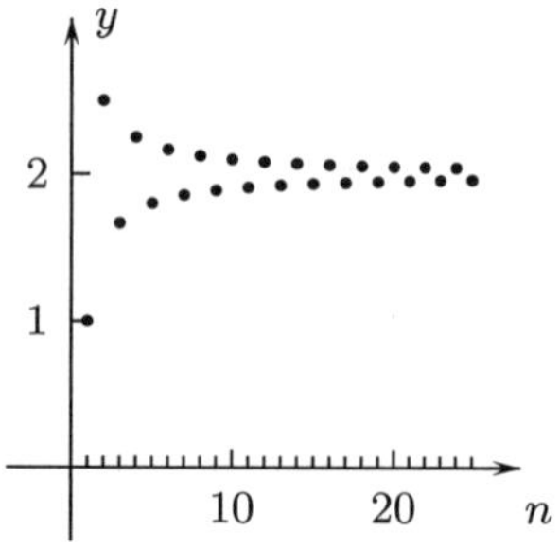

Figure T1

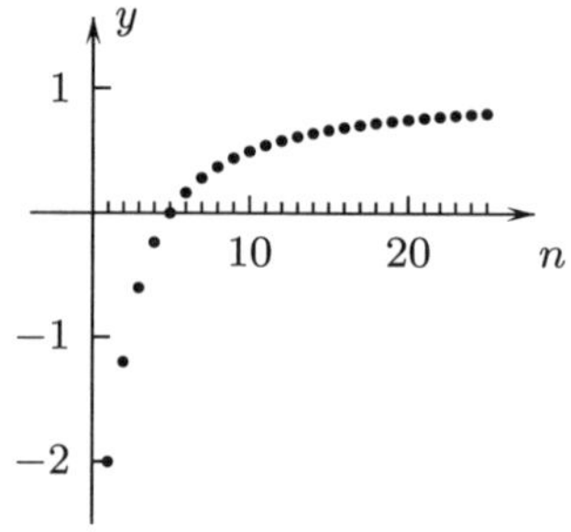

Figure T2

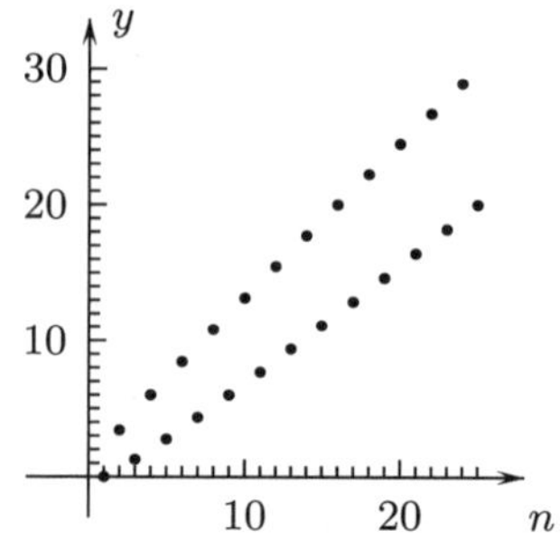

Figure T3

T2. **(a)** $a_n = \frac{n^2 - 5n}{n^2 + 1}$ • The table is below.

n	1	2	3	4	5	6
$a_n = \frac{n^2 - 5n}{n^2 + 1} \doteq$	−2	−1.2	−0.6	−0.235	0	0.162

(b) Figure T2

(c) $\lim_{n\to\infty} \frac{n^2 - 5n}{n^2 + 1} = 1$ • The sequence converges to 1.

T3. **(a)** $a_n = n + (-1)^n\sqrt{n}$ • The table is below.

n	1	2	3	4	5	6
$a_n = n + (-1)^n\sqrt{n} \doteq$	0	3.4142	1.2679	6	2.7639	8.4495

(b) Figure T3

(c) $\lim_{n\to\infty} (n + (-1)^n\sqrt{n}) = \infty$ • The sequence diverges to ∞.

T4. **(a)** $a_n = n(-1)^n$ • The table is below.

n	1	2	3	4	5	6
$a_n = n(-1)^n \doteq$	−1	2	−3	4	−5	6

(b) Figure 4

(c) $\lim_{n\to\infty} (n(-1)^n)$ diverges without a limit because $n(-1)^n$ equals n for even n and $-n$ for odd n.

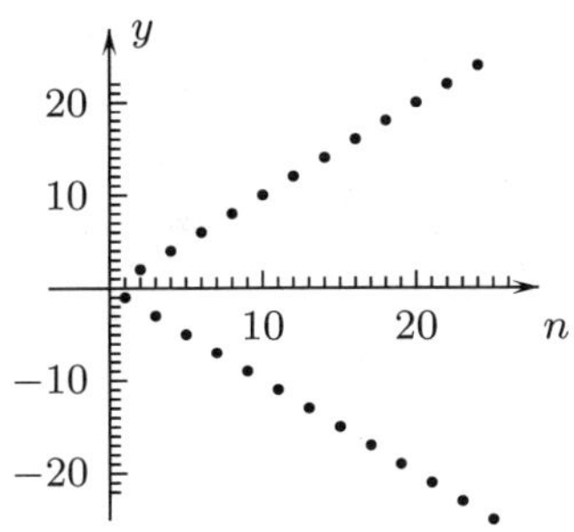

Figure T4

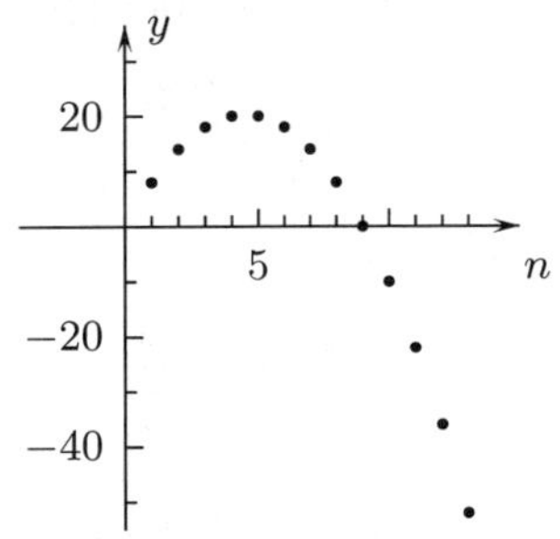

Figure T5

T5. **(a)** Figure T5 **(b)** $f(x) = 9x - x^2$ • $f'(x) = 9 - 2x = 2(4.5 - x)$ is positive for $x < 4.5$ and negative for $x > 4.5$ • Increasing for $1 \le n \le 4$ • Nondecreasing for $1 \le n \le 5$ • Decreasing for $n \ge 5$ • Nonincreasing for $n \ge 4$ **(c)** Least upper bound = 20 • There are no lower bounds

T6. **(a)** $x_0 = 1,\ x_1 = 1 + \frac{1}{1} = 2,\ x_2 = 1 + \frac{1}{2} = \frac{3}{2} = 1.5,\ x_3 = 1 + \dfrac{1}{3/2} = \frac{5}{3} \doteq 1.666667,\ x_4 = 1 + \dfrac{1}{5/3} = \frac{8}{5} = 1.6,\ x_5 = 1 + \dfrac{1}{8/5} = \frac{13}{8} = 1.625,\ x_6 = 1 + \dfrac{1}{13/8} = \frac{21}{13} \doteq 1.615385,\ x_7 = 1 + \dfrac{1}{21/13} = \frac{34}{21} \doteq 1.619048$
(b) The sequence appears to converge.

T7. **(a)** $a_n = 4(a_{n-1})^2$

n	1	2	3	4	5	6
$a_n \doteq$	1	4	64	16,384	1.07×10^9	4.6×10^{18}

(b) The sequence appears to diverge to ∞.

Problems 8.1

1. **(a)** $a_n = \dfrac{n^2+1}{5n}$ • The table is below.

n	1	2	3	4	5	6
$a_n = \dfrac{n^2+1}{5n} \doteq$	0.4	0.5	0.6667	0.85	1.04	1.2333

(b) Figure A1 **(c)** $\lim_{n\to\infty} \left(\dfrac{n^2+1}{5n}\right) = \lim_{n\to\infty} \left(\dfrac{n}{5} + \dfrac{1}{5n}\right) = \infty$ • Diverges to ∞

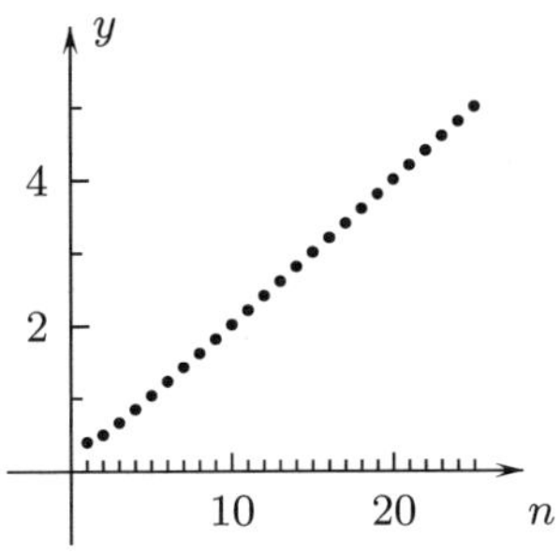

Figure A1

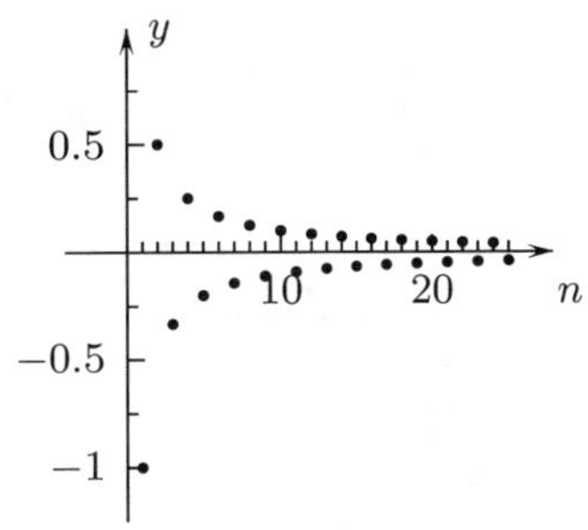

Figure A2

2. **(a)** $a_n = \dfrac{(-1)^n}{n}$ • See the table below

n	1	2	3	4	5	6
$a_n = \dfrac{(-1)^n}{n} \doteq$	-1	0.5	-0.3333	0.25	-0.2	0.1667

(b) Figure A2 **(c)** $\lim_{n\to\infty} (\dfrac{(-1)^n}{n}) = 0$ • Converges to 0

7a. Increasing

8. $\left\{n + \dfrac{36}{n}\right\}_{n=1}^{\infty}$ is decreasing for $1 \le n \le 6$ and increasing for $n \ge 6$.

10. Since $\left\{1 - e^{-n/10}\right\}_{n=1}^{\infty}$ is increasing for $n \ge 1$ and $\to 1$ as $n \to \infty$, the least upper bound of $\left\{(-1)^n(1 - e^{-n/10})\right\}_{n=1}^{\infty}$ is 1 and its greatest lower bound is -1.

11. The least upper bound is $\frac{3}{2}$ and the greatest lower bound is -1

16. The least upper bound is $\frac{1}{2}\pi$ and the greatest lower bound is $\frac{1}{4}\pi$.

20. **(a)** $\lim_{n\to\infty} a_n = \lim_{n\to\infty} (\dfrac{6}{2 + 1/n}) = 3$ **(b)** $\lim_{n\to\infty} b_n = \lim_{n\to\infty} (\dfrac{3}{2/n - 1}) = -3$
(c) $\lim_{n\to\infty} (a_n + b_n) = 3 - 3 = 0$

22. **(a)** $x = 0,\ 8$ **(b)** The table is below.

n	1	2	3	4	5	6	7	8	9
$a_n = 4(a_{n-1})^{1/3} \doteq$	4	6.349	7.407	7.7972	7.9318	7.9772	7.9924	7.9975	7.9992

(c) The fixed point $x = 8$ appears to be asymptotically stable.

23. **(a)** The positive fixed point is $x = 2$. **(b)** The table is below.

n	1	2	3	4	5	6	7	8	9
$a_n = 1 + \dfrac{2}{a_{n-1}} \doteq$	3	1.6667	2.2	1.9091	2.0476	1.9767	2.0118	1.9942	2.0029

(c) $x = 2$ seems to be an asymptotically stable fixed point.

27. $\lim_{n\to\infty} (\dfrac{1}{n})^n = \lim_{n\to\infty} \dfrac{1}{n^n} = 0$

28. $\lim_{n\to\infty} (3 - \dfrac{1}{n})^n = \infty$ because $(3 - \dfrac{1}{n}) \ge 2$ for $n \ge 1$.

29. $\lim_{n\to\infty} \ln(1+n)^{1/n} = \lim_{n\to\infty} \dfrac{\ln(1+n)}{n} = \lim_{x\to\infty} \dfrac{\ln(1+x)}{x} = \lim_{x\to\infty} \dfrac{1}{1+x} = 0$ • $(1+n)^{1/n} \to 1$

31. $\lim_{n\to\infty} n\sin(\dfrac{1}{n}) = \lim_{n\to\infty} \dfrac{\sin(\frac{1}{n})}{\frac{1}{n}} = \lim_{x\to 0+} \dfrac{\sin x}{x} = \lim_{x\to 0+} \dfrac{\cos x}{1} = 1$ by L'Hopital's rule. • Converges to 1.

38. $\lim_{n\to\infty} \ln(1 - \dfrac{2}{n})^n = \lim_{n\to\infty} n\ln(1 - \dfrac{2}{n}) = \lim_{x\to 0+} \dfrac{\ln(1-2x)}{x} = \lim_{x\to 0+} \dfrac{-2}{1-2x} = -2$ •
$(1 - \dfrac{2}{n})^n$ converges to e^{-2}

40. **(a)** $P_n = 2nR\sin\left(\dfrac{\pi}{n}\right)$

43. One answer: $n = 700$ since $(700)^{1/700} \doteq 1.0094$ is between 1 and 1.01 and greater than 1.001

44. For $n \ge 1$ and $\epsilon > 0$, $\left|\dfrac{1}{n^2}\right| < \epsilon \iff n^2 > \dfrac{1}{\epsilon} \iff n > \sqrt{1/\epsilon}$ • Take $N \ge \sqrt{1/\epsilon}$

46. $1 < e^{1/n} < 1+\epsilon \iff \ln(e^{1/n}) < \ln(1+\epsilon) \iff \dfrac{1}{n} < \ln(1+\epsilon) \iff n > \dfrac{1}{\ln(1+\epsilon)}$ • Choose $N \geq \dfrac{1}{\ln(1+\epsilon)}$

52. There exist $n > N$ for which $|a_n - L| \geq \epsilon$

53. Set $\epsilon = 0.00005$. Then for $n > 20000$, $1 + \dfrac{1}{n} < 1.00005$ so $\left|1 + \dfrac{1}{n} - 1.0001\right| > 0.00005$

54. Set $\epsilon = 1$. Then for every odd $n, (-1)^n = -1$ and $|(-1)^n - 1| > 1$

Tune-Up Exercises 8.2

T1.^{O} **(a)** $\displaystyle\sum_{j=0}^{100}(0.95)^j = \frac{1-(0.95)^{101}}{1-0.95} \doteq 19.8875$

(b) The Geometric Series converges because $0.95 < 1$. • $\displaystyle\sum_{j=0}^{\infty}(0.95)^j = \frac{1}{1-0.95} = 20$

T2.^{O} $3^3+3^4+3^5+3^6+3^7+3^8+3^9 = 3^3(1+3+3^2+3^3+3^4+3^5+3^6)$

$$= 3^3\sum_{j=0}^{6}3^j = 27\left(\frac{1-3^7}{1-3}\right) = 29,511$$

T3.^{O} $\displaystyle\sum_{j=3}^{\infty}(-\tfrac{3}{4})^j = (-\tfrac{3}{4})^3 + (-\tfrac{3}{4})^4 + (-\tfrac{3}{4})^5 + \cdots = (-\tfrac{3}{4})^3[1 + (-\tfrac{3}{4}) + (-\tfrac{3}{4})^2 + \cdots]$

$$= (-\tfrac{3}{4})^3\sum_{j=0}^{\infty}(-\tfrac{3}{4})^j = (-\tfrac{3}{4})^3\left(\frac{1}{1+\frac{3}{4}}\right) = \frac{-\frac{27}{64}}{\frac{1}{4}} = -\frac{27}{112}$$

T4.^{O} **(a)** $\displaystyle\sum_{j=0}^{\infty}(-4)^j$ diverges without limit because $-4 \leq -1$

(b) $\displaystyle\sum_{j=0}^{10}(-4)^j = \frac{1-(-4)^{11}}{1-(-4)} = 838,861$ • $\displaystyle\sum_{j=0}^{11}(-4)^j = \frac{1-(-4)^{12}}{1-(-4)} = -3,355,443$

Problems 8.2

1. $\displaystyle\sum_{j=0}^{12}(\tfrac{1}{2})^j = \frac{1-(\frac{1}{2})^{13}}{1-\frac{1}{2}} \doteq 1.999755859$

2. $\doteq 0.800003051$

3. $\tfrac{3}{4}(1+\tfrac{3}{4}+(\tfrac{3}{4})^2+\cdots+(\tfrac{3}{4})^{30}) = \tfrac{3}{4}\displaystyle\sum_{j=0}^{30}(\tfrac{3}{4})^j = \tfrac{3}{4}\left(\frac{1-(\frac{3}{4})^{31}}{1-\frac{3}{4}}\right) \doteq 2.999598190$

5. 6.66666

11. 1

13. Converges to $\frac{7}{6}$

14. Diverges to ∞

15. Diverges without a limit

16. Converges to 98.01

21. Converges to $\frac{15}{4}$

23. $3.4+0.02121212\cdots = 3.4+21(10^{-3}+10^{-5}+10^{-7}+\cdots) = 3.4+21\times 10^{-3}(1+10^{-2}+10^{-4}+\cdots) = 3.4+0.021\sum_{j=0}^{\infty} 10^{-2j} = 3.4+\dfrac{0.021}{1-10^{-2}} = 3\frac{139}{330}$

24. $\frac{1373}{333}$

28. $\sum_{j=0}^{\infty}(\frac{3}{4})^j - \sum_{j=0}^{n}(\frac{3}{4})^j = \sum_{j=n+1}^{\infty}(\frac{3}{4})^j = (\frac{3}{4})^{n+1}\sum_{j=0}^{\infty}(\frac{3}{4})^j = (\frac{3}{4})^{n+1}\dfrac{1}{1-\frac{3}{4}} = 4(\frac{3}{4})^{n+1}$ • $4(\frac{3}{4})^{n+1} < 0.004 \iff (\frac{3}{4})^{n+1} < 0.001 \iff (n+1)\ln(\frac{3}{4}) < \ln(0.001) \iff n+1 > \dfrac{\ln(0.001)}{\ln(\frac{3}{4})}$ • $\dfrac{\ln(0.001)}{\ln(\frac{3}{4})} \doteq 24.0118$ • $n+1 = 25$ • $n = 24$

30. $\doteq \$247,114.61$

Tune-Up Exercises 8.3

T1.^{O} $\dfrac{9^j}{5+10^j} < \dfrac{9^j}{10^j}$ • The series converges absolutely by the Comparison Test with convergent geometric series $\sum_{j=0}^{\infty}(\frac{9}{10})^j$.

T2.^{O} Because $\dfrac{25}{\sqrt{j^2+9}} = \dfrac{25}{j\sqrt{1+9/j^2}}$ behaves like $\dfrac{25}{j}$ for large j, we compare with $\sum_{j=1}^{\infty}\dfrac{1}{j}$ • $\lim_{j\to\infty}\dfrac{\frac{25}{\sqrt{j^2+9}}}{\frac{1}{j}} = \lim_{j\to\infty}\dfrac{25}{\sqrt{1+9/j^2}} = 25$ • The series diverges by the Limit-Comparison Test with the divergent Harmonic Series $\sum_{j=1}^{\infty}\dfrac{1}{j}$.

T3.^{O} $f(x) = \dfrac{1}{x\ln x}$ is positive and decreasing for $x > 1$. • Use $u = \ln x$, $du = (1/x)\,dx$: For $X > 2$, $\displaystyle\int_2^X \frac{1}{x\ln x}\,dx = \int_{x=2}^{x=X}\frac{1}{u}\,du = \Big[\ln|u|\Big]_{x=2}^{x=X} = \Big[\ln|\ln x|\Big]_2^X = \ln(\ln X) - \ln(\ln 2)$ • $\ln(\ln X) - \ln(\ln 2) \to \infty$ as $X\to\infty$ • $\displaystyle\int_2^X \frac{1}{x\ln x}\,dx = \infty$ • $\sum_{j=2}^{\infty}\dfrac{1}{j\ln j}$ diverges to ∞ by the Integral Test.

T4.^{O} $\tan^{-1} j \to \frac{1}{2}\pi$ as $j\to\infty$ • $(-1)^j\tan^{-1} j$ does not $\to 0$ as $j\to\infty$ • $\sum_{j=1}^{\infty}(-1)^j\tan^{-1} j$ diverges.

T5.^{O} Because $\dfrac{k^2}{\sqrt{k^5+7}}$ behaves like $\dfrac{k^2}{\sqrt{k^5}} = \dfrac{1}{k^{1/2}}$ for large k, we compare with $\sum_{k=1}^{\infty}\dfrac{1}{k^{1/2}}$. • $\dfrac{\frac{k^2}{\sqrt{k^5+7}}}{\frac{1}{k^{1/2}}} = \dfrac{k^{2+1/2}}{\sqrt{k^5+7}} = \dfrac{1}{\sqrt{1+k^5/7}} \to 1$ as $k\to\infty$ • The series diverges by the Limit-Comparison

Test with the divergent p-series $\sum_{k=1}^{\infty} \frac{1}{k^{1/2}}$.

T6.⁰ $\lim_{j\to\infty} \frac{\left|\frac{(-1)^j}{2j^2-15}\right|}{\frac{1}{j^2}} = \lim_{j\to\infty} \frac{1}{2-15/j^2} = \frac{1}{2}$ • The series converges absolutely by the Limit-Comparison Test with the convergent p-series $\sum_{j=1}^{\infty} \frac{1}{j^2}$.

Problems 8.3

1. $\left|\frac{\cos j}{j^{3/2}}\right| \le \frac{1}{j^{3/2}}$ • The series converges absolutely by the Comparison Test with the convergent p-series $\sum_{j=1}^{\infty} \frac{1}{j^{3/2}}$.

2. The series diverges because $\frac{4^j}{3^j - 10} = (\frac{4}{3})^j \frac{1}{1-10/3^j} \to \infty$ as $j \to \infty$ and the terms do not $\to 0$.

3. For $j \ge 2$, $\frac{1}{\ln(\ln j)} > \frac{1}{\ln(j)} > \frac{1}{j}$ • The series diverges by the Comparison Test with the divergent Harmonic Series $\sum_{j=1}^{\infty} \frac{1}{j}$.

4. The series converges.

18. $\doteq 1.539767731$

Tune-Up Exercises 8.4

T1. **(a)** $s_n = \sum_{j=0}^{n} \frac{3^j}{j!}$ • The table is below.

n	0	1	2	3	4	5	6	7
$s_n \doteq$	1	4	8.5	13	16.375	18.4	19.413	19.846

(b) Figure T1 **(c)** $a_j = \frac{3^j}{j!}$ • $\frac{|a_{j+1}|}{|a_j|} = \frac{\frac{3^{j+1}}{(j+1)!}}{\frac{3^j}{j!}} = \frac{j!}{(j+1)!}(3) = \frac{3}{j+1}$ • $\frac{3}{j+1} \to \rho = 0$ as $j \to \infty$ • The series converges by the Ratio Test because $\rho < 1$.

(d) $\sum_{j=0}^{16} \frac{3^j}{j!} \doteq 20.085536488$ • $\sum_{j=0}^{17} \frac{3^j}{j!} \doteq 20.085536851$ • One explanation: These partial sums are very close because the series converges rapidly.

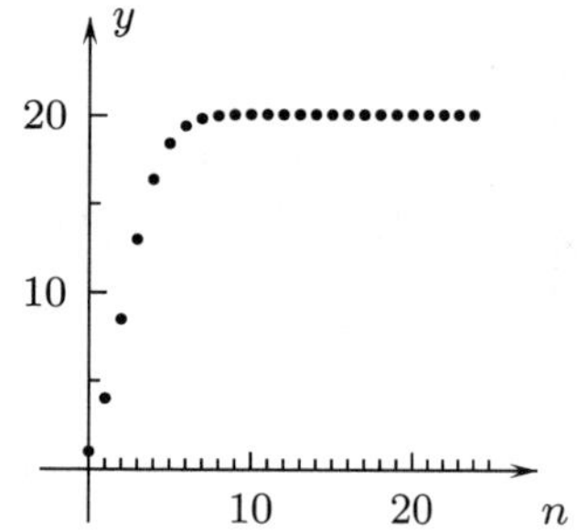

Figure T1

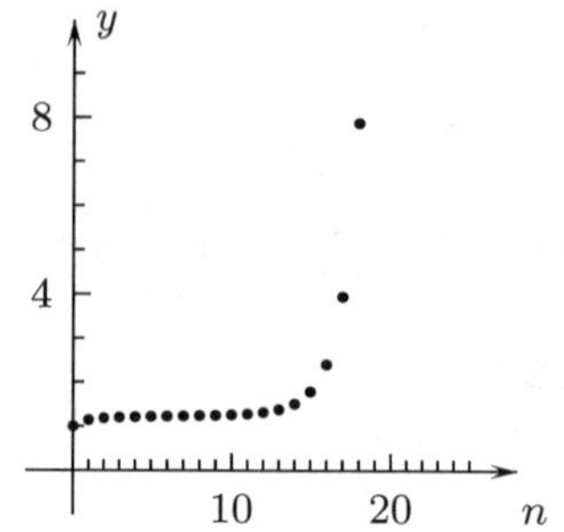

Figure T2

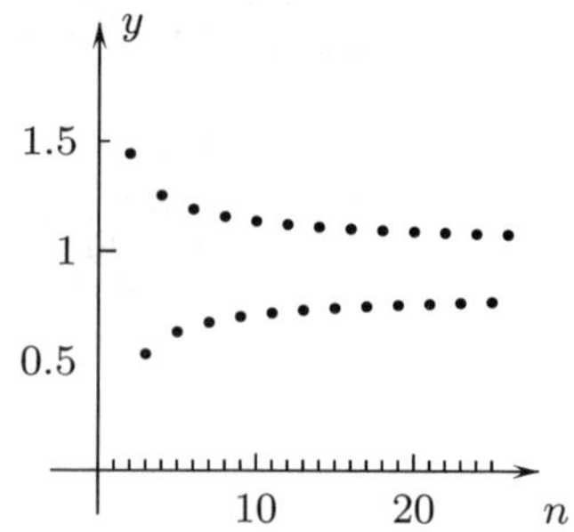

Figure T3

T2. **(a)** $s_n = \sum_{j=0}^{n} \frac{j!}{7^j}$ • The table is below.

n	0	1	2	3	4	5	6	7
$s_n \doteq$	1	1.1429	1.1837	1.2012	1.2112	1.2183	1.2244	1.2305

(b) Figure T2 **(c)** $a_j = \frac{j!}{7^j}$ • $\frac{|a_{j+1}|}{|a_j|} = \frac{\frac{(j+1)!}{7^{j+1}}}{\frac{j!}{7^j}} = \frac{1}{7}\frac{(j+1)!}{j!} = \frac{j+1}{7}$ • $\frac{j+1}{7} \to \rho = \infty$ as $j \to \infty$ • The series diverges by the Ratio Test because $\rho > 1$. • **(d)** $\sum_{j=0}^{24} \frac{j!}{7^j} \doteq 4611.13$ • $\sum_{j=0}^{25} \frac{j!}{7^j} \doteq 16{,}177.44$ • One explanation: The second partial sum is much larger than the first because the series diverges rapidly to ∞.

T3. **(a)** $s_n = \sum_{j=2}^{n} \frac{(-1)^j}{\ln j}$ • The table is below.

n	2	3	4	5	6	7	8	9
$s_n \doteq$	1.4427	0.5325	1.2538	0.6325	1.1906	0.6767	1.1576	0.7025

(b) Figure T3 **(c)** $\frac{1}{\ln j} > 0$ and $\frac{1}{\ln(j+1)} < \frac{1}{\ln j}$ for $j \geq 2$ • $\frac{1}{\ln j} \to 0$ as $j \to \infty$ • The series converges by the Alternating-Series Test. • $\sum_{j=2}^{\infty} \frac{1}{\ln j}$ diverges by the Comparison Test with the divergent Harmonic Series $\sum_{j=1}^{\infty} \frac{1}{j}$ because $\frac{1}{\ln j} > \frac{1}{j}$ for $j \geq 2$. • The series converges conditionally. **(d)** $\sum_{j=2}^{20} \frac{(-1)^j}{\ln j} \doteq 1.089812500$ • $\sum_{j=2}^{21} \frac{(-1)^j}{\ln j} \doteq 0.761353759$ • $\sum_{j=2}^{22} \frac{(-1)^j}{\ln j} \doteq 1.1084869212$ • Two explanations: The partial sum with $n = 22$ is between those for $n = 20$ and 21 because the series satisfies the conditions of the Alternating-Series Test, and these partial sums are not very close because the series converges relatively slowly.

T4. **(a)** $s_n = \sum_{j=1}^{n} \frac{10j(-1)^j}{j+5}$ • The table is below.

n	1	2	3	4	5	6	7	8
$s_n \doteq$	−1.6667	1.1905	−2.5595	1.8849	−3.1151	2.3395	−3.4939	2.6600

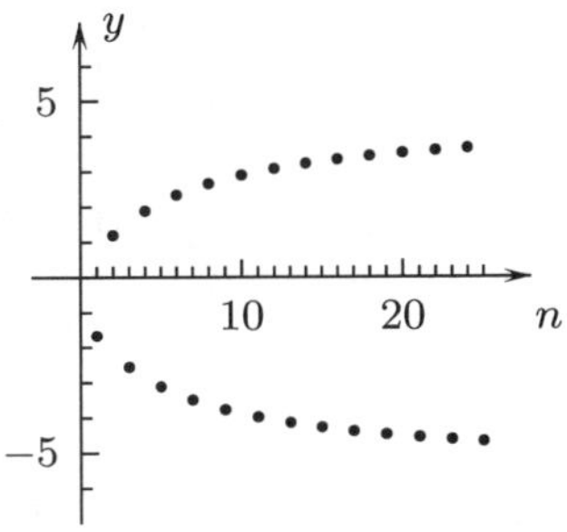

Figure T4

(b) Figure T4 **(c)** $\dfrac{10j}{j+5} = \dfrac{10}{1+5/j} \to 10$ as $j \to \infty$ • The series diverges because its terms do not tend to 0. •
(d) $\displaystyle\sum_{j=0}^{19} \frac{10j(-1)^j}{j+5} \doteq -4.470708310$ • $\displaystyle\sum_{j=0}^{20} \frac{10j(-1)^j}{j+5} \doteq 3.529291690$ • One explanation: The second partial sum is 8 more than the first because the 20th term is $\dfrac{10(20)}{20+5} = 8$.

Problems 8.4

1.[A] **(a)** The series converges absolutely. • Figure A1

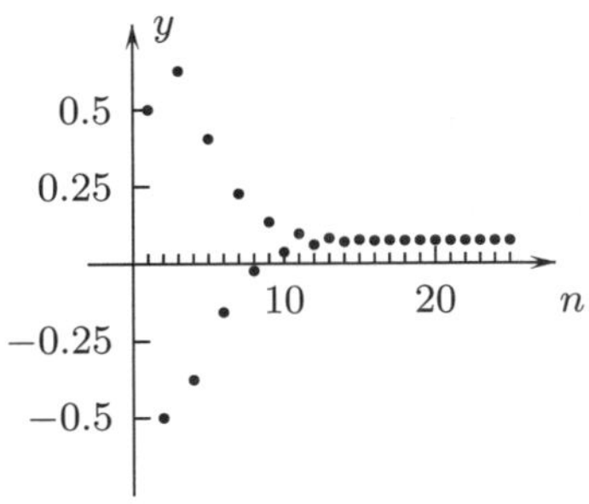

Figure A1

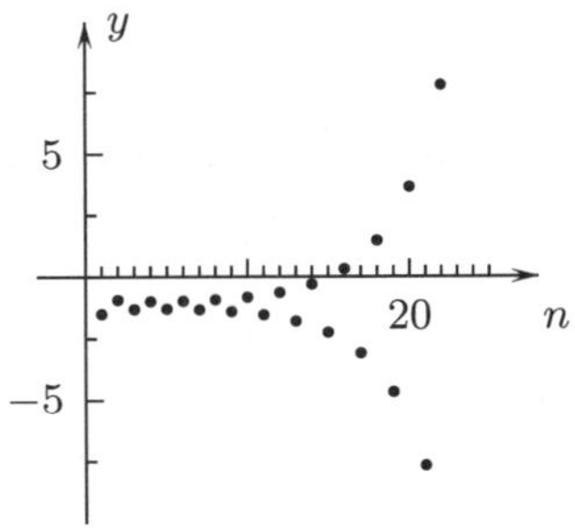

Figure A2

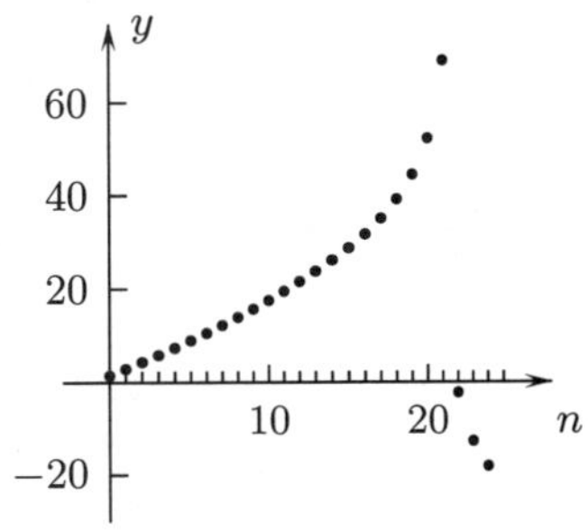

Figure A3

2.[A] **(a)** The series diverges. • Figure A2

6.[A] The series converges absolutely. • Figure A6

7.[A] The series converges absolutely.

10.[A] Diverges

14.[A] Converges absolutely

Tune-Up Exercises 8.5

T1. **(a)** $f(x) = 2 + xe^x, f'(x) = xe^x + e^x, f''(x) = xe^x + 2e^x, f^{(3)}(x) = xe^x + 3e^x, f^{(4)}(x) = xe^x + 4e^x$ • $f(0) = 2, f'(0) = 1, f''(0) = 2, f^{(3)}(0) = 3, f^{(4)}(0) = 4$ •
$P_4(x) = 2 + x + \frac{1}{2}(2x^2) + \dfrac{1}{3!}(3x^3) + \dfrac{1}{4!}(4x^4) = x + x^2 + \frac{1}{2}x^2 + \frac{1}{6}x^3$ **(b)** Figure T1
(c) The values are in the table below. Possible descriptions: $P_4(x)$ approximates xe^x much more closely at $x = \pm 0.1$ than at ± 1 or ± 2 • $P_4(x)$ is greater than xe^x for $x = -2, -1, -0.1$ and is less than xe^x for $x = 0.1, 1$, and 2.

x	-2	-1	-0.1	0.1	1	2
$2 + xe^x - P_4(x) \doteq$	-0.9373	-0.0345	-4.08×10^{-7}	4.25×10^{-7}	0.0516	2.1114

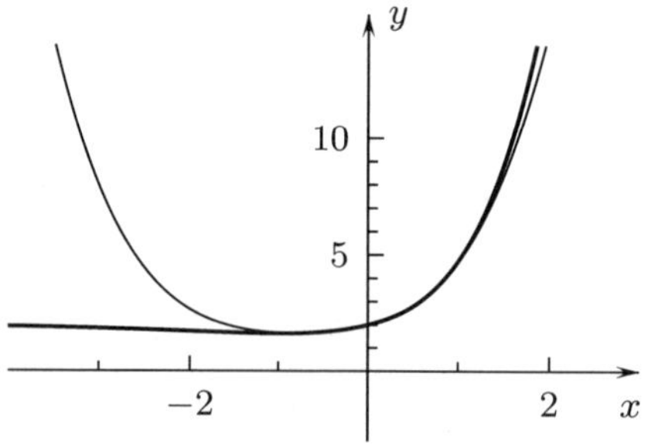

Figure T1

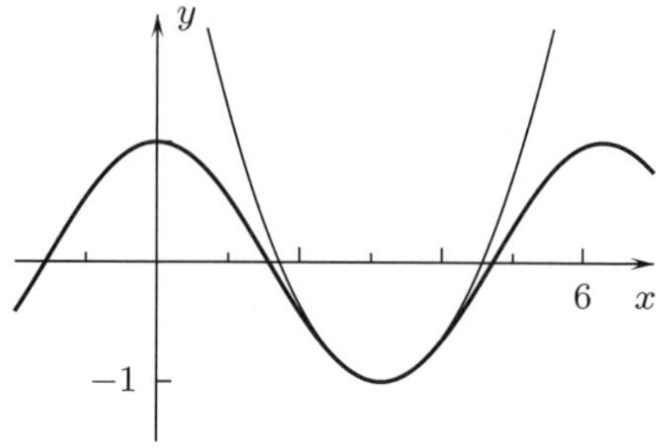

Figure T2

T2. **(a)** $f(x) = \cos x, f'(x) = -\sin x, f''(x) = -\cos x, f^{(3)}(x) = \sin x$ •
$f(3) = \cos(3), f'(3) = -\sin(3), f''(3) = -\cos(3)$ •
$P_3(x) = \cos(3) - \sin(3)(x-3) - \frac{1}{2}\cos(3)(x-3)^2 + \frac{1}{6}\sin(3)(x-3)^4$ **(b)** Figure T2
(c) The values are in the table below. Possible descriptions: $P_3(x)$ approximates $\cos x$ much more closely at $x = 2.9$ and 3.1 than at $1, 2, 4$ or 5. • $P_3(x)$ is greater than $\cos x$ for $x = 1, 2, 2.9, 3.1, 4$ and 5.

x	1	2	2.9	3.1	4	5
$\cos x - P_3(x) \doteq$	-0.5438	-0.0388	-4.11×10^{-6}	-4.14×10^{-6}	-0.041	-0.6123

T3. **(a)** $f(x) = 1 + 2x + 3x^2 + 4x^3 + 5x^4 + 6x^5$
$f'(x) = 2 + 6x + 12x^2 + 20x^3 + 30x^4$
$f''(x) = 6 + 24x + 60x^2 + 120x^3$
$f^{(3)}(x) = 24 + 120x + 360x^2$
$f(0) = 1, f'(0) = 2, f''(0) = 6 f^{(3)}(0) = 24$
$P_3(x) = 1 + 3x + \frac{1}{2}(6x^2) + \frac{1}{3!}(24x^3) = 1 + 2x + 3x^2 + 4x^3$.
(b) $P_3(x)$ is the first four terms of the polynomial $f(x)$.

T4. **(a)** $f^{(j)}(x) = e^x$ for all $j \geq 0$ • $f^{(j)}(0) = 1$ for all $j \geq 0$ •
$P_4(x) = 1 + x + \frac{1}{2}x^2 + \frac{1}{3!}x^3 + \frac{1}{4!}x^4$ •
[Maximum of $|f^{(5)}(x)| = |e^x|$ for $0 \leq x \leq 0.5$] $= e^{0.5}$ •
[Error estimate] $= \frac{1}{5!}e^{0.5}(0.5)^5 \doteq 4.29 \times 10^{-3}$
(b) $|e^{0.5} - P_4(0.5)| \doteq 2.84 \times 10^{-4}$

Problems 8.5

1. $f(x) = \cos x,\ f'(x) = -\sin x,\ f''(x) = -\cos x,\ f'''(x) = \sin x,\ f^{(4)}(x) = \cos x,$
$f^{(5)}(x) = -\sin x,\ f^{(6)}(x) = -\cos x$ • $f(0) = 1, f'(0) = 0, f''(0) = -1, f^{(3)}(0) = 0,$
$f^{(4)}(0) = 1, f^{(5)}(0) = 0, f^{(6)}(0) = -1$ • $P_6(x) = 1 - \frac{1}{2!}x^2 + \frac{1}{4!}x^4 - \frac{1}{6!}x^6$ • Figure A1

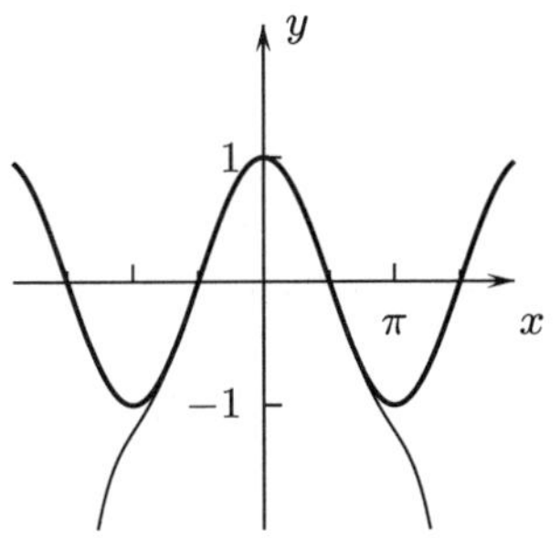

Figure A1

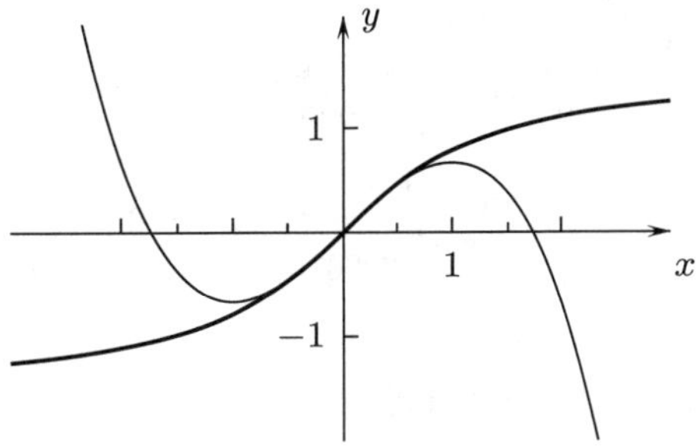

Figure A2

2. $P_4(x) = x - \frac{1}{3}x^3$ • Figure A2

3. $f(x) = x^{1/2}$, $f'(x) = \frac{1}{2}x^{-1/2}$, $f'(x) = -\frac{1}{4}x^{-3/2}$ • $f(9) = 9^{1/2} = 3$, $f'(9) = \frac{1}{2}9^{-1/2} = \frac{1}{6}$, $f''(9) = -\frac{1}{4}9^{-3/2} = -\frac{1}{108}$ • $P_2 = 3 + \frac{1}{6}(x-9) - \frac{1}{216}(x-9)^2$ • Figure A3

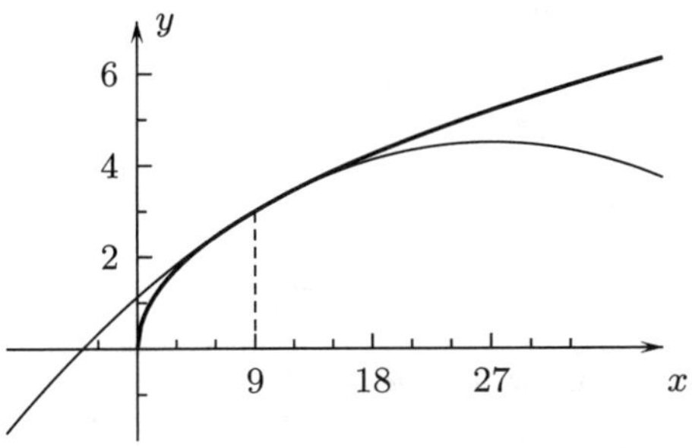

Figure A3

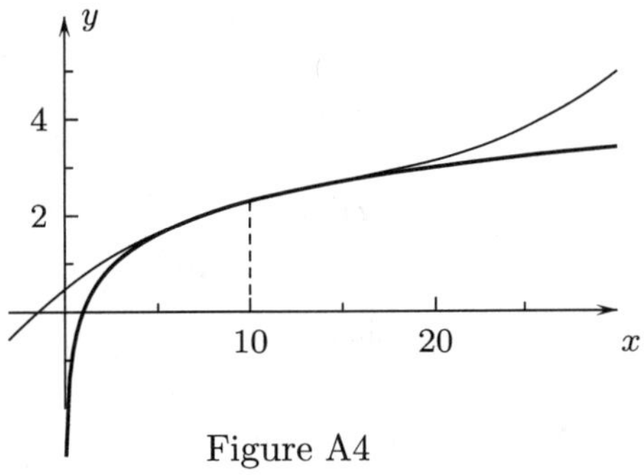

Figure A4

4. $P_3(x) = \ln(10) + \frac{1}{10}(x-10) - \frac{1}{2!}\frac{1}{100}(x-10)^2 + \frac{1}{3!}\frac{1}{500}(x-10)^3$ • Figure A4

6. $P_2(x) = x$ • Figure A6

Figure A6

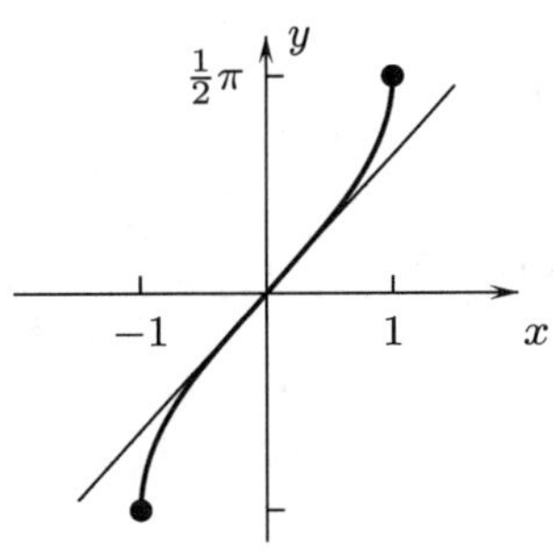

7. $P_4(x) = -\frac{1}{2}x^2 - \frac{1}{12}x^4$

13. $P_2(x) = 1 + x^2$

17. $P_3(\frac{1}{2})$ is closer to $e^{1/2}$ than is $Q_3(\frac{1}{2})$, where $P_3(x)$ is the third-degree Taylor polynomial at $x = 0$ and $Q_3(x)$ is the third-degree Taylor polynomial at $x = 1$.

24. **(a)** e^x **(b)** $\cos x$ **(d)** $\sinh x$

25. **(a)** $f'(0) = 1$ **(b)** $f''(0) = 0$ **(c)** $f^{(4)}(0) = -96$ **(d)** $f^{(5)}(0) = 360$

27. $R_2(1.03) = \frac{1}{3!}(\frac{8}{27}c^{-7/3})(0.03)^3 > 0$ so $P_2(1.03) < (1.03)^{2/3}$ • $P_2(1.03) = 1.0199$ is less than $(1.03)^{2/3} = 1.01990131$

29. $R_5(\frac{1}{10}\pi) = \dfrac{1}{6!}[-\sin c](\frac{1}{10}\pi)^6 < 0$ since $0 < c < \frac{1}{10}\pi$ • $|R_5(\frac{1}{10}\pi)| < \dfrac{1}{6!}\sin(\frac{1}{10}\pi)(\frac{1}{10}\pi)^6$ • $\sin(\frac{1}{10}\pi) \doteq 0.309016994$ • $P_5(\frac{1}{10}\pi) \doteq 0.309017054$

Tune-Up Exercises 8.6

T1. $f(x) = \displaystyle\sum_{j=0}^{\infty} \frac{1}{j!} f^{(j)}(0)x^j = \sum_{j=0}^{\infty} \frac{1}{j!}(5)^j x^j = \sum_{j=0}^{\infty} \frac{1}{j!}(5x)^j$

$\left|\dfrac{a_{j+1}}{a_j}\right| = \dfrac{\frac{1}{(j+1)!}|5x|^{j+1}}{\frac{1}{j!}|5x|^j} = \dfrac{j!}{(j+1)!}|5x| = \dfrac{1}{j+1}|5x| \to \rho = 0$ as $j \to \infty$ • The series converges absolutely for all x; its radius of convergence is ∞.

T2. Add the series $e^x = \displaystyle\sum_{j=0}^{\infty} \frac{1}{j!}x^j$ and $e^{2x} = \displaystyle\sum_{j=0}^{\infty} \frac{1}{j!}(2x)^j = \sum_{j=0}^{\infty} \frac{1}{j!}(2)^j x^j$ • $e^x + e^{2x} = \displaystyle\sum_{j=0}^{\infty} \frac{1}{j!}(1+2^j)x^j$

T3. $e^x \cos x = (1 + x + \frac{1}{2}x^2 + \frac{1}{6}x^3 + \frac{1}{24}x^4 + \cdots)(1 - \frac{1}{2}x^2 + \frac{1}{24}x^4 - \ldots) =$
$1(1 - \frac{1}{2}x^2 + \frac{1}{24}x^4 - \ldots) + x(1 - \frac{1}{2}x^2 + \cdots) + \frac{1}{2}x^2(1 - \frac{1}{2}x^2 + \cdots) + \frac{1}{6}x^3(1 - \cdots) + \frac{1}{24}x^4(1 - \cdots) =$
$1 - \frac{1}{2}x^2 + \frac{1}{24}x^4 + x - \frac{1}{2}x^3 + \frac{1}{2}x^2 - \frac{1}{4}x^4 + \frac{1}{6}x^3 + \frac{1}{24}x^4 + \cdots = 1 + x - \frac{1}{3}x^3 - \frac{1}{6}x^4 + \cdots$

T4. For $x \neq 0$, $\dfrac{\sin x}{x} = \dfrac{1}{x}(x - \frac{1}{3!}x^3 + \frac{1}{5!}x^5 - \cdots$ • This is the series because its value is 1 at $x = 0$ • With summation notation $\dfrac{1}{x}\displaystyle\sum_{j=0}^{\infty} \frac{(-1)^j}{(2j+1)!}x^{2j+1} = \sum_{j=0}^{\infty} \frac{(-1)^j}{(2j+1)!}x^{2j}$

T5. $\dfrac{1}{1-\sin x} = 1 + \sin x + \sin^2 x + \sin^3 x + \sin^4 x + \cdots = 1 + (x - \frac{1}{6}x^3 + \cdots) + (x - \frac{1}{6}x^3 + \cdots)^2 + (x - \frac{1}{6}x^3 + \cdots)^3 + (x - \cdots)^4 \cdots = 1 + [x - \frac{1}{6}x^3 + \cdots] + [x^2 - 2x(\frac{1}{6}x^3) + \cdots] + [x^3 - \cdots] + [x^4 - \cdots] \cdots =$
$1 + x - \frac{1}{6}x^3 + x^2 - \frac{1}{3}x^4 + x^3 + x^4 + \cdots = 1 + x + x^2 + \frac{5}{6}x^3 + \frac{2}{3}x^4 + \cdots$

T6. Set $u = -t^2$ in $\dfrac{1}{1-u} = \displaystyle\sum_{j=0}^{\infty} u^j$: For $|t| < 1$, $\dfrac{1}{1+t^2} = \displaystyle\sum_{j=0}^{\infty}(-t^2)^j = \sum_{j=0}^{\infty}(-1)^j t^{2j}$ • For $|x| < 1$, $\tan^{-1} x = \displaystyle\int_0^x \frac{1}{1+t^2}\,dt = \sum_{j=0}^{\infty}(-1)^j \int_0^x t^{2j}\,dt = \sum_{j=0}^{\infty}(-1)^j \frac{1}{2j+1}x^{2j+1}$

T7. $e^u = 1 + u + \frac{1}{2}u^2 + \frac{1}{6}u^3 + \cdots$ • $\sin x = x - \frac{1}{6}x^3 + \cdots$ •
$e^{\sin x} = 1 + (x - \frac{1}{6}x^3 + \cdots) + \frac{1}{2}(x - \frac{1}{6}x^3 + \cdots)^2 + \frac{1}{6}(x - \frac{1}{6}x^3 + \cdots)^3 + \cdots$
$= 1 + [x - \frac{1}{6}x^3 + \cdots] + \frac{1}{2}[x^2 + \cdots] + \frac{1}{6}[x^3 - \cdots] + \cdots = 1 + x + \frac{1}{2}x^2 + \cdots$ (The coefficient of x^3 is 0.)

Problems 8.6

1. **(a)** $R = 1$ • Ratio Test: $\dfrac{|a_{j+1}|}{|a_j|} = \dfrac{j+1}{j}|x| = (1+\dfrac{1}{j})|x| \to |x| = \rho$ as $j \to \infty$ • $\rho < 1 \iff |x| < 1, \rho > 1 \iff |x| > 1$. • The series converges absolutely for $-1 < x < 1$ diverges for $x < -1$ and $x > 1$. **(b)** For $x = 1$, the series is the harmonic series $\sum_{j=1}^{\infty} \dfrac{1}{j}$, which diverges. • For $x = -1$, the series is $\sum_{j=1}^{\infty} \dfrac{(-1)^j}{j}$ which converges by the Altenating Series Test since $\dfrac{1}{j+1} < \dfrac{1}{j}$ for $j \geq 1$ and $\dfrac{1}{j} \to 0$ as $j \to \infty$, the series converges conditionally. **(c)** Prediction for $x = \frac{1}{2}$: $\sum_{j=1}^{\infty} \dfrac{1}{j}(\frac{1}{2})^j \approx 0.69315$ since $\sum_{j=1}^{n} \dfrac{1}{j}(\frac{1}{2})^j \doteq 0.69315$ for $n = 15, 20, 25$ • $x = 2$: $\sum_{j=1}^{n} \dfrac{1}{j}(2)^j \doteq 4749.80$ for $n = 15$ and $\doteq 8845.80$ for $n = 16$

2. **(a)** $R = 1$ **(b)** The series diverges at $x = -1, 1$ **(c)** $\sum_{j=1}^{\infty} \sqrt{j}(\frac{3}{4})^j \approx 5.5432$ • $\sum_{j=1}^{n} \sqrt{j}(10)^j \doteq 245,490$ for $n = 5$

9. $f(x) = \sum_{j=0}^{\infty} \dfrac{1}{j!} f^{(j)}(0)x^j = \sum_{j=0}^{\infty} \dfrac{j^3}{j!}x^j$ • Prediction: $\sum_{j=0}^{\infty} \dfrac{j^3}{j!} \approx 13.591$ since $\sum_{j=0}^{\infty} \dfrac{j^3}{j!} \approx 13.591$ for $n = 10, 20, 30$

12. The series converges in $[1, 5)$ • The series diverges for $x \geq 5$ and $x < 1$.

18. $f(x) = (1+x)^5$ • $f'(x) = 5(1+x)^4$ • $f''(x) = 20(1+x)^3$ • $f'''(x) = 60(1+x)^2$ • $f^{(4)}(x) = 120(1+x)$ • $f^{(5)}(x) = 120$ • $f^{(j)}(x) = 0$ for $j \geq 6$ • $f(0) = 1,\ f'(0) = 5,\ f''(0) = 20,\ f'''(0) = 60,\ f^{(4)}(0) = 120,\ f^{(5)}(0) = 120,\ f^{(j)}(0) = 0$ for $j \geq 6$ • The MacLaurin series converges because it is finite and it equals $(1+x)^5$ because the remainder $R_n(x)$ in Taylor's Formula is zero for $n \geq 5$ • $(1+x)^5 = 1 + 5x + \dfrac{1}{2!}(20)x^2 + \dfrac{1}{3!}(60)x^3 + \dfrac{1}{4!}(120)x^4 + \dfrac{1}{5!}(120)x^5 = 1 + 5x + 10x^2 + 10x^3 + 5x^4 + x^5$

19. $f(x) = \dfrac{1}{1-x}$

21. $\ln\left(\dfrac{1+x}{1-x}\right) = \ln(1+x) - \ln(1-x) = \sum_{j=1}^{\infty} \dfrac{(-1)^{j+1}}{j}x^j - \sum_{j=1}^{\infty} \dfrac{(-1)^{j+1}}{j}(-x)^j = \sum_{j=1}^{\infty} \dfrac{[(-1)^{j+1}+1]}{j}x^j = \sum_{j=1}^{\infty} \dfrac{2}{2j-1}x^{2j-1}$

22. $\sum_{j=0}^{\infty} \dfrac{[3(2)^j + 4(5)^j]}{j!}x^j$

26. $\dfrac{d}{dx}(1-x)^{-1} = (1-x)^{-2}$ • $\dfrac{d^2}{dx^2}(1-x)^{-1} = 2(1-x)^{-3}$ • $(1-x)^{-3} = \frac{1}{2}\dfrac{d^2}{dx^2}\sum_{j=0}^{\infty} x^j = \frac{1}{2}\sum_{j=0}^{\infty} \dfrac{d^2}{dx^2}(x^j) = \sum_{j=2}^{\infty} \dfrac{j(j-1)}{2}x^{j-2}$

27. $\sum_{j=2}^{\infty} \dfrac{2j(2j-1)(2j-2)}{j!}x^{2j-3}$

31. $x^{30}\sin x = \sum_{j=0}^{\infty} \frac{(-1)^j}{(2j+1)!} x^{2j+31}$ • $\frac{d^2}{dx^2}(x^{30}\sin x) = \frac{d^2}{dx^2}\sum_{j=0}^{\infty}\frac{(-1)^j}{(2j+1)!}x^{2j+31}$
$= \sum_{j=0}^{\infty} \frac{(-1)^j(2j+31)(2j+30)}{(2j+1)!} x^{2j+29}$

32. $\sum_{j=0}^{\infty} (j+10)(j+9)(j+8)x^{j+7}$

34. $\int_0^x e^{t^2}\, dt = \sum_{j=0}^{\infty} \frac{1}{j!}\int_0^x t^{2j}\, dt = \sum_{j=0}^{\infty} \frac{1}{j!(2j+1)} x^{2j+1}$

35. $\sum_{j=0}^{\infty} \frac{(-1)^j}{(2j+1)!(6j+4)} x^{6j+4}$

37. $e^x \sin x = (1 + x + \frac{1}{2}x^2 + \frac{1}{6}x^3 + \frac{1}{24}x^4 + \cdots)(x - \frac{1}{6}x^3 + \frac{1}{120}x^5 + \cdots) =$
$(x + x^2 + \frac{1}{2}x^3 + \frac{1}{6}x^4 + \frac{1}{24}x^5 + \cdots) + (-\frac{1}{6}x^3 - \frac{1}{6}x^4 - \frac{1}{12}x^5 + \cdots) + (\frac{1}{120}x^5) =$
$x + x^2 + \frac{1}{3}x^3 - \frac{1}{30}x^5 + \cdots$

38. $x^2 + x^3 + \frac{11}{12}x^4 + \frac{5}{6}x^5 + \cdots$

41. $\sec x = \frac{1}{\cos x} = 1/(1 - \frac{1}{2!}x^2 + \frac{1}{4!}x^4 - \frac{1}{6!}x^6 + \cdots) = 1 + \frac{1}{2}x^2 + \frac{5}{24}x^4 + \frac{61}{760}x^6 + \cdots$

42. $x - \frac{1}{2}x^2 + \frac{5}{6}x^3 - \frac{1}{2}x^4 + \cdots$

44. $\cos(\sin x) = 1 - \frac{1}{2!}(\sin x)^2 + \frac{1}{4!}(\sin x)^4 - \frac{1}{6}(\sin x)^6 + \cdots$
$= 1 - \frac{1}{2!}(x - \frac{1}{3!}x^3 + \frac{1}{5!}x^5 + \cdots)^2 + \frac{1}{4!}(x - \frac{1}{3!}x^3 + \frac{1}{5!}x^5 + \cdots)^4 - \frac{1}{6!}(x + \cdots)^6 = 1 - \frac{1}{2}x^2 + \frac{5}{24}x^4 - \frac{37}{720}x^6 + \cdots$

45. $f^{(10)}(0) = \frac{9}{2}(10)!$

CHAPTER 9
CONIC SECTIONS AND POLAR COORDINATES

This chapter contains topics in analytic geometry that we will need in Chapters 10 and 11, where we will study curves, surfaces, and functions of two variables. The first three sections deal with CONIC SECTIONS—circles, ellipses, parabolas, and hyperbolas. In Section 9.1 we derive the equations of these curves from their definitions in terms of distances in an xy-plane in cases where the axes of the curves are parallel to the coordinate axes. In Section 9.2 we show how conic sections whose axes are not parallel to coordinate axes can be studied by using rotated coordinates. In Section 9.3 we show that parabolas, ellipses, and hyperbolas are, in fact, cross sections of cones. Then we derive focus-eccentricity definitions of ellipses and hyperbolas. In Section 9.4 we discuss polar coordinates, which we will use for examples of curves and to study planetary motion in Chapter 10 and with integrals of functions of two variables in Chapter 12.

Section 9.1

Parabolas, ellipses, and hyperbolas

OVERVIEW: *In this section we derive equations for parabolas, ellipses, and parabolas in xy-planes in cases where the axes of the curves are parallel to the x- or y-axes and some of their applications.*

Topics:

- ***Equations of parabolas in standard position***
- ***Translation of axes***
- ***Equations of ellipses with horizontal and vertical axes***
- ***Equations of hyperbolas with horizontal and vertical axes***
- ***Parabolic trajectories***
- ***Optical properties of conic sections***
- ***Hyperbolas and navigation***

Cross sections of cones

Parabolas, ellipses, and hyperbolas are called CONIC SECTIONS because they can be obtained as the intersections of planes with double right circular cones (Figure 1). Such a cone consists of all the lines through a point and the circumference of a circle. The point is the VERTEX of the cone and lies on its AXIS, which is a line through the center of the circle and perpendicular to the plane containing the circle. The lines that form the cone are its ELEMENTS.

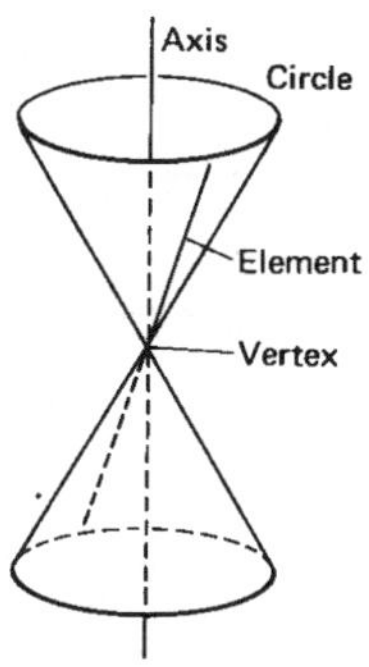

A double cone
FIGURE 1

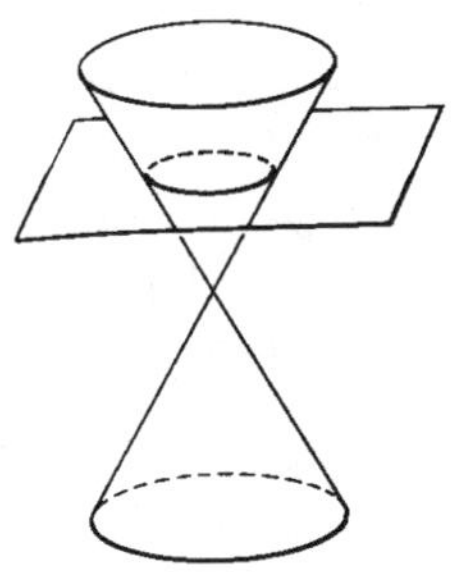

A circle
FIGURE 2

♦ SUGGESTIONS TO INSTRUCTORS: The main topics in this section can be divided into three parts, as follows:
Part 1 (Parabolas; Tune-up Exercises T1 through T3; Problems 1–3, 11, 16, 17, 21, 23, 24, 30, 37) Have students begin work on Questions 1, 2, and 8 before class. Problems 2, 3, 17, 30, and 37 are good for in-class discussion.
Part 2 (Ellipses; Tune-up Exercises T4 and T5; Problems 4–6, 10, 14, 18, 20, 22, 27–29, 32, 33, and 36) Have students begin work on Questions 3 through 6 before class. Problems 4, 10, 22, 27, 32, 33, and 36 are good for in-class discussion.
Part 3 (Hyperbolas and general conic sections; Tune-up Exercises T6 and T7; Problems 7–9, 12, 13, 15, 19, 25, 26, 31, 34, 35, 38) Have students begin work on Question 7 before class. Problems 7, 9, 13, 15, 31, 34, and 38 are good for in-class discussion.

Visualize a plane that intersects the cone in Figure 1 but does not pass through its vertex. As you can imagine, the intersection of the plane with the the cone is a circle if the plane is perpendicular to the axis of the cone, as in Figure 2. We will see in Section 9.2 that the intersection is an ellipse if the plane is tilted and cuts across one half of the cone (Figure 3); the intersection is a parabola if the plane is parallel to one of the elements of the cone (Figure 4); and the intersection is a hyperbola if the plane intersects both halves of the cone (Figure 5).

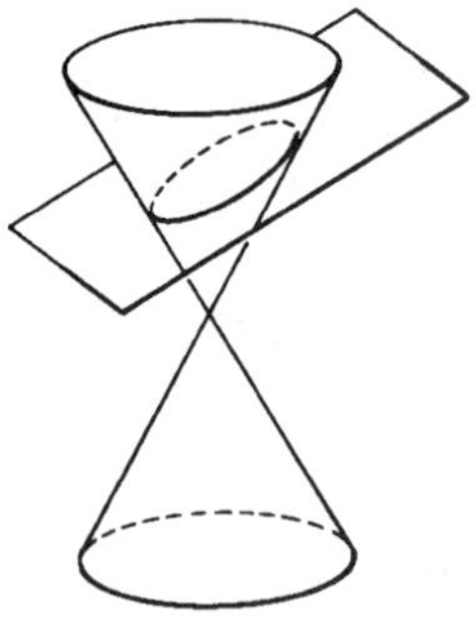

An ellipse
FIGURE 3

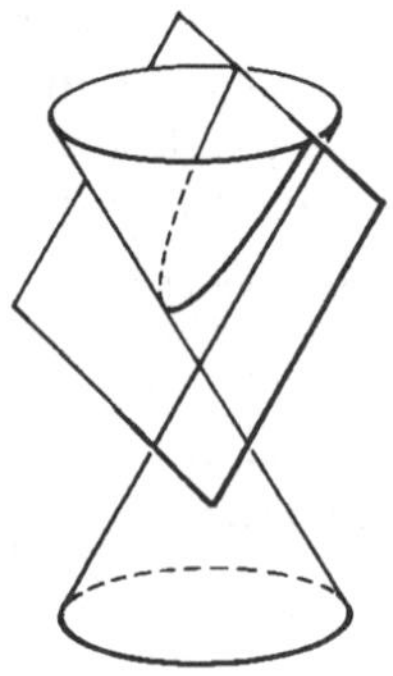

A parabola
FIGURE 4

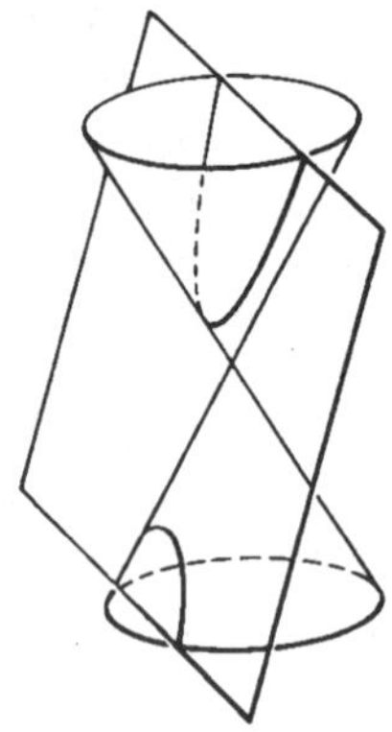

A hyperbola
FIGURE 5

Conic sections were studied as early as the fourth century B.C., when the Greek geometer Menaechmus used them to solve the problem of constructing a cube of volume twice that of a given cube. Later Greek mathematicians, notably Euclid (ca. 287-212 B.C.), Apollonius (ca. 225 B.C.), and Pappus (ca. 300 A.D.) developed the theory of conic sections with a style of proof that is similar to that used in modern plane geometry.

Now conic sections are studied through analytic geometry, an approach Descartes and Fermat initiated in the seventeenth century. During the 2000 years from Menaechmus to Descartes and Fermat, conic sections were investigated primarily for their own sake, as a topic in geometry, but in the seventeenth century, their importance to astronomy and physics began to emerge. At that time Galileo and others showed that if there were no air resistance, projectiles such as cannon balls would have parabolic trajectories; Kepler discovered that the planets move in elliptical orbits; and Newton showed that objects under the force of gravity can also have parabolic or hyperbolic paths. Newton also invented the reflecting telescope, using a mirror with parabolic cross sections. Many large modern telescopes use mirrors with hyperbolic cross sections in addition to those with parabolic cross sections, and, until recently, navigational systems were based on properties of hyperbolas.

Conic sections also arise in pure mathematics. We have seen, for instance, that the derivatives of logarithms are the functions k/x, whose graphs are hyperbolas, and in Chapter 11 we will use conic sections to study maxima and minima of functions of two variables.

Parabolas

A PARABOLA is the locus (set) of points P in a plane that are equidistant from a line and a point in the plane that is not on the line. The line is the DIRECTRIX and the point is the FOCUS of the parabola (Figure 6). The line through the focus and perpendicular to the directrix is the AXIS of the parabola. The point where the curve intersects its vertex is its VERTEX.

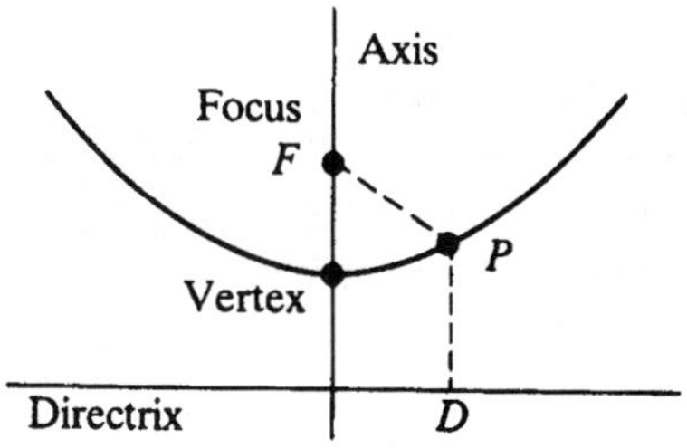

FIGURE 6
$\overline{PF} = \overline{PD}$

A parabola in a coordinate plane has the simplest equation if its focus is on a coordinate axis and its vertex is at the origin as in Figures 7 and Figure 8.

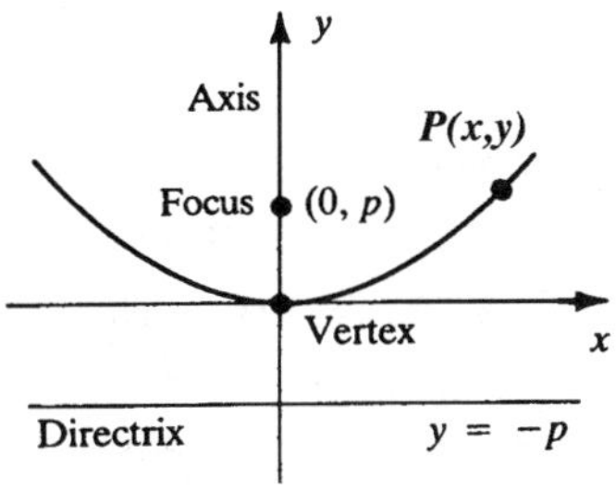

$$y = \frac{1}{4p}x^2 \ (p > 0)$$

FIGURE 7

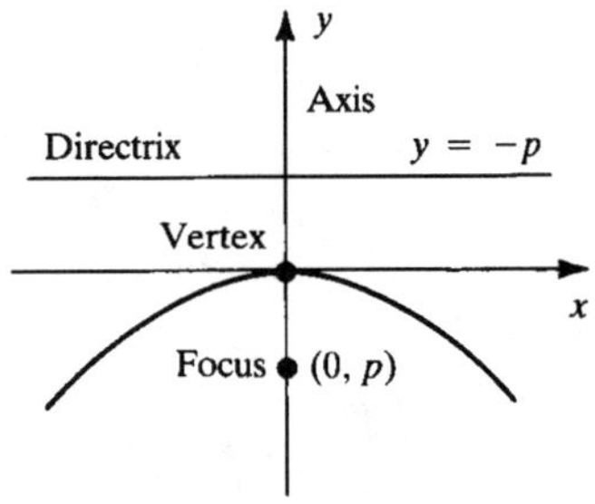

$$y = \frac{1}{4p}x^2 \ (p < 0)$$

FIGURE 8

Question 1 Figure 7 shows a variable point $P(x, y)$ on the parabola with focus $F(0, p)$ and directrix $y = -p$ with a positive constant p. **(a)** Express the (positive) distances $\overline{PF}$ and $\overline{PD}$ in terms of x, y, and p. **(b)** The point P is on the parabola if and only if $\overline{PF} = \overline{PD}$. Put the formulas from part (a) in this equation. Then square both sides and solve for y to obtain an equation of the curve.

The derivation in Question 1 of the equation $y = \dfrac{1}{4p}x^2$ with a positive p for the parabola in Figure 7 would give the same equation for the parabola in Figure 8 if p were negative. Interchanging the roles of x and y gives the equation $x = \dfrac{1}{4p}y^2$ for the parabolas in Figures 9 and 10.

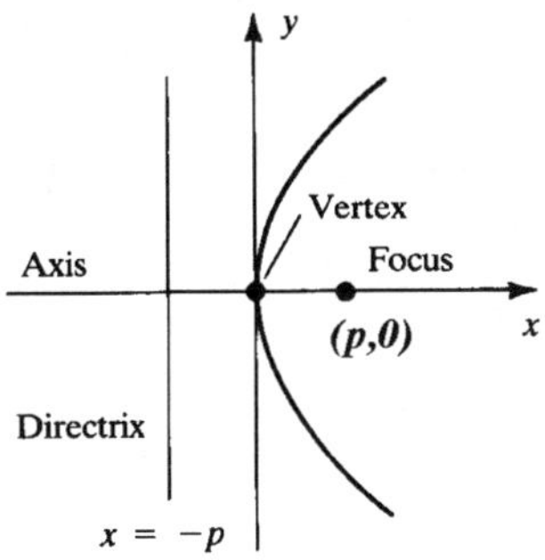

$$x = \frac{1}{4p}y^2 \ (p > 0)$$

FIGURE 9

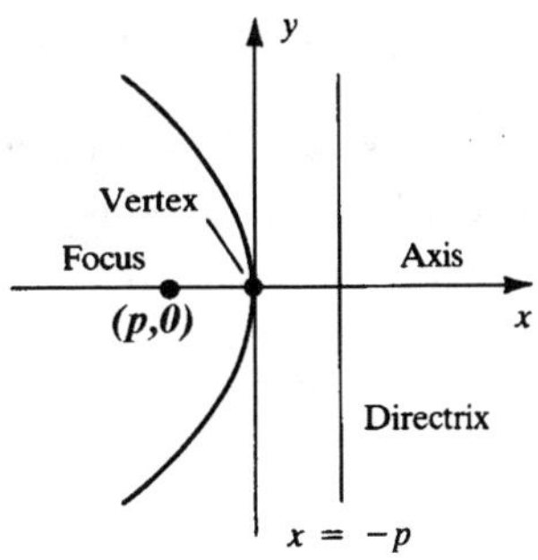

$$x = \frac{1}{4p}y^2 \ (p < 0)$$

FIGURE 10

We summarize these results as a rule.

Rule 1 *The parabola with its vertex at the origin in an xy-plane, with focus $(0, p)$, and with directrix $y = -p$ (Figures 7 and 8) has the equation,*

$$y = \frac{1}{4p}x^2. \tag{1}$$

If the vertex of the parabola is at the origin, the focus is $(p, 0)$, and the directrix is $x = -p$ (Figures 9 and 10), then the equation is

$$x = \frac{1}{4p}y^2. \tag{2}$$

The parabolas in Figures 7 through 10 are said to be in STANDARD POSITION relative to the x- and y-axes, and their equations **(1)** and **(2)** are said to be in STANDARD FORM.

Example 1 Find the focus and directrix of the parabola $x = -y^2$ and then sketch the curve with the focus and directrix.

SOLUTION The equation $x = -y^2$ is of the form **(2)** with $p = -\frac{1}{4}$. Therefore, the focus of the parabola is the point $(p, 0) = (-\frac{1}{4}, 0)$ and the directrix is the line $x = -p$, which is $x = \frac{1}{4}$. The parabola, focus, and directrix are drawn in Figure 11. □

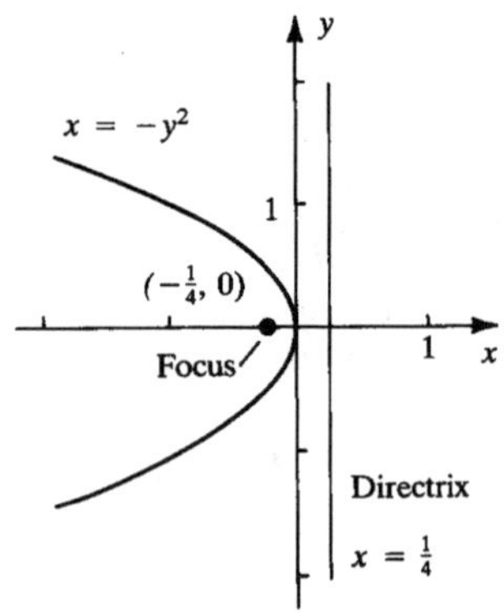

FIGURE 11

Translation of axes

If a parabola has a horizontal or vertical axis but, like the parabola in Figure 12, its vertex is not at the origin, we can put its equation in standard form by introducing new $x'y'$-axes that are parallel to the original xy-axes. This choice of new coordinates is called a TRANSLATION of axes. If the origin of the $x'y'$-axes is at the point with xy-coordinates $x = h$ and $y = k$, as in Figure 13, then the new and old coordinates are related by the equations†

$$\begin{cases} x' = x - h \\ y' = y - k. \end{cases} \tag{3}$$

†Notice that the symbols x' and y' in **(3)** indicate new coordinates and not derivatives of functions x and y.

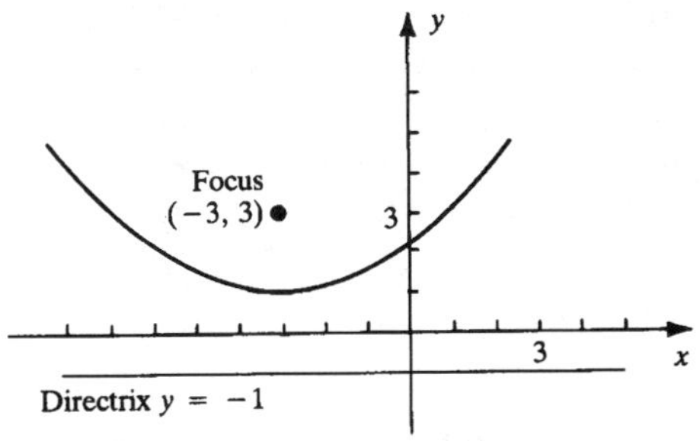

FIGURE 12

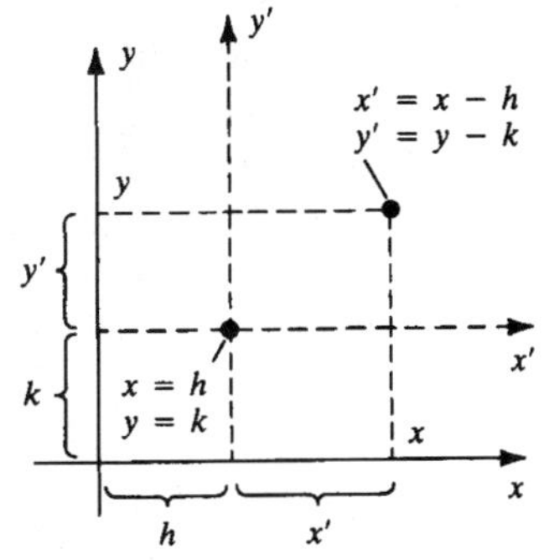

FIGURE 13

To find an equation for a parabola in Figure 12, we choose the constants h and k in **(3)** so that the point with xy-coordinates (h, k) is the vertex of the parabola.

Question 2 The focus of the parabola in Figure 12 is the point $(-3, 3)$ and its directrix is the line $y = -1$. Where is its vertex?

Example 2 Find an equation of the parabola in Figure 12.

SOLUTION To introduce new $x'y'$ coordinates, as in Figure 14 with origin at the vertex $(-3, 1)$, we set $h = -3$ and $k = 1$ in **(3)** and define

$$\begin{cases} x' = x + 3 \\ y' = y - 1. \end{cases} \tag{4}$$

Comparing Figure 14 with Figure 7 shows that the parabola has an equation of the form

$$y' = \frac{1}{4p}(x')^2 \tag{5}$$

where p is the distance from the vertex to the focus and to the directrix. In this case $p = 2$, so formulas **(4)** and **(5)** give the equation we want:

$$y - 1 = \tfrac{1}{8}(x + 3)^2. \ \square$$

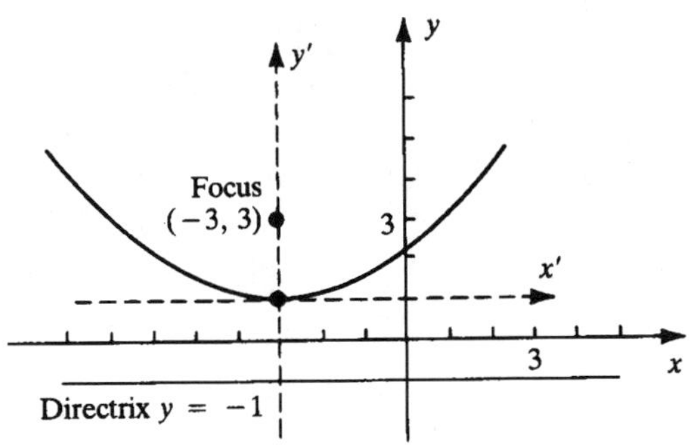

FIGURE 14

In the next Example, we find the vertex, focus, and directrix of a parabola by completing a square in its equation.

Example 3 Draw the parabola $y^2 + 2y + 8x - 15 = 0$. Show its focus and directrix.

SOLUTION We isolate the terms involving y on the left by adding $-8x + 15$ to both sides of the equation. This gives $y^2 + 2y = -8x + 15$. Then we complete the sum $y^2 + 2y$ into a square by adding the square $1^2 = 1$ of half of the coefficient of y to both sides of the equation and obtain $y^2 + 2y + 1 = -8x + 16$. Then, writing the left side as a square and factoring -8 from the two terms on the right yields $(y+1)^2 = -8(x-2)$. We divide by -8 to have

$$x - 2 = -\tfrac{1}{8}(y+1)^2. \qquad \textbf{(6)}$$

This is the type of parabola in Figure 10 with $p = -2$ and with x and y replaced by x' and y', where

$$\begin{cases} x' = x - 2 \\ y' = y + 1. \end{cases} \qquad \textbf{(7)}$$

Consequently, the vertex is at $x = 2, y = -1$; the focus is two units to the left of the vertex at $x = -, y = -1$; and the directrix is the vertical line $x = 4$, two units to the right of the vertex. To obtain the sketch of the parabola in Figure 15, we plot its vertex and its y-intercepts, $y = 3$ and $y = -5$, which we find by setting $x = 0$ in **(6)** and solving for y. □

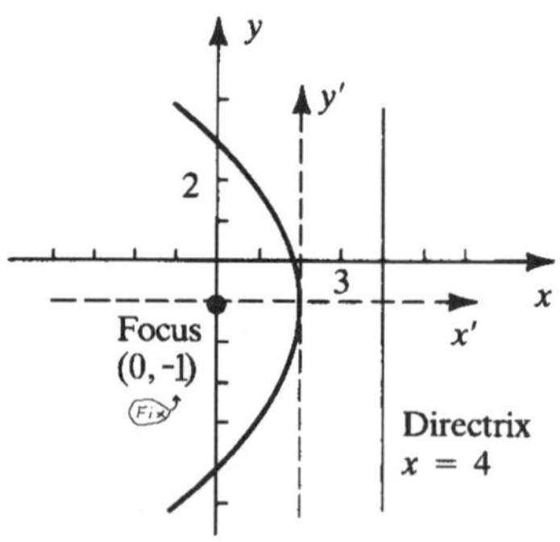

FIGURE 15

Ellipses

The ELLIPSE in Figure 16 is the locus of points P such that the sum $\overline{PF_1} + \overline{PF_2}$ of the distances between P and two fixed points F_1 and F_2 is equal to a constant that is greater than the distance between the fixed points. To see why the ellipse has its particular shape, imagine that the solid lines in the drawing represent a string whose ends are fastened at F_1 and F_2 and which is held taut by a pencil at the point P. If we move the pencil around F_1 and F_2, keeping the string taut, we would draw the ellipse.

The points F_1 and F_2 are called the FOCI of the ellipse and the point midway between them is its CENTER.

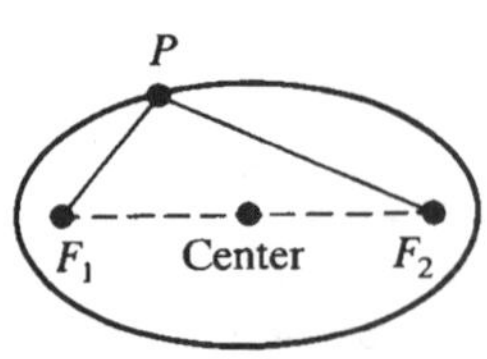

$\overline{PF_1} + \overline{PF_2} = \text{Constant}$

FIGURE 16

To find an equation for the ellipse in Figure 17, whose foci are $F_1(-c, 0)$ and $F_2(c, 0)$ on the x-axis, we pick the positive constant a such that the sum of the distances from the point $P(x, y)$ on the curve equals $2a$. Then the ellipse is determined by the equation

$$\overline{PF_1} + \overline{PF_2} = 2a. \tag{8}$$

Expressed in xy-coordinates, equation **(8)** reads

$$\sqrt{(x-c)^2 + (y-0)^2} + \sqrt{(x+c)^2 + (y-0)^2} = 2a.$$

Moving the second square root to the right side of the last equation and squaring both sides gives

$$(x-c)^2 + y^2 = (x+c)^2 + y^2 - 4a\sqrt{(x+c)^2 + y^2} + 4a^2.$$

This equation simplifies to $cx + a^2 = a\sqrt{(x+c)^2 + y^2}$. Squaring both sides again and rewriting the results gives

$$\frac{x^2}{a^2} + \frac{y^2}{a^2 - c^2} = 1.$$

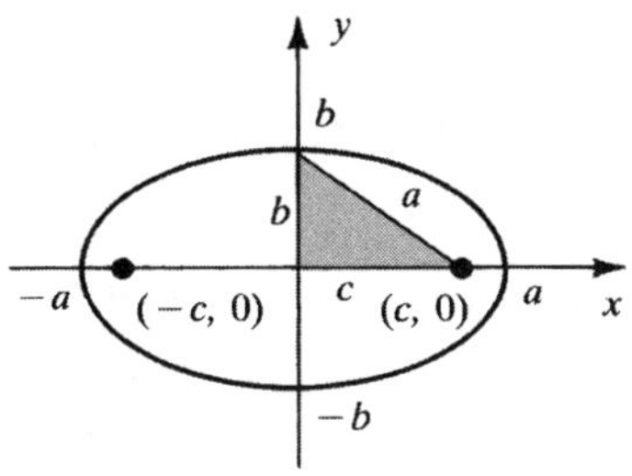

$$\frac{x^2}{a^2} + \frac{y^2}{b^2} = 1$$

$(a > b)$

FIGURE 17

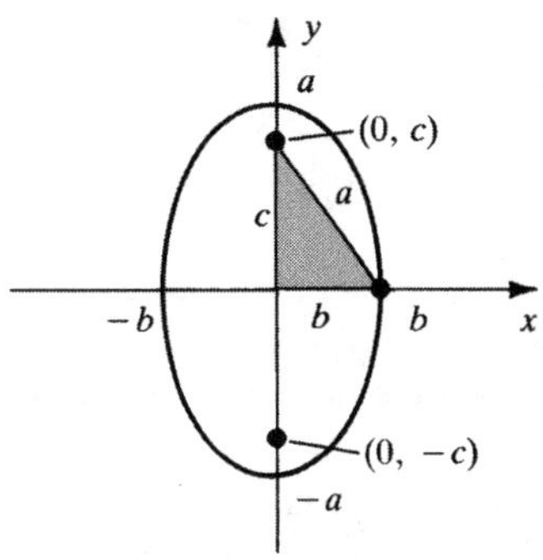

$$\frac{x^2}{b^2} + \frac{y^2}{a^2} = 1$$

$(a > b)$

FIGURE 18

To put the last equation in a better form, we set $b = \sqrt{a^2 - c^2}$ to give the ellipse the equation $\dfrac{x^2}{a^2} + \dfrac{y^2}{b^2} = 1$. This establishes part (a) in the next rule. Part (b) follows by interchanging x and y.

Rule 2 **(a)** *Suppose the foci of an ellipse are $(c, 0)$ and $(-c, 0)$ on the x-axis, as in Figure 17, and that the sum of the distances from an arbitrary point on the ellipse to the foci is $2a$ $(0 \le c < a)$. Then the ellipse has the equation*

$$\frac{x^2}{a^2} + \frac{y^2}{b^2} = 1 \tag{9}$$

where

$$b = \sqrt{a^2 - c^2}. \tag{10}$$

(b) *If the foci of the ellipse are $(0, c)$ and $(0, -c)$ on the y-axis, as in Figure 18, then the equation is*

$$\frac{x^2}{b^2} + \frac{y^2}{a^2} = 1. \tag{11}$$

In the next Question, we will see how the numbers a and b are related to the ellipses in Figure 17 and 18.

Question 3 What are the x- and y-intercepts **(a)** of the ellipse $\dfrac{x^2}{a^2} + \dfrac{y^2}{b^2} = 1$ and **(b)** of the ellipse $\dfrac{x^2}{b^2} + \dfrac{y^2}{a^2} = 1$?

The intercepts are shown in Figures 17 and 18. The line segment across the ellipse through its foci is called its MAJOR AXIS. Its length in both cases is $2a$. The perpendicular line segment across the ellipse through its center is its MINOR AXIS, and its length is $2b$. If the ellipse is a circle, then $a = b$. Otherwise b is less than a, as can be seen from its definition **(10)**. If the ellipse is not a circle, then the ends of its major axis are its VERTICES.

Question 4 Use the definition $\overline{PF_1} + \overline{PF_2} = 2a$ of the ellipse to show that the point at an end of the minor axis in Figure 17 or Figure 18 is the distance a from either focus, as shown in Figures 17 and 18.

The result of Question 4 shows that c and b are the lengths of the legs and a is the length of the hypotenuse of the shaded right triangles in Figures 17 and 18. This provides a way to remember equation **(10)** that relates these numbers.

Question 5 Sketch the ellipse $\dfrac{x^2}{4} + \dfrac{y^2}{9} = 1$, showing its foci.

Translating coordinates

If the axes of an ellipse are parallel to the coordinate axes but its center is not at the origin, we complete the squares to determine auxiliary $x'y'$-axes such that the new origin is at the center of the ellipse.

Example 4 Sketch the ellipse $x^2 + 4y^2 - 4x + 8y + 4 = 0$. Show its foci and vertices.

SOLUTION We first group together the terms in x and the terms in y and subtract 4 from both side of the equation to obtain

$$(x^2 - 4x) + (4y^2 + 8y) = -4.$$

Factoring the coefficient of y^2 from the two terms involving y gives

$$(x^2 - 4x) + 4(y^2 + 2y) = -4.$$

We complete the square inside the first pair of parentheses by adding and subtracting the square $(-2)^2 = 4$ of half the coefficient of x, and complete the square inside the second pair of parentheses by adding and subtracting the square $1^2 = 1$ of half the coefficient of y. We obtain

$$(x^2 - 4x + 4 - 4) + 4(y^2 + 2y + 1 - 1) = -4$$

and then

$$(x - 2)^2 - 4 + 4(y + 1)^2 - 4 = -4.$$

Next, we add 8 to both sides. This gives $(x - 2)^2 + 4(y + 1)^2 = 4$ and then

$$\frac{(x-2)^2}{2^2} + \frac{(y+1)^2}{1^2} = 1.$$

We can put this equation in the form

$$\frac{(x')^2}{2^2} + \frac{(y')^2}{1^2} = 1 \tag{12}$$

by introducing the auxiliary coordinates

$$\begin{cases} x' = x - 2 \\ y' = y + 1. \end{cases} \tag{13}$$

The center of the ellipse is at the origin of the new coordinate system. This is where $x' = 0$ and $y' = 0$, which, by **(13)**, has the original coordinates $x = 2$ and $y = -1$. Then equation **(12)** shows that the x'- and y'-intercepts are $x' = \pm 2$ and $y' = \pm 1$. The ellipse has a horizontal major axis of length $2(2) = 4$ and a vertical minor axis of length $2(1) = 2$.

Equation **(12)** shows that the ellipse is of the type in Figure 17, relative to the $x'y'$-coordinates, with $a = 2$ and $b = 1$. Formula **(7)** then gives $c = \sqrt{2^2 - 1^2} = \sqrt{3}$, so the foci are $\sqrt{3}$ units to the left and right of the center. They are at $x = 2 - \sqrt{3} \doteq 0.268, y = -1$ and at $x = 2 + \sqrt{3} \doteq 3.732, y = -1$. The ellipse is drawn in Figure 19. □

FIGURE 19

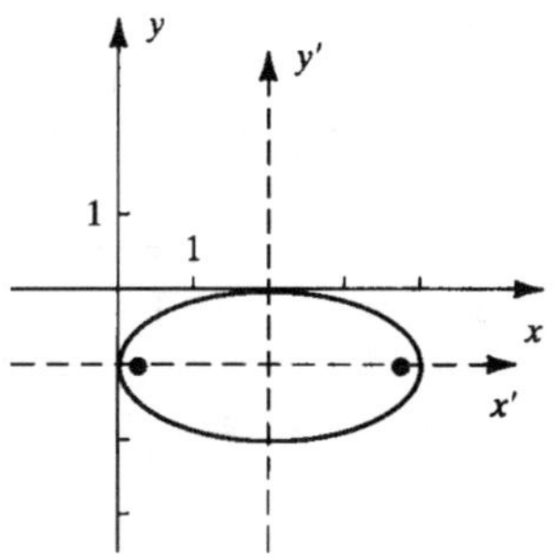

Question 6 An ellipse is the locus of points P such that the sum of the distances from P to $(-1, -2)$ and to $(-1, 4)$ is 8. **(a)** Is its major axis horizontal or vertical? **(b)** Where is its center? **(c)** What are the constants a and c in an equation for the ellipse? **(d)** What is b?

Example 5 Give an equation of the ellipse in Question 6, and then sketch it.

SOLUTION Because the major axis of the ellipse is vertical, its equation is

$$\frac{(x')^2}{b^2} + \frac{(y')^2}{a^2} = 1$$

relative to translated $x'y'$-coordinates with origin at the center of the ellipse. In Question 6 you determined that the center is $(-1, 1)$, that $a = 4$, and that $b = \sqrt{7}$. Therefore, $x' = x + 1, y' = y - 1$, and the ellipse (Figure 20) is

$$\frac{(x+1)^2}{7} + \frac{(y-1)^2}{16} = 1. \square$$

$$\frac{(x+1)^2}{7} + \frac{(y-1)^2}{16} = 1$$

FIGURE 20

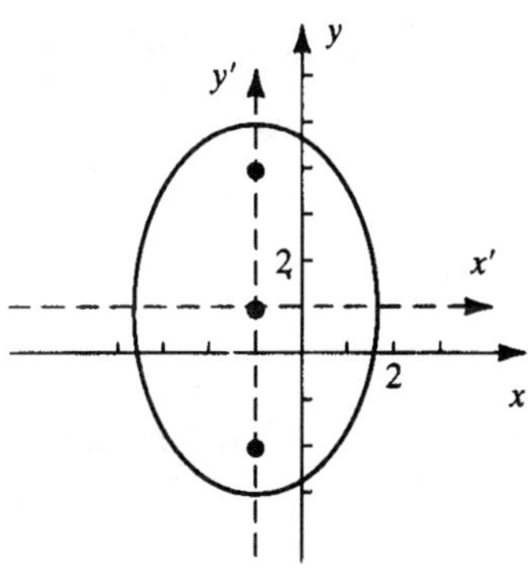

Hyperbolas

A HYPERBOLA is the locus of points P in a plane such that the absolute value of the difference $|\overline{PF_1} - \overline{PF_2}|$ between the distances from P to two points F_1 and F_2 is a constant (Figure 21).

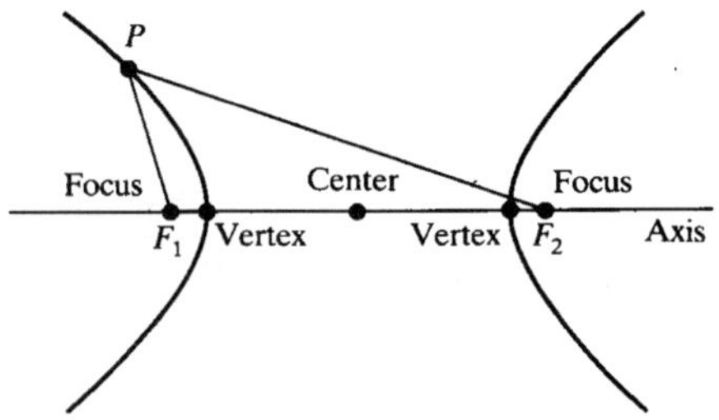

$|\overline{PF_1} - \overline{PF_2}| = \text{Constant}$

FIGURE 21

The points F_1 and F_2 are the FOCI of the hyperbola, the line through the foci is the AXIS, the point halfway between the foci is the CENTER, and the points where the axis intersects the hyperbola are the VERTICES.

Hyperbolas, like ellipses, have the simplest equations if their centers are at the origin and their foci are either on the x- or y-axis. Calculations similar to those used in deriving the equations for ellipses in Rule 2 give the following results.

Rule 3 **(a)** *If the foci of a hyperbola are the points $(c, 0)$ and $(-c, 0)$ on the x-axis (Figure 22) and the absolute value of the difference between the distances from points on the hyperbola to the foci is $2a$ $(0 < a < c)$, then the hyperbola has the equation*

$$\frac{x^2}{a^2} - \frac{y^2}{b^2} = 1 \tag{14}$$

with

$$b = \sqrt{c^2 - a^2}. \tag{15}$$

(b) *If the foci of the hyperbola are $(0, c)$ and $(0, -c)$ on the y-axis (Figure 23), the equation is*

$$\frac{y^2}{a^2} - \frac{x^2}{b^2} = 1. \tag{16}$$

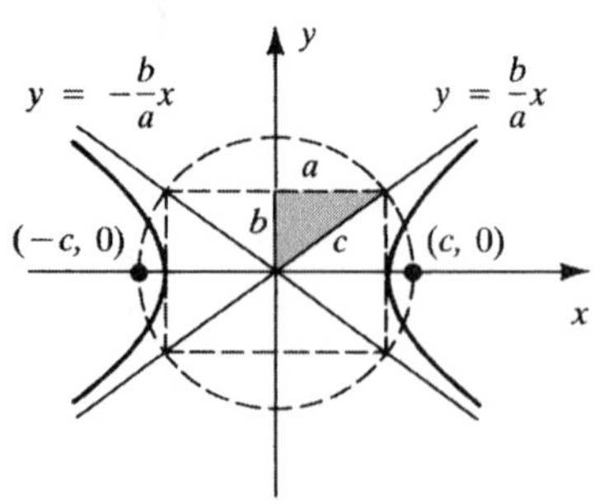

$\frac{x^2}{a^2} - \frac{y^2}{b^2} = 1$

FIGURE 22

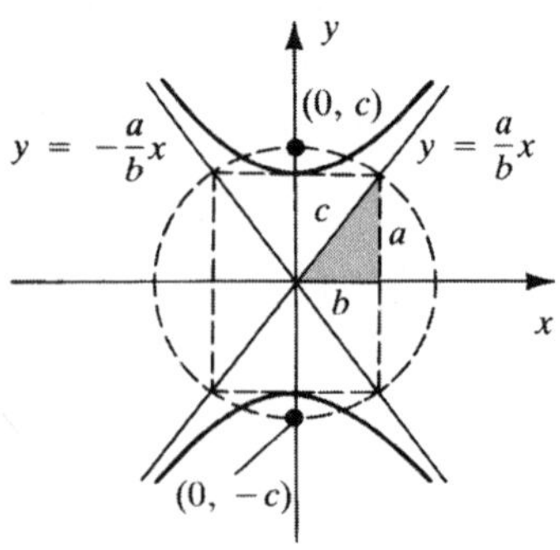

$\frac{y^2}{a^2} - \frac{x^2}{a^2} = 1$

FIGURE 23

In the equations for ellipses that were given in Rule 2, the constants a and b were labeled so that $a \geq b$. In the equations for hyperbolas, the letter b goes with the term that has the minus sign; either a or b may be the larger number or they might be equal.

Hyperbolas, unlike parabolas and ellipses, have ASYMPTOTES. Imagine that a point (x, y) is moving out the upper right branch of the hyperbola in Figure 22. Then x and y satisfy equation **(14)** as they both tend to ∞. We factor **(14)** to obtain

$$\left(\frac{x}{a} + \frac{y}{b}\right)\left(\frac{x}{a} - \frac{y}{b}\right) = 1$$

and then

$$\frac{x}{a} - \frac{y}{b} = \frac{1}{\dfrac{x}{a} + \dfrac{y}{b}}.$$

This equation shows that $\dfrac{x}{a} - \dfrac{y}{b} \to 0$, and consequently, $y - \dfrac{bx}{a} \to 0$ as the point (x, y) moves out the curve on the upper right and x and y tend to ∞. Consequently, the line $y = \dfrac{bx}{a}$ through the origin with slope b/a is an asymptote of the hyperbola. A similar argument would show that $y = -\dfrac{bx}{a}$ is also an asymptote as is shown in Figure 22. The asymptotes in the case of Figure 23 are $y = \pm\dfrac{ax}{b}$.

Example 6 Sketch the hyperbola $9x^2 - 16y^2 = 144$. Show its foci, vertices, and asymptotes.

SOLUTION We divide both sides of the equation by 144 to obtain

$$\frac{x^2}{4^2} - \frac{y^2}{3^2} = 1.$$

To determine whether this is an equation for a hyperbola of the type in Figure 22 or in Figure 23, we first set $y = 0$ in the equation and solve for the x- intercepts $x = \pm 4$. Then we set $x = 0$ and find that the curve has no y-intercepts. The hyperbola is of the type in Figure 22 with $a = 4$ and $b = 3$. Its center is at the origin and its vertices are $(4, 0)$ and $(-4, 0)$.

To draw the hyperbola, as in Figure 24, we first draw the rectangle of width $2a = 8$ and height $2b = 6$ with its center at the origin, as shown with dashed lines in Figure 24. The vertices of the hyperbola are at the intersections of this rectangle with the x-axis.

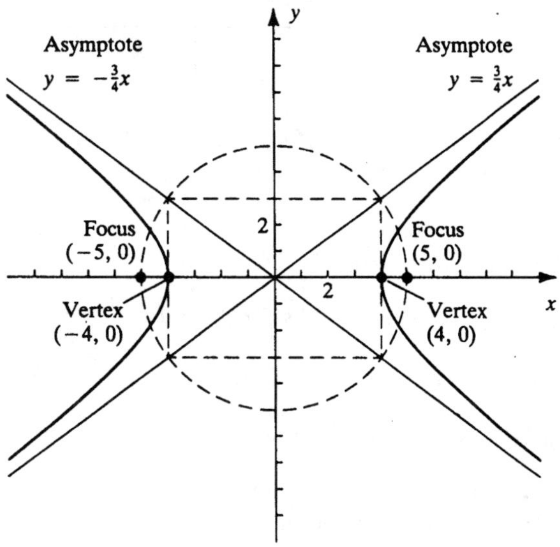

$$\frac{x^2}{4^2} - \frac{y^2}{3^2} = 1$$

FIGURE 24

Next, we draw the asymptotes through the corners of the rectangle. They have the equations $y = \pm\frac{3}{4}x$. Then we draw the hyperbola through its x- intercepts and so it approaches the asymptotes as x and y tend to ∞ or to $-\infty$.

To plot the foci, we draw the circle through the corners of the rectangle. The foci are at the intersection of this circle with the x-axis. They are at $(\pm c, 0)$ with $c = \sqrt{a^2 + b^2} = \sqrt{4^2 + 3^2} = 5$. □

Translation of coordinates

If the axis of a hyperbola is horizontal or vertical but its center is not at the origin, we find its center, foci, vertices, and asymptotes by translating coordinates. We complete the squares, just as for ellipses, to find $x'y'$-coordinates with origin at the center of the hyperbola.

Question 7 Sketch the hyperbola $y^2 - x^2 - 10y + 6x = 0$. Show its center, foci, and asymptotes.

A hyperbola for which a and b are equal, as in Question 7, is called a RECTANGULAR HYPERBOLA because its asymptotes are perpendicular to each other.

Example 7 Give an equation of the hyperbola that is the locus of points that are either four units closer to $(1, 0)$ than to $(-5, 0)$ or four units closer to $(-5, 0)$ than to $(1, 0)$.

SOLUTION The locus is a hyperbola with foci $(1, 0)$ and $(-5, 0)$, so its equation is of the form

$$\frac{(x')^2}{a^2} - \frac{(y')^2}{b^2} = 1$$

with translated coordinates x' and y'. The origin of the new coordinates is at the center $(-2, 0)$ of the hyperbola midway between the foci. Therefore, $x' = x + 2$ and $y' = y$. The constant a is 2 because the absolute value of the difference between the distances from points on the hyperbola to the foci is 4. The constant c is the distance 3 from the center to the foci. Since $b = \sqrt{c^2 - a^2} = \sqrt{3^2 - 2^2} = \sqrt{5}$, the hyperbola has the equation

$$\tfrac{1}{4}(x + 2)^2 - \tfrac{1}{5}y^2 = 1$$

It is drawn in Figure 25. □

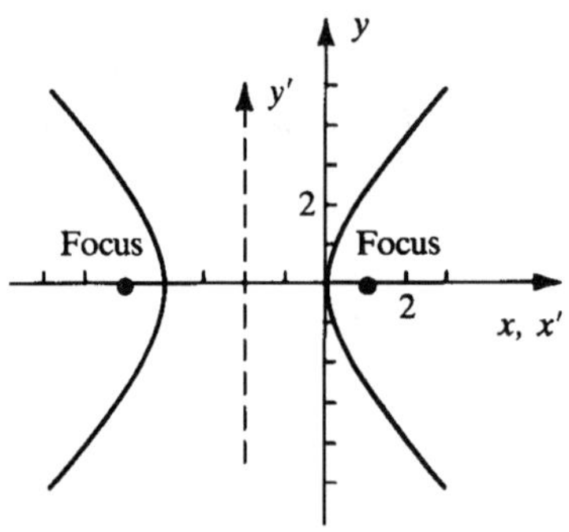

FIGURE 25

Generating conic sections

The parabolas with vertical axes and equations $y - k = \dfrac{1}{4p}(x - h)^2$ can be generated on a calculator or computer with function-graphing algorithms by rewriting the equation in the form $y = k + \dfrac{1}{4p}(x - h)^2$ so it is the graph of a function $y(x)$. These algorithms can also be used for parabolas $x - h = \dfrac{1}{4p}(y - k)^2$ with horizontal axes and for ellipses and hyperbolas with horizontal and vertical axes by solving for y to obtain, in each case, formulas for two functions $y = y(x)$ whose graphs form the conic section. It is easier, however, in the latter cases, to use algorithms for parametric curves, $x = x(t), y = y(t)$.

To generate the parabola $x-2 = \frac{1}{8}(y+1)^2$ of Example 3, for instance, you can set $x = 2+\frac{1}{8}(t+1)^2$ and $y = t$ and have the t-range be the same as the y-range in the window you choose for the plot.

Ellipses and hyperbolas can also be generated using trigonometric functions. It follows from the definition of $\sin t$ and $\cos t$ that, for positive r, the circle $\dfrac{x^2}{r^2} + \dfrac{y^2}{r^2} = 1$ of radius r with its center at the origin is given by $x = r\cos t, y = r\sin t$ for $0 \le t \le 2\pi$ (Figure 26). These formulas are called PARAMETRIC EQUATIONS of the ellipse.

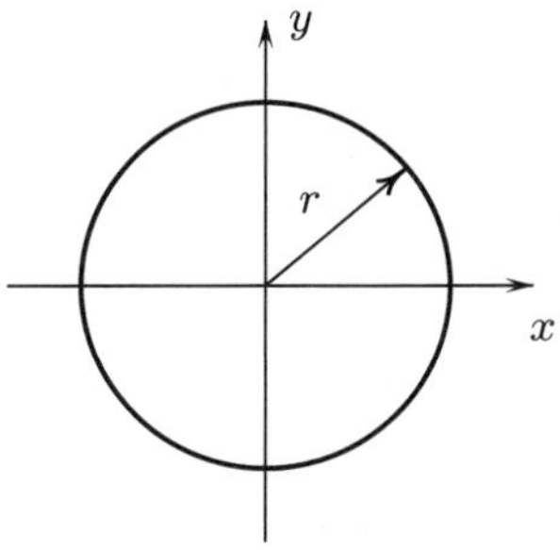

$\begin{cases} x = r\cos t \\ y = r\sin t \end{cases}$

FIGURE 26

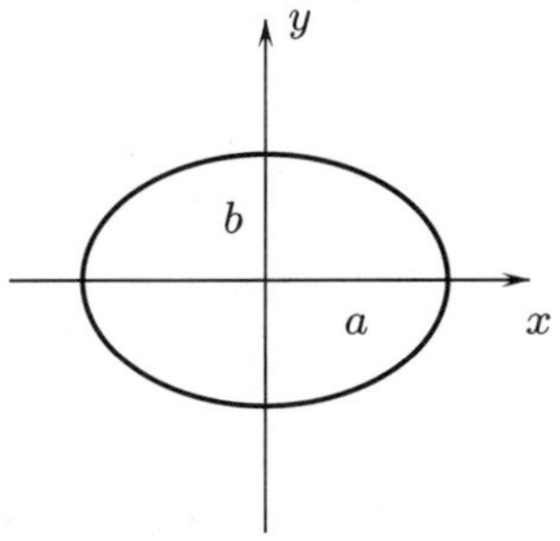

$\begin{cases} x = a\cos t \\ y = b\sin t \end{cases}$

FIGURE 27

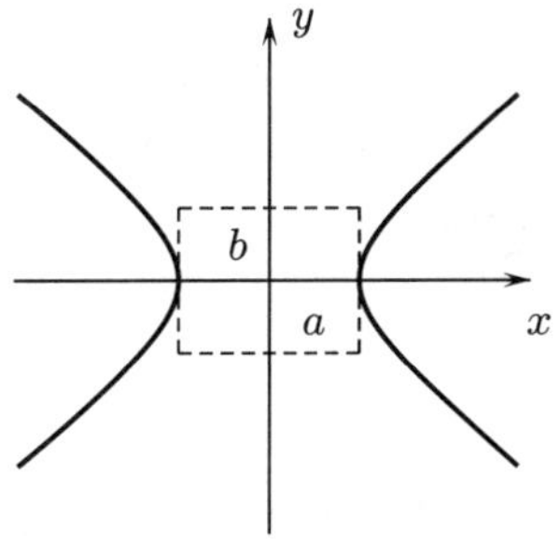

$\begin{cases} x = a\sec t \\ y = b\tan t \end{cases}$

FIGURE 28

The ellipse $\dfrac{x^2}{a^2} + \dfrac{y^2}{b^2} = 1$ with positive a and b in Figure 27 is given by the similar equations

$$x = a\cos t,\ y = b\sin t \quad \text{for } 0 \le t \le 2\pi. \tag{17}$$

These formulas give the ellipse because the point $(a\cos t, b\sin t)$ goes around the origin once as t goes from 0 to 2π, and for each t

$$\frac{(a\cos t)^2}{a^2} + \frac{(b\sin t)^2}{b^2} = \cos^2 t + \sin^2 t = 1$$

so the point is on the ellipse.

The hyperbola $\dfrac{x^2}{a^2} - \dfrac{y^2}{b^2} = 1$ in Figure 28 is given by the parametric equations

$$x = a\sec t,\ y = b\tan t \quad \text{for } -\tfrac{1}{2}\pi < t < \tfrac{1}{2}\pi, \tfrac{1}{2}\pi < t < \tfrac{3}{2}\pi. \tag{18}$$

The hyperbola $\dfrac{y^2}{a^2} - \dfrac{x^2}{b^2} = 1$ is given by

$$x = a\tan t,\ y = b\sec t \quad \text{for } -\tfrac{1}{2}\pi < t < \tfrac{1}{2}\pi, \tfrac{1}{2}\pi < t < \tfrac{3}{2}\pi. \tag{19}$$

Formulas **(18)** and **(19)** give points on the respective hyperbolas because $\sec^2 t - \tan^2 t = \dfrac{1-\sin^2 t}{\cos^2 t} = 1$.

You may have to enter $\sec t$ in your calculator or computer as $1/\cos t$. Also, you can distinguish between equations **(18)** and **(19)** by noting that $\tan t$ is zero at $t = 0$ but $\sec t$ is never 0, so the hyperbola **(18)** does not cross the y-axis and the hyperbola **(19)** does not cross the y-axis.

If an ellipse or hyperbola has horizontal and vertical axes but its center is a point (h, k) other than the origin, you can use **(17)**, **(18)**, or **(19)** with h added to the formula for x and k added to the formula for y.

Parabolic trajectories

If an object is moving on a vertical line, we measure its position by introducing a y-axis along the line. Then, if there is no air resistance or other force on the object other than the downward force of gravity and the positive side of the y-axis points up, then the object's upward acceleration $\dfrac{d^2y}{dt^2}$ equals $-g$, where g is the constant downward acceleration due to gravity (32 feet/second^2, 9.6 meters/second^2, or 960 centimeters/second^2). In this case, the object's upward velocity is

$$\frac{dy}{dt} = \int \frac{d^2y}{dt^2}\,dt = \int (-g)\,dt = -gt + v_1 \tag{20}$$

with a constant v_1 that can be determined from the value of dy/dt at one particular time.

Integrating again, we find that the object's y-coordinate at time t is

$$y = \int (-gk + v_1)\,dt = -\tfrac{1}{2}gt^2 + v_1t + s_1 \tag{21}$$

with another constant s_1 that could be found from the object's position at a given time.

If the only force is gravity but the object does not move on a vertical line, then it moves in a vertical plane, which we choose to be an xy-plane with its y-axis pointing up. Then the object's upward acceleration is $-g$, so its upward velocity and y-coordinate are given by **(20)** and **(21)**. With gravity being the only force, there are no horizontal forces on the object and its horizontal acceleration d^2x/dt^2 is zero. Its horizontal velocity is constant,

$$\frac{dx}{dt} = \int \frac{d^2x}{dt^2}\,dt = \int 0\,dt = v_2 \tag{22}$$

and its x-coordinate is linear:

$$x = \int \frac{dx}{dt}\,dt = \int v_2\,dt = v_2t + s_2. \tag{23}$$

The constants v_2 and s_2 can be determined from the object's horizontal velocity and x-coordinates at any particular time.

We can solve equation **(23)** for t in terms of x and substitute the result in **(18)** to express y in terms of x. This yields an equation of the form

$$y = ax^2 + bx + c \tag{24}$$

with constants a, b and c for the object's path. Since the curve **(24)** is a parabola, we have shown that projectiles subject only to the force of gravity have parabolic paths. We will study paths of projectiles further in Chapter 10.

An optical property of parabolas

The simplest reflecting telescopes have one mirror with parabolic cross sections. The parallel rays of light from the stars are reflected by the mirror to an eyepiece or camera at the focus of the parabolas (Figure 29). This occurs because the tangent line at point P on a parabola makes equal angles with the line parallel to the axis and with the FOCAL RADIUS, which is the line from P to the focus shown in Figure 30. In the next Question we will derive this property in the case of the parabola $y = x^2$.

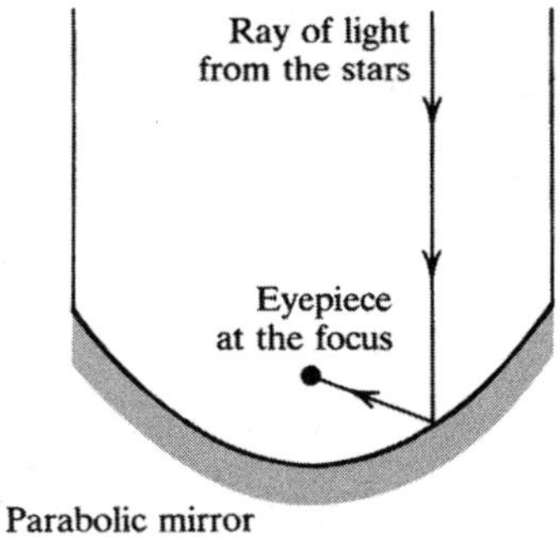

FIGURE 29

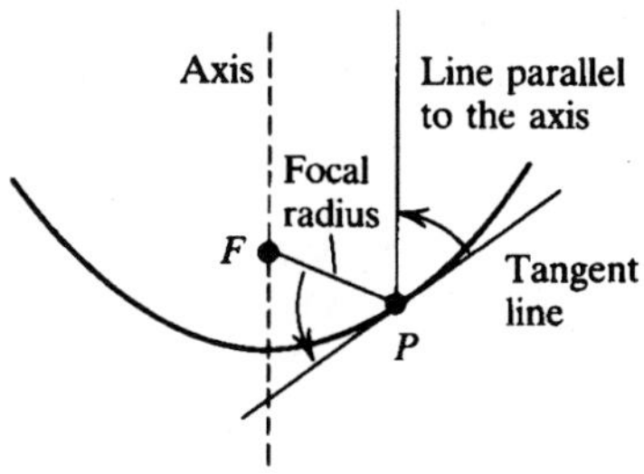

FIGURE 30

Question 8 Figure 31 shows the parabola $y = x^2$, its focus $F(0, \frac{1}{4})$, and the tangent line at the point $P(a, a^2)$ on it for an nonzero constant a. **(a)** Find the y-coordinate of the point Q, where the tangent line intersects the y-axis. **(b)** Show that the triangle PQF is isosceles. **(c)** Show that the tangent line makes equal angles with the line $x = a$ and the focal radius FP.

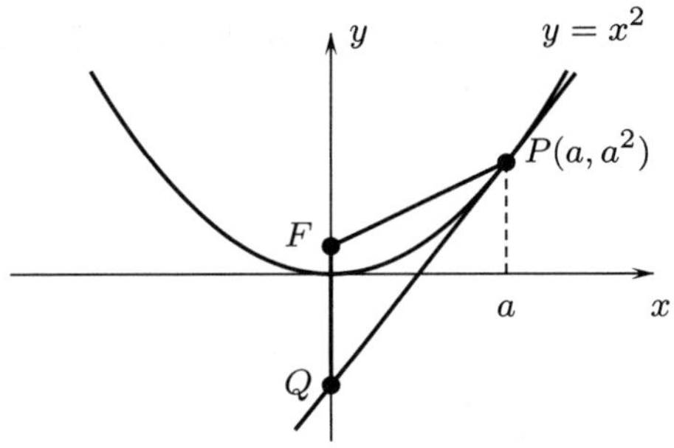

FIGURE 31

A geometric derivation of the reflecting property of parabolas is described in Problem 37.

Optical properties of ellipses and hyperbolas

If a ray of light passes through one focus of an ellipse and reflects off the ellipse, then it passes through the other focus, as shown in Figure 32. This is because the tangent line at a point P on an ellipse makes equal angles with the focal radii at that point (see Problem 37).

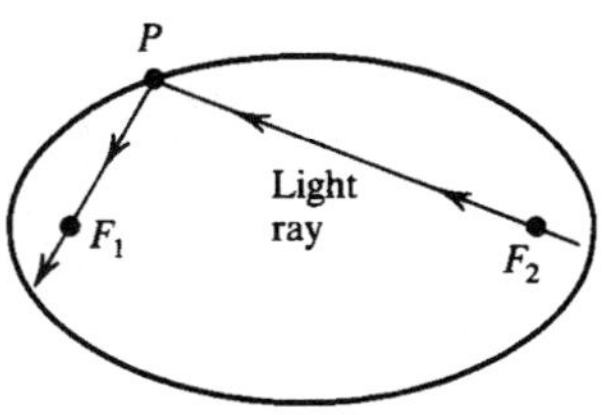

FIGURE 32

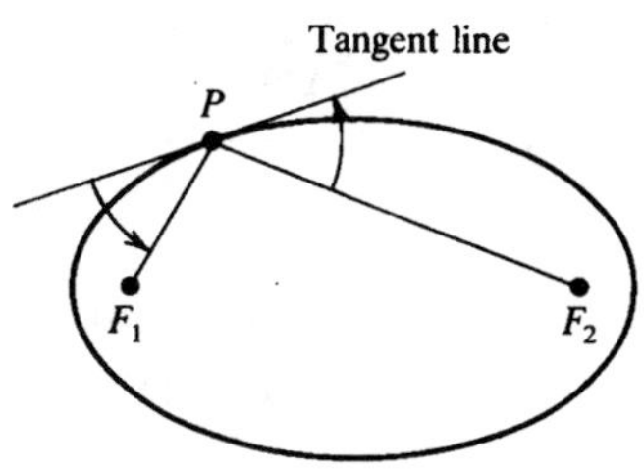

FIGURE 33

The corresponding acoustic property of ellipses is used in "whispering rooms," which have ceilings with cross sections that are arcs of ellipses. A person standing at one focus can hear someone whispering at the other focus because all of the sound from one focus that hits the ceiling is reflected to the other focus. Reflectors with elliptical cross sections are also used to focus ultrasound signals in medical treatments.

Hyperbolas have a similar optical property. A ray of light that approaches a hyperbola from the side opposite a focus and in a line through that focus reflects off the hyperbola in a line through the other focus (Figure 34). This is because the tangent line at a point P on the hyperbola bisects the angle made by the focal radii at P as shown in Figure 35. (See Problem 38.)

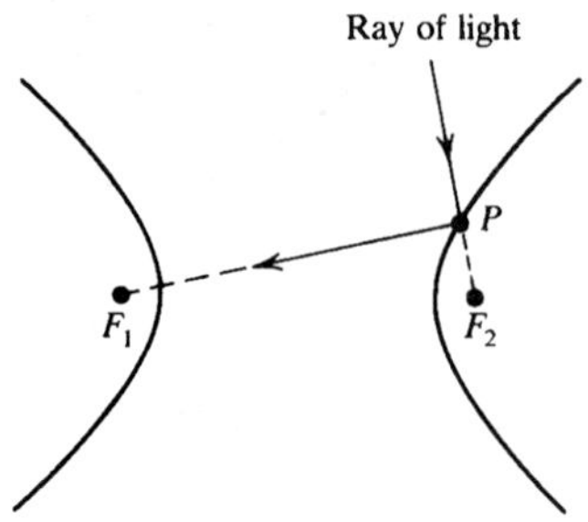

FIGURE 34

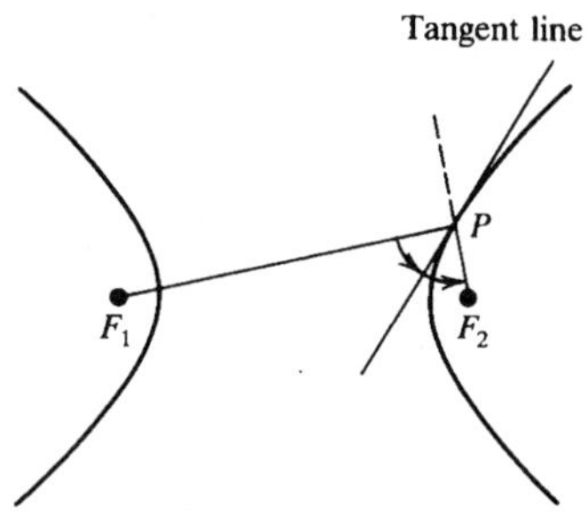

FIGURE 35

This property of hyperbolas is employed in the Cassegrain telescope, which uses a mirror with hyperbolic cross sections and a mirror with parabolic cross sections (Figure 36). One focus of the hyperbolic mirror is at the focus of the parabolic mirror and the other is at the vertex of the parabolic mirror, where there is an eyepiece or camera. The parallel rays of light from the stars reflect off the parabolic mirror towards its focus and then off the hyperbolic mirror toward the eyepiece or camera.

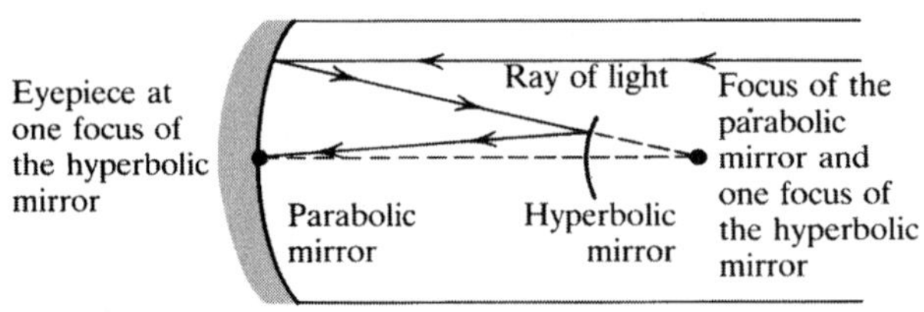

A Cassegrain telescope
FIGURE 36

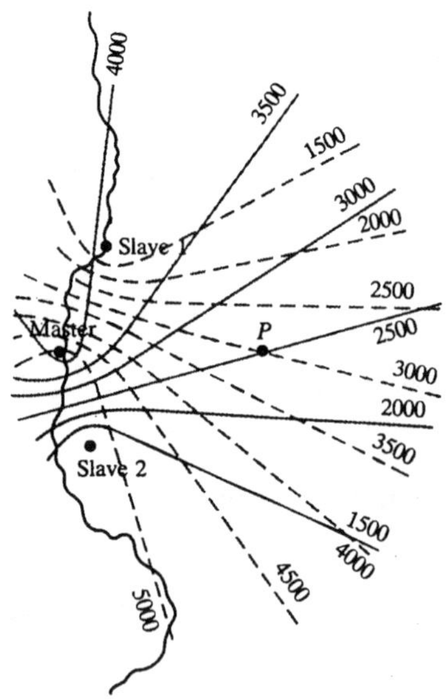

FIGURE 37

Hyperbolas in navigation

The long range navigation system LORAN was used before the current systems of satelite navigation were developed. LORAN enabled ships to determine their location by radio. Typically, a“master” station transmited pulses at 0.05 second intervals and “slave” stations transmitted pulses 0.001 seconds later. A ship had equipment that measured the delay between reception of signals from the two stations. By using one master and two slaves, as in Figure 37, the ship found its location as an intersection of hyperbolas. The numbers on the hyperbolas drawn with solid lines show how much longer (microseconds) it takes signals from slave 1 to reach points on it than signals from the master. The numbers on the hyperbolas drawn with broken lines indicate how much longer it takes signals from slave 2 to reach points on the curve than signals from the master. A ship at point P in Figure 34, for example, would receive signals from the master station 3000 microseconds (0.003 seconds) before the signals from slave 1 and 2500 microseconds before the signals from slave 2. Its position would be the intersection of the hyperbolas along which these time delays occur.

In the currently prevelant Global Positioning System an object's location is found as the intersections of four spheres by using laser beams to measure the distances from the objects to four orbiting satelites.[(1)]

Principles and procedures

- If you have an equation, $y = \frac{1}{4p}x^2$ or $x = \frac{1}{4p}y^2$, of a parabola in standard position, you can tell its orientation by noting that x^2 or y^2 is never negative. The orientation of the curve then determines how the focus and directrix are related to the vertex at the origin. Position them $|p|$ units from the vertex.
- The distinction between the two equations, $\frac{x^2}{a^2} + \frac{y^2}{b^2} = 1$ and $\frac{x^2}{b^2} + \frac{y^2}{a^2} = 1$, of ellipses in standard form is that in each case $a \geq b$ so that the length of the major axis is $2a$, the length of the minor axis is $2b$ and the foci are on the major axis the distance $c = \sqrt{a^2 - b^2}$ from the vertex.
- You can recall the equation $a^2 = b^2 + c^2$ relating the parameters a, b and c for an ellipse by visualizing the right triangles in Figures 17 and 18 with vertices at the center of the ellipse, at one end of the minor axis, and at one focus.
- In the standard equations, $\frac{x^2}{a^2} - \frac{y^2}{b^2} = 1$ and $\frac{y^2}{a^2} - \frac{x^2}{b^2} = 1$, of a hyperbola there are no restrictions on the relative sizes of the positive parameters a and b; a or b may be larger or they might be equal. What distinguishes the two equations is that the denominator a^2 always goes with the positive term and b^2 goes with the negative term.
- To determine the orientation of a hyperbola given by a standard equation, look for its x- and y-intercepts. The hyperbola $\frac{x^2}{a^2} - \frac{y^2}{b^2} = 1$ has x-intercepts $x = \pm a$, as you can see by setting $x = 0$ and solving for y, but does not cross the y-axis, since there are no solutions with $x = 0$. Similarly, $\frac{y^2}{a^2} - \frac{x^2}{b^2} = 1$ has y-intercepts $y = \pm a$ and does not cross the x-axis.
- To sketch the hyperbola $\frac{x^2}{a^2} - \frac{y^2}{b^2} = 1$, draw the rectangle whose sides are a units to the right and left of and b units above and below the origin. For $\frac{y^2}{a^2} - \frac{x^2}{b^2} = 1$, draw the rectangle whose sides are a units above and below and b units to the right and left of the origin. Then draw the circle centered at the origin that passes through the corners of the rectangle. The asymptotes pass through the origin and opposite corners of the rectangle, and the foci are at the points where the circle crosses the axis of the curve.
- To recall the equation $c^2 = a^2 + b^2$ relating the parameters a, b and c for a hyperbola, visualize the right triangles in Figures 22 and 23 that occupy one-fourth of the rectangle described above.
- If the axis or axes of a conic section are vertical and/or horizontal and the center is at a point (h, k) other than the origin, then translated coordinates $x' = x - h, y' = y - k$ are introduced so that the curve has a standard equation relative to these new coordinates. If an equation for the curve is given, then the new coordinates are found by factoring or completing squares in the equation. If the curve is described geometrically, perhaps by giving the focus and directrix of a parabola or the foic and vertices of an ellipse or hyperbola, then the new coordinates are determined by locating the center of the curve from the description of the curve.

[(1)] "Satellite navagation systems", *McGraw-Hill Encyclopedia of Science and Technology*,Volume 16, New York NY: McGraw Hill, 1997, pp. 53–59.

Tune-Up Exercises 9.1[♦]

[A]*Answer provided.* [O]*Outline of solution provided.*

T1.[O] Give an equation of the parabola that is the locus of points equidistant from its focus $(1, 1)$ and directrix $y = 3$. Then draw the curve with the focus and directrix.

♦ Type 1, equation of a parabola

T2.[O] Give an equation of the parabola passing through $(0, -2)$ that has vertex $(-1, 2)$ and axis $y = 2$. Draw the curve with its focus and directrix.

♦ Type 1, equation of a parabola

T3.[O] Sketch the parabola $x^2 + 6x - 2y + 5 = 0$. Show its focus and directrix.

♦ Type 1, sketching a parabola

T4.[O] Sketch the ellipse $25x^2 + 9y^2 - 50x + 36y = 164$ with its foci.

♦ Type 3, sketching an ellipse

T5.[O] Give an equation of the ellipse that has foci $F_1(0, 2)$ and $F_2(4, 2)$ and vertices $V_1(-1, 2)$ and $V_2(5, 2)$. Then sketch it with its foci.

♦ Type 1, equation of an ellipse

T6.[O] Sketch the hyperbola $x^2 - 9y^2 - 4x - 18y = 14$. Show its foci and asymptotes.

♦ Type 3, sketching a hyperbola

T7.[O] Give an equation for the locus of points that are either four units closer to $(0, -2)$ than to $(0, 4)$ or four units closer to $(0, 4)$ than to $(0, -2)$. Then draw the curve with its foci and directrices.

♦ Type 3, equation of a hyperbola

Problems 9.1

[A]*Answer provided.* [O]*Outline of solution provided.*

Give equations of the parabolas in conic sections in Problems 1 through 14. Then sketch the curves, showing all foci, directrices of parabolas, and asymptotes of hyperbolas.

1.[O] The parabola with focus $(0, 4)$ and directrix $y = 0$

♦ Type 1, equation of a parabola

2.[A] The parabola with focus $(3, 3)$ and directrix $x = 1$

♦ Type 1, equation of a parabola

3. The parabola with focus $(4, -5)$ and directrix $y = 1$

♦ Type 1, equation of a parabola

4.[O] The ellipse with foci $(-3, 0)$ and $(3, 0)$ and vertices $(-5, 0)$ and $(5, 0)$

♦ Type 1, equation of an ellipse

5.[A] The ellipse with foci $(0, 4)$ and $(0, -4)$ and minor axis of length six.

♦ Type 1, equation of an ellipse

6. The locus of points such that the sum of their distances from $(1, 1)$ and $(1, 3)$ is $2\sqrt{2}$

♦ Type 1, equation of an ellipse

7.[O] The hyperbola with foci $(-4, 0)$ and $(4, 0)$ and vertices $(-3, 0)$ and $(3, 0)$

♦ Type 1, equation of a hyperbola

8. The hyperbola with foci $(0, 2)$ and $(0, -2)$ and vertices two units apart

♦ Type 1, equation of a hyperbola

9. The hyperbola with vertices $(1, 3)$ and $(9, 3)$ and foci ten units apart

♦ Type 1, equation of a hyperbola

10. The locus of points such that the sums of their distances to $(0, -1)$ and $(0, -5)$ is twelve

♦ Type 1, equation of an ellipse

11. The locus of points that are equidistant from the point $(-3, 0)$ and the line $y = 3$

♦ Type 1, equation of a parabola

12.[A] The locus of points such that the absolute value of the differences of their distances from $(3, 6)$ and $(3, 0)$ is 4

♦ Type 1, equation of a hyperbola

13. The locus of points P such that $|\overline{PF_1} - \overline{PF_2}| = 6$, where $F_1 = (-4, 1)$ and $F_2 = (6, 1)$

♦ Type 1, equation of a hyperbola

14. The locus of points P such that $\overline{PF_1} + \overline{PF_2} = 12$, where $F_1 = (2, -1)$ and $F_2 = (2, 3)$

♦ Type 1, equation of an ellipse

[♦]The Tune-up Exercises and Problems are classified by type and content. The types are (1) basic, reactive; (2) basic reflective; (3) intermediate, reactive; (4) intermediate, reflective; (5) advanced, reactive; (6) advanced, reflective; and (7) advanced, theoretical.

Sketch the conic sections in Problems 15 through 22, showing all foci, directrices of parabolas, and asymptotes of hyperbolas.

15.[O] $x^2 - y^2 + 2x + 6y = 12$

♦ Type 3, graph of a conic section

16.[O] $y^2 - 6y + 2x + 13 = 0$

♦ Type 3, graph of a conic section

17. $x^2 - 2x - y = 0$

♦ Type 3, graph of a conic section

18.[O] $9x^2 + 16y^2 = 144$

♦ Type 3, graph of a conic section

19.[A] $x^2 - y^2 + 2x + 2y = 1$

♦ Type 3, graph of a conic section

20. $x^2 + 4y^2 = 16$

♦ Type 3, graph of a conic section

21.[A] $y = \frac{1}{2}x^2 - 3$

♦ Type 3, graph of a conic section

22. $9x^2 + 25y^2 + 18x + 50y = 191$

♦ Type 3, graph of a conic section

23.[A] Find equations for and draw the two parabolas with focus $(0, 0)$ and axis $y = 0$ that pass through the point $(3, 4)$.

♦ Type 3, equation of a parabola

24. Find an equation for the parabola with axis $x = 0$ that passes through the points $(1, 4)$ and $(2, 7)$.

♦ Type 3, equation of a parabola

25.[A] Find an equation for the hyperbola with foci $(3, 0)$ and $(-3, 0)$ and directrices $y = \pm x$.

♦ Type 3, equation of a hyperbola

26. Find an equation for the hyperbola with foci $(0, 5)$ and $(0, -5)$ and directrices $y = \pm\frac{3}{4}x$.

♦ Type 3, equation of a hyperbola

27.[A] Find the area of the region inside the ellipse $\dfrac{x^2}{a^2} + \dfrac{y^2}{b^2} = 1$.

♦ Type 3, an area

28. The region inside the ellipse of Problem 27 with $a > b$ generates an OBLATE SPHEROID when it is rotated about the y-axis and a PROLATE SPHEROID when its is rotated about the x-axis. Find the volumes of these two solids.

♦ Type 3, a volume

29. Give an integral that equals the length of the ellipse of Problem 27. Why is it improper? (This integral for $a \neq b$ is called an ELLIPTIC INTEGRAL.)

♦ Type 3, length of a curve

30.[A] A headlight reflector has parabolic cross sections and a circular face. The face has a radius of three inches and the reflector is three inches deep. What is its volume?

♦ Type 4, a volume

31. Find the volumes of the solids generated when the region bounded by the hyperbola $y^2 - x^2 = 1$ and by the lines $x = \pm 1$ is rotated **(a**[A]**)** about the x-axis and **(b)** about the y-axis.

♦ Type 4, a volume

32. Explain why you see an ellipse when you look at a circle from an angle. (Imagine that the circle is in an xy-plane and that your line of sight is perpendicular to the x-axis and makes an acute angle α with the y-axis.)

♦ Type 4, circles and ellipses

33. **(a)** Show that for any constant $k > -1$ the curve $\dfrac{x^2}{10 + k} + \dfrac{y^2}{1 + k} = 1$ is an ellipse with foci $(3, 0)$ and $(-3, 0)$. **(b)** Does this equation give all ellipses with these foci? **(c)** What is the graph of this equation for other values of k?

♦ Type 6, ellipses

34. Prove that the ellipse $\frac{1}{169}x^2 + \frac{1}{144}y^2 = 1$ and the hyperbola $\frac{1}{9}x^2 - \frac{1}{16}y^2 = 1$ have the same foci and intersect at right angles.

♦ Type 4, orthogonal ellipses

35. Coordinates can be chosen so that any conic section has an equation of the form $y^2 = Lx + kx^2$ with constants $L > 0$ and $k \geq -1$. The number L is the LATUS RECTUM of the curve. **(a)** For what values of k is the curve a parabola, an ellipse, or a hyperbola? **(b)** Show that for any $k \geq -1$, L is the width of the conic section at a focus.

♦ Type 6, latus rectum of a conic section

36. **(a)** According FERMAT'S PRINCIPLE in optics, a ray of light from point Q in Figure 38 that reflects off a line and then goes to point R will reflect at the point P such that the distance $\overline{QP} + \overline{PR}$ is a miniumum. Express this distance as a function $D(x)$ of the distance x in the drawing. Then show that the global minimum of $D(x)$ occurs for P such that $\sin\alpha = \sin\beta$ and, consequently, the angles α and β in the drawing are equal. **(b)** Figure 39 shows the tangent line at a point P on an ellipse with foci F_1 and F_2. Because the ellipse is the locus of points P such that $\overline{PF_1} + \overline{PF_2} = k$ for a positive constant k, and because every point Q on the tangent line other than P is outside the ellipse, $\overline{QF_1} + \overline{QF_2} > \overline{PF_1} + \overline{PF_2}$ for all $Q \neq P$ on the tangent line. Use this fact with the result of part (a) to show that the tangent line makes equal angles with the focal radii PF_1 and PF_2.

◆ Type 4, the reflective property of an ellipse

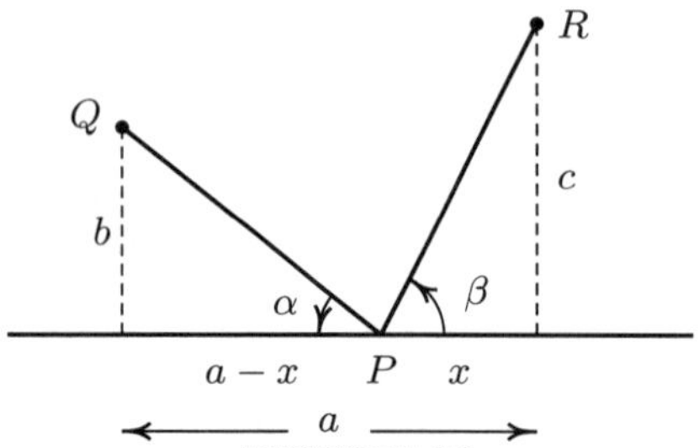

FIGURE 38

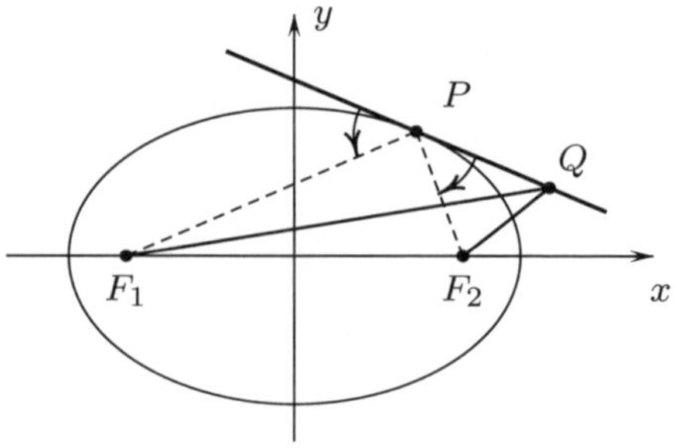

FIGURE 39

37. Figure 40 shows the tangent line at a point P on a parabola with focus F and directrix as labeled in the drawing. Because the parabola is the locus of points P that are equidistant from the focus and the directrix, $\overline{PF} = \overline{PD_1}$, and because every point Q on the tangent line other than P is closer to the directrix than to the focus, $\overline{QF} > \overline{QD_2}$ for all $Q \neq P$ on the tangent line. Let R, as in the drawing, be a point on the vertical line extending up from P. **(a)** Explain why $\overline{RQ} + \overline{QD_2} > \overline{RD_1}$ and consequently, $\overline{RQ} + \overline{QD_2} > \overline{RP} + \overline{PF}$. **(b)** Use the results of part (a) of this Problem and part (a) of Problem 36 to show that the tangent line makes equal angles with the focal radius PF and the vertical line segment PR.

◆ Type 6, the reflective property of a parabola

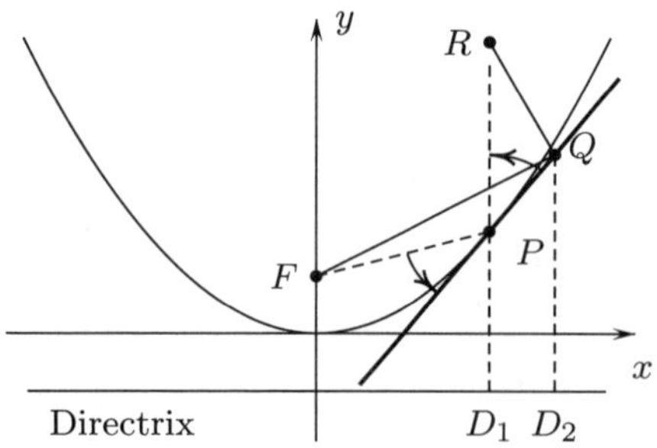

FIGURE 40

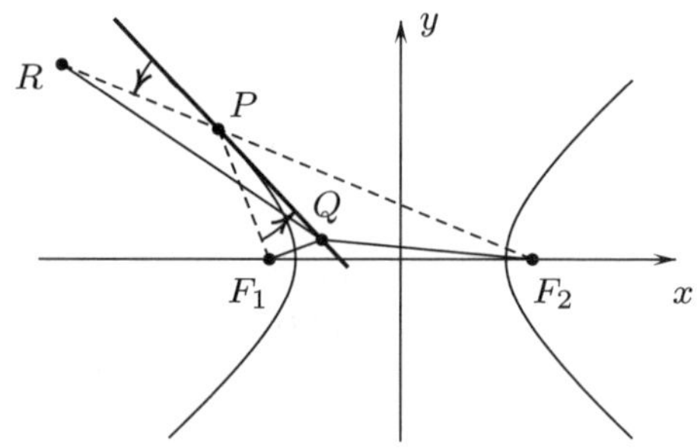

FIGURE 41

38. Figure 41 shows the tangent line at a point P on an hyperbola with foci F_1 and F_2 and a point R on the focal radius PF_2 extended to the left. **(a)** The left half of the hyperbola is the locus of points P such that $\overline{PF_2} - \overline{PF_1} = k$ for a positive constant k. Moreover, for every point Q on the tangent line other than P, $\overline{QF_2} - \overline{QF_1}$ is less than k and consequently $\overline{QF_2} - \overline{QF_1} < \overline{PF_2} - \overline{PF_1}$. Use this inequality to show that $\overline{RQ} + \overline{QF_1} > \overline{RQ} + \overline{QF_2} + \overline{PF_1} - \overline{PF_2}$ and then that this quantity is greater than $\overline{RF_2} + \overline{PF_1} - \overline{PF_2}$, which equals $\overline{RP} + \overline{PF_1}$. **(b)** Use the result of part (a) of this Problem and part (a) of Problem 36 to show that the tangent line bisects the angle between the focal radii PF_1 and PF_2.

◆ Type 6, the reflective property of a hyperbola

Section 9.2

Rotating axes, discriminants, and sections of cones

OVERVIEW: *We begin this section by showing how, by rotating coordinates, equations for conic sections with axes that are not parallel to the xy-axes can be put in forms that can be analyzed by the techniques of Section 9.1. Next, we define the* DISCRIMINANT *of an equation of a conic section and explain how it is used to determine whether the curve is a parabola, ellipse, or hyperbola without going through the procedure of rotating axes. Then we discuss* DEGENERATE CONIC SECTIONS, *which are curves with equations of the form* $Ax^2+Bxy+Cy^2+Dx+Ey+F=0$ *that are not actual parabolas, ellipses, or hyperbolas. We close the section with a discussion of a proof that the parabolas, ellipses, and hyperbolas are, in fact, sections of cones. This proof also provides focus-eccentricity definitions of ellipses and hyperbolas that are similar to the definition of a parabola. These alternate definitions will be used in Chapter 10, where we study planetary motion*

Topics:

- ***Rotating coordinate axes***
- ***Classifying conic sections by their discriminants***
- ***Degenerate conic sections***
- ***Sections of cones and eccentricity***

Rotating coordinate axes

The parabolas, ellipses, and hyperbolas we studied in Section 9.1 all had equations of the form $Ax^2+Cy^2+Dx+Ey+F=0$ with no xy-terms. Conic sections whose axes are not parallel to the coordinate axes have equations $Ax^2+Bxy+Cy^2+Dx+Ey+F=0$ with nonzero B. At the beginning of this section we show how new rotated $x'y'$-coordinates can be chosen so that relative to the new coordinates the equation of the curve has no $x'y'$-term. The new equation can then be analyzed with the techniques by completing squares and translating axes if necessary. We begin by deriving the necessary formulas for rotating coordinate axes.

Imagine that new $x'y'$-axes are obtained by rotating xy-axes through a positive acute angle α, as in Figure 1. If the point P with xy-coordinates (x,y) is positioned as in Figure 1, then its coordinates relative to the new axes are the positive numbers x' and y' in the drawing. We need formulas for this change of coordinates.

Question 1 **(a)** Express the distances $\overline{OV}, \overline{UV}, \overline{OR}$, and $\overline{RQ}$ in Figure 1 in terms of $x', y', \cos\alpha$, and $\sin\alpha$. **(b)** Use the results of part (a) to give formulas for x and y in terms of $x', y', \cos\alpha$, and $\sin\alpha$.

FIGURE 1

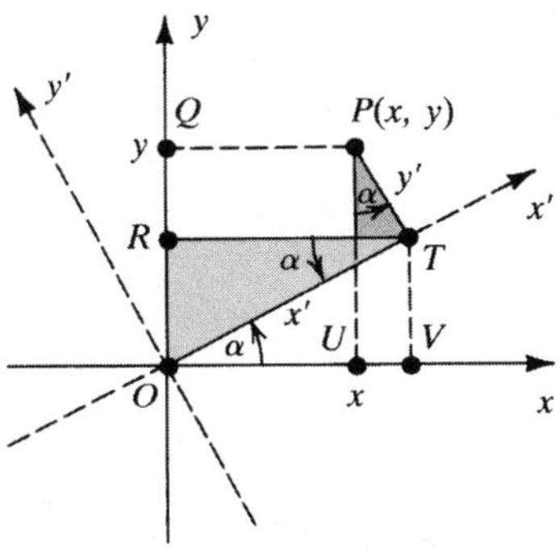

♦ SUGGESTIONS TO INSTRUCTORS: The main topics in this section can be divided into three parts, as follows:
Part 1 (Analyzing conic sections by rotating coordinates; Tune-up Exercises T2–T4; Problems 7–15, 17, 32–44, 50, 51) Have students begin work on Questions 1 and 2 before class. Problems 7–9 and 41 are good for in-class discussion.
Part 2 (Classifying conic sections by their discriminants and degenerate conic sections; Tune-up Exercises T1 and T5; Problems 1–6, 16, 21, 22, 41, 46) Have students begin work on Questions 3 and 4 before class. Problems 1–3, 16, 21, and 22 are good for in-class discussion.
Part 3 (Sections of cones and eccentricity; Tune-up Exercises T6 and T7; Problems 18-20, 23–31, 45, 47–49) Have students begin work on Questions 5 through 8 before class. Problems 23, 27, and 48 are good for in-class discussion.

The formulas you derived in Question 1 for a positive acute angle α and P as in Figure 1,

$$\begin{cases} x = x' \cos\alpha - y' \sin\alpha \\ y = x' \sin\alpha + y' \cos\alpha \end{cases} \tag{1}$$

are valid for any α and for any location of the point P.[†]

To transform an equation

$$Ax^2 + Bxy + Cy^2 + Dx + Ey + F = 0 \tag{2}$$

of a conic section relative to xy-coordinates into an equation in rotated $x'y'$-coordinates, we substitute equations **(1)** for x and y in **(2)** to obtain

$$A(x'\cos\alpha - y'\sin\alpha)^2 + B(x'\cos\alpha - y'\sin\alpha)(x'\sin\alpha + y'\cos\alpha) + C(x'\sin\alpha + y'\cos\alpha)^2 \\ + D(x'\cos\alpha - y'\sin\alpha) + E(x'\sin\alpha + y'\cos\alpha) + F = 0.$$

Then, when we expand the squares and the products and rearrange terms, we have

$$\begin{aligned} &\big[A\cos^2\alpha + B\cos\alpha\sin\alpha + C\sin^2\alpha\big](x')^2 \\ &\quad + \big[-2A\cos\alpha\sin\alpha + B(\cos^2\alpha - \sin^2\alpha) + 2C\cos\alpha\sin\alpha\big]x'y' \\ &\quad + \big[A\sin^2\alpha - B\cos\alpha\sin\alpha + C\cos^2\alpha\big](y')^2 \\ &\quad + \big[D\cos\alpha + E\sin\alpha\big]x' + \big[-D\sin\alpha + E\cos\alpha\big]y' + F = 0. \end{aligned}$$

which can be written as

$$A'(x')^2 + B'x'y' + C'(y')^2 + D'x' + E'y' + F = 0$$

with

$$A' = A\cos^2\alpha + B\cos\alpha\sin\alpha + C\sin^2\alpha \tag{3}$$

$$B' = B(\cos^2\alpha - \sin^2\alpha) + 2(C - A)\cos\alpha\sin\alpha \tag{4}$$

$$C' = A\sin^2\alpha - B\cos\alpha\sin\alpha + C\cos^2\alpha. \tag{5}$$

If the B in the original equation **(2)** is not zero, we want to choose α so the number B' in **(4)** is zero and, consequently, there is no $x'y'$-term in the new equation for the conic section. To achieve this, we set the right side of **(4)** equal to zero and then divide by $\cos^2\alpha$ to obtain

$$B\tan^2\alpha - 2(C - A)\tan\alpha - B = 0.$$

This equation must be satisfied by the angle α in order to have $B' = 0$. By the Quadratic Formula, this equation holds if

$$\tan\alpha = \frac{C - A}{B} \pm \sqrt{\left(\frac{C - A}{B}\right)^2 + 1}.$$

Taking the plus sign gives the formula for $\tan\alpha$ in the next rule.

Rule 1 *If the constant B is not zero in the equation $Ax^2 + Bxy + Cy^2 + Dx + Ey + F = 0$ for a conic section, then an equation in rotated coordinates* **(1)** *with no $x'y'$-term can be obtained by choosing α such that*

$$\tan\alpha = \frac{C - A}{B} + \sqrt{\left(\frac{C - A}{B}\right)^2 + 1} \tag{6}$$

and then making the substitutions

$$\begin{cases} x = x' \cos\alpha - y' \sin\alpha \\ y = x' \sin\alpha + y' \cos\alpha. \end{cases} \tag{7}$$

[†]These formulas are used extensively for rotating pictures in computer algorithms for graphic design and animation.

Rotating axes: the case of A = C

If $A = C$ and $B \neq 0$ in the equation $Ax^2 + Bxy + C^2 + Dx + Ey + F = 0$ of a conic section, as in the next Example, then formula **(6)** gives $\tan\alpha = 1$ and we can use the angle $\alpha = \frac{1}{4}\pi$, for which the x'-axis is rotated $\frac{1}{4}\pi$ radians counterclockwise from the x-axis. Then $\sin\alpha = 1/\sqrt{2}$ and $\cos\alpha = 1/\sqrt{2}$, so that formulas **(7)** become

$$\begin{cases} x = \dfrac{1}{\sqrt{2}}(x' - y') \\[2ex] y = \dfrac{1}{\sqrt{2}}(x' + y') \end{cases} \tag{8}$$

Example 1 Find the foci, vertices, and asymptotes of the hyperbola $y = 1/x$ in Figure 2.

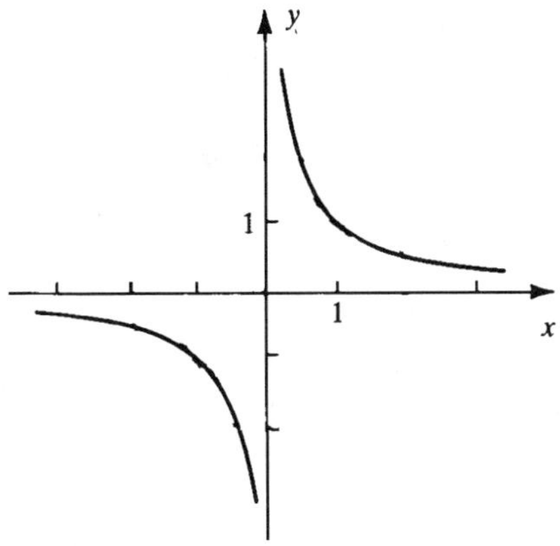

FIGURE 2

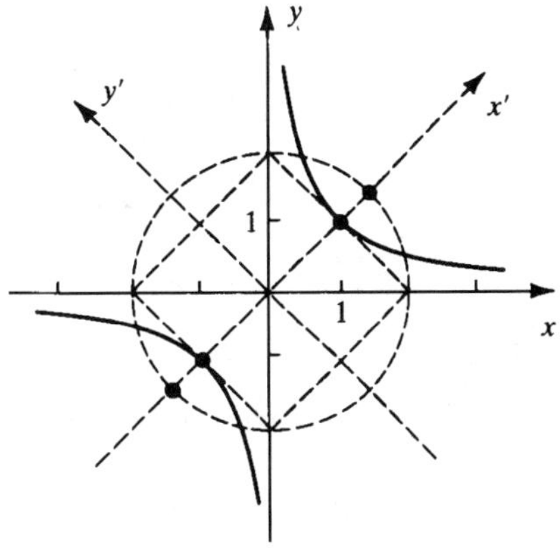

FIGURE 3

SOLUTION We write the equation $y = 1/x$ as $xy = 1$, which is of the form **(2)** with $A = 0, B = 1$, and $C = 0$. Because A and C are equal, we can use $\alpha = \frac{1}{4}\pi$ and formulas **(8)**. Making these substitutions in $xy = 1$ gives the equation

$$\left(\frac{1}{\sqrt{2}}x' - \frac{1}{\sqrt{2}}y'\right)\left(\frac{1}{\sqrt{2}}x' + \frac{1}{\sqrt{2}}y'\right) = 1.$$

Expanding the product yields

$$\frac{1}{2}(x')^2 - \frac{1}{2}(y')^2 = 1. \tag{9}$$

This equation, relative to the $x'y'$-coordinates, is of the form covered in Rule 3 in Section 9.1 with $a = \sqrt{2}$ and $b = \sqrt{2}$. We do not have to translate axes in this case because there are no first-degree terms in equation **(9)**.

To sketch the curve, we draw auxiliary $x'y'$-axes, rotated 45° counterclockwise from the xy-axes, as in Figure 3. The curve does not intersect the y'-axis because x' cannot be zero in **(9)**. To find where it crosses the x'-axis, we set $y' = 0$ in **(9)** and solve for $x' = \pm\sqrt{2}$. These are the coordinates of the vertices of the hyperbola on the x'-axis. To find the xy-coordinates of the vertices, we set $x' = \pm\sqrt{2}$ and $y' = 0$ in **(9)**. This shows that the vertices are at $x = 1, y = 1$ and $x = -1, y = -1$.

We draw the square through the vertices with sides parallel to the x'- and y'-axes, as in Figure 3. The asymptotes of the hyperbola pass through the corners of this square and, consequently, are the x- and y-axes.

Next, we draw the circle through the corners of the square. The foci are at the intersections of this circle with the x'-axis, as shown in Figure 3. To find the coordinates of the foci, we apply the formula $c = \sqrt{a^2 + b^2}$ from Rule 3 of Section 9.1. We conclude that $c = \sqrt{2+2} = 2$ and the foci are at $x' = \pm 2, y' = 0$ on the x'-axis. Substituting these numbers in the change of variables formulas **(8)** shows that the coordinates of the foci, relative to the original xy-axes, are $x = \sqrt{2}, y = \sqrt{2}$ and $x = -\sqrt{2}, y = -\sqrt{2}$. □

Rotating axes: the case of $\mathbf{A \neq C}$

If $B \neq 0$ and $A \neq C$ in the equation $Ax^2 + Bxy + C^2 + Dx + Ey + F = 0$, we can calculate $\tan\alpha$ from **(6)** and then find $\sin\alpha$ and $\cos\alpha$, without finding α, from a right triangle as in Figure 4.

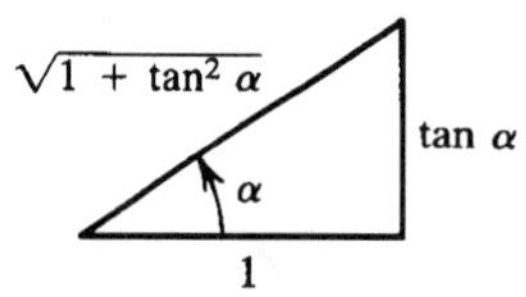

FIGURE 4

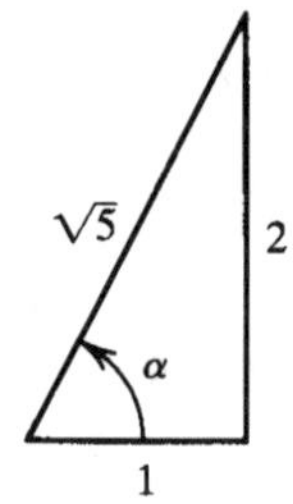

FIGURE 5

We will illustrate this procedure with the equation

$$17x^2 - 12xy + 8y^2 = 20. \tag{10}$$

Here $A = 17, B = -12$ and $C = 8$, so that by **(6)**, we can obtain an equation in $x'y'$-coordinates with no $x'y'$-term if we rotate the axes an angle α such that

$$\begin{aligned}\tan\alpha &= \frac{C-A}{B} + \sqrt{\left(\frac{C-A}{B}\right)^2 + 1} = \tfrac{3}{4} + \sqrt{\left(\tfrac{3}{4}\right)^2 + 1} \\ &= \tfrac{3}{4} + \sqrt{\tfrac{25}{16}} = \tfrac{3}{4} + \tfrac{5}{4} = 2.\end{aligned}$$

We draw such an angle by drawing the right triangle in Figure 5 with base of length 1, height 2, and hypotenuse of length $\sqrt{1^2 + 2^2} = \sqrt{5}$.

Question 2 What are $\sin\alpha$ and $\cos\alpha$ in Figure 5?

Example 2 Sketch the conic section with equation **(10)**.

SOLUTION With the values from Question 2, the changes of coordinates **(7)** are

$$\begin{cases} x = \dfrac{1}{\sqrt{5}}(x' - 2y') \\ y = \dfrac{1}{\sqrt{5}}(2x' + y'). \end{cases}$$

Making these substitutions in **(10)** yields

$$\tfrac{17}{5}(x' - 2y')^2 - \tfrac{12}{5}(x' - 2y')(2x' + y') + \tfrac{8}{5}(2x' + y')^2 = 20$$

which simplifies, after several lines of calculations, to

$$\tfrac{1}{4}(x')^2 + (y')^2 = 1. \tag{11}$$

By Rule 2 of Section 9.1, equation **(11)** is that of an ellipse with its center of the origin of the $x'y'$-plane, whose major axis runs from $x' = 2$ to $x' = -2$ on the x'-axis, and whose minor axis runs from $y' = 1$ to $y' = -1$ on the y'-axis. The distance c from the center of the foci is $\sqrt{a^2 - b^2} = \sqrt{2^2 - 1^2} = \sqrt{3}$, so the foci are at $x' = \pm\sqrt{3}$ on the x'-axis.

To obtain the drawing of the ellipse in Figure 6, we draw the x'-axis parallel to the hypotenuse of the right triangle in Figure 5, and mark the ends of the major and minor axes on the x'- and y'-axes. Then we draw the ellipse through those points and plot the foci a distance $c = \sqrt{3}$ from the origin on the x'-axis. □

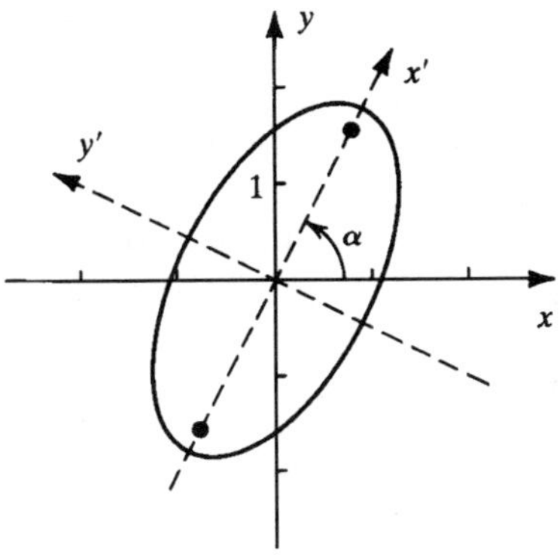

FIGURE 6

Classifying conic sections by their discriminants

We will now show how to determine whether a conic section

$$Ax^2 + Bxy + Cy^2 + Dx + Ey + F = 0$$

is a parabola, an ellipse, or a hyperbola by calculating its DISCRIMINANT, which is the number

$$B^2 - 4AC. \tag{12}$$

We start by looking at some special cases.

Example 3 What are the discriminants of **(a)** the parabola $y = x^2$, **(b)** the ellipse $\frac{1}{4}x^2 + \frac{1}{9}y^2 = 1$, and **(c)** the hyperbola $x^2 - y^2 = 1$?

SOLUTION **(a)** We write $y = x^2$ in the form $x^2 - y = 0$ to see that here $A = 1$, $B = 0, C = 0$, and the discriminant is $B^2 - 4AC = 0^2 - 4(1)(0) = 0$.
(b) We write the equation as $\frac{1}{4}x^2 + \frac{1}{9}y^2 - 1 = 0$ to see that in this case $A = \frac{1}{4}$, $B = 0, C = \frac{1}{9}$, and $B^2 - 4AC = 0^2 - 4(\frac{1}{4})(\frac{1}{9}) = -\frac{1}{9}$.
(c) We write the equation as $x^2 - y^2 - 1 = 0$, for which $A = 1, B = 0, C = -1$, and $B^2 - 4AC = 0^2 - 4(1)(-1) = 4$. □

Notice in this example, the discriminant is zero for the parabola, is negative for the ellipse, and is positive for the hyperbola. We will now show that these criteria always hold. To accomplish this we need to calculate the discriminant of a conic section

$$A'(x')^2 + B'x'y' + C'(y')^2 + D'x' + E'y' + F' = 0$$

in rotated $x'y'$-coordinates in terms of the discriminant of the same curve in the original coordinates.

Formulas **(3)**, **(4)**, and **(5)** give

$$\begin{aligned}(B')^2 &- 4A'C' \\ &= [B(\cos^2\alpha - \sin^2\alpha) + 2(C - A)\sin\alpha\cos\alpha]^2 \\ &- 4[A\cos^2\alpha + B\sin\alpha\cos\alpha + C\sin^2\alpha][A\sin^2\alpha - B\sin\alpha\cos\alpha + C\cos^2\alpha].\end{aligned}$$

Expanding the square and the products on the right side of this equation yields the following identity, after a lengthy calculation and repeated use of the Pythagorean Identity $\sin^2\alpha + \cos^2\alpha = 1$:

$$B^2 - 4AC = (B')^2 - 4A'C'. \qquad \textbf{(13)}$$

We describe equation **(13)** by saying that the discriminant is INVARIANT when coordinates are rotated.

Now we can see how the discriminant is used to classify conic sections. We suppose the axes have been rotated, if necessary, using Rule 1, so the equation is

$$A'(x')^2 + C'(y')^2 + D'x' + E'y' + F' = 0 \qquad \textbf{(14)}$$

in $x'y'$-coordinates with no $x'y'$-term.

If the curve with equation **(14)** is a parabola, then either $A' = 0$ or $C' = 0$ and $B^2 - 4AC = 0 - 4A'C' = 0$. If the curve is an ellipse, then A' and C' are either both positive or both negative and $B^2 - 4AC = 0 - 4A'C' < 0$. Finally, if the curve is a hyperbola, then one of A' or C' is positive and the other negative and $B^2 - 4AC = 0 - 4A'C' > 0$. This leads us to the following rusult.

Theorem 1 *If the conic section $Ax^2 + Bxy + Cy^2 + Dx + Ey + F = 0$ is a parabola, then its discriminant $B^2 - 4AC$ is zero, if it is an ellipse, then $B^2 - 4AC$ is negative, and if it is a hyperbola then $B^2 - 4AC$ is positive.*

Example 4 The curve $x^2 - xy + 5y^2 - 6x - y - 4 = 0$ is either a parabola, an ellipse, or a hyperbola. Which is it?

SOLUTION The discriminant $B^2 - 4AC = (-2)^2 - 4(1)(5) = -16$ is negative, so the curve is an ellipse. □

Question 3 Use the discriminant to determine whether $x^2 - xy - 1 = 0$ is a parabola, an ellipse, or a hyperbola. Then generate the curve in the window $-3 \le x \le 3, -3 \le y \le 3$ by first solving the equation for y.

Degenerate conic sections

The graph of an equation of the form $Ax^2 + Bxy + Cy^2 + Dx + Ey + F = 0$ that is not a parabola, ellipse, or hyperbola is called a DEGENERATE CONIC SECTION.

Example 5 Describe the degenerate conic sections with equations
(a) $x^2 - 2xy + y^2 = 0$, **(b)** $x^2 - 4 = 0$, and **(c)** $y^2 + 1 = 0$.

SOLUTION **(a)** The equation $x^2 - 2xy + y^2 = 0$ can be written $(x - y)^2 = 0$, so its graph is the line $y = x$. **(b)** The equation $x^2 - 4 = 0$ holds if and only if $x = 2$ or $x = -2$ so its graph is the the vertical lines with these equations. **(c)** The equation $y^2 + 1 = 0$ has no solutions, so its graph is the EMPTY SET □

A degenerate conic section is called a degenerate parabola, ellipse, or hyperbola, according to whether its discriminant is zero, negative or positive.

Example 6 Are the three curves of Example 5 degenerate parabolas, degenerate ellipses, or degenerate hyperbolas?

SOLUTION The discriminant of $x^2 - 2xy + y^2 = 0$ is $(-2)^2 - 4(1)(1) = 0$, the discriminant of $x^2 - 4 = 0$ is $0^2 - 4(1)(0) = 0$, and the discriminant of $y^2 + 1 = 0$ is $(0^2) - 4(0)(1) = 0$, so all three curves are degenerate parabolas. □

Question 4 Describe and classify the degenerate conic sections **(a)** $x^2 + y^2 = 0$, **(b)** $x^2 + y^2 + 1 = 0$, and **(c)** $x^2 - xy = 0$.

Cross sections of cones

We begin our study of parabolas, ellipses, and hyperbolas as cross sections of cones by looking at the case of an ellipse. We consider a double, right circular cone with a vertical axis and a plane that cuts across its top, as in Figure 7.

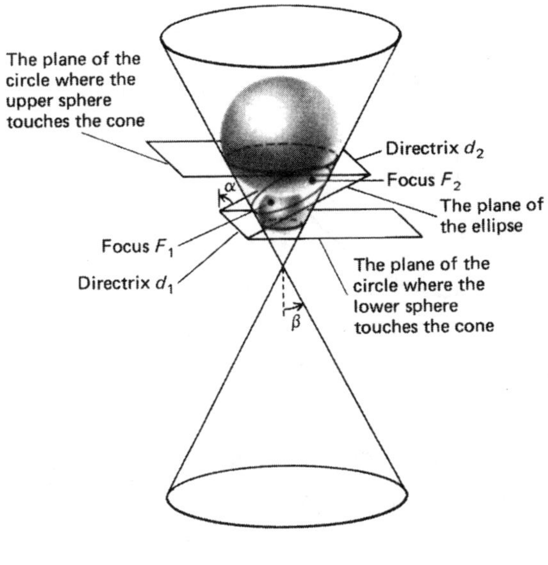

FIGURE 7

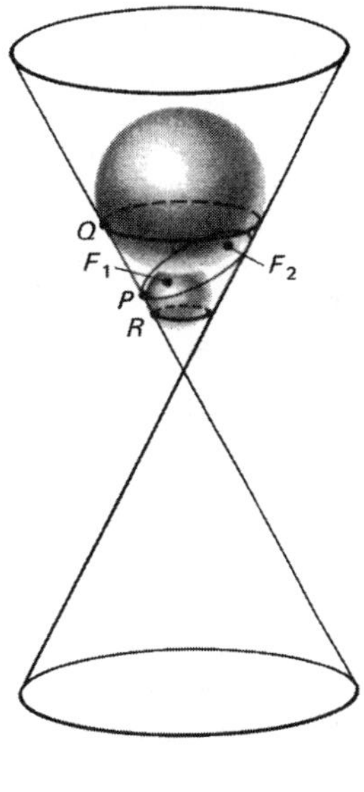

FIGURE 8

The shape of the cone is determined by the angle β in Figure 7 between its axis and its elements (the lines through the vertex that form the cone). The smallest angle between vertical lines and lines in the plane that slices the cone is called the angle BETWEEN vertical lines and the plane. We denote it with the letter α. It measures how much the plane is tilted. Because the plane slices the top of the cone, the angles α and β satisfy the inequalities $\beta < \alpha \leq \frac{1}{2}\pi$.

Question 5 What is the angle α in Figure 7 if the plane is horizontal, and what sort of curve is the cross section in this case?

Now suppose that $\beta < \alpha < \frac{1}{2}\pi$, so the plane cuts across the top of the cone but the cross section is not a circle. Imagine that spheres have been placed in the top half of the cone, above and below the plane, as in Figure 7, with the sizes of the two spheres chosen so that the lower sphere touches the plane at the point F_1 and touches the sides of the cone in a horizontal circle, and so that the upper sphere touches the plane at a point F_2 and touches the cone in another horizontal circle.[†]

We want to show that the curve that is the intersection of the plane with the cone is the ellipse with foci F_1 and F_2. To this end, we let P denote an arbitrary point on the curve and let Q be the point where the element of the cone through P intersects the upper circle, as in Figure 8. The line through P and Q and the line through P and F_1 are both tangent to a circular cross section of the sphere. To use this fact, we need the result stated in the next question.

Question 6 Show that $\overline{PR} = \overline{PF_1}$ in Figure 8. Use the fact that a tangent line to a circle is perpendicular to the radius at the point of tangency.

The reasoning in Question 6 can also be applied to the upper sphere in Figure 8 to show that $\overline{PQ} = \overline{PF_2}$. Therefore,

$$\overline{PF_1} + \overline{PF_2} = \overline{PR} + \overline{PQ}.$$

Then because $\overline{PR} + \overline{PQ}$ is the distance between the two horizontal circles in Figure 8, measured along an element of the cone, it is the same for all points P on the curve. Thus, the curve is the locus of

[†]This approach to analyzing cross sections of cones was first given by a Belgian mathematician, G. P. Dandelin (1794–1847).

points P such that $\overline{PF_1} + \overline{PF_2}$ equals a constant and, by Definition 2 of Section 9.1, the curve is, in fact, an ellipse with foci F_1 and F_2.

A focus-directrix definition of an ellipse

In Section 9.1, we defined a parabola to be the locus of points P that are equidistant from its focus F and its directrix. We let D be the foot of the perpendicular line from P to the directrix, as in Figure 9. Then the parabola is defined by the equation,

$$\overline{PF} = \overline{PD}. \tag{15}$$

FIGURE 9

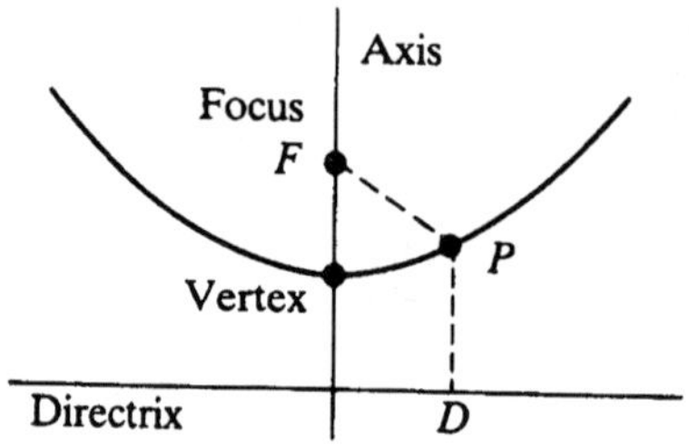

Every noncircular ellipse and every hyperbola has two directrices, one for each focus (Figures 10 and 11), which can be used to give these curves definitions similar to **(15)**. We will derive this result for an ellipse and leave the case of a hyperbola as Problem 45.

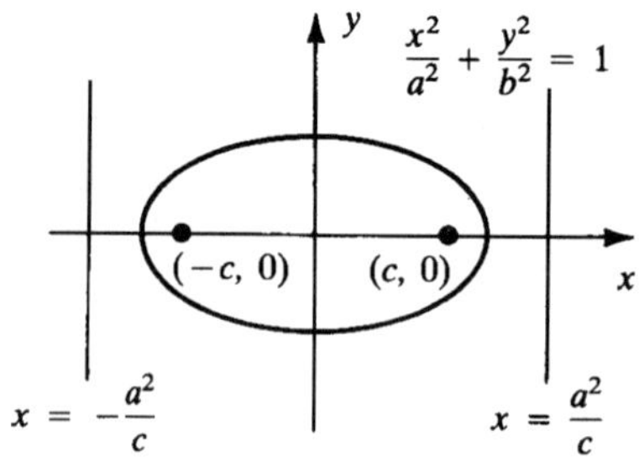

FIGURE 10

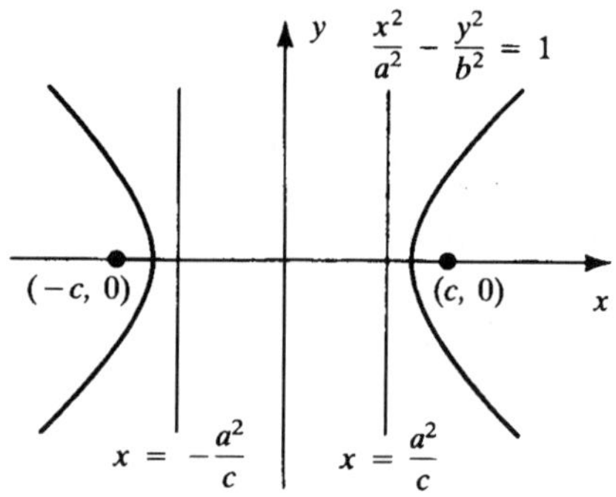

FIGURE 11

The two spheres inside the cone in Figure 7 touch the cone along horizontal circles, and the horizontal planes through those two circles intersect the plane of the ellipse in two lines, which we define to be the directrices. Figure 12 shows the forus F and directrix for the lower sphere.

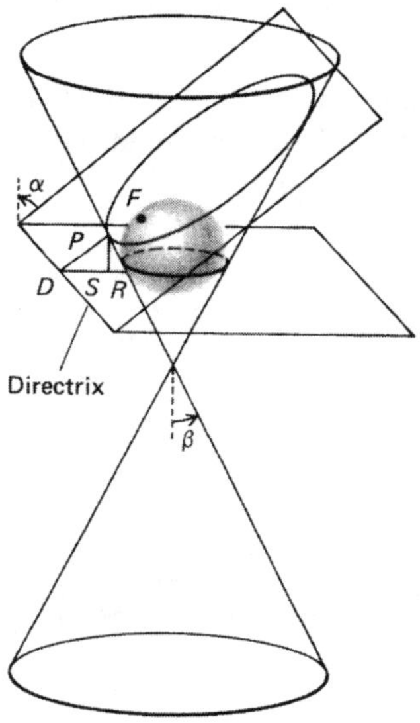

FIGURE 12

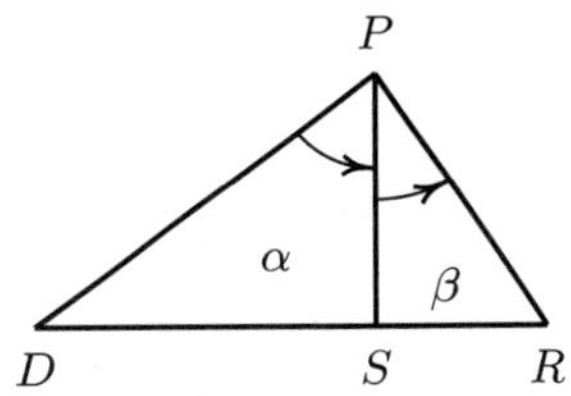

FIGURE 13

We let D be the foot of the perpendicular line in Figure 12 from the point P on the curve to the directrix and let R be the intersection of the horizontal plane through the lower circle with the element through P. These points form the triangle in Figure 13.

Question 7 **(a)** Express $\overline{PD}$ in Figure 13 in terms of $\overline{PS}$ and α. Then express $\overline{PR}$ in terms of $\overline{PS}$ and β. **(b)** Use the results of part (a) to express $\overline{PR}$ in terms of $\overline{PD}, \alpha$, and β.

As before, $\overline{PR} = \overline{PF}$, so the result of Question 7b gives

$$\overline{PF} = \left(\frac{\cos\alpha}{\cos\beta}\right)\overline{PD}. \tag{16}$$

We define the ratio $\cos\alpha/\cos\beta$ to be the ECCENTRICITY e of the ellipse, and obtain the following result for the case of an ellipse.[†]

Theorem 2 **(a)** *Consider a double right circular cone whose elements make an angle β with its vertical axis and a plane that does not pass through the vertex of the cone and makes an angle α with vertical lines. The intersection of the plane with the cone is a circle if $\alpha = \frac{1}{2}\pi$, is an ellipse if $\beta < \alpha < \frac{1}{2}\pi$, is a parabola if $\alpha = \beta$, and is a hyperbola if $0 \le \alpha < \beta$.*
(b) *Except in the case of a circle, the conic section may be characterized as the locus of points P in the plane such that*

$$\overline{PF} = e\overline{PD} \tag{17}$$

where $\overline{PF}$ is the distance from P to a focus, $\overline{PD}$ is the distance from P to the corresponding directrix, and the constant e is given by

$$e = \frac{\cos\alpha}{\cos\beta}. \tag{18}$$

Formula **(18)** is also used in the case of a circle, for which $\alpha = \frac{1}{2}\pi, \cos\alpha = 0$, and consequently the eccentricity e is zero.

[†]The symbol e here represents a parameter that can have any nonnegative value and is not the base of the natural logarithm.

In the case of $\alpha = \beta$, for which, $e = 1$, we use only one sphere, tangent to the cone and to the plane that forms the cross section., as in Figure 14. The derivation of **(17)** is the same as for an ellipse and gives the usual equation, $\overline{PF} = \overline{PD}$ for a parabola.

In the case of $\alpha < \beta$, we use two spheres as in Figure 15. The derivation of **(17)** applies to each focus and the corresponding directrix, and minor modifications of the argument given above for the case of an ellipse show that the cross section is a hyperbola (see Problem 45).

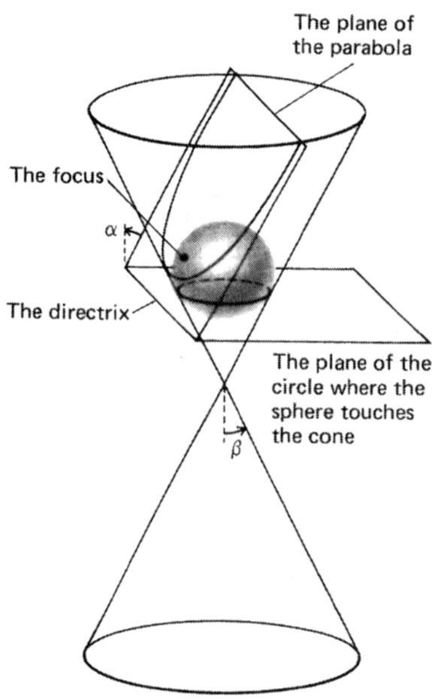

FIGURE 14

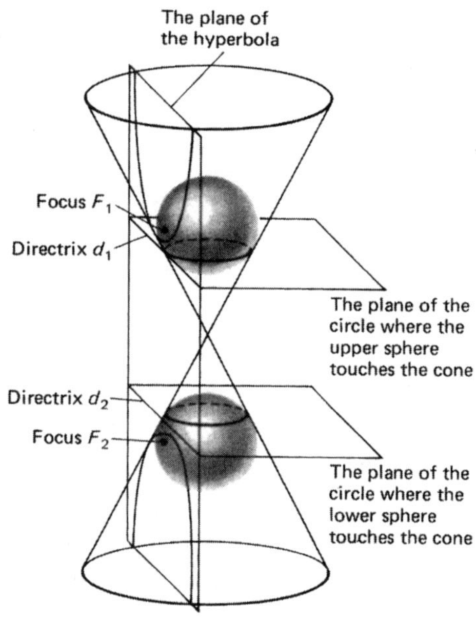

FIGURE 15

Directrices of ellipses and hyperbolas in standard position

Recall from Section 9.1 that if a and b are positive constants with $a > b$, then the curve

$$\frac{x^2}{a^2} + \frac{y^2}{b^2} = 1 \tag{19}$$

is a noncircular ellipse with foci at $x = \pm c$ on the x-axis with $c = \sqrt{a^2 - b^2}$ (Figure 10).

Question 8 Suppose that the focus on the right of the ellipse in Figure 10 is at $x = c$, as shown in the drawing and the directrix on the right is $x = k$. **(a)** Give the equations obtained from $\overline{PF} = e\overline{PD}$ with $F = (c, 0), D = (k, 0)$ with P first on one end and then the other end of the major axis of the ellipse. **(b)** Use the equations from part (a) to find formulas for e and k in terms of a, b, and c.

The calculations in the response to Question 8 could be carried out for the hyperbola in Figure 11, to give the next result.

Theorem 3 *The noncircular ellipse in Figure 10 or the hyperbola in Figure 11 has eccentricity*

$$e = \frac{c}{a} \tag{20}$$

and its directrices are the vertical lines

$$x = \pm\frac{a^2}{c}. \tag{21}$$

Example 7 Give an equation of the conic section with foci $(0, 2)$ and $(0, -2)$ and eccentricity $\frac{2}{3}$. Sketch the curve with its foci and directrices.

SOLUTION Since $e = \frac{2}{3}$ is less than 1, the curve is an ellipse. The constant c is 2 because the foci are at $(0, \pm 2)$. The equation $e = c/a$ reads $\frac{2}{3} = 2/a$, so that $a = 3$. Then, by Rule 2 of Section 9.1, $b = \sqrt{a^2 - c^2} = \sqrt{5}$.

Because the foci are on the y-axis, the equation of the ellipse is $\dfrac{x^2}{b^2} + \dfrac{y^2}{a^2} = 1$, which in this case is $\frac{1}{5}x^2 + \frac{1}{9}y^2 = 1$. Since $a^2/c = \frac{9}{2}$, the directrices are $y = \pm\frac{9}{2}$ (Figure 16). □

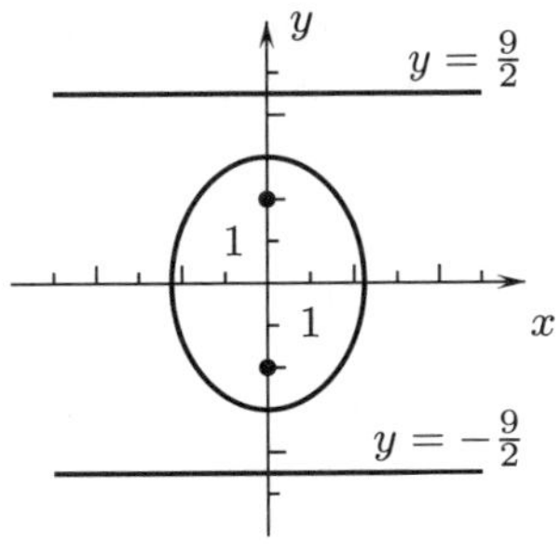

$\frac{1}{5}x^2 + \frac{1}{9}y^2 = 1$
FIGURE 16

Example 8 What is the eccentricity of the hyperbola $x^2 - 2y^2 = 8$?

SOLUTION We divide the given equation by 8 to put it in the standard form $\dfrac{x^2}{8} - \dfrac{y^2}{4} = 1$. Then a^2 goes with the positive term and b^2 with the negative term, so $a = \sqrt{8}$ and $b = 2$. By the formula in (3) of Section 9.1, $c = \sqrt{a^2 + b^2} = \sqrt{12}$ and by formula **(6)** above, $e = c/a = \sqrt{1.5}$. □

Principles and Procedures

- Every second-degree equation, $Ax^2 + Bxy + Cy^2 + Dx + Ey + F = 0$, with variables x and y, is either the graph of a regular or degenerate parabola, ellipse, or hyperbola and we can determine whether the graph is a parabola, ellipse, or hyperbola, by calculating the discriminant $B^2 - 4AC$. The graph is an ellipse if the discriminant is negative, a parabola if it is zero, and a hyperbola if it is positive. These facts will be used to classify surfaces given by second-degree equations in Chapter 10 and to study maxima and minima of functions of two-variables in Chapter 11.
- To remember the formula for the discriminant, notice that it is the same as the expression inside the square-root sign in the Quadratic Formula, $x = \dfrac{-b \pm \sqrt{b^2 - 4ac}}{2a}$, for the solutions of the Quadratic Equation $ax^2 + bx + c = 0$.
- To recall the discriminant test for classifying conic sections, calculate the discriminant for a circle, such as $x^2 + y^2 = 1$ (where $B^2 - 4AC = 0^2 - 4(1)(1) = -4$ is negative), for a familiar parabola, such as $x^2 - y = 0$ (where $B^2 - 4AC = 0^2 - 4(1)(0) = 0$), and for a familiar hyperbola, such as $x^2 - y^2 = 1$ (where $B^2 - 4AC = 0^2 - 4(1)(-1) = 4$ is positive).
- To remember where the minus sign goes in formulas **(1)** for rotating axes, notice that for a positive acute angle α, as in Figure 1, y is positive if x' and y' are positive.
- It takes some practice to carry out correctly the algebra that is required to convert an equation of a conic section to rotated coordinates—whether you do the calculations by hand or use a calculator or computer algebra system. If there is a nonzero $x'y'$-term in your final equation, then you made a mistake either in the formulas for changing coordinates or in their application.

- Some degenerate conic sections are also cross sections of a double cone. In particular, a plane through the vertex is a single point (a degenerate ellipse), one line (a degenerate parabola), or two intersecting lines (a degenerate hyperbola), depending on the angle of the plane.
- The focus-directrix definitions of ellipses and hyperbolas and the definition of eccentricity will be used in Section 9.3 to derive a formula for conic sections. This formula will be used in Chapter 11 to derive Kepler's Law that planets have elliptical orbits from Newton's Law of gravity.
- The definition of eccentricity relates to the terms "eccentric" (deviating from a standard), "ellipsis" (an omission), "parable" (a comparison or short narrative), and "hyperbole" (an exaggerated statement) in everyday English. "Ellipse" and "ellipsis" both derive from Greek words for "falling short"; "parabola" and "parabole" both come from Greek works for "throwing beside"; and "hyperbola" and "hyperbole" both come from Greek words for "overshooting." Circles have zero eccentricity and are not eccentric. Noncircular ellipses, with eccentricity between 0 and 1, are somewhat eccentric compared to circles, but fall short of being as eccentric as parabolas or hyperbolas. The eccentricity of all parabolas is exactly 1, and the eccentricity of hyperbolas exceeds 1.

Tune-Up Exercises 9.2[♦]

[A]*Answer provided.* [O]*Outline of solution provided.*

Use equal scales on the axes in all drawings of xy-planes in the answers to Tune-up Exercises and Problems in this section.[†]

T1.[O] Use the discriminant to classify the conic sections **(a)** $3x^2 - 4xy + y^2 + 2x - y = 4$, **(b)** $2x^2 - 4xy + 2y^2 + 3x + 5y = 10$, and **(c)** $x^2 + xy + y^2 = 4$.

♦ Type 1, classifying conic sections by the discriminant

T2.[O] Sketch the ellipse $5x^2 + 6xy + 5y^2 = 8$. Plot its foci on the x' or y'-axis without finding their xy-coordinates.

♦ Type 3, sketching an ellipse by rotating axes

T3.[O] Sketch the parabola $x^2 - 2xy + y^2 - \sqrt{2}\,x - \sqrt{2}\,y = 0$. Plot its focus and draw its directrix relative to the $x'y'$-axes.

♦ Type 3, sketching a parabola by rotatng axes

T4.[O] Sketch the hyperbola $x^2 + 4xy + y^2 = 1$. Plot its foci relative to $x'y'$-coordinates.

♦ Type 3, sketching a hyperbola by rotating axes

T5.[O] Sketch and classify the degenerate conic sections **(a)** $x^2 - 2xy + y^2 = 0$ and **(b)** $(x-1)(y-1) = 0$.

♦ Type 2, sketching and classifying degenerate conic sections

T6.[O] What is the eccentricity of the ellipse $\frac{1}{25}x^2 + \frac{1}{16}y^2 = 1$? Sketch it with its foci and directrices.

♦ Type 3, eccentricity, foci, and directrices

T7.[O] What is the eccentricity of the hyperbola $x^2 - y^2 = 1$? Where are its foci? What are its directrices? Sketch the curve with the foci and directrices.

♦ Type 3, eccentricity, foci, and directrices

[♦]The Tune-up Exercises and Problems are classified by type and content. The types are (1) basic, reactive; (2) basic reflective; (3) intermediate, reactive; (4) intermediate, reflective; (5) advanced, reactive; (6) advanced, reflective; and (7) advanced, theoretical.

[†]To obtain approximately equal scales in the window on a Texas Instruments TI-82 or TI-83 calculator use ($x$max $-$ $x$min) $= \frac{3}{2}$($y$max $-y$min). On a TI-85 or TI-86, use ($x$max $-$ $x$min)$= \frac{7}{4}$($y$max $-$ $y$min). On a TI-89, use ($x$max $-$ $x$min) $= 2$($y$max $-$ $y$min). On a TI-92. use ($x$max $-$ $x$min) $= \frac{5}{2}$($y$max $-$ $y$min).

Problems 9.2

[A]*Answer provided.* [O]*Outline of solution provided.*

Use the discriminant to classify the conic sections in Problems 1 through 6.

1.[O] $x^2 + 3xy - y^2 + 6x - 6 = 0$

♦ Type 1, classifying conic sections by the discriminant

2.[O] $3x^2 - xy + 4y^2 - 16x - 3y = 0$

♦ Type 1, classifying conic sections by the discriminant

3.[A] $-x^2 + 4xy - 4y^2 + x = 0$

♦ Type 1, classifying conic sections by the discriminant

4.[A] $x^2 + xy + y^2 = x + y$

♦ Type 1, classifying conic sections by the discriminant

5. $x^2 + 2xy + y^2 = x - y$

♦ Type 1, classifying conic sections by the discriminant

6. $x^2 + 3xy + y^2 = -x + y + 2$

♦ Type 1, classifying conic sections by the discriminant

Sketch the conic sections in Problems 7 through 14. Plot the vertices and foci on the x'- and/or y'-axes without finding their xy-coordinates.

7.[O] $13x^2 - 10xy + 13y^2 = 72$

♦ Type 3, sketching conic sections by rotating axes

8.[O] $x^2 + 2xy + y^2 + 8x - 8y = 0$

♦ Type 3, sketching conic sections by rotating axes

9.[A] $5x^2 + 26xy + 5y^2 = -72$

♦ Type 3, sketching conic sections by rotating axes

10. $5x^2 + 6xy + 5y^2 = 8$

♦ Type 3, sketching conic sections by rotating axes

11. $x^2 - 6xy + y^2 = 4$

♦ Type 3, sketching conic sections by rotating axes

12. $x^2 + 2xy + y^2 - 4x + 4y = 2$

♦ Type 3, sketching conic sections by rotating axes

13. $7x^2 + 8xy + 7y^2 = 0$

♦ Type 3, sketching conic sections by rotating axes

14. $x^2 - 2xy + y^2 = 1$

♦ Type 3, sketching conic sections by rotating axes

15. Find the asymptotes of the hyperbola $xy + x = 1$ without rotating coordinates.

♦ Type 4, asymptotes of a hyperbola

16. For what values of the constant B is the conic section $x^2 + Bxy + y^2 = 1$ **(a)** a circle, **(b)** a noncircular ellipse, **(c)** a degenerate parabola, and **(d)** a hyperbola?

♦ Type 1, classifying conic sections by the discriminant

17. Archimedes found a geometric solution of the equation which in modern notation reads $x^3 - x^2 + 1 = 0$ by finding the intersection of the hyperbola $xy - y + 1 = 0$ and the parabola $y = x^2$. Show that his reasoning was correct and use your calculator or computer to find the exact or approximate value of the solution.

♦ Type 3, solving an equation with conic sections

18. Find equations for and sketch the conic sections with foci at $(-1, 0)$ and $(1, 0)$ in an xy-plane with eccentricities **(a**[A]**)** $e = \frac{1}{3}$, **(b**[A]**)** $e = 3$, **(c)** $e = \frac{12}{13}$, and **(d)** $e = \frac{13}{12}$.

♦ Type 3, eccentricity of conic sections

19. Find the eccentricities of the conic sections **(a**[A]**)** $x^2 + y^2 = 1$, **(b**[A]**)** $x^2 + 9y^2 = 1$, **(c**[A]**)** $x^2 - 10y^2 = 100$, **(d)** $4x^2 + y^2 = 36$, **(e)** $y = 97x^2$, and **(f)** $x^2 - 25y^2 = 100$.

♦ Type 3, eccentricity of conic sections

20. Show that the conic sections with eccentricity $\sqrt{2}$ are the rectangular hyperbolas (those with perpendicular asymptotes).

♦ Type 4, eccentricity of rectangular hyperbolas

21. The degenerate hyperbola $y^2 - 3xy + 2x^2 = 0$ consists of two intersecting lines. Find their equations either by factoring the equation or using the Quadratic Formula to solve for y.

♦ Type 3, degenerate conic sections

22.[A] Describe the degenerate conic section $x^2 + \frac{1}{4}y^2 - 2x - y + 2 = 0$.

♦ Type 4, a degenerate conic section

23.[A] The orbit of the earth is an ellipse of eccentricity 0.0167 with the sun at one focus, and with a major axis of length 185.8 million miles. What are **(a)** the earth's APHELION (farthest distance from the sun) and **(b)** the earth's PERIHELION (closest distance from the sun)?

♦ Type 3, eccentricity

24. The eccentricity 0.2486 of Pluto's orbit is the greatest of the orbits of all of our sun's planets. What is the ratio of the lengths of the major and minor axes of Pluto's orbit?

♦ Type 3, eccentricity

25.[A] The Russian satellite Lunik 3 had an APOGEE (farthest distance from the surface of the earth) of 250,000 miles and a PERIGEE (closest distance to the surface of the earth) of 25,000 miles. What was the eccentricity of the ellpitical orbit of Lunik 3? (Use 4000 miles as the radius of the earth.)

♦ Type 3, eccentricity

26. The U. S. satellite Explorer 6 had eccentricity 0.75 and apogee 26,400 miles. What was its perigee? (See Problem 25.)

♦ Type 3, eccentricity

27. The asteroid Icarus (Figure 17) has a perihelion of 17 million miles (inside the orbit of Mercury) and an aphelion of 183 million miles (outside the orbit of Mars). What is the eccentricity of its elliptical orbit? (See Problem 23.)

♦ Type 3, eccentricity

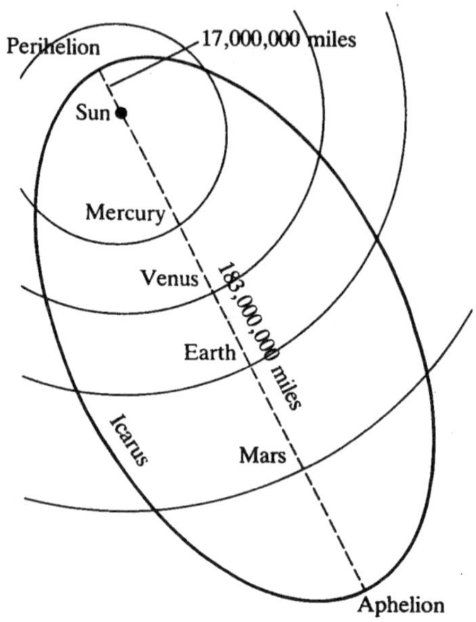

FIGURE 17

In Problems 28 through 31 sketch the ellipses and hyperbolas with their foci and directrices.

28. $\frac{1}{9}x^2 + \frac{1}{4}y^2 = 1$

♦ Type 3, finding foci and directrices

29. $x^2 - y^2 = 16$

♦ Type 3, finding foci and directrices

30. $\frac{1}{25}x^2 + \frac{1}{169}y^2 = 1$

♦ Type 3, finding foci and directrices

31. $\frac{1}{16}x^2 - \frac{1}{9}y^2 = 1$

♦ Type 3, finding foci and directrices

Sketch the conic sections in Problems 32 through 39. Plot the vertices and foci on the x'- and/or y'-axes without finding their xy-coordinates.

32. $x^2 + 2xy + y^2 = 4\sqrt{2}\,y$

♦ Type 3, drawing a conic section by rotating axes

33. $41x^2 - 24xy + 9y^2 = 45$

♦ Type 3, drawing a conic section by rotating axes

34. $32x^2 - 60xy + 7y^2 = 52$

♦ Type 3, drawing a conic section by rotating axes

35. $4x^2 - 4xy + y^2 - \sqrt{5}\,x - 2\sqrt{5}\,y = 0$

♦ Type 3, drawing a conic section by rotating axes

36. $7x^2 - 6\sqrt{3}\,xy + 13y^2 = 16$

♦ Type 3, drawing a conic section by rotating axes

37. $x^2 - 2\sqrt{3}\,xy - y^2 = 8$

♦ Type 3, drawing a conic section by rotating axes

38. $4x^2 - 6xy - 4y^2 = 5$

♦ Type 3, drawing a conic section by rotating axes

39. $-7x^2 + 4\sqrt{8}\,xy + 7y^2 = 81$

♦ Type 3, drawing a conic section by rotating axes

40. Find an equation for the hyperbola $4xy - 3x^2 = 20$ in rotated $x'y'$-coordinates such that the new equation has no $x'y'$-term.

♦ Type 3, rotating coordinates

41.[A] Use the fact that the distance from (x, y) to the line $y = x$ is $\dfrac{|y - x|}{\sqrt{2}}$ to find an equation in the form $Ax^2 + Bxy + Cy^2 + Dx + Ey + F = 0$ for the parabola with focus $(0, 3)$ and directrix $y = x$.

♦ Type 4, finding an equation of a conic section

42. Use the fact that the distance from (x, y) to the line $y = 2x$ is $\dfrac{|y - 2x|}{\sqrt{5}}$ to find an equation in the form $Ax^2 + Bxy + Cy^2 + Dx + Ey + F = 0$ for the parabola with focus $(0, -4)$ and directrix $y = 2x$.

♦ Type 4, finding an equation of a conic section

43.[A] Give an equation in the form $Ax^2 + Bxy + Cy^2 + Dx + Ey + F = 0$ for the locus of points such that the sum of the distances from $(1, 0)$ and $(0, 1)$ is 6.

♦ Type 4, finding an equation of a conic section

44. Give an equation in the form $Ax^2 + Bxy + Cy^2 + Dx + Ey + F = 0$ for the locus of points that are either 6 units closer to $(3, -4)$ than to $(-3, 4)$ or are 6 units closer to $(-3, 4)$ than to $(3, -4)$.

♦ Type 4, finding an equation of a conic section

45. Prove Theorem 2 in the case of a hyperbola, for which $0 \le \alpha < \beta$.

♦ Type 7, a proof

46. Explain why the discriminant of an equation $Ax^2 + Bxy + Cy^2 + Dx + Ey + F = 0$ is not changed when the coordinates are translated.

♦ Type 7, a proof

47. What is the greatest eccentricity of the cross sections of a right circular cone whose elements make an positive acute angle β with its axis?

♦ Type 6, eccentricity

48. The curve $x^2 + \dfrac{y^2}{1-k} = 1$ is an ellipse for $k < 1$ and a hyperbola for $k > 1$. What happens to its foci and directrices as k crosses 0? What happens to them when k crosses 1?

♦ Type 6, dependence offoci and directrices on a parameter

49. A conic section has focus $(0, 0)$ corresponding to the directrix $x = k$ with $k > 0$. What happens to the other focus as k increases through all positive values?

♦ Type 6, dependence offoci and directrices on a parameter

50. Descartes' first work using analytic geometry was in the solution of problems such as the following posed by Pappus. Let PA, PB, and PC be the line segments from a point P to three given lines L_1, L_2, L_3 that make prescribed angles with the lines. Find the locus of points P such that $\overline{PA}\,\overline{PB} = \overline{PC}^2$. Find the locus where the three lines are the x-axis, the y-axis, and the line $x = 1$, and the three angles are right angles.

♦ Type 6, finding a conic section

51. In Fermat's first treatise on analytic geometry *Ad Locos Planoa et Solidos Isagoge*, he solved the following problem. Given any number of lines and an angle for each line, find the locus of points P such that the sum of the squares of the lengths of the line segments from P to the given lines at the given angles is a given constant. Explain why in each case the locus is a conic section.

♦ Type 6, finding a conic section

Section 9.3

Polar Coordinates

OVERVIEW: *In this section we discuss* POLAR COORDINATES, *which describe points by giving their distances and directions from the origin in a coordinate plane. We also look at equations of curves in polar coordinates and derive polar equations for conic sections. Then we show how areas of regions can be calculated with polar coordinates.*

Topics:

- ***Polar coordinates***
- ***Curves in polar coordinates***
- ***Polar equations of conic sections***
- ***Finding areas in polar coordinates***

Polar coordinates

Many problems are simplified if the points in a plane are located by giving as their POLAR COORDINATES their distances and directions from a fixed point or "pole." The motion of a planet around the sun, for example, is studied by using polar coordinates with the fixed point the center of the sun (Figure 1). The planet's position is described by giving its distance r to the origin and an angle θ as in Figure 1 because then, by Newton's law of gravity, the sun's force on the planet is equal to a constant divided by r^2.

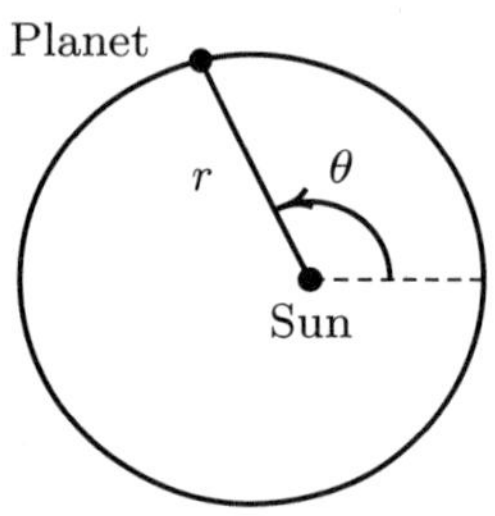

FIGURE 1

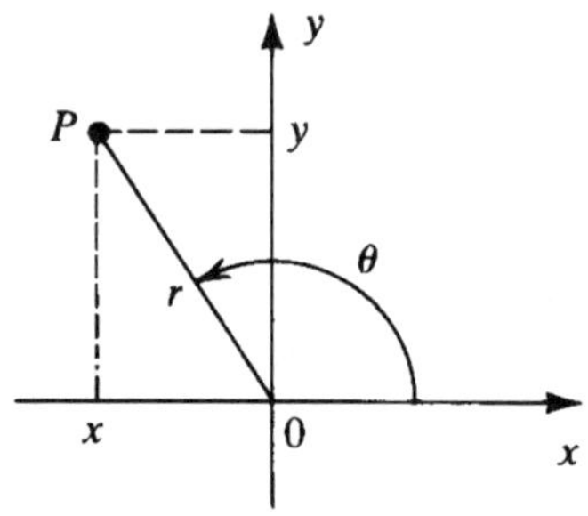

$P = [r, \theta]$

FIGURE 2

To find polar coordinates of a point P other than the origin in an xy-plane with equal scales on the axes, we draw a line segment from the point to the origin, as in Figure 2, measure the distance r from P to the origin, and let θ be an angle from the positive x-axis to the line segment. This gives one pair $[r, \theta]$ of polar coordinates of the point.[†] The point also has polar coordinates $[r, \theta + 2n\pi]$ for any integer n because the angle $\theta + 2n\pi$ has the same terminal side as the angle θ for any integer n. The origin has polar coordinates $[0, \theta]$ with any angle θ.

In certain contexts, such as the discussion of polar equations of curves, we allow negative values of r. To locate the point $[r, \theta]$ where r may be positive, negative, or zero, we draw an r-axis with its origin at the origin in the xy-plane and with its positive side at an angle θ from the positive x-axis, as in Figure 3. The point at r on this r-axis has polar coordinates $[r, \theta]$ or $[r, \theta + 2n\pi]$ with any integer n.

[†]Square brackets are used for polar coordinates to distinguish them from rectangular coordinates (x, y).

♦ SUGGESTIONS TO INSTRUCTORS: Have students begin work on Questions 1 through 4 before class. Problems 17, 20, 25, 28, 29, 34, and 42 are good for in-class discussion.

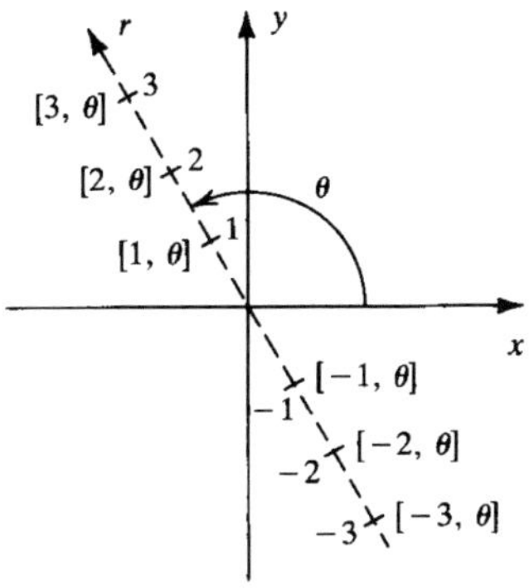

FIGURE 3

If a point other than the origin has polar coordinates $[r, \theta]$, then, by definition of the cosine and sine, its rectangular coordinates (x, y) satisfy $\cos\theta = x/r$ and $\sin\theta = y/r$, so that

$$x = r\cos\theta, \quad y = r\sin\theta. \tag{1}$$

These formulas are used to transform polar coordinates to rectangular coordinates.

Example 1 Plot the point with polar coordinates $[5, \frac{2}{3}\pi]$. Then find its xy-coordinates.

SOLUTION Because the value 5 of r in the polar coordinantes is positive, we draw the terminal side of an angle of $\frac{2}{3}\pi$, as in Figure 4 and plot the point a distance 5 from the origin on that line. By formulas **(1)** the xy-coordinates of the point are $x = 5\cos\left(\frac{2}{3}\pi\right) = -\frac{5}{2}$ and $y = 5\sin\left(\frac{2}{3}\pi\right) = \frac{5}{2}\sqrt{3}$. □

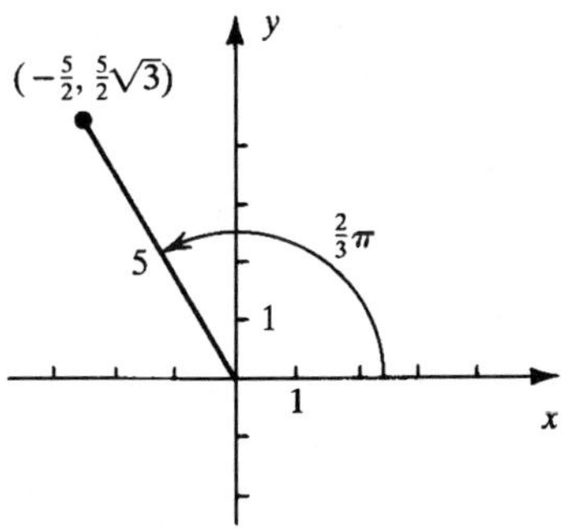

The point with polar coordinates $[5, \frac{2}{3}\pi]$

FIGURE 4

If a point has rectangular coordinates (x, y), then it has polar coordinates $[r, \theta]$ and $[-r, \theta]$ with

$$r = \sqrt{x^2 + y^2} \tag{2}$$

the point's distance to the origin and suitable angles θ. In most cases the suitable angles can be found from a drawing that shows the point in an xy-plane.

Example 2 Give all possible polar coordinates of the point with rectangular coordinates $(-4, 4)$.

SOLUTION We plot the point and then draw an r-axis with its origin at the origin of the xy-plane and its positive side passing through the point, as in Figure 5. The distance from the point to the origin is $\sqrt{4^2 + 4^2} = 4\sqrt{2}$ and the r-axis is at an angle $\frac{3}{4}\pi$ radians from the positive x- axis, so the point has polar coordinates $[4\sqrt{2}, \frac{3}{4}\pi]$. Then we direct the r-axis in the opposite direction, as in Figure 6, to see that the point also has polar coordinates $[-4\sqrt{2}, \frac{7}{4}\pi]$. All possible coordinates of the point are given by $[4\sqrt{2}, \frac{3}{4}\pi + 2n\pi]$ and $[-4\sqrt{2}, \frac{7}{4}\pi + 2n\pi]$ with integers n. □

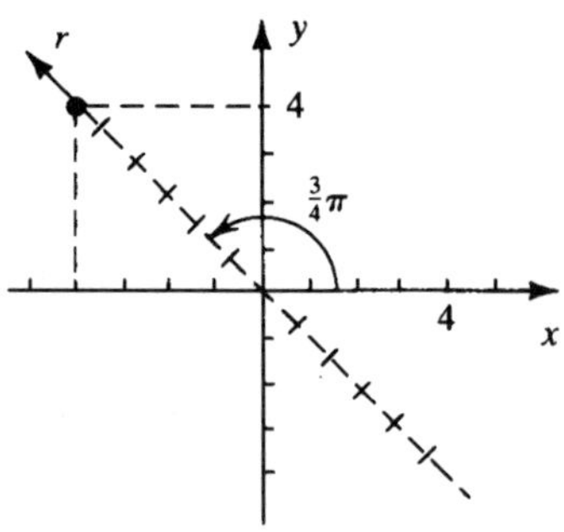

$r = 4\sqrt{2}$, $\theta = \frac{3}{4}\pi$

FIGURE 5

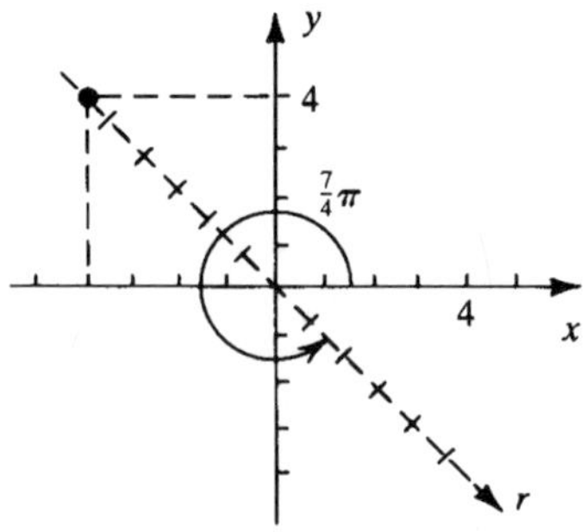

$r = -4\sqrt{2}$, $\theta = \frac{7}{4}$

FIGURE 6

Question 1 Use the inverse tangent function to give exact expressions for all polar coordinates $[r, \theta]$ of $(7, 5)$.

Polar equations of curves

The GRAPH of an equation in polar coordinates is the set of points such that at least one of its possible polar coordinates satisfies the equation.

In many cases, not all possible polar coordinates of a point on a curve satisfy a particular polar equation for the curve. The circle $x^2 + y^2 = 1$ in Figure 7, for instance, has the equation $r = 1$. The points on the circle have polar coordinates $[1, \theta]$, which satisfy this equation, and polar coordinates $[-1, \theta]$ which do not.

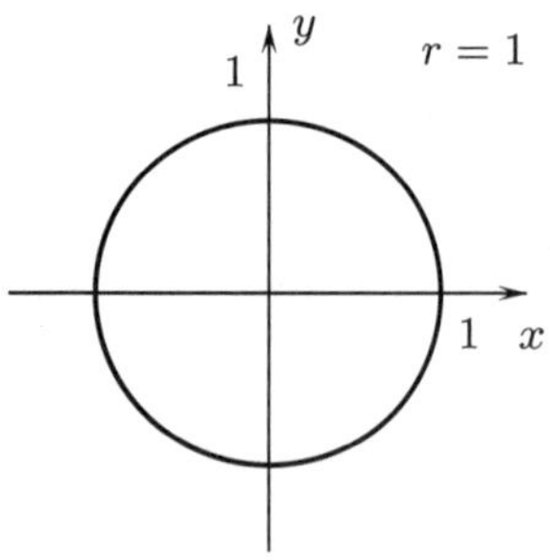

$x^2 + y^2 = 1$

FIGURE 7

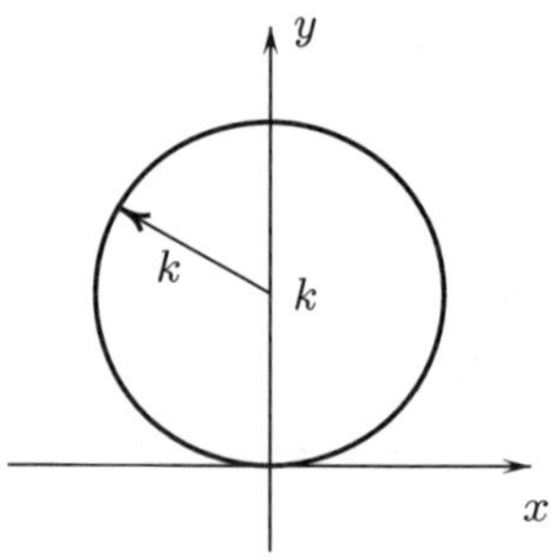

$x^2 + (y - k)^2 = k^2$

FIGURE 8

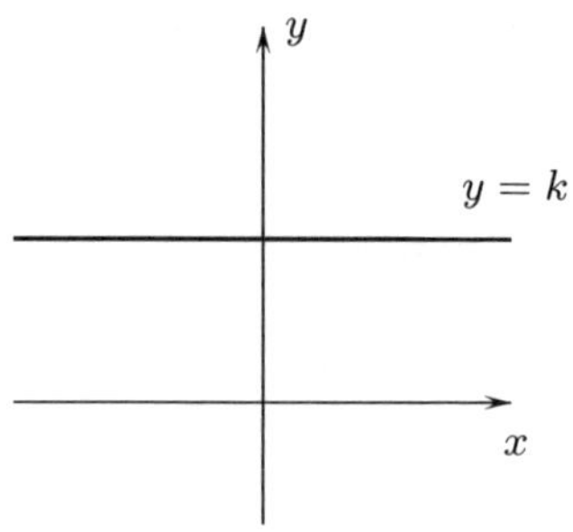

$y = k$

FIGURE 9

Example 3 The circle of radius $k > 0$ with its center at $y = k$ on the y-axis in Figure 8 is the set of points a distance k from $(0, k)$ and, consequently, has the equation $x^2 + (y - k)^2 = k^2$ in xy-coordinates. Use formulas **(1)** to convert this to an equation for the circle in polar coordinates.

SOLUTION **(a)** Substituting formulas **(1)** into $x^2 + (y - k)^2 = k$ gives $(r\cos\theta)^2 + (r\sin\theta - k)^2 = k^2$, which expands to $r^2\cos^2\theta + r^2\sin^2\theta - 2rk\sin\theta + k^2 = k^2$. Subtracting k^2 from both sides and using the Pythagorean Identity $\cos^2\theta + \sin^2\theta = 1$ simplifies this equation to $r^2 = 2rk\sin\theta$. We can divide both sides by r for $r \neq 0$ to obtain the equation $r = 2k\sin\theta$. Since $2k\sin\theta = 0$ for $\theta = 0$, we did not lose any solutions by dividing by r and the polar equation is $r = 2k\sin\theta$. □

Example 4 Give a polar equation for the horizontal line $y = k$ in Figure 9.

SOLUTION Substituting $y = r\sin\theta$ in the equation $y = k$ gives $r\sin\theta = k$, which we rewrite as $r = k\csc\theta$. □.

Question 2 Find polar equations for **(a)** the circle $(x-k)^2+y^2=k^2$ in Figure 10 and **(b)** the vertical line $x=k$ in Figure 11.

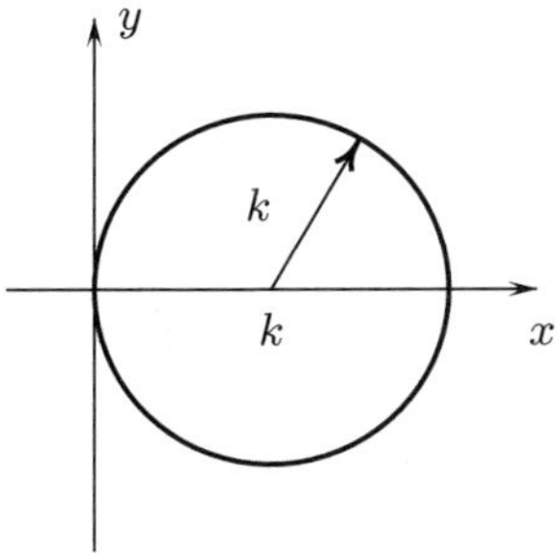

$(x-k)^2+y^2=k^2$

FIGURE 10

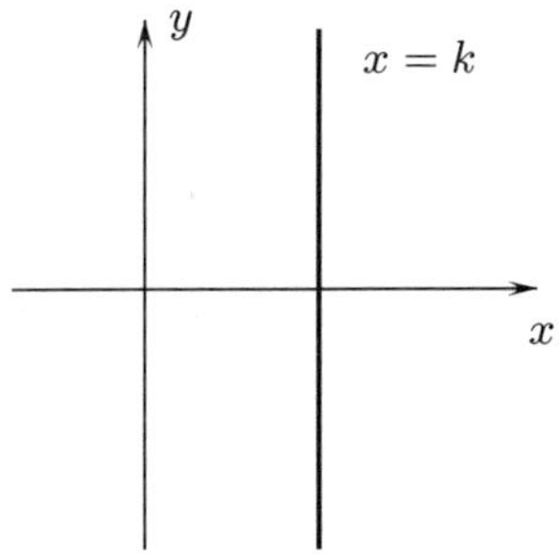

$x=k$

FIGURE 11

In determining the shape of a curve that is given by an equation in polar coordinates, it often helps to first study the graph of the equation in an a θr-plane.

Example 5 Generate in an xy-plane the curve with the polar equation $r=3\cos(2\theta)$.† Then use the equation to explain the shape of the curve.

SOLUTION Figure 12 shows the curve $r=3\cos(2\theta)$ in an xy-plane. To explain its shape, we generate the graph of $r=3\cos(2\theta)$ in a θr-plane as in Figure 13. The shape of the latter curve is explained by the folowing table, where values of θ have been chosen to cover a full revolution $0\le\theta\le 2\pi$ and to include the values where $3\cos(2\theta)$ is zero and has its greatest and least values.

2θ	0	$\frac{1}{2}\pi$	π	$\frac{3}{2}\pi$	2π	$\frac{5}{2}\pi$	3π	$\frac{7}{2}\pi$	4π
θ	0	$\frac{1}{4}\pi$	$\frac{1}{2}\pi$	$\frac{3}{4}\pi$	π	$\frac{5}{4}\pi$	$\frac{3}{2}\pi$	$\frac{7}{4}\pi$	2π
$r=3\cos(2\theta)$	3	0	-3	0	3	0	-3	0	3

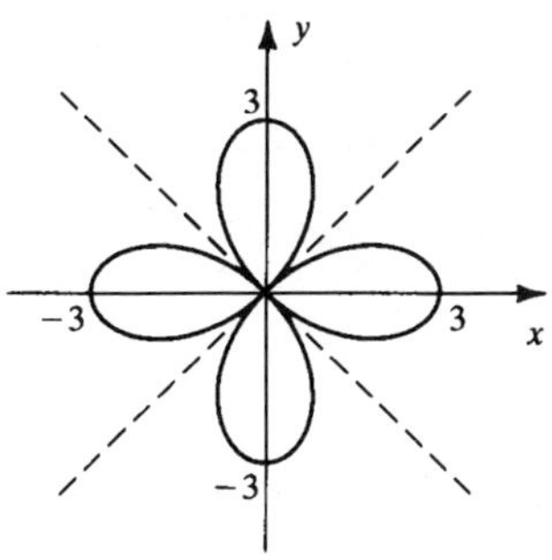

$r=3\cos(2\theta$

FIGURE 12

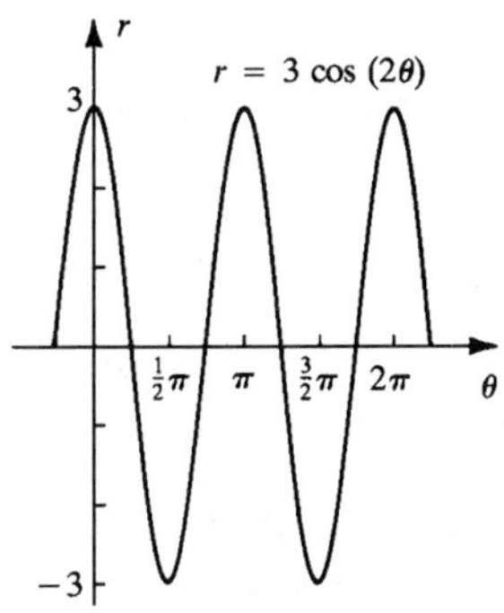

FIGURE 13

†Your calculator or computer software may have a polar-coordinate mode that enables you to enter the formula $r=3\cos(2\theta)$ directly. If not, use the equations $x=3\cos(2t)\cos t$, $y=3\cos(2t)\sin t$ in the algorithm for generating parametric curves.

Now imagine that the curve in Figure 12 is drawn by a marker that moves up and down an r-axis as the r-axis rotates counterclockwise about the origin. The marker is at $3\cos(2\theta)$ when $\theta = 0$. As θ increases from 0 to $\frac{1}{4}\pi$, $r = 3\cos(2\theta)$ decreases from 3 to 0, and the marker draws the portion of the curve in Figure 14 that goes from $r = 3, \theta = 0$ on the x-axis to the origin..

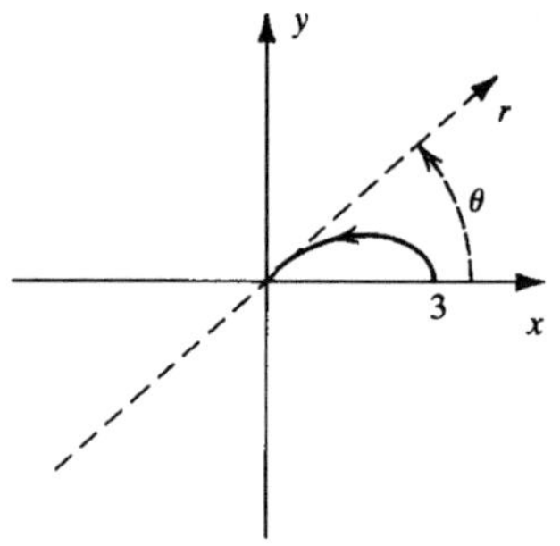

$0 \le \theta \le \frac{1}{4}\pi$, $r > 0$

FIGURE 14

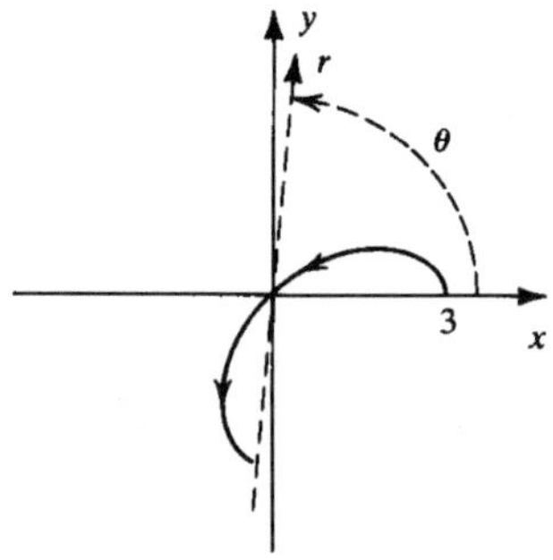

$\frac{1}{4}\pi \le \theta \le \frac{1}{2}\pi$, $r < 0$

FIGURE 15

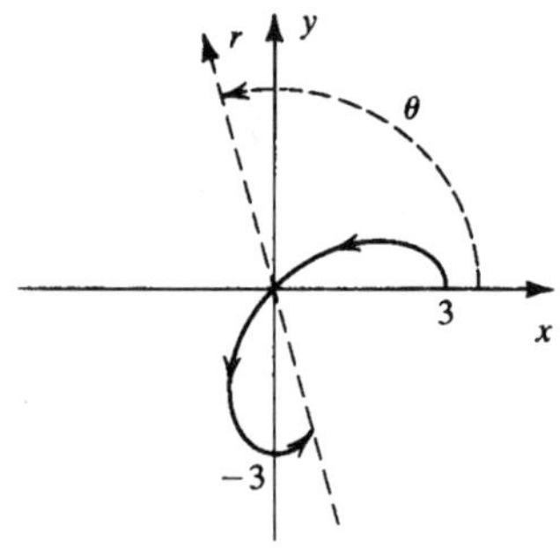

$\frac{1}{2}\pi \le \theta \le \frac{3}{4}\pi$, $r < 0$

FIGURE 16

Next, as θ increases from $\frac{1}{4}\pi$ to $\frac{1}{2}\pi$, the value $3\cos(2\theta)$ of r decreases from 0 to -3 , and the marker draws the portion of the curve in Figure 15 that goes from the origin to the point $r = -3, \theta = \frac{1}{2}\pi$ on the negative y-axis. Figures 17 and 18 shows how the next portions of the curve are generated. The graph in Figure 12 is competed as θ increases to 2π. This curve is often called a FOUR-PETALED ROSE. □

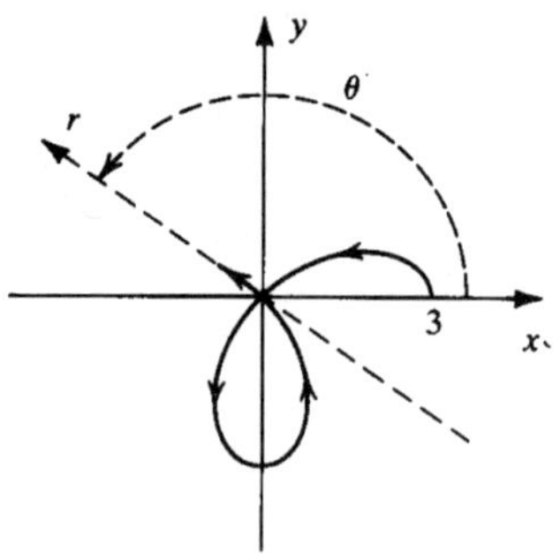

$\frac{3}{4}\pi \le \theta \le \pi$, $r > 0$

FIGURE 17

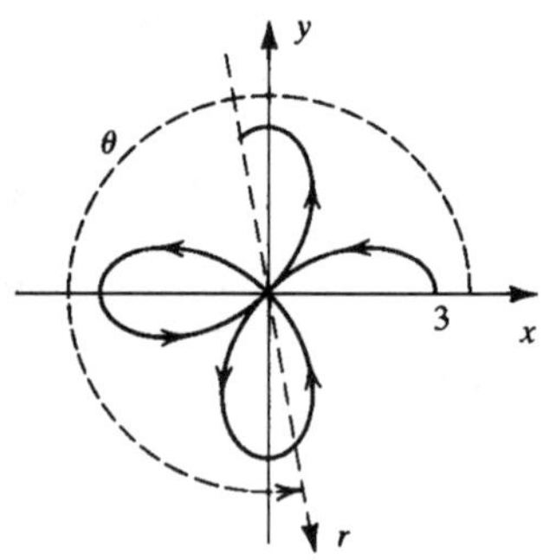

$\pi \le \theta \le \frac{7}{4}\pi$, $r > 0$

FIGURE 18

Question 3 Generate in an xy-plane the CARDIOID with the polar equation $r = 1 + \cos\theta$. Then use the formula to explain the shape of the curve.

Polar equations of conic sections

In Theorem 2 of Section 9.2 we saw that every noncircular conic section has an equation of the form

$$\overline{PF} = e\overline{PD} \qquad \textbf{(3)}$$

where P is a point on the curve, F is its focus, and D is the closest point from P to the directrix of the curve. If we put the focus at the origin in an xy-plane and have the directrix be the line $x = k$, as in Figure 19, then the x-axis is the axis of the conic section, and $\overline{PF} = r, \overline{PD} = k - r\cos\theta$, so that **(3)** reads $r = e(k - r\cos\theta)$. Solving for r yields the following polar equation of the conic section:

$$r = \frac{ke}{1 + e\cos\theta} \qquad \textbf{(4)}$$

FIGURE 19

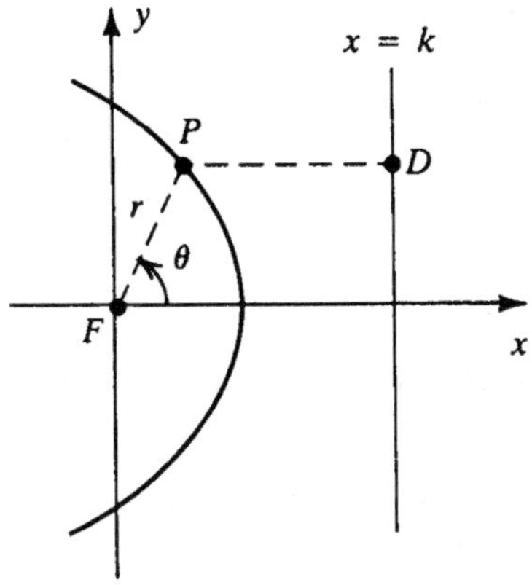

We will need formula **(4)** when we study Kepler's laws of planetary motion in Chapter 10.

Areas in polar coordinates

Imagine that a region is bounded by lines from the origin at $\theta = \alpha$ and $\theta = \beta$ and by the curve with the polar equation $r = r(\theta)$, where $r(\theta)$ is positive for $\alpha \le \theta \le \beta$, as in Figure 20. To find the area of the region, we use a partition

$$\alpha = \theta_0 < \theta_1 < \theta_2 < \cdots < \theta_N = \beta$$

of the interval $\alpha \le \theta \le \beta$ into N subintervals of equal width $\Delta\theta = (\beta - \alpha)/N$.

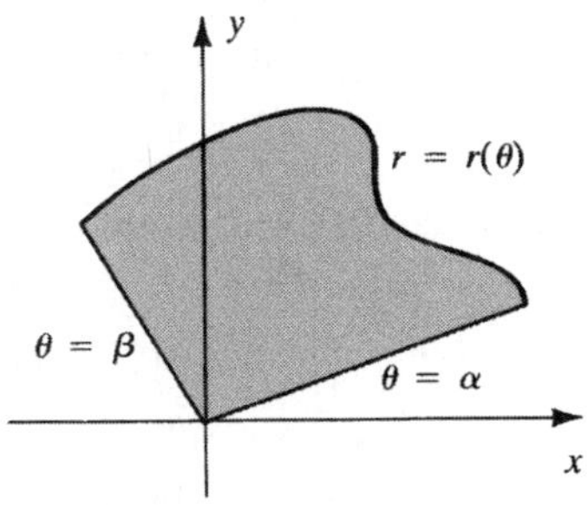

FIGURE 20

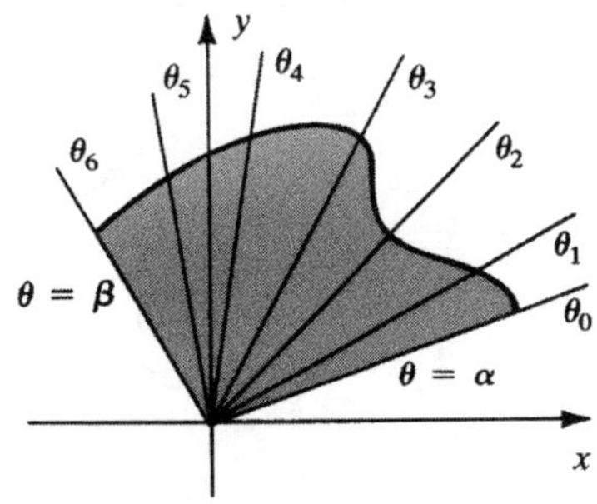

FIGURE 21

The lines from the origin at the angles θ_j slice the region into N pieces, as shown in Figure 21 for $N = 6$. We approximate the jth piece by the segment of the circle of radius $r(\theta_j)$ for $\theta_{j-1} \le \theta \le \theta_j$ (Figure 22). Because the angle in the segment is $\Delta\theta$, the segment is the fraction $\Delta\theta/(2\pi)$ of the entire circle and, consequently, has area

$$\left(\frac{\Delta\theta}{2\pi}\right)\left(\pi[r(\theta_j)]^2\right) = \tfrac{1}{2}[r(\theta_j)]^2\Delta\theta.$$

FIGURE 22

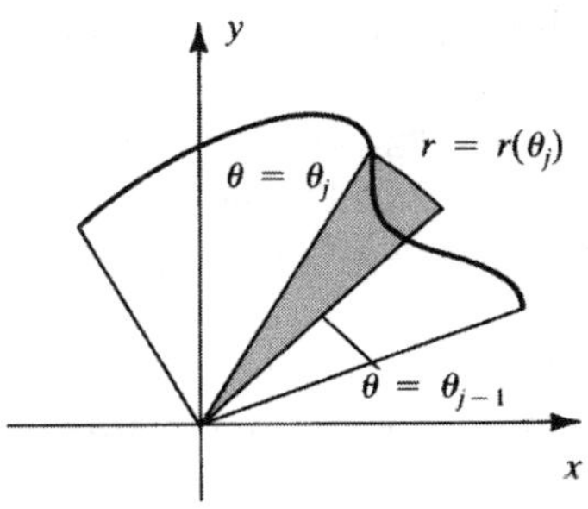

The total area of all N pieces is

$$\sum_{j=1}^{N} \tfrac{1}{2}[r(\theta_j)]^2 \Delta\theta$$

which is a Riemann sum for the integral $\int_{\alpha}^{\beta} \frac{1}{2}[r(\theta)]^2\, d\theta$ in the case of positive $r(\theta)$. This leads us to the following rule, which also applies if $r(\theta)$ has zero or negative values.

Rule 1 *If a region is bounded by the curve with polar equation $r = r(\theta)$ and by the lines from the origin at angles $\theta = \alpha$ and $\theta = \beta$ with $0 \le \beta - \alpha \le 2\pi$, then the area of the region is*

$$\int_{\alpha}^{\beta} \tfrac{1}{2}[r(\theta)]^2\, d\theta. \qquad \textbf{(5)}$$

Because of the square in formula **(5)**, areas of regions bounded by curves with polar equations involving $\sin\theta$ and $\cos\theta$ are frequently given by integrals involving $\sin^2\theta$ or $\cos^2\theta$. As was explained in Section 6.3, such integrals can often be evaluated with the double-angle formulas,

$$\sin^2\theta = \tfrac{1}{2}[1 - \cos(2\theta)] \qquad \textbf{(6)}$$

$$\cos^2\theta = \tfrac{1}{2}[1 + \cos(2\theta)]. \qquad \textbf{(7)}$$

Example 6 Find the area of the region bounded by the cardioid from Question 3 with polar equation $r = 1 + \cos\theta$.

SOLUTION Since the curve is traversed once as θ runs from 0 to 2π, it bounds a region of area

$$\begin{aligned}\int_0^{2\pi} \tfrac{1}{2}(1+\cos\theta)^2\, d\theta &= \int_0^{2\pi} \tfrac{1}{2}(1 + 2\cos\theta + \cos^2\theta)\, d\theta \\ &= \int_0^{2\pi} [\tfrac{1}{2} + \cos\theta + \tfrac{1}{4}(1+\cos(2\theta)]\, d\theta \\ &= \int_0^{2\pi} [\tfrac{3}{4} + \cos\theta + \tfrac{1}{4}\cos(2\theta)]\, d\theta \\ &= \left[\tfrac{3}{4}\theta + \sin\theta + \tfrac{1}{8}\sin(2\theta)\right]_0^{2\pi} = \tfrac{3}{2}\pi. \;\square\end{aligned}$$

Question 4 Find the area of the region inside one petal of the four-petaled rose $r = 3\cos(2\theta)$ of Example 5.

Principles and procedures

- Rectangular coordinates (x, y) can be calculated from polar coordinates with the formulas $x = r\cos\theta, y = r\sin\theta$.
- Polar coordinates $[r, \theta]$ of a point with $r > 0$ can be found from its rectangular coordinates (x, y) by plotting the point in an xy-plane, calculating its distance r to the origin, and determining an angle θ from the positive x-axis to the line segment from the origin to the point, using the inverse tangent function as necessary. The other possible polar coordinates of the point are $[r,\ \theta + 2n\pi]$ and $[-r,\ \theta + (2n+1)\pi]$ with integers n. The origin $(0, 0)$ has polar coordinates $[0, \theta]$ with any angle θ.
- In most applications, only one pair of values of r and θ are needed, but multiple values must be considered in determining the graphs in xy-planes of equations given in polar equations.

- To recall the formula for calculating areas in polar coordinates, find the area of a sector of a circle of radius r that is subtended by an angle $\Delta\theta$ at the center of the circle. The sector is the fraction $\dfrac{\Delta\theta}{2\pi}$ of the entire circle and, consequently, has area $\dfrac{\Delta\theta}{2\pi}(\pi r^2) = \frac{1}{2}r^2\Delta\theta$. This leads to the expression $\frac{1}{2}r^2 d\theta$ in the integral $\displaystyle\int_\alpha^\beta \frac{1}{2}r^2\, d\theta$ for area in Rule 1.
- To remember which of the double-angle identities **(6)** and **(7)** has the plus sign and which has the minus sign, use the facts that $\sin(0) = 0$ and $\cos(0) = 1$ to check the formulas at $\theta = 0$.

Tune-Up Exercises 9.3♦

[A]*Answer provided.* [O]*Outline of solution provided.*

Use equal scales on the xy-axes in all of the Tune-up Exercises and Problems of this section.

T1.[O] Plot the point with polar coordinates $[-2, \frac{3}{4}\pi]$ and give its rectangular coordinates.

♦ Type 1, polar coordinates

T2.[O] Give all possible polar coordinates of the point with rectangular coordinates $(-5\sqrt{3}, -5)$.

♦ Type 1, polar coordinates

T3.[O] Sketch the curve in an xy-plane with polar equation $r = -3\sec\theta$.

♦ Type 1, a curve in polar coordinates

T4.[O] Generate the LIMAÇON in an xy-plane with polar equation $r = 1 - 2\sin\theta$ and copy it on your paper. Then use the equation to explain its shape.

♦ Type 1, a curve in polar coordinates

T5.[O] **(a)** Find an equation in rectangular coordinates for the curve with the polar equation $r = 4\cos\theta$.
(b) Draw the curve from part (a) and use the polar equation to calculate its area.

♦ Type 2, a curve in polar coordinates

Problems 9.3

[A]*Answer provided.* [O]*Outline of solution provided.*

In Problems 1 through 6 give rectangular coordinates of the points with the given polar coordinates. Plot the points in Problems 1 through 6. (Use equal scales on the xy-axes in all of the Problems of this section.)

1.[O] $[3, \frac{3}{4}\pi]$

♦ Type 1, polar coordinates

2.[O] $[2, \pi]$

♦ Type 1, polar coordinates

3.[A] $[-\frac{1}{2}, \frac{1}{6}\pi]$

♦ Type 1, polar coordinates

4 $[0, \pi]$

♦ Type 1, polar coordinates

5.[A] $[-2, \frac{15}{6}\pi]$

♦ Type 1, polar coordinates

6. $[\pi, 0]$

♦ Type 1, polar coordinates

In Problems 7 through 12 give all possible polar coordinates of the points with the given rectangular coordinates.

7.[O] $(-3, -3)$

♦ Type 1, polar coordinates

8.[A] $(0, -4)$

♦ Type 1, polar coordinates

9. $(7, 0)$

♦ Type 1, polar coordinates

10.[A] $(-1, \sqrt{3})$

♦ Type 1, polar coordinates

11.[A] $(3, 4)$

♦ Type 1, polar coordinates

12.[A] $(-3, -6)$

♦ Type 1, polar coordinates

♦The Tune-up Exercises and Problems are classified by type and content. The types are (1) basic, reactive; (2) basic reflective; (3) intermediate, reactive; (4) intermediate, reflective; (5) advanced, reactive; (6) advanced, reflective; and (7) advanced, theoretical.

Give equations in polar coordinates for the curves in Problems 13 through 16.

13.[A] $x = 2$

♦ Type 1, a curve in polar coordinates

14. $y = 3$

♦ Type 1, a curve in polar coordinates

15.[A] $x = 3y$

♦ Type 1, a curve in polar coordinates

16.[A] $x^2 = y^2$

♦ Type 1, a curve in polar coordinates

Generate the curves with the polar equations in Problems 17 through 24 and copy them on your paper. Then use their equations to explain their shapes.

17. $r = -1 - 3\cos\theta$

♦ Type 2, a curve in polar coordinates

18. $r = 4\sin(3\theta)$

♦ Type 2, a curve in polar coordinates

19. $r = 4 - 4\cos\theta$

♦ Type 2, a curve in polar coordinates

20. $r = \sin(2\theta)$

♦ Type 2, a curve in polar coordinates

21. $r = 5 + 3\cos\theta$

♦ Type 2, a curve in polar coordinates

22. $r = 5$

♦ Type 2, a curve in polar coordinates

23. $r^2 - 5r + 4 = 0$

♦ Type 2, a curve in polar coordinates

24. $\theta = 1$

♦ Type 2, a curve in polar coordinates

25. Generate the curve with the polar equation $r = \dfrac{1}{\sqrt{1+\cos^2\theta}}$ and copy it on your paper. Then show the curve is an ellipse by changing the equation to rectangular coordinates.

♦ Type 4, a curve in polar coordinates

26. Generate the curve with the polar equation $r = \dfrac{1}{\sqrt{\cos^2\theta - \sin^2\theta}}$. Then show the curve is a hyperbola by changing the equation to rectangular coordinates.

♦ Type 4, a curve in polar coordinates

27. **(a)** Give an equation of the hyperbola $r = \sqrt{\sec\theta\csc\theta}$ in rectangular coordinates. **(b)** Find numbers a, b, c, and d such that the portion of the hyperbola with positive x and y is traversed for $a < \theta < b$ and the portion with negative x and y is traversed for $c < \theta < d$.

♦ Type 4, a curve in polar coordinates

28. Describe the curve with polar equation $r = |\cos\theta|$.

♦ Type 4, a curve in polar coordinates

29.[A] Find the area of the region bounded by the LIMAÇON $r = 5 + 4\sin\theta$.

♦ Type 3, area in polar coordinates

30.[O] Find the area of one petal of the THREE-PETALED ROSE $r = 2\cos(3\theta)$.

♦ Type 3, area in polar coordinates

31.[A] Find the area of the region bounded by the x- and y-axes and by the portion of the LOGARITHMIC SPIRAL $r = e^{\theta}$ for $0 \le \theta \le \frac{1}{2}\pi$.

♦ Type 3, area in polar coordinates

32. Describe how the HYPERBOLIC SPIRAL $r = 2\pi/\theta$, defined for $\theta > 0$, the ARCHIMEDIAN SPIRAL $r = \theta/(2\pi)$, defined for $\theta \ge 0$, and the LOGARITHMIC SPIRAL $r = e^{\theta/(2\pi)}$, defined for $\theta \ge 0$, differ in shape.

♦ Type 2, a curve in polar coordinates

33. How does the shape of the LEMNISCATE OF BERNOUILLI $r^2 = a^2\cos(2\theta)$ depend on the value of the positive constant a?

♦ Type 2, a curve in polar coordinates

34. Find the values of θ_0 such that $\lim_{\theta\to\theta_0} |r| = \infty$ for the conic sections **(a)** $r = \dfrac{1}{1+\cos\theta}$, **(b)** $r = \dfrac{1}{1+\sqrt{2}\cos\theta}$, and **(c)** $r = \dfrac{1}{1+\frac{1}{2}\cos\theta}$. **(d)** Which of the curves in parts (a), (b), and (c) is a parabola, which is an ellipse, and which is a hyperbola? How do these facts relate to the answers of parts (a), (b), and (c)?

♦ Type 4, asymptotes of conic sections

35. Show that the curve with polar equation $r = \dfrac{ke}{1 + e\cos\theta}$ with a positive constant k is an ellipse if $0 \le e < 1$, a parabola if $e = 1$, and a hyperbola if $e > 1$ by transforming the equation to rectangular coordinates and calculating its discriminant.

♦ Type 4, a curve in polar coordinates

36. Figure 23 shows the CONCHOID OF NICOMEDES, which has the polar equation $r = 2 + \sec\theta$. Find its vertical asymptote by switching the equation to rectangular coordinates and solving for $\sqrt{x^2 + y^2}$.

♦ Type 4, a curve in polar coordinates

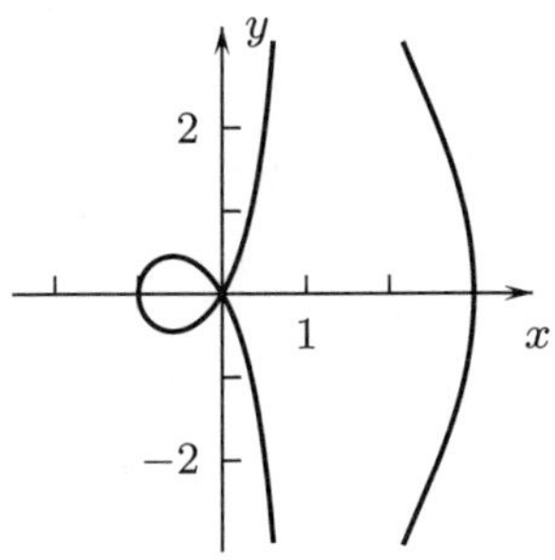

FIGURE 23

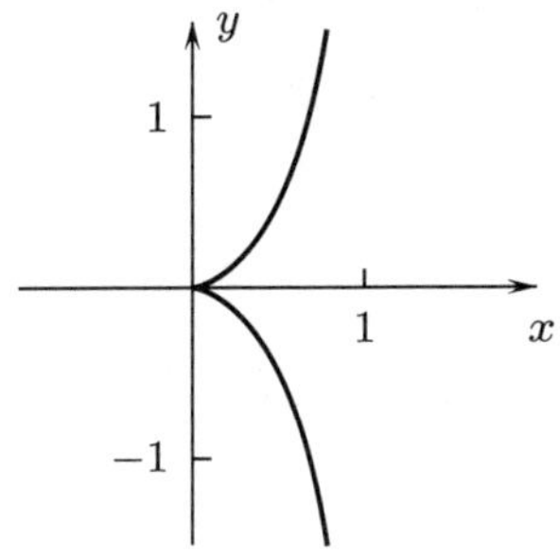

FIGURE 24

37. The CISSOID OF DIOCLES in Figure 24 has the polar equation $r = \sin\theta\tan\theta$. Show that it extends only for $0 \le x < 1$ and that it has $x = 1$ as a vertical asymptote by switching the equation to rectangular coordinates and solving for y^2.

♦ Type 4, a curve in polar coordinates

38. Figure 25 shows the graph of $r = \sin\left(\frac{1}{3}\theta\right)$. What is the smallest value of b such that the entire curve is traversed for $0 \le \theta \le b$?

♦ Type 4, a curve in polar coordinates

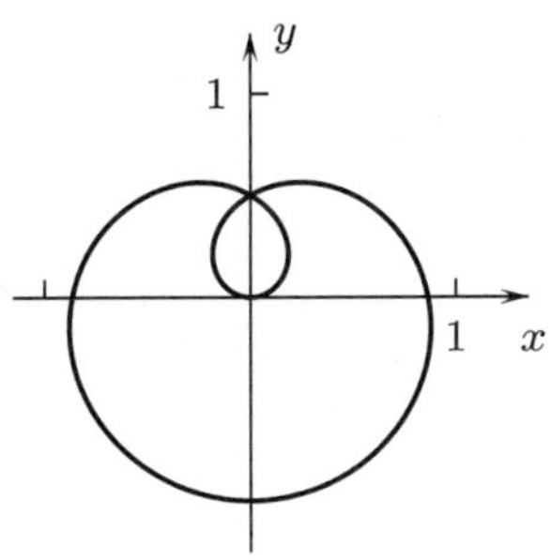

FIGURE 25

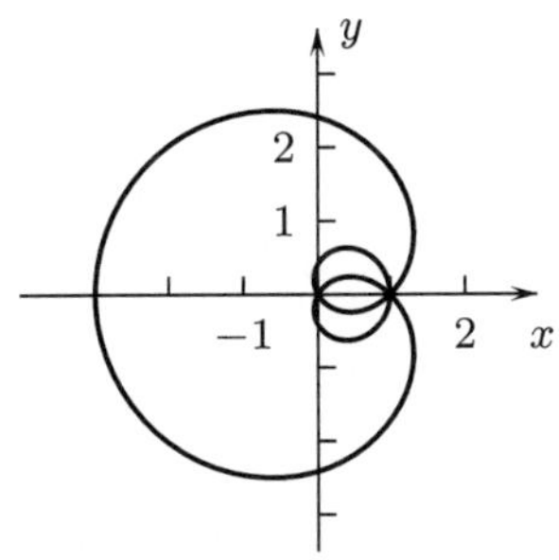

FIGURE 26

39. **(a)** Which part of FREETH'S NEPHROID $r = 1 + 2\sin\left(\frac{1}{2}\theta\right)$ in Figure 26 is traversed for $0 \le \theta \le 2\pi$ and which part for $2\pi \le \theta \le 4\pi$? **(a)** Use the results of part (a) to prove that the curve is symmetric about the x-axis.

♦ Type 4, a curve in polar coordinates

40. Use the fact that any triangle inscribed in a semicircle with its hypotenuse along a diameter is a right triangle to show that the graph of $r = 2k\cos\theta$ with $k > 0$ (a one-petaled rose) is a circle.

♦ Type 4, a curve in polar coordinates

41. Show that $r = k\sec(\theta - \alpha)$ with $k \ne 0$ and $0 < \alpha < \pi$ is a polar equation of the line perpendicular to the ray $\theta = \alpha$ and passing through the point with polar coordinates $[k, \alpha]$.

♦ Type 4, a curve in polar coordinates

42. According to POISEUILLE'S LAW, the velocity of a fluid at a distance r feet from the center of a pipe of radius R feet is $k(R^2 - r^2)$ feet per minute with a positive constant k. **(a)** What is the velocity next to the wall of the pipe? **(b)** Give a formula for the rate of flow through the pipe in terms of R and k.

♦ Type 3, Poiseuille's Law

RESPONSES TO QUESTIONS IN CHAPTER 9

Responses 9.1

Response 1 (a) $\overline{PF} = \sqrt{(x-0)^2+(y-p)^2}$, $\overline{PD} = |y+p|$ • $\sqrt{(x-0)^2+(y-p)^2} = |y+p|$ • $x^2+(y-p)^2 = (y+p)^2$ • $x^2+y^2-2py+p^2 = y^2+2py+p^2$ • $4py = x^2$ • $y = \frac{1}{4p}x^2$

Response 2 The vertex is the point $(-3,1)$ midway between the focus and directrix.

Response 3 (a) The x-intercepts of $\frac{x^2}{a^2}+\frac{y^2}{b^2}=1$ are the solutions $x=\pm a$ of $\frac{x^2}{a^2}+\frac{0^2}{b^2}=1$. • The y-intercepts of $\frac{x^2}{a^2}+\frac{y^2}{b^2}=1$ are the solutions $y=\pm b$ of $\frac{0^2}{a^2}+\frac{y^2}{b^2}=1$.
(b) The x-intercepts of $\frac{x^2}{b^2}+\frac{y^2}{a^2}=1$ are the solutions $x=\pm b$ of $\frac{x^2}{b^2}+\frac{0^2}{a^2}=1$. • The y-intercepts of $\frac{x^2}{b^2}+\frac{y^2}{a^2}=1$ are the solutions $y=\pm a$ of $\frac{0^2}{b^2}+\frac{y^2}{b^2}=1$.

Response 4 For the point $P=(0.b)$ at the top of the minor axis of the ellipse in Figure 17, $\overline{PF_1}=\overline{PF_2}$ and $\overline{PF_1}+\overline{PF_2}=2a$, so that $\overline{PF_1}=a$. •
For the point $P=(b,0)$ at the right of the minor axis of the ellipse in Figure 18, $\overline{PF_1}=\overline{PF_2}$ and $\overline{PF_1}+\overline{PF_2}=2a$, so that $\overline{PF_1}=a$.

Response 5 The equation of the ellipse is in standard form. • Setting $y=0$ and solving for x in $\frac{x^2}{2^2}+\frac{y^2}{3^2}=1$ shows that the x-intercepts are $x=\pm 2$. • Setting $x=0$ and solving for y shows that the y-intercepts are $y=\pm 3$. • $a=3, b=2$ • $c=\sqrt{a^2-b^2}=\sqrt{3^2-2^2}=\sqrt{5}$. • Draw the ellipse as in Figure R5 with foci at $(0,\pm\sqrt{5})$.

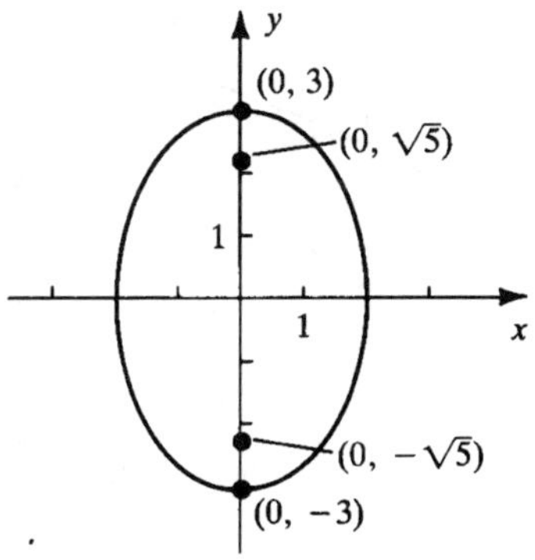

Figure R5

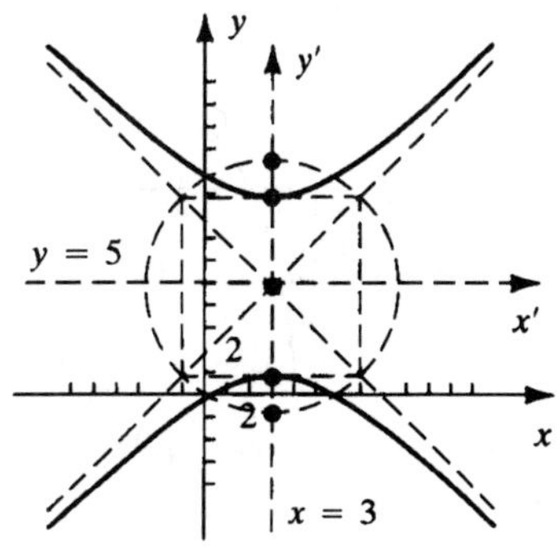

Figure R7

Response 6 (a) The foci are $(-1,2)$ and $(-1,4)$, and the major axis is on the vertical line $x=-1$ through them.
(b) The center is $(-1,1)$ midway between the foci.
(c) $2a=8$ • $a=4$ • c is half the distance between the foci: $c=\frac{1}{2}[4-(-2)]=3$.
(d) $b=\sqrt{a^2-c^2}=\sqrt{4^2-3^2}=\sqrt{7}$.

Response 7 Group the terms in x and y: $(y^2-10y)-(x^2-6x)=0$ •
Complete the two squares by adding $5^2=25$ inside the first set of parentheses and $3^2=9$ inside the second, and add the same numbers on the right of the equation: $(y^2-10y+25)-(x^2-6x+9)=25-9$ •
$(y-5)^2-(x-3)^2=16$ •
$\frac{(y-5)^2}{4^2}-\frac{(x-3)^2}{4^2}=1$ • Axis: $x=3$, Center: $(3,5)$ • $a=4, b=4$ •

Draw the square with its sides four units above, below, to the left and to the right of the center and the circle through its corners; draw the asymptotes through the corners of the square; draw the hyperbola; and plot the foci at the intersection of the circle with the y'-axis. • Figure R7 • ($c = \sqrt{a^2 = b^2} = \sqrt{32}$, but we do not need this number to plot the foci.)

Response 8 **(a)** $\frac{dy}{dx} = 2x$, so the slope of the tangent line at $x = a$ is $2a$. • Tangent line: $y = a^2 + 2a(x - a)$ or $y = 2ax - a^2$ • Set $x = 0$ to obtain the y-coordinate $-a^2$ of Q.
(b) $\overline{PF}^2 = a^2 + \left(a^2 - \frac{1}{4}\right)^2 = a^2 + a^4 - \frac{1}{2}a^2 + \frac{1}{16} = a^4 + \frac{1}{2}a^2 + \frac{1}{16} = \left(a^2 + \frac{1}{4}\right)^2$ • $\overline{PF} = a^2 + \frac{1}{4} = \overline{FQ}$
(c) The angle FPQ equals the angle PQF because the triangle FPQ is isosceles, and the angle PQF equals the angle between QP and the vertical line $x = a$ because the y-axis is also vertical. • The angle FPQ equals the angle between QP and the vertical line $x = a$, so the tangent line makes equal angles with the line $x = a$ and the focal radius.

Responses 9.2

Response 1 **(a)** $\overline{OV} = x'\cos\alpha$ • $\overline{UV} = y'\sin\alpha$ • $\overline{OR} = x'\sin\alpha$ • $\overline{RQ} = y'\cos\alpha$
(b) $x = \overline{OV} - \overline{UV} = x'\cos\alpha - y'\sin\alpha$ • $y = \overline{OR} + \overline{RQ} = x'\sin\alpha + y'\cos\alpha$

Response 2 $\sin\alpha = 2/\sqrt{5}$ • $\cos\alpha = 1/\sqrt{5}$

Response 3 $A = 1, B = -1, C = 0$ • The discriminant $B^2 - 4AC = (-1)^2 - 4(1)(0) = 1$ is positive. • Hyperbola • $y = x - 1/x$ • Figure R3

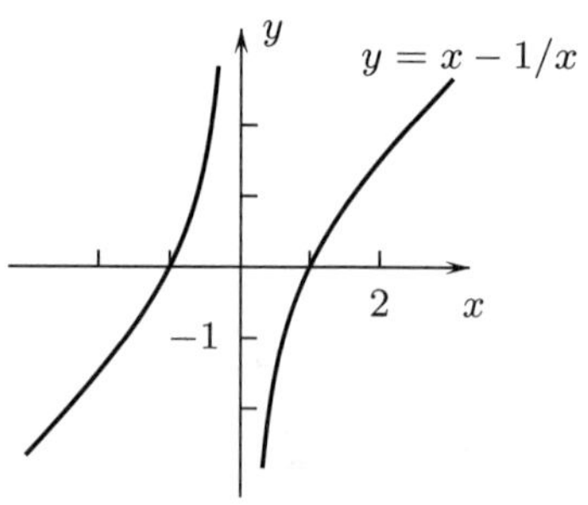

Figure R3

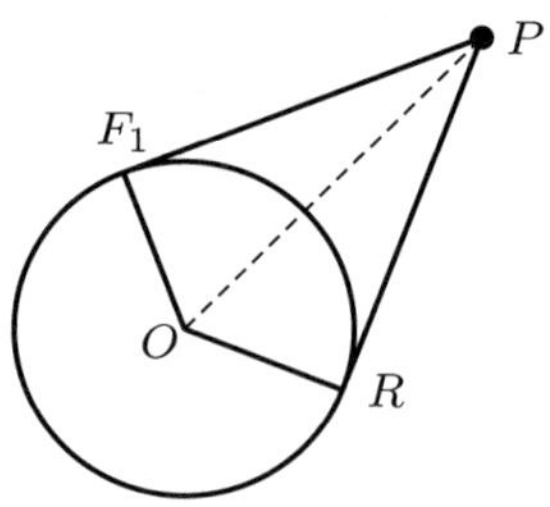

Figure R6

Response 4 **(a)** The graph of $x^2 + y^2 = 0$ consists of the single point $(0, 0)$. • $A = 1, B = 0, C = 1$ • [Discriminant] $= 0^2 - 4(1)(1) = -4$ • Degenerate ellipse
(b) $x^2 + y^2 + 1 = 0$ has no solutions; its graph is empty. • [Discriminant] $= 0^2 - 4(1)(1) = -4$ • Degenerate ellipse
(c) $x^2 - xy = 0 \iff x(x - y) = 0 \iff x = 0$ or $y = x$ • The graph consists of the y-axis and the line $y = x$. • [Discriminant] $= (-1)^2 - 4(1)(0) = 1$ • Degenerate hyperbola

Response 5 If the plane is horizontal, then $\alpha = \frac{1}{2}\pi$ and the cross section is a circle.

Response 6 The line segment PR is tangent to the lower sphere because it is on the cone, and the line segment PF_1 is tangent to the lower sphere because it is in the plane of the ellipse. • PR and PF_1 are tangent to a circular cross section of the lower sphere, as in Figure R6. • The triangles OPR and OPF_1 have right angles at R and F_1 and the same hypotenuse and, consequently, are congruent, so that $\overline{PR} = \overline{PF_1}$.

Response 7 $\dfrac{\overline{PS}}{\overline{PD}} = \cos\alpha$ and $\dfrac{\overline{PS}}{\overline{PR}} = \cos\beta$ • $\overline{PS} = \cos\alpha\,\overline{PD}$ and $\overline{PS} = \cos\beta\,\overline{PR}$ • $\overline{PR} = (\cos\alpha/\cos\beta)\overline{PD}$

Response 8 **(a)** $a-c = e(k-a)$ and $a+c = e(k+a)$ • Solve for $e = c/a$ and $k = a/e = a^2/c$.

Responses 9.3

Response 1 Draw an r-axis through the origin with the point $(7,5)$ on its positive side (Figure R1) • The positive side of the r-axis is at an angle $\theta = \tan^{-1}\left(\frac{5}{7}\right)$ from the positive x-axis. • The distance from $(7,5)$ to the origin is $\sqrt{7^2+5^2} = \sqrt{74}$. • Polar coordinates $\left[\sqrt{74}, \tan^{-1}\left(\frac{5}{7}\right) + 2n\pi\right]$ and $\left[-\sqrt{74}, \tan^{-1}\left(\frac{5}{7}\right) + (2n+1)\pi\right]$ with integers n

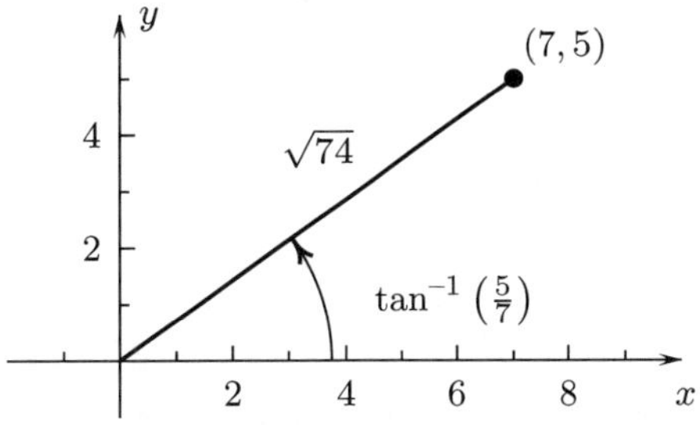

Figure R1

Response 2 **(a)** $(r\cos\theta - k)^2 + (r\sin\theta)^2 = k^2$ • $r^2\cos^2\theta - 2rk\cos\theta + k^2 + r^2\sin^2\theta = k^2$ • $r^2 = 2rk\,\cos\theta$ • $r = 2k\,\cos\theta$.
(b) $r\cos\theta = k$ • $r = k\,\sec\theta$

Response 3 Figure R3a shows the curve in an xy-plane. • As is shown by the graph of $r = 1+\cos\theta$ in the θr-plane of Figure R3b, $r(0) = 1+\cos(0) = 2$, so the point $(2,0)$ is on the polar curve. • r decreases to 0 as θ increases from 0 to π, generating the top half of the polar curve. • The bottom half is generated as θ increases from π to 2π and r increases from 0 back to 2.

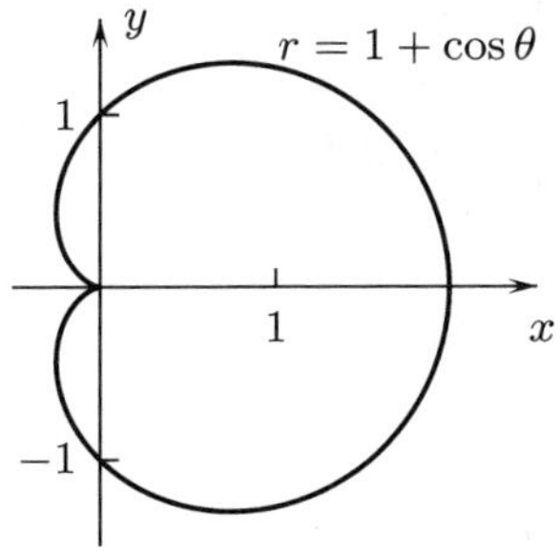

Figure R3a

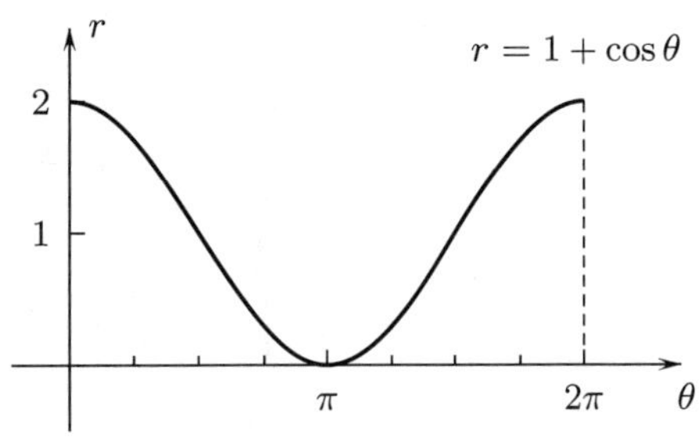

Figure R3b

Response 4 Figure 14 shows that half a petal is generated for $0 \le \theta \le \frac{1}{4}\pi$. •
Use the double-angle formula **(7)** and the substitution $u = 4\theta$, $du = 4\,d\theta$:

$$[\text{Area of one petal}] = 2\int_0^{\pi/4} \tfrac{1}{2}[3\cos(2\theta)]^2\,d\theta = 9\int_0^{\pi/4} \cos^2(2\theta)\,d\theta$$

$$= \tfrac{9}{2}\int_0^{\pi/4} [1+\cos(4\theta)]\,d\theta = \tfrac{9}{8}\int_{\theta=0}^{\theta=\pi/4} (1+\cos u)\,du$$

$$= \tfrac{9}{8}\Big[u + \sin u\Big]_{\theta=0}^{\theta=\pi/4} = \tfrac{9}{8}\Big[4\theta + \sin(4\theta)\Big]_{\theta=0}^{\theta=\pi/4} = \tfrac{9}{8}\pi$$

CHAPTER 9 ANSWERS

Tune-up exercises 9.1

T1. The vertex $(1, 2)$ is midway between the focus and the directrix. • Figure T1 • $x' = x - 1, y' = y - 2$ • Because the vertex is 1 unit from the focus and directrix, $p = -1$ • $y - 2 = -\frac{1}{4}(x - 1)^2$

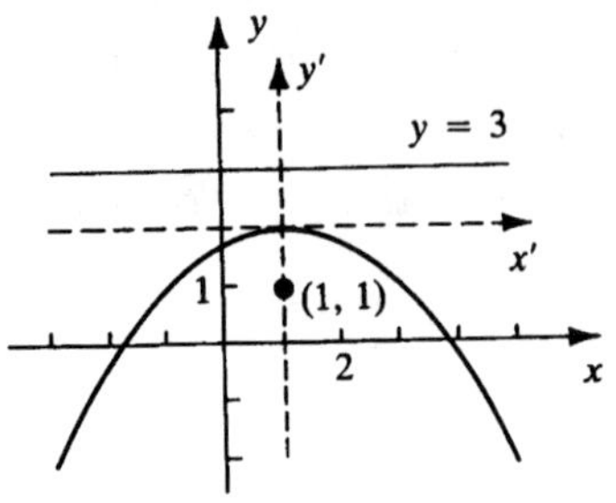

Figure T1

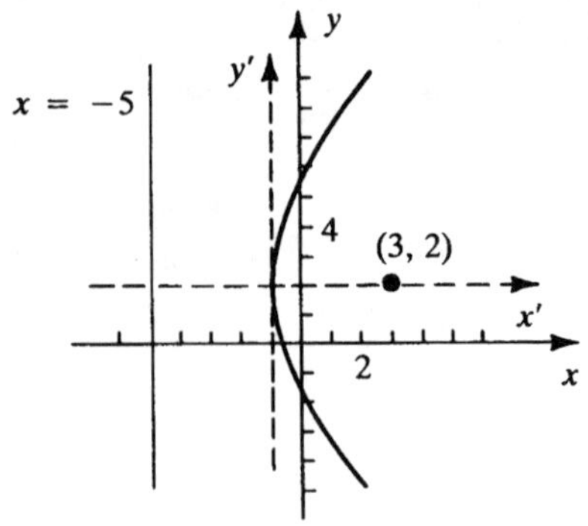

Figure T2

T2. The vertex is $(-1, 2)$. • Figure T2 • $x' = x + 1, y' = y - 2$ • $x + 1 = \dfrac{1}{4p}(y - 2)^2$ • Set $x = 0, y = -2 : 1 = \dfrac{4^2}{4p}$ • $p = 4$ • $x + 1 = \frac{1}{16}(y - 2)^2$ • The focus is at $(3, 2)$, four units to the right of the vertex, and the directrix is $x = -5$, four units to the left.

T3. $x^2 + 6x = 2y - 5$ • Add $[\frac{1}{2}(6)]^2 = 9$ to both sides of the equation to complete the square on the left: $x^2 + 6x + 9 = 2y + 4$ • $(x + 3)^2 = 2(y + 2)$ • $y' = \frac{1}{2}(x')^2$ with $y' = y + 2, x' = x + 3$ • Figure T3 • $p = \frac{1}{2}$ • The vertex is $(-3, -2)$, the focus is $(-3, -\frac{3}{2})$, and the directrix is $y = -\frac{5}{2}$.

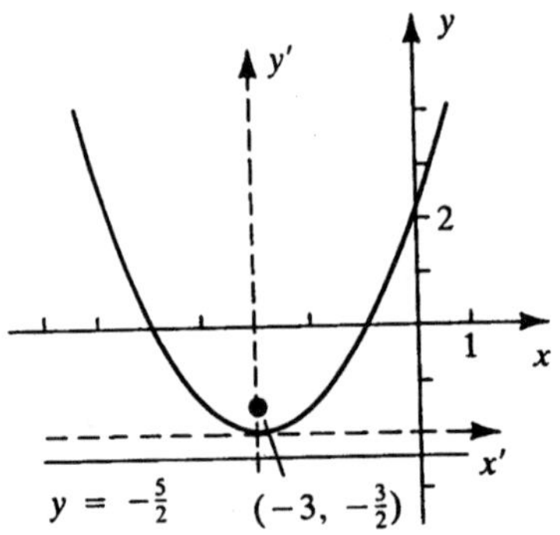

Figure T3

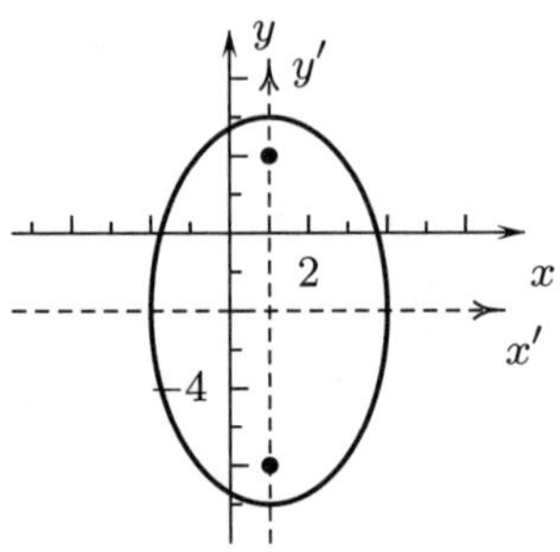

Figure T4

T4. Collect terms: $25(x^2 - 2x) + 9(y^2 + 4y) = 164$ • Complete the squares: $25(x^2 - 2x + 1) + 9(y^2 + 4y + 4) = 164 + 25(1) + 9(4)$ • $25(x - 1)^2 + 9(y + 2)^2 = 225$ • Divide by 225: $\dfrac{(x-1)^2}{3^2} + \dfrac{(y+2)^2}{5^2} = 1$ • $x' = x - 1, y' = y + 2$ • Center: $(1, -2)$ • $a = 5, b = 3$ • The major axis is of length 10 and the minor axis is of length 6. • Figure T4 • $c = \sqrt{a^2 - b^2} = 4$ • Foci: $(1, -6)$, $(1, 2)$

T5. Plot the foci and vertices, as in Figure T5 • Center: $(2,2)$ • $x' = x - 2, y' = y - 2$ • $a = \frac{1}{2}\overline{V_1V_2} = 3$, $c = \frac{1}{2}\overline{F_1F_2} = 2$ • $b = \sqrt{a^2 - c^2} = \sqrt{5}$ • $\dfrac{(x')^2}{a^2} + \dfrac{(y')^2}{b^2} = 1$ • $\dfrac{(x-2)^2}{9} + \dfrac{(y-2)^2}{5} = 1$ • Draw the ellipse through the vertices with its minor axis of length $2\sqrt{5}$ • Figure T5

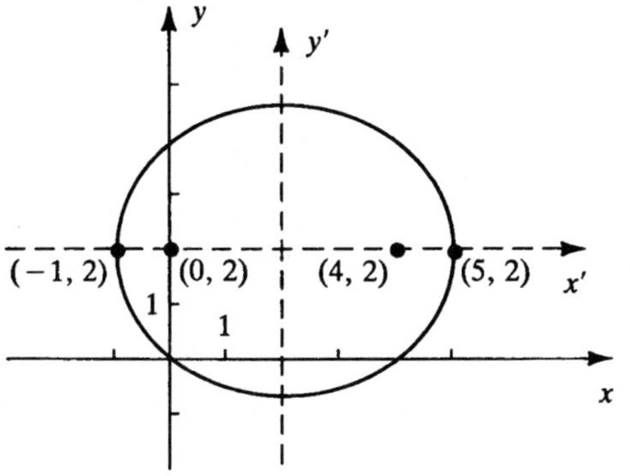

Figure T5

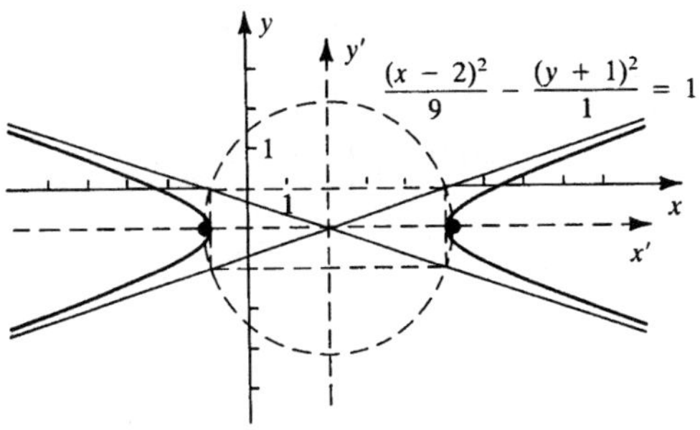

Figure T6

T6. Group terms and complete the squares: $(x^2 - 4x) - 9(y^2 + 2y) = 14$ • $(x^2 - 4x + 4) - 9(y^2 + 2y + 1) = 14 + 4 - 9(1)$ • $(x-2)^2 - 9(y+1)^2 = 9$ • Divide by 9: $\dfrac{(x-2)^2}{3^2} - \dfrac{(y+1)^2}{1^2} = 1$ • $x' = x - 2, y' = y + 1$ • $a = 3, b = 1, c = \sqrt{a^2 + b^2} = \sqrt{10}$
Draw the new axes with center at $(2, -1)$, the rectangle with sides of lengths $2a = 6$ and $2b = 2$, and the circle through the corners of the rectangle. Then draw the hyperbola with the foci at the intersections of the circle and the x'-axis, as in Figure T6.

T7. The center is the point $(0, 1)$ midway between the foci. • $x' = x, y' = y - 1$ • $2a = 4$, so $a = 2$ • $c = \frac{1}{2}\overline{F_1F_2} = 3$ • $b = \sqrt{c^2 - a^2} = \sqrt{5}$ • $\dfrac{(y')^2}{a^2} - \dfrac{(x')^2}{b^2} = 1$ • $\dfrac{(y-1)^2}{4} - \dfrac{x^2}{5} = 1$ • Figure T7

Figure T7

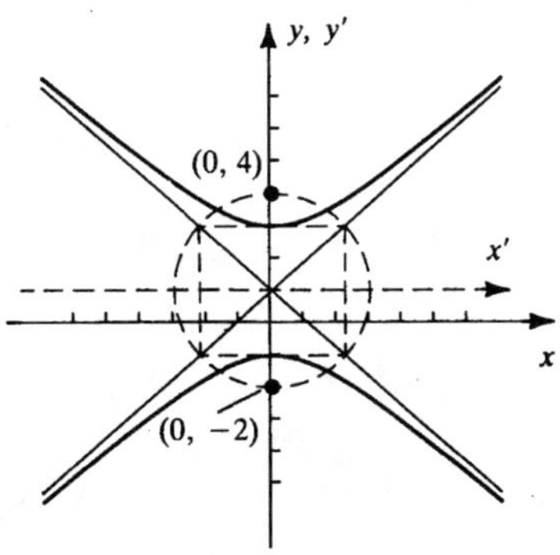

Problems 9.1

1. The vertex $(0, 2)$ is midway between the focus and the directrix. • Figure A1 • $x' = x, y' = y - 2$ • The distance from the vertex to the focus and directrix is 2, so $p = 2$. • $y' = \dfrac{1}{4p}(x')^2$ • $y - 2 = \frac{1}{8}x^2$

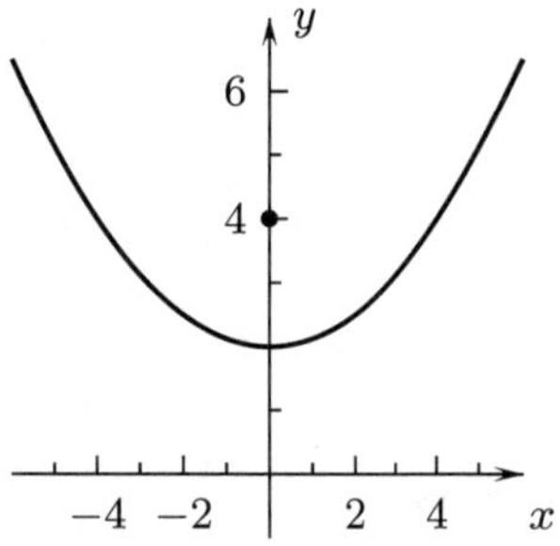

Figure A1

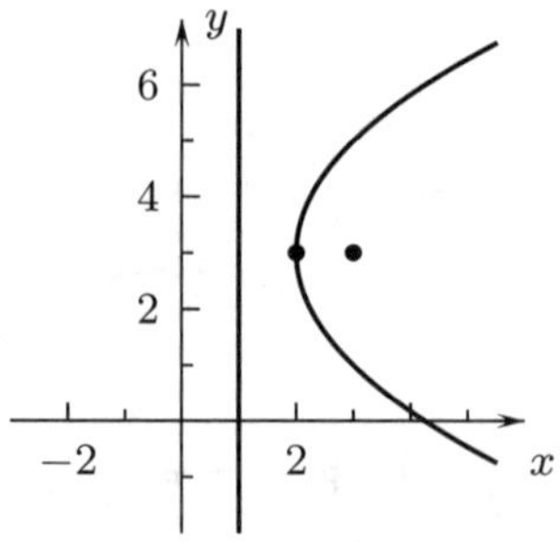

Figure A2

2. $x - 2 = \frac{1}{4}(y-3)^2$ • Figure A2

4. The center is at the origin. • $c = 3$ • $a = 5$ • $b = \sqrt{a^2 - c^2} = 4$ • $\dfrac{x^2}{25} + \dfrac{y^2}{16} = 1$ • Figure A4

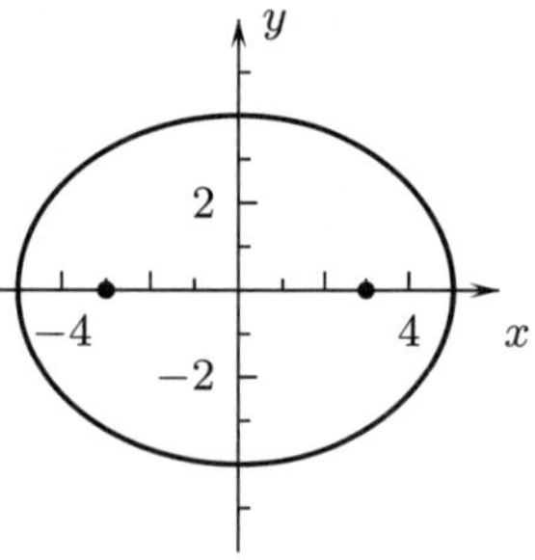

Figure A4

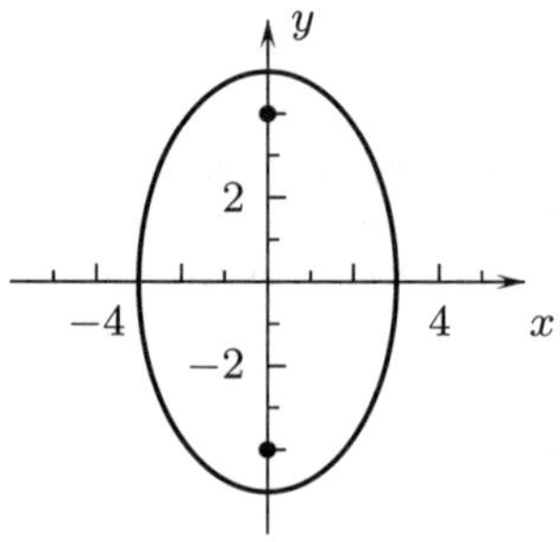

Figure A5

5. $\dfrac{x^2}{9} + \dfrac{y^2}{25} = 1$ • Figure A5

7. The center is at the origin. • $a = 3, c = 4$ • $b = \sqrt{c^2 - a^2} = \sqrt{7}$ • $\dfrac{x^2}{9} - \dfrac{y^2}{7} = 1$ • Figure A7

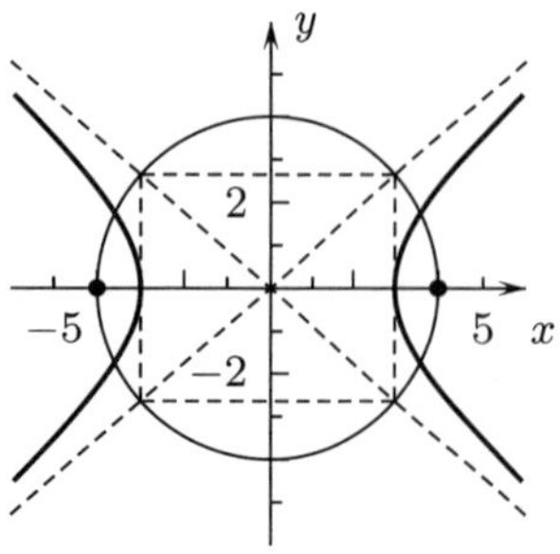

Figure A7

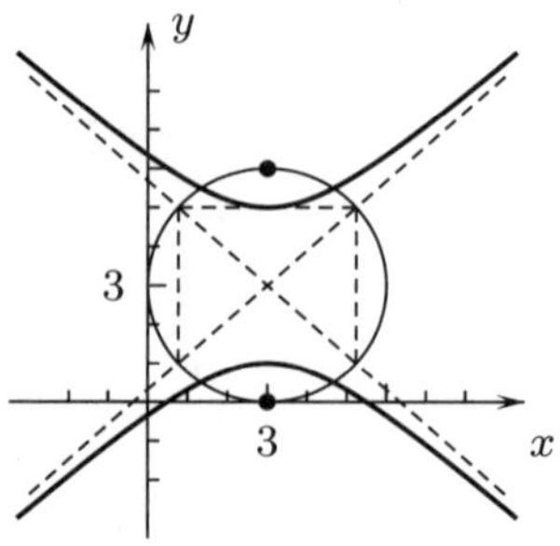

Figure A12

12. $\dfrac{(y-3)^2}{4} - \dfrac{(x-3)^2}{5} = 1$ • Figure A12

15. $(x+1)^2-(y-3)^2=4$ • $\dfrac{(x+1)^2}{2^2}-\dfrac{(y-3)^2}{2^2}=1$ • $a=2, b=2, c=2\sqrt{2}$ • Center: $(-1,3)$ • Foci: $(-1\pm 2\sqrt{2},3)$ • Vertices: $(1,3)$ and $(-3,3)$ • Asymptotes: $y=3\pm(x+1)$ • Figure A15

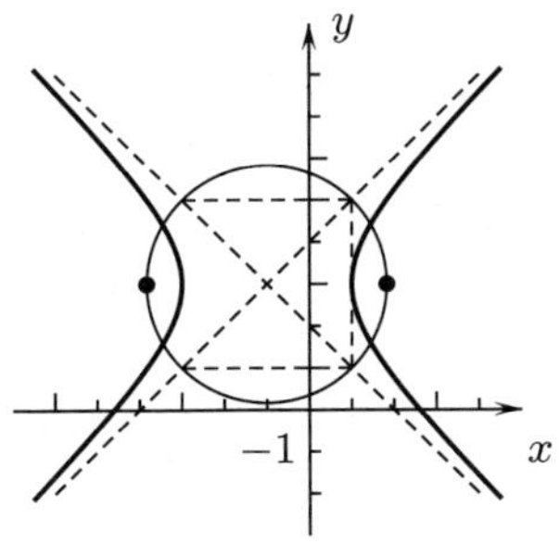

Figure A15

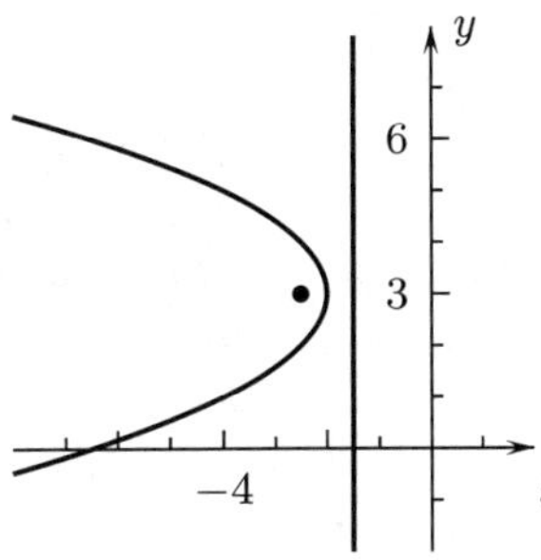

Figure A16

16. $(y-3)^2=-2x-4$ • $x+2=-\frac{1}{2}(y-3)^2$ • $p=-\frac{1}{2}$ • Vertex: $(-2,3)$ • Focus: $(-\frac{5}{2},3)$ • Directrix: $x=-\frac{3}{2}$ • Figure A16

18. $\dfrac{x^2}{4^2}+\dfrac{y^2}{3^2}=1$ • $a=4, b=3, c=\sqrt{a^2-b^2}=\sqrt{7}$ • Center: $(0,0)$ • Foci: $(\pm\sqrt{7},0)$ • Vertices: $(\pm 4,0)$ • Figure A18

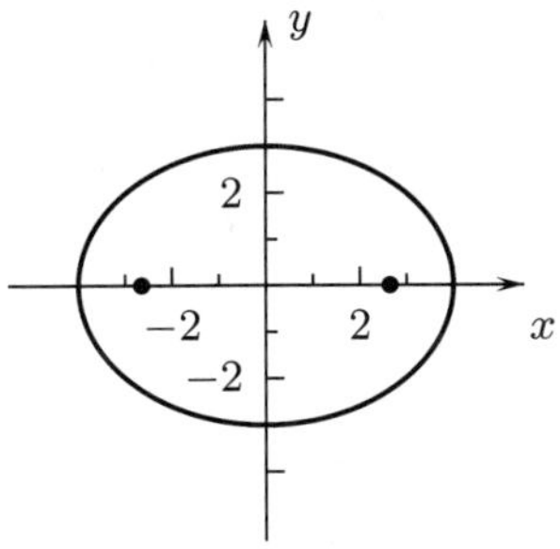

Figure A18

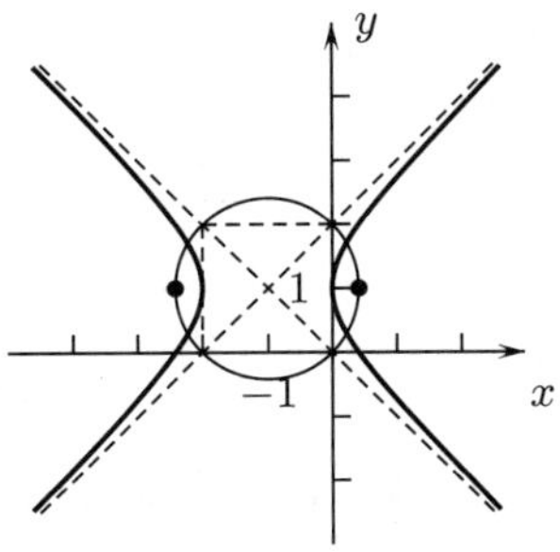

Figure A19

19. Center: $(-1,1)$ • Foci: $(-1\pm\sqrt{2},1)$ • Vertices: $(0,1)$ and $(-2,1)$ • Asymptotes: $y-1=\pm(x+1)$ • Figure A19

21. Vertex: $(0,-3)$ • Focus: $(0,-\frac{5}{2})$ • Directrix: $y=-\frac{7}{2}$ • Figure A21

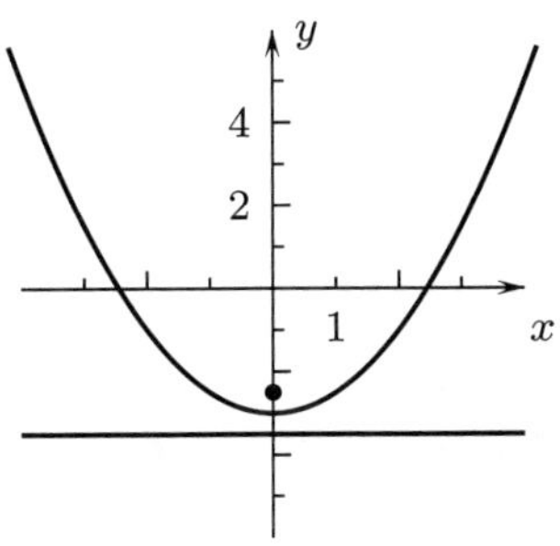

Figure A21

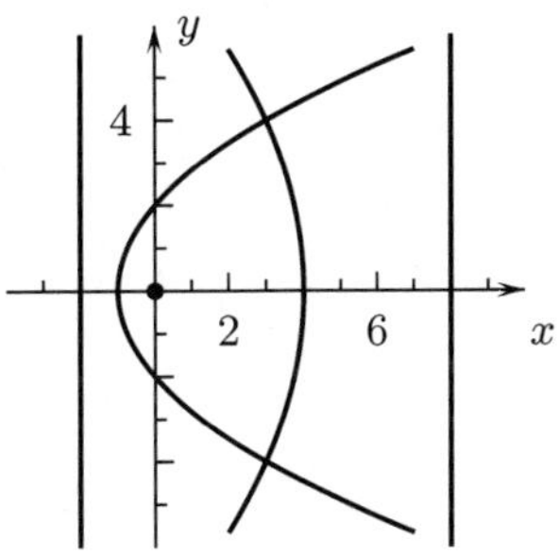

Figure A23

23. $x+1=\frac{1}{4}y^2$ and $x-4=-\frac{1}{16}y^2$ • Figure A23

25. $\frac{2}{9}x^2-\frac{2}{9}y^2=1$

27. πab

30. $\frac{27}{2}\pi$ cubic inches

Tune-up exercises 9.2

T1. **(a)** $B^2-4AC=(-4)^2-4(3)(1)=4$ • Hyperbola
(b) $B^2-4AC=(-4)^2-4(2)(2)=0$ • Parabola
(c) $B^2-4AC=1^2-4(1)(1)=-3$ • Ellipse

T2. $A=C \implies \alpha=\frac{1}{4}\pi$ •

$$\begin{cases} x=\dfrac{1}{\sqrt{2}}(x'-y') \\ y=\dfrac{1}{\sqrt{2}}(x'+y') \end{cases} \qquad \textbf{(22)}$$

$\frac{5}{2}(x'-y')^2+3(x'-y')(x'+y')+\frac{5}{2}(x'+y')^2=8$ • $8(x')^2+2(y')^2=8$ • $\dfrac{(x')^2}{1^2}+\dfrac{(y')^2}{2^2}=1$ • $a=2, b=1, c=\sqrt{a^2-b^2}=\sqrt{3}$ • • Foci: $x'=0, y'=\pm\sqrt{3}$ • Figure T2

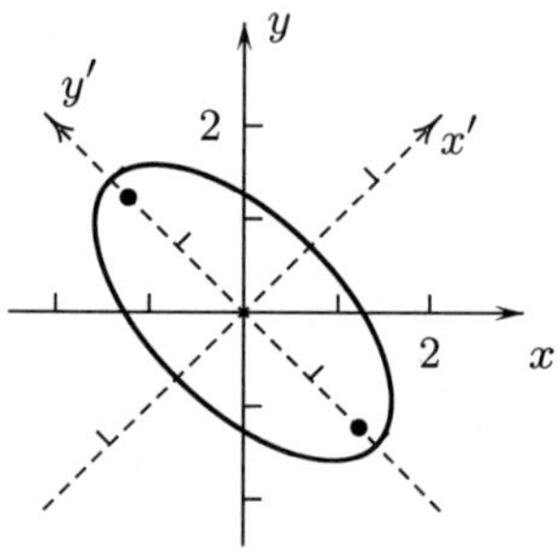

Figure T2

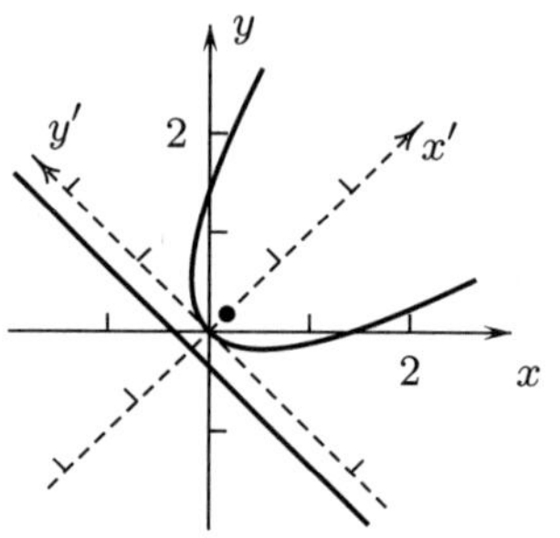

Figure T3

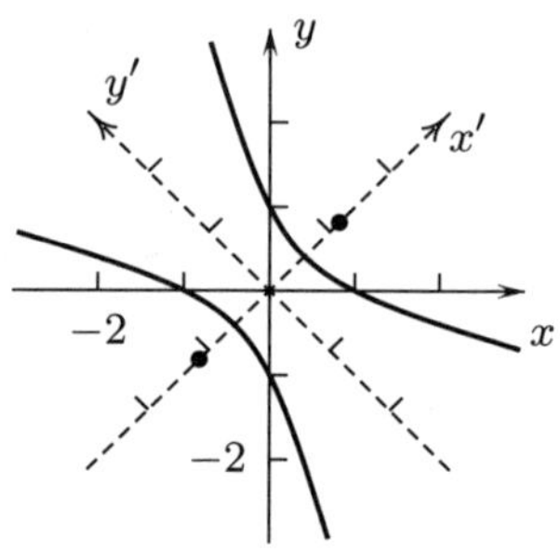

Figure T4

T3. $A=C \implies \alpha=\frac{1}{4}\pi$ • Use formulas **(22)**.
$\frac{1}{2}(x'-y')^2-(x'-y')(x'+y')+\frac{1}{2}(x'+y')^2-(x'-y')-(x'+y')=0$ •
$2(y')^2-2x'=0$ • $x'=(y')^2$ • $p=\frac{1}{4}$ Focus: $x'=\frac{1}{4}, y'=0$ • Directrix: $x'=-\frac{1}{4}$ • Figure T3

T4. $A=C \implies \alpha=\frac{1}{4}\pi$ • Use formulas **(22)**.
$\frac{1}{2}(x'-y')^2+2(x'-y')(x'+y')+\frac{1}{2}(x'+y')^2=1$ •
$3(x')^2-(y')^2=1$ • $a=\frac{1}{3}\sqrt{3}, b=1, c=\sqrt{a^2+b^2}=\frac{2}{3}\sqrt{3}$ • Figure T4

T5. **(a)** $(x-y)^2=0$ • The line $y=x$ • Figure T5a • $B^2-4AC=(-2)^2-4(1)(1)=0$ • Degenerate parabola
(b) The lines $x=1$ and $y=1$ • Figure T5b • $xy-x-y+1=0$ $B^2-4AC=1^2-4(0)(0)=1$ • Degenerate hyperbola

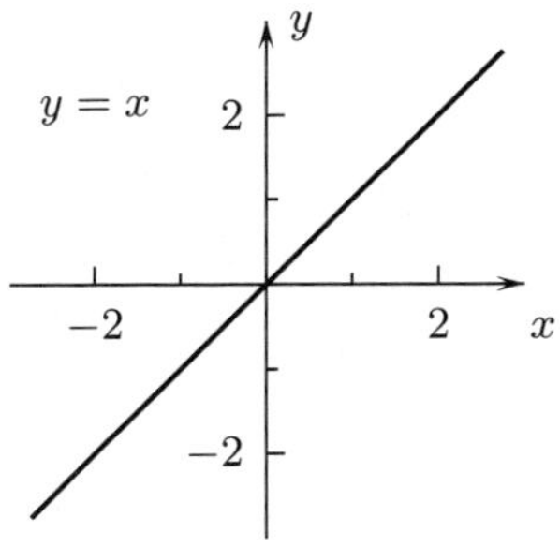

Figure T5a

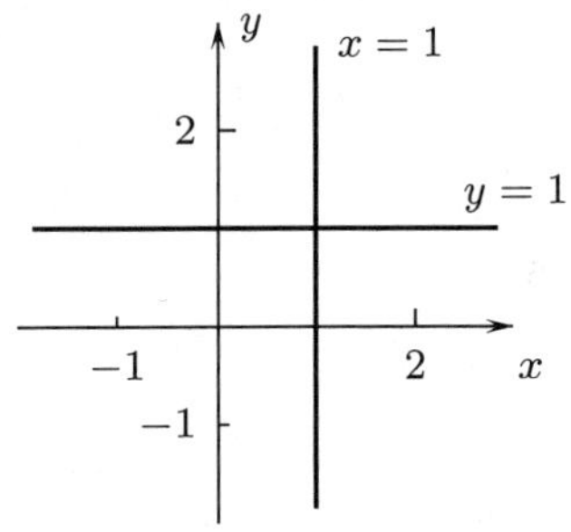

Figure T5b

T6. $a = 5, b = 4, c = \sqrt{a^2 - b^2} = 3$ • $e = c/a = \frac{3}{5}$ • Foci: $(\pm c, 0) = (\pm 3, 0)$ • $a^2/c = \frac{25}{3}$ • Directrices: $x = \pm\frac{25}{3}$ • Figure T6

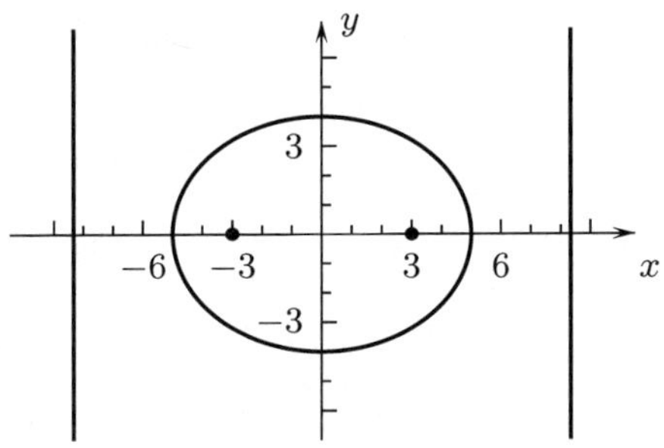

Figure T6

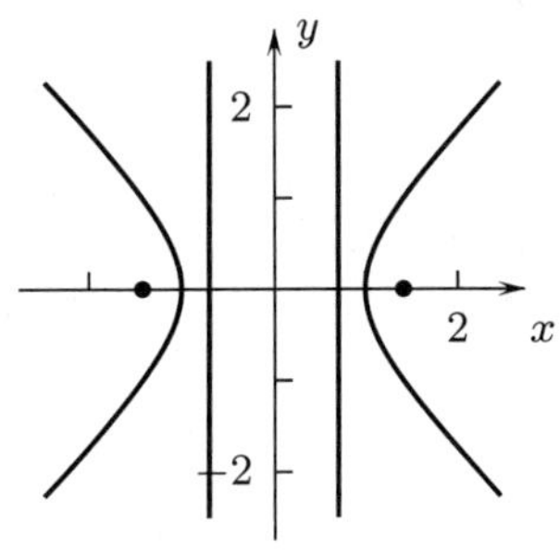

Figure T7

T7. $a = 1, b = 1, c = \sqrt{a^2 + b^2} = \sqrt{2}$ • $e = c/a = \sqrt{2}$ • Foci: $(\pm c, 0) = (\pm\sqrt{2}, 0)$ • $a^2/c = 2/\sqrt{2}$ • Directrices: $x = \pm 1/\sqrt{2}$ • Figure T7

Problems 9.2

1. $B^2 - 4AC = 3^2 - 4(1)(-1) = 13$ • Hyperbola

2. $B^2 - 4AC = 1^2 - 4(3)(4) = -47$ • Ellipse

3. Parabola

4. Ellipse

7. $A = C \implies \alpha = \frac{1}{4}\pi$ • Use $x = \dfrac{1}{\sqrt{2}}(x' - y'),\ y = \dfrac{1}{\sqrt{2}}(x' + y')$ •
$\frac{13}{2}(x' - y')^2 - 5(x' - y')(x' + y') + \frac{13}{2}(x' + y')^2 = 72$ •
$8(x')^2 + 18(y')^2 = 72$ • $\dfrac{(x')^2}{3^2} + \dfrac{(y')^2}{2^2} = 1$ •
$a = 3, b = 2, c = \sqrt{a^2 - b^2} = \sqrt{5}$ • Foci: $x' = \pm\sqrt{5}, y' = 0$ • Figure A7

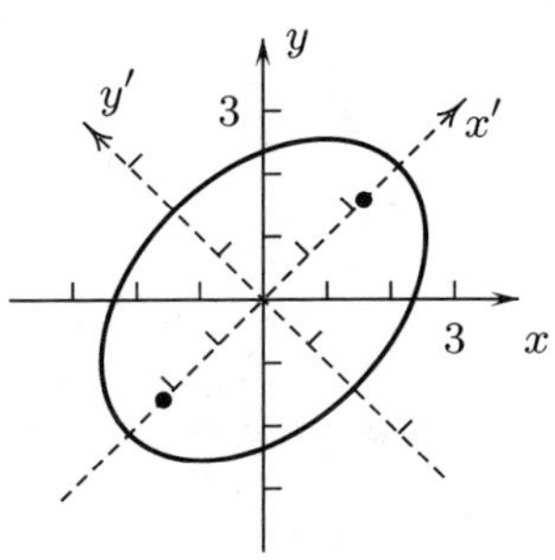

Figure A7

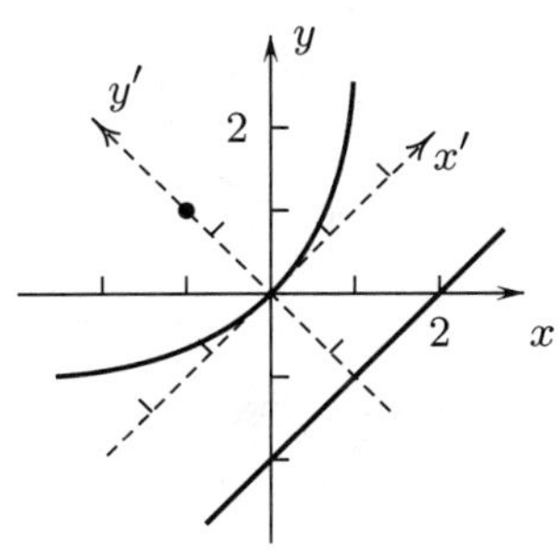

Figure A8

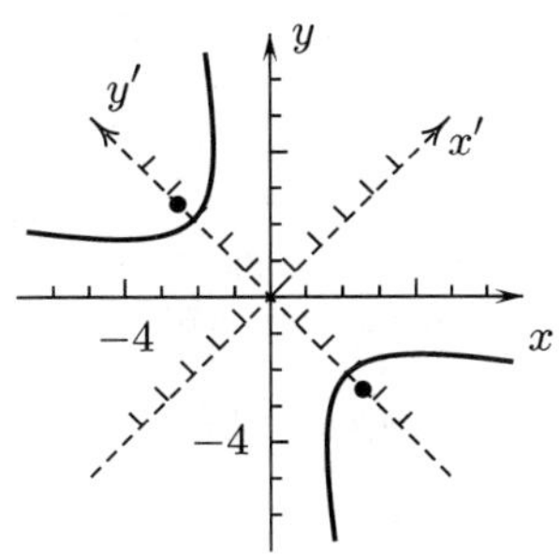

Figure A9

8. $A = C \implies \alpha = \frac{1}{4}\pi$ • Use $x = \dfrac{1}{\sqrt{2}}(x' - y'),\ y = \dfrac{1}{\sqrt{2}}(x' + y')$ •
$\frac{1}{2}(x' - y')^2 + (x' - y')(x' + y') + \frac{1}{2}(x' + y')^2 + \dfrac{8}{\sqrt{2}}(x' - y') - \dfrac{8}{\sqrt{2}}(x' + y') = 0$ •
$2(x')^2 - 8\sqrt{2}(y') = 0$ • $y' = \dfrac{1}{4\sqrt{2}}(x')^2$ • $p = \sqrt{2}$ • Focus: $x' = 0, y' = \sqrt{2}$ • Directrix: $y' = -\sqrt{2}$ • Figure A8

9. $A = C \implies \alpha = \frac{1}{4}\pi$ • Use $x = \dfrac{1}{\sqrt{2}}(x' - y'),\ y = \dfrac{1}{\sqrt{2}}(x' + y')$ •
$\frac{5}{2}(x' - y')^2 + 13(x' - y')(x' + y') + \frac{5}{2}(x' + y')^2 = -72$ •
$18(x')^2 - 8(y')^2 = -72$ • $\dfrac{(y')^2}{3^2} - \dfrac{(x')^2}{2^2} = 1$ $a = 3, b = 2, c = \sqrt{a^2 + b^2} = \sqrt{13}$ • Foci: $x' = 0, y' = \pm\sqrt{13}$ • Figure A9

22. The point $(1, 2)$

23. **(a)** [Aphelion] $\doteq$ 94.45 million miles **(b)** [Perilion] $\doteq$ 91.35 million miles

25. [Eccentricity] $\doteq 0.795$

41. $x^2 + 2xy + y^2 - 12y + 18 = 0$

43. $35x^2 + 2xy + 35y^2 - 36x - 36y - 288 = 0$

Tune-up exercises 9.3

T1. Draw the r-axis at and angle $\frac{3}{4}\pi$ from the x-axis and plot the point $r = -2$ on it. • Figure T1 • $x = -2\cos(\frac{3}{4}\pi) = \sqrt{2}, y = -2\sin(\frac{3}{4}\pi) = -\sqrt{2}$

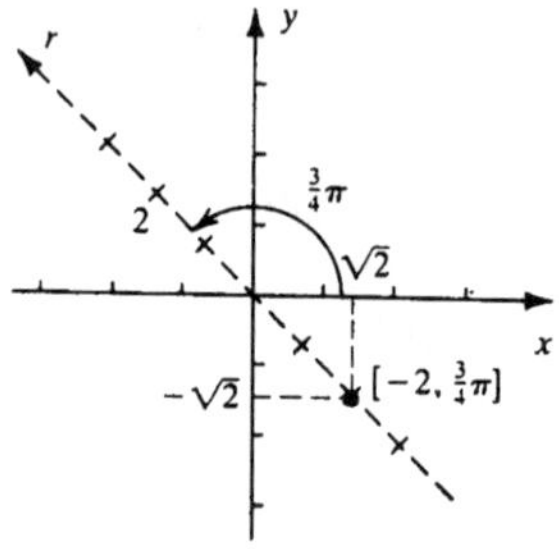

Figure T1

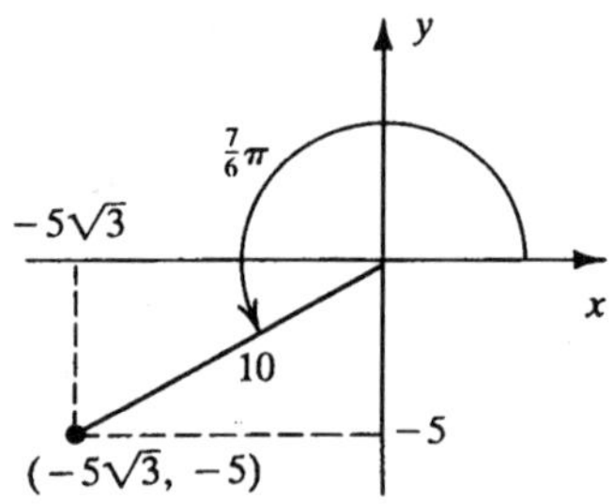

Figure T2

T2. The positive r is $[(-5\sqrt{3})^2 + (-5)^2]^{1/2} = 10$, and one angle for this r is $\theta = \frac{7}{6}\pi$ • Figure T2 • Polar coordinates: $[10, (\frac{7}{6} + 2n)\pi]$ and $[-10, (\frac{1}{6} + 2n)\pi]$

T3. $r = -3\sec\theta \iff r\cos\theta = -3 \iff x = -3$. Figure T3

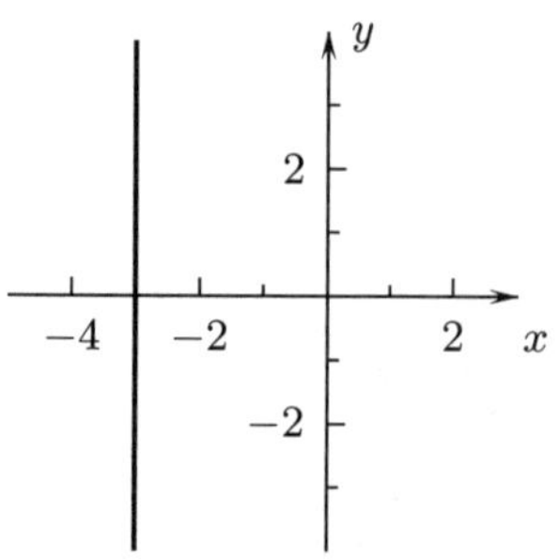

$r = -3\cos\theta$

Figure T3

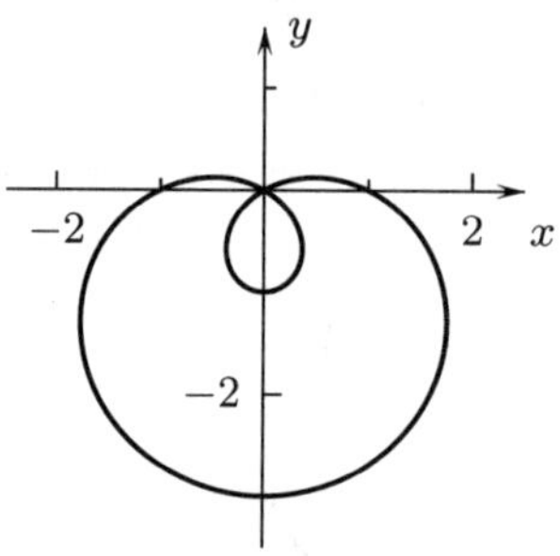

$r = 1 - 2\sin\theta$

Figure T4a

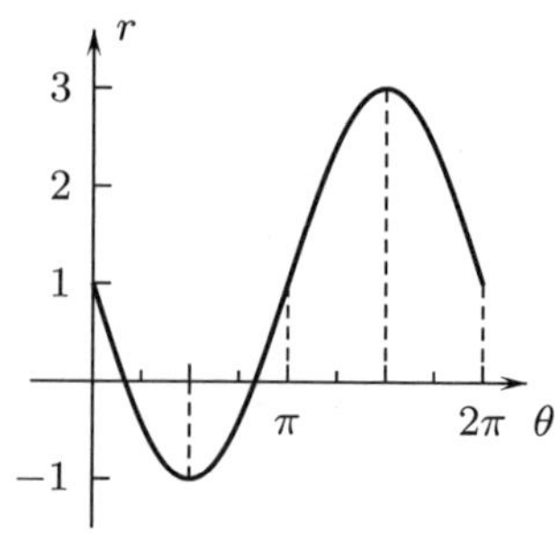

$r = 1 - 2\sin\theta$

Figure T4b

T4. Figure T4a • To explain the shape, generate $r = 1 - 2\sin\theta$ in an θr-plane (Figure T4b). • The curve is generated as θ increases from 0 to 2π. • $r = 0 \iff \sin\theta = \frac{1}{2}$ • $r = 0$ at $\theta = \frac{1}{6}\pi$ and $\theta = \frac{5}{6}\pi$. • r is 1 at $\theta = 0$ and decreases to 0 as θ increases to $\frac{1}{6}\pi$; this generates the piece of the outer loop above the positive x-axis. • r decreases to -1 as θ increases to $\frac{1}{2}\pi$; this generates the left side of the inner loop. • r increases to 0 as θ increases to $\frac{5}{6}\pi$; this generates the right side of the inner loop. • r increases to 1 as θ increases to π; this generates the portion of the outer loop above the negative x-axis. • r increases to 3 as θ increases to $\frac{3}{2}\pi$; this generates the rest of the left side of the outer loop. • r decreases to 1 as θ increases to 2π to draw the rest of the right side of the outer loop.

T5. Multiply $r = 4\cos\theta$ by r: $r^2 = 4r\cos\theta$ • This is the same curve because $\cos\theta$ is 0 at $\theta = \frac{1}{2}\pi$. • Switch to rectangular coordinates: $x^2 + y^2 = 4x$. • Complete the square: $(x^2 - 4x + 4) + y^2 = 4$ • $(x-2)^2 + y^2 = 4$ • The graph is the circle in Figure T5.
(b) The circle is traversed once for $-\frac{1}{2}\pi \le \theta \le \frac{1}{2}\pi$ •

$$[\text{Area}] = \frac{1}{2}\int_{-\pi/2}^{\pi/2} (4\cos\theta)^2\,d\theta = \frac{1}{2}\int_{-\pi/2}^{\pi/2} 16(\tfrac{1}{2})[1+\cos(2\theta)]\,d\theta$$ • Use $u = 2\theta, du = 2d\theta$:

$$[\text{Area}] = 4\int_{-\pi/2}^{\pi/2} d\theta + 2\int_{\theta=-\pi/2}^{\theta=\pi/2} \cos u\,du = 4\pi + 2\Big[\sin u\Big]_{\theta=-\pi/2}^{\theta=\pi/2} = 4\pi + 2\Big[\sin(2\theta)\Big]_{\theta=-\pi/2}^{\theta=\pi/2} = 4\pi$$

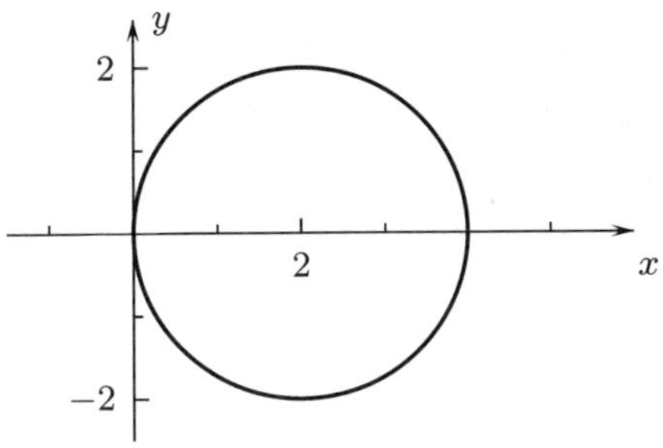

Figure T5

Problems 9.3

1. Figure A1 • $(x,y) = (3\cos(\frac{3}{4}\pi), 3\sin(\frac{3}{4}\pi)) = (-\frac{3}{2}\sqrt{2}, \frac{3}{2}\sqrt{2})$

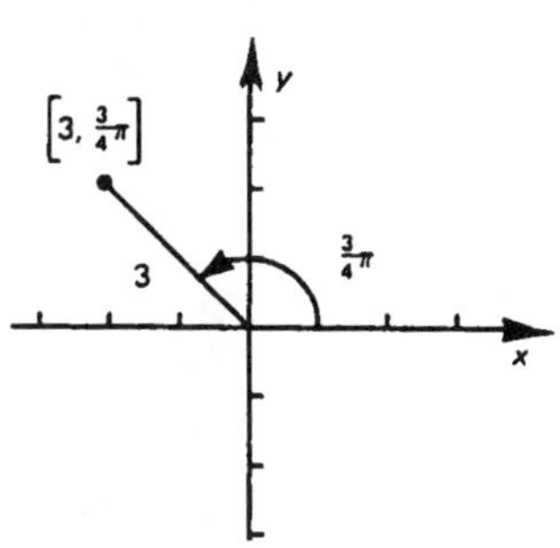

Figure A1

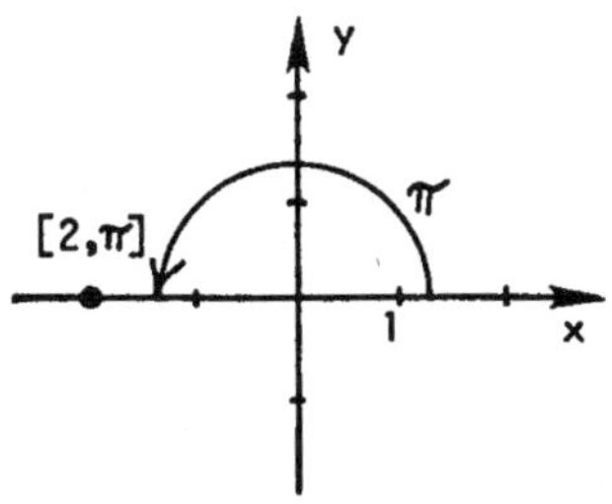

Figure A2

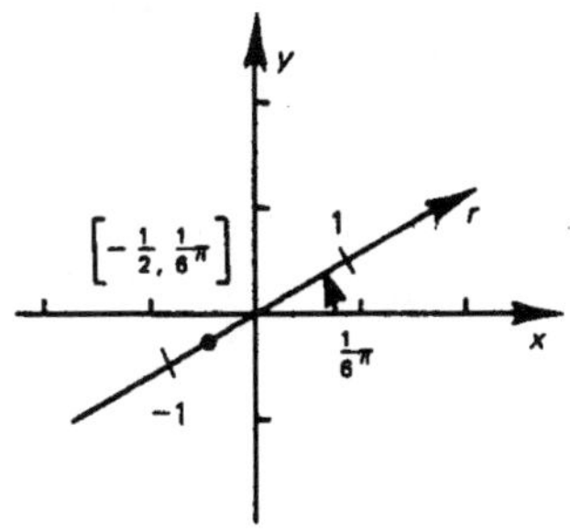

Figure A3

2. Figure A2 • $(x,y) = (2\cos(\pi), 2\sin(\pi) = (-2, 0)$

3. Figure A3 • $(x,y) = (-\frac{1}{2}\cos(\frac{1}{6}\pi), -\frac{1}{2}\sin(\frac{1}{6}\pi) = (-\frac{1}{4}\sqrt{3}, -\frac{1}{4})$

5. Figure A5 • $(x,y) = (-2\cos(\frac{15}{6}\pi), -2\sin(\frac{15}{6}\pi) = (0, -2)$

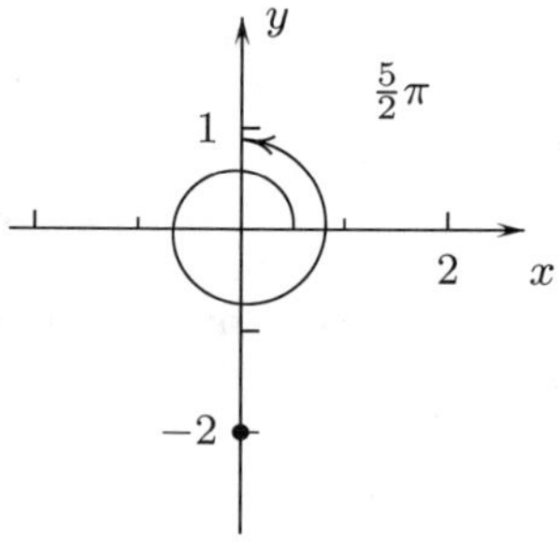

Figure A5

7. $r = \sqrt{(-3)^2 + (-3)^2} = 3\sqrt{2}$ and $\theta = \frac{5}{4}\pi + 2n\pi$ or $r = -3\sqrt{2}$ and $\theta = \frac{1}{4}\pi + 2n\pi$ with integers n

8. $r = 4$ and $\theta = \frac{3}{2}\pi + 2n\pi$ or $r = -4$ and $\theta = \frac{1}{2}\pi + 2n\pi$ with integers n

10. $r = 2$ and $\theta = \frac{2}{3}\pi + 2n\pi$ or $r = -2$ and $\theta = -\frac{1}{3}\pi + 2n\pi$ with integers n

11. $r = 5$ and $\theta = \tan^{-1}(\frac{4}{3}) + 2n\pi$ or $r = -5$ and $\theta = \tan^{-1}(\frac{4}{3}) + (2n+1)\pi$ with integers n

12. $r = 3\sqrt{5}$ and $\theta = \tan^{-1}(2) + (2n+1)\pi$ or $r = -3\sqrt{5}$ and $\theta = \tan^{-1}(2) + 2n\pi$ with integers n

13. $r = 2\sec\theta$

15. $\theta = \tan^{-1}(\frac{1}{3})$

16. $\theta = \pm\frac{1}{4}\pi$

29. 33π

30. $\frac{1}{3}\pi$

31. $\frac{1}{4}(e^{\pi} - 1)$

Appendix 1

Using your graphing calculator or computer

OVERVIEW: *Your graphing calculator or computer is a powerful and flexible tool, which you would probably be able to use fairly well without reading any instructions. It is important, however, to learn how to take advantage of some of its less features and how to avoid making errors using it. Study this section carefully and be sure you can work the tune-up exercises at the end. If you are using a Texas Instruments TI-82, TI-83, TI-85, or TI-86 calculator, read the operating instructions with the set of responses to questions keyed to that calculator.*

Topics:

- ***Priority of operations in calculations***
- ***The need to use * and proper parentheses***
- ***Extraneous lines in graphs***
- ***Graphs, intersections, zeros, maxima, and minima***
- ***Decimals, exact values, and pixel coordinates***

Priority of operations

The meaning of a formula involving functions, powers, sums, differences, products, and quotients depends on how the formula is interpreted to determine the order in which the operations are performed. Calculators and computers generally interpret formulas with the rules that are used in manual calculations, but with some exceptions, depending on the type of calculator or computer software being used. To explain the possible exceptions, we need to look first at the order of operations in manual calculations.

Question 1 Calculating $5 + 2\sqrt{9}$ involves addition, multiplication, and the taking of a square root. In what order would you do these operations?

Question 2 In what order would you add, subtract, multiply, and take the cube to evaluate $1 + 2(11 - 2^3) - 4$?

Question 3 What steps would you use to evaluate $3(4)/\sqrt{36} - 10/(8 - 3)$?

These questions illustrate the following principles that are generally used in manual calculations and on calculators and computers.

Rule 1 *Operations are performed from left to right, except as described in Rules 2 through 5 below.*

Rule 2 *Expressions inside parentheses are evaluated as soon as they are reached.*

Rule 3 *Addition and subtraction have the lowest priority. If an addition or subtraction is followed by multiplication, division, a power, or a function, the addition or subtraction is postponed until another addition or subtraction or the end of the expression is reached.*

Rule 4 *Multiplication and division have medium priority. If a multiplication or division is followed by a power or a function, the multiplication or division is postponed until the power or function has been evaluated.*

Rule 5 *The taking of powers and evaluation of functions have the highest priority and are performed as soon as they are reached.*

Look over your calculations in Questions 1 through 3 to see how these rules apply in those cases.

The quantities in Questions 1 through 3 can be entered in most graphing calculators as

$$5 + 2 * \sqrt{\ } 9, \qquad 1 + 2 * (11 - 2 \wedge 3) - 4, \qquad \text{and} \qquad 3 * 4/\sqrt{\ } 36 - 10/(8 - 3) \tag{1}$$

with $*$ for multiplication and $\wedge$ for exponentiation. Expressions **(1)** could also be used on most computers with perhaps some changes in notation. For example, a symbol such as sqrt(9) might be needed in place of $\sqrt{\ }$ for the square root and $**$ instead of $\wedge$ for exponentiation.

Most graphing calculators allow for the representation of multiplication by juxtaposing symbols without the multiplication symbol $*$. On these machines the quantities in Questions 1 through 3 could also be entered as

$$5 + 2\sqrt{\ } 9, \qquad 1 + 2(11 - 2 \wedge 3) - 4, \qquad \text{and} \qquad 3(4)/\sqrt{\ } 36 - 10/(8 - 3). \tag{2}$$

The need to use $*$ and proper parentheses

Most graphing calculators and some computer software systems interpret certain expressions in unexpected ways because they use one or more of the following modifications of Rules 3 through 5 above. Answer the questions following these rules to see whether they apply to your machine.†

Rule 6 *The taking of powers has priority over the evaluation of functions that appear before their variables, such as the trigonometric functions, logarithms, e^x, $\sqrt{\ }$, and negation. Also, the parentheses in expressions such as $\sin(2)$ and $e \wedge (2)$ are ignored. (See Question 4.)*

Rule 7 *Multiplication by juxtaposition has priority over division and multiplication represented by $*$. (See Question 5.)*

Rule 8 *Multiplication represented by juxtaposition, where the second term is a number or a variable, has priority over the evaluation of functions that appear before their arguments. (See Questions 6 and 7.)*

Question 4 With your calculator or computer in radian mode, attempt to evaluate $\sin^3(2)$ by entering $\sin(2) \wedge 3$.‡ Do you obtain $(\sin 2)^3$, $\sin(2^3)$, or an error message? Explain.

Question 5 Attempt to evaluate $\frac{1}{5}(10) = 2$ by entering $1/5(10)$. Explain the result.

Question 6 Evaluate $\sqrt{\ }(4)\pi$. Do you obtain $(\sqrt{4}\,) * \pi$, $\sqrt{(4 * \pi)}$, or an error message? Explain.

Question 7 When you evaluate $\sin(5)(10)$, do you obtain $(\sin 5) * 10$, $\sin(5 * 10)$, or an error message? Explain.

Any difficulties that arise from calculations as in Questions 4 through 7 can be avoided by using multiplication signs or by putting parentheses around expressions that are to be evaluated first. To find the value of $\sin^3(2)$, enter $(\sin(2)) \wedge 3$; to calculate $(\sqrt{4}) * \pi$, enter $(\sqrt{\ }4)\pi$ or $\sqrt{\ }4 * \pi$; to calculate $(\sin(5)) * (10)$, enter $(\sin(5))(10)$ or $\sin(5) * 10$; and to calculate $\frac{1}{5}(10)$, enter $(1/5)(10)$ or $1/5 * 10$.

The next question shows how using the negation symbol $^-$ for subtraction can lead to an error message or give an incorrect result because a product is calculated instead of a difference. (If your machine does not use a negation symbol, skip this question.)

Question 8 Evaluate the expressions $2\pi - \pi$, $2\pi\,^-\pi$, $8 - 5$, and $8\,^-5$, where $-$ is the subtraction symbol and $^-$ is the negation symbol, and explain the results.

---

†The Texas Instruments TI-83 calculator uses Rules 3 through 5 without the modifications described in Rules 6 through 8. It does not use Rule 7, and to avoid the ambiguities in notation that lead to Rules 6 and 8, the symbols for built-in functions are accompanied by open parentheses when they are entered. The expression for the value of a function must then be terminated either by an ending parenthesis symbol or the end of the expression. The TI-86 generally uses Rule 6 but not Rules 7 or 8.

‡$\sin^3 x$ denotes $(\sin x)^3$.

Graphs, intersections, zeros, maxima, and minima

A graph $y = f(x)$ is generated on a calculator or computer by plotting points (x, y), where x is the horizontal coordinate of a pixel (dot) on the screen and y is the vertical pixel coordinate that is closest to the corresponding value $f(x)$ of the function.

The viewing window—the portion of the xy-plane that is shown by the calculator or computer—is selected to show whatever properties of the function are currently of interest, and does not generally show all of the domain of definition of the function. Also, the intervals of x and y that are used in the viewing window are often called the "ranges" of x and y, although usually neither of these intervals is the range of values of the function.

Question 9 Generate† the curves $y = x^2 - x - \frac{1}{2}$ and $y = \dfrac{1}{x^2+1}$ in the window $-2 \le x \le 2$, $-1 \le y \le 1.5$ as in Figure 1. Use your calculator or computer to find the approximate coordinates of their intersection that is to the right of the y-axis.

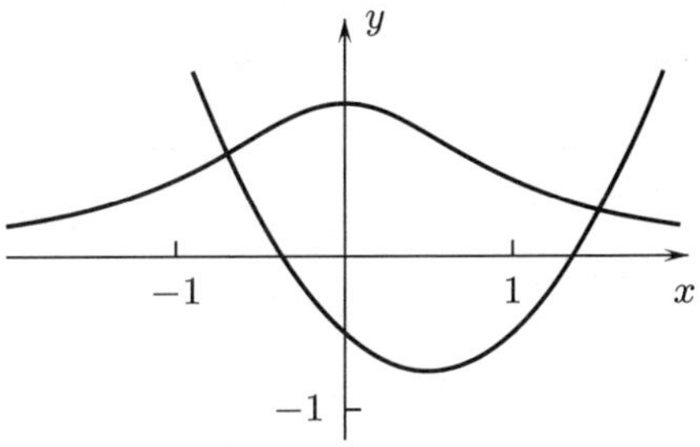

FIGURE 1

Question 10 Use your calculator or computer to find the approximate x-intercept of $y = x^2 - x - \frac{1}{2}$ that is to the right of the y-axis.

Question 11 Use your calculator or computer to find the approximate minimum value of $x^2 - x - \frac{1}{2}$ and the approximate value of x where it occurs.

Extraneous lines in graphs

In many cases calculators and computers connect pieces of graphs by vertical lines that are not part of the actual graphs.

Question 12 Generate $y = \dfrac{1}{x - \pi}$ and $y = \dfrac{x - \pi}{|x - \pi|}$ in the window $-5 \le x \le 10$ $-2 \le y \le 2$, and compare the results with the actual curves in Figures 2 and 3.

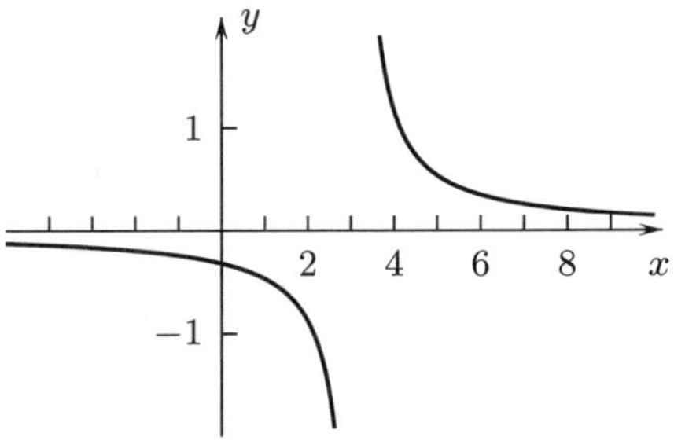

FIGURE 2

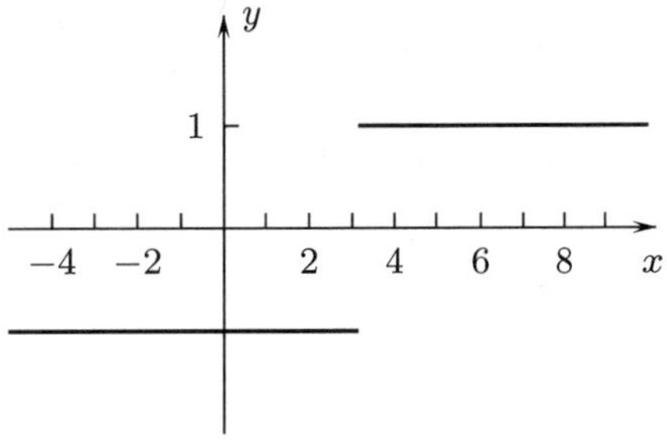

FIGURE 3

†Whenever you are asked in this text to generate a curve, generate it on your calculator or computer.

Decimals, exact values, and pixel coordinates

Not all numbers can be represented exactly as finite decimals, and it is important to distinguish exact expressions, such as $\frac{1}{3}$ and π, and decimal approximations, such as 0.33333 and 3.14159. You also need to recognize when coordinates obtained from graphs generated by calculators and computers are approximations.

Question 13 Use your calculator or computer to complete the table below of ten-digit values of $5x^{1/3} = 5\sqrt[3]{x}$ at $x = -27, -30, 4, 6, 8,$ and 10. The value $5(-27)^{1/3} = 5(-3) = -15$ is exact, but -15.53616253 is only a decimal approximation of $5(-30)^{1/3}$, which cannot be represented by a finite decimal. Its value to 20 decimal places, for example, is -15.53616252976929433439. Which y-values in the table in addition to -15 do you recognize as exact?

x	$y = 5x^{1/3} \doteq$	x	$y = 5x^{1/3} \doteq$
-27	-15	-30	-15.53616253
4		6	
8		10	

Question 14 (TI-82, TI-83, TI-85, and TI-86 calculators only) The calculator screen is formed of square or rectangular pixels that are darkened to display printing and graphs. To see how the choice of graphing window affects the coordinates of the pixels, set the window to $-47 \le x \le 47, -31 \le y \le 31$ on a TI-82 or TI-83 and $-63 \le x \le 63, -31 \le x \le 31$ on a TI-85 or TI-86 with x-scale $= 10$ and y-scale $= 10$ and unselect all functions in the Y= menu so no graphs are drawn. Press [GRAPH] on a TI-82 or TI-83 and select <GRAPH> on a TI-85 or TI-86, and use [▲] and the other cursor keys to see that in this case the pixels' coordinates are multiples of 1 and the pixels are 1 unit apart in the x- and y-directions. (Press [CLEAR] to remove the menu at the bottom of the TI-85 or TI-86 screen.) **(a)** Find choices of x-min, x-max, y-min, and y-max so that the pixels' coordinates are multiples of 100 and the pixels are 100 units apart in the x- and y-directions. **(b)** Find choices of x-min, x-max, y-min, and y-max so that the pixels' coordinates are multiples of 0.01 and the pixels are 0.01 units apart in the x- and y-directions.

Question 15 (TI-82, TI-83, TI-85, and TI-86 calculators only) Generate the curve $y = 5x^{1/3}$ with the window from the beginning of Question 14 so the pixels are at integer values of x and y. Use the <Trace> operation to trace values of the function along the graph. The calculator displays the x-coordinates of the pixels, as in Question 14, but not the y-coordinates of the pixels. What does it display?

Tune-Up Exercises

[A]*Answer provided.* [O]*Outline of solution provided.*

Use your calculator or computer to find the approximate decimal values of the expressions in Exercises T1 through T8. Do not simplify the expressions before entering them and be sure your machine is in radian mode for the trigonometric function in Exercise T1. In some cases extra parentheses are needed to express numerators, denominators, and exponents. The outlines of solutions are keyed to TI-82, TI-83, TI-85, and TI-86 calculators.

T1.[O] **(a)** $\sqrt{6}\ \cos(9/7)$ **(b)** $6\cos(\pi/7^3)$ (Edit the expression from part (a).)

T2.[O] $(-5 - 1.63 \times 10^{-2})^{-1}$

T3.[O] $\dfrac{2+8}{4-6} - 3^{5-1}$

T4.[O] $A + BC^D$ with $A = 7$, $B = 6$, $C = 5$, and $D = 4$ (Store the values first.)

T5.[O] $\frac{1}{2}\log_{10}(7)$

T6.[O] $\dfrac{1.34 \times 10^6 - 4 \times 10^5}{7.12 \times 10^{-8}}$

T7.[O] $\sqrt{4 + 7^{8-10}}$

T8.[O] $(-32)^{4/5}$

Exercises T9 through T16 illustrate types of errors that can arise from using improper expressions for functions when generating their graphs. Attempt to generate the curves in the window $-3 \le x \le 3, -2 \le y \le 2$ by entering the given expressions in your calculator or computer. You might obtain the correct curves, incorrect curves, or error messages, depending on what machine you are using. Explain the results.†

T9. Attempt to use $y = 1/2x$, $y = (1/2)x$, and $y = 1/2 * x$ to generate the line $y = \frac{1}{2}x$ in Figure 4.

T10. Try using $y = e\hat{}(-1)x$, $y = (e\hat{}(-1))x$, and $y = e\hat{}(-1) * x$ to generate the line $y = e^{-1}x$ in Figure 5.

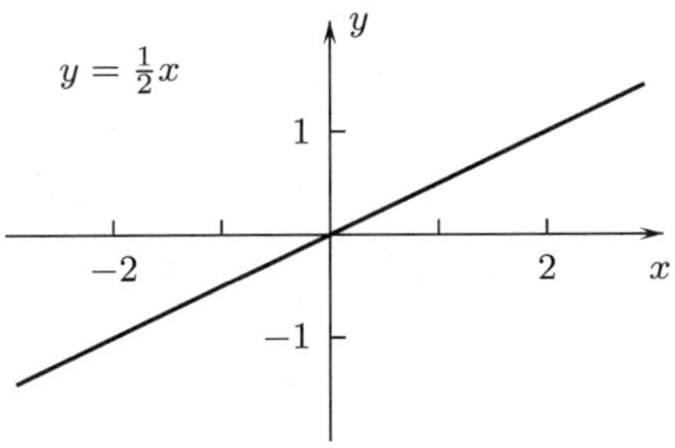

FIGURE 4

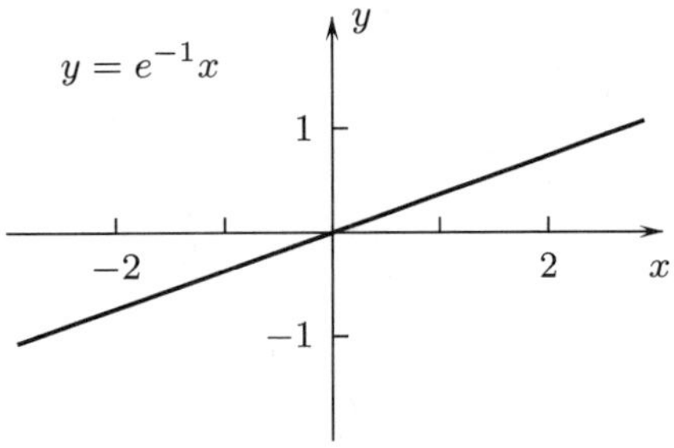

FIGURE 5

T11. Use $y = \sin(3)x^2$, $y = (\sin 3)x^2$, and $y = \sin(3) * x^2$ to generate the parabola $y = [\sin(3)]x^2$ in Figure 6.

T12. Use $y = \sin(x)^2$ and $y = (\sin x)^2$ to generate $y = \sin^2 x$ in Figure 7. (Recall that $\sin^2 x$ represents $(\sin x)^2$.)

†The ambiguities of notation that cause the types of errors illustrated in Exercises T9 through T13, T15, and T16 do not occur with a TI-83 calculator, and most of the ambiguities do not arise with the TI-86. (See the footnote before Rule 6 above).

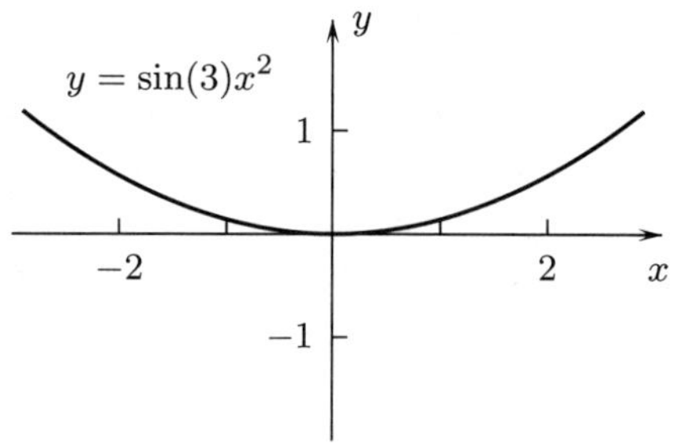

FIGURE 6

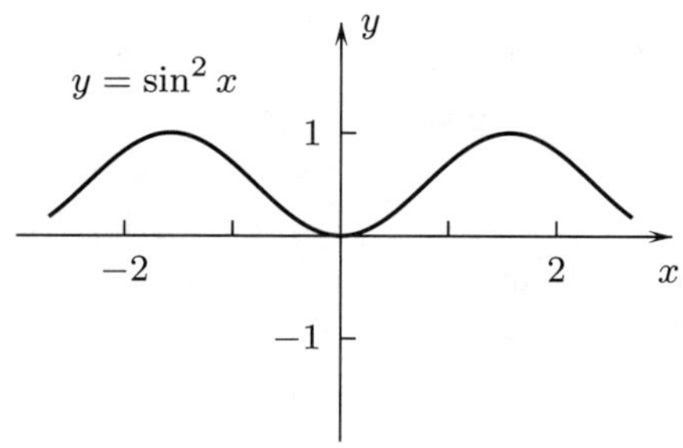

FIGURE 7

T13. Try using $y = \sqrt{\ }2x$, $y = \sqrt{\ }(2)x$ and $y = (\sqrt{\ }2)x$ to generate the line $y = (\sqrt{2})\,x$ in Figure 8.

T14. Attempt to use $y = x^{2}\,{}^{-}1$, with ${}^{-}$ the negation symbol and $y = x^2 - 1$, with $-$ the subtraction symbol, to generate the parabola $y = x^2 - 1$ in Figure 9.

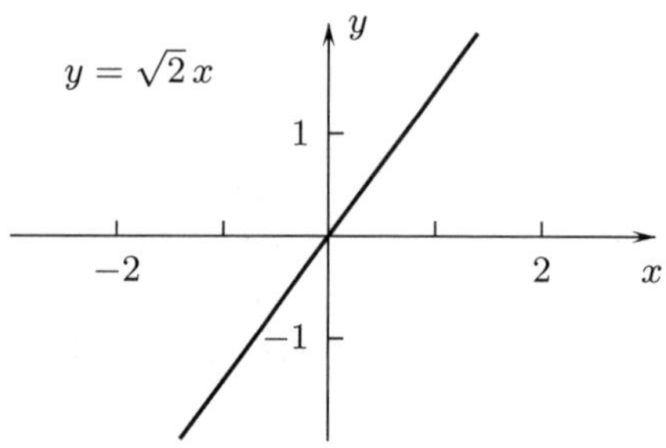

FIGURE 8

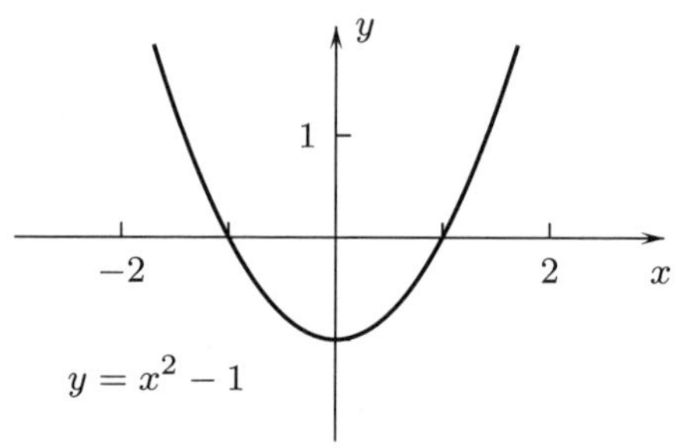

FIGURE 9

T15. Try using $y = 0.5\sin(x)^{-1}$ and $y = 0.5(\sin x)^{-1}$ to generate $y = \dfrac{0.5}{\sin x}$ in Figure 10.

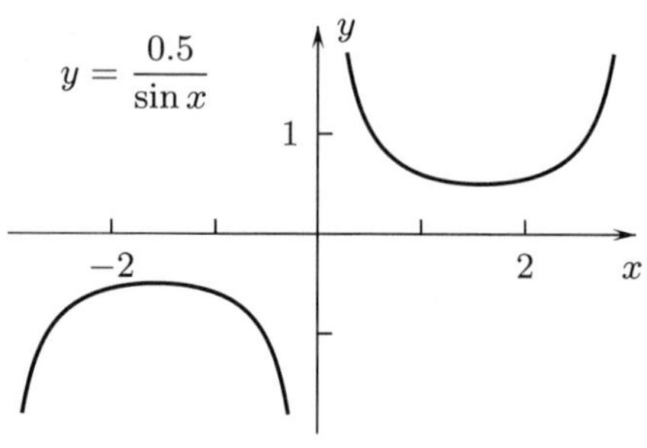

FIGURE 10

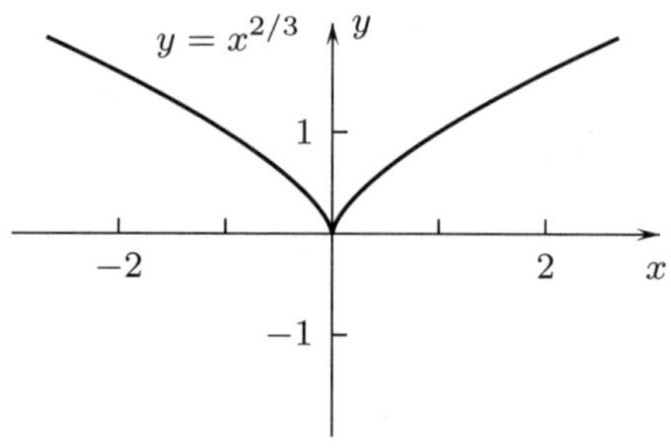

FIGURE 11

T16.[A] Use $y = x \wedge (2/3)$ to generate $y = x^{2/3}$ in Figure 11.

T17.[A] Generate the graph of $4x^2 - x^6$ in the window $-2 \leq x \leq 2, -2 \leq y \leq 4$. Then use the graph, a trace operation, or a procedure for finding maxima on your calculator or computer to find the approximate maximum value of $4x^2 - x^6$ (the y-coordinate of the highest point or points on its graph) and the approximate values of the two x's where the maximum occurs.

T18.[A] Generate the curves $y = 3\sqrt{x} - x$ and $y = x^2$ in the window $-0.5 \leq x \leq 2, -0.5 \leq y \leq 4$ and find the approximate coordinates of their intersection.

Problems

[A] *Answer provided.* [O] *Outline of solution provided.*

1.[A] The parentheses in the expression $1+((y/x)-(3(w^2)))$ determine the order of operations without using the priority conventions described in Rules 1 through 5. Rewrite the expression without parentheses by employing these conventions.

2.[A] Write the following expressions without parentheses by using built-up fractions $\frac{A}{B}$ and superscripts A^B: **(a[A])** $2/(x+3)$, **(b[A])** $2/x+3$, **(c)** $x \wedge 4+y$, **(d)** $x \wedge (4+y)$, **(e[A])** $x/y * z$, **(f)** $x/(y * z)$.

3.[A] Write $x \wedge (y \wedge z)$ and $(x \wedge y) \wedge z$ using superscripts A^B for exponents.

4. Which of **(a)** $(x * y) \wedge z$, **(b)** $x \wedge (y+z)$, **(c)** $(x \wedge y) \wedge z$, **(d)** $(x \wedge y) * (x \wedge z)$, **(e)** $(x \wedge z) * (y \wedge z)$, and **(f)** $x \wedge (y * z)$ have the same values for every choice of positive x, y, and z?

5.[A] Find the approximate maximum value of $2^x - \frac{1}{5}(3^x)$. Use the window $-1 \le x \le 5$, $-5 \le y \le 5$.

6. Find the approximate minimum value of $x^2 - \sin(4x)$. Be sure your calculator or computer is in radian mode, and use the window $-2.5 \le x \le 3$, $-2 \le y \le 5$.

7.[A] Find the approximate coordinates of the intersections of $y = 5 - x^2$ and $y = x^6$. Generate the curves in the window $-2.5 \le x \le 2.5$, $-1 \le y \le 7$.

8.[A] Find the approximate coordinates of the intersections of $y = \cos x$ and $y = x^2$. Be sure your calculator or computer is in radian mode, and generate the curves in the window $-3 \le x \le 3$, $-1.5 \le y \le 2$.

9. Find the approximate solutions of $\frac{1}{x} + x^2 = 3$. Generate $y = \frac{1}{x} + x^2$ and $y = 3$ in the window $-2.5 \le x \le 2.5, -2 \le y \le 5$.

10. Find the approximate solutions of $\sqrt{x} + \frac{1}{\sqrt{x}} = 2.5$. Generate $y = \sqrt{x} + \frac{1}{\sqrt{x}}$ and $y = 2.5$ with $-1 \le x \le 6, -0.5 \le y \le 4$.

RESPONSES TO QUESTIONS

Responses 1–3 in Appendix 1

Response 1 You would first calculate the square root of 9 to obtain $5 + 2(3)$, then multiply the 2 and 3 to have $5 + 6$, and then perform the addition for the result 11.

Response 2 You could start with the expression inside the parentheses by taking the cube of 2 to have $1 + 2(11 - 8) - 4$, and then subtracting 8 from 11 to obtain $1 + 2(3) - 4$. You could calculate the product next, yielding $1 + 6 - 4$. Then, adding the 1 and 6 would give $7 - 4$, and subtracting 4 from 7 would give the final result 3.

Response 3 You could multiply the 3 and 4 first to have $12/\sqrt{36} - 10(8 - 3)$. Then you could evaluate the square root, yielding $12/6 - 10/(8 - 3)$. Dividing 12 by 6 would give $2 - 10/(8 - 3)$. Next you would perform the subtraction inside the parentheses to have $2 - 10/5$. Finally, you would divide 5 into 10 to obtain $2 - 2$ and subtract for the final answer 0.

Responses 4–15 in Appendix 1

Texas Instruments TI-82 calculators

To turn the calculator on, press the [ON] key. Press [2nd] followed by the up cursor key [▲] to increase display contrast and by [▼] to decrease it. Change the four AAA batteries as soon as the screen dims when graphs are generated.

Press the [MODE] key. The screen should show

Normal Sci Eng
Float 0123456789
Radian Degree
Func Par Pol Seq
Connected Dot
Sequential Simul
FullScreen Split

with the words that are printed in bold type here highlighted. If the wrong item is highlighted on any line, use the cursor keys to move the flashing box to the correct item and press [ENTER]. <Normal> denotes normal notation, <Sci> is scientific notation, and <Eng> is engineering notation for decimals. <Float> causes decimals to be printed with 10 digits. Choosing a nonnegative integer instead of <Float> causes that many digits be shown after decimal points. <Radian> is for radians and <Degree> for degrees. <Func> is selected to generate graphs $y = f(x)$ of functions. <Par> is used with parametric equations, <Pol> is for polar coordinates, and <Seq> is for plotting sequences. Points on graphs are connected by lines if <Connected> is selected; graphs are generated one after another with <Sequential>; and the entire screen is used for graphs with <FullScreen>.

Press [WINDOW] and [▶] to highlight <FORMAT>. The screen should show

RectGC PolarGC
CoordOn CoordOff
GridOff GridOn
AxesOn AxesOff
LabelOff LabelOn

with the words in bold highlighted. Rectangular coordinates are used if <RecGC> is selected and polar coordinates with <PolarGC>. Coordinates of the cursor are displayed with <CoordOn>. You might occasionally want to choose <GridOn>, so that dots are placed on the screen at the points whose coordinates correspond to the tickmarks on the axes. The axes would not be shown with <AxesOff> and labels would be displayed with <LabelOn>.

Press [2nd] [STAT PLOT]. The word "Off" should appear under "Plot 1", "Plot 2", and "Plot 3" so no statistical plots will be drawn. If the word "On" is shown, scroll down to 4 and press [ENTER] twice.

The [CLEAR] key either clears an entry or returns you to a previous screen. [2nd] activates the blue commands above the keys. [2nd] [QUIT] returns you to the home screen where calculations are made. If you make an error in a command or calculation, the type of error is given and you have the choice of pressing [1] to go to the error to fix it, or [2] to cancel the incorrect command.

The key [ALPHA] puts the calculator in alpha mode, activating the yellow letters and other symbols above the keys. Pressing [2nd] [ALPHA] locks it in alpha mode and pressing [ALPHA] or [ENTER] takes it out of alpha-lock mode. Entering a number followed by [STO▶], [ALPHA], the letter, and [ENTER] assigns that number to the letter. The number can then be recalled with [ALPHA] followed by the letter.

In the home screen, [2nd] [ENTRY] recalls the last expression that was evaluated so it can be edited, if necessary, and used again. The [ON] key stops the generation of graphs, the running of programs, and other operations. The [ENTER] key can be used to interrupt and resume the generation of graphs.

Refer to the owner's manual for further information.

Response 4 The TI-82 generally uses Rules 6 through 8 in Appendix 1. In this case it probably uses Rule 6 and calculates the cube before the sine to yield 0.9893582466, which is the approximate decimal value of $\sin(2^3) = \sin(8)$. (If you obtained 0.1391731010, your calculator is using degrees instead of radians. Press [MODE], put the cursor on <Radian> and press [ENTER] to select radian mode. Press [2nd] [QUIT] [2nd] [ENTRY] to return to the home screen and recall the last typed line, and then [ENTER] for the correct answer.)

Response 5 Your calculator probably evaluates $1/5(10)$ as $1/(50) = 0.02$ because, by Rule 7, it multiplies the 10 and the 5 before performing the division.

Response 6 Your calculator probably interprets $\sqrt{}(4)\pi$ as $\sqrt{(4 * \pi)} \doteq 3.544907702$ using Rule 8: It multiplies 4 and π before taking the square root.

Response 7 Your calculator probably gives $\sin(50) \doteq -0.2623748537$ because, by Rule 8, it does the multiplication before evaluating the sine.

Response 8 Your calculator gives $3.141592654 \doteq \pi$ for $2\pi - \pi$, which it obtains by subtracting π from 2π. It probably gives the value $-19.73920880 \doteq -2\pi^2$ for $2\pi\,{}^-\pi$, which it interprets as 2π multiplied by $-\pi$. The expression $8-5$ equals 3, and you probably get an error message with $8\,{}^-5$.

Response 9 To use the required graphing window, press [WINDOW] [▼] and enter the values x-min $= -2$, x-max $= 2$, x-scl $= 1$, y-min $= -1$, y-max $= 1.5$ and y-scl $= 1$. Setting x-scl and y-scl equal to 1 causes the tickmarks to be one unit apart on both axes. Use [▲] [▼] [◀] and [▶] to move the cursor and [DEL] and [2nd] [INS] to delete and insert symbols.

Next press [$Y=$]. Press [CLEAR] to remove any previous formula for Y_1 and then [X,T,θ] [x^2] [$-$] [X,T,θ] [$-$] [1] [$\div$] [2] to have the first line read $Y_1 = X^2 - X - 1/2$. ([X,T,θ] is the same as [ALPHA] [X] in <Function> mode.) Move the cursor to after the equal sign on the second line, press [CLEAR] and then [1] [$\div$] [(] [X,T,θ] [x^2] [$+$] [1] [)], so the second line reads $Y_2 = 1/(X^2 + 1)$. Be sure only the first and second equal signs are highlighted.

Press [GRAPH]. Because the first two equal signs are highlighted, the parabola $y = x^2 - x - \frac{1}{2}$ will be drawn, followed by "Agnesi's versiera" $y = 1/(x^2 + 1)$, as in Figure 1 of Appendix 1.

Press [2nd] [CALC], move the cursor down to <intersect>, and press

ENTER. The screen should show the prompt "First curve?" and a flashing × should be on one of the curves. Press ENTER to select it. The × should move to the other curve and the prompt "Second curve?" should appear. Press ENTER to select the second curve. When the prompt "Guess?" appears, press ▶ or ◀ to trace the curve until the × is near the intersection at the right of the screen, and press ENTER. The calculator then does some extensive calculations to find the most accurate coordinates $(1.5250303, 0.300068702)$ of the intersection that it can determine.

Response 10 With the graphs from Question 9 displayed, press 2nd CALC, move the cursor down to <root> (for "zero"), and press ENTER. The screen should read "Lower bound?" If the × is not on the parabola $y = x^2 - x - \frac{1}{2}$, press ▲ so it is. We want the x-intercept that is near the right of the screen. Press ▶ or ◀ until the × is just to the left of that intercept and press ENTER. "Upper Bound?" should appear. Move the × to the right of the intercept and press ENTER. When "Guess?" appears, move the × near the intercept and press ENTER again. The approximate value 1.3660254 of the x-intercept should be displayed.

Response 11 With the graphs from Question 9 displayed, press 2nd CALC, move the cursor down to <minimum> and press ENTER. The screen should read "Lower bound?" If the × is not on the parabola, press ▲ so it is. Press ▶ or ◀ until the × is to the left of the lowest point on the parabola and press ENTER. "Upper Bound?" should appear. Move the × to the right of the lowest point and press ENTER. When "Guess?" appears, put the × near the lowest point and press ENTER again. The exact or approximate decimal values of the minimum $y = -0.75$ and of the value $x = 0.5$ where it occurs should be displayed.

Response 12 Generate $Y_1 = 1/(x - \pi)$ first and then $Y_1 = (x - \pi)/\text{abs}(x - \pi)$ with the given ranges of x and y. (ABS is on the key with x^{-1}.) Be sure only the equal sign on the first line is highlighted. The two pieces of each graph will be connected by an extraneous vertical line if <Connected> has been chosen in the <Mode> menu and not if <Dot> has been selected.

Response 13 To do these calculations quickly, set $Y_1 = 5X \wedge (1/3)$ in the Y= menu. Press 2nd TblSet for <TABLE SETUP> and select <Ask> after <Indpnt>, so you will be able to give any values you want for x and <Auto> after <Depend> so the values of the function will be calculated automatically. Press 2nd TABLE and then (–) 2 7 ENTER. The value -27 of x will appear in the first column and the value -15 of Y_1 in the second column. Press (–) 3 0 ENTER to have -30 in the first column and -15.54 in the second. To display more digits of this value of Y_1, move the cursor to the second row of the second column. To complete the table below, move the cursor to other rows of the first column and enter the values of x. Only the values of Y_1 at -27 and $x = 8$ are exact because only -27 and 8 of the x-values are perfect cubes. (To clear the table, select <Auto> in the <TABLE SETUP> menu, display the table, and select <Ask> again.)

x	$y = 5x^{1/3} \doteq$	x	$y = 5x^{1/3} \doteq$
-27	-15	-30	-15.53616253
4	7.9370052598	6	9.085602964
8	10	10	10.772173450

Response 14 **(a)** One answer: $-4700 \leq x \leq 4700, -3100 \leq y \leq 3100$. Use x-scale $= 1000$ and y-scale $= 1000$ and check with the cursor. **(b)** One answer: $-0.47 \leq x \leq 0.47$, $-0.31 \leq y \leq 0.31$ with x-scale $= 0.1$ and y-scale $= 0.1$

Response 15 Figure R15. The trace procedure gives the x-coordinates of the pixels and the values of the function at those points, correct to eight digits, rather than the y-coordinates of the pixels as when no graph is being traced.

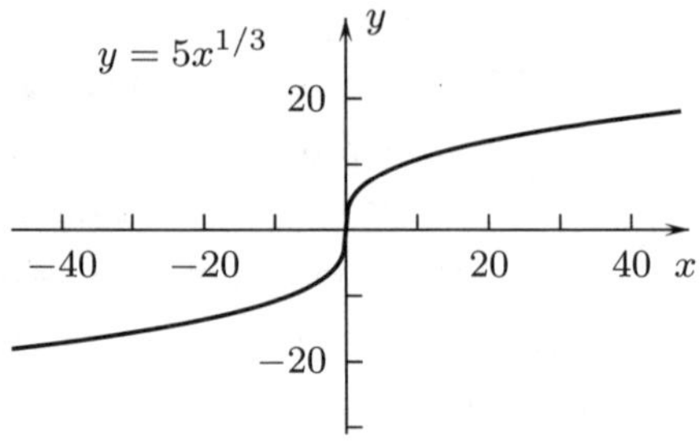

Figure R15

Responses 4–15 in Appendix 1

Texas Instruments TI-83 calculators

To turn the calculator on, press the ON key. Press 2nd followed by the up cursor key ▲ to increase display contrast and by ▼ to decrease it. Change the four AAA batteries as soon as the screen dims when graphs are generated.

Press the MODE key. The screen should show

Normal Sci Eng
Float 0123456789
Radian Degree
Func Par Pol Seq
Connected Dot
Sequential Simul
Real a+bi re^θi
Full Horiz G–T

with the words that are printed in bold type here highlighted. If the wrong item is highlighted on any line, use the cursor keys to move the flashing box to the correct item and press ENTER. <Normal> denotes normal notation, <Sci> is scientific notation, and <Eng> is engineering notation for decimals. <Float> causes decimals to be printed with 10 digits. Choosing an integer instead of <Float> causes that many digits be shown after decimal points. <Radian> is for radians and <Degree> for degrees. <Func> is selected to generate graphs $y = f(x)$ of functions. <Par> is used with parametric equations, <Pol> is for polar coordinates, and <Seq> is for plotting sequences. Points on graphs are connected by lines if <Connected> is selected; graphs are generated one after another with <Sequential>. With <Real> selected, only real numbers are used; with <a+ib> complex numbers are given in rectangular coordinates; and with <re^θi> complex numbers are displayed in polar coordinates. The entire screen is used for graphs with <Full> and the top half with <Horiz>. With <G–T> the left half of the screen shows the graphs and the right side the table.

Press 2nd FORMAT. The screen should show

RectGC PolarGC
CoordOn CoordOff
GridOff GridOn
AxesOn AxesOff
LabelOff LabelOn
ExprOn ExprOff

with the words in bold highlighted. Rectangular coordinates are used if <RecGC> is selected and polar coordinates with <PolarGC>. Coordinates of the cursor are displayed with <CoordOn>. You might occasionally want to choose <GridOn>, so that dots are placed on the screen at the points whose coordinates correspond to the tickmarks on the axes. The axes would not be shown with <AxesOff> and labels would be displayed with <LabelOn>. With <ExprOn> selected, the formula for a curve is displayed when it has been selected for tracing or other operations.

Press [2nd] [STAT PLOT]. The word "Off" should appear under "Plot 1", "Plot 2", and "Plot 3" so no statistical plots will be drawn. If the word "On" is shown, scroll down to 4 and press [ENTER] twice.

The [CLEAR] key either clears an entry or returns you to a previous screen. [2nd] activates the yellow commands above the keys. [2nd] [QUIT] returns you to the home screen where calculations are made. If you make an error in a command or calculation, the type of error is given and you have the choice of pressing [1] to cancel the incorrect command, or [2] to go to the error to fix it.

The key [ALPHA] puts the calculator in alpha mode, activating the green letters and other symbols above the keys. Pressing [2nd] [A-LOCK] locks it in alpha mode and pressing [ALPHA] or [ENTER] takes it out of alpha-lock mode. Entering a number followed by [STO▶], [ALPHA], the letter, and [ENTER] assigns that number to the letter. The number can then be recalled with [ALPHA] followed by the letter.

In the home screen, [2nd] [ENTRY] recalls the last expression that was evaluated so it can be edited, if necessary, and used again. The [ON] key stops the generation of graphs, the running of programs, and other operations. The [ENTER] key can be used to interrupt and resume the generation of graphs.

Refer to the owner's manual for further information.

Response 4 You do not have to worry about Rules 6 through 8 in Appendix 1 with the TI-83. It gives multiplication by juxtaposition the same priority as multiplication with a multiplication symbol, and automatically inserts parentheses after functions that appear before the values of their variables. You select the variables by closing the parentheses. In this case sin(2) ∧ 3 gives the expected $(\sin 2)^3 \doteq 0.7518269447$. (If you obtained $4.2150670996 \times 10^{-5}$, your calculator is using degrees instead of radians. Press [MODE], put the cursor on <Radian> and press [ENTER] to select radian mode. Press [2nd] [QUIT] [2nd] [ENTRY] to return to the home screen and recall the last typed line, and then [ENTER] for the correct answer.)

Response 5 The TI-83 gives the expected value 2.

Response 6 The TI-83 gives the expected value $(\sqrt{4})\pi = 2\pi \doteq 3.544907702$.

Response 7 The TI-83 gives the expected value $(\sin 5) \times 10 \doteq -9.589242747$.

Response 8 Your calculator gives $3.141592654 \doteq \pi$ for $2\pi - \pi$, which it obtains by subtracting π from 2π. It probably gives the value $-19.73920880 \doteq -2\pi^2$ for $2\pi\,{}^-\pi$, which it interprets as 2π multiplied by $-\pi$. The expression $8-5$ equals 3, and you probably get an error message with $8\,{}^-5$.

Response 9 To use the required graphing window, press [WINDOW] and enter the values Xmin = −2, Xmax = 2, Xscl = 1, Ymin = −1, Ymax = 1.5 and Yscl = 1. Setting Yscl and Yscl equal to 1 causes the tickmarks to be one unit apart on both axes. Have Xres = 1 so graphs are drawn with maximum resolution. Use [▲] [▼] [◀] and [▶] to move the cursor and [DEL] and [2nd] [INS] to delete and insert symbols.

Next press [$Y=$]. Press [CLEAR] to remove any previous formula for Y_1 and then press [X,T,θ,n] [x^2] [−] [X,T,θ,n] [−] [1] [÷] [2] to have the first line read $Y_1 = X^2 - X - 1/2$. ([X,T,θ,n] is the same as [ALPHA] [X] in <Function> mode.) Move the cursor to after the equal sign on the second line, press [CLEAR] and then [1] [÷] [(] [X,T,θ,n] [x^2] [+] [1] [)], so the second line reads $Y_2 =$

$1/(X^2+1)$. Be sure only the first and second equal signs are highlighted.

Press GRAPH . Because the first two equal signs are highlighted, the parabola $y = x^2 - x - \frac{1}{2}$ will be drawn, followed by "Agnesi's versiera" $y = 1/(x^2+1)$, as in Figure 1 of Appendix 1.

Press 2nd CALC , move the cursor down to <intersect>, and press ENTER . The screen should show the prompt "First curve?" and a flashing $\times$ should be on one of the curves. Press ENTER to select it. The $\times$ should move to the other curve and the prompt "Second curve?" should appear. Press ENTER to select the second curve. When the prompt "Guess?" appears, press ▶ or ◀ to trace the curve until the $\times$ is near the intersection at the right of the screen, and press ENTER . The calculator then does some extensive calculations to find the most accurate coordinates $(1.5250303, 0.30068702)$ of the intersection that it can determine.

Response 10 With the graphs from Question 9 displayed, press 2nd CALC , move the cursor down to <zero>, and press ENTER . The screen should read "Left bound?" If the $\times$ is not on the parabola $y = x^2 - x - \frac{1}{2}$, press ▲ so it is. We want the x-intercept that is near the right of the screen. Press ▶ or ◀ until the $\times$ is just to the left of that intercept and press ENTER . "Right Bound?" should appear. Move the $\times$ to the right of the intercept and press ENTER . When "Guess?" appears, move the $\times$ near the intercept and press ENTER again. The approximate value 1.3660254 of the x-intercept should be displayed.

Response 11 With the graphs from Question 9 displayed, press 2nd CALC , move the cursor down to <minimum> and press ENTER . The screen should read "Left bound?" If the $\times$ is not on the parabola, press ▲ so it is. Press ▶ or ◀ until the $\times$ is to the left of the lowest point on the parabola and press ENTER . "Right Bound?" should appear. Move the $\times$ to the right of the lowest point and press ENTER . When "Guess?" appears, put the $\times$ near the lowest point and press ENTER again. The exact or approximate decimal values of the minimum $y = -0.75$ and of the value $x = 0.5$ where it occurs should be displayed.

Response 12 Generate $Y_1 = 1/(x-\pi)$ first and then $Y_1 = (x-\pi)/\text{abs}(x-\pi)$ with the given ranges of x and y. (For abs press MATH ▶ 1 .) Be sure only the equal sign on the first line is highlighted. The two pieces of each graph will be connected by an extraneous vertical line if <Connected> has been chosen in the <Mode> menu and not if <Dot> has been selected.

Response 13 To do these calculations quickly, set $Y_1 = 5X \wedge (1/3)$ in the Y= menu. Press 2nd TBLSET for <TABLE SETUP> and select <Ask> after <Indpnt>, so you will be able to give any values you want for x and <Auto> after <Depend> so the values of the function will be calculated automatically. Press 2nd TABLE and then (–) 2 7 ENTER . The value -27 of x will appear in the first column and the value -15 of Y_1 in the second column. Press (–) 3 0 ENTER to have -30 in the first column and -15.54 in the second. To display more digits of this value of Y_1, move the cursor to the second row of the second column. To complete the table below, move the cursor to other rows of the first column and enter the values of x. Only the values of Y_1 at -27 and $x = 8$ are exact because only -27 and 8 of the x-values are perfect cubes. (To clear the table, select <Auto> in the <TABLE SETUP> menu, display the table, and select <Ask> again.)

x	$y = 5x^{1/3} \doteq$	x	$y = 5x^{1/3} \doteq$
-27	-15	-30	-15.53616253
4	7.93700525984	6	9.08560296416
8	10	10	10.7721734502

Response 14 **(a)** One answer: $-4700 \leq x \leq 4700, -3100 \leq y \leq 3100$. Use x-scale $= 1000$ and y-scale $= 1000$ and check with the cursor. **(b)** One answer: $-0.47 \leq x \leq 0.47$, $-0.31 \leq y \leq 0.31$ with x-scale $= 0.1$ and y-scale $= 0.1$

Response 15 Figure R15. The trace procedure gives the x-coordinates of the pixels and the values of the function at those points, correct to eight digits, rather than the y-coordinates of the pixels as when no graph is being traced.

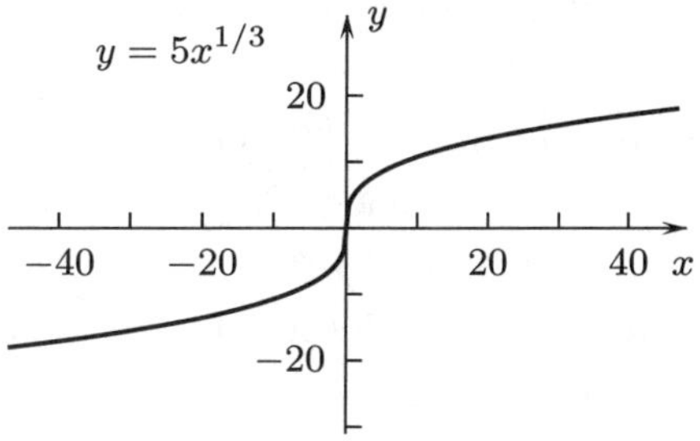

Figure R15

Responses 4–15 in Appendix 1

Texas Instruments TI-85 and TI-86 calculators

Read the parts of the owner's manual dealing with operating the calculator and on generating graphs of functions. To turn the calculator on, press the ON key. Press 2nd followed by the up cursor key ▲ to increase display contrast and by ▼ to decrease it. Change the four AAA batteries as soon as the screen dims when graphs are generated.

Press the 2nd and MODE keys. The screen should show

Normal Sci Eng
Float 012345678901
Radian Degree
RectC PolarC

Func Pol Param DifEq
Dec Bin Oct Hex
RectV CylV SphereV
dxDer1 dxNDer

with the words that are printed in bold here highlighted. If the wrong item is highlighted on any line, use the cursor keys to move the flashing box to the correct item and press ENTER. <Normal> denotes normal notation, <Sci> is scientific notation, and <Eng> is engineering notation for decimals. <Float> causes decimals to be printed with twelve digits. Choosing a nonnegative integer instead of <Float> causes that many digits be shown after decimal points. (Use the second 0 for ten digits and the final 1 for eleven.). <Radian> is for radians and <Degree> is for degrees. The other selections will be explained as needed.

The key 2nd activates the yellow commands above the keys. EXIT is used to return to a previous screen and to remove menus. 2nd QUIT returns you to the home screen where calculations are made. CLEAR with the cursor on a blank line of the home screen clears the screen. In other cases it clears the line with the cursor or removes a menu.

Press EXIT or 2nd QUIT to return to the home screen and then GRAPH for the first row of the graph menu. Press MORE to see the second row of the menu and then F3 for <FORMT>. The screen should read

RectGC PolarGC	**GridOff** GridOn
CoordOn CoordOff	**AxesOn** AxesOff
DrawLine DrawDot	**LabelOff** LabelOn
SeqG SimulG	

with the words in bold highlighted. With <RecGC> and <CoordOn> selected, rectangular coordinates are used and the coordinates of the cursor are displayed with graphs. Points on graphs are connected if <DrawLine> is chosen and not with <DrawDot>. Use <SeqG> (sequential graphs) to have two or more graphs drawn one after the other, and <SimulG> (simultaneous graphs) to have them drawn at the same time. If <GridOn> were selected, dots would be placed on the screen at the points whose coordinates correspond to the tickmarks on the axes. The axes would not be shown with <AxesOff> and labels would be displayed with <LabelOn>.

The key ALPHA puts the calculator in upper-case alpha mode, activating the blue letters and other symbols above the keys. Pressing ALPHA ALPHA locks the calculator in upper-case alpha mode and pressing ALPHA or ENTER takes it out of upper-case alpha-lock mode. 2nd ALPHA puts it in lower-case alpha mode. Entering a number followed by STO▶, one or more letters, and ENTER assigns that number to the letter or letters. The number can then be recalled by entering the letter or letters. The calculator is locked in alpha mode after STO▶ is pressed.

If you make an error in a command or calculation, the type of error is given and a menu appears. Select <GOTO> to go to the error to correct it or <QUIT> to cancel the incorrect command.

In the home screen, 2nd ENTRY recalls the last expression that was evaluated so it can be edited, if necessary, and used again. The ON key stops the generation of graphs, the running of programs, and other operations. The ENTER key can be used to interrupt and resume the generation of graphs.

Response 4 The TI-85 and TI-86 generally use Rule 6 in Appendix 1. In this case they probably calculate the cube before the sine to yield 0.989358246623, which is the approximate decimal value of $\sin(2^3) = \sin(8)$. (If you obtained 0.13917310096, your calculator is using degrees instead of radians. Press 2nd MODE, put the cursor on <Radian> and press ENTER to select radian mode. Press 2nd QUIT 2nd ENTRY to return to the home screen and recall the last typed line, and then ENTER for the correct answer.)

Response 5 The TI-85 probably evaluates $1/5(10)$ as $1/(50) = 0.02$ because it uses Rule 7 and multiplies the 10 and the 5 before performing the division. The TI-86 probably gives $(1/5)(10) = 0.02$, as expected, because it does not use Rule 7.

Response 6 The TI-85 calculator probably gives $\sqrt{(4 * \pi)} \doteq 3.54490770181$ because is uses Rule 8 and multiplies the 4 and the π before taking the square root. The TI-86 probably yields $(\sqrt{4})\pi = 2\pi \doteq 6.28318530718$, as expected, because it does not use Rule 8.

Response 7 The TI-86 probably gives $\sin(50) \doteq -0.262374853704$, as expected, because it uses Rule 8 and does the multiplication before evaluating the sine. The TI-86 probably yields $10\sin(5) \doteq -9.58924274663$ because it does not use Rule 8.

Response 8 Your calculator gives $3.14159265359 \doteq \pi$ for $2\pi - \pi$, which it obtains by subtracting π from 2π. It probably gives the value $-19.7392088022 \doteq -2\pi^2$ for $2\pi\ {}^{-}\ \pi$, which it interprets as 2π multiplied by $-\pi$. The expression $8 - 5$ equals 3, and you probably get -40 or an error message with $8\,{}^{-}5$, which the calculator either interprets as the product of 8 and -5 or does not accept.

Response 9 Press [EXIT] or [2nd] [QUIT] to go to the home screen and then [GRAPH] for the graph menu. Press [F2] for <RANGE> on a TI-85 and <WIND> (for "window") on a TI-86. Enter the values x-Min $= -2$, x-Max $= 2$, x-Scl $= 1$, y-Min $= -1$, y-Max $= 1.5$, and y-Scl $= 1$. Setting x-Scl and y-Scl equal to 1 causes the tickmarks to be one unit apart on both axes. Use [▲] [▼] [◄] and [►] to move the cursor and [DEL] and [2nd] [INS] to delete and insert symbols.

Next press [$F1$] for < $y(x)$ => and [CLEAR] to erase any previous definition of $y1$. Press [x-VAR] [x^2] [−] [x-VAR] [−] [1] [÷] [2] to have the first line read $y1 = x^2 - x - 1/2$. (You could also select x from the menu or enter it as a lower case variable with [2nd] [alpha] [X].) If the first equal sign is not highlighted, move the cursor to that line and press [F5] for <SELCT> to highlight it. Move the cursor to after the equal sign on the second line, press [CLEAR] and then [1] [÷] [(] [x-VAR] [x^2] [+] [1] [)], so the second line reads $y2 = 1/(x^2 + 1)$. Be sure the first and second equal signs are highlighted and any other equal signs are not.

Press [F5] or [2nd] [M5] for <GRAPH>. Because the first two equal signs are highlighted, the parabola $y = x^2 - x - \frac{1}{2}$ will be drawn, followed by "Agnesi's versiera" $y = 1/(x^2 + 1)$, as in Figure 1 of Appendix 1.

Press [MORE] to display more of the graph menu and [F1] for <MATH>. Press [MORE] and [F5] for <ISECT>. The screen should show a flashing × on one of the curves. Press [►] or [◄] to trace the curve until the × is near the intersection at the right of the screen, and press [ENTER] to select the first curve. The × should move to the other curve. Press [ENTER] to select the second curve. On a TI-86 press [ENTER] again to enter the x-coordinate of the × as an initial guess of the solution. The calculator will carry out extensive calculations to find the most accuratare coordinates $(1.5250302511, 0.30068701569)$ of the intersection that it can determine.

Response 10 With the graphs from Question 9 displayed, press [GRAPH] [MORE], [F1] for <MATH>, and select <ROOT> ("zero"). If the × is not on the parabola $y = x^2 - x - \frac{1}{2}$, press [▲] so it is. On a TI-85 press [►] or [◄] until the × is near the x-intercept at the right of the screen and press [ENTER]. On a TI-86 position the × to the left of the intersection and press [ENTER] to choose a lower bound for the search for the solution, move the × to the right of the intersection and press [ENTER] to choose a upper bound, and put the × near the intersection and press [ENTER] for an initial guess. The approximate value 1.3660254038 of the x-intercept should be displayed.

Response 11 With the graphs from Question 9 displayed, press [GRAPH] [MORE], [F1] for <MATH>, and select <FMIN>. If the × is not on the parabola $y = x^2 - x - \frac{1}{2}$, press [▲] so it is. On a TI-85 press [►] or [◄] until the × is near the lowest point on the curve and press [ENTER]. On a TI-86 position the × to the left of the intersection and press [ENTER] to choose a lower bound for the search for the minimum, move the × to the right of the intersection and press [ENTER] to choose a upper bound, and put the × near the lowest point and press [ENTER] for an initial guess. The exact or approximate decimal values of the minimum $y = -0.75$ and of the value $x = 0.5$ where it occurs should be displayed.

Response 12 Generate $y1 = 1/(x - \pi)$ first and then $y1 = (x - \pi)/\text{abs}(x - \pi)$ with the given ranges of x and y. Press [2nd] [CATALOG] (and [CATLG] on the TI-86) [ENTER] or [2nd] [alpha] [*A*] [2nd] [alpha] [*B*] [2nd] [alpha] [*S*] for <abs>. The two pieces of each graph will be connected by extraneous lines if <DrawLine> has been chosen in the FORMAT menu (found with <GRAPH> <MORE>) and not if <DrawDot> has been selected.

Response 13 To do these calculations quickly, set $y1 = 5x \wedge (1/3)$ in the $y=$ menu. Press [2nd] [QUIT] and [(−)] [2] [7] [STO▶] [x-VAR] [2nd] [:] [2nd] [alpha] [Y] [ALPHA] [ALPHA] [1] to have the screen read $-27 \to x : y1$. The two [ALPHA] keys are needed because [STO▶] locks the calculator in alpha mode. Press [ENTER] for the value -15 of the function. Press [2nd] [ENTRY] and edit the expression to read $-30 \to x : y1$. Press [ENTER] for the second value -15.5361625298, which rounded to ten digits is the number in the table. Repeat the procedure for each other x to complete the table. Only the values of y at $x = -27$ and $x = 8$ are exact because only -27 and 8 of the x-values are perfect cubes.

x	$y = 5x^{1/3} \doteq$	x	$y = 5x^{1/3} \doteq$
-27	-15	-30	-15.53616253
4	7.9370052598	6	9.085602964
8	10	10	10.772173450

Response 14 **(a)** One answer: $-6300 \le x \le 6300, -3100 \le y \le 3100$. Use x-scale $= 1000$ and y-scale $= 1000$ and check with the cursor.
(b) One answer: $-0.63 \le x \le 0.63$, $-0.31 \le y \le 0.31$ with x-scale $= 0.1$ and y-scale $= 0.1$

Response 15 Figure R15. The trace procedure gives the x-coordinates of the pixels and the approximate decimal values of the function at those points, rather than the y-coordinates of the pixels as in Question 14.

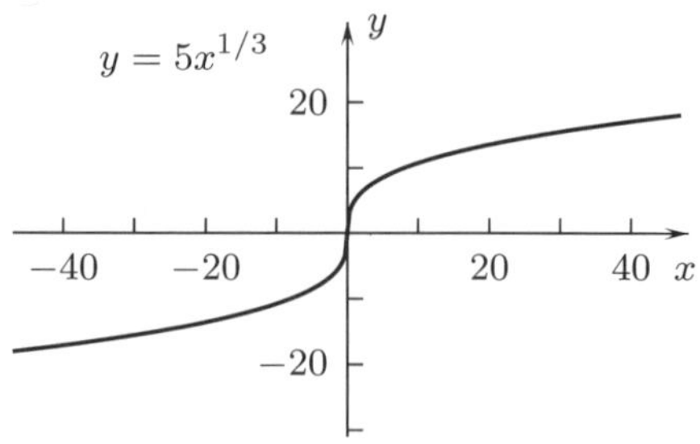

Figure R15

ANSWERS AND OUTLINES OF SOLUTIONS

Tune-up exercises for a TI-82 calculator

T1a. 0.6888851432 • (If your result is 2.448873047, your calculator is not in radian mode.) • Press [2nd] [√] [6] [cos] [(] [9] [÷] [7] [)] to have the screen read $\sqrt{\ } 6\cos(9/7)$. Then press [ENTER] for the answer.

T1b. 5.999748331 • If your last operation was the calculation of $\sqrt{\ } 6\cos(9/7)$, press [2nd] [ENTRY] to put it back on the screen. If you performed other calculations, type $\sqrt{\ } 6\cos(9/7)$ again. Press [◀] until the cursor is over the square root sign and press [DEL] to delete it. Move the cursor to the 9 and press [2nd] [π] to replace the 9 with π. Put the cursor on the close parenthesis and press [2nd] [INS]. Press [∧] [3] to insert ∧ 3 before the close parenthesis, so the screen reads $6\cos(\pi/7 \wedge 3)$, and press [ENTER] for the answer.

T2. -0.1993501186 • Press [(] [(–)] [5] [–] [1] [.] [6] [3] [2nd] [EE] [(–)] [2] [)] [x^{-1}] to display $(^{-}5 - 1.63\text{E}^{-}2)^{-1}$. ($A\,\text{E}\,n$ stands for $A \times 10^n$.) Press [ENTER] for the answer. Notice that [(–)] is for negation, [–] is for subtraction, and [x^{-1}] is for taking reciprocals.

T3. -86 • Use [(] [2] [+] [8] [)] [÷] [(] [4] [–] [6] [)] [–] [3] [∧] [(] [5] [–] [1] [)] to display $(2+8)/(4-6) - 3 \wedge (5-1)$. Press [ENTER] for the answer.

T4. 3757 • To put the commands on one line for possible later editing, press [7] [STO▶] [ALPHA] [A] [2nd] [:] [6] [STO▶] [ALPHA] [B] [2nd] [:] [5] [STO▶] [ALPHA] [C] [2nd] [:] [4] [STO▶] [ALPHA] [D] [2nd] [:] [ALPHA] [A] [+] [ALPHA] [B] [ALPHA] [C] [∧] [ALPHA] [D] so the screen reads $7 \to A : 6 \to B : 5 \to C : 4 \to D : A + BC \wedge D$. Press [ENTER] to store the values of A, B, C and D and calculate the answer.

T5. 0.422549020 • Use [(] [1] [÷] [2] [)] [LOG] [(] [7] [)] [ENTER].

T6. $1.320224719 \times 10^{13}$ • Use [(] [1] [.] [3] [4] [2nd] [EE] [6] [–] [4] [2nd] [EE] [5] [)] [÷] [(] [7] [.] [1] [2] [2nd] [EE] [(-)] [8] [)] [ENTER].

T7. 2.00509555 • Use [2nd] [√] [(] [4] [+] [7] [∧] [(] [8] [–] [1] [0] [)] [)] [ENTER].

T8. 16 • The TI-82 calculator will evaluate $x^{1/n}$ for negative x and odd integers n, but will not accept $x^{m/n}$ with negative x, odd n, and m an integer > 1. Consequently, it is generally best to write $x^{m/n}$ as $(x^m)^{1/n}$ for odd n. The symbols [(-)] [3] [2] [∧] [(] [4] [÷] [5] [)] [ENTER] give the wrong result since the negation is perfomed after the power. The symbols [(] [(-)] [3] [2] [)] [∧] [(] [4] [÷] [5] [)] [ENTER] probably result in an error message since the calculator cannot evaluate the four-fifths power of the negative number. The expressions [(] [(-)] [3] [2] [)] [∧] [4] [∧] [(] [1] [÷] [5] [)] [ENTER] and [(] [(-)] [3] [2] [)] [∧] [(] [1] [÷] [5] [)] [∧] [4] [ENTER] give the correct result, since for any positive odd integer n and negative X, the calculator interprets $X \wedge (1/n)$ as the negative nth root of X.

T9. $y = 1/2x$ does not work because the multiplication by juxtaposition is performed before the division, giving $y = 1/(2x)$. • $y = (1/2)x$ and $y = 1/2 * x$ work because the expression inside parentheses is evaluated first and multiplication with $*$ is performed after the division.

T10. $y = e \wedge (-1)x$ does not work because the multiplication by juxtaposition is performed before the exponentiation, giving $y = e^{-x}$. • $y = (e \wedge (-1))x$ and $y = e \wedge (-1) * x$ work because the expression inside parentheses is evaluated first and multiplication with $*$ is performed after the exponentiation.

T11. $y = \sin(3)x^2$ does not work because the multiplication by juxtaposition is performed before the sine is calculated, giving $y = \sin(3x^2)$. • $y = (\sin 3)x^2$ and $y = \sin(3) * x^2$ work because the expression inside parentheses is evaluated first and multiplication with $*$ is performed after the sine is evaluated.

T12. $y = \sin(x)^2$ does not work because the square is calculated before the sine, giving $y = \sin(x^2)$. • $y = (\sin x)^2$ works because the expression inside parentheses is evaluated first.

T13. $y = \sqrt{\ } 2x$ and $y = \sqrt{\ }(2)x$ do not work because the product is calculated before the square root, giving $y = \sqrt{2x}$. • $y = (\sqrt{\ } 2)x$ works because the expression inside parentheses is evaluated first.

T14. $y = x^2\,{}^{-}1$ does not work because it gives x^2 multiplied by -1. • $y = x^2 - 1$ works.

T15. $y = 0.5\sin(x)^{-1}$ does not work because it gives $0.5\sin(x^{-1})$. • $y = 0.5(\sin x)^{-1}$ works.

T16. If you obtain only the right half of the curve in Figure 11 of Section 1, try $y = (x \wedge (1/3)) \wedge 2$ or $y = (x^2) \wedge (1/3)$.

T17. See the response to Question 11 of Appendix 1. • Maximum ≈ 3.079 at $x \approx \pm 1.075$

T18. See the response to Question 9 of Appendix 1. • $x \approx 1.472, y \approx 2.168$

Tune-up exercises for a TI-83 calculator

T1a. 0.6888851432 • (If your result is 2.448873047, your calculator is not in radian mode.) • Press [2nd] [√] [6] [)] [cos] [9] [÷] [7] [)] to have the screen read √(6) cos (9/7). Then press [ENTER] for the answer.

T1b. 5.999748331 • If your last operation was the calculation of √(6) cos (9/7), press [2nd] [ENTRY] to put it back on the screen. If you performed other calculations, type √(6) cos (9/7) again. Press [◀] until the cursor is over the square root sign and press [DEL] to delete it. Move the cursor to the close parentheses after the 6 and delete it. Put the cusor on the 9 and press [2nd] [π] to replace the 9 with π. Put the cursor on the close parenthesis and press [2nd] [INS]. Press [∧] [3] to insert ∧ 3 before the close parenthesis, so the screen reads $6\cos(\pi/7 \wedge 3)$, and press [ENTER] for the answer.

T2. −0.1993501186 • Press [(] [(−)] [5] [−] [1] [.] [6] [3] [2nd] [EE] [(−)] [2] [)] [x^{-1}] to display $(^-5 - 1.63\text{E}^-2)^{-1}$. ($A\,\text{E}\,n$ stands for $A \times 10^n$.) Press [ENTER] for the answer. Notice that [(−)] is for negation, [−] is for subtraction, and [x^{-1}] is for taking reciprocals.

T3. −86 • Use [(] [2] [+] [8] [)] [÷] [(] [4] [−] [6] [)] [−] [3] [∧] [(] [5] [−] [1] [)] to display $(2+8)/(4-6) - 3 \wedge (5-1)$. Press [ENTER] for the answer.

T4. 3757 • To put the commands on one line for possible later editing, press [7] [STO▶] [ALPHA] [A] [ALPHA] [:] [6] [STO▶] [ALPHA] [B] [ALPHA] [:] [5] [STO▶] [ALPHA] [C] [ALPHA] [:] [4] [STO▶] [ALPHA] [D] [ALPHA] [:] [ALPHA] [A] [+] [ALPHA] [B] [ALPHA] [C] [∧] [ALPHA] [D] so the screen reads $7 \to A : 6 \to B : 5 \to C : 4 \to D$: $A + BC \wedge D$. Press [ENTER] to store the values of A, B, C and D and calculate the answer.

T5. 0.422549020 • Use [(] [1] [÷] [2] [)] [LOG] [7] [)] [ENTER].

T6. $1.320224719 \times 10^{13}$ • Use [(] [1] [.] [3] [4] [2nd] [EE] [6] [−] [4] [2nd] [EE] [5] [)] [÷] [(] [7] [.] [1] [2] [2nd] [EE] [(-)] [8] [)] [ENTER].

T7. 2.00509555 • Use [2nd] [√] [4] [+] [7] [∧] [(] [8] [−] [1] [0] [)] [)] [ENTER].

T8. 16 • Use [(] [(-)] [3] [2] [)] [∧] [(] [4] [÷] [5] [)] [ENTER].

T9. All three expressions work on the TI-83. (The first does not work on most other calculators.)

T10. All three expressions work on the TI-83.

T11. The first and third expressions work. $(\sin 3)x^2$ cannot be entered on the TI-83 because sin always comes with an open parentheses.

T12. $y = \sin(x)^2$ and $y = (\sin(x))^2$ work, but $y = (\sin 3)^2$ cannot be entered.

T13. You can use $y = \sqrt{}\,(2)x$.

T14. $y = x^2\,^-1$ does not work because it gives x^2 multiplied by −1. • $y = x^2 - 1$ works.

T15. Use $y = 0.5\sin(x)^{-1}$.

T16. If you obtain only the right half of the curve in Figure 11 of Section 1, try $y = (x \wedge (1/3)) \wedge 2$ or $y = (x^2) \wedge (1/3)$.

T17. See the response to Question 11 of Appendix 1. • Maximum ≈ 3.079 at $x \approx \pm 1.075$

T18. See the response to Question 9 of Appendix 1. • $x \approx 1.472, y \approx 2.168$

Tune-up exercises for a TI-85 or TI-86 calculator

T1a. 0.688885143177 • (If your result is 2.44887304686, your calculator is not in radian mode.) • Press [2nd] [√] [6] [cos] [(] [9] [÷] [7] [)] to have the screen read $\sqrt{}\,6\cos(9/7)$. Then press [ENTER] for the answer.

T1b. 5.999748331 • If your last operation was the calculation of $\sqrt{}\,6\cos(9/7)$, press [2nd] [ENTRY] to put it back on the screen. If you performed other calculations, type $\sqrt{}\,6\cos(9/7)$ again. Press [◀] until the cursor is over the square root sign and press [DEL] to delete it. Move the cursor to the 9 and press [2nd] [π] to replace the 9 with π. Put the cursor on the close parenthesis and press [2nd] [INS]. Press [∧] [3] to insert ∧ 3 before the close parenthesis, so the screen reads $6\cos(\pi/7 \wedge 3)$, and press [ENTER] for the answer.

T2. −0.199350118613 • Press [(] [(−)] [5] [−] [1] [.] [6] [3] [EE] [(−)] [2] [)] [2nd] [x^{-1}] to display $(^{-}5 - 1.63\text{E}^{-}2)^{-1}$. ($A\,\text{E}\,n$ stands for $A \times 10^n$.) Press [ENTER] for the answer. Notice that [(−)] is for negation, [−] is for subtraction, and [x^{-1}] is for taking reciprocals.

T3. −86 • Use [(] [2] [+] [8] [)] [÷] [(] [4] [−] [6] [)] [−] [3] [∧] [(] [5] [−] [1] [)] to display $(2+8)/(4-6) - 3 \wedge (5-1)$. Press [ENTER] for the answer.

T4 3757 • To put the commands on one line for possible later editing, press [7] [STO▶] [A] [ALPHA] [2nd] [:] [6] [STO▶] [B] [ALPHA] [2nd] [:] [5] [STO▶] [C] [ALPHA] [2nd] [:] [4] [STO▶] [D] [ALPHA] [2nd] [:] [ALPHA] [A] [+] [ALPHA] [B] [×] [ALPHA] [C] [∧] [ALPHA] [D] so the screen reads $7 \to A : 6 \to B : 5 \to C : 4 \to D : A + B * C \wedge D$. Press [ENTER] to store the values of A, B, C and D and calculate the answer. The colon separates two commands on one line. The [ALPHA] keys are needed because [STO▶] locks the calculator in upper-case alpha mode.

T5. 0.42254902007 • Use [(] [1] [÷] [2] [)] [LOG] [(] [7] [)] [ENTER].

T6. $1.3202247191 \times 10^{13}$ • Use [(] [1] [.] [3] [4] [EE] [6] [−] [4] [EE] [5] [)] [÷] [(] [7] [.] [1] [2] [EE] [(-)] [8] [)] [ENTER].

T7. 2.00509554966 • Use [2nd] [√] [(] [4] [+] [7] [∧] [(] [8] [−] [1] [0] [)] [)] [ENTER].

T8. 16 • The TI-85 and TI-86 calculatosr evaluate $x^{1/n}$ for negative x and odd integers n, but does not always give the correct, real value of $x^{m/n}$ with negative x, odd n, and m an integer > 1. It is generally best to write $x^{m/n}$ as $(x^m)^{1/n}$ for odd n. The symbols [(-)] [3] [2] [∧] [(] [4] [÷] [5] [)] [ENTER] give the wrong result(−16) since the negation is perfomed after the power. The symbols [(] [(-)] [3] [2] [)] [∧] [(] [4] [÷] [5] [)] [ENTER] probably yield, on a TI-85, (−12.94427191, 9.40456403668) representing $-12.94427191 + 9.40456403668\,i$, which is the approximate decimal value of one of the complex four-fifth roots of −32, rather than the real four-fifth root 16. These symbols, however, give the correct result on a TI-86. The expressions [(] [(-)] [3] [2] [)] [∧] [4] [∧] [(] [1] [÷] [5] [)] [ENTER] and [(] [(-)] [3] [2] [)] [∧] [(] [1] [÷] [5] [)] [∧] [4] [ENTER] give the correct result, since for any positive odd integer n and negative X, the calculator interprets $X \wedge (1/n)$ as the negative nth root of X.

T9. $y = 1/2x$ does not work on a TI-85 because the multiplication by juxtaposition is performed before the division, giving $y = 1/(2x)$, but does work on a TI-86. • $y = (1/2)x$ and $y = 1/2 * x$ work on both calculators because the expression inside parentheses is evaluated first and multiplication with $*$ is performed after the division.

T10. $y = e \wedge (-1)x$ does not work on a TI-85 because the multiplication by juxtaposition is performed before the exponentiation, giving $y = e^{-x}$, but does work on a TI-86. • $y = (e \wedge (-1))x$ and $y = e \wedge (-1) * x$ work because the expression inside parentheses is evaluated first and multiplication with $*$ is performed after the exponentiation.

T11. $y = \sin(3)x^2$ does not work on a TI-85 because the multiplication by juxtaposition is performed before the sine is calculated, giving $y = \sin(3x^2)$, but does work on a TI-86. • $y = (\sin 3)x^2$, and $y = \sin(3) * x^2$ work because the expression inside parentheses is evaluated first and multiplication with $*$ is performed after the sine is evaluated.

T12. $y = \sin(x)^2$ does not work because the square is calculated before the sine, giving $y = \sin(x^2)$. • $y = (\sin x)^2$ works because the expression inside parentheses is evaluated first.

T13. $y = \sqrt{}2x$ does not work on a TI-85 because the product is calculated before the square root, giving $y = \sqrt{2x}$, but does work on a TI-86. • $y = (\sqrt{}2)x$ works because the expression inside parentheses is evaluated first.

T14. $y = x^{2^{\,-}}1$ does not work because it gives x^2 multiplied by -1. • $y = x^2 - 1$ works.

T15. $y = 0.5\sin(x)^{-1}$ does not work because it gives $0.5\sin(x^{-1})$. • $y = 0.5(\sin x)^{-1}$ works.

T16. If you obtain only the right half of the curve in Figure 11 of Section 1, try $y = (x \wedge (1/3)) \wedge 2$ or $y = (x^2) \wedge (1/3)$.

T17. Max ≈ 3.079 at $x \approx \pm 1.075$

T18. $x \approx 1.472, y \approx 2.168$

Problems

1. $1 + y/x - 3w^2$

2. (a) $\dfrac{2}{x+3}$ (b) $\dfrac{2}{x} + 3$ (e) $\dfrac{xz}{y}$

3. x^{y^z} • $(x^y)^z$

5. Maximum ≈ 2.6307 at $x \approx 2.833$ • Figure A5

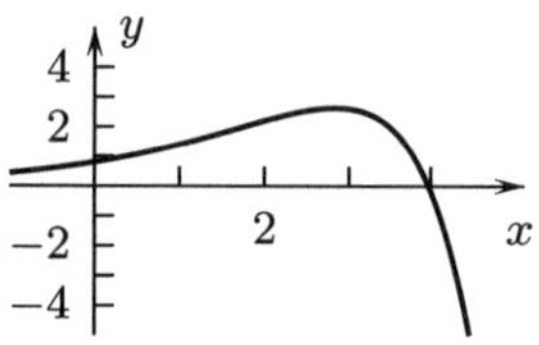

Figure A5

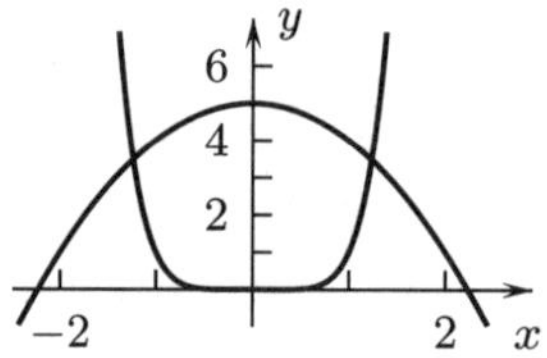

Figure A7

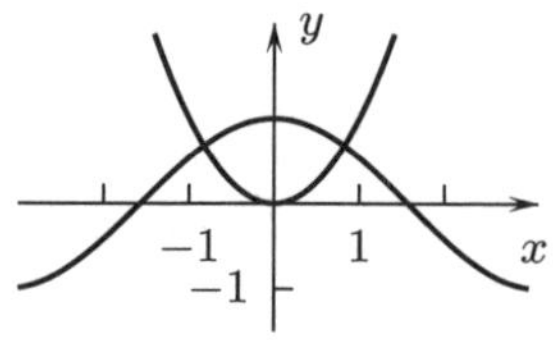

Figure A8

7. The intersections are approximately $(1.231, 3.484)$ and $(-1.231, 3.484)$. • Figure A7

8. The intersections are approximately $(\pm 0.8241, 0.6919)$. • Figure A8

Appendix 2

Exponential functions and logarithms

Topics:

- ***The exponential functions* $\mathrm{b^x}$ *and their graphs***
- ***Algebraic properties of* $\mathrm{b^x}$**
- ***Compound interest and the number* e**
- ***Logarithms* $\log_\mathrm{b}\mathrm{x}$ *and their graphs***
- ***Algebraic properties of* $\log_\mathrm{b}\mathrm{x}$**
- ***The natural logarithm* lnx**

The function $\mathrm{b^x}$

For any positive number b and positive integers p and q, $b^{p/q}$ represents the pth power of the q root of b, which equals the qth root of the pth power of b, and $b^{-p/q}$ represents the reciprocal of that number:

$$b^{p/q} = \sqrt[q]{b^p} = \left[\sqrt[q]{b}\right]^p \quad \text{and} \quad b^{-p/q} = \frac{1}{\sqrt[q]{b^p}} = \frac{1}{\left[\sqrt[q]{b}\right]^p}.$$

For an irrational number x, b^x can be defined to be the limit as $n \to \infty$ of b^{x_n}, where x_n is the rational number (fraction) obtained by taking only the first n digits after the decimal point in the decimal expansion of x. If $x = \sqrt{2} = 1.41421356237\ldots$, for example, we would use $x_1 = 1.4, x_2 = 1.41, x_3 = 1.414, x_4 = 1.4142$, etc. To have the formulas and identities involving b^x apply with any exponents, b^0 is defined to be 1.

If $b = 1$, then $b^x = 1$ for all x and its graph is a horizontal line (Figure 1). For $b > 1$, b^x is a positive function that tends to ∞ as $x \to \infty$ and tends to 0 as $x \to -\infty$ (Figure 2). For $0 < b < 1$, b^x is an positive function that tends to 0 as $x \to \infty$ and tends to ∞ as $x \to -\infty$ (Figure 3).

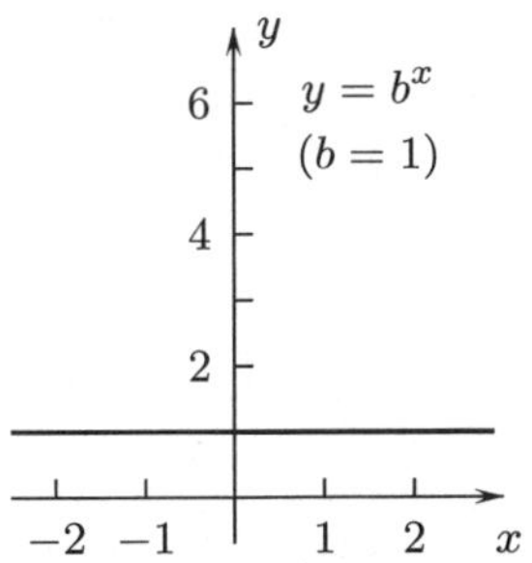

FIGURE 1

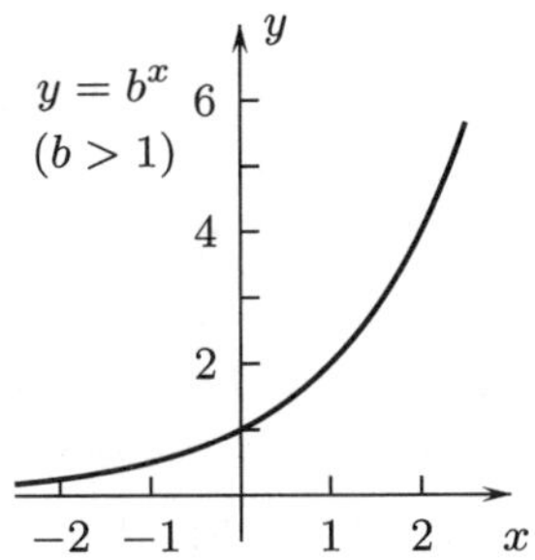

FIGURE 2

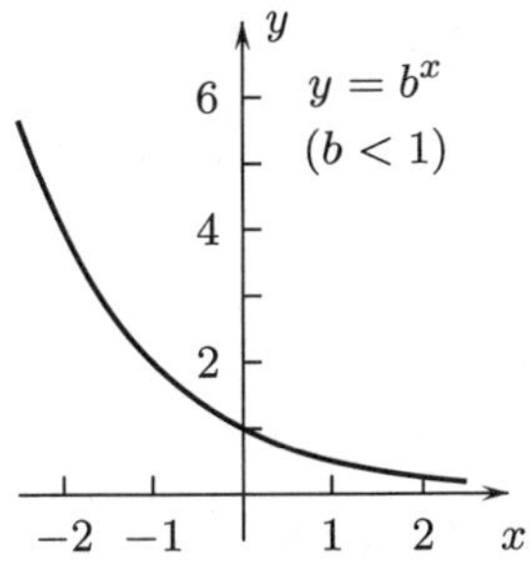

FIGURE 3

Laws of exponents

For any real number b, b^2 represents $b \times b$, which we also write as bb; b^3 represents $b \times b \times b$, which we also write bbb; and so forth. Consequently,

$$b^3b^2 = (bbb)(bb) = bbbbb = b^5 = b^{3+2}$$
$$(b^3)^2 = (bbb)^2 = (bbb)(bbb) = b^6 = b^{3(2)}$$
$$(bc)^3 = (bc)(bc)(bc) = (bbb)(ccc) = b^3c^3.$$

These are special cases of the following identities, which hold for positive b and c and any x and y:

$$b^x b^y = b^{x+y} \tag{1}$$

$$(b^x)^y = b^{xy} \tag{2}$$

$$(bc)^x = b^x c^x \tag{3}$$

$$b^{-x} = \frac{1}{b^x} \tag{4}$$

The advantage of exponential notation is illustrated in the next example.

Example 1 Simplify the formula $y = \dfrac{\sqrt{b}}{\sqrt[3]{b}}$ **(a)** without using fractional or negative exponents by taking the sixth power of both sides, simplifying, and taking the sixth root, and **(b)** by using fractional and negative exponents and rules **(1)** through **(4)**.

SOLUTION **(a)** Taking sixth powers gives

$$y^6 = \frac{(\sqrt{b})^6}{(\sqrt[3]{b})^6} = \frac{(\sqrt{b}\sqrt{b})(\sqrt{b}\sqrt{b})(\sqrt{b}\sqrt{b})}{(\sqrt[3]{b}\sqrt[3]{b}\sqrt[3]{b})(\sqrt[3]{b}\sqrt[3]{b}\sqrt[3]{b})} = \frac{bbb}{bb} = b.$$

Then taking sixth roots yields $y = \sqrt[6]{b}$ since $y > 0$.

(b) With fractional exponents we have a more direct calculation:

$$y = \frac{\sqrt{b}}{\sqrt[3]{b}} = b^{1/2}b^{-1/3} = b^{1/2-1/3} = b^{1/6} = \sqrt[6]{b}. \;\square$$

Zero and negative bases

Rules **(1)** through **(4)** often do not apply or have to be modified if the bases b and c can have zero or negative values, because (i) we cannot divide by zero; (ii) every positive number k has two square roots $\pm k^{1/2}$ and, in general, two nth roots $\pm k^{1/n}$ for any even positive integer n; and (iii) we cannot take irrational (nonfractional) powers or even roots of negative numbers if we use only real numbers. These restrictions are illustrated by the following statements.

(i) The equation $b^{1-1} = b/b$ is true for any $b \neq 0$ but is meaningless if $b = 0$.

(ii) Solving $b^2 = 9$ by taking the half power of both sides of the equation and using **(2)** gives $(b^2)^{1/2} = 9^{1/2} = 3$, but if we allow negative values of b, there is a second solution $b = -3$.

(iii) Using **(2)** to solve $b^2 = -9$ by writing $b = (b^2)^{1/2} = (-9)^{1/2}$ does not make sense if we consider only real numbers because there is no real number whose square is -9. The solutions of $b^2 = -9$ are the "imaginary" numbers $b = \pm 3i$, where i is the symbol used for a square root of -1.

Compound interest and the number e

Suppose that a bank account has an initial balance of C dollars, that the account pays $P\%$ annual interest compounded n times a year, where n is a positive integer, and that no further depositis or withdrawals are made. We let $r = P/100$ denote the fraction or decimal corresponding to $P\%$. Then the interest rate (measured as a fraction instead of a percent) is r/n for each nth of a year. Consequently, if the balance is B dollars just before an interest payment is made, then the account earns $(r/n)B$ dollars interest and increases to $(1 + r/n)B$ dollars.

After t years, with $t = 1/n, 2/n, 3/n, \ldots$, interest has been paid nt times, and the initial deposit of C dollars has grown to

$$B = C\left(1 + \frac{r}{n}\right)^{nt} \text{ dollars.} \tag{5}$$

The limit of the quanity **(5)** as the number n tends to ∞ is the balance when the interest is COMPOUNDED CONTINUOUSLY. It can be expressed in terms of the special number $e = 2.718281828\ldots$, which is defined by

$$e = \lim_{n\to\infty}\left(1+\frac{1}{n}\right)^n. \qquad \textbf{(6)}$$

The exponential function e^x can be given for any real x by

$$e^x = \lim_{n\to\infty}\left(1+\frac{x}{n}\right)^n. \qquad \textbf{(7)}$$

Consequently, an initial balance of C dollars that earns $P\%$ annual interest compounded continuously grows after t years to

$$\lim_{n\to\infty}\left[C\left(1+\frac{r}{n}\right)^{nt}\right] = C\left\{\lim_{n\to\infty}\left[\left(1+\frac{r}{n}\right)^n\right]\right\}^t = C(e^r)^t = Ce^{rt} \text{ dollars.}$$

Figure 4 shows the graph of the balance in an account containing \$100 initially that earns 50% annual interest compounded annually ($n = 1$) and the graph of the balance $100e^{0.5t}$ with interest compounded continuously. Figure 5 shows the graphs of the balances with interest compounded semi-annually ($n = 2$) and continuously. Notice that the graph of the step function in Figure 5 is closer to the curve for interest compounded continuously than is the graph of the step function in Figure 4. The graph for interest compounded more than two times a year would consist of more, shorter line segments and be even closer to the exponential curve.

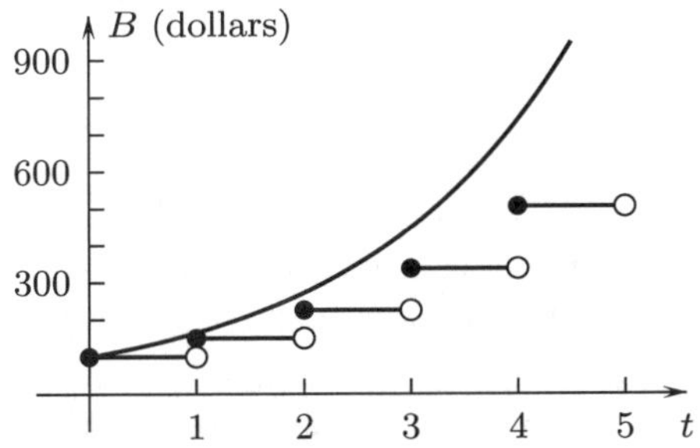

FIGURE 4

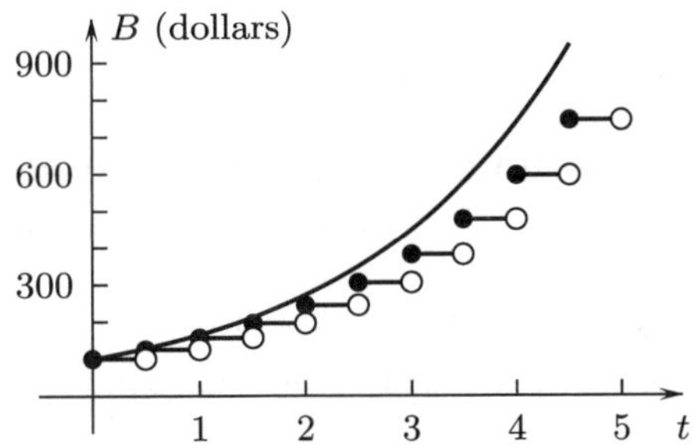

FIGURE 5

Logarithms

For $b > 1$, the exponential function $y = b^x$ is an increasing function, defined for all x (Figure 6), and its inverse is the logarithm $y = \log_b x$ to the base b, which is defined for $x > 0$ (Figure 7). The graphs of these functions are mirror images about the line $y = x$ in Figures 6 and 7 because the scales on the axes are equal. Notice that $b^x \to \infty$ as $x \to \infty$, that $b^x \to 0$ as $x \to -\infty$, that $\log_b x \to \infty$ as $x \to \infty$, and that $\log_b x \to -\infty$ as $x \to 0^+$.

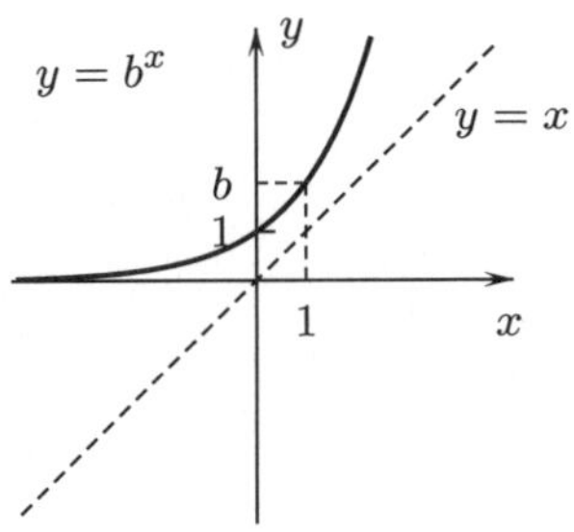

FIGURE 6

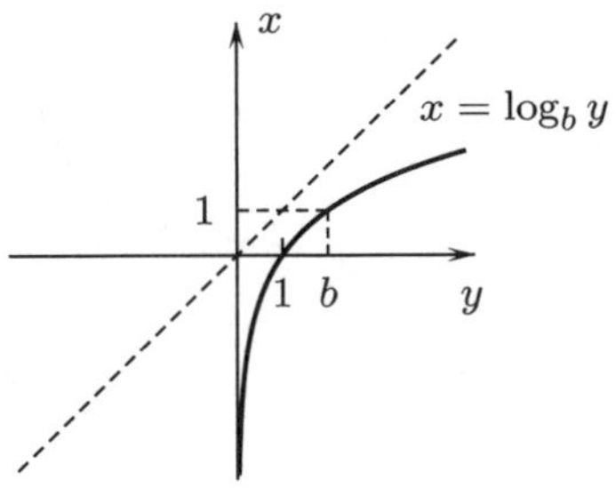

FIGURE 7

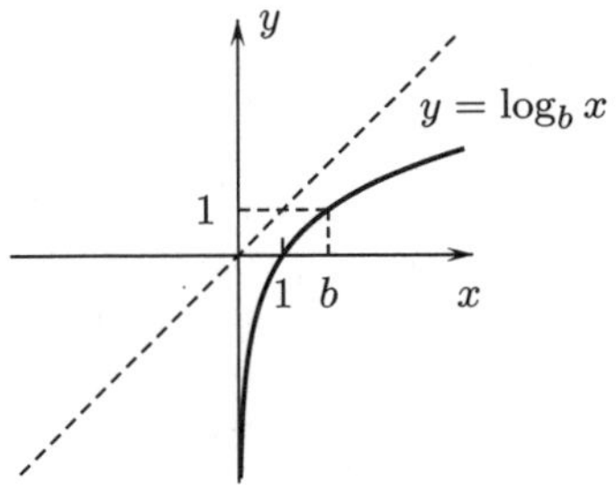

FIGURE 8

Interchanging x and y in Figure 7 gives the graph $y = \log_b x$ in Figure 8 of the logarithm with independent variable x and dependent variable y.

Because b^x is defined for all x and its range is the interval $y > 0$, $\log_b x$ is defined for $x > 0$ and its range is the set of all numbers y. The fact that b^x and $\log_b x$ are inverse functions is also expressed in the two equations,

$$\log_b(b^x) = x \quad \text{for all } x \tag{7}$$

$$b^{\log_b x} = x \quad \text{for all } x > 0. \tag{8}$$

Notice that, while these formulas are visually quite complex, all they state is that b^x and $\log_b x$ undo each other.

The basic properties of $\log_b x$ follow from properties of b^x listed above: For any $b > 1$, any n, and any positive x and y,

$$\log_b(1) = 0 \tag{9}$$

$$\log_b(b) = 1 \tag{10}$$

$$\log_b(xy) = \log_b x + \log_b y \tag{11}$$

$$\log_b\left(\frac{x}{y}\right) = \log_b x - \log_b y \tag{12}$$

$$\log_b(x^n) = n \log_b x. \tag{13}$$

Logarithms with different bases, $b > 1$ and $c > 1$, are related by the formula

$$\log_b x = \left[\log_b c\right]\left[\log_c x\right] = \frac{\log_c x}{\log_c b}. \tag{14}$$

The common and natural logarithms

The logarithm to the base 10, $\log_{10} x$, is called the COMMON LOGARITHM and is often denoted $\log x$ with no subscript. It was employed to simplify calculations involving products and powers before calculators and computers were readily available, and is still used in applications.

The logarithm that is most convenient for calculus is $\log_e x$, with base equal to the base e of the natural exponential function e^x. This logarithm is called the NATURAL LOGARITHM and is usually denoted $\ln x$:

$$\ln x = \log_e x \qquad \text{for all } x > 0. \tag{15}$$

Since e^x and $\ln x$ are inverse functions,

$$e^{\ln x} = x \qquad \text{for all } x > 0 \tag{16}$$

$$\ln(e^x) = x \qquad \text{for all } x. \tag{17}$$

The graphs of e^x and $\ln x$ are shown together in Figure 9.[†]

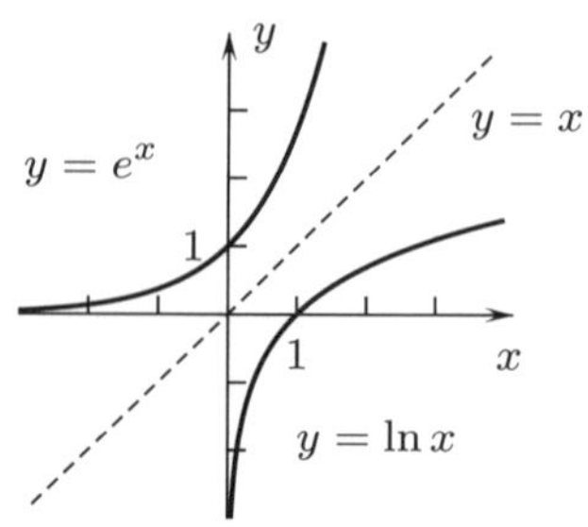

FIGURE 9

[†]The natural logarithm is denoted $\log x$, rather than lnx, in some calculus computer software.

The natural logarithm can be used to find logarithms to any base $b > 1$ by using **(14)** with $c = e$:

$$\log_b x = \frac{\ln x}{\ln b} \qquad \text{for all } x > 0. \tag{18}$$

The natural logarithm can also be used to change bases in formulas for exponential functions. In particular, if we want to express b^t with $b > 1$ as an exponential function with base e, we can write $b = e^{\ln b}$, and then, since $(e^{\ln b})^t = e^{t \ln b}$,

$$b^t = e^{t \ln b}. \tag{19}$$

Appendix 3

Trigonometric and inverse trigonometric functions

Topics:

- ***Angles and radians***
- ***Circular motion***
- ***Trigonometric functions and right triangles***
- ***Isosceles-right and equilateral triangles***
- ***Trigonometric functions of arbitrary angles***
- ***Some trigonometric identities***
- ***A direct definition of the tangent function***
- ***The inverse sine, cosine, and tangent functions***
- ***The Law of Cosines and the Law of Sines***

Angles and radians

In analytic geometry and calculus and in applications to many other fields the trigonometric functions are defined without direct reference to triangles, as they are in trigonometry. Instead, we think of an angle as a rotation of one of its sides, which we call its initial side, to its other side, which we call the terminal side. By convention, angles corresponding to counterclockwise rotations are considered to be positive (Figure 1) and those corresponding to clockwise rotations are considered to be negative (Figure 2).

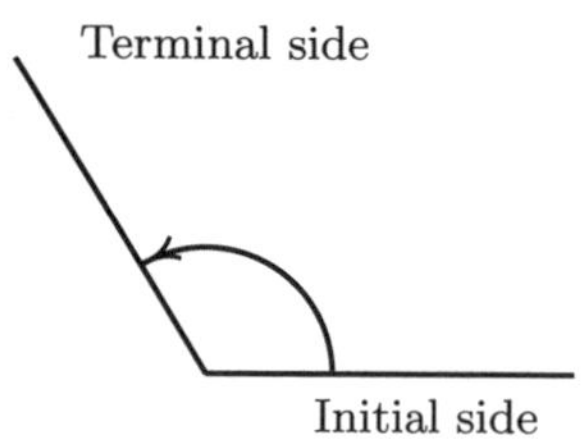

A positive angle
FIGURE 1

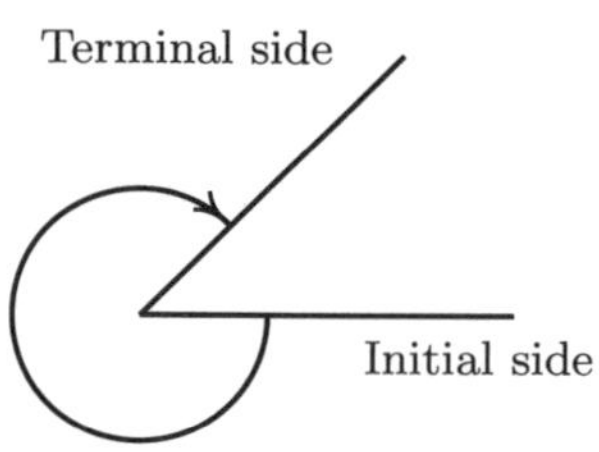

A negative angle
FIGURE 2

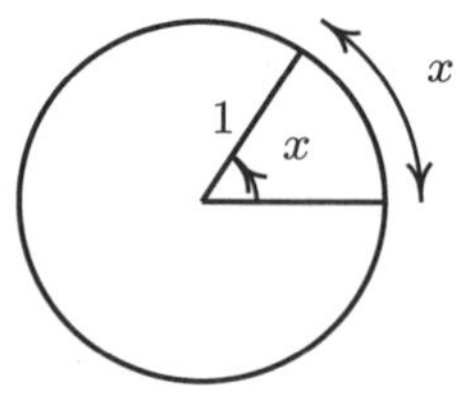

FIGURE 3

In calculus, angles are generally measured in RADIANS instead of degrees. The radian measure of an angle is determined by placing the vertex of the angle at the center of a circle of radius 1. The radian measure of the angle equals the length of the arc of the circle subtended (marked off) by the angle if the angle is given by a counterclockwise rotation, as in Figure 3. The angle is negative and equals the negative of the length of the arc if the angle is given by a clockwise rotation.

Since the circumference of a circle of radius 1 is 2π, a full counterclockwise revolution is 2π radians and corresponds to 360 degrees, and a half counterclockwise revolution is π radians, which corresponds to 180 degrees. Consequently, degrees can be converted to radians and radians to degrees with the formulas,

$$\theta \text{ degrees} = [\theta \text{ degrees}]\left[\frac{\pi \text{ radians}}{180 \text{ degrees}}\right] = \left[\frac{\pi}{180}\right]\theta \text{ radians} \tag{1}$$

$$x \text{ radians} = [x \text{ radians}]\left[\frac{180 \text{ degrees}}{\pi \text{ radians}}\right] = \left[\frac{180}{\pi}\right] x \text{ degrees.} \tag{2}$$

If a circle has radius r, then a positive angle x with its vertex at the center of the circle subtends an arc of length rx (Figure 4). Moreover, because the sector inside the angle x is the fraction $\dfrac{x}{2\pi}$ of the circle, the area of the sector is $\left(\dfrac{x}{2\pi}\right)(\pi r^2) = \frac{1}{2}xr^2$.

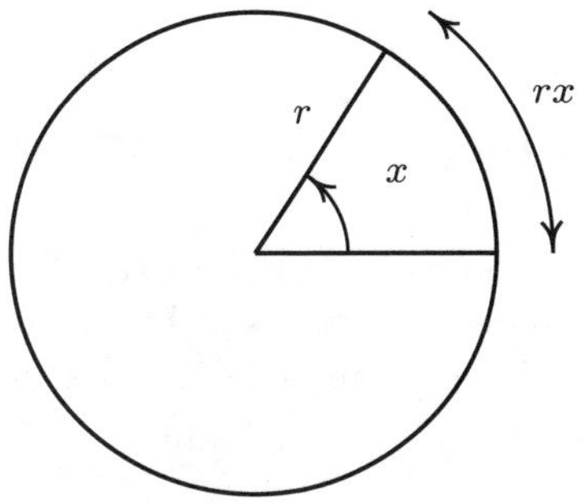

Arclength $= rx$

Area $= \frac{1}{2}xr^2$

FIGURE 4

We will generally give values of angles in radians and we will always assume that the trigonometric functions are given relative to radians. If we want to give degree measure of an angle, we will put a degree sign after its value, as in the expression 90° for 90 degrees.

Trigonometric functions and right triangles

If all you know about a triangle is its three angles, you cannot determine the lengths of its sides. You would need to know how large the triangle is. You can, however find the ratios of the lengths of its sides from its angles. These ratios for right triangles are given by the SINE, COSINE, TANGENT, COTANGENT, SECANT and COSECANT functions.†

The sides adjacent to the right angle in a right triangle are called its LEGS. The side opposite the right angle is the HYPOTENUSE. If θ (theta) is an acute angle in the triangle, we write "Opp" for the length of the leg opposite θ, "Adj" for the length of the adjacent leg, and "Hyp" for the length of the hypotenuse, as in Figure 5.

The values of the sine, cosine, and tangent of the angle θ are the ratios‡

$$\sin\theta = \frac{\text{Opp}}{\text{Hyp}}, \quad \cos\theta = \frac{\text{Adj}}{\text{Hyp}}, \quad \tan\theta = \frac{\sin\theta}{\cos\theta} = \frac{\text{Opp}}{\text{Adj}}. \qquad \textbf{(3)}$$

The secant, cosecant, and cotangent are the reciprocals of the cosine, sine, and tangent, respectively:

$$\sec\theta = \frac{1}{\cos\theta} = \frac{\text{Hyp}}{\text{Adj}}, \quad \csc\theta = \frac{1}{\sin\theta} = \frac{\text{Hyp}}{\text{Opp}}, \quad \cot\theta = \frac{1}{\tan\theta} = \frac{\text{Adj}}{\text{Opp}}. \qquad \textbf{(4)}$$

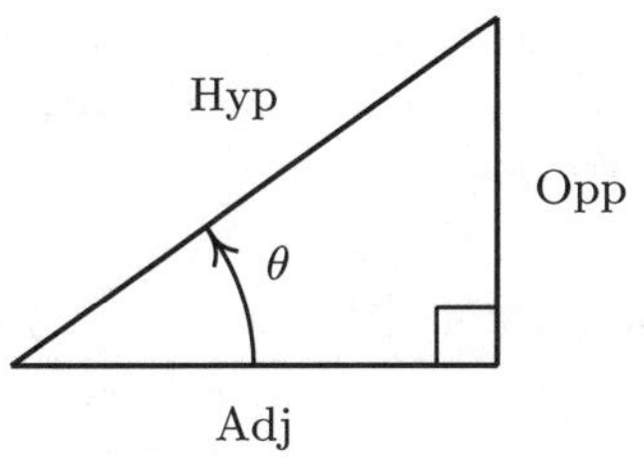

FIGURE 5

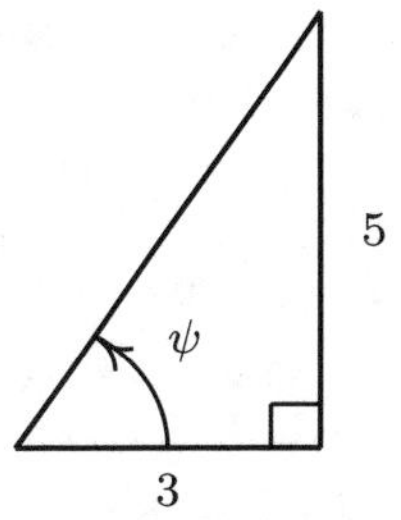

FIGURE 6

†A right angle is 90°, a right triangle is a triangle with a right angle, and an acute angle is a positive angle that is smaller than a right angle.

‡Some students remember formulas **(3)** with the mnemonic "SOHCAHTOA" for "Sine equals Opposite over Hypotenuse; Cosine equals Adjacent over Hypotenuse; Tangent equals Opposite over Adjacent."

Example 1 Find the sine, cosine, tangent, secant, cosecant, and cotangent of the angle ψ (psi) in the right triangle of Figure 6.

SOLUTION The length of the hypotenuse of the triangle is $\sqrt{3^2+5^2} = \sqrt{34}$, so

$$\sin\psi = \frac{5}{\sqrt{34}}, \quad \cos\psi = \frac{3}{\sqrt{34}}, \quad \tan\psi = \frac{5}{3}$$
$$\sec\psi = \frac{\sqrt{34}}{3}, \quad \csc\psi = \frac{\sqrt{34}}{5}, \quad \cot\psi = \frac{3}{5}. \ \square$$

Two special triangles and three special angles

Figures 7 and 8 show an isosceles-right triangle and a 30°-60°-right triangle, both with hypotenuse of length 1. These triangles can be used to recall the values of the trigonometric functions at $\frac{1}{4}\pi$ radians (45°), $\frac{1}{6}\pi$ radians (30°), and $\frac{1}{3}\pi$ radians (60°).

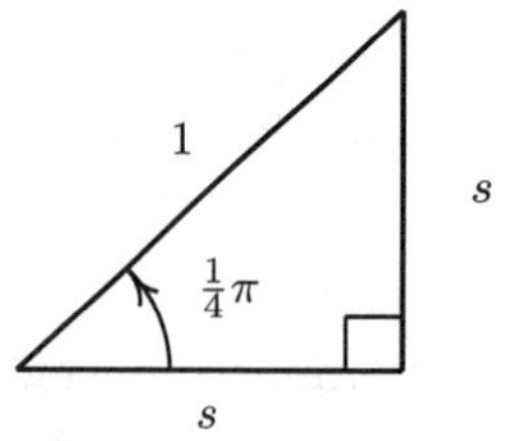

$s = 1/\sqrt{2}$

FIGURE 7

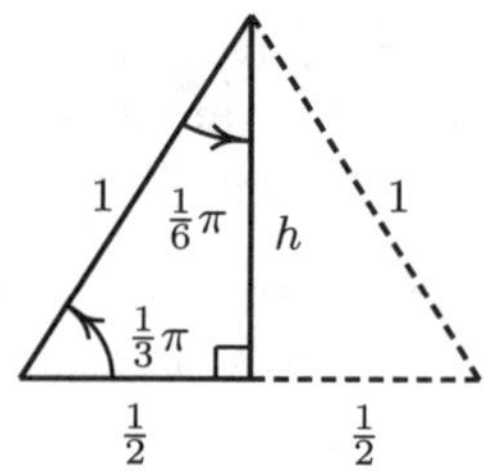

$h = \frac{1}{2}\sqrt{3}$

FIGURE 8

The Pythagorean Theorem applied to the triangle in Figure 7 gives $s^2 + s^2 = 1$, so that $s^2 = \frac{1}{2}$, $s = 1/\sqrt{2}$, and

$$\sin\left(\tfrac{1}{4}\pi\right) = \frac{1}{\sqrt{2}}, \quad \cos\left(\tfrac{1}{4}\pi\right) = \frac{1}{\sqrt{2}}, \quad \tan\left(\tfrac{1}{4}\pi\right) = 1. \tag{5}$$

Similarly, the 30°-60°-right triangle in Figure 8 is half of an equilateral triangle with sides of length 1. Consequently, the length of its base is $\frac{1}{2}$, and, again by the Pythagorean Theorem, its height h satisfies $\left(\frac{1}{2}\right)^2 + h^2 = 1$, so that $h^2 = \frac{3}{4}$, $h = \frac{1}{2}\sqrt{3}$, and

$$\sin\left(\tfrac{1}{6}\pi\right) = \tfrac{1}{2}, \quad \cos\left(\tfrac{1}{6}\pi\right) = \tfrac{1}{2}\sqrt{3}, \quad \tan\left(\tfrac{1}{6}\pi\right) = \frac{1}{\sqrt{3}} \tag{6}$$
$$\sin\left(\tfrac{1}{3}\pi\right) = \tfrac{1}{2}\sqrt{3}, \quad \cos\left(\tfrac{1}{3}\pi\right) = \tfrac{1}{2}, \quad \tan\left(\tfrac{1}{3}\pi\right) = \sqrt{3}. \tag{7}$$

Trigonometric functions of arbitrary angles

To define the trigonometric functions of an arbitrary angle of x radians, we place the vertex of the angle at the origin of a uv-plane with equal scales on the coordinate axes and with the initial side of the angle along the positive u-axis, as in Figures 9 and 10. (We use uv-planes, rather than xy-planes, because we are using x as an angle.)

We pick any point (u, v), other than the origin, on the terminal side of the angle and calculate its distance $r = \sqrt{u^2+v^2}$ to the origin. In the case of a positive acute angle x as in Figure 9, x is an angle in a right triangle such that u is the length of the side adjacent to x, v is the length of the opposite side, and r is the length of the hypotenuse. Hence,

$$\begin{aligned} \sin x &= \frac{v}{r}, & \cos x &= \frac{u}{r}, & \tan x &= \frac{v}{u} \\ \sec x &= \frac{r}{u}, & \csc x &= \frac{r}{v}, & \cot x &= \frac{u}{v}. \end{aligned} \tag{8}$$

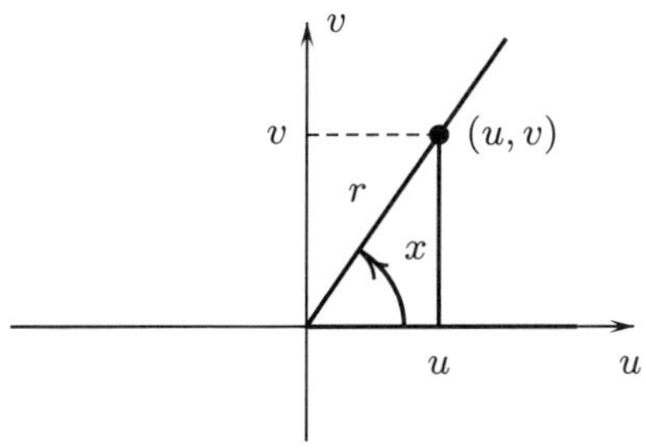

FIGURE 9

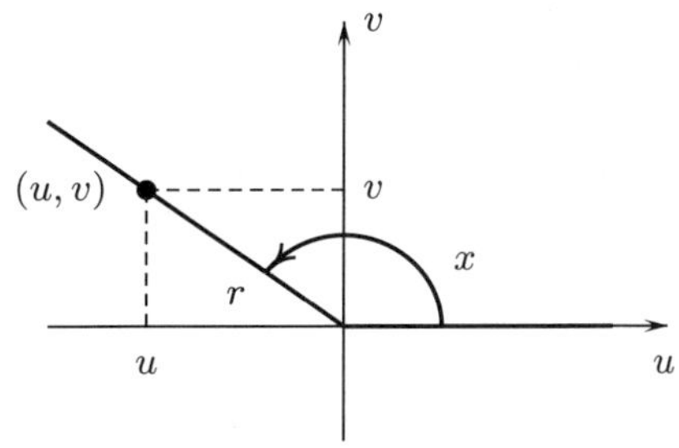

FIGURE 10

Formulas **(8)** also define the trigonometric functions for all other values of the variable x except where a denominator is zero. If x is not a positive acute angle, then the coordinates u and v, which take the place of the lengths of the legs of the right triangle in definitions **(3)** and **(4)**, can be positive, negative, or zero, depending on the choice of x. In Figure 10, for instance, x is an angle between $\frac{1}{2}\pi$ and π, and consequently u is negative and v is positive. The number r, which takes the place of the length of the hypotenuse of the triangle in the definitions, is always positive because it is a length and not a coordinate.

The unit circle

We can simplify definitions **(8)** of the trigonometric functions by setting $r = 1$. We draw the UNIT CIRCLE—the circle of radius 1 with its center at the origin—in a uv-plane (Figures 11 and 12). The point where the terminal side of the angle x intersects the unit circle is a distance $r = 1$ from the origin. Using it in **(8)** gives $\sin x = v/r = v$ and $\cos x = u/r = u$ so that $\sin x$ is the vertical v-coordinate of the point and $\cos x$ is its horizontal u-coordinate.

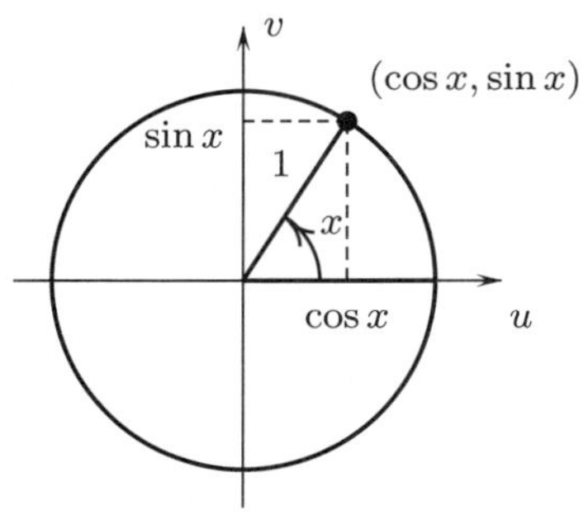

FIGURE 11

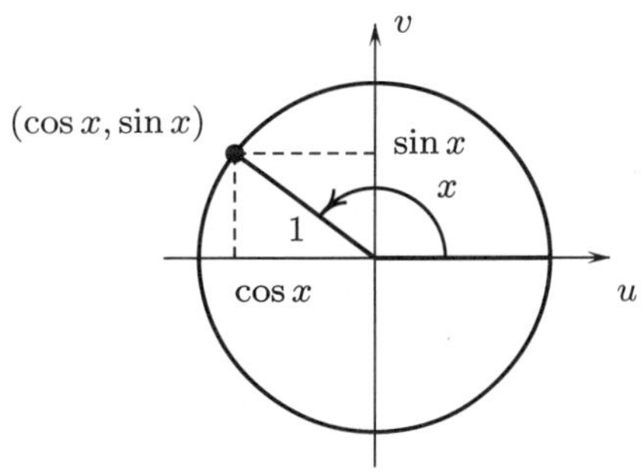

FIGURE 12

You can see from Figures 11 and 12 that $\sin x$ is 0 at $x = 0$, increases to 1 as x increases from 0 to $\frac{1}{2}\pi$, decreases back to 0 as x increases from $\frac{1}{2}\pi$ to π, decreases to -1 as x increases from π to $\frac{3}{2}\pi$, and increases to 0 as x increases from $\frac{3}{2}\pi$ to 2π. This behavior then repeats every 2π as x increases, and gives the graph of $\sin x$ the shape in Figure 13.

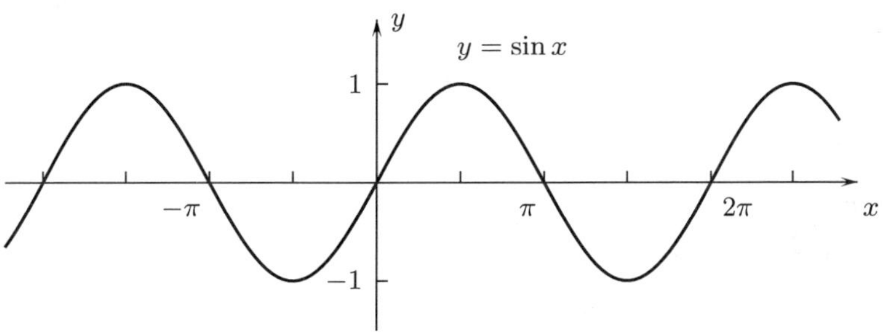

FIGURE 13

Similarly, $\cos x$ is 1 at $x = 0$, decreases to 0 as x increases to from 0 to $\frac{1}{2}\pi$, decreases to -1 as x increases from $\frac{1}{2}\pi$ to π, increases to 0 as x increases from π to $\frac{3}{2}\pi$, and increases to 1 as x increases from $\frac{3}{2}\pi$ to 2π. This behavior repeats every 2π as x increases, and gives the graph of $\sin x$ the shape in Figure 14.

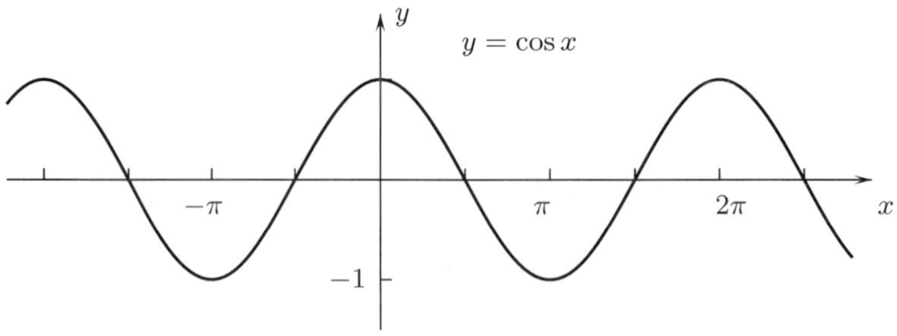

FIGURE 14

Some trigonometric identities

Increasing the angle x by a full revolution gives an angle with the same terminal side and the same values of the sine and cosine. Consequently,

$$\sin(x + 2\pi) = \sin x \quad \text{and} \quad \cos(x + 2\pi) = \cos x \quad \text{for all } x. \tag{9}$$

We say that $\sin x$ and $\cos x$ are PERIODIC of period 2π.

The PYTHAGOREAN IDENTITY

$$\sin^2 x + \cos^2 x = 1 \quad \text{for all } x \tag{10}$$

holds because the point $(\cos x, \sin x)$ is on the unit circle and its distance $\sqrt{(\cos x)^2 + (\sin x)^2}$ to the origin is 1.

Figures 15 through 17 yield, in the case of positive, acute angles x, the following identities, which are valid for all x:

$$\cos(-x) = \cos x \quad \text{and} \quad \sin(-x) = -\sin x \tag{11}$$

$$\cos(x + \pi) = -\cos x \quad \text{and} \quad \sin(x + \pi) = -\sin x \tag{12}$$

$$\cos(x + \tfrac{1}{2}\pi) = -\sin x \quad \text{and} \quad \sin(x + \tfrac{1}{2}\pi) = \cos x. \tag{13}$$

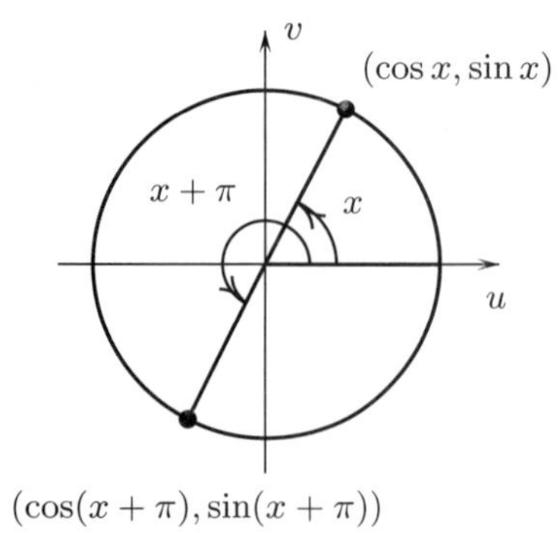

FIGURE 15

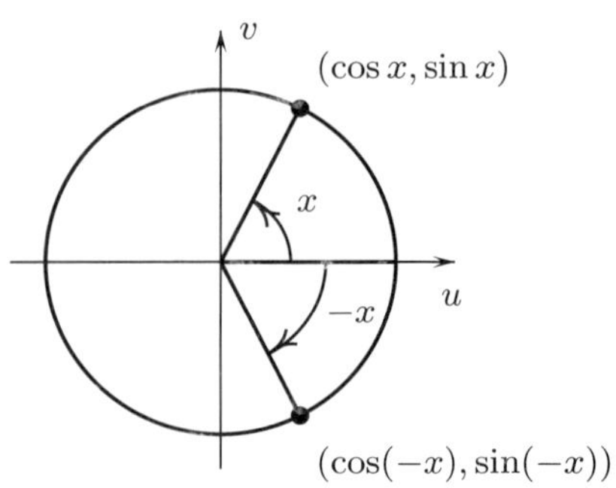

FIGURE 16

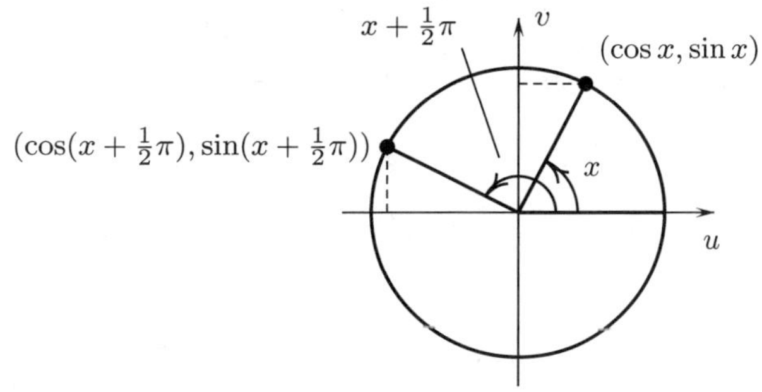

FIGURE 17

Identities **(11)** and **(12)** then show that

$$\tan(-x) = \frac{\sin(-x)}{\cos(-x)} = \frac{-\sin x}{\cos x} = -\tan x \qquad \textbf{(14)}$$

$$\tan(x + \pi) = \frac{\sin(x + \pi)}{\cos(x + \pi)} = \frac{\sin x}{\cos x} = \tan x. \qquad \textbf{(15)}$$

Thus, $\tan x$ is an odd function that is periodic of period π.

To derive other identities, we use Figure 18, which shows positive angles α and β with $\alpha + \beta < \frac{1}{2}\pi$. Because α is an acute angle in the right triangle OCD whose hypotentuse has length 1, we have

$$\overline{OC} = \cos\alpha$$
$$\overline{CD} = \sin\alpha.$$

The angle β is the same in the right triangles OBC and CDE because the legs of the triangles are perpendicular. Therefore,

$$\overline{BC} = \overline{OC}\sin\beta = \cos\alpha\sin\beta$$
$$\overline{ED} = \overline{CD}\cos\beta = \sin\alpha\cos\beta.$$

Finally, because $\alpha + \beta$ is an acute angle in the right triangle OAD we obtain

$$\sin(\alpha + \beta) = \overline{AD} = \overline{AE} + \overline{ED} = \overline{BC} + \overline{ED} = \cos\alpha\sin\beta + \sin\alpha\cos\beta. \qquad \textbf{(16)}$$

Similar reasoning would yield **(16)** and

$$\cos(\alpha + \beta) = \overline{OA} = \overline{OB} - \overline{AB} = \overline{OB} - \overline{EC} = \cos\alpha\cos\beta - \sin\alpha\sin\beta \qquad \textbf{(17)}$$

for any α and β.[†]

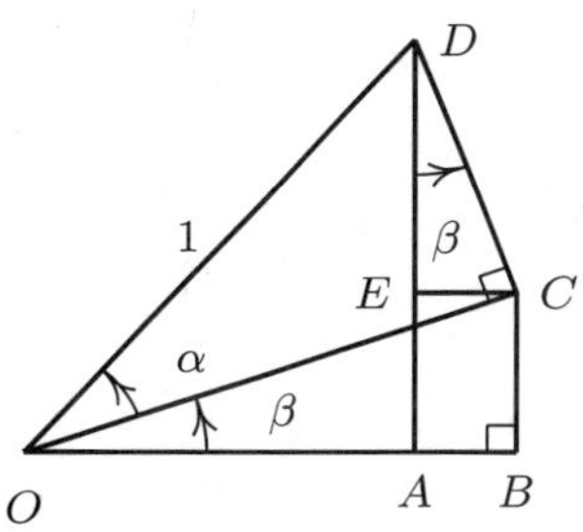

FIGURE 18

[†]These identities can also be derived by using differential equations or complex exponential functions, as we will see in Section 13.2.

Identities **(16)** and **(17)** can be used to derive the following:

$$\sin\alpha\sin\beta = \tfrac{1}{2}[\cos(\alpha-\beta) - \cos(\alpha+\beta)] \tag{18}$$

$$\cos\alpha\cos\beta = \tfrac{1}{2}[\cos(\alpha-\beta) + \cos(\alpha+\beta)] \tag{19}$$

$$\sin\alpha\cos\beta = \tfrac{1}{2}[\sin(\alpha-\beta) + \sin(\alpha+\beta)] \tag{20}$$

$$\sin(2\alpha) = 2\sin\alpha\cos x \tag{21}$$

$$\cos(2\alpha) = \cos^2\alpha - \sin^2\alpha = 2\cos^2\alpha - 1 = 1 - 2\sin^2\alpha \tag{22}$$

$$\sin^2\alpha = \tfrac{1}{2}\big[1 - \cos(2\alpha)\big] \tag{23}$$

$$\cos^2\alpha = \tfrac{1}{2}\big[1 + \cos(2\alpha)\big] \tag{24}$$

$$\sin\alpha - \sin\beta = 2\sin\left(\tfrac{1}{2}(\alpha-\beta)\right)\cos\left(\tfrac{1}{2}(\alpha+\beta)\right) \tag{25}$$

$$\cos\alpha - \cos\beta = 2\sin\left(\tfrac{1}{2}(\alpha-\beta)\right)\sin\left(\tfrac{1}{2}(\alpha+\beta)\right) \tag{26}$$

$$\tan(\alpha+\beta) = \frac{\tan\alpha + \tan\beta}{1 - \tan\alpha\tan\beta}. \tag{27}$$

There is no need to memorize identities **(16)** through **(27)**. They are used occasionally in calculus and other fields, but you can look them up whenever you need them.

A direct definition of the tangent function

We defined the tangent function above in terms of the sine and cosine by

$$\tan x = \frac{\sin x}{\cos x}$$

for all x such that $\cos x \neq 0$. All properties of the tangent function can be determined from properties of the sine and cosine using this formula. It is useful, however, to have a direct geometric definition of the tangent using a unit circle in a uv-plane, as in the definitions of the sine and cosine.

We draw the angle x and the vertical tangent line, $u = 1$, at the right of the unit circle. If the terminal side of the angle is to the right of the vertical axis, as in Figures 19 and 20, we extend it until it intersects the tangent line. The vertical, v-coordinate of the intersection is $\tan x$ because the horizontal coordinate is 1 and by definition **(8)**, $\tan x = v/u = v$.

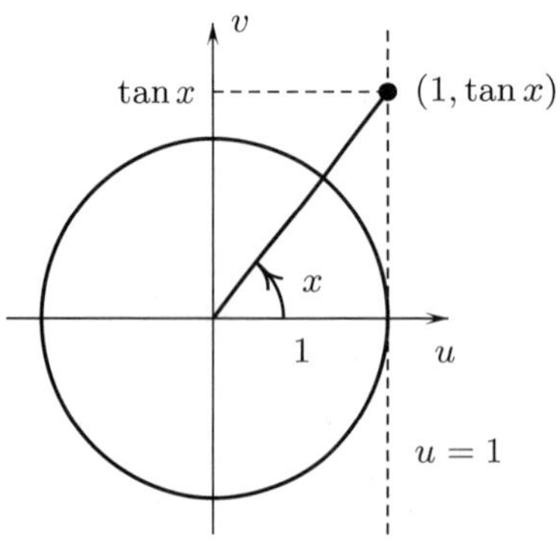

FIGURE 19

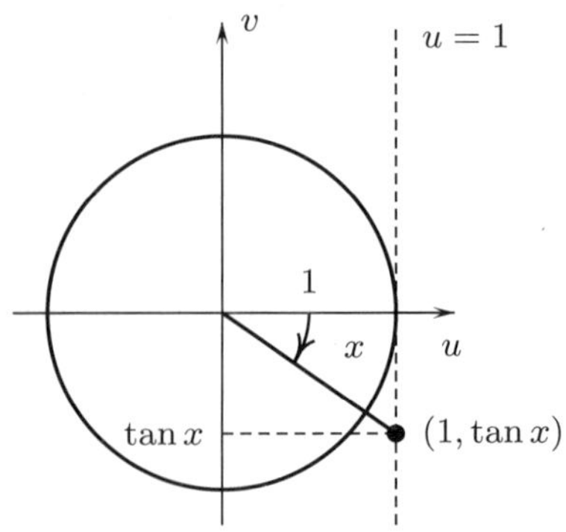

FIGURE 20

If the terminal side of the angle is to the left of the vertical axis, as in Figures 21 and 22, we extend it backwards until it intersects the tangent line $u = 1$ at the right of the unit circle. The vertical, v-coordinate of the intersection equals $\tan x$ in this case because $\tan x$ is periodic of period π.

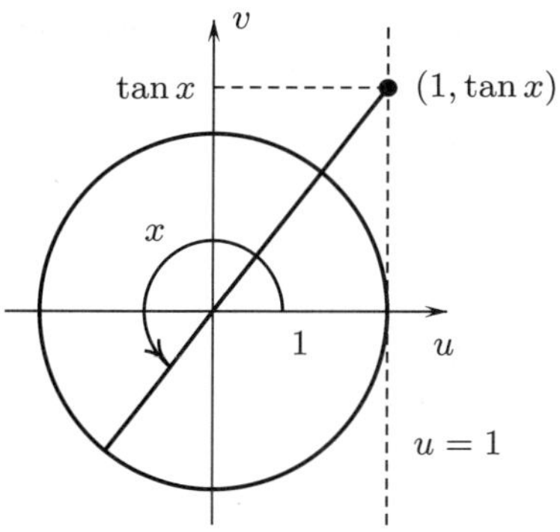

FIGURE 21

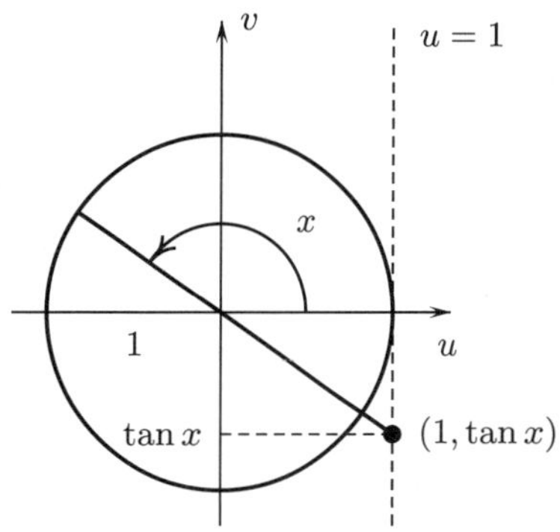

FIGURE 22

Figures 19 through 22 explain the shape of the graph of the tangent function in Figure 23: $\tan x$ is increasing for $-\frac{1}{2}\pi < x < \frac{1}{2}\pi$; $\tan 0 = 0$; $\tan x \to \infty$ as $x \to \frac{1}{2}\pi^-$; $\tan x \to -\infty$ as $x \to -\frac{1}{2}\pi^+$; and the graph repeats every π units to the right and left.

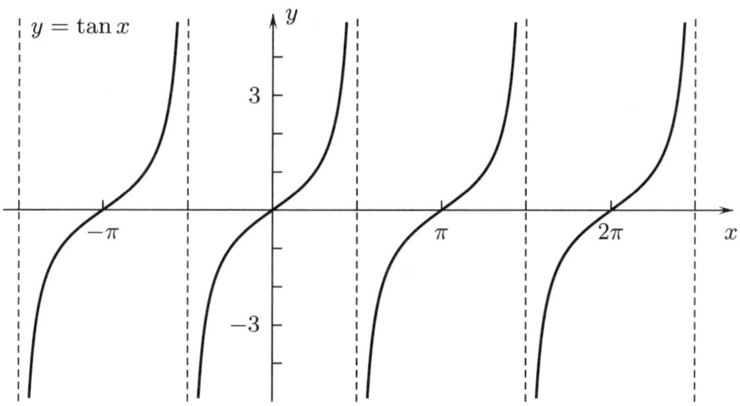

FIGURE 23

The secant, cosecant, and cotangent functions

The secant, cosecant, and cotangent are used to simplify formulas involving the recriprocals of the cosine, sine, and tangent. They can be defined by

$$\sec x = \frac{1}{\cos x}, \qquad \csc x = \frac{1}{\sin x}, \qquad \cot x = \frac{1}{\tan x} = \frac{\cos x}{\sin x} \tag{28}$$

and their graphs are shown in Figures 24 through 26.

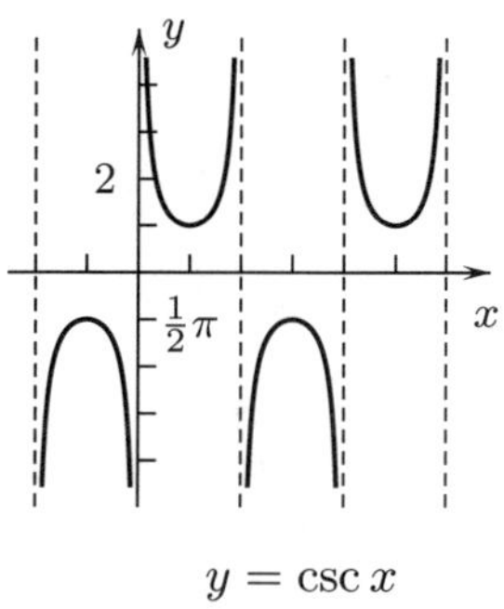

FIGURE 24

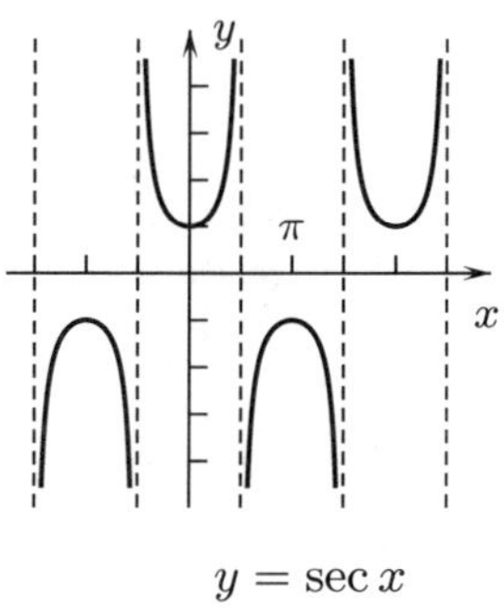

FIGURE 25

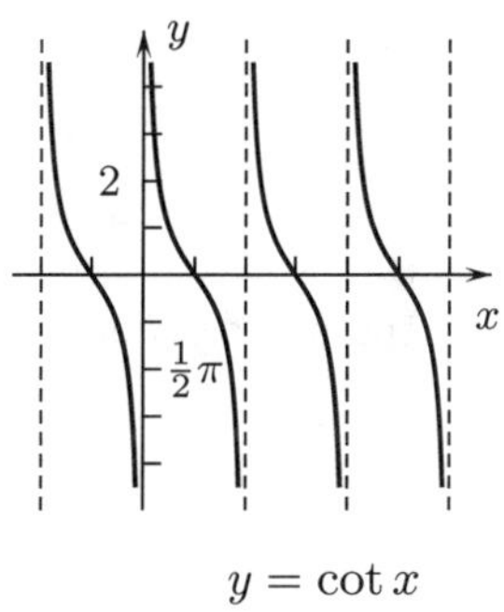

FIGURE 26

The inverse sine, tangent, cosine, and cotangent

The sine function, defined for all x, does not have an inverse because there are an infinite number of solutions x of $\sin x = y$ for each y with $-1 \le y \le 1$. To obtain an inverse, we restrict the domain of $\sin x$ to $-\frac{1}{2}\pi \le x \le \frac{1}{2}\pi$, as in Figure 27. The inverse of this function is the INVERSE SINE, which is denoted $\sin^{-1} x$ or $\arcsin x$. Its graph is the curve in Figure 28. For $-1 \le x \le 1$, $\sin^{-1} x$ is the angle y between $-\frac{1}{2}\pi$ and $\frac{1}{2}\pi$ whose sine is x:

$$y = \sin^{-1} x \iff x = \sin y \quad \text{and} \quad -\tfrac{1}{2}\pi \le y \le \tfrac{1}{2}\pi. \tag{29}$$

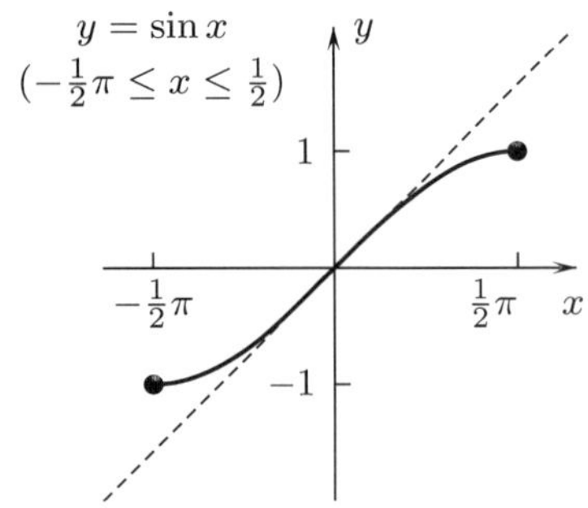

FIGURE 27

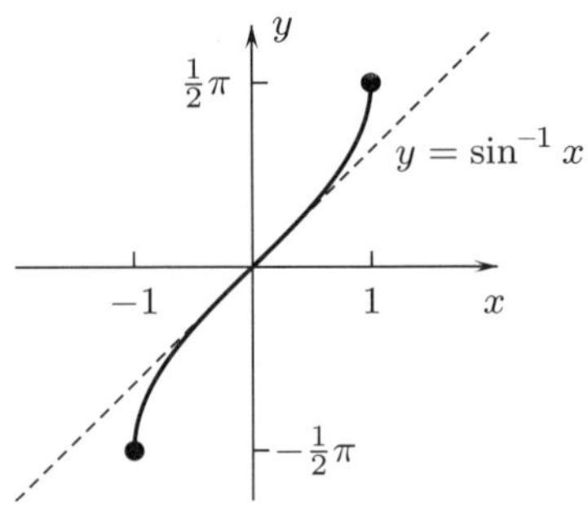

FIGURE 28

For the inverse tangent, we restrict the domain of $\tan x$ to $-\frac{1}{2}\pi < x < \frac{1}{2}\pi$ (Figure 29). The inverse is denoted $\tan^{-1} x$ or $\arctan x$ (Figure 30). It is defined for all x, and its value at x is the angle y between $-\frac{1}{2}\pi$ and $\frac{1}{2}\pi$ whose tangent is x:

$$\tan^{-1} x = y \iff x = \tan y \quad \text{and} \quad -\tfrac{1}{2}\pi < x < \tfrac{1}{2}\pi. \tag{30}$$

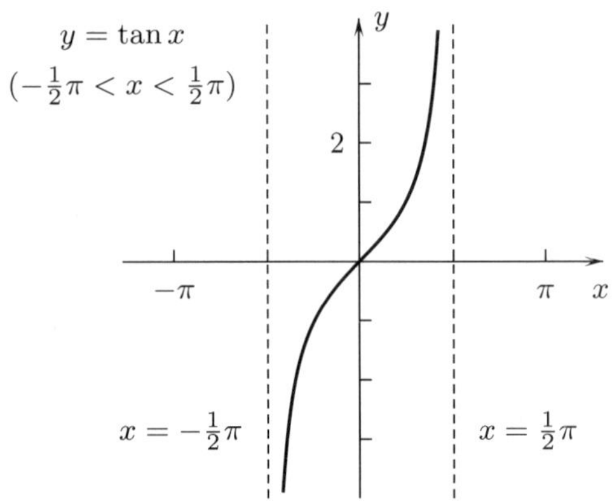

FIGURE 29

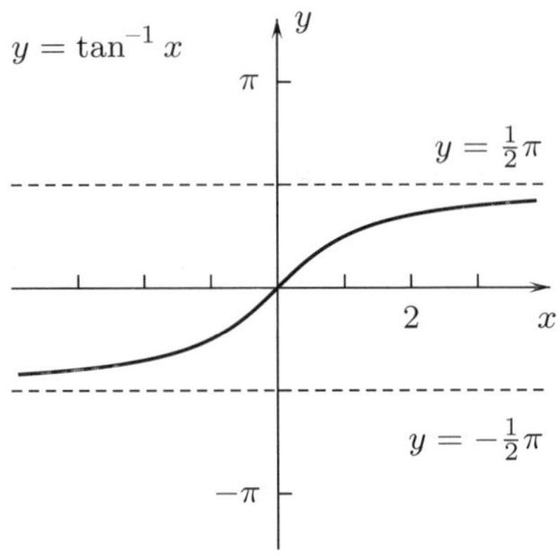

FIGURE 30

The inverse cosine is denoted $\cos^{-1} x$ or $\arccos x$. It is the inverse of $\cos x$ with its domain restricted to $0 \le x \le \pi$ (Figure 31). Its graph is in Figure 32. It is defined for $-1 \le x \le 1$, and its value at x is the angle y between 0 and π whose cosine is x:

$$y = \cos^{-1} x \iff x = \cos y \quad \text{and} \quad 0 \le y \le \pi. \tag{31}$$

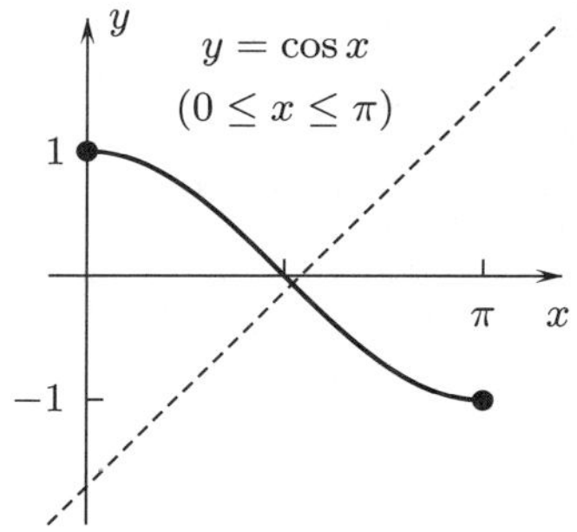

FIGURE 31

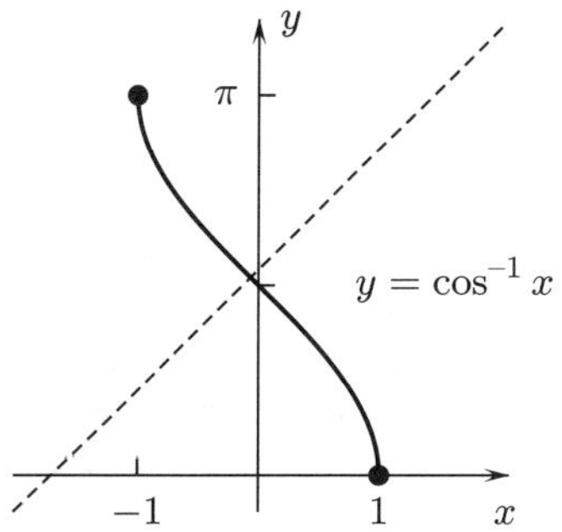

FIGURE 32

The inverse cosine can also be defined, for $-1 \leq x \leq 1$, by

$$\cos^{-1} x = \tfrac{1}{2}\pi - \sin^{-1} x. \tag{32}$$

This formula is easy to remember because for positive x, as in Figure 33, $\sin^{-1} x$ and $\cos^{-1} x$ are the acute angles in a right triangle and the sum of the acute angles in a right triangle is $\frac{1}{2}\pi$ radians.

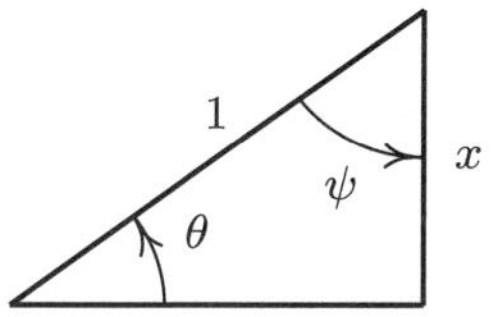

$\theta = \sin^{-1} x$

$\psi = \cos^{-1} x$

FIGURE 33

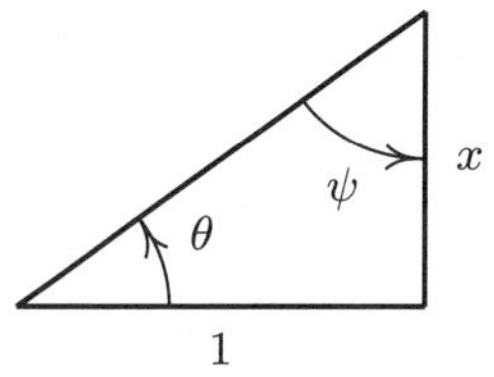

$\theta = \tan^{-1} x$

$\psi = \cot^{-1} x$

FIGURE 34

The inverse cotangent function can be defined by a similar formula (Figure 34):

$$\cot^{-1} x = \tfrac{1}{2}\pi - \tan^{-1} x. \tag{33}$$

The next examples show how angles in right triangles can be found from lengths of the sides by using $\sin^{-1} x$, $\cos^{-1} x$, or $\tan^{-1} x$.

Example 2 Express the angle ψ of Figure 6 on the second page of this appendix in terms of the lengths of the sides of the triangle by using the inverse sine, inverse cosine, and inverse tangent. Check that your formulas are consistent by calculating the approximate decimal value of ψ with all three formulas.

SOLUTION Since $\sin\psi = 5/\sqrt{34}$ and $0 < \psi < \frac{1}{2}\pi$, we have $\psi = \sin^{-1}(5/\sqrt{34}) \doteq 1.03$ radians. Also $\cos\psi = 3/\sqrt{34}$ and $0 < \psi < \frac{1}{2}\pi$, so $\psi = \cos^{-1}(3/\sqrt{34}) \doteq 1.03$ radians, or $\tan\psi = 5/3$ and $0 < \psi < \frac{1}{2}\pi$, so $\psi = \tan^{-1}(5/3) \doteq 1.03$ radians. □

Example 3 Juliet is standing outside Romeo's house looking up at his bedroom window 7 meters above the ground (Figure 35). She is 8 meters from the house and her eyes are 1.6 meters ($\doteq$ 63 inches) above the ground. What is the angle θ between her line of sight and the horizontal?

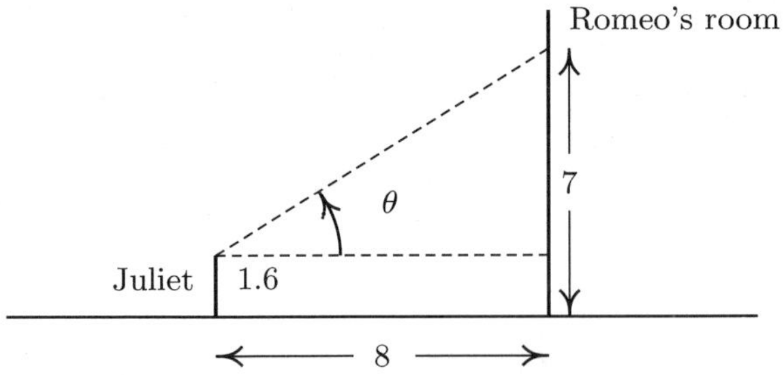

FIGURE 35

SOLUTION Because θ is the positive acute angle such that $\tan\theta = \dfrac{7-1.6}{8} = 5.4$, it is given by $\theta = \tan^{-1}(5.4) \doteq 1.39$ radians. □

The Law of cosines

Suppose that two sides of a triangle are of lengths a and b and the angle between them is θ, as in Figure 36. Then the coordinates of A are $(b, 0)$ and the coordinates of B are $(a\cos\theta, s\sin\theta)$, so that

$$\begin{aligned} c^2 = \overline{AB}^2 &= (a\cos\theta - b)^2 + (a\sin\theta - 0)^2 = a^2\cos^2\theta + 2ab\cos\theta + b^2 + a^2\sin^2\theta \\ &= a^2(\cos^2\theta + \sin^2\theta) + b^2 - 2ab\cos\theta. \end{aligned}$$

Then, since $\cos^2\theta + \sin^2\theta = 1$, we obtain $c^2 = a^2 + b^2 - 2ab\cos\theta$ and finally

$$c = \sqrt{a^2 + b^2 - 2ab\cos\theta}. \tag{34}$$

This formula is called the LAW OF COSINES.

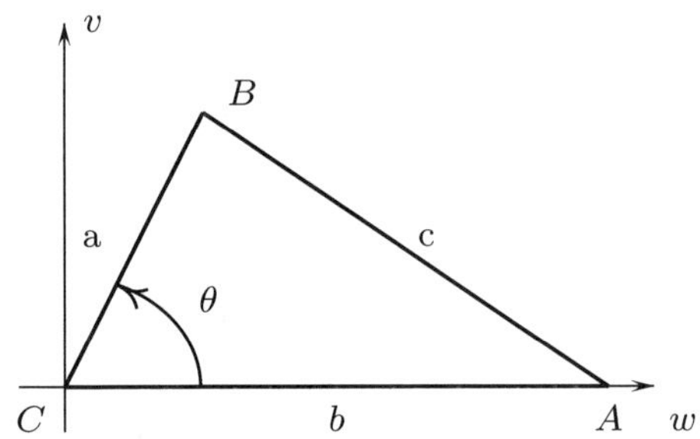

FIGURE 36

Example 4 Two sides of a triangle are 8 and 10 meters long and the angle between them is $\frac{4}{5}\pi$ radians. How long is the other side?

SOLUTION The length of the third side is $\sqrt{8^2 + 10^2 - 2(8)(10)\cos(\frac{4}{5}\pi)} \doteq 17.13$ meters. □

The Law of Sines

The Law of Cosines is frequently used, as in Example 4, to study a nonright triangle when we know the lengths of two sides and the angle between them. Now we look at the case where you need to relate two angles α and β and the lengths a and b of the sides opposite them, as in Figure 37. We let h denote the altitude of the triangle as in the drawing. Then $h = a \sin \beta$ and $h = b \sin \alpha$ so that

$$\frac{\sin \alpha}{a} = \frac{\sin \beta}{b}.$$

We can apply this reasoning to all three angles in a triangle to obtain

$$\frac{\sin \alpha}{a} = \frac{\sin \beta}{b} = \frac{\sin \gamma}{c} \tag{35}$$

where, as in Figure 38, c is the length of the side oppose the angle γ. This is the LAW OF SINES.

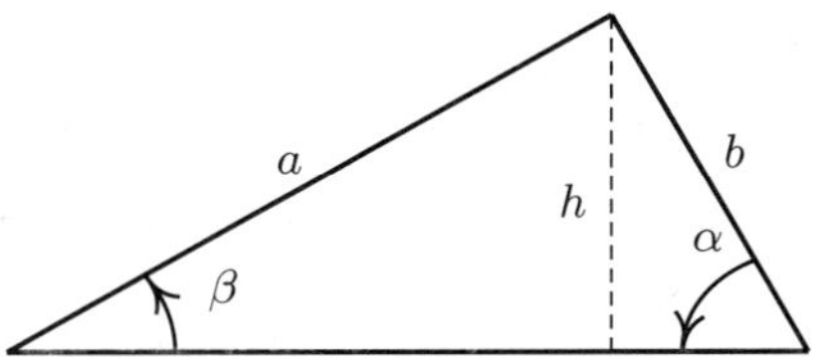

FIGURE 37

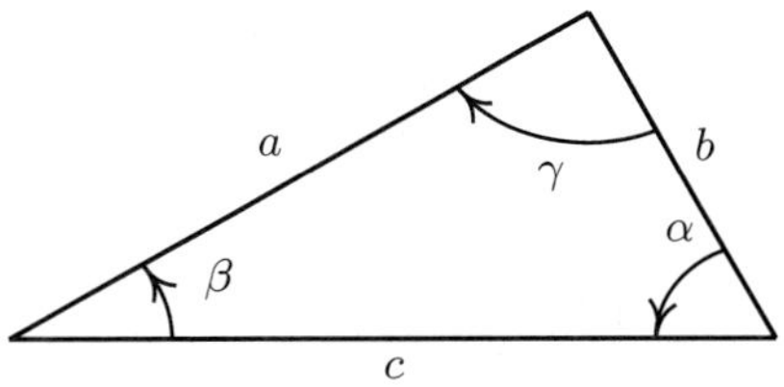

FIGURE 38

Example 5 Two angles in a triangle are 0.3 and 0.5 radians. The side opposite the angle of 0.5 radians is 10 yards long. How long is the side opposite the angle of 0.3 radians?

SOLUTION With s the length of the side opposite the angle of 0.3 radians,

$$\frac{\sin(0.3)}{s} = \frac{\sin(0.5)}{10}$$

so that $s = 10 \left[\frac{\sin(0.3)}{\sin(0.5)}\right] \doteq 6.16$ yards. □

Index